The Proterozoic Biosphere is the first major study of the paleobiology of the Proterozoic earth. It is a multidisciplinary work dealing with the evolution of the earth, the environment, and life during the 40% of earth's history that extended from the close of Precambrian Archean Eon (2500 Ma ago) to the widespread appearance of invertebrate metazoan fossils (about 550 Ma ago). *The Proterozoic Biosphere* includes a vast amount of new data on Proterozoic organisms and their modern analogues. Prepared by the Precambrian Paleobiology Research Group, a multidisciplinary consortium of forty-one scientists from eight countries, this monograph will serve as a benchmark in the development of the science of the paleontology, the organic geochemistry, and the paleoecology of Proterozoic sediments.

The three main goals of the study are:

(1) to amass, evaluate, and synthesize the large body of paleobiologic data available from previous studies, eliminating mistakes so that future investigations will not be encumbered by them;

(2) to generate new data and new analyses based on the re-examination of previous studies and on new investigations within an interdisciplinary framework;

(3) to build toward the future by placing special emphasis on new or relatively neglected aspects of paleobiologic study and by highlighting major unsolved problems in the field.

The Proterozoic Biosphere

FRONTISPIECE. Evidence of the Proterozoic biosphere and its environment (clockwise from upper left):

Core sample of banded iron-formation from the Hotazel Formation (2300 Ma), Kuruman region, northern Cape Province, South Africa.

Vertical section of finely layered, flat-laminated, modern microbial mat from "Area 5" (main PPRG study site) at Guerrero Negro, Baja California Sur, northern Mexico ($\times$ 2.1).

Holotype specimen of *Tribrachidium heraldicum* Glaessner from the Rawnsley Quartzite, Ediacara Member (560 Ma), central Flinders Ranges, South Australia, Australia ($\times$ 1.8).

Reef-like bioherm of columnar to domical, solitary and interconnected carbonate stromatolites, about 50 cm in height, in the upper intertidal zone of Shark Bay (Hamelin Pool), Western Australia.

Holotype specimen of *Eoentophysalis arcata* Mendelson and Schopf from the Sukhaya Tunguska Formation (1000 Ma), Turukhansk region, Siberia, USSR ($\times$ 990).

Proterozoic crude oil from seeps in the Nonesuch Shale (1055 Ma) at White Pine Copper Mine, White Pine, northern Michigan, USA.

Holotype specimen of *Cucumiforma vanavaria* Mikhailova from the Kamovsk Formation (675 Ma), Krasnoyarsk region, Siberia, USSR ($\times$ 540).

(Upper center) Vertical section of columnar stromatolites (*Coloniella*) from the Albanel Formation (2000 Ma), Lake Albanel region, central Quebec, Canada ($\times$ 0.3).

(Lower center) Sclerites of *Chancelloria* from the lowermost Cambrian (Tommotian Stage, 550 Ma) of the eastern Siberian Platform, USSR ($\times$ 10).

Photos by S. Bengtson, D. J. Des Marais, H. J. Hofmann, C. V. Mendelson, B. N. Runnegar, J. W. Schopf, G. Vidal.

THE PROTEROZOIC BIOSPHERE

A Multidisciplinary Study

Edited by
J. WILLIAM SCHOPF AND CORNELIS KLEIN

CAMBRIDGE
UNIVERSITY PRESS

Published by the Press Syndicate of the University of Cambridge
The Pitt Building, Trumpington Street, Cambridge CB2 1RP
40 West 20th Street, New York, NY 10011, USA
10 Stamford Road, Oakleigh, Melbourne 3166, Australia

© Cambridge University Press 1992

First published 1992

Printed in the United States of America

Library of Congress Cataloging-in-Publication Data

The proterozoic biosphere: a multidisciplinary study / edited by
 J. William Schopf and Cornelis Klein.
   p.   cm.
 Includes bibliographical references and index.
 ISBN 0-521-36615-1
  1. Paleontology – Proterozoic.   I. Schopf, J. William, 1941–
 II. Klein, Cornelis, 1937–
 QE724.5.P76 1992
 560'.171–dc20   91-15085
                                             CIP

ISBN 0-521-36615-1 hardback

THIS MONOGRAPHIC WORK IS DEDICATED TO
ROBERT M. GARRELS (1916–1988),
COLLEAGUE, FRIEND, TEACHER.

A contribution of the Center for the Study of
Evolution and the Origin of Life (CSEOL),
University of California, Los Angeles.

TERRA BOOKS

THE PROTEROZOIC BIOSPHERE—A MULTIDISCIPLINARY STUDY

J.W. Schopf and C. Klein (eds)

Cambridge University Press, Cambridge, UK. 1992, XXIV + 1348pp., US$195.00 (hbk). ISBN 0-521-36615-1.

Like its precursor volume *Earth's Earliest Biosphere* (1983), this second release by the UCLA-based 'Precambrian Paleobiology Research Group' is a remarkable erudition that is unrivalled in its field. As in the case of its predecessor, it owes much to the vision and drive of J. William Schopf who has pioneered the quest for an interdisciplinary approach to early biological and environmental evolution. Drawing from an impressive collation of published and newly generated data, a galaxy of authors have brought their expertise to bear on subjects ranging from early crustal evolution to molecular phylogenetics, the overall effort aiming at a holistic picture of life's evolution during the 0.5–2.5 Ga time interval. Particular attention is given to the Precambrian–Cambrian transition that marks a cardinal quantum step in the evolution of terrestrial life.

Having to cope with a huge pile of paperwork whose printed condensate just falls short of 1400 pages, the editors found it useful to divide the book into two parts. Part 1 starts with a summary of 'Geology and Paleobiology of the Archean Earth', followed by twelve chapters which address the whole spectrum of questions pertinent to the palaeontological and biogeochemical record of Proterozoic life: Chapter 5 presents the microfossil record of prokaryotes and protists; Chapters 7 and 8 deal with records of carbonaceous relics, trace and body fossils, and of metaphytes and metazoans; Chapters 9 and 11 present Proterozoic molecular phylogenetics, biotic diversity and rates of evolution; Chapters 10 and 12 consider Proterozoic biostratigraphy, and palaeobiogeography. Chapter 6 deserves special mention for its excellently compiled and highly educational treatise on recent mat-forming microbial ecosystems as modern analogues of ancient stromatolites, thus illustrating superbly the utility of the present as a key to the geological past. Part 1 concludes with an overview by Schopf on 'Evolution of the Proterozoic Biosphere: Benchmarks, Tempo and Mode'.

Part 2 combines elements of a data catalogue with textbook-like accounts of the principal analytical and other working procedures employed in the study; for instance, a flow chart of the processing techniques applied to the rock samples (Chapter 15), whole rock, kerogen and extractable organic matter analysis (Chapters 16, 17 and 18), and the sophisticated microelectrode techniques used in the study of modern mat-building microorganisms (Chapter 20). Further prominent chapters are the 'Atlas of Representative Proterozoic Microfossils' (Chapter 24), an 'Informal Revised Classification of Proterozoic Microfossils' (Chapter 25), and 'Described Taxa of Proterozoic and Selected Earliest Cambrian Carbonaceous Remains, Trace and Body Fossils' (Chapter 23), that follows up the metazoan connection of the Late Proterozoic biosphere.

Combining broad coverage and datastore, the book is more of a handbook than a topically focused text. It represents a gigantic undertaking (with 80 pages of references) that aims to be comprehensive rather than streamlined—there may be differences of opinion over the necessity of including some of the data. A mildly disturbing feature of the book is a certain lack of systematic rigidity in the arrangement of the individual chapters, with topically different presentations sometimes randomly intermingled and subjects torn apart by different (albeit complementary) treatments in Parts 1 and 2. Reference to "*Soviet* type specimens" (see pp. 1057 and 1059) in a 1992 book is anachronistic and possibly insulting to contemporaneous Russian palaeontology.

Such minor flaws cannot detract from the fact that this first systematic attempt to explore the interaction of biological and environmental evolution during Proterozoic times is certainly a benchmark in its field, and is indispensible reading for any student of Precambrian life. While not substituting for the wealth of specialist literature currently available, the *Proterozoic Biosphere* will probably serve as the principal introductory and standard text for the next decade.

Reviewed by Manfred Schidlowski, Max-Planck Institut für Chemie (Otto Hahn Institut), D-55020 Mainz, Germany.

Contents

List of Contributors   xxi

Preface   xix

PART 1

1. Geology and Paleobiology of the Archean Earth   5
J. William Schopf, Sherwood Chang,
W. Gary Ernst, Heinrich D. Holland,
James F. Kasting, Donald R. Lowe

1.1 Introduction   7
J. William Schopf

1.2 Formation of the Earth and the Origin of Life   9
James F. Kasting, Sherwood Chang
    1.2.1 Origin and Development of the Early Atmosphere   9
    1.2.2 Timing and Environmental Setting of the Origin of Life   10
    1.2.3 Chemical Evolution and the Origin of Life   10
    1.2.4 Paleoenvironmental Considerations   11
    1.2.5 Geologic and Paleobiologic Considerations   11

1.3 The Archean Geologic Record   13
Donald R. Lowe, W. Gary Ernst
    1.3.1 Formation and Evolution of the Pre-Archean Earth   13
    1.3.2 Principal Archean Geologic Terranes   14
    1.3.3 Archean Crustal Evolution   16
    1.3.4 Conclusions   18

1.4 The Environment of the Archean Earth   21
Heinrich D. Holland, James F. Kasting
    1.4.1 Introduction   21
    1.4.2 Evidence for Liquid Water and the Early Archean Surface Temperature   21
    1.4.3 Archean Ocean Chemistry   21
    1.4.4 The Archean Atmosphere   23
    1.4.5 The Archean Terrigenous Environment   23
    1.4.6 The Environmental Influence of Archean Impact Events   23

1.5 Paleobiology of the Archean   25
J. William Schopf
    1.5.1 Introduction   25
    1.5.2 Recent paleobiological studies of the Early Archean Fig Tree and Onverwacht Groups (Swaziland Supergroup), South Africa   25
    1.5.3 Controversial Microfossils Previously Described from the Early Archean Warrawoona Group, Western Australia   27
    1.5.4 Newly Discovered Microfossils from the Early Archean Apex Basalt (Warrawoona Group), Western Australia   28
    1.5.5 The Early Archean Microbiota   37
    1.5.6 Conclusions   39

1.6 Geology and Paleobiology of the Archean Earth: Current Status and Future Research Directions   41
J. William Schopf
    1.6.1 Archean Paleobiology: Problems and Perspectives   41

2. Geological Evolution of the Proterozoic Earth   43
Donald R. Lowe, Nicolas J. Beukes,
John P. Grotzinger, Raymond V. Ingersoll,
Joseph L. Kirschvink, Cornelis Klein,
Ian B. Lambert, and Ján Veizer

2.1 Introduction   45
Donald R. Lowe

2.2 Proterozoic Sedimentary Basins 47
    John P. Grotzinger, Raymond V. Ingersoll
    2.2.1 Introduction 47
    2.2.2 Preservation of the Proterozoic Record 47
    2.2.3 Extensional Settings 47
    2.2.4 Compressional Settings 49
    2.2.5 Strike-Slip Basins 49
    2.2.6 Conclusions 50

2.3 Late Proterozoic Low-Latitude Global
    Glaciation: the Snowball Earth 51
    Joseph L. Kirschvink
    2.3.1 Introduction 51
    2.3.2 Mechanisms Responsible for Low-Latitude
          Glaciation 51
    2.3.3 Implications of the Global Snowball Model 52

2.4 The Proterozoic Sedimentary Record 53
    Donald R. Lowe
    2.4.1 Introduction 53
    2.4.2 Greenstone Association 53
    2.4.3 Cratonic Association 54
    2.4.4 Secular Development of Proterozoic
          Supracrustal Sequences 54

2.5 Proterozoic Mineral Deposits Through Time 59
    Ian B. Lambert, Nicolas J. Beukes,
    Cornelis Klein, Ján Veizer
    2.5.1 Introduction 59
    2.5.2 Greenstone Belt Stage 59
    2.5.3 Cratonization Stage 60
    2.5.4 Stable Craton-Rifting Stage 61

2.6 Recycling and Preservation Probabilities of
    Sediments 63
    Ján Veizer
    2.6.1 Introduction 63
    2.6.2 Preservation Probabilities for Tectonic
          Domains 63
    2.6.3 Preservation Probabilities for Sediments 65

2.7 Major Events in the Geological Development
    of the Precambrian Earth 67
    Donald R. Lowe
    2.7.1 Introduction 67
    2.7.2 Principal Precambrian Crust-Forming
          Episodes 67
    2.7.3 Secular Variation in the Patterns of Crustal
          Formation 72
    2.7.4 Rates of Crustal Growth 74
    2.7.5 Conclusions 74

2.8 Summary and Conclusions 77
    Donald R. Lowe

2.9 Unresolved Problems and Future Research
    Directions 79
    Donald R. Lowe

3. **Proterozoic Biogeochemistry** **81**
    J. M. Hayes, David J. Des Marais, Ian B. Lambert,
    Harald Strauss, Roger E. Summons

3.1 Principles of Molecular and Isotopic
    Biogeochemistry 83
    Roger E. Summons, J. M. Hayes
    3.1.1 Preservation and Syngenicity 83
    3.1.2 Formation of Kerogen 84
    3.1.3 Biomarkers: Origin and Significance 84
    3.1.4 Rudiments of Isotopic Biogeochemistry 88
    3.1.5 Isotopes in Natural Systems 89

3.2 Concentrations of Organic Carbon and Maturities
    and Elemental Compositions of Kerogens 95
    Harald Strauss, David J. Des Marais, J. M. Hayes,
    Roger E. Summons
    3.2.1 Review of Previous Studies 95
    3.2.2 Results of This Study 96
    3.2.3 Trends in Abundance and Preservation of
          Organic Matter 96

3.3 Abundance and Composition of Extractable
    Organic Matter 101
    Roger E. Summons
    3.2.1 Preliminary Analysis of Kerogen and
          Bitumen 101
    3.3.2 Intensively Studied Rock Units and
          Petroleums 104

3.4 The Carbon-Isotopic Record 117
    Harald Strauss, David J. Des Marais, J. M. Hayes,
    Roger E. Summons
    3.4.1 Total Organic Carbon and Whole-Rock
          Carbonate 117
    3.4.2 Isotopic Compositions of Closely Defined
          Phases 118
    3.4.3 The Proterozoic Carbon Cycle 125

3.5 The Sulfur-Isotopic Record 129
    J. M. Hayes, Ian B. Lambert, Harald Strauss
    3.5.1 Results of New Isotopic Analyses of Sulfides
          and Sulfates 129
    3.5.2 An Overview of the Record and its
          Interpretation 130

3.6 Unsolved Problems and Conclusions 133
    J. M. Hayes, David J. Des Marais,
    Ian B. Lambert, Harald Strauss,
    Roger E. Summons
    3.6.1 The Subject 133
    3.6.2 The Proterozoic 133

4. **Proterozoic Atmosphere and Ocean** **135**
    Cornelis Klein, Nicolas J. Beukes,
    Heinrich D. Holland, James F. Kasting,
    Lee R. Kump, Donald R. Lowe

Contents

4.1 Introduction Cornelis Klein	137	
4.2 Time Distribution, Stratigraphy, and Sedimentologic Setting, and Geochemistry of Precambrian Iron-Formations Cornelis Klein, Nicolas J. Beukes	139	
4.2.1 Introduction	139	
4.2.2 Distribution of Iron-Formations Throughout the Precambrian	139	
4.2.3 Stratigraphic Setting and Sedimentology of the Depositional Basins of Iron-Formations	139	
4.2.4 Average Chemistry of Several Major Iron-Formations	141	

4.3 Models for Iron-Formation Deposition 147
Nicolas J. Beukes, Cornelis Klein
    4.3.1 Introduction 147
    4.3.2 Paleoenvironmental Interpretation of Iron-Formation Deposition in the Transvaal Supergroup, South Africa 147
    4.3.3 Paleoenvironmental Interpretation of Iron-Formation Deposition in Later Proterozoic Time 151

4.4 Distribution and Paleoenvironmental Interpretation of Proterozoic Paleosols 153
Heinrich D. Holland
    4.4.1 Introduction 153
    4.4.2 Implications for the Proterozoic Atmosphere 153

4.5 Other Geological Indicators 157
Donald R. Lowe
    4.5.1 Glacial Deposits 157
    4.5.2 Red Beds 157
    4.5.3 Detrital Uraninites 158

4.6 Atmospheric Evolution: the Rise of Oxygen 159
James F. Kasting, Heinrich D. Holland, Lee R. Kump
    4.6.1 Introduction 159
    4.6.2 The Three-Stage Model for the Rise of O_2 159
    4.6.3 Development of the Modern Aerobic (Stage III) Atmosphere 160
    4.6.4 Biogeochemical Cycles and Their Effects on O_2 161
    4.6.5 Modern and Ancient Controls on O_2
    4.6.6 Ozone and Other Trace Gases 162

4.7 Proterozoic Climates: the Effect of Changing Atmospheric Carbon Dioxide Concentrations 165
James F. Kasting
    4.7.1 The Climatic Record 165
    4.7.2 Solar Luminosity and Possible Greenhouse Gases 165
    4.7.3 Proterozoic CO_2 Levels from Climate Model Calculations 166
    4.7.4 The Late Precambrian Icehouse 167

4.8 Chemistry and Evolution of the Proterozoic Ocean 169
Heinrich D. Holland
    4.8.1 Introduction 169
    4.8.2 The Influence of Atmospheric Oxygen Concentrations on the Proterozoic Ocean 169
    4.8.3 An Evaluation of the Oceanic Oxygen Flux to Oxidize Hydrothermal Reductants 171

4.9 Conclusions and Unsolved Problems 173
Cornelis Klein, Nicolas J. Beukes, Heinrich D. Holland, James F. Kasting, Donald R. Lowe

5. Proterozoic and Selected Early Cambrian Microfossils: Prokaryotes and Protists 175
Carl V. Mendelson, John Bauld, Robert J. Horodyski, Jere H. Lipps, Toby B. Moore, J. William Schopf
5.1 Introduction 177
Carl V. Mendelson

5.2 Historical Development of Proterozoic Micropaleontology 179
J. William Schopf
    5.2.1 Introduction 179
    5.2.2 Phases of Development of Proterozoic Micropaleontology 180

5.3 Preservation of Prokaryotes and Organic-Walled and Calcareous and Siliceous Protists 185
Robert J. Horodyski, John Bauld, Jere H. Lipps, Carl V. Mendelson
    5.3.1 Introduction 185
    5.3.2 Prokaryotes 185
    5.3.3 Eukaryotes (Primarily Acritarchs) 189
    5.3.4 Preservation Potential of Naked, Agglutinated, Calcareous, and Siliceous Protists 192
    5.3.5 Bias in the Microfossil Record: Prokaryotes and Eukaryotes as a Function of Paleoenvironment and Lithology 192

5.4 Proterozoic Prokaryotes: Affinities, Geologic Distribution, and Evolutionary Trends 195
J. William Schopf
    5.4.1 Introduction 195
    5.4.2 Morphometric Effects of Preservational Compression of Microfossils in Shales 195
    5.4.3 Methods for Inferring the Probable Affinities of the Principal Categories of Proterozoic Microfossils 202
    5.4.4 Coccoidal Microorganisms and Proterozoic Microfossils 204

5.4.5 Ellipsoidal Microorganisms and Proterozoic Microfossils — 210
5.4.6 Non-septate Filamentous Microorganisms, Prokaryotic Sheaths, and Proterozoic Microfossils — 211
5.4.7 Septate Filamentous Microorganisms and Proterozoic Microfossils — 215
5.4.8 Summary: Inferred Affinities and Evolutionary Trends of the Principal Categories of Proterozoic Prokaryotic Microfossils — 218

5.5 Proterozoic and Early Cambrian Acritarchs — 219
Carl V. Mendelson, J. William Schopf
5.5.1 Introduction — 219
5.5.2 Major Types of Proterozoic and Early Cambrian Acritarchs — 219
5.5.3 Milestones in Morphological Evolution in Acritarchs — 228
5.5.4 Closing Remarks — 231

5.6 Distinctive Problematical Proterozoic Microfossils — 233
Toby B. Moore, Robert J. Horodyski, Jere H. Lipps, J. William Schopf
5.6.1 Introduction — 233
5.6.2 Distinctive Proterozoic Micro-problematica — 233

5.7 Proterozoic and Cambrian Skeletonized Protists — 237
Jere H. Lipps
5.7.1 Introduction — 237
5.7.2 Proterozoic Skeletonized Protists — 237
5.7.3 Cambrian Skeletonized Protists — 237
5.7.4 Summary — 239

5.8 Unsolved Problems and Future Research Directions — 241
Carl V. Mendelson, J. William Schopf
5.8.1 Introduction — 241
5.8.2 Major Problem Areas — 241

5.9 Summary and Conclusions: The Current Status of Proterozoic Micropaleontology — 243
Carl V. Mendelson, J. William Schopf

6. Modern Mat-Building Microbial Communities: a Key to the Interpretation of Proterozoic Stromatolitic Communities — 245
Beverly K. Pierson, John Bauld, Richard W. Castenholz, Elisa D'Amelio, David J. Des Marais, Jack D. Farmer, John P. Grotzinger, Bo Barker Jørgensen, Douglas C. Nelson, Anna C. Palmisano, J. William Schopf, Roger E. Summons, Malcolm R. Walter, David M. Ward

6.1 Introduction
Beverly K. Pierson — 247

6.2 Proterozoic Stromatolites — 253
Malcolm R. Walter, John P. Grotzinger, J. William Schopf
6.2.1 Stromatolite Morphology and Taxonomy — 253
6.2.2 Microbiotas of Proterozoic Stromatolitic Communities — 255
6.2.3 Paleoecology of Proterozoic Stromatolites — 255
6.2.4 Growth and Lithification of Proterozoic Stromatolites — 257
6.2.5 Temporal Distribution, Taxonomic Diversity, and Abundance of Proterozoic Stromatolites — 258
6.2.6 Interpretation of the Stromatolitic Fossil Record — 259

6.3 Modern Microbial Mats — 261
John Bauld, Elisa D'Amelio, Jack D. Farmer
6.3.1 Morphology of Modern Microbial Mats — 261
6.3.2 Biotic and Environmental Factors Controlling Mat Morphology, Lamination, Fabric, and Morphogenesis — 266
6.3.3 Feedback Relationships Between Biotic and Environmental Factors — 267

6.4 The Microenvironment Within Modern Microbial Mats — 271
Bo Barker Jørgensen, Richard W. Castenholz, Beverly K. Pierson
6.4.1 Microenvironmental Parameters — 271
6.4.2 The Physical Microenvironment — 271
6.4.3 The Chemical Microenvironment — 273
6.4.4 Microbial Movements — 274

6.5 Photosynthetic Activity in Modern Microbial Mat-Building Communities — 279
Richard W. Castenholz, John Bauld, Beverly K. Pierson
6.5.1 Types of Photosynthetic Activity — 279
6.5.2 Photosynthetic Activity in Microbial Mats — 280
6.5.3 Pigments in Mats — 284
6.5.4 Productivity in Mat Ecosystems — 285

6.6 Chemotrophy and Decompsition in Modern Microbial Mats — 287
Bo Barker Jørgensen, Douglas C. Nelson, David M. Ward
6.6.1 Decompsition and Element Cycling — 287
6.6.2 Heterotrophic Processes — 287
6.6.3 Chemolithotrophic Processes — 291

6.7 Grazing and Bioturbation in Modern Microbial Mats — 295
Jack D. Farmer
6.7.1 Introduction — 295
6.7.2 Grazing in Microbial Mats of Restricted, Hypersaline Environments — 295
6.7.3 Grazing in Thermal Spring Mats — 297
6.7.4 Implications of Faunal Studies of Microbial Mats for Proterozoic Paleobiology — 297

Contents

6.8 The Biogeochemistry of Carbon in Modern
    Microbial Mats 299
    David J. Des Marais, John Bauld,
    Anna C. Palmisano, Roger E. Summons,
    David M. Ward
    6.8.1 Introduction 299
    6.8.2 Elemental Abundances 299
    6.8.3 Stable Carbon Isotopes 299
    6.8.4 Biomarkers 303

6.9 Modern Phototrophic Microbial Mats: 309
    Anoxygenic, Intermittently Oxygenic/Anoxygenic,
    Thermal, Eukaryotic, and Terrestrial
    David M. Ward, John Bauld,
    Richard W. Castenholz, Beverly K. Pierson
    6.9.1 Introduction 309
    6.9.2 Anoxygenic Mats 309
    6.9.3 Intermittently Oxygenic/Anoxygenic Mats 312
    6.9.4 Thermal Mats 313
    6.9.5 Eukaryotic Microbial Mats 321
    6.9.6 Terrestrial Mats 324

6.10 Case Study of a Modern Microbial Mat-Building
    Community: the Submerged Cyanobacterial Mats
    of Guerrero Negro, Baja California Sur, Mexico 325
    David J. Des Marais, Elisa D'Amilio,
    Jack D. Farmer, Bo Barker Jørgensen,
    Anna C. Palmisano, Beverly K. Pierson
    6.10.1 Introduction 325
    6.10.2 The Benthic Mats: Trends Along the Salinity
           Gradient 325
    6.10.3 Trends With Depth in a Mat 328
    6.10.4 Comparison of Guerrero Negro and Solar
           Lake Mat Communities 331
    6.10.5 Characteristics of Stromatolites Predicted
           from Studies of Microbial Mats 332

6.11 A General Comparison of Microbial Mats and
    Microbial Stromatolites: Bridging the Gap
    Between the Modern and the Fossil 335
    Malcolm R. Walter, John Bauld,
    David J. Des Marais, J. William Schopf
    6.11.1 Introduction 335
    6.11.2 Analogies and Homologies Between
           Extant Microbial Mats and Ancient
           Stromatolites 335
    6.11.3 General Geochemical Comparisons 337
    6.11.4 Closing Remarks 338

6.12 Unsolved Problems and Future Research
    Directions 339
    Beverly K. Pierson

6.13 Summary and Conclusions: The Current
    Status of Studies of Modern Microbial
    Mat-Building Communities and Their Relevance
    to Interpretation of Proterozoic Stromatolites 341
    Beverly K. Pierson

7. **Proterozoic and Earliest Cambrian Carbonaceous
   Remains, Trace and Body Fossils** 343
   Kenneth M. Towe, Stefan Bengtson,
   Mikhail A. Fedonkin, Hans J. Hofmann,
   Carol Mankiewicz, Bruce N. Runnegar

7.1 Introduction 345
    Kenneth M. Towe

7.2 Criteria for Acceptance of Reported
    Proterozoic Megafossils 347
    Kenneth M. Towe

7.3 Proterozoic Carbonaceous Films 349
    Hans J. Hofmann
    7.3.1 Introduction 349
    7.3.2 Informal Categories of Megascopic
          Carbonaceous Films 349
    7.3.3 Morphological Characteristics, Distributions,
          and Possible Affinities 354
    7.3.4 Summary 357

7.4 Proterozoic and Early Cambrian Calcareous
    Algae 359
    Carol Mankiewicz
    7.4.1 Introduction 359
    7.4.2 Classification 359
    7.4.3 Diversity 361
    7.4.4 Paleoecology 363
    7.4.5 Stratigraphic Range 363
    7.4.6 Summary 367

7.5 Proterozoic Metazoan Body Fossils 369
    Bruce N. Runnegar, Mikhail A. Fedonkin
    7.5.1 Introduction 369
    7.5.2 The Ediacara Faunas 369
    7.5.3 Conclusions 388

7.6 Proterozoic Metazoan Trace Fossils 389
    Mikhail A. Fedonkin, Bruce N. Runnegar
    7.6.1 Introduction 389
    7.6.2 Reported Pre-Vendian Trace Fossils 389
    7.6.3 Biostratigraphy of Vendian Trace Fossils 390
    7.6.4 Conclusions 395

7.7 Proterozoic and Earliest Cambrian Skeletal
    Metazoans 397
    Stefan Bengtson
    7.7.1 Introduction 397
    7.7.2 Representativity of the Skeletal Fossil Record 397
    7.7.3 Geographical Distribution of Early Skeletal
          Fossils 398
    7.7.4 Appearances of Skeletal Types 399
    7.7.5 Phylum-level Taxa in the Latest Proterozoic
          and Early Cambrian 402

7.8 Megascopic Dubiofossils Hans J. Hofmann	413
7.8.1 Introduction	413
7.8.2 Horizontal Spindles and Ropes	414
7.8.3 Vertical Cylindrical Structures	416
7.8.4 Discoidal Structures	416
7.8.5 Other Megascopic Dubiofossils	418
7.8.6 Conclusions	419
7.9 Unsolved Problems and Future Research Directions Kenneth M. Towe	421
7.10 Summary and Conclusions Kenneth M. Towe	423

8. The Proterozoic-Early Cambrian Evolution of Metaphytes and Metazoans
Stefan Bengtson, Jack D. Farmer,
Mikhail A. Fedonkin, Jere H. Lipps,
Bruce N. Runnegar

8.1 Introduction Stefan Bengtson, Jere H. Lipps	427
8.2 Origins of Multicellular Individuality Jack D. Farmer	429
8.2.1 Introduction	429
8.2.2 The Concept of Individuality	429
8.2.3 The Nature of Multicellularity	429
8.2.4 The Evolution of Multicellularity	431
8.3 The Major Biotas of Proterozoic to Early Cambrian Multicellular Organisms Stefan Bengtson, Mikhail A. Fedonkin, Jere H. Lipps	433
8.3.1 Introduction	433
8.3.2 Evolutionary Relationships of the Three Biotas	433
8.3.3 Environmental and Taphonomic Settings of the Biotas	433
8.4 Ecology and Biogeography Jere H. Lipps, Stefan Bengtson, Mikhail A. Fedonkin	437
8.4.1 Trophic Relationships and Life Modes	437
8.4.2 Paleobiogeography	438
8.5 The Evolution of Metazoan Body Plans Stefan Bengtson, Jack D. Farmer	443
8.5.1 Introduction	443
8.5.2 Genealogies and Body Plans—Two Perspectives on Phyla	443
8.5.3 The Evolution of Body Plans	444
8.5.4 Recognizing Body Plans in the Fossil Record	445
8.6 Origins of Biomineralization in Metaphytes and Metazoans Stefan Bengtson, Bruce N. Runnegar	447
8.6.1 Introduction	447
8.6.2 Carbonates	447
8.6.3 Phosphates	448
8.6.4 Opal	449
8.6.5 Origin of Phosphate *Versus* Carbonate	449
8.6.6 Origins of Skeletal Biomineralization	450
8.7 The Precambrian-Cambrian Evolutionary Transition Jere H. Lipps, Stefan Bengtson, Jack D. Farmer	453
8.7.1 Introduction	453
8.7.2 Hypotheses on the Origin and Early Evolution of Metazoans and Metaphytes	453
8.7.3 Evaluating the Hypotheses	454
8.7.4 What Type of Hypothesis are We Looking For?	457
8.8 Unsolved Problems and Future Research Directions Stefan Bengtson	459
8.9 Summary and Conclusions Jere H. Lipps, Stefan Bengtson	461

9. Molecular Phylogenetics, Molecular Paleontology, and the Proterozoic Fossil Record
Bruce N. Runnegar, David J. Chapman,
Walter M. Fitch — 403

9.1 Introduction Bruce N. Runnegar	465
9.2 Methods of Molecular Phylogenetics Walter M. Fitch	467
9.2.1 Introduction	467
9.2.2 Sequence Alignment of Homologous Amino Acids or Nucleotides	467
9.2.3 Methods for Inferring the Ancestral Relationships Between Sequences	468
9.2.4 Use of Molecular Information for Dating Biological Events in Geological Time	469
9.3 The Tree of Life Bruce N. Runnegar	471
9.3.1 Life's Deepest Branches	471
9.3.2 Endosymbiosis and the Origin of Organelled Eukaryotes	471
9.3.3 Rooting the Tree of Life in the Fossil Record	472
9.3.4 Radiation of the Eubacteria and Archaebacteria	474
9.3.5 Origin of the Eukaryotic Kingdoms	475

Contents

9.4 Origin and Divergence of Protists ... 477
David J. Chapman
9.4.1 What is a "Protist"? ... 477
9.4.2 The Ancestral Host for Organelle-Generating Endosymbioses ... 477
9.4.3 The Endosymbiotic Origin of Chloroplasts and Mitochondria ... 478
9.4.4 The Evolutionary Radiation of Protists ... 480

9.5 Origin and Diversification of the Metazoa ... 485
Bruce N. Runnegar

10. **Biostratigraphy and Paleobiogeography of the Proterozoic** ... 487
Hans J. Hofmann, Stefan Bengtson, J. M. Hayes, Jere H. Lipps, J. William Schopf, Harald Strauss, Roger E. Summons, Malcolm R. Walter

10.1 Introduction ... 489
Hans J. Hofmann

10.2 Proterozoic Biostratigraphy: Problems and Perspectives ... 491
Hans J. Hofmann
10.2.1 Introduction ... 491
10.2.2 Major Problems ... 491

10.3 Stratigraphic Distribution of Prokaryotes and Acritarchs ... 497
J. William Schopf
10.3.1 Proterozoic Prokaryotes ... 497
10.3.2 Proterozoic Acritarchs ... 497

10.4 Stratigraphic Distribution of Skeletonized Protists ... 499
Jere H. Lipps

10.5 Stratigraphic Distribution of Megafossils ... 501
Hans J. Hofmann, Stefan Bengtson
10.5.1 Carbonaceous Compressions ... 501
10.5.2 Soft-Bodied Metazoans ... 504
10.5.3 Trace Fossils ... 504
10.5.4 Early Skeletonized Metazoans ... 505
10.5.5 Camasid Dubiofossils ... 506

10.6 Stratigraphic Distribution of Stromatolites and Allied Structures ... 507
Malcolm R. Walter

10.7 Proterozoic Distribution of Biogeochemical Fossils ... 511
Harald Strauss, J. M. Hayes, Roger E. Summons
10.7.1 Stable Isotope Records ... 511
10.7.2 Molecular Fossils ... 512

10.8 Summary: Current Status of Proterozoic Biostratigraphy ... 513
Hans J. Hofmann

10.9 Proterozoic Paleobiogeography ... 515
Hans J. Hofmann

10.10 Unsolved Problems and Future Research Directions ... 517
Hans J. Hofmann

10.11 Summary and Conclusions ... 519
Hans J. Hofmann

11. **Biotic Diversity and Rates of Evolution During Proterozoic and Earliest Phanerozoic Time** ... 521
J. John Sepkoski, Jr., J. William Schopf

11.1 Introduction ... 523
J. John Sepkoski, Jr.

11.2 The Proterozoic Fossil Record: Special Problems in Analyzing Diversity Patterns ... 525
J. John Sepkoski, Jr., J. William Schopf
11.2.1 General Considerations ... 525
11.2.2 Taxonomic Problems ... 525
11.2.3 Problems of Correlation and Dating ... 526
11.2.4 Problems of Biased Sampling ... 526
11.2.5 A Note of Optimism ... 527

11.3 Patterns of Proterozoic Microfossil Diversity: An Initial, Tentative, Analysis ... 529
J. William Schopf
11.3.1 Introduction ... 529
11.3.2 Methods of Analyses ... 529
11.3.3 Assessment of the Components of the Analyses ... 530
11.3.4 Analyses of Diversity ... 537
11.3.5 Evaluation of Detected Diversity Patterns ... 548
11.3.6 Interpretive Summary ... 550
11.3.7 Overview ... 552

11.4 Proterozoic-Early Cambrian Diversification of Metazoans and Metaphytes ... 553
J. John Sepkoski, Jr.
11.4.1 Introduction ... 553
11.4.2 Data ... 553
11.4.3 General Patterns of Early Metazoan Diversification ... 554
11.4.4 Faunal Heterogeneity in the Vendian-Cambrian Diversification ... 557
11.4.5 Exponential Diversification in the Proterozoic-Phanerozoic Transition ... 559
11.4.6 Diversification of Metaphytes ... 559

11.5 Proterozoic and Earliest Phanerozoic Biotic
     Diversity: Unsolved Problems and Future
     Research Directions   563
     J. John Sepkoski, Jr. and J. William Schopf

11.6 Biotic Diversity During Proterozoic and Earliest
     Phanerozoic Time: Summary and Conclusions   565
     J. John Sepkoski, Jr. and J. William Schopf

12. A Paleogeographic Model for Vendian and Cambrian Time   567
Joseph L. Kirschvink

12.1 A Paleogeographic Model for Vendian and
     Cambrian Time   569
     12.1.1 Introduction   569
     12.1.2 Paleomagnetic Data   570
     12.1.3 Tectonic Reconstructions   574
     12.1.4 Configuration and Breakup of Late
            Proterozoic Super(?) Continents   580

13. Evolution of the Proterozoic Biosphere: Benchmarks, Tempo, and Mode   583
J. William Schopf

13.1 Prologue   585

13.2 Times of Origin and Earliest Evidence
     of Major Biologic Groups   587
     13.2.1 Prokaryotes   587
     13.2.2 Eukaryotes   591

13.3 Tempo and Mode of Proterozoic Evolution   595
     13.3.1 Introduction   595
     13.3.2 Hypobradytely and the Evolution of
            Proterozoic Life   595

13.4 A Synoptic Comparison of Phanerozoic and
     Proterozoic Evolution   599

PART 2

14. Geographic and Geologic Data for PPRG Rock Samples   603
Toby B. Moore, J. William Schopf

14.1 Introduction and Numerical Listing of Geologic
     Samples Included in the PRRG Collections   605

14.2 Geographic and Geologic Data for Rock
     Samples Included in the PPRG Collections   614
     14.2.1 Africa   614
     14.2.2 Asia   620
     14.2.3 Australia   628
     14.2.4 Europe   652
     14.2.5 North America   655
     14.2.6 South America   687

Table 14.5 Estimated Ages of Geologic Units
           Represented in the PPRG Collections   688

15. Flow Chart and Processing Procedures for Rock Samples   695
Harald Strauss, David J. Des Marais,
J. M. Hayes, Toby B. Moore, J. William Schopf

15.1 Summary of Processing Procedures   697

16. Procedures of Whole Rock and Kerogen Analysis   699
Harald Strauss, David J. Des Marais,
J. M. Hayes, Ian B. Lambert, Roger E. Summons

16.1 Carbon: Whole Rock Analysis   701
     16.1.1 Organic Carbon   701
     16.1.2 Rock-Eval Analysis   701
     16.1.3 Carbonate Carbon   702

16.2 Sulfur: Whole Rock Analysis   703
     16.2.1 Sulfide Sulfur   703
     16.2.2 Sulfate Sulfur   703

16.3 Notation and Precision of Isotopic
     Measurements   705
16.4 Kerogen   707
     16.4.1 Kerogen Isolation   707
     16.4.2 Elemental Analysis of Carbon, Hydrogen,
            and Nitrogen   707
     16.4.3 Kerogen Color   707

17. Abundances and Isotopic Compositions of Carbon and Sulfur Species in Whole Rock and Kerogen Samples   709
Harald Strauss, Toby B. Moore

17.1 Introduction   711
17.1 Explanation of Symbols   711
     Table 17.1 New Analyses of the Abundance and
                Isotopic Composition of Organic
                Carbon in Whole Rock Samples   712
     Table 17.2 New Analyses of the Isotopic
                Composition of Carbonates   721
     Table 17.3 Previously Published Analyses of the
                Abundance and Isotopic Composition
                of Carbon in Whole Rock Samples   721

Contents

Table 17.4	Summary of All Results for Abundance and Isotopic Composition of Carbon in Whole Rock Samples	748
Table 17.5	New Analyses of the Elemental and Carbon Isotopic Composition of Kerogen Samples	759
Table 17.6	New Analyses of Sulfur Abundance and Isotopic Composition of Sulfides and Sulfates	762
Table 17.7	Previously Published Analyses of the Isotopic Composition of Sulfides and Sulfates	764
Table 17.8	Summary of All Results for Isotopic Composition of Sulfides and Sulfates	787
Table 17.9	Previously Published Analyses of Elemental and Carbon Isotopic Composition of Kerogen Samples	791
Table 17.10	Summary of Elemental and Carbon Isotopic Composition of Kerogen Samples	796

18. Procedures for Analysis of Extactable Organic Matter 799
Roger E. Summons and Harald Strauss

18.1 Sample Selection and Handling 801

18.2 Extraction Procedures 803
  18.2.1 Extraction and Chromatographic Separation of Bitumens 803
  18.2.2 Extraction and Fractionation of Signature Lipids from Extant Microbes and Modern Sediments 803

18.3 Analyses of Extracts 805
  18.3.1 Gas Chromatographic Analysis of Hydrocarbons and Other Lipids 805
  18.3.2 Gas Chromatography-Mass Spectrometry 805

18.4 Kerogen Analyses 807
  18.4.1 Kerogen Hydrogen Pyrolysis 807
  18.4.2 Kerogen Pyrolysis-Gas Chromatography 807

18.5 Urea Adduction 809

19. Composition of Extractable Organic Matter 811
Roger E. Summons, Harald Strauss

19.1 Explanation of Tables Summarizing the Bitumen Content and Composition of the Analyzed Samples 813

19.2 Discussion of the Data 815
  Table 19.1 Rock-Eval Data for PPRG Samples 815
  Table 19.2 Organic Geochemical Data for PPRG Samples 816
  Table 19.3 Extract Data for BMR Samples 817
  Table 19.4 Organic Geochemical Data for BMR Samples 817

20. Modern Mat-Building Microbial Communities: Methods of Investigation and Supporting Data 821
Richard W. Castenholz, Elisa D'Amelio, Jack D. Farmer, Bo Barker Jørgensen, Anna C. Palmisano, Beverly K. Pierson, David M. Ward

20.1 Methods of Light Microscopy 823
Jack D. Farmer, Beverly K. Pierson
  20.1.1 Preserved and Sectioned Material of Whole Mat 823
  20.1.2 Living Material and Wet Mounts 823

20.2 Methods of Electron Microscopy 825
Elisa D'Amelio, Beverly K. Pierson
  20.2.1 Stereoscopic Electron Microscopy (SEM) 825
  20.2.2 Transmission Electron Microscopy (TEM): Special Techniques for Mats 825

20.3 Culture Methods 827
Richard W. Castenholz, Beverly K. Pierson
  20.3.1 Filamentous Anoxygenic Phototrophs 827
  20.3.2 Purple Bacteria and Green Sulfur Bacteria 827
  20.3.3 Cyanobacteria 827
  20.3.4 Other Conspicuous Bacteria of Microbial Mats 827

20.4 Methods of Pigment Study 829
Anna C. Palmisano, Richard W. Castenholtz, Beverly K. Pierson
  20.4.1 Organic Extracts of Chlorophyll *a* and Bacteriochlorophylls *a, b, c, d, e, g* 829
  20.4.2 Analysis of Lipophilic Pigments from Phototrophic Microbial Mats by High Performance Liquid Chromatography 829
  20.4.3 "*In Vivo*" Analysis of Pigments by Ultrasonic Disruption in Buffer 831

20.5 Methods of Light Measurement 833
Bo Barker Jørgensen, Beverly K. Pierson, Richard W. Castenholz
  20.5.1 Definitions 833
  20.5.2 Total Irradiance Within Mats 833
  20.5.3 Spectral Distribution of Radiation Within Mats 833
  20.5.4 Photoacoustic Methods 835

20.6 Microelectrode Measurements 837
Bo Barker Jørgensen

20.7 Meiofauna: Strategies for Field Studies   839
      Jack D. Farmer

20.8 Rate Measurements of Anaerobic Processes
      in Hot Spring Mats   841
      David M. Ward

20.9 Chemical Analyses   843
      Bo Barker Jørgensen

21. Construction and Use of Geological, Geochemical, and Paleobiological Databases   855
J. M. Hayes, Stefan Bengtson, Hans J. Hofmann, Jere H. Lipps, Donald R. Lowe, Carol Mankiewicz, Carl V. Mendelson, Toby B. Moore, Bruce N. Runnegar, Harald Strauss

21.1 Sample Inventory and Curation   857

21.2 Results and Correlative Information   859
      21.2.1 Geochemistry and Lithology   859
      21.2.2 Paleontology   859

21.3 Bibliographic System   861

21.4 Summary   863

22. Proterozoic and Selected Early Cambrian Microfossils and Microfossil-Like Objects   865
Carl V. Mendelson, J. William Schopf

22.1 Introduction   867
      22.1.1 Coverage   867
      22.1.2 Content   867
      22.1.3 Caveats   867
      22.1.4 Conventions   868
      22.1.5 Abbreviations   868
      Table 22.1 Proterozoic and Selected Early Cambrain Micrononfossils and Micropseudofossils   869
      Table 22.2 Proterozoic and Selected Early Cambrian Microdubiofossils   876
      Table 22.3 Proterozoic and Selected Early Cambrain Microfossils   884

23. Described Taxa of Proterozoic and Selected Earliest Cambrian Carbonaceous Remains, Trace and Body Fossils   953
Kenneth M. Towe, Stefan Bengtson, Mikhail A. Fedonkin, Hans J. Hofmann, Carol Mankiewicz, Bruce N. Runnegar

23.1 Proterozoic and Selected Cambrian Megascopic Carbonaceous Films   957
      Hans J. Hofmann

23.2 Proterozoic and Early Cambrian Calcareous Algae   981
      Carol Mankiewicz

23.3 Proterozoic Fossils of Soft-Bodied Metazoans (Ediacaran Faunas)   999
      Bruce N. Runnegar

23.4 Proterozoic Metazoan Trace Fossils   1009
      Bruce N. Runnegar

23.5 Proterozoic and Earliest Cambrian Skeletal Metazoans   1017
      Stefan Bengtson

23.6 Proterozoic and Selected Cambrian Megascopic Dubiofossils and Pseudofossils   1035
      Hans J. Hofmann

24. Atlas of Representative Proterozoic Microfossils   1055
J. William Schopf

24.1 Introduction   1056

24.2 Illustrated Type Specimens   1057

24.3 Illustrated Specimens   1062

25. Informal Revised Classification of Proterozoic Microfossils   1119
J. William Schopf

25.1 Introduction   1121

25.2 Informal Revised Classification   1123
      25.2.1 Non-Septate Filaments   1123
      23.2.2 Septate Unbranched Filaments   1124
      25.2.3 Solitary Coccoidal Cells and Sphaeromorph Acritarchs   1159
      25.2.4 Colonial Coccoidal and Ellipsoidal Cells   1161
      25.2.5 Miscellaneous Prokaryotes   1162
      25.2.6 Miscellaneous Eukaryotes   1163

25.3 Temporal Distribution of Benthic and Planktonic Proterozoic Microfossils   1167

Contents

26. Models for Vendian-Cambrian Biotic Diversity and for Proterozoic Atmospheric and Ocean Chemistry   1169
J. John Sepkoski, Jr., James F. Kasting

26.1 Stratigraphic Ranges of Vendian and Cambrian Animal Families   1171
J. John Sepkoski, Jr.
   26.1.1 Taxonomy and Stratigraphic Ranges of Animal Families in the Vendian and Cambrian: Data and Analytical Results for Section 11.4   1171

26.2 Models Relating to Proterozoic Atmospheric and Oceanic Chemistry   1185
James F. Kasting
   26.2.1 Box Models Relating to the Rise in Atmospheric Oxygen (Section 4.6)   1185
   26.2.2 One-Dimensional Climate Modeling of Past CO_2 Concentrations (Section 4.7)   1187

27. Glossary of Technical Terms   1189
J. William Schopf, Cornelis Klein

References Cited   1205

Subject Index   1287

Index to Geologic Units   1297

Taxonomic Index   1309

List of Contributors

JOHN BAULD
Groundwater Branch, Bureau of Mineral Resources, G.P.O. Box 378, Canberra City, A.C.T. 2601, Australia

STEFAN BENGTSON
Institute of Palaeontology, Uppsala Universitet, Norbyvagen 22 (Box 558), S-751 22, Uppsala, Sweden

NICOLAS J. BEUKES
Department of Geology, Rand Afrikaans University, Johannesburg, 2000, South Africa

RICHARD W. CASTENHOLZ
Department of Biology, University of Oregon, Eugene, Oregon, 97403, USA

SHERWOOD CHANG
Planetary Biology Branch, National Aeronautics and Space Administration, Ames Research Center, Moffett Field, California, 94035, USA

DAVID J. CHAPMAN
Department of Biology, University of California, Los Angeles, California, 90024, USA

ELISA D'AMELIO
Planetary Biology Branch, National Aeronautics and Space Administration, Ames Research Center, Moffett Field, California, 94035, USA

DAVID J. DES MARAIS
Planetary Biology Branch, National Aeronautics and Space Administration, Ames Research Center, Moffett Field, Califonria, 94035, USA

W. GARY ERNST
School of Earth Sciences, Stanford University, Stanford, California, 94305, USA

JACK D. FARMER
Department of Earth and Space Sciences, University of California, Los Angeles, California, 90024, USA

MIKHAIL A. FEDONKIN
Palaeontological Institute, USSR Academy of Sciences, Profsoyuznaya ulitsa 113, Moscow, 117868, USSR

WALTER M. FITCH
Department of Ecology and Evolutionary Biology, University of California, Irvine, California, 92717, USA

ROBERT M. GARRELS*
Department of Marine Sciences, University of South Florida, St. Petersburg, Florida, 33701, USA

JOHN P. GROTZINGER
Department of Earth, Atmosphere, and Planetary Sciences, Massachusetts Institute of Technology, Cambridge, Massachusetts, 02139, USA

JOHN M. HAYES
Biogeochemical Laboratories, Departments of Geological Sciences and of Chemistry, Geology Building, Indiana University, Bloomington, Indiana, 47405, USA

HANS J. HOFMANN
Department of Geology, University of Montreal, Québec, H3C 3J7, Canada

HEINRICH D. HOLLAND
Department of Earth and Planetary Sciences, Hoffman Laboratory, Harvard University, Cambridge, Massachusetts, 01238, USA

ROBERT J. HORODYSKI
Geology Department, Tulane University, New Orleans, Louisiana, 70118, USA

RAYMOND V. INGERSOLL
Department of Earth and Space Sciences, University of California, Los Angeles, California, 90024, USA

BO BARKER JØRGENSEN
Institute of Ecology and Genetics, University of Aarhus, Ny Munkegade, DK-8000, Aarhus C., Denmark

*deceased

JAMES F. KASTING
Department of Geosciences, Pennsylvania State University, University Park, Pennsylvania, 16802, USA

JOSEPH L. KIRSCHVINK
Division of Geological and Planetary Science, California Institute of Technology, Pasadena, California, 91125, USA

CORNELIS KLEIN
Department of Geology, University of New Mexico, Albuquerque, New Mexico, 87131, USA

LEE R. KUMP
Department of Geosciences, Pennsylvania State University, University Park, Pennsylvania, 16802, USA

IAN B. LAMBERT
Secretariat, Resource Assessment Commission, Queen Victoria Terrace, Canberra City A.C.T., 2600, Australia

JERE H. LIPPS
Museum of Paleontology and Department of Integrative Biology, University of California, Berkeley, California, 94720, USA

DONALD R. LOWE
Department of Geology, Stanford University, Stanford, California, 94305, USA

CAROL MANKIEWICZ
Departments of Biology and Geology, Beloit College, Beloit, Wisconsin, 53511, USA

CARL V. MENDELSON
Department of Geology, Beloit College, Beloit, Wisconsin, 53511, USA

TOBY B. MOORE
Department of Earth and Space Sciences, University of California, Los Angeles, California, 90024, USA

DOUGLAS C. NELSON
Department of Microbiology, University of California, Davis, California, 95616, USA

ANNA C. PALMISANO
Environmental Safety Department, Procter and Gamble Company, Ivorydale Technical Center, Cincinnati, Ohio, 45217, USA

BEVERLY K. PIERSON
Department of Biology, University of Puget Sound, Tacoma, Washington, 98416, USA

BRUCE N. RUNNEGAR
Department of Earth and Space Sciences and Institute of Geophysics and Planetary Physics, University of California, Los Angeles, California, 90024, USA

J. WILLIAM SCHOPF
Department of Earth and Space Sciences and Center for the Study of Evolution and the Origin of Life, Institute of Geophysics and Planetary Physics, University of California, Los Angeles, California, 90024, USA

J. JOHN SEPKOSKI, JR.
Department of the Geophysical Sciences, University of Chicago, Illinois, 60637, USA

HARALD STRAUSS
Institut für Geologie, Ruhr-Universität Bochum, Postfach 102148, 4630 Bochum 1, Germany

ROGER E. SUMMONS
Bureau of Mineral Resources, Geology and Geophysics, G.P.O. Box 378, Canberra City, A.C.T. 2601, Australia

KENNETH M. TOWE
Department of Paleobiology, U.S. National Museum, Smithsonian Institution, Washington, D.C., 20560, USA

JÁN VEIZER
Institut für Geologie, Ruhr-Universität Bochum, Postfach 102148, 4630 Bochum 1, Germany and Derry Laboratory, Ottawa-Carleton Geoscience Center, University of Ottawa, K1N 6N5, Canada

MALCOLM R. WALTER
M. R. Walter Pty. Ltd., P.O. Box 258, Northbridge, 2063, Australia

DAVID M. WARD
Department of Microbiology, Montana State University, Bozeman, Montana, 59717, USA

Preface

J. WILLIAM SCHOPF   CORNELIS KLEIN

The 41 contributors to this monograph are members of the Precambrian Paleobiology Research Group (PPRG), a multidisciplinary, international consortium of physical and biological scientists actively investigating the Precambrian history of the earth, and of life. Initially organized more than a decade ago, the PPRG includes geologists and biologists of diverse backgrounds and expertise: sedimentologists, petrologists, geochemists, and atmospheric scientists, microbiologists, phycologists, molecular phylogenists, and paleontologists. In 1979—80—supported by NASA and by funds provided by the National Science Board's 1977 Alan T. Waterman Award to J. W. Schopf—the group worked together at UCLA for 14 months on problems centered on the paleobiology of the Archean, the earlier half of Precambrian time encompassing the initial 40% of the history of the planet (viz., extending from 4500 to 2500 Ma ago). This first PPRG project culminated with publication in 1983 of *Earth's Earliest Biosphere* (J. W. Schopf, ed.; Princeton University Press), a monographic work concerned with the origin and evolution of the Archean biota and the development of the Archean environment.

Stemming from what was viewed by the group as the success of this initial venture—a remarkable educational experience that both broadened and deepened the perspectives of the two dozen participants—in the mid-1980s the group elected to undertake a follow-on study, a sequel to be devoted to the paleobiology of the Proterozoic, the latter half of Precambrian time extending from the close of the Archean, 2500 Ma ago, to the first appearance in the geologic record of skeletonized invertebrate metazoans, about 550 Ma ago, that marks the beginning of the richly fossiliferous (and paleobiologically well-documented) Phanerozoic Eon of earth history. This monographic work—like its forerunner, also encompassing some 40% of total earth history—is the result of that effort.

From the time of inception of this Proterozoic project, there has been enthusiastic agreement within the PPRG that the field is ripe for such a venture. As an area of active scientific effort, Proterozoic paleobiology is little more than two decades old. During this period, the field has grown, markedly but more or less willy-nilly. A monographic treatise bringing together in one place a summary of past accomplishments in the field, contributing new data to the present development of the science, and providing a firm foundation for future growth and advance, seemed very much in order. The project was thus designed to address the past, present, and future of Proterozoic paleobiological studies:

The past. As research plans were formulated, it was agreed early that a principal goal of the venture would be to amass and synthesize the large body of relevant paleobiologic data that has become available over the years from previous studies. As part of that task, effort would be devoted, as feasible, to "separate the wheat from the chaff" so that future investigators might be less encumbered than those of the present by mistakes of the past.

The present. Detailed analyses of the enormous body of data available seemed certain to provide new insight and new understanding (e.g., of the patterns and rates of Proterozoic biotic evolution; of Proterozoic trends in the carbon- and sulfur-isotopic ratios of sedimentary materials; of cyclicity in Proterozoic geologic processes and their relation to environmental and biologic change). Moreover, the project presented a cardinal opportunity to make new measurements and to undertake new investigations—in short, to contribute new data. A second objective was thus to generate new data and new analyses bearing on important questions in the field.

Building toward the future. Finally, a venture such as this would be of limited value were it to ignore the certainty of the future growth of this multidisciplinary science, a future that is bound to incorporate many of the new approaches and new tools currently being developed in ancillary fields. A third goal was thus to build toward the future, by (*i*) placing special emphasis on new or relatively neglected aspects of the science that present promising avenues for future advance (e.g., studies of "biomarkers" and the extract organic geochemistry of Proterozoic sediments; of the molecular phylogenetics of extant early-evolving lineages; of geomagnetic-based Proterozoic paleogeography; of the ecology of modern microbiological analogues of Proterozoic stromatolitic communities); and (*ii*) explicitly highlighting major, current, unsolved problems in the field.

This two-part monograph is structured to meet each of these

goals. Syntheses of past progress in the field are presented largely in Part 2, a series of Chapters that include tabulations of the organic geochemical and isotopic data now available for Proterozoic organic matter (Chapters 17 and 19) and that summarize and evaluate previously reported occurrences of Proterozoic microfossils (Chapters 22 and 24) and of various types of Proterozoic megascopic trace and body fossils (Chapter 23), with the sources of these and related data specified in a bibliography containing some 3,500 entries. New data and new analyses are presented in Part 1, a series of research papers that spans virtually the entire gamut of this newly emergent science. And important unsolved problems in the field are specifically addressed in the concluding discussions of each section of Part 1, discussions that it is hoped will serve as a "primer" for talented graduate students in search of significant research problems.

In addition to the goals noted above, a "hidden agenda item" must be acknowledged. For those actively investigating the paleobiology of the Precambrian, it is a striking, and somewhat discouraging fact that the progress of recent years has remained largely ignored in college-level textbooks and coursework. It is perhaps true that in some sense the Precambrian has only rather recently been "discovered," a post-World War II development spurred by the advent of improved radiometric dating techniques. Nevertheless, the Precambrian is no longer *terra incognita*; there is a notable gap between that which is known, and that which is taught. Indeed, although together the Archean and Proterozoic span more than four-fifths of the history of the planet and some 85% of the known history of life—and although the development of the present-day atmosphere, oceans, crustal regimes, ecology, and basic biologic makeup of the planet all date from this early, enormously long phase of earth history—the geology, biology, and evolution of the Precambrian world continue to occupy a minuscule fraction of relevant college curricula. Part of this neglect may be due to the absence of a handy "source-book" to provide the relevant information. But a major part almost certainly stems from a widespread lack of appreciation of the overall evolutionary continuum that unites the Precambrian with the more recent history of the planet, and of life. Coupled with the earlier PPRG monograph on Archean paleobiology, it is hoped that this treatise, devoted to *The Proterozoic Biosphere*, will begin to fill this gap.

As this Proterozoic project began to take shape, it soon became apparent that the initial, Archean, PPRG consortium would have to be substantially broadened—additional expertise would be required if the goals set by the group were to be achieved. Moreover, generation of new data would require substantial field work in order to augment the pooled sample collections available for study; this, in turn, would require significant financial support. Thus, over the 1983–87 period, PPRG membership was rapidly expanded to include the 41 contributors to this monograph; grant proposals were submitted, and support was obtained from NASA, the NSF and, especially, from the National Geographic Society which provided generous funding for numerous geological and biological field excursions; geological field work was carried out by subsets of the group worldwide (viz., in northwestern and northern Australia; South Africa; northern China; eastern Siberia; southern Sweden; southeastern Spain; southern Ontario, arctic, and north-central Canada; and in Michigan, Minnesota, Idaho, Montana, and Arizona, USA); a paleontology contingent traveled to Moscow, USSR, to study and photograph type specimens of Soviet Proterozoic microfossils; and biological field excursions, to investigate extant mat-forming microbial communities, were carried out in numerous locales (viz., in northern and southern Baja California, Mexico; North Island, New Zealand; Iceland; Christmas, Turks, and Caicos Islands; and in Wyoming and Utah, USA). During this five-year period, the entire group met yearly at UCLA, hosted by the Center for the Study of Evolution and the Origin of Life (CSEOL), where it carried out week-long scientific symposia and planning sessions.

Beginning in July, 1987, the data-generating phase of the project began in earnest: nearly half of the contributors to this book were in residence at CSEOL, gathering data and carrying out collaborative research, for periods ranging from 2–14 months; and the entire PPRG met quarterly, to assess progress, plan on-going studies, and to construct the extensive outline on which this monograph is based. Leadership was shared (K. M. Towe and J. W. Schopf each serving as PPRG Director for half of this phase of the project), and decisions were reached by group consensus. Among such decisions, it was determined that prime responsibility for each portion of this monograph would be delegated to the principal author of that Chapter, and that all other co-authors, regardless of their relative contributions, would be listed alphabetically. This intensive, collaborative, 14 month-long phase culminated in late August, 1988, with a three-day, CSEOL-sponsored, public symposium on "The Proterozoic Biosphere" at which preliminary results of the endeavor were first presented.

Table P-1 summarizes various measures comparing the first, Archean, PPRG project (1979–80) with the subsequent, Proterozoic study (1987–88) on which this book is based. Clearly, this Proterozoic project has been an enormous task, in comparison with the Archean venture involving twice as many participants, three times as many geologic samples, the generation of roughly three times as much geochemical data, and evaluation of nearly an order of magnitude more micropaleontological reports. Those involved in these efforts have benefited greatly from the experience, and it is particularly gratifying to note that of the 41 PPRG participants contributing to this Proterozoic monograph, six are present or former students of individuals involved in the initial, Archean, project. There is a continuity from the past, to the present, and into the future of this science to which the PPRG, via its activities and this book hopes to contribute.

Acknowledgments

A venture of this magnitude could not have occurred without the assistance and considerable cooperation of a large number of individuals, many of whom went well beyond the bounds of their usual duties in helping the PPRG accomplish its tasks. For many of the geological and biological field trips enumerated above, for example, local scientists and their parent research organizations played pivotal roles in assisting the

Table P-1 *Comparison of Archean and Proterozoic PPRG projects.*

Aspect of the Study	PPRG 1979–80	PPRG 1987–88
Principal Focus:	Archean (and Early Proterozoic)	Proterozoic (and Early Cambrian)
Geologic Time Considered:	4.5 to ~2.0 Ga	2.5 to ~0.5 Ga
Number of Contributors:	19	41
		(10 previous plus 31 new contributors)
Countries Represented:	4	8
	(Australia, Canada, Germany, USA)	(Australia, Canada, Denmark, Germany, South Africa, Sweden, USA, USSR)
PPRG Geologic Samples:	542	1,781
Analyses:		
del^{13}C$_{organic}$ and TOC	43 geologic units	152 geologic units
Kerogen Composition	23 geologic units	49 geologic units
Rock-Eval Characteristics	—	8 geologic units
Extractable Organics	—	5 geologic units
del^{13}C$_{carbonate}$	23 geologic units	4 geologic units
del^{34}S$_{sulfide}$	—	31 geologic units
del^{34}S$_{sulfate}$	—	2 geologic units
"Microfossiliferous" Units Evaluated	61 geologic units (21 Archean)	465 geologic units
Authentic Microfossils	333 taxonomic occurrences in 24 geologic units (6 Archean occurrences in 2 geologic units)	2,835 taxonomic occurrences in 395 geologic units
Resulting Product:		
Monograph Title	*Earth's Earliest Biosphere, its Origin and Evolution* (J. W. Schopf, Ed., 1983)	*The Proterozoic Biosphere, A Multidisciplinary Study* (J. W. Schopf and C. Klein, eds., 1991)
Contents	154 figs.; 67 tables; 1,700 references cited	318 figs.; 116 tables; 3,480 references cited
Subjects Addressed	(1) History of Precambrian paleobiology	(1) History of subdisciplines of Proterozoic paleobiology
	(2) Archean geology, plate tectonics	(2) Archean and Proterozoic geology, plate tectonics
	(3) Evolution of Precambrian atmosphere and oceans	(3) Evolution of Precambrian atmosphere and oceans
	(4) Archean organic and isotopic geochemistry	(4) Proterozoic organic and isotopic geochemistry
	(5) Archean (and Early Proterozoic) microfossils	(5) Proterozoic (and Early Cambrian) microfossils
	(6) Archean stromatolites	(6) Proterozoic stromatolites
	(7) Biochemical evolution	(7) Molecular phylogeny
	(8) Major evolutionary events	(8) Major evolutionary events
	(9) Formation of planet/origin of life	(9) Microbiology of modern mat-building microbial communities
		(10) Environmental models of banded iron-formation deposition
		(11) Proterozoic (and Early Cambrian) carbonaceous films, calcareous algae, metazoan body and trace fossils, and skeletal protists and metazoans
		(12) Late Proterozoic-Early Paleozoic paleogeography
		(13) Proterozoic biostratigraphy and paleobiogeography
		(14) Proterozoic (and Early Cambrian) patterns of biotic diversity and major extinctions

group, a generosity and hospitality that is deeply appreciated. Similarly, at UCLA, the technical and office staffs both of the Department of Earth and Space Sciences and of the Institute of Geophysics and Planetary Physics markedly contributed to the success of the project and, repeatedly and with much kindness, to the well-being of visiting PPRG members. As host of the project, CSEOL had principal responsibility for the welfare of these numerous visitors, a charge ably shouldered by the CSEOL staff, particularly by Administrative Assistant Richard Mantonya; and as unofficial hostess (and historian) of the group, Jane Shen-Schopf deserves heartfelt kudos. In addition, numerous UCLA students helped to carry out a wide variety of tasks, from the cutting of thin sections to the cataloguing of rock specimens, from library research to computer data entry. All of these individuals participated in this venture; all played a role; without their help, this project would not have come to fruition.

Finally, financial support making this project possible was provided by the National Aeronautics and Space Administration (NASA Grants NAGW-252 and NAGW-825, to the PPRG, and NGR 05007407, to J. W. Schopf), by the National Science Foundation (NSF Grants EAR-8516509, to the PPRG, and BSR-8613583, to J. W. Schopf), by the National Geographic Society (NGS Grants 302085 and 329386 to the PPRG) and, most generously, by the UCLA Center for the Study of Evolution and the Origin of Life. Acknowledgment of other grants and assistance is provided in many of the Chapters of this book. Additionally, the research of D. R. Lowe (Chapters 1, 2, and 4) was supported by NSF Grants EAR-8611686 and EAR-8896218, and by NASA Grants NAG-9344 and NCA-2332; the research of C. Klein on iron-formations, especially in South Africa (Chapter 4), was supported by NSF Grants EAR-8419161, EAR-845681, and EAR-8617805; and that portion of the editing for publication carried out during 1989 by J. W. Schopf at The Natural History Museum, London, U.K., was supported by a Fellowship from the John Simon Guggenheim Memorial Foundation.

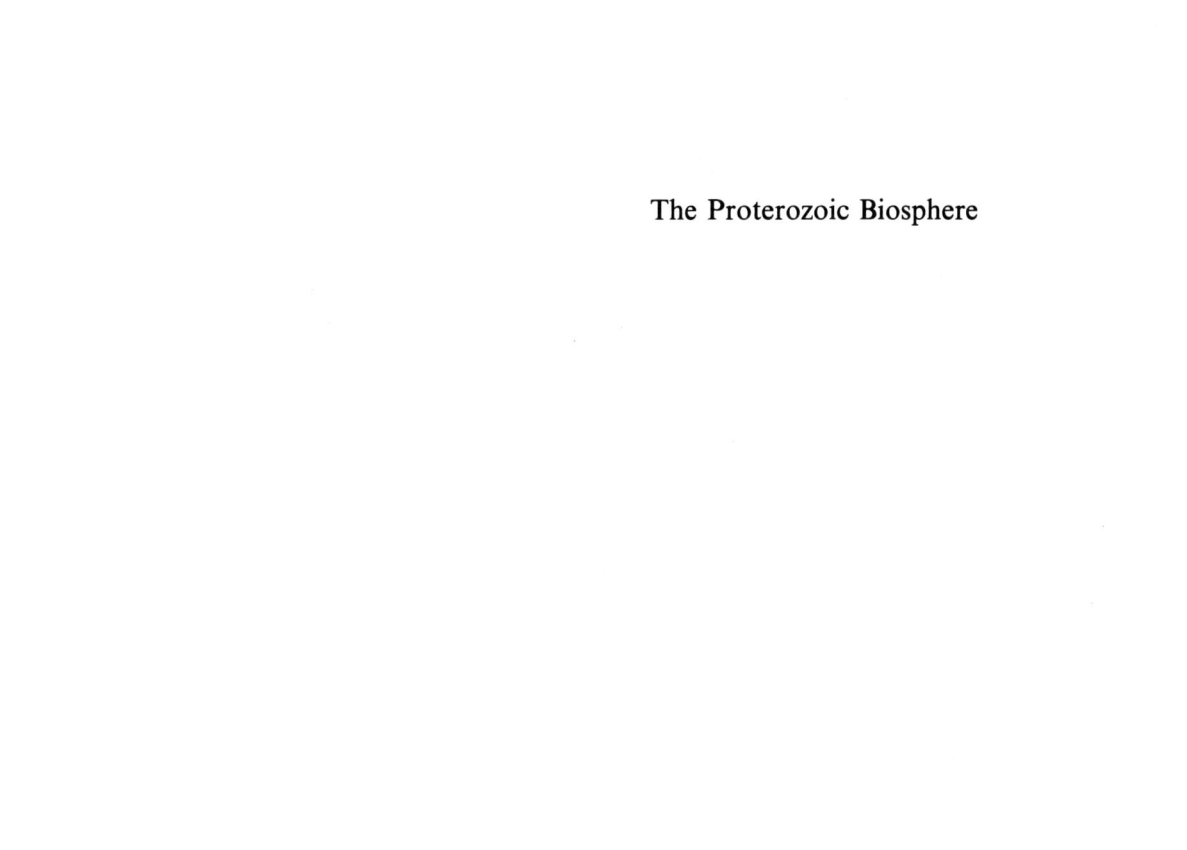

The Proterozoic Biosphere

PART I

1

Geology and Paleobiology of the Archean Earth

J. WILLIAM SCHOPF   SHERWOOD CHANG   W. GARY ERNST
HEINRICH D. HOLLAND   JAMES F. KASTING   DONALD R. LOWE

1.1

Introduction

J. WILLIAM SCHOPF

This monographic, two-part work is devoted to multifaceted consideration of the nature and evolutionary development of the Proterozoic biosphere and its environment, that is, of the evolution of life and the interrelated history of the atmosphere, oceans, and surficial planetary environment from the beginning of the Proterozoic Eon of geologic time, 2500 Ma ago, to the beginning of the Phanerozoic, a period spanning more than 40% of all of earth history.

Time divisions adopted in this work, largely after Harrison and Peterman (1980) and similar to those employed in the previous collaborative monograph of the Precambrian Paleobiology Research Group (Schopf 1983a), are summarized in Table 1.1.1. The boundary dates and age-related names there specified will be familiar to most readers. It should be noted, however, that considerable uncertainty exists regarding precise dates both of the Proterozoic-Cambrian "boundary" and of the base

Table 1.1.1. *Subdivisions of the Precambrian.*

Age (Ma)	Subdivisions Here Used			Age (Ma)	Subdivisions Proposed by the IUGS Subcommission on Precambrian Stratigraphy (Cowie et al. 1989)			Age (Ma)
	Eon	Era	Period		Period	Era	Eon	
	Phanerozoic	Paleozoic	Cambrian		Cambrian	Paleozoic	Phanerozoic	
550				550				
				?570				?570
			Vendian		"Neoproterozoic III"			
650		Late Proterozoic		650				650
					Cryogenian	Neoproterozoic		
				850				850
900				900	Tonian			
				1000				1000
					Stenian			
	Proterozoic	Middle Proterozoic		1200				1200
					Ectasian	Mesoproterozoic	Proterozoic	
				1400				1400
					Calymmian			
1600				1600				1600
					Statherian			
				1800				1800
		Early Proterozoic			Orosirian			
				2050		Palaeoproterozoic		2050
					Rhyacian			
				2300				2300
					Siderian			
2500				2500				2500
	Archean						Archean	

of the Vendian; throughout this work, the date of the former is assumed (rather arbitrarily) to be 550 Ma, whereas the base of the Vendian is placed at 650 Ma. For comparison with the subdivisions of the Proterozoic here used, Table 1.1.1 also lists subdivisions recently proposed by the IUGS Subcommission on Precambrian Stratigraphy (Cowie et al. 1989). At the time of this writing, this newly proposed system for Proterozoic subdivision has not been formally accepted by the appropriate IUGS body; to avoid potential confusion, and because the newly coined Period and Era names there listed are not yet in common usage, this proposed IUGS system has not been followed here.

Proterozoic earth history, of course, is part of a continuum that extends back through geologic time to the beginnings of the planet. The purpose of this Chapter is to consider this earlier, Archean, history of the planet and of life in order to set the stage for those Proterozoic developments addressed in some detail in following Chapters of the work. These introductory discussions need not be exhaustive; the geology and paleobiology of the Archean have been considered rather recently and at length in the earlier PPRG monograph *Earth's Earliest Biosphere* (Schopf 1983a). For the most part, therefore, the following portions of this Chapter present abbreviated, updated discussions of topics covered in greater depth in the earlier volume; however, new data are also included, particularly with regard to Archean paleobiology and the oldest cellularly preserved microscopic fossils now known (Section 1.5). Considered sequentially in the following discussions are the formation of the planet and the origin of life, the Archean geologic record, the environment of the Archean earth, and Archean paleobiology.

1.2

Formation of the Earth and the Origin of Life

JAMES F. KASTING   SHERWOOD CHANG

1.2.1 Origin and Development of the Early Atmosphere

Theories of the formation of earth and of the development of its atmosphere and oceans have changed markedly over the past half century. As a consequence, ideas about the chemical pathway leading to the origin of life, which are dependent on models of the early earth environment, have also been subject to change. More than sixty years ago, Oparin (1924) proposed that life originated under a strongly reducing atmosphere rich in hydrogen (H_2), ammonia (NH_3), methane (CH_4), and other saturated and unsaturated hydrocarbons. Reactions between these compounds produced additional, more complex organic compounds which eventually rained into the primordial ocean. There they underwent further transformations ultimately giving rise to the first living systems. A few years later Haldane (1928) also speculated on the early conditions necessary for the origin of life, but he gave carbon dioxide (CO_2) the dominant role as the atmospheric carbon source. Oparin's model was given a more solid theoretical foundation by Urey (1952) and received additional support from the successful laboratory synthesis of amino acids and other organic compounds under such conditions (Miller 1953; Miller and Urey 1959).

The concept of a highly reducing primitive atmosphere was challenged by Rubey (1951, 1955). Based on his analysis of the geological history of seawater, Rubey concluded that the volatile compounds making up the atmosphere and oceans came from degassing of the planet's interior. The volcanic gases released on the early earth, like those today, would have consisted primarily of relatively oxidized compounds: H_2O rather than H_2, CO_2 rather than CO, and nitrogen (N_2) rather than NH_3. Small amounts of more highly reduced compounds would have been present, but the bulk of the atmosphere would have had an almost neutral oxidation state.

Holland (1962) attempted to reconcile the two opposing viewpoints by suggesting that the primitive atmosphere went through two stages: a highly reducing Stage I; and more neutral, but still anoxic, Stage II. In Holland's model, as in Urey's, the earth accreted cold and core formation took place gradually as the interior of the planet was heated by the decay of radioactive elements. The volcanic gases released prior to core formation would have been highly reduced as a consequence of the presence of metallic iron in the upper mantle. Stage I could therefore have lasted for perhaps half a billion years, if core formation was indeed a slow process.

Today, all of these previous ideas about the origin and evolution of the early atmosphere appear unrealistic, because the earth is thought to have formed much more rapidly than in Urey's model. The currently accepted theory of accretion is that of Safronov (1969). His estimate for the length of time required for this process to occur is between 10^7 and 10^8 years. If the earth accumulated this fast, its interior would have been heated to several thousand degrees celsius by gravitational energy released by infalling planetesimals (Kaula 1979). The core would probably have formed as the planet was accreting (Stevenson 1983). The atmosphere should have begun to accumulate once the planet had reached a sufficient size (0.3 of its final radius) for infalling planetesimals to have been devolatilized on impact (Benlow and Meadows 1977; Lange and Ahrens 1982). Any atmosphere of impact origin would have contained H_2, H_2O, CO, CO_2, and N_2 in proportions dependent on the overall redox state of the gas cloud generated by vaporization of the target and projectile. Water, the most abundant volatile compound, was initially present as steam. The infrared blanketing (or "greenhouse") effect of this dense steam atmosphere would have raised the earth's surface temperature to over 1200°C, above the solidus for typical silicate rocks (Matsui and Abe 1986; Zahnle et al. 1988). This steam atmosphere and accompanying magma ocean should have persisted for several tens of millions of years, until the frequency of impacts dropped off sufficiently to allow water vapor to condense. Alternatively, if the impactors were large and relatively infrequent, as suggested by Wetherill (1985), the oceans may have rained out and been revaporized several times during the latter stages of accretion. Very large impacts, such as the one thought to have formed the Moon (Hartman et al. 1986), may have stripped off the atmosphere and oceans altogether, forcing the entire process of atmospheric formation to begin anew.

1.2.2 Timing and Environmental Setting of the Origin of Life

The considerations outlined above seem to rule out the first hundred million years of planetary history as a likely time for life to have originated. It would appear that the first biological systems must have post-dated the main accretionary period. As discussed further in Section 1.4, the post-accretionary environment may still have been a difficult place for life to gain a foothold. Intense bombardment apparently continued until at least 3.8 Ga ago, based on the lunar record (Shoemaker 1982). Alternatively, the clustering of lunar crater ages around 3.8 Ga may represent a period of late, heavy bombardment caused by the breakup of some large, inner-solar-system object. Estimates of the cratering frequency can be used in conjunction with educated guesses about the time scale for prebiotic synthesis to speculate about the timing of the origin of life. Maher and Stevenson (1988) suggest that life could have originated between 4.0 and 4.2 Ga at the mid-ocean ridges, or 3.7 to 4.0 Ga at the surface. (The ocean floor is protected from all but the very largest thermal insults by the insulating layer of water above it.) A related approach is to estimate when the last impact capable of vaporizing the entire ocean would have occurred. Such an event would almost certainly have sterilized the surface of the planet. According to Sleep et al. (1989), at least one ocean-vaporizing impact should have occurred after 4.3 Ga, and the probability for the occurrence of such an event remains significant until as late as 3.8 Ga. Thus, life may not have originated until just before Isua time. (The oldest known sedimentary rocks, found at Isua, west Greenland, were formed about 3.8 Ga ago.) In this case, the development of living systems from chemical precursors would have taken less time than previously thought.

Where life originated remains unknown, but it is almost certain that it occurred in an environment maintained far from equilibrium, perhaps at a geophysically active boundary region where fluxes of matter and energy established steep gradients in physical and chemical properties. These regions include fumaroles and volcanic vents on continents and continental shelves, deep-sea plate spreading centers, island-arc volcanic vents, the land-air interface, and the sea-air interface. On the prebiotic earth, a multitude of self-generated dynamic structures must have existed, differing in physical and chemical composition, size, distribution, abundance, and lifetime against dissipation. Then, as today, such structures would have encompassed a range from those as small and ephemeral as bubbles and vesicles at the sea-air interface, to ones as large and persistent as tectonic plates. The ability of a planet to produce and sustain a complex hierarchy of such systems over geologic time may be a prerequisite for the origin and maintenance of life (Chang 1988).

Chemical evolution may have begun in the atmosphere. Yet despite the importance of the composition of the atmosphere in determining the nature and abundances of the organic compounds that could have been formed in it, the oxidation state at the time when life originated remains a matter of dispute (Chang et al. 1983). Although Rubey's N_2-CO atmosphere is favored by geochemists (e.g., Walker 1977; Holland 1984), some organic chemists have difficulty in reconciling the apparent inefficiency of prebiotic synthesis in such a gas mixture with the perceived need to produce a high abundance of organic matter for chemical evolution. What production rates were actually needed of what kinds of compounds are, of course, unknown. The expected yields in some synthetic processes, for example, in electric discharges, (Miller and Schlesinger 1984; Stribling and Miller 1987), would have been considerably smaller than in the Oparin/Urey model atmosphere. Nonetheless, because Rubey's model atmosphere lacked free O_2, prebiotic syntheses of organic compounds could still have occurred, although perhaps involving different starting materials, energy sources, and locales than supposed in the Oparin/Urey model. At least one important biological precursor molecule, formaldehyde (H_2CO), can be produced from CO_2 and H_2O in a Rubey-type atmosphere, either photochemically (Pinto et al. 1981) or in lightning discharges (Miller and Schlesinger 1984). Hydrogen cyanide (HCN) can be formed photochemically, given a trace amount of methane in the atmosphere (Zahnle 1986); it could also have been generated by lightning if the atmospheric C:O ratio was unity or higher (Chameides and Walker 1981). This latter situation might have been realized if atmospheric CO_2 was reduced to CO by reactions with metallic iron brought in by late impacts.

Other sources of organic matter could have offset some of the supposed inefficiencies in atmospheric synthesis mechanisms. Cometary or asteroidal impacts might have delivered intact organic matter to the primitive earth (Oro 1961; Chang 1979). Perhaps more importantly, organic synthesis need not have taken place in the atmosphere; it could have occurred in the ocean. Despite uncertainties regarding the early composition of seawater, it is almost certain that the ocean's reducing capacity in the form of dissolved ferrous iron would have been orders of magnitude higher than that of the atmosphere. In this context it is significant that near-UV irradiation of bicarbonate in solutions of ferrous iron at neutral pH has been shown to provide an efficient synthesis of formaldehyde and other organic compounds (Borowska and Mauzerall 1988). This photochemical synthesis, simulating reactions at the seawater-air interface, may be viewed as the first step in the chemical evolution of photosynthetic organisms.

1.2.3 Chemical Evolution and the Origin of Life

Although it is not possible to specify a universally accepted sequence of transformations, the chemical evolution of living systems is usually assumed to have occurred in stages beginning with molecular simplicity and leading to complexity of function (Steinman and Kenyon 1969; Miller and Orgel 1974; see also Shapiro 1986, and Thaxton et al. 1984, for discussions of the limitations of the chemical evolution hypothesis). In a typical chemical evolution model, simple organic compounds formed in the atmosphere or oceans are transformed into biomonomers (e.g., amino acids, nucleotides) in the oceans. Condensation reactions convert the biomonomers to oligomers and polymers (e.g., peptides and polynucleotides) with rudimentary catalytic properties. Encapsulation of these polymers within a primitive membrane limits matter and energy exchange with the environment. Further development within the membrane-bound environment eventually affords many polymeric subunits (i.e., enzymes and nucleic acids) acting cooperatively to achieve the complex functions of metabolism and self-replication. The end product of such chemical evolution is generally thought to have been a heterotrophic ancestral organism. Alternatively, life may have begun along protocellular lines with synthesis, growth, and replication of membranes preceding synthesis of proteins and nucleic acids. In this model (Morowitz et al. 1988), the membranes provide sites for development of energy transduction mechanisms and photosynthetic growth processes and as well as microenvironments for synthesis of biopolymers. This protocellular model would have led to a photoautotrophic primordial organism. Other primitive bioenergetic systems have been suggested by Baltscheffsky (1986) and Weber (1978). In addition to organic life forms, crystal genes and clay mineral systems (Cairns-Smith 1982) as well as mineral-organic systems (Orgel 1986) have also been proposed as the first organisms.

Perhaps the most significant new insight bearing on theories for the origin of life is the discovery that ribonucleic acids (RNA) can function as catalysts as well as carriers of genetic information (Cech et al. 1981; Cech 1985). Catalysis by RNA lends added significance to the non-enzymatic template-directed RNA polymerase model of Inoue and Orgel (1983). Prior to this discovery, it was presumed that the primitive self-replication process required both proteins and nucleic acids, as in all modern cells. The recognition of RNA catalysis means that RNA may be endowed with the dual capacity to carry genetic information as well as catalyze the replication of the information. This simplification of the replication process means that RNA in principle can replicate itself (Pace and Marsh 1985; Orgel 1986a). A demonstration of this capability would be a remarkable achievement.

1.2.4 Paleoenvironmental Consideration

Numerous paleoenvironmental questions regarding the origin of life are as yet unresolved. One of the most fundamental is whether the early earth was hot or cold. Arguments can be made for either alternative. Proponents of a cold early earth point out that solar luminosity was at least 25 to 30% lower at this time (Gough 1981). A cold environment is favored by some organic chemists as being most favorable for chemical evolution (Miller and Orgel 1974). On the other hand, large amounts (10 to 20 bars) of carbon dioxide may have been present in the atmosphere at the close of accretion (Holland 1984) and could conceivably have remained there for hundreds of millions of years on an ocean-covered early earth (Walker 1985). The mean surface temperature under these conditions would have been 80° to 100°C (Kasting and Ackerman 1986). Although the oceans would have been warm, boiling would have been prevented by the large vapor pressure of water in the overlying atmosphere. (Liquid water would be stable up to its critical temperature, 374°C or even higher if one takes into account the effect of salinity.) Thus, such an atmosphere would not have evolved into a runaway greenhouse of the type that may have occurred on Venus (Goody and Walker 1972). Whether such a warm, dense, CO_2 atmosphere actually existed is uncertain, but it cannot be ruled out on the basis of any obvious physical principles. Clearly, the pH and concentrations of CO_2, bicarbonate, metal ions, and other solutes in the oceans would have been influenced by such an atmosphere, but how they in turn would have affected abiotic organic syntheses and the process of chemical evolution remains to be determined. Whether life could have originated under those conditions is unknown.

Another environmental parameter that might have influenced the origin and early evolution of life was the flux of solar ultraviolet (UV) radiation. In the absence of atmospheric ozone, or some equivalent UV absorber, radiation between 200 and 300 nm would have reached the surface virtually unattenuated. Whether this would have posed a serious problem for early life is a matter of debate. Some researchers (e.g., Margulis et al. 1976; Rambler and Margulis 1980) have argued that a UV screen was not necessary: organisms could have protected themselves by living at a sufficient depth underwater, by forming mats, or by developing internal UV-protection mechanisms. On the other hand, living at depth underwater might not have been an acceptable strategy if the earliest organisms were photosynthetic. It is not known whether communal lifestyles and biological protection mechanisms might have taken a long time to develop (Sagan 1961, 1973); or whether life was communal at the outset and already endowed with UV-protection mechanisms. Sagan suggests that life may have formed under a UV screen composed of organic compounds (primarily aldehydes) formed from methane photolysis. This is an unlikely hypothesis because the methane abundance was probably low and because highly soluble aldehydes would have been rapidly rained out of the early atmosphere. Another candidate for a UV screen is elemental sulfur vapor produced from photochemical reactions of volcanic SO_2 and H_2S (Kasting et al. 1989). Sulfur ring molecules (S_8, in particular) exhibit high absorption between 200 and 300 nm and should also be relatively stable against photolysis. In order for a sulfur UV screen to have been effective, however, the surface temperature of the early earth must have been at least several tens of degrees higher than today; otherwise, most of the sulfur vapor would have condensed to form particles. A CO_2-rich atmosphere with a surface pressure of several bars could have provided the necessary warmth.

1.2.5 Geologic and Paleobiologic Considerations

The biogeological record of early life does not extend far enough into the past to provide much insight into the unresolved issues addressed above. As is discussed in Section 1.5, sediments 3.3 to 3.5 Ga old from South Africa and Western Australia contain the earliest evidence of stromatolites (reviewed in Walter 1983) and microfossils (Schopf and Walter 1983; Schopf and Packer 1987). Analogies between ancient and present-day stromatolitic microbial communities suggest that the early ecosystems included both heterotrophic and autotrophic organisms, and that some of the latter were phototactic and probably capable of using sunlight directly as their energy source. Sedimentological studies (Byerly et al. 1986; D. R. Lowe pers. comm. to J.F.K.) indicate that stromatolites grew in narrow, shallow water zones along shorelines of volcanic platforms where they were subject to periodic agitation by waves or currents and occasional rapid burial by lava flows.

Localization of stromatolitic or microbial mat growth may have been influenced by hot springs or hypersaline conditions. The overall impression is of a shallow, marine, hydrothermal environment dominated by episodic volcanism. It is highly probable that similar environments existed earlier than even 3.8 Ga ago. Hydrothermal and volcanic environments have been suggested as the loci for the origin of life by a number of workers (e.g., Fox and Dose 1982; Sylvester-Bradley 1976; Corliss et al. 1981). It is especially noteworthy that the contemporary microorganisms with the most ancient lineages based on molecular phylogenies are anaerobic, thermophilic, sulphur-metabolizing, archaebacteria (Woese 1987). These organisms were isolated from hot springs and hydrothermal vents where they thrive at temperatures up to 105°C. Their habitats are consistent with notions about the early locales where life may have arisen, but whether life actually originated in such environments remains an open question.

Clearly, more research is needed to diminish the large gaps that exist in our understanding of the relationship between models of early planetary conditions, models for the origin of life, and the biological and geological records of early life.

1.3

The Archean Geologic Record

DONALD R. LOWE   W. GARY ERNST

1.3.1 Formation and Evolution of the Pre-Archean Earth

Isotopic data from meteorites and the Moon indicate that the solar system is approximately 4.6 to 4.5 Ga old (Patterson 1956; Papanastassiou and Wasserburg 1971). Accretion of planetesimals in progressively more peripheral regions of the solar nebula (Safronov 1972; Cameron 1978) gave rise to condensed bodies orbiting the newly formed sun. Although initial accumulates may have been rather cold, burial of heat during successive impacts probably resulted in increasingly hot planetary accretion (Wetherill 1976; Kaula 1979). This primordial phase of formation of the inner solar system probably was completed by about 4.5 Ga, at which time earth differentiation began.

The relative abundance of long-lived radioactive isotopes of elements such as U, K, and Th (Lee 1970), and the initial presence of short-lived radioisotopes, would have resulted in additional planetary self-heating during and directly following the accretionary stage. Because iron melts at lower temperatures than silicates under high confining pressure, and because silicate and Fe(+Ni) liquids are immiscible, a dense metal-rich melt must have formed within the earth. Due to gravitational instability, the iron-rich liquid would have migrated downward, displacing silicates upward, producing a molten metallic core overlain by a largely solid silicate mantle (Elsasser 1963). Infall of core material evidently accompanied later stages of planetesimal accretion (Stevenson 1983). The conversion of potential energy to heat would have liberated substantial amounts of additional thermal energy, contributing to the fusion and more complete separation of core and mantle (Birch 1965; Flaser and Birch 1973), and, possibly, partial melting of near-surface portions of the latter (Ringwood 1979). Slight cooling since this earliest stage of planetary differentiation has resulted in solidification of the inner core. By analogy with the Moon, the separation of Fe(+Ni) core and encompassing, largely solid, silicate mantle, and possibly the generation of a partially molten upper mantle, were completed by about 4.4 Ga at the latest.

On earth, no crustal rocks have survived from this primeval stage of terrestrial evolution; the most ancient rocks now known, from the Slave Province of northern Canada (Bowring et al. 1989), are no older than about 4.0 Ga, and 3.8 Ga-old rocks have long been known from the Isua Supracrustal Belt of western Greenland (Moorbath et al. 1972; Moorbath 1975). Froude et al. (1983), Myers and Williams (1985), and Compston and Pidgeon (1986) have reported 4.1 to 4.2 Ga old zircons from Precambrian metasediments in the Narryer Complex, Western Australia, so (now-obliterated) continental crust or silicic volcanic complexes had begun to form by this time.

Considerably more evidence of the initial stages of crustal formation are preserved on the Moon. The ferroan anorthositic impact breccias of the lunar highlands contain rock fragments at least as old as 4.4 Ga, and the highlands are strongly pulverized, attesting to continued, intense meteorite bombardment there, and presumably on the other inner planets, to about 3.9 Ga (Wetherill 1972; Burnett 1975; Shoemaker 1984). The less heavily cratered lunar maria are floored by Early Archean basalts. Thus, the thermal budget of the Moon allowed local production of mafic magma within the lunar mantle as late as 3.8 to 3.3 Ga ago. The more ancient highlands appear to have formed from the accumulation of calcic plagioclase which rose buoyantly toward the surface of the hypothesized lunar magma ocean (e.g., Longhi 1978; Smith 1979).

If a terrestrial magma ocean ever existed, its crystallization, except near the surface, would have been dominated by higher pressure phase equilibria than on the Moon because of the greater mass of the earth. Early crystallization of garnet rather than calcic plagioclase would have resulted in the settling of dense aluminous crystalline phases instead of flotation. The thin, transitory, mafic crust that would have solidified at the upper cooling surface, lying upon less-dense, partially molten material, would have foundered in a rapidly convecting mantle, undoubtedly aided by continued meteoritic bombardment. Therefore, dense, refractory ferromagnesian constituents should have been stabilized in relatively deeper portions of the gradually thickening, largely solid mantle of the primitive earth, with more fusible LIL (large ion lithophile), incompatible, and volatile elements concentrated toward the surface in a relatively silicic, globe-encompassing rind and dense primitive atmosphere (Hargraves 1976; Shaw 1976; Fyfe 1978). More permanent continental crust probably did not form during this early, hot stage because any sialic concentrates reaching or formed at the earth's surface would have continued to be reincorporated into the rapidly convecting, rehomogenizing mantle (Jacobsen and Wasserburg 1981). Outgassing must have been intense, with a considerable portion of the volatiles transferred from the mantle to the hydrosphere and atmosphere. By about 4.2 Ga at the latest, a solid lithospheric skin was in existence, portions of which were silicic, as required to explain the occurrence of 4.2 Ga-old zircons in Western Australia. However, these zircons could have been derived as easily from ancient felsic volcanic rocks as from ancient blocks of continental crust; they do not, in such sparse abundance within quartzites 3.4 to 3.2 Ga-old, provide solid evidence for the existence of pre-4.0 Ga old continental crust.

At 3.8 Ga and probably earlier, water-laid sediments were being deposited on the earth. The surface temperature, therefore, was less than the boiling point of water. Thus, however the initial cooling and chemical differentiation of the earth was accomplished, an outer rind of crust and overlying watery and gassy envelopes existed by 3.8 Ga (Jacobsen 1984). From this point in earth history onward, the evolution of the continental crust and its sedimentary and biological systems can be traced by direct reference to the extant, if fragmentary, geologic record.

1.3.2 Principal Archean Geologic Terranes

Archean rocks form cratonic nuclei, dispersed enclaves, and remobilized remnants within the Precambrian basements of all continents (Fig. 1.3.1). They represent three principal petrologic suites or terrane types (Windley 1977): (*i*) granite-greenstone terranes; (*ii*) craton cover and craton margin sequences; and (*iii*) high-grade gneiss terranes. In this Section are reviewed the supracrustal associations characterizing each of these suites and the conditions under which they formed.

1.3.2A Granite-Greenstone Terranes

Granite-greenstone metasedimentary supracrustal greenstone successions are intruded by coeval and younger granitoid plutons. Early intrusives are commonly tonalites, trondhjemites, and granodiorites, with later, post-tectonic units dominated by potash-rich granitic rocks. Greenstone assemblages formed throughout the Archean, but appear to represent three principal episodes of crustal growth. The oldest known greenstone belt is the Early Archean 3800 Ma Isua supracrustal sequence in west Greenland. The association originally included thick units of mafic volcanic rock, felsic conglomerate and breccia, banded iron-formation, chert, and immature clastic sediments (Allaart 1976; Nutman et al. 1984).

A younger and areally separate generation of late Early and

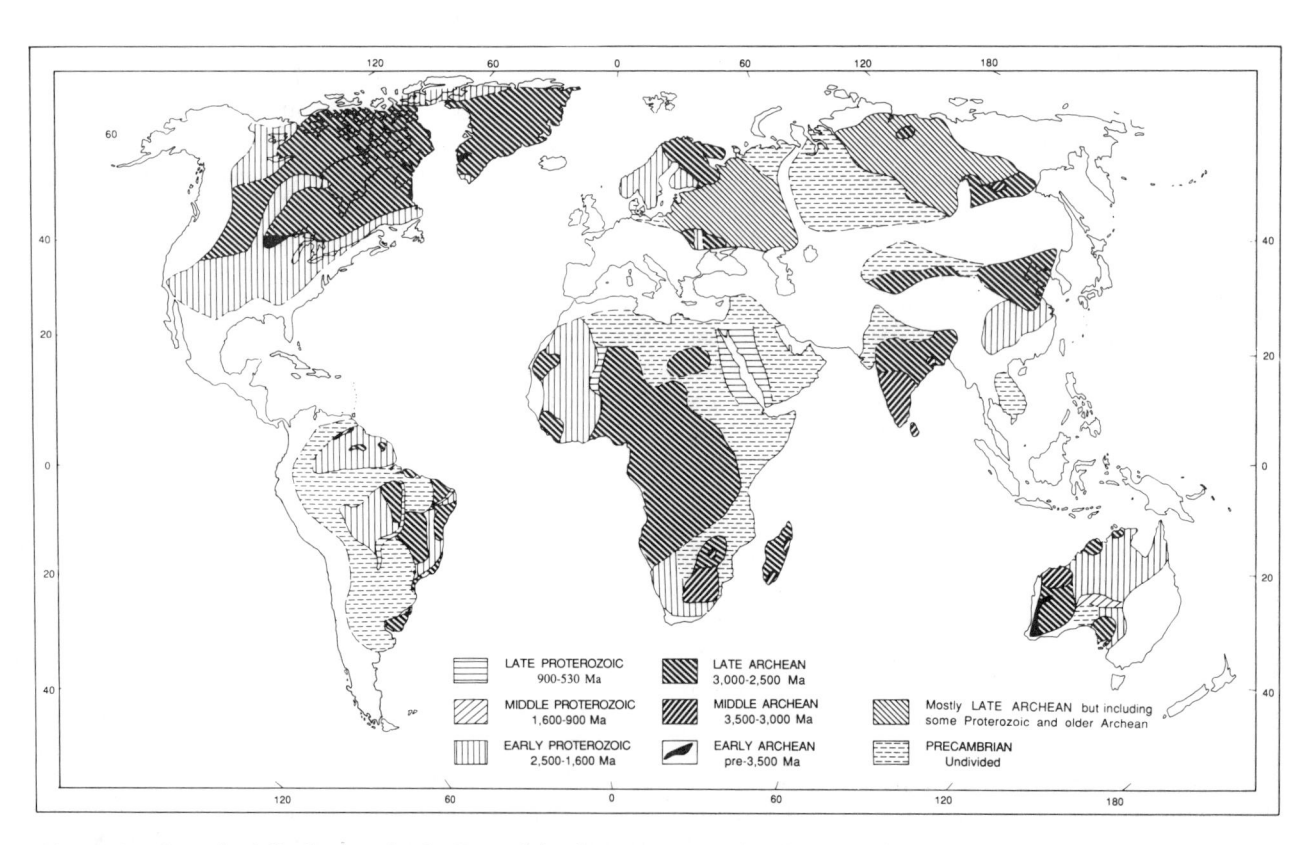

Figure 1.3.1. Generalized distribution of major Precambrian basement terranes based on the estimated ages of the initial greenstone and intrusive stage of crust formation. Most Middle and Late Proterozoic orogenic or "mobile" belts represent zones of older crust that have been either reworked during later igneous and metamorphic events or buried beneath younger fold and thrust belts.

Middle Archean, 3500 to 3200 Ma-old greenstone belts, termed the Barberton Cycle (Section 2.7), includes the Barberton Belt (Viljoen and Viljoen 1969a, b; Annhaeusser 1973) and related rocks in the Kaapvaal Craton, South Africa, and belts in the eastern Pilbara Block, Western Australia (Hickman 1983). Less studied greenstones of this age include the Sebakwian of Zimbabwe (Wilson et al. 1978) and supracrustal sequences in the Dharwars Craton of peninsular India (Naqvi and Rogers 1987). Although structurally complex, Middle Archean greenstone belts include stratigraphic sections from 10 to 15 km thick comprising lower, predominantly volcanic, and upper, predominantly sedimentary, divisions. The volcanic divisions, represented by the Onverwacht Group of the Barberton Belt (Viljoen and Viljoen 1969a, b) and the Warrawoona Group in the eastern Pilbara Block (Hickman 1983), make up the bulk of most sections. They consist of tholeiitic basalt, komatiitic basalt, and komatiite; subordinate felsic volcanic and volcaniclastic rocks; and thin layers of sedimentary rock. Sedimentary units, commonly silicified and/or carbonated, include tuff, current-worked volcaniclastic debris, carbonaceous chert, and evaporites (Lowe 1980a, 1982). As discussed in Section 1.5, these rocks contain the oldest reported stromatolites (Lowe 1980b; Walter et al. 1980; Byerly et al. 1986) and carbonaceous microfossils (Awramik et al. 1983; Walsh and Lowe 1985; Schopf and Packer 1987). They were deposited on low-relief simatic volcanic platforms under mainly low-energy, shallow-water, probably marine conditions during breaks in eruptive activity (Lowe and Knauth 1977; Barley et al. 1979; Lowe 1980a, 1982; Groves and Batt 1984).

The upper divisions of these greenstone sequences are composed largely of orogenic detrital and volcaniclastic sedimentary rocks, represented by the Fig Tree, Moodies, and Gorge Creek Groups. They were deposited under active tectonic regimes in environments ranging from alluvial to deep-marine (Eriksson 1977, 1978, 1979; Barley 1987). There is considerable evidence that many of these units were deposited in foreland basins developed adjacent to active fold and thrust belts (e.g., Jackson et al. 1987).

Late Archean greenstone belts, 3100 to 2600 Ma old, are present on all continents. This cycle of greenstone and crust formation has been termed the Superior Cycle (Section 2.7). The best studied belts are those in the Superior and Slave Provinces of Canada; the Yilgarn Block, Western Australia; and the Zimbabwe Craton. All contain thick mafic to locally komatiitic volcanic sequences interbedded with felsic volcanic and volcaniclastic units, calc-alkaline flow rocks, and sedimentary layers (Gee et al. 1981; Jensen 1985; Hallberg 1986; Card 1986). Sedimentary units are dominated by immature graywacke, shale, and chemically deposited chert, jasper, and iron-formation (Lowe 1980a, 1982; Groves and Batt 1984; Ojakangas 1985). In contrast to Middle Archean belts, thick detrital sedimentary units, although more abundant toward the top of most sections, are interstratified throughout the volcanic sequences (Hyde 1980; Ojakangas 1985; Blackburn et al. 1985; Hallberg 1986). These belts record volcanism and sedimentation under tectonically active conditions, mainly in deep-water, commonly turbiditic environments. Alluvial and, locally, shallow water fringes accumulated around emergent tectonic uplifts and volcanic islands. Stromatolites occur widely near or at the tops of the volcanic sequences in units representing transitory shallow-water carbonate-evaporite facies deposited mainly on subsiding felsic volcanic complexes (Henderson 1975; Hofmann et al. 1985b), and, less commonly, at the tops of and along unconformities within mafic units (Martin et al. 1980).

1.3.2B Craton Cover and Craton Margin Deposits

Craton cover and craton margin deposits accumulate, respectively, upon and around buoyant blocks of sialic crust. They represent a variety of basinal settings including stable cratonic platforms; anorogenic intracratonic basins; intracratonic extensional, strike-slip, transgressional, and foreland basins; and continental margin rises, terraces, and embankments (Ingersoll 1988). All are characterized by sediments derived by erosion of the underlying or adjacent sialic crust. Units deposited in tectonically active basins are generally dominated by clastic sediments. Coarser units contain abundant quartz and, if immature, K-feldspar. Stable cratonic sequences are dominated by quartzite-carbonate-shale associations.

Unmetamorphosed Archean cratonic sedimentary rocks are well-developed only as Late Archean sequences on Early and Middle Archean crustal blocks, including the Kaapvaal Craton of southern Africa and the Pilbara Block of Western Australia. The well-studied Archean and Early Proterozoic Kaapvaal sequence includes the Pongola (3000 Ma), Witwatersrand (2800 to 2700 Ma), Ventersdorp (2700 to 2500 Ma), and Transvaal (2500 to 2100 Ma) Supergroups, and Waterburg Group (2000 to 1700 Ma) and related strata (Tankard et al. 1982). This sequence records over 1000 Ma of predominently shallow-water and terrestrial sedimentation of quartzose clastic sediments, shale, carbonate, iron-formation, volcanic rocks, and, locally, evaporites. The tectonic settings were diverse, possibly including rifted shelf during Pongola sedimentation (Armstrong et al. 1982; Burke et al. 1985b); foreland basin deposition during Witwatersrand time (Winter 1987); an unstable shelf undergoing rifting (Tankard et al. 1982) or extension related to subduction (Winter 1987) or continental collision (Burke et al. 1985a), during accumulation of the Ventersdorp Supergroup; stable shelf, slope, and rise sedimentation in the Transvaal Basin (Beukes 1983); and unstable shelf and alluvial sedimentation in Waterburg time (Tankard et al. 1982), possibly within a foreland basin. A very similar, although less well resolved, sequence along the southern margin of the Pilbara Craton includes the Fortescue (ca. 2700 to 2600 Ma), Hamersley (ca. 2600 to 2400 Ma), and Turee Creek (ca. 2400 to 2300 Ma) Groups (Goode 1981; Morris and Horwitz 1983). The Fortescue contains thick basaltic and volcaniclastic units, many reflecting terrestrial sedimentation, and appears to document a regional episode of crustal fracturing and rifting, but the nature of the cratonic basin in which the overlying groups were deposited is controversial (Trendall and Blockley 1970; Morris and Horwitz 1983). The bulk of the sediments, however, reflects clastic-starved sedimentation of iron, carbonate, and volcanic ash under quiet, deep-water conditions (Morris and Horwitz 1983).

1.3.2C High-Grade Gneiss Terranes

Relatively large areas of complexly deformed, amphibolite- to granulite-grade gneisses occur in all Archean basement blocks. These include the Early to Middle Archean rocks of west Greenland, Labrador, and northern Scotland (Bridgwater et al. 1976); the Limpopo Mobile Belt of southern Africa (Robertson and du Toit 1981; Fripp 1983); the Narryer Gneiss Complex of Western Australia (Myers and Williams 1985; Myers 1988); the Peninsular Gneisses of southern India (Naqvi and Rogers 1987); the Napier Complex in Antarctica (Black et al. 1983); and large gneiss terranes in the Aldan Shield, USSR (Kazansky and Moralev 1981). These high-grade terranes commonly contain Early Archean as well as Middle and Late Archean supracrustal components intruded by granitoid plutons and overprinted by several generations of tectonism and metamorphism. The most common rocks are varieties of orthogneiss interpreted to be metamorphosed tonalitic, trondhjemitic, and granodioritic intrusive rocks (Windley 1977). Supracrustal units occur as deformed enclaves of amphibolite and mafic granulite, representing mafic volcanic rocks; mica schist, generally interpreted as metapelitic sediments; marbles and impure altered carbonate sediments; quartzites, formed by the recrystallization of chert and/or quartzitic sandstones; and quartz-magnetite iron-formation.

Two important questions are whether these high-grade supracrustal rocks represent fundamentally different types of sedimentary sequences than the lower-grade Archean terranes or, if similar to other Archean sequences, whether they represent mainly greenstone or cratonic rocks, or both (Windley 1977). The presence of the Isua supracrustals within the west Greenland terrane indicates unambiguously that the supracrustal protoliths included greenstone sequences, and petrologic, geochemical, and isotopic arguments suggest the widespread presence of greenstone components (Windley 1977; McLennan et al. 1984; Taylor et al. 1986). However, the proportion, thickness, and extent of carbonate bodies in some high-grade terranes exceed those of carbonate units in greenstone belts (Windley 1977), and REE (rare earth element) data (McLennan et al. 1984; Taylor et al. 1986) suggest a cratonic derivation for some detrital protoliths. Thick cross-stratified orthoquartzites are locally present and imply a cratonic depositional setting. Such units are generally poorly developed in greenstone belts but can be locally prominent, as in the Moodies Group of South Africa (Eriksson 1977). Zircons, both fresh and detrital, are abundant in felsic volcanic and epiclastic portions of greenstone belts as well as in craton-derived quartzites and sandstones, and, hence, in the absence of complementary isotopic studies, do not provide unambiguous means of discriminating between cratonic and greenstone source terranes and depositional settings. The weight of available evidence suggests that high-grade terranes include both greenstone and cratonic components differing only in metamorphic grade from their less-altered equivalents (Windley 1977; Taylor et al. 1986). This conclusion is supported by transitions in peninsular India and other shield areas from amphibolite-grade granite-greenstone belts to granulite facies tonalitic gneiss terranes (e.g., Mason 1973; Hubregtse 1980; Condie and Allen 1984; Hansen et al. 1984; Glikson 1984; Srikantappa et al. 1984).

1.3.3 Archean Crustal Evolution

The setting, processes, history, and influence of the formation of the earth's continental crust have been the focus of an enormous amount of recent research and discussion (see reviews in Moorbath and Windley 1981; Kröner 1981b, 1987; Taylor and McLennan 1985). Although many issues remain controversial, most investigators agree that components now making up the continental crust have been withdrawn magmatically from the mantle and assembled into cratons through volcanism, intrusion, sedimentation, and tectonism during relatively discrete crust-forming cycles. A considerable part of that activity occurred during the Archean.

Fundamental questions remain regarding: (*i*) the ensimatic versus ensialic setting of Archean as well as Proterozoic crustal formation; (*ii*) the nature and evolution of tectonic processes accompanying crustal generation and, specifically, the role of modern, mobilistic plate tectonic regimes; (*iii*) the rate of crustal formation over time; and (*iv*) the extent of crustal recycling, including both surficial gardening of older crust by later igneous, tectonic, and metamorphic events and complete destruction of continental blocks through re-homogenization within the mantle as a consequence of subduction or analogous processes. In this discussion, we will focus briefly on only the first two of these problems, those concerning the setting and the processes of crustal formation.

Although the geologies and histories of many Archean terranes remain poorly resolved, available data suggest that preserved Archean rocks record three major crust-forming cycles (Section 2.7):

(1) Early Archean, 3800 to 3500 Ma (Isua Cycle);
(2) Early to Middle Archean, 3550 to 3200 Ma (Barberton Cycle); and
(3) Late Archean (3100?) 3000 to 2600 Ma (Superior Cycle).

Each of these episodes can be correlated with formation of extensive tracts of silalic crust and, with the exception of the earliest event, with the metamorphic and magmatic overprinting of older terranes. Each cycle of crustal generation can generally be divided into four, in part overlapping stages: (*i*) an initial greenstone stage dominated by mafic volcanism and plutonism; (*ii*) an orogenic stage of deformation and metamorphism, commonly including calcalkaline volcanism and orogenic clastic sedimentation and generally culminating in a post-orogenic, post-metamorphic interval of erosion and cooling of the new crust; (*iii*) one or more post-orogenic periods of intracrustal melting and intrusion of late-stage potassic magmas; and (*iv*) eventual subsidence, submergence, and cratonic sedimentation.

1.3.3A Early Archean (Isua) Cycle

Early Archean crust survives in a number of high-grade terranes, including the Nain Province in Labrador and west Greenland, the Slave Province in the Northwest Territories, Canada, and the Western Gneiss Terrane in the Yilgarn Block, Western Australia. The Central Zone of the Limpopo Belt may also contain remnants of Early Archean crust, but the stratigraphic and geochronologic history of this area remains controversial. The oldest elements in Canada, west Greenland, and Western Australia include greenstone and/or intrusive rocks about 3900 to 3600 Ma-old. In the Western Gneiss Terrane, early intrusive and metamorphic events were evidently followed by erosion and deposition of cratonic sediments on the newly formed cratonic block 3500 to 3200 Ma (Myers 1988). No similar cratonic sediments are present in Greenland, but the Malene greenstone-like supracrustals accumulated between about 3550 and 3100 Ma and may include components derived from older continental rocks (Beech and Chadwick 1980; McLennan et al. 1984; Appel 1988). Early Archean gneisses in the Slave Province are in contact with younger gneisses, about 3100 to 3000 Ma old, and with Late Archean greenstones (Bowring et al. 1989).

1.3.3B Early to Middle Archean (Barberton) Cycle

The Middle Archean Barberton Cycle, represented by the Kaapvaal and Pilbara Cratons, involved greenstone volcanism and sedimentation (3550 to 3200 Ma), tonalitic and trondhjemitic intrusion (3500 to 3200 Ma), and late granitoid intrusion (3050 to 2700 Ma). The last regional metamorphic event was at about 3050 Ma in Barberton, roughly coincident with deposition of the first clear craton-cover sediments of the Pongola Supergroup. The Late Archean Superior Cycle included greenstone volcanism, sedimentation, and intrusion (ca. 3100 to 2650 Ma) and late intrusive activity (2600 to 2400 Ma). Between 15% and 40% of the present continental mass was probably in existence by 3000 Ma, and between 40% and 80% by 2500 Ma (Veizer 1976b; Lambert 1976; Jacobsen and Wasserburg 1981; DePaolo 1981; Taylor and McLennan 1985), although extreme estimates suggest that the volume of the Early Archean continental crust could have equaled or exceeded that of continents on the present earth (Armstrong 1981).

Continuing controversies concern the tectonic setting(s) of greenstone belts and the nature of underlying basement. The case for an ensialic or rift-related setting turns in part on the abundance of basalt and dacite/rhyolite bimodal volcanism, not only in Early (Nutman et al 1984), Middle, and Late Archean (Thurston et al. 1985) greenstone belts but also in younger, Proterozoic, greenstone or greenstone-like terranes (e.g., Boardman and Condie 1986; Ethridge et al. 1987). Bimodal volcanism alone, however, is ambiguous; felsic magmas can originate both by the anatectic melting of deep crustal rocks, including metamorphosed greenstone sequences, and by fractional crystallization of mafic magmas. Details of geochemistry and inferred magma-chamber dynamics also do not provide firm constraints on tectonic setting (Boardman and Condie 1986; compare discussions of Thurston et al. 1985, and Condie and Shadel 1984). In the final analysis, arguments are commonly based on the presumed difficulty of generating large volumes of felsic magmas through fractional crystallization alone (Thurston et al. 1985; Boardman and Condie 1986; Ethridge et al. 1987).

Overall, there is little or no direct evidence that older continental blocks served as basement or nuclei for the construction of Early or Middle Archean crustal units. The volcanic portions of the Middle Archean greenstone terranes record volcanism and sedimentation on tectonically quiescent, simatic volcanic platforms (Lowe 1980b, 1982; Groves and Batt 1984). Volcanic, intrusive, and sedimentary components carry strong mantle geochemical and isotopic signatures, and there is no isotopic, geochronological, or sedimentological evidence within the belts for the existence of substantially older underlying crust. The oldest gneisses in both the Barberton and Pilbara Blocks yield ages comparable to or only slightly older than the oldest ages within the supracrustal sequences (Barton et al. 1983; Cooper et al. 1982). Adjacent gneiss terranes containing somewhat older ages, such as the Ancient Gneiss Complex, are in structural, not stratigraphic contact with the main greenstone terranes.

These relationships suggest that both supracrustal sequences and the deep intrusive and metamorphic roots of the Middle Archean greenstone platforms formed as parts of the same cycle of crustal generation, possibly in part by metamorphism, melting, and mobilization of the lowest, hence oldest, parts of the volcanic piles themselves (O'Nions and Pankhurst 1978). These platforms appear analogous in tectonic style and depositional setting to modern oceanic islands, such as Hawaii, Iceland, or the Galapagos Platform, but the presence of calc-alkaline volcanic rocks and other aspects of the igneous geochemistry (e.g., Barley et al. 1984) suggest that all or parts of the sequences may represent subduction-related volcanic arcs. Their preservation can be attributed to the greater buoyancy and resistance to recycling of massive volcanic/sedimentary/intrusive edifices relative to surrounding, thinner oceanic crust. Only during the post-platform orogenic stage of its evolution was the Barberton Belt clearly influenced, probably through collision, by other crustal blocks, including the coeval to slightly older Ancient Gneiss Complex (Jackson et al. 1987).

1.3.3C Late Archean (Superior) Cycle

The best-studied Late Archean shields, the Yilgarn Block and Superior Province, are complex mosaics of elongate, commonly parallel greenstone belts, extensive plutonic tracts, and, in Canada, linear metasedimentary provinces. The parallelism of these domains to each other and to inferred syndepositional volcanic and structural trends has been interpreted to reflect craton formation through either ensialic rifting (Groves and Batt 1984; Davis et al. 1986; Hallberg 1986) or arc accretion (Langford and Morin 1976; Card 1986). In both blocks, the ages of initial greenstone volcanic/plutonic activity tend to decrease more or less regularly across the shields, normal to the strike of the belts. The oldest belts show initial volcanism and sedimentation at about 3100 to 3000 Ma and the youngest from about 2800 to 2650 Ma. Clastic units in the younger belts locally contain detritus eroded from adjacent, slightly older belts (Davis et al. 1986; Fletcher et al. 1984). Plutonism marked individual belts concurrently with or slightly younger than the volcanic and sedimentary development, but the terminal stages of intense plutonism and orogenesis, about 2700 Ma in the Yilgarn and 2700 to 2670 in the Superior Province, were apparently short, intense, and craton-wide, affecting belts of all ages throughout the blocks (de Laeter et al. 1981; Card 1986; Davis et al. 1986).

In a number of Late Archean terranes, including the Slave Province of Canada (Bowring et al. 1989), the Yilgarn Block of Western Australia (Gee et al. 1981; Fletcher et al. 1984), and the Zimbabwe Craton of southern Africa (Wilson et al. 1978), Late Archean greenstones are in contact with significantly older, commonly gneissic rocks. Some of these contacts have been interpreted, at least locally, as unconformities (Wilson et al. 1978; Kusky 1988), indicating deposition of greenstone volcanic rocks directly on older basement. In other areas, an early juxtaposition of older rocks and Late Archean greenstones is suggested by the presence of older xenocrysts and xenoliths in greenstone volcanic rocks or associated plutons (Compston et al. 1985; Davis et al. 1986). Although these occurrences indicate the existence of older rocks during granite-greenstone development, they do not provide unambiguous means of distinguishing among older greenstone-related felsic volcanic rocks, older continental basement, and blocks of older crust accreted or subducted during greenstone volcanism as the zircon sources.

Some areas show evidence that Late Archean tectonic, metamorphic, and/or intrusive events have affected both older rocks and the adjacent granite-greenstone terranes. In the western Yilgarn Block, Western Australia, both Late Archean greenstones and Early to Middle Archean Western Gneiss Terrane have been affected by 2700 Ma-old deformation and plutonism (Myers 1988). Although the Western Gneiss Terrane has been interpreted to represent a regional east-dipping sialic basement underlying the entire Yilgarn Block (Gee et al. 1981), recent studies suggest that the central and eastern parts of the Yilgarn Block are not underlain by large tracts of substantially older Archean crust (Fletcher et al. 1984; Gee et al. 1986) and that the Western Gneiss Terrane was accreted to the Late Archean granite-greenstone belts during or only slightly before the 2700 Ma terminal orogenic event (Myers 1988). Other Late Archean terranes, although less well studied, appear to exhibit similar age relationships (Hoffman 1986; Lobach-Zhuchenko et al. 1986; Nisbet 1987), and it seems likely that older Archean rocks juxtaposed with Late Archean greenstones commonly represent accreted terranes rather than underlying basement.

The common presence of calc-alkaline volcanic components, especially within the upper parts of the greenstone sequences (Jensen 1985; Hallberg 1986), the abundance of volcanic and intrusive suites bearing a strong chemical similarity to those developed along Phanerozoic convergent margins (Windley 1981; Kröner 1981; Jensen 1985), and the widespread evidence for compressional, nappe-forming events during deformation (Lamb 1984; Myers 1984; Stowe 1984), are consistent with the inference that Late Archean greenstone belts evolved mainly along convergent plate margins.

Most if not all Late Archean granite-greenstone belts probably formed by arc accretion and the progradation of accretionary complexes (Langford and Morin 1976; Blackburn 1980; Hoffman 1986). Some volcanic arcs probably developed as continental arcs along the margins of older Early and Middle Archean continental or microcontinental blocks or on continent-like crust made up of older, deformed and metamorphosed arc volcanic-intrusive complexes. The resulting crustal blocks are mosaics of heavily deformed, intruded, and metamorphosed arc volcanic-intrusive suites and, locally, older basement to continental arcs; intra-arc, backarc, and forearc basin sedimentary assemblages; subduction complexes including seamounts and older autochthonous cratonic and arc fragments; and deformed post-accretionary orogenic sediment wedges. There is little evidence that any of the major Archean greenstone terranes developed as ensialic volcanic sequences or within continental rifts.

Archean crustal evolution was probably controlled by the same basic thermally driven plate tectonic regime that governed lithospheric evolution throughout the Phanerozoic. Contrasts between Archean and younger terranes in terms of preserved rock types, magmatic and depositional styles, and tectonic development, probably reflect differences in the expression of tectonic processes rather than fundamental differences in style. The volcanic portions of the best studied Middle Archean belts formed as large, oceanic volcanic islands and arcs that were delaminated from subducting oceanic lithosphere and accreted to growing, juvenile crustal blocks because of their thickness and buoyancy. Younger, Late Archean belts were constructed mainly from arc and adjacent sedimentary/volcanic terranes through an extended process of accretion. Once formed and removed from sites of active arc magmatism and accretion, the new cratons underwent erosion and, generally, episodic intracrustal melting (Taylor and McLennan 1985). After intervals of 100 to 300 Ma, they had cooled, subsided, and been reduced by erosion sufficiently to become platforms and shelves hosting a subsequent generation of cratonic sediments.

The changing patterns of crustal evolution from Early to Late Archean probably reflects the increasing scale of upper mantle circulation as a consequence of the declining thermal budget of the earth (Ernst in press), as schematically shown in Figure 1.3.2.

1.3.4 Conclusions

Because advective heat transfer has almost certainly existed since formation of the earth's core 4.5 to 4.4 Ga ago, mantle circulation has probably guided if not driven the evolution of the earth's crust and lithosphere throughout geologic history. The existence of mineral grains at least 4.2 Ga old suggests that a peripheral solid rind, probably composed of

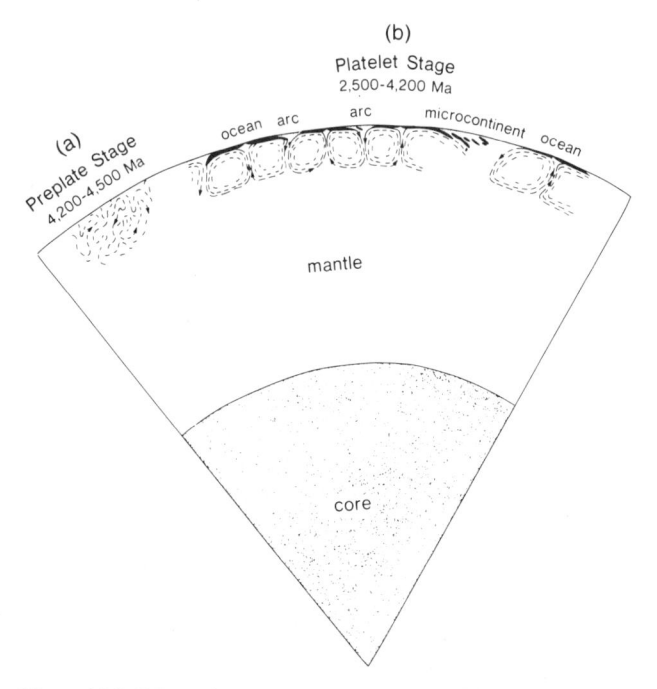

Figure 1.3.2. Schematic cross-section of the Earth showing microplate arrangement of crustal blocks above small but vigorous Archean convection cells. Waning heat flow accounts for the enlarging scale of decelerating advective transport (Ernst in press).

small, thin platelets, must have bounded the earth's surface quite early in its history, a supposition consistent with experimentally determined phase equilibrium relationships (Ernst in press). During the pre-Archean (Hadean), mantle circulation apparently was so vigorous that virtually all sialic and buoyant quasi-sialic crustal platelets formed at the earth's surface were returned to the mantle because of viscous drag (Condie 1980; Hargraves 1981). Reworking as a consequence of meteorite impacts was also significant until about 3.9 Ga, when planetesimal sweep-up lessened appreciably (Wetherill 1976).

During the Archean, from about 3.9 to 2.5 Ga, the stabilization of a globe-encircling mosaic of lithospheric platelets capped by thin simatic and, locally, sialic crust resulted in increasing preservation of the products of platelet tectonics as a consequence of the diminishing return of buoyant crustal blocks to the mantle. Extremely rapid and initially, perhaps, poorly organized mantle overturn would have driven platelets against and beneath one another, but because of the overall high global geothermal gradient and the small magnitude of the lithospheric/asthenospheric density inversion and consequent minor gravitational instability, the platelets would not have been subducted to appreciable depths before rising temperature caused their transformation to asthenosphere. This accounts for the lack of Archean and Early Proterozoic high-pressure metamorphic rocks such as blueschists and eclogites, the absence of early alkalic igneous suites, and the occurrence in Archean and some Early Proterozoic greenstone belts of highly refractory komatiites and related magnesian lavas.

The earliest preserved sialic blocks were formed as thick, buoyant simatic platforms that were partially stabilized and "cratonized" by the intrusion of massive volumes of tonalitic, trondhjemitic, and granodioritic magmas. The progressive accretion of similar terranes along convergent margins resulted in the assembly of enormous tracts of Archean crust. The evolution of such subduction-resistant sialic microcontinents in the Archean was accompanied by the development of the first craton cover and marginal sequences.

1.4

The Environment of the Archean Earth

HEINRICH D. HOLLAND   JAMES F. KASTING

1.4.1 Introduction

The Archean geologic record and early solar evolution are sufficiently well understood so that one can at least begin to reconstruct the nature of earth surface environments during the Middle and Late Archean. The Early Archean is still largely terra incognita. The presence of 4.1 to 4.2 Ga old zircons in the Mount Narryer Quartzite, Western Australia (Section 1.3; Kinny et al. 1988) indicates the presence of continental rocks more than 4.0 Ga before present, the 3.96 Ga zircons in tonalites of the Slave Province in Canada demonstrate the presence of continental crust at that time (Bowring 1989), and the Nd and Sr systematics of clastic sediments from Isua, west Greenland (Jacobsen and Dymek 1988) suggest that the amount of pre-3.8 Ga continental crust was very large. However, the nature of this crust and conditions in the atmosphere, in the oceans, and at their interface with the lithosphere are still poorly understood.

1.4.2 Evidence for Liquid Water and the Early Archean Surface Temperature

The earliest firm environmental data come from the nature of the supracrustal rocks at Isua, west Greenland. Nutman et al. (1984) have suggested that these metasedimentary rocks were deposited in a predominantly submarine volcanic area. Intermittent volcanic activity and associated instability produced clastic sediments which alternated with more slowly deposited chemical sediments. The lithologic and geochemical characteristics of the clastic components of the supracrustal sequences are consistent with derivation from felsic and basic volcanic rocks, and do not require a continental source.

The presence of chemical sediments in the Isua sequence demands the presence of liquid water on the earth 3.8 Ga ago. This, in turn, implies temperatures in excess of 0°C, at least in the Isua region, and strongly suggests that the hydrologic cycle was already in full swing. All these conclusions are consistent with the rare gas evidence for the degassing history of the earth (see, for example, Allègre et al. 1983; Holland 1984, Ch. 3; and Azbel and Tolstikhin 1990). Much of the earth's inventory of volatiles was probably released to the atmosphere during the first 500 Ma of earth history or, indeed, during the accretion process itself (see Section 1.2). This would have ensured the presence of ocean-sized bodies of water on the surface of the earth by at least 4.0 Ga ago, and probably earlier. The simultaneous release of a large fraction of the earth's CO_2 inventory is consistent with surface temperatures well in excess of 0°C during Isua time (see Section 4.7), with weathering that involved the acid titration of surface rocks, with oceans that were saturated with respect to quartz (SiO_2), calcite ($CaCO_3$) and dolomite ($CaMg(CO_3)_2$), and with sediments that contained these minerals as chemical precipitates. There is every reason to believe that 3.8 Ga ago the exogenic cycle was operating much as it does today.

As noted in Section 1.3, the endogenic cycle must have been more active in the Archean than it is today. The rate of heat generation due to the decay of long-lived radionuclides (^{238}U, ^{235}U, ^{232}Th, ^{40}K, ^{87}Rb) was several times greater then than now, and it is difficult to imagine an alternative to mantle convection as the dominant means for dissipating this heat (see, for example, Nisbet 1987, Ch. 8). If the Archean was not dominated by modern plate tectonics, then plates must have been created by some other very vigorous process. The large scale geology of Archean terranes does not look like that of Proterozoic terranes, and Nisbet (1987, p. 171) has wondered whether perhaps the earth in Early Archean time would have had "a few small continental nuclei scattered across a global ocean like water-beetles; by the end of the Archean, after a splendid Late Archean pandemonium, the bulk of the continental crust would be fractionated and sailing as huge stately floes across the simatic sea." Perhaps, but the Early Archean continents may well have been too large and too sluggish to be described as water beetles.

1.4.3 Archean Ocean Chemistry

Little is known of the composition of the oceans through which the Archean continents moved. Chloride was almost certainly the dominant anion in seawater, but its concentration can only be inferred by indirect means. Today, most of the chloride in the upper 30 km of the earth is dissolved in seawater. Evaporites probably contain less than one-half of the marine complement of chloride. The ratio of Cl to H_2O in the oceans plus evaporites is largely determined by the average $HCl:H_2O$ ratio in volatiles that have been released from the earth's interior. There is no good reason to believe that this ratio has varied drastically with time, and there are good reasons to believe that most of the earth's inventory of volatiles was released early in the Archean. Both arguments suggest that the $Cl:H_2O$ ratio in the oceans plus sediments has not varied greatly with time, and that the salinity of the Archean oceans was not very different from that of present-day seawater.

The composition of the chemical metasediments at Isua (see above), and the composition of later Archean sediments, demonstrate beyond reasonable doubt that the Archean oceans were saturated or – more likely – supersaturated with respect to $CaCO_3$ and $CaMg(CO_3)_2$ (see, e.g., Gibbs et al. 1986; and Holland 1984, Ch. 5). Unfortunately, this does not fix the concentration of the individual constituent ions of these minerals in Archean seawater. They are, however, related in an interesting fashion. In solutions saturated with respect to $CaCO_3$, equilibrium obtains in the reaction

$$CaCO_3 + CO_2 + H_2O \rightleftharpoons Ca^{+2} + 2HCO_3^-. \quad (1.4.1)$$

Thus, the activities of Ca^{+2} and HCO_3^- in sea water are related to P_{CO_2}, the CO_2 pressure in the atmosphere, by

$$\frac{a_{Ca^{+2}} \cdot a^2_{HCO_3^-}}{P_{CO_2}} = K_{(1.4.1)}. \quad (1.4.2)$$

In seawater today, the product of the concentration of these species is

$$\frac{m_{Ca^{+2}} \cdot m^2_{HCO_3^-}}{P_{CO_2}} \, 1.6 \times 10^{-4} \frac{(mol/kg)^3}{atm}. \quad (1.4.3)$$

In Archean seawater this product was probably somewhat larger, because there is good evidence for a greater degree of supersaturation of Archean sea water with respect to $CaCO_3$ (e.g., Grotzinger 1989). The product could not have been very much greater, however, since it is limited by the value at which the inorganic precipitation of aragonite becomes geologically rapid.

P_{CO_2} was almost certainly higher in the Archean than it is today. If Kasting's (1987) estimates are correct (see Section 4.7), P_{CO_2} was on the order of 1 atm some 3.5 Ga before present; hence, at that time,

$$m_{Ca^{+2}} \cdot m^2_{HCO_3^-} \approx 2 \times 10^{-4} (mol/kg)^3. \quad (1.4.4)$$

This product cannot be decomposed uniquely, but some reasonable limits can be set on $m_{Ca^{+2}}$ and $m_{HCO_3^-}$. If the ratio of $m_{Ca^{+2}}:m_{HCO_3^-}$ in Archean seawater was the same as today (i.e., 4.2), $m_{HCO_3^-}$ was approximately 35 mmol/kg and $m_{Ca^{+2}}$ was about 150 mmol/kg. This is probably a strong upper limit for $m_{Ca^{+2}}$. Saturation of seawater with respect to dolomite as well as to aragonite demands that $m_{Mg^{+2}} > m_{Ca^{+2}}$. If the earlier argument for a salinity close to that of the present-day ocean is correct, then essentially all of the Cl^- would have been needed to balance Ca^{+2} and Mg^{+2}, and Archean seawater would have contained very little Na^+. This seems unlikely.

At the other extreme, we can explore the consequences of an ocean in which $m_{Ca^{+2}}$ was equal to that of modern seawater. In that case, $m_{Ca^{+2}} = 10$ mmol/kg and $m_{HCO_3^-} \approx 140$ mmol/kg in seawater at 3.5 Ga. The evaporation of such seawater would have generated mineral assemblages rich in bicarbonate minerals but without gypsum and anhydrite, because Ca^{+2} would have been exhausted by the precipitation of calcite. Pseudomorphs of barite after gypsum in an evaporite sequence from the 3.5 Ga-old Warrawoona Group of Western Australia have been reported by Buick and Dunlop (1987). It is not sure, however, that this sequence is the product of the evaporation of seawater. A lacustrine origin is hard to rule out (Grotzinger 1989). Other evidence for Archean gypsum and anhydrite is scarce. This could mean that the Archean oceans were "soda oceans," as advocated by Kempe and Degens (1985), but gypsum and anhydrite are so soluble that their absence from Archean sedimentary assemblages may well be due to later dissolution rather than to non-precipitation. Alternatively, dissolved sulfate concentrations in the Archean oceans may have been low as a consequence of the absence of oxidative weathering of pyrite on an anoxic early earth (Walker and Brimblecombe 1985).

Even if the occurrence of gypsum in the Warrawoona sequence tells us little about the composition of seawater 3.5 Ga ago, it is very strong evidence that the temperature at that time and in that location was not much greater than 57°C, the transformation temperature of gypsum to anhydrite in water (Hardie 1967). This temperature decreases rapidly with increasing salinity of the solutions from which gypsum crystallizes. The occurrence of gypsum in the Warrawoona sequence therefore indicates that the temperature of the earth need not have been very different from that of the present day and could not have been as high as 100°C. Inconclusive arguments based on weathering rates suggest temperatures cooler than those of the last 100 Ma. However, the absence of evidence for glaciation older than ~2.5 Ga (Frakes 1979) suggests just the opposite. In short, we do not presently know whether the Archean climate was warm or cold.

The permitted range of the HCO_3^- concentration in Archean seawater is large, but it does set some useful bounds on the pH of Archean seawater. In the ocean-atmosphere system today,

$$\frac{a_{H^+} \cdot m_{HCO_3^-}}{P_{CO_2}} = 5.4 \times 10^{-8}. \quad (1.4.5)$$

This ratio depends only on the temperature and on the activity coefficient of HCO_3^- in seawater. It has probably changed rather little with time. If so, the indicated pH range is 5.8 to 6.4 for all likely values of $m_{HCO_3^-}$. If Kasting's (1987) estimate of 1 atm for P_{CO_2} 3.5 Ga ago turns out to be too high, the pH of average seawater at that time will have been closer to its present day value of 8.1.

The high proposed P_{CO_2} in the atmospheres would have driven silicate-carbonate reactions toward the carbonate side, and the low pH of seawater would have slowed or prevented reverse weathering reactions. This is consistent with the intensity of the degree of acid titration of Archean sedimentary rocks (Holland 1984, Ch. 5).

1.4.4 The Archean Atmosphere

The oxidation state of the atmosphere during the Archean is still uncertain. Little information has been added during the last few years that would alter the assessments by Walker et al. (1983), Holland (1984), and by Kasting (1987). Life surely began early in the Archean (Section 1.5), and green-plant photosynthesis could have been invented more than 3.5 Ga ago (cf. Schopf and Packer 1986, 1987). However, the evidence for a significant level of free O_2 in the Archean is limited and equivocal. In the absence of O_2 in the atmosphere, Fe^{+2} released from solid phases during weathering tends to be flushed out of soils. In the presence of abundant O_2, Fe^{+2} in soils is oxidized rapidly and quantitatively, and is retained in soils as ferric oxide or hydroxide (see Section 4.4). The loss and retention of iron in Archean paleosols is therefore an interesting semi-quantitative indicator of the oxidation state of the Archean atmosphere. At present there are no well-documented and well-studied Archean paleosols. Those that have been studied suggest that a small amount of O_2 was present in the atmosphere 3.0 Ga ago.

1.4.5 The Archean Terrigenous Environment

Evidence from the Fe^{+2} content of Archean limestones and dolomites is consistent with the foregoing conclusion, but it is also inconclusive. In the absence of O_2, much of the Fe^{+2} released during weathering can be expected to enter the oceans as a major constituent of river waters. Fe^{+2} is apt to behave like Mg^{+2} under these conditions, and the ratio of $Fe^{+2}:Mg^{+2}$ in rivers is apt to be close to that in average crustal rocks, i.e., about unity. The potential mechanisms for removing Fe^{+2} from the oceans are quite diverse. Oxides, carbonates, silicates, and sulfides come to mind; all of these mineral groups are represented in Archean banded iron-formations (see Section 4.2), but only oxides are common in "normal" Archean sediments. This is somewhat curious. The thermochemical data for assessing the equilibrium constant of the reaction

$$\underset{\text{greenalite}}{Fe_6Si_4O_{10}(OH)_8} + 6CO_2 \rightleftharpoons \underset{\text{siderite}}{6FeCO_3} + \underset{\text{quartz}}{4SiO_2} + 4H_2O$$

(1.4.6)

are rather poor (Klein and Bricker 1977) but they suggest that siderite and amorphous SiO_2 are stable with respect to greenalite at 25°C and at CO_2 pressures >1 atm. Siderite and/or ankerite might therefore be expected to occur as major components of Early and Middle Archean marine carbonates. The available evidence (e.g., Veizer 1983; Holland 1984, Ch. 8; Gibbs et al. 1986) strongly indicates that this is not the case. Near-surface seawater seems to have been poor in Fe^{+2} except in unusual areas where upwelling mid- or deep-ocean water brought Fe^{+2} into shallow water environments.

Sulfides are not attractive candidates for the removal of major portions of river-borne Fe^{+2} under anoxic conditions, because the ratio of $Fe^{+2}:S^{-2}$ in crustal rocks is very much larger than unity. That leaves oxides as the most likely of the dominant marine iron precipitates, but the precipitation of both magnetite and hematite requires the prior oxidation of Fe^{+2} to Fe^{+3}. This is consistent with the presence of O_2 in the Archean atmosphere. Unfortunately these observations cannot be taken as proof of its presence. Braterman et al. (1983) have pointed out that solar ultraviolet light can oxidize Fe^{+2} to Fe^{+3} in solutions in the absence of O_2. However, in the presence of organic matter, visible sunlight can also reduce Fe^{+3} to Fe^{+2} (McKnight et al. 1988). To complicate matters further, free oxygen could have been present in the surface waters of some ocean basins and yet still have been absent from the atmosphere (Section 4.6). The overall effects of aqueous photochemistry on the marine geochemistry of iron during the Archean are therefore difficult to assess.

Atmospheric photochemistry may also have played a role in removing Fe^{+2} from the oceans. Kasting, Holland, and Pinto (1985) have shown that oxidants and reductants are both produced photochemically in low-O_2 atmospheres. Some of these compounds are highly soluble in water. These are washed out of the atmosphere very efficiently with rainwater. H_2O_2 is the most important of the oxidants. Its flux to the Archean oceans could well have exceeded the river flux of Fe^{+2}, and Fe^{+2} could have been oxidized to Fe^{+3} in the shallow oceans by H_2O_2 alone.

The marine budget of iron must have included both river inputs and hydrothermal inputs. It has been suggested that the chemistry of the Archean oceans was essentially controlled by hydrothermal inputs (Fryer et al. 1979), but this is unlikely. Mg^{+2} is removed quantitatively from seawater which reacts with ocean floor basalts at elevated temperatures (Von Damm et al. 1985). Seawater cycling through ocean floor basalt during the Archean at a rate much greater than today would therefore have assured that the ocean floor served as a much more important sink for Mg^{+2} than today. This is difficult to believe, since dolomite is abundant in the few extensive Archean carbonate sequences which have been described to date (e.g., Gibbs et al. 1986; Grotzinger 1989). If a large fraction of the river flux of Mg^{+2} had been transferred to the ocean floor during the hydrothermal cycling of seawater, Archean carbonate sequences should be dominated by limestone.

Even if the ratio of the river input of Mg^{+2} to the hydrothermal output of Mg^{+2} during the Archean was not very different from the present-day, the hydrothermal input of iron might have been considerably greater (see Section 4.6). The concentration of Fe^{+2} in these hydrothermal solutions increases rapidly with increasing temperature (Mottl and Holland 1978). The abundance of komatiites and komatiitic basalts in Archean greenstone belts suggests that temperatures in the upper mantle were higher then than now. Black smokers may therefore have been hotter; if so, the input of iron per kilogram of average hydrothermal solution would have been greater, perhaps by as much as an order of magnitude. The implications of such a rapid hydrothermal flux of iron for the formation of banded iron-formations are discussed in Section 4.3.

1.4.6 The Environmental Influence of Archean Impact Events

The influx of extraterrestrial material must also have been more rapid in the Archean than it is today. An estimate of the mass influx rate around 3.8 Ga ago can be derived from analysis of the lunar cratering record and the composition of the lunar crust (Sleep et al. 1989; Kasting 1990). For reasonable assumptions concerning impactor velocities, the loss of material to space during impacts, and the time dependence of the impact flux, the predicted terrestrial infall rate is 8×10^{14} g yr^{-1}, or 1.6×10^{-4} gm cm^{-2} yr^{-1} (Kasting 1990). This is about one-tenth of the rate of supply of sediment to the oceans today (Milliman and Meade 1983). By 3.5 Ga, the infall rate was probably lower by a factor of 10 to 50 (Wilhelms 1987). Extraterrestrial material should therefore have been a relatively minor component of Early and Middle Archean sediments. This expectation is borne out by recent data on the iridium content of Archean sediments. The Ir:Cr ratio of sediments is a sensitive indicator of a meteoritic component. Preliminary measurements of the Ir and Cr content of the 3.8 Ga old metasediments at Isua, west Greenland, have shown that their Ir:Cr ratio is terrestrial (R. Dymek pers. comm. to H.D.H.). The input of extraterrestrial Ir and Cr seems to have been minor compared to the terrestrial input of these elements. If the Ir:Cr ratio of the Isua metasediments turns out to be representative of sediments deposited 3.8 Ga ago, then the rate of infall of extraterrestrial material at that time was already much smaller than the rate of erosion and sedimentation by purely terrestrial processes.

This does not mean that the impact of extraterrestrial objects was trivial. The 100-km diameter objects that formed the lunar basins around 3.8 Ga (Sleep et al. 1989) would have impacted the earth with an energy of some 2×10^{33} ergs, assuming a collision velocity of about 20 km s^{-1}. If 25% of this energy went into production of steam, this would have been sufficient to boil off the topmost 35 m of ocean water. Objects several times this size may well have hit the earth during this period; these could have evaporated the entire photic zone (Sleep et al. 1989). The best place to live during Isua times may have been the floor of the deep ocean, especially the regions in and around hydrothermal systems at midocean ridges (Maher and Stevenson 1988).

Even at 3.5 Ga, after the period of heavy bombardment had ended, impacts would still have been a force to be reckoned with. Lowe has found 3.5 Ga-old spherule deposits on two continents (Lowe et al. 1989). One of the beds contains a 10 cm-thick layer of pure spherules with an iridium content equal to 30% of the chondritic value. These spherules appear to have been deposited in the aftermath of a giant tsunami. The absence of coarse ballistic ejecta in these beds suggests that the impact was not local, so it seems likely that this material was deposited globally. The iridium values indicate that an approximately 3 cm-thick layer of impactor material was distributed around the earth, implying an impactor mass of about 5×10^{19} gm. This is smaller than the lunar basin-forming objects 3.8 Ga ago, but some 20 to 50 times larger than the bolide that wreaked havoc on the earth some 65 Ma ago (Alvarez 1986). The energy released during this event was about 10^{32} ergs – enough to boil off perhaps the top 2 m of the oceans and to destroy organisms living near the surface. Yet, stromatolite deposits are found directly above the spherule bed, indicating that benthic prokaryotic communities, if they were affected at all, were able to recover from the devastation in a geologically short period of time. The Early Archean environment was evidently sufficiently benign for microbial life, but it would not have been healthy for most higher organisms.

1.5
Paleobiology of the Archean
J. WILLIAM SCHOPF

1.5.1 Introduction

Virtually all aspects of Archean paleobiology have been recently and thoroughly reviewed in the previous collaborative publication of the Precambrian Paleobiology Research Group (PPRG), *Earth's Earliest Biosphere* (Schopf 1983a). There are included detailed discussions of paleobiologic aspects of the origin of life (Chang et al. 1983); the Archean organic geochemical (Hayes et al. 1983) and carbon, sulfur, hydrogen, and nitrogen isotopic records (Schidlowski et al. 1983; see also Chapter 3 herein); Archean biochemistry (Gest and Schopf 1983; Chapman and Schopf 1983); and Archean stromatolites (Walter 1983), microfossils (Schopf and Walter 1983), and environmental (Walker et al. 1983) and ecological evolution (Hayes 1983a; Schopf et al. 1983).

The following discussion, therefore, deals only with those developments in Archean paleobiology that post-date publication of the earlier PPRG volume (Schopf 1983a), in particular, with recent discoveries of Archean stromatolites and microfossils that provide new insight into the nature and biologic composition of the Archean biota. Included are formal descriptions of three new taxa of cellularly preserved filamentous microbes from bedded cherts of the Early Archean Apex Basalt (Warrawoona Group) of Western Australia.

1.5.2 Recent Paleobiologic Studies of the Early Archean Fig Tree and Onverwacht Groups (Swaziland Supergroup), South Africa

In comparison with the Proterozoic fossil record, that known from the Archean is minuscule: although literally hundreds of microfossiliferous formations, containing more than 2800 occurrences of bona fide microfossils, have been discovered in the Proterozoic (Sections 5.4, 5.5, Table 22.3), fewer than 30 such units (containing 43 categories of putative microfossils, the vast majority of which are of questionable authenticity) have been reported from the Archean (Schopf and Walter 1983). Similarly, whereas stromatolites are abundant, widespread, and in fact essentially ubiquitous in Proterozoic carbonate terranes – being represented in the known fossil record by hundreds of taxonomic occurrences reported from a large number of Proterozoic basins (Section 6.2.5) – only about a dozen occurrences of Archean stromatolites have as yet been reported (Walter 1983).

Among the notably few Archean units to have been investigated paleobiologically in some detail, those of the Early Archean (~ 3300 to ~ 3500 Ma-old) Fig Tree and Onverwacht Groups (Swaziland Supergroup) of the eastern Transvaal, South Africa, have received particular attention. Beginning with the studies of Pflug (1966), and of Barghoorn and Schopf (1966), more than 30 publications appeared prior to 1983 reporting discovery of microfossil-like objects in these strata (for discussion and relevant references, see Schopf and Walter 1983). Although most of the objects thus reported are now regarded as nonfossils (viz., bubbles and similar artifacts, modern biologic contaminants, mineralic dendrites, and nonbiogenic aggregates of amorphous organic matter; Schopf and Walter 1983, p. 216), certain of the simple, unicell-like, organic spheroids reported from these units are possibly of biologic origin (Muir and Grant 1976; Knoll and Barghoorn 1977; Schopf and Walter 1983, Photo 9-2).

Recently, additional, biologically more convincing structures have been reported from Fig Tree and Onverwacht Group sediments. In particular, Byerly et al. (1986) have described the low-relief, laterally linked, wavy laminated, stratiform to short-columnar stromatolites shown in Figure 1.5.1E from a thin chert unit of the Fig Tree Group southwest of Barberton, South Africa. Although these structures are not known to contain cellularly preserved remnants of the microorganisms responsible for their formation, the evidence presented by Byerly et al. (1986) seems wholly consistent with their inferred biogenicity. Moreover, from the same general area, Walsh and Lowe (1985) have reported the occurrence of threadlike filamentous microfossils in chert units both of the Hooggenoeg and Kromberg Formations of the stratigraphically underlying Onverwacht Group. As shown in Figure 1.5.2, the Kromberg filaments occur in association with wavy, flat-laminated, stromatolite-like laminae, composed chiefly of amorphous kerogen. Within this stromatolite-like fabric, and commonly oriented parallel or subparallel to the component laminae, are thin (1.0 to 2.5 μm-diameter), straight to gently curved, solid or hollow (Fig. 1.5.2F, at arrow), kerogenous (and possibly

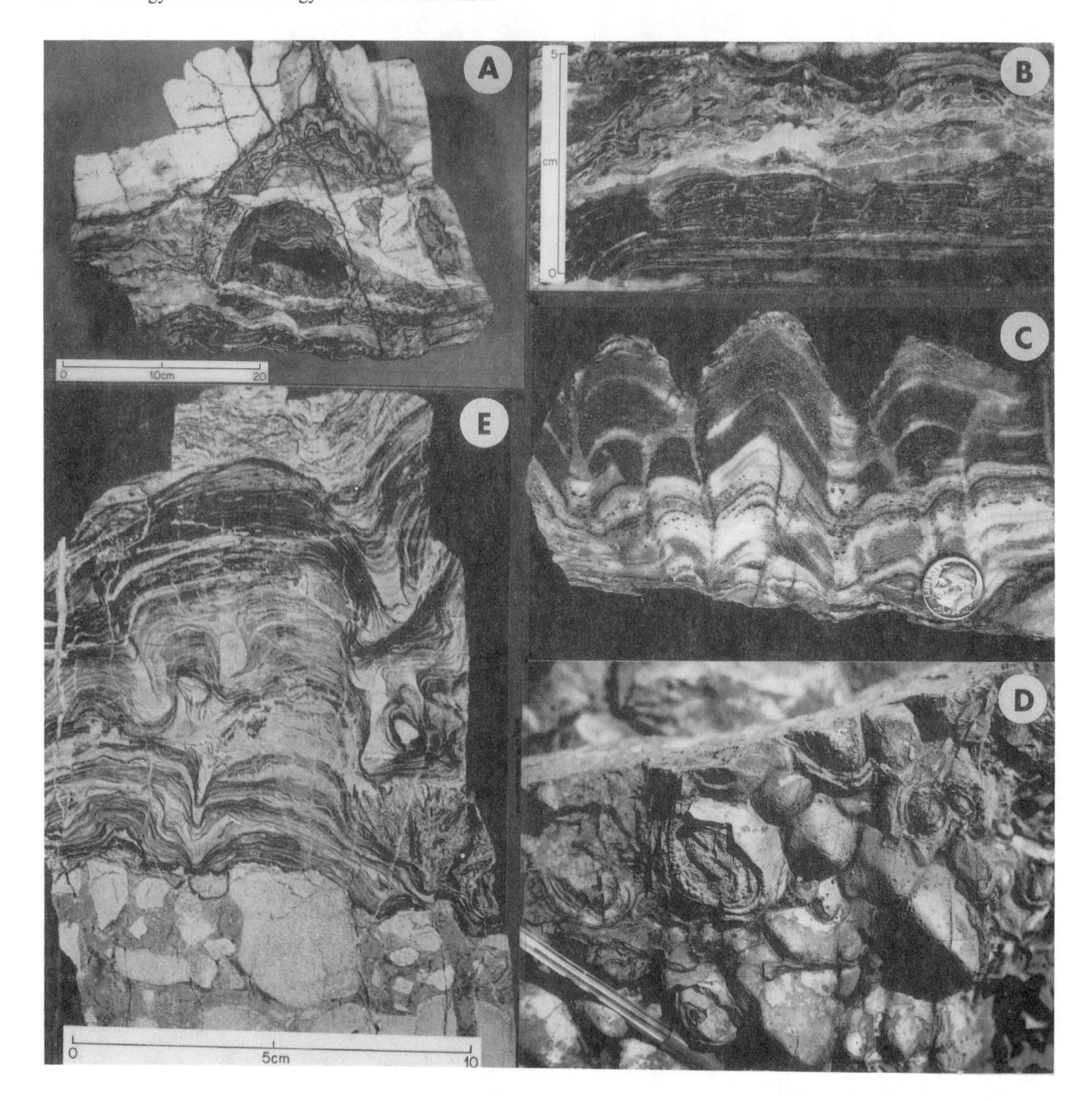

Figure 1.5.1. Domical (A), stratiform (B), conical (C, D), and laterally linked short-columnar Early Archean stromatolites from the ~3300 to ~3600 Ma-old Warrawoona Group of Western Australia (A–D) and the ~3300 to ~3400 Ma-old Fig Tree Group of South Africa (E). (A, B) Vertical sawn slab faces showing domical (A) and stratiform (B) portions of a dolomitic chert stromatolite from the Towers Formation (Table 1.5.1), 2.5 km southwest of Panorama Homestead, in the northeastern part of the North Pole Dome, eastern Pilbara Block, northwestern Western Australia (Walter et al. 1980; Walter 1983, Photos 8-1, 8-2; Schopf 1983, Frontispiece). (C, D) Vertical sawn slab face (C; coin for scale is 17 mm in diameter) and eroded surface morphology (D; pen for scale is 6 mm wide) of concentrically laminated, conical, chert stromatolites from the "Strelley Pool chert," near the top of the Warrawoona Group, about 30 km west of the North Pole Dome, northwestern Western Australia (Lowe 1980). (E) Vertical sawn slab face showing, from base upward, brecciated top of a silicified komatiitic lava flow; small, basal, domical stromatolites; laterally linked, somewhat asymmetrical, short-columnar stromatolites; and very low-relief stromatolitic laminae with high inheritance, from a thin, gray to black chert bed about 20 km southwest of Barberton, South Africa (Byerly et al. 1986).

mineral-encrusted), microbe-like filaments (Fig. 1.5.2B–G). Not uncommonly the opaque, solid, filaments are broken and disrupted (e.g., Fig. 1.5.2C–E), and although such breaks are in some examples rather regularly distributed, suggesting possible cellularity (Fig. 2.5.1G), well-defined, unquestionable cells have not been detected. The exact affinities of these filaments are uncertain. In general morphology, they are similar to many extant prokaryotes; among these, they seem more comparable to filamentous bacteria (or to the tubular sheaths of such bacteria) than to most modern cyanobacteria (Section 5.4).

Together with available organic geochemical data (Chapter 3; Hayes et al. 1983; Schidlowski et al. 1983), the recently

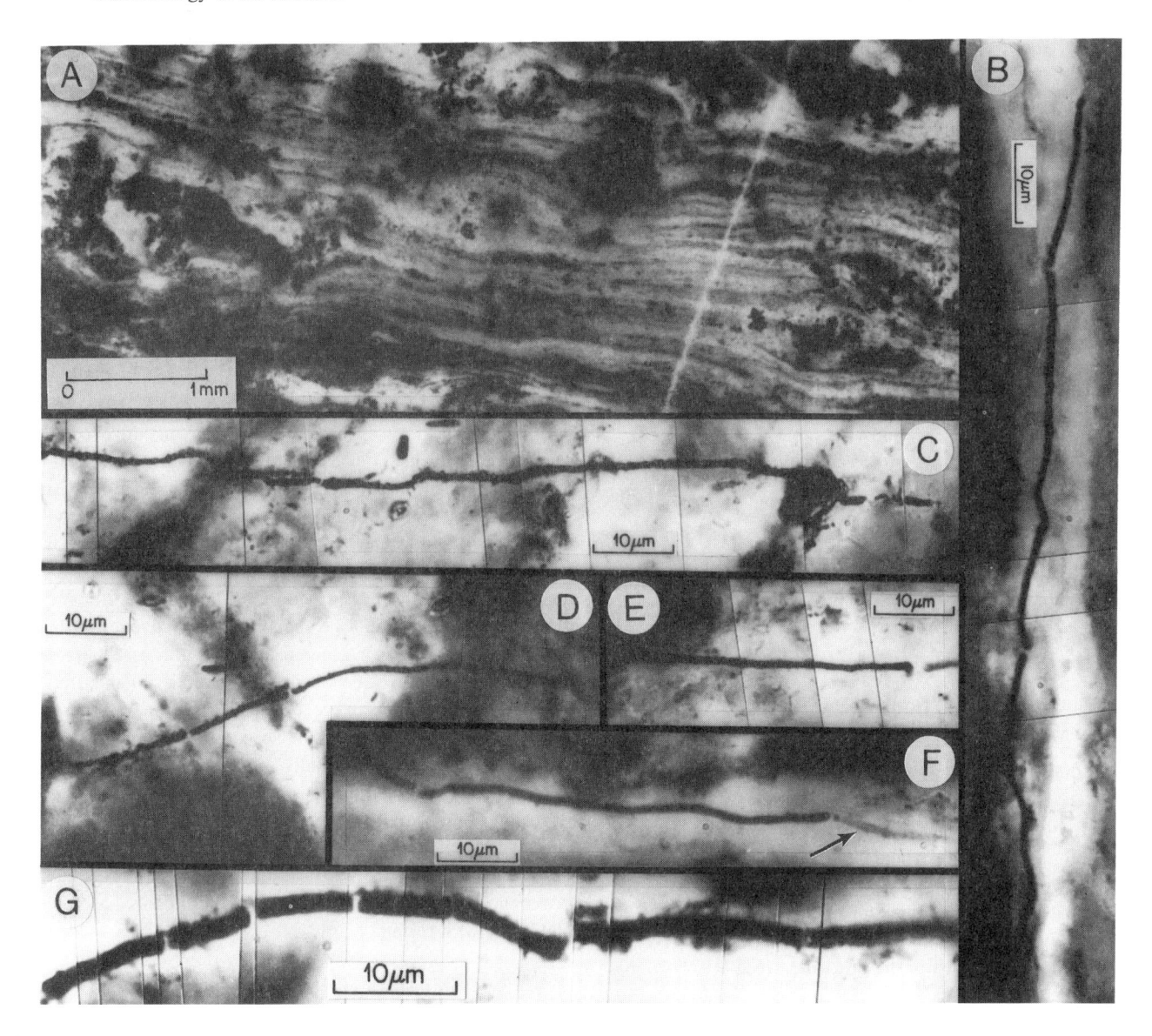

Figure 1.5.2. Optical photomicrographs showing wavy, kerogenous, stromatolite-like laminae (A) and unnamed, threadlike, filamentous microfossils (B–G) in petrographic thin sections of black and white banded chert from the "stratiform stromatolitic facies" described by Walsh and Lowe (1985) from the lower third of the Early Archean Kromberg Formation, Onverwacht Group, in the Komati River valley, Transvaal, South Africa (locality cf. PPRG 1394); B–E, G, show composite photomicrographs; arrow in F points to tubular, sheath-like, hollow filament.

reported stromatolites (Fig. 1.5.1E; Byerly et al. 1986) and filamentous microfossils (Fig. 1.5.2; Walsh and Lowe 1985) discussed above and, possibly, certain of the unicell-like spheroids previously reported from other cherts of the Fig Tree and Onverwacht Groups (Muir and Grant 1976; Knoll and Barghoorn 1977; Schopf and Walter 1983, Photo 9-2), provide convincing evidence of Early Archean biologic activity.

1.5.3 Controversial Microfossils Previously Described from the Early Archean Warrawoona Group, Western Australia

Paleobiologic studies of the ~3300 to ~3600 Ma-old (Table 1.5.1) Early Archean Warrawoona Group of Western Australia were initiated in the late 1970s by Dunlop, Muir, Milne, and Groves (1978) who reported discovery of carbonaceous spheroids interpreted as fossil unicells. These microfossil-like objects, however, were later reinterpreted by Schopf and Walter (1980; 1983, p. 226) as "most likely to be nonfossils, solid carbonaceous globules of apparently nonbiologic origin." Shortly after the initial Dunlop et al. (1978) report, flat-lying, domical, and conical stromatolites were also discovered in Warrawoona strata (Fig. 1.5.1A–D; Lowe, 1980; Walter et al. 1980; Walter 1983); like the globular microstructures, however, and because these stromatolites do not contain cellularly preserved microfossils, they too have been regarded as being of unproven biogenicity (Buick et al. 1981).

In 1983, what appeared to be a major advance in the search for evidence of Early Archean life was reported by Awramik et al. (1983): discovery of cellularly preserved filamentous microbes in cherty, finely laminated, stromatolite-like strata of the Towers Formation of the Warrawoona Group. Filamentous microfossils and possible microfossils were described from

Table 1.5.1. *Stratigraphy and geochronology of the Pilbara region, Western Australia.*

Stratigraphic Unit	Approximate Age (Ga) and Geochronologic Technique[a]
MOUNT BRUCE SUPERGROUP	
Hamersley Group	2.5 (Sm-Nd, U-Pb)
Fortescue Group	2.8 (Pb)
PILBARA SUPERGROUP	
Whim Creek Group	3.0 (Sm-Nd, Pb)
Gorge Creek Group	
Warrawoona Group	
Wyman Formation	
Euro Basalt	
Panorama Formation	3.3 (Pb)
*Apex Basalt	
*Towers Formation	3.4–3.5 (Pb)
Duffer Formation	3.4–3.5 (Sm-Nd, U-Pb, Pb)
Mt. Ada Basalt	
McPhee Formation	3.6–3.7 (Sm-Nd)
North Star Basalt	

[a]Data from: Blake and McNaughton (1984); Glickson (1986)
*Indicates microfossiliferous formation (Figs. 1.5.3 through 1.5.7, Table 1.5.2)

two localities, both in the North Pole Dome area of the Pilbara Block of northwestern Western Australia:

(1) Rosette-like aggregates of fine radiating filaments, interpreted by Awramik et al. (1983, pp. 366, 371) as "possible microfossils" and regarded by Schopf and Walter (1983) as "dubiomicrofossils" (Schopf and Walter 1983, p. 215, Photo 9-6), "microstructures that are of possible, but as yet unproven biogenicity" (Schopf and Walter 1983, p. 226), were reported from "Locality B" of Awramik et al. (1983), about 5 km southeast of the other described Warrawoona locality.

(2) Four species of filamentous microfossils (viz., *Eoleptonema australicum*, *Primaevifilum septatum*, *Archaeotrichion contortum*, and *Siphonophycus antiquus*; Table 1.5.2) were described by Awramik et al. (1983) from "Locality A," northwest of Locality B; these morphotypes, organic in composition and in part demonstrably cellular in organization, were regarded both by Awramik et al. (1983) and by Schopf and Walter (1983) as authentic microfossils.

Controversy has swirled about both of these sets of microstructures. Substantial caution was exercised by Awramik et al. (1983) and by Schopf and Walter (1983) in reporting and interpreting the rosette-like aggregates from Locality B. As expressed by Schopf and Walter (1983, p. 230), "it seems most appropriate to regard these rosette-like aggregates as only possible fossils; they may represent a previously undescribed variety of Precambrian microbe, or they may be non-biogenic, aggregated microstructures of solely non-biologic origin." Essentially the same interpretation was reached by Buick (1984) who concluded that these aggregated "filaments are too simple in form for their origin to be determined, so they should be regarded as dubiofossils, perhaps biogenic, perhaps inorganic." Thus, and despite differences in emphasis (Awramik et al. 1988; Buick 1988), all of the workers involved are in basic agreement: the rosette-like aggregates are properly regarded as dubiomicrofossils; because their biogenicity has not been established, they cannot be regarded as firm evidence of Archean life.

There also has been considerable discussion regarding the four taxa of microfossils described from "Locality A" (see Awramik et al. 1988). There seems ample evidence that these fossil filaments are indigenous to the rock samples in which they occur, are syngenetic with deposition of those rocks, and are of biologic origin (Awramik et al. 1983, 1988; Schopf and Walter 1983) – that is, that they are authentic microfossils. However, what *is* in question is the exact source of the microfossiliferous samples: despite repeated visits to the outcrop, both by the original collector (Awramik et al. 1988) and by others (Buick 1988), the precise layer in the Towers Formation from which the fossiliferous cherts were collected in 1977 has not been relocated. Clearly, this is a tantalizing set of observations, one suggesting that it must be very probable that cellularly preserved evidence of Early Archean life exists in Warrawoona Group sediments. Just as clearly, however, the available data fall short of providing the critical "hard-nosed" proof that properly should be demanded of discoveries of this sort, discoveries that alter generalizations (in this case, regarding the oldest unequivocal evidence of life on earth) and, thus, become incorporated in college textbooks and even in common parlance. The harsh reality is that because this discovery has not been reconfirmed, the results of this study, despite their presumed validity, are of limited value. Fortunately, as discussed below, other studies recently have shown that unquestionably biogenic microfossils, of demonstrable Early Archean age, are indigenous to Warrawoona strata; moreover, the provenance of these newly discovered microfossils has been established with certainty.

1.5.4 Newly Discovered Microfossils from the Early Archean Apex Basalt (Warrawoona Group), Western Australia

Discovery of microfossils at two new localities in two formations of the Warrawoona Group (Schopf and Packer 1986, 1987) obviates many of the difficulties of the earlier reports discussed above. As expressed by Buick (1988), "Schopf and Packer (1987) recently reported the discovery of some intriguing microstructures that appear to be genuine microbial fossils in authentic Archaean sediments of the Warrawoona Group. When a full description of the host rocks becomes available, these may well prove to be incontrovertible evidence of the existence of prokaryotic life in the early Archaean, rendering this current discussion [of the controversy surrounding earlier reports] redundant except as a cautionary exercise."

The locations in Western Australia of the two recently reported fossiliferous localities of the Warrawoona Group are shown in Figure 1.5.3. Microfossils occur in chert units both of the Apex Basalt immediately south of Chinaman Creek about 12 km west of Marble Bar, Western Australia (Locality 2 of Fig. 1.5.3), and of the stratigraphically immediately underlying Towers Formation near Strelley Pool, Western Australia (Locality 1 in Fig. 1.5.3). Available geochronologic data in-

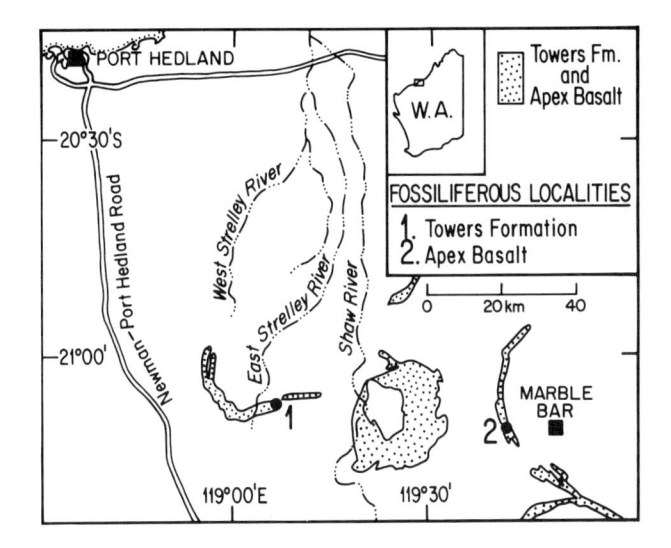

Figure 1.5.3. Location in Western Australia of microfossiliferous localities of the Apex Basalt ("Locality 2," Figs. 1.5.4, 1.5.5, 1.5.6; PPRG 1458) and the Towers Formation ("Locality 1," Fig. 1.5.7; PPRG 2007) of the Early Archean Warrawoona Group.

dicate that both of these formations are on the order of 3300 to 3500 Ma in age (Table 1.5.1).

Filamentous, very dark brown to black (opaque), kerogenous microfossils (Figs. 1.5.4 through 1.5.6) occur in a 10 m-thick, bedded, dark gray to black chert unit of the Apex Basalt along Chinaman Creek, west of Marble Bar, in an area well-mapped by Hickman and Lipple (1978). The filaments are cellularly preserved (e.g., Figs. 1.5.4C, D, F, G, H; 1.5.5A–E), permineralized within subangular to rounded, brown to very dark-brown, kerogen-rich, sedimentary siliceous clasts one to a few millimeters in diameter (Fig. 1.5.4A), members of a population of such clasts ranging from less than 1 mm to more than 20 cm in size. Laterally, over a distance of less than 100 m from the fossiliferous locality, this fossil-bearing chert unit merges into a continuous, brecciated, bedded, gray chert unit, 20 to 30 m in thickness, that is concordant and interfingering with associated volcanics. Fossiliferous samples have been collected on two occasions from this locality: in June, 1982 (Figs. 1.5.4A–D, F–H; 1.5.5A, C–E; 1.5.6A–D) and in August, 1986 (Figs. 1.5.4E; 1.5.5B; 1.5.6E–G).

The occurrence of microfossils within lithic clasts but not in the surrounding matrix in this unit (Fig. $1.5.4A_2$) indicates that they pre-date deposition of the bedded chert, having been preserved initially in an older unit that was subsequently eroded, the clasts being transported and redeposited to form a clastic component of the bedded chert of the Apex Basalt. The clasts, in turn, are a primary, integral component of the bedded cherts as indicated by numerous petrographic relationships (e.g., the occurrence of secondary quartz-filled veinlets that transect both the clasts and their encompassing bedded chert matrix; Fig. $1.5.4A_1$). Thus, the microfossiliferous clasts seem clearly to be syngenetic with deposition of the bedded cherts, and because the subangular to rounded siliceous clasts were transported and redeposited to become a component of this unit, the fossils themselves are evidently older (although whether they are greatly older than, or essentially penecontemporaneous with, the bedded chert unit cannot be ascertained from the evidence at hand).

The occurrence of these microfossils in petrographic thin sections (Figs. 1.5.4 through 1.5.6), their presence in the clasts but not in the surrounding chert matrix (Fig. $1.5.4A_2$), and their similarity in color and texture to dark brown to black kerogenous particles finely distributed throughout the clasts (Fig. 1.5.4A), indicate that they are indigenous to the clasts in which they occur. The microfossils demonstrably are not of secondary, post-depositional, geologic origin, nor are they modern contaminants.

The biogenicity of these filaments is firmly established by their morphological complexity, being composed of well-defined, barrel-shaped, discoidal or quadrate cells (e.g., Figs. 1.5.4C, D, F–H; 1.5.5A–E), and exhibiting rounded or conical terminal cells (e.g., Figs. 1.5.4C, D; 1.5.5A) and medial bifurcated cells and paired half-cells (Fig. 1.5.6F, G) that reflect the original presence of partial septations and, thus, the occurrence of prokaryotic binary cell division. Their biological origin similarly is indicated by their organic, carbonaceous composition; by the degree of regularity of cell shape and dimensions in numerous specimens of particular species (e.g., compare Figs. 1.5.5A through 1.5.5E) and the limited, definable, cell size ranges of such taxa (Table 1.5.2); and by their obvious similarity in salient morphological characteristics to cellular, unbranched, uniseriate, prokaryotic trichomes, both extant (Fig. 5.4.2) and Proterozoic (see Chapter 24).

Finally, and in notable contrast with the controversial filamentous microfossils earlier reported from Warrawoona sediments (Section 1.5.3), there is no question as to the provenance of these microfossils: as noted above, photomicrographs are here included of filaments detected in cherts collected at the Chinaman Creek locality on two occasions (the latter, in August, 1986, by a fledgling graduate student who had not previously visited the outcrop and was armed solely with maps, notes, and field photographs from the previous, June, 1982, PPRG collecting trip).

Thus, the cellular filaments recently discovered in bedded cherts of the Apex Basalt at Chinaman Creek meet all of the criteria required of bona fide, authentic, Archean microfossils (Schopf and Walter 1983): (*i*) they occur in rocks of unquestionably Archean age; (*ii*) they are demonstrably indigenous to these Archean sediments; (*iii*) they occur in clasts that are assuredly syngenetic with deposition of this sedimentary unit (with the fossils themselves pre-dating deposition of the bedded cherts in which the clasts occur); and (*iv*) they are certainly biogenic. Moreover, to these criteria may now be added a fifth: (*v*) as demonstrated by replicate sampling of the fossiliferous outcrop, the provenance of these microfossils is firmly established.

1.5.4A Systematic Paleontology

At the times of announcement of their discovery (Schopf and Packer 1986) and of their initial description (Schopf and Packer 1987), too few data were available to justify taxonomic treatment of the filaments of the Apex Basalt. Since then, however, additional studies have been carried out; based on size data for about 840 cells (occurring in 85 filaments), six fila-

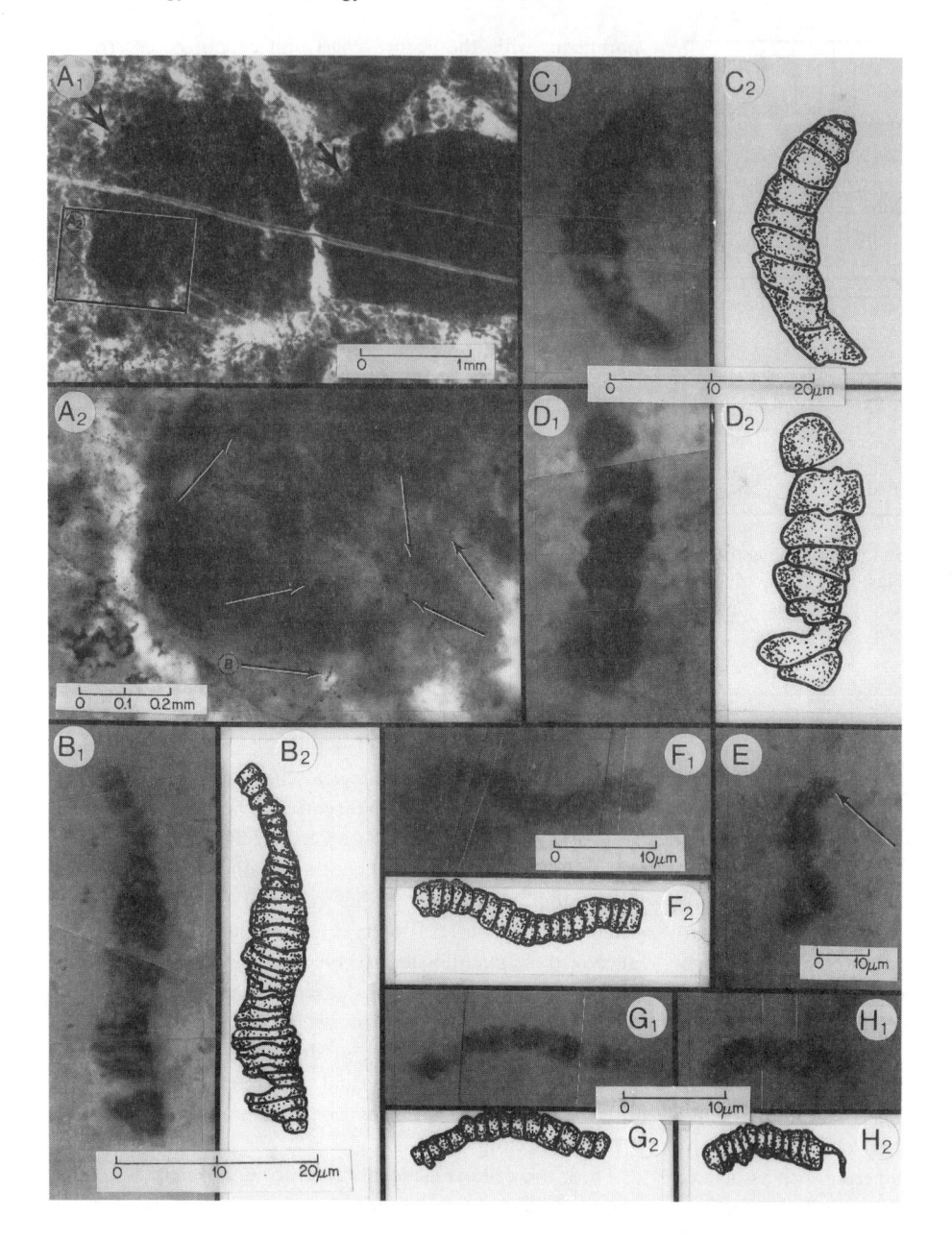

30   Geology and Paleobiology of the Archean Earth

Figure 1.5.4. Optical photomicrographs showing specimens in petrographic thin sections of gray, carbonaceous, bedded chert from the Early Archean Apex Basalt, Warrawoona Group, collected in 1982 (A–D, F–H) and 1986 (E) along Chinaman Creek near Marble Bar, Western Australia. [Thin section identification number, stage coordinates on Leitz Ortholux II microscope BM-NH #907464, and England Finder Slide coordinate are listed for each specimen; acquistion numbers are listed for specimens deposited in the collections of the British Museum-Natural History, London; $B_1, C_1, D_1, F_1, G_1, H_1$ show composite photomicrographs; $B_2, C_2, D_2, F_2, G_2, H_2$ show interpretive tracings of photomicrographs.] (A_1) microfossiliferous, subangular to rounded, light- to dark-brown, organic-rich, sedimentary lithic clasts (broad arrows), transected by a secondary quartz-filled veinlet, encompassed by the surrounding unfossiliferous chert matrix; rectangle at lower left delineates area shown in A_2 (thin section 4 of 6/15/82-1-D, slide label front left, 110.8×46.4, England slide V47). (A_2) a portion of a lithic clast shown in A_1 containing numerous very dark brown to black (opaque), carbonaceous, filamentous microfossils (arrows) permineralized in fine-grained quartz; arrow denoted by the circled "B" shows the location of the filament illustrated in B_1, B_2 (thin section 4 of 6/15/82-1-D, slide label front left, 110.8×46.4, England slide V47). (B_1, B_2) degraded cellular filament (or wrinkled sheath?) occurring near the upper surface of the thin section at the location denoted by the arrow with the circled "B" in A_2 (thin section 4 of 6/15/82-1-D, slide label front left, 111.2×45.7, England slide V47; BM-NH Number V.63165[1]). (C_1, C_2) *Primaevifilum conicoterminatum* Schopf, n. sp., HOLOTYPE; note conically tapered terminus (thin section 4 of 6/15/82-1-B, slide label front right, 113.1×23.9, England slide X24; BM-NH Number V.63164[1]). (D_1, D_2) *Primaevifilum conicoterminatum* Schopf, n. sp., PARATYPE, showing barrel-shaped medial cells (thin section 4 of 6/15/82-1-B, slide label front right, 112.5×24.5, England slide W25; BM-NH Number V.63164[2]). (E) *Primaevifilum conicoterminatum* Schopf, n. sp., PARATYPE; note (at arrow) conically tapered terminus (thin section PPRG 2006-1-A, slide label back right, 115.8×36.5, England slide Z37). (F_1, F_2) *Primaevifilum delicatulum* Schopf, n. sp., PARATYPE (thin section 4 of 6/15/82-1-B, slide label front right, 113.5×25.3, England slide X26; BM-NH Number V.63164[3]). (G_1, G_2) *Primaevifilum delicatulum* Schopf, n. sp., HOLOTYPE, exhibiting well defined, quadrate to discoidal medial cells (thin section 4 of 6/15/82-1-D, slide label front left, 111.3×45.6, England slide V47; BM-NH Number V.63165[2]). (H_1, H_2) *Primaevifilum delicatulum* Schopf, n. sp., PARATYPE, showing well defined medial cells (thin section 4 of 6/15/82-1-B, label front right, 113.6×25.2, England slide X26; BM-NH Number V.63164[4]).

Paleobiology of the Archean

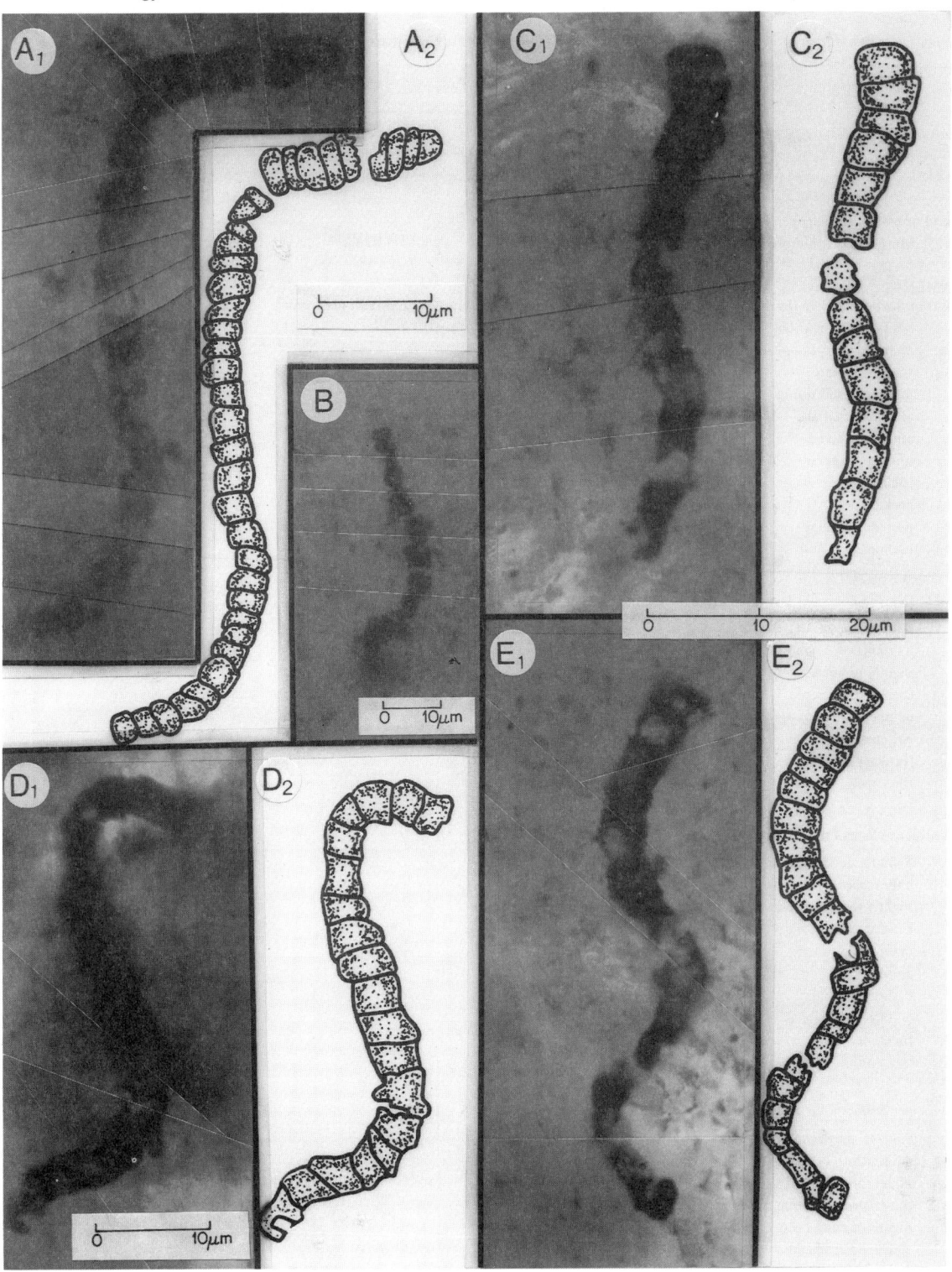

Figure 1.5.5. Optical photomicrographs showing permineralized, cellularly preserved specimens of *Primaevifilum amoenum* Schopf, n. sp., in petrographic thin sections of carbonaceous, bedded, gray chert from the Early Archean Apex Basalt, Warrawoona Group, collected in 1982 (A, C–E) and 1986 (B) along Chinaman Creek near Marble Bar, Western Australia. [Thin section identification number, stage coordinates on Leitz Ortholux II microscope BM-NH #907464, and England Finder Slide coordinate are listed for each specimen; acquisition numbers are listed for specimens deposited in the collections of the British Museum-Natural History, London; A_1, B, C_1, D_1, E_1 show composite photomicrographs; A_2, C_2, D_2, E_2 show interpretive tracings of photomicrographs.] (A_1, A_2) *Primaevifilum amoenum* Schopf, n. sp., HOLOTYPE, showing discoidal to quadrate medial cells and rounded, hemispheroidal terminal cell (thin section 4 of 6/15/82-1-F, slide label back right, 114.7 × 7.4, England slide Z7; BM-NH Number V.63166[1]). (B) PARATYPE (thin section PPRG 2006-1-A, slide label back right, 112.4 × 50.9, England slide X52). (C_1, C_2) PARATYPE (thin section 4 of 6/15/82-1-B, slide label front right, 112.0 × 25.0, England slide W26; BM-NH Number V. 63164[5]). (D_1, D_2) PARATYPE (thin section 4 of 6/15/82-1-D, slide label front left, 113.1 × 45.5, England slide X46; BM-NH Number V.63165[3]). (E_1, E_2) PARATYPE (thin section 4 of 6/15/82-1-B, slide label front right, 111.9 × 25.4, England slide W26; BM-NH Number V.63164[6]).

Table 1.5.2. *Filamentous microfossils from carbonaceous cherts of the Early Archean Warrawoona Group (Apex Basalt and Towers Formation), Western Australia.*

Taxon Morphotype [Formation, Locality] Figured Specimens	Filaments Morphology [Interpretation]	Maximum Observed Length (μm)	Cell Shape	Medial Portion of Filaments Cell/Filament Diameter (μm) Range	Avg.	Cell Length (μm) Range	Avg.	Number Measured Cells	Filaments	Terminal Cell Shape
I. SEPTATE FILAMENTS:										
I.1 *Eoleptonema australicum* Schopf 1983 (in Awramik, Schopf and Walter 1983) [Towers Formation, North Pole Dome]										
HOLOTYPE: Fig. 3e (Awramik et al. 1983), Photo 9-5C (Schopf and Walter 1983); TOPOTYPES: Photo 9-5A, B (Schopf and Walter 1983)	Solitary or gregarious, unbranched, cylindrical; possibly septate, uniseriate, and not at all constricted at septa; untapered toward apices [filamentous prokaryote, probably cellular trichome]	340	if septate, quadrate to elongate	0.8–1.1	~0.9	~1.0	~1.0	10	15 (~260 observed)	flat to slightly rounded
I.2 Unnamed "Type C" Narrow Filament [Apex Basalt, Chinaman Creek]										
Fig. 1.5.6C–E	Solitary, uniseriate, unbranched, cylindrical; slightly to not at all constricted at septa; untapered toward apices; not ensheathed [prokaryote cellular trichome]	40	elongate to nearly quadrate	1.5–2.5	2.0	2.0–3.0	2.5	46	5	?blunt-rounded
I.3 *Primaevifilum delicatulum* Schopf n. sp. [Apex Basalt, Chinaman Creek]										
HOLOTYPE: Fig. 1.5.4G, Fig. 2A (Schopf and Packer 1987); PARATYPES: Figs. 1.5.4F, H; TOPOTYPES: Figs. 2D, 2G (Schopf and Packer 1987)	Solitary, uniseriate, unbranched, cylindrical; slightly to not at all constricted at septa; untapered toward apices; not ensheathed [prokaryote cellular trichome]	45	discoidal to nearly quadrate	2.0–3.2	2.5	0.7–2.0	1.5	445	34	flat to slightly rounded
I.4 *Primaevilfilum amoenum* Schopf n. sp. [Apex Basalt, Chinaman Creek]										
HOLOTYPE: Fig. 1.5.5A; PARATYPES: Figs. 1.5.5B–E; TOPOTYPE: Fig. 2F (Schopf and Packer 1987)	Solitary, uniseriate, unbranched, cylindrical; slightly to not at all constricted at septa; slightly tapered near apices; not ensheathed [prokaryote cellular trichome]	80	discoidal to quadrate	2.0–4.5	3.8	1.5–3.8	2.5	252	27	rounded, hemispheroidal
I.5 *Primaevifilum septatum* Schopf 1983 (in Awramik, Schopf and Walter 1983) [Towers Formation, North Pole Dome]										
HOLOTYPE: Fig. 3a (Awramik et al. 1983) Photo 9-8 (Schopf and Walter 1983)	Solitary, uniseriate, unbranched, cylindrical; not at all to slightly constricted at septa; slightly tapered toward apices; not ensheathed [prokaryote cellular trichome]	120	quadrate	4.0–6.0	~5.0	4.5–7.0	~5.7	28	4 (10 observed)	unknown

Table 1.5.2 (*Continued*)

Taxon/Morphotype [Formation, Locality] Figured Specimens	Morphology [Interpretation]	Filaments Maximum Observed Length (μm)	Cell Shape	Medial Portion of Filaments						Terminal Cell Shape
				Cell/Filament Diameter (μm)		Cell Length (μm)		Number Measured		
				Range	Avg.	Range	Avg.	Cells	Filaments	
I.6 *Primaevifilum conicoterminatum* Schopf n. sp. [Apex Basalt, Chinaman Creek]										
HOLOTYPE: Fig. 1.5.4C, Fig. 2C (Schopf and Packer 1987); PARATYPES: Figs. 1.5.4D, E; TOPOTYPE: Fig. 2E (Schopf and Packer 1987)	Solitary, uniseriate, unbranched, cylindrical; somewhat to markedly constricted at septa; tapered toward apices; not ensheathed [prokaryote cellular trichome]	30	barrel-shaped to discoidal	4.5–7.0	5.9	3.0–4.5	3.9	83	16	conical to blunt-rounded
I.7 Unnamed "Type A" Broad Filament [Apex Basalt, Chinaman Creek]										
Fig. 1.5.6A	Solitary, uniseriate, unbranched, cylindrical; not at all constricted at septa; evidently not ensheathed [prokaryote cellular trichome]	45	short cylinder	7.0–8.5	8.0	4.5–5.0	4.8	12	2	unknown
I.8 Unnamed "Type B" Broad Filament [Apex Basalt, Chinaman Creek]										
Fig. 1.5.6B	Solitary, uniseriate, unbranched, cylindrical; not at all constricted at septa; evidently not ensheathed [prokaryote cellular trichome]	30	short cylinder	9.5–11.0	10.5	6.0–7.5	7.0	4	1	unknown
II. NON-SEPTATE FILAMENTS:										
II.1 *Archaeotrichion contortum* Schopf 1968 [Towers Formation, North Pole Dome (Awramik et al. 1983; Schopf and Walter 1983)]										
HYPOTYPES: Fig. 3h, i (Awramik et al. 1983), Photo 9-4 A–I Schopf and Walter 1983)	Solitary or irregularly entangled, non-septate (tubular), unbranched, untapered, cylindrical [non-septate, threadlike, filamentous prokaryote]	180	—	0.3–0.7	—	—	—		~40 (~1200 observed)	rounded, hemispheroidal
II.2 *Siphonophycus antiquus* Schopf 1983 (in Awramik, Schopf and Walter 1983) [Towers Formation, North Pole Dome]										
HOLOTYPE: Fig. 3b (Awramik et al. 1983), Photo 9-7C (Schopf and Walter 1983); TOPOTYPES: Photos 9–5D, 9–7A, B, D, E (Schopf and Walter 1983)	Solitary, non-septate (tubular), unbranched, untapered, cylindrical [tubular sheath of filamentous prokaryote]	600	—	3.0–9.5	~6.0	—	—		9 (~180 observed)	—

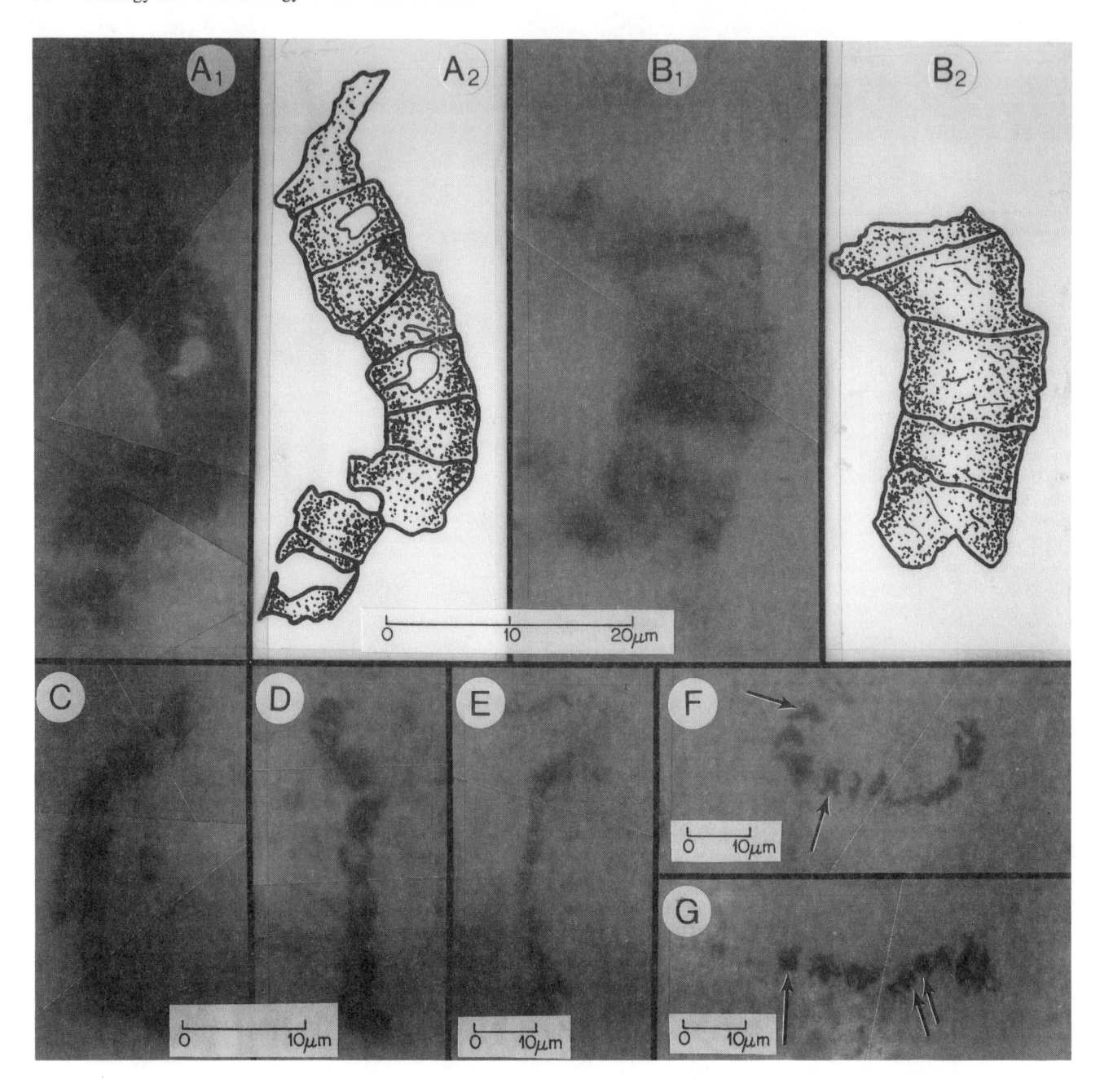

34 Geology and Paleobiology of the Archean Earth

Figure 1.5.6. Optical photomicrographs showing unnamed, permineralized, filamentous microfossils in petrographic thin sections of carbonaceous, bedded, gray chert from the Early Archean Apex Basalt, Warrawoona Group, collected in 1982 (A–D) and 1986 (E–G) along Chinaman Creek near Marble Bar, Western Australia. [Thin section identification number, stage coordinates on Leitz Ortholux II microscope BM-NH #907464, and England Finder Slide coordinate are listed for each specimen; acquisition numbers are listed for specimens deposited in the collections of the Bristish Museum-Natural History, London; A_1, B_1, C–G show composite photomicrographs; A_2, B_2 are interpretive tracings of photomicrographs.] (A_1, A_2) unnamed "Type A" broad filament (thin section 4 of 6/15/82-1-B, slide label front right, 111.4 × 23.2, England slide V24; BM-NH Number V.63164[7]). (B_1, B_2) unnamed "Type B" broad filament (thin section 4 of 6/15/82-1-D, slide label front left, 110.5 × 47.8, England slide V49; BM-NH Number V.63165[4]). (C, E) unnamed "Type C" narrow filaments (C, thin section 4 of 6/15/82-1-F, slide label front left, 115.6 × 65.8, England slide Z68, BM-NH Number V.63166[2]; D, thin section 4 of 6/15/82-1-B, slide label front right, 113.1 × 24.4, England slide X25, BM-NH Number V.63164[8]; E, thin section PPRG 2006-1-A, slide label back right, 115.8 × 36.5, England slide Z37). (F, G) degraded cellular filaments showing (arrows) bifurcated cells and paired half-cells reflecting the occurrence of partial septations (F, thin section PPRG 2006-1-A, slide label back right, 110.5 × 46.8, England slide U48; G, thin section PPRG 2006-1-A, slide label front left, 99.5 × 21.6).

mentous morphotypes have been recognized in the deposit (Table 1.5.2) of which the three most abundant (viz., *Primaevifilum delicatulum* Schopf n. sp.; *P. amoenum* Schopf n. sp.; and *P. conicoterminatum* Schopf n. sp.) are formally described below. Because of the exceptionally great age of these filaments, about 2000 Ma older than known Proterozoic trichomic species of comparable morphology (see discussions of fossil and modern morphological analogues, below), and because it remains to be established whether these Early Archean microfossils are members of evolutionary lineages represented by younger morphological analogues, these morphotypes have not been referred to Proterozoic taxa and the newly defined species are regarded as "prokaryotes *incertae sedis*."

Type locality. The three species formally diagnosed below are described based on permineralized specimens studied in petrographic thin sections of bedded carbonaceous chert from the Early Archean Apex Basalt (Warrawoona Group) collected from outcrop by D. Blight, D. J. Chapman, A. H. Hickman, C. Klein, and J. C. G. Walker at Schopf locality 4 of 15 June 1982; as specified by "locality 2" in Figure 1.5.3, the microfossiliferous cherts occur in northwestern Western Australia, about 12 km west of the town of Marble Bar, immediately south of Chinaman Creek (viz., at grid reference number 799558 on the Marble Bar 1:100,000 Australian National Topographic Map No. 2855, and at 21°11'4" S longitude and 119°42'36" E latitude within Archean map unit "Acj" of the Marble Bar structural belt on the Geological Survey of Western Australia 1:250,000 Marble Bar Geological Map Sheet SF 50-8).

Location and repository of type specimens. Captions for Figures 1.5.4, 1.5.5, and 1.5.6 list thin section identification numbers, stage coordinates on BM-NH (British Museum–Natural History) Leitz Ortholux II microscope #907464, and the England Finder Slide coordinate for all illustrated specimens; acquisition numbers are listed for holotype specimens (and for all other figured specimens occurring in the same thin sections) deposited in the permanent collections of the British Museum (Natural History), Cromwell Road, London.

Description of new taxa.
Kingdom Procaryotae Murray 1968,
Incertae Sedis
Genus *Primaevifilum* Schopf 1983 (in Awramik et al. 1983).
Type species: *Primaevifilum septatum* Schopf 1983 (in Awramik et al. 1983).

Species *Primaevifilum delicatulum* Schopf n. sp. (Figure 1.5.4F–H; Table 1.5.2).

Diagnosis: trichomes solitary, uniseriate, unbranched, and cylindrical, slightly to not at all constricted at septa, not tapered toward apices, not sheath enclosed, up to 45 μm long (incomplete specimen), composed of discoidal to nearly quadrate cells 2.0 to 3.2 μm in diameter and 0.7 to 2.0 μm in length, with average dimensions of 2.5 μm wide and 1.5 μm long ($n = 445$ cells measured in 34 trichomes), terminated by flat to slightly rounded terminal cells.

Etymology: with reference to delicate appearance.

Type specimen: trichome shown in Figure 1.5.4G (thin section 4 of 6/15/82-1-D; with slide label to front left, stage coordinates on BM-NH microscope #907464 111.3 × 45.6 and England Finder Slide coordinate V47; BM-NH Number V.63165 [2]).

Fossil and modern morphological analogues: among Proterozoic septate filaments, *P. delicatulum* is similar in cellular dimensions and terminal cell shape to microfossils here included in informal species 3N21 of Table 25.3 (*Contortothrix vermiformis* Schopf 1968) and earliest represented in the known Proterozoic fossil record by three species of *Oscillatoriopsis* described by Xu (1984a,b) from the ~1425 Ma-old Gaoyuzhuang Formation of northern China. Among modern taxa, as is noted in Table 25.3, informal species 3N21 seems particularly similar in morphology to certain leucothricacean (viz., *Thiothrix nivea*) and gliding bacteria (viz., *Achroonema inequale*).

Species *Primaevifilum amoenum* Schopf, n. sp. (Figure 1.5.5A–E; Table 1.5.2)

Diagnosis: trichomes solitary, uniseriate, unbranched, and cylindrical, slightly to not at all constricted at septa, slightly tapered near apices, not sheath enclosed, up to 80 μm long (incomplete specimen), composed of discoidal to quadrate cells 2.0 to 4.5 μm in diameter and 1.5 to 3.8 μm in length, with average dimensions of 3.8 μm wide and 2.5 μm long ($n = 252$ cells measured in 27 trichomes), terminated by rounded, hemispheroidal, terminal cells.

Etymology: with reference to pleasing, well-proportioned appearance.

Type specimen: trichome shown in Figure 1.5.5A (thin section 4 of 6/15/82-1-F; with slide label at back right, stage coordinates on BM-NH microscope #907464 114.7 × 7.4 and England Finder Slide coordinate Z7; BM-NH Number V.63166[1]).

Fossil morphological analogues: among Proterozoic septate filaments, *P. amoenum* is similar in cellular dimensions and terminal cell shape to microfossils here included in informal species 3N29 of Table 25.3 (*Filiconstictosus diminutus* Schopf and Blacic 1971) and earliest represented in the known Proterozoic fossil record by unnamed septate filaments reported by Diver (1974) from the ~1364 Ma-old Bungle Bungle Dolomite of Western Australia.

Species *Primaevifilum conicoterminatum* Schopf, n. sp. (Figure 1.5.4C–E; Table 1.5.2)

Diagnosis: trichomes solitary, uniseriate, unbranched, and cylindrical, somewhat to markedly constricted at septa, tapered toward conical apices, not sheath enclosed, up to 30 μm long (incomplete specimen), composed of barrel-shaped to discoidal cells 4.5 to 7.0 μm in diameter and 3.0 to 4.5 μm in length, with average dimensions of 5.9 μm wide and 3.9 μm long ($n = 83$ cells in 16 trichomes), terminated by conical to blunt-rounded terminal cells.

Etymology: with reference to tapered conical terminus.

Type specimen: trichome shown in Figure 1.5.4C (thin section 4 of 6/15/82-1-B; with slide label at front right, stage coor-

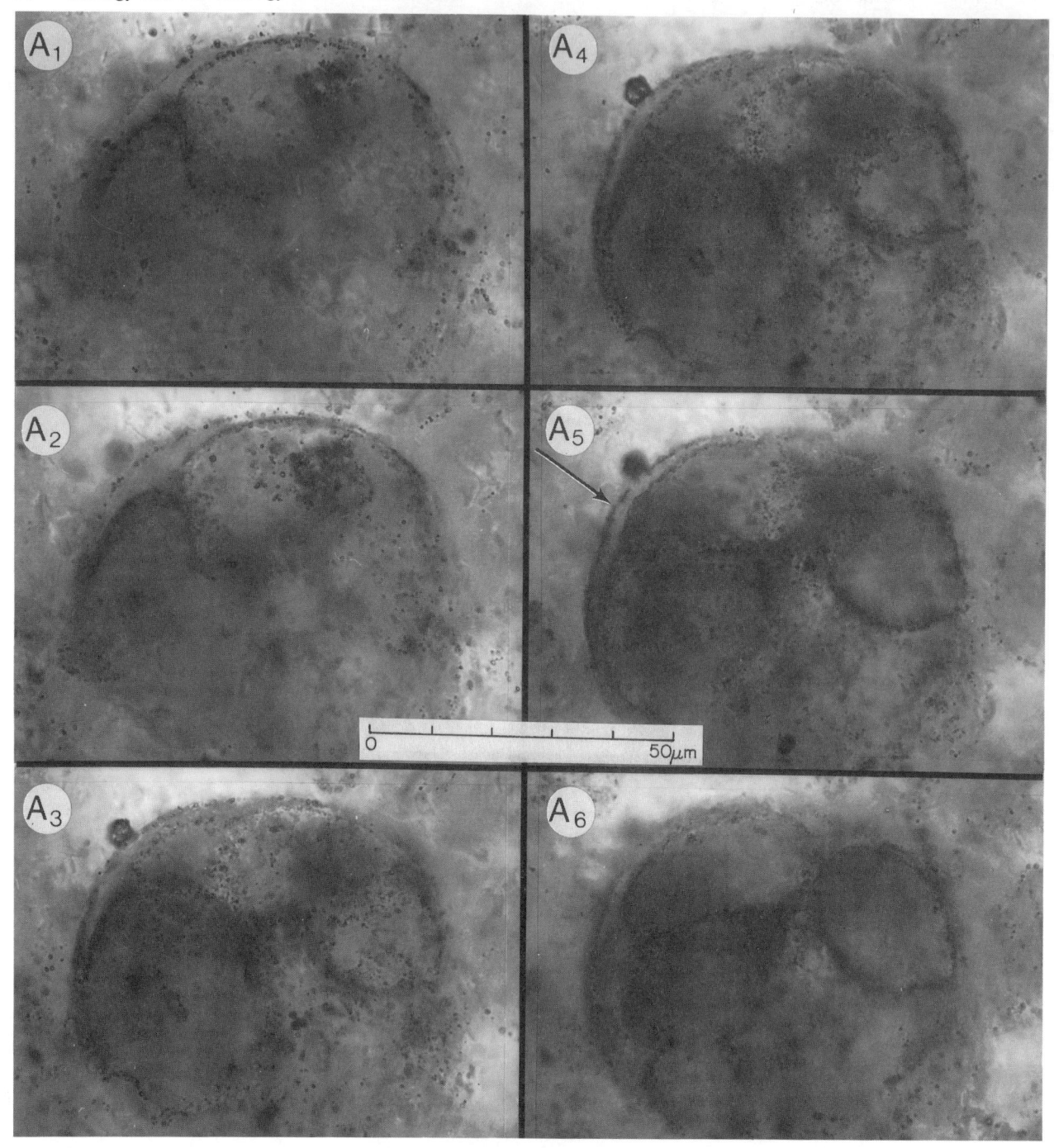

Figure 1.5.7. Optical photomicrographs showing an unnamed, chroococcalean-like, sheath-enclosed, spheroidal colony of closely adpressed ellipsoidal cells in a petrographic thin section of black, bedded, carbonaceous chert from the "middle chert horizon" of the Early Archean Towers Formation, Warrawoona Group, near Strelley Pool, Western Australia (North Shaw 1:100,000 Topographic Map Number 2755, at grid reference number 219636). (A_1, A_6) serial optical sections at progressively greater depth beneath the surface of the thin section; arrow in A_5 points to well defined, colony-encompassing, lamellated sheath (thin section 4 of 6/13/82-2-A, label front right, Leitz Ortholux II microscope BM-NH #907464 stage coordinates 17.1 × 109.1, England Finder Slide coordinate T17).

dinates on BM-NH microscope #907464 113.1 × 23.9 and England Finder Slide coordinate X24; BM-NH Number V.63164[1]).

Fossil and modern morphological analogues: among Proterozoic septate filaments, *P. conicoterminatum* is similar in terminal cell shape and more or less similar in cellular dimensions to microfossils here included in informal species 3N41 of Table 25.3 (*Halythrix nodosa* Schopf 1968) and earliest represented in the known Proterozoic fossil record by *Filiconstrictosus* sp. reported by Horodyski and Donaldson (1980) from the ~1400 Ma-old Dismal Lakes Group of northwestern Canada. Among modern taxa, as is noted in Table 25.3, informal species 3N41 seems particularly similar in morphology to the oscillatoriacean cyanobacterium *Oscillatoria acuta*.

1.5.5 The Early Archean Microbiota

In addition to the cellular filaments preserved in bedded cherts of the Apex Basalt discussed above, sheath-enclosed colonial unicells have recently been detected in cherts of the stratigraphically immediately underlying Towers Formation of the Warrawoona Group (Schopf and Packer 1986, 1987) at Strelley Pool, west of the Chinaman Creek locality (viz., Locality 1 in Fig. 1.5.3). Serial optical sections of the most distinctive of these sheath-enclosed colonies are shown in Figure 1.5.7, a specimen compared by Schopf and Packer (1986, 1987) with chroococcalean cyanobacteria. Indeed, comparison of this colony with fossil and extant *Chroococcus*-like chroococcaleans seems plausible: as shown in Figure 1.5.8, the ranges of cell diameter and sheath thickness of this Early Archean colony are comparable to those of numerous Proterozoic chroococcaleans having similarly prominent lamellated sheaths; and, as shown in Figure 5.4.14, both this Towers colony and the 17 *Chroococcus*- and *Gloeocapsa*-like Proterozoic taxa here recognized (Table 25.3) are comparable in their ranges of cell size and sheath thickness to modern *Chroococcus* and *Gloeocapsa*. However, and although other sheath-enclosed possible chroococcaleans have also been detected in the Towers cherts at Strelley Pool (Schopf and Packer 1987, Figs. 2H-J), only a single specimen of the large-celled variety shown in Figure 1.5.7 has been discovered; thus, and despite its distinctive morphology, this morphotype has not been formally described.

Morphological characteristics of the six morphotypes of filamentous microfossils now known to occur in the Warrawoona cherts at Chinaman Creek are listed in Table 1.5.2, as are such characteristics of the four types of fossil filaments previously described from Warrawoona cherts of the Towers Formation by Awramik et al. (1983). As discussed above, the former are definitely, and the latter are "very probably," authentic Archean microfossils. In addition, as noted above, sheath-enclosed colonies, comparable to extant and Proterozoic chroococcaleans, are also known to occur in Warrawoona Group cherts. What can be inferred regarding the probable biological affinities and metabolic characteristics of these Warrawoona microfossils? More generally, what levels of evolutionary development and ecological complexity had been attained by Early Archean time?

As is discussed in some detail by Schopf and Walter (1983, pp. 236-237), relevant paleontological and organic geochemical data seem to indicate that the Warrawoona biota was composed solely of prokaryotes, probably including anaerobes, autotrophs and heterotrophs, and various photoresponsive, including phototactic, microorganisms that at least in part occurred as members of microbial, stromatolitic, biocoenoses. These interpretations, based on data available in 1983 and evidently applicable to the Early Archean biota in general, are consistent with all additional evidence gathered to date, including the discovery of filamentous microfossils (Fig. 1.5.2; Walsh and Lowe 1985) and excellently preserved microbial stromatolites (Fig. 1.5.1E; Byerly et al. 1986) of comparable age in the Onverwacht and Fig Tree Groups of South Africa (Section 1.5.2).

Because of its evolutionary and environmental significance, the cardinal ecological question to be asked is whether oxygen-producing (as opposed to solely anoxic, bacterial) photosynthesizers were represented in the Early Archean biota. Not inconsistent with this supposition are (*i*) the well-established occurrence in the Early Archean environment (Chapter 4) of the reactants required for oxygenic photosynthesis, CO_2 and H_2O; (*ii*) the presence of kerogen (Chapter 3) and of oxidized iron minerals (e.g., hematite) in Early Archean sediments (Chapter 4), materials that might evidence the products of oxygenic photosynthesis, cellular organic matter and free oxygen; (*iii*) the occurrence of Early Archean stomatolites (Fig. 1.5.1; Section 1.5.2) which, like the vast majority of their modern morphological counterparts, might have been produced by cyanobacterium-dominated microbial communities (Chapter 6); and (*iv*) the carbon isotopic composition of Early Archean organic matter (Chapter 3; Schidlowski et al. 1983) which is evidently indicative of enzymatic CO_2-fixation similar to that occurring in exant oxygen-producing photoautotrophs.

However, none of these lines of evidence is at all conclusive: most would be equally consistent with the Early Archean presence of solely anoxic, bacterial, stromatolite-building photoautotrophs (Section 6.11; Schopf and Walter 1983), the sole exception being the occurrence of Early Archean oxidized sedimentary minerals for which UV-induced photodissociation of water vapor would seem a plausible non-biological source of the required free oxygen.

What relevance, if any, do the newly detected Warrawoona microfossils have to the solution of this problem? As indicated above, the colony illustrated in Figure 1.5.7 is notably similar in morphology both to Proterozoic (Fig. 1.5.8) and extant (Fig. 5.4.14) chroococcalean cyanobacteria. Moreover, the cells of this Early Archean colony are nearly an order of magnitude larger than those of any morphologically comparable, lamellated sheath-enclosed, extant non-oxygen-producing prokaryote (e.g., *Siderocapsa*; Buchanan and Gibbons 1974, p. 464). In short, if this colony were found in a modern microbial community, and if its affinities were to be determined solely on the basis of morphology, it would almost certainly be referred

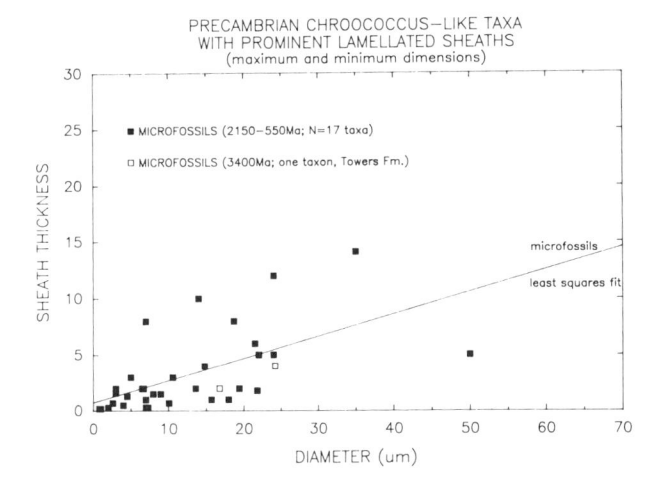

Figure 1.5.8. Cell diameters and sheath thicknesses (maximum and minimum dimensions) for Archean and Proterozoic *Chroococcus*-like microfossils with prominent lamellated sheaths.

to the Chroococcales and, as such, be inferred to be an oxygen-producing cyanobacterium. It is important to note, however, that only a single specimen of this distinctive morphotype has been detected in the Warrawoona cherts; it would thus seem inappropriate to stake a major generalization (viz., a minimum date for the time of origin of oxygenic photosynthesis) on this single known example.

Do the newly detected filamentous fossil microorganisms shed light on this problem? Among modern filamentous microbes, quadrate, discoidal, and barrel-shaped medial cells like those of the fossil filaments are exhibited both by bacteria and cyanobacteria, as are rounded (Fig. 1.5.5A) and conical terminal cells (Fig. 1.5.4C, E). Similarly, bifurcated cells and paired half-cells like those of the fossil filaments (Fig. 1.5.6F, G), indicative of prokaryotic organization (and significant because they reflect the presence of partial septations, thus providing direct evidence that cell division in these Early Archean trichomes occurred via the same mechanism as that in extant filamentous prokaryotes), are also exhibited both by modern bacteria and cyanobacteria.

Perhaps significantly, however, the cell sizes of the majority of the fossil trichomes seem more suggestive of cyanobacterial, than of bacterial, affinity. As tabulated in Table 1.5.2, cell widths of the eight trichomic morphotypes reported from the Warrawoona cherts range from a minimum of $0.8 \mu m$ to a maximum of $11.0 \mu m$, with an average of $4.8 \mu m$ (with the six septate morphotypes from the Chinaman Creek locality, microfossils of unquestionable Early Archean provenance, ranging in diameter from 1.5 to $11.0 \mu m$, with an average of about $5.5 \mu m$); and diameters of the two tubular, non-septate, sheath-like morphotypes range from $0.3 \mu m$ to $9.5 \mu m$ with an average of about $3.3 \mu m$. As discussed in Sections 5.4.6A and 5.4.7A, and based on morphometric analyses of more than 800 relevant extant prokaryotic species and varieties, analogy with extant filamentous microbes suggests that fossil cellular trichomes $< 1.5 \mu m$ in width and tubular sheaths $< 2.0 \mu m$ in diameter can be regarded as probably representing bacteria, whereas fossil trichomes or sheaths $> 3.5 \mu m$ in diameter can be interpreted as probably being of cyanobacterial origin. Applying these criteria, one of the reported Warrawoona trichomic morphotypes (viz., entry I.1 in Table 1.5.2) would be regarded as a "probable bacterium"; two of the trichomic morphotypes (I.2 and I.3) as "prokaryotes," either bacteria or cyanobacteria; and the remaining five morphotypes (I.4 through I.8) would appear to represent "probable cyanobacteria" (with two of the morphotypes from the Chinaman Creek locality, I.2 and I.3, being interpreted as either bacteria or cyanobacteria; and the remaining four, I.4, I.6, I.7, and I.8, as "probable cyanobacteria"). Similarly, one of the non-septate, tubular morphotypes (entry II.1 in Table 1.5.2) would be regarded as "probably bacterial," and the other (II.2) as "probably cyanobacterial".

Therefore, by analogy with extant microorganisms, the majority of the filamentous morphotypes now known from the Warrawoona cherts seem more likely to be of cyanobacterial, than of bacterial, affinities. Although some extant cyanobacteria are capable of facultative anoxygenic photosynthesis (Section 6.9), the ability to carry out oxygen-producing photoautotrophy is a universal characteristic of the group. Based on these observations, it thus seems a plausible inference that cyanobacteria, and therefore oxygen-producing photosynthesizers, are likely to have been represented in the Warrawoona biota and, hence, that this relatively advanced level of physiological evolution may have been attained at least as early as ~ 3500 Ma ago (cf. Schopf and Packer 1986, 1987).

This inference can be questioned on several grounds. In the first place, although uncommon among extant taxa, bacterial trichomes with widths in excess of $11.0 \mu m$ (the broadest of the Warrawoona cellular filaments; Table 1.5.2) are known to occur in the modern biota (Fig. 5.4.29A); indeed, some 16% of modern bacterial taxa have widths greater than the $3.5 \mu m$ "limit" here used to define "probable cyanobacteria" (Fig. 5.4.31). Thus, even assuming the validity of the "inference by analogy" here applied, the differentiation of "probable bacterial" from "probable cyanobacterial" fossil trichomes is by no means absolute. Interestingly, this is not true of the Warrawoona tubular sheaths: the broadest tubular sheaths of modern filamentous bacteria are about $7.5 \mu m$ in diameter (Fig. 5.4.23A) whereas the largest such fossils known from the Warrawoona cherts are $9.5 \mu m$ (Table 1.5.2); sheaths of modern bacteria, unlike those of extant cyanobacteria (Figs. 5.4.23B, 5.4.27A), do not attain the dimensions of the largest, comparable, Warrawoona microfossils.

Also embedded in this "inference by analogy" are a number of key assumptions. To apply this reasoning, for example, it must be assumed that the Warrawoona microfossils are members of still extant lineages, part of an unbroken evolutionary continuum that extends from the Early Archean to the present, rather than being members of extinct lineages only remotely related to living microbes. Although such an inference is evidently valid for Proterozoic microorganisms, as documented in Section 5.4, because of the paucity of relevant fossil data it is an assumption that cannot yet be evaluated for the much older microorganisms of the Archean. Moreover, this approach assumes that the morphology of fossil microorganisms can be used validly as an indicator of their physiological capabilities, in short, that such microbes have not been affected markedly by the "Volkswagen Syndrome" (Schopf 1977; Schopf et al. 1983), a lack of external morphological evolution masking important changes in internal, biochemical, machinery.

Further studies and additional data will be required before the question of the possible Early Archean appearance of oxygen-producing photosynthesis is finally resolved. However, it seems evident that if the range of morphotypes now known from Warrawoona cherts were to be found in a microbial community today, and if morphology were the only criterion available on which to base inferences of the biologic affinities of such microbes, the majority of these forms would be regarded as cyanobacteria rather than as non-oxygen-producing prokaryotes. And this observation raises an intriguing possibility: either the majority of the known Warrawoona morphotypes were, indeed, cyanobacteria (and, thus, presumably either obligately or facultatively oxygen-producing) or they were entirely anaerobic, in which case the "cyanobacterium-like" morphotypes would represent atypical and possibly now extinct lineages, varieties of very early evolving prokaryotes that heretofore have gone unnoticed in the fossil record.

1.5.6 Conclusions

As documented by previous, thorough, reviews (see Schopf 1983a), and here updated by brief consideration of significant recent developments, it is evident that as early as ~3500 Ma ago microbial communities were extant, morphologically varied, and possibly physiologically advanced, and that Archean microfossils, both filamentous and colonial, are notably similar in morphological detail to extant prokaryotes, a similarity evidently extending even to the mechanisms of cell division. As discussed in Sections 5.4 and 13.3.2, Precambrian (Proterozoic and, evidently, Archean) prokaryotes seem clearly to have been "hydrobradytelic" (Schopf 1987), exhibiting exceptionally slow rates of morphological evolutionary change.

However, and although substantial progress thus has been made, the deficiencies in the known Archean fossil record are striking. Indeed, only three assuredly microfossiliferous units have as yet been reported from the entire Archean (Schopf and Walter 1983) – cherts of the Onverwacht and Warrawoona Groups, here discussed, and stromatolitic cherty carbonates of the ~2800 Ma-old Fortescue Group of Western Australia from which cyanobacterium-like filaments have been reported and illustrated by Schopf and Walter (1983, pp. 237–238, Photos 9-9, 9-10). Much remains to be learned about the nature, composition, distribution, and evolutionary development of the Archean biota. Nevertheless, it is abundantly clear that relatively complex microbial ecosystems became established early in the history of the earth, and that the evolutionary roots of the Proterozoic biosphere, the principal subject of this monographic effort, extend far into the Archean past.

Acknowledgments

Fieldwork by the PPRG of June, 1982, on which this work is based, was supported by NASA grant NGW-825; I much appreciate the help of the other participants (K. J. Armstrong, D. Blight, D. J. Chapman, J. M. Hayes, A. H. Hickman, C. Klein, D. D. Radke, J. C. G. Walker, and M. R. Walter). I also thank B. M. Packer for collecting additional chert samples from the Chinaman Creek locality of the Apex Basalt in August, 1986, during the course of her field studies of the Fortescue Group which were supported by NASA Grant NGR 05-007-407 (to J.W.S.), with the generous assistance of the Western Australia Geological Survey, Perth, and the Bureau of Mineral Resources, Canberra; the scientific staff of the British Museum (Natural History) where this research was carried out in 1989; and Toby B. Moore for discussion and helpful review of an earlier version of this paper. The research here reported was supported by NASA Grant NGR 05-007-407, NSF Grant BSR 86-13583, and by a Guggenheim Fellowship.

1.6

Geology and Paleobiology of the Archean Earth: Current Status and Future Research Directions

J. WILLIAM SCHOPF

1.6.1 Archean Paleobiology: Problems and Perspectives

(1) The foregoing discussions of the formation of the earth and the origin of life, and of the geology, environment, and paleobiology of the Archean earth are, in large measure, abbreviated, updated summaries of topics addressed in the previous PPRG collaborative work, *Earth's Earliest Biosphere* (Schopf 1983a), to which the reader is referred for more detailed consideration of these and related aspects of Archean earth history.

(2) As based on direct evidence from the geologic record, understanding of the origin of life, of Archean geology, of the early development of the atmosphere, oceans, and the surficial environment, and of the nature and distribution of earth's earliest ecosystems is hampered chiefly by (*i*) a paucity of preserved, ancient, sedimentary rock sequences; (*ii*) the predominantly biochemical, intracellular nature of early evolution and the difficulties inherent in the preservation of evidence of such evolutionary innovations in the fossil record; and (*iii*) the relative newness of studies of these problem areas and the resulting absence of an extensive base of data and experience on which reliable interpretations can be firmly grounded.

(3) With regard to the origin of life, there is need for (*i*) detection and characterization of pre-biotic organic matter in ancient sediments to provide direct confirmation of the chemical evolution hypothesis; (*ii*) improved models of the energetics and dynamics of earth's accretion and core formation, of the origin and early evolution of the atmosphere, and of the influence of impactors in the early environment, to provide tighter constraints on the physical, chemical, and environmental conditions on the pre-biotic earth and on the possible contribution of extraterrestrial organic matter to the early carbon budget; and, perhaps most importantly, (*iii*) in laboratory experiments designed to simulate organic syntheses under early earth conditions, improved consonance between the experimental conditions used and the atmospheric compositions suggested by refined models of the early environment.

(4) Geologically, there is need for (*i*) additional, detailed, petrotectonic analyses of relatively well-preserved ancient rock sequences; (*ii*) new, more refined data regarding, especially, the paleomagnetic and geochronologic framework of early crustal evolution; and, above all, (*iii*) discovery and characterization of additional Archean (and Hadean), relatively unmetamorphosed, geologic terranes, units that might contain evidence by which to resolve the numerous uncertainties regarding the earliest history of life and of the planet.

(5) At present, and despite the progress of recent years, numerous, biologically important, Archean environmental parameters remain poorly constrained. Habitable environments, in many respects like those of the Proterozoic and even the present-day, appear to have existed since early in the Archean. For example, since the beginning of the known rock record, liquid water oceans and an atmosphere containing N_2, CO_2, and water vapor have been extant; the chemistry of sea water has evidently remained relatively unchanged; and shallow water, open ocean, and land surface environments have existed. Nevertheless, there is pressing need for direct evidence to constrain current models of: (*i*) solar evolution and the related temperature regime of the Archean environment; (*ii*) the P_{CO_2} of the Archean atmosphere; (*iii*) the P_{O_2}, P_{O_3}, and ambient UV flux in the early environment; and (*iv*) the evolution of the Earth–Moon system and of interrelated day-length on the Archean earth.

(6) The Archean fossil record, even in comparison with that known from the later (Proterozoic) Precambrian, is minuscule. Nevertheless, sufficient data are available to establish that by $\sim 3.5\,\text{Ga}$ ago prokaryotic biocoenoses were extant, forming laminar, stromatolitic, microbial mats at the sediment-water interface in shallow water to intermittently exposed settings; that these microbiological communities contained both heterotrophs and autotrophs; that at least some of these microorganisms were anaerobes; and that the major mat-building taxa were filamentous, phototactic, and presumably photoautotrophic (although not necessarily oxygen-producing). Major unsolved paleobiologic questions include (*i*) the time of origin of oxygenic photosynthesis (possibly, but not assuredly, established by $\sim 3.5\,\text{Ga}$); (*ii*) the possible contribution of methane-producing and -utilizing microorganisms and of sulfate-reducing microbes to early ecosystems (metabolic processes suggested by available isotopic data to have been extant during the Late Archean to Early Proterozoic); (*iii*) the time of first

appearance of widespread phytoplankton (possibly related to the development of an effective UV-absorbing atmosphere); (*iv*) the possible importance of thermoacidophile-dominated fumarolic and vent-system microbial communites in the early biosphere (as suggested by molecular phylogenies); and (*v*) the evolutionary relationships of Archean prokaryotes (perhaps most fruitfully analyzed by combined study of cellularly preserved microfossils and the molecular biology of extant microbes).

(7) From this brief analysis, and despite the numerous deficiencies of the Archean rock and fossil records as now known, it is abundantly evident that the antecedents of the Proterozoic biosphere, the subject addressed in the remaining chapters of this interdisciplinary monograph, were already well established during the Archean, at a very early stage in the history of the earth.

2
Geological Evolution of the Proterozoic Earth

DONALD R. LOWE   NICOLAS J. BEUKES   JOHN P. GROTZINGER
RAYMOND V. INGERSOLL   JOSEPH L. KIRSCHVINK
CORNELIS KLEIN   IAN B. LAMBERT   JÁN VEIZER

2.1

Introduction

DONALD R. LOWE

The Proterozoic Eon extends from 2500 to 550 Ma, from the close of the Archean to the beginning of the Cambrian. It includes three principal geochronologic divisions: Lower or Proterozoic I (2500 to 1600 Ma), Middle or Proterozoic II (1600 to 900 Ma), and Upper or Proterozoic III (900 to 550 Ma). These definitions are consistent with previous usage (Schopf 1983a) and with recommendations of the Subcommission on Precambrian Stratigraphy of the International Union of Geological Sciences (Plumb and James 1986). Although some criticism has been voiced at the use of absolute ages rather than stratigraphic or paleontologic events for subdividing Precambrian time (Cloud 1987), we find that the lack of well-developed, widespread, narrowly constrained, isochronous Precambrian biostratigraphic markers, equivalent to Phanerozoic faunal successions, presents an as yet insurmountable barrier to the establishment of globally useful Precambrian biostratigraphic subdivisions.

Systematic treatment of the geological evolution of the Proterozoic earth and similar long-term or large-scale aspects of Proterozoic history is complicated at present by our incomplete knowledge of existing Proterozoic rocks, selective preservation/obliteration of certain types of terranes, and uncertain geochronology and correlation. Many Proterozoic sequences remain essentially unstudied, especially in parts of central and northern Africa, South America, and Asia, and their stratigraphies, ages, and tectonic settings are unresolved. Information from these sequences is essential to evaluation of global patterns of Proterozoic geologic evolution, sediment recycling, and tectonics. The selective recycling of sediments deposited in non-cratonic settings has left few if any examples of Proterozoic oceanic domains and only sparse, deformed and metamorphosed representatives of most arc, oceanic island, and eugeoclinal deposits. Most Proterozoic sedimentological and paleobiological research has focused on well-preserved, unstrained, relatively unmetamorphosed cratonic or craton-margin sequences, perhaps leading to the impression that these represent the dominant styles of Proterozoic sedimentation and tectonism. However, enormous areas of basement on all continents are composed of deformed, metamorphosed, and intruded Proterozoic sedimentary sequences, most reflecting sedimentation in tectonically and magmatically active non-cratonic settings.

Although high-precision dating is providing increasingly accurate estimates of the ages of rock units, available ages are still sparse. Many important sedimentary sequences lack materials for precise dating of the time of deposition. Rb-Sr and K-Ar ages are suspect and must generally be regarded as minimum ages. Recent studies have demonstrated that even the most reliable age dating techniques, including U-Pb dating of zircons, must be approached with care to understand the geologic meaning of the derived ages. The absence of reliable, precise ages for many sequences and of useful, narrow biostratigraphic subdivisions makes regional correlation problematic.

There can be little doubt that some form of plate tectonics provided the driving force for Proterozoic tectonism and crustal development. There is still considerable controversy, however, as to the degree to which Proterozoic plate systems may have differed from those of the Phanerozoic, including the sizes of individual plates, rates of plate motion, and relative roles of rifting versus accretionary or collisional tectonics in crustal generation.

In the following discussion we will develop an overview of Proterozoic geologic regimes and attempt to construct a synthetic picture of Proterozoic crustal development that can serve as a basis for evaluating interrelationships among crustal, oceanic, atmospheric, and biological evolution. In discussing crustal evolution, it is important to note what is meant by the "age" of individual crustal blocks. Three particularly important ages need to be carefully distinguished: (*i*) Sm-Nd and Nd-isotopic model ages of the oldest juvenile components of a crustal block, which are related, often complexly because of contamination, to the time original melts became independent of the mantle reservoir; (*ii*) the eruption and crystallization ages of the oldest juvenile rocks, not older recycled mineral grains or terranes, within the cratonic block; these are usually greenstone volcanic and sedimentary units and closely associated plutons, many of which record events preceding the existence of a stable sialic block; and (*iii*) the ages of the last major deformational/plutonic/metamorphic events leading to the formation of a stable sialic block and followed by deposition of cratonic successions unconformably on deeply eroded, mainly igneous and metamorphic basement. While all three record various "ages of formation," commonly separated by tens to hundreds of millions of years, the following discussion will focus principally on the second and third types of ages as the ages of crustal formation.

2.2

Proterozoic Sedimentary Basins

JOHN P. GROTZINGER   RAYMOND V. INGERSOLL

2.2.1 Introduction

Sedimentary basins are the storehouses of fossilized organisms, their fragments, and biochemical markers. The sediments that entomb these fossil remnants record the environmental conditions in which early life evolved. Useful information may be extracted from the sediments concerning the composition of seawater, paleoenvironmental zonation including water depth estimates and paleoecologic belts, and the evolution of these environments through time. However, the successions that fill sedimentary basins are the products of several influences including tectonic setting and subsidence history, sea-level history, the supply of sediments and nutrients, and the effects of organism growth, among other factors. These elements combine to strongly influence sequences of sediments and paleoenvironments in basins, and the history of burial that invariably promotes post-depositional alteration of the sediments and their biologic components.

It is useful to examine settings, styles, and the evolution of Proterozoic sedimentary basins as part of a comprehensive assessment of factors that may have controlled the evolution of early life. In large part, the origin and development of Proterozoic basins are similar to those developed in Phanerozoic time and have been preserved in extensional settings (rifts and passive margins), compressional settings (foreland basins and strike-slip basins), and hybrid settings resulting from combinations of the end-members. In addition, thin sequences are preserved in some areas as cratonic veneers.

In a general sense, the development of Proterozoic sedimentary basins can be understood in terms of actualistic geodynamic concepts involving interactions among lithospheric plates of both continental and oceanic affinity. On a more specific level, however, there are differences which perhaps can be related to the secular evolution of the earth's lithosphere.

2.2.2 Preservation of the Proterozoic Record

In any discussion of the Proterozoic stratigraphic record, it is essential to remember that only a small proportion of the original deposits and basins of Proterozoic history is preserved. Today's record reflects the vagaries of preservation inherent in multiple erosional and tectonic recycling events that have occurred continuously through earth history (Veizer and Jansen 1979, 1985; Veizer 1988a). In general, oceanic realms have higher recycling rates and lower preservation potential, whereas continental realms have higher preservation potential, due to faster creation and destruction of oceanic lithosphere during seafloor spreading. Most major sedimentary basins form along continental margins of convergent, divergent, and transform character and have intermediate recycling rates and preservation potential (Ingersoll 1988; Veizer 1988c).

Ingersoll (1988) described 23 different basin types, based on modern settings. The majority of these basin types have not been identified in the Proterozoic record. For example, deep-marine Proterozoic deposits formed on oceanic crust are presently unknown. The largest sediment accumulation in the modern world are huge submarine fans shed from areas of continental collision and deposited in remnant ocean basins (e.g., the Bengal-Nicobar and Indus fans). Comparable Proterozoic sequences have not been identified; thus, major areas of sediment accumulation (and potential fossil records) are unrepresented.

The following discussion summarizes actualistic geodynamic constraints on Proterozoic histories for several basin types. Paleobiologists should realize that the Proterozoic stratigraphic record is limited. The restricted types and numbers of Proterozoic basins remaining should illustrate the biases of the record.

2.2.3 Extensional Settings

2.2.3A Rift Basins

Rift basins are known on a global basis from throughout the Proterozoic. They developed in isolation with no subsequent history of further extension or deformation (e.g., the Mid-continental Rift of the USA) or, more commonly, as an initial phase in the evolution of the world's Proterozoic orogenic belts. The rifts are dominated by immature siliciclastic sediments of fluvial and marine origin, interspersed with thin units of carbonate and possible evaporites. Mafic to felsic volcanic rocks and intrusive sills also occur. In some cases, the rifts are dominated by thick sedimentary sequences (e.g., the

Ocoee Supergroup of the eastern USA; the Pahrump Group of the western USA; the Upper Proterozoic of the North Atlantic region) whereas in others, voluminous volcanism was common (e.g., the Mid-continent Rift of the USA). Rift-related lithologies typically overlie older Proterozoic or Archean basement.

Substantial differences between Phanerozoic and Proterozoic rifts are not observed. In both, attenuation of continental lithosphere led to block faulting of the upper crust to produce basins that filled with sediments derived from the uplifted margins of the blocks and shoulders of the rifts. In some cases, extension led to major mafic volcanism with (?MORB-like, i.e., possible mid-ocean basalt-like) affinities (e.g., the Mid-continent Rift), that, perhaps, resulted from lower-mantle upwelling, crustal underplating, and ponding of melts (cf. Hoffman 1989). In the Cape Smith thrust-fold Belt of northern Canada, volcanics show increasing MORB-like signatures with increasing height in the structural thrust wedge. Palinspastic restoration of the thrust sheets combined with documentation of ophiolitic stratigraphy suggests that the uppermost slices of largely mafic volcanics were extruded through highly attenuated continental crust (Hynes and Frances 1982; St. Onge et al. 1988).

It is clear that Proterozoic rift assemblages represent a continuum of extensional regimes, ranging from limited (e.g., the Mid-continent Rift) to extreme (the Cape Smith Belt). Substantial thicknesses of strata (10 to 15 km) are present in some rifts (e.g., the Belt Basin, USA; the Pine Creek Belt, Australia; the Mid-continent Rift, USA; the Upper Proterozoic of east-central Greenland), suggesting crustal and lithospheric thicknesses not substantially different from modern values. Under the condition of limited stretching, as shown by examples such as the Mid-continent Rift, lithospheric plates on the order of 125 km are required in order to produce rift accumulations greater than 10 km in thickness (McKenzie 1978). This is, perhaps, not surprising given the evidence for a 150- to 200-km-thick lithosphere for the Kaapvaal Craton (South Africa) as early as the Late Archean (Boyd et al. 1985). The results of Boyd et al. (1985), from analysis of peridotitic inclusions in diamonds, indicate that a cold, thick, basalt-depleted root has been present beneath the Kaapvaal Craton since 3.0 Ga.

2.2.3B Passive Margins

This Section discusses basins in which the primary driving force for subsidence was thermal cooling of formerly extended continental lithosphere. Therefore, it necessarily includes passive margins adjacent to oceanic lithosphere, as well as intracontinental rifts that may record stretching great enough to markedly thin the lithosphere, but less than that required to generate oceanic lithosphere (unless a component of transform movement is involved during rifting). As Ethridge et al. (1987) point out, the distinction between these two situations is one of a matter of degree, and the end products, sediments accumulating through passive thermally driven subsidence, may be very similar. This subtle distinction, however, is central to the major question of whether or not continental fragmentation and dispersal occurred during the Proterozoic. An analysis of the sequences that may have formed as a result of thermal cooling of extended continental lithosphere can place some new constraints on this subject.

To date, no quantitative test has been made to show that Proterozoic extensional basins formed as a result of thermal cooling (cf. Steckler and Watt 1982; Bond and Kominz 1984). However, on the basis of qualitative comparisons, the similarity of Proterozoic and Phanerozoic basins is striking, both in terms of the positions of inferred thermal-phase sediments between documented rift and foredeep sediments, and the onlap geometry and internal stratigraphy of the inferred thermal-phase sediments (Hoffman and Bowring 1984; Grotzinger 1986a; Ethridge et al. 1987; Grotzinger and McCormick 1988). Thermal-phase sediments overlie rift-phase sediments and step over their growth faults, eventually onlapping unfaulted cratonic basement. Cratonic onlap may be due to flexural loading of the margin during sedimentation (Watts et al. 1982; Bond et al. 1989).

Several stratigraphic features are consistent with these sequences having formed during cooling phases following extension of lithosphere. First, all sequences face toward their associated orogenic hinterlands, and progradation of the major stratigraphic units occurred in that direction (Hoffman 1973; Ricketts and Donaldson 1981; Ethridge et al. 1987; Grotzinger and McCormick 1988). Sequences are parallel to the trends of overlying foreland-basin axes, and perpendicular to tectonic indicators of vergence where thrusting and/or folding subsequently occurred. Furthermore, sequence development commonly involves the formation of carbonate platforms that overstep basal siliciclastic sediments and directly onlap cratonic basement (Grotzinger and McCormick 1988). Such stratigraphic onlap is typical of Phanerozoic passive margins (Vail et al. 1984; Bond et al. 1989) and is controlled by sedimentary loading on the increasingly rigid lithosphere as it cools (Watts et al. 1982).

Thus, in a general way, Proterozoic thermal-phase sequences can be directly compared to, and interpreted in terms of, actualistic geodynamic principles. It can be inferred that Proterozoic continental lithosphere behaved similarly to Phanerozoic lithosphere, responding to the effects of rifting (attenuation) by cooling and adjusting to new equilibrium thermal thicknesses. In the process, basins were created through the combined effects of local isostatic compensation of the attenuated region as well as flexural subsidence of the adjacent craton by sediment loading.

There is a substantial difference, however, between Phanerozoic and Proterozoic passive-margin sequences. Paleozoic through modern passive margins commonly have had durations as long as 140 Ma and sedimentary sequences of up to 6 or 7 km thick (Vail et al. 1977; Steckler and Watts 1982). Some were as long-lived as 200 Ma. In contrast, Proterozoic passive margin sequences are rarely over a few kilometers thick. Although there are several possible ways to explain this observation, the simplest is that thermal phases were of short duration and thus precluded the development of thicker sediment prisms. This interpretation is supported by new geochronological data from three separate basins, which indicate that thermal-phase sediments spanned less than 80 Ma for the Epworth Group and less than 50 Ma for the Kimerot Group in northern Canada (Grotzinger 1989; Bowring and Grotzinger 1989), and only a few tens of Ma for the Mount Partridge Group (Ethridge et al. 1987).

If thermally driven passive subsidence phases were of shorter duration, then continents probably were not fragmented and dispersed for long periods of time (so as to generate broad ocean basins between fragments). The thin passive margins of probably shorter duration suggest that if other factors were constant (i.e., spreading rates), then ocean basins between continental fragments were not as large as they have been subsequently. However, other factors, such as spreading rates, may not have been similar. For example, it has been theorized that the increased heat production of the early earth (McKenzie and Weiss 1980) may have been accommodated by holding the number of plates constant and increasing spreading/subduction rates or, alternately, keeping spreading/subduction rates constant and increasing the number of plates (Bickle 1978; Stacey 1981). The interpretation of short-lived thermal phases is consistent with either model but does not discount the possibility that extension during rifting may not have been followed by emplacement of the oceanic crust in some cases.

2.2.4 Compressional Settings

2.2.4A Foreland Basins

Foreland basins originated primarily in response to flexure of the lithosphere during emplacement of tectonic loads. They are particularly well suited for analyzing the rheology of the Proterozoic lithosphere because their geometry is a function of the mechanical properties and thermal structure of the lithosphere (Beaumont 1981; Jordan 1981; Watts et al. 1982; Kominz 1986). Consequently, the architecture of foreland-basin stratigraphy and the distribution of related unconformities potentially can be used to extract information on the thermal age and structure of continental lithosphere at the time of basin development. It follows that long-term secular evolution of any parameters that affect lithospheric rheology may be manifested in the geologic history of foreland basins. Several Proterozoic foreland basins have been analyzed using geologic and geochronological data, combined with geophysical modeling, in order to help place constraints on the thermal age and structure of the lithosphere at the time of the development of those basins (Hoffman and Bowring 1984; Ethridge et al. 1987; Hoffman 1987; Grotzinger and McCormick 1988; Grotzinger et al. 1988).

Proterozoic foreland basins share many of the distinctive characteristics of other well-documented, but younger analogues (Hoffman 1973; Ricketts and Donaldson 1981; Le Gallais and Lavoie 1982; Hoffman and Bowring 1984; Ethridge et al. 1987; Grotzinger and McCormick 1988). In foreland basins, deep water, axially deposited turbidites commonly overlie shallow-water platform carbonate and siliciclastic sediments and grade up into shallow-marine and alluvial deposits that were shed toward the foreland. Alluvial deposits include conglomerates containing clasts derived by reworking of earlier deposited lithofacies equivalents, thus indicating migration of the depocenter toward the foreland with time.

"Passive-margin" platforms were covered tectonically during lithospheric flexure as indicated by drowning and mass sliding of portions of their outer parts, and synchronous uplift and erosion of their inner parts (e.g., Grotzinger and McCormick 1988). Flexural arches ("peripheral bulges") were intermittently reactivated throughout deposition of foreland-basin axes, and their trends are generally parallel to associated orogenic belts. Similarly, they are approximately perpendicular to the trend of finite-stretching lineations in thrust nappes. These relationships indicate that basin development occurred by flexure of the lithosphere during emplacement of thrust nappes rooted in the hinterland of each belt, resulting in subsidence of the foreland and activation of related arches.

2.2.4B Development of Early Proterozoic Foreland Basins on Thermally Young Lithosphere

An implication of the model for short-lived "passive margins" discussed above is that foreland basins would have formed on thermally young continental lithosphere following collapse of the inferred, immature, thermally subsiding basins. The apparently "warm" behavior of continental lithosphere during development of some Early Proterozoic foreland basins (e.g., the Bear Creek foredeep; Grotzinger and McCormick 1988) might be considered as evidence for an elevated average conductive geotherm for Precambrian continental lithosphere, thus arguing against the widely accepted theory that heat loss through continental lithosphere was not significantly greater than at present (cf., Bickle 1978; Boak and Dymek 1982). Alternately, if it is true that foreland-basin development occurred on thermally young lithosphere due to collapse of immature passive continental margins, then the apparently "warm" behavior of the lithosphere can be attributed to local thermal anomalies associated with continental rifting.

Several features of Early Proterozoic foreland-basin evolution are consistent with their development on thermally young lithospheres. These include: (*i*) the thinness of underlying "passive-margin" sequences; (*ii*) the areally restricted occurrence of regional unconformities related to passage of the peripheral bulge on the tops of some passive-margin carbonate platforms; (*iii*) narrow transition zones between the thick axial turbidite sequences and equivalent thin shelf sequences; (*iv*) pervasive low-grade metamorphism of basins; and (*v*) high ductile strain in thrust nappes (Grotzinger and McCormick 1988).

2.2.5 Strike-Slip Basins

Strike-slip deformation and basin formation occur where adjacent crustal blocks move laterally with respect to each other (Christie-Blick and Biddle 1985). Most strike-slip basins are hybrids and are associated with regions of crustal extension or compression, which alternate in time and space along strike-slip systems. Sedimentary basins that develop along strike-slip faults have various shapes and sizes and differ in the details of their tectonic and depositional history (Crowell 1974; Nilsen and McLaughlin 1985). These differences are related in part to the geometry and orientation of the principal bounding faults. Because the principal faults may be curved, braided, or *en echelon*, areas of compression (or transpression) may develop adjacent to or even within areas of extension (or transtension).

In certain situations, rotation of crustal blocks about vertical axes may occur along broad zones of strike-slip deformation (transrotation), such as in southern California (Luyendyk and Hornafius 1987; Ingersoll 1988). Several aspects of the basin architecture provide clues to their origin (see Christie-Blick and Biddle 1985).

Basins of pull-apart origin are common along strike-slip faults and may be filled with thick sedimentary sequences. Strike-slip basins may form during local extension along the zone of principal displacement. Basins vary greatly in size, but are generally smaller than those produced by regional extension (e.g., intracontinental rifts) or compression (e.g., foreland basins). Furthermore, any given strike-slip basin may experience alternate extension and shortening on timescales varying from thousands to millions of years.

Descriptions of Proterozoic strike-slip basins are uncommon. Notable exceptions include the Early Proterozoic Et Then Basin (Hoffman 1973) and Nonacho Basin (Aspler and Donaldson 1985), both of northwest Canada. The basins are dominated by dextral and sinistral fault systems, respectively, and occur within 150 km of each other. These basin-bounding fault systems have recently been related to tectonic escape of crustal blocks during continental collision and indentation (Hoffman 1988).

The Et Then Basin contains over 4 km of alluvial conglomerate and sandstone that locally interfingers with lacustrine siltstones. Interestingly, the siltstones contain unusual pedestal-shaped stromatolite mounds formed predominantly of carbonate sediment, and completely encased in siliciclastic sediment. Sedimentation is related to subsidence along a braided fault system in which individual strands exceed 100 km (Hoffman 1973).

The Nonacho Basin contains a sequence of alluvial-fan, braided-stream, fan-delta, beach, and lacustrine facies. Although true stratigraphic thicknesses are unknown, longitudinal cumulative thicknesses are in excess of 40 km. Deposition occurred in several sub-basins covering an area over $12,000 \text{ km}^2$, formed by block faulting along a regionally extensive system of sinistral strike-slip faults. Areas that are proximal to the major basin-bounding faults are dominated by high-gradient alluvial fan and braid plain facies, that pass abruptly into deeper water lacustrine facies. Sub-basins that were more distal with respect to the major basin-bounding faults are characterized by low-gradient alluvial-plain sediments containing ephemeral pond facies.

2.2.6 Conclusions

Overall, the setting, style, and evolution of Proterozoic sedimentary basins closely resemble those of younger, Phanerozoic basins, reflecting the fundamental similarity of lithospheric dynamics throughout the last 2500 million years of earth history. Possible second order differences, such as the thinness of Early Proterozoic passive margin sequences and the development of Proterozoic foreland basins mainly on thermally young lithosphere, are consistent with inferred models of the earth's thermal evolution.

2.3

Late Proterozoic Low-Latitude Global Glaciation: the Snowball Earth

JOSEPH L. KIRSCHIVINK

2.3.1 Introduction

A fundamental question of earth history concerns the nature of the Late Proterozoic glaciogenic sequences that are known from almost all of the major cratonic areas, including North America, the Gondwana continents, and the Baltic Platform. A major controversy involves the probable latitude of formation for these deposits – were they formed at relatively high latitudes, as were those of the Permian and our modern glacial deposits, or were many of them formed much closer to the equator? Arguments supporting a low depositional latitude for many of these units have been discussed extensively for the past 30 years (e.g., Harland 1964), beginning with the field observations that some of the diamictites had a peculiar abundance of carbonate fragments, as if the ice had moved over carbonate platforms. Indeed, many of these units, such as the Rapitan Group of the Canadian Cordillera, are bounded above and below by thick carbonate sequences which, at least for the past 100 Ma, are only known to have been formed in the tropical belt within about 33° of the equator (Ziegler et al. 1984). Other anomalies include dropstones and varves in the carbonates, as well as evaporites (for a complete review, see Williams 1975). Either the earth was radically different during the late Precambrian glacial episode(s), or the major continental land masses spent an extraordinary amount of time traversing back and forth between the tropics and the poles.

Paleomagnetic data have been invoked to both support and attack the low-latitude interpretations of these deposits (e.g., Harland 1964; Crawford and Daily 1971; Tarling 1974; McElhinny and Embleton 1976b; Morris 1977; Stupavsky et al. 1982; Embleton and Williams 1986). As discussed in Chapter 12, however, most of the earlier reported paleomagnetic data are not as convincing as they should be, particularly in terms of using geological tests to constrain the time at which remanent magnetization was acquired. Two important exceptions to this are the data of Embleton and Williams (1986) and of Sumner et al. (1987 and in prep.) for the varved sediments of the Elatina Formation of South Australia (discussed further in Chapter 12). During the uppermost Marionan glaciation in Australia, it now seems clear that these extensive, sea-level deposits (including varves and dropstones) were formed by widespread continental glaciers which were within a few degrees of the equator. The data are difficult to interpret in any fashion other than that of a widespread, equatorial glaciation.

A global climatic mechanism that could lead to such a widespread, low-latitude glaciation is not yet available. Williams (1975) suggested that, if the earth's obliquity reached angles higher than about 54°, the relative annual heat balance would shift so as to warm the poles more than the equator. Several arguments can be marshaled against this. First, although the obliquity of the earth varies by a few degrees with a period of a few tens of thousands of years (and hence is a component of Melankovitch cycles), the physical basis for the changes are fairly well understood. No mechanism is yet known that would lead to much larger oscillations of the sort proposed by Williams (1975). Although lack of a known mechanism should never be used alone to argue against the reality of an effect (e.g., continental drift), such an absence demands that the hypothesis receive especially critical scrutiny. Second, a redistribution of the radiant energy balance to polar latitudes should also move the carbonate belts from equatorial latitudes to the poles, where the glaciers (in Williams' model) should not encounter them. Finally, Vanyo and Awramik (1982, 1985), Awramik and Vanyo (1986), and Vanyo et al. (1986) argue convincingly that the obliquity 800 Ma ago was in the range of the present values, based on detailed studies of both modern and ancient heliotropic stromatolites.

2.3.2 Mechanisms Responsible for Low-Latitude Glaciation

It is necessary to consider an alternate, equally speculative mechanism that might yield widespread, low-latitude, sea-level glaciations, and perhaps help to interpret the last major episode of banded iron-formation deposition associated with it (Summer et al. in prep., and Sections 4.2 and 4.3). As discussed in Chapter 12, large portions of the continental land masses probably were within middle to low latitudes during the late Precambrian glacial episode, a situation that has not been encountered at any subsequent time in earth history. In a qualitative sense, this could have had a fundamental impact on global climate, as most of the solar energy adsorbed by the earth today is trapped in the tropical oceans (in contrast to the continents which are relatively good reflectors) and in high-latitude oceans which often have fog or other cloud cover. Furthermore, if extensive areas of shallow, epicontinental seas were within the tropics, a slight drop in sea level would convert large areas of energy-absorbing oceanic surface to highly reflective land surface, perhaps enhancing the glacial tendency. Escape from the "ice house" would presumably be through the gradual buildup of the greenhouse gas, CO_2, contributed to the air through volcanic emissions. The presence of ice on the continents and pack ice on the oceans would inhibit both silicate weathering and photosynthesis, which are the two major sinks for CO_2 at present (Section 4.7). Hence, this would be a rather unstable situation with the potential for fluctuating rapidly between the "ice house" and "greenhouse" states. A major question, of course, is whether the planet could get cold enough to permit the glaciers to advance to equatorial zones, without the poles reaching temperatures low enough to freeze the atmospheric CO_2 into dry ice, robbing the planet of the greenhouse rescue and yielding a permanent ice catastrophe.

Whatever the triggering mechanism, if the earth had normal obliquity during an equatorial glaciation we would expect that areas of high latitude would be at least as cold, if not colder, than the equator. A reasonable inference from this would be the presence of floating pack-ice over most of the ocean surface at middle to high latitudes, as well as glaciers on those land areas with larger net precipitation than sublimation or evaporation. Thus, the earth would have resembled a highly reflective "snowball". In this model, however, it is not clear what fraction of the equatorial oceans in deep water would form pack ice, as these zones would still absorb large amounts of the incident solar radiation, perhaps enough to prevent ice formation. Hence, we might expect to find some warm tropical "puddles" in the sea of ice, shifting slightly from north to south with the seasons. In turn, this should produce extreme climatic shifts in some local areas as envisioned by Williams and Tonkin (1985) and Williams (1986).

2.3.3 Implications of the Global Snowball Model

This global snowball model has several implications which might lend themselves to geological tests. First, it implies that the glacial units should be more or less synchronous (Harland 1964), and standard radiometric, paleomagnetic, paleontologic, or geochemical techniques should be capable of testing this concept. Second, Late Proterozoic strata from widely separate areas which preserve a record of these climatic fluctuations might bear an overall similarity in lithologic character, which would be a result of the global scale of the climatic fluctuations. Third, the presence of floating pack ice should reduce evaporation, act to decouple oceanic currents from wind patterns and, by inhibiting oceanic to atmosphere exchange of O_2, would enable the oceanic bottom waters to stagnate and become anoxic. Over time, ferrous iron generated at the mid-oceanic ridges or leached from the bottom sediments would build up in solution and, when circulation became reestablished toward the end of the glacial period, the iron could oxidize to form a "last-gasp" blanket of banded iron-formation deposition in upwelling areas. Iron-rich deposits of this sort are known from several late Precambrian glacial units in Canada, Brazil, Australia, and South Africa (Section 4.2). The banded iron-formations in the Rapitan Group of northern Canada are interbedded with tillites and contain occasional dropstones.

In closing, it is perhaps worth noting again that this Late Proterozoic glacial episode marks a major turning point in the evolution of life. Although preceded by abundant evidence of the presence of protists and prokaryotes (Sections 5.4, 5.5), it is followed by the first clear record of metazoan animals (the Ediacaran Fauna) and shortly thereafter by the appearance of mineralized fauna in the Cambrian (Chapters 7 and 8). It is tempting to extend the snowball earth speculation to suggest that these evolutionary changes were made possible by the glaciations – the periodic removal of all life from higher latitudes would create a series of post-glacial sweepstakes, perhaps allowing novel forms to establish themselves, free from the competition of a preexisting biota.

Acknowledgments

Supported by the PPRG and NSF grants EAR-8721391 and PYI-8351370, and contributions from the Chevron Oil Field Research Company and the Arco Foundation. Contribution no. 4807 of the Division of Geological and Planetary Sciences of the California Institute of Technology.

2.4

The Proterozoic Sedimentary Record

DONALD R. LOWE

2.4.1 Introduction

Whereas the Archean sedimentary record is dominated by the deformed volcanic-sedimentary associations of greenstone belts, the Proterozoic is noted for its thick, mature, often nearly undisturbed cratonic sedimentary deposits. In fact, however, Proterozoic supracrustal sequences present a rich and enormously diverse assemblage of sedimentary and volcanic rocks in which greenstone-type lithologies are abundantly represented. To emphasize the co-evolution of the continental crust and the global sedimentary record and to attempt to discriminate between ocean- (mantle-) and continent-dominated sedimentary successions, we here divide Proterozoic supracrustal rocks into two major lithologic associations: greenstone and cratonic (Section 1.3). The following discussion outlines the characteristics and distribution of these associations.

2.4.2 Greenstone Association

The greenstone volcanic-sedimentary association consists of thick sequences of mafic to intermediate volcanic rocks containing subordinate silicic/felsic volcanic units and interflow sedimentary layers. Although greenstone belts are often considered characteristic of and perhaps restricted to the Archean, based on the presence of komatiitic volcanic rocks, greenstones, as a general mafic to intermediate volcanic/sedimentary association, are abundant as the oldest recognizable components of most Proterozoic continental basement blocks. They occur as metamorphosed enclaves within granite-greenstone basement terranes, such as the 2.2 to 2.0 Ga Guiana Shield of South America (Gibbs and Barron 1983; Gruau et al. 1985), the 1.9 to 1.6 Ga basement of the southwestern and central United States (Van Schmus et al. 1987), and the 1.9 to 1.7 Ga Svecofennian, and 1.7 to 1.55 Ga Southwestern Provinces of Scandinavia (Gorbatschev 1985). Smaller belts occur within the internal zones of orogenic provinces, such as the 1.95 to 1.85 Ga Great Bear magmatic zone in the Wopmay Orogen, Canada (Hildebrand et al. 1987; Hoffman 1988), the 1.9 to 1.8 Ga Flin Flon, La Ronge, and related predominantly volcanic supracrustal belts of the Trans-Hudson Orogen, Canada (Hoffman 1988), and portions of the 950 to 550 Ma Pan-African basement of northeastern Africa and the Arabian Peninsula (Stoeser and Camp 1985).

Proterozoic greenstone sequences are more diverse in makeup than their Archean counterparts. Most are thick, highly deformed and metamorphosed accumulations of submarine tholeiitic basalts, andesitic to rhyolitic calc-alkaline volcanic rocks, and fragmental volcaniclastic units. Associated immature sedimentary rocks include turbiditic mudstones and lithic to feldspathic graywackes, generally of volcanic provenance. In some areas, however, greenstones developed adjacent to or upon older basement blocks and include detrital sediments derived from cratonic sources (e.g., Condie and Martel 1983). Orthochemical and biogenic sedimentary layers are rare but occur locally, principally as iron-formation and cherty rocks.

Individual stratigraphic sequences are enormously variable, even within single basement blocks, and no single greenstone stratigraphic "model" depicts the evolution of all or even most Proterozoic greenstone belts. In many areas, such as the Guiana Shield (Gibbs and Olszewski 1982) and the southeastern United States (Boardman 1986), the lower parts of the greenstone sequences are dominated by tholeiitic extrusive units and the upper by thick intermediate to felsic calc-alkaline volcanic and pyroclastic deposits co-magmatic with extensive granitoid intrusive rocks.

Although Proterozoic greenstone terranes have been extensively investigated petrologically and geochemically, they have provoked little sedimentological study, probably because metamorphism and deformation combined to obscure many essential sedimentological features and stratigraphic relationships.

The largest Proterozoic greenstone tracts, like their Archean equivalents, form the oldest recognizable elements of large continental blocks. The greenstones represent pre- and syn-cratonization stages of Proterozoic crustal development and are dominated by mantle-derived volcanic and intrusive rocks. The prevailing view is that these sequences formed through volcanism and sedimentation in predominantly non-cratonic or marginal cratonic settings including oceanic spreading centers, intraplate oceanic volcanic islands, back-arc basins, and oceanic and continental volcanic arcs (e.g., Hoffman 1988). Most are interpreted as arc-related volcanic-sedimentary complexes. Their late-stage development was characterized by compression; intrusion of tonalitic, trondhjemitic, and granodioritic plutons; and by metamorphism, so that most are now heavily altered and deformed remnants of more extensive fold and thrust belts. Deformation commonly accompanied progradation of accretionary complexes above subduction zones or involved back-arc or retro-arc thrusting and obduction onto adjacent foreland cratonic blocks during suturing (Karlstrom and Houston 1984; Stoeser and Camp 1985; Kröner 1985; Van Schmus et al. 1987; Hoffman 1988).

Minor Proterozoic greenstones formed within cratonic rifts, such as the Keeweenawan lavas of the Mid-continent Rift (Van Schmus and Hinze 1985). Rift greenstones may include sediments derived by erosion of clearly older cratonic blocks, unusually voluminous felsic eruptives and plutons, and igneous suites marked by evidence of anatexis of older crustal components. Where adjacent sialic blocks dominate sedimentation and magmatism, the mafic greenstone components may be greatly subordinate to felsic eruptives and craton-derived sediments. This general setting has been inferred during deposition of supracrustal sequences making up much of the exposed basement of northern Australia (Ethridge et al. 1987; Page 1988) although there is a clear absence of detritus derived from substantially older Proterozoic or Archean basement (McCulloch 1987; Page 1988). In this and other areas of bimodal volcanism and extensive felsic volcanism, where dated older detrital components and clear geochemical signatures of the older basement rocks are absent or ambiguous, the tectonic setting remains controversial.

2.4.3 Cratonic Association

Proterozoic cratonic sediment accumulated as flanking aprons along continental margins and as cover and basin fill sequences upon continental blocks. They divide generally into the same major petrologic facies as their Phanerozoic analogues (Dickinson and Suczek 1979; Ingersoll 1989). Exceptions include units such as Early Proterozoic banded ironformations, which bear strongly on environmental contrasts between the Proterozoic and Phanerozoic earth (see Chapter 4). The thickest sequences were deposited in basins formed by dynamic subsidence resulting from intra-cratonic extension and rifting and later passive subsidence during cooling of newly formed continental margins. Some sequences were deposited in failed rifts, such as the Early to Middle Proterozoic McArthur Basin sequence (Jackson 1987), the Middle Proterozoic Belt Supergroup (see discussion in Winston 1986), the Late Proterozoic Adelaidean succession in the Adelaide Geosyncline (Preiss 1987), and the Early to Late Proterozoic sequences around Jixian in northern China (Yang et al. 1986). In many areas, however, rifting led to the formation of oceanic crust with well-developed fringing passive margin sequences.

Most rift sequences include thick but laterally limited sections of immature, often arkosic, craton-derived, clastic sedimentary rocks deposited under alluvial, fluvial, lacustrine, or shallow-marine conditions. Overlying, thick, passive margin sections are typically more mature and characterized by quartzite-shale-carbonate associations. Basal siliciclastic formations commonly become more mature upward, and higher units include carbonate formations with stromatolite buildups as well as local evaporite facies (Karlstrom and Houston 1984; Grotzinger 1986b; Preiss 1987).

Proterozoic foreland basins, many of late Early Proterozoic age, contain thick predominantly clastic sequences (Hoffman 1987; Grotzinger et al. 1987), although Hoffman (1987) has recently suggested that many Early Proterozoic ironformations were also deposited in foreland basins. Foredeep development commonly marked the closing phases of tectonic cycles. Depositional settings ranged from fluvial to deepwater turbiditic systems. Many basins contain initially deep-water deposits grading upward into shallow-water and terrestrial sections (Hoffman 1987), reflecting progradation of the frontal thrusts.

With the exception of rift and foreland basin deposits, craton cover sediments, although widespread, occur mainly as thin remnants of thin, more extensive sheets. Some cratons, however, like the Siberian and Eastern European Cratons, are largely covered by Proterozoic and younger cratonic sediments (Khain 1985).

The ratio of preserved cratonic to greenstone deposits appears to increase generally with time reflecting the progressive growth of the continental crust and the greater preservation potential of its cover and flanking sedimentary sequences. To some extent, however, the apparent abundance of Proterozoic cratonic deposits relative to greenstones is an artifact. Vast tracts of Proterozoic basement make up many continental blocks, but the included greenstone remnants are often overlooked in assessments of overall Proterozoic sediment abundance. In fact, the volume of Proterozoic supracrustal rocks in greenstone enclaves may exceed that of the better preserved and more extensively studied cratonic sequences.

2.4.4 Secular Development of Proterozoic Supracrustal Sequences

2.4.4A Early Proterozoic (2500 to 1600 Ma)

Late Archean orogenesis formed large blocks of continental crust, portions of which are preserved within most present-day Precambrian basement terranes. During the Early Proterozoic, these and older Archean continents served as nuclei around and upon which thick sedimentary sequences accumulated.

There are relatively few preserved sedimentary rocks representing the interval 2600 to 2300 Ma except those that are parts of cratonic sequences that began accumulating during the Late Archean on the older, Early to Middle Archean Pilbara and Kaapvaal Cratons. Exceptions include the 2.4 to 2.1 Ga Huronian Supergroup of the eastern Great Lakes region, which overlies Late Archean granite greenstone basement of the Superior Province (Young 1983), and lower parts of the Karelian succession, especially the Sumian-Sariolan and Jatulian Groups (Gorbatshev and Gaal 1987), deposited on the Late Archean Karelian Craton.

Beginning diachronously at about 2200 Ma and continuing

locally to the close of the Early Proterozoic at 1600 Ma, the Archean cratons became the loci of rifting and, subsequently, accretionary activity that led to the formation of thick bodies of Early Proterozoic supracrustal rocks and culminated in the formation and stabilization of large tracts of new continental crust. As well as adding immense new areas of sialic crust to the continental inventory, large areas of older Archean crust were reworked during widespread and intense structural, intrusive, and metamorphic events.

Large crustal blocks formed during this interval include: (i) the 1.9 to 1.6 Ga Precambrian basement of much of the southwestern and central United States (Van Schmus et al. 1987; Bickford 1988; Grambling et al. 1988; Karlstrom and Bowring 1988); (ii) the 2.3 to 2.0 Ga Amazonian Craton of South America (Gibbs and Barron 1983; Gruau et al. 1985); (iii) the 1.9 to 1.75 Ga Svecofennian Province adjacent to the Archean Karelian Craton of Scandinavia (Gorbatschev 1985; Gorbatschev and Gaal 1987); (iv) large parts of the West African Craton (Wright et al. 1985); and (v) major portions of the northern and southern Australian basement (Ethridge et al. 1987; Page 1988). Narrower, heavily deformed, late Early Proterozoic orogenic belts fringe and separate many present Archean blocks. These include (i) belts ringing the Late Arachean Superior Craton (Baragar and Scoates 1981; Hoffman 1988) including the Labrador Trough or New Quebec Belt, the Cape Smith Fold Belt, the Belcher Island Fold Belt, the Trans-Hudson Orogen, and the 2.1 to 1.85 Ga Lake Superior region; (ii) the 1.95 to 1.9 Ga Wopmay Orogen, Kilohigok Basin, and Thelon Orogen surrounding the Late Archean Slave Craton, northern Canada (Hoffman 1988); (iii) the multiply deformed and metamorphosed Namaqua Province west and south of the Early to Middle Archean Kaapvaal Craton (Kröner 1978; Botha 1983); (iv) the Nabberu Basin lying along the northern margin of the Late Archean Yilgarn Block, Western Australia (Goode 1981); (v) the Pine Creek Geosyncline surrounding a Late Archean basement block in northernmost northern Australia (Needham et al. 1988); (vi) thick basinal units in the Broken Hill or Olary Block, Mount Painter Province, and Peake and Denison Ranges adjacent to the Late Archean Gawler Craton, southern Australia (Fanning et al. 1988; Stevens et al. 1988); and (vii) the 2.2 to 2.0 Ga Magondi Mobile Belt flanking the northern margin of the Early to Late Archean Zimbabwe Craton (Treloar 1988).

Individual local Early Proterozoic tectonic cycles, which commenced and ended diachronously over the earth's surface, commonly involved a sequence of events resembling part or all of a Wilson cycle: (i) an initial rifting stage involving older Archean cratonic blocks accompanied by the deposition of immature rift-related sediments; (ii) the subsequent development of a subsiding continental margin and the accumulation of quartzite-carbonate-shale marginal rise, terrace, and embankment sequences; (iii) greenstone volcanism and sedimentation concurrent with and/or following (ii) in areas removed from the craton margin; and (iv) one or more orogenic stages involving accretion of the volcanic terrane(s) along the continental margin and thrusting of the marginal sediments onto the adjacent Archean foreland.

Sediments representing the early rift stage are locally well preserved, such as the Akaitcho Group in the Wopmay Orogen adjacent to the Late Archean Slave Craton (Hoffman 1980; Easton 1981; Grotzinger 1986a); the Seward Group in the Labrador Trough developed along the margin of the Late Archean Superior Province (Wardle and Bailey 1981; LeGallais and Lavoie 1982); and the Deep Lake Group in the Medicine Bow Mountains along the southern edge of the Archean Wyoming Province (Karlstrom and Houston 1984). They consist largely of thick sections of immature fluvial to locally turbiditic clastic sediments.

Rift sequences are succeeded by thick marginal quartzite-shale-carbonate sections deposited in environments ranging from shallow-shelf (miogeoclinal) to deep water (eugeoclinal). Well preserved examples, again from North America, include the 1.95 to 1.84 Ga Epworth Group in the Wopmay Orogen (Hoffman 1988; Grotzinger 1986b); the Knob Lake Group in the Labrador Trough (Wardle and Bailey 1981; LeGallais and Lavoie 1982); and the Libby Creek Group in the Medicine Bow Mountains (Karlstrom and Houston 1984). These and similar predominantly miogeoclinal sequences are some of the thickest and most extensive cratonic deposits of Early Proterozoic age. Their wide extent attests to the global nature of late Early Proterozoic rifting and passive margin development.

The craton margin sequences, when traced toward the internal portions of the orogenic zones, commonly pass into or are in tectonic contact with coeval and younger greenstone terranes such as the Great Bear Magmatic Zone in the western portion of the Wopmay Orogen (Hildebrand et al. 1987); the Doublet Group of the Labrador Trough (Wardle and Bailey 1981); and 1.9 to 1.6 Ga greenstones of the central and southwestern USA, developed south of the Archean Wyoming Province (Van Schmus et al. 1987; Bickford 1988; Grambling et al. 1988; Karlstom and Bowring 1988).

During deformation, most Early Proterozoic marginal successions were thrust onto adjacent Archean foreland and overriden by the deformed and metamorphosed greenstone sequences. The initial stages of deformation were commonly marked in the craton margin sequences by the upward, generally unconformable transition from stable margin quartzite-carbonate-shale units to immature or submature clastic foreland basin deposits (e.g., Hoffman 1988).

In most areas, accretion and orogenic activity was completed prior to 1800 Ma, although in the southwestern USA and Scandinavia arc accretion continued until 1600 Ma or less (Gorbatschev 1985; Gorbatschev and Gaal 1987; Van Schmus et al. 1987). Many Early Proterozoic orogenic belts are overlain unconformably by late Early Proterozoic craton cover or foreland basin sequences. In other areas, latest Early Proterozoic rifting affected portions of the newly formed Early Proterozoic blocks. Post-orogenic craton cover sequences include the approximately 1731 Ma-old Roraima Formation (Rogers et al. 1984) of the Guiana Shield and the 1.76 to 1.65 Ma Matzatzal, Ortega, Sioux, and Baraboo Quartzites of the central United States (Dott 1983; Soegaard and Eriksson 1985). Examples of latest Early Proterozoic rifts include the McArthur Basin (Jackson 1987) and possibly the Mount Isa Basin (Derrick 1982) of Australia.

2.4.4B Middle Proterozoic (1600 to 900 Ma)

Late Early Proterozoic craton formation was completed diachronously, but mid-Proterozoic depositional styles prevailed in most areas by 1750 to 1600 Ma. The late Early and Middle Proterozoic represents a period of prolonged, almost exclusively craton-related sedimentation. There are few greenstones of this age and few areas underlain by continental crust formed during this interval, although large areas of older crust were reworked during Middle Proterozoic tectonic and magmatic episodes. Local greenstone sequences were deposited in rifts, such as the Keeweenawan volcanics of the 1200 to 1100 Ma-old Mid-continent Rift system (Van Schmus and Hinze 1985), and in volcanic arcs now preserved locally in the interior portions of orogenic belts, such as the Central Metasedimentary Belt of the Grenville Province (Condie and Moore 1977).

Major Middle Proterozoic cratonic deposits include early postorogenic foreland sequences, rift deposits, craton cover units, and in some Late Proterozoic orogenic belts, heavily deformed craton margin sequences. In some areas, late Early and early Middle Proterozoic sedimentation involved the deposition of thick alluvial or shallow-water clastic sequences in what were probably late foreland or successor basins to Early Proterozoic orogenic terranes. Overall, however, this interval is represented by shallow-water to terrestrial sedimentation in three main depositional settings: (*i*) cratonic rifts; (*ii*) broad cratonic basins and cratonic platforms; and (*iii*) craton margins.

The predominantly cratonic control on preserved mid-Proterozoic supracrustal sequences is emphasized by the development of extensive rhyolite-granite extrusive-intrusive provinces. These reflect large-scale intracrustal melting and possibly underplating events (Van Schmus et al. 1987; Hoffman 1989b). The 1500 to 1340 Ma-old Granite-Rhyolite Province of the southern mid-continent region consists of rhyolitic to dacitic volcanic rocks and related shallow intrusive units, possibly emplaced during extension rifting of underlying, late Early Proterozoic crust (Van Schmus et al. 1987) or through large-scale mantle upwelling and crustal underplating (Hoffman 1989). Similar provinces include the latest Early Proterozoic granites in the Mount Isa Inlier (Wyborn et al. 1988), the 1750 to 1550 Ma Trans-Scandinavian granite-porphyry Belt (Gorbatschev 1985), and the 1592 Ma Gawler Range Volcanics in southern Australia (Fanning et al. 1988).

A number of major tectonic events occurred during the Middle Proterozoic, over wide areas gardening older Proterozoic and, in some instances, Archean crust. Examples include the Kibaran and Irumide Belts in central Africa (Robertson et al. 1981), which involve extensive remobilization of adjacent Archean blocks, and the Grenville Province of North America (Moore et al. 1986). In central Africa, the Kibaran Orogeny occurred at about 1350 to 1250 Ma, and a more widespread period of deformation was roughly correlative with the Grenville Orogeny in North America (1100 to 900 Ma). Although the Grenville Province occupies an exposed area of nearly 700,000 square km in eastern North America, it is not clear that this represents a fundamentally new tract of Middle Proterozoic continental crust. The dominance of craton-related intrusive suites and craton-derived metasedimentary assemblages, and the areal paucity of greenstones, are among a variety of features suggesting that the Grenville Province represents a late Early or early Middle Proterozoic craton margin succession and arc terrane that was thrust onto remobilized older Proterozoic and Archean crust (Dewey and Burke 1973; Baer 1983; Windley 1986; Dickin and McNutt 1989). Although tectonism and magmatism were widespread during the Middle Proterozoic, there were few if any discrete basement blocks formed during this interval. Exceptions may include portions of the Albany-Fraser Province and Musgrave Block of Australia.

2.4.4C Late Proterozoic (900 to 550 Ma)

The Late Proterozoic, 950 to 550 Ma, was an interval of widespread rifting, possibly related to the fragmentation and dispersal of a supercontinent assembled at 1400 to 1000 Ma (Piper 1976; Bond et al. 1984). Rifting was diachronous, beginning in some areas as early as 1200 Ma and in others possibly as late as 750 Ma (Wright et al. 1985; Cahen et al. 1984; Hanson et al. 1988a). Thick, well-preserved post-1000 Ma cratonic rift and overlying passive margin deposits are present on most continents. They include sequences in the Adelaide Geosyncline, Amadeus Basin, and Kimberley region of Australia (Preiss and Forbes, 1981); along the Cordilleran margin of North America from the southern Great Basin to the McKenzie Mountains of northwestern Canada (Stewart 1972); in the Appalachian Orogen of eastern North America and related belts in east Greenland, the British Isles, Norway, and Spitzbergen (Winchester 1988); and in the Ural Mountains of the Soviet Union (Khain 1985). Less well-studied sections occur in orogenic belts around the North China Platform, as in the Shaanxi region, China, and in Braziliano Belts of South America. Widely preserved rift sequences are exemplified by the Ocoee Supergroup of the southern Appalachians (King 1964), Valdres and Hedmark Groups in southern Norway (Nysteun and Siedlecka 1988), the lower part of the Warrina Supergroup in the Adelaide Geosyncline (Preiss 1987), and the Nosib Group in the Damara Orogen (Martin and Porada 1977; Behr et al. 1983). Although highly variable in makeup, the rift sequences are commonly thick but laterally restricted basinal successions composed largely of immature craton-derived feldspathic sandstone, siltstone, and conglomerate, and bimodal basalt-rhyolite volcanic suites. Glaciogenic units are widespread (Harland 1983). Deposition generally occurred under rapidly subsiding alluvial to shallow-water conditions, although some basins accumulated deeper water successions, including thick sequences of turbiditic feldspathic sandstone in the Ocoee Supergroup of the southern Appalachians (King 1964). Local shallow-marine and continental evaporite units reflect the restricted nature of many of these early rift basins (Rowland et al. 1980; Behr et al. 1983).

The overlying craton margin sequences also developed diachronously and some are the equivalents of rift sequences elsewhere. Most are predominantly clastic quartzite-shale-carbonate associations composed largely of mature quartzose sandstone units deposited in current- and wave-active shallow marine waters, shelf and basinal shales, and platformal carbonates. Periods of rapid, post-rifting cooling and subsidence are reflected in the development of deeper-water conditions accompanied by outer shelf, slope, and basinal eugeoclinal sedimentation of both clastic and carbonate sediments (e.g., Porada and Witting 1983). Evaporites were deposited locally, such as gypsum in the Coppercap and Redstone River Formations of the MacKenzie Mountains Supergroup (Ruelle 1982), and most carbonate sequences contain major stromatolitic units. Glaciogenic units, including diamictites, fluviatile sandstones and conglomerates, and varved siltstone-mudstone deposits, are widespread and mark the major Late Proterozoic glacial advances between about 900 and 650 Ma (Harland 1983). Some recent work suggests that these glaciations may have involved low-latitude ice sheets (Williams 1975; Embleton and Williams 1986; Section 2.3 and Chapter 12). Iron-formation constitutes a minor but widespread facies of Late Proterozoic rift and marginal sequences, including iron-rich sediments in the glaciogenic Rapitan Group in the MacKenzie Mountains of northwestern Canada (Young 1976; Yeo 1981), the Pan-African belts in West Africa (Wright 1985), and the Damara Orogen of southern Africa (Breitkopf 1988; Sections 4.2 and 4.3). More seaward facies of these sequences include deepwater eugeoclinal associations, although these are more restricted in outcrop and commonly heavily deformed and metamorphosed.

In many areas, rifting was aborted prior to the development of wide ocean basins, such as in the Adelaide Geosyncline and Amadeus Basin, Australia, and the Damara Orogen of southern Africa. In others, it preceded a full Wilson cycle involving the formation of an open ocean and the deposition of thick passive margin sequences, as in the North American Appalachian and Cordilleran Orogens.

Greenstone associations of Late Proterozoic age were deposited locally during rifting, but most underlie limited tracts of land and are areally subordinate to associated cratonic rift and marginal sequences. Post-rift Late Proterozoic greenstones, associated with the development of magmatic arcs, are widespread in the interior zones of many orogens, but most are parts of allochthonous thrust complexes and generally do not constitute significant tracts of new crust comparable to the vast arc accretionary complexes of the Archean and Early Proterozoic.

Perhaps the most extensive Late Proterozoic greenstone terrane is that in northeastern Africa and the Arabian Peninsula that developed between about 1200 and 650 Ma. Initial rifting before 950 Ma, and possibly as old as 1200 to 1000 Ma (Jackson 1980), was followed by accumulation of a craton margin sequence (Kröner et al. 1987) that was deformed during subsequent arc collision, obduction, and magmatism from 950 to 630 Ma. This accretionary event led to the formation of a substantial tract of new, largely Late Proterozoic or Pan-African continental crust of the Arabian-Nubian Shield (Stoeser and Camp 1985; Kröner 1985). The accretionary complex includes thick sequences of mafic to intermediate volcanic rocks and local immature volcaniclastic sediments and ophiolites containing cherts.

Over most of Africa, South America, and central Asia, compression and deformation followed rifting closely in the latest Precambrian and earliest Paleozoic Pan-African, Braziliano, Baikalian orogenic cycles. In much of North America and Australia, the main deformation was later, during Paleozoic Caledonian or younger orogenic events.

Late Proterozoic platform deposits are widely developed. They include the Nama Group of southern Africa, which contains both platform and foreland basin facies of the Damara Orogen (Germs 1983); 1050 Ma to Paleozoic rocks of the Taoudini Basin covering a vast area of the West African Craton (Wright et al. 1985); portions of the Sinian sequence on the Archean North China Craton and Proterozoic Yangtse Platform of southern China (Yang et al. 1986); and cover sequences on the Siberian Platform in the Soviet Union (Khain 1985). All represent stable to locally active shelf settings dominated by mature craton-derived detritus and stromatolitic carbonate sedimentation.

2.5

Proterozoic Mineral Deposits Through Time

IAN B. LAMBERT   NICOLAS J. BEUKES   CORNELIS KLEIN   JÁN VEIZER

2.5.1 Introduction

The temporal distributions of important types of mineral deposits are summarized in Figure 2.5.1. During the Precambrian, there were major changes in the styles of mineralization and these have been related to evolutionary trends for the lithosphere, atmosphere, hydrosphere, and biosphere (Veizer 1976a; Cloud 1976b; Lambert and Groves 1981; Meyer 1981, 1988).

The present-day pattern of mineralization reflects the effects of recycling superimposed on evolutionary trends. This has been emphasized by Veizer (Section 2.6) and others who simulated the overall trends in terms of growth of continental crust until 1.8 to 1.5 Ga ago, when a near present-day steady state was established. In this model, cumulative metal reserves reached essentially their present levels by the Middle Proterozoic, and the preservation potentials of different ore types are related to the rates of generation/destruction of their host tectonic domains.

For the purpose of this review of Proterozoic metallogeny, the Precambrian is divided into three major evolutionary stages, which are referred to as the Greenstone Belt Stage, the Cratonization Stage, and the Stable Craton-Rifting Stage, in order of decreasing age. These are similar to the four stages in the simulation by Veizer, except that it is convenient in the present discussion to combine his third and fourth stages. It is noteworthy that the various evolutionary stages can have different age ranges in different crustal segments. For example, the Cratonization Stage commenced in the Kaapvaal Craton of South Africa around 3.0 to 2.8 Ga ago, when greenstone belt development was at its height in the Yilgarn Craton of Australia and the Superior Province of Canada. Also, greenstone belt development had ceased in most terranes prior to the end of the Archean, 2.5 Ga ago, but it continued until 2.0 to 1.8 Ga in some areas (Section 2.4).

2.5.2 Greenstone Belt Stage

A brief outline of Greenstone Belt Stage mineralization is necessary to set the scene for discussion of the Cratonization and Stable Craton-Rifting Stages which dominated the Proterozoic.

Archean crust consists essentially of granite-greenstone and gneiss terranes. The mineralization reflects high levels of igneous and metamorphic activity, and the most significant deposits occur in the volcanosedimentary greenstone belts. For more information on Archean metallogeny, the reader is referred to Lambert and Groves (1981) and Meyer (1981, 1988).

Several types of stratabound deposits, which occur in association with felsic to mafic volcanic rocks, formed at or near the rock-water interface. The most widespread are banded iron-formations (BIF; Section 4.2), which are commonly referred to as Algoma-type after deposits in North America. These can be oxide, carbonate, or sulfide types, in decreasing order of overall abundance, and they have variable proportions of interbedded sedimentary strata. Sulfide and carbonate types can be auriferous, particularly in strongly deformed zones. Volcanogenic massive sulfide deposits of copper (and zinc) are particularly prevalent in Canada. Bedded barite, at least some of which has replaced evaporitic gypsum/anhydrite, occurs in Australia, South Africa, Canada, and India.

Vein and shear-zone gold deposits occur on all continents and formed at favorable structural/lithological sites, apparently derived mainly from metamorphic fluids. Ultramafic-mafic flows and intrusives have magmatic segregations that are locally of considerable economic significance, particularly in Western Australia where sulfide concentrations are mined for nickel (platinoids and copper), and in Zimbabwe where chromite deposits are exploited. Hydrothermal deposits of tin, tungsten, lithium, tantalum, beryllium, and molybdenum are quite widely distributed in association with late-stage granitoids and pegmatites, but are generally minor.

The oxide type iron-formations and the bedded barite in the Archean are of particular interest in the present context because they imply a degree of oxidation by oxygen-producing photosynthetic microorganisms and/or by photodissociation of water. They do not prove that the Archean atmosphere was oxygenated (Section 4.6) or that sulfate was abundant throughout the hydrosphere. In fact, generally low sulfate contents are implied for the hydrosphere by the general absence of sulfate minerals in Archean volcanogenic sulfide deposits, and by the sulfur isotope data discussed in Section 3.5.

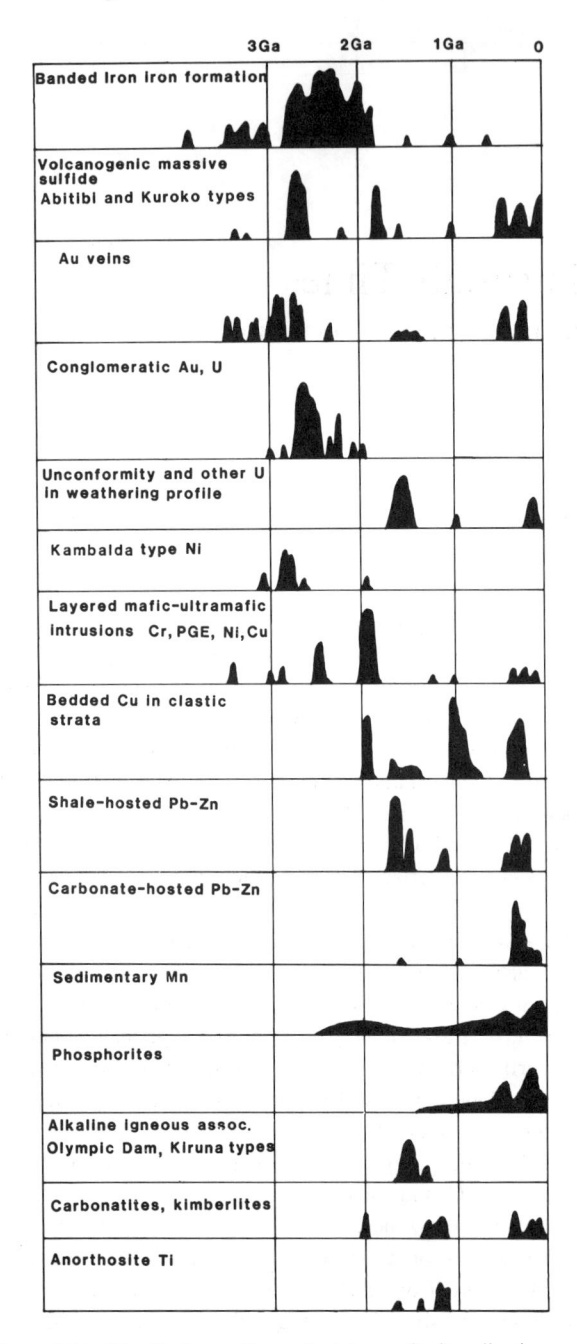

Figure 2.5.1. Distributions of important types of mineralization through geologic time, adapted from Lambert and Groves (1981) and Meyer (1981, 1988). Heights of the curves represent the relative quantities of mineralization known to occur.

2.5.3 Cratonization Stage

This commenced 3.0 to 2.5 Ga ago, and dominated the Early Proterozoic. It was heralded by widespread emplacement of potassic granitoids, and characterized by generally lower levels of igneous activity than the Greenstone Stage, by widespread epicratonic sedimentation, and by the emplacement of large layered mafic-ultramafic intrusives. There is also evidence for increased oxidation of the hydrosphere/atmosphere and proliferation of sulfate-reducing bacteria in the earliest Proterozoic (Section 3.5). The main mineralization of this evolutionary stage contrasts markedly with that of the Greenstone Stage, being predominantly in epicratonic sedimentary sequences and large layered intrusives.

Important gold and/or uranium deposits in pyritic quartz conglomerates, as exemplified by the huge Witwatersrand deposits (Pretorius 1976, 1989) and lower grade deposits in Brazil and Canada, are restricted to Late Archean and Early Proterozoic sequences. There is sedimentological evidence in these deposits for transport of gold, uraninite, and pyrite in high-energy fluvial systems, and for concentration of these heavy minerals by physical and biological processes in nearshore environments. The stability of uraninite and pyrite during extensive physical reworking favors low oxygen levels in the early atmosphere (Cloud 1976b; Section 4.6). However, there is also evidence of hydrothermal addition/redistribution of gold, uranium, iron, and sulfur within the Witwatersrand reefs, possibly reflecting the importance of metamorphic fluids (Phillips 1987) or of igneous related fluids at a magmatic arc/back-arc basin interface (Pretorius 1989). This hydrothermal activity could well have played a critical role in producing the higher grade deposits.

Early Proterozoic iron-formations, which are found on all continents, occur within predominantly sedimentary sequences and are generally larger than Archean examples (Section 4.2). These have been termed Hamersley-type after the extensive Western Australian deposits. They are generally only mined where they have been enriched by secondary processes. Studies of the Hamersley Basin deposits indicate that "shales" interbedded with BIF are fine pyroclastic and chemical deposits and that there was very little terrigenous input into what is considered to have been a marginal marine depositional basin (Morris and Horwitz 1983). Before the evolution of a largely oxidized hydrosphere, Fe^{+2} would have become enriched in deep, reduced, sulfide-poor waters as a result of hydrothermal processes and seafloor weathering and, in the absence of silica-secreting organisms, these waters would also have been enriched in silica. The iron-formations could have precipitated as upwelling reduced waters entered oxygenated, possibly evaporative shelf environments (Section 4.3).

There are widespread examples of stratiform manganese oxide and carbonate deposits in Proterozoic and, less commonly, Late Archean sequences (Roy 1980, 1981). These are of particular importance in India, South Africa, the USSR, and Brazil. One major type is associated with iron-formation and thus is restricted to the Late Archean-Early Proterozoic and the Late Proterozoic. A second major type is associated with carbonaceous shales and their metamorphic equivalents; this occurs in sequences from Late Archean to Late Proterozoic age. In addition, deposits associated with non-carbonaceous sericitic meta-pelites and chert breccias are locally important. The deposits of the carbonaceous shale association, which are commonly dominated by rhodochrosite pods, are seldom rich enough to be mined but they constitute protore to major supergene enriched deposits. Both these associations appear to have formed in deeper water basinal facies of progradational sequences deposited on continental platforms following major transgressive events, and the Late Proterozoic examples are commonly in glaciomarine diamictite sequences. In the iron-formation association, manganese was separated from iron because the former metal is less easily converted from the soluble reduced form to the insoluble oxidized form, so the appearance of the first significant manganese deposits in latest Archean-Early Proterozoic sequences is consistent with increasing oxidation levels from that time. In the carbonaceous shale association, the precipitation of iron as pyrite was probably the main process involved in its separation from manganese. The cessation of extensive manganese and iron-formation deposition in the Middle Proterozoic is probably a reflection of a combination of global depletion of these metals in the hydrosphere, because of the extent of earlier precipitation, and the paucity of reduced waters. The reappearance of such deposits in the Late Proterozoic may have been related to prolonged and extensive oceanic anoxic events, which are implied by the carbon isotopic data for that period (Lambert and Donnelly, 1989), and which again could have resulted in significant solution of these metals in deep waters.

Volcanogenic massive sulfide deposits occur in several approximately 2.0 to 1.8 Ga-old sequences, mainly in North America and Scandinavia, but mineralization of this type is uncommon in younger Proterozoic sequences (Gale 1983). These deposits formed at and near the submarine surface in volcanic and volcanosedimentary sequences. Some are in settings similar to those of Archean massive sulfide deposits and appear to have formed during Early Proterozoic greenstone belt development, for example Flin Flon and some other copper (zinc) deposits in southern Canada and the northern USA. Others are in sequences having higher proportions of felsic volcanic and sedimentary rocks, for example the Cu-Pb-Zn deposits at Skellefte and elsewhere in Scandinavia, and these more resemble deposits in Phanerozoic volcanic arc settings.

The large layered mafic-ultramafic intrusives of the latest Archean and Early Proterozoic were products of crustal stability and mantle tapping fracture zones, and they contain some very important deposits of magmatic origin (MacDonald 1987). The 2.0 Ga-old Bushveld Complex in South Africa has vast chrome and platinum group element deposits. The 2.75 Ga-old Stillwater Complex in the USA, and the 2.55 Ga-old Great Dyke of Zimbabwe, are also enriched in these metals. Magma mixing and wall-rock assimilation could have played critical roles in the concentration of the metals. The 1.7 Ga-old Sudbury Irruptive in Canada is the world's major source of nickel and a significant copper producer; there is active debate as to whether emplacement of this body was triggered by impact of a large extraterrestrial object.

2.5.4 Stable Craton-Rifting Stage

This stage commenced in the Early Proterozoic. Interpreted rift sequences were deposited prior to the major global orogeny which occurred roughly 1.8 Ga ago (Etheridge et al. 1988; Section 2.4). During the ensuing Middle and Late Proterozoic, there was widespread deposition of ensialic rift sequences, but it is noteworthy that active continental margin sequences are less common in the geological record for this stage than for the Phanerozoic Stage. These observations, supported by paleomagnetic data, and the widespread development of ^{34}S-rich sedimentary sulfides (Section 3.5), suggest the existence of a supercontinent during this period (Lambert and Donnelly 1989). Although extensional tectonic settings appear to have been common in this stage, these do not appear to have resulted in major separations of cratonic blocks, and many of the intracratonic basins were partly or wholly closed off from the open ocean. Break-up of the supercontinent occurred at about the Proterozoic-Cambrian boundary, following a prolonged period of widespread oceanic anoxia (Lambert and Donnelly 1989).

Vein-type uranium deposits formed close to Early-Middle Proterozoic unconformities (Ferguson and Goleby 1980; Marmont 1987). Major examples occur in the Pine Creek Geosyncline of northern Australia and the Athabasca Basin of western Canada. The main metal sources were probably Late Archean high-level granites enriched in potassium and uranium, and Early to Middle Proterozoic sedimentary and igneous rocks derived from these. The deposits appear to have formed at moderate to low temperatures in multiple solution-precipitation events mainly during the Middle Proterozoic. They can be related to the evolution of significant levels of oxygen in the atmosphere and hydrosphere because hexavalent uranium is very soluble in oxidized waters. Such waters, probably dominated by intraformational fluids, evidently migrated through faults, fractures, and permeable zones and the uranium was precipitated by reductants, commonly organic matter and ferrous iron minerals. Important noble metal mineralization is associated with uranium in the fault-controlled Coronation Hill and Jabiluka deposits in the Pine Creek Geosyncline.

A number of very large stratiform lead zinc deposits formed in Proterozoic extensional basins (Gustafson and Williams 1981; Lambert 1983). The oldest major examples are the 1.7 to 1.65 Ga-old Broken Hill, Mount Isa, and McArthur River deposits in Australia. The other major examples are the Sullivan deposits in Canada (ca. 1.5 Ga; Ethier et al. 1976), and the Gamsbert-Aggeneys deposits in South Africa (probably ca. 1.3 Ga old; Köppel 1980; Rozendaal 1980). No major stratiform lead-zinc deposits are known in Late Proterozoic sequences. Upward enrichment of lead in the crust appears to have been a prerequisite for the formation of such deposits, and the major tectonothermal event approximately 1.8 Ga ago could have played an important role in this enrichment; the major potassic granite intrusives in the Late Archean could also have contributed. The stratiform lead-zinc deposits formed at and near the sediment-water interface in depressions in ensialic troughs, well above basement. The ore-forming fluids were probably essentially heated basinal brines that leached metals from the trough sequences. The sulfide appears to have been partly biogenic and partly hydrothermal.

Alkaline and alkali-rich igneous rocks are a feature of the Middle to Late Proterozoic extensional basins (Wyborn et al. 1987). Important associated mineralization includes hydrothermal uranium deposits, such as those in the ~1.7 Ga old sequence at Mary Kathleen and the ~1.6 Ga old Olympic Dam sequence, both in Australia, and the Late Proterozoic sequence at Rössing in South Africa. Magmatic segregations of ilmenite occur in North America, Scandinavia, and the USSR in late Early to Middle Proterozoic anorthositic intrusives, which are commonly associated with alkali-rich rapikivi granitoids.

Other alkaline rocks of economic importance in Precambrian cratons are diamondiferous kimberlites, which are most abundant in southern Africa and Siberia, and carbonates that can have associated deposits of copper, phosphate, and rare earths, the richest being at Palabora in South Africa. The oldest of these mantle-derived rocks formed ~2 Ga ago, and many are of Phanerozoic age.

Non-bedded hydrothermal copper mineralization is unusually well represented in Australian Early to Middle Proterozoic sequences (Lambert et al. 1985). The largest example, Olympic Dam (which also contains uranium, rare earth, and gold mineralization), formed from fluids which were probably generated in association with alkaline magmatic activity in an ensialic rift, whereas the Mount Isa and Tennant Creek deposits apparently originated during late stages of regional metamorphism. Porphyry copper deposits are not a feature of the Proterozoic, presumably reflecting the paucity in that Eon of high level, hydrous intrusives of the types associated with the numerous examples of such deposits in late Phanerozoic continental margin and volcanic arc environments.

Major bedded copper (uranium) mineralization became abundant in the Late Proterozoic, as exemplified by the African Copper Belt deposits (Fleischer et al. 1976) and by White Pine, Canada (Brown 1978). However, the important Udokan deposits of the USSR are in Early Proterozoic strata and there are many sub-economic examples in Early and Middle Proterozoic sequences. The moderately oxidizing conditions favoring copper transport in low temperature brines (Haynes and Bloom 1987) would have become widespread only after the evolution of an oxygenous atmosphere. While there may have been syngenetic copper sulfide deposition in some cases, most deposits exhibit evidence for introduction of moderately oxidized cupriferous fluids into pyritic sediments during diagenesis (Gustafson and Williams 1981). Red bed sandstones, which are well developed in the Late Proterozoic, are commonly present close beneath and/or beside the deposits. The red beds are widely considered to have been the major sources of the copper, and other potential sources are basic volcanics and metalliferous basement rocks. The sulfide was generated by bacterial sulfate reduction in the host sediments, but hydrothermal sulfide is probably also present in the high grade African deposits.

Evaporitic gypsum and anhydrite beds and/or nodules occur in Late Proterozoic sequences of Australia, Canada, Zambia, and Zaire. Pseudomorphic evaporitic calcium sulfates have been recognized in shallow water, Early to Late Proterozoic sequences. In contrast to Archean examples, there is abundant evidence for the former existence of halite in most of the Proterozoic evaporitic strata, and this may reflect generally increased concentrations of chloride in the Proterozoic hydrosphere.

Phosphorites first became important at about the Proterozoic-Cambrian boundary, but some phosphorus-enriched beds in India may be of Middle Proterozoic age (Cook and Shergold 1984). The phosphate deposits are considered to have formed from deep organic-rich waters welling up onto shallow shelves. The latter would have been abundant during continental dispersion events, a major example of which was the break up of the supercontinent near the Proterozoic-Cambrian boundary. The carbon isotope data from carbonates of Late Proterozoic age (Lambert and Donnelly 1989; Section 3.4) support greater than normal levels of burial of organic carbon in sediments during this period, which was characterized by major changes in the biota (Sections 5.4, 5.5, 11.3, 11.4, Chapters 7, 8).

Despite evidence for abundant microbial activity in Proterozoic sedimentary environments, few economic deposits of fossil fuels are known in basins of this age. There are "coal" (Shungite) deposits of probable microbial origin in Early Proterozoic strata of northern Europe and elsewhere; and oil and gas have been recovered from several unmetamorphosed Late Proterozoic sequences (Murray 1965), for example in the Amadeus Basin of central Australia, and in Siberia and Oman (see Chapter 3). Recently, indigenous oil was encountered in late Early to Middle Proterozoic strata of the McArthur Basin of northern Australia (Jackson et al. 1986). These discoveries, from very limited exploration in unmetamorphosed Proterozoic basins, coupled with the evidence for prolonged oceanic anoxia in the Late Proterozoic, attest to the potential existence of major Proterozoic oil and gas fields.

CHAPTER 2.6

Recycling and Preservation Probabilities of Sediments

JÁN VEIZER

2.6.1 Introduction

Throughout the Precambrian, the record of past life becomes progressively more attenuated with increasing geological age (Sections 5.4, 5.5, 11.3). This may be a consequence of (*i*) a lesser abundance and diversity of life forms; (*ii*) obliteration of the fossil record by post-depositional phenomena; or (*iii*) some combination of both. Available carbon isotopic data (Section 3.4) indicate that the reservoir of reduced carbon and, hence, the extant biomass, have been substantial as far back as the existing rock record extends. It is, therefore, unlikely that the observed paucity of fossils in older rocks is primarily the result of a rarity of primitive life forms. Although the appearance of skeletal life forms in the Phanerozoic enables their easier preservation and recognition, this factor alone cannot be held responsible for the diminishing number of fossils with increasing age of the rocks. Geological phenomena, such as recrystallization, ultimately reflecting the increase in pressure and temperature during the diagenetic and epigenetic history of the rocks, are a significant culprit. Statistically, the chance that these agents erase the record increases with the time available for their action. Nevertheless, the major impact of such agents is confined to the times of sediment consolidation into rocks. Past this stabilizing stage, of perhaps $\sim 10^{7 \pm 1}$ years duration, the effects are much less pronounced, particularly for platform sediments. These, as is discussed below, represent the bulk of the Proterozoic sedimentary record. If so, the preservation of the sediments themselves is the main constraining factor for survival of the entombed fossils. It is, therefore, essential to calibrate the existing statistical databases of fossil frequency to the preservation probabilities of their host sedimentary rocks.

2.6.2 Preservation Probabilities for Tectonic Domains

The tenets of plate tectonics demand that tectonic evolution follows the following sequence: mantle → oceanic crust → volcanic arc → orogeny → craton (including its non-sedimentary basement). As material moves along this chain, it undergoes progressive fractionation and refinement. For example, the melt generated in the mantle forms the oceanic crust. The latter, in turn, is mostly resorbed back into the mantle, but a subordinate portion feeds the successor tectonic domain, the island arc. This pattern continues along the mantle → craton succession. As a consequence, each tectonic domain is involved in a continuous generation/destruction ("birth/death") cycle, with the rate of recycling decreasing downstream due to the necessity to process ever diminishing quantities of material per unit time.

The above systematics are amenable to quantitative treatment analogous to that of the populations dynamics in living systems. The "birth/death" cycle for a given population – in this case, for a tectonic domain – imposes an internal age structure on its constituent units (rocks). The proportion of progressively older units diminishes, usually exponentially, and the rate of this decay is inversely proportional to the rate of recycling (Fig. 2.6.1). Knowing the age structure of the rocks making up a given geological population – information decipherable from the geologic record – it is possible to calculate the rate of recycling and, hence, the theoretical preservation potential of its constituent rocks. This approach has been systematized by Veizer and Jansen (1985) and discussed in detail by Veizer (1988a). For steady-state systems, the calculated theoretical half-lives (or 50% probabilities of destruction) for the rocks of global tectonic domains are the following: active margin basins ~ 27 Ma; oceanic intraplate basins ~ 51 Ma; oceanic crust ~ 59 Ma; passive margin basins ~ 75 Ma; immature orogenic belts ~ 78 Ma; mature orogenic belts (roots) ~ 355 Ma; and platforms ~ 361 Ma. For continental basement, the half-life depends on the relative ease of resetting of ages established by different radiometric techniques and therefore varies between ~ 673 and ~ 1728 Ma limits. The generation/destruction cycle of the oceanic crust is a good illustration of this concept. The calculated steady-state rate of recycling of ocean crust is ~ 3.5 km^2 (or 20 km^3) per year. The theoretical half-life of newly generated crust is ~ 59 Ma and its "oblivion age" (or 95% probability of destruction) should be about 4.5 half-lives, that is, ~ 260 Ma. In reality, geological populations deviate somewhat from a simple exponential decay law, and the observed empirical oblivion ages attain only 3.0 to 3.5 half-lives (Veizer 1988a). Note that these are only statistical probabilities valid for large populations and are not necessarily applicable to specific geologic domains (this is a situation reminiscent of

64 Geological Evolution of the Proterozoic Earth

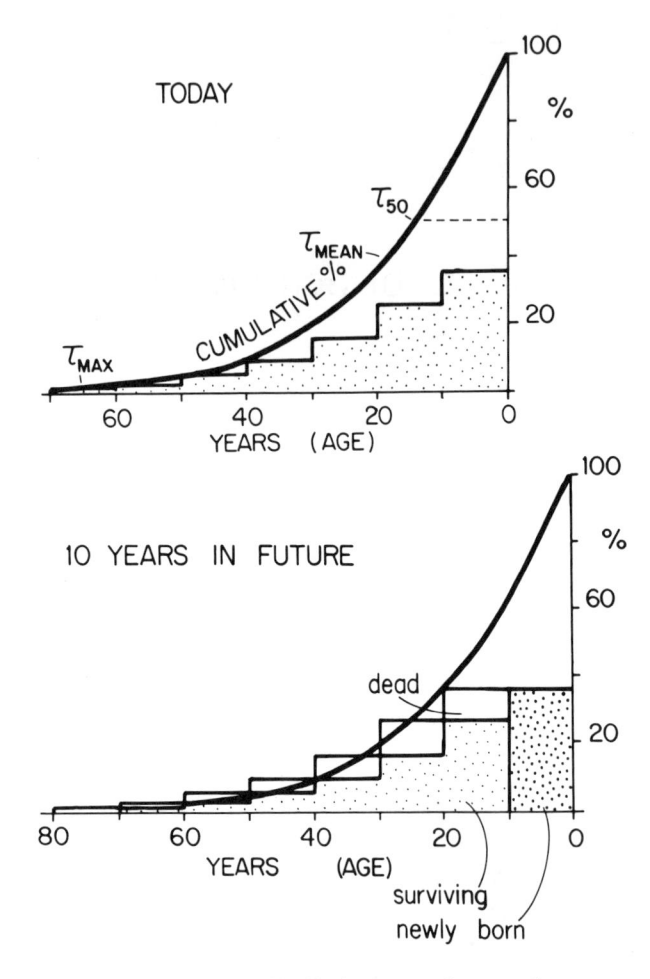

Figure 2.6.1. Simplified age distribution pattern for a steady-state extant population. In this example, the recycling rate is 35% of the total population for a 10 year interval (35×10^{-3} yr^{-1}). τ_{50} is the half-life or 50th percentile. τ_{max}, defined as the 95th percentile, is the "oblivion age," the age at which the resolution of the database becomes essentially indistinguishable from the background (see Veizer and Jansen 1985).

living populations in which the life-span of an individual does not necessarily follow the prediction of the actuarial tables). All these caveats not withstanding, this concept implies that all tectonic domains have predictable life-spans and that their geologic records therefore extend only as far back as their oblivion ages. Consequently, the probability of preserving tectonic domains with half-lives of $\leqslant 100$ Ma, such as active margin basins and active orogenic belts, into the Precambrian, is remote (Fig. 2.6.2). If so, Precambrian terranes should encompass mostly rocks of the metamorphic continental basement, roots of former mountains (mature orogenic belts), and platformal sedimentary sequences. This is not to say that one cannot recognize vestiges of, for example, former oceanic crust in the assemblage of rocks of a basement of an orogenic root. However, these are only reprocessed remnants of *former* oceanic crust, now incorporated into a downstream tectonic domain.

The above probabilities of preservation of tectonic domains are those for a steady-state system. The continental crust probably reached its near-present-day size at $\sim 1.75 \pm 0.25$ Ga ago (Veizer 1988a,b,c; Section 2.4). Considering that this represents more than 15 half-lives for fast cycling tectonic domains, the steady-state inferences are directly applicable. Even for platforms and orogenic roots, the above time interval represents ~ 5 half-lives, a duration of recycling sufficient to obliterate most of the "memory" of a continental growth phase. In contrast, for the slowly cycling continental basement, the steady-state preservation probabilities predict vestiges of early continents considerably in excess of their observed size (Fig. 2.6.3). This discrepancy reflects the growth pattern of continents, which follows a logistic function (Veizer 1988a,c). The nucleation of continents, which commenced ~ 4 Ga ago, was followed by an exponential growth phase throughout the Archean, curtailment of growth rates in the Early Proterozoic, and attainment of near-steady-state or "carrying capacity" of the system at $\sim 1.75 \pm 0.25$ Ga years ago. With little new continental crust generated, still less could have been preserved, no matter how low the rate of recycling.

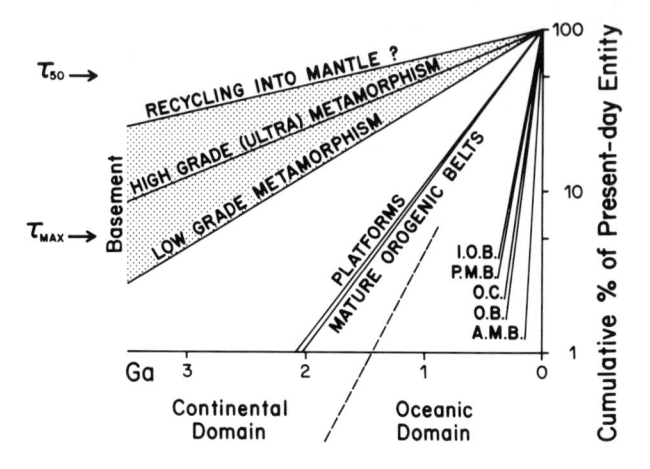

Figure 2.6.2. Preservation probabilities for major global tectonic domains. A.M.B. denotes active margin basins; O.B., oceanic intraplate basins; O.C., oceanic crust; P.M.B., passive margin basins; I.O.B., immature orogenic belts. Note the semi-logarithmic scale used in this and in Figures 2.6.3 through 2.6.5. See text and Veizer (1988b) for discussion.

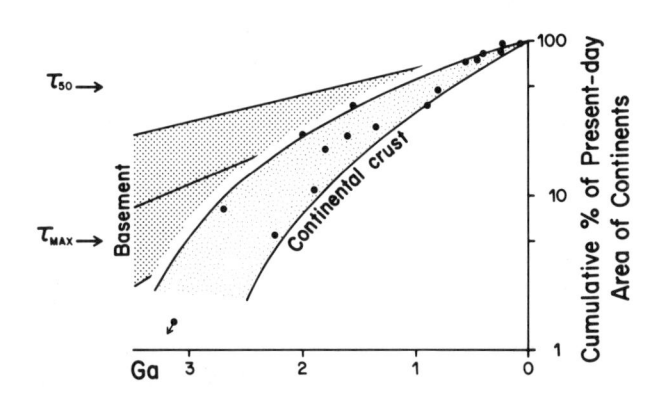

Figure 2.6.3. The observed present-day age distributions of continental and oceanic crusts, compared with the theoretical probabilities of preservation summarized in Figure 2.6.2 (cf. Veizer 1988a).

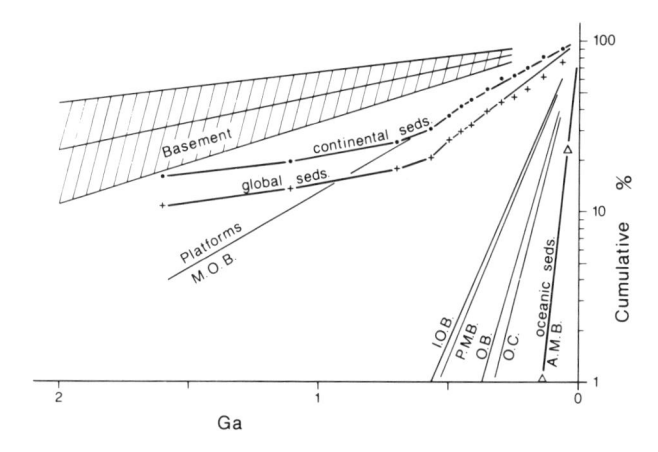

Figure 2.6.4. The observed present-day age distributions for global, continental, and oceanic sedimentary masses. These patterns are superimposed on preservational probabilities for tectonic domains. M.O.B. denotes mature orogenic belts; other abbreviations are as in Figure 2.6.2. The data on which these plots are based and their sources are listed in Veizer and Jansen (1985); the shallowing of the age slopes from recycling rates typical of platforms to those comparable to basement may reflect progressively larger proportions of metamorphosed sediments included in the data derived from Ronov (1982).

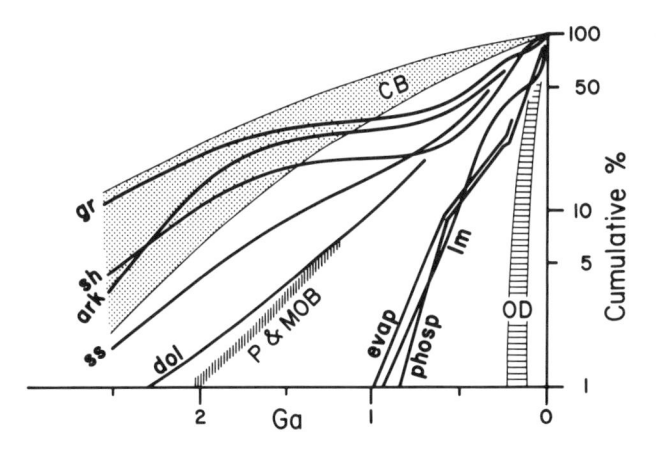

Figure 2.6.5. Observed cumulative mass-age distributions of major sedimentary lithological types (cf. Veizer 1988a). Data for carbonates and clastics after Ronov (1964), for evaporites from Holser (1984), and for phosphorites after Cook and McElhinny (1979). Explanation of abbreviations: gr denotes graywackes; sh, shales; ark, arkoses; ss, sandstones; dol, dolostones; evap, evaporites; lm, limestones; phosp, phosphorites; CB, continental basement (Fig. 2.6.3); P, platforms; MOB, mature orogenic belts; OD, oceanic domain (viz., active to passive margin basins of Fig. 2.6.2).

2.6.3 Preservation Probabilities for Sediments

The age structure for the global, continental, and oceanic sedimentary masses are summarized in Figure 2.6.4. These estimates, particularly for the pre-1.6 Ga masses, are somewhat uncertain. Nevertheless, it appears that continental and global sediments represent polymodal distributions, with age slopes becoming progressively shallower with increasing antiquity. If representative, such patterns could result from progressively older sediments being recycled at every slower rates. This could be a consequence of their increasing stability, such stability reflecting either physicochemical attributes or geological settings. The first alternative, termed differential recycling, has been advocated by Garrels and Mackenzie (1971). They proposed that the respective half-lives for clastic:carbonate:evaporitic rocks have been ~600:300:200 Ma, reflecting principally their differential susceptibility to chemical dissolution. As a consequence, the older sequences contain diminishing proportions of carbonates and particularly salts, and are thus less susceptible to recycling. If this were the case, the time-dependent distribution of different lithologies should form a pattern of progressively shallower slopes from salts to clean sandstones. Although the available data are poor, this is hardly the case (Fig. 2.6.5). Lithologies with widely differing solubilities (evaporites, limestones, phosphorites) appear to have comparable probabilities of preservation. In addition, graywackes and arkoses are surely not the most resistant of all lithologies.

In my view, the destruction of sediments proceeds mostly en masse. The principal agents are physical weathering and transport, both controlled by the rate of uplift and, hence, tectonic setting (Veizer 1988a). With increasing geological age, the continental (or global) sedimentary mass becomes dominated by sediments associated with progressively more stable tectonic domains, each domain having a characteristic package of sediment lithologies. These complex interplays of packages have been responsible for the apparent time-dependent lithological evolution of sediments summarized in the work of Ronov (1964) and modelled as differential recycling by Garrels and Mackenzie (1971). The existing inventories are as yet insufficient to model effectively such a tectonically controlled secular evolution of the sedimentary system. Nevertheless, the preservation probabilities of sediment should approximate those of their host tectonic settings. In the absence of better data, the probabilities summarized in Figure 2.6.2 may serve as a first-order calibrating tool for preservation of sediments and their entombed fossils. Because the dominant tectonolithologic package of the Proterozoic consists of platform sediments (shales-sandstones-dolostones; Fig. 2.6.5), the probability of preservation of such rocks and fossils is ~30% for the Late Proterozoic, ~15% for 1 Ga old sequences, and ~2% for 2 Ga old rocks. Sediments incorporated into the basement have a somewhat better chance of survival, but the nearly ubiquitous occurrence of metamorphism usually obliterates morphological remnants of former life. The probabilities of unperturbed survival from the Precambrian for the remaining tectonic domains are negligible. If encountered, they represent statistical anomalies.

2.7

Major Events in the Geological Development of the Precambrian Earth

DONALD R. LOWE

2.7.1 Introduction

The evolution of Proterozoic supracrustal rocks was strongly influenced by the existence of large blocks of continental crust, of Archean age in the Early Proterozoic but including large Early Proterozoic basement blocks during the Middle and Late Proterozoic. Proterozoic sedimentary rocks show a marked increase in craton-derived components, cratonic depositional styles, and cratonic geochemical and isotopic signatures relative to those of Archean age. At the same time, greenstone remnants in most Proterozoic basement blocks preserve an extensive but highly altered and deformed record of non-cratonic depositional settings.

As advocated by many previous investigators (Burke and Dewey 1973; Burke et al. 1976; Windley 1977, 1983; Hoffman 1988) and as emphasized elsewhere in this volume (Sections 1.3, 2.2, and 2.4), the basic mechanics of terrestrial heat loss, the compositions and distribution of Proterozoic igneous and sedimentary rocks, the importance or horizontal compression in major orogenic events throughout geologic history, and the inferred dominance of accretion along convergent plate boundaries during the assembly of both Archean and Proterozoic cratons argue strongly that basic lithospheric dynamics have remained essentially the same throughout formation of the preserved geologic record. Plate tectonics provides a geologically consistent framework for interpreting earth history and the growth and evolution of the continental crust throughout the Precambrian as well as the Phanerozoic.

Major controversies remain, however, regarding the application of plate tectonic models to the interpretation of specific aspects of Precambrian crustal history, such as the ensialic versus ensimatic nature of orogenic zones (e.g., Kröner 1977). The following discussion traces the formation of the earth's continental crust through a series of discrete Archean and Proterozoic crust-forming events and evaluates evolution of the style and rate of crust formation during the Precambrian in terms of evolving lithospheric dynamics.

2.7.2 Principal Precambrian Crust-Forming Episodes

Major orogenies and episodes of crust formation are non-uniformly distributed over geologic time (Stockwell 1961; Hurley and Rand 1969; Windley 1977; Baer 1983). Major periods of activity mark especially the Late Archean and Early Proterozoic (Windley 1976). However, orogenesis and crust formation are not equivalent processes, and the distribution of orogenic episodes, whether measured by the secular distribution of radiometric dates of igneous and metamorphic rocks (Stockwell 1961) or by the spatial distribution of orogenic zones (Hurley and Rand 1969), does not provide a precise estimate of the ages or geographic extent of crust-forming events. Many orogenic events profoundly altered enormous tracts of older crust without producing any new, first cycle, continental domains.

In order to determine the distribution of primary crust-forming events in the Precambrian, we have estimated the area and cratonization ages of the major Precambrian basement domains making up the present-day continental crust. Cratonization ages have been taken as the initial crystallization ages of the greenstone and/or early tonalitic-trondhjemitic-granodioritic intrusive suites. Where possible, we have avoided (*i*) Sm-Nd model ages (Nelson and DePaolo 1985), which bear a close relationship to, but commonly pre-date, eruption/crystallization ages and which thus reflect mantle rather than crustal events; (*ii*) ages of late potassic granites and related intrusive units, which commonly represent postcratonization rifting or basement remobilization; (*iii*) Rb-Sr and K-Ar ages, which also commonly reflect overprinting and remobilization of older crust; and (*iv*) ages of detrital minerals, which are readily recycled from older blocks. This analysis differs from that of Hurley and Rand (1969) inasmuch as postcratonization metamorphic and intrusive events are not counted as crust-forming episodes.

The approach is semi-quantitative at best, because of incomplete knowledge of many parts of the earth's crust: (*i*) available radiometric ages and geologic studies are insufficient to characterize the makeup and evolution of many exposed Precambrian terranes; (*ii*) many large basement blocks are deeply buried beneath younger covering sediments, such as the Siberian and Eastern European Cratons (Khain 1985); (*iii*) many terranes have been dated using Rb-Sr and K-Ar techniques, which are commonly insufficient to distinguish primary

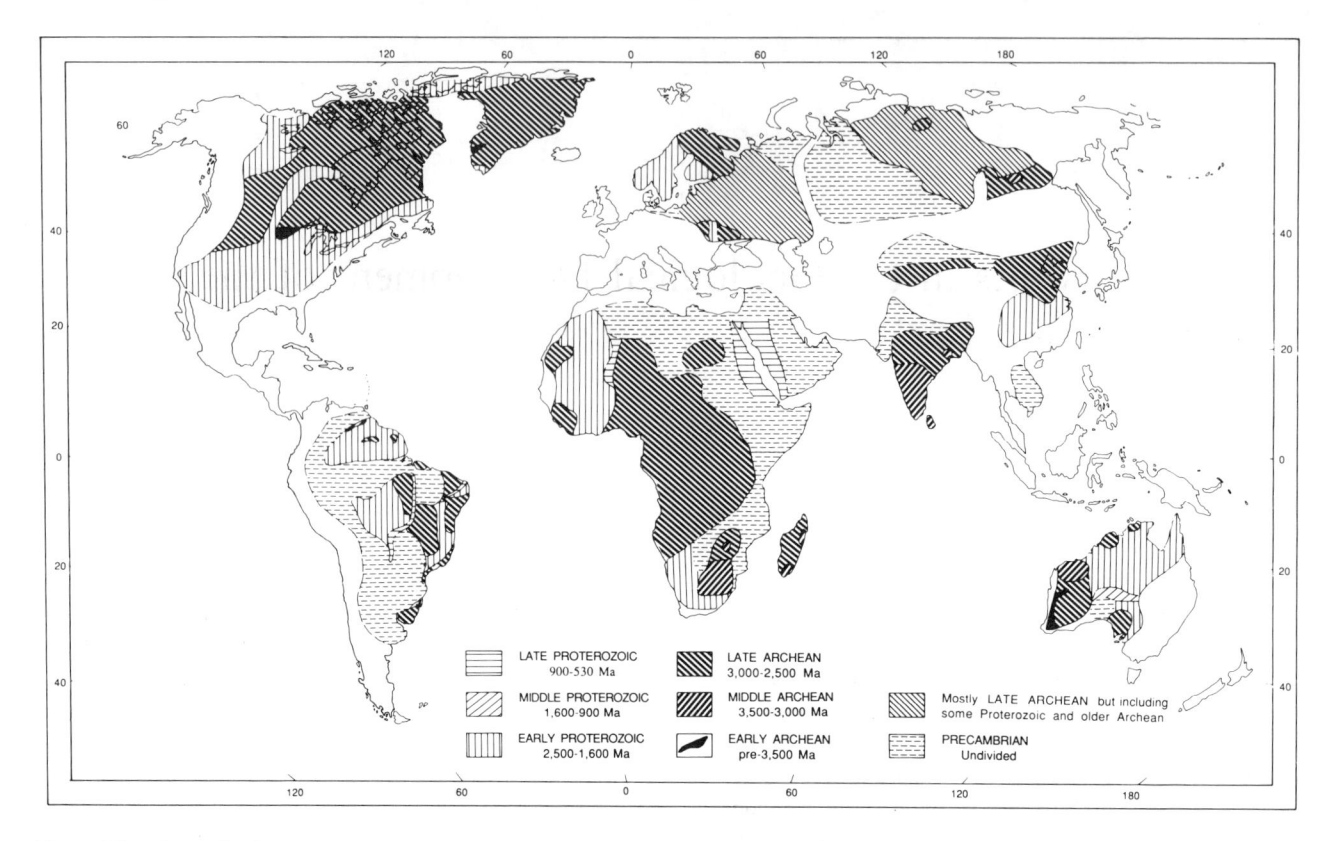

Figure 2.7.1. Generalized age distribution of major Precambrian basement terranes based on the estimated ages of the initial crust-forming events. Most Middle and Late Proterozoic orogenic or "mobile" belts represent zones of older crust that has been either reworked during the later igneous and metamorphic events or buried beneath younger fold and thrust belts.

crust-forming events from later metamorphic and tectonic overprinting; and (iv) the nature of basement beneath many orogenic zones remains controversial. Granite-greenstone terranes and heavily deformed mobile zones commonly contain enclaves of older continental material. It is often ambiguous whether such enclaves reflect the presence of older basement unconformably or structurally beneath the deformed successions or whether they are exotic slices of older continental crust caught within a new crustal block during its formation.

A particularly difficult and important problem is encountered when attempting to estimate the amount of new continental basement represented by Middle and Late Proterozoic mobile belts. Many Proterozoic mobile zones developed in part as fold and thrust belts structurally emplaced over older basement. The Grenville Province of eastern North America, for example, underlies an area of between 1 and 2 million km². If the bulk of the province represents juvenile late Middle Proterozoic continental crust, it clearly marks a major crust-forming event. If, however, Grenville metasedimentary and meta-igneous rocks are entirely allochthonous and rest structurally on older Archean or Early Proterozoic basement (e.g., Baer 1983; Dickin and McNutt 1989), the province represents a tract of buried Archean or Early Proterozoic continental crust, not Middle Proterozoic crust. Similarly, Middle Proterozoic Kibaran and Irumide and Late Proterozoic Pan-African Belts in central Africa underlie enormous areas, but, if developed as ensialic basins (Kröner 1983) or as allochthonous fold and thrust belts, they do not represent new Middle or Late Proterozoic crust and must be inventoried as older crustal domains. In the present analysis, the areas of mobile belts compressed between sutured older cratons and lacking prominent granite-greenstone terranes have been divided more or less equally among the adjacent older cratonic blocks. Where large granite-greenstone terranes are present, as in the Late Proterozoic Arabian-Nubian Shield of northeastern Africa and western Arabia and the Touareg Shield of north-central Africa (Caby et al. 1981), these have been included as primary new continental domains.

The results of the present survey (Tables 2.7.1 through 2.7.3 and Figs. 2.7.1 to 2.7.3) strongly suggest that the preserved Precambrian continental crust was formed mainly during five principal crust forming episodes or cycles (see also Section 1.3):

(1) pre-3.5 Ga (represented by less than 1% of the present Precambrian crust), here termed the Isua Cycle;
(2) 3.5 to 3.2 Ga (less than 6%), the Barberton Cycle;
(3) 3.0 to 2.6 Ga (50% to 60%), the Superior Cycle;
(4) 2.2 to 1.6 Ga (30% to 40%), the Colorado Cycle; and
(5) post-1.0 Ga (less than 6%), the Nubian Cycle.

Of these, the Superior and Colorado Cycles account for more than 80% of the preserved continental crust of Precambrian age. If there has been substantial loss of continental crust through recycling into the mantle (Armstrong 1981; Veizer and Jansen 1979), the relative proportions of crust formed during the Archean and Early Proterozoic may have been even greater.

Table 2.7.1. *Principal Archean and Proterozoic basement terranes used in calculation of crustal growth curves.*

Cycle/Geologic Age	Basement Block	Area[a] (km^2)
Isua Cycle: Pre-3.5 Ga	West Greenland	50,000
	Minnesota River Valley, USA	104,000
	Slave Province, Canada	50,000
	Labrador, Canada	50,000
	Western Gneiss Province, Australia	151,000
	Enterby Land, Antarctica	50,000
	Total Isua Cycle:	455,000
Barberton Cycle: Middle Archean 3.5 to 3.2 Ga	Pilbara Block, Western Australia	204,000
	Kaapvaal Craton, southern Africa	737,000
	Zimbabwe Craton, southern Africa	50,000
	Madagascar	134,000
	Southern India	490,000
	South America	253,000
	Siberian Craton[b]	800,000
	East European Craton[b]	400,000
	Total Barberton Cycle:	3,068,000
Superior Cycle: Late Archean 3.0 to 2.5 Ga	North America:	
	Northern Canada	3,644,000
	Western Canada and USA	1,216,000
	Eastern Canada	380,000
	Superior Province	2,357,000
	Greenland	1,700,000
	Africa:	
	Central Africa[c]	9,524,000
	Northeastern Sahara	760,000
	Liberian Craton	295,000
	Northwestern Africa	400,000
	Zimbabwe Craton	395,000
	Madagascar	465,000
	Australia:	
	Yilgarn Block	646,000
	Gawler Craton	250,000
	Kimberley Block	154,000
	Pine Creek Inlier	50,000
	Tennant Creek region	10,000
	South America	1,370,000
	Siberian Craton	3,600,000
	East European Craton	4,500,000
	India	1,650,000
	North China Craton	1,430,000
	Total Superior Cycle:	34,796,000
Colorado Cycle: Early Proterozoic 2.2 to 1.6 Ga	North America:	
	Wopmay Orogen and Northwestern Canada	907,000
	Trans-Hudson Orogen	610,000
	Southwwest, central, eastern USA, and eastern Canada	4,741,000
	Africa:	
	West African Craton	3,330,000
	Namaqua Province	700,000
	Svecofennian and South West Province, Scandinavia	1,250,000
	East European Craton	1,100,000
	South America	3,780,000
	Australia	3,922,000
	Total Colorado Cycle:	20,340,000
Middle Proterozoic	Musgrave Block, Australia	141,000
	Total Middle Proterozoic:	141,000
Nubian Cycle: Late Proterozoic 900 to 600 Ma	Nubian-Arabian Craton	1,000,000
	Pharusian Belt	250,000
	South America	535,000
	Total Nubian Cycle:	1,785,000
	Total Measured Precambrian Crust:	60,585,000

[a]All areas were measured on 1:10,000,000 scale geologic maps; areas in Australia and South America were determined using a digitizer; all other areas were determined using a grid overlay method.
[b]Geology and geochronology of the Siberian and East European Cratons are poorly constrained. Estimated basement age domains are here based on recorded ages and geologies of exposed shield areas.
[c]Includes Congo Craton; Tanzanian and Zambian Cratons; Kibaran, Irumide, and portions of the Pan-African Belts in central Africa; and areas east of the West African Craton and Pharusian Belt.

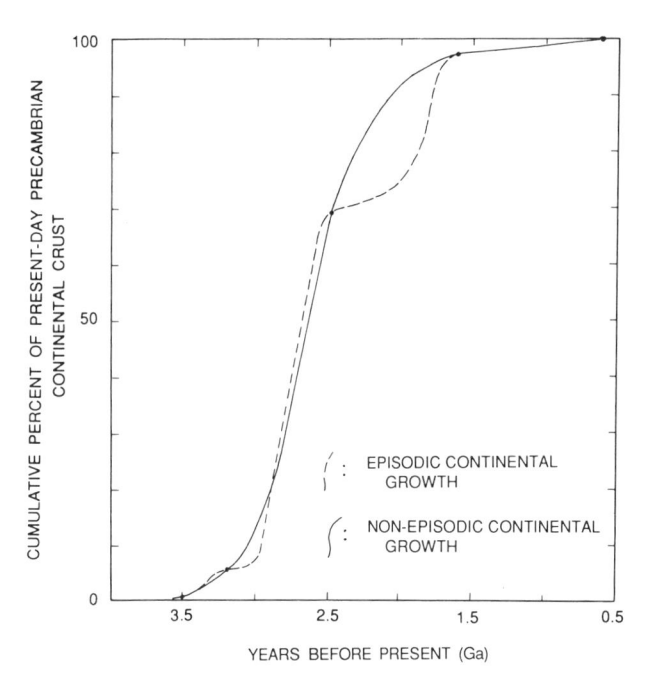

Figure 2.7.2. Growth patterns of the present-day continental crust based on observed initial ages of formation of major basement blocks (Fig. 2.7.1). Actual distribution of present-day crustal ages suggests that the crust has formed episodically (dashed curve) rather than continuously (solid curve).

Table 2.7.2. *Estimated proportions of continental crust formed during each of the major crust-forming cycles and the Middle Proterozoic.*

Cycle	I	II	III	IV
Isua	0.7	0.8	0.7	0.6
Barberton	5.1	3.8	5.0	
Superior	57.5	53.0	55.5	59.5
Colorado	33.6	39.4	32.2	39.9
Middle Proterozoic	0.2	0.3	1.2	
Nubian	3.0	3.7	5.4	
	100.1	100.0	100.0	100.0

I. From Table 2.7.1.
II. From Table 2.7.1 but excluding large unexposed portions of Siberian and East European Cratons and central African area.
III. From Table 2.7.1 but area estimates modified on presumption that some of the interior portions of the larger Middle and Late Proterozoic mobile belts represent juvenile crust. Modifications include considering 100 km-wide portions of the Damara, Zambesi, and Mozambique Belts as Late Proterozoic, and a portion of the Grenville Belt as Middle Proterozoic. This increases the Middle Proterozoic area to 740,000 km² and the Nubian Cycle area to 3,295,000 km² while decreasing the Superior Cycle area to 34,105,000 km² and the Colorado Cycle to 19,740,000 km².
IV. North American data from Table 2.7.1.

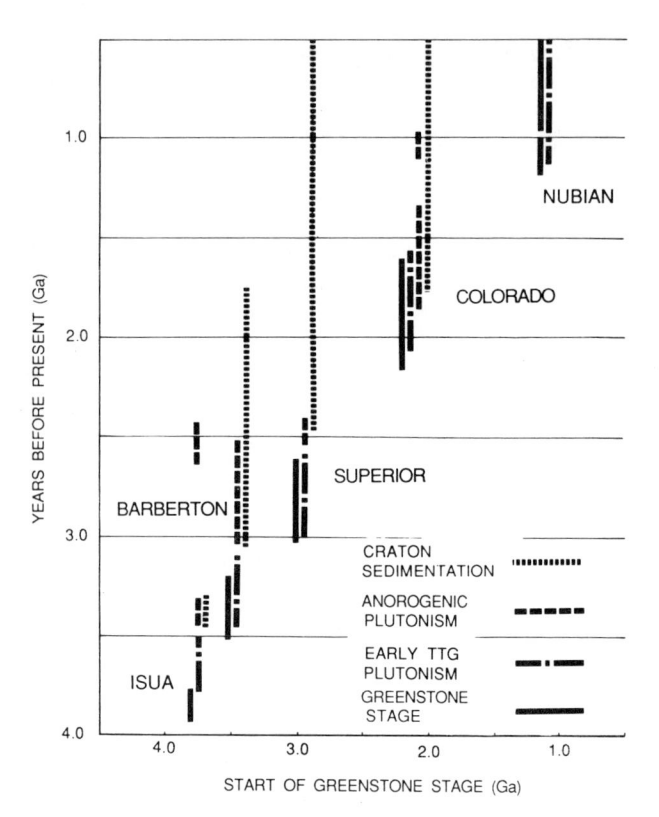

Figure 2.7.3. Principal Precambrian crust-forming episodes. Episodes are spaced along the abscissa based on the inferred time of beginning of the greenstone stage. Not all anorogenic plutonic and craton sedimentation events are shown.

2.7.2A Isua Cycle: pre-3.5 Ga

Small crustal blocks, lithosomes, or mineral grains with maximum ages older than 3.6 to 3.5 Ga occur widely in Precambrian terranes. Some of the larger domains of this age include the 4.0 to 3.9 Ga-old gneiss terrane in the Slave Province, Canada (Bowring et al. 1989); the 3.8 to 3.7 Ga-old Isua greenstone sequence and the intruding 3.7 to 3.6 Ga-old Amitsoq orthogneisses of west Greenland (Bridgewater et al. 1976; Nutman et al. 1984); the 3.6 Ga Minnesota River Valley gneisses and related rocks (Goldich and Wooden 1980); and the 3.6 Ga Narryer Complex in the Western Gneiss Belt, Western Australia (Myers and Williams 1985; Myers 1988). These blocks have or can be inferred to have experienced a greenstone stage of development before about 3.7 Ga, a concurrent and/or subsequent early intrusive stage and, at least locally, a post-cratonization interval of cratonic sedimentation, such as that represented by the 3.6 Ga-old quartzites of the Narryer Complex (Myers 1988).

In general, these old blocks of continental crust have been strongly affected by younger intrusive, metamorphic, and structural events and their areas and histories, except locally, are poorly constrained. In many cases, the greenstone stages have been completely obliterated and only orthogneiss marking early intrusive events remains; remaining metagreenstone remnants may have not been dated (Goldich and Wooden 1980; Bowring et al. 1989).

2.7.2B Barberton Cycle: 3.5 to 3.2 Ga

Latest Early to Middle Archean blocks of continental

Table 2.7.3. *Estimated crustal growth during major Precambrian crust-forming cycles and the Middle Proterozoic.*

Cycle	Time (Ga) (greenstone age)	Duration (Ma)	Area (km²/10³)	Percent	Cumulative Percent	Growth Rate (km²/a)
Isua*	4.0–3.6	400	455	0.7	0.7	0.07
Barberton*	3.5–3.2	300	3,068	5.1	5.8	0.5
Superior	3.0–2.5	500	34,796	57.5	63.3	3.5
Colorado	2.2–1.6	600	20,340	33.6	96.9	1.7
Middle Proterozoic	1.6–0.9	700	141	0.2	97.1	0.01
Nubian	0.9–0.6	300	1,785	3.0	100.1	0.3

*Early, low crustal growth rates almost certainly reflect the effect of later Archean and Proterozoic recycling of Early and Middle Archean crust (Section 2.6).

crust include the Kaapvaal Craton, South Africa; the Pilbara Block, Western Australia; portions of the Zimbagwe Craton, Zimbabwe; and areas in India. Small cratonic domains and isolated rock and mineral ages representing this interval are widespread in younger basement terranes on most continents (Fig. 2.7.1). Greenstone activity in the best dated blocks commenced 3.55 to 3.5 Ga ago and ended probably not later than about 3.2 Ga ago. Tonalitic and trondhjemitic intrusive events were essentially coeval with volcanism, but anorogenic granites continued to be emplaced during the latest Archean. The oldest cratonic sequences overlying these blocks are generally about 3.0 Ga in age, including the Pongola Supergroup, South Africa, and the Whim Creek Group, Western Australia.

2.7.2C 3.2 to 3.0 Ga

There are virtually no crustal blocks having formative granite-greenstone activity between 3.2 and 3.0 Ga ago. Intrusive activity continued locally in older crustal blocks, but little or no new crust was formed during this interval.

2.7.2D Superior Cycle: 3.0 to 2.6 Ga

The Late Archean was perhaps the most important crust-forming interval in the history of the earth. Major continental domains of this age occur in every continent (Table 2.7.1), and probably over 60% of the extant continental crust of Precambrian age formed at this time. Major greenstone development occurred from about 3.0 to 2.6 Ga. Intrusive activity coincided with or lagged slightly behind volcanism, culminating generally at 2.7 to 2.6 Ga. The oldest craton cover units overlying these Late Archean blocks are generally 2.4 to 1.9 Ga old.

During this same interval, thick cratonic sequences continued to accumulate on the older, Early to Middle Archean cratons.

2.7.2E 2.6 to 2.2 Ga

The interval from 2.6 to about 2.2 Ga ago is represented by an extremely sparse supracrustal record. Exceptions include craton cover successions accumulating on the stabilized Early to Middle Archean Kaapvaal and Pilbara Cratons and local rift sequences, including the Huronian Supergroup (ca. 2.4–2.1 Ga) of the Lake Superior region (Young 1983), and the Sumian, Sariolan, and Lapponian Groups (2.5–2.3 Ga) of the Baltic Shield (Grobatschev and Gaal 1987).

2.7.2F Colorado Cycle: 2.2 to 1.6 Ga

A major episode of diachronous greenstone development and craton formation commenced about 2.2 Ga ago and climaxed between 1.9 and 1.7 Ga ago with the formation of the Penokean and central and southwestern USA Proterozoic Provinces, the Svecofennian and Southwestern Provinces of Scandinavia, and much of the exposed and buried crust of Australia. The earliest documented granite-greenstone activity of this interval led to the formation of the Amazonian and West African Cratons at about 2.2 to 2.0 Ga (Gibbs and Olszewski 1982; Gruau et al. 1985; Cahen et al. 1984). The ages of greenstones in the Guiana Shield, however, are based on Sm-Nd model ages (Gruau et al. 1985) and crystallization ages may be somewhat younger. Many slightly younger Early Proterozoic cratons record this early activity in 2.4 to 2.0 Ga Sm-Nd model ages of supracrustal suites having crystallization ages of 2.0 to 1.6 Ga (Nelson and DePaulo 1985; Wilson et al. 1985; Page 1988; Barovich et al. 1989). Cratonic blocks of the Colorado Cycle continued to form diachronously until about 1.65 to 1.6 Ga, represented by the youngest granite-greenstone terranes in the southcentral USA (Van Schmus et al. 1987).

2.7.2G 1.6 to 0.95 Ga

The Middle Proterozoic, from about 1600 to 900 Ma ago, represents the longest interval of Precambrian time showing little greenstone activity and is represented by few if any new crustal blocks. Tectonism, especially rifting, was recurrent and widespread. Rifting of newly formed Early Proterozoic cratons began in the latest Early Proterozoic, as in the McArthur Basin and Mount Isa region of Australia (Jackson et al. 1987; Beardsmore et al. 1988); a major, perhaps global rifting event commenced about 1200 to 1100 Ma (Sawkins 1976; Hanson et al. 1988b); and a major interval of rifting and basinal development was initiated at about 1000 Ma, preceding the Pan-African and related Late Proterozoic tectonic cycles.

Diachronous orogenic events during the Middle Proterozoic include the widespread Irumide and Kibaran events in Africa (1.4 to 1.2 Ga; Cahen et al. 1984; Borg 1988; Klerkx et al. 1987), Urucuano tectonism in South America (Bernasconi 1987), formation of the Albany-Fraser and Musgrave Provinces in Australia (1.4 to 1.1 Ga), and the Grenville Orogeny in North America (1.1 to 1.0 Ga; Easton, 1986). Characteristically, Middle Proterozoic orogeny formed elongate zones of intense deformation or mobile belts. Most Middle Proterozoic mobile belts are made up of early rift deposits, voluminous craton margin sequences, and minor volcanic suites that were severely deformed, metamorphosed, and obducted onto adjacent older cratonic foreland areas during Middle to Late Proterozoic suturing. These histories probably reflect more or less complete Wilson cycles of ocean formation and closure. Mafic and calc-alkaline volcanic units occur widely as minor components of the interior portions of orogenic belts. In spite of an enormous amount of rifting, sedimentation, magmatism, metamorphism, and tectonism and the wide development of mobile belts during the Middle Proterozoic, however, it is unclear whether any substantial areas of new continental crust formed during this interval.

2.7.2H Nubian Cycle: post-1.0 Ga

The Late Proterozoic Nubian Cycle commenced with widespread diachronous continental rifting, some as early as 1200 Ma ago (Hansen et al. 1988) but with much between about 1000 and 900 Ma, probably involving the breakup of a supercontinent (Piper 1976). Greenstone volcanism and sedimentation was widespread but, as in most Middle Proterozoic orogenic belts, greenstone rocks are greatly subordinate to craton-derived sedimentary sequences. Only in northeast Africa and Arabia was a substantial accretionary tract of Late Precambrian crust formed. There, early volcanism and deformation occurred between about 920 and 600 Ma, with later events continuing until 450 Ma (Reymer and Schubert

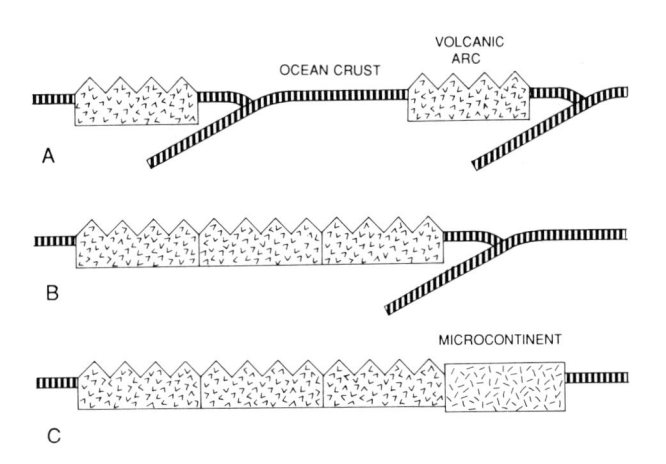

Figure 2.7.4. Schematic diagram depicting the sequential (A through C) early development of large cratons by the build-up of accretionary complexes. In many instances, as shown in C, accretion was terminated by collision with a microcontinent.

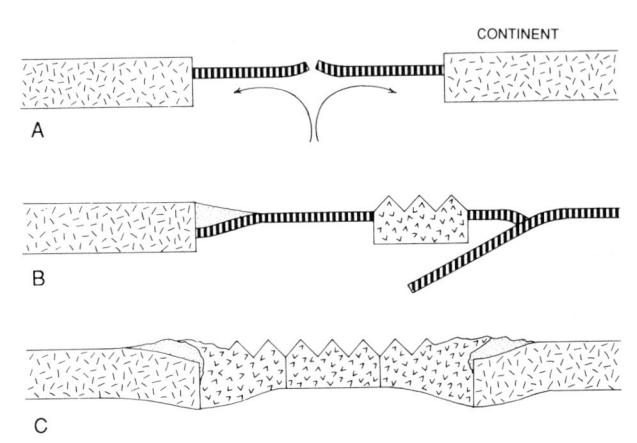

Figure 2.7.5. Schematic diagram depicting the sequential (A through C) development of Early Proterozoic cratons through initial rifting (A) followed by the growth of accretionary complexes. Craton growth was terminated by collision of the accretionary complexes with Archean continents. Some Early Proterozoic belts were involved in collision before significant accretionary growth occurred whereas others evolved into large, Archean-style, continental blocks. Symbols as for Figure 2.7.4.

1984; Kröner 1984; Kröner et al. 1987; Dixon and Golombek 1988).

2.7.3 Secular Variation in the Patterns of Crustal Formation

Comparison of the patterns of crust formation during each major crust-forming cycle suggests that large blocks of continental crust form through a common sequence of development stages (Figs. 2.7.4 through 2.7.6):

(1) an early granite-greenstone stage dominated by basaltic to calc-alkaline volcanism, by tonalitic, trondhjemitic, and/or granodioritic plutonism, and by volcaniclastic sedimentation in areas around magmatic arcs and subduction zones (*pre-cratonization stage*);

(2) a deformational stage involving horizontal compression, thickening, and metamorphism of the early volcanic and sedimentary sequences and the deposition of thick units of orogenic clastic sediments, commonly in foreland basins (*early cratonization stage*);

(3) an accretionary stage derived from the foregoing and marked by the initial formation of quasi-sialic components, their assembly into large crust-like masses through accretion, and partial stabilization by intrusion and magmatism (*late-cratonization stage*); and

(4) a prolonged interval of erosion and craton-related sedimentation on the newly formed craton (*post-cratonization stage*). Intracrustal and subcrustal intrusion and underplating were almost certainly acting in concert with near-surface magmatic and tectonic events to thicken the developing blocks of continental crust during stages 1 through 3; stages 3 and 4 commonly overlap in time.

These four stages appear to constitute the entire histories of some Archean cratons. Post-Archean crust-forming cycles, however, were generally more complex and display an increasingly prominent early rift stage, decreasing importance of greenstone magmatism and arc accretion, and a well-developed terminal stage of continental convergence or suturing (Figs. 2.7.5 and 2.7.6). The following discussion will explore some of the secular variations in the nature of crust-forming cycles.

2.7.3A Early to Middle Archean Domains (Isua and Barberton Cycles)

Known blocks of continental crust formed during the Isua Cycle are too few and fragmentary to compare meaningfully at the present time with younger crustal units. The geologic development of the Kaapvaal and Pilbara Cratons formed during the Barberton Cycle are reasonably well documented and serve as a basis for comparison with later crust-forming events.

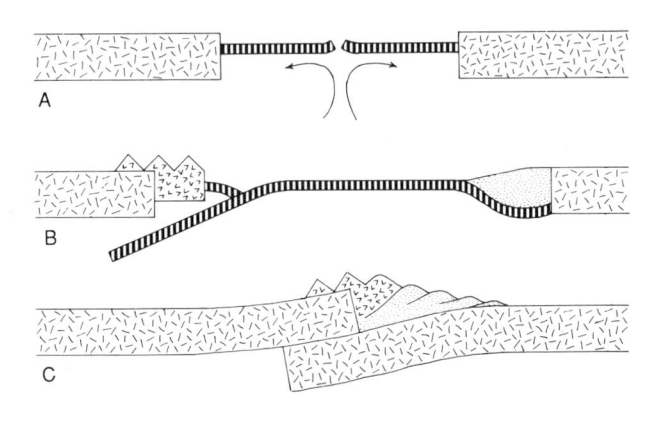

Figure 2.7.6. Schematic diagram depicting the sequential (A through C) early development of Late Proterozoic orogenic belts. Although some new Late Proterozoic crustal blocks were assembled through Early Proterozoic-style accretion (Fig. 2.7.5), most Late Proterozoic mobile belts contain only minor accretionary complexes and represent metamorphic zones and fold and thrust belts developed on older Archean or Early Proterozoic crust. Symbols as for Figures 2.7.4 and 2.7.5.

Both the Kaapvaal and Pilbara Cratons developed over about 300 million years, between about 3.5 and 3.2 Ga ago, by greenstone volcanism and sedimentation, concurrent tonalitic and trondhjemitic intrusion, and a culminating orogenic stage, all probably within magmatic arcs of accretionary complexes (Section 1.3). Neither craton shows evidence of a pre-greenstone basement or a pre-greenstone rift stage and, although both experienced compression during their orogenic stages, neither lies adjacent or or upon substantially older crustal blocks and there is no evidence that continent collision and suturing played a role in their evolution. Fold and thrust complexes and foreland basins commonly developed during late orogenesis by the collision of penecontemporaneous, quasi-continental arc complexes (Jackson et al. 1987; Lowe and Byerly, in press).

2.7.3B Late Archean domains (Superior Cycle)

Among the Late Archean cratons, reasonably detailed pictures have emerged for the evolution of the Superior and Slave Provinces, Canada (Percival and Williams, 1989; Kusky 1989), and the Yilgarn Block, Western Australia (Fletcher et al. 1984; Hallburg 1986; Gee et al. 1981). Both the Slave and Superior Provinces are mosaics of granitoid plutons and greenstone terranes representing volcanic arcs, associated arc-derived sedimentary successions and, possibly, oceanic islands assembled as accretionary complexes above subduction zones. The development of the Yilgarn Block is considerably more controversial, but its overall makeup and evolution suggest that it, too, represents a largely arc-related accretionary complex.

Small Early to Middle Archean blocks are associated with many of these Late Archean domains, such as the 3.6 Ga Western Gneiss Province along the western edge of the Yilgarn Block (Myers and Williams 1985; Myers 1988), the 3.6 Ga gneiss block along the southern edge of the Superior Province (Goldich and Wooden 1980), and a remnant of Early and Middle Archean crust along the western margin of the Slave Province (Bowring and Van Schmus 1984; Kusky 1989; Bowring et al. 1989). In the Slave Province, greenstone belt development followed rifting of an Early to Middle Archean microcontinent (Kusky 1989), but in the Yilgarn Block (Myers 1988) and possibly the Superior Province, there is no evidence that the older domains were associated with the developing accretionary complexes prior to the terminal orogenic events.

Unfortunately, little detail is available on the development of some of the largest continental blocks, including the Congo, Siberian, and East European Cratons.

Available evidence suggests that Late Archean crustal blocks formed by the buildup and accretion of arc and arc-related volcanic and sedimentary sequences above subduction zones. Ages within the cratons suggest that volcanism and intrusion commonly occurred over intervals of 200 to 400 million years, although individual greenstone successions probably developed over much shorter intervals (e.g., Ludden et al. 1986). Locally, there is evidence that greenstone development was preceded by rifting of older microcontinents (Kusky 1989) and a number of microcontinents were incorporated into the growing accretionary tracts. Microcontinent suturing locally formed high-grade belts, such as the Limpopo Belt (e.g., Burke et al. 1985a), but did not destroy the large accretionary tracts. There is no evidence that the enormous Late Archean granite-greenstone terranes are allochthonous or were obducted onto older crust as Phanerozoic-style fold and thrust belts. However, in both the Yilgarn and Superior Provinces, the climactic intrusive stage may have coincided with collision between the accretionary tract and older microcontinental fragments.

2.7.3C Early Proterozoic Domains (Colorado Cycle)

Late Archean patterns of crustal evolution characterize many of the larger Early Proterozoic shields. The enormous 1.9 to 1.6 Ga-old Proterozoic granite-greenstone province of the central and southwestern USA and the Svecofennian Province of the Baltic region were both constructed by the assembly of ancient arc and arc-derived sedimentary terranes (Karlstrom and Bowring 1988; Grambling et al. 1988; Bickford 1988). The Amazonian Craton has not been shown to be accretionary in nature, but it might be suspected that this large craton developed in a manner similar to other large tracts of early Precambrian crust.

These large blocky Early Proterozoic accretionary complexes, however, although fundamentally like their Late Archean counterparts, differ in showing prominent early, ocean-forming rift stages involving older continental blocks. The central USA Province developed along the southern margin of the Archean Wyoming Province, and greenstone accretion along this margin was preceded by an interval of rifting and craton margin sedimentation (Karlstrom and Houston 1984). Formation and accretion of the Svecofennian and related Early Proterozoic belts in Scandinavia was also preceded by rifting and craton margin sedimentation along the Archean Karelian Craton (Gorbatschev and Gaal 1987). In both of these areas, however, initial rifting was followed by the construction of broad accretionary complexes that incorporated few or no older crustal blocks.

Other Early Proterozoic domains are preserved as elongate fold and thrust belts between blocks of Archean crust. The Trans-Hudson Orogen in Canada, part of a major circum-Superior accretionary province, is a broad Early Proterozoic accretionary tract caught between converging Late Archean continental blocks (Hoffman 1988). Other Early Proterozoic orogens survive as narrow fold and thrust belts obducted onto the bounding Archean blocks during suturing (Hoffman 1988).

These relationships suggest that Early Proterozoic orogenesis and crust formation were transitional between Archean styles (Fig. 2.7.5), dominated by the formation of large arc-dominated accretionary tracts with little or no continental influence, and Late Proterozoic and Phanerozoic styles, in which early continental rifting and late continental suturing were prominent and accretionary belts generally small.

2.7.3D Late Proterozoic Domains (Nubian Cycle)

From 1200 to 950 Ma ago, rifting initiated an orogenic cycle that climaxed with the Nubian Cycle of crust formation, represented by Pan-African, Brasiliano, and Baikalian orogenesis. In virtually all areas, the juvenile components of Late Proterozoic orogenic belts consist largely of heavily deformed supracrustal sequences, the formation of which was dominated by craton and craton-margin sedimentation. Greenstone sequences and arc accretion played a minor role in the evolution of most Late Proterozoic belts. Only in northeastern Africa and Arabia was rifting followed by the development of an accretionary complex that forms the basement of much of the Arabian-Nubian Shield (Kröner 1985; Kröner et al. 1987). The formation of the Late Proterozoic Arabian-Nubian crust was initiated by rifting and deposition of early rift and craton-margin sequences in eastern Egypt. These sediments were subsequently deformed and metamorphosed by collision with an eastward-growing accretionary complex. Growth of the accretion/obduction complex involved volcanism and deformation over about 300 million years and was terminated by the welding of another block of older continental crust against its eastern margin (Raymer and Schubert 1984; Kröner et al. 1987; Dixon and Golombek 1988).

The Arabian-Nubian Shield is one of the few large accretionary granite-greenstone tracts of post-Early Proterozoic age. Still younger, Phanerozoic granite-greenstone terranes occur locally, such as the Klamath Mountains and western Paleozoic and Mesozoic Provinces of California (Miller and Harwood 1989), but their ultimate fate, as new blocks of continental crust or terranes to be obliterated during later tectonism, has yet to be decided.

Other Late Proterozoic mobile belts are widely preserved, principally as fold and thrust belts and metamorphic zones marking sutures between older continental blocks. Examples include the Paraguay-Araguiaia Belt between the Archean Sao Francisco and Early Proterozoic Amazonian Cratons in South America (Almeida and Hasui, 1984; Bernasconi 1987), and the Damara, Katangan, Zambesi, Mozambique, Togo, Rokelide-Mauritanide, and related fold belts in Africa (Kröner 1977; Wright et al. 1985). Each of these belts is marked by (*i*) a fold and thrust belt where craton-margin or basinal sequences have been obducted onto adjacent Archean or Early Proterozoic foreland blocks during suturing; and (*ii*) commonly, a broad zone of intense metamorphism and intrusion on one or both of the adjacent older crustal blocks marking the site of a Late Proterozoic continental volcanic arc. Greenstones are present locally within the fold and thrust belts but large tracts of new continental crust built up by the growth of accretionary wedges are lacking.

In summary, the Late Proterozoic Nubian Cycle probably involved the formation of little new preserved continental crust. Late Proterozoic orogens (Fig. 2.7.6) are composed of (*i*) allochthonous cratonic rift and craton-margin sequences obducted onto adjacent, older cratonic foreland areas during suturing, and (*ii*) metamorphic and intrusive complexes developed on older crust and marking continental arc root zones. No large, open accretionary granite-greenstone tracts, comparable to those of Archean and Early Proterozoic age, developed at this time. Only one relatively large accretionary complex in northeast Africa and western Arabia formed and was preserved between converging older cratons.

2.7.4 Rates of Crustal Growth

Studies by Reymer and Schubert (1984, 1987) have estimated crustal addition rates based on a present-day arc model. Recasting the areal data from the present study into volumetric form using a constant crustal thickness of 50 km, equivalent to that used by Reymer and Schubert (1984), crustal growth rate estimates are obtained for each of the major Precambrian crust-forming intervals (Table 2.7.3). This approach almost certainly overemphasizes crustal development during the early greenstone and intrusive stages, however, because major crustal thickening occurred through anorogenic intrusion and underplating long after initial cratonization (Taylor and McLennan 1985). This may, in fact, represent a major process of deep crustal growth during intervals, such as 2.6 to 2.2 Ga and 1.6 to 0.9 Ga, represented by little greenstone activity and little apparent near-surface crustal growth.

The results (Table 2.7.3) argue for enormous variations in crustal growth rates, ranging from less than $0.05 \text{ km}^3/\text{a}$ during the Middle Proterozoic to rates of $3.4 \text{ km}^3/\text{a}$ during the 500 million-year-long Superior cycle and $1.65 \text{ km}^3/\text{a}$ over the 600 million-year-long Colorado Cycle. Intervals between major crust-forming cycles must have involved more or less continuous plate activity and heat loss, but little arc activity that produced new continental crust. Evidently, not all arc magmatism translates directly into the formation of preservable continental crust, making it difficult to assess long-term crustal accretion rates represented by modern arc magmatism.

The enormous width of Archean and Early Proterozoic accretionary tracts, commonly up to 1000 km, and the estimated high crustal accretion rates, suggest that the rapid growth of the continental crust during these intervals was related to the following three main factors:

(1) During the Early and Middle Archean, subduction processes were insufficient to recycle all newly formed and older, perhaps impact-modified, crustal components back onto the mantle, permitting the long-term preservation of continental rocks at the surface.
(2) High spreading and subduction rates promoted the rapid growth of arcs and the accumulation of buoyant arc terranes within accretionary complexes.
(3) The paucity of continental blocks allowed the assembly of large accretionary tracts before accretion was terminated by collision with older crustal blocks, even when spreading and subduction rates were high.

During later geologic time, not only did spreading rates probably decrease but the size and number of continental blocks increased. These factors resulted in greatly reduced rates of arc growth and accretion and the increasing dominance of continental collision and suturing in late-stage orogenesis.

The minor role played by continental blocks during the evolution of the Archean crust argues strongly that little continental crust existed at that time. Scenarios suggesting formation of large volumes of continental crust before about 2.6 Ga ago (Fyfe 1978; Armstrong 1981; Reymer and Schubert 1984) are probably incorrect.

2.7.5 Conclusions

The evolution of the earth's continental crust can be divided into three principal stages based on the patterns of crustal development: (*i*) an Archean microcontinental or permobile stage characterized by the existence of a few small continental blocks, rapid plate movements, and high arc-growth and accretion rates; (*ii*) an Early Proterozoic transitional stage during which continental blocks influenced but did not control continent growth; and (*iii*) a Late Precambrian and Phanerozoic stage in which lithospheric dynamics were dominated by the existence of large continental blocks and relatively slower spreading rates.

Archean continental growth during the Isua, Barberton, and Superior Cycles occurred primarily through a modified Wilson cycle that lacked prominent continental rifting and continent suturing stages. Archean continents developed initially as large granite-greenstone accretionary complexes and commonly reached thousands of kilometers in width before undergoing cratonization and stabilization. This unimpeded growth reflects both the paucity of older continental blocks and higher accretion rates, permitting the assembly of enormous accretionary complexes. It is not clear, however, whether these complexes accreted gradually over long intervals of time, were constructed over short intervals by the rapid assembly of extant active and fossil arc terranes, or whether both or intermediate accretionary rates prevailed. The climactic intrusive and orogenic events seem to have coincided with collision of small blocks of older continental crust with the growing accretionary tracts.

The Early Proterozoic Colorado Cycle marks a clear transition in the history of the earth's crust. Archean styles of crustal evolution prevailed in many areas, with accretionary complexes developing into immense tracts of new continental crust, such as the central USA Province. In other areas, however, accretion was terminated by collision of Archean continental blocks with the growing accretionary tracts. Preserved accretionary belts of this type, including the 300-km-wide Trans-Hudson Orogen, are clearly transitional into much narrower mobile zones representing fully allochthonous fold and thrust belts obducted onto older basement blocks during suturing without the formation of fundamentally new crustal domains. Most Early Proterozoic accretionary tracks developed adjacent to older cratonic blocks and show well-developed craton-margin sequences that have been thrust onto the Archean foreland basement during continent-accretionary complex collisions.

Little new continental crust was generated during the Middle and Late Proterozoic. The Middle Proterozoic, 1600 to 900 Ma, was characterized by widespread orogenesis, and mobile belts of this age are well developed in Africa, North America, and other continents. There are no major granite-greenstone terranes of this age, however, and no corresponding large blocks of continental crust formed during this interval.

A major cycle of rifting and orogenesis, the Nubian Cycle, was initiated and terminated during the late Middle and Late Proterozoic, 1200 to 600 Ma ago. In contrast to Archean and many Early Proterozoic crust-forming cycles, Late Proterozoic orogens are dominated by craton-derived rift and craton and craton-margin deposits. Accretionary tracts developed locally, as in the Pan-African belts of northeastern Africa and Arabia, but these are more comparable to Early Proterozoic granite-greenstone tracts preserved between continental blocks. There appear to be no preserved analogues of the enormous Archean and Early Proterozoic accretionary tracts. Late Proterozoic orogenesis involved complete Wilson cycles initiated by continental rifting and terminated with the obduction of the craton margin sequences and some eugeoclinal arc and accretionary belts onto the adjacent older continental foreland areas during suturing. Very little new crust was formed at this time.

The dominance of large blocks of continental crust on the modern earth, and the relatively slow spreading and subduction rates suggest that large accretionary tracts are unlikely to develop in the future and that the continental inventory has probably reached a more or less steady state.

2.8

Summary and Conclusions

DONALD R. LOWE

(1) The setting, style, and evolution of Proterozoic sedimentary basins closely resemble those of Phanerozoic basins, reflecting the fundamental similarity of lithospheric dynamics throughout the last 2,500 million years of earth history. Some differences, such as the thinness of Early Proterozoic passive margin sequences and the development of Proterozoic foreland basins mainly on thermally young lithosphere, are consistent with inferred models of the earth's thermal evolution.

(2) Current understanding of the age composition of the continental crust indicates that Precambrian continental crust formed during five major crust-forming cycles: (*i*) the pre-3.5 Ga Isua Cycle; (*ii*) the 3.55 to 3.3 Ga Barberton Cycle; (*iii*) the 3.0 to 2.6 Ga Superior Cycle; (*iv*) the 2.2 to 1.6 Ga Colorado Cycle; and (*v*) the post-1.0 Ga Nubian Cycle. Each of these cycles involved one or more shorter intervals of greenstone activity, mainly within subduction-related volcanic arcs, followed by assembly of large accretionary complexes that were stabilized by tonalite-trondhjemite-granodiorite intrusive suites and by later granitic intrusions to form thick blocks of continental crust.

(3) The changing nature of crust formation during the Precambrian reflects the increasing abundance of continental crust at the earth's surface. Archean cycles involved the rapid construction of enormous accretionary complexes that were little influenced by blocks of older continental crust. Individual orogenic-accretionary cycles did not show prominent initial rift or terminal suturing stages because of the paucity and small size of continental blocks. The Early Proterozoic Colorado Cycle was intermediate in character between Archean and Late Proterozoic styles. The formation of large accretionary complexes and vast new continental blocks was widely preceded by the rifting of Archean continents. Locally, accretion was terminated by suturing without formation of accretionary complexes. Little new crust formed during the Late Proterozoic as continental margin and volcanic arc sequences were widely obducted onto adjacent older continental blocks during orogenesis and suturing.

2.9

Unresolved Problems and Future Research Directions
DONALD R. LOWE

Having approached the problem of the geologic evolution of the Proterozoic earth, we must recognize the incompleteness of the data upon which our interpretations are based. The stratigraphy, structure, and evolution of enormous blocks of continental crust in Africa, South America, Asia, and Antarctica remain poorly constrained. High-precision zircon geochronology, which represents the key to resolving the histories of Precambrian crustal blocks, has to date been applied to only a few terranes, mainly in North America and Australia. The future development of a more detailed and accurate picture of the evolution of the Precambrian earth and its crust require systematic geological and geochronological dissection of these enormous areas of virtually unexplored Precambrian crust.

We have presented a model that suggests that the Precambrian continental crust formed discontinuously during a number of major crust-forming cycles. These were separated by intervals of little new crustal growth but considerable tectonic and magmatic activity within older continental blocks. The concept of large-scale cycles of orogenic, magmatic, or crustal activity is an old one and by no means non-controversial. The criteria by which cyclicity should be measured, however, remain poorly defined. We have chosen the crystallization ages of granite-greenstone terranes, and available, often imprecise age data from a limited number of areas seem to support cyclic, discontinuous continental growth. If, however, extensive granite-greenstone terranes with intermediate ages exist in some of the as-yet poorly studied cratons, the apparent cyclicity could vanish.

The unknown nature of the crust beneath many Middle and Late Proterozoic orogenic belts presents another stumbling block to evaluating the amount of crust formed during major periods of plate activity. Do these orogens represent fold and thrust belts developed on top of sutured Early Proterozoic and Archean continents, or are there significant intervening domains of new, later Precambrian crust?

Clearly, major goals of Precambrian geologic research over the next decades must include (*i*) detailed stratigraphic and structural interpretation of the poorly studied portions of the earth's crust; (*ii*) the development of accurate histories of these areas by combining geological studies with high-precision zircon and Sm-Nd geochronology; and (*iii*) the construction of an accurate map of the age-of-formation of the earth's continental crust.

Many other important problems remain concerning the nature of plate evolution during the Precambrian, the mechanics of growth and stabilization of accretionary complexes, the role of underplating and the origin of anorogenic granite provinces, and the changing role of continental crust in lithospheric dynamics. Hypotheses and partial solutions already exist for many of these problems. The enormous effort currently directed at understanding the evolution of Precambrian terranes should insure significant additional progress in the future.

3
Proterozoic Biogeochemistry

J. M. HAYES   DAVID J. DES MARAIS   IAN B. LAMBERT
HARALD STRAUSS   ROGER E. SUMMONS

3.1
Principles of Molecular and Isotopic Biogeochemistry
ROGER E. SUMMONS   J. M. HAYES

Biogeochemistry encompasses the study of chemical fossils. It includes and draws on knowledge of the biochemical activities of contemporary organisms in modern sedimentary environments, including their roles in the biogeochemical cycling and isotopic fractionation of important elements such as carbon, oxygen, sulfur, and nitrogen, and their production of taxonomically distinctive organic compounds. This Section deals with the chemical entities preserved in the Proterozoic sedimentary record that may carry information about the biology and evolution of early life.

Chemical fossils can be discerned at the atomic level, in the occurrence of anomalous concentrations of a particular element or an isotope; at a molecular level, in the structure and stereochemistry of hydrocarbons derived from membrane lipids or pigments; and at a macromolecular level by way of the preservation of detailed chemical structures in kerogen and morphologically distinct microfossils. Paleobiochemical information is encoded in the nucleic acids of extant organisms and in their comparative biochemistry; this topic is treated in Chapter 9. Here we examine and discuss the occurrence of isotopic and molecular fossils. A considerable and consistent body of information derived, in part, from techniques developed during exploration for petroleum and minerals is now available. Rapid expansion of this knowledge is presently taking place, particularly with regard to chemical processes in early preservation of organic matter, structures of kerogen, isotopic composition of individual biomarkers, and global secular variations in organic and inorganic isotopic abundances.

3.1.1 Preservation and Syngenicity

The earliest organic geochemical studies of Proterozoic sediments held much promise as a means of understanding early evolution through the study of fossil lipids (e.g., Eglinton et al. 1964; Barghoorn et al. 1965; Burlingame et al. 1965: Oró et al. 1965; Johns et al. 1966; Hoering 1967, 1976, 1981; McKirdy 1974). Diverse biogenic and potentially informative molecular classes were identified in bitumens from rocks of Proterozoic and Archean age. These biological marker compounds (biomarkers) included hydrocarbons of various types (see McKirdy and Hahn 1982, for a review), fatty acids (e.g., Han and Calvin 1969; Smith et al. 1970), amino acids (Barghoorn 1957), and porphyrins (Barghoorn et al. 1965; Kvenvolden et al. 1969). However, as basic knowledge and new technical and instrumental skills developed during the study of Phanerozoic sediments and petroleums, it became evident that some early studies of Proterozoic bitumens were beset with uncertainty regarding the origins of the organic matter. Specifically, the metamorphic history of many of the sediments under study precluded the indigenous occurrence of certain of the molecular compound classes which had been detected (Leventhal et al. 1975; Wedeking and Hayes 1983; Hayes et al. 1983; Janks et al. 1985). The problems were particularly severe when solvent extracts from rocks containing very low concentrations of organic carbon had been analyzed. If present, genuine Proterozoic molecular fossils would be confined to the most thermally recalcitrant molecules such as hydrocarbons, and would be restricted to sediments with low levels of thermal alteration (McKirdy and Hahn 1982; Hoering and Navale 1987) and relatively high total organic matter contents. The presence of significant amounts of immobile, well-preserved kerogen, as proven by elemental H/C ratio or some other criterion, and rigorous geological and sedimentological characterization of samples are prerequisites to reliable study of Proterozoic organic matter (e.g., McKirdy 1974; McKirdy and Hahn 1982; Hayes et al. 1983b; Janks et al. 1985; Hoering and Navale 1987; Summons et al. 1988b).

As well as kerogen quantity and quality, the presence of indigenous petroleum deposits in Proterozoic sediments gives credence to some studies of primitive organic matter. Petroleum fields occur continuously throughout the Phanerozoic. In geologically straightforward situations, information on kerogen and bitumen type, abundance, and thermal maturity, as well as on the stratigraphic position of porous reservoirs and impervious seals in the associated sedimentary sequence, can be used to define the petroleum distributions on a regional scale. The factors that establish the likelihood of petroleum deposits in such situations should, logically, have application to older rocks and it is now established that significant oil deposits occur in geological settings indicative of derivation from

Proterozoic source rocks. Large oil fields of the Siberian Platform are reservoired in Late Proterozoic and Cambrian sediments in association with source rocks of the same age (Meyerhoff 1980; Aref'ev et al. 1980). Similarly, in Oman, extensive oil fields are reservoired in Late Proterozoic and younger rocks where the known and likely sources are within sediments of the Proterozoic Huqf Group (Alsharhan and Kendal 1986; Al-Marjeby and Nash 1980; Hughes Clarke et al. 1988). As is discussed below, these petroleums have unique compositional and isotopic features indicative of Proterozoic origins.

Deposits of primitive organic matter are known from the Middle Proterozoic, where study of the Nonesuch Formation (ca. 1050 Ma) of the North American Mid-continent Rift over a number of years has revealed that significant concentrations of organic matter (kerogen and bitumen) are found in a relatively narrow horizon within the basal section, just above the Copper Harbor Conglomerate. The organic rich sediments, located at the White Pine Copper Mine, Michigan, have generally experienced moderate thermal alteration (Murray et al. 1980; Dickas 1986; Pratt et al. 1989; this study) and are geologically situated where migration of large amounts of petroleum from younger rocks is not feasible. Thus, it seems likely that the numerous petroleum seeps within the formation have been sourced from within the host rocks. A more extensive, but similarly well defined situation is found in the McArthur Basin of northern Australia, where a thick sequence of flat lying and unmetamorphosed Proterozoic sediments (ca. 1300 to 1500 Ma) contains several organic-rich units, some of which have oil seeps and "live" oil (Muir et al. 1980; Jackson et al. 1986, 1988). An appreciation of the overall abundance of organic matter can be gained by noting that within the Roper and McArthur Groups there are several tens of meters thickness of organic rich sediments extending laterally over hundreds of kilometers. The very significant contents of organic carbon in the Barney Creek Formation (0.2 to 10.4% of total organic carbon, "TOC"), the Yalco Formation (0.8 to 5.4% TOC), the Velkerri Formation (0.9 to 7.2% TOC), and the McMinn Formation (0.7 to 2.9% TOC) illustrate the magnitude of these deposits of fossil organic matter, while the stratigraphy, regional geology, and correlation of thermal maturity parameters within this sedimentary sequence indicate that the petroleum originates in the host rocks (Jackson et al. 1988; Crick et al. 1988). In all these examples, the co-occurrence of oils and rich petroleum source rocks of appropriate thermal maturity provides strong evidence for Proterozoic origins of the associated organic matter and for its preservation over the time interval since deposition.

Besides the above considerations, there are internal tests that can be applied to establish the integrity of isolated but locally abundant deposits of sedimentary bitumen and kerogen. Immobile kerogen of appropriately low thermal alteration should predominate. It may contain recognizably Proterozoic microfossils. Carbon isotopic signatures of the carbonate, kerogen, and bitumen subfractions should be consistent with relatedness (e.g., Galimov 1980; Stahl 1978; Knoll et al. 1986; Burwood et al. 1988). Furthermore, the hydrocarbons may include some of the unusual biomarker assemblages found in sediments of comparable age, and inappropriate molecules, such as those related exclusively to vascular plants, should be absent.

3.1.2 Formation of Kerogen

It is now established that many of the most robust components of fossil organic matter are derived from the original lipid constituents of living organisms. The precursors are mainly the components of cell membranes, but also include pigments such as carotenoids and chlorophylls and storage or protective lipids such as triglycerides, wax esters, resins, and hydrocarbons. Another important contribution comes from the structural material including the resistant biopolymers constructed from polysaccharides, sporopollinen, and polymethylenic matter like *polymère résistante de Botryococcus* (PRB) (Berkaloff et al. 1983; Largeau et al. 1986; Boon and De Leeuw 1987; Chalansonnet et al. 1988; Kadouri et al. 1988). These biopolymers commonly occur in vegetative cells as well as reproductive cysts of protists and in the sheaths of cyanobacteria.

As the remains of organisms are deposited in the sediments, many of the labile and easily assimilated components, particularly compounds containing oxygen, phosphorus, and nitrogen, will be removed by bacterial attack. Thus, the biochemical components which are normally rapidly metabolized and recycled in a living cell will be consumed by detritivores, leaving a product enriched in resistant components. If the sediments remain anoxic, the terminal stages in decomposition, sulfate reduction and methanogenesis, will add new lipid signatures to the complex of organic matter. Erosion and transport may lead to exposure to oxygen and renewed episodes of oxidation followed by anaerobic reworking. Preservation of high contents of organic matter will generally only occur below stable anoxic water columns. These conditions are enhanced during extended periods of thermally or salinity-induced stratification. In these cases, evidence of input from halophiles and methanogens may also be detected.

In the marine environment, or in other instances of elevated sulfate content, sulfate-reducing bacteria and, possibly, associated sulfur-oxidizing bacteria will be active. Copious amounts of sulfide, polysulfide, and elemental sulfur can be produced. Provided the supply of metal ions is insufficient to trap all sulfide, residual HS will be available to react with substrates such as unsaturated lipid in a process of natural "vulcanization". This will lead to polymerization and condensation, and eventually to the generation of protokerogen. The totality of processes occurring during the early stages of burial, under mild conditions of temperature and pressure, is termed diagenesis. The principal organic chemical reactions include hydrogenation, clay-catalyzed rearrangements, aromatization, metal complexation, and condensation. The geopolymer that results is termed kerogen and amounts of unbound lipids, including free hydrocarbon (Hunt 1979; Durand 1980; Tissot and Welte 1984), may persist as a residue.

3.1.3 Biomarkers: Origin and Significance

Alfred Treibs' deduction (1934) that the metallated porphyrins in ancient sediments and petroleum derived from the chlorophyll of plants marked the recognition of the first biological marker compounds, or "biomarkers". Subsequent research has led to the generally accepted view that the bulk of petroleum and dispersed organic matter now present in the lithosphere is biogenic in origin. These materials comprise the preserved remnants of primary carbon fixation by photosynthetic organisms, minor inputs from heterotrophs higher in the food chain, and the products of bacteria which repeatedly attack, degrade, and recycle the debris of higher organisms. Biomarkers are distinctive molecules that can be identified in this organic matter and which, by their presence and abundance, inform us of some aspect of these processes. Thus n-alkanes (e.g., 1 of Fig. 3.1.1), usually the simplest and most abundant components of petroleum, may be useful as biomarkers through consideration of their isotopic compositions, chain lengths, and preferences for even or odd numbers of carbon atoms.

Most biomembranes are constructed from glycerol-based phospholipids containing non-rigid, n-acyl carbon chains and formed into a lipid bilayer. Modifications which thicken or disrupt an otherwise homogeneous bilayer structure and allow an organism to control the fluidity of its membranes usually have some phylogenetic significance because their details are dependent on the level of biochemical sophistication within a particular group. Primitive biota such as bacteria and cyanobacteria utilize branched-chain monomethyl, cyclopropyl, ω-cyclohexyl or monounsaturated fatty acids in addition to the normal saturated species (e.g., Kaneda 1977; Suzuki et al. 1981; Fulco 1983; Dasgupta et al. 1984; Parkes and Taylor 1986; Dowling et al. 1986). Eukaryotic algae and plants introduce polyunsaturation and longer carbon chains into the polar lipid moieties. Besides the non-rigid components, lipid bilayers have rigid inserts which have further and profound phylogenetic significance which is discussed below. Archaebacteria constitute a major exception in that their lipid bilayers are composed of isoprenoid chains, sometimes containing pentacyclic rings, and which are bound to the glycerol head groups by ether rather than ester linkages. The phytanyl and biphytanyl glycerol ethers of the archaebacteria show some exclusivity in their distribution within the kingdom and appear to constitute both the fluid and rigid partners of the biomembrane (Kates 1978; Langworthy 1985; De Rosa et al. 1986), because neither hopanoids nor steroids have been detected in archaebacteria (Rohmer et al. 1984; Taylor 1984).

Unsaturation in membrane lipid acyl chains is not preservable since it does not survive oxidation, contact with strongly reducing conditions, or exposure to reactive sulfur species during sedimentary diagenesis. However, other aspects, including isotopic composition and structural details, will be preserved, albeit to varying degrees. The diverse fossil forms of simple branched and cyclic alkanes are shown in Figure 3.1.1, in which representative examples of methyl branching (e.g., 2-4), cyclicity (e.g., 5, 6), chain length, and style of isoprenoid linkage (compare 7 to 10) are illustrated. These hydrocarbons, so ubiquitous in sedimentary bitumens and chemically bound in macromolecular kerogens, constitute informative molecular fossils for Proterozoic organisms.

Strong evidence now exists for a phylogenetic, and evolutionarily determined, hierarchy in the types and structures of the rigid components of biomembranes (Nes 1974; Ourisson et al. 1979, 1987; Rohmer et al. 1979, 1984; Nes and Nes 1980; Taylor 1984). These "rigidifiers" or strengtheners comprise a variety of molecules constructed from isoprene (C_5) units, usually in chains of 30 (sometimes 35, 40, 45 or 50) carbon atoms and known collectively as polyterpenoids or polyisoprenoids. Their structures are shown in Figure 3.1.2. The C_{30} compounds (triterpenoids), derived from squalene (11), can be cyclized into pentacyclic hopanoids (e.g., 13) in a sequence of enzyme-mediated steps that does not require the presence of molecular oxygen. As discussed in more detail below, production of hopanoids is characteristic of many classes of bacteria as well as cyanobacteria. In an alternate process that *does* require molecular oxygen at several stages, squalene can be converted to tetracyclic steroids (e.g., 16). This process is found almost exclusively in eukaryotes (e.g., Ourisson et al. 1987). Within the steroids themselves there is an evolutionary hierarchy, postulated on the basis of proposed improvements in function (Bloch 1983), progressing along the biosynthetic steps from lanosterol (15) to cholesterol (16). Structural variety with phylogenetic significance is associated with the patterns of unsaturation and alkylation in the sterol rings and side chain (e.g., Nes and McKean 1977; Djerassi 1981; Schmitz 1983), and the alkylation patterns have most potential for preservation in sediments. While few prokaryotes are known to biosynthesize

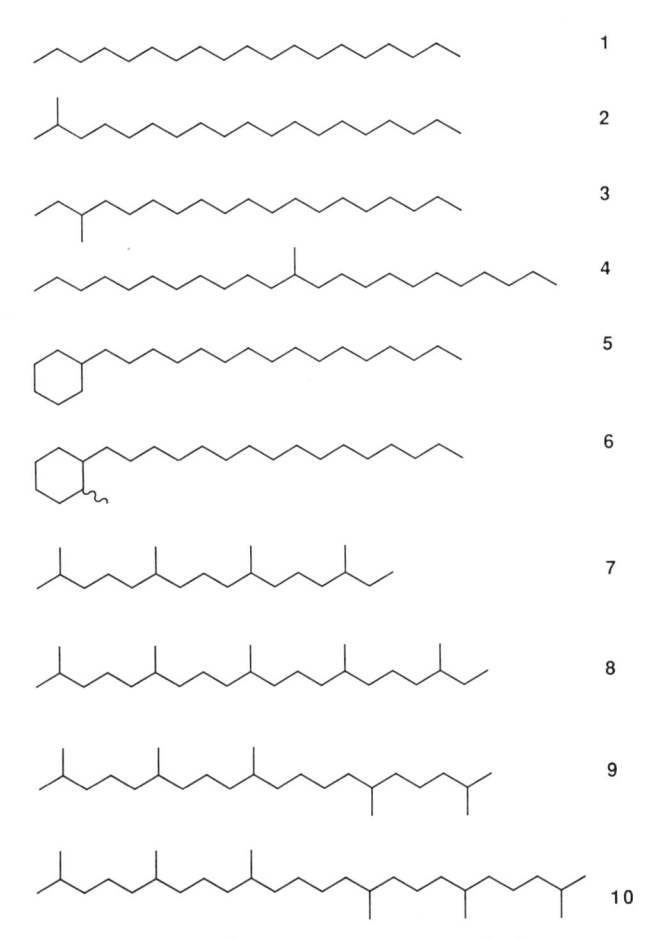

Figure 3.1.1. Structures of simple branched and cylic alkane biomarkers found in sediments and petroleum.

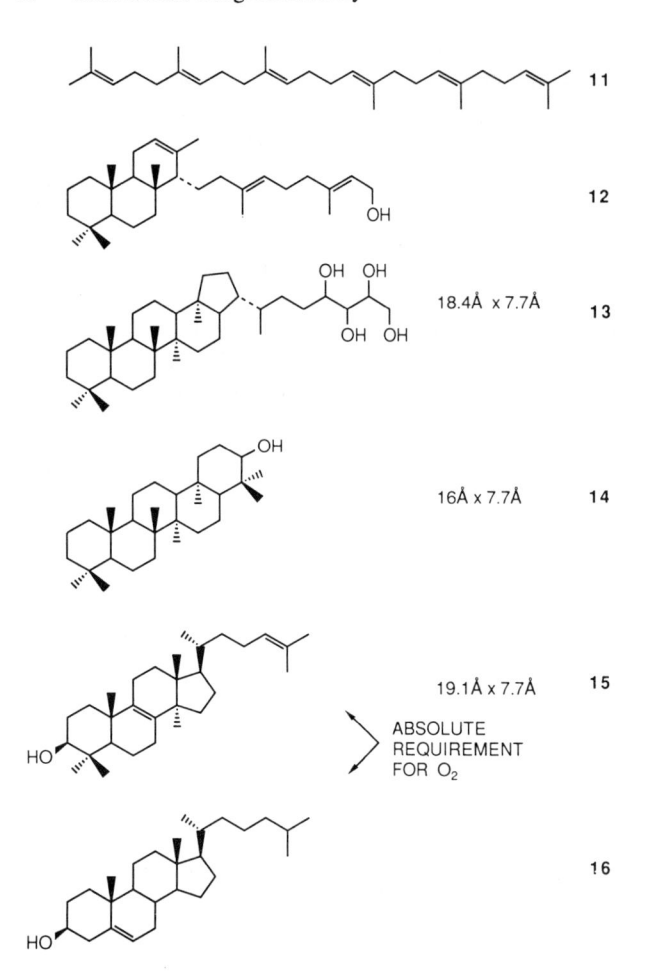

Figure 3.1.2. Structures of triterpenoids which are, or may be, rigidifying constituents of cell membranes of extant organisms. Squalene (**11**) is the biosynthetic precursor of the hopanoids (e.g., **13**), tetrahymanol (**14**), and the steroids (e.g., **15**, **16**). Tricyclic terpenoids (e.g., **12**) are postulated to exist in some extant species on the basis of their ubiquitous fossil record.

sulfur bacteria examined, as well as in a diverse selection of Gram-positive and Gram-negative chemoheterotrophs. Hopanoids have not been detected in archaebacteria, purple sulfur bacteria, or Chlorobiaceae, nor in various other Gram-positive and Gram-negative bacteria. There appear to be no rigid taxonomic distinctions which can be made on the basis of hopanoid content (Rohmer et al. 1984; Taylor 1984). The precise nature of membrane rigidifying components in the many eubacteria and cyanobacteria which lack hopanoids is not generally known but may include both cyclic and acyclic C_{30}, C_{40}, and C_{50} carotenoids (Rottem 1979; Taylor 1984; Ourisson et al. 1987).

The pentacyclic triterpenoids, although first known as natural products from higher plants and particularly ferns, were recognized early as abundant and ubiquitous hydrocarbon components of petroleum (Hills and Whitehead 1966). Subsequent work showed that the hopanes (e.g., **17**, **18**; Fig. 3.1.3) occurred in oils and sediments with geological ages predating the macrofossil record of higher plants. At about the same period, hopanoid hydrocarbons such as diploptene were recognized as components of living cyanobacteria and bacteria (Gelpi et al. 1970; De Rosa et al. 1971; Bird et al. 1971) and an "extended" C_{35} pentacyclic triterpenoid was isolated from *Acetobacter xylinum* (Foster et al. 1973). As a direct consequence of these observations, it was realized (Ensminger et al. 1974; Van Dorsselaer et al. 1974) that the great abundance and variety of petroleum triterpenoids owed their origins to bacteria sterols as major cell components, there is a notable exception in the methylotrophic bacterium *Methylococcus capsulatus*. *Methylococcus* is known to biosynthesize both hopanoids and sterold (Bird et al. 1971; Bouvier et al. 1976), the latter being dominated by the 4,4,14-trimethyl, 4,4-dimethyl, and 4-methyl sterols which are generally found as intermediates in the biosynthesis of cholesterol in eukaryotes. In *Methylococcus*, these compounds are produced by a primitive and non-specific squalene epoxide cyclase, the biosynthetic pathway is not fully expressed to cholesterol, and it does not extend to alkylation of the sterol side-chain at C-24 (Bouvier et al. 1976).

Many bacteria have been carefully screened for their content of, and ability to biosynthesize hopanoids or steroids (Rohmer et al. 1984). Reports of steroid occurrences (e.g., De Souza and Nes 1968) supported by unambiguous evidence of biosynthesis are rare and include *Methylococcus capsulatus* and *Nannocystis excedans* (Kohl et al. 1983; Rohmer et al. 1987). Several groups of prokaryotes, including mycoplasmas, utilize steroids in membrane construction and obtain them from other organisms by sequestration. Hopanoids have been detected in most cyanobacteria, obligate methylotrophs, and all purple non-

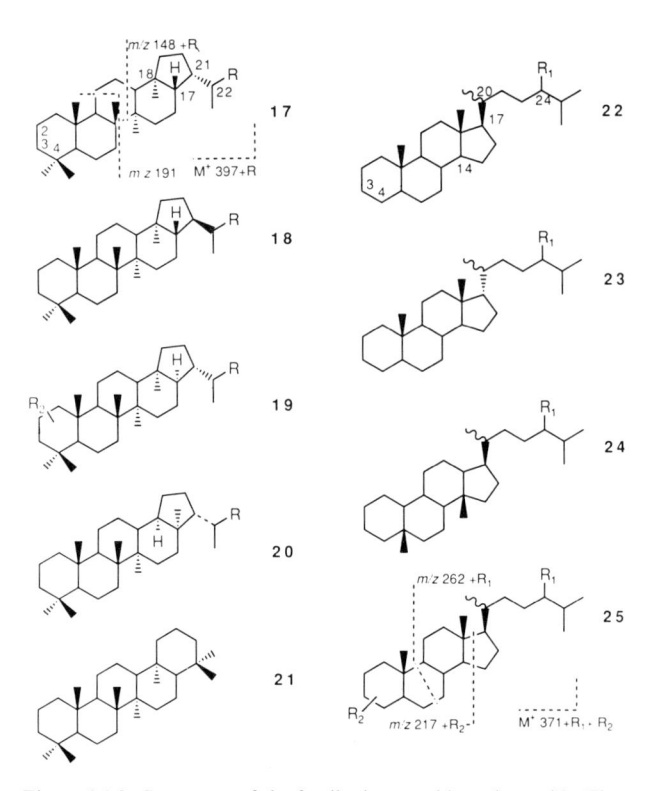

Figure 3.1.3. Structures of the fossil triterpenoids and steroids. These compounds differ from their biological precursors by the absence of unsaturation and other functionalities and by stereochemical changes induced during diagenesis. Also shown are the origins of mass spectral fragment ions used for SIR and MRM analyses of these hydrocarbons.

and only rarely to plants. When Nes (1974), Ourisson et al. (1979, 1987), Rohmer et al. (1984), and others demonstrated that bacteriohopane polyols (e.g., **13**) and sterols (e.g., **16**) were essential components of the membranes of prokaryotes and eukaryotes, respectively, sedimentary hopanes (**17–19**), steranes (**22–25**), and other distinctive terpenoids (e.g., **20**, **21**) were exposed as biomarkers of major significance with potential to establish the nature of primitive microbes. Different patterns of alkylation in the A-ring of steranes and hopanes (e.g., **19** and **25**; R_2 = Me) add to the structural variety of these biomarkers.

Subtle but readily established differences in three dimensional structure (i.e., stereochemistry) distinguish the steroids and hopanoids of living organisms from their fossil counterparts. Some of these stereochemical changes take place during early sediment diagenesis and are influenced by water-column chemistry and sediment lithology, including the presence of acidic clays and sulfide. Further significant stereochemical changes occur during deep burial when concomitant heating alters relative abundances of stereoisomers. These changes encode information on the relative duration and thermal stress of these events (Seifert and Moldowan 1978; Mackenzie et al. 1980, 1982; Mackenzie and McKenzie 1983; Grantham 1986b; ten Haven et al. 1986; Rullkötter et al. 1986). Thus, further aspects of sedimentary evolution are preserved in the stereochemical aspects of the biomarker record (depicted in Fig. 3.1.3) and this record should show consistency with other geological data. In precursor sterols, for example, the biologically determined stereochemistry is normally (20R)-5α(H), 14α(H), 17α(H), whereas the principal fossil steranes comprise mixtures, all partly epimerized at C-20, of (i) 5α(H), 14α(H), 17α(H), as in structure **22** where R = H, CH_3, C_2H_5; (ii) 5α(H), 14β(H), 17β(H), as in structure **23**; and (iii) the rearranged steranes (diasteranes), 13α(H), 17β(H), as in structure **24**. In hopanoids isolated from living organisms, biological stereochemistry is normally (22R)-17β(H), 21β(H) (**17**). On deep burial, this single form is progressively replaced by mixtures, partially epimerized at C-22, of the more stable 17β(H), 21α(H) moretanes (**18**) and the 17α(H), 21β(H) (**19**) hopane isomers. At advanced levels of maturity, only the latter will remain.

Based on their ubiquity in the fossil record, it has been proposed that tricyclic polyisoprenoids constitute another structural variant of the triterpenoid membrane rigidifiers, but one which has not yet been recognized in a living system (Ourisson et al. 1982, 1987; Renoux and Rohmer 1986). An analogous situation existed before the hopanoids were recognized in bacteria. The postulate of an evolutionary hierarchy proceeding from acyclic polyprenoids such as squalene (**11**), to tricyclic polyprenoids (e.g., **12**), pentacyclic hopanoids (e.g., **13**), tetrahymanol (**14**), the steroids lanosterol (**15**) and cycloartenol, and culminating in cholesterol (**16**) is based on the demonstrated requirements for increasingly sophisticated and substrate-specific enzyme systems for their biosynthesis (Ourisson et al. 1987). Because they are presently known only for their fossil analogues (Anders and Robinson 1971; Aquino-Neto et al. 1982; Moldowan et al. 1983; Ekweozor and Strausz 1983; Zumberge 1983), the tricyclic terpenoids have been described as "orphan" biomarkers (Ourisson et al. 1982). However, independent evidence for a primitive microbial source exists in the action of the squalene cyclases of the protozoon *Tetrahymena pyriformis*, which is able to cyclize hexaprenyl methyl ether to a variety of tri-, tetra-, and pentacyclic triterpenoids including **12** (Renoux and Rohmer 1986).

Lipophilic pigment derivatives constitute another group of biomarkers carrying phylogenetic information and with potential for preservation over long periods of geological time (e.g., Barghoorn et al. 1965). Phytane and pristane have long been recognized as likely degradation products of the phytyl side chain of chlorophyll, although other sources such as tocopherols (Goossens et al. 1984) and the lipids of archaebacteria are now recognized and may contribute in certain environments (reviewed by Volkman and Maxwell 1986). The porphyrin moieties derived from chlorophylls are not as refractory as the alkyl chains, but constitute more versatile biomarkers by virtue of their structural diversity and the great variety of diagenetic reaction pathways open to them. As a result, compounds in this class are useful as markers for paleoenvironment and are sensitive indicators of the thermal history of sediments. Extractable porphyrins are relatively more abundant in immature sediments and, unlike many sedimentary lipids, some can be traced to primary photosynthetic sources. Recent work exploits this fact to establish the isotopic composition of the CO_2 sources used by algae and bacteria (Hayes et al. 1987). The other major class of pigments found in photosynthetic organisms and possessing a record of geological preservation is the carotenoids. These act as light-harvesting "antennae" or as photoprotective agents and, as discussed above, may also have a role in strengthening membranes in some classes of bacteria. Large amounts of β-carotene occur in many cyanobacteria and also in halophilic algae such as *Dunaliella*. The fossil analogue, perhydro-β-carotene, is one of the most abundant individual hydrocarbons in the Green River Formation (Murphy et al. 1967), thus demonstrating potential for preservation over significant periods of geological time. Aromatic carotenoids such as isorenieratene, which are specifically produced by green photosynthetic bacteria of the genus *Chlorobium*, indicate not only the presence of a particular organism, but also the existence of specific paleoenvironmental conditions, in this case the presence of sulfide-rich water within the photic zone. Biomarkers derived from these *Chlorobium*-specific pigments have a record in the Mesozoic (Schaefle et al. 1975), the Paleozoic (Summons and Powell 1986, 1987), and the Proterozoic (see below).

Extended (C_{21}-C_{40}) acyclic isoprenoids are found as free sedimentary hydrocarbons and as substituents bound into the macromolecular complex of kerogen (e.g., Haug and Curry 1974; Waples et al. 1974; Michaelis and Albrecht 1979; Chappe et al. 1982, 1983; reviewed by Volkman and Maxwell 1986). There is useful structural variety because of different arrangements of the isoprene (C_5) units. These can be regular head-to-tail, as exemplified by phytane (**7**), or irregular tail-to-tail as in squalane (**10**). These features have direct structural relationships to the alkyl chains of distinctive ether lipids of specific groups of archaebacteria (e.g., methanogens, halophiles, and thermoacidophiles; reviewed by Kates 1978; Langworthy 1985). In the particular case of the unusual C_{25} hydrocarbon

2,6,10,15,19-pentamethyleicosane (9), a C_{15} unit and a C_{10} unit are joined tail-to-tail and can be distinguished from the regular structure (8) found in halophiles (Summons 1987). Hydrocarbon 9 has been isolated from cultures of the methanogen *Methanosarcina barkerii*. Its recognition in sediments (Brassell et al. 1981), sometimes in association with copious quantities of biogenic methane, clearly indicates the presence of methanogens. The sedimentary C_{40} hydrocarbon biphytane (Moldowan and Seifert 1979) has a distinctive head-to-head linkage of two C_{20} units and shares this characteristic with ether lipids of some methanogens.

3.1.4 Rudiments of Isotopic Biogeochemistry

Natural variations in the abundances of the stable isotopes of some elements of biogeochemical interest – carbon, hydrogen, nitrogen, oxygen, and sulfur – encode information about the development over time of the biogeochemical cycles of those elements. Because those cycles are mediated by biological processes, the isotopic records also carry a great deal of information about the evolution of the biota. Finally, because the importance of various biochemical processes (e.g., sulfate reduction, oxygenic photosynthesis) can be related to the availability or production of specific reactants or products (SO_4^{2-}, O_2), the isotopic records can also provide information about the development of the environment. In this Section, we discuss the mechanisms by which natural isotopic variations are created and the logic of their interpretation.

3.1.4A Notation, Mass Balances

Natural isotopic variations are small, but can be observed with great precision by techniques of differential measurement. The delta notation is employed for reporting isotopic abundances. In the case of carbon, for example,

$$\delta^{13}C_{PDB}(x) \equiv [(R_x - R_{PDB})/R_{PDB}]10^3 \quad (3.1.1)$$

where R is the ratio of isotopic abundances ($^{13}C/^{12}C$), x refers to a particular carbon sample, and PDB refers to the standard on which differential measurements are based. Quantitatively, δ is the relative difference, expressed in parts per thousand (or "permil" units, designated by ‰), between two isotope ratios. The accepted value for R_{PDB} is 0.0112372, so, for carbon, each 1‰ change in δ corresponds to 0.0011% ^{13}C.

Delta values for all elements are defined as indicated in equation 3.1.1. The zero point of the scale of isotopic abundances must always be specified by identification of an isotopic standard. Values of R are computed with the minor isotope placed in the numerator of the ratio. For hydrogen, the standard is "Standard Mean Ocean Water" (SMOW); for nitrogen, the standard is atmospheric nitrogen (AIR); for sulfur, the standard is sulfur in troilite nodules from the Canyon Diablo meteorite (CDT). Oxygen isotopic abundances in sedimentary carbonates are reported relative to oxygen in PDB; other oxygen-isotopic abundances are reported relative to SMOW. More detailed discussions can be found in reviews or texts by Hoefs (1987), Valley et al. (1986), Hayes (1983a, b), or Arthur et al. (1983).

A critical concept in most considerations of isotopic abundances is that of mass balance. For example, if the flow of an element in a natural system is divided, the compositions of the streams (input a, outputs b and c) will be related by the following equations:

$$\varphi_a = \varphi_b + \varphi_c \quad (3.1.2)$$

$$\varphi_a F_a = \varphi_b F_b + \varphi_c F_c \quad (3.1.3)$$

where φ represents the flux of the element in moles per unit time and F represents isotopic composition in terms of fractional abundance. In the case of carbon, for example, $F = {}^{13}C/({}^{12}C + {}^{13}C)$. Equation 3.1.3 is exact as written. Within the range of most natural variations, use of equation 3.1.3a (an approximate form, shown below), will not introduce serious errors.

$$\varphi_a \delta_a = \varphi_b \delta_b + \varphi_c \delta_c \quad (3.1.3a)$$

3.1.4B Isotope Effects, Fractionation

The distribution of an isotope among species at equilibrium will be nonuniform if the chemical bonding of the isotope is not the same in all reacting species. Oxygen atoms (O*, for example) can be exchanged between water and carbon dioxide:

$$H_2O^* + CO_2 \leftrightarrows H_2CO_2O^* \leftrightarrows H_2O + COO^* \quad (3.1.4)$$

As a result of this exchange and of differences in the chemical bonding of O in CO_2 or H_2O, oxygen-18 will accumulate in carbon dioxide relative to water such that

$$^{18}\alpha(CO_2/H_2O) \equiv {}^{18}R_d/{}^{18}R_w = (1000 + \delta_d)/(1000 + \delta_w)$$

$$= 1.0412 \quad (3.1.5)$$

where $^{18}\alpha(CO_2/H_2O)$ is the fractionation factor that quantifies the equilibrium isotope effect for exchange of oxygen between water and carbon dioxide. The subscripts d and w refer to carbon dioxide and water, respectively, and $^{18}R = {}^{18}O/{}^{16}O$. If a fractionation factor is expressed in terms of the delta notation, it takes the form of the second quotient shown in equation 3.1.5, with the requirement that both δ values must be referred to the same standard. Thus, for CO_2 and H_2O in equilibrium at 25°C, oxygen in CO_2 will be enriched in ^{18}O by 41.2‰ relative to that in H_2O. Fractionation factors often vary significantly within environmental temperature ranges. When the oxygen-isotopic compositions of coexisting water and carbonate minerals can be measured or estimated, temperatures of ancient environments can be determined. Equilibrium isotope effects have determined the relative isotopic abundances of many inorganic constuents of the sedimentary record.

Isotopic differences between species not related by chemical equilibria can be controlled by kinetic isotope effects. For example, the enzyme, ribulose bisphosphate carboxylase oxygenase ("rubisco"), catalyzes the reaction in which carbon dioxide and ribulose bisphosphate combine to form a six-carbon product that decomposes to yield two molecules of phosphoglyceric acid ("PGA"). Each of the molecules of PGA contains three carbon atoms, and the pathway of carbon fixation in which this reaction is the first step is termed the C-3 pathway. The abundance of ^{13}C in organic material deriving from C-3 carbon fixation is depleted relative to that in the source CO_2 due, in part, to a kinetic isotopic effect associated with the action of rubisco. The magnitude of the isotope effect is dependent upon the precise mechanism of the reaction. Rubisco from modern green plants displays a normal kinetic isotope effect of 2.9%, or 29‰ (Roeske and O'Leary 1984, 1985); that is, the carbon fixed at any instant is depleted by 29‰ relative to carbon dioxide at the reaction site. Magnitudes of kinetic isotope effects can also be expressed in terms of fractionation factors. In the present case, for example, one might write either

$$^{13}\alpha(CO_2/\text{fixed carbon}) = 1.029 \quad (3.1.6)$$

or

$$^{13}\alpha(\text{fixed carbon}/CO_2) = 0.9710 \quad (3.1.7)$$

Finally, the fractionation can be expressed in terms of an ε value, defined as follows

$$\varepsilon \equiv (\alpha - 1)1000 \quad (3.1.8)$$

Thus, for rubisco, $\varepsilon(\text{fixed carbon}/CO_2) = -29‰$.

3.1.5 Isotopes in Natural Systems

3.1.5A Carbon Fixation

In nature, processes of two kinds are involved in the synthesis of organic material carbon by autotrophs: (*i*) transport of inorganic carbon to the C-fixing enzyme (e.g., rubisco); and (*ii*) formation of a chemical bond ("fixation"). O'Leary (1981) and Farquhar et al. (1982, 1989) have shown that isotopic compositions of most plants can be understood in terms of a combination of isotope effects associated with these processes. The simplified scheme below depicts the relevant carbon flows:

$$CO_2(\text{environment}) \underset{\varphi_o,\delta_o}{\overset{\varphi_i,\delta_i}{\rightleftarrows}} CO_2 \text{ (at site of fixation)}$$

$$\xrightarrow{\varphi_f,\delta_f} R\text{-}CO_2. \quad (3.1.9)$$

The flows and isotopic compositions denoted by i, o, and f refer to carbon flowing **in**to the cell, **o**ut of the cell, and being **f**ixed. Mass balance at the site of fixation requires

$$\delta_i = (\varphi_o/\varphi_i)\delta_o + (\varphi_f/\varphi_i)\delta_f. \quad (3.1.10)$$

In the absence of active transport, $\varphi_i = p_a/r$ and $\varphi_o = p_i/r$, where p_a and p_i are the atmospheric and internal partial pressures of CO_2 and r is the resistance to diffusion of CO_2 along the pathway linking the atmosphere and the site of fixation. Rearrangement of equation 3.1.10 in terms of fractionation factors and pressure ratios yields, without approximation,

$$\varepsilon_P = \varepsilon_t + (\varepsilon_f - \varepsilon_t)p_i/p_a \quad (3.1.11)$$

where ε_P is the overall isotope effect (fixed carbon/source CO_2) associated with autotrophic carbon fixation, ε_t is the isotope effect associated with mass transport of inorganic carbon, and ε_f is the isotope effect associated with carbon fixation. For diffusion of CO_2 in water, $\varepsilon_t = -0.7‰$ (O'Leary 1984). For modern C-3 plants, the effective value of ε_f is about $-27‰$ (Farquhar et al. 1982, 1989). This value is 2‰ less than that associated with rubisco because a portion of carbon fixation in modern plants takes place by alternate mechanisms (e.g., carboxylation of phosphoenolpyruvate) that have isotope effects much smaller than $-29‰$ (Farquhar and Richards 1984).

In experiments with modern algae, it has been observed that ε_P becomes more negative as the pressure at which CO_2 is supplied increases (Mizutani and Wada 1982; McCabe 1985). In the terms of equation 3.1.11, as p_a increases, $p_i/p_a \to 1.0$ and $\varepsilon_P \to \varepsilon_f$. Ancient values of ε_P can be estimated from observed isotopic compositions of carbonate minerals and sedimentary organic matter (Hayes et al. 1989). When a record of ε_P values extending over the past 300 Ma is compiled, significant variations are apparent, and it has thus been suggested (Popp et al. 1989) that these can be interpreted in terms of variations in p_a, with larger values of ε_P indicating larger values of p_a. To whatever extent this hypothesis can be sustained, isotopic compositions of organic carbon may provide a sort of "paleobarometer" for atmospheric carbon dioxide.

Oxygen-producing green plants, algae, and cyanobacteria, are, of course, not the only autotrophs (or even photoautotrophs) of biogeochemical interest, particularly in the Precambrian. Different enzymatic mechanisms of carbon fixation, with different isotopic fractionation factors, are employed by green photosynthetic bacteria (Sirevåg et al. 1977), by photosynthetic bacteria of the genus *Chloroflexus* (Holo and Sirevåg 1986), and by acetogenic and methanogenic bacteria (Wood et al. 1986).

3.1.5B Carbonate and Organic Carbon in Sedimentary Rocks

A schematic view of the critical branch point in the biogeochemical cycle of carbon is shown in Figure 3.1.4. An isotopic mass balance like that specified in equations 3.1.2 and 3.1.3 must prevail. Specifically, we can write

$$\delta_{\text{input}} = f_{\text{org}}\delta_{\text{org}} + (1 - f_{\text{org}})\delta_{\text{carb}} \quad (3.1.12)$$

where δ_{input} is the isotopic composition of the carbon input entering the reaction chamber, $f_{\text{org}} (= \varphi_{\text{org}}/\varphi_{\text{input}})$ is the fraction of input carbon leaving the reaction chamber in the form of organic carbon, δ_{org} is the isotopic composition of that carbon, and δ_{carb} is the isotopic composition of carbonate carbon leaving the reaction chamber. The "reaction chamber" is the global environment. Equation 3.1.12 pertains to the operation of the *global* carbon cycle, and indicates that much can be learned about its history by analysis of the isotopic compositions of organic and carbonate phases in sedimentary rocks (Broecker 1970; Sundquist and Broecker 1985; Holser et al.

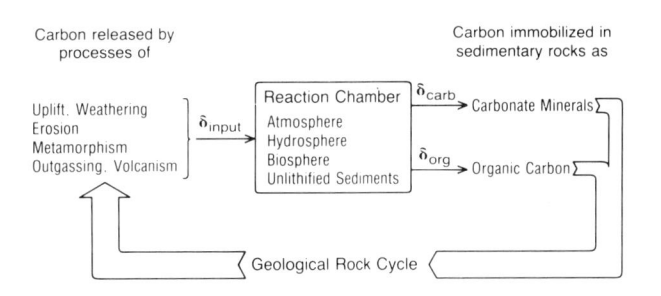

Figure 3.1.4. A schematic view of the biogeochemical cycle of carbon.

1988). Specifically, if δ_{input}, δ_{org}, and δ_{carb} can be determined for a point in geologic time, f_{org} can be calculated for that same time.

The significance of f_{org} merits special attention. Two points are of particular importance. First, it is defined strictly in terms of the global mass balance. It expresses a global average that includes diverse environments. In some, depositional processes are yielding shales rich in organic carbon and poor in carbonate; in others, sediments are rich in carbonate and poor in organic material. It is the net effect of these processes that controls f_{org} at any point in time. Second, even though it relates to organic carbon, f_{org} is independent of biological productivity. It refers explicitly to the burial, *not* merely the synthesis, of organic material. A system with large f_{org} might be characterized by high productivity and average levels of preservation of organic material *or* by moderate to low levels of productivity and high levels of preservation (e.g., Bralower and Thierstein 1984). Whatever the case, knowledge of f_{org} is of great value in reconstruction of the crustal redox budget (e.g., Berner 1989). In order to work toward that goal, we must consider in more detail the isotopic terms in equation 3.1.12.

As noted above (Fig. 3.1.4), the sources of input carbon are diverse. A value for δ_{input} cannot be measured precisely, but some limiting cases can be considered. Mantle carbon has an isotopic composition in the range -5 to $-6‰$ vs. PDB (Des Marais and Moore 1984). Estimates of the global average isotopic composition for total crustal carbon (both organic and inorganic) fall between -4.4 and $-5.2‰$ vs. PDB (Holser et al. 1988). Therefore, whether it is assumed that all carbon entering the reaction chamber derives from the mantle, *or* that it derives from the recycling of average crust, the value assigned to δ_{input} will be near $-5‰$. This assignment will be in error whenever the recycling component is not "average" crust but is, instead, enriched in either organic or carbonate carbon. The former circumstance might arise, for example, during rapid erosion of shales at the close of a glacial interval. During such an inverval, δ_{input} will be less than $-5‰$, and values of f_{org} determined by insertion of observed values of δ_{org} and δ_{carb} in equation 3.1.12 will be in error if δ_{input} is not adjusted. We choose not to attempt recognition of variations in δ_{input}, but instead to remember the uncertainties imposed by our ignorance of those variations.

If δ_{input} is adequately constrained, accurate treatment of the mass balance requires only knowledge of globally representative values of δ_{org} and δ_{carb}. In particular, reconstruction of primary isotopic compositions (i.e., those that might have prevailed at some point prior to the immobilization of carbon in sinks effectively removed from the reaction chamber) is neither required nor desirable. With this in mind, we can usefully define Δ_B, the isotopic difference between organic and inorganic carbon phases undergoing burial

$$\Delta_B = \delta_{carb} - \delta_{org} \tag{3.1.13}$$

and write (accepting $-5‰$ as an estimate of δ_{input})

$$f_{org} = (\delta_{carb} + 5)/\Delta_B^* \tag{3.1.14}$$

where Δ_B^* refers to the globally averaged value of Δ_B. For any environment, the magnitude of Δ_B depends not only on isotope effects associated with primary production of organic carbon, but also on those associated with all processes following carbon fixation and preceding the immobilization of organic carbon in sediments. As a result, Δ_B can ultimately differ significantly from isotopic differences expected on the basis of the fixation process alone. A systematic approach to this subject has been described by Hayes et al. (1989), and several relevant examples have been studied in detail (Hayes et al. 1987, 1989; Freeman et al. 1990).

It is by design that equation 3.1.14 is made dependent on the isotopic record of carbonate, rather than organic, carbon. Because it depends primarily on the isotopic composition of marine bicarbonate, the isotopic composition of carbonate entering the sedimentary record at any point in time is relatively uniform (geographically). Isotopic compositions of organic phases can be more strongly affected by local conditions and, especially, by thermal alteration of sedimentary rocks long after burial. Variations due to the latter cause are particularly to be excluded in attempts to reconstruct the development of the carbon cycle, and it appears most practical to attempt estimation of globally representative values of Δ_B based on isotopic differences observed in settings minimally, if at all, affected by post-depositional thermal alteration.

3.1.5C Sulfur, an Isotopic Comparison with Carbon

The biosphere depends absolutely on capture of energy through catalysis, by "primary producers," of the photochemically driven reduction of carbon and oxidation of water to produce organic material and O_2. From that starting point, energy is recovered within the global ecosystem through catalysis, by consumers and secondary producers, of a remarkable variety of exergonic redox processes. In these, the most important reactants and products involve not only carbon and oxygen, but also sulfur, nitrogen, iron, and manganese. Sulfur is of particular interest here. It is, quantitatively, very important; much of the oxidation of organic matter in the modern sea bed is directly coupled, not to the reduction of O_2 but to the reduction of SO_4^{2-} (Jørgensen, 1982b; Henrichs and Reeburgh 1987). There is a potentially informative isotopic record; for sulfur as for carbon, readily available sedimentary constituents record isotopic compositions of reactants *and* products in ancient redox cycles.

There are interesting contrasts between natural fractionations of carbon and sulfur isotopes. In each case, a reductive process is responsible for the principal fractionation – fixation of CO_2 as a carbon source by autotrophs in the case of carbon; utilization of SO_4^{2-} as an electron acceptor by anaerobic heterotrophs in the case of sulfur. A normal kinetic isotope effect is associated with each of these processes and, as a result, abundances of heavy isotopes in the products, organic carbon and sulfide, are lower than those in the reactants, inorganic carbon and sulfate. The maximum isotopic difference expected between reactants and products is given by the isotope effect itself (i.e., $29‰$ in the case of fixation of carbon by rubisco). Isotope effects accompanying sulfate reduction have not been studied nearly as intensively as those involved in carbon fixation. Isotopic abundances of sulfate and sulfide in a wide variety of natural environments can, however, be consistently interpreted in terms of an isotope effect approaching $60‰$

(Chambers and Trudinger 1979; Sweeney and Kaplan 1980). Accepting this estimate, and continuing to draw a parallel between carbon and sulfur, we might expect that the typical isotopic difference between sulfate and sulfide would approximate 60‰, just as Δ_B, the isotopic difference between carbonate and organic material, can usually be related to the isotope effect accompanying carbon fixation (see Hayes et al. 1989, for a detailed discussion).

In fact, we would find that the isotopic difference between sulfate and average sulfide is commonly much less than 60‰. Reconstructing the carbon and sulfur isotopic records for the Phanerozoic, Veizer et al. (1980) found that adoption of 40‰ as the average isotopic difference between sulfate and sulfide yielded good agreement between predicted and observed isotopic variations. This marked attenuation of isotopic differences occurs because most reduction of sulfate takes place in a system that is subject to closure, while that of carbon takes place in a system virtually certain to remain open. In plainer words, the supply of sulfate in sedimentary muds is likely to be limited by slow diffusion of sulfate from the overlying water column. In the extreme case, transfer of sulfate into the region in which sulfate reduction is occurring will cease, the system will be "closed" (i.e., to transport of matter across its boundaries), and, as all the sulfate is reduced, the isotopic composition of the pooled product sulfide will become equal to that of the starting sulfate (individual microscopic samples of sulfide will be depleted in ^{34}S if they formed early in the process of sulfate reduction, strongly enriched if they formed late). In contrast, fixation of CO_2 occurs mainly "where the sun shines," in gas and liquid phases where mass transport is much less strongly hindered. Concentrations of CO_2 may be diminished near sites of active fixation, but most natural systems do not approach closure. As a result, fractionations of carbon isotopes between carbonate and organic matter are quite consistently related to the isotope effect associated with the mechanism of carbon fixation, whereas fractionations of sulfur isotopes depend on the mechanism of reduction *and* on the supply of sulfate at the site of reduction.

The supply of sulfate depends on both the initial concentration – if it is high, it will take longer for utilization to approach completion and for sulfate-sulfide differences to approach zero – and on the "connection" between the site of sulfate reduction and the major sulfate reservoir; if transport of sulfate occurs easily, concentrations of sulfate will never drop to zero and isotopic fractionations will approach maximum values. Occurrence of sulfides with isotopic compositions averaging to that of coeval marine sulfate might represent a system (*i*) in which biologically catalyzed redox reactions did not occur; (*ii*) from an environment in which concentrations of sulfate were so low that reduction approached completion and any trace of isotopic fractionation associated with biological processes was erased; or (*iii*) from an environment in which initial concentrations of sulfate were high, but in which isolation from any sources of sulfate replenishment led eventually to quantitative reduction of sulfate and no net isotopic fractionation.

Isotopic compositions of carbonate and well-preserved organic carbon are generally well separated and carry information about the structure of the global carbon cycle at any point in time, but interpretation of the isotopic difference between sulfate and sulfide can be much more complex. The graphs shown in Figure 3.1.5 are illustrative. In the upper portion of each, the sloping line represents the isotopic composition of oxidized material (carbonate, sulfate) being buried in sediments. For carbon, the equation of that line is obtained in its simplest form by rearrangement of equation 3.1.14:

$$\delta_{\text{carb}} = \Delta_B^* f_{\text{org}} - 5. \qquad (3.1.15)$$

For sulfur, a derivation paralleling that of equations 3.1.12 to 3.1.15 yields

$$\delta_{\text{sulfate}} = \Delta_S^* f_{\text{sulfide}} + 2 \qquad (3.1.16)$$

where $\Delta_S \equiv \delta_{\text{sulfate}} - \delta_{\text{sulfide}}$ and Δ_S^* is the globally averaged value of that isotopic difference. In each case, (*i*) the isotopic composition of the oxidized portion of the redox couple is a linear function of the fractional division of the element between its oxidized and reduced forms, and (*ii*) the slope of that relationship is determined by Δ^*.

These Δ^* terms are concepts, not constants. For any point in time, there *will* be some average isotopic fractionation for carbon or sulfur, and, to the extent that it can be accurately estimated, an f value characteristic of the global mass balance can be calculated. But isotopic compositions of organic carbon and sulfide depend strongly on conditions prevailing in local environments. Secondary processes affecting organic material will be controlled by community structure which will, in turn, depend on the physical environment. Variation in Δ_B of several permil can easily result, as indicated in Figure 3.1.5a. Variations in Δ_S can be much larger, with values ranging from 60 to 0‰. For any global value of f_{sulfide}, δ_{sulfate} will take the value specified by the line in Figure 3.1.5b and values of δ_{sulfide} will range between the lines marked "minimal reduction" and "maximal reduction." Even within a single environment, sulfide produced in areas of low sedimentary porosity and/or with high ratios of organic matter to sulfate will tend to yield average δ values near the maximal reduction line (individual microsamples of sulfide will yield both high and low δ values, depending on their times of formation). Areas well connected to the sulfate supply and/or with low ratios of organic matter to sulfate will tend to yield δ values near the minimal reduction line. A world dominated by environments in which the extent of reduction of sulfate was low would yield $\Delta_S^* \to 60$; one dominated by environments in which the extent of reduction of sulfate was high (either because initial concentrations were low or because replenishment of sulfate did not occur) would yield $\Delta_S^* \to 0$.

Variations observed in Δ_B^* during the Precambrian have been discussed by Hayes (1983a), who showed that a gradual decrease in Δ_B^* provided an explanation for the otherwise surprising observation that δ_{carb} values changed little, if at all, while average δ_{org} values changed from ~ -35‰ in the Archean to ~ -29‰ in the Late Proterozoic. Likely variations in Δ_S^* have not been previously considered, but can readily be predicted, given the principles introduced here. Prior to the rise of atmospheric O_2, both concentrations of sulfate and levels of oxidation of sedimentary materials are likely to have been low.

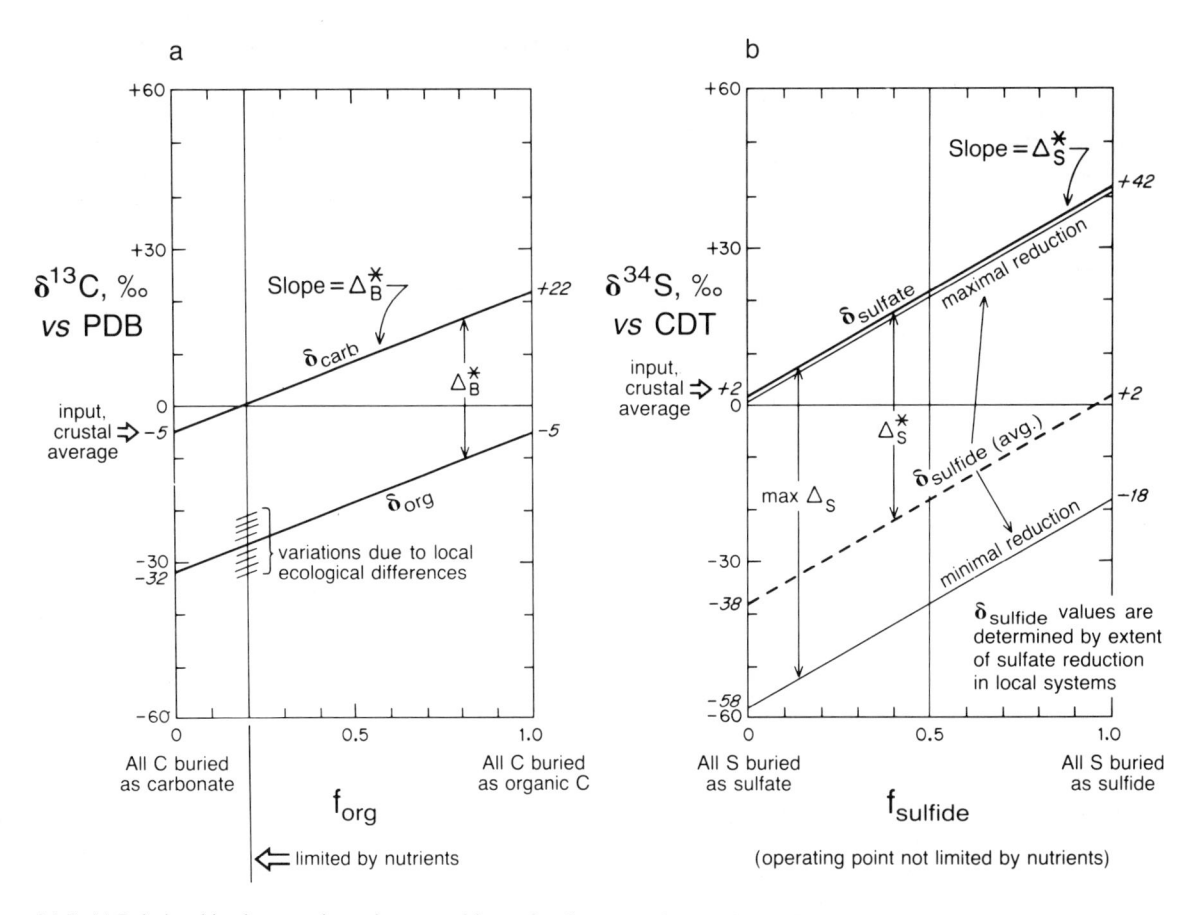

Figure 3.1.5. (a) Relationships between isotopic compositions of carbonate carbon and organic carbon and the fraction of carbon buried in the form of organic material. Very high values of δ_{carb} corresponding to $f_{org} > 0.5$ are not observed, presumably because the global inventory of phosphate is not large enough to sustain the synthesis of such large quantities of biomass. (b) A similar graph representing isotopic relationships within the biogeochemical cycle of sulfur. For reasons discussed in the text, values of $\delta_{sulfide}$ vary more widely than those of δ_{org}. Moreover, the operating point of the cycle with respect to $f_{sulfide}$ is limited only by redox balances, not by supplies of scarce nutrients.

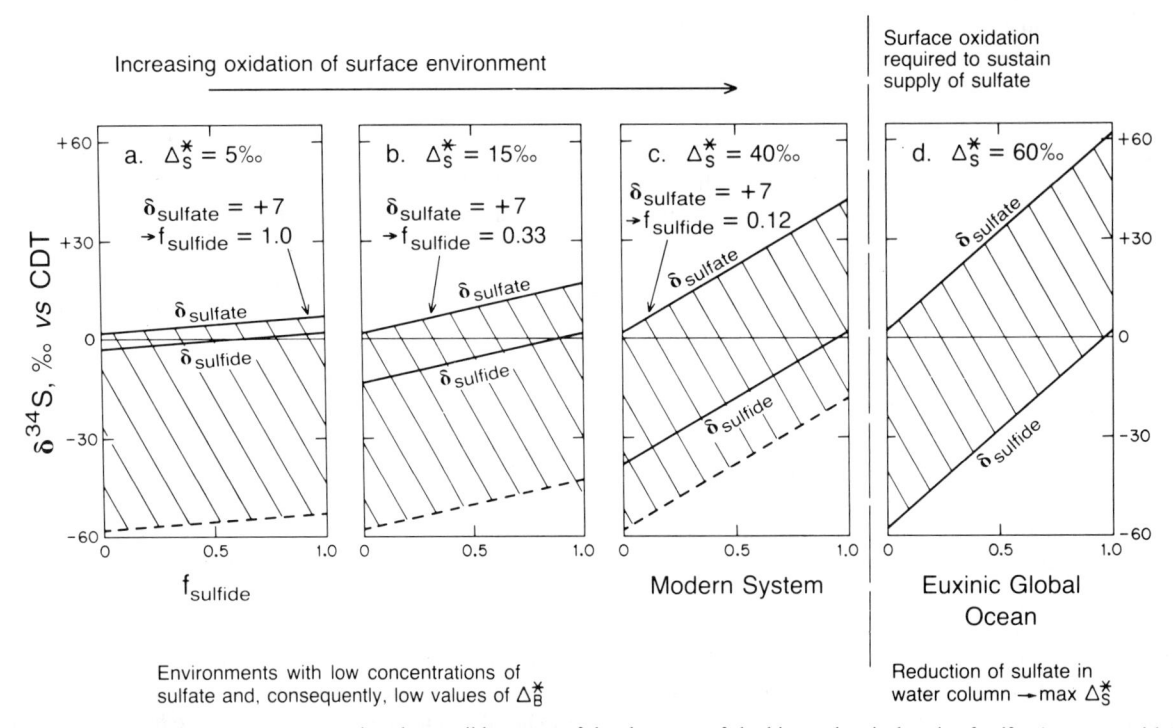

Figure 3.1.6. Fractionation diagrams suggesting the possible course of development of the biogeochemical cycle of sulfur (compare with Fig. 3.1.5). In each case, the shaded areas indicate the range of $\delta_{sulfide}$ values that might be observed. Solid lines within the $\delta_{sulfide}$ fields indicate global average values.

Wherever reduction of sulfate occurred, it is likely to have proceeded nearly to completion and yielded low values of Δ_S. Accordingly, Δ_S^* will have been low, and the fractionation diagram representative of the global sulfur cycle will be something like Figure 3.1.6a. As levels of O_2 increased, concentrations of sulfate and levels of oxidation of sedimentary materials will have risen. The globally averaged extent of reduction of sulfate in sedimentary environments will have decreased, Δ_S^* will have increased, and the fractionation diagram will progress through that shown in Figure 3.1.6b as it approaches its modern state (Fig. 3.1.6c). Finally, if the global cycle of sulfur were ever dominated by environments in which the isotopic difference between sulfide and sulfate approached 60‰, the maximum possible, a fractionation diagram like that shown in Figure 3.1.6d would result. Extreme isotopic differences are characteristic of euxinic environments, represented, for example, by bottom waters in the modern Black Sea (Sweeney and Kaplan 1980). In such settings, reduction of sulfate occurs in the water column. Diffusional barriers to replenishment of sulfate are practically absent and, because supplies of sulfate are essentially infinite, the extent of reduction of sulfate is very low, leading to maximal isotopic fractionation.

Figure 3.1.6d represents, therefore, the ultimate "Oceanic Anoxic Event" (Schlanger and Jenkyns 1976), one in which *all* deep waters are anoxic and sulfidic. The impact on the sulfur cycle is clear; Δ_S^* approaches 60‰ because nearly all burial of sulfur takes place in regions of maximal fractionation. But attainment of $\Delta_S^* = 60$ as the culmination of a sequence of increasing levels of global oxidation appears paradoxical until it is recognized as a signal of *two* conditions: global anoxia in the deep ocean *and* abundant supplies of sulfate.

Evidently, care must be taken in the interpretation of the Precambrian record of sulfur-isotopic abundances. In particular, the modern fractionation diagram (Fig. 3.1.6c) must not be allowed to dominate consideration of the significance of observed δ values. Consider, for example, $\delta_{\text{sulfate}} = +5‰$. For some time at which fractionation in the sulfur cycle was described by Figure 3.1.6a, that observation would indicate burial of essentially all sulfur in the form of sulfide. If changes in Δ_S^* were not taken into account and the $+5‰$ value were interpreted in terms of Figure 3.1.6c, it would be wrongly concluded that most sulfur was being buried in the form of sulfate.

3.2

Concentrations of Organic Carbon and Maturities and Elemental Compositions of Kerogens

HARALD STRAUSS   DAVID J. DES MARAIS   J. M. HAYES
ROGER E. SUMMONS

Information regarding the abundance, composition, and distribution of organic matter guides analyses of isotopic and molecular fossils, allows comparison of rates of burial of organic carbon in various depositional environments, and facilitates comparisons between Proterozoic and Phanerozoic biogeochemical processes.

3.2.1 Review of Previous Studies

There has been no systematic inventory of organic carbon in Proterozoic sedimentary rocks. The existing record derives to a large extent from analyses of samples specifically selected for their contents of organic matter and/or fossils. The predominant lithologies are shaley and are characterized by rather high contents of total organic carbon ("TOC"); lithologies likely to be low in organic carbon are poorly represented. Individual reports have been tabulated in Chapter 17 and are summarized here in Table 3.2.1. Abundances of total organic carbon are similar to those in comparable Phanerozoic sediments (Ronov 1958); no systematic variation through time is indicated.

Assessment of post-depositional alteration and preservation is critical for any interpretation of the organic carbon record. Elemental compositions of sedimentary organic material can indicate the extent of thermally induced remobilization of organic material. Results of prior studies are summarized in Table 3.2.2. Given that the H/C ratio of fresh sedimentary organic material is commonly 1.4 or higher, it is evident that all of these results are indicative of significant levels of thermal maturation through expulsion of hydrocarbons.

Further relevant data appear in Table 3.2.3, which summarizes results of "Rock-Eval" pyrolytic analyses of Proterozoic samples in the collection of the Bureau of Mineral Resources, Canberra, Australia. Entries in this Table include "T_{max}", the temperature at which the pyrolytically driven release of volatile hydrocarbons is maximized; "S_2", the yield of hydrocarbons produced by pyrolysis of kerogen; "HI", the "hydrogen index", a measure of the hydrocarbon yield relative to the quantity of organic carbon (HI $\equiv S_2$/TOC); an *estimate* of the H/C ratio; and "EOM", the yield of extractable organic matter. The estimates of H/C in Table 3.2.3 are based on empirical correlations between H/C and HI (Espitalié et al. 1977; Tissot and Welte 1984). For the most part, reasonable agreement is observed where samples of a given unit appear in both Tables 3.2.2 and 3.2.3. The Tapley Hill Formation is a notable exception, in that chemical determinations of H/C are uniformly lower than the single Rock-Eval result. If quantities of organic carbon were low ($\sim$ 1 mg/g), this might be attributed to enhancement of the hydrocarbon yield due to the presence of H_2O or some other source of H. This is not the case, and it appears likely that the Rock-Eval study utilized a sample differing significantly from those examined by conventional chemical analyses. Many of the entries in Table 3.2.3 derive from extensive and detailed studies of the McArthur Basin and its hydrocarbon potential (Powell et al. 1987; Crick et al. 1988). Well preserved organic matter is associated with a number of units in the Middle Proterozoic McArthur Basin of Australia and many Late Proterozoic sediments with wide geographic distribution. Some of these contain extractable organic matter that has yielded significant new biogeochemical information (Section 3.3).

Table 3.2.1. Previous reports of the abundance of total organic carbon in Proterozoic sediments.

	Lithology	TOC mg/g	N
Late Proterozoic	Shale	16.6	26
	Carbonate	3.2	154
	Chert	0.7	2
Middle Proterozoic	Shale	14.9	362
	Carbonate	5.0	294
	Chert	1.8	49
Early Proterozoic	Shale	10.5	1,184
	Carbonate	2.1	812
	Chert	2.1	120
Total Proterozoic	Shale	11.6	1,572
	Carbonate	2.9	1,260
	Chert	2.0	171

Table 3.2.2. *Previous reports of H/C ratios for Proterozoic kerogens.*

Unit	Age (Ma)	H/C ratio Avg.	Range	N	Refs[a]
Wonoka Fm.	585	0.47		1	1
Pertatataka Fm.	610	0.66		1	2
Brachina Fm.	650	0.11	0.06–0.15	3	3
Umberatana Fm.	675	0.49		1	1
Tapley Hill Fm.	750	0.13	0.01–0.25	31	1,2,3,4
Skillogalee Dolomite	770	0.22	0.10–0.33	11	1,2,3,5
Rhynie Fm.	785	0.85		1	5
Burra Group	800	0.17		1	6
Black Mudstone, Utah	830	0.1		1	7
Bitter Springs Fm.	850	0.46	0.10–0.82	6	1,6,7
Nonesuch Fm.	1,055	0.5		1	7
McMinn Fm.	1,340	0.7	0.36–1.24	10	1,6,8
Lansen Creek Shale	1,350	0.93	0.60–1.24	4	8
Velkerri Fm.	1,350	0.81	0.65–1.01	7	8
Bungle Bungle Dolomite	1,364	0.39	0.31–0.45	5	6
Yalco Fm.	1,485	0.99	0.89–1.19	6	8
Lynott Fm.	1,490	0.94		1	8
Marimo Slate	1,500	0.05		1	3
Balbirini Dolomite	1,500	0.99		1	8
HYC Pyritic Shale	1,500	0.51	0.39–0.64	2	1,6
Barney Creek Fm.	1,500	1.08	0.38–2.02	21	8
Paradise Creek Fm.	1,650	0.2		1	7
Urquhart Shale	1,670	0.12	0.10–0.13	2	1,7
Earaheedy Group	1,800	0.27		1	6
Rove Fm.	1,900	0.41		1	6
Golden Dyke Fm.	1,900	0.04		1	3
Duck Creek Dolomite	2,000	0.18		1	6
Gunflint Fm.	2,090	0.35		1	6
Upper Albanel Fm.	2,100	0.14		1	6
McLeary Fm.	2,100	0.26		1	6
Malmani Dolomite	2,400	0.16	0.14–0.22	3	6
Mt. McRae Shale	2,500	0.11		2	6
Wittenoom Dolomite	2,500	0.14		1	6

[a] 1, McKirdy 1976. 2, McKirdy et al. 1980. 3, McKirdy and Powell 1974. 4, McKirdy et al. 1975. 5, Dungworth and Schwartz 1972. 6, Hayes et al. 1983. 7, Leventhal et al. 1975. 8, Powell et al. 1987.

3.2.2 Results of This Study

"Survey level" analyses, described in Chapter 16, included determination of the abundance of total organic carbon. An attempt was made to select a representative suite of Proterozoic sediments with different lithologies, facies, and ages, and from different areas, but initial sampling was to a degree biased toward samples containing abundant organic material. A principal objective was the identification of samples for further, more detailed studies of extractable organic material and kerogen.

Detailed results of 731 analyses are given in Chapter 17; a summary is presented in Table 3.2.4. Overall, the TOC values are in good agreement with previously published Precambrian data (Hayes et al. 1983). Differences between the data summarized in Table 3.2.4 and those presented in Table 3.2.1 can best be explained in terms of a more representative selection during initial acquisition of samples. The tendency toward higher concentrations of organic carbon in shales, as opposed to carbonates and cherts, is obvious, and has also been observed in Phanerozoic sediments (Ronov 1958, 1980). This trend must be related to differences in initial contents of organic matter and to subsequent degradational processes, both being related to the environment of deposition.

Results of elemental analyses of 121 kerogens prepared from samples in the PPRG collection are summarized in Table 3.2.5. A full record of individual results can be found in Chapter 17. Low H/C ratios (<0.5), indicative of substantial losses of hydrocarbons, are observed for many of these kerogens. Samples with higher H/C ratios tend to be shales or shaley carbonates, possibly because preservation is enhanced when higher quantities of organic material are present.

Samples containing more than 5 mg/g organic carbon were subjected to Rock-Eval pyrolysis. Results of analyses confirmed extensive alteration of the organic matter to subgraphitic and graphitic levels in many of the Early Proterozoic units. Notable exceptions are mostly confined to the Late Proterozoic. Results are tabulated in detail in Chapter 17, and summarized in Table 3.2.6. In particular, sediments from different stratigraphic levels within the Late Proterozoic Chuar Group were found to contain well-preserved organic matter presumably originating from the diverse assemblage of microfossils in these same strata (e.g., Bloeser 1980; Horodyski and Bloeser 1983; Bloeser 1985). Extractable organic matter from these same samples is described in a later portion of this Section.

3.2.3 Trends in Abundance and Preservation of Organic Matter

All presently available data regarding the distribution of organic carbon in Proterozoic sedimentary rocks are summarized in Table 3.2.7, which combines results of the present study (Table 3.2.4) with those of all earlier studies (Table 3.2.1). Parallel results for Phanerozoic sediments (Ronov 1958, 1980) are given for comparison. While it appears that abundances of organic carbon are similar, it must be borne in mind that the Proterozoic data do not derive from systematic sampling designed to reflect accurately the entire volume of the rock record.

The relationship between ages and H/C ratios of kerogens is summarized in Figure 3.2.1. As previously observed (Hayes et al. 1983), *some* younger materials are less altered and *all* older materials have been significantly affected by loss of hydrocarbons. As with biological debris generally, young is no guarantee of preservation; great age virtually ensures significant degradation.

Abundances of hydrogen and nitrogen in Proterozoic kerogens are compared in Figure 3.2.2. The distribution of results suggests that N/C ratios in immature kerogens were in the range of 0.02 to 0.03 and that losses of H and N are correlated for N/C < 0.025. Essentially identical results have been obtained in studies of Phanerozoic kerogens. Waples (1977) found that 189 samples with H/C ratios generally between 0.6 and 1.0 yielded N/C ratios in the range of 0.017 to 0.033. At H/C ratios

Table 3.2.3. Previous reports of Rock-Eval data for Proterozoic sediments[a].

Unit	Age[b] (Ma)	TOC (mg/g)	T_{max} (°C)	S_2 (mg HC/g)	HI (mg HC/g C_{org})	"H/C"	Maturity	EOM (mg/g)	N	Ref[c]
Rodda Beds	610	5.9–7.9	437–443	0.7–0.96	112–143	0.75	margin. mature	0.03–0.23	5	1
Pertatatala Fm.	610	1.1–8.6	331–493	0.01–0.32	9–291	0.4–1.0	mature	0.02–0.38	16	1
Svalbard (Vendian)	700	1.6–3.7	276–424	0.00–0.04	0–11	≤0.4	data inconcl.	0.01–0.20	6	2
Tapley Hill Fm.	750	15	424	1.2	84	0.7	data inconcl.	0	1	1
Areyonga Fm.	760	3.8	428	0.62	163	0.8	data inconcl.	0.03	1	1
Visingso Group	775	0.4–2.6	278–437	0.01–0.13	33–129	0.5–0.75	data inconcl.	0.08–0.16	8	3
Chuar Group	850	12–17	446–449	1.6–1.9	116–139	0.7–0.8	mature	0.18–0.76	2	4
Bitter Springs Fm.	850	0.07–0.45	220–458	0.00–0.2	0–100	0.4–0.7	mature	0.04–1.5	20	1
Victor Bay Fm.	1,000	12–25	465–495	0.15–0.67	10–32	0.45	overmature	0.04–0.32	5	2
Nonesuch Fm.	1,055	4.2–20	436–445	0.8–11	117–426	0.6–0.8	mature	1.04	6	1,6
McMinn Fm.	1,340	9.1–30	430–441	1.4–13	91–451	0.7–1.2	margin. mature	0.09–2.3	23	5
Velkerri Fm.	1,350	34–87	435–449	15–37	266–498	1.0–1.3	mature	1.9–30	23	5
Lansen Creek Shale	1,350	9.0–43	436–452	1.6–16	173–416	0.85–1.1	mature	0.21–4.2	17	5
Corcoran Fm.	1,360	5.9	454	0.44	75	0.7	mature	0	1	5
Mainoru Fm.	1,470	5.2–5.6	445–460	0.17–0.19	33–34	0.55	mature	0	2	5
Balbarini Dolomite	1,482	6.1	437	2.2	361	1.0	mature	1.6	1	5
Yalco Fm.	1,485	32–54	429–434	17–32	523–588	1.3	mature	0.45–3.2	2	5
Lynott-Carabirini Mbr.	1,490	16–46	432–483	0.19–3.5	7–125	0.45–0.75	mature	2.0	7	5
Lynott-Donnegan Mbr.	1,490	7.6–8.1	464–474	0.11–0.16	14–20	≤0.5	overmature		2	5
Lynott-Hot Spring Mbr.	1,490	1.0–12	502–551	0.19–0.43	27–300	0.5–1.0	overmature		6	5
Reward Dolomite	1,495	6.6–18	425–459	0.11–0.91	14–87	0.45–0.7	mature	1.9	11	5
Barney Creek Fm.	1,500	7.4–110	432–451	0.2–66	24–649	0.55–1.3	mature	0.08–5.0	15	5
Teena Dolomite	1,510	21.4	454	1.2	54	0.5	overmature		1	5

[a]See text for discussion of method.
[b]BMR preferred age assignments for the McArthur Basin sediments exceed these. See Jackson et al. 1988.
[c]1, Summons, unpublished. 2, Summons, Knoll, and Swett, unpublished. 3, Summons and Vidal, unpublished. 4, Summons et al. 1988. 5, Crick et al. 1988. 6, Pratt et al. 1989.

below 0.6, Waples (1977) observed that losses of N and H were correlated as shown in Figure 3.2.2.

An attempt to evaluate differences in composition and preservation of kerogens from different paleoenvironments was

Table 3.2.4. *New observations of the abundance of total organic carbon in Proterozoic sediments.*

	Lithology	TOC (mg/g)	N
Late Proterozoic	Shale	2.30	128
	Carbonate	0.35	52
	Chert	0.44	83
Middle Proterozoic	Shale	3.94	59
	Carbonate	0.30	28
	Chert	0.47	35
Early Proterozoic	Shale	5.59	40
	Carbonate	0.56	64
	Chert	1.29[a]	64
Total Proterozoic	Shale	3.31	227
	Carbonate	0.43	144
	Chert	0.75[a]	182

[a]Anthraxolitic cherts from the Gunflint Formation have been excluded.

made by analyzing a suite of stromatolitic samples. Remarkably, a large percentage of these cluster in an area of low H/C and N/C ratios. This is, however, almost certainly not a paleoenvironmental signal. These same samples have very low TOC and, consequently, organic material is poorly preserved. As shown in Figure 3.2.2, some samples with N/C > 0.04 were observed both in the present study and in some prior investigations (for details, see Table 17.9). In most instances, equilibration of such materials with aqueous solutions buffered at high pH removes most of this "organic" nitrogen, indicating that it was, instead, NH_4^+ bound at exchangeable sites, presumably in clays not removed during preparation of the kerogen. Accordingly, these "high-N" results are regarded as artifacts.

Concentrations of organic carbon in Proterozoic sedimentary rocks do not differ significantly from those in Phanerozoic sedimentary rocks. Caution is in order. A truly systematic approach to sampling would remove biases inherent in the Proterozoic database, which may be dominated by materials *selected for* organic analysis. But while that bias may have given undue weight to samples rich in organic carbon, a competing factor is clearly evident. In contrast with the Phanerozoic record, that of the Proterozoic is dominated by units with low H/C ratios. *Initial* (post-diagenetic, post-lithification) concentrations of TOC in these units must have been considerably

Table 3.2.5. New analyses of H/C ratios and carbon-isotopic compositions for Proterozoic kerogens.

Unit	Age (Ma)	Lithology[a]	H/C Avg.	H/C Range	$\delta^{13}C$, ‰ vs. PDB Avg.	$\delta^{13}C$, ‰ vs. PDB Range	N
Khatyspyt Fm.	620	SH	1.30		−36.2		1
Boonall Dolomite	640	CARB	0.23		−30.4		1
Sheepbed Fm.	650	SH	0.23	0.22–0.25	−29.5	−29.9–−28.5	5
Twitya Fm.	700	SH	0.44	0.39–0.55	−29.3	−32.1–−28.1	4
Luoquan Fm.	700	SH	0.31	0.26–0.36	−29.5	−29.9–−28.9	3
Biri Fm.	700	CARB	0.57	0.47–0.69	−31.6	−32.4–−30.5	4
Tindelpina Shale	750	SILT	0.18		−23.9		1
Woocalla Dolomite	750	SH	0.60		−29.3		1
River Wakefield Subgp.	775	CH	0.82		−24.1	−31.7–−16.4	2
Kingston Peak Fm.	850	CARB	0.27		−26.0		1
Beck Spring Dolomite	850	CARB	0.14		−26.5		1
Love's Creek Mbr.	850	CH	0.35		−26.1		1
		CARB	0.39		−29.7		1
Walcott Member	850	SH	0.84	0.77–0.91	−26.9	−27.8–−26.3	3
		CH	0.79		−27.2		1
		CARB	0.78	0.69–0.87	−26.5	−26.5–−26.5	2
Awatubi Mbr.	862	SH	0.70	0.60–0.79	−29.8	−32.2–−27.4	2
Carbon Canyon Mbr.	900	SH	0.74		−28.4		1
		CARB	0.64		−26.3		1
Jupiter Mbr.	930	SH	0.57	0.49–0.60	−24.0	−30.0–−19.6	4
Red Pine Shale	950	SH	0.68	0.62–0.68	−17.4	−17.7–−17.1	2
Tanner Mbr.	950	SH	0.53	0.51–0.55	−24.1	−28.1–−20.9	4
Allamoore Fm.	1,050	CH	0.31		−32.4		1
Nonesuch Shale	1,055	SILT	0.53		−33.5		1
Hongshuizhuang Fm.	1,250	SH	0.67	0.62–0.71	−33.2	−33.4–−33.0	2
Wumishan Fm.	1,325	CH	0.54		−30.1		1
McMinn Fm.	1,340	SILT	0.96	0.30–1.19	−32.6	−33.2–−32.1	5
Yangzhuang Fm.	1,350	CH	0.49		−28.4		1
Bungle Bungle Dolomite	1,364	CARB	0.30	0.26–0.34	−23.9	−25.1–−22.6	2
Dismal Lakes Gp.	1,400	CH	0.67		−28.7		1
Greyson Shale	1,420	SH	0.37		−28.3		1
Gaoyuzhuang Fm.	1,425	CH	0.45	0.38–0.52	−32.1	−31.4–−31.0	2
Newland Limestone	1,440	SH	0.34	0.26–0.42	−29.7	−31.8–−27.5	2
Urquhart Shale	1,670	SH	0.11		−28.3		1
Tuanshanzi Fm.	1,750	CARB	0.57		−31.4		1
Chuanglinggou Fm.	1,850	SH	0.41	0.33–0.53	−32.3	−33.0–−31.3	3
Koolpin Fm.	1,885	CH	0.17		−30.4		1
Fontano Fm.	1,900	SH	0.14		−30.1		1
Union Island Fm.	1,900	SH	0.23		−38.9		1
Rove Fm.	1,900	SH	0.47	0.42–0.51	−31.8	−32.1–−31.6	3
Rocknest Fm.	1,925	CARB	0.20	0.11–0.45	−21.7	−24.7–−12.6	5
Tyler Fm.	1,950	SH	0.29	0.26–0.32	−31.7	−31.8–−31.7	2
Gunflint Fm.	2,090	CH	0.41	0.32–0.54	−32.8	−33.1–−31.7	4
		CARB	0.44		−31.7		1
Reivilo Fm.	2,300	SH	0.26	0.13–0.35	−35.9	−36.7–−34.5	3
Klipfonteinheuwel Fm.	2,325	CH	0.13		−37.9		1
Gamohaan Fm.	2,450	CARB	0.26	0.25–0.28	−33.7	−34.6–−32.1	5
		SH	0.18		−34.9		1
Jeerinah Fm.	2,650	SH	0.42	0.36–0.48	−43.2	−45.1–−41.4	2
Tumbiana Fm.	2,750	SH	0.47		−58.7		1
		CARB	0.30	0.28–0.32	−53.4	−53.8–−52.9	2
		CH	0.22		−51.8		1
Kromberg Fm.	3,300	CH	0.14	0.12–0.16	−29.7	−32.6–−26.2	3
Figtree Group	3,350	CH	0.19	0.18–0.19	−28.6	−28.6–−28.5	2
Hooggenoeg Fm.	3,450	CH	0.20		−36.7		1
Theespruit Fm.	3,550	CH	0.06		−15.1		1

[a]CARB = carbonate, CH = chert, SH = shale, SILT = siltstone.

Table 3.2.6. *New Rock-Eval analyses of Proterozoic sediments.*

Unit	Age (Ma)	TOC (mg/g)	T_{max} (°C)	S_2 (mg HC/g)	HI (mg HC/g C_{org})	"H/C"	N
Woocalla Dolomite	750	4.7–5.8	459–485	0.13–0.19	28–33	0.45	2
Chuar Group	850	3.0–17	423–470	0.32–1.9	48–160	0.5–0.8	8
McMinn Fm.	1,340	14–20	430–437	4.8–8.2	306–403	0.9–1.1	5
Barney Creek Fm.	1,500	5.6	447	0.17	30	0.45	1

Table 3.2.7. *Abundance of organic carbon in Proterozoic sediments*[a].

Eon/Era	Lithology	TOC (mg/g)	N
Phanerozoic[b]	Shale	9.4	
	Carbonates	2.5	
Late Proterozoic	Shale	4.7	154
	Carbonate	2.5	206
	Chert	4.5	85
Middle Proterozoic	Shale	13.4	421
	Carbonate	4.6	322
	Chert	1.3	84
Early Proterozoic	Shale	10.3	1,224
	Carbonate	2.0	876
	Chert	1.8	184
Total Proterozoic	Shale	10.5	1,799
	Carbonate	2.7	1,404
	Chert	2.3	353
All Proterozoic lithologies		6.6	3,556

[a] Data of Tables 3.2.1 and 3.2.4.
[b] Ronov (1958, 1980).

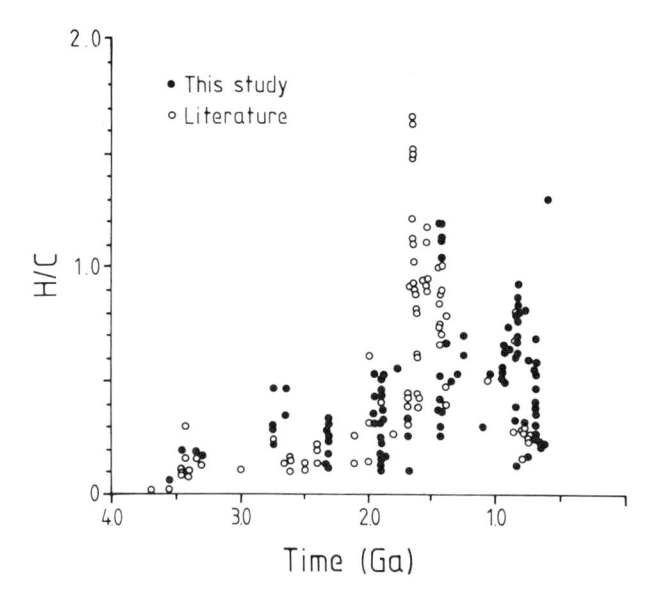

Figure 3.2.1. Atomic ratio of hydrogen to carbon in kerogens prepared from rocks of Proterozoic and Archean age. For numerical tabulation of data, see Tables 17.5 and 17.9.

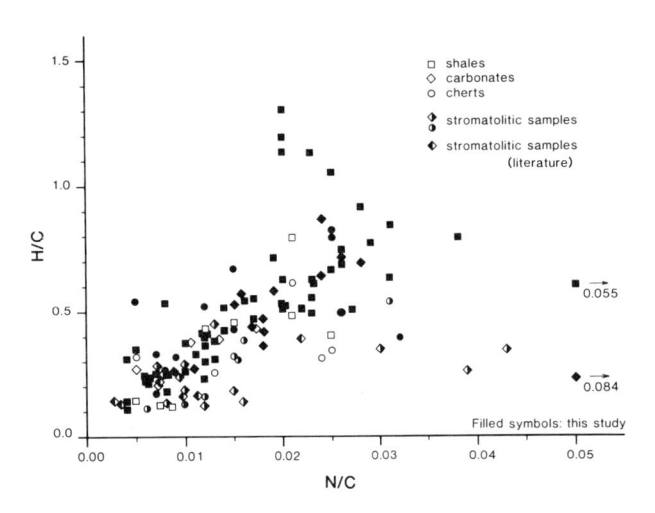

Figure 3.2.2. Atomic ratio of hydrogen to carbon as a function of atomic ratio of nitrogen to carbon in kerogens prepared from rocks of Proterozoic and Archean age. For numerical tabulation of data, see Tables 17.5 and 17.9.

higher than present concentrations. Consider organic material with an initial H/C ratio of 1.6 and a present ratio of 0.4. Even if the expelled hydrocarbons had an average H/C ratio of 3 (this high value being chosen to allow for losses of CH_4 at the high temperatures required to produce H/C $\sim$ 0.4), the initial quantity of organic carbon would have exceeded that now found in the rock by a factor of two. This is shown by the hydrogen mass balance:

$$C_1H_{1.6} \rightarrow C_{1-x}H_{0.4(1-x)} + C_xH_{3x}. \quad (3.2.1)$$

In this equation, the reactant is represented as kerogen with an H/C ratio of 1.6 and the products as residual kerogen with an H/C ratio of 0.4 and a mobile hydrocarbon fluid with an average H/C ratio of 3. The fractional carbon loss is represented by x, and its value can be determined by solution of the equation

$$1.6 = 0.4(1 - x) + 3x. \quad (3.2.2)$$

The result, $x = 0.46$, indicates that 46% of the carbon initially present was lost as a result of thermolysis of the kerogen. On this basis, confidence may be increased in the hypothesis that average contents of organic carbon are equal in Phanerozoic and Proterozoic sediments. Unfortunately, the trend to low H/C ratios in older sediments is paralleled by a loss of biochemical information.

3.3

Abundance and Composition of Extractable Organic Matter

ROGER E. SUMMONS

3.3.1 Preliminary Analysis of Kerogen and Bitumen

A combination of ubiquity, high relative abundance, and tractability in routine chemical analysis has determined that sedimentary alkanes would be the initial and continuing focus of geochemical study. As a consequence, the emphasis of this Section is on hydrocarbon composition and its interpretation. Other components of sedimentary organic matter, such as the aromatic, nitrogen-, oxygen-, and sulfur-containing (NSO) compounds and kerogen itself, are understudied but will be important in the future as new techniques expand the range of analyses that are routinely conducted. Recently recognized thiolane and thiophene biomarkers are particularly abundant in many oils and sediment extracts and are a source of well preserved and significant structural detail (Brassell et al. 1986; Cyr et al. 1986; Payzant et al. 1986; Schmid et al. 1987; Sinninghe Damste et al. 1987). Methodology for artificial and controlled degradation of kerogen by pyrolysis (e.g., Larter and Senftle 1985; Lewan et al. 1986), by catalytic hydrogenolysis (Mycke and Michaelis 1986), and by cleavage of specific types of bonds involved in cross linking (Chappe et al. 1980) is rapidly improving and may provide an important focus for future analyses of Proterozoic kerogens.

Kerogen, because of its immobility and relative stability, constitutes a valuable means of evaluating syngenicity in associated extractable organic matter (EOM). Kerogen type, as defined by Tissot and Welte (1984), while commonly determined from measurements of elemental H/C and O/C ratios, can also be conveniently estimated from Rock-Eval pyrolysis data (Espitalié et al. 1977). During the progressive heating of a whole rock sample to 550°C, volatile bitumen is first driven off, combusted and quantified (S_1 peak, kg/tonne). This is followed by combustion and quantification of the kerogen (S_2 peak, kg/tonne), measurement of the CO_2 generation from kerogen (S_3), and recording of the temperature at which the pyrolytic yield of hydrocarbons maximizes (T_{max}, °C). The production index ($S_1/S_1 + S_2$), therefore, records the proportion of bitumen in the sample and the T_{max} provides an indication of the maturity of the kerogen because it increases as the H/C ratio declines. Ratios of Hydrogen Index (S_2/organic carbon) and Oxygen Index (S_3/organic carbon) provide further qualitative information on the elemental composition of the kerogen because they have been shown to correlate well with elemental H/C and O/C, respectively (Espitalié et al. 1977).

Van Krevelen plots of elemental H/C vs. O/C, such as shown in Figure 3.3.1e for the McArthur Basin sediments, Barney Creek Formation, confirm the presence of immature Type I kerogen in the upper part of the unit and its progressive maturation in more deeply buried samples. Where burial is deep, the organic matter has passed through the main phase of oil generation, the kerogen is overmature (Fig. 3.3.1b; $T_{max} > 460°C$), and the abundance and composition of the associated EOM reflects this (Fig. 3.3.1c). Petrographic examination of organic matter *in situ* in thin section, particularly in regard to reflectance (Fig. 3.3.1c), and the relationships to lithological features should not be overlooked as valuable tools for the study of organic matter (e.g., Crick et al. 1988).

Comparison of the isotopic compositions of kerogen and bitumen subfractions provide further evidence for genetic relationship. Because the lipid components of living organisms are slightly depleted in ^{13}C relative to other components and to total biomass, the ultimate collection of their alkane products in the saturated hydrocarbon fractions of petroleum or sedimentary EOM means that they, too, will be depleted relative to the total bitumen and to the residual kerogen. Correlated isotopic curves have been described by Stahl (1978) and Schoell (1984). Table 3.3.1 summarizes the carbon-isotopic compositions of kerogen and bitumen subfractions for many of the samples discussed below. These data conform to the expected trends within subfractions and also highlight those situations where the organic matter is unusually heavy or light due to global secular variation or environmental factors in the immediate vicinity of sediment deposition. Many kerogen assemblages are quite heterogeneous, with some components maturing to yield liquid hydrocarbons at lower temperatures than others. Isotopic (Burwood et al. 1988) and hydrocarbon analyses (Larter and Senftle 1985; Lewan et al. 1986) of the pyrolysis products can produce further valuable information. There is limited, but useful application of kerogen degradation to Proterozoic kerogens (e.g., Summons et al. 1988a).

When subjected to the thermal stress of burial, the kerogen

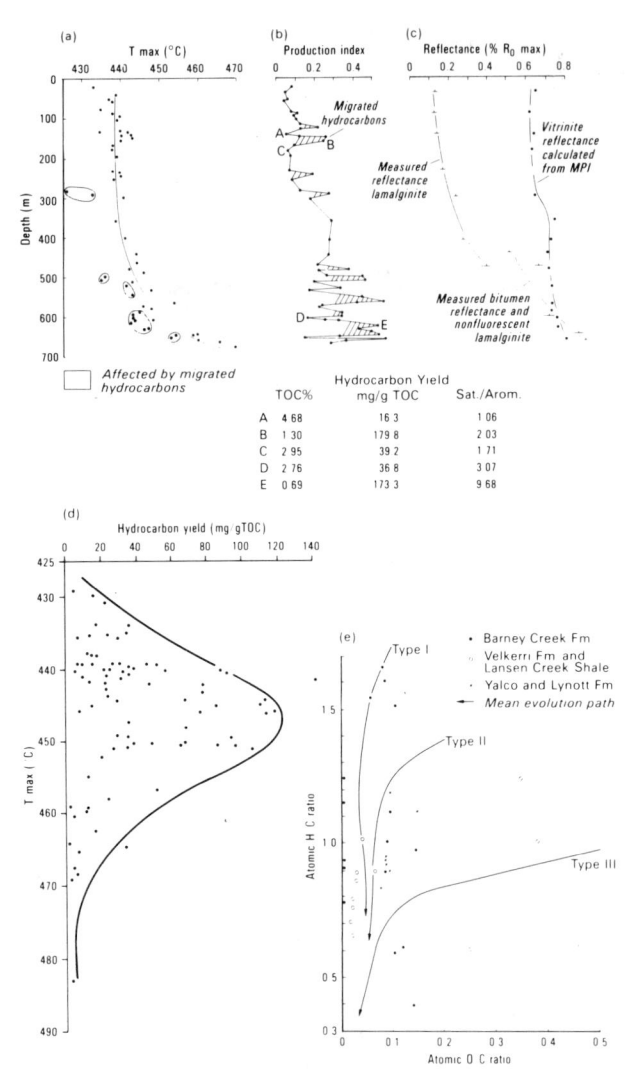

Figure 3.3.1. Comparison of maturity-dependent parameters for McArthur Basin sediments. (a) and (b) show Rock-Eval T_{max} and Production Index variation with depth within the Barney Creek Formation in the GR10 borehole; (c) shows the measured reflectances of organic matter in the same samples and compares these to the vitrinite reflectances calculated from the Methyl Phenanthrene Indices; (d) is a composite plot of samples from all McArthur units showing maximum hydrocarbon yield occurs in the T_{max} range 445–450°C; (e) shows variations in atomic H/C and O/C ratios in selected McArthur Basin samples. Reduction in atomic H/C corresponds to increasing maturity as measured by other parameters.

ted to smaller and smaller molecules by cleavage of carbon-carbon bonds (Hunt 1979; Tissot and Welte 1984). These products gradually lose their structural resemblances to the original biogenic precursors. As the residual kerogen becomes more graphitic, it also becomes devoid of informative chemical structure. When the kerogen is demonstrably of low to moderate maturity, the bulk characteristics of the EOM are similar to those of Phanerozoic bitumens.

An example of a sedimentary sequence which has been examined in considerable detail is that in the McArthur and Roper Groups from the McArthur Basin in northern Australia. Examination of core samples from multiple boreholes through several of the organic-rich units shows that EOM content and composition vary with depth according to the expected maturation trends (Crick et al. 1988). For example, saturate to aromatic ratios of less than one are found with marginally mature Type I kerogen and increase to more than three in more deeply buried samples. There are strong correlations with other maturity indicators including Rock-Eval T_{max}, organic-matter reflectance, fluorescence, color, and the Methyl Phenanthrene Index (MPI-1), a molecular parameter based on the relative abundances of isomers of methyl phenanthrene (Radke and Welte 1983; Boreham et al. 1988a). Organic petrographic analysis, applied to core samples through the Barney Creek Formation, demonstrated an abundance of *in situ*, highly fluorescent and poorly reflective lamalginite (mean max $R_0 \approx 0.2\%$) in marginally mature shallow samples. With increased burial depth, fluorescence was progressively lost to the point where Rock-Eval pyrolysis and EOM data indicated maximum hydrocarbon yields. Fluid inclusions containing liquid hydrocarbon were evident. In deeper samples, there was a systematic increase in reflectance to a mean max $R_0 \approx 1.0\%$ and a decrease in hydrocarbon yields on pyrolysis (Crick et al. 1988). Figure 3.3.1a–c compares Rock-Eval, extract, and petrographic data from a borehole (GR7) from the Barney Creek Formation, showing where different parameters identify the zone of maximal hydrocarbon generation. Data on the abundance and composition of EOM components from the McArthur Basin sediments are summarized in Table 3.3.2.

Gas chromatographic analysis of the saturated hydrocarbons, now carried out using high resolution, fused-silica capillary columns, reveals the first details of biomarker content and composition. It also shows any evidence for alteration by biodegradation or thermal cracking. Unaltered Proterozoic samples usually display an envelope of abundant C_{12} to C_{19} n-alkanes, each main peak being separated by a cluster of well resolved peaks for the simple branched and cyclic components. A large continuous hump of unresolved components underlying the main sharp peaks, together with a reduction in the abundance of the low molecular weight (low MW) n-alkanes, is usually an indication of later bacterial or thermal alteration (e.g., Connan and Restle 1984). At low carbon numbers (C_{10} to C_{20}) the chromatogram will show separately resolved peaks for monomethyl alkanes, cyclohexanes, and acyclic isoprenoids and these will generally be identifiable by their relative retention times. Very high relative abundances of the monomethyl alkanes, without strong preference for mid-chain isomers, is often associated with very low abundances of isopren-

cross-linking bonds rupture progressively, some intramolecular transfer of hydrogen occurs, and free hydrocarbons are liberated into the rock matrix. Recent, immature sediments generally have 1 to 10 mg hydrocarbon per g of organic carbon while concentrations in ancient, mature, non-reservoir sediments are in the range of 10 to 150 mg/g. As the breakdown of kerogen (the process is termed catagenesis) proceeds, there will be a decrease in the proportions of polar (NSO) compounds and higher molecular weight components (asphaltenes), an increase in the ratio of saturated to aromatic hydrocarbons, and a progressive decrease in the elemental H/C ratio of the residual kerogen. The final stages of cracking, termed metagenesis, occur when the mobile components themselves are conver-

Table 3.3.1. Carbon-isotopic compositions of bitumen subfractions and kerogens from a selection of Proterozoic sediments.

Unit	Age (Ma)	$\delta^{13}C$, ‰ vs. PDB				Sample Number[a]
		Sat.	Arom.	Polar	Ker.	
Rodda Beds	610	−31.8	−32.1	−31.9		4044
		−30.7	−30.7	−31.3	−29.4	3581
		−30.0	−30.5	−30.5	−28.7	3355
		−30.0	−30.6	−30.6	−29.4	3356
Pertatataka Fm.	610		−28.4	−29.3		3610
Khatyspyt Fm.	620	−36.6	−35.9	−35.7	−37.3	4385
V.-Chonskaya Oil	620	−34.8	−34.5	−33.3		319
Danilovskaya Oil	620	−34.7	−34.0	−33.0		320
Duliominskaya Oil	620	−33.9	−33.9	−32.9		321
Kwagunt Fm.	850	−26.5	−26.4	−25.6	−25.7	3239
		−26.8	−26.5	−25.0	−26.0	3240
		−27.5	−26.4	−26.7	−26.5	4216
		−27.7	−26.6	−27.1	−26.5	4217
		−27.4	−26.5	−26.6		4418
Bitter Springs Fm.	850	−27.6	−29.0			2449
		−27.4				2452
			−26.1	−26.6	−25.4	2462
		−27.2	−25.7	−26.0		3340
		−28.3	−26.1	−26.8		3341
				−26.9		3592
				−25.4		3594
		−23.7		−27.7		3346
Nonesuch Shale	1,055	−32.7	−32.9	−32.8	−32.2	3354
Nonesuch Seep Oil	1,055	−32.0	−32.2	−32.5		306
McMinn Fm.	1,340	−31.2	−32.4	−32.0		3131
Velkerri Fm.	1,350	−33.3	−33.2	−33.4		3813
		−32.1	−31.1	−31.8		3819
Barney Creek Fm.	1,500	−31.1	−31.6	−30.6	−31.7	1810
		−30.1	−30.1			1830

[a]Bureau of Mineral Resources, Canberra, Australia.

oids such as pristane and phytane. Strong mid-chain preference for branching in longer ($>C_{20}$) monomethyl alkanes is notable for its association with abundant acyclic isoprenoids in some environments including the Proterozoic oils of Oman and the Siberian Platform (e.g., Fig. 3.3.2).

Combined gas chromatography-mass spectrometry (GC-MS) is required to detect and correctly identify the more complex biomarker structures such as the extended ($>C_{20}$) acyclic isoprenoids, steranes, and triterpanes. In immature samples, where these components may be very abundant, complete scans of the mass spectrometer may be used to generate spectra for correlation with the spectra of reference materials. However, it is usually observed that most oils and mature sediments have such low levels of biomarkers that they can only be detected by more sensitive techniques. In the method known as selected ion recording (SIR), only the most

Table 3.3.2. Contents of organic carbon and extractable organic matter and ratio of saturated to aromatic hydrocarbons in McArthur Basin sediments.

Unit	Age (Ma)	TOC (mg/g)		EOM (mg/g C_{org})		Sat./Arom.		N
		Avg.	Range	Avg.	Range	Avg.	Range	
McMinn Fm.	1,340	19.5	13.8–29.6	21.0	14.6–29.8	1.54	0.47–3.63	4
Velkerri Fm.	1,350	47.3	6.7–72.0	86.8	30.5–186	1.38	0.59–2.95	5
Lansen Creek Shale	1,350	20.6	9.0–29.8	82.6	19.9–205	1.89	0.63–3.16	4
Yalco Fm.	1,485	41.7	31.5–53.7	17.7	9.7–33.5	0.31	0.09–0.46	3
Barney Creek Fm.	1,500	24.3	3.1–51.7	44.2	10.9–110	2.12	1.06–5.82	15

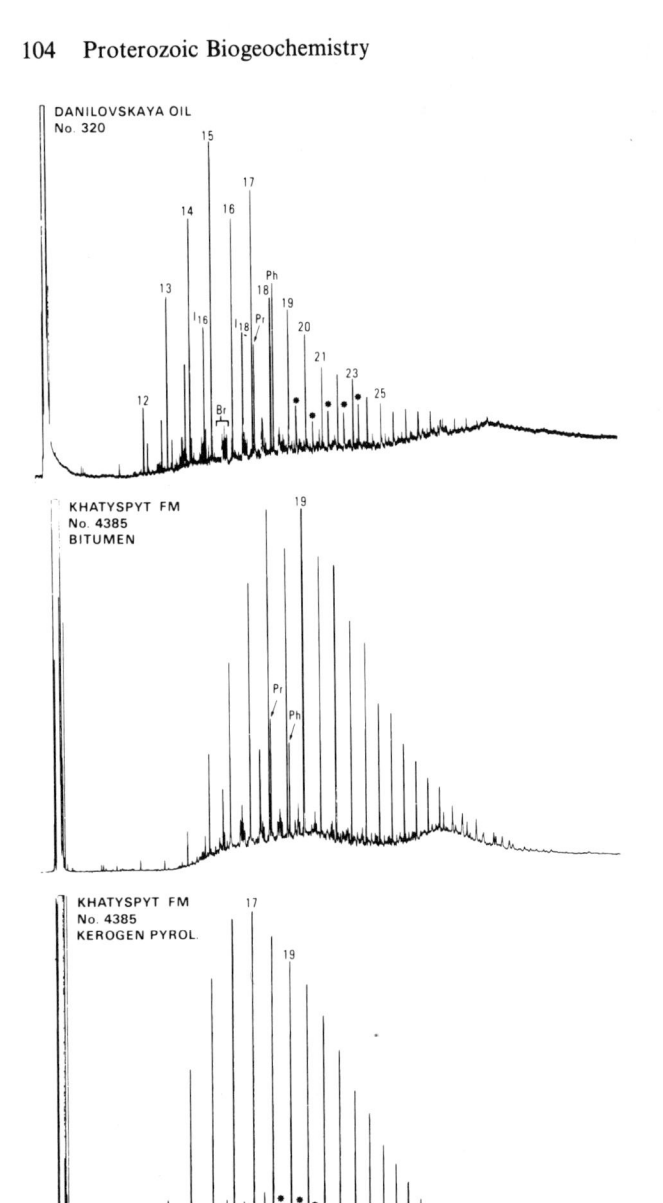

Figure 3.3.2. Gas chromatographic traces (FID response) showing saturated hydrocarbon distributions in some Proterozoic organic matter from the Siberian Platform. BMR no. 320 is an oil from the Lena-Tunguska region; BMR no. 4385 (PPRG no. 2723) is a bitumen extracted from sediment of the Khatyspyt Formation. The lower trace shows hydrocarbons released by hydrous pyrolysis of the Khatyspyt Formation kerogen. Numbered peaks are n-alkanes designated according to their carbon chain length; Pr = pristane; Ph = phytane; I16, I18, etc., refer to the carbon numbers of other isoprenoid alkanes; mid-chain branching in monomethyl alkane components is denoted by ∗; Br denotes a suite of monomethyl alkanes with predominance of 2-Me and 3-Me isomers.

abundant fragment ions of a molecule are detected and measured; the relative retention times of each chromatographic peak are used to aid identification (e.g., Seifert and Moldowan 1978). With the relatively new GC-MS techniques of multiple metastable reaction monitoring (MRM) (Warburton and Zumberge 1983; Moldowan et al. 1985; Summons 1987), and GC-MS-MS, the instrument is programmed to only detect specific parent ions decomposing to characteristic daughter ions. The combination of significantly enhanced selectivity with improved signal to noise ratio leads to reliable and sensitive analyses of biomarkers because only the target compounds are detected. MRM analyses are especially useful in studies of Proterozoic sediments, where low biomarker concentrations and the presence of many co-eluting components otherwise yields complex chromatograms which are difficult to interpret (e.g., Summons et al. 1988a, b).

3.3.2 Intensively Studied Rock Units and Petroleums

This Section describes specific details of the hydrocarbon compositions of extractable bitumens from sediments which, by the criteria described above, contain Proterozoic organic matter. Much of the work (cited as "this study") is preliminary and results from recent and continuing studies utilizing improved analytical techniques. Full and definitive accounts will appear in reviewed journals.

3.3.2A Lena-Tunguska Oils, Siberian Platform

Deposits of petroleum from the Lena-Tunguska region of eastern Siberia have been known for many years (Petrov et al. 1977; Meyerhoff 1980) and have been recognized for their unusual hydrocarbon compositions (Makushina 1978; Aref'ev et al. 1980; Fowler and Douglas 1987). Oils in commercial quantities are reservoired in marine shelf carbonates and terrigenous clastic rocks of Cambrian and Vendian age. They may have been sourced by either deeply buried basinal shales of Riphean age, or by the overlying carbonates and carbonate evaporites of Vendian to Cambrian age, as reviewed by Meyerhoff (1980). While a definitive oil to source rock correlation has not yet been made, certain aspects of the biomarker composition, on which sediment lithology exerts significant control, suggest the most likely origins are in the Vendian to Cambrian carbonates (Fowler and Douglas 1987; Summons and Powell 1989).

Striking features evident from the GC and GC-MS (MRM) data obtained in a recent study (Summons et al. 1988b; Summons and Powell 1989) are shown in Figures 3.3.2, 3.3.3 and 3.3.4. The GC data (Fig. 3.3.2) reveal high relative abundances of acyclic isoprenoids, Pr/Ph (Pristane/Phytane) ratios of less than one, and the presence of C_{20+} monomethyl alkanes. The GCMS data for steranes and triterpanes are notable for the virtual absence of rearranged steranes (diasteranes), high relative abundances of $\alpha\beta\beta$ steranes, and high abundances of 30-norhopanes including norhopane and 29,30-bisnorhopane. The presence of aryl isoprenoids in the aromatic hydrocarbon fraction suggests a contribution from the Chlorobiaceae family of green sulfur bacteria. Combinations of these features have been reported for Phanerozoic oils and their marine carbonate or carbonate evaporite source rocks (e.g., McKirdy et al. 1983, 1984; Zumberge 1984; Summons and Powell 1987; Price et al. 1988). Other unusual features of the oils include the very strong preference for C_{29} steranes (Aref'ev et al. 1980; Fowler and Douglas 1987) and the presence of 3β-methyl and 2α-methyl steranes (Summons and Capon 1988; Summons et al. 1988b) and 3β-ethyl steranes with the same 24-ethyl substitution and apparent stereoisomer preferences as the

al. 1987), with a slight to moderate predominance of C_{27} steranes, a feature probably reflecting a high input of cholesterol from metazoan fecal pellets combined with some algal sterol sources. The absence of body and trace fossils from all but a few sequences of latest Vendian age indicates that the organic matter contributions from metazoans are likely to have been minor in these source environments. Thus, preservation of algal steroids, possibly dominated by input from Chlorophyceae, could lead to the pattern found in these cases.

Monomethyl alkanes have been reported in the Siberian Platform oils by Makushina (1978) and found to have mid-chain branching preference by Fowler and Douglas (1987). This feature is evident in the isomers of the C_{20} to C_{30} homologues, whereas a preference for terminally branched 2-methyl and 3-methyl isomers is seen at lower carbon numbers (Fig. 3.3.1). Similar features are also evident in the Late Proterozoic oils of Oman (e.g., Klomp 1986; Grantham et al. 1988). The variation in branch-site preference with carbon chain length provides strong evidence for a distinctive biological, probably bacterial, origin for these compounds (e.g., Summons et al. 1988b). Bacterial lipids are often distinguished by the presence of fatty acid carbon chains with 2-methyl and 3-methyl branching (Fulco 1983) and by occasional mid-chain branching (Dowling et al. 1986). The former style of branching accounts for the high abundance of 2-methyl and 3-methyl alkanes in some geolog-

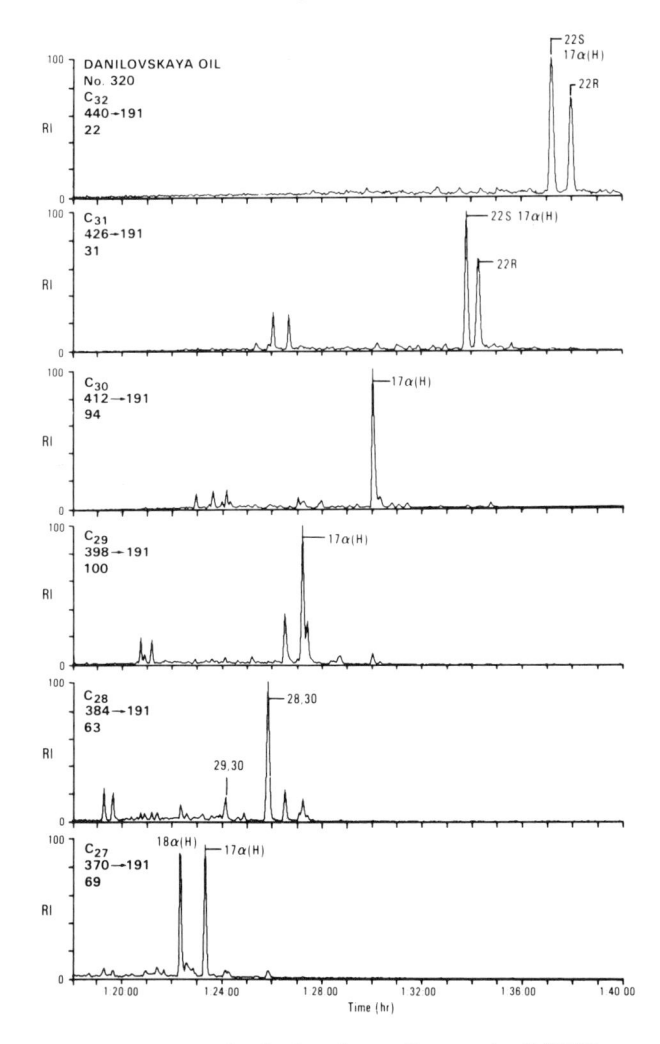

Figure 3.3.3. Sterane distributions from a Proterozoic oil (BMR no. 320) from the Lena-Tunguska region of the Siberian Platform. The chromatograms were obtained by metastable reaction monitoring (MRM) in GC-MS of molecular ions fragmenting to daughter ions at m/z 217 or m/z 231. The trace identifiers show relative abundance normalized to the most intense reaction. GC-MS was carried out on Ultra-1. Assignments provide details of C-20 stereochemistry (R or S) and $\beta\alpha$, $\alpha\alpha\alpha$, and $\alpha\beta\beta$ denote 13β(H),17α(H)-diasteranes, 5α(H),14α(H),17α(H)-, and 5α(H),14β(H),17β(H)-steranes, respectively.

desmethyl steranes. A high relative abundance of the biomarker 28,30-bisnorhopane, which is found in these oils, is also distinctive and associated with a limited number of Phanerozoic marine sediments with very high organic matter contents (e.g., Grantham et al. 1980; Katz and Elrod 1983; Volkman et al. 1983).

Sterols with 24-ethyl substitution are often reported to be most abundant in organic matter produced by terrestrial higher plants and chlorophycean algae (e.g., Patterson 1971; Volkman 1986). Thus, the association of 24-ethyl substituted steranes with otherwise marine indicators in the Siberian Platform oils is enigmatic. It is also significant that the same association has been observed in the Proterozoic oils from Oman (Grantham 1986a; Grantham et al. 1988) and is present in several other sediments of probable Vendian age (see below). Marine oils from the Phanerozoic normally display suites of C_{26} to C_{30} sterane pseudohomologues (Moldowan et al. 1985; Brassell et

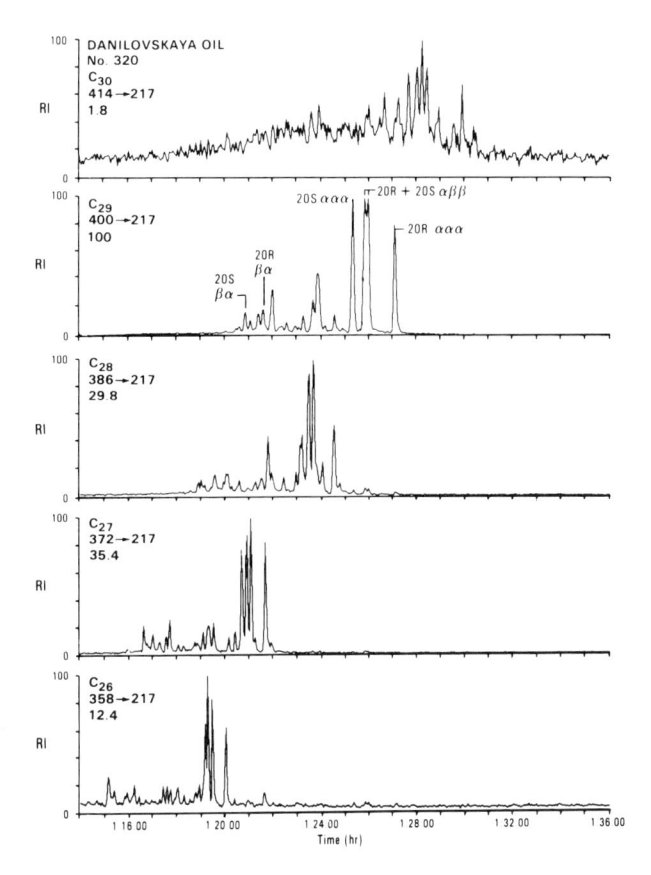

Figure 3.3.4. MRM chromatograms (e.g., $M^+ \to 191$) for C_{31}-C_{27} triterpanes of a Siberian Platform oil (BMR no. 320). The 17α(H), 21β(H)-hopanes are denoted $\alpha\beta$. R and S refer to the stereochemistry at C-22. 28,30 denotes 28,30-bisnorhopane, and 29,30 denotes 29,30-bisnorhopane.

ically younger oils (e.g., Tissot and Welte 1984), particularly those derived from bacterially degraded, non-marine organic matter (e.g., Powell et al. 1989).

The carbon isotopic composition of Siberian Platform oils is unusually light (-34 to $-35‰$), and data for individual subfractions are given in Table 3.3.1. Similar light carbon isotopic characteristics have been reported for the Huqf Oils from Oman (Grantham et al. 1988), some sediments from Spitzbergen and East Greenland (Knoll et al. 1986), and are also evident in the Khatyspyt Formation.

3.3.2B Khatyspyt Formation, Siberian Platform

The present study of Proterozoic organic matter included three sediment samples from the Khatyspyt Formation, Khorbusuonka River, in the Olenyok Uplift of the Siberian Platform. They comprised thinly laminated silty carbonates and associated bituminous organic films present on bedding planes. Rock-Eval analyses showed the presence of abundant kerogen of low to moderate maturity, particularly in the organic films. The age of samples has been estimated at 620 Ma by Fedonkin (pers. comm. 1988), and the bitumens are interesting because of the extremely close resemblance of their hydrocarbon biomarker and carbon isotopic compositions to those of the Lena-Tunguska oils. Bitumen extracted from the organic films lack the high abundance of mid-chain monomethyl alkanes, but hydrocarbons released by pyrolysis of the residual kerogen have increased relative abundances of mono-

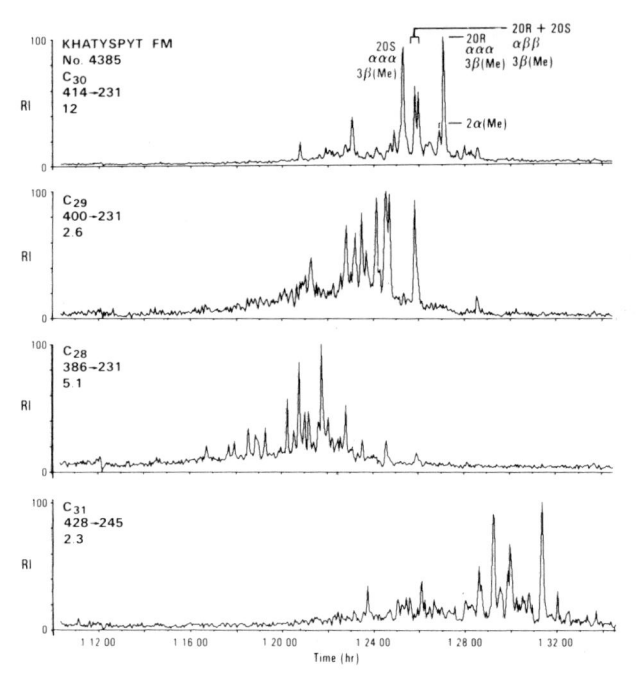

Figure 3.3.6. MRM chromatograms showing the methyl sterane distribution of a sediment extract from the Khatyspyt Formation (BMR no. 4385 = PPRG no. 2723). Assignments are as in Figure 3.3.3 and intensities are normalized to the C_{29} desmethyl sterane. The distribution of homologues and stereoisomers closely resembles that of the desmethyl steranes.

methyl alkanes. Other features, which include low Pr/Ph ratio, the dominance of C_{29} (24-ethyl) steranes, and uncommonly light carbon isotope signature (near $-37‰$), are the same as those found in the Lena-Tunguska oils. We have not detected 4-methyl steranes in any Siberian Platform samples (cf. Fowler and Douglas 1987), but instead a high abundance of 3β-methyl, 2α-methyl, and 3β-ethyl isomers with C-24 alkylation and stereoisomer preferences similar to those of the desmethyl steranes (compare Figs. 3.3.5 and 3.3.6). The origin of these 3β- and 2α-alkylated steroids is not presently known (Summons and Capon 1988), but their distribution patterns would be consistent with formation via biological methylation of sedimentary steroids during early diagenesis. One of the earliest steps in the diagenesis of any sterol is dehydration to a Δ^2 sterene (e.g., Brassell and Eglinton 1981; Mackenzie et al. 1982). Biological methylation, with preference for C-3, of a proportion of the mixture of sterenes (and concomitant reduction of the remainder) could lead to a mixture of 3-methyl and 2-methyl steranes having side-chain and stereoisomer preferences which closely match those of the desmethyl steranes. Steranes with two extra carbons attached to ring A (i.e., 3β-ethyl steranes) have not been reported previously, but their close association here with 3β-methyl steranes indicates that their formation involved sequential methylations. There is a direct analogy in prokaryote biochemistry in that the sites of methylation of the pentacyclic ring system in 3β-methyl and 2β-methyl hopanoids produced by some extant bacteria and cyanobacteria (Bisseret et al. 1985; Zundel and Rohmer 1985a, b) are identical to those found in these steranes. Many of the subsequent diagenetic and thermally induced modifications to sedimentary sterenes and

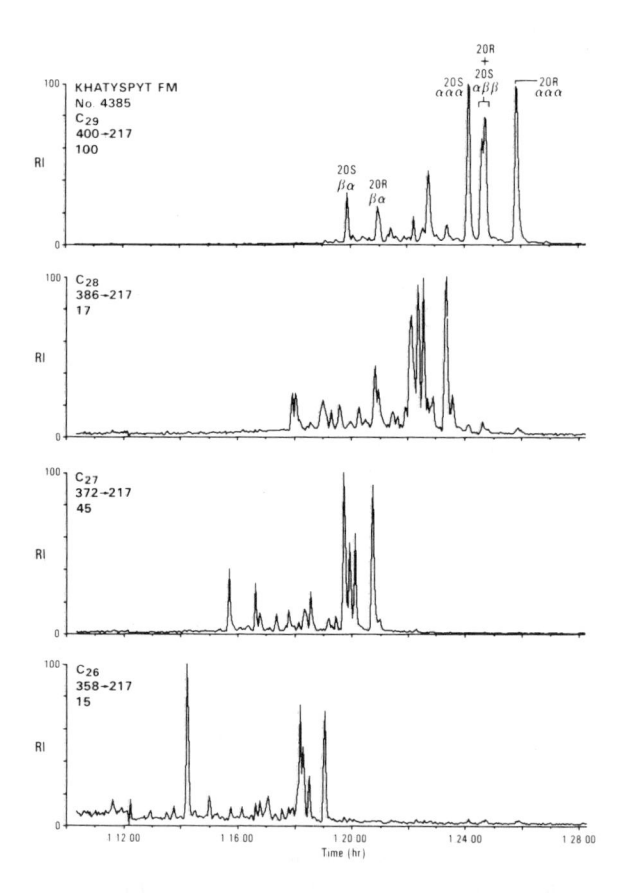

Figure 3.3.5. MRM chromatograms showing the sterane distribution of a sediment extract from the Khatyspyt Formation (BMR no. 4385 = PPRG no. 2723). Assignments are as in Figure 3.3.3.

steranes, that is, those which lead to the complex mixtures of stereoisomers (Mackenzie et al. 1982; ten Haven 1986), would take place unaffected by the additional ring-A alkyl group. This would give rise to similar mixtures of isomers at C-5, C-14, C-17, and C-20, and also to a preference for the thermodynamically stable equatorial orientation of the alkyl groups [3β(Me), 3β(Et), and 2α(Me)]. It is therefore likely that the anomalous 3- and 2-alkyl steroids, unlike conventional 4-methyl steranes, are not representative of original and distinctive eukaryotic inputs but, rather, are signals for microbial reworking.

3.3.2C The Huqf Group Oils, Oman

Two distinctly different families of Proterozoic petroleum are presently known from the Sultanate of Oman. One family, known as the Huqf oils, has been chemically and isotopically correlated with rich source rocks within the Huqf Group carbonates and evaporites (Grantham 1986b; Grantham et al. 1988; Hughes Clarke 1988). Crystalline basement rocks dated radiometrically at >650 Ma, evidence for glaciogenic deposition (ca. 615 Ma) in the Abu Mahara Formation, and stratigraphic correlations suggest Vendian to Early Cambrian ages for the Huqf source rocks (Gorin et al. 1982; Hughes Clarke 1988). A second family, known as the "Q" oils, which have no clearly defined source rock, has compositional features and stratigraphic position suggesting a Late Proterozoic source (Grantham et al. 1988).

The characteristics of the Huqf oils closely resemble those of the Siberian Platform oils discussed above, and include high abundances of monomethyl alkanes (Klomp 1986), low pristane to phytane ratio, predominance of C_{29} steranes (24-ethyl cholestanes) with $\alpha\beta\beta$ stereochemistry, low relative abundances of rearranged steranes (Grantham 1986b), and an extremely light carbon isotope signature (ca. $-36‰$ vs. PDB; Grantham et al. 1988). The "Q" oils have moderate concentrations of monomethyl alkanes, pristane to phytane ratios of just greater than one, a strong preference for C_{27} steranes (cholestanes), and $\delta^{13}C$ values of -30 to $-31‰$ (Grantham 1986b; Grantham et al. 1988). Geological data and the presence of monomethyl alkanes have been used to postulate a source for the "Q" oils deep within the Huqf Group sediments (Grantham et al. 1988).

Information on the triterpenoid distributions of the Oman oils (Grantham, 1986b) suggests that the normal series of 17α(H), 21β(H)-hopanes is predominant in both Huqf and "Q" oils with the latter having high levels of tricyclic terpenoids and many unidentified components. The relative abundance of norhopane relative to hopane in the m/z 191 chromatograms of both oils (Grantham, 1986b) suggests that an homologous series of 30-norhopanes may be present, as has been observed in the Siberian Platform samples. This unusual feature of hopane distributions is often found alongside a low abundance or absence of rearranged steranes. It is considered to be a powerful indicator for source rocks with carbonate lithologies (e.g., Price et al. 1988) and is, thus, consistent with origin of these oils from within the carbonates of the Huqf Group.

3.3.2D Rodda Beds, Officer Basin, South Australia

Detailed studies have been made of the organic matter from Lower Cambrian sediments and oils from the Observatory Hill Formation, eastern Officer Basin (McKirdy et al. 1983, 1984). Below this and the Relief Sandstone are the Rodda Beds, a thick (>1200 m) sequence of marine laminated calcareous and dolomitic siltstones, black shales, and sands with medium to coarse conglomerate lenses. The Rodda Beds have been assigned a Late Proterozoic age on the basis of facies and stratigraphic correlations within this basin and with radiogenically dated sediments of the Amadeus Basin to the north, and the Adelaide Geosyncline to the south (Brewer et al. 1987). Sedimentological and facies analysis suggests the unit might be correlative with the Pertatataka Formation which has been dated at approximately 610 Ma (Preiss and Forbes 1981; Zang and Walter 1988). Organic matter contents range from 0.3 to 1.5% TOC through several hundred meters of the sediment which has been cored in several boreholes (Ungoolya-1, Karlaya-1, and Marla-9; Brewer et al. 1987). Rock-Eval pyrolysis parameters and the presence of oil bleeds confirm the presence of mature kerogen.

Analyses of saturated hydrocarbons from the extractable bitumens of the three boreholes reveal patterns of alkanes resembling those of the Late Proterozoic samples from Siberia and Oman. Bitumens from the Marla-9 borehole are described here. Monomethyl alkanes are present, albeit in low relative abundance, and there are significant concentrations of acyclic isoprenoids with Pr/Ph = 1. Most distinctive, however, are the compositions of the steroids and triterpenoids, with 24-ethyl steranes predominating in all samples examined (Fig. 3.3.7). The hopanoids (Fig. 3.3.8) comprise two homologous series. One corresponds to the normal 17α(H), 21β(H)-hopanes. On the basis of relative retention times, members of the other series,

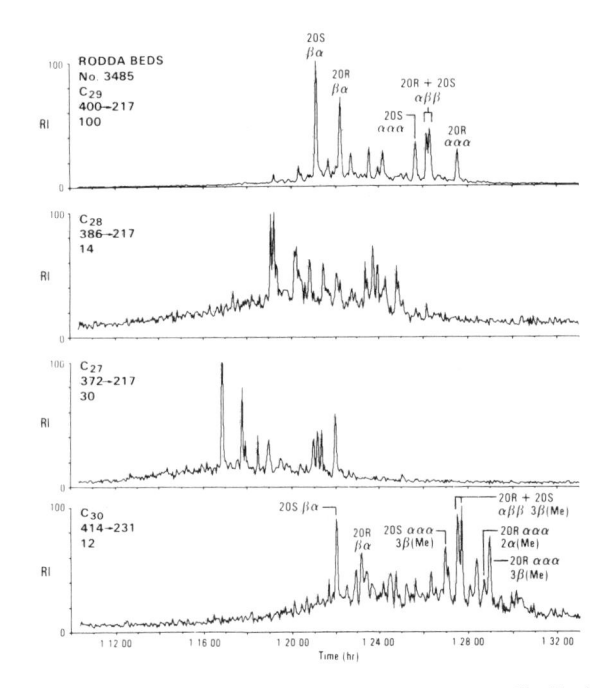

Figure 3.3.7. MRM chromatograms showing the sterane distribution of a sediment extract from the Rodda Beds, Officer Basin, South Australia (BMR no. 3485). Assignments are as in Figure 3.3.3. The distribution of isomers of the 3β-methyl steranes closely resembles that of the desmethyl steranes, including the high relative abundance of 13β(H),17α(H)-diasteranes.

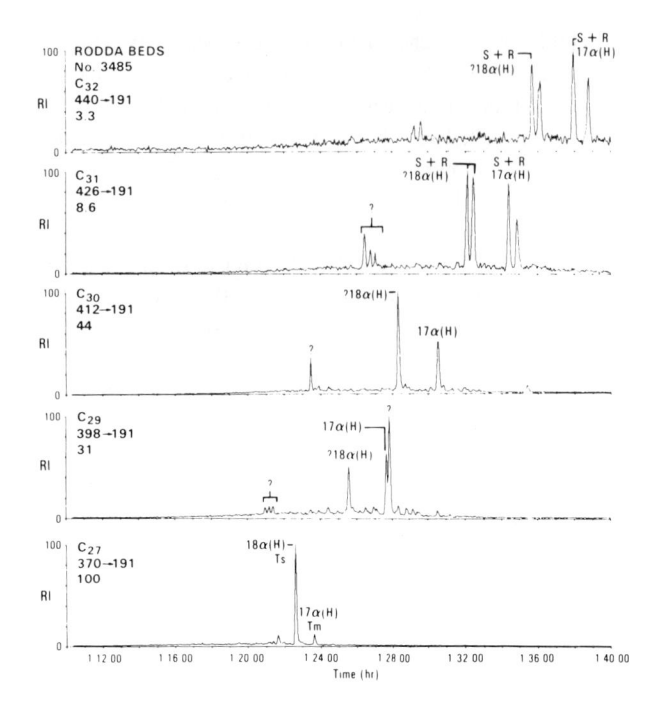

Figure 3.3.8. MRM chromatograms showing the triterpane distribution of a sediment extract from the Rodda Beds. Officer Basin, South Australia (BMR no. 3485). The elution positions of 17α(H), 21β(H)-hopanes and putative 18α(H)-neohopanes are designated 17α(H) and 18α(H), respectively. S and R refer to C-22 stereochemistry in the hopanoids and it should be noted that it is not yet possible to make a definitive structural or stereochemical assignment for probable 18α(H)-neohopanes.

eluting earlier, appear to be 18α(H)-neohopanes (e.g., Fig. 3.1.3, **20**). Although their precise structures have not yet been established by comparison with authentic standards, the recognition of a new suite of distinctive triterpenoids with abundances similar to those of the ubiquitous 17α(H), 21β(H)-hopanes is significant. The new compounds most likely stem from a variation in bacteriohopane biosynthesis and, accordingly, represent a biochemically novel class of bacteria.

3.3.2E Bitter Springs and Pertatataka Formations, Amadeus Basin, Central Australia

The distribution of organic matter in Amadeus Basin sediments has been discussed in the context of petroleum prospectivity by Jackson et al. (1984). Generally, organic matter is present as bitumen in the Gillen Member of the Bitter Springs Formation, and as kerogen and bitumen in the Pertatataka Formation. However, its abundance is quite low, usually being less than 0.5% TOC, and, when present, the kerogen is mature with respect to petroleum generation. The sediments are discussed here in detail because of their geological significance, the interest associated with the rich microfossil assemblages of both formations (e.g., Schopf 1968b; Schopf and Blacic 1971; Schopf and Oehler 1976; Zang and Walter 1988), and other features of paleobiological significance (e.g., Southgate 1987). Two organic facies types are evident in both units. One is associated with the more organic-rich shales which have abundant monomethyl alkanes and alkyl cyclohexanes and low abundances of isoprenoids and polycyclic biomarkers (Fig. 3.3.9). The other facies, associated with carbonates and the bituminous evaporites of the Gillen Member, has lower contents of monomethyl alkanes

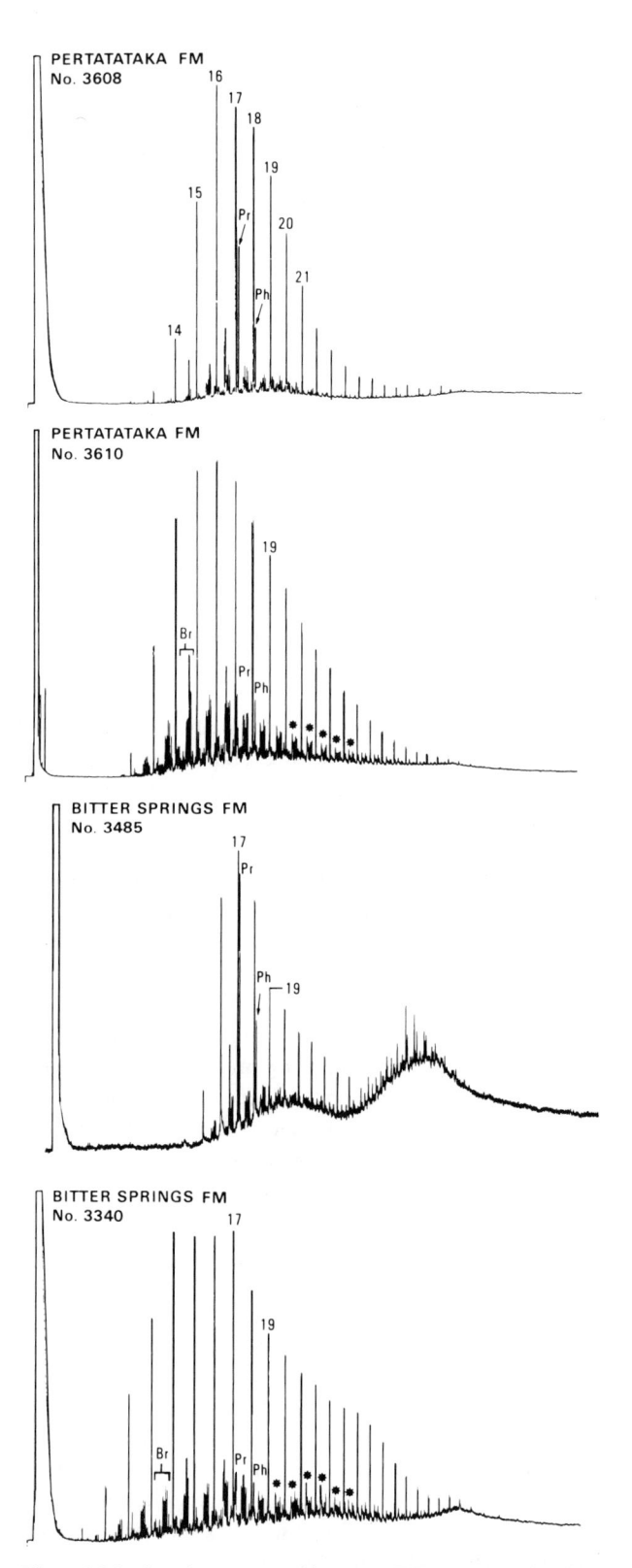

Figure 3.3.9. Gas chromatographic traces (FID response) showing saturated hydrocarbon distributions in sediment extracts from the Bitter Springs (Gillen Member) and Pertatataka Formations, Amadeus Basin, central Australia. Peak identifiers are as in Figure 3.3.2.

Table 3.3.3. *Characteristics of a suite of sediment samples from the Chuar Group, Grand Canyon, Arizona.*

Unit	Age (Ma)	Lithol.	TOC (mg/g)	H/C	N/C	$\delta^{13}C$, ‰ vs. PDB TOC	$\delta^{13}C$, ‰ vs. PDB Ker.	T_{max} (°C)	S_2	HI	Sample Nos. PPRG	Sample Nos. BMR
Walcott Mbr.	850	SH	29.6	0.84	0.031	−27.7	−27.8				1088	
		SH	0.5			−26.1					1093	
		CARB/SH	17.2	0.87	0.024	−26.5	−26.5	447	1.81	105	1095	4216
		CARB/SH	19.7	0.91		−26.5					1097	
		SH	11.8	0.77	0.029	−26.4	−26.4				1112	4419
		CARB/CH	18.2	0.69	0.026	−26.5	−26.5	470	1.05	57	1123	4217
		SH	13.1	0.79	0.038	−27.0	−27.4	423	0.89	68	1087	4215
		CARB/SH	11.2	0.92			−25.7	446	1.61	139		3239
		CARB/SH	16.6	0.93			−26.0	449	1.92	116		3240
		CARB	3.0					453	0.32	106		4203
Awatubi Mbr.	862	SH	1.7	0.60	0.055	−32.05	−32.2				1083	4214
		SH	12.9					426	0.62	48		4229
Carbon Canyon Mbr.	900	SILT	17.6	0.74	0.026	−28.7	−28.4				1126	4421
		SH	0.2			−24.5					1153	
		CARB	0.3			−29.1					1156	
Jupiter Mbr.	930	SH	7.7	0.50	0.027	−25.5	−25.7				1062	
		SH	7.6	0.66	0.025	−19.7	−20.6				1063	
		CARB	0.1			−21.9					1064	
		SH	2.9	0.49	0.023	−18.6	−19.6	436	0.02		1067	4218
		SILT	11.4			−28.9					1124	
		SH	3.5	0.63	0.031	−30.4	−30.0				1125	4420
Tanner Mbr.	950	SH	4.8	0.51	0.020	−24.8	−24.8				1055	
		SH	3.4			−23.2					1058	
		SH	2.2	0.55	0.023	−21.8	−22.4				1059	
		SH	5.0	0.53		−19.5					1060	
		SH	1.9	0.55	0.017		−28.1				1061	
		SH	4.9	0.51	0.022	−20.7	−20.9				1080	

and high contents of isoprenoids, steranes, and triterpanes. The steroidal hydrocarbon compositions of both units are similar, but anomalous with respect to all other Late Proterozoic materials studied to date. In a study of over 20 samples of cored sediments from five boreholes penetrating both units, it has been found that isomers of a C_{30} desmethyl sterane, C_{28} to C_{30} 4-methyl, 3β-methyl, and 2α-methyl steranes, and C_{29} to C_{31} 3β-ethyl steranes are abundant, and that neither C_{27} nor C_{29} desmethyl steranes are predominant (Summons, this study). The desmethyl steranes do not show the strong carbon-number preferences that characterize the Oman, Siberian Platform, and Rodda Bed samples, and the identification of C_{30} desmethyl homologues is of interest because it predates other recognized occurrences by several hundred million years (e.g., Moldowan et al. 1985). There are significant concentrations of the 4-methyl steranes, including 24-ethyl-4α-methylcholestane and 4α,23,24-trimethylcholestane (dinosterane). These have not otherwise been detected in sediments older than mid-Triassic (e.g., Summons et al. 1987; Goodwin et al. 1988), corresponding in age to the first generally recognized dinoflagellate cysts. Although it is very unlikely that these hydrocarbons could have migrated from younger rocks, these potentially important findings should be considered tentative until verified by further work.

3.3.2F Chuar Group, Grand Canyon, Arizona

The Chuar Group sediments of the Grand Canyon are notable as the site of exceptional microfossil preservation (e.g., Walcott 1883; Ford and Breed 1973; Schopf et al. 1973a; Bloeser et al. 1977; Bloeser 1980, 1985; Horodyski and Bloeser 1983). Samples of dolostone from the Walcott Member, Kwagunt Formation, have high contents of mature kerogen and extractable bitumens bearing distinctive hydrocarbon signatures (Summons et al. 1988a). Further analyses of sediments from the Kwagunt and underlying Galeros Formations (this study) confirm the wide occurrence kerogen and bitumens, and there is a consistent trend of increasing maturity down through the sequence. Table 3.3.3 summarizes the organic geochemical parameters of a suite of samples, all from surface outcrop, from this sequence. Preservation of kerogen, as determined by elemental H/C ratios and Rock-Eval pyrolysis, is best in the Walcott Member and declines progressively with depth. The sediments are reported to range in age from ca. 950 Ma to ca. 825 Ma (Elston 1979).

Distributions of hydrocarbons from sediments of the basal section of the Walcott Member and several underlying units are shown in Figure 3.3.10, but only the former is discussed in detail here. Variability within the major alkanes, as in other sediments of this age, relates to the relative abundance of acyclic isopren-

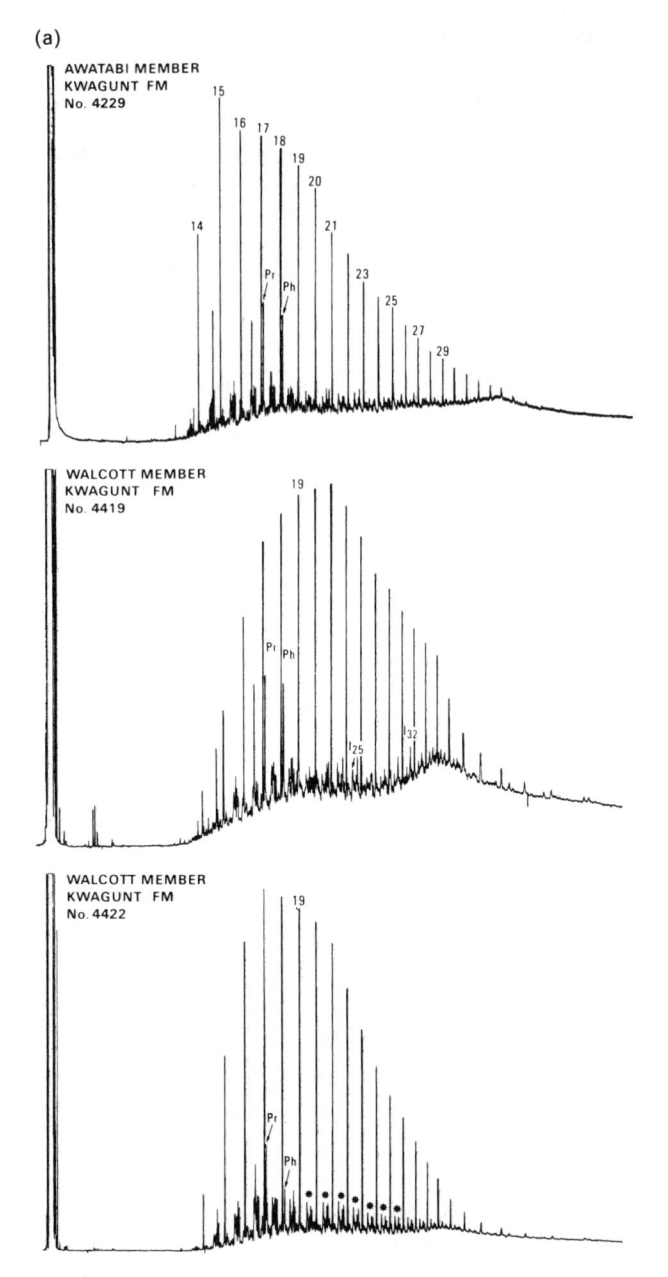

Figure 3.3.10. Gas chromatographic traces (FID response) showing saturated hydrocarbon distributions in samples from the Kwagunt (a) and Galeros (b) Formations, Chuar Group, Grand Canyon, Arizona. Peak identifiers are as in Figure 3.3.2.

oids and monomethyl branched alkanes and also to the chain length and carbon number predominances. Sedimentological features at the base of the Walcott Member include the presence of minor evaporite minerals and solution collapse structures indicating likely episodes of hypersalinity during deposition. This is consistent with aspects of the hydrocarbon distributions, including the slight even over odd predominance of the n-alkanes, low Pr/Ph ratio, and the high relative abundances of extended acyclic isoprenoids seen in many of the samples, and with observations made on numerous Phanerozoic sediments (e.g., McKirdy et al. 1983; ten Haven et al. 1985, 1988). C_{20+} n-alkanes are particularly abundant, and kerogen and bitumen fractions are somewhat enriched in ^{13}C (about $-24‰$).

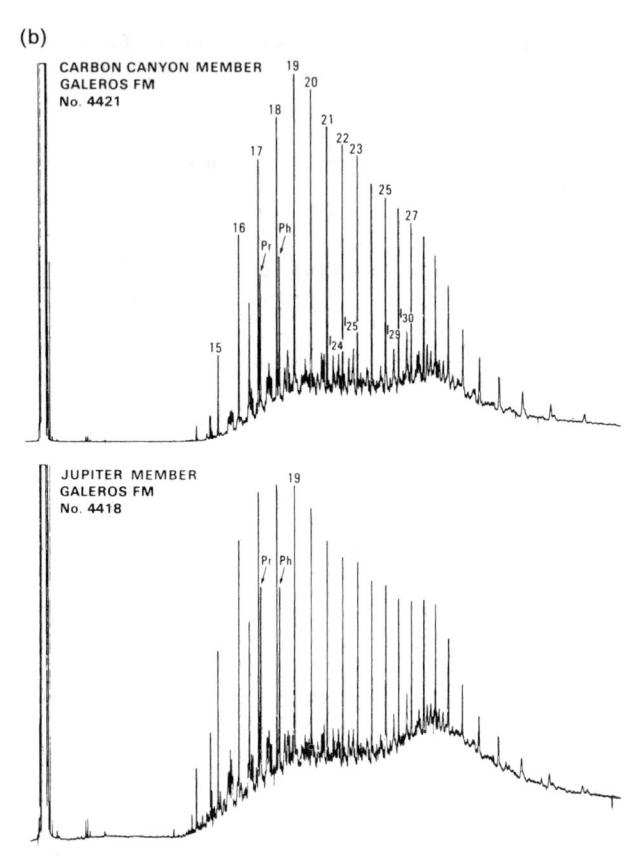

The most striking features of the hydrocarbons from the base of the Walcott Member are found in the compositions of steranes and triterpanes (Summons et al. 1988a; this study). The steranes, which are in high relative abundance in several of the samples, are unlike any previously reported assemblage. Side-chain alkylation at C-24 is restricted to isomers of 24-methyl cholestane, present in only 5% of the abundance of the equivalent cholestanes. The 24-ethyl steranes, which are the dominant pseudohomologues in most of the sediments described above, were present only in trace amounts in several samples from this horizon. Moreover, steranes with additional methyl groups in ring A show the same characteristic C-24 alkylation patterns as observed in the desmethyl steranes. They comprise mostly 3β-methyl steranes, a minor component of 2α-methyl steranes, and with 4α-methyl steranes absent. These features are shown in Figure 3.3.11, where the main compounds are isomers of cholestane and a C_{26} sterane with probable contraction of the side-chain. Such an assemblage of steroids is unlikely to arise from a diverse community of organisms. The only extant algae likely to show an analogous pattern of desmethyl sterols belong to the Rhodophyta (e.g., Patterson 1971; Ferezou et al. 1974; Schmitz 1983). As mentioned above, the origins of 3β-methyl and 2α-methyl steranes are not presently known. However, the observations made here, where there is a predominance of C_{27} steranes, and in those sediments with predominances of C_{29} steranes, confirm that the C-24 alkylation patterns of the 3β-methyl steranes are strongly aligned to those of the desmethyl steranes.

Pentacyclic triterpanes consist of the common pseudohom-

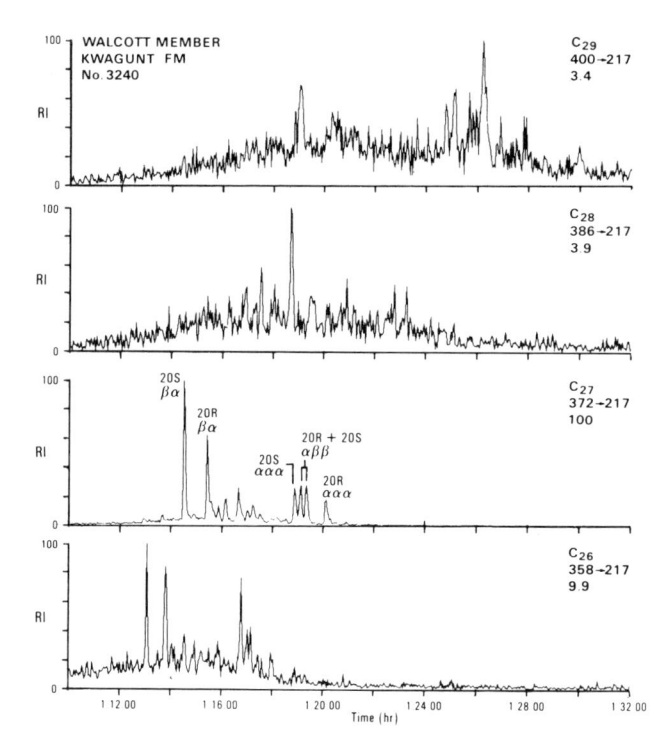

Figure 3.3.11. MRM chromatograms showing the sterane and methyl sterane distribution of a sediment extract (BMR no. 3240 = PPRG no. 1095) from the base of the Walcott Member, Kwagunt Formation, Chuar Group. Peak assignments are as in Figure 3.3.3.

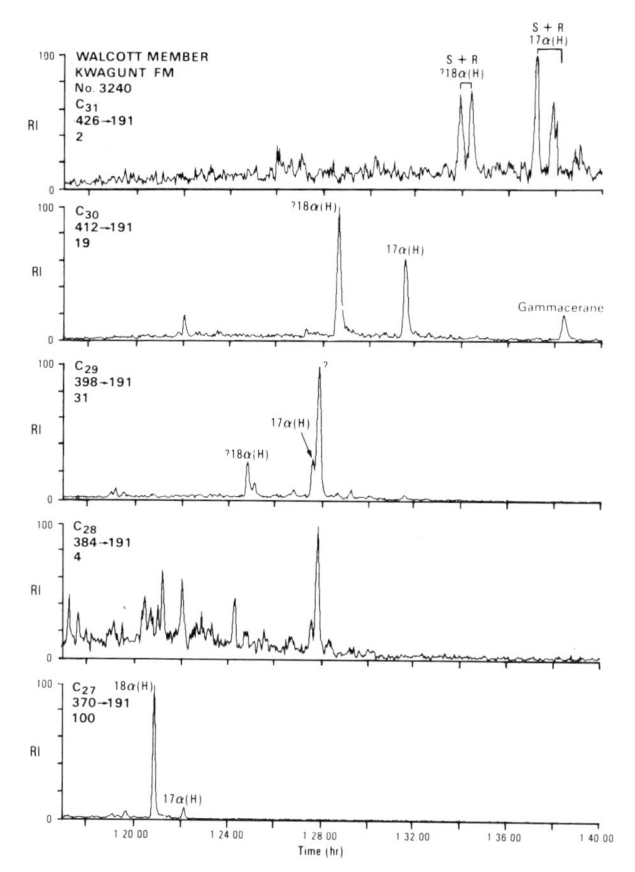

Figure 3.3.12. MRM chromatograms showing the triterpane distribution of a sediment extract (BMR no. 3240 = PPRG no. 1095) from the base of the Walcott Member, Kwagunt Formation, Chuar Group. Peak assignments are as in Figure 3.3.4.

ologous series of 17α(H), 21β(H)-hopanes and a series of putative 18α(H)-neohopanes as shown in Figure 3.3.12. The latter compounds had not been reported in ancient sediments prior to their recognition in high abundance in this unit, although they have since been found as the major hopane series in samples from the Rodda Beds (see above) and as significant components of the Nonesuch Formation and McArthur and Roper Group sediments. The same compounds have now been shown to occur in low abundance in some Phanerozoic sediments (Summons, unpubl.), where their presence is revealed by the high specificity of MRM analysis. The structural difference between 17α(H), 21β(H)-hopanes (e.g., Fig. 3.1.3, **19**) and the putative 18α(H)-neohopanes is thought to correspond to transposition of a single methyl substituent from C-18 to C-17. Although further work is needed to confirm this point, the difference between these hopanoid types is almost certainly biosynthetic as opposed to diagenetic in its origins. Such a structural modification has not yet been reported in hopanoids from any specific contemporary organism, but it seems likely that they are the products of bacteria, and could conceivably arise by a simple change in the mode of cyclization of squalene during hopanoid biosynthesis. Some of the Walcott Member sediments also contain gammacerane, a probable derivative of the protozoan triterpenoid tetrahymanol, which itself results from another modification to the biosynthetic pathway from squalene to the polycyclic triterpenoids (Ourisson et al. 1987; Rohmer et al. 1984) and is often observed in ancient sediments from highly saline environments (e.g., Shi Ji-Yang et al. 1982).

3.3.2G Nonesuch Formation, Mid-Continent Rift, Michigan

Analysis of the seep oil and cored sediments from the Nonesuch Formation (ca. 1000 Ma) within the region of the White Pine Mine (this study) has confirmed many of the characteristics reported in earlier studies (e.g., Eglinton et al. 1964; Barghoorn et al. 1965; Hoering 1967b; Imbus et al. 1988). Kerogen is in low abundance and mature according to Rock-Eval and elemental H/C determination. Saturated hydrocarbons from the bitumens resemble other Proterozoic samples in having a predominance of low MW n-alkanes and abundant n-alkylcyclohexanes and monomethyl alkanes (Fig. 3.3.13). GC-MS with MRM has confirmed that the sediments also contain low levels of triterpanes and steranes. The latter compounds lack the strong carbon number predominances evident in Late Proterozoic sediments and comprise mainly C_{27} to C_{29} diasteranes, as well as ααα and αββ steranes in a pattern consistent with the shale lithology of the unit. Samples of core from the base of the Nonesuch Formation, but remote from the White Pine mineralization, show better preservation of kerogen and a sterane composition (Fig. 3.3.14) similar to that of the oil and shale from the White Pine Mine (Pratt et al. 1989). The pentacyclic triterpenoids in these latter sediments contain a significant proportion of the extended series of putative C_{27} to C_{35} 18α(H)-neohopanes (Fig. 3.3.15), distinguishing them from

112 Proterozoic Biogeochemistry

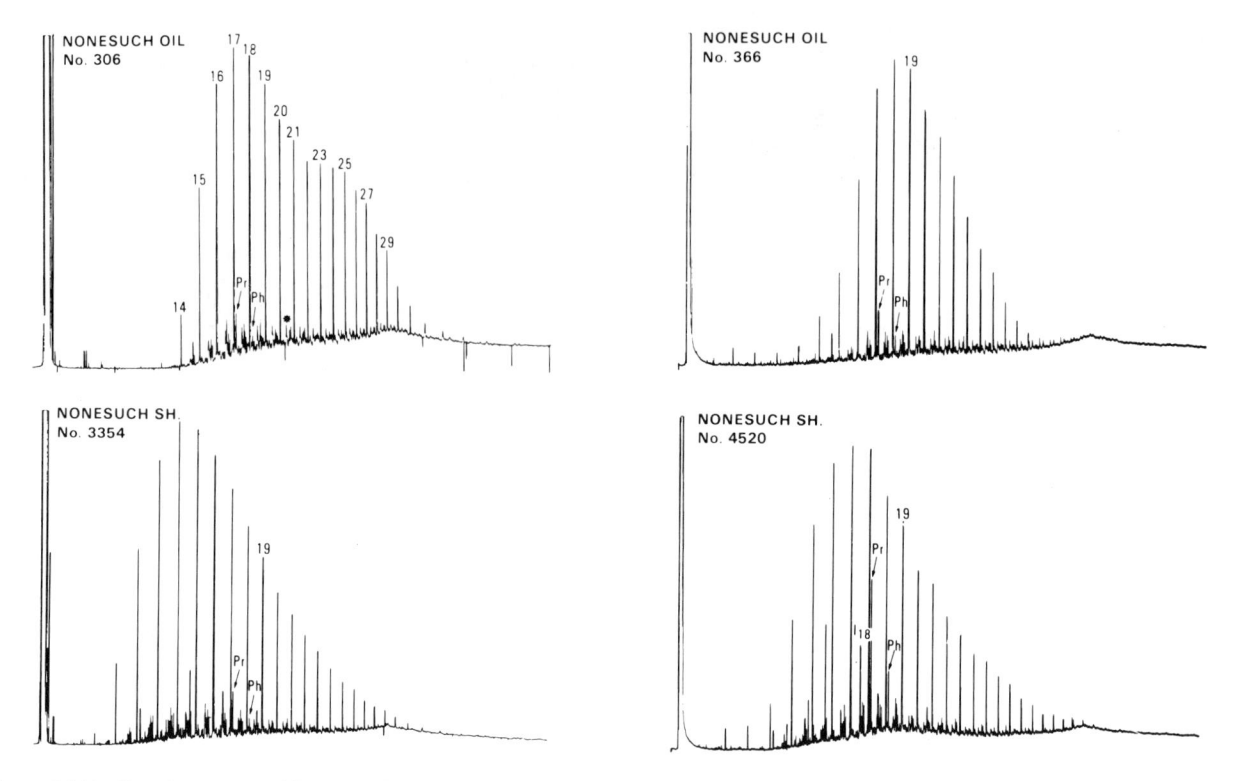

Figure 3.3.13. Gas chromatographic traces (FID response) showing saturated hydrocarbon distributions from sediment and seep oils from the Nonesuch Formation.

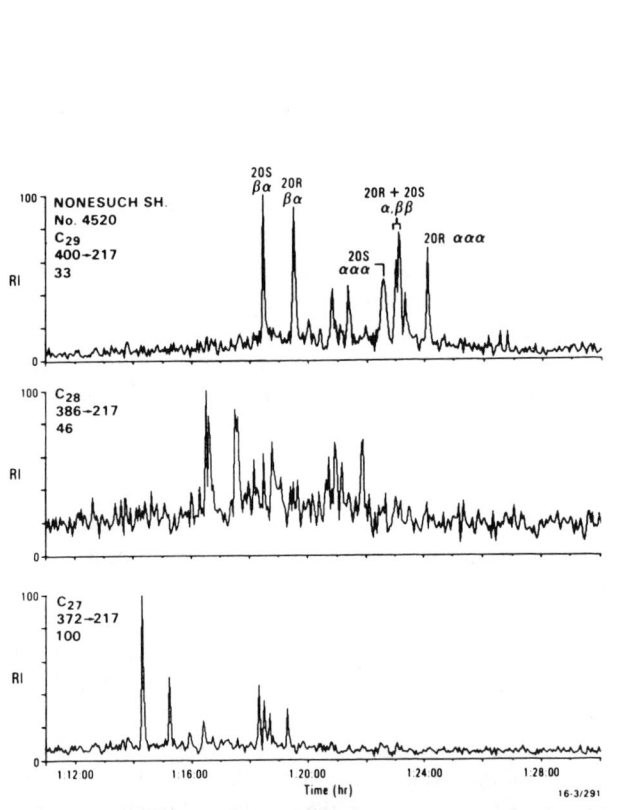

Figure 3.3.14. MRM chromatograms showing the steranes of sample BMR no. 4520 from the basal Nonesuch Formation. Peak assignments are as in Figure 3.3.3.

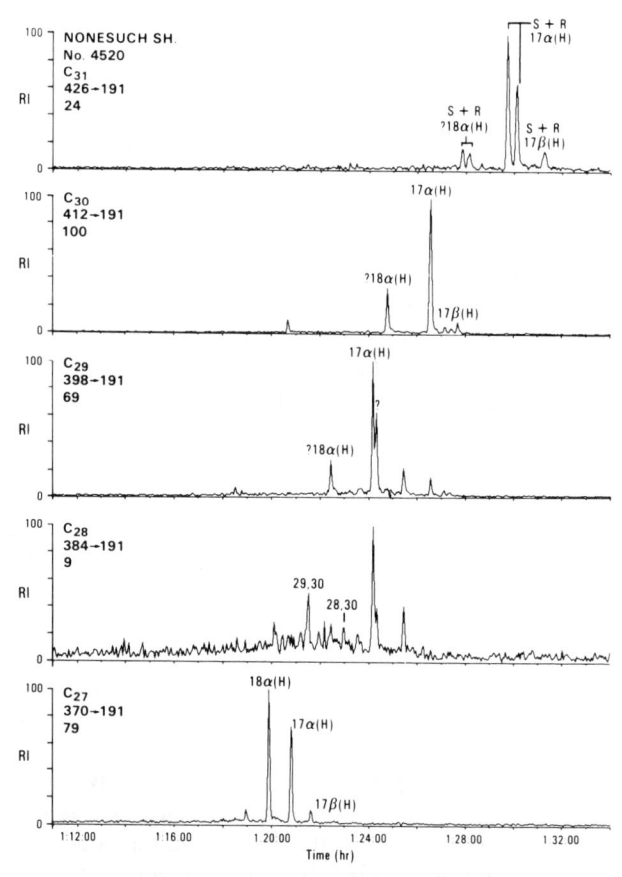

Figure 3.3.15. MRM chromatograms showing the triterpane distribution of sample BMR no. 4520 from the basal Nonesuch Formation. Peak assignments are as in Figure 3.3.4.

There are five potential petroleum source rocks at various stages of maturity (Jackson et al. 1986; Jackson et al. 1988; Crick et al. 1988) with extensive regional distribution, and representing a wide variety of paleoenvironmental settings. Some sediments of the Barney Creek Formation (ca. 1500 Ma) and the McMinn Formation (ca. 1300 Ma) are immature with respect to petroleum generation. These ages are conservative compared to those preferred by Jackson et al. (1988) and based on the data of Kralik (1982). Nevertheless, the sediments themselves represent the oldest presently known deposit of organic matter with a sufficiently mild thermal history in which we may expect significant preservation of syngenetic hydrocarbon biomarkers. The extent and degree of this preservation, summarized by the parameters in Table 3.2.3, is unprecedented in rocks of this age.

Sediments of the Barney Creek, Yalco, Velkerri, and

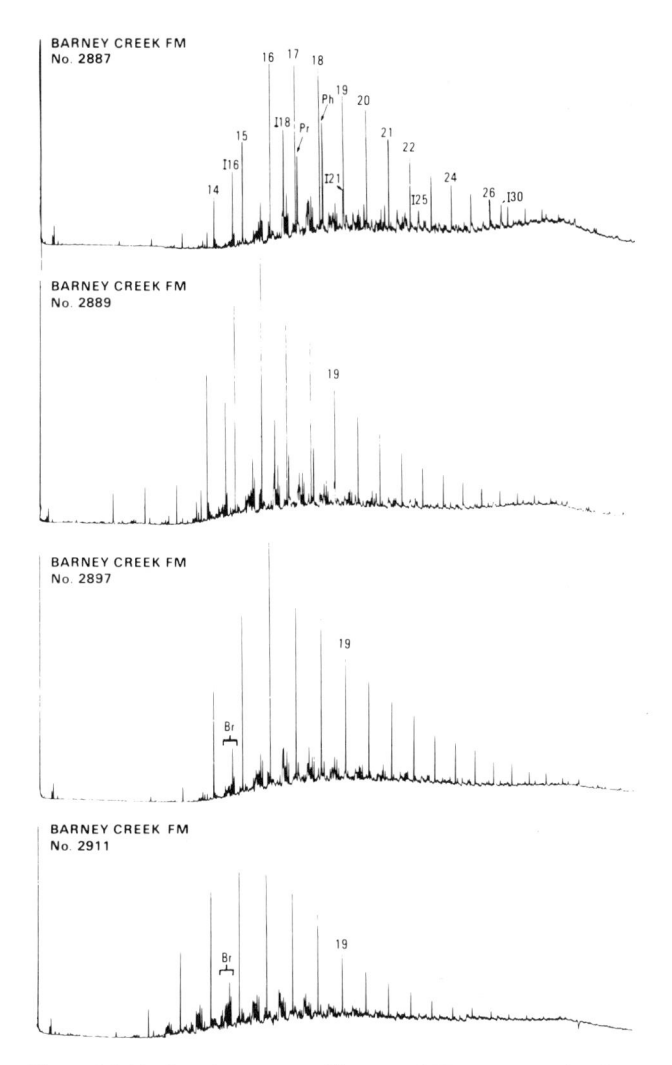

Figure 3.3.16. Gas chromatographic traces (FID response) showing saturated hydrocarbon distributions in a selection of samples from the Barney Creek Formation of the McArthur Group, McArthur Basin, northern Australia. The samples BMR no. 2887, 2889, 2897, and 2911 are from a single borehole GR 7, and correspond to increasing burial depth.

those of seep oil and shale from the White Pine Mine and suggesting lateral organic facies variations.

3.3.2H McArthur and Roper Groups, Northern Australia

The sedimentary sequence in the McArthur Basin ranges in age from approximately 1800 to 1400 Ma and comprises four units with exceptional preservation of syngenetic organic matter. The geology (e.g., Plumb et al. 1980), micropaleontology (e.g., Muir 1974, 1983; Oehler 1977, 1978; Oehler et al. 1976; Peat et al. 1978), and petroleum potential (e.g., Muir et al. 1980) of these deposits were the subjects of limited but active research interest prior to a major interdisciplinary study carried out at the Bureau of Mineral Resources, Australia. This work comprehensively reviewed the stratigraphy and sedimentology, examined the distribution and preservation of organic matter and its potential as a petroleum resource, and established the presence of distinctive patterns of biomarkers in the sedimentary bitumens.

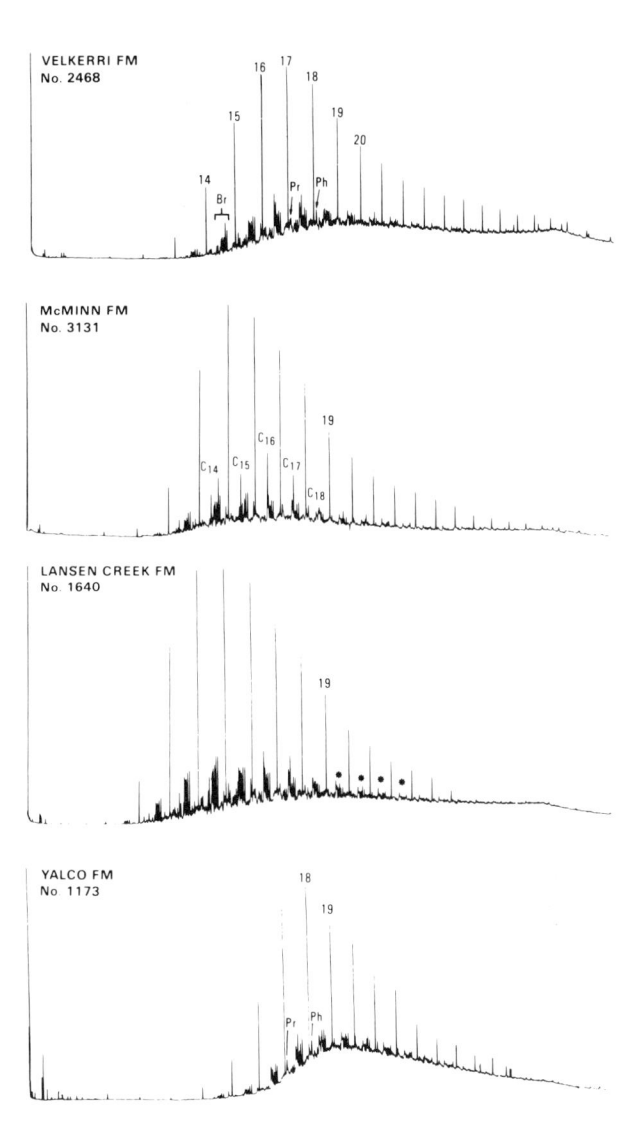

Figure 3.3.17. Gas chromatographic traces (FID response) showing saturated hydrocarbon distributions in a selection of samples from the Yalco Formation (BMR no. 1173), McArthur Group, and from the Velkerri (BMR no. 2468), McMinn (BMR no. 3131), and Lansen Creek (BMR no. 1640) Formations, Roper Group, of the McArthur Basin. The samples come from different boreholes and demonstrate the ubiquity of the pattern of abundant low molecular weight alkanes and monomethyl alkanes in this sequence.

McMinn Formations, despite their variety of lithologies and depositional environments and the 200 Ma range in their ages, show hydrocarbon distributions that are strikingly similar, both within themselves and to the other Proterozoic sediments discussed above (Summons et al. 1988b). The main sources of variation relate to the relative abundances of monomethyl alkanes and acyclic isoprenoids. Shallowly buried sections of the Barney Creek Formation, which comprise dolomitic siltstones, stromatolitic cherty dolostones, tuffs, and occasional evaporites, show abundant acyclic isoprenoids including the regularly branched C_{24} and C_{25} isomers and squalane, and have a Pr/Ph ratio of 1 or less (Fig. 3.3.16). These compounds may be the remains of archaebacterial membrane lipids. Most samples from elsewhere in the McArthur and Roper Groups show the characteristics of low abundances of isoprenoids and high relative abundances of n-alkylcyclohexanes and monomethyl alkanes (Fig. 3.3.17). Samples of the mature Velkerri Formation, which has the highest abundances of the latter compounds, lacked the mid-chain predominances of the equivalent compounds from the Siberian Platform oils. Low carbon number predominances were also characteristic of the n-alkylcyclohexanes, methyl-n-alkylcyclohexanes, and n-alkanes, and there is presently no known specific source for

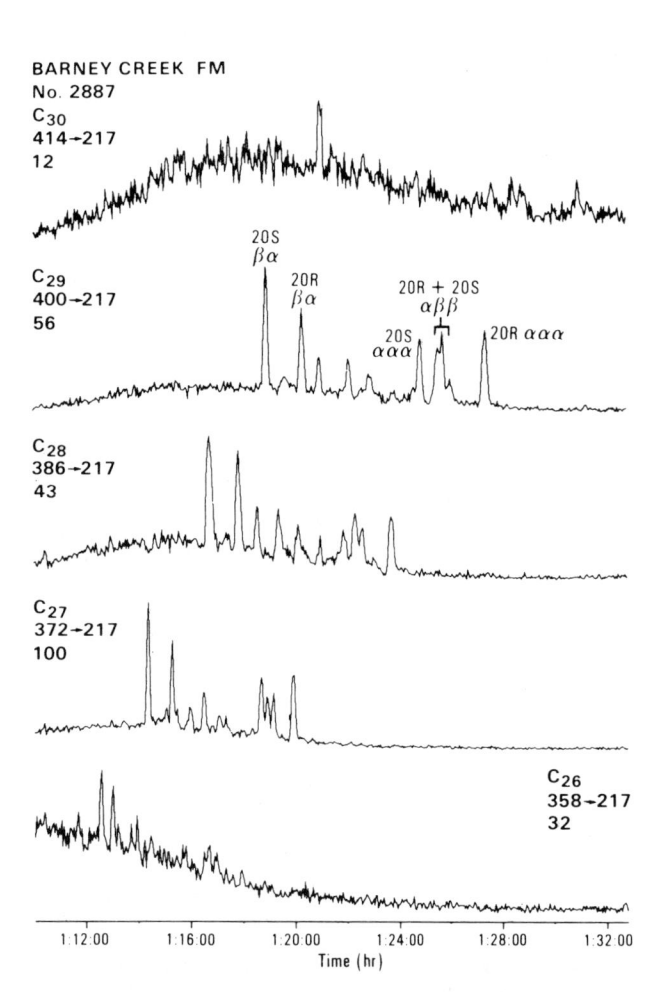

Figure 3.3.18. MRM chromatograms showing the sterane distribution of a sediment sample (BMR no. 2887) from the Barney Creek Formation. Peak assignments as in Figure 3.3.3.

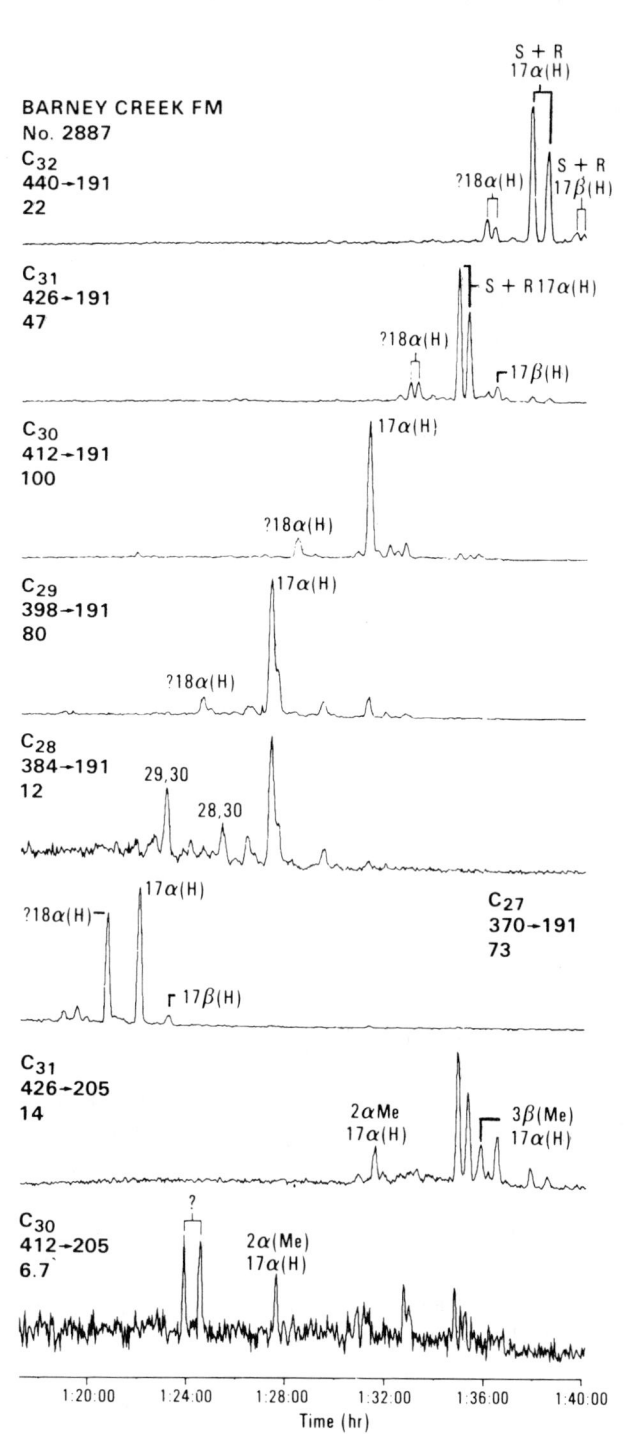

Figure 3.3.19. MRM chromatograms showing the triterpane distribution of a sediment sample (BMR no. 2887) from the Barney Creek Formation. Peak assignments as in Figure 3.3.4. Additional peaks show probable $2\alpha(Me)$ and $3\beta(Me)$ hopanes and unidentified C_{30} triterpanes.

these compounds, although their presence in high relative abundance seems to be characteristic of many ancient sediments and oils (e.g., Fowler and Douglas 1984; Fowler et al. 1986; Reed et al. 1986; Hoffmann et al. 1987). Sealed-tube and Curie-point pyrolysis experiments have shown that they can be produced directly from free fatty acids (Hoffmann et al. 1987) and, more significantly, from the resistant biopolymers of algae and cyanobacteria (Chalansonnet et al. 1988; Kadouri et al. 1988).

Pentacyclic triterpanes and steranes were present at all stratigraphic levels, demonstrating the inputs of membrane lipids from organisms with both prokaryotic and eukaryotic affinities. The triterpanes of the Barney Creek Formation, shown in Figure 3.3.18, comprised minor 2α-methyl hopanes, putative 18α(H)-neohopanes, and some previously unencountered species in addition to the normal series of extended 17α(H), 21β(H)-hopanes. Steranes, the normal series of C_{27} to C_{29} isomers (Fig. 3.3.19), with a distribution similar to that seen in the Nonesuch Formation, and without strong carbon number predominances as is seen in many of the Late Proterozoic units described above, were usually in low abundance, even in the least mature samples.

3.4

The Carbon-Isotopic Record

HARALD STRAUSS   DAVID J. DES MARAIS   J. M. HAYES
ROGER E. SUMMONS

The abundances of ^{13}C in marine carbonate and in fresh organic material are determined by (i) relative rates of immobilization of oxidized and reduced carbon in sediments and (ii) isotopic fractionations imposed by the biological carbon cycle. The record of variations in these abundances is, therefore, of great interest. Much would be learned about the development of the global environment and the evolution of the biota if ancient isotopic compositions could be reconstructed with perfect accuracy. Present interpretations, however, must be based on available data. Isotopic analyses of crudely defined phases such as "total organic carbon" are plentiful. Although of varying significance, these results provide the foundation of the carbon-isotopic record in Proterozoic materials, and will be considered first. Results of analyses of petrographically defined carbonate phases or of specific organic materials are more dependably informative, but rarer. These will be used to evaluate the record estimated from cruder measurements, and it will be shown that broad features are clear enough that significant conclusions can be drawn.

3.4.1 Total Organic Carbon and Whole-Rock Carbonate

3.4.1A Organic Carbon

Results of 687 analyses described in the literature are summarized in the middle columns of Table 3.4.1. For comparison, results of 541 new analyses undertaken in the course of this work are summarized in the last columns of Table 3.4.1.

The relative abundance of ^{13}C in organic carbon is generally lower in shales than in carbonates. This may be due to paleoenvironmental factors. For example, Lewan (1986) has suggested that ^{13}C is depleted in the organic matter of many Phanerozoic black shales because stratified water bodies developed in their environments of deposition, resulting in fixation of ^{13}C-depleted CO_2. Conversely, many carbonates may have formed under conditions of enhanced salinity, with biological factors favoring decreased isotopic fractionation (see Section 6.8). However, because concentrations of organic carbon average three- to ten-fold higher in shales than in carbonates (Tables 3.2.1 and 3.2.4), it is likely that preservational factors are at least partly responsible for the apparent depletion of ^{13}C in Proterozoic shales. Independent of lithology, Precambrian rocks with higher contents of organic carbon are systematically depleted in ^{13}C relative to organic-carbon-poor rocks (Hayes et al. 1983). This trend is likely due to effects of secondary processes in the depositional environment, which *can* result in enrichment of ^{13}C in residual organic matter (Hayes et al. 1989), and of thermal alteration, which generally *does* (Section 3.1.6).

In spite of their variety, the data in Table 3.4.1 are not yet so well balanced and globally inclusive that they do not give special weight to a few, heavily studied units. For example, the notable depletion of ^{13}C in Middle Proterozoic shales studied by others (middle columns, Table 3.1.4) is associated primarily with numerous reports of analyses of exceptionally well preserved organic material from the McArthur Basin, Australia. Table 3.4.1 is, thus, a scoreboard reflecting progress on recon-

Table 3.4.1. *Isotopic composition of total organic carbon in sedimentary rocks of Proterozoic Age*[a].

		Prior Reports[b]		This Work[c]	
Era	Lithology	δ^{13}C ‰ vs. PDB	N	δ^{13}C ‰ vs. PDB	N
Late	Shale	-27.1 ± 5.0	124	-26.4 ± 4.2	110
Proterozoic	Carbonate	-23.3 ± 6.3	111	-26.3 ± 4.1	71
	Chert	-24.8 ± 4.1	14	-24.7 ± 5.4	76
Middle	Shale	-30.4 ± 3.0	60	-27.1 ± 5.0	55
Proterozoic	Carbonate	-24.6 ± 4.2	15	-26.9 ± 2.2	23
	Chert	-28.1 ± 1.9	6	-28.9 ± 1.9	43
Early	Shale	-27.5 ± 4.6	175	-29.8 ± 4.3	32
Proterozoic	Carbonate	-25.0 ± 7.5	114	-27.0 ± 7.0	49
	Chert	-25.9 ± 7.3	53	-31.0 ± 4.7	82
	BIF[d]	-29.6 ± 7.7	15		

[a]Indicated uncertainties are standard deviations of populations of N observations.
[b]Detailed results and citations are given in Table 17.3.
[c]Detailed results are given in Table 17.1.
[d]BIF = banded iron-formation.

struction of the Proterozoic isotopic record, not a definitive summary.

A more detailed summary of new results, separated by stratigraphic unit and arranged in order of increasing age, is provided in Table 3.4.2, which also includes results of 81 new analyses of *Archean* samples. All of these results will be discussed in more detail below, after additional information relevant to their interpretation has been introduced.

3.4.1B Carbonate Carbon

There have been many reports of isotopic analyses of Proterozoic carbonates. Only a few new analyses have been undertaken in the course of this project, and their results are summarized in Table 3.4.3. Bedded limestones from the Early Cambrian part of the Chapel Island Formation, Newfoundland, Canada (Members 2 and 4), have $\delta^{13}C$ values near $-1‰$ (mode 1, Table 3.4.3). Calcareous nodules from the same unit (mode 2) are more strongly depleted in ^{13}C, with some δ values as low as $-23‰$, and clearly reflect incorporation of CO_2 derived from oxidation of organic matter. Values of $\delta > 0$ are common in Late Proterozoic carbonates (e.g., Kaufman et al. 1990) and this phenomenon was observed in samples of the Reed Dolomite from the Precambrian-Cambrian boundary section in the White Inyo Mountains, California. Samples of kutnahorite, $Ca(Mn, Mg)(CO_3)_2$, from the Early Proterozoic Hotazel Formation, South Africa, were strongly depleted in ^{13}C. Chemical and mechanical separation of individual carbonate phases and subsequent isotopic analysis revealed no systematic patterns of isotopic abundance, suggesting that all phases incorporated CO_2 derived from oxidation of organic material (N. J. Beukes, pers. comm. to H. S.). Isotopic compositions observed for carbonates from two "S bands" in the Dales Gorge Member of the Early Proterozoic Hamersley Banded Iron Formation, Western Australia, were similar to those previously observed for other beds in that unit (Becker and Clayton 1972; Baur et al. 1985).

Results of previous isotopic analyses of Proterozoic carbonates are listed in detail in Table 17.3.

3.4.1C Overview of the Record

Results of the present work and of prior analyses are compiled by stratigraphic unit in Table 17.4 and graphically summarized in Figure 3.4.1. The appearance of the record may be distorted by temporal misplacements. The assignment of ages to sedimentary rock units usually depends on stratigraphic correlations that may be in error and which are always subject to refinement. For nearly all Precambrian units, the lack of biostratigraphic control means that relative ages of widely separated units are, at present, determined entirely by geochronometric techniques. Materials suitable for application of these techniques might be distant–both temporally and geographically–from the sedimentary units of interest. In some cases, very large uncertainties may result. Post-depositional redistribution of parent or daughter isotopes can lead to inaccurate assignment of ages, a point of particular concern wherever ages have been assigned on the basis of a single geochronometric measurement or where discordant ages have been obtained. Conceivable refinements of ages are, however, not likely to change the overall form of this record. In particular, the observed distribution of isotopic abundances, in which any given time interval is likely to be characterized by a *range* of δ values, immediately suggests that multiple processes have shaped this record of whole-rock isotopic compositions.

The isotopic variations shown in Figure 3.4.1 represent multiple signals corrupted by unknown levels of noise. The noise includes all of the temporal uncertainties noted above as well as errors of measurement. The signals carry information about relative rates of burial of organic and inorganic carbon (f_{org}, Section 3.1.5), isotopic fractionation in the biological carbon cycle (Δ_B, Section 3.1.5), and –unfortunately–post-depositional alteration of initial isotopic compositions. Three signals must be resolved, but only two lines of information, the carbon-isotopic records in carbonates and in organic material, are evident. Additional information is required. For organic materials, isolation of kerogen and determination of the H/C ratio allows assessment of the extent of degradation and, therefore, recognition of samples in which the signal of alteration is absent. For carbonates, the extent of diagenetic alteration of primary compositions can be estimated from patterns of the distribution of ^{13}C, ^{18}O, and trace metals among petrographically distinct phases.

3.4.2 Isotopic Compositions of Closely Defined Phases

3.4.2A Carbonates

Processes of diagenetic stabilization can alter isotopic compositions of carbonates. Such alterations have been extensively studied in Phanerozoic materials, in which Veizer and coworkers (e.g., Veizer et al. 1986) have examined correlations between changes in carbon- and oxygen-isotopic compositions and changes in concentrations of trace metals. Similarly, Lohmann and coworkers (e.g., Given and Lohmann 1986) have reconstructed the isotopic evolution of limestones by combining petrographic techniques with isotopic analyses of samples prepared by micro-drilling. From studies like these has come recognition that post-depositional shifts in carbon-isotopic compositions can be substantial, even in the absence of thermal effects. When special care has been taken to reconstruct primary isotopic compositions, significant revisions of the Phanerozoic record of isotopic abundances in carbonates have resulted (e.g., Popp et al. 1986).

The first selective analyses of petrographically distinct phases in Precambrian materials were reported by Tucker (1982), who observed that sparry phases were depleted in ^{13}C by 2 to 3‰ relative to algal laminites and micrites in samples of the Late Proterozoic Beck Springs Dolomite (Death Valley, California). More recently, considerable evidence has accumulated that diagenetically produced carbon-isotopic shifts in Proterozoic carbonates tend to be markedly smaller than those observed in Phanerozoic materials. For example, Fairchild and Spiro (1987) have found that $\delta^{13}C$ values of nonluminescent micrites in Late Proterozoic carbonates from northeastern Spitsbergen are generally within 1‰ of whole-rock values earlier reported by Knoll et al. (1986). Aharon et al. (1987) observed that whole-rock carbon-isotopic compositions of Late Proterozoic dolomites from the southern margin of the Himalayan Belt did

Table 3.4.2. *Abundance and isotopic composition of total organic carbon in Proterozoic stratigraphic units.*

Unit	Age (Ma)	Lithol.[a]	TOC (mg C/g) mean	TOC (mg C/g) range	δ¹³C, ‰ vs. PDB mean	δ¹³C, ‰ vs. PDB range	N
Chulaktau Fm.	550	CH	2.6	1.8–3.7	−36.0	−36.2 − −35.8	3
Wyman Fm.	550	SILT	1.7	0.93–2.3	−13.1	−13.4 − −12.9	4
Chapel Island Fm.	560	SILT	0.11	0.02–0.45	−26.5	−32.2 − −17.9	45
Julie Mbr.	600	CARB	0.17	0.14–0.19	−26.7	−28.3 − −24.3	3
Yudoma Fm.	600	CH	1.4		−32.6		1
Cijara Fm.	610	SH	1.2		−30.0		1
Pertatataka Fm.	610	CH	0.11		−27.9		1
Estenilla Fm.	620	CARB	1.0	0.80–1.2	−19.9	−27.1 − −12.7	2
Fuentes Fm.	620	SH	0.78		−27.6		1
Khatyspyt Fm.	620	SH	210.		−36.4		1
		CARB	4.9		−37.1		1
Pusa Fm.	620	SH	1.6		−31.0		1
Timperley Shale	630	SH	0.29	0.28–0.30	−25.9	−26.0 − −25.7	2
Boonall Dolomite	640	CARB	0.09		−31.2		1
Egan Fm.	640	CARB	0.24		−29.5		1
Chichkan Fm.	650	CH	0.16	0.15–0.16	−28.9	−29.0 − −28.9	2
Sheepbed Fm.	650	SH	5.5	1.4–10.2	−28.9	−30.0 − −26.1	10
Jarrad Sandstone	675	CARB	0.07		−25.7		1
Biri Fm.	700	CARB	23.	16.0–34	−29.3	−32.1 − −27.5	4
Keele Fm.	700	CARB	0.54	0.06–1.4	−23.8	−26.7 − −22.2	10
Loquan Fm.	700	SH	1.3	0.30–3.8	−27.4	−30.0 − −18.9	11
Twitya Fm.	700	SH	1.0	0.08–2.4	−28.7	−32.2 − −26.8	12
Min'yar Fm.	740	CH	0.78	0.13–1.4	−29.4	−30.6 − −28.1	6
Coppercap Fm.	750	CARB	2.2	0.02–14	−28.6	−33.2 − −21.8	19
Tapley Hill Fm.	750	SILT	5.8	2.6–11	−23.7	25.1 − −22.5	3
Woocala Dolomite	750	SILT	3.7	2.0–5.0	−30.4	−31.1 − −29.3	4
Areyonga Fm.	760	CH	0.10	0.08–0.15	−26.8	−29.2 − −23.8	6
Shezal Fm.	770	QUAR	0.12		−26.6		1
Skillogalee Dolomite	770	CH	0.46	0.08–1.2	−20.1	−29.1 − −16.6	21
River Wakefield Subgp.	775	CH	0.88	0.34–1.5	−14.3	−17.6 − −11.0	3
		SH	0.31	0.30–0.33	−19.1	−19.8 − −18.4	3
		CARB	0.33	0.27–0.38	−18.1	−19.6 − −16.6	2
Visingsö Beds	775	SH	1.6	0.38–5.4	−26.5	−28.7 − −22.0	8
		CARB	1.3	0.46–2.2	−28.1	−28.3 − −27.9	2
Auburn Dolomite	780	CARB	0.55		−19.3		1
Rhynie Sandstone	785	CH	3.1		−17.6		1
Jiudingshan Fm.	800	CH	0.55		−23.6		1
Little Dal Fm.	800	CARB	0.34	0.02–1.1	−26.4	−33.0 − −20.2	11
		CH	0.40	0.11–0.69	−26.1		2
		SH	7.1	0.37–6.9	−25.6	−26.6 − −24.8	5
Sayunei Fm.	800	SILT	0.06	0.06–0.06	−16.7	−17.0 − −16.4	2
Torridon Group	800	CONGL	0.08	0.04–0.10	−26.1	−27.0 − −25.6	3
Beck Spring Dolomite	850	CARB	0.35		−25.7		1
Bitter Springs Fm.	850	CARB	0.18	0.13–0.28	−28.8	−30.2 − −25.5	7
		CH	0.16	0.06–0.45	−25.8	−29.7 − −21.7	29
Kingston Peak Fm.	850	CARB	11.	6.6–16	−22.3	−24.2 − −19.9	4
Walcott Mbr.	850	SH	20.	12–30	−26.9	−27.7 − −26.4	3
		CH	3.1	0.52–8.1	−26.7	−27.0 − −26.1	3
		CARB	18.	17.0–18	−26.5	−26.5 − −26.5	2
Zhangqu Fm.	850	CH	0.52		−29.4		1
Awatubi Mbr.	862	SH	7.4	17–18	−29.5	−32.0 − −27.0	2
Carbon Canyon Mbr.	900	SH	0.17		−24.4		1
		CARB	0.32		−29.1		1
		CH	0.96		−26.1		1
Shorikha Fm.	900	CH	0.38		−29.1		1
Jupiter Mbr.	930	SH	6.6	2.9–11	−24.6	−30.4 − −18.5	5

Table 3.4.2. (Continued).

Unit	Age (Ma)	Lithol.[a]	TOC (mg C/g) mean	TOC (mg C/g) range	$\delta^{13}C$, ‰ vs. PDB mean	$\delta^{13}C$, ‰ vs. PDB range	N
Jupiter Mbr.	930	CARB	0.09		−21.9		1
Katav Fm.	940	CARB	0.55	0.13–0.97	−28.8	−29.1−−28.5	2
Red Pine Shale Fm.	950	SH	1.8	1.5–2.2	−16.9	−17.1−−16.8	2
Tanner Mbr.	950	SH	3.7	1.9–5.0	−22.0	−24.8−−19.5	6
Bederysh Mbr.	1,000	SH	1.3		−30.0		1
Sukhaya Tunguska Fm.	1,000	CH	0.18		−26.5		1
Allamoore Fm.	1,050	CH	0.44		−26.8		1
Nonesuch Shale	1,055	SH	1.9	0.09–6.9	−29.9	−33.9−−26.6	7
		CARB	7.9	1.1–15	−31.9	−32.5−−31.3	2
Tieling Fm.	1,185	SH	0.97		−31.7		1
		CARB	0.14		−24.5		1
Fengjiawan Fm.	1,190	CARB	0.08		−26.1		1
Duguan Fm.	1,200	CARB	0.09		−26.8		1
Avsyan Fm.	1,240	SH	3.5	0.86–7.9	−23.9	−29.5−−21.6	4
Hongshuizhuang Fm.	1,250	SH	8.3	7.3–9.4	−28.5	−33.4−−23.6	2
Xunjiansi Fm.	1,250	CARB	0.15		−25.7		1
		CH	0.05	0.05–0.05	−28.2	−28.3−−28.2	2
Revet Mbr.	1,260	CH	0.43		−30.1		1
Mount Shields Fm.	1,320	CARB	0.05		−26.3		1
Wumishan Fm.	1,325	SH	0.19	0.05–0.33	−28.4	−28.8−−28.0	2
		CARB	0.17	0.05–0.30	−28.3	−29.2−−27.4	4
		CH	0.42	0.12–1.7	−28.4	−30.4−−26.3	15
Rossport Fm.	1,339	CARB	0.42	0.10–0.85	−26.4	−27.3−−25.4	4
McMinn Fm.	1,340	SILT	11.	2.0–21	−32.8	−33.5−−31.7	9
Tukan Fm.	1,350	SH	1.6	1.6–1.7	−24.9	−25.6−−24.3	2
Yangzhuang Fm.	1,350	CH	0.21	0.13–0.29	−27.5	−28.3−−26.3	4
Bungle Bungle Dolomite	1,364	CARB	0.16	0.08–0.23	−24.6	−25.3−−24.0	2
		CH	0.08	0.08–0.08	−21.6	−25.2−−17.9	2
Helena Fm.	1,380	CARB	0.08		−26.1		1
Dismal Lakes Group	1,400	CH	0.65	0.11–1.4	−28.5	−30.5−−27.2	6
Greyson Shale	1,420	SH	1.2	0.82–1.6	−29.8	−31.1−−28.3	2
Gaoyuzhuang Fm.	1,425	CH	1.2	0.30–3.2	−30.9	−32.0−−29.9	8
Newland Limestone	1,440	CARB	0.51	0.21–1.0	−27.3	−31.0−−24.3	7
		SH	0.93	0.28–1.6	−24.7	−32.3−−13.6	15
Altyn Fm.	1,440	CARB	0.07		−26.4		1
Chamberlain Shale	1,450	SH	0.79		−32.0		1
Prichard Fm.	1,450	SILT	0.30		−23.9		1
		QUAR	0.93		−23.1		1
Barney Creek Fm.	1,500	SH	4.8		−32.6		1
Paradise Creek Fm.	1,650	CH	0.09		−28.8		1
Urquhart Shale Fm.	1,670	SH	9.2	0.27–16	−26.1	−28.5−−20.2	6
Vempalle & G.R. Fm.	1,700	CH	0.08	0.05–0.12	−27.8	−29.2−−26.8	3
Tuanshanzi Fm.	1,750	CARB	0.86	0.13–1.9	−30.5	−31.5−−29.5	3
Chuanlinggou Fm.	1,850	SH	3.5	0.76–5.8	−32.1	−33.1−−30.2	3
Koolpin Fm.	1,885	CH	1.2		−30.6		1
Fontano Fm.	1,900	SH	6.6	4.0–9.1	−28.3	−30.2−−26.4	2
Rove Fm.	1,900	SH	15.	0.84–23	−30.8	−32.0−−27.6	4
		CARB	0.30	0.08–0.51	−28.6	−31.7−−25.6	2
Spartan Group	1,900	SH	6.5	3.5–12	−29.9	−30.5−−29.4	3
Union Island Group	1,900	SH	21.		−39.4		1
Odjick Fm.	1,925	SH	0.22		−25.6		1
Rocknest Fm.	1,925	CARB	0.49	0.06–1.4	−19.2	−25.8−−12.0	13
		CH	0.71	0.53–1.0	−19.1	−24.7−−12.3	3
Tyler Fm.	1,950	CH	5.2	5.1–5.4	−31.4	−31.5−−31.2	2
		CARB	11.		−32.3		1
Kona Dolomite	2,000	CARB	0.06	0.03–0.10	−23.7	−27.7−−14.9	7
Lookout Schist	2,000	SH	100.		−20.6		1

The Carbon-Isotopic Record

Table 3.4.2. (*Continued*).

Unit	Age (Ma)	Lithol.[a]	TOC (mg C/g) mean	TOC (mg C/g) range	$\delta^{13}C$, ‰ vs. PDB mean	$\delta^{13}C$, ‰ vs. PDB range	N
Biwabik Iron-Fm.	2,090	CH	0.15	0.08–0.32	−29.0	−32.4 – −24.6	4
		CARB	0.14		−28.2		1
Gunflint Iron-Fm.	2,090	CARB	6.	0.15–29	−32.7	−34.4 – −30.9	10
		CH	1.6	0.07–11	−32.6	−35.2 – −18.3	57
		SH	16.		−31.9		1
Pokegama Quartzite	2,090	QUAR	0.15	0.08–0.22	−30.9	−32.7 – −29.1	3
Bar River Fm.	2,250	QUAR	0.07		−28.8		1
Bruce Fm.	2,250	SILT	0.04		−28.3		1
		CARB	0.04		−21.4		1
Espanola Fm.	2,250	SILT	0.17		−27.0		1
		CARB	0.16		−25.9		1
Gowganda Fm.	2,250	SILT	0.04	0.04–0.05	−27.6	−28.4 – −26.2	3
		CONGL	0.10		−28.4		1
Lorrain Fm.	2,250	QUAR	0.06		−28.9		1
Mississagi Fm.	2,250	QUAR	0.06		−29.0		1
Pecors Fm.	2,250	SILT	0.17		−36.2		1
Ramsay Lake Fm.	2,250	CONGL	0.18		−36.6		1
Serpent Fm.	2,250	QUAR	0.07		−29.3		1
Gordon Lake Fm.	2,259	QUAR	0.07		−26.7		1
Reivilo Fm.	2,300	SH	4.8	1.3–11	−37.0	−38.1 – −35.1	4
		CARB	0.67		−31.2		1
Rooinekke Fm.	2,300	CARB	0.05	0.03–0.07	−22.6	−26.7 – −18.5	2
Klipfonteinheuwel Fm.	2,325	CH	0.83	0.80–0.86	−37.6	−38.5 – −36.8	2
Kuruman Iron-Fm.	2,350	CH	0.38		−34.1		1
Malmani Subgp.	2,400	CH	0.22		−32.0		1
Matinenda Fm.	2,400	SILT	0.03		−26.1		1
		QUAR	0.07	0.05–0.09	−30.2	−31.4 – −29.0	2
McKim Fm.	2,400	QUAR	0.11	0.08–0.14	−31.9	−35.1 – −28.7	2
		SILT	0.10	0.06–0.14	−27.7	−30.5 – −25.2	4
Gamohaan Fm.	2,450	SH	22.	21.0–24	−39.1	−40.5 – −37.5	3
		CARB	1.5	0.45–2.6	−33.8	−36.3 – −32.2	11
		CH	0.65	0.52–0.82	−33.2	−36.2 – −26.5	4
Keewatin Group	2,500	CH	0.45	0.40–0.50	−19.7	−24.1 – −15.3	2
Jeerinah Fm.	2,650	SILT	5.23	1.3–9.2	−45.8	−48.6 – −42.9	2
Rietgat Fm.	2,650	CH	0.69		−43.1		1
Soudan Iron-Fm.	2,650	SH	0.23		−20.5		1
		CH	0.08	0.06–0.10	−27.6	−29.5 – −24.6	3
Ventersdorp Spgp.	2,650	SILT	0.27		−36.5		1
		CARB	0.08	0.07–0.08	−29.4	−30.9 – −27.8	2
Venterspost Fm.	2,670	CONGL	0.09		−31.9		1
		SH	0.08		−23.7		1
Nymerina Basalt	2,700	CARB	0.86		−53.2		1
Kameeldorns Fm.	2,705	CH	0.38		−32.5		1
Carbon Leader Mbr.	2,710	ANTHR	120.	97.0–130	−37.9	−38.8 – −37.0	2
		SILT	0.16	0.09–0.24	−27.9	−30.6 – −25.2	2
Government Subgp.	2,710	SH	1.4		−36.7		1
North Leader Mbr.	2,710	SILT	0.20	0.08–0.32	−28.4	−32.4 – −24.3	2
Witwatersrand Spgp.	2,710	SH	0.05		−24.9		1
Jeppestown Shale	2,720	SH	2.8		−37.3		1
Maddina Basalt	2,735	CH	0.54	0.18–0.89	−46.3	−46.7 – −45.9	2
Kuruna Siltstone	2,740	CARB	0.81		−55.8		1
Kimberley Shale	2,750	SH	0.08		−23.7		1
Tumbiana Fm.	2,750	SH	1.2	0.63–1.8	−48.6	−60.9 – −36.2	2
		CARB	0.36	0.05–0.49	−46.1	−53.5 – −28.5	7
		CH	0.43	0.26–0.56	−50.2	−52.2 – −47.8	3
Kylena Basalt	2,768	CARB	0.38		−37.4		1
Insuzi Group	3,000	CH	0.18	0.05–0.31	−22.9	−28.1 – −17.8	2

Table 3.4.2. (Continued).

Unit	Age (Ma)	Lithol.[a]	TOC (mg C/g) mean	TOC (mg C/g) range	$\delta^{13}C$, ‰ vs. PDB mean	$\delta^{13}C$, ‰ vs. PDB range	N
Gorge Creek Group	3,200	CH	0.08		−29.8		1
Cleaverville Fm.	3,300	CH	0.10		−26.3		1
		SH	0.56		−28.6		1
Kromberg Fm.	3,300	CH	2.03	0.15–4.6	−31.5	−36.5 − −26.2	9
Fig Tree Group	3,350	CH	0.75	0.14–2.2	−24.5	−35.4 − −26.9	8
Apex Basalt Fm.	3,400	CH	0.10	0.07–0.13	−27.0	−30.0 − −22.5	3
Towers Fm.	3,435	CH	0.75	0.08–2.1	−30.5	−36.9 − −18.1	9
Hooggenoeg Fm.	3,450	CH	2.64	1.8–3.4	−36.0	−38.6 − −32.4	5
Theespruit Fm.	3,550	CH	6.15	3.3–9.0	−15.5	−16.3 − −14.6	2

[a] ANTHR = Anthraxolite, CARB = Carbonate, CH = Chert, CONGL = Conglomerate, QUAR = Quartzite, SH = Shale, SILT = Siltstone.

not differ significantly from those of related dolomicrites. Micro-sampling of carbonates from the Rocknest Formation showed that fine-grained carbonate phases were generally enriched in ^{13}C relative to sparry phases, but the difference was only 0.1 to 0.3‰, and whole-rock δ values were close to those of the fine-grained phases (J. T. Hayes and J. M. Hayes, unpublished). Finally, Kaufman et al. (1990) have observed that carbon-isotopic compositions of nonluminescent micrites and of whole rock samples from Late Proterozoic Namabian carbonates are very well correlated and that, although $\delta^{13}C$ values for nonluminescent micrites are generally greater than those for whole-rock carbonates, the average difference is only 0.3‰.

It is not clear *why* diagenetic shifts of carbon-isotopic abundances should have been systematically lower for Proterozoic carbonates. Kaufman et al. (1990) speculate that, in the absence of biogenic carbonates, initial porosities of Proterozoic carbonates were lower than those of Phanerozoic materials. In that case, void-filling phases, even if they were isotopically exotic (i.e., Zempolich et al. 1988), might not make up a consequential fraction of whole-rock carbon.

Caution is in order. In spite of the existence of numerous reports of analyses, the carbon-isotopic record in Proterozoic carbonates has not been examined nearly as intensively or critically as that in Phanerozoic carbonates. The Proterozoic covers a time interval four times longer than that of the Phanerozoic. Sampling densities are low and there are numerous gaps in the record, some of which can never be filled. Very recent results, for example, suggest the possibility of a significant isotopic excursion in the Early Proterozoic (Baker and Fallick 1989a, 1989b). Thermally altered and isotopically unrepresentative samples (e.g., carbonates not formed in equilibrium with open marine carbonate) must be recognized and excluded. If it proves to be general, however, the tendency for diagenetic shifts to be smaller in Proterozoic carbonates offers hope that the whole-rock isotopic record is reasonably accurate.

3.4.2B Organic Carbon

If primary phases can be recognized and selectively analyzed in order to provide accurate values for initial isotopic compositions of carbonates, some analogous procedure might be possible for organic materials. The task would not be to select and micro-sample a petrographically defined mineral phase, but to identify a particular organic compound that preserved initial isotopic compositions. This approach has met with some success in studies of Phanerozoic sections (Hayes et al. 1987, 1989), but has not yet been attempted with Precambrian samples. It will certainly be of very restricted

Table 3.4.3. *New determinations of the isotopic compositions of Proterozoic carbonates.*

Unit	Age (Ma)	Lithology[a]	$\delta^{13}C$, ‰ vs. PDB Avg.	$\delta^{13}C$, ‰ vs. PDB Range	$\delta^{18}O$, ‰ vs. PDB Avg.	$\delta^{18}O$, ‰ vs. PDB Range	N
Chapel Island Fm. (Mode 1)	560	LS	−0.8	−3.1–0.6	−14.6	−17.5 − −11.2	4
(Mode 2)	560	LS	−19.3	−13.9 − −23.1	−18.8	−21.5 − −17.6	8
Reed Dolomite	570	DOL	1.7	1.0–2.6	−12.3	−13.4 − −11.7	5
Rocknest Fm.	1,925	DOL	1.2	0.8–1.7	−9.9	−11.0 − −8.1	9
Hotazel Fm.	2,250	LS	−9.3	−11.1 − −7.5	−12.8	−14.4 − −11.2	2
		DOL	−11.3		−12.0		1
Dales Gorge Mbr.	2,500	ND	−6.2	−7.3 − −5.1	−12.0	−12.7 − −11.2	2

[a] LS = Limestone, DOL = Dolomite, ND = Not Determined.

The Carbon-Isotopic Record

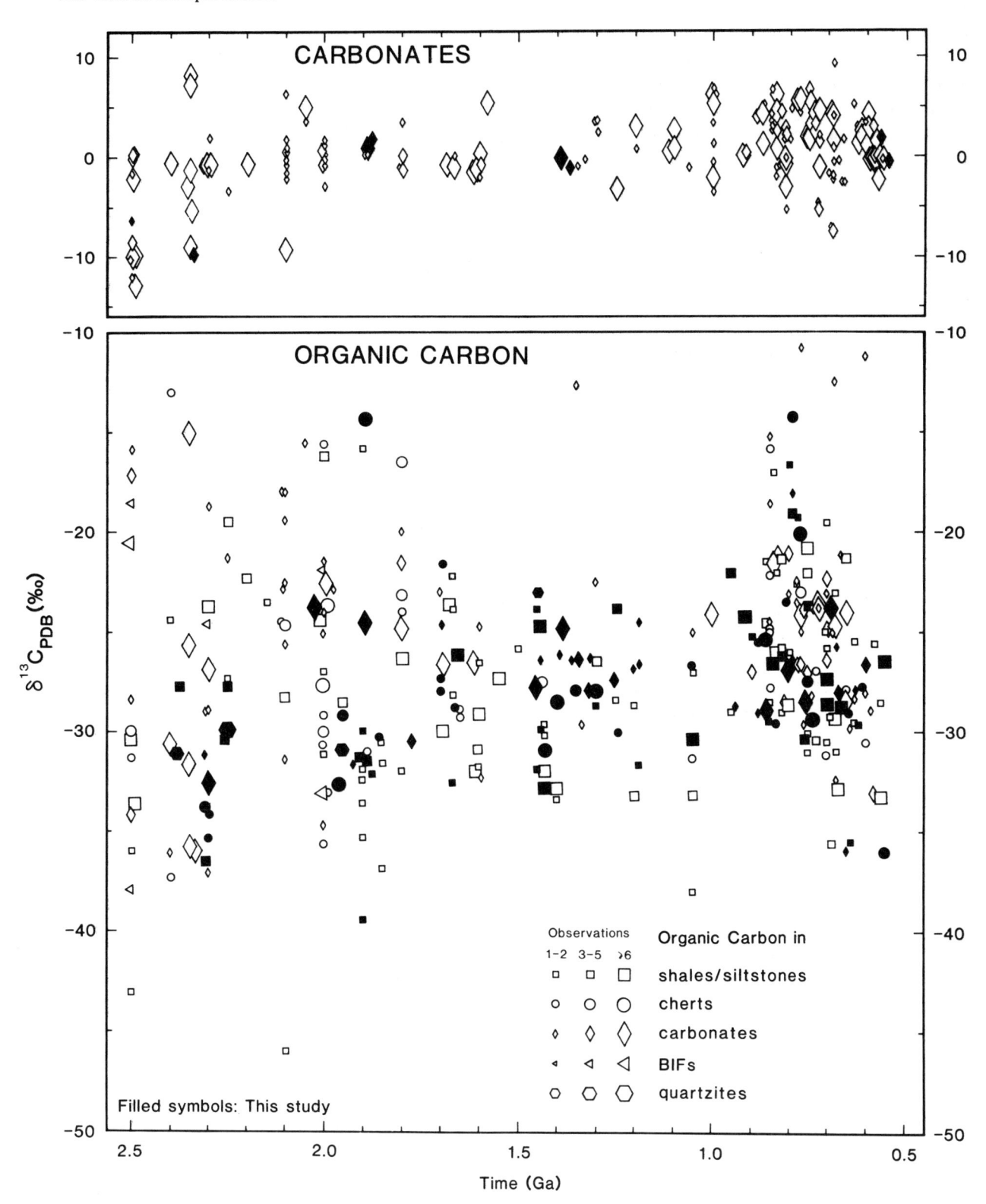

Figure 3.4.1. Average isotopic compositions of carbonate carbon and organic carbon in stratigraphic units of Proterozoic age. For listing of points and sources of data, see Table 17.4.

applicability in the Precambrian, where extractable organic matter is either confined to a few organic-rich, well preserved units, or is of uncertain origin, possibly representing younger materials that have migrated into the rock long after initial deposition. Specific organic molecules produced by chemical or pyrolytic degradation of kerogen may provide useful alternatives.

Although the organic-carbon isotopic record cannot be perfected, it can, at least, be edited. The task is to exclude from consideration organic materials that (i) may not be in place and/or (ii) may have been so severely altered that their isotopic compositions have been changed significantly. The first objective can be approached through analysis of kerogen rather than total organic carbon. Kerogens are immobile. A few may have accumulated increments of imported carbon at some point in their histories, but isotopic compositions of most should be minimally affected by addition of migrated materials. The second objective can be approached by noting *elemental*, as well as isotopic, compositions of kerogens. The carbon-isotopic composition of a kerogen can change only through loss or addition of carbon. In most settings, thermally driven loss of hydrocarbons is by far the most important process. Because loss of hydrocarbons also strongly affects the H/C ratio of the residual kerogen, driving it eventually to zero, the extent of hydrocarbon loss and, thus, of isotopic alteration, can be estimated from the H/C ratio. Based on earlier observations of relationships between isotopic shifts and H/C ratios (Hayes et al. 1983a, b), we will exclude from the kerogen isotopic record δ values derived from kerogens with H/C ratios of 0.2 or lower.

Results of new determinations of H/C ratios and of carbon-isotopic compositions in Proterozoic kerogens are summarized in Table 3.2.5. Results of prior studies are summarized in Table 17.9. Data from all sources are combined in Figure 3.4.2, which provides two views of the record of isotopic abundances in organic carbon. In Figure 3.4.2A, an outer envelope encloses all points representative of total organic carbon and an inner envelope encloses points representative of kerogens with H/C $\geq$ 0.2. The number of points within the latter (shown in Figure 3.4.2B) is small enough that its shape is likely to change as the variety of samples examined increases, but two points are unmistakable. First, the range of isotopic compositions in well-preserved kerogens is generally much smaller than that in total organic carbon. Second, for most of the Proterozoic, the δ values for well-preserved kerogens are near the minima observed for total organic carbon.

The observed contraction and shift of the isotopic range indicates that the last of the three signals described earlier – post-depositional alteration – has been resolved. For clarity, we will oversimplify the interpretation, focusing, for the moment, on the Early and Middle Proterozoic: because all well-preserved kerogens have relatively low δ values, we conclude that, initially, all sedimentary organic material had low δ values. Qualifications should be added. It is not required that *all* sedimentary organic material was depleted in ^{13}C. Estimates of the relative rates of burial of carbonate and organic carbon will not be seriously in error provided the δ value in well-preserved kerogens is representative of the global average for organic carbon being immobilized in sediments during a particular time interval. When geochemical accounts are being reckoned, "organic carbon being immobilized in sediments" is what matters, and that is certainly not the same thing as "all organic carbon in all conceivable environments." If organic material with δ values considerably higher or lower than those within the kerogen envelope existed, our view of the Proterozoic carbon cycle will be *incomplete*, but our interpretation need not be *erroneous*. In dealing with any geologic record, we must always consider that some evidence might be missing. If we knew that some pool of organic carbon with δ values well

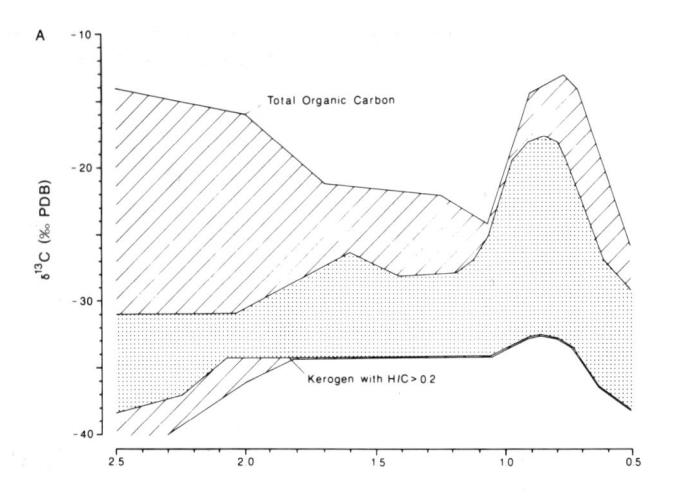

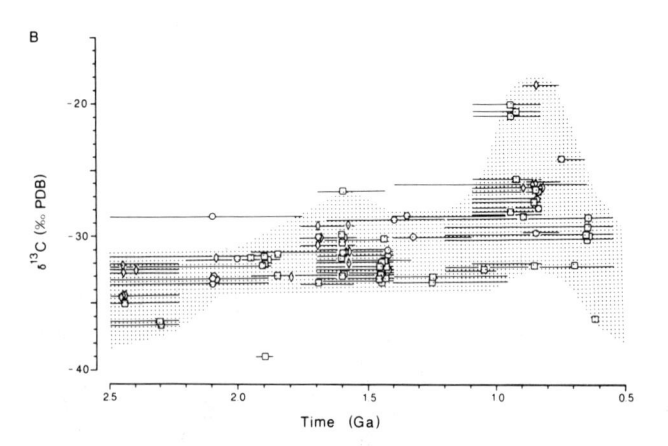

Figure 3.4.2. A. Ranges of isotopic compositions of samples of organic carbon from rocks of Proterozoic age. Outer envelope encloses δ values for all analyses of total organic carbon (Table 17.4). Inner envelope encloses only δ values for kerogens with H/C $\geq$ 0.20 (Table 17.10). B. View showing placement of points within kerogen envelope. Horizontal lines indicate maximum range of age.

outside the kerogen envelope had existed, that might provide some clues regarding carbon flows in Proterozoic environments. But we would still have the task of deducing biogeochemical pathways and mechanisms of isotopic fractionation relating abundant kerogens with δ values near $-32‰$, and lower, to carbonates with δ values near 0‰.

As shown in Figure 3.4.2A, the relationship between the envelopes enclosing δ values for total organic carbon and for well-preserved kerogens changes in the Late Proterozoic. Evidence for secular changes in isotopic compositions during this interval is well established. Knoll et al. (1986) observed coupled variations in the isotopic compositions of organic and carbonate carbon in rocks with ages of 900 to 550 Ma and reviewed numerous other reports indicating that the isotopic variations must have been global. Results of more recent analyses (summarized in Table 17.4) are consistent with that conclusion. Particularly notable is an enrichment of ^{13}C in marine carbonates, with δ values rising from ~ 0 in the Middle Proterozoic to $+5‰$ and higher at points within the Late Proterozoic. The shapes of the envelopes enclosing both the

TOC and the kerogen data sets are consistent with that change, and a broadening of the kerogen envelope is not unexpected. Shifts in the isotopic composition of marine carbonate occurred rapidly (Knoll et al. 1986). Because ages for the units included in Figure 3.4.2 are not known precisely, the envelope enclosing δ values must broaden to include the range of isotopic maxima and minima throughout the Late Proterozoic.

The fit of the kerogens inside the TOC envelope varies systematically. In the Early Proterozoic, the highest δ values in the TOC data set are 15‰ greater than those for coeval well-preserved kerogens. The gap decreases to about 6‰ in the Middle Proterozoic, and to almost nothing at about 850 Ma. The overall trend can be understood in terms of preservational effects. Rocks with longer crustal residence times are more likely to have been deeply buried for longer intervals, with consequent severe alteration of organic matter. The greatest enrichments of ^{13}C are expected – and observed – in the oldest rocks. The close agreement between maximum δ values for TOC and well-preserved kerogens in the Late Proterozoic cannot be readily explained, but is based on only a few points.

We conclude that the differences between the kerogen and total organic carbon isotopic records shown in Figure 3.4.2 can be understood in terms of preservational effects. In that case, the best view of isotopic abundances in Proterozoic organic carbon is that provided by well-preserved kerogens. We turn now to interpretation of that record.

3.4.3 The Proterozoic Carbon Cycle

3.4.3A Isotopic Fractionation and Development of the Ecosystem

The Early Proterozoic. If samples from iron-formations are excluded, δ values for carbonates in the time interval from 2.5 to 2.0 Ga are near 0 or +1‰. During this same interval, well-preserved kerogens have δ values ranging between -31 and -40‰. Accordingly, values of Δ_B are in the range 32 to 40‰. To consider the significance of this observation, we can begin by thinking in terms of modern, open marine environments. Biogeochemical relationships between carbonates and organic carbon in such systems have recently been considered by Hayes et al. (1989). The following points are critical:

(1) Fixation and reduction of carbon occur through the C_3 pathway, in which the key enzyme is ribulose bisphosphate carboxylase oxygenase ("rubisco").
(2) The dissolved CO_2 and bicarbonate utilized by primary producers as their carbon source are in isotopic equilibrium with the carbonate that is incorporated in the bulk of carbonate minerals.
(3) Isotopic fractionations associated with remineralization of organic carbon are small in comparison to those imposed by primary production.

Given the isotope effects associated with enzymatic carbon fixation catalyzed by rubisco ($\varepsilon = -27$‰) and with chemical equilibria mediating isotopic exchange between carbonates and dissolved CO_2 ($\varepsilon \approx 11$‰ at 25°), the maximum fractionation possible between carbonates and biomass of primary producer organisms would be 38‰ (for a discussion of isotopic relationships, see Hayes et al. 1989). Therefore, the largest values of Δ_B observed in the Early Proterozoic ($\Delta_B > 38$‰) could not be associated simply with burial of organic carbon not fractionated isotopically after primary biosynthesis.

In modern environments, primary photosynthate is not simply buried. Although processes of carbon fixation in primary producers exert initial control over the isotopic composition of organic carbon, that material is subsequently reworked by consumer communities that have evolved to utilize primary material as completely and efficiently as possible. The residues of those communities enter the sedimentary record along with whatever traces of primary material have escaped utilization, and isotope effects associated with consumer metabolism can thus play a significant role in controlling the ^{13}C content of the organic material eventually immobilized in sedimentary rocks. Hayes et al. (1989) have shown that isotope effects associated with consumer communities can *decrease* the isotopic difference between carbonates and residual organic material in environments apparently dominated by respiratory metabolism (respiration $\equiv$ transfer of electrons from organic material to inorganic electron acceptors such as O_2, NO_3^-, SO_4^{2-}). Conversely, there is evidence (Hayes et al. 1987; Boreham et al. 1989) that the isotopic difference between carbonates and organic material can be *increased* in environments where fermentative processes dominate remobilization of sedimentary carbon (fermentation $\equiv$ transfer of electrons between carbon compounds). Because they act to increase Δ_B, fermentative processes are of particular interest here.

Methanogenesis is an important component of terminal metabolism in fermentative systems. Isotopically, biogenic methane is depleted in ^{13}C by 30 to 50‰ relative to the carbon source being utilized by the fermentative community (Games and Hayes 1976; Whiticar et al. 1986). In one case where sedimentary total organic carbon is depleted in ^{13}C relative to photosynthetic carbon sources (the Messel Shale; Hayes et al. 1987), there is now compelling evidence that recapture of methane carbon was at least partly responsible for the depletion of ^{13}C in sedimentary TOC relative to primary inputs (Freeman et al. 1990). Earlier, Hayes (1983a) hypothesized that recapture of isotopically depleted methane could explain the prevalence of Δ_B values as large as 59‰ in the Late Archean, and outlined isotopic budgets and microbial communities capable of producing the observed depletions. Although biological oxidation of methane can occur anaerobically (Alperin and Reeburgh 1985), there is no evidence that this process leads to efficient incorporation of methane-derived carbon in biomass. Because such incorporation is required to produce the very large Δ_B values in Late Archean systems, that isotopic signal was interpreted as indicative of aerobic utilization of CH_4 beginning about 2.8 Ga ago (Hayes 1983a). An aerobic process is clearly indicated in the Messel Shale, which contains isotopically depleted hydrocarbon biomarkers derived from aerobic methylotrophs (Freeman et al. 1990). There, secondary inputs, prominently including ^{13}C-depleted methylotrophic biomass, yielded sedimentary TOC with a δ value 8‰ lower than that of the biomass of oxygenic algae in the water column (Hayes et al. 1987). Such depletions increase the isotopic difference between sedimentary TOC and the carbon source utilized by primary producers and are characterized by higher values of Δ_B, easily exceeding the 32 to 35‰ maximum noted above and providing one possible explanation for the large fractionations observed in Early Proterozoic sediments.

The proposed recycling of methane requires stratification, or separation, of anaerobic and aerobic zones. More generally, any environment in which distributions of oxidants and reductants are stratified or separated can support processes that might increase Δ_B. Chemoautotrophy and bacterial (anaerobic) photosynthesis are of particular interest. Organisms in the first category, sulfide- and ammonia-oxidizing bacteria, for example, can thrive at boundaries between redox zones. As they obtain energy through catalysis of an exergonic chemical reaction, such bacteria fix carbon by means of the C_3 pathway, building biomass for which dissolved inorganic carbon serves as the carbon source. If that carbon is depleted in ^{13}C relative to dissolved inorganic carbon in the aerobic photic zone—and this is often the case in natural environments (e.g., Takahashi et al. 1968; Deuser 1970)—the chemoautotrophic biomass will be depleted in ^{13}C relative to that of primary producers in the same environment. If that material contributes significantly to sedimentary TOC, Δ_B will be enhanced. Significant levels of sulfide oxidation are observed in many modern marine sediments (Jørgensen 1987), and carbon-fixation associated with that process has been recognized and quantified in some modern systems (Kepkay et al. 1979). Biomarkers with isotopic compositions possibly associated with chemoautotrophic inputs have been recognized in the Messel Shale (Freeman et al. 1990). If Proterozoic sediments systematically accumulated greater proportions of chemoautotrophic biomass than do modern marine sediments, that might also explain the observed Δ_B values.

Photosynthetic bacteria are important producers in some modern stratified lakes (e.g., Fry 1986), and Summons and Powell (1986, 1987) have shown that they contributed significantly to organic matter in some Paleozoic marine sediments. The same conditions that stabilize redox stratification and, thus, the creation of an anaerobic photic zone, may lead to stratification of the distribution of ^{13}C in dissolved inorganic carbon. Some photosynthetic bacteria (the Chlorobiaceae, or green photosynthetic bacteria, and *Chloroflexus*) do not utilize the C_3 pathway for fixation of carbon, but instead employ reaction mechanisms that do not discriminate as strongly against ^{13}C (Sirevåg et al. 1977). As a result, their biomass may be *enriched* in ^{13}C relative to that of oxygenic primary producers (this was observed by Summons and Powell 1986), but ^{13}C-depleted organic matter would still result if isotopic depletion in the anaerobic zone exceeded ~10‰. Other photosynthetic bacteria (the Chromatiaceae, or purple photosynthetic bacteria) utilize the C_3 pathway for carbon fixation. If their inorganic carbon source is depleted in ^{13}C, they are likely to produce organic material with a δ value significantly lower than that of aerobic primary producers (e.g., Fry 1986). It is, therefore, possible that inputs from photosynthetic bacteria contributed to the larger values of Δ_B observed in Early Proterozoic materials.

Notably, all of the mechanisms discussed here involve stratification of the Early Proterozoic ocean.

The Middle Proterozoic. In Figure 3.4.2, the envelope enclosing δ values for well-preserved kerogens trends upward while average δ values for carbonates continue to vary between 0 and +1‰. A decrease in Δ_B results. Values as low as 28.5‰, observed by Knoll et al. (1986) as characteristic of the extensive Late Proterozoic succession in Svalbard and east Greenland, are first observed in strata with ages near 1500 Ma and become increasingly frequent thereafter. Because the fractionation between dissolved CO_2 and carbonate minerals is approximately 11‰, the fractionation then attributable to biologically mediated processes is only 17.5‰, a level that *could* be associated entirely with operation of the C_3 pathway and related biosynthetic processes. For reference, the comparable value in modern marine environments is 10‰; during most of the Phanerozoic, the comparable value was 16‰ (calculated from modern marine organic δ values of −21‰ and pre-Eocene values of −27‰; see Degens 1969).

Two factors are likely responsible for the decrease in Δ_B, and a third must be considered. First, the processes outlined in the preceding paragraphs (methane recycling, chemoautotrophy, anaerobic photosynthesis) may have declined in importance. As these contributions to sedimentary carbon pools declined, average δ values for organic carbon increased. Second, ε_P, the fractionation factor for photosynthetic carbon fixation, may have become less negative. In that case, even if secondary processes retained their prominence, lower overall fractionations would be observed. Noting that algae grown at high P_{CO_2} were depleted in ^{13}C relative to those grown at normal levels, Mizutani and Wada (1982) suggested that this phenomenon might explain the low $\delta^{13}C$ values characteristic of Precambrian organic materials. More recently, Popp et al. (1989) reviewed evidence for a link between ε_P and P_{CO_2}, the partial pressure of carbon dioxide. Following McCabe (1985), they suggested that, for $-1.7 > \varepsilon_P > -27$‰, each 3‰ change in ε_P corresponds to a 1.5-fold change in P_{CO_2}. If that same sensitivity to P_{CO_2} characterized photosynthetic fractionation of carbon isotopes during the Early and Middle Proterozoic, and if the observed 5‰ decrease in Δ_B were ascribed entirely to changes in the value of ε_P, a two-fold decrease in P_{CO_2} would be indicated. Extrapolations involved in this estimate are enormous. There is no reason to place much confidence in the quantitative result, but the idea that P_{CO_2} declined appears reasonable. The global cooling indicated by the Late Proterozoic record of low latitude glaciation might have been due at least in part to a weakening of the terrestrial greenhouse (see Section 4.7).

Consideration of differences between modern and Proterozoic primary producers raises physiological and biochemical questions that suggest the third possible cause for the decrease in Δ_B. It is, for example, quite conceivable that Proterozoic rubisco was different enough from the modern enzyme that it imposed larger or smaller isotopic fractionations on photosynthetically fixed carbon. It is known that rubisco isozymes from various plant, algal, and bacterial species differ in substrate specificities for carboxylase and oxygenase functions (Jordan and Ogren 1981), and it is possible that these differences are evidence of significant evolutionary changes in rubisco. As global levels of O_2 increased (see below) and the oxygenase activity of rubisco took an increasing toll on carbon fixation, there would have been strong pressure to improve the efficiency of rubisco's carboxylase function. Any such developments might have changed the associated isotope effect. It is also possible that physiological changes affecting mass transport of inorganic carbon to the intracellular site of fixation could affect photosynthetic fractionation of carbon isotopes. Badger and Andrews (1987) have presented convincing arguments for coevolution of CO_2-concentrating mechanisms and changes to the kinetic properties of rubisco, both adaptations occurring in response to a declining CO_2/O_2 ratio in the external environment.

The Late Proterozoic. Because of relatively rapid secular changes in the isotopic composition of marine carbonate during the Late Proterozoic, it is important that values of Δ_B be estimated from closely associated carbonate-organic pairs. Approximately 45 such pairs can be found in Table 17.4. On average, they yield $\Delta_B = 29.2‰$, the standard deviation of the population being 2.4‰. The record of secular changes in δ values indicates significant changes in the global carbon cycle, and the occurrence of intermittent global cooling suggests greenhouse fluctuations. Because ε_P ought to vary in response to changes in P_{CO_2}, it will be important to search for secular variations in Δ_B during this interval.

As shown in Table 3.4.2, organic carbon in the Chulaktau and Khatyspyt Formations (550 and 620 Ma, respectively) is strongly depleted in ^{13}C, reaching levels not seen since nearly two billion years earlier. Similar depletion has been observed ($-37‰$, Grantham et al. 1988) in the Late Proterozoic Huqf Group, Oman. It is evident that carbon flows and pathways within the communities supplying organic carbon to these units must have differed significantly from those in communities generating the organic carbon that makes up the bulk of the sedimentary record. These units contain well preserved organic carbon, and it is likely that more detailed studies will yield useful information.

3.4.3B Burial of Organic Carbon and the Global Redox Balance

As outlined in Section 3.1.5, the requirement for mass balance in the global carbon cycle makes it possible to calculate f_{org}, the fraction of recycling carbon being buried in the form of organic material, whenever globally representative values are known for the isotopic compositions of organic and carbonate carbon. Specifically, we can write

$$f_{org} = (\delta_{carb} + 5)/\Delta_B^*. \quad (3.4.1)$$

Solution of this equation using values characteristic of the Early Proterozoic yields $f_{org} \approx 0.17$. This value cannot be interpreted in terms of productivity of the Early Proterozoic ecosystem. It merely indicates that approximately 17% of carbon passing through the exogenic cycle was preserved in organic form long enough to be immobilized in accumulating sediments. The *rate* of burial (moles/time) can be determined only if the throughput of carbon in the exogenic cycle is known. Moreover, a high rate of organic-carbon burial might derive from an ecosystem characterized by low productivity but efficient preservation of organic material, and a low rate of organic-carbon burial might derive from an ecosystem with high productivity and very aggressive recycling mechanisms. But *any* burial of organic material releases oxidizing power in the surface environment. If that supply of oxidant exceeds the requirements of contemporary sinks (oxidation, much of it biologically catalyzed, of exposed minerals and organic material), oxidizing power will accumulate in the surface environment. That oxidizing power may be in the form of O_2 or Fe^{3+} or SO_4^{2-} (or other minor oxidants). Of these, only O_2 can increase without limits imposed by solubility in the ocean or finite supplies of precursors.

We can describe as "redox neutral" the point at which the supply of oxidants exactly balances the sinks. The value of f_{org} characteristic of redox neutrality for the modern carbon cycle is not known, and, of course, still less is known about ancient carbon cycles. From independent geochemical evidence regarding redox conditions at earth's surface (Section 4.6), we can judge that levels of O_2 were not rising significantly during the Early Proterozoic. On that basis, we might conclude that $f_{org} = 0.17$ corresponded to redox neutrality, but even that might be an overestimate because crustal inventories of ferric iron and sulfate probably *were* increasing during this interval.

As Δ_B^* decreased during the Middle Proterozoic, f_{org} increased even though δ_{carb} remained roughly constant. Increased burial of organic carbon might be seen as reflecting the advent of eukaryotic primary producers, with resultant improved production of organic carbon, but the higher values of f_{org} should also steadily create a more aggressive oxidizing environment, making the preservation of organic material less likely.

Problems associated with interpretation of f_{org} blossom dramatically during the Late Proterozoic, where high values of δ_{carb} lead to f_{org} values approaching 0.40. Estimating fluxes, durations of isotopic excursions, and the strength of sinks for O_2, Knoll et al. (1986) calculated that the quantities of organic material buried during the Late Proterozoic were so large that the oxidizing power released could not possibly have been consumed by oxidation of recycling sulfides and ferrous iron *and* that quantities of O_2 released would have easily exceeded the present atmospheric inventory. This is a pleasing result inasmuch as it is consistent with numerous suggestions that a rise in atmospheric O_2 might have initiated the development of complex organisms (e.g., Runnegar 1982c), but the apparent over-supply of oxygen points to a significant problem. Relationships between f_{org} and the global redox balance are apparently not adequately described by the present model and, therefore, our understanding of the linkage between the carbon cycle and the development of the environment is fundamentally deficient. The problem would be solved if there were a category of O_2 sinks that came into play only as O_2 levels rose, but the identity of those sinks has not been described. Crust-mantle interactions may hold the key.

Periodic sharp declines of δ_{carb} during the Late Proterozoic suggest intervals during which f_{org} dropped to levels below 0.1. Correlations between these features, glacial maxima, and possible extinctions await exploration.

3.5

The Sulfur-Isotopic Record

J. M. HAYES   IAN B. LAMBERT   HARALD STRAUSS

The global average δ value for all forms of sulfur in the sedimentary shell is about $+2‰$ vs. CDT (Holser et al. 1988). The zero point of the sulfur-isotopic scale is fixed by the $^{34}S/^{32}S$ ratio in troilite from the Cañon Diablo meteorite. If the meteorite carried sulfur representative of materials from which the earth accreted, the alignment between its isotopic composition and the crustal average indicates that geochemical partitioning of sulfur between the core, mantle, and crust has caused only minimal fractionation of the isotopes of sulfur. If processes delivering sulfur to the crustal inventory have imposed such small fractionations, then the global δ value, averaged over all forms of sedimentary sulfur, has *always* been near zero. This widely accepted hypothesis provides a basis for consideration and interpretation of the sulfur isotopic record.

Though sulfur isotopes are apparently not fractionated by processes of planetary differentiation, they *are* subject to fractionation at earth's surface. In the modern world, processes mediating the distribution of sulfur among its oxidized and reduced forms in the atmosphere, hydrosphere, and biosphere result in significant isotopic fractionation (Section 3.1.5). For processes occurring at earth-surface temperatures, the notable fractionations are associated with biologically catalyzed reactions. To probe the evolution of the biogeochemical cycle of sulfur and its interactions with other elemental cycles, we can ask when the distribution of sulfur isotopes in sediments first records the effects of these biological processes and how the distribution of sulfur among its oxidized and reduced forms reflects development of global redox levels.

3.5.1 Results of New Isotopic Analyses of Sulfides and Sulfates

The existing, fragmentary record of the isotopic composition of marine sulfate has been supplemented by analyses of samples from North America, Africa, and Australia (Table 3.5.1). Results of analyses of new samples from the Early Archean Towers Formation, Pilbara Block, Western Australia, are consistent with earlier reports (Lambert et al. 1978). The Middle Proterozoic Newland Limestone, Belt Supergroup, western United States, consists of shales and carbonates. The interbedded barite horizon analyzed here appears to have formed by early diagenetic replacement of anhydrite (Schieber 1988). Results of analyses of later Proterozoic sulfates consistently yield values near $+20‰$. Among these, sulfates from the Late Proterozoic Bitter Springs Formation in the Amadeus Basin of central Australia had previously been analyzed by Solomon et al. (1971). In view of the evidence for remobilization, dehydration, and rehydration of these minerals during burial and uplift, it was considered warranted to conduct a systematic Sr and S isotopic study of the various forms of sulfates from a drillhole through the evaporites. The $^{87}Sr/^{86}Sr$ values, 0.70568–0.70636, are in the unaltered marine range, regardless of mineralogy or form. The range of $\delta^{34}S$ values is similar to that of the previous data; they show no systematic trends with depth, mineralogy, or form of sulfate. The upper Redstone River Formation of the Mackenzie Mountain Supergroup of western Canada contains a succession of evaporites and redbeds considered to have formed in coastal playa environments (Ruelle 1982). The Bakoye Group, Taoudenni

Table 3.5.1. *New analyses of sulfur-isotopic compositions of sulfate in Precambrian sedimentary rocks.*

| Unit | Age (Ma) | $\delta^{34}S$, ‰ vs. CDT | | Mineral | N |
		Avg.	Range		
Bakoye Group	620	27.6		Barite	1
Redstone River Fm.	770	20.2	15.6–25.5	Anhydrite	3
Bitter Springs Fm.[a]	850	18.6	16.3–20.7	Gypsum	9
		18.0	17.6–18.4	Anhydrite	2
Zambian Copperbelt[b]	1,000	20.7	20.3–21.0	Anhydrite	2
Newland Limestone	1,440	15.4	13.6–18.3	Barite	3
Towers Fm.	3,435	5.4	4.5–6.3	Barite	2

[a]Unpublished results (I. B. Lambert) of analyses of samples from DDH Alice Springs 3, depths 131'8" to 824'6". See text for discussion of Sr isotopes in these samples.
[b]Unpublished results (I. B. Lambert) of analyses of samples from cores NN23 (1482') and NN17 (1851'), Nkana North Limb, Zambian Copperbelt.

Basin, West Africa, contains carbonates, tillites, and a massive barite bed (Deynoux and Trompette 1976; Clauer et al. 1982).

Results of new analyses of Precambrian sulfides are summarized in Table 3.5.2. Sulfides from Late Archean and Early Proterozoic units mainly yield δ values within ten permil of zero. Much higher values are commonly observed during the Middle and Late Proterozoic, with δ occasionally exceeding +50‰.

3.5.2 An Overview of the Record and its Interpretation

3.5.2A The Early Proterozoic

Results of new analyses are plotted in Figure 3.5.1 together with results of prior studies of the antiquity of sulfate reduction, paleoenvironmental conditions, and ore genesis. The variable δ ranges for Early Proterozoic sulfides and sulfates contrast with the marked concentration of Archean δ values around 0‰. The Archean data have been interpreted in terms of generally low concentrations of sulfate in the hydrosphere (Lambert and Donnelly 1989). The interpretation is based on

Table 3.5.2. *New analyses of sulfur-isotopic compositions of sulfide in Precambrian sedimentary rocks.*

Unit	Age (Ma)	$\delta^{34}S$, ‰ vs. CDT Avg.	Range	Mineral	N
Chapel Island Fm.	560	−18.0		Pyrite	1
Wlodawa Fm.[a]	560	20.5	20.1–20.9	Pyrite	2
Denying Fm.[b]	570	7.7		Pyrite	1
Almodovar del Rio Gp.	600	−2.1		Pyrite	1
Cijara Fm.	610	8.2	−2.9–19.4	Pyrite	2
Lublin Fm.[a]	625	16.5	−8.3–50.2	Pyrite	3
Doushantou Fm.[b]	650	22.7	11.4–26.7	Pyrite	8
Sheepbed Fm.	650	2.8	−12.6–22.6	Pyrite	4
Datanpo Fm.[b]	680	30.4		Pyrite	1
Backlundtoppen Fm.[c]	700	−7.0		Pyrite	1
Luoquan Fm.	700	23.0		Pyrite	1
Twitya Fm.	700	30.4	26.2–33.7	Pyrite	4
Svanbergfjellet Fm.[c]	750	15.1		Pyrite	1
Tapley Hill Fm.[c]	750	31.9	14.0–41.0	Pyrite	18
Aralka Fm.[e]	750	47.3	39.4–54.5	Pyrite	9
Coppercap Fm.	750	6.3		Cu sulfides	1
Little Dal Group	800	−28.0		Pyrite	1
Kingbreen Fm.[c]	830	−13.1		Pyrite	1
Walcott Mbr.	850	11.7	7.8–18.6	Pyrite	3
Tarcoola Fm.[f]	850	28.0	17.6–37.8	Pyrite	6
Kortbreen Fm.[c]	860	28.3		Pyrite	1
Awatubi Mbr.	862	−18.8	−38.9–1.4	Pyrite	2
Jupiter Mbr.	930	16.7	13.9–19.5	Pyrite	2
Tanner Mbr.	950	15.7		Pyrite	1
Belt Supergroup[g]	1,400	−4.8	−14.0–18.1	Pyrite	8
Greyson Shale	1,420	5.0		Pyrite	1
Newland Limestone	1,440	19.3	14.7–31.2	Pyrite	6
Prichard Fm.	1,450	14.3		Pyrite	1
Urquhart Shale	1,670	28.5	22.7–36.3	Pyrite	3
Gunflint Iron-Fm.	2,090	6.6	5.6–7.6	Pyrite	2
Gamohaan Fm.	2,450	−2.7	−13.9–1.8	Pyrite	5
Dales Gorge Mbr.	2,500	0.9	−4.2–6.0	Pyrite	6
Jeerinah Fm.	2,650	7.9	6.4–9.4	Pyrite	2
		6.0[h]	−2.5–19.2	Pyrite	13
Venterspost Fm.	2,670	1.1		Pyrite	1
Kameeldorns Fm.	2,705	−0.9		Pyrite	1
Carbon Leader Mbr.	2,710	3.3		Pyrite	1
Kuruna Siltstone	2,740	0.3		Pyrite	1

[a]Samples from Moczydlowska (1988)
[b]Samples from Lambert et al. (1987)
[c]Samples from Knoll et al (1986)
[d]DDH Eukaby 8, MD003, PD004, and PD005
[e]DDH Illogwa Creek 6
[f]Wilgena 1 core
[g]Pyritic, laminated shales
[h]multiple micro-samples from PPRG samples no. 375 to 377

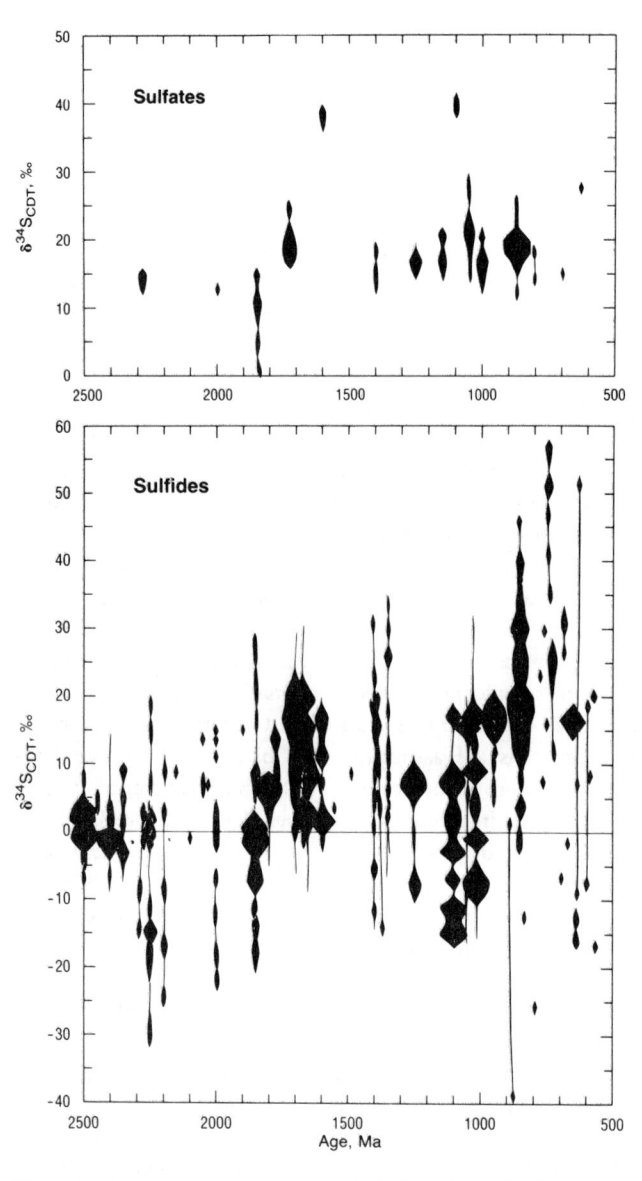

Figure 3.5.1. Isotopic compositions of sulfide sulfur and sulfate sulfur in stratigraphic units of Proterozoic age. Vertical spreads of spindles indicate ranges of δ values. Widths indicate frequency of observation. Numbers of points contributing to each spindle vary. For listing of points and sources of data, see Tables 3.5.1, 3.5.2, 17.6, and 17.7.

the common absence of evidence for large isotopic fractionations like those which accompanied sulfide formation by bacterial sulfate reduction or seafloor hydrothermal processes in younger environments. It envisages that most of the sulfides in Archean sedimentary strata (commonly tuffaceous) formed directly or indirectly from sulfurous emanations which were abundant because of extremely high levels of igneous activity in this Eon, and which would have reacted rapidly with Fe^{2+} in the hydrosphere. Clearly, there were some environments with appreciable concentrations of sulfate during the Archean. Possibly as a result of locally high levels of microbial photosynthesis which oxidized exhalative H_2S, sulfate minerals formed in certain shallow-water evaporative environments as much as 3.45 Ga ago (Lambert et al. 1978).

Wide ranges of δ values in some Archean sulfides, particularly in the Michipicoten and associated iron-formations in Canada, have been interpreted as evidence for bacterial reduction of sulfate (Goodwin et al. 1976; Schidlowski et al. 1983). Similarly, Ripley and Nicol (1981) suggested that bacterial reduction of sulfate was responsible for formation of pyrite with δ values between -2.3 and $+11.1‰$ in Late Archean metasediments from northern Minnesota. However, there is textural and other evidence that these sulfides, which include abundant vein and massive varieties, are mainly of hydrothermal origin. It follows that thermochemical redox reactions could have caused the observed isotopic fractionations, and this is supported by the close association with volcanics and the predominance of δ values between -4 and $+4‰$.

A different model for the Archean (Ohmoto and Felder 1987) proposes that marine concentrations of sulfate could have been moderately high, and bacterial reduction of sulfate widespread, with very little isotopic fractionation occurring because of high temperatures in the hydrosphere. This model, which should be tested by looking for small fractionations associated with bacterial reduction of sulfate in modern solar ponds, offers no explanation for the marked change in isotopic trends in the Early Proterozoic.

The sustained change in the sulfur isotope record in the Early Proterozoic, to relatively large δ ranges and to mean values well removed from 0‰, can be accounted for by creation of a significant global pool of sulfate and its enrichment in ^{34}S as a result of bacterial and hydrothermal reduction processes (see Section 3.1.5). Stabilization of sulfate as a mobile and ubiquitous constituent of the oceanic mixed layer is likely to have paralleled the stabilization of oxygen in the atmosphere, and may have been accompanied by the proliferation of sulfate-reducing bacteria. Prior to oxygenation of the surface environment and the general reduction in concentrations of dissolved iron (marked by deposition of the great iron-formations during the interval 2.8–1.8 Ga), sulfur may have been so immobile and concentrations of sulfate so low that redox processes in any locality tended to completion and isotopic fractionations remained small.

As oxidation advanced, the lifetime of sulfate against reductive removal from the hydrosphere approached the stirring time of the ocean and an isotopically homogenous pool of marine sulfate was created. Because concentrations of sulfate were low, opportunities for removal of sulfate in evaporites were minimal. As the global cycle of sulfur was established, patterns of isotopic fractionation may have resembled those indicated by Figures 3.1.6a and 3.1.6b, with values of $f_{sulfide}$ approaching 1.0. Values of $\delta_{sulfate}$ observed during the Early Proterozoic range between $+1.5$ and $+24.5‰$ and may represent global cycles with Δ_S^* in that range. It is well to remember that, if concentrations of sulfate were low, values of $\delta_{sulfate}$ were weakly buffered. In particular, as sulfate reduction proceeded, values of $\delta_{sulfate}$ in any system would rise rapidly unless supplies of sulfate were replenished by mass transport. If reduction of sulfate were incomplete and the residue formed sulfate minerals, their isotopic compositions would *not* reflect the marine δ value, possibly exceeding it substantially. This is not likely to be a problem with massive evaporites, but could significantly affect isotopic compositions of residual sulfates disseminated in chemical sediments. Especially for such materials, observed values of $\delta_{sulfate}$ represent maxima for isotopic compositions of marine sulfate. In some unusual circumstances (e.g., Lambert et al. 1978) sulfate deriving from oxidation of sulfide sulfur may contribute to formation of sulfate minerals. It is, therefore, possible for observed values of $\delta_{sulfate}$ to fall below the isotopic composition of marine sulfate.

3.5.2B Middle and Late Proterozoic

Overall, Proterozoic sulfides in sedimentary strata exhibit a total δ range of -32 to $+58‰$, but there is a concentration of values between -10 and $+20‰$, considerably above the common range for modern marine sulfides. Such values can result from reduction of sulfate in hydrothermal systems, but most of the Proterozoic sulfides studied do not exhibit features characteristic of hydrothermal activity. They are also expected from bacterial reduction of marine sulfate that is (i) enriched in ^{34}S, and (ii) present at low concentrations in the hydrosphere. Both of these factors favor development of bacterially produced sulfides enriched in ^{34}S, the first simply by providing a favorable starting point for the process of enrichment, the second by creating a situation in which reduction would more readily approach completion and, thus, maximal enrichment. Because Figure 3.5.1 provides some evidence of isotopic enrichment of marine sulfate, and because low environmental concentrations of sulfate are highly plausible, the elevation of δ values relative to modern marine sulfides may seem unexceptional.

From a global perspective, however, the proportion of positive δ values among the data available for Late Proterozoic sulfides is enigmatically high. At steady state, the mass-balanced average δ for all forms of sulfur leaving the atmosphere + hydrosphere + biosphere must be equal to that of sulfur entering the same pool. If the overall δ value of the input is, as expected, $+2$, then that of the output must also be $+2$. Given the paucity of sulfate minerals in the record and the expectation that they were much less abundant in the not-yet-fully oxygenated global environment than in the Phanerozoic, we need not be too concerned with their impact on the mass balance. But the sulfides are a different matter. For every gram of sulfur with a δ value of $+20$ removed from the atmosphere + hydrosphere + biosphere, mass balance requires removal of another with $\delta = -16$ (or of ten grams with $\delta = -1.6‰$, etc.).

The latter deposits are not abundant in the record summarized in Figure 3.5.1. Is it possible that they existed long enough to fulfill the requirements of mass balance, but have not been preserved?

The isotopically heavy sulfides may reflect in part a bias in sampling, resulting from the relative ease of obtaining samples from mineralized areas where hydrothermal sulfides could be present in strata for considerable distances from known ore deposits. Alternatively, there is sedimentological evidence (see Chapter 2) for the existence of numerous major intracratonic troughs in the Proterozoic, possibly on supercontinents, and it is considered that restricted marine or non-marine environments in these resulted in generation of isotopically heavy sulfides as bacterial sulfate reduction proceeded. The occurrence of some very high δ_{sulfide} values may indicate that δ_{sulfate} approached levels characteristic of globally euxinic conditions in the deep ocean (see Fig. 3.1.6d). Indeed, given the compelling evidence for burial of large quantities of organic material during this interval (see Section 3.4), the occurrence of an "oceanic anoxic event" of extreme proportions is very likely.

Under conditions of marine stagnation and high rates of burial of organic material, the large quantities of O_2 that would otherwise ventilate the deep ocean must accumulate in the atmosphere. Globally, deposition of sulfides would take place under two very different sets of conditions. In the deep ocean, reduction of sulfate would occur in the water column and Δ_S would tend to very high values, approaching 60‰. If most burial of sulfur occurred in the deep ocean, Δ_S^* would also approach 60‰ and, if tectonic and climatic conditions did not favor removal of sulfate via formation of evaporites, the resulting high values of f_{sulfide} would drive δ_{sulfate} to values near +60‰. In relatively shallow, epicratonic basins, penetration of abundant atmospheric O_2 into the water column would ensure that reduction of sulfate occurred below the sediment-water interface. There, particularly in the absence of bioturbation, Δ_S would tend to low values due to restricted diffusion of sulfate from the water column. Accordingly, δ_{sulfide} in such basins would tend to very high values, approaching the value of δ_{sulfate} (up to +60‰). Fractionation would thus be minimized in restricted marine environments, which tended toward closure, at the same time it was maximized in the deep ocean. Continental breakup near the end of the Proterozoic apparently did not result in widespread preservation of deep marine sequences on the plate margins; it is feasible that many were eventually subducted, giving rise to the marked decreases in δ observed in post-Cambrian marine sulfate (Claypool et al. 1980).

3.5.2C Linkage of the Carbon and Sulfur Cycles

Another interesting observation concerns the apparent breakdown in relationships between isotopic abundances of sulfur and carbon around the Proterozoic-Phanerozoic boundary (Lambert et al. 1987). Through the Phanerozoic, there is a general inverse correlation at a time scale of 10^7 years between $\delta^{13}C$ values for carbonates and $\delta^{34}S$ values for evaporitic sulfates. This has been explained in terms of an overall redox balance between the carbon and sulfur cycles (Veizer et al. 1980). Increasing $\delta^{13}C$ values for carbonates signal a release of oxidizing power by the carbon cycle. If sulfide is available, a portion of that oxidizing power will be utilized in the oxidation of sulfide and the $\delta^{34}S$ values of sulfates will decrease, establishing the inverse correlation. From a Phanerozoic perspective, therefore, the general enrichment of ^{13}C in Late Proterozoic carbonates (see Section 3.4) is expected to be complemented by a depletion of ^{34}S in evaporitic sulfates during that same interval. The sulfate record is so incomplete that it provides no strong evidence for or against such a depletion. As noted above, however, the sulfide record suggests that sulfate was enriched in ^{34}S, not depleted. If so, the carbon-sulfur linkage must be "turned on" only later, near the beginning of the Cambrian.

Ultimately, the establishment of the linkage may be traced to the major rise in atmospheric oxygen that appears to have occurred in the latest Proterozoic (Knoll et al. 1986). It is likely that O_2 is the indispensable, mobile intermediate that transfers electrons between the cycles of carbon and sulfur. For example, burial of organic matter and release of O_2 will eventually create an aggressively oxidizing environment in which the extent of oxidation of sulfur will, inevitably, increase. According to this view, Phanerozoic variations in sulfur redox levels are responses to forcing imposed by the carbon cycle and *transmitted by* O_2. If levels of O_2 were low, carbon-sulfur redox linkages were likely to be weak to nonexistent. As levels of O_2 increased, the C-S linkage "evolved" just as multicellular animals that required high levels of O_2 also evolved.

3.6

Unsolved Problems and Conclusions

J. M. HAYES   DAVID J. DES MARAIS   IAN B. LAMBERT
HARALD STRAUSS   ROGER E. SUMMONS

3.6.1 The Subject

Biogeochemistry is that point on the surface of science toward which all lines converge without seeming ever to meet. Our major unsolved problem is the recognition, let alone the integration, of all of the boundary conditions and lines of evidence that bear on the reconstruction of the development of earth's environment. Biogeochemists must be physicists and oceanographers as well as biologists, geologists, and chemists. The development of such diversity runs counter to classical scientific practice. The problem is probably physiological as well as psychological: the development of minds *capable of* such diversity may be a task for evolution.

3.6.2 The Proterozoic

Concentrations of organic carbon in Proterozoic sedimentary rocks are equal to those in Phanerozoic sediments. Moreover, many units, some as old as 1700 Ma, contain organic materials so well preserved that molecular structures within them carry information of undoubted evolutionary significance.

Hydrocarbons present in the extractable organic matter from large suites of samples of well-preserved Proterozoic sediments are consistent with the presence of membrane lipids of archaebacteria, eubacteria, and eukaryotes as far back as 1690 Ma. The low carbon number predominances of the alkanes and the recognition of high concentrations of monomethyl branched alkanes in most of the samples distinguish them from most Phanerozoic bitumens, and suggest eubacteria with unusual lipid compositions were important precursor organisms. Many samples that have independent evidence for hypersalinity in their depositional environments contain well preserved suites of acyclic isoprenoid alkanes, including extended C_{21} to C_{30} isomers, and are consistent with inputs of archaebacterial lipid.

Steranes, probably derived from eukaryotic membrane lipids, and hopanes derived from the bacteriohopane polyols of eubacterial membranes, are present in rocks as old as 1700 Ma but are in very low concentrations in the oldest and in the more thermally mature samples. Late Proterozoic sediments have the same high relative abundances of steranes as Phanerozoic sediments but sometimes show very strong predominances of either cholestane or 24-ethyl cholestane isomers. It is possible that the increase in steroid abundances at this time correlates with the well established radiation of planktonic algae (Vidal and Knoll 1982, 1983; Knoll 1983; Sections 5.5, 11.3) which preceded the appearance of the metazoan fauna (Cloud and Glaessner 1982; Cloud 1985). Also, the presence of C_{21} to C_{30} monomethyl branched alkanes with predominance of centrally branched isomers and moderately light carbon isotope signatures are unusual features which, in combination, would also distinguish them from most Phanerozoic bitumens. Anomalous triterpane distributions are commonly encountered and include the presence of high abundances of putative 18α(H)-neohopanes and 3β-methyl and 2α-methyl steranes. Some of the distinctive features of the Proterozoic bitumens, including the high abundances of 3β-methyl and 2α-methyl steranes and of monomethyl alkanes, do extend to Paleozoic sediments and oils. Thus, hydrocarbon biomarkers might encode a loose hierarchy of structures which reflect trends in the evolution of lipid biosynthetic pathways.

Correlations between isotopic abundances and elemental compositions of kerogens indicate that post-depositional processes have altered the carbon-isotopic compositions of many samples of Proterozoic organic material. When altered materials are excluded, a sequence of systematic shifts in isotopic abundances and fractionations can be observed. Isotopic differences between Early Proterozoic organic materials and coeval carbonates are generally larger than those in Phanerozoic materials. That enhanced fractionation of carbon isotopes might be ascribed to (i) generally different carbon pathways within the global ecosystem, possibly associated with the minimal oxygenation of the environment; or (ii) enzymological and/or physiological differences in primary producers.

During the Middle and Late Proterozoic, isotopic differences between carbonates and organic materials decrease, slowly approaching Phanerozoic values. This trend may be associated with changes in either of the factors noted above or with a decline, caused by decreasing abundances of CO_2, in the isotopic fractionation associated with photosynthetic carbon fixation. During the Late Proterozoic, a global pattern of secular variations in carbon-isotopic abundances is evident. Elevation of ^{13}C contents of carbonates and organic materials is indicative of enhanced levels of burial of organic carbon. This can be qualitatively correlated with an increase in oxygenation of the surface environment, but detailed redox balances have not yet been constructed.

The sulfur-isotopic record is consistent with the occurrence of significant levels of biological reduction of sulfate throughout the Proterozoic. Unusually high contents of ^{34}S in many Late Proterozoic sulfides suggest that marine sulfate must have been highly enriched in ^{34}S, a development plausibly associated with global euxinicity in the deep ocean. Details of the sulfur-isotopic mass balance during the Late Proterozoic cannot, at present, be resolved.

Foremost among future studies must be detailed studies of single basins. Geological, paleontological, and sedimentological evidence should be utilized to identify suites of samples likely to represent contrasting environments within ancient ecosystems. Closely spaced stratigraphic successions and apparently different facies should be sampled and studied using molecular and isotopic techniques. The goal should be not only to extend characterizations of biomarkers and details of isotopic records, but to reconstruct flow pathways of both matter and energy within biological communities, and to recognize interactions between communities and their environments. There is every reason to expect success in this work and, as a result, a greatly improved view of the development of the global environment.

4
Proterozoic Atmosphere and Ocean

CORNELIS KLEIN   NICOLAS J. BEUKES   HEINRICH D. HOLLAND
JAMES F. KASTING   LEE R. KUMP   DONALD R. LOWE

4.1 Introduction

CORNELIS KLEIN

In this Chapter an overview is developed of aspects of the Proterozoic atmosphere and oceans based as much as possible on geologic evidence, but supplemented by theory, whenever such evidence is indirect, incomplete, or lacking. Much of the theoretical treatment is rather oversimplified and speculative. Several biologically important aspects of the Proterozoic environment are addressed, namely, the partial pressures of oxygen and carbon dioxide in the atmosphere and possible changes in their partial pressures as a function of Precambrian time. Aspects of the chemistry and evolution of the Proterozoic ocean are discussed as well.

Banded iron-formations (BIFs) are the most abundant chemical sediments found throughout much of Precambrian time. Because they are generally devoid of clastic components, their chemistry, their oxidation state, and their temporal distribution provide important clues about the chemistry and the chemical evolution of the Precambrian ocean and atmosphere. Section 4.2 provides a synopsis of the average major element chemistry of banded iron-formations throughout the Precambrian; all iron-formations older than about 1.9 Ga represent very similar chemical systems. Iron-formations formed between 0.8 and 0.6 Ga are distinctly different and are more highly oxidized. Few iron-formations are younger than about 1.8 Ga; a minor resurgence in BIF deposition occurred between 0.8 and 0.6 Ga. After about 1.85 Ga, the atmosphere and oceans became rather highly oxygenated and the oceans as a whole became depleted in iron. The overall REE (rare earth element) patterns, and positive Eu and negative Ce anomalies in banded iron-formations ranging from 3.8 to 1.9 Ga in age, strongly support the notion that iron-formations were deposited from marine systems with input from deep sea hydrothermal sources. The hydrothermal input into the ocean system has apparently decreased with time, and at the time of the deposition of iron-formations of about 0.8 to 0.6 Ga it is essentially absent from the system.

In Section 4.3 a new model for iron-formation deposition is developed; this is based mainly on recent geochemical and sedimentological studies of iron-formation sequences in the Transvaal Supergroup of South Africa. Because shallow-water stromatolitic limestones and shales in this sequence have geochemical signatures that are distinctly different from those of the associated, microbanded iron-formations, the iron-formations are concluded to be the result of chemical deposition in deep marine environments, with the stromatolitic limestones and shales having been formed in overlying shallow water. This leads to a model of a density-stratified ocean system during much of Precambrian time, up to about 1.8 Ga, in which the surface waters are somewhat oxic and rich in organic matter, and the deeper waters are anoxic and enriched in dissolved ferrous iron and probably silica. Iron-formation precipitation would have taken place along the chemocline, in response to Eh/pH gradients and fluxes of organic matter from the shallow water photic zone and ferrous iron from the deeper waters. The analytical results suggest a decoupling of primary organic productivity and the deposition of iron-formation. Furthermore, combined petrographic and geochemical data lead to the conclusion that many of the very fine-grained siderites are primary precipitates formed in the water column. These siderites show a ^{13}C depletion of about 4‰ with respect to modern marine carbonates. This ^{13}C depletion of the siderites appears to be a primary signature of the Early Proterozoic deep ocean waters. It is likely that the ocean system at that time was stratified with regard to the isotopic composition of carbon, unlike present-day oceans in which the δ^{13}C gradient is small.

Inferences about the O_2/CO_2 ratio in the Proterozoic atmosphere are largely based on studies of Proterozoic paleosols. The identification of a Proterozoic rock as a paleosol is far from straightforward. It can be done, however, and in Section 4.4 it is concluded that, on the basis of paleosols spanning the age range between 2.2 to 1.8 Ga, there was a significant increase about 1.9 Ga ago in the ratio of P_{O_2}/P_{CO_2} in the Proterozoic atmosphere.

The distribution of red beds supplies additional evidence for a rise in the oxygen content of the atmosphere during the Proterozoic, as discussed in Section 4.5. It is suggested that the period between 2.2 and 2.0 Ga anticipates the abundant appearance of red beds worldwide after 2.0 Ga, and that there was an increase in the atmospheric oxygen content after about 2.0 Ga. The major occurrences of detrital uraninite- and pyrite-rich rock types prior to 2.0 Ga, and the general absence of these rock types from younger sediments also suggest that P_{O_2} increased soon after 2.0 Ga.

Section 4.6 discusses the rise of oxygen during the Precambrian on the basis of a three-box model of the atmosphere-ocean system. During stage I, termed "reducing," the atmosphere was essentially devoid of free oxygen. Atmospheric O_2 levels predicted during Stage I (prior to ?2.4 Ga) are on the order of 10^{-14} PAL (times the present atmospheric level) or lower. Independent calculations based on the survival of uraninite imply that P_{O_2} must have been less than $10^{-3.2}$ bar (less than 0.0032 PAL) during the Archean. During Stage II (from 2.4 to about 1.85 Ga), surface environments were oxidized and the deep ocean was reduced. Models for banded iron-formation deposition suggest an upper limit of approximately 0.03 PAL for atmospheric P_{O_2} during Stage II; calculations involving paleosol data yield P_{O_2} values of less than 0.05 PAL, consistent with the model predictions. In Stage III, beginning approximately 1.85 Ga, the deep ocean became oxic and therefore depleted in iron. The reasons for the increase in P_{O_2} levels in Late Proterozoic time, and the various factors that might have been responsible for increases in the O_2 content of the atmosphere at various stages in Precambrian time, are still rather obscure.

Aspects of Proterozoic paleoclimates are reviewed in Section 4.7 and approximate limits on atmospheric CO_2 concentrations during Proterozoic time are derived from climate models. CO_2 levels must have declined during the Proterozoic as solar luminosity increased. P_{CO_2} in the Early Proterozoic was probably about 60 times greater than the present level; in the Late Proterozoic, during global glaciation, P_{CO_2} values were perhaps three times greater than the present. Possible reasons why P_{CO_2} levels in the late Precambrian (during the major periods of glaciations) did not rise to high enough levels to prevent such glaciations, are discussed.

Section 4.8 addresses the chemistry and evolution of the Proterozoic ocean, especially the effect of changes in atmospheric P_{O_2} levels on the chemistry of seawater. The oxidation of organic matter in the upper parts of the oceans was probably the major cause for O_2-depletion in downwelled water in the Proterozoic ocean.

4.2

Time Distribution, Stratigraphy, and Sedimentologic Setting, and Geochemistry of Precambrian Iron-Formations

CORNELIS KLEIN   NICOLAS J. BEUKES

4.2.1 Introduction

This Section deals with an overview of several Precambrian iron-formations ranging in age from 3.8 Ga (Isua, west Greenland) to about 0.8 Ga (the approximate age for iron-formation in the Rapitan Group, Yukon and Northwest Territories, Canada). The term iron-formation as used here is similar to that of James (1954) with some modifications as suggested by Trendall (1983a). As such, iron-formation is defined as: *a chemical sediment, typically thin-bedded or laminated, whose principal chemical characteristic is an anomalously high content of iron, commonly but not necessarily containing layers of chert.* Major mineralogical variations, reflecting different bulk chemistries, are reflected in the nomenclature of four different types: (*i*) *oxide iron-formation;* (*ii*) *carbonate iron-formation;* (*iii*) *sulfide iron-formation* and; (*iv*) *silicate iron-formation.* In James' original definition of 1954, a quantitative lower limit of iron content (of 15 weight percent or more iron) was incorporated. This arbitrary lower limit is commonly too restrictive in the evaluation of rock types that reflect a range in iron content (from ferruginous to iron-rich) all of which show the pertinent characteristics of iron-formation.

Although many Precambrian iron-formations have undergone various metamorphic conditions, most of this overview is based on data derived from iron-formations that have undergone only low grade metamorphic (lower greenschist) conditions. The mineralogy of these iron-formations consists mainly of chert, magnetite, various carbonates (siderite, members of the dolomite-ankerite series, and calcite), hematite, and silicates such as greenalite, stilpnomelane, minnesotaite and riebeckite. The mineralogy of iron-formations as a function of diagenesis and metamorphism is reviewed in Klein (1983).

4.2.2 Distribution of Iron-Formations Throughout the Precambrian

The assessment of the distribution of iron-formations in the Precambrian is based on recent compilations by James (1983) and Walker et al. (1983). The ages for specific BIFs (banded iron-formations) in these tabulations appear not to have changed significantly since 1983. For example, Walker et al. (1983) report an age for the Labrador Trough sequence as 1.87 Ga (from Fryer 1972); Hoffman (1988) gives a date for the Wishart-Sokoman-Menihek sequence of the Labrador Trough (the New Quebec Orogen) as 1.88 Ga; and Chevé and Machado (1988) report a date of 1.88 Ga (± 2 Ma) for the Sokoman Formation. In short, in an overview of iron-formation abundances throughout Precambrian time, the dates of many BIFs are now probably better known than their relative abundances.

Interpretations of relative abundances versus age of iron-formations have been made by James (1983), James and Trendall (1982), Walker et al. (1983), and Gole and Klein (1981), with results that differ mainly in the estimated abundance of Archean iron-formations. There is general agreement that the largest iron-formations appear to be those of the Hamersley Range of Western Australia (dated at 2.5 Ga) and the Transvaal Supergroup, with an age of approximately 2.5 to 2.3 Ga (Klein and Beukes 1989). There is additional agreement on a sharp decline in major iron-formation in the Precambrian record at or before 1.8 Ga, except for several that show approximate ages between 0.8 and 0.6 Ga (such as iron-formations in the Rapitan Group, N.W.T., Canada; in the Urucum-Mutun region, Brazil-Bolivia; and in the Damara Supergroup of Namibia). There is, however, less agreement in the evaluation of the relative abundances of the generally smaller, discontinuous, and commonly tectonically deformed iron-formations in the Archean. Gole and Klein (1981) conclude that the size and extent of Archean iron-formations have commonly been underestimated. The schematic relative abundance curve shown in Figure 4.2.1 is adapted from that given by Gole and Klein (1981) with some minor modifications as well as the names added for several well-studied and major iron-formations. This curve is in essence a "best estimate" of total volume of iron-formation (not iron ore) for all iron-formations tabulated by James (1983), James and Trendall (1982), and Walker et al. (1983), relative to a maximum represented by the total iron-formation content of the Hamersley Range of Western Australia.

4.2.3 Stratigraphic Setting and Sedimentology of the Depositional Basins of Iron-Formations

The stratigraphic sequences in which Precambrian iron-formations occur are highly variable (Gole and Klein 1981) with wide diversity in lithological associations. In general, the

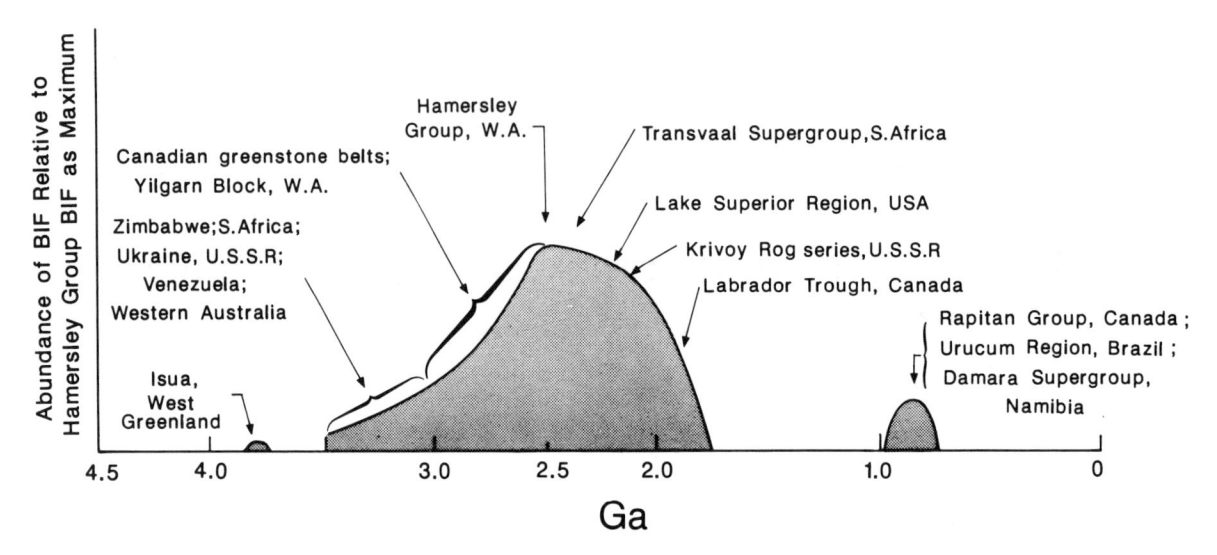

Figure 4.2.1. Highly schematic diagram showing the relative abundance of Precambrian banded iron-formations versus time, with several of the major iron-formations or major iron-formation regions identified. Estimated abundance values are relative to the Hamersley Group banded iron-formation volume taken as a maximum (adapted from Goles and Klein 1981).

stratigraphic associations of Archean iron-formations appear somewhat less variable than those of Proterozoic age, because volcanic rocks dominate many Archean banded iron-formation sequences. In Proterozoic iron-formations, volcanic rocks tend to be much less abundant, but materials of volcaniclastic parentage occur in the form of stilpnomelane-rich bands with still clearly recognizable shard structures (LaBerge 1966a, b). Such are described from two major iron-formations (both of which have undergone only very low grade metamorphism and some local tectonic deformation) for which the stratigraphy is extremely well known, namely, those of the Hamersley Range of Western Australia (Trendall and Blockley 1970) and of the Kuruman and Griquatown iron-formations in the Transvaal Supergroup of South Africa (Beukes 1983).

An evaluation of the sedimentologic setting and subsequent interpretation of the original basin in which iron-formation was deposited is generally a much more difficult task than measurement and compilation of the stratigraphy of an iron-formation. Furthermore, iron-formation sequences have commonly been primarily studied by mineralogists, metamorphic petrologists, and economic geologists with lesser input from sedimentologists. The assessment of the sedimentary setting of metamorphosed iron-formations, be they of Archean or Proterozoic age, is a difficult task. Major metamorphosed iron-formation sequences that show extensive regions of very low grade metamorphic conditions are found in the Lake Superior region of the USA and Canada, and in the Labrador Trough sequence in Canada. Sedimentologic models for the Animikie Basin in the Lake Superior region are found in the work of James (1954) and extensive subsequent analysis by Morey (see Morey 1983, for a review). Although a broad sedimentologic picture is now available for the Lake Superior region, Morey (1983) notes that "understanding the detailed sedimentologic history of a particular iron-formation in the basin is still an important objective." The sedimentology of the iron-formations and associated lithologies in the Labrador Trough have been extensively studied by Dimroth and colleagues, by Zajac (see Gross and Zajac 1983, for a review), and more recently by Simonson (1985a). Gross and Zajac (1983) provide a broad picture of chemical precipitation of iron-formation in high-energy sedimentary environments of shallow troughs, layered basins, and tidal flats. They do not support the replacement origin for these iron-formations as proposed by Dimroth and Chauvel (1973). However, many details of the reconstructed sedimentary basins are uncertain.

Although the stratigraphy of the Hamersley Range in Western Australia (see Trendall and Blockely 1970, and a more recent review by Trendall 1983b), which has undergone only greenschist facies (burial) metamorphism, is extremely well documented, the sedimentologic reconstruction of this vast iron-formation basin is only in its infancy (Simonson 1989). Trendall (1983b) notes that the "lack of detailed information on the lateral extent, thickness variations, and depositional environment of each unit of the Fortescue Group (which underlies the major iron-formation sequences) precludes any definite conclusion concerning the tectonic development of the basin during deposition."

In terms of major, well-documented, only slightly metamorphosed and essentially non-deformed iron-formation sequences, the most complete stratigraphic and sedimentologic reconstruction of the basinal sequence is available from the Griqualand West region of the Transvaal Supergroup in South Africa. This is almost totally based on the work of N. J. Beukes since 1973 (see Beukes 1983, for a review; see also Beukes 1984). This basinal sequence involves a facies transition from underlying limestone, dolomite, and shale to an overlying sequence of siderite-rich to oxide-rich iron-formations. Because it has undergone only a very low grade metamorphic overprint (within an estimated temperature range of 110° to 170°C) and is essentially undeformed, it provides a unique opportunity for evaluation of the geochemistry of the various lithologies as a function of the depositional basin constraints; this has led to much new information on chemical deposition processes and seawater chemistry (e.g., Klein and Beukes 1989). These matters are addressed below.

The very late Proterozoic iron-formations, among them iron-formations in the Rapitan Group of the Yukon and Northwest Territories, Canada (with an approximate age of about 0.7 to 0.8 Ga), those of the Morro du Urucum region of western Brazil and eastern Bolivia (approximately 0.8 Ga-old; Dorr 1973a), and the iron-formations in the Damara Supergroup of Namibia (with an approximate age of about 0.8 Ga; Kröner 1981a), all occur in stratigraphic sequences associated with deposits of glacial origin (Young 1976; Yeo 1986; Henry et al. 1986). The sedimentologic setting of these Late Proterozoic deposits is thus very different from all of the other Precambrian iron-formation sequences and their mineralogy tends to be mainly hematite and chert (see below for aspects of the Rapitan sequence).

4.2.4 Average Chemistry of Several Major Iron–Formations

The averages for the major oxide components as well as sulfur and carbon contents for a range of Precambrian iron-formations are given in Table 4.2.1 on an H_2O- and CO_2-free basis. All original analyses were recalculated in this way in order to make the analyses of unmetamorphosed and metamorphosed iron-formations properly comparable, assuming that metamorphic reactions are essentially isochemical except for the loss of H_2O and CO_2. The chemical similarities between, for example, the unmetamorphosed and highly metamorphosed Labrador Trough iron-formations (Table 4.2.1, columns 13 and 14) suggest that this is a reasonable scheme for the major element data. There is a relatively wide range between individual analyses as shown by the ranges, but the averages are remarkably similar for most of the analysis columns (except number 15 for a Rapitan Group iron-formation) even though iron-formations are very inhomogeneous and some of the averages and ranges in Table 4.2.1 are based on a relatively small number of samples. Even so, the general similarity of all of the averages, except those for the Rapitan iron-formation, is well shown in Figure 4.2.2. In the selection of these analyses, every effort was made to exclude iron-formation samples that had undergone any type of secondary alteration, such as oxidation or leaching, thus excluding any materials that might be considered iron ore, or in the process of becoming ore. This is reflected in the values for total iron, ranging from a minimum of 23 weight percent (column 7, S bands) to a maximum of 34 weight percent (column 6, BIF of the Brockman Iron Formation), as well as in the $Fe^{3+}/(Fe^{2+}+Fe^{3+})$ values. The latter range from 0.05 (column 10, siderite BIF in the Kuruman Iron Formation) to 0.58 (column 3, metamorphosed iron-formation in Montana), except for one value of 0.97 (column 15, Rapitan). It is instructive to compare the 0.05 to 0.58 range with the values for two of the iron oxides, magnetite [$Fe^{3+}/(Fe^{2+}+Fe^{3+}) = 0.67$] and hematite [$Fe^{3+}/(Fe^{2+}+Fe^{3+}) = 1$]. This illustrates that the iron in most iron-formations is in an average oxidation state between that of wüstite (FeO) and that of magnetite (Fe_3O_4), reflecting the very common association of magnetite with Fe^{2+}-containing minerals such as carbonates (siderite and ankerite), silicates (such as greenalite, in essentially unmetamorphosed iron-formations; minnesotaite;

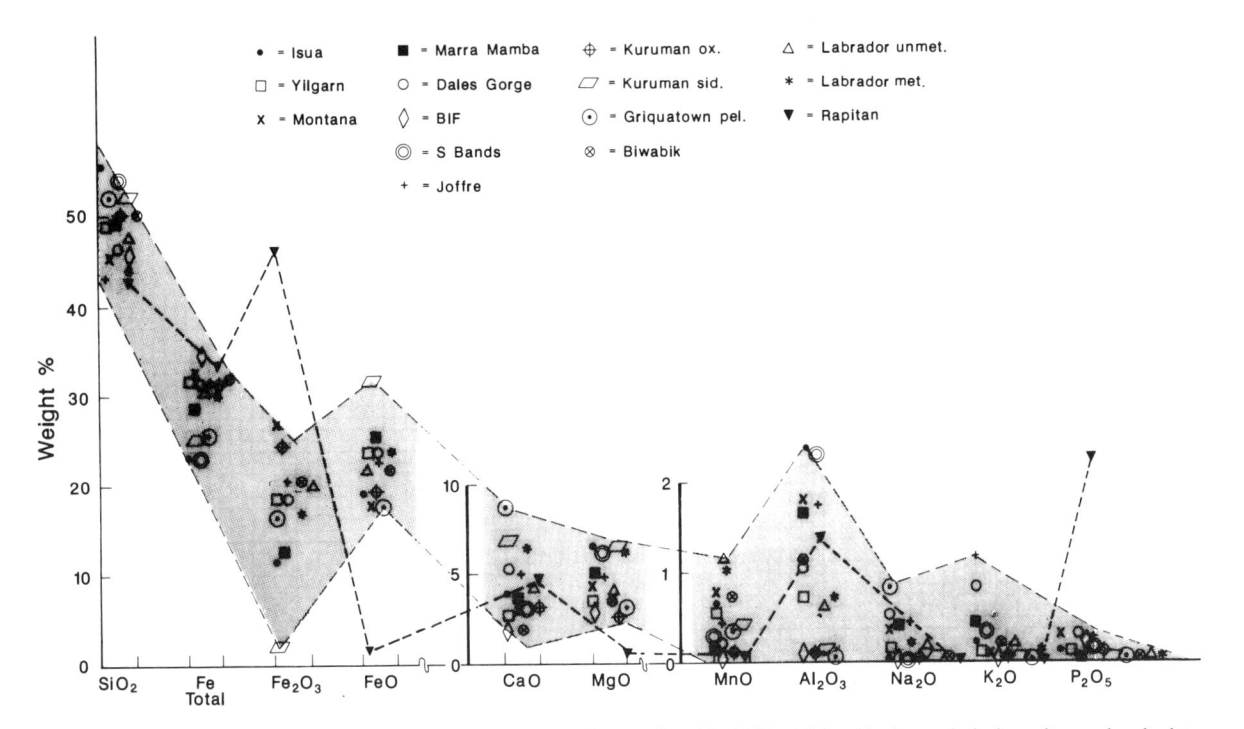

Figure 4.2.2. Plot of the major chemical components of iron-formations as listed in Table 4.2.1, with the analytical results recalcualted to 100% on an H_2O- and CO_2-free basis. Area enclosed by thin dashed lines brackets the overall range of all values except those of the Rapitan Group iron-formations which are denoted by solid triangles connected by a thick dashed line. Note that data for the various components are presented relative to three different (vertical) weight percentage scales.

Table 4.2.1. *Averages and ranges of major components in bulk analyses of Archean and Proterozoic banded iron-formations (recalculated to 100% on an H_2O- and CO_2-free basis).*

wt.%	1 Isua n = 28	2 Yilgarn Block n = 35	3 Montana n = 8	4 Hamersley Basin — Marra Mamba n = 9	5 Hamersley Basin n = 21	6 Hamersley Basin — Dales Gorge Member BIF	7 Hamersley Basin — Dales Gorge Member S bands	8 Joffre Member n = 17
SiO_2	55.86	49.07	45.53	49.16	46.20	45.76	54.01	43.31
	6.1–88.0	21.5–68.3	33.6–59.5	26.8–67.4	24.8–60.2			7.7–59.1
TiO_2	0.06	0.04	0.06	0.18	0.04	0.01	0.11	0.7
	<0.01–0.80	0–0.18	0–0.18	0–0.88	0–0.15			0.01–0.28
Al_2O_3	1.36	0.70	1.80	1.64	1.03	0.09	2.41	1.72
	0.03–12.4	0.01–3.51	0.1–5.86	0–6.00	0–5.12			0.04–7.54
Fe_2O_3	11.76	18.98	26.91	12.93	18.40	(49.17)[a]	(32.76)[a]	20.16
	0.13–33.4	1.7–37.6	3.4–38.3	0–39.5	0.1–35.9			3.1–40.1
FeO	19.25	23.65	17.51	25.49	23.88	n.d.	n.d.	22.53
	2.97–54.6	14.4–66.6	4.5–27.3	13.6–38.8	13.3–45.6			14.3–33.6
MgO	6.69	3.46	3.82	5.03	3.15	2.85	6.15	4.86
	1.01–16.5	1.0–9.63	1.25–5.72	2.14–9.42	1.5–6.66			1.78–14.81
MnO	0.62	0.55	0.64	0.13	0.18	0.05	0.28	0.36
	0.01–3.45	0.04–2.76	0.01–3.26	0.03–0.36	0–0.50			0.04–2.68
CaO	3.96	2.68	3.01	3.79	5.22	1.75	3.36	4.97
	0.03–20.3	0.32–7.61	0.51–11.48	0–10.47	0.74–39.6			0.05–37.4
Na_2O	0.01	0.11	0.34	0.38	0.50	0.04	0.02	0.39
	<0.01–0.62	0.02–0.57	0.01–0.97	0.01–0.91	0–4.97			0.01–2.11
K_2O	0.20	0.10	0.07	0.43	0.81	0.02	0.33	1.15
	<0.01–0.89	0–0.61	0.01–0.21	0.10–0.91	0–4.28			0–2.85
P_2O_5	0.16	0.16	0.29	0.08	0.31	0.22	0.19	0.25
	<0.01–0.41	0.02–0.45	<0.01–0.55	0.02–0.17	0.17–0.57			0.10–0.60
S	0.17	0.81	0	0.35	0.14	0.03	0.38	0.11
	0–3.54	0–5.20	0–0.01	0–2.11	0–1.36			0–0.84
C	0.63	0	0.02	0.50	0.22	n.d.	n.d.	0.15
	0–2.98	0–0.30	0–0.10	0–1.97	0–1.58			0–0.55
Total Fe	23.20	31.65	32.43	28.85	31.43	34.39	22.91	31.61
$\frac{Fe^{3+}}{(Fe^{2+}+Fe^{3+})}$	0.55	0.42	0.58	0.31	0.41	n.d.	n.d.	0.45

n.d. = not determined
[a] Fe_2O_3 values in parentheses means that total iron was recalculated as Fe_2O_3
[b] Mn-rich banded iron-formation excluded from averages

Sources:
1: Dymek and Klein (1988)
2: Gole (1981)
3: Immega and Klein (1976)
4: Klein and Gole (1981)
5: Trendall and Blockley (1970); Trendall and Pepper (1977); Klein and Gole (in prep.)
6: Samples from 103.78m of core of 17 BIF bands, Ewers and Morris (1981)
7: Samples from 18.65m of core of 16 S bands, Ewers and Morris (1981)
8: Trendall and Blockley (1970); Trendall and Pepper (1977); Klein and Gole (in prep.)
9 and 10: Klein and Beukes (1989)
11: Beukes and Klein (1990)
12: Lepp (1966)
13: Klein (1974); Klein and Fink (1976); Klein (1978); and Lesher (1978)
14: Klein (1978)
15: Klein and Beukes (1989)

Table 4.2.1. (Continued).

wt.%	9 Kuruman IF Oxide BIF n = 9	10 Kuruman IF Siderite BIF n = 16	11 Kuruman and Griquatown IF Peloidal n = 8	12 Biwabik n = 9	13 Labrador Trough unmet. n = 22	14 Labrador Trough met. n = 22	15 Rapitan n = 4
SiO_2	50.30	52.45	52.27	50.62	47.81	44.33	43.19
	37.0–56.38	11.90–65.50	21.3–69.22	46.9–53.4	16.1–87.2	15.2–63.6	30.85–62.32
TiO_2	<0.04	<0.04	0.04	0.06	0.04	0.10	0.20
	<0.04	<0.04–0.04	<0.04–0.05	0.02–0.13	0–0.25	0–0.59	0.03–0.55
Al_2O_3	0.08	0.13	0.06	1.13	0.62	0.74	1.41
	0.03–0.18	0.02–0.29	0.01–0.38	0.33–2.28	0.05–1.98	0–4.23	0.12–3.98
Fe_2O_3	24.59	1.67	16.99	20.28	19.96	16.87	46.62
	14.48–30.93	0.56–2.94	0.8–31.38	12.0–26.3	0–65.0	0–77.9	33.03–50.25
FeO	19.01	31.30	17.75	21.43	21.69	23.68	1.43
	8.77–21.85	15.26–37.58	10.48–23.24	17.4–26.7	2.1–46.8	1.1–40.9	0.50–3.47
MgO	2.55	6.60	2.82	3.17	4.00	6.25	1.20
	1.49–3.40	2.74–7.82	0.77–4.54	2.59–4.08	0–15.04	0.48–13.66	0.19–3.36
MnO	0.10	0.42	0.34	0.72	1.15[b]	1.01[b]	0.06
	0.06–0.19	0.08–0.54	0.09–0.44	0.41–1.02	0.06–2.52	0.12–2.66	0.03–0.09
CaO	3.13	7.07	8.76	1.98	4.30	6.55	4.40
	0.72–6.23	1.93–12.60	1.66–18.33	1.02–3.17	0–22.3	0–22.4	1.13–6.01
Na_2O	0.02	0.03	0.84	0.06	0.17	0.21	0.04
	0.01–0.07	0.01–0.03	0.01–2.19	0.02–0.09	0.01–3.05	0–1.49	0.01–0.06
K_2O	0.03	0.03	0.03	0.17	0.20	0.10	0.04
	0–0.08	0–0.09	0.01–0.31	0.07–0.31	0.03–0.46	0.02–0.54	<0.01–0.1
P_2O_5	0.12	0.08	0.07	0.09	0.04	0.06	1.26
	0.04–0.34	0.01–0.11	<0.01–0.26	0.05–0.17	0–0.15	0–0.13	0.49–2.16
S	0.01	0.07	<0.01	n.d.	0.01	0.06	n.d.
	0–0.05	0.01–0.15	0.001–0.03		0–0.05	0–0.44	
C	0.01	0.10	0.02	0.29	<0.01	0.06	0.14
	0.01–0.02	0.05–0.20	0.01–0.04	0.07–0.58	0–0.09	0–0.63	0.13–0.16
Total Fe	31.99	25.50	25.69	30.85	30.83	30.21	33.76
$\frac{Fe^{3+}}{(Fe^{2+}+Fe^{3+})}$	0.54	0.05	0.46	0.46	0.45	0.39	0.97

Table 4.2.2. *Averages of trace elements and rare earth elements (REE) for several Precambrian iron-formations as well as data for some single iron-formation samples (REE determined by neutron activation analysis at Washington University, St. Louis, Missouri, using procedures described by Korotev 1987).*

ppm	1 Isua n = 28	2 Brockman BIF #18-2 n = 1	3 Kuruman IF sid BIF n = 16	4 Kuruman IF mag-sid BIF n = 4	5 Kuruman IF Ouplaas Member n = 4	6 Sokoman IF hem-oolitic n = 1	7 Rapitan IF granular n = 1	8 Rapitan IF nodular n = 1
Sc	2.45	0.20	0.21	0.098	0.141	0.056	5.22	1.39
V	33.2	—[d]	<150	<150	<150	—[d]	—[d]	—[d]
Cr	31.6	3.7	3.23	2.75	3.6	3	25	18
Co	11.0	0.39	0.59	0.37	0.49	8.08	4.27	1.3
Ni	58.2	<40	29.3	29.5	46	13	14	9
Cu	27.5	—[d]	<15	<15	<15	25	93	67
Zn	83.7	—[d]	25.5	25	<20	44	53	41
As	0.72[a]	1.6	11.7	<0.3	0.85	42.7	10.5	3.43
Se	1.24[b]	<1	<0.2	<0.2	—[d]	<1	5.6	2.1
Br	0.58[c]	<1	2.83	2.14	0.27	0.22	0.29	0.74
Rb	<18	20	18.9	23	25	17	22	19
Sr	9.7	68	9.3	7.25	45	25	96	98
Y	6.0	—[d]	3.8	4.2	2	4	7	8
Zr	10.0	<40	31.8	18.5	26	<5	30	<5
Nb	0.63	—[d]	7.4	7	7	8	8	5
Sb	0.15	0.16	0.16	0.13	0.05	0.09	0.33	0.04
Cs	0.06	0.45	0.27	0.11	0.43	0.09	0.47	0.1
Ba	<5	30	<30	<30	11	60	24	334
La	2.34	1.76	1.421	1.02	0.78	5.58	7.56	2.32
Ce	3.92	2.24	2.06	1.41	0.94	9.48	17.1	4.32
Nd	2.03	—[d]	—[d]	—[d]	—[d]	5.4	9.8	4.4
Sm	0.510	0.258	0.228	0.177	0.110	0.86	2.26	0.715
Eu	0.369	0.114	0.088	0.069	0.033	0.298	0.604	0.189
Tb	0.118	0.046	0.042	0.041	0.019	0.147	0.34	0.169
Yb	0.585	0.495	0.242	0.223	0.150	0.491	1.34	1.21
Lu	0.102	0.091	0.052	0.052	0.031	0.066	0.218	0.208
Hf	0.25	0.06	0.040	<0.06	0.067	<0.1	0.56	<0.1
Ta	0.04	<0.4	0.025	0.05	0.019	<0.05	0.083	0.039
Pb	16.8	—[d]	—[d]	—[d]	—[d]	—[d]	—[d]	—[d]
Th	0.21	0.093	0.102	<0.04	0.042	0.034	0.7	0.099
U	0.08	<0.15	0.035	<0.2	0.086	0.09	0.18	0.05

[a] average of 14 determinations
[b] average of 5 determinations
[c] average of 17 determinations
[d] — = not determined

Sources: 1 – Dymek and Klein (1988); all BIF samples
          2 – Beukes and Klein (in preparation); microbanded BIF sample
          3 – Klein and Beukes (1989), Table 5; siderite-ankerite iron-formation samples
          4 – Klein and Beukes (1989), averaged samples are 7, 8, 9, and 10; magnetite-siderite iron-formation samples
          5 – Beukes and Klein (1990), samples 10, 11, 12, and 13; the Ouplaas Member is the uppermost part of the Kuruman Iron Formation
          6 – Beukes and Klein (in preparation); sample number 5-35, from the Upper Red Cherty Member
          7 – Klein and Beukes (in preparation); sample Y5-168.7, granular hematite-chert
          8 – Klein and Beukes (in preparation); sample Y7-142, laminated and nodular hematite-chert

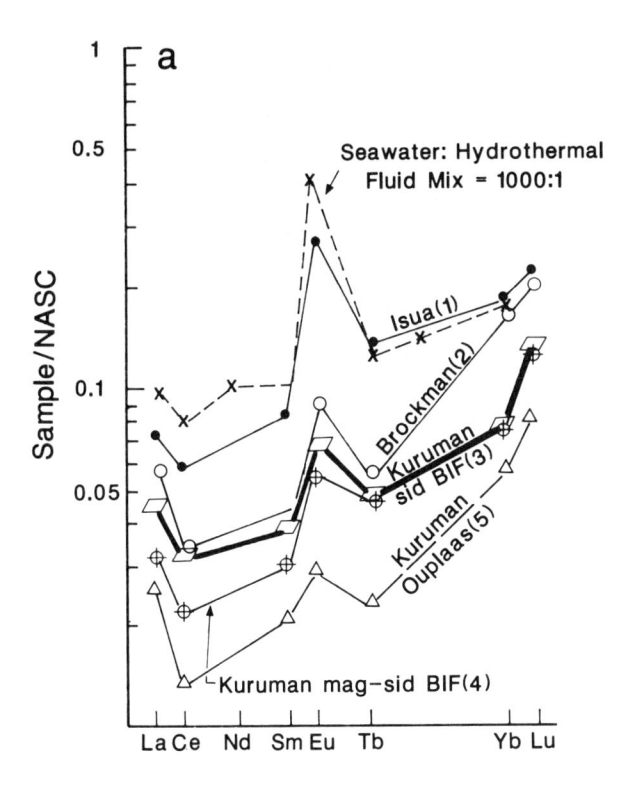

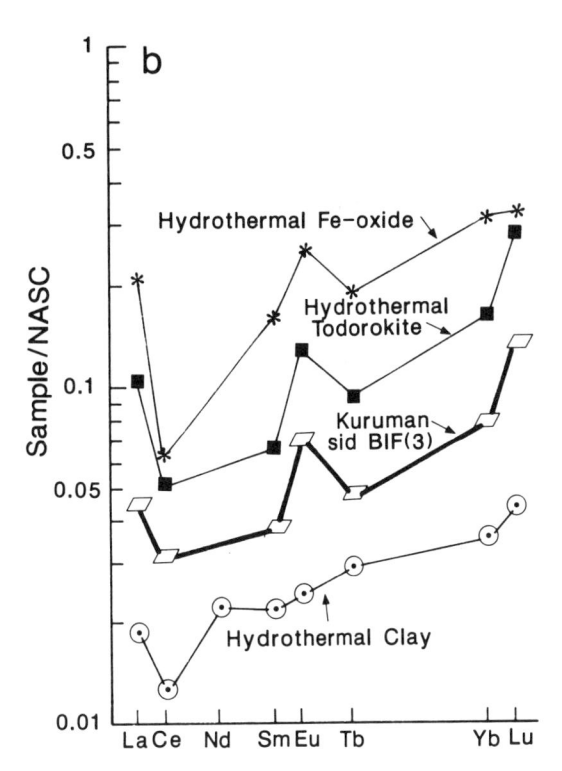

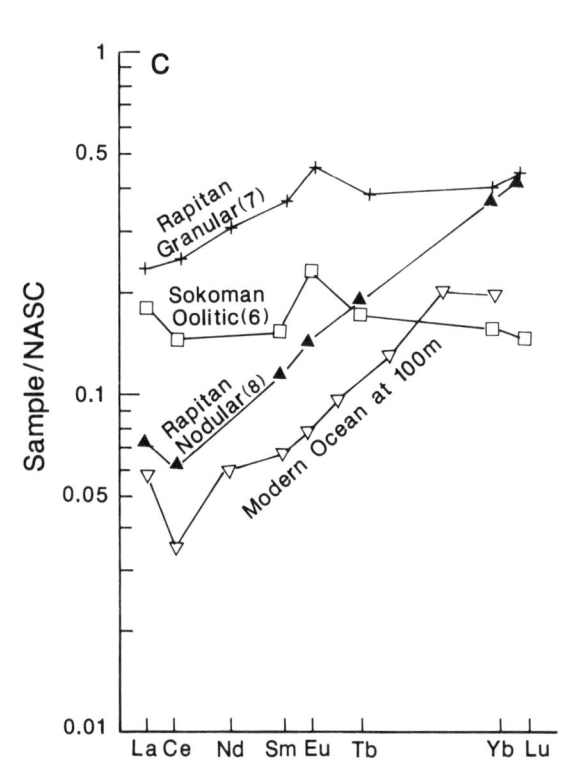

Figure 4.2.3. REE (rare earth element) patterns, normalized to the North American Shale Composite (NASC), for (a) several BIFs, (b) deep marine hydrothermal deposits and siderite-rich iron-formation from the Kuruman BIF, and (c) several hematite-rich BIF samples. (a) Patterns for analyses 1 through 5 listed in Table 4.2.2 compared with that for a calculated mixture of 1000 parts of North Atlantic Seawater from 100 m depth (Elderfield and Greaves 1982) and one part of hydrothermal fluid from the East Pacific Rise (Michard et al. 1982). Among the iron-formation patterns, there is a clear decrease in overall REE abundance, as well as a decrease in the positive Eu anomaly, with decreasing geologic age. (b) Patterns for some deep marine hydrothermal deposits from the Galapagos mounds (Corliss et al. 1978) compared with the average pattern for siderite-rich iron-formation from the Kuruman sequence (analysis 3 in Table 4.2.2). (c) Patterns for an oolitic hematite-rich sample from the Sokoman Iron Formation (analysis 6 in Table 4.2.2) and two hematite chert-rich samples from the Rapitan Group (Klein and Beukes in prep.), a hematite-rich, granular, volcaniclastic iron-formation (analysis 7 in Table 4.2.2; field sample no. Y5-168.7) and a very hematite-rich, laminated and nodular lutite (analysis 8 in Table 4.2.2; field sample number Y7-142).

members of the cummingtonite-grunerite series; and members of the orthopyroxene series in metamorphosed assemblages), and locally pyrite. The average $Fe^{3+}/(Fe^{2+} + Fe^{3+})$ value for 12 determinations in Table 4.2.1 (excluding the Rapitan value in column 15) is 0.42, which recalculates to an average oxidation state for iron in iron-formations of $Fe^{2.4+}$. This means that any models for iron-formation precipitation require considerably less oxygen input than would be needed if iron-formations are assumed to consist mainly of hematite, Fe_2O_3, as has been commonly done in the literature. The only iron-formation that is radically different from any of the others is the Rapitan iron-formation (column 15) which consists almost totally of chert and hematite. This difference is clearly shown in Figure 4.2.2.

Table 4.2.2 gives a listing of averages for trace and rare earth elements (REE) for several iron-formations in which these determinations were made. The REE patterns of analyses given in Table 4.2.2, normalized to the North American Shale Composite (NASC; Gromet et al. 1984) are given in Figure 4.2.3. Figure 4.2.3a shows REE patterns for the Archean Isua iron-formations and the very early Proterozoic Brockman and Kuruman microbanded iron-formations. All patterns are essentially the same showing pronounced positive Eu anomalies, negative Ce anomalies, and depletion in the light REE. Such patterns, also described by Fryer (1983) for other Archean and Proterozoic iron-formations, are very similar to those of some modern deep sea hydrothermal deposits (Figure 4.2.3b) and that which is obtained when modern deep sea hydrothermal fluids are mixed with typical North Atlantic Seawater. The pattern of such a mixture, consisting of 1000 parts of North Atlantic Seawater from 100 meters depth (Elderfield and Greaves 1982) and one part of hydrothermal fluid from the East Pacific Rise (Michard et al. 1983), is shown at the top of Figure 4.2.3a. Such mixing calculations, originally made by Dymek and Klein (1988), produce REE patterns with a striking similarity to those of the various iron-formations. This similarity leads to the interpretation that Precambrian iron-formations ranging in age from those at Isua (3.8 Ga) to those about 1.9 Ga-old, (e.g., the Soloman Iron Formation) are the result of chemical precipitation from solutions that represent mixtures of seawater and hydrothermal input.

Figure 4.2.3b shows the similarity of the iron-formation patterns to those of some modern deep sea hydrothermal deposits, and Figure 4.2.3c shows the fairly flat REE pattern for a sample of oolitic hematite-rich iron-formation of the Sokoman Iron Formation in which a positive Eu anomaly is still clearly present. For comparison, this diagram also shows a pattern for modern North Atlantic Seawater (at 100 m depth), obtained from Elderfield and Greaves (1982), which in contrast totally lacks the Eu anomaly, and exhibits a considerably stronger negative Ce anomaly. The two iron-formation samples of the Rapitan sequence (about 0.7 to 0.8 Ga in age) are considerably different from any of the patterns in Figure 4.2.3a or of that given in Figure 4.2.3c for the Sokoman iron-formation sample. A nodular and laminated hematite-chert sample (number 8) shows a pattern that completely lacks an Eu anomaly with a general REE pattern distribution similar to that of the modern seawater pattern at 100 meter depth. The other sample (number 7; granular with a volcaniclastic component) does show an Eu anomaly, probably reflecting the proximity of this volcaniclastic iron-formation sample to hydrothermal source terranes, as in rift areas of modern oceans. The patterns of Figures 4.2.3a and 4.2.3c show a continuous decrease in the size of the positive Eu anomaly from older to younger iron-formations, with the anomaly completely absent in an iron-formation of Rapitan age (see also Fryer 1983). This is verified by Moeller and Danielson (1988) who, on the basis of chondrite-normalized REE patterns for iron-formations in the Hamersley Range, conclude that positive Eu anomalies start to disappear in iron-formations of middle Early Proterozoic age. Very similar conclusions on the basis of the Nd and Sr isotopic compositions of a range of Proterozoic iron-formations, are reported by Derry and Jacobsen (1988). They state that the transition from Archean mantle-dominated systematics to modern seawater behavior was established by approximately 0.7 to 0.8 Ga. It appears, therefore, that over time the hydrothermal component in Archean-Early Proterozoic ocean water decreased, and that at the time of deposition of the Late Proterozoic Rapitan iron-formation it was generally absent from the system.

Further support for the notion of hydrothermal input into ocean waters from which iron-formations precipitated can be obtained from the chemical distinctions between REE-rich deep sea sediments and REE-poor hydrothermal deposits such as studied by Bonnot-Courtois (1981) in the FAMOUS area and the Galapagos mounds. The hydrothermal deposits consist in large part of the Fe-rich layer silicate nontronite, whereas their bulk compositions are remarkably similar to those of typical BIFs. Figure 4.2.4 illustrates the nature of this distinction in terms of (Co + Ni + Cu) abundances versus total REE content. All the iron-formation data points fall in or near the field of the hydrothermal deposits.

In short, the striking similarities in the REE patterns of iron-formations to seawater dominated-hydrothermal fluid mixtures, and the trace element distribution plots in Figure 4.2.4, suggest that similar processes with hydrothermal inputs have been responsible for the origin of iron-formations throughout Precambrian time. This conclusion is supported by sulfur (Goode et al. 1983) and Nd (Jacobsen and Pimental-Klose 1988) isotopic compositions of Archean and very early Proterozoic iron-formations, and Sr^{87}/Sr^{86} ratios in early Precambrian carbonates (Veizer et al. 1982). ^{18}O depletion in chert and carbonates from this time period (Perry and Tan 1972; Veizer and Hoefs 1976; Perry and Ahmad 1983; Beukes et al., 1990) may also be related to enhanced hydrothermal activity in the ocean system.

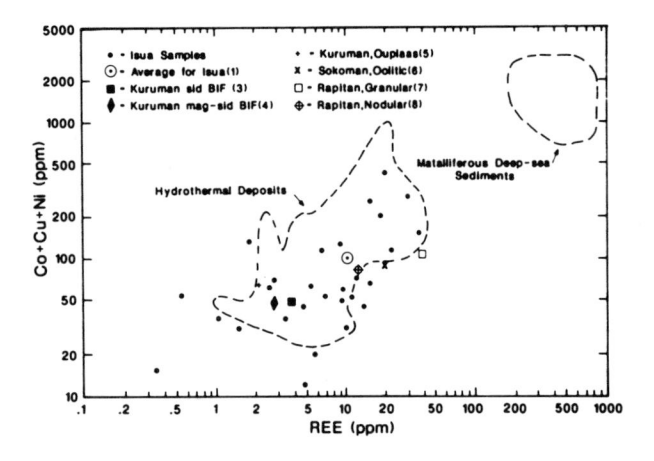

Figure 4.2.4. A plot of (Co + Ni + Cu) abundances versus total REE content (La + Ce + Nd + Sm + Eu + Tb + Yb + Lu) for analyses listed in Table 4.2.2 as well as data for all analyzed BIF samples from the Early Archean Isua Supracrustal Group (Dymek and Klein 1988). The field labeled "Hydrothermal Deposits" encloses data for deposits from the FAMOUS and Galapagos regions, which are mostly green muds/nontronite, whereas the field labeled "Metalliferous Deep-Sea Sediments" represents mostly DSDP samples from East Pacific sites (see Bonnot-Courtois 1981 for an extended discussion of these data). Data points for all the iron-formation samples are contained almost entirely within the "hydrothermal" field, suggesting a possible common origin.

4.3

Models for Iron-Formation Deposition

NICOLAS J. BEUKES   CORNELIS KLEIN

4.3.1 Introduction

Over the last 15 years a number of authors have specifically addressed the chemical and/or sedimentological aspects of iron-formation deposition. In chronological order, these include: Cloud (1973); Eugster and Chou (1973); Holland (1973b); Drever (1974); Button et al. (1982); Ewers (1983); Holland (1984); Garrels (1987); and Morris and Trendall (1988). In most of these models, the chemical sediment (iron-formation) is precipitated from a relatively open stratified Precambrian ocean in which an upper oxic layer overlies a much larger volume of anoxic water. Garrels' (1987) model, in contrast, is based on the evaporation of stream water in restricted basins. Because of Garrels' attempt to explain the microbanded aspects of the iron-formations in the Hamersley region in such a restricted basin, this model has been questioned by Morris and Trendall (1988).

In most of the marine models, the necessary concentrations of iron (and, possibly, silica and other oxide components) in solution are stored in the large volume of essentially anoxic deep water, which through upwelling comes into contact with the more oxygenated shallower water. Where these two water masses interact, through upwelling, is generally considered to be the depositional environment for the various types of iron-formation, with the oxygen content of the upper waters responsible for any precipitation of iron hydroxides and/or oxides.

In several models (e.g., Cloud 1973), microorganisms are considered to have played a direct role in precipitation of the iron. Furthermore, Perry et al. (1973), Walker (1984, 1987), and Baur et al. (1985) explain the isotopically light carbonate carbon in iron-formation as resulting from the oxidation of organic matter, by a reaction such as: $6 Fe_2O_3 + C \rightarrow 4 Fe_3O_4 + CO_2$, in which primary hematite is converted to magnetite. There are several problems with this reaction mechanism: (i) magnetite appears commonly as a primary precipitate in iron-formations, (Klein 1974; Klein and Bricker 1977); (ii) productivity of organic matter in the water column during iron-formation deposition may well have been low, as deduced from very low organic carbon values in BIF (Klein and Beukes 1989); and (iii) the scarcity of well-preserved organic remains (microfossils or stromatolites) in banded (non-peloidal) iron-formations (Walter and Hoffman 1983).

The fine-scale alternation of iron-rich and iron-poor microbands, mainly in iron-formations of the Hamersley Range, Western Australia and the Kuruman Iron Formation sequence in South Africa, is interpreted as the result of deep water (below wave base) deposition linked with evaporation. The interpretation of microbands as varves (Trendall and Blockley 1970) led Garrels (1987) to develop a quite realistic chemical precipitation scheme for alternating hematite-chert-carbonate microbands. However, the evaporative origin proposed for such varves is not without problems as pointed out by Kaufman et al. (1990). These authors conclude that mixing of waters of different isotopic compositions is a more likely mechanism for the production of primary isotopic compositions in microbanded Brockman iron-formations of the Hamersley Range of Western Australia, than is evaporation.

The non-microbanded iron-formations exhibiting granules, ooids, and cross bedding are considered the shallow water equivalents of the microbanded BIFs of deeper water origin.

In general, in the above models for BIF deposition, the predicted paleoenvironment is based upon modern oceanic conditions and present-day shallow water carbonate deposition (Drever 1974). They also have in common a lack of sedimentologic constraint of any basinal configurations, because, in general, detailed sedimentologic interpretations of BIF sequences have not been available. Joint geochemical and sedimentologic studies of the iron-formation sequences in the Transvaal Supergroup by Klein and Beukes (1989), Beukes and Klein (1990), and Beukes et al. (1990), over the last seven years, are providing a much-needed correlation between detailed geochemistry and the well-established sedimentologic setting (see Beukes 1983, for review) of these iron-formations. The following Section provides a synopsis of these studies which lead to a new interpretation of the depositional setting of iron-formations as based on the South African occurrences.

4.3.2 Paleoenvironmental Interpretation of Iron-Formation Deposition in the Transvaal Supergroup, South Africa

The geochemistry and sedimentology of the transition from underlying interbedded carbonate and shale to banded

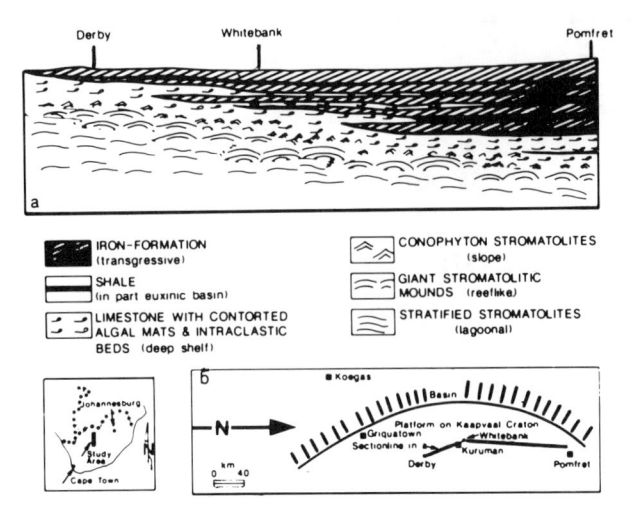

Figure 4.3.1. (a, above) Schematic reconstruction of the transgressive environment of iron deposition on top of the Kaapvaal Craton which consists mainly of limestone, and lesser dolomite, containing abundant stromatolites and lithified microbial mats; the carbonate and shale units are interpreted as representing regressive increments of sedimentation and the iron-formation as transgressive units (Klein and Beukes 1989). (b, below) Location of the schematic cross section shown in a, above.

iron-formation in the Transvaal Supergroup (with an age of approximately 2.5 to 2.3 Ga) of South Africa is documented in detail by Klein and Beukes (1989). Figure 4.3.1 is a schematic reconstruction of a transgressive iron-formation sequence (the Kuruman and Griquatown Iron Formations) on top of a stromatolitic carbonate platform sequence on the Kaapvaal Craton. The stromatolitic carbonates and associated shale beds represent regressive increments of sedimentation, whereas the iron-formation units were deposited immediately following transgressive events. At such times, siliciclastic input was inhibited and primary organic matter and carbonate production shifted shoreward (Klein and Beukes 1989). This process of alternating regressive and transgressive sedimentation is well illustrated by geochemical signatures in the sequence.

The shallow water stromatolitic limestones and associated carbonaceous shales have geochemical signatures that are distinct from those of the interbedded, deeper water, micro-banded iron-formations. This is well illustrated in Figure 4.3.2 with the limestones and shales displaying REE patterns similar to those of modern shallow marine surface waters with terrigenous detrital influx, and the iron-formation patterns being similar to deep marine water (with no terrigenous input) with a pronounced hydrothermal component admixed, as evidenced by the positive Eu anomalies. A cross-plot of Al_2O_3 versus organic carbon contents (Fig. 4.3.3a) further supports the concept of the iron-formation having been deposited from water masses devoid of detrital input, as opposed to that of the association of limestone and shale. The geochemistry and sedimentologic data support the concept of the iron-formations having been precipitated in deep marine environments, far removed from weathered continental sources. The iron, and probably silica as well, were derived from a hydrothermal source (as discussed in Section 4.2).

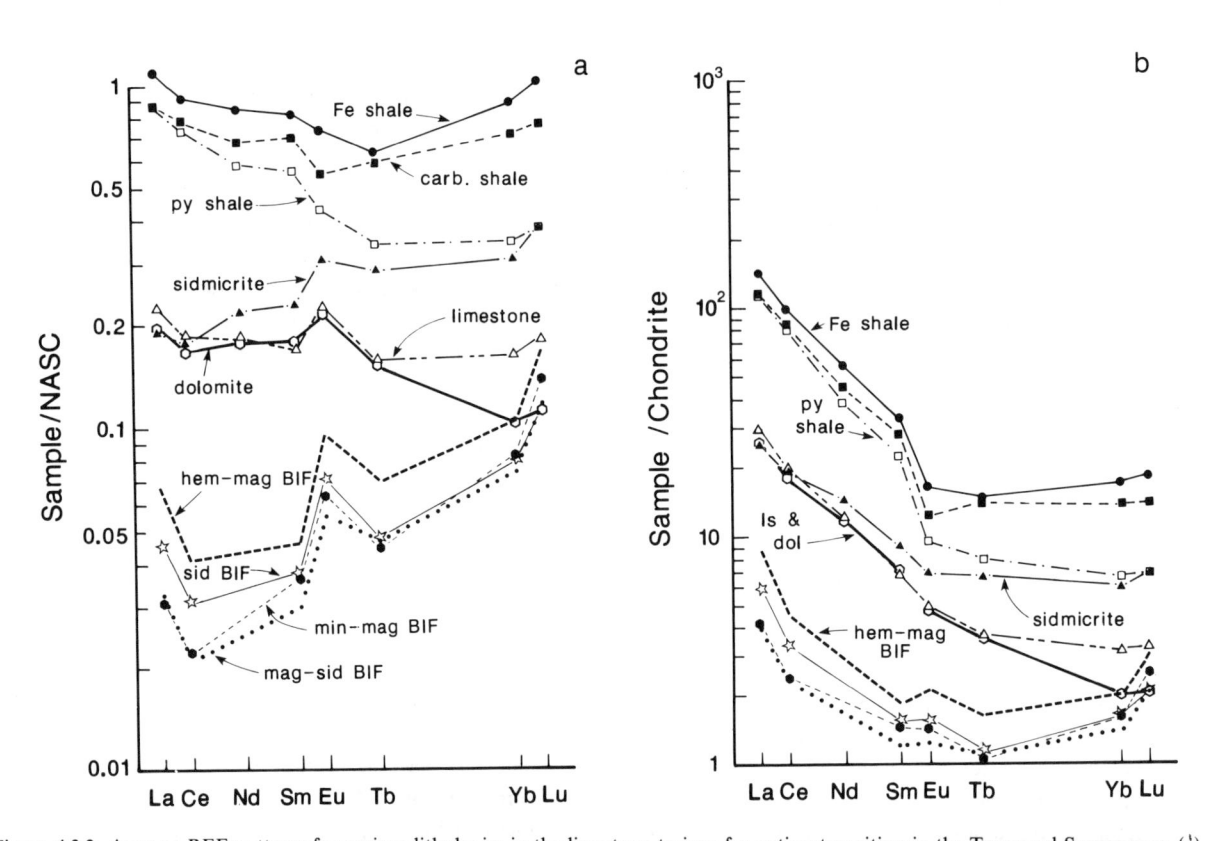

Figure 4.3.2. Average REE patterns for various lithologies in the limestone to iron-formation transition in the Transvaal Supergroup, (a) normalized to the North American Shale Composite (NASC), and (b) normalized to chondrites (Klein and Beukes 1989).

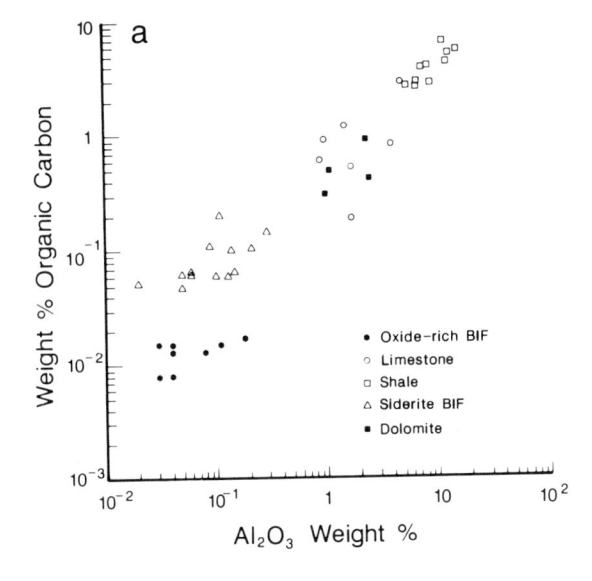

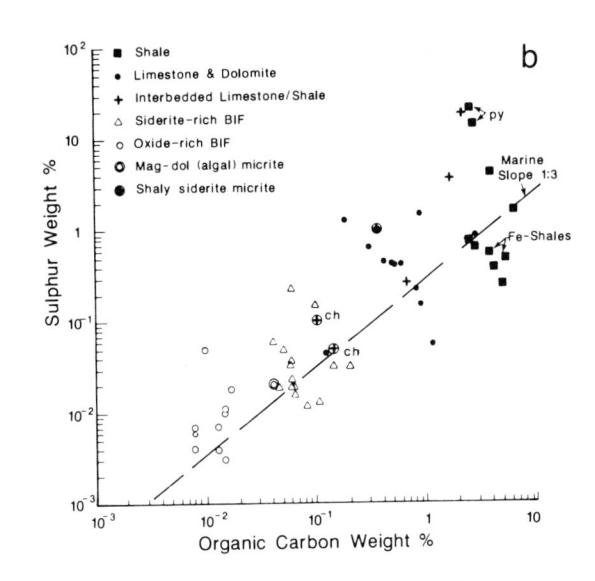

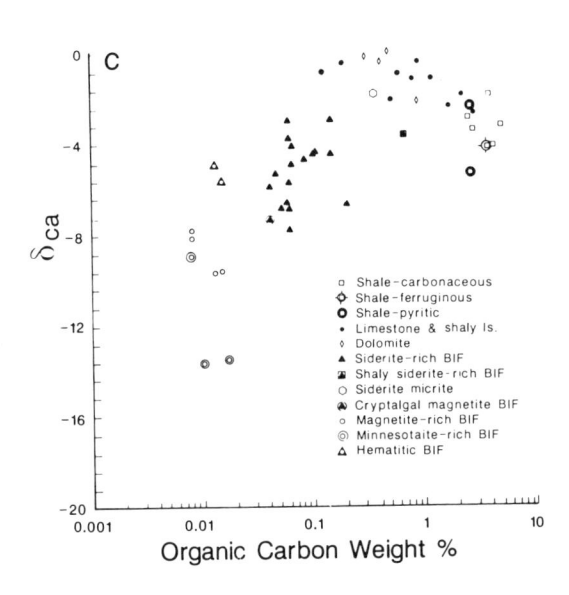

Figure 4.3.3. Geochemical aspects of the transition from limestone/shale to microbanded iron-formation in the Transvaal Supergroup (cf. Fig. 4.3.2). (a) Total organic carbon as a function of weight percent Al_2O_3 (Klein and Beukes 1989). (b) Sulfur as a function of weight percent of total organic carbon. (c) Carbon isotopic composition of carbonates (δ_{ca}) as a function of weight percent of total organic carbon (Beukes et al. 1990).

A further aspect of the studies by Klein and Beukes (1989), Beukes and Klein (1990), and Beukes et al. (1990) is the insight they provide into the relationship of the primary production of organic (carbon) matter and iron-formation deposition. Figure 4.3.3a suggests that most of the organic (carbon) matter in the depositional basin was derived from the shallow, near-shore stromatolitic limestone/shale lithofacies with very little organic carbon reaching the deeper and distal parts of the iron-formation depositional system. This notion is further supported by the carbon-sulfur-iron relationship shown in Figure 4.3.3b for the various lithofacies in the limestone to iron-formation transition. The organic carbon/sulfur ratios are very similar to the present-day open marine curve, suggesting that sulfate reduction was taking place at that time (Berner 1984; Raiswell and Berner 1985; Leventhal 1987). The age of the transition zone (from limestone to BIF deposition) studied by Klein and Beukes (1989) conforms to that of a sequence in which Cameron (1982) established positive evidence for sulfate reduction, as based on sulfur isotopic studies. Figure 4.3.3b also depicts the very low sulfur contents of the iron-formations which would correlate with low initial organic carbon contents. Other evidence presented by Klein and Beukes (1989) and Beukes and Klein (1990) in favor of low primary organic productivity in the locale of iron-formation deposition are: (i) high estimated rates of sedimentation for microbanded iron-formations which would have been favorable to the preservation of organic matter (Müller and Mangini 1980) if it had originally been present; (ii) very low phosphorus and low barium contents in most iron-formations; and (iii) little preservation of organic carbon in siderite-rich iron-formations (see Fig. 4.3.3a) which must have been deposited under reducing conditions. It is therefore concluded that iron-formation deposition and primary organic productivity were decoupled as first hinted at by Towe (1983). This is in contrast to models that couple deposition of iron-formation to either aerobic (Cloud 1968; Button et al. 1982) or anaerobic (Hartman 1984; Walker 1987) photosynthesis.

Decoupling of iron-formation deposition and primary pro-

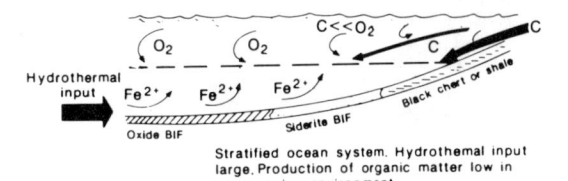

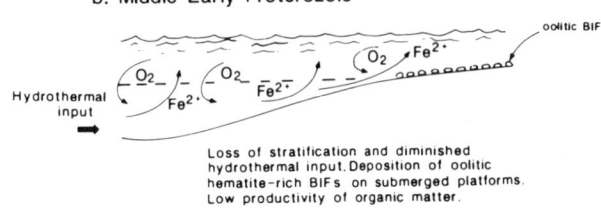

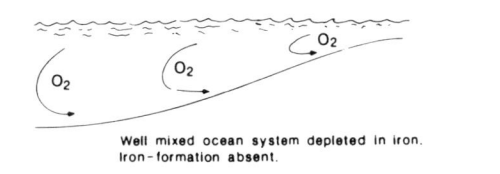

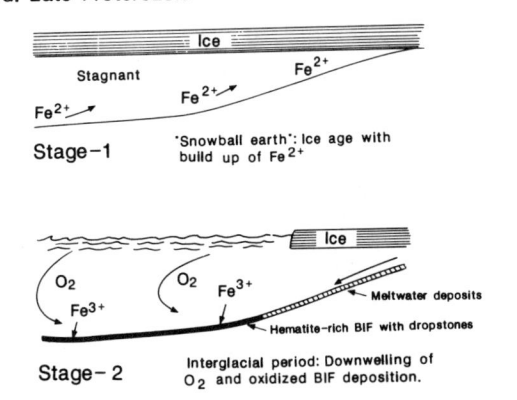

Figure 4.3.4. Paleoceanographic models for iron-formation deposition from the Archean to the Late Proterozoic. (a) Archean to Early Proterozoic: stratified ocean system with predominantly deep water deposition of microbanded iron-formation (after Klein and Beukes 1989). (b) Middle Early Proterozoic: breakdown of the stratified ocean system and deposition of hematite-rich oolitic iron-formations. (c) Middle Proterozoic: iron-depleted, well mixed ocean system with no deposition of iron-formations. (d) Late Proterozoic: "snowball Earth," with build-up of ferrous iron in solution in deeper water during glacial periods (Stage 1) and deposition of iron oxides during interglacial periods (Stage 2).

1968; Walker 1987) or abiotic oxidation (Cairns-Smith 1978; Francois 1986) oxide-rich iron-formations would have been deposited in shallow water environments. However, there is no geological evidence for this because typical Archean iron-formations are very similar in texture, bulk composition (major and trace elements), and mineralogy to the giant microbanded Early Proterozoic iron-formations (Gole and Klein 1981; Klein and Beukes 1989; see also Figs. 4.2.2 and 4.2.3) suggesting similar deep water environments of deposition. Archean shallow water chemical sediments, such as Early Archean stromatolitic cherts (Lowe 1980; 1983; Walter et al. 1980; Byerly et al. 1986) and stromatolitic Middle (Mason and Von Brunn 1977; Beukes and Lowe, in press) and Late (Henderson 1975; Martin et al. 1980) Archean carbonates, are concluded to have formed in the presence of oxygen-producing cyanobacteria (Schopf and Packer 1986, 1987; Section 1.5) and are depleted in iron. Especially in the Middle Archean Pongola stromatolites (Mason and Von Brunn 1977) there is good evidence that they were deposited in tidal marine environments (Von Brunn and Mason 1977; Beukes and Lowe, in press) which would imply that Archean ocean surface waters were depleted in dissolved iron (Towe 1983). This is similar to the Early Proterozoic environment for the deposition of iron-poor stromatolitic limestones in a shallow water platform and that of microbanded iron-formation under deeper basinal conditions (Beukes 1987). The interpretation of such conditions does not favor the coupling of iron-formation deposition to any photochemical process, be it biogenic or abiogenic.

Possibly a more satisfactory model is one of a more or less permanently density-stratified ocean system with surface waters that were somewhat oxic and deeper waters that were anoxic and enriched in dissolved ferrous iron. Iron-mineral precipitation would have taken place along the chemocline with the mineral assemblages reflecting variable Eh/pH conditions, determined by the fluxes of organic matter from the shallow photic zone and of ferrous iron from the deeper waters (Fig. 4.3.4a). In regions of high organic matter supply, black shales or carbonaceous chert would have been deposited. Siderite-rich iron-formations would have precipitated in locales of intermediate supply. In regions of low organic input, the presence of some oxygen in the surface waters would have caused oxidation of ferrous iron and the subsequent precipitation of oxide-rich iron-formations along the chemocline (Fig. 4.3.4a). Oversaturation with respect to silica, due to the absence of silica-secreting microbes (Button et al. 1982; Siever 1988), could have caused almost continuous precipitation of chert, interrupted by the deposition of iron minerals due to seasonal or other effects on circulation, especially in the shallow surface water layer.

Another implication of decoupling primary organic productivity from iron-formation deposition is that ^{13}C depletion in the carbonates of iron-formations in general (Becker and Clayton 1972; Perry et al. 1973; Baur et al. 1985), and of siderite-rich iron-formations in particular (Beukes et al. 1990), may not be the result of the degradation of organic matter. Siderite-rich iron-formations in the Transvaal Supergroup are depleted by about 4 permil in ^{13}C with respect to their shallow water stromatolitic limestone equivalents which display a modern marine $\delta^{13}C$ value of close to zero permil

ductivity implies transfer of some oxygen from sites of its (photosynthetic) production to sites of the most oxidized (oxide type) iron-formation deposition. In turn, this would imply the presence of some oxygen in the atmosphere and surface ocean waters even in Archean times. If the Archean atmosphere and surface ocean were anoxic as suggested by Walker et al. (1983), Walker (1987), and Kasting (1987), dissolved ferrous iron would have been present in it and either through biotic (Cloud

(Fig. 4.3.3c; see also Beukes et al. 1990). This translates to a $\delta^{13}C$ difference of about 9 permil if effects of mineral fractionation (Golyshev et al. 1981) are taken into consideration. Klein and Beukes (1989) and Beukes and Klein (1990) present petrographic and geochemical evidence indicating that many of the siderite microsparites are primary sedimentary precipitates formed in the water column. This is in contrast to the suggestion by Walker (1984) that siderite in iron-formations is the diagenetic product of a ferric oxide precipitate having reacted with organic matter as a reductant. The oxide-rich iron-formations in the Kuruman sequence consistently contain less carbon than do siderite-rich iron-formations (see Fig. 4.3.3), suggesting that the latter could not have been derived from the former by making use of organic carbon as a reductant. This process is also suggested as the cause for ^{13}C depletion in the siderites (Walker 1984), yet carbonates in the oxide-rich iron-formation are generally more depleted in ^{13}C than those in siderite-rich iron-formations (Fig. 4.3.3c). Similarly, the isotopic composition of organic carbon in oxide-rich iron-formations is more enriched in ^{13}C than that of siderite-rich units and not the other way around as would be the case if the siderite had been derived from a reaction between hematite and organic matter (Beukes et al. 1990). It is therefore concluded that the ^{13}C depletion in siderite microsparites represents a primary signature of Early Proterozoic deep ocean water, and it appears that the ocean system at that time was most probably also stratified with regard to the ^{13}C composition of total dissolved carbon in contrast with present-day oceans (Schidlowski et al. 1983).

4.3.3 Paleoenvironmental Interpretation of Iron-Formation Deposition in Later Proterozoic Time

In middle Early Proterozoic time, the stratified ocean system may have started to break down (Fig. 4.3.4b) with the development of abundant oolitic iron-formations such as those of the Lake Superior area (Goodwin 1956; Shegelski 1982; Simonson 1985b; Ojakangas 1988), the Sokoman Iron Formation in the Labrador Trough (Dimroth and Chauvel 1973; Chauvel and Dimroth 1974), and of the Nabberu Basin of Western Australia (Goode et al. 1983) in shoal areas. A mechanism for the transport of iron to surface ocean water must have developed. This may have been the result of a declining chemical density stratification due to lesser hydrothermal input as indicated by REE results (Fig. 4.2.3). Following this period of time (ca. 1.9 Ga), the oceans may have become completely mixed, oxygenated, and depleted in iron (Fig. 4.3.4c) as indicated by the absence of iron-formations in late Early and Middle Proterozoic times (Fig. 4.2.1).

In Late Proterozoic time, however, iron-formations are again part of the geologic record. These iron-formations are intimately associated with glaciomarine deposits and may also contain interbedded manganese deposits (Dorr 1973b; Walde et al. 1981; Roper 1956). In the Rapitan sequence, the iron-formation beds, composed essentially of hematite femicrite, occur immediately above planes of transgression at the base of progradational, shallowing-upward, glaciomarine sequences. A distinctive feature of the glaciations is that they apparently took place in near equatorial settings at ocean level as indicated by paleomagnetic data (Frakes 1979; Walter 1979) and abundant glaciomarine deposits (Yeo 1986). This may call for a "snowball earth" at certain periods of time (Section 2.3). A combination of all of these features may provide an explanation for the iron and manganese deposits of this period. In a "snowball-type" earth situation, sea level stand would have been very low and the ocean highly stagnant such that reducing conditions could have developed for accumulation of dissolved iron and/or manganese either from hydrothermal sources or from dissolution of material along basin floors. The onset of interglacial or postglacial stages would have resulted in transgressions, restoration of ocean circulation, and precipitation of iron and manganese-rich sediments at the base of prograding glacial meltwater deposits (Fig. 4.3.4d). Hydrothermal activity, generally thought to have been instrumental in the formation of the Late Proterozoic iron and manganese deposits (Yeo 1986; Breitkopf 1988), may then not have played such an important role. This is suggested by REE concentrations displaying patterns similar to modern deep marine water without any clear-cut hydrothermal signature in the form of a positive Eu anomaly (Fig. 4.2.3c).

4.4

Distribution and Paleoenvironmental Interpretation of Proterozoic Paleosols

HEINRICH D. HOLLAND

4.4.1 Introduction

Proterozoic paleosols are potential indicators of the O_2/CO_2 ratio in the atmosphere at their time of formation. In the absence of oxidants in the atmosphere, the cations that are released by the attack of weathering acids on the parent rocks of soils are flushed out of soils. Some of these cations are removed entirely; others tend to be reprecipitated at or below the water table. Fe^{+2} is perhaps the best example of this class of cations. In reduced soils, Fe^{+2} is flushed out of the upper soil layers but may be reprecipitated at depth as a Fe^{+2}-carbonate or -silicate. In contrast, Fe^{+2} in oxygenated soils is oxidized quantitatively to Fe^{+3} and is precipitated rapidly as one or more of the oxides or hydroxides of Fe^{+3}. The air and water in most soils today are highly oxidizing, because the O_2/CO_2 ratio in the atmosphere is very high. The only modern soils that are reducing are those of low permeability and that contain large quantities of organic reductants.

No vascular land plants existed during the Proterozoic. Thin mats of terrestrial algae and bacteria may have covered part of the land surface, but their effect on the oxidation state of soils was certainly much less than that of vascular land plants today. Their effect may have been sufficiently small, so that it can be neglected in the interpretation of paleosol chemistry (Pinto and Holland 1988). If so, the behavior of cations such as Fe^{+2} in Proterozoic soils was determined by the rate of supply of oxidants from the atmosphere and by their rate of use in oxidizing ions released by the attack of weathering acids. The functional relationships between the pertinent variables have been explored by Holland and Zbinden (1988) and by Pinto and Holland (1988). These authors have shown that the behavior of iron in paleosols can be used to make semi-quantitative estimates of the O_2/CO_2 ratio in the atmosphere at the time of paleosol formation. The estimates are only semi-quantitative, because they depend on a knowledge of the permeability of the paleosol at the time of soil formation. This parameter is difficult to reconstruct. The permeability of most modern soils is sufficiently high so that the diffusive transport of O_2 and CO_2 through pore spaces in soils is much more important quantitatively than the advection of these gases into soils with rain water. It is likely that this was also true in many Proterozoic soils. However, it is possible that the permeability of Proterozoic soils was significantly lower than that of modern soils, and that advection of O_2 and CO_2 with rain water was as important or more so than the diffusive transport of these gases from the atmosphere into Proterozoic soils. The difference is important for two reasons. Firstly, the diffusive transport of O_2 and CO_2 in soils depends on their diffusion constants, which are very similar; the advective transport of the two gases depends on their solubilities in water, which are quite different. Secondly, the relative effect of oxidants and reductants that are produced photochemically in the atmosphere depends on the mechanism of gas transfer into soils. Diffusive transport in most modern soils is so rapid that photochemical products such as H_2O_2, which are washed out of the atmosphere with rain water, have little effect on the redox balance of soils. It has been shown, however, that in low-O_2 atmospheres ($P_{O_2} < 0.01$ PAL) the concentration of H_2O_2 can exceed that of O_2 in rain water (Kasting et al. 1985). In highly impermeable soils that developed under such an atmosphere, the net concentration of advected oxidants could therefore have been influenced strongly by the presence of H_2O_2 and other photochemical products.

It could be argued that all of these uncertainties combine to render paleosols an unattractive source of information regarding the evolution of the atmosphere. However, other approaches to atmospheric evolution are even more burdened with uncertainties; the semi-quantitative estimates of the O_2/CO_2 ratio in the Precambrian atmosphere that are potentially extractable from paleosol data are therefore useful. To extract these estimates, we must establish that the rock units which we are studying are indeed paleosols, and that the pertinent chemical parameters of the paleosols have not been compromised by diagenesis and metamorphism. Holland and Zbinden (1988) have developed some useful criteria for identifying paleosols. These criteria have been applied to several Paleozoic and Precambrian paleosols (Zbinden et al. 1988; Holland et al. 1989; Feakes et al. 1989), and work is in progress on several additional Precambrian poleosols.

4.4.2 Implications for the Proterozoic Atmosphere

The results to date are encouraging, but not yet conclusive. As predicted by Pinto and Holland (1988), the behavior of iron in paleosols depends on the ratio of $(D_{O_2})_e$, the effective oxygen demand of the parent rock, to $(D_{CO_2})_e$, the effective acid demand of the parent rock (see Feakes et al. 1989). The effective R value of the parent rocks, R_e, is defined by the expression

$$R_e \equiv \frac{(D_{O_2})_e}{(D_{CO_2})_e}. \quad (4.4.1)$$

For most igneous rocks, R_e is approximately equal to

$$R_e \equiv \frac{\Delta M_{FeO}}{8[\Delta M_{CaO} + \Delta M_{MgO} + \Delta M_{Na_2O} + \Delta M_{K_2O} + \Delta M_{MnO}]}. \quad (4.4.2)$$

In this expression, the individual ΔM_i values are the differences between the number of moles of i in the parent rock and in its most weathered equivalent. When the most weathered equivalent contains essentially no FeO, CaO, MgO, Na$_2$O, K$_2$O, and MnO, the value of R_e is equal the value of R which is defined elsewhere (Holland 1984, Ch. 7) in terms of the concentration of these oxides in parent rocks alone.

Figure 4.4.1 summarizes much of the currently available data for the behavior of iron in paleosols older than ca. 200 Ma. The locations of the poleosols are listed in Table 4.4.1. Well-documented paleosols are starred in Figure 4.4.1. Data for paleosols that have been identified in drill cores alone are considered doubtful, and have been denoted by a question mark. In all of the paleosols that are younger than ca. 1.8 Ga, all or nearly all of the "FeO" present in the parent rocks was oxidized to Fe$_2$O$_3$ and retained in the weathering profile. In the two high-R paleosols which are more than 2.0 Ga-old, a large fraction of the "FeO" in the parent rocks was apparently lost during weathering. The concentration of total iron in the upper part of these paleosols is much less than that of their parent rocks. The loss of so much "FeO" from these paleosols implies that O$_2$ and other oxidants were exhausted within the paleosols

Table 4.4.1. *Paleosols included in the data plotted in Figure 4.4.1.*

Number	Paleosols and Location	References
1.	Denison Paleosol, southern Ontario, Canada	Gay and Grandstaff 1979; G-Farrow and Mossman 1988
2.	Pronto Paleosol, southern Ontario, Canada	Gay and Grandstaff 1979; G-Farrow and Mossman 1988
3.	Hekpoort Paleosol, Transvaal, South Africa	Button 1979; Hart 1986; Retallack 1986
4.	Dominion Paleosol, South Africa	Grandstaff et al. 1986
5.	Pre-Pongola (Jerico Dam) Paleosol, South Africa	Grandstaff et al. 1986
6.	Abitibi Paleosol, Canada	Vogel 1975
7.	Drakenstein Paleosol, Griqualand West, South Africa	Beukes 1985
8.	Flin Flon Paleosol, Manitoba, Canada	Goetz 1980; Holland et al. 1989
9.	Cape Wrath Paleosol, northwest Scotland	Williams 1968
10.	Sturgeon Falls Paleosol, Michigan, USA	Zbinden et al. 1988
11.	Squaw Creek Paleosol, Texas, USA	Capo 1984
12.	Butler Paleosol, Saint Francois Mountains Missouri, USA	Blaxland 1974
13.	Flagstaff Mountain Paleosol, Boulder, Colorado, USA	Wahlstom 1948
14a,b.	Arisaig Paleosols, Arisaig, Nova Scotia, Canada	Feakes et al. 1989

during weathering, that Fe^{+2} was not oxidized to Fe^{+3}, and that iron was therefore lost from the upper portions of the paleosols.

The Hekpoort Paleosol is known from outcrops along a strike distance of several hundred kilometers. Its most completely described outcrop is near Watervaal Onder, South Africa (Button 1979). The compositional transition there from the paleosol to the Hekpoort Basalt is obscured by modern weathering of the basalt, and Retallack (1986) has suggested that the upper portion of the paleosol is actually a shale. We believe that this is not the case, because a very similar profile, not as severely obscured by modern weathering, is exposed at the entrance to the Daspoort Tunnel near Pretoria (Hart 1986), and because metamorphosed equivalents of closely related profiles are exposed further to the west. It is possible that iron was lost from the Hekpoort Paleosol not during weathering but later, during diagenesis and/or mild metamorphism. This is, however, unlikely. Thin, contorted veinlets containing iron-rich chlorite are common in the uppermost, iron depleted part of the paleosol. These veinlets were probably desiccation cracks formed during weathering, and were filled with iron-rich clays while they were open. It is therefore very difficult to explain this

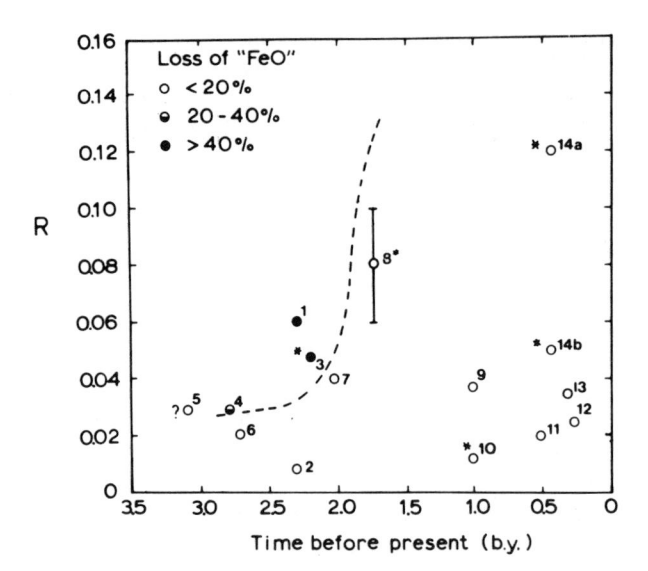

Figure 4.4.1. Summary plot of data for the oxidation state of paleosols during the last 3.5 Ga. As discussed in the text, R is the ratio of the oxygen demand to the CO$_2$ demand of the parent rocks of these paleosols.

juxtaposition of iron-rich and iron-poor material in the paleosol if iron loss from the paleosol occurred after the deposition of the overlying sediments (Holland and Feakes 1989).

The extensive "FeO" loss from the Hekpoort (Table 4.4.1, no. 3) and the Denison (Table 4.4.1, no. 1) Paleosols, and the apparent reduction of iron in paleosols developed on roughly contemporaneous low-R parent rocks can be used to set limits on the probable O_2/CO_2 ratio in the atmosphere between 2.4 and 2.2 Ga ago. As demonstrated by Pinto and Holland (1988), the change in P_{O_2} and P_{CO_2} in soils unaffected by biological processes is given the expression

$$\frac{P^\circ_{O_2} - P^L_{O_2}}{P^\circ_{CO_2} - P^L_{CO_2}} = \phi R_e, \qquad (4.4.3)$$

where

$P^\circ_{O_2} = P_{O_2}$ in the atmosphere;
$P^L_{O_2} = P_{O_2}$ in the soil air at a depth L;
$P^\circ_{CO_2} = P_{CO_2}$ in the atmosphere;
$P^L_{CO_2} = P_{CO_2}$ in soil air at a depth L;
R_e = effective R value of soil system; and
ϕ = a parameter whose value depends on the relative importance of the supply of O_2 and CO_2 via diffusion through soil air and via advection with rain water.

The value of R_e at which O_2 was apparently just exhausted in paleosols prior to 2.0 Ga ago is approximately 0.04. If the permeability of these paleosols was comparable to those of most modern soils, the value of ϕ was close to 1. The ratio of $P^L_{CO_2}$ at a depth of 10 m to $P^\circ_{CO_2}$ is on the order of 0.6 (Pinto and Holland 1988). Substituting these values into equation 4.4.3, we obtain

$$\frac{P^\circ_{O_2}}{P^\circ_{CO_2}} = (0.04)(0.4) = 0.016. \qquad (4.4.4)$$

This value of $P^\circ_{O_2}/P^\circ_{CO_2}$ is rather uncertain, but it is clearly much lower than the current value of 600. It seems likely that $P^\circ_{CO_2}$ was significantly higher 2.2 Ga ago than now. If we adopt the likely range of 0.02 to 0.25 atm given in Section 4.7, we obtain an atmospheric O_2 pressure of ca. 3×10^{-4} to 4×10^{-3} atm. This range is consistent with other indicators of atmospheric O_2 pressure, but the agreement should not be taken as proof that the value of $P^\circ_{O_2}$ during the Early Proterozoic has indeed been bracketed.

The data in Figure 4.4.1 suggest that the ratio $P^\circ_{O_2}/P^\circ_{CO_2}$ increased significantly between 2.2 and 1.8 Ga. This also agrees with other lines of geologic evidence, including the extensive oxidation of carbonate and silicate facies Kuruman Iron Formation exposed below a ca. 1.9 Ga erosion surface in western South Africa (Holland and Beukes 1990). No upper limit on the value of the $P^\circ_{O_2}/P^\circ_{CO_2}$ ratio during the past 1.8 billion years is set by the data in Figure 4.4.1. The paleosol data for the Archean suggest that molecular O_2 was present in the atmosphere as early as 3.0 Ga ago. These data are still very fragmentary, but studies that are currently planned and under way should go far to test the validity of this proposition.

4.5 Other Geological Indicators

DONALD R. LOWE

In addition to BIFs and paleosols, a number of geological materials are thought to be particularly useful indicators of ancient environments. The following discussion briefly reviews the implications of three of these indicators: glaciogenic deposits, red beds, and detrital uraninites.

4.5.1 Glacial Deposits

Glacial and periglacial sediments provide important indicators of paleoclimates, although the recognition of glaciogenic sediments and their discrimination from the products of mass flowage and even tidal processes (Williams 1987; Kvale 1989) can be problematic.

The distribution and characteristics of the principal Late Archean and Proterozoic glacial deposits have been reviewed by Harland (1983). Possible Late Archean Witwatersrand (2800 to 2700 Ma) glacial deposits, examined by the author, remain controversial. These diamictites were deposited under tectonically active conditions in environments that would have been favorable to the generation of debris flows.

Huronian glaciation (2400 to 2100 Ma) is well documented from the Lake Superior region of Canada and other areas in North America (Young 1970), and possible glacial strata have been described in the 2500 to 2300 Ma Transvaal Supergroup of South Africa (Visser 1971).

Middle Proterozoic glacial deposits have been reported locally (Harland 1983). Probably the major glacial episode of the Proterozoic was the Late Proterozoic glaciation (Harland 1964; Frakes 1979; Harland 1983; Chapter 12). Some recent studies suggest that Late Proterozoic glaciation may have involved extensive low-latitude ice sheets (Williams 1975; Embleton and Williams 1986; Chapter 12).

Overall, glaciation was sporadic throughout the Proterozoic, although the only episode that was demonstrably global in extent was that toward the end of the Eon. The development of thick stromatolitic carbonates and locally evaporites throughout the Proterozoic (Grotzinger 1986a) indicates the wide existence of warmer climatic conditions. Available evidence suggests that the Proterozoic earth's surface temperature and climatic regime ranged between extremes that were not radically different from those characterizing the Phanerozoic.

4.5.2 Red Beds

The formation of modern red beds has been documented in both desert (Walker 1967) and tropical (Walker 1974) environments in response to the slow oxidation of detrital minerals, particularly iron silicates. Although considerable debate has surrounded the detrital versus authigenic origins of red beds (van Houten 1973), extensive work on both modern and ancient red beds suggests that most formed authigenically under alluvial or less commonly shallow marine conditions where iron-bearing detrital minerals are subject to frequent wetting and drying by oxygenated pore waters (Walker 1967, 1974; Walker et al. 1978).

The geologic distribution of red beds is decidedly nonuniform. Although extensive alluvial sequences are developed in Archean and Early Proterozoic greenstone and cratonic deposits, classical red beds are absent. Reported occurrences of early Precambrian red sediments or early-formed red pigmentation, such as in the Late Archean of Canada (Shegelski 1980; Dimroth and Kimberley 1976; Dimroth and Lichtblau 1978), are local and clearly anomalous against the overall background of unoxidized terrestrial Archean and Early Proterozoic clastic sediments. The availability of some oxygen is required by the widespread deposition of Archean and Early Proterozoic ironformation (Section 4.2) and the concurrent development of local red clastic units might be expected. Some red beds may have been deposited between about 2.4 and 2.1 Ga ago (e.g., Cannon 1965; Salop 1977; Roscoe 1969, 1973), but the ages of most of these units are poorly constrained and probably lie toward the younger part of this interval.

The oldest reasonably well-dated red beds closely preceding the major interval of red bed deposition after 2000 Ma are in the 2224 to 2090 Ma-old Rooiberg Group in South Africa (Twist and Cheney 1986). This terrestrial, volcanic and sedimentary sequence includes a lower succession of dark brown, green, and black rhyolitic volcanic units interbedded with buff or gray sedimentary units, and an upper sequence of red rhyolites interbedded with red hematitic sediments. Twist and Cheney (1986) suggest that this transition, which parallels a similar upward change from gray to red sediments in the approximately correlative Huronian Supergroup of Canada (Roscoe 1969,

1973), may mark the transition from a non-oxic to an oxic atmosphere. Abundant red sediments appear in the geologic record after 2000 Ma, especially within foreland basins developed in association with late Early Proterozoic orogenic belts (Knight and Morgan 1981; Ramaekers 1981; and many others).

4.5.3 Detrital Uraninites

Like red beds, the presence of abundant detrital uraninites in Late Archean and pre-2.0 Ga Early Proterozoic sedimentary units has been considered by many to be indicative of an early oxygen-deficient atmosphere (Roscoe 1969; Grandstaff 1976, 1980; Langford 1983). Major Precambrian uraninite occurrences include those in the Witwatersrand Supergroup of Late Archean age (2800 to 2700 Ma) on the Kaapvaal Craton of South Africa (Simpson and Bowles 1977), and those in the 2400 to 2100 Ma Huronian Supergroup deposited on Late Archean basement of the Great Lakes region of Canada (Roscoe 1969; Robertson 1978). These deposits consist of thick auriferous quartz-pebble conglomerate and sandstone containing both detrital uraninite and pyrite. Both were deposited in braided alluvial systems draining older, exposed, Archean granite-greenstone basement terranes.

Cloud (1976a) has emphasized the absence of economic grade uraniferous and pyritiferous conglomerates after about 2000 Ma. Although isolated occurrences of easily oxidized detrital minerals, such as uraninite and pyrite, have been reported from modern environments (Zeschke 1951; Steacy 1953; Davidson and Cosgrove 1955), they differ fundamentally from early Precambrian deposits by (*i*) containing relatively few, isolated, uraninite or pyrite grains; and (*ii*) generally representing areas of rapid erosion and transport, containing immature recently eroded sediments instead of largely quartzitic, extensively weathered and transported sediments such as those of the Witwatersrand.

Langford (1983) has further pointed out that Witwatersrand- and Huronian-type placer uraninites contain both uranium and thorium. After about 2000 Ma, this type of uranium deposit is replaced by vein-, sandstone-, and unconformity-type deposits in which uranium minerals, lacking thorium (which is left behind as insoluble oxides at the sites of weathering), are deposited by meteoric or hydrothermal solutions after sediment deposition. He related these contrasts in uranium ore types to the appearance of oxygen in the atmosphere at about 2000 Ma.

Grandstaff (1976, 1980) calculated that oxygen partial pressures must have been below 10^{-2} to 10^{-6} times present atmospheric levels for detrital uraninites to have survived under environmental conditions like those prevailing during deposition of the Witwatersrand Supergroup.

4.6

Atmospheric Evolution: the Rise of Oxygen

JAMES F. KASTING   HEINRICH D. HOLLAND
LEE R. KUMP

4.6.1 Introduction

The rise of oxygen in the earth's atmosphere has been discussed extensively over the last 30 years (Rubey 1955; Holland 1962; Berkner and Marshall 1965b; Brinkman 1969; Cloud 1972; Walker 1977; Walker et al. 1983; Holland 1984; Kasting 1987). However, the subject remains highly controversial. The geologic record includes a variety of different types of evidence that can be used to infer paleo-O_2 levels. Unfortunately, this evidence is often ambiguous and the interpretations occasionally conflict (e.g., Clemmey and Badham 1982). Our discussion of the rise of oxygen parallels that of Walker et al. (1983) and Kasting (1987) in that it relies on a simple, three-box model of the atmosphere-ocean system. This model is described in Chapter 26. The discusion also provides a number of reasonable alternatives to the Walker/Kasting scenario that reflect the diversity of opinions of the co-authors of this Section and of other researchers in the field. Some assessment of Archean oxygen levels is provided to place the subsequent evolution of O_2 levels in its proper context.

4.6.2 The Three-Stage Model for the Rise of O_2

Kasting (1987), following Walker et al. (1983), conceptualized the evolution of atmospheric O_2 as progressing in three stages. In Stage I, which Walker et al. termed "reducing", the entire atmosphere/ocean system was essentially devoid of free oxygen. Prior to the development of oxygenic photosynthesis, the supply rate of reduced substances (H_2, CO, and H_2S from volcanoes, and Fe^{2+} from seawater/basalt interaction) would almost certainly have exceeded the production of O_2 from photolysis followed by the escape of hydrogen to space. Atmospheric O_2 levels predicted during Stage I are of the order of 10^{-14} PAL (times the present atmospheric level) or lower. Even after biological photosynthesis had been "invented", atmospheric P_{O_2} should have remained at these low levels until the net production rate of oxygen exceeded the supply of reduced gases to the atmosphere. Highly productive regions of the surface ocean could, however, have developed substantial concentrations of dissolved O_2, creating localized "oxygen oases". Fisher (1965) coined this term to describe oxygen-rich microenvironments in stromatolitic mats. We suggest here that oxidizing conditions could have been maintained throughout entire lakes or ocean basins, even though the atmosphere itself was almost totally anoxic. As discussed in Chapter 26, dissolved O_2 concentrations approaching 10% of the present are theoretically possible under such conditions. Such oxygen oases could have provided an environment conducive to BIF deposition in an "upside-down" Archean biosphere (Walker 1987).

At some point during the Archean or the Early Proterozoic, the supply of reduced substances decreased, or the oxygen production rate increased, and oxygen began to accumulate in the atmosphere. The most visible evidence of this increase is the widespread occurrence of red beds after approximately 2.0 Ga ago (Section 4.5). The near-surface ocean during this stage would also have been oxygenated globally, and not just locally. Except in areas of deep water formation, the deep ocean remained anoxic and carried ppm quantities of iron in solution. Upwelling, anoxic, Fe-enriched deep waters appear to have been responsible for the formation of banded iron sediments until mid-Proterozoic time (Section 4.2). The simultaneous presence of an oxidized surface environment and a reduced deep ocean identifies the Early Proterozoic system as being in Stage II. The three-box model of Chapter 26 can be used to derive an upper limit of ~ 0.03 PAL for the atmospheric O_2 concentration during Stage II (see also Kasting 1987). This limit is not absolute but, rather, is directly proportional to the organic carbon burial rate and is inversely proportional to the rate of deep ocean mixing.

Stage II ended when the deep ocean became oxic and, therefore, depleted in iron. At least 0.002 PAL of atmospheric O_2 was required for this transition to have occurred (Chapter 26). Walker et al. (1983) proposed that Stage III was initiated by a decrease in the supply of ferrous iron to the deep sea or by an increase in the rate of photosynthesis. Other possible causes include changes in ocean circulation (e.g., the establishment of polar deep-water formation) or in other factors besides photosynthetic rate that determine organic carbon burial rates (e.g., the reactivity of the organic matter). It seems likely that the timing of this change coincided with the disappearance of the Superior-type iron-formations around 1.85 Ga. This interpretation fails to account for the occurrence of some Late

Proterozoic banded iron deposits in the Rapitan Group of the Yukon and Northwest Territories and in the Urucum area of Brazil. As discussed in Section 4.2, however, these latter BIFs are composed almost totally of hematite and chert and are invariably associated with glacial deposits. Hence, their formation appears to involve a separate cause and should not be interpreted as signalling a return to Stage II conditions (see Section 4.3).

The arguments given above can be used to construct a tentative time history of atmospheric oxygen (Fig. 4.6.1). The shaded region in the Figure represents permissible values of P_{O_2}; the vertical dashed lines show suggested transition times between the different evolutionary stages. In drawing this curve we have followed Kasting (1987) in placing the transition from Stage I to Stage II at 2.4 Ga ago. This interpretation is consistent with the data on red beds and detrital mineral deposits presented in Walker et al. (1983), and reviewed in Section 4.5, and with the relatively broad range of sulfur isotope ratios in 2.3 to 2.2 Ga-old sulfide deposits from Canada and South Africa (Cameron 1982; Lambert and Donnelly 1989; see also Chapter 3). A wide spread in sulfur isotope ratios is thought to indicate the presence of abundant dissolved sulfate in seawater which, in turn, implies the presence of some atmospheric oxygen (Walker and Brimblecombe 1985). If the date for the first appearance of red beds is pushed forward to about 2.0 Ga (Section 4.5.2). Stage II might be postponed for another 400 million years. Indeed, Stage II may have been relatively short because the age of the youngest large Early Proterozoic iron-formation, the Sokoman, has been pushed back from 1.7 to 1.88 Ga (Hoffman 1988; Holland and Beukes 1990; and Section 4.2).

Evidently, the date at which the atmosphere first became oxygenated is still a matter of dispute. To complicate matters further, paleosol data (Section 4.4) imply measurable quantities of free O_2 back to at least 3.0 Ga ago. An upper limit on P_{O_2} during the Archean can be derived from the observation that detrital uraninite grains were not oxidized during this time. Holland (1984, p. 331) has shown that preservation of uraninite requires that $P_{O_2} \cdot P_{CO_2} \leq 10^{-4.9 \pm 0.8}$ bar^2. A lower limit of 0.02 bar for P_{CO_2} prior to 2.5 Ga can be derived from climate models (see Section 4.7). If we ignore the uncertainties in both estimates, this implies that P_{O_2} must have been less than $10^{-3.2}$ bar, or 0.003 PAL, throughout the Archean.

The solid vertical bar in Figure 4.6.1 shows an estimate of Early Proterozoic P_{O_2} derived from the paleosol data presented in Section 4.4 in which it is suggested that the ratio P_{O_2}/P_{CO_2} at this time was approximately 0.03 ± 0.01. The CO_2 partial pressure during the Early Proterozoic, based on climate models, was between 0.25 bar and 0.02 bar. Combining these estimates with the O_2/CO_2 ratios obtained from the paleosols yields: 4×10^{-4} bar $< P_{O_2} < 0.01$ bar. Encouragingly, this range is consistent with the O_2 partial pressures predicted from the simple three-box model of the atmosphere/ocean system.

4.6.3 Development of the Modern Aerobic (Stage III) Atmosphere

The three-box model, unfortunately, tells little about the rate at which oxygen levels rose once the atmosphere had entered Stage III. Walker (1977) has argued that, once the deep oceans were cleared of ferrous iron, P_{O_2} would have rapidly approached its present level because the planetary system would have been operating much as it does today. This argument is not very convincing, however, because the carbon cycle must have evolved significantly since mid-Proterozoic time. For example, the appearance and diversification of multicellular organisms during the Vendian and Cambrian should have facilitated carbon burial by producing forms of organic matter (e.g., collagen, fecal pellets) that were more resistant to degradation than simple unicellular plants and animals. The appearance of land plants during the Late Silurian would have created materials (e.g., lignin) that were even more resistant to degradation. An increase in atmospheric P_{O_2} may have been required to keep the organic carbon burial rate approximately constant in the face of these biological innovations (Kump 1988).

Evidence that P_{O_2} was increasing during the Late Proterozoic is provided by the carbon isotopic record (Section 3.4). The average $\delta^{13}C$ value of marine carbonates in the late Riphean (800 to 700 Ma ago) was about $+5‰$ (Knoll et al. 1986). If the overall rate of carbon cycling was the same as today, this implies a doubling of the rate of organic carbon burial during this period. The excess O_2 produced would be ~ 10 times the present atmospheric O_2 reservoir. Although the bulk of this oxygen must have been consumed by reactions with reduced minerals or with volcanic gases (Section 4.8), it seems likely that at least part of it accumulated in the atmosphere. A second, shorter interval of rapid organic carbon deposition occurred near the end of the Proterozoic (Tucker 1986; Margaritz et al. 1986; Aharon et al. 1987). By the same reasoning, it seems likely that this event was also associated with an increase in atmospheric P_{O_2}.

These considerations suggest that atmospheric P_{O_2} did not increase abruptly to its modern level, but was still substantially below its present level during the latter part of the Proterozoic.

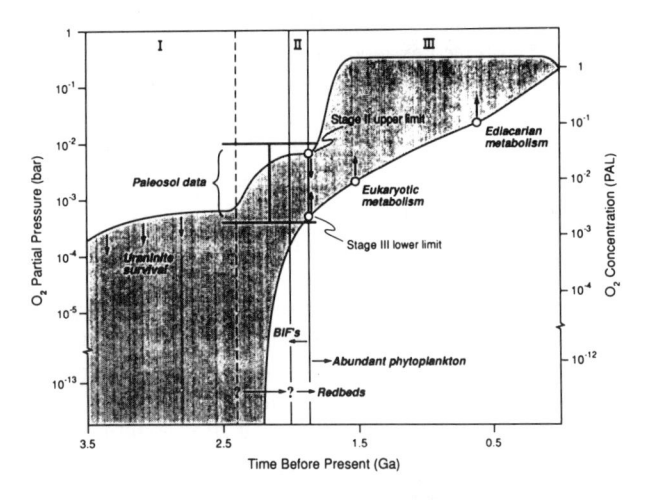

Figure 4.6.1. Tentative time history of atmospheric oxygen. The shaded region shows the range of O_2 concentrations that is consistent with the specified geological and biological constraints and with the predictions of the simple three-box model of the evolution of the atmosphere/ocean system discussed in the text.

As indicated in Figure 4.6.1, lower limits on P_{O_2} can be estimated by considering the oxygen requirements of the evolving biota. For example, an O_2 level of at least 0.01 PAL was probably required by 1.5 Ga to sustain the metabolism of eukaryotic organisms that appeared at or somewhat earlier than this time (Section 5.5). (This argument is not entirely convincing, because an effective dissolved O_2 level of 0.01 PAL might have been achieved in the local oxygen oases of Stage I. However, the high O_2 levels in such oases may not have been steady enough to support widespread, planktonic, eukaryotes.) Approximately 10 times this much oxygen (0.1 PAL) was required by 0.6 Ga to satisfy the oxygen requirements of the Ediacaran "worm" *Dickinsonia* (Runnegar 1982c). Runnegar's criterion is essentially the same as that derived from the study of shelled metazoans by Rhoads and Morse (1971). The latter approach has been reevaluated, however, by Thompson et al. (1985), who find that some classes of shelled organisms could have survived at substantially lower oxygen levels. Runnegar further suggests that the wide, flat body shapes of *Dickinsonia* and other members of the Ediacaran Fauna were evolved so as to maximize their uptake of oxygen in a low-P_{O_2} environment.

The idea that atmospheric oxygen levels increased at the end of the Proterozoic is not new. It was proposed by Nursall (1959) and was the basis for the model of atmospheric evolution of Berkner and Marshall (1965b). These authors went one step further and suggested that the rise in P_{O_2} at that time triggered the entire metazoan radiation. While this assertion remains speculative and may well be too simplistic, it is consistent with the carbon isotopic record and with theoretical ideas about the evolution of the carbon cycle.

4.6.4 Biogeochemical Cycles and Their Effects on O_2

The discussion above suggests when the rise of oxygen may have taken place, but it does not describe why it occurred. To do so requires that one consider the geochemistry of carbon, sulfur, and iron and the effects of the evolving biota on the cycling of these elements. This is not the place for a comprehensive treatment of this subject; detailed analyses have been given by Holland (1978, 1984). A few observations will be made here, however, to illustrate the nature of the questions involved and to point out some of the issues that remain to be resolved.

One long-standing question concerns the level of atmospheric oxygen during the Archean. We suggested above that the Archean atmosphere was in Stage I, that is, free oxygen was essentially absent. But this does not have to have been the case. Life had already originated by 3.5 Ga and so, too, in all likelihood, had photosynthesis (Section 1.5). The earliest photosynthetic organisms may have synthesized organic compounds via photosystem I, i.e., by bacterial photosynthetic reactions such as

$$CO_2 + 2H_2 \rightarrow CH_2O + H_2O \qquad (4.6.1)$$

and

$$CO_2 + 2H_2S \rightarrow CH_2O + H_2O + 2S. \qquad (4.6.2)$$

Free oxygen is not produced in these reactions; their main effect would have been to draw down atmospheric hydrogen concentrations by providing an additional sink for reduced volcanic gases. However, it is possible, even likely, that organisms quickly discovered how to use photosystem II to carry out cyanobacterial ("green plant") photosynthesis:

$$CO_2 + H_2O \rightarrow CH_2O + O_2. \qquad (4.6.3)$$

Here, the hydrogen from water is used to reduce carbon dioxide, and molecular oxygen is the unavoidable byproduct of the process.

What would have been the effect on the atmosphere of the invention of Photosystem II? If the burial rate of organic carbon then was the same as it is now, the net oxygen production rate would have been of the order of 10^{13} moles yr^{-1} (Holland 1978). This is more than enough to overwhelm the reduced gases (H_2, CO, and SO_2) released from present-day volcanoes. [The present volcanic emission rate of reduced gases is equivalent to an oxygen sink of some (1 to 10) $\times 10^{11}$ moles yr^{-1}, depending on whether one accepts the lower estimate of Walker (1977) or the higher estimate of Holland (1978)]. However, if the organic carbon burial rate in the Archean was somewhat lower, and the volcanic outgassing rate higher, the flux of reduced gases would have overwhelmed the photosynthetic source of oxygen, and the atmosphere would have remained anoxic.

The situation is further complicated by the participation of iron in the oxygen cycle. We know that ferrous iron was being oxidized in some localities during the Archean because hematite and magnetite occur in Archean iron-formations. One can estimate the amount of oxygen involved by once again comparing with conditions on the modern earth. The amount of iron that is presently eroded from the continents and carried to the oceans by rivers is about 7×10^{14} g Fe yr^{-1} (Martin and Meybeck 1979). This iron is currently oxidized and is transported in particulate form; however, on an anoxic early earth, at least 25% would have been present as dissolved, ferrous iron (Garrels 1987). If all of this ferrous iron were oxidized to magnetite, then the rate of oxygen consumption would have been about 6×10^{11} moles yr^{-1}, roughly the same as the amount consumed by volcanic gases. Oxidation of the ferrous iron contained in hydrothermal fluids would consume an additional 4×10^{11} moles O_2 yr^{-1} (Section 4.8), or possibly more if the ridge axes were hotter in the past. Thus, the available sinks for O_2 could have been comparable in magnitude to the present O_2 production rate.

The preceding discussion points out why it is difficult to determine from a purely theoretical standpoint when the atmosphere should have become oxygen-rich. Although we have couched our discussion in terms of the Archean, it is relevant to the Early Proterozoic as well. Many of the largest BIFs (e.g., those of the Hamersley Basin) were deposited around the time (2.5 Ga) when the transition from Stage I to Stage II may have occurred. One would like to know whether the atmosphere contained free oxygen at that time. The first good evidence for bacterial sulfate reduction, namely, an increase in the spread of isotopic ratios of both sulfides and sulfates, is found at around 2.3 to 2.2 Ga (Cameron 1982; Lambert and Donnelly 1989; see also Chapter 3). One possible explanation is that oceanic sulfate levels increased at this time as a consequence of an increase in oxidative weathering of pyrite on the continents (Walker and Brimblecombe 1985). If so, an increase in atmospheric P_{O_2} is implied. Indeed, this rise in P_{O_2} could have been caused by the increased rate of microbial sulfate reduction and pyrite burial in the global sulfur cycle. This is because the burial of pyrite can be considered an oxygen source, if the organic carbon utilized by the bacteria replaces that which would otherwise have undergone aerobic decomposition (unlikely) or fermentation and methanogenesis (more likely).

On the other hand, the Archean oceans may have been rich in dissolved sulfate, but bacterial sulfate reduction had either not yet been invented (Schidlowski 1979) or produced a much smaller degree of isotopic fractionation because of higher ocean temperatures at that time (Ohmoto and Felder 1987). The current controversy could be settled by determining the relationship between the concentration of reduced sulfur and elemental carbon in highly carbonaceous Archean shales. If bacterial sulfate reduction became a globally established mode of organic matter decomposition in the Early Archean, the concentration of pyrite and C_{org} should be as well correlated in Archean sediments as they are in more modern sediments. If bacterial sulfate reduction is a Proterozoic invention, highly carbonaceous Archean sediments should be either pyrite-free, or should contain pyrite the concentration of which is poorly or not at all correlated with C_{org} content.

4.6.5 Modern and Ancient Controls on O_2

One would also like to know how the atmospheric oxygen level was regulated during Stage III and what controls it at present. The conventional view (see, e.g., Holland 1978, 1984) is that the concentration of dissolved O_2 in ocean water affects the production rate of oxygen by altering the fraction of photosynthetically produced organic matter that gets buried in sediments. That fraction is currently very low ($\sim 0.1\%$) as a consequence chiefly of the operation of efficient aerobic decay mechanisms. If atmospheric P_{O_2} were to decrease, dissolved O_2 concentrations would likewise decrease, and a larger fraction of the organic matter would be preserved. This, in turn, would generate more oxygen and would thereby tend to counteract the original decrease in O_2. Thus, the carbon/oxygen cycle includes a stabilizing negative feedback loop.

The present oxygen control system may actually be more complicated than described here for reasons alluded to above. A large fraction (perhaps half) of the organic matter that becomes buried is terrestrially derived material that is deposited in river deltas (Ittekott 1988). The preservation of organic matter in these sediments is apparently affected more by its burial rate than by ambient O_2 levels (Berner and Canfield 1988). This terrestrial material is more resistant to degradation and has a much higher C:P ratio than does typical marine organic matter (Likens et al. 1981). Thus, the presence of terrestrial vegetation probably enhances atmospheric oxygen levels by increasing the efficiency of usage of the phosphate derived from rock weathering. This opens up the possibility that strictly terrestrial processes (e.g., forest fires) may play a role in present-day oxygen regulation (Kump 1988).

Changes in phosphate availability could also have had an important influence on oxygen levels in the past. Roughly half of the present influx of phosphorus from rivers is removed from the modern oceans by co-precipitation with $Fe(OH)_3$. A significant amount of Fe^{+2} (1 to 10 mmol/kg) is present in the hydrothermal solutions that flow from black smokers at the crest of mid-ocean ridges. This Fe^{+2} is oxidized and precipitates as P-rich $Fe(OH)_3$ (R. Feely pers. comm. to H. D. H. 1988). This precipitate ultimately coats mineral grains in marine sediments (see, for example, Sherwood et al. 1986). If, as seems likely, the hydrothermal flux of Fe^{+2} to the oceans was considerably greater in the Archean and Early Proterozoic than it is today, the rate of phosphate removal as a constituent of iron-phosphate hydroxides may well have been more efficient then than it is today. If so, the oceans were operating at a lower phosphate concentration than today, and the biologic productivity of the oceans was almost certainly lower. However, the burial rate of organic matter with sediments seems to have been comparable to that during more recent times (Holland 1978, Ch. 5; Schidlowski et al. 1983). It follows that the fraction of the primary organic matter that was buried during the Archean and Early Proterozoic was greater then than now. This is best explained in terms of a lower O_2 content in the atmosphere prior to ca. 1.7 Ga which, in turn, is consistent with the evidence provided by the BIFs and paleosols. There are, however, so many uncertainties in the rather long chain of reasoning which connects the chemistry of paleosols, BIFs, and the level of atmospheric O_2 in the Precambrian, that it would be unwise to place a great deal of faith in the resilience of this chain.

4.6.6 Ozone and Other Trace Gases

Our discussion to this point has dealt strictly with the rise of atmospheric oxygen. Other biologically significant changes should have accompanied this process. Perhaps the most important of these would have been a corresponding increase in atmospheric ozone and the establishment of an effective screen against solar ultraviolet (UV) radiation. The critical wavelength region for life is between 210 and 290 nm, where absorption by water vapor and CO_2 is negligible, but where absorption by DNA is strong (Sagan 1961, 1973; Berkner and Marshall 1965b). In the absence of O_2 and O_3, radiation at these wavelengths could have reached the ground virtually unattenuated. (See Section 1.2 for a discussion of other possible primitive UV screens.) This may not have posed much of a problem for many early prokaryotes which could have protected themselves in a variety of different ways: by living under a sufficient depth of water, by forming mats, and by developing efficient UV-repair mechanisms (Margulis et al. 1976; Rambler and Margulis 1980). Other organisms, however, may have been at greater risk. Eukaryotic phytoplankton, for example, would have been periodically brought up to the top of the water column by advection and were probably much more sensitive than prokaryotes to the damaging effects of UV radiation (Margulis et al. 1976).

The relationship between atmospheric P_{O_2}, ozone, and solar UV radiation has been explored by a number of investigators (Ratner and Walker 1972; Levine et al. 1980; Kasting et al. 1985, and references therein; Kasting 1987). There now seems to be general agreement among the various models. Figure 4.6.2 (from Kasting 1987) summarizes the results. The UV screen

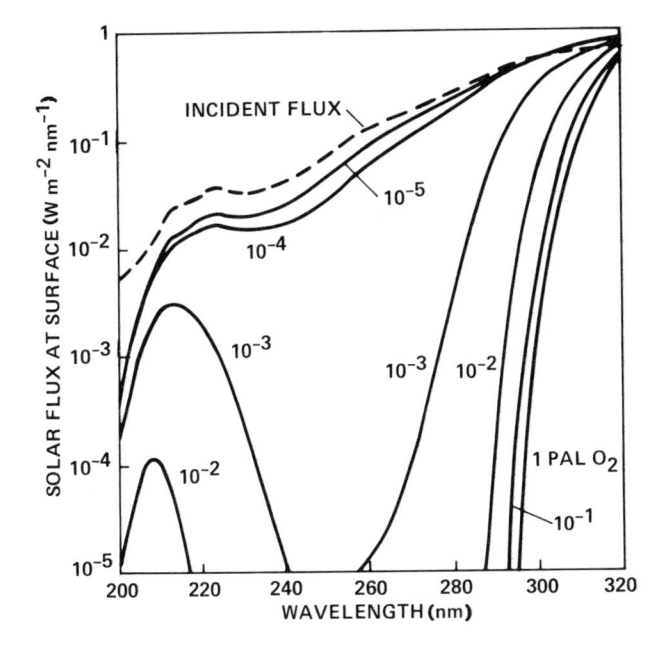

Figure 4.6.2. Ultraviolet energy flux at the earth's surface for the specified atmospheric O_2 levels (daytime values, for a solar zenith angle of 50°); the uppermost dashed curve shows the UV flux incident at the top of the atmosphere (Kasting 1987).

eukaryotic phytoplankton (at about 1.8 to 1.5 Ga; Section 5.5). However, it is conceivable that the evolution of eukaryotic phytoplankton may have created a positive feedback, whereby an increase in oxygen and ozone levels, caused by increased phytoplankton productivity and organic carbon burial, promoted further increases in productivity and oxygen production by generating an increasingly effective UV shield (Holland 1984; Kasting 1987).

The rise in atmospheric O_2 should also have affected the concentrations of other biogenic trace gases (Kasting et al. 1985; Kasting 1987). If the production rate of nitrous oxide (N_2O) was fixed, this trace gas should have become progressively more abundant as P_{O_2} increased because of a decrease in its rate of photolysis. Methane, which might have exceeded 100 ppm in the Archean and Early Proterozoic atmospheres (Kasting et al. 1983; Zahnle 1986), should have fallen to very low concentrations as P_{O_2} began to rise, perhaps reaching a minimum (0.1 ppm), and then increasing to somewhere near the present value (1.6 ppm) in the Late Proterozoic to early Phanerozoic as tropospheric OH densities declined. These predictions are sensitive to the assumed production rates of the various gases, and thus to understanding of how the biota evolved. For this reason it is doubtful that much will ever be known about how the trace gases fluctuated through time.

A possible exception to the foregoing might occur in the case of carbon dioxide (CO_2). CO_2 affects the earth's climate by the greenhouse effect. It also affects the rate of rock weathering directly, and evidence of these effects might be preserved in paleosols. It is theoretically possible, therefore, to infer past CO_2 concentrations by examining the earth's climate record and by studying the chemistry of paleosols. The next Section describes one such attempt to do so.

would have begun to form at an O_2 level of 10^{-3} PAL and should have been firmly entrenched by the time P_{O_2} reached 10^{-2} PAL (2×10^{-3} bar). Given the range of plausible oxygen concentrations inferred for the Early Proterozoic, an effective UV screen should have been in place by this time. Establishment of the screen may have preceded the appearance of

4.7

Proterozoic Climates: the Effect of Changing Atmospheric Carbon Dioxide Concentrations

JAMES F. KASTING

4.7.1 The Climatic Record

The history of the earth's climate during the Proterozoic is not understood as well as that of the more recent past; nevertheless, a great deal is known about it compared to the climate of the Archean Eon (Section 1.4). The Proterozoic was marked by two well-documented glacial periods, separated by a long span of apparently ice-free conditions (Frakes 1979; Crowley 1983; Crowell 1983). The earlier, Huronian, glaciation, named after the rock units in Canada in which it was first identified, occurred about 2.3 Ga ago. The later glacial period actually represents a series of glacial episodes, collectively termed the Late Precambrian glaciations, which lasted from about 0.9 or 0.8 to 0.6 Ga ago. As discussed in Chapter 12, these Late Proterozoic glaciations seem to have been extraordinarily severe: evidence for ice cover is found on all the continents, extending in most cases to paleolatitudes below 30° (Frakes 1979; Walter 1979; Chapter 12). The fact that the glaciers occurred at low latitudes was in one respect unavoidable, as the continents appear at this time to have been strung out along the equator (Section 2.3 and Chapter 12). But equatorial ice cover also implies that the Late Proterozoic must have been a time of intense cold.

Other empirical methods for inferring paleoclimates have been suggested. Knauth and Epstein (1976) and Karhu and Epstein (1986) have attempted to determine ancient temperatures using oxygen isotopes in cherts and chert-phosphate pairs. Both studies yielded very high ocean temperatures (60° to 120°C) throughout the Precambrian and into the Phanerozoic. In the latter study, the authors attempted to remove the effects of possible changes in the isotopic composition of seawater with time. Unfortunately, the temperatures inferred from the isotopic data exhibit no correlation with the climate history suggested by the glacial record. It therefore seems likely that the isotopic data reflect temperatures during diagenesis rather than during deposition. The apparent secular trend in the isotope ratios might then be explained by a decline in the geothermal gradient over time. Alternatively, the low $\delta^{18}O$ of Archean and Proterozoic oceans may reflect increased rates of isotopic exchange during low-temperature alteration of oceanic crust (Perry et al. 1978; Veizer et al. 1982).

Until more convincing methods for measuring paleotemperatures are developed, a less direct approach must suffice. Some useful constraints on past surface temperatures can be obtained from straightforward theoretical arguments. Because there is no evidence that the oceans were ever completely frozen, an absolute lower limit on tropical surface temperatures is $-2°C$, the freezing point of seawater. Actual surface temperatures were almost certainly higher than this throughout the geologic past. The mean surface temperature during the peak of the Pleistocene glaciation is thought to have been about 10°C (Schneider and Londer 1984), some 5 degrees lower than at present. Following Kasting (1987), 5°C is adopted here as a reasonable lower limit for the mean surface temperature during the Precambrian. Upper limits on surface temperature are more difficult to establish. Kasting (1987) has suggested that 20°C is a reasonable upper limit for the mean temperature during glacial periods. This is the approximate mean surface temperature of the earth during the Miocene, when the current polar ice caps are thought to have first appeared (Frakes 1979). Much of the Proterozoic was evidently ice-free, however, so mean surface temperatures could conceivably have been significantly higher than 20°C. One relatively loose constraint that would apply during ice-free periods stems from the existence throughout the rock record of sulfate minerals that appear to have been deposited originally as evaporitic gypsum ($CaSO_4 \cdot 2H_2O$). Because gypsum converts to anhydrite ($CaSO_4$) above 58°C, earth's mean surface temperature has on average been below this value since at least 3.5 Ga ago (Walker 1982).

4.7.2 Solar Luminosity and Possible Greenhouse Gases

To determine what the climatic record implies about paleo-atmospheric composition, one is forced to rely on climate theory. Stellar evolution models predict that the Sun was approximately 30% less luminous than today when it entered the main sequence, and that its brightness has increased more or less linearly with time (Newman and Rood 1977; Gough 1981). According to Gough, the Sun's output was some 18% less than today at the beginning of the Proterozoic (2.5 Ga ago) and 6.5% less during the Late Proterozoic (0.8 Ga ago). The earth's surface temperature at these times was determined by the intensity of the incident solar radiation, the planetary albedo (which is affected strongly by clouds), and by the greenhouse effect of its atmosphere. One of the difficulties in simulating past climates arises from uncertainty in how to model clouds. If one simply assumes that the amount of cloud cover has remained constant, it is possible to use climate models to estimate the concentrations of infrared-active (or "greenhouse") gases in the paleo-atmosphere. Given a fixed planetary albedo, some enhancement in the greenhouse effect is needed to compensate for the reduced solar liminosity during the Precambrian.

Greenhouse gases that may have been present in the atmosphere of the early earth include ammonia (NH_3), methane (CH_4), and carbon dioxide (CO_2) (Sagan and Mullen 1972; Kiehl and Dickinson 1987; Owen et al. 1979; Kasting 1987, and references therein). The first two of these gases are reduced species, the concentrations of which probably fell precipitously when free oxygen began to accumulate in the atmosphere. Ammonia is photochemically unstable even in a reducing atmosphere and could only have been abundant if it was somehow shielded from photolysis (Kuhn and Atreya 1979; Kasting 1982). It is suggested in Section 4.6 that free O_2 had reached appreciable levels by the Early Proterozoic. Thus, the greenhouse gas most likely to have been an important constituent of the Proterozoic atmosphere was carbon dioxide. Carbon dioxide is a good candidate for other reasons as well. The process by which CO_2 is lost from the atmosphere/ocean system over long time periods involves weathering of silicate minerals on the land, followed by deposition of carbonate minerals on the sea floor. Because rates of weathering are highly dependent on temperature, the carbonate-silicate cycle provides a natural feedback mechanism that should have enhanced the atmospheric CO_2 concentration when the solar luminosity was lower (Walker et al. 1981; Berner et al. 1983; Kasting et al. 1988).

4.7.3 Proterozoic CO_2 Levels from Climate Model Calculations

If changes in cloudiness are ignored, approximate limits on atmospheric CO_2 concentrations during the Early and Late Proterozoic can be derived from climate models, using the solar luminosity estimates given above. A brief description of the model used here is given in Chapter 26. These calculations are slightly different from those of Kasting (1987), because different absorption coefficients are used here, and because oxygen and ozone are included in the model. Together, O_2 and O_3 account for about 6 degrees Celsius of warming in the present atmosphere. O_2 concentrations have been estimated based on the discussion in Section 4.6. The assumed limits are 0 to 0.03 PAL (times the present atmospheric level) during the Early Proterozoic and 0.03 to 1 PAL during the Late Proterozoic. Ozone profiles corresponding to these oxygen levels are taken from the model of Kasting et al. (1985). For each combination of O_2 concentration and solar flux, the climate model was used to calculate surface temperature as a function of the atmospheric CO_2 partial pressure. [Actual partial pressures were used, as opposed to the concentration units employed by Kasting (1987).] The results are shown in Figure 4.7.1. The result of each set of calculations appears as a shaded band, the width of which

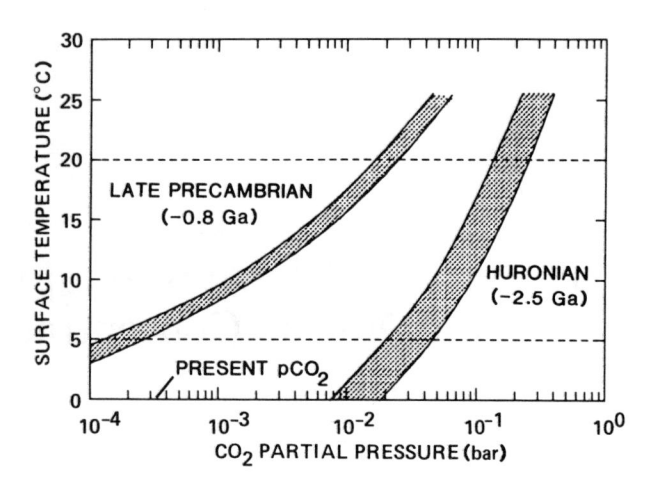

Figure 4.7.1. Surface temperature as a function of the partial pressure of atmospheric CO_2 during two different Proterozoic glacial periods; shaded areas show values calculated for a range of atmospheric O_2 and O_3 concentrations, using a one-dimensional (globally averaged) climate model with zero cloud feedback.

shows the effect of the assumed variations in O_2. The dashed horizontal lines show the estimated limits of the mean surface temperature during glacial periods.

This calculation shows that the earth's climatic record is consistent with a broad range of atmospheric CO_2 levels, even under relatively restrictive assumptions concerning the effect (or lack thereof) of changes in cloud cover. Permissible CO_2 partial pressures range from 0.02 bar to 0.25 bar during the Early Proterozoic, and 10^{-4} bar to 0.025 bar during the Late Proterozoic. These values can be compared with the 1988 CO_2 partial pressure of 3.5×10^{-4} bar, or the estimated "pre-industrial" value of 2.8×10^{-4} bar. In spite of the large range in possible CO_2 concentrations, one trend is clear: atmospheric CO_2 levels should have declined during the Proterozoic as solar luminosity increased. The minimum estimate for P_{CO_2} during the Early Proterozoic is some 60 times the present level. In contrast, if the Late Proterozoic was as cold as the glacial record suggests, P_{CO_2} at this time was probably within a factor of three of its present value.

These calculations can be combined with arguments presented in Sections 1.2 and 1.4 to estimate how atmospheric carbon dioxide levels have varied throughout the earth's history (Fig. 4.7.2). The shaded area in this Figure shows the range of CO_2 partial pressures that are consistent with the predictions of this one-dimensional climate model. The curve is constrained at four time periods: the present, the two Precambrian glacial periods, and the period shortly after the earth was formed. Estimated CO_2 partial pressures at 4.5 Ga ago range from 0.1 to 10 bars. The lower limit represents the amount of CO_2 needed to keep the surface temperature above the freezing temperature (Kasting et al. 1984); the upper limit is the CO_2 pressure predicted for an ocean-covered early earth (Section 1.2). The shape of the curve between the tie points is dictated by the need to keep the earth relatively warm during ice-free periods. The results shown here are similar to those of Kasting (1987), except that the present model allows somewhat lower CO_2 levels

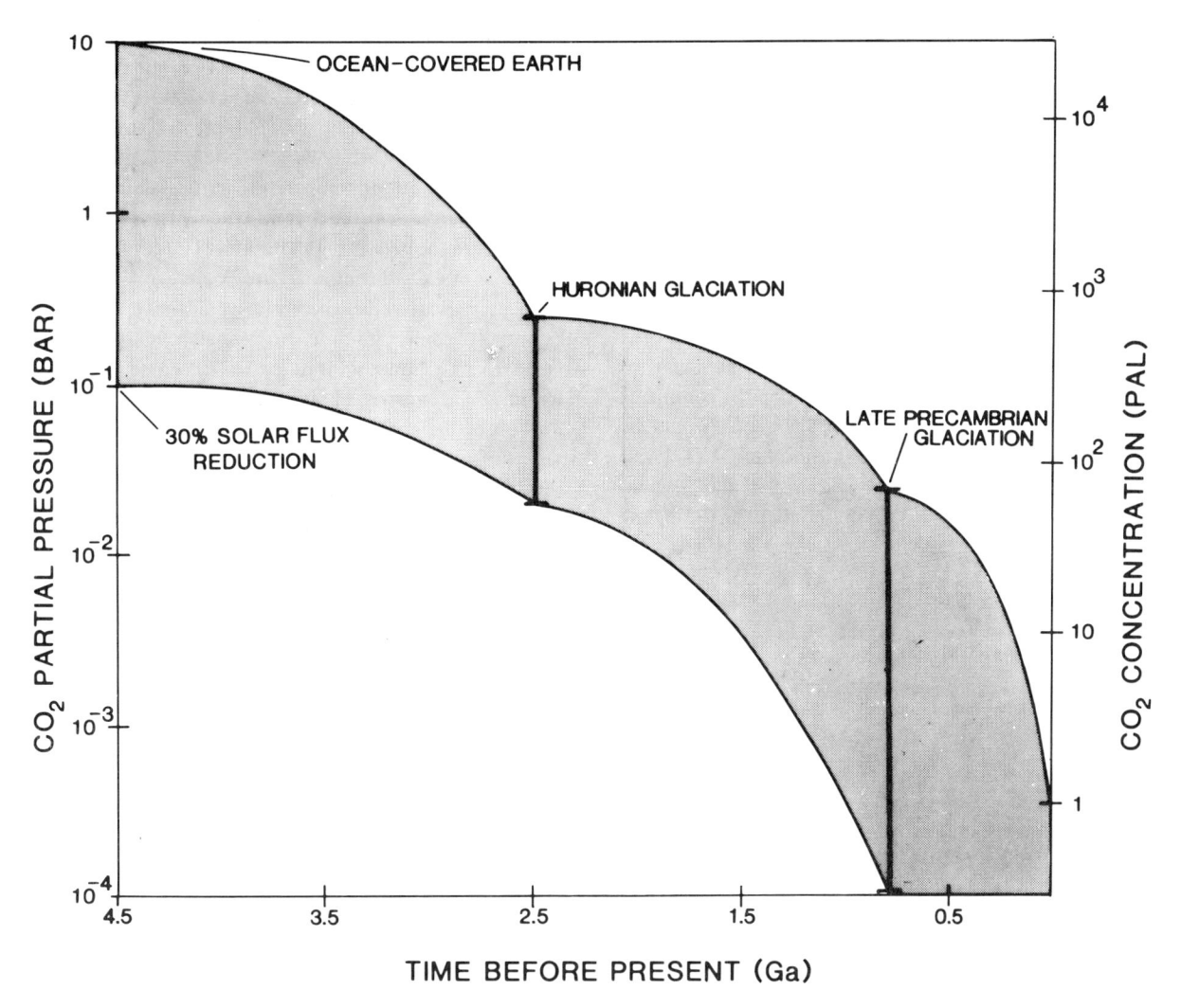

Figure 4.7.2. The range of atmospheric CO_2 partial pressures (shaded area) consistent with results of the climatic model calculations graphically summarized in Figure 4.7.1 and with other considerations discussed in Sections 1.2 and 1.4.

throughout the Proterozoic because it includes consideration of the effects of warming by oxygen and ozone.

The prediction that P_{CO_2} declined during the Proterozoic could be upset if the amount of cloud cover was much less toward the beginning of this Eon, or if high-altitude cirrus clouds (which have a net warming effect) were more prevalent at that time. A large decrease in cloudiness had been proposed by Henderson-Sellers (1979) and Rossow et al. (1982) as a means of offsetting the faint young sun. If cloud effects are eliminated entirely from the model, a mean temperature of 8°C during the Early Proterozoic can be sustained at the present CO_2 concentration. This would imply that P_{CO_2} need not have changed at all during the last 2.5 Ga. A negative cloud feedback of this magnitude seems unlikely, however, particularly during the mid-Proterozoic or the Early Archean when the climate was, if anything, warmer than today. Furthermore, increases in snow and ice, which have not been considered here, should have tended to offset any decrease in cloud albedo during glacial periods. Thus, there is good reason to expect that the decline in P_{CO_2} shown in Figure 4.7.2 actually occurred. This prediction is consistent with the observed trend in the ^{13}C content of well-preserved kerogens (Section 3.4). The generally heavier isotopic values that one finds in the Early Proterozoic could have resulted from a lower degree of biological fractionation in a CO_2-rich environment.

4.7.4 The Late Precambrian Icehouse

Although the preceding analysis illustrates the general relationship between past atmospheric CO_2 levels and climate, the details of the Proterozoic climatic record require additional explanation. The Late Proterozoic glaciations, in particular, present an intriguing problem for climatologists. If the evidence for ice in the tropics is accepted, the earth was colder during this period than at any other time in its history. What factor or combination of factors might have caused the climate to go into such a deep freeze?

The answer to this puzzle may involve the long-term carbon cycle and its effect on atmospheric P_{CO_2}. This idea was anticipated by Roberts (1971, 1976), who suggested that the Late Proterozoic glaciations were caused by a draw-down of CO_2, or an "anti-greenhouse" effect. Roberts, however, failed to consider changes in solar output, which make the anti-greenhouse theory much easier to understand. If P_{CO_2} was equal to or lower than its present level during the Late Proterozoic, there would be little difficulty in explaining the cold climate, since the Sun was still some six to eight percent dimmer at that time. Climate models that include ice albedo feedback (but no CO_2 feedback) predict that the earth would become totally glaciated were a few percent decrease in solar luminosity to occur (North 1975; Wang and Stone 1980). The real problem with the Late Proterozoic is not why it was cold but, rather, why atmospheric CO_2 did not rise to high enough levels to prevent such a deep glaciation? As discussed by Walker et al. (1981), if the present earth were suddenly to become colder, the rate of silicate weathering on the continents would decrease, and fewer Ca^{+2} and Mg^{+2} ions would be available to form carbonate minerals. Atmospheric CO_2 levels would rise because the metamorphic source of CO_2 would exceed the weathering sink. Why did this not happen during the Late Proterozoic?

A tentative answer to this question has been suggested by Marshall et al. (1988). As mentioned above (and in Section 2.3 and Chapter 12), the continents were at this time situated much closer to the equator than they are at present. If the tropics were as warm then as they are today, the rate of silicate weathering (which depends both on temperature and rainfall) would have been faster than it is at the present when large portions of the continents are located at high latitudes. Thus, if the metamorphic source of CO_2 during the Late Proterozoic was comparable to its present value, the earth would have to have been substantially cooler in order to keep the carbon cycle in balance. Exactly how cold it might have become is not easy to determine. The answer should depend on the precise configuration of the continents, the distribution of exposed rock types, the rate of carbonate metamorphism, and on atmospheric circulation and precipitation patterns. Qualitatively, however, this explanation seems capable of accounting for the anomalously cold Late Proterozoic climate.

4.8

Chemistry and Evolution of the Proterozoic Ocean

HEINRICH D. HOLLAND

4.8.1 Introduction

The foregoing Sections have summarized the somewhat fragmentary evidence for the evolution of near-surface environments during the Proterozoic Eon. The evolution of seawater during this period was in part caused by, and in part the cause of, these changes. Seawater continued to be supersaturated with respect to calcite, dolomite, and quartz (see Section 1.4), but its pH almost certainly rose during the course of the Proterozoic in response to decreasing atmospheric P_{CO_2} (see Section 4.7). The salinity of Proterozoic seawater is still poorly constrained by observational data, but the arguments developed in Section 1.4 for circumscribing the salinity of Archean seawater apply with even greater force to the salinity of Proterozoic seawater; it is likely, therefore, that the salinity of Proterozoic seawater did not differ greatly from that of modern seawater.

The bicarbonate concentration and the ratio of the bicarbonate to the calcium concentration in Archean seawater is still uncertain (see Section 1.4). This is also true of Proterozoic seawater until late in the Eon. In the Late Proterozoic, about 900 Ma-old marine evaporites are known from the Amadeus Basin in central Australia (Wells et al. 1970; Stewart 1979; Lindsay 1987a); the sequence of evaporite minerals is the same as that in Phanerozoic marine evaporites. Carbonates are followed by gypsum ± anhydrite, and these minerals are followed by halite. This sequence is difficult to understand (see Holland 1984, Ch. 9) if the oceans were a "soda ocean" as proposed by Kempe and Degens (1985). The occurrence of widespread anhydrite in parts of the highly metamorphosed Grenville carbonates in the northeastern United States is suggestive of a similar evaporite sequence, and indicates that a "soda ocean," if it ever existed, only did so more than ca. 1.4 Ga ago.

4.8.2 The Influence of Atmospheric Oxygen Concentration on the Proterozoic Ocean

Perhaps the most interesting changes that took place in the chemistry of seawater during the Proterozoic are related to the evolution of atmospheric oxygen (see Sections 4.3, 4.4, 4.5, and 4.6). Free O_2 was probably present in the atmosphere at the beginning of the Proterozoic. The paleosol data, and evidence from the oxidation of carbonate facies Kuruman Iron Formation below a ca. 1.9 Ga-old unconformity in Griqualand West (Holland and Beukes 1990), indicate that the O_2 content of the atmosphere rose dramatically about 1.90 to 1.85 Ga ago, probably to levels $\geqslant 15\%$ PAL. This rise coincides with the end of deposition of the large Archean and Early Proterozoic iron-formations (see Section 4.2). The reappearance of BIFs in the Late Proterozoic may signal a temporary return to low-O_2 conditions. However, by the end of the Proterozoic, P_{O_2} must have been at least 6 to 10% of its present level in order to have supported the contemporary biota (see, for example, Runnegar 1982c; Knoll in press).

It is clear that P_{O_2} has continuously adjusted itself, so that the net rate of O_2 production by green plant photosynthesis has much higher O_2 demand. Several lines of evidence point in this rocks exposed to weathering, of volcanic gases, and of reductants injected hydrothermally into the oceans. The high value of P_{O_2} today (Table 4.8.1) is demanded by the intense photosynthetic activity of the flora. The low value of P_{O_2} at the beginning of the Proterozoic (Table 4.8.1) is probably due to a much lower rate of photosynthetic O_2 production rather than due to a much higher O_2 demand. Several lines of evidence point in this direction. The degree of acid titration of Proterozoic rocks is similar to that of rocks of all ages. The ratio of the rate of volcanic CO_2 emission to the rate of erosion of igneous and high grade metamorphic rocks has therefore been roughly constant during the past 3.5 Ga (Holland 1984, Ch. 5). The rate of erosion has probably decreased somewhat during this period, but almost certainly by less than an order of magnitude. The redox state of volcanic gases determines in large part the ratio of elemental carbon to carbonate carbon, and of sulfide sulfur to sulfate sulfur, in sedimentary rocks (Holland 1978, Ch. 6). Highly reducing volcanic gases give rise to sediments containing a greater abundance of $C°$ and S^{-2} than of CO_3^{-2} and SO_4^{-2}. The ratios of $C°:CO_3^{-2}$ and of $S^{-2}:SO_4^{-2}$ are reflected in the isotopic composition of carbon and sulfur in sedimentary rocks. The $\delta^{13}C$ values of marine carbonates and of organic carbon in sedimentary rocks have varied somewhat with time, but not unidirectionally (see, for example, Schidlowski et al. 1983), and the $\delta^{13}C$ value of Phanerozoic marine

Table 4.8.1. *Comparison of redox processes in a model Early Proterozoic ocean with those in the modern oceans.*

Redox Process	Modern Ocean	Early Proterozoic Model Ocean
O_2 pressure in atmosphere	0.2 atm (1 PAL)	0.002 atm (1% PAL)
O_2 concentration in surface seawater	250 mol/kg	2.5 mol/kg
Mass of oceans	1.4×10^{21} kg	1.4×10^{21} kg
Mixing time of oceans	1000 yr	1000 yr
O_2 supply to deep sea	3.5×10^{14} mol/yr	3.5×10^{12} mol/yr
Hydrothermal input of Fe^{+2}	2×10^{12} g/yr	2×10^{12} g/yr
O_2 required to oxidize input of hydrothermal reductants	4×10^{11} mol/yr	4×10^{11} mol/yr
O_2 use in deep sea by oxidation of C_{org}	1.7×10^{14} mol/yr	3.5×10^{12} mol/yr
Rate of marine photosynthesis	4×10^{15} mol/yr	8×10^{13} mol/yr
Preservation rate of C_{org}	0.2%	10%
C_{org} burial rate with marine sediments	1×10^{14} g/yr	1×10^{14} g/yr
Mean sedimentation rate	2×10^{13} kg/yr	2×10^{13} kg/yr
C_{org} content of average sediment	0.5%	0.5%
Phosphate content of average seawater	2 μmol/kg	0.04 μmol/kg

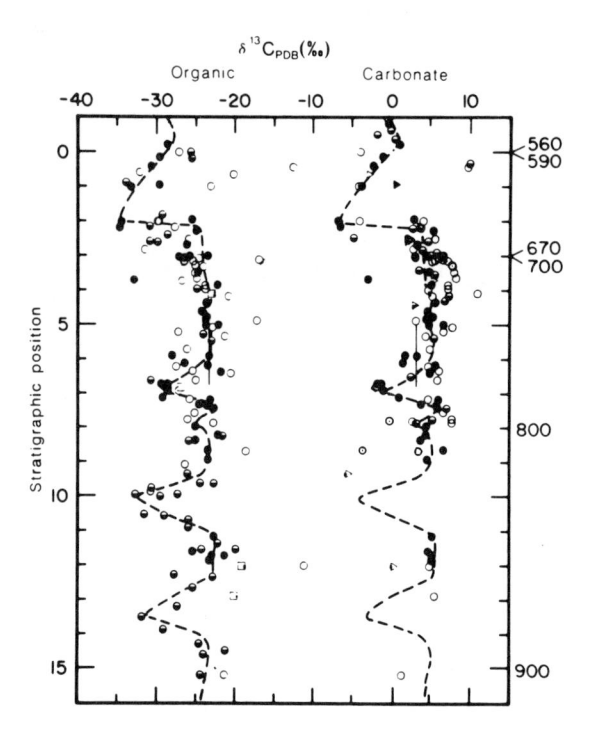

Figure 4.8.1. Secular variation of carbonate carbon and organic carbon isotopic values in Upper Proterozoic sequences of northeastern Spitsbergen (circles), Nordaustlandet (triangles), and central east Greenland (squares). [After Knoll et al. (1986).]

carbonates is roughly correlated with the $\delta^{34}S$ value of sulfur in anhydrites of marine evaporites in a manner that suggests the occurrence of only small deviations in the overall redox state of the crust, and hence of volcanic gases, during the last 600 Ma (Veizer et al. 1980). The isotopic history of carbon and sulfur during the last 300 Ma of the Proterozoic falls somewhat outside this pattern. The data in Figure 4.8.1 show the pattern of $\delta^{13}C$ variations within the Upper Proterozoic to Lower Paleozoic Lomfjorden and Hinopenstretet Supergroups of Spitsbergen and correlative successions from Nordaustlandet and east Greenland (Knoll et al. 1986). These three sections represent different portions of a single, tectonically dissected, sedimentary basin. Long sequences of isotopically heavy carbonate carbon ($\delta^{13}C = +4$ to $+7‰$ PDB) are interrupted by intervals of lighter carbonate carbon (-2 to $0‰$ PDB). The $\delta^{13}C$ values of organic carbon are ca. 28.5‰ lighter than, and covary smoothly with, the $\delta^{13}C$ values of the carbonates.

This pattern is similar to that observed in several other Vendian (Ediacaran) sequences (Magaritz et al. 1986; Lambert et al. 1987; Wang et al. 1987; Kaufman et al. 1988), and may therefore be representative of the $\delta^{13}C$ history of HCO_3^- in the world ocean during Late Proterozoic time (Knoll, in press).

The history of $\delta^{34}S$ in marine sulfate during the last 300 Ma of the Proterozoic is much more poorly documented. The data in Figure 4.8.2 are taken largely from the critical review by Lambert and Donnelly (1988). There are too few data for a meaningful comparison of the $\delta^{13}C$ to $\delta^{34}S$ relationships during the Late Proterozoic and the Phanerozoic. The isotopically heavy carbonates at ca. 820 Ma ago seem to be coeval with isotopically relatively heavy sulfates, but the dating is somewhat uncertain, and it is not clear whether this departure from the Phanerozoic $\delta^{13}C$ to $\delta^{34}S$ relationship is real. As

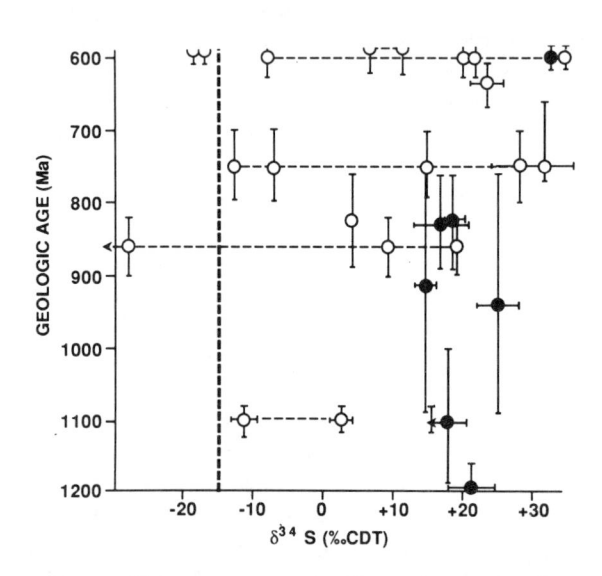

Figure 4.8.2. $\delta^{34}S$ for Upper Proterozoic sulfates (solid circles) and sulfides (open circles). [After Knoll in press (data largely from Lambert and Donnelly 1988).]

discussed in Section 4.6, the isotopically heavy carbon in Late Proterozoic carbonates could signal an increase in atmospheric O_2 between 850 and 630 Ma ago. However, other explanations can also be advanced for the relatively large positive values of $\delta^{13}C$ in these carbonates. Volcanic gases could, for instance, have been somewhat more reducing and/or the ratio of $(CO + CO_2)$ to $(H_2O + H_2)$ could have been somewhat smaller (Holland 1989).

4.8.3 An Evaluation of the Oceanic Oxygen Flux to Oxidize Hydrothermal Reductants

During the deposition of the major Early Proterozoic banded iron-formations (see Sections 4.2 and 4.3), P_{O_2} must have been sufficiently low so that a significant amount (perhaps several ppm) of Fe^{+2} could be present in intermediate depth and/or deep ocean water. This requirement imposes some interesting constraints on the composition and circulation of Proterozoic seawater (Kasting 1987). The flux of seawater through mid-ocean ridges today is ca. 2×10^{16} g/yr (Mottl 1983). The mean Fe^{+2} content of the high temperature solutions is ca. 100 ppm. The input of Fe^{+2} to the oceans is therefore ca. 2×10^{12} g/yr. This rate may well have been higher during the Proterozoic if the exit temperatures at the vents were significantly greater than today, but it seems unlikely that the flux of Fe^{+2} was greater than ca. 2×10^{13} g/yr. Fluxes of seawater through oceanic crust much larger than the present-day flux are ruled out by the prevalence of dolomite in Proterozoic carbonate sequences. This is also suggested by the time variation of the $^{87}Sr:^{86}Sr$ ratio of Proterozoic carbonate sediments. The evolution of the isotopic composition of Sr in the Proterozoic oceans is only known adequately during the last few hundred million years of that Eon (Veizer 1980); during this period the record looks very much like that of Phanerozoic marine carbonates.

The Fe^{+2} from hydrothermal vents was certainly accompanied by other reduced substances. Among these, Mn^{+2}, H_2S, H_2, and CH_4 are and were quantitatively the most important. The composition of hydrothermal vent solutions from the East Pacific Rise at 21°N, 13°N, and 11°N (Bowers et al. 1988; Campbell et al. 1988; and Welhan et al. 1984), and from the Juan de Fuca Ridge (Von Damm and Bischoff 1987), shows that some 10 to 30 mmol O_2 are required to oxidize the reduced components in 1 kg of these solutions. H_2S is usually the dominant oxygen sink; the O_2 demand is not a strong function of the Fe^{+2} concentration, because the concentration of this cation is negatively correlated with the concentration of H_2S.

The quantity of O_2 needed to oxidize the current input of hydrothermal reductants is approximately

$$(20 \pm 10) \times 10^{-3} (\text{mol/kg}) \times 2 \times 10^{13} (\text{kg/yr})$$
$$= (4 \pm 2) \times 10^{11} \text{ mol } O_2/\text{yr}. \quad (4.8.1)$$

The current rate of supply of O_2 to the deeper parts of the oceans is about

$$250 \times 10^{-6} \frac{\text{mol } O_2}{(\text{kg})} \times \frac{1.4 \times 10^{21} \text{ kg s.w.}}{1000 \text{ yrs}} = 3.5 \times 10^{14} \text{ mol/yr} \quad (4.8.2)$$

i.e., roughly 1000 times the rate required to oxidize the flux of

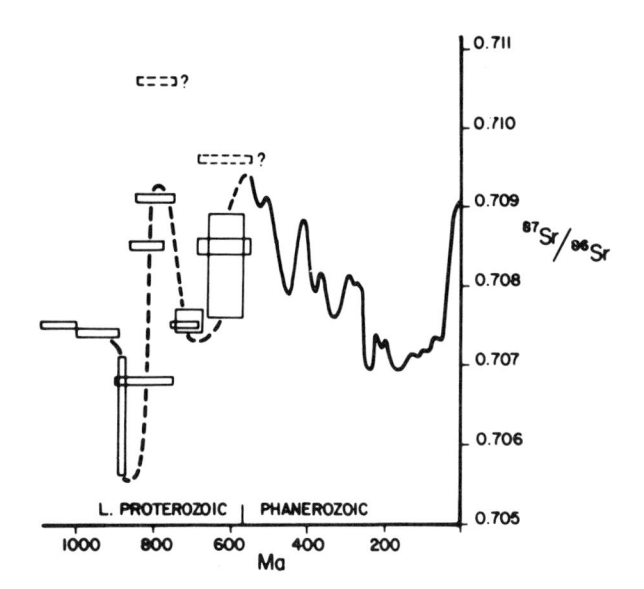

Figure 4.8.3. The isotopic composition of Sr in Late Proterozoic marine carbonates, as summarized by Veizer et al. (1983).

hydrothermal reductants. This agrees with the observation that the hydrothermal reductants are oxidized completely and relatively quickly without significantly decreasing the O_2 content of intermediate depth seawater. Even if the level of atmospheric O_2 were 100 times lower than at present, the supply of O_2 to the deep oceans would exceed the O_2 demand of hydrothermal reductants. The preservation of Fe^{+2} in intermediate and/or deep ocean water for upwelling and precipitation in banded iron-formations therefore demands either (i) a much higher rate of reductant input; or (ii) a very much lower O_2 content of the atmosphere; or (iii) a much slower mixing of the oceans; or (iv) the presence of other reductants to remove excess O_2 from downwelling seawater.

The first of these alternatives is unattractive, although not impossible. The second is permitted by the paleosol data. The third is possible within the limits imposed by ocean overturn due to the input of heat through the ocean floor. The fourth is very likely. Today, the O_2 content of downwelled seawater decreases progressively as it moves southward from the North Atlantic into the South Atlantic and thence into the Indian or Pacific Oceans (see, for example, Broecker and Peng 1982, Ch. 6). The oxidation of organic matter is the major cause for this loss of O_2; the same process was probably responsible for a part of the O_2 loss in the Proterozoic oceans. A simple calculation shows that this is reasonable.

If P_{O_2} in the atmosphere in the Early Proterozoic was 1% PAL, the rate of O_2 downwelling was ca. 3.5×10^{12} mol/yr for oceans with a turnover time of 1000 years (see equation 4.8.2). If this quantity of O_2 was used up in the oxidation of organic matter, the rate of CH_2O oxidation was also ca. 3.5×10^{12} mol/yr. If the input or organic matter into the deep oceans was at a much slower rate, the hydrothermal input of reductants would have been oxidized by the O_2 remaining in seawater after completion of C_{org} oxidation. On the other hand, if organic matter sank into the deep oceans at a much more rapid rate, seawater sulfate would have been reduced. Most or all of the hydrothermal flux of iron would then have been precipitated as pyrite in the deep ocean, and would not have been available for precipitation as a constituent of banded iron-formations.

Today, approximately 4% of the total quantity of marine organic matter is oxidized in the intermediate and deep water column. If we take this figure for the Proterozoic oceans as well, we obtain a total rate of photosynthesis of ca. 1×10^{15} gC°/yr. This is about 2% of the present rate of marine photosynthesis. Since P_{O_2} is so low in our model Proterozoic earth, the degree of preservation of organic carbon must have been much higher than today. If we take what is probably a reasonable preservation rate of 10%, we obtain a C° burial rate of 1×10^{14} g, which corresponds to a C° content of ca. 0.5% in average Proterozoic sediments.

The above calculation is obviously unsatisfactory; it involves too many unproven and easily questioned assumptions. It does have the virtue, however, of showing that we can construct a self-consistent model of the redox balance of the Early Proterozoic oceans which fits current views of the composition of the Early Proterozoic atmosphere and of the origin of banded iron-formations.

4.9

Conclusions and Unsolved Problems

CORNELIS KLEIN NICOLAS J. BEUKES HEINRICH D. HOLLAND
JAMES F. KASTING DONALD R. LOWE

Well-dated Precambrian rock sequences are required for evaluating Precambrian environmental conditions and their changes over time. During the last ten years, the ages of Precambrian iron-formations, paleosols, red beds, and glacial sequences have become better constrained. The timing of changes in environmental conditions, such as P_{O_2} levels of the atmosphere and ocean systems, is therefore becoming better known. The sharp drop in the abundance of major Precambrian iron-formation sequences at about 1.85 Ga suggests that at that time the deeper parts of the oceans became oxygenated and depleted in iron. Precambrian paleosol data suggest that the ratio $P_{O_2}:P_{CO_2}$ of the atmosphere increased significantly between 2.2 and 1.8 Ga ago. This is corroborated by the extensive oxidation of carbonate- and silicate-rich iron-formation assemblages in the Kuruman Iron Formation, South Africa, which are exposed below an erosional surface dated at about 1.9 Ga. The abundance of red beds in known Precambrian sequences increases significantly after 2.0 Ga. Such evidence is based on data from a wide range of Precambrian rock types and suggests strongly that there was a considerable increase in atmospheric oxygen about 1.9 Ga.

Recent geochemical and sedimentologic studies of the Early Proterozoic Kuruman and Griquatown iron-formation sequence in South Africa provide some clear insights into the environment from which these chemical sediments were precipitated. The overall REE patterns of these iron-formations, and especially their pronounced positive Eu anomalies (as well as small negative Ce anomalies), when compared with the REE patterns of modern ocean water and modern deep sea hydrothermal deposits, lead to the conclusion that iron-formations are the result of chemical precipitation from a seawater-dominated hydrothermal fluid mixture.

Petrographic and geochemical evidence indicate that the very fine-grained siderites which are the major carbonate components of micro- and mesobanded siderite-rich iron-formations in the South African iron-formations are probably primary precipitates formed in the water column. They are depleted by about 4 permil in ^{13}C with respect to shallow water stromatolitic limestones (which display a modern marine $\delta^{13}C$ signature close to zero). The difference is thought to be a primary signature of Early Proterozoic deep ocean water.

Detailed reconstruction of the sedimentologic environment of the South African iron-formations and associated limestones and shales, coupled with much geochemical information on the various lithologies, leads to the conclusion that the Early Proterozoic ocean was strongly density-stratified; surface waters were somewhat oxic whereas deeper waters were anoxic and enriched in dissolved ferrous iron. It appears likely that the ocean was also stratified with respect to the isotopic composition of total dissolved carbon.

Most of the data underlying the above conclusions on iron-formations are derived from the Early Proterozoic iron-formation sequence in South Africa. The concept of a stratified ocean should be tested further on the basis of additional Precambrian sequences in which deep and shallow water marine lithofacies can be clearly distinguished, and in which the paragenetic history of the carbonate mineral assemblages and kerogen distribution can be well established.

It is not clear why the ocean was stratified in Early Proterozoic time. Salinity stratification created by thermohaline circulation patterns offers the most likely explanation. The consequences of this type of stratification for the isotopic and elemental composition of the deeper oceans could have been quite dramatic if the mixing time of the oceans was much longer than at present. Our view of the Proterozoic oceans will be shaped considerably by the reality and the intensity of such stratification.

Other aspects of the Precambrian environment that raise important questions are related to the timing of glaciations, the history of the rise of atmospheric oxygen, and the history of atmospheric CO_2.

Although 2.7 Ga-old glacial deposits are reported in the Witwatersrand Supergroup, the oldest well documented glaciation is that of Huronian age in eastern North America. These deposits are possibly correlative or nearly correlative with diamictites in the Transvaal Supergroup in South Africa. It is important to search for older glacial episodes because they are powerful indicators of climate.

Although it is now well-documented that there was a relatively rapid rise in the O_2 content of the atmosphere ca.

1.9 Ga ago, it is not clear whether this apparent change was due to biological innovations leading to an increase in the biological productivity of the oceans, to changes in physical oceanography, or to geochemical changes, perhaps involving an increase in the availability of phosphate.

The history of atmospheric oxygen is a complicated problem that is not likely to be solved to anyone's satisfaction in the near future. Two very important questions related to the rise of O_2 ought, however, to be tractable at the present time. One concerns Archean O_2 levels: did the Archean earth have a reducing, Stage I atmosphere ($P_{O_2} \sim 10^{-12}$ PAL), or a mildly oxidizing, Stage II atmosphere ($P_{O_2} \sim 10^{-3}$ PAL)? The paucity of red beds during the Archean argues for the former, while the retention of iron in granitic paleosols argues for the latter. Since both red beds and paleosols are generated by surficial weathering processes, it seems peculiar that they should be telling us different things. Additional detailed studies of Archean paleosols, and improved models for understanding the formation of paleosols and red beds are clearly needed.

The second question that ought to be answered in the near future is whether O_2 rose quickly or slowly during the Middle Proterozoic, that is, during the early part of Stage III. As discussed in Section 4.6, it is attractive to associate the rise of the metazoans near the end of the Proterozoic with an increase in atmospheric O_2, especially because the carbon isotopic record indicates rapid burial of organic carbon during that time. However, recent work on paleosols described in Section 4.4 and by Holland et al. (1990) indicates that P_{O_2} may have been within a factor of two or three of its present level as early as 1.9 Ga ago. If this is true, then biological evolution was probably not limited by O_2 availability, and the 1.3 Ga-long "waiting period" before metazoans evolved must be attributed to other factors. This question could be answered definitively by the identification and analysis of additional mid-Proterozoic paleosols, particularly ones developed on rocks that have high O_2 demands.

Serious questions also remain regarding the history of atmospheric CO_2. The evolutionary curve presented in Section 4.7 is based entirely on theoretical calculations using a one-dimensional climate model. Two key assumptions are made in deriving this curve: (i) CO_2 and H_2O have been the major greenhouse gases throughout earth history; and (ii) cloudiness has not changed in any dramatic fashion over time. Either or both of these assumptions could be wrong. Clearly, some independent method of estimating P_{CO_2} is needed to test the climate model predictions. One possibility is to analyze carefully the extent of weathering in paleosols; another is to try to develop a relationship between atmospheric P_{CO_2} and the isotopic composition of sedimentary organic carbon. Unfortunately, both approaches are fraught with uncertainty. Nevertheless, it is clear that additional information is required to establish the history of atmospheric CO_2 with a significant degree of confidence.

At this time we are just beginning to define important aspects of the evolution of the Proterozoic earth. The environment of the early earth will become better understood by a close interplay of theoretical model calculations and carefully based geological observations.

5

Proterozoic and Selected Early Cambrian Microfossils: Prokaryotes and Protists

CARL V. MENDELSON   JOHN BAULD   ROBERT J. HORODYSKI
JERE H. LIPPS   TOBY B. MOORE   J. WILLIAM SCHOPF

5.1

Introduction

CARL V. MENDELSON

This Chapter considers the preserved, known record of Proterozoic and selected Early Cambrian microfossils: the microbiology of the middle eon of earth history. The evolutionary changes (evidenced morphologically) that took place during the Proterozoic were somewhat transitional between those of the preceding Archean (Section 1.5) and succeeding Phanerozoic Eons. The Early Proterozoic record is dominated by simple bacterial and cyanobacterial prokaryotes, some of which exhibit a significant degree of morphological complexity by about 2 Ga (Section 5.4); by the Late Proterozoic, various types of eukaryotic phytoplankters had arisen, including "giant" sphaeromorph and acanthomorph acritarchs as well as the enigmatic melanocyrillids. The evolutionary fabric of the Proterozoic is a complex one, and holds the key to the evolution of significant grades in microbiological organization. Here we attempt to dissect that fabric so that we can study it with critical and (we hope) open eyes.

Although the amount of information available for the task is less than one might prefer, it nevertheless is immense, even overwhelming; included in this mass of data are many uncritical reports of microfossils that must be filtered out before meaningful interpretations can be made. We might compare this dataset to that available for recent reviews of the Archean (Schopf and Walter 1983) and Early Proterozoic (Hofmann and Schopf 1983) microbiotas. The Archean compilation included 43 categories of microfossils and microfossil-like objects from 28 geologic units; two of these categories were accepted as representing true microfossils. The Early Proterozoic compilation yielded 421 occurrences of microfossils and microfossil-like objects from 40 geologic units; 290 of these were considered unequivocal microfossils (see Hofman and Schopf 1983, Table 14.2). Although the format of data summary in the two 1983 compilations differs from that employed here (Chapter 22), a rough comparison with the present study is instructive. The Proterozoic database used here consists of 3,277 records (3,494 if unfigured specimens are included) of occurrences of individual types of microfossils from 470 geologic units; 2,841 of these occurrences are here interpreted as being of authentic microfossils (Table 22.3), whereas 231 are regarded as dubiofossils (Table 22.2) and 205 as nonfossils and pseudofossils (Table 22.1). The sheer bulk of these data dictated, in part, the two-part format of the present monograph.

The coverage of this Chapter begins with the historical development of Proterozoic micropaleontology (Section 5.2), the initial stages of which determined the subsequent dichotomization of the science into two methodological schools: the palynologic (shale-facies) school, and the thin-section (chert-facies) school. Only recently have these two paths intermeshed productively. Section 5.3 treats the preservational aspects of Proterozoic and Early Cambrian microfossils, and their modern counterparts; the recent surge in interest in taphonomy has permitted more informed evaluation of putative microfossils and has thus helped to clear some of the haze that obstructed earlier views of early evolutionary patterns. The preservation discussion leads to an analysis of the microfossils themselves: bacteria and cyanobacteria (Section 5.4); organic-walled microfossils of uncertain systematic position, known as acritarchs (Section 5.5); distinctive, problematical, Proterozoic microfossils, including the enigmatic vase-shaped vesicles called melanocyrillids (Section 5.6); and the very rare mineralized protists (Section 5.7). This material is synthesized into a broad-brush overview of major events in the Proterozoic evolution of microorganisms (Section 5.8).

For biostratigraphic applications of the microfossils considered here, see Chapter 10. Diversity studies based on Chapters 5, 22, and 25 are presented in Section 11.3.

5.2

Historical Development of Proterozoic Micropaleontology

J. WILLIAM SCHOPF

5.2.1 Introduction

Despite important mid- to late-nineteeth century antecedents and a series of ground-breaking studies during the early 1900s, major success in Proterozoic micropaleontology – like that in Precambrian paleobiology generally – is a product chiefly of the past two decades. Indeed, of the more than 2800 taxonomic occurrences of authentic Proterozoic microfossils now known (Table 22.3), the vast majority, whether discovered in shales or cherts, have been reported since 1970 (Fig. 5.2.1). This recent progress, however, has not been devoid of setbacks. In particular, as is shown in Figure 5.2.2, and despite the nearly asymptotic rise since 1970 in the number of publications reporting occurrences of bona fide Proterozoic microfossils, a substantial subset of the literature has included significant errors of interpretation, reports of "microfossil-like" objects here regarded (Table 22.1) as nonfossil contaminants and pseudofossils. Based on comparison of cumulative curves showing the number of reported occurrences of authentic microfossils and of "microfossil-like" nonfossils reported from Proterozoic strata over the past century (Fig. 5.2.3), the historical development of the field can be subdivided into four, more or less discrete, successive phases, each of which is characterized in comparison with its predecessor by increases both in the quantity and the quality of available micropaleontological data (Fig. 5.2.4):

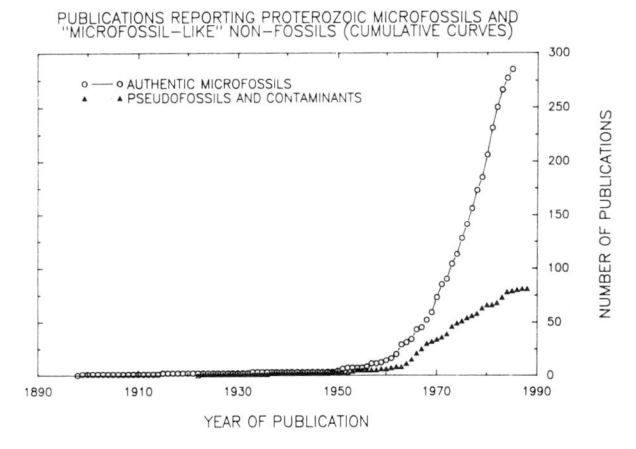

Figure 5.2.2. Number of publications reporting Proterozoic microfossils and "microfossil-like" nonfossils, 1890–1990.

Figure 5.2.1. Number of microfossil taxa reported from Proterozoic shales and cherts, 1890–1990.

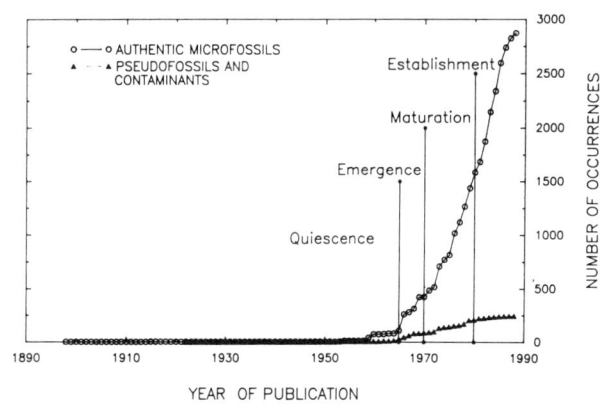

Figure 5.2.3. Number of reported occurrences of Proterozoic microfossils and "microfossil-like" nonfossils, 1890–1990.

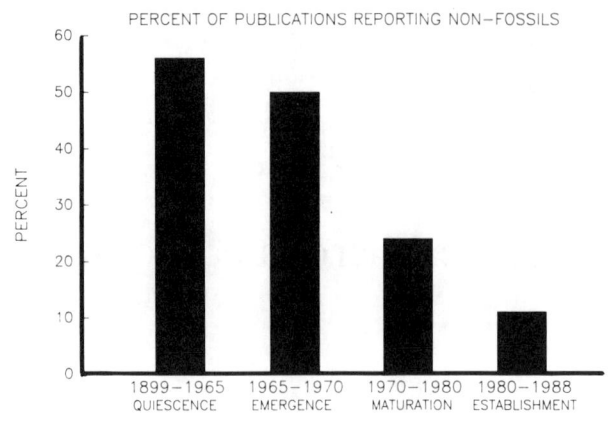

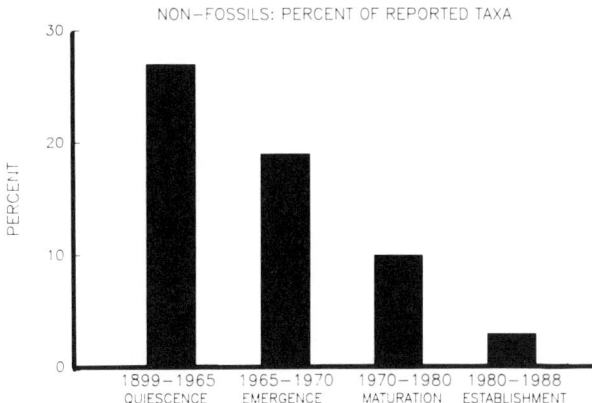

Figure 5.2.4. Historical development of Proterozoic micropaleontology: percent of publications reporting nonfossils, and nonfossils as a percent of reported taxa, 1899–1988.

(1) 1899–1965, a long period of *quiescence*, the formative phase in the development of the field;
(2) 1965–1970, a brief phase of *emergence* characterized by solid discoveries but also by use of new techniques that led to flawed interpretations;
(3) 1970–1980, a *maturation* phase during which sound studies markedly increased while erred reports became progressively less numerous; and
(4) 1980–present, a period of *establishment* of the field and its increasing acceptance as a significant subdiscipline of paleontological science.

5.2.2 Phases of Development of Proterozoic Micropaleontology

5.2.2A Quiescence: 1899–1965

The roots of Proterozoic paleobiology, like so many other aspects of natural science, extend to the work of Charles Darwin and publication in 1859 of *The Origin of Species*; as Darwin then wrote in assessing problems confronted by his theory of evolution:

"There is another...difficulty, which is much more serious. I allude to the manner in which species belonging to several of the main divisions of the animal kingdom suddenly appear in the lowest known fossiliferous rocks....if the theory be true it is indisputable that before the lowest Cambrian stratum was deposited long periods elapsed...and that during these vast periods the world swarmed with living creatures...[However], to the question why we do not find rich fossiliferous deposits belonging to these assumed earliest periods prior to the Cambrian system, I can give no satisfactory answer. The case at present must remain inexplicable; and may be truly urged as a valid argument against the views here entertained.

C. Darwin. 1859. *The Origin of Species*. Chapter X: On the Imperfection of the Geological Record. [John Murray: London; 6th edition; 1902; pp. 446–449].

Thus, Darwin stated the problem. Not surprisingly, a series of scientists took up the challenge, among them such influential workers as the Canadian geologist-paleontologist J. W. Dawson (1875) and the French micropaleontologist M. L. Cayeux (1894a, 1894b, 1895), but probably the first to make a lasting contribution to the field was the American paleontologist Charles Doolittle Walcott, the acknowledged "Father of Precambrian Paleobiology" (Schopf 1970; Cloud 1983). Following extensive field work beginning in 1882 in the Grand Canyon of the Colorado River, Walcott published a series of papers (Walcott 1883, 1895, 1899) reporting discovery both of stromatolites and of microfossils (viz., the "megasphaeromorph" acritarch *Chuaria*) in units of the Late Proterozoic Chuar Group. Other Proterozoic microfossil-like objects, less convincing primarily because of their minute size and inadequate illustration, were also reported by Walcott (1915), Moore (1918), Gruner (1923, 1924), and Ashley (1937); but skepticism was widespread (e.g., Hawley 1926; Seward 1931) – for members of a paleontological fraternity reared on studies of fossil organisms with "preservable hard-parts," the preservation of minute "soft-bodied" microbial fossils over periods of hundreds of millions of years seemed clearly to be a logical impossibility.

Proterozoic stromatolites were also reported during this period, most notably by Matthew (1890), Walcott (1883), and C. L. and M. A. Fenton (1931, 1937), but their biological interpretation similarly met with skepticism, in some quarters verging on derision. As the noted British paleobotanist Sir Albert Charles Seward wrote in 1931:

"The general belief among American geologists and several European authors in the organic origin of [the stromatolite] *Cryptozoon* is, I venture to think, not justified by the facts... [such forms] are precisely the same in their series of concentric shells as many concretions which are universally assigned to purely inorganic agencies... It is clearly impossible to maintain that all such concentrically constructed bodies are even in part attributable to algal activity... It is claimed that sections of a Pre-Cambrian limestone from Montana show bodies similar in form and size to cells and cell-chains of [bacteria]... These and similar contributions... are by no means convincing... We can hardly expect to find in Pre-Cambrian rocks any actual proof of the existence of bacteria... Cyanophyceae or similar primitive algae may have flourished in Pre-Cambrian seas...but to regard these hypothetical plants as the creators of reefs of *Cryptozoon* and allied structures is to make a demand upon imagination inconsistent with Wordsworth's definition of that quality as 'reason in its most exalted mood'."

A. C. Seward. 1931. *Plant Life Through the Ages*. [Cambridge University Press: Cambridge, England; pp. 86, 87, 92].

Seward's words had impact. The Precambrian putative fossils, whether algal stromatolites or bacterial cells, fell within the purview of paleobotany. And Seward, a Cambridge don, Fellow of the Royal Society, and author of numerous paleobotanical texts, was the most influential paleobotanist of his generation worldwide. If A. C. Seward expressed skepticism, doubt was certified as well-founded; if A. C. Seward thought such Precambrian "fossils" had been misinterpreted, well, they no doubt had been. Despite Seward's other accomplishments – and they were innumerable, for he truly was a major contributor to the development of paleobotanical science – his disbelief in the potential preservability of minute Precambrian plant fossils stifled the field for more than two decades.

Indeed, it was not until after World War II, in part spurred by technological advancements made during that dark period, that the current outlines of Precambrian studies in general, and of Proterozoic micropaleontology in particular, began to take shape. And it is particularly interesting that this progress was stimulated not by one or a few breakthrough discoveries but, rather, by an extensive series of interconnected, synergistic, developments. Foremost among these were the rapidly improving techniques available for dating of ancient rocks, advances in geochronology that stemmed from the discovery of radioactivity near the turn of the century but which profited greatly from improvement of mass spectrometric equipment during the war years. For the first time (and because, at least in the United States – via the newly established National Science Foundation, a direct descendant of the WWII Office of Naval Research – funding had now become available to academia for purchase of such equipment), widespread dating of Precambrian terranes became an increasingly practical possibility. During the 1940s, the science of pre-Quaternary palynology was becoming increasingly well established, a subdiscipline of paleobotany involving study of plant microfossils isolated from clastic sediments by acid maceration and a field that was proving notably useful in the search for new sources of petroleum (and, thus, a development also spurred by the war effort). The search for pollen- and spore-like "palynomorphs" in Proterozoic deposits was a logical extension of successful Phanerozoic studies, one pursued actively in the post-war years both in Europe and, especially, in the Soviet Union. In particular, the early studies of B. V. Timofeev in Leningrad merit kudos (e.g., Timofeev 1958, 1960), not so much because of their scientific accuracy (his early reports of Precambrian "trilete spores" having subsequently been shown to have been in error) but, rather, because of the prescience of these studies, laying the groundwork for an approach that has since paid enormous dividends. During this period (but similarly dating from pre-war years), the Soviets were also active in studies of Proterozoic stromatolites and megascopic algae, led by V. P. Maslov (1939), A. G. Vologdin (1962a, 1966), and by Vologdin's colleagues, K. B. Korde (e.g., Vologdin and Korde 1965) and N. A. Drozdova (e.g., Vologdin and Drozdova 1964). Vologdin was the most prolific, and evidently the most influential of the group. And although Vologdin regarded stromatolites as individual megascopic fossil algae, a view later shown to be erroneous (with discovery in 1960 of living stromatolites at Shark Bay, Western Australia; reviewed in Walter 1976b), his work nevertheless documented the distribution of numerous Proterozoic stromatolitic occurrences, establishing the framework for subsequent studies of stromatolite-based biostratigraphy.

In Australia in the late 1940s, R. C. Sprigg (1947, 1949) published papers that would forever change the outlook on the search for Precambrian life. Sprigg is credited as discoverer of the now famous latest Proterozoic (Vendian-age) fauna of the Ediacara Hills of South Australia, an assemblage of fossil metazoans that by the late 1950s (Glaessner 1958a, 1958b; Termier and Termier 1960) had established beyond all doubt that fossils of "soft-bodied" animals were, indeed, preserved, and could, indeed, be discovered, in Proterozoic deposits. And, based on field work in southern Canada in the early 1950s, S. A. Tyler of the University of Wisconsin and E. S. Barghoorn of Harvard University published what was later to prove an epoch-making report describing for the first time a diverse assemblage of microscopic organisms well preserved in Proterozoic cherty stromatolites (Tyler and Barghoorn 1954). Although affinities suggested by Tyler and Barghoorn for various of these fossils were later discounted (among others, identifications of Early Proterozoic "fungi" and "discoasters"), this first report of the discovery of microfossils in the now famous mid-Precambrian (viz., Early Proterozoic) Gunflint Iron Formation paved the way for numerous, later, highly successful investigations of fossil, Proterozoic, microbial communities.

Thus, by the end of the post-war decade or soon thereafter, coupled with coincident advances in geochronometric techniques, important progress had been made in all aspects of Proterozoic paleontology: palynological studies of shale microfloras were beginning to flourish, especially in the Soviet Union; modern stromatolites had been discovered in Western Australia, and Soviet scientists had begun intensive studies of fossil, Proterozoic, stromatolitic assemblages; fossil metazoans, of increasingly accepted Proterozoic age, had been discovered and described from the Ediacara Hills of South Australia; and a diverse assemblage of fossil microorganisms had been reported from cherty Proterozoic stromatolites in southern Canada. Nevertheless, few scientific papers were as yet published (Fig. 5.2.2) and, thus, few fossil occurrences had as yet been documented (Fig. 5.2.3). Moreover, studies of the early history of life have since evolved into far more than the simple cataloguing of various types of fossils detected in Proterozoic strata at various locales. Such fields as isotopic and organic geochemistry, aspects of microbiology, biochemistry and molecular biology, and studies of the evolution of earth's atmosphere, oceans, and crustal regimes, to name just a few, have become integral parts of the total Proterozoic paleobiologic mix. Germinal studies leading to development of these related aspects of the science, studies resulting in the current, hallmark, interdisciplinary character of the field, also appeared during the 1950s and early 1960s.

Among the earliest of such studies was the reported synthesis by S. L. Miller (1953, 1955) of amino acids and other simple organic compounds under "possible primitive earth conditions." Except for those (few) individuals concerned with the earliest history of life, these studies were presumably regarded by paleontologists of the day as having little obvious immediate relevance. As has since been shown, however, the early history of life is characterized more by biochemical, than by morphological innovation, and with the later addition of organic geochemistry as a tool with which to uncover fossil evidence of such biochemical developments (e.g., Hayes et al. 1983; Schidlowski et al. 1983), the results of Miller's experiments could be viewed as providing at least a potential pathway toward discovering in the fossil record direct evidence of the timing and nature of the beginnings of life itself (Chang et al. 1983).

At about the same time as Miller's work, studies were published by Rubey (1951, 1955), soon to be followed by the contributions of Holland (1962) and of Berkner and Marshall (1964a), dealing with the Precambrian history of earth's atmosphere and oceans. These analyses made it obvious that early biochemical innovations, especially the origin of oxygen-producing photosynthesis, played a crucial role in the evolutionary development of earth's environment. Finally, based on studies of enzymatic fractionation of the stable isotopes of carbon during green-plant photosynthesis (Park and Epstein 1960), on studies of the isotopic values of organic and inorganic carbon sequestered in Precambrian sediments (Hoering 1961), and on studies establishing the Precambrian relevance of the isotopic biochemistry of sulfur and its fractionation during microbial sulfate reduction (Thode and Monster 1965), it became apparent during this period that organic geochemistry could play a major role in understanding the early history of life (Abelson 1959). Indeed, for the field to prosper – for plausible answers to be obtained to the pressing problems in the science – it had become evident that a broad-based, multidisciplinary approach should be adopted, one involving not only studies of fossils and "normal" paleontology but also based on related studies of the history of the early environment and the evolution of biochemical capabilities. Perhaps the first manifest evidence of this realization was the convening of a Woodring Conference on "Major Biologic Innovations and the Geologic Record," held in Virginia, USA, in June 1961, a meeting that brought together interested scientists from disparate fields, several of whom became leading contributors to the study of Proterozoic paleobiology (Cloud and Abelson 1961; Cloud 1983). With these developments, the long quiescence of Proterozoic paleobiology drew to a close; the time was now ripe for the emergence of a new subdiscipline of paleontological science.

5.2.2B Emergence: 1965–1970

In retrospect, the brief five-year span between 1965 and 1970 appears to have been the period of "make or break" – the emergence phase – of Proterozoic micropaleontology. During this period appeared:

(1) the first major publication discussing and describing a diverse assemblage of three-dimensionally preserved Proterozoic microorganisms, that of the 2000 Ma-old Gunflint Iron Formation of southern Ontario, Canada (Barghoorn and Tyler 1965), a paper that was followed within two months by a confirmatory report dealing with the same assemblage from the same locality (Cloud 1965) and that within a scant five months was followed by a third report illustrating even better preserved microfossils from the Late Proterozoic of central Australia (Barghoorn and Schopf 1965);

(2) the first lengthy, truly interdisciplinary study of a Proterozoic fossiliferous formation, that of the 1000 Ma-old Nonesuch Formation of northern Michigan, a study involving not only geology and micropaleontology but, in addition, organic geochemical analyses of hydrocarbons, porphyrins, and the carbon isotopic composition of the preserved organic matter (Barghoorn et al. 1965);

(3) the first monograph describing a Proterozoic microbial stromatolitic community, that of the 850 Ma-old Bitter Springs Formation of central Australia (Schopf 1968a), a work that established an approach and format that have been followed by the great majority of such studies to the present-day;

(4) a series of major taxonomic studies, by European and Soviet workers, of predominantly planktonic microfossils occurring in Proterozoic shales and siltstones (e.g., Roblot 1964; Timofeev 1966, 1969);

(5) the first comprehensive microbiological and sedimentological analyses of stromatolite-building communities of modern microorganisms designed to relate such biocoenoses to the formative agents of Proterozoic fossil stromatolites (e.g., Monty 1967; Gebelein 1969), and an influential study of the potential use of fossil stromatolites for intrabasinal and interregional correlation of Proterozoic strata (Cloud and Semikhatov 1969);

(6) new data and excellent summaries and analyses of the status of Proterozoic organic geochemistry and of the known carbon isotopic fossil record (Hoering 1967a, 1967b; Degens 1969);

(7) significant papers dealing with latest Proterozoic metazoans (e.g., Cowie 1967), including the first broad-based critical evaluation of various non-fossil structures that had previously been regarded as "possible Precambrian animal fossils" (Cloud 1968b); and

(8) overviews of the Proterozoic geologic, environmental, and paleontologic records intended to tie together the data as then available and to increase awareness of significant, but as yet unsolved, problems (Cloud 1968a; Schopf 1970).

Despite this progress, all was not sanguine. Mineralic pseudofossils and modern contaminants continued to be reported incorrectly as fossil Proterozoic microorganisms (Figs. 5.2.3, 5.2.4, Table 22.1); use of transmission electron microscopy, a new technique in Precambrian micropaleontology (e.g., Cloud and Hagen 1965; Schopf et al. 1965; Oberlies and Prashnowsky 1968), led to flawed interpretations (Table 22.1); a seemingly promising report of the discovery of Precambrian amino acids (Schopf et al. 1968) was soon shown to be mistaken, the amino acids being contaminants decidedly younger than the sediments from which they had been extracted (Kvenvolden et al. 1969); and other studies established that virtually all categories of soluble, extractable, organics presented similar problems – clearly, the possibility of post-depositional contamination presented a major difficulty for organic geochemistry (Hoering 1967b).

Nevertheless, this brief five-year period was probably the most remarkable, and most positive, in the history of the science. Discoveries of lasting import had been made. The first major studies in the field had been completed. Problems of interpretation had been confronted, and for the most part had been resolved. A new, multidisciplinary research strategy had been devised and found effective. New scientific questions, prompted by the newly available data, had been posed. And tentative generalizations regarding the early history of life, based for the first time on a fairly extensive body of direct evidence from the fossil record, had been drawn. With these now sound underpinnings, the field was primed to enter a new phase of development, one in which gains of the previous few years could be consolidated and built upon.

5.2.2C Maturation: 1970–1980

As is evident from Figures 5.2.1 through 5.2.4, the decade of the 1970s was a period of gradual maturation of the field. Not only did the number of relevant publications and of reported occurrences of Proterozoic microfossils markedly increase during the decade but, of equal significance, the number of published errors (i.e., of misidentified nonfossils) began to level out (Figs. 5.2.3, 5.2.4). Microfossil-based understanding of the early history of life was aided substantially during this period by an impressive set of ancillary developments, among them publication of a major treatise on the evolutionary origins of eukaryotic cells (Margulis 1970); the appearance of monographic works on the bioenergetics of early-evolving microorganisms (Broda 1975), the evolutionary development of the earth's rock record (Garrels and Mackenzie 1971), and on the nature, distribution, and paleoecological significance of modern and fossil stromatolites (Walter 1976); and by publication of various shorter works dealing with such subjects as the ecologic restriction by metazoans of stromatolite growth (Garrett 1970), the importance of assessing preservational and degradational variants in studies of Proterozoic microbial communities (Hofmann 1976), the distribution and significance of hydrocarbons extractable from Proterozoic sediments (Smith et al. 1970), the evolutionary development of the early ocean-atmosphere system (Holland 1973a) and the environmental implications of Precambrian banded iron-formations (Trendall 1973), and the effect on earth's ambient surface temperature of a gradual increase in solar luminosity over Precambrian time (Sagan and Mullen 1972). Reviews of these and earlier developments have been published by Schopf (1970, 1974b, 1975, 1977, 1978) and by Cloud (1974, 1976a, 1976b, 1980, 1983).

5.2.2D Establishment: 1980–Present

By the beginning of the 1980s, the field of Precambrian paleobiology, including studies of Proterozoic micropaleontology, had become firmly established as a sound scientific discipline. Among the most voluminous relevant publications to appear during the past decade was the PPRG-authored compendium on Archean paleobiology (Schopf 1983a). And a large number of excellent workers, all of whom deserve much credit (and many of whom are listed in Table 10.2.1), are making significant contributions to development of the science. In Proterozoic micropaleontology, for example, publications continue to rise; new fossils continue to be discovered; and errors continue to diminish (Figs. 5.2.1 through 5.2.4). Much of the remainder of this Chapter is devoted to consideration of the Proterozoic microfossil record as now known, paleontological data amassed during the historical development of the science.

5.3

Preservation of Prokaryotes and Organic-Walled and Calcareous and Siliceous Protists

ROBERT J. HORODYSKI   JOHN BAULD   JERE H. LIPPS
CARL V. MENDELSON

5.3.1 Introduction

One of the goals of Proterozoic paleontology is to determine the nature of the Proterozoic biosphere as a function of time. In order to accomplish this, we must ask the question: How representative of Proterozoic life are the preserved fossils? Insight into this can be attained through studies of the composition, degradation, and ecologic distribution of modern prokaryotes and plant protists, and through taphonomic studies of the fossil material.

5.3.2 Prokaryotes

5.3.2A Biochemistry and Preservation Potential of Prokaryotic Cell Walls and Sheaths

The contents of prokaryotic cells are completely enclosed by a thin, flexible cytoplasmic membrane composed of phospholipid and protein. This membrane is, in turn, encased in a thick, porous cell wall which confers rigidity and maintains cell shape (Beveridge 1981). An outermost, extracellular component, the glycocalyx, is almost universally present in natural populations of prokaryotes (Costerton et al. 1981). Murein (=peptidoglycan), a repeating network of parallel polysaccharide chains covalently cross-linked by many short peptide bridges (Beveridge 1981), provides the rigid framework for the eubacterial cell wall. Within the eubacteria, murein comprises approximately 90% of Gram-positive cell walls but only about 5 to 20% of lipid-rich Gram-negative cell walls. In contrast, archaebacterial cell walls lack murein and are characterized by considerable chemical diversity (Kandler and König 1985). Cyanobacteria are considered to have a Gram-negative cell wall organization but to possess a thicker murein layer (Jürgens and Weckesser 1985).

The prokaryotic glycocalyx ranges in consistency from that of a diffluent, water-soluble, amorphous mucilage to that of a firm, water-insoluble, typically visibly layered structure. The latter is characteristic of sessile cyanobacterial populations or of those constructing microbial mats. In general, the prokaryotic glycocalyx is a highly hydrated (ca. 99% water) and highly ordered polymeric matrix composed of polyanionic polysaccharide or glycoprotein (Costerton et al. 1981). Cyanobacterial sheaths, which consist of heteropolysaccharide-protein complexes (see also Section 6.8.4C), dominate assemblages preserved in many Precambrian stromatolites (Golubić and Hofmann 1976; Awramik et al. 1985; Knoll 1985a) and the early diagenetic stages of their recent microbial mat analogues (Horodyski et al. 1977; Boon et al. 1983; Klok et al. 1984). In contrast, well-preserved trichomes and/or cells tend to be rare (e.g., Knoll 1985a) suggesting that sheaths are preserved preferentially relative to cells of the same organism (see Bauld 1981, for detailed discussion).

Golubić and Hofmann (1976) have speculated that bound metals, accumulated in the sheath, may inhibit its decomposition. Despite reports that bacterial exopolysaccharides absorb heavy metals, thus protecting the enveloped cells against the toxic effects of such metals (Bitton and Freihofer 1978; Rudd et al. 1984), and that once bound, metals inhibit the decomposition of microbial exopolysaccharides in soils (Martin et al. 1972), it is yet to be experimentally demonstrated that heavy metals bound to the sheaths of cyanobacteria (Weckesser et al. 1988) inhibit sheath decomposition. Furthermore, prokaryotic cell walls are also known to avidly bind heavy metals (Hoyle and Beveridge 1984, and references therein). Since, in a single deposit, the diagenetic milieu is identical for sheaths and cells, preferential preservation of sheaths would also require that *pre-mortem* sheaths preferentially accumulate heavy metals relative to cell walls (assuming a concentration-dependent inhibition). Alternatively, sheath recalcitrance to enzymatic attack may reside in its higher order chemical structure; for example, the configuration of polysaccharide-protein interaction may inhibit sheath decomposition.

5.3.2B Degradational and Preservational Artifacts in Modern and Fossil Microbial-Mat Communities

Microbial-mat microbiotas preserved in Proterozoic strata most commonly occur in silicified stratiform (flat) stromatolites, silicified microbial peats, and silicified intraclasts and mat fragments. These microbiotas, which generally consist largely of microfossils of apparent or presumed prokaryotic affinity, are among the best preserved Proterozoic assemblages and have played a major role in shaping early ideas concerning the development of Proterozoic life (e.g., Schopf 1974a, 1974b,

1975). Although degradation of the mat-building microbes was recognized in early studies (Barghoorn and Tyler 1965; Schopf 1968; Schopf and Barghoorn 1969), among the first to address specifically the topic of degradationally produced artifacts were Awramik et al. (1972), who noted that modern coccoid cyanobacterial sheaths often resist *post-mortem* degradation better than cytoplasm, and that cellular contents may degrade to form a dense granule within the sheath. These results were supported and expanded through subsequent laboratory and field studies (Knoll and Barghoorn 1975; Horodyski and Vonder Haar 1975; Golubić and Hofmann 1976; Golubić and Barghoorn 1977). In some fossil taxa (viz., *Eoentophysalis*) the preferential preservation of cyanobacterial gelatinous envelopes over the more readily degradable cells may be due to the accumulation of the iron-chelating pigment scytonemine within the envelopes (Golubić and Hofman 1976). Preferential preservation of prokaryotic cell walls over sheaths (although less likely) should also be considered, and has been reported by Zhang (1988) in microfossils from the Dahongyu Formation (ca. 1700 Ma) of Hebei Province, North China. Zhang found that the presumed vegetative cells of the endolithic colonial cyanobacterium *Eohyella campbellii* showed a minimum of shrinkage deformation, whereas the surrounding sheaths were highly degraded or absent. Perhaps details of physiology and biochemistry, as well as microenvironmental conditions, significantly determine specific degradational sequences (Zhang 1988).

Dense organic-rich accumulations, commonly referred to as "spots," are not uncommon in coccoid microfossils preserved in some Proterozoic cherts. They were first described from the Late Proterozoic Bitter Springs Formation and were interpreted initially as possible "eye spots" and later as preserved nuclei or other organelles, particularly pyrenoids, of eukaryotic algae (Barghoorn and Schopf 1965; Schopf 1968; Schopf and Blacic 1971; Schopf 1974; Schopf and Oehler 1976; Oehler 1977). Subsequent workers contended that similar dark inclusions within coccoid microfossils from other Proterozoic units could be explained as degradational features of prokaryotes (Golubić and Hofmann 1976; Hofman 1976; Knoll et al. 1978; Horodyski and Donaldson 1980). Degradation of cytoplasm in eukaryotic algae also can produce "spots" (Knoll 1983). The early interpretation of the Bitter Springs "spots" as preserved nuclei, other organelles, or starch granules has met with some criticism (Awramik et al. 1972; Knoll and Barghoorn 1975; Knoll et al. 1978; Golubić and Hofmann 1976; Knoll 1983); however, Schopf (1974a) argued that they represent original structures as evidenced by their size, shape, and texture, as well as their absence from microfossils (viz., filaments) known to be prokaryotic. He further noted that they are too small and regular to be cytoplasmic remnants and that cytoplasmic remnants occur in some unicells, but differ in color and texture. Furthermore, transmission electron microscopy illustrates a close similarity between the Bitter Springs "spots" and pyrenoids in extant eukaryotic micro-algae as well as the presence of internal membranous sacs difficult to explain if they are prokaryotic (Schopf and Oehler 1976; Oehler 1977). It is generally agreed that dark inclusions in some coccoid microfossils may represent collapsed cytoplasm within a prokaryotic sheath; however, a consensus has not been reached for the Bitter Springs specimens. Although this problem was of importance during the 1970s when the Bitter Springs fossils were cited as representing the oldest assured eukaryotes (Schopf 1974a), today it is of much less importance because older eukaryotic fossils have been recognized, particularly among the acritarchs and certain carbonaceous compressions (see Sections 5.5 and 7.5).

Although the selective preservation of cyanobacterial sheaths over cellular contents and the degradational condensation of cytoplasm are perhaps the most significant observations that have been made on degradational and preservational artifacts in silicified-mat microbiotas, a number of other observations are also pertinent. Studies on a colonial coccoid cyanobacterium from a coastal sabkha in Baja California have shown that the thick sheaths surrounding groups of cells are more resistant to degradation than thinner sheaths surrounding individual cells and cellular contents (Horodyski and Vonder Haar 1975). If such outer envelopes were preserved and cellular contents and envelopes surrounding individual cells were not preserved, then these outer envelopes could be misinterpreted as single cells, which, owing to their large size, might be considered to be eukaryotic. Kinks and folds in filament sheaths, particularly narrow sheaths, could be misinterpreted as septa (Hofmann 1976; Knoll 1977); and desiccation of filamentous cyanobacteria has been shown to result in shrinkage of the cells and the production of artifacts, which could be misinterpreted as primary attributes (Awramik et al. 1972; Hofmann 1976; Golubić and Barghoorn 1977; Horodyski 1977; Horodyski et al. 1977).

When studying silicified Proterozoic microbial-mat microbiotas it is important to distinguish authentic biologic attributes from those produced during *post-mortem* degradation. In many cases such a distinction can be made on the basis of an understanding of degradationally produced features in modern microorganisms in combination with information that is available in the microfossil assemblage, where most taxa are represented by numerous specimens within which the variability of attributes can be studied.

5.3.2C Preservation-Related Taxonomic Problems

Most of the microfossils in silicified Proterozoic microbial mats have relatively simple coccoid or filamentous morphologies and possess a limited number of attributes available for taxonomic characterization (see Section 5.4). The coccoids can be characterized by the size, shape, thickness, and microfabric of the cell-like envelope; the presence or absence and the nature of material within the envelope; the geometric relationships of cell-like envelopes; the nature of enclosing sheath-like envelopes; and, if colonial, the geometry of the colony. The filaments can be characterized by the dimensions of the cells; the presence or absence of constrictions between cells; the shape of the terminal portion of the trichome; the presence and geometry of specialized cells; the presence or absence and thickness of a sheath; the number of layers composing the sheath; and the number of trichomes and their arrangement within a sheath. It is well established that some of these attributes vary within a cyanobacterial clone as a function of environment or age of the microbe or colony, a taxonomic complication that is resolved for modern cyanobacteria by comparing cultures grown under identical conditions. In addition to this intraspecific variability, some of these attributes can be modified, in some cases substantially, during *post-mortem* degradation, thereby making taxonomic treatment of fossil microbiotas difficult. For example, studies of the colonial coccoid cyanobacterium *Entophysalis*, the modern morphologic counterpart of the relatively common Proterozoic mat-building fossil *Eoentophysalis*, have shown that substantial shrinkage accompanies *post-mortem* degradation, with cells first becoming polyhedral, then stellate or lunate, and in advanced stages forming granules one-tenth or less of their original size or completely disappearing (Golubić and Hofman 1976; Golubić and Barghoorn 1977). Furthermore, the gelatinous envelopes surrounding cells contract uniformly with water loss. Similar diminution in cell size has been documented in both coccoid and filamentous cyanobacteria (Hofmann 1976; Golubić and Barghoorn 1977; Horodyski et al. 1977), and it is a feature that must carefully be considered in Precambrian studies as cell size is typically a heavily weighted taxonomic attribute for "chert-facies" microfossils. Other studies have shown that the terminal portion of trichomes may shrink and attain a variety of shapes upon desiccation (Hofmann 1976; Golubić and Barghoorn 1977; Horodyski 1977) and that differential shrinkage within a cylindrical trichome can produce heterocyst-like features (Hofmann 1976; Golubić and Barghoorn 1977). Such features could be mistaken for orginal morphologic attributes in fossil specimens and be used both as taxonomic characters and as a basis for inferring biologic affinities. It also has been shown that differential degradation can result in the preferential preservation of filament sheaths with an absence of cellular contents (Golubić and Barghoorn 1977; Horodyski et al. 1977).

The taxonomy of microfossils preserved in silicified microbial mats can be approached in two ways. At one extreme, one can regard all of the fossils as form taxa and define them solely on the basis of their morphologic attributes and not upon their presumed affinity. This is an approach that by necessity is commonly followed for empty tubular cylinders, which could represent sheaths of filamentous cyanobacteria or bacteria or even the walls of fungi, and for the morphologically simple, isolated coccoid forms, which could represent bacteria, cyanobacteria, or unicellular eukaryotic algae. This approach can be expanded to include degradational artifacts as is illustrated by Hofmann's (1976) recognition in the Early Proterozoic Belcher Islands assemblage of five form genera of moderate-sized spheroidal microfossils based upon the nature of their internal contents, which he interpreted as being degradational products.

The second approach is to take consideration of morphologic variability and degradation patterns within a population of specimens. Such an approach can work well when studying fossils from the same locality, and particularly from a single horizon, where individual populations can be recognized and taphonomic noise can be filtered out. It may be more difficult to follow such an approach when comparing microfossils from different localities or different basins where the microbiotas have been subjected to different degradational histories as well as originating under different environmental conditions.

Both of these approaches have merit. The form taxa approach simplifies the taxonomy of silicified-mat microbiotas, it eliminates much of the personal biases of the investigator, and it may yield a clearer description of the fossil assemblage. The second approach, which attempts to define biologically distinct entities in order to identify or reconstruct those microorganisms most representative of the original biota and to describe the morphologic variability within each taxon, is particularly useful for paleoecologic studies of Precambrian assemblages (see Knoll 1984) and for comparing Precambrian microbiotas. Although these two approaches seem mutually exclusive, many Precambrian paleontologists take an intermediate approach, treating isolated, morphologically simple spheroids and empty tubular envelopes as form taxa, commonly with the understanding that their affinities even at the kingdom level are uncertain, and treating fossils having more taxonomic information as biologically distinct entities.

5.3.2D Ecological Distribution of Modern Prokaryotes

There are very few habitats on earth so severe as to exclude prokaryotes, which are ubiquitous, often cosmopolitan, in aquatic and terrestrial environments. The *potential* distribution of any prokaryote is determined by the full range of physicochemical (nonbiologic) factors to which it is exposed. Its *actual* distribution (realized potential) will be a subset of this environmental envelope, within which the organism can reproduce, constrained by factors such as competition and predation, and enhanced by survival strategies such as metabolic dormancy to cope with transitory unfavorable conditions. As a group, extant prokaryotes are reported to have the capacity to grow (reproduce) at temperatures as low as $-12°C$ and as high as $+110°C$; at salinities between that of distilled water and saturated halite (but having an absolute requirement for free, liquid water); both at dissolved oxygen concentrations ranging from zero to supersaturated and at dissolved sulfide concentrations up to ca. 10 mM; at pH ranging from 0.5 to 11.0; in darkness and under full incident sunlight; at redox potentials between ca. -400 mV and $+800$ mV, and under high hydrostatic pressure (Knoll and Bauld 1989 and references therein).

Individual species, genera, or physiologically distinguishable groups are generally more restricted with respect to their environmental ranges or envelopes. Cyanobacteria provide a particularly apt illustration of this point – Proterozoic fossils commonly show striking morphological similarity to extant benthic cyanobacteria (see Table 25.2), a similarity echoed in the taxonomic assemblages and community structures of modern microbial mats and their stromatolitic analogues (Knoll and Bauld 1989; see Section 6.12). The recent demonstration that chroococcoid cyanobacteria are an important component of modern oceanic (Johnson and Sieburth 1979) and limnetic (Caron et al. 1985) phytoplankton communities suggests a comparable, if not greater, role for cyanobacteria in Proterozoic plankton. Presumptive paleoenvironmental limits can sometimes be inferred from cyanobacterial fossil remains by analogy with their modern counterparts (Knoll and Bauld 1989 and references therein).

The upper temperature limit for cyanobacteria (ca. 72–74°C; Castenholz 1979) falls far short of the current upper limit for prokaryotes (110°C for the archaebacterium *Pyrodictium occultum*; Stetter 1986). The lower limit for cyanobacteria, though poorly defined, appears likely to be higher than the purported prokaryotic limit (ca. −12°C to −10°C; Michener and Elliott 1964; Mazur 1980). Likewise, cyanobacteria are not represented among the acidophilic prokaryotes, being unable to grow at pH less than about 4 (Brock 1973), thus leaving an ecological niche to be filled by eukaryotic phototrophs. In fact, it appears that these two boundaries (temperature ⩽ ca. 72–74°C and pH ⩾ 4) hold for all phototrophic prokaryotes (Knoll and Bauld, in press). In contrast, cyanobacterial representatives are conspicuously alkalophilic, even producing blooms in highly alkaline (pH = 11.3) carbonate lakes (Ciferri 1983).

Although cyanobacteria can grow, albeit slowly, at salinities approaching halite saturation, they are restricted in nature to somewhat lower salinities by competition from halophilic eukaryotic micro-algae (Brock 1976). Filamentous cyanobacteria appear to have a lower salinity tolerance than coccoid species, generally being unable to grow at salinities greater than ca. 150–200 parts per thousand (Ehrlich and Dor 1985). The ability of certain cyanobacteria to carry out facultative anoxygenic photosynthesis using sulfide (⩽ 2–10 mM) makes them formidable competitors in fluctuating habitats or those in which oxygen and sulfide coexist (Padan 1979; Cohen et al. 1986).

5.3.2E How Representative are the Known Fossils Preserved in Proterozoic Microbial Mats?

A considerable amount of current understanding of Proterozoic microbial life is based on studies of microbial mat assemblages preserved in Proterozoic cherts; therefore, it is important to ask how representative are these microfossils? Two aspects of this question are addressed here: How representative are these fossils of the microorganisms that originally composed a particular mat? And how representative are these fossils of Proterozoic microbial-mat communities in general. A third aspect – How representative are these fossils of Proterozoic microbial life? – is discussed in Section 5.3.5.

The question of how representative these fossils are of the microorganisms that originally composed a particular mat can be addressed by studying both fossil material and modern microbial mats. The abundance of microfossils varies considerably in silicified Proterozoic microbial mats: in some cherts microfossils are quite abundant, whereas in others they are quite rare. Unfortunately, the volumetric abundance of microfossils is rarely reported in publications (even though this can be determined relatively easily through standard petrographic point-counting techniques applied to the surface of a thin section, or by determining the cross-sectional area of microfossils at the surface of a thin section for numerous fields of view and comparing the sum of these areas to the total area viewed). Consequently, there are little available quantitative data on microfossil abundance in Proterozoic cherts. However, in many microfossiliferous Proterozoic cherts, the fossils constitute only a few percent to a fraction of a percent of the volume of the chert. For example, in the most fossiliferous interval found in a number of microfossil-bearing cherts from the Middle Proterozoic Dismal Lakes Group of Canada, the fossils constitute only 0.1% of the volume of the laminae in which the microfossils were considered to be "well preserved" (Horodyski and Donaldson 1980). Studies conducted on modern microbial mats have shown that cyanobacterial sheaths commonly are more resistant to degradation than are trichomes or cells (Section 5.3.2B); thus, very thin-sheathed microorganisms would be expected to have a relatively low preservation potential. In the thick, well-laminated microbial mats at the Laguna Mormona sabkha, Baja California, the thick-sheathed oscillatoriacean cyanobacterium *Lyngbya aestuarii* has a high preservation potential, with sheaths of this species being readily identifiable at depth in the mat. In contrast, remains of another oscillatoriacean in this same mat, *Microcoleus chthonoplastes*, which is much more abundant than *L. aestuarii* in these mats but has a much thinner sheath, are very difficult to recognize at depth in the mat (Horodyski et al. 1977). In addition, very small or very narrow microorganisms can be expected to have a very low preservation potential because diagenesis and low-grade metamorphism would tend to reorganize the organic matter of such taxa to a degree that they would commonly be unrecognizable as microfossils.

It seems reasonable to conclude that the microfossil assemblages preserved in Proterozoic cherts are considerably biased inasmuch as bacteria, for example, and particularly the smaller and narrower bacteria, are almost certainly underrepresented. In some cherts, and particularly in those in which microfossils constitute only a few percent or less of the silicified microbial mat, the fossils may represent only the most degradationally resistant component of the original microbial mat.

The second question to be considered is how representative are the fossils known from Proterozoic silicified microbial mats of all Proterozoic stromatolite-forming microbiotas? As discussed above, the fossil record of such organisms is expected to be considerably biased inasmuch as very small and very narrow microorganisms, as well as thin-sheathed and sheathless microorganisms, are evidently more susceptible to degradation and to diagenetic modification. Thus, many bacteria and the small, narrow, or thin-sheathed and sheathless cyanobacteria would have a low to negligible probability of being preserved as fossils. In addition, it is necessary to evaluate whether these silicified assemblages reflect environmentally induced biases. Microorganisms in mats in certain environments, such as ephemeral streams, flood plains, braided streams, and ephemeral pools in bedrock, clearly would have a negligible chance of being preserved. Inasmuch as most of the Proterozoic microbiotas from silicified microbial mats occur in strata that generally are considered to be marine, one should ask whether there is an environmentally induced bias among the microfossils occurring in chert in marine strata. Some workers have proposed that the record could be biased towards microorganisms that are tolerant of elevated salinities. This argument is based on the discovery of an 8,000-year-old algal peat containing partially preserved cellular structures in a Persian Gulf sabkha (Golubić 1976), with the preservation evidently being due to the pickling effects of saline brines. Hypersaline environments, such as sabkhas and saline lagoons, could thus represent settings where organic material could have had a greater chance of being preserved during the interval preceding silicification. A number of Proterozoic microfossil-bearing cherts have been shown to have been closely associated with hypersaline environments (see Knoll 1985a). For example, microfossil-bearing laminae in a chert from the Middle Proterozoic Dismal Lakes Group exhibit a displacive fabric that has been attributed to precipitation of an evaporite mineral, possibly a sulfate (Horodyski and Donaldson 1983), and a microfossil-bearing chert in the Late Proterozoic Chuar Group is very closely associated with pseudomorphs after lensoidal gypsum and dissolutional collapse breccias (Horodyski, unpublished data). However, evidence for hypersaline conditions is not presented in most publications reporting microfossils from Proterozoic chert. This lack of information regarding associated evaporite minerals could reflect the absence of such association, or it could be the result of insufficient study. This is not a simple question to answer as evidence of hypersaline conditions is often subtle and evaporitic environments are not necessarily evaporite-preserving environments.

5.3.3 Eukaryotes (Primarily Acritarchs)

5.3.3A Biochemistry and Preservation Potential of Nonmineralized Eukaryotic Plant Protists: Vegetative Cells and Cysts

The vegetative cell walls of nonmineralized eukaryotic plant protists are composed of a variety of compounds (see Tappan 1980, and references therein, for a complete inventory of such materials). The biochemistry of these cell-wall components is, to a significant degree, correlative with the preservation potential of such organisms in the fossil record. Most cell walls are evidently dominated by water-soluble polysaccharides. For example, certain red algae (e.g., many florideophycids) construct their walls of cellulose, although the cell wall of coralline algae is noncellulosic, consisting of a pectinlike material. Other rhodophytes (e.g., some bangiophycids) have a rather complex wall of xylan fibrils embedded in mucilage surrounded by a cuticle of mannan (Tappan 1980). Many xanthophytes and chrysophytes have cellulosic cell walls, but others have walls composed of noncellulosic polysaccharides. Some chrysophytes lack a cell wall entirely.

The divisions Pyrrhophyta (dinoflagellates) and Chlorophyta (green algae) perhaps have the closest systematic affinities to Proterozoic and Early Cambrian phytoplankters. Cellulose is very common as a cell-wall component in both of these groups; however, exceptional cell-wall biochemistries, and the presence of resistant organic cysts in some, make a more detailed discussion fruitful.

In *dinoflagellates*, the cell wall (or amphiesma) may be quite complex, having four distinct layers. But because the wall is composed primarily of cellulose, vegetative cells of dinoflagellates are thought to have an extremely low preservation potential (Tappan 1980; Edwards 1987). Many workers argue that the vegetative dinoflagellate theca is virtually impossible to preserve; consequently, all fossil dinoflagellates are probably represented by the resistant cyst stage (Evitt 1985).

Dinoflagellate life cycles may include a stage during which a resistant cyst is produced. Most cysts that have been studied appear to be hypnozygotes (zygotic cells in a resting stage), and thus represent a nonmotile portion of the life cycle that follows sexual reproduction. Evitt (1985) pointed out that only a small minority of modern dinoflagellates produce resting cysts. Moreover, not all of these cysts are preservable – in fact, a few are made of cellulose! Edwards (1987) estimated that only about 10% of modern taxa produce preservable cysts.

Preservable dinoflagellate cysts are commonly regarded as being composed of "sporopollenin," a complex polymer or group of polymers of the general type that makes up the resistant outer coating of spores and pollen grains of vascular plants. The structure and composition of sporopollenin is still a subject of research; it is reported to be a high molecular weight and acid-resistant polymer of carbon, hydrogen, and oxygen (Brooks et al. 1971). In its general properties, sporopollenin is essentially indistinguishable from the insoluble, particulate, organic fraction in sedimentary rocks that geochemists call kerogen (see Chapter 3). The fossil record of dinoflagellates, then, would appear to be a very biased one consisting of the small subset of dinoflagellates that produce cysts made of sporopollenin (Evitt 1985). This generalization may be premature: recently, sporopollenin was detected in the *vegetative* cell wall of a few dinoflagellate species (Morrill and Loeblich 1981).

Typically, the cell walls of *green algae* have an inner cellulosic layer and an outer pectic layer; other chlorophytes may use mannans and xylans instead of true cellulose (Tappan 1980). The great majority of Proterozoic and Early Cambrian chlorophytes, however, are probably allied to the enigmatic and noncellulosic prasinophytes (see Section 5.5). This group of green flagellates is characterized by having a scaly quadriflagellate (or biflagellate) stage; the flagella also typically bear distinctive scales (Norris 1980). Some oceanic prasinophytes, such as *Halosphaera* and *Pachysphaera*, encyst during their life cycles, producing a highly resistant (sporopollenin-containing) nonmotile stage called the phycoma. The phycoma actually grows considerably during its existence; therefore, many workers avoid using the term cyst (see Norris 1980; Colbath 1983). Evidence that certain acritarchs are the remains of prasinophyte phycomata is very compelling. For example, Wall (1962) showed convincingly the morphologic similarities between *Tasmanites* and *Pachysphaera*, and between *Leiosphaeridia* and *Halosphaera*. The latter genus of each pair is a modern, phycoma-producing oceanic prasinophyte. *Tasmanites* is a common Paleozoic microfossil (and it has also been reported from the Proterozoic; see Section 5.5), and *Leiosphaeridia* is probably the most common microfossil recovered from Proterozoic shales.

The preservation discussion above argues strongly that many acritarchs may represent the nonmotile, acid-resistant stage (cyst or phycoma) of phytoflagellates related to, and perhaps including, dinoflagellates and prasinophytes. But the *vegetative cells* of some chlorophytes (as well as certain dinoflagellates discussed previously) have recently been shown to resist acetolysis – that is, some of them contain sporopollenin or other refractory organic polymers. For example, the modern freshwater, marine, and soil alga *Chlorella fusca* (Chlorococcales) has an outer trilaminar layer that has been determined to contain sporopollenin. After acetolysis treatment, the inert remains bear a striking resemblance to the small, smooth-walled sphaeromorph acritarchs known as leiospheres (see Atkinson et al. 1972, Fig. 23; Pickett-Heaps 1975, Fig. 3.11). Species of at least two other chlorococcaleans, *Pediastrum* and *Scenedesmus*, probably also contain sporopollenin-like compounds because their cell walls resist acetolysis (at least partially; Pickett-Heaps 1975). Finally, sporopollenin has been documented to occur in the zygote walls of charophycean chlorophytes, such as *Chara* and *Nitella* (see Blackmore and Barnes 1987), as well as those of *Coleochaete* (Delwiche et al. 1989). The fact that most of these chlorophytes are typical freshwater forms may temper enthusiasm about relating them to the fossil record, in which marine sedimentary rocks yield most microfossils. Still, these examples are important: they suggest that a more systematic study of algal vegetative cells may yield additional examples of potentially preservable species.

5.3.3B Degradational and Preservational Artifacts

Proterozoic acritarchs are acid-resistant, organic-walled microfossils that most commonly are studied in macerations derived by digesting fine-grained siliciclastic rocks in hydrofluoric acid (Vidal 1976a), although they also have been studied in macerations of carbonates (Vidal 1976a) and in petrographic thin sections of shale (Horodyski 1980) and chert (Knoll and Calder 1983; Knoll 1984). The hollow, empty nature of some specimens probably is the result of physical loss of internal contents via tears or openings in the wall, structures including those which have been attributed to excystment (Vidal 1976a, 1981a); however, in other specimens the lack of internal contents could be the result of their susceptibility to degradation.

Acritarchs can be modified by a variety of *postmortem* processes. Vidal (1979a, 1981a, 1981b) noted a direct correlation between the state of preservation of Proterozoic acritarchs and the energy of the environment and the associated potential for reworking; acritarchs in nearshore settings commonly were reported to be corroded and worn, whereas those from low-energy settings, such as prodeltaic or distant-delta environments, were regarded as relatively well preserved. In part this appears to be the result of abrasion, but it also could be caused by chemical or biochemical degradation under oxidizing or alkaline conditions (Vidal 1979a, 1981a; Vidal and Siedlecka 1983). Reworking and degradation prior to final deposition also would facilitate the separation of double-walled acritarchs into two components or the loss of the outer wall by abrasion (Vidal 1981a, 1981b).

In siliciclastic rocks, acritarchs typically are compressed, and some of their morphological features clearly are the result of sediment compaction. As discussed in Section 5.4.2, such compressional features include folds, such as those in *Kildinosphaera* (Vidal 1981b; Vidal and Ford 1985), and cracks, such as those in *Trachysphaeridium laufeldii* (Vidal and Ford 1985). Although such features clearly are artifacts, they provide information on the strength and rigidity of the vesicle and may be useful for recognizing certain acritarch taxa (see Vidal and Ford 1985). The acritarch vesicle may be further modified by imprints of detrital grains and by disruption caused by growth of diagenetic and metamorphic minerals (Halpern 1988).

5.3.3C Taxonomic Problems

Historically, acritarchs occurring in Proterozoic mudstones generally have received different taxonomic treatment from large spheroidal microfossils preserved in chert. In part, this was because of the division of Precambrian microfossil workers into two schools: those that primarily study microfossils in thin sections of chert, and those that primarily study microfossils in macerations obtained by acid dissolution of fine-grained siliciclastic rocks. However, part of this difference in taxonomic treatment is because microfossils preserved in different lithologies and treated by different processes offer different attributes useful for their taxonomic characterization. Large spheroidal microfossils in chert typically are three-dimensionally preserved and are characterized foremost by their size, the thickness and nature of their envelope, the nature of their internal contents, and the geometric relationship to other specimens. Recently, workers have successfully identified, in chert, acritarchs that originally were described from siliciclastic rocks (e.g., Knoll and Calder 1983; Knoll 1984), and progress is now being made in establishing a single taxonomy for acritarchs occurring in chert and siliciclastic rocks.

Proterozoic acritarchs preserved in shale typically are flattened [note, however, that three-dimensionally preserved acritarchs have been reported from siliciclastic rocks (Vidal 1976a)] and are characterized by the shape, thickness, and sculpture of the vesicle, the shape and ornamentation of processes, and the nature of excystment structures. Some diagnostic features of Proterozoic acritarch taxa are very fine and delicate, and their recognition may be dependent upon the lithology. For example, fine details of the walls of certain acritarchs may be of micrometer size and best observed by interference-contrast or phase-contrast microscopy of specimens separated from their mineralic matrix. Such details can be studied in well-preserved acritarchs removed from mudstone by acid dissolution; however, these details may be lost in chert due to taphonomic reorganization of the organic material. Furthermore, acritarchs in chert commonly cannot be viewed with the high degree of optical resolution that is obtainable in acid-resistant residues, and many of the specimens preserved in chert are not of sufficient structural integrity to survive acid dissolution intact. In addition, some attributes of acritarchs preserved in mudstone, such as compactionally produced features that may reflect the rigidity of the wall (Vidal 1981b), are not present in the three-dimensionally preserved envelopes occurring in chert. Finally, low-grade metamorphism may be responsible for poor preservation of acritarchs, and it may render them difficult or impossible to identify (e.g., Hofmann et al. 1979; Horodyski 1981); however, readily identifiable acritarchs have been recovered from folded rocks metamorphosed to prehnite-pumpellyite facies (Vidal 1976b).

5.3.3D Distribution of Modern Nonmineralized Eukaryotic Plant Protists

Modern organic-walled eukaryotic protists exhibit an impressive range of environmental tolerances, accounting for their broad distribution in fresh, brackish, and marine waters, and in a variety of subaerial habitats. For many environmental factors these eukaryotic protists are more restricted than many prokaryotes (see Knoll and Bauld 1989), but a full evaluation of their environmental tolerances must await a more complete understanding of their life cycles, which may include highly resistant cysts or other dormant stages. Because microbiologists have tended to concentrate on the physiologically more active vegetative stages, environmentally tolerant encysted stages are relatively poorly known (see Tappan 1980).

Proterozoic and Early Cambrian nonmineralized eukaryotes seem to have closest affinities with modern and fossil dinoflagellates and green algae; accordingly, the modern ecology of these two groups is considered briefly in the following paragraphs.

Dinoflagellates are second in importance only to diatoms as primary producers in the oceans. The majority of modern dinoflagellates are photoautotrophs; the remainder display a remarkable array of physiologies: some are myxotrophic (obtaining nutrients through both photosynthesis and ingestion); some are holozoic (grazing or predating other organisms); and some live on decaying organic matter as saprophytes. Still other dinoflagellates live as parasites or as symbionts (zooxanthellae) within other organisms, and some host their own symbionts. It is not surprising, therefore, that dinoflagellates exploit a wide range of habitats in coastal and oceanic marine waters, fresh-, salt-, and brackish water lakes, snow, and even the interstices of sand and silt grains on beaches (reviewed in Tappan 1980; Edwards 1987).

Individual dinoflagellate species typically are either cosmopolitan or are restricted to the warm waters of tropical seas. Photoautotrophic dinoflagellates prefer the upper 25 meters of the ocean, but some shade species live at depths of up to 200 meters (Tappan 1980). Many dinoflagellates that produce cysts during their life cycles provide a "benthic" population, largely in neritic waters (Dale 1983). According to Tappan (1980), such meroplanktonic forms (having both planktonic and benthic stages) are morphologically simple. Holoplanktonic dinoflagellates, those that lack a benthic stage, commonly are found in the open ocean; many of these exhibit more complicated morphologies that increase their surface area-to-volume ratio, and that may be adaptations to retard sinking or to facilitate access to nutrients. But studies of the functional morphology of phytoplankton form have not reached a consensus (see, for example, Smayda 1970; Taylor 1980; Vogel 1981; and Reynolds 1984).

Tappan (1980) reviewed the salinity and temperature tolerances of dinoflagellates. Both eury- and stenohaline marine forms exist, but most species grow best at a salinity of about 20 parts per thousand (compared to the normal oceanic value of 35 parts per thousand), partly accounting for the abundance of dinoflagellates in less saline neritic waters. The temperature optimum of most dinoflagellates lies between 18 and 25°C. Some species of *Ceratium* remain physiologically active at $-1°C$ (Nordli 1957), whereas *Crypthecodinium cohnii* can live at 35°C.

Tappan (1980, p. 825) characterized the *green algae* as having the "greatest diversity in morphology, habit, and habitat of any of the algal divisions." But only about 10% of the approximately 5,500 species of green algae occur in marine waters (Dawes 1981). Of these marine forms, the most important group in terms of the Proterozoic biosphere are the prasinophytes.

Prasinophytes are most common in neritic waters of the temperate zones, typically as plankton occurring at depths of $\geqslant 10$ meters or, during blooms, in the surface film of the sea (Norris 1980; Tappan 1980; Richardson 1984). *Halosphaera* and *Pachysphaera*, among the best studied of modern genera, produce a resistant, nonmotile phycoma during their life cycles. But, unlike the cysts of many meroplanktonic dinoflagellates, these phycomata are not restricted to shelf sediments; perhaps a higher content of oily substances allows them to remain in the upper water column for longer periods of time, thus allowing them to inhabit habitats away from the shelf and farther into the open ocean. Richardson (1984) argued that the fossil record substantiates this broader distribution, as fossil prasinophytes may occur in profusion in sediments interpreted to be of deeper water origin. Tappan (1980) suggested that prasinophytes may be "disaster species" because their fossils are found in profusion in sediments just above extinction horizons (e.g., the end-Frasnian extinction, which decimated many of the acritarchs). Many of the larger sphaeromorph acritarchs are probably prasinophytes; a detailed examination of their history may help identify *Proterozoic* extinction events (see Sections 5.5.2A, 5.5.3D, 5.8).

5.3.3E How Representative is the Known Biota of Proterozoic Acritarchs?

In contrast to Proterozoic silicified microbial-mat microbiotas, it is difficult to evaluate how representative the known record is of Proterozoic acritarchs. The relatively few data available that bear upon this question (Section 5.5.2) suggest that the known record is probably representative of Late Proterozoic acritarchs, and there are few data indicating the contrary. [Lower and Middle Proterozoic acritarchs are excluded from this discussion because they are known from fewer localities (Section 5.5.2), and organic-walled animal protists also are excluded as they are poorly represented in the Proterozoic fossil record. The single moderately convincing example of a possible organic-walled animal protist of Late Proterozoic age is the vase-shaped fossil *Melanocyrillium*; however, it alternatively could represent the sporangium of a eukaryotic alga (Section 5.6).]

Evidence suggesting that the known record is representative of Late Proterozoic acritarchs includes:

(1) Late Proterozoic depositional environments appear to have been adequately sampled for acritarchs. Although acritarchs of this age are most diverse in strata deposited in open-shelf settings (Vidal and Knoll 1983), they also have been reported from deeper water settings as well as nearshore and coastal settings (Ivanovskaya and Timofeev 1971; Vidal 1976a, 1979b, 1981b; Knoll and Vidal 1983). There may, however, be a tectonically indiced bias in that deeper water sediments are less commonly preserved and are more likely to have been later metamorphosed.

(2) Acritarchs that have been reported from chert deposited in Late Proterozoic open-coastal environments – the same en-

vironments that contain diverse acritarch assemblages in Upper Proterozoic siliciclastic rocks – can be compared to acritarch taxa known from siliciclastic rocks (Knoll and Calder 1983; Knoll 1984). Significantly, these cherts do not contain fossils of organic-walled protists that are not known from siliciclastic rocks. However, localities identified as containing fossiliferous Proterozoic open-coastal chert are few in number.

(3) Upper Proterozoic cherts from other environments do not contain fossils of *large* organic-walled protists that differ substantially from acritarchs preserved in siliciclastic rocks. However, some of these cherts do contain small- to medium-sized, thin-walled spheroidal fossils that are not commonly reported from siliciclastic rocks (Figs. 5.4.15, 5.5.6, 5.5.7, 5.5.10). As discussed in Sections 5.4.4 and 5.5.2, these fossils could represent eukaryotic algal unicells, and their distribution could be environmentally controlled as indicated by Knoll and Swett (1985), who noted the presence of "chert-facies" microfossils in lagoonal shales from the Late Proterozoic Veteranen Group in Spitsbergen. However, the scarcity or apparent scarcity of these relatively small, thin-walled forms in shales could be due to other factors. They probably are more susceptible to *post-mortem* changes than the more robust acritarchs, they may not be recognized owing to separation techniques (maceration and sieving) commonly used in the preparation of acritarchs, and they simply may be overlooked by workers who concentrate on the larger forms and particularly on those forms that exhibit surface ornamentation or other attributes that make those specimens useful time-stratigraphic indicators.

Obviously, more work needs to be done to evaluate the organic-walled protist record. In particular, Middle Proterozoic (and even Lower Proterozoic) siliciclastic rocks need to be sampled more thoroughly for microfossils, and cherts and other lithologies such as phosphates that could preserve degradationally susceptible organisms should be studied from offshore environments.

5.3.4 Preservation Potential of Naked, Agglutinated, Calcareous, and Siliceous Protists

In general, only those protists with hard parts have a chance of being preserved in the fossil record. These include organisms with agglutinated, calcareous, and siliceous skeletons or cell elements.

Protist cells are covered by an envelope or pellicle which functions in a variety of ways but chiefly to exchange gases and fluids between the environment and the cytoplasm and to maintain the shape of the organism. It may be a single membrane, or it may be thicker and more complex. The pellicle consists of proteins and degrades rapidly upon death of the protist. Only a few occurrences of fossil naked protists are known, almost entirely from organic-rich lignitic or bituminous shales (Tappan 1980). It is conceivable that naked protists might be preserved in cherts that contain permineralized prokaryotes.

Agglutinated skeletons are common among foraminifera, thecamoebians, and some tintinnids. These skeletons are made either by utilizing available sedimentary clasts or by selecting a particular subset of such particles from the sediment. Unless selection results in a skeleton of a single mineralogy that is particularly susceptible to dissolution or diagenesis (e.g., agglutinated siliceous sponge spicules), fossils of such protists are likely to be preserved as part of the sedimentary rock.

Calcareous protists are unknown in either Proterozoic or Cambrian rocks, although other calcareous fossils occur in rocks of both ages. In the Cambrian, megascopic calcareous skeletons are widespread, diverse, and abundant, yet calcareous protists are unknown. This observation suggests that selective destruction of calcareous protists did not occur, but that Cambrian protists simply may not have secreted calcareous skeletons.

Siliceous protists are known from the Cambrian, but not earlier. Although of uncommon occurrence, they have been reported from fine-grained clastic rocks, limestones, and cherts. In modern seas, silica is rapidly dissolved and lightly silicified skeletons are easily destroyed. However, the relative scarcity of Cambrian siliceous protists may not be a result of such dissolution; other siliceous fossils (e.g., sponge spicules) are found commonly in the Cambrian, but fossil radiolaria and other siliceous protists are rare. Had these protists been a prominent component of the early biota, they likely would be of more common occurrence.

5.3.5 Bias in the Microfossil Record: Prokaryotes and Eukaryotes as a Function of Paleoenvironment and Lithology

As discussed above (Sections 5.2.2E, 5.3.3E), there is reason to believe that the known Proterozoic fossil record is considerably biased. For example, fossil bacteria have rarely been reported in Proterozoic rocks, even though they must have been abundant considering the evidence for degradation in preserved Proterozoic microbial mats and by analogy to modern microbial mats. Furthermore, there may be a strong environmentally induced bias in microfossil associations preserved in chert. Some workers (Horodyski and Donaldson 1983; Knoll 1985a, 1985b) contend that hypersaline conditions enhanced the preservation potential of Proterozoic microorganisms, and that some of the microfossils preserved in Proterozoic cherts originated in hypersaline environments. Inasmuch as the predominantly prokaryotic biotas of modern sabkhas and hypersaline lagoons provide a limited sample of the modern microbial and micro-algal biosphere, fossils preserved in Proterozoic rocks deposited in similar settings may not be representative of the Proterozoic biosphere. There might also be an environmental bias introduced by preferential silicification in environments where organic matter was abundant in near-surface sediments (Knoll 1985b).

Proterozoic microfossil-bearing cherts largely originated in very shallow to intertidal settings and consist largely of fossils of benthic, mat-building microorganisms. In contrast, microfossil-bearing siliciclastic rocks originated in a variety of settings and consist largely of fossils of planktonic microorganisms. There are substantial differences between microfossils in cherts and those in siliciclastic rocks; however, microfossils in the limited number of Proterozoic fossiliferous cherts regarded as representing open-coastal settings are similar to those occurring in siliciclastic rocks (Knoll and Calder 1983; Knoll 1984).

Table 5.3.1. *Type of Proterozoic and Early Cambrian microfossil detected as a function of the primary lithology reported. Total occurrences of unequivocal microfossils in the PPRG microfossil database is 2,841; 2,721 of these were used to construct this table [minor lithologies and highly problematic microfossils (Section 5.7) were excluded]. Assignment of microfossils to prokaryotes and eukaryotes is based on Chapter 25. The table is subject to many errors (e.g., some microfossils have been reported from many lithologies in the same paper; and the prokaryote/eukaryote determination may not always be accurate).*

Primary Lithology	Prokaryote Occurrences (and percent)	Eukaryote Occurrences (and percent)
chert[a]	716 (26.3%)	89 (3.3%)
carbonate[b]	37 (1.4%)	54 (2.0%)
sandstone[c]	23 (0.8%)	22 (0.8%)
siltstone	25 (0.9%)	61 (2.2%)
shale[d]	468 (17.2%)	1226 (45.1%)

[a] includes jaspilite
[b] includes limestone, dolostone
[c] includes graywacke, quartzite
[d] includes argillite, claystone, mudstone

Can the question of how representative is the Proterozoic and Early Cambrian microfossil record be addressed in a quantitative fashion? In terms of representation of environments of deposition, a quantitative approach is not fruitful. Few microfossiliferous Proterozoic deposits have been studied in a detailed sedimentologic manner; consequently, the published literature includes only minimal environmental information. Recently, a trend toward more detailed sedimentologic characterization of Proterozoic microfossil-bearing rocks has been established (e.g., Vidal 1981b; Knoll 1984; Knoll et al. 1989).

However, if lithology can be regarded as providing at least a rough index of environment, it is possible to perform a *crude* quantitative analysis of environmental bias in the record of Proterozoic microfossil-bearing strata as now known. Table 5.3.1 conveys a very general sense of the environmental bias of fossil-yielding strata. The table presents little that is new: it simply confirms (in a crude quantitative manner) that most prokaryotic microfossils (in terms of occurrences) have been described from cherts, whereas the majority of eukaryotes have been described from siliciclastic rocks, primarily shales. Analysis of the database used to construct Chapter 22 reveals that about 33% of all unequivocal microfossils (i.e., excluding nonfossils, pseudofossils, and dubiofossils) have been studied in thin section; about 64% have been studied in maceration; and less than 1% have been studied as carbonaceous films. Only about 2 to 3% have been studied by both thin section and maceration. The correspondence between technique and lithology is not surprising; clearly, few prokaryotes have been reported from siliciclastics, and few eukaryotes have been studied in cherts and carbonates. However, recent work (by Hofmann, Horodyski, Knoll, and Vidal, among others) is showing that these results may be exaggerated, for the pace at which eukaryotes are being discovered in cherts, and prokaryotes are being reported from shales, is accelerating. This development has been spurred by the gradual breakdown of separation between the two "facies-based" schools of Precambrian micropaleontology (Section 5.2).

5.4

Proterozoic Prokaryotes: Affinities, Geologic Distribution, and Evolutionary Trends

J. WILLIAM SCHOPF

5.4.1 Introduction

The following is a discussion of the affinities, temporal distribution, and inferred evolutionary trends of Proterozoic prokaryotes; these matters regarding Proterozoic eukaryotes are addressed in Section 5.5. Of the 297 informal species of fossil prokaryotes here recognized (Table 25.1, 25.3, Fig. 25.4), all but 22 morphologically distinctive taxa, comprising four categories of Proterozoic prokaryotes, are encompassed by the following discussion. Collectively, these 22 species of easily identifiable but relatively uncommon fossils comprise less than 3% of all taxonomic occurrences of prokaryotes now known from the Proterozoic fossil record (Table 25.3). The four categories of Proterozoic prokaryotes not here considered include:

(1) helical, *Spirulina*-like morphotypes (code "TUC" in Tables 25.1, 25.3, and Fig. 25.4, represented in the Proterozoic fossil record by five species; e.g., Chapter 24: Pl. 19, C; Pl. 22, H; Pl. 31, B);
(2) branched, tubular morphotypes (code "TB," three fossil species; e.g., Chapter 24: Pl. 39, A; Pl. 46, E);
(3) *Polybessurus* sp. (code "POLY," one fossil species; e.g., Chapter 24: Pl. 35, A-E; Pl. 36, A-G; Pl. 37, A, C, F-J; Pl. 38, A-H); and
(4) unusual "bizarre" morphotypes (code "BZ," 13 fossil species; e.g., Chapter 24: Pl. 1, K, L; Pl. 2, A-D, O-R; Pl. 4, C; Pl. 42, E, I).

As is discussed below, the vast majority of known Proterozoic prokaryotes appear to be of cyanobacterial affinity. Among solitary and colonial coccoidal and ellipsoidal fossils, most are inferred to be of chroococcalean (predominantly chroococcacean) affinity, comparable in salient morphological characteristics to the representative modern chroococcaceans shown in Figure 5.4.1. The great preponderance of known filamentous Proterozoic prokaryotes are comparable to modern nostocalean (viz., oscillatoriacean) cyanobacteria. Both sheath-enclosed (cf. Fig. 5.4.2B) and naked cellular trichomes (cf. Fig. 5.4.2A) are common components of many Proterozoic assemblages, as are empty, tubular, prokaryotic sheaths, the occurence of which reflects either the preferential preservation of sheaths as opposed to trichomes (Section 5.3) or the vacating of sheaths by trichomes capable of photo- and chemotactic gliding mobility.

5.4.2 Morphometric Effects of Preservational Compression of Microfossils in Shales

As is discussed in Section 5.2 (see also, Section 11.3 and Chapter 25), historically, micropaleontological studies of Proterozoic strata have progressed along two parallel, but largely independent, pathways: (*i*) investigations of plankton-dominated microfloras preserved as two-dimensional compressions and studied in acid-resistant macerations of shales/siltstones; and (*ii*) studies of benthic-dominated microbial communities preserved via three-dimensional permineralization and studied in petrographic thin sections of

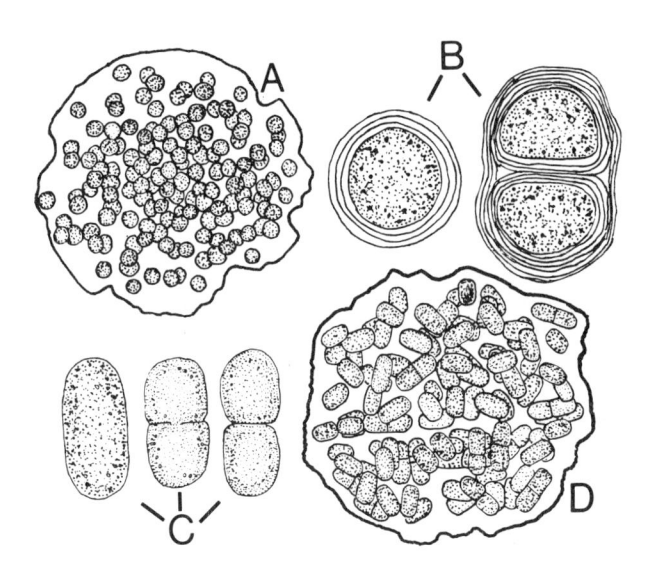

Figure 5.4.1. Representative modern coccoidal and ellipsoidal chroococcalean (Chroococcaceae) cyanobacteria. (A) cf. *Microcystis-Aphanocapsa*, unordered colonies of coccoidal cells enclosed in an amorphous mucilage. (B) cf. *Gloeocapsa-Chroococcus*, coccoidal to ellipsoidal cells enveloped by distinct, commonly multilamellated, sheaths. (C) *Synechococcus* sp., ellipsoidal cells, single, paired or in small clusters, not enclosed by a common mucilage. (D) *Aphanothece* sp., ellipsoidal colonial cells enclosed in an amorphous mucilage.

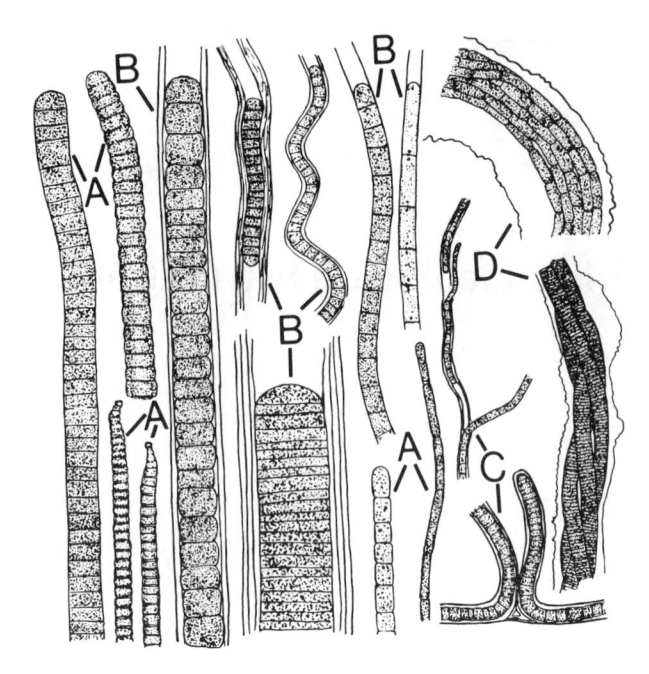

Figure 5.4.2. Representative modern filamentous nostocalean cyanobacteria illustrating the range of cell shape and organization characteristic of the Oscillatoriaceae. (A) *Oscillatoria* spp., unbranched, uniseriate, cylindrical to somewhat tapered trichomes, typically with discoidal to quadrate (or less commonly elongate) medial cells and rounded (or less commonly globose or attenuated) end cells, lacking a well-defined encompassing sheath. (B) *Lyngbya* spp., monotrichomic filaments composed of unbranched cylindrical trichomes enclosed within a well-defined, unlamellated or lamellated cylindrical sheath. (C) Filaments cf. *Symploca* spp. exhibiting false-branching. (D) Polytrichomic filaments cf. *Microcoleus-Hydrocoleum* composed of numerous trichomes enclosed within a common sheath.

carbonates/cherts. Because of the differing types of preservation occurring in these two lithologies and the differing expertise thus required for preparation and study of the preserved microfossils, few workers (and even fewer publications) have dealt with Proterozoic microfossils occurring both in shales/siltstones and in carbonates/cherts. Indeed, as is discussed in Chapter 25 (see also Section 11.3), much of the current incoherence in the systematics of Proterozoic microfossils stems from the differing taxonomic conventions adopted by these two contrasting, facies-based schools. With but few exceptions, taxa described from one facies have not been compared with coeval taxa reported from the other, chiefly because of a lack of data that would permit realistic morphometric comparison of microfossils preserved as two-dimensional compressions in shales/siltstones with uncompressed modern microorganisms and with microfossils preserved three-dimensionally in cherts/carbonates. The studies discussed below were carried out in order to redress this deficiency.

5.4.2A Compression and Radial Cracking of Organic-Walled Ellipsoidal and Spheroidal Microfossils

As is shown in Figures 5.4.3 and 5.4.4, fossils of originally ellipsoidal (Fig. 5.4.3A–D) and spheroidal (Figs. 5.4.3E, 5.4.4A, B, E, F) microorganisms preserved as flattened, two-dimensional, ellipses and discs in shales/siltstones typically exhibit compression-produced, central and tangential, folds, pleats, and flexures (Figs. 5.4.3A–E, 5.4.4E, F), and peripheral radial cracks (Figs. 5.4.3D, E, 5.4.4A, B). In order to compare morphometrically such fossils with uncompressed microfossils or with modern microorganisms, the distortion of original morphology resulting from such compression need be assessed. Because of their morphological simplicity and excellent preservation, specimens of *Macroptycha* spp. from the Late Proterozoic Miroedikha Formation of Siberia (Fig. 5.4.3A, B) are ideal candidates for such an assessment. Discovered and first described by Timofeev (in Timofeev and German 1976), *Macroptycha* is a genus composed of originally ellipsoidal vesicles, subdivided into three species based on the occurrence of one, two, or three, distinctive, symmetrically distributed, pleat-like longitudinal folds (pleats that were originally interpreted by Timofeev as representing the walls of internal, cellular, chambers). Table 5.4.1 summarizes the results of a series of measurements and calculations made on seven specimens of *Macroptycha*. As is there shown: (*i*) measurement of the preserved, flattened, ellipse-shaped specimens (#1) has been used to calculate the area of each compressed ellipse (#3); (*ii*) for each specimen, addition of this area (#3) to the measured area occupied by pleat-like folds (#2) yields the area of an equivalent "unfolded" ellipse (#4); (*iii*) calculation of the dimensions of these "mathematically unfolded" ellipses (#5) permits calculation of the length and width of the original, uncompressed, ellipsoidal vesicles (#6); and (*iv*) comparison of these dimensions (#6) with those of the preserved, compressed, ellipses (#1) provides means to determine the percent change in length and width resulting from preservational compression (#7).

Despite the difficulty in measuring precisely the dimensions of such microfossils (even in greatly enlarged photomicrographs), and especially the uncertainty in determining accurately the total area of all regions occupied by folds (including minute pleats and flexures; e.g., Fig. 5.4.3B), it is notable that the measured dimensions of the preserved, compressed microfossils (#1) and the calculated dimensions of the original uncompressed vesicles (#6) are in close agreement. Such agreement is not unexpected: unless the ellipsoids had been ruptured during preservation, the preserved ellipses could not have lengths or widths greater than those of the original vesicles, and if compressed symmetrically, as in these and most Proterozoic specimens, the lengths and widths of the compressed vesicles would be expected to approximate the original dimensions. It seems evident, therefore, that the measured dimensions of flattened, originally ellipsoidal microfossils preserved by compression in shales/siltstones provide a reasonably accurate indication of the original, uncompressed, dimensions of such vesicles and, thus, an appropriate basis for morphometric comparison with uncompressed microfossils and modern microorganisms.

Similarly, and for much the same reasons, the diameters of compressed, flattened, originally spheroidal microfossils preserved in shales/siltstones, if not altered by radial cracking (see below), tend to closely approximate the uncompressed diameters of such vesicles. Such flattened spheroids rather commonly exhibit central folds (Fig. 5.4.4E), prominent tangential

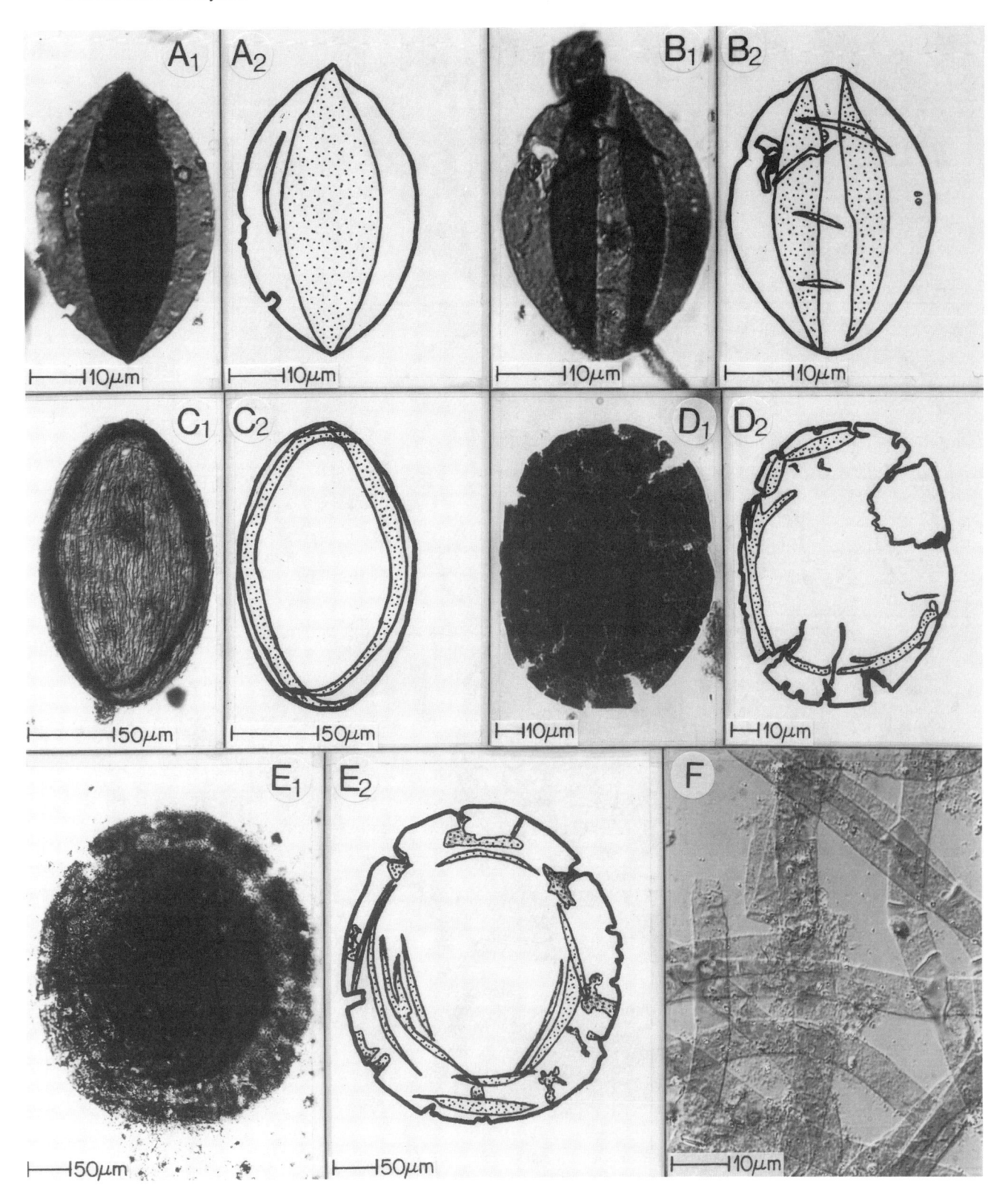

Figure 5.4.3. Effects of preservational compression on originally ellipsoidal (A–D), spheroidal (E), and cylindrical-tubular microfossils (F) detected in acid-resistant residues of Proterozoic shales from the Miroedikha (A, B), Kamarovsk (C), and Lakhanda Formations (E, F) of Siberia, and the Redkino Formation of Bashkiria, USSR (D). [A_1, B_1, C_1, F, interference contrast; D_1, E_1, transmitted light; A_2, B_2, C_2, D_2, E_2 show interpretive tracings of photomicrographs with stippling indicating folded areas.] (A_1, A_2) *Macroptycha uniplicata* Timofeev 1976, exhibiting a single, medial-longitudinal, pleat-like fold. (B_1, B_2) *Macroptycha biplicata* Timofeev 1976, with two, prominent, symmetrically distributed, longitudinal pleat-like folds. (C_1, C_2) *Cucumiforma vanavaria* Mikhailova 1986, with peripheral, tangential, pleats. (D_1, D_2) *Satka granulosa* Jankauskas 1980, with peripheral folds and radial cracks. (E_1, E_2) A greatly compressed, tangentially folded and radially cracked specimen of *Pterospermopsimorpha* sp. (F) *Leiothrichoides tenuitunicatus* Hermann 1981, flattened ribbon-like microfossils resulting from compression of the originally cylindrical, tubular, empty sheaths of prokaryotic filaments.

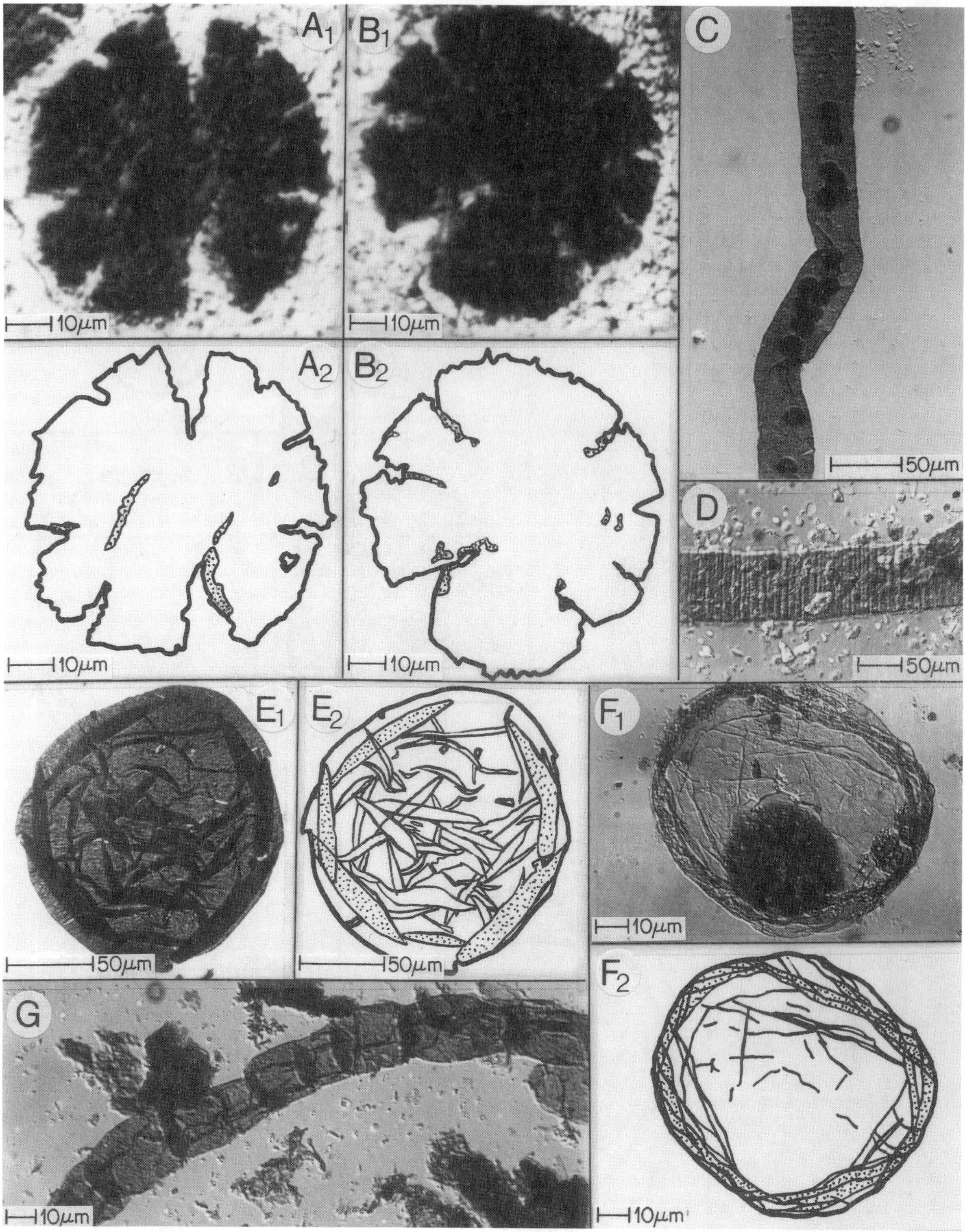

Figure 5.4.4. Effects of preservational compression on originally spheroidal (A, B, E, F), cylindrical-tubular (C), and cylindrical-septate microfossils (D, G) detected in acid-resistant residues of Proterozoic shales from the Lakhanda (A, B) and Miroedikha Formations (C, G) of Siberia, and the Redkino (D, F) and Akberdin Formations (E) of Bashkiria, USSR. [A_1, B_1, E_1, transmitted light; C, D, F_1, G, interference contrast; A_2, B_2, E_2, F_2 show interpretive tracings of photomicrographs with stippling indicating folded areas.] (A_1 and A_2 and B_1, B_2) *Turuchanica ternata* (Timofeev) Jankauskas in press, exhibiting deeply incised radial cracks. (C) *Palaeolyngbya sphaerocephala* Hermann 1986, the flattened, ribbon-like, originally cylindrical-tubular sheath of a filamentous cyanobacterium; note the occurrence within the flattened sheath of numerous disarticulated discoidal cells. (D) *Striatella coriacea* Assejeva 1983, the compressed, flattened remnants of an originally cylindrical-septate cyanobacterial trichome; note the occurrence of relatively well-preserved, regularly spaced, transverse septa. (E_1, E_2) a highly compressed specimen of the thick-walled sphaeromorph *Kildinella lophostriata* Jankauskas 1979, exhibiting numerous central folds and prominent, tangential, pleats. (F_1, F_2) a highly compressed specimen of the thin-walled sphaeromorph *Leiosphaeridia incrassatula* Jankauskas 1980, exhibiting numerous, peripheral, tangential, accordion pleats and flexures. (G) *Arctacellularia doliiformis* Hermann 1976, the compressed, strap-like remnants of an originally cylindrical-septate filament; note the more or less regular spacing of flattened transverse septa.

Table 5.4.1. *Calculated size changes in ellipsoidal microfossils resulting from preservational compression in shales.*

Taxon	Compressed Shape	Compressed Ellipsoidal Microfossils							
		Measured Dimensions		Calculated Dimensions					
		(#1A)×(#1B) Compressed Ellipse Dimensions (μm) Length × Width	(#2) Folded Region Area (μm^2)	(#3) Compressed Ellipse Area (μm^2)	(#4) Unfolded Ellipse Area (μm^2)	(#5A)×(#5B) Unfolded Ellipse Dimensions (μm) Length × Width	(#6A)×(#6B) Uncompressed Ellipsoid Dimensions (μm) Length × Width	(#7A) Calculated Size Change Due To Compression Length	(#7B) Width
Macroptycha uniplicata Timofeev (Timofeev & German 1976, Pl. 10, Fig.7)	Folded Ellipse	42.2 × 23.0	395	762	1,157	42.2 × 34.9	42.2 × 22.2	0%	+4%
M. uniplicata Timofeev (Fig. 5.4.3A)	Folded Ellipse	53.0 × 34.0	651	1,415	2,166	53.0 × 52.0	53.0 × 33.0	0%	+3%
M. uniplicata Timofeev (Timofeev & German, 1976, Pl. 10, Fig. 6)	Folded Ellipse	40.0 × 26.1	366	820	1,186	40.0 × 37.8	40.0 × 24.0	0%	+9%
Macroptycha biplicata Timofeev (Fig. 5.4.3B)	Folded Ellipse	40.5 × 30.5	341	970	1,311	40.5 × 41.2	40.5 × 26.3	0%	+16%
M. biplicata Timofeev (Timofeev & German 1976, Pl. 10, Fig. 8)	Folded Ellipse	42.9 × 31.6	456	1,065	1,521	42.9 × 45.1	42.9 × 28.7	0%	+10%
M. biplicata Timofeev (Timofeev & German 1976, Pl. 10, Fig. 9)	Folded Ellipse	47.8 × 27.3	649	1,025	1,674	47.8 × 44.6	47.8 × 28.4	0%	−4%
Macroptycha triplicata Timoveev (Timofeev & German 1976, Pl. 10, Fig. 12)	Folded Ellipse	55.3 × 32.9	771	1,429	2,200	55.3 × 50.7	55.3 × 32.3	0%	+2%

#3 = (3.14)(#1A/2)(#1B/2)
#4 = (#2) + (#3) = (3.14)(#5A/2)(#5B/2)
#5A = (#1A) = (#6A)
#5B = (1.27)(#4/#5A)
#6B = (3.14/2)(#5B)
#7B = [(#1B/#6B)(100)] − 100

pleats (Fig. 5.4.4E), and peripheral accordion pleats and flexures (Fig. 5.4.4F). The abundance and complexity of these folds (Figs. 5.4.4E, F) preclude measurements of the type discussed above for flattened ellipsoids. Nevertheless, like the ellipsoids, such spheroids – *whether thin-walled* (Fig. 5.4.4F) *or thick-walled* (Fig. 5.4.4E – typically have been symmetrically compressed as though they were minimally elastic spherical "balloons". If not ruptured, and if disc-shaped (rather than being crumpled and folded over), the diameters of such fossils closely approximate those of the original uncompressed vesicles, providing an appropriate basis for comparison with uncompressed microfossils and modern microorganisms.

If highly compressed in shales/siltstones, originally ellipsoidal (Fig. 5.4.3D) and spheroidal microfossils (Figs. 5.4.3E, 5.4.4A, B), especially relatively thick-walled specimens, typically exhibit peripheral, radial cracking. Table 5.4.2 summarizes the results of measurements and calculations made on four such specimens in order to assess the degree of morphometric change resulting from preservational radial cracking. As is there shown (Table 5.4.2): (*i*) measurement of the actual circumference (not including gaps due to radial cracking) of the flattened, ellipse- or disc-shaped microfossils (#3) was used to calculate the original, uncracked, dimensions of the flattened vesicles (#4, #5); (*ii*) measurement of the dimensions of the microfossils as preserved, dimensions affected by size increases due to radial cracking (#1, #2), were used to calculate preserved circumferences (#6); and (*iii*) comparison of measured, actual, circumferences (#3) and calculated, preserved, circumferences (#6) were used to calculate percent size change due to radial cracking (#7).

As is summarized in Table 5.4.3, observed size increases due to radial cracking range from a minimum of 10% to a maximum of nearly 40%, with an average increase of 23%. Although the morphometric effects of radial cracking on highly compressed ellipsoidal and spheroidal microfossils can thus be substantial, taxa exhibiting such effects (principally *Turuchanica* spp.) are of uncommon occurrence in the Proterozoic fossil record, accounting for less than 2% of all known occurrences of Proterzoic spheroidal and ellipsoidal microfossils (Table 25.3). As indicated in Table 5.4.4, a correction factor of 20% has been here used for comparison of radially cracked microfossils with uncracked microfossils and modern microorganisms.

Table 5.4.2. *Calculated size changes in ellipsoidal and spheroidal microfossils resulting from preservational radial cracking in shales.*

		Compressed Radially Cracked Ellipsoidal and Spheroidal Microfossils						
		Measured Dimensions			Calculated Dimensions			
		(#1A)×(#1B) Cracked Ellipse	(#2) Cracked Disc	(#3e, #3d) Uncracked Ellipse/Disc	(#4A)×(#4B) Uncracked Ellipse	(#5) Uncracked Disc	(#6e, #6d) Cracked Ellipse/Disc	(#7e, #7d) Calculated Change
Taxon	Compressed Shape	Dimensions (μm) Length × Width	Diameter (μm)	Circumference (μm)	Dimensions (μm) Length × Width	Diameter (μm)	Circumference (μm)	In Diameter due to Radial Cracking
Satka granulosa Jankauskas (Fig. 5.4.3D)	Radially Cracked Ellipse	103 × 82		258	90 × 71		292	+13%
Turuchanica ternata (Timofeev) Jankauskas (Fig. 5.4.4A)	Radially Cracked Disc		68	162		52	214	+32%
Turuchanica ternata (Timofeev) Jankauskas (Fig. 5.4.4B)	Radially Cracked Disc		90	204		65	283	+39%
Pterospermopsimorpha sp. (Fig. 5.4.3E)	Radially Cracked Disc		352	1,010		321	1,106	+10%

#3e = #3 ellipse
#3d = #3 disc
#4A = (1 − #7e)(#1A)
#4B = (1 − #7e)(#1B)
#5 = (#3d/3.14)
#6e = #6 ellipse = $(2)(3.14)(\sqrt{[\#1A/2]^2/2 + [\#1B/2]^2/2})$
#6d = #6 disc = (3.14)(#2)
#7e = #7 ellipse = [(#6e/#3e)(100)] − 100
#7d = #7 disc = [(#6d/#3d)(100)] − 100

Table 5.4.3. *Summary of size changes resulting from compression and radial cracking of ellipsoidal and spheroidal microfossils preserved in shales.*

Compressed Shale of Microfossils	Specimens Measured	Percent Size Change Range	Average	Source of Data
Folded Ellipses:	7	−4% to +16%	+6%	Table 5.4.1
Radially Cracked Ellipses and Discs:	4	+10% to +39%	+23%	Table 5.4.2

Table 5.4.4. *Summary of correction factors used for morphometric comparison of compressed (two-dimensional) microfossils preserved in shales with modern microorganisms and with uncompressed (three-dimensional) microfossils.*

Type of Microfossil Preserved in Shale	Correction Factor* Diameter	Cell Length	Cell Width
Flattened Tubular Sheaths:	1.57	—	—
Flattened Cellular Trichomes:	—	—	1.57
Discs with Folds:	—	—	—
Radially Cracked Discs:	1.20	—	—
Ellipses with Folds:	—	—	—
Radially Cracked Ellipses:	—	1.20	1.20

*Uncompressed Dimensions (of modern microorganisms or of three-dimensionally preserved microfossils) = Compressed Dimensions (of two-dimensionally preserved shale microfossils) divided by the "Correction Factor."

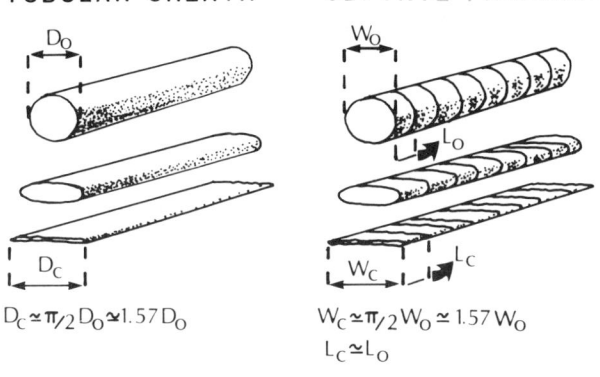

Proterozoic Prokaryotes

TUBULAR SHEATH

$D_C \simeq \pi/2 \, D_O \simeq 1.57 D_O$

SEPTATE FILAMENT

$W_C \simeq \pi/2 \, W_O \simeq 1.57 W_O$
$L_C \simeq L_O$

Figure 5.4.5. Correction factors for morphometric comparisons of uncompressed cylindrical-tubular microbial sheaths or cylindrical-septate microbial trichomes with compressed fossil analogues preserved in shales.

5.4.2B Compression of Cylindrical-Tubular and Cylindrical-Septate Microfossils

As studied in acid-resistant residues of shales/siltstones, the carbonaceous, originally cylindrical-tubular sheaths of bacterial and cyanobacterial filaments have typically been flattened due to preservational compression into two-dimensional ribbon-like strands (Figs. 5.4.3F, 5.4.4C). Such flattened sheaths are relatively common components of assemblages reported from shales/siltstones, accounting for about 40% (114) of the 277 occurrences of such fossils now known from the Proterozoic and Early Cambrian fossil records (Tables 22.3, 25.3). As shown in Figure 5.4.5, if such cylindrical-tubular sheaths have been affected by compression in a manner analogous to that occurring during flattening of a thin-walled garden hose, the compressed diameter of such microfossils ("Dc") – the diameter that would be measured in acid-resistant residues of shales/siltstones – would be related to the original, uncompressed, diameter ("Do") by the following equation:

$$Dc \simeq pi/2 Do \simeq 1.57 Do. \qquad (5.4.1)$$

Are the "garden hose analogy" and the resultant equation valid? Figure 5.4.6A shows the size distribution of the 114 occurrences of tubular sheaths as reported from macerations of shales/siltstones 1600 to 530 Ma; the broadest such sheaths are nearly 160 μm in diameter. For comparison, these same data, but corrected by use of the above equation for the effects of preservational compression, are plotted in Figure 5.4.6B. The corrected data (Fig. 5.4.6B), with a maximum diameter of about 100 μm, exhibit a size range and pattern of side distribution that closely mimic those of modern cyanobacterial (oscillatoriacean) sheaths (Fig. 5.4.27A). The size range and pattern of size distribution of corrected data for the 114 occurrences of fossil sheaths reported from shales/siltstones (Fig. 5.4.6B) coincides with those of uncorrected data for the 163 occurrences of sheaths reported from thin sections of stromatolitic cherts, and the two sets of data plotted together (277 occurrences) also closely mimics size data for modern oscillatoriacean sheaths (Fig. 5.4.7A). Finally, Figure 5.4.7B compares the size distribution of the 64 taxa of tubular microfossils currently known from Proterozoic and Lower Cambrian shales/siltstones and carbonates/cherts (Table 25.3) with that for sheaths of 199 species and varieties of modern oscillatoriacean cyanobacteria. These plots (cf. Fig. 5.4.27 in which these data for modern and fossil sheaths are plotted separately) also show a marked degree of detailed similarity (e.g., in their distribution of peaks, patterns of size distribution, and overall size ranges). These comparisons, of the size distribution of modern cyanobacterial (oscillatoriacean) sheaths (Fig. 5.4.27A) with (*i*) the corrected data for fossil sheaths occurring in shales/siltstones (Fig. 5.4.6B), (ii) in the combined data from shales/siltstones (corrected) and from carbonates/cherts (uncorrected) for all occurrences of fossil sheaths (Fig. 5.4.7A), and (*iii*) data for all recognized taxa of fossil sheaths (Figs. 5.4.7B, 5.4.27), substantiate, quite effectively, the validity of the correction factor (cf. Hofman 1979, p. 156).

Originally cylindrical-septate filaments occurring in shales/siltstones have similarly been affected by preservational compression. However, as is shown in Figure 5.4.4D, G, although the original diameters, and thus the cell widths, of such flattened septate fossils have evidently been altered by

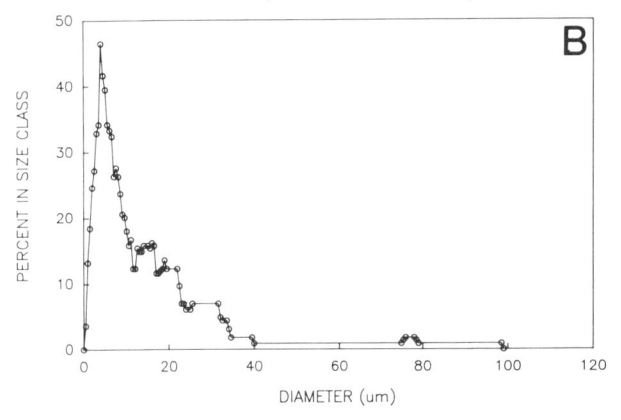

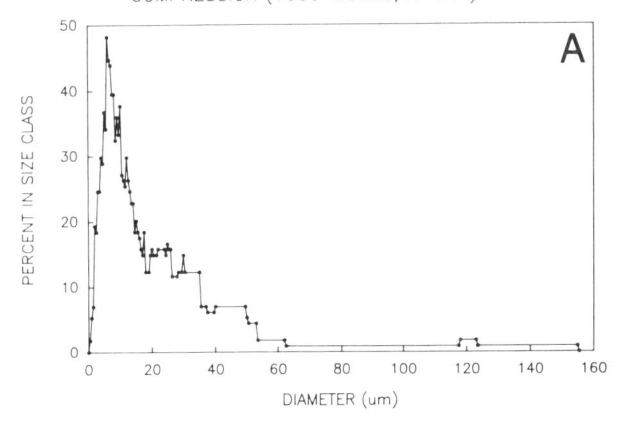

Figure 5.4.6. Distribution of diameters for 114 occurrences of tubular sheaths isolated by acid maceration from Proterozoic and Early Cambrian (1600–530 Ma) shales from geologic units listed in Table 22.3. (A) Sheath diameters *not* corrected for the effects of preservational compression. (B) Sheath diameters corrected for the effects of preservational compression.

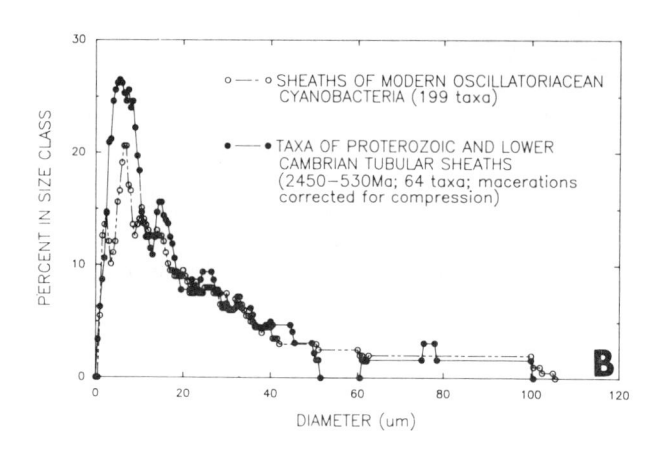

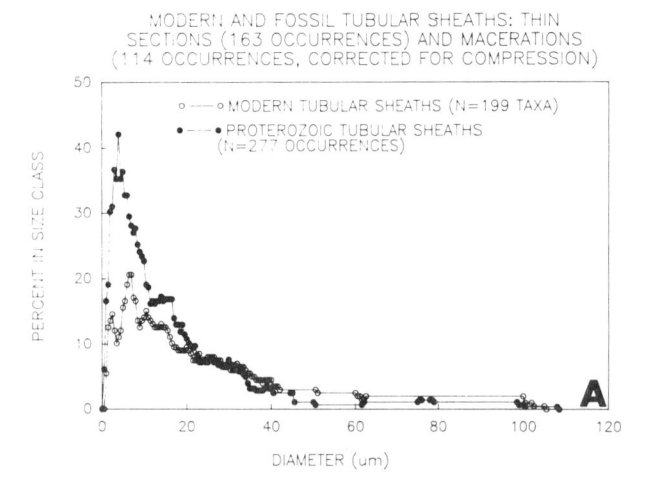

Figure 5.4.7. Comparison of the distribution of diameters of sheaths of 199 taxa of modern oscillatoriacean cyanobacteria with that of (A) 277 occurrences of fossil tubular sheaths and with that of (B) 64 taxa of fossil sheaths listed in Table 25.3, with diameters (both for occurrences and taxa) of fossils isolated from shales/siltstones by acid maceration corrected for preservational compression.

compression, the location and regular distribution of transverse septa, demarcating cell lengths, are little changed. Thus, as shown in Figure 5.4.5, for comparison of compressed septate filaments with uncompressed cellular microfossils or modern trichomes, the cell width of the flattened microfossils ("Wc") is related to the original, uncompressed width ("Wo") by the following equation:

$$Wc \simeq pi/2 Wo \simeq 1.57 Wo, \quad (5.4.2)$$

whereas the length of compressed cells ("Lc") and of original, uncompressed cells ("Lo") are approximately equivalent (viz., Lc ≃ Lo).

5.4.2C Correction Factors for Morphometric Comparison of Compressed Microfossils with Uncompressed Microfossils and Modern Microorganisms

Based on the foregoing analyses, Table 5.4.4 summarizes the set of correction factors here used for morphometric comparison of compressed (two-dimensional) organic-walled microfossils preserved in shales/siltstones with modern microorganisms and with uncompressed (three-dimensional), permineralized microorganisms preserved in carbonates/cherts. Size data for fossils preserved in shales/siltstones belonging to the various categories of microfossils discussed in the remainder of this Section, and the data listed in Table 25.3 (the classification of Proterozoic microfossils used both below and for the analyses of Proterozoic biotic diversity presented in Section 11.3), have been corrected by use of these factors for the effects of preservational compression.

5.4.3 Methods for Inferring the Probable Affinities of the Principal Categories of Proterozoic Microfossils

As is discussed elsewhere in this work (e.g., Section 11.3, Chapter 25), determination of the probable affinities of Proterozoic microfossils is fraught with difficulty. Taxonomy-related deficiencies in the relevant literature are numerous and varied (Section 11.3.3C), and even such a seemingly straightforward problem as determining whether a particular microfossil is of prokaryotic or of eukaryotic affinity can, in practive, be difficult (or even impossible) to resolve. For example, in comparison with extant microorganisms, small Proterozoic fossil unicells (e.g., spheroidal microfossils 5 μm in diameter) are morphologically similar to relatively large coccoidal prokaryotic bacteria, "normal-sized" spheroidal prokaryotic cyanobacteria, and small eukaryotic green or red unicellular algae, but because many diagnostic characters are almost never preserved (e.g., nuclei, other intracellular organelles, chlorophylls and accessory pigments) – and even if it is assumed that the microfossils are not members of an extinct lineage, only remotely related to members of the modern biota – in practice, commonly it is simply not possible to demonstrate unequivocally that such microfossils are related to one rather than another of these modern groups.

5.4.3A Morphometric Comparison of Fossil and Modern Microorganisms

In view of such difficulties as those discussed above, it would be unrealistic to assume that the biological affinities of all, or perhaps even most Proterozoic microfossils can be determined unequivocally. Such interpretations are constrained by the fossil data actually available and, thus, by the potential preservability of diagnostically useful characters. The approach here used, therefore, is to concentrate on those characters that are demonstrably preserved in Proterozoic microfossils, and to infer "probable affinity" by comparing this necessarily limited and morphologically based suite of characters with the same characters as exhibited by extant microorganisms. Seven such characters, measured for the Proterozoic taxa here considered and collectively relevant to the great majority of all prokaryotes now known, are illustrated in Figure 5.4.8. Despite recent progress toward understanding patterns of microbial phylogeny as revealed by (unpreservable) biochemical characteristics (e.g., ribosomal RNA), this morphologically based approach is valid: among modern microbes, morphology plays a prominent role for differentiation of many bacterial (Buchanan and Gibbons 1974) and virtually all cyanobacterial species (Desikachary 1959).

Proterozoic Prokaryotes

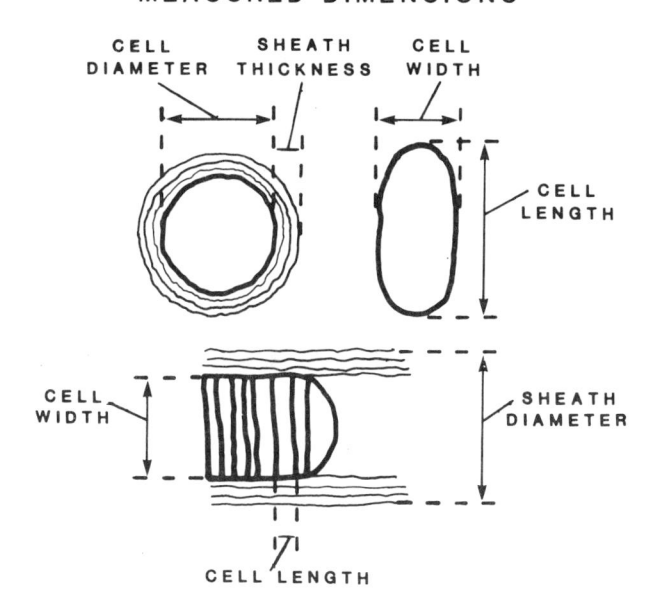

Figure 5.4.8. Dimensions measured for morphometric comparisons of modern microorganisms and comparable Proterozoic microfossils.

5.4.3B Inferences of Affinity Based on the Size Ranges of Modern Microorganisms

A principal technique here used for inferring the "probable affinity" of a Proterozoic taxon is one of comparing fossil and modern size ranges (e.g., as illustrated in Fig. 5.4.8, of cell diameters, cells lengths and widths, thicknesses of encompassing sheaths). Embedded in this strategy is the assumption that most Proterozoic prokaryotes are members of still extant lineages (evidently a valid assumption, as documented below) and that their patterns of size distribution can therefore be expected to be comparable to those of modern analogues. Thus, data for modern microorganisms have been used to define the size ranges characteristic of extant taxonomic groups, size distributions which provide a basis for inferring the affinity of morphologically comparable fossil taxa. In simple cases, those in which the size ranges of extant taxonomic groups do not significantly overlap (e.g., Fig. 5.4.9A), definition of the characteristic patterns of size distribution – and, thus, inference of the affinity of morphologically comparable fossils – is relatively straightforward.

Unfortunately, however, the occurrence of non-overlapping size ranges (for virtually all of the characters measured) is the exception, rather than the rule. A typical situation is illustrated in Figure 5.4.9B which shows three hypothetical modern taxa with significantly overlapping size ranges. If the size range of the diameters of the living "Type A" organism (Fig. 5.4.9B) were typical of micro-algal eukaryotes, that of "Type B" were characteristic of bacteria, and the size range of the "Type C" organism were representative of cyanobacteria, and if this distribution pattern were mimicked by a Proterozoic population of morphologically comparable, simple spheroidal microfossils, how would the "probable affinites" of the fossils be assessed? The approach here used is one of assuming analogy when analogy seems justified, but of asserting ignorance when

ignorance is unavoidable; that is, in this particular example, those ("Type B") fossils smaller than about 2.0 μm in diameter would be inferred to be of "probable bacterial affinities"; those 2.0 to 2.5 μm in diameter, occupying the area of overlap between the bacterial and cyanobacterial size ranges, would be regarded as "probable prokaryotes" (viz., non-nucleated bacterial or cyanobacterial microbes); fossils 2.5 to 8.0 μm in diameter (in the range typical of "Type C") would be regarded as "probable cyanobacteria"; those 8.0 to 9.5 μm in diameter, occurring in the overlap region between prokaryotes and eukaryotes, would be regarded as "prokaryotes or eukaryotes"; and those greater than 9.5 μm in diameter (in the range characteristic for "Type A") would be regarded as "probable eukaryotes."

A non-hypothetical example of the effect of such overlapping size ranges may help to clarify this approach. As is shown in Figure 5.4.10 (and discussed below in Section 5.4.4A), although some degree of overlap occurs at smaller and at larger cell sizes, the patterns of size distribution of modern coccoidal bacteria and cyanobacteria overlap markedly for those taxa 1.5 to 2.5 μm in diameter. In the approach here used, fossil taxa

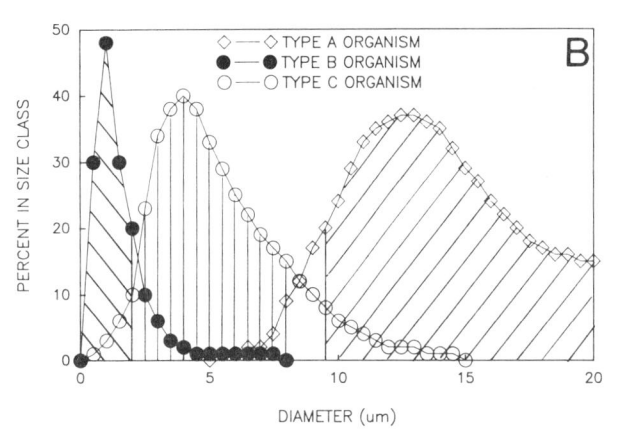

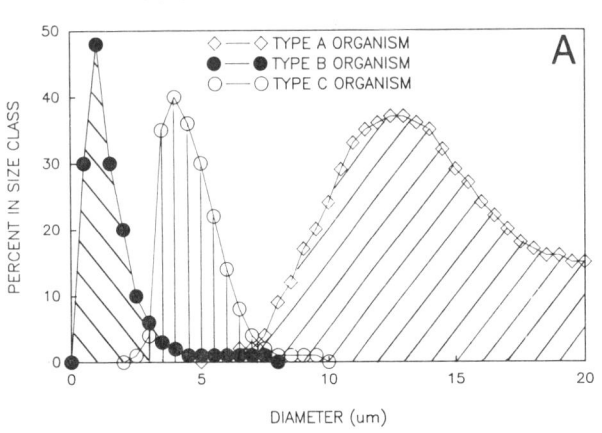

Figure 5.4.9. Hypothetical size distributions for three coccoid microorganisms. (A) Three components that lack a significant overlap of size ranges. (B) Three components with significantly overlapping size ranges.

0 to 1.5 μm in diameter are regarded as "probable bacteria"; those 1.5 to 2.5 μm in diameter as "prokaryotes"; and those ⩾ 2.5 μm as "probable cyanobacteria."

5.4.3C Interpretive Key to the Inferred Probable Affinities of the Principal Categories of Proterozoic Microfossils

Table 5.4.5 presents an interpretive key to the affinities here inferred for the principal categories of prokaryotic and eukaryotic Proterozoic microfossils. As noted above (Section 5.4.1), the prokaryotic categories apply to all but a few percent of reported occurrences of putative Proterozoic prokaryotes. With regard to probable eukaryotes, the key applies to all recognized taxa (Table 25.3) with the exception of eight, non-acritarch (mostly filamentous), eukaryote-like species (code "EU" in Tables 25.1, 25.3, and Fig. 25.4; e.g., Chapter 24: Pl. 11, C; Pl. 12, A; Pl. 13, D; Pl. 26, E; Pl. 28, C; Pl. 40, A; Pl. 45, A).

5.4.4 Coccoidal Microorganisms and Proterozoic Microfossils

Data discussed in this Section are relevant to the following four broad categories of Proterozoic, solitary and colonial, coccoidal prokaryotes and eukaryotes:

(1) solitary, coccoidal unicells ⩽ 60 μm in diameter (code "S" in Tables 25.1, 25.3, and Fig. 25.4, 24 fossil species; e.g., Chapter 24: Pl. 2, E–G; Pl. 3, D–N; Pl. 10, N; Pl. 33, F, G, K, L);

(2) sphaeromorph acritarchs > 60 μm in maximum diameter (code "MS," 10 fossil species, e.g., Chapter 24: Pl. 14, A, B; Pl. 19, A; and code "MG," 15 fossil species, e.g., Chapter 24:

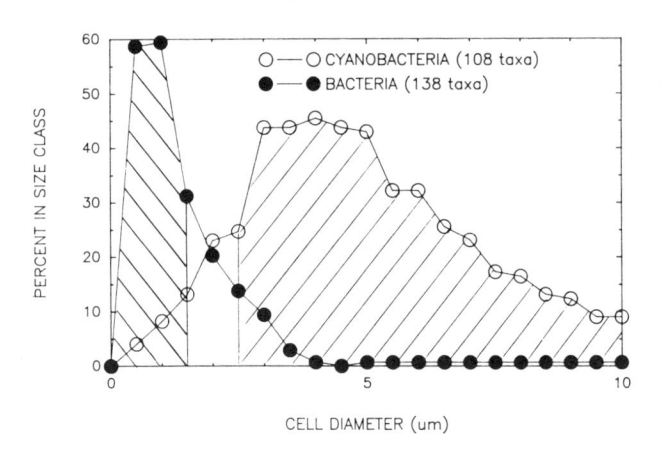

Figure 5.4.10. Overlapping size distributions of modern coccoid free-living bacteria (138 taxa) and modern spheroidal cyanobacteria ⩽ 10 μm in diameter (108 taxa) used to infer the probable affinities of fossil analogues: < 1.5 μm = "bacteria"; 1.5–2.5 μm = "prokaryotes"; 2.5–10 μm = "cyanobacteria" (see Tables 5.4.5 and 5.4.6 for details). [Bacterial data from Buchanan and Gibbons (1974), Laskin and Lechevalier (1977), and Starr et al. (1981). Cyanobacterial data from Desikachary (1959).]

Pl. 24, D; Pl. 42, A, D; for discussion of the affinities, geologic distribution and evolutionary trends inferred for these fossils, see Section 5.5);

(3) unordered colonies of coccoidal cells (code "CIR," 17 fossil species, e.g., Chapter 24: Pl. 10, J; Pl. 32, H, I; Pl. 45, C; code "CLI," three fossil species; code "CSF," nine fossil species,

Table 5.4.5. *Interpretive key to the inferred probable affinities of the principal categories* of Proterozoic microfossils.*

1. Microfossil preserved as kerogenous, two-dimensional compression in clastic sediment, commonly shale ———————————————————————— 2
1. Microfossil three-dimensionally preserved by permineralization, commonly in chert ————— 3
2. Apply "Correction Factors" (Table 5.4.4) to dimensions of microfossils preserved as compressions, measured in acid resistant residues, for morphometric comparison with three-dimensionally permineralized microfossils, measured in petrographic thin sections, and with modern microorganisms:
   Measured width of compressed, originally cylindrical-tubular, non-septate filament × 0.64 (i.e., × 2/pi);
   Measured cell width of compressed, originally cylindrical-septate filament × 0.64 (i.e., × 2/pi);
   Measured diameter of compressed, originally spheroidal, peripherally cracked disc × 0.83 (i.e., × 1/1.20);
   Measured width and length of compressed, originally ellipsoidal, peripherally cracked ellipse × 0.83 (i.e., × 1/1.20) ———————————————————— 3
3. Filamentous microfossil ————————————————————————— 4
3. Ellipsoidal or spheroidal microfossil, simple or ornamented ————————————— 5
3. Other (e.g., prismatic, polygonal, or fusiform microfossil) ————————————— 17
4. Cylindrical-tubular non-septate filament ————————————————————— 6
4. Cylindrical-septate uniseriate cellular filament, with or without encompassing tubular sheath ——————————————————————————————— 8
5. Unornamented ellipsoid or spheroid, solitary or colonial ——————————————— 9
5. Solitary ornamented spheroid or ellipsoid ———————————————————— 16
6. Tubular lamellated filament ———————————————————— Tubular Sheath of Filamentous Cyanobacterium
6. Tubular unlamellated filament ————————————————————————— 7

Table 5.4.5. (Continued)

7. Tubular filament diameter <2.0 μm	Bacterium (?Tubular Sheath of Filamentous Bacterium or ?Unsheathed, Elongate, "Thread Cell" Bacterium)
7. Tubular filament diameter 2.0 to 3.5 μm	Tubular Sheath of Filamentous Prokaryote (?Bacterium or ?Cyanobacterium)
7. Tubular filament diameter >3.5 μm	Tubular Sheath of Filamentous Cyanobacterium
8. Cellular filament width <1.5 μm	Septate Filamentous Bacterium
8. Cellular filament width 1.5 to 3.5 μm	Septate Filamentous Prokaryote (?Bacterium or ?Cyanobacterium)
8. Cellular filament width >3.5 μm	Septate Filamentous Cyanobacterium
9. Ellipsoidal cell or vesicle, solitary or colonial	10
9. Spheroidal cell or vesicle, solitary or colonial	11
10. Vesicle (cell) dimensions <3.5 × 5.0 μm	Ellipsoidal Bacterium
10. Vesicle (cell) dimensions ⩾3.5 × 5.0 μm	Ellipsoidal Cyanobacterium
11. Vesicle (cell) diameter <1.5 μm	Coccoid Bacterium
11. Vesicle (cell) diameter 1.5 to 2.5 μm	Coccoid Prokaryote (?Bacterium or ?Cyanobacterium)
11. Vesicle (cell) diameter >2.5 μm, <10.0 μm	Coccoid Cyanobacterium
11. Vesicle (cell) diameter ⩾10.0 μm	12
12. Vesicle (cell) and/or colony encompassed by prominent, commonly lamellated, sheath	13
12. Vesicle (cell) not sheath enclosed	14
13. Vesicle (cell) diameter ⩽58.0 μm	Chroococcacean Cyanobacterium cf. *Chroococcus* or *Gloeocapsa*
13. Vesicle (cell) diameter >58.0 μm	Eukaryotic Sphaeromorph Acritarch (?Chlorophycean)
14. Vesicle (cell) diameter ⩾10.0 μm, ⩽25.0 μm	15
14. Vesicle (cell) diameter >25.0 μm, ⩽45.0 μm	Eukaryotic Sphaeromorph Acritarch (?Chlorophycean or ?Rhodophycean)
14. Vesicle (cell) diameter >45.0 μm	Eukaryotic Sphaeromorph Acritarch (?Chlorophycean)
15. Vesicle (cell) solitary	Prokaryotic or Eukaryotic Sphaeromorph Acritarch (?Bacterium cf. *Thiovulum* or ?Cyanobacterial Prokaryote, or ?Chlorophycean or ?Rhodophycean Eukaryote)
15. Vesicle (cell) colonial, with or without evident encompassing sheath	Coccoid Cyanobacterium
16. Vesicle ornamented with branched or unbranched, solid or hollow spiny processes	Acanthomorph Acritarch
16. Vesicle ornamented at one pole only	Oömorph Acritarch[#]
16. Vesicle ornamented with crests dividing surface into approximately equidimensional polygonal fields	Herkomorph Acritarch
16. Vesicle ornamented with equatorial flange which may be supported by processes or radial folds	Pteromorph Acritarch[#]
17. Vesicle elongate to fusiform, with or without polar spines	Netromorph Acritarch
17. Vesicle prismatic, commonly tetrahedral or octahedral	Prismatomorph Acritarch
17. Vesicle polygonal	18
18. Vesicle with equatorial flange which may be supported by processes or radial folds	Pteromorph Acritarch[#]
18. Vesicle with crests dividing surface into approximately equidimensional polygonal fields	Herkomorph Acritarch
18. Vesicle without equatorial flange or surficial crests	Polygonomorph Acritarch[#]

*Including coccoidal, ellipsoidal, and filamentous bacterial and cyanobacterial microfossils and major groups of Proterozoic acritarchs, but not including numerous atypical, morphologically distinctive fossil taxa (e.g., *Eosphaera*, *Archaeorestis*, *Eoastrion*, *Kakabekia*, *Polybessurus*, *Palaeosiphonella*)

[#] Reported pre-Lower Cambrian occurrences of these groups are here regarded as questionable (see Fig. 5.5.13) e.g., Chapter 24: Pl. 44, K, M; Pl. 48, A, B; Pl. 54, J; code "CRCF," seven fossil species, e.g., Chapter 24: Pl. 5, A, C; Pl. 48, F; and code "CFL," eight fossil species, e.g., Chapter 24: Pl. 24, A); and

(4) ordered colonies of coccoidal cells (code "G," six fossil species, e.g., Chapter 24: Pl. 4, F; Pl. 10, K, L; Pl, 45, B, D–I; code "CHR," 11 fossil species, e.g., Chapter 24: Pl. 10, L; Pl. 32, J; Pl. 33, A; code "MYX," one fossil species, e.g., Chapter 24: Pl. 39, C; Pl. 41, B, D; code "PLEU," four fossil species, e.g., Chapter 24: Pl. 50, B, C; code "TET," six fossil species, e.g., Chapter 24: Pl. 50, A; Pl. 54, H; code "CPL," two fossil species, e.g., Chapter 24: Pl. 41, A; and code "CCU," four fossil species, e.g., Chapter 24: Pl. 4, A, B; Pl. 10, E).

5.4.4A Morphometric Characteristics of Modern Coccoidal Bacteria, Cyanobacteria, and Eukaryotic Micro-algae Used to Infer the Affinities of Comparable Proterozoic Microfossils

Figure 5.4.11 summarizes the size distributions of modern, coccoidal, free-living bacteria (138 species and varieties), chroococcalean cyanobacteria (121 species and varieties), and chlorophycean (226 species) and rhodophycean (eight species), eukaryotic micro-algae. As shown in Figure 5.4.12, the size distributions of these groups overlap substantially, particularly for cells less than about $10\,\mu m$ in diameter. However, as is summarized graphically in Figure 5.4.13 (and tabulated in Table 5.4.6), about 75% of modern coccoid bacteria and only about 11% of extant coccoid cyanobacteria are less than $1.5\,\mu m$ in diameter, whereas only about 12% of such bacteria and about 85% of such cyanobacteria are greater than $2.5\,\mu m$ in diameter. Thus, coccoid Proterozoic microfossils less than $1.5\,\mu m$ in diameter are here regarded (Tables 5.4.5, 5.4.6) as "probable bacteria"; those 1.5 to $2.5\,\mu m$ in diameter as "prokaryotes"; and those greater than $2.5\,\mu m$ as either "probable cyanobacteria" or as various types of "probable eukaryotes."

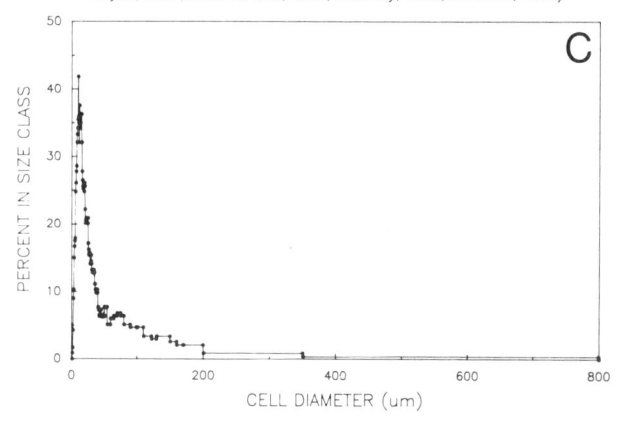

About 89% of modern coccoid cyanobacteria and about 29% of eukaryotic micro-algae are less than $10\,\mu m$ in diameter (Table 5.4.6, Fig. 5.4.13); most extant, coccoid, cyanobacteria larger than $10\,\mu m$ are morphologically distinctive chroococcaceans (viz., *Chroococcus* spp. and *Gloeocapsa* spp.; Fig. 5.4.1B) that are encompassed by thick, commonly multilamellated, prominent sheaths (Fig. 5.4.12). The largest modern coccoid bacterium (*Thiovulum majus* Heinze) is about $25\,\mu m$ in diameter (Buchanan and Gibbons 1974, p. 463), and the largest extant chroococcaceans (*Chroococcus giganteus* West, *C. macrococcus* [Kutzing] Rabenhorst, and *C. turgidus* var. *maximus* Nygaard) have cells $50\,\mu m$ to nearly $60\,\mu m$ in diameter (Desikachary 1959, pp. 99–102). Thus, Proterozoic coccoidal microfossils 2.5 to $10\,\mu m$ in diameter are here regarded (Tables 5.4.5, 5.4.6) as "probable cyanobacteria"; solitary cells 10 to $25\,\mu m$ in diameter that are not encompassed by prominent sheaths are regarded as "sphaeromorph acritarchs" (viz., of either prokaryotic or eukaryotic affinity); sheath-enclosed, *Chroococcus*-like microfossils $10\,\mu m$ to about $60\,\mu m$ in diameter are regarded as "probable chroococcacean cyanobacteria"; and non-colonial, non-sheath-enclosed vesicles larger than $25\,\mu m$ in diameter are regarded as "probable eukaryotic sphaeromorph acritarchs."

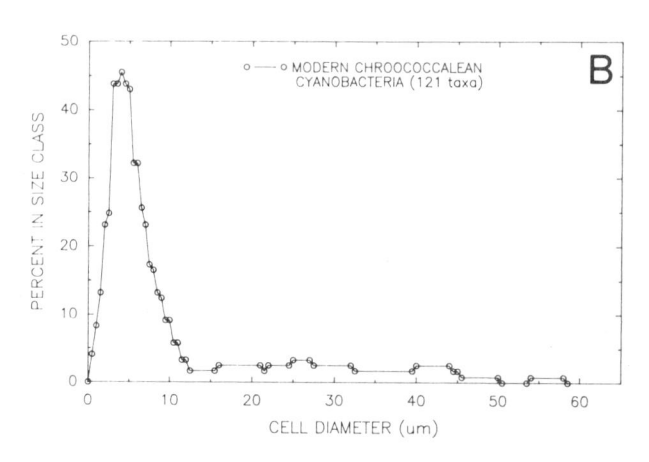

5.4.4B Proterozoic (and Archean) Chroococcus-like Microfossils with Prominent Lamellated Sheaths

Gloeocapsa-like microfossils (Chapter 24: Pl. 4, F; Pl. 10, K, L; Pl. 45, B, D–I) and *Chroococcus*-like colonies (Chapter 24: Pl. 10, L; Pl. 32, J; Pl. 33, A), composed of spheroidal to ellipsoidal cells that are typically encompassed by thick, com-

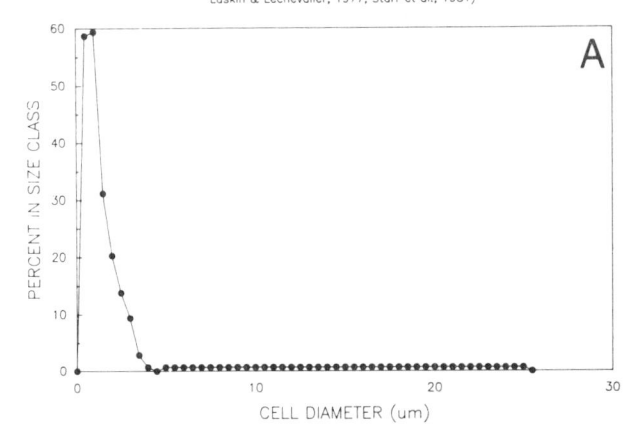

Figure 5.4.11. Size distributions of modern coccoid microorganisms. (A) Free-living bacteria (138 taxa). (B) Chroococcalean cyanobacteria (121 taxa; data from Desikachary, 1959). (C) Chlorophycean (226 taxa) and rhodophycean (8 taxa) eukaryotic algae.

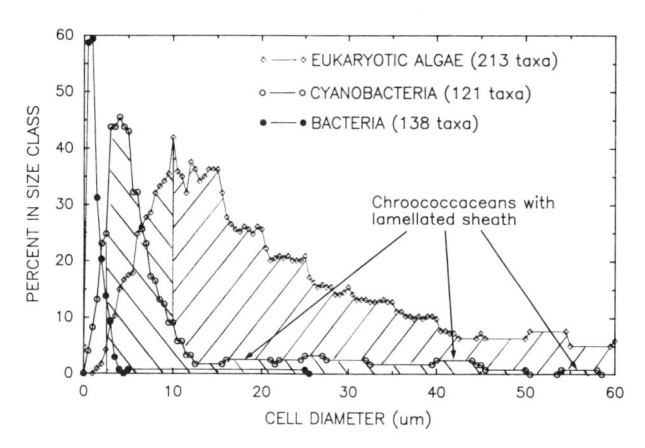

Figure 5.4.12. Overlapping size ranges of modern coccoid bacteria, cyanobacteria, and eukaryotic algae ≤60 μm in diameter illustrating the definitions here used (Tables 5.4.5, 5.4.6) for inferring the probable affinities of fossil analogues. [Sources of data cf. Fig. 5.4.11.]

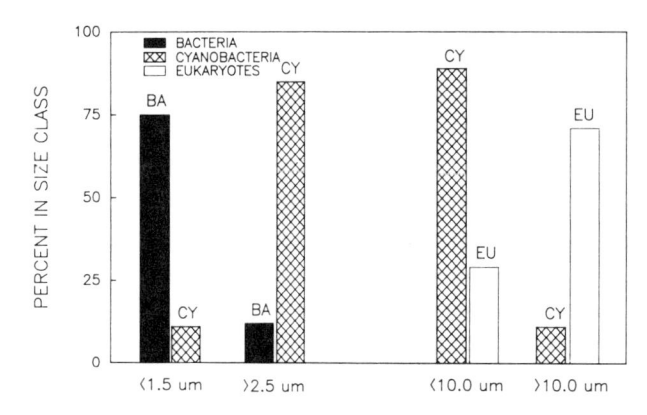

Figure 5.4.13. Histograms summarizing the size distributions of modern coccoid microorganisms used to infer the probable affinities of fossil analogues: coccoids <1.5 μm = "bacteria"; 1.5–2.5 μm = "prokaryotes"; 2.5–10 μm = "cyanobacteria"; >10 μm = "sphaeromorph acritarchs" (see Tables 5.4.5 and 5.4.6 for details).

Table 5.4.6. *Morphological characteristics of modern coccoid prokaryotes and eukaryotic micro-algae used to infer the probable affinities of comparable fossil taxa.*

	Modern Coccoid Microorganisms						
	Free-living Spheroidal Bacteria		Chroococcacean and Entophysalidacean Cyanobacteria		Chlorophycean and Rhodophycean Eukaryotes (Vegetative Cells)		Inferred Probable Affinities
Cell Diameter	Percent	Number	Percent	Number	Percent	Number	of Fossil Analogues
>0.1 μm	100%	138	—	—	—	—	<1.5 μm = "Bacterium"
>0.5 μm	80%	110	100%	121	—	—	——————————
>1.5 μm	25%	35	89%	108	100%	234	1.5–2.5 μm = "Prokaryote"
>2.5 μm	12%	17	85%	103	98%	230	——————————
>5.0 μm	1%	1	47%	57	94%	220	2.5–10.0 μm = "Cyanobacterium"
>10.0 μm	1%	1	11%	13	71%	165	——————————
>12.0 μm	1%	1	6%	7	63%	147	10.0–25.0 μm = "Sphaeromorph
>25.0 μm	0%	0	5%	6	29%	68	Acritarch"[2]
>58.0 μm	—	—	0%	0	9%	21	——————————
>200.0 μm	—	—	—	—	1%	2	>25.0 μm = "Eukaryotic
>800.0 μm	—	—	—	—	0%	0	Sphaeromorph Acritarch"[2]
Number of Species and Varieties[1]:	138		121		234		
Median Diameter:	0.9 μm		4.0 μm		12.5 μm		
Average Diameter:	1.4 μm		5.8 μm		21.0 μm		
Diameter, Range:	0.2–25.0 μm		0.5–58.0 μm		1.5–800.0 μm		
Modal Size Class:	0.5–2.5 μm		2.0–8.0 μm		5.0–20.0 μm		

[1]Bacterial data from: Buchanan and Gibbons 1974; Laskin and Lechevalier 1977; Starr et al. 1981. Cyanobacterial data from: Desikachary 1959. Micro-algal data from: West and Fritsch 1927; Prescott 1954; Taylor 1957; Smith and Bold 1968; Bourrelly 1970; La Rivers 1978.
[2]If colonial and/or enclosed in lamellated sheath = "Chroococcacean Cyanobacterium"

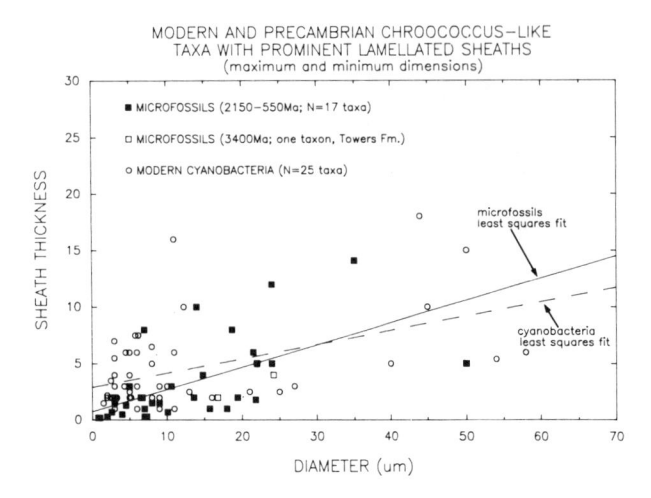

Figure 5.4.14. Cell diameter versus sheath thickness (maximum and minimum dimensions) for *Chroococcus*-like taxa with prominent lamellated sheaths: one unnamed taxon from the Early Archean Towers Formation, Warrawoona Group, of Western Australia (Fig. 1.5.5); 17 Proterozoic (2150–550 Ma) taxa listed in Table 25.3; and 25 modern chroococcacean taxa (Desikachary 1959).

monly multilamellated sheaths (cf. Fig. 5.4.1B), are particularly distinctive components of the Proterozoic biota. A total of 17 such taxa, having the size distribution summarized in Figure 5.4.14, are now known from the Proterozoic fossil record (Tables 25.1, 25.3, Fig. 25.4). Interestingly, the ranges of cell size and sheath thickness exhibited by these fossil taxa are quite similar to those of living morphological analogues and, like modern *Gloeocapsa* and *Chroococcus*, the cell diameters of the fossil taxa are rather strongly and positively correlated with the thickness of their encompassing sheaths (Fig. 5.4.14). As is shown in Figure 5.4.14, the range of cell size and sheath thickness exhibited by one *Chroococcus*-like morphotype (Fig. 1.5.5) reported from cherts of the Early Archean (~3.4 Ga) Towers Formation of Western Australia (Schopf and Packer 1987) is comparable both to those of the Proterozoic taxa and to those of modern species, supporting interpretation of this fossil colony as a probable chroococcacean and, thus, presumably an oxygen-producing photoautotroph (for related discussion, see Section 1.5).

5.4.4C Solitary and Colonial Coccoidal Proterozoic Microfossils ≤60 μm in Diameter

Figure 5.4.15 summarizes the size distributions for all occurrences of non-*Chroococcus*-like solitary and colonial coccoidal microfossils ≤60 μm in diameter currently known from Early, Middle, and Late Proterozoic shales/siltstones and stromatolitic carbonates/cherts (Chapter 22). As is clearly illustrated, the modal size class for such fossils reported from stromatolitic cherty units consistently falls toward the smaller cell size end of the spectrum, and the cell size ranges in these units tend to be decidedly narrower, than those for fossils reported from shales/siltstones. The majority of such fossils reported from chert thin sections are colonial, small-celled, benthic coccoids (cf. Fig. 5.4.1A), comparable in size range to many modern solitary and colonial chroococcacean cyanobacteria (e.g., Fig. 5.4.16). The apparent rarity of similar-sized

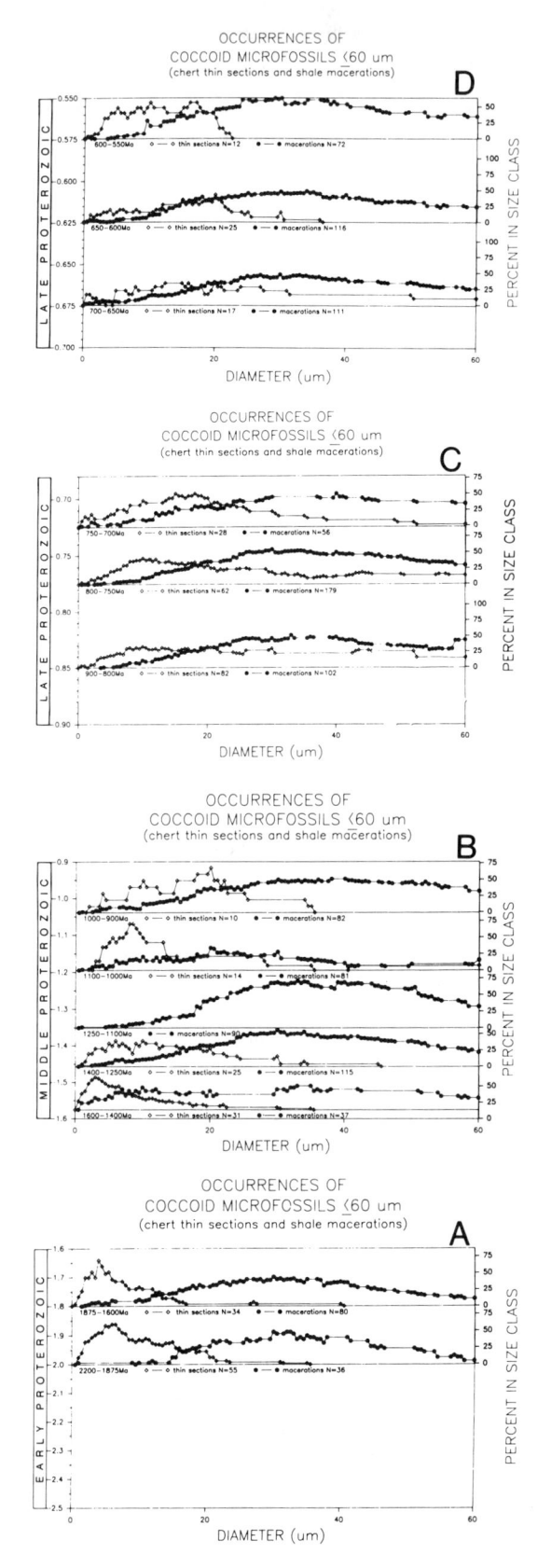

Figure 5.4.15. Size distributions of known Proterozoic occurrences of coccoid microfossils ≤60 μm in diameter reported from thin sections of cherts and macerations of shales from geologic units listed in Table 22.3. (A) Early Proterozoic occurrences (2200–1600 Ma). (B) Middle Proterozoic occurrences (1600–900 Ma). (C) Late Proterozoic occurrences (900–700 Ma). (D) Late Proterozoic occurrences (700–550 Ma).

Proterozoic Prokaryotes

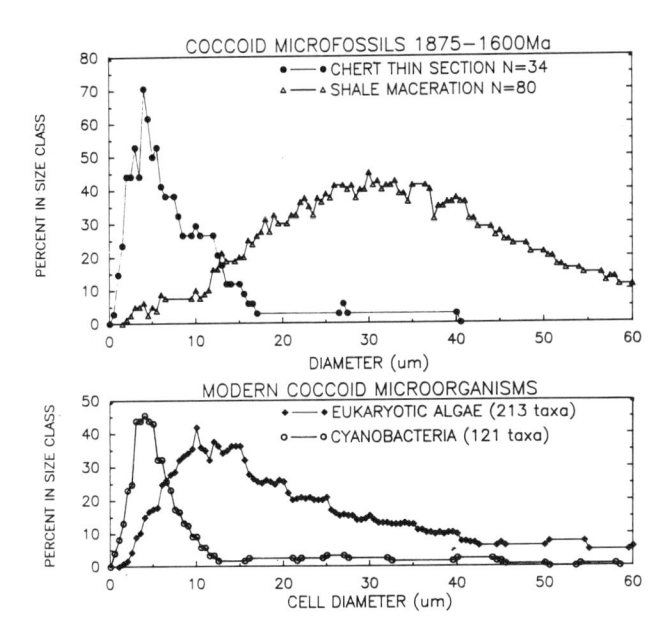

Figure 5.4.16. Above: size distributions of coccoid microfossils ≤ 60 μm in diameter reported from thin sections of cherts (34 occurrences) and macerations of shales (80 occurrences) from Early Proterozoic (1875–1600 Ma) geologic units listed in Table 22.3. Below: size distributions of 121 taxa of modern coccoid cyanobacteria (Desikachary 1959) and 213 taxa of modern eukaryotic algae ≤ 60 μm in diameter (sources of data listed in Fig. 5.4.11C).

coccoids in shales/siltstones presumably reflects, in part, differential preservation, with small-celled taxa, comparable in diameter to fine (8 to 16 μm) and very fine (4 to 8 μm) silt-size clastic particles, having been preferentially destroyed. (Fossils of this size may also be lost during palynological preparation, and it thus seems clear that absence of evidence of such small-celled coccoids in Proterozoic shales/siltstones cannot be taken as "evidence of absence" – cyanobacteria such as *Synechococcus* spp., ≤ 1 μm in diameter and a major component of the modern marine ecosystem, may have thrived in the Proterozoic seas but be undetectable in Proterozoic clastic sediments).

Large-celled coccoids ≤ 60 μm in diameter are abundant in shales/siltstones and are relatively rare in carbonates/cherts (Fig. 5.4.15). This disjunct pattern of distribution (attributable neither to differential preservation nor to techniques of sample preparation) suggests that the biological populations sampled by the two lithotypes significantly differ. Moreover, if all species from cherts and shales are considered together, there is a distinct lack of similarity between the size ranges of modern coccoid cyanobacterial taxa and fossil species either of Proterozoic solitary coccoids (Fig. 5.4.17) or of most Proterozoic colonial coccoids (Fig. 5.4.18). Indeed, as exemplified in Figure 5.4.16, the size distribution of assemblages of such fossils reported from shale macerations, which commonly include colonial taxa but which are dominated by solitary planktonic species, seems decidedly more similar to that of modern eukaryotic micro-algae than to that of cyanobacterial prokaryotes.

Read literally, these data could be interpreted as indicating that Proterozoic assemblages from "nearshore," stromatolitic carbonates/cherts were dominated by benthic, cyanobacterial prokaryotes, whereas those from "offshore" shales/siltstones, throughout virtually all of the sampled Proterozoic (Fig. 5.4.15), included a major component of eukaryotic plankton. Although other data suggest that this interpretation may well be correct (Section 5.5), extant coccoidal cyanobacteria 50 μm to nearly 60 μm in diameter are known (Section 5.5.4A) and it is at least conceivable, therefore, that the spheroidal vesicles of this size occurring in shale microfloras could be prokaryotic. Clearly, the affinities of these relatively large, predominantly planktonic, coccoidal Proterozoic taxa are difficult to determine. Moreover, because of the evolutionary significance of the origin of the eukaryotic cell, it seems important that particularly rigorous criteria be met for such an affinity to be accepted without reservation. For this reason, and in constrast with the

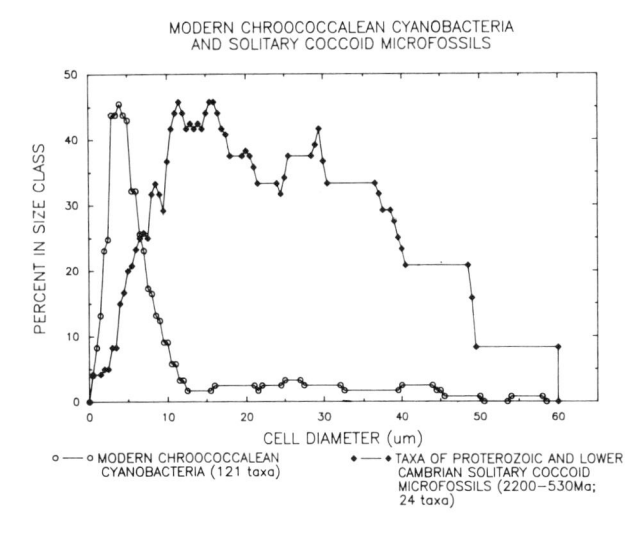

Figure 5.4.17. Comparison of the size distributions of 121 taxa of modern chroococcalean cyanobacteria (Desikachary 1959) and of 24 taxa of Proterozoic and Early Cambrian (2200–530 Ma) solitary coccoid microfossils ≤ 60 μm in diameter listed in Table 25-3.

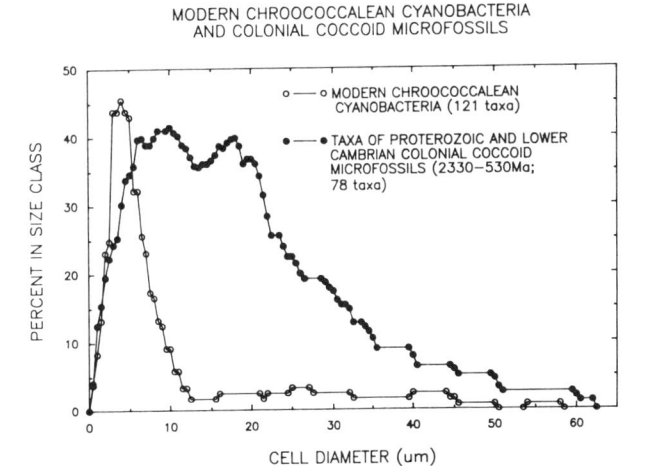

Figure 5.4.18. Comparison of the size distributions of 121 taxa of modern chroococcalean cyanobacteria (Desikachary 1959) and of 78 taxa of Proterozoic and Early Cambrian (2330–530 Ma) colonial coccoid microfossils ≤ 60 μm in diameter listed in Table 25-3.

"probable affinities" summarized in the interpretive key discussed above (Section 5.4.3C, Table 5.4.5), in the classification system here used (Table 25.3) and in discussions of patterns of biotic diversity evidenced by the known Proterozoic fossil record (Section 11.3), only those taxa with maximum cell diameters greater than 60 μm (and, thus, larger than all known extant prokaryotes) have been interpreted as assuredly eukaryotic.

5.4.4D Inferred Evolutionary Trends

It seems reasonable to expect that consideration of the Proterozoic size distributions summarized in Figure 5.4.15, and of the temporal distributions of the various subgroups of coccoidal taxa here classified as prokaryotes (Fig. 25.4), should provide a basis for interpretation of Proterozoic evolutionary trends. Significantly, however, over the entire Proterozoic, there appear to be no major differences or detectable trends in the patterns of size distribution exhibited by coccoid microfossils ≤ 60 μm in diameter, whether from shales/siltstones or from carbonates/cherts (Fig. 5.4.15). Similarly, subgroups of these fossils that are especially well-represented in the known Proterozoic record (e.g., solitary coccoid cells ≤ 60 μm in diameter, code "S" in Fig. 25.4; and irregular colonies of coccoid cells, code "CIR") exhibit no obvious time-related trends other than gradual increases in the detected number of taxa (Fig. 25.4), increases that are evidently a function of a gradual increase in the cumulative number of formations sampled (Section 11.3). And highly ordered colonies of such cells (e.g., colonial planar tetrads of coccoid cells, code "TET" in Fig. 25.4; and cuboidal colonies of coccoid cells, code "CCU") were demonstrably already extant relatively early in the Proterozoic (Fig. 25.4); available data provide no evidence of the evolution of colonial organization. Thus, coccoidal microfossils ≤ 60 μm in diameter, whether solitary or colonial, appear to have exhibited little evolutionary change over much of the Proterozoic; evidently, if fossil evidence of the phylogeny and evolutionary development of such organisms exists, it is to be found only in older, presumably in Archean-age, deposits.

5.4.5 Ellipsoidal Microorganisms and Proterozoic Microfossils

Data discussed in this Section relate to Proterozoic, solitary or irregularly colonial, ellipsoidal prokaryotes (code "EIC" in Tables 25.1, 25.3, and Fig. 25.4; e.g., Chapter 24: Pl. 9, G, H; Pl. 10, 0; cf. Fig. 5.4.1C; cf. Fig. 5.4.1C, D)

5.4.5A Morphometric Characteristics of Modern Ellipsoidal Bacteria and Cyanobacteria Used to Infer the Probable Affinities of Comparable Proterozoic Microfossils

Figure 5.4.19 summarizes the distribution of cell sizes exhibited by modern ellipsoidal bacteria (45 species and varieties) and modern ellipsoidal cyanobacteria (47 species and varieties). As is evident from these plots of the maximum and minimum cell sizes of each taxon, the majority (about 78% of species and varieties) of modern ellipsoidal bacteria are smaller than 3.5×5.0 μm, whereas most ellipsoidal cyanobacteria (about 70%) have cells equal to or larger than this size. Thus, as

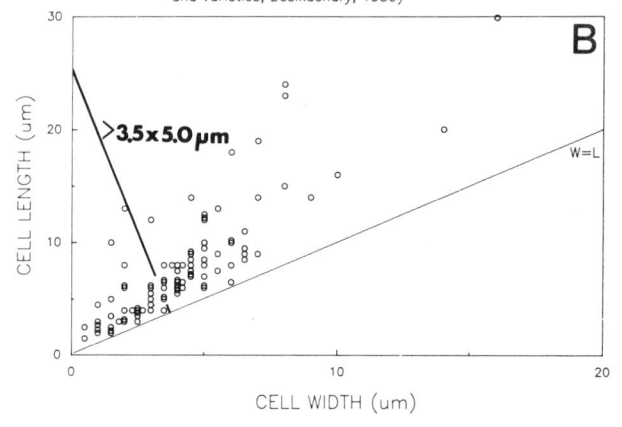

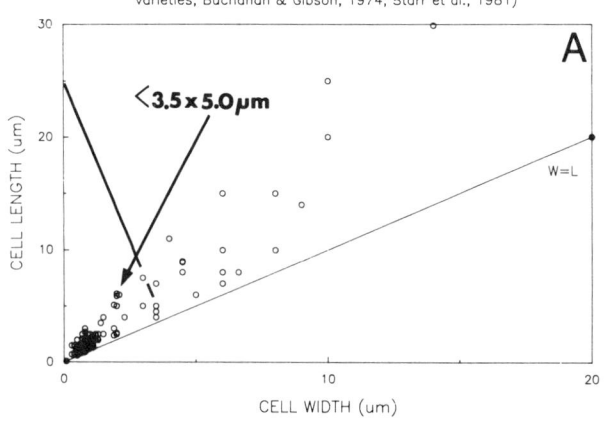

Figure 5.4.19. Cell length versus cell width for modern ellipsoidal microoganisms. (A) Maximum and minimum cell sizes for 45 taxa of modern ellipsoidal bacteria; the majority are smaller than 3.5×5.0 μm (Fig. 5.4.20). (B) Maximum and minimum cell sizes for 47 taxa of modern ellipsoidal cyanobacteria; the majority are larger than 3.5×5.0 μm (Fig. 5.4.20).

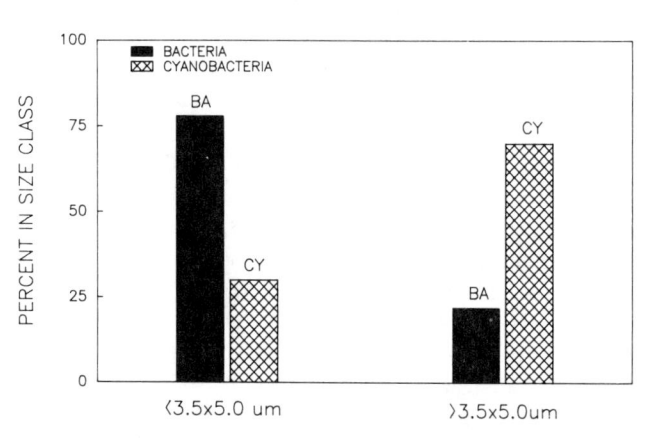

Figure 5.4.20. Histograms showing the size distributions of modern ellipsoidal microorganisms used to infer the probable affinities of fossil analogues: $< 3.5 \times 5.0$ μm = "bacteria"; $> 3.5 \times 5.0$ μm = "cyanobacteria" (see Tables 5.4.5 and 5.4.7 for details).

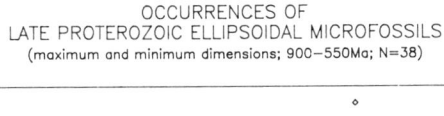

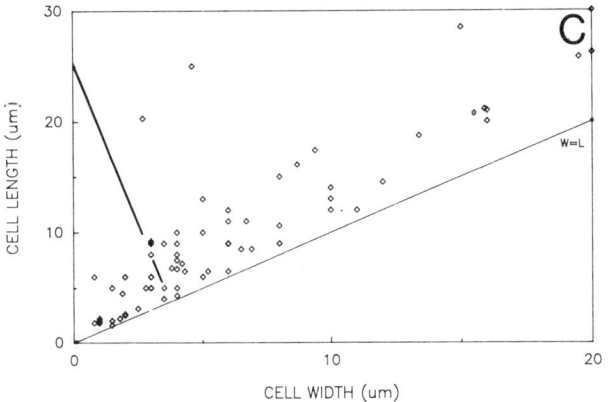

is summarized graphically in Figure 5.4.20 (and tabulated in Table 5.4.7), solitary or colonial ellipsoidal Proterozoic microfossils smaller than $3.5 \times 5.0\,\mu m$ are here regarded as "probable bacteria," and those $\geqslant 3.5 \times 5.0\,\mu m$ as "probable cyanobacteria" (Table 5.4.5).

5.4.5B Solitary and Colonial Ellipsoidal Proterozoic Microfossils

Figure 5.4.21 summarizes the size distributions (maximum and minimum dimensions) for the 69 occurrences of ellipsoidal microfossils now known from Early, Middle, and Late Proterozoic sediments (Table 22.3). Fourteen species of such ellipsoidal microfossils are here recognized (Table 25.3), the size ranges of which are plotted in Figure 5.4.22.

5.4.5C Inferred Evolutionary Trends

As is shown in Figures 5.4.21 and 5.4.22, both "probable bacteria" and "probable cyanobacteria" are represented in the Proterozoic record of ellipsoidal microfossils as now known, and the size of such microfossils increases (resulting in increased occurrences of "probable cyanobacteria") between the Early (Fig. 5.4.21A) and Middle Proterozoic (Fig. 5.4.21B). However, the apparent paucity of "probable cyanobacteria" in Early Proterozoic assemblages may simply be a function of the relatively small number of fossiliferous units yet sampled (a total of only 11 occurrences of ellipsoidal microfossils being currently known between 2350 and 1600 Ma); available data are too few to provide firm evidence of major evolutionary trends.

5.4.6 Non-septate Filamentous Microorganisms, Prokaryotic Sheaths, and Proterozoic Microfossils

Data discussed in this Section are relevant to the following two categories of Proterozoic microfossils:

(1) unlamellated, unbranched, tubular prokaryotic sheaths (code "TU" in Tables 25.1, 25.3, and Fig. 25.4, 55 fossil

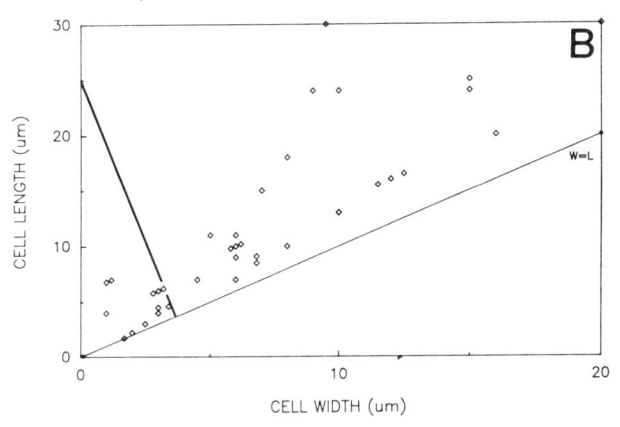

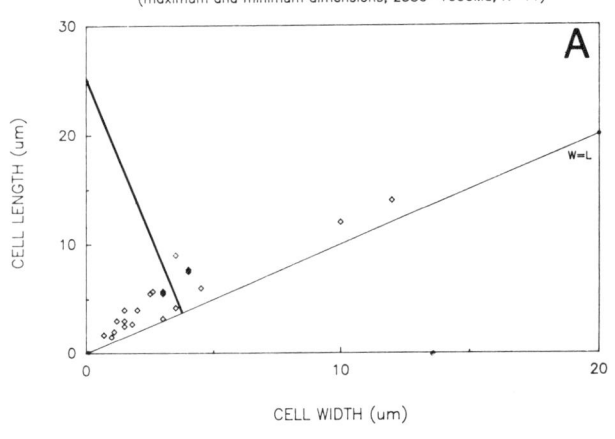

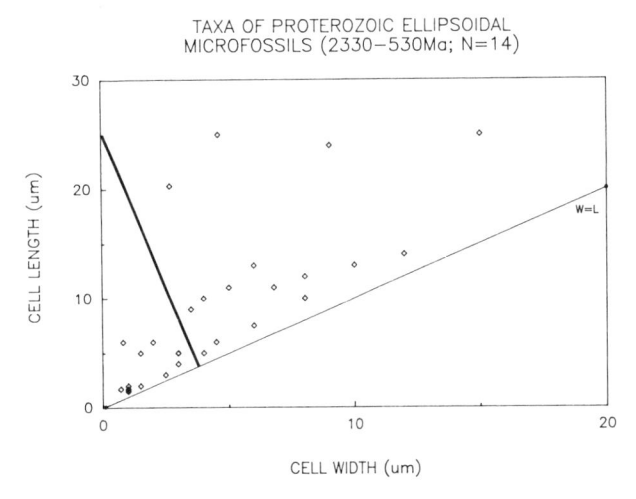

Figure 5.4.21. Cell lengths and widths for occurrences of ellipsoidal Proterozoic microfossils reported from geologic units listed in Table 22.3; diagonal line perpendicular to W=L separates plotted points into those less than, and those greater than, $3.5 \times 5.0\,\mu m$. (A) Maximum and minimum dimensions for 11 Early Proterozoic (2350–1600 Ma) occurrences. (B) Maximum and minimum dimensions for 20 Middle Proterozoic (1600–900 Ma) occurrences. (C) Maximum and minimum dimensions for 38 Late Proterozoic (900–550 Ma) occurrences.

Figure 5.4.22. Maximum and minimum cell lengths and widths for 14 taxa of ellipsoidal Proterozoic and Early Cambrian (2350–530) microfossils listed in Table 25.3; diagonal line perpendicular to W=L separates plotted points into those less than, and those greater than, $3.5 \times 5.0\,\mu m$.

Table 5.4.7. *Morphological characteristics of modern ellipsoidal prokaryotes used to infer the probable affinities of comparable fossil taxa.*

	Modern Ellipsoidal Prokaryotes				
	Free-living Bacteria		Chroococcalean Cyanobacteria		Inferred Probable Affinities of Fossil Analogues
Dimensions	Percent	Number	Percent	Number	
$>0.2 \times 0.6$ μm	100%	45	—	—	$<3.5 \times 5.0$ μm =
$>0.4 \times 1.4$ μm	91%	41	100%	47	"Bacterium"
$>3.5 \times 5.0$ μm	22%	10	70%	33	————————
$>14.0 \times 30.0$ μm	0%	0	2%	1	$>3.5 \times 5.0$ μm =
$>16.0 \times 30.0$ μm	—	—	0%	0	"Cyanobacterium"
Number of Species and Varieties[1]:	45		47		
Median Dimensions:	0.7×1.0 μm		2.5×4.0 μm		
Average Dimensions:	4.4×8.4 μm		8.4×15.9 μm		
Cell Dimensions, Range:	0.3×0.7 μm – 14.0×30.0 μm		0.5×1.5 μm – 16.0×30.0 μm		
Modal Size Class:	0.5×1.0 μm – 5.0×9.0 μm		1.0×2.5 μm – 7.0×13.0 μm		

[1]Bacterial data from: Buchanan and Gibbons 1974; Starr et al. 1981. Cyanobacterial data from Desikachary 1959.

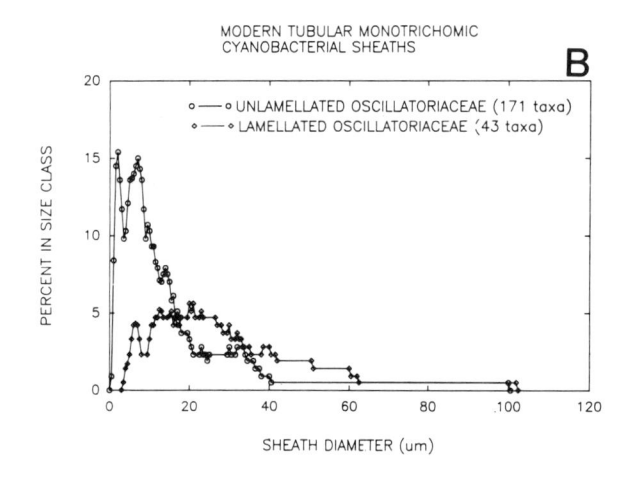

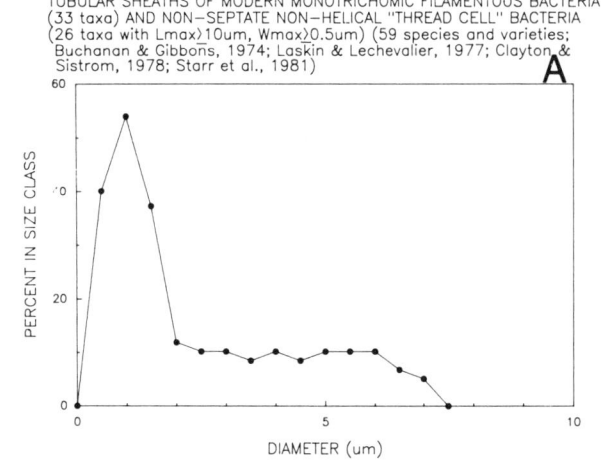

Figure 5.4.23. Distribution of diameters of modern non-septate microorganisms and unbranched tubular sheaths. (A) Non-septate, non-helical, "thread cell" bacteria (26 taxa with maximum lengths $\geqslant 10$ μm and maximum widths $\geqslant 0.5$ μm) and unbranched tubular sheaths of modern filamentous bacteria (33 taxa). (B) Unlamellated (171 taxa) and lamellated (43 taxa), unbranched, tubular sheaths of oscillatoriacean cyanobacteria (data from Desikachary 1959).

species; e.g., Chapter 24: Pl. 1, A–F, M, N; Pl. 7, A–D; Pl. 26, A; Pl. 27, B; Pl. 30, A, B, H, I, S; Pl. 31, J; Pl. 32, G; Pl. 46, A–D, F, G; cf. Fig. 5.4.2B); and

(2) lamellated, unbranched, tubular prokaryotic sheaths (code "TUL," five fossil species; e.g., Chapter 24: Pl. 6, B–E; Pl. 49, A, E; cf. Fig. 5.4.2B).

5.4.6A Morphometric Characteristics of Modern Non-septate "Thread Cell" Bacteria and Unbranched Prokaryotic Tubular Sheaths Used to Infer the Probable Affinities of Comparable Proterozoic Microfossils

Figure 5.4.23 summarizes the size distributions of the tubular sheaths of modern monotrichomic filamentous bacteria (33 species and varieties) and modern, non-septate, non-helical, "thread cell" bacteria with maximum lengths greater than 10 μm and maximum widths greater than 0.5 μm (26 species and varieties; Fig. 5.4.23A); of the tubular unlamellated sheaths of modern monotrichomic oscillatoriacean cyanobacteria (171 species and varieties; Fig. 5.4.23B); and of the tubular lamellated sheaths of modern monotrichomic oscillatoriacean cyanobacteria (43 species and varieties; Fig. 5.4.23B). As is shown in Figure 5.4.24, there is a substantial overlap in the 2.0 to 3.5 μm size range between bacterial "sheaths" (including "thread cell" bacteria) and unlamellated cyanobacterial sheaths. However, as is summarized graphically in Figure 5.4.25 (and tabulated in Table 5.4.8), about 80% of modern bacterial sheaths and non-helical "thread cell" bacteria, and only about 9% of cyanobacterial sheaths, are less than 2.0 μm in diameter; whereas only about 12% of modern bacterial sheaths and "thread cell" bacteria, and about 71% of cyanobacterial sheaths, are greater than 3.5 μm in diameter. Thus, unbranched tubular microfossils less than 2.0 μm in diameter are here regarded (Tables 5.4.5, 5.4.8) as "probably bacterial"; those 2.0 to 3.5 μm in diameter as "prokaryotic"; and those greater than 3.5 μm in diameter as "probably cyanobacterial."

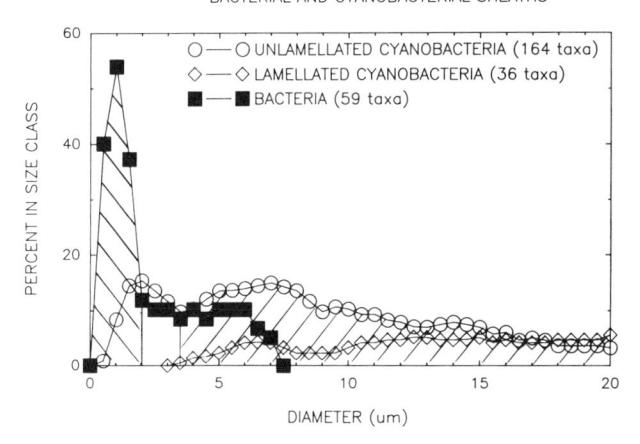

Figure 5.4.24. Overlapping distributions of the diameters of modern "thread cell" bacteria and unbranched tubular bacterial sheaths (59 taxa), and the unlamellated (164 taxa) and lamellated (36 taxa), unbranched, tubular sheaths of modern oscillatoriacean cyanobacteria ⩽20 μm in diameter, used to infer the probable affinities of fossil analogues: non-septate filaments <2.0 μm = "bacteria"; 2.0–3.5 μm = "prokaryotes"; >3.5 μm = "cyanobacteria" (see Tables 5.4.5 and 5.4.8 for details; sources of data cf. Fig. 5.4.23).

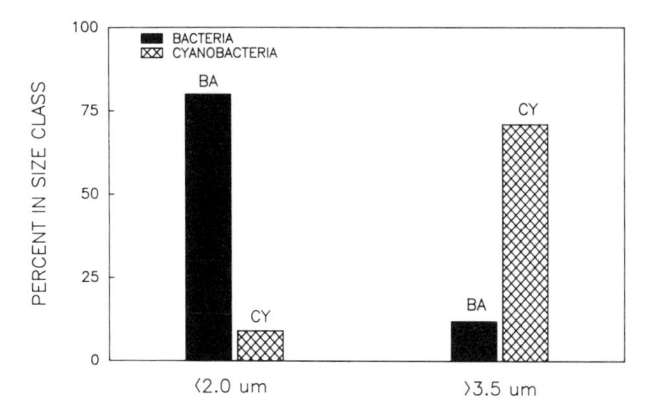

Figure 5.4.25. Histograms summarizing the distribution of diameters of non-septate "thread cell" bacteria and the unbranched tubular sheaths of modern filamentous bacteria ("BA"), and of the unbranched tubular sheaths of modern oscillatoriacean cyanobacteria ("CY"), used to infer the affinities of fossil analogues: non-septate filaments <2.0 μm = "bacteria"; 2.0–3.5 μm = "prokaryotes"; >3.5 μm = "cyanobacteria" (see Tables 5.4.5 and 5.4.8 for details; sources of data cf. 5.4.23).

Table 5.4.8. *Morphological characteristics of unbranched monotrichomic sheaths of modern prokaryotes used to infer the probable affinities of comparable fossil taxa.*

	Modern Unbranched Monotrichomic Sheaths								
	Bacterial Sheaths*		Oscillatoriacean Cyanobacteria						Inferred Probable Affinities of Fossil Analogues
			All Sheaths		Unlamellated		Lamellated		
Diameter	Percent	Number	Percent	Number	Percent	Number	Percent	Number	
>0.2 μm	100%	59	—	—	—	—	—	—	<2.0 μm = "Bacterium"
>0.7 μm	59%	35	100%	200	100%	157	—	—	————————
>2.0 μm	20%	12	91%	182	89%	139	—	—	2.0–3.5 μm =
>3.5 μm	12%	7	71%	142	63%	99	100%	43	"Prokaryote"
>6.0 μm	8%	5	60%	119	50%	78	95%	41	————————
>7.2 μm	0%	0	52%	104	42%	66	88%	38	>3.5 μm =
>100.0 μm	—	—	1%	1	0%	0	1%	2	"Cyanobacterium"
>102.0 μm	—	—	0%	0	—	—	0%	0	————————
Number of Species and Varieties[1]:	59		200		157		43		
Median Diameter:	1.1 μm		6.5 μm		5.7 μm		13.5 μm		
Average Diameter:	1.5 μm		9.6 μm		7.5 μm		18.0 μm		
Diameter, Range:	0.3–7.2 μm		0.8–102.0 μm		0.8–100.0 μm		3.6–102.0 μm		
Modal Size Classes:	0.5–2.0 μm		1.0–3.0 μm, 4.0–18.0 μm		1.0–3.0 μm, 4.0–13.0 μm		5.0–8.0 μm, 9.0–20.0 μm		

* = Including elongate bacterial "thread cells" with maximum width ⩾0.5 μm and maximum length ⩾10.0 μm.
[1] Bacterial data from: Buchanan and Gibbons 1974; Laskin and Lechevalier 1977; Clayton and Sistrom 1978; Starr et al. 1981. Cyanobacterial data from Desikachary 1959.

5.4.6B Non-septate Filamentous Proterozoic Microfossils

Figure 5.4.26 summarizes the size distribution of all occurrences of unbranched, tubular microfossils now known from units of Early, Middle and Late Proterozoic age (Table 22.3). As is shown in Figure 5.4.27 (see also Fig. 5.4.7

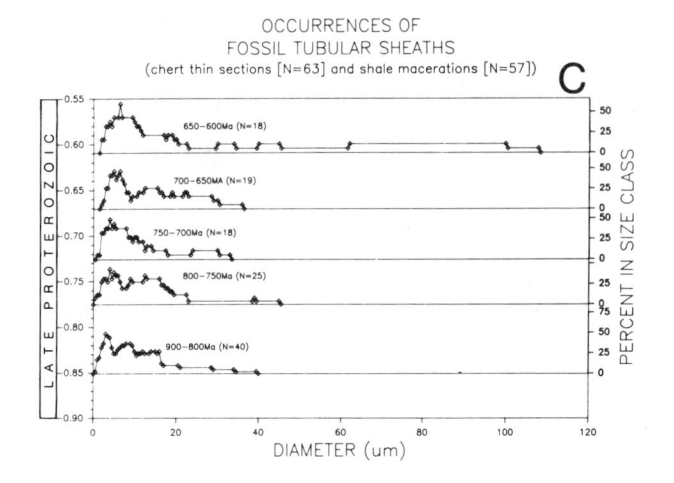

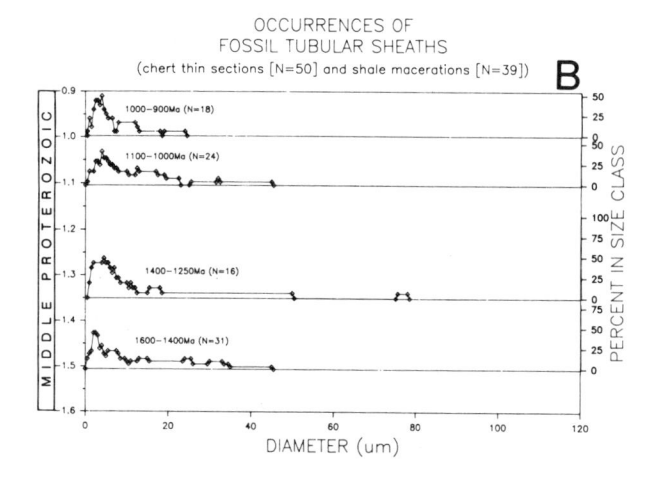

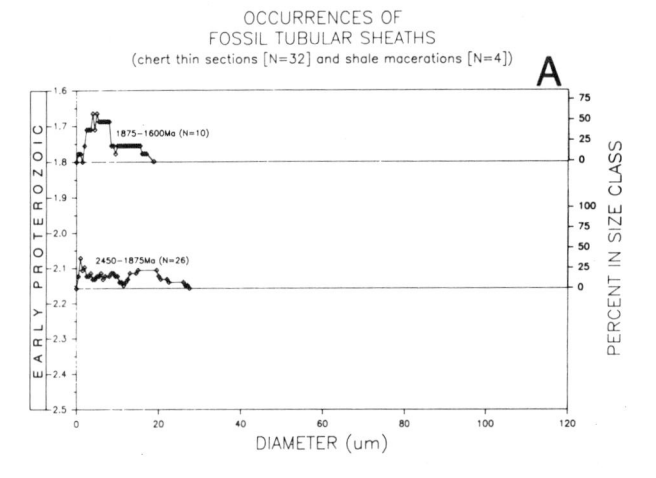

Figure 5.4.26. Distributions of diameters of known Proterozoic occurrences of non-septate filamentous microfossils reported from thin sections of cherts and macerations of shales from geologic units listed in Table 22.3. (A) Early Proterozoic occurrences (2450–1600 Ma). (B) Middle Proterozoic occurrences (1600–900 Ma). (C) Late Proterozoic occurrences (900–550 Ma).

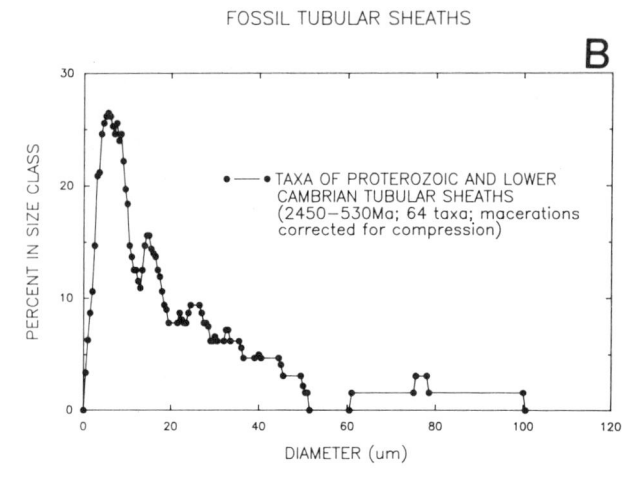

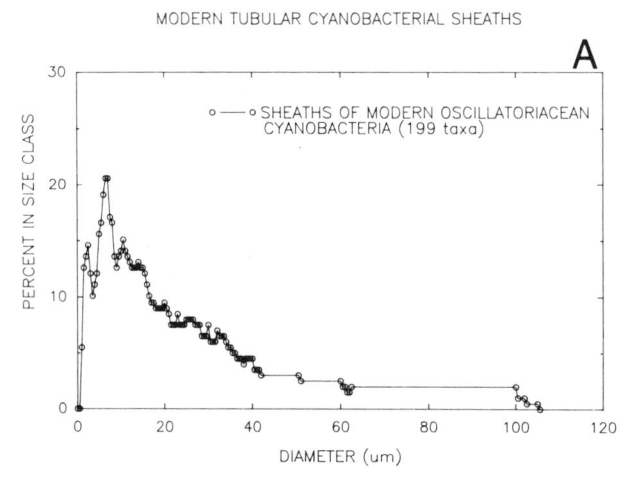

Figure 5.4.27. Distribution of diameters of taxa of modern and fossil tubular sheaths. (A) Sheaths of modern oscillatoriacean cyanobacteria (199 taxa; Desikachary 1959). (B) Diameters for 64 taxa of Proterozoic and Early Cambrian (2450–530 Ma) tubular sheaths listed in Table 25.3.

in which the data plotted in parts A and B of Fig. 5.4.27 are superimposed), there is a marked similarity in the patterns of size distribution of Proterozoic tubular microfossils and the sheaths of modern oscillatoriacean cyanobacteria. The relatively few lamellated Proterozoic tubular microfossils now known (Table 25.3) are also comparable in size distribution (Fig. 5.4.28) to the lamellated sheaths of extant oscillatoriaceans (e.g., *Lyngbya* spp.).

5.4.6C Inferred Evolutionary Trends

Like Proterozoic coccoidal prokaryotes ⩽60 μm in diameter (Section 5.4.4D), and despite earlier interpretations to the contrary (Schopf 1977), no obvious evolutionary trends are reflected in the known Proterozoic fossil record of non-septate filamentous microfossils. As is shown in Figure 25.4, the reported number of such taxa gradually increases over the Proterozoic, but this increase is strongly coupled to, and is evidently a function of, the number of assemblages sampled (Section 11.3). Non-septate microfossils known from the Early Proterozoic exhibit substantial diversity (being represented by a wide range of taxa including relatively large species, 15 μm to more than 20 μm in diameter; Fig. 5.4.26 and Chapter 24: Pl.

Proterozoic Prokaryotes

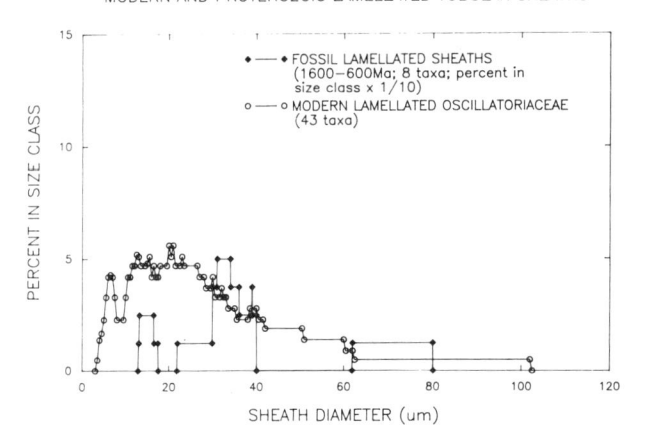

Figure 5.4.28. Comparison of the distribution of diameters of the lamellated tubular sheaths of 43 taxa of modern oscillatoriacean cyanobacteria (Desikachary 1959) and of 8 taxa of Proterozoic (1600–600 Ma) lamellated sheaths listed in Table 25-3 (plotted at 1/10 times the percent in size class).

1, A–F, M, N), and no evident trends in patterns of size distribution can be discerned over the remainder of the Proterozoic (Fig. 5.4.26). As noted above, the pattern of size distribution of such Proterozoic fossils is markedly similar to that of modern cyanobacterial (oscillatoriacean) sheaths (Figs. 5.4.7, 5.4.27). Thus, the majority of such fossils appear to be of oscillatoriacean affinity, with available fossil evidence suggesting that such cyanobacteria were both extant and relatively diverse early in the Proterozoic, and that they exhibited little morphological evolution over subsequent geologic time; if preserved in the geologic record, evidence of the origin and evolutionary diversification of this group is presumably sequestered in older (Archean) deposits.

5.4.7 Septate Filamentous Microorganisms and Proterozoic Microfossils

Data discussed in this Section relate to the following two categories of Proterozoic microfossils:

(1) naked, cellular, uniseriate, unbranched, prokaryotic trichomes (code "3N" in Tables 25.1, 25.3, and Fig. 25.4, 69 fossil species; e.g., Chapter 24: Pl. 5, I; Pl. 7, F, G; Pl. 10, F; Pl. 20, B–F; Pl. 22, C–F; Pl. 30, J–R; Pl. 31, A, C–I, K, L; Pl. 32, A–D, F; Pl. 40, C; Pl. 49, B, D; cf. Fig. 5.4.2A); and
(2) sheath-enclosed, cellular, uniseriate, unbranched, prokaryotic trichomes (code "3S," 24 fossil species; e.g., Chapter 24: Pl. 10, D; Pl. 28, A; cf. Fig. 5.4.2B).

5.4.7A Morphometric Characteristics of Modern Septate Filamentous Bacteria and Cyanobacteria Used to Infer the Probable Affinities of Comparable Proterozoic Microfossils

Figure 5.4.29 summarizes the distributions of cell widths of modern, septate, free-living, filamentous bacteria (83 species and varieties) and modern septate filamentous cyanobacteria (447 species and varieties). As is shown in Figure 5.4.30, there is a substantial overlap in size ranges between these two

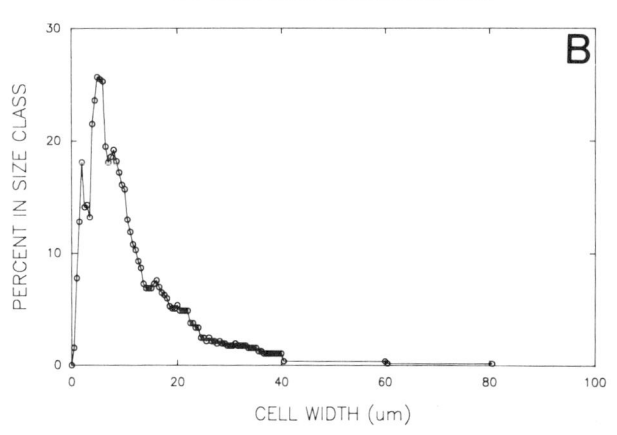

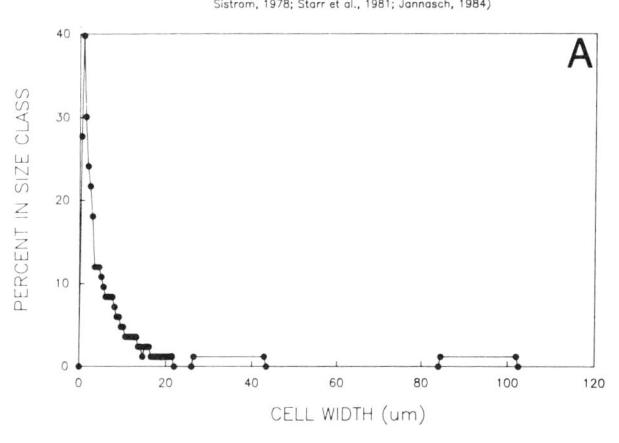

Figure 5.4.29. Distributions of cell widths of modern septate filamentous microorganisms. (A) 83 taxa of filamentous bacteria. (B) 447 taxa of oscillatoriacean cyanobacteria.

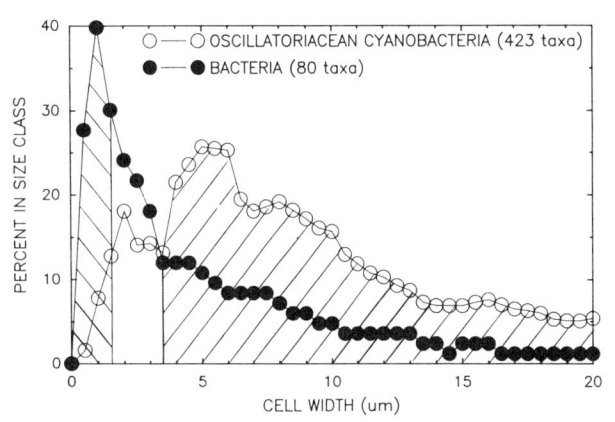

Figure 5.4.30. Overlapping distributions of cell widths of modern septate filamentous bacteria (80 taxa) and of oscillatoriacean cyanobacteria (423 taxa) $\leqslant 20\,\mu m$ in diameter used to infer the probable affinities of fossil analogues: $<1.5\,\mu m =$ "bacteria"; 1.5–$3.5\,\mu m =$ "prokaryotes"; $>3.5\,\mu m =$ "cyanobacteria" (see Tables 5.4.6 and 5.4.9 for details; sources of data cf. Fig. 5.4.29).

Table 5.4.9. *Morphological characteristics of modern septate filamentous prokaryotes used to infer the probable affinities of comparable fossil taxa.*

	Modern Septate Filamentous Prokaryotes				
	Free-living Bacteria		Oscillatoriacean Cyanobacteria		Inferred Probable Affinities of Fossil Analogues
Cell Width	Percent	Number	Percent	Number	
>0.2 μm	100%	83	—	—	<1.5 μm = "Bacterium"
>0.4 μm	100%	83	100%	447	————————
>1.5 μm	39%	32	85%	380	1.5–3.5 μm = "Prokaryote"
>3.5 μm	16%	13	69%	309	————————
>80.0 μm	1%	1	0%	0	>3.5 μm = "Cyanobacterium"
>102.0 μm	0%	0	—	—	————————
Number of Species and Varieties[1]:	83		447		
Median Cell Width:	1.5 μm		5.3 μm		
Average Cell Width:	3.6 μm		6.7 μm		
Cell Width, Range:	0.3–102.0 μm		0.5–80.0 μm		
Cell Width, Modal Size Class:	0.5–3.5 μm		1.5–12.0 μm		

[1] Bacterial data from: Buchanan and Gibbons 1974; Laskin and Lechevalier 1974; Clayton and Sistrom 1978; Starr et al. 1981; Jannasch 1984b. Cyanobacterial data from Desikachary 1959.

categories of modern microorganisms for filaments 1.5 to 3.5 μm in diameter. However, as is summarized graphically in Figure 5.4.31 (and tabulated in Table 5.4.9), about 61% of such bacteria and about only 15% of such cyanobacteria are less than 1.5 μm in diameter, whereas only 16% of bacteria and 69% of cyanobacteria are greater than 3.5 μm in diameter. Thus, Proterozoic septate microfossils less than 1.5 μm in diameter are here regarded (Tables 5.4.5, 5.4.9) as "probable bacteria"; those 1.5 to 3.5 μm in diameter as "prokaryotes"; and those greater than 3.5 μm in diameter as "probable cyanobacteria."

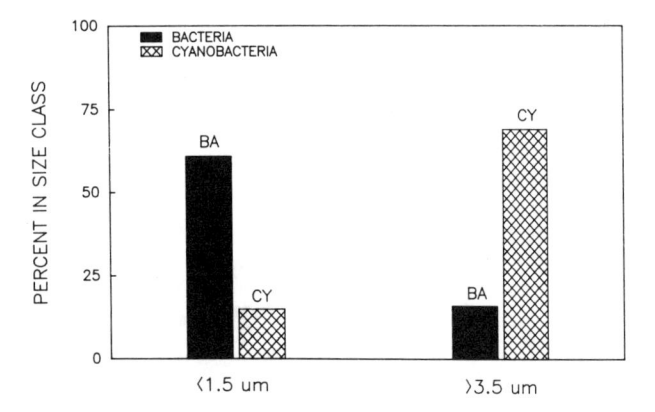

Figure 5.4.31. Histograms summarizing the distributions of cell widths of modern septate filamentous bacteria and cyanobacteria used to infer the probable affinities of fossil analogues: diameter <1.5 μm = "bacteria"; 1.5–3.5 μm = "prokaryotes"; >3.5 μm = "cyanobacteria" (see Tables 5.4.5 and 5.4.9 for details; sources of data cf. Fig. 5.4.29).

5.4.7B Septate Filamentous Prokaryotic Microfossils

Figure 5.4.32 summarizes the size distributions of all occurrences of septate filamentous microfossils now known from shales/siltstones and carbonates/cherts of Early, Middle, and Late Proterozoic age (Table 22.3). Figure 5.4.33 compares the size distributions (maximum and minimum medial cell widths and lengths) of all taxa (Table 25.3) of naked and sheath-enclosed trichomes now known from Proterozoic strata (Fig. 5.4.33D) with those of modern septate bacteria (Fig. 5.4.33A, C) and modern oscillatoriacean cyanobacteria (Fig. 5.4.33B, C). As is evident from this comparison, the majority of Proterozoic species are more similar in medial cell size and shape to modern cyanobacteria (viz., oscillatoriaceans) than they are to extant filamentous bacteria.

Like the sheath-enclosed, *Chroococcus*- and *Gloeocapsa*-like microfossils discussed above (Section 5.4.4B), cellularly preserved trichomes exhibit sufficient morphological complexity to permit multifactor comparison of modern and fossil taxa, in particular, comparison of the dimensions (widths and lengths) and shapes of intercalary cells occurring in trichomes characterized by distinctive terminal cells. As is shown in Figure 5.4.34A, modern (46 oscillatoriacean taxa) and Proterozoic trichomes with rounded terminal cells (22 taxa) exhibit similar distributions of intercalary cell sizes and shapes. Similarly (Fig. 5.4.34B), the dimensions of the intercalary cells of modern (12 oscillatoriacean taxa) and Proterozoic trichomes having globose terminal cells (7 taxa) are essentially indistinguishable (viz., with such cells occurring exclusively in the field defined by cell width ⩽15 μm and cell length ⩽10 μm), but their range of cell size differs significantly from that of intercalary cells exhibited by trichomes with rounded terminal cells. Thus, both for trichomes with rounded terminal cells and with globose

Proterozoic Prokaryotes

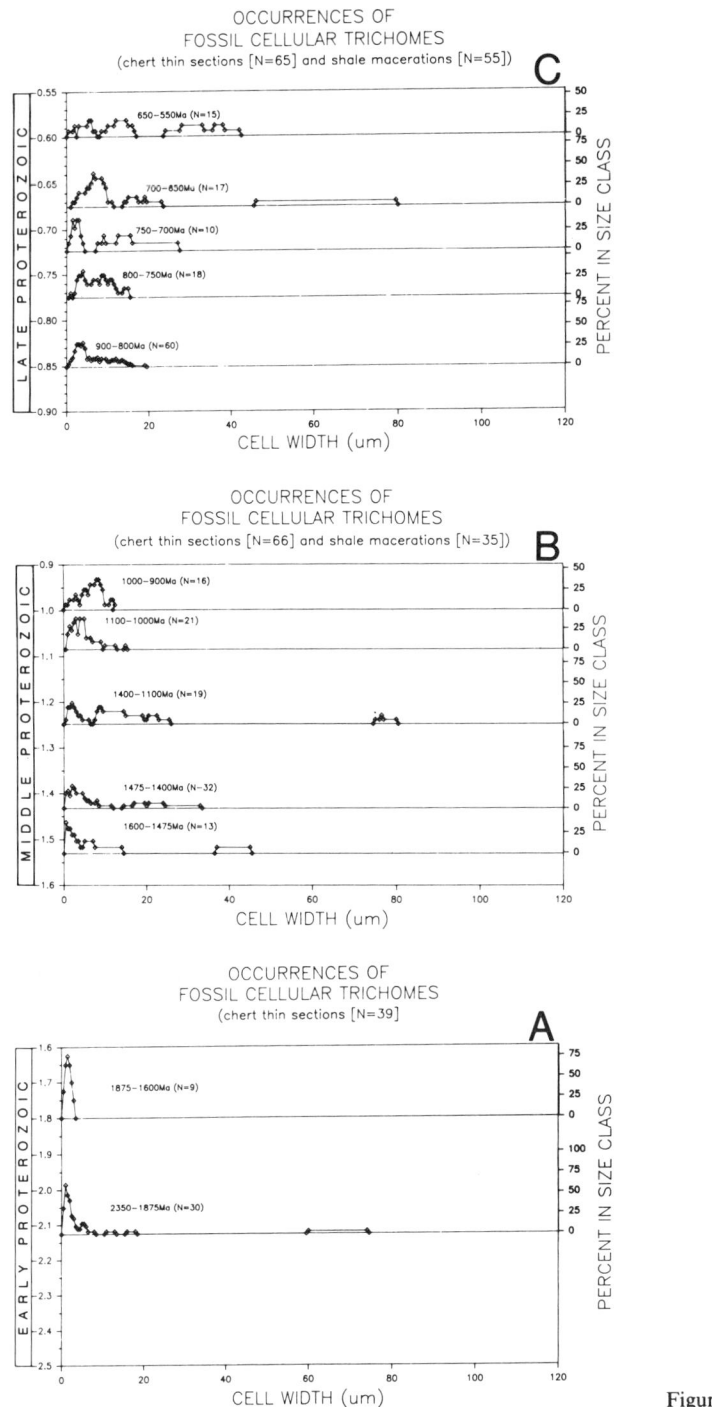

Figure 5.4.32. Distributions of cell widths of known Proterozoic occurrences of septate filamentous microfossils reported from thin sections of cherts and macerations of shales from geologic units listed in Table 22.3. (A) Early Proterozoic occurrences (2350–1600 Ma). (B) Middle Proterozoic occurrences (1600–900 Ma). (C) Late Proterozoic occurrences (900–550 Ma).

terminal cells, there appears to be a strong correlation between the type of terminal cell and the range of intercalary cell sizes, a correlation exhibited both in modern oscillatoriacean cyanobacteria and in Proterozoic septate microfossils, with the modern and fossil trichomes exhibiting notably similar ranges of intercalary cell sizes and shapes.

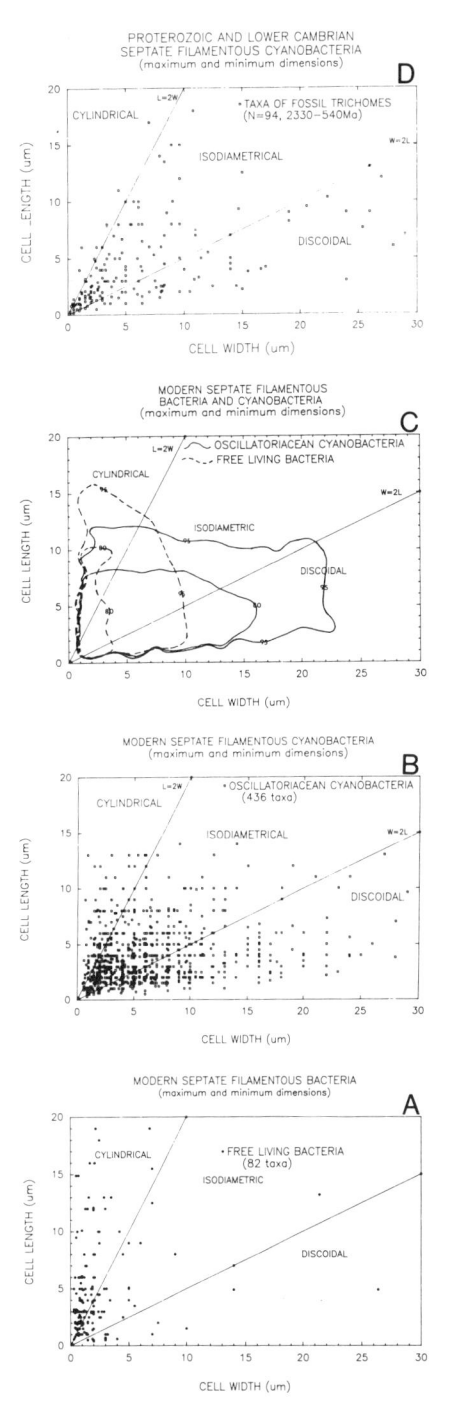

Figure 5.4.33. Widths versus lengths (maximum and minimum dimensions) of medial cells for taxa of modern and fossil septate filaments with medial cell widths $\leqslant 30\,\mu m$ and lengths $\leqslant 20\,\mu m$. (A) 82 taxa of modern free-living filamentous bacteria (source of data cf. Fig. 5.4.29A). (B) 436 taxa of modern oscillatoriacean cyanobacteria (Desikachary 1959). (C) 95% and 80% size-distribution envelopes for modern filamentous bacteria (dashed line; derived from Fig. 5.4.33A) and for oscillatoriacean cyanobacteria (continuous line; derived from Fig. 5.4.33B); e.g., 95% of the 436 taxa of oscillatoriacean cyanobacteria plotted in Figure 5.4.33B (viz., all taxa except for the largest 5%) are included within the area delineated by the cyanobacterial 95% size-distribution envelope. (D) 94 taxa of Proterozoic and Early Cambrian (2330–540 Ma) septate filamentous microfossils listed in Table 25.3, the majority of which have non-cylindrical medial cells and are thus more "cyanobacterium-like" (compare with Figs. 5.4.33B and 5.4.33C) than "bacterium-like" (Figs. 5.4.33A and 5.4.33C).

notable similarity of tubular Proterozoic microfossils to the cylindrical sheaths of modern oscillatoriaceans (Figs. 5.4.27, 5.4.28); (*ii*) the comparability of the ranges of cell sizes and shapes of most Proterozoic septate microfossils (Fig. 5.4.33D) to those of modern oscillatoriacean cyanobacteria (Fig. 5.4.33B); and (*iii*) the detailed morphometric similarity between fossil and modern trichomes characterized by the same type of distinctive terminal cells (Fig. 5.4.34). Thus, like the other categories of fossil prokaryotes discussed in this Section, the majority of Proterozoic septate filaments appear to be of cyanobacterial affinity, with the evolutionary diversification of such prokaryotes having evidently occurred early in geologic time, pre-dating deposition of the Proterozoic sediments here investigated, and the group having exhibited minimal morphological evolutionary change over Proterozoic time.

5.4.8 Summary: Inferred Affinities and Evolutionary Trends of the Principal Categories of Proterozoic Prokaryotic Microfossils

The following generalizations emerge from the foregoing survey of the Proterozoic microbial fossil record as now known:

(1) With the exception of "Proterozoic microbial(?) problematica," uncommon microfossils of particularly unusual morphology discussed below (Section 5.6), all informal species of prokaryotic microfossils here recognized (Table 25.1) are comparable in salient morphological characteristics to extant prokaryotes, predominantly, but not exclusively, to modern cyanobacteria.

(2) Dominant among such fossils are solitary and colonial coccoids referrable to the Chroococcales (chiefly, members of the Chroococcaceae with subsidiary, but locally abundant entophysalidaceans, pleurocapsaceans, and hyellaceans) and unbranched, uniseriate, filaments referrable to the Nostocales (predominantly, to the Oscillatoriaceae).

(3) At least as early as 2100 Ma (the beginning of the relatively continuous fossil record as now known; Section 11.3), numerous bacterial lineages and oscillatoriacean, chroococcacean, and probably entophysalidacean cyanobacteria were well established, evidently exhibiting a substantial degree of morphological and taxonomic diversity.

(4) Many Proterozoic prokaryotes are comparable in morphometric detail to members of the modern biota (Figs. 5.4.14, 5.4.34, Tables 25.2, 25.3), with available data indicating that prokaryotes in general, and cyanobacteria in particular, are "hypobradytelic" (Schopf 1987), having exhibited little morphological evolution over exceptionally long periods of geologic time (for discussion, see Section 13.3.2). Thus, and despite earlier, hopeful interpretations to the contrary (Schopf 1977), the temporal distribution of such fossils cannot be expected to provide a basis for meaningful biostratigraphic subdivision of the Proterozoic rock record.

(5) If evidence of the origin and evolutionary deversification of the major prokaryotic groups (including the ecologically crucial origin of oxygen-producing photoautotrophic cyanobacteria) exists in the early fossil record, it must occur in sediments decidedly older (viz., of Archean age) than those here investigated.

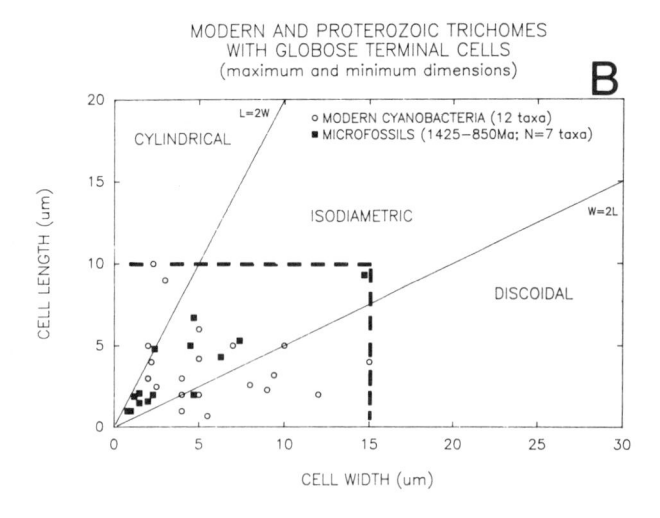

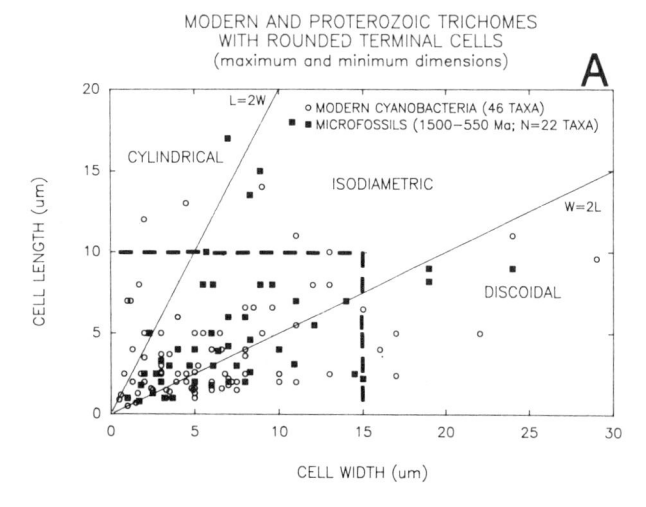

Figure 5.4.34. Widths versus lengths (maximum and minimum dimensions) of medial cells for taxa of modern and Proterozoic trichomes with medial cell widths $\leq 30\,\mu m$ and lengths $\leq 20\,\mu m$ and having distinctive terminal cells. (A) Trichomes with rounded terminal cells: 46 representative taxa of modern oscillatoriacean cyanobacteria (Desikachary 1959) compared with 22 taxa of Proterozoic (1500–550 Ma) trichomes listed in Table 25.3. (B) Trichomes with globose terminal cells: 12 taxa of modern oscillatoriacean cyanobacteria (Desikachary 1959) compared with 7 taxa of Proterozoic (1425–850 Ma) trichomes listed in Table 25-3. Note that for both the modern and fossil trichomes, those with rounded terminal cells (Fig. 5.4.34A) exhibit a substantially broader range of medial cell sizes than those with globose terminal cells (Fig. 5.4.34B).

5.4.7C Inferred Evolutionary Trends

Not surprisingly, like non-septate fossil filaments (the vast majority of which originally encompassed cellular trichomes like those here considered), the Proterozoic fossil record of septate prokaryotic trichomes reveals no discernible evolutionary trends. For example, the size distribution of such fossils known from the Early Proterozoic (Fig. 5.4.32A) does not differ significantly from those known from Middle (Fig. 5.4.32B) or Late Proterozoic sediments (Fig. 5.4.32C). Such fossils appear to be predominantly of cyanobacterial, oscillatoriacean affinity, an interpretation consistent with (*i*) the

5.5

Proterozoic and Early Cambrian Acritarchs

CARL V. MENDELSON J. WILLIAM SCHOPF

5.5.1 Introduction

Acritarchs are "small microfossils of unknown and probably varied affinities consisting of a central cavity enclosed by a wall of single or multiple layers and of chiefly organic composition..." (Evitt 1963, p. 300). These organic-walled microfossils exhibit a variety of shapes and sizes and are important components of Proterozoic and Paleozoic ecosystems.

Many Paleozoic acritarchs bear striking resemblance to living and fossil dinoflagellate cysts (see Evitt 1985, pp. 42–43). Others, including many Proterozoic and Early Cambrian forms, may be confidently assigned to the prasinophyte green algae (although such forms are still commonly treated among the acritarchs in the literature). The phylogenetic placement of prasinophytes, however, is subject to debate. For example, Tappan (1980) and Round (1984) review the case for placing these algae in the Division Prasinophyta; Norris (1980) argued for a Class Prasinophyceae within the Division Chlorophyta; and Mattox and Stewart (1984), on the basis of comparative cytology, allocated prasinophyte genera to other classes of green algae (i.e., they rejected the concept of a natural group of prasinophyte algae). Finally, the prasinophytes as a natural grouping are absent from a preliminary cladistic analysis of green algae (Mishler and Churchill 1985). The preservation potential and modern ecology of both dinoflagellates and prasinophytes are discussed in Section 5.3.3.

Most acritarchs probably represent the remains of a thick-walled resting stage, or cyst, in the life cycle of planktonic eukaryotic algae. But the definition of the term "acritarch" does not exclude vegetative cells, some of which are known to be highly resistant to degradation (Section 5.3.3). Lindgren (1981, p. 7) concluded that "there seems to be no possibility of discriminating between vegetative and resting (cyst) stages in fossil material." Finally, some simple spheroidal acritarchs may represent the sheaths of cyanobacterial unicells or the outer common sheath of a colonial cyanobacterium (reviewed in Horodyski 1980; see Section 5.3). The morphologically distinctive melanocyrillids (Bloeser 1985) fit the definition of acritarch; because they are so distinctive, however, they are here treated separately (Section 5.6.2B).

5.5.2 Major Types of Proterozoic and Early Cambrian Acritarchs

For convenience, we follow the simple morphological classification of acritarchs proposed by Downie et al. (1963), in which the acritarchs are organized into subgroups on the basis of general morphology. As pointed out by Downie et al. (1963), this system lies completely outside the jurisdiction of any codes of nomenclature. Moreover, it is not a phylogenetic statement; rather, it is a classification of convenience. It is conceivable, however, that the Downie et al. subgroups may have phylogenetic significance at least for the *early* stages of acritarch evolution (viz., during the Middle and Late Proterozoic), when experiments in different *Baupläne* may have occurred.

The most common Proterozoic and Early Cambrian acritarchs are the simple sphaeromorphs (e.g., Chapter 24: Pl. 19, G; Pl. 42, D; Pl. 55, C), forms that have a spherical to ellipsoidal vesicle and which lack processes. Some sphaeromorphs bear distinctive surface ornamentation (see Section 5.5.3, below). Less common are the morphologically complex acritarchs, which include the acanthomorphs, herkomorphs, netromorphs, prismatomorphs, pteromorphs, and many minor subgroups (for a complete inventory of such forms, see Tappan 1980). The morphologically complex types are here grouped together as "nonsphaeromorphs."

5.5.2A Sphaeromorph Acritarchs

As discussed above (Section 5.4.4), three categories of sphaeromorph acritarchs are here recognized (Table 25.3): (*i*) coccoid microfossils $\leq 60\,\mu m$ in maximum diameter; (*ii*) "mesosphaeromorph" acritarchs, 60 to $200\,\mu m$ in maximum diameter; and (*iii*) "megasphaeromorph" acritarchs $>200\,\mu m$ in maximum diameter.

The affinities and geologic distribution of coccoid microfossils $\leq 60\,\mu m$ in diameter are addressed in Section 5.4.4C. As is there discussed, although many of the larger microfossils of this category are similar in size to extant eukaryotic micro-algae (Figs. 5.4.12, 5.4.16, Tables 5.4.5, 5.4.6), coccoidal prokaryotes as large as 50 to $60\,\mu m$ in diameter are known from the modern flora. Thus, by analogy with members of the extant biota, coccoidal microfossils ≤ 60 μm in diameter cannot be regarded as unequivocally eukaryotic.

In contrast – and if their cellular nature can be clearly established – because of their decidedly larger cells, "mesosphaeromorphs," 60 to 200 μm in maximum diameter, and "megasphaeromorphs," 200 μm to more than 7,500 μm in diameter, seem assuredly to be eukaryotes. It is important to note, however, that large size, by itself, cannot be considered an absolute indicator of eukaryotic affinity. Some large sphaeromorph-like fossils may represent the preserved spheroidal envelopes (sheaths) of colonial, coccoid, prokaryotes (e.g., chroococcaceans) or the thick, resilient, originally mucilagenous matrix of colonial, filamentous, nostocalean cyanobacteria (e.g., so-called *Nostoc* balls). Indeed, some colonial cyanobacteria (viz., *Nostoc* spp.) even produce ribbon-like megascopic "thalli," a few to several millimeters in width and several centimeters in length, that easily could be confused with *Vendotaenia* and similar Late Proterozoic putative eukaryotic metaphytes. However, unlike the vegetative cells of eukaryotic micro-algae (for a relevant discussion of the "Divisional Dispersion Index," see Chapter 25), spheroidal colonies produced by colonial prokaryotes tend to be highly variable in size, due to the occurrence of fragmentation at various stages during colony formation. Moreover, large colonies of this type also tend to be of variable, and not uncommonly of highly irregular shape. Studies of fossil populations of sphaeromorphs can thus be used to differentiate true cells from sphaeromorph-like prokaryotic envelopes. In addition, and perhaps most importantly, it is commonly possible to identify preserved cell walls (characterized in some groups by their surficial texture – the occurrence of pits, pores and the like – and more generally by their relatively uniform thickness and their robust, dense, commonly granular, well-defined nature), a certain indicator of cellularity.

Although demonstrably cellular "mesosphaeromorphs" and "megasphaeromorphs" thus seem assuredly eukaryotic, because of the absence of preservable definitive taxonomic characters (e.g., accessory pigments) and the simple morphology of such sphaeromorphs, their phylogenetic relations within the Eukaryota are uncertain. Among extant microorganisms they seem most similar to various unicellular, planktonic, chlorophycean (including prasinophycean) and some rhodophycean micro-algae (cf. Table 5.4.5). However, little is known of the life cycles of these fossils, and although no systematic attempt has been made to distinguish between the preserved (relatively thin-walled?) vegetative cells and (relatively thick-walled?) reproductive cysts of such taxa, it seems likely, as discussed above, that both cell types are included among reported fossil occurrences. Thus, "mesosphaeromorph" and "megasphaeromorph" acritarchs are probably best regarded as planktonic, eukaryotic, micro-algae *incertae sedis*. The remainder of this discussion deals with these two categories of large, organic-walled, spheroidal eukaryotes, planktonic microfossils that played a significant role in the Proterozoic biosphere.

"*Mesosphaeromorph*" acritarchs (code "MS" in Tables 25.1, 25.3, Fig. 25.4; Chapter 24: Pl. 14, A, B; Pl. 19, A), sphaeromorphs of a size well known from the Phanerozoic fossil record (Tappan 1980), occur also throughout much of the Proterozoic.

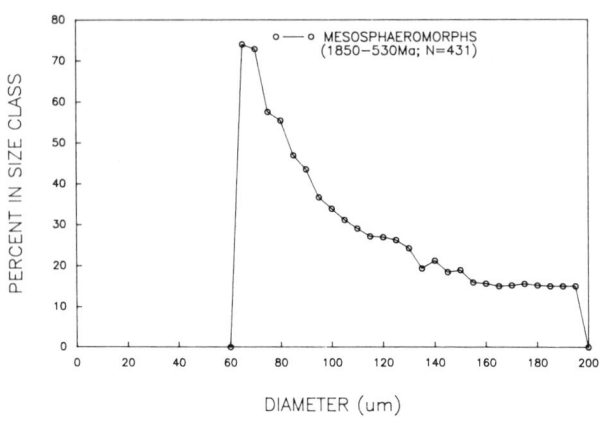

Figure 5.5.1. Size distribution for 431 occurrences of Proterozoic and Early Cambrian (1850–530 Ma) "mesosphaeromorphs" (spheroidal acritarchs with maximum diameters > 60 μm, < 200 μm) reported from geologic units listed in Table 22.3.

Figure 5.5.1 summarizes the size distribution for 431 reported occurrences, ranging from the Early Proterozoic (ca. 1850 Ma) into the Early Cambrian. Such occurrences are particularly numerous in Proterozoic shales/siltstones (Fig. 5.5.2), but "mesosphaeromorphs" have also been reported from Proterozoic cherts/carbonates (Fig. 5.5.3). Figure 5.5.4 summarizes the size distribution of the ten species of "mesosphaeromorphs" here recognized (Table 25.3). Relatively small taxa (less than 140 μm in diameter) predominate in assemblages older than about 1100 Ma, with taxa up to 200 μm in diameter being abundant throughout much of the Late Proterozoic (Figs. 5.5.2, 5.5.3). As is shown in Figure 11.3.9, "mesosphaeromorph" species diversity remained relatively low throughout the Early Proterozoic, increased during the Middle Proterozoic to reach a maximum at about 950 Ma, and gradually declined throughout the remainder of the era; three of the ten Proterozoic taxa are particularly comparable in morphology to living micro-algal species (Table 25.2, Fig. 25.4).

Unlike "mesosphaeromorphs," "*megasphaeromorph*" acritarchs (code "MG" in Tables 25.1, 25.3, Fig. 25.4; Chapter 24: Pl. 24, D; Pl. 42, A, D), sphaeromorphs > 200 μm in maximum diameter, are of rare occurrence in Phanerozoic strata, with taxa larger than about 600 μm being unknown in the Phanerozoic fossil record (Tappan 1980). Nevertheless, "megasphaeromorphs" were a significant component of Proterozoic ecosystems: as shown in Figure 5.5.5, more than 100 occurrences of "megasphaeromorphs" have been reported from Proterozoic and Early Cambrian strata, occurring both in shales/siltstones and in carbonates/cherts (Fig. 5.5.6). As is noted in Figure 5.5.6 (and indicated in Fig. 11.3.9), only four (questionable) "megasphaeromorph" occurrences have been reported from the Early Proterozoic; the established, relatively continuous fossil record of the group extends only from near the beginning of the Middle Proterozoic (Figs. 5.5.6, 11.3.9). The size range of the 14 Proterozoic "megasphaeromorph" species here recognized (Table 25.3) is summarized in Figure 5.5.7.

Proterozoic and Early Cambrian Acritarchs

OCCURRENCES OF MESOSPHAEROMORPH ACRITARCHS >60.0um, <200.0um (shale macerations)

[Figure B: Late Proterozoic size distributions, age bins 600–550 Ma (N=30), 650–600 Ma (N=42), 700–650 Ma (N=39), 750–700 Ma (N=19), 800–750 Ma (N=87), 900–800 Ma (N=54)]

OCCURRENCES OF MESOSPHAEROMORPH ACRITARCHS >60.0um, <200.0um (shale macerations)

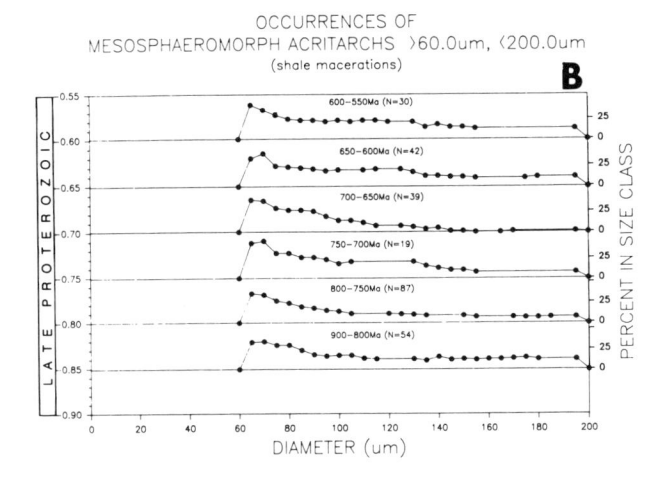

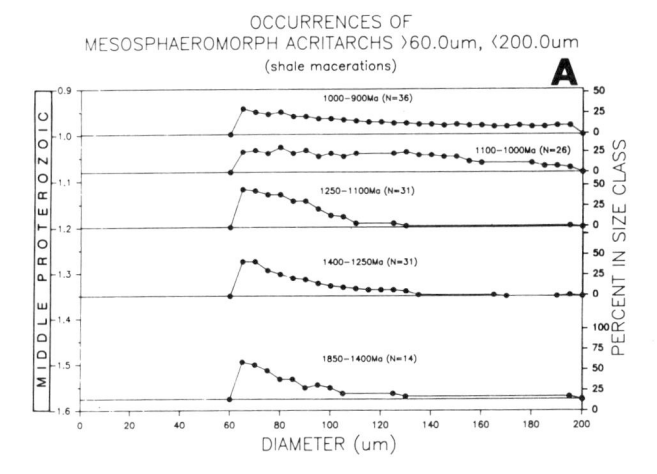

[Figure A: Middle Proterozoic size distributions, age bins 1000–900 Ma (N=36), 1100–1000 Ma (N=26), 1250–1100 Ma (N=31), 1400–1250 Ma (N=31), 1850–1400 Ma (N=14)]

Figure 5.5.2. Size distributions for occurrences of Proterozoic "mesosphaeromorphs" (spheroidal acritarchs with maximum diameters >60 μm, <200 μm) reported from macerations of shales from geologic units listed in Table 22.3. (A) Middle Proterozoic (1850–900 Ma) occurrences. (B) Late Proterozoic (900–550 Ma) occurrences.

TAXA OF PROTEROZOIC AND LOWER CAMBRIAN MESOSPHAEROMORPHS (maximum diameter >60.0um, <200.0um)

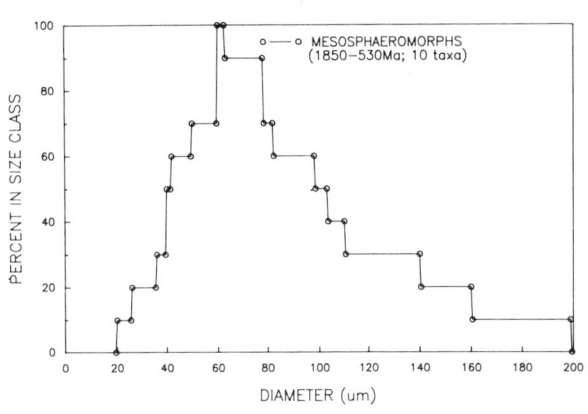

[Plot: MESOSPHAEROMORPHS (1850–530 Ma; 10 taxa), percent in size class vs diameter (μm)]

Figure 5.5.4. Size distribution for 10 taxa of Proterozoic and Early Cambrian (1850–530 Ma) "mesosphaeromorphs" (spheroidal acritarchs with maximum diameters >60 μm, <200 μm) listed in Table 22.3.

As is shown in Figure 5.5.6 (and discussed in Section 5.5.3D, Fig. 5.5.12), the size ranges exhibited by reported occurrences of these fossils show a well-defined time-correlative pattern of distribution: occurrences reported throughout much of the Middle Proterozoic include only relatively small "megasphaeromorphs" (less than 1,500 μm in diameter); those between 1100 and 900 Ma include fossils of increasingly large size (more than 2,000 μm to more than 5,000 μm in diameter); occurrences reported from strata 900 to 800 Ma include the largest "megasphaeromorphs" now known (individuals larger than 7,500 μm in diameter); and those reported from the remainder of the Proterozoic exhibit decreasing size ranges and an increasing relative abundance of small individuals, culminating with all "megasphaeromorphs" younger than 600 Ma being less than 1,000 μm in diameter. This time-correlative pattern of size distribution is paralleled by Proterozoic "megasphaero-

OCCURRENCES OF MESOSPHAEROMORPH ACRITARCHS >60.0um, <200.0um (chert thin sections)

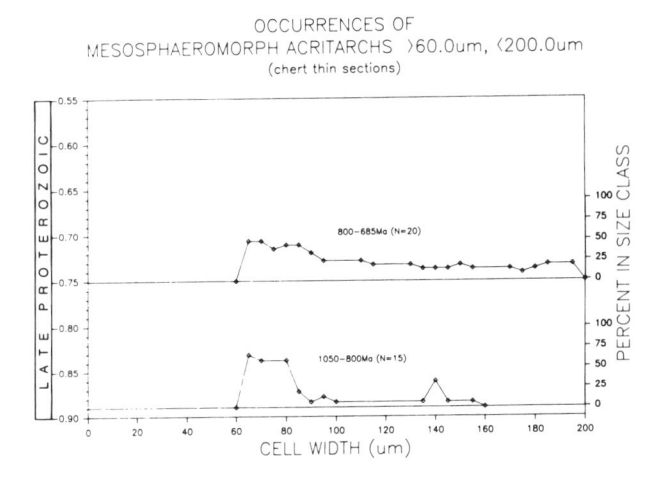

[Plot: Late Proterozoic size distributions, age bins 800–685 Ma (N=20), 1050–800 Ma (N=15)]

Figure 5.5.3. Size distributions for occurrences of "mesosphaeromorphs" (spheroidal acritarchs with maximum diameters >60 μm, <200 μm) reported from thin sections of cherts from Proterozoic (1050–685 Ma) geologic units listed in Table 22.3.

PROTEROZOIC AND LOWER CAMBRIAN MEGASPHAEROMORPH ACRITARCHS (>200.0 um)

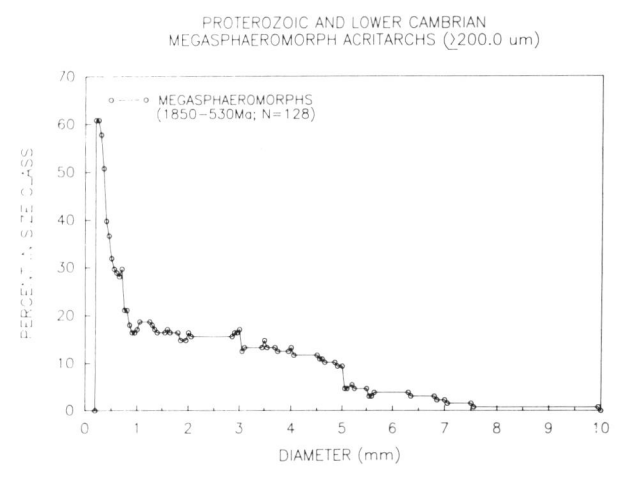

[Plot: MEGASPHAEROMORPHS (1850–530 Ma; N=128), percent in size class vs diameter (mm)]

Figure 5.5.5. Size distribution for 128 occurrences of Proterozoic and Early Cambrian (1850–530 Ma) "megasphaeromorphs" (spheroidal acritarchs with maximum diameters ≥200 μm) reported from geologic units listed in Table 22.3.

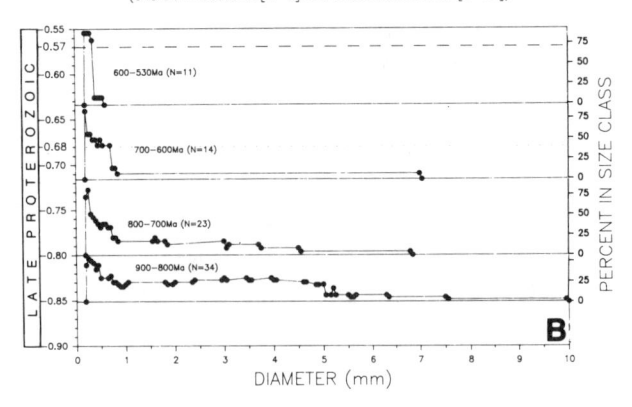

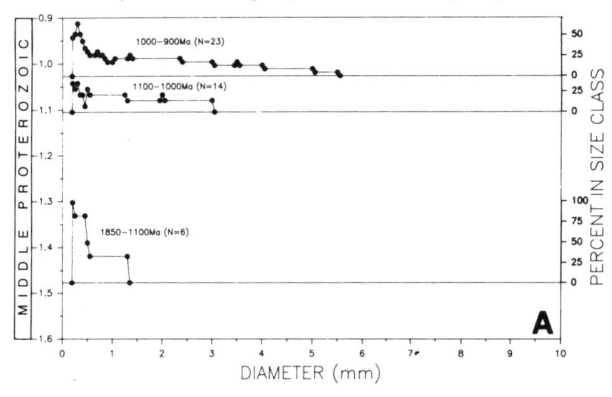

Figure 5.5.6. Size distributions for occurrences of Proterozoic and Early Cambrian "megasphaeromorphs" (spheroidal acritarchs with maximum diameters ≥ 200 μm) reported from thin sections of cherts and macerations of shales from geologic units listed in Table 22.3. (A) Middle Proterozoic (and four Early Proterozoic, 1850 Ma) occurrences. (B) Late Proterozoic occurrences.

morph" species diversity (Fig. 11.3.9): diversity was low prior to 1200 Ma, increased to reach a maximum at about 900 Ma, and declined, gradually but markedly, throughout the remainder of the Proterozoic.

5.5.2B Nonsphaeromorph Acritarchs

The nonsphaeromorph acritarchs of the Proterozoic and Early Cambrian vary greatly in size, shape, wall structure, and surface sculpture. Analysis of the available literature (coverage estimated at 90% of publications on *nonsphaeromorphs*) reveals 577 occurrences of morphologically complex acritarchs in about 125 geologic units of this time period (Table 5.5.1). Acanthomorphs (e.g., Figs. 5.5.8C, 5.5.9, 5.5.11) exhibit prominent simple or branched processes; they account for about 67% of the occurrences. Herkomorphs (e.g., Fig. 5.5.10, H, I), contributing about 7.1%, have a spherical to subpolygonal vesicle that is divided into polygonal fields by raised crests. Netromorphs (5.4%) have elongate to fusiform vesicles that may be produced into processes or spines at one or both poles. Oömorphs (2.8%) have a unique egg-shaped vesicle, commonly heavily ornamented on one end. Polygonomorphs (e.g., Fig.

5.5.10, F), accounting for about 3.3% of occurrences, encompass vesicles having a polygonal outline, and which typically have simple processes. The vesicle of prismatomorphs (3.5%, e.g., Chapter 24: Pl. 56, C) is polyhedral, typically resembling an octahedron or tetrahedron. Pteromorphs (Fig. 5.5.10, G) contribute another 3.5%; this subgroup is characterized by an equatorial flange that may be supported by radial folds or processes. Additional examples of these subgroups and others are illustrated in Chapter 24.

As is apparent in Table 5.5.1, these major subgroups of complex acritarchs occur predominantly in the Early Cambrian. Analysis of questionable occurrences discussed below reinforces this impression (Fig. 5.5.13). Note, however, that the acanthomorphs have a significant Late Proterozoic fossil record. In the following paragraphs, each of the major subgroups of nonsphaeromorph acritarchs is discussed in turn.

Acanthomorphs are among the most interesting of the several subgroups of complex acritarchs. They are common in lower Paleozoic sedimentary rocks worldwide, where they are used for detailed biostratigraphy (including stratotype definitions). Their Proterozoic record, however, is very limited. Acanthomorphs are unknown from the Archean (Schopf and Walter 1983); similarly, unquestionable acanthomorphs are unknown from Lower and Middle Proterozoic strata (Hofmann and Schopf 1983; see Chapter 22). Evidently, and with but one exception (the remarkable biota of the Lakhanda Formation of Siberia, which is poorly dated at 950 Ma), well-documented acanthomorph occurrences are confined to strata of Late Proterozoic (900–550 Ma) and younger age.

Reports of four occurrences of acanthomorphs older than 1000 Ma are very dubious (Fig. 5.5.13). Among these reports, Timofeev (1966, pl. 76, fig. 14) published a drawing of (?) *Acantholigotriletum* sp. from "undifferentiated Sinian deposits" of northern China. However, this report is questionable because (*i*) the age is very poorly constrained; (*ii*) the drawing is interpretive (perhaps partly illustrating diagenetic features); and *iii*) only one specimen is figured, and it is not discussed in the text. Published illustrations of the other three putative pre-

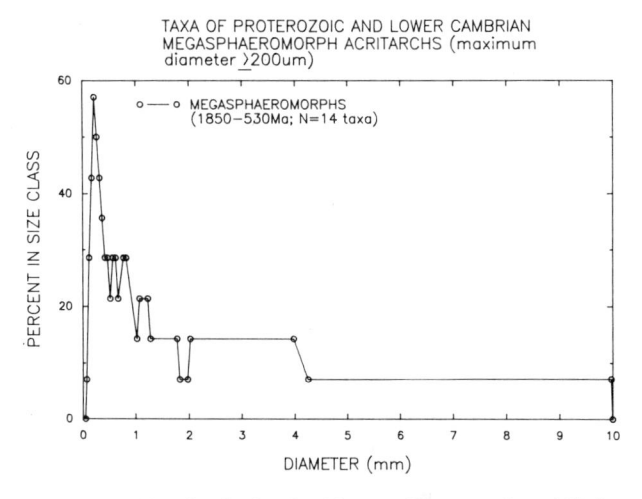

Figure 5.5.7. Size distribution for 14 taxa of Proterozoic and Early Cambrian (1850–530 Ma) "megasphaeromorphs" (spheroidal acritarchs with maximum diameters ≥ 200 μm) listed in Table 25.3.

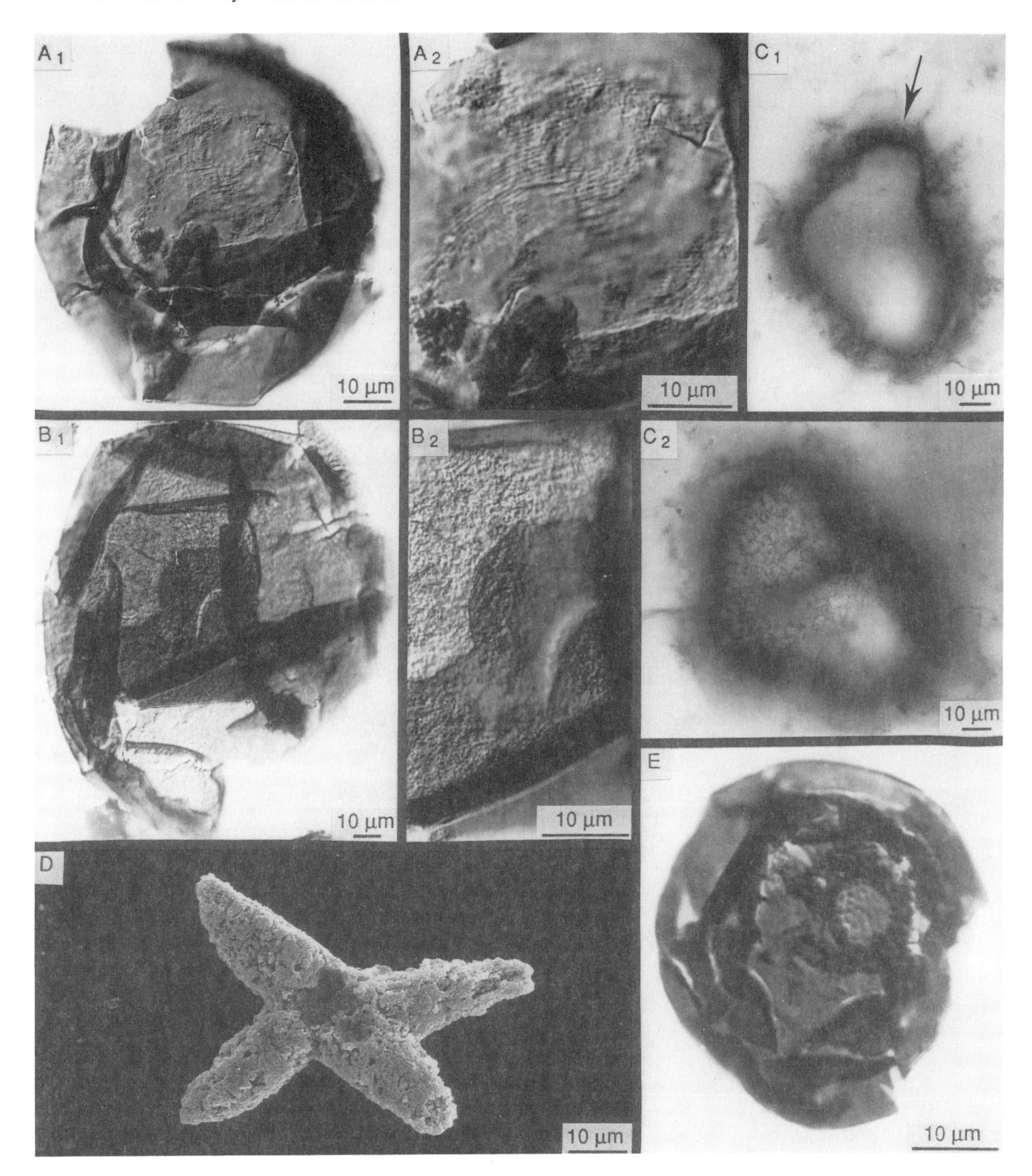

Figure 5.5.8. Late Proterozoic microfossils from the Awatubi Member, Kwagunt Formation, Chuar Group of northern Arizona, USA (A, B, E; Vidal and Ford 1985); the Hunnberg Formation, Roaldtoppen Group of Nordaustlandet, Svalbard (C; Knoll 1984); and the Båtsfjord Formation, Barents Sea Group of the Varanger Peninsula, East Finnmark, northern Norway (D; Vidal and Siedlecka 1983). [A, B, D, E, acid-resistant residues; C, petrographic thin section. A–C, E, light photomicrographs; D, scanning electron micrograph. A_1, A_2, B_1-B_2, and C_1-C_2 each show single specimens at different magnifications and/or focal depths.] (A_1, A_2) *Kildinosphaera lophostriata* (Jankauskas 1979) Vidal 1983; note distinctive surface sculpture of concentric striae. (B_1, B_2) *Kildinosphaera chagrinata* Vidal 1983; B_1 shows compressional folds; B_2 shows granular ornamentation on surface of vesicle. (C_1, C_2) *Cymatiosphaeroids kullingii* Knoll 1984; arrow in C_1 identifies thin, solid processes that arise from inner wall and support smooth outer wall or membrane; C_2 shows surface view. (D) *Podolina minuta* Hermann 1976 emend. Vidal 1983; distinctive star-shaped form (central part of specimen indicates where "process" may have been detached). (E) *Leiosphaeridia* sp. A of Vidal and Ford (1985); small hemispherical granular structure may be operculum covering excystment opening.

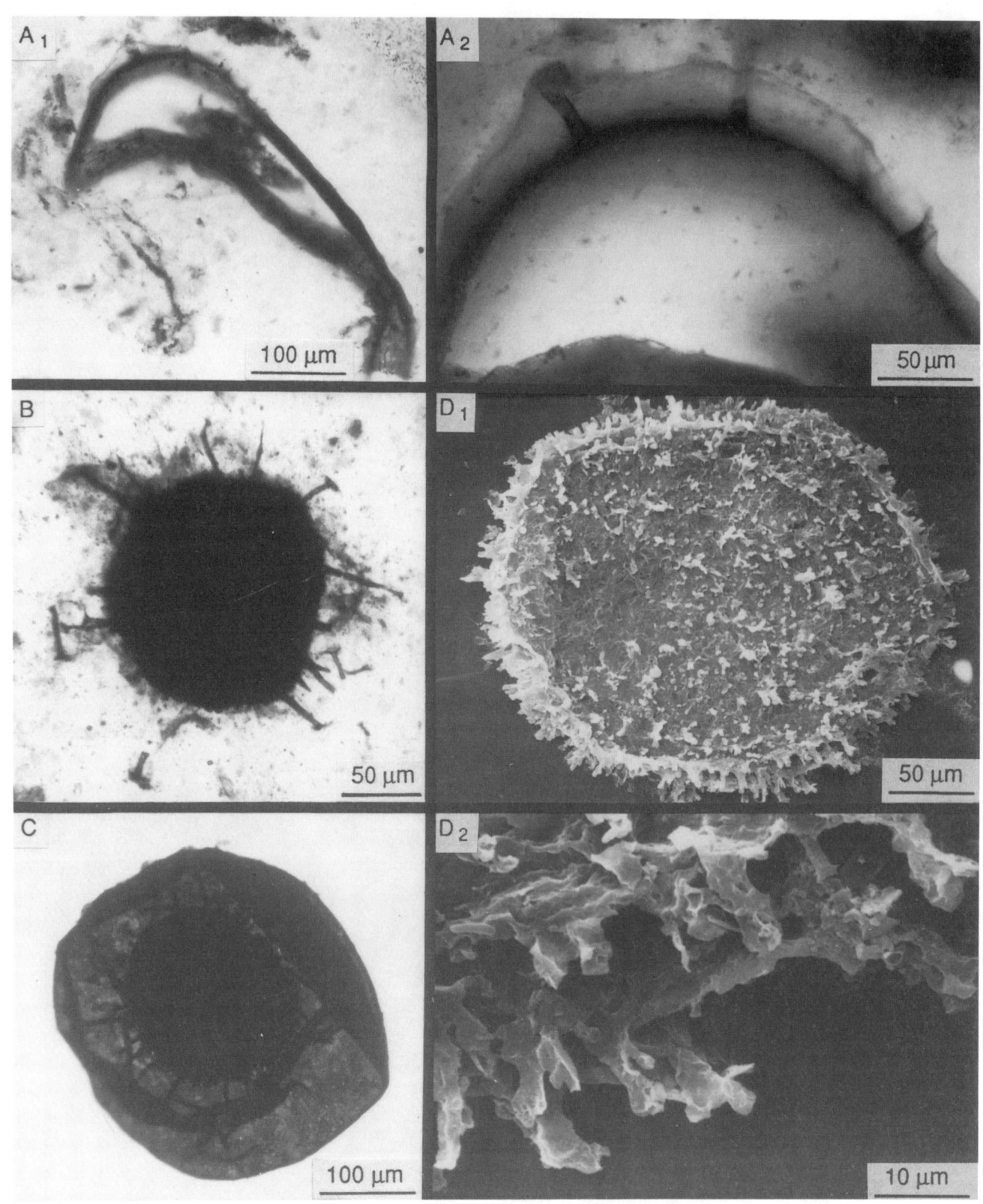

Figure 5.5.9. Late Proterozoic acritarchs from the Hunnberg Formation, Roaldtoppen Group of Nordaustlandet, Svalbard (A; Knoll 1984); the Svanbergfjellet Formation, Akademikerbreen Group of northeastern Spitsbergen, Svalbard (B, C; Butterfield et al. 1988); and the Pertataka Formation of central Australia (D). [A, B, petrographic thin sections; C, D, acid-resistant residues. A–C, light photomicrographs; D, scanning electron micrograph. A_1, A_2 and D_1, D_2 each show single specimens at different magnifications.] (A_1, A_2) *Trachyhystrichosphaera vidalii* Knoll 1984; note collapsed double-walled vesicle (A_1) and hollow, cylindrical processes arising from inner wall and supporting outer wall or membrane (A_2). (B) Large acanthomorph of Butterfield et al. (1988); note long (up to 60 μm) processes that flare distally. (C) Acanthomorph-containing unornamented vesicle of Butterfield et al. (1988); microfossil similar to that in "B," above, is preserved within a larger, unornamented vesicle. (D_1, D_2) Large acanthomorph (cf. Zang and Walter 1989); note high density of short, fragile processes (photograph courtesy of Zang Wenlong).

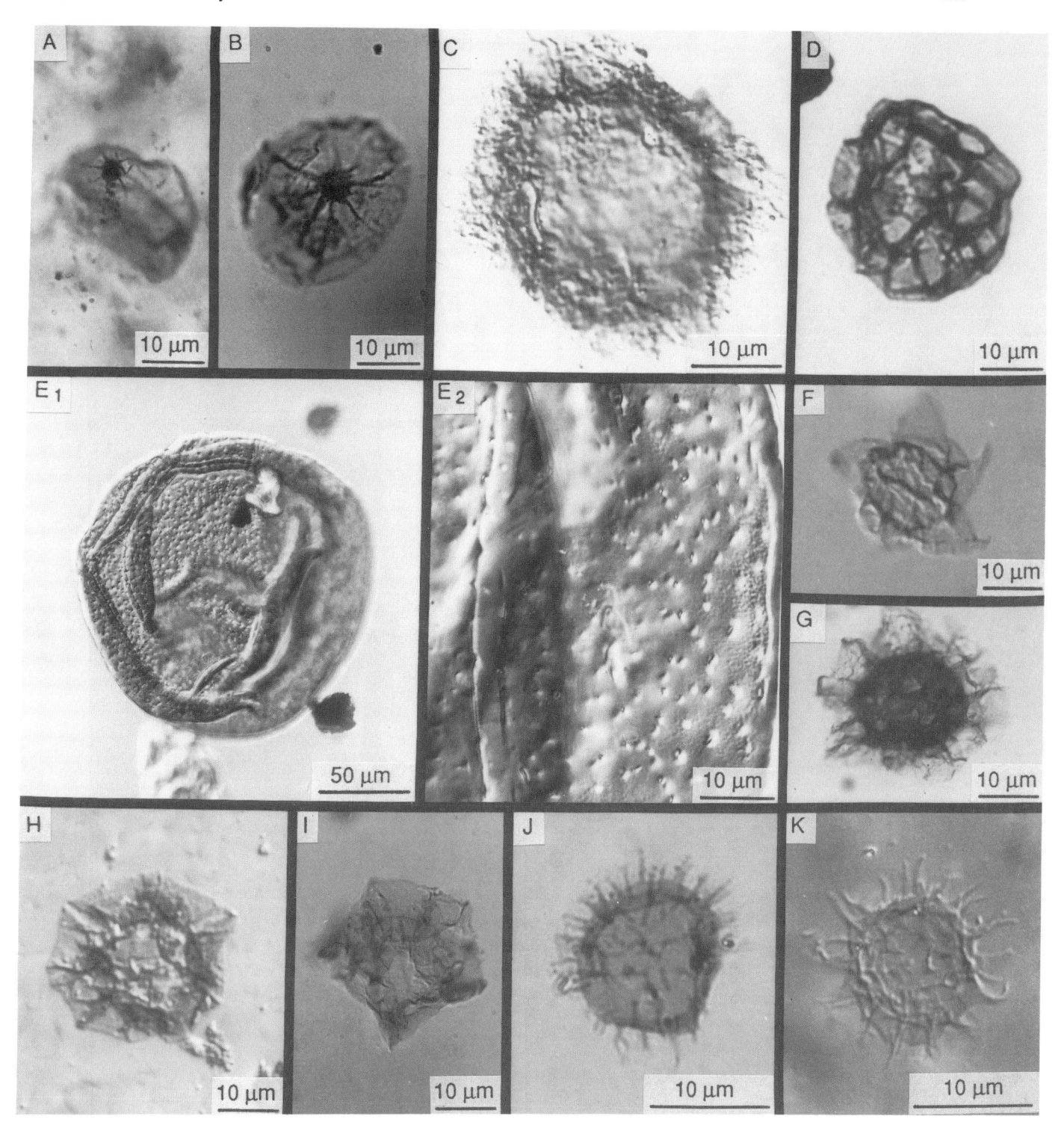

Figure 5.5.10. Light photomicrographs showing Early Cambrian acritarchs in acid-resistant residues of clastic sediments from the Tokammane Formation of northeastern Spitsbergen, Svalbard (A, F, G, I, K; Knoll and Swett 1987); the *Mickwitzia* Sandstone of Västergötland, Sweden (B, E; Moczydlowska and Vidal 1986); the Mazowsze Formation, Drill-Hole Parczew IG-10 of the Lublin Slope of east-central Poland (C; Moczydlowska and Vidal 1986); and the Fucoid Beds of northwestern Scotland, U.K. (D, H, J; Downie 1982). [E_1, E_2 show a single specimen at different magnifications.] (A, B) *Archaeodiscina umbonulata* Volkova 1968; enigmatic spheromorph with a distinctive, dark organic body displaying radiating "processes." (C) *Granomarginata squamacea* Volkova 1968; spongy corona surrounds the vesicle of this sphaeromorph. (D) *Retisphaeridium dichamerum* Staplin, Jansonius & Popcock 1965; perhaps a herkomorph or a sphaeromorph; the reticulate pattern of folds may be due to compression (see Downie 1982). (E_1, E_2) *Tasmanites volkovae* Kirjanov 1974; a thick-walled prasinophyte displaying well-preserved pores; the right portion of E_1 is shown at higher magnification in E_2. (F) *Estiastra minima* Volkova 1969; distinctive polygonomorph displaying five broad deltoid processes. (G) *Pterospermella* cf. *P. solida* (Volkova 1968) Volkova 1979; prasinophyte bearing a distinctive equatorial flange (i.e., a pteromorph acritarch). (H) *Cymatiosphaera postii* Jankauskas 1979; prasinophyte bearing crests that divide the vesicle surface into polygonal fields (i.e., a herkomorph acritarch). (I) *Cymatiosphaera* sp. of Knoll and Swett (1987). (J) *Micrhystridium ordense* Downie 1982; note rather blunt terminations of processes. (K) *Micrhystridium dissimilare* Volkova 1969; note thin conical processes.

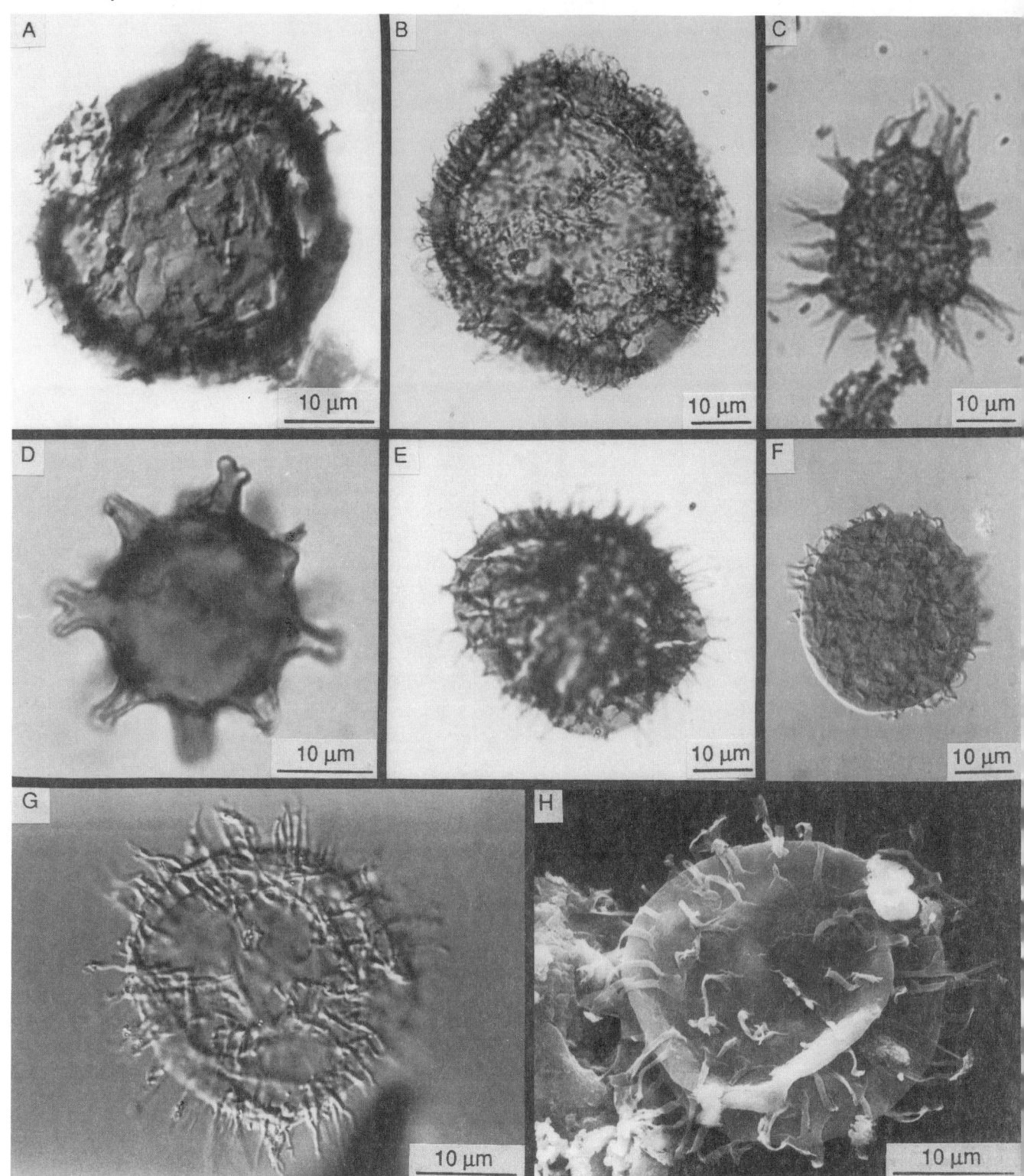

Figure 5.5.11. Late Proterozoic (A) and Early Cambrian (B–H) acritarchs in acid-resistant residues of clastic sediments from the Awatubi Member, Kwagunt Formation, Chuar Group of northern Arizona, USA (A; Vidal and Ford 1985); the Hardeberga Sandstone of Skåne, southern Sweden (B; Moczydlowska and Vidal 1986); the Fucoid Beds of northwestern Scotland (C; Downie 1982); the Lingulid Sandstone (D; Vidal 1981) and the *Mickwitzia* Sandstone (G; Moczydlowska and Vidal 1986) of Västergötland, Sweden; shales of the *Holmia kjerulfi* Zone, Lake Mjøsa region, southern Norway (E; Moczydlowska and Vidal 1986); and the Tokammane Formation of northeastern Spitsbergen, Svalbard (F, H; Knoll and Swett 1987). [A–G, light photomicrographs; H, scanning electron micrograph.] (A) *Vandalosphaeridium walcottii* Vidal & Ford 1985; note distinctive processes. (B) *Comasphaeridium brachyspinosum* (Kirjanov 1974) Moczydlowska and Vidal 1988; note dense covering of thin, hairlike processes. (C) *Goniosphaeridium implicatum* (Fridrichsone 1971) Downie 1982; vesicle has a somewhat polyonal outline. (D) *Multiplicisphaeridium dactilum* Vidal 1988; note sturdy, unornamented processes that bifurcate at tips. (E) *Skiagia compressa* (Volkova 1968) Downie 1982; tips of processes are funnel-shaped. (F, H) *Skiagia ciliosa* (Volkova 1969) Downie 1982; note flaring at tips of processes. (G) *Baltisphaeridium cerinum* Volkova 1968.

Table 5.5.1. *Reported occurrences (including questionable occurrences) of Proterozoic and Early Cambrian nonsphaeromorph (morphologically complex) acritarchs. Table should be studied in conjunction with Figure 5.5.13 and text because many Proterozoic occurrences are problematic. A few very important additional occurrences (for which details are unavailable) have been reported by Butterfield and Rainbird (1988) and Knoll and Butterfield (1989). [Total occurrences covered by Table = 524. This number differs from that in text (577) due to the exclusion of 11 highly doubtful occurrences older than 1 Ga, and of a few relatively minor subgroups.]*

Subgroup	Age (Ma)									
	999–950	949–900	899–850	849–800	799–750	749–700	699–650	649–600	599–550	549–530
acanthomorphs	6	1	6	—	7	2	4	15	19	321
herkomorphs	—	—	—	—	1	—	2	—	1	37
netromorphs	1	—	—	—	—	—	1	1	1	27
oömorphs	—	—	—	—	—	—	—	—	—	16
polygonomorphs	—	—	—	—	—	—	—	—	—	19
prismatomorphs	1	—	—	—	1	3	—	6	—	9
pteromorphs	—	4	—	—	2	—	—	—	1	9

Data based on Chapter 22 and the following additional references: Akul'cheva et al. (1981), Brasier (1977), Butterfield et al. (1988), Downie (1982), Fajzulina et al. (1973, 1982, 1984), Fajzulina and Treshchetenkova (1979), Fridrikhsone (1971), German et al. (1989), Il'yasova and Lysova (1959), Kir'yanov (1974), Knoll and Ohta (1988), Korolev and Ogurtsova (1982), Mikhajlova and Podkovyrov (1987), Moczydlowska (1980, 1988a), Moczydlowska and Vidal (1986, 1988), Muir et al. (1979), Naumova (1968), Ogurtsova (1975, 1985), Pashkyavichene (1980), Pyatiletov (1976, 1978, 1980a, 1980b), Pyatiletov and Karlova (1980), Pyatiletov and Rudavskaya (1985), Rudavskaya (1965, 1973), Rudavskaya and Timofeev (1963), Smith (1977), Song et al. (1984), Stanevich (1986), Timofeev (1957, 1959, 1973), Treshchetenkova et al. (1982), Tynni (1978), Vanguestaine (1974), Volkova (1968, 1969, 1981), Voronin et al. (1982), Wang and Chen (1987), Wang et al. (1983, 1984), Ważyńska (1967), Wood and Clendening (1982), Xing (1982), Yankauskas (1975, 1976), Yankauskas and Posti (1976), Yin (1985, 1987), and Zang and Walter (1989).

1000-Ma acanthomorphs (all assigned to *Baltisphaeridium* sp.), one from the Mbuji Mayi (=Bushimay) Supergroup (Maithy 1975, pl. 4, fig. 37) and the other two from the Semri Group of the Vindhyan Supergroup (Nautiyal 1983)], do not permit confident interpretation.

The earliest well-documented acanthomorphs occur in the Lakhanda Formation of Yakutia, in eastern Siberia (Timofeev and German 1976; German 1981; German et al. 1989). This formation is poorly dated at about 950 Ma. Sporadic occurrences of well-documented acanthomorphs persist through the Late Proterozoic, becoming abundant in the Early Cambrian (Fig. 5.5.13). These acanthomorphs are discussed in detail below.

Herkomorphs, composing a distinctive subgroup of acritarchs, are extremely rare in the Proterozoic. An ornamented specimen considered to be similar to the genus *Acrum* Fombella 1977, from the Svanbergfjellet Formation (ca. 750 Ma) of northeastern Spitsbergen (Butterfield et al. 1988), is probably the earliest well-documented herkomorph. A few similar forms occur in older deposits, and are typically assigned to the genus *Satka* Jankauskas 1979 (see Yankauskas 1982; German et al. 1989; Chapter 24: Pl. 5, A; Pl. 42, B, C). Also noteworthy are the "polygonally segmented and bumpy surfaced sphaeromorphs" detected by Horodyski (1980, pl. 2, Fig. 8-15, 21) in the 1450 Ma-old Chamberlain Shale of the Belt Supergroup of Montana, USA. If these specimens have not been significantly altered by diagenesis, they may represent very early occurring herkomorphs.

Most reports of Proterozoic *netromorphs* are not convincing (but see German et al. 1989, for a different view). As Peat et al. (1978) noted, thin-walled sphaeromorphs may become folded or rolled up, in some cases mimicking typical netromorph morphology. Some good candidates for early netromorphs have been illustrated by Pyatiletov (1980a), Pyatiletov and Karlova (1980), and Fajzulina et al. (1982). Well-documented netromorphs are known from strata just below the Vendian-Cambrian boundary (e.g., Pashkyavichene 1980).

Although *oömorphs* are fairly common and diverse in the Lower Cambrian (Timofeev 1957; Yankauskas 1975; Downie 1982), they are evidently unknown in the Proterozoic.

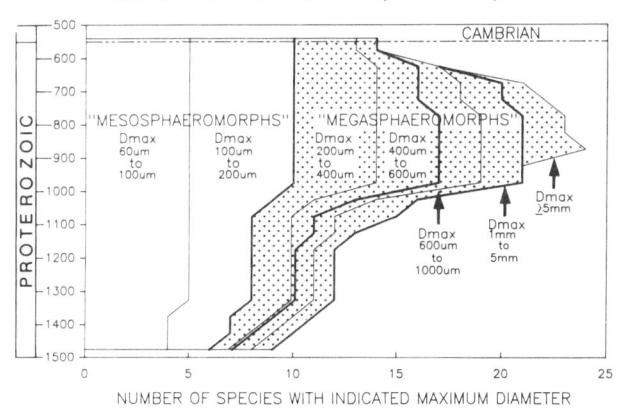

Figure 5.5.12. Temporal distributions (first and last occurrences; Fig. 25.4), from 1500 Ma to the Early Cambrian, of 24 "mesosphaeromorph" and "megasphaeromorph" species (Table 25.3) having the indicated maximum diameters.

Well-documented *polygonomorphs* seem also to be confined to Early Cambrian and younger strata (e.g., Downie 1982).

Prismatomorphs, enigmatic polyhedral acritarchs, are earliest known from the Lakhanda Formation (ca. 950 Ma; Timofeev and German 1976; see Chapter 24: Pl. 56, C). Other good Proterozoic examples of this subgroup occur in the Visingsö Group (ca. 775 Ma; Vidal 1976a), the Båtsfjord Formation of the Barents Sea Group (ca. 730 Ma; Vidal and Siedlecka 1983), the Olkha Formation of Siberia (ca. 700 Ma; Akul'cheva et al. 1981), and the Derlo Formation of Podilia (ca. 600 Ma; Timofeev 1973). A few occurrences are known from the Early Cambrian (Table 5.5.1).

Pteromorphs are among the more difficult subgroups to evaluate because poorly preserved sphaeromorphs may crudely mimic their morphology. Figure 5.5.13 shows the distribution of a number of questionable occurrences illustrated by Timofeev (1969), Yankauskas (1982), and Yin (1987); one of Yankauskas' specimens, assigned to *Pterospermella simica* (= *Pterospermopsis simicus*), is illustrated here (Chapter 24: Pl. 18, E, F). Additional pteromorphs, probably of Late Proterozoic age, have been illustrated by German et al. (1989). Very convincing pteromorphs are found in strata of Early Cambrian age (e.g., Knoll and Swett 1987; Volkova et al. 1979; see Fig. 5.5.10, G).

5.5.3 Milestones in Morphological Evolution in Acritarchs

Too few data are available to permit a definitive statement regarding the times of origin of various important morphological features in acritarchs. The following discussion, therefore, should be regarded as a status report, a summary of current knowledge that will undoubtedly change, perhaps markedly, over the next few years. Because of the vagaries of preservation (Section 5.3), only the most obvious morphologic features may be discussed with some degree of confidence. These features include processes, excystment structures, distinctive vesicle ornamentation, and relatively large size (Fig. 5.5.14).

5.5.3A Processes

Simple cylindrical processes characterize the earliest acanthomorphs from the Lakhanda Formation (e.g., Chapter 24: Pl. 14, E). Such processes dominate acanthomorph taxa of the Proterozoic and Early Cambrian. The first occurring significant complex processes typify *Vandalosphaeridium walcottii* Vidal and Ford 1985 (Fig. 5.5.11, A), from the Kwagunt Formation of the Chuar Group, northern Arizona (ca. 850 Ma); these thick, sturdy processes flare distally, and apparently support a smooth membrane (Vidal and Ford 1985). In *V. reticulatum* (Vidal 1976) Vidal 1981, from the Visingsö Group of southern Sweden (ca. 775 Ma), the distal portion of the processes branch and form a net, or reticulum (Vidal 1976a, fig. 14). See Figure 5.5.9 for additional examples of complex processes in Proterozoic acritarchs. Typical process-bearing Early Cambrian forms, such as *Micrhystridium* and *Skiagia*, are illustrated in Figures 5.5.10 and 5.5.11.

Figures 5.5.9, 5.5.10, and 5.5.11 show that process length varies significantly. Might length be correlative with vesicle diameter? The data for Proterozoic and Early Cambrian acritarchs argue against this notion: scatter plots of process length (both minimum and maximum) vs. vesicle diameter (both minimum and maximum) reveal no discernible correlation.

5.5.3B Excystment Structures

Because the majority of acritarchs are considered to represent the resistant cyst stage of eukaryotic phytoplankters, these microfossils should show evidence of the excystment process. Phanerozoic acritarchs exhibit a wide variety of excystment structures that commonly typify specific taxa (Tappan 1980). Only limited data are available, however, regarding the excystment structures of Proterozoic acritarchs.

Partial rupture of the vesicle is the simplest, perhaps the most primitive, and probably the most common type of excystment structure exhibited by Proterozoic acritarchs. The vesicle simply ruptures, leaving a gap that may vary in its degree of development. Accidental (e.g., preservational) breaks, however, may be indistinguishable from the results of such simple excystment.

Some acritarchs break into two equal parts, as a result of a median split of the vesicle. Because partial ruptures, if equatorially located, may mimic median splits, documenting early examples of this mode of excystment is highly problematic. Yankauskas (1982) illustrated specimens of *Leiosphaeridia bicrura* from the Zigazino-Komarovsk Formation (ca. 1350 Ma) of the southern Ural Mountains, which he interpreted as exhibiting a median split. Peat et al. (1978, Fig. 5m) reported similar examples from the McMinn Formation of the Roper Group (ca. 1340 Ma) of north-central Australia. Although these examples bear a resemblance to the classic median split common in Paleozoic acritarchs (Loeblich and Tappan 1969), their morphology could be a result simply of the diagenetic extension of a simple rupture. The "median split" reported to occur in *Kildinosphaera chagrinata* (see Vidal and Siedlecka

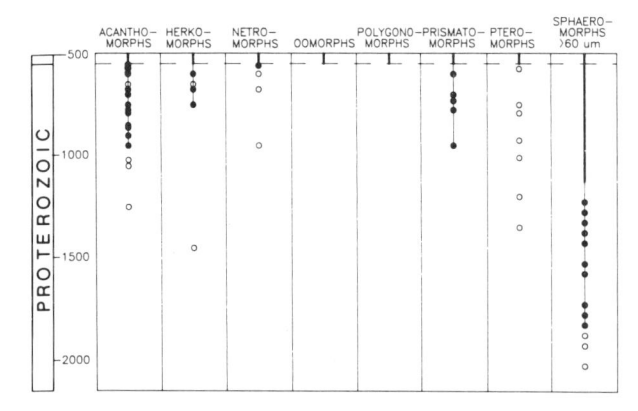

Figure 5.5.13. Temporal distribution of major morphological subgroups of Proterozoic and Early Cambrian acritarchs as determined from the illustrated literature. Degree of confidence in assignment: questionable (open symbols), well documented (solid symbols). Based on Chapter 22 and additional references listed in Table 5.5.1.

1983, fig. 4A–C), from the Båtsfjord Formation of the Barents Sea Group of East Finnmark (ca. 730 Ma), may have a similar explanation.

Lowermost Cambrian strata also contain acritarchs with putative median splits. Examples include *Leiosphaeridia* spp. from the Tokammane Formation of Svalbard (Knoll and Swett 1987) and *Leiosphaeridia dehisca* from the Tommotian of the East European Platform (Volkova et al. 1979). However, when compared with the classic Paleozoic forms, these examples, like those discussed above, are also not completely convincing. Probably the most likely candidate for an Early Cambrian acritarch bearing a true median split is a specimen of *Baltisphaeridium ciliosum* Volkova 1969 (Volkova et al. 1979, pl. 2, Fig. 1). Clearly, any history of the median split as a mode of excystment must be based on a complete analysis of actual specimens, preferably of many examples within large populations.

Some acritarchs excysted via a circular to subpolygonal opening called the pylome (or cyclopyle), which commonly is plugged with an operculum that may be highly ornamented. Such pylomes resemble somewhat the archeopyle, the polygonal excystment opening occurring in many living and fossil dinoflagellates (see Tappan 1980; Evitt 1985).

An early, questionable, pylomate acritarch has been reported from the McMinn Formation of (ca. 1340 Ma; Peat et al. 1978, Fig. 3b) of northern Australia; the ragged outline of the "opening" in this specimen suggests that diagenetic factors may have been responsible for its formation. Yankauskas (1980) illustrated well-developed, nearly circular openings in *Leiosphaeridia kulgunica*, from the Podinzer Formation (ca. 925 Ma) of the southern Ural Mountains (Chapter 24: Pl. 19, G). In the absence of a definitive operculum, however, it is difficult to evaluate the putative pylomes in these specimens; nevertheless, they are among the most promising early candidates. *L. kulgunica* has also been repoeted from the older Zil'merdak Formation (ca. 1000 Ma) of the southern Urals, but this occurrence was not illustrated (Yankauskas 1980b).

The most impressive early pylomate acritarchs are those described by Vidal and Ford (1985) from the Kwagunt Formation of the Chuar Group (ca. 860 Ma), northern Arizona. Distinct circular openings are evident in *Trachysphaeridium laminaritum* Timofeev 1969, *T. laufeldii* Vidal 1976, and *Leiosphaeridia* sp. A (Fig. 5.5.8, E). Specimens of the latter two taxa retain distinctive echinate opercula, which convincingly document that their pylomes were original, biologic, features. Specimens of *T. laufeldii* exhibiting echinate opercula have also been recovered from the slightly younger Visingsö Group (ca. 775 Ma) of southern Sweden (Vidal 1976a). Vidal and Ford (1985) also reported (but did not illustrate) similar complex excystment structures from even older strata of the Chuar Group (Galeros Formation, ca. 930 Ma) and Uinta Mountain Group (ca. 950 Ma). Distinctive polygonal openings characterize many of the enigmatic melanocyrillids (Section 5.6.2B; Chapter 24: Pl. 29), which are known from sediments as old as 950 Ma (Hofmann 1987).

5.5.3C Distinctive Vesicle Ornamentation

Palynologists use a variety of terms to describe surface sculpture in acritarchs (Tappan and Loeblich 1971); examples include laevigate (featureless), granulate (bearing tiny bumps, or grana), echinate (bearing short, spinelike grana), and costate (bearing parallel ridges). Because preservational artifacts commonly mimic a variety of these ornaments, the following discussion is limited to highly distinctive (and perhaps significant) types of surface ornament.

The well-preserved organic-walled microfossils from the Kwagunt Formation of the Chuar Group (Vidal and Ford 1985) include excellent examples of primary surface features. For example, *Kildinosphaera chagrinata* Vidal 1983 displays a finely granulate, shagreen-like surface (Fig. 5.5.8, $B_1 - B_2$). Fine, concentric striae characterize the surface of *K. lophostriata* (Jankauskas 1979) Vidal 1983 (Fig. 5.5.8, A_1-A_2). Careful examination of such surfaces by Vidal, Yankauskas, and others has led to recognition of distinct types of sphaeromorphs that are otherwise nondescript.

Relatively simple sphaeromorphs may also display a porate wall structure; some examples from the Proterozoic and Early Cambrian resemble the common Paleozoic prasinophyte genus *Tasmanites* Newton 1875 (see Wall 1962). Recognition of a primary porate structure is rendered difficult by various diagenetic processes which may result in the development of holes in the vesicle wall (e.g., pyrite growth, impressions of detrital grains; see Section 5.3.3B). The specimens of *Trematosphaeridium holtedahlii* Timofeev 1966, illustrated by Timofeev (1969, Pl. 5, Fig. 1; pl. 10, Fig. 6, 14; pl. 19, Fig. 11; pl. 30, Fig. 9, 10) and Yankauskas (1982, pl. 43, Fig. 4), have probably suffered diagenetic alteration: the porous nature of these vesicles seems unlikely to be primary. *Tasmanites vindhyanensis* Maithy and Shukla 1977 may have similarly been altered diagenetically, although its apparent porous nature seems more regular than that in Timofeev's specimens (Maithy and Shukla 1977, pl. 4, Fig. 33). *Tasmanites rifejicus* Jankauskas 1978, from the Podinzer Formation (ca. 925 Ma) of the southern Ural Mountains, was originally described as having a primary porous wall. The taxon was later rejected because the porosity of the wall was reinterpreted as a secondary feature (Yankauskas et al. 1987; German et al. 1989; see Chapter 24: Pl. 19, A_1-A_2).

Specimens assigned to *T. rifejicus* by Vidal and Ford (1985, Fig. 7E) may be the earliest examples of acritarchs bearing a primary, porous, tasmanitid-like wall. They report (but do not figure) this taxon from the 950-Ma-old Red Pine Shale of the Uinta Mountain Group; and they illustrate a specimen from the Chuar Group [from either the Galeros Formation (ca. 930 Ma), according to the description; or the Kwagunt Formation (ca. 860 Ma), according to the figure caption]. The Chuar specimen bears pores that are fairly constant in dimensions and may have raised rims. Vidal and Siedlecka (1983) figured a similar specimen from the Klubbnes Formation, Vadsø Group (ca. 810 Ma), of East Finnmark. Finally, Knoll and Swett (1985) figured this taxon from the Glasgowbreen Formation, Veteranen Group (ca. 850 Ma) of Spitsbergen. Because diagenetic factors have not been excluded as possibly responsible for the porate structure of the walls of these vesicles, these Proterozoic specimens are here regarded as being questionably tasmanitid-like (Fig. 5.5.14). The occurrence of tasmanitid walls

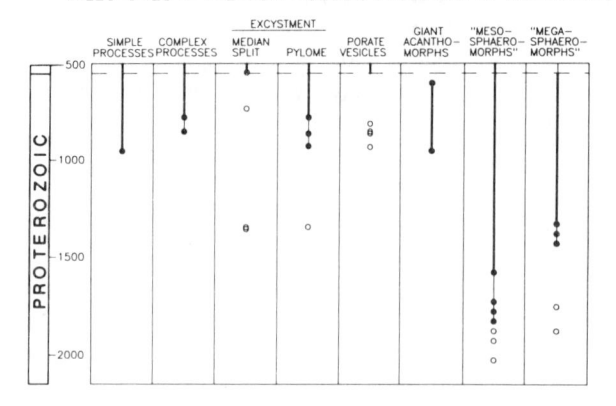

Figure 5.5.14. Milestones in the morphologic evolution of Proterozoic and Early Cambrian acritarchs as determined from the illustrated literature. Degree of confidence in assignment: questionable (open symbols), well documented (solid symbols). Based on Chapter 22 and additional references listed in Table 5.5.1.

appears to be fairly well established by the Early Cambrian. Specimens reminiscent of later Paleozoic, assured tasmanitids are illustrated in a number of papers on Early Cambrian acritarchs (e.g., Vidal 1979b; Volkova et al. 1979; Moczydłowska and Vidal 1986; Knoll and Swett 1987; see Fig. 5.5.10, E).

5.5.3D Large Size

Relatively few morphologic trends can be discerned in the evolutionary development of acritarchs and prasinophytes (Fig. 5.5.14). An exception to this generalization is the remarkable disparity in vesicle size between many Proterozoic and Early Cambrian acritarchs. Trends in vesicle size are particularly striking among the larger sphaeromorphs and the acanthomorphs.

With regard to *sphaeromorph acritarchs*, as is discussed above (Section 5.5.2A), the size ranges of reported occurrences of Proterozoic "mesosphaeromorph" and "megasphaeromorph" acritarchs both exhibit distinctive time-correlative patterns of distribution. Prior to 1100 Ma, reported occurrences of both groups are dominated by individuals having relatively small vesicles; between 1100 and 900 Ma, occurrences for both groups exhibit progressively increasing size distributions; and between 900 and 800 Ma, occurrences for both groups include the largest individuals known from the Proterozoic fossil record (Figs. 5.5.2, 5.5.6). Correlative with this Middle to Late Proterozoic pattern of increasing size distribution, both groups exhibit major increases in total number of species (Fig. 25.4) and in the average number of species occurring per sampled plankton-containing assemblage (Fig. 11.3.9).

In contrast with the notably similar early development of these two groups, their Late Proterozoic histories markedly differ. Subsequent to attainment of maximum diversity near the beginning of the Late Proterozoic, the patterns of size distribution of "mesosphaeromorphs" remain relatively constant throughout the remainder of the era (Fig. 5.5.2) whereas size ranges of reported "megasphaeromorphs" become progressively and markedly diminished (Fig. 5.5.6). Evidently, during the Late Proterozoic, "megasphaeromorph" acritarchs underwent

major extinction, an episode that appears to have occurred not as a single event but, rather, in a continuum of phases – a long-term, progressive "wave" of sequential size-related extinction. As summarized graphically in Figure 5.5.12 (and tabulated in Table 5.5.2), the largest "megasphaeromorphs" (viz., taxa with maximum diameters $>5,000\,\mu m$) are evidently both the latest to have originated (known first from the fossil record at about 975 Ma) and among the earliest to have become extinct (known last at about 675 Ma). Intermediate-size "megasphaeromorphs" (those 600 to 5,000 μm in maximum diameter) originated earlier and became extinct later (at about 675 to 625 Ma), but did not survive into the Phanerozoic. Similarly, two-thirds of the Proterozoic informal taxa 400 to 600 μm in maximum diameter here recognized (Table 25.3), and one-quarter of such "megasphaeromorphs" 200 to 400 μm in maximum size, are evidently unknown in Early Cambrian deposits. As is discussed below, although comparable data for Proterozoic acanthomorph (spiny) acritarchs are far less numerous than those available for large sphaeromorphs, it is notable that the size distribution of acanthomorphs also markedly decreases over the Late Proterozoic and, like sphaeromorph taxa, only small-diameter species of acanthomorphs are known to have survived into the earliest Phanerozoic (Fig. 5.5.15).

Perhaps the most intriguing features of Proterozoic *acanthomorph acritarchs* are that many of them are big – some are giant even by Paleozoic standards – and many display rather complex morphologies. For example, some specimens are double-walled; short, hollow, cylindrical processes arise from the inner wall and seemingly support the thin outer wall or membrane (Fig. 5.5.8, C; Fig. 5.5.9, A). The processes borne on some of these acritarchs may be quite distinctive, ranging from short and flimsy (Fig. 5.5.9, D) to types that are sturdily built and flare distally (Fig. 5.5.9, B). And most impressively, a few of these acanthomorphs have been preserved within what appear

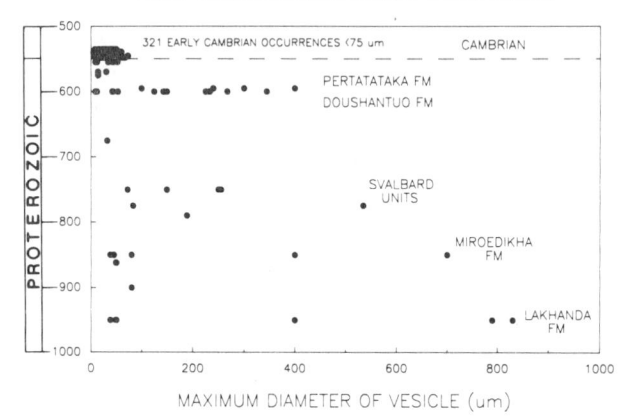

Figure 5.5.15. Temporal distribution of all Proterozoic and Early Cambrian acanthomorph acritarch occurrences ($n = 373$) as a function of maximum vesicle diameter reported (or measured from photomicrograph). [Total does not agree with Table 5.5.1 ($n = 381$) because size data not available for all occurrences.] Important formations yielding giant acanthomorphs (see text) appear at their approximate ages. Compilation based on Chapter 22 and additional references cited in Table 5.5.1.

Table 5.5.2. *Sequential size-related extinction of large sphaeromorph acritarchs.*

Size Category (Maximum Diameter)	Number of Taxa	First Noted Occurrence	Period of Maximum Diversity	Last Noted Occurrence	Percent Not Surviving Into Cambrian
≥5000 μm	3	~975 Ma	~875 Ma	~675 Ma	100%
1000–5000 μm	3	>1500 Ma	~1075 Ma	~675 Ma	100%
600–1000 μm	2	>1500 Ma	~1025 Ma to ~675 Ma	~625 Ma	100%
400–600 μm	3	~1125 Ma	~1025 Ma to ~775 Ma	<550 Ma	67%
200–400 μm	4	>1500 Ma	~975 Ma to ~625 Ma	<550 Ma	25%
100–200 μm	5	>1500 Ma	~925 Ma to <550 Ma	<550 Ma	0%
60–100 μm	5	>1500 Ma	~1325 Ma to <550 Ma	<550 Ma	0%

to be large, simple sphaeromorphs. Butterfield et al. (1988) illustrated such a microfossil from the Svanbergfjellet Formation (ca. 750 Ma; see Fig. 5.5.10, C) of Svalbard, and German et al. (1989, pl. 2, Fig. 1) documented a similar form in the Miroedikha Formation (ca. 850 Ma) of the Turukhansk region of central Siberia. Such preservation strengthens the argument that certain acritarchs formed as endocysts, similar to many fossil and living dinoflagellates (Evitt 1985).

Only 64 total possible occurrences of Proterozoic acanthomorphs are now known (Table 22.3); of these, 11 are here regarded as questionable. Significantly, of the 53 remaining, acceptable, occurrences, at least 15 are giant, "mega-acanthomorphs," having a maximum vesicle diameter exceeding 200 μm (Fig. 5.5.15); and 21 of these occurrences are of vesicles that exceed 100 μm in maximum diameter. It is likely that known occurrences will markedly increase in the next few years; among recently discovered, diverse, Proterozoic microbiotas are several that contain large "mega-acanthomorphs" [Butterfield and Rainbird 1988; Knoll and Butterfield 1989 (see especially their analysis of the thesis of Zang Wenlong); Zang and Walter 1989; A. H. Knoll, pers. commun. to C.V.M. 1988].

Large acanthomorphs (maximum vesicle diameter exceeding 100 μm) have been detected in one possibly Middle Proterozoic unit and several Upper Proterozoic units (see Knoll and Butterfield 1989). These include the Lakhanda Formation of eastern Siberia (ca. 950 Ma; Timofeev and German 1976; German 1981; German et al. 1989; Chapter 24: Pl. 14, D, E); the Miroedikha Formation of the Turukhansk region, north-central Siberia (ca. 850 Ma; German et al. 1989); the Wynniatt Formation of Victoria Island, arctic Canada (700–1200 Ma; Butterfield and Rainbird 1988; Knoll and Butterfield 1989); strata in the Kabakovo-62 drillhole of the southern Ural Mountains (ca. 790 Ma; Yankauskas 1982); the Hunnberg (ca. 775 Ma; Knoll 1984; see Fig. 5.5.9, A), Ryssö (ca. 750 Ma; Knoll and Calder 1983), and Svanbergfjellet Formations (ca. 750 Ma; Butterfield et al. 1988; see Fig. 5.5.9, B, C) of Svalbard; the Yudoma Formation of eastern Siberia (ca. 600 Ma; Pyatiletov 1980b); the Doushantuo Formation of Hubei, China (ca. 600 Ma; Awramik et al. 1985; Yin 1985, 1987); metasediments from Prins Karls Forland, western Svalbard (ca. 600 Ma; Knoll and Ohta 1988); and the Pertatataka Formation of north-central Australia (ca. 600 Ma; Zang and Walter 1989; see Fig. 5.5.9, D).

As Knoll and Butterfield (1989) point out, the Late Proterozoic seems to be a time of large acanthomorphs. Such large forms evidently became extinct during the Vendian (Fig. 5.5.15), but not necessarily correlative with, or as a consequence of, the Late Proterozoic Varanger glacial episode: the large acanthomorphs of the Pertatataka Formation post-date deposition of the uppermost Proterozoic tillites of the Amadeus Basin (Zang and Walter 1989). Large acanthomorphs do not seem to be unusually constrained environmentally (Knoll and Butterfield 1989), although solid knowledge of the environmental contexts of these microfossils must await more detailed sedimentological studies of many of their associated deposits.

On the basis of available data, the large acanthomorphs that survived through the Varanger glaciation soon disappeared (Zang and Walter 1989): such large forms are *entirely absent* from the relatively well-sampled Lower Cambrian deposits of the world (Fig. 5.5.15). Indeed, Early Cambrian acanthomorphs are remarkably small (e.g., Fig. 5.5.10, J, K; Fig. 5.5.11, B, D–H). An analysis of the available literature (references listed in Table 5.5.1) yields 321 occurrences of acanthomorphs in the Lower Cambrian. All of these forms have maximum vesicle diameters ≤73 μm; only 20 (6.2%) have a maximum diameter between 50 and 73 μm. Indeed, about 50% of the occurrences of Lower Cambrian acanthomorphs are ≤25 μm in maximum vesicle diameter. Later Paleozoic assemblages, for example those of the Ordovician, typically include much larger acanthomorphs (e.g., Tappan 1980). It seems clear (even considering the low density of acanthomorph data for the Late Proterozoic) that, like the Proterozoic sphaeromorphs discussed above, many Late Proterozoic acanthomorphs were remarkably large and that, in comparison, Early Cambrian acanthomorphs (and sphaeromorphs) were notably small.

5.5.4 Closing Remarks

In their discussion of the affinities of the Pertatataka microbiota, Zang and Walter (1989) stress the continuationist tradition in taxonomy (see Gould 1983): they compare some of their giant acanthomorphs to members of much smaller Early Cambrian genera, such as *Skiagia* and *Gorgonisphaeridium*. As suggested by Knoll and Butterfield (1989), however, it seems more appropriate to place emphasis on comparison of such forms with other Late Proterozoic acanthomorphs. The large acanthomorphs, "megasphaeromorphs" (some of which may initially have enclosed acanthomorphs), and other morphologically complex Late Proterozoic acritarchs document an early evolutionary phase of experimentation for eukaryotic phytoplankters; the composition of this assemblage markedly changed as various lines of acritarchs became extinct during the Late Proterozoic, some lines survived into the Cambrian (e.g., Fig. 5.5.12), and still others seeded an early Phanerozoic renewal. Understanding of the evolutionary dynamics of that renewal must await careful analysis of Middle and Upper Cambrian acritarch assemblages.

5.6

Distinctive Problematical Proterozoic Microfossils

TOBY B. MOORE   ROBERT J. HORODYSKI   JERE H. LIPPS
J. WILLIAM SCHOPF

5.6.1 Introduction

In this section, three groups of particularly distinctive problematical Proterozoic microfossils are considered, grouped together because of their unusual, but easily identifiable morphologies and their uncertain biological affinities. These three groups include:

(1) morphologically distinctive ("bizarre") microbial(?) problematica (code "BZ" in Tables 25.1, 25.3, and Fig. 25.4; e.g., Chapter 24: Pl. 1, K, L; Pl. 2, A–D, O–R; Pl. 4, C; Pl. 42, E, I);

(2) vase-shaped ("melanocyrillid") eukaryotic problematica (codes "A-20" through "A-23"; e.g., Chapter 24: Pl. 29, A–H); and

(3) three types of dubiofossils that are somewhat similar in morphology to skeletonized protists.

5.6.2 Distinctive Proterozoic Micro-problematica

5.6.2A Proterozoic microbial(?) problematica

Listed in Table 5.6.1 are sixteen taxa and unnamed forms of Proterozoic micro-problematica having "bizarre," atypical, morphology. Included among these are morphotypes such as *Germinosphaera* (Chapter 24: Pl. 42, E, I), possibly remnants of reproductive cells with attached vegetative filaments, or perhaps distorted sphaeromorph acritarchs; *Metallogenium*, a radiating colonial prokaryote similar to *Eoastrion*; *Redkinia* (Chapter 24: Pl. 52, B–D), probably the remnant of an invertebrate (arthropod?) carapace (see Chapters 7 and 23); and an "unnamed organic tube with a globose end" (Knoll 1982), possibly a distorted cyanobacterial sheath. More than one-half of such reported Proterozoic bizarre morphotypes occur in the Early Proterozoic Gunflint Iron Formation of southern Ontario, Canada.

The well-known microbiota of the approximately 2.0 Ga-old Gunflint Iron Formation contains seven bizarre genera. These include *Archaeorestis* (Barghoorn and Tyler 1965), a branched tubular filament, possibly a cyanobacterial sheath; *Eoastrion* (Chapter 24: Pl. 2, C–D; Pl. 4, C), radiating, filamentous, presumably prokaryotic colonies; *Eomicrhystridium*, probably a distorted prokaryotic unicell cf. *Huroniospora*; and *Eosphaera* (Chapter 24: Pl. 1, K–L), a distinctive, spheroidal, evidently planktonic microfossil that according to Kazmierczak (1976), who compared *Eosphaera* to *Eovolvox*, and Tappan (1976), who compared *Eosphaera* to the extant red alga *Porphyridium*, may be of eukaryotic affinity. Other bizarre genera from the Gunflint include *Exochobrachium* (of which only two specimens have been described), a morphotype characterized by an unusual tri-radiate morphology (Awramik and Barghoorn 1977, fig. 5g) that may be a degradational artifact; *Kakabekia* (Chapter 24: Pl. 2, O–R), an unusual form consisting of an umbrella-like "crown" connected via a filamentous stipe to a spheroidal bulb; *Veryhachium*? (Hofmann 1971a) described from a single excellently preserved specimen that most likely is unrelated to Phanerozoic *Veryhachium*; and *Xenothrix*, a morphotype similar to *Archaeorestis* and probably the sheath or envelope of a colonial prokaryote. Some of these morphotypes, for example *Eoastrion* and *Kakabekia*, also occur in other units of similar age, such as the Duck Creek Dolomite, and the Barney Creek and Frere Formations of Australia (Table 5.6.1).

Because the biological affinities of these problematic microfossils are uncertain or unknown, their evolutionary relationships can only be guessed. As suggested above, many appear to represent "normal" taxa that have been unusually preserved, and therefore are degradational variants described solely because of their bizarre, atypical, morphologies. Some, however, may represent extinct forms that lack modern analogues. Common to all these bizarre morphotypes is their rarity in the known fossil record. As indicated in Table 5.6.1., several of these forms have been reported to occur in approximately coeval units on more than one continent; thus, they may have biostratigraphic potential (Schopf 1977), although this potential seems clearly limited by their relative scarcity.

The relatively common occurrence of these bizarre micro-problematica in geologic units about 2.0 Ga in age (Table 5.6.1, Fig. 25.4) may be of some significance. As discussed in Chapter 4, free atmospheric oxygen evidently became relatively abundant at about this time. Thus, as suggested by Awramik and Barghoorn (1977), certain of these bizarre morphotypes may represent Early Proterozoic "experiments" in evolution,

234 Prokaryotes and Protists

Table 5.6.1. *Proterozoic and Early Cambrian problematic microfossils.*

Taxon	Unit	Age	References	Interpretation*
1. *Archaeorestis*	Gunflint Fm.	2090	Barghoorn & Tyler 1965 Awramik & Barghoorn 1977	[BZ1], sheath/envelope of a prokaryote
2. *Ceratophyton*	(Lower Cambrian)	(540)	Volkova et al. 1979	[BZ2], conical bodies, possibly of animal origin
3. *Eoastrion*	Gunflint Fm. Duck Creek Dol. Barney Creek Fm. Frere Fm.	1500–2090	Barghoorn & Tyler 1965 Knoll & Barghoorn 1976 Oehler 1977 Walter et al. 1976 Knoll et al. 1988	[BZ3; Chapter 24: Pl. 2, C, D; Pl. 4, C], colonial radiating filamentous prokaryote
4. *Eomicrhystridium*	Gunflint Fm.	2090	Hofmann 1971	distorted *Huroniospora*
5. *Eosphaera*	Gunflint Fm.	2090	Barghoorn & Tyler 1965	[BZ4; Chapter 24: Pl. 1, K, L]. colonial prokaryote *Incertae Sedis*, probably planktonic
6. *Exochobrachium*	Gunflint Fm.	2090	Awramik & Barghoorn 1977	[BZ5], degraded colonial coccoid prokaryote
7. *Germinosphaera*	Dashka Fm.	750	Mikhajlova 1986	[BZ6, BZ7; Chapter 24: Pl. 42, E, I], degraded reproductive body with attached vegetative filaments
8. *Kakabekia*	Gunflint Fm. Tyler Fm. Frere Fm.	1875–2090	Barghoorn & Tyler 1965 Cloud & Morrison 1980 Walter et al. 1976	[BZ8; Chapter 24: Pl. 2, O–R], prokaryote *Incertae Sedis*, probably planktonic
9. *Metallogenium*	Tyler Fm.	1950	Cloud & Morrison 1980	[BZ3; Chapter 24: Pl. 2, C–D; Pl. 4, C], probably *Eoastrion* sp.
10. *Redkinia*	Yaryshev Fm.	610	Aseeva in press	[BZ9; Chapter 24: Pl. 52, B–D], (?)invertebrate carapace
11. *Vervhachium*?	Gunflint Fm.	2090	Hofmann 1971	[BZ11], (?)prokaryote *Incertae Sedis*
12. *Xenothrix*	Gunflint Fm.	2090	Awramik & Barghoorn 1977	[BZ10], envelope of colonial prokaryote
13. (Unnamed Type 4)	Belcher Gp.	2150	Hofmann & Jackson 1969	[BZ4; Chapter 24: Pl. 1: K–L], *Eosphaera* sp.
14. (Unnamed Form A)	Draken Congl.	750	Knoll 1982	[BZ12], "unnamed organic tube with globose end," *Incertae Sedis*, possibly sheath of filamentous prokaryote
15. (Branched tubes)	Yudoma Fm.	600	Lo 1980	[BZ13], *Incertae Sedis*
16. (radiate structure)	Gunflint Fm.	2090	Cloud 1965	[BZ3; Chapter 24: Pl. 2: C–D; Pl. 4: C], *Eoastrion* sp.

*code in brackets "[ ]" refers to the classification in Chapter 25.

ecophenotypic responses to increased levels of atmospheric oxygen. If the occurrence of such bizarre forms reflects such a drastic, and evidently time-restricted, environmental change, their potential for biostratigraphic correlation of units of this age would be significantly enhanced. However, it also should be noted that the relative abundance of such forms in units approximately 2.0 Ga may be facies-controlled. The majority of these forms are associated with, or preserved within, shallow-water, stromatolitic and commonly oolitic, banded iron-formations, a somewhat unusual facies essentially restricted to the Early Proterozoic (Chapter 4).

5.6.2B Proterozoic Vase-Shaped Microfossils

Vase-shaped fossils (Chapter 24: Pl. 29, A–H) are one of the morphologically most distinctive types of microfossils known to occur in Proterozoic rocks. First noted by Ewetz (1933) in phosphate nodules from the Visingsö Beds in Sweden, they since have been reported from at least fifteen localities of Late Proterozoic age (Hofmann 1987). In recent years, excellently preserved specimens have been reported by Bloeser et al. (1977) from the Chuar Group in northern Arizona; these forms were subsequently described as the genus *Melanocyrillium* (Bloeser 1985) in reference to their opacity, black color, and juglike morphology. Vase-shaped microfossils are generally robust and thick-walled; they most commonly range from about 40 to 200 μm in length, although specimens as short as 20 μm and as long as 400 μm have been detected in the Chuar Group (Horodyski 1987), and specimens 1,000 μm long have been reported from the Dolomite Series in eastern Greenland (Vidal 1979b). Scanning electron microscopy of the Chuar specimens reveals a complex structure with the organic envelope having a pylome with a round or polygonal opening, which may be bordered by a smooth, triangular or hexagonal collar (Bloeser et al. 1977; Bloeser 1985; Chapter 24: Pl. 29, A–H). Vase-shaped microfossils have been reported from mudstone (Bloeser et al. 1977; Bloeser 1985), chert (Knoll and Calder 1983; Horodyski 1987), carbonate (Binda and Bokhari 1980; Horodyski 1987), and phosphate (Ewetz 1933; Knoll and Vidal 1980).

Bloeser et al. (1977) interpreted specimens from the Chaur Group as chitinozoans; but Bloeser (1980, 1985) later noted that they are distinctly different from any previously reported chitinozoan. She classified them as "microfossils incertae sedis" and tentatively interpreted them as the encystment stage of an unknown alga. Vase-shaped microfossils from other Proterozoic units have been interpreted as possible animal protists, particularly tintinnids (Fairchild et al. 1978; Knoll and Vidal 1980; Knoll and Calder 1983), an interpretation that is consistent with the occurrence of gammacerane, a biomarker thought to be derived from protozoan tetrahymenol, in associated strata of the Chuar Group (Summons et al. 1988). However, Bloeser (1985) noted substantial differences in external morphology between vase-shaped microfossils from the Chuar Group and tintinnids. New material from the Chuar Group reveals the presence of abundant 10-μm-sized organic-walled spheroids within some specimens (Horodyski 1987), suggesting that *Melanocyrillium* may represent the sporangium of a megascopic, eukaryotic alga.

5.6.2C Proterozoic Skeletonized Protist-Like Problematica

Licari (1978) figured three types of distinctive, skeletonized protist-like objects (viz., *Bullasphaera variegata*, unnamed spindle-shaped objects, and *Protegocista*, a nomen nudum) from the Late Proterozoic Beck Spring Dolomite of southeastern California. All of these objects are here regarded as dubiofossils. *Bullasphaera variegata* is spheroidal, ranging from 1.9 to 8.1 μm in diameter, with a covering of short spines or with a single larger spine. *Protegocista*, a morphotype not formally described by Licari (1978), is the name he used in reference to cylindrical or ellipsoidal to ovoidal, or rarely spherical, objects ranging in size from 5.1 to 21.6 μm. Some of these specimens have radial spines. Both *Bullasphaera* and *Protegocista* are interpreted here as most probably remnants of mineralic crystals that have been partially dissolved. The spindle-shaped objects figured by Licari (1978) resemble pennate diatoms in overall shape, but no morphologic structures characteristic of diatoms are discernable. Moreover, fossil pennate diatoms are otherwise known only from sediments of Tertiary age. The Beck Spring spindles are interpreted here as probably being partially dissolved or corroded inorganic crystals.

5.7

Proterozoic and Cambrian Skeletonized Protists

JERE H. LIPPS

5.7.1 Introduction

Protists with mineralized or agglutinated skeletons or elements on their cell walls include foraminifera, radiolaria, and a host of algal and minor protozoan groups. Many such protists have been reported to occur in Proterozoic and Cambrian rocks, although the validity of most of these reports is questionable. The following discussion deals solely with skeletonized protists; organic-walled algal protists (viz., acritarchs; Section 5.5) and enigmatic vase-shaped protists (*Melanocyrillium*; Section 5.6.2B) specifically have been excluded. Although organic-walled acritarchs were evidently extant as early as 1.8 Ga (Section 5.5), skeletonized protists did not appear until more than a billion years later.

5.7.2 Proterozoic Skeletonized Protists

Although objects interpreted as foraminifera, radiolaria, tintinnids, and other skeletonized protists have been reported from Proterozoic sediments, most such objects are regarded here as contaminants, artifacts, or results of similar misinterpretation. Pflug (1965), for example, described tiny, possibly chambered objects (*Scaniella*) from Proterozoic units of Sweden which he interpreted as foraminifera. Glaessner (1978) considered them to be Cretaceous contaminants, but they are too small to be foraminifera. They are most likely silicified ooids and catagraphs; they are certainly not foraminifera. Pflug (1965) also figured tubes about 0.15 mm long, from the Middle Proterozoic Belt Supergroup of Montana, that he interpreted as foraminifera; these are of algal origin. Similarly, Hovasse (1956) and Hovasse and Couture (1961) described *Birrimarnoldi antiqua* as a foraminferan from the Precambrian of the Ivory Coast. Loeblich and Tappan (1987) considered these to be minute siliceous and iron-oxide globules.

Tintinnids were reported from the Precambrian of the USSR by Koshevoj (1987), but none of this material exhibits characteristics of tintinnids. Indeed, these putative fossils cannot be compared with any known group of protists, prokaryotes, plants, or animals; they are prabably inorganic pseudofossils. Rao and Mohan (1954) illustrated a hexagonal pattern within a circular outline, regarded by them as a radiolarian, from the Proterozoic Dogra Slates of Kashmir. This material needs restudy for proper determination, but probably these objects are nonbiogenic.

Among the more unusual and important discoveries in recent years is that of tiny organic or siliceous scales preserved in chert beds or nodules in laminated limestones of the Tindir Group of northwestern Canada (Allison and Hilgert 1986). Although Allison and Hilgert (1986) considered these fossiliferous strata to be lowermost Cambrian, they are probably uppermost Proterozoic based on the carbon isotopic signature of the enclosing rocks and the occurrence higher in the section of Early Cambrian shelly fossils. So-called sponge spicules reported to occur in association with the scale fossils (Allison and Hilgert 1986) are probably nonbiogenic (A. H. Knoll, pers. comm., 1990).

Allison and Hilgert (1986) described 26 species and six varieties referred to 17 genera, all new, and none clearly assignable to known protists. The fossils were separated into three groups and compared with known major kinds of protists. One group contains simple, imperforate scales of monocrystalline silica not unlike scales possessed by the chrysophyte alga *Ochromonas* or the rhizopod protozoa *Sphenoderia*, *Tracheleuglypha*, and *Trinema*. Taxa in the second group are disks with prominent pores arranged in hexagonal or radial patterns. Some have spines or raised rims. One cluster of overlapping scales suggests that they covered a single cell. These scales also are composed of monocrystalline silica. The third group contains scales of greater variety, both in morphology and size. In this group, more than one type of scale appears to have been derived from a single organism, further complicating assignment of these fossils to a particular biologic group. Some of the scales in this group resemble those of prymnesiophytes, others resemble chrysophyte scales, and still others do not resemble scales of any known organism. Regardless of the affinities of these Tinder Group fossils, they indicate that a wide variety of protistan grade organisms was present in the Late Proterozoic and that knowledge of life at this time is still very incomplete.

5.7.3 Cambrian Skeletonized Protists

Authentic skeletonized protists are earliest known from

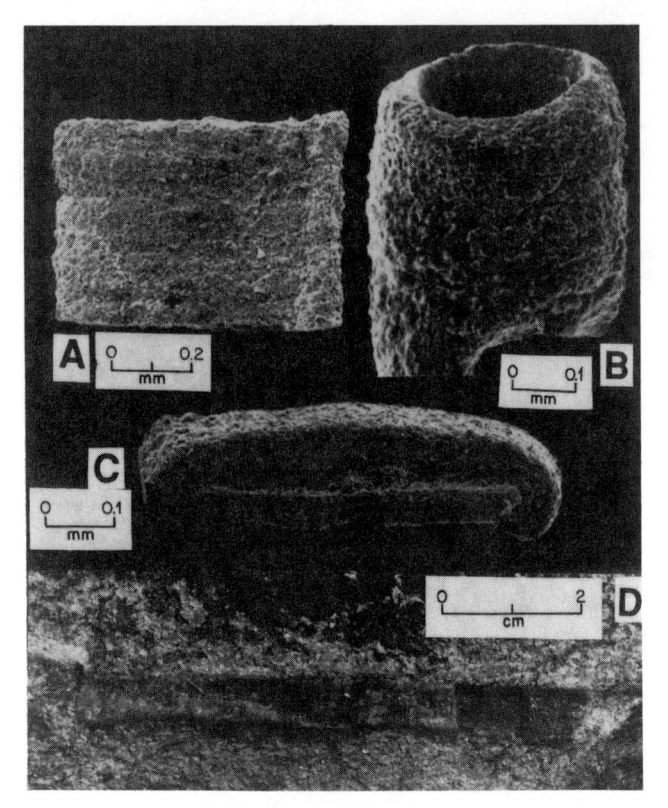

Figure 5.7.1. Scanning electron micrographs (A–C) and optical photomicrograph (D) showing *Platysolenites antiquissimus* Eichwald from the Lower Cambrian Lontova Horizon exposed on the left bank of the Tosna River, near the mouth of the Sablinka River, southeast of Leningrad, USSR (A–C), and from the Lower Cambrian Chapel Island Formation, 148 m above its lowest exposure at Little Dantzic Cove (see Landing et al. 1989), Burin Peninsula, Newfoundland, Canada (D). (A) Side view of a flattened fragment with constrictions (University of California Museum of Paleontology no. 38639). (B) Apertural view of a small unflattened fragment with two constrictions (UCMP 38640). (C) End view of a flattened specimen (UCMP 38641). (D) Broken specimen exposed on the rock surface (Royal Ontario Museum no. 46151).

the Cambrian. Foraminifera, radiolaria, and diatoms have been reported from the Cambrian; all must be evaluated.

5.7.3A Foraminifera

Three genera, *Platysolenites*, *Spirosolenites*, and *Yanichevskyites*, interpreted by various workers either as foraminifera or as worm tubes, have been described from the Lower Cambrian of the Baltic Platform. *Platysolenites* (Fig. 5.7.1) was first described by Eichwald (1860) from the Lower Cambrian Blue Clay near Leningrad and interpreted to be a worm tube. It is agglutinated, ranging from small fragments a fraction of a centimeter in length to specimens as long as 10 cm. The tube is most commonly flattened, and some specimens may have a central groove, both due to compression. Numerous transverse swollen growth bands and constrictions occur on most specimens.

Soviet and some other authors have considered *Platysolenites* to be a worm tube, but Glaessner (1978) restudied specimens from the Baltic Platform and concluded that they are identical to the agglutinated foraminiferan *Bathysiphon*. Rozanov (1979, 1983) and Rozanov and Zhuravlev (in press) state that the specimens they studied from the Blue Clay exhibit original, biogenically secreted silica tubes, and hence cannot be foraminifera but are worms. Brasier (1989) has also concluded that *Platysolenites* is not a foraminiferan. However, no living worm is known to secrete a tube composed of silica, whereas members of at least one suborder of foraminifera do (Resig et al. 1980; Loeblich and Tappan 1987). Moreover, *Platysolenites* has no characters that would distinguish it as a worm of any kind, and *P. antiquissimus*, the type species, is identical to fossil *Bathysiphon* (Danner 1955; Miller 1988a, 1988b), the descriptions and illustrations of which are indistinguishable from those of *Platysolenites*. Like *Platysolenites*, fossils of *Bathysiphon* are flattened, have numerous transverse constrictions, and possess a medial furrow caused by compression.

Bathysiphon was first described by Sars (1872) from modern sediments in Hardangerfjord on the west coast of Norway, with *B. filliformis* as the type species. The neotype of *B. filliformis* (see Gooday 1988) is finely agglutinated, as long as the longest *Platysolenites*, has transverse rings, and easily fragments. In all of these characters and in overall morphology, including its collapsed, flattened, tubular form, *Platysolenites* is closely similar to *Bathysiphon*. *Bathysiphon* ranges in age from Late Triassic to Recent (Loeblich and Tappan 1987) and is known chiefly from deep-water deposits.

Because of their general form and morphological simplicity, agglutinated worm tubes and agglutinated foraminiferal tests share many characters and can be difficult to differentiate. Nevertheless, *Platysolenites* is here considered a foraminiferan because that is the most parsimonious conclusion based on its shared characters with *Bathysiphon*.

Platysolenites is known throughout the Baltic Platform, from the Blue Clay in the Leningrad area (Timofeev 1955) to Poland (Rozanov 1979, 1983) and into Scandinavia (Føyn and Glaessner 1979; Glaessner 1978; Hamar 1967). It also has been reported from England (Brasier 1986, 1989), Wales (Rushton, in Allen and Jackson 1978), New Brunswick (Matthew 1889), Newfoundland (Landing et al. 1989), and California (Firby-Durham 1974; Onken and Signor 1988b).

Spirosolenites Glaessner and *Yanichevskyites* Sokolov (the latter, a nomen nudem; Sokol 1965), were considered by Rozanov (1979, 1983) to be synonyms of *Platysolenites*. Forms described as *Yanichevskyites* simply represent small juvenile specimens of *Platysolenites* (Rozanov 1979, 1983), as do the nearly cylindrical tubes described by Brasier (1984) as *Hyperammina?* (see Brasier 1986).

Although Rozanov (1979, 1983) considered the spiral tubes referred to *Spirosolenites* as varieties of *Platysolenites*, Glaessner (1978) pointed out that the spiral types occur only at a single Lower Cambrian locality in Finnmark originally described by Hamar (1967). Later, Glaessner (in Føyn and Glaessner 1979) described these spiral tubes as a new foraminiferal genus which was subsequently recognized by Loeblich and Tappan (1987) as a valid genus and, chiefly because of its coiled morphology, was referred by them to the superfamily Ammodiscacea rather than to the superfamily Astrorhizacea where they placed *Platysolenites*. Because such spiral specimens are not known throughout the geographic or geologic range of

Platysolenites, and because they exhibit a distinctive spiral morphology, the genus *Spirosolenites* is retained here for spiral agglutinated tubes considered to be foraminifera.

Conway Morris and Fritz (1980) illustrated two types of microfossils that might be foraminifera, a simple cap-shaped silicified one and an agglutinated one, from the Lower Cambrian of the Mackenzie Mountains of northwestern Canada. These microfossils both resemble known, very simple, living foraminifera. However, because these kinds of agglutinated foraminifera have virtually no diagnostic characteristics, comparison of the microfossils to foraminifera must remain uncertain. Similarly, Waern (1971) described and illustrated brownish disks, 0.4 to 0.9 mm in diameter, from a Lower Cambrian core sample from Bodahamm, Sweden, that may be foraminifera. Although flattened by compression, the disks appear to have been originally spheroidal. Kobluk and James (1979) illustrated an encrusting coelobiont referred to the foraminifera (*Wetheredella*) from Lower Cambrian reefs in Labrador, but this genus is considered to be an alga (Loeblich and Tappan 1987). This occurrence needs detailed study and description.

Indisputable foraminifera have been reported from Middle Cambrian rocks in Sardinia and the northwestern USA. *Psammosphaera*, *Hemisphaerammina*, possibly two ammodiscids, and straight astrorhizid-like tubes occur in the Cabitza Formation in southwest Sardinia (Cherchi and Schroeder 1985), and an ammodiscid and members of an undescribed foraminiferan genus occur in the Spence Shale of Idaho (Lipps 1985). The Sardinian ammodiscids differ from most known types in having fewer whorls and a relatively thicker tube; however, they do resemble *Spirosolenites* from the Lower Cambrian. The remaining Sardinian fossils closely resemble known foraminifera.

5.7.3B Radiolaria

From Cambrian deposits, both nonbiologic objects and true fossils have been described as radiolaria. Peng (1984) listed and illustrated microscopic objects, interpreted to be radiolaria, from a Lower Cambrian ophiolite sequence in Mongolia that consists mostly of metamorphosed volcanogenic sedimentary rocks. His listing includes six families of radiolaria, but only one specimen, which is most likely a volcanic shard, is figured (Peng 1984). Assuming that this rock sequence is not considerably younger than reported, the remaining reported "radiolaria" may also be volcanic shards; no comparably diverse radiolarian fauna is known from Cambrian rocks anywhere else in the world.

Radiolaria were described from the Early Cambrian (Atdabanian of the Batenevsky Ridge) by Nazarov (1973, 1975). These siliceous microfossils closely resemble radiolaria, although Nazarov (pers. comm., 1986) has suggested that they might possibly be sponge spicules. Nazarov and Ormiston (1985) included them as radiolarians; Rozanov and Zhuravlev (in press) regard them as demosponges.

Conway Morris and Chen (1990) described *Blastulospongia polytreta* from the Lower Cambrian Shuijingtuo Formation in the Taishanmiao geologic section of Hubei Province, China. These fossils closely resemble unnamed microfossils from the Middle Cambrian of Utah, USA (White 1986), *Blastulospongia monothalamos* from the Middle Cambrian of New South Wales, Australia (Pickett and Jell 1983), and *Blastulospongia mindyallica* from the Upper Cambrian of Queensland, Australia (Bengtson 1986c). All of these fossils are siliceous spheres perforated with as many as 600 pores. They have been interpreted as sphinctozoan sponges (Pickett and Jell 1983), radiolarians (Bengtson 1986c, White 1986), or as being of uncertain affinity (Conway Morris and Chen 1990).

In its porous nature and spherical shape, *Blastulospongia* resembles the much larger *Jawonya* (Kruse 1987) from Australia, which is most likely not a radiolarian but a sphinctozoan. However, it differs from *Jawonya* by being about 130 times smaller and lacking an extended, nonporous rim encircling an opening into the sphere. Conway Morris and Chen (1990) described two specimens of *Blastulospongia* that may have been benthic, attached to other skeletons, although they do not rule out diagenetic bonding between the microfossils and the associated debris. The Australian fossils of Pickett and Jell (1983) and of Bengtson (1986c) have indentations that may have served as sites of attachment. The possible benthic habit of these fossils does not rule out radiolarian affinity; radiolaria may have originated as benthic organisms (Petrushevskaya 1977). *Blastulospongia* spp. cannot be shown definitely to be radiolaria, but that conclusion seems most consistent with currently available data.

Bengtson (1986c) also described *Echidnina runnegari* from the Upper Cambrian of Queensland. It is a spherical microfossil composed of numerous siliceous spicules that resembles some radiolarians although the spicules are typical hexactinellid (Bengtson 1986c). At present, it is best considered a sponge. Upper Cambrian radiolaria occur in the Cow Head Group of western Newfoundland (Iams and Stevens 1988); none of these fossils was described or illustrated in the initial report, but two groups were recognized among them.

5.7.3C Algae

Difficult to interpret are "diatom-like" objects described from the Cambrian of the Tannu-Ola Mountains in Tuva in the USSR (Vologdin 1962b). These objects formed the basis for description of a new family, six new genera, and nine new species. The specimens are polygonal, trigonal, or irregular in shape, and all have a porous appearance. The presence of the pores suggested to Vologdin (1962b) that the objects were remains of diatoms or diatom-like organisms. Judging from the published data, however, they are more likely nonbiologic, possibly particles of volcanic ash.

5.7.4 Summary

Near the beginning of the Late Proterozoic, organic-walled acritarchs had already become abundant and diverse (Sections 5.5, 11.3), testifying to the relatively early development of unmineralized protistan phytoplankton. By the time of appearance of Vendian-Ediacaran metazoans, phytoplankton had declined in diversity and complexity, but later radiated in the Early Cambrian into many new types (Section 5.5). At this time the first skeletal protists appear in the fossil record, with the fragmentary evidence now available suggesting that they were much more diverse than the known record indicates. A number of groups of protists are known to have been extant in the Cambrian, and both foraminifera and radiolaria increased in diversity throughout the period becoming relatively diverse in the Early Ordovician, a pattern resembling that of early metazoan diversification (Section 11.4; Sepkoski 1978). Skeletonized protists, however, are still not well enough studied in the Cambrian to document adequately their early evolutionary trends.

The overall early Phanerozoic evolutionary pattern of single-celled eukaryotes thus resembles that of the Metazoa. Both protists and metazoans developed skeletons near the beginning of the Cambrian, exhibited some increase of diversity through the Middle Cambrian, and a renewed radiation in the Ordovician. Although the documented diversity of early Phanerozoic protists is considerably less than that of metazoans, the similarity in the patterns of development of the two groups seems to indicate that whatever induced the metazoan radiation had effects extending well beyond multicellular megascopic animals to affect much, and probably all, of the marine ecosystem.

5.8

Unsolved Problems and Future Research Directions

CARL V. MENDELSON   J. WILLIAM SCHOPF

5.8.1 Introduction

Understanding of the Proterozoic biosphere has dramatically increased over the past two decades. Major patterns of Proterozoic biotic change and the timing and nature of early evolutionary innovations are becoming increasingly better defined. As in most scientific endeavors, however, greater understanding does not mean that there are fewer questions; on the contrary, improved knowledge has allowed questions to be asked that had either not been considered, or that could not have been articulated, even a few years ago. Unsolved problems are being generated faster than they can be answered (the mark of a healthy science); future research directions are expanding rapidly. In the following paragraphs, a number of these currently pressing problems are organized into a series of categories and briefly discussed.

5.8.2 Major Problem Areas

5.8.2A Taphonomy

An understanding of taphonomic processes greatly improves the ability to interpret fossil microorganisms of the types here considered (Section 5.3). The fruitful beginnings of the past decade must be followed up. In particular, and in addition to continued study of natural communities, experiments should be designed to answer specific, paleontologically relevant questions. For example, what is the fate of modern prasinophyte phycomata that are artificially compressed in different types of sediment? Do consistent fold patterns develop as a function of wall thickness? Do thin-walled and thick-walled vesicles consistently deform differently? Is it true that unlike the cellulosic cell walls of vascular plants, those of phytoplankters are not preservable? Under what conditions are the sheaths of prokaryotes preserved preferentially to cell walls (and vice versa)? And, if the sheaths of Proterozoic prokaryotes were composed of carbohydrates, as are those of their modern analogues, why are they commonly well preserved whereas the carbohydrate (cellulosic) cell walls of phytoplankters are evidently not? Such fossilization studies might yield taxonomic characters that would improve classification of many types of Proterozoic microfossils (including the morphologically nondescript sphaeromorphs), a classification that would have biological meaning.

5.8.2B Paleoecology

What are the environmental constraints on large acanthomorphs and "megasphaeromorphs"? Taxa of these types have been found in a wide variety of sediments, ranging from shallow-water chert assemblages to deposits interpreted to be deep-water turbidites (see Knoll and Butterfield 1989). Does their environmental distribution change through the Late Proterozoic?

Many freshwater algae have been documented to incorporate sporopollenin-like substances in their cell walls. Which microfossiliferous Proterozoic formations might be fresh water in origin? And to what extent is the occurrence of cellular microfossils in silicified Proterozoic stromatolites a result of their preservation under evaporitic conditions?

5.8.2C Modern Microbiology

In recent years, an increasing number of studies have been carried out on modern microbial mats and the vertical and horizontal distribution of their contained microbes; such studies form a basis for direct comparison with fossilized, Proterozoic, microbial-mat communities (Chapter 6). Comparable studies by phycologists on extant plankton are in their nascent stages. Which types of algae are characterized by geologically preservable cell walls? Which phytoplankters produce cysts, under what environmental conditions, and what is the composition of these cysts? How commonly does cyst formation reflect the presence of a sexual stage in an algal life cycle, and are there morphologically distinctive cysts that could be used in the Proterozoic fossil record as firm evidence of an alternation of haploid and diploid generations? To what extent does such information constrain interpretations of putative cysts known from the Proterozoic?

5.8.2D Taxonomy

At present, the field is far from achieving stability in the systematics of Proterozoic microfossils. A concerted program of taxonomic standardization (e.g., Culver et al. 1987) seems in order. A significant step in this direction has been taken by

German et al. (1989) for Precambrian microfossils from the USSR, and effort has been made here along similar lines (Chapter 25). The informal revised classification of Proterozoic microfossils presented in Chapter 25 is, however, just that: informal. This or some other comprehensive revision needs to be formalized, a daunting task indeed. A large number of type specimens, and of populations of such microfossils, need to be studied carefully, and appropriate nomenclature must be employed (a task requiring an enormous amount of bibliographic work). Until such standardization is carried out, the current nomenclatural chaos will continue unabated, greatly hindering communication and attempts to extract meaningful evolutionary information from the known Proterozoic fossil record. Standardization is needed in a published format similar to that in the *Treatise on Invertebrate Paleontology*.

5.8.2E Evolutionary Innovations

How old are the eukaryotes? This and similar questions are familiar to Proterozoic micropaleontologists. Sections 5.4, 5.5, and Chapters 22 and 25 provide well-documented data that permit a current, tentative, answer. But improved environmental control is greatly needed. For example, what is the size distribution of modern eukaryotic phytoplankters in open-coastal waters? Might exceptionally large planktonic prokaryotes have existed during the Proterozoic? And if so, how could they be distinguished in the fossil record from co-existing planktonic eukaryotes? Similarly, what is the size distribution of filamentous prokaryotes in the upper few millimetres of microbial mats? In filamentous microorganisms, is cell size a solid criterion for differentiation of prokaryotes and eukaryotes, or must some additional criterion (e.g., true branching) be applied? The answers to such questions, based at least in part on detailed microenvironmental data from the modern biota, should provide a much improved basis for assignment of microfossil taxa to one or the other of these two great categories of organisms.

5.8.2F Diversity Studies

Even if it is assumed that the taphonomic haze can be cleared, and that the nomenclatural chaos can be put in order, major problems will still confront those trying to extract meaningful diversity information from the Proterozoic microfossil record. Many of these problems are addressed in Section 11.3.3. Additional biologically based problems, however, also exist. For example, the morphological similarity of spiny acritarchs to dinoflagellates suggests that the two groups may be related. If so, how many biological species do two morphologically different acritarchs represent? The answer may be one or it may be many. Seemingly identical dinoflagellate cysts can germinate into more than one type of vegetative cell; and apparently identical vegetative cells may give rise to a number of morphologically distinct cysts (reviewed in Edwards 1987; also see Evitt 1985). To some extent, this situation is analogous to that in ichnology (Section 7.6): different traces can be made by a single animal, and different animals can make nearly indistinguishable traces. Seeing through such *biological* filters presents a challenging problem both for studies of Proterozoic acanthomorph diversity in particular, and of Proterozoic biotic diversity in general.

5.9

Summary and Conclusions: the Current Status of Proterozoic Micropaleontology

CARL V. MENDELSON   J. WILLIAM SCHOPF

The relatively youthful science of Proterozoic micropaleontology is healthy: unsolved research problems are being generated faster than they can be answered (Section 5.8.2). What has been accomplished? How has the material discussed here (Chapter 5) informed our understanding of early microbial evolution? Some conclusions are well documented, whereas many others lack significant constraints and require further study for validation; all of these conclusions speak to fundamental aspects of Proterozoic micropaleontology.

(1) Proterozoic micropaleontology is a relatively new discipline. The historical development of the field can be divided into four phases (Section 5.2.2): a long period of *quiescence* (1899–1965), a brief period of *emergence* (1965–1970), a decade of *maturation* (1970–1980), and the current period of *establishment* (1980–present). Much progress has been made, and each succeeding phase shows a marked decrease in the number of errors published. Still, the number of practitioners of this science remains small, a situation that should gradually change as appreciation of the relevance and potential of Proterozoic micropaleontology becomes more widespread and the history of Precambrian life becomes increasingly incorporated into biology and geology textbooks and courses.

(2) Studies of the composition, degradation, and ecologic distribution of modern prokaryotes and protists, coupled with taphonomic analyses of fossil assemblages, have allowed more informed interpretations of the nature of the known fossil record and, thus, of the history of the Proterozoic biosphere (Section 5.3). The question of how representative is the known Proterozoic fossil record cannot yet be answered definitively, but at least that question can now be articulated clearly.

(3) The historical dichotomization of the field into two "facies-based" traditions is becoming somewhat tempered. The results and limitations of the chert-facies (thin section) school are now being actively compared with those of the shale-gacies (acid maceration) school (Sections 5.2, 5.3), and quantification, albeit rudimentary, of the morphometric effects of preservational compression in shales can now be systematized (Section 5.4.2).

(4) Proterozoic microfossils exhibit only a very limited array of diagnostic taxonomic characters. Indeed, many such fossils are simple organic tubes or smooth, unornamented, spheroids that might be allied to members of any of a number of extant taxonomic groups, whether at the class, phylum or division, or even the kingdom level. The biologic relationships of Proterozoic microfossils can now be assessed more rigorously through morphometric comparison of fossil and modern taxa, particularly via comparative analyses of size ranges (Section 5.4.3); a relatively crude – but nonetheless useful – interpretive key to the inferred "probable affinities" of such fossils, application of which may help to make the processs of identification more systematic, is now available (Table 5.4.5).

(5) Most reported Proterozoic prokaryotes resemble modern members of the cyanobacteria, including solitary and colonial chroococcaleans and uniseriate, unbranched, nostocaleans. Archaebacterial and many eubacterial (including cyanobacterial) lineages seem to have been well established by the Early Proterozoic (Section 5.4.8), with the morphometric similarity of such Proterozoic fossils to living taxa supporting the hypothesis that prokaryotes are "hypobradytelic" (Schopf 1987), having exhibited little morphological evolution over exceptionally long periods of geologic time (see Section 13.3.2). The major branches of the prokaryotic evolutionary bush must have diverged during the earlier, Archean, eon of the Precambrian.

(6) In contrast, available data seem to indicate that the major branches of the acritarch evolutionary bush are to be found within the Proterozoic (Section 5.5.3). Both sphaeromorph and acanthomorph acritarchs exhibit pronounced increases in taxonomic diversity, and correlative increases in vesicle size, during the Middle to Late Proterozoic. The relative "giants" of the Proterozoic, hundreds of micrometers in diameter, apparently became extinct toward the close of the eon; indeed, early Phanerozoic forms are typically much smaller in size, with all reported Early Cambrian acanthomorph taxa, for example, being less than 75 μm in diameter (Fig. 5.5.15).

(7) The Late Proterozoic seems to have been a time of major evolutionary change in the eukaryotic phytoplankton, primarily represented by acritarchs. Beginning in the earliest Cambrian, a radiation of comparable importance occurred in such phytoplankton and, also, evidently among eukaryotic skeletonized protists, although confirming data for the latter

are sorely limited (Section 5.7). An equally important episode may have occurred among prokaryotes during the Early Proterozoic: *Lagerstätten,* such as the Gunflint Iron Formation, preserve a virtual menagerie of unusual and problematic forms that cannot be placed confidently within any existing classification (Section 5.6). Do such bizarre forms identify a stage of early experimentation in physiology or *Baupläne* among prokaryotes? Or, do they reflect the relatively special environmental conditions required for deposition of banded iron-formations, within which they typically occur?

Finally, the relatively complex vase-shaped microfossils known as melanocyrillids (Section 5.6.2B) seem to be restricted to Upper Proterozoic strata. These distinctive fossils might represent the sporangia of eukaryotic algae, or they may be remnants of the organic-walled cysts of heterotrophic protists, an interpretation consistent with their co-occurrence with tetrahymenol-derived gammacerane (Summons et al. 1988b). The fact that these enigmatic microfossils were first well characterized only during the past decade suggests that additional surprises are probably in store as the fabric of Proterozoic evolution is increasingly scrutinized.

(8) In broad-brush outline, major aspects of the evolutionary development and potential biostratigraphic usefulness of known Proterozoic microorganisms can be summarized as follows: (*i*) numerous prokaryotic lineages, most notably including photoautotrophic cyanobacteria, had become established and relatively diverse prior to or shortly after the beginning of the Early Proterozoic; (*ii*) because of the evidently hypobradytelic tempo of prokaryotic evolution in general, and of cyanobacterial evolution in particular, and the relative uniformity of sampled prokaryotic assemblages throughout the Proterozoic, fossil prokaryotes (with the possible exception of Early Proterozoic micro-problematica) cannot be expected to provide a basis for biostratigraphic subdivision of Proterozoic strata; (*iii*) eukaryotic phytoplankters, evidenced by the occurrence both of relatively large-diameter sphaeromorph acritarchs and of eukaryote-derived sterane hydrocarbons (Summons et al. 1988a), were evidently extant as early as 1800 to 1700 Ma; (*iv*) between 1800 and 1200 Ma, the taxonomic diversity (and evidently the abundance) of such eukaryotes gradually increased, followed by a shorter period of major, relatively rapid diversification, prior to about 900 Ma (Fig. 5.5.12), which more or less coincided with the appearance of probable excystment structures in sphaeromorphs (Fig. 5.5.14) and which therefore can be hypothesized as possibly reflecting the advent of meiosis, syngamy, and the resultant genetic variability provided by eukaryotic sexuality (cf. Schopf et al. 1973b); (*v*) near the beginning of the Late Proterozoic, the planktonic flora included "giant" sphaeromorphs and acanthomorphs, hundreds (and for sphaeromorphs, thousands) of micrometers in diameter, which gradually disappear from the known fossil record throughout the remainder of the Era; (*vi*) the temporal distribution of such "giant" acritarchs, perhaps coupled with that of other morphologically distinctive, coeval, organic-walled microfossils (e.g., *Melanocyrillium* spp.), holds promise for biostratigraphic zonation of the Late Proterozoic; and (*vii*) unquestionable skeletonized protists, presumably descendants of unmineralized Late Proterozoic ancestors, occur earliest in Cambrian strata, exhibiting a pattern of early Phanerozoic diversity change that broadly parallels that of early evolving, skeletonized, Metazoa.

(9) Proterozoic micropaleontology has progressed markedly during the past two decades; accordingly, numerous new research agendas can now be envisioned. Future investigations can be expected to follow many paths – paths that are sinuous and commonly interwoven. These paths will lead to increased understanding of microbial taphonomy and paleoecology, and of related aspects of modern microbiological communities; to new taxonomic approaches, perhaps based in part on chemical analyses of individuals or populations of microfossils; to increased interplay between molecular phylogenies and relationships suggested by the known fossil record, and between biochemical evolution and geochemically detectable organic biomarkers and isotopic signatures; to better recognition, and temporal resolution, of early evolutionary innovations; and to a more complete and better documented assessment of prokaryote and protist diversity through time.

Acknowledgments

We are grateful to the following individuals who generously supplied photomicrographs or negatives for the illustrations in Section 5.5: Nicholas J. Butterfield and Andrew H. Knoll (Harvard University); Charles Downie (University of Sheffield); Małgorzata Moczydłowska and Gonzalo Vidal University of Lund); and Zang Wenlong (Institute of Geology and Palaeontology, Nanjing).

6

Modern Mat-Building Microbial Communities: A Key to the Interpretation of Proterozoic Stromatolitic Communities

BEVERLY K. PIERSON   JOHN BAULD   RICHARD W. CASTENHOLZ
ELISA D'AMELIO   DAVID J. DES MARAIS   JACK D. FARMER
JOHN P. GROTZINGER   BO BARKER JØRGENSEN   DOUGLAS C. NELSON
ANNA C. PALMISANO   J. WILLIAM SCHOPF   ROGER E. SUMMONS
MALCOLM R. WALTER   DAVID M. WARD

6.1 Introduction

BEVERLY K. PIERSON

6.1.1 Introduction

Modern microbial mats are structurally coherent macroscopic accumulations of microorganisms. Mats are widely distributed on earth. They are found in a surprisingly large number of diverse environments from the equatorial zones to both polar regions. They vary in size from extensive terrestrial and hypersaline mats that cover areas several square kilometers in extent to minute mats only a few square centimeters in area found in small thermal springs. They vary in thickness from massive accumulations measured in meters, such as those in the Persian Gulf and the Red Sea region, to thin films less than a few millimeters in thickness. In addition to being highly varied in size, modern microbial mats are also very diverse in morphology, community structure, and physiological characteristics. What do such mats have in common? Under what conditions do they form? What is the basis of their diversity? What insight do they provide, if any, to the interpretation of the widespread stromatolites of the Proterozoic?

6.1.1A Terminology

Microbial mats are accretionary cohesive microbial communities which are often laminated and found growing at the sediment-water (occasionally sediment-air) interface. Most mats stabilize unconsolidated sediment. The mats are comprised of the various microorganisms that accumulate along with their metabolic products. The most conspicuous of these products is usually a copious amount of extracellular polysaccharide which helps hold the cells together to form a cohesive structure. Very thin cohesive microbial communities (e.g., 100 to 200 μm thick) are called biofilms (Hamilton 1987). The term "mat" is usually applied to thicker (greater than one millimeter) more complex assemblages of microorganisms. However, there is no specific separation between the two terms; conceptually they are very similar and even thick microbial mats could properly be referred to as biofilms (Hamilton 1978). Mats form at interfaces and most commonly are benthic aquatic communities found in stationary or flowing systems. Mats also form on the surface of soil (forming desert crusts) covering rocks, and even occur within rocks as endolithic communities. They may also occur on the surfaces of trees or other vegetation at a solid-air interface. Such mats formed at an air interface are here referred to as "terrestrial mats" (Section 6.9.6). An important part of the definition of microbial mats is their coherent structure which distinguishes them from settled, thick accumulations of originally suspended (e.g., planktonic) microorganisms. The degree of cohesion in microbial mats varies, as does the degree of stability of unconsolidated sediment. Some mats accumulate substantial amounts of inorganic matter, whether by sedimentation or precipitation, within their organic matrix of polysaccharide and cells. Not all mats do so, however, and the presence of sedimentary or precipitated matter is not essential to the definition. Depending on the stability of the environment, the presence of grazers, and numerous other factors, mats may accrete to great thicknesses and show a history of recent decomposition and/or preservation. Metabolic activity is usually confined to the uppermost few millimeters (occasionally centimeters), although some inactive organisms may remain viable at greater depths.

Many mats are laminated, and many factors contribute to the presence of recognizable layers. Metabolically distinct groups of microorganisms contribute to visually distinct layers recognized by differences in pigmentation and/or degree of cohesion. Seasonal or annual periods of growth may alternate with periods of deposition resulting in layers of primarily organic matter alternating with inorganic accumulations. The accretionary process and accompanying decomposition and preservation result in the accumulation of layers in various states of preservation capped by a layer of relatively thin, living, metabolically active mat.

Microbial mats existed throughout much of the Precambrian (see Section 1.5), and were particularly widespread and abundant during the Proterozoic (Section 6.2). Some of these became lithified and preserved in the fossil record as stromatolites. Controversy surrounds the definition of stromatolites. The original use of the term by Kalkowsky (1908) has been modified and re-modified (for example, Awramik and Margulis 1974; Walter 1976a). A thorough discussion of the terminology and variations of definitions has been presented by Krumbein (1983). The definition of Walter (1976a), modified from Awramik and Margulis (1974), is widely used: "Stromatolites are organosedimentary structures produced by sediment trapping, binding, and/or precipitation as a result of the growth and metabolic activity of microorganisms, principally cyanophytes (cyanobacteria)." Walter's definition leaves the possibility open to call contemporary microbial mats stromatolites, if they contain sediment or precipitates or are lithified to any degree. Krumbein's (1983) definitions distinguish among the fossil remains of ancient microbial communities and contemporary living structures. Krumbein (1983) defines stromatolites as "...laminated rocks, the origin of which can clearly be related to the activity of microbial communities, which by their morphology, physiology, and arrangement in space and time interact with the physical and chemical environment to produce a laminated pattern which is retained in the final rock structure." He uses the term "potential stromatolite" to describe living microbial communities involved in sediment trapping or precipitation. Potential stromatolites are specifically defined as "unconsolidated laminated systems, clearly related to the activity of microbial communities." Thus, the term "potential stromatolite" is suggested to replace "recent" or "living" stromatolites. Walter's definition, which includes "potential stromatolites," is more encompassing.

In an attempt to impose order on the discrepancies in these definitions, Burne and Moore (1987) proposed a new term – "microbialites." Microbialites are defined as "organosedimentary deposits that have accreted as a result of a benthic microbial community trapping and binding detrital sediment and/or forming the focus of mineral precipitation." This term excludes many contemporary microbial mats and implies no particular laminated structure. The internal structure of microbialites is further described by use of a specific subset of terms (Burne and Moore 1987): stromatolites, for example, are "laminated microbialites"; thrombolites are "clotted (unlaminated) microbialites"; and oncolites are "concentrically laminated microbialites."

In this chapter, the term "stromatolite" is used as defined by Walter (1976a) with reference to the fossil remains of microbial mats. Living microbial systems, in contrast, are here referred to as "microbial mats" and the terms stromatolite (Walter 1976a), potential stromatolite (Krumbein 1983), or microbialite (Burne and Moore 1987) are thus avoided in discussing contemporary microbial ecosystems. "Microbial mat" is a term that is biological in origin, encompassing a much wider range of cohesive interface-inhabiting microbial communities than that of the competing terminology. While many of these communities are laminated organosedimentary structures, others are not laminated and have little or no sediment trapping or precipitation of minerals within them. Some of these mats, therefore, may not be appropriate analogues for interpretation of organosedimentary Proterozoic stromatolites. Taken as a whole, however, their study can play an important role in understanding the major evolutionary events that occurred in the Proterozoic biosphere.

6.1.1B Environments Where Modern Mats Occur

Microbial mats occur primarily in environments that are at least somewhat restrictive to grazing fauna. In general, these restrictive environments include thermal springs; hypersaline lakes and lagoons; supratidal, intertidal and subtidal marine sediments; ice-covered Antarctic lakes; freshwater lakes and streams; deep sea hydrothermal vents; and soil and rock surfaces exposed to high insolation and/or desiccation. It is a formidable task to attempt to catalogue all microbial mat sites currently known worldwide. Figure 6.1.1 shows the geographic distribution of many significant contemporary microbial mat communities. Although the locations of many such mats are illustrated on this map, others, of necessity, have been omitted. Figure 6.1.1 shows the locations of, and cites relevant references for, most mats that have been subjected to physiological and ecological study or that are otherwise particularly significant. Many mats that have been described only in terms of species composition or gross structure have been omitted. Thus, the mats selected for inclusion in the figure reflect the emphasis of this chapter on understanding the physiology, ecology, and biogeochemistry of modern microbial mat communities.

The most extensively studied hot spring mats occur in Yellowstone National Park, Wyoming, USA (1; numbers refer to locations on the map in Fig. 6.1.1; references are cited in the caption for Fig. 6.1.1). Work pioneered by Brock (1978) and continued by many others has been focused on phototrophic mats in primarily alkaline hot springs. Yellowstone National Park includes many acidic environments as well, however, and examples of many different types of microbial mats. Many mats show the effects of different degrees of grazing and have been used to study the role of movement of phototrophic bacteria in forming analogues of fossil stromatolites. Mat communities occurring in several other hot springs in the western USA have been subjected to significant physiological analysis including those at Hunter's Hot Springs (3), Stinky Springs (2), and Kahneeta Springs (4). Hot springs in New Zealand (5) and Iceland (6) have also been studied extensively, and two mats were selected from these environments for study by the PPRG because of their potential significance in understanding the evolution during the Precambrian of oxygenic mat communities from anoxygenic precursor communities (see Section 6.9). Other hot spring mats that have been studied include those in the USSR (7, 8, 10), Israel (9), and Japan (61). While the ecosystems of the totally aphotic deep sea hydrothermal vents figure prominently in some ideas of early evolution (Baross and Hoffman 1985; Section 1.2), the microbial mats (or biofilms) at these sites are composed entirely of chemotrophs and are very thin. Relatively little is known about them. The best studied mats are from the Guaymas Basin on the East Pacific Rise (60).

The most widely distributed and intensely studied of all microbial mats are those found in the marine and hypersaline marine environments. These environments contain some of the most extensive microbial mats in the world today.

Worldwide, mats occurring in the intertidal zones of the temperate regions are all very similar, ranging from purely cyanobacterial mats to cyanobacterial mats with underlying, brightly colored layers of anoxygenic phototrophs, primarily purple sulfur bacteria. Such mats are abundant in intertidal sand flats and salt marshes, where sufficient biogenically produced sulfide creates anaerobic zones suitable for massive development of the purple bacteria. Such mats have been

Modern Mat—Building Microbial Communities

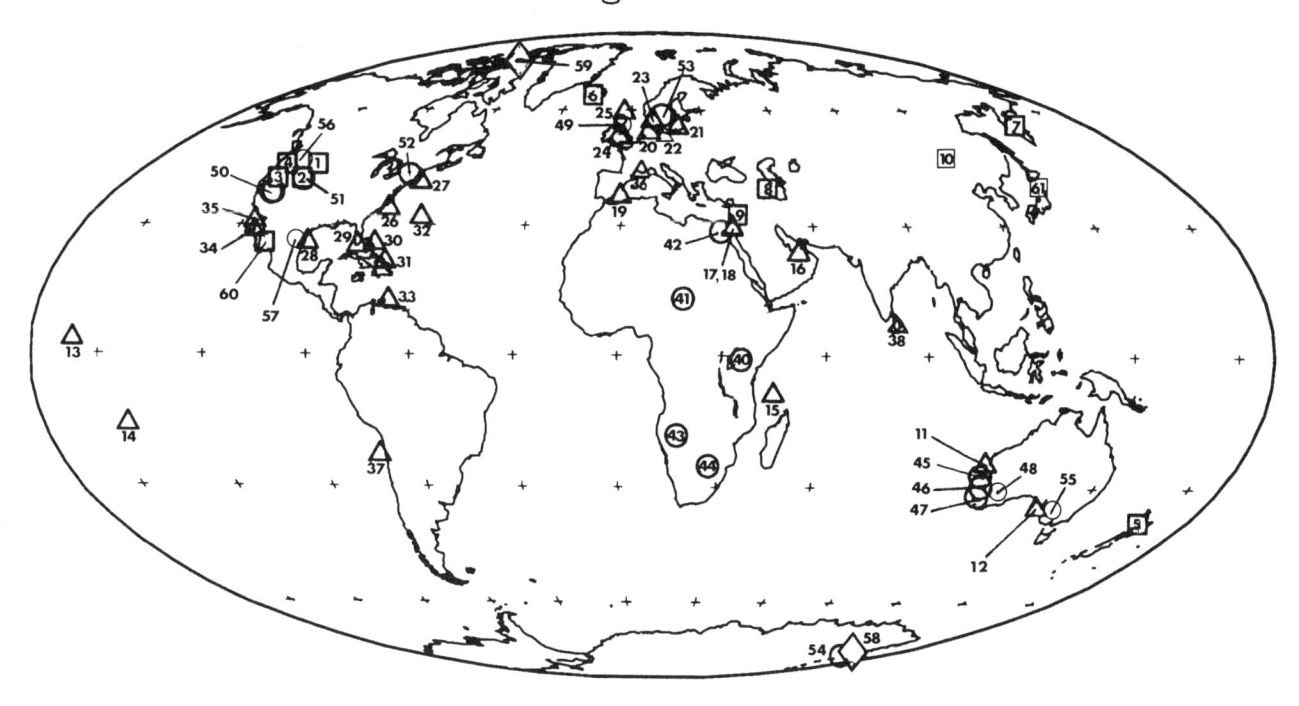

Figure 6.1.1. Geographic distribution of modern mat-building microbial communities: □, Hot Springs; △, Marine and Hypersaline Marine; ○, Inland Waters; ◇, Terrestrial.

□ Hot Springs: (1) Yellowstone National Park, Wyoming, USA (Walter et al. 1972; Doemel and Brock 1974; Walter et al. 1976; Castenholz 1977; Doemel and Brock 1977; Brock 1978; Sandbeck and Ward 1981; Revsbech and Ward 1984; Ward et al. 1984; Vanyo et al. 1986; Anderson et al. 1987; Giovannoni et al. 1987; Bateson and Ward 1988). (2) Stinky Springs, Utah, USA (Cohen 1984a; Ward et al. 1989). (3) Hunter's Hot Springs, Oregon, USA (Castenholz 1973a; Wickstrom and Castenholz 1978; Wickstrom and Castenholz 1985; Richardson and Castenholz 1987). (4) Kahneeta Springs, Oregon, USA (Castenholz 1968; Pierson et al. 1984). (5) North Island, New Zealand (Brock and Brock 1971; Castenholz 1976, 1988). (6) Iceland (Castenholz 1969, 1973b, 1976; Jørgensen and Nelson 1988). (7) Kamchatka Peninsula, USSR (Orleanskii and Gerasimenko 1982; Gorlenko et al. 1985). (8) Caucasus, USSR (Gorlenko et al. 1985). (9) Dead Sea Area, Israel (Oren 1989). (10) Bolsherechensky Thermal Springs, Northeast Shore of Lake Baikal, USSR (Gorlenko et al. 1985; Gorlenko and Yurkov 1988). (60) Hydrothermal vents, Guaymas Basin (E. Pacific) (Jannasch 1984; Belkin and Jannasch 1989; Nelson et al. 1989). (61) Japan (Maki 1986, 1987).

△ Marine and Hypersaline Marine: (11) Shark Bay, Western Australia (Playford and Cockbain 1976; Bauld et al. 1979; Bauld 1984; Awramik and Riding 1988). (12) Spencer Gulf, South Australia (Bauld et al. 1980; Skyring et al. 1983; Bauld 1984). (13) Christmas Island, Pacific Ocean (Yehuda Cohen, pers. comm. to B. K. P.). (14) Rangiroa Island, Tuamoty Archipelago, Pacific Ocean (Defarge, et al. 1985). (15) Aldabra Island, Seychelles, Indian Ocean (Potts and Whitton 1977, 1979, 1980). (16) Abu Dhabi Coast, United Arab Emirates, Persian Gulf (Kinsman and Park 1976; Cardoso et al. 1978). (17) Solar Lake, Sinai, Red Sea (Krumbein et al. 1977; Krumbein and Cohen 1977; Jørgensen et al. 1983; Revsbech et al. 1983; Cohen 1984b). (18) Gavish Sabkha, Sinai, Red Sea (Krubein et al. 1979; Gerdes et al. 1985). (19) Ebro Delta, Spain (Mediterranean Sea) (Esteve et al. 1988). (20) Dutch Wadden Sea, The Netherlands (Cadee and Hegeman 1974). (21) Puck Bay, Poland (Baltic Sea) (Witkowski 1986). (22) Island of Mellum, Federal Republic of Germany (North Sea) (Stal et al. 1985). (23a) Limfjørden, Denmark (Jørgensen 1977; Jørgensen and Revsbech 1983). (23b) Øresund, Denmark (Gargas 1970; Fenchel and Straarup 1971). (24) Menai Strait, Wales (Irish Sea) (Whale and Walsby 1984). (25) Orkney Islands (North Sea) (van Gemerden et al. 1989). (26) Shackleford Banks, North Carolina, USA (Bebout et al. 1987). (27) Great Sippewissett Salt Marsh, Massachusetts USA (Nicholson et al. 1987; Pierson et al. 1987). (28) Laguna Madre and Baffin Bay, Texas, USA (Sorensen and Conover 1962; Dalrymple 1965). (29) Southern Florida, USA (Gebelein 1974, 1976; Golubić and Focke 1978; Simmons et al. 1985). (30) Bahamas (Atlantic Ocean) (Monty 1965, 1967; Gebelein 1974, 1976; Monty and Hardie 1976; Golubić and Focke 1978; Dill et al. 1986). (31) Turks and Caicos Islands (Atlantic Ocean) (Yehuda Cohen, pers. comm. to B. K. P.). (32) The Bermuda Islands (Atlantic Ocean) (Gebelein 1969, 1976; Goulubić and Focke 1978). (33) Bonaire, The Netherlands Antilles (Caribbean Sea) (Golubić and Focke 1978; Lyons et al. 1984a). (34) Guerrero Negro, Baja California Sur, Mexico (Javor and Castenholz 1981, 1984a, b; Jørgensen and Des Marais 1986, 1988; Jørgensen et al. 1987). (35) Laguna Figueroa (Mormona), Baja California Sur, Mexico (Horodyski 1977; Horodyski et al. 1977; Margulis et al. 1980; Stolz and Margulis 1984; Stolz 1984). (36) Camargue, Rhone Delta, France (Caumette et al. 1988). (37) Coasts of Peru and Chile (Gallardo 1977; Maier and Gallardo 1984b). (38) Mannar Lagoon, Northwest Sri Lanka (Gunatilaka 1975).

○ Inland Waters: (40) East African Rift Valley Lakes (Lake Magadi) Kenya (Eugster and Jones 1968; Beadle 1974; Bauld 1981). (41) Lake Yao (and other saline lakes), Chad (Round 1961; Bauld 1981). (42) Wadi Natrun, Egypt (Imhoff et al. 1979; Bauld 1981). (43) Etosha Pan, Namibia (Yehuda Cohen, pers. comm. to B. K. P.). (44) Transvaal Province, South Africa (Gomes 1985). (45) Hutt Lagoon, Western Australia (Bauld 1986). (46) Lake Thetis, Western Australia (Bauld 1986). (47) Lake Clifton, Western Australia (Moore 1987). (48) Lake Cowan, Western Australia (Clarke and Teichert 1946). (49) Malham Tarn, North Yorkshire, England (Pentecost 1987). (50) Walker Lake, Nevada, USA (Osborne et al. 1982). (51) Great Salt Lake, Utah, USA (Halley 1976). (52) Green Lakes, New York, USA (Eggleston and Dean 1976). (53) Lake Gardsjon, Sweden (Lazarek 1982). (54) Dry Valley Lakes and Streams, S. Victoria Land, Antarctica (Love et al. 1983; Wharton et al. 1983; Parker and Wharton 1985; Simmons et al. 1985; Howard-Williams et al. 1986; Vincent and Howard-Williams 1986). (55) Coorong Lagoon Region, South Australia (Walter et al. 1973; vonder Borch 1976). (56) Pine Creek Lake outlet, Absaroka Mtns., Montana, USA (Haack and McFeters 1982). (57) Stream near Cuatro Cienegas, Mexico (Winsborough and Golubić 1987).

◇ Terrestrial Mats: (58) Dry Valleys and Ross Island Ice Shelf, Antarctica (Fogg and Stewart 1968; Friedman and Ocampo-Friedmann 1984; McKay and Friedman 1985). (59) Truelove Lowland, Devon Island, Northwest Territories, Canada (Leal Dickson, pers. comm. to B. K. P.).

studied extensively in northern Europe (20–25) and the USA (26–27).

In the subtropical and tropical zones, where high evaporation rates lead to relatively high salinities and where seasonal fluctuations in temperature tend to be minimal, extensive accumulations of microbial mats occur, particularly in sheltered bays and lagoons. The composition of the mats varies tremendously in these environments. Mats may be composed almost entirely of cyanobacteria, but commonly they include surface layers of cyanobacteria covering lower layers of purple sulfur bacteria and associated organisms when biogenic sulfide is available. Variations also occur with salinity, a factor particularly well illustrated in the hypersaline lagoons of Guerrero Negro (34; see Section 6.10). By far the most extensively studied microbial hypersaline system is that of Solar Lake on the shore of the Red Sea in the Sinai (17). Within this small "lake" occur several different mat types that exhibit substantial seasonal variation. Shark Bay, Western Australia, is another intensively studied hypersaline region that includes several distinct environments. At Shark Bay, extensive inter- and supratidal cyanobacterial mats occur, as well as large subtidal stromatolites constructed by precipitation and sedimentation within mats composed largely of diatoms and other eukaryotic algae in addition to prokaryotes. Other, less well-studied marine marginal areas include Spencer Gulf, South Australia (12); Christmas Island (13) and Rangiroa Atoll (14) in the Pacific; Aldabra Island (15) in the Indian Ocean; Abu Dhabi (16) in the Persian Gulf; the Gavish Sabkha (18) near Solar Lake; the Ebro (19) and Rhone Deltas (36) of the Mediterranean Sea; and the Gulf Coast, Caribbean, and Atlantic Ocean regions (28–33). Included in this latter region are the recently described large lithified subtidal stromatolites off the Bahamas (3; Dill et al. 1986). Mats very similar to the well-studied mats of Solar Lake also occur in the hypersaline lagoons of Baja California Sur, Mexico (34, 35), and the mats within the salt evaporation ponds at Guerrero Negro (34) were selected as particularly well-suited for intensive study by the PPRG because of their potential similarity to the stromatolite-building communities of some Late Proterozoic units (e.g., the Bitter Springs Formation of central Australia; see Sections 6.10 and 6.11).

One of the least studied, yet possibly very extensive marine mat systems is that composed of the chemotrophic sulfur-oxidizing bacteria of the genus *Thioploca* found off the coasts of Chile and Peru (37). Descriptions of these organisms are based solely on collections; the component species have not yet been isolated in culture (Maier and Gallardo 1984b). Although little *in situ* work has as yet been carried out, evidence suggests that mats formed by these bacteria may contribute significantly to vast areas of the marine benthos (Gallardo 1977).

Microbial mat communities found within inland water habitats, ranging from hypo- to hypersaline and including lacustrine and fluvial environments, show greater variation than any of those noted above. Because few of these mats have been studied ecologically or physiologically in great detail, they are reviewed here only briefly. For most such mats, sedimentological and organismal descriptions are available, and in some cases growth and productivity have been measured. Of particular interest among these widely scattered environments are the large thrombolites recently described in hyposaline Lake Clifton, Western Australia (47). Calcification in freshwater environments is widespread and can result in rather substantial thrombolitic bioherms such as those in Green Lake, New York (52). The participation of cyanobacteria in such freshwater calcification has been observed in many different localities (Monty 1976), and Pentecost (1987) has recently studied the effects of various environmental factors on calcification and the growth rate of *Rivularia* colonies in several localities in North Yorkshire, England (49).

The role of diatoms in mat and stromatolite formation has gained recent attention. Diatoms commonly form benthic and periphyton communities which can become quite thick and mat-like. The seasonal occurrence of such gelatinous accumulations of diatoms is so common in freshwater streams and lakes that no particular locations have been placed on the map shown in Figure 6.1.1. A general review of the ecology of diatoms in various habitats (Patrick 1977) points to their ubiquitous distribution. The mats that form in streams are usually complex communities that are transient and rarely accrete. Dense gelatinous mats of complex communities of diatoms and heterotrophic bacteria have been studied in oligotrophic freshwater streams (56). The significance of benthic diatom mats in the lithification process in freshwater fluvial stromatolites (57) has recently been described (Winsborough and Golubić 1987). In these thick lithifying mats, calcification occurs within the gelatinous layers of both diatoms and cyanobacteria.

Even in the extreme environment of Antarctica, the dry valleys of South Victoria Land harbor ice-covered lakes receiving sufficient light to support thick benthic mats of algae and cyanobacteria (43).

Terrestrial mats are widely distributed but relatively poorly studied. Vast areas of the arid western USA have soils and rock surfaces covered by crusts of varying thickness held together primarily by cyanobacteria (Campbell 1979). No particular locations have been placed on the map (Fig. 6.1.1) for this type of mat except to note the terrestrial mats and endolithic microbial communities of the dry valleys of South Victoria Land (58) and the cyanobacterial mats in the Arctic (59).

6.1.1C Approaches to the Study of Modern Microbial Mats

The study of microbial mats has a recent history with recognizable phases. Intensive study of a large variety of microbial mats initially focused on analysis of the structure of the mats, description of the microbial components, and studies of their sedimentology. Analysis of the microbial components and structural features was by light microscopy. Most of the significant work on structural and sedimentological analysis which occurred before the mid-1970s is referenced in Walter (1976b). Beginning in the 1960s and continuing through the 1980s, classical and novel techniques for microbial ecology were applied to microbial mats to assess the role of particular organisms in the mat community. Mat microorganisms were isolated and studied in pure culture. Physiological ecology

Introduction

became the focus of study of microbiologists who were more interested in the activity of the biotic components than their structure and sedimentology. The application of microautoradiography and electron microscopy and, most significantly, the use of microelectrodes permitted investigation of the orientation and activity of microorganisms *in situ* at a scale of resolution comparable to the dimensions of the biotic components themselves. Many of the significant advances in the study of the microbial ecology of mats are discussed or referenced in two recent volumes: Cohen et al. (1984a), and Cohen and Rosenberg (1989). Application of these new techniques has revealed that microbial mats are dynamic systems at a microscale. Questions can now be posed regarding evolutionary responses to environmental parameters at the level of the cell *in situ* in microbial mats. Environmental parameters such as oxygen, hydrogen sulfide, pH, and light are significantly altered within mats on a sub-millimeter scale. The significance of a particular mat in providing answers to a particular evolutionary question may not become clear until such high resolution analysis has been applied to it (for example, see Jørgensen and Nelson 1988).

The purpose of Chapter 6 is to apply the study of contemporary microbial mats to interpretation of the evolution of microbial life during the Proterozoic. This approach includes the following four components:

(1) An analysis of the structural features of modern microbial mats (Section 6.3) is necessary to provide a basis for interpretation of the structural information preserved in fossil stromatolites (Section 6.2).

(2) An analysis of the biomarkers and isotopic ratios of modern mats exhibiting various activities and stages of preservation (Section 6.8) is necessary to identify and interpret organic and isotopic signals in the geochemical record (Section 6.11.3).

(3) An understanding of the organisms, their distributions, and their activities in various microbial mats (i.e., the microbial ecology of mats; Section 6.3 through 6.7) is necessary to interpret successfully what types of organisms might have contributed to the mat communities of the Proterozoic.

(4) Finally, a holistic view of certain significant mats is necessary to understand what entire communities might have looked like and how they might have functioned in the Proterozoic; Section 6.9 surveys several such significant contemporary mat communities, and Section 6.10 summarizes results of a detailed investigation of one particular community, the hypersaline flat mats of Guerrero Negro, Baja California Sur, Mexico. In light of these studies, Section 6.11 synthesizes the broad parallels that can be drawn between modern microbial mats and fossil, Proterozoic, stromatolites.

6.2

Proterozoic Stromatolites

MALCOLM R. WALTER   JOHN P. GROTZINGER   J. WILLIAM SCHOPF

6.2.1 Stromatolite Morphology and Taxonomy

The element of stromatolite morphology that records the original microbial mat most directly is the lamina. The geometry of stacking of the laminae determines the gross morphology. All stromatolites may be described as having stratiform, conical, or convex laminae, a very simple classification that has the merit of having at least a coarse-grained biological significance for extant microbial mats. Stromatolites with clotted, unlaminated fabrics are called thrombolites (Kennard and James 1986).

Almost all little-metamorphosed limestones, dolomites, and magnesites of Proterozoic age contain stromatolites. They also occur in phosphorites, iron-formations and, rarely, in sandstones. Usually only structured stromatolites (i.e., those with more than a few millimetres of growth relief on their laminae) are conspicuous, but stratiform stromatolites are very abundant. Stratiform types are often very difficult to recognize both in the field and in the laboratory, but lateral and vertical gradation into more obvious stromatolites is one of the more useful features that allows identification of a finely laminated rock as a stromatolite. The problems of distinguishing stromatolites from similar but abiogenic sediments are discussed by Walter (1983). Stromatolites are described by sedimentologists, whose interest is to interpret paleoenvironments, and biostratigraphers, who have used them to determine the relative ages of rock sequences (see Section 10.6). Only the biostratigraphers need to describe stromatolites in detail, so it is the biostratigraphic literature that contains most information on morphology. Detailed descriptions of fossil stromatolites have concentrated on structured forms with a columnar morphology because they have the greatest array of features that can be used for making taxonomic distinctions. Figure 6.2.1 illustrates the range of features that characterize stromatolites.

Stromatolite taxonomy has been a subject of much controversy. Krylov (1976) was able to document twelve different taxonomic systems which overlap to varying extents. Most of these use formal, binomial, Linnean nomenclature to communicate taxonomic distinctions. Some stromatolite workers, however, prefer to avoid use of this formal nomenclature, considering it inappropriate for objects that are only partly biogenic and that were constructed by consortia of microorganisms rather than single species. Such objections have come largely from geologists, perhaps especially from those with little experience in the use of biological systematics, and despite this criticism the use of binomial systematics in the study of stromatolites continues to grow. The reason is simple: it provides a rigorous system for the storage and communication of large amounts of information about complex objects (Bertrand-Sarfati and Walter 1981). More than a century of use demonstrates its utility. It can communicate fine distinctions and its results are reproducible. Although arguments over taxonomic principles and practice will no doubt continue into the future, at present there is sufficient general agreement among stromatolite workers for this system of nomenclature to work better than any other yet devised.

The various approaches to taxonomy utilize different combinations of the features shown in Figure 6.2.1 to define taxa. Groups ("form-genera") and forms ("form-species") are the only categories in general use. It is now well established that a plexus of environmental and biological influences determines the form and fabric of stromatolites (e.g., Walter 1977; Semikhatov et al. 1979; Burne and Moore 1987), although only the most obvious of these influences have been studied to any significant extent. The small scale features of stromatolites – that is, the stromatolite "fabric" (lamina shape and microstructure) – reflect the general shapes of the constructing microbial colonies and frequently also record some information on the size, shape, and orientation of the cells in those colonies (e.g., Semikhatov et al. 1979; Walter 1983). As a consequence, taxonomic treatments of stromatolites that give weight to fabric features may provide evidence of the past distribution in time and space of different microbial communities. Often, however, too little regard is given to the effects of the diagenetic alteration of such microstructures, with the result that spurious taxa are defined. Difficulties in dealing with literature in foreign languages, and inadequate descriptions, have both resulted in a proliferation of synonyms. There is not yet a treatise of stromatolite taxonomy in which the systematics is rationalized; this is becoming an urgent requirement. Despite these serious limitations, a great deal of progress has been made in the last 30 years and

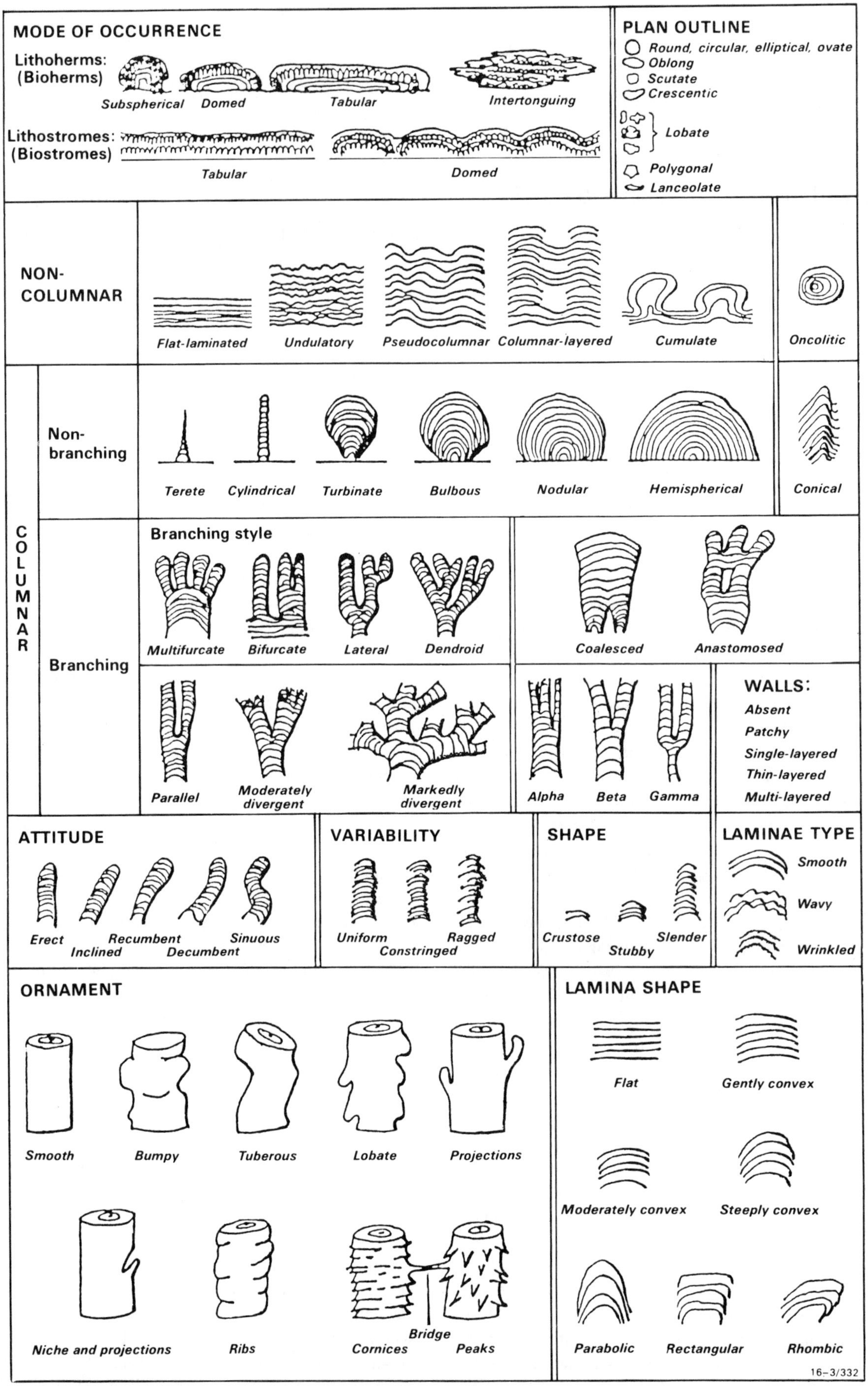

Figure 6.2.1. Morphological features used to characterize stromatolites.

stromatolites are being used successfully in biostratigraphy (see Section 10.6). As is discussed below (Section 6.2.5), it is possible to examine patterns of relative stromatolite diversity through time because the database is relatively large (over 800 forms have been described from about 200 basins), and the taxonomic problems can reasonably be assumed to have a uniform effect throughout the data.

6.2.2 Microbiotas of Proterozoic Stromatolitic Communities

As is summarized in Chapter 22 (Table 22.3), microfossiliferous stromatolitic units are now known to be both widespread and abundant in Proterozoic terranes. Over the past two decades, more than 1100 total occurrences of benthic microfossils have been reported from 190 Proterozoic formations (Fig. 11.3.1A); many, but not all, of these fossil microorganisms have been preserved by permineralization in cherty portions of carbonate stromatolites.

Microfossiliferous Proterozoic stromatolites commonly include various planktonic components, especially small coccoid unicells that are evidently of chroococcacean, cyanobacterial affinity (e.g., Figs. 5.4.15, 5.4.16), as well as non-cyanobacterial prokaryotes. With but few (generally quite local) exceptions, however, such assemblages tend to be dominated by benthic, colonial, chroococcalean (viz., chiefly chroococcacean and entophysalidacean) and, particularly, nostocalean cyanobacteria, the latter of which are represented by both cylindrical tubular sheaths (Section 5.4.6; Fig. 5.4.26) and fossil cellular trichomes (Section 5.4.7; Fig. 5.4.32) that are essentially indistinguishable from the sheaths and trichomes of extant oscillatoriaceans (Figs. 5.4.7, 5.4.33, 5.4.34).

Since the early 1970s, numerous stromatolitic microbiotas have been studied in considerable detail. Because of their significance in the historical development of the field (see Section 5.2.2) and the subsequent studies prompted by their rather detailed description in the mid- to late 1960s, the microbiological communities of the Early Proterozoic Gunflint Iron Formation of southern Canada (Barghoorn and Tyler 1965; Cloud 1965), and the Late Proterozoic Bitter Springs Formation of central Australia (Schopf 1968a; Schopf and Blacic, 1971), have been particularly intensively investigated, but many other such assemblages, spanning virtually all of the Proterozoic, are now known (e.g., those of the Dismal Lakes Group and the Kasegalik and McLearly Formations of Canada; the Gamohaan Formation of South Africa; the Duck Creek, Amelia, and Skillogalee Dolomites, and the Frere and Paradise Creek Formations of Australia; the Tyler, Beck Spring, Galeros, and Kwagunt Formations of the USA; the Vempalle Formation of India; the Gaoyuzhuang and Wumishan Formations of China; and the Zil'merdak, Sukhaya Tunguska, Shorikha, Min'yar, Olkha, Chichkan, and Yudoma Formations of the USSR; see Table 22.3 for details and reference citations). Photomicrographs of representative microorganisms from these and other Proterozoic stromatolitic units are included in Chapter 24.

Most studies of Proterozoic stromatolitic communities have been concerned with documenting the range of morphological and taxonomic variability represented in such assemblages.

Although some such studies have also addressed related taphonomic problems (see Section 5.3.2), virtually no detailed analyses have been made in stromatolitic communities of variability in lamina-by-lamina biotic composition, of other possible trends with depth, or of lateral biotic variability, studies of the type carried out on modern mat-building communities (Sections 6.4, 6.5, 6.9, 6.10).

Despite the deficiencies in available data outlined above, it is notable that there are numerous well-documented similarities, both in ecology and, moreover, in morphology – at scales ranging from that of gross morphology to that reflected by minute features of small-scale fabric (Sections 6.2.3, 6.3, 6.11) – between modern microbial mats and Proterozoic stromatolites. Clearly, such similarities are consistent with formation of both types of structure by the same or very similar set of biotic agents. Further, in both the modern mats (Section 6.3.2) and the fossil stromatolites (Chapter 24: Pl. 3, A; Pl. 6, A; Pl 21, H), microbial filaments (both sheaths and trichomes) tend to be oriented parallel or subparallel to, and commonly partially or wholly comprise, the characteristic accreted laminae, similarly suggesting a common mode of lamina formation. Finally, the validity of this supposition seems confirmed by the dominance of cyanobacteria and, in particular, of members of the same two cyanobacterial families (viz., the Oscillatoriaceae and Chroococcaceae), in both the modern and the fossil biocoenoses. Much remains to be learned regarding Proterozoic stromatolitic communities; for example, although photosynthetic bacteria must surely have played a role in such assemblages, that role, and even their presence in Proterozoic stromatolitic communities, has yet to be adequately documented. Nevertheless, the known stromatolitic microfossil record seems unequivocal: like their modern microbial mat counterparts, the vast majority of Proterozoic stromatolites were evidently formed by, and produced as a result of the activities of, cyanobacterium-dominated microbial communities.

6.2.3 Paleoecology of Proterozoic Stromatolites

Stromatolites occupied every major ecological niche known to be important in the construction of Phanerozoic platforms by more complex and environmentally sensitive organisms (Grotinger 1989a). However, no firm relationship has yet been established between the inferred variety of microbial benthic communities and their potential effect on development of specific facies in Proterozoic carbonate platforms. Although it is now generally accepted that gross morphology of stromatolites is determined mainly by environmental factors, and that the fabric is controlled primarily through biological interactions (e.g., Semikhatov 1976), few studies attempt to relate specific stromatolite types to specific settings. Some exceptions include work by Donaldson (1976), Hoffman (1976), Grey and Thorne (1985), Grotzinger (1986a,b) and Southgate (1989). Few generalizations can be made at this point, but it appears that certain stromatolite types recur in specific settings (Fig. 6.2.2). These types include conical columnar stromatolites and asperiform or "microdigitate" stromatolites. The development of conical and microdigitate stromatolites probably had locally important effects on the evolution of particular carbonate platforms where these forms were present, especially

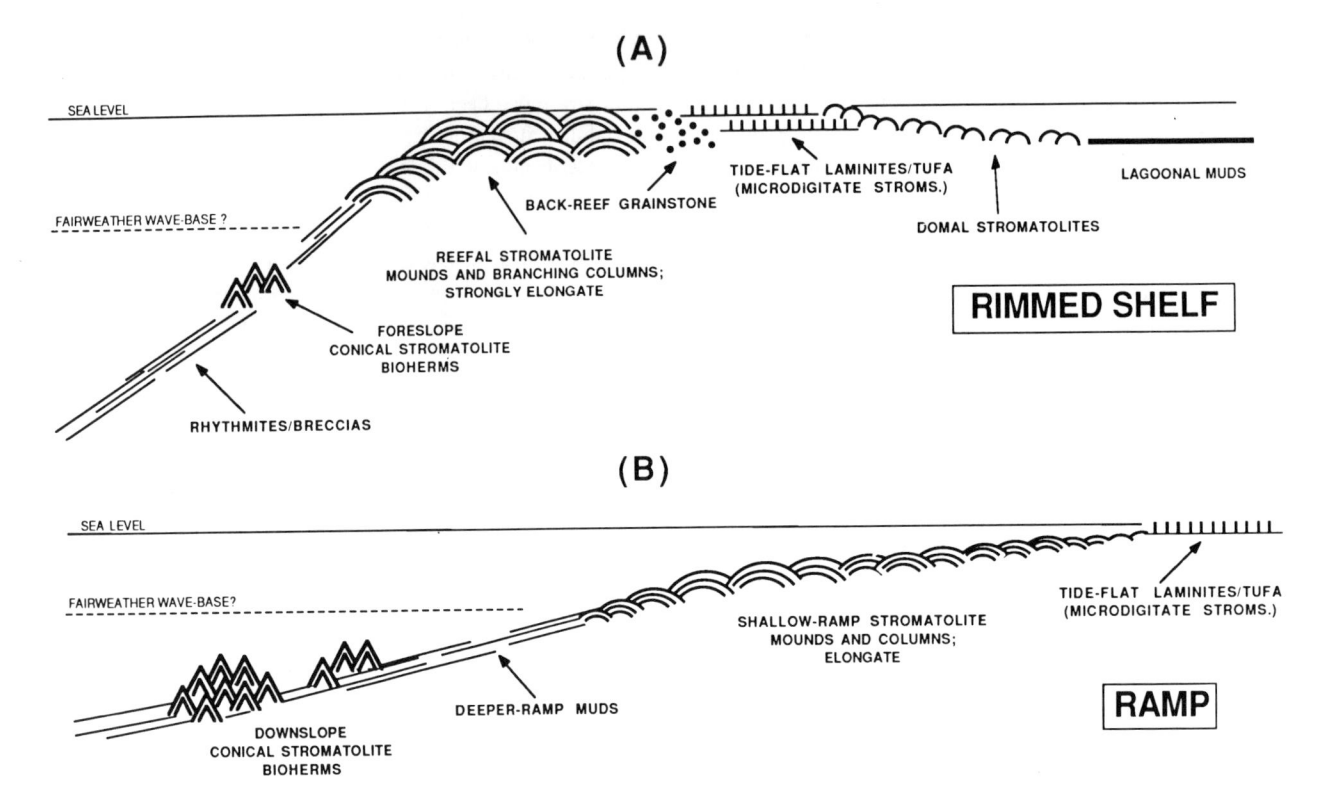

Figure 6.2.2. General distribution of specific stromatolite types across Proterozoic carbonate platforms. (A) Rimmed shelf as a high-energy (windward) margin characterized by a barrier reef complex of strongly elongate stromatolite mounds and columns. Conical stromatolites may form below wave-base as foreslope deposits; domal stromatolites of the inner-shelf lagoon are weakly elongate to non-elongate as a result of their restricted, low-energy setting: tufas (including microdigitate stromatolites) form by precipitation of tidal flats. (B) Ramp as a moderate-energy platform. Elongation of stromatolite mounds and columns is dependent on the relative amount of wave surge and/or tidal strength; in low-energy settings, elongation is minimal. Note the decrease of stromatolite mound size toward both deeper and shallower settings; elongation may show a similar relationship (Grotzinger 1989a).

with regard to carbonate production. It is noteworthy that these two stromatolite groups represent paleoenvironmental end-members, and that they also contain the most clear-cut evidence for a precipitational origin. The great majority of stromatolite types occur somewhere between these end members, including domes, linked domes, most columnar stromatolites, and their elongate equivalents; these latter types are the dominant Proterozoic reef builders.

6.2.3A Conical Columnar Stromatolites

Conical columnar stromatolites apparently developed in deep to shallow subtidal paleoenvironments. The potential subtidal restriction of *Conophyton* was first recognized by Donaldson and Taylor (1972). Subsequently, this stromatolite group has been recognized as the dominant and often exclusive stromatolitic component of basinal and slope environments (Fig. 6.2.2), and as a transitional facies in incipient to terminally drowned platform sequences (e.g., Donaldson 1976; Hoffman 1976; Grotzinger 1986a; Beukes 1987). Additionally, conical columnar stromatolites occur in probable deep ramp facies in the Middle to Late Proterozoic Beck Spring Dolomite (Pahrump Group, western USA; Grotzinger, unpublished data), and also in the Late Proterozoic Bambui Group of Brazil (Cloud and Dardene 1973).

In the Pethei platform (Hoffman 1974, 1989a), conical stromatolites were important sources of carbonate precipitation in basinal and deep to shallow slope and ramp facies where carbonate production might not have occurred otherwise. This may have been an important control in maintaining the relatively gentle gradients of the Pethei platform (as evidenced in the Taltheilei and Wildbread Formations), in contrast to other platforms where the lack of contemporaneous carbonate production in basin and slope environments resulted in rapid slope steepening (e.g., the Rocknest platform; Grotzinger 1986a). This relationship is apparently supported by the lack of evidence for steep platform slopes in another two platforms where conical stromatolites had a significant role in slope and basinal carbonate production (viz., the Cambellrand and Dismal Lakes platforms; Beukes 1987; Kerans and Donaldson 1989).

6.2.3B Microdigitate Stromatolites

Asperiaform or "microdigitate" columnar stromatolites appear to be almost exclusively restricted to peritidal environments (Fig. 6.2.2). These stromatolites are generally less that one centimeter wide, less that 10 cm high, and tend to form units up to a few meters thick that extend laterally for hundreds of kilometers. They are particularly well-developed as capping facies in shallowing-upward tidal flat cycles (e.g., Hoffman 1975; Kerans 1982; Grey and Thorne 1985; Grotzinger 1986a, b;

Grotzinger and Gall 1986). They also tend to occur as encrusting layers that drape unconformity-related erosional surfaces (Grotzinger et al. 1987).

The likely precipitate origin of microdigitate stromatolites may have had a significant effect on the evolution of tidal flat sequences. The ability to directly produce carbonate probably enabled tidal flats to shallow and prograde more efficiently during the development of shallowing-upward peritidal cycles (Grotzinger 1986b), rather than having to depend on shoreward transport of subtidally produced carbonate (cf., James 1984).

6.2.3C Other Stromatolite Types

Most other stromatolite types probably formed in a diverse range of deeper subtidal to intertidal reef, open shelf, lagoonal, and tidal flat settings (Fig. 6.2.2). Examples of lacustrine stromatolites are also known (e.g., Southgate 1989). In general, inferred low energy coastlines are characterized by non-elongate, domal or linked-domal stromatolites with less common non-elongate columnar stromatolites. These facies are typically developed behind barrier reef complexes of rimmed shelves in protected inner-shelf peritidal or lagoonal environments (Fig. 6.2.2a), or as deeper subtidal to intertidal facies of low energy ramps.

In contrast, inferred high-energy coastlines typically had non-linked to partially linked columnar stromatolites with elongation aspect ratios of >5:1. Non-elongate domes and linked domes are rare. Most shoal-water reef complexes are also characterized by strongly elongate columnar stromatolites (Fig. 6.2.2). These commonly are basic components of much larger, strongly elongate mounds on a scale of 10 to 100 meters in length. Stromatolite mounds may interfinger with lenses of grainstone to form "mound-and-channel" reef belts along platform margins of rimmed shelves (Fig. 6.2.2a). Buildups of strongly elongate stromatolites are characteristic of many platform positions including barrier reefs of rimmed shelves, and deeper subtidal to intertidal settings of high energy ramps (Fig. 6.2.2b).

The variation in form of these stromatolite types as a function of platform position is generally limited to changes in the synoptic profiles of mounds and/or individual stromatolites, and in their elongation aspect ratios. There is an excellent correlation between inferred paleowater depth and stromatolite form such that shallow-water forms often have lower synoptic relief (e.g., Grotzinger 1986a; Hoffman 1989a; Ricketts and Donaldson, in press). The correlation between inferred paleowater depth and elongation aspect is less clear, owing to other external factors such as wave interference patterns that also influence elongation (cf., Logan et al. 1974).

6.2.3D Stromatolite Reefs

A reef, as defined by Lowenstam (1950) in terms of ecologic principles, is "...the product of the actively building and sediment-binding biotic constituents, which, because of their potential wave resistance, have the ability to erect rigid, wave-resistant topographic structures." Lowenstam also stated that those principles applied to both modern and ancient reefs, independent of time, and allowed for specific evolutionary adaptation. It is clear that this definition is independent of the specific organism responsible for reef growth. The culminating stage of reef development, following growth to sea level, was lateral expansion of the reef habitat to "...influence its surroundings, locally affecting circulation and salinity and affecting sedimentation actively through contribution (of sediment) and passively through creating a zone of turbulence."

Lowenstam's ecologic definition of a reef has major significance for stromatolite-constructed buildups. Because most stromatolites show evidence of having been produced primarily through the trapping and binding and/or precipitation-inducing activity of benthic microbial communities, they can be regarded as having had the same function as did individual metazoans during development of Phanerozoic reefs – they are directly responsible for the vertical accretion of the structure upward into the zone of physical destruction (via wave-action). In many cases, reef development subsequently caused strong paleoenvironmental zonation of carbonate platforms. Stromatolites and stromatolite bioherms are generally streamlined, having become elongated during their buildup into the zone of intense wave action.

Many Proterozoic stromatolite buildups (e.g., Grotzinger 1989a, b) demonstrate that stromatolites did form true reefs that are directly comparable, in a paleoecological context, to Phanerozoic, metazoan-constructed counterparts. These stromatolite buildups exhibit all of the critical criteria cited by Lowenstam to be indicative of a true reef, including evidence for growth from deeper to shallower water wave-agitated settings, and continued growth and flourishing within the zone of active physical destruction. Stromatolite reefs occupied a variety of different niches, similar to younger counterparts. These include major barrier reefs adjacent to large seaways, patch reefs that formed in restricted lagoons located behind such barriers, patch reefs located on gentle ramps facing open seaways, pinnacle reefs analogous to the Silurian reefs of the Great Lakes region, USA, and even downslope bioherms that grew entirely within a deeper, quieter-water setting.

6.2.4 Growth and Lithification of Proterozoic Stromatolites

The single most important difference between Proterozoic and Phanerozoic carbonate platforms is the absence in the former of carbonate secreting Metazoa and calcifying algae. Consequently, all Proterozoic carbonate muds and coarser allochems (generally intraclasts and ooids) must have precipitated either abiotically, or by precipitation regulated indirectly by microbial activities (such as fixation of CO_2 during photosynthesis). Surprisingly, this difference seems to have had little effect on the gross facies mosaics and morphologies of Proterozoic platforms, which closely mimic those of the Phanerozoic platforms (Grotzinger 1989a). This relationship demonstrates that zones of Proterozoic carbonate production and accumulation must have been nearly identical to those of their Phanerozoic counterparts.

Proterozoic stromatolitic sediment probably accumulated through one or more of three possible mechanisms: (*i*) *in situ* precipitation as fibrous (marine) cement; (*ii*) *in situ* precipitation of micrite that was passively accreted by settling from suspension, or was actively accreted through "trapping and binding" by microbes; and (iii) precipitation of micrite imported from laterally adjacent environments. Historically, the latter model for stromatolite growth has been favored. However, this previously prevalent view was based largely on early observations of modern stromatolites; more recent studies are beginning to recognize the importance of *in situ* precipitation in stromatolite development.

6.2.4A Stromatolites and Proterozoic Carbonate Production.

Buildups. Stromatolites probably represent prolific *in situ* carbonate production in the form of major buildups and reefs, some of which may have been comparable in scale to the largest of Phanerozoic barrier and pinnacle reefs. Stromatolites may have also been responsible for the production of "intraclast" grainstone that occurs adjacent to many buildups and can be accounted for through erosion and degradation.

Thus, emphasis can be placed on stromatolitic buildups not only as carbonate "depositories" but also as actual "factories." Other evidence for this model includes locally abundant marine cementation in the vicinity of stromatolites, as cements between component particles of stromatolites, and also in the direct precipitation of stromatolitic laminae (e.g., Kerans 1982). The presence of substrate-parallel layers of neomorphic fibrous cement that coat and form adjacent to stromatolites is evidence for *in situ* precipitation of accretionary layers. The preservation of radial fibrous fabrics in neomorphic carbonates (usually dolomite) that constitute "microdigitate" stromatolites (microbial tufas) of early Proterozoic tidal flat facies is similarly evidence for *in situ* sediment accumulation by precipitation. Other, more indirect arguments also apply, but with a correspondingly lower degree of confidence. For example, the minimal incorporation of sand grains into stromatolitic laminae where stromatolites are strongly admixed with siliciclastics is a common field observation (e.g., Donaldson 1963; Serbryakov and Semikhatov 1974). Additionally, the fine lamination of Proterozoic stromatolites has been cited as evidence of possible precipitation, in contrast to Phanerozoic "trapping and binding" stromatolites which often show poor lamination (e.g., Cloud and Semikhatov 1969; Walter 1972a). In general, however, the degree of autochthoneity of most stromatolitic carbonate has not been demonstrated.

Carbonate muds. A related feature that also serves to distinguish Proterozoic from Phanerozoic platforms is the dominance of fine-grained carbonate in Proterozoic platforms. Most Proterozoic carbonate mud production was apparently aided by the presence of stromatolites, in that many Proterozoic carbonate mud units are associated with stromatolitic units.

Based on the demonstrable involvement of ancient microbial communities in stromatolite construction (Walter 1983; Klein et al. 1987), and because the volume of stromatolitic sediment is generally equal to or greater than the volume of non-stromatolitic carbonate (e.g., Grotzinger 1986a), it seems likely that micritic carbonate was indirectly precipitated through the activity of stromatolite-forming benthic microorganisms. This may have occurred during photosynthesis, when CO_2 was extracted from the water column producing an increase in pH and a resulting decrease in carbonate solubility. Such a mechanism might have been effective on a platform-wide scale, inducing carbonate pricipitation wherever stromatolitic buildups and reefs were developed. On many platforms, stromatolites probably formed veneers spanning deeper subtidal to upper intertidal environments, and it is conceivable that the "factory" spanned the entire platform, was operative on a diel basis, and that its performance was dependent only on the rate of turnover of bank water.

It is also possible that during the Proterozoic, "whitings" (Cloud 1942) may have been a common or even diel phenomenon. Although it is reasonable to expect that this may have occurred inorganically, the effect would have been greatly enhanced through the activities of benthic and planktonic photosynthetic microbiotas.

6.2.4B Diagenesis of Proterozoic Stromatolites

Microbial components of Proterozoic stromatolites were best preserved during early silicification of sediments. Although the mechanisms by which such silicification occurs are poorly understood, it is likely that Proterozoic seawater had a much higher concentration of dissolved silica than in younger times owing to the absence of organisms with siliceous tests (Holland 1984). Due to the excellent fabric preservation observed in Proterozoic cherts that have replaced primary carbonates (cf. Hofmann and Jackson 1988), it is likely that replacement occurred before inversion of metastable carbonate minerals occurred. Such conditions would have been very favorable for the preservation of delicate microorganisms.

Most modern marine carbonate sediments consist of aragonite and high-magnesium calcite, both of which are metastable at surface conditions and invert readily to low-magnesium calcite during diagenesis. Aragonite and high-magnesium calcite appear to have been important components of Proterozoic carbonates as well (Grotzinger and Read 1983; Grotzinger 1989a). During neomorphism of aragonite and high-magnesium calcite to low-magnesium calcite (or dolomite; Grotzinger and Read 1983), recrystallization occurs. Micrites (with grains $<2\,\mu m$) become somewhat coarser grained (2 to $10\,\mu m$), and cements also increase in crystal size. This process often results in fabric destruction of primary elements such as crystal fibers, leaving only "ghosts" of their former presence; in the process, most microorganisms are obliterated (see Hofmann and Grotzinger 1986 for a rare exception). These fabrics do, however, preserve larger scale elements in much more detail than would result from other diagenetic processes such as coarse, sucrosic dolomitization, or dissolution of sediment and reprecipitation of void-filling cement. Future work on the carbonate petrology and diagenesis of stromatolites will probably provide much important information on the relative roles of "trapping and binding" versus precipitation in the formation of stromatolites.

6.2.5 Temporal Distribution, Taxonomic Diversity, and Abundance of Proterozoic Stromatolites

First documented by Awramik (1971), Proterozoic stromatolite diversity was later analyzed by Walter and Heys (1985) after there had been a five-fold increase in the number of described taxa. The latter authors also attempted to analyze Proterozoic stromatolite abundance. The geological abundance of thrombolites has been documented by Kennard and James (1986). There is now a huge literature, especially in Russian, Chinese, English, and French, on the time ranges of stromatolite taxa, and it is beyond the scope of this discussion to present a comprehensive compilation of these data. Earlier compilations made when the task was not overwhelming (e.g., Walter 1972a; Semikhatov 1976; Preiss 1976b; Krylov and Semikhatov 1976) remain useful. More recently, these summaries have been supplemented by compilations for particular regions (e.g., Semikhatov 1980; Chumakov and Semikhatov 1981; Raha and Sastry 1982; Zhu 1982; Golovenok 1985; Liang et al. 1985) or for particular sets of taxa (e.g., Grey and Thorne 1985) or geological times (Bertrand-Sarfati and Walter 1981). A recent brief summary is presented by Hofmann (1987).

Figure 6.2.3 presents data on the diversity and abundance of stromatolites through the Proterozoic and Cambrian and, for thrombolites, for all of the Phanerozoic. The data are replotted from Walter and Heys (1985) and Kennard and James (1986); details of the methods of data collation and presentation can be found in those papers. Diversity was scored as the number of forms in each time interval. In estimating abundance, ideally one would want to know thicknesses of stromatolite-containing sedimentary sections in large numbers of sequences of different ages. Few such data are available, especially for stratiform stromatolites. For this reason, abundance is scored as the number of formations from which thrombolites have been recorded for each time interval, and the number of forms within each time interval multiplied by the number of basins from which each form has been recorded. Both of these are measures of frequency of occurrence of particular types of stromatolites. The available data are unevenly distributed in two ways: first, the time intervals to which the stromatolites are referred are of differing durations; and second, there are more rocks to sample in some time intervals as compared with others (i.e., the number of basins or of formations, or the rock mass are not uniformly distributed through time). To compensate for both of these effects, the raw counts were divided separately by the number of basins in each time interval and by the function $(1 + T/R)$, where T is the duration of the time interval and R is the average time range of all taxa; the two results were then averaged. On Figure 6.2.3 the results are referred to as "relative abundance" and "relative diversity." These corrections were not used on the counts of thrombolites, and those data are presented as they were by the original authors (Kennard and James 1986). The taxonomic problems referred to above are clearly one source of error in these analyses, but the effect should be minimal because only relative, not absolute, diversity is being considered. Dating of the occurrences is the other major source of error; because of the imprecision imposed by present geochronometric methods, only long time intervals have been used (in particular, the time intervals here used are those of the standard stratigraphic scheme used in the Soviet Union, the source of most of the plotted data).

6.2.6 Interpretation of the Stromatolitic Fossil Record

As discussed in Section 10.2, the Proterozoic stromatolite record reveals some significant temporal patterns, the most outstanding of which are the following:

(1) Ministromatolites with a radial fibrous fabric (Hofmann and Jackson 1987), characteristic of peritidal environments, were abundant in the Early and Middle Proterozoic, with a marked decline thereafter (Grey and Thorne 1985; Grotzinger, in press).
(2) *Conophyton* and related stromatolites, characteristic of quiet subtidal environments, were abundant in the Early and Middle Proterozoic, with a marked decline thereafter (Komar et al. 1965; Zhu 1982; Walter and Heys 1985).
(3) The maximum recorded diversity of stromatolites, represented by more than 600 taxa, was reached during the Middle and Late Proterozoic (Walter and Heys 1985).
(4) The abundance and diversity of all stromatolites declined toward the end of the Proterozoic (Awramik 1971; Walter and Heys 1985).
(5) Thrombolites, which became abundant during the Early Cambrian (Walter and Heys 1985; Kennard and James 1986), were absent or nearly absent during the Proterozoic.

Within this coarse framework, many finer patterns of temporal distribution have been recognized, at least in the Middle and Late Proterozoic, on a time scale of 100 to 300 Ma. The major features of the record listed above all seem to be real and not just artifacts of a limited data set. Walter and Heys (1985) found no correlation between their measures of abundance and diversity and the numbers of authors that had published on each time period represented in their analysis. The total number of data points is large enough (Fig. 6.2.3) to have some confidence to the interpretations. Most significantly, Awramik's (1971) analysis of only 20% of the data later available to Walter and Heys (1985) produced essentially the same result. Nonetheless, there have been many fewer monographic studies of Early Proterozoic stromatolites than of younger ones, so caution is appropriate when interpreting the apparently low abundance and diversity of stromatolites of that age.

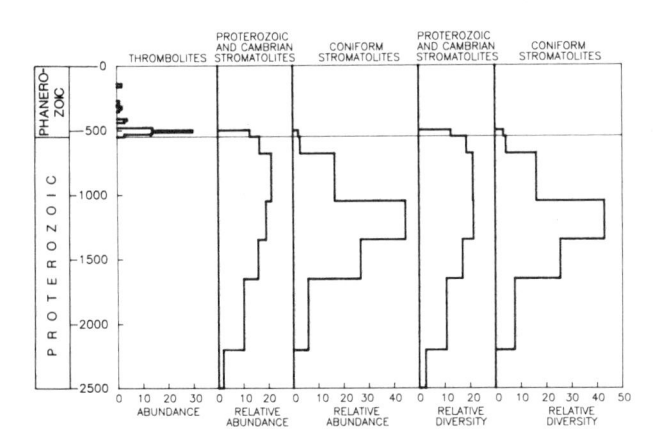

Figure 6.2.3. Abundance and diversity of Proterozoic and Phanerozoic thrombolites and of Proterozoic and Cambrian stromatolites. Data from: Walter and Heys (1985) and Kennard and James (1986).

Although it is reasonable to postulate a link between microbial evolution and stromatolite change (see Figure 11.3.10), such a link cannot at present be firmly demonstrated. Other explanations are proffered and cannot yet be refuted. For example, Gebelein (1974, 1976) suggested that major low-stands in sea level would have caused substantial changes in the composition of benthic microbial communities independent of biotic evolution, and that this could be sufficient explanation for any observed changes in stromatolite microbial assemblages. A similar proposal was made by Kennard and James (1986) to account for the temporal distribution of thrombolites (with all of these authors also allowing some role for evolution). Awramik (1971) and Walter and Heys (1985) postulated a direct link between the decline in the abundance and diversity of stromatolites in the Late Proterozoic and the ecological effects of newly evolved grazing and burrowing metazoans. No one disputes that grazing and burrowing metazoans disrupt the fabric and restrict the distribution of extant microbial mats (Section 6.7), but unequivocal evidence of these same processes in fossil stromatolites is not found until the Cambrian. Burrows are abundant in thrombolites of Cambrian and younger ages. Microscopic and mesoscopic animals, such as nematodes and tiny annelids, are abundant in many modern mats, including those at Shark Bay (Section 6.7). A comparable meiofauna probably inhabited Proterozoic stromatolites, and a search for evidence for this fauna has begun, so far without success. An alternative hypothesis attributes the Late Proterozoic stromatolite decline to the effects of the addition of eukaryotic algae to the cyanobacterial communities, suggesting that this would have produced less coherent mats, or that macroscopic algae would have outcompeted the microorganisms for space and nutrients. This seems unlikely, as evidence for the existence of eukaryotic algal plankton now extends back to 1.7 Ga (Sections 5.4, 5.5), long before the Late Proterozoic decline in stromatolite abundance and diversity. Finally, recent studies suggest that there may have been a decline in the diversity (and presumably abundance) of both benthic and planktonic cyanobacteria during the Late Proterozoic (Section 11.3), leading to the suggestion that another extrinsic factor, deteriorating climate, for example, might have been the root cause of some major changes in such communities. Grotzinger (in press) has proposed that an abundance of ministromatolites with a radial fibrous fabric which record *in situ* precipitation of calcium carbonate in Early and Middle Proterozoic sequences is consistent with other evidence suggesting that seawater was greatly oversaturated with respect to calcium carbonate and perhaps also had a high HCO_3^- to Ca^{2+} ratio at that time. All of these explanations for changes in stromatolite diversity and abundance during the Proterozoic are best regarded as working hypotheses in various stages of development, and in any event they are not mutually exclusive.

Finally, the revelation that some of the stromatolites and most of the described microfossils of the Late Proterozoic Bitter Springs Formation of central Australia are lacustrine (Southgate 1987), rather than marine as was long assumed, indicates the danger of interpreting the Proterozoic fossil record without the support of a sedimentological framework. That framework is still very limited, particularly in the Late Proterozoic. As a result, it would be foolish to be too emphatic in insisting on any particular interpretation of detected trends in the record of stromatolites. But the regularities mentioned here, and many others, exist and indicate that the stromatolite record will continue to be a rich source of information about the history of benthic microbial communities.

6.3

Modern Microbial Mats

JOHN BAULD   ELISA D'AMELIO   JACK D. FARMER

6.3.1 Morphology of Modern Microbial Mats

Microbial mats are usefully defined as "accretionary, cohesive, microbial communities, which are often laminated, and found growing at solid-aqueous interfaces; most can stabilize unconsolidated sediment" (see Section 6.1). In addition to autochthonous organic matter produced by the constructing (and associated) microbes, mats may comprise both organic (live and/or dead) and inorganic allochthonous sediment. In addition, some mats may incorporate autochthonous chemical precipitates important for penecontemporaneous cementation and lithification and, hence, preservation in the rock record (see Section 6.3.3).

The morphological attributes of microbial mats are usefully considered at several levels of scale ranging from the macroscopic to the microscopic, i.e., from gross external shape (cm to m), through internal fabric of the organosedimentary structure (mesoscopic; mm to cm), to the form and arrangement of constructing microbes (μm to mm); the latter may require high quality light, or electron, microscopy for satisfactory resolution. Discriminatory, characteristic morphological attributes appear to be the product of both environmental and biotic factors (see Section 6.3.2) but, in general, gross macroscopic features are most frequently the consequence of physical, environmental processes (Horodyski 1976, 1977a; Playford 1980; see Walter et al. 1973 for probable exceptions) while micro- and some mesoscopic features result from biotic controls (Gebelein 1974; Awramik et al. 1976). The presence of microbial communities able to exert such control is, in turn, a consequence of physicochemical constraints associated with the overall environment which may be further modified by complex interactions and feedback effects (see Section 6.3.3).

Despite what is considered to be the extreme variation in gross morphology of ancient stromatolites (Hofmann 1973), and the fact that the geological literature is replete with arcane descriptive terminology (see Preiss 1976a, b), the morphology of modern microbial mats is restricted to several fundamental and relatively simple shapes which can be modified by such contemporaneous or post-depositional events as desiccation, erosion, burial, compression, and diagenesis. Table 6.3.1 presents a representative sample of modern microbial mats together with information about their macrostructure and microstructure, including synoptic profile, fabric, and constructing microbes. Synoptic relief of ancient stromatolites ranges from several millimeters to several (ca. 5 to 15) meters (Hofmann 1973). Until recently, modern microbial mats giving rise to lithified structures of significant relief were known only from a few localities, mostly, though not exclusively, in such hypersaline environments as Hamelin Pool (Logan et al. 1974), the Great Salt Lake (Carozzi 1962), and Green Lake (Dean and Eggleston 1975). Recently, there has been a series of discoveries from both marine and lacustrine habitats of near-normal sea water salinities (e.g. Dravis 1983; Moore et al. 1983; Dill et al. 1986). Generally these occurrences exhibit a maximum synoptic relief in the range 0.5 to 1.0 m, penecontemporaneous cementation and lithification, and a poorly laminated fenestrate fabric.

Other mats, lacking significant lithification, produce less synoptic relief (frequently 2 to 4 cm, rarely as much as 10 cm; Table 6.3.1), with domes and cones being the most common morphologies. In both of these cases, comparable morphology can be produced by taxonomically distinct, though morphologically analogous, microbial constructing agents. (Domes generated through gas bubbles or substrate conformation are excluded from consideration in this discussion.) Even domes of similar dimensions and general appearance may be separable on the basis of their fabric; for example, "*Scytonema*" domes exhibit a radial fabric as a consequence of the coarseness of the radially oriented *Scytonema* trichomes (Monty 1976), whereas the narrower *Phormidium* trichomes, despite their periodic vertical orientation, produce only a minimal (or zero) radial overprint in the fabric of domal structures which they construct or in which they reside (Gebelein 1969; Monty and Hardie 1976; Monty 1976; Golubić and Focke 1978).

Distinctive conical stromatolites (*Conophyton*), common during the Proterozoic (Walter and Heys 1985), have two possible modern analogues from disparate environments – geothermal (Walter et al. 1972; 1976) and marginal marine (Horodyski 1977b; Javor and Castenholz 1981) – each constructed by a distinctive suite of filamentous prokaryotes and under differing geochemical regimes (Table 6.3.1). Neither of these analogues is totally satisfactory, however, because the

Table 6.3.1. *Representative selection of modern microbial mats illustrating the range of macrostructure and microstructure together with relevant biotic and environmental information. Sources of data are given in parentheses, with number corresponding to the list at the bottom of the table.*

Mat Type (Reference #)	Habitat	Gross Morphology & Synoptic Profile	Mat Topography	Asymmetry	Dominant Constructors (Diameter)
Smooth/Stratiform (3, 4, 15, 16, 21)	Marginal marine: intertidal semi-permanent standing water; salinities ca. 50–120‰ (low energy)	Flat, no relief (low amplitude undulations inherited from original surface)	Smooth, flat; may develop gas bubbles and polygonal desiccation cracks	Absent	*Microcoleus chthonoplastes* (trichome ca. 5 μm, filament ca. 30 μm) with other cyanobacteria (incl. presumptive *Chloroflexus*); filamentous, gliding, ensheathed
Domal (7, 8, 18)	(Low) intertidal to subtidal; latter can be high energy; salinities ca. normal marine	Hemispheroidal domes; relief 0.4–3.0 cm	Smooth, curved; (occasionally knobby crusts)	Absent/rare to ellipsoidal as current speed increases	*Phormidium hendersonii* (ca. 1.5 μm) essentially monospecific; filamentous, gliding, ensheathed
Tufted (4, 11, 13, 16)	Marginal marine: intertidal, intermittently emergent; salinities ca. 30–100‰ (low to moderate energy)	Conical (occasionally subspherical) superimposed on stratiform; relief $\leqslant$1 cm, occasionally 3–4 cm	Pinnacles, cones, and ridges arise from smooth mat platform (see above); these may form reticulate or anastomosing patterns on bedding plane	Axis may be inclined to vertical	*Lyngbya aestuarii* (ca. 15–20 μm), other filamentous cyanobacteria at varying, usually low levels; filamentous, gliding, ensheathed
Columnar/-Conical (24, 25)	Alkaline geothermal effluents (pH 7–9; 32–59 °C); submerged or with intermittent subaerial exposure	Columnar to conical, emerging from stratiform sheets; relief mostly 2–3 cm but range is from <0.1–10 cm	Surface of mat and columns/cones smooth, but with vertical ridges and spines common	May be subcylindric in transverse section; axis may be inclined to vertical	*Phormidium tenue* var. *granuliferum* (1.0 μm), *Chloroflexus aurantiacus* (0.5–1.0 μm) subdominant; filamentous, gliding, ensheathed
Colloform (1, 2, 3, 16, 21, 23)	Subtidal to (very low) intertidal; marginal marine and saline lacustrine (salinity max. ca. 65–70‰); environment erosive	Cylindrical to club shaped columns (occas. small cones) arise from lithified substrate; relief $\leqslant$0.5 m to 0.75 m with bioherms occasionally to 1 m or greater	Mat 2–5 mm thick follows lithified surface; mat with soft smooth surface	Bioherms may be ellipsoidal, e.g., 3 m $\times$ 0.6 m at 0.6 m relief; compound structures show irregular branching	(Stalked) diatoms with filamentous/coccoid cyanobacteria and other eukaryotic micro-algae; coccoid cyanobacteria in carbonate "mush" above lithified zone; stalked diatoms
Domal pancakes (17, 19)	Freshwater marshes and mud flats bordering freshwater lakes; intermittent marine incursions bring storm deposits; normally low energy	Domes and pancakes $\leqslant$30 cm diameter having relief $\leqslant$6–10 cm (or arising from sheet ca. 0.5–3.0 cm thick)	Smooth "pincushions" (spongy)	?	*Scytonema mychrous* (ca. 12–15 μm) (V) and *Schizothrix calcicola* (ca. 1.5 μm) (H); filamentous, gliding (*S.c.* only), ensheathed
Pustular/Mammill-ate (2, 3, 9)	(Upper) intertidal marine to hypersaline; sustained subaerial exposure	("Stratiform") sheets with $\leqslant$2–4 cm relief; low domal and ridge/rill undulations usually $\leqslant$5 cm	Irregular cauliflower-like hemispheroidal pustules	Some ridge/rill structures – current controlled?	*Entophysalis major* (ca. 5–6 μm) essentially monospecific; coccoid, colonial, ensheathed

1. J. Bauld, unpubl.; 2. J. Bauld in prep.; 3. Bauld (1984); 4. Davies (1970); 5. Eriksson and Truswell (1974); 6. Fairchild and Subacius (1986); 7. Gebelein (1969); 8. Golubić and Focke (1978); 9. Golubić and Hofmann (1976); 10. Hofmann (1969); 11. Horodyski (1977b); 12. Jackson et al. (1987); 13. Javor and Castenholz (1980); 14. Knoll and Golubić (1979); 15. Krumbein et al. (1977); 16. Logan et al. (1974a, b); 17. Monty (1976); 18. Monty (1979); 19. Monty and Hardie (1976); 20. Park (1977); 21. Playford and Cockbain (1976); 22. Upfold (1984); 23. Walter and Bauld (1986); 24. Walter et al. (1972); 25. Walter et al. (1976a).

Fabric	Accretion Process and Rate	Penecontemporaneous Lithification	Grazing and/or Bioturbation	Probable Proterozoic Analogue(s) (Reference #)
(Usually) fine lamination, often striking; laminae 1–10 mm thick, frequently sedimentation-event determined	Sediment trapping and building; laminations commonly persist to 30–50 cm depth; accretion ca. 1–10 mm/y	Absent or rare	Grazing and burrowing recorded where low salinity waters inundate mats; in-mat meiofauna present (e.g., nematodes)	Transvaal Dolomite (5) Balbirini Dolomite (12)
Finely laminated; alternating thick (ca. 600 μm) and thin (ca. 30 μm) laminae–former with erect trichomes, latter with prostrate trichomes	Sediment trapping and binding; accretion seasonal, avg. rate 0.5 mm/d to max. ca. 3.0 cm/d	Filaments calcify but structure is not lithified	Not recorded but potential grazers are usually present in habitat	Bambui Group (6) (note: laterally linked; the only microfossils present are coccoid)
Finely laminated (generally in low sediment influx areas)	Sediment trapping and binding; accretion to ca. 1 cm common	Absent/rare	Grazing by cerithid gastropods and corixid insects	Stoer Group (22) Lynott Formation (12)
Distinctly laminated to diffusely streaky; most laminae 5–15 μm thick; macrolaminations 35–60 μm; columnar laminae are irregularly conical (axial laminae comparable to *Lyngbya* tufts axes)	Contemporaneous silicification, biomass, minor detrital; accretion estimated at 0.5–3.0 mm/y	Silicification; aragonite formation in carbonate-rich effluents	Potential up to 51°C; grazing by ephydrid flies and potamocyprid ostracods	Sibley Group (10)
Lithified fabric is coarsely fenestral to (irregularly) laminoid; laminae 1–3 mm thick; may be unlaminated (thrombolitic); lamination of unlithified surface mat is poor/coarse	Trapping and building with contemporaneous cementation; accretion rates estimated at ≤1 mm/y	Carbonate cementation (aragonite) with subsequent lithification	Forams, serpulid worms, boring pelecypods, meiofauna (incl. nematodes), crustaceans, gastropods, fish	No exact equivalents known; similarities to Paleozoic thrombolites (see Kennard and James 1986)
Laminated, radial fabric; laminations derive from alternating erect (V) and prostrate (H) filament orientation, consequent upon alternating populations of cyanobacteria	Sediment trapping and binding; filaments become calcified; accretion to ≤6–10 cm over 10–15 cm of subsurface accretion; accretion <2.5 cm over "several years"	Cementation and lithification does not occur despite rapid calcification of filaments	Grazing and bioturbation occurs; insect larvae, ostracods and other small crustaceans, flatworms, crabs, gastropods, plant roots	?
Unlaminated to (rarely) poorly laminated; irregular fenestral, clotted fabric (low sediment influx); subsurface cells tend to form columnar stacks	Sediment trapping and binding; accretion estimated at ≤2 mm/y	Poor	Nematodes	Bitter Springs Formation (14) Belcher Group (9)

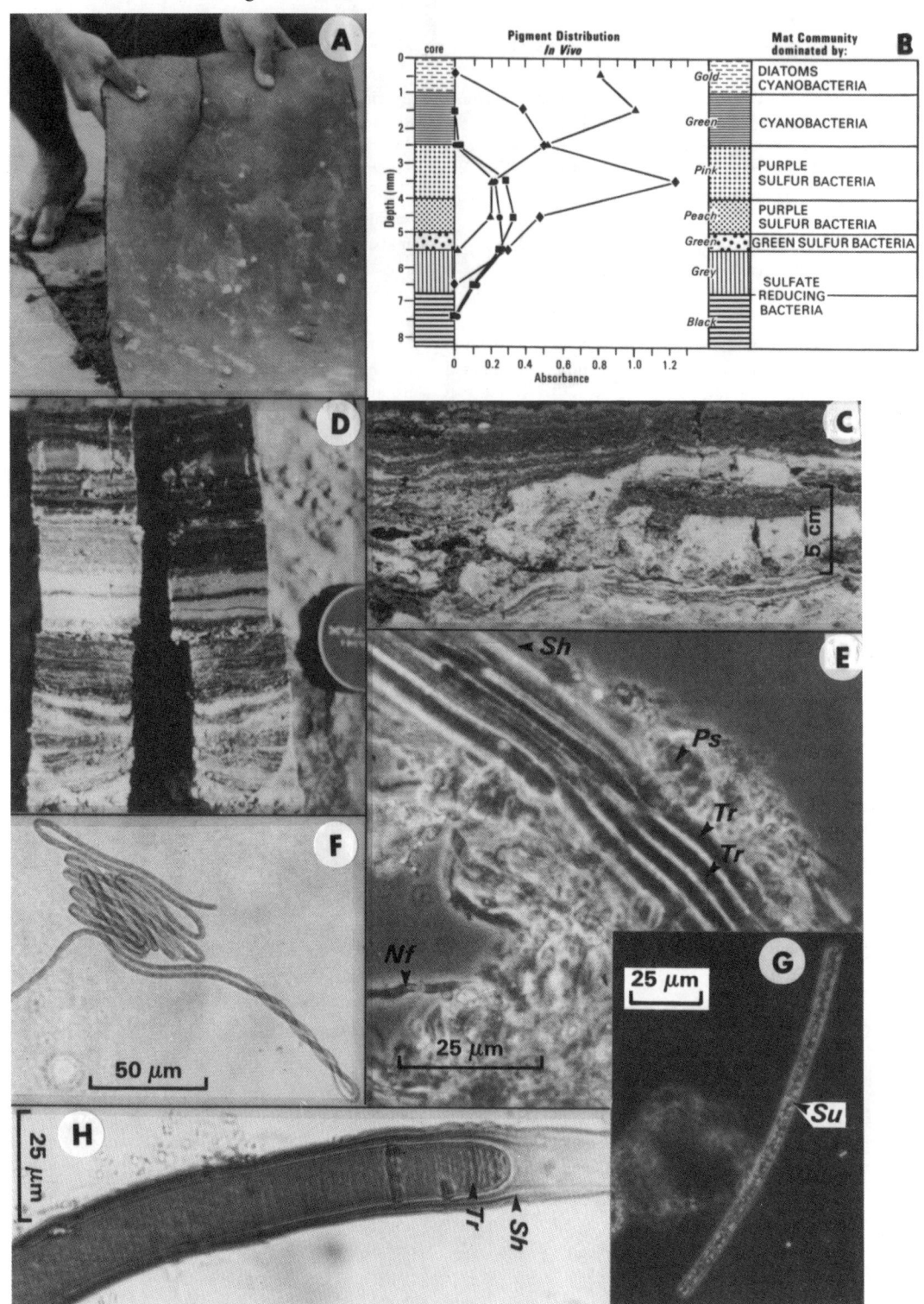

264   Modern Mat-Building Microbial Communities

Figure 6.3.1. Representative prokaryote-dominated microbial mats and their agents of construction (for eukaryote-containing mats, see Fig. 6.9.11). (A) A water-saturated example of a classic "smooth" mat from the high intertidal zone of Spencer Gulf, South Australia, illustrating the cohesive strength of mats constructed by ensheathed, filamentous, gliding cyanobacteria (here, primarily *Microcoleus chthonoplastes*; see [E], below). (B) Vertical distribution of photosynthetic pigments and dominant microbial components in a multi-layered mat from Great Sippewissett Marsh, Massachusetts, USA, illustrating the characterisitic millimeter-scale vertical stratification of microbial populations and their light-absorbing pigments. (C) Upper portion of a box core showing a vertical profile through a finely laminated mat colonizing mangrove sediments at Crane Key, Florida, USA. Laminations are produced as the mat accretes vertically by the interaction of the constructing agents (filamentous cyanobacteria and associated biomass, the darker areas in the photograph) with deposited carbonate sediment (lighter areas). Sediment delivery rates may fluctuate from those which closely match the trapping and binding capacity of the microbial community (fine laminations) to storm deposits which greatly exceed this capacity. (D) Vertical profile through a core from laminated "smooth" mat at Gladstone Embayment, Shark Bay, Western Australia. This habitat is characterized by almost continuous water coverage with infrequent tidal flooding. The freshly cut faces of the split core show alternating dark (organic-rich microbial sediment including amorphous FeS_2) and light bands (fine-grained carbonate sediment). The primary agent of construction is the gliding, filamentous cyanobacterium *Microcoleus chthonoplastes* (see [E], below). The lens cap is 54 mm in diameter. (E) Phase-constrast photomicrograph of the gliding, filamentous cyanobacterium *Microcoleus chthonoplastes* which constructs "smooth" mat. Several cellular trichomes (Tr) are present within the hyaline sheath (Sh) to which adhere organic and inorganic detrital particles, and microbes such as purple sulfur bacteria (Ps) and narrow filamentous prokaryotes (Nf). (F) Phase-contrast photomicrograph of a filamentous cyanobacterium (cf. *Phormidium*) isolated into culture from "smooth" mat collected at Spencer Gulf (South Australia), where it is a minor but constant component. The organism glides actively and, as here shown, exhibits a propensity to become entangled, thus promoting mat cohesion. (G) Phase-contrast photomicrograph of *Beggiatoa* sp. from "smooth" mat collected at Gladstone Embayment, South Australia (see [D], above). Gloubules of elemental sulfur (Su), present in the periplasmic space, are visible in this short trichome. This organism occurs in the micro-niche where O_2 and H_2S coexist. (H) Phase-contrast photomicrograph of the gliding, filamentous cyanobacterium *Lyngbya aestuarii*, responsible for constructing "tufted" mats (see [I]–[K], below). This example, from Fisherman Bay, Spencer Gulf, South Australia, shows a lamellated, deeply pigmented, tubular sheath (Sh) enclosing a single trichome (Tr) composed of uniseriate, disc-shaped cells. *L. aestuarii* produces recognizably related but distinctive mat morphologies, the distribution of which may be environmentally controlled (see [I]–[K], below). (I) Photograph showing an anastomosing network of microbially produced low relief ridges (Ri) colonizing coarse, shelly sediment (Si) in a relatively high energy and frequently inundated environment (Mambray Creek, Spencer Gulf, South Australia). When the mat is subaerially exposed, pinnacles are poorly developed and not obvious, behaving in a manner analogous to camel-hair paint brushes. Coin is 28 mm in diameter. (J) Oblique view of the well-developed network of ridges (Ri) and moderate-relief pinnacles (Pi) formed in a low energy environment subject to infrequent but sustained inundation (Lharidon Bight, Shark Bay, Western Australia). Coin in richt foreground is 28 mm in diameter. (K) Exceptionally strong development of pinnacles (Pi) in a well protected environment (margin of a salt concentration pond protected by a levee) where inundation is most commonly due to wind-driven seiching of the water mass (Concentrator Pond No. 4, Exportadora de Sal SA, Guerrero Negro, Baja California Sur, northern Mexico). (L) Vertical view of a pustular mat surface showing cauliflower-like "pustules" of *Entophysalis major* which form an incomplete cover over the sediment. Also visible are shells (Sl) deposited between pustules and small (white) crystals of gypsum and halaite which coat the mat surface after subaerial exposure and evaporation (Hamelin Pool, Shark Bay, Western Australia). Coin is 28 mm in diameter. (M) Phase-contrast photomicrograph of *Entophysalis major*, the coccoid, colonial cyanobacterium which constructs pustular mats (cf. [L], above, and [N], below). The firm, multi-layered gelatinous sheaths (Sh), which enclose cells (Ce) and groups of cells, are deeply pigmented at the colony surface (Pg) but only slightly pigmented at depth (Po), consistent with the suggestion that sheath pigment may be produced in response to high intensities of incident solar radiation. (N) Pustular (also referred to as "mammillate") mat, constructed by *Entophysalis major* (see [M], above), occurs most commonly in coarse, well-drained sediments (here shown at Hutchison Embayment, Shark Bay, Western Australia). The morphology and growth habit of colonies of *Entophysalis major* is expressed in the irregular topography and unlaminated fabric of the mat. Scale, divided into 1 cm squares, is 10 cm long. [All photographs and photomicrographs are by J. Bauld; Fig. 6.3.1B is modified after Pierson et al. (1987b, Fig. 4) and Nicholson et al. (1987, Fig. 2).]

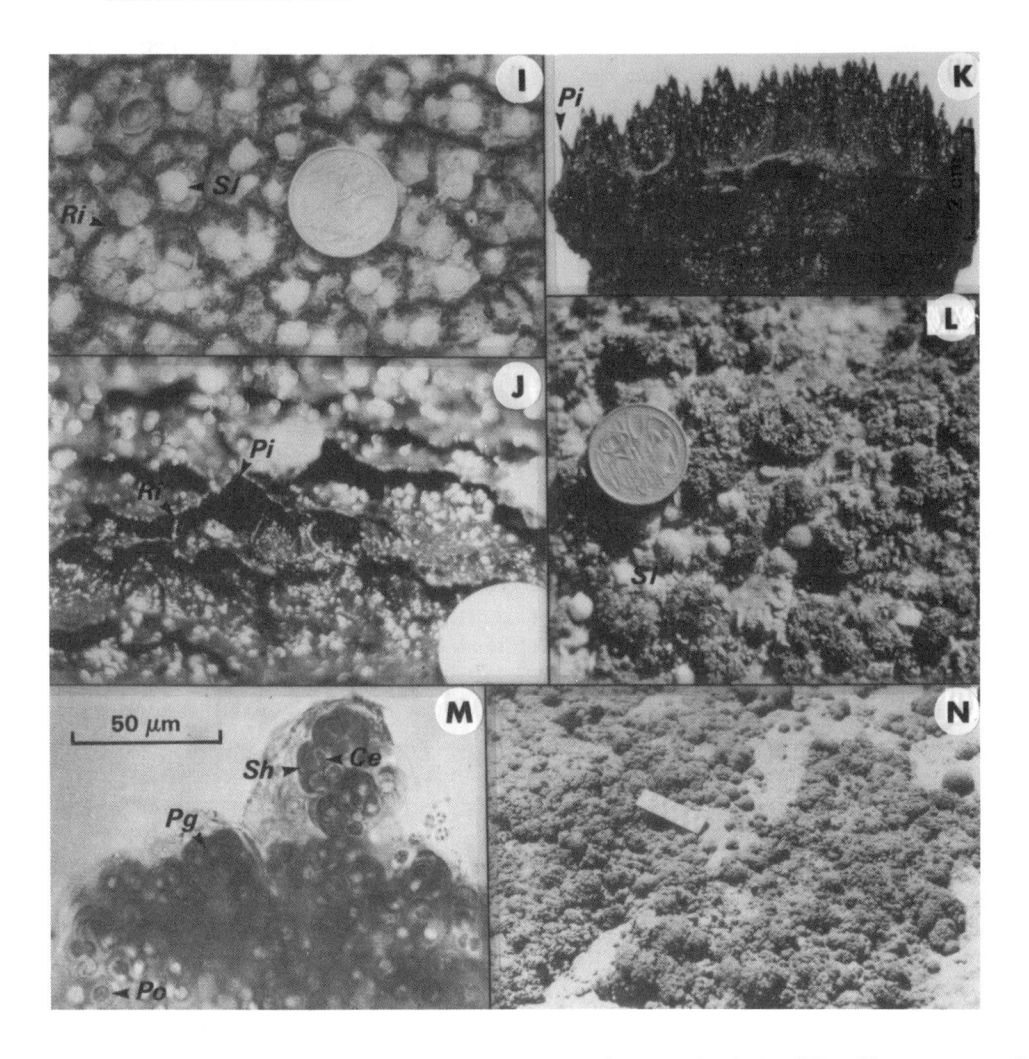

modern examples have a considerably lower synoptic relief than those known from the Proterozoic (Donaldson 1976; Walter et al. 1976a), despite convincing similarities in laminar fabric and morphogenesis (see Section 6.3.2) to each other and to that of the fossil forms (Donaldson 1976; Walter et al. 1976a; Horodyski 1977b).

Laminations are usually most striking in mats constructed solely or predominantly by filamentous microbes. The visibility of laminae may be augmented by inorganic deposition and/or biotic pigmentation (see Bauld 1986). Sediment deposition followed by trapping and binding may enhance, or even cause, laminations. However, laminations may be readily evident in the absence of such phenomena; for example, diel laminations are evident in *Phormidium* domes as a consequence of alternating vertical and prostrate filament orientation of the constructing cyanobacterium, whether or not sediment is entrapped (Golubić and Focke 1978; Monty 1979). In the absence of fabric obliteration by grazing, bioturbation, or intruding lithification (see Section 6.3.3), the degree of lamination is strongly correlated with microbial morphology and growth habit (see Table 6.3.1). Poor lamination may result directly from cell morphology, such as is the case with pustular mat (having irregular cauliflower-like topology) which is constructed by the colonial coccoid cyanobacterium *Entophysalis*, or in colloform mat which is constructed by a consortium of prokaryotic and eukaryotic phototrophs including stalked (branching) diatoms (Fig. 6.9.11). Unlaminated flat mats, the end-member of this progression, are constructed in geothermal effluents by *Chlorobium*, a small, non-motile, rod-shaped prokaryote (Castenholz et al. submitted-a). A representative suite of prokaryote-dominated microbial mats and the agents of their construction is shown in Figure 6.3.1.

6.3.2 Biotic and Environmental Factors Controlling Mat Morphology, Lamination, Fabric, and Morphogenesis

Physicochemical (environmental) factors ultimately proscribe the growth of those microbes which have the potential to construct mats; these, together with biotic factors, determine the extent to which such potential is realized (see Knoll and Bauld 1989). For example, salinity, temperature, and pH severely constrain which potential mat-builders can grow, but exert minimal control over morphology or fabric. Environmental factors such as light, dissolved oxygen, sulfide, and sedimentation may have additional, direct effects on lamination and fabric. Biotic controls include bioturbation and grazing, the levels of which are proscribed by environmental factors, an indication of the degree to which interactive feedback between biotic and abiotic factors may exert control over mat construction (see Section 6.3.3).

Physical environmental factors for which there is evidence of control over mat macrostructure (gross morphology) include prevailing wind direction, water depth, current velocity and direction, rates of sediment transport and deposition, and the nature of the substrate (Hoffman 1967; Gebelein 1969; Logan et al. 1974; Horodyski 1976; Semikhatov et al 1979; Playford 1980). For example, the morphology of subtidal cyanobacterial biscuits and domes is considered (Gebelein 1969; Fig. 19) to be directly controlled by current velocity and the rate of sediment movement. As current velocity increases, morphology changes from irregular and substrate-conformable to streamlined ellipsoidal. At a smaller scale, assymetric laminations reflect non-uniform sediment supply.

Where the agents of mat construction are phototrophs, organismal responses to light regime are important. Control on fabric by microbial response to light has been demonstrated in several instances (Gebelein 1969, 1974; Walter et al. 1976; Golubić and Focke 1978; Monty 1979). The detailed studies of Golubić and Focke (1978) show that domal mats built by *Phormidium hendersonii* exhibit prominent fine laminations. These are the result of alternating sediment-poor (hyaline) and sediment-rich (opaque) layers comprising erect and prostrate filaments, respectively, repeating on a diel cycle. Vertical filament orientation is produced by positive phototaxis during daylight hours. Overnight, the cyanobacterial filaments revert to a prostrate orientation on the mat surface without further vertical migration. In contrast, laminations in stratiform geothermal mats constructed by the anoxygenic filamentous phototroph *Chloroflexus aurantiacus* have been shown by Doemel and Brock (1974) to arise by nocturnal vertical migration through the overlying oxygenic cyanobacteria to the mat surface (for further discussion see Section 6.3.3).

The well-documented role of light in laminae production leads to consideration of the possibility that light may play a significant role in morphogenesis. Walter et al. (1976), as a consequence of their studies of extant mats which produce *Conophyton*-like conical mats in geothermal effluents, proposed that phototaxis by the gliding filamentous agents of construction is the mechanism underlying both lamination and morphogenesis. The formation of tufts, and later cones, was considered to be initiated on an otherwise flat mat by knotted entanglements of randomly gliding filaments. Subsequently, positive phototopotaxis is initiated in filaments deflected upward by the knot nucleation with concomitant selection of the most actively responding gliding filaments, resulting in maximal accretion in the tips of the tufts or cones. A similar mechanism was invoked by Horodyski (1977b) for comparable morphogenesis of analogous *Conophyton*-like "tufts" by *Lyngbya aestuarii*.

The occurrence of inclined Proterozoic columnar stromatolites has led to the proposal that growth direction is itself influenced by the direction of incident sunlight (i.e., via heliotropism; Nordeng 1963). More recently, Awramik and Vanyo (1986) reported observations on modern microbial mats that they considered to support this proposal. They examined the two presumed modern analogues of *Conophyton* (see Section 6.3.1 and Table 6.3.1) and subtidal columnar stromatolites (colloform mat) and found examples consistent with inclination due to heliotropism. However, occurrences were not ubiquitous, and Awramik and Vanyo (1986) thus suggested that the heliotropic response could readily be suppressed by other environmental factors. Walter et al. (1972; 1976a) did not find convincing examples of heliotropism during their investigations of morphogenesis in modern *Conophyton* from geothermal habitats, nor did Playford and Cockbain (1976) observe such inclinations on columnar stromatolites at Hamelin Pool.

The decline in stromatolite abundance and diversity toward

the end of the Proterozoic (Section 6.2.5) has been attributed to the advent of widespread grazing and burrowing fauna. The advent of metaphytes and their role in competitive exclusion may also be significant (see Sections 6.2.5 and 6.7). The activities of grazing and burrowing fauna prevent stable mat accumulation, or the preservation of their laminated remains, through non-competitive exclusion (Garrett 1970a; Awramik 1971; Mazzullo and Friedman 1977; Walter and Heys 1985). This paradigm has its origins in the studies of Garrett (1970a) on intertidal mat communities of Andros Island. Walter and Heys (1985) further speculated that the Late Proterozoic decline of subtidal stromatolites was due to the subtle effects of grazing and burrowing by meiofauna. Modern meiofaunal species such as nematodes and small annelids are capable of disturbing sedimentary fabric by moving grains (Gebelein 1977; Farmer and Richardson 1988). Active grazing on, or within, modern microbial mats is documented for ephydrid flies (Brock et al. 1969; Jørgensen and Nelson 1988), corixid insects (Javor and Castenholz 1984a), potamocyprid ostracods (Wickstrom and Castenholz 1985), nematodes (Schwarz et al. 1975; Farmer and Richardson 1988; Bauld 1984), and cerithid gastropods (Garrett 1970a; Gunatilaka 1975; Schwarz et al. 1975; Gerdes and Krumbein 1984; Javor and Castenholz 1984a; Bauld 1984). Cerithid gastropods actively graze upon, and limit the extent of colonization of microbial mats in Laguna Guerrero Negro (*Cerithidea californica*; Javor and Castenholz 1984a), Sabkha Gavish (*Pirenella conica*; Gerdes and Krumbein 1984), and Andros Island *Batillaria minima*; Garrett 1970a). Other fauna observed within or on mats include ciliates, rotifers, annelids, staphilinid beetles, and crabs (Gerdes and Krumbein 1987; Schwarz et al 1975). Reworking by crabs (Schwarz et al. 1975) and surface-contouring by grazing fish (Davies 1970; Schwarz et al. 1975) have also been reported.

Investigations of the origins of laminated fabrics within mat communities of Concentrator Pond 5 at Guerrero Negro, Baja California Sur, Mexico (see Section 6.10) have been aided by the restriction of faunal components due to the high prevailing salinity; bioturbation agents are meiofauna, mainly nematodes. Mat structure and microbial composition at this locality are discussed in Section 6.10. Diel movements (vertical migrations) of the gliding filamentous cyanobacteria *Microcoleus chthonoplastes* and *Phormidium molle* in response to microgradients of light, oxygen, and sulfide are the factors most important in defining fabrics which are expressed by the orientation of intact filaments, empty sheaths, and newly sheathless trichomes. Farmer and Richardson (1988) recognized four distinctive (micro) laminated fabrics in thin sections of these mats; only the first of these, a "homogeneous" fabric characterized by random orientations of sheaths and filaments and by "floating" detrital grains, is interpreted to be a consequence of microbioturbation by gliding filamentous prokaryotes (especially *Microcoleus* and *Beggiatoa*) and grazing nematodes. Microlaminated fabrics characterized by vertical, or by prostrate, orientations of sheaths and trichomes are interpreted by Farmer and Richardson (1988) to originate under conditions of moderate, continuous sedimentation, or of catastrophic (storm generated) burial, respectively. The fourth fabric type identified by these workers is often associated with "storm" laminae and is characterized by dense concentrations of organic matter and associated authigenic carbonate. Although authigenic carbonate is most abundant in these layers, its presence does not appear to lead to significant lithification.

6.3.3 Feedback Relationships Between Biotic and Environmental Factors

Feedback interactions among and between biotic and abiotic factors influence microbial mats (and by inference, stromatolites) at all scales of morphology. Feedback interactions between microbial components, expressed through changing physicochemical environmental parameters, are exemplified by light, oxygen, and sulfide which exert direct control on the type and level of physiological activity and, indirectly, on fabric. These factors at the same time induce behavioral responses in mat-building microbes, are resources for which they compete and, in the case of the latter two, are also products of microbial metabolism ultimately controlled by light intensity, quality and duration (see Section 6.4; Krumbein et al. 1979; Bauld 1986; Jørgensen and Nelson 1988). Although there is evidence that light alone, mediated via phototactic responses, can produce laminated fabric (Golubić and Focke 1978; see Section 6.3.2), the nocturnal upward migration of *Chloroflexus* (Doemel and Brock 1974) was found to be due to positive aerotaxis (Doemel and Brock 1977) in response to reduced oxygen levels. The latter was attributable to continued heterotrophic respiration in the nocturnal absence of oxygenic activity by the overlying cyanobacteria.

It appears not uncommon for filamentous cyanobacteria to undergo nocturnal migrations to the surface of microbial mats, a phenomenon usually interpreted as resulting from positive aerotaxis (Pentecost 1984; Whale and Walsby 1984). Many of these mats undergo anaerobic decomposition, via bacterial sulfate reduction with copious H_2S production (see Skyring and Bauld 1990), and the consequent possibility of negative sulfidotaxis has not been excluded. In contrast, nocturnal downward gliding into an anoxic reducing microenvironment has been reported for *Oscillatoria terebriformis* (Richardson and Castenholz 1987a), a response attributable to its inability to survive dark oxic conditions (Richardson and Castenholz 1987b). Where geothermal sulfide is present in overlying effluent waters, laminations may also arise from inversions of mat zonation such that oxygenic cyanobacteria are sandwiched between a sulfide-rich surface layer of *Chloroflexus* and an anoxic zone of decomposition (Jørgensen and Nelson 1988).

In the absence of sediment input, regularly laminated fabrics are produced by the diel response of constructing microbe(s) to light (and thereby to biogenic oxygen and sulfide). Similar laminations may be produced under conditions of constant sediment input as a result of the activity of one organism (Golubić and Focke 1978) or the alternating activities of two organisms (Gebelein 1969, Fig. 14). However, deposition (and hence trapping and binding) of allochthonous sediment, which also contributes to laminated fabric, may be highly irregular in frequency, volume, and distribution, particularly in intertidal environments or as a result of storm activity (Monty 1976; Park 1976). The resultant interaction between biotic and physical processes controls the periodicity, or lack thereof, of laminations evinced in the mat fabric. The physical dimensions of the microbial components, their behavior, and their capacity for rapid cementation, may control fabric via the size of sediment particles which can be incorporated into the accreting structure. Microbial mats constructed by cyanobacteria preferentially sequester fine-grained sediment (e.g., Gebelein 1969; Golubić and Focke 1978), while those constructed by mixed cyanobacterial-eukaryotic micro-algal communities, in which the main agents of construction are often diatoms (see Figs. 6.3.1 and 6.9.11; John and Bauld, in prep.) but may also include filamentous green algae (Section 6.9.5), contain relatively coarse sediment particles (Neumann et al. 1970; Logan et al. 1974; Playford and Cockbain 1976; Dravis 1983; Dill et al. 1986; Awramik and Riding 1988).

Interactions between constructing microbes and other biota also illustrate the feedback nature of the relationships and their mediation via environmental parameters. Cerithid gastropods can sustain long periods at salinities up to ca. 70‰ (Javor and Castenholz 1984a; Gerdes and Krumbein 1984), but their distribution is primarily regulated by their requirement for "havens" of normal 35‰ sea water from which they make feeding excursions to less habitable areas. *Cerithidea. californica* was demonstrated to be capable of grazing on all mats tested, and to restrict mat colonization of most lower intertidal sediments. However, under normal field conditions, only tufted (*Lyngbya*-constructed) mats were subjected to heavy grazing. *Calothrix* mats escaped grazing because the high intertidal areas they colonized were too dry, and the *Microcoleus* mats were too remote from "fresh" seawater havens. To some extent, grazing may be self-limiting; Javor and Castenholz (1984a) observed that *C. califirnica*, rather than grazing on intact mats, shows a preference for the soft, organic-rich sediments which it had previously destroyed but preferred to rework, thus preventing reestablishment of a stable mat.

The anomalous distribution of a nodular geothermal mat (constructed by *Calothrix thermalis* and *Pleurocapsa minor*) with respect to seasonal and diel thermal gradients was found to be due to grazing pressure exerted by the ostracod *Potamocypris* sp. (Wickstrom and Castenholz 1985). Selective grazing of *Calothrix* by *Potamocypris* prevented competitive exclusion of *Pleurocapsa* by the more rapidly growing *Calothrix*. Further, persistence of the nodular mat at its upper and lower temperature boundaries was due to non-competitive exclusion, by ostracod grazing, of competing mat communities (Wickstrom and Castenholz 1985). In addition to affecting mat morphology (either directly, by surface grazing, or indirectly, by controlling the distribution of a particular morphology), fauna may interact in a positive, literally constructive way. For example, columnar microbial mats, superficially resembling small *Conophyton*, are produced by an association between a mound-building polychaete worm and filamentous cyanobacteria (Shinn 1972). Polychaete worm-tubes also contribute to the mechanical strength of subtidal Bahamian mats (Neumann et al. 1970), as do the encrusting sponges and the lithified tubes of serpulid worms to subtidal colloform mats of Hamelin Pool (Logan et al. 1974a; Playford and Cockbain 1976; Walter and Bauld 1986). The latter are also known to form living reefs on their own (Boscence 1973), and reefs built by tube-dwelling spirorbid worms are known from the Lower Triassic (Peryt 1974).

Although mats may accrete by incorporating, through trapping and binding, allochthonous sediment particles, the preservation of gross morphology, characteristic fabric, and microfossils over geological time requires penecontemporaneous lithification, a process initiated as autochthonous inorganic cements precipitate from mat-sediment pore waters. Primary lithification of Proterozoic stromatolites was most commonly by calcium or calcium-magnesium carbonates (Section 6.2), but occurrences incorporating silicate (Knoll and Simonson 1981), phosphate (Chauhan 1979), and ferruginous oxides (Walter and Hofmann 1983) are also known. Infiltration of silica-rich pore waters appears to be most effective in preserving organic matter and, hence, microfossils (e.g., Hofmann and Schopf 1983). Penecontemporaneous lithification by carbonate (e.g., Playford and Cockbain 1976; Moore 1987) and silica (e.g., Walter et al. 1976a) is reported to occur in modern microbial mats. Such occurrences often (but not exclusively; see Gomes 1985; Burne and Moore 1987) exhibit poorly laminated to unlaminated, fenestrate to clotted fabrics more closely analogous to Cambrian and younger thrombolites (Kennard and James 1986) than to Proterozoic stromatolites.

The process of carbonate cementation (calcification) in lithifying modern microbial mats remains poorly understood (see summary by Lyons et al. 1984b). Such mats are unattractive experimental systems because the radioisotopic tracer ^{14}C binds to, and exchanges with, particulate carbonates and problems of physical damage prevent the use of microelectrodes, ehich are very powerful probes in non-lithified mats (see Chapter 20). Although frequently associated with cyanobacterial components in tufaceous microbial mats (see Golubić 1973; Pentecost and Riding 1986), nucleation and crystal growth can be demonstrated to occur on empty sheaths of cyanobacteria not normally associated with calcification (Pentecost and Bauld 1988) suggesting that the interaction of porewater chemical composition and biomolecular configuration may play a significant role (Bauld, unpublished observation). Porewater geochemistry of mat sediments from Solar Lake was interpreted by Lyons et al. (1984b) to indicate that degradation via bacterial sulfate reduction was a major process leading to carbonate precipitation and, presumably, later lithification.

Microbial decomposition of original phototrophic biomass is often nearly complete (see Skyring and Bauld 1990), but it may proceed without destroying fabric or morphology. In contrast, the depredations of grazing, bioturbation, desiccation, and erosion modify or obliterate primary fabric and characteristic morphology while constructing microbes remain recognizable. Lithification can be a powerful preservative, but diagenetic crystal growth and evaporite formation (e.g., of gypsum, anhydrite, and halite) tend to destroy not only fabric but also microfossils (Park 1977; Hofmann and Schopf 1983). A significant proportion of mats (see Table 6.3.1) does not undergo such lithification, suggesting that analogous Proterozoic mats may have been excluded from the fossil record (see Section 5.3.2). On the other hand, there is evidence to support the view that, during the Early and Middle Proterozoic, sea water was oversaturated with respect to calcium carbonate (Section 6.2; Grotzinger 1989), a phenomenon which may have resulted in more widespread calcification. Despite this, it seems probable that knowledge of Proterozoic benthic microbial communities and their ecology is distorted by selective preservation (Section 5.3.2), thus somewhat constraining use of the components of modern microbial mats as analogues of Proterozoic stromatolitic microfossils.

With regard to comparison of modern mats and fossil stromatolites, it seems significant that although bioherms with columnar branching appear to be widespread in the Proterozoic (see Walter 1976b; Horodyski 1977a; Grey and Thorne 1985; Jackson et al. 1987), they are relatively rare in modern environments (Logan et al. 1974a, b; Grey et al. 1990). Similarly, modern analogues of the small digitate columnar branching stromatolites so common in the Proterozoic (Grey and Thorne 1985; Hofmann and Jackson 1987) remain essentially unknown (see Grey et al. 1990, for a possible exception). On the other hand, as discussed above, two possible modern analogues of *Conophyton* have been recognized, a stromatolite morphology common only during the Proterozoic (see Section 6.3.2). These analogues are constructed by morphologically comparable, although taxonomically distinct, microbial communities which, however, occur in totally disparate habitats. Flat (stratiform), finely laminated microbial mats are widespread in a diversity of modern environments and have a multitude of Proterozoic counterparts. Their modern analogues are almost invariably constructed by filamentous phototrophs which, though taxonomically diverse and of variable diameter, are capable of gliding motility. Such flat mats are known from marine, lacustrine, and geothermal habitats with widely varying sedimentary regimes (see, e.g., Cohen et al. 1984). Finally, there appears to be a clear lineage from modern mats of poorly laminated to fenestral fabric, constructed by the coccoid, colonial cyanobacterium *Entophysalis*, to the Proterozoic analogue *Eoentophysalis*, both at the microfossil and the environmental levels (Golubić and Hofmann 1976; Knoll and Golubić 1979).

6.4

The Microenvironment Within Modern Microbial Mats

BO BARKER JØRGENSEN   RICHARD W. CASTENHOLZ
BEVERLY K. PIERSON

6.4.1 Microenvironmental Parameters

Modern microbial mats range in thickness from less than one millimeter to over one meter. The mats are built by microorganisms that range from one to a few micrometers in cell width. What is the scale in space and time that is important to the life and function of the microorganisms? At what resolution is it relevant or necessary to study their microenvironment?

Phototrophic mats, with the exception of highly gelatinous types, are actively growing only in the uppermost few millimeters into which light penetrates. To resolve a distribution over one mm within such euphotic zones, the size of the measuring instruments should ideally be less than 0.1 mm. Physical and chemical parameters vary rapidly within thin euphotic zones. The chemical pools, for example of oxygen, can turn over within less than a minute, and a one second time resolution is therefore necessary to analyze their dynamics. Several microsensors are currently available which meet these requirements; use of these instruments in studies of microbial mats is discussed in this Section.

Some environmental factors, such as the temperature of hot springs or the salinity of salt ponds, determine the type and general distribution of microbial mats that can occur, but do not exhibit steep gradients at a scale less than one mm. Other factors, such as light, oxygen, sulfide, pH, nutrients, and various metabolites, exhibit dramatic variations within the euphotic zone and are the cause of microzonations and migrations of organisms at and near the mat surface. The diel cycle of light and dark is particularly important for the chemical microenvironment. As this cycle changes through the seasons, zonation changes; diel cycles are therefore a key to formation of the characteristic lamination of thick microbial mats (Doemel and Brock 1974; Brock 1978; see Sections 6.3, 6.4.4).

6.4.2 The Physical Mocroenvironment

The fluid mechanical forces that determine the water movement in the microzone at the surficial mat-water interface are quite different from the dynamic forces in the bulk water above. At a small distance (of a few mm to one cm) from the mat surface, viscous forces dominate over inertial forces. The internal friction of the water creates a viscous boundary layer, that is, a film of water which tends to "stick" to the mat surface and does not participate in the general flow or circulation of the overlying water (Wimbush 1976; Boudreau and Guinasso 1982). As the surface is approached within this boundary layer, eddy diffusion gradually becomes insignificant relative to molecular diffusion which is the principal mechanism for mass transport across the solid-water interface (Vogel 1981; Santchi et al. 1983; Jørgensen and Revsbech 1985; Jørgensen and Des Marais in press).

By use of oxygen microelectrodes, the thickness of the diffusive boundary layer has been found to vary from 0.2 to more than 1 mm (Jørgensen and Des Marais in press). This thickness depends on the flow velocity of the overlying water and the roughness of the mat surface. Although molecular diffusion over such short distances is a relatively efficient transport process, the boundary does provide a transport resistance that delays chemical communication between the mat and its environment. The mean diffusion time of a small molecule like O_2 over a 0.3 to 1.0 mm thick boundary layer is 1 to 10 minutes, that is, of the same order of magnitude as the turn-over time of O_2 within the mat. The thicker the boundary layer, the more extreme are the chemical fluctuations that can occur at the mat surface (Jørgensen and Revsbech 1985). In mats of some filamentous bacteria (viz., *Beggiatoa* sp.), the surface structure varies with flow velocity and thereby with the thickness of the diffusive boundary layer. Under stagnant conditions, when the diffusive boundary layer is relatively thick, long ends and loops of filaments stretch up from the loose and fluffy mat surface. Under flowing water, the mat surface is smooth. This change of structure results from chemotactic responses by individual filaments to oxygen microgradients within the diffusive boundary layer (Møller et al. 1985).

In contrast with chemical transport, sunlight penetrates "instantaneously" into microbial mats and distributes according to the scattering and absorption properties of the mat. Solar spectral radiation is like that of a black body of approximately 6000°K. Before light reaches the earth's surface, however, this spectrum is modified due to absorption by atmospheric gases. As shown in Figure 6.4.1, within the spectral range of 350 to

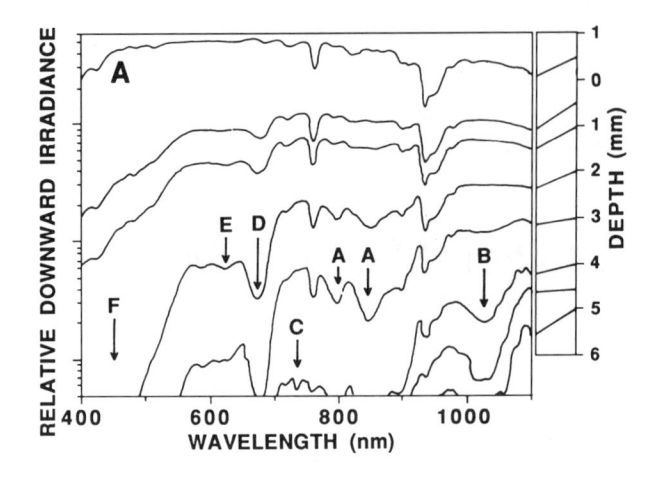

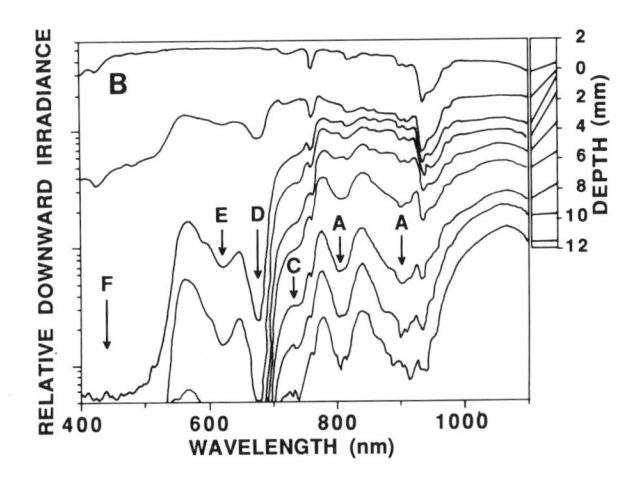

Figure 6.4.1. Spectra of relative downward irradiance plotted on a logarithmic scale at different depths in two microbial mats: (A) A "sand mat" from Sippewissett Salt Marsh, Massachusetts and (B) "Rabbit Creek spouter mat" from Yellowstone National Park, Wyoming. Spectra were recorded in bright sunlight using a battery-operated recording spectroradiometer connected to a tapered fiber optic tip inserted into the mat. The depths in mm above or below the mat surface are indicated for each spectrum. The solar spectrum just above the surface of the mat (+0.5 mm) shows little detail due to the logarithmic scale; however, sharp minima of the solar spectrum are recognized at all depths. Absorption maxima of predominant pigments in the mats are indicated on the figure with arrows (A, Bchl a; B, Bchl b; C, Bchl c; D, Chl a; E, phycocyanin; F, carotenoids) and are recognized as minima of residual radiation, especially in the deeper mat layers. (B. K. Pierson, unpublished data.)

1050 nm, the range of radiation potentially useful for photosynthetic microorganisms, sharp troughs develop in the light spectrum due to absorption by O_2 (at about 765 and 690 nm) and by H_2O vapor (at about 725, 825, and 935 nm). Below and above this range, strong absorption is caused mostly by O_3, in the ultraviolet, and by CO_2, in the infrared. As light penetrates through a water column to a microbial mat surface, the spectral range is narrowed further due to absorption by water, especially of infrared light. A one-meter-deep water column thus absorbs about 99% of 750 to 900 nm light. The near-infrared (NIR) absorption bands of bacteriochlorophylls therefore play a role for photosynthesis only under a minimum of water coverage. In deeper (substantially greater than 1 m) waters, chlorophyll and the phycobilins, and especially the blue-green absorbing carotenoids, serve as the principal light-harvesting pigments.

The light flux, which reaches different layers within the mats, can be measured either as (i) the total downwelling light passing a horizontal surface, i.e., "*downward irradiance*"; (ii) downward directional light within a defined spherical angle, i.e., "*downward radiance*"; or (iii) the integrated light from all sides which reaches a defined point in the mat, i.e., "*scalar irradiance.*" The definitions of these light parameters and the applicable measuring techniques are described in Chapter 20. Recent measurements of radiance distributions are based on use of fiber-optic probes which can be made with tip sizes as small as a few μm (Vogelmann and Bjorn 1984; Jørgensen and Des Marais 1986a, b; Vogelmann et al. 1988).

Figure 6.4.1 shows two examples of the downward spectral irradiance measured in mats of different fabric and pigment composition: a "sand mat" from an intertidal salt marsh (Pierson et al. 1987) and an air-exposed gelatinous mat sprinkled by spray from a hot spring (Castenholz 1984). Measurements were done with a tapered fiber-optic bundle connected to a scanning spectroradiometer (see Chapter 20). As light penetrated through the different microbial layers of the mats, its spectral composition changed according to the absorption bands of the dominant pigments. In the "sand mat" (Fig. 6.4.1A). the blue light (<500 nm) fell below the detection limit at a depth of 2 mm; only NIR radiation (>700 nm) was detectable at a depth of 4 mm, where a BChl b-containing purple bacterium, *Thiocapsa pfennigii*, utilized light of 1000 to 1050 nm. This exceptional range falls within a window of the water absorption spectrum and comprises the longest wavelengths at which photons have sufficient energy to drive photosynthesis. In the gelatinous mat (Fig. 6.4.1B), blue light absorption due to carotenoids was also intensive, but NIR radiation penetrated much deeper than in the sand mat; troughs at 800 to 810 nm and 910 to 920 nm were presumably due to BChl a absorption by an undescribed filamentous bacterium (see Section 6.5).

Scaler irradiance describes the radiance flux (quantum flux) that reaches individual phototrophic microorganisms from all sides and which may be quite different from either the downward radiance or downward irradiance. Figure 6.4.2 shows scalar irradiance spectra of visible light at different depths in two cyanobacterial mats from hypersaline ponds (Jørgensen and Des Marais 1988). Scalar irradiances were calculated from radiance measurements done at 50 to 100 μm depth intervals with a fiber-optic microprobe consecutively pointing toward different directions. The dense populations of cyanobacteria in the mat from Pond 5 (Fig. 6.4.2, left) had extinguished the usable light by 0.25 mm below the mat surface; photosynthesis thus takes place only in a 250-μm-thick layer. The mat from Pond 8 (Fig. 6.4.2, right) was more gelatinous, had less intensive light absorption, and relatively more light scattering. As a result, the scalar irradiances at the mat surface were dramatically affected; there was much more light available than expected, especially in the orange-red, 550 to 650 nm range where absorption by photosynthetic pigments was mininal. These optical properties of microbial mats are important for the

The Microenvironment

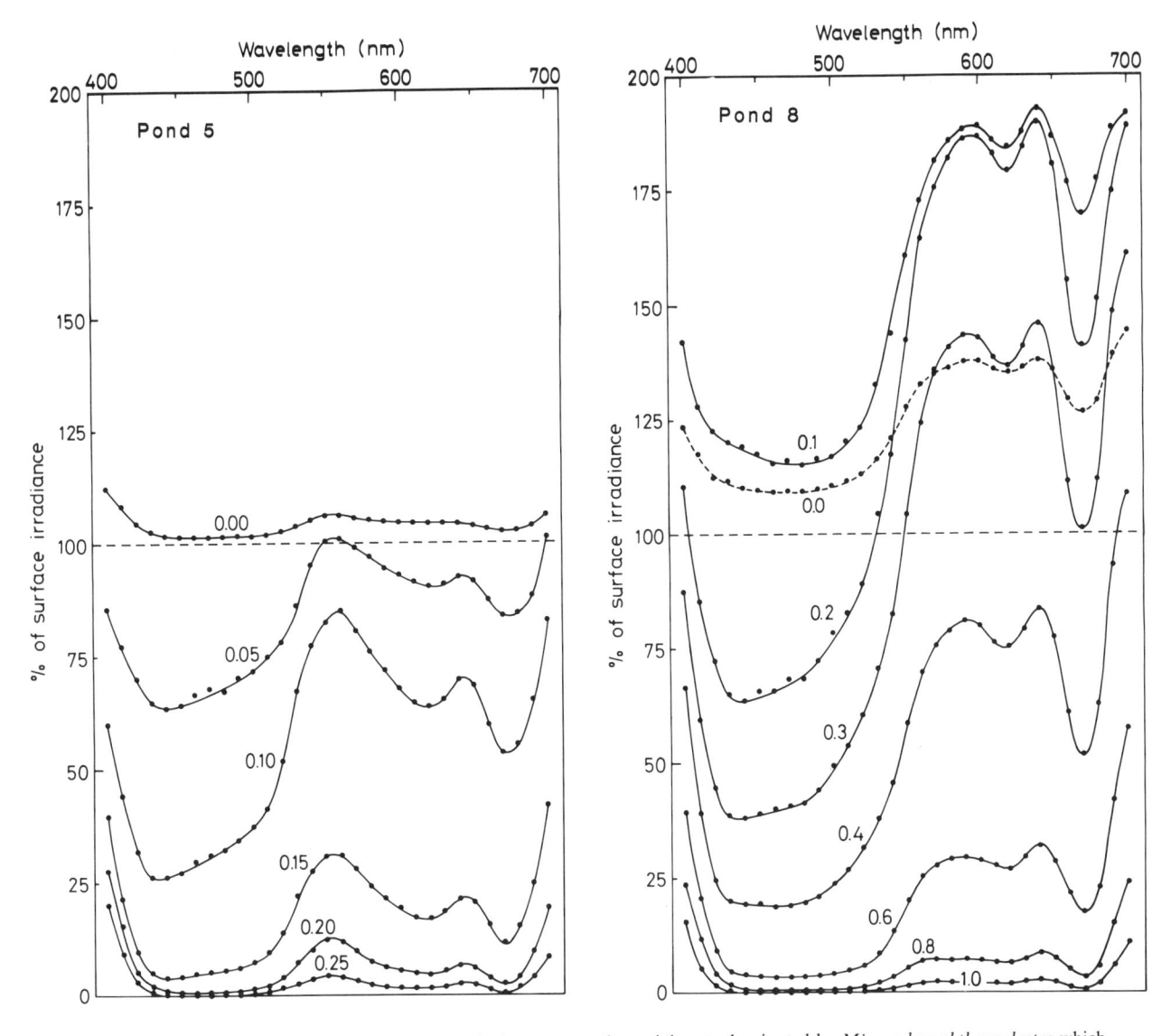

Figure 6.4.2. Spectra of scalar irradiance at different depths in two cyanobacterial mats dominated by *Microcoleus chthonoplastes* which develop in hypersaline ponds at Guerrero Negro, Baja California Sur, Mexico. The one meter-deep ponds have salinities of 70 to 90‰ (Pond 5) and 110 to 130‰ (Pond 8) and seasonal temperatures of 15° to 23°C. For each depth and wavelength, the scalar irradiances are expressed as percent of the downward scalar irradiance at the surface, i.e., of the light intensity at which the mats were illuminated. The depts in mm below the mat surface are indicated on the spectra. Notice how the total quantum flux available to the phototrophic cells differs spectrally from the light source, even at the mat surface (Jørgensen and Des Marais 1988).

zonation and spectral utilization of light. The organisms living on the surface of cyanobacterial mats utilize a broad range of the visible spectrum, while underlying oxygenic phototrophs utilize mostly the orange-red light. Phototrophic bacteria living below the oxygenic zone utilize the remaining NIR radiation (Jørgensen and Des Marais 1988; Jørgensen et al. 1987; Jørgensen 1989; Pierson et al., in press).

6.4.3 The Chemical Microenvironment

The presence or absence of oxygen is generally the most important chemical factor determining the distribution and metabolism of microorganisms in mats. The oxic zone is a chemically unstable environment which often extends only a millimeter below the mat surface. The oxic-anoxic interface thus adjusts within minutes or seconds to changes in photosynthetic oxygen production. This can particularly affect micro-aerophilic bacteria living at the oxic-anoxic interface. Thus, for the sulfide oxidizing gradient bacteria, *Beggiatoa* s.pp., even changing cloud cover perturbs oxygen production sufficiently to move their preferred habitat up or down (Jørgensen and Revsbech 1983). Aerobic organisms must therefore be adapted to survive anoxic periods (e.g., during the night), and anaerobic organisms in this habitat must survive oxic periods during the day. Diel, cyclic variations of oxygen turn on and off such anaerobic processes as nitrogen fixation at the mat surface (Stal et al. 1985; Bebout et al. 1987), whereas other processes, such as sulfate reduction, seem to persist through the oxic period (Y. Cohen and D. E. Canfield pers. comm. to B. B. J.). Strictly anaerobic methanogenic bacteria, isolated from the mat surface, have also been found to survive exposure to atmospheric levels of oxygen (N. P. Revsbech pers. comm. to B. B. J.). On the other hand, aerobic, phototrophic cyanobacteria may survive for months within a dark, anoxic mat (Jørgensen et al. 1988), probably by carrying out fermentative metabolism or anaerobic respiration with elemental sulfur (Oren and Shilo 1979; Richardson and Castenholz 1987b).

Through use of microelectrodes, the dynamic balance between photosynthesis and respiration, as well as the rapid cycling of oxygen and sulfide, have been demonstrated. Figure 6.4.3 shows steady state distributions of O_2, H_2S, and pH during day and night in a cyanobacterial mat from Solar Lake, Sinai. The oxic zone was only 0.5 mm deep at night and H_2S, produced from intensive sulfate reduction, diffused almost to the mat surface. Due to oxidation of H_2S to H_2SO_4, and to heterotrophic CO_2 production, pH fell by almost one unit in the mat. A high rate of oxygenic photosynthesis during the day caused oxygen to increase to levels up to 7 times air saturation in the top saturation in the top millimeter. Autotrophic CO_2 consumption correspondingly resulted in a rise of pH to 9.2, a value which may have resulted in limitation of CO_2-uptake by the phototrophs (see Section 6.8). The O_2-H_2S interface moved vertically between day and night, as did the sulfide oxidizing bacteria *Beggiatoa* s.pp. occupying this zone. These bacteria catalyze such rapid oxidation of the sulfide that the residence time of O_2 and H_2S is only 20 seconds in the 0.2-mm-deep zone where oxygen and hydrogen sulfide coexist in dynamic steady state (Revsbech et al. 1983).

The diel variations of oxygen, sulfide, and pH in this Solar Lake mat illustrate the dynamic nature of the microenvironment. Figure 6.4.4 shows the details of the chemical zonation, over a 28 h period, in relation to changing light and temperature. Throughout the night, the mat was anoxic and rich in H_2S. The pH changed only slowly relative to O_2 and H_2S because of the more than ten-fold larger pool of the bicarbonate system and, thus, its correspondingly higher buffering capacity. At dawn the H_2S was rapidly depleted from the photic zone, and O_2 built up to supersaturation. The pH followed with a distinct time lag. At sunset this sequence was reversed.

The chemical zonations shown in Figures 6.4.3 and 6.4.4 seem to be typical of oxygenic mats, although the depth scale may vary ten-fold depending on the amount of polysaccharide sheath material present (and, thus, on the density of phototrophic organisms) and the depth of light penetration. The database of these chemical microgradients includes cyanobacterial mats from marine coasts of Germany, Denmark, and North Carolina, USA (Stal et al. 1985; Revsbech et al. 1986; Bebout et al. 1987), from hypersaline ponds of Sinai and Mexico (Revsbech et al. 1983; Jørgensen et al. 1983; Jørgensen and Des Marais 1986a), and from hot springs of Wyoming, USA (Yellowstone National Park), Iceland, and New Zealand (Revsbech and Ward 1984a; Jørgensen and Nelson 1988; Castenholz et al. submitted-b).

In hot springs with anoxic and highly sulfide spring water, the predominant mats at the higher temperatures are partly or completely anoxygenic (Sections 6.5 and 6.9). Mats composed solely of green or purple sulfur bacteria are sparse. The few microelectrode studies made in mats of *Chloroflexus* sp. in Yellowstone National Park, and of *Chlorobium* sp. in New Zealand, have detected no traces of oxygen in the mats. Instead, there is a dynamic variation of H_2S in the photic zone, with

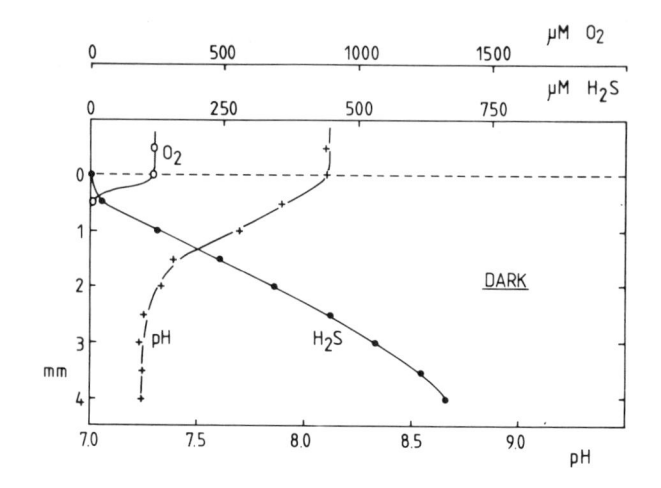

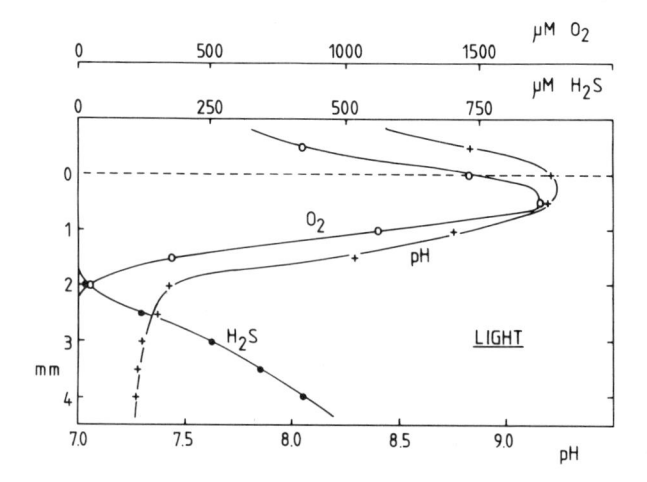

Figure 6.4.3. Distribution of oxygen, sulfide, and pH in a *Microcoleus chthonoplastes* cyanobacterial mat from Solar Lake, Sinai, during day and night. Measurements were done by microelectrodes in a core of intact mat covered by oxic brine from the lake. (Redrawn from Revsbech et al. 1983.)

rapid depletion in the light due to phototrophic consumption, and accumulation in the dark due to sulfate or elemental sulfur reduction within the mat and to diffusion of H_2S from above (Giovannoni et al. 1987a; Castenholz et al. submitted-b). An interesting transition between the anoxygenic and oxygenic mats has recently been discovered in a sulfidic hot spring in Iceland (Jørgensen and Nelson 1988). A small population of photoautotrophic cyanobacteria (*Mastigocladus* sp.), occurring within a predominantly *Chloroflexus* mat, produced at 1 mm depth a 0.5-mm-thick oxic zone wedged into the sulfide gradient and having no diffusional contact with the spring water above the mat; the photosynthetically produced oxygen was thus completely recycled within the mat.

6.4.4 Microbial Movements

Mobile responses of microorganisms to environmental parameters include phototaxis, photophobic responses, photokinesis, chemotaxis, thermotaxis, thigmophobic responses, and rheotaxis.

Photosensory responses include phototaxis *per se*, photophobic reactions, and photokinesis. Phototaxis is a directed

The Microenvironment

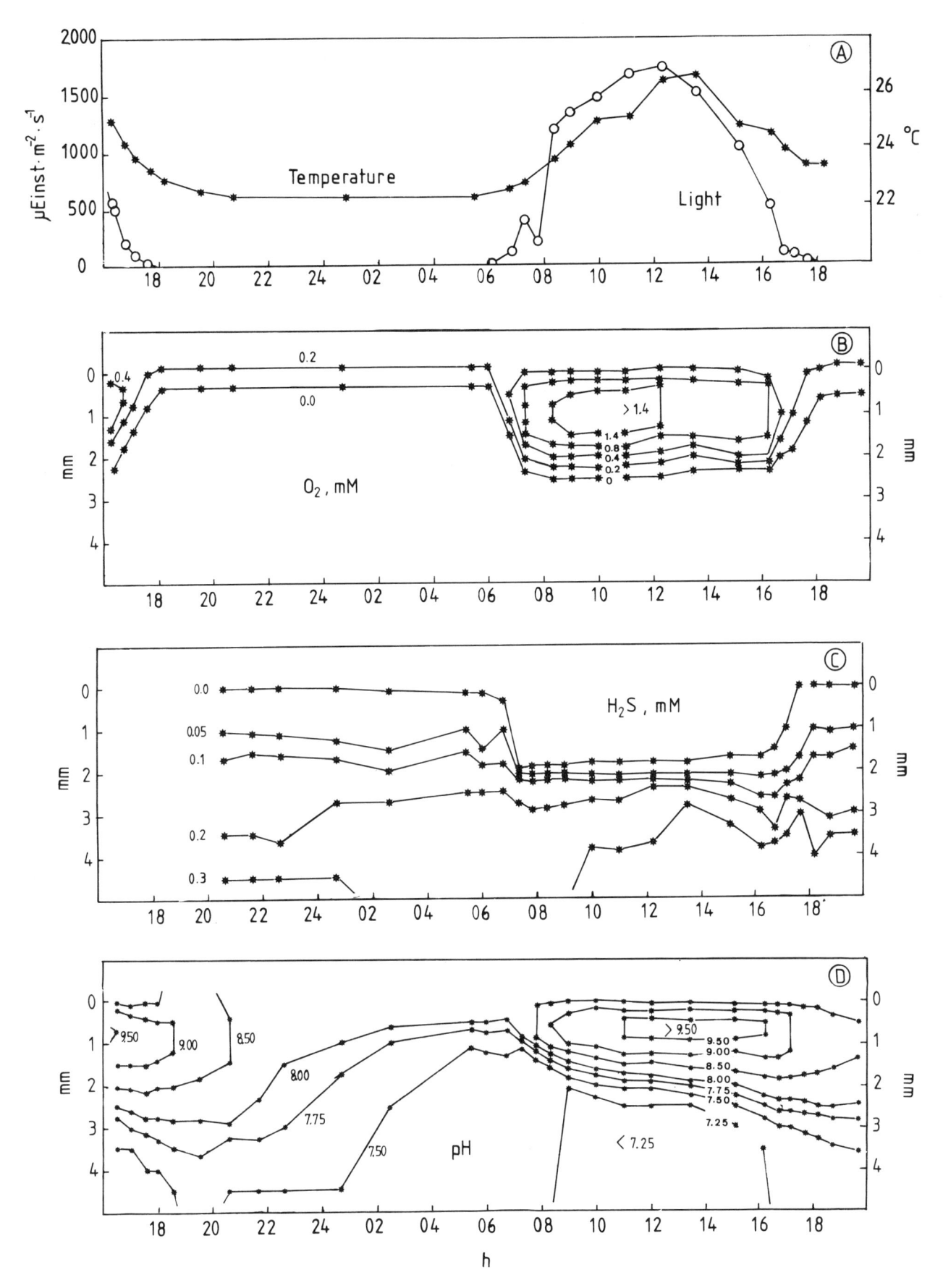

Figure 6.4.4. Detailed chemical zonation in a *Microcoleus chthonoplastes* cyanobacterial mat from Solar Lake, Sinai, during a diel cycle. Temperature and solar illumination at the mat surface are shown together with isopleths of O_2, H_2S, and pH. (Redrawn from Revsbech et al. 1983).

movement with respect to light direction. Many gliding cyanobacteria exhibit this response, doing so by various means including a lessening of reversal frequency when gliding toward light or, at the other extreme, by actual steering toward a source of light (see Castenholz 1982; Häder 1987a, b). The response is "positive" when the net movement is toward the light, and "negative" when away from the light. Photophobic responses are independent of light direction. Instead, they are induced by sudden changes in light intensity (whether temporal or spatial) and are usually manifested by a stopping and reversal of direction. They are referred to either as "step-up reversals" (when the light change is from low to high) or "step-down reversals." Step-down reversals tend to keep the organisms in illuminated regions, whereas step-up reversals accomplish the opposite. Photophobic reactions are known in cyanobacteria, flagellated photosynthetic bacteria, and in some non-photosynthetic bacteria such as *Beggiatoa* (Nelson and Castenholz 1982). Photokinesis merely describes the dependence of motility rate on light intensity. Many cyanobacteria show this response. In a microbial mat environment, the upward or downward movement of cyanobacteria, for example, could involve all three of these responses in a single organism. Mainly filamentous, gliding microorganisms are involved in vertical movements of populations in mats; speeds may range from less than 1 to approximately 5 μm/sec.

Of the other "tactic" responses that may be involved in vertical movements in mats, chemotaxis may be the most significant. Little is known about chemotaxis in cyanobacteria (Malin and Walsby 1985; Richardson and Castenholz 1987a). Either directed movements or phobic responses may occur. Chemotaxis in flagellated bacteria has been studied more extensively (Glagolev 1984; Armitage 1988). Thermotaxis, with respect to mat organisms, has been tentatively identified in *Oscillatoria terebriformis* and appears to be a "step-up phobic" response occurring when supra-optimal temperatures are contacted by filament tips (Castenholz 1968). Thigmophobic responses in mats involve the reversal of gliding direction when obstacles such as a subtle change in substrate viscosity are met (R. W. Castenholz unpublished data). Rheotaxis is a directed movement in response to a current movement, that is, either moving upstream (positive) or downstream (negative). Although positive rheotaxis is known in eukaryotic organisms, it has been reported only once in a prokaryote, in gliding mycoplasmas (Rosengarten et al. 1988). Rheotaxis may be important in horizontal dispersal on mat surfaces.

Within the microscale in which the steep gradients of light,

Table 6.4.1. *Vertical position of motile microorganisms in microbial mats during darkness or low or high irradiance. (M = marine, FW = freshwater, FW-T = freshwater, thermal)*

| Organism | Type of Organism and Metabolism | Position in Mat | | | References |
		darkness	low light	high light	
Oscillatoria margaritifera (M)	cyanobacterium, oxygenic-anoxygenic?	up	up	down	R. Castenholz unpubl.
Oscillatoria terebriformis (FW-T)	cyanobacterium, oxygenic	down	up	down	Castenholz 1968; Richardson and Castenholz 1987a.
Oscillatoria princeps (FW)	cyanobacterium, oxygenic	up	up	up	R. Castenholz unpubl.
Oscillatoria boryana (FW-T)	cyanobacterium, oxygenic-anoxygenic	up	up	down	Castenholz et al. 1989b
Spirulina subsalsa (M)	cyanobacterium, oxygenic-anoxygenic?	up	up	down	R. Castenholz unpubl.
"*Microcoleus lyngbyaceus*" (FW)	cyanobacterium, oxygenic-anoxygenic?	up	up	down	Pentecost 1984
Microcoleus chthonoplastes (M)	cyanobacterium, oxygenic-anoxygenic	up	up	down	Whale and Walsby 1984
Chloroherpeton thalassium (M)	Chlorobiaceae, anoxygenic	up (if anoxic)	up (if anoxic)	down	R. Castenholz unpubl.
Chloroflexus aurantiacus (FW-T)	Chloroflexaceae, anoxygenic	up (in part)	?	down	Doemel and Brock 1974
Chloroflexus sp. (M)	Chloroflexaceae anoxygenic	up	up	down	Mack and Pierson 1988
Chromatium spp. (M, FW-T)	Chromatiaceae, anoxygenic	up (in part)	down	down	Jǫrgensen 1982; R. Castenholz unpubl.
Beggiatoa sp. (M)	non-photosynthetic, autotroph	up	down	down	Jørgensen 1982
Beggiatoa sp. (FW)	non-photosynthetic, heterotroph	up	down	down	Nelson and Castenholz 1982

The Microenvironment 277

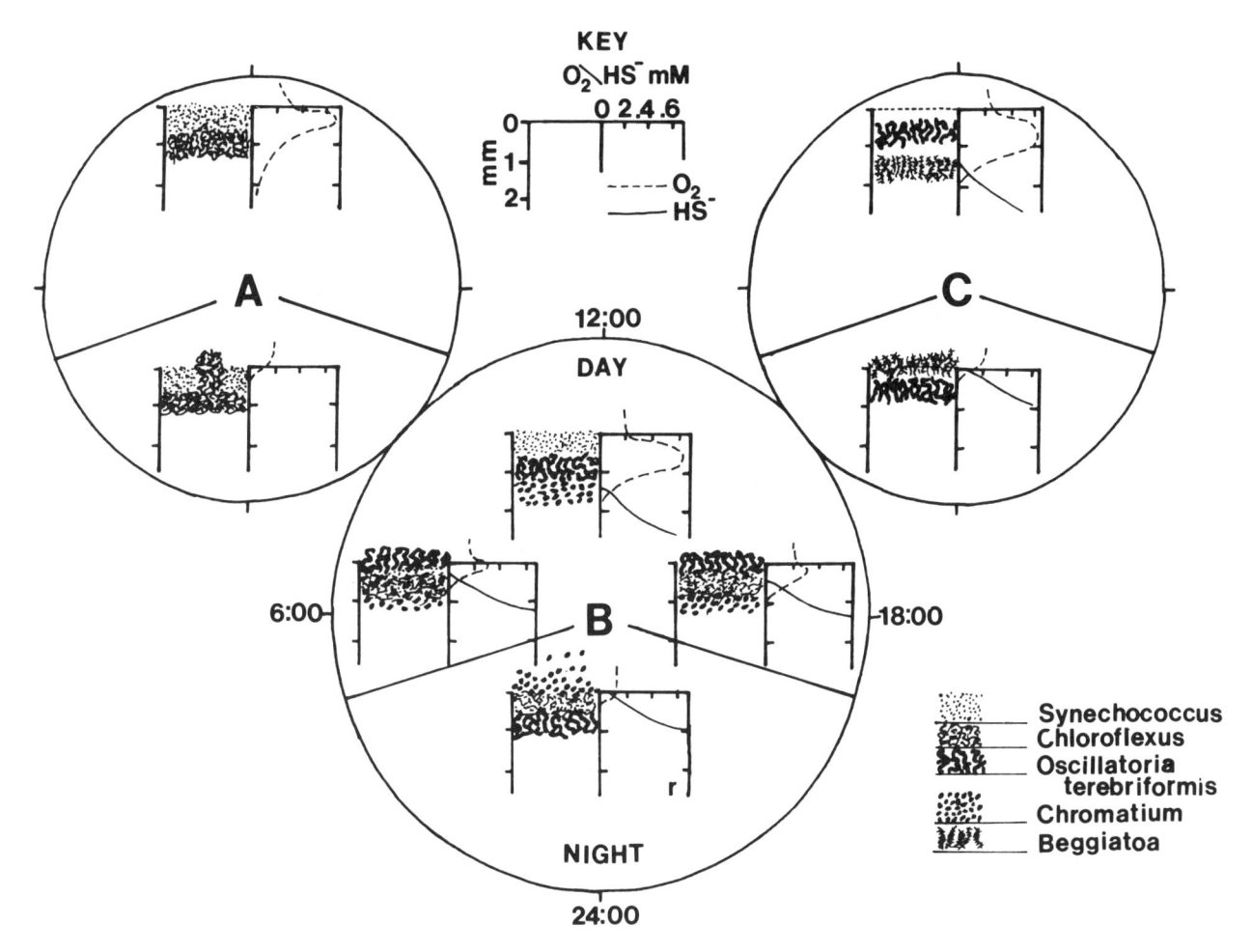

Figure 6.4.5. General pattern of microbial vertical movements in mats of Hunter's Hot Springs, Oregon, USA, in three temperature zones over a 24 h period. In all cases, clear skies and high light intensity of summer are assumed. Approximate vertical distances are in millimeters; O_2(---) and sulfide (——) levels are in mM. (A) 55–68°C; (B) 45–55°C; (C) 35–45°C (from Castenholz 1968, 1973a; Richardson and Castenholz 1987b; and Castenholz, unpublished data). (A), from 68° to 55°C, is a gelatinous mat with two phototrophs, the unicellular cyanobacterium *Synechococcus lividus* on top with the anoxygenic photo/chemo-heterotroph, filamentous *Chloroflexus aurantiacus* generally below. At night some of the *Chloroflexus* filaments migrate to the surface. This might be positive aerotaxis. (B) represents the zone from 55° to 45°C; during a summer morning the filamentous, oxygenic cyanobacterium *Oscillatoria terebriformis* migrates by gliding motility to the mat surface (probably positive phototaxis). During a bright midday it migrates downward again. It resurfaces in the moderate light of late afternoon (Castenholz 1968). After dark it migrates downward again (cues or causes unknown) (Richardson and Castenholz 1987b). The photosynthetic, anoxygenic purple sulfur bacterium, *Chromatium* sp., in contrast, moves in darkness to the surface and into the water above the mat by swimming motility (possibly positive aerotaxis). At night *Chromatium* may metabolize as a microaerophilic chemotroph. At dawn it descends into the mat (possibly negative phototaxis) and remains there all day. (C) represents a lower temperature range (about 45°–35°C). In this zone the two conspicuous components are the *Oscillatoria terebriformis* and the non-photosynthetic filamentous *Beggiatoa* sp. The *Oscillatoria* again descends at night, while the *Beggiatoa* glides to the surface (negative sulfidotaxis or positive aerotaxis?). During daylight *Beggiatoa* descends into the gelatinous mat while light intensity determines the position of *Oscillatoria* (Jørgensen 1982).

O_2, pH, and sulfide occur, vertical movements of some microbial components (filaments capable of gliding motility and some flagellated cells) take place. The vertical movements tend to follow diel schedule (Table 6.4.1), but less predictable cues may result in more erratic behavior. The most universal cue appears to be light intensity (Castenholz 1968; Whale and Walsby 1984; Pentecost 1984; Richardson and Castenholz 1987a; Castenholz et al. submitted-b). Other likely cues are substances such as O_2, sulfide, and CO_2 which may establish steep gradients in mats and act as chemotactic attractants or repellents (Malin and Walsby 1985; Møller et al. 1985; Richardson and Castenholz 1987a).

The adaptive significance of movements for phototrophs is to position themselves in optimal light intensities during daytime and in favorable chemical environments at night. Although the vertical distance of movement is usually less than one mm, the changes in the light and chemical environments within that distance can be dramatic. The response time for movement from one extreme to another is usually less than an hour (Castenholz 1968; Richardson and Castenholz 1987a, b). For some microorganisms, particularly non-phototrophs such as *Beggiatoa*, movements may be entirely phobic responses to avoid the damaging effects of high light (Nelson and Castenholz 1982) and of moderate or high O_2 (Møller et al. 1985). *Beggiatoa* positions itself in its optimal environment, normally the transition zone of low O_2 and low sulfide (Nelson et al.

1986). The above remarks regarding the adaptive significance of microbial movements are speculative deductions based on behavioral patterns but on little experimental work.

An example of the complex patterns of movements seen within microbial mats is illustrated in Figure 6.4.5 in which the microbial mat system within a single hot spring outflow is represented. The composition of organisms changes at different points in the thermal gradient (see Section 6.9).

Phototactic responses of cyanobacteria and possibly other filamentous phototrophic bacteria (e.g. *Chloroflexus*), together with cohesive interactions, may be responsible for the initiation and construction of "*Conophyton*"-like conical, columnar, and pinnacle-shaped mats (e.g., Walter et al. 1972, 1976a; Brock 1978; Section 6.3). Heliotropic orientation of these structures has also been noted and discussed (Awramik and Vanyo 1986). Although phototactic orientation may also be involved in the formation of mat laminations, this has not been confirmed by experimentation or satisfactorily explained (Monty 1967; Gebelein 1969).

In microbial mat systems, horizontal surface movements by random gliding, swimming, or passive flotation, or in some cases by directed tactic movements, are certainly involved in dispersal and colonization but have been little studied. A series of papers on the hydrophobicity and adhesion properties of cyanobacteria has resulted in the conclusion that some non-gliding cyanobacteria attach to substrates by the combined effects of co-flocculation and hydrophobic interaction (e.g., Bar-Or and Shilo 1987, 1988; Fattom and Shilo 1985). Although the release of gliding hormogonia that lack the surface properties of their "parent" filaments may be one form of dispersal, these authors believe that a regulated decrease in the two surface characteristics during "late" growth phases of non-motile cyanobacteria may enable such gliding hormogonia to detach and, by passive means, to colonize new areas.

6.5

Photosynthetic Activity in Modern Microbial Mat-Building Communities

RICHARD W. CASTENHOLZ   JOHN BAULD   BEVERLY K. PIERSON

6.5.1 Types of Photosynthetic Activity

Photosynthesis is defined as the conversion of radiant energy into chemical energy. This may include reduction of CO_2, a process referred to as *autotrophy* but not limited to photosynthetic organisms. If CO_2 is the principal source of carbon, and radiant energy the principal source of energy, the organism is a "photoautotroph." If organic carbon is mainly used as the carbon source and light is the source of energy, the organism is a "photoheterotroph."

The quantum of energy contained in a photon of light can be converted into chemical energy only if absorbed. Pigments absorb radiant energy. When a photon is absorbed by a pigment, the pigment is energized to an excited state having an energy distribution in its electrons different from that of its ground state. In the abundant light-harvesting (antenna) pigments (see Table 6.5.2), this absorbed energy will be transferred by excitation resonance to another light-harvesting pigment and, ultimately, to a reaction center (RC) pigment (see Häder and Tevini 1987).

Reaction center chlorophyll (RC-Chl) and bacteriochloro-

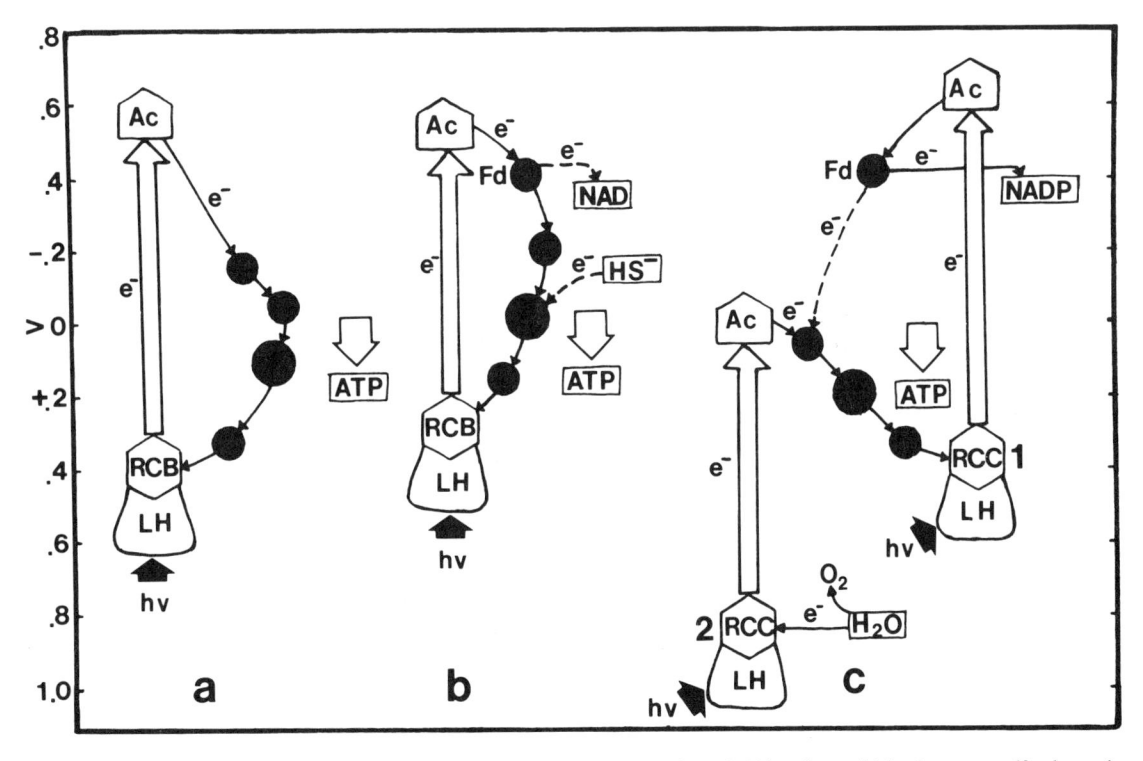

Figure 6.5.1. Schematic drawings of the photosynthetic process in (a) purple bacteria and *Chloroflexus*, (b) in the green sulfur bacteria, and (c) in cyanobacteria or the chloroplasts of eukaryotic organisms. V = voltage of E_0' (standard mid-point redox potential), LH = light harvesting pigments, Ac = primary electron (e^-) acceptor, RCB = reaction center bacteriochlorophyll, RCC = reaction center chlorophyll a, dark circles = electron carriers (e.g., cytochromes, iron sulfur proteins, quinones, ferredoxin [Fd], and others) (modified from Gottschalk 1986; and Dawes 1986).

phyll (RC-BChl), whether they enter the excited state by energy transfer or by direct absorption of photons, are the only pigment complexes capable of performing photochemistry. The excited state of RC-Chl or RC-BChl transforms these pigments into strong reductants able to expel an electron from the pigment molecule and reduce a relatively low potential acceptor against the thermodynamic gradient (Fig. 6.5.1). The reduction of an acceptor constitutes the trapping of chemical energy, there being a 800 to 1000 mV difference between the reduced acceptor and the oxidized RC-BChl or RC-Chl. The electron can be returned, to reduce the RC-pigment molecule, via a sequential series of intermediate electron carriers. Since reaction center complexes are arranged in membrane systems, the electron transfer reactions are used to generate a proton-motive force across the membrane. This can be used directly for ATP synthesis (Fig. 6.5.1a). Thus, ATP is a principal product of photosynthesis whether in a photoautotroph or photoheterotroph, and in the latter, it is the only product. The production of ATP as a result of the electron transfer described above is part of a cyclic process (cyclic photophosphorylation) in which there is no net input or output of electrons (Fig. 6.5.1a).

Most photosynthetic purple bacteria and known types of *Chloroflexus* have ATP as the only direct product of photosynthesis (Fig. 6.5.1a). By the use of this ATP and various reductants (when required), either organic carbon or CO_2 is incorporated into cell matter. This type of photosynthesis is anoxygenic (i.e., non-O_2 producing).

In the green sulfur bacteria (also anoxygenic), in addition to cyclic electron flow and ATP synthesis, there is a net inflow of electrons from exogenous sulfide (H_2S, HS^-, or S^{2-}) or other reduced sulfur compounds (e.g., $S°$, $S_2O_3^{2-}$) to the reaction center, and a net outflow of electrons in the form of reduced ferredoxin or NADH which can reduce CO_2 (Fig. 6.5.1b).

The reaction center BChl's of all of the bacteria discussed above are part of a single membrane-contained complex called a "photosystem" (Thornber 1986). In the cyanobacteria, prochlorobacteria, and in eukaryotic chloroplasts, there are two photosystems (PS I and PS II) which operate in series. Thus, there are two reaction centers. The RC-Chl a PS II, with a very positive redox potential, can oxidize H_2O to O_2 with the catalysis of a manganoprotein in PS II (Fig. 6.5.1c). Water is the source of electron inflow ($H_2O \rightarrow 2e^- + 2H^+ + \frac{1}{2}O_2$) and O_2 is evolved – hence, this is oxygenic photosynthesis. The net inflow of electrons from water results in a net outflow of electrons in the form of NADPH (capable of reducing CO_2), after passing through several electron carriers and requiring two quanta of energy for each electron (one quantum from each reaction center) (Fig. 6.5.1c). ATP is generated as a result of non-cyclic electron transport between PS II and PS I. In the absence of photooxidation of water, ATP can be produced by cyclic electron transport in PS I only.

Some species of cyanobacteria are able to temporarily dispense with photosystem II (and H_2O as a source of electrons) by using electrons from sulfide and the RC Chl a of PS I; these species are thus capable of reverting, temporarily, to anoxygenic photosynthesis.

6.5.2 Photosynthetic Activity in Microbial Mats

6.5.2A Oxygenic Photosynthesis in Mats

Oxygenic photosynthesis occurs in the surface layer of most microbial mats. The organisms responsible are usually cyanobacteria, but some eukaryotic algae (particularly diatoms) are often present (see Sections 6.3, 6.9, 6.10). When sulfide levels are high or persistent, the sole or predominant type of photosynthesis may be anoxygenic (Fig. 6.5B).

In low sulfate (that is, non-sulfide generating) mats, oxygenic photosynthesis may be the only type occurring. This is often the case in hot springs, or in newly colonized areas where undermats of anaerobic sulfide-producing bacteria have not become established (e.g., Stal et al. 1985; de Wit and van Gemerden 1987b). When sulfate concentration is high, such as in marine, some hypersaline, and some hot spring waters, undermats of sulfide-using anoxygenic phototrophs develop, often as brightly colored red layers (purple sulfur bacteria) or as green layers (green sulfur bacteria) (Javor and Castenholz 1981; Stal et al. 1985; Jørgensen and Des Marais 1986a; Nicholson et al. 1987).

The depth to which oxygenic photosynthesis occurs depends primarily on the steepness of the light gradient in those spectral regions required by light-harvesting pigments of cyanobacteria (see Figs. 6.4.1 and 6.4.2) and by those of diatoms, green algae, or other but rarer eukaryotic algae (see Table 6.5.2). The depth of oxygenic photosynthesis ranges from less than one mm to several mm (Section 6.4) and, less commonly, to depths of 1 to 2 cm if the mats are quite translucent (Jørgensen et al. 1983, 1987, 1988; Revsbech et al. 1983; Revsbech and Ward 1984a; Castenholz 1984; Richardson and Castenholz 1987a; Pierson et al. 1987; Jørgensen and Nelson 1988; Pierson and Castenholz, in prep.). In addition to the availability of visible radiation, the depth to which oxygenic photosynthesis occurs in compact mats is limited by the release and diffusion rate of biogenic sulfide generally from deeper parts of the mat. Sulfide, depending on concentration, partially or completely inhibits oxygenic photosynthesis (see Section 6.9; Oren et al. 1979; Cohen et al.

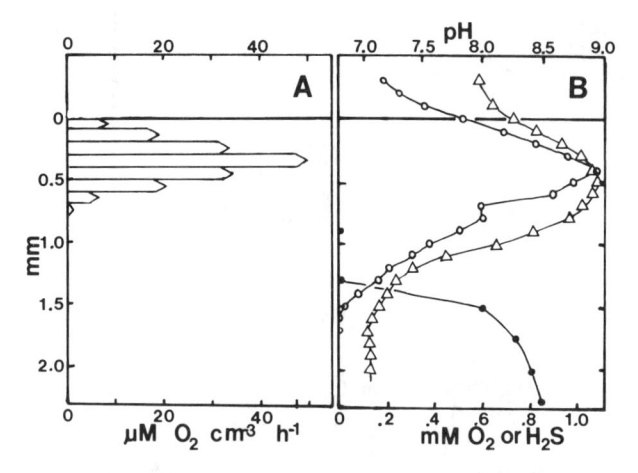

Figure 6.5.2. Rates of oxygenic photosynthesis (A) and profiles of O_2 (–○–○–), sulfide (–●–●–), and pH (–△–△–) (B) in a "shallow flat mat" of saline Solar Lake, Sinai, as measured by microelectrodes. The upper 0.2 mm of mat was dominated by diatoms, from 0.2 to 0.5 mm by the cyanobacterium *Microcoleus chthonoplastes*, and from 0.5 to 0.8 mm and below by the cyanobacterium *Phormidium* sp. together with anoxygenic *Chloroflexus* (see Jørgensen et al. 1983).

1986). The lower the pH, the greater the proportion of sulfide in the protonated form. H_2S being more membrane-permeable, is also more toxic to oxygenic photosynthesis (Howsley and Pearson 1979; Cohen et al. 1986). During darkness, generally the entire mat becomes anoxic, with or without sulfide biogenesis, whereas photosynthetic conditions can result in supersaturated levels of O_2 in the mat (Section 6.4 and Fig. 6.5.2).

6.5.2B Anoxygenic Photosynthesis of Cyanobacteria in Mats

Some of the most common cyanobacteria of the upper layers of mats are capable of carrying out sulfide-dependent anoxygenic photosynthesis. Two representatives of such cyanobacteria (*Oscillatoria boryana* and *Microcoleus chthonoplastes*) and their mats are described in more detail in Sections 6.4 and 6.9. There appear to be three tactics employed by sulfide-tolerant cyanobacteria:

(1) *Oscillatoria limnetica*, which forms floccose mats in Solar Lake, Sinai (e.g., Cohen 1984a, b), temporarily loses the capacity for oxygenic photosynthesis and converts to sulfide-dependent anoxygenic photosynthesis during exposure to sulfide levels as low as 80 μM. Photosystem II is 50% inhibited even by 10 μM sulfide (see Cohen et al. 1986).

(2) Some species of cyanobacteria (e.g., marine *Microcoleus chthonoplastes*, and freshwater *Oscillatoria boryana* and *O. amphigranulata*) are able to carry out oxygenic photosynthesis in concert with anoxygenic photosynthesis at relatively high sulfide levels (Castenholz and Utkilen 1984; Cohen et al. 1986; Jørgensen et al. 1986a; de Wit and van Gemerden 1987a; Castenholz et al. submitted-b).

(3) *Oscillatoria terebriformis* from hot springs, and *Oscillatoria* sp. from a saline hot spring (Wilbur Springs, California, USA), are able to carry out little or no anoxygenic photosynthesis in the presence of sulfide, but are able to sustain oxygenic photosynthesis when exposed to sulfide levels of over 1 mM and 0.3 mM, respectively (Castenholz 1977; Cohen et al. 1986).

There are also many species that show little or no tolerance to sulfide (e.g., Castenholz 1976, 1977; Garlick et al. 1977; Cohen et al. 1986).

6.5.2C Photoheterotrophy of Cyanobacteria in Mats

Cyanobacteria, once considered to be "classical" obligate photoautotrophs (Smith et al. 1967), are now known to exhibit several modes of growth. In laboratory culture, a substantial number of species are capable of photoheterotrophic growth using exogenous organic carbon sources in the absence of CO_2; in the dark, some of these can sustain chemoheterotrophic growth by aerobic respiration (Rippka et al. 1979; Smith 1983). However, photoheterotrophic growth rates of free-living cyanobacteria appear to be significantly slower than those obtained under photoautotrophic conditions, and aerobic chemoheterotrophic rates may be more than an order of magnitude slower (see Section 6.6) (Rippka 1972; Joset-Espardellier et al. 1978; Smith 1983; see also Table 1 of Richardson and Castenholz 1987b). However, phototrophic growth rates of isolated symbiotic *Anabaena azollae* were enhanced two- to three-fold with the addition of 8 mM fructose (Rozen et al. 1986). Very few organic substrates are known to be used by cyanobacteria, the most common being glucose, fructose, or sucrose (Rippka et al. 1979; Smith 1983).

Cyanobacterial photoheterotrophy and its prerequisite, photoassimilation (e.g., Gibson 1981), have been little studied in natural habitats. Usually, photoheterotrophy is estimated by measuring the photoassimilation of ^{14}C-labelled organic carbon substrates (with and without DCMU [3-(3,4-dichlorophenyl)-1, 1-dimethylurea]). Photoassimilation of organic substrates has been demonstrated for cyanobacterium-dominated microbial mat communities in geothermal (Castenholz et al. submitted-b), lacustrine (Bauld 1987), and marginal marine (Bauld 1987; Bauld, unpubl.) habitats. In these benthic communities, photoheterotrophic activity comprises as much as 50% of total heterotrophic activity, comparable to that reported for freshwater planktonic cyanobacterial communities (30 to 50%: McKinley and Wetzel 1979; Ellis and Stanford 1982). The design and interpretation of such experiments should be approached with some caution since it has been shown, for example, that in *Oscillatoria boryana*, about 50% of apparent photoheterotrophic uptake in fact takes place via heterotrophic release of CO_2 of the carboxyl group of acetate and its subsequent fixation by DCMU-sensitive oxygenic photosynthesis (Bauld 1987; Castenholz et al. submitted-b).

Just how important is heterotrophy to the overall carbon economy of cyanobacteria in their various benthic habitats? Self-shading and sediment burial are characteristic of benthic microbial communities, and the consequent rapid vertical diminution of light intensity (e.g., Jørgensen et al. 1987, 1988) constrains activity and survival. The meager light energy available under such conditions, while insufficient to support photoautotrophic growth, might support slow photoheterotrophic growth (e.g., Van Baalen et al. 1971). Natural populations of *Oscillatoria agardhii*, a common inhabitant of the light-limited metalimnion of lakes, are reported to assimilate organic substrates from approximately natural concentrations under chemoheterotrophic, mixotrophic, and photoheterotrophic conditions (Saunders 1972; Ellis and Stanford 1982). Microbial mat pore waters may contain substantial concentrations of dissolved organic carbon (Lyons et al. 1982), but it is not clear how much of this is reasily assimilable by cyanobacteria.

6.5.2D Anoxygenic Phototrophs With Single Photosystems in Mats

Most microbial mats contain one or more species of anoxygenic single-photosystem bacteria: green sulfur bacteria, purple sulfur bacteria, purple nonsulfur bacteria, *Chloroflexus*, *Heliothrix*, *Erythrobacter*, "heliobacteria," and others that are incompletely described. Some mats are composed entirely of single-photosystem organisms (see Section 6.9). All of these bacteria contain reaction center and light-harvesting bacteriochlorophylls (Bchl), and all lack chlorophyll *a*, *b*, *c*, and *d*. The diversity of pigments and phototrophic metabolism among the anoxygenic phototrophs is great, as is the diversity of habitats in which they occur (Table 6.5.1).

Since single-photosystem anoxygenic photosynthesis is prob-

Table 6.5.1. *Occurrences of anoxygenic phototrophic bacteria in various mat environments.*

Organism	Environment	Type of Mat Layer	Major Phototrophic Metabolism	References
Chloroflexaceae	(a) Hot spring (40–72°C) no or low sulfide alkaline-neutral	Distinct layer under cyanobacteria	Photoheterotrophy	1, 2, 5
	(b) Hot spring (40–66°C) high sulfide alkaline-neutral	Distinct surface layer above cyanobacteria or monospecific laminated mat	Photoautotrophy	3, 4, 5
	(c) Marine intertidal high sulfide alkaline	Distinct migrating layer below layer of purple sulfur bacteria	Photoheterotrophy and/or photoautotrophy?	6
	(d) Marine hypersaline fluctuating sulfide alkaline	Intimately associated with or just below cyanobacteria or as deep layer below cyanobacteria and purple sulfur bacteria	Photoautotrophy and/or photoheterotrophy?	7, 8, 9
Heliothrix oregonensis	(a) Hot spring (35–55°C) no sulfide alkaline	Distinct surface layer above cyanobacteria	Photoheterotrophy	5, 10
Unidentified Bchl *a*-filaments	(a) Hot spring (30–50°C) no sulfide alkaline	Distinct deep mat layer below cyanobacteria and *Chloroflexus*	Photoheterotrophy	5, 9
	(b) Hypersaline mat sulfide alkaline	Free-living and intra-sheath dwelling with *Microcoleus*	Photoautotrophy(?)	11
Chlorobiaceae	(a) Hot springs (45–55°C) high sulfide slightly acidic-neutral	Unispecific, non-layered mat	Photoautotrophy	12
	(b) Marine intertidal sulfide neutral-alkaline	Distinct deep layer below purple sulfur bacteria	Photoautotrophy	13, 14
Chromatiaceae	(a) Hot spring (up to 57°C) sulfide	Distinct surface or under layer often associated with *Chloroflexus*	Photoautotrophy and/or photoheterotrophy?	15, 20
	(b) Marine intertidal sulfide alkaline	Distinct layer below cyanobacterial surface layer; also seen as a surface layer perhaps growing partially as chemolithoautotroph	Photoautotrophy	13, 14, 16, 17
	(c) Marine hypersaline	May be distinct lower layer beneath cyanobacteria or may be mixed with *Chloroflexus* in lower cyanobacterial zone	Photoautotrophy and/or photoheterotrophy?	18, 19

1. Bauld and Brock 1973; 2. Pierson and Castenholz 1974; 3. Giovannoni et al. 1987; 4. Jørgensen and Nelson 1988; 5. Castenholz 1984; 6. Mack and Pierson 1988; 7. Jørgensen and Revsbech 1983; 8. Cohen 1984; 9. Pierson unpublished; 10. Pierson et al. 1984; 11. D'Amelio et al. 1987; 12. Castenholz 1988; 13. Nicholson et al. 1987; 14. Pierson et al. 1987; 15. Madigan 1986; 16. Fenchel and Straarup 1971; 17. Stal et al. 1985; 18. Javor and Castenholz 1984; 19. Jørgensen and Des Marais 1986; 20. Castenholz unpublished.

ably more ancient than dual-photosystem oxygenic photosynthesis, mat layers composed of anoxygenic organisms may be close analogues to ancient mats that existed before oxygenic cyanobacteria evolved. Because species of the filamentous anoxygenic phototroph *Chloroflexus* appear to be derived from the deepest lineage among all the phototrophs (Oyaizu et al. 1987), mats composed of these organisms are of particular interest in the study of Archean and, possibly, Early Proterozoic evolution (see Sections 1.5, 6.9).

Only one species of *Chloroflexus* has been described, *Chloroflexus aurantiacus*, a thermophile growing in alkaline hot springs. Most commonly, *Chloroflexus* grows photoheterotrophically under a layer of oxygenic cyanobacteria and utilizes organic substrates excreted or lost by the cyanobacteria, thereby cycling carbon within the mat (Section 6.9; Bauld and Brock 1973; Pierson and Castenholz 1974; Anderson et al. 1987; Bateson and Ward 1988). At higher sulfide levels, *Chloroflexus* may occur as a sulfide-dependent photoautotroph (Section 6.9; Giovannoni et al. 1987a; Jørgensen and Nelson 1988).

Mesophilic *Chloroflexus*-like organisms have been identified by pigment analysis and electron microscopy in mats from a large variety of marine intertidal and hypersaline environments (Section 6.10; Jørgensen and Revsbech 1983; Stolz 1984; Cohen 1984b; Mack and Pierson 1988). None of these mesophilic strains has been successfully isolated, although relatively pure populations have been grown and maintained in enrichment cultures.

Other filamentous anoxygenic phototrophs occur as major components of laminated mats in both thermal and marine hypersaline environments. *Heliothrix oregonensis* forms a prominent surface layer, covering oxygenic cyanobacteria, and grows as an active motile photoheterotroph in some alkaline hot springs lacking sulfide (Pierson et al. 1984, 1985). An unidentified Bchl *a*-containing filamentous phototroph has been found both free-living and as an intimate co-inhabitant within *Microcoleus* sheaths in hypersaline sulfide-containing laminated mats (D'Amelio et al. 1987). An ultrastructurally identical Bchl *a*-containing filamentous photoheterotroph has been found in highly translucent laminated thermal mats at Yellowstone National Park, Wyoming, USA (Castenholz 1984; Pierson and Castenholz, in preparation). The presence of all of these diverse filamentous anoxygenic phototrophs as prominent members of several mat communities suggests that ancient mats may have been built by several different types of phototrophs before the evolution of cyanobacteria.

Although purely anoxygenic mats are formed by unicellular purple and green sulfur bacteria in some hot springs, the most ubiquitous occurrence of these phototrophs in mats is in marine intertidal and hypersaline environments (see Section 6.9). Several species of purple bacteria and green bacteria form conspicuous, colorful layers commonly underlying a layer of cyanobacteria (Fenchel and Straarup 1971; Javor and Castenholz 1984b; Stal et al. 1985; Jørgensen and Des Marais 1986a; Nicholson et al. 1987; Pierson et al. 1987). Prominent accumulations of purple sulfur bacteria often occur on the surface of sandy intertidal sediments (de Wit and van Gemerden 1987b). Sulfide must be present for development of these mat layers.

Other anoxygenic single-system phototrophic bacteria are notable in their absence as major constituents of microbial mats. The photoheterotrophic purple non-sulfur bacteria can be enriched from mats, but they do not appear to be major

Table 6.5.2. *Pigments occurring in mat-forming microorganisms.*

Chlorophylls and Phycobilins from Microorganisms found in Microbial Mat Communities[a].

Pigment	Function	Major *in vivo* Abs. Max.	Organisms	References
Chl *a*	RC & LH	435, 670–680	Cyanobacteria Bacillariophyceae (diatoms) Chlorophyceae	1, 9, 10
Chl *b*	LH only	480, 650	Chlorophyceae	9, 10
Chl *c*	LH only	---, 645	Bacillariophyceae	9, 10
Bchl *a*	RC & LH	375, 590, 790–812, 830–920	Chromatiaceae Rhodospirillaceae Chlorobiaceae Chloroflexaceae *Heliothrix oregonensis* unidentified filaments	1, 2
Bchl *b*	RC & LH	390–400, 600–610 800–850, 1015–1040	Chromatiaceae Rhodospirillaceae	1, 3
Bchl *c*	LH only	335, 450–460, 740–760	Chlorobiaceae Chloroflexaceae	1, 4, 5
Bchl *d*	LH only	320–330, 440–460, 725–745	Chlorobiaceae	1, 5
Bchl *e*	LH only	340–360, 450–460, 715–725	Chlorobiaceae	1, 5
Bchl *g*	RC & LH	420, 575, 670[b], 788	*Heliobacterium chlorum* and related organisms	6
Allophycocyanin	LH only	645–650	Cyanobacteria	7
Phycocyanin	LH only	610–625	Cyanobacteria	7
Phycoerythrocyanin	LH only	570, 590–595	Cyanobacteria	7
c-Phycoerythrin	LH only	560–565	Cyanobacteria	7
r-Phycoerythrin	LH only	490–500, 545–560	Cyanobacteria	7, 8

[a]See Chapter 20 for absorption coefficients from the chlorophylls in organic solvents. LH = light-harvesting; RC = reaction center.
[b]The absorption maximum at 670 nm is probably due to a Chl *a*-like pigment (lacking phytyl) formed naturally from Bchl *g* and which accumulates in aging cells.

1. Trüper and Pfennig 1981; 2. Garcia et al. 1986; 3. Trüper and Imhoff 1981; 4. Brune et al. 1987; 5. Gloe et al. 1975; 6. Beer-Romero and Gest 1987; 7. Cohen-Bazire and Bryant 1982; 8. Waterbury et al. 1987; 9. Glazer 1983; 10. Govindjee and Braun 1974.

components of any of the mat layers. The occurrence of the Bchl *g*-containing photoheterotrophs ("heliobacteria") within microbial mats has yet to be demonstrated. Species of *Erythrobacter*-like organisms have been cultured from mats but have not yet been demonstrated as major components (J. Hawkins pers. comm. to B. K. P.).

6.5.3 Pigments in Mats

Microbial mats are colorful and often startlingly so (see Frontispiece). The most obvious distinguishing characteristics of the different layers in laminated mats are their different colors which are due to a diverse array of pigments present in the various groups of phototrophic prokaryotes (Table 6.5.2). Yellow pigments that probably serve a protective function are present in the sheaths of some species of cyanobacteria. Most of the pigments contributing to the color of mats, however, are intracellular, membrane bound, protein complexes. The diversity in wavelengths of light absorbed by different microorganisms, and hence their color, is determined both by the presence of differing chromophores and by the association of the same chromophores with differing proteins. The functions of these pigments include photochemistry, light-harvesting (LH), and protection. Most of the chlorophylls function to harvest light energy (Thornber 1986). There are eight different LH chlorophylls in various phototrophic prokaryotes (Chl *a* and *b*, Bchl *a*, *b*, *c*, *d*, *e*, *g*). Eukaryotic diatoms also contain Chl *c*. Phototrophs contain several carotenoids which function both as LH pigments and for protection against photooxidative damage (see Chapter 20 and Section 6.10). Cyanobacteria also contain phycobiliprotein complexes which function as light-harvesting systems.

The effect of these pigments on the light-environment within the mat is very significant (see Section 6.4). The broad overlapping absorption maxima of all the carotenoids (from 400 to 550 nm), and the intense Soret peaks of the chlorophylls within this same region (see Table 6.5.2) contribute to rapid extinction of light energy below 600 nm within the upper zones of most mats. In order for phototrophs to form dense mats of any appreciable thickness (greater than 2 mm), diversity of light-harvesting chlorophylls with absorption maxima in the red and near infrared (greater than 600 nm) is necessary. It seems likely that diversification of pigment systems evolved very early in

Table 6.5.3. *Productivity in modern microbial mats.*

Mat (location)	Method	Productivity (gC/m^2/d)	References
MARINE AND HYPERSALINE			
Solar Lake (Sinai)	^{14}C, suspension	8–12.0	1
Spencer Gulf (South Australia)	^{14}C, cores[a] (*in situ*)	0.15–3.07[b]	2
Salt Marsh (Tijuana Estuary, CA)	O$_2$, cores (*in situ*)	0.51–0.93[c]	3
Long Island Sound (CT)	^{14}C, suspension	0.88–9.52	4
Coral reef lagoon, French Polynesia	O$_2$, cores	2.3	5
Enewetak Atoll	O$_2$, cores	3.35[b]	6
Coral Reef Algal Turf (Tague Bay, St. Croix)	O$_2$, *in situ*[d]	2.1–3.1	7
Salt Marsh (Sippewissett Salt Marsh, MA)	^{14}C, suspension	0.4–0.8[b]	8
Hamelin Pool (West, Australia)	^{14}C, core	0.09–0.48	9
Sabkha Gavish (Sinai)	^{14}C, core	3.1–3.8[b]	10
GEOTHERMAL			
Hengill-Ölfus Hot Springs, Iceland (winter)	^{14}C, core	0.26–1.33	11
Drakesbad Hot Springs (CA)	^{14}C, core	7–12	12
Ohanepecosh Hot Springs WA (summer)	O$_2$, *in situ*	2.2–4.2	13
Tecopa Bore, Mojave Desert, CA	^{14}C, cores	1.83–4.53	14
Lake Baikal Hot Springs	^{14}C, core	0.012–0.9	15
ANTARCTIC LAKE			
Algal Lake, Ross Island	^{14}C, cores	1.91–3.63	16
STROMATOLITE (Proterozoic)			
Transvaal Supergroup	estimated	0.190–3.6	17

[a]"Cores" – refers to an intact subsample
[b]Calculated from an hourly rate (multiplied by 5)
[c]Calculated from annual rate (divided by 365)
[d]*In situ* measurements on 18–24 month colonized plates

1. Krumbein et al. 1977; 2. Bauld 1984; 3. Zedler 1980; 4. Burkholder et al. 1965; 5. Sournia 1976; 6. Johannes et al. 1972; 7. Carpenter 1985; 8. Pierson unpublished; 9. Bauld et al. 1979; 10. Krumbein et al. 1979; 11. Sperling 1975; 12. Lenn 1966; 13. Stockner 1968; 14. Naiman 1976; 15. Gorlenko and Yurkov 1988; 16. Goldman et al. 1972; 17. Lanier 1986.

mat communities due to competition for limiting light. The red and near infra-red wavelengths penetrate relatively well in compact mats but not in water, so that aquatic environments over 20 cm in depth would be essentially devoid of radiation in these wavelength regions (Section 6.4; Wetzel 1983; Pierson and Olson 1989).

The methods used for analysis of pigments from mats, and the absorption characteristics of such pigments in organic solvents, are given in Table 20.1.

6.5.4 Productivity in Mat Ecosystems

The primary productivity of freshwater and marine phytoplankton in a large number of aquatic communities has been measured over the past 40 years. Problems of methodology and biological variances have led to a considerable range in the values obtained (Legendre et al. 1983). The application of similar techniques to microbial mats has been far more limited (Bauld 1981a). Variations in results are great, however, and the problems of methodology are even more significant than in water column systems. Table 6.5.3 summarizes data on primary productivity obtained from several different microbial mat communities. Values range over three orders of magnitude and no consistent generalizations can be made. Problems in methodology, which make comparisons of the data from different mats nearly impossible, fall into two main categories: (i) methods of incubation of the cells; and (ii) calculation of the rate of primary production. Primary productivity can be determined in mats by measuring the rate either of O_2-production using microelectrodes, or that of ^{14}C-HCO_3-fixation (see Chapter 20).

The data in Table 6.5.3 are reported as $gC/m^2/d$. A few investigators measured primary productivity at several times during the day and integrated these values to yield the total reported here. Others made determinations for a fixed period of time under optimal conditions and reported their data as $gC/m^2/h$. We have arbitrarily multiplied these values by a factor of five to convert them to $gC/m^2/d$. This conversion factor was determined by a dawn-to-dusk experiment for Spencer Gulf data (Bauld unpubl.). The productivity values reported for mats in Table 6.5.3 are within the range reported for water column ecosystems. Two values (those for Solar Lake and for Drakesbad Hot Springs) are extremely high, but are comparable to the highest rates reported for phytoplankton production in the Peru Current ($10 gC/m^2/d$; Steemann Nielsen 1975).

In cases where the anoxygenic primary productivity in mats has been measured separately from oxygenic productivity, it has been found that the anoxygenic phototrophs can contribute as much as 90% of the total productivity in the mat system (Pierson unpubl.; Gorlenko and Yurkov 1988). Such values are in the same range as those calculated for planktonic anoxygenic phototrophs in holomictic and meromictic lakes (Biebl and Pfennig 1979).

Values obtained for primary productivity in modern microbial mats cannot be applied readily to determination of the productivity of Proterozoic stromatolitic biocoenoses. One recent attempt to estimate primary productivity in such stromatolites suggested values in the range of $0.19-3.6 gC/m^2/d$ (Lanier 1986). However, these values are based on such a large number of unproven (and unprovable) assumptions that although they fall well within the range of the values most commonly obtained for modern mats, their comparison to modern values is of questionable significance.

6.6

Chemotrophy and Decomposition in Modern Microbial Mats

BO BARKER JØRGENSEN   DOUGLAS C. NELSON   DAVID M. WARD

6.6.1 Decomposition and Element Cycling

Microbial mats, in which phototrophs such as cyanobacteria constitute the dominant mat-building organisms, often have a very high organic productivity (Section 6.5). Yet, the net accretion of microbial biomass at the mat surface is low and is often balanced by decomposition in the deeper layers (Doemel and Brock 1977). Even in the 1-m-thick mats, which have grown in the hypersaline Solar Lake in Sinai for over 1000 years, the net accretion constitutes only about 1% of the gross primary production of organic matter (Krumbein et al. 1977; Jørgensen and Cohen 1977).

In view of the foregoing, it is evident that rapid decomposition and recycling of the microbial biomass can occur in mats. A direct recycling of photosynthate occurs by the dark respiration of phototrophic organisms and by their exudation of organic molecules such as glycolate. Glycolate is formed by the oxidase activity of the key enzyme for CO_2-fixation, ribulose bisphosphate carboxylase/oxidase. Glycolate formation is particularly enhanced by conditions which prevail in the euphotic zone of phototrophic mats: high light intensity, high O_2 concentration, and high pH (i.e., low CO_2 concentration; Section 6.4). In a hot spring mat dominated by the coccoid cyanobacterium *Synechococcus lividus*, up to 12% of the photosynthetically fixed carbon was found to be photo-excreted (Bauld and Brock 1974). Up to 60% of this exudate was glycolate, which within the mat was readily assimilated by *Chloroflexus*-like filamentous bacteria (Bateson and Ward 1988; Section 6.9).

In contrast with the direct, rapid recycling discussed above, a slower recycling of biomass takes place upon burial and death of the mat-building phototrophs. Microbial remains are gradually hydrolyzed, assimilated, and respired or fermented by a variety of heterotrophic organisms (Fig. 6.6.1). In the aerobic surface layers of the mat, the assimilated organic compounds are completely oxidized to CO_2 and H_2O within the aerobically respiring cells. Deeper in the mat, organic molecules are degraded through anaerobic microbial food chains in which inorganic nitrogen and sulfur compounds are also reduced (Fenchel and Jørgensen 1977).

The heterotrophic components of microbial mats are poorly characterized because they cannot be microscopically distinguished from many non-heterotrophs. Modern techniques involving recognition of heterotrophs based on biochemical markers are now improving this situation. Groups of bacteria can be identified based on their characteristic membrane lipids (Section 6.8), while individual species can be identified by use of recombinant DNA techniques which recognize nucleotide sequences of individual 16S ribosomal RNA genes (Pace et al. 1986b; Ward 1989). By more traditional enrichment and isolation methods, more than ten different aerobic and anaerobic heterotrophs have been identified in a single cyanobacterial mat from Octopus Spring, Yellowstone National Park, Wyoming, USA (see Ward et al. 1987; Section 6.9). Among these species are the aerobic eubacteria *Isosphaera pallida* and *Thermus aquaticus*; the anaerobic fermentative *Thermobacteroids acetoethylicus*; the sulfate-reducing *Thermodesulfobacterium commune*; and the methane-producing archaebacterium *Methanobacterium thermoautotrophicum*. As the names imply, these are thermophilic organisms, especially adapted to a hot spring environment.

6.6.2 Heterotrophic Processes

6.6.2A Aerobic Respiration

Microbial mats growing in oxic environments generally have only a thin oxic surface layer in which aerobic respiration can take place. During the night, oxygen penetrates into this zone by diffusion from the overlying water. During the day, the oxic zone of phototrophic mats is highly expanded due to internal (photosynthetic) O_2-production. Within the surface layer, there is thus a diel cycle between aerobic metabolism during the day, and anaerobic metabolism during the night (Section 6.4). In some hot springs, anoxic conditions prevail during the day at the surface due to high sulfide concentration in the overlying source water. In these springs the aerobes may be restricted to only a thin oxic zone wedged into the mat during the daytime (Jørgensen and Nelson 1988; Section 6.9). The presence of free O_2 in microbial mats is probably the most important chemical factor determining species composition, physiology, and geochemistry. Close coupling between aerobic

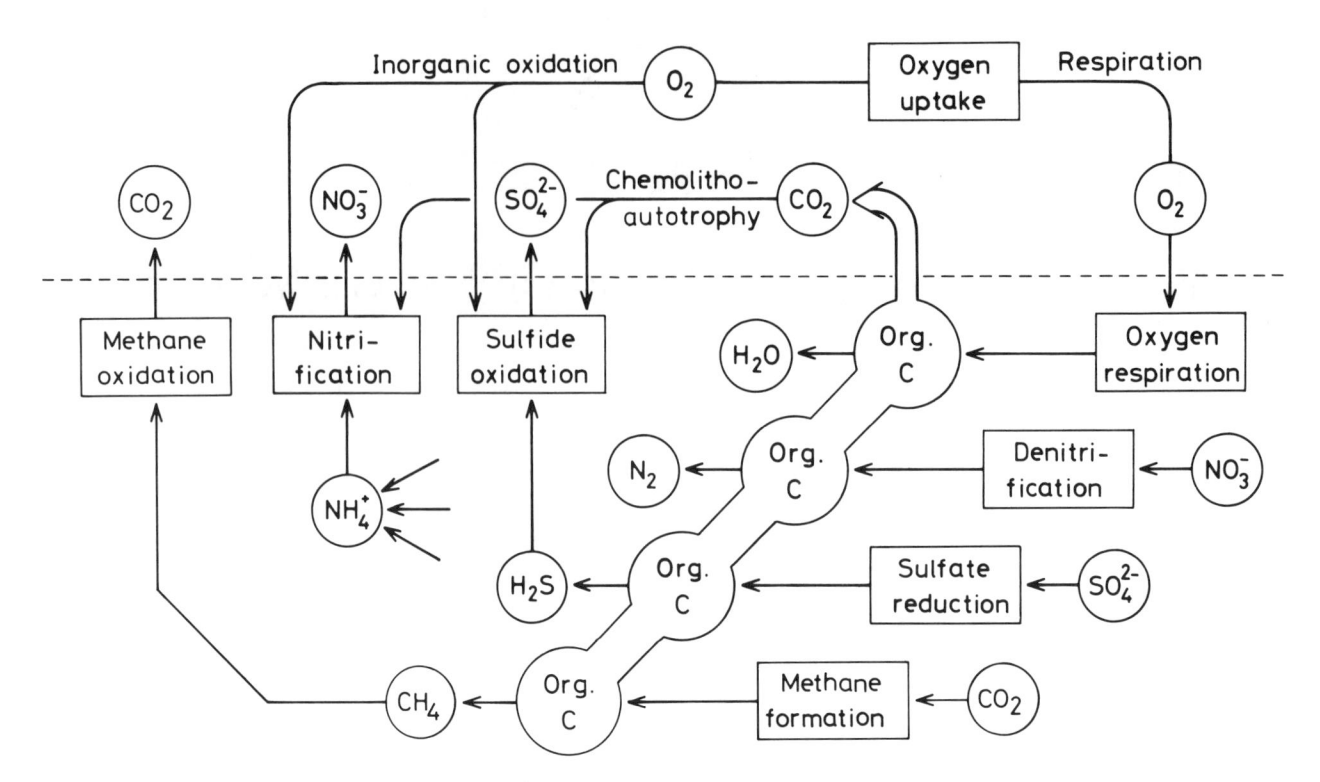

Figure 6.6.1. Respiration and element cycling associated with the degradation of organic material in microbial mats. The organic matter is oxidized to CO_2 through a vertical sequence of O_2, NO_3^-, and SO_4^{2-} respiration processes. Of the respiration products, only H_2S carries useful energy, which can be utilized by chemolithotrophic sulfide oxidizing bacteria. Ammonia is released by degradation of nitrogenous compounds and may be oxidized to NO_3^- by lithoautotrophic, nitrifying bacteria. A significant fraction of the oxygen consumption in mats is used for these lithotrophic processes. If available electron acceptors for respiration become depleted deeper in the mat, methane becomes the end product of the anaerobic fermentations (Jørgensen 1980).

respiration and light conditions has been demonstrated by use of microelectrodes in cyanobacterial mats from hypersaline ponds of Guerrero Negro, Baja California Sur, Mexico (Fig. 6.6.2). The thickness of the oxic zone was observed to change with light intensity from 0.25 mm in the dark to 2.0 mm under near-daylight illumination, that is, an eight-fold increase in total depth. The rate of aerobic respiration per unit area of the mat also increased, from 4 to 18 nmol O_2/cm^2/min, that is, a 4.5-fold increase. Although these mats were covered by fully aerated water, the respiration rate during the night was thus strongly limited by the availability of O_2 through diffusion.

In spite of the thinness of the oxic zone, it has been possible by use of microelectrodes and computer modelling to estimate the distribution of aerobic respiration at $< 100\,\mu m$ resolution (Revsbech et al. 1986). Figure 6.6.3 shows a typical, but highly compressed, zonation in a cyanobacterial mat growing along the shore of a protected, brackish lagoon in Denmark. The euphotic zone is only 0.3 mm deep; O_2 diffuses from the zone of its production to a depth of about 0.8 mm. Aerobic respiration is high in the zone of active photosynthesis, possibly due to exudation of glycolate or of other organic substrates from the phototrophs. A second peak of O_2 respiration occurs at the oxic-anoxic interface where O_2 reacts with H_2S and other reduced compounds diffusing up from deeper layers. Chemolithotrophic sulfur bacteria such as *Beggiatoa* are concentrated at this interface and are probably responsible for the intensive consumption of O_2 used for H_2S-oxidation.

An overall budget of the O_2 balance in the cyanobacterial mat discussed above shows that respiration consumes 76% of the O_2 produced by the total photosynthesis of 62 nmol/cm^2/min. The remaining 24% is lost by diffusion into the water along the steep surface gradient. The mat ecosystem is thus autotrophic during the day when it builds up biomass. Most of this biomass is again respired during the night or after burial into anoxic layers. The result is thus a very small, or perhaps even zero, net growth. Anoxic degradation by sulfate reduction or methanogenesis is generally included in the O_2 respiration due to reoxidation of reduced end-products (e.g., H_2S, CH_4, NH_4^+, etc.) at the oxic-anoxic interface. In the mat illustrated in Figure 6.6.3, these interface products account for 21% of the total respiration.

6.6.2B Fermentation and Methanogenesis

Anaerobic degradation takes place both in deeper layers of the mats and near their upper surface. It is interesting that methanogenic bacteria and other strict anaerobes, which are normally killed by oxygen, seem to survive prolonged periods of exposure to high O_2 at or near the mat surface (Revsbech and Ward 1984a). Sulfate reduction in the euphotic surface layer can even be stimulated in the light in spite of high O_2 concentrations (Cohen 1984a). Conversely, the aerobic phototrophs and heterotrophs must survive dark anoxic periods and even high H_2S concentrations. The cyanobacteria have physiological adaptations to such an anaerobic metabolism; in the

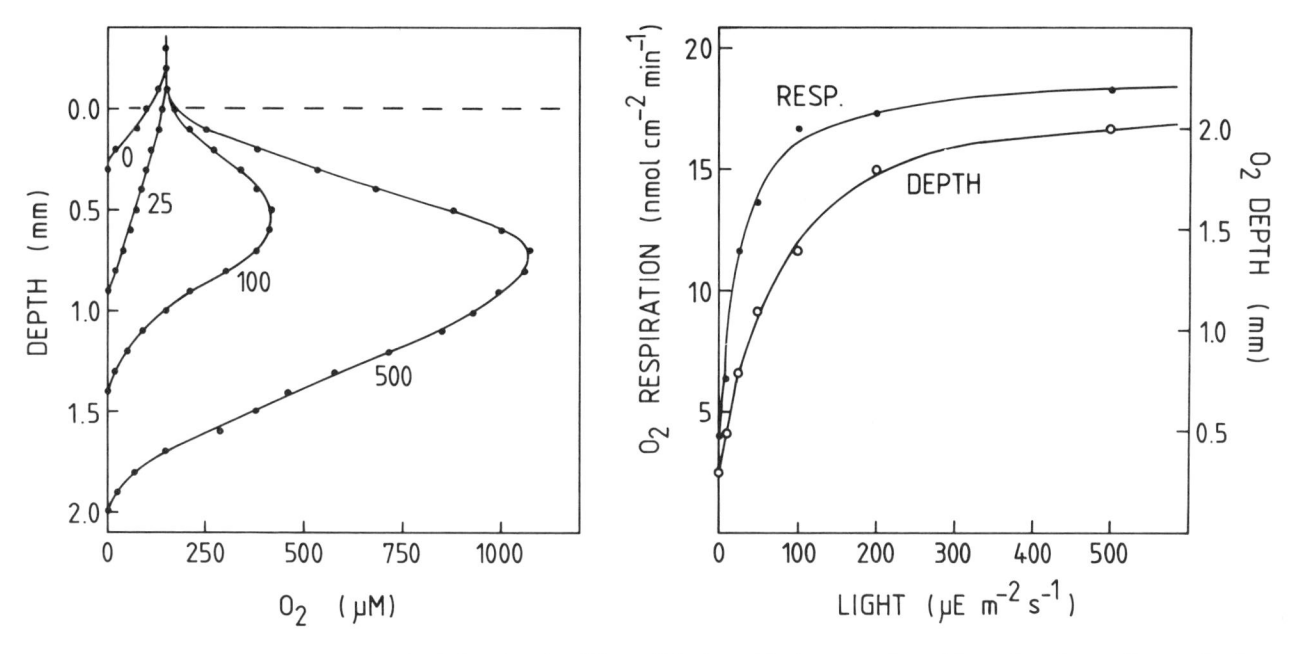

Figure 6.6.2. Left: Oxygen distributions in the dark and under different light intensities in a cyanobacterial mat from hypersaline ponds of Guerrero Negro, Baja California Sur, Mexico. Numbers on curves indicate illumination in μEinsteins/m^2/s. Right: Depth of the oxic zone (O_2 depth) and total rates of respiration in the same mat under different light intensities. (Unpublished data of B. B. Jørgensen).

absence of O_2, several species are known to switch to fermentation, or to anaerobic respiration using elemental sulfur as the electron acceptor (Oren and Shilo 1979). Cyanobacteria can even occur deep within an anoxic mat, possibly surviving here for months or years by a fermentative metabolism (Richardson and Castenholz 1987b; Jørgensen et al. 1988).

Fermentable substrates such as carbohydrates and proteins are abundant in the upper layers of microbial mats (Krumbein et al. 1977; Doemel and Brock 1977; Boon et al. 1983; Aizenshtat et al. 1984; de Leeuw et al. 1985). Such substrates presumably include both cellular components of mat organisms and extracellular components (e.g., sheaths and other exopolymers). The substrates are fermented by different anaerobic heterotrophs producing a wide array of possible end-products (Fig. 6.6.4). Cyclic oxidation and reduction of the intermediate electron carrier, NADH, plays a central role in these fermentations.

The energy yield from anaerobic formation of "electron sink products" is low. A higher yield of ATP per substrate molecule is gained from fermentation to acetate, H_2, and CO_2. This fermentation, however, requires efficient removal of H_2 (i.e., an "interspecies H_2 transfer") by sulfate-reducing, methanogenic, or acetogenic bacteria (Ward et al. 1984; Ward and Winfrey 1985). There is thus a tight coupling between the fermentation pathways and the terminal processes of anaerobic decomposition. The H_2 in microbial mats seems mostly to be derived from these bacterial fermentations (Oremland 1983; Skyring et al. 1988), but it may also be produced directly by cyanobacteria during anoxygenic photosynthesis (Belkin and Padan 1978a, b, 1983).

Because of competition between methanogens and sulfate-reducers for common substrates, especially acetate and H_2, methane formation in anaerobic environments is markedly influenced by the availability of sulfate. In low-sulfate hot spring mats, intensive methanogenesis occurs with H_2 and CO_2 as the main substrates; in this setting, acetate and other volatile fatty acids are mostly photoassimilated by filamentous bacteria, probably *Chloroflexus* sp. (Section 6.9; Sandbeck and Ward 1981; Anderson et al. 1987). In non-thermal mats, which are mostly high in sulfate, sulfate reduction predominates and methanogenesis proceeds at low rates via degradation of methylated amines and other "non-competitive" substrates (Giani et al. 1984; King 1988).

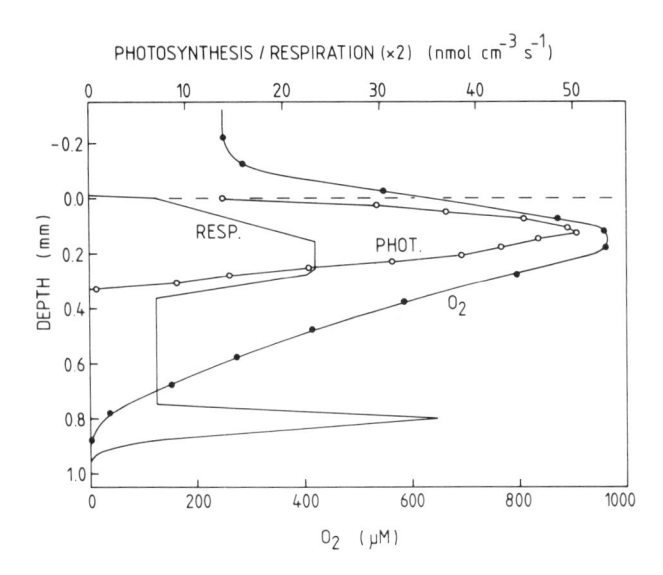

Figure 6.6.3. Depth distribution of O_2, photosynthesis, and respiration in a cyanobacterial mat from a coastal sediment in Limfjorden, Denmark. The O_2 and photosynthesis distributions were measured by oxygen microelectrodes. The distribution of O_2 respiration was calculated by computer simulation of the O_2 diffusion and dynamics during light-dark transitions. (Redrawn from Revsbech et al. 1986).

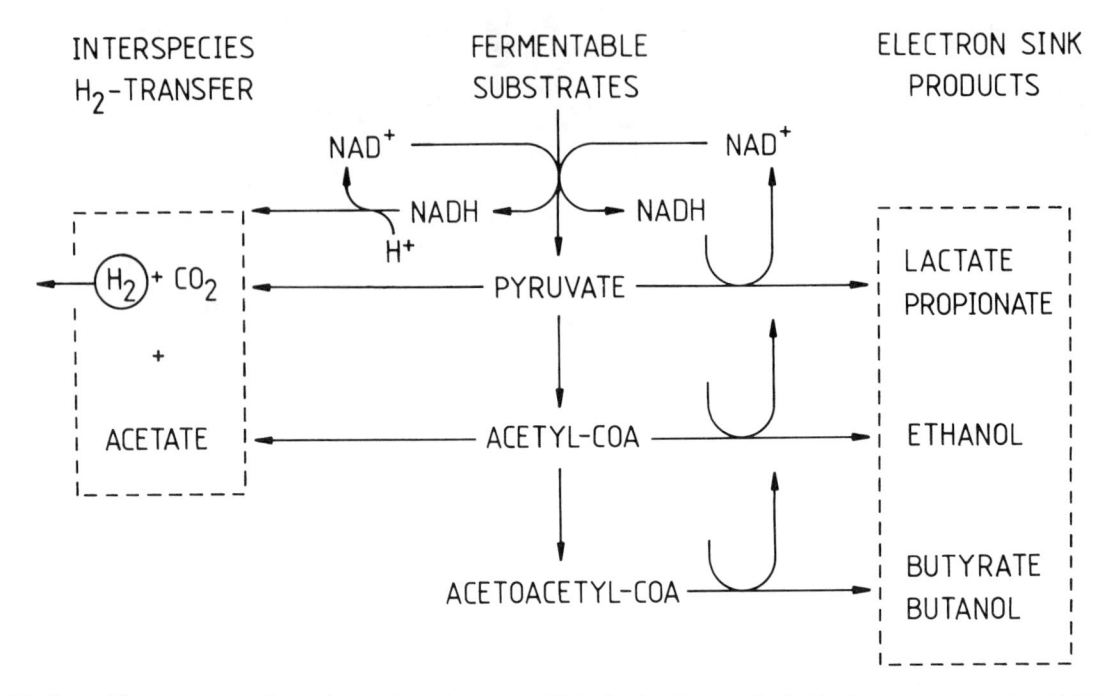

Figure 6.6.4. General fermentation pathways in anoxic environments. Right: In the absence of suitable electron acceptors for NADH, the reduced fermentation products (lactate, ethanol, etc.) serve as electron sinks for the regeneration of oxidized NAD^+. Left: In the presence of H_2-utilizing sulfate reducers and methanogens, the fermentation is shifted toward the production of free H_2, CO_2, and acetate.

6.6.2C Anaerobic Respiration

The role of nitrate respiration by denitrifying bacteria in microbial mats has not been determined experimentally. Based on the steep chemical microgradients found in most mats, however, a zone of denitrification would be expected to be very thin, <1 mm. Only recently have N_2O and NO_3^- microsensors been developed which allow measurement of denitrification at 0.1 mm resolution (Revsbech et al. 1988; Sweerts and de Beer 1989).

Sulfate reduction has been measured in both hot spring and hypersaline mats by radiotracer techniques (e.g., Jørgensen and Cohen 1977; Bauld et al. 1979; Ward and Olsen 1980; Skyring 1984, 1987). By application of special $^{35}SO_4^{2-}$-impregnated glass rods (Skyring et al. 1983) or silver wires (Cohen 1984a) it has been possible to measure the activity at 1 mm depth resolution. The reduction rates determined from surface layers of microbial mats of Solar Lake, Sinai, are among the highest recorded for any natural environment, 5,400 nmol $SO_4^{2-}/cm^3/d$ (Fig. 6.6.5; Jørgensen and Cohen 1977). Due to the intensive photosynthetic productivity and the lack of bioturbating invertebrates (which in other sedimentary environments transport organic matter into deeper layers), sulfate reduction is strongly concentrated toward the mat surface. Thus, 50% of the sulfate reduction takes place at 0 to 5 mm depth where counts of viable, lactate-utilizing, sulfate-reducing bacteria reach 2×10^6 per cm^3.

A high proportion of the organic matter in high-sulfate mats is degraded and oxidized through bacterial sulfate reduction. In the *Microcoleus chthonoplastes* mats at Solar Lake, Sinai, total reduction rates in the entire mat column were 6.7 μmol $SO_4^{2-}/cm^2/d$ (Jørgensen and Cohen 1977). The rate of gross photosynthesis measured seven years later was 15.6 μmol $O_2/cm^2/d$ (Revsbech et al. 1983). Since the stoichiometric ratio between sulfate and oxygen is 1:2, this latter rate would suggest that sulfate reducers respired 86% of the total organic carbon produced, an estimate that for a variety of reasons seems to be too high.

Thermophilic sulfate reducers such as *Thermodesulfobacterium commune* are well adapted to life in hot spring mats

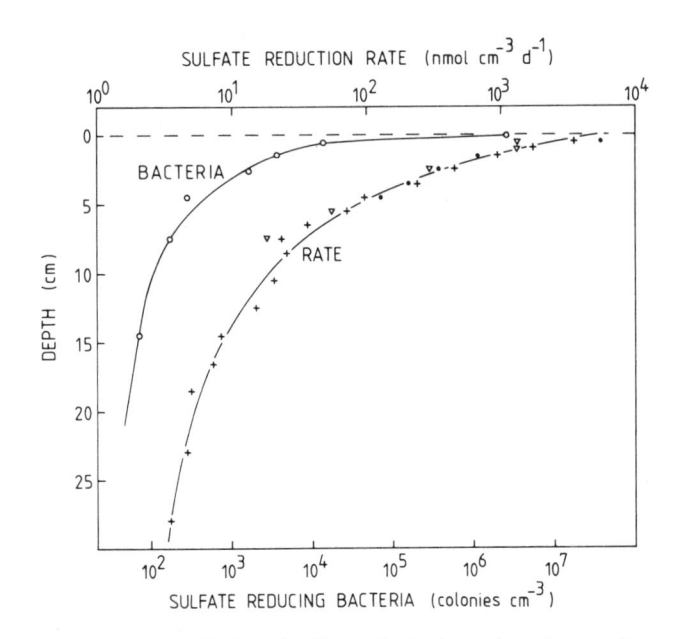

Figure 6.6.5. Distribution of sulfate reducing bacteria and rates of sulfate reduction in the deep cyanobacterial mats of Solar Lake, Sinai. Sulfate reduction was measured by radiotracer technique while the viable counts of the bacteria were done in agar shake cultures. (Redrawn from Jørgensen and Cohen 1977).

(Zeikus et al. 1983), but the upper temperature limit for such microbes has not been determined. In mats from hypersaline ponds, however, sulfate reduction appears to be inhibited by the high salinities (Section 6.10), although rapid sulfate reduction has nevertheless been recorded at 180‰ (Jørgensen and Cohen 1977).

6.6.3 Chemolithotrophic Processes

Table 6.6.1 lists the main types of known chemolithotrophic bacteria, all of which can generate the energy needed for their growth and maintenance by the oxidation of inorganic compounds (e.g., NH_4^+, H_2S, $S°$, H_2). Most lithotrophic bacteria, including all those listed in Table 6.6.1, can use CO_2 for the biosynthesis of their cell constituents, they are therefore referred to as "lithoautotrophs." The first four groups in the Table are lithoautotrophs which use the Calvin cycle for CO_2-fixation, as do photoautotrophic algae and cyanobacteria. They consequently also fractionate the stable carbon isotopes during their growth, as do the phototrophs (Section 6.8). The remaining three groups listed in Table 6.6.1 employ other pathways of CO_2-fixation which fractionate carbon isotopes differently. The principal oxidants shown to be used by pure cultures of lithoautotrophs include O_2, NO_3^-, CO_2, and $S°$. Furthermore, a few of the sulfate-reducing bacteria oxidize H_2 with SO_4^{2-}, and Fe^{3+} serves as an oxidant for Sulfolobus (Brock 1985). There is growing evidence, mostly based on biogeochemical field data, that oxidized manganese and iron may serve as electron acceptors in a variety of environments (e.g., Jørgensen 1989). Although some chemolithotrophs in aquatic environments have been studied (e.g., nitrifying bacteria), little is known of their activity in mats. Colorless sulfur bacteria (and certain iron bacteria) are the only chemolithotrophs known to form mats in which they are the dominant components.

Understanding is emerging of the diverse niches occupied by sulfur bacteria (Table 6.6.2). Thermothrix is restricted to flowing freshwater thermal springs (Fig. 6.9.5) where it forms long filaments in response to micro-oxic conditions (Brannan and Caldwell 1986), concentrating streamer growth in the region which contains both O_2 and H_2S. Other genera of sulfur bacteria, although found in freshwater habitats, are more frequently encountered in marine environments where the sulfur cycle is more vigorous as a result of the much higher concentration of sulfate and of its biological reduction. Thiothrix dominates in flowing water systems where H_2S and O_2 are presumably high in concentration but temporally variable (Larkin and Strohl 1983). Thioploca apparently occurs where sulfide exists principally or exclusively as metal precipitates rather than as free H_2S (Henrichs and Farrington 1984); whether this environment is anoxic but oxidized (sensu Jørgensen 1988), or micro-oxic, has yet to be established (Maier and Gallardo 1984b; Fossing and Nelson unpubl.). Extensive and dense mats of Thioploca (up to 1 kg/m², wet weight including sheaths) have been inferred to have played a role in oil production associated with the Cenozoic Monterey Formation of California, USA (Williams and Reimers 1983).

Beggiatoa provides an unparalleled example of a microorganism dependent on the microenvironment of opposing diffusion gradients. Beggiatoa mats are often found at the sediment-water interface of organic-rich coastal sediments. Around deep-sea hydrothermal vents, notably in the Guaymas Basin of the Gulf of California, Mexico, mats have been found of giant Beggiatoa with filament widths of 100 μm (Jannasch

Table 6.6.1. Types of chemolithotrophs and related bacteria.

Bacterial group	Energy substrate(s)	Oxidant(s)	Representative genera
Hydrogen bacteria	H_2	O_2, sometimes NO_3^-	14 diverse genera[a]
Nitrifying bacteria	NH_4^+, NO_2^-	O_2	Nitrosomonas, Nitrobacter
Colorless sulfur bacteria	H_2S, $S°$, $S_2O_3^{2-}$	O_2, sometimes NO_3^-	Thiobacillus, Thiomicrospira, Beggiatoa[b], Thermothrix[b], Thioploca[b], Thiothrix[b]
Iron bacteria	Fe^{2+}	O_2	Thiobacillus ferrooxidans, Gallionella ferruginea[b]
Methanogens	H_2	CO_2	Methanobacterium, Methanococcus, and 8 other genera[c]
Autotrophic acetogens	H_2	CO_2	Acetobacterium woodii, Clostridium aceticum, Desulfovibrio barsii[d]
Sulfur-dependent thermophilic archaebacteria	H_2	$S°$	Thermoproteus tenax, Pyrodictium brockii[e]
	$S°$	O_2, Fe^{3+}	Sulfolobus spp.[e]

[a] See Stainer et al. (1986) for more detailed description of these genera.
[b] These bacteria are known to be major constituents of microbial mats or biofilms.
[c] See Jones et al. (1987) for more detailed description of the other lithoautotrophic representatives of this group.
[d] See Ljungdahl (1986) for a discussion of the approximately 20 species comprising this recently identified group.
[e] See Brock (1985).

Table 6.6.2. *Chemolithotrophic mat-forming microorganisms.*

Bacterial genus	Bacterial Morphology	Habitat	Mat Description
Beggiatoa spp.	Cylindrical cells with length 0.5 to 2 × diameter; cells occur in filaments up to 1 cm long and contain internal S° globules; filaments glide (1–5 μm/s); filament width ranges from 1 to 100 μm (strain invariant).	Typically aquatic sediments (especially marine) with high H_2S flux and some O_2 present (e.g., sediments of salt marshes, eutrophic bays, and harbors; sulfur springs, including deep-sea vents).	In sediments, mats are dense networks of filaments typically <1 mm thick which "track" the O_2/H_2S interface using gliding motility; deep-sea vent mats are composed of 30–100 μm diameter filaments, may be centimeters thick, occurring in an uncharacterized microenvironment.
Thioploca spp.	Single filaments (1 to >40 μm wide) superficially identical to *Beggiatoa* except that several occur in parallel or braided within a common sheath.	Sediment regions containing little or no free H_2S; known from sediments of a few freshwater lakes; occur principally off the coast of S. America (10–30°S lat.) in surface sediments below water column O_2 minimum; typical depth of overlying water is 100–500 m; biomass densities can exceed 1 kg/m².	Fairly uniform distribution throughout upper 10 cm of marine sediment; this sediment is anoxic but oxidized with the possible exception of top few mm which may be micro-oxic.
Thiothrix spp.	Nearly cylindrical cells forming filaments often several cm long; one end of filament attaches to solid surfaces via a "holdfast"; cells contain S° globules and only glide in dispersal stage.	Flowing water containing both H_2S and O_2 (e.g., tidal creeks and both freshwater and deep-sea vent sulfur springs); mats attach to solid surfaces including rocks, vegetation, shells, and animals.	Filaments form rosette patterns of "holdfasts," resulting in dense "streamers" moved by flowing water.
Thermothrix thioparus	Individual cells 0.5–1.0 × 3–5 μm forming chains up to 1 cm long; encrusted with external S° granules.	Flowing sulfidic hot springs (55°–80°C, 30–600 μM H_2S); not observed in marine environments.	Filaments form bundles several cm long which attach to solid surfaces and form "streamers" in flowing water; O_2 or NO_3^- may serve as e^- acceptor.
Gallionella sp.	Crescent- or bean-shaped "apical" cells ca. 0.5 × 1–2 μm; cells excrete numerous organic fibrils (10–20 nm diameter) which form a twisted "stalk" (up to 50 μm long) of unknown composition.	Freshwater and occasionally marine environments which are very low in organic matter and neutral in pH; slightly reduced environment (+330 to +200 mV) but some O_2 present as well as abundant Fe^{2+} (e.g., mineral springs, wells and drainage).	Individual cells are attached by stalk to substrate; when bacteria divide by binary fission, organic fibrils result in branched stalks; variable $Fe(OH)_3$ encrustation; numerous other bacteria may contribute to the "mat."

1984). The 0.5-mm-thick *Beggiatoa* mat from Aarhus Bay, Denmark, depicted schematically in Figure 6.6.6, is typical of the mat-behavior of this organism. Even though the overlying water is aerated and stirred, a diffusive boundary layer exists which reduces the O_2 concentration at the mat surface to 10 μM (4% of air saturation). H_2S, the product of sulfate reduction in the sediment, diffuses from below attaining a concentration of ca. 100 μM at the lower mat boundary. H_2S-consumption takes place within the 100 μm vertical zone in which H_2S and O_2 co-occur, essentially a steady-state situation because both of these nutrients are constantly renewed by diffusion along their respective concentration gradients. In mats that also contain oxygenic phototrophs (e.g., marine and hypersaline cyanobacterial mats), O_2 can shift dramatically within the upper centimeter on a diel basis, and *Beggiatoa* has been observed to follow the vertically moving H_2S-O_2 interface. Documented negative tactic responses to excess O_2 and light (Møller et al. 1985; Nelson and Castenholz 1982), along with a postulated negative response to excess H_2S, explain this behavior. It has been further demonstrated (Jørgensen and Des Marais 1986a) that anoxygenic photosynthetic sulfur bacteria (e.g., *Chromatium*) also compete for this H_2S, the outcome being dependent on whether light or O_2 has deeper contact with H_2S.

Iron-oxidizing bacteria of the genus *Gallionella*, which are probably capable of chemoautotrophy, are characteristically found in macroscopic masses. Such masses occur in aquatic environments which are virtually devoid of dissolved organic compounds, slightly reduced (+200 to +320 mV), rich in Fe^{2+}, and micro-oxic, conditions under which Fe^{2+} is relatively stable and biological iron oxidation can compete with non-biological oxidation. *Gallionella* mats are composed of bacteria, ferric hydroxide, and excreted bacterial polymers; their participation in sedimentary formation of iron ore has been suggested (Hanert 1981; Ghiorse 1984).

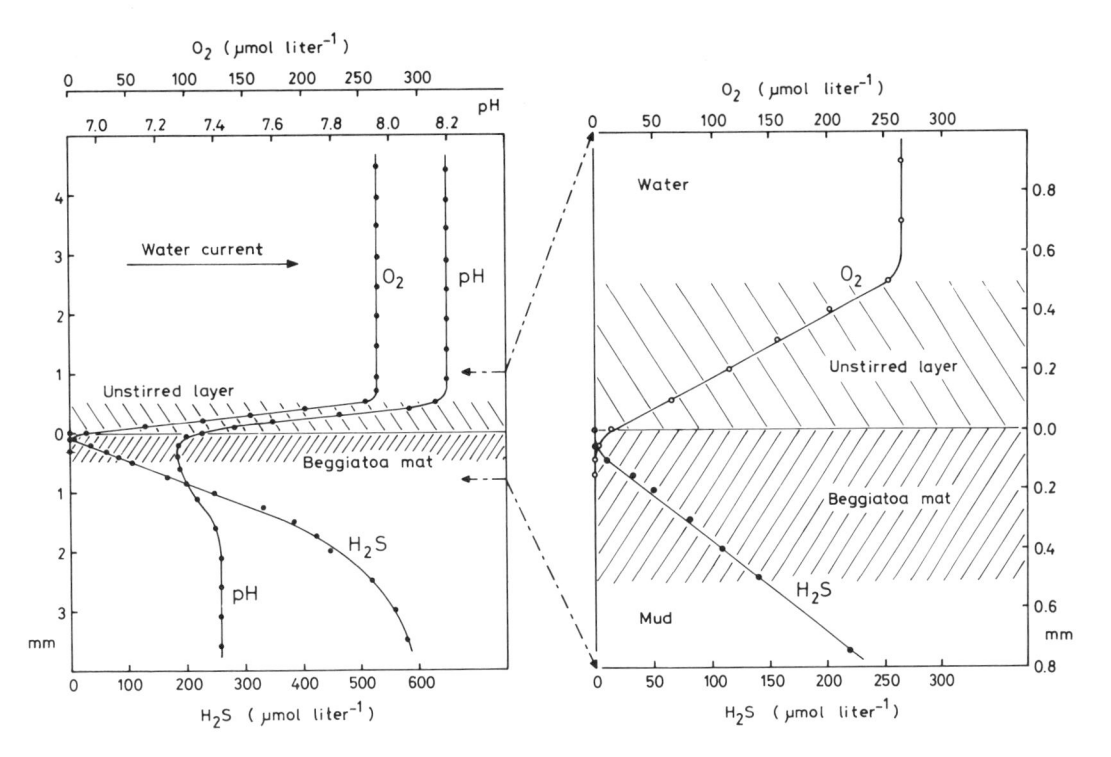

Figure 6.6.6. The H_2S/O_2 microenvironment of a *Beggiatoa* mat found at a coastal sediment-water interface in Aarhus Bay, Denmark (Jørgensen and Revsbech 1983).

6.7
Grazing and Bioturbation in Modern Microbial Mats
JACK D. FARMER

6.7.1 Introduction

The primary consumers, or "grazers" of microbial mats are significant to Proterozoic paleobiology in several important respects. Faunal studies of microbial mats provide important paleobiological constraints for interpreting the adaptive pathways followed during the early evolution of the benthic Metazoa and for understanding the innovations that have shaped the history of microbial systems in general. For example, grazing and bioturbation by metazoans has been suggested as a causal factor in the Late Proterozoic decline of stromatolite diversity and abundance (Walter and Heys 1985; Awramik 1971, 1981; Garrett 1970a, b). This hypothesis is based on the general tendency for modern, potentially stromatolite-forming, microbial mats to be restricted to extreme environments from which metazoan grazers are largely excluded. Studies in both intertidal marine (Robles and Cubit 1981; Lubchenco and Cubit 1980; Castenholz 1961) and freshwater environments (Hill and Knight 1988, 1987; Hart 1985; Murphy 1984; McAuliffe 1984; Lamberti and Moore 1984; Gregory 1983) demonstrate the importance of grazers in altering the biomass, primary productivity, and structure of algal communities encrusting hard substrates. In marine sedimentary systems, Decho and Castenholz (1986) have shown that the distribution of species of meiofaunal copepods is coupled closely to their preferred microbial food sources. However, comparatively little is known about the interactions of most grazers in laminated mat systems. Important questions to be resolved include trophic preferences, participation in nutrient cycles, and controls on distribution.

6.7.2 Grazing in Microbial Mats of Restricted, Hypersaline Environments

Ecological studies of primary consumers, and their effects on microbial mats in hypersaline environments, have focused on the coastal sabkhas marginal to the Gulf of Aquaba, Sinai (Gerdes and Krumbein 1987; Gerdes et al. 1985b; Friedman and Krumbein 1985; Gerdes and Krumbein 1984), and similar environments or salterns associated with Laguna Guerrero Negro and Laguna Ojo de Libre on the Vizcaino Peninsula, central Baja California Sur, Mexico (Javor and Castenholz 1984a, 1981; Section 6.10). The following brief synopsis is intended to serve as a general framework for identifying major grazing strategies and their physical and biological effects in modern microbial mats.

6.7.2A Sabkhas, Southern Sinai Peninsula

The Gavish Sabkha and Solar Lake are coastal-margin brine pans located on the plain of the southeastern Sinai Peninsula, bordering the Gulf of Aquaba. These two areas share many important physical and biological similarities, discussed in detail elsewhere (Gerdes and Krumbein 1987; Gerdes et al. 1985a, b). Major differences between the two areas relate to seasonal variations in cycles of salinity stratification, and to the occurrence of winter flooding of marginal littoral habitats at Solar Lake. Both ecosystems are markedly similar in faunal composition, and both are characterized by low diversity, grazer-dominated food chains comprised of primarily cosmopolitan genera and species. Faunal compositions and trophic relationships in the two areas are summarized in Gerdes et al. (1985b) and in Gerdes and Krumbein (1987). The grazers of mats in both areas are largely restricted to marginal wetlands or the shallow, permanently submerged areas of lagoons. Faunal diversity and abundance are highest over a salinity range of 70 to 140‰, and for mats having heterogenous surface textures, such as nodular, flocculose, blister, or pinnacle mats (Gerdes et al. 1985b).

The focus of this discussion is the more diverse of the two faunas, that from the Gavish Sabkha. Thirty-one species of animals, belonging to ten supraspecific categories, have been recognized in this fauna, not including several unidentified species of ciliate protozoans (Gerdes and Krumbein 1987, Table 4). Of the 31 species, at least 19 are known to be consumers of diatoms and coccoid cyanobacteria, two are detritus feeders, and one is a species that feeds on filamentous cyanobacteria. At all salinities, the fauna is dominated by species of insects (coleopterans and dipterans), most of which are grazers.

Within the wetlands habitat of both sabkhas, salinities range from 50 to 70‰. In this zone, the aspect dominants are terrestrial arthropods, which include adults and larvae of two species of the salt beetle, *Bledius*. Adults of *Bledius capra* excavate chambered burrows to depths of 10 to 12 cm in which they store and feed on lumps of cyanobacterial mat (Gerdes and Krumbein 1987; Bro Larsen 1936). Also common are feeding coeleopterans including adults of *Georyssus* sp. and larvae of *Lophyridia aulica*. The amphipod *Halophiloscia* sp., and larvae of the brine fly *Musca drassirostris*, prefer shallowly submerged to seasonally exposed mats of the littoral zone transition, where they feed on decaying organic detritus. The nematode *Enoplus communis*, common at the higher salinities within this transition, feeds primarily on coccoid cyanobacteria and diatoms. At Solar Lake, *Monhystera* sp. is the common nematode in this habitat, where it feeds on filamentous cyanobacteria (Gerdes et al. 1985b).

The littoral zone of both sabkhas is characterized by periodic or seasonal flooding. Over the salinity range from 65 to 140‰, the faunas are dominated by diatom and coccoid cyanobacterial feeders, including the ostracods *Cyprideis torosa* and Paracyprideinae sp.; the copepod *Robertsonia salsa*; the turbellarian *Macrostomum* sp.; five species of coleopterans; and the nematodes *Oncholaimus* sp. and *Adoncholaimus okyris*. Also present is the copepod *Nitocra* sp. which feeds on the filamentous cyanobacterium *Microcoleus* (Gerdes and Krumbein 1987).

Within the littoral zone marginal to the coast in both sabkhas, seeps maintain a constant influx of normal seawater. In higher areas, characterized by a cover of the green alga *Enteromorpha*, the euryhaline gastropod *Pirenella conica* is common, as are several marine annelids. *P. conica*, however, achieves its highest densities (2000/m² for the 5 to 10 mm size class) in lower areas that are covered by shallow water films of salinity 70‰ or less (Gerdes and Krumbein 1984). Where abundant, *P. conica* completely inhibits surficial microbial mat development by plowing the sediment surface, although within sediments microbial growth rates may be enhanced by grazing and the production of feces (Fry 1982; Fenchel and Jørgensen 1977). High densities of *P. conica* are sustained by the prolific growth of microbial films on sediment grains and fecal pellets (Gerdes and Krumbein 1984).

The Gavish Sabkha includes several wide-ranging species that tolerate salinities up to 250‰. A coccoid chaeopteran, *Ochthebius auratus*, is found on the spongy and flocculose mats of the submerged littoral zone, beneath salt crusts, and in low oxygen environments of the deeper pinnacle mats on the upper lagoon slope where it feeds on cyanobacteria and diatoms. Larvae and adults of the dipteran *Hecamede grisescens* are also common beneath salt crusts where they feed on diatomaceous and cyanobacterial films. Adults of the coleopteran *Anacaena* sp. occupy a similar range of habitats, including flocculent mats of the peripheral wetland and littoral zone, and the mat-coated underfaces of salt crusts, where they feed on filamentous cyanobacteria.

6.7.2B Lagoons, Sabkhas, and Salterns, Vizcaino Peninsula, Central Baja California Sur, Mexico

Javor and Castenholz (1984a) documented the effects of grazing on microbial mats in intertidal pond and tidal creek environments marginal to Laguna Guerrero Negro, Baja California Sur, Mexico. A variety of mat types and associated invertebrate species are found in intertidal and supratidal zones along the lagoon margin. However, the most conspicuous grazers include a waterboatman (*Trichocorixa* sp.) and a marine gastropod (*Cerithidea californica*), both of which are associated with *Microcoleus* mats in the lower intertidal zone and with *Lyngbya*-*Microcoleus* mats in the middle intertidal zone. The waterboatman is common in intertidal ponds over a salinity range of 35 to 70‰; it produces a fecal floc that partially covers he mat surface, but that does not prevent mat development (Javor and Castenholz 1984a). This species is tolerant of hypersalinity, but the eggs can hatch only at lower salinities (<43‰; see Davis 1966). *Cerithidea californica*, the marine gastropod, tolerates salinities up to 88‰, but is restricted to areas that receive regular influxes of 35‰ seawater (Javor and Castenholz 1984a).

The effects of grazing by *Cerithidea californica* at Laguna Guerrero Negro are comparable to those noted above for *Pirenella conica* in the Gavish Sabkha. Where cerithids are abundant (800 to 1100 individuals/m²), they completely inhibit mat development. This species shows a preference for algal-rich sediments produced by grazing and burrowing (Javor and Castenholz 1984a). The major limitation on mat distribution by *Cerithidea californica* is its restriction to environments having a frequent tidal exchange with the lagoon.

Extensive *Lyngbya* and *Calothrix* mats occur on intertidal flats around the margins of Laguna Ojo de Libre (Javor and Castenholz 1981). Javor and Castenholz (1984a) suggest that cerithids do not colonize these areas due to restricted circulation and hypersalinity. Subtidal, flat-laminated benthic mats are well developed within the salt evaporation ponds of Exportadora de Sal, S.A., over a salinity range of about 60 to 130‰ (Section 6.10). Cerithids and other macroinvertebrate grazers are absent within the saltern, due to relatively high and constant salinities (Section 6.10). Small epifaunal and meiofaunal grazers are abundant in the mats and appear to have little impact on their distribution. A species of waterboatman (Corixidae) is common in Ponds 4 and 5 in the flocculent surface layer of shallow water mats over a salinity range of 65 to 85‰. For the same salinity range, the meiofauna is dominated by several species of crustaceans and nematodes. In the diatom-rich surface layer of the mat (viz., the uppermost 0.5 to 1.0 mm), the harpacticoid copepod *Cletocampus dietersi* is abundant (200 individuals/cm²), along with an unidentified cladoceran species (100 individuals/cm²). These species appear to feed primarily on diatoms and/or organic detritus. At a depth of 1 to 2 mm within the mat, the meiofauna is dominated by the nematodes *Monhystera* sp., *Microlaimus* sp., and *Oncholaimus* sp., and by two unidentified species of chromadorids. *Oncholaimus* is predatory on other nematodes, but also consymes diatoms. *Microlaimus* sp. and the chromadorids possess buccal armature and appear to feed on diatoms, coccoid cyanobacteria, and organic detritus (Tietjen and Lee 1973). *Monhystera* lacks buccal armature and is probably capable of consuming only smaller coccoid cyanobacteria or bacteria (Tietjen 1967). In plastic-embedded thin sections, nematodes are seen to be concentrated in a zone immediately above the chemocline (based on the shallowest occurrence of *Beggiatoa*), and below the dense surficial layer of *Microcoleus*. Estimates of nematode abundance, for all species combined, average 1600 individuals/cm^2.

At higher salinities (90 to 120‰, Ponds 6 and 8), saltern mats are dominated by coccoid cyanobacteria (Section 6.10), and meiofaunal diversity is lower. The meiofauna characteristic of this salinity range is dominated by nematodes (*Monhystera* sp. and an unidentified species of Microlaimidae) and the larvae of the brine fly *Ephydra* sp.. The surface of Pond 8 mat is tunneled with shallow horizontal galleries, often occupied by the larvae of an unidentified chaeopteran.

6.7.3 Grazing in Thermal Spring Mats

Wickstrom and Castenholz (1985) have documented the effects of grazing by a thermophilic ostracod, *Potamocypris*, on hot spring mats in Oregon. The temperature range tolerated by the ostracod is 28° to 53.5°C (Wickstrom and Castenholz 1973). Where abundant, it is associated with a nodular cyanobacterial mat composed of *Calothrix thermalis* and *Pleurocapsa minor* (Wickstrom and Castenholz 1978). *Calothrix* is faster growing than *Pleurocapsa*, and thus dominates in ungrazed areas. The unique nodular mat texture in this setting reflects a dynamic equilibrium between the two competing cyanobacterial species, the coexistence of which depends on selective grazing by the ostracod. Nodular mats also occur in other thermal springs in western North America where *Potamocypris* is abundant (Wickstrom and Castenholz 1985). Diel and seasonal changes in temperature are reflected by range extensions or retractions of the nodular mat in response to grazing. Persistence of the nodular mat also depends on continual upstream grazing on the faster-growing *Synechococcus-Chloroflexus* mat, the boundary between the two systems being maintained at the 51°C isotherm, near the upper temperature tolerance of the ostracod.

Several studies (Wickstrom and Wiegert 1980; Gorden and Wiegert 1977; Collins et al. 1976; Wiegert and Mitchell 1973; Brock et al. 1969) have described successional cycles in microbial mats in hot springs at Yellowstone National Park, Wyoming, USA, cycles that are controlled by larval grazing of the brine fly *Paracoenia turbida*. Optimal growth of these mats occurs between 40°C and 73°C. Above 40°C, biomass is added to the surface faster than it is degraded within the anerobic zone of the mat (Wickstrom and Wiegert 1980). With continued accretion, the mat surface shallows and eventually diverts the stream flow, producing islands of cooler, exposed mat which attract *Paracoenia*. The brine flies feed on the exposed part of the mat and deposit eggs. At temperatures less than 40°C, the upper few millimeters of mat hosts *Paracoenia* larvae which actively graze the mat. Maceration of the mat, and the mixing of larval feces, speeds the cycling of nutrients within the system. Bioturbation expands the oxygenated region of the mat, accelerating decay by aerobic bacteria. Older larvae pupate and emerge, but those remaining die as the cooler islands are destroyed by grazing and become flooded with hot water, thus repeating the successional cycle (Gorden and Wiegert 1977).

6.7.4 Implications of Faunal Studies of Microbial Mats for Proterozoic Paleobiology

The studies discussed above demonstrate the importance of primary consumers in microbial ecosystems. Together with other relevant data, such studies provide a potentially useful framework for assessing aspects of the Proterozoic stromatolite record.

As illustrated in the above discussion, recent studies in marginal marine environments reveal moisture and salinity to be major factors in controlling the distribution of metazoan grazers. In some sedimentary systems lacking mats, macrofaunal grazers utilize the microbial films on sand grains and fecal pellets as a major food resource. Intense grazing and bioturbation by macroinvertebrates with body sizes exceeding a few millimeters prevent the development of microbial mats.

When grazers have small body sizes (less than a few millimeters), microbial mats and grazers can coexist. The great majority of small grazers are arthropods (insects and crustaceans) that feed on bacteria, coccoid cyanobacteria, and diatoms. Comparatively few smaller species possess the necessary adaptations for consuming filamentous cyanobacteria, the major constructors of most highly laminated mats and stromatolites. The effects of microfaunal grazing and bioturbation are expressed as a variety of mat textures and fabrics that reflect various equilibria between competing algal species that are mediated by selective grazing. Examples include nodular mats and successional changes in thermal spring mat systems, and flocculose textures in heavily grazed mats of sabkha environments.

Grazing in thermal spring systems is known to promote the cycling of nutrients and to increase the depth of the aerobic zone, enhancing the decay of organic matter. Grazing of bacterial systems by meiofaunal protozoans is known to decrease bacterial biomass and the rate of mineralization, increase the rate of decomposition, and regenerate biologically usable phosphorus that would otherwise accumulate in the bacterial biomass (Fenchel 1977). However, in soil ecosystems where predation rates are low, nematodes can attain population densities high enough to cause net immobilization of nutrients (Coleman et al. 1978). Predator abundances decline rapidly with increasing salinity in microbial communities (Gerdes et al. 1985a, b); a similar limitation may prevail in the microbial ecosystems of hypersaline environments where food chains are short and populations of some predator species (e.g., nematodes) can be very high.

The dominance of arthropods (insects and crustaceans) as grazers of modern microbial mats challenges the applicability of some of these observations to Proterozoic stromatolites. Proterozoic grazers must have differed significantly from these dominant grazers of modern mats, and their effects may have been quite different as well. Nevertheless, modern mat-grazer interactions do suggest a lack of causal connection between the appearance of small grazers and the Late Proterozoic decline of stromatolites (Section 6.2.5). In current ecosystems, small grazers do not inhibit development of microbial mats, although they may affect mat fabric and surface texture. Comparative microfabric or biomarker studies may eventually demonstrate the presence of small grazers in Proterozoic stromatolites, although this is problematical at present. In any case, microfaunal studies in modern mats suggest that other factors (e.g., the evolution of algal metaphytes) were probably of more importance than grazing by metazoans as a causal factor for the pre-Vendian decline of microbial stromatolites.

6.8

The Biogeochemistry of Carbon in Modern Microbial Mats

DAVID J. DES MARAIS   JOHN BAULD   ANNA C. PALMISANO
ROGER E. SUMMONS   DAVID M. WARD

6.8.1 Introduction

Organic matter occurring in Proterozoic stromatolites contains morphologic, molecular, and isotopic information regarding the original biota (see Sections 3.2, 3.3, 3.4, 5.4). Studies of modern microbial mat communities help to interpret this record of ancient stromatolitic organisms and their environments. Accordingly, this section deals with the biogeochemistry of the diverse types of microbial mats that have been examined to date (see also Sections 6.1, 6.9, Tables 6.8.1, 6.8.2).

6.8.2 Elemental Abundances

Microcoleus mat from Solar Lake, Sinai, was analyzed for its elemental content (Aizenshtat et al. 1984; Table 6.8.1). Such mat organic matter has a very high organic oxygen content (O/C values between 0.41 and 0.66) relative to organic matter from other contemporary environments (which has O/C values typically less than 0.3; see Tissot and Welte 1984). This is because prokaryotic sheaths are quite abundant in microbial mats (see Section 6.8.4C) and consist mainly of heteropolysaccharides (mostly neutral sugars with some O-methyl sugars and uronic acids), a protein moiety, but lack lipids (Schrader et al. 1982; Adhikary et al. 1986; Tease and Walker 1987). Because sheaths tend to be better preserved in mat sediments than are many other prokaryotic constituents (Section 5.3.2; Golubić and Hofmann 1976; Klok et al. 1984), this high oxygen content might be expected to be reflected in the stromatolitic fossil record. Indeed, McKirdy (1976) has reported high oxygen content in stromatolitic organic matter, evidently a legacy of the polysaccharide-rich sheaths of ancient mat-building cyanobacteria and related prokaryotes.

Modern mat organic matter also tends to exhibit a high sulfur content, a characteristic which has been attributed (Aizenshtat et al. 1983) to the sulfide-rich, iron-poor environment of many such mats. Sulfate-reducing bacteria produce sulfide, which, in environments low in ambient iron concentrations, forms highly labile polysulfides. These polysulfides react further to form a diverse array of organosulfur compounds.

Commonly, mat organic matter is also rich in nitrogen, having an N/C value similar to the value of 0.15 measured for marine plankton (Redfield et al. 1963). However, despite the relatively high nitrogen content of sheaths and the resistance of sheath material to degradation, this initially high nitrogen content has not resulted in Proterozoic stromatolitic kerogens being more nitrogen-rich than other kerogens of comparable age (see Section 3.2).

6.8.3 Stable Carbon Isotopes

In many modern mats, especially those from coastal marine environments (see Table 6.8.2), the $\delta^{13}C$ of the organic matter is markedly higher than that typically observed both in modern aquatic organisms (e.g., Deines 1980; Schidlowski et al. 1983) and in Proterozoic stromatolites (Section 3.4). It is useful to explore this isotopic contrast between ancient stromatolites and modern mats, because it might reflect the evolution either of the biota or of their environment.

The carbon isotopic composition of the organic matter in a microbial mat reflects isotopic discrimination by the biochemical processes which sustain the network of carbon pathways that occur within that mat (Fig. 6.8.1). If a branch point in this network leads to two reactions in which the degree of discrimination between ^{13}C and ^{12}C differ, then the $\delta^{13}C$ values of the resultant products also differ (see Monson and Hayes 1980).

One key branch point in a mat accompanies the uptake of dissolved inorganic carbon (DIC) from the overlying water column (process A in Fig. 6.8.1). Carbon entering the mat is either fixed into biological material or it diffuses back out. Because this process is analogous to the diffusion of gaseous

Table 6.8.1. *Elemental analyses of Solar Lake Microcoleus mat (Aizenshtat et al. 1984).*

Depth (cm)	Weight Percentage					Molar Ratio		
	C	H	N	S	O	H/C	O/C	N/C
0–5	46	5.4	7.6	1.4	40	1.4	0.65	0.14
5–10	45	6.0	7.7	2.2	39	1.6	0.66	0.15
10–45	53	6.3	8.0	3.9	29	1.4	0.41	0.13

Table 6.8.2. *Carbon data for mats from marine, freshwater, and hot spring sites.*

Locality and dominant flora	$\delta^{13}C_{mat}$[a]	$\Delta\delta^{13}C$[b]	DIC (mM)	pH	Temp. (°C)
Cyanobacteria (Behrens and Frishman 1971)					
Baffin Bay, Texas	−14 to −16				
Cyanobacteria (Barghoorn et al. 1977)					
Hypersaline pool, Red Sea	−11.7				
Supratidal mat, Cape Cod	−15.7				
Intertidal mats, Shark Bay, Australia					
Entophysalis	−19.1				
Lyngbya/Entophysalis	−10.6				
Schizothrix	−8.4				
Bahia Honda, Bahamas	−8.4				
Crane Key, Florida	−11.1				
Green Lake, New York	−21.1				
Cyanobacteria (Schidlowski et al. 1984)					
Solar Lake, Egypt	−4 to −8				
Gavish Sabkha, Egypt	−6 to −17				
Cyanobacteria (Schidlowski et al. 1985)					
Gavish Sabkha, Egypt	−6 to −17	7 to 15[c]			
Solar Lake, Egypt	−3 to −16	6 to 13[c]			
Cyanobacteria (Aizenshtat et al. 1984)					
Solar Lake, Egypt	−6 to −16				
Guerrero Negro, Mexico, salt pond					
(Des Marais et al. 1989)					
Microcoleus		−1 to 1	1.7—2.2	8.4—8.8	25
Microcoleus, Synechococcus		1 to 3	2.1–2.8	8.3–8.7	25
Shark Bay, Western Australia (this work)					
Diatom (colloform) mat					
Booldah	−12.1	17.9[c]			
Carbla Point	−10.0(−19.5)[d]	15.0[c]			
Playford's Landing	−12.8(−15.9)[d]	17.7[c]			
Entophysalis (pustular) mat					
Booldah	−19.2	24.8[c]			
Carbla Point	−12.9(−19.7)[d]	18.2[c]			
Playford's Landing	−10.7(−21.8)[d]	15.9[c]			
Lyngbya (tufted) mat					
Booldah	−10.4	15.6[c]			
Carbla Point	−11.3(−16.6)[d]	16.2[c]			
Gladstone Embayment	−7.1(−14.4)[d]	9.6[c]			
Playford's Landing	−10.1(−15.0)[d]	15.2[c]			
Microcoleus (smooth) mat					
Carbla Point	−11.6(−16.1)[d]	16.9[c]			
Gladstone Embayment	−8.2(−13.2)[d]	11.4[c]			
Gladstone Embayment	−11.5	15.4[c]			
Playford's Landing	−8.3(−18.9)[d]	13.7[c]			
Yellowstone National Park, USA (Estep 1984)					
Synechococcus-Chloroflexus					
Fairy Creek Spring 2		17 to 22	70–80	7.2–8.3	40–74
Clearwater Springs	−5 to −17		2–6	4.5–6.0	28–67
Octopus Spring		7 to 15		7.4	55–62
Phormidium-Chloroflexus					
Mammoth Hot Springs		22 to 24		7.0–8.0	40–54
Boulder Spring	−10 to −12			7.2	60
Chlorogleopsis-Chloroflexus					
Serendipity Sp.-Weigert's Ch.		12 to 15	35–45	7.0	56–57
Cyanobacteria					
Fairy Creek Meadows		9 to 14		7.5–9.0	34–48
Queen's Laundry		12 to 15	40–60	6.8–8.0	48–50
Cyanidium					
Nymph Creek	−9 to −14		1.3–6.0	3.5–4.5	37–42

Table 6.8.2. (Continued)

Locality and dominant flora	$\delta^{13}C_{mat}$[a]	$\Delta\delta^{13}C$[b]	DIC (mM)	pH	Temp. (°C)
Zygogonium					
Obsidian Creek	−8 to −15			2.5–3.7	25–32
Yellowstone National Park, USA					
(Barghoorn et al. 1977)					
Cyanobacteria					
Fairy Meadows	−18.5				
Yellowstone National Park, USA					
(Seckbach and Kaplan 1973)					
Nymph Creek					
Cyanidium caldarium	−11 to −24			2.7	44
Yellowstone National Park, USA					
(Madigan et al. 1989)					
Chromatium tepidum					
Roland's Well	21.6(24.0)[e]	21.7		6.7	52
New Pit	(24)[e]	15.2		6.5	55
New Spring	(24.8)[e]	10.7		6.5	53
New Zealand (Seckbach and Kaplan 1973)					
Orakei Korako spring					
cyanobacteria	−18.5			8.4–9.8	30–64
New Zealand thermal areas (this work)					
Orakei Korako spring stream					
Chlorogleopsis		20.4	2.5	7.1	60
Chlorogleopsis		12.5	2.3	7.7	53
Chlorogleopsis		6.8	2.0	8.2	47
Chlorogleopsis		4.7	1.9	8.4	42
Rotorua, along lake					
Chlorobium		17.0	4.8	5.85	44
Oscillatoria amphigranulata		12.3	7.9	8.9	48
Oscillatoria (Govt. Vent Pool)			5.3	7.1	18
Tokaanu pools					
Phormidium, O. terebriformis		6.7	0.64	7.25	43
Phormidium		11.9	0.70	7.52	46
Chlorogleopsis, Synechocystis		21.2	0.91	5.41	45
Chlorogleopsis, Phormidium		16.1	1.5	5.83	51
Chlorogleopsis		14.8	0.59	6.24	59
Waimongu stream					
Chlorogleopsis		20.3	0.68	4.4	53

[a] Values are $\delta^{13}C$ of mat organic matter. The $\delta^{13}C$ of dissolved inorganic carbon (DIC) were not reported.
[b] Values are for $\delta^{13}C_{CO_2} - \delta^{13}C_{mat}$.
[c] Values are for $\delta^{13}C_{carbonate} - \delta^{13}C_{mat}$. The carbonate is likely several permil more ^{13}C-enriched than the coexisting DIC which was available to mat autotrophs. Thus, the value for $\delta^{13}C_{CO_2} - \delta^{13}C_{mat}$ was likely several permil smaller.
[d] Numbers in parenthesis give $\delta^{13}C$ values for fatty acid fractions of mats.
[e] Values are for $\delta^{13}C_{CO_2} - \delta^{13}C_{phytol}$, where phytol was extracted from *Chromatium tepidum*.

CO_2 into a photosynthesizing leaf, it can be modelled using an equation developed for isotope discrimination by plants (Farquar et al. 1982). The equation relates isotope discrimination to the relative rates of diffusion and carboxylation reactions, and can be adapted to mats as follows:

$$C_e/C_w = (\delta^{13}C_m - \delta^{13}C_w + a)/(a + b) \qquad (6.8.1)$$

where $\delta^{13}C_m$ and $\delta^{13}C_w$ are the $\delta^{13}C$ of the mat and of the CO_2 in the overlying water, respectively; "a" is the $\delta^{13}C$ shift associated with CO_2 diffusion to the site of CO_2 fixation within the mat; "b" is the $\delta^{13}C$ shift due to CO_2 fixation; "C_w" is the CO_2 concentration in the overlying water; and "C_e" is the CO_2 concentration in the vicinity of the CO_2-fixing enzyme. Discrimination during aqueous CO_2 diffusion is small; the term "a" is assigned the value −1 (Roeske and O'Leary 1984). The fixation of CO_2 by cyanobacteria, diatoms, and purple sulfur bacteria occurs principally via the enzyme ribulose 1,5 bisphosphate carboxylase (e.g., see Schidlowski et al. 1983). For enzyme purified from green plants, the value for "b" is −29 (Roeske and O'Leary 1984); this value is used for the illustration that follows. For a pure culture of a species of the cyanobacterium *Microcoleus*, Des Marais et al. (1989) measured a discrimina-

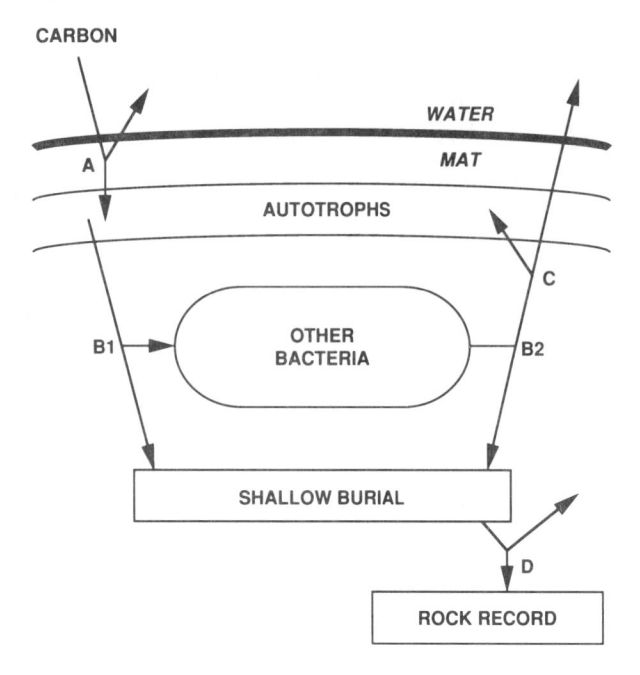

Figure 6.8.1. Carbon flow in a microbial mat. Key branch points include: A – imported carbon is either fixed or it diffuses back out; B_1 – organic carbon from autotrophs is either buried or used by bacteria; B_2 – bacterial organic carbon is either buried or converted to either DIC or to low molecular weight organics; C – degraded carbon is either refixed or it exits the mat; D – organic matter at greater depths is either degraded thermally and released or it enters the geologic rock record.

Table 6.8.3. *Carbon in a Guerrero Negro pond and Microcoleus mat*[a].

Table entry #	Sample	DIC conc., mM	pH	$\delta^{13}C$	Flux, mM C/m²/d
	Pond water:				
1	DIC[b]	2.0	8.5	−6	
2	DOC[b]	2.0		−12	
3	Total mat organics				
	Net DIC fluxes:				
4	Day (water to mat)			−6	100
5	Night (mat to water)			+1	80
	Mat porewater:				
6	DIC	20.0	7.0	−5	
7	DOC	10.0	7.0	−13	

[a]Growth conditions: salinity = 85‰, T_{day} = 28°C, T_{night} = 22°C, midday illumination = 2000 µE/m²/s, water depth was a constant 70 cm. Mat was from Pond #5 (see Section 6.10).
[b]DIC and DOC are dissolved inorganic and organic carbon, respectively. DOC analyzed by method of Bauer et al. (1988).

tion of −26. If CO_2 diffusion between water and the cyanobacterial carboxylase enzyme is rapid, C_e/C_w approaches unity and $\delta^{13}C_m$ becomes very negative. If the CO_2 uptake rate is limited by diffusion, C_e/C_w becomes small and $\delta^{13}C_m$ becomes more positive and approaches $\delta^{13}C_w$.

Natural microbial mats are, of course, more complex than implied in the above discussion, because cyanobacteria can actively transport HCO_3^- (e.g., Badger and Andrews 1982), and because additional branch points for carbon flow occur between each of the several membranes in an organism (Kerby and Raven 1985). Moreover, mat organisms other than cyanobacteria fix CO_2. Also, following fixation, organic carbon is subsequently altered by bacterial diagenesis (Fig. 6.8.1).

The $\delta^{13}C_m$ values in the literature (Table 6.8.2) are rather positive, suggesting that carbon uptake by most such modern mats, particularly those in marine environments, is largely diffusion-limited. Diffusion limitation can be assessed more precisely by examining CO_2 uptake by a typical marine mat in daylight. This approach minimizes other sources of discrimination which can alter $\delta^{13}C_m$, but which are not directly involved with primary carbon fixation. In the study here described, the abundance and isotopic composition of the assimilated carbon was determined by estimating the carbon withdrawn from a benthic chamber placed over the mat (entry #4, Table 6.8.3). The $\delta^{13}C$ value of this assimilated carbon is identical to $\delta^{13}C_w$, indicating that *no* discrimination occurs during DIC uptake. Applying the above equation, diffusion is essentially 100% responsible for limiting the rate of CO_2 fixation. With pronounced diffusion limitation, C_e/C_w (and thus C_e) is small. Low values of C_e are also indicated by the high pH values observed in the daytime in the mat photic zone. This finding also shows that carbon to be fixed enters the mat as HCO_3^-, not CO_2. If selective withdrawal into the mat of CO_2 is substantial, the DIC in the benthos should become more ^{13}C-enriched, because CO_2 in isotopic equilibrium with HCO_3^- is ^{13}C-depleted by several permil at ambient temperatures.

Larger isotopic discriminations by mat communities occur in hot springs (Estep 1984; see Table 6.8.2). Due to the thermal source of DIC, the waters associated with these mats have DIC concentrations elevated above those of the surface waters, which are equilibrated with the atmosphere, and thus enhance DIC diffusion rates. A clear correlation exists (Table 6.8.2, Fig. 6.8.2) between dissolved CO_2 concentration and isotopic discrimination. Higher Proterozoic atmospheric CO_2 levels (see Section 4.7) would have increased such discrimination; this effect may have caused ancient stromatolitic organic matter to be more ^{13}C-depleted than are modern mats.

Such processes as excretion, heterotrophy, and decomposition modify mat organic matter (processes B_1 and B_2 in Fig. 6.8.1), yet they apparently alter $\delta^{13}C_m$ values minimally. Dissolved organic carbon (DOC) in porewaters near the mat surface has a $\delta^{13}C$ value essentially equal to $\delta^{13}C_m$ (compare entries 3 and 7 in Tables 6.8.3), suggesting that little discrimination occurs during the decomposition of cellular material to DOC. Organic matter buried in mat sediments is isotopically similar to that at the mat surface (Aizenshtat et al. 1984), similarly indicating little diagenetic alteration of $\delta^{13}C_m$. This might be due in part to the abundant mucilage produced by mat phototrophs, because such mucilage is relatively resistant to decomposition, thus preserving the isotopic signature created by the near-surface community.

The carbon cycle of a microbial mat is completed when remineralized carbon is reassimilated by autotrophs (process C in Fig. 6.8.1). If this carbon assimilation is essentially 100%

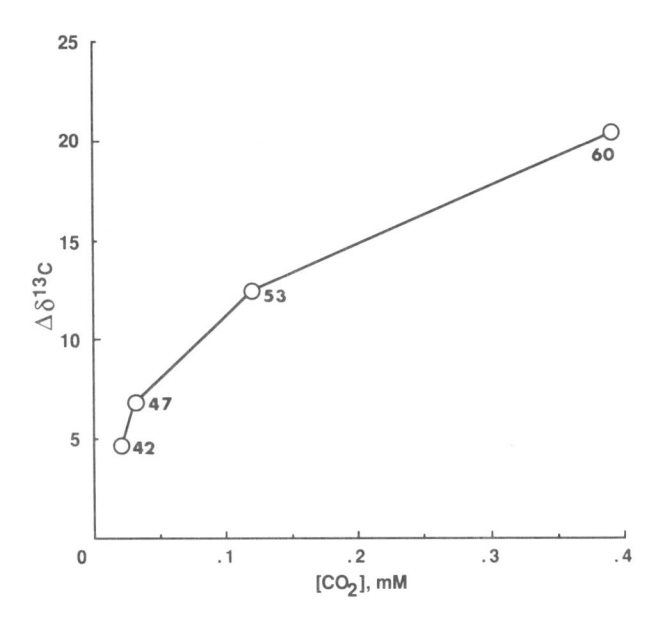

Figure 6.8.2. Carbon isotope discrimination during mat growth versus the ambient concentration of dissolved carbon dioxide in a spring outflow at Orakei Korako thermal area, New Zealand. The term $D\delta^{13}C$ equals $d^{13}C_{CO_2}$ minus $\delta^{13}C_{mat}$. The mat was dominated by *Chlorogleopsis* cyanobacteria. Water temperatures (°C) at the sampling points are indicated by the numbers adjacent to the data points.

complete, with none of this carbon reentering the overlying water column (as would be expected, for example, during the daytime, total reassimilation being due to high rates of photosynthesis), then no isotopic fractionation can occur. However, such mats do lose carbon to the overlying water at night (entry 5, Table 6.8.3). If this "lost" carbon were to be partially recovered by mat autotrophs, selective retention of ^{12}C could occur, as would be evidenced by a ^{13}C-enriched carbon component escaping into the water column. Such ^{13}C enrichment is indeed observed (compare entries 5, 6 and 1, Table 6.8.3), suggesting that isotopically selective carbon uptake occurs at night; sulfide-oxidizing bacteria are very likely responsible for this uptake (see Section 6.10).

The processes principally responsible for creating the $\delta^{13}C$ signature of mat organic matter are those which control the movement of carbon across the mat-water interface. Isotopic discrimination varies as a function of the relative rates (per unit area of mat surface) of carbon diffusion versus enzymatic carbon uptake. Diffusion rates of DIC are influenced by carbon concentrations in the water, as well as by the permeability of the mat. Schidlowski et al. (1984) proposed that the high salinities and high temperatures associated with hypersaline mats result in lower CO_2 solubility and, thus, should augment diffusion limitation and suppress isotopic discrimination. However, Des Marais et al. (1989) observed (Fig. 6.10.5) that discrimination does not change appreciably with salinity. Evidently, no such change occurs, because increasing salinity lowers rates of enzymatic CO_2-fixation at least as much as it lowers CO_2 solubility; the ratio of the DIC supply rate to the DIC fixation rate thus remains relatively constant, and isotopic discrimination is not altered.

Enzymatic fixation rates are also controlled by the amount of energy available for autotrophy (e.g., light energy, or chemical energy such as that from sulfide oxidation), and by the ambient temperature. The fixation rate per unit area of mat also varies as a function of the density of active autotrophic organisms. This density can be lowered by physical or chemical processes such as sedimentation or authigenic mineral precipitation. For this reason, studies of carbon isotopic fractionation by carbonate-precipitating microbial mats are warranted. Calcifying mats are perhaps better analogues of many stromatolite-building benthic Proterozoic communities than are purely organic microbial mats.

6.8.4 Biomarkers

Many modern microbial mats have been studied (see Table 20.5) with the aim of interpreting the origins of specific "chemical fossils" which are likely to provide an indication of the types of microorganisms which lived in Proterozoic environments. From the microbiological perspective, can mats (or stromatolites) built by physiologically or phylogenetically distinctive microorganisms (e.g., photosynthetic bacteria, cyanobacteria, eukaryotic algae) be identified by the occurrence of distinctive biomarkers? From the geochemical perspective, are surviving biomarker signatures or individual biomarkers (e.g., monomethylalkanes, steranes, and hopanes) representative of specific microbial groups?

The general problem of relating organic constituents of modern mats to their fossil counterparts is complicated by many factors, as shown in Figure 6.8.3. Mats are comprised of diverse microbial species (see Section 6.9); relating a suite of extracted lipids to particular component microorganisms requires more knowledge of community composition and structure, and of regulation of biomarker synthesis, than is currently available. The organic components of extant microorganisms are commonly chemically complex, and their analysis often requires degradative methods, reducing the value (and increasing the difficulty of interpretation) of the information gained. Finally, both diagenesis (including microbial degradation within mats) and catagenesis (see Chapter 3) may alter the original structures and stereochemistries of biomarkers. Studies of the vertical distributions in sediments of pigments (discussed below) and lipids (Cardoso et al. 1976; Philp et al. 1978; Boon et al. 1983; Edmunds and Eglinton 1984; Boudou et al 1986a, b; Dobson et al. 1988) have revealed evidence of diagenesis, as well as demonstrating the survival potential of various types of biomarkers.

6.8.4A Sheath and Cell Exterior Components as Biomarkers

Microbes that construct modern mats are commonly ensheathed within a firm, gel-like, mucilaginous material having a sharply defined outer boundary (e.g., Figs. 5.4.1, 5.4.2). The amorphous extracellular slimes and mucilages of some prokaryotes are chemically well-documented (e.g., Wolk 1973; Costerton et al. 1981), but cyanobacterial sheaths, which are often lamellated and sometimes pigmented, have been analyzed only recently. Sheath fractions from both coccoid (Schrader et al. 1982; Adhikary et al. 1986; Tease and Walker 1987) and

ORGANIC COMPONENTS IN LIVING MICROORGANISMS
- CELL WALL POLYMERS
- LIPOPOLYSACCHARIDE
- COMPLEX MEMBRANE LIPIDS
- FREE LIPIDS
- PIGMENTS

⬆

COMPLEXITY OF MICROBIAL MAT COMMUNITIES
- UNKNOWN SPECIES COMPOSITION
- UNKNOWN COMMUNITY STRUCTURE
- REGULATION OF BIOMARKER SYNTHESIS

ANALYTICAL LIMITATIONS
- EXTRACTION BIASES
- LOSS OF INFORMATION FROM DEGRADATION OF POLYMERS

DIAGENETIC AND CATAGENIC ALTERATIONS
- SELECTIVE PRESERVATION
- ALTERED STRUCTURE/STEREOCHEMISTRY
- KEROGEN FORMATION

⬇

BIOMARKERS IN THE GEOLOGIC RECORD

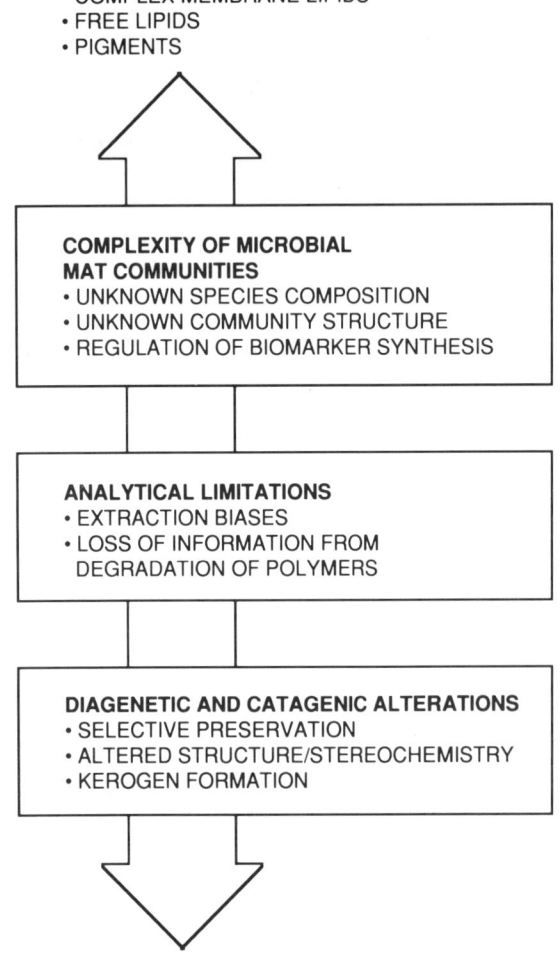

Figure 6.8.3. Difficulties of relating the organic chemical composition of modern microbial mat communities to biomarkers in the geochemical record.

filamentous (Weckesser et al. 1988) cyanobacteria contain mainly heteropolysaccharides (mostly neutral sugars but with some O-methyl sugars and uronic acids), together with a protein moiety, but no lipid. Sheath preparations from about twenty cultures of *Lyngbya aestuarii*, a filamentous mat-constructing cyanobacterium, contain a similar array of neutral sugars and uronic acids (Bauld and Dudman, unpubl.). Sheaths from other prokaryotes may be proteinaceous (e.g., *Methanospirillum hungatei*, Beveridge et al. 1985) or consist of protein-polysaccharide-lipid complexes (e.g., *Sphaerotilus natans*, Romano and Peloquin 1963; *Leptothrix discophora*, Emerson and Ghiorse 1987). Evidence from both ancient and modern mats suggests that sheaths can be preserved preferentially relative to the cells which produce them (Section 5.3.2; Golubić and Hofmann 1976; Doemel and Brock 1977; Klok et al. 1984; Boon et al. 1983). The extent to which such preferential preservation is attributable to chemical composition (and/or structure), environmental factors, or both, remains unanswered.

Studies of reducing sugars and pyrolysis products of deep mat layers represent an initial approach to understanding the chemistry of preserved sheath remains (Boon et al. 1983; de Leeuw et al. 1985). Reactivity of immune sera prepared against specific mat-forming bacteria (Tayne et al. 1987) or intact mat samples (Muyzer et al. 1986), with deep mat layers, suggests the survival of antigenic determinants for at least 10 years to perhaps 2000 years.

6.8.4B Pigment Biomarkers

Chlorophylls, sources of porphyrin biomarkers, reflect the presence of different phototrophic microorganisms in modern mats (see Section 6.5). Most mats contain both cyanobacteria and photosynthetic bacteria, and thus contain both cyanobacterial chlorophyll *a* and bacteriochlorophylls derived from purple and/or green bacteria (e.g., Javor and Castenholz 1981; Pierson et al. 1987b; Palmisano et al. 1989a). Mats built exclusively by green and/or purple bacteria contain only bacteriochlorophylls *a* and/or *c* (Castenholz 1973b; Ward et al. 1989b). Mats containing diatoms may contain chlorophyll *c* (Palmisano et al. 1989a). Within mats, the vertical distributions of the various chlorophylls appear to reflect the stratification of different phototrophs (Bauld and Brock 1973; Pierson et al. 1987; Palmisano et al. 1989b), although the survival of chlorophylls at depths well below the photic zone complicates direct association of pigments with specific living phototrophs. Variable relative abundances of chlorophylls and their phaeophytin derivatives suggest that degradation occurs at differing rates within mat communities (Palmisano et al. 1989a).

In general, the distribution within mats of phycobilins, mainly phycocyanin, evidences the distribution of cyanobacteria (although phycobilins also occur in the eukaryotic alga *Cyanidium caldarium*, an inhabitant of acid thermal mats; Ward et al. 1989b).

Carotenoid pigments, precursors to some distinctive hydrocarbon biomarkers occurring in Proterozoic sediments (Chapter 3), have been studied in several microbial mats (see Table 20.5). Although some carotenoids (e.g., β-carotene) are produced by many microorganisms, others may be useful biomarkers for specific groups of photosynthetic taxa. For example, the occurrence of γ-carotene and chlorobactene may reflect the presence of the green bacteria *Chloroflexus* and *Chlorobium*, respectively; myxoxanthophyll, zeaxanthin, canthaxanthin, and echinenone may reflect cyanobacterial contributions; and fucoxanthin, diadinoxanthin, and diatoxanthin may be derived from diatoms in modern mats (Palmisano et al. 1989b, and in prep.). Carotenoids, like chlorophylls, can survive at depth in mats in the absence of oxygen and light, but differential preservation alters their initial composition, as is shown by the preferential survival of carotenoids in a Guerrero Negro cyanobacterial mat (Fig. 6.8.4).

6.8.4C Lipid Biomarkers
Complex Lipids.
Although microbial lipids are usually dominated by such complex polar lipids as glyco- and phospholipids and lipopolysaccharides, most studies have considered only the free lipids of modern mats (Table 20.5). There is clearly need for more and

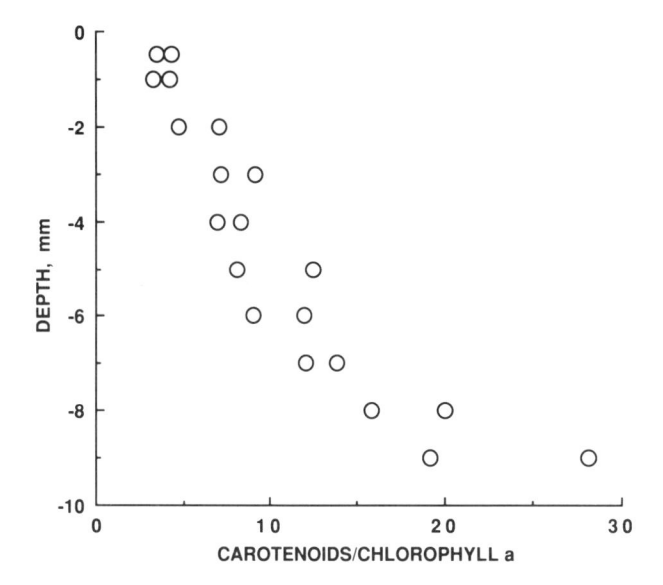

Figure 6.8.4. Selective preservation of carotenoids relative to chlorophyll *a* during burial in the Laguna Guerrero Negro Pond 8 cyanobacterial mat (Palmisano et al. 1989a).

better analyses of complex lipids in modern mats, because of their quantitative significance (Table 6.8.4) and, especially, because many hydrocarbon geolipids are thought to be of diagenetic origin. Degradative studies (e.g., via saponification, acid hydrolysis, or ether cleavage) of a few modern mats have revealed mainly short-chain ($<C_{20}$) alkyl moieties in whole mat or total lipid extracts; some products are distinctive (e.g., archaebacterial ethers: Ward et al. 1985; and fatty acids, such as $10\,MeC_{16}$, characterizing sulfate-reducing bacteria: Boon et al. 1983; de Leeuw et al. 1985). More specific information is obtained by degradative study of specific molecular fractions such as glyco- and phospholipids. For example, when compared with a suite of lipids produced by bacteria isolated from the Octopus Spring (Yellowstone National Park, Wyoming) cyanobacterial mat, the glyco- and phospholipid methanolysis products appear to reflect the trophic structure of the mat community (i.e., phototrophs greater than heterotrophic consumers greater than such terminal anaerobic consumers as methanogenic and sulfate-reducing bacteria; Ward et al. 1989a). Maximum information can, of course, be obtained from studies

Table 6.8.4. *Abundances of lipid fractions in the Octopus Spring 55°C cyanobacterial mat.*

Compound Class	µg/g dry wt.
glycolipid fraction	6100
phospholipid fraction	6200
neutral lipid fraction	4000
free hydrocarbons	72
free wax esters	2405
free alcohols	401
free fatty acids	65

[from Ward et al. 1989b]

of undegraded complex lipids. For example, in the Octopus Spring cyanobacterial mat, predominant complex lipids resembled the complex lipids of the mat-forming phototrophic microorganisms (pers. comm. from G. Dobson to D. M. W.).

Hydrocarbons. Hydrocarbon distributions in mats dominated by cyanobacteria tend to be generally quite similar, including mainly C_{15} to C_{19} normal alkenes and alkanes, monomethylalkanes, and phytenes, the latter presumably derived from degradation of the chlorophyll *a* phytol side chain (Table 20.5). In Figure 6.8.5, hydrocarbons of a salt pond cyanobacterial mat, and of a cyanobacterium isolated from the same mat, are compared with those extracted from a Middle Proterozoic shale. The dominance of short-chain hydrocarbons typical for sediments >1000 Ma (Chapter 3), smoothed by mixed inputs and diagenetic and catagenic alteration products, is comparable to the dominance of short-chain hydrocarbons (and complex lipid alkyl moieties) of cyanobacteria and cyanobacterial mats which typically contain only minor amounts of higher molecular weight hydrocarbons (e.g., $>C_{20}$) like those commonly derived from vascular plants. The similarity of hydrocarbons in different cyanobacterial mats is illustrated by comparison of hydrocarbons from the Guerrero Negro hypersaline mat (Fig. 6.8.5b) with those from hot spring cyanobacterial mats (Figs. 6.8.6A, 6.8.7). However, the spectrum of hydrocarbons detected differs for mats built by physiologically and phylogenetically distinct phototrophs. For example, C_{29} to C_{33} di- and triunsaturated alkenes, produced by *Chloroflexus*, dominate in anoxygenic mats built by this green bacterium (Fig. 6.8.6B). Mats produced mainly by eukaryotic algae and dominated by diatoms (e.g., certain of those at Hamelin Pool, Lake Clifton, and Lake Thetis, Western Australia; at Lake Fellmongery, South Australia; and at Lee Stocking Island, Bahamas) are characterized by highly branched C_{20} and C_{25} isoprenoid alkanes and alkenes, as well as sterenes (Fig. 6.8.6C).

Of particular interest are the mid-chain branched monomethylalkanes which occur in Proterozoic sediments (Fig. 6.8.5a). Such compounds also occur in all modern cyanobacterial mats (Figs. 6.8.5b, 6.8.6A, 6.8.7) and in many natural cyanobacterial samples (Table 20.6), but are absent from mats which lack cyanobacteria. Although they are not always detected in cyanobacteria isolated from mats (e.g., Fig. 6.8.5c), a range of different mid-chain branched monomethylalkanes has been found in nearly two-thirds of all cyanobacteria investigated for their hydrocarbon content (Table 20.6), and evidence for the *de novo* biosynthesis of these compounds by cyanobacteria is excellent (Han et al. 1969; Fehler and Light 1970, 1972). Though mid-chain monomethylalkanes are found in many insects, a few plants, and in wool (Nelson 1978), they have not been reported to occur in microorganisms other than cyanobacteria and, thus, they may be useful biomarkers for cyanobacteria in the Precambrian geochemical record. Alternatively, diagenetic conversion of mid-chain branched monomethyl fatty acids (e.g., $10\,MeC_{16}$ from sulfate-reducing bacteria: Dowling et al. 1986) might explain the suites of monomethylalkane isomers found in Proterozoic sediments and oils (Summons 1987). With regard to such a possibility, the recent discovery of suites of monomethylalkanes in modern

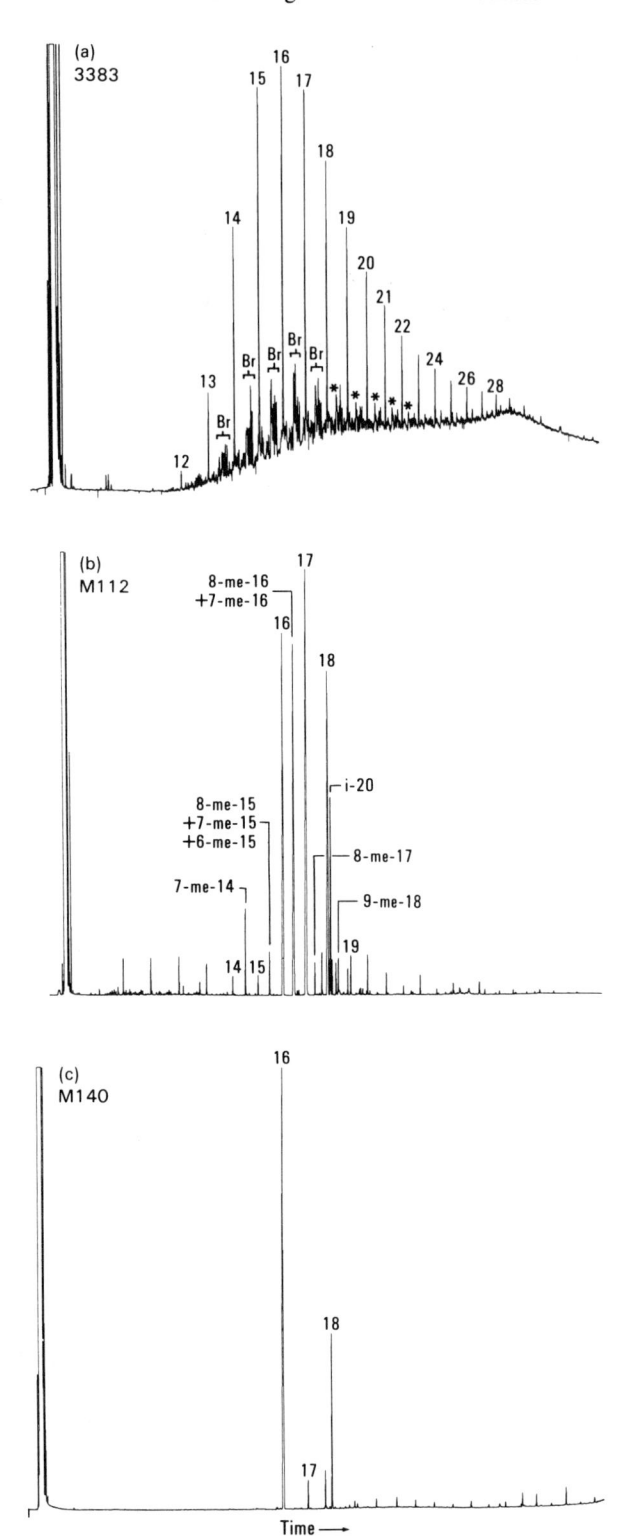

Figure 6.8.5. Comparison of hydrocarbon distributions in (a) an extract of a sediment from the Middle Proterozoic Velkerri Formation, McArthur Basin, northern Australia; (b) a modern cyanobacterial mat dominated by *Microcoleus chthonoplastes* (Guerrero Negro salt pond 5, hydrogenated sample); and (c) a cyanobacterium (possibly *Microcoleus*) isolated from the same mat. Numbers signify alkane chain length; Br = all-isomer suite of monomethyl alkanes; * = predominance of centrally branched isomers at high carbon numbers; n-me-m indicates monomethylalkane with methylation of a straight-chain alkane of m carbon atoms at carbon number n. (Previously unpublished results of R. S.).

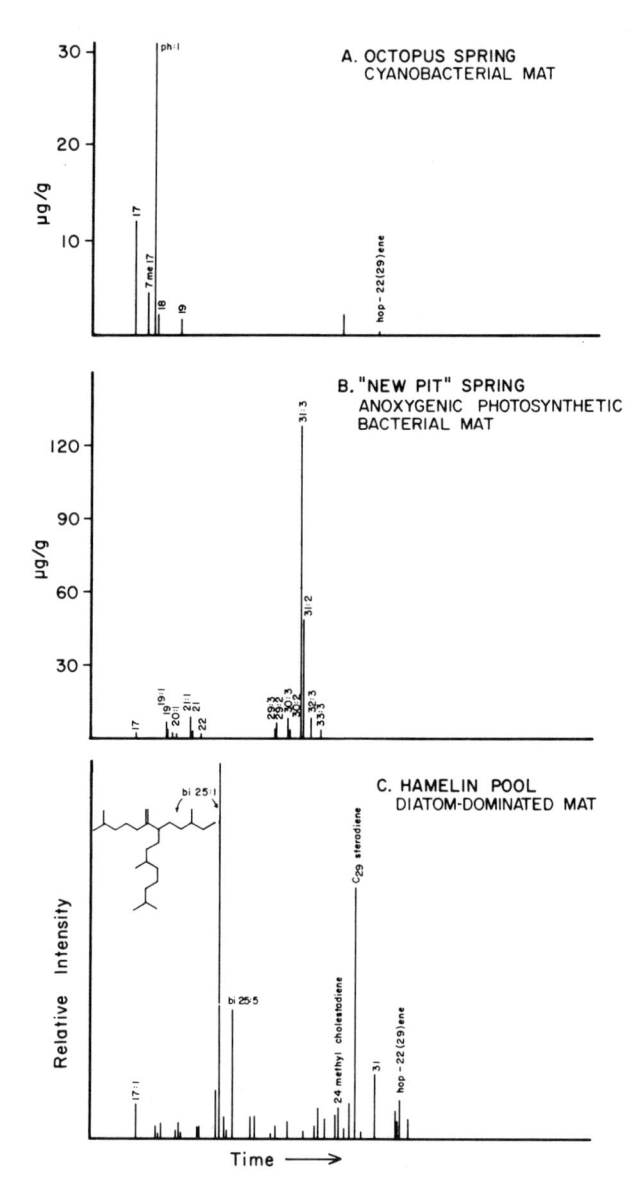

Figure 6.8.6. Varying hydrocarbon composition of microbial mats built predominantly by physiologically and phylogenetically distinct photosynthetic microorganisms. (A) Octopus Spring 55°C cyanobacterial mat (see Dobson et al. 1988). (B) "New Pit" Spring anoxygenic photosynthetic bacterial mat (see Ward et al. 1989b). (C) Subtidal colloform mat from Hamelin Pool, Western Australia, built mainly by diatoms (R. S. unpubl.). Labelling as in Figure 6.8.5; Sq = squalene; :n refers to the number of double bonds; bi = branched isoprenoid.

cyanobacterial mats (Fig. 6.8.7; Shiea et al. 1990; Robinson and Eglinton 1990) suggests that the direct production of a variety of structural isomers within the mat environment (perhaps by cyanobacteria) is possible.

The source of dimethylalkanes reported from cyanobacterial mats (de Leeuw et al. 1985; Shiea et al. 1990; Robinson and Eglinton 1990) is of interest relative to their suggested occurrence among Proterozoic geolipids (Klomp 1986).

Wax esters. Wax esters (C_{28}–C_{37}) are major components of hot spring microbial mats (Tables 6.8.4 and 20.5; Dobson et al. 1988; Ward et al. 1989a; Shiea et al. submitted), and are known

The Biogeochemistry of Carbon

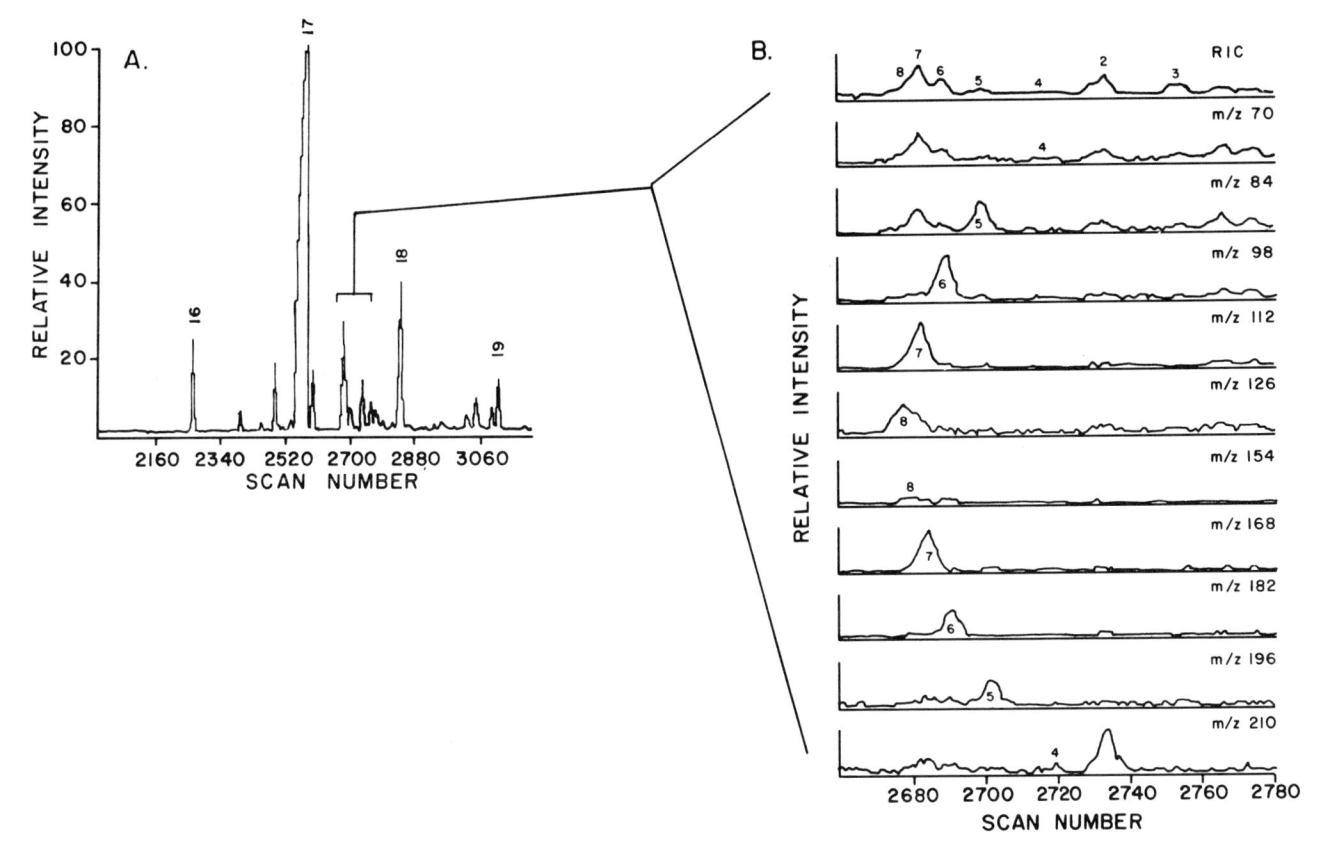

Figure 6.8.7. Suites of mid-chain branched monomethylalkanes in the "Weigert's Channel" hot spring cyanobacterial mat (see Shiea et al. 1990). (A) Predominant hydrocarbons. (B) Mass fragmentograms for key ions of structural isomers of methylheptadecanes (numbers within peaks correspond to site of methylation). Labelling as in Figure 6.8.5.

to occur in *Chloroflexus* and *Chlorobium* which inhabit such mats (Knudsen et al. 1982; Beyer et al. 1983; Shiea et al. submitted).

Alcohols. Nonsteroidal alcohols appear to be decomposition products of the predominant pigments occurring in various mat types. Thus, phytol dominates in mats with chlorophyll *a*-containing phototrophs (cyanobacteria and eukaryotic algae) (Fig. 6.8.8 and Table 20.5), whereas *n*-octadecanol, esterified to bacteriochlorophyll c_s (Gloe and Risch 1978), is dominant in anoxygenic *Chloroflexus* mats. Nonsteroidal alcohols may also be derived from the decomposition of wax esters in mats (Dobson et al. 1988; Shiea et al. submitted).

The distribution of sterols in modern mats is especially relevant to the interpretation of their hydrogenated derivatives, steranes, which occur in Proterozoic sediments and oils. Most modern mats contain sterols (Table 20.5), compounds that have been suggested by some workers to have been derived from cyanobacteria (e.g., Volkman 1986). However, a critical review of the literature reveals that virtually all investigations interpreted as suggesting the presence of sterols in cyanobacteria and other prokaryotes (Table 20.7) are fraught with problems related principally to the incorrect classification of the microorganisms studied, a lack of culture and medium purity, the reporting of relative rather than of absolute sterol abundance, and the use of inadequate methodology (Ward et al. 1989a). With the exception of methylotrophic bacteria (and *Nan-*

nocystis excedens), there is no convincing evidence of novel sterol structure or of *de novo* sterol biosynthesis in cyanobacteria or other prokaryotes.

In mat communities, the abundance and distribution of sterols differs markedly depending on the presence of eukaryotic organisms (Fig. 6.8.8). Hot spring microbial mats provide a test case regarding the presence or absence of sterols in prokaryotes, because eukaryotes do not inhabit neutral-alkaline thermal mats. Only in mats of acid hot springs containing or built by the eukaryotic alga *Cyanidium caldarium* are sterols found to be abundant and diverse. The low levels of a few common sterols reported from some neutral-alkaline mats (especially, from completely anaerobic mats where molecular oxygen-requiring sterol biosynthesis should be impossible) appear likely to be non-indigenous. Considering that many (probably all) mats in marine, lagoonal, or hypersaline environments contain diatoms as well as eukaryotic infauna (see Section 6.7), the abundant and diverse sterol assemblages reported from such mats seem more likely to be derived from eukaryotes than from cyanobacteria or other prokaryotes.

Hopanols (as well as hopenes and hopanoic acids) have been detected in several microbial mats (Table 20.5). Cyanobacteria and many other prokaryotes are known to produce hopanoid alcohols and hydrocarbons, possibly used to stabilize membranes as is thought to be the case with sterols in eukaryotes (Taylor 1984; Rohmer et al. 1984; Ourrison et al. 1987). Because only aerobic or facultatively aerobic prokaryotes are known to

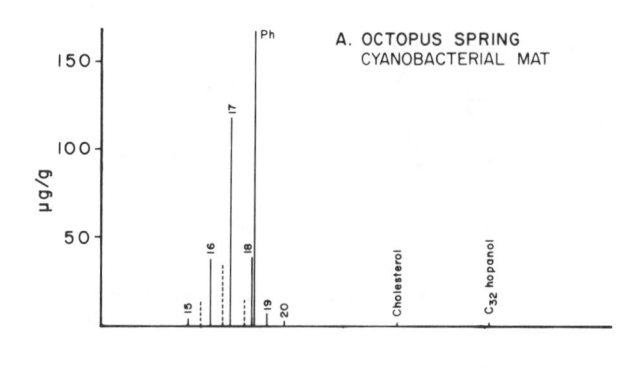

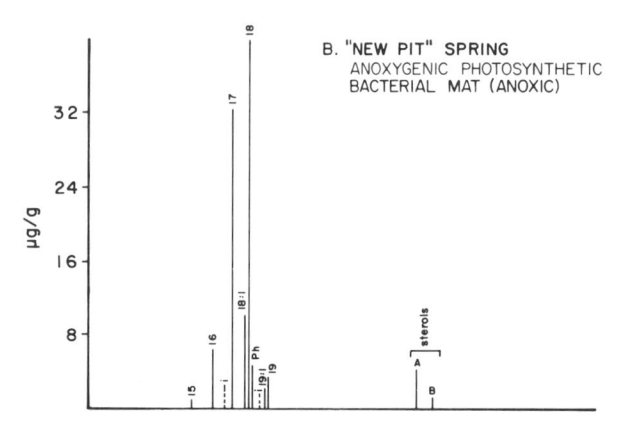

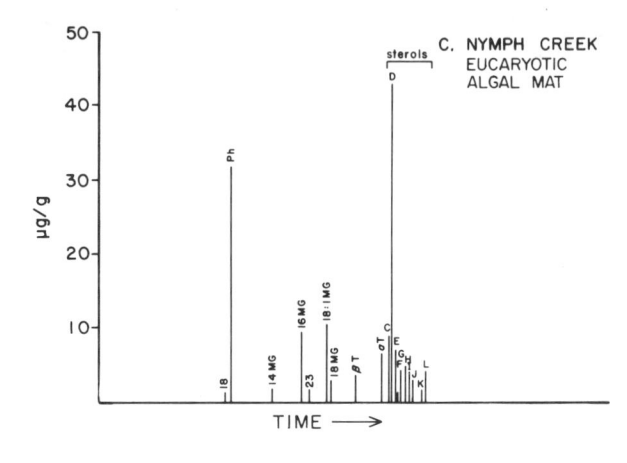

Figure 6.8.8. Comparison of free alcohols in hot spring microbial mats built predominantly by physiologically and phylogenetically distinctive phototrophic microorganisms. (A) Octopus Spring 55°C cyanobacterial mat (see Dobson et al. 1988). (B) "New Pit" Spring anoxygenic photosynthetic bacterial mat, and (C) Nymph Creek 47°C eukaryotic algal mat (see Ward et al. 1989b). Numbers correspond to carbon number of normal alcohols; dashed lines represent branched components of carbon number equivalent to that of the normal alcohol which follows; :1 = monounsaturated; ph = phytol; MG = monoglyceride; T = tocopherol; A, B, and D, are cholesterol, 24-ethyl-cholesterol, and ergosterol, respectively (see Ward et al. 1989b for other sterols).

produce hopanoids (Taylor 1984), a better understanding of their distribution in modern mats would be of considerable interest.

Fatty acids. Free fatty acids, of strikingly similar composition, are minor components of all microbial mats (Table 20.5). Where comparison is possible, a similarity of these free acids to complex-lipid fatty acids has been noted, suggesting a common biosynthetic origin (Dobson et al. 1988; Ward et al. 1989a).

Protokerogen. Degradation of the insoluble organic residues of several modern cyanobacterial mats (Table 20.5) by various techniques (e.g., via oxidation, saponification, or pyrolysis) releases alkyl moieties which in composition resemble the lipid fractions of such mats (i.e., predominantly composed of $<C_{20}$ normal, unsaturated and hydroxy fatty acids and alkanes, and including phytol, phytene, phytanic acids). Modern mats may provide controlled system in which to investigate the pathways by which physically and chemically resilient cell polymers (Berkaloff et al. 1983; Largeau et al. 1984, 1986; Derenne et al. 1988; Kadouri et al. 1988; Laureillard et al. 1988; Chalansonnet et al. 1988) become entrapped or condensed into a "protokerogen" matrix.

6.9

Modern Phototrophic Microbial Mats: Anoxygenic, Intermittently Oxygenic/Anoxygenic, Thermal, Eukaryotic, and Terrestrial

DAVID M. WARD   JOHN BAULD   RICHARD W. CASTENHOLZ
BEVERLY K. PIERSON

6.9.1 Introduction

Considerable effort has been made to understand the modern microbial mats which occur in marine or hypersaline settings similar to those in which most Proterozoic stromatolites are thought to have formed (Section 6.10). However, to understand the range of possible microbial communities which could have formed fossil stromatolites, the variety of different environments in which modern mats are known to occur also needs to be examined. As a reflection of the physiology of the principal mat-building primary producers, microbial mats can be classified as *oxygenic, anoxygenic,* or *intermittently oxygenic/anoxygenic.* Typical *oxygenic* mats, encompassing most well-studied, coastal mats, include as major components cyanobacteria and, in some cases, eukaryotic algae (both of which evolve O_2; Section 6.5), overlying a layer of photosynthetic bacteria. The total range of known oxygenic mats is far greater, however, and includes thermal mats of simple biotic composition, some with atypical vertical organization, and mats formed predominantly or even exclusively by eukaryotic algae. *Anoxygenic* mats, in which photosynthetic bacteria, incapable of oxygenic photosynthesis, are the primary producers, may have been especially important as evolutionary forerunners of mats built principally by oxygenic phototrophs. *Intermittently oxygenic/anoxygenic* microbial mats, here defined as those constructed by cyanobacteria that shift daily from oxygenic to anoxygenic photosynthesis, might be representative of the type of mat that was transitional in evolutionary history between solely anoxygenic and dominantly oxygenic mats. Interestingly, however, not all mats are formed by photoautotrophs: indeed, the occurrence of mats formed by chemoautotrophs as primary producers in H_2S/O_2 gradients suggests the possibility that some ancient stromatolites might have been built by microbes other than photosynthetic microorganisms. And mats which form in terrestrial environments illustrate an alternative to the aquatic setting in which most mats occur.

This section addresses the various uncommon extant mat types noted above, the relatively rare occurrence of which seems chiefly to be a reflection of the earth's present environment, in particular, of the predominantly oxic nature of the setting in which most mats form, the photic zone sediment-water interface. However, given the dramatic changes in the environment known to have occurred over the geologic past, and especially in light of the well-documented onset during Early Proterozoic time of widespread oxic conditions in shallow water settings (Chapter 2), it is easily conceivable that certain of these "a typical" mat types may have been decidedly more abundant, and ecologically far more important, early in earth history.

6.9.2 Anoxygenic Mats

Purely anoxygenic mats (those lacking cyanobacteria or other O_2-producing phototrophs) are relatively rare and are known at present from hot spring outflows only (Table 6.9.1). It is possible that they were far more abundant when anoxic, sulfide- or H_2-rich environments were more widespread.

6.9.2A "Green Chloroflexus" Mats

The best-developed laminated mats of this type are the "green *Chloroflexus*" (GCF) mats known from some of the Mammoth Hot Springs of Yellowstone National Park, Wyoming, USA. The highly calcareous springs of that region deposit travertine and spring sources or aquifers commonly become plugged, so that GCF mats vary in abundance from year to year. During the years of greatest water flow, the total combined areas of these mats commonly cover less than a few square meters. "Pure" GCF mats occur only in spring waters that, at their source, are in the range between 50° and 65°C and that carry primary sulfide (H_2S and HS^-) in the range of 30 to 130 μM (Castenholz 1977; Giovannoni et al. 1987a). The pH is usually between 6.2 and 6.5 where the water emerges, but with loss of CO_2 it rises to 6.8 to 7.0 downstream where the spring water has cooled to 50°C. The apparent cause of GCF dominance under these conditions is that the high sulfide concentration inhibits growth of cyanobacteria that otherwise would grow between 50° and 74°C (Castenholz 1976). Springs with similar sulfide levels, but that at their source are less than 50°C, are usually dominated by the sulfide-tolerant cyanobacterium *Spirulina labyrinthiformis* (Castenholz 1977). Temperatures in the range of 55° to 65°C permit growth of *Chloroflexus*

Table 6.9.1. *Locations and characteristics of purely anoxygenic microbial mats.*

Location	Anoxygenic Phototroph	Top Mat Color	Chlorophyll[a] Type	Mat Structure	Temperature (°C)	pH	Sulfide in Water (μM)	Reference
Mammoth Springs,[b] Yellowstone Nat. Park, U.S.A.	*Chloroflexus*	dull green	BChl *c*	laminated mat	50–65	6.2–6.8	30–130	1, 2, 3, 4
Mammoth Springs, Y.N.P., U.S.A.	*Chloroflexus*	orange-salmon	BChl *c*	biofilm	60–65	6.5–7.0	<70	1, 5
River Group, Lower Geyser Basin, Y.N.P., U.S.A.	*Chloroflexus*	dull brown-green	BChl *c*	biofilm	50–64	8.3	70	5
Steamboat Springs, Reno, Nevada, U.S.A.	*Chloroflexus*	dull green	BChl *c*	biofilm + laminated mat	<63	6.3	60	6
Hvergerdi Springs, Iceland	*Chloroflexus*	orange	BChl *c*	biofilm + streamers	<66–68	7.5–8.6	20–80	5, 7, 8
Ystihver, Husavik, Iceland	*Chloroflexus*	orange	BChl *c*	biofilms + streamers	<65	8.4	present	4, 6
Krisuvik, Iceland	*Chloroflexus*	orange	BChl *c*	biofilm + crust	45–56	9.1	40–300	5
Maruia Springs, S. Island, New Zealand	*Chloroflexus*	dull green	BChl *c*	biofilm + laminae	62–63	8.8	1300	5
Rotorua Springs, N. Island, New Zealand	*Chloroflexus*	dull brown-green	BChl *c*	biofilm + laminae	<65	7.3–8.3	20–1400	5, 6
Mammoth Springs,[c] Y.N.P., U.S.A.	*Chromatium* + *Chloroflexus*	dull red	BChl *a* + *c*	laminated mat	48–55	6.2–6.8	50–60	1, 4, 9
Thermopolis Springs, Wyoming, U.S.A.	*Chromatium*	dull red	BChl *a*	biofilm	<57	6.2–6.8	30–160	6
Wilbur Hot Springs, N. California, U.S.A.	*Chromatium*	dull pink to red	BChl *a*	powdery biofilm	<44	7.4–7.6	2700–3800	6
	Chlorobium	green	BChl *c*	dense slime	<44	7.4–7.6	2700–3800	6
Rotorua Springs,[d] N. Island, New Zealand	*Chlorobium*	green	BChl *c*	dense slime	40–55	4.5–6.2	300–1800	3, 10
Lake Rotoiti Springs, N. Island New Zealand	*Chlorobium*	green	BChl *c*	dense slime	44–50	4.8–5.7	350–400	3, 10

[a]BChl = bacteriochlorophyll
[b]Includes "New Pit Spring" and Painted Pool mats
[c]Includes "Roland's Well" mat
[d]Includes "Travel Lodge Stream" mat

1. Castenholz (1977); 2. Giovannoni et al. (1987a); 3. Castenholz (1988); 4. Ward et al. (1989b); 5. Castenholz (1973b); 6. R. Castenholz (unpublished); 7. Castenholz (1976); 8. Jørgensen and Nelson (1988); 9. Madigan (1986); 10. Castenholz et al. (submitted a).

but not of other known anoxygenic phototrophs such as *Chromatium tepidium* and thermophilic *Chlorobium*, taxa which grow only below about 55°C.

GCF mats are a dull- to deep-green with a satin-like surface. The surficial green layer is less than 1 mm thick and is composed of a tight fabric of "interwoven" *Chloroflexus*-type filaments (Figure 6.9.1A, B). Below this uppermost layer are numerous orange to brown colored layers (collectively up to 1 cm thick) of gel-like material which also contains *Chloroflexus*, presumably remnants of those originally in the top layer. The organisms comprising the green top layer consume sulfide in the light (Fig. 6.9.2A). CO_2 incorporation is light- and sulfide-dependent; acetate incorporation is greatly stimulated by light and is even more enhanced by the presence of light and sulfide (Giovannoni et al. 1987a). *Chloroflexus* strains isolated from certain of these mats differ from the originally described *Chloroflexus aurantiacus* by their inability to grow heterotrophically with O_2; in addition, they appear to be better adapted to use sulfide as photoautotrophs (Giovannoni et al. 1987a). However, anaerobic photoheterotrophic growth of the GCF strains is possibly more rapid. Very little usable light penetrates the top green layer (Pierson and Castenholz unpubl.). Because the gel-like undermat is anoxic (Giovannoni et al. 1987a), and light penetration is minimal, it is unlikely that the *Chloroflexus* within these deeper layers is actively growing even though the organisms in these layers show the same light-

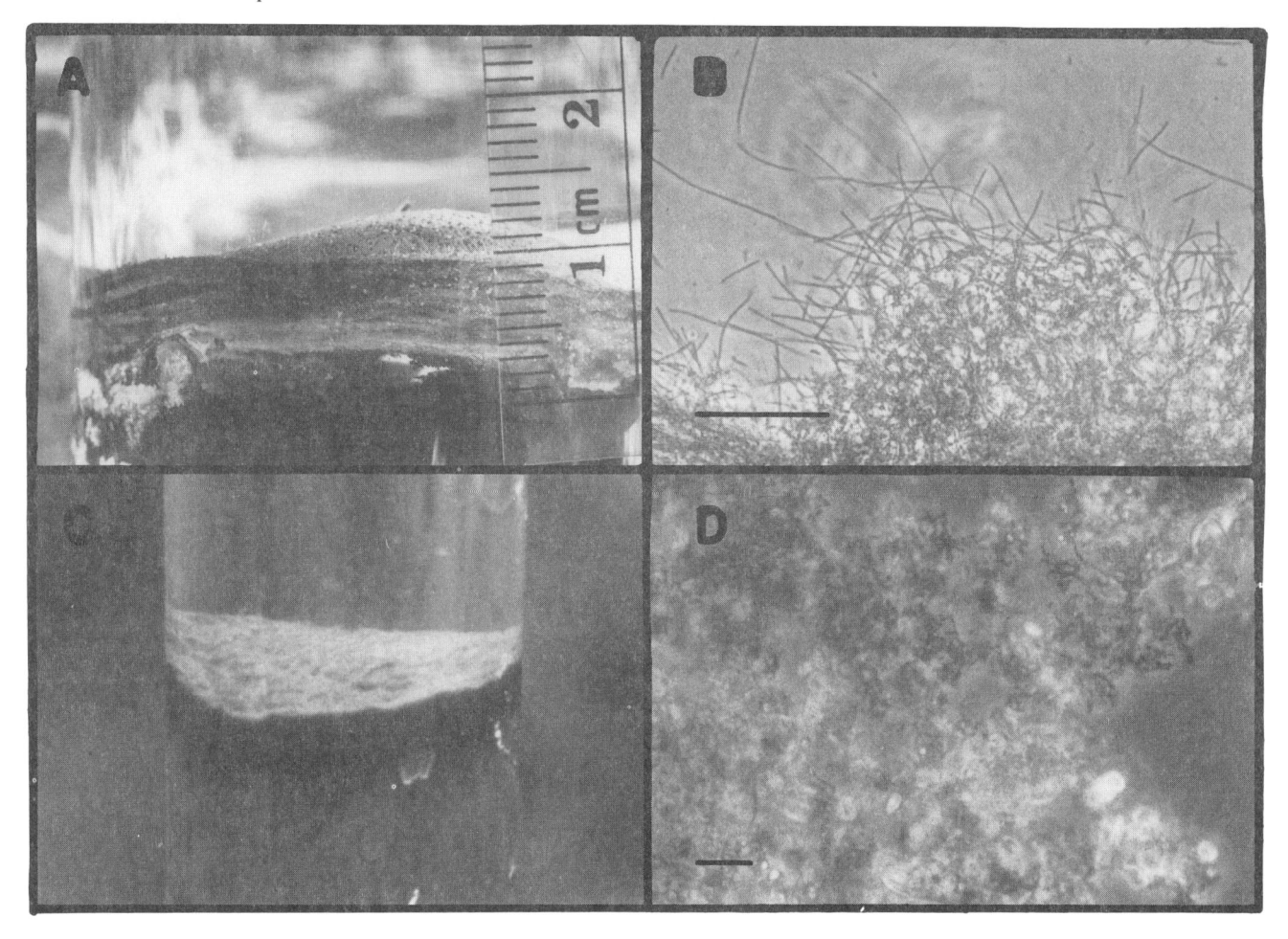

Figure 6.9.1. Well-studied anoxygenic microbial mats (Ward et al. 1989b). [Lines for scale in B and D represent 50 μm.]. (A) "New Pit" Spring, Mammoth Terraces, Yellowstone National Park, Wyoming, USA; top ~1 mm is dark green; undermat laminations are orange. (B) Photomicrograph of top layer of "New Pit" Spring mat showing predominating *Chloroflexus* sp. filaments. (C) "Travel Lodge" Stream, Rotorua, North Island, New Zealand; note nonlaminated green mat, 2 to 3 mm thick, above mud substrate. (D) Photomicrograph of mat from "Travel Lodge" Stream showing predominating *Chlorobium* sp. cells.

stimulated activities as those in the surface green layers (Pierson unpubl.). Dark anaerobic metabolism is unknown in *Chloroflexus*. However, the occurrence of laminations within this thick undermat, probably a result of episodic changes in the local environment and dependent on the filamentous nature of the microorganisms, shows that purely anoxygenic photosynthetic bacteria are capable of producing layered structures that, if preserved in the geologic record, would be similar to many flat-laminated Proterozoic stromatolites.

6.9.2B Chromatium/Chloroflexus Mats

In similar, but rarer, lower temperature springs in the Mammoth area, the purple sulfur bacterium *Chromatium* sp. (probably *C. tepidum*; Madigan 1986) forms a reddish top layer that overlies a laminated undermat consisting of *Chloroflexus* (Castenholz 1977; Ward et al. 1989b). It is unknown why *Chromatium* rather than "green *Chloroflexus*" is dominant in these few springs. Light-dependent sulfide uptake in this system resembles that observed in the "pure" *Chloroflexus* mats discussed above (N. P. Revsbech and D. M. Ward unpubl.). However, action spectra for the photoassimilation of acetate and bicarbonate shows activity only with wavelengths absorbed by the *Chromatium* and not the *Chloroflexus* (Pierson unpubl.).

6.9.2C Chlorobium Mats

Mats in which the unicellular green sulfur bacterium *Chlorobium* sp. is the sole phototroph have recently been discovered in New Zealand. The conditions which appear to promote *Chlorobium* dominance are temperatures between 40° and 55°C, high sulfide, and pH levels generally between 4.5 and 6.2 (Castenholz et al. submitted-a). Under similar conditions but at higher pH, either photoautotrophic *Chloroflexus* (>56°C) or the sulfide-tolerant cyanobacterium *Oscillatoria amphigranulata* (<56°C) is normally dominant in New Zealand hot springs (Castenholz 1976; Castenholz 1988). The newly found *Chlorobium* mat occurs as a dense green slime cover; it is unlaminated, like all other known mats that lack filamentous components (Fig. 6.9.1C, D). Metabolically, however, this mat appears to be similar to "green *Chloroflexus*" mats, being characterized by rapid light-dependent depletion of sulfide within the uppermost 1.2 mm, the entire photic zone of the most studied mat (Fig. 6.9.2b).

diffusible and potentially toxic H$_2$S is present at levels of only 25 to 75 µM (Fig. 6.9.3A). Biogenic sulfide formed within the undermat permeates the *Microcoleus* layer (top 0 to 1 mm) during the night. With increasing light intensity after sunrise, sulfide is initially consumed anoxygenically. Via the anoxygenic process, *M. chthonoplastes* is able to locally deplete sulfide within the mat to less than 300 µM (ca. 75 µM H$_2$S), at which concentration oxygenic photosynthesis begins. Thus, a narrow band of O$_2$ can be built up and sustained within the mat even if sulfide levels below and above are very high (Figs. 6.9.3a, 6.9.4). Due to its phototrophic flexibility, this species is well adapted to actively carry out photosynthesis throughout the day under a variety of changing environments having high, low, or fluctuat-

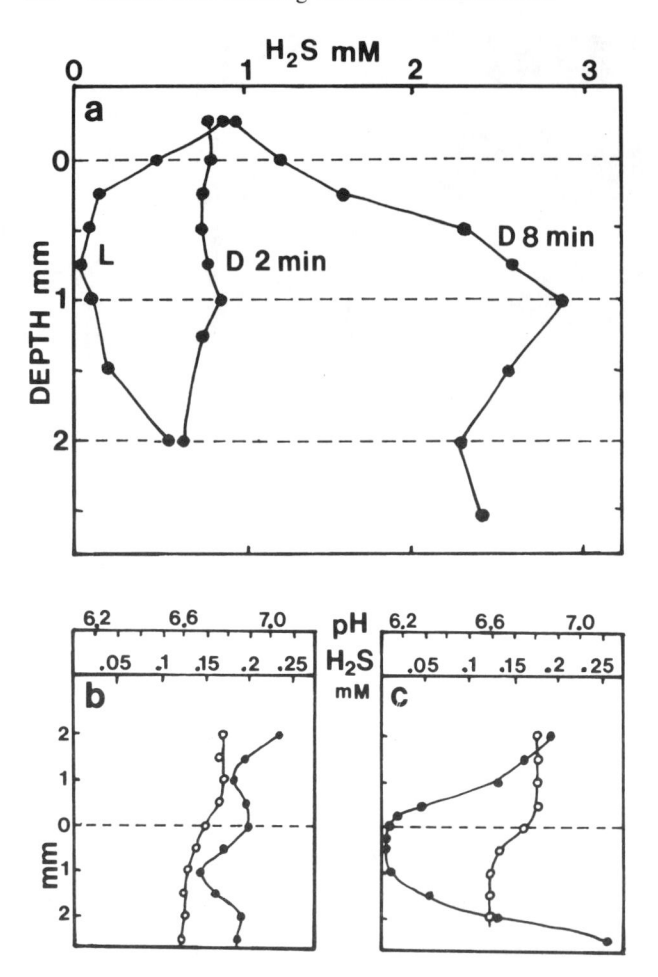

Figure 6.9.2. Microprofiles within purely anoxygenic mats. (a) Sulfide profiles within a 64°C "green *Chloroflexus*" mat at Painted Pool, Mammoth Springs, Yellowstone National Park, Wyoming, USA (Giovannoni et al. 1987a); pH 6.3; incident light (visible only) = 1990 µE/m^2/s; L = steady state in light; D = after 2 min and 8 min of shift to darkness. (b) Profile of sulfide (●) and pH (○) in darkness within the "Travel Lodge" Stream *Chlorobium* mat at Rotorua, New Zealand, at 45°C; no O$_2$ was present (Castenholz et al. submitted-a). (c) Profile of the same *Chlorobium* mat as shown in "b" (at the left) after 10 min of light (270 W/m^2); sulfide in upper 1 mm of mat was depleted to "zero" within 5 min (Castenholz, submitted-a).

6.9.3 Intermittently Oxygenic/Anoxygenic Mats

Mats of this type occur where the primary or biogenic sulfide supply is abundant and cyanobacteria capable of shifting from oxygenic to anoxygenic photosynthesis are present.

The photosynthetic reactions of *Microcoleus chthonoplastes* have been detailed by Jørgensen et al. (1986), Cohen et al. (1986), de Wit and van Gemerden (1987b), and de Wit et al. (1988). Well developed *Microcoleus* mats have been studied in Solar Lake, Sinai (see Section 6.4), and Baja California Sur, Mexico (see Section 6.10). This *Microcoleus* species is capable of alternately (or perhaps even concurrently) carrying out oxygenic and anoxygenic photosynthesis. Anoxygenic photosynthesis is dominant in environments containing about 500 µM sulfide, whereas the oxygenic process dominates if sulfide concentrations are less than about 100 to 300 µM. During photosynthetic periods, the pH is 8.0 to 9.0, and the more

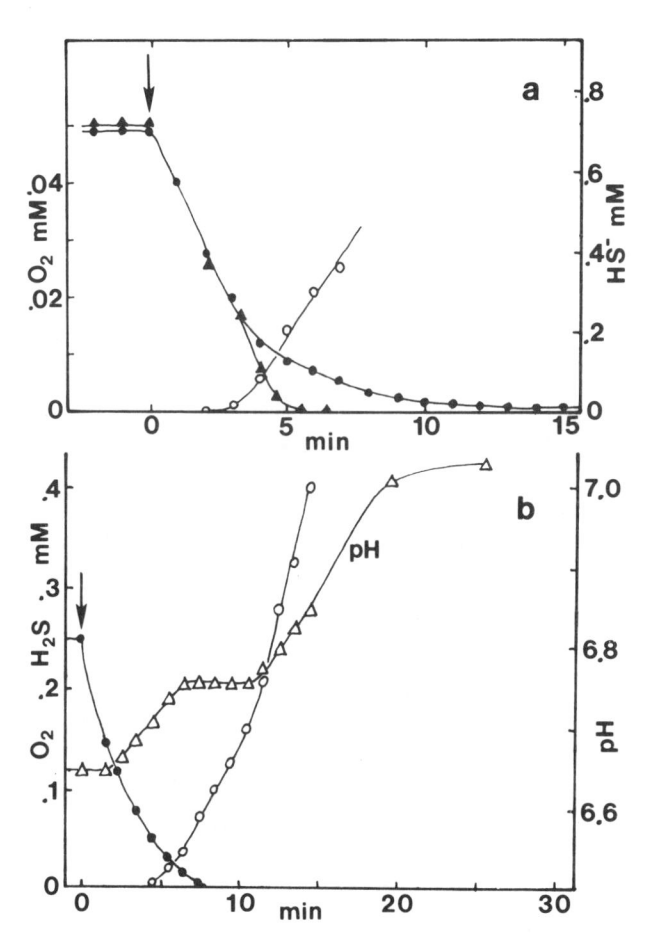

Figure 6.9.3. Time courses of sulfide depletion (anoxygenic photosynthesis) and of O$_2$ production (oxygenic photosynthesis) in two types of intermittently oxygenic/anoxygenic mats. (a) An isolated fragment of *Microcoleus chthonoplastes* mat from Solar Lake, Sinai (Jørgensen et al. 1986); (△) without 3-(3,4-dichlorophenyl)-1,1-dimethylurea (DCMU) to inhibit oxygenic photosynthesis, HS$^-$ was rapidly used and O$_2$ (○) built up; with DCMU (●), HS$^-$ persisted after lower concentrations were reached and O$_2$ (not shown) was not evolved; arrow indicates shift from dark to 400 µE/m^2/s (150 W halogen lamp with optic light guide), 25°C, pH 8.5 to 8.7. (b) Surface 0.1 mm of *Oscillatoria boryana* mat core from "Government Vent Pool," Rotorua, North Island, New Zealand (Castenholz et al. submitted b); sulfide (●) was rapidly used (anoxygenic photosynthesis) and O$_2$ (○) built up; pH (△) rose through steps with photosynthetic activity; arrow indicates shift from dark to 155 W/m^2 (halogen lamp with fiber optic light guide), 20° to 22°C.

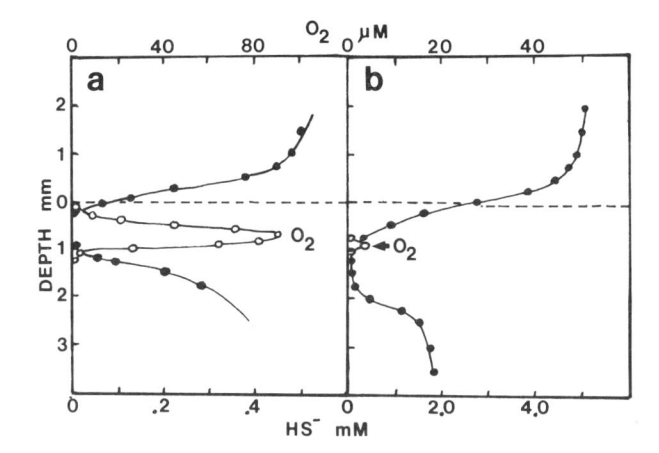

Figure 6.9.4. Profile of O_2 (○) and sulfide (●) in the intermittently oxygenic/anoxygenic Solar Lake *Microcoleus* mat after exposure to high sulfide in the dark. (a) A substantial O_2 maximum developed under 600 μM initial sulfide in light (400 μE/m^2/s) between 0.5 and 1.0 mm depth (*Microcoleus*). (b) A very small O_2 maximum developed under 5 to 6 mM initial sulfide at 1,950 μE/m^2/s illumination, 25°C; note the broad region of HS$^-$ depletion below the O_2 maximum, resulting mainly from anoxygenic photosynthesis of cyanobacteria and photosynthetic bacteria (cf. Jørgensen et al. 1986).

ing sulfide levels within the mat. However, because *Microcoleus chthonoplastes* requires trace amounts of O_2 for biosynthetic purposes, sustained growth under continuous anoxic conditions is impossible (de Wit et al. 1988).

In a tepid pond in North Island, New Zealand ("Government Vent Pool," Rotorua), that has both primary and biogenic sulfide input, extensive surface mats of dark brown *Oscillatoria boryana* develop (Castenholz et al. submitted-b). This species is very similar to *M. chthonoplastes* in its ability to shift from anoxygenic to oxygenic photosynthesis (Fig. 6.9.3b). The sulfide level that allows the shift to oxygenic photosynthesis appears to be lower for *Oscillatoria* than for *Microcoleus*, but because pH is generally between 6.6 and 7.0, the H$_2$S values associated with the shift are roughly similar (i.e., about 50 μM). *O. boryana* does not appear to be able to carry out oxygenic photosynthesis in the presence of higher sulfide levels even under high light, and it seems likely that both modes of photosynthesis alternate closely in time (or perhaps even operate in concert) only at very low sulfide levels. The principal difference between the two cyanobacteria, however, is the ability of *O. boryana* to rapidly move to different depths within the mat by gliding in response to changes in light intensity. This behavioral flexibility allows this species to avoid unfavorable (photo-oxidative) conditions of high light intensity during the day by migrating to an environment of optimal light and anaerobic conditions where anoxygenic photosynthesis is performed.

It has been suggested that the earliest stages in the evolution of oxygenic photosynthesis during the Precambrian may have begun with cyanobacterial prototypes that were able to make use of H$_2$O as a photoreductant alternating with (or in concert with) sulfide-requiring anoxygenic photosynthesis (Jørgensen et al. 1986). A possible selective value for this may have been severe localized competition for sulfide (in microbial mats)

between sulfidophilic anoxygenic prototypes of cyanobacteria and anoxygenic photosynthetic bacteria, such as prototypes of green bacteria or of *Chloroflexus*. Thus, the earliest biogenic production of O_2 may have been "sandwiched" within mats but enveloped above and below by highly reducing, sulfide-rich environments (see the discussion of "inverted mats" below, Section 6.9.4B).

6.9.4 Thermal Mats

Diverse microbial communities exist in thermal environments as a result of varying temperature, pH, H$_2$S, and O_2 (see Table 6.9.2 and references therein regarding the following discussion). In acid hot springs (pH 2 to 4), mats are built by eukaryotic algae. In neutral to alkaline hot springs, however, many different microbial communities are found, even within a single spring. This is clearly illustrated in typical hot spring effluents. For example, the neutral sulfidic spring illustrated in Figure 6.9.5 contains an anoxygenic *Chloroflexus* mat (see Section 6.9.2A) in the anoxic, sulfide-rich source pool, but within centimeters of this mat, because of the H$_2$S/O$_2$ gradient, the cascading effluent is inhabited by chemolithotrophic *Thermothrix* streamers (see Section 6.6); cyanobacterial mats form only in the distal pool where the effluent is sulfide-depleted. In nonsulfidic neutral-alkaline springs, such as Octopus Spring (Yellowstone National Park, Wyoming, USA), temperature is the main factor controlling the distribution of microorganisms (Fig. 6.9.6). Source pools contain living bacteria (up to 100°C), taxa which have never been cultured but the nature of which has been partially revealed by 5S rRNA analysis (Stahl et al. 1985). As yet uncharacterized nonphotosynthetic filamentous bacteria commonly form streamers at 74° to 89°C (Fig. 6.9.7A, B). The upper temperature limit of thermophilic cyanobacteria varies from about 64°C (e.g., *Chlorogloeopsis* in Iceland and New Zealand hot springs) to 74°C (e.g., *Synechococcus lividus* in Octopus Spring). Cyanobacteria, together with a diverse collection of other prokaryotes, form laminated mats which extend to lower temperatures (e.g., from ca. 70° to 40°C in Octopus Spring). Mat community diversity increases with decreasing temperature as the upper temperature limits of different organisms are encountered. Below about 42°C (ca. 49°C in Oregon hot springs), the mats are grazed by metazoans. In springs having temperatures that prohibit growth of eukaryotes, the biota of microbial mats may be especially relevant as a model of Precambrian stromatolitic communities.

Temperature-dependent community changes affect the thickness and texture of mats. In Octopus Spring, for example, the *Synechococcus* mat is flat and only a few millimeters thick near the upper temperature limit (ca. 70°C), but at 50° to 55°C well-laminated flat mats may accumulate to a thickness of several centimeters (Fig. 6.9.7C, D). Below 50° to 57°C, streamers and vertical structures (e.g., cones and columns in quiescent pools) are often found (Fig. 6.9.7E, F, G), correlating with the presence of the cyanobacterium *Phormidium tenue* which may play a role in their formation (Walter et al. 1976a). The relationship between conical mats and their stromatolitic counterparts (*Conophyton*) has been observed previously (see Section 6.3; Walter et al. 1972; Doemel and Brock 1974; Awramik and Vanyo 1986). At lower temperatures, tufted mats are formed by

Table 6.9.2. *The variety of microbial communities in thermal environments as determined by environmental features.*

Microbial Community	Best-studied Example	Predominant Genera				
		eukaryotic algae	cyanobacteria	photosynthetic bacteria	chemolithotrophic bacteria	other[a] organisms
I. Neutral-alkaline Hot Springs Without Primary Sulfide						
source pool	Octopus Spring, Yellowstone National Park (YNP)	—	—	—	—	*Thermus*-like eubacteria; unknown archaebacterium
streamers	Octopus Spring, YNP	—	—	—	—	unknown filamentous bacteria
cyanobacterial mats	Octopus Spring, YNP	—	*Synechococcus*	—	—	—
		—	*Synechococcus*	*Chloroflexus*	—	see Table 6.9.3
		—	*Synechococcus, Phormidium*	*Chloroflexus*	—	—
		—	*Calothrix*	—	—	ephydrid flies, mites
	Hunter's Hot Spring, OR	—	*Synechococcus*	—	—	—
		—	*Synechococcus*	*Chloroflexus*	—	—
		—	*Oscillatoria*	*Chloroflexus*	—	—
		—	*Calothrix, Pleurocapsa*	—	*Beggiatoa*	ostracods
"inverted" cyanobacterial mats	Kah-nee-ta Springs, OR	—	mixed cyanobacteria	*Heliothrix, Chloroflexus*	—	—
translucent cyanobacterial splash-zone mats	Rabbit Creek Spouter, YNP	—	*Phormidium, Pseudoanabaena, Chlorogloeopsis, Synechococcus*	*Chloroflexus, Heliothrix*-like filaments	—	—
II. Near-neutral Hot Springs With Primary Sulfide						
chemolithotrophic streamers	Jamez Spring, NM	—	—	—	*Thermothrix*	—
anoxygenic photosynthetic bacterial mats	"New Pit" Spring, YNP	—	—	*Chloroflexus*	—	—
	"Travel Lodge Stream", NZ	—	—	*Chlorobium*	—	—
	"Roland's Well", YNP	—	—	*Chromatium Chloroflexus*	—	—
intermittently oxygenic/ anoxygenic cyanobacterial mats	Stinky Springs, UT	—	*Oscillatoria*	*Chlorobium*	—	—
"inverted" mats with *Chloroflexus* above cyanobacteria	Iceland "Spring A"	—	*Chlorogloeopsis*	*Chloroflexus*	—	—
III. Acid Hot Springs						
eukaryotic algal mats	Nymph Creek, YNP	*Cyanidium*	—	—	—	*Dactylaria gallopava* (fungus), unknown archaebacterium
	Norris Junction, YNP	*Zygogonium*	—	—	—	
IV. Deep-sea Hydrothermal Vents						
possibly chemolithotrophic biofilms and mats	Galapagos vents, Guaymas Basin	—	—	—	*Beggiatoa*, various morphologies	–

[a]Natural microbial mats are not pure cultures of the predominant microorganisms. Thus, in all cases other microorganisms which have not been cultured are likely to be present.

1. Brock 1978; 2. Stahl et al. 1985; 3. Castenholz 1973a; 4. Ward et al. 1987; 5. Revsbech and Ward 1984b; 6. Wiegert and Mitchell 1973; 7. Revsbech and Ward 1984a; 8. Wickstrom and Castenholz 1985; 9. Pierson et al. 1985; 10. Castenholz 1988; 11. Castenholz 1984; 12. Caldwell et al. 1976; 13.. Ward et al. 1989b; 14. Giovannoni et al. 1987a; 15. Jørgensen and Nelson 1988; 16. Ward et al. 1985; 17. Revsbech and Ward 1983; 18. Jannasch and Wirsen 1981; 19. Jannasch 1984; 20. Karl 1987.

Environmental Features				
°C	pH	μM H_2S	O_2	References
92–100	7–9	—	?	1, 2
75–85	7–9	—	low	1
70–74	7–9	—	present	3
57–70	7–9	very low	superoxic	4, 5
40–57	7–9	—	superoxic	4
<42	7–9	—	—	6
70–74	8.4	—	present	3
54–70	8.4	biogenic	superoxic	3, 7
49–54	8.4	biogenic?	superoxic	3
<49	8.4	biogenic?	superoxic	3, 8
35–55	alkaline	absent	superoxic	9, 10
35–50	neutral-alkaline	absent	aerated	11
55–75	7.0	30	O_2/H_2S interface	12
52–58	6.3	34	absent	13, 14
45	5.9	250	absent	10
52–55	6.3	34	absent	13
42–46	6.1	1200	biogenic where H_2S depleted	13, 15
55–61	7.5	50–80	biogenic in cyanobacterial layer	15
20–57	2–4	—	superoxic	1, 16, 17
20–31	2–4	—	—	1
10–20° above seawater	seawater pH	+	possibly O_2/H_2S intersect	18, 19, 20

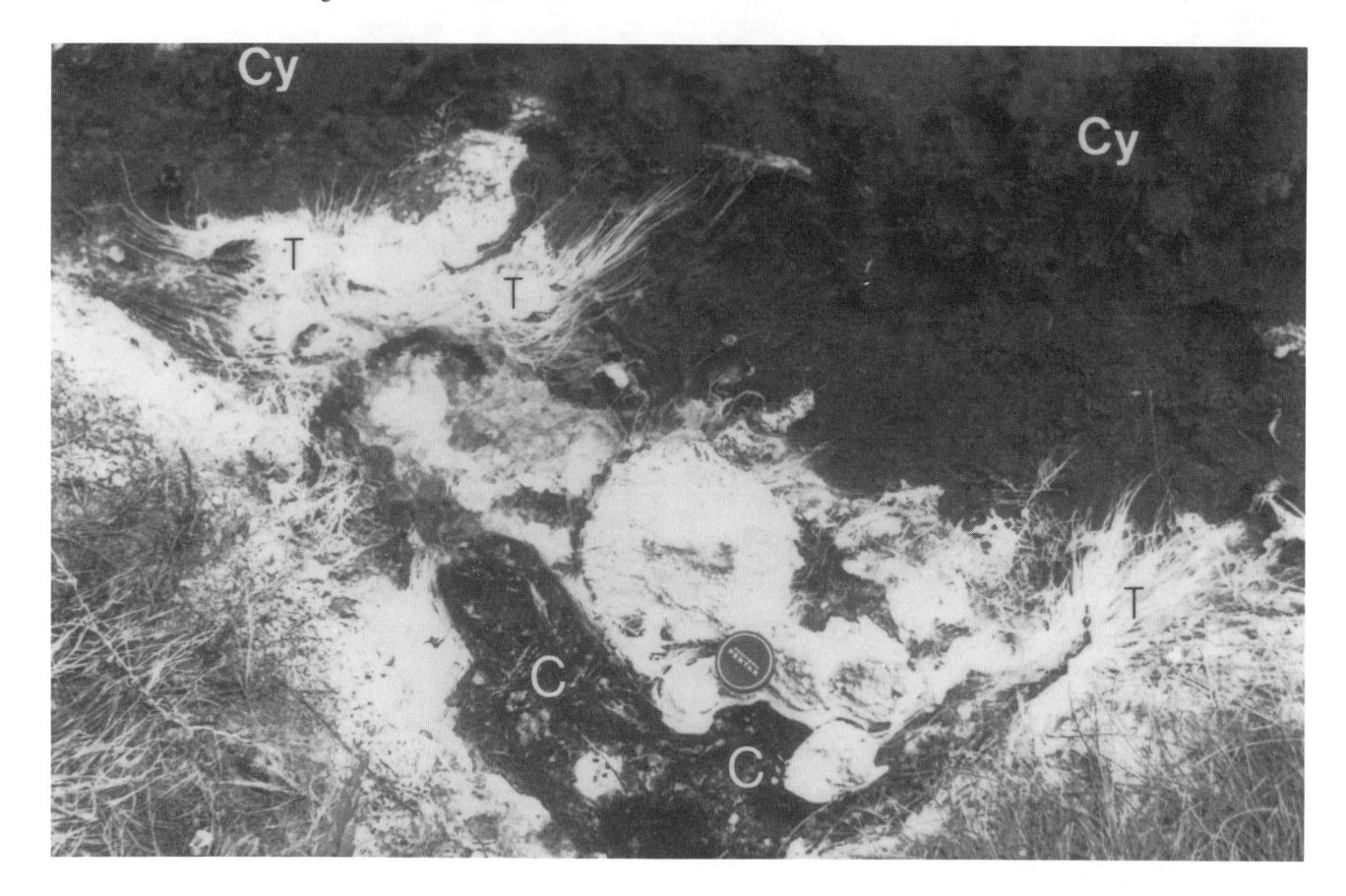

Figure 6.9.5. A typical neutral sulfidic hot spring setting (Painted Pool, Mammoth Hot Springs, Yellowstone National Park; now defunct). C, "green *Chloroflexus* mat"; T, *Thermothrix* streamers at O_2/H_2S interface; Cy, cyanobacterial mat. [Modified after Castenholz 1988].

the cyanobacterium *Calothrix*, in some cases associated with *Pleurocapsa* (Fig. 6.9.7H).

In addition to commonly occurring submerged mats, relatively rare translucent mats are known from geyser spray zones (e.g., Rabbit Creek Spouter, Yellowstone National Park; see Castenholz 1984). The distribution of thermophilic cyanobacteria and the mats they form has been reviewed previously (Castenholz 1969a, 1973a, 1984; Brock 1978). In the following discussion, a well-studied example of a typical hot spring cyanobacterial mat is compared with two other mats having atypical vertical organization.

6.9.4A A Typical Hot Spring Cyanobacterial Mat: Octopus Spring

The 50° to 55°C cyanobacterial mat of Octopus Spring (Fig. 6.9.7C, D) has been studied in detail (Ward et al. 1987), providing understanding of the structure and function of a modern mat that may be analogous in these respects to some Proterozoic stromatolites.

Community structure. The community in this mat is complex and phylogenetically diverse (Table 6.9.3). Microbes isolated from the mat include members of six eubacterial "phyla" and one archaebacterium, and there is direct evidence that the community is even more diverse than is summarized in Table 6.9.3 (Ward et al 1985, 1989b, in press; Tayne et al. 1987). This high inferred diversity is not unexpected; culture-dependent diversity estimates have been long suspected to, and have now been proven to underestimate the true diversity of natural microbial communities (Ward 1989, 1990). The bacteria cultured from the mat represent all trophic levels in the totally microbial food chain of the mat (Fig. 6.9.8). Lipid biomarkers suggest that phototrophs are more abundant than aerobic and anaerobic heterotrophic consumers, with the terminal members of the food chain being least abundant (Section 6.8).

Community function. The cyanobacterial mat of Octopus Spring is thought to be produced by *Synechococcus lividus* (Fig. 6.9.8). The intense oxygenic photosynthesis of this cyanobacterium results in superoxic (ca. 6 times air saturation), alkaline conditions within the photic zone (Fig. 6.9.9B) which promote photorespiratory production of glycolic acid (Bateson and Ward 1988). At high O_2 concentrations, the absence of sulfide from the photic zone forces the photosynthetic bacterium *Chloroflexus aurantiacus* to be mainly photoheterotrophic, assimilating glycolic acid and other low molecular weight organic compounds that are photoexcreted by the cyanobacterium (Bauld and Brock 1974) or are produced as a result of anaerobic decomposition (Anderson et al. 1978). The rate of photosynthetic mat accretion (18 to 45 μm/day) is balanced by that of decomposition so that there is little net mat accumulation (Doemel and Brock 1977). Aerobic bacteria have been

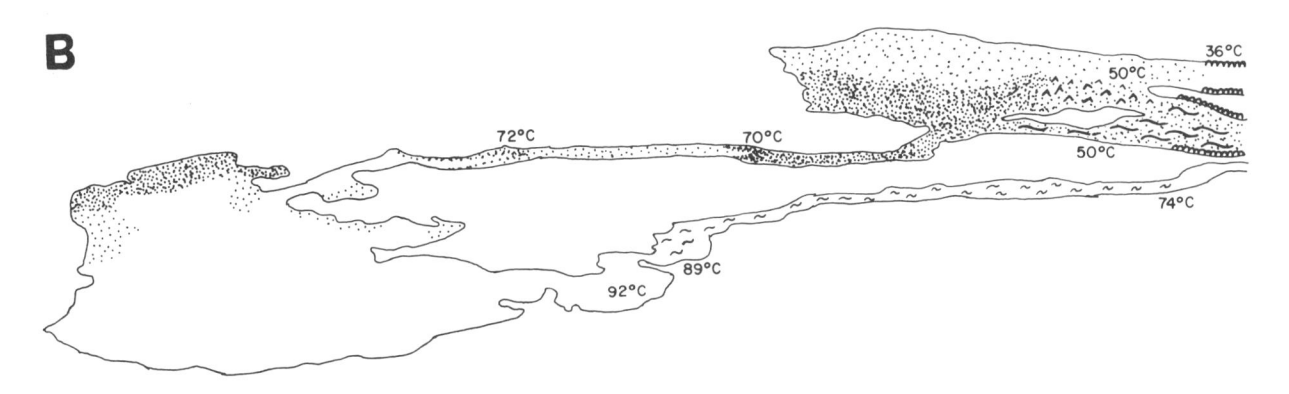

Figure 6.9.6. A typical nonsulfidic neutral-alkaline hot spring setting. (A) Octopus Spring, Lower Geyser Basin, Yellowstone National Park. (B) Schematic showing locations of different microbial communities along the thermal gradients of the spring and its effluent channels. Dots indicate distribution of photosynthetic microbial mats (see Fig. 6.9.7C, D); density of dots indicates mat thickness: short wavy lines indicate nonphotosynthetic filamentous bacterial streamers (see Fig. 6.9.7A, B); long wavy lines indicate streamers containing photosynthetic microorganisms above mats (see Fig. 6.9.7E); inverted "v" indicate columnar and conical stromatolites within photosynthetic mats (see Fig. 6.9.7F, G); scallop symbol indicates distribution of nodular *Calothrix* and *Pleurocapsa* mats (see Fig. 6.9.7H). In the cooler waters downstream of the extreme right edge of the effluent channel in the foreground, the distribution of microbial communities is similar to that of the effluent channel detailed in the background.

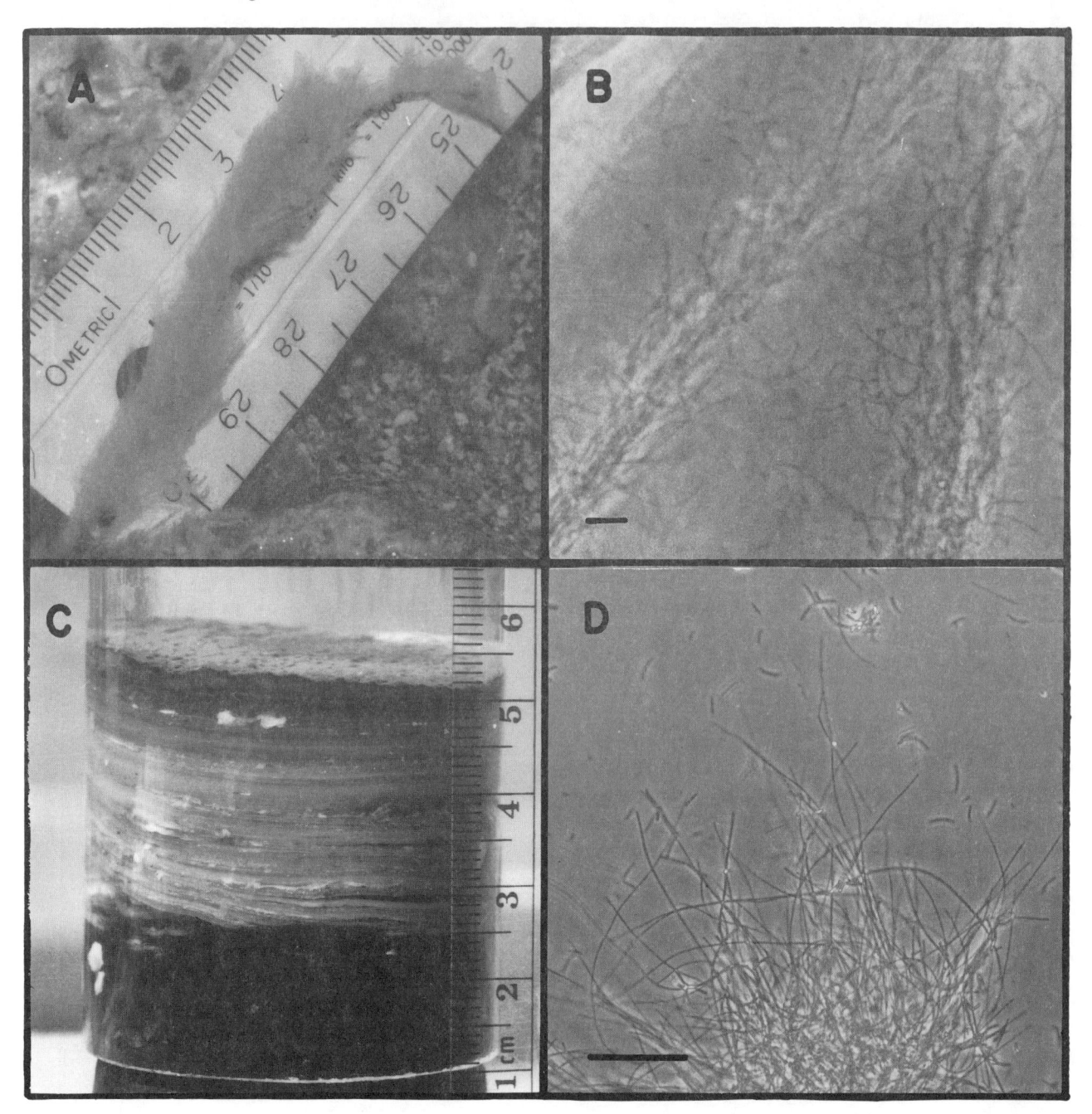

Figure 6.9.7. Microbial communities within typical nonsulfidic neutral-alkaline hot springs. Lines for scale in B and D represent 50 μm. (A) Streamers of nonphotosynthetic filamentous bacteria inhabiting the Octopus Spring channel at ca. 74° to 89°C. (B) Photomicrograph showing nonphotosynthetic filamentous bacteria inhabiting the streamer community shown in "A," at left. (C) 50°–55°C Octopus Spring laminated cyanobaterial mat; top ca. 1 mm is green; undermat laminations are orange (cf. Ward et al. 1989). (D) Photomicrograph of the top green layer of the 50°–55°C Octopus Spring cyanobacterial mat (see Ward et al. 1989) showing sausage-shaped cyanobacterial cells (presumably *Synechoccus lividus*) and filamentous bacteria (presumably *Chloroflexus aurantiacus*). (E) Streamers containing *S. lividus*, *C. aurantiacus*, and *Phormidium tenue* occurring above the microbial mat at ca. 40° to 57°C in Octopus Spring. (F) Stromatolite-like columns (each ca. 1 cm wide) at Column Spouter Spring, Fairy Creek Meadows, Yellowstone National Park, Wyoming, USA; diameter of nail head is ca. 1 cm. (G) Conical stromatolite-like structures at Octopus Spring, Yellowstone National Park, showing heliotropism as reported by Awramik and Vanyo (1986); diameter of coin is 2.4 cm. (H) *Calothrix* and *Pleurocapsa* nodular mat, reported by Castenholz (1973a) from Hunter's Hot Spring, Oregon, with grazing ostracods (at arrows).

cultured from the mat, but little is known quantitatively about the importance of aerobic decomposition. Under anaerobic conditions, fermentation of cellular organopolymers occurs (Anderson et al. 1987). Methanogens, the predominant terminal anaerobes of this system (Ward 1978), consume the H_2 produced during fermentation (Sandbeck and Ward 1981, 1982). Because of this interspecies H_2-transfer (see Section 6.6), the methanogens cause fermenters to produce mainly acetate, H_2, and CO_2. Acetate and minor organic fermentation products are recycled by their photoheterotrophic uptake by *C. aurantiacus*.

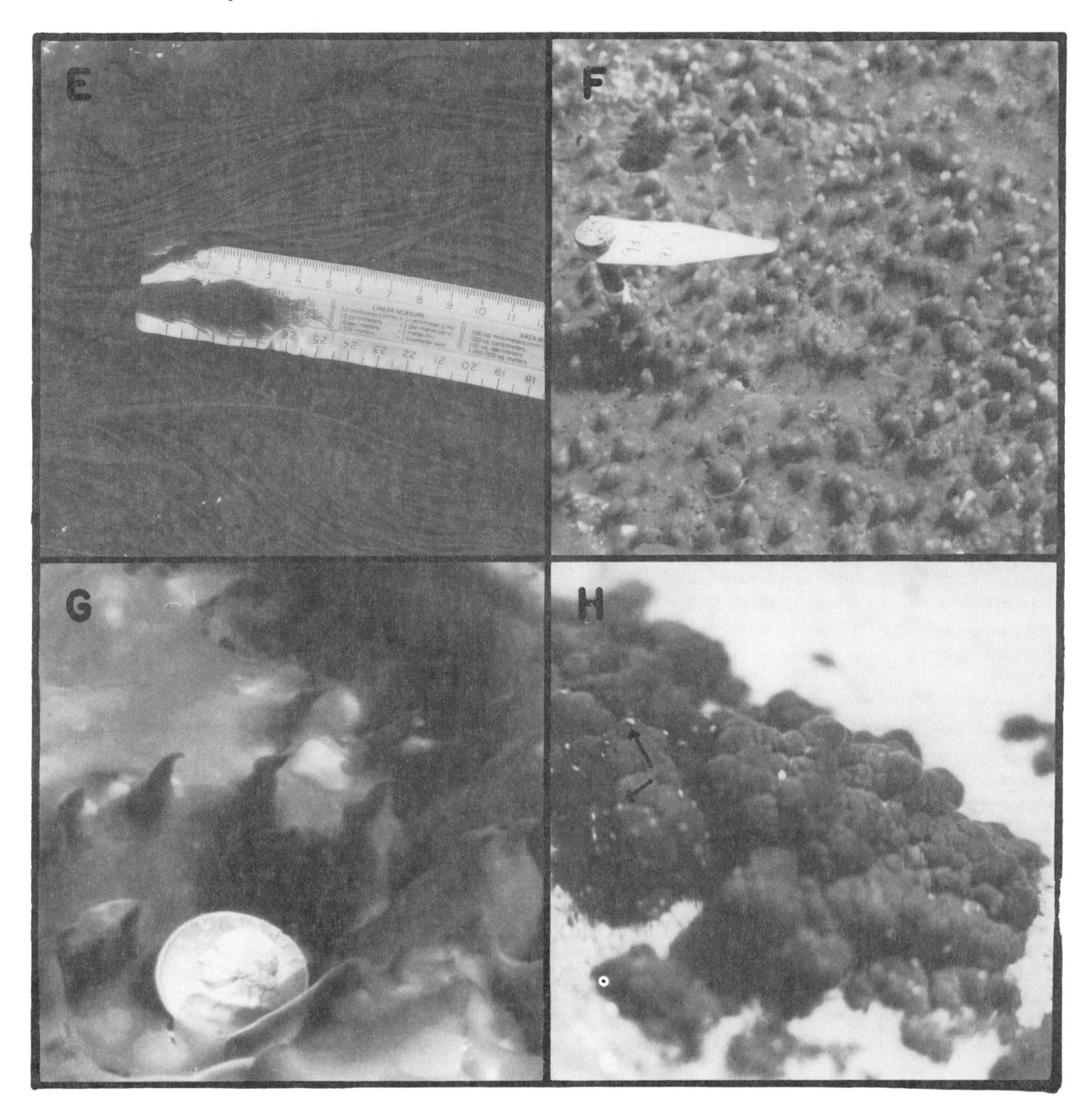

Figure 6.9.7. (Continued)

It is interesting to note that in high sulfate hot springs, the situation differs: sulfate reduction dominates over methane production (Ward and Olsen 1980; Ward et al. 1984), and biogenic sulfide production may increase the importance of anoxygenic photosynthesis by purple and green bacteria as well as chemolithotrophic sulfide oxidation (Revsbech and Ward 1984b).

Vertical distribution of organisms and processes. The metabolic processes outlined above occur at different vertical positions within the cyanobacterial Octopus Spring mat, as illustrated in Figure 6.9.9. Oxygenic photosynthesis is restricted to the uppermost 1 mm of the mat which contains both cyanobacteria and *Chloroflexus*. The latter organism and other anoxygenic filamentous phototrophs predominate at depths 1 to 3 mm beneath the mat surface. Anaerobic decomposition processes (fermentation and methanogenesis) occur principally within the top 0 to 5 mm of the mat. Thus, most of the mat thickness (all material below a depth of ca. 5 mm) apparently represents mainly recalcitrant remains of the active microbial layer. At the morphological level, the mat-forming organisms show differential preservation during decomposition. For example, moribund filaments of *Chloroflexus* can be detected in relatively deep mat layers (Tayne et al. 1987), where morphologic remains of the cyanobacterium *S. lividus* do not occur (Doemel and Brock 1977). The vertical distributions of O_2 and pH (and H_2S,

Table 6.9.3. *Known or suspected members of the 50° to 55°C Octopus Spring cyanobacterial mat community.*

Organism	Phylogenetic Type[a]	Physiologic Type	References
I. Phototrophs			
Synechococcus lividus	eubacterium (cyanobacterium)	cyanobacterium	1
Chloroflexus aurantiacus	eubacterium (green non-sulfur group)	photosynthetic bacterium	1, 2
II. Heterotrophic Consumers			
Thermus aquaticus	eubacterium (radioresistant cocci)	aerobic heterotroph	3
Isosphaera pallida	eubacterium (planctomycete group)	aerobic heterotroph	4, 5
Thermomicrobium roseum[b]	eubacterium (green-nonsulfur group)	aerobic heterotroph	6
Thermobacteroides acetoethylicus	eubacterium (Gram-positive group)	anaerobic fermenter	7, 8
Thermoanaerobium brockii	eubacterium (Gram-positive group)	anaerobic fermenter	8, 9
Thermoanaerobacter ethanolicus	eubacterium (Gram-positive group)	anaerobic fermenter	10
Clostridium thermohydrosulphuricum	eubacterium (Gram-positive group)	spore-forming anaerobic fermenter	8, 11
Clostridium thermosulfurogenes	eubacterium (Gram-positive group)	spore-forming anaerobic fermenter	12
Clostridium thermoautotrophicum[b]	eubacterium (Gram-positive group)	spore-forming anaerobic fermenter	13
III. Terminal Anaerobic Food Chain Consumers			
Thermodesulfobacterium commune	eubacterium (unique group)	sulfate-reducer	14
Methanobacterium thermoautotrophicum	archaebacterium	methanogen	8, 15

[a]Phylogenetic groupings based on Woese (1987) and data from Bateson et al. (1989), Bateson, Thibault and Ward (1990), and Ward et al. (in press).
[b]*T. roseum* and *C. thermoautotrophicum* have never actually been isolated from the Octopus Spring mat, but might be mat inhabitants; distinctive diol lipids of the former have been observed in the mat (see Ward et al. 1989a).
1. Bauld and Brock (1973). 2. Tayne et al. (1987). 3. Brock and Freeze (1969). 4. Giovannoni et al. (1987b). 5. Doemel and Brock (1977). 6. Jackson et al. (1973). 7. Ben-Bassat and Zeikus (1981). 8. Zeikus et al. (1980). 9. Zeikus et al. (1979). 10. Wiegel and Ljungdahl (1981). 11. Wiegel et al. (1979). 12. Schink and Zeikus (1983). 13. Wiegel et al. (1981). 14. Zeikus et al. (1983). 15. Sandbeck & Ward (1982).

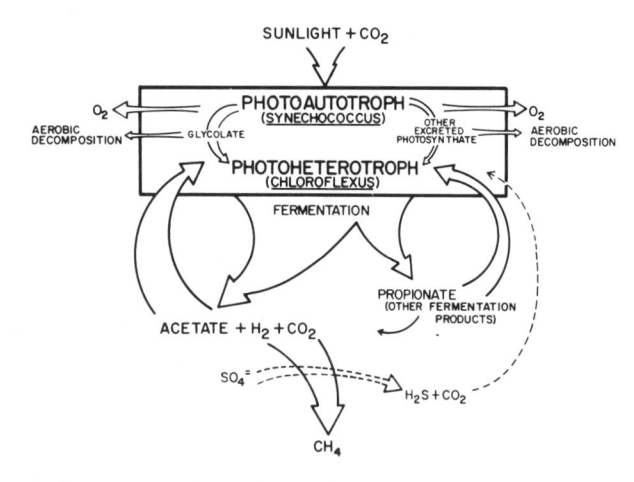

Figure 6.9.8. Model of carbon and electron flow through the 50°–55°C cyanobacterial mat community of Octopus Spring (see Ward et al. 1987). Widths of arrows approximate importance in overall carbon and electron flow of the processes indicated.

see Revsbech and Ward 1984b) vary with light intensity (Fig. 6.9.10). Over a diel cycle, the vertical positioning and occurrence of various metabolic processes may change. For example, anaerobic decomposition reactions and bacteriochlorophyll synthesis by *Chloroflexus* are likely to be inhibited by O_2 and thus may occur only at night when the upper portions of the mat are almost completely anoxic.

6.9.4B Mats With Atypical Vertical Sequences

Two exceptions are known to the "normal" positioning of cyanobacteria above anoxygenic phototrophic bacteria in hot spring mats, systems that are commonly referred to as "inverted mats." In some sulfidic (ca. 240 μM H_2S) Icelandic hot springs, cyanobacteria (*Chlorogloeopsis* and *Phormidium*) are confined to a layer underlying *Chloroflexus* sp. (Jørgensen and Nelson 1988). In these inverted mats, the level of sulfide above the cyanobacteria (65 to 100 μM H_2S) inhibits cyanobacterial growth at temperatures greater than 55°C (Castenholz 1976)

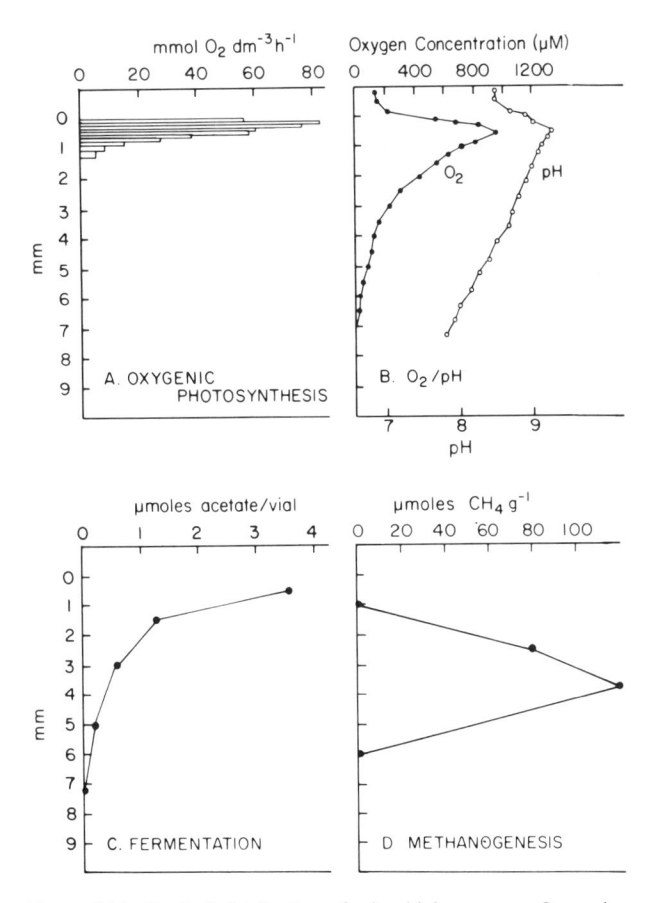

Figure 6.9.9. Vertical distribution of microbial processes, O_2, and pH within the 50°–55°C Octopus Spring cyanobacterial mat community (see Ward et al. 1987). (A) Oxygenic photosynthesis in full sunlight (1,720 $\mu E/m^2/s$). (B) O_2 and pH microprofiles in full sunlight (1,720 $\mu E/m^2/s$). (C) Fermentation. (D) Methanogenesis.

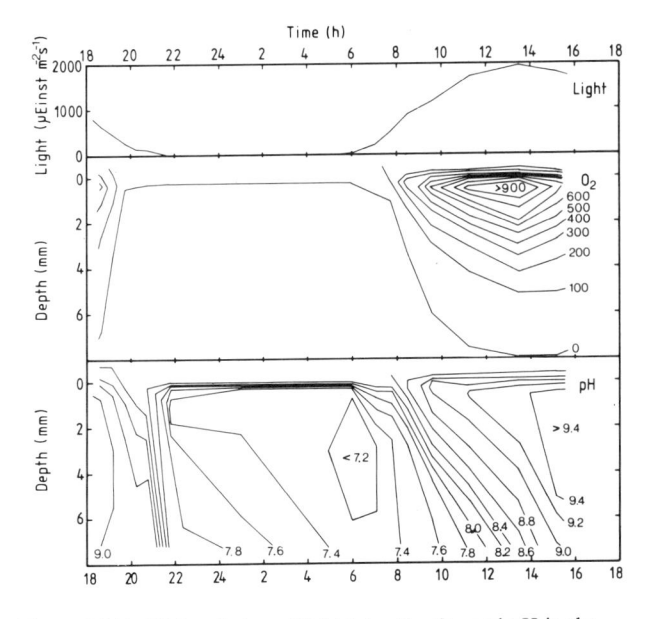

Figure 6.9.10. Diel variation of light intensity, O_2, and pH in the 50°–55°C Octopus Spring cyanobacterial mat (see Revsbech and Ward 1984a).

but does not inhibit the growth of *Chloroflexus* sp. which forms a surface mat at temperatures up to 68°C (Castenholz 1973b; Jørgensen and Nelson 1988). Under these conditions, *Chloroflexus* probably grows as a sulfide-dependent photoautotroph independent of the cyanobacteria (see Section 6.9.2A). Presumably, the cyanobacteria can develop only as an undermat below *Chloroflexus* which reduces the sulfide concentration to tolerable levels (Jørgensen and Nelson 1988). Oxygen and oxygenic photosynthesis are confined to the narrow subsurface *Chlorogloeopsis* layer in these mats, reaching maximum levels at a depth of 1.1 mm beneath the surface (Jørgensen and Nelson 1988). No O_2 is detected at depths of less than 0.8 mm, and rapid consumption of oxygen within the mat is inferred.

Production and confinement of oxygen to a subsurface layer of cyanobacteria within an otherwise anoxygenic mat may be of particular relevance regarding the origin of oxygen production in mat-dwelling cyanobacterial ancestors. As do intermittently oxygenic/anoxygenic mats (Section 6.9.3), the Icelandic inverted mats suggest the possibility that the first oxygen-producing phototrophs may have developed within, rather than on the surface of, an otherwise anoxygenic, phototrophic community. Thus, ancestral cyanobacteria may have produced oxygen long before production exceeded consumption within such mats, and the advent of oxygen-producing photosynthesis need not necessarily have resulted in the escape of free oxygen into the surrounding environment.

A second type of extant inverted mat is found in some nonsulfidic alkaline springs in central Oregon, USA, and in Yellowstone National Park, occurring over a temperature range of 35° to 55°C. In these mats, the surface layer consists of an anoxygenic filamentous phototroph, *Heliothrix oregonensis*, which contains Bchl *a* as its only chlorophyll (Pierson et al 1984, 1985). The filaments glide and form characteristic puffs and tufts above a mixed assemblage of cyanobacteria. The mat contains oxygen throughout (Castenholz 1988b), and organisms of its surface layer are capable of photoheterotrophic activity despite the saturated levels of oxygen.

6.9.5 Eukaryotic Microbial Mats

6.9.5A Marine and Lacustrine Eukaryotic Mats

Although most mats in marine or hypersaline environments are principally cyanobacterial (e.g., Gebelein 1969), mats built predominantly by eukaryotic microorganisms are cosmopolitan in shallow marine, lacustrine, and flowing waters (Table 6.9.4). Following the earlier work of Hommeril and Rioult (1962, 1965), the significance of eukaryotic micro-algae in subtidal microbial mats was emphasized by Bathurst (1967). He observed that relatively thin (ca. 0.25-cm-thick), gelatinous coherent mats, composed of both filamentous cyanobacteria and mucilage-producing diatoms, were widespread on the Great Bahama Bank and, in his view, responsible for sediment stability and resistance to hydrodynamic erosion. These mats show best development at depths down to 10 m (Neumann et al. 1970; Scoffin 1970; Gebelein 1976; Dravis 1983; Dill et al. 1986), although in the very clear waters around Bermuda, mats possibly produced by eukaryotes are observed to depths of 50 m (Gebelein 1976). Filamentous, commonly branching,

Table 6.9.4. *Examples of marine and lacustrine microbial mats constructed principally by eukaryotic algae.*

Locality	Habitat	Salinity (‰)	Mat Description	Constructing Organisms	Other Organisms	Comments	Reference
Rock Harbour Cays, Little Bahama Bank (Stn. F3)	submerged <2 m	33–38	Thick (1–2 cm), rigid resilient mat with porous surface; stratiform not associated with lithified stromatolites	*Cladophoropsis membranacea*, *Gracilaria blodgettii*, *Thalassiothrix woodii*	*Nitzschia closterium*, *Schizothrix calcicola*	No stalked diatoms; several benthic diatoms abundant but only *Th. woodii* and *N. closterium* secrete significant mucilage; highly erosion resistant	1
Bimini Lagoon, Great Bahama Bank	submerged ca. 1 m avg. 26°C	31–45 avg. 37	Dense mat up to ca. 1 cm thick with erect filaments extending 0.5–1.0 cm above mat-sediment; more effective sediment binding than local cyanobacterial mats	*Enteromorpha prolifera*	*Chaetomorpha* sp., *Ulothrix* sp., *Cladophora* sp., *Oscillatoria* sp.	Capable of rapid seasonal colonization; subaerial exposure is detrimental; can grow through 1–2 cm sediment cover	2
Walker Lake, Nevada (highly siliclastic lake)	shallow, turbulent shoreline	ca. 9	Thin crust-like "base" mat, with seasonally abundant outer layer; coats cm-scale carbonate stromatolites	Outer Community: *Cladophora glomerata*, *Ulothrix* cf. *aequalis*; Understory: *Gongrosira* sp., *Amphithrix janthina*	*Schizothrix* sp., *Homeothrix* sp., *Anabaena* sp.	Stalked/sessile diatoms are seasonally abundant	3
Hamelin Pool, Shark Bay	moderately hypersaline; submerged to ca. 4 m	60–70	Colloform mat; coats low intertidal and subtidal stromatolites and hard-grounds; softly cohesive surface pale- to orange-brown; bright, blue-green layer between surface community and lithified carbonate	*Mastogloia halophila* (?), *Mastogloia reimeri*, *Brachysira aponina*, (other *Mastogloia* spp.?)	*Amphora coffeaeformis*, *Amphora ventricosa* (other *Amphora* spp., *Mastogloia* spp., *Navicula* spp.)	More than 60 species of diatoms identified; only 5 species are centric forms never significant quantitatively; species with mucilaginous stalks are dominant agents of construction	4, 5

1. Neumann et al. (1970). 2. Scoffin (1970). 3. Osborne et al. (1982). 4. J. John (pers. comm. to J.B.). 5. Bauld, D'Amelio, John, and Pierson (unpublished results).

green algae provide the coherent framework in Bahamian subtidal mats (Neumann et al. 1970; Scoffin 1970). In contrast to the Hamelin Pool mats discussed below, Bahamian mats appear to lack structural input from stalked diatoms (Fig. 6.9.11). The algal floras of mats coating subtidal lithified stromatolites (Dravis 1983; Dill et al. 1986) have not been described in detail, although macroscopic eukaryotic algae such as the dasyclad *Batophora* (Dravis 1983), and "a complex consortium of blue-green algae (and) specially adapted diatoms" (Dill et al. 1986), are reported. Eukaryotic mats are not confined to warm waters; the filamentous diatom *Amphipleura rutilans* constructs dense mats in shallow (ca. 0.5-m-deep) nearshore areas of the Chukchi Sea above the Arctic Circle (Matheke and Horner 1974).

Eukaryotic micro-algae are conspicuous components of microbial mats associated with the lithified stromatolites and hardgrounds of the sublittoral platform of Hamelin Pool, Shark Bay, Western Australia (Fig. 6.9.11; Playford and Cockbain 1976; Bauld et al. 1979; Bauld 1984; Golubić 1985; Awramik and Riding 1988). Taxonomic (John and Bauld unpubl.) and pigment analyses (Palmisano et al. 1989; Pierson unpubl.) of these mats, have clearly established their diatomaceous nature and the role that these eukaryotic micro-algae play in mat construction and cohesion (Bauld, D'Amelio, John, and Pierson unpubl.). These and other submerged mats from Australian coastal saline lakes (see below), and also those coating the Lee Stocking Island stromatolites (Dill et al. 1986), are characterized by distinctive lipid biomarkers (see Section 6.8.4). More than 60 species of diatoms (in comparison with only 6 to 8 species of cyanobacteria) have been identified, of which all but five are pennate. *Mastogloia* (23 species) and *Amphora* (11 species) are the most abundant genera. Many of the pennate diatoms are representatives of genera frequently found in marine and/or saline lacustrine benthic microbial communities. For example, *Amphora coffeaeformis*, quantitatively dominant in Hamelin Pool colloform mats, has a cosmopolitan distribution in peripyton of widely varying salinities (e.g., Ehrlich 1978; Ehrlich and Dor 1985; J. John pers. comm. to J. B.). *A. coffeaeformis* and many other pennate diatoms produce polysaccharide-containing mucilage

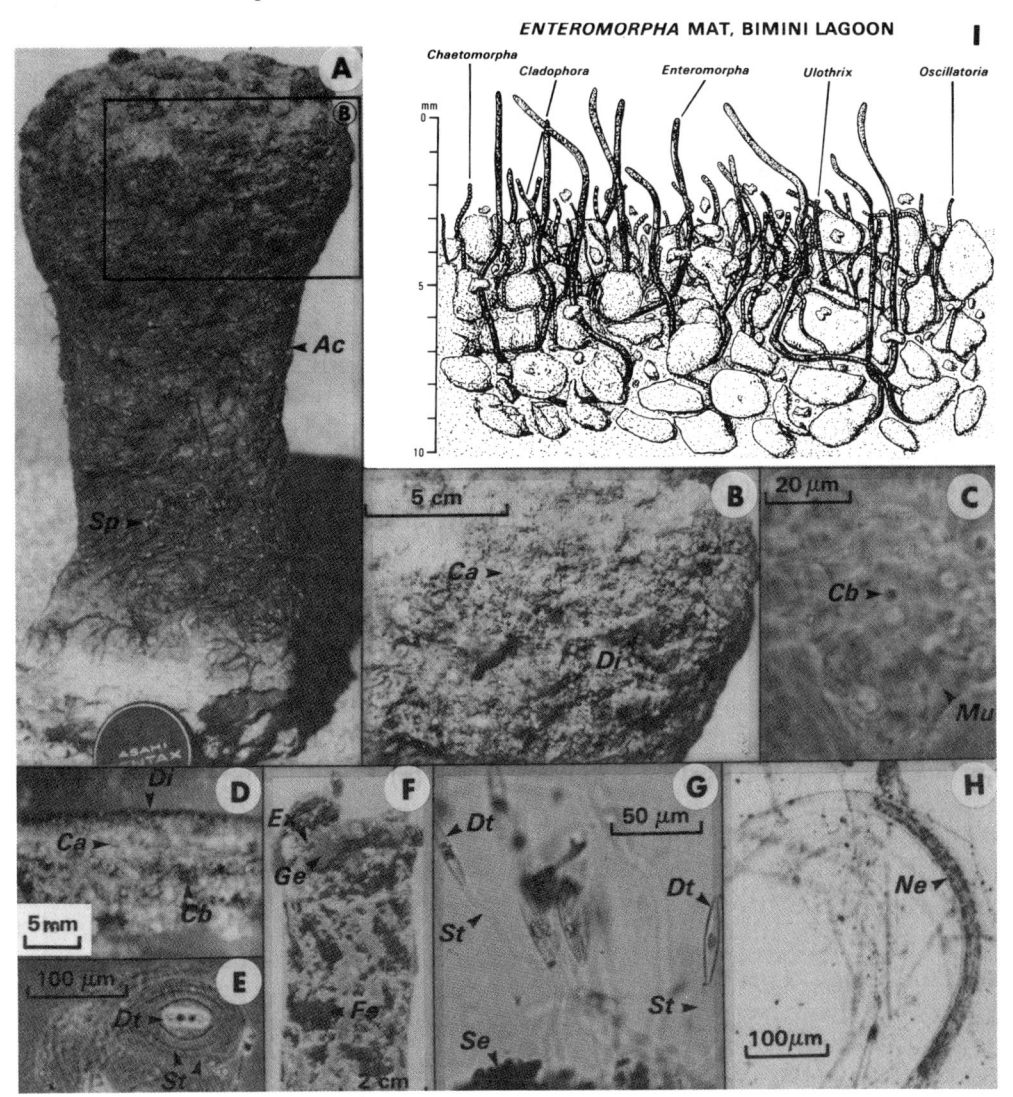

Figure 6.9.11. Illustrative examples of eukaryote-containing microbial mats and their component organisms (for examples of prokaryote-dominated mats, see Fig. 6.3.1). (A) Subtidal stromatolite from Hamelin Pool, Shark Bay (Western Australia), in growth orientation but removed from water. The lithified columnar structure has a thin outer "rind," the surface of which is still intact in some areas (e.g., upper portion of the structure, enclosed by the rectangle and shown in [B], below) despite abrasive damage during collection. The middle and lower sides of the column are colonized by a "beard" of the chlorophyte *Acetabularia* sp. (Ac; arrow points to an individual stalk) and serpulid worm tubes (Sp; at arrow, visible as white spots). Lens cap is 54 mm in diameter. (B) Area enclosed by rectangle in [A], above, showing smooth diatomaceous outer layer (Di; see also [D], below). Much of the surrounding area has been abraded during collection, exposing the immediately underlying "crunchy" carbonate mush (Ca) which contains unicellular cyanobacteria. (C) Phase-contrast photomicrograph of the firm gel removed from the cavity in [F], below, showing the copious production of extracellular mucilage (Mu) in relation to the biomass of cyanobacterial cells (Cb). (D) Vertical section through diatomaceous mat removed from the top of a columnar stromatolite head (collected at a water depth of ca. 2 m at Carbla Point, Hamelin Pool, Shark Bay, Western Australia). The surface layer of diatoms (Di) immediately overlies a crunchy carbonate layer (Ca) that contains dark areas where unicellular cyanobacteria (Cb) are numerous; below the bottom of the sample shown, the carbonate is too firmly cemented to permit collection. (E) Phase-contrast photomicrograph of *Mastogloia* cf. *halophila*, one of the major agents of construction of the diatomaceous mats shown in [B] and [D], above; the diatom cells (Dt) are borne on characteristic extracellular stalks (St) providing cohesion for these somewhat fragile mats. (F) Portion of a small core removed from the side of a subtidal Hamelin Pool stromatolite (cf. [A], above) oriented so that the outer surface of the core is at the top of the photograph. The outer diatomaceous layer has been lost during collection and manipulation. The lithified portion of the stromatolite here shown is characterized by a combination of strong cementation and large fenestrae (Fe). The external cavity (Ex) is colonized by coccoid cyanobacteria which produce extracellular gel (see [C], above) giving the mass (Ge) a firm, rubbery resilience. (G) Phase-contrast photomicrograph of the stalked diatom *Brachysira aponina* from shallow subtidal mats at Flagpole Landing, Hamelin Pool, Shark Bay, Western Australia. Diatom cells (Dt) are borne at the tips of firmly gelatinous stalks (St). Note the occurrence of trapped sediment particles (Se). (H) Additional eukaryotic components of the Shark Bay stromatolites include meiofauna such as nematodes (Ne) which move freely through the porous interstices of the diatom-dominated mats. (I) Vertical profile through a subtidal eukaryotic mat from Bimini Lagoon, Bahamas, constructed primarily by non-motile filamentous chlorophytes such as *Enteromorpha* and *Cladophora*; gliding filamentous cyanobacteria (e.g., *Oscillatoria*) are also present but play a subsidiary role. Note the coarse nature of the trapped and bound sediment. [All photographs and photomicrographs by J. Bauld, except Fig. 6.9.11F which is provided courtesy of G. W. Skyring; Fig. 6.9.11I is modified after Fig. 22 of Scoffin (1970)].

(D'Amelio and Bauld unpubl.); some produce branched stalks. Together, these serve to trap and bind sediment particles in a manner analogous to that of intermeshed, filamentous, mat-forming cyanobacteria, producing a soft but coherent surficial diatomaceous mat. The two most important stalked diatoms in the Hamelin Pool mats are *Brachysira aponina* and *Mostogloia halophila* (?), the former being the dominant cohesive agent in low intertidal to high subtidal habitats, and the latter in deeper subtidal localities (John and Bauld unpubl.).

Similar eukaryote-prokaryote consortia produce mats colonizing lacustrine sediments. The mats coating stromatolites in hyposaline Walker Lake, Nevada, USA, are dominated by the filamentous green alga *Gongrosira* sp., together with the cyanobacterium *Schizothrix* sp., with stalked diatoms being seasonally abundant (Osborne et al. 1982). Diatoms are very common inhabitants both of lacustrine periphyton (Roemer et al. 1984) and of microbial mats (Bauld 1981b). Although diatoms play an important cohesive role in, for example, mats of certain Australian coastal saline lakes (Bauld 1986 and references therein; Grey et al. 1990; J. John pers. comm. to J. B.) and those of the hypersaline Gavish Sabkha (Ehrlich and Dor 1985), many quantitatively important species are small and rapidly motile, and thus fill "residential" (e.g., Solar Lake, Ehrlich 1978) rather than structural niches. Mucilage- and stalk-producing diatoms are also reported to construct mats and carbonate stromatolites in flowing freshwater environments (Winsborough and Golubić, 1987).

6.9.5B Geothermal Eukaryotic Mats

Mats built exclusively by eukaryotic algae are common in effluents of acid hot springs (Table 6.9.2). Because cyanobacteria (and photosynthetic bacteria) cannot tolerate extreme acidity (Brock 1973), these mats are unique in composition and are useful for observing eukaryotic influences on mat biomarkers (Section 6.8.4). In such settings and over the temperature range of approximately 30° to 57°C, *Cyanidium caldarium* (and perhaps related algae; De Luca and Moretti 1983) is the principal mat-building phototroph (Ward et al. 1989a). The cohesiveness of *Cyanidium*-produced nonlaminated mats may be provided by the filamentous fungus *Dactylaria gallopava*, which is cross-fed by *Cyanidium* (Belly et al. 1973). *Cyanidium* mats resemble other mats built by oxygenic phototrophs with respect to the distribution and intensity of oxygenic photosynthesis (Revsbech and Ward 1983). At temperatures below about 31°C, the filamentous alga *Zygogonium* sp. forms nonlaminated purple-colored mat-like accumulations.

6.9.6 Terrestrial Mats

Terrestrial (or subaerial) mats are those which cover substrate that is not permanently inundated by standing or flowing water. Unless located in humid tropical areas, most terrestrial mats, therefore, undergo long periods of desiccation. Surface, particle-binding mats of cyanobacteria or of green or yellow-green eukaryotic algae are not restricted to moist regions. "Desert crusts" on soils of vegetated high and low deserts are composed largely of sheathed filamentous oscillatorian cyanobacteria that are capable of gliding motility, a feature that may be required for mat spreading (see Campbell 1979; Shields et al. 1957). Many other sheathed unicellular, colonial, and filamentous heterocystous cyanobacteria are present in "desert crusts." Sheaths usually show yellowish or other pigmentation. Growth may be limited to a few weeks or months of the year; most cyanobacterial inhabitants are rapidly reactivated upon wetting (e.g., Campbell 1979; Potts and Friedmann 1981; Potts et al. 1983). Some terrestrial mats contain cyanobacteria that appear to be morphologically identical to marine species, such as *Microcoleus chthonoplastes* (Campbell 1979) and *Chroococcus* sp. (Potts et al. 1983). Because desiccation in desert rocks may also involve elevated salinities and low water potential, this apparent commonality may be more than fortuitous. It has been suggested that adaptation of marine cyanobacteria to terrestrial habitats during the Precambrian may have been influential in the formation of paleosols (see Campbell 1979). Cyanobacteria are also well known as pioneers on juvenile fields of lava and tephra, either as free-living mats or as nitrogen-fixing components of lichens (see Englund 1978; Fritz-Sheridan 1987).

In exceptionally harsh modern environments (e.g., the Negev Desert and the dry valleys of Antarctica), extreme desiccation or wind scouring restricts "mat" formation (by cyanobacterial and algal/fungal lichen-like assemblages) to the inside of porous or semi-porous rocks. These "endolithic mats" occur as pigmented laminae one to a few millimeters below the rock surface (see Friedmann 1982; Friedmann and Ocampo-Friedmann 1984).

Extensive, thick, gelatinous, terrestrial or semi-terrestrial mats form in polar regions in melt water areas that remain moist during the short summer (Davey 1983; Davey and Marchant 1983). Consisting mainly of cyanobacteria (oscillatorians and *Nostoc* sp.), these mats naturally freeze-dry with the end of the growing season and may then disappear largely through wind disruption (W. Vincent pers. comm. to R. W. C.).

Some of the best known terrestrial "mats", composed largely of cyanobacteria, are the vertically oriented, darkly colored covers on steep faces of mountains and cliffs, usually in seeps or where rain or melt waters overflow. These mats, commonly referred to as "tintenstrichen," are composed primarily of coccoid cyanobacteria, and are well described by Jaag (1945) and by Golubić (1967).

6.10

Case Study of a Modern Microbial Mat-Building Community: the Submerged Cyanobacterial Mats of Guerrero Negro, Baja California Sur, Mexico

DAVID J. DES MARAIS ELISA D'AMILIO JACK D. FARMER
BO BARKER JØRGENSEN ANNA C. PALMISANO
BEVERLY K. PIERSON

6.10.1 Introduction

A diverse population of microbial mats has developed within a solar saltern operated by Exportadora de Sal, S. A. (ESSA), at Guerrero Negro, Baja California Sur, Mexico. The presence of mats in diverse but stable environments made this site particularly attractive for study. These mats are useful analogues for interpretation of shallow water, Proterozoic, benthic microbial communities (see Sections 6.2, 6.10.5, 6.11).

The site lies on the Pacific coast at 28°N latitude, about 700 km south of the Mexican–USA border. Annual precipitation ranges between 15 and 120 mm/yr., although in 1978 and 1983 it was 305 mm/yr and 330 mm/yr, respectively (ESSA, pers. comm. to D. J. D.). Winds are west to west-northwesterly and average 5 m/sec. Daily air temperatures range from 20° to 32°C in summer, and from 8° to 24°C in winter. Water temperatures range from 22° to 29°C in summer, and 14° to 22°C in winter. Seawater is pumped from the lagoon into Pond 1 (Fig. 6.10.1) then flows to Ponds 2, 3, 4, 5, 6, 8, etc. No groundwater enters these evaporation ponds; they lie above the water table. Runoff has minimal effect on the brines, except during wet years. This saltern is the world's largest, yielding a sodium chloride harvest that exceeds 6×10^6 tons/yr.

In the salinity range from 50 to 60‰ (Ponds 1, 2, 3, and part of 4), the ponds are dominated by a seagrass, *Ruppia sp.*, and a green alga, *Enteromorpha* (Javor 1983b). The flora also includes diatoms (*Navicula* sp., *Grammatophora* sp., *Striatella* sp., and *Licmophora* sp.) and cyanobacteria (*Entophysalis*). The well-developed permanently submerged mats, which are described in greater detail below, occur in Ponds 4 through 8, in the salinity range from 65 and 125‰. The cyanobacterium *Aphanothece halophytica* occurs in environments with salinities up to 200‰. *Dunaliella* sp., usually common in such environments, is not abundant here.

Levels of phosphate, nitrate, and ammonia are lower than typical for a saltern (Javor 1983a), perhaps due to negligible runoff and the absence of wastewater. Under these conditions of relatively low levels of biologically available nitrogen and phosphorus, benthic communities are favored over planktonic ones because they retain nutrients more efficiently (Javor 1983).

6.10.2 The Benthic Mats: Trends Along The Salinity Gradient

6.10.2A Field and Microscope Observations

Ponds 4 through 8 contain mats having a range of structures and biotic compositions (Fig. 6.10.1). This range is attributed to the salinity gradient noted above, the nature of the competing organisms (e.g., plankton, seagrass), water turbulence, sedimentation, and the saltern's seasonal management. Within Ponds 4 through 8, water depth was between 0.5 m and 0.8 m during the course of this study.

The low-salinity (southeast) end of Pond 4 (Fig. 6.10.1) had a substrate composed of homogeneous, gray-black, sulfide-rich mud overlain by 1 to 5 mm of yellowish, diatom-derived organic detritus. Dense stands of the seagrass *Ruppia maritima* covered about 70% to 85% of the surface in some areas. To the northwest, *Ruppia* became less abundant, and patches of thin microbial mats appeared; farther to the northwest, the mat thickened rapidly and became well-laminated.

The lower salinity limit for mats within Pond 4 apparently reflected a combination of hydrographic controls on sedimentation and competition with *Ruppia*. In the absence of such limitations, *M. chthonoplastes* mats can grow at salinities as low as that of normal seawater (Bauld 1984). Salinities in Pond 4 were evidently sufficiently high to exclude macroinvertebrate grazers; none were observed. Meiofaunal grazers (nematodes, copepods, cladocerans, and ciliates) and small surface grazers (water boatmen and larval coleopterans) were common, but did not influence the area distribution of the mats. *Ruppia* created an environment unfavorable for mat growth by shading the substrate, by bioturbation due to root growth and, most importantly, by baffling wave action and enhancing sedimentation of suspended particulate organic matter (POM). Mats developed at salinities above 60‰; *Ruppia* tolerates salinities from zero up to 60‰ (Den Hartog 1970; Setchell 1924).

In the northwestern end of Pond 4, waves eroded mats and thus limited their thickness. Mat fragments broke up rapidly and constituted a major source of POM in the ponds. These

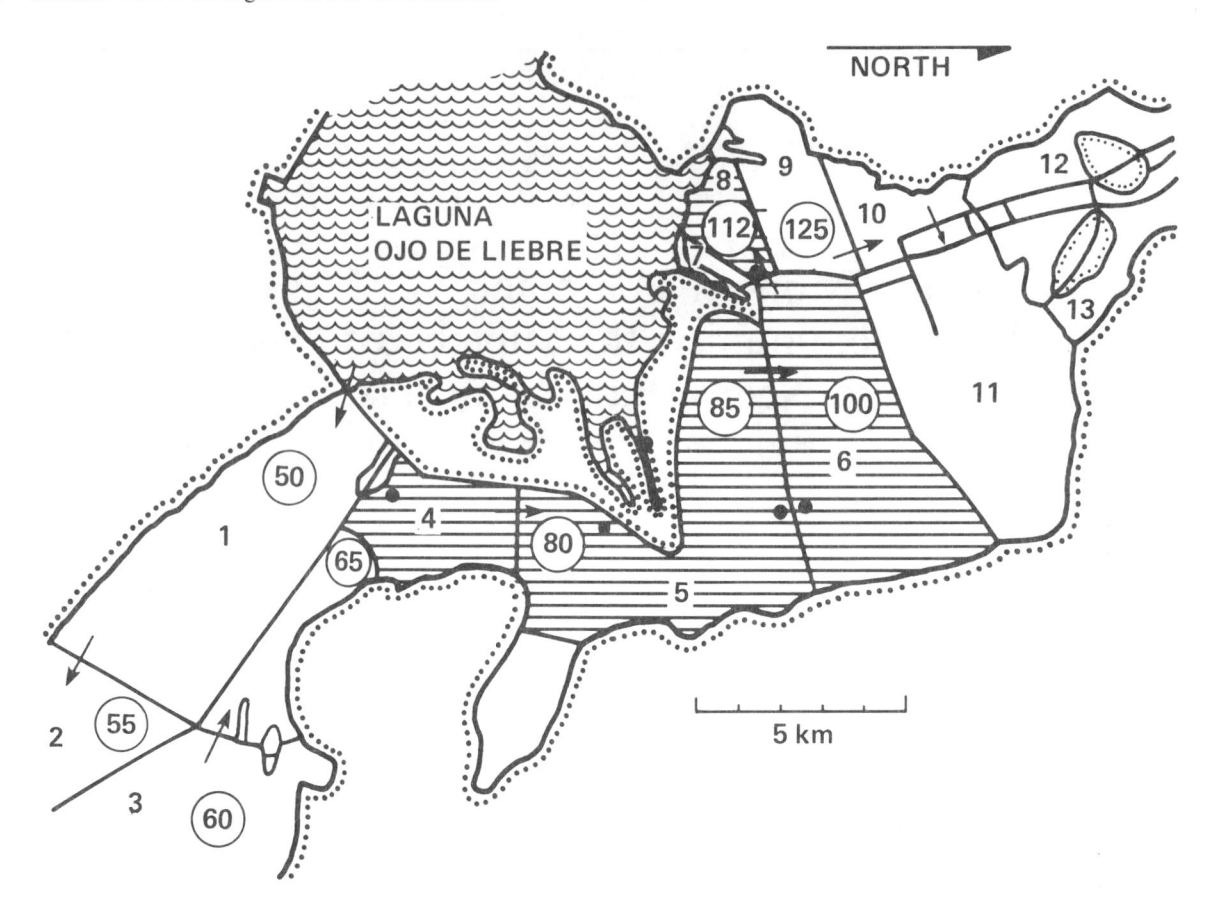

326 Modern Mat-Building Microbial Communities

Figure 6.10.1. Map of concentrator ponds of the company Exportadora de Sal, S.A., at Guerrero Negro, Baja California Sur, Mexico. Arrows indicate direction of flow of evaporating seawater. Circled numbers give average salinities (‰) for each pond. Areas in Ponds 4 through 8 which sustain mats perenially are delineated by horizontal hachures. Mat sampling sites in Ponds 4, 5 (southern and northern sites), 6, and 8 are indicated by black dots.

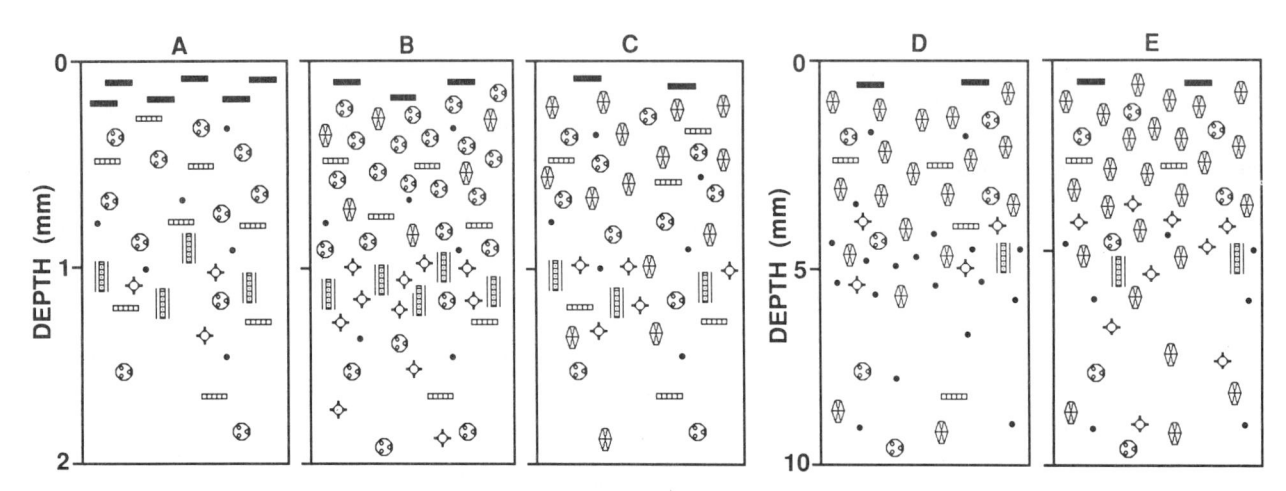

Figure 6.10.2. Schematic representation of the distribution of microorganisms with depth in the mats at several sites along the salinity gradient in the concentrating ponds at Guerrero Negro, Baja California Sur, Mexico. The relative abundances of the symbols indicate the relative densities of particular organisms. Depths (mm) into the mats are indicated on the vertical scales. Symbols represent the following – solid horizontal bar: diatoms; hachured horizontal bar: *Oscillatoria* spp.; circles with smaller interior circles: *M. chthonoplastes* bundles; hexagons with interior lines: unicellular cyanobacteria; black dots: purple sulfur bacteria; vertical bars bounded by lines: *Beggiatoa* sp.; circles with four projections: *Chloroflexus*. Capital letters designate the sites as follows: A, Pond 4; B, Pond 5, southern site; C, Pond 5, northern site; D, Pond 6; E, Pond 8. Parts A through C represent a depth range from approximately 0 to 2 mm; parts D and E represent a depth range of approximately 0 to 10 mm; depths of mat chemoclines are indicated by the presence of *Beggiatoa* sp. (vertical bars bounded by lines).

rip-up areas were renewed by lateral growth of the adjacent mat.

The uppermost 1 mm of mat in Pond 4 (Fig. 6.10.2A) was typically a loose fabric of organic detritus and diatoms. Most bundles of *M. chthonoplastes* were oriented at steep angles within this layer, suggesting migration to the mat surface due to sedimentation (Farmer and Richardson 1988). The filamentous cyanobacteria *Oscillatoria* and *Phormidium* also occurred within the upper 2 mm of the mat. Nematodes were numerous. Between 1 mm and 2 mm depth, abundant spherical cells resembling *Chromatium* sp. contained sulfur granules. *Beggiatoa alba* also occupied this depth interval during the day; it migrated upwards to the mat surface at night. Pond 4 mat showed distinct colored laminations at the mm scale, but stratification of organisms at the 100 μm scale was not nearly as well-developed as in mats growing at higher salinities. Rapid sedimentation apparently disrupted this stratification. Also, the abundant organic detritus decayed and produced sulfide, which sustained large populations of *Chromatium* and *Beggiatoa*.

Pond 5 was uniformly underlain by well-developed mats (see Frontispiece), the constituents and structure of which are depicted schematically in Figure 6.10.3. The biological composition of these mats varied locally, perhaps due to differences in sedimentation and salinity. This contrast is illustrated by comparing two mats from different sites in Pond 5 (Fig. 6.10.2). The southern site (Fig. 6.10.2B) was the closer of the two to the inlet from Pond 4; therefore, its brine had a lower salinity and a higher POC content. Mats at this site accreted at about 1 cm/yr, due to bacterial growth and organic sedimentation. These mats have been described in detail (D'Amelio et al. 1989). Diatoms (*Nitzschia* sp. and *Navicula* sp.) populated the surface. The mats were dominated by *M. chthonoplastes*, which formed flat-laminated layers. The trichomes of *M. chthonoplastes* were most concentrated at a depth between 300 and 700 μm. The cyanobacteria *Oscillatoria*, *Phormidium*, *Spirulina*, and *Aphanothece* occurred commonly within the upper 2 mm of mat. A newly discovered filamentous purple sulfur bacterium (D'Amelio et al. 1987) inhabited the uppermost 1 mm to 1.5 mm of mat, both inside and outside *M. chthonoplastes* sheaths. The chemocline, the depth interval in which measurable (by microelectrode) concentrations of oxygen and sulfide coexisted, occurred at 0.7 mm to 1.2 mm during conditions of full daylight (midday, 1000 to 2000 μE/m²/sec). This interval was dominated by *Beggiatoa* sp. and *Chloroflexus* sp. As in Pond 4 and elsewhere, *Beggiatoa* sp. followed the chemocline as it moved to the mat surface at night.

At the northern Pond 5 site (Fig. 6.10.2C), mats contained fewer diatoms and more unicellular cyanobacteria near the mat-water interface than did mats at the southern Pond 5 site. Unicellular cyanobacteria imparted a reddish brown appearance to the northern site mats. The southern site mats, where filamentous cyanobacteria dominated, were olive-green.

The differences between the two Pond 5 sites discussed above perhaps reflect lower local rates of sedimentation (0.4 cm/yr versus 1.0 cm/yr) and somewhat higher (by as much as 5 to 10‰) salinities at the northern site. At the southern site, faster sedimentation may have favored the more motile filamentous cyanobacteria, and the more abundant decaying organic detritus led to higher sulfide concentrations and larger populations of sulfide-dependent bacteria (e.g., *Beggiatoa* sp. and associated *Chloroflexus* sp.).

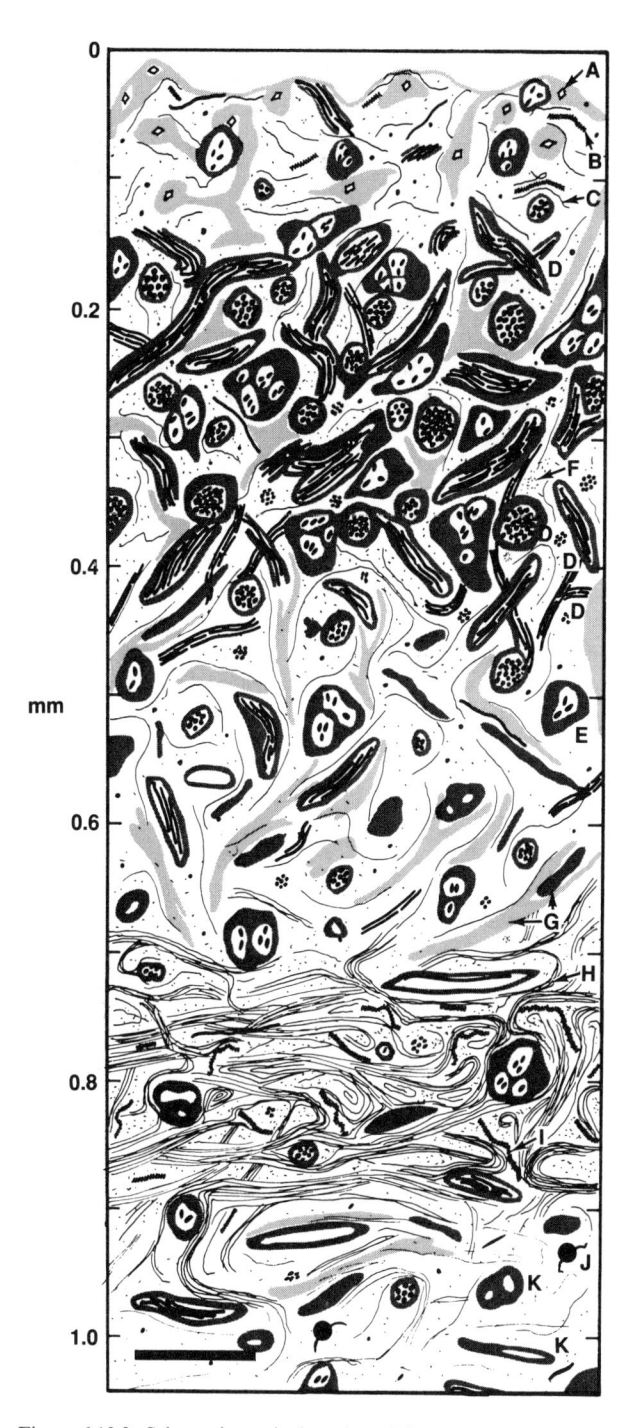

Figure 6.10.3. Schematic vertical section of the topmost 1 mm of mat in Pond 5. Horizontal bar at lower left represents 100 μm. Letters along the right margin indicate the following: A, diatoms; B, *Spirulina* spp.; C, *Oscillatoria* spp.; D, *Microcoleus chthonoplastes*; E, nonphotosynthetic bacteria; F, unicellular cyanobacteria; G, fragments of bacterial mucilage; H, *Chloroflexus* spp.; I, *Beggiatoa* spp.; J, meiofauna; K, abandoned cyanobacterial sheaths.

Mats in Pond 6 were more variable in biotic composition and more easily disrupted than those in Pond 5. Pond 6 mats were more gelatinous, reflecting a dominance of slime-producing unicellular cyanobacteria over sediment-binding filamentous forms. Where moderate to high organic sedimentation occurred (see site location in Fig. 6.10.1), the mats had a 5-mm-thick, orange-brown, partially translucent, surface layer which consisted primarily of low-density populations of coccoid (*Aphanothece*) and filamentous (*Spirulina*, *Phormidium*) cyanobacteria (Jørgensen and Des Marais 1986a) enclosed within a slime matrix. Small tufts of diatoms (*Nitzschia*) grew on the surface. *M. chthonoplastes* formed scattered but thick bundles at a depth of 3 to 5 mm. A distinctive, 0.3-mm-thick, pink band of purple sulfur bacteria (*Chromatium* sp.) occurred at a depth of 5.5 mm. *Chloroflexus* was locally abundant at a depth of several millimeters.

Mats in Pond 8 were subject to smaller salinity variations and less wave-driven rip-up and sedimentation than were Pond 6 mats. Pond 8 mats contained numerous unicellular cyanobacteria (*Synechococcus*, *Aphanothece*), although *M. chthonoplastes* and *Phormidium* were locally abundant. *Beggiatoa* occurred at the chemocline at depths ranging from 1.5 mm to several millimeters. The mats were well layered at the submillimeter scale. Purple sulfur bacteria and *Chloroflexus* were locally prominent, especially where organic sedimentation was highest.

Mat growth in Pond 9 competed with pervasive gypsum precipitation. Thin *Aphanothece* mats developed but were continually disrupted by waves. *Aphanothece* and purple sulfur bacteria formed endolithic communities within the gypsum.

6.10.2B Biogeochemical Trends

As discussed above, the biotic composition of the various mats studied showed trends that were correlative with the salinity gradient. Such trends can also be reflected by biogeochemical (biomarker) indicators of changing microbial populations. The distribution of lipophilic pigments such as carotenoids, for example, can reflect the relative abundances of the organisms producing the pigments. Interpretation of such distributions can be complicated, however, by biotic responses to environmental changes, for although a decrease in fucoxanthin content between two sites might reflect a decrease in the abundance of diatoms, an increase in such pigments (e.g., β-carotene) can occur as a physiological response to increased salinity (Ben-Amotz and Avron 1983).

Studies of pure cultures of cyanobacteria isolated from Guerrero Negro mats illustrate the effects of salinity on pigment biomarkers resulting from (*i*) differential growth of species, and (*ii*) physiological responses to salinity (Palmisano et al. 1989a). *M. chthonoplastes* and *Synechococcus* sp. were grown over a range of salinities. *M. chthonoplastes* shows an increase in the ratio of carotenoids:chlorophyll *a* (μg/μg), ranging from 0.35 to 1.02 over a salinity increase of 30 to 120‰. *Synechococcus* sp. exhibits no trend in this ratio as a function of salinity (and has a 5- to 10-fold greater carotenoid:chlorophyll *a* ratio than *M. chthonoplastes* at any given salinity). As noted above, unicellular cyanobacteria are relatively more abundant than filamentous forms at higher salinities. Thus, salinity-related changes in carotenoid biomarkers reflect both species abundance and metabolic responses (Palmisano et al. 1989a).

Light penetration into mats is affected by the relative densities of microflora as well as by their pigment contents. At higher salinities, where unicellular cyanobacteria dominate, mats are reddish-brown and exhibit a lower density of organisms. Consequently, these mats are more transparent than the dark green mats in which filamentous cyanobacteria dominate. Specific aspects of the photic zone of the mats are discussed in Section 6.4 (and illustrated in Fig. 6.4.2).

Salinity-related changes in the level of biological activity also occur (Des Marais et al. 1989). Figure 6.10.4 shows trends in pH, dissolved inorganic and organic carbon concentrations (DIC and DOC, respectively), and gross rates of oxygenic photosynthesis. In Ponds 1 through 9, values for DIC and DOC range from 2 to 4 mMolal and 0.3 to 6 mMolal, respectively. In Figure 6.10.4B and C, the concentrations of dissolved carbon species have been normalized to the halide content of the brines in order to compensate for concentration increases due to evaporation. Between the lagoon inlet and Pond 4, brine pH increases by approximately one unit whereas DIC declines substantially. These trends reflect CO_2 uptake by blooms of planktonic algae. As the brine passes through Pond 4 and continues beyond, brine pH no longer increases and DIC no longer decreases, indicating that net inorganic carbon uptake by the biota has slowed.

Trends in DOC support the observations made for DIC (Fig. 6.10.4C; Des Marais et al. 1989). A rapid DOC increase at lower salinities perhaps reflects organic carbon produced both by photoautotrophs and by decomposition of organic matter. Net addition of DOC to the brine slows considerably at Pond 4 and beyond. The small net DOC flux from the mats reflects very efficient recycling of DOC and/or a dramatic decline in the rate of DOC production.

Gross rates of oxygenic photosynthesis decline with increasing salinity (Fig. 6.10.4D). The decline in oxygen production with salinity could be caused by the increasing proportion of unicellular cyanobacteria, which form a more gelatinous, dispersed mat. Alternatively, higher salinities might slow photosynthesis in all cyanobacteria.

The $\delta^{13}C$ values both of the dissolved inorganic carbon and the mat organic matter do not vary appreciably within the salinity range where mats are well developed (Fig. 6.10.5; Des Marais et al. 1989). The relationship between carbon isotopic composition and the carbon budget was examined in mats from Pond 5, as is discussed in Section 6.8. To summarize the key points there developed: (*i*) in such mats, little isotopic discrimination accompanies the photosynthetic uptake of inorganic carbon from the overlying brine, because the supply of this carbon is diffusion-limited and the isopopic fractionation associated with such aqueous diffusion is small; and (*ii*) a small amount (approximately 5 to 7‰) of discrimination occurs at night, perhaps reflecting isotopic fractionation during inorganic carbon uptake by chemoautotrophic bacteria.

6.10.3 Trends With Depth in a Mat

The biological, chemical, and isotopic composition of a 1 m^2 area of smooth *Microcoleus chthonoplastes* mat from the Pond 5 northern site (Fig. 6.10.2C) was examined in order to understand the relationship between this benthic community and the buried organic matter. Two sets of these interdisciplinary observations were made at two scales of depth. The first

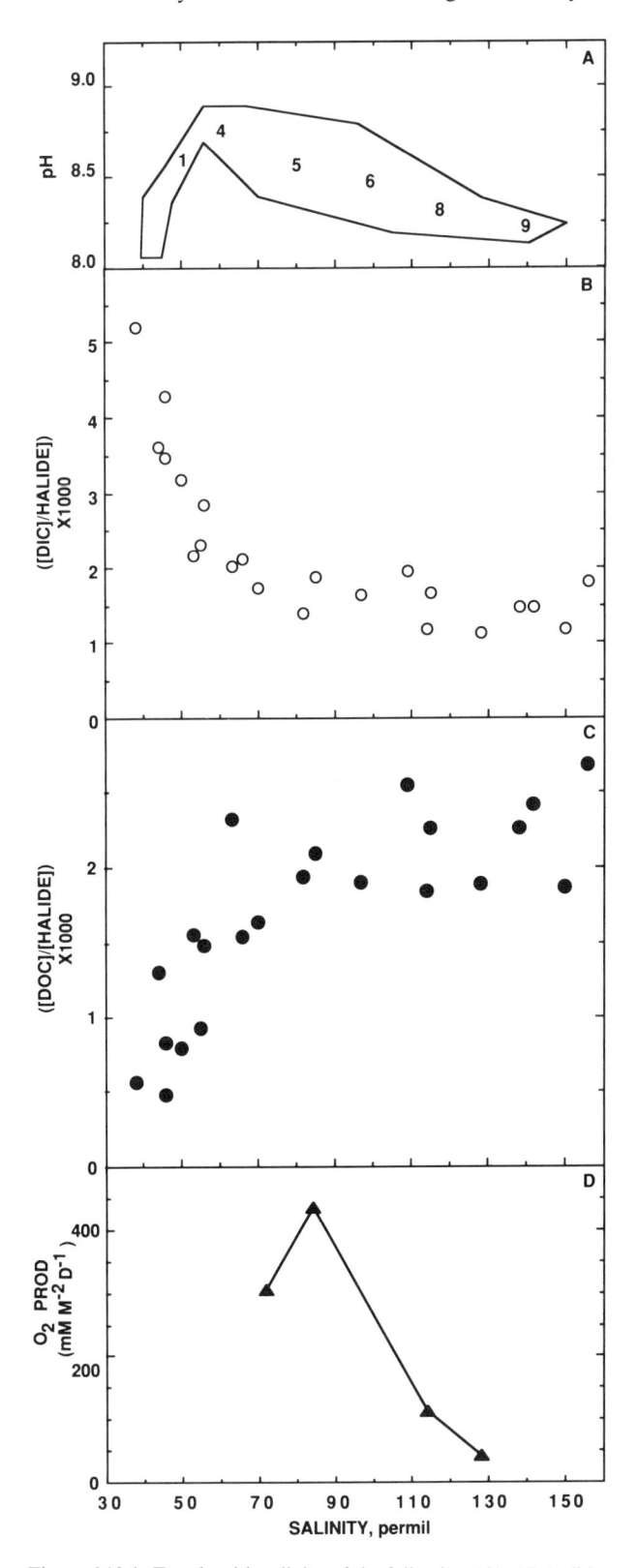

Figure 6.10.4. Trends with salinity of the following: (A) pH, (solid line shows boundary which encloses all data points; numbers indicate ponds). (B) Concentration ratios of dissolved inorganic carbon (DIC) over total halide for three sampling times (November 1985; January 1986; and June 1986). (C) Concentration ratios of dissolved organic carbon (DOC) over total halide (sampling times are same as for B). (D) Rates of oxygenic photosynthesis (filled triangles). [Data are from Des Marais et al. 1989.]

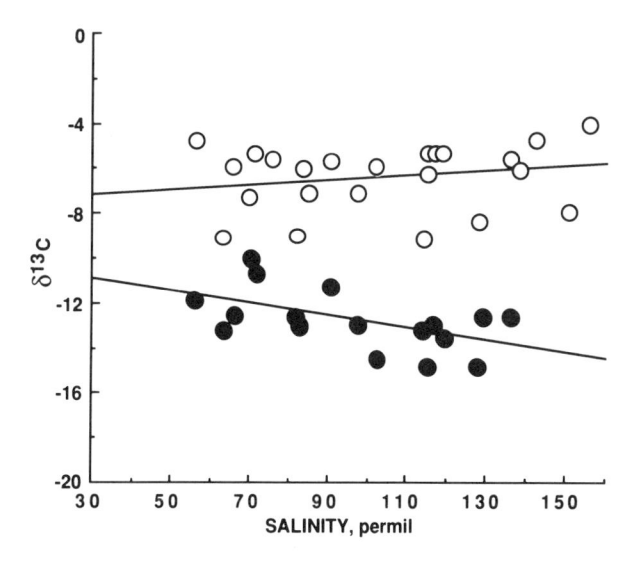

Figure 6.10.5. Carbon isotopic composition of total DIC (open circles) and total organic carbon in top 2 mm of mats (filled circles) over a range of salinities in the concentrating ponds (lines are least squares fits of the DIC and mat carbon).

set addresses the topmost 2 to 3 mm of mat and includes the communities of the photic zone, the chemocline, and the region immediately below. The second set extends to 4 cm depth and examines some of the geologically potentially preservable features of the mat.

6.10.3A The Photic Zone and Chemocline

In this mat, the photic zone and chemocline comprised a thickness of only a few millimeters, as illustrated in Figures 6.10.6A and 6.10.6D. Thinness of this zone can be attributed to a series of physical, chemical, and light-related properties, as described in Section 6.4. The particular stratification of the organisms (seed Fig. 6.10.2C) is due both to optical properties of the mat (Jørgensen and Des Marais 1988) and to chemical gradients. In the mat photic zone, visible wavelengths of light (400 to 700 nm) are absorbed by pigments more strongly than are infrared wavelengths (700 to 1100 nm; see Fig. 6.10.6A). This observation is consistent with the measured distribution both of chlorophylls (Fig. 6.10.6B) and of carotenoids (Fig. 6.10.6C). Absorption of visible wavelengths 430 nm to 530 nm is attributed (Palmisano et al. 1989a) to the carotenoids fucoxanthin (in diatoms) and myxoxanthophyll (in cyanobacteria) and to absorption by the Soret band of chlorophyll a (present both in diatoms and in cyanobacteria). Chlorophyll a also absorbs strongly at 677 nm. Infrared radiation penetrates more deeply, providing energy to green and purple sulfur bacteria. For example, bacteriochlorophyll a (e.g., present in purple sulfur bacteria and in the green bacterium *Chloroflexus*) absorb radiation at 800 nm and 866 nm. Bacteriochlorophyll c (in *Chloroflexus*) absorbs at 752 nm.

Chemical gradients also influenced stratification of the microorganisms in ways which were consistent both with the microscopic observations and with the light-related data obtained for photosynthesizing mats. Diatoms occupied the topmost layers, generating the first peak of oxygen production (Fig. 6.10.6D); they were also the source of the fucoxanthin, and

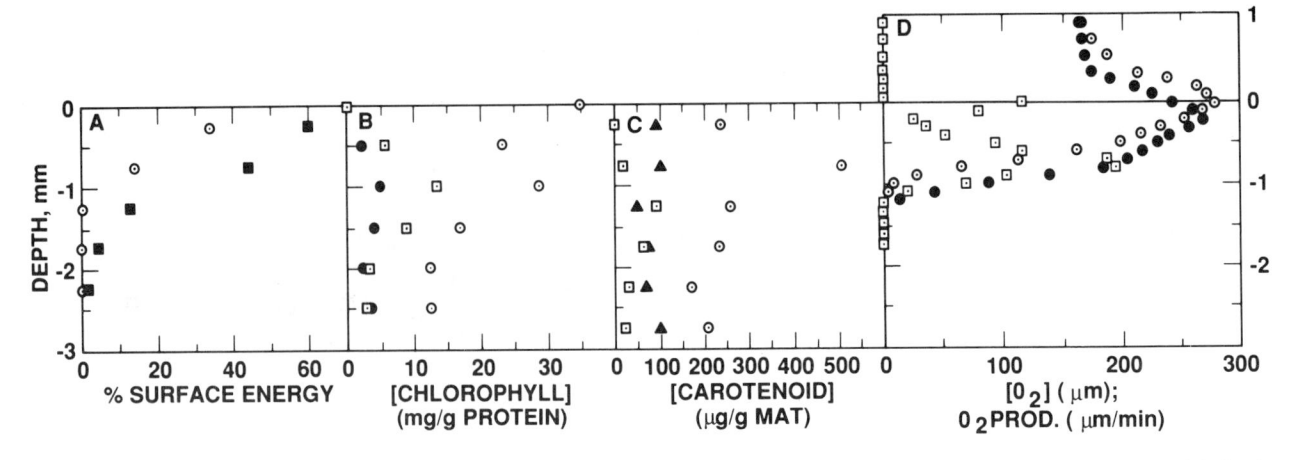

Figure 6.10.6. Light, pigment and oxygen data for Pond 5 mat at the northern site, April 1988. (A) Distribution of radiant energy with depth (circles: 400 to 700 nm; squares: 700 to 1100 nm). (B) Concentrations (mg/g protein) of chlorophylls with depth (open circles: chlorophyll a; filled circles: bacteriochlorophyll a; squares: bacteriochlorophyll c), obtained by methods described in Chapter 20. (C) Concentrations (μg/g mat) of carotenoids with depth (filled triangles: fucoxanthin; circles: myxoxanthophyll; squares: γ-carotene), obtained by method of Palmisano et al. (1989). (D) Oxygen concentrations (replicate profiles shown by filled and open circles) and gross rates of oxygenic photosynthesis (squares) as a function of depth in the mat, obtained by method of Revsbech and Jørgensen (1986).

contribute to the chlorophyll a observed near the mat surface (Fig. 6.10.6B, C). The restriction of diatoms to or near the mat surface was consistent with their inability to tolerate sulfide. The dominant photosynthetic population, situated immediately below the diatoms, consisted of cyanobacteria, and was evidenced by a large peak of oxygen production at a depth of 0.8 mm (Fig. 6.10.6D) which coincided with the highest concentrations of chlorophyll a and myxoxanthophyll detected in the mat (Figs. 6.10.6B and C). Below a depth of 0.8 mm, oxygen concentrations and oxygen production rates both dropped steeply (Fig. 6.10.6D) to the chemocline, at a depth of 1.2 mm, the thin zone where oxygen and sulfide coexist (D'Amelio et al. 1989). That *Beggiatoa* is observed here (Fig. 6.10.2C) is consistent with its requirement for both oxygen and sulfide. The presence of *Chloroflexus* at the chemocline suggests that this green bacterium utilizes both sulfide and near-infrared radiation, consistent with the requirements of some *Chloroflexus* strains found in sulfidic hot springs (Giovannoni et al. 1987a).

6.10.3B Properties of Organic Matter Buried at Depth

Organic matter deeply buried in mats such as these potentially could enter the geologic record. The nature of such preserved organic matter is determined both by the original composition of organics derived from the surface microbial community and which survive diagenesis, and by the composition of organics synthesized at depth, perhaps by photosynthetic organisms surviving anaerobically in the dark (e.g., cyanobacteria; Jørgensen and Cohen 1987) and by other bacteria (e.g., anaerobic heterotrophs and sulfate reducers; Jørgensen and Cohen 1977).

Recent studies have shown that most photosynthetic pigments survive well-below the photic zone of a *M. chthonoplastes* mat from the southern site (Fig. 6.10.2B) in Pond 5 (Palmisano et al. 1989a). This survival is illustrated by comparing pigment compositions in the mat near-surface layers (0 to 1 mm depth) with those in deeper layers (7 to 10 mm depth; Fig. 6.10.7A). Pigments in this deeper interval had been buried below the photic zone for up to 7 months. The relative concentrations of pigments reflect the relative abundances of their source organisms. The diatom-derived carotenoid fucoxanthin is detected near the mat surface, but disappears at depth due to its relatively low chemical stability. The relatively abundant occurrence of myxoxanthophyll reflects the dominance of cyanobacteria in this mat, both in the photic zone and at depth. The minimal amounts of γ-carotene within the top 0.05 cm of the mat, and the higher concentrations of this carotenoid at greater depths, correlate with the distribution of *Chloroflexus* sp.; indeed, concentrations of bacteriochlorophyll a, produced by *Chloroflexus* and purple sulfur bacteria, actually increase at depths up to 1.0 cm.

These observations are consistent with those of Edmunds and Eglinton (1984) who noted good carotenoid preservation for hundreds of years in a microbial mat from Solar Lake, Sinai. Preservation of carotenoids is enhanced by neutral pH, anoxic conditions, and darkness.

Elemental and isotopic abundances were measured in the 1 m^2 portion of mat (Pond 5, northern site) selected for interdisciplinary study. The data presented in Figure 6.10.7B and C, and in Figure 6.10.6 are from this mat. The sedimentation rate of the mat was estimated from carborundum layering experiments to be 0.5 cm/yr. Thus, mat material at a depth of 4 cm had been buried for approximately 8 years. Both elemental C/N values and δ^{13}C were remarkably constant with depth and were approximately equal to values measured at the mat surface. The C/N values closely approximate the Redfield ratio, the elemental ratio measured for organic matter in living marine plankton. Results of these chemical analyses are consistent with microscopic observations indicating that major components of the mat (viz., cyanobacterial colonies and sheath material) persist at depth and retain the biogeochemical signatures of the surface community.

6.10.3C Preservation in the Rock Record

Microbial mats in the Guerrero Negro saltern were not lithified by carbonate, silica, or other mineralic material. The organic mats attained thicknesses of 5 to 10 cm. Because the mats are known to have been growing in this area for approximately 18 years at rates in the range 0.5 to 1.4 cm/yr, simple extrapolation would suggest that they should have accumulated to thicknesses of between 9 and 25 cm, depending on the locality. However, because the mats were not lithified, they were readily disrupted by wind action, they decayed relatively rapidly during (winter) periods of low salinity, and they decomposed gradually at depth.

These mats provide a promising opportunity to examine the effects of (both biological and abiotic) diagenesis over a 20-year period in a system with excellent environmental controls and which lacks the complications attendant to lithification. The original biogeochemical features of the mats have been shown to be well-preserved several years after deposition, a fidelity of preservation evidently due to the resistance of sheath material to bacterial degradation, to the chemical environment at depth in the mat (viz., hypersalinity and the absence of light and oxygen), and perhaps also to the survival of various microorganisms at depth. If such a mat had been lithified within months or a few years of its formation, and if the lithification process did not markedly affect the chemistry of the organic components, its organic matter should be well-preserved. For example, if a mat similar in all other respects to those of Guerrero Negro were to become lithified within a period of three months, its degree of preservation could mimic that observed in a three month-old layer within a Guerrero Negro mat. Once a mat becomes lithified, its rate of organic degradation can be expected to slow, because lithification limits the exchange of pore fluids with the overlying brine. Among other things, this serves to restrict the influx of sulfate, and thus the rate of sulfate-supported, biologically induced diagenesis within the mat (Jørgensen and Cohen 1977). Further studies of diagenesis in lithifying mats are needed; both systems precipitating silica and those precipitating carbonate merit attention.

6.10.4 Comparison of Guerrero Negro and Solar Lake Mat Communities

The permanently submerged mats at Guerrero Negro are remarkably similar to those at Solar Lake, Sinai (D'Amelio et al. 1989). At the two localities, vertical distributions of organisms follow similar patterns (see Fig. 6.10.3): diatoms and unicellular cyanobacteria, commonly *Synechococcus* spp., occupy the mat surface; ensheathed bundles of *M. chthonoplastes* extend throughout almost the entire photic zone; the chemocline is populated by *Chloroflexus* spp. and *Beggiatoa spp.*; and several other cyanobacteria, purple photosynthetic bacteria, and nonphotosynthetic bacteria occur. At Guerrero Negro, the proportion of unicellular cyanobacteria increases in ponds having relatively higher salinities. At Solar Lake, relatively deep mats occurring near the halocline are also subjected to high salinities and contain relatively large populations of unicellular cyanobacteria.

Mats from these two sites are dissimilar in respects which likely reflect environmental differences. In the Solar Lake mat, the vertical zonation of organisms (e.g., of *M. chthonoplastes* and unicellular cyanobacteria) and of authigenic minerals (e.g., of carbonates and gypsum) is more pronounced over several millimeters of depth than are such features at Guerrero Negro. Similarly, seasonal salinity variations in Solar Lake exceed those in the ponds at Guerrero Negro. Such environmental variations at Solar Lake favor the development of an *M. chthonoplastes* mat in the winter (when salinity is as low as 45‰), and mineral precipitation in late summer (having salinities up to 180‰). Of the two mats, those at Solar Lake typically

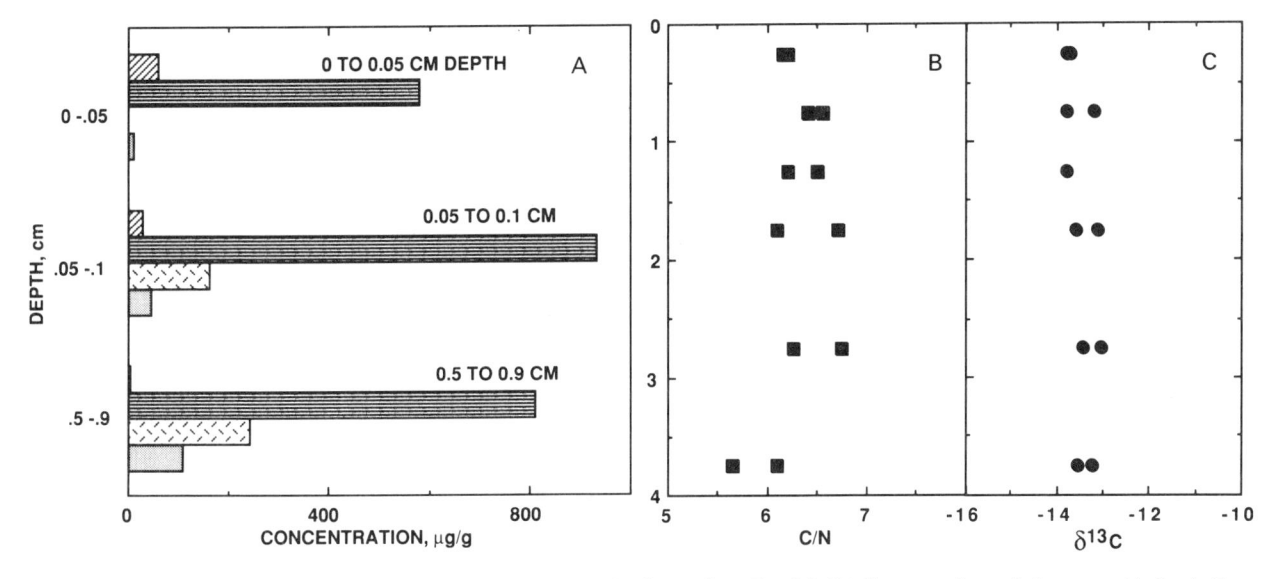

Figure 6.10.7. Chemical and carbon isotopic composition with depth of mats from Pond 5. (A) Concentrations of pigments with depth (dot-filled bars: fucoxanthin; coarse horizontal hachures: chlorophyll *a*; diagonal hachures: myxoxanthophyll; staggered lines: γ-carotene; fine horizontal hachures: bacteriochlorophyll *a*), data from southern Pond 5 site (Palmisano et al. 1989). (B) Elemental carbon to nitrogen ratio of total organic matter with depth from mat at northern Pond 5 site, April, 1988. (C) Carbon isotopic composition of total organic carbon with depth from mat at northern Pond 5 site, April, 1988, obtained by method of Des Marais et al. (1989).

have a denser fabric. Bundles of *M. chthonoplastes* are more densely packed and their sheaths are thicker. These differences likely reflect a slower rate of accretion in the Solar Lake mat, which typically ranges from 0.08 to 0.1 cm/yr in comparison with a rate of 1.0 to 1.5 cm/yr for the Guerrero Negro *M. chthonoplastes* mats (D'Amelio et al. 1989). The slower rate at Solar Lake is due both to the much higher summer salinities, which slow microbial growth, and to smaller inputs of organic detritus. Thus, upward migration of organisms in the Solar Lake mat is not as rapid as that observed at Guerrero Negro, and the cyanobacteria can develop denser fabrics and thicker sheaths.

The similarities of the mats at these localities suggest that their biogeochemistry should be similar. Indeed, depth profiles of oxygen, sulfide, and of oxygenic photosynthesis for *M. chthonoplastes* mats from both localities are virtually identical (D'Amelio et al. 1989). *M. chthonoplastes* mats from other coastal marine environments are also similar (see Section 6.3), indicating that this mat type is widely distributed and seems a likely modern analogue of at least some ancient mats preserved in the Proterozoic rock record as stratiform stromatolites.

6.10.5 Characteristics of Stromatolites Predicted from Studies of Microbial Mats

Although, unlike stromatolites, the microbial mats in the saltern at Guerrero Negro are not lithified, they exhibit trends with sedimentation rate and silinity, and offer insights about the processes of burial and preservation of organic matter, that are relevant to interpretation of their fossil stromatolitic counterparts.

6.10.5A Sedimentation Rate

At Guerrero Negro, the influx of detritus has both physical and chemical effects on the accreting mats. Physically, the microflora compete with incoming detritus in order to occupy the sediment surface at a density sufficient to maintain a coherent mat. Increasing sedimentation rates were observed to produce several responses in the Guerrero Negro mats. Most notably, populations of filamentous cyanobacteria increased, relative to those of unicellular cyanobacteria, presumably because of the selective advantage of gliding motility; moreover, as their populations increased, the filaments imparted increased coherence to the mat surface. Also correlative with increases in sedimentation rates, filaments tend to be oriented more perpendicular to the growth surface (Farmer and Richardson 1988), due to phototaxis and the required increased rate of vertical migration, and cyanobacterial sheaths tended to be thinner, because the trichomes occupied a given depth for a shorter interval of time and therefore deposited less sheath. If a mat surface was buried catastrophically, the trichomes did not migrate upward en masse; they retained a horizontal orientation. With rapid sedimentation, the mat became less stable against physical disruption, because the network of filaments and sheaths which bind the mat became more dispersed by the detritus.

Increases in the rate of organic sedimentation were observed to influence mats chemically by resulting in an increased rate of organic decomposition relative to that of oxygenic photosynthesis. The enhanced decomposition fueled higher rates of bacterial sulfate reduction, resulting in increased sulfide production. Sulfide-dependent microorganisms such as *Beggiatoa* sp., anoxygenic photosynthetic bacteria, and cyanobacteria of the genus *Oscillatoria* tended to become more abundant relative to sheath-forming cyanobacteria. The relative abundance of non-ensheathed cyanobacteria increased, and the mat became less stable against physical disruption.

6.10.5B Salinity

The abundance of unicellular cyanobacteria in Guerrero Negro mats was observed to increase with increasing salinity, relative to filamentous forms, and the resulting more gelatinous mats of the unicellular cyanobacteria were more easily disrupted. Such an increase in unicellular cyanobacteria is consistent with observations made in stratified solar ponds (Dor and Paz 1989). As salinity increased and the mats at Guerrero Negro became progressively more translucent, the photic zone of these mats deepened, enlarging zones for anoxygenic photosynthetic bacteria. Rates of gross photosynthesis declined. However, the net rate of mat accretion did not change with salinity in the range 80 to 120‰. Salinity variations exert no major effect on the magnitude of carbon isotopic discrimination during mat growth (Des Marais et al. 1989).

6.10.5C Burial and Preservation

Although these mats did not form lithified stromatolites, their study suggests that the abundance of photosynthetic microorganisms preserved in Proterozoic stromatolites (Section 5.4) is controlled by several factors, including the initial abundance of such organisms in the surface community and their ability to survive at depth. For example, with increasing depth in the Guerrero Negro mats, populations of ensheathed cyanobacteria sustained themselves better than did non-ensheathed taxa and other microbial components having cell diameters larger than 1 μm. Rapid burial resulted in surficial communities being buried beneath the photic zone but with their gross structure being essentially unaltered.

Numerous chemical and isotopic features of buried surface mat communities were also observed to persist at depth. Such characteristics could be preserved in a stromatolite. The elemental C/N and $\delta^{13}C$ values of mat organic matter were observed to not change greatly with depth, perhaps due to the occurrence both of highly preservable, abundant cyanobacterial sheath material and of surviving cyanobacteria. Thus, that organic matter which survives early diagenesis and lithification, well records the elemental and isotopic characteristics of the original photosynthetic community. The photosynthetic pigments reflect both the relative abundances of organisms and the survivability of these pigments in the subsurface. For example, in the Guerrero Negro mats, cyanobacterial pigments predominate. Carotenoids evidently survive better than chlorophylls, but diatom carotenoids are relatively unstable (Palmisano et al. 1989a).

6.10.5D The Origin of Stromatolitic Laminae

The characteristic laminated fabric of stromatolites can

be formed by, and records chemical aspects of, cyclic variations in the accretion of the mat. Such variations include changes in sedimentation rate (whether total or organic) and those due to tactic responses of the mat-forming microbiota (Sections 6.3, 6.4; Monty 1976; Gerdes and Krumbein 1987). Laminations can also reflect changes in the relative abundance of components of the microflora due to changes of season, salinity, water depth, etc. (Monty 1976; Krumbein et al. 1977). Periodic water turbulence and mat disruption can also be important. At Guerrero Negro, mat lamination reflects most strongly the disruption and sedimentation due to wind. Periodic fluctuations in salinity, although minor relative to those occurring in less controlled environments, were also observed to influence rates of mat decomposition and dispersal.

6.11

A General Comparison of Microbial Mats and Microbial Stromatolites: Bridging the Gap Between the Modern and the Fossil

MALCOLM R. WALTER   JOHN BAULD   DAVID J. DES MARAIS
J. WILLIAM SCHOPF

6.11.1 Introduction

Ever since the work of Walcott, Black, Mawson, and Pia it has been assumed that extant microbial mats are analogous to ancient stromatolites (Glaessner 1972b). For just as long, however, it has been known that stromatolites were far more abundant in the past than they are at present, and that they were particularly abundant during the Proterozoic. It is widely stated that during the Proterozoic, stromatolites were also morphologically much more diverse than they are at present. This may not be correct; the relative rarity of extant stromatolite-building microbial mats makes it difficult to gain a comprehensive view of their morphological diversity, particularly because their morphology is of little interest to most microbiologists and, thus, is often not thoroughly described. To date, no comprehensive survey exists of the morphological diversity of extant mats that would allow systematic comparison with the myriad of described fossil stromatolites. However, all of the most basic morphological categories of stromatolites that are known in the Proterozoic are also represented among extant mats: columnar stromatolites, widespread in Proterozoic terranes, are morphologically comparable to the subtidal bioherms of the Caribbean and of Shark Bay, Western Australia, as well as numerous modern lacustrine lithified mat deposits; stromatolites of the distinctive conical *Conophyton* group are comparable to much smaller forms known from extant hot springs; various types of Proterozoic stratiform stromatolites closely resemble flat-laminated mats of modern intertidal zones; and domical fossil stromatolites, and domical modern mats, are both well-known. Most of the microfossils in stromatolites are remarkably similar in morphology to modern mat-building taxa; where stromatolitic microfossils are preserved, they rarely contain any surprises for microbiologists familiar with the microorganisms of extant mat systems (see Section 5.4 and Chapter 24). In this regard, however, it is worth emphasizing that although the stromatolite record demonstrates that microbial mats covered vast tracts of the marine and lacustrine realms during Proterozoic times, most stromatolites are devoid of cellularly preserved microfossils; indeed, cellular preservation is of rare occurrence and seems possibly biased toward those fossil mats that lived in hypersaline environments, where rates of microbial degradation are minimal.

6.11.2 Analogies and Homologies Between Extant Microbial Mats and Ancient Stromatolites

For the reasons outlined above, it seems plausible to consider that many extant microbial mats are at least analogous, and probably homologous, with ancient stromatolites. However, this is a comparison that is sometimes used very superficially: in particular, because the most conspicuous and readily accessible microbial mats are both cyanobacterial and intertidal, it has been widely assumed that stromatolites must similarly be cyanobacterial and intertidal. Over the past 20 years, however, the intertidal myth has largely been put to rest, chiefly as a result of the description of extant, lithified, subtidal and lacustrine mats, and of numerous examples of ancient stromatolitic reefs and banks which certainly formed under entirely subaqueous, rather than intertidal and intermittently exposed, conditions (with many of these examples having been known for decades, but having been largely ignored in western literature because of the dominating influence of studies of Holocene intertidal mats).

The cyanobacterial interpretation of ancient stromatolites is more recalcitrant, and for good reason: evidence indicates that cyanobacteria played a significant role in the formation of many, and probably even most, Proterozoic stromatolites (Sections 5.4, 6.2.2). However, the common assumption that *all* such stromatolites are of cyanobacterial origin is challenged by the following observations: (*i*) eukaryotic micro-algae appear to have been extant at least as early as 1.7 Ga (Sections 5.4, 5.5) with some micro-algae, especially diatoms, being prominent in the mats constructing extant subtidal bioherms (Section 6.9.5) that closely resemble fossil stromatolites, especially thrombolites; and (*ii*) modern mats are constructed not only by cyanobacterial oxygenic phototrophs, but also by chemotrophs and anoxygenic phototrophs (Sections 6.9.2, 6.9.3), and an appreciable fraction of known stromatolitic Proterozoic microfossils are just as similar morphologically to extant chemotrophs and anaerobic phototrophs as they are to modern cyanobacteria (Section 5.4). Section 6.9 well illustrates the known diversity of extant mats; morphometric data for Proterozoic stromatolitic microfossils are presented in Sections 5.4 and 5.5; and fossil microorganisms from many Proterozoic stromatolitic assemblages are illustrated in Chapter 24.

The view that all fossil stromatolites are cyanobacterial has been a useful simplifying assumption, but it is almost certainly wrong. Stromatolites younger than 1.7 Ga would have been built by mats that contained eukaryotic micro-algae (at least as planktonically derived, allochthonous components, such as the "meso-" and "megasphaeromorphs" known to occur in a number of Middle and Late Proterozoic stromatolitic assemblages; Section 5.5, Chapter 22), and some ancient stromatolites may have been dominantly algal in origin (e.g., possibly some Cambro-Ordovician thrombolites). Moreover, stromatolites built by anoxygenic phototrophs may well have formed in Proterozoic anoxic settings, such as areas of active bacterial sulfate reduction, and were presumably even more widespread during the Archean, prior to the development of a stable oxic atmosphere and of widespread oxic conditions within the photic zone at and near the sediment-water interface (see Sections 1.5, 4.6). Ancient equivalents of the extant mats formed by sulfide-dependent prokaryotes (e.g., *Chloroflexus*, *Chlorobium*, *Chromatium*; Section 6.9.2) and, possibly, mats formed by sulfur-oxidizing bacteria such as *Thioploca* and *Begiatoa*, may have been locally abundant and were possibly widespread during the Proterozoic. Evidence for these should be sought, for example, in mat-stabilized, fine-grained clastic rocks, such as those represented by some, organic-rich, black or gray Proterozoic shales. As yet there are no means of identifying unequivocally stromatolites of these types, although biomarker techniques offer some promise. However, such mats exist now and must be expected to have occurred in the past. The assumption that *all* stromatolites are necessarily cyanobacterial inhibits appreciation of the probability that a diverse range of mat types occurred in the Proterozoic.

The foregoing comments are intended as a caveat regarding the blanket usage of oversimplified analogies. In this vein, it is of interest to note that although there are now hundreds of examples of the successful use of conceptual models derived from studies of the bioherms and mats of Shark Bay, the Persian Gulf, the Bahamas, and elsewhere as applied to the interpretation of ancient stromatolitic sequences, even here the analogies may be less than perfect. In particular, Awramik and Riding (1988) have suggested that the commonly made comparison between the subtidal lithified bioherms of Shark Bay and the Caribbean, for example, and the columnar and domical stromatolites of the Proterozoic, might not be valid because of differences of fabric resulting from the abundance of bioclastic debris in Shark Bay and the involvement of diatoms in the formation of the modern structures (Section 6.9.5A). It is certainly true that in several respects these extant bioherms more closely resemble Paleozoic thrombolites than they do older stromatolites, but their comparison with older structures, if done judiciously, is still instructive.

The abundance and patterns of temporal distribution of Proterozoic stromatolites documented in Sections 6.2 and 10.7 indicate that these structures have potential to yield a wealth of paleobiological and paleoenvironmental information. As discussed in Sections 6.2.5 and 6.7.4, the Late Proterozoic decline in the abundance and diversity of stromatolites has been interpreted speculatively in several ways; among these, the decline has been explained in terms of the evolution of the constructing organisms, a decrease in the abundance of microbial communities resulting from major changes in sea level or of large-scale climatic change, the competition for nutrients and space between Late Proterozoic mat communities and early evolving macroscopic algae, and the disruption of microbial mats due to grazing and burrowing by Late Proterozoic metazoans. The ecological studies of mat communities in Baja California Sur, Mexico, discussed above (Section 6.10), have provided new insights that bear on these interpretations. In particular, as is discussed at some length in Section 6.7.4, microfaunal studies of the Baja mats suggest that factors other than metazoan grazing are likely to have been responsible for the decline. Grazers smaller than a few millimeters in size do not prevent development of extant mats (Section 6.7.4). They can, however, affect mat fabric and thus could, for example, have caused a decline of specific fabric types, such as that of coniform (e.g., *Conophyton*) stromatolites (see Section 6.2.5). On balance, however, it does seem that other factors, such as the evolution of algal metaphytes or the occurrence of major changes in the Late Proterozoic environment, are likely to have been more important than metazoan grazing in accounting for the pre-Vendian stromatolite decline.

The use in recent years of microelectrode and comparably refined techniques has provided a vast amount of new information on the structure and function of microbial mats (e.g., Section 6.4). In addition to the numerous complexities and variations thus revealed, it is significant that such studies establish that the range of organization exhibited by the majority of modern mats can be characterized by a relatively small number of frequently occurring patterns such as those shown in Figure 6.10.2. Most commonly, at the upper surface of the mat a millimeter or two of oxygenic phototrophs overlies a zone containing anoxygenic phototrophs (such as *Chloroflexus* and purple sulfur bacteria) and chemotrophs (e.g., *Beggiatoa*), below which are such anaerobes as sulfate-reducers, methanogens, and other, predominantly heterotrophic, prokaryotes. These microlayered ecosystems accrete to produce laminated sediments, the laminae of which result from various combinations of biological and environmental processes (Sections 6.3.2, 6.10,5D). Four principal types of laminar fabric are described above (Section 6.3.2; Farmer and Richardson 1988) from mats constructed primarily by the cyanobacterium *Microcoleus chthonoplastes* in Baja California Sur, Mexico. Of these, the "homogenized" fabric resulting from microbioturbation has not yet been recognized in fossil stromatolites. However, fossil examples of microlaminated fabrics with vertically and horizontally oriented filaments (the second and third fabric types described from Baja) are known from Proterozoic stromatolites. As noted in Section 6.2.2, for example, cellular trichomes and abandoned sheaths in numerous stratiform and domical Proterozoic stromatolites commonly are parallel or subparallel to the stromatolitic lamination. The fourth fabric type described from Baja, also represented in the fossil record, is a derivative of the former types in which mineral precipitation has apparently been induced as a result of microbial decomposition of mat organic matter.

Although scores of examples of cellularly preserved Proterozoic mat communities are now known (virtually all of which have been preserved by permineralization in fine-grained silica; Section 5.4, Chapter 22), such preservation was a rare event, as is demonstrated by the vastly greater abundance in Proterozoic strata of carbonate stromatolites that are entirely devoid of microfossils. When microorganisms were preserved, however, preservation was commonly accompanied by loss of a significant assortment of useful taxonomic characters (see Section 5.4). Studies of preservation in extant mats (Sections 5.3.2, 6.10.5) show that it also is selective, and that whole populations of cells can be lost. For example, as described in Section 6.9.4, in hot springs there are mats constructed by the unicellular cyanobacterium *Synechococcus* and the phototrophic bacterium *Chloroflexus*; decomposition within the uppermost 5 mm of the mats destroys all of the *Synechococcus* but leave remnants of the *Chloroflexus*. The sheaths of cyanobacteria and of other prokaryotes seem to be particularly resistant to decomposition; they thus form a major component of both extant and fossil mats (Section 5.4). In addition, as mats accrete, the finely structured, layered microbial ecosystems described above migrate upward; previously separate laminae can become superimposed as the upwardly migrating living community intermingles with remnants (abandoned sheaths, moribund cells, etc.) of the previous community. Catastrophically buried laminae are at least partly exempted from this process; comparison of such laminae with those in accreting mat systems may therefore be expected to provide useful insight into the dynamics of mat communities, a possibility that has yet to be exploited in the study of fossil mats.

6.11.3 General Geochemical Comparisons

Recent biogeochemical studies, both of microbial mats (see Sections 6.8 and 6.10) and of stromatolites (Chapter 3), generally support the earlier observations of McKirdy (1976). The preservable features of mat and stromatolite organic matter which warrant comment and comparison include their abundance, their elemental and isotopic compositions, and their solvent-extractable constituents.

In Proterozoic stromatolites, the abundance of organic carbon (viz., that in kerogen plus extractable compounds) typically is low. All but a very few of the stromatolites analyzed in the studies summarized in Chapter 3 contain less than 1 mg/gC, the average TOC (total organic carbon) value being between 0.3 and 0.6 mg/gC (see data for carbonate stromatolites, Table 3.2.4). Values typical of modern microbial mats (on a wet weight basis, free of inorganic detritus) are decidedly higher, generally between 10 and 20 mg/g C (Des Marais et al. 1989). Such processes as sedimentation, compaction, diagenesis, lithification, and thermal maturation have evidently affected the organic carbon content of fossil stromatolites.

Relative to non-stromatolitic kerogens, both the kerogen of stromatolites and the organic matter of modern mats have high contents of organic oxygen (Fig. 6.11.1). This high oxygen content presumably reflects the abundance of heteropolysaccharide-rich sheaths in the mat- and stromatolite-

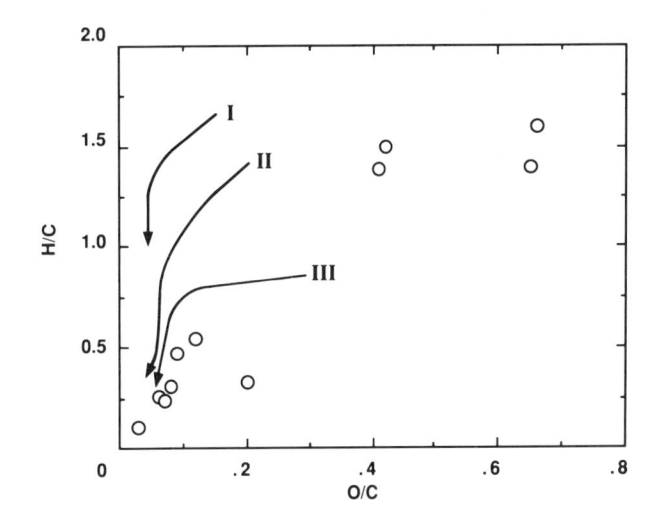

Figure 6.11.1. Elemental O/C versus H/C for organic matter from cyanobacterial mats (data points at upper right; Aizenshtat et al. 1984) and for kerogen from Proterozoic stromatolites (data points at lower left; McKirdy et al. 1980). Lines I, II, and III represent "van Krevelen plots" (see Tissot and Welte 1984) illustrating the typical changes that occur during thermal maturation of the three types of kerogen common in Phanerozoic sediments (viz., Type I = "algal"; Type II = of mixed origin; Type III = "coaly," derived from vascular plants). Because vascular plants did not exist during the Proterozoic, Type III kerogens should not occur in Proterozoic sediments.

forming microbial communities (see Section 6.8.2), thus supporting analogies both between the modern and ancient communities and between the layered, megascopic structures in which they occur.

The diverse, extractable, biomarker compounds in modern mats (Section 6.8.4) are demonstrably derived from, and thus indicative of the presence of, particular components of the mat-building communities. Study of such compounds in Proterozoic stromatolites can therefore be expected to be fruitful. Unfortunately, the very low abundance of extractable organic matter in virtually all stromatolites thus far examined (Chapter 3; McKirdy 1976), and in silicified stromatolites in particular (the facies in which exceptionally well-preserved cellular microfossils commonly occur), makes the possibility of postdepositional contamination such a serious concern that it has effectively precluded any extensive research to date. In well-preserved stromatolites, where the extractable compounds are most likely to be syngenetic, *n*-alkanes range from C15 to C23, with C18 hydrocarbons being most abundant (McKirdy 1976). Such a pattern is typical of organic matter derived from extant cyanobacteria (e.g., Winters et al. 1969; Gelpi 1970). As noted in Section 3.3, certain methyl alkanes are perhaps even more diagnostic of cyanobacteria; a search for these particular alkanes and for other key biomarkers in well-preserved stromatolites is warranted.

The δ^{13}C values of organic matter from modern microbial mats are markedly higher (i.e., the organic matter is less enriched in the lighter isotope, ^{12}C) than are values of Proterozoic stromatolitic kerogens (Fig. 6.11.2). This contrast indicates that change of some sort has occurred, even though benthic prokaryotic photosynthetic communities appear to

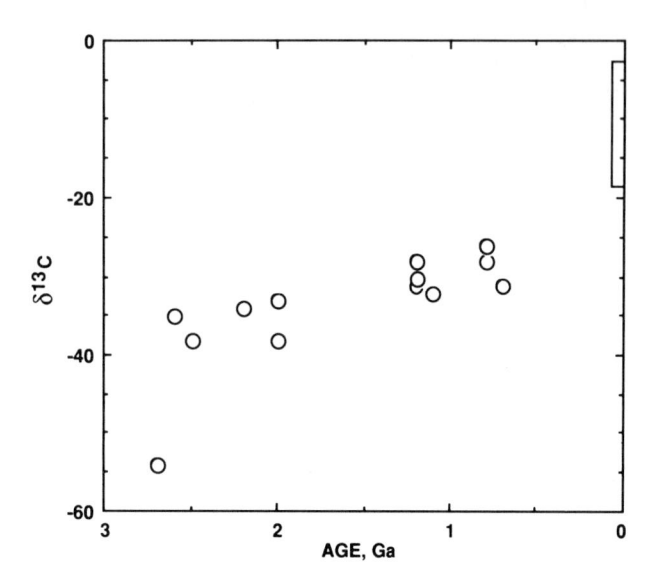

Figure 6.11.2. Carbon isotopic composition ($\delta^{13}C_{org}$ vs. PDB, ‰) of stromatolite-derived kerogens (open circles; data from section 3.4, plotted only for kerogens with H/C > 0.14) and marine microbial mat-derived organic matter (box, upper right; data from Section 6.8) versus age.

have evolved little or not at all over Proterozoic time (Section 5.4 and Chapter 13). This isotope trend apparently represents a continuation of the trend of increasing $\delta^{13}C$ characteristic of well-preserved Proterozoic organic matter (Fig. 6.11.2, Section 3.4). It is conceivable that the organic matter of stromatolites has recorded a global decline in the inorganic carbon content of the ocean-atmosphere system (see Sections 3.4, 4.7, 6.8). If a long-term, global, environmental change such as this in fact occurred, it should perhaps be most clearly interpretable from evidence sequestered in the preserved organic constituents of microbial communities of well-established biologic composition, members of which have exhibited exceptionally slow (viz., "hypobradytelic"; Schopf 1987) morphological and presumably biochemical/physiological evolution (Section 13.3.2).

6.11.4 Closing Remarks

The question of what biological information can be obtained from stromatolites has been discussed in some detail by Walter (1983). They, together with stromatolitic microfossils, are a potential source of information on paleoecology, cell sizes and orientations, cell division patterns, microbial phylogeny and physiology, pathways of carbon-flow, tropic responses (particularly phototropism), and ecosystem evolution. As discussed above (Section 6.11.3), the younger and better preserved Proterozoic stromatolites are also a potential source of biomarkers that could eventually provide significant taxonomic information regarding the biotic composition of Proterozoic stromatolitic communities. The patterns of variation in the abundance and diversity of stromatolites through time no doubt encode significant biological and environmental information, but interpretations of these patterns are at present speculative. It is clear that microbial mats covered vast areas in a wide range of shallow-water, photic environments during the Proterozoic. The majority, but certainly not all, were built by cyanobacterium-dominated biocoenoses. Future studies of stromatolites and of fossil stromatolitic microbiotas, if based on firm understanding of the morphology and ecology of modern mat communities and, especially, an appreciation of the impressive range of physiologies, biotic compositions, and environments represented among such communities, hold promise for providing greatly improved understanding of the Proterozoic biosphere and its development through time.

6.12

Unsolved Problems and Future Research Directions

BEVERLY K. PIERSON

The success of any attempt to answer questions of early evolution can be measured in its development of new (or substantiation of old) explanations for observed phenomena (see Section 6.13) and in its identification of new questions and clarification of unsolved problems for new directions of future research. This section briefly identifies some of the specific questions and gaps in knowledge that have emerged from the foregoing discussion that we believe are worth pursuing in future research.

Studies on microbial mats such as those reported here reveal that their contemporary diversity is tremendous, but that most study has focused on relatively few types of mats, primarily those of the marine and hypersaline marine environments. To enhance the value of knowledge of modern mats to the interpretation of ancient stromatolites, more detailed morphologic study is needed on the modern systems. Detailed studies of mat fabrics are often ignored by microbial ecologists. Particularly useful would be microscopic studies of histological thin sections of embedded mat from a large variety of habitats to determine fabric and changing orientations of microorganisms. Further detailed descriptions of more obvious meso- and megascopic structures are also needed. Studies of sections of various mats in different stages of diagenesis would also be useful. Such studies would enable construction of an expanded catalogue of mat morphologies more representative of the diversity of modern mats.

Parallel to this gap in knowledge of modern mats is the lack of adequate examples of preserved mat morphologies in Proterozoic stromatolites that reflect ancient mats constructed by primarily anoxygenic phototrophs or by chemotrophic bacteria. Such mats exist today and their counterparts were surely abundant in the Proterozoic. Because detail of community structure is lacking in most stromatolites, search for those formed after catastrophic burial might provide a more reliable record of original community structure due to their better preservation. Relatively little is known about the effects of the process of lithification on the various components of diverse mat communities. Because most study of lithified modern mats has been on thrombolitic bioherms, more study is needed on stratified prokaryotic mats that are in the process of lithification. All of the above studies taken together would strengthen the use of modern mats as structural analogues to the mat communities contributing to Proterozoic stromatolites.

One of the major emphases of microbiologists in their studies of modern non-lithifying mats is the identification of community members and their activities within the distinct layers of laminated mats. Certain physiologically distinct taxa of prokaryotes contribute to different layers of mats. The unanswered questions are: What microorganisms were responsible for laminate in Proterozoic stromatolites? Was the vertical distribution of microorganisms the same in the Proterozoic as it is today? There is little indication of what stratified communities existed in Proterozoic mats. The best evidence to determine if parallels for such modern stratification existed in Proterozoic mats may be from mats preserved after catastrophic burial. Although evidence of layering of physiologically distinct communities may not have been well-preserved in stromatolites, careful examination with this question in mind would be worth pursuing.

Except for exceptionally well-preserved cyanobacterial colonies, trichomes, and sheaths in stromatolites (Section 5.4, Chapter 24), we have relatively little evidence for the identification of other organisms that lived within them. How diverse and complex were the Proterozoic mat microbiota? In the absence of structural preservation, biochemical fossils may provide insight. The use of biomarkers to identify community members is rapidly expanding in the study of modern mats but is still limited by the lack of a large enough database. There is a need to expand our catalogue of specific biomarkers for distinct taxa, or at least physiologically distinct groups, of microbes. Furthermore, it is essential to identify biomarkers that would have left a preservable chemical signal in the fossil record. Potentially fruitful classes of molecules to study for their significance as biomarkers are the complex lipids and hopanoids. Although problems of level of detection and syngenicity plague the use of biomarkers to identify community members from Proterozoic stromatolites, efforts to improve resolution in this direction hold promise. As difficult as it is, this approach may hold the best promise for identifying the community composition of ancient mats.

It is possible that the application of stable carbon isotope analysis to organic matter preserved in stromatolites could also provide information on the metabolic origin of such carbon. There are significant gaps, however, in the database of $\delta^{13}C$ analyses in modern mats. Such analyses have just begun on the rarer but very diverse mats constructed by anoxygenic phototrophs and chemotrophic communities in thermal environments. Also needed are such data from modern calcifying mats.

While much progress has been made in recent years in the study of basic physiological activities in modern mats, several areas in need of further research can be identified. In comparison to well-studied photosynthetic activities, little is known about nitrogen cycling in modern mats. Very little is known about the activities of chemolithotrophs (many of which are autotrophs) in modern mats and how significant their metabolism is to the overall turnover of carbon and oxygen in these communities. Relatively little is known about the role of specific chemo- and photoheterotrophs in modern mat environments. Even though much more is known about photosynthesis, few studies have been done to estimate primary productivity in diverse mat environments or to distinguish among the relative contributions of anoxygenic and oxygenic photoautotrophs and chemoautotrophs to mat primary production.

Motility of diverse microorganisms within modern microbial mats is vigorous and in response to changing environmental cues. Oxygenic and anoxygenic phototrophs, as well as chemotrophs, respond with vertical migrations to diel changes in light, oxygen, sulfide, and probably many other factors. Clarification of the cues for these movements in modern mats may better aid our interpretation of orientations preserved in stromatolites. Motility studies are becoming more sophisticated and are revealing much greater complexity than was previously thought. Migrations of phototrophic organisms are influenced not only by light, but also by oxygen and sulfide, making interpretation of evidence of past taxis very difficult.

Finally, what caused the decline of stromatolites in the Late Proterozoic? This question remains unanswered. The significance of competition from emerging metaphytes appears to have been strengthened, however, and the search for more evidence of such competition in both modern mat environments and in Late Proterozoic environments should continue.

6.13

Summary and Conclusions: The Current Status of Studies of Modern Microbial Mat-Building Communities and Their Relevance to Interpretation of Proterozoic Stromatolites

BEVERLY K. PIERSON

Modern microbial mats are geographically widely distributed and surprisingly abundant, particularly in marine marginal habitats. Modern microbial mats are also biologically exceptionally diverse, a fact that may be overlooked because of the relative inaccessibility and very limited distribution of many types of mats. Furthermore, while significant diversity is seen in morphological characteristics of modern mats expressed at the micro-, meso-, and megascopic levels, this diversity is greatly expanded when microbiological taxa comprising mats are identified along with their notably diverse physiologies. The Proterozoic stromatolitic record is also abundant and diverse, indicating the abundance and probable diversity of microbial mat communities during this era. Both abundance and diversity declined during the Late Proterozoic, the diversity of Proterozoic stromatolites being expressed chiefly in morphological characteristics. Due to the generally poor and highly selective preservation of the constructing microbial taxa, relatively little can be concluded about the potentially high microbial diversity that probably existed in Proterozoic mat-building communities. Even less can be inferred about the potentially diverse physiologies that had developed in structurally very similar microorganisms. The microfossil evidence that is available, however, strongly supports the interpretation that most of the stromatolites were constructed at least in part by cyanobacterial communities. This inference is further supported by both ancient and modern evidence for similar selective preservation of sheaths during early diagenesis in mats and in microfossils associated with stromatolites (e.g., the microscopic evidence and the high O/C ratio). The carbon isotopic discrimination patterns ($\delta^{13}C$) seen in modern microbial mats and Proterozoic organic carbon are consistent with this interpretation, when tempered with the consideration of possible effects of relative environmental availability of inorganic carbon, diel fluctuations of activity within mats, and potential metabolic diversity within the communities.

The objective of studies such as this is to improve understanding of the origin and evolution of the diversity seen in modern microbial mat communities. Studies of the diversity among different modern microbial mats and the diversity (both temporal and spatial) within a particular mat, have been the focus of much of this section. Some important generalizations that have significant implications for interpretation of the Proterozoic mats preserved as stromatolites, and for speculations about those that were not preserved, can be made. Modern microbial mats encompass the highly oxic mats in which cyanobacteria may be the sole or major constructors accompanied by eukaryotic microorganisms, the relatively rare totally anoxic mats devoid of cyanobacteria, and the widespread and abundant "cyanobacterial mats" which contain dynamic and fluctuating oxic and anoxic zones and are comprised of a very diverse assemblage of prokaryotes. These diverse prokaryotes interact with, and alter the microenvironments within, the mat, often migrating up or down in response to changing environmental parameters. It seems likely that all of the diverse types of mats mentioned above existed during the Proterozoic, although unequivocal identification of any particular mat type with particular stromatolites is generally lacking.

Our knowledge of the fluctuating nature of the mat environment has advanced most rapidly with the advent of microelectrodes and other microprobes. Refinement of measurements of the microenvironment at the scale of individual cells might at first seem irrelevant to interpreting Proterozoic stromatolites on a much larger scale. However, our thinking about mat communities, and the evolution of the microorganisms inhabiting them, is markedly influenced by these measurements. First, it is now clear that microgradients of light, oxygen, sulfide, and pH are the controlling factors for migrations of organisms within mats and that they are the prime environmental factors for turning off and on various physiological activities within mat microorganisms on a diel basis. Second, only with the use of microelectrodes have we become aware of the incredible magnitude of these variations in oxygen, sulfide, and pH experienced by many microorganisms within mats. Many abundant mat-building cyanobacteria are very tolerant of these changes and switch metabolism from oxygenic to anoxygenic photosynthesis, and even to chemotrophy, in response to changing parameters. Other microorganisms, such as *Beggiatoa* spp., exhibit substantial diel migrations to remain in the very narrow micro-zone of overlapping oxygen and sulfide required for their metabolism. Even anaerobic processes such as sulfate reduction occur at unexpectedly high rates in upper layers of mats in the vicinity of oxygenic activity, and are not restricted to deeper, strictly anoxic layers. Within the small distance occupied by a few prokaryotic cells, the environment can change dramatically. In some mats (e.g., the *Chloroflexus* mats in thermal sulfide springs in Iceland), very narrow biogenic oxic zones have been found where cyanobacteria reside. The oxygen they produce is confined within this narrow zone of an otherwise anoxic mat.

Given these wide-ranging environmental fluctuations occurring daily within extant dynamic mats, we can infer that mats of high diversity are significantly regulated internally to withstand certain environmental changes occurring externally. As long as light is available to sustain an ecosystem based on phototrophy, these mats remain highly stable but internally dynamic communities. Since present microbial diversity is the result of evolutionary changes occurring within microorganisms that sense significant environmental changes occurring on a very small spatial scale (perhaps too small a scale to be preserved in the fossil record), major evolutionary events, such as the development of oxygenic photosynthesis, could thus have occurred within mats without significantly altering their overall morphology or perhaps even their stability.

Some major event(s) did occur, however, that resulted in the decline of both abundance and diversity of stromatolites near the end of the Proterozoic. These events were not buffered by the high degree of metabolic diversity assured to have been established by that time within the mat communities. Since the small grazers found within mats today are quite compatible with persistance of these mats, and only the larger arthropod grazers truly destroy mats, it seems unlikely that the evolution of the meiofauna was significant in this regard. Evolution of competing metaphytes and environmental changes resulting in a deterioration of the environment or perhaps a decrease in lithification, and hence preservation of mats, may be more significant factors.

7

Proterozoic and Earliest Cambrian Carbonaceous Remains, Trace and Body Fossils

KENNETH M. TOWE STEFAN BENGTSON MIKHAIL A. FEDONKIN
HANS J. HOFMANN CAROL MANKIEWICZ
BRUCE N. RUNNEGAR

7.1

Introduction

KENNETH M. TOWE

This Chapter deals with the records of all of the Proterozoic fossil finds which are not included among the prokaryotic or protistan fossils dealt with in Chapter 5. In general, therefore, Chapter 7 deals with the earliest fossil records of higher organisms on earth. However, where systematic assignment is subject to debate there is unavoidable overlap with the prokaryotes and protists.

In historical perspective, publications describing presumed Precambrian megafossils of various kinds go back more than 100 years. Reports of carbonaceous films appeared as early as 1854 (Eichwald 1854). "Trace fossil" descriptions date from 1866 (Dawson 1866), and "body fossils" from 1872 (Billings 1872a, b). Related to the now-famous Ediacaran faunas, the first unequivocal megafossil to be described was that of *Rangea schneiderhöhni* reported by G. Gürich in 1930 from rocks in southwest Africa. Since these early reports, hundreds of widely, if not universally accepted Proterozoic megafossils have been described from around the world. This record of Proterozoic megafossil remains is therefore not without its share of problems which are similar to those associated with the phylogenetically lower organisms described in Chapter 5. These problems include decisions regarding biogenicity and fossil syngenicity, as well as doubts about geologic age. There are disparate taxonomic judgments, including the differing environmental and/or evolutionary interpretations such judgments may engender. In some instances the reasoning may be circular. In others, the taxonomic judgments may be influenced more by existing biostratigraphic and paleobiological thinking than by independent systematic thought. In many ways such judgments are "model dependent," with the systematic assignments reflecting what an individual feels ought to be present at some given time in the Proterozoic rather than what the fossils might themselves represent, independent of their geological age. The geological age assignment of the rocks may be questioned if the contained fossils do not seem to "fit" their presumed stratigraphic position. The limited availability and study of potentially fossiliferous sites with increasing geologic age tends to compound these problems and continually raises the question of how representative the known Proterozoic fossil record of the higher taxa really is.

The fossils listed and discussed herein, and in Chapter 8, include mostly megascopic remains, but a number of microscopic fossils are included as well. The subject includes the records of such categories as Proterozoic carbonaceous films (7.3); calcareous algae, including earliest Cambrian forms (7.4); soft-bodied metazoans (7.5); metazoan trace fossils (7.6); and small shelly fossils, also including earliest Cambrian forms (7.7). In addition, there is discussion of some of the more enigmatic remains – the megascopic dubiofossils (7.8).

7.2

Criteria for Acceptance of Reported Proterozoic Megafossils

KENNETH M. TOWE

It is important to emphasize that, like the microfossils described in Chapter 5, many Proterozoic megafossils reported in the literature, especially the early literature, are not widely accepted by the paleontological community. In what follows in this Chapter, certain criteria were applied by the respective authors in their evaluation of the various "fossils" in question. In addition to the obvious requirement that the geological age be correct and that the preservation and paleoenvironment be plausible, syngenicity with deposition and the biogenicity of the object were critically examined insofar as possible.

Biogenicity. An acceptably biogenic fossil (real fossil) is one which can be related by a competent investigator to conventional (or at least plausible) biological morphology and/or biological processes. To be acceptable, a fossil cannot be explained by either purely inorganic processes (e.g., sedimentation, diagenesis, tectonic deformation) or by processes of sample preparation (artifacts and contaminants). "Fossils" which deviate from demonstrable acceptability may be placed into one of two categories: pseudofossil or dubiofossil (Hofmann 1972).

A "pseudofossil" is a fossil-like object that resembles a real fossil and was once thought to be a fossil, but which has been interpreted to be the result of a purely inorganic, nonbiologic process. In discussions and tabulations of such objects, the presumed origin of the pseudofossil is commonly indicated (e.g., "soft-sediment slump structure") rather than simply the less helpful notation of its assignment to the pseudofossil category (see, e.g., Table 23.1).

"Dubiofossil" is a term used to express uncertainty as to the biogenicity of the object in question. In general, dubiofossils lack sufficient, distinctive, biologic characteristics, or exhibit one or more possibly nonbiogenic features, either of which makes a confident judgment as to their biogenicity difficult. The term is therefore reserved for fossil-like structures of possible, but unproven, biological origin. Because dubiofossil assignments are concerned foremost with biogenicity, the term differs from "problematica," usage of which usually refers to an authentic fossil of unknown or uncertain systematic position.

As more data or new insights become available, a dubiofossil may be reassigned to the pseudofossil or fossil category. To this extent, then, the term is a taxonomic "waiting room" for fossil materials. Ideally, a dubiofossil should not carry a Linnean designation. Yet subsequent investigators reevaluating primary assignments must unavoidably place previously named fossils into this category, pending further more definitive study.

Syngenicity. Fossils which are acceptably syngenetic are those which can be demonstrated to have been formed in or entombed and preserved contemporaneously in the sediments which now contain them. The sediments which enclose the fossil must be well-dated and "in place." Obviously, a body fossil (or a rock containing a body fossil) which has been reworked into a sediment from significantly older rocks will certainly bias the biostratigraphic utility and may bias environmental or evolutionary interpretations made from it. Organisms which lived in one place but were removed by processes of sedimentation (e.g., a turbidity flow) to be preserved in another locale will be depositionally syngenetic, but may not be environmentally so.

7.3
Proterozoic Carbonaceous Films
HANS J. HOFMANN

7.3.1 Introduction

Carbonaceous remains comprise fossils preserved as micrometer-thick films on bedding planes and, exceptionally, as envelopes of three-dimensional structures. These millimeter- to centimeter-sized bodies are generally simple films with or without distinct ornamentation; variably oriented folds are best regarded as compactional artifacts. Their outlines range from ribbon-like to regular and irregular, round and angulate shapes. Such films were first reported from the Vendian of Russia in the mid-18th century under the name *Laminarites* (Eichwald 1854). Other early reports are those of carbonaceous discs, from the Late Proterozoic of India (King, 1872, pp. 68–69), the Grand Canyon (Powell 1876), and southern Sweden (Nathorst 1879; Wiman 1894), all now referred to *Chuaria*. Walcott (1899) was the first to describe a diversity of new taxa from the Middle and Late Proterozoic. Sporadic reports during the following six decades, widely scattered in the literature, slowly raised the number of known occurrences that ultimately came to include compressions as old as Early Proterozoic.

Several assemblages of morphologically distinct remains are now known, although, due to a lack of distinctive characteristics other than gross shape, the biological affinities of the elements of these biotas are to a large extent still uncertain. (For reviews in the English language, see Gnilovskaya, 1979, and Hofmann 1985b.) Further detailed microscopic and chemical work is required before a better understanding is possible.

About 73 form genera and 80 form species have now been proposed, but these include many synonyms and *nomina nuda*; numerous other occurrences have been reported without Linnean nomenclature (Table 23.1; Figs. 7.3.1 to 7.3.8). Several attempts at suprageneric biological classification of the remains have been made, but none is comprehensive, nor entirely satisfactory, inasmuch as each is purely morphological, referring only to discoidal and/or filamentous remains of which the biological affinities remain obscure. Among groups proposed by various authors are the following: Fermoriidae (Sahni 1936), Chuariidae (Wenz 1938), Megasphaeromorphida (Timofeev 1970), Vendotaenides (Gnilovskaya 1971b), Chuariamorphida (Sokolov 1976a, p. 137), Chuariaceae (Duan 1982), Huaiyuanellidae (Xing 1984b), Longfengshanides (Duan et al., 1985), Cyphomegacritarchs (Fu 1986), and Vendophyceae (Gnilovskaya 1988a). The most recent monograph is edited by Gnilovskaya (1988a), and contains the most comprehensive suprageneric classification to date.

7.3.2 Informal Categories of Megascopic Carbonaceous Films

To accomodate all morphologic types of megascopic carbonaceous films, Hofmann (1985b, 1987) used a tentative informal system of categories for genus-level taxa, named after the dominant genus. The scheme is here amended to include additional categories, and to add or exclude forms. For a selection of representative taxa, see Figures 7.3.9 and 7.3.10.

Chuarid remains: spheroids and discs, normally with concentric or oblique wrinkles (*Chuaria* and others, including synonyms [*Fermoria*, *Huainania*, ?*Ljadlovites*, *Ovidiscina*, *Protobolella*, some *Kildinella* and *Trachysphaeridium*, and ?*Vindhyanella*]).

Tawuid remains: rectilinear or curvilinear tomaculate structures, normally not twisted, and without transverse elements (*Tawuia* and synonyms and possible synonyms [*Bagongshanella*, *Bipatinella*, *Conicina*, *Eurycyphus*, *Fengyangella*, ?*Fermoria*, *Lakhandinia*, *Linguiformis*, *Liulaobeia*, ?*Mezenia*, *Pumilibaxa*, *Sicyus*, *Stenocyphus*]).

Ellipsophysid remains: ovate to spatulate forms morphologically intermediate between chuarids and tawuids (*Ellipsophysa*, ?*Glossophyton*, *Nephroformia*, *Phascolites*, *Shouhsienia*).

Grypanid remains: slender, curvilinear structures with pronounced spiraliform tendency and occasionally with coarse transverse markings, giving a bead-like internal appearance (*Grypania* and synonyms ["*Helminthoidichnites*", *Sangshuania*], and possibly *Katnia* and *Loriforma*).

Longfengshanid remains: round to oblong structures with narrow stipe or appendage (?*Glossophyton*, ?*Krishnania*, *Longfengshania*, *Paralongfengshania*).

Moranid remains: irregular round forms without wrinkles (?*Morania*, ?*Ljadlovites*, ?*Vindhyavasinia*).

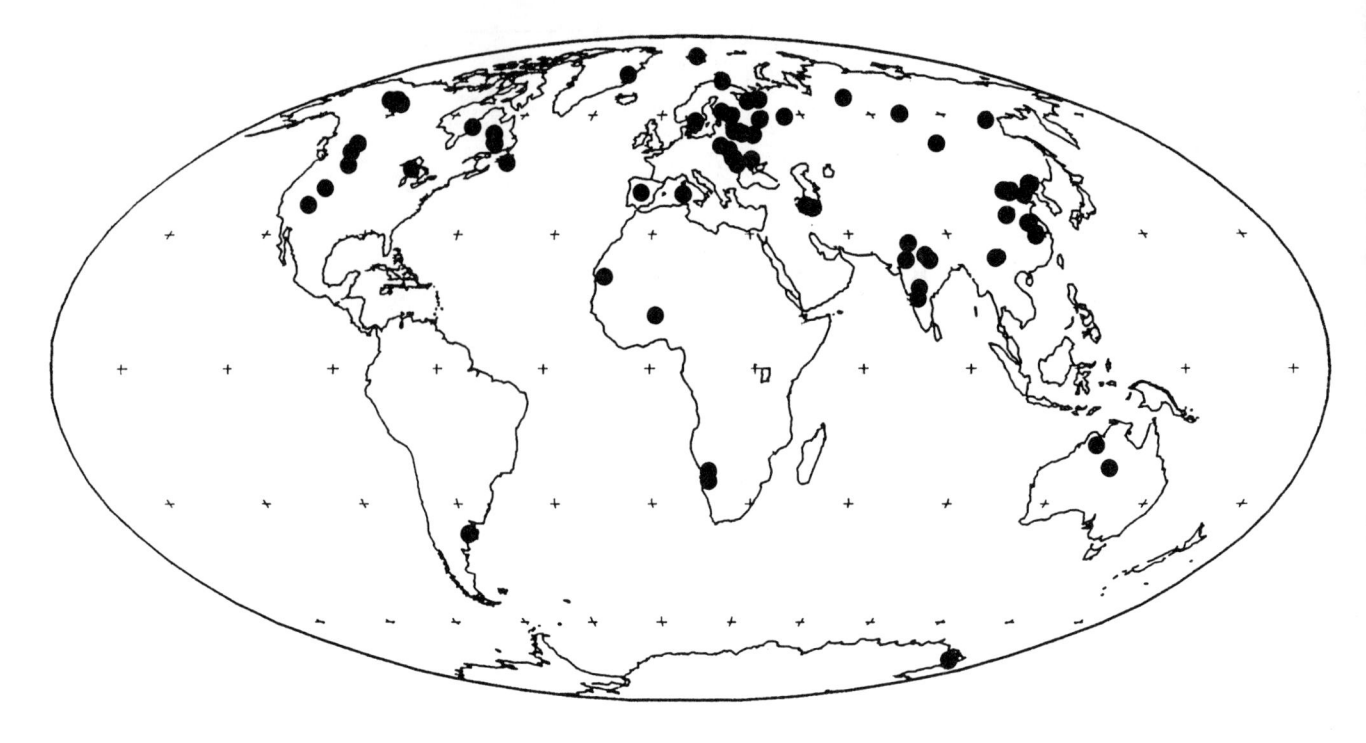

Figure 7.3.1. Distribution of carbonaceous films for all of the Proterozoic.

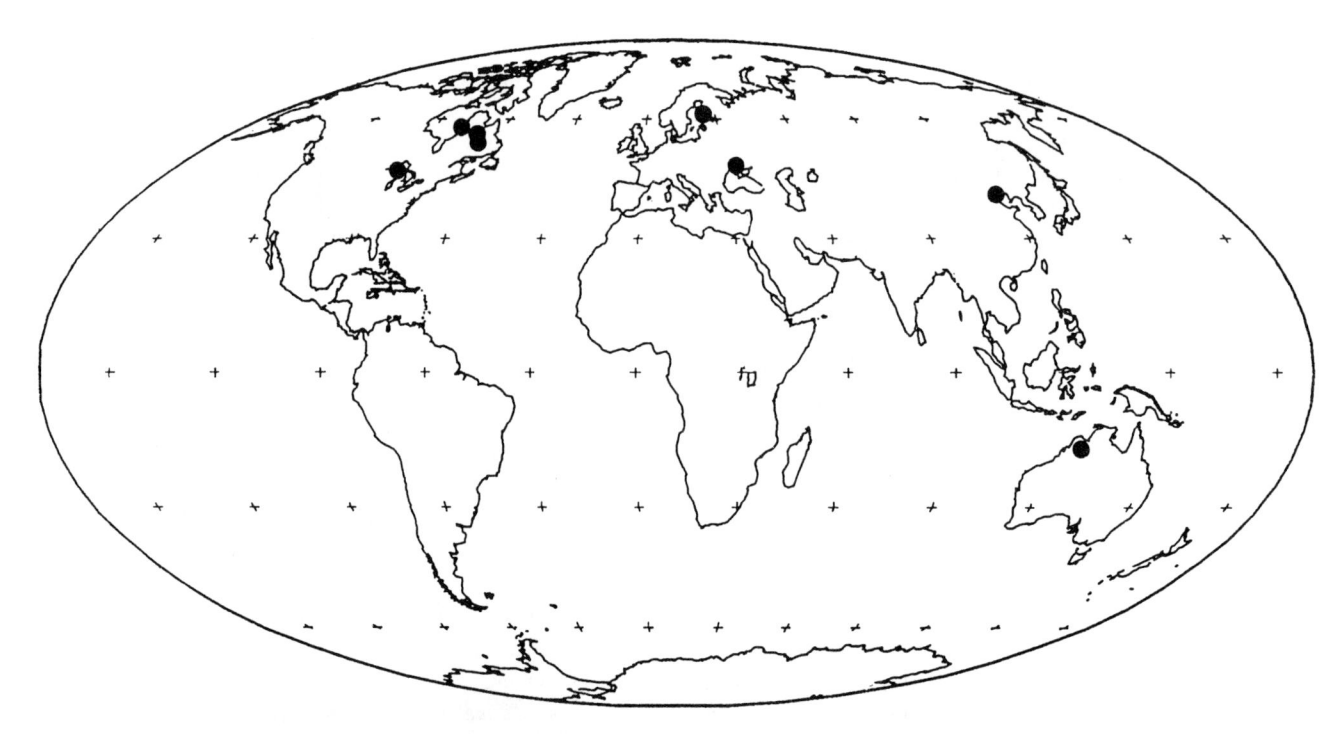

Figure 7.3.2. Distribution of carbonaceous films for the Early Proterozoic.

Proterozoic Canbonaceous Films 351

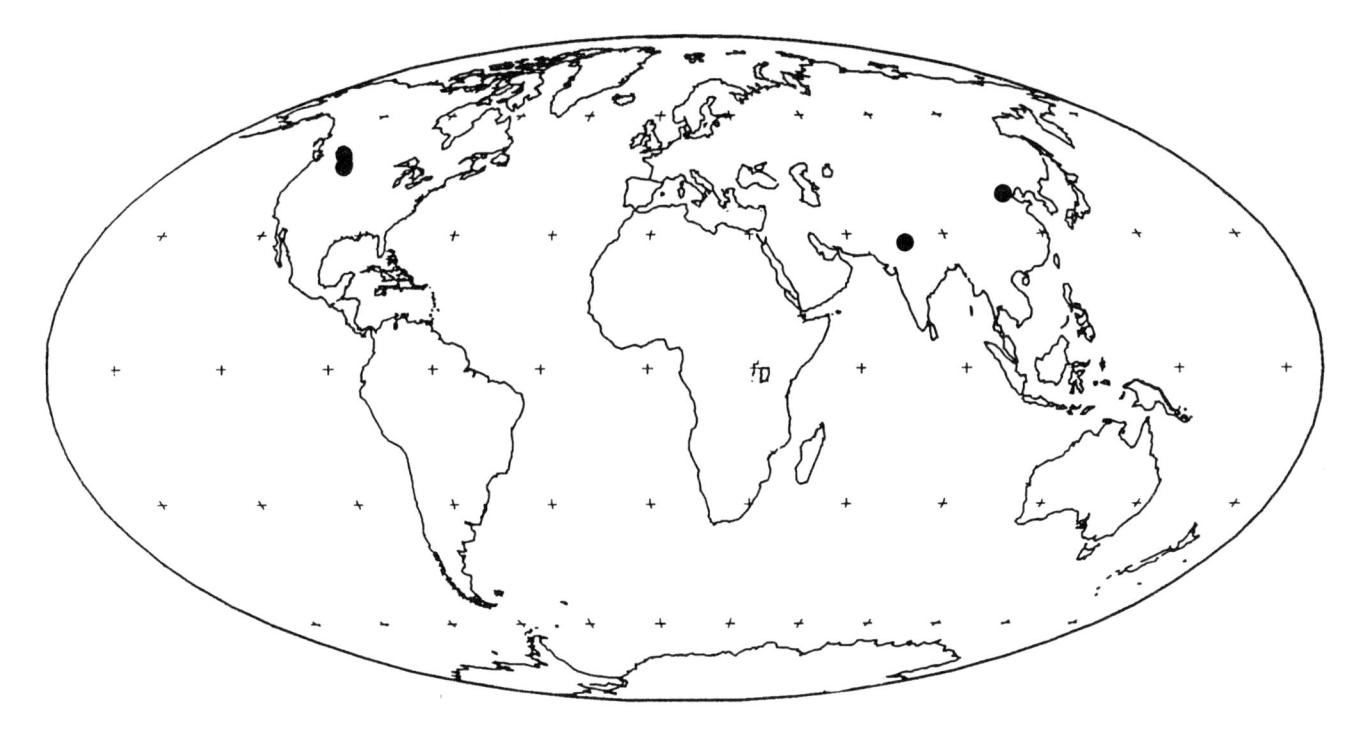

Figure 7.3.3. Distribution of carbonaceous films for the Middle Proterozoic.

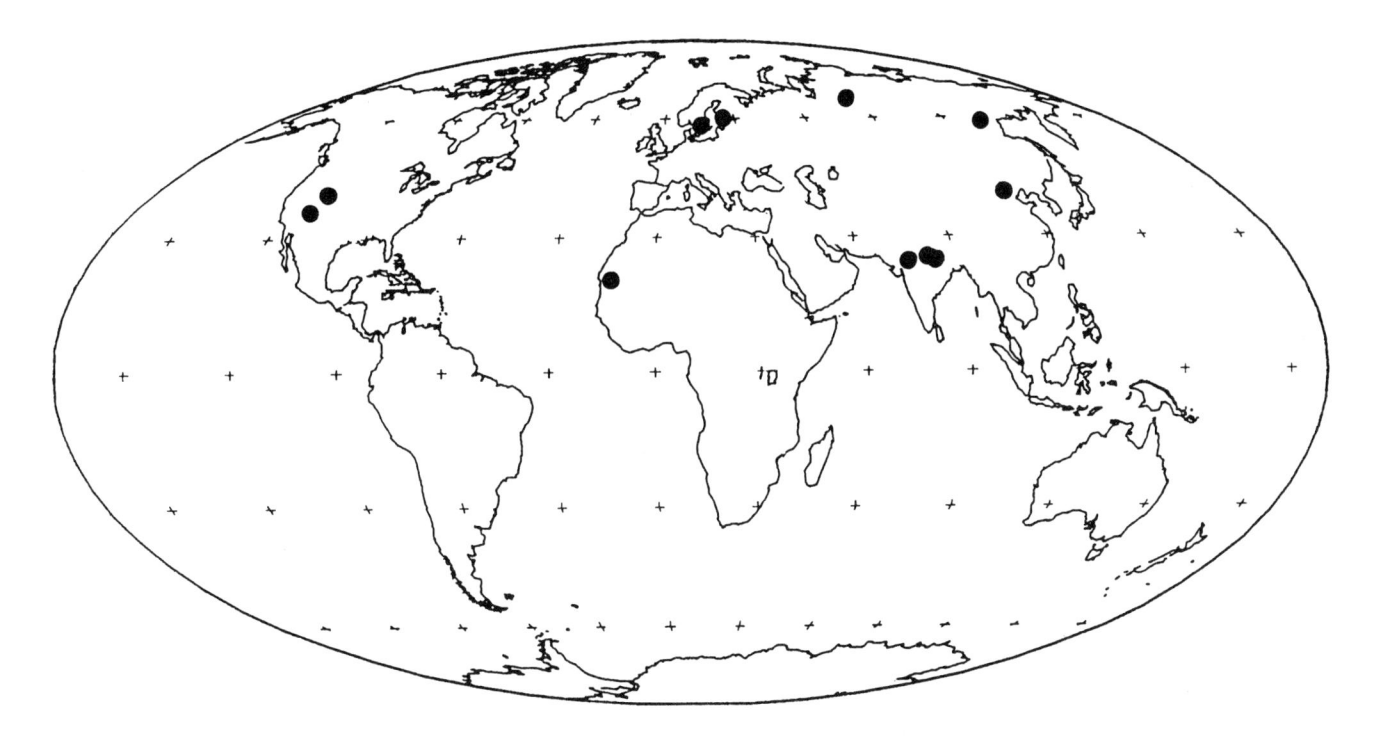

Figure 7.3.4. Distribution of carbonaceous films for the Middle to Late Proterozoic.

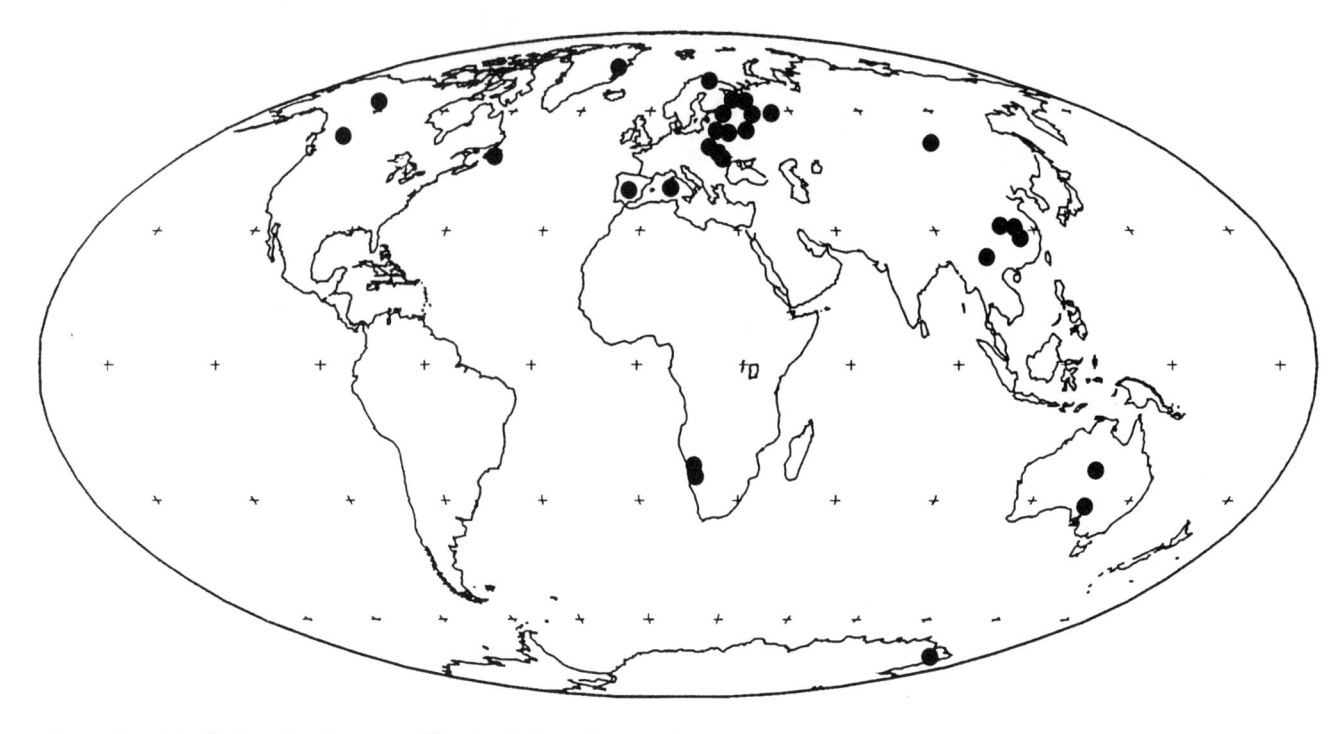

352 Proterozoic Remains, Trace and Body Fossils

Figure 7.3.5. Distribution of carbonaceous films for the Late Proterozoic.

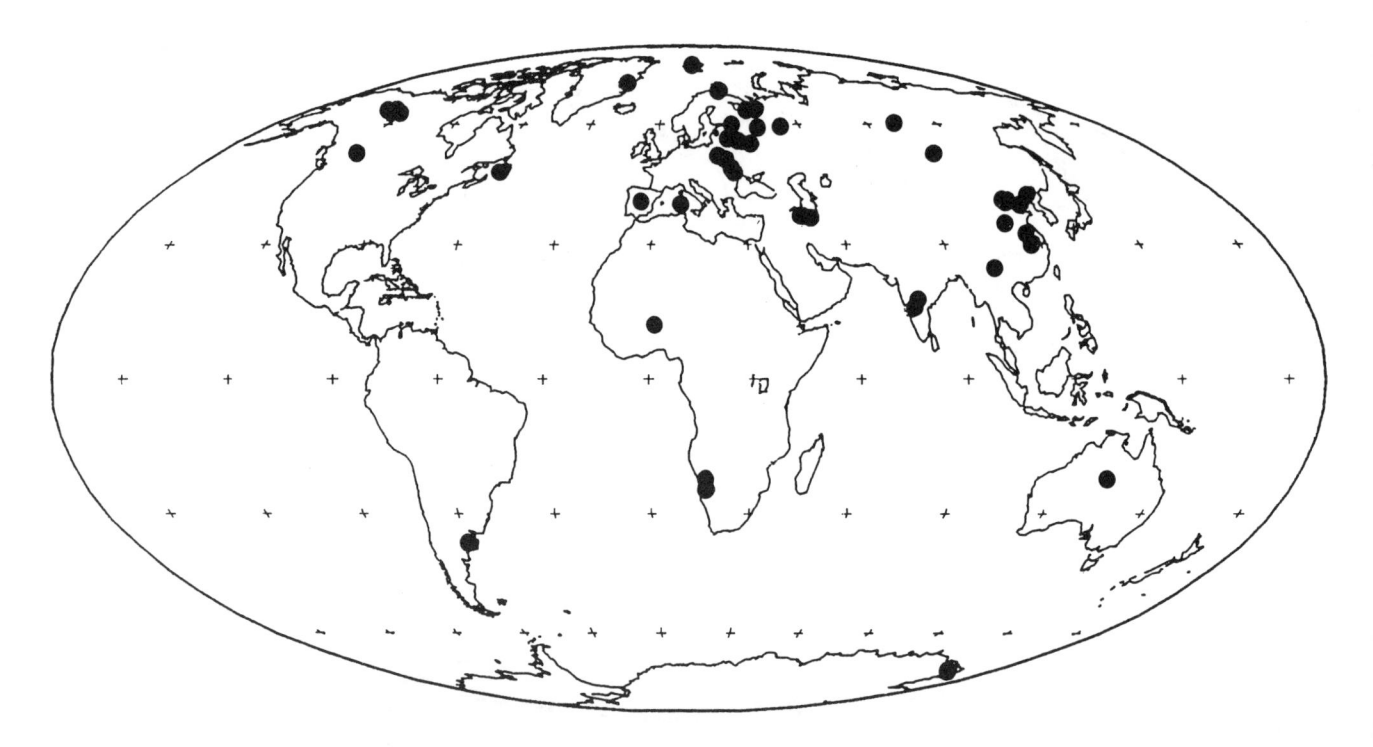

Figure 7.3.6. Distribution of carbonaceous films for the Vendian.

Proterozoic Carbonaceous Films

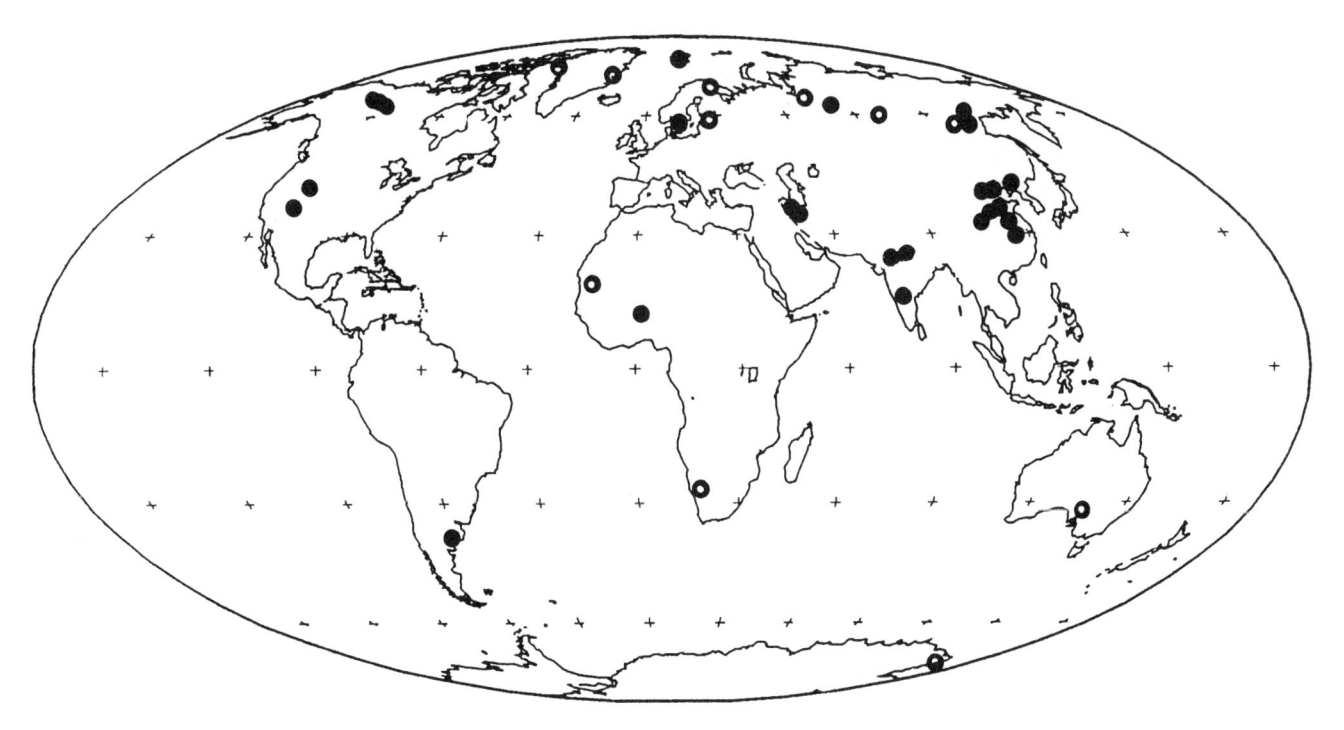

Figure 7.3.7. Occurrences of *Chuaria*, with macroscopic forms shown by "●" and microscopic forms attributed to *Chuaria* denoted by "○".

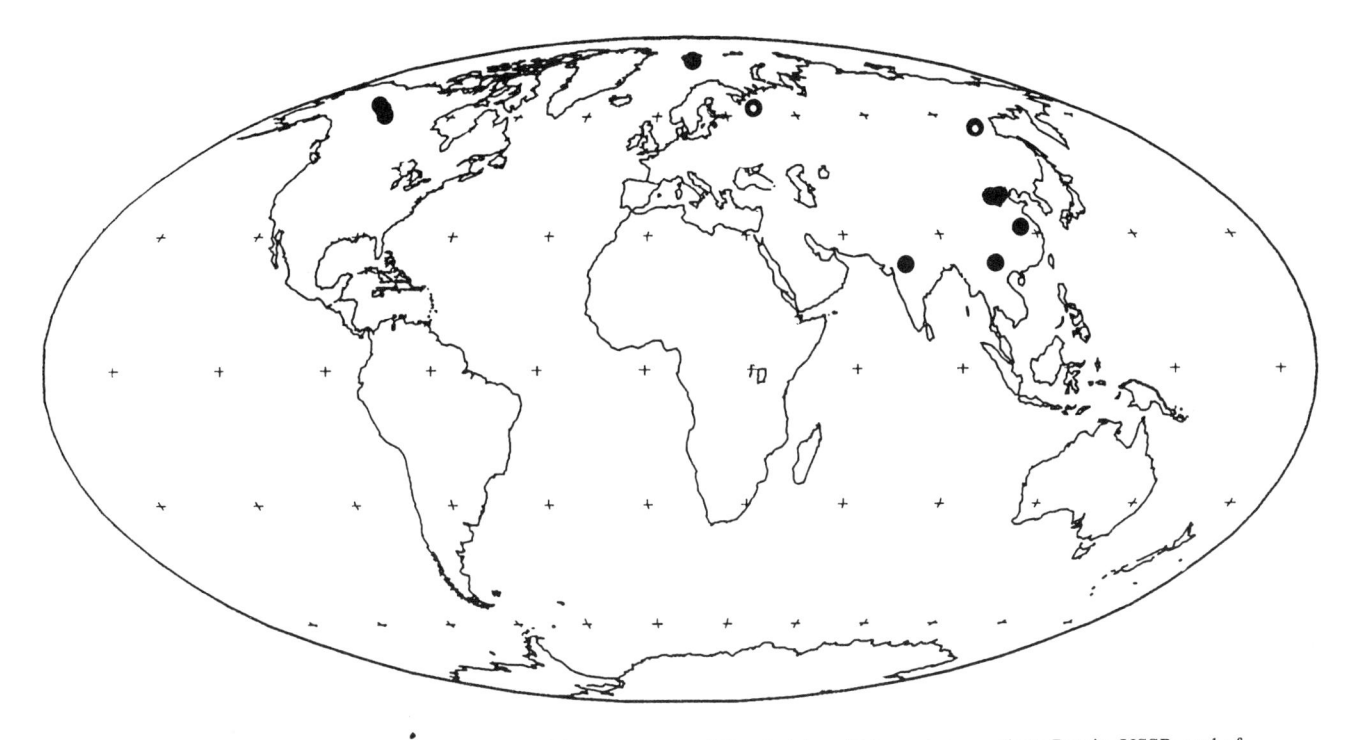

Figure 7.3.8. Occurrences of *Tawuia* shown by "●"; possible occurrences of *Tawuia* (viz., of *Menzenia* in northern Russia, USSR, and of *Lakhandinia* in eastern Siberia, USSR) are denoted by "○".

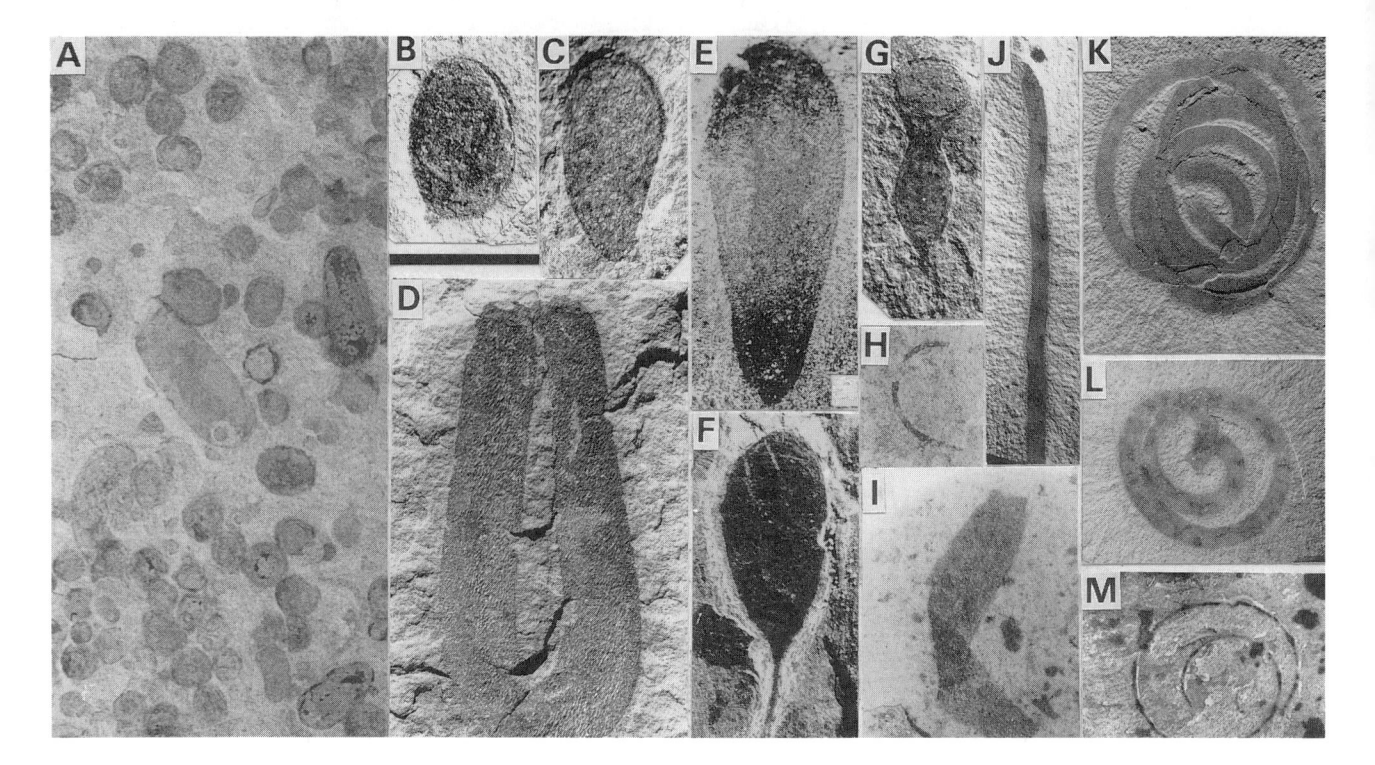

Figure 7.3.9. Representative Proterozoic carbonaceous films. Bar for scale (above D) represents 10 mm for A, D, H, I, K, L, and M; 8 mm for B, C, F, and G; 7 mm for E; and 5 mm for J. (A) *Chuaria circularis* and short specimens of *Tawuia dalensis*, Little Dal Group, Mackenzie Mountains, N.W.T., Canada, Geological Survey of Canada (GSC) no. 77191 (Hofmann 1985a, Pl. 35, Fig. 5). (B) *Shouhsienia shouhsienensis*, Changlongshan Formation, Hebei, China (Du and Tian 1985a, Pl. 1, Fig. 13). (C) *Ovidscina longa*, Changlongshan Formation, Hebei, China (Du and Tian 1985a, Pl. 1, Fig. 20). (D) *Tawuia dalensis*, U-shaped specimen, Little Dal Group, Mackenzie Mountains, N.W.T., Canada, GSC no. 57893 (Hofmann and Aitken 1979, Fig. 13E). (E) *Glossophyton mucronatus*, Changlongshan Formation, Hebei, China (Duan et al. 1985, Pl. 17, Fig. 5; Du and Tian, 1985a, Pl. 1, Fig. 24). (F) *Longfengshania ovalis*, Changlongshan Formation, Hebei, China (Du and Tian 1985a, Pl. 2, Fig. 9; Duan et al. 1985, Pl. 17, Fig. 9). (G) *Paralongfengshania sicyoides*, Changlongshan Formation, Hebei, China (Duan et al. 1985, Pl. 17, Fig. 10; Du and Tian 1985a, Pl. 2, Fig. 13). (H) *Grypania spiralis*, Little Dal Group, Mackenzie Mountains, N.W.T., Canada, GSC no. 77204 (Hofmann 1985a, Pl. 39, Fig. 4). (I) Holotype of *Lanceoforma striata*, Greyson Shale, Montana, USA, U.S. National Museum (USNM) no. 210908 (Walter et al. 1976, Pl. 2, Fig. 2). (J) *Sangshuania linearis*, Gaoyuzhuang Formation, Jixian, northern China. (K) *S. sangshuanensis*, Gaoyuzhuang Formation, Jixian, northern China. (L) *S. sangshuanensis* with bead-like markings, Gaoyuzhuang Formation, Jixian, northern China (specimen Jg-j126 of Du Rulin). (M) *Grypania* ("*Helminthoidichnites?*") *spiralis*, Greyson Shale, Deep Creek Canyon, Montana, USA, USNM no. 33794(210905) (Walcott 1899, Pl. 24, Fig. 5; Walter et al. 1976C, Pl. 2, Fig. 9; GSC photograph 200881-A). Photographs in B, C, E–G, J, K, courtesy of Du Rulin; photograph in I, L, courtesy of M. R. Walter.

Beltinid remains: irregular angulate forms (*Beltina* and synonyms and possible synonyms [?*Lanceoforma, Radicula,* ?*Vindhyavasinia*]).

Vendotaenid remains: slender filaments, generally unbranched and twisted, smooth or patterned (*Vendotaenia* and others, including synonyms and possible synonyms [*Aataenia,* ?*Caudina, Fasciculella, Fusosquamula, Kanilovia,* ?*Katnia, Laminarites,* ?*Lanceoforma,* ?*Lorioforma,* ?*Pilitella, Proterotainia,* ?*Sarmenta, Serebrina, Sinotaenia, Tyrasotaenia, Vindhyania*]).

Eoholynid remains: noticeably branched aggregates of filamentous forms (?*Daltaenia, Enteromophites, Eoholynia, Kalusina*).

Sinosabelliditid remains: tomaculate forms with regularly spaced, narrow transverse annulations (*Sinosabellidites* and similar forms [*Anhuiella, Huainanella, Huaiyuanella,* ?*Katnia, Paleorhyncus, Pararenicola, Protoarenicola,* ?*Ruedemannella*]).

Sabelliditid remains: slender tubes or ribbons with regular transverse annulations [segmentation?] of narrow, funnel-shaped rings (*Calyptrina, Paleolina, Saarina, Sabellidites,* ?*Shaanxilihes*).

Other remains: poorly characterized forms (e.g., *Corycium, Misraea, Orbisiana*).

7.3.3 Morphological Characteristics, Distributions, and Possible Affinities.

Chuarids. Aside from the vendotaenids, chuarid remains comprise the most widespread and most commonly identified films in Late Proterozoic sequences, having been noted at about 50 localities worldwide (Fig. 7.3.7; Hofmann 1985b, p. 22). Although typical *Chuaria* is megascopic (greater than about 0.2 to 0.3 mm), there is a continuous size gradation to microscopic spheroids, making it impossible to delimit a lower size range objectively. As a result, there are many reported occurrences of sphaeromorphs attributed to *Chuaria* in the size range of 0.06 to 0.2 mm, without associated megascopic specimens. All such occurrences are plotted as a separate category in Figure 7.3.7.

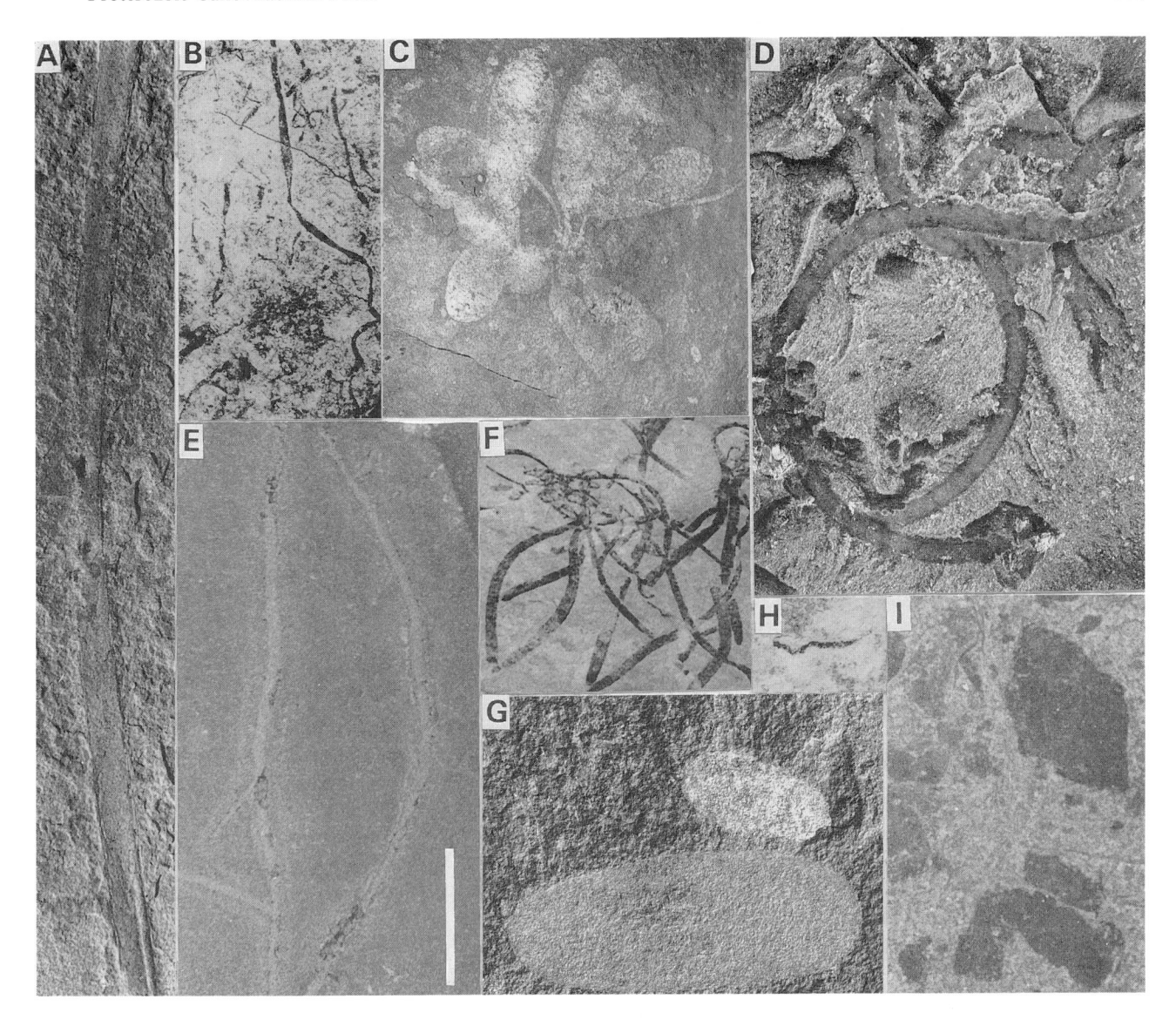

Figure 7.3.10. Representative Proterozoic (and Cambrian) carbonaceous films. Bar for scale in (E) represents 10 mm for A–G and I; and 2 mm for H. (A) Paratype of *Tawuia dalensis*, a long fragment, Little Dal Group, Mackenzie Mountains, N.W.T., Canada, GSC paratype no. 57890 (Hofmann and Aitken 1979, Fig. 13A). (B) Holotype of *Tyrasotaenia podolica*, Kanilov Formation, Kitaigorod, USSR, specimen no. 6931/1 (Gnilovskaya 1971b, Pl. 11, Fig. 1). (C) *Longfengshania stipitata*, Little Dal Group, Mackenzie Mountains, N.W.T., Canada, GSC no. 77200 (Hofmann 1985a, Pl. 38, Fig. 4). (D) "*Helminthoidichnites*" (= *Grypania*) *meeki*, Greyson Shale, Deep Creek Canyon, Montana, USA, USNM no. 33793 (Walcott 1899, Pl. 27, Fig. 7; Walter et al. 1976C, Pl. 2, Fig. 12; GSC photograph 200881-C). (E) Holotype of *Daltaenia mackensiensis*, Little Dal Group, Mackenzie Mountains, N.W.T., Canada, GSC no. 77203 (Hofmann 1985a, Pl. 39, Fig. 2). (F) *Sabellidites* sp., Chapel Island Formation, Member 2 (Cambrian part), Fortune Head, Newfoundland, Canada. (G) *Morania*? sp., Attikamagen Formation, Schefferville, Quebec, Canada (B. L. Stinchcomb collection; GSC photograph 200810-D). (H) Hypotype of *Tyrasotaenia* sp., Little Dal Group, Mackenzie Mountains, N.W.T., Canada, GSC no. 77211 (Hofmann 1985a, Pl. 39, Fig. 11). (I) *Beltina danai*, Little Dal Group, Mackenzie Mountains, N.W.T., Canada.

Chuaria has been variously regarded as a problematicum, brachiopod, gastropod, hyolithid operculum, trilobite egg, medusoid, chitinous foraminifer, green alga, and megascopic acritarch (for a comprehensive annotated bibliography, see Spamer 1988). The most recent interpretation of the genus is by Sun (1987), who regards the remains as probable colonies of filamentous cyanobacteria, comparable to modern *Nostoc* colonies; the interpretation is identical to that first proposed for *Morania*-like structures from the Belt Supergroup (Walcott 1919, p. 226) and later for similar structures in the Early Proterozoic of Michigan (Tyler et al. 1957, p. 1300). Sun based his conclusion on the presence of a few structures interpreted as filaments associated with some specimens of *Chuaria* from China. However, it is unclear from his illustrations whether these are actually inside the envelope or attached on the outside. *Chuaria* with associated external filamentous structures occurs in northwestern Canada (e.g., Hofmann 1985a, Pl. 35, Fig. 8). Filaments are not known from the type material of *Chuaria circularis*, making a rediagnosis (Sun 1987b, p. 115) of the genus somewhat premature. While the interpretation of the genus as a nostocalean cyanobacterial colony is certainly plausible and attractive, some doubt remains; the presence of internal filaments needs to be confirmed for at least some other *Chuaria* occurrences. Also, the absence of *Chuaria* in the post-Proterozoic geologic record needs explanation, inasmuch as one would expect it to be present, given the existence of the modern analogue. Thus, for the time being, we still regard *Chuaria* as a megascopic acritarch. The larger discs referred to *Beltanelloides* and *Beltanelliformis*, and earlier included in the chuarids, are here excluded; their large size and lack of carbonaceous test indicate that they are more likely to be related to metazoan structures such as *Nemiana* and *Bergaueria*. Two other occurrences of reported *Chuaria* (Table 23.1, items 3, 61) are excluded here for similar reasons.

Tawuids. Tawuid remains are represented by *Tawuia* (Figs. 7.3.9A, D; 7.3.10A), which was originally regarded as probably algal, but possibly metazoan (Hofmann and Aitken 1979), and placed into the Vendotaenides of Gnilovskaya (1971). As new occurrences became known (Fig. 7.3.8), *Tawuia* was found invariably associated with *Chuaria*, while the converse is not true. This led to suggestions that both genera may be closely related systematically (Hofmann 1981a, b; Duan 1982; Sun 1987b), an opinion further supported by data on allometric growth curves for *Tawuia* (Knoll 1982a, Fig. 3; Hofmann 1985a, p. 336), and by small circular discs attached to (or contained within) *Tawuia* specimens (Hofmann 1985a, Pl. 35, Figs. 6–7) and interpreted as reproductive structures. *Tawuia* is mostly preserved as smooth, I-, J-, C-, U-, and S-shaped compressions of variable length. An interesting morphometric study, characterizing the form of *Tawuia* as parabolic, is presented by Fu (1986).

Where preserved three-dimensionally in laminated argillaceous dolostone, the fossils have a carbonized wall about 1 μm thick, surrounding a pure, equigranular (microsparitic) calcite filling. This indicates that the living structure was a cylindrical object with hemispherical termini, and had a distinct, resistant envelope around a more readily decomposed interior of different makeup. An elemental and isotopic analysis of the wall of one specimen from the Mackenzie Mountains (locality 1 of Hofmann, 1985a, p. 382) gave the following results: C = 80.56%; H/C = 0.162; N/C = 0.028; $\delta^{13}C = -24.48$ (J. M. Hayes 1986, pers. comm.). The axial stipe present in some flattened specimens suggests that it might have had internal heterogeneity, perhaps an organ like an alimentary canal, but this has been rejected for lack of other, compelling evidence (no opening or other specialized structures). One line of support for a metazoan relationship is provided by the coetaneous occurrence, in China, of the worm-like fossil *Sinosabellidites*, the gross tomaculate shape and size-range of which are strikingly similar to those of *Tawuia*; the chief observable difference is the presence of very distinct, closely spaced, fine transverse ornamentation in *Sinosabellidites*. Timofeev and German (1979) described a microscopic form under the name *Lakhandinia*, which may be similar to *Tawuia*.

Sun (1987b, p. 124, Pl. 4, Fig. 6) reported finding numerous obscure filamentous impressions on the surface of one specimen of *Tawuia*, leading to the interpretation that, like *Chuaria*, *Tawuia* may represent nostocalean colonies. Once again, such an interpretation is quite plausible, because modern counterparts of sausage-shaped colonies exist (e.g., *Wollea*). However, the illustration of supposed filaments is not convincing; also, typical *Tawuia* so far has not shown preserved filaments. Thus, the nature of *Tawuia* and its relation to *Chuaria* and *Sinosabellidites* remains unresolved.

Ellipsophysids. Ellipsophysid remains resemble very short specimens of tawuids, and appear to intergrade with them. They have been given various names such as *Shouhsientia* (Fig. 7.3.9B); some, such as *Nephroformia*, *Pumilibaxa*, and possibly *Glossophyton* (Fig. 7.3.9E), may be regarded as short forms of *Tawuia* (Hofmann 1985a, b). The fossil *Phascolites* probably belongs to the ellipsophysids; the medial constriction appears to be due to fortuitous superposition of two elliptical *Shouhsienia* specimens. It is possible that ellipsophysids are also related to the longfengshanids, representing structures that emerged or became liberated from originally stalked bodies.

Grypanids. Grypanid remains are curved and spiraliform ribbons with, in some specimens, a uniseriate, beaded, internal pattern. These include fossils first reported under the name *Helminthoidichnites meeki* (Walcott 1899) from the Middle Proterozoic of Montana, which are not trace fossils, but body fossils (Fig. 7.3.10D). They can all be placed in *Grypania* (Fig. 7.3.9M), following Walter et al. (1976c). Identical fossils in coeval rocks in China are named *Sangshuania* (Du et al. 1986; = junior synonym for *Grypania*). The ribbons are of millimetric width and centimetric length, and are preserved as regular spirals (*S. sangshuanensis*; Fig. 7.3.9K, L), sinusoidal structures, or as drawn-out-and-kinked spirals (*S. linearis*; Fig. 7.3.9J). In both Montana and China, well-preserved specimens clearly show mutually contiguous elliptical internal structures arranged in a uniseriate row that is as wide as the surrounding envelope, thus giving the fossil the appearance of a giant coiled trichome with megascopic cells inside a sheath (Fig. 7.3.9K, L). The consistent spiral nature of these remains suggests the existence of unidirectionally coiled structural elements inside, such as, perhaps, fibrous spiral strands within the envelope.

Similar, but three-dimensionally preserved, spiraliform fossils are known from India (Beer 1919; Mathur 1983, pp. 112–113) and Siberia (Sokolov 1975), and may be present in the Late Proterozoic in Newfoundland (Billings 1872a, p. 478) and Brittany (Lebesconte 1887), but these are not carbonaceous, and may have quite different affinities. The grypanid fossils are interpreted as possibly bead-like algal colonies within an envelope, composed of oblate spheroidal or short cylindrical cells of megascopic size (like *Chuaria* and *Tawuia*) peculiar to the Proterozoic. They appear to differ from tawuid remains by their greater filament length/width ratio, distinctive content, and greater tendency for coiling.

Longfengshanids. The longfengshanid films are oval to oblong structures with a single stipe or appendage. These fossils have thus far been found only in China and northwestern Canada (*Longfengshania*; Fig. 7.3.10C), and possibly in India (*Krishnania*, Sahni and Shrivastava 1954). As with *Tawuia*,

Longfengshania is only found associated with *Chuaria*. Indeed, some specimens of *Longfengshania* have a main body whose shape closely resembles that of *Chuaria* and short specimens of *Tawuia* (*Shouhsienia*, *Nephroformia*, *Pumilibaxa*), suggesting a link between these genera (see also Du and Tian 1985b). *Paralongfengshania* differs from *Longfengshania* by a medial constriction (Fig. 7.3.9E). *Glossophyton* may be a fragmentary *Longfengshania*. The longfengshanid organisms appear to have been epiplanktonic or epibenthonic, as shown in the model of Chen and Zheng (1986, p. 227), and *Longfengshania* may represent the sessile stage for the planktonic *Chuaria* and *Tawuia*. More recently, Zhang (1988) has compared *Longfengshania* to the Bryophyta. Smaller structures of similar morphology, at the lower limit of resolution by the unaided eye, and possibly belonging to the longfengshanid category, are Gnilovskaya's (1979) genera *Sarmenta*, *Caudina*, and *Primophlagella*, and the "*Phycomycetes*" of Timofeev (1969, Pl. 34, Figs. 3, 5, 6), which are included in the Section on microfossils (Section 5.5).

Moranids. Moranid remains are relatively large, ellipsoidal to irregular, round compressions without wrinkles (e.g., Fig. 7.3.10G); they appear to be distinct from chuarids in that a physically resistant wall or envelope was either less well developed or absent, as in essentially homogeneous gelatinous masses. The fossils have been attributed to the genus *Morania*, and compared to modern free-floating globoidal colonies of the cyanobacterium *Nostoc* (Walcott 1919, p. 226, Pls. 46, 53; Tyler et al. 1957, Pl. 1, Fig. 4). Sun (1987a,b) has made the same interpretation for chuarid remains, though this needs confirmation by demonstrating the presence of filaments in other *Chuaria* occurrences.

Beltinids. Beltinid fossils are angulate films of kerogenous or carbonized matter, sometimes folded, that have been placed in *Beltina* Walcott (e.g., Fig. 7.3.10I); other taxa include *Laceoforma* (Fig. 7.3.9I) and ?*Radicula*. Originally referred to the eurypterids, *Beltina* probably includes diverse remains, such as structureless biogenic matter, or fragments of prokaryotic mats and colonies, or eukaryotic algae. The latter affinities cannot be totally excluded, given potentially close modern analogues among the Chlorophyta (*Monostroma*) and Rhodophyta (*Porphyra*). Beltinids are so scrappy looking that they have attracted little detailed investigation: further microscopic study may be rewarding.

Vendotaenids. Vendotaenid remains are generally found on bedding planes as black, twisted and untwisted fragments of unbranched ribbons. Some of them are reported to bear longitudinal striations and/or microscopic discoids which are used to suggest algal (metaphyte) affinities for the fossils, although the primary nature of such structures is not always clearcut, nor is the distinction between the main constituent genera (*Vendotaenia*, *Tyrasotaenia* [Fig. 7.3.10B, H], *Kanilovia*). It is likely that vendotaenid remains, the stratigraphic range of which extends back to 1.8 Ga (Hofmann and Chen 1981), but which are most abundant in the late Vendian, comprise phylogenetically diverse organisms, and include filamentous prokaryotic (chiefly cyanobacterial?) aggregates as well as eukaryotic elements.

Eoholynids. Eoholynid structures are aggregates of noticeably branched macroscopic filaments. They include small taxa (*Eoholynia*) as well as large ones (*Entomorphites* and, possibly, *Daltaenia*, shown in Fig. 7.3.10E). As with other categories of carbonaceous films, their morphologic detail and biological affinities are poorly known; algal affinities are considered likely, particularly for the larger ones. Eoholynids are of Late Proterozoic and younger age, the questionably oldest form being *Daltaenia* from the Little Dal Group of northwestern Canada.

Sinosabelliditids. Sinosabelliditids are intriguing black films with more or less regular transverse markings. *Sinosabellidites*, which has the size and gross morphology of *Tawuia* except for the transverse pattern, and co-occurs with it, has been regarded as having affinities with worms. It is unclear whether this pattern is primary or taphonomic, although its regular nature suggests that it is primary. Similar transverse features, here interpreted as secondary, are known from microscopic filaments assigned to *Plicatidium* (e.g., see Yankauskas 1980a, Pl. 12, Fig. 15). There is thus a need to determine whether *Sinosabellidites* is a preservational variant of *Tawuia*, or whether it is more like *Sabellidites* (Fig. 7.3.10F) and other Sabelliditidae. Other sinosabelliditid remains (*Paleorhynchus*, *Protarenicola*, *Pararenicola*, etc.) may be more poorly preserved individuals of *Sinosabellidites*.

Sabelliditids. The sabelliditids are slender, rhythmically annulated black films, occuring chiefly in Lower Cambrian deposits, but ranging into the latest Vendian. Ultrastructural investigations by Urbanek and Mierzejewska (1983) to verify Sokolov's attribution of sabelliditids to the "worm" phylum Pogonophora Johannson 1938 (e.g., Sokolov 1967; Glaessner 1978, p. A107) proved inconclusive; their affinities still remain undetermined.

7.3.4 Summary.

In summary, carbonaceous films are known from sediments as old as Early Proterozoic, and are increasingly important elements of biotas of the Middle and Late Proterozoic. They indicate the existence of megascopic entities in the oceans of that remote past. Although their affinities still need to be more precisely determined, their probable affinities lie with both prokaryotic and eukaryotic organisms. Of special paleobiologic and biostratigraphic significance are genera that are morphologically complex, geographically widespread, and stratigraphically restricted, such as *Grypania*, *Longfengshania*, *Sinosabellidites*, *Tawuia*, and *Vendotaenia*. Much further work is needed to clarify the nature and distribution of these remains.

Acknowledgments

Thanks are due Du Rulin and M. R. Walter for providing photographs of some specimens. J-J. Chauvel provided photographs of *Montfortia* used for comparing this genus with *Grypania*. Financial support from the Natural Sciences and Engineering Research Council of Canada (Grant no. A7484) is gratefully acknowledged.

7.4 Proterozoic and Early Cambrian Calcareous Algae

CAROL MANKIEWICZ

7.4.1 Introduction

"Calcareous algae" are those algae and cyanobacteria that precipitate $CaCO_3$ on or within the thallus (plant body) while living. Today, this artificial, heterogeneous group includes representatives from the Cyanobacteria, Chlorophyta, Rhodophyta, Phaeophyta, and Chrysophyta; in the Cambrian, perhaps only Cyanobacteria and possible members of the Rhodophyta and Chlorophyta calcified their thalli (e.g., Chuvashov and Riding 1984). Note that stromatolites, organosedimentary structures, are not included. The degree of control over precipitation of carbonate (in the form either aragonite or calcite) is poorly understood and varies among the "algae" from highly active (best exemplified by the coccolithophorids of the Chrysophyta) to passive (e.g., many cyanobacteria; see Leadbeater and Riding 1986).

Calcareous algae first flourished around the transition between the Proterozoic and Cambrian. Pyatiletov et al. (1981) noted that only noncalcified forms of the genera *Obruchevella*, *Proaulopora*, and *Glomovertella* occur prior to the Cambrian; the same might be said of *Girvanella* and possibly *Renalcis*. Thus, calcification of the "algae" seemingly paralleled that in the metazoans (see Section 8.5 for discussion).

During the Early Cambrian, calcareous algae attained virtually world-wide distribution (Fig. 7.4.1) and sedimentologic importance. They were the primary constructors of bioherms throughout the Early Cambrian, rarely relinquishing dominance to the archeocyaths (James and Debrenne 1980; Rowland and Gangloff 1988).

In 1878, Nicholson and Etheridge described the calcareous alga *Girvanella problematica* from the Silurian of Great Britain; Chapman (1908) extended the range of this species to the Early Cambrian of South Australia. In 1886, Bornemann published the first taxonomic study of Early Cambrian calcareous algae in which *Epiphyton flabellatum* was described. Soviet scientists, however, have dominated the taxonomic study of Early Cambrian calcareous algae. Vologdin began work in the early 1930s, and was followed by Kordeh, Voronova, Luchinina, and others. Kordeh alone named about 40% of all species of Early Cambrian algae; the effect of her work, which commenced in the early 1950s, is evident in cumulative curves of described species versus year of publication (Fig. 7.4.2).

Nearly 400 species of Early Cambrian calcareous algae have been described. Of these, about 68% are interpreted here as fossils, 22% as dubiofossils, and 10% as pseudofossils. Of the fossils, however, only about 40% represent occurrences from more than a single geographic locality; only about 15% of the dubiofossils have multiple occurrences (Fig. 7.4.3).

7.4.2 Classification

7.4.2A Adapting to modern classification schemes.

Historically, modern methods of classification of calcareous algae have been adapted to fossil calcareous algae in an attempt to impart an evolutionary perspective to the study. Such an approach has proved fruitful in some cases. For example, detailed studies of modern calcifying cyanobacteria suggest that the Cambrian genera *Girvanella* (Riding 1977), *Angulocellularia* (Riding and Voronova 1982a), and *Tubomorphophyton* (Riding and Voronova 1982b) have cyanobacterial affinities. In addition, similarities in general morphology between the long-ranging genus *Solenopora* and the more recent coralline (rhodophyte) algae suggest that they may be related (e.g., Wray 1977), but Brooke and Riding (1987) proposed that solenoporids are actually a heterogeneous group that includes rhodophytes, cyanobacteria, and metazoans.

Unfortunately, few Cambrian taxa have obvious modern analogues, rendering classification highly subjective, equivocal, and variable. For example, the common genus *Epiphyton* has been compared to chlorophytes (Bornemann 1886), rhodophytes (Johnson 1966; Kordeh 1961, 1973; Voronova 1976), and cyanobacteria (Riding and Voronova 1982b); Saltovskaya (1975) inferred affinity with filamentous cyanobacteria for *Epiphyton* as well as for *Renalcis*, whereas Pratt (1984) suggested a colonial coccoid cyanobacterial affinity for the same two genera. In addition, some workers have classified some calcareous algae as foraminifers, as for *Obruchevella* (Rejtlinger 1948), *Wetheredella* (Wood 1948; Kobluk and James 1979), and *Renalcis* (Riding and Brasier 1975).

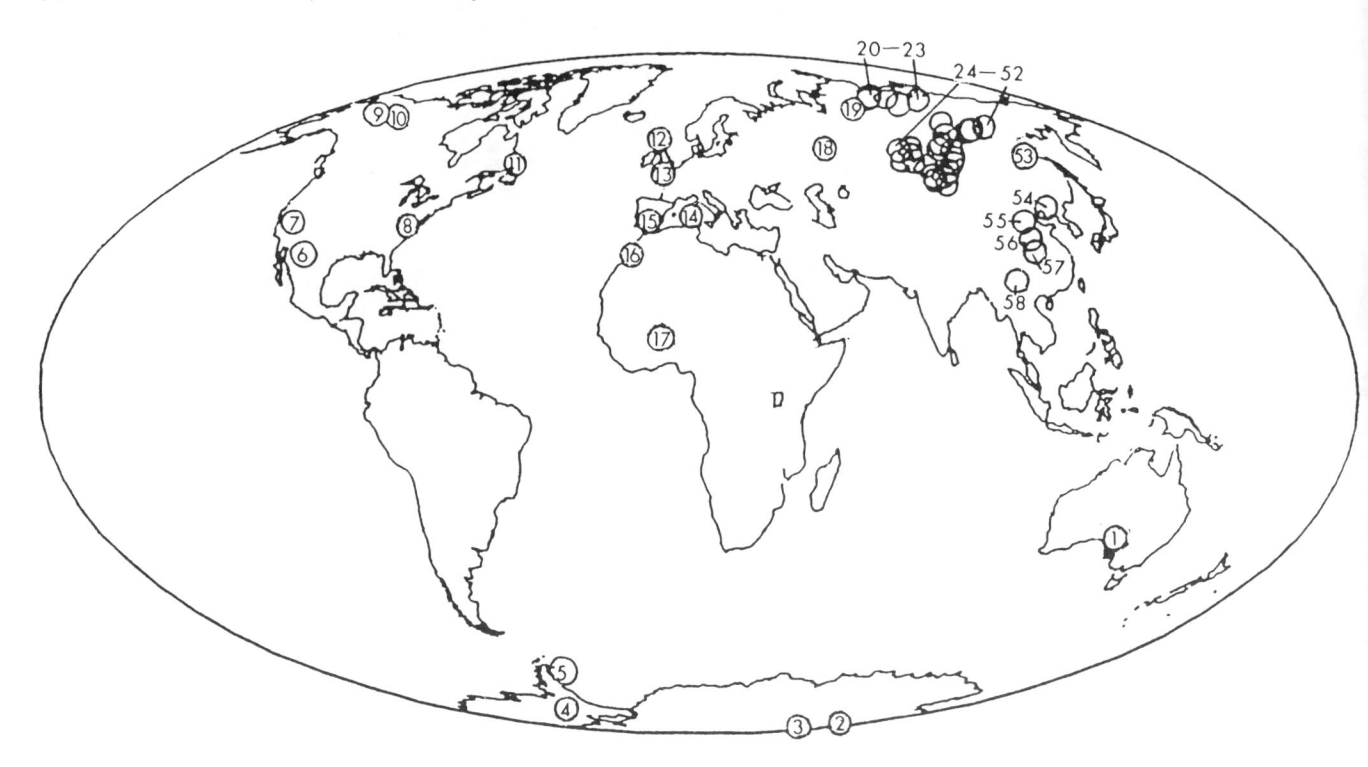

Figure 7.4.1. Distribution of Early Cambrian calcareous algae. Note the density of points in the Soviet Union and the absence of reported occurrences in South America. Numbers represent localities as follows: 1 Australia, South Australia, Flinders Ranges; 2 Antarctica, Transantarctic Mtn.; 3 Antarctica, Antarctic Expedition; 4 Antarctica, Ellsworth Mtn., Mt. Lymburner; 5 Antarctica, Weddell Sea; 6 Mexico, Caborca; 7 United States, Nevada (W) and California (E); 8 United States, Virginia (SW); 9 Alaska, Yukon; 10 Canada, NW Territories, Mackenzie Mtn.; 11 Canada, Labrador (S); 12 United Kingdom, Scotaland (NW), Inchnadamph; 13 France, Manche, Carteret; 14 Italy, Sardina (SW); 15 Spain, near Córdoba, Las Ermitas; 16 Morocco, Anti-Atlas Mtn.; 17 Mali, Gourma; 18 Soviet Union, Ural Mtn. (S); 19 Siberia, Olenek River; 20 Siberia, Yakutia, Lena River (lower reaches); 21 Siberia, Kotuy River; 22 Siberia, Anabar Massiv, Bol'shaya Kuonamka River; 23 Siberia, Ehriechka River, near Nemakit Daldyn River mouth; 24 Siberia, Yakutia, Chara River; 25 Siberia, Yakutia, Amga R, near Khomuskakh settlement; 26 Siberia, Yakutia, Lena River (middle reaches), near Kuchuguj-Keteme River mouth; 27 Siberia, Yakutia, Botoma River (right tributary of Lena River); 28 Siberia, Yakutia, Sinyaya River (left tributary of Lena River); 29 Siberia, Yakutia, Mukhatta River (left tributary of Lena River); 30 Siberia, Yakutia, Lena River, Oj-Muran village, opposite Buary River; 31 Siberia, Yakutia, Lena River, Zhurinskij mys (headlands); 32 Siberia, Yakutia, Lena River, near Zhura village; 33 Siberia, Yakutia, Lena River, Negyurchyuneh stream; 34 Siberia, Yakutia, Lena River, Krestyakh village; 35 Siberia, Yakutia, Lena River, Elovka settlement; 36 Siberia, Yakutia, Lena River, mouth of Peleduj River; 37 Siberia, Irkutsk amphitheater, Lena River, Markovskaya area; 38 Siberia, Irkutsk amphitheater, Ilim River; 39 Siberia, Eastern Sayan, Khara-Zhalga area; 40 Siberia, Yangud River basin; 41 Siberia, Kuznetsk Alatau, Tom' River; 42 Siberia, Kuznetsk Alatau, Batenevsk ridge; 43 Siberia, Irkutsk, Manzurka River; 44 Siberia, Altai Sayan, Lebed River; 45 Siberia, Western Sayan, Kyzas River; 46 Siberia, Irkut River; 47 Siberia, Tuva (Central), Bayan-Kol River; 48 Siberia, Tsaganolom uplift; 49 Mongolia, Ehgijn-Gol River (upper reaches); 50 Mongolia, Sartantuin Mtn., Idehr River basin; 51 Mongolia, Sehr' Range; 52 Mongolia, Khara-Usu Lake; 53 Siberia, Dzhagdy Range, Udy River (tributary of Mel'kan River); 54 China, Liaoning Province, South Manchuria; 55 China, Shanxi Province (N); 56 China, Henan Province, Yiyang County; 57 China, Hubei Province (W), Yichang County; 58 China, Yunnan Province, Jinning County, Meishucun area.

Affinity might be inferred from preservation styles. Today, all marine calcified chlorophytes are aragonitic in composition and characteristically are preserved as molds. Moldic preservation, on the other hand, is not typical of the Early Cambrian algae, thus strengthening the interpretation of rhodophyte or cyanobacteria affinites for the early algae. On the basis of fabric preservation of dense microcrystalline calcite and the relative enrichment of magnesium, James and Klappa (1983) inferred an original magnesian-calcite composition for *Renalcis* in Cambrian reef limestones from Labrador; such a composition is consistent with a cyanobacterial or rhodophyte affinity.

Study of calcification fabric also may prove fruitful in determining affinity. The microcrystalline calcite of modern coralline algae displays two orientations: an inner layer of calcite oriented tangential to the cell wall, and an outer perpendicular layer (Cabioch and Giraud 1986). The aragonite crystals of most chlorophytes lack orientation (Borowitzka et al. 1986) and the crystals of cyanobacteria either show no orientation or are perpendicular to the sheath (Pentecost and Riding 1986).

In summary, comparisons of Cambrian forms with modern taxa, and analysis of preservation fabrics, are most consistent with cyanobacterial affinities for many of the Cambrian "calcareous algae." Some show similarities to the rhodophytes; few (none?) seem to be related to the chlorophytes.

7.4.2B Classification by morphotypes.

Realizing the problems of determining the biological affinity of Early Cambrian calcareous algae, Riding and

Proterozoic and Early Cambrian Calcareous Algae

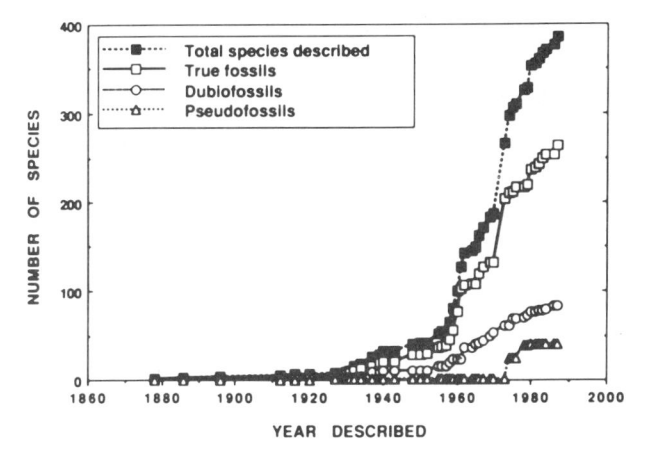

Figure 7.4.2. Cumulative plot of the number of described species of Early Cambrian calcareous algae versus year of description. As with many other groups, taxonomic studies of calcareous algae proliferated post-1950. The rise in reports of pseudofossils in the mid-1970s reflects species described as rhodophytes (e.g., referred to the family Manicosiphoniaceae) from the Proterozoic (Sinian) of China interpreted by Glaessner (1980) and by Hofmann and Jackson (1987) to be nonbiological radial-fibrous mineralic structures.

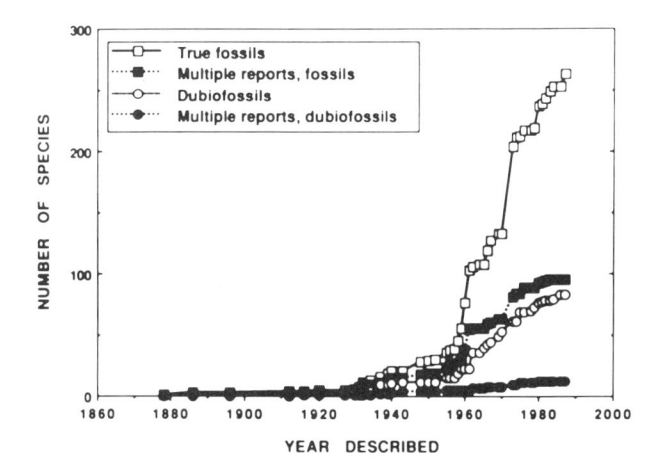

Figure 7.4.3. Cumulative plot of described species of Early Cambrian calcareous algae interpreted to be fossils (squares) and dubiofossils (circles) versus year of description. Open symbols represent all fossil species; solid symbols represent species that have been figured from more than one locality. The difference between the two "true-fossil" curves (open and solid squares) emphasizes the effect of monographic biases. However, excessive splitting of taxa has not been factored out, because although reported from several localities, many species are recognized by but a single worker. The large discrepancy between the two "dubiofossil" curves (open and solid circles) perhaps stresses the doubtful nature of these taxa.

Voronova (1985) proposed a morphological classification that was adopted with minor alteration for this study (see Table 23.2). They group the algae into six primary morphological series: spherical, botryoidal, dendritic, tubiform, tuberous, and cup-like. Most Early Cambrian calcareous algae display botryoidal, dendritic, or tubiform forms (Figs. 7.4.4 through 7.4.7). Most of the spheroidal forms are dubiofossils and may be altered oöliths (Riding and Voronova 1985, p. 60) and those in the tuberous group include Kordeh's dasycladacean (chlorophyte) algae that have been questioned by Bassoulet et al. (1979, p. 432); the tuberous group comprises the problematic radiocyathans (Section 7.7.5), which were not included in this study.

As with all classification schemes, some forms defy classification and appear transitional. Riding and Voronova recognized the possibility of morphological convergence and cautioned against equating a morphologic series with taxonomic affinity.

7.4.3 Diversity

Strong geographic (Siberia) and monographic (e.g., Kordeh 1961, 1973) biases hinder evaluation of diversity. The value of inferred species diversity is especially questionable due to extreme splitting or lumping biases, probable plasticity of form, and diagenetic alteration.

7.4.3A Extreme splitting

Most described genera of Early Cambrian calcareous algae are represented by only one to three species, probably reflecting unnecessary splitting of genera. On the basis of inferred (but unconvincingly demonstrated) reproductive structures, Kordeh (1961, 1973) classified numerous simple forms as several genera of eukaryotic algae. The genus *Epiphyton*, on the other hand, comprises about 90 Early Cambrian species (about one-third of all Early Cambrian calcareous algae species), rendering recognition of species virtually impossible.

7.4.3B Extreme lumping

Soviet paleontologists commonly recognize 10 to 20 genera of calcareous algae. Perusal of the non-Soviet literature, however, gives the impression that only three genera of calcareous algae occur in the Cambrian: *Girvanella*, *Renalcis*, and *Epiphyton* (Figs. 7.4.6, 7.4.7). This discrepancy probably cannot be attributed to provinciality of the Soviet genera, but rather to the lack of recognition and/or acceptance of many genera by non-Soviet workers. For example, many illustrations of *Girvanella* from North American deposits display the growth form of *Razumovskya* (e.g., James 1981, Fig. 12A, 14C; Read and Pfeil 1983, Fig. 5A, 7A; Kobluk 1985, Fig. 7); many *Epiphyton* might be called *Gordonophyton* (as recognized by James 1981, Fig. 9D) or *Kordephyton* (Fig. 7.4.7; Read and Pfeil's flat-lying *Epiphyton*?, 1983, Fig. 6E).

7.4.3C Plasticity of form

Today, calcareous algae characteristically display marked plasticity of form as a function of environmental variables as demonstrated for the foliose, lightly calcified brown alga, *Padina jamaicensis* (Lewis et al. 1987), and for the coralline alga, *Lithophyllum congestum* (Steneck and Adey 1976). Plasticity of form in the Cambrian genera would lead to a proliferation of specific, and possibly generic, names. Many of the numerous species of *Epiphyton* might be explained by plasticity of this branching form.

7.4.3D Diagenetic alteration

Calcareous algae, and carbonates in general, are especially sensitive to diagenetic alteration that can easily confuse taxonomists. Voronova described the species *Microcodium laxus* from the Nemakit-Daldyn Horizon of Siberia (Voronova

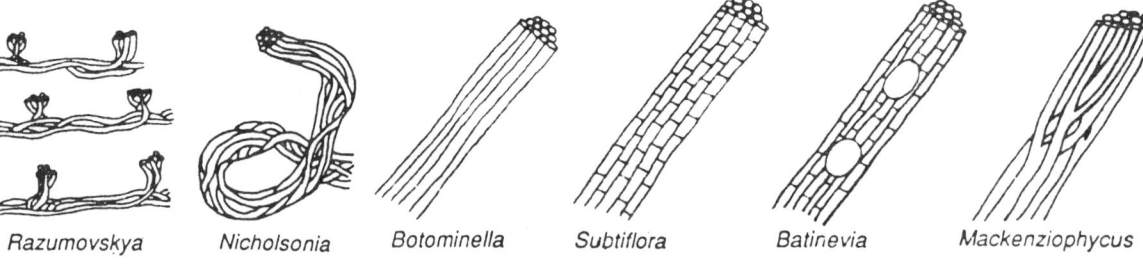

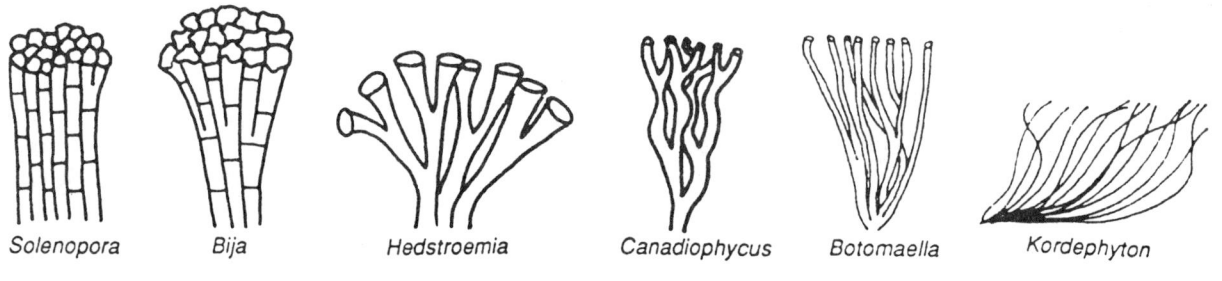

Figure 7.4.4. Interpretive simplified drawings showing variety of morphologies characteristic of Early Cambrian calcareous algae. As discussed in the text, some of these genera are probably diagenetic variants of others. Botryoidal forms can be differentiated by wall structure: micritic (*Renalcis*), fibrous (*Tarthinia, Acanthina*), or peloidal (*Gemma*). Dendritic forms may display dichotomous branching (*Epiphyton, Gordonophyton, Tubomorphophyton, Epiphytonoides, Chabakovia, Cambrina, Serligia*), irregular branching (*Sajania, Bajanophyton, Korilophyton*), or a micritic bushlike appearance (*Angulocellularia*). Internal structure (e.g., hollow, solid, or chambered) further divides dentritic forms. Tubiform morphologies can be (1) single straight (*Proaulopora*) or curved (*Obruchevella*) tubes; (2) irregular masses (*Wetheredella, Girvanella*); (3) bundles (*Razumovskya, Nicholsonia, Botominella, Subtiflora, Batinevia, Mackenziophycus*); or (4) fans (*Solenopora, Bija, Hedstroemia, Canadiophycus, Botomaella, Kordephyton*).

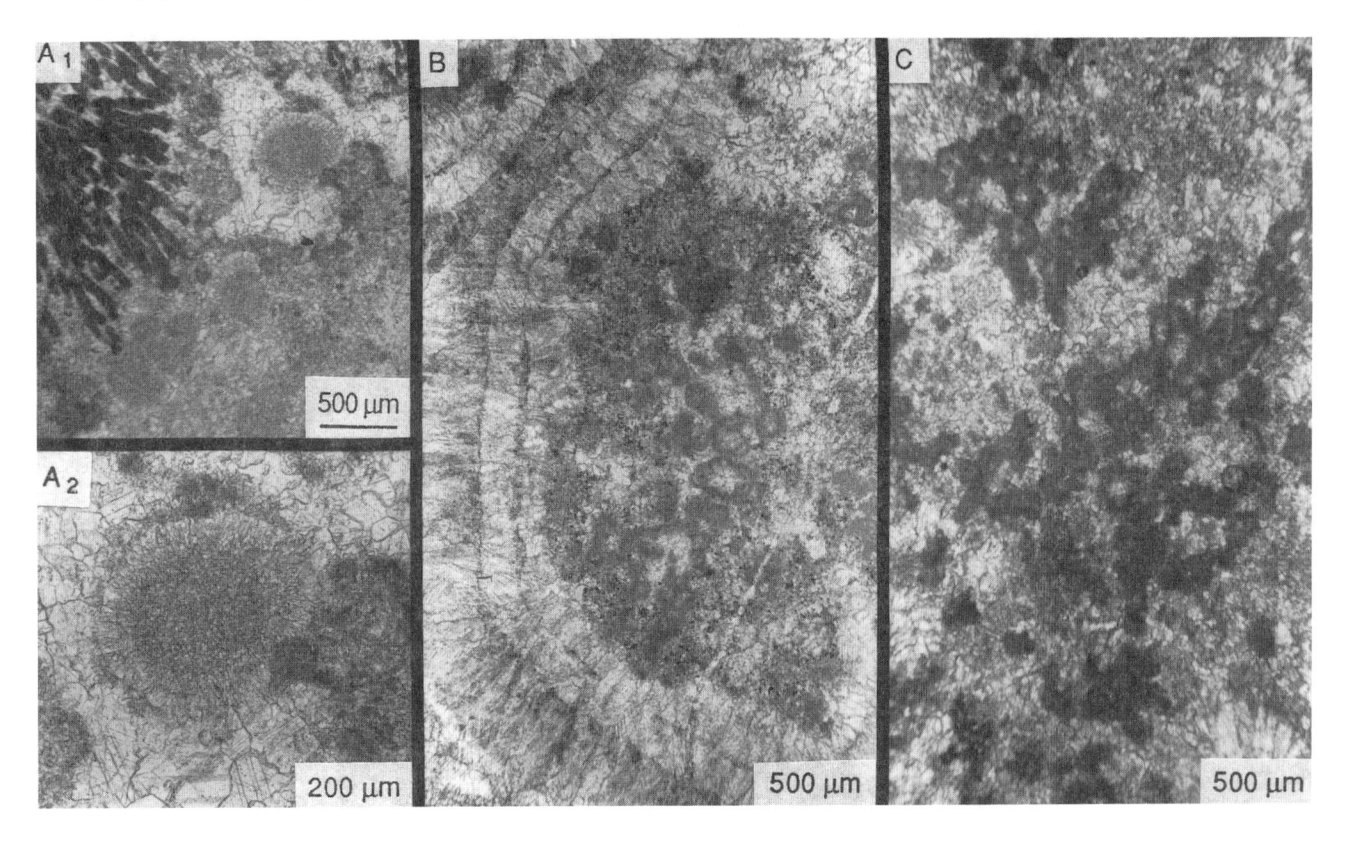

Figure 7.4.5. Optical photomicrographs (transmitted light) showing spheroidal (A) and botryoidal (B, C) calcareous algae in petrographic thin sections of carbonate rocks from the Atdabanian Stage (Lower Cambrian) of the Sehr' Range of western Mongolia (A_1, A_2, B) and from the Lower to Middle Cambrian Shady Dolomite of Virginia, USA (C). (A_1, A_2) *Acanthina* (a single specimen shown at two magnifications) showing circular outline and fibrous rim (PPRG sample no. 2787). (B) Irregular bubble-like morphology of *Renalcis* and thick isopachous cements (PPRG sample no. 2786). (C) *Renalcis* showing greater organization of bubble-like bodies into branches, approaching the morphology of *Chabakovia*.

and Missarzhevskij 1969) but later (Voronova 1976) decided the specimens were recrystallized forms of the common genus *Renalcis*; Riding and Voronova (1984, p. 207) hint that the same may be true of the genus *Panomninella*. Many of Kordeh's genera, such as *Cambrina*, *Filaria*, *Kyzassia*, *Potentilina*, and *Sporinula*, may be diagenetic variants of the genus *Epiphyton*; *Epiphytonoides* may be altered *Gordonophyton* or *Chabakovia*.

Saltovskaya (1975) and Pratt (1984) suggested that *Epiphyton* and *Renalcis* represent endpoints of a diagenetic continuum that encompasses genera such as *Chabakovia* and *Gordonophyton*. Pratt proposed that the different forms resulted from variations in environmental conditions (mainly turbulence) and synsedimentary or postmortem calcification of coccoid cyanobacterial colonies. Riding and Voronova (1985, p. 74) and Cherchi and Schroeder (1985a, p. 144), however, argued rather convincingly in favor of the uniqueness of *Epiphyton* and *Renalcis* taxa.

7.4.4 Paleoecology

Relatively few accounts of the paleoecology of calcareous algae have been published and unfortunately rarely accompany detailed taxonomic studies. Whereas most taxonomic studies are published in the Soviet literature, most paleoecologic/sedimentologic studies come out of the non-Soviet world. An exception has been the study of cavity dwellers by Kobluk and James (1979) and Kobluk (1981, 1985). Other examples of paleoecologic/sedimentologic studies include those of James and Kobluk (1978), James (1981), James and Klappa (1983), Read and Pfeil (1983), Coniglio and James (1985), and Selg (1986).

7.4.5 Stratigraphic Range

In comparison with the range chart here included (Fig. 7.4.8), previously published range charts for Early Cambrian taxa are more restricted in terms of taxa shown (e.g., Wray 1977) or in terms of geographic distribution (Riding and Voronova 1984). Although the present study was not intended to be a taxonomic revision, some general decisions on taxonomic validity had to be made in order to construct a range chart. Difficulties arise due to the inability to correlate stratigraphic units from one continent to another with confidence. Given the restrictions, a "best-guess" range chart was constructed for genera listed in Chapter 23 interpreted to be Proterozoic or Early Cambrian calcareous algae (Fig. 7.4.8).

Most Vendian or older occurrences are questionable, representing either single occurrences, poorly illustrated and/or described taxa, or taxa from poorly dated strata. Klein et al. (1987) reported calcified sheaths referred to the genus *Siphonophycus* from 2.5- to 2.3-Ga-old strata of the Transvaal Supergroup in South Africa, but the carbonate associated with the sheaths could be diagenetic (N. J. Beukes, pers. comm. 1988). Other possible Vendian or older occurrences of calcareous

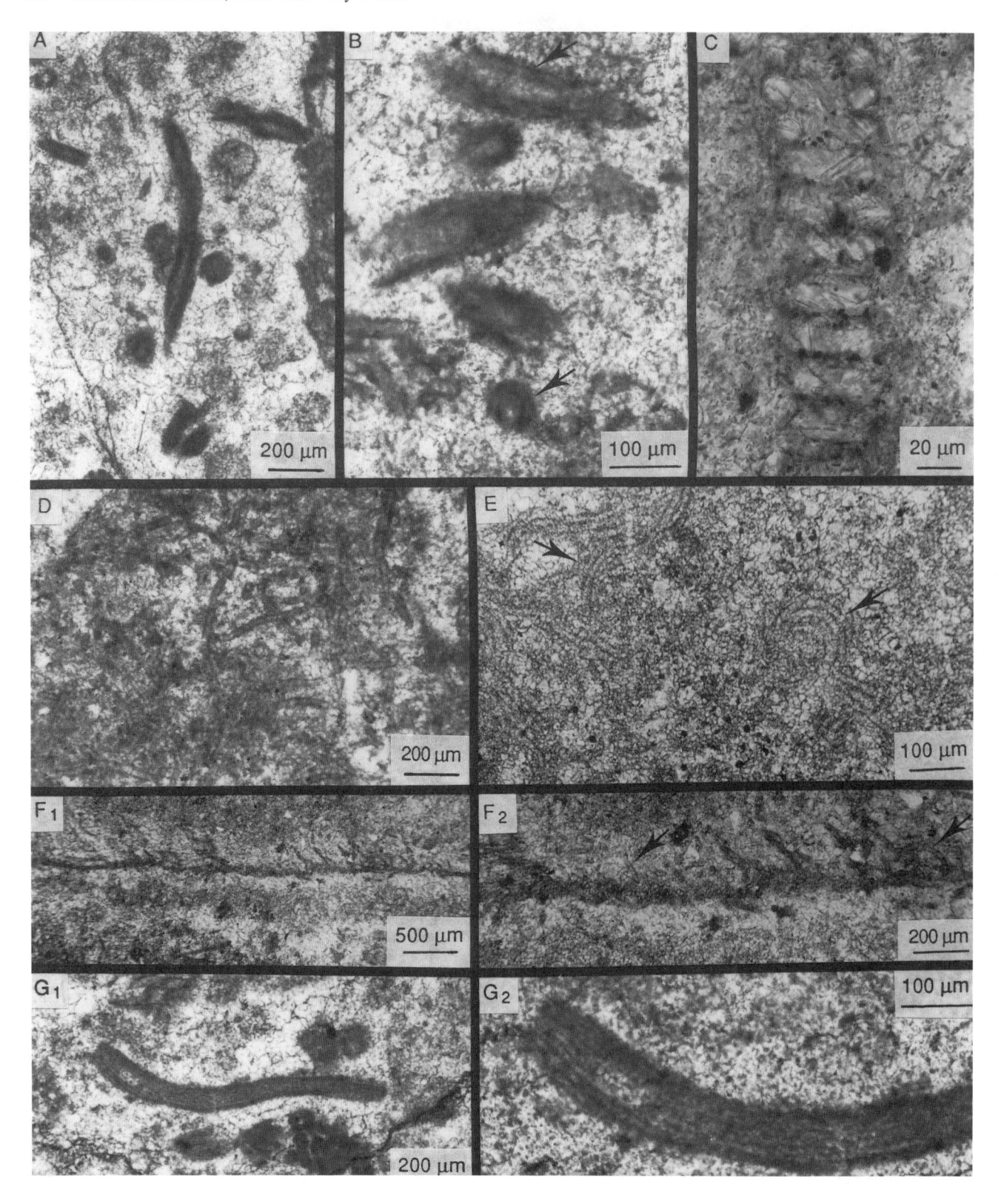

Figure 7.4.6. Optical photomicrographs (transmitted light) showing tubiform calcareous algae in carbonate rocks from the Atdabanian Stage (Lower Cambrian) of the Sehr' Range (A, B, E, G) and Tsaganolomsk Uplift (D) of western Mongolia; the Middle Cambrian Burgess Shale of British Columbia, Canada (C); and the Lower to Middle Cambrian Shady Dolomite of Virginia, USA (F). (A) Thick-walled tubes of *Proaulopora* (PPRG sample no. 2789). (B) Layered-wall character (at arrows) of segments of *Proaulopora* (PPRG sample no. 2789). (C) Characteristic helical coils of *Obruchevella* (Smithsonian Institution, Washington, D.C., no. 444294). (D) Tangled filaments of *Girvanella* (PPRG sample no. 2785). (E) *Nicholsonia*, differentiated from *Girvanella* by the occurrence of bundles of parallel filaments (at arrows; PPRG sample no. 2788). (F_1, F_2) *Razumovskya* showing "felted" fabric; higher magnification (F_2) reveals the tubiform nature (at arrows) of the genus. (G_1, G_2) Filament bundles and oval-shaped area devoid of filaments typical of *Batinevia* (PPRG sample no. 2789).

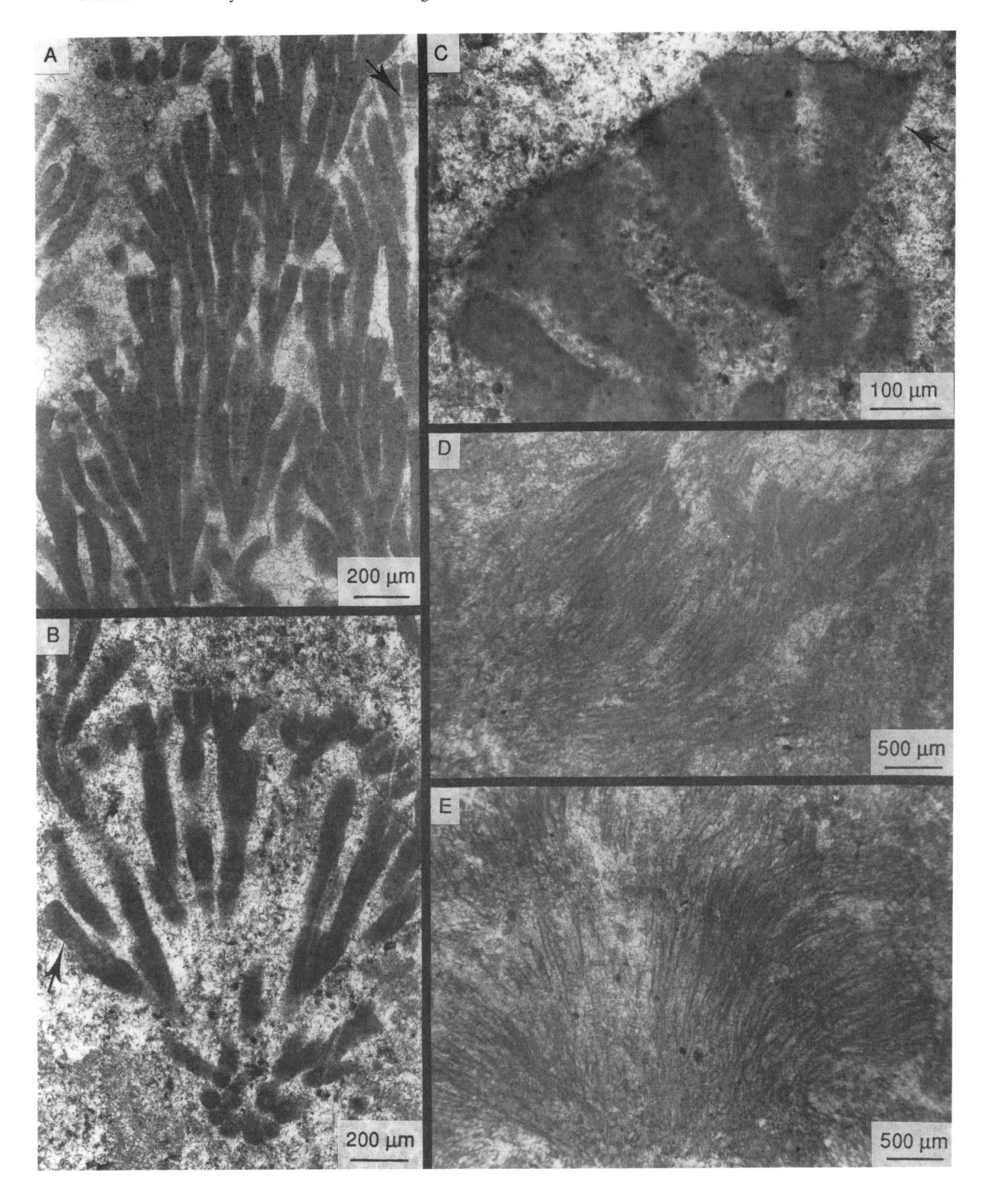

Figure 7.4.7. Optical photomicrographs (transmitted light) showing tubiform and dentritic calcareous algae from the Atdabanian Stage (Lower Cambrian) of western Mongolia (A–C) and the Lower to Middle Cambrian Shady Dolomite of Virginia, USA (D, E). All of these forms might be referred to the genus *Epiphyton*. However, note the barely visible segmentation (arrows in A–C) that is characteristic of *Gordonophyton*; such segmentation is a taxonomic feature potentially lost during diagenetic alteration. In D and E, the thin wispy filaments typify the relatively rarely occurring genus *Kordephyton*.

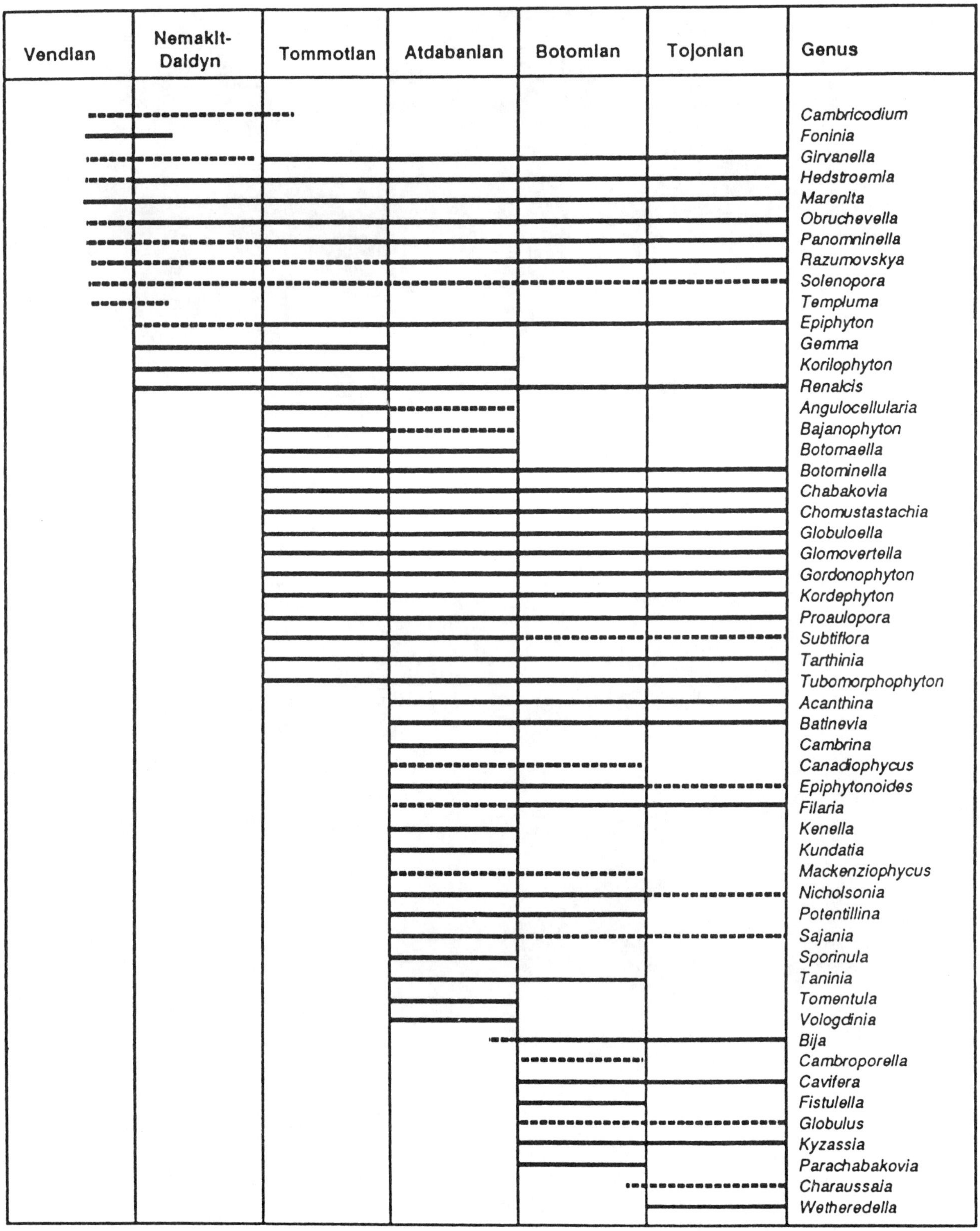

Vendian	Nemakit-Daldyn	Tommotian	Atdabanian	Botomian	Tojonian	Genus
	----	----				Cambricodium
	—					Foninia
----	----	————————————————				Girvanella
----	————————————————————					Hedstroemia
	————————————————————					Marenita
----	————————————————————					Obruchevella
----	————————————————————					Panomninella
----	----	————————————————				Razumovskya
----	----	----	----	----	----	Solenopora
----	----					Templuma
	----	————————————————				Epiphyton
	————————————————————					Gemma
	————————————————————					Korilophyton
	————————————————————					Renalcis
			----	----		Angulocellularia
			----	----		Bajanophyton
		————————————————				Botomaella
		————————————————				Botominella
		————————————————				Chabakovia
		————————————————				Chomustastachia
		————————————————				Globuloella
		————————————————				Glomovertella
		————————————————				Gordonophyton
		————————————————				Kordephyton
		————————————————				Proaulopora
		————————	----	----		Subtiflora
		————————————————				Tarthinia
		————————————————				Tubomorphophyton
			————————			Acanthina
			————————			Batinevia
			————————			Cambrina
			----	----		Canadiophycus
			————————	----		Epiphytonoides
			----	————		Filaria
			————————			Kenella
			————————			Kundatia
			----	----		Mackenziophycus
			————————	----		Nicholsonia
			————————			Potentillina
			————————	----		Sajania
			————————			Sporinula
			————————			Taninia
			————————			Tomentula
			————————			Vologdinia
			----			Bija
				----		Cambroporella
				————		Cavifera
				————		Fistulella
				----	----	Globulus
				————		Kyzassia
				————		Parachabakovia
				----	----	Charaussaia
				————		Wetheredella

Figure 7.4.8. Stratigraphic ranges of calcareous algae from the latest Vendian through the Early Cambrian.

algae include a possible rhodophyte from the Nama Group of Namibia (Grant et al. 1987) and a possible alga from the Pahrump Group of California, USA (R. J. Horodyski, pers. comm. 1988).

7.4.6 Summary

Serious monographic and geographic biases plague the study of calcareous algae. Simple morphologies of the Early Cambrian algae, and the vagaries of diagenesis, complicate taxonomic classification. Taxonomic revision is necessary if we ever hope to use calcareous algae to their full potential for paleoenvironmental study and for deciphering the early evolution of calcareous metaphytes.

Acknowledgments

I thank J. F. Read, Virginia Polytechnic Institute and State University, for loaning me thin sections from the Shady Dolomite for study and photography. I thank M. A. Fedonkin for providing thin sections of calcareous algae from N. A. Drozdova's collection from western Mongolia.

7.5

Proterozoic Metazoan Body Fossils

BRUCE N. RUNNEGAR   MIKHAIL A. FEDONKIN

7.5.1 Introduction

The search for the oldest animal fossils (Fig. 7.5.1) has produced a large number of objects of pre-Vendian age that are generally considered to be metazoan dubiofossils or pseudofossils (Sections 7.6, 7.8; Chapter 23). They include well-understood sedimentary structures such as the sinuous mudcracks that may form between parallel ripple crests (Fig. 7.8.3) – often interpreted as worm trails (e.g., Frarey and McLaren 1963; Angelucci 1970; Glaessner 1969) – as well as more enigmatic "metazoans" that are often known only from a single specimen (Fig. 7.5.2A; see Section 7.6).

It may not be a simple matter to decide whether a complicated structure is or is not a fossil. To some, the unique specimen of *Brooksella canyonensis* (Fig. 7.5.2A) is as convincing an early metazoan as is the mediocre "medusoid" from the Ediacara Member of the Rawnsley Quartzite shown in Figure 7.5.2B. However, the South Australian "medusoid" is regarded as a true fossil because it has features that are found on numerous other better specimens. So one important test of authenticity is the presence of similar structures at the same horizon or locality. Another is the degree of complexity; none of the purported pre-Vendian fossil metazoans has features which could not have been produced by inorganic processes.

7.5.2 The Ediacara Faunas

Nearly all of the known Vendian metazoans are components of the Ediacara fauna, reported first from south west Africa in the 1920s (Gürich 1929; Jenkins 1985) and discovered independently at Ediacara in South Australia about 20 years later (Sprigg 1947, 1988). This distinctive association, comprising the impressions of sizable, soft-bodied organisms, has since been found on all continents except South America and Antarctica (Sprigg 1946, 1947; Glaessner 1984). The most informative sites are in Namibia (Gürich 1933; Richter 1955; Pflug 1966, 1970a, b, 1972, 1973; Germs 1972a, 1973a; Hahn and Pflug 1985, 1988; Jenkins 1985); southern and central Australia (Sprigg 1947, 1948; Glaessner and Daily 1959; Glaessner and Wade 1966; Wade 1969, 1972a, b; Jenkins and Gehling 1978; Glaessner 1984; Gehling 1987, 1988; Walter et al. 1989); England and Wales (Ford 1958, 1963; Cope 1977, 1983); southeastern Newfoundland (Anderson and Misra 1968; Misra 1969; Anderson 1978; Anderson and Conway Morris 1982); eastern Europe (Zaika-Novatsky et al. 1968; Palij 1969, 1976; Palij et al. 1979; Velikanov et al. 1983); northern Russia (Keller et al. 1974; Keller and Fedonkin 1977; Fedonkin 1981; Sokolov and Fedonkin 1984; Fedonkin 1985); the Olenyok Uplift and other parts of Siberia (Sokolov and Fedonkin 1984; Fedonkin 1985; Vodanyuk 1989); northwest Canada (Hofmann 1981; Narbonne and Hofmann 1987; Aitken 1989); and North Carolina, USA (Gibson et al. 1984). Other probable or possible occurrences of this biota include discoidal and other structures reported from Norway, Iran, British Columbia, India, and Liaoning, China (Føyn and Glaessner 1979; Hahn and Pflug 1980; Hofmann et al. 1985; Sun 1986a; Xing et al. 1989; Mathur and Shanker 1989), the "pennatulid" *Paracharnia* from China (Sun 1986b), and enigmatic worm-like fossils from North Carolina (Cloud et al. 1976).

In addition to the foregoing, there are coeval trace fossils (reviewed in Section 7.6, and by Crimes 1987) and at least three

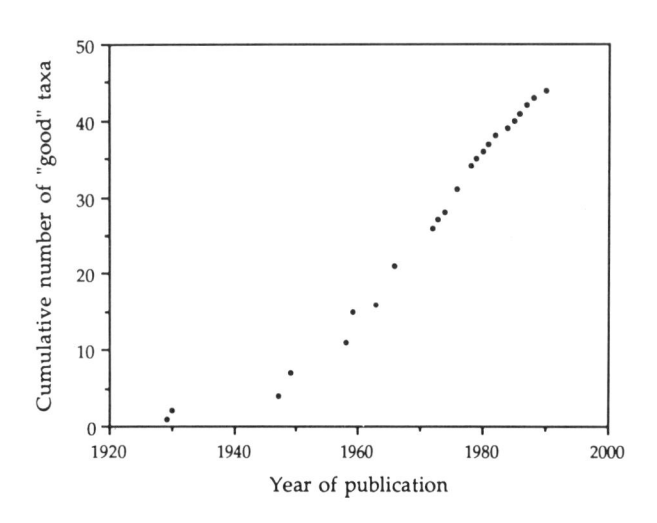

Figure 7.5.1. Rate of publication of distinctive, identifiable taxa from the Vendian Ediacara faunas. Almost all taxa are monospecific genera. The diagram indicates that, on average, about one new taxon is described each year.

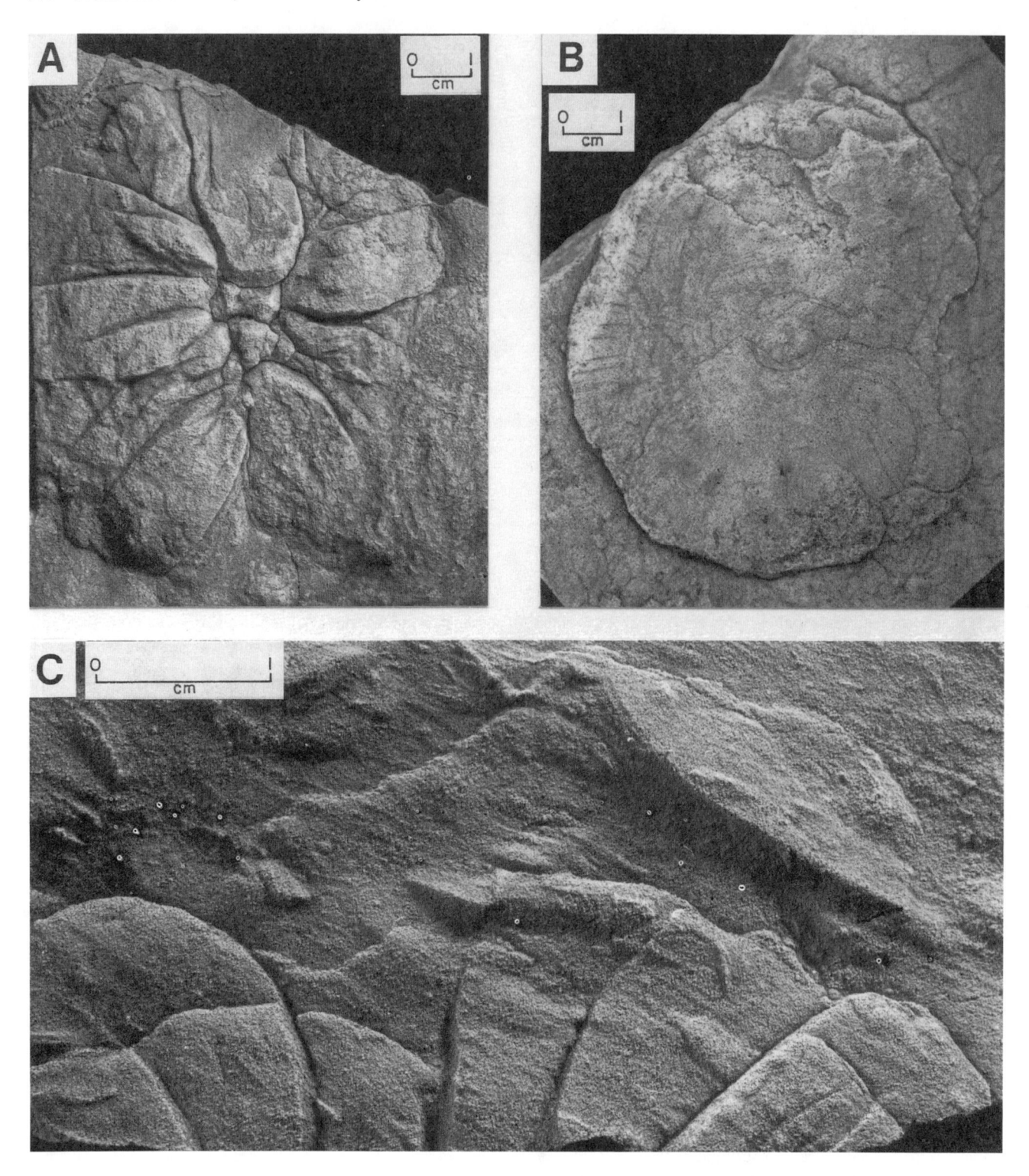

Figure 7.5.2. Vendian pseudofossils (A, C) and an Ediacaran "medusoid" (B). Lines for scale are approximate. (A) *Brooksella canyonensis* Bassler 1941; U.S. National Museum (USNM) no. 99438, Nankoweap Group, Grand Canyon, Arizona, USA. (B) Cast of unidentifiable "medusoid" on the base of a slab of sandstone from the Rawnsley Quartzite, Mt. Scott Range, northern Flinders Ranges, South Australia; note the "old-elephant-skin texture" of the adjacent sandstone. (C) Supposed second specimen of *Brooksella canyonensis* discovered at the original locality by G. Webers and P. Cloud in 1962; note the tool marks that are truncated by post-depositional fractures; USNM no. 399543.

other kinds of fossils that seem to be of metazoan origin. The latter include: (i) *Cloudina hartmannae* (Germs 1972b), a complex, tubular, calcareous structure (Fig. 7.5.11A) that occurs just above[1] typical members of the Ediacara fauna in Namibia (Germs 1972a; Glaessner 1976); (ii) *Redkinia spinosa* (Sokolov 1976, 1985), microscopic, comb-like objects of organic composition from mid-Vendian strata of the Russian Platform (Fig. 7.5.11C; see also Chapter 24, Pl. 52B-D); and (ii) *Saarina* sp., an organic-walled, annulated tube from a subsurface part of the rock unit that contains the Ediacara fauna in the Archangel region of northern Russia (Fig. 7.5.11B).

Each of these taxa is poorly understood. *Cloudina* appears to have been constructed by a solitary, sessile organism that inhabited a thin-walled, funnel-shaped tube made originally of aragonite or magnesian calcite. As the organism grew vertically, it created new funnel-shaped chambers that were often attached eccentrically to inner walls of the previously formed ones. The open space between successive walls was filled with carbonate cement during diagenesis (Germs 1972b; Glaessner 1976a; S. Grant pers. comm. to B. N. R.). *Cloudina* is best known from the Vendian of Namibia, but it is also found in Brazil (Hahn and Pflug 1985; Zaine and Fairchild 1987), Oman (Simon Conway Morris pers. comm. to B. N. R.) and, perhaps, in other parts of the world including Antarctica and North America. It appears to be the oldest organism known to have formed a mineral skeleton.

Sokolov (1976) first thought that *Redkinia* might be related to an early onychophoran like *Aysheaia* from the Burgess Shale (Whittington 1985), but its jaw-like morphology led to it being regarded as a new kind of scolecodont (Redkiniida; Sokolov 1985). However, the delicate nature of the secondary spines of *Redkinia* (Chapter 24, Pl. 52B-D) make it unlikely that the whole comb functioned as a mandible. Instead, *Redkinia* may be a fragment of the edge of an arthropod carapace; there are also some superficial similarities to the much larger "arthropod" arms of *Anomalocaris* (Briggs 1979; Dzik and Lendzion 1988). For the present, *Redkinia* is best regarded as a core member of the Problematica (Bengtson 1986b).

Saarina is one of several similar genera of Saarinidae and Sabellidititidae from Vendian and Early Cambrian strata of the Soviet Union which Sokolov (1965, 1967, 1968) has suggested are the tubes of early Pogonophora. Urbanek and Mierzejewska (1983) attempted to test this hypothesis by comparing the ultrastructure of Cambrian sabelliditids with that of modern pogonophoran tubes, but the results of their study were inconclusive. Excellent specimens from a borehole in the Archangel region, identified as *Palaeolina* sp. in Sokolov and Ivanovski (1985), are here referred to *Saarina* Sokolov because they appear to be constructed from a series of funnel-shaped increments (Fig. 7.5.11B). *Palaeolina*, like other sabelliditids, has transverse markings (growth lines?) but lacks the regularly annulated structure of the saarinids (Sokolov 1965, 1967, 1968).

Corumbella werneri (Hahn et al. 1982; Zaine and Fairchild 1987) is a distinctive annulated tube from the ?Vendian Corumbá Group, Brazil, that may be allied to the saarinids. Known only from casts and molds, *Corumbella* has regular longitudinal marks that have been interpreted as short, equally spaced septa (Hahn et al. 1982).

Another suite of annulated tubes has been described from the Late Proterozoic Liulaobei and Jiuliqiao Formations of Anhui, China, by Zheng (1980), Wang (1982), Sun et al. (1986), and Chen (1988). Their interpretation of at least some of these fossils, and *Pararenicola huaiyuanensis* in particular, as the remains of worm-like metazoans has resulted in criticisms of both the biological interpretation and the assigned age (Cloud 1986; Vidal and Moczydłowska 1987). Despite Sun's (1986, 1987) effective replies, the matter remains unresolved.

In contrast to the Middle Cambrian rhabdopleurid described by Bengtson and Urbanek (1986), none of these Late Proterozoic, annulated, organic-walled tubes shows growth increments that could be construed to be fuselli of the hemichordate type. When the problem of determining the phylogenetic relationships of a much more morphologically complex and taxonomically diverse group of organisms such as the graptolites is appreciated (Urbanek 1986), it becomes clear that it may never be possible to classify *Saarina*, *Sabellidites*, *Corumbella*, and *Pararenicola* in any biologically meaningful way. On the other hand, if the similarities in age and morphology are indicative of a common heritage among this set of taxa, then the combination of characters displayed by the group may provide an answer to the question of their affinities.

7.5.2A Age of the Ediacara Faunas

All known occurrences of the Ediacara fauna are either in strata that underlie those containing the earliest skeletal fossils or in locations where other evidence indicates a latest Proterozoic age. The stratigraphic distance from the youngest Ediacaran fossil to the oldest Cambrian skeletal fossil or trace fossil varies from as little as 10 meters at Ediacara (Goldring and Curnow 1967; Jenkins et al. 1983) to some thousands of meters in Newfoundland (Anderson 1978; Anderson and Conway Morris 1982). Stratigraphic studies in Australia had suggested that a substantial hiatus separates the Ediacara fauna from the overlying Cambrian succession (Daily 1972, 1973), but this interpretation is being challenged by regional sequence stratigraphy (Lindsay 1987b; von der Borch et al. 1988) and local analysis (Mount 1989a; Walter et al. 1989; but see also Jenkins 1989). It now seems likely that the successions in the Flinders Ranges and Amadeus Basin are reasonably continuous and that the Ediacara fauna pre-dates the beginning of the Cambrian by as little as ten to twenty million years. A recent U-Pb date of 566 ± 5 Ma on zircon from a tuff interbedded with the Ediacara fauna in Newfoundland (Benus 1989) is consistent with this interpretation, as is the age of 540 to 590 Ma estimated for the beds containing *Pteridinium* in North Carolina (Harris and Glover 1988; Gibson 1989), and also a set of mediocre Rb-Sr whole-rock isochrons from south Australia (Compston et al. 1987). However, the presence of the Zone III trace fossils *Diplocraterion*, *Plagiogmus*, and *Rusophycus* (Fig. 7.6.1) in the oldest Cambrian horizons of the Flinders Ranges (Daily 1976) is evidence for an hiatus in the South Australian

---

[1] The only published illustration of a well-characterized member of the Ediacara fauna that is said to come from a horizon above *Cloudina* is an excellent specimen of *Rangea* from the "Basal Clastic Member of the Schwarzrand Formation" on the farm Chamis, near Helmeringhausen, Namibia (Germs 1972a, Pl. 20, Fig. 7; 1973a, Fig. 1E).

succession. The unpalatable alternatives are that the Ediacara fauna is as young as Cambrian in its type area (Crimes 1987; Mount 1989a), or that *Diplocraterion, Plagiogmus,* and *Rusophycus* appear earlier in South Australia than elsewhere. On taphonomic grounds the former is more likely than the latter; the Ediacaran body fossils are confined to a distinctive rock unit within the Rawnsley Quartzite (Pound Subgroup) known as the Ediacara Member (Jenkins et al. 1983). This unusual facies represented a taphonomic window that may well be diachronous, being older in the south than in the north (Mount 1989b). The recent discovery of many Middle Cambrian Burgess Shale taxa in the Early Cambrian Chengjiang fauna of southern China, and the possible occurrence of an "Ediacaran" sea-pen in the Burgess Shale (Zhang and Hou 1985; Conway Morris 1989a), should remind us that it is necessary to view stratigraphic ranges derived from *Fossil-Lagerstätten* with considerable circumspection.

The younger isotopic ages suggested recently for the base of the Cambrian (Jenkins 1981, 1984b; Odin et al. 1983; Conway Morris 1989b), and hence the Ediacara fauna, are at odds with dates that have been widely used in the Soviet Union and China (Keller 1979; Sokolov and Fedonkin 1984; Xing et al. 1984; Odin et al. 1985). This is a matter that will be addressed by the newly formed IUGS Working Group on the Terminal Proterozoic System; most of the available published dates are shown on the Vendian-Early Cambrian correlation chart (Table 7.5.2).

7.5.2B Extinction of the Ediacara Fauna

Cloud (1948) and Schindewolf (1955) argued forcefully for a period of "eruptive" evolution in the history of the Metazoa at the beginning of the Phanerozoic, thereby displacing the established notion that the perceived abruptness of the Precambrian-Cambrian boundary event was largely due to the inadequacy of the Late Proterozoic fossil record. Although this idea of a major metazoan radiation is generally accepted in one guise or another (Towe 1970; Sepkoski 1978; Runnegar 1982a; Valentine 1988; Bergström 1989), the scenario has now been enlarged to incorporate a mass extinction of the pre-existing Ediacara fauna (Seilacher 1983, 1984, 1989). According to Lewin (1984), it was "the first major extinction of many that have punctuated the long history of multicellular animals, and it removed a form of life that was alien to all that followed."

Reports of the discovery of *Pteridinium* and *Dickinsonia* in Cambrian strata (Cloud and Nelson 1966; Borovikov 1976) have proved to be incorrect, but there are some typical Ediacaran taxa that may have direct Phanerozoic descendants. These taxa, and lines of additional relevant evidence, include:

(1) A possible "pennatulid" from the Burgess Shale (Conway Morris 1989).

(2) The previously unemphasized but possibly significant similarities among various kinds of Vendian and Cambrian discoidal fossils such as *Aspidella hatyspytia* Vodanyuk, *Bonata* Fedonkin, *Eldonia* Walcott, *Eomedusa* Popov, *Rotadiscus* Sun & Hou, *Tateana* Sprigg, *Spriggia* Southcott, *Stellostomites* Sun & Hou, *Velumbrella* Stasinska, *Yunnanomedusa* Sun & Hou (Stasinska 1960; Popov 1967, 1968; Durham 1974; Fedonkin 1980; Jenkins 1984a; Sun 1986a; Sun and Hou 1987; Vodanyuk 1989).

(3) The close proximity in time between the triradial body plan of the Vendian trilobozoans *Tribrachidium* (Fig. 7.5.9A), *Skinnera, Albumares* (Fig. 7.5.9C), and *Anafesta* (Fig. 7.5.9D), and the three-fold symmetry of Early Cambrian anabaritids (Conway Morris and Chen 1990), which may imply a phylogenetic connection (Fedonkin 1985b).

(4) The artificial separation of Vendian "medusoids" from comparable Phanerozoic "trace fossils," such as the forms placed in *Bergaueria* by Arai and McGugan (1969).

(5) The similarity of *Ovatoscutum* from Ediacara (Fig. 7.5.6E) and the White Sea (Fig. 7.5.7F) to Phanerozoic fossils such as *Plectodiscus* Reudemann and *Silurovelella* Fisher, which are regarded as chondrophorine cnidarians (Fisher 1957; Yochelson and Stanley 1981; Yochelson et al. 1983; Stanley 1986 Stanley and Yancey 1986).

(6) The possibility, discussed below, that *Spriggina* may be a stem-group arthropod and *Arkarua* (Gehling 1987) an ancestral echinoderm.

(7) The fact that the associated trace fossils demonstrate the existence of otherwise unknown, "invisible," Vendian animals that may well have Paleozoic descendants.

Given the number of disparate metazoan clades that may have survived, albeit in reduced numbers, from the Vendian to the Cambrian, the case for an end-Proterozoic mass extinction is weak. The Ediacara fauna – like most Precambrian fossil assemblages – represents a trivial sample of the contemporaneous biota, and so it is difficult to use its disappearance to make a case for anything more than "background" attrition.

7.5.2C Preservation of the Fossils

At Ediacara and other localities in the Flinders Ranges,

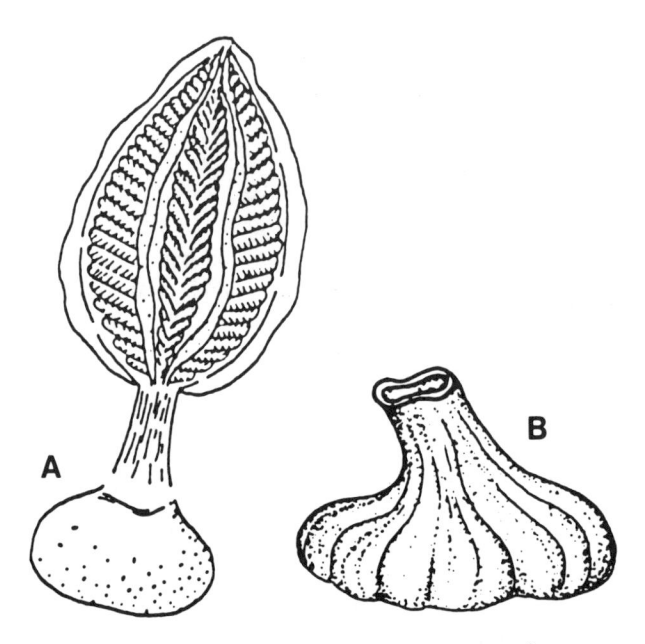

Figure 7.5.3. Reconstructions of two sessile animals from the Ediacara fauna. (A) *Rangea schneiderhoehni* Gürich 1929, Nama Group, Namibia (after Jenkins 1985, Fig. 8; republished with permission). (B) *Inaria karli* Gehling 1988, a possible actinian-grade cnidarian (after Gehling 1988, Fig. 5B; republished with permission).

Table 7.5.1. *Classification of the Vendian Metazoa* (B. N. R.)

Phylum Petalonamae Pflug 1972
    Class Erniettamorpha Pflug 1972
    Family Erniettidae Pflug 1972
1 *Ernietta* Pflug 1970 (= *Namalia* Germs 1968; *Erniobaris*, *Erniobeta*, *Erniocarpis*, *Erniocentris*, *Erniocoris*, *Erniodiscus*, *Erniofossa*, *Erniograndis*, *Ernionorma*, *Erniopelta*, *Erniotaxis* Pflug 1972; ? = *Baikalina* Sokolov 1972)
    Family Pteridiniidae Richter 1955
2 *Pteridinium* Gürich 1933 (= *Inkrylovia* Fedonkin 1979)
3 ?*Phyllozoon* Jenkins & Gehling 1978
    ?Family Dickinsonidae Harrington & Moore 1955
4 *Dickinsonia* Sprigg 1947 (? = *Praecambridium* Glaessner & Wade 1966)
5 ?*Vendia* Keller 1969
6 ?*Vendomia* Keller 1976
    Class Rangeomorpha Pflug 1972
    Family Rangeidae Pflug 1970
7 *Rangea* Gürich 1929
8 Spindle-shaped form (Newfoundland)
9 Bush-like form (Newfoundland)
    Family Charniidae Glaessner 1979
10 *Charnia* Ford 1958 (= *Glaessnerina* Germs 1973)
11 ?*Charniodiscus* Ford 1958 (= *Arborea* Glaessner & Wade 1966)

Phylum Trilobozoa Fedonkin 1985
    Family Albumaresidae Fedonkin 1985
12 *Albumares* Fedonkin 1976
13 *Anfesta* Fedonkin 1984
14 *Skinnera* Wade 1969
    Family Tribrachididae nov.
15 *Tribrachidium* Glaessner 1959
    [?Family Anabaritidae Glaessner 1979
00 *Anabarites* Missarzhevsky 1969]

Phylum Cnidaria Hatschek 1888 (?Phylum Vendiata Gureev 1985)
    ?Class Cyclozoa Fedonkin 1983
    Family Cyclomedusidae Gureev 1987
16 *Tateana* Sprigg 1949
17 *Ediacaria* Sprigg 1947
18 *Tirasiana* Palij 1976
    Class Hydrozoa Owen 1843
19 *Ovatoscutum* Wade 1971
20 ?*Spriggia* Southcott 1958 (= *Madigania* Sprigg 1949)
21 ?*Eoporpita* Wade 1972
22 ?*Wigwamiella* Runnegar 1991
    Class Anthozoa Ehrenberg 1834
23 *Inaria* Gehling 1988
24 ?*Beltanelliformis* Menner 1974 (= *Nemiana* Palij 1976; *Hagenetta* Hahn & Pflug 1988)
25 *Hiemalora* Fedonkin 1982

Phylum "Vermes" (including Mollusca and extinct phyla?)
26 ?*Archaeichnum* Glaessner 1963
27 ?*Cloudina* Germs 1972
28 *Cochlichnus* Hitchcock 1858
29 ?*Didymaulichnus* Young 1972
30 *Gordia* Emmons 1884
31 *Harlaniella* Sokolov 1972
32 *Helminthoidichnites* Fitch 1850
33 "Large sinuous trails" (tube) Glaessner 1969
34 *Neonereites* Seilacher 1960
35 *Palaeopascichnus* Palij 1976
36 *Planolites* Nicholson 1873
37 *Sellaulichnus* Jiang 1982
38 *Yelovichnus* Fedonkin 1985

Phylum Arthropoda Siebold & Stannius 1845
    ?Family Sprigginidae Glaessner 1958
39 *Spriggina* Glaessner 1958 (= *Marywadea* Glaessner 1976; ? = *Praecambridium* Glaessner & Wade 1966)
    Incertae sedis
40 ?*Parvancorina* Glaessner 1958
41 ?*Onega* Fedonkin 1976
42 ?*Redkinia* Sokolov 1976

Phylum Echinodermata
43 ?*Arkarua* Gehling 1987

Incertae sedis
44 *Ausia* Hahn and Pflug 1985
45 *Bomakiella* Fedonkin 1985
46 *Bonata* Fedonkin 1980
47 *Lorenzinites* Glaessner & Wade 1966
48 *Paracharnia* Sun 1986
49 Pectinate form (Newfoundland)
50 Star-shaped form (Newfoundland)
51 *Wigwamiella* Runnegar 1991

---

the fossils are usually found as impressions on the bases of thin (2 to 5 cm), ripple-marked, sandstone flags (Wade 1968; Gehling 1988). If the animals were "resistant" they formed a normal concave external mold on the base of the bed (Figs. 7.5.6D, 7.5.8C, 7.5.9C). Less resistant animals collapsed and disappeared prior to lithification, allowing the sand to fall and fill the volume they had once occupied (Figs. 7.5.2B, 7.5.5B). A good analogy is the formation of sandstone casts of halite crystals in arid-zone playas; the salt dissolves quickly and is cast by subsequently deposited sand (Richter 1985). Composite molds and casts were formed if the interface on which the dead or dying organism lay were either a thin algal film (Gehling 1986) or a veneer of clay on the surface of an underlying sandy bed. Following the terminology for composite molds introduced by McAlester (1962) and modified by Wade (1968), it has become conventional to refer to convex impressions formed from below as "counterpart casts" and concave impressions formed in the same way as "counterpart molds" (Fig. 7.5.6; Gehling 1988). In such situations, part and counterpart may be separated by an almost infinitesimal thickness.

Gehling's (1986) proposal (see also Seilacher 1989; and McMenamin and Schulte McMenamin 1989) that the preservation of the Ediacara fauna was facilitated by the widespread occurrence of algal films receives support from the nature of the bases of the fossiliferous sandstone beds. It is as if the sand were laid down on a variety of plastic "cling wrap." Fedonkin refers to these characteristic surfaces (Figs. 7.5.2B, 7.5.7F) as "old-elephant-skin texture." Apparently, if decomposition gases built up beneath such a film, it was breached here and there to produce the structure known as *Pseudorhizo-*

Table 7.5.2. *Vendian-Early Cambrian correlation chart*

Ma	SIBERIAN STAGES AND EQUIVALENTS	CHINESE STAGES	NAMIBIA S of OSIS RIDGE	SOUTH AUSTRALIA FLINDERS RANGES	CENTRAL AUSTRALIA AMADEUS BASIN	Ma
530	L E N I A N	TOJONIAN (ELANKIAN)	LONGWANGMIAOIAN		WIRREALPA LIMESTONE BILLY CREEK FORMATION	GILES CREEK DOLOMITE	530
		BOTOMIAN	TSANGLANGPUAN		ORAPARINNA SHALE BUNKERS SANDSTONE PARARA LIMESTONE (AJAX LST)		
540		ATDABANIAN	QIONGZHUSIAN CHIUNGCHUSSUIAN) MEISHUCUNIAN	GROSS AUB FM	WILKAWILLINA LIMESTONE	TODD RIVER DOLOMITE	540
		TOMMOTIAN	MEISHUCUNIAN	NABABIS FORMATION STOCKDALE FM NOMTSAS FORMATION	PARACHILNA FM URATANNA FM	ARUMBERA FM UNIT 4	
550	NEMAKIT-DALDYN	MEISHUCUNIAN			ARUMBERA FM UNIT 3	550
				TOTEM OR URUSIS FM	RAWNSLEY		
				NASEP FORMATION	*Charnia* (*Glaessnerina*)	ARUMBERA FM UNIT 2	
				NUDAUS FORMATION	EDIACARA MEMBER		
560				ZARIS FORMATION	*Pteridinium*		560
		(EDIACARAN)		*Pteridinium* DABIS FORMATION unconformity	QUARTZITE		
570					BONNEY SANDSTONE	ARUMBERA FM UNIT 1	570
580					WONOKA FORMATION	JULIE FORMATION	580
590		VENDIAN	SINIAN		588 ± 35 Rb-Sr BUNYEROO FORMATION	PERTATATAKA	590
600					ABC RANGE QTZITE BRACHINA		600
610		(VARANGERIAN)		JBELIAT GROUP WEST AFRICA	*Bunyerichnus* 609 ± 64 Rb-Sr SUBGROUP NUCCALEENA FM	FORMATION	610
620					ELATINA FORMATION	OLYMPIC FORMATION	620
					TREZONA FORMATION		

Table 7.5.2. (Continued)

Ma	WESTERN AUSTRALIA KIMBERLEY REGION	CHINA YANGTZE GORGE HUBEI PROVINCE	CHINA YUNNAN PROVINCE	WESTERN MONGOLIA SALANY-GOL	SOVIET UNION OLENEK UPLIFT SIBERIAN PLATFORM	Ma
530	ANTRIM PLATEAU BASALTS	SHIHLUNGDONG FM	LONGWANGMIAO FM		KUONAMKA FORMATION	530
		TIENHEBAN FM SHIPAI FM	TSANGLANGPU FM	KHAIRKHAN FM		
540		SHUIJINTUO FORMATION 573 ± 7 Rb-Sr	YUANSHAN FM QIONGZHUSI FM BADAOWAN MBR	SALANYGOL FM	ERKEKET FORMATION	540
			DAHAI MEMBER	BAYANGOL FORMATION	KESSYUSE FM 583 ± 20 K-Ar(G)	
550		HUANGSHANDONG MEMBER	ZHONGYICUN MBR		600 ± 20 K-Ar(G)	550
	FLAT ROCK FORMATION		BAIYANSHAO MBR	TSAGANOLOM	TURKUTUT FORMATION	
560	NYULESS SANDSTONE					560
		DENGYING	DENGYING	FORMATION	*Charnia* KHATYSPYT	
570	TIMPERLEY SHALE				*Charnia* FORMATION 646 ± 20 K-Ar	570
		DOLOMITE *Paracharnia*	DOLOMITE		MAASTAKH FORMATION	
580	BOONALL DOLOMITE				650 ± 20 K-Ar(G) unconformity	580
590	639 ± 48 Rb-Sr ELVIRE FORMATION	DOUSHANTUO	DONGLONGTAN			590
600	MOUNT FORSTER SANDSTONE					600
610	RANFORD FORMATION	FORMATION	FORMATION			610
620	"CAP DOLOSTONES" MOONLIGHT VALLEY, FARGOO and EGAN TILLITES unconformity	NANTUO FORMATION LIANTUO FM unconformity	NANTUO FORMATION CHENGJIANG FORMATION unconformity			620

Table 7.5.2. (Continued)

Ma	SOVIET UNION ALDAN-LENA RIVERS SIBERIAN PLATFORM	SOVIET UNION PODOLIA UKRAINE	SOVIET UNION ALTAI-SAYAN FOLD BELT	SOUTHERN NORWAY	NORTHERN NORWAY FINNMARK	Ma
	ELANKOE FM					
	TITARY FM		OBRUCHEV FM			
530	KETEME FM					530
	KURTOGINA FM					
	SINSK FM		SANASHTYKGOL FM			
	PEREKHOD			1b	DUOLBASGAISSA	
	FORMATION		KAMESHKI FM			
540		SAMETS FM	KIYA FM		FORMATION	540
	PESTROTSVET		NATALEVKA FM	1a		
	FORMATION					
	588 ± 20 K-Ar(G)	ZUBRUCH	USTKUNDAT	RINGSAKER QTZITE		
	533 ± 20 K-Ar(G)	FORMATION	FORMATION	VANGSÅS	BREIVIK FORMATION	
550		KHMELNITSK FM		FORMATION		550
		STUDIENTSA FM				
	YUDOMA	KRUSHANOVKA FM	BELKIN	EKRE SHALE		
560				>630 Rb-Sr		560
		ZHARNOVKI FM				
		DANILOVKA FM			STAPOGIEDDE	
		NAGORIANY FM				
570		550 ± 20 K-Ar(G)				570
		Redkinia	HORIZON			
	GROUP	YARYSHEV FM			FORMATION	
		577 ± 20 K-Ar(G)				
		MOGILEV FM				
580		*Pteridinium*				580
	unconformity	GRUSHKA FM				
		unconformity				
590						590
					MORTENSNES	
				MOELV TILLITE	TILLITE	
600				unconformity	654 ± 23 Rb-Sr	600
610						610
620						620

Table 7.5.2. (*Continued*)

Ma	SOVIET UNION BALTIC SHIELD EASTERN EUROPE	POLAND	SPITZBERGEN	UNITED STATES NORTH CAROLINA	UNITED KINGDOM ENGLISH MIDLANDS	Ma
530						530
540	TISKRE & LYUKATI FMS	RADZYN & KAPLONOSY FORMATIONS MAZOWSZE FORMATION	TOKAMMANE FORMATION BLÅREVBREEN MBR	540 ± 7 Rb-Sr MILLINGPORT FORMATION	HARTSHILL FORMATION	540
	LONTOVA FM 592 ± 16 Rb-Sr					
550	ROVNO HORIZON	WLODAWA FM			540 ± 58 Rb-Sr CALDECOTE VOLCANIC FORMATION	550
	RESHMA FM (KOTLIN SERIES)			*Pteridinium* CID FORMATION		
560	LYUBIM FM 627 ± 15 Rb-Sr	LUBLIN FM			BRAND GROUP MAPLEWELL	560
570	584 ± 20 K-Ar(G) UST'-PINEGA FM *Charnia*	SIEMIATYCZE-BIALOPOLE FORMATION	DRACOISEN	TILLERY FORMATION	*Charnia* GROUP BLACKBROOK GROUP	570
580	(REDKINO SERIES) *Redkinia* *Pteridinium* PLETENYOV FM *Pteridinium*					580
590	SVISLOCH FORMATION	SLAWATYCZE FM	FORMATION	554 ± 50 Rb-Sr 586 ± 10 U-Pb UWHARRIE FORMATION		590
600	VILCHITSY FM BLON FM LAPICHY FM POLES'E GROUP	POLESIE FORMATION				600
	unconformity	unconformity		unconformity VIRGILIANA FORMATION		
610				AARON FORMATION		610
620			WILSONBREEN FORMATION	*Vermiforma* 620 ± 20 U-Pb HYCO FORMATION		620

Table 7.5.2. (*Continued*)

Ma	CANADA NEWFOUNDLAND SW BURIN PENINSULA	CANADA NEWFOUNDLAND AVALON PENINSULA	CANADA YUKON WERNECKE MTNS	CANADA NORTHWEST TERR. MACKENZIE MTNS	UNITED STATES CALIFORNIA-NEVADA	Ma
					MULE SPRING FM	
530	BRIGUS FORMATION	BRIGUS FORMATION			SALINE VALLEY FM	530
					HARKLESS FM	
540		BONAVISTA GROUP	SEKWI FORMATION		POLETA FM CAMPITO FORMATION	540
			VAMPIRE FORMATION	BACKBONE RANGES FORMATION	DEEP SPRING FM	
	RANDOM FORMATION CHAPEL ISLAND FM	RANDOM FORMATION			REED DOLOMITE	
550	RECONTRE FM			INGTA FORMATION	WYMAN FORMATION concealed base	550
		SIGNAL HILL GROUP	RISKY FORMATION	RISKY FORMATION		
		ST JOHN'S GROUP				
560			SILTSTONE UNIT 1	BLUEFLOWER		560
	MARYSTOWN	MISTAKEN POINT FM *Charnia*	CARBONATE UNIT			
		566 ± 5 U-Pb				
570		BRISCAL FORMATION	SILTSTONE UNIT 2	FORMATION		570
	GROUP			*Pteridinium* (*Inkrylovia*)		
580		DROOK FORMATION	GAMETRAIL FM "Goz Siltstone"	GAMETRAIL FM		580
			SHEEPBED FM	SHEEPBED FM		
		GASKIERS FORMATION				
590			KEELE FORMATION	KEELE FORMATION		590
		MALL BAY FORMATION				
600						600
		606 ± 4 U-Pb	TWITYA FORMATION	TWITYA FORMATION		
		HARBOUR MAIN GROUP				
610						610
			RAPITAN			
620			GROUP			620

stomites howchini (Fig. 7.5.5D; Glaessner and Wade 1966, Pl. 103; Wade 1968, Figs. 4–6). As the same texture is found on Late Devonian sandstone bedding surfaces that contain imprints of "by-the-wind-sailors" (Gutschick and Rodriguez 1990, Figs. 6–7), it must indicate a particular and unusual taphonomic environment.

Some of the Ediacaran animals which were initially "resistant" collapsed under the weight of overlying sediments. *Tribrachidium*, for example, is found beneath small craters that have formed when the sand overlying the animal has fallen downward into the space previously occupied by the body. These "temporarily resistant" animals are preserved as foreshortened concave external molds on the lower surfaces of sandstone beds (Runnegar 1991b).

In a second kind of preservation, which is rare in south Australia (Glaessner and Wade 1966, Bl. 101, Figs. 1–3; Wade 1971) but common elsewhere, the fossils are embedded in massive sandstones (commonly orthoquartzites). A deposit of this nature in the Dabis Formation, Namibia, has been well described by Jenkins (1985; Fig. 7.5.02). Similarly, Fedonkin (1985a, Bl. 11, Fig. 1) has illustrated a vertically oriented specimen of *Pteridinium carolinaensis* that was found *in situ* in glauconitic sandstone of the White Sea coast.

The richly fossiliferous surfaces of the Mistaken Point Formation, Newfoundland, represent a third mode of preservation. There, the fossils lie on hemipelagic sediments immediately beneath volcanic ash beds which may have buried the animals *in situ* (Benus 1989). The sequence containing the fossiliferous layers is thought to have been deposited in deep water, below normal storm wave base. The sites in North Carolina (Gibson et al. 1984; Gibson 1989) and England (Ford 1980) may represent comparable deep-water conditions. In northwestern Canada, fossiliferous upper slope deposits of the Mackenzie Mountains can be traced laterally into shallower water shelf sediments in the Wernecke Mountains (Narbonne 1989). This area promises to be of fundamental importance in the development of an understanding of the palaeoecology of the Ediacaran animals. The preservation is somewhat different at other sites, but, by and large, the fauna is found in sandstones or siltstones rather than claystones or carbonates. As a result, fine details are difficult to resolve and interpretation becomes both difficult and controversial.

7.5.2D Taxonomy and affinities of the Ediacara fauna.

The nature of the preservation of the soft-bodied metazoans has had an important effect on the taxonomic practices used by those working with these fossils. Forms such as *Dickinsonia* (Fig. 7.5.7A–D) and *Pteridinium* (Fig. 7.5.9E) that were "resistant" and morphologically complex lie at one end of a continuum. They are known from many specimens, have a wide geographic distribution, are easily recognized even when poorly preserved, and exist in two or more varieties that are widely regarded as different species. Unfortunately, only a tiny fraction of the named Ediacaran taxa has these "three star" properties (Table 7.5.1; Table 23.3); most belong to monospecific genera and many are known only from the holotype. To make matters worse, the preservation may vary from individual to individual and from site to site. As a result, published names are difficult to use and many specimens are hard to identify. For example, the holotype of *Ediacaria flindersi* (Fig. 7.5.4D) was the first Ediacaran fossil described from South Australia (Sprigg 1947). The name has the considerable advantage of priority over all other Vendian "medusoids" except *Paramedusium* Gürich 1930, yet is rarely used even at the generic level, because the combination of characters it represents is not found on many other specimens. Ironically, many of the fossils that have been referred to *E. flindersi* do not resemble the holotype (Harrington and Moore 1956, Fig. 60; Glaessner and Wade, Bl. 99, Fig. 6; Wade 1972, Bl. 43, Fig. 1; Fedonkin 1985a. Bl. 1, Fig. 5 [herein Fig. 7.5.4F]), and specimens identified as *E. flindersi* by different authors may be totally dissimilar. This is a common situation in the literature and the result is a jumble of useless and misapplied names.

It is even difficult for students of these fossils to agree on what is being observed. The now-famous holotype of *Mawsonites spriggi*, which appeared on the cover of *Science* (Cloud and Glaessner 1982), is interpreted as a cast of the knobbly exumbrella surface of a cnidarian medusa by Glaessner and Wade (1966) and Sun (1986c), but as a complex metazoan burrow by Seilacher (1984, 1989). It may simply be a cast of the mesogloeal core of a jellyfish that has flowed under load pressure (Runnegar 1991c). On the other hand, some of the "medusoids" may merely be the sandstone casts of holes left by the holdfasts of "sea-pens" that were washed away by currents.

As a result of differences in observation, description, and interpretation it has not been possible to achieve a consensus view of nature of the Vendian metazoans. Those who have worked in South Australia have largely attempted to relate the Ediacaran animals to the modern biota (e.g., Glaessner 1984). In contrast, Pflug (1970, 1972b) recognized that the Namibian fossils were not referable to living taxa and he therefore proposed a new phylum (Petalonamae) to accommodate them. This idea was not well received because it was accompanied by complex, even fanciful reconstructions of the animals. A third strategy has been adopted by Fedonkin (1984, 1985a, b, 1987); he used differences in symmetry to identify and categorize major taxonomic groups that in general are regarded as extinct higher taxa of phylum- or class-grade. A model for the early history of the Metazoa based on symmetry-related changes in gross body form resulted from this work.

It was Seilacher (1983, 1984, 1985, 1989) who invigorated the study of Ediacaran metazoans with the proposal that few, if any, of the fossils could be referred to the living phyla. Reassessed in this way, it is clear that few of the described taxa can be unequivocally referred to known living or extinct animal groups. For example, some of the discoidal "medusoids" may be the remains of cnidarian jellyfish but none has the diagnostic scyphozoan characters – including four-fold symmetry – that are visible in younger Mazon Creek and Solnhofen examples (Nitecki 1979; Barthel 1978). On the other hand, some of the assignments made in the past may be correct but unprovable. The recognition of many critical features is hampered by the nature of the preservation.

Given these caveats, what can be said at present about the Ediacara faunas? The dominance of cnidarians as measured by the abundance of "medusoids" and other discoidal fossils is no

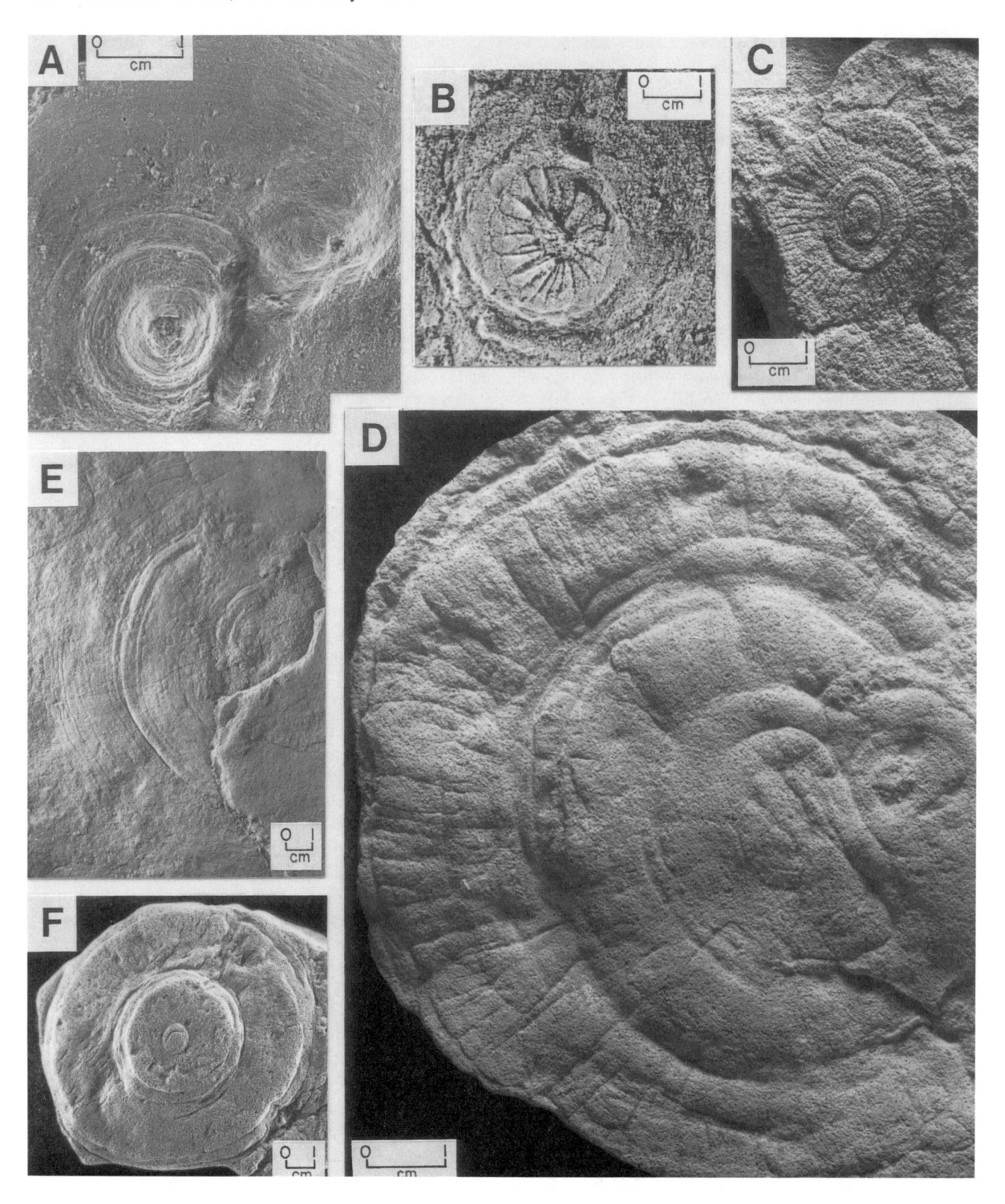

Figure 7.5.4. Various Vendian "medusoids." Lines for scale are approximate. (A) *Cyclomedusa delicata* Fedonkin 1981, Palaeontological Institute, USSR Academy of Sciences, Moscow (PIN) no. 3992/335, Ust'-Pinega Formation, White Sea, USSR. (B) *Bonata septata* Fedonkin 1980, Geological Institute, USSR Academy of Sciences, Moscow (GIN) no. 4482/52-7, Ust'-Pinega Formation, White Sea, USSR. (C) *Tateana inflata* Sprigg 1949, South Australia Museum (SAM) no. 2032, Rawnsley Quartzite, Ediacara, Flinders Ranges, South Australia. (D) Holotype of *Ediacaria flindersi* Sprigg 1947, SAM no. 2058(Tl), Rawnsley Quartzite, Ediacara, Flinders Ranges, South Australia. (E) *Kaisalia mensae* Fedonkin 1984, PIN no. 3393/284, Ust'-Pinega Formation, White Sea, USSR. (F) *Tirasiana disciformis* Palij 1976, PIN no. 4482/153, Ust'-Pinega Formation, White Sea, USSR.

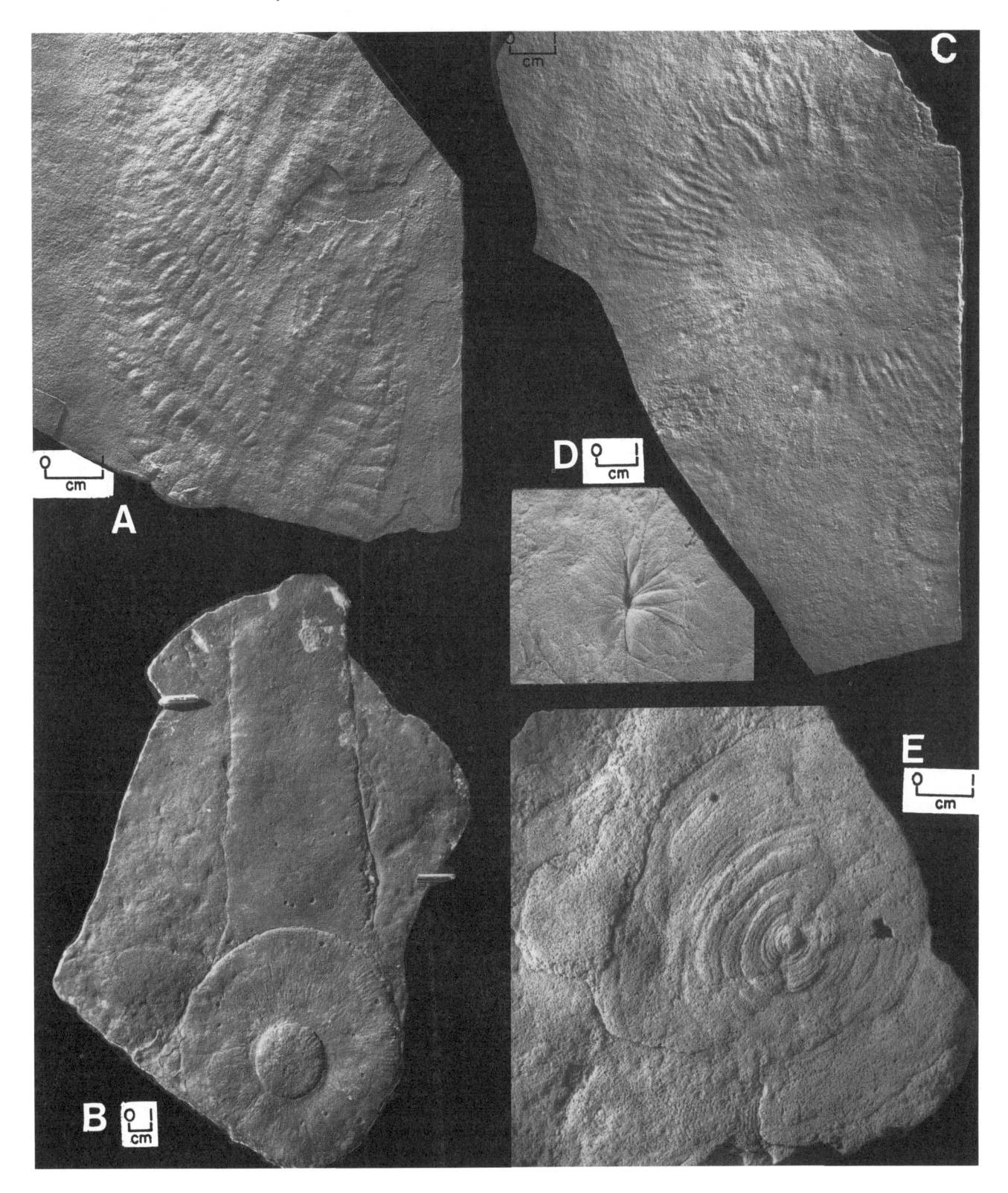

Figure 7.5.5. Vendian "sea-pens" (A, B) and medusoids (C–E). Lines for scale are approximate. (A) Frond of *Charnia masoni* Ford 1958, PIN no. 3995/125, Khatyspyt Formation, Khorbusuonka River, Olenyok Uplift, USSR. (B) Base of stem and holdfast of *Charniodiscus arboreus* (Glaessner) 1959, uncatalogued specimen, Rawnsley Quartzite, Ediacara, Flinders Ranges, South Australia. (C) *Hiemalora pleiomorphus* Vodanyuk 1989, PIN no. 3995/00, Khatyspyt Formation, Olenyok Uplift, USSR. (D) *Pseudorhizostomites howchini* Sprigg 1949, GIN no. 4482/27, Ust'-Pinega Formation, White Sea, USSR. (E) Holotype of *Spriggia annulata* (Sprigg) 1949, SAM no. 2031(T30), Rawnsley Quartzite, Ediacara, Flinders Ranges, South Australia.

Figure 7.5.6. Various Vendian metazoans. Lines for scale are approximate. (A) Spindle-shaped form (Misra 1969) photographed in the field by B. R. Pratt, Mistaken Point Formation, Avalon Peninsula, Newfoundland, Canada. (B) *Dickinsonia costata* Sprigg 1947, rubber cast of uncatalogued specimen, Rawnsley Quartzite, Ediacara, Flinders Ranges, South Australia. (C, D) *Dickinsonia costata* Sprigg 1947, rubber cast (C) and natural external mold (D) of juvenile specimen, same slab as SAM no. P13794, Rawnsley Quartzite, Ediacara, Flinders Ranges, South Australia. (E) *Ovatoscutum concentricum* Glaessner & Wade 1966, SAM no. P24300, Rawnsley Quartzite, Bunyeroo Gorge, Flinders Ranges, South Australia.

Figure 7.5.7. (A) Expanded and contracted specimens of *Dickinsonia costata* Sprigg 1947, University of Adelaide, South Australia (UA) no. F17462F, Rawnsley Quartzite, Brachina Gorge, Flinders Ranges, South Australia. (B) *Dickinsonia costata* Sprigg 1947, rubber cast of dorsal surface, SAM no. P18888, Rawnsley Quartzite, Flinders Ranges, South Australia. (C) *Dickinsonia tenuis* Glaessner & Wade 1966, rubber cast of folded specimen, SAM no. P13768, Rawnsley Quartzite, Flinders Ranges South Australia. (D) *Dickinsonia elongata* Glaessner & Wade 1966, natural mold of anterior half of right side of large individual, SAM no. P27980, Rawnsley Quartzite, Brachina Gorge, Flinders Ranges, South Australia. (E) *Praecambridium sigillum* Glaessner & Wade 1966, UA no. F17033, Rawnsley Quartzite, Ediacara, Flinders Ranges, South Australia. (F) *Ovatoscutum concentricum* Glaessner & Wade 1966, PIN no. 3993/381, Ust'-Pinega Formation, White Sea, USSR. Lines for scale are approximate.

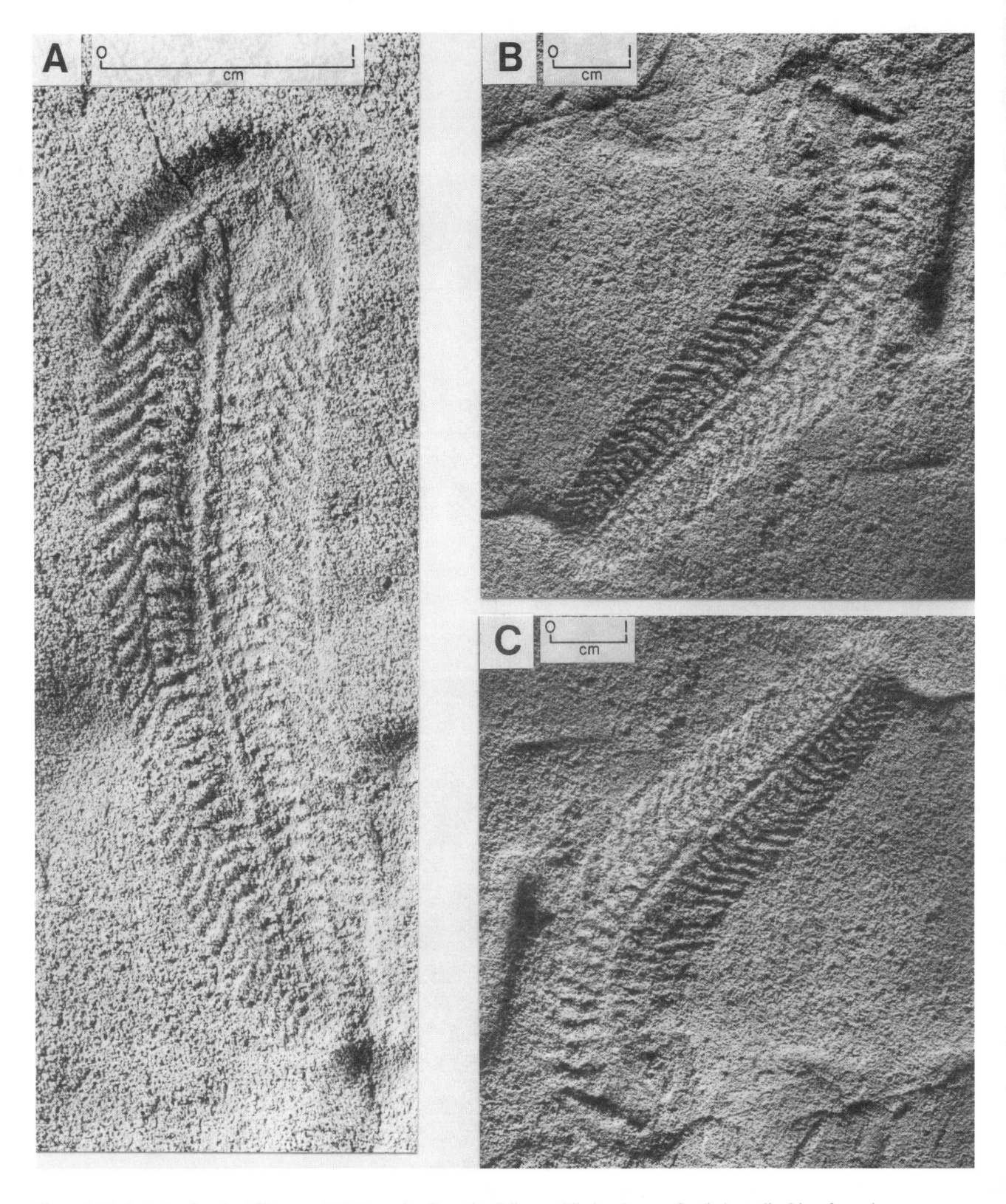

Figure 7.5.8. *Spriggina flounderi* Glaessner 1958, Rawnsley Quartzite, Ediacara, Flinders Ranges, South Australia. Lines for scale are approximate. (A) Dorsal surface, UA no. F17354 (photograph courtesy of J. Pojeta, Jr.). (B, C) Holotype; note that the lighting makes (B) appear to be a positive cast of the natural external mold (C); SAM no. 00000.

Proterozoic Metazoan Body Fossils

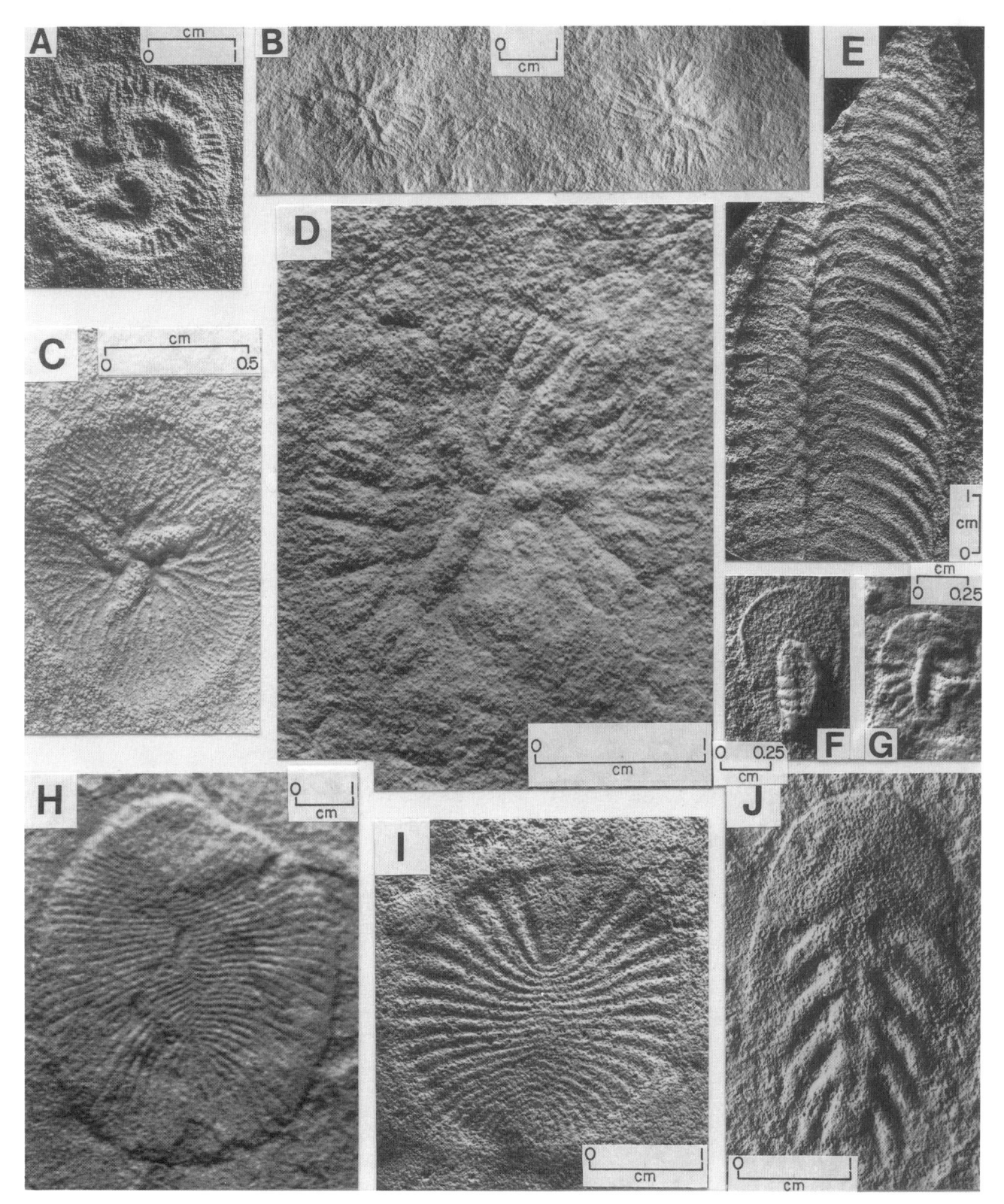

Figure 7.5.9. (A) Rubber mold of the holotype of *Tribrachidium heraldicum* Glaessner 1959, SAM no. P12898, Rawnsley Quartzite, Ediacara, Flinders Ranges, South Australia (see also Frontispiece to Volume 1). (B) *Anfesta stankovskii* Fedonkin 1984, PIN no. 3993/260A, Ust'-Pinega Formation, White Sea, USSR. (C) *Albumares brunsae* Fedonkin 1976, PIN no. 3992/510, Ust'-Pinega Formation, White Sea, USSR. (D) *Anfesta stankovskii* Fedonkin 1984, PIN no. 3993/260A, Ust'-Pinega Formation, White Sea, USSR. (E) *Pteridinium carolinaense* (St. Jean) 1973, PIN no. 3992/400, Ust'-Pinega Formation, White Sea, USSR. (F) *Onega stepanovi* Fedonkin 1976, PIN no. 3992/401, Ust'-Pinega Formation, White Sea, USSR. (G) Holotype of *Vendomia menneri* Menner 1976, GIN no. 4464/57A, Ust'-Pinega Formation, White Sea, USSR. (H) *Dickinsonia costata* Sprigg 1947, PIN no. 3993/1305, Ust'-Pinega Formation, White Sea, USSR. (I) *Dickinsonia costata* Sprigg 1947, SAM uncatalogued specimen, Rawnsley Quartzite, Ediacara, Flinders Ranges, South Australia. (J) Holotype of *Vendia sokolovi* Keller 1969, Ust'-Pinega Formation, Yarensk Borehole, USSR. Lines for scale are approximate.

Figure 7.5.10. (A) Holotype of *Charnia masoni* Ford 1958, University of Leicester, England (UL) no. 2382, Woodhouse Beds, Leicestershire, U.K. (photograph courtest of R. J. F. Jenkins). (B) *Beltanelloides sorichevae* Sokolov 1972 (= *Beltanelliformis brunsae* Menner 1974), PIN no. 3992/501, Ust'-Pinega Formation, Yarnema Borehole (118.4 m depth), White Sea, USSR. (C) "Form A" of Glaessner (1969), a tubular fossil that might be the discarded, originally mucilagenous coat of an echiuroid-like worm, Rawnsley Quartzite, Mt. Scott Range, northern Flinders Ranges, South Australia. Lines for scale are approximate.

longer certain. Some may be the burrows of anemones or the holdfasts of "sea-pens" and others extinct animals with no living counterparts. One way to approach this problem is to classify the animals as if there were to be no Phanerozoic. A biologist of the time would probably have grouped all of the foliate forms – those Seilacher (1989) has described as having the construction of an "air-mattress" – into a single higher taxon. Thus, Pflug's phylum Petalonamae may be the best repository for genera such as *Ernietta*, *Pteridinium*, *Rangea* (Fig. 7.5.3A, and even *Dickinsonia* (Fig. 7.5.7A–D). However, the genus *Ovatoscutum* (Figs. 7.5.6E, 7.5.7F) is sufficiently similar to Phanerozoic fossils interpreted as the floats of chondrophores (Fisher 1957; Yochelson and Stanley 1981; Yochelson et al. 1983; Stanley and Yancey 1986; Stanley 1986) to warrant its inclusion in the Cnidaria. Some of the foliate forms, such as *Charnia* (Figs. 7.5.5A, 7.5.10A) and *Charniodisus* (Fig. 7.5.5B), resemble living pennatulacean octocorals (Jenkins 1985) and could belong to this phylum, as may rare fossils displaying impressions of tentacles (Fig. 7.5.5C; Fedonkin 1985a, Bl. 6, Fig. 5; Jenkins 1989, Fig. 2A). On the other hand, the presence of triradiate symmetry in the discoidal forms *Tribrachidium*, *Albumares*, *Anfesta*, and *Skinnera* provides a tentative basis for grouping these genera with the otherwise dissimilar Cambrian taxon *Anabarites* in an extinct higher taxon, the Trilobozoa (Table 7.5.1).

The existence of vermiform mobile animals is best demonstrated by the trace fossils (Section 7.6). Sinuous surface trails up to several millimeters in width (Fig. 7.6.4c) and strings of fecal pellets (Fig. 7.6.4G) point to the existence of soft-bodied bilateria of molluscan to annelid grade. Large diameter tubes, now filled with sand, may be the mucilagenous coats of echiuroid-like worms (Fig. 7.5.10C).

Even though several annelids and arthropods have been identified in the Ediacaran faunas, the presence of higher metazoans is difficult to prove. The best candidate is probably *Spriggina* (Fig. 7.5.8) which appears to have many of the characters that might be expected in a primeval trilobite.

One possible scenario for the origin of the Arthropoda is that the characteristic exoskeleton evolved independently in two different lines (Cisne 1974). This model, which recognizes two arthropod phyla (Uniramia and "Schizoramia" or Trilobita plus Crustacea plus Chelicerata), points to an animal like *Spriggina* as the direct common ancestor of all non-uniramian arthropod lineages. Interpreted as an arthropod, *Spriggina* has a crescent-shaped head-shield followed by about 40 segments, each bearing narrow, jointed walking legs and, presumably, also gill branches (Birket-Smith 1981b). The mouth is assumed to have been underneath the head-shield.

Glaessner steadfastly rejected suggestions that *Spriggina* might be a primitive arthropod because he believed that the appendages were unjointed and bore straight acicular setae at their ends (Glaessner 1984, pp. 61–62; pers. comm. to B.N.R.). He considered *Spriggina* to be an errant polychaete, a view not shared by Conway Morris (1979) in a review of the Burgess Shale and other Cambrian annelids. In stark contrast, Seilacher (1989) has inverted the animal so that the "head shield" becomes a "holdfast" and the "appendages" a foliate body. This reorientation (Fig. 7.5.8C) was mandatory if *Spriggina* were to

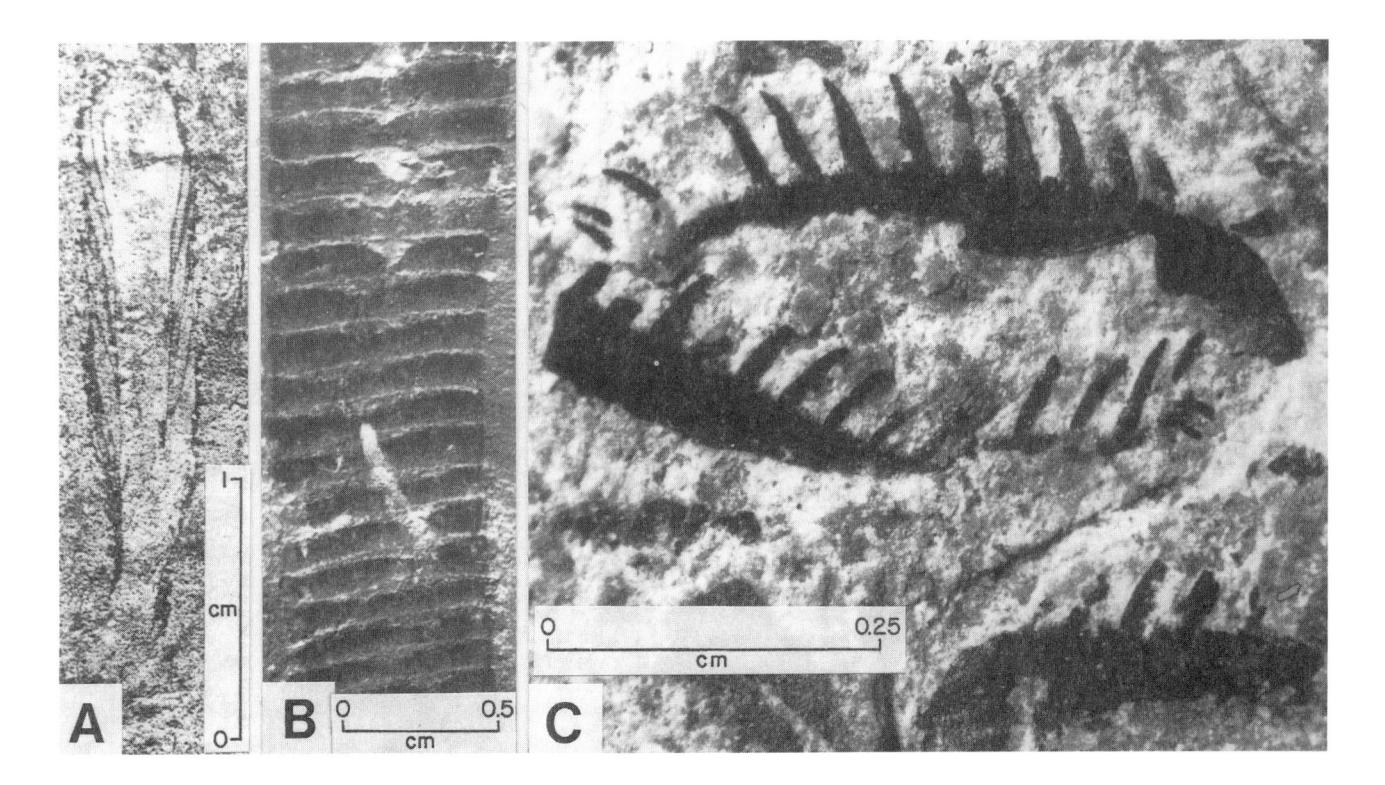

Figure 7.5.11. Vendian problematica. Lines for scale are approximate. (A) *Cloudina hartmannae* Germs 1976, Zaris Formation, Nama Group, Namibia (after Glaessner 1976, Pl. 2, Fig. 1; republished with permission). (B) *Saarina* sp., PIN no. 3993C/1, Ust'-Pinega Formation, Malinovka PGO Archangelskgeologia Borehole (360 m depth), USSR. (C) *Redkinia spinosa* Sokolov 1976, PIN no. 3993C/5, Redkino Formation, Nepeitsino Borehole (1417 to 1426 m depth), USSR.

be accommodated within Seilacher's extinct kingdom, the "Vendozoa."

7.5.3 Conclusions

The ecological significance of the Ediacaran animals remains enigmatic. Their large size alone makes it unlikely that they were pioneers in metazoan evolution. Most had a high ratio of surface to volume that may reflect low levels of ambient oxygen (Runnegar 1982a, b, 1991c), the presence of photosynthetic endosymbionts, or the ability to absorb nutrients through the body wall (Seilacher 1984, 1989; McMenamin 1986; McMenamin and Schulte McMenamin 1989). "Medusoids" of apparently comparable construction are known from the Early Cambrian of China (Sun and Hou 1987) and perhaps the Late Cambrian of the Siberian Platform (Popov 1967, 1968), but most Ediacaran body-plans disappeared from the fossil record before the beginning of the Cambrian. Perhaps their peculiar mode of construction (Seilacher 1989) became obsolete in the face of competition from other newly evolved animal groups. In any case, it was the "invisible" soft-bodied Precambrian animals and not the Ediacaran ones that gave rise to the Cambrian faunas.

7.6

Proterozoic Metazoan Trace Fossils

MIKHAIL A. FEDONKIN   BRUCE N. RUNNEGAR

7.6.1 Introduction

Trace fossils, or "ichnofossils," are structures produced in soft sediments and hard substrates as a result of the living activities of organisms. They include surface tracks and trails, subsurface burrows and borings, as well as fecal material and the marks produced by dying animals (Häntzschel 1975). To a large extent, the factors that control the distribution of different kinds of trace fossils are environmental rather than temporal (Seilacher 1964a, b; Osgood 1970; Häntzschel 1975; Fedonkin 1977; Savrda and Bottjer 1986), so most ichnofossil taxa have long stratigraphic ranges. Therefore, the abrupt appearance of may different kinds of trace fossils at the end of the Proterozoic is regarded as a singular biological event. It represents either the evolutionary origin of soft-bodied metazoans (Cloud 1948; Gould 1977; Sepkoski 1978) or the first appearance in the fossil record of animals capable of leaving preservable traces (Towe 1970; Durham 1978; Runnegar 1982a, b).

7.6.2 Reported Pre-Vendian Trace Fossils

There are many reports of pre-Vendian trace fossils in the literature but none has withstood close scrutiny (Cloud 1968; Häntzschel 1975; Glaessner 1979a). Here, all are considered to be pseudofossils or, at best, unsubstantiated dubiofossils (Section 7.8). However, this view is not universally accepted for a few structures listed with the authentic Precambrian trace fossils in Table 23.4. Briefly:

(1) In 1935, C. E. van Grady collected a lobate, radiating structure (Fig. 7.5.2A) from the mid-Proterozoic Nankoweap Group in the Grand Canyon, Arizona. Initially thought to be a jellyfish, it was assigned to a new species of *Brooksella* Walcott by Bassler (1941). Since then, *Brooksella canyonensis* has been considered to be a siphonophore (Caster 1957), a cnidarian medusa (Harrington and Moore 1956; Glaessner 1962, Vologdin 1966), a sub-radial fracture system (Cloud 1960, 1968), a fluid escape structure (McMenamin and Schulte McMenamin 1989), fecal mounds (Glaessner 1969; Kauffman and Steidtmann 1981), and an advanced metazoan burrow (Kauffman and Fürsich 1983; Glaessner 1984; Fürsich and Bromley 1985). A restudy of the holotype of *B. canyonensis* shows no trace of the "Spreite-like laminae" reported by Kauffman and Fürsich (1983); instead, fine irregular layers continue across the slab beneath the "fossil" (Runnegar and Stait, in prep.). The structure appears to be the sandstone cast of a fortuitous aggregation of mud-rolls or *Tongallen* (*sensu* Hati and Hayasaka 1961). A second specimen (Fig. 7.5.2C), collected at the original locality in 1962 (Cloud 1968, p. 27), turns out to be a microfaulted, tool-marked bed base quite different in detail from the holotype of *Brooksella canyonensis*.

(2) Well-formed burrows from the main ore horizon in the Zambian Copperbelt were described by Clemmey (1976) as one billion-year-old metazoan trace fossils. It was subsequently pointed out by Cloud, Gustafson, and Watson (1980) that these structures are likely to be modern termite galleries.

(3) Structures from the Early Proterozoic Medicine Peak Quartzite of Wyoming, which were documented by Kauffman and Steidtmann (1981, 1983) as possible metazoan burrows, seem to be largely of metamorphic origin. Some may be fluid evasion pathways (Cloud 1983) that have been accentuated by metamorphic recrystallization and vug formation. However, almost identical "burrows" can be found in cross-cutting quartz veins in the same blocks of cross-bedded quartzite as these "dubiofossils" (B. Runnegar field observations, 1988).

(4) An authentic metazoan burrow, *Rugoinfractus ovruchensis*, described originally be Palij (1974) from Riphean-age strata in the Ukraine, is now thought to have come from rocks of Paleozoic age (M. A. Fedonkin, unpubl.; Hofmann et al. 1983).

(5) The only known structures that might conceivably be megascopic pre-Vendian metazoan trace fossils are burrow-like features described by Faul (1949, 1950) from the Early Proterozoic of Michigan, and puzzling, bead-shaped, bedding-plane markings (Figs. 7.6.3D, 7.8.5D) reported by Horodyski (1982) and Grey and Williams (1990) from the mid-Proterozoic of Montana and Western Australia, respectively. Both of these structures remain in the literature as unexplained dubiofossils; neither provides any positive evidence for the existence of pre-Vendian metazoans (see also Section 7.8). Furthermore, it is known that some complex inorganic sedimentary structures may mimic marks made by metazoans (Figs. 7.6.4A, 7.8.3A).

(6) Microscopic pellets from a variety of Proterozoic rocks

were considered by Robbins, Porter, and Haberyan (1985) to be the fecal pellets of minute metazoans. Although this conclusion is consistent with the hypothesis that most Precambrian animals were tiny and therefore inconspicuous (Towe 1970, 1981; Runnegar 1982a), there is little evidence that the pellets extracted by Robbins et al. had passed through the guts of metazoans. This investigation stands as a promising, but unconvincing, attempt to search for hints of pre-Vendian animals in the microfossil size-range.

(7) It has been suggested that the occurrence of well-defined, delicate laminae in mid-Proterozoic open-shelf sediments is evidence that metazoans had not yet evolved (Byers 1976). This may be true, and more studies of this kind are to be encouraged, but it is also clear that trace fossils can be preserved in apparently undisturbed, finely laminated sediments (Goldring and Seilacher 1971).

(8) The identification of authentic Precambrian metazoan trace fossils is complicated by the lack of a widely accepted and easily-recognized younger limit to the Proterozoic. This problem is particularly acute in those Precambrian-Cambrian boundary sequences that contain many trace fossils but no body fossils (Crimes 1987). In such situations it becomes difficult to avoid circular arguments for the age of the strata and their contained trace fossils.

Nevertheless, it is clear from stratigraphic sections in Australia, Canada, and the Soviet Union that there are some distinctive trace fossils that appear within the Vendian, at or just below the level of the Ediacara faunas (Glaessner 1969; Fedonkin 1985a; Crimes 1987; Narbonne and Hofmann 1987; Gibson 1989; Aitken 1989; Walter et al. 1989). As might be expected, the diversity is low; of the six kinds of trace fossils described by Glaessner (1969) from the fossiliferous horizon at Ediacara, two may be body fossils (Table 23.4; Figs. 7.5.10C, 7.6.3B), and the others are various kinds of horizontal trails and marks. However, even these and other simple structures (Fig. 7.6.3C, E) have great importance, for they provide unequivocal evidence that mobile, bilateral animals not found as body fossils existed with more typical members of the Ediacara fauna.

An estimate of the age and affinities of a number of distinctive Vendian and very Early Cambrian trace fossils is presented in Figure 7.6.1 and Table 23.4. Given the difficulty of correlating within the Vendian (Fig. 7.5.2), the precise age and order of appearance of many of the taxa remain uncertain. Nevertheless, it is still possible to make some general statements about the nature of the early trace fossil succession and its biological significance.

7.6.3 Biostratigraphy of Vendian Trace Fossils

Crimes (1987) identified three successive global fossil zones in rocks ranging from late Vendian to middle Atdabanian. In this scheme, Zone I is late Vendian in age and Zones II and III belong to the Early Cambrian. In a similar fashion, Narbonne et al. (1987) described and named three ichnofossil zones within the Chapel Island Formation of southeastern Newfoundland, and Walter et al. (1989) have identified three successive trace fossil assemblages in the Precambrian-Cambrian boundary sequences of central Australia. However, as the oldest Australian "assemblage" is the occurrence of a single species, Assemblages 2 and 3 of Walter et al. correspond, respectively, to Zones I and II + III of Crimes. The oldest Newfoundland zone, the *Harlaniella podolica* Zone, is thought to be equivalent to the second half of Crimes' Zone I (Narbonne et al. 1987), and the two younger zones to his Zones II and III (Fig. 7.6.1).

On the Russian Platform, *Harlaniella podolica* occurs in the late Vendian Kotlin Horizon well above representatives of the Ediacara fauna (Sokolov and Fedonkin 1984; Fedonkin 1985a).

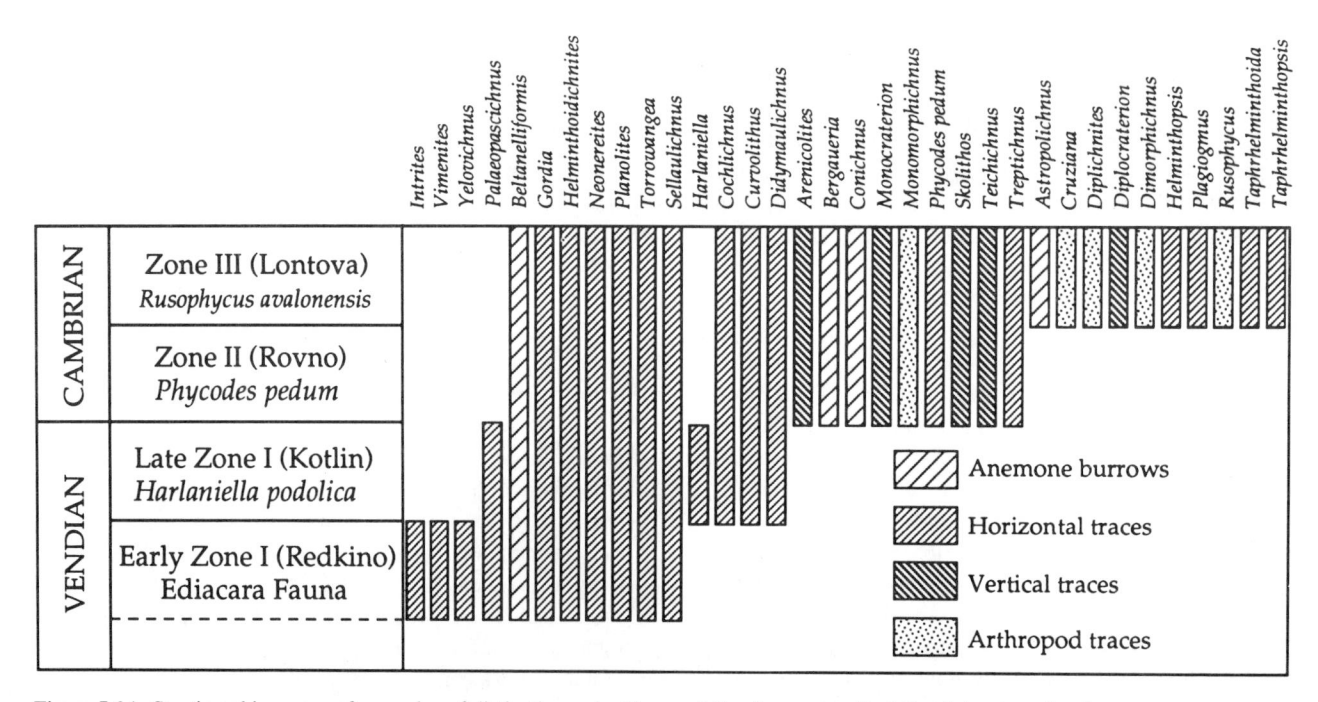

Figure 7.6.1. Stratigraphic ranges of a number of distinctive and widespread Vendian and earliest Cambrian trace fossils.

The overlying Rovno Horizon yields typical Zone II traces and is regarded by Crimes (1987), but not Sokolov and Fedonkin (1984), as Tommotian in age. An even more diverse suite of trace fossils is found in the overlying Lontova Horizon (Fedonkin 1977, 1981). These relationships are summarized in Figure 7.6.1.

7.6.3A Taxa Found with or Below the Ediacara fauna

Horizontal burrows from the Elkera Formation of central Australia identified as *Planolites ballandus* (Glaessner 1984, p. 70; Walter et al. 1989) may be the oldest known metazoan trace fossils, possibly older than the Ediacara fauna. However, the estimate of their age depends upon a correlation of the Georgina Basin sequence with the Amadeus Basin and Flinders Ranges successions. *Planolites* and another kind of burrow called *Torrowangea* occur with and below Ediacaran metazoans in the Vendian sequences of the Wernecke Mountains (Narbonne and Hofmann 1987; Aitken 1989) and with or close to *Pteridinium* in North Carolina (Gibson 1989). Moreover, Crimes (1987) concluded that *Planolites* appeared before any other trace fossil in four of eleven well-studied Precambrian-Cambrian boundary sequences from various parts of the world. Thus, it seems that one of the oldest kinds of trace fossil is a simple, narrow, irregular, horizontal burrow. Following Clark (1964) and Trueman (1975), it is reasonable to suppose that such burrows were formed by the peristaltic contraction of fluid-filled bodies of pseudocoelomate or coelomate worms. In addition, the presence of uniserial (*Neonereites uniserialis*) and biserial (*Neonereites biserialis*) strings of fecal pellets in mid-Vendian strata of the White Sea coast (Fig. 7.6.2G; Fedonkin 1981, 1985) and North Carolina (Gibson 1989) is good evidence for the existence of triploblastic sediment-feeders with a one-way gut.

Another kind of Vendian organism must have produced the structures known variously as *Beltanelloides sorichevae* Sokolov 1972, *Beltanelliformis brunsae* Menner 1974, *Nemiana simplex* Palij 1976, or *Hagenetta aarensis* Hahn and Pflug 1988 (Fig. 7.5.10B). Narbonne and Hofmann (1987, p. 666) concluded that "*Beltanelliformis* (= *Beltanelloides*) and *Nemiana* are best explained as preservational variants of a single globular biological taxon," and they rejected the idea that these structures were formed by the hemispherical bases of sea anemones, as is thought to be the case for the Cambrian trace fossil *Bergaueria* (Arai and McGugan 1968; Alpert 1973; Boyd 1974). However, well-preserved trace fossils from the Devonian of Poland, which show many of the features of *Beltanelliformis*, have been interpreted as anemone burrows (*Alpertia santacrucensis*; Orlowski and Radwanski 1986); the sequence of events postulated by Orlowski and Radwanski for the formation of *Alpertia* (Fig. 7.6.2) would explain most of the properties of *Beltanelliformis* described by Narbonne and Hofmann (1987). Consequently, it is possible that *Beltanelliformis* and *Alpertia*, like the cylindrical structures placed in *Bergaueria* (e.g., Palij 1976, Bl. 28, Figs. 1–2), were cnidarian resting traces. Modern anemones may live gregariously (Doubilet and Kohl 1980) and could be expected to produce the close packing seen in slabs of *Beltanelloides* (Fig. 7.5.10B), "*Nemiana*" (Palij 1976, Bl. 21, Fig. 5), *Bergaueria* (Arai and McGugan 1968, Bl. 36, Fig. 12), *Alpertia* (Orlowski and Radwanski 1986, Pls. 5–6), and *Hagenetta* (Hahn and Pflug 1988, Pl. 2, Figs. 1–6).

Fedonkin (1985, Pl. 22, Fig. 2) has illustrated a specimen of *Bergaueria* or *Beltanelloides* which seems to have left a trail (Spreite) as it moved sideways throught the sediment. However, even the inert bodies of stranded jellyfish may leave trails of this type (Branagan 1976; Vialov et al. 1977) so this unusual specimen fails to provide a unique solution to the problem of the affinities of *Beltanelloides*. The structures from the Nama Group illustrated by Crimes and Germs (1982, Pl. 1, Fig. 1) as *Bergaueria* and considered by Crimes (1987) to be "remarkably similar" to *Intrites punctatus* from the White Sea (Fig. 7.6.4B) may also be cnidarian resting traces. Comparable objects occur with the probable Ediacaran anthozoan *Inaria* in south Australia (Gehling 1988, Fig. 8E).

The common trace fossils at Ediacara are simple, one millimeter-wide surface trails that may be identified as *Helminthoidichnites tenuis* (*sensu* Hofmann and Patel 1989). Abrupt bends and acute forks in these structures indicate that the producers may have been equidimensional rather than vermiform, so an organism such as a tiny mollusc is a likely source. In contrast, the sinusoidal ichnofossil *Cochlichnus serpens*, well-illustrated by Webby (1970) from Precambrian-Cambrian boundary beds in western New South Wales, is similar in shape and size to modern and fossil nematode tracks (Moussa 1970; Fordyce 1980; Runnegar 1982b; Tabuse, Nishiwaki and Miwa 1989; but see also Metz 1987).

The closely spaced meanders of *Yelovichnus* (Fig. 7.6.3A; Glaessner 1969, Fig. 5C; Cope 1983, Pl. 2, Fig. 2) are more difficult to interpret. They may be the grazing traces of a sizable animal that could move its head from side to side, or the burrow of a vermiform animal that periodically reversed its direction of motion. The former interpretation is more likely if *Palaeopascichnus*, and perhaps *Harlaniella*, are considered homologous to *Yelovichnus*. *Palaeopasichnus* is the cast of a series of shallow subparallel grooves that seem to have lain at right angles to the principal direction of motion (Fig. 7.6.3F). *Nenoxites* (Fig. 7.6.3D) may have a comparable origin. On the other hand, *Vimenites bacillaris* (Fedonkin 1985a) and *Syringomorpha ?nilssoni* (Gibson 1989) are clusters of shallow excavations that appear to be aligned in the direction of motion.

Some structures are hard to categorize. Sandstone casts of parallel-sided tubes about 2 to 3 cm in diameter are relatively common in the Ediacara Member of Rawnsley Quartzite (Fig. 7.5.10C; Glaessner 1969, Fig. 5E). They were considered by Glaessner (1969) to be surface trails left by sizable sand-dwellers. However, they are now known to be three-dimensional in form and oval in cross-section, and so are more likely to be either body fossils or the casts of mucilagenous tubes like those formed by echiuroids and enteropneusts (Fisher and MacGintie 1928; Paul et al. 1978). On the other hand, Vendian fossils from North Carolina which Cloud, Wright, and Glover (1976) regarded as vermiform body fossils may be early horizontal trails. The repeated pattern of loops, which led Cloud et al. to suggest that several similar specimens on one bedding plane had been aligned by currents, could also be interpreted as the repetitive programmed motion of a foraging organism. Field photographs of the Spanish Early Cambrian

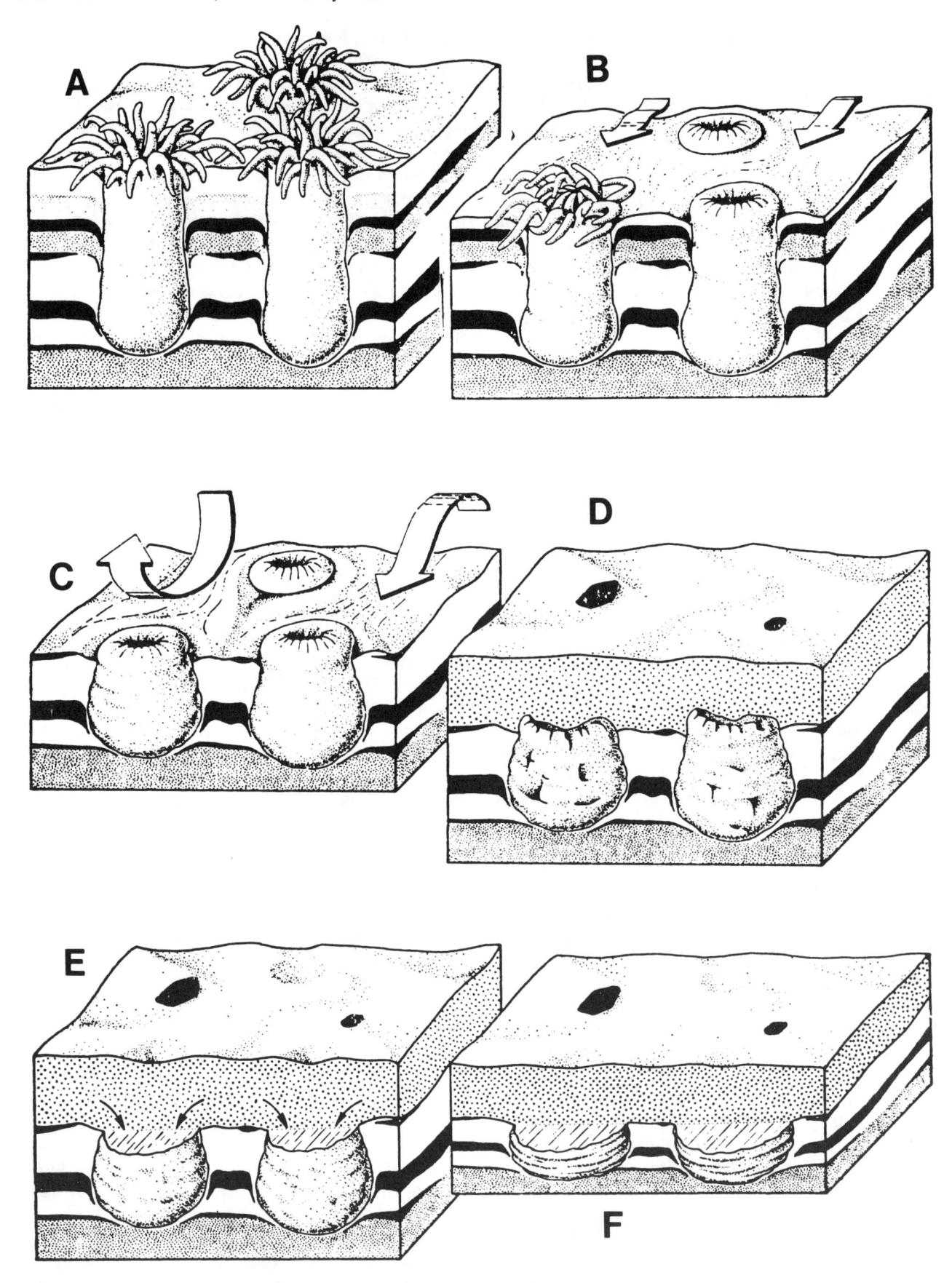

Figure 7.6.2. Interpretive diagram showing stages (A through F) in the preservation of the Devonian trace fossil, *Alpertia santacrucensis*, from Poland (after Orlowski and Radwanski 1986, Fig. 2; republished with permission). (A) Normal conditions. (B–D) Burial during a storm. (E, F) Casting and compaction of emptied burrows. The result (F) resembles Vendian specimens of *Beltanelliformis brunsae* (Fedonkin 1981; Narboone and Hofmann 1987).

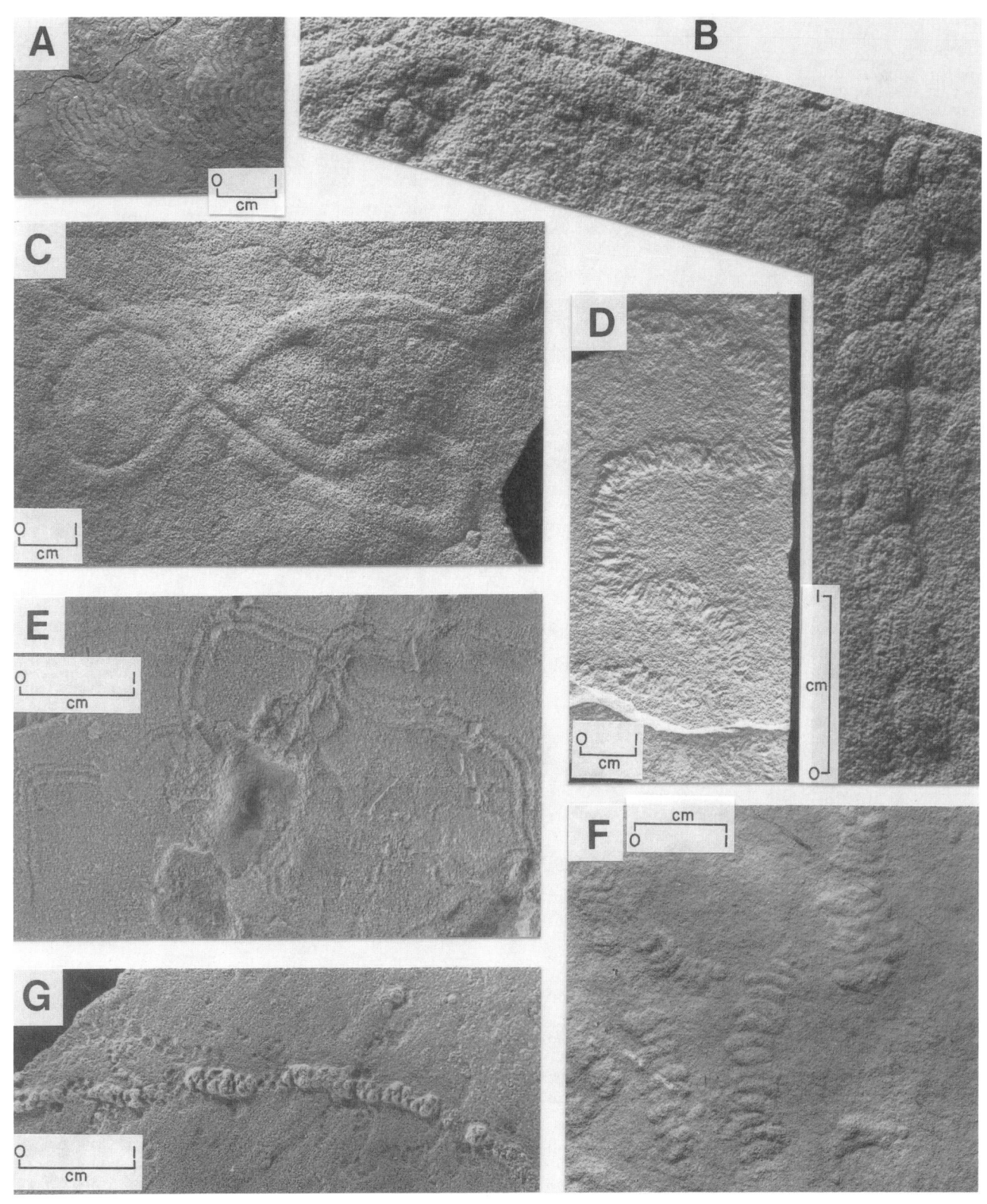

Figure 7.6.3. Vendian trace fossils. Lines for scale are approximate. (A) Holotype of *Yelovichnus gracilis* Fedonkin 1985, Palaeontological Institute, USSR Academy of Science, Moscow (PIN) no. 3993/1309, Ust'-Pinega Formation, Valdai Series (100 m below top of section), Winter Shore, White Sea, USSR. (B) Structure identified by Glaessner (1969) as *Cylindrichnus* Bandel 1967 (= *Margaritichus* Bandel 1973, non *Cylindrichnus* Toots 1966) and interpreted as a row of fecal pellets; interpreted here as probably twisted tissue; South Australia Museum (SAM) no. 12893, Ediacara Member, Rawnsley Quartzite, Ediacara, South Australia. (C) *Gordia* sp., on the base of a sandstone flag, SAM no. P27977, Ediacara Member, Rawnsley Quartzite, Ediacara, South Australia. (D) Holotype of *Nenoxites curvus* Fedonkin 1976, Geological Institute, USSR Academy of Science, Moscow (GIN) no. 4310/202-4, Ust'-Pinega Formation, Valdai Series, Summer Shore, White Sea, USSR. (E) *Sellaulichnus meishucunensis* Jiang 1982, GIN no. 4482/66, Ust'-Pinega Formation, Winter Shore, White Sea, USSR. (F) *Palaeopascichnus delicatus* Palij 1976, PIN no. 3994/295, Mogilev Formation, Podolia, USSR. (G) *Neonereites biserialis* Seilacher 1960, GIN no. 4464/180, Ust'-Pinega Formation, Valdai Series, Winter Shore, White Sea, USSR.

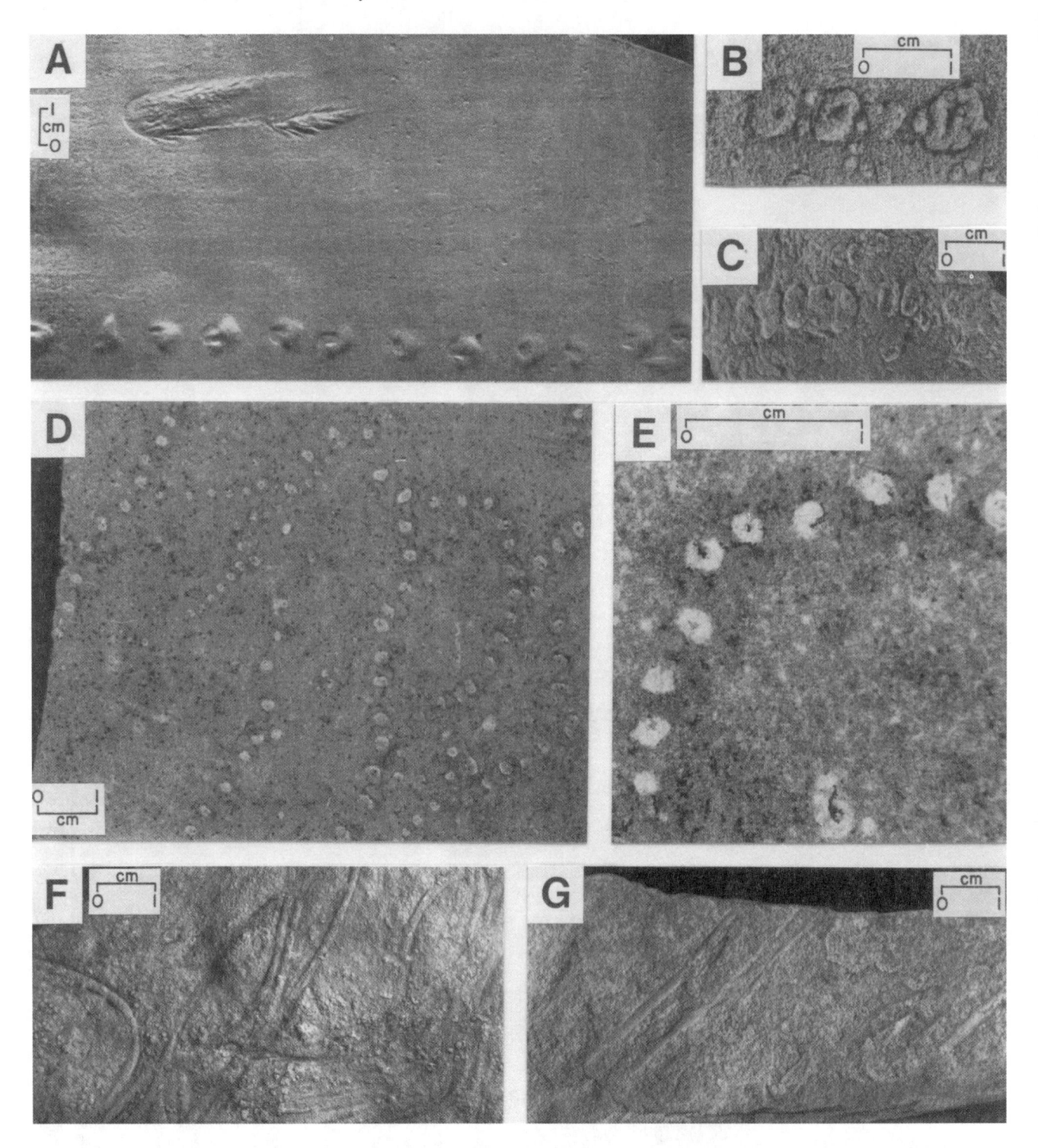

Figure 7.6.4. Proterozoic trace fossils (B, C, F, G), dubiofossils (D, E), and a pseudofossil (A). Lines for scale are approximate. (A) Sandstone cast of a series of depressions caused by a rolling object, Nama Group, Namibia (after Glaessner 1963, Pl. 2, Fig. 4; republished with permission). (B) Holotype of *Intrites punctatus* Fedonkin 1980, GIN no. 4482/81, Ust'-Pinega Formation, Valdai Series, Winter Shore, White Sea, USSR. (C) Holotype of *Neonereites renarius* Fedonkin 1985, GIN no. 4482/163, Ust'-Pinega Formation, Valdai Series, White Sea, USSR. (D, E) Puzzling bedding-plane markings (cf. Fig. 7.8.5D), Appekunny Argillite, Montana, USA (after Horodyski 1982, Pl. 1; republished with permission). (F) *Helminthoida* sp., PIN no. 3993/874, Ust'-Pinega Formation, Valdai Series (top of section), Winter Shore, White Sea, USSR. (G) Holotype of *Skolithos declinatus* Fedonkin 1985, PIN no. 3993/640, Ust'-Pinega Formation, Valdai Series (top of section), Winter Shore, White Sea, USSR.

bilobate trail *Taphrhelminthopsis circularis* (Crimes et al. 1977, Pl. 8) show some similarity in this respect to *Vermiforma antiqua* from North Carolina (Cloud et al. 1976, Fig. 3).

7.6.3B Taxa Found Above the Ediacara Fauna in Vendian or Earliest

Cambrian strata. Many different kinds of trace fossils appear abruptly with the earliest skeletal fossils. The situation in southern China is typical: no trace fossils are known from the underlying Precambrian dolomites, but several distinctive types appear with the earliest small skeletal fossils in the basal Cambrian Zhongyicun Member of the Dengying Formation (Jiang et al. 1982; Crimes and Jiang 1986). A similar phenomenon is seen in a different environmental setting in central Australia; there the Arumbera Sandstone spans the Proterozoic-Cambrian boundary with no discernable break, and the trace fossil diversity increases markedly at a horizon considered to mark the beginning of the Cambrian (Walter et al. 1989). In central Australia, and elsewhere, the Cambrian beds contain complex horizontal traces (*Didymaulichnus, Helminthopsis, Phycodes pedum, Plagiogmus, Taphrhelminthoida, Taphrhelminthipsis, Treptichnus*) and closely-spaced vertical burrows (*Diplocraterion, Monocraterion, Skolithos*) that are rare in (Fig. 7.6.4G) or absent from Vendian strata. Arthropod crawling and resting traces appear somewhat later in the Early Cambrian, a little below the first trilobites (Crimes 1987).

7.6.4 Conclusions

The first metazoan trace fossils are no older than the early Vendian. Those that occur with or below the Ediacara fauna are mostly simple horizontal trails or burrows and more complicated meanderform grazing structures. Nevertheless, these fossils provide good evidence for the existence in Vendian times of mobile, bilateral, triploblastic animals of the pseudocoelomate or coelomate grade of organization. None of these animals is preserved in any other way, so the Vendian traces provide some indication of the limits of the fossil record. A major increase in trace fossil diversity occurred at the beginning of the Cambrian in concert with the abrupt appearance of organisms with mineral skeletons.

7.7
Proterozoic and Earliest Cambrian Skeletal Metazoans
STEFAN BENGTSON

7.7.1 Introduction

The advent of organisms with mechanically and chemically resistant skeletons signifies the end of the impoverished Proterozoic fossil record and the beginning of the abundantly fossiliferous Phanerozoic. This Section discusses the paleontological evidence for the times of appearance of the various types of skeletons and the organisms that formed them. These data provide a minimum age for each known skeleton-forming phylum. The taxonomic data on which the presentation is based are presented and referenced in Table 23.5. Broader ecological and evolutionary aspects of the data are dealt with in Chapters 8 and 11.

"Skeleton" is here taken to imply both inorganic (organic tissues stiffened by biominerals; cf. Section 8.6) and purely organic (mainly scleroproteins and chitins) skeletons in metazoans. Prokaryotes, protists, and multicellular algae contain some groups that also have mineralized skeletons. These are treated separately in Chapter 5 and Section 7.4.

7.7.2 Representativity of the Skeletal Fossil Record

The stratigraphic distribution of skeleton-forming phylum-level clades, as seen in the fossil record of the Vendian-Cambrian, is given in Figure 7.7.1 (see discussion of phyla, below). All the well-mineralized phyla living today are present in this record, which suggests that most or all of the living "soft" phyla were also in existence at that time. Some of these have indeed been found as fossils, due to exceptional preservational circumstances (Conway Morris 1985c), although at least 10 living soft-bodied phyla lack a fossil record and several others are represented by only one or a few unique occurrences. Information on soft-bodied fossils can be used to assess how representative the record of skeletal fossils is in terms of total diversity.

Soft-body records, such as can be derived from the Ediacaran fossil assemblages (Section 7.5), trace fossils (Section 7.6), and the unique Cambrian *Lagerstätten* represented by the Burgess and Chengjiang biotas and other occurrences (Conway Morris and Robison 1982; Müller and Walossek 1985; Whittington 1985, Zhang 1987; Conway Morris et al. 1987; Hou and Sun 1988), give rich information on the non-skeletal metazoans present at that time. Even so, the available record shows only a fraction of the taxa living at the time.

The number of phylum-level taxa was probably substantially greater than now (Chapter 8), but because phyla are largely arbitrarily defined, the percentage of recovery can only be estimated in a very general sense. On lower taxonomic levels and for shorter stratigraphic intervals, the record becomes increasingly poor, particularly when the data derive only from skeletal fossils. About 30 to 40% of genera in modern intertidal environments have a fossil record (T. J. M. Schopf 1978). About 30% of the species living in modern marine communities produce any kind of fossilizable skeleton (Johnson 1964), and the corresponding number for all living species is less than 10% (Nicol 1977). Well-mineralized taxa in well-represented and well-studied sequences may have up to 77% recovery (Valentine 1989b). In the Middle Cambrian Burgess Shale only about 20% of the metazoan species had normally preservable hard parts (Conway Morris 1986), so the recovery percentage is not likely to exceed 15%. Recovery of higher taxa is strongly dependent on diversity (e.g., Raup 1975), and assuming that present distributions of mineral skeletons were about the same in the Cambrian, the probability of recovering phyla (or any other taxa) in which most members produce skeletons reaches 0.95 at a sample size of about a dozen species (Bengtson and Conway Morris, in press). The prerequisite is, however, that recovery of species has reached a saturation point, and it is clear from the cumulative curve of published genera (Fig. 7.7.2) that we are still far from this condition.

The first appearances of the skeletal phyla plotted against time (Fig. 7.7.1) form an undulating line, with 90% of the points lying within an interval of about 10 Ma. The resolution of this line depends on the correlation scheme selected (see Chapter 10), but errors due to correlation difficulties are small compared to those resulting from biases of the record. The first appearances of Precambrian-Cambrian skeletal fossils are strongly tied to sedimentary facies (Brasier 1982; Mount and Signor 1988). Selective preservation and sampling further play a major role – most of the earliest skeletal fossils have been extracted from the rocks by dissolution of calcareous rocks, which means that the record is biased towards non-calcareous biominerals,

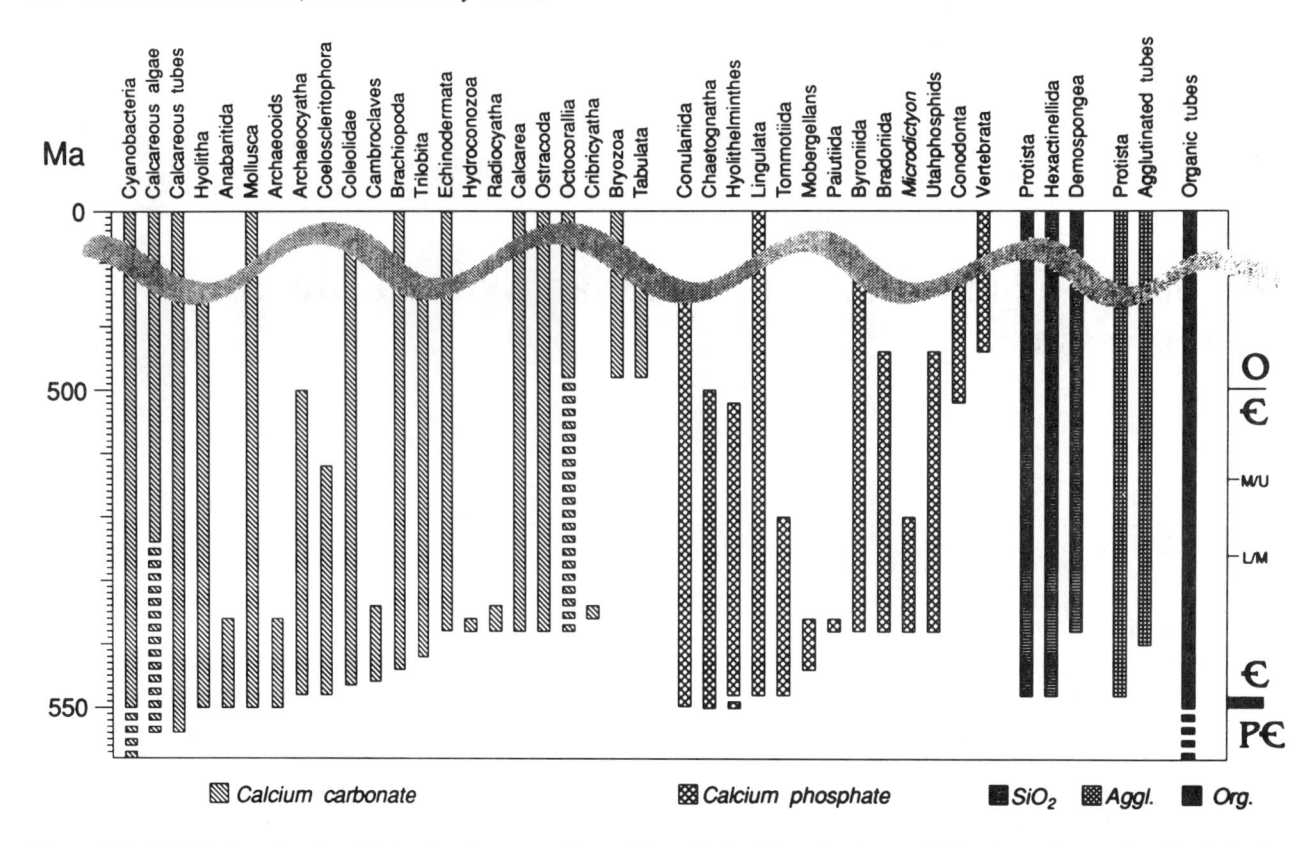

Figure 7.7.1. Distribution of major skeleton-forming groups through time. Vertical bars denote age distribution as known from fossils; broken bars indicate problems of identification or of age assessment. Time scale according to correlation chart shown in Table 7.5.2; the as yet undefined Precambrian-Cambrian boundary is shown as a transitional interval spanning the Nemakit-Daldynian (550 to 548 Ma), and a missing time interval from the Lower Ordovician to the Upper Tertiary is indicated by a gray undulating band. Note that not all phylum-level taxa listed in Section 7.7. and Table 23.2 have been included, that some of the taxa on the diagram are subgroups of such "phyla," and that a number of problematic tubicolous organisms have been lumped into non-taxonomic categories as "calcareous tubes," "agglutinated tubes," and "organic tubes."

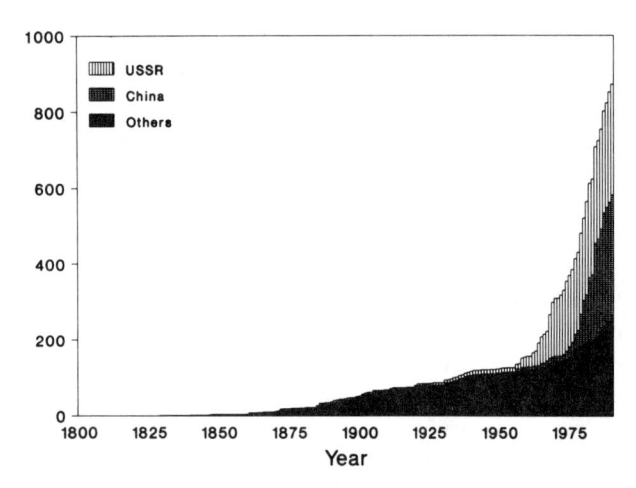

Figure 7.7.2. Cumulative plots of published genera of Early Cambrian fossils (exclusive of archaeocyathans and trilobites) from 1800 to 1990. Note the major contributions by Soviet and Chinese specialists from the early 1960s and the mid-1970s, respectively. Also note that there is no indication that the curve is leveling off (data for 1990 are incomplete).

and in particular, calcium phosphate. Also, there is a bias in favor of environments where calcareous skeletons have been secondarily replaced or where shell cavities have been filled with non-calcareous minerals.

Although the interval during which the metazoan phyla appeared in the fossil record spans tens of millions of years, it is a razor-sharp line on the time scale generally applied in Proterozoic biostratigraphy. At face value, this line marks the sudden appearance of readily preservable multicellular organisms. Fossils with mineralized skeletons constitute the overwhelming majority of these organisms, but the almost simultaneous appearance of many kinds of multicellular organisms that are preserved because of size and resistant non-mineralized tissue, or that leave traces in the sediment, makes it impossible to view biomineralization as the sole event responsible for the appearance of metazoan fossils.

7.7.3 Geographical Distribution of Early Skeletal Fossils

Faunas of the earliest periods (up to and including the Atdabanian) of biomineralizing metazoans have been found on all continents, but at present there is a strong dominance of data from the Siberian Platform and China. These two areas and Australia, all at low latitudes during this time, represent the highest known diversities (on the generic level) of early skeletal fossils. Higher-latitude regions, such as the East European Platform and western North America, exhibit lower diversities in the earliest Cambrian (Section 8.4).

7.7.4 Appearances of Skeletal Types

The following types of metazoan skeletons are all represented in the earliest skeletal biotas:

(1) Calcareous and siliceous spicular skeletons.
(2) Internal and external calcareous reinforcements of soft bodies.
(3) Tubular skeletons of differing structure and composition.
(4) Shells, including the calcareous conchs of molluscs or mollusc-like fossils and the calcareous or phosphatic bivalved shells of brachiopods or brachiopod-like fossils.
(5) Calcareous or phosphatic sclerites forming parts of a composite dermal exoskeleton.
(6) Phosphatic teeth or grasping spines.
(7) Cone-shaped, cap-shaped, or plate-shaped skeletal structures of uncertain function and position in the body but probably primarily protective in nature; most of the plate-shaped structures are phosphatic.

The distinction between some of these groups, in particular that between tubes and shells, and between groups 5, 6, and 7, is somewhat arbitrary. Each category includes two or more major taxonomic groups that evolved their skeleton separately.

Diagenetic effects have obscured much of the original shell mineralogy, but the preservation of the original mineral (commonly in the case of calcium phosphate and calcite) and cases of replication of crystal morphology on the surface of steinkerns have provided reliable data on the detailed mineralogy of at least some early skeletal fossils. For others, at least the approximate original composition (calcareous, phosphatic, or siliceous) may be known from circumstantial evidence. The distribution of various skeletal minerals among the early biomineralizing phyla is presented in Figure 7.7.1.

7.7.4A Calcareous and Siliceous Spicular Skeletons

Spicules are minute mineralized elements formed within living tissues. They are widely distributed among living phyla and are formed from a variety of minerals (Rieger and Sterrer 1975; Lowenstam and Weiner 1983). In the fossil record they are often poorly preserved because of their size and mineralogical composition. An intensified search among the earliest skeletal faunas suggests that a considerable diversity of metazoan taxa may be represented by spicules. However, the identification of taxa based on spicules is difficult, because the spicular morphology of metazoans is only occasionally related to larger anatomical features, and a particular morphology is often not confined to a single clade.

Spicules have been reported from Proterozoic rocks, but in no case has their presence been unequivocally demonstrated. The oldest unquestionable spicules appear to be hexactinellid-type remains in the first skeletal assemblages of the Siberian Platform (Zhuravleva 1983) and China (Ding and Qian 1988). From the Atdabanian and onwards, spicules are common constituents of Early Cambrian rocks, although they have been only sporadically reported in the literature (e.g., Mostler 1985).

A spicule assemblage from the late Atdabanian of South Australia includes representatives of hexactinellids, demosponges, calcareous sponges, possible octocorals, and tunicates, as well as unknown groups (Fig. 7.7.3A-H; Bengtson et al. 1990).

Calcareous reinforcements of bodies. This type of skeleton is often massive and has good fossilization potential. Archaeocyathans and echinoderms, characterized by internal such skeletons, have an excellent fossil record which may closely correspond to their actual distribution in time and space. In both these groups the skeleton appears to be integrated with the basic function of the animals, rather than, e.g., an armor added for protection. There are also a number of Cambrian fossils that appear to represent massive basal skeletons of sponge-like or cnidarian-like animals (Jell 1984).

7.7.4B Tubular Skeletons

A large number of genera of early skeletal fossils are represented by tubes of various composition (Fig. 7.7.4; see Section 8.2 for a discussion of the ecological and evolutionary significance of the common occurrence of tubes). Their composition includes phosphatic (hyolithelminths), calcareous (anabaritids, cribricyathans, coleolids), siliceous (platysolenitids), agglutinating (*Onuphionella*), and purely organic (sabelliditids) tubes. Calcareous tubes occur together with some Ediacaran fossils of the Nama Group in Namibia (Germs 1972). Although the type locality of *Cloudina hartmannae* Germs 1972, does not have a clear stratigraphic relation with the Ediacaran fauna, fossils which resemble *Cloudina* are also known from carbonate rocks that interfinger with the clastic rocks bearing the Ediacaran fossils (Glaessner 1976a, 1984). The organic tubes of sabelliditids, occurring in late Vendian deposits, consist of a laminated middle zone lined by homogenous layers (Urbanek and Mierzejewska 1977). The chemical composition is not known.

Tubular fossils account for a large number of the problematic taxa, because soft-part anatomy of the animals that constructed the tubes is not intimately related to the morphology and structure of the tubes. There are at least seven phylum-level taxa represented by tubular fossils (trilobozoans, coleolids, cribricyathans, sabelliditids, agmatans, hyolithelminths, paiutiids, conulariids, ?mobergellans but no living phylum has been shown to be represented among these, notwithstanding claims to the contrary (e.g., Poulsen 1963; Sokolov 1965, 1967; Glaessner 1976a; Grigor'eva and Zhegallo 1979).

7.7.4C Shells

Mineralized shells – hardened external coverings consisting of one or a few units covering most of the body – appear independently among at least four groups (arthropods, molluscs, hyoliths, brachiopods) at the Precambrian-Cambrian boundary. They are a good source of information on body size and shape, but many important anatomical features may not be reflected at all in the shells. Of the living phyla, molluscs and brachiopods unquestionably occur in the Tommotian and Meishucunian biotas (Pel'man 1977; Runnegar 1983), and arthropods become abundant from the Atdabanian and

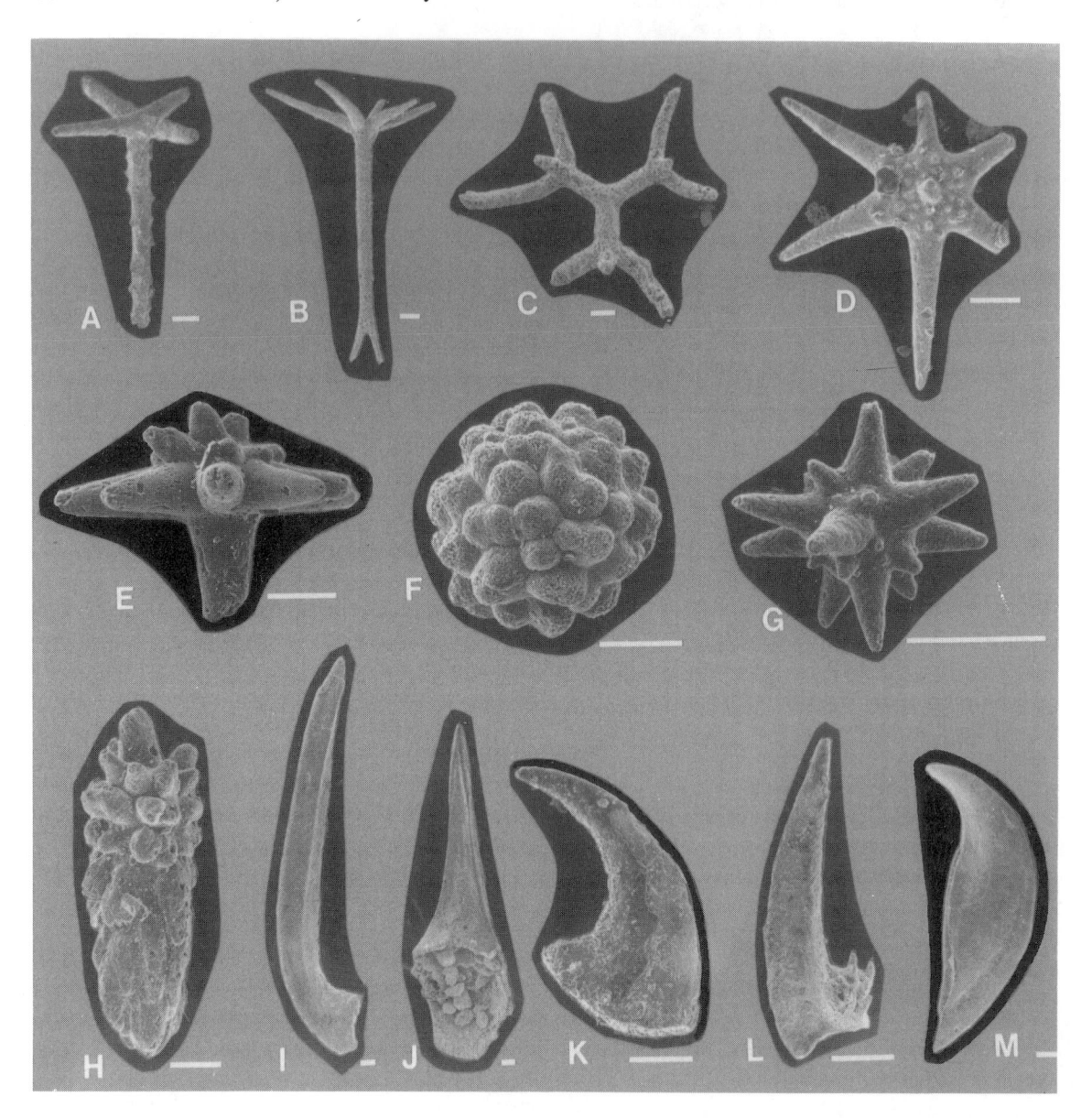

Figure 7.7.3. Early Cambrian skeletal fossils: siliceous (A, B) and calcareous (C, D) sponge spicules, calcareous spicules of organisms of unknown affinity (E–G), structural element possibly of an octocoral (H), and tooth-shaped objects of phosphatic (I–K) or unknown (L, M) composition. Lines for scale represent 100 μm. (A) Hexactinellid sponge, Ajax Limestone, Flinders Ranges, South Australia (Bengtson et al. 1990). (B) Demosponge(?), *Taraxaculum volans* Bengtson 1990, Ajax Limestone, Flinders Ranges, South Australia (Bengtson et al. 1990). (C) Calcareous sponge, *Dodecaactinella cynodontota* Bengtson & Runnegar 1990, Ajax Limestone, Flinders Ranges, South Australia (Bengtson et al. 1990). (D) Heteractinid sponge, *Eiffelia araniformis* (Missarzhevsky 1981), Ajax Limestone, Flinders Ranges, South Australia (Bengtson et al. 1990). (E–G) Unknown organisms, possibly sponges or (F) ascidians, Kulpara Limestone, Yorke Peninsula, South Australia (Bengtson et al. 1990). (H) Octocoral(?), *Microcoryne cephalata* Bengtson 1990, Ajax Limestone, Flinders Ranges, South Australia (Bengtson et al. 1990). (I, J) Chaetognaths(?), *Protohertzina unguliformis* Missarzhevsky 1973 (I) and *P. anabarica* Missarzhevsky 1973 (J), lower Meishucunian Stage, Meishucun, Yunnan, China (Qian and Bengtson 1989). (K) Chaetognath(?), *Mongolodus* cf. *rostriformis* Missarzhevsky 1977, Ajax Limestone, Flinders Ranges, South Australia (Bengtson et al. 1990). (L) *Yunnanodus dolerus* Wang & Jiang 1980, middle Meishucunian Stage, Meishucun, Yunnan, China (Qian and Bengtson 1989). (M) *Crytochites pinnoides* Qian 1984, middle Meishucunian Stage, Xianfeng, Yunnan, China (Qian and Bengtson 1989).

Qiongzhusian, but a large number of the cap-shaped fossils in the earliest Cambrian are of unknown affinity, and a few appear to represent extinct phyla (Hyolitha, etc.).

7.7.4D Calcareous or Phosphatic Sclerites

A sizable fraction of the earliest skeletal assemblages consists of isolated skeletal elements from composite exoskeletons. Some of these belong to arthropods, the most common of which are trilobites. These are sufficiently often found as complete articulated exoskeletons, and their interpretation is straightforward. But the majority belong to short-lived groups of uncertain affinity, and finds of articulated skeletons are extremely rare. The problems of interpretation have led to taxonomic oversplitting in that the nature of the sclerites as

Figure 7.7.4. Early Cambrian skeletal fossils of tube-dwelling organisms: anabaritids (A–C, calcareous tubes), hyolithelminths (D–F, phosphatic tubes), hyoliths (G–K, calcareous conchs and lids), and organisms of possible hyolith affinity (L, M, calcareous tubes). Lines for scale represent 100 μm. (A) *Anabarites trisulcatus* Missarzhevsky 1969, lower Meishucunian Stage, Meishucun, Yunnan, China (Qian and Bengtson 1989). (B, C) *Anabarites sexalox* Conway Morris & Bengtson 1990, Ajax Limestone, Flinders Ranges, South Australia (Bengtson et al. 1990). (D) *Hyolithellus filiformis* Bengtson 1990b, Parara Limestone, Yorke Peninsula, South Australia (Bengtson et al. 1990). (E) *Hyolithellus* cf. *micans* Billings 1871, Parara Limestone, Yorke Peninsula, South Australia (Bengtson et al. 1990). (F) *Torellella* sp. Parara Limestone, Yorke Peninsula, South Australia (Bengtson et al. 1990). (G, H) Operculum and conch of *Parkula bounites* Bengtson 1990, Parara Limestone, Yorke Peninsula, South Australia (Bengtson et al. 1990). (I) Conch of *Hyptiotheca karraculum* Bengtson 1990, Parara Limestone, Yorke Peninsula, South Australia (Bengtson et al. 1990). (J, K) Operculum and conch of *Conotheca australiensis* Bengtson 1990, Parara Limestone, Yorke Peninsula, South Australia (Bengtson et al. 1990). (L) *Paragloborilus subglobosus* He 1977, middle Meishucunian Stage, Xianfeng, Yunnan, China and Bengtson 1989). (M) *Actinotheca holocyclata* Bengtson 1990, Parara Limestone, Yorke Peninsula, South Australia (Bengtson et al. 1990).

parts of disarticulated skeletons has not been recognized (Bengtson 1985). About 6% of the genera in the Early Cambrian belong to such poorly known groups of sclerite-bearing organisms; they represent at least four phylum-level taxa (coeloscleritophorans, paracarinachitids-cambroclaves, tommotiids, utahphosphids), including forms with calcareous (e.g., coeloscleritophorans) and phosphatic (e.g., tommotiids) sclerites. None of these taxa is known after the Cambrian.

7.7.4E Phosphatic Teeth or Grasping Spines

Almost any collection of skeletal microfossils from Lower Cambrian rocks contains spine-shaped fragments. Many of these are pieces of larger, spiny sclerites, shells, or tubes, but a few are discrete elements having a recognizable morphology, composition, and structure (Fig. 7.7.3I-M). Most commonly they are composed of calcium phosphate. They may represent several phylum-level taxa, including the living Chaetognatha (Szaniawski 1982; Bengtson 1983; Qian and Bengtson 1989).

7.7.4F Mineralized Skeletons of Uncertain Nature

Many of the earliest skeletal fossils are plate- or cap-shaped structures that have not yet been adequately explained in terms of functional morphology. Most of these are of phosphatic composition (*Fomitchella, Mobergella, Tumulduria, Microdictyon*, and others). Although their biology and affinities are poorly understood, they may be quite distinct forms that are known in considerable detail (e.g., *Microdictyon*, Fig. 7.7.8K-L).

7.7.5 Phylum-level Taxa in the Latent Proterozoic and Early Cambrian

The use of the phylum as a natural unit is not clearcut (see Section 8.5). Present-day phyla are generally clearly circumscribed, but this is not the case during the early phases of metazoan radiation. In phyla of the late Vendian-Early Cambrian, evolutionary distances to the common ancestor were smaller, many short-lived lineages existed, and the limited number of data available from fossils further obscures phylum divisions. The extinct "phyla" recognized herein share no known synapomorphic characters with other known groups and show basic features of skeleton formation and body organization not known in any other group. They are also sufficiently well understood that their lack of association to other groups is not likely to be due solely to our ignorance. This latter criterion for acceptance of an extinct phylum weeds out a large number of potential phylum-level taxa that are not sufficiently well-known, but makes the number of phyla difficult to assess because of the temporary and somewhat arbitrary nature of the cut-off point. Examples of genera that may or may not be closely related to some of the "phyla" are given in Figures 7.7.5L, 7.7.6A-B, and 7.7.8I-J.

The list below includes also non-skeletal body fossils in the Early Cambrian. The genera referred to the respective "phyla" (and those currently not referred to any of the latter) are listed in Table 23.5. Geographic and stratigraphic ranges as given below refer to published occurrences of named taxa.

Poriferans (Fig. 7.7.3A-D, E?, F?, G?). Poriferans, or sponges, are well-represented throughout the Phanerozoic fossil record and are widely distributed in modern marine and fresh-water environments. They are, without exception, sessile filter-feeders.

Most modern poriferans have a spicular skeleton of calcite or opal, and an increasing number of forms are now known to produce a non-spicular calcareous skeleton (cf. comments on archaeocyathans below). There have been a considerable number of Precambrian sponge spicules reported in the literature, but none of these is convincing or even probable, although only some of the reports have been subject to closer scrutiny and reinterpretation (Cloud 1968; Miller 1987). The earliest hexactinellid spicules occur in the lower Meischu-

cunian (Ding and Qian 1988, and the first definitive spicules of calcareous sponges in the upper Atdabanian (Bengtson 1990). Demosponge spicules are probably present in the upper Atdabanian as well (Bengtson 1990).

Many of the earliest sponges should presently be left without formal assignment to suprageneric taxa. Nevertheless, a workable hierarchical classification exists for a number of them, mainly due to the plentiful information available on Middle Cambrian sponges (Walcott 1920; Rigby 1986). This system has been used wherever practicable in the present context.

Although most probably metazoans (Bergquist 1985; Chapter 9), poriferans lack the individuality of most metazoans, and their "bodies" rather behave like colonies of individual cells in many aspects of growth and function (Fry 1970).

Range – Nemakit-Daldyn to Recent.

Geographic distribution (P€–€) — North America, Europe, Asia, Africa, Australia, Antarctica.

Key references – Sdzuy 1969; Finks 1970, 1983; Rigby 1983, 1986; Mostler 1985, 1986; Bengtson 1990.

Archaeocyathans. This group is characteristic of many Lower Cambrian carbonate sequences. They had a porous calcareous skeleton commonly consisting of an outer and an inner cup with radiating walls crossing the space between the cups. They were sedentary filter-feeders, in many cases reef-building.

Archaeocyathans were long thought to be an extinct phylum confined to the Early Cambrian. They have recently been shown to have survived at least into the Late Cambrian (Debrenne, Rozanov and Webers 1984), and recent research on living and fossil coralline sponges (e.g., Vacelet 1985) suggests that the archaeocyathan type of skeleton may well have been formed by a sponge. Until this question has been clarified, the archaeocyathans are best treated as a separate clade of problematic affinity.

Range. Lower Tommotian to Upper Cambrian.

Geographic distribution (PC–C). North America, southern Europe, Asia, Australia, North Africa, Antarctica.

Key references. Zhuravleva 1960; Hill 1972 (and references therein); Debrenne et al. 1984; Rowland and Gangloff 1988; Zhuravlev 1989.

Radiocyathans. This is a small group of sponge-like fossils with walls reinforced by dumbbell-shaped calcareous sclerites having a central shaft and a rosette at the outer and inner end. They have an exclusively Lower Cambrian range and are primarily known from Australia and Siberia.

The affinities of radiocyathans are uncertain. They have been referred to sponges or archaeocyathans, or been treated as a separate group with affinities either to these phyla or to receptaculitid algae. At present they are too poorly known to enable a more definitive placement.

Range. Upper Tommotian-Botomian.

Geographic distribution. Australia, Mongolia, Siberia, Canada, Morocco.

Key references. Bedford, R., and Bedford, W. R., 1934; Bedford, R., and Bedford, J., 1936; Debrenne 1970, 1971; Nitecki and Debrenne 1979; Zhuravleva and Myagkova 1981; Zhuravlev 1986.

Figure 7.7.5. Early Cambrian skeletal fossils: molluscs (A, B), mollusc-like fossils (C–G), and phosphatic (H–K) or calcareous (L) shells of probable opercular nature. Line for scale in (A) represents 1 mm; all other lines for scale represent 100 μm. (A) Gastropod, *Aldanella attleborensis* (Shaler & Foerste 1888), Chapel Island Formation, Burin Peninsula, Newfoundland, Canada (Bengtson and Fletcher 1983). (B) Monoplacophoran, *Archaeospira ornata* Yu 1979, middle Meishucunian Stage, Xianfeng, Yunnan, China (Qian and Bengtson 1989). (C) *Xianfengella prima* He & Yang 1982, middle Meishucunian Stage, Xianfeng, Yunnan, China (Qian and Bengtson 1989). (D) *Ocruranus trulliformis* (Jiang 1980b), middle Meishucunian Stage, Xianfeng, Yunnan, China (Qian and Bengtson 1989). (E) *Maikhanella pristinis* (Jiang 1980b), lower Meishucunian Stage, Meishucun, Yunnan, China (Qian and Bengtson 1989). (F) *M. cambrina* (Jiang 1980), middle Meishucunian Stage, Xianfeng, Yunnan, China (Qian and Bengtson 1989). (G) *Proplina*? sp., Parara Limestone, Yorke Peninsula, South Australia (sample no. L1852 of Bengtson et al. 1990). (H) *Mobergella? bella* (He & Yang 1982), middle Meishucunian Stage, Xianfeng, Yunnan, China (Qian and Bengtson 1989). (I) *Mobergella holsti* (Moberg 1892), Kalmarsund Sandstone, Sweden (Bengtson 1968). (J, K) *Mobergella turgida* Bengtson 1968, Kalmarsund Sandstone, Sweden (Bengtson 1968). (L) *Archaeopetasus excavatus* Conway Morris & Bengtson 1990, Parara Limestone, Yorke Peninsula, South Australia (Bengtson et al. 1990).

Cnidarians (Fig. 7.7.3H?). Given the great importance of cnidarians as biomineralizers during most of the Phanerozoic and their presumed early origins, it is somewhat paradoxical that skeletalized cnidarians are not known for certain from the Lower Cambrian. In addition to the possible cnidarians listed in Table 23.5, there are trace fossils referred to cnidarians (cf. Chapter 7.3). Regarding the proposed cnidarians in the Ediacara fauna, see Section 7.3 and Table 23.5.

Key reference. Jell 1984.

Priapulids are a small phylum of worm-like predatorial coelomates that burrow in modern sandy or muddy environments.

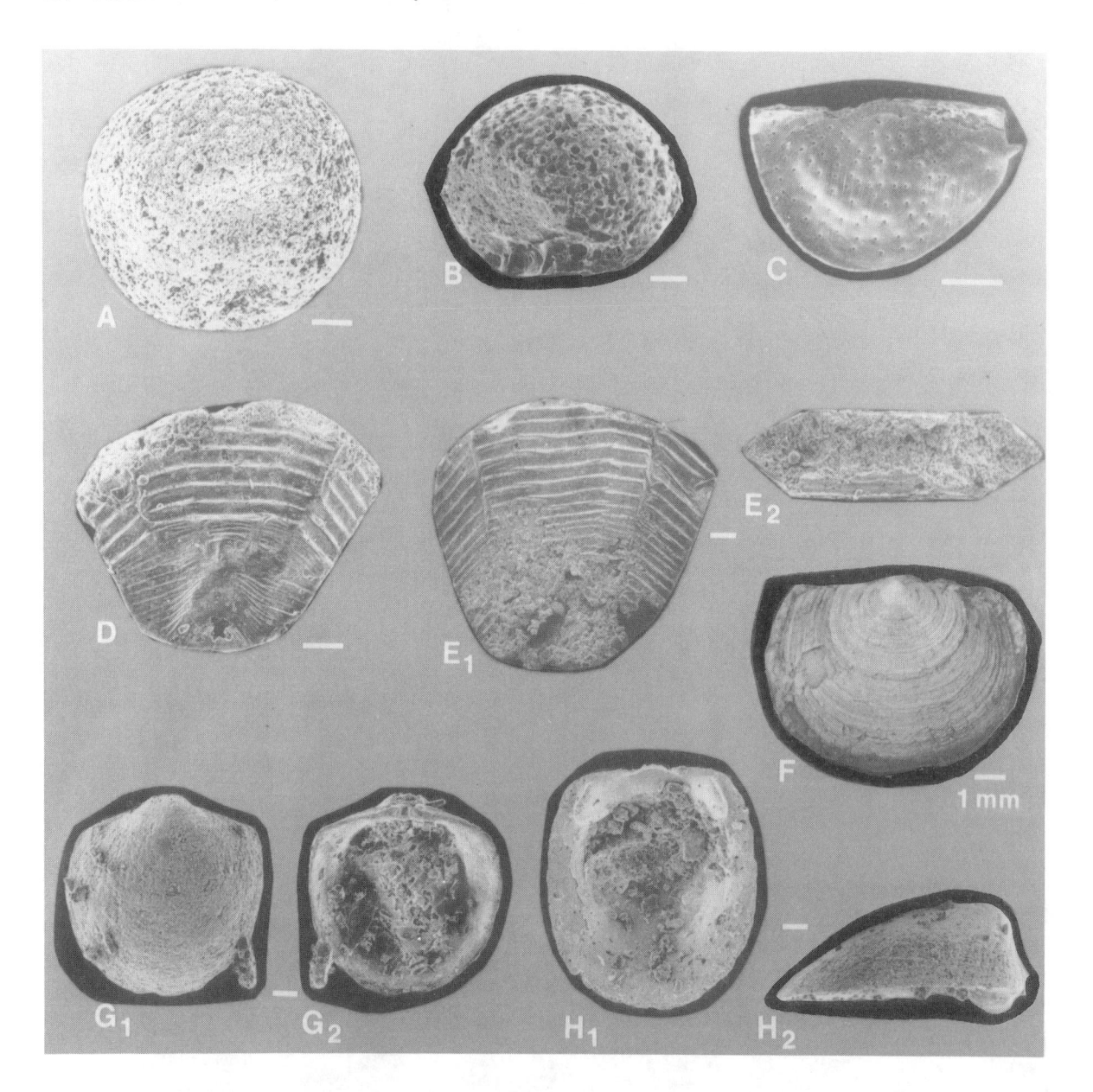

404 Proterozoic Remains, Trace and Body Fossils

Figure 7.7.6. Early Cambrian skeletal fossils. Line for scale in (F) represents 1 mm; all other lines for scale represent 100 μm. (A) *Archaeooides* sp., a spherical fossil of uncertain affinity, lower Meishucunian Stage, Nanjiang, Sichuan, China (Yang et al. 1983). (B) *Aetholicopalla adnata* Conway Morris 1990, an encrusting hemispherical fossil of possible algal origin, Parara Limestone, Yorke Peninsula, South Australia (sample no. L1763b of Bengtson et al. 1990; scanning electron micrograph courtesy of S. Conway Morris). (C) Ostracod, *Epactridion portax* Bengtson 1990, Parara Limestone, Yorke Peninsula, South Australia (Bengtson et al. 1990). (D, E) Conulariid, *Hexangulaconularia formosa* He 1984, lower Meishucunian Stage, Nanjiang, Sichuan, China (Yang et al. 1983). (F) Brachiopod, *Micromitra undosa* (Moberg 1892), Kalmarsund Sandstone, Sweden (Åhman and Martinsson 1965). (G) *Aroonia seposita* Bengtson 1990, the "pit valve" of a brachiopod-like animal, Ajax Limestone, Flinders Ranges, South Australia (Bengtson et al. 1990). (H) *Apistoconcha apheles* Conway Morris 1990, the ventral valve of a brachiopod-like animal, Parara Limestone, Yorke Peninsula, South Australia (sample no. L1762 of Bengtson et al. 1990).

They have a poor fossil record, with the splendid exception of the Middle Cambrian Burgess Shale, where they show a morphological diversity surpassing that of the living members of the phylum (Conway Morris 1977a). Representatives of this early assemblage have recently been discovered also in the Lower Cambrian Chengjiang fauna (Hou and Sun 1988). Some burrows of the same age may well be attributed to priapulids (Jensen 1990).

Range. Atdabanian to Recent.
Geographic distribution (PC–C). China.

Key references. Conway Morris 1977a; van der Land and Nørrevang 1985; Sun and Hou 1987b; Hou and Sun 1988.

Annelids. Many references to Late Precambrian and Early Cambrian annelids exist in the literature (e.g., Glaessner 1976a), but most of them are based on tubular fossils or trace fossils lacking characters reflecting on the anatomy, or on worm-shaped impressions with more or less distinct annulation. True scolecodonts, i.e. polychaete jaws, are not known from this time.

The Middle Cambrian Burgess Shale has preserved a diverse fauna of polychaetes (Conway Morris 1979), and probable annelids in similar preservation have also been found in Lower Cambrian rocks (Glaessner 1979C; Hou and Sun 1988).

Range. Atdabanian-Recent.

Geographic distribution (PE–E). Australia, China.

Key references. Walcott 1911; Howell 1962; Glaessner 1979C; Conway Morris 1979; Hou and Sun 1988.

Trilobozoans (Figs. 7.5.9A-D, 7.7.4A-C). Anabaritids are characteristic components of the first rich assemblages of skeletal fossils to appear around the Precambrian-Cambrian boundary. Their calcareous tubular skeletons have a consistent triradial symmetry, most commonly expressed in three equally developed longitudinal sulci dividing the tube into three equal lobes.

There is considerable variation in divergence angle (from near 0° to about 80°), cross-sectional shape (round, triangular, more or less deeply sulcate, also with an extra set of shallower sulci dividing each lobe into two equal parts; some forms are also provided with external longitudinal flanges along the shell), surface sculpture (from smooth to distinctly annulated), internal structures (the insides may be smooth or provided with longitudinal rows of spine- or plate-like processes), and general growth habits (the tubes may be straight, sinuously curved, or regularly bent; longitudinal twisting around the axis is commonly present). It appears that much of this variation is intraspecific, and the establishment of a number of genera and families on the basis of morphs (see list of taxa in Table 23.5) is weakly founded. Nevertheless, a number of distinct species appear to be present.

Due to their radial symmetry and their sharing of the unusual triradial symmetry with certain Vendian/Ediacaran medusoid-like forms (*Tribrachidium*, *Albumares*, *Anafesta*, *Skinnera*), anabaritids have been proposed to belong to the Cnidaria (Missarzhevsky 1974; Fedonkin 1985a), specifically the Scyphozoa (Val'kov 1982). Proposals of annelid affinity have also been made (Glaessner 1976a). As the triradial symmetry appears to be a stable and fundamental character of the group, and the presence of internal projections suggests a firmer attachment to the tubes than is known in annelids, the latter proposal appears weakly founded. The scyphozoan hypothesis is also problematic, particularly as even the Ediacaran triradiate forms are not demonstrably scyphozoans.

Range. Nemakit-Daldyn to Atdabanian.

Geographic distribution. Siberian Platform, Kazakhstan, China, south Australia, northwestern Canada, Sweden.

Key references. Voronova and Missarzhevsky 1969; Rozanov 1969; Missarzhevsky 1974; Val'kov and Sysoev 1970; Abaimova 1978; Val'kov 1982; Fedonkin 1985a,d; Bengtson 1990.

Coleolids. The family Coleolidae Fisher (1962) includes a variety of thick-walled calcareous tubes ranging from the Lower Cambrian to the Carboniferous. There may be some doubt whether it constitutes a monophyletic group, but at least the Lower Cambrian representatives are a distinct group of tubular fossils having a very low angle of divergence and commonly an ornament of longitudinal ribs. Although mostly found fragmented on bedding planes, occurrences of vertically embedded tubes show that the life attidude of at least some coleolids was that of a sedentary animal sitting vertically in muddy bottoms.

Range. Tommotian to Carboniferous.

Geographic distribution (PE–E). North America, Europe, Asia, Australia.

Key references. Fisher 1962; Brasier and Hewitt 1979.

Cribricyathans are a group of problematic microfossils having calcareous tubes of complex microstructure, commonly a few millimeters in length. There is an outer wall composed of short stacked concentric lamellae, apparently representing incremental growth, and sometimes a thin inner porous wall. They are almost exclusively associated with archaeocyathans.

The cribricyathans appear to be restricted to the Lower Cambrian and have mostly been reported from the Altaj-Sayan Region, Tuva, USSR. It has been suggested (Germs 1972b; Glaessner 1976) that the Precambrian tubular fossil *Cloudina* (cf. Sections 7.7, 8.5) belongs to this group, but the similarities are only of a very general nature and do not prove a close relationship. Cribricyathans have been regarded as planktic larval stages (Vologdin 1932) or other life stages (Zhuravleva and Okuneva 1981) of archaeocyathans, but a number of workers on the group have interpreted them as a separate problematic group (Hill 1972), sometimes at the level of an independent phylum (Jankauskas 1972).

Range. Atdabanian to Tojonian.

Geographic distribution (PE–E). Altaj-Sayan Region (USSR).

Key references. Vologdin 1932, 1964, 1966a; Hill 1972; Jankauskas 1965, 1972, 1973; Zhuravleva and Okuneva 1981.

Sabelliditids occur in Vendian-Cambrian boundary deposits as long organic tubes, up to a few millimeters wide, with more or less distinct transverse annulations on the outer side. They have been compared with pogonophorans (Sokolov 1967). Investigations into the ultrastructure of sabelliditid tubes (Urbanek and Mierzejewska 1977) show them to consist of fine organic laminae, but there are no convincing similarities with the structures of pogonophoran tubes.

Range. Nemakit-Daldyn to Tommotian.

Geographic distribution (PE–E). North America, Europe, Asia.

Key references. Sokolov 1965, 1967, 1968; (Urbanek and Mierzejewska 1977).

Agmatans are a small group of Early Cambrian fossils having tubes with or without an external calcareous wall and with the interior filled with agglutinated material deposited in a funnel-shaped pattern around a central tube. They have been compared with a number of animal groups, including annelids and protozoans (see Lipps and Sylvester 1968, and Yochelson 1977, for reviews), but none of the assignments to living phyla have been well supported.

Range. Atdabanian to Tojonian.

Geographic distribution (PE–E). North America, Europe.

Key reference. Yochelson 1977.

Hyolithelminths (Fig. 7.7.4D-F). Early skeletal fossils represented by elongate phosphatic tubes are commonly referred to the Hyolithelminthes. Although these have often been thought to carry opercula with paired muscle scars, it is now recognized that the alleged opercula belong to a different organism (cf. mobergellids). The tubes are built up of phosphatized laminae. The hyolithelminths were sedentary animals, and at least some may have been infaunal.

Hyolithelminths have been suggested to be of pogonophoran (Poulsen 1963) and annelid (Glaessner 1976a) affinities. Although it has been claimed that *Hyolithellus* has a microstructure similar to that of serpulid polychaetes (Grigor'eva and Zhegallo 1979), this claim has not been sufficiently substantiated and does not seem to be in agreement with other observations (Hurst and Hewitt 1977; Hewitt 1980; Bengtson et al. 1990).

Range. Nemakit-Daldyn(?)-Tommotian-Middle Cambrian (upper boundary somewhat arbitrary, as phosphatic tubes of uncertain affinity occur throughout most of the Paleozoic).

Geographic distribution (PЄ–Є). North America, Europe, Asia, West Africa(?), Australia, Antarctica.

Key references. Fisher 1962; Poulsen 1963; Rozanov 1969; Glaessner 1976a; Hurst and Hewitt 1977; Grigor'eva and Zhegallo 1979; Bengtson 1990.

Paiutiids. The paiutiids are a small group of problematic Early Cambrian organisms with phosphatic tubular skeletons having up to seven longitudinal septa arranged in bilateral symmetry. They have been found only in a stratigraphically restricted interval in the North American Cordillera.

Tynan (1983) regarded the paiutiids to be related to anthozoan corals, on the basis of the internal septation. This proposal does not find support in the structure and composition of the tubes, however. The laminated wall structure and the phosphatic composition prompt comparisons with the Hyolithelminthes, but the fine structure of both these groups needs to be studied in detail before this possibility can be evaluated. The paiutiids should presently be considered to be of unknown affinity.

Range. Atdabanian.
Geographic distribution (PЄ–Є). Western North America.
Key reference. Tynan 1983.

Conulariids (Fig. 7.7.6D-E). Conulariids are tubular or pyramidal fossils, typically with a quadriradiate (in fact, double bilateral) symmetry and quadratic to rectangular transverse cross-section (Babcock and Feldmann 1986). Their phosphatic conchs consist of densely set transverse rods, often forming a chevron pattern on the flat faces, embedded in a presumably flexible integument. The flat faces are usually delimited by longitudinal corner furrows.

The group is well known from Ordovician to Triassic rocks, but recent finds from the Lower Cambrian of China (e.g., Chen 1982; He 1987; Qian and Bengtson 1989) show the salient characteristics and may also be referred to the conulariids.

Conulariids have been interpreted as sessile scyphozoans due to their alleged tetraradiate symmetry and the preservation of tentacles in the Ordovician presumed relative *Conchopeltis*. However, the latter relationship does not withstand scrutiny (Oliver 1984), and the conulariids may be regarded as a separate clade of uncertain affinities (Babcock and Feldman 1986).

Range. Nemakit-Daldyn to Triassic.
Geographic distribution (PЄ–Є). China.
Key references. Kiderlen 1937; Moore and Harrington 1956; Babcock and Feldman 1986; He 1987; Qian and Bengtson 1989.

Mobergellans (Fig. 7.7.5H-K). Mobergellans are a small and exclusively Lower Cambrian group represented by rounded phosphatic shells with distinctly displayed paired muscle impressions on the inside. The shells are usually convex, but in some forms they are flat discs that may have only a restricted convex part in the center.

Mobergellans have been interpreted as monoplacophorans, due to their paired muscle prints, but the phosphatic composition is unknown among monoplacophorans, and the often flat shape of the shells suggests that they were opercula of a larger animal rather than conchs.

Range. Upper Tommotian to Upper Atdabanian.
Geographic distribution (PЄ–Є). North America, Europe, Asia.
Key references. Hedström 1923, 1930; Fisher 1962; Bengtson 1968.

Arthropods (Fig. 7.7.6C). Although by many zoologists considered to encompass several phyla (e.g., Uniramia, Crustacea, Chelicerata), arthropods are here treated together as a phylum-level taxon. Trace fossils attributable to arthropods occur in probable Tommotian-equivalent strata, but definite body fossils of arthropods are not known until the Atdabanian.

Range. ?Tommotian to Recent.
Geographic distribution (PЄ–Є). North and South America, Europe, Africa, Asia, Australia, Antarctica.

Coeloscleritophorans (Fig. 7.7.7A-G). The coeloscleritophorans are an exclusively Cambrian group of metazoans characterized by calcareous, probably aragonitic, hollow sclerites which were periodically molted during ontogenetic growth. They include vagrant, bilaterally symmetrical forms with scale- and spine-shaped sclerites (halkieriids and wiwaxiids); probably sedentary forms with bag-shaped bodies and covered with spiny, usually composite, sclerites (chancelloriids); and several apparently intermediate forms with spine-shaped sclerites and unknown body shape (sachitids, siphogonuchitids).

Coeloscleritophoran taxonomy is in a confused state due to the establishment of a large number of taxa on the basis of individual sclerites, often of poor preservation. The list of taxa in Table 23.5 is based on the revision by Qian and Bengtson (1989) and Bengtson (1990).

Coeloscleritophorans appear to constitute a major Cambrian clade without obvious relations to other contemporaneous clades, i.e., they could be regarded as an extinct phylum. Conway Morris (1985b) proposed a relationship to a stock of metazoans of platyhelminth grade that may have been close to the molluscs (cf. also Chapter 9).

Range. Tommotian to Upper Cambrian.

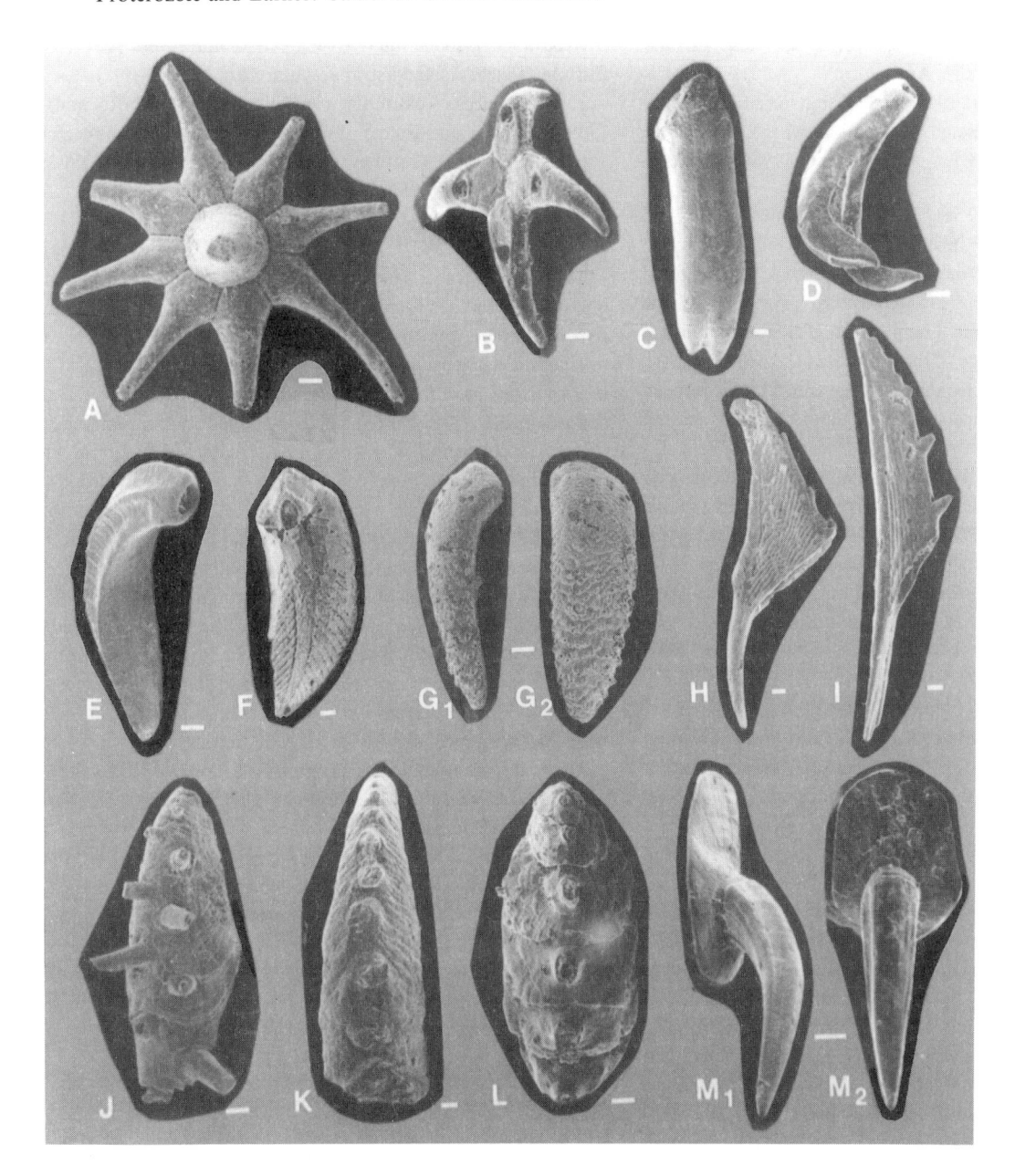

Figure 7.7.7. Early Cambrian skeletal fossils: calcareous sclerites of coeloscleritophorans (A–G), paracarinachitids (H?, I?, J–L), and a cambroclave (M). Lines for scale represent 100 μm. (A) *Chancelloria* sp., Ajax Limestone, Flinders Ranges, South Australia (Bengtson et al. 1990). (B) *Archiasterella hirundo* Bengtson 1990, Parara Limestone, Yorke Peninsula, South Australia (Bengtson et al. 1990). (C) *Eremactis conara* Bengtson & Conway Morris 1990, Ajax Limestone, Flinders Ranges, South Australia (Bengtson et al. 1990). (D) *Drepanochites dilatatus* Qian & Jiang 1982, middle Meishucunian Stage, Xianfeng, Yunnan, China (Qian and Bengtson, 1989). (E) *Siphogonuchites triangularis* Qian 1977, middle Meishucunian Stage, Xianfeng, Yunnan, China (Qian and Bengtson 1989). (F) *Thambetolepis delicata* Jell 1981, Parara Limestone, Yorke Peninsula, South Australia (Bengtson et al. 1990). (G) *Hippopharangites dailyi* Bengtson 1990, Parara Limestone, Yorke Peninsula, South Australia (Bengtson et al. 1990). (H, I) *Scoponodus renustus* Jiang 1982, middle Meishucunian Stage, Xianfeng, Yunnan, China (Qian and Bengtson 1989). (J) *Paracharinachites parabolicus* Qian & Bengtson 1989, middle Meishucunian Stage, Xianfeng, Yunnan, China (Qian and Bengtson 1989). (K) *P. sinensis* Qian & Jiang 1982, middle Meishucunian Stage, Xianfeng, Yunnan, China (Qian and Bengtson 1989). (L) *P. spinus* Yu 1984, middle Meishucunian Stage, Meishucun, Yunnan, China (Qian and Bengtson 1989). (M) *Cambroclavus undulatus* Mambetov 1979, Shabakty Formation, Malyj Karatau, Kazakhstan, USSR (Mambetov and Repina 1979).

Geographic distribution (PЄ–Є). North America, Europe, Asia, West Africa?, Australia, Antarctica.

Key references. Walcott 1911, 1920; Bengtson and Missarzhevsky 1981; Jell 1981; Bengtson and Conway Morris 1984; Conway Morris 1985b; Qian and Bengtson 1989; Bengtson 1990.

Paracarinachitids (Fig. 7.7.7H?-I?, J-L). Paracarinachitids are an enigmatic group of organisms with a restricted distribution in the Early Cambrian. They are characterized by elongated sclerites with a row of spines along the convex side, usually one spine per growth increment. The mineralogical composition is unknown. The morphology of the animals is unknown, but the narrow shape of the sclerites suggests that they may have been radial elements in a cone-shaped structure similar to the capitulum of cirrepedes.

Paracarinachitids have been intepreted as polyplacophoran molluscs (Yu 1984, 1987), but this interpretation is not tenable (Qian and Bengtson 1989). They may be related to the cambroclaves, but the further affinities are unknown.

Range. Meishucunian, Atdabanian(?).

Geographic distribution (PЄ–Є). China, France.

Key references. Yu 1984, 1987; Qian 1984; Kerber 1988; Qian and Bengtson 1989.

Cambroclaves (Fig. 7.7.7M) are a little known and exclusively Lower Cambrian group. The group is mostly represented by disarticulated sclerites, although a few more complete sets of articularted sclerites have been found. The sclerites have a circular to elongated basal shield and an elongate spine, and were probably of calcareous composition. They formed a coat of interlocking sclerites, but the shape of the complete skeleton is not known.

A relationship with the endoparasitic acanthocephalans has been proposed (Qian and Yin 1984b), but this is based only on the morphological similarities of individual sclerites and hooks. There is no evidence that these similarities are anything but convergent, and the nature of the articulated cambroclave scleritome does not suggest any further similarities. The cambroclaves are currently regarded as a clade of uncertain affinities, but they may be closely related to the paracarinachitids.

Range. Meishucunian, Atdabanian.

Geographic distribution (PЄ–Є). Kazakhstan, China, Australia.

Key references. Qian 1978; Mambetov and Repina 1979; Qian and Yin 1984b; Conway Morris and Chen *in press*; Conway Morris *in* Bengtson 1990.

Molluscs (Fig. 7.7.5A-B, C?-G?). A rich assemblage of molluscs and mollusc-like fossils appears at the Precambrian-Cambrian boundary. Representatives of three living classes (monoplacophorans, gastropods, and bivalves) were present before the end of the Atdabanian, in addition to one extinct class (rostroconchs) and a number of forms of more uncertain affinity. Many cap-shaped shells have been questionably assigned to molluscs. These are listed under ?Phylum Mollusca in Table 23.5.

Range. Nemakit-Daldyn to Recent.

Geographic distribution (PЄ–Є). North and South America, Europe, Africa, Asia, Australia, Antarctica.

Key references. Vostokova 1962; Yochelson 1979; Rozanov 1969; Runnegar and Pojeta 1974, 1985; Pojeta 1975; Runnegar and Jell 1976, 1980; Golubev 1976; Runnegar 1978, 1981, 1983; Yu 1979, 1987; Jell 1980; Runnegar and Bentley 1983.

Hyoliths (Fig. 7.7.4G-K, L?-M?) are bilaterally symmetrical metazoans with a cone-shaped aragonitic conch carrying an operculum and, in some forms, a pair of helens, curved appendages that protrude through slits between conch and operculum. They usually have a flatter, presumed ventral, and a more convex, presumed dorsal side, and were probably benthic deposit feeders. A sediment-filled U-shaped gut is sometimes preserved, consisting of a convoluted ventral and a straight dorsal limb. The two main groups, orthothecids and hyolithids, differ most conspicuously in that the latter have a ventral lip, ligula, that is expressed in both conch and operculum.

Hyoliths are present in the earliest skeletal assemblages. They attained a world-wide distribution and high diversity in the Cambrian, and survived until the end of the Paleozoic. Their true diversity in the Early Cambrian assemblages is difficult to assess, however, because a large number of taxa have been based on very imperfectly preserved conchs, so that a considerable amount of synonymy may be present.

Although often linked with molluscs, hyoliths have been suggested to constitute an extinct phylum related to the Sipunculida (Runnegar 1975).

Range. Nemakit-Daldyn to Permian.

Geographic distribution (PЄ–Є). North and South America, Europe, Africa, Asia, Australia, Antarctica.

Key references. Yochelson 1969, 1974; Fisher 1962; Marek 1963b, 1967; Sysoev 1958, 1968, 1973; Rozanov 1969; Marek and Yochelson 1964, 1976; Runnegar 1975.

Microdictyon (Fig. 7.7.8K-L). The genus *Microdictyon* is widespread in the Lower Cambrian and ranges into the Middle Cambrian. *Microdictyon* sclerites are phosphatic netlike plates with a crudely hexagonal meshwork. They did not grow once they were formed, and are always delimited by a peripheral girdle. Recent discoveries of soft-body preservation in the Chengjiang fauna of Yunnan (Chen 1989a) suggest that the *Microdictyon* animal was a bizarre worm-like organism with tentacle-shaped appendages along one side and paired sclerites situated on both sides of the body near the attachment point of each appendage.

Microdictyon does not offer comparisons with any known group of organism. There is a certain resemblance to the equally problematic fossil *Hallucigenia* Conway Morris, 1977 from the Middle Cambrian Burgess Shale.

Range. Atdabanian-lower Middle Cambrian, Silurian?

Geographic distribution (PЄ–Є). North America, Europe, Asia, Australia.

Key references. Missarzhevsky and Mambetov 1981; Bengtson et al. 1986; Hinz 1987; Chen 1989a.

Tommotiids (Fig. 7.7.8A-G) are an exclusively Cambrian group

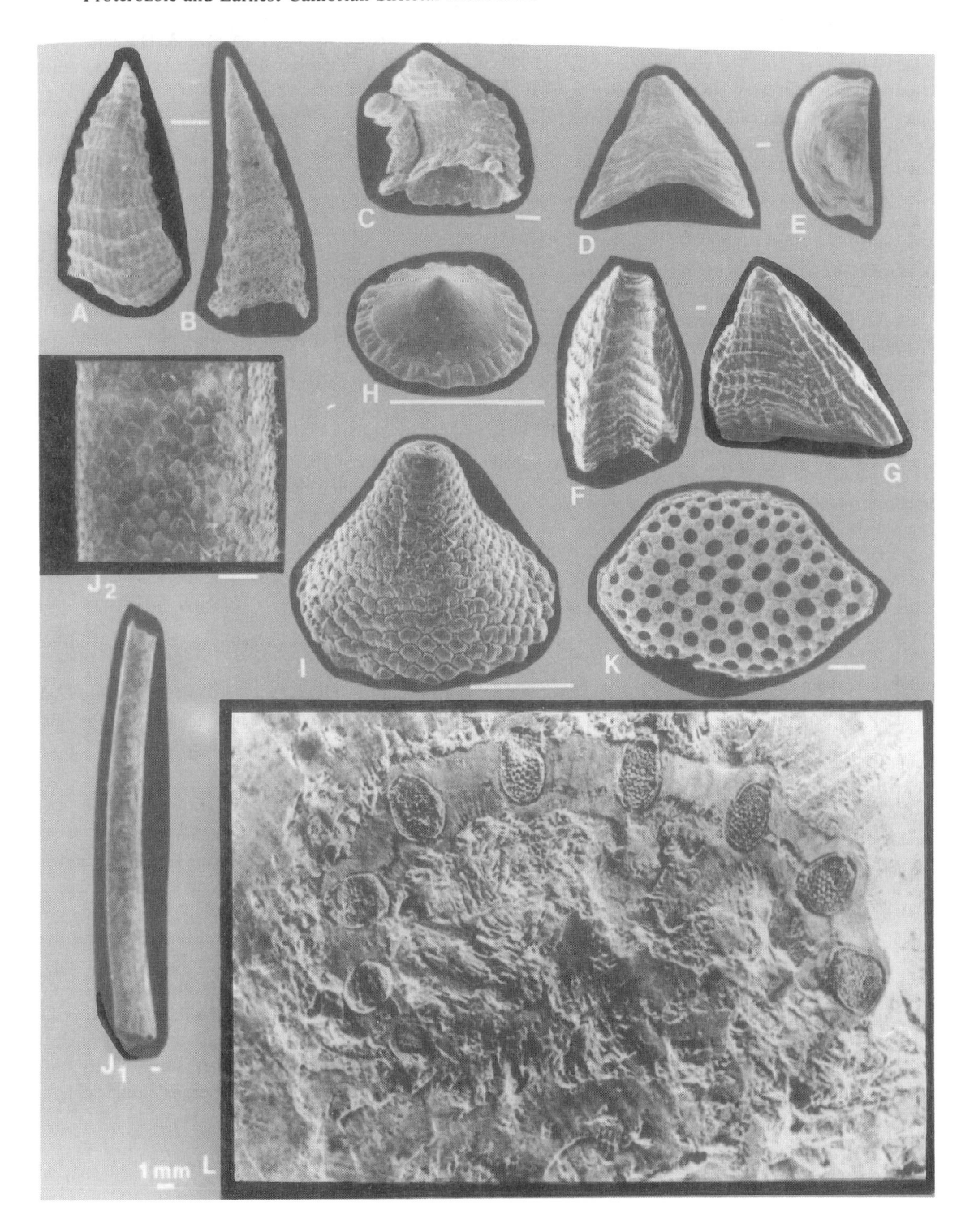

Figure 7.7.8. Early Cambrian skeletal fossils: tommotiids (A–G), a utahphosphid (H), and other phosphatic sclerites (I–L). Line for scale in (L) represents 1 mm; all other lines for scale represent 100 μm. (A) *Lapworthella fasciculata* Conway Morris & Bengtson 1990, Ajax Limestone, Flinders Ranges, South Australia (Bengtson et al. 1990). (B) *L. tortuosa* Missarzhevsky 1966, middle Tommotian Stage, River Lena, Siberian Platform, USSR. (C) *Eccentrotheca guano* Bengtson 1990, Kulpara Limestone, Yorke Peninsula, South Australia (Bengtson et al. 1990). (D, E) Sellate (D) and mitral (E) sclerites of *Tannuolina zhangwentangi* Qian & Bengtson 1989, upper Meishucunian Stage, Meishucun, Yunnan, China (Qian and Bengtson 1989). (F, G) Sellate (F) and mitral (G) sclerites of *Camenella parilobata* Bengtson 1986a, Khairkhan Formation, Salany-Gol, Mongolia (Bengtson 1986). (H) Sclerite of *Hadimopanella apicata* Wrona 1982, upper Atdabanian, Olenyok Uplift, Siberian Platform, USSR (sample no. 63e, collected by N. P. Lazarenko in 1958). (I) *Stoibostrombus crenulatus* Conway Morris & Bengtson 1990, Parara Limestone, Yorke Peninsula, South Australia (Bengtson et al. 1990). (J) *Mongolitubulus* sp., Shabakty Formation, River Uchbas, Malyj Karatau, Kazakhstan, USSR. (K) *Microdictyon rhomboidale* Bengtson et al. 1986, Shabakty Formation, River Uchmas, Malyj Karatau, Kazakhstan, USSR (Bengtson et al. 1986). (L) *Microdictyon sinicum* Chen et al. 1989a, a complete(?) specimen with sclerite pairs in place, lower Qiongzhusian, Chengjiang, Yunnan, China (Chen et al. 1989).

of metazoans carrying conical phosphatic sclerites, usually distinctly ornamented with growth lines. The sclerotome may be poorly organized (sunnaginiids, lapworthellids), or the sclerites may be clearly differentiated into usually two sclerite types occurring in left and right symmetry forms (mitrosagophorans). Tommotiids are particularly characteristic of the Tommotian-Atdabanian, but a few forms range into the Middle Cambrian.

Although tommotiid sclerites, being phosphatic, are commonly easily retrieved in large numbers and good preservation from carbonate rocks using weak organic acids, their taxonomy is presently unstable due to the establishment of a number of taxa without a proper sclerotome analysis. The family assignments used in the list of genera (Table 23.5) is based on Landing's (1984) scheme, but should be regarded as temporary.

Tommotiids are a short-lived group having no obvious relationships with other phyla. They may be regarded as benthic metazoans, possibly segmented. A relationship of the mitrosagophorans with the annelid-like Paleozoic Machaeridia has been proposed (Bengtson 1970, 1977a; Jell 1979).

Range. Tommotian to Middle Cambrian.

Geographic distribution (P€–€). North America, Europe, Asia, Australia, Antarctica.

Key references. Missarzhevsky 1966; Rozanov and Missarzhevsky 1966; Fonin and Smirnova 1967; Rozanov 1969; Bengtson 1970, 1977a, 1986a; Matthews 1973; Bischoff 1976; Landing 1984; Laurie 1986; Bengtson 1990.

Brachiopods (Fig. 7.7.6F, G?, H?) are present in the earliest Tommotian assemblages and appear to increase slowly in diversity throughout the Early Cambrian. A number of brachiopod-like fossils occur that suggest a great initial variability of the "inarticulate" type. In many of these cases, however, their affinity to brachiopods is dubious, and some of the shells have not even been conclusively shown to be bivalved. Although there was a predominance of phosphatic-shelled forms, calcareous shells are by no means uncommon.

Range. Lower Tommotian to Recent.

Geographic distribution (P€–€). North and South America, Europe, Africa, Asia, Australia, Antarctica.

Key references. Walcott 1912; Williams 1965; Pel'man 1977.

Echinoderms appeared in the Atdabanian and produced a number of short-lived Cambrian classes. Living echinoderms are all characterized by pentaradial symmetry, a water vascular system, and a mesodermal skeleton consisting of calcite in a stereom meshwork. Among the early forms, the carpoids lacked visible symmetry. These were probably an early offshoot of the echinoderm branch. Some carpoids have been proposed as stem groups of chordates (Jefferies 1986 and earlier papers referred to therein), but this proposed origin of chordates is controversial (e.g., Philip 1979; Parsley 1988). Helicoplacoids also lacked symmetry, but they have a three-rayed ambulacral system that appears to correspond to the three main branches of the five-rayed system in other echinoderms (Derstler 1981; Paul and Smith 1984).

Range. Upper Atdabanian to Recent.

Geographic distribution (P€–€). North and South America, Europe, Africa, Asia, Australia, Antarctica.

Key references. Durham and Caster 1963; Durham 1967, 1978a; Sprinkle 1973, 1976; Paul 1979; Jefferies 1986; Derstler 1981; Paul and Smith 1984.

Chaetognaths (Fig. 7.7.3I-K). The slender, spine-shaped, phosphatic elements known as protoconodonts are a characteristic component of many of the earliest skeletal faunas around the world. They are characterized by a lamellar structure indicating accretionary growth on the inner side and basal margin, which differentiates them from most true conodonts. The elements have been interpreted as grasping hooks of predatorial animals.

The genus *Protohertzina*, together with *Anabarites*, is generally regarded as an index fossil for Precambrian-Cambrian boundary deposits, although both genera have lately been shown to range in to the late Atdabanian (Bengtson 1990).

The protoconodont elements show a striking resemblance to grasping hooks of recent chaetognaths, and Szaniawski (1982) has shown that the similarities extend to the fine- and ultrastructural levels. His proposal of chaetognath affinities for the protoconodonts appears well supported. The relationship of protoconodonts to the para- and euconodonts (cf. Bengtson 1983) is obscure.

Range. Nemakit-Daldyn to Recent.

Geographic distribution (P€–€). North America, Europe, Asia, Australia.

Key references. Missarzhevsky 1973; Bengtson 1976, 1977a, 1983; Szaniawski 1982, 1983; Nowlan et al. 1985; Brasier and Singh 1987; Qian and Bengtson 1989; Bengtson 1990.

Other tooth-shaped fossils (Fig. 7.7.3L-M). Cambrian rocks abound with tooth-like fossils that may or may not be related to conodonts. The taxonomy of these forms is mostly unmanageable, as most named taxa are based on fragmented spines with unknown structure and composition. Some forms are better known, however, such as the distinctly ornamented *Rhombocorniculum*. Some have a morphology that strongly suggests a function as jaws in a predatorial animal, for example the mandible-like hooks of *Cyrtochites* (Fig. 7.7.3M). Others are of more obscure function. Several major taxa are probably represented among these fossils, but they are presently only listed alphabetically (Table 23.5), as there is insufficient information on the suprageneric taxonomy.

Key references. Walliser 1958; Bengtson 1983; Qian and Bengtson 1989.

Utahphosphids (Fig. 7.7.8H) are usually found as minute, button-shaped phosphatic sclerites, about 50 to 250 µm in diameter. They consist of a fibrous core and an enamel-like capping, and are interpreted as dermal sclerites. Occasional specimens preserved several sclerites set in a matrix of finely granular apatite; in Upper Cambrian representatives of the genus *Utahphospha*, the sclerites are fused into polygonal platelets with no intervening matrix, forming a hollow cone with an open tip. No complete specimens have been found, and the shape of the whole body is unknown.

Bengtson (1977b) pointed out that the sclerites show certain histological similarities with vertebrate dermal skeletons, and the proposal of vertebrate affinity of utahphospids has been followed by Wrona (1982, 1987) and Dzik (1986). Comparisons with tunicates have also been made (Bendix-Almgreen and Peel 1988). Although most arguments point towards a chordate affinity, the evidence is not compelling, and the utahphosphids should be regarded as being of uncertain systematic affinity.

Range. Atdabanian to Lower Ordovician.

Geographic distribution (PЄ–Є). North America, Greenland, Spitsbergen, Europe, Siberian Platform, Turkey, Antarctica.

Key references. Müller and Miller 1976; Gedik 1977, 1981; Bengtson 1977; Repetski 1981; Wrona 1982, 1987; Bendix-Almgreen and Peel 1988.

7.8

Megascopic Dubiofossils

HANS J. HOFMANN

7.8.1 Introduction

The search for evidence of the earliest megascopic forms of life on earth has resulted in an extensive literature on dubiously or incorrectly identified Precambrian "fossils": there are large numbers of reports of "burrows," "worm casts," "jellyfish," "medusoids," and many other questionable remains (Fig. 7.8.1). The most famous of the Precambrian pseudofossils is "*Eozoon*," reported from marbles in the Grenville structural province of Canada (Fig. 7.8.2A, B), which generated an intense and long-standing controversy in the 19th century and volumes of literature (see O'Brien 1970, and Hofmann 1971, pp. 6–12). Claims for the biogenic nature of many remains have not withstood scrutiny, and some others have been shown to be post-Vendian. Inventories and critical reevaluations have appeared from time to time, such as the reviews of Walcott (1899), Raymond (1935), Seilacher (1956), Schindewolf (1956), Häntzschel (1962, 1965, 1975). Cloud (1968), Hofmann (1971b), and Glaessner (1979a, 1984).

Given the wide variety of objects encountered, and the various degrees of confidence with which they are interpreted, a three-fold classification has been found useful in referring to Precambrian remains: those that are demonstrably biogenic (real *fossils*), those that are questionably biogenic (*dubiofossils*), and others that are demonstrably nonbiogenic (*pseudofossils*); crtieria for aiding attribution of remains to these categories are given by Hofmann (1972) (see also Section 7.2). As discussed in Section 7.2, assignment to the dubiofossils should be considered an interim measure, pending new information or new specimens that would allow such remains to be relegated to either the fossils or the pseudofossils. The list of "taxa" in the dubiofossil and pseudofossil categories is long (Table 23.6), and inasmuch as most have been discussed in the more recent of the reviews

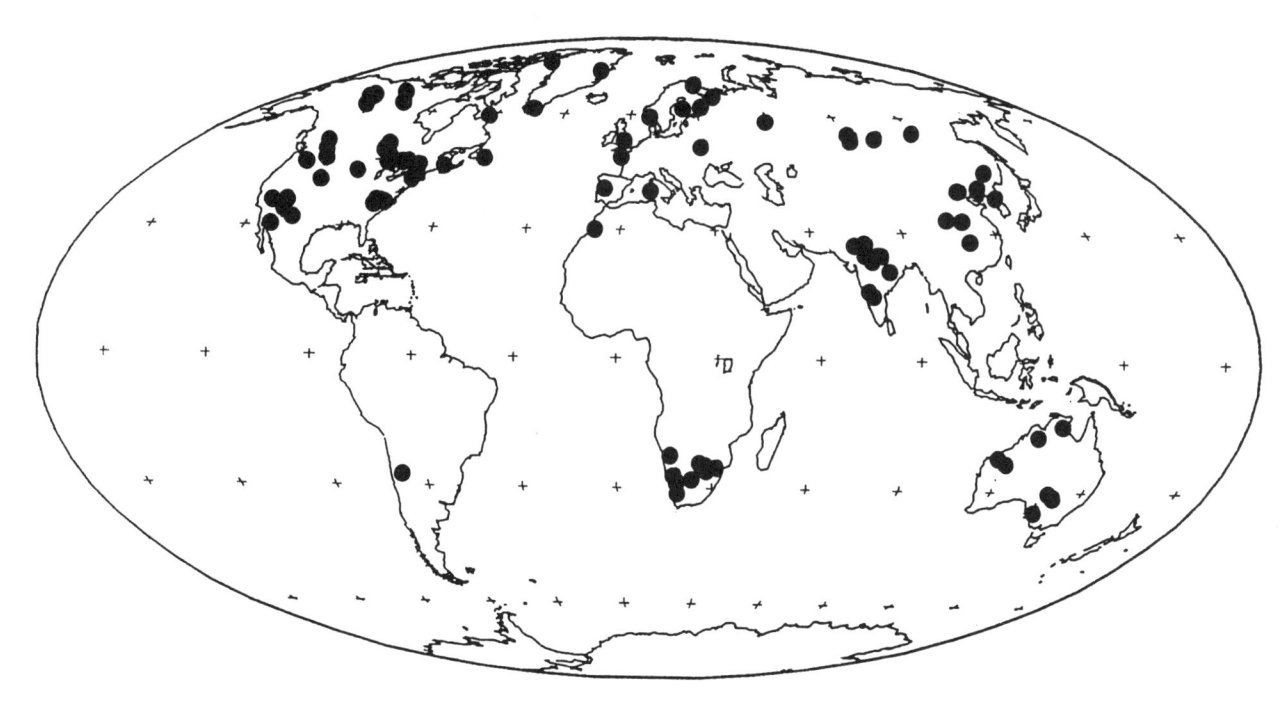

Figure 7.8.1. Geographic distribution of Proterozoic megascopic dubiofossils.

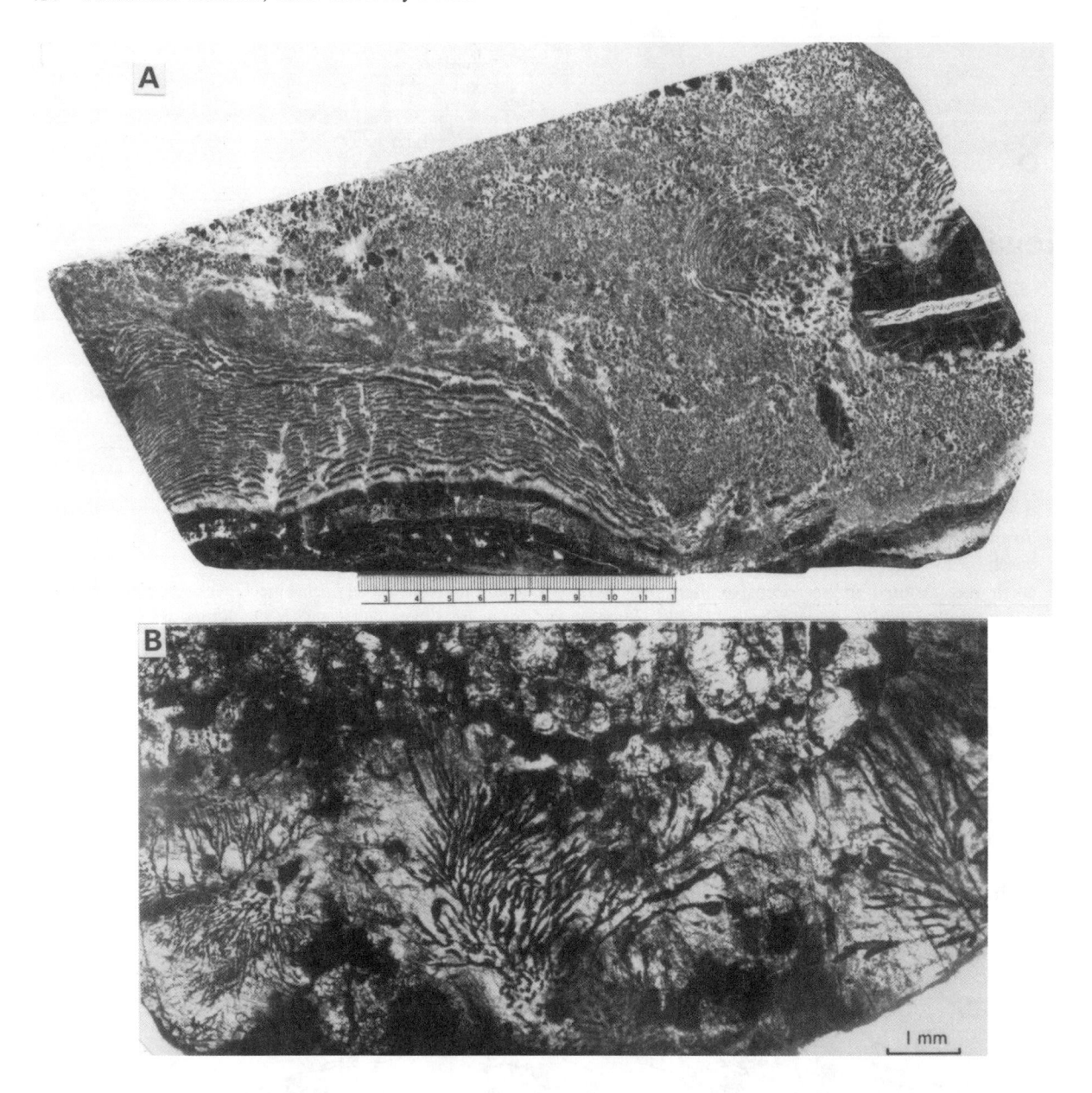

414   Proterozoic Remains, Trace and Body Fossils

Figure 7.8.2. *Eozoon canadense* and *Archaeospherina*, Côte St. Pierre, Quebec, Canada. (A) Large polished slab of topotype specimen, showing the rhythmic banding ("*Eozoon*") of dark serpentine and light colored calcite that, in the mid-1850s, gave rise to the interpretation of this material as biogenic; the serpentine grains in the central and upper portions of the evenly mottled ophicalcite were later named "*Archaeospherina*" and considered to be solitary chambers or distinct organisms; Geological Survey of Canada (GSC) type no. 24368; GSC photograph 200446-C (Hofmann 1971, Pl. 2). (B) Petrographic thin section of a calcitic layer, showing "canals"; these dendritic features are composed of serpentine and usually occur only within the thickest calcite layers; GSC type no. 152; GSC photograph 200329-P, −Q (Hofmann 1971b, Pl. 3, Fig. 1).

cited above, we here concern ourselves only with some examples of those that are probably or possibly biogenic and of Precambrian age, and thus may have some relevance in the discussion of the Precambrian biosphere.

7.8.2 Horizontal Spindles and Ropes

Several groups of remains are represented. The first comprises curved horizontal spindles and ropes of sand. Although some relatively simple forms such as "*Manchuriophycus*" can readily be explained as compacted mudcrack or syneresis fillings (e.g., Häntzschel 1949, 1975), some more complex curved spindles with small lateral corrugations associated with ripple-marked sandstones appear animate ("*Rhysonetron*", Fig. 7.8.3A). However, these, too, can be ascribed to compaction (Hofmann 1971, pp. 36–38). The sinuous pattern of these structures is a response to varying tensional stress directions associated with the propagation of a crack within a drying mud in ripple troughs. The corrugations represent a lateral rippling effect involving axial (longitudinal) transport of the sand-filling with respect to the surrounding mud layer during the terminal stages of sediment compaction. Orthogonal junctions of spindles are developed as a result of stress reorientation in the mud after a primary crack has relieved the original stress field.

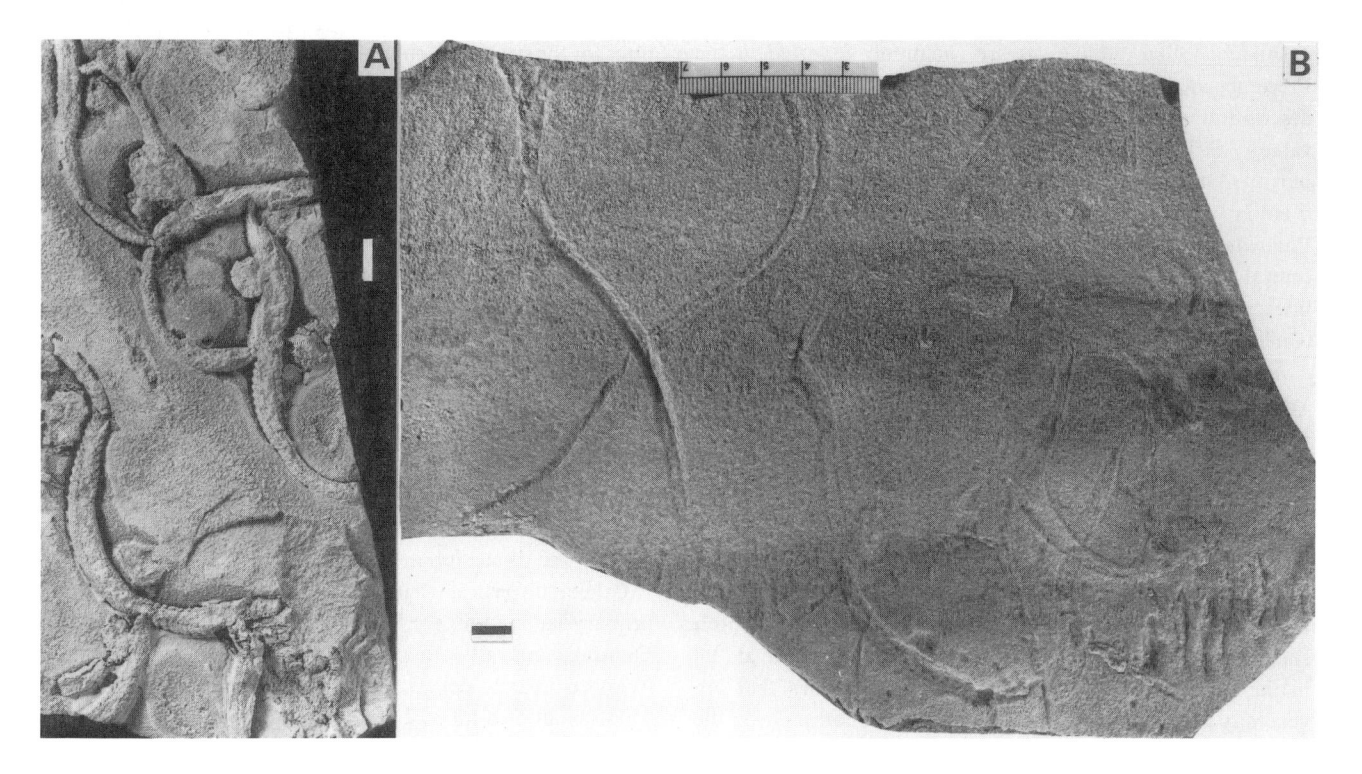

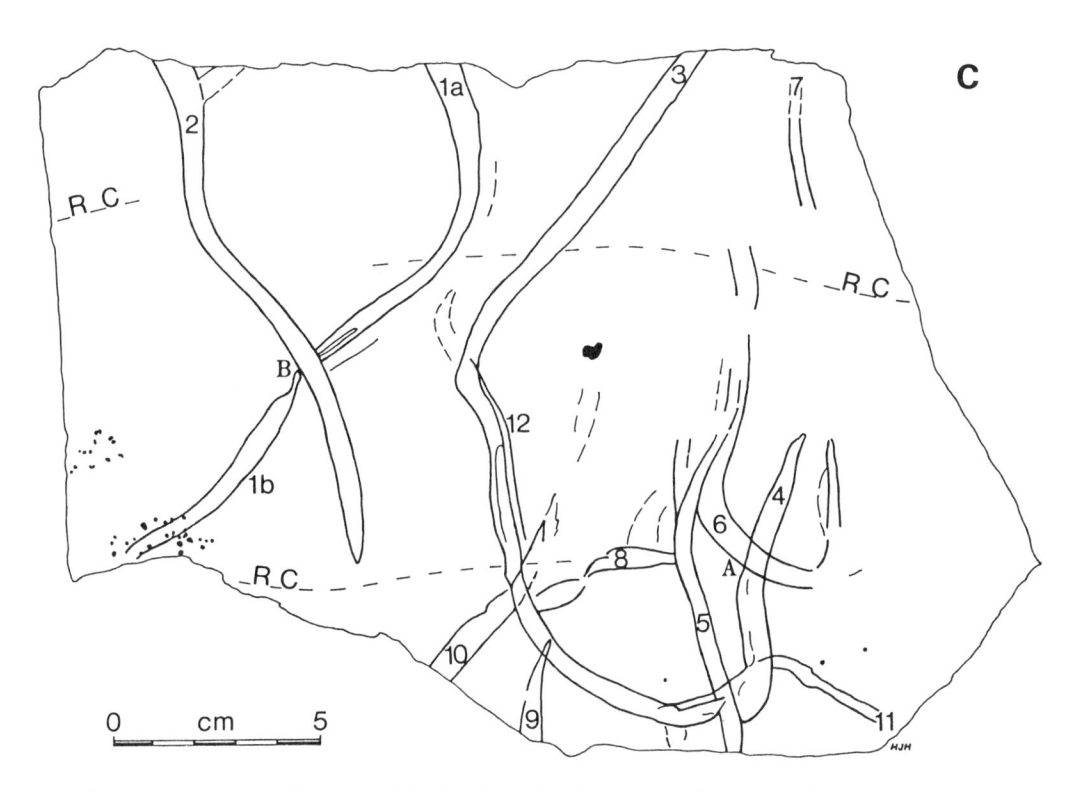

Figure 7.8.3. Early Proterozoic pseudofossils (A) and dubiofossils (B, C). (A) Upper bedding surface view of corrugated spindles ("*Rhysonetron*") in quartz arenite from the Huronian-age Bar River Formation, Flack Lake, Ontario, Canada; bar for scale represents 1 cm; GSC type no. 9876; GSC photograph 200090-C (Hofmann 1971b, Pl. 13, Fig. 1). (B) Spindle-shaped ropes of sand on lower surface of slab from the Ajibik Quartzite near Ishpeming, Michigan, USA; bar for scale represents 1 cm; Museum of Comparative Zoology, Harvard University (MCZ) no. 1017; GSC photograph 200329-E (Faul 1949, Fig. 1). (C) Analysis of vermiform markings (specimens labeled 1 through 12) shown in (B), above; "RC" denotes ripple crests; "A" and "B" indicate sites where two spindles cross; small black patches represent coarse quartz grains. The spindles are not confined to ripple troughs, as are normal sinuous mudcrack fillings associated with rippled sand. These structures are regarded here as dubiofossils.

Mudcrack-like dubiofossils are commonly reported in the Precambrian literature (again, only in passing), with little or no description or illustration, making such references of little value. An interesting exception are the "flexuous annelid burrows" reported from the Belt Supergroup of Montana (Fenton and Fenton 1937, p. 1951). These resemble the *Planolites* from the Late Proterozoic Elkera Formation in central Australia (Glaessner 1984, p. 70). The most recent interpretation, however, is that they are shrinkage crack-fillings (Cuffey 1988).

Another striking exception is the occurrence of sinuous, sand-filled ropes from the Early Proterozoic Ajibik Quartzite in Michigan (Fig. 7.8.3B, C; Faul 1949, 1950), for which a purely mechanical explanation presents certain problems. The structures, in plan-view, exhibit a pattern that seems difficult to reconcile with a simple mudcrack hypothesis: the long, curved ropes of sand are not confined to the ripple troughs; the ropes are continuous across one another and appear to be superposed, as shown by the continuity of the raised margins of both spindles at "A" in Figure 7.8.3C, and by the relationship at "B" in the same figure; moreover, they cross at angles that are generally not orthogonal and certainly not uniform. Given the orientation of stresses in a cracking mud layer, it is possible, but unlikely, that a crack could propagate in continuity across an already existing crack. If the mudcrack idea is to apply, it would require that the spindles represent fillings of at least two successive mud layers, the first having become cracked and filled before the second, and both subsequently compressed, so that two different sets of crack-fillings became superposed. The different amounts of relief of different structures may be related to variable thickness of sand accumulation in the ropes and spindles. Alternatively, the first crack may have been healed before a subsequent, discordant cracking episode. A burrow origin is untenable, because of the tapering form of the structures, and because burrowing activity would have removed fill from an earlier formed burrow at the crossing point (quite apart from any theoretical consideration of the presumed age of origin of the Metazoa). An explanation as sand casts of spindle- or rope-like objects may be appropriate. As discussed in Section 7.3, *Nostoc*-like colonies have been used to explain other Proterozoic remains, and one may speculate whether they could not also provide the explanation for the Ajibik structures. A lack of associated carbonaceous material would argue against this.

7.8.3 Vertical Cylindrical Structures

A second group of dubiofossils encompasses vertical or steeply inclined cylindrical structures resembling *Skolithos* or root casts. These also occur in Early Proterozoic rocks. The most interesting are probably those reported from the 2.4 to 2.0 Ga-old Medicine Peak Quartzite in Wyoming (Fig. 7.8.4A; Kauffman and Steidtmann 1981); nine different morphologies of tubes were recognized in sediments for which tidal flat, bar, and beach environments were inferred. They include simple to rarely branching sinuous, mostly subvertical, tubes with elliptical to ovate cross sections, 0.2 to 3 cm wide and up to 25 cm long; others are 2 to 10 cm wide and 1 to 5 cm deep pits. The authors of that paper considered several alternative explanations for the structures, ruling out mud-cracking and gas-escape mechanisms, but retaining water-escape as a possibility. Finding no close analogues among nonbiologic sedimentary structures, they concluded that, although no unequivocal evidence for biogenicity is present, the tubes most closely resemble Phanerozoic ichnogenera, particularly *Palaeophycus*, *Macanopsis*, *Bergaueria*, *Thalassinoides*, *Diplopichnus*, *Skolithos*, *Monocraterion*, and *Chondrites*. The range of morphology represented by these taxa is indicative of just how variable the Medicine Peak shapes are. To accept them as metazoan burrows would require a three-fold extension backward in time from the oldest generally accepted evidence for metazoans. Apart from a discussion by Cloud (1983) and a reply by Kauffman and Steidtmann (1983), no new conclusive information has come to light in the past few years to warrant reassignment of these structures to either fossils or pseudofossils (but see discussion in Section 7.6).

Other subvertical structures (Fig. 7.8.4C), cutting across primary lamination, are found in the Espanola Formation in Ontario, Canada (Hofmann et al. 1980, pp. 1352–1353), which is approximately coeval with the Medicine Peak Quartzite. These sand-filled tubes, 2 mm wide and up to 2.5 cm long, and originally ascribed to dewatering, might otherwise readily be referred to *Skolithos* were it not for their great age. Other *Skolithos*-like tubes from the Early Proterozoic Sosan Group of Great Slave Lake can be dismissed because sedimentary laminae are not truncated (Fig. 7.8.4B; see also Hofmann 1971, pp. 39–40). Yet others still (e.g., *Sabellarites* of Dawson 1890) remain dubiofossils because of the inadequate nature of the specimens.

Structures from the Middle Proterozoic Greyson Shale in Montana were used by Walcott (1899, p. 236–237, Pl. 24, Figs. 8–9) to erect two new species: *Planolites superbus* (Fig. 7.8.5A, B) and *P. corrugatus* (Fig. 7.8.5E). Some palichnologists have accepted them as trace fossils (e.g., Seilacher 1956, p. 165; Häntzschel 1965, p. 72), whereas Cloud (1968, p. 55) thought they may possibly be algal. A cross-section view of *P. superbus* shows the structure to be a deformed sandy lens impressed into the underlying mud (Fig. 7.8.5B). The "corrugations" on *P. corrugatus*, which appears to be branching, are secondary features due to irregular breakage of the argillite around the spindle-shaped structure, the internal composition of which has not yet been ascertained. Both ichnospecies were questionably put into synonymy with *P. beverleyensis* in a monographic review of the genus (Pemberton and Frey 1982, p. 867, 868). The present view is that they should be removed from this synonymy and attributed to the dubiofossils.

7.8.4 Discoidal Structures

A third group of dubiofossils includes various discoidal markings, sometimes with indistinct radial pattern, resembling bona fide fossils. An early occurring example is *Aspidella* from the Late Proterozoic of Newfoundland (Fig. 7.8.5C). It has been variously considered biologic (either a mollusc or a crustacean) or as inorganic (of concretionary origin, a crater associated with a gas escape vent, etc.). Although most authors have treated it as nonbiologic (for a tabular summary, see Hofmann 1971b, p. 16), the evidence now available remains inconclusive, partic-

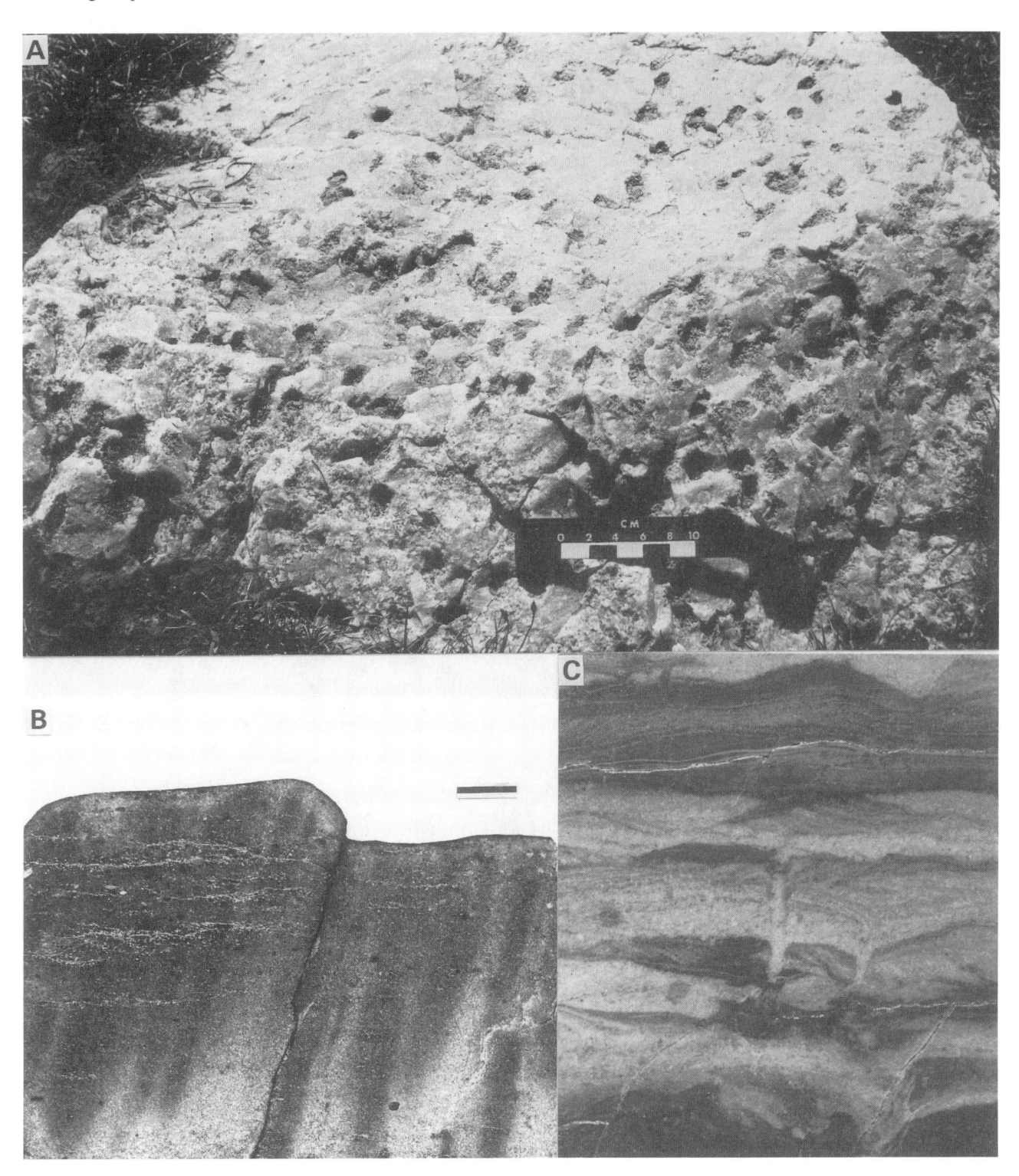

Figure 7.8.4. Early Proterozoic tubular pseudo- and dubiofossils. (A) Field view of block with short, inclined, cylindrical structures resembling *Skolithos* or root casts, from the Medicine Peak Quartzite, Medicine Bow Mountains, Wyoming, USA (Kauffman and Steidtmann 1981, text-Fig. 9; photograph courtesy of E. Kauffman). (B) Polished vertical section of glauconitic, carbonate-cemented, quartz arenite with inclined *Skolithos*-like structures from the Sosan Group, Great Slave Lake, northwestern Canada; the light colored glauconitic laminae pass across the cylinders without disruption, indicating that the structures are not burrows; bar for scale represents 1 cm; GSC type no. 24969; GSC photograph 201117 (Hofmann 1971, Pl. 25, Fig. 3). (C) Polished vertical section of argillaceous quartz arenite with *Skolithos*-like vertical structures from the Espanola Formation, Quirke Lake, Ontario, Canada; the vertical cylinders have been regarded previously as possible dewatering structures; magnification shown by bar for scale in (B).

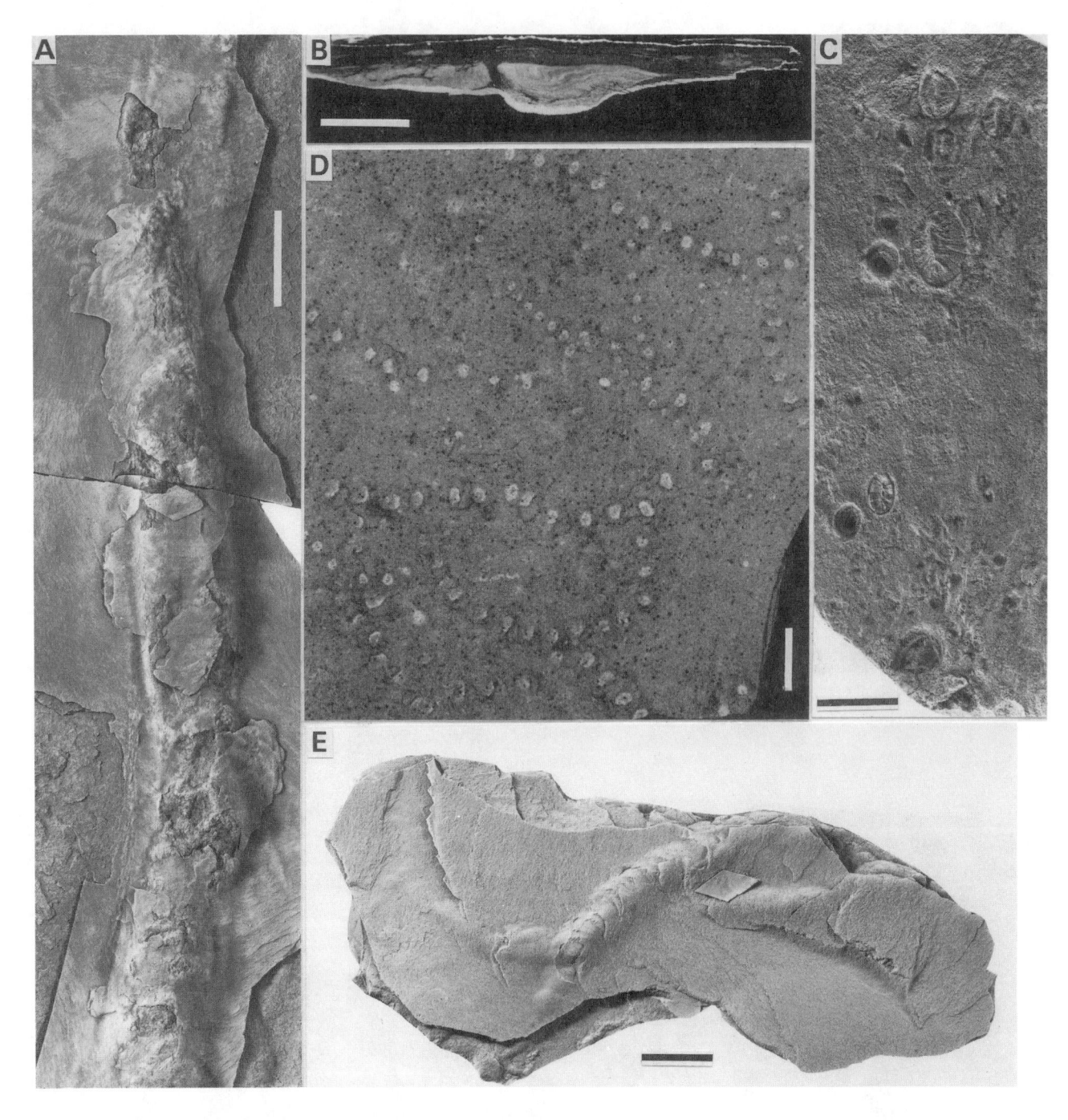

Figure 7.8.5. Middle and Late Proterozoic dubiofossils. Bars for scale represent 1 cm. (A, B) *Planolites superbus* Walcott (1899, Pl. 24, Fig. 9) from the Greyson Shale, Montana, USA; U.S. Geological Survey (USGS) holotype no. 33797; (A) view of lower surface (GSC photograph 200883); (B) cross-sectional view showing deformed lenticular sandy layer protruding into underlying shale. (C) Upper bedding surface with several specimens of *Aspidella terranovica*, showing preferred orientation, from the Late Proterozoic St. John's Group, St. John's, Newfoundland, Canada; GSC type no. 24371; GSC photograph 200329-K (Hofmann 1971b, Pl. 5, Fig. 5). (D) Bedding plane view of *Neonereites*-like problematica (cf. Fig. 7.6.4D, E) from the Middle Proterozoic Appekunny Argillite, Montana, USA; U.S. National Museum (USNM) no. 311313 (Horodyski 1982, Pl. 1, Fig. 3a; photograph courtesy of R. J. Horodyski). (E) *Planolites corrugatus* Walcott (1899, Pl. 24, Fig. 8) from the Greyson Shale, Neihart, Montana, USA; corrugations are deflected cleavage planes in shale; USNM holotype no. 33796; GSC photograph 200882.

ularly in light of discoveries of similar structures (*Irridinitus*) in the Vendian of the White Sea region (Fedonkin, 1981) and in the Miette Group of the Rocky Mountains (Hofmann et al. 1985). *Aspidella* needs restudy. Larger *Cyclomedusa*-like discs with concentric structure in China (but without radial patterns), originally thought to be cnidarians, have also been reinterpre-ted as fluid escape structures, based on the presence of a central conduits attached to the disc (Sun 1986, pp. 348–356).

7.8.5 Other Megascopic Dubiofossils

An additional controversial dubiofossil is the radial-lobate *Brooksella canyonensis* from the Nankoweap Formation of the Grand Canyon (Fig. 7.5.2A, B; Bassler 1941). Originally presumed to be a medusoid, it was subsequently regarded as inorganic (e.g., Cloud 1968), but later referred to the trace fossils (to *Asterosoma*?, by Glaessner 1969a; and to *Dactyloidites*, by Fürsich and Bromley 1985; see also Kauffman and Fürsich 1983). A cross-section is said to show spreiten (Kauffman and Fürsich 1983), although recent restudy suggests that the "Spreite-like laminae" may have been misidentified (Section 7.6.2). Interpretation of this dubiofossil is hampered by a lack of new material that could provide more information on morphologic variability, as well as allowing for serial sectioning to determine the internal structure more closely.

Structures interpreted as metazoan fecal pellets were described from the Upper Riphean Zilmerdak Group of the Urals (Sabrodin 1971; see also Glaessner 1984, Fig. 1.7 and p. 25). They are convoluted in section, but it is not clear from the two-dimensional view whether the structures are cylindrical or platy. One of the contorted bodies illustrated has uniform width over most of its length, an unlikely situation for a section of a contorted cylinder, but more probable for a folded plate. They may be deformed intraclasts.

Still other groups of dubiofossils include forms with more complex morphology. Particularly intriguing forms were reported by Horodyski (1982); these are strings of flattened, randomly oriented beads from the mid-Proterozoic Apekunny Formation in Montana (Figs. 7.6A-D, E, 7.8.5D). Similar structures are found in the coeval Manganese Group of Western Australia (Grey and Williams 1987, 1990). The latter are serially arranged subspherical or discoidal structures 1 to 3 mm in diameter, apparently linked by a fine thread; they are thought to have green or brown algal affinities. As both forms also resemble *Neonereites uniserialis* among the ichnofossils, and some inorganic structures (strings of "millet-seed" gypsum casts) in the Balbirini Dolomite of the McArthur Basin (Jackson et al. 1987, p. 139) of northern Australia, their affinities remain undetermined.

Finally, there are the peculiar camasid dubiofossils in mid-Proterozoic carbonate rocks in Montana and Siberia (*Camasia, Copperia, Newlandia, Saralinskia*, and a few others), reported by Walcott (1914), Vasiliev et al. (1968), Sosnovskaya (1980, 1981), and Sosnovskaya and Shipitzyn (1984a, b). Walcott attributed his taxa to microbial activity, but Fenton and Fenton (1936) and later authors have considered such structures to be diagenetic pseudofossils (see Häntzschel 1965, 1975). Different types of phenomena are represented, and in some specimens of *Newlandia* fine internal rhythmic lamination is preserved, suggesting the need for further study.

7.8.6 Conclusions

This brief sampling of what is an enormous pool of dubio- and pseudofossils, demonstrates the urgent need for much further work and more systematic reporting of these enigmatic remains. There is particular scope for the experimental approach to simulate and duplicate such structures in the laboratory or on the computer. On the one hand, the present dogma stipulates that there is no conclusive evidence for the existence of metazoan life before the Late Proterozoic, making it therefore difficult to explain biologically any structure that resembles Phanerozoic body or trace fossils; mechanical or chemical processes are thus the alternative explanations sought for. On the other hand, when conclusive evidence for such a non-biological process is not forthcoming, one is left with the dilemma of still having to explain the structures, and reconsidering possible biological alternatives. Removal of a structure from the dubiofossils will be facilitated when new specimens, new data, new concepts or ideas, and new techniques are brought to bear on the problem, and more emphasis is placed on the experimental approach to simulate or duplicate the remains. Certainly, one aspect will need improvement generally, and that is a much greater thoroughness in documenting a new occurrence on the part of the author(s) who first report such problematic structures. That means a minimum of information on size, shape, composition, *detailed* stratigraphic and geographic information (coordinates, locality map), photographs (of thin or polished sections also), and repositories of specimens in collections, to make a report verifiable. Such essential information is almost completely lacking in many reports of dubiofossils. Rules, codes, procedures, and recommendations exist for mineralogical, chemical, stratigraphical, zoological, and biological nomenclature, but not yet for sedimentary structures or dubiofossils. Why, in keeping with the aims of those codes, not report new finds in a more systematic way henceforth?

Acknowledgments

Financial support from the Natural Sciences and Engineering Research Council of Canada is gratefully acknowledged.

7.9

Unsolved Problems and Future Research Directions

KENNETH M. TOWE

The record of megascopic fossil remains extends back into the Early Proterozoic, and although it is extensive and varied (Chapter 23), the distribution in time is markedly skewed toward the more recent end of the Proterozoic (Chapter 11). This is expecially true for the record of (bio)mineralized megascopic organisms. Thus, in comparison with the total fossil record of megascopic remains, that of skeletal metazoans is even more compressed into the latter portion of the Proterozoic, so much so that discussion of mineralized remains inevitably invites, if not requires, comparison with the earliest Cambrian forms in order to be meaningful (e.g., Sections 7.4, 7.7).

However, the skeletal fossil record, while not without its taxonomic and functional morphologic enigmas, is significantly less subject to doubt than is the record of early trace fossils and older carbonaceous remains where many more problems invite inquiry. The search for the "world's oldest..." has created a tortuous path leading toward the goal of deciphering the early history of the Metazoa. This path is strewn with argument, controversy, and pseudofossils (Section 7.6). As the age of the enclosing sediments increases, so does the likelihood that an object will be classified as a dubiofossil, questioned as to its true age or its biological affinity, or placed into the limbo of *incertae sedis*. Acceptance of fossils purported to represent evidence of the earliest metazoans (like the controversy surrounding the early eukaryotes) has been fraught with the difficulties that model dependence may engender, and honest differences of opinion have kept the subject from consolidating toward consensus (see, e.g., Cloud 1986; Sun 1986).

What are the true affinities of the Grypanids?, the Longfengshanids?, the Sinosabelliditids? (Section 7.3). The megascopic fossil record may be the "court of last resort" on questions regarding the history of metazoan life, but this record is of much less value where the "jury," failing to reach a verdict, remains hung. For the future, therefore, the continuation of interdisciplinary biochemical approaches to estimating the time(s) of appearance of different taxa (Chapter 9) looms as a major step toward resolution of these questions. It is at least as important as further, more critical and sophisticated field searches for megafossils in appropriately dated sediments around the world.

7.10

Summary and Conclusions

KENNETH M. TOWE

(1) The Proterozoic history of life as recorded in megascopic remains is significant and varied. The known fossil record can be divided into several categories which include: carbonaceous films, remains of calcareous algae, body fossils and trace fossils of metazoans, and small shelly fossils.

(2) Because of the poor preservation and simplicity of form of many Proterozoic megascopic remains, the category of *dubiofossil* is appropriately applied to those for which biogenicity (or syngenicity or Precambrian age of the enclosing sediments) is open to controversy. With a history of publications dating back more than 100 years, close to 300 megascopic "Proterozoic organisms" have been restudied and relegated to this category pending further research and additional evidence.

(3) *Carbonaceous films* of megascopic dimensions occupy an important place in the fossil record, extending back into the Early Proterozoic. Published reports date from the mid-18th century. Although a number of formal suprageneric classifications have been proposed for these remains, their obscure biological affinities are better served by a more informal, tentative, working classification. These enigmatic films have therefore been treated here in the following eleven informal categories: chuarids, tawuids, ellipsophysids, grypanids, longfengshanids, moranids, beltinids, vendotaenids, eoholynids, sinosabellidilitids, and sabelliditids. Among these, chuarids and vendotaenids are the more common and widespread in Proterozoic strata. Although both tawuids and longfengshanids are invariably found associated with chuarids, suggesting links between them, the reverse is not true. The distinctive, and intriguing, ribbon-like grypanids, commonly spiralled, have now been reported from Middle Proterozoic rocks of widely differing litholgies in Montana (USA), India, China, and the USSR.

(4) The approximately 400 species of *calcareous algae* described from the Early Cambrian (dominated, notably, by but three genera: *Girvanella*, *Renalcis*, and *Epiphyton*) contrast strikingly with the limited, and commonly questionable, fossil record of calcareous algal materials in rocks of Vendian or older age.

(5) The now well-known, if enigmatic, Ediacara fauna is the keystone of the Proterozoic metazoan *body fossil* record. Occurrences of these soft-bodied impressions have been reported from numerous localities around the world in rocks of Vendian age. The fossils provide clear evidence of multicellular diversification at the tissue-level of organization, but the basic body plans of these organisms were mostly replaced in the Cambrian by those developed and evolved from unknown, as yet "invisible," unmineralized Proterozoic animals.

(6) The metazoan ichnofossil (*trace fossil*) record decreases markedly in its diversity and abundance from the Early Cambrian back into the Late Proterozoic. It extends with certainty only into the early Vendian. Pre-Vendian trace fossils reported in the literature have proven to be either pseudofossils or, in a few cases, dubiofossils. The preservation of horizontal burrows in units stratigraphically equivalent to those containing the Ediacara assemblage provides clear evidence that the fauna of this age included bilateral, mobile animals, capable of peristaltic motion.

(7) *Small shelly fossils* from latest Proterozoic and earliest Cambrian sediments, including calcareous, phosphatic, and siliceous skeletal remains, have been reported from all continents, but are particularly well known from the Siberian Platform, China, and south Australia. Their distribution in time indicates an essentially penecontemporaneous evolution of biomineralization across several mineralogical boundaries. These early skeletal biotas include an astonishing variety of spicules, internal and external skeletal reinforcements, shells, sclerites, and teeth- and spine-like processes, together with an assortment of cone-, cap-, and plate-shaped structures of more uncertain function. The evidently rather "abrupt" appearance in numerous clades of these diverse types of mineralized structures, and their apparent absence from members of the older, Ediacara fauna, are obviously striking, and require explanation.

(8) The Proterozoic Eon was a time of lengthy development and testing of various megascopic forms of organization, ranging from prokaryotic aggregates to metaphytic and metazoan tissues. Despite its obvious deficiencies, the Proterozoic fossil record is now much better known than ever before. Critical attention has been paid to separating the megascopic pseudofossil "chaff" from the bona fide fossil "wheat."

Nevertheless, considerable caution must be exercised in interpreting these important remains, in terms both of over-interpreting (this *might* be the "earliest," let's report it!) and under-interpreting (this doesn't fit; forget it!). The tasks of establishing the correct biological affinities of a great many of these fossils, of understanding their significance in the long sweep of evolutionary history and, above all, of determining their impact on the development of the earth's early ecosystem, will occupy interested paleobiologiss for years to come.

8

The Proterozoic-Early Cambrian Evolution of Metaphytes and Metazoans

STEFAN BENGTSON   JACK D. FARMER   MIKHAIL A. FEDONKIN
JERE H. LIPPS   BRUCE N. RUNNEGAR

8.1

Introduction

STEFAN BENGTSON   JERE H. LIPPS

The fossil record of the later Proterozoic through the Early Cambrian is marked by extraordinary change. This change indicates a fundamental reorganization of the biosphere from the exclusively single-celled prokaryotic and protistan ecosystems that prevailed during much of the Proterozoic, to ecosystems characterized by complex multicellular plants and animals of the latest Proterozoic and Early Cambrian. The first recorded events in this transition took place about 900 Ma and the last about 550 Ma, a period of time exceeding that since the end of the Paleozoic. But the final and most dramatic phase, the "Cambrian Explosion," occurred over a few tens of Ma at the onset of the Cambrian.

The glaring contrast between the Precambrian and the Phanerozoic has long been recognized as a major problem in the history of life. Darwin (1859) attempted to explain the sudden appearance of the Cambrian fauna by inadequacies of the rock record, and Walcott (1910) used a similar concept in his "Lipalian Interval" at the base of the Cambrian. Certainly the abrupt appearance in some local areas (for example, in the contact between Yudomian dolomites and Tommotian limestones in the Aldan-Lena region of Yakutia; Rozanov et al. 1969; Khomentovskij and Karlova 1986) may still be explained by incompleteness of the record. Yet the current knowledge of the fossil occurrences of multicellular and other organisms from the later Proterozoic to the Early Cambrian (Chapter 7) fully supports the view pioneered by Cloud (1948, 1968) and Schindewolf (1955) that the biotic changes are indeed real, not due to artifacts of preservation.

As the fossil record near the Precambrian-Cambrian boundary became better known in more recent years, various hypotheses were proposed to account for the observations, chiefly focusing on aspects of the origin, radiation, or skeletonization of metazoans. At least 25 hypotheses have been proposed, but none has received widespread acceptance.

The hypotheses can be separated into those based either on biotic or abiotic factors. Biotic factors are those that are related to the biology of the organisms involved; abiotic ones are those that are related to environmental factors beyond the organisms themselves. Biotic factors drive evolution from within the system while abiotic factors drive evolution from outside the system, and this difference is important in evaluating the fossil evidence. Table 8.1.1 lists the major contrasting families of hypotheses, grouped according to the dominant factor invoked. The list does not include less tenable ideas or slight variations of those listed.

Some of these hypotheses attempt only to explain the record of metazoans or even only that of their skeletons. They commonly omit reference to metaphytes, protists, or prokaryotes. Some hypotheses implicitly or explicitly include the origination of metazoans within their framework, while others are concerned only with the radiation or skeletonization of Precambrian-Cambrian metazoans.

Many of the mechanisms invoked in the various hypotheses summarized in Table 8.1.1 are in themselves plausible, or even necessary, components of any evolutionary scenario. But a satisfactory explanation of the unique course of evolution deduced from the empirical data must specifically identify those factors that had a decisive influence on the pattern and timing. This can also be expressed as the problem of finding the limiting factors during various stages of evolution. The complexity of the diversification, including biotic interactions, indicates that multiple, interacting causes are responsible for the Proterozoic-Phanerozoic radiation, and that the answers to the problem are themselves complex. An acceptable hypothesis must take into account the history of groups other than metazoans, and evolving ecological interactions is more than likely to be a major component of any answer. Stanley's "cropping hypothesis" (1973a), for example, shows that it is possible to explain major features of the late Precambrian to Early Cambrian evolution simply with the introduction of new trophic levels in the ecosystem.

We find it useful to separate the complex of questions into six topics: (*i*) Why and when did multicellular individuality evolve? (*ii*) What were the evolutionary relationships between the major biotas of the Late Proterozoic to Early Cambrian? (*iii*) What ecological interactions were involved in the radiations of protists, metazoans, and metaphytes? (*iv*) What was the timing and pattern of the radiation of protist and multicellular phyla? (*v*) What role did biomineralization play in the "Cambrian Explosion"? (*vi*) What environmental "triggers" (glaciations,

Table 8.1.1. *Major hypotheses proposed to account for the observed pattern of appearance of metazoans in the fossil record*

BIOTIC FACTORS:	
Size	*Precambrian metazoans did not leave a fossil record because they were small.* They lived in the plankton (Brooks 1894; Raymond 1939), interstitially (Boaden 1975), or on the sediment surface (Clarke 1964). Larger organisms evolved with the acquisition of sedentary (Nicol 1966) or burrowing (Clarke 1964) habit.
Skeletons	*Precambrian metazoans did not leave fossil record because they were soft* (Rutten 1971; Glaessner 1972, 1984; Durham 1971, 1978b). Skeletonization evolved secondarily (after the evolution of calcium regulation; Lowenstam and Margulis 1980a, b; Kaz'mierczak et al. 1985; Degens et al. 1985) as defense against predators (Evans 1912; Sollas 1912), for biomechanical reasons (Raymond 1939; Nicol 1966), or primarily as excretion of excess calcium (Kaz'mierczak et al. 1985; Degens et al. 1985) or phosphorus (Rhodes and Bloxam 1971).
Migration	*Metazoans invaded marine shallow waters from habitats lacking a recoverable fossil record*, for example the shoreline (Axelrod 1958) or deeper waters (Snyder 1947).
Ecology	*Advent of predators and/or croppers provided selection mechanisms for diversification* (Stanley 1973a). Extinction of Ediacara biota provided the ecological space for the "Cambrian explosion" (Hsü et al. 1985; McMenamin and Schulte McMenamin 1989).
Genetics	*Evolution of regulatory genes made rapid appearances of new body plans possible* (Valentine and Campbell 1975). Evolution of sexuality promoted diversification (Schopf et al. 1973b; Stanley 1975a).
ABIOTIC FACTORS:	
Oxygen	*Accumulation of sufficient oxygen in the atmosphere and oceans allowed the origination and diversification of larger organisms* with complex musculature and metabolism (Berkner and Marshall 1964a, b, 1965a, 1967; Cloud 1968, 1976a; Nursall 1959; Runnegar 1982a; Berry and Wilde 1987; Wilde 1987). Higher oxygen levels permitted the synthesis of structural proteins, particularly collagen (Towe 1970, 1981).
Calcium	*Increased calcium levels in the oceans permitted the development of calcareous skeletons* (Daly 1907; Kaz'mierczak et al. 1985).
Phosphorus	*Increased availability of phosphorus led to an increase in biomass and diversity* (Cook and Shergold 1984, 1986). Abundance of phosphorus accounts for early predominance of phosphatic skeletons (Rhodes and Bloxam 1971; Lowenstam and Margulis 1980a, b; Cook and Shergold 1984, 1986; Brasier 1986).
Glaciation	*Late Proterozoic glaciation of the earth changed paleoceanography and trophic resource supply* so that all groups could radiate in concert (Harland 1964; Rudwick 1964).
Land/sea patterns	*Continental reconfiguration and sea-level changes increased the extension and diversity of habitable shallow marine environments*, causing a general biotic diversification (Valentine and Moores 1972; Brasier 1982, 1985, in press).
Cosmic causes	*A bolide impact* (Hsü et al. 1985) *or an increase in cosmic radiation* (Schindewolf 1954, 1955, 1958) *triggered the restructuring of the biosphere.*

oxygen levels, breakup of a supercontinent, etc.) were significant for the timing of the radiations? There is no final and satisfactory answer to any of these questions today, but we can look at each one in light of available evidence from the geologic record and from the biology of living organisms to see how far the present evidence can be taken. Sections 8.2–8.6 deal with each one of the topics in turn, and in Section 8.7 we discuss the possible components of an integrated model for the biological revolution marking the end of the Proterozoic.

8.2

Origins of Multicellular Individuality

JACK D. FARMER

8.2.1 Introduction

The development of multicellular individuality laid the foundation for the extraordinary evolutionary advances that restructured the biosphere during the Late Proterozoic. With the evolution of complex morphogenesis and of more inclusive levels of functional integration, a new level of individuality emerged. Multicellularity opened the door to new ways of occupying the environment, and for acquiring and retaining resources. Ecosystems underwent rapid change under the new regime, and with longer food chains, and increasingly complex trophic interactions, evolution was cannalized in new directions marked by the emergence of a new global ecology.

Direct fossil evidence of the steps that led to multicellular organisms does not exist; those steps must therefore be inferred from biological theory and observations of currently living systems. The objective of this Section is to characterize the features of multicellular individuality, and to identify the important factors that may have contributed to its evolution.

8.2.2 The Concept of Individuality

The definition of individuality presents a number of conceptual difficulties (Buss 1987; Rosen 1979; Beklemishev 1969; Mackie 1963) that are rooted in subjective, anthropocentric arguments that tend to pervade such discussions. Behavioral definitions are particularly prone to such problems, because individual behavior is often difficult to specify outside of human experience.

An alternative to a behavioral definition of individuality identifies the individual as a unit possessing evolutionary significance. The essential feature of individuality is autonomy (Rosen 1979, p. xix). However, individuality and coloniality represent endpoints of a spectrum of possible types and degrees of integration, and the individual as an autonomous evolutionary unit has meaning only within the context of a particular organizational level (Rosen 1979). This problem is exemplified by highly integrated colonial organisms that consist of organically linked modules. In such cases, it is difficult to specify units that were once capable of independent existence, but which now function within a broader corporate individuality (see discussion by Rosen 1979).

At higher levels of morphological and behavioral integration, evolutionary criteria for individuality are difficult to apply and it becomes easier to identify different levels of functional integration than to specify units that possess evolutionary significance (Rosen 1979). A definition of individuality that is based on an objective evaluation of the function and biological significance of organizational units has advantages over the use of strictly behavioral, morphological, or evolutionary criteria. This approach is especially relevant when considering the origin of multicellular individuality, which is generally thought to have evolved from highly integrated aggregations of unicellular organisms (Buss 1987).

For purposes of discussion, multicellular individuality is herein characterized as an obligate association of cells that exhibits functional specialization and differentiation of cell types to form tissues, coordination of metabolic activity by systems of intercellular communication, preservation of individual integrity by self/non-self recognition, and patterned morphogenesis with temporal and spatial control of cellular division, differentiation, and movement. Interestingly, most of these characteristics are also exhibited to some degree by colonial unicellular forms. The evolution of multicellular individuality was foreshadowed by the development of coloniality in unicellular organisms. The following discussion will explore some of the important biological features of multicellular organisms, and the development of these characteristics among unicellular colonial organisms. This approach is intended to provide an actualistic conceptual framework for evaluating hypotheses about the origin of multicellular individuality.

8.2.3 The Nature of Multicellularity

8.2.3A Large Size

A general advantage of multicellularity is the adaptive value of increased size (Bonner 1965; 1974). The ecological advantages of large size are numerous, and have in many cases provided the basis for exploiting new adaptive zones. In higher organisms, large size has been shown to have advantages in predator-prey interactions, and is also correlated with enhanced reproductive success (Stanley 1973b).

Among unicellular colonial forms, large size has been shown to share some of these same advantages. For example, the bacterium *Myxococcus xanthus* forms spherical colonies that enhance success in predation by providing pockets which entrap prey, while maintaining concentrations of digestive enzymes and released cellular contents (Shapiro 1988, pp. 84–85). Borass (1984) showed that aggregation may also be important in the avoidance of predation in species of *Chlorella*.

8.2.3B Cellular Differentiation

As noted above, it is easier to identify different levels of functional integration than to specify levels of "autonomously defined individuality" (Rosen 1979, pp. xx–xix). In the transition from coloniality to multicellularity, evolution apparently attained more inclusive levels of organization in stepwise fashion by the morphological differentiation of cells, followed by increasing degrees of functional integration (Buss 1987, pp. 171–172). Nevertheless, cellular differentiation within what can be regarded as autonomous individuals is, with few exceptions, restricted to three groups: plants, fungi, and animals.

Buss (1987, p. 103) identified cellular differentiation as an important evolutionary "trigger" that led to the emergence of multicellular individuality. Within the somatic environment of a multicellular organism, variant cells that replicate faster are more likely to pass on heritable traits, particularly if they arise in cell lineages that lead to the germ line, or that undergo somatic transformation to form germinative cells. With the differentiation of reproductive cells, and the sequestering of a cellular germ line from somatic functions, a synergism developed between selection at the cellular level for optimal growth and replication among the competing cell lineages during ontogeny, and morphogenetic patterns that were favorable to the whole organism (Buss 1987). This interaction between selection at different levels of organization marked a transition to a new mode of selection, and probably accounts for the high levels of functional integration and complex morphogenesis exhibited by living metaphytes and metazoans.

Multicellularity, accompanied by differentiation of reproductive cells, is well known in species of myxobacteria (Reichenbach 1984). These "social prokaryotes" produce fruiting bodies and desiccation-resistant myxospores. New colonies are generated from a dispersive spore that consists of a resting body of cells. Simple systems of cellular differentiation are also present in species of acrasid and dictyostelid slime molds and in the colonial phytoflagellate protozoan *Volvox* (see Buss 1987, pp. 70–75). In *Volvox*, new colonies arise from germ cells within the host colony that are liberated upon its death. In the sessile colonial ciliate *Zoothamnium geniculatum*, both somatic and reproductive cells are present. Colonies apparently have a restricted lifespan, and new colonies are established from a cluster of dispersive cells, or "telotrochs" (Fenchel 1987, p. 6). *Pseudomonas* colonies also exhibit cellular differentiation and a common extracellular "skin" (Shapiro 1988). Extracellular secretions appear to enhance the resistance of cell aggregates to antibiotics, and may also be important in the maintenance of microenvironments.

8.2.3C Intercellular Coordination

Cells or groups or groups of cells communicate by two basic types of chemical messengers: hormones and neurotransmitters (see review by Snyder 1985). Dobzhansky et al. (1977) suggest that intercellular inductors, such as steroids and peptides, originally functioned in the *intra*cellular regulation of cell growth, shape, and size. Microenvironments created within aggregations of cells may have favored the adaptation of such systems to fulfill *inter*cellular functions. Many new biosynthetic pathways appear to have originated by the addition of steps to established pathways, rather than by the "invention" of entirely new pathways (Schopf 1978). This probably happened frequently during the evolution of multicellularity, a view which finds support in studies of the beta-ketoadipate pathway in bacteria; this work suggests that even complex and specialized pathways that seem to be homologous may have had polyphyletic origins, their development having been narrowly constrained by physicochemical laws (Canovas et al. 1967).

Simple systems of *inter*cellular communication are also present in some colonial unicellular organisms. Cellular "flares" in *Myxococcus xanthus* show purposeful movement toward prospective prey, suggesting that cells in aggregation communicate by an as yet unspecified biochemical system (Reichenbach 1984). Similarly, "swarm" behavior in aggregates of *Proteus mirabilis* is characterized by synchronous flagellar activity which seems to be regulated by biochemical, possibly hormonal, gradients. Growth in colonies of *Pseudomonas* is a highly regulated process involving both the temporal control of cellular growth and of cellular differentiation to produce unique and heritable surface textures and pigmentation patterns (Shapiro 1988).

8.2.3D Maintenance of Microenvironments

Another advantage of multicellularity is enhanced efficiency in enzyme production and greater control of concentration gradients and nutrient losses by diffusion (Reichenbach 1984, pp. 4–5). In unicellular organisms, aggregation confers similar ecological advantages by allowing for the creation and maintenance of favorable microenvironments. For example, Richardson et al. (1988) demonstrated that naturally occurring aggregates of *Microcystis* maintain internal pH and oxygen levels significantly higher than the surrounding external environment. These conditions are favorable for the oxidation of manganese, a required nutrient, and Mn oxides were observed to precipitate both within cellular aggregates and near their surfaces. Similarly, maintenance of microenvironments has been shown to be important in nitrogen fixation by nonheterocystous cyanobacteria such as *Trichodesmium* (Paerl and Bebout 1988).

By promoting selection for a division of labor within groups of cells, cellular aggregation has been suggested to be a key factor in the evolution of multicellularity, a process which may have led to cellular differentiation (Buss 1987; Nursall 1962). It is apparent from the examples discussed above that cellular aggregation confers important ecological advantages to unicellular organisms, even to those which do not exhibit cellular differentiation. It is reasonable to assume that during the evolutionary development of multicellularity, cellular differentiation may have been importantly influenced by microenvironmental factors that were actively regulated by aggregations of cells. In this scenario, the proximal cause or "trigger" in the development of multicellularity is the improved regulation of important microenvironmental factors made possible by cellular aggregation. This may have played a fundamental role in the development of coloniality in many groups of unicellular organisms, and could have set the stage for subsequent developments (cellular differentiation, intercellular communication, and control of morphogensis) which led up to multicellular individuality.

8.2.3E Morphogenesis

Cellular differentiation is governed by regulatory genes that control the timing of expression of other genes. Differentiation of cells follows the activation and selective transcription of large numbers of genes that encode structural proteins. In contrast, morphogenesis is governed by regulatory genes that control the frequency and direction of cellular division, and the movement of cells to specified locations in the developing embryo, prior to cellular proliferation.

Particularly exciting discoveries in the field of regulatory genetics have come from the study of homeotic genes known to regulate cellular differentiation in the nematode *Caenorhabditis elegans* (see Wood et al. 1988), and spatial organization in the fruitfly *Drosophila melanogaster* (see Gehring 1985). Mutations at the regulatory sites are known to have pleiotropic effects on the integration and expression of numerous structural genes. Such changes can lead to radical alterations in patterns of development (Wilkins 1986, and references therein).

Simple regulatory genes which control cellular differentiation and morphogenesis appear to also be present in some unicellular colonial forms. Recent studies of the "homeobox" (a short region of DNA encoding a polypeptide domain 60 amino acids in length) support this idea (Wilkins 1986, p. 465). Interestingly, the homeobox is very similar to certain DNA-binding regions in prokaryotic regulatory genes (Laughn and Scott 1984). The homeobox is present in several genes believed to regulate segmentation patterns in *Drosophila* (McGinnis et al. 1984a, b) and is homologous with gene sequences in other insects and in annelids, vertebrates, and even yeast. Although the homeobox has been suggested to have a general function in regulating morphogenesis in these groups (McGinnis et al. 1984b), Wilkins (1986, p. 466) cautions that it is premature to ascribe a general developmental function to genes with homeoboxes: they appear to be absent in echinoderms and nematodes.

8.2.4 The Evolution of Multicellular Individuality

Comparative studies of cellular ultrastructure, biochemistry, and genetics indicate that multicellularity evolved from unicellular precursors many times during the history of life (Dobzhansky et al. 1977). Cellular aggregation is a common feature of unicellular prokaryotes and eukaryotes (Shapiro 1988). Of 23 monophyletic protist groups, at least 17 have colonial representatives (Buss 1987, p. 70). The Protozoa include colonial species that originate by incomplete cell division, as well as multinucleate forms that originate by endomitosis (larger ciliates, amoebas, or foraminiferans) or by fusion to form a syncytium (e.g., acellular slime molds) (Fenchel 1987, p. 5). These observations suggest several possible alternative evolutionary pathways leading to multicellular individuality, some of which have been outlined in the previous Section.

In the absence of direct fossil evidence, the events that led to the emergence of multicellular higher organisms must be inferred from living systems. Many unicellular colonial organisms exhibit simple systems of cellular differentiation, intercellular communication, and regulation of morphogenesis, which are analogous to those found in extant multicellular forms. In addition, cellular aggregates also benefit from the advantages of larger size in predator-prey interactions, and in greater control over the extracellular microenvironment. It is highly unlikely that any of these simpler extant systems are the direct precursors of those found in complex multicellular organisms. However, they serve to suggest the possible adaptive pathways that may have been followed in the development of multicellularity.

The features that most clearly define multicellular individuality among living metaphytes and metazoans are complex morphogenetic development and highly integrated systems for intercellular communication. Buss (1987, p. 103) suggests that cellular differentiation resulting in the isolation of a cellular germ line from somatic functions was an important evolutionary "trigger" which permitted the development of high levels of integration and complex morphogenesis in multicellular organisms. However, it is likely that this step was preceeded by the development of coloniality among unicellular precursors, based on the advantages of cellular aggregation noted above. The evolution of the first simple systems of cellular differentiation may have been directed in large part by microenvironmental factors within cellular aggregates that were under direct biological control. This may have been the most fundamental and proximal cause in the developments which led to multicellularity.

The intrinsic biological factors that have been suggested to account for the apparently rapid diversification of multicellular life at the end of the Proterozoic include specialization for alternating haploid and diploid generations resulting from eukaryotic sexuality (Schopf et al. 1973b), the expansion of heterotrophy (Stanley 1973a), and advances in the regulatory genome (Valentine and Campbell 1975). Extrinsic physical factors which may account for this rapid diversification include the buildup of oxygen in the atmosphere to a critical threshold required for the synthesis of collagen (Towe 1970), and climatic change regulated by plate tectonics (Valentine and Moores 1971). The extent to which these or other factors are responsible for the rapid diversification of multicellular life at the end of the Proterozoic cannot be decided on the basis of present evidence. However, it is likely that all of the basic precursor systems needed for the transition to multicellular individuality had evolved much earlier, among unicellular colonial organisms.

8.3

The Major Biotas of Proterozoic to Early Cambrian Multicellular Organisms

STEFAN BENGTSON   MIKHAIL A. FEDONKIN   JERE H. LIPPS

8.3.1 Introduction

On a broad scale, three successive assemblages of multicellular organisms can be identified during the Late Proterozoic-Early Cambrian:

(1) the pre-Ediacaran assemblage represented by carbonaceous films;
(2) the Ediacaran assemblage of large, soft-bodied multicellular organisms and small, simple trace fossils; and
(3) the diverse skeletal fossils and large, complex trace fossils of the Early Cambrian radiation.

Each of these assemblages represents a biota that also contained prokaryotes and protists. The biotas are accounted for in detail in Chapters 7 and 23, and only a very brief summary follows here.

The pre-Ediacaran macroscopic fossils occur as compressed and uncompressed carbonaceous films, e.g., *Grypania* (a helically or spirally coiled fossil), *Tawuia* (an elongate sausage-like fossil), *Chuaria* (a spheroidal fossil), and *Longfengshania* (an oval to oblong structure with a single stalk or appendage). *Grypania*, *Tawuia*, and *Longfengshania* may represent multicellular organisms, whereas *Chuaria* is regarded as an acritarch (Section 5.5). The Anhui biota of sinosabelliditids is probably also pre-Ediacaran and possibly represents early metazoans.

The Ediacaran biota contains simple trace fossils and body fossils of medusoid, frond-shaped, and other organisms (Chapter 7). The small trace fossils are preserved on the surface of horizontal bedding planes. Strings of fecal pellets indicate the presence of metazoans with a one-way gut. The highest diversity of these trace fossils occurs in the middle Vendian of the Russian Platform (Fedonkin 1976, 1981, 1987); at higher stratigraphical levels in the Vendian, the number of taxa is generally lower. At the same time that these animals existed, acritarchs were reasonably diverse, stromatolites abundant, and metaphytes were present (Zhang 1989).

The earliest Cambrian assemblages include all major groups of organisms. Acritarchs diversify after a decline in the latest Proterozoic. Protozoans are represented by foraminiferans and radiolarians. Metazoans include practically all phyla with a fossil record as well as numerous forms not attributable to known phyla. Calcareous algae diversify. Trace fossils appear in higher diversity and larger size and complexity. The rapid appearance of this multitude of fossils is called the "Cambrian Explosion".

Although the rise of metazoans and metaphytes is commonly said to have been rather sudden, the total time involved in the whole transition was enormous. At least 450 Ma passed between the appearance of the pre-Ediacaran *Tawuia* biota and the "Cambrian explosion". Thus, the biotas involved in the transition from ecosystems dominated by prokaryotes and protists to those dominated by metazoans and metaphytes endure for a period longer than from the end of the Paleozoic to the Recent. The transition is *not* sudden, although the final phase, involving the establishment of complex modern-type ecosystems, i.e., the "Cambrian explosion," was so.

8.3.2 Evolutionary Relationships of the Three Biotas.

The mutual relationship of the three biotas in the Late Proterozoic to Early Cambrian is of critical concern. They may be seen as essentially distinct, the multicellular members of one biota not giving rise to those of succeeding biotas; or they may be seen to demonstrate evolutionary sequences of multicellular organisms. These alternative possibilities are to a certain extent possible to test in the fossil record.

The hypothesis that the lineages in each biota are separate from those of the next can be disproven by demonstrating lineages that cross from one biota to another. Reconstructing lineages in early fossil multicellular organisms is difficult because the number of reliable characters is small and the biology of the organisms is unknown. A number of alternative hypothetical phylogenies can be constructed that are possible or even plausible, but are not testable with the known fossils. The only reliable information for the identification of phylogenetic lineages is the distribution of precisely defined characters; only such characters are accepted here.

Of the sparse pre-Ediacara macroscopic fossils, only *Grypania*, *Tawuia*, *Longfengshania*, *Sinosabelliditides*, *Pararenicola*, and *Protoarenicola* are sufficiently characteristic to be considered possible early metazoans or metaphytes. The spirally or helically coiled *Grypania* may be the most significant of these, both because of its age (~1.4 Ga) and its regular morphology. Although a eukaryote interpretation of *Grypania* seems most likely, a prokaryotic origin cannot be excluded (Section 7.3); possibly it represents large colonies of filamentous cyanobacteria. A similar but considerably smaller form, the cyanobacterium *Obruchevella*, which forms helically coiled filaments (see Section 7.4; Song 1984), and some closely related Cambrian fossils like *Spirellus* (cf. Peel 1988; Qian and Bengtson 1989), represent a lineage that can be traced across the Vendian-Cambrian boundary.

The worm-like fossils (*Sinosabellidites*, *Pararenicola*, and *Protoarenicola*) described from deposits below rocks interpreted as Vendian tillites in Anhui, China, may represent compressed tubes, although they have been considered to be cuticles of worm-like animals (Sun et al. 1986; see Section 7.5). If they are tubes, they may be compared with the late Vendian-Early Cambrian sabelliditids. Vidal and Moczydłowska (1987) suggested that the Chinese fossils are indeed sabelliditids and that the deposits may have been misdated (see also Cloud 1986). Their stratigraphical position appears to support an older age, however (Sun 1987a).

In the Ediacaran biota, the most suggestive similarities with later occurring phyla exist for the cnidarian-like fossils. As discussed in Section 7.5, hydrozoans (chondrophores) and anthozoans (octocorals and actinians) may be present. These cnidarian groups are also known from the Cambrian (Stasinska, 1960; Stanley 1986; Crimes 1987; Bengtson et al. 1990). The presence of other extant metazoan phyla in the Ediacaran biota has not been sufficiently substantiated to warrant acceptance, but a case can be made for animals of arthropod-grade organization (*Spriggina*, cf. Section 7.5).

Among extinct forms, the triradially symmetrical trilobozoan lineage apparently survived from the Ediacaran into the Cambrian. The trilobozoans are a distinct component of the Ediacaran biota (see Section 7.5) and the first Cambrian assemblages contain the triradially symmetrical, calcareous tubes of the anabaritids (see Table 23.5). The living chamber of the anabaritids was likewise triradially symmetrical, indicating that the whole body was so too. While this symmetry is a simple one, and conceivably may have arisen convergently in a time of high diversity of "body plans," a consistently triradial symmetry of the whole body is not known in any other multicellular animals, fossil or Recent. The triradiate symmetry in Ediacaran and Early Cambrian forms may be a shared derived character of a separate lineage of early sessile metazoans (note also that the three vanes of *Pteridinium* might fit into this concept).

Conway Morris (1989) has drawn attention to similarities between the Burgess Shale problematicum *Mackenzia costalis* and the Ediacarian fossil *Platypholinia pholiata* Fedonkin 1985. Such a comparison may also be made between the Ediacaran *Onega stepanovi* Fedonkin 1976 and the unnamed arthropod-like fossil figured by Conway Morris et al. (1987, Fig. 1a) from the Lower Cambrian Buen Formation of north Greenland. Both these potential cases of Cambrian survivors of the Ediacaran biota need to be further investigated, but they do suggest that some of the short-lived Cambrian clades had their independent origin at least as early as in the Ediacaran.

Even though some phylum-level taxa may thus be represented both in the Ediacaran and the Early Cambrian biotas, there is no suggestion of a direct ancestor-descendant relationship. Such relationships are notoriously difficult to identify in fossils, especially in this case of rare fossils of poorly understood anatomy. If the Ediacaran taxa would largely represent ancestors of Cambrian forms, more Ediacara-type fossils might be expected in the Cambrian where soft-bodied preservation is well known. The Ediacaran biota, as known, most likely became extinct before the Cambrian, and the hypothesis of a disparate origin for each known biota remains viable.

The lack of clearly identifiable lineages extending through the pre-Ediacaran, Ediacaran and Cambrian biotas indicates that direct phylogenetic connections between the known fossils were weak at best, and that the hypothesis of distinct origins for them is supportable based on present evidence. But unless we postulate a polyphyletic origin for metazoans, the ancestors of the succeeding biotas must have been present in previous ones. Why then is the evidence lacking? Although several possible reasons exist, differential taphonomy may be sufficient to account for this apparent absence of ancestors to the Cambrian taxa in the early biotas.

8.3.3 Environmental and Taphonomic Settings of the Biotas

Sedimentologic and topologic data indicate that the Ediacaran biota is found in shallow-water (Australia) to deep-water (Newfoundland) deposits, onshore and offshore. It occurs in sandstones (Australia), mudstones (western Soviet Union), carbonate rocks (Siberia), and turbiditic sequences (Newfoundland). These occurrences are widespread throughout the world. Similarly, the skeletal fossils of the Lower Cambrian are known from mudstones and turbiditic deposits as well as carbonate rocks and near-shore clastics. Thus, both of these biotas were capable of occupying similar if not identical habitats. The earlier *Tawuia* assemblage may also have occupied a variety of habitats, but its occurrence does not justify any conclusions regarding habitat.

In a gross sense, all of the three biotas represent similar taphonomic settings. Each one occurs in several different sedimentologic and geologic conditions over long time periods, indicating that they were capable of preservation (and discovery) in a variety of conditions. Also, soft-bodied preservation is well known in the Cambrian, although Ediacaran fossils are not. Thus, different taphonomies cannot be invoked to account for the three different biotas in their entirety; the differences between observable fossil assemblages are real.

Yet taphonomy probably significantly biased the record connecting the three biotas. The evidence for this bias is circumstantial: the generally small size of the earliest Cambrian organisms implies that their Ediacaran ancestors were also minute, and minute forms lacking skeletons are unlikely to be preserved in most sedimentary environments. The metazoan fossil record is biased towards skeletal forms or large, soft-bodied forms. Meiofaunal non-skeletal metazoans are unknown as fossils.

The Ediacaran fossils seem remarkably large – they may be of meter size, although most are of centimeter or decimeter size (see Section 7). Contrarily, the earliest Cambrian skeletal fossils are often of millimeter size or smaller, although evidence from both trace and skeletal fossils shows that centimeter sizes are by no means uncommon. In the clastic depositional environments of the Ediacaran, non-mineralized organisms of millimeter size and smaller, or their traces, are not likely to be preserved. Animals of this size commonly rely on ciliated surfaces for locomotion. Metazoans living in today's interstitial habitats belong to various minor and major phyla, and are typically worm-like, with a body length down to tens of micrometers (see Chapter 6). Moving between sand grains without disturbing the sedimentary particles, they are not likely to produce recognizable trace fossils. Small nematodes might conceivably leave traces in fine-grained sediments because they move by undulatory body motions. However, in spite of their thick cuticle, extreme abundance, and ubiquitous presence in modern biotas, fossil nematodes have not been positively identified in Phanerozoic rocks (the so-called nematode trace fossils from Eocene non-marine sediments [Moussa 1970] are now attributed to insect larvae [Metz 1987]). Even though metazoan cell walls have low preservational potential, a resistant cuticle, such as the highly cross-linked collagen in nematodes (McBride and Harrington 1967), or the mineralized spicules incorporated in some meiofauna body walls (Reiger and Sterrer 1975) might be expected to have a reasonable chance of fossilization, especially in cherts. Yet no records exist.

Thus, taphonomic bias against meiofaunal elements makes any inference about a lack of connections between the pre-Ediacaran, Ediacaran, and Cambrian multicellular organisms suspect. Likewise, the even greater puzzle of the discrepancy between the ca. 600 Ma age of the earliest definitive metazoan fossils and the ca. 1000 Ma age that has been suggested for oldest branching of metazoan lineages based on molecular studies of modern metazoans (Runnegar 1982b; see also Chapter 9) may be attributed to this bias as well.

The major biotas spanning the Proterozoic-Phanerozoic transition are unique, as far as the fossil evidence reveals. The strong taphonomic bias against non-skeletal meiofaunal organisms throughout the entire fossil record indicates that a major part of early metazoan history may be unrecorded in the rocks. While that bias is not so significant in the Phanerozoic, most of the first Cambrian skeletal fossils are close to meiofaunal size, indicating that the meiofauna was of great importance in the early history of multicellular life. The meiofaunal bias is so strong that further speculation based on the fossil record about the relationship between the three biotas or possible pre-Ediacaran metazoans is unwarranted. At this time, the evidence concerning the origin of metazoans must be sought in disciplines other than paleontology or in new paleontological techniques.

8.4

Ecology and Biogeography

JERE H. LIPPS   STEFAN BENGTSON   MIKHAIL A. FEDONKIN

8.4.1 Trophic Relationships and Life Modes

Ecological structure is commonly bound by trophic interactions, and trophic relationships of organisms have probably always been a fundamental property of ecologically co-occurring species. These interactions are mostly not directly observable in the fossil record, and interpretations of them depend on understanding of the autecology of the fossil organisms. This is particularly precarious in the older parts of the fossil record.

The pre-Ediacaran biota shows little to indicate trophic interactions involving multicellular organisms. The decline in stromatolite diversity and abundance beginning about 1 Ga ago has been suggested to be caused by grazing organisms disrupting the microorganism-bound laminations (Garrett 1970b; Awramik 1971; Walter and Heys 1985). Grazing or predating pressures is normally expected to raise diversities rather than lower them (Paine 1966), but this argument is not applicable to stromatolites, because the morphological diversities of the latter do not primarily reflect taxonomic diversities but rather the possibilities for microorganisms in various environmental settings to build coherent mats unchecked by grazers. Nevertheless, direct evidence for heterotrophic multicellular organisms is scarce or lacking. Tawuiids had bag-like morphologies without any evidence of attachment structures – they may be interpreted as thin-walled floating photosynthesizers, either colonial protists or multicellular individuals. Longfengshaniids also were bag-like, but with a stalk that suggests attachment to a substrate. The sinosabelliditids are currently the most promising evidence for pre-Ediacaran heterotrophic metazoans.

The Ediacaran biota is composed chiefly of passively feeding organisms, such as the medusoids and frond animals, and detritivores represented by trace fossils. Possibly some elements of the fauna were photoautotrophs harboring algal endosymbionts (Fischer 1965; Conway Morris 1985a; McMenamin 1986a; McMenamin and Schulte McMenamin 1989), although no evidence exists for such an interpretation other than the flattened morphology of some species. Such a morphology has many other functional possibilities, including uptake of dissolved organic matter (McMenamin 1986a), diffusion of various gases (Runnegar 1982C), feeding on algal mats (Seilacher 1984, 1989), and suspension feeding; any of these are difficult to demonstrate in the fossil record. The Ediacaran organisms may have taken phytoplankton by suspension feeding as their chief dietary item, although smaller unfossilizable zooplankton and suspended detritus could have been an important food source as well. The trace fossils are chiefly horizontal shallow tracks made on the surface of the sea floor, and so they probably indicate detrital and bacterial feeders. They appear to have been formed by vagile benthic metazoans occupying a relatively narrow zone near the sediment-water interface and using various types of peristaltic locomotion. Possibly the organisms represented by these trace fossils also fed on algal mats covering the sea floor in the euphotic zone. The vagile metazoans were capable of ingesting the sediment and passing fecal pellets (as in *Neonereites*) and in efficiently covering the sediment surface (as in *Palaeopascichnus*, *Nenoxites*, *Yelovichnus*, and *Helminthoida*). Indications of predation on any of the larger metazoans is wanting. Thus, no complex trophic interactions between members of the Ediacaran biota are indicated by the fossil evidence. The Ediacaran multicellular biota may thus have been structured differently than any subsequent biota in the fossil record, each element functioning relatively independently of the others.

The earliest Cambrian fossils show evidence of more complex trophic interactions and life postures, most significantly the dramatic increase of suspension and deposit feeders and the advent of predation.

Suspension feeders were represented by the infaunal and epifaunal tube-dwelling animals, and various epifaunal sponges, archaeocyathans, brachiopods, and other taxa. Tubular fossils are a particularly characteristic component of the early skeletal faunas. Tubes are a simple kind of external protective skeletons that may be formed in several different ways, by secretion of a mucus cover analogous (or even homologous) to the lining of a burrow, by formation of a cuticle or shell from a mantle-like epithelium, by secretion from a local gland in the growing region, or by external assembly by a protruding organ. Functionally, the tubes serve as protection (mainly against predators) for an animal of limited mobility, and the passive mode of life usually implies that the animal was feeding on matter suspended or dissolved in the water. Less commonly, tubicolous animals may be predators (catching prey drifting by) or deposit feeders, but at least the latter feeding mode requires some measure of mobility.

The sclerite-bearing metazoans typical of this time (tommotids, coeloscleritophorans, cambroclaves, and others) in most cases represent sluggish benthic deposit feeders or grazers on algal mats, as understood from the anatomy observable in one of the few Middle Cambrian survivors of this type (Bengtson and Conway Morris 1984; Conway Morris 1985b). Hyoliths were probably also deposit feeders, as indicated by preserved gut fillings (Runnegar et al. 1975). The diverse groups of molluscs present probably included suspension feeders, deposit feeders, and grazers. The separation between these trophic groups is not particularly clear even in modern ecosystems but, in general, the incoming of the earliest Cambrian faunas indicates a sharp rise in the utilization of organic matter near the sediment-water interface.

Early Cambrian predation is difficult to assess relative to the later Phanerozoic because of unique and poorly understood anatomies; however, a variety of predators were present. (The term "predation" is here restricted to predation on multicellular animals.) Tooth-shaped elements practically identical in shape to those of modern predators occur in the fossil assemblages (Figs. 8.4.1D–E, 7.7.3I–M). The giant Early-Middle Cambrian *Anomalocaris* was probably a portentous predator on larger metazoans (Briggs and Mount 1982; Whittington and Briggs 1985). Many of the exoskeletal tubes, conchs, and sclerites, as well as the spicules, had a morphology that would make them effective in deterring predation (Figs. 7.7.3 through 7.7.8). The latter, however, constitute only indirect evidence of predation. More direct evidence exists in the occurrence of shells drilled by other organisms (Fig. 8.4.1A–C). Shell-boring predators, mostly molluscs, are common from the Mesozoic onwards (e.g., Vermeij 1987); they were, however, very rare in the Cambrian and throughout the rest of the Paleozoic. Early Cambrian trace fossil evidence (Fig. 8.4.1F) shows that other arthropods, probably trilobites, occasionally fed on infaunal worms (Bergström 1973; Jensen 1990). Trilobites themselves show evidence of predation by other animals (Pocock 1974; Alpert and Moore 1975; Conway Morris and Jenkins 1985; Babcock and Robison 1989).

Plots of the Early Cambrian metazoan genera, grouped into their most probable life modes and feeding habits (Figs. 8.4.2, 8.4.3), suggest that suspension feeders dominated until the latest part of the Early Cambrian, when they lost ground mostly due to the decline of archaeocyathans and the rising dominance of trilobites. (Trilobites are grouped in their own trophic category in the plots, as they are a dominant group that could probably feed in several ways, including deposit feeding, suspension feeding, and predation.) Predators and herbivores throughout this time seem to have been of low diversity, although they were present already in the Nemakit-Daldyn. Even though generic diversity does not directly reflect the quantitative importance of various life modes in the ecosystems, the genera are currently our best taxonomic measure, and the diversity at this level gives at least a rough picture of the relative ecological diversity. (Note, however, that in the earlier part of the time period the number of problematica with unknown life habits is high, and for the Nemakit-Daldyn the total number of genera is low, making the curves in this part of the diagrams very uncertain.) The bias towards large skeletal taxa in these data leads to an under-representation of ecological types such as infauna, meiofauna, and plankton. Nevertheless, the data seem to suggest that the basic characters of a modern marine ecosystem were present already from the beginning of the Cambrian, although subsequent events led to an increase in the number of ecological niches as reflected in increasing taxonomic diversity.

The history of life modes and feeding strategies through the later Proterozoic and Early Cambrian thus indicates the development of increasingly complex trophic interactions. The pre-Ediacaran biota was composed largely of primary producers: there is little evidence of utilization by other organisms than bacteria. The Ediacaran biota contained suspension and deposit feeding animals that most likely utilized single-celled primary producers in the water column or on the bottom. No evidence of predation exists in the Ediacaran fossil record, the fauna being passive in its feeding habits. The Cambrian is marked by the initiation of an essentially modern type of marine ecological structure with several trophic levels and a complex food web.

The development and diversification of these various trophic strategies is the single most important aspect of the evolutionary history of these biotas. No other factor can by itself account for the simultaneous diversification of protists, metaphytes, and metazoans. On the other hand, this factor is so all-explaining as to be almost trivial for an explanatory model: trophic and taxonomic diversification are two sides of the same coin. Independent evidence is needed to explain the timing and pattern of this event.

The indication that the trophic web was significantly extended through the advent of predation and herbivory (i.e., utilization of multicellular algae, not including grazing on prokaryote or eukaryote mats) in the latest Proterozoic is one important piece of such evidence. Similarly, the evidence for suspension feeding in the Ediacaran and Early Cambrian also suggests extension of the food web by direct utilization of planktic organisms. The incoming of predators, herbivores, and suspension feeders can be seen as the most likely proximal cause for the explosive radiation of eukaryotes at the Precambrian-Cambrian boundary, including the massive development of mineralized skeletons. The earlier, more slowly progressing, diversification during the preceding 400 to 450 Ma (or more) may reflect evolution of various multicellular lineages either themselves autotrophic or living off the prokaryotic and protist producers, but not actively interacting.

8.4.2 Paleobiogeography

The paleobiogeography of each biota is difficult to determine with a similar degree of confidence because reliable reconstructions of continental configurations cannot be accomplished for each time and the distribution of fossiliferous deposits is limited. The present distribution of pre-Ediacaran megafossils bears no relationship to their biogeography when they were alive, nor necessarily to the subsequent biotas, because of the long times between them. This biota contains so

Ecology and Biogeography

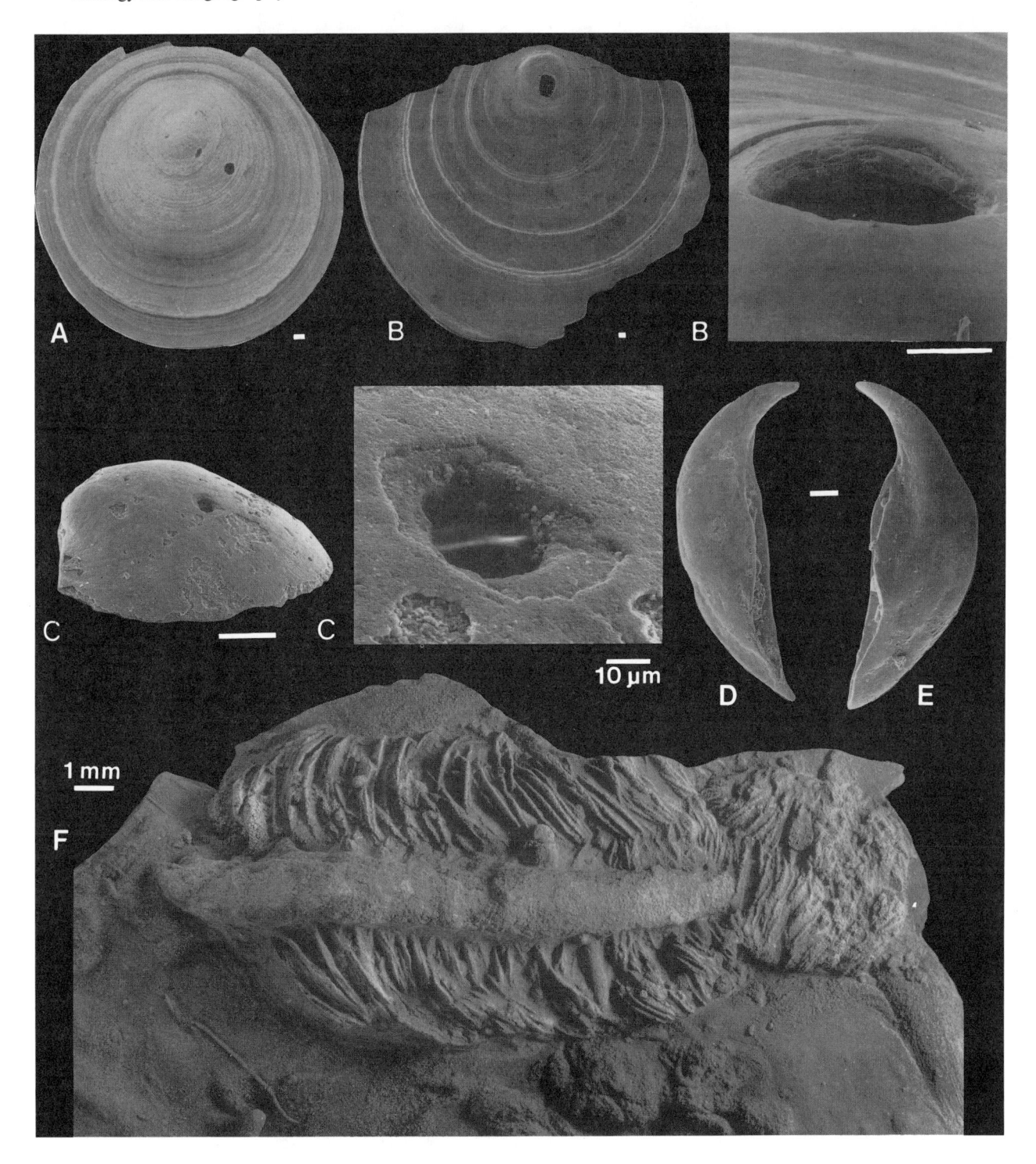

Figure 8.4.1. Evidence for predation (A–C, F) and predators (D, E) in the Early Cambrian. Except as otherwise indicated (C_2, F), lines for scale represent 100 μm. (A, B_1 and B_2) Shells of *Mobergella holsti* (Molberg 1892) with drill holes, from south Sweden (Bengtson 1968). (C_1, C_2) Shell of *Aroonia seposita* Bengtson 1990, with drill hole, from the Flinders Ranges of South Australia (Bengtson et al. 1990). (D, E) Mandible-like elements of *Cyrtochites pinnoides* Qian 1984, from Yunnan, southern China (Qian and Bengtson 1989). (F) *Rusophycus dispar* hunting burrow, from central Sweden (Jensen 1990).

few taxa that biogeographic patterns may not be meaningful in any case. Nevertheless, *Tawuia* and *Chuaria*, a common co-occurring form, are known from Canada, Spitzbergen, India, and China (Hofmann 1987), places that were on different crustal plates (Chapter 12), indicating that the assemblage was at least widespread.

The Ediacaran biota is known from all continents except Antarctica and questionably from South America (Hofmann 1987). Not all elements of the biota, which now consists of over 100 species, are found everywhere together, and again reliable continental reconstructions for this interval of time are wanting (see, however, Chapter 12). The Ediacara biota appears wide-

440   The Proterozoic-Early Cambrian Evolution

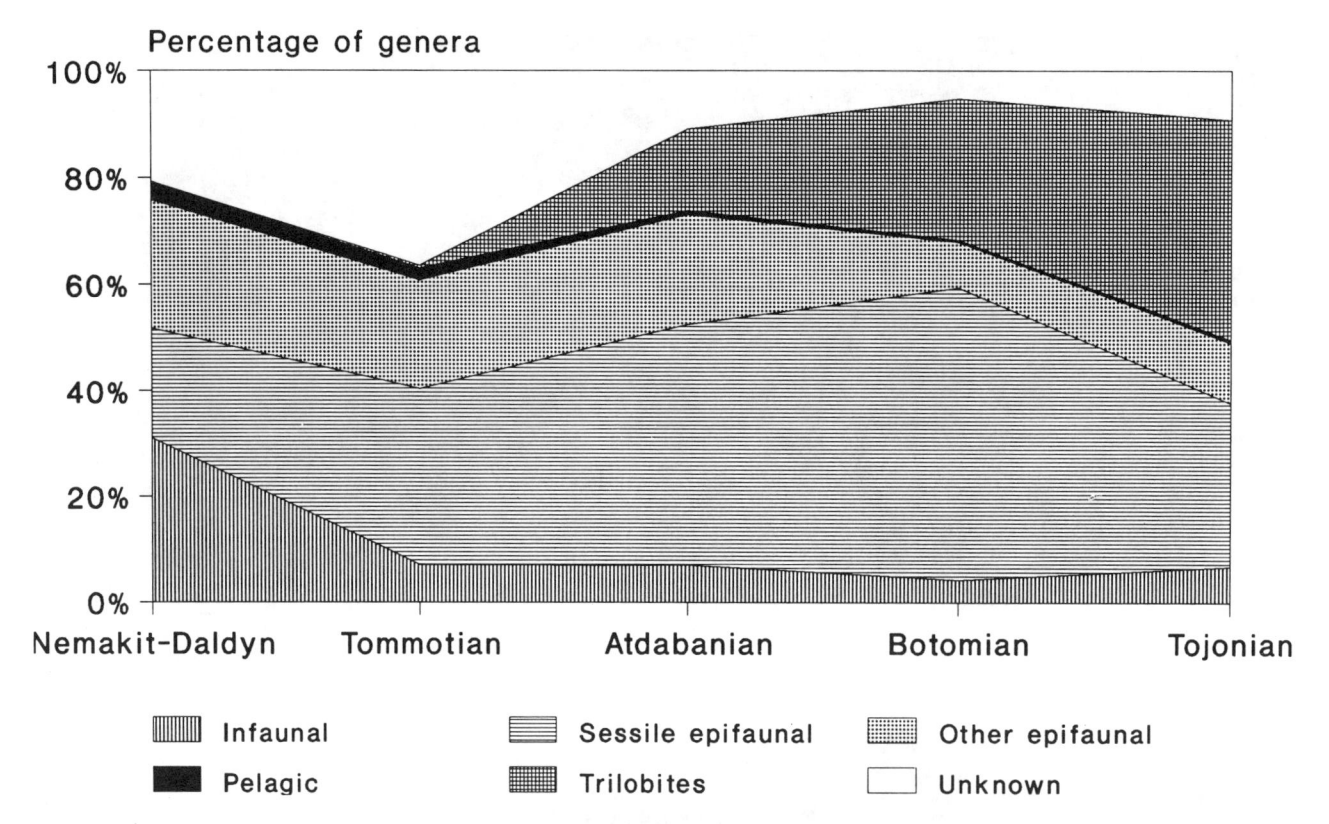

Figure 8.4.2. Percentages of genera of Early Cambrian skeletal fossils assigned to probable life attitudes. Total number of genera: Nemakit-Daldyn, 29; Tommotian, 252; Atdabanian, 511; Botomian, 610; Tojonian, 211. Taxonomic and stratigraphic data from Table 23.5 (see also Section 7.7).

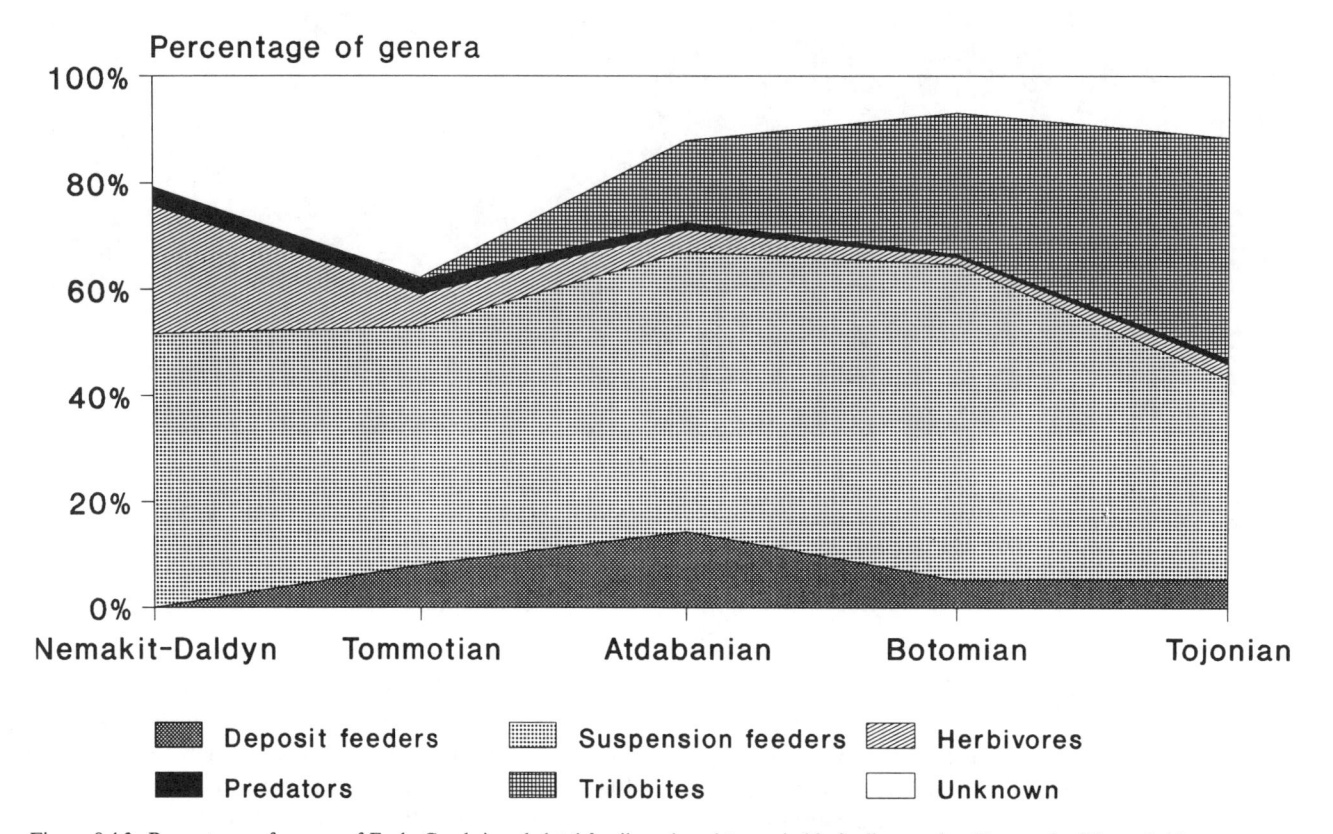

Figure 8.4.3. Percentages of genera of Early Cambrian skeletal fossils assigned to probable feeding modes. Data as for Figure 8.4.2.

Ecology and Biogeography 441

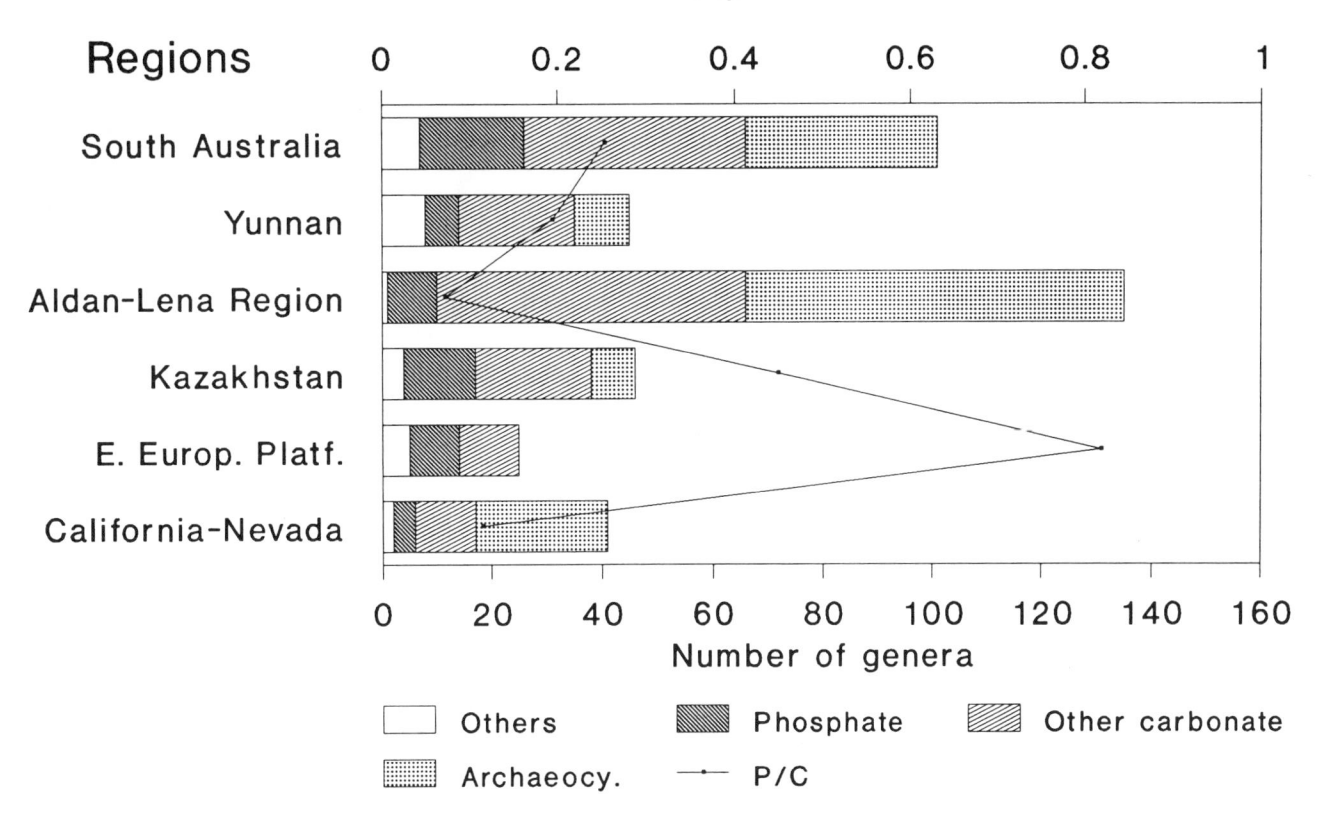

Figure 8.4.4. Generic diversities and shell minerals of early Early Cambrian (Namakit-Daldyn to Atdabanian) skeletal fossils (except trilobites) from the six key regions indicated (data compiled from Table 23.5 and literature referred to in Section 7.7). P/C curve (solid line) shows the ratio of genera with phosphatic skeletons to those with carbonatic skeletons.

spread nevertheless because many taxa can be found at several Ediacaran occurrences.

The Early Cambrian biota of archeocyathans and small shelly fossils shows geographic changes in total species diversity (Fig. 8.4.4). The most diverse small shelly faunas (100 genera) occur on the Siberian Platform (Aldan-Lena region) and in Australia. Paleomagnetic data (see Chapter 12) indicate that these areas were on or near the equator. Regions of somewhat higher paleolatitudes (about 20–30°), including South China, Kazakhstan, and the southwestern United States and northwestern Mexico, have an intermediate diversity (40 to 50 genera). The highest-latitude regions, Scandinavia and the East European Platform, show lower diversities (about 25 genera). This diversity pattern is enhanced by archaeocyathan distributions. They, too, are abundant and diverse in reconstructed low-latitude areas where the main reefs occur during the Tommotian-Atdabanian, and sparse or absent at higher latitudes. Reef-building archaeocyathans also appear from the latest Atdabanian in California-Nevada (Gangloff 1976;

Rowland and Gangloff 1988) as well as in other regions on the North American Plate. This changing distribution of archaeocyathans may be related to northward continental drift during the course of the Early Cambrian (Chapter 12).

The Nemakit-Daldyn biota, characterized mainly by *Anabarites* and *Protohertzina*, had a wide distribution, whereas the subsequent Tommotian biotas were considerably more provincial. The geographic distribution of diversity apparently varies subsequently, however, although the patterns are obscured by current taxonomic uncertainties.

The paleobiogeography of the Proterozoic and Early Cambrian remains tenuous at best, although a general biogeographic partitioning appears to have developed through this period of time. The earliest Cambrian faunas show a typical latitudinal diversity gradient whereas no such pattern can yet be detected in Proterozoic biotas. Even among the acritarch protists, paleobiogeographic patterns are not distinguishable, other than onshore offshore ones (Vidal and Knoll 1983).

8.5

The Evolution of Metazoan Body Plans

STEFAN BENGTSON   JACK D. FARMER

8.5.1 Introduction

There is a great mass of knowledge on anatomy and embryology of multicellular animals, but the evolutionary history of basic metazoan structures continues to be poorly comprehended. Part of the reason for this is the lack of reliable data on the interrelationships of phyla. Comparative anatomy, so successful in unravelling relationships at lower taxonomic levels, yields ambiguous results at the phylum level, as indicated by the many contradictory schemes of metazoan evolution (e.g., Nursall 1962; Hadzi 1963; Jägersten 1968; Salvini-Plawen 1978; Nielsen 1985). Molecular analyses of phylogeny still have too low a resolution to discriminate relationships between phyla (see Chapter 9). Bergström (1986), however, has used molecular data to challenge some of the traditional views on metazoan phylogeny.

The earliest metazoan fossils are very different from each other but not so different from later, even extant, representatives of the same groups. Why are these differences pronounced so early in the evolutionary history? Why is there no evidence of a gradual evolution of such groups? How long did it take to make, for example, a flatworm, a polychaete, or a snail, and from what were they made? What are the phylogenetic relationships between the major divisions of the animal kingdom? In what environments did the early stages of metazoan evolution take place? What types of body plans evolved in which order of events? And which adaptational, historical, and biomechanical constraints governed their evolution? These questions are most commonly couched in terms of the origin of phyla or body plans, which thus is a real problem of great importance for the Late Precambrian biotic evolution.

8.5.2 Genealogies and Body Plans – Two Perspectives on Phyla

There are two major approaches to the diversification of metazoans into phylum-level taxa. One is purely genealogical, in that it centers on phylogenetic evolution, looking at phyla as major clades of lineages. The other concentrates on the question why body plans become established and fixed to serve as bases for subsequent radiations of animals.

These two approaches mirror the phylogenetic/phenetic dichotomy in general taxonomy, often construed as a conflict between classification methods based on character distributions (phylogenetic systematics, or cladistics) and those based on similarities (phenetic systematics). Although this methodological conflict is real, it can be resolved for practical purposes (Fortey and Jefferies 1982). The two perspectives on the diversification of phyla are equally valid, and both are indispensable, but it is important to be clear about which one is being used.

In the genealogical perspective, phyla are major clades into which taxa group only by virtue of common ancestry. Taxonomic rank of a particular group may be least ambiguously defined by phylogenetic distance to the last common ancestor with sister taxa. Hennig (1966, pp. 154–193) proposed that absolute time of divergence be used to define higher taxonomic categories, placing, for example, the divergence of phyla in the late Precambrian. An effectively identical measure of taxonomic rank was also adopted by Raup (1983) in an analysis to show that the early appearance of major taxa in a radiation is an artifact of the geometry of the evolutionary tree. Using Hennig's definition of phyla, it is impossible to avoid this conclusion, but this may be of little significance, as the definition presupposes that which is to be explained.

The genealogical aspect of phyla may, however, be profitably used to analyze questions of phylum diversities through time (e.g., Sepkoski 1978; Strathmann and Slatkin 1983; Raup 1983). Assuming constant and equal rates of speciation and extinction after the initial establishment of a would-be phylum, the probability of survival for small (tens of species) such clades over 600 Ma is low, even with very sluggish rates of species turnover (Fig. 8.5.1). This supports the intuitively obvious interpretation that a "phylogenetic tree" is more likely to resemble a bush, i.e., the number of extinct twigs at the base should greatly exceed the number of branches surviving at the top. Using calculations based on the number of surviving low-diversity phyla, and making different assumptions about rates of species turnover, Strathmann and Slatkin (1983) suggest that there may have been hundreds or even thousands of pre-Ordovician phyla. This may be compared with lower estimates of up to a hundred Early Cambrian phyla based on extrap-

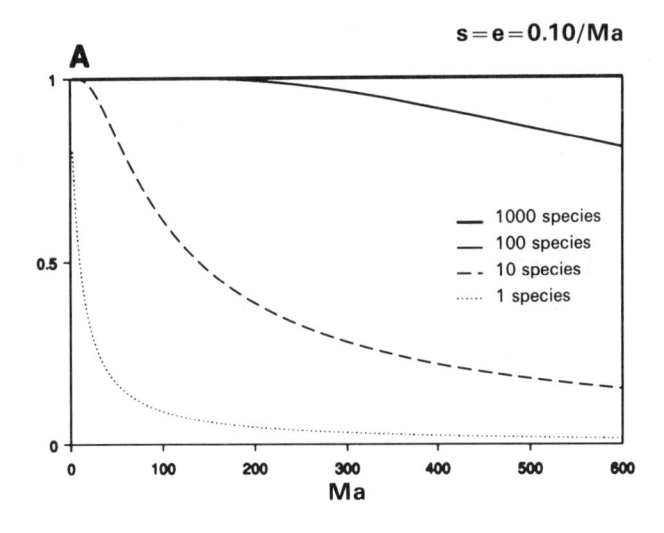

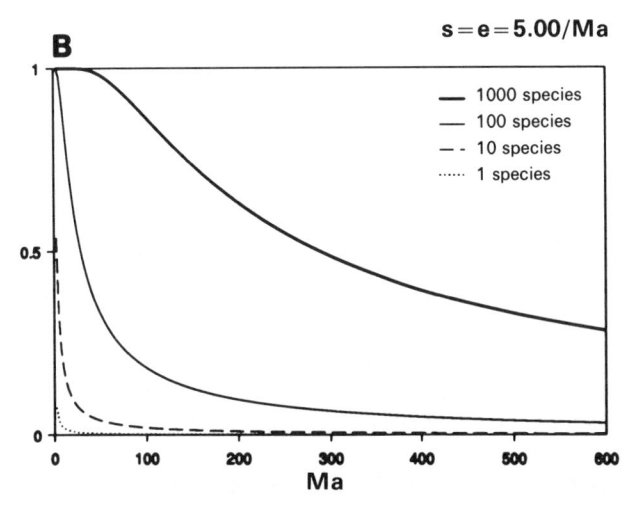

Figure 8.5.1. Plots of $p = st/(1 + et)$, the survival probabilities (p) for clades of various initial sizes (1 to 1000 species) over time (t), assuming constant speciation (s) and extinction (e) rates, for slow rates (A), with $s = e = 0.10/\text{Ma}$; and for rapid rates (B), with $s = e = 5.00/\text{Ma}$.

olation from skeletal fossils (Valentine 1987), but since a genealogical phylum definition in fact extends down to the species level, the difference could be regarded as being due to the difference in the phylum concepts.

The phenetic approach to phyla groups taxa that share a set of fundamental characters, a "body plan," inherited from a common ancestor (e.g., Simpson 1950). There is a considerable measure of arbitrariness in such a definition, due to the lack of objective criteria for "body plans" and "fundamental characters". In fact, "body plan" often carries with it an unfortunate teleological implication; as used herein the term strictly implies only a shared character complex. Traditionally, phyla are groupings of *Recent* animals that cannot be shown to be closely related to any other such group (Bengtson 1986b); the existence of any significant numbers of extinct phyla has been implicitly or even explicitly (Glaessner 1984, p. 135) denied by paleontologists. Most traditional phyla are recognized through a unique character complex, and there are only few living groups that do not fit easily into established phyla (e.g., Rieger 1980). The early fossil record has many such groups, however (Hoffman and Nitecki 1986; Chapters 7 and 23), and it is clear that for present purposes we need a different criterion for what – if anything – constitutes a phylum.

8.5.2A A Pragmatic Phylum Definition

A strict genealogical definition of a phylum as a holophyletic clade (monophyletic in the cladistic sense, i.e., containing the founding species and all its descendants; cf. Ashlock 1971) that diverged from its sister group in the late Precambrian (Hennig 1966) avoids the body-plan problem in the definition. Although it has a measure of arbitrariness, the criterion of divergence in the late Precambrian in fact follows established practice – there is no evidence of traditionally recognized phyla originating in the Phanerozoic. A strict adherence to the holophyly criterion in classification, as insisted on by cladists, easily becomes impractical, but in this case it can be defended also from a phenetic point of view: there is an "ignorance criterion" in the traditional phylum definition (a phylum is a group of organisms with unknown phylogenetic relationships; cf. Bengtson 1986) which presupposes that no close relatives can be clearly identified. Adult synapomorphies are rarely shared between two phyla; whatever shared characters exist are generally regarded as symplesiomorphies at the phylum level. This leads to an acceptance of the holophyly criterion for recognized phyla. Further additions to the list, in the form of extinct phyla according to the proposed genealogical definition, will then cause the least disruption of existing usage.

This phylum definition may be seen as a practical measure. It is biologically meaningful if the lineages that make their stage appearance at the Proterozoic-Phanerozoic transition have evolved their morphological characters independently during a burst of massive early radiation, rather than having evolved from each other in a sequential evolution of traditionally presumed type.

8.5.3 The Evolution of Body Plans

Given the severe constraints imposed by regulatory systems in extant organisms, we would expect the origination of novel body plans (in the sense of sets of shared characters that define the basic functional physiology, morphology, and anatomy of animal groups) and higher taxa to be an extremely rare event in evolution. The establishment of new biochemical pathways requires more than the development of novel structural genes. Such changes must be accompanied by modifications of the regulatory genome and the integration of control substances into pre-existing physiological systems.

Yet there remains the classical problem of the widespread and explosive appearance of multicellular life representing a large number of body plans at the end of the Proterozoic. Valentine and Campbell (1975) suggest that this was the result of rapid advances in the regulatory genome. Vermeij (1973) views the versatility of a given phenotype in terms of the number and range of independent variables controlling form, which increase the number of possible solutions to adaptive problems. Once morphogenetic control systems had appeared, versatility and adaptive innovation could drive rapid and complex evolutionary change. Buss (1987) believes that the epigenetic programs and regulatory systems characterizing the ontogeny of higher organisms represent altruistic compromises that developed between competing cell lineages that were of benefit to the individual. It can be argued that the very presence of such complex interactions between cell lineages during development constitutes *prima facie* evidence that they arose as variants in the course of ontogeny.

The other side of the story, namely the question why new body plans did not continue to evolve at more or less the same pace during the Phanerozoic, has also been addressed with reference to both intrinsic and extrinsic mechanisms. In the first category, the possibility that developmental constraints became too severe once the adult organization of the metazoan phyla had evolved (Strathmann 1978), is in good congruence with current understanding of animal development. Hypotheses of the second category generally center on ideas of lack of competition in the earliest ecosystems involving multicellular organisms. The adaptive space theory (e.g., Valentine 1980; Conway Morris 1989a) says that ecologic niches were largely unoccupied at the time so that competition was minimized and a large number of body plans were possible. Erwin et al. (1987) suggest that the reason why the Cambrian radiation produced new body plans, whereas the Triassic one did not, is to be found in the fact that at the start of the latter radiation a diversity of families were already present in the not quite empty habitats.

But the concept of empty ecological niches allowing for unrestrained development of body plans is flawed by the fact that there is no direct relationship between life mode and body plan. Suspension feeders, for example, may use a variety of techniques involving many different organs to gather their food, and burrowers similarly use different techniques and body parts to dig into the sediment. Selection does not work on body plans, but on their effects. So whereas the low occupation of ecological space at the close of the Proterozoic may have facilitated radiation as such, it does not suffice to explain the difference to the Triassic radiation.

Given the distinctness of metazoan phyla even at their first appearances in the fossil record and the fact that they all appear to have originated very early in metazoan history, the evolution of body plans appears uniquely connected with the earliest phases in the evolution of the Metazoa. It is difficult to envisage any unique features of the late Precambrian environment (low oxygen levels, glaciations, etc.) that would be likely to have influenced this phenomenon and that were not at least locally recurring in the Phanerozoic; thus, the early establishment of the major aspects of metazoan organization is more easily explained as a result of greater genetic and developmental flexibility of the early multicellular organisms.

However, it is likely that most if not all physiological prerequisites for the transition to multicellular individuality were present long before the final radiation of phyla at the Precambrian-Cambrian transition. This is suggested by the molecular (and possibly also paleontological) indications of an earlier divergence of main metazoan lineages. It is supported by the fact that the phylum radiation took place simultaneously with protist and metaphyte radiations, which points to an extrinsic trigger of the different radiations or to close ecological interdependance between them. In either case it suggests that all groups were ready for the big leap in advance, otherwise the timing of their respective radiation would not have been so good. The most likely such trigger may have been oxygen levels (Section 8.4).

8.5.4 Recognizing Body Plans in the Fossil Record

Although the geometry of the phylogenetic bush and evidence from the fossil record indicate that the early metazoan radiation contained a considerably higher number of phyla than are alive today, this cannot be readily translated into diversity of body plans. Except in a few cases, soft-part anatomy of extinct phyla is not known. Certain basic anatomical features can be deduced: for example, a bilaterally symmetrical animal with a dorsal exoskeleton is likely to have had a locomotory sole, and if the animal was larger than millimeter size the means of propagation were probably muscular rather than ciliary. But most details of anatomy – and, obviously, physiology, biochemistry, and molecular biology – of the kind that are now yielding so much information on various living metazoans, are beyond direct observation in fossils.

Another limitation of the fossil record is that preserved larvae are seldom available for study. An analysis of the evolution of metazoan morphologies must take the whole life-cycle into account. This is particularly important in view of the ideas that heterochronic processes played a significant role in the rapid evolution of different body plans (e.g., McNamara 1982; Bergström 1986).

An animal's size largely determines its anatomy. The cells of a small metazoan (millimeter size or below) can interact directly with the environment by virtue of their large surface/volume ratio. Such an organism does not need particular respiratory, digestive, circulatory, and neural systems, and its locomotory system may be very simple, consisting of a ciliated epithelium. Many small taxa now living in the meiobenthos have secondarily lost organs that are attributes of larger relatives, and their evolution probably involved neoteny in larval forms that repeatedly invaded the interstitial environment (Westheide 1984). Thus, a number of supposedly primitive and even ancestral metazoans, such as "pseudocoelomates" and "archiannelids," can be shown to be secondarily derived from larger taxa (Lorenzen 1985; Westheide 1985).

Proceeding from such general principles, it is possible to construct plausible interpretations of fossil animals regarding the presence or absence of gills, circulatory system, alimentary canal, and so forth. However, such reconstructions based on actualistic principles have their main value in limiting and giving direction to speculations. Only exceptionally can they be tested against observable features in the fossils themselves. Thus they are seldom a source of information on anatomical features.

As in the case of the question of how to define a phylum, we may state that the concept of body plan has biological meaning in that it signifies a fundamental organization shared by members of phylum and not present in other phyla, but it is of little use when dealing with fossil remains of what may be extinct phyla, because not enough of the anatomy is preserved to determine the body plan. The existence of soft-body preservation of Burgess Shale type (e.g., Whittington 1985) makes little difference in this regard, as even such exquisitely preserved fossils leave too much of the anatomy and ontogenetic development unknown. We can "dissect" the animals only when we already have a living template (e.g., Conway Morris 1977a). A case in point is the Ediacaran biota, where soft structures are preserved but can only be interpreted within the framework of very specific hypotheses (Section 7.5).

What is left, then, of the body plans in the fossil record is the fossilized small subset of the complete set of characters present in the phyla. The crucial question here is whether these characters may be subject to normal systematic analyses with the same possibilities and limitations as lower-level taxa later in the fossil record. The answer would be yes, if the radiation of phyla was no different from later radiations of lower-level taxa, only occurring earlier. But we know that the phylum level is effectively where ordinary systematic analysis breaks down. As Cambrian molluscs, crustaceans, hemichordates, etc., have a modern appearance with regard to what could be regarded as their body plan, chances would seem slim even if we had access to live animals from this time. The presence of additional lineages representing possible intermediate forms, and the possibility that some phyla, such as chordates, had not yet diverged fully from their sister group, give us some hope, but convincing examples are still lacking. Precambrian metazoans and metaphytes have so far mostly provided material to weave stories and breed controversies around.

8.6

Origins of Biomineralization in Metaphytes and Metazoans

STEFAN BENGTSON   BRUCE N. RUNNEGAR

8.6.1 Introduction

The appearance of skeletons was an important development in the history of multicellular life, both practically in terms of fossilization potential and biologically in terms of increased functional potential. Section 7.7 and Table 23.5 summarize the fossil evidence for the appearances of different types of skeletons. Hard skeletons may be formed by organic substances, such as scleroproteins and polysaccharides, and through agglutination of externally derived mineral particles. The most common and diverse types of hard skeletons, however, are those that employ biologically formed minerals. The ability to produce biominerals and to use them for constructional purposes has opened new paths in the evolution of metazoans and metaphytes.

Biologically induced mineralization may have occurred as far back as 2.5 to 2.3 Ga, in connection with the photosynthetic activity of cyanobacteria (Klein et al. 1987). Calcified cyanobacteria have their mucilagenous sheaths impregnated with crystals, perhaps as a byproduct of the photosynthetic removal of CO_2 from the water in which they lived (Riding 1977). But if the term biomineralization is restricted to imply direct biological feedback on the mineralization process, the phenomenon is of more recent origin. Chains of magnetite crystals, similar to those produced by magnetotactic bacteria, have been identified in the Gunflint chert, and well-preserved magnetite of apparent biogenic origin is known from the Late Proterozoic Nama Group (Chang and Kirschvink 1989). Modern organisms are known to produce minerals of considerable diversity; about 60 such biominerals have been identified (Lowenstam and Weiner 1989). Many of these, however, are unstable under normal diagenetic conditions, and most of them lack a record as fossils. Given the widespread occurrence of minute mineral units, such as spicules, in living tissues (Rieger and Sterrer 1975; Lowenstam and Weiner 1989), they are likely to have been present already in Proterozoic organisms. Apart from the magnetite mentioned above, however, definitive Proterozoic examples are not known.

The evolution of biominerals as structural materials is likely to be reliably registered in the fossil record, because of the good fossilization potential of mineralized tissues. The record shows that the various skeletal biominerals appeared almost instantaneously at the Proterozoic-Phanerozoic transition (Fig. 7.7.1). Mineralized skeletons seem to have evolved separately in at least 38 lineages (Fig. 7.7.1; Sections 7.7 and Table 23.5).

Biominerals used for skeleton-building purposes are esssentially limited to calcite, magnesian calcite, aragonite, apatite, and opal. The original mineralogy of the various groups of late Precambrian and Cambrian fossils is not always well known. In most groups, the skeletal composition is only known from preserved mineralogy in different types of rocks, or inferentially through comparisons with known related taxa. A few studies have yielded more detailed information from petrographic and geochemical studies of fossils and surrounding rocks (e.g., James and Klappa 1983) and from studies of replicated crystal morphologies (Runnegar 1985).

8.6.2 Carbonates

The most common skeleton-forming minerals are calcium carbonates, mainly calcite, magnesian calcite, and aragonite. They appear to have been dominant already from the beginning of skeleton biomineralization. Calcite (and magnesian calcite) often preserves the original crystallographic structure and can thus be clearly identified, whereas few cases of preserved aragonite are known, and none from the Precambrian-Cambrian transition.

The tubular fossil *Cloudina* is the earliest known example of a mineralized skeleton (Sections 7.5, 7.7, 10.6, Table 23.5). Other early carbonate tube-building animals include the triradially symmetrical anabaritids (Sections 7.7, 8.4, Table 23.5) that appear to have formed their tubes of aragonite. Later faunas contain more diverse tubular fossils of carbonate composition, some resembling, for example, protective structures built by certain modern annelids. The shells of the hyoliths (Section 7.7, Table 23.5) were most probably composed of aragonite. In younger Paleozoic members of the group a structure resembling molluscan crossed-lamellar fabric has been observed.

Aragonite is above all characteristic of the early molluscs (Runnegar 1985). The degree of biological control of the mineralization process varied greatly. Some Cambrian mollusc shells seem to have consisted of a single layer of spherulitic

447

aragonite prisms beneath an organic periostracum. This structure is similar to inorganically precipitated calcium carbonate and need not have been mediated by a protein substrate. The shape of the spherulitic "prisms" is molded by surface forces rather than chemical bonds. Nacre, on the other hand, is formed by flat aragonitic tablets in which growth is very slow on the [001] face due to mediation by a protein matrix. The resulting layered microstructure is much stronger than fibrous aragonite. Nacreous linings appeared at least by the Middle Cambrian and may have been present already in Early Cambrian time. Other molluscan ultrastructures, such as tangentially arranged fibrous aragonite, crossed-lamellar aragonite, and foliated calcite, had evolved by the Middle Cambrian.

A number of Cambrian solitary and colonial animals with basal skeletons of calcium carbonate have been referred to cnidarians (Jell 1984; Section 7.7; Table 23.5), but undoubted skeleton-forming cnidarians are not known until the Ordovician. Coral skeletons are built up of spherulitic tufts (trabeculae) formed of fibrous calcite or aragonite. As with the spherulitic "prisms" of some mollusc shells, the process of formation appears to involve little matrix-mediated control of crystal shape (Constantz 1986), but nucleation of the fibrous trabeculae may be under more direct biochemical control.

The cups of archaeocyathans (Section 7.7; Table 23.5) are preserved as microgranular calcite, interpreted to represent original magnesian calcite (James and Klappa 1983). Calcium carbonate (argonite or calcite) skeletons are also formed by several groups of sponges ("sclerosponges" and "sphinctozoans") from the Middle Cambrian until the Recent (Vacelet 1985).

In the skeleton types described above, growth occurs by addition of material to earlier formed growth stages, which puts strong geometrical constraints on the morphology. Ways of avoiding these problems are (*i*) periodical molting of the exoskeleton, or (*ii*) continuous construction and destruction of the mineral phase by intimately associated living tissue.

Trilobites had periodically molted exoskeletons composed of low-magnesian calcite (Wilmot and Fallick 1989), often with well-preserved crystallographic fabrics. Other examples are the Cambrian coeloscleritophorans (Section 7.7; Table 23.5). The ubiquitous recrystallization and occasionally preserved fibrous structure suggest that their sclerites were aragonitic (Bengtson et al. 1990).

In the calcareous endoskeleton of echinoderms there is close interaction of mineral and living tissue. The skeleton consists of a meshwork (stereom) of magnesian calcite containing minor amounts of organic matter (Lowenstam and Weiner 1989). Each skeletal element behaves optically like a large single crystal.

Spicules of magnesian calcite are characteristic of calcareous sponges and octocorals (Section 7.7; Table 23.5). In both groups the spicules are formed by specialized cells (sclerocytes), sometimes originating intracellularly and only later erupting from the cell membrane to be further enlarged by enveloping sclerocytes. Sponge spicules grow in crystallographic continuity, so that each behaves optically as a single crystal of calcite. In contrast, octocoral spicules typically are composed of smaller acicular crystals (Majoran 1987). As the echinoderm plates and sponge and octocoral spicules are made of magnesian calcite, it has been suggested that magnesium is used to shape the crystals by selectively poisoning appropriate parts of the lattice (O'Neill 1981). Calcitic sponge spicules have been found in the late Atdabanian 535 Ma (Fig. 7.7.3C–D), and possible octocoral spicules also appear in beds of the same age (Fig. 7.7.3H). The fossil sponge and octocoral spicules have the same crystallographic properties as modern spicules of these groups.

Calcified cyanobacteria become common in the Early Cambrian. One group including the helically coiled filamentous genus *Obruchevella* is present as uncalcified filaments in rocks of Vendian age, but is frequently calcified after the beginning of the Cambrian (Peel 1988). Structures interpreted as aragonitic encrusting algae have been reported from the Late Proterozoic *Cloudina*-bearing beds in Namibia (Grant et al. 1987), and possible calcareous algae are also known from the 550 Ma Nemakit-Daldyn beds of the northern Siberian Platform (Riding and Voronova 1984). Indubitable examples are first known from the Middle Cambrian (Section 7.4).

8.6.3 Phosphates

Skeleton-forming apatite occurs today only in vertebrates and inarticulate brachiopods. Amorphous calcium phosphate, however, is produced by members of at least 20 recent phyla (Lowenstam and Weiner 1989) and in the radular teeth of chitons this may later crystallize into apatite (Lowenstam and Weiner 1985). Among the earliest skeletal organisms apatite appears to have been more widespread than today.

Tubular fossils of phosphatic composition are a common constituent of Cambrian faunas (Section 7.7; Table 23.5). Hyolithelminth tubes consisted of lamellae with fibrous elements in adjacent lamellae oriented at angles, producing a force-resistant structure similar to that of arthropod cuticles (Bengtson et al. 1990). Conulariids, with some representation in the Cambrian (Fig. 7.7.6D–E), had tetraradially symmetrical cones built up of more or less strongly phosphatized lamellae forming a flexible integument reinforced by stiff transverse rods (Babcock and Feldman 1986).

Phosphatic conchs or shells were also widespread. In addition to phosphatic inarticulate brachiopods, there are also a number of problematic organisms with phosphatic shells that grew by successive accretion of growth laminae. A characteristic Cambrian example is the tommotiids (Section 7.7; Fig. 7.7.8A–G; Table 23.5). Examples of periodically molted exoskeletons of calcium phosphate are rare, but the valves of the ostracod-like bradoriids are commonly preserved as phosphate. Although some of them appear to have been flexible, they were most probably to varying degrees impregnated with apatite crystallites. Whether or not the ecdysis involved resorption of mineral matter is not known, but resorption may explain the common occurrence of collapsed or buckled valves.

The net-like sclerites of *Microdictyon* (Fig. 7.7.8K–L) were constructed of two or three distinct layers of apatite and show no evidence of incremental growth.

The phosphatic bone of vertebrates is intimately associated

with fibrillar collagen, which does not seem to be the case in other phosphatic skeletons. Vertebrates have a plastic mode of skeleton formation as a result of a constant physiological exchange between mineralized and cellular tissues, similar to the condition in echinoderms. Although undoubted vertebrate remains are not known until the Ordovician, certain Cambrian phosphatic fossils show a fine structure suggesting association with fibrous organic matter that may be homologous with vertebrate collagen. The small button-shaped sclerites of the utahphosphans (Fig. 7.7.8H) consist of a thin dense apatite layer covering a porous core; the latter has fine tubules or fibrils perpendicular to the lower surface (Bengtson, 1977b; Wrona 1982). The "buttons" are more or less densely set in an integument that is impregnated with smaller apatitic crystallites. Similarly, tooth-shaped conodonts had a fibrous organic matrix in which the apatite crystallites were embedded (Szaniawski 1987). In both of these cases, a chordate affinity has been proposed using partly independent lines of evidence.

Interpretation of other suggested Cambrian biomineralizing chordates (*Palaeobotryllus*, *Anatolepis*) is even more problematic (Müller 1977; Bockelie and Fortey 1976; Repetski 1978; Peel 1979).

Several of the other Cambrian fossils of phosphatic composition are spine- or tooth-shaped objects (Section 7.7; Fig. 7.7.3I-K; Table 23.5). This may reflect the fact that apatite is a hard mineral suitable for wear-resistant surfaces.

8.6.4 Opal

Biogenic opal (a mineral gel consisting of packed spheres of hydrated silica) has had limited potential as a skeletal material except in very small organisms. This is probably because morphogenetic control of opaline skeletons is essentially limited to the intracellular environment (Simpson and Volcani 1981, Table 1-1); the opal is formed within a cytoplasmic membrane (silicalemma). Opal is most widespread among protists (Round 1981; Bovee 1981), but occurs as well in some metazoans and higher plants (Simpson and Volcani 1981). The only metazoans known to form it are hexactinellid sponges and demosponges, which use it for spicule formation (e.g., Hartman 1981). Most biogenic opal formed today is either dissolved in the water column before it is incorporated in the sediment or dissolved during early diagenesis (Hurd 1972, 1973; Johnson 1976), but under certain circumstances opaline skeletons may be preserved, usually as microcrystalline quartz or as replacements by other minerals.

The distribution of opal among the earliest skeletal fossils differs significantly from that of calcium carbonates and phosphates. Only four groups of silica-producing organisms are known from the time period under consideration: hexactinellids, demosponges, radiolarians, and, possibly, chrysophytes (Table 23.5). All are still living. Whether this apparent immortality of opal-producing lineages is a chance effect due to the small number of clades involved or whether it has a more profound meaning, the pattern differs considerably from that seen in the carbonate and phosphatic groups. In the latter two, the Cambrian radiation appears to have produced a large number of taxa of which only a few survived. A possible explanation for this pattern is that the radiation of the main branches of opal-producing taxa (protists and sponges) considerably predated that of the other skeleton-forming taxa, so that the phylogenetic bush had already been well-pruned in the Cambrian. This is supported by molecular phylogeny (Chapter 9), but there is no unequivocal evidence in the fossil record for an earlier origin of silica biomineralization. In fact, the present abundance of opal-producing organisms in the seas is largely due to massive Phanerozoic radiations of taxa such as radiolarians, diatoms, and dinoflagellates.

8.6.5 Origin of Phosphate *Versus* Calcite

The diversity of fossils with apatitic skeletons in the earliest Phanerozoic has prompted the idea that this mineral has a particular significance among the earliest skeletal invertebrates (e.g., Rhodes and Bloxam 1971; Lowenstam and Margulis 1980a, b; Cook and Shergold 1984, 1986; Lowenstam and Weiner 1989). A commonly held view is that animals using phosphate, rather than carbonate or silica, diversified first. Phosphate has been stated to be the dominating (Lowenstam and Margulis 1980a; Cook and Shergold 1984, 1986) or even exclusive (Jiang 1985) mineral of the earliest skeletal faunas. This has been explained with reference to the major phosphogenic event at the Precambrian-Cambrian boundary (e.g., Rhodes and Bloxam 1971; Cook and Shergold 1984). The subsequent replacement of phosphates by carbonates is presumed to have occurred both within phylogenetic lineages such as the Ostracoda, Brachiopoda, and Cnidaria, and also by the replacement through extinction of organisms with phosphatic skeletons by organisms with carbonate hard parts. The data presented here on the distribution of clades displaying different biomineralizing habits through time (Fig. 7.7.1), as well as the taxonomic diversity of animals with carbonatic rather than phosphatic skeletons (Fig. 8.6.1; Table 23.5), give little support to these views.

First, the relative amount of phosphate versus carbonate bound up in biominerals in the Cambrian has been exaggerated by sampling biases: (*i*) most early skeletal fossils are of millimetric size, and the chemical extraction of microfossils is more likely to destroy carbonates than phosphates; and (*ii*) unrecognized cases of secondary phosphatization are common (the Cambrian was a time of extensive deposition of phosphatic sediments). As seen in Figure 7.7.1, the number of higher taxa forming their skeletons of calcite or aragonite was approximately twice that for calcium phosphate. Seen at the generic level, the dominance of the calcium carbonates was even stronger – about 4 to 10 times more genera had carbonatic skeletons than phosphatic ones throughout the earliest Cambrian (Fig. 8.6.1). Even without a correction for the sampling biases, these relations between the two major skeletal materials are not radically different from those existing today.

Second, the widespread distribution of phosphate skeletons among different clades in the Early Cambrian as compared to the Recent, is parallelled by that shown by carbonate ones. Among the animal clades shown in Figure 7.7.1, 39% of the carbonate ones ("calcareous tubes" are not included) survive until the present, as compared to 23% of the phosphatic ones (protoconodonts are regarded as chaetognaths with mineralized grasping spines). Both categories include clades that are

Percent of genera

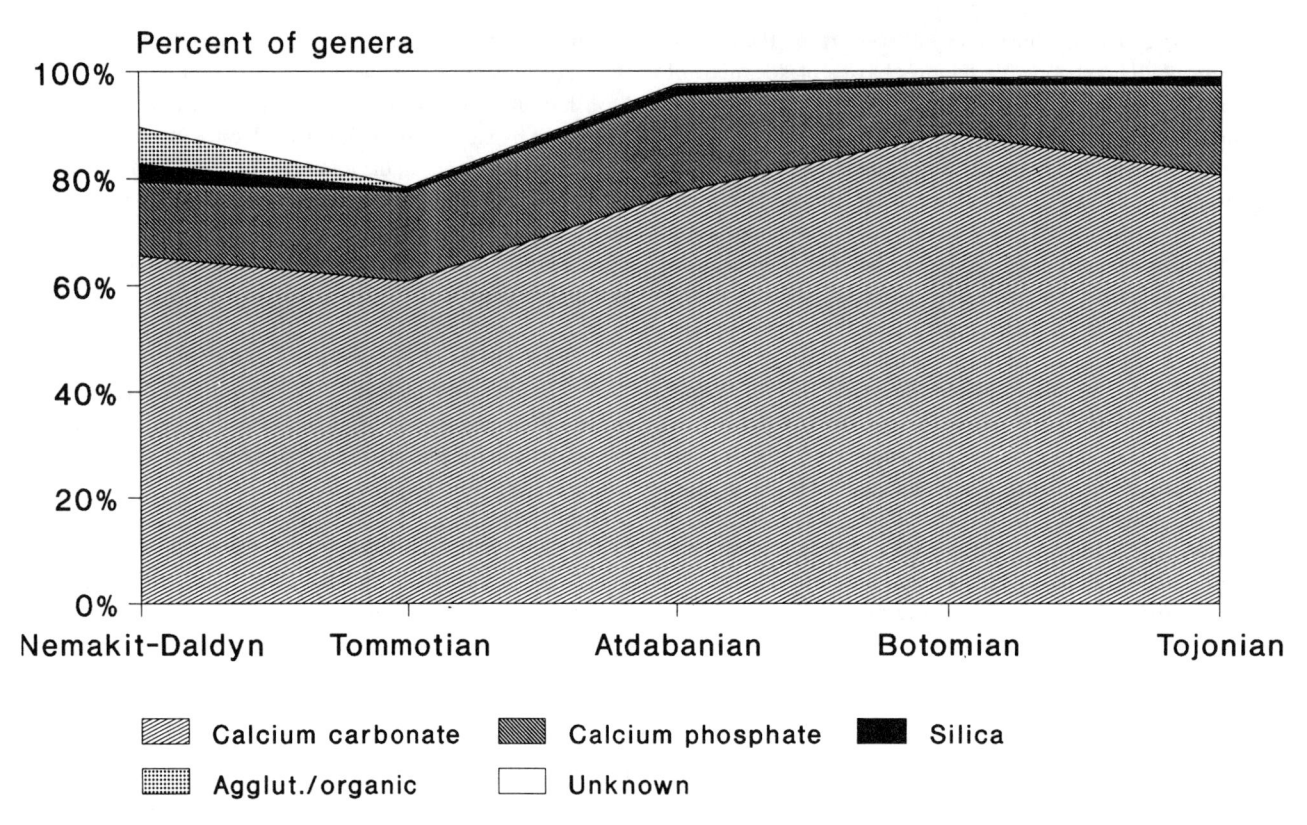

[Chart showing stacked area graph with categories: Calcium carbonate, Calcium phosphate, Silica, Agglut./organic, Unknown; x-axis stages: Nemakit-Daldyn, Tommotian, Atdabanian, Botomian, Tojonian]

Figure 8.6.1. Percentages of genera of Early Cambrian skeletal fossils assigned to skeletal composition. Data as for Figure 8.4.2.

today very successful and diverse. The restriction of phosphate minerals to two major clades today may simply be the result of different evolutionary success of various early lineages. Nothing in the history of vertebrates, however, suggests that their skeletal mineralogy put them at an evolutionary disadvantage, and there is no reason to assume that the shell mineral was the particular factor that decided the survival of each of the early lineages.

Third, there are no convincing examples of a phylogenetic transition from phosphate to carbonate in the history of lineages. For example, a suggested evolutionary succession from phosphate to carbonate hard parts within the cnidarians depends upon the taxonomic decision to place the extinct conulariids within the Cnidaria, a decision questioned, for example, by Oliver (1984) and by Babcock and Feldmann (1986). The proposed secondary origin of carbonate brachiopods from phosphate ones and the derivation of carbonate ostracods from preexisting phosphate forms have the merit of linking groups that are clearly closely related, but the proposal of a mineralogical transition is nevertheless weakly founded. In neither case has a strict phylogenetic analysis been able to demonstrate that the carbonate forms are derived from the phosphate ones. The phosphate-carbonate division may well be original in both groups, as suggested for the brachiopods by Gorjansky and Popov (1986; cf. also Holmer 1989, pp. 67–68).

Another suggestion (e.g., Cook and Shergold 1984, 1986) is that the appearance of phosphatic skeletons was connected to regions of rich phosphate sedimentation. The distribution of these skeletal types in different areas, as seen in Figure 8.4.4, provides some support for this argument. During the period covered by this diagram (Nemakit-Daldyn to Atdabanian), Yunnan and Kazakhstan regions were the sites of extensive phosphate deposition (Yeh et al. 1986; Eganov et al. 1986), and the relative diversity of genera there with phosphatic rather than carbonatic skeletons is relatively high (30 to 70%). The Siberian Platform had low rates of phosphate deposition, and the corresponding percentage is very low (about 10%). However, the highest phosphate/carbonate percentage (more than 80%) occurs on the East European Platform, which does not represent an area of major phosphate deposition during this time.

8.6.6 Origins of Skeletal Biomineralization

Present knowledge of the fossil record confirms that mineralized skeletons of many different kinds and compositions appeared very rapidly in a number of clades at the beginning of the Phanerozoic. Analysis of the precise pattern is still difficult, because in many cases the original mineralogy is insufficiently known and the taxonomic understanding of the various enigmatic early skeletal fossils is incomplete.

The early Phanerozoic radiation cannot be seen only as a radiation of biomineralizing taxa, however. The trace fossil record shows a similar rapid diversification of behavior in non-biomineralizing organisms (Seilacher 1956; Crimes 1987; Section 7.6), and the appearance at the same time of resistant organic structures and agglutinating tubular fossils (Fig. 7.7.1) shows that the key event is not biomineralization as such. To a certain extent, the appearance of mineralized skeletons may be seen as one of many aspects of the early radiation of multicellular organisms. Nevertheless, the apparent absence of skeletal biominerals in the Ediacaran fauna and the nearly simultaneous "skeletalization" of cyanobacteria (notwithstanding reports of earlier sporadic cases of mineralized cyanobacterial sheaths; Klein et al. 1987), algae, heterotrophic protists (foraminiferans and radiolarians), and metazoans, seems to call for specific explanations.

Attempts to explain the appearance of skeletons have often foundered on the lack of universality. For example, models involving calcium availability or regulation (e.g., Lowenstam and Margulis 1980a, b; Degens 1984; Kaz'mierczak et al. 1985; Degens et al. 1985) do not explain the simultaneous appearance of opaline skeletons. Similarly, the proposal that biomineralization began as a phosphate-excreting process in a time of high phosphate availability (Rhodes and Bloxam 1971; Cook and Shergold 1984, 1986) is not consistent with the pattern of appearance of various biominerals as discussed above. Several other models, for example those attempting to explain the utilization of apatite by vertebrates (Pautard 1961; Ruben and Bennett 1987), are even more restricted in their applicability. Models based on increasing pO_2 may have more explanatory power, as an increasing availability of oxygen would make it easier for organisms to form skeletal minerals and proteins and would also make mineralized exoskeletons less of a respiratory disadvantage (Towe 1970). (There is a general but not perfect correlation between distribution of mineralized skeletons and oxygen levels in modern marine faunas; Rhoads and Morse 1971; Savrda et al. 1984; Thomas et al. 1985.)

The synecologically based explanation discussed in Section 8.4 – implying that biomineralization in animals and plants arose primarily in response to selection pressures induced by grazers and predators – may be sufficient to account for the rapid radiation of skeleton-bearing forms. Such an explanation stresses the view that the early evolution of skeletons was a complex event that was integrated with other aspects of the rapid biotic diversification at this period.

8.7

The Precambrian-Cambrian Evolutionary Transition

JERE H. LIPPS   STEFAN BENGTSON   JACK D. FARMER

8.7.1 Introduction

The Precambrian-Cambrian evolutionary events were truly unique in the history of life, making their interpretation difficult as well. The fossil record and theory indicate that these events were complex, including far more than the visible proliferation of metazoan fossils near the base of the Cambrian. The events are now known to have involved all groups of organisms, from protists (like acritarchs) through multicellular plants and animals, non-mineralizing as well as mineralizing organisms, autotrophs as well as heterotrophs. All this implies profound changes in biogeography and trophic relationships. Furthermore, the record indicates that this development of more complex biotas took place over a very long period of time starting perhaps at least 1000 Ma ago and lasting about 450 million years. The three major biotas that can be discerned during this time reflect an evolution of the ecological structure of the whole biosphere over a long geological time.

Any hypothesis to explain the Precambrian-Cambrian history by reference to singular events, such as skeletonization or global transgressions, is at best incomplete. The search for Phanerozoic analogues is highly problematic as the global environment and ecological structure of the biotas were different from Phanerozoic conditions and probably varied significantly at different times. Among the important differences from the Phanerozoic were (*i*) the common occurrence of sediment-binding microbial mats; (*ii*) the absence or near-absence of bioturbating organisms; (*iii*) the absence or near-absence of macrophagous heterotrophs; and (*iv*) the low diversity of phytoplankton. In addition, different global levels of oxygen, carbon dioxide, and calcium, among other possibilities, may have helped to sustain life modes that are poorly or not at all represented today. The general pattern also must have been altered by phenomena such as the worldwide Late Proterozoic glaciation and the break-up of the Proterozoic supercontinent (Chapters 2, 4, 12).

All of these factors combine to indicate that testing of hypotheses will be difficult at best. In addition, modern theory does not easily explain many of the ecological structures observed even in today's biotas. Thus, explanations of the Precambrian-Cambrian events described here may remain unsatisfactory and certainly simplistic for some time. Nevertheless, the hypotheses that have been proposed to date can be measured against the known fossil record, and that measure eliminates many of them. In the following discussion, we separate the questions of origin and subsequent diversification of multicellular organisms, and test applicable hypotheses against the record and the theory.

8.7.2 Hypotheses on the Origin and Early Evolution of Metazoans and Metaphytes

None of the previously proposed hypotheses based on the fossil record (Table 8.1.1) specifically addresses the origin of multicellular organisms. Such hypotheses are based on biological observations of modern organisms. A large portion of the history of early multicellular life is unknown in the rock record because plausible initial and intermediate forms are missing.

The first biota that might include multicellular organisms, the pre-Ediacaran *Tawuia* assemblage, contains simple possible metaphytes resembling also in some respects aggregates of prokaryotes. These observation may not be irreconcilable because the origin of metaphytes is traced through molecular studies of 18S RNA to cyanobacteria. Molecular data indicate that the separation of the major branches of metaphytic algae was part of the same radiation that involved the separation and early branching of plants and metazoans. Yet metaphytes remained sparse in the Proterozoic fossil record, and althouth they diversified somewhat in the 900 to 700 Ma interval, they probably were not a major component of marine ecosystems until the Ediacaran or Cambrian, much like the metazoans.

The Ediacaran assemblages suggest that at this time multicellular heterotrophs were already reasonably diversified in morphology and life habits, indicating an unknown prehistory. The next fossil assemblage (Early Cambrian) confirms that diversification was on-going and still did not leave a record. The unrecorded prehistory of metazoans conceals their origin(s), body-plan diversification, and ecological radiation. Molecular-clock methods of determining the age of divergence of metazoans, subject as they are to many uncertainties, indicate that this prehistory extends back in time to about 1000 Ma (Runnegar 1982b; Chapter 9), leaving some 400 million years for the earliest evolution, of which we have no direct fossil evidence. This date corresponds to indirect evidence from stromatolite morphologies that grazing organisms may have been active from about 1000 Ma (see Sections 6.2 and 11.3). Not all groups that appear in either the Ediacaran or Early Cambrian biotas necessarily extend far back in time, and the development of these biotas may have happened soon before or at the time they are recorded in the fossil record. But the Metazoa are most likely much older than their first fossils, and their poor early record is due to their being soft and tiny (see Section 8.5).

The Precambrian-Cambrian fossil record has been interpreted in two ways: (*i*) that the unrecorded history of metazoans and metaphytes conceals a long period of diversification that resulted in the biotas observed in the fossil record (Walcott 1910; Rutten 1971; Glaessner 1972a, 1984; Durham 1971, 1978b); or (*ii*) that the fossil occurrences accurately record the diversification events (Cloud 1948, 1968; Schindewolf 1955). Both views are probably correct – the metazoans and metaphytes *did* have a long unrecorded history, and their known fossil record *does* reveal major events in that history. There is no evidence that the diversification of metazoans and metaphytes was gradual or exponential (Sepkoski 1978; Stanley 1976a, b; Section 11.4).

We interpret the fossil record of the Late Proterozoic-Early Cambrian to be truly reflective of events in metazoan-metaphyte evolution, and not just a partial record of gradual evolution. This interpretation is based on the observations that the three biotas were distinct, the latter two appeared in a geologically abrupt way, the Ediacaran biota appears to have become extinct, and there are few connections between them. The appearance of these biotas in the fossil record implies that once multicellular organisms originated, their subsequent development was marked by separate events of major importance in their evolution.

The fossil record does not, however, provide any evidence of the origin of metazoans or of metaphytes. They are far too well developed at their first fossil appearance to support any discussion of their origins. Such inferences must come from non-paleontological evidence.

8.7.3 Evaluating the Hypotheses

Any hypotheses proposed to account for early metazoan-metaphyte evolution must be evaluated in terms of these events, either separately or jointly. The previously proposed hypotheses, listed in Table 8.1.1, are discussed below in light of the preceding Sections, and a general conclusion follows. Some of them can be eliminated for one of two reasons: they may be incompatible with known facts, or they may invoke a mechanism that is plausible but in the context trivial, i.e., it fails to identify any limiting conditions. Some of the hypotheses may again be both plausible and non-trivial, but currently untestable. The latter hypotheses should not be discarded, but retained as a part of one or several alternative scenarios, in the hope that we shall eventually find means of testing them.

8.7.3A Biotic Factors
Size. The suggestion that Precambrian metazoans were not visible in the fossil record because they were too small is an old one. To some extent it is unavoidable and even trivial: the evolution from protist to metazoan must have involved a size increase, an inescapable effect of an evolutionary increase in variance (Stanley 1973b; Gould 1988). But early metazoans need not necessarily have been small, and evolutionary innovation may even be favored by the reproductive strategies associated with centimeter rather than millimeter size (Olive 1985). Section 8.3 presents the case for the ancestors of the Cambrian metazoans being of meiofaunal size, but as argued in that Section the taphonomic bias against preservation of meiofauna is so strong that the fossil record currently cannot be used to support any general hypothesis about Precambrian meiofauna. Comparative studies of well-preserved (e.g., in phosphorites and cherts) Phanerozoic and Proterozoic sediments with regard to metazoan-induced fabric, meiofaunal body fossils, and presence of spicules and cuticles of the types characteristic of meiofauna, may be a fruitful avenue of future research. At present, however, the size argument must be classified as plausible but untestable.

If the Metazoa had a long Precambrian history as small creatures without fossilization potential, as suggested by some bodies of evidence, the sudden size increase at the Proterozoic-Phanerozoic transition needs an explanation. The hypothesis that attainment of sedentary habits by planktic or meiofaunal predecessors led to larger sizes can be eliminated on the basis that all groups, from protists to metazoans, both free-living and sedentary, benthic and planktic, underwent changes at this time. Furthermore, the Ediacaran biota includes medusoids, trace fossils, and some others that were not sedentary. More acceptable mechanisms for a size increase would be increasing O_2 levels (see below) and selective advantage in predator-prey interactions (Section 8.2). But a special mechanism is not necessary to explain the sudden size increase of the metazoans: it may be seen as one aspect of a rapid diversification event starting with small organisms.

Skeleton. The idea that the "Cambrian explosion" is essentially an event of biomineralization, an "explosion" of fossils rather than of organisms, can be effectively disproven through evidence from non-mineralized fossils and trace fossils (Chapter 7, and Sections 8.3, 8.6). Hypotheses to explain the almost simultaneous appearance of skeletons in the geological record through reference to an abiotic trigger, such as increasing calcium levels, fail as general theories, as do those based on the evolution of calcium-regulating mechanisms (Section 8.6). Whereas the physiological regulation of various ions is a prerequisite for controlled biomineralization, it is unlikely to be a trigger mechanism for biomineralization.

The hypothesis that skeletonization followed from increasing size is also flawed by the simultaneous diversification in many other marine groups than metazoans, from small radiolarians and foraminiferans to larger benthic algae and phytoplankton. There is no clear relationship between size and skeletalization in the fossil record.

As discussed in Section 8.6, a synecologically based hypothesis for the appearance of skeletons seems best to account for their multifarious nature. This allows for various functions of hard skeletons, but in concordance with the proposed major importance of predators and herbivores in driving evolutionary diversification (Section 8.4), prime weight is given to the protective functions of skeletons. This latter suggestion, branded as naive by Glaessner (1984, p. 174), is supported by mounting evidence for predation in Cambrian ecosystems (cf. Section 8.4). In this view, skeletonization is only one aspect of the major ecology-driven diversification of the total biota (cf. Stanley 1976b).

Migration. This old hypothesis simply accounts for the rather sudden appearance of certain groups (originally the Early Cambrian biota, but applicable to any of the three biotas now known). The hypothesis suggests that a biota resided in habitats that failed to leave a fossil record, for example, the intertidal zones, and suddenly moved into fossilizable habitats. It does not account for the diversification seen in marine phytoplankton for which a long record exists. The fossil record shows that the marine habitats have been inhabited continuously for over 1000 million years. Although regional migration and facies shifts are likely to account for the sharp lower boundaries of many local fossil assemblages, the idea that the abruptness of the *global* appearance of any of the three early assemblages is facies-related can now be discounted as a serious consideration.

Ecology. The hypothesis that the advent of predators and herbivores prompted the general diversification is attractive, because trophic interactions have powerful selective effects that can be transferred through the food chain and thus affect many groups. Trophic proliferation, in a general sense, is supported by the fossil record. The pre-Ediacaran and Ediacaran biotas appear not to have contained active predators, although croppers working the benthic substratum, and suspension feeders collecting plankton, radiated. The Early Cambrian biota was more diversified trophically, including representatives at all levels of the food chain. As discussed in Section 8.4, there is good evidence that macrophagous predators and herbivores were present in the earliest Cambrian ecosystems. The hypothesis needs additional development, however, by the further documentation of the predators and croppers so that the trophic interactions might be traced.

Although trophic interactions are clearly necessary for the diversification observed in all types of organisms, it is far from clear whether this was a decisive factor for the timing of the diversification at the beginning of the Cambrian. Such a hypothesis would require the trophic linkage between a wide variety of marine groups occupying widely divergent habitats in order to account for the simultaneous changes in diversity of many different kinds of organisms. Predators/croppers would have had to appear nearly simultaneously in a number of habitats not normally linked by predation or cropping (i.e., open ocean and shallow neritic zone). Thus, trophic interactions provide a necessary framework for biological diversification, but with regard to the events around the Precambrian-Cambrian boundary a hypothesis based solely on trophic interactions is in the category of plausible but trivial.

A common observation with regard to the early metazoans is that they radiated into what was essentially empty ecospace. Such a situation may well have been the result of the flooding of continental shelves in the late Precambrian. An additional hypothesis, that the extinction of a late Precambrian biota paved way for the subsequent Precambrian-Cambrian radiation, is more problematic in that the known diversity of the late Precambrian biota is so much lower than that of even the earliest Cambrian assemblages (Section 11.4). It should be stressed that ecospace is not primarily defined by the physical environment, but that evolving organisms define their own ecological niches and create new ones through biotic interactions.

Genetics. Valentine and Campbell (1975) pointed to the importance of regulatory genes in the rapid development of new body plans, but did not ascribe the Cambrian evolutionary event to the evolution of gene regulation. If the changes in the Precambrian-Cambrian biotas resulted from the appearance of new gene structures, then these must have appeared in all groups from protists through metazoans at about the same time. Such a hypothesis is thus highly unlikely to account for the appearance and diversification of the Cambrian biota, and with regard to the Ediacaran and pre-Ediacaran biotas it is untestable. The evolution of an integrated system of regulatory genes is obviously a prerequisite for the evolution of complex organisms, but there is no indication that absence of a such was ever a limiting factor that delayed the appearance of multicellular organisms or their diversification.

On the other hand, the subsequent evolution of developmental pathways, regulated by a genome that was ever increasing in complexity, may be seen as a factor that restricted evolutionary variability at subsequently higher levels of organization. This could account for the observation that the evolution of features such as metazoan body plans (as defined in Section 8.5) appears to have a heyday restricted to a short initial period, and that subsequently no new types of multicellular organisms built from unicellular "scratch" are known to have evolved.

Sexuality would confer a strong selective advantage on organisms that attained it, and would most likely promote diversification (Schopf et al. 1973b; Stanley 1975a). Schopf et al. (1973b) regarded the advent of meiosis and the switchover from a haploid-dominant to a diploid-dominant life cycle as key innovations that dramatically increased genetic variability within populations, thus triggering an escalation in evolutionary rates (see Sections 5.5, 11.3 and 13.3).

The metazoans and metaphytes of the Ediacaran and Cambrian biotas probably were sexual for a long prior period, because sexual reproduction appears to be a heritable character of ancient origin. Furthermore, acritarchs, although they underwent significant changes in diversity and morphology in the Late Proterozoic (Section 5.5), had preexisted since the mid-Proterozoic, possibly with a sexual stage in their life cycle. If sexuality played a role in metazoan and metaphyte evolution, it occurred in the earlier unrecorded history of the groups. As with the evolution of regulatory genes, sexuality is a prerequisite for the later evolutionary events, but there is currently no way of testing the hypothesis that it was a triggering factor.

8.7.3B Abiotic Factors

Oxygen. The idea that an increase of oxygen in the atmosphere and oceans allowed the development and radiation of metazoans is an attractive one based on the requirements for collagen, metabolic processes of larger organisms, and the geological record. In particular, collagen, which is found only in the Metazoa, is essential in providing a skeletal framework and is considered prerequisite for the attainment of large body size and mobility in animals. Towe (1970) has pointed out that collagen requires oxygen for its synthesis, and suggests that the explosive diversification of megascopic animal life may have been triggered by the buildup of oxygen to an undetermined critical threshold required for collagen synthesis.

The geological record shows that oxygen was probably absent in the early atmosphere and that until various elements in the primitive oceans were oxidized, free O_2 could not accumulate in the oceans and atmosphere, without which metazoans could not function. The chief problem with the hypothesis of metazoan diversification being triggered by O_2 is that there is no direct method of determining what the levels of oxygen were in either the oceans or the atmosphere at any particular time (see Section 4.6). In general, three lines of indirect evidence are used to circumscribe the timing of an oxic ocean – the termination of the deposition of banded iron-formations about 1.8 Ga; the occurrence of the Ediacarian fossil *Dickinsonia*, surmised by Runnegar (1982C) to have had a relatively low oxygen requirement based on its thin, flattened morphology; and the burial of abundant carbon between 900 and 700 Ma (see Chapters 2 and 3).

The geological and evolutionary events and rising oxygen levels inferred by Runnegar (1982C) or, in a more general way, by Cloud (1968), are attractive because they make a logical sequence. However, there are alternative explanations for each of these observations; for example, that banded iron-formations formed under unique oceanographic conditions (i.e., upwelling) and that *Dickinsonia* was thin and flat for reasons other than oxygen uptake (i.e., passive feeding on suspended food particles or dissolved organic matter, or possession of symbionts). This evidence and other secondary evidence has been debated extensively in the past. At this time, however, no direct evidence defines when oxic conditions became prevalent; current opinions suggest that they arose some time between 2000 and 500 Ma. The abundant burial of carbon in sediments between 900 and 700 Ma may be strong evidence for the production of high levels of oxygen (Chapter 3).

Clearly, metazoans require free oxygen, and so it must be a prerequisite for their evolution although oxic conditions need not have occurred simultaneously with that evolution. Yet other organisms, the acritarchs, that predate the appearance of metazoans by almost a billion years also show significant changes near the Precambrian-Cambrian boundary. Acritarchs, generally assumed to be the cysts of phytoplankton (Section 5.5), may themselves be the best indication of the time of the appearance of significant O_2 because photosynthesis is its only probable source. Phytoplankton, under eutrophic oceanographic conditions, could produce enough oxygen to attain the present atmospheric level in a very short geologic time, and even under poor oceanographic conditions would probably require very little geologic time. Cyanobacteria are another oxygen source, but they first appeared much earlier in the fossil record, and making them the producers of O_2 in the oceans could push the timing for oxic oceans and atmospheres even further back in time (for discussion, see Sections 4.4 and 4.6).

Another problem with O_2 levels as an evolutionary mechanism is that diversity is not O_2-dependent, and there is no reason that diversity should respond to O_2 levels. It could have acted as a trigger mechanism, allowing preexisting small and diverse metazoans to become large and skeletonized enough to fossilize. The case for increasing oxygen levels as a trigger for the appearance of large metazoans and their subsequent radiation remains circumstantial. There is no direct means to determine oxygen levels at any specific geologic time, and until such a method is found, oxygen levels cannot be constrained. In any case, the diversification of life forms and trophic strategies can not be explained by oxygen levels, so that even if oxygen is a trigger, the proximal causes of diversification must be sought elsewhere. As discussed above, these causes are most likely found in ecological interactions.

Calcium. Increase in calcium levels has been invoked to explain the sudden skeletonization of invertebrates in the earliest Cambrian. While calcium must be available in sea water to provide for some kinds of skeletons, skeletonization involved a wide variety of materials. Siliceous, organic, and agglutinated skeletons in particular probably required no increase in calcium levels. In addition, some aspects of the Cambrian diversification involved groups that predated this event (acritarchs, for example), and that would have been fully functional both before and after the event. As an all-encompassing hypothesis, an increase in calcium levels can be rejected on the basis of the phytoplankton and skeletal diversity record.

Phosphorus. As discussed in Section 8.6, the early predominance of phosphatic skeletons has been greatly exaggerated in the literature. Hypotheses to suggest that skeletonization started as a detoxification response to increasing levels of phosphorus in the seas are hardly viable. As in the case of calcium levels, changes in the levels of phosphates will not account for the skeletal diversification, radiation of the entire marine biota, or the changes recorded among the phytoplankton.

But phosphorus, essential for all life, is a limiting element in today's biosphere and may have been so during substantial periods of earth history. Against this background, a general increase in the availability of phosphorus may boost biomass and thereby diversification. This may have been the situation associated with the Late Proterozoic-Early Cambrian phosphogenic event (Cook and Shergold 1986). Although no direct measure of ancient biomass exists, the dramatic increase in various evidences for biotic activity around this time is consistent with a significant rise in biomass. The main problem with this hypothesis is that phosphogenesis is generally associated with high organic productivity, and the causal relationship is generally very complex. The concentration of phosphorus can be both a cause *for* and a result *of* organic productivity. It is thus very difficult to devise critical tests for the hypothesis.

Glaciation. Continental glaciation induces many oceanographic and atmospheric changes that could affect the marine biota, but they remain difficult or impossible to test. In particular, periods of glaciation are usually associated with increased oceanographic circulation because the thermal and salinity gradients that drive currents are intensified. But the Late Proterozoic glaciations appear to have been equatorial (see Chapter 12), and circulation patterns are impossible to infer without evidence from higher latitude regions. With equatorial glaciation on continents arrayed along the equator, circulation would most likely include currents divergent away from the continents, thus promoting upwelling and eutrophication of the continental shelf waters and perhaps encouraging new feeding strategies and life modes among organisms living there. Such a model is complex and cannot now be tested, although it remains a viable alternative.

Land/Sea Patterns. Sea level changes and continental reconfigurations have been favored explanations for Phanerozoic extinction and radiation events because of the expansion or retraction of habitable area, or because of climatic changes induced by more, or less, sea surface area. Diversity has little or no correlation with large marine regions (although it might on small islands), but rather with habitat diversity. If an area contains a variety of habitats, diversity can be high; fewer habitats, lower diversity. Climates can be ameliorated or intensified by proximity of oceans and their extent.

The breakup of a supercontinent in the Late Proterozoic (Chapter 12) would cause partitioning of sea and atmosphere, resulting in a more heterogeneous world. At the same time, the extension of shallow-sea habitats would increase as the water displaced by active ridges flooded old and new continental shelves. In these kinds of models, marine habitats, nutrient supply, productivity, and trophic resource availability, among others, would change significantly, thus affecting the marine biota. A number of observations in the geological record from this time (sea level changes, sedimentary patterns, isotopic data, igneous activity, continental configuration, and climatic and oceanographic changes), can be brought to bear on such a model (see discussions by, e.g., Brasier 1979, 1982, 1985, in press; Conway Morris 1987; and Knoll 1987, 1989). Testing is difficult, however, because no single factor stands alone in promoting evolutionary change.

The chief problem with these hypotheses is that they explain everything, yet explain nothing. The changes in continental configuration would surely have effects on the marine biota, but the cause and effects remain unspecified and cannot be tested by data or comparison to modern ecologic situations. As a result, these hypotheses are scenarios, like many paleontologic explanations, that fail rigorous scientific testing but which could in fact be correct.

Cosmic Causes. No convincing case has as yet been made for an extraterrestrial trigger of the Late Proterozoic-Cambrian evolutionary events. Hsü's et al. (1985) proposal that the "Cambrian explosion" was triggered by a major impact event was based on inconclusive geochemical data from a "wrong" stratigraphical horizon (Awramik 1986; Conway Morris and Bengtson 1986). As discussed above, there is no lack of known physical events that are likely to have influenced evolution, and there seems currently to be no need to invoke unknown ones.

8.7.4 What Type of Hypothesis are We Looking For?

As discussed in the preceding Sections, there are a number of conditions necessary for the evolution of the Phanerozoic biosphere, and most of them have been addressed by more or less elaborate hypotheses. But these are difficult or impossible to test, as they deal with very complex relationships where a number of factors combine to influence the course of evolution. A testable hypothesis would identify factors that have been limiting for evolution at any one period of time.

For the concordant radiation it is not necessary to assume an extrinsic trigger: an ecological event such as the introduction of croppers/predators is sufficient, and diversification thereafter drives itself. This diversification would affect all organisms, also prokaryotes, because of new types of interactions (parasitism, etc.) within an ever-changing biota. The astounding speed of the process is not remarkable in light of comparable incidents of radiation in the Phanerozoic; it can be explained by diversification into ecological "empty space" by organisms with suitable genetic mechanisms (sexuality, regulatory genes, heterochronic development) for rapid reorganization of genome and body plan.

But the delay of the event by hundreds of millions of years from the probable origin of multicellularity can be explained by an abiotic limiting factor. As discussed above, the prime candidate for this is atmospheric oxygen levels. As pointed out by Knoll (1987), the Late Proterozoic record of carbon isotopes suggests a period of massive burial of organic carbon that would have released major amounts of O_2 into the atmosphere. This could be due to the presence of narrow ocean basins with anoxic bottom water (Knoll 1987), or to an increased productivity of phytoplankton that initially was not countered by increasing consumption, because of restricted benthic life and the absence of bioturbation.

8.8

Unsolved Problems and Future Research Directions

STEFAN BENGTSON

The last decades have seen enormous advances with regard to elucidating the biological and geological events leading from the Proterozoic to the Phanerozoic world. Nevertheless, understanding of the causative processes are in many ways as speculative as they were in the beginning of the century. We have identified further possible mechanisms and eliminated a few others, but on the whole there is too little constraint on the hypotheses for them to be really informative. The following questions stand out as important for further clarification of the problem:

(1) What are the true ages and branching sequences of the main lineages of multicellular organisms? The fossil record may help to calibrate the answer, but the solution is likely to come from molecular biology.
(2) What was the mode of life of the first metazoans: planktic, epibenthic, meiobenthic? This is a question that has plagued zoology just as much as the question of the phylogeny of the main metazoan branches, and there is no solution in sight from the zoological side. A determined search for meiobenthic fossils (spicules, cuticles, etc.) in suitable facies of both Proterozoic and Phanerozoic age may help to answer at least the last possibility in the affirmative or negative.
(3) What is the relationship between phytoplanktonic, benthic algal, and benthic metazoan diversification? There still remains a lot of work on the taxonomy of all groups and on their biostratigraphical correlation in order to find out whether the diversification patterns of these categories mirror each other, if one precedes the others, or if they are essentially unconnected. Diversity data must also be seen in relation to facies types and taphonomic biases.
(4) Is there any correlation in the Late Proterozoic between biotic diversity and (*i*) sea level, (*ii*) extent of shelf seas, (*iii*) splitting up of shelf seas, (*iv*) temperature, and (*v*) nutrient supply? Physical data that would pertain to these questions are likely to come from further sedimentologic, tectonic, stratigraphic, and geochemical work in this interval.

Of all the factors necessary for the successful evolution of the Phanerozoic biota during the Late Proterozoic, which factors were limiting, at which time, and in what way? Which were independent of the others, and which were interconnected? These questions are central to the test of any specific model, but they are only partly open to analysis.

8.9

Summary and Conclusions

JERE H. LIPPS   STEFAN BENGTSON

We conclude that no hypothesis yet proposed can account for the Late Proterozoic-Early Cambrian biotic events clearly and unambiguously. All hypotheses are flawed by being so general that they are untestable in a rigorous fashion or they fail on the basis of documented evidence. We believe that these events were complex, like life and the earth themselves, and that therefore the underlying causes were complexly interrelated as well.

The fossil biotas of the Late Proterozoic and Early Cambrian do indicate major ecological restructuring in the marine biosphere. The first multicellular biota contains only probable metaphytes such as *Tawuia* and was, like the preexisting prokaryote and protist assemblages, dominantly photosynthetic. No evidence exists that these organisms were eaten by any animal large enough to leave marks, although bacterial decay and protistan scavenging no doubt occurred. The Ediacaran biota shows utilization of trophic resources in the water column and on the surface of the sea floor. Again, no evidence exists to support interpretations that predation was significant on the fauna. The Early Cambrian biota, however, shows a complete range of trophic types. In a general way, these developments indicate that trophic specialization was an important aspect of Late Proterozoic-Early Cambrian evolution. It may not have been the key to the events reflected in the fossil record, but trophic interactions are strongly subject to selection and the fossil record indicates that they played an important role. Why the marine biota should have become more trophically complex is yet another question. One explanation might be that once multicellular organisms, especially metazoans, evolved, evolution simply proceeded toward more complexity, of which trophic specialization is a major component. Contrarily, trophic interactions may have created selection pressures that led to further diversification. Trophic diversification might be related to increased primary productivity, for which evidence exists in the fossil record at about the right time. This increased productivity would have resulted from changed oceanographic conditions that promoted the redistribution of nutrients from deeper water to the euphotic zone, presumably due to increased upwelling intensity.

The early evolution of metazoans and metaphytes seems linked to a complex of geologic, oceanographic, and biologic factors, none of which can be clearly isolated as a primary causal mechanism. Historical developments involve sequences of events, each influencing later ones. Narrowly constructed hypotheses are therefore probably unlikely to account satisfactorily for these events even though they may be rigorously testable. The real explanation lies buried in the complexity of the biologic and geologic systems, and may never be clearly identified because it involves many interacting factors. Explanations involving such systems are usually scenarios that cannot be tested with scientific rigor; that, however, does not invalidate their possible truth. It means that we may never be able to test models. The search should go on with imagination and originality!

9
Molecular Phylogenetics, Molecular Paleontology, and the Proterozoic Fossil Record

BRUCE N. RUNNEGAR   DAVID J. CHAPMAN   WALTER M. FITCH

9.1

Introduction

BRUCE N. RUNNEGAR

Twenty years ago, living microorganisms were named and classified on the basis of their morphology and physiology. This worked well for closely related taxa, but the higher level relationships between very distant lineages remained obscure because too few homologous characters were available for analysis. This situation changed dramatically in the 1980s following the invention of rapid and efficient methods for sequencing proteins and nucleic acids (both DNA and RNA). Suddenly, large numbers of homologous characters (single amino acids or single nucleotides) became available for comparison between closely and distantly related taxa. A concomitant development of computer-based methods for analyzing the new data (Section 9.2) has led to the production of "phylogenetic trees" which are designed to display, in a graphical way, the genealogical relationships between taxa at the tips of their branches. The same molecular methods have now been applied to representatives of most kinds of living organisms and they complement formal cladistic analyses based primarily upon shared-derived (synapomorphic) morphological characters.

These new methods have enabled a second approach to historical biology. The molecular comparisons may, in principle, produce information about the history of organisms with no fossil record. For example, colorful, tropical, soft-bodied molluscs known as nudibranchs are unknown as fossils. When did they first evolve? Molecular methods can provide an answer.

For Precambrian paleobiology the applications are obvious. However, the proportion of species with living descendants diminishes as one moves further backwards in time. It therefore becomes vital to integrate the historical information that is available in the chromosomes of living organisms with the paleontological data available from the stratigraphic record. In particular, it becomes critical to attempt to identify and date in years (Ma or Ga) significant dichotomies in the history of life (e.g., the origin of the first terrestrial tetrapod from a lobe-finned fish). These key dichotomies are called "divergences," their ages are known as "divergence times," and they may be "shallow" or "deep" depending on whether they occurred relatively recently or in the distant past.

The idea of using differences in the amino acid sequences of related proteins from different organisms as an independent "molecular clock" was suggested by Zuckerkandl and Pauling (1962), and was profitably exploited by early work on common water-soluble proteins such as cytochomes c and the globins (Dickerson 1971). The hypothesis remains controversial, even though it is widely agreed that sequence differences may be used to construct believable phylogenetic trees. In fact, these two uses of the sequence data are not entirely separable because the depth of divergences measured from the present contains an implication of age as well as relationship.

A quite different approach to the use of organic molecules in historical biology falls under the umbrella of "molecular paleontology" (Section 3.3). The precise characterization of distinctive organic molecules from ancient sedimentary rocks was revolutionized by the development of gas chromatography-mass spectroscopy (GC-MS) in the 1960s. By 1967 these chemical fossils were already becoming known as "biological markers" – a term that is now usually contracted to *biomarkers* — and their great significance to the petroleum industry was just beginning to be appreciated (Eglinton and Calvin 1967). Thirty years later, study of the organic geochemistry of sedimentary rocks has become a major enterprise (Philp and Lewis 1987; Summons 1988) and the importance of this central field of molecular paleontology continues to grow as analytical techniques improve and the taxonomic and environmental significance of individual biomarkers becomes well understood (Mackenzie et al 1982; Ourisson et al. 1979, 1982; Brassell et al. 1986; Summons 1988). In particular, the recent introduction of metastable reaction monitoring (MRM) and analogous methods as a refinement of GC-MS has allowed trace amounts of distinctive biomarkers to be unambiguously identified in rocks as old as Middle Proterozoic (Section 3.3).

The ultimate goal these techniques is an understanding of the evolution of life at both the organismal and the molecular levels. By using the genealogical information gleaned from universally distributed homologous molecules such as the ribosomal RNAs (rRNAs), it has been possible to derive a phylogenetic classification of all living organisms (Section 9.3). In a similar way, the sequences of homologous genes may be used to explore the evolutionary history of gene families and their products. Finally, if the approximate times of origin of some critical molecules and biochemical pathways can be deduced from the fossil record, these data may be used to relate the evolution of the biosphere to the environment it inhabited (Schopf 1978).

9.2
Methods of Molecular Phylogenetics
WALTER M. FITCH

9.2.1 Introduction

There are many kinds of biochemical data that can be used to infer phylogenetic relationships of organisms, and there are many methods that can be applied in that effort. This Section gives a brief description of those methods that are appropriate for organisms that last shared a common ancestor in the Precambrian.

The first thing to note is that few biochemicals survive in rocks for more than 600 Ma (Chapter 3); therefore, this topic is constrained by the need to obtain data from living organisms. In addition, some biochemical characters may change too slowly or too rapidly for the task. Others contain too little information. A lipid molecule present in *all* eukaryotes is of no value in estimating relationships *within* the Eukaryota. Neither are pigments that differ from one species to the next. In addition, one wants to consider many characters so as to avoid the need to rely on one or a few pieces of evidence. Consequently, the only practical sources of biochemical/genetic information for Precambrian studies are protein and nucleic acid sequences.

There are, however, many methods for obtaining information about such sequences. They include differential migration in gels, antibody cross-reactivity, amino acid compositions, DNA-DNA hybridization, gene mapping, and direct sequencing. This Section discusses only sequenced proteins and nucleic acids on the grounds that they appear to be the only data currently available that are useful for lineages of organisms whose most recent shared ancestors lived in the Precambrian. Other methods – such as maps of gene order or chromosome structure – may become useful in the future.

9.2.2 Sequence Alignment of Homologous Amino Acids or Nucleotides

After the sequences are obtained from suitable organisms, they must be correctly aligned so that homologous amino acids/nucleotides are opposite one another. This may prove to be impossible if they are not derived from a common ancestor (i.e., if they are non-homologous) or if they have evolved so far that they no longer show any reliable trace of their ancestry. Obtaining an acceptable alignment may be difficult because evolution frequently inserts into or deletes from the gene a series of nucleotides. Such insertions or deletions require a gap in one or more of the sequences being aligned. To prevent indiscriminate placement of gaps, a penalty must be paid in any algorithm used to generate an alignment (Needleman and Wunsch 1970). The penalty will be different for different kinds of comparisons (Fitch and Smith 1983). As the number of sequences and their evolutionary distance increases, the selection of sites for gaps becomes more and more difficult. Ultimately, the prevalence of gaps may deprive the aligned sequences of any value in determining phylogenetic relationships.

Having achieved an acceptable alignment, it should be true that the *n*-th nucleotide or amino acid in each sequence derives from the same ancestral nucleotide or set of three nucleotides (codon), that is, that the positions are *homologously* aligned. Thus, each position in the sequence is a single *character* and its *character state* is the nucleotide (adenine, guanine, etc.) or amino acid (alanine, glycine, etc.) that occupies that position in any given sequence.

The set of aligned sequences may be analyzed by any of the methods discussed below, but one caveat needs to be made. Even if the data and the method were good enough to infer the correct genealogical relationships of the sequences, that genealogy need not match the correct phylogeny of the organisms from which the proteins of nucleic acids were isolated. This is because genes have a way of duplicating themselves. Thus, the two common components of vertebrate red blood cells, the *a* and *b* hemoglobins, are only two of a larger family of homologous proteins. A phylogeny of the *a* or *b* hemoglobins may well represent accurately the phylogeny of the organisms sampled since the alpha hemoglobins represent a subset of homologous sequences that are said to be *orthologous* because there is an exact correspondence between the history of the gene and the history of the organism. The *a* and *b* hemoglobins are *paralogous* because, after gene duplication, both genes and their products evolve in parallel in the genome. It is therefore necessary to distinguish between orthologous and paralogous genes when making sequence comparisons. Errors will be made if paralogous genes are used as if they were orthologous.

9.2.3 Methods for Inferring the Ancestral Relationships Between Sequences

Given an alignment, one must choose a method for inferring the ancestral relationships between sequences. Available methods fall into two classes called distance procedures and parsimony procedures.

9.2.3A Distance Procedures

Distance procedures involve making a table for pairwise comparison of distances. For aligned orthologous sequences, one examines every possible pair of sequences and counts the number of mismatches. These mismatches may each be given a weight of one or some other value. If they are nucleotide sequences, they usually are weighted one, but if they are amino acid sequences they may be weighted according to the number of nucleotides that must differ in their underlying codons or according to the frequencies with which they have been inferred to replace one another during evolution. This (possibly weighted) count of differences is the observed *distance*.

When the distance measure is a simple count of the number of differences, one may make a Poisson correction – assuming that all positions in the sequence are always equally likely to be the locus of the next change – to convert the number of differences into the number of amino acid replacements (or nucleotide substitutions) that one would have expected to occur in order to see the number of differences observed (Cantor and Jukes 1969). If the differences occur at less than twenty percent of the sites, the corrections are usually negligible. The distinction between differences, replacements (or substitutions), and mutations should be kept in mind. Although a mutation must underlie each of the other events, replacements and substitutions represent only a tiny fraction of the mutations that have occurred in related lineages.

Distance data are used in the the following way. Starting with a distance matrix, one constructs from it a tree that serves as a hypothesis to describe the evolutionary relationships of the taxa under consideration. There are a number of ways to approach this problem (Sokal and Sneath 1973) but a principal procedure involves an operation to find the pair of sequences that is least distant from each other and hence the closest relatives. The joined pair is then treated as a unit and the process is repeated. With each interation, the number of unjoined units is reduced by one until all of the sequences are united into a single tree.

The next step is to assign distances to the branches of the tree and to sum the lengths of all of the branches connecting each pair of sequences to obtain phyletic distances. Obviously, there should be some similarity between phyletic and observed distances, and the degree of similarity becomes the criterion for testing the goodness-of-fit of the hypothesized tree to the observed distances.

An alternative procedure for assigning branch lengths involves minimization of the sum of the lengths of all branches of the tree – the *tree length* – with the proviso that no phyletic distance can be less than the observed distance. This constraint results from the observation that the actual number of changes cannot be less than the observed number of differences.

All reasonable methods of analysis allow for the exploration of alternative trees. This is because no practical algorithm can be guaranteed to find the "best" tree on the first attempt. It should be noted that there are more than one million unrooted trees for ten sequences and that their number increases by more than an order of magnitude with each additional sequence. So one cannot, at present, explore them all.

The exploration of alternative trees requires that there be a criterion to decide which is best. A criterion appropriate to the first method of assigning branch lengths is that the tree that is best has the minimum sum of the squares of the differences between all pairs of observed and phyletic distances. An alternative, appropriate to the second method of assigning branch lengths, is that the tree of shortest length is best.

9.2.3B Parsimony Procedures

The second general class of analytical methods – *parsimony* – tries to find the tree that permits the observed sequences to have been derived with the fewest number of replacements or substitutions from a common ancestor (Fitch 1971). This is not because evolution is intrinsically parsimonious – it is not – but instead results from the following reasoning.

Consider a group of organisms (e.g., human, chimpanzee, gorilla, and orangutan) and a set of characters showing two character states (e.g., the amino acids alanine and glycine). Those character states divide the organisms into two subsets (of which one is, for example, gorilla and orangutan). Then, in the absence of any other information, it is a better bet that one of the two subsets – orangutan and gorilla, or chimpanzee and human – is a *monophyletic* group (having an ancestor not shared with any of the other taxa) than is the bet that any one of the other four pairs is monophyletic. Of course, it is possible that the chimpanzee and human lineages each had a parallel amino acid replacement from alanine to glycine, but the two specific parallel events required by the hypothesis of gorilla-chimp monophyly is less probable *a priori* than the single event required by the hypothesis that humans and chimpanzees are monophyletic.

The most parsimonious fit of sequence data to a prescribed tree is a simple well-known procedure. Finding the tree with the most parsimonious length from all possible trees is one of the problems that mathematicians describe as "hard," meaning that after decades of work no solution for this enormous class of problems has been found. Thus, both distance and parsimony procedures require heuristic approaches to search for the best tree and there is no guarantee that it will be located. As in the distance procedure, not every difference needs to be given a weight equal to one in parsimony methods.

It is well recognized that the manner in which a character evolves determines which method of phylogenetic reconstruction is most likely to give the correct tree (Felsenstein 1979). Thus, any one method of analysis is unlikely to be appropriate for hundreds or thousands of characters evolving in different ways. Unfortunately, it is not easy to determine which characters should be analyzed by any particular method. Some of the reasons for this ambiguity may be seen if we examine the circumstances under which various methods tend to work best.

9.2.3C Relative Usefulness of Distance and Parsimony Procedures

Distance methods, in which least distant sequences are considered to be most closely related, make an assumption about the general uniformity of evolutionary rates. Clearly, sequence A may be more closely related phylogenetically to sequence B than either A or B is to sequence C, but at the same time be more similar to C than either A or C is to sequence B. This situation will occur if A and C have changed little from their common ancestral condition while B has changed markedly since its more recent common ancestry with A. Nevertheless, such methods frequently seem to give good results, especially if large numbers of characters are used and if characters that evolve rapidly in one lineage are compensated for by other characters that evolve rapidly in the other lineages.

Parsimony methods attempt to overcome the loss of information that in distance methods results from reducing the differences between subsets of sequences to a single average number. And by using all the data, parsimony procedures are no longer subject to the rate problem, at least not in the same way. However, for parsimony to work properly it is necessary that amino acid replacements and nucleotide substitutions occur rarely at each site. An ideal situation – one that would always yield the same unique (and correct!) answer – is one where each character had changed only once during the course of evolution. As more changes take place at previously altered sites the results of a parsimony analysis become more and more ambiguous. It follows that the best solution would be to analyze those characters that evolved at a uniform rate by a distance method and those that changed rarely by parsimony. If both methods produce the same result the tree may be considered "robust."

It should be noted that none of the methods necessarily determines where the root of the tree is situated. Those who use distance methods often opt for the first or "deepest" junction. Users of parsimony frequently take the point within the tree that, on average, is farthest from the tips of the branches. But this is a secondary result in that the goodness-of-fit of the tree to the original data does not depend upon the location of the root.

9.2.4 Use of Molecular Information for Dating Biological Events in Geological Time

Finally, there is the problem of using molecular information for dating biological events in geological time. Given the existence of an acceptable tree, one has ancestral *nodes* (branch-points) that are successively more remote from the sequences representing the branch tips. Whenever all of the characters have evolved at a reasonably uniform average rate, then they are displaying clock-like behavior, and the distance of any ancestral node from the tips connected to it becomes an approximate measure of the time of divergence.

Many tests have been performed which show that collections of related sequences have evolved in a clock-like manner (Sarich and Wilson 1973; Fitch and Langley 1976). These tests do not require information from the fossil record and hence they only demonstrate that the process is clock-like without specifying a particular rate. All such tests have shown, however, that the clock is "sloppy" compared, for example, to the processes of radioactive decay (Fitch 1976). If replacements and substitutions are regarded as the analogues of nuclear disintegrations, molecular methods require about four times as much data to achieve a comparable accuracy. Moreover, there are some clear exceptions to general rates. For example, rodent genes seem to evolve at a faster rate than the comparable genes of other mammals but not nearly as fast as might have been estimated given the small generation times involved (Li et al. 1987).

If a clock-like behavior can be demonstrated independently of the fossil record, then paleontological evidence and isotopic dates may be used to calibrate the molecular clock. Nevertheless, there are severe, if obvious, limitations: (*i*) the clock-like behavior is frequently not demonstrated for the sequences employed; (*ii*) the paleontological dates are subject to considerable error; (*iii*) the whole process depends heavily on the correctness of the topology of the tree and the assigned branch lengths; and (*iv*) there is a considerable but dangerous tendency to extrapolate divergence times, sometimes to dates many times greater than the calibration date. Thus, in summary, dates based on molecular methods are useful but the technique requires caution in its use.

9.3

The Tree of Life

BRUCE N. RUNNEGAR

9.3.1 Life's Deepest Branches

In order to develop a phylogenetic tree of all living things, it was necessary to find a molecule that existed before the last common ancestor of all living organisms and that had been inherited by each living species. Ideally, such a molecule would be a linear polymer of a small, finite number of kinds of subunits (e.g, 4 nucleotides or 20 amino acids), and it would be abundant in cells, unique to each species, and easily sequenced in the laboratory. These specifications are fulfilled by the ribosomal RNA molecules (rRNAs) that constitute part of the protein-translating machinery of all living cells. These molecules (or DNA copies) can be sequenced directly, in whole or in part, rapidly, and efficiently (Field et al. 1988). Most of the recent tree construction has come from comparisons of nucleotide sequences of the so-called "small subunit" rRNAs (16S to 18S, about 1,500 to 1,900 nucleotides in length), thus supplanting earlier comparative studies of the tiny (120-nucleotide) 5S rRNA (Hori and Osawa 1987) and even smaller 16-18S rRNA pieces called "oligonucleotides" (Woese 1987). Results from the "large subunit" rRNAs (23S to 28S, about 2,500 to 3,500 nucleotides in length) tend to match those obtained from the 16 to 18S rRNAs, and there is some evidence that the two parts of the RNA operon have been evolving at approximately the same rate.

Studies, first of oligonucleotide catalogues (Fox et al. 1980) and then of complete 16S and 18S rRNA molecules (Woese 1987), allowed Woese and his colleagues to produce a universal tree of life (Fig. 9.3.1). Because the most distantly related living organisms were incorporated in the tree, its root could not be determined through the use of an outgroup, and it therefore required some ingenuity to determine which part of the tree represents the most recent common ancestor of all living things (Woese et al. 1990).

One early suggestion was that the triple junction between the three main branches (Eubacteria, Archaebacteria, Eukaryota) might be close to the last common ancestor. This possibility was explored by Lake (1983, 1988), who argued on the basis of both ribosome structure and small subunit sequence comparisons[1] that the deepest divergence within the "Archaebacteria" divided that group into two lineages, such that the extremely thermophilic Archaebacteria – which Lake calls the "Eocytes" – lie with the Eukaryotes and not with a the rest of the Archaebacteria. Thus, Lake (1988, 1989) envisaged two primary clades ("Parkaryotes" and Karyotes") that diverged within but near the base of the Archaebacterial branch of the Woese et al. tree (Fig. 9.3.2; Table 9.3.1). By contrast, Van Valen and Maiorana (1980), Hori and Osawa (1987), Cavalier-Smith (1987a, b), and Woese et al. (1990) root the tree partway along the eubacterial branch (Fig. 9.3.2). In this interpretation, the Archaebacteria and Eukaryota are sister groups derived from a (presumably) bacterial lineage that last shared a common ancestor with the Eubacteria some considerable time before that dichotomy. Cavalier-Smith (1987b) has formulated a precise model for a sequential evolution along the archaebacterial/eukaryotic pathway; a critical step was the loss, in the line leading to the Archaebacteria, of the ability to manufacture the peptidoglycan murein. Murein forms a covalently bonded network around eubacterial cells and it both maintains and constrains cell shape. Freed of constraining network, archaebacterial and eukaryotic cells have been able to diversify in ways that the eubacterial cells could never achieve.

These new developments in understanding have necessitated novel names for some of the highest level taxa. Unfortunately, each major hypothesis has generated its own vocabulary, so there are now competing terms for many of the newly identified higher taxonomic groups; a summary of the principal alternatives is given in Table 9.3.1. In this context it should be noted that "Metabacteria" is a name that has also been proposed as a neutral alternative for the interpretation-rich word "Archaebacteria" (Hori et al. 1982; Hori and Osawa 1987).

9.3.2 Endosymbiosis and the Origin of Organelled Eukaryotes

One of the great triumphs of modern molecular biology has been the demonstration that eukaryotic chloroplasts and mitochondria are ultimately of endosymbiotic origin. This idea was championed by Margulis (1970) using cytological evidence, but it was the subsequently discovered sequence similarites between chloroplast 16S rDNA and the 16S rRNA from *E. coli* (Schwarz and Kössel 1980; Gray 1988) which, for many people,

---

[1] Using a technique that accounts for unequal rates on different branches (Lake 1987b).

472   Molecular Phylogenetics

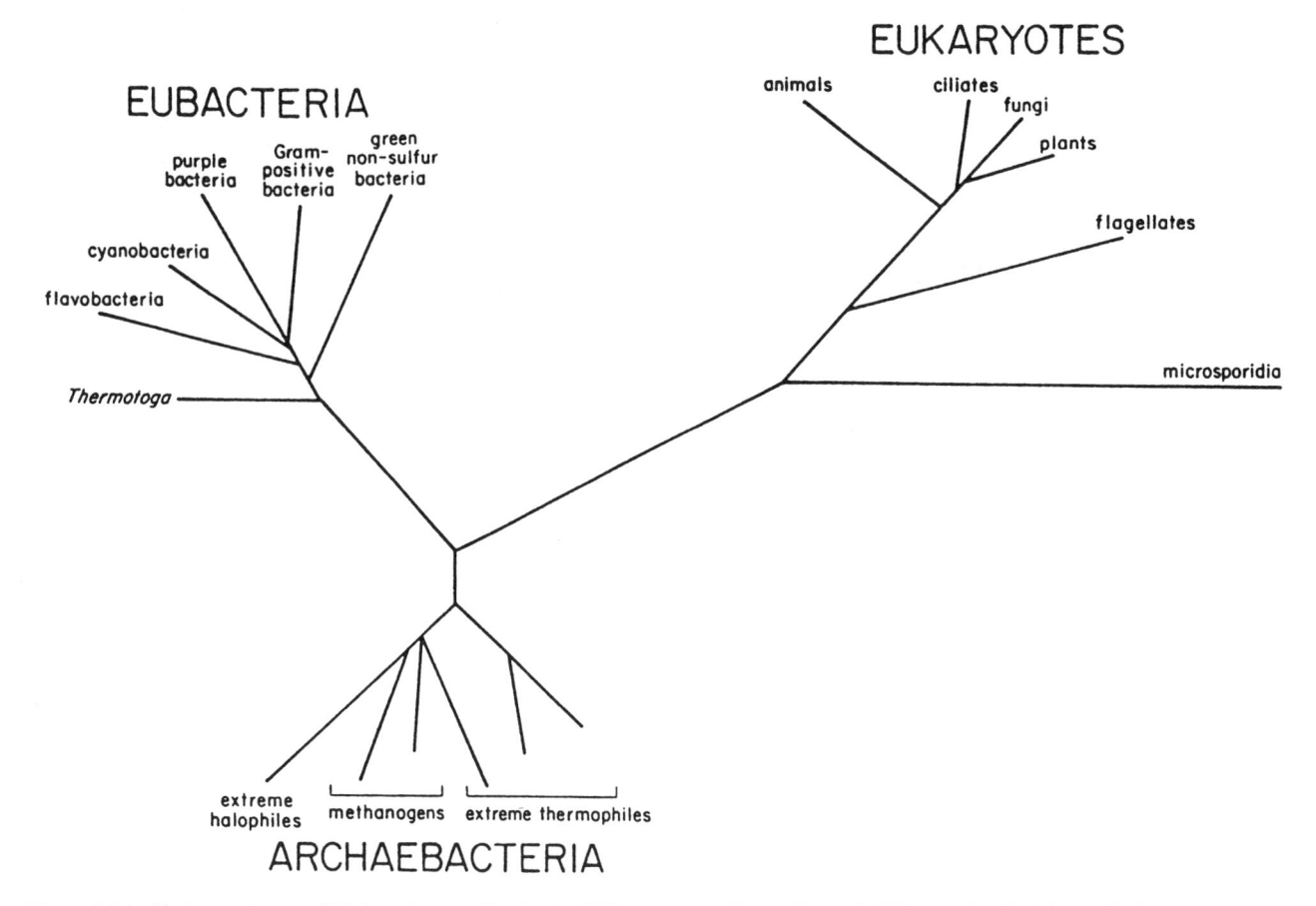

Figure 9.3.1. Phylogenetic tree of life based on small subunit rRNA sequences. From Woese (1987), reproduced with permission.

confirmed the endosymbiont hypothesis. There is also now equally good evidence for a distant relationship between the mitochondria and the living purple bacteria (Rhodobacteria; Küntzel and Köchel 1981; Yang et al. 1985; Gray 1988), so this group of Eubacteria is believed to be the source for the ancestral mitochondrion (or mitochondria; Gray et al. 1989).

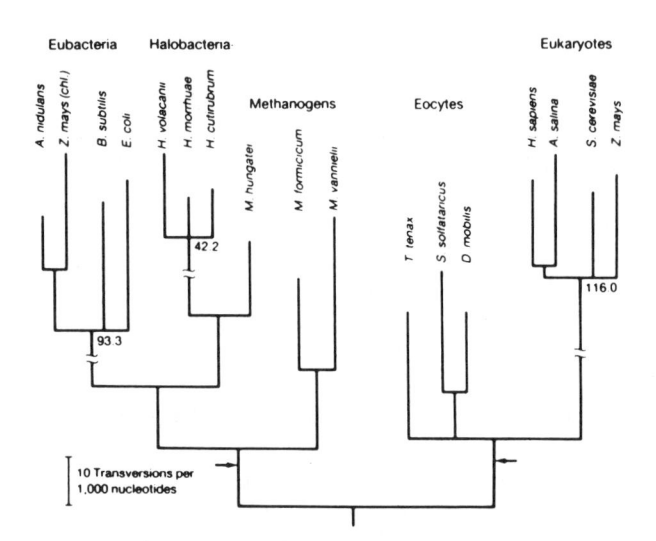

Figure 9.3.2. Rooted tree of life based on Lake's method of "evolutionary parsimony." In this interpretation, the Archaebacteria are a stem group divided between the eubacterial and eukaryotic branches. From Lake (1988), reproduced with permission.

The implications of these discoveries are that the eukaryotic cell is a kind of ultimate ectoparasite. Eukaryotes never evolved the ability to carry out either aerobic respiration or autotrophic photosynthesis. Instead, eukaryotes have, on more than one occasion, co-opted eubacterial cells for just these purposes.

Some living unicellular eukaryotes (e.g., the diplomonad *Giardia*) have neither chloroplasts nor mitochondria. Inasmuch as the 16S-like rRNA of *Giardia* has retained many of the features that might have been present in the common ancestor of eukaryotes and prokaryotes (Sogin et al. 1989a), it is almost certain that the earliest eukaryotes lacked these organelles. This means that the earliest eukaryotic unicells were heterotrophic anaerobes; the anaerobic fungi which inhabit the rumen of cattle and sheep (Bauchop 1979) are modern-day analogues.

9.3.3 Rooting the Tree of Life in the Fossil Record

Doolittle et al. (1989) have attempted to use the principle of the molecular clock to date the "prokaryote-eukaryote divergence time." They first used known or assumed divergence times to calibrate the rates of evolution of proteins such as cytochrome *c*, Cu-Zn superoxide dismutase, and lactate dehydrogenase, and then extrapolated these rates to obtain estimates of the times of divergence of typical eukaryotic proteins from one or more of their eubacterial homologues. A similar procedure was used by Runnegar (1982) to obtain a date for the origin of the animal phyla.

The ten proteins used by Doolittle et al. (1989) yielded an

Table 9.3.1. *List of some of the names that have been proposed recently for the higher taxonomic groups of living organisms. Taxa with a rank HIGHER THAN KINGDOM are listed in capital letters;* **Kingdom-level Taxa** *are in bold type; "Phyla" are in normal type*

Cavalier-Smith (1987a,b)	Lake (1988, 1989)	Woese et al. (1990)
EMPIRE BACTERIA	SUPERKINGDOM PARKARYOTES	DOMAIN BACTERIA
Kingdom Eubacteria	**Eubacteria**	DOMAIN ARCHAEA
Cyanobacteria		
Prochlorobacteria		
Chlorobacteria		
Rhodobacteria		
Heliobacteria		
Kingdom Archaebacteria		**Kingdom Euryarchaeota**
Halobacteria	**Halobacteria**	
Methanobacteria	**Methanogens**	
	SUPERKINGDOM KARYOTES	
Sulfobacteria	**Eocytes**	**Kingdom Crenarchaeota**
EMPIRE EUKARYOTA	**Eukaryotes**	DOMAIN EUCARYA
Kingdom Protista		
Kingdom Plantae		
Kingdom Fungi		
Kingdom Animalia		

average age of 1.8 ± 0.4 Ga for the eukaryote-eubacterial split. However, if the tree of life is correctly rooted along the eubacterial branch, this date corresponds to the time of existence of the most recent common ancestor of all living things. Intuitively, this seems unlikely, but it is not so easy to find evidence which would refute this conclusion. It could even be argued that the mass extinction which may have resulted from the major impact that produced the Sudbury Complex 1.8 ± 0.02 Ga ago (Faggart et al. 1985) could have so reduced the diversity of life that the likelihood of more than one clade surviving the following 1.8 Ga was small (Raup and Valentine 1983, give some estimates of the probability of survival from small beginnings). Whether such speculations have any meaning remains to be seen; for the present, the best approach is to try to refute the Doolittle et al. date by demonstrating the existence of archaebacterial and/or eukaryotic life from evidence in rocks older than about 1.8 Ga.

The discovery of trace amounts of $5\alpha(H)14\alpha(H)17\alpha(H)$-24-cholestane and related steranes in 1.7 Ga-old organic-rich sediments of the McArthur Basin (Section 3.3) has been taken to be strong evidence for the existence of eukaroytic organisms at that time (Summons et al. 1988b). However, although the steranes certainly seem to have required an oxygen-dependent step in their manufacture (Schopf 1978; Ourisson et al. 1987). it is not so clear that they could not be of prokaryotic origin (e.g., Weeks and Francesconi 1978).

The key to relating the tree of life to the fossil record seems to lie in a study of the small subunit rRNAs from cyanobacteria and green plant chloroplasts published by Giovannoni et al. (1988). Surprisingly, the modern cyanobacterial groups, which traditionally have been identified on the basis of morphology, converge to a point that is relatively high up on the eubacterial branch of the tree of life (Fig. 9.3.1). Even more astonishing was the discovery that the chloroplast sequences indicate that chloroplasts have been eukaryotic endosymbionts since about the time of the cyanobacterial radiation.

If we accept these data at face value, how can they be reconciled with the Precambrian fossil record? Schopf and Packer (1987) provided limited evidence for cyanobacterial diversification about 3.4 Ga ago (for additional data and discussion, see Section 1.5); however, the inferior preservation of the fossils (shown in Figs. 1.5.4 through 1.5.7) and the simple morphology of the cells makes this conclusion, as they clearly indicated, "suggestive" rather than "compelling." On the other hand, it is difficult to doubt the presence of representatives of modern cyanobacterial groups in the cherts of the Gunflint Iron Formation (Chapter 24, Pl. 2, A–R; Lanier 1989) and Belcher Supergroup (Chapter 24, Pl. 1, G–O; Hofmann 1976), each about 2.0 Ga in age. Given the stratigraphic continuity between these Early Proterozoic stromatolitic biotas and those of the Late Proterozoic and Phanerozoic (Section 5.4), the radiation of the Cyanobacteria can be conservatively judged to have occurred prior to 2.0 Ga ago.

The oldest known, chloroplast-bearing, megascopic eukaryote is probably *Grypania spiralis* (Walcott), found in rocks about 1.4 Ga in age in the USA, China, and India (Figs. 7.3.9, 7.3.10; Walter et al. in press). However, sizable, organic-walled, spheroidal unicells such as those illustrated by Peat et al. (1978) from the Middle Proterozoic McArthur Basin of Australia are considered to be the remains of photosynthetic planktic eukaryotes (Section 5.5). As these kinds of cells are also found in rocks as old as about 2.0 Ga (Section 5.5), the endosymbiotic origin of eukaryotic chloroplasts may have occurred prior to that time.

It is customary to regard the Cyanobacteria as "derived" rather than "primitive" members of the Eubacteria. To a large extent, this conclusion depends upon the belief that the water-splitting photosynthetic apparatus (Photosystem II) of cyanobacteria and the plant chloroplasts is more advanced than the comparable apparatus (Photosystem I) used by other living photosynthetic eubacteria (Pierson and Olson 1990). It should be noted, however, that some authors regard the cyanobacterial

474 Molecular Phylogenetics

photosynthetic pigments and reaction centers as primitive. For example, Cavalier-Smith (1987b, Fig. 8) derives all extant organisms from a cell having most of the properties of a modern cyanobacterium. Inasmuch as the rRNA sequence comparisons give a genealogy which is independent of cell physiology, the resulting tree provides little indication of the nature of the primitive condition.

Procholorophytes (*Prochloron* and *Prochlorothrix*) are free-living photosynthetic prokaryotes that are thought to be closer to the ancestors of green plant chloroplasts rather than to Cyanobacteria because they contain chlorophylls *a* and *b* and lack phycobilins (Lewin and Cheng 1989). The available molecular data are ambiguous about the exact relationships between these three groups (Morden and Golden 1989; Turner et al. 1989), but the prochlorophytes (or Prochlorobacteria) are a second major group of oxygenic eubacteria that might be expected to have a long geological history.

There is no living chloroplast-bearing eukaryote which does not also use mitochondria for aerobic respiration. This implies that the acquisition of mitochondria by unicellular eukaryotes preceeded the engulfment of the ancestral chloroplast(s). Thus, *Grypania spiralis* was probably "fully organelled" in the sense that it had both mitochondria and chloroplasts. The same argument may be applied to smooth, large-celled, unicellular acritarchs interpreted as planktic eukaryotes (Section 5.5), but this interpretation is weakened by the uncertainty which accompanies the taxonomic placement of these featureless fossils.

Biogeochemical evidence – principally measurements of the stable isotopes of carbon and sulfur – for the nature of Archean and Proterozoic bacterial life is discussed in Sections 3.4 and 3.5. Briefly: (*i*) the $\delta^{13}C$ values of organic matter (kerogen) from rocks as old as about 3.5 Ga indicate that carbon-fixing photosynthesis has been operating since that time (Schidlowski 1988); (*ii*) an increase in the variance of $\delta^{34}S$ in the Late Archean (ca. 2.7 Ga) or Early Proterozoic (2.4 to 2.2 Ga) may mark the origin of dissimilatory sulfate reduction (Lambert and Donnelly in press); and (*iii*) extraordinarily light $\delta^{13}C$ values ($<40‰$ vs. PDB) in the Late Archean (ca. 2.7 Ga) and Early Proterozoic (ca. 2.1 Ga) may indicate that methylotrophs were recycling methane that had been generated by methanogenic archaebacteria (Hayes 1983). This last point is the only known possible physical evidence for the existence of the Archaebacteria in the Archean.

9.3.4 Radiation of the Eubacteria and Archaebacteria

Ribosomal RNA sequences are now available from a large number of living Eubacteria and Archaebacteria and their documentation and analysis has been treated in a series of reports that are summarized by Woese (1987). The deepest divergence so far documented is between the thermophilic species *Thermotoga maritima* (Thermotogales) and all other Eubacteria; this has led some authors to regard thermophily as a primitive condition for all Eubacteria and possibly for all bacteria (Achenbach-Richter et al. 1987).

Other major eubacterial clades (cyanobacteria, green non-sulfur bacteria, purple bacteria, Gram-positive bacteria etc.) appear at various levels higher up the tree (Fig. 9.3.3). In general, low branches such as the green non-sulfur bacteria (*Chloroflexus* and relatives) are regarded as having primitive phenotypes, but, as pointed out above, this is not necessarily the case.

The Archaebacteria are distributed in a more dispersed clade with the extreme thermophiles (Sulfobacteria or Crenarch-

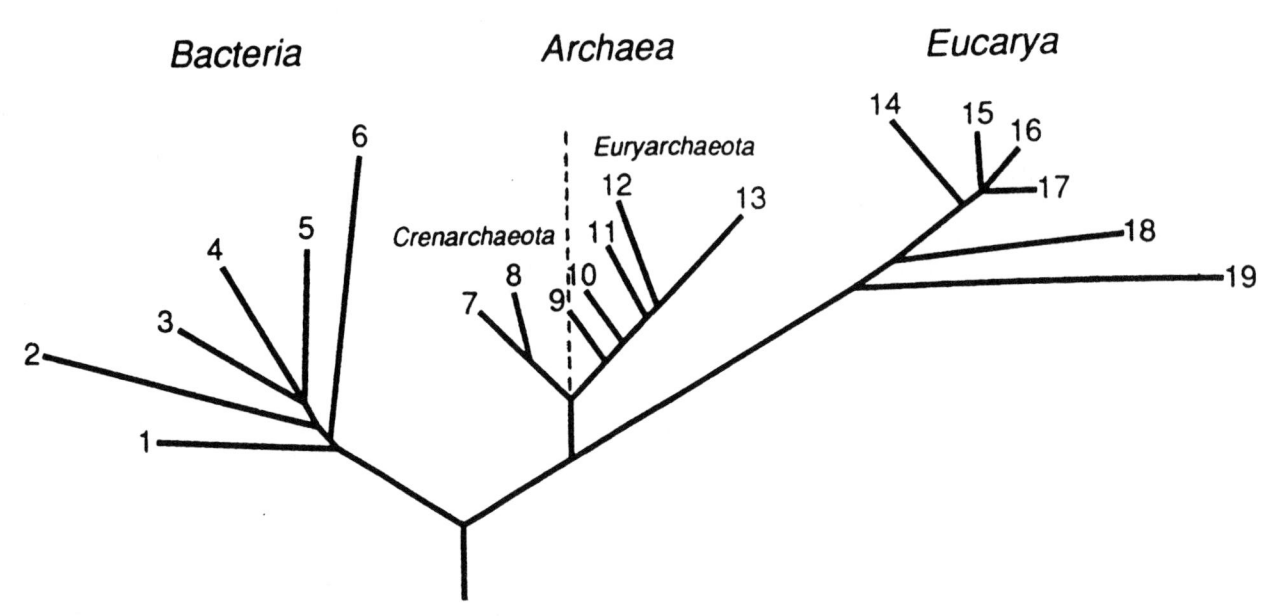

Figure 9.3.3. Rooted tree of life from Woese et al. (1990), reproduced with permission. In this interpretation, the root lies along the eubacterial branch. *Bacteria* – 1, the Thermotogales; 2, the flavobacteria and relatives; 3, the cyanobacteria; 4, the purple bacteria; 5, the Gram-positive bacteria; and 6, the green non-sulfur bacteria. *Archaea* – the kingdom Crenarchaeota: 7, the genus *Pyrodictium*; and 8, the genus *Thermoproteus*; and the kingdom Euryarchaeota: 9, the Thermococcales; 10, the Methanococcales; 11, the Menthobacteriales; 12, the Methanomicrobiales; and 13, the extreme halophiles. *Eucarya* – 14, the animals; 15, the ciliates; 16, the green plants; 17, the fungi; 18, the flagellates; and 19, the microsporidia.

aeota) lying on one side of a deep divergence (Fig. 9.3.3). The other side of the clade, which Woese et al. (1990) unify into the Kingdom Euryarchaeota, comprise the "methanogens" (Methanobacteria = Methanococcales plus Methanobacterales plus Methanomicrobiales) and the "extreme halophiles" (Halobacteria). Lake's (1988) "Eocytes" correspond to the Sulfobacteria/Crenarchaeota branch of the archaebacterial tree (Table 9.3.1).

9.3.5 Origin of the Eukaryotic Kingdoms

As more and more rRNA sequence data have appeared, it has become increasingly obvious that the history of the eukaryotes is both long and complex. A simple "five kingdom" classification of life of the kind introduced by Whittaker (1959) is no longer workable, even within the eukaryotic branch of the universal tree (Sogin et al. 1989; Woese et al. 1990). One alternative, advocated by Sogin et al. (1989a, b), is to lump all eukaryotes into a single kingdom; others prefer a more traditional approach with green plants, fungi, animals, and some unicellular groups (e.g., the ciliates) each being given kingdom rank.

This topic is beyond the scope of this brief review. Instead, we shall focus on the origin and evolution of two kinds of eukaryotes, namely the unicellular protists and the multicellular Metazoa. These are treated in Sections 9.4 and 9.5, respectively.

9.4

Origin and Divergence of Protists

DAVID J. CHAPMAN

9.4.1 What is a "Protist"?

It is instructive before beginning this Section to outline what is meant by "Protists," Most importantly, they are not a natural taxonomic (or phylogenetic) grouping. The term has been used as one of convenience to describe those eukaryotic organisms that belong to the three groups commonly known as algae, fungi, and protozoa. This is the same delineation (less the Eubacteria and Actinomycetes) used by Ragan and Chapman (1977), whose review can serve as the baseline for the last 12 years of research and discussion in protistan evolution. While the algae, fungi, and protozoa are familiar to most, it must be noted that these groups, themselves, are not natural assemblages, but rather conglomerates of convenience whose systematic origin is buried in early natural history. Many workers now exclude the fungi from the protists, perhaps implying a systematic integrity for the protists, something that does not in fact exist. "Protist" is a term of convenience to accommodate organisms that do not fit neatly into other easily recognizable taxa such as metazoa, insects, fish, birds, and flowering plants. One is thus reduced to a simple operational description. Protists are eukaryotes with the simplest gross morphological organization in the vegetative state. Unicells, colonies of unicells, and simple multicells lacking tissue differentiation (with some notable exceptions such as the brown algae) are the predominant morphologies represented. Even this definition is relative. Certainly, protists lack the morphological complexity of metazoa, either invertebrates or vertebrates, and higher plants, but this does not translate into biochemical simplicity. Cellular, physiological, biochemical, and molecular diversity and complexity are hallmarks of the protists. Although there is no *a priori* reason for the idea, all the evidence suggest that the relatives of modern extant protists were the earliest eukaryotes.

Acceptance of the proposal that protists represent the ancestral eukaryote stock focuses much of the evolutionary discussion on three aspects: the nature of the ancestral organism that provided the "source-stock" for the host in the symbioses that subsequently generated organelles; the origins of mitochondria and chloroplasts and the question of probable multiplicity of evolutionarily significant endosymbioses; and the evolutionary radiation of protists themselves.

9.4.2 The Ancestral Host for Organelle-Generating Endosymbioses

The identification of the ancestral/original organism, from which derived the nucleocytoplasmic component of the eukaryote, has received limited attention. Two principal schools of thought exist. There are proponents (Lake et al. 1984; Zillig et al. 1985; Wolters and Erdmann 1986; Zillig 1987; Lake 1987a, 1988, 1989a, b) who suggest that this ancestral entity is closely related to the sulfur-dependent archaebacteria in the orders Sulfolobales and Thermoproteales ("Eocytes" of Lake et al. 1984) or who suggest (Searcy 1986, 1987) that these two orders show features that were suited to give rise to the nucleocytoplasmic component of eukaryotes. The alternative proposal (Stackebrandt and Woese 1981; Woese 1982; Olsen et al. 1985; Pace et al. 1986a; Woese and Olsen 1986; Olsen 1987; Woese 1987; Olsen and Woese 1989; Woese 1989; Gouy and Li 1989a) regards the archaebacteria as a distinct and coherent phylogenetic group, quite separate from the eubacteria and eukaryotes, with all three major groups perhaps deriving from a universal common ancestor: the "progenote" theory (Woese and Fox 1977). Alternative suggestions, e.g., for a eukaryotic origin from an aerobic "posibacterium" (Cavalier-Smith 1981, 1987a, b), have received little support. Part of the problem lies with disagreement over the interpretation of data of nucleotide sequences; the various methods for constructing evolutionary trees; and the choice of representative organisms and sequence alignments (see, e.g., Achenbach-Richter et al. 1988, Olsen 1988). Furthermore, studies of comparative biochemistry have not yet provided any compelling evidence to support a particular assertion. Wolters and Erdmann (1989) have correctly pointed out that one way out of the dilemma would be an enlargement of the dataset of large subunit rRNA sequences and inclusion of data from a greater variety of organisms. This has been started with analyses by Leffers et al. (1988) of 23S rRNA sequences. Very recent analyses, using independent data such as derived amino acid sequences for RNA polymerase (Zillig et al. 1989), H^+-ATPase (Gogarten et al. 1989), and glyceraldehyde-3-phosphate dehydrogenase (Hensel et al. 1989), while a move in the right direction, have not served to clarify the situation but rather, in some instances, to raise more questions and concerns. This is in contrast to the question of organellar origins, where comparative biochemical studies were one of the early driving forces for theories of endosymbiosis.

9.4.3 The Endosymbiotic Origin of Chloroplasts and Mitochondria

The theory of the endosymbiotic origin of organelles had a checkered history until 1970. The publication of the synthesis by Margulis (1970) served to focus attention on the theory and to provide impetus for much of the research that followed. The overwhelming body of available evidence now supports the idea that the original chloroplasts and mitochondria were derived from endosymbioses by eubacteria, notwithstanding that chloroplasts and mitochondria of some cells may be derived by secondary symbioses (Taylor 1987). For a full discussion, the reader is referred to key reviews by Whatley (1981), Whatley and Whatley (1981), Doolittle and Bonen (1981), Gray and Doolittle (1982), George et al. (1983), Gray (1983), Cavalier-Smith (1986), Taylor (1980b, 1987), Cattolico (1986), Cavalier-Smith (1987b, d), Gray (1988), Giovannoni et al. (1988), Gray (1989), and the literature cited therein.

9.4.3A Chloroplasts

Delihas and Fox (1987) and Van den Eynde et al. (1988), based on consideration of 5S rRNA sequences, discuss a polyphyletic origin of chloroplasts. Stackebrandt and Woese (1981) have argued that the chloroplasts of higher plants and *Euglena* are phylogenetically deeper than cyanobacteria, suggesting that the ancient phenotype may not have been a cyanobacterium but rather a related organism. The dendrogram of Seewaldt and Stackebrandt (1982), illustrates a deeper branching, assuming no fast clock, for chloroplasts of higher plants, *Chlamydomonas*, and *Euglena*. In this regard, the reader is referred to the work by Chapman and Schopf (1983) proposing the need for a protocyanobacterium in early evolution. In support of multiple origins of chloroplasts, Margulis and Obar (1985) proposed the eubacterium *Heliobacterium* or a relative as the ancestral source for chloroplasts of the Chrysophyceae (yellow-brown algae). Witt and Stackebrandt (1988), using 16S rRNA catalogues and reverse transcriptase sequencing, have shown conclusively that the chrysophyte chloroplast cannot be traced back to a *Heliobacterium*-type ancestor. Studies with 16S rRNA by these authors and by Doolittle and Bonen (1981) and by Kössel et al. (1983) suggest that the phylogenetic position of the chloroplast of such divergent protists as *Ochromonas* (Chrysophyceae), *Euglena* (Euglenophyceae), and even *Chlamydomonas* (Chlorophyceae) is far from resolved and may represent a case of multiple origins from some, as yet unknown, prokaryotes. (See also the conclusions on the *Euglena* chloroplast by de Wachter et al. 1985, and Van den Eynde et al. 1988, based on 5S rRNA sequences.)

Upon its discovery some 14 years ago, the chlorophyll *b*-containing prokaryote, *Prochloron*, acquired immediate notoriety in the chloroplast-origin hypotheses, cautionary advice (Chapman and Trench 1982) notwithstanding. Catalogues from 16S rRNA now make it clear that *Prochloron* is typically cyanobacterial, with no special relationship to chlorophyll *b*-containing chloroplasts (Seewaldt and Stackebrandt 1982; Stackebrandt et al. 1982; Stackebrandt 1983; Giovannoni et al. 1988; Reichenbach et al. 1988). A similar lack of a close relationship is apparent from dendrograms based on 5S rDNA sequences (Van den Eynde et al. 1988). The original analysis of *Prochloron* 5S rRNA (MacKay et al. 1982) did not demonstrate any specific relationship to chloroplasts. Burger-Wiersma et al. (1986) have recently isolated a filamentous counterpart to *Prochloron* possessing chlorophyll *b*, which has been designated *Prochlorothrix*. Turner et al. (1989), employing 16S rRNA sequence phylogeny, have shown that *Prochlorothrix*, like *Prochloron*, shows no close relationship to green chloroplasts of the *Chlamydomonas* and *Euglena* genre [The different conclusion reached by Morden and Golden (1989a), from the nucleotide sequence of the gene encoding the photosystem II protein Dl, *psbA* gene, suggesting a common ancestry of "prochlorophytes" and chloroplasts, has since been shown to be dependent upon the analytical consideration given to a missing 7-amino acid sequence (Morden and Golden 1989b).] Indeed, but for chlorophyll *Prochloron* and thus *Prochlorothrix* would never have been an issue (Chapman and Trench 1982). One must ask if there are O_2-producing photosynthetic prokaryotes with a biochemistry different from cyanobacteria, that have become extinct or have yet to be discovered, which gave rise through endosymbiosis to chloroplasts whose ancestors exist today? The answer is almost certainly affirmative; as Doolittle and Bonen (p. 253, 1981) have earlier pointed out, "...the suggestion...that plastids arose on many different occasions from different oxygenic-photosynthetic prokaryotes, not all of which we would now call cyanobacteria, seems likely to be correct."

Endosymbiotic theories invoking only a single evolutionary origin of chloroplasts are perhaps more convenient and tidy, and simpler to incorporate into "trees." This thinking has perhaps prejudiced our attitudes and willingness to entertain possibilities different from a "single origin" hypothesis. Twenty years ago Raven (1970), and then later Whatley and Whatley (1981), proposed multiple origins. Many of the arguments, either for or against, were based upon comparative biochemistry (for summaries of the earlier data, see Ragan and Chapman 1978, and more recently, Rothschild and Heywood 1987); indeed, it has been difficult until now to build a persuasive argument for one or other scenario. Taylor (1978, 1979, 1987) and Cavalier-Smith (1981, 1982, 1987a, b), for example, have supported the single-origin approach but not to the exclusion of subsequent secondary symbioses. *Prochloron*, the chlorophyll *b*-containing prokaryote (summarized by Lewin 1984; Chapman and Trench 1982; and Raven 1987), has had a disproportionate influence on endosymbiotic theories. The same criticism can be leveled at cyanelles, in particular those of *Cyanophora* and *Glaucocystis*. Superficially they represent a nice scenario for endosymbiosis, but a more careful analysis of the data summarized in the review by Trench (1982), and more recent work by Maxwell et al. (1986), Breiteneder et al. (1988), Wassman et al. (1988), Evrard et al. (1990), and work referenced therein, suggests that cyanelles may not be the beautiful intermediate between cyanobacteria and chloroplasts that many have suggested. Burnap and Trench (1989, p. 100) have suggested that the *Cyanophora*-cyanelle complex is a case of convergent evolution and "not necessarily a replay of the events hypothesized as occurring in the evolution of chloroplasts." Certainly, there would be no argument that at least some chloroplasts were derived directly from endosymbioses (multiple or single) of cyanobacteria or close relatives (e.g., Bonen and Doolittle 1975; Stackebrandt and Woese 1981).

In considering the likely multiplicity of the origins of chloroplasts it is instructive to note the diversity of these organelles at the molecular level. Sequence analysis of the small subunit 16S rRNA (Witt and Stackebrandt 1988) reveals a red algal plastid (*Porphyridium*) remote from plastids of the yellow-brown algae (*Ochromonas*), euglenoids (*Euglena*), and green algae as represented by *Chlamydomonas*. These latter three and higher plants (e.g., *Nicotiana*) appear to share a distant common ancestor, but the evidence would not suggest that their immediate plastidial progenitor was the same. The *Cyanophora* cyanelle shows a close relationship to the liverwort chloroplast (Giovannoni et al. 1988).

Immunological comparisons of light harvesting fucoxanthin-chlorophyll protein (17 to 21 KDa) suggest a close relationship between the Prymnesiophyceae and diatoms (Fawley et al. 1987, Hiller et al. 1988). In turn, the diatom complex did cross react with that from brown algae but not with that from green algae (*Chlamydomonas*), euglenoids, dinoflagellates, cryptomonads, or yellow-brown (Chrysophyte) algae as represented by *Olisthodiscus* (Friedman and Alberte 1987). As expected, similar comparisons have shown cross-reactivity of prymnesiophytes with diatoms but not with dinoflagellates or cryptomonads (Hiller et al. 1988). Mapping of chloroplast DNA from a brown alga, a euglenoid, and a cyanelle (Loiseaux-de Goer et al. 1988; Markowicz et al. 1988a,b) has revealed significant difference from the "green algal" lineage. In the chromophytic *Olisthodiscus* (Reith and Cattolico 1986), as well as in the cyanelle and in the red algae *Cyanidium* and *Porphyridium* (Steinmuller et al. 1983, Valentin and Zetsche 1989), the small subunit of ribulose-1,5-bisphosphate carboxylase (Rubisco) is encoded on the chloroplast genome, not on the nuclear genome as it is in higher plants and green algae. A detailed structural and functional analysis of Rubisco from a variety of algae (Newman et al. 1989) further supports an evolutionary divergence of "chlorophytic" chloroplasts from rhodophytic and chromophytic counterparts. Plastid DNA configuration (Coleman 1985) shows two basic types not out of keeping with delineations observed from other data. The concensus of data now available leaves a series of inescapable conclusions; namely, that (*i*) the red algal chloroplast probably derived from a cyanobacterium by direct assimilation; that (*ii*) the "Prochlorophyta" do not represent the ancestral stock of chlorophyll *a*, *b* chloroplasts, and have been grossly overrated in evolutionary significance; that (*iii*) cyanelles may not be the intermediates which they have previously been considered; that (*iv*) chromophyte chloroplasts of diatoms, brown algae, chrysophytes, prymnesiophytes, and possibly of dinoflagellates, probably have a common ancestral origin; and that (*v*) algal chloroplasts almost certainly have multiple evolutionary origins through both primary and secondary symbioses. These and other data (see below) also serve as a reminder that the taxonomic dichotomy built on chlorophyll *a*, *b* versus chlorophyll *a*, *c* is now irrelevant.

In recent years, ultrastructural studies combined with existing biochemical data have been used to propose a eukaryotic (secondary symbiotic) origin (i.e, through phagocytosis of an existing separate eukaryote) for the chloroplasts of the euglenoids from green algae (Gibbs 1978, 1981a,b); of cryptomonads from red algae (Ludwig and Gibbs 1985, 1987); and of Chlorarachniophyta from green algae (Hibberd and Norris 1984; Ludwig and Gibbs 1987, 1989) or some other group (Raven 1987). The well-documented occurrence in dinoflagellates of "chrysophytan"-type chloroplasts (Tomas and Cox 1973; Jeffrey and Vesk 1976, Kite and Dodge 1985, 1988), of presumptive "chlorophytan" chloroplasts (Watanabe et al. 1987), and of a second, non-dinoflagellate nucleus (Dodge 1971, 1983; Kite et al. 1988), together with the occurrence of the *Mesodinium*/cryptomonad assemblage (Hibberd 1977; Lindholm *et al.* 1988), the *Amphidinium* and *Gymnodinium*/cryptomonad system (Wilcox and Wedemayer 1984, 1985), the *Dinophysis*/cryptomonad system (Schnepf and Elbrächter 1988), and of dinoflagellates with apparent phycobiliproteins (Hu et al. 1980; Zhang et al. 1982; Geider and Gunter 1989), are relevant to this issue.

A series of focused investigations on the division of the nucleomorph in cryptomonads (McKerracher and Gibbs 1982; Morrall and Greenwood 1982) and the localization of both DNA and RNA in this body (Hansmann et al. 1986; Hansmann 1988), combined with the results from dinoflagellates, Chlorarachniophyta, and euglenoids are persuasive arguments for secondary symbioses (Taylor 1979, 1987). This method for the acquisition of chloroplasts in the evolution of photosynthetic protists is now as important as direct assimilation of prokaryotes and is almost certainly a continuing evolutionary occurrence.

It is important, however, to distinguish between the organismic origin of the chloroplast within the various eukaryotes (i.e., the source prokaryotic endosymbiont for the future chloroplast); the phylogeny of those individual endosymbionts (or chloroplasts-to-be) which may well have had a common ancestor; chloroplasts deriving from secondary symbioses; and any subsequent change in an established chloroplast, regardless of origin or ancestry.

Wolters and Erdmann (1988) have suggested that 5S rRNA sequences define a monophyletic group for the Cyanobacteria, the plastids of *Euglena*, and the cyanelle *Cyanophora*. There is also the suggestion of Witt and Stackebrandt (1988), who provided additional evidence that the common ancestor of cyanobacteria and chloroplasts predated the chloroplast ancestor itself (see also Bonen and Doolittle 1975; Zablen et al. 1975). Additional sequences, especially of small subunit rRNA from a wider array of chloroplasts and hosts (nucleocytoplasmic), are obviously needed to unravel the puzzle. Giovannoni et al. (1988, p. 3591) have asked the similar questions: "Do green chloroplasts share a common ancestor with modern prochlorophytes? Do rhodophyte, chlorophyte, chrysophyte and cryptomonad chloroplasts have monophyletic or polyphyletic origins?"

The origin of chloroplasts thus appears to have been a complex of events involving both primary and secondary symbioses and both eukaryotes and prokaryotes, and not a single event from which all other chloroplasts evolved simply by intra-host change and evolution.

9.4.3B Mitochondria

The origin of the mitochondrion raises the same questions of multiplicity. There would now appear to be general agreement that the eubacterium giving rise through endosymbiosis to the mitochondrion originated in the α-subdivision of the purple non-sulfur photosynthetic bacteria (Villaneuva et al. 1985; Yang et al. 1985; Delihas and Fox 1987). This group includes *Paracoccus denitrificans* which had been postulated by John and Whatley (1975, 1977) as a mitochondrial candidate even before the advent of nucleotide sequence data (see John 1987, for a concise summary of the biochemical and cytochrome sequence data used in support of this hypothesis).

Superficially, this would appear to be a clear-cut case for a monophyletic origin of the mitochondrion, because no other serious contenders have been proposed. Ultrastructural observations, however, have suggested that the scenario might not be that simple, and Stewart and Mattox (1980, 1984) have proposed a polyphyletic origin for mitochondria.

Gray et al. (1984) and Gray (1988) have provided nucleotide sequence data in support of a multiple origin, noting that plant mitochondria branch separately from animal and fungal mitochondria. It must be emphasized, however, that Stewart and Mattox (1984) did not suggest in their proposal an origin from outside of the purple non-sulfur bacterial group. All of their comparison bacteria belonged to either the α- or β-subdivisions. This would suggest multiple origins with all the endosymbionts or promitochondria being closely related to each other. Most recently Boer, quoted in Gray and Boer (1988), has reported that nucleotide sequences of both the small subunit and large subunit mitochondrial rRNA of the green alga *Chlamydomonas reinhardtii* do not support branching with other plant mitochondria, but instead suggest an early offshoot to the lineage leading to animal mitochondria. Gray et al. (1989) have now provided a more detailed analysis of this problem, showing that the mitochondrial lineages of *Chlamydomonas* and higher plants are quite separate, in constrast to a close relatedness of the nuclear-cytoplasmic branches. In considering a number of explanations, these workers have presented a hypothesis that the rRNA genes of the plant mitochondria were acquired after the separation of *Chlamydomonas* and higher plants from their last ancestor. The possibility of such an event lies in the assumption that the higher plant ancestor (a green alga already with a mitochondrion) ingested a eubacterium and that there followed transfer of genes to the endosymbiont's mitochondrion. They have described this scenario as "mitochondrial succession." This is not outside the realm of possibility (cf. the chloroplast problem) and more extensive data from within the various chlorophytan lineages, together with additional data from within the α- (and possibly β-) subdivisions, could produce a resolution (Gray et al. 1989). The complexity and problems observed in elucidating chloroplast origins may well be repeated with mitochondria. Currently, the situation appears simpler than for chloroplasts, but this may simply be a reflection of fewer data and the fact that the superficial uniformity of mitochondria has prejudiced thinking more strongly towards monophyly.

One of the most intriguing questions is that regarding the origin of the protistan cilia or flagella. They are quite different from prokaryotic flagella. For a number of years, Margulis (Sagan 1967) has championed an evolutionary origin of cilia/flagella from spirochaetes through a symbiotic relationship. It is an idea that has not received wide acceptance. However the recent discovery of a circular chromosomal genome (6 to 9Mba DNA) associated with the basal body of *Chlamydomonas* flagella (Hall et al. 1989), while not proving the spirochaete relationship, certainly adds a new dimension to the hypothesis.

9.4.4 The Evolutionary Radiation of Protists

9.4.4A Search for the "Rosetta Stone"

The desire to elucidate evolutionary relationships has prompted researchers, perhaps unwittingly, to look for a single criterion that would provide the answer, as if, as Ragan and Chapman (1978, p. 212) noted "... phylogeneticists were gazing heavenward toward the Phylogeny hovering somewhere near the fourth level of the Divided Line." Rothschild et al. (1986) have cautioned against looking for such a rosetta stone, reminding that a phylogeny should be built on all appropriate data, not just one set. A similar caution has been expressed by Jupe et al. (1988, p. 229) who questioned whether "... enthusiasm for molecular approaches overwhelm prudence which cautions against any system of single character taxonomy." Penny (1989, p. 305), in discussing the *Prochloron-Prochlorothrix* question (*vide supra*) has added his voice to this chorus: "It is always safer to reconstruct trees by more than one method, whenever possible using more than a single molecule. One method with one coding sequence seems insufficient to reconstruct unambiguously the whole of evolution." These caveats are particularly appropriate to the protists where the evolution of organelles appears to be anything but tightly integrated with the nucleocytoplasmic (host) evolution. Within protistan phylogenetics there are but a few instances (Gunderson et al. 1987; Sogin et al. 1986b, c, 1988, 1989; Edman et al. 1988; Lynn and Sogin 1988; Ragan 1988, 1989) where nucleotide sequence results have been considered in the light of other biochemical or ultrastructural data. Gouy and Li (1989b), in attempting to build a global phylogeny, have in fact used a multiple data base, using sequence data from both large and small subunit rRNA and six proteins, including cytochrome *c*, Cu/Zn superoxide dismutase, glyceraldehyde-3-phosphate dehydrogenase, and (α) ATPase. It is a highly commendable approach which suffers, through no fault of Gouy and Li, from a less than desirable suite of organisms with which to work. Cedergren et al. (1989), also using both large and small subunit rRNA but from a greater selection of organisms, have also attempted to produce a global phylogeny, again with a good measure of success. To be sure, there are problems and inconsistencies, but these may well be merely a reflection of patchiness or the current lack of relevant information. The foregoing discussion has emphasized the complexity and problems involved in the elucidation of mitochondrial and chloroplast origins. Any single character will provide only a phylogeny of that character or component to which it belongs. For example, chloroplast 16S rRNA nucleotide sequences allow determinations about chloroplast phylogenies, but not about host or mitochondrion; nuclear 5S, 16S, 23S sequences permit phylogenetic construction of the nucleocytoplasmic genome or host, but not of the chloroplast and mitochondrion. In the same vein, Nei (1987), and Pamilo and Nei (1988), have emphasized the difference between a gene tree (constructed from sequences for a single gene) and a species tree (evolution of a group of species reflecting actual pathways). But in the final analysis, a phylogeny of protists involves the whole organism: the host or nucleocytoplasmic entity *and* the endosymbiotically derived organelles.

Early phylogenetic schemes were drawn without the availability of nucleotide sequences, relying on criteria further removed from the genome and, thus, except in the case of proteins (e.g., Hunt et al. 1985, who provide a series of protein trees), lacked quantitative precision. Without providing an explanation, it should be noted that ferredoxin (Schwartz and Dayhoff 1981; Matsubara et al. 1983; Minami et al. 1985, Uchida et al. 1988) and cytochrome c_6 sequences (Schwartz and Dayhoff 1981; Okamoto et al. 1987) from chloroplasts produce some unusual alignments (e.g., red algal and brown algal chloroplast relatedness) that are not in keeping with data from other arenas. The mitochondrial cytochrome data (Dayhoff and Schwartz 1981) would appear less controversial and more in keeping with other data, even to the point of providing an early divergent origin for *Euglena* and the kinetoplastid *Crithidia* (see comments below). Plastocyanin (Guss and Freeman 1983; Yoshizaki et al. 1989) and ribulose-1,5-bisphosphate carboxylase (Sailland et al. 1986; Keen et al. 1988) have too few known sequences, and from too divergent groups, to be of quantitative use. On the other hand, the sequences of the photosynthetic biliproteins (found only in cyanobacteria, Rhodophyceae, Cryptophyceae, and cyanelles) all suggest a common origin (Glazer and Apell 1977) for these proteins, a scenario not unexpected and in keeping with other data on cyanelle and the appropriate chloroplast origin. Other proteins (e.g., the superoxide dismutases, ATPases, glyceraldehyde-3-phosphate dehydrogenase) have yet to be fully tapped for information.

9.4.4B Phylogenetic Reconstructions

By far, the greatest number of sequences have originated from 5S rRNA. With only approximately 120 bases, 5S rRNA was conducive to sequencing, and the total number of sequences now determined is over 500. Different tree drawing methodologies have been employed, the merits, disadvantages, strengths, and weaknesses of which are discussed in Section 9.2. Most of the dendrograms thus produced have dealt with specific taxonomic units, for eukaryotes usually at the level of phylum, class, or order (and at the family level in the case of prokaryotes). A number of global or protistan phylogenies have been produced, examples of which, using cladistic analysis, are shown in Küntzel et al. (1981), Kumazaki et al. (1983), Hori et al. (1985), Walker (1985b), Hori and Ozawa (1986, 1987), Erdmann et al. (1987), Manske and Chapman (1987), and Wolters and Erdmann (1986, 1988); Mannella et al. (1987) have used correspondence analysis to produce taxonomic clusters. However, use of 5S rRNA, containing a relatively small number of bases, does have drawbacks (e.g., Fox and Stackebrandt 1987; Woese 1982) which can be further compounded by the different treeing methods. Any usefulness of 5S rRNA is probably restricted to identifiably monophyletic groups whose taxa are neither too closely nor too distantly related. The small subunit 16 to 18S rRNA, with approximately 1500 nucleotides, is a more reliable molecular character. Phylogenetic reconstructions with a wide array of protistan organisms, using 16S rRNA, have been made extensively by Sogin and coworkers (McCarroll et al. 1983; Sogin et al. 1986a; Gunderson et al. 1987; Sogin and Gunderson 1987; Lynn and Sogin 1988; Medlin et al. 1988; Edman et al. 1989; Sogin et al. 1989; Bhattacharya et al. 1990) and others (Wolters and Erdmann 1986, 1988; Vossbrinck et al. 1987; Bhattacharya and Druehl 1988; Baverstock et al. 1989; Hendriks et al. 1989). Specific questions of importance, for example, the origin of chloroplasts (Bonen and Doolittle 1976, Doolittle and Bonen 1981; Witt and Stackebrandt 1988; Giovannoni et al. 1988), the origin of mitochondria (Yang et al. 1985), and the phylogeny of *Prochloron* (Seewaldt and Stackenbrandt 1982; Stackebrandt 1983) have likewise been addressed through use of 16S rRNA sequences or catalogues. An excellent example to illustrate this is provided by Figure 9.4.1, taken from Gunderson et al. (1987).

9.4.4C The Protist Phylogenetic Tree

If one assumes that the tree presented by Gunderson et al. (1987) is one of the most precise phylogenetic trees of protists currently available, a number of important and significant conclusions can be drawn from this and subsequent work. The protists demonstrate extreme diversity and very early divergence. A number of workers have commented that the diversity of these early eukaryotes exceeds that of plants and animals, and that their divergence occurred long before that of plant and animals. As shown in Figure 9.4.1, the microsporidia (represented by *Vairimorpha*) and metamonads (*Giardia*) are the earliest, most deeply branched eukaryotes (Vossbrinck et al. 1987, Sogin et al. 1989), preceding even the kinetoplastids and euglenoids. The divergence of *Giardia* precedes that of *Vairimorpha*. The microsporidia and metamonads lack mitochondria and possess other presumably primitive features (Canning 1988; Cavalier-Smith 1987c), such as the lack of golgi and endoplasmic reticula, indicating that organelle appearance was not in lockstep with the development of a nucleus in eukaryotic evolution and that the earliest eukaryotes were simple nucleated cells. In this regard, one should also note Cavalier-Smith's (1987a) positioning of his eukaryotic Archezoa (microsporidia plus diplomonads) prior to the appearance of mitochondria, a position that he has recently reemphasized (Cavalier-Smith 1989).

The red algae, in particular, and dinoflagellates, have often been considered primitive eukaryotes. The red algae (see Bhattacharya et al. 1990, and comment in Gunderson et al. 1987, p. 5827) do not now appear to be very primitive eukaryotes, a statement that can also apply to the dinoflagel-

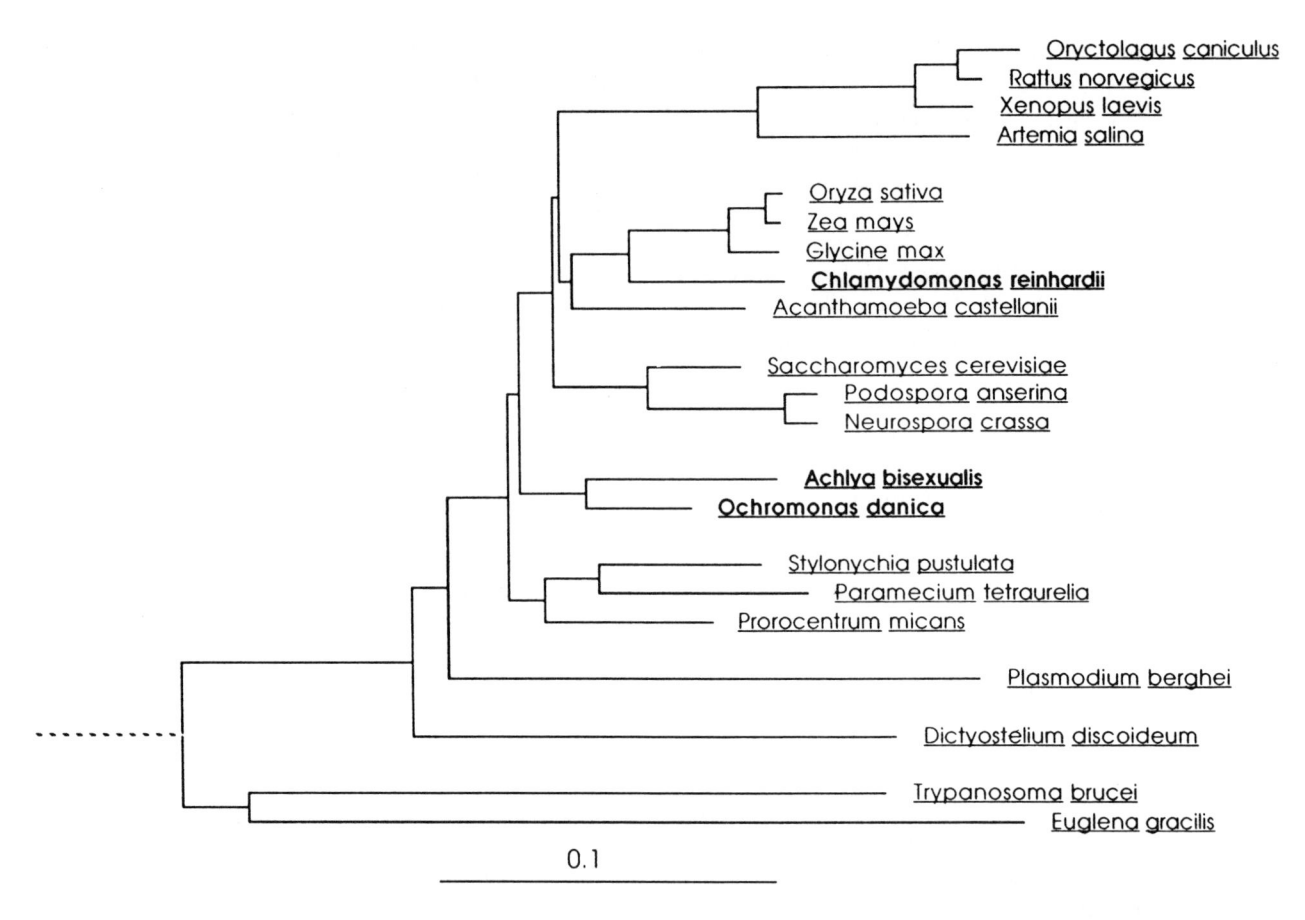

Figure 9.4.1. Eukaryotic phylogeny inferred from small subunit rRNA sequence similarities. The analysis is limited to approximately 1530 positions. The evolutionary distance between nodes of the tree is represented in the horizontal component of their separation in the figure. From Gunderson et al. (1987), reproduced with permission.

lates whose relationship to the ciliates is now apparent and which has been confirmed by large subunit (23S) rRNA sequence analysis (Nanney et al. 1989; Preparata et al. 1989; Lenaers et al. 1989). Herzog and Maroteaux (1986) suggest an early branching for dinoflagellates from the eukaryotic lineage, but this conclusion is obviously prejudiced by the too few (only three) protists examined. A similar conclusion on dinoflagellate relationships has been observed by Baroin et al. (1988) and Qu et al. (1988) using large subunit 28S rRNA. The latter authors also comment on a more recent origin for red algae, and in a recent publication (Perasso et al. 1989) provide evidence for this. Most interesting, however, is the close relationship suggested for red algae and cryptomonads. Classical phylogenies would not have predicted this, and it also should be remembered that these are only partial sequences; more definitive conclusions must await complete sequences a 16S rRNA sequences. Despite the claims by Demoulin (1974, 1985) for a red algal-fungal relationship, the data (Perasso et al. 1989; Bhattacharya et al. 1990) from the nucleic acid field do not support it; hopefully, this idea can be put to rest. Kwok et al. (1986) have found no evidence from DNA hybridization experiments to support the relationship.

An extensive series of data from 5S rRNA sequences exists for the various fungal groups (Walker and Doolittle 1982a, b, 1983; Walker 1984a, b, 1985a; Huysmans et al. 1983; Chen et al. 1984;

Gottschalk and Blanz 1984, 1985). As might be expected, the data provide mixed signals about the relationships of the various groups of fungi to each other. Even allowing for the problems inherent in 5S rRNA phylogenetics, there is no suggestion for a red algae-higher fungi relationship. It is worth noting that Stöcklein et al. (1983), using 18S rRNA catalogues, place the yeast, *Saccharomyces*, as a deep branching organism, and that Baharaeen and Vishniac (1984) support the idea, based on 25S rRNA hybridizations, that the primitive higher fungi were yeastlike. Edman et al. (1988, 1989) and Stringer et al. (1989) have also assigned the ancestral rooting of the fungal stock to the yeasts, with the enigmatic *Pneumocystis* being allied with the fungi and not the protozoa (see also Hendriks et al. 1989). The exact relationships of all of the true fungi are still problematic. More sequences, including complete ones from the large subunit rRNA should resolve this question.

Interestingly the most deeply branched protists, on the bases of small subunit (16S) rRNA sequence data and apart from the microsporidia, are the euglenoids and kinetoplastids, a situation supported by the partial 28S rRNA sequences (Perasso et al. 1989). These data indicate a relationship between these two latter groups, and it should be noted that Kivic and Walne (1984) had proposed a similar relationship between euglenoids and kinetoplastids, independent of sequence data. The other early branching groups are represented by acellular (*Physarum*) slime molds and certain amoebae (e.g., *Naegleria*) now postulated (Clark and Cross 1988; Baverstock et al. 1989) to be polyphyletic. A similar conclusion has been reached from analysis of 28S (large subunit) rRNA (Baroin et al. 1988; Qu et al. 1988). The Oömycetes or water molds (*Achlya*), and chrysomonads or yellow-brown algae (*Ochromonas*), show a close relationship to each other and, in turn, only a distant relationship to the fungi (*Podospora*, *Neurospora*), confirming what has been anticipated from a knowledge of comparative biochemistry. An unusual feature, with no obvious explanation at this time, is the close phylogenetic proximity of *Chlamydomonas* (green algae) to *Acanthamoeba* (Rhizopoda). It is instructive to compare this dendrogram (Fig. 9.4.1), admittedly representing the nucleocytoplasmic phylogeny, with those based on 5S rRNA and with phylogenies based on cladistic analysis (Wolters and Erdmann 1986, 1988) and cladistic analysis lacking any nucleotide input (Lipscomb 1985).

9.4.4D Inferred Geologic Times of Divergence

With the exception of the work of Hori and Osawa (1987) with 5S rRNA, geological time frames have not been incorporated into protist dendrograms. Ragan (1988) has discussed individual events within a time frame. Hasegawa et al. (1988) have also provided some divergence times, although their data for the *E. coli*-eukaryote divergence, ca. 17 Ga (more than three times the age of the earth!), obviously illustrates some of the problems inherent in these calculations, something these authors did discuss. Gouy and Li (1989b) suggest a time frame of 0.8 to 1.1 Ga (large subunit data) for the fungal divergence. For the green algal (*Chlamydomonas*) to higher plant divergence, Amati et al. (1989) have calculated a time of 700 to 750 million years ago. Sogin and Gunderson (1987) noted that their data suggested an extensive radiation, over a short period of time, giving rise to the plant, animal, fungi, and some protistan lineages. They suggested that this radiation occurred within a 50 to 100 million year time span, corresponding "to the concurrent acquisition of mitochondria and an environmental shift to an oxidizing environment" (p. 137). This is geologically a short time, and it is tempting to speculate that this occurred during the mid-Proterozoic. However, one must caution that this is based upon analyses of only a few taxa, and that additional sequences may well produce additional lineages and a longer time span. Examination of the fossil record (Sections 5.4, 5.5) suggests a time frame of 2.0 to 1.7 Ga for the origin of cellular eukaryotes. If the majority of protists appeared during this time, what does the deep branching of the kinetoplastids, euglenoids, slime molds, and especially microsporidia suggest? If we accept a constant rate of change in substitution, then eukaryotes (e.g., microsporidia) emerged much earlier than we have believed, based on the fossil record. Vossbrinck et al. (1987) have suggested the divergence of the microsporidia occurred as long ago as 2.0 to 2.8 Ga. Hasegawa et al. (1986) calculated (but see earlier cautionary note on *E. coli*-eukaryote divergence calculation) the divergence time for *Dictyostelium discoideum* at 1.83 ± 0.22 Ga (large subunit rRNA) and 2.21 ± 0.23 Ga (small subunit rRNA), while *Physarum polycephalum* gave a time of 3.03 ± 0.39 Ga (large subunit rRNA). In this regard, it should be remembered that dendrograms produced from large subunit rRNA (Baroin et al. 1988; Qu et al. 1988) and small subunit rRNA sequences (McCarroll et al. 1983; Gunderson et al. 1987; Elwood et al. 1985; Lynn and Sogin 1988; Clark and Cross 1988; Johansen et al. 1988; Sogin et al. 1988) demonstrate a deep divergence for *Physarum* and *Crithidia* (see also Schnare et al. 1986), but less for *Dictyostelium*. Regardless of any particular time of origin, especially for the metamonads (*Giardia*), microsporidia, (*Vairimorpha*), amoeba (as represented by *Naegleria*), acellular (*Physarum*) and cellular (*Dictyostelium*) slime molds, the euglenoids (*Euglena*), and kinetoplastids (*Trypanosoma*), the data suggest an evolution of eukaryotes much further back in time than previously suspected. It is now possible to speculate on the times for the major benchmarks in protistan evolution: ca. 2.5 Ga for the origin of eukaryotes; and 1.0 to 1.5 Ga for the massive radiation of fungi, photosynthetic protists, and animals. These are not outside the realm of possibility if the earliest (3.5 Ga) fossil prokaryotes (Section 1.5; Schopf and Packer 1986, 1987) were indeed O_2-producing photosynthesizers. Indeed, relating the earliest currently presumed eukaryotes from the fossil record (from approximately 1.8 Ga) to protistan groups on the phylogenetic tree takes on added importance, because it is questionable whether some of the protists implicated in the molecular phylogenetic trees as being among the earliest would leave a fossil record.

In summary, the great divergence and antiquity of the protists is documented. Their exact age, and thus an establishment for the time of appearance of eukaryotes, will require further interaction between studies of paleobiology and of molecular evolution.

Acknowledgments

It is not easy to establish and interpret the evolutionary history of the protists, the oldest and most diverse eukaryote conglomerate, but one plagued by a minimal early fossil record. Objectivity can frequently be muddied by the vagaries of personal prejudice and interpretation. I am indebted to a number of colleagues who have reviewed this manuscript and reminded me of my own prejudices and errors: Drs. Cliff Brunk, Mark Ragan, Bill Schopf, Mitchell Sogin, and Bob Trench. If there is objectivity here, it is to the credit of these reviewers; the opinions and errors remain mine. Drs. Ragan and Sogin are due additional thanks for alerting me to their papers in advance of publication. The author's evolutionary research was supported by NSF grant BSR-83-14838.

9.5

Origin and Diversification of the Metazoa

BRUCE N. RUNNEGAR

The paleontological evidence for the early evolution of the Metazoa is treated in Chapters 7 and 8. Here we shall deal briefly with molecular data that pertain to this problem.

The traditional approach to understanding the relationship between the metazoan phyla and the history of their evolution has been based largely on the meager information that is available from comparative anatomy and developmental biology of living animals (Anderson 1982). Unfortunately, there are far too few homologous characters that can be recognized in different phyla, and many of those that have been used are either primitive features or are polyphyletic. As a result, there are many hypotheses about how the principal phyla are related to one another and whether or not they constitute a monophyletic clade. Because most of the phyla appear abruptly in the Cambrian (Section 8), it is even uncertain whether multicelled animals had a relatively short (50 to 100 Ma) or relatively long (200 to 500 Ma) Precambrian history. The field is in a state of flux at present (Ax 1989; Conway Morris 1989a, b; Bergström 1989; Briggs and Fortey 1989; Gould 1989; Kauffman 1989; Lake 1990; Valentine 1989a; Patterson 1989, 1990).

Fortunately, new molecular data now make these questions far more tractable. By far the best available dataset is a series of partial 18S rRNA sequences assembled by Field et al. (1988). These data have been interpreted using distance methods (Field et al. 1988; Raff et al. 1989), a cladistic approach (Ghiselin 1988, 1989), and parsimony (Patterson 1989; Smith 1989; Lake 1990). The final answers are not yet in, but a general consensus is beginning to emerge (Fig. 9.5.1).

Despite an original suggestion that the Metazoa might be polyphyletic (Field et al. 1988), there now seems to be overwhelming evidence that all metazoans belong to a single monophyletic clade (Runnegar 1986; Walker 1989; Bode and Steele 1989). Within this tree, the cnidarians are, as expected, most distant from the other "higher" metazoans. The traditional deep division between the proterostomes on one hand, and the deuterostomes on the other, shows up very clearly in the rRNA tree, with the Echinodermata plus Chordata (deuterostomes) well separated from the other bilaterally symmetrical phyla (Fig. 9.5.1). Unexpectedly, arthropods appear as a kind of "stem group" from which other segmented (Annelida) and non-segmented phyla (Mollusca, Sipunculida) are derived (Lake 1990). Thus, the rRNA data seem to settle, once and for all, the question of whether the molluscs are closer to flatworms or to annelids; the sequence comparisons clearly point to a close relationship with the Annelida. Similarly, brachiopods – which have, at times, been regarded as deuterostomes – are much closer to annelids and arthropods than they are to chordates or echinoderms.

The question of whether the Metazoa had a short or long Precambrian history is more difficult to settle with existing molecular data. The estimates of several hundred million years given by Runnegar (1982b, 1986) based on protein sequence comparisons have been criticized by Erwin (1989). On the other hand, the distances between echinoderm and chordates measured by Field et al. (1988) are substantially less than the distances between either of these phyla and the cnidarians. Inasmuch as the echinoderm-chordate divergence can be no younger than the beginning of the Cambrian, proportionality would suggest a relatively long Precambrian history for the Metazoan clade. No doubt this question will be settled as more metazoan rRNA sequences become available in the not too distant future.

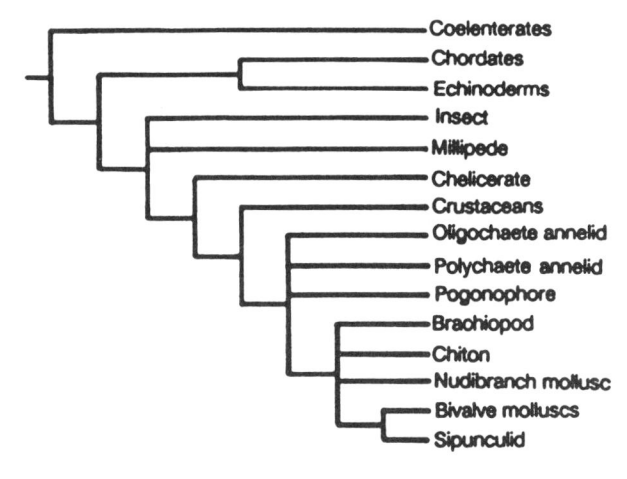

Figure 9.5.1. Relationships of major metazoan classes and phyla derived from the rRNA sequence data of Field et al. (1988) using parsimony analysis. From Patterson (1990), reproduced with permission.

10

Biostratigraphy and Paleobiogeography of the Proterozoic

HANS J. HOFMANN STEFAN BENGTSON J. M. HAYES
JERE H. LIPPS J. WILLIAM SCHOPF HARALD STRAUSS
ROGER E. SUMMONS MALCOLM R. WALTER

10.1

Introduction

HANS J. HOFMANN

Biostratigraphy deals with bodies of rock defined or characterized by their fossil content. Biogeography is concerned with the geographic distribution of organisms. The basic biostratigraphic principles and concepts now in use were developed in the early- to mid-nineteenth century by pioneers such as William Smith (1769–1839), Georges Cuvier (1769–1832), Alcide d'Orbigny (1802–1857), and Albert Oppel (1831–1865) who divided the stratigraphic record into successions of distinct faunal assemblages; the fundamental biostratigraphic unit still in use is the *biozone*, which usually is named after a dominant or a characteristic species. Fossils were unknown in pre-Cambrian rocks in 1835, when Adam Sedgwick introduced the concept of the Cambrian System; in fact, this interval was subsequently given names that referred to the presumed nonexistent or primitive paleontologic record (Agnotozoic, Archeozoic, Azoic, Eozoic, Protozoic, etc.).

Precambrian paleontology started in the 1850s, with the discovery of remains thought to be organic (for an historical summary, see Section 5.2 and Hofmann 1982, pp. 246–247). Although many of the early reported forms later were shown to be pseudofossils, some were true fossils. The number of accepted fossil occurrences increased slowly over the next 100 years, but only after the Second World War did Proterozoic biotic abundance and diversity become established by discoveries in various parts of the world (see Section 5.2). By the late 1950s, data were sufficient to be put to use in subdividing and correlating sequences locally and regionally, principally in the Soviet Union, giving rise to the subdiscipline of Precambrian biostratigraphy. This new field has rapidly expanded, particularly in the past three decades, as geologists have become increasingly interested in the early history of life, and in the practical application of their work. However, the quality of the time resolution for Precambrian biozones is still much less than that obtainable with Phanerozoic fossils, which may be of the order of 1 Ma, as compared to the approximately 100 Ma resolution, at best, for most of the Proterozoic. Nevertheless, fossils can serve for the global recognition of broad intervals, divisions such as those variously referred to as Ediacaran (or Ediacarian), Vendian, and Sinian (based on a peculiar assemblage of soft-bodied metazoans), or the even broader divisions of the Proterozoic (e.g., early, middle, and late Riphean), characterized by different fossil assemblages.

A promising recent adjunct in the study of the Precambrian biosphere is the systematic organic geochemical analysis of stratigraphically continuous intervals. Molecular fossils and biologically cycled stable isotopes of closely spaced samples have provided useful new perspectives on both short- and long-term trends of isotope fractionation of carbon, oxygen, and sulfur.

Proterozoic biostratigraphy is not without difficulties, and much work must be done before the various local and regional schemes can be unified into a coherent, generally accepted global system. A discussion of some of these problems follows.

10.2

Proterozoic Biostratigraphy: Problems and Perspectives

HANS J. HOFMANN

10.2.1 Introduction

Proterozoic fossils are abundant and exhibit a considerable degree of diversity, and successions of similar assemblages are found in different regions. Therefore, the same principles and methods can be used for Proterozoic sequences that are used for the Phanerozoic record. Similarity of fossil form is inferred to mean either similar process or similar stage of evolution (synchroneity), or both. In addition to the general problems in Phanerozoic sequences, the older units have additional diverse problems unique or peculiar to the pre-Phanerozoic record. It may be recalled that attributes of good index fossils include distinctive appearance (morphological complexity), wide geographic dispersal, short stratigraphic range, abundance of individuals, and presence in a wide variety of lithofacies.

10.2.2 Major Problems

A major problem in Proterozoic biostratigraphy is the nature of the remains. When compared to the Paleozoic, the Proterozoic fossils exhibit a relatively low biotic diversity and a primitive evolutionary stage of development: they have less morphologic complexity and less variety, resulting in fewer characters that can be used for circumscribing taxa or making correlations. For example, among the microfossils, how many attributes can be used to establish taxa for the most cosmopolitan of Proterozoic organisms, the simple spheroids and tubular forms? Even among the carbonaceous megafossils, relatively few morphological characters are usable in a diagnosis. The situation is considerably improved for the metazoans, but even here common forms such as the cyclomedusoids have simple, concentric construction. Moreover, some of the "taxa" used in Proterozoic biostratigraphy are not even the remains of organisms, but are biosedimentary structures: their formation involves sedimentary and diagenetic processes in addition to biological processes (e.g., stromatolites and oncolites). Still others are structures for which biogenicity is even more doubtful (e.g., catagraphs and camasids), yet they, too, have been used to make correlations. In addition to these shortcomings, the relatively slow rates of evolution of the early biotas allow for only rather broad biozones in the Proterozoic.

A second major problem is taxonomy. Just what is the best way to diagnose and name the diverse remains? The answer to this question depends to a great extent on such factors as the type of fossil under consideration, the state of its preservation, the method used to study the fossil, and the background and experience of the investigator. A useful biostratigraphic scheme is one with efficient presentation of distinct morphologic data. For better or for worse, paleontologists decided long ago to use Linnean binomial nomenclature to convey such information; rules were eventually set up to guide practitioners in matters such as formally erecting new taxa, priority, requirements for basic data, depositories, and so forth. Unfortunately, paleontologists use two major codes for biological nomenclature (one zoological, the other botanical; the bacterial code is inapplicable, because metabolism, which is the basis for bacterial systematics, cannot be determined for fossil specimens). While most provisions of the codes coincide, certain peculiar differences make some taxonomic problems difficult. Neither code deals specifically with the most widely used elements in Proterozoic biostratigraphy, the stromatolites, because they are neither animal nor vegetal nor wholly biologic. Despite this, many taxonomically oriented stromatolite researchers doing biostratigraphic work have attempted to adhere to the provisions of the codes.

Acritarchs present difficulties too. Mostly, acritarchs are the remains of phytoplankton and are treated under the botanical code, but they may also include heterotrophic protists (or their parts) that could be considered under the zoological code. Trace fossils which, like stromatolites, are biosedimentary structures, are specifically included in the most recent version of the zoological code.

Regardless of the type of biologic remains considered, and despite the existence of the codes, many Proterozoic taxa are either so inadequately diagnosed, described, or illustrated as to provide only limited information for other workers. This situation is made worse at times by the lack of appreciation for degradational and preservational variability of the morphology of both microscopic and megascopic fossils, or neglect of a sedimentary perspective for the matrix in which the fossils are enclosed. For microfossils, accidental irregularities such as folds or taphonomic perforations have too often been used as a basis for new taxa. Some stromatolite taxa have been based on small specimens taken from one part of a bioherm, without consideration of the full range of morphologic variability within the bioherm. On a more general level, stromatolite taxonomy is even more complicated, because fabric (microstructure) enters into the diagnosis of some taxa (= Spongiostromidae of Gürich 1906) while it does not in others; there is no unanimity on the need for separate but complementary taxonomic systems to accommodate both the stromatolite (gross morphology) and spongiostrome (microstructure) concepts. This situation may in part be the reason why stromatolite biostratigraphy is still strictly empirical and is questioned by some. If the systematics of stromatolites is not systematic, that is, if the sets of attributes are not used consistently for groups ("genera") and forms ("species") so as to allow tracking of attributes with time, how can evolutionary trends based on such taxa be determined? Stromatolitology is still in need of a theory that explains in what way stromatolites have changed with time.

A third major problem is the dissemination and accessibility of data, both physically and linguistically. Early work in Proterozoic biostratigraphy was done in the Soviet Union, and publications resulting from this work, precise locality data, and specimens for comparison, were often not available elsewhere, even if the literature was linguistically accessible; visits by foreigners to outcrops to view field relationships commonly

Table 10.2.1. *Contributors to the database on Proterozoic biostratigraphy. Abbreviations listing domicile are 2-letter country codes used by UNESCO. Abbreviations for regions are continent and country codes. Codes for specialties are: B = Body fossils; C = Chemofossils; S = Stromatolites; K = Catagraphs; M = Microfossils; I = Ichnofossils; O = Oncolites; D = Dubiofossils*

	Domicile	Specialties	Region
Aceñolaza, F. G.	AR	I	SA AR
Aharon, P.	US	C	AS IN
Aseeva, E. A.	SU	M	EU SU
Awramik, S. M.	US	M	NA US
Banerjee, S. M.	IN	S	AS IN
Barghoorn, E. S.	US	M	NA CA
Barman, G.	IN	S	AS IN
Basu, P. C.	IN	S	AS IN
Brasier, M. D.	GB	B	WORLD
Butin, R. V.	SU	S	EU SU
Cao R-J.	CN	S	AS CN
Chauhan, S. S.	IN	S	AS IN
Cloud, P. E., Jr.	US	B M S D	WORLD
Crimes, T. P.	GB	I	WORLD
Diver, W. L.	GB	M	AU WA
Dol'nik, T. A.	SU	S	AS SU
Du, R-L.	CN	B	AS CN
Fairchild, T. R.	BR	S	SA BR
Fedonkin, M. A.	SU	B I	WORLD
Fenton, C. L. & M. A.	US	S	NA US
Ford, T. D.	GB	B	NA US
German (Hermann), T. N.	SU	M	SU
Glaessner, M. F.	AU	B	WORLD
Gnilovskaya, M. B.	SU	B	EU SU
Golovanov, N. P.	SU	S	EU SU
Golovenok, V. K.	SU	S	AS SU
Germs, G. J. B.	ZA	B I	AF NA
Grey, K.	AU	S	AU WA
Hayes, J. M.	US	C	WORLD
Hofmann, H. J.	CA	B M I S D	WORLD
Horodyski, R. J.	US	M S	NA US
Jenkins, R. J. F.	AU	B	AU SA
Kiryanov, V. V.	SU	M	SU
Knoll, A. H.	US	M C	WORLD
Komar, V. A.	SU	S	SU
Korolyuk, I. K.	SU	S	SU
Krylov, I. N.	SU	S	SU
Kumar, S.	IN	S	AS IN
Lambert, I. B.	AU	C	AU
Liang, Y-Z.	CN	S	AS CN
Luo, Q-L.	CN	M	AS CN

Table 10.2.1. (*Continued*)

	Domicile	*Specialties*	Region
Magaritz, M.	IL	C	AS SU
Maithy, P. K.	IN	M S O K	AS IN
Makarikhin, V. V.	SU	S O	EU SU
Maslov, V. P.	SU	S O	SU
Mendelson, C. V.	US	M D	WORLD
Mikhaylova, N. S.	SU	M	SU
Milshtein, V. E.	SU	O K	SU
Muir, M. D.	GB	M	AU NT
Narbonne, G. M.	CA	B I	NA CA
Naumova, S. N.	SU	M	SU
Paczesna, J.	PO	I	EU PO
Paliy, V. M.	SU	B I	EU SU
Peat, C. J.	GB	M	AU NT
Pflug, H. D.	DE	B D	WORLD
Preiss, W. V.	AU	S	AU SA
Pyatiletov, V. G.	SU	M	SU
Raaben, M. E.	SU	S	SU
Raha, P. K.	IN	S	AS IN
Rudavskaya, V. A.	SU	M	SU
Runnegar, B. N.	AU	B C	AU SA
Sahni, M. R.	IN	B	AS IN
Sarfati (Bertrand-Sarfati), J.	FR	S O	WORLD
Schidlowski, M.	DE	C	WORLD
Schopf, J. W.	US	M C D	WORLD
Semikhatov, M. A.	SU	S	WORLD
Serebryakov, S. N.	SU	S	SU
Sergeev, V. N.	SU	M	SU
Shapovalova, I. G.	SU	S	AS SU
Shenfil', V. Yu.	SU	S	AS SU
Sokolov, B. S.	SU	B I	SU
Sprigg, R. C.	AU	B	AU SA
Summons, R. E.	AU	C	AU
Sun, W-G.	CN	B	AS CN
Timofeev, B. V.	SU	M	SU
Tucker, M. E.	GB	C	AF MA
Valdiya, K. S.	IN	S	AS IN
Verma, K. K.	IN	S	AS IN
Vidal, G.	SE	M	WORLD
Viswanathiah, M. N.	IN	S	AS IN
Vlasov, F. Ya.	SU	S	EU SU
Volkova, N. A.	SU	M	SU
Vologdin, A. G.	SU	S	AS SU
Vorontsova, G. A.	SU	O K	AS SU
Wade, M.	AU	B	AU SA
Walcott, C. D.	US	B M S D	NA US
Walter, M. R.	AU	B M I S D	WORLD
Webby, B. D.	AU	I	AU
Xing, Y-Sh.	CN	B M	AS CN
Yakshin, M. S.	SU	O K	AS SU
Yankauskas (Jankauskas), T. V.	SU	M	EU SU
Yin, L-M.	CN	M	AS CN
Zabrodin, V. E.	SU	K	SU
Zhang, P-Y.	CN	M	AS CN
Zhang, Y.	CN	M	AS CN
Zhang, Z-B.	CN	M	AS CN
Zheng, W-W.	CN	B	AS CN
Zhu, Sh-X.	CN	S	AS SU
Zhuravleva, Z. A.	SU	O K	AS SU

494   Biostratigraphy and Paleobiogeography

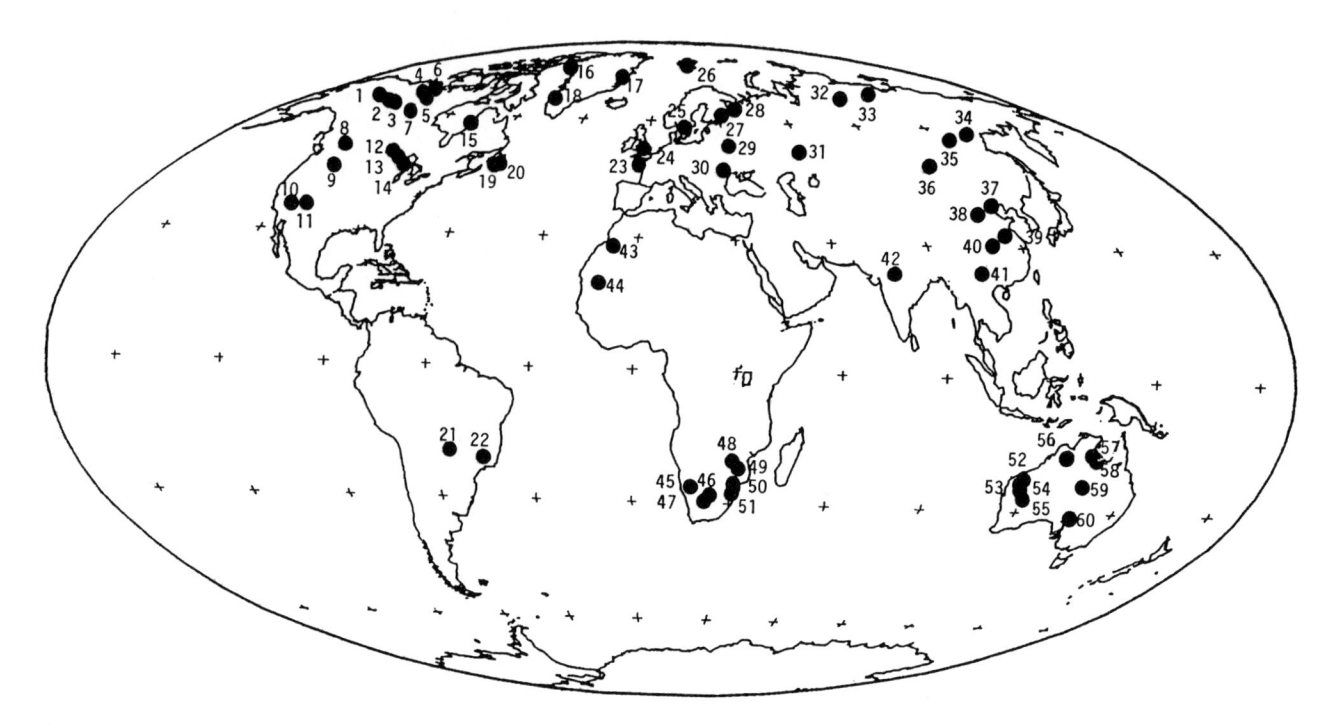

Figure 10.2.1. Geographic distribution of paleontologically and biostratigraphically significiant Archean and Proterozoic geologic sections. (A = Archean, >2500 Ma; L = Lower Proterozoic, 2500–1600 Ma; M = Middle Proterozoic, 1600–900 Ma; U = Upper Proterozoic (not including Vendian), 900–670 Ma; V = Vendian (Uppermost Proterozoic), 670–550 Ma; C = Cambrian.)

Number	Geologic Unit	Geologic Age
1.	Tindir Group	V C
2.	Wernecke Mountains: Windermere (Ekwi) Supergroup	V C
3.	Mackenzie Mountains: Little Dal Group; Windermere Supergroup	U V
4.	Dismal Lakes Group	M
5.	Epworth Group (Rocknest Formation)	L
6.	Snofield Lake Carbonate	A
7.	Great Slave Supergroup	L
8.	Windermere Supergroup	U V
9.	Belt (Purcell) Supergroup	M
10.	California: Pahrump Group and younger formations	M
11.	Grand Canyon, Arizona: Nankoweap Formation; Chuar Group	M U
12.	Steeprock Group	A
13.	Animikie Group; Gunflint Formation	L
14.	Keewenawan; Nonesuch Formation	M
15.	Belcher Supergroup	L
16.	Thule Group	M U
17.	Eleonore Bay Group	U V
18.	Isua Group	A
19.	Burin Peninsula: Chapel Island Formation	V C
20.	Avalon Peninsula: Conception Group; St. John's Group	V
21.	Corumbá Group	V
22.	Bambui Group	U
23.	Brioverian	V
24.	Charnian	V
25.	Visingsö Formation	U V
26.	Murchinsonfjorden Supergroup	U V
27.	Jatulian	L
28.	White Sea Coast: Valdaj "Series"	V
29.	East European Platform: Vendian	V
30.	Podolia: Vendian	V
31.	Southern Urals: Riphean	M U
32.	Anabar Massif: Riphean; Yudomian; Nemakit-Daldyn Formation	M U V C
33.	Olenyok Uplift: Riphean; Yudomian	M U V
34.	Yudoma-Maya region: Riphean; Yudomian	M U V
35.	Aldan Shield: Riphean; Yudomian	M U V
36.	Lake Baikal region: Riphean; Yudomian	M U V
37.	Jixian region: Changcheng "System"; Jixian "System"; Qingbaikou "System"	L M U
38.	Hutuo Group	L
39.	Huainan region: Feishui Group; Huainan Group	U
40.	Yangtze Gorge: Sinian	U V
41.	Meishucun region: Precambrian-Cambrian transition	V C
42.	Vindhyan Supergroup	U V
43.	Anti-Atlas: Infracambrian; Adoudounian; Taliwinian	U V C
44.	Hank Supergroup; Atar Group	U
45.	Nama Group	V C
46.	Transvaal Supergroup	L
47.	Ventersdorp Supergroup	A
48.	Bulawayo Group	A
49.	Belingwe Group	A
50.	Swaziland Supergroup (Onverwacht Group; Fig Tree Group)	A
51.	Pongola Supergroup	A
52.	Warrawoona Group	A
53.	Fortescue Group	A
54.	Wyloo Group (Duck Creek Dolomite)	L
55.	Earaheedy Group (Frere Formation)	L
56.	Bungle Bunble Dolomite	L M
57.	Roper Group	M
58.	McArthur Group	L M
59.	Bitter Springs Formation	U
60.	Adelaidean; Ediacaran	M U V

were not possible. Much recent work has been published in Chinese, and while generally available (often with English summaries), it is linguistically not accessible for most non-Chinese without translation. Precambrian biostratigraphy and biogeography have global perspectives, and the literature is multilingual; proficiency in Russian, Chinese, and English (or access to translations) are now essential for evaluation of activity in these fields. The linguistic requirement poses a considerable hindrance to a better understanding of the biotas, as well as a deterrent to individuals trying to do biostratigraphic work with Proterozoic fossils on any basis other than local or regional.

At present, the taxonomy of Precambrian fossils is untidy if not chaotic, characterized by a proliferation of redundant (synonymous) taxa that constitute an extensive excess baggage in Precambrian paleontology. Of approximately 1250 genus-level Precambrian taxa reported up to 1986, about half are either synonyms, inorganic objects, contaminants, or younger than Precambrian (Hofmann 1987, p. 136). The present situation is even worse for species-level taxa. Given the current state of relative flux in the taxonomy of the biostratigraphically used groups, particularly the acritarchs (Yankauskas et al. 1987), but also the stromatolites, and the uncertain biologic significance of oncolites and catagraphs, the quality of some of the detailed biostratigraphic schemes now used and based on these groups is questionable.

A fourth problem besetting Proterozoic biostratigraphy is the lack of geochronologic control. Many biostratigraphically important rock sequences, such as those represented by the Vindhyan of India, the Belt and Windermere of North America, the Adelaidean of Australia, and even the major biostratigraphic reference sections in the Soviet Union and China, are poorly dated radiometrically. Without more precise dating, the time-spans of taxa cannot be adequately constrained, and the rates of taxonomic change, if not evolution, of Proterozoic taxa cannot be determined.

Despite all these major difficulties, knowledge today of the stratigraphic and geographic distributions of Proterozoic fossils is greatly superior to what it was three decades ago. The advances have been made through systematic investigations of many thick, well-exposed, fossiliferous sections by specialists studying selected groups of remains, principally acritarchs and stromatolites. A number of empirical biostratigraphic schemes have been developed on different continents for the different groups of fossils, and similarities between assemblages of different regions have been used to make correlations. Only a few studies have attempted to integrate data for all groups. Important syntheses or compilations of biostratigraphic data on a regional basis include the following: Khomentovskiy et al. (1972), Sokolov (1979), Keller and Rozanov (1979; for translation, see Urbanek and Rozanov 1983), Wang et al. (1980), Keller (1982), Schopf (1983a), Velikanov et al. (1983), Sokolov and Ivanovskiy (1985), Xing et al. (1985), and Zhao et al. (1985).

In Table 10.2.1 are listed many of the workers who have contributed in varying amounts and in different ways to the database of Precambrian biostratigraphy; the location of biostratigraphically significant Proterozoic sections is summarized in Figure 10.2.1. The stratigraphic distributions of the different fossil groups are treated individually in the following sections of this Chapter.

10.3

Stratigraphic Distribution of Prokaryotes and Acritarchs

J. WILLIAM SCHOPF

10.3.1 Proterozoic Prokaryotes

The known geochronologic ranges of each of the 297 informal species of Proterozoic prokaryotes here recognized (Chapter 25) are summarized graphically in Table 25.4. The known temporal distributions of occurrences of these various types of Proterozoic microfossils are summarized in some detail in Section 5.4. In particular:

(1) The size distribution (of diameters, for more than 1500 occurrences) of solitary and colonial coccoid microfossils $\leqslant 60\,\mu m$ in diameter, interpreted as probable or possible prokaryotes and occurring in Early, Middle, and Late Proterozoic shales and cherts (and divided into 13 temporal subunits, 2200 to 550 Ma in age), is summarized in Figure 5.4.15.
(2) The size distribution (of widths versus lengths, maximum and minimum dimensions, for about 70 occurrences) of known Early, Middle, and Late Proterozoic solitary and colonial ellipsoidal microfossils, interpreted as bacterial or cyanobacterial prokaryotes, is summarized in Figure 5.4.21.
(3) The size distribution (of widths, for about 250 occurrences) of tubular, non-septate microfossils, interpreted as sheaths of filamentous prokaryotes and reported from Early, Middle, and Late Proterozoic shales and cherts (and divided into 11 temporal subunits, 2450 to 550 Ma), is summarized in Figure 5.4.26.
(4) The size distribution (of widths, for about 260 occurrences) of filamentous septate microfossils, interpreted as prokaryotic trichomes and known from Early, Middle, and Late Proterozoic shales and cherts (and divided into 12 temporal subunits, 2350 to 550 Ma in age), is summarized in Figure 5.4.32.

As is discussed in Section 5.4, none of these categories of Proterozoic prokaryotes appears to exhibit discernable, time-related, morphologic trends of potential biostratigraphic usefulness. However, available data do suggest that prokaryotic taxonomic diversity may have changed over Proterozoic time (see Section 11.3); such changes, if substantiated in the future by additional data, might ultimately provide a useful basis for a coarse-grained biostratigraphic zonation, especially of the Late Proterozoic.

10.3.2 Proterozoic Acritarchs

The known geochronologic ranges of each of the 76 informal species of Proterozoic microscopic eukaryotes here recognized (Chapter 25), including 68 taxa of acritarchs, are summarized in Table 25.4. The known temporal distributions of occurrences of these acritarch taxa are summarized in some detail in Section 5.5. In particular:

(1) The size distribution (of diameters, for more than 400 occurrences) of "mesosphaeromorph" acritarchs (having maximum diameters between 60 and $200\,\mu m$), occurring in Early, Middle, and Late Proterozoic shales and cherts (and divided into 13 temporal subunits, 1850 to 550 Ma), is summarized in Figures 5.5.2 and 5.5.3.
(2) The size distribution (of diameters, for about 130 occurrences) of "megasphaeromorph" acritarchs (having maximum diameters ranging from $200\,\mu m$ to more than $7,500\,\mu m$), reported from Early, Middle, and Late Proterozoic, and Early Cambrian shales and cherts (and divided into seven temporal subunits, 1850 Ma to 530 Ma), is summarized in Figure 5.5.6.
(3) The temporal distribution of the largest diameter sphaeromorph acritarchs reported from Proterozoic and Early Cambrian sediments 1500 to 530 Ma in age is summarized in Figure 5.5.12 and Table 5.5.2.
(4) The temporal distribution of the largest diameter acanthomorph (spiny) acritarchs reported from Proterozoic and Early Cambrian sediments 1000 to 530 Ma in age is summarized in Figure 5.5.15.
(5) The known temporal distribution of eight, major, Proterozoic acritarch subgroups is summarized in Figure 5.5.13.
(6) The known temporal distribution of important milestones in the morphological evolution of Proterozoic acritarchs (e.g., times of first known appearance of various types of processes, wall structure, and excystment mechanisms) are summarized in Figure 5.5.14.

In comparison with data available regarding the Proterozoic distribution of other types of acritarchs, those for large sphaeromorphs (viz., having vesicles with diameters in excess of $60\,\mu m$) are relatively numerous. As discussed in Section 5.5, the largest category of such sphaeromorphs, "megasphaeromorphs," exhibit relatively well-documented, time-correlative trends of potential biostratigraphic usefulness. Occurrences of these acritarchs are particularly numerous in units 1000 to 800 Ma in age (Fig. 5.5.6), a period that coincides with their greatest known taxonomic diversity and the presence of the largest diameter taxa (Fig. 5.5.12). As summarized in Table 5.5.2, available data suggest that the occurrence of sphaeromorphs larger than 5 mm in diameter may be restricted to strata ~ 975 to ~ 675 Ma; those larger than 1000 μm, to strata older than about ~ 675 Ma; and those larger than 600 μm, to the Proterozoic.

Although relevant data for morphologically more complex acritarchs are less numerous than those for sphaeromorphs, certain of these forms also appear to exhibit time-related trends of potential biostratigraphic utility. In particular, as shown in Figure 5.5.15, "giant acanthomorph" acritarchs (400 μm to more than 800 μm in diameter) may be restricted to units between about 950 and 600 Ma in age; all Early Cambrian acanthomorphs are evidently smaller than 75 μm; and the limited data available suggest that, like sphaeromorphs, the maximum diameters of acanthomorphs may have gradually decreased over the Late Proterozoic.

Finally, although the temporal distribution of other (non-sphaeromorph, non-acanthomorph) Proterozoic acritarch subgroups is at present poorly documented (Fig. 5.5.13), and although much remains to be learned regarding the temporal distribution of major milestones in the morphological evolution of early acritarchs (Fig. 5.5.14), such distributions also hold promise for biostratigraphic zonation, particularly of the Late Proterozoic.

10.4
Stratigraphic Distribution of Skeletonized Protists
JERE H. LIPPS

Skeletonized protists include all single-celled eukaryotes with an agglutinated or secreted mineral skeleton. While a great variety of these are known from Phanerozoic rocks (Lipps 1981), only melanocyrillids have a Proterozoic fossil record (Section 5.6). Melanocyrillids possess a vase-like, unilocular test, presumably composed of a complex kerogenous, organic polymer similar to "sporopollenin" (Bloeser 1985, p. 745). They occur world-wide in rocks ranging between 900 and 700 Ma (Hofmann 1987), and they are thus useful biostratigraphically. Their environmental tolerances are unknown. They have been considered as protists by some, but are now best regarded as micro-problematica (see Section 5.6). All reports of Proterozoic skeletonized protists are based on contamination, pseudofossils, or errors of interpretation (see Section 5.7).

True skeletonized protists (e.g., Fig. 5.7.1), such as foraminifera, radiolaria, and those with mineralized scales, first appear in the fossil record near the base of the Cambrian, in rocks containing the first skeletonized metazoans. Their diversity remains low throughout the Cambrian, but increases markedly in the Ordovician. Thus, protists do not become biostratigraphically useful until long after the end of the Proterozoic.

10.5

Stratigraphic Distribution of Megafossils

HANS J. HOFMANN   STEFAN BENGTSON

Following the convention adopted in Chapter 7, in which megafossils preserved as carbonaceous compressions are distinguished from trace fossils and fossils of soft-bodied and skeletonized Metazoa, the ranges of the taxa belonging to these various groups are here treated separately.

10.5.1 Carbonaceous Compressions

Macroscopic carbonaceous films (Figs. 7.3.9, 7.3.10) are known from the Proterozoic of most continents, dating back to about 2 Ga. In fact, one of them (the vendotaenid "*Laminarites*" *antiquissimus*) was one of the first bona fide fossils to be reported from what is now classed as Precambrian (Eichwald 1854). The stratigraphic and geographic distribution of these remains was summarized by Hofmann (1985b, Figs. 2, 3), and is here brought up to date in Table 10.5.1, Table 23.1, and in Figures in Section 7.3. (The numbering system used in the earlier compilation [Hofmann 1985b] is maintained in Tables 10.5.1 and 23.1 to facilitate cross-referencing; new localities are identified by numbers higher than 64, and some previously listed localities [e.g., 3, 61] have been removed from the inventory because the remains are no longer considered to belong to the category of carbonaceous films.)

Several observations can be made from the compilation. To begin with, rounded and angulate films without ornamentation and folds (that is, moranids and beltinids) have the longest stratigraphic ranges. They are known from rocks as old as about 2 Ga, occurring in euxinic Early Proterozoic (Aphebian) shales surrounding, on three sides, the Superior Structural Province of the Canadian Shield. Other occurrences are in Middle and Late Proterozoic units.

Both moranids and beltinids are of undetermined affinities, but most likely they represent whole and fragmented colonies and mats of prokaryotes and amorphous organic matter. Such remains therefore tend to be of very little or no use in biostratigraphy. Some late Early Proterozoic beds have been reported to contain chuarids and vendotaenids (Hofmann and Chen 1980), but these too may be bacterial or cyanobacterial in origin.

The first biostratigraphically significant event after the Early Proterozoic is the appearance of more complex megascopic compressions in the mid-Proterozoic, with the development of the spiraliform grypanids and other remains in the Belt Supergroup of Montana (Walcott 1899; Walter et al. 1976) and the coeval Gaoyuzhuang Group in northern China (Du et al. 1986). The fossils from these two regions are so similar to one another in morphology as to indicate generic identity, and they suggest a basis for the eventual recognition of a *Grypania* biozone. These two fossil-bearing sequences also have yielded similar radiometric ages in the 1.5 to 1.2 Ga range, which reinforces the potential value of *Grypania* as a zone fossil for the oldest biozone recognizable on the basis of such megafossils. However, with only two certain occurrences, additional data are necessary to confirm the utility of this taxon as a zone fossil. A third occurrence may be poorly preserved specimens attributed to *Grypania*, from the somewhat younger Little Dal Group (Hofmann 1985a, p. 349).

Spiraliform fossils that are three-dimensionally preserved elsewhere in Proterozoic rocks ("*Arenicolites*" Billings 1872a; "*Montfortia*" Lebesconte 1891; "*Helminthoidichnites*" Sokolov 1975; *Spiroichnus* Mathur 1983) do not appear to be carbonaceous, are of younger age, and may have affinities different from *Grypania* (see Section 7.3).

The next younger carbonaceous fossils of potential use in biostratigraphy are those in the *Tawuia* assemblage that has been recognized in five main areas on three continents (Hofmann 1987, Fig. 15). *Tawuia*, together with associated taxa like *Longfengshania*, *Sinosabellidites*, and other forms can serve as a basis for a broad *Tawuia* biozone characterizing the pre-Vendian part of the Upper Proterozoic. This taxon occurs in both basinal and platform sequences. *Sinosabellidites*, so far reported only from China, is similar in size and gross morphology to *Tawuia*, differing mainly by the presence of closely spaced transverse annulations the origins of which still need to be determined; this genus may be related to either *Tawuia* or to *Protoarenicola*. *Protoarenicola* and associated *Pararenicola* occur in the upper part of the *Tawuia* zone (Jiuliqiao Formation) in Anhui Province in China. These taxa form shiny black ribbons with regularly spaced annulations; although they have been regarded as worms (Sun et al. 1986, Fig. 2), such an attribution needs to be confirmed.

Table 10.5.1. *Stratigraphic ranges of Proterozoic carbonaceous films (Compiled by H. J. Hofmann).*

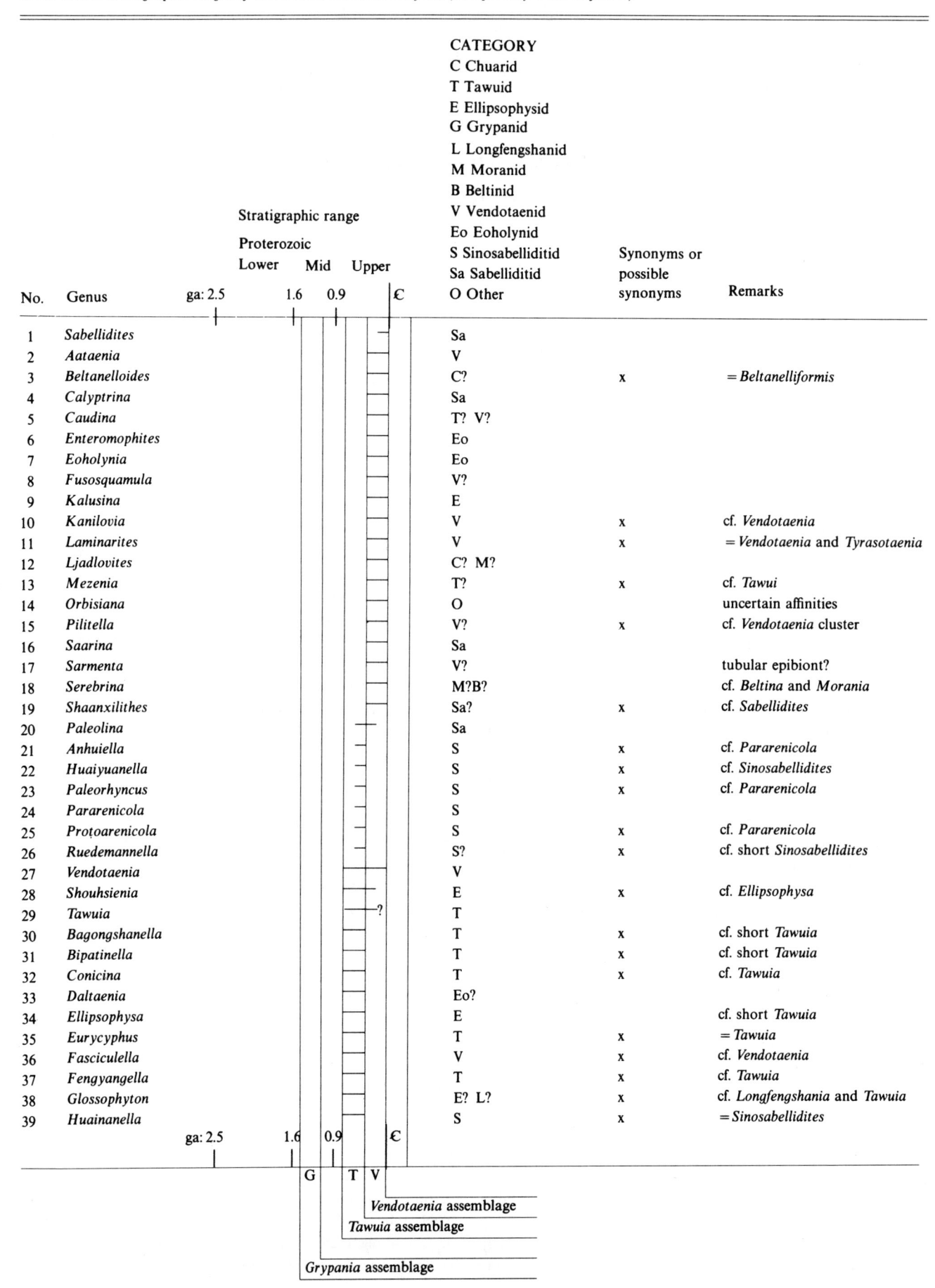

No.	Genus	Stratigraphic range Proterozoic	Category	Synonyms or possible synonyms	Remarks
1	Sabellidites	Upper–℃	Sa		
2	Aataenia	Upper	V		
3	Beltanelloides	Upper	C?	x	= *Beltanelliformis*
4	Calyptrina	Upper	Sa		
5	Caudina	Upper	T? V?		
6	Enteromophites	Upper	Eo		
7	Eoholynia	Upper	Eo		
8	Fusosquamula	Upper	V?		
9	Kalusina	Upper	E		
10	Kanilovia	Upper	V	x	cf. *Vendotaenia*
11	Laminarites	Upper	V	x	= *Vendotaenia* and *Tyrasotaenia*
12	Ljadlovites	Upper	C? M?		
13	Mezenia	Upper	T?	x	cf. *Tawui*
14	Orbisiana	Upper	O		uncertain affinities
15	Pilitella	Upper	V?	x	cf. *Vendotaenia* cluster
16	Saarina	Upper	Sa		
17	Sarmenta	Upper	V?		tubular epibiont?
18	Serebrina	Upper	M?B?		cf. *Beltina* and *Morania*
19	Shaanxilithes	Upper	Sa?	x	cf. *Sabellidites*
20	Paleolina	Upper	Sa		
21	Anhuiella	Mid–Upper	S	x	cf. *Pararenicola*
22	Huaiyuanella	Mid–Upper	S	x	cf. *Sinosabellidites*
23	Paleorhyncus	Mid–Upper	S	x	cf. *Pararenicola*
24	Pararenicola	Mid–Upper	S		
25	Protoarenicola	Mid–Upper	S	x	cf. *Pararenicola*
26	Ruedemannella	Mid–Upper	S?	x	cf. short *Sinosabellidites*
27	Vendotaenia	Mid–Upper	V		
28	Shouhsienia	Mid–Upper	E	x	cf. *Ellipsophysa*
29	Tawuia	Mid–Upper?	T		
30	Bagongshanella	Mid	T	x	cf. short *Tawuia*
31	Bipatinella	Mid	T	x	cf. short *Tawuia*
32	Conicina	Mid	T		cf. *Tawuia*
33	Daltaenia	Mid	Eo?		
34	Ellipsophysa	Mid	E		cf. short *Tawuia*
35	Eurycyphus	Mid	T	x	= *Tawuia*
36	Fasciculella	Mid	V	x	cf. *Vendotaenia*
37	Fengyangella	Mid	T	x	cf. *Tawuia*
38	Glossophyton	Mid	E? L?	x	cf. *Longfengshania* and *Tawuia*
39	Huainanella	Mid	S	x	= *Sinosabellidites*

CATEGORY
C Chuarid
T Tawuid
E Ellipsophysid
G Grypanid
L Longfengshanid
M Moranid
B Beltinid
V Vendotaenid
Eo Eoholynid
S Sinosabelliditid
Sa Sabelliditid
O Other

ga: 2.5 — 1.6 — 0.9 — ℃

G | T | V
Vendotaenia assemblage
Tawuia assemblage
Grypania assemblage

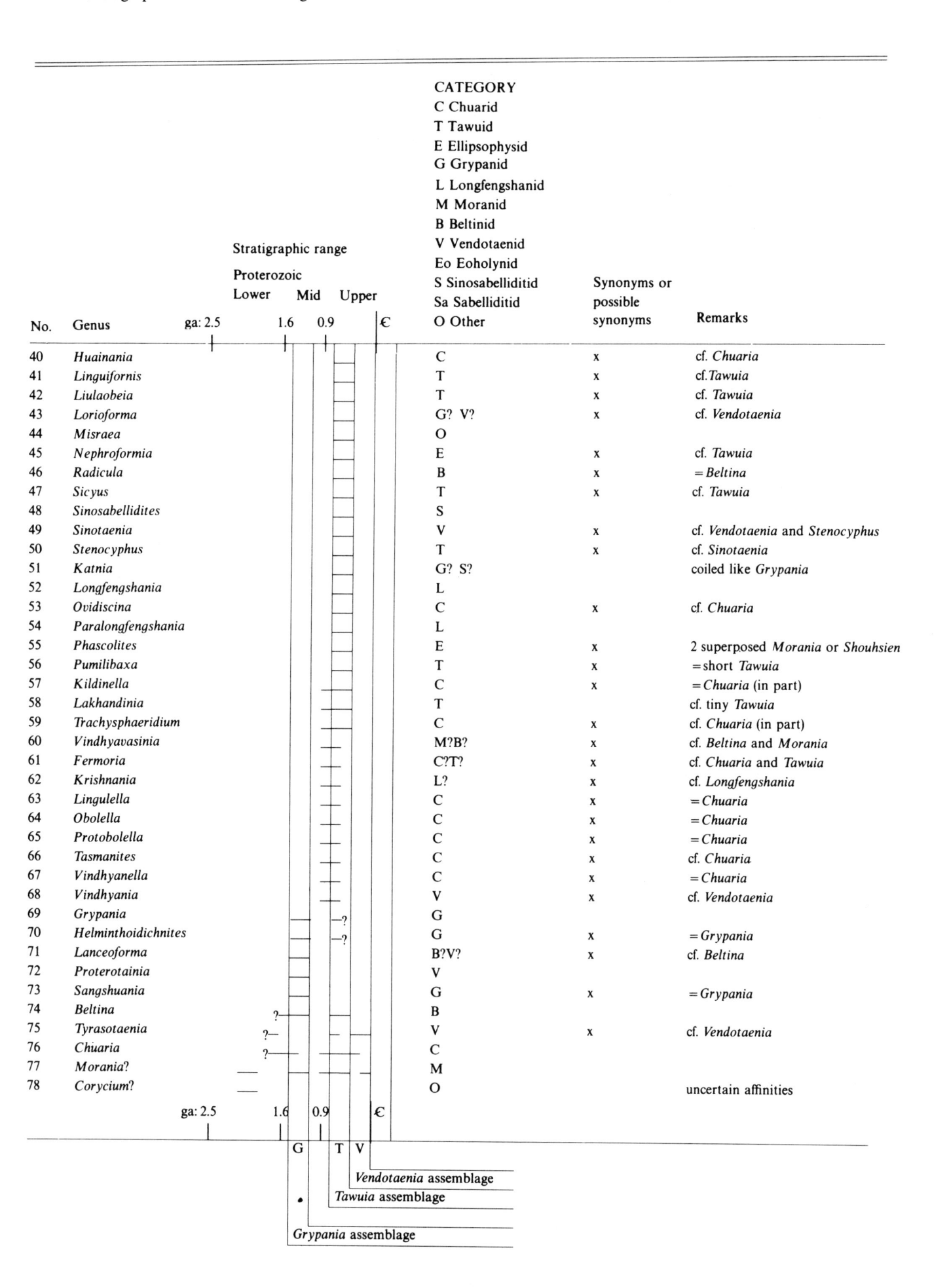

Stratigraphic Distribution of Megafossils

CATEGORY
- C Chuarid
- T Tawuid
- E Ellipsophysid
- G Grypanid
- L Longfengshanid
- M Moranid
- B Beltinid
- V Vendotaenid
- Eo Eoholynid
- S Sinosabellditid
- Sa Sabellditid
- O Other

Stratigraphic range Proterozoic: Lower (ga: 2.5) – Mid (1.6) – Upper (0.9) – €

No.	Genus	Category	Synonyms or possible synonyms	Remarks
40	Huainania	C	x	cf. Chuaria
41	Linguifornis	T	x	cf. Tawuia
42	Liulaobeia	T	x	cf. Tawuia
43	Lorioforma	G? V?	x	cf. Vendotaenia
44	Misraea	O		
45	Nephroformia	E	x	cf. Tawuia
46	Radicula	B	x	= Beltina
47	Sicyus	T	x	cf. Tawuia
48	Sinosabellidites	S		
49	Sinotaenia	V	x	cf. Vendotaenia and Stenocyphus
50	Stenocyphus	T	x	cf. Sinotaenia
51	Katnia	G? S?		coiled like Grypania
52	Longfengshania	L		
53	Ovidiscina	C	x	cf. Chuaria
54	Paralongfengshania	L		
55	Phascolites	E	x	2 superposed Morania or Shouhsien
56	Pumilibaxa	T	x	= short Tawuia
57	Kildinella	C	x	= Chuaria (in part)
58	Lakhandinia	T		cf. tiny Tawuia
59	Trachysphaeridium	C	x	cf. Chuaria (in part)
60	Vindhyavasinia	M?B?	x	cf. Beltina and Morania
61	Fermoria	C?T?	x	cf. Chuaria and Tawuia
62	Krishnania	L?	x	cf. Longfengshania
63	Lingulella	C	x	= Chuaria
64	Obolella	C	x	= Chuaria
65	Protobolella	C	x	= Chuaria
66	Tasmanites	C	x	cf. Chuaria
67	Vindhyanella	C	x	= Chuaria
68	Vindhyania	V	x	cf. Vendotaenia
69	Grypania	G		
70	Helminthoidichnites	G	x	= Grypania
71	Lanceoforma	B?V?	x	cf. Beltina
72	Proterotainia	V		
73	Sangshuania	G	x	= Grypania
74	Beltina	B		
75	Tyrasotaenia	V	x	cf. Vendotaenia
76	Chuaria	C		
77	Morania?	M		
78	Corycium?	O		uncertain affinities

G – Grypania assemblage
T – Tawuia assemblage
V – Vendotaenia assemblage

Although vendotaenids have been reported from the Early and Middle Proterozoic, only in the Late Proterozoic, particularly the Vendian, do they become abundant in many regions, where they are preserved as large, often twisted carbonaceous ribbons ("*Laminarites*") that are found mainly in terrigenous units. They have been assigned to several genera, all considered to be metaphytes (Gnilovskaya 1971, 1979). However, evidence indicates that some of the morphologic attributes used in defining *Vendotaenia* and *Tyrasotaenia* may not be primary, and also that these ribbons may be the remains of bacterial colonies (Vidal 1989). Three successive biotas have been recognized on the Russian Platform (Gnilovskaya 1985, p. 125): a lower assemblage with *Eoholynia*, *Leiothrichoides*, and *Orbisiana*; a middle one with *Aataenia*, *Vendotaenia*, and *Leiothrichoides*; and a latest Vendian-Early Cambrian assemblage with *Dvinia* and *Tyrasotaenia*.

Lastly, the sabelliditids are characteristically found in latest Vendian and Cambrian beds of eastern Europe (Baltic Stage) and Siberia (e.g., the Platonov and Sukharikha Formations). They have also been cited as occurring in rocks as old as 740 Ma (viz., in the Jiuliqiao Formation of China: Wang 1982; Wang et al. 1984; Chen and Zheng 1986). These latter occurrences, however, could also be interpreted as sinosabelliditids, which are abundant at this level.

10.5.2 Soft-bodied Metazoans

Metazoans are the most common biostratigraphic tools used for the Phanerozoic, and they offer one of the best ways for correlating Vendian post-tillite sequences. Representatives of this fauna of soft-bodied organisms, commonly called the Ediacara(n) Fauna, were originally collected in the Nama Group in the period from 1908 to 1914 by P. Range and H. Schneiderhöhn (Richter 1955, p. 244). They were first described, however, only many years later (Gürich 1930a, 1933) when they were regarded as probably Cambrian (with a pre-Cambrian assignment being considered, but rejected). A probable Cambrian age was also attributed to the faunally diverse assemblage subsequently discovered in the Ediacara Hills of South Australia (Sprigg 1947, 1949). Opinions seem to have changed after a third assemblage of such metazoans was discovered in the Charnwood Forest in England (Ford 1958), where a pre-Cambrian assignment was indicated by field relationships with overlying Cambrian rocks in a locality relatively close to the type area of the Cambrian System. This find provided further data and focused interest on early metazoans, by then known from three continents. The question of the stratigraphic position of the two other occurrences was taken up, and a pre-Cambrian age gradually became accepted. Further finds in the 1960s in Eurasia (Zaika-Novatskiy 1965) and North America (Anderson and Misra 1968) augmented the available data which have continued since to grow spectacularly. The most diverse assemblage presently known occurs in the Soviet Union, on the southeastern shore of the White Sea (Fedonkin 1981). Over one hundred species are now known (see Section 7.5), and the widespread distribution of particularly characteristic taxa at many localities on at least five continents demonstrates the global nature of this fauna and its biostratigraphic utility. In fact, the recognition of a separate biozone (Termier and Termier 1960), or of a geologic system and period based on this fauna, has been proposed (Jenkins 1981; Cloud and Glaessner 1982; Sokolov and Fedonkin 1984; Xing 1984), although no formal international agreements on either the name (Ediacar(i)an/Vendian/Sinian) or the boundaries have yet been concluded.

In addition to the now numerous reports of bona fide early metazoans are also many reports of remains of pseudofossils or, at best, dubiofossils (see Table 23.6), and some that are Cambrian or younger. Authentic fossils that support long-distance correlations are treated in Section 7.5. The quality of such correlations varies greatly, depending on the number of taxa in common and the type of fossils preserved. Little difficulty is encountered when distinctive fossils are present, for example, *Charnia*, *Charniodiscus*, *Dickinsonia*, *Pteridinium*, *Rangea*, or *Tribrachidium* (Figs. 7.5.5 through 7.5.10). If only a few cyclomedusoids or other morphologically indistinct taxa are present, as for some occurrences, the confidence in the correlation is greatly reduced.

Thus, although many endemic genera and species occur in the White Sea area, Podolia, and at Ediacara, the fauna in each region, preserved in shallow water sediments of stable platform settings, is so abundant that their use for biostratigraphic correlation can be considered firm.

Similarly, the faunas of the Avalon Peninsula in Newfoundland and of Charnwood Forest in England have much in common, although they differ somewhat in taxonomic composition from those at Ediacara and the White Sea. Such differences probably reflect differences in environment, the Avalon and Charnian sediments representing deeper marine depositional environments characterized by active margin or island arc sedimentary tectonics with turbidite accumulation and an abundance of volcanogenic detritus.

Reports of metazoans older than tillites at the base of the Vendian are difficult to evaluate, because the material on which the metazoan interpretations are based is poor and sparse. Sinosabelliditids, known only from eastern China, may be worms (Pogonophora), but the evidence is inconclusive and they may be algal.

10.5.3 Trace Fossils

Interest in ichnofossil-based stratigraphic correlation is growing. Indeed, ichnofossils have been proposed to define the base of the Cambrian (Crimes 1987; Narbonne et al. 1987). Trace fossils are autochthonous, formed and preserved within the sedimentary environment soon after deposition; they reflect ambient physical and chemical conditions, and the gross morphology and behavioral patterns of animals that can be traced through time. The widespread preservation of undisturbed, finely laminated mudstones throughout most of the Proterozoic is testimony to the general absence of burrowing animals for that interval, at least those that are macroscopic. Traces that are reported to be pre-Vendian are here regarded as dubious, or as trace fossils of uncertain age (Section 7.6, Table 23.4).

In general terms, Late Proterozoic trace fossils (Figs. 7.6.3, 7.6.4) exhibit evolutionary trends both of increasing size and of complexity of pattern. Precambrian traces tend to be small, simple, of low diversity and abundance, and to show preferential development parallel to bedding (*Helminthoidichnites* [*Gordia*], *Planolites*); the three-dimensionally preserved spiraliform structures (*Spiroichnus*, "*Arenicolites*" *spiralis*, "*Montfortia*" *filiformis*, "*Helminthoidichnites*" *spiralis*) reported from various Late Proterozoic rocks may also be included here. By comparison, traces in Cambrian and younger ichnofaunas are larger, more diverse, more complex, and show vertical development (for graphic summaries, see Fedonkin 1981, Figs. 5, 12; Crimes and Anderson 1985; Paceśna 1986, Fig. 3; Crimes 1987, Figs. 1–3.). Some ichnogenera are restricted to the Precambrian, but relatively few such sections are well-studied (see Crimes 1987) so that the geographic distribution of such genera is poorly known. The greatest known diversity is that reported for the Redkino Group of the Russian Platform, from which 14 ichnogenera have been described (Fedonkin 1985b, p. 116), many of which range into the Cambrian; the immediately overlying Kotlin Formation has a much reduced diversity, with only eight ichnogenera present of which six continue on into the Cambrian.

Genera that appear to be restricted to the uppermost Proterozoic are *Intrites*, *Harlaniella*, *Medvezhichnus*, *Nenoxites*, *Palaeopascichnus*, *Stelloglyphus*, *Suzmites*, *Vendichnus*, and *Vimenites* (compare with the list in Alpert 1977, reproduced in Byers 1982). Many new taxa make their appearance in the Cambrian (Baltic Stage and equivalents), including those of the Rovno Formation which contains the trace fossil *Phycodes pedum* that has been used as an index for the Cambrian (Crimes 1987; Narbonne et al. 1987). In the candidate stratotype section for the Precambrian-Cambrian boundary on the Burin Peninsula in Newfoundland, the lowermost occurrence of this species is only centimeters above the highest observed occurrence of *Harlaniella podolica*, providing one of the reasons for selecting and proposing the location of the boundary at this level (Narbonne et al. 1987).

10.5.4 Early Skeletonized Metazoans

Skeletal fossils abound in the Phanerozoic, where they provide the main framework for biostratigraphy. The lack of such readily preserved fossils in the Proterozoic is a major reason for the comparatively lower correlation potential of Proterozoic rocks (among other reasons being the higher biological diversities and the better preservation potential, typical of the Phanerozoic rocks). The transition to the Cambrian thus signifies a considerable sharpening of biostratigraphic resolution.

Nevertheless, the global correlation of the earliest skeletal biotas is fraught with many difficulties. Current international efforts to define the Precambrian-Cambrian boundary have so far floundered on the problems of correlating with sufficient precision the boundary interval in different key regions, such as Siberia, China, and the Avalon Platform (Cowie 1985). Other potential means of long-distance correlation, such as organic microfossils (Moczydłowska and Vidal 1988) and trace fossils (Crimes 1987; Narbonne et al. 1987), along with data from, for example, stable isotopes (Hsu et al. 1985; Magaritz et al. 1986; Tucker 1986; Knoll et al. 1986; Lambert et al. 1987; Aharon et al. 1987), magnetostratigraphy (Kirschvink and Rozanov 1984), and geochronometry (Odin et al. 1983; Cowie and Johnson 1985), have not as yet led to greater precision, although they may provide grounds for more successful integrated approaches.

Not unlike the situation in Proterozoic biostratigraphy, a major impediment in using skeletal fossils in earliest Cambrian biostratigraphy is the prevailing poor state of taxonomy. About four-fifths of the taxa of early skeletal fossils recognized today were established during the last two decades (see Section 7.7). Much of this taxonomic work has been of a preliminary nature, and the often obscure biology of the various groups involved has prevented stable taxonomic practices from evolving. Most such taxa have been erected on imperfectly preserved and illustrated material, and consequently their usefulness for biostratigraphic correlation is limited.

The tubular fossil *Cloudina* (Fig. 7.5.11) is currently regarded as the oldest fossil with a mineralized skeleton, being reported from rocks that interfinger with beds containing elements of the Ediacara fauna in Namibia (Germs 1972b; Glaessner 1976, 1984). *Cloudina*-like fossils have been reported from Argentina (Yochelson and Herrera 1974), Brazil (Hahn and Pflug 1985) and Spain (Palacios 1987), but the stratigraphic and taxonomic relationships to the Namibian occurrences are not clear.

In a number of geographical areas, the first more diverse biota of skeletal fossils to appear is an *Anabarites-Protohertzina* assemblage (see Figs. 7.7.3, 7.7.4). It characterizes the Nemakit-Daldyn beds of northern Siberia (Voronova and Missarzhevsky 1969; Rozanov et al. 1969; Missarzhevsky 1973, 1982, 1983; Shishkin 1974; Val'kov and Sysoev 1970; Val'kov 1975, 1982) and the lower part of the Meishucunian Stage in China (Zhong [Chen] 1977; Qian 1977, 1978; Qian et al. 1979; Qian and Yin 1984a; Yin et al. 1980; Luo et al. 1980, 1982, 1984; Chen 1982). This assemblage has also been found to initiate the succession of skeletal fossils in the Lesser Himalayas, India (Azmi et al. 1981; Azmi and Pancholi 1983; Bhatt et al. 1985; Brasier and Singh 1987; Kumar et al. 1987), the Elburz Mountains, Iran (B. Hamdi, pers. comm., to S. B. 1986), and western Canada (Conway Morris and Fritz 1980; Nowlan et al. 1985).

The occurrence of *Anabarites trisulcatus* and *Protohertzina anabarica/unguliformis* has long been considered indicative of Vendian-Cambrian boundary beds, but their stratigraphic co-occurrence with trace fossils of somewhat younger aspect in the Vampire Formation of the Yukon Territory (Nowlan et al. 1985; Crimes 1987), and the discovery of representatives of both key genera in late Atdabanian strata in South Australia (Bengtson et al. in prep.), suggest that their range may be longer than assumed. Nevertheless, the geographically widespread occurrence of an *Anabarites-Protohertzina* fauna in a consistent position in the faunal succession makes it possible to recognize it as a biostratigraphic unit. In want of a globally acceptable name for this unit, it is here referred to as the "Nemakit-Daldynian." This unit is to be regarded as an approximate equivalent of the Manykayan of Missarzhevsky (1982) and the Lower Meishucunian of Chinese workers.

The skeletal biotas immediately succeeding the *Anabarites-Protohertzina* assemblage, in particular the Tommotian of the Siberian Platform and the Middle Meishucunian of China, show a higher degree of endemism, and their global correlation is generally problematic. In the Atdabanian, diversity increases and endemism decreases, resulting in more secure correlation.

The range chart of the major groups of skeletal fossils (Section 7.7) is based on the known occurrences of genera as listed in Table 23.5. The stratigraphic framework of this Table comprises the Siberian Lower Cambrian Stages (Spizharskij et al. 1983) plus the "Nemakit-Daldynian", tentatively correlated and assigned absolute ages in Table 7.5.2. Although the pattern of first appearances of skeletal fossils is strongly facies-dependent (e.g., Brasier 1982; Mount and Signor 1985), this does not change the large-scale picture of an almost instantaneous appearance in the fossil record of most major groups of skeletal fossils.

10.5.5 Camasid Dubiofossils

Dubiofossils known as camasids have been used by some Soviet workers to correlate between the Poludennaya, Tjurim, and Aramon Formations in central Asia and the Belt Supergroup in North America (Vasiliev et al. 1968a, 1968b; Sosnovskaya 1980, 1981). These structures, referred to "*Newlandia*" and "*Camasia*", apparently are similar to structures in the Newland Limestone of Montana. A large number of other, endemic "taxa" are also reported. The correlations noted above were not accompanied by systematic paleontologic descriptions or photographic illustrations and, thus, the documentation for the correlations are not available. Moreover, the remains described by Walcott (1914) under such names as *Copperia* and *Greysonia*, reinterpreted by Fenton and Fenton (1936) and others as inorganic, are at best dubiofossils (with a possible microbial contribution to their origin). Their apparent restriction to the mid-Proterozoic, like that of molar-tooth structure, is intriguing and worthy of further investigation. If not biogenic, perhaps unique physicochemical or environmental conditions existed, or were more widespread, to promote formation of such structures at this particular interval of geologic time. Such conditions might have chronostratigraphic significance.

10.6

Stratigraphic Distribution of Stromatolites and Allied Structures

MALCOLM R. WALTER

In this discussion, the term "stromatolite" is taken to include oncolites and thrombolites (i.e., it is used synonymously with "microbialite" as defined by Burne and Moore 1987).

It is now well established that a plexus of environmental and biological influences determine the form and fabric of stromatolites (e.g., Walter 1977; Semikhatov et al. 1979; Burne and Moore 1987, although only the most obvious of these influences have been studied to any significant extent. The small-scale features of stromatolites, the "fabric" (lamina shape and microstructure) record the shapes of the constructing microbial colonies and frequently also record some information on the size, shape, and orientation of the cells in those colonies (e.g., Semikhatov et al. 1979; Walter 1983). As a consequence, taxonomic treatments of stromatolites that give weight to fabric features will have the best chance of reflecting the past distribution in time and space of different microbial communities.

Some confidence in the use of stromatolites in biostratigraphy can be gained by first considering what might have produced the succession of assemblages that has been described. Work of the last 30 years has documented direct fossil evidence of microbial evolution through the Proterozoic, and this is now being importantly supplemented by various geochemical techniques (see Chapter 3 and Section 10.7). Microfossil studies are only possible, of course, where suitable preservation has occurred, and this represents a significant limitation to the record. Many of the monographically documented fossil stromatolitic microbiotas (Section 6.2.2; Table 25.7) are from stratiform stromatolites (e.g., the Bitter Springs Formation and the Belcher Islands units) that formed in littoral environments where even today microbiotas are of low diversity and are dominated by cyanobacteria (Knoll 1985a). The small number of well-documented fossil microbiotas from structured (non-stratiform) stromatolites (e.g., Schopf and Sovietov 1976; Awramik and Semikhatov 1979) show community structures different from those in stratiform stromatolites. Even with all the recent advances (Sections 5.4, 5.5), the microfossil record, as currently known, samples a smaller range of paleoenvironments at any one time than does the extraordinarily abundant record of stromatolites. The significant point for this discussion is that even if the relatively limited Proterozoic microfossil record documents microbial evolution, then we can, nevertheless, expect a fuller record of evolutionary change in benthic communities to be encoded in stromatolites.

Another observation reinforces the view that is reasonable to expect to be able to document substantial changes in stromatolites over time. It is now well established, though still poorly documented, that extant subtidal stromatolites are inhabited and probably constructed by diverse microbiotas including abundant micro-algae (Dravis 1983; Bauld 1984; Awramik and Riding 1988). The few known examples have a distinctive fabric which might be a result of an ability of these algae to trap and bind coarse sediment (Awramik and Riding 1988). Whatever the explanation of the fabric features, the advent and diversification of eukaryotic algae during the Proterozoic must have increased the range of fabrics and chemical microenvironments within benthic microbial mats previously dominated by cyanobacteria, and therefore inevitably increasing also the diversity of stromatolites.

Given that it is reasonable to postulate a link between microbial evolution and stromatolite change, it is still a long way from being able to demonstrate that observed stromatolite changes are due necessarily to biotic evolution. Nevertheless, there seems ample evidence of a major decline in the abundance and diversity of stromatolites during the Late Proterozoic (Fig. 6.2.3; Awramik 1971; Walter and Heys 1985), a decline that appears to correlate with a decrease in the apparent diversity of Late Proterozoic stromatolite-building benthos (Fig. 11.3.10). As discussed in some detail in Section 6.2.6, alternative, non-biologic explanations for the stromatolite decline have also been presented, and all current explanations are best regarded as working hypotheses in various stage of development that in any event are not mutually exclusive. So, although there is not yet an accepted "theory of stromatolite change," enough is known so that it is no surprise that stromatolites did change.

For the sake of completeness, it is here necessary to mention "catagraphs." These are carbonate grains with a range of fabrics. They have been interpreted as biogenic and have been used in biostratigraphy (see Hofmann 1987, for discussion and references). Many are diagenetically altered ooids and intraclasts; others are coated grains seemingly identical to modern "grapestone" and similar such grains. There is no satisfactory explanation as to why these are likely to be of any stratigraphic value; their use is empirical. However, as is often suggested (but has yet to be documented), some catagraphs might be fecal pellets, in which case they would be of considerable interest in providing new evidence on the early history of the metazoans. Similarly, some may partly result from microbial boring of carbonate grains and so may provide an interesting record of this ecological niche (Knoll 1985b).

Already by Early Proterozoic time, carbonate platforms had developed which show the range of reef and carbonate clastic facies once thought typical only of the Phanerozoic (Section 6.2). Throughout the Proterozoic, stromatolites formed reefs of many types, from huge barriers to much smaller patch and pinnacle reefs. Coherent benthic microbial communities covered vast tracts in both marine and lacustrine environments, forming a wide range of stromatolite morphologies each characteristic of its own paleoenvironment. Against this background, the temporal distribution of stromatolite types can be examined for any regularities that might encode information on environmental or biological evolution. The Proterozoic record does reveal some interesting temporal patterns (Fig. 6.2.3), the most outstanding of which are:

(1) An abundance of ministromatolites with a radial-fibrous fabric (Hofmann and Jackson 1987), characteristic of peritidal environments, in the Early and Middle Proterozoic, with a marked decline thereafter (Grey and Thorne 1985; Grotzinger in press).
(2) An abundance of *Conophyton* and related conical stromatolites, characteristic of quiet subtidal environments, in the Early and Middle Proterozoic, with a marked decline thereafter (Komar et al. 1965; Zhu 1982; Walter and Heys 1985).
(3) The decline in abundance and diversity of all stromatolites just prior to or about 700 Ma ago (Awramik 1971; Walter and Heys 1985).
(4) The absence or near absence of Proterozoic thrombolites, which became abundant during the Early Cambrian (Walter and Heys 1985; Kennard and James 1986).

The foregoing represents the coarse framework of stromatolite distribution, familiar to all biostratigraphers who use these fossils. It is this framework that allows an experienced biostratigrapher to walk through a stromatolitic carbonate sequence in the field and make a first rough assessment of the age of the rocks. Within this framework many finer patterns have been recognized, at least in the Middle and Late Proterozoic, but the time resolution is still only on the order of 100 to 300 Ma. Nonetheless, this approach has proven very useful, commonly because it is better than any alternative. The use of stromatolites in biostratigraphy has a history extending back to the early part of this century (reviewed in Walter 1972), but its first substantial use was in the USSR by V. P. Maslov during the 1930s and it is there that stromatolite distribution is most thoroughly documented (e.g., Semikhatov 1976, 1980 and references therein). During the last 30 years, stromatolites have also been extensively used biostratigraphically in China (e.g., Liang et al. 1985), India (e.g., Raha and Sastry 1982), Australia (e.g., Grey 1984), North Africa (Bertrand-Sarfati 1972), and to a lesser extent elsewhere. The results are summarized and reviewed in English by Preiss (1976), Semikhatov (1976), Bertrand-Sarfati and Walter (1981), and by Hofmann (1987). During the early years of the development of this field, most observations were referred to standard successions of assemblages recognized first in the USSR. Predictably, more work has modified the known ranges of many taxa and many new taxa have been described that are considered to be endemic to particular regions. Apparent endemism could be a product of variable taxonomic practices, but it is described from the USSR (Semikhatov 1980; Golovenok 1985) where this field has reached considerable maturity and there is some agreement about systematics. Because of this, and equally because of language barriers, recent work concentrates on developing local and regional biostratigraphic schemes.

There is no need to repeat here the information from the detailed reviews of stromatolite biostratigraphy referred to above (see, for example, papers in Walter 1972), but it can be briefly summarized in the context of the subdivisions of the Proterozoic proposed by the Subcommission on Precambrian Stratigraphy (Plumb and James 1986). Distinctive stromatolites are known from the Early Proterozoic; forms of *Pilbaria* Walter and *Pseudogymnosolen* (*Asperia* Semikhatov) are particularly characteristic of this Era (Grey and Thorne 1985; Grotzinger in press). The latter is an example of the ministromatolites with a radial-fibrous fabric discussed in Section 6.2. Over one hundred forms of stromatolites have been described from Early Proterozoic strata (Bertrand-Sarfati and Walter 1981; Zhu et al. 1987); although the latter authors are able to recognize five distinct assemblages, more work is required to demonstrate the utility of these and other assemblages for subdividing and correlating within the Early Proterozoic.

Much more is known about Middle and Late Proterozoic stromatolites: over 600 taxa have been described. The maximum recorded diversity of stromatolites was reached during these Eras (Fig. 6.2.3; Walter and Heys 1985). In China, four (Zhu 1982) or seven (Liang et al. 1985) successive assemblages have been recognized, with further subdivision into "subassemblages." In the USSR there are four such major assemblages, within which some subdivision is possible (Chumakov and Semikhatov 1981), although there are many local complexities (Semikhatov 1980). Comparable successions are recognized elsewhere, including Australia (e.g., Grey 1982; Preiss 1987; Walter et al. 1987, and references therein). The papers of Semikhatov (1976) and Preiss (1976) still effectively summarize the known basic patterns of stromatolite distribution, although there has been much subsequent work. The early part of the Middle Proterozoic (the Early Riphean) is characterized by *Kussiella* Krylov, *Omachtenia* Nuzhnov, *Conophyton cylindricum* Maslov, and others. The later Middle Proterozoic (the Middle Riphean) contains a diverse array of forms of *Conophyton* (17% of all stromatolite taxa in any basin; Walter and Heys 1985), with *Baicalia* Krylov and other groups. The early part of the Late Proterozoic (the Late Riphean) contains far fewer forms of *Conophyton*, but exhibits a high diversity of small columar branching stromatolites, particularly those with "walls" (in which laminae envelop columns); examples are forms of *Gymnosolen* Steinmann, *Boxonia* Korolyuk, and *Minjaria* Krylov. By the later Late Proterozoic (in the USSR, the Kudash and the Vendian), *Conophyton* is both rare and of low diversity, and among the more distinctive stromatolites are forms of *Linella* Krylov. This youngest Proterozoic assemblage spans the Vendian-Riphean boundary, defined to include the glaciogenic deposits of the Baltic Platform in the basal Vendian.

Stromatolite biostratigraphy continues to be widely used successfully, with some quite unexpected results, as in Western Australia where application of this method recently contributed to the recognition and dating of a previously unknown sedimentary basin (Grey 1987). It is most effective and convincing when used in conjunction with all other available techniques, as advocated by Golovenok (1985) and demonstrated by Plumb (1985) with an example from the McArthur Basin in northern Australia. However, taxonomic problems continue to plague the field; it is time for monographic revisions of the more significant taxa, taking full account of modern understanding of the morphogenesis of microbial mats (e.g., Section 6.3) and the diagenetic alteration of carbonate sediments.

10.7

Proterozoic Distribution of Biogeochemical Fossils

HARALD STRAUSS   J. M. HAYES   ROGER E. SUMMONS

10.7.1 Stable Isotope Records

The stable isotope compositions of carbon, hydrogen, nitrogen, and sulfur, chemically bound in kerogen, constitute one of the more inert and geologically stable indicators of biological processes. The isotopic composition of total organic carbon, for example, resists thermal alteration during burial to depths in excess of those which adversely affect morphological detail in microfossils and corrupt molecular fossils by categenesis. The distribution of Proterozoic organic carbon and analyses of its isotopic compositions have been reviewed by Hayes et al. (1983), and Schidlowski et al. (1983). These and new results are treated in detail in Chapter 3 of this monograph.

The most significant finding to emerge from new work is the detection of an apparent global secular variation in the $\delta^{13}C$ content of organic matter during Proterozoic time (Fig. 10.7.1). Samples of organic matter with the highest (>0.2) elemental H/C ratios, and thus the least alteration, show a trend from light values of about $-34‰$ in the Early Proterozoic toward heavier values of about $-31‰$ in the Middle Proterozoic. Relative to an observed constant isotopic composition of around zero permil for carbonate carbon, there appears to be a decrease in the isotopic separation between inorganic and organic carbon phases. The causes of this trend are presently unknown, but may include changes in P_{CO_2} or in the metabolism or physiology of photosynthetic organisms. During the Late Proterozoic, where there is a significantly better record of preservation, dramatic fluctuations in the isotopic composition of organic carbon have been recorded (Hsu et al. 1985; Margaritz et al. 1986; Tucker 1986; Knoll et al. 1986; Lambert et al. 1987; Aharon et al. 1987). Closely spaced sampling and analyses of both organic and inorganic carbon phases from the same samples in continuous sections from Svalbard and east Greenland (Knoll et al. 1986) reveal dramatic and tightly coupled secular trends to heavier than normal values (Fig. 4.8.1). This phenomenon has been interpreted as due to burial (averaged globally) of larger than normal amounts of organic carbon, thereby generating an equivalent amount of excess oxidizing power, most likely oxygen. The trends accurately documented in the Greenland section correlate well with contemporaneous sections in many other localities (Knoll et al. 1986). Because carbon in fossil organic matter is continually returned to the biosphere through erosion and tectonism, such isotopic fluctuations, even in a limited sample set, are a better permanent record burial of organic carbon than studies of the distribution of organic matter in sediments. The Late Proterozoic carbon isotope variations show major fluctuations between 900 and 700 Ma ago, a major shift to extremely light values at about 650 Ma and a gradual increase in ^{13}C contents at the base of the Cambrian (Fig. 4.8.1). These patterns cannot be classified as biostratigraphic markers in the strictest sense. They are indicative not only of the biosynthesis of organic carbon, but of variations in conditions regulating its preservation (see discussion in Knoll et al. 1986). Consequently, these patterns encode information about oxygenation of bottom waters and rates of sedimentation and, secondarily, about oceanographic and global tectonic processes.

The sulfur isotopic record and its biogeochemical significance have recently been reviewed by Schidlowski et al. (1983), and by Lambert and Donnelly (1987) (see Chapter 3). A very narrow range of isotope ratios characterizes sulfide sulfur during the Late Archean and Early Proterozoic. At 2.4 to 2.2 Ga, a much broader range and $^{34}S/^{32}S$ values clearly removed from zero (evidence coming from the Transvaal Group, 2.3 Ga: Cameron 1982; and later from the Gunflint Formation: Shegelski 1985), indicate that the biogeochemical cycle of sulfur had begun to approach its modern form. Middle, and particularly Upper Proterozoic sequences are often characterized by sulfides enriched in ^{34}S. A few preserved Proterozoic sulfates have yielded ^{34}S-enriched (but not extremely heavy) isotope values. Evidence for restricted marine to lacustrine environments suggests the operation of a Rayleigh distillation, giving rise to unusually heavy isotope values for sulfides. The apparent "imbalance" of the exogenic sulfur cycle might in part result from a selective preservation of intracratonic basins (partially closed with respect to sulfate replacement), and a biased sampling toward mineralized areas.

Very limited work has been done on organic hydrogen and nitrogen and their significance in Proterozoic strata (Schidlowski et al. 1983). Their relative abundances and isotope ratios in kerogen have been applied to evaluate the thermal alteration

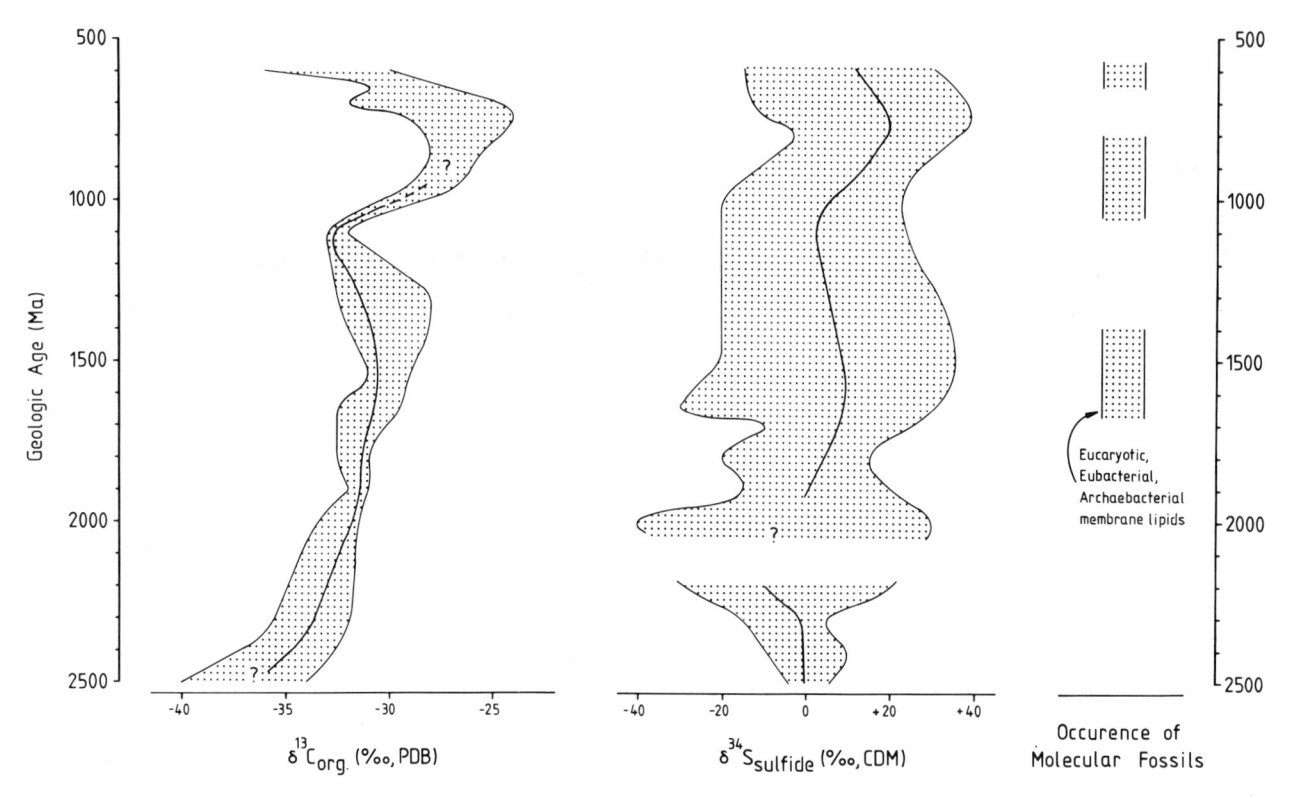

Figure 10.7.1. Proterozoic distribution of biochemical fossils: extractable molecular fossils and isotopic values of organic carbon and sulfide sulfur indicative of biological activity.

and postdepositional history of a given sediment. The isotopic records for both elements are rather sparse throughout the Proterozoic and their interpretation is thus necessarily inconclusive.

10.7.2 Molecular Fossils

Much of the organic matter produced by Early and Middle Proterozoic microorganisms has been lost through processes of crustal recycling or has been severely altered by metamorphism. However, the recent work detailed in Chapter 3 shows that there is an abundant and significant record of well-preserved organic-rich sediments beginning with the McArthur sediments at 1.7 Ga and continuing to the base of the Cambrian. These sediments and some oils have had mild to moderate thermal histories and, as a direct result, contain comprehensive suites of molecular fossils or biomarkers. None of the signatures thus far encountered in Proterozoic bitumens constitute unequivocal biostratigraphic markers, although most hydrocarbon distributions display one or more features that distinguish them from Phanerozoic materials. These features include the following: (i) strong predominance of low molecular weight (C_{14} to C_{19}) alkanes and low abundance of waxes; (ii) high abundances of monomethyl alkanes, sometimes with mid-chain predominances in the C_{20} to C_{25} homologues; (iii) high relative abundances of cyclohexyl alkanes and methyl-cyclohexyl alkanes; (iv) very low contents of polycyclic biomarkers such as steranes and hopanes in the Middle Proterozoic, followed by an increase in their abundance in the Late Proterozoic, usually accompanied by strong preferences for either C_{27} or C_{29} pseudohomologues; (v) presence of unusual triterpenoids including putative C_{29} to C_{35} neo-hopanes; and (vi) absence of vascular plant-derived diterpanes and triterpanes.

10.8

Summary: Current Status of Proterozoic Biostratigraphy

HANS J. HOFMANN

Proterozoic biostratigraphy has developed in the past three decades and found its widest application in Eurasia; Russian and Chinese language references dominate the literature. Zonations have been chiefly based on stromatolites and microphytolites for most of the Proterozoic; acritarchs, metazoans, and carbonaceous compressions of as yet undetermined affinities, form the basis for broad Late Proterozoic schemes. The stratigraphic distribution and relative importance of various categories are compiled in Figure 10.8.1. There is a general trend of increasing size and morphologic complexity of organisms throughout the Proterozoic. Stromatolite reefs, however, are generally more imposing in the Early and Middle Proterozoic than at later times.

The stromatolite record begins at about 3.5 Ga (see Section 1.5). Both in terms of abundance and diversity, it is marked by two peak developments (Fig. 6.2.3), a smaller one in the late Early Proterozoic, and a major one near the end of the Middle Proterozoic. Broad evolutionary trends are evident. The record between about 2.6 and 1.0 Ga shows an abundance of mini-stromatolites containing a radial-fibrous microfabric (pseudogymnosolenids), found in beds formed in peritidal environments. Stromatolites with conical and ridge-like laminae (conophytonids and thyssagetids) were characteristic of quiet, subtidal environments during the same time interval. A marked decline in both abundance and diversity of all stromatolite types occurs after the Middle Proterozoic, a decline which has been attributed to numerous causes (for discussion, see Section 6.2.6).

Assemblages of stromatolite groups and forms are used to recognize broad biozones that correspond to the major stratigraphic divisions established in the areas of the major continental stratotypes (e.g., the Lower, Middle, and Upper Riphean, and Vendian = Yudomian, of the Soviet Union; the Hutuo, Changcheng, Jixian, Qingbaikou, and Sinian "systems" of China, etc.). Microphytolite zonations often accompany those based on stromatolites. Such microphytolite-based zones should be considered suspect, however, because the biogenicity of these remains has not been demonstrated, and many such forms are known to have longer ranges than claimed earlier. Stromatolite biostratigraphy needs to be refined considerably, after thorough revision of stromatolite taxonomy.

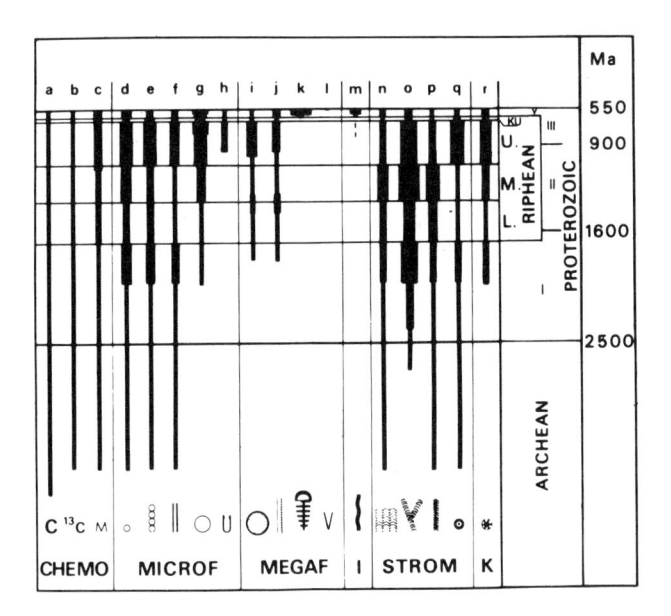

Figure 10.8.1. Stratigraphic distribution and relative importance of various categories of Proterozoic remains. [Modified from Hofmann 1985b, Fig. 3.] a – free carbon, kerogen; b – fractionated isotopes of C and S; c – molecular fossils; d – coccoid microfossils < 30 μm; e – filamentous microfossils; f – tubular microfossils, sheaths; g – coccoid microfossils > 30 μm; h – vase-shaped microfossils; i – megascopic carbonaceous films; j – megascopic carbonaceous films; k – soft bodied metazoan fossils; l – shelly fossils; m – ichnofossils; n – stratiform stromatolites; o – branching columnar stromatolites; p – conical stromatolites; q – oncolites; r – catagraphs.

Microfossils show a trend of increasing size with time, especially for the coccoid forms, and in both basinal and shelf environments (Sections 5.4, 5.5). Shale-facies planktonic forms are generally larger than coeval benthic forms in cherty carbonate biotas (Fig. 5.4.15). Early Proterozoic microfossils preserved in chert are already morphologically diverse, but simple coccoid and filamentous forms predominate; "bizarre" forms without good modern homologues are minor elements in the biotas (Section 5.6).

Among Proterozoic shale-facies microfossils, known as far back as the Early Proterozoic, sphaeromorphs are the most abundant, followed by acanthomorph (spiny) acritarchs. A first period of acritarch diversification occurred near the beginning of the Late Proterozoic, with the development of double-walled forms bearing short cylindrical processes, acritarchs with pylomes, and prismatomorphs (Figs. 5.5.12, 5.5.14). Enigmatic vase-shaped fossils (melanocyrillids), which have been regarded as heterotroph protists but may be algae or algal cysts (Section 5.6.2B), characterize the period between approximately 900 and 700 Ma ago. An apparent decline in diversity coincided with the Early Vendian glaciations; this was followed by rapid morphological diversification around the Precambrian-Cambrian transition (Vidal and Knoll 1983, Fig. 4). Genera of acritarchs thought to have relatively good biostratigraphic value for the pre-tillite Late Proterozoic include *Octoedrixium*, *Podolina*, *Pterospermopsimorpha*, and *Vandalosphaeridium*. *Granomarginata squamosa* has been considered a good index marking the lowermost Cambrian (Vidal and Knoll 1983).

Although carbonaceous films have been found in strata as old as about 2.0 Ga, the first biostratigraphically significant forms are spiraliform ribbons and associated forms found in 1.5 to 1.2 Ga old assemblages in Montana and China. These films may eventually form the basis for recognizing a "*Grypania* biozone." A second assemblage with *Tawuia*, *Longfengshania*, *Chuaria*, and other taxa, characterizes 950 to 700 Ma-old sequences, and serves to define a "*Tawuia* assemblage zone." Possibly, an upper subzone may be recognized within the *Tawuia* zone, based on *Sinosabellidites*, *Protarenicola*, and similar taxa, but these taxa will first have to be found in areas in addition to that of their type area in China. *Vendotaenia* and other vendotaenids are typical of, but not restricted to, the Vendian; abundant long, twisted or irregularly curved filaments serve to characterize a "*Vendotaenia* assemblage zone."

Metazoans are known with certainty only from the middle and upper Vendian (Ediacaran), and form the basis for the "Ediacara Fauna" having worldwide distribution (Section 7.5; Table 23.3). This fauna includes very large, soft-bodied organisms with high surface-to-volume ratios; many of the taxa are morphologically quite distinct and are therefore readily recognizable. The Ediacara Fauna is biostratigraphically the most useful of all Proterozoic biotas yet recognized.

Unquestioned trace fossils are also restricted to the upper Vendian, and include mainly simple, horizontal traces. Diversity is low, and only five genera are known to be restricted to the pre-Cambrian and therefore of potential use in biostratigraphy. Larger and more complex traces, and abundance of vertical burrows are evidently indicative of Cambrian and younger sequences.

Calcareous algae, such as *Renalcis*, *Girvanella*, and *Obruchevella*, and skeletonized protists appear to have developed concurrently with skeletonization in the Metazoa at the beginning of the Cambrian.

Among the chemofossils of potential biostratigraphic interest are stable isotope data and extractable molecular fossils. As discussed in Chapter 3 and Section 10.7, recent studies of well-preserved Proterozoic sediments, samples with abundant kerogen and mild thermal histories (with H/C >0.2), have dispelled the notion that Precambrian biogeochemical studies are generally unproductive. Contrary to earlier views, isotopic fractionation of carbon in Proterozoic kerogen does show a time-dependent change, with a general trend toward isotopically heavier values with the progress of time (Fig. 3.4.2). These results may be interpreted in terms of a corresponding decrease in atmospheric CO_2 abundance over the same time interval. Moreover, extraction of the kerogens and associated organic matter has yielded compounds that can be related to particular groups of organisms. Thus, steranes, characteristic of eukaryotic membranes, extended (C_{20}) acyclic isoprenoids consistent with an input from archaebacteria, and pentacyclic triterpanes derived from bacteria, have been shown to occur in rocks as old as 1690 Ma (viz., the Barney Creek Formation of northern Australia), providing chemical evidence of the existence of these particular organisms during the Early Proterozoic. Quantitatively, the relative abundances of steranes and triterpanes exhibit significant increases in the Late Proterozoic.

10.9

Proterozoic Paleobiogeography

HANS J. HOFMANN

In comparison with biostratigraphy, biogeographic analysis of the Proterozoic has been slow to develop. A number of global syntheses of lithofacies distributions have been prepared, of which the most elaborate and useful is the set of colorful maps for the Middle and Late Proterozoic by Ronov et al. (1984) and Wang (1985). However, relatively few maps showing the distribution of Precambrian biotas exist, and those are chiefly for the latest Proterozoic. Most use present-day geographic outlines and positions of the continents.

A principal drawback in using a paleogeographic base for plotting Precambrian distributions has been the paucity of data that would permit the reliable palinspastic reconstruction of the ancient continents during various times of the Precambrian. Although reconstructions based on paleomagnetic data have been prepared by several authors (e.g., Morel and Irving 1978; Piper 1982, 1987), such maps differ from one another in too many important ways to allow full confidence in the method. A new set of reconstructions for the Late Proterozoic and Cambrian is presented and discussed by Kirschvink in Chapter 12.

Although paleomagnetic data, combined with climatically sensitive sedimentologic and paleontologic features can be used to obtain paleolatitudes, paleolongitude is not determinable by present methods. Given the mobilism of plate tectonics, the elaborate mosaic resulting from the multiple fragmentation and reassembly of Precambrian continental masses poses great challenges to the paleogeographer.

The task of deciphering early geologic history and paleogeography is greatly aided by modern precise radiometric dating techniques (e.g., Pb/U isotope studies on single mineral grains), which allow resolution of some Precambrian geologic events at a scale of plus or minus only a few million years. For example, the chronology of events associated with the Early Proterozoic Wopmay Orogen in northwestern Canada (Hoffman 1988), which is among the best documented and understood of Precambrian orogens, would not have been possible without precise geochronology. The combination of detailed geologic, paleomagnetic, and radiometric studies of many more Precambrian regions is essential for more precise determination of the ancient paleogeography. Development of methods for establishing paleolongitudes also would be of great assistance in preparing more accurate maps.

As a consequence of the difficulties in making accurate palinspastic maps for the Precambrian, most paleontologic data are plotted on maps showing the present-day continental alignments. Yet some geologists have ventured beyond this limit. As one such example, Jux (1977, Fig. 3) has prepared a map showing the distribution of particular Precambrian fossil groups on a palinspastic base that suggests a tropical position (<30° paleolatitude) for *Chuaria* occurrences. However, this map, which also locates the coeval magnetic south pole at about 42°S, is for the Eocambrian (= Vendian) interval, whereas *Chuaria* is mainly pre-Vendian (see Section 5.5).

Three alternate maps showing the distribution of the Ediacara Fauna were presented by Donovan (1987), one for each of the late Precambrian or Early Cambrian reconstructions of Morel and Irving (1978), Piper (1982, 1987), and Bond et al. (1984). Another map for the Ediacara Fauna, using a reconstruction of Smith et al. (1981) for the Early Cambrian, is given by Conway Morris (1987, Fig. 5). The wide discrepancies among these various maps indicate the still quite speculative nature of this method, and the imprecise stable of knowledge of Proterozoic paleobiogeography for even the youngest part of the Precambrian (that portion which presumably is also the best known, because some extrapolation from the richly fossiliferous Cambrian is possible). At best, the different configurations proposed for the same slice of time afford the opportunity to test multiple working hypotheses.

Like paleogeographic studies, paleoenvironmental analysis is also lagging behind taxonomic and biostratigraphic studies. In this respect, more work has been done on stromatolites than on other Proterozoic fossil groups. Stromatolites are most commonly developed in relatively shallow water platform or shelf-carbonate environments (Walter 1977), and different facies can be recognized. A peritidal carbonate tufa facies, represented by the Pseudogymnosolenaceae, is common in 2.6 to 1.0 Ga old rocks (Grey 1984), and stromatolites of the *Conophyton* group are characteristic of facies free of terrigenous clastics in which little or no wave or current activity occurs. Oncolites and catagraphs are almost exclusively high-energy, shallow water features. Carbonaceous remains occur in both lagoonal and offshore basinal environments, and metazoans and trace fossils are found in littoral and basinal settings as well. Microfossils occur both in benthic and planktonic assemblages, and may show distinct lateral zonation. Vidal and Knoll (1983) have distinguished inshore from offshore acritarch biotas in which assemblages of stratigraphically long-ranging, morphologically simple forms of low taxonomic diversity dominate the inshore environments, and contemporaneous, morphologically complex biotas, with higher diversity and greater biostratigraphic significance, occur in offshore settings. Knoll (1985, Fig. 5) presents an interesting graphic summary diagram of the paleoecological settings for 17 biotas including microfossils, stromatolites, oncolites, and carbonaceous megafossils, spanning supratidal to open shelf settings. Unfortunately, however, as discussed in Section 11.3, data of this type are currently available for only a few Proterozoic microfossiliferous deposits.

It is evident from a survey of the literature that relatively little effort has yet been devoted to the world-wide paleobiogeographic perspective as compared with that applied to biostratigraphy and taxonomy. Many local and regional syntheses have appeared, sometimes accompanied by maps showing the distribution of one or another fossil group, but relatively few global compilation maps have been prepared (for a compilation map of all reported Precambrian fossils a decade ago, see Murray 1980). Aspects of this deficiency have been addressed in the present volume, in particular, the geographic distribution of known occurrences of (*i*) paleontologically and biostratigraphically significant Archean and Proterozoic geologic sections is summarized in Figure 10.2.1; (*ii*) Proterozoic carbonaceous films, in Figure 7.3.1 (and for various portions of the Proterozoic, in Figs. 7.3.2 through 7.3.6); (*iii*) Proterozoic microfossils, in Figure 11.3.2; (*iv*) the "megasphaeromorph" acritarch *Chuaria*, in Figure 7.3.7; (*v*) *Tawuia* and similar Proterozoic possible metaphytes, in Figure 7.3.8; (*vi*) Proterozoic trace fossils, dubiofossils, and pseudofossils, in Table 23-4.1; (*vii*) Ediacaran metazoan body fossils, in Table 23.3.1; (*viii*) other Proterozoic putative metazoan body fossils, in Table 23.3.2; (*ix*) Proterozoic megascopic dubiofossils, in Figure 7.8.1; (*x*) Early Cambrian small shelly fossils, in Table 23.5; and (*xi*) Early Cambrian calcareous algae, in Figure 7.4.1. The abundance and density of the points on the various maps reflect the positions of the ancient biotopes as well as the locus of paleontologic interest and activity in an area: the greater the number of paleontologists, the greater the chance for reports and description of taxa and assemblages.

10.10

Unsolved Problems and Future Research Directions

HANS J. HOFMANN

On a general level, a priority need in Proterozoic biostratigraphy is, without doubt, comprehensive studies that would provide relief from the taxonomic chaos that presently afflicts the discipline, particularly with respect to acritarchs, stromatolites, and microphytolites, the principal groups currently used for biostratigraphic subdivision. Redundant synonyms abound and need to be eliminated; as discussed in Section 11.3, this situation directly affects the closely related topic of taxonomic diversity in the Proterozoic. The daunting task of taxonomic revision would be helped by agreement among the principal workers as to which attributes are diagnostic biological features and which, for example, are diagenetic, taphonomic, or artifactual. In order to accomplish this, a way needs to be found to bridge the language barriers; quantification, and the presentation of data in graphic and photographic form may be at least a partial solution (for example, see Zhang and Hofmann 1981). Certainly, the data are now so voluminous and so diverse that taxonomy needs to become much more computer-oriented to be efficient. Diagnostic fossil attributes need to be identified in such a way that they can be handled electronically, as are X-ray diffraction data or chemical compositions in mineralogy and petrology. Technology is available to capture, analyze, manipulate, and store images, although such equipment has requirements for extensive memory capabilities and is costly.

High quality, scaled photographs, including orthogonal views of non-uniform taxa like stromatolites, are essential, as is precise information on stratigraphic position, age, and geographic coordinates. Numerous publications fail to provide sufficient data identifying the exact locality from which the fossils were derived. Inclusion of universally accessible information, such as latitude and longitude, would obviate such problems and would reduce the chance of introducing errors later.

Ways need to be found to make type specimens more accessible than they commonly are now. New taxa should be described only in widely circulated publications. Although personal examination of type specimens is most desirable, it is commonly not feasible for financial or other reasons (including the loan policies of the repository institutions). In such cases, scaled color photographs of holotype and other type material may have to serve as a substitute (with such photographs made available at minimal cost, or on an exchange basis).

Other unsolved problems include the nature of dubiofossils. All objects referred to this category need to be restudied so that they can be established as belonging either to the fossils or pseudofossils. Such restudy is particularly important for dubiofossils with potentially great implication for early evolution (such as structures purported to represent Early Proterozoic "metazoans").

More attention needs to be given to the sedimentological and taphonomic context of Proterozoic fossils, and tools need to be developed to determine paleolongitude. Sampling of unexplored sedimentary facies, of minimally studied regions (such as South America, northern Canada, Antarctica, etc.), and of under-investigated stratigraphic intervals (particularly the Lower Proterozoic), need to be greatly expanded. More biochemical and stable isotope studies of closely spaced samples in extensive, continuous, stratigraphic sections can help provide information regarding ecosystem changes over time. New methods for obtaining precise radiometric dates directly from particular sedimentary rock types could improve time resolution: for instance, the Pb/U or other isotopic systems could be applied to rocks containing authigenic phosphates (Pouliot and Hofmann 1981).

A host of more specific Proterozoic paleontologic problems await solution. A partial list might include such items as the origin and evolution of pylomes in acritarchs; the time of origin of eukaryotes, metazoans, the burrowing habit, and of land-surface and freshwater biotas; and the frequency, amplitude, and causes of fluctuations in taxonomic diversity in Proterozoic biotas.

10.11

Summary and Conclusions

HANS J. HOFMANN

During the past 30 years, various biostratigraphic schemes have been constructed for the Proterozoic. The divisions created are crude in comparison with zonations for the Phanerozoic, and are based on assemblages of stromatolites, acritarchs, metazoans, and carbonaceous films.

Early Proterozoic sequences are the oldest containing extensive accumulations of stromatolitic platform carbonates; large columnar structures with second-order ministromatolites are common. Sphaeromorph acritarchs are present in Early Proterozoic basinal sediments, but are relatively small and nondescript. Various peculiar and enigmatic microfossils, of possible biostratigraphic utility, are found associated with Early Proterozoic banded iron-formations.

Early and Middle Proterozoic platform sequences commonly contain characteristic peritidal ministromatolites with radial-fibrous fabric (pseudogymnosolenids), and deeper water carbonates of this age have stromatolites with conical and ridge-like surfaces (conophytonids and thyssagetids).

Middle and Upper Proterozoic sequences are reported to be typified by characteristic assemblages of columnar branching stromatolites, such as *Kussiella* and *Omachtenia* in the Lower Riphean, *Baicalia* and *Tungussia* in the Middle Riphean; *Gymnosolen*, *Inzeria*, and *Minjaria* in the Upper Riphean; and *Boxonia* in the Vendian (= Yudomian).

Crude biostratigraphic zonation based on carbonaceous films now appears possible, with a "*Grypania* assemblage zone" about 1.5 to 1.2 Ga; a "*Tawuia* assemblage zone" about 0.9 to 0.7 Ga; and a "*Vendotaenia* assemblage zone" in the Vendian. The Ediacara Fauna of middle to late Vendian age is the most distinct and most precisely identified of all recognized Proterozoic biozones.

Large sphaeromorph acritarchs, with vesicles several hundred to a few thousand micrometers in diameter, appear to have been particularly characteristic of the 1.1 to 0.8 Ga interval, and other microfossils, particularly ornamented acritarchs, have been used for subdividing the remainder of the Upper Proterozoic. Vase-shaped microfossils (melanocyrillids) and more elaborately ornamented vesicles appear characteristic of the 0.9 to 0.7 Ga interval, whereas acritarchs with complex processes (baltisphaerids) are typical of late Vendian and younger sequences.

Organisms with hard parts, metazoans, calcareous algae, and skeletonized protists, appear near the beginning of the Cambrian.

Molecular fossils from sediments with mild thermal histories can be related to specific groups of organisms and traced back to about 1.7 Ga; they show an increase in abundance and diversity with time, beginning at about 1 Ga. Carbon isotope ratios for coexisting kerogen and carbonate, from closely sampled intervals in long continuous stratigraphic sections, exhibit considerable fluctuation in values, providing a basis for regional, and possibly broader, correlations, and for tracking of major events in the development of the Late Proterozoic biosphere and ecosystems.

Few studies have concentrated on the global paleobiogeography of Proterozoic biotas, although separate planktonic and benthic realms have generally been distinguished. Paleomagnetically supported palinspastic reconstructions for the Proterozoic mostly suggest that the land masses (and, hence, known fossiliferous Proterozoic strata) occupied low to moderate paleolatitudinal positions.

Unresolved problems abound; there is no shortage of projects for further study.

Acknowledgments

The contribution by Hofmann was in part financially supported by the Natural Sciences and Engineering Research Council of Canada. Searches for skeletal protists in Proterozoic rocks by Lipps was supported by NSF grant BSR-8509301.

11

Biotic Diversity and Rates of Evolution During Proterozoic and Earliest Phanerozoic Time

J. JOHN SEPKOSKI, JR.   J. WILLIAM SCHOPF

11.1

Introduction

J. JOHN SEPKOSKI, Jr.

"Diversity" is commonly defined as the number of kinds of organisms within an ecologic unit. "Kinds" are usually species but can also be genera, families, or higher taxa or even morphotypes. "Ecologic units" can be environmental patches, local communities, or whole ecosystems; in paleontology, the unit is often a complex of ecosystems or even the whole world ocean.

Diversity is interesting because it varies among taxa, places, and times, and study of this variation can lead to understanding of the processes that control the relative success of different groups of organisms. In the investigation of life's history, analyses of changing patterns of diversity have provided insight into the macroevolutionary dynamics of evolving clades and ecosystems and into the physical processes on earth that govern or limit evolutionary change. Because diversity represents the sum of all organisms in a clade or place, it provides a convenient index of evolutionary activity when mapped over time: when diversity is increasing, rates of evolution generally are high and new morphotypes, adaptations, and interactions often are appearing; when diversity is stable, rates of evolution usually are lower and changes in taxic structure of the ecosystem tend to be slow; when diversity is declining or fluctuating erratically, physical (or "extrinsic") changes on earth, or even beyond earth, may be exerting dominant influence over evolutionary activity. Even when macroevolutionary interpretations are not possible because of inaccurate information or overwhelming biases, compilation of diversity patterns produces a synthesis of all data on fossil organisms within a standardized stratigraphic framework and thus provides an overview of current knowledge of the fossil record. Gaps in knowledge thus can be identified and agenda for future research can be devised.

There has been considerable success in documenting, analyzing, and interpreting patterns of diversity for animals and plants in the Phanerozoic (e.g., Newell 1952, 1967; Valentine 1969, 1973; Sepkoski 1981, 1984; Niklas et al. 1980, 1983; Tiffney 1981; Benton 1985). Analogous studies of older, principally unicellular organisms previously have not been attempted, however, because of the youth of Precambrian paleobiology and the difficulties of taxonomy, stratigraphy, and international synthesis (although see Awramik 1971; Schopf 1977; Vidal and Knoll 1982; Walter and Heys 1985). The purpose of this Chapter is to attempt to apply methods of analyzing Phanerozoic diversity to the fossil record of the Proterozoic and of the Proterozoic-Phanerozoic transition. The goals are to (*i*) produce a more precise summary of the large-scale nature of the Proterozoic fossil record; (*ii*) document changing patterns of complexity and biotic dominance in the earth's early ecosystems in order to identify, in particular, times of radiation and decline in diversity; (*iii*) as appropriate, link patterns of biotic complexity to changes in the earth's physical and chemical environments; and (*iv*) document the initial radiations of metazoans and metaphytes and search for patterns during the Proterozoic microbial record that might herald these radiations.

Below, problems of analyzing diversity in the Proterozoic fossil record are considered first. Then, patterns of microbial diversity in the known Proterozoic record are documented and interpretations offered. This is followed by more detailed documentation of the initial diversification of metazoan and metaphytic genera during the Proterozoic-Cambrian transition. Finally, problems and programs for future research are discussed.

11.2

The Proterozoic Fossil Record: Special Problems in Analyzing Diversity Patterns

J. JOHN SEPKOSKI, JR.   J. WILLIAM SCHOPF

11.2.1 General Considerations

Biologically meaningful estimates of fossil diversity are dependent upon a plethora of conditions, including (*i*) stable taxonomy and uniform application of taxonomic names; (*ii*) accurate correlation and dating of fossiliferous formations; (*iii*) reasonably uniform preservation of organisms among clades and through time; and (*iv*) fairly even, or unbiased, sampling of fossils, formations, and environments in space and time (see also Raup 1972, 1976, 1979). As discussed in Chapter 7 and below, none of these conditions is met to any great degree for the Proterozoic. Precambrian paleobiology is still a young science with relatively few specialists worldwide, and sampling of the Proterozoic record and analysis of its fossils are still extremely incomplete. However, this does not necessarily preclude analysis of diversity patterns. As stated above, at minimum such analyses will provide a synthetic picture of current knowledge. And even if this knowledge is very limited, major biotic events such as large-scale radiations may be discernable in the data. Indeed, even with very few detailed data, Darwin (1859) was able to recognize the Proterozoic-Cambrian transition as a pivotal point in the evolution of life.

11.2.2 Taxonomic Problems

Most studies of global fossil diversity, such as those conducted for Phanerozoic groups, have been based upon taxonomic compilations taken from the published literature. These are lists of taxa (mostly species, genera, or families) with ages of their first and last documented fossil occurrences. Analyses of such lists assume that published taxa are valid and diagnosable, definitely biogenic, belong to the major taxon in question, and have been properly identified (i.e, the name has been correctly applied) in each fossil occurrence.

There are many examples of violations of these assumptions in the published record for the Proterozoic. As discussed below (Section 11.3.3; see also Chapter 22), numerous abiogenic structures have been formally described and named as microbial taxa. (These as well as dubiofossils – objects of possible but unproven biogenicity – have been culled from the data analyzed here wherever possible.) For material that is definitely biogenic, there are problems in the published literature of few exhaustive monographic studies and too many cursory descriptions, inadequate diagnoses, and indistinct photographs (especially in the pre-1980 literature); these are often compounded by limited availability of type or reference material. (Taxa of questionable validity generally have been either ignored or informally synonymized in the data analyzed here; see Chapter 25.) Finally, among valid microfossil taxa there frequently remain questions of proper taxonomic assignment, as discussed in Section 5.4. For some Proterozoic carbonaceous megafossils, assignments to kingdom are also uncertain: some may be fragments of prokaryotic mats, some may be multicellular algae, and some arguably may even be animals (e.g., *Tawuia*, *Sinosabellidites*).

Analyses of taxonomic lists also make the assumption that synonyms, including life stage and preservational variants, have been culled or lumped. This is a particular problem in Proterozoic paleobiology which has developed somewhat independently in the English, Russian, and Chinese literatures with different methodologies, inadequate communication, and little transfer of type or reference material (with regard to megafossils, see Chapter 7). Only now is a uniform style of erecting new taxa being developed, and every attempt has been made here to eliminate redundant names that have evolved among the independent traditions (see Chapter 25).

A more insidious problem in Proterozoic taxonomy is the question of what constitutes biologically (or taxonomically) meaningful characters. Microbial fossils generally are simple, with the dual problems that distinct species, or even higher taxa, often are lumped because of similar morphologies whereas other are split because of complex life stages or variable preservation (Section 5.4; Chapter 25). Preservation is a particular problem in correctly identifying and naming microbial fossils preserved as three-dimensional permineralized bodies in cherts versus those preserved as two-dimensional compressions in shales; different techniques (i.e., thin sections versus macerations) are used to study these fossils, and different systematic conventions have developed for naming them (often without adequate reference to living prokaryotes), matters that are addressed in some detail in Section 5.4. These problems are not confined to microbial fossils: in the Ediacaran fauna, some taxa may be preservational variants, especially among simple animals with few characters (e.g., medusoids); and in the earliest skeletal fossils, some named sclerites probably belong to single scleritomes.

As noted below in Section 11.3.3, concerted attempts have been made within the analyzed data to synonymize chert and shale microfossil taxa, eliminate preservational variants, and separate taxa on the basis of meaningful characters, including gross size and shape, cell wall structure, ornamentation, cellular organization, and others (see Chapter 25). This has eliminated obvious synonyms from the taxonomic lists but has not solved the problem of lumping morphologically similar but phylogenetically distinct organisms.

11.2.3 Problems of Correlation and Dating

Taxonomic lists require reasonably reliable ages for first and last appearances of taxa in order to compile global diversity patterns. However, time scales and correlations within the Proterozoic are still primitive (Chapter 7 and Section 11.3.3E), with problems including the following: (*i*) low diversity and evolutionary rates among Proterozoic organisms, providing only very generalized biostratigraphic intervals (often of regional rather than global utility); (*ii*) few reasonably synchronous marker events (e.g., glacial episodes); and (*iii*) questionable reliability of many radiometric dates, particularly those in the older literature or based on whole-rock shale analyses or metamorphic minerals. In Section 11.3, the conservative assumption is made that most reported ages for fossiliferous formations are within 10% of true ages. This translates to intervals of around 200 Ma for the Early Proterozoic and 100 Ma for the Middle. The situation is somewhat better for the Late Proterozoic where there are more means of correlating formations, including more biostratigraphic indices (e.g., "ornamented" acritarchs) and marker horizons (e.g., the Varangian tillites). Still, intervals range from 50 Ma (late Riphean) to 25 Ma (late Vendian) at best.

Thus, intervals for compiling global diversity in the Proterozoic are very crude compared to the 5 to 10 Ma-long stages used in the Phanerozoic. This in itself, however, does not preclude analyses of diversity, since narrow intervals are necessary only when rates of diversification are rapid (e.g., the Proterozoic-Cambrian transition). All evidence indicates extremely low rates of morphologic and ecologic evolution through most of the Proterozoic (the "hypobradytely" of Schopf 1987; see Sections 5.4, 55, and 13.3). Therefore, very long time intervals, even up to 200 Ma duration, may still permit identification of major changes and trends in microbial diversity.

11.2.4 Problems of Biased Sampling

All lists of fossil taxa are incomplete because of imperfect sampling and preservation. This in itself will not induce false patterns of diversity unless sample sizes are extremely small. However, if the lists are biased because of non-uniform sampling of taxa, time intervals, environments, or geographic regions, false patterns may be produced. As is discussed in Chapter 2, the temporal and spatial distribution of Proterozoic sedimentary rocks is far from uniform in time; because of the

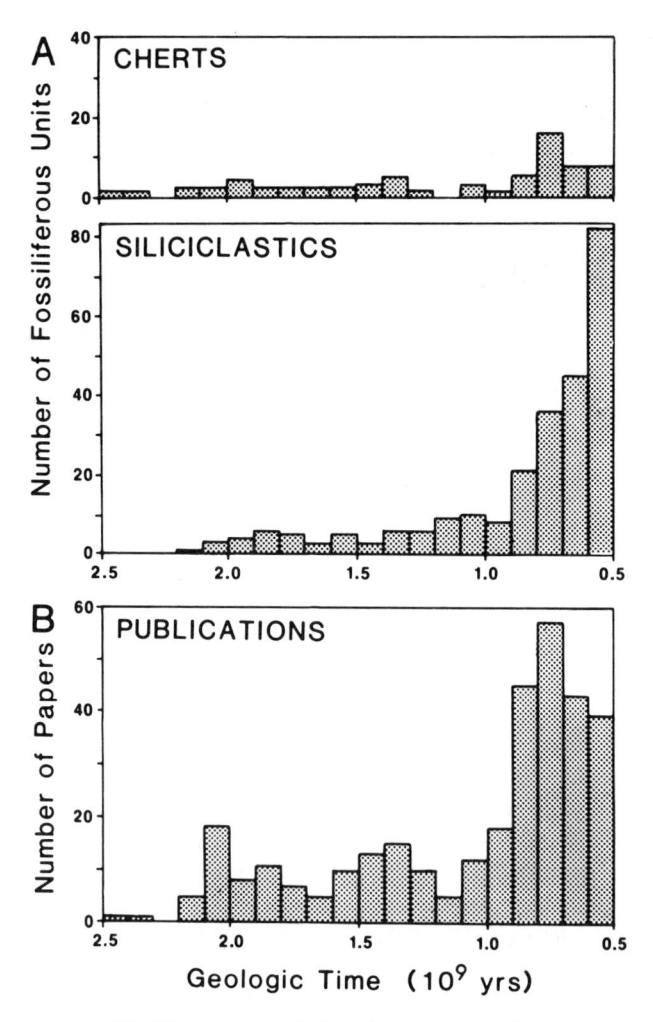

Figure 11.2.1. Histograms displaying the frequencies of microfossiliferous rock units (A) and taxonomic publications (B) as a function of time through the Proterozoic. Fossiliferous rock units (stratigraphic formations) increase markedly in frequency in the Upper Proterozoic, especially among siliciclastics (shales and siltstones). With more units to sample, more paleontologic investigations have been conducted, resulting in more taxonomic papers and monographs. Frequencies were compiled for 100 Ma-long intervals from the PPRG microfossil data base (Chapter 22).

recycling of sediments by tectonism and erosion, there is a far greater volume and exposure area for sampling of younger formations (Garrels and MacKenzie 1971). Figure 11.2.1A illustrates the number of microfossiliferous rock units listed in the analyzed database (Chapter 22) as a function of Proterozoic time. As expected, Late Proterozoic units are more numerous than are those from the Early Proterozoic, especially among siliciclastic sediments. Fossiliferous cherts are slightly more uniform in time, although the numbers are small and do increase slightly toward the Late Proterozoic. With more rock to sample, there has been more extensive paleontologic effort on the Late Proterozoic, as shown in Figure 11.2.1B, which illustrates the number of taxonomic papers and monographs published on microfossils in each 100 Ma-long interval of the Proterozoic.

The skewed sampling of the Proterozoic record means that any increase in observed diversity toward younger time intervals is ambiguous: such increases could be real, or they could reflect uneven sampling, and their relationship to the units analyzed must therefore be carefully assessed (cf. Section 11.3.4). This problem has also been encountered in Phanerozoic studies (Raup 1972, 1976). Bambach (1977), however, showed that such problems could be at least partially circumvented if both global and local diversities were examined. For the marine biota, global diversity is the total number of taxa in the world ocean as implied by assuming each taxon is present at all times between its first and last documented occurrences and then summing all taxa in each time interval; more taxa will be known worldwide and temporal ranges of taxa will be most complete where the record is best sampled and, therefore, diversity will tend to be highest. Local, or "alpha," diversity, on the other hand, is the average number of taxa in fossil assemblages of a given time interval; it does not assume that any taxon is present where it has not been sampled and therefore is less sensitive to biases of non-uniform sampling. Thus, as discussed in Section 11.3.4, average local diversity can be used as a check of global diversity; if both show similar trends, it can be inferred with some confidence that the patterns are in part real (cf. Sepkoski et al. 1981). This is shown below for Proterozoic microfossils in Section 11.3.

However, there are still some problems using local diversity. It must be assumed that the sampling of environments through time is reasonably uniform (see Section 11.3.3B and Knoll 1985). Unfortunately, this is a difficult matter to evaluate critically for Proterozoic microfossils: different environments have different characteristic diversities, as seen in both modern and fossil situations in onshore-offshore and pole-to-equator diversity gradients. When analyzing local diversity, it also must be assumed that the proportion of fossilizable taxa has not changed radically through time. This assumption is virtually impossible to evaluate, although for the Phanerozoic at least, there is no evidence to suggest the contrary, as indicated by similar diversity patterns among skeletal and trace fossils (Sepkoski et al. 1981).

11.2.5 A Note of Optimism

The problems and biases involved in the current sampling of the Proterozoic record might suggest that studies of diversity, especially among microbial taxa, may have limited value: although compiled diversity patterns may document current knowledge of the Proterozoic record, they could be of little value for identifying and interpreting evolutionary histories. This is certainly the most cautious and conservative viewpoint, but it may not be the most reasonable. Small samples and noisy data certainly limit the precision with which patterns can be ascertained; however, they need not mask the largest changes or longest trends in the fossil record.

The first diversity curve for Phanerozoic taxa was compiled in the mid-nineteenth century by John Phillips (1860). This was done at a time when only several thousands of fossil species had been described (mostly from western Europe), stratigraphic correlations were very imperfect, and time intervals were highly generalized. Yet, Phillips was able to perceive a number of

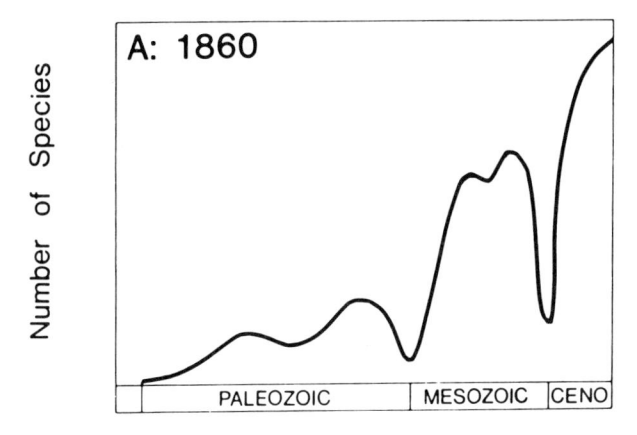

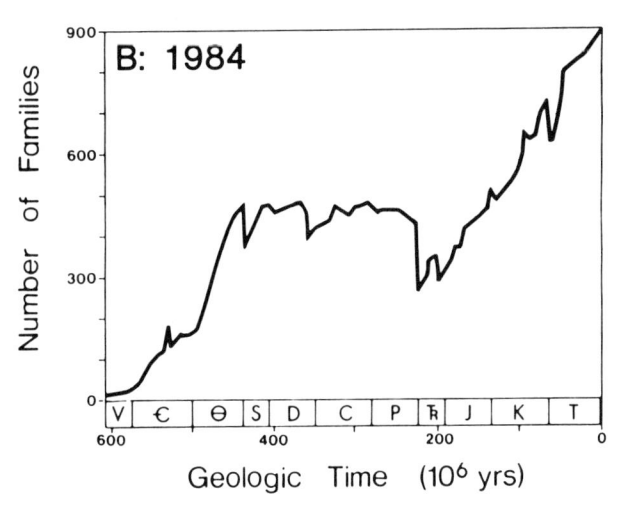

Figure 11.2.2. A comparison of Phanerozoic diversity curves from the nineteenth and twentieth centuries, illustrating the effects of increased sampling, greater geographic coverage, and enhanced temporal control. (A) The first published diversity curve for species, redrafted and modified from Phillips (1860). (B) A diversity curve for marine animal families, modified from Sepkoski (1984). Although the two curves differ considerably in detail, the qualitative, coarse-level patterns are similar: increased diversity from the Paleozoic to Mesozoic to Cenozoic, with mass extinctions (major declines in diversity) separating the eras.

large-scale changes in the Phanerozoic record: the faunal differences between the Paleozoic, Mesozoic, and Cenozoic Eras (which he named), the irregular increase in diversity toward the Recent, and two of the major mass extinctions. His diversity curve is illustrated in Figure 11.2.2 along with a curve from the current era for comparison. The two diversity curves certainly differ in detail, reflecting the greater precision in sampling, correlation, and taxonomy that a century's work has provided. But the broad-scale patterns are the same.

At present, the science is at the John Phillips stage of analyzing Proterozoic diversity patterns. As discussed below (Section 11.3), one must be very wary of overinterpreting the coarse data that are currently available. But one can be confident that the early stage of investigation does not necessarily exclude the possibility of discerning true patterns in the history of life.

11.3

Patterns of Proterozoic Microfossil Diversity: An Initial, Tentative, Analysis

J. WILLIAM SCHOPF

11.3.1 Introduction

As noted in Section 11.2, assessment of patterns of biotic diversity evidenced by the known Proterozoic microfossil record is a difficult, complex, multifaceted problem. Despite the progress of recent years, available fossil data are as yet limited; data that are available show strong time-related biases; ages of most fossiliferous units are poorly constrained; the current classification and taxonomy of Proterozoic microfossils are confused and incoherent; evidence regarding the specific paleoecologic setting of the vast majority of Proterozoic microfossil assemblages is unavailable; and relatively few exhaustive (monographic) studies of such assemblages have as yet been carried out. Given this situation, it may be regarded as either courageous or foolish (or perhaps both) to even attempt to assess Proterozoic biotic diversity. Any generalizations drawn seem certain to be controversial, chiefly because such an assessment has not previously been carried out and it remains to be established – indeed, *a priori*, it is not possible to establish – whether currently available data are of sufficient quality and quantity to justify such generalizations. At most, the following initial, tentative, analysis can be expected to yield only a coarse-grained, broad-brush description of major patterns of Proterozoic biotic diversity. At least, this attempt should provide a reasonably thorough evaluation of diversity patterns reflected by the fossil record as now known, an analysis that may serve as a foundation for further work in future years. One conclusion seems certain: generalizations here drawn will be tested, repeatedly, against new discoveries and additional data, in the years to come; the final arbiter of the success of this exercise, or lack thereof, will be the test of time.

11.3.2 Methods of Analyses

Of the 2800 occurrences of bona fide microfossils now known from Proterozoic sediments (Chapter 22), more than 85% have been reported during the past two decades (Section 5.2). Preserved in shales/siltstones (62% of occurrences) and stromatolitic carbonates/cherts (38%), both planktonic (1580 occurrences in 248 formations) and benthic taxa (1219 occurrences in 208 formations) are now well known from Proterozoic units worldwide (Figs. 11.3.1 through 11.3.3, Chapter 22).

To investigate patterns of Proterozoic biotic diversity potentially reflected by these data:

(1) An extensive database, summarized in Chapter 22 and estimated to include perhaps 90% of all putative Proterozoic microfossils reported through mid-1988, was amassed from literature sources.

(2) Authentic, well-established microfossil occurrences were separated from questionable reports (Chapter 22), with only the former being considered in analyses of diversity.

(3) Following the approach outlined in Section 5.4, appropriate morphometric data for shale microfloras were corrected for the effects of preservational compression.

(4) As discussed in Section 5.4 and Chapter 25, relevant size data for 1450 varieties and species of living prokaryotes and eukaryotic micro-algae were used to design a system of species-level classification encompassing morphologically comparable Proterozoic microfossils.

(5) 373 taxonomically informal Proterozoic fossil species were thus recognized (Chapter 25), including 91 "Lazarus taxa" (29% of prokaryotes, 5% of eukaryotes) known from the Proterozoic and the present-day but evidently not reported from the intervening Phanerozoic (Table 25.2).

(6) Fossil species were grouped and subgrouped based on cellular organization, habitat (281 benthic and 92 planktonic taxa), and inferred affinity (297 prokaryotic and 76 eukaryotic taxa) into the categories summarized in Table 25.1, and their Proterozoic distributions were tabulated in sequential 50 Ma-long intervals (Fig. 25.4).

(7) For each group, "global diversity" (the total observed or inferred Proterozoic species richness) was analyzed using data for first known and last known (for inferred global diversity, including present-day) occurrences.

(8) 50 Ma-long intervals containing ≥20 reported occurrences of benthic species (Table 25.5) or of planktonic species (Table 25.6) were selected for analyses of "mean alpha diversity" (the average number of species per assemblage), "maximum alpha diversity" (the maximum number of benthic species and maximum number of planktonic species reported from any of the assemblages within an interval),

530 Biotic Diversity and Rates of Evolution

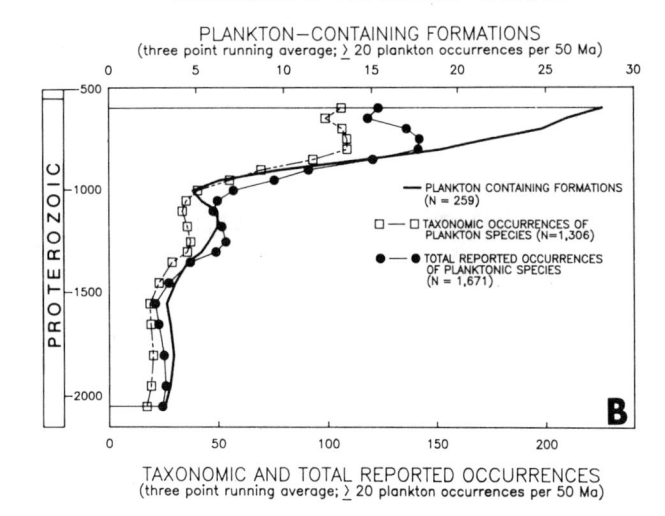

11.3.3 Assessment of the Components of the Analyses

11.3.3A General Characteristics of the Available Data

In view of the relative youth of Proterozoic paleobiologic studies (see Section 5.2), it is not surprising that available data are decidedly less numerous and less comprehensive than would be preferred for the analyses here attempted. As noted above, and despite major progress during the past two decades, the known Proterozoic fossil record pales in comparison with that of later geologic time. Moreover, as may be noted from inspection of Chapter 22, the vast majority (70% to 80%) of the data now available are based on first ("discovery") reports only, a much smaller percentage ($\leqslant 25\%$) on follow-up, more detailed, studies, and only a small fraction ($\sim 5\%$) on exhaustive, monographic, analyses; very few studies include detailed sedimentological-paleoecological analyses of the microfossiliferous strata; and many "early" (pre-1980) reports contain significant errors of interpretation.

The current taxonomy of reported Proterozoic microfossils similarly leaves much to be desired. As a result of the historical development of the field (Section 5.2; see also Chapter 25), studies of Proterozoic microfossils have developed along two largely independent pathways based on investigations of two lithologically differing suites of samples (viz., shales/stiltstones and stromatolitic carbonates/cherts) in which the reported microfossils have been preserved by differing mechanisms (via compression in shales/siltstones and via permineralization in carbonates/cherts) and have been detected and studied in differing types of preparations (in acid-resistant macerations of shales/siltstones and in petrographic thin sections of carbonates/cherts). As a result, microfossils reported from shales/silstones have been classified according to systematic conventions different from those used for fossils from carbonates/cherts (Chapter 25). At least in part because of the resulting plethora of synonymies, the taxonomy of Proterozoic microfossils is confused and incoherent.

Finally, because of time-related variations in the preserved Proterozoic rock record, there are significant temporal variations in both the quality and quantity of available paleontolog-

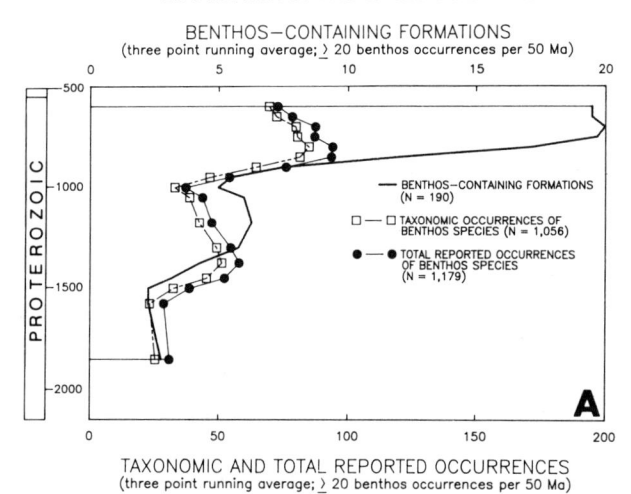

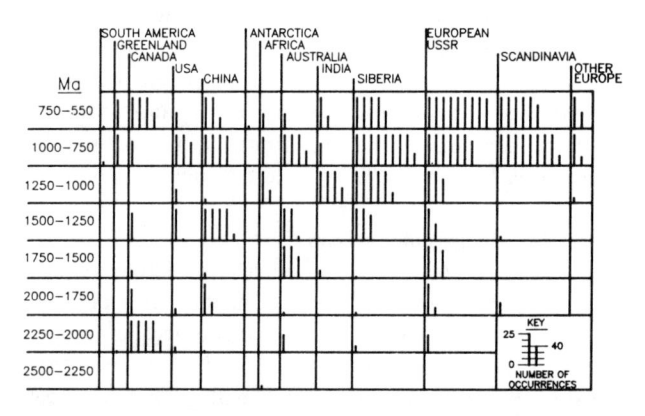

Figure 11.3.1. Temporal distribution (running average ± 75 Ma of estimated formation ages) of occurrences of Proterozoic microfossils and fossiliferous formations. (A) Benthos-containing formations (Tables 22.3 and 25.5), taxonomic occurrences of benthic species (Tables 25.3 and 25.5), and total reported occurrences of benthic species (Table 25.5). (B) Plankton-containing formations (Tables 22.3 and 25.6), taxonomic occurrences of planktonic species (Tables 25.3 and 25.6), and total reported occurrences of planktonic species (Table 25.6).

and "beta diversity" (a measure of between-assemblage species richness).

The geologic ages of most of the Proterozoic assemblages considered in these analyses are known only approximately; in many of the figures included with this discussion, therefore, data from consecutive 50 Ma-long intervals have been averaged with the resulting plots showing running means ± 50 Ma, ± 75 Ma, or $\pm 8\%$ to 10% of estimated formation ages. These analyses are thus necessarily coarse-grained, especially in comparison with diversity studies of Phanerozoic megafossils for which a much larger set of far more precisely dated assemblages is available for study.

Figure 11.3.2. Histograms summarizing the geographic distribution of total reported occurrences of Proterozoic microfossils (Tables 22.3, 25.5, and 25.6) grouped in 200 Ma- or 250 Ma-long intervals.

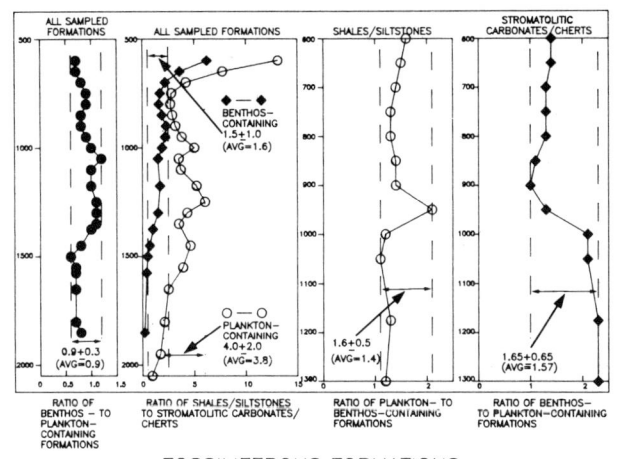

SAMPLED PROTEROZOIC LITHOLOGIES

Figure 11.3.3. Temporal distribution (running average ±75 Ma of estimated formation ages) of sampled Proterozoic lithologies. The ratio of sampled benthos-containing (Table 25.5) to plankton-containing formations (Table 25.6), 1850 to 600 Ma, is more or less uniform, ranging from 0.6 to 1.2 (mean = 0.9). The ratio of sampled benthos-containing "offshore" shales/siltstones to "nearshore" stromatolitic carbonate/cherts (Table 22.3), 1500 to 700 Ma, is 0.5 to 2.5 (mean = 1.6; N = 118 formations); this ratio for sampled plankton-containing units (Table 22.3), 1800 to 700 Ma, is 2.0 to 6.0 (mean = 3.8; N = 127 formations). Sampled "offshore" and "nearshore" lithologies from 1300 to 600 Ma (Table 22.3) contain a more or less uniform mix of analyzed assemblages, a ratio of plankton-to benthos-containing shales/siltstones ranging from 1.1 to 2.1 (mean = 1.4) and of benthos- to plankton-containing stromatolitic carbonate/cherts ranging from 1.0 to 2.3 (mean = 1.57).

ical data: in general, Upper Proterozoic deposits tend to be less metamorphosed than Middle and Lower Proterozoic units; moreover, the area of surface outcrop, volume of potentially fossiliferous strata, number of named formations per unit volume, and degree of paleontological research effort all increase over the Proterozoic correlative with decreasing geologic age. As a result of this "pull of the Phanerozoic," data available from the later Proterozoic (viz., ≤1000 Ma) are decidedly more numerous (Fig. 11.2.1), of higher quality and, hence, presumably of greater reliability, than those available from older sediments.

The deficiencies in quality, quantity, and overall coverage of available data outlined above, coupled with numerous other evident deficiencies in the available database (see below), suggest that the known fossil record can be used to support only the most broad-brush, coarse-grained generalizations regarding Proterozoic biotic diversity, and that such generalizations should be accepted only after critical and thorough evaluation.

11.3.3B Coverage of the Database

As is detailed in Chapter 22, *bibliographic coverage* of the Proterozoic microfossil database used in these analyses is extensive; certainly, and despite its numerous deficiencies, it is the most comprehensive such database now available. In compiling the database, effort was made to include all of the major (and much of the minor) micropaleontological literature published worldwide prior to mid-1988 for the entire Proterozoic and, to the extent feasible, to avoid significant non-random omission of relevant papers. Overall, this effort appears to have been reasonably successful: the database is estimated to include more than 90% of all major Proterozoic micropaleontological publications and on the order of ≥85% of all relevant papers published prior to mid-1988.

It should be recognized, however, that bibliographically the database is incomplete, and it is thus conceivable that significant non-random data omission may have occurred. Although obviously difficult to evaluate critically, evidence suggests that such omission has probably been minimal, especially with regard to the later Proterozoic (≤1000 Ma) where omission of significant data seems unlikely to have played a major role in distorting the diversity pattern detected:

(1) The apparent trends in Late Proterozoic diversity here inferred (Sections 11.3.5, 11.3.6) are based on a relatively large number of microfossil assemblages (Fig. 11.3.1) and, hence, a relatively large number of publications.
(2) The assemblages described in these publications become progressively less diverse over time, producing a pattern of gradual, long-term, Late Proterozoic decline.
(3) For this pattern to be negated by evidence available in the literature would require that numerous other papers reporting literally scores of additional, but exceptionally diverse, assemblages had been inadvertently omitted from the database (Section 11.3.5; Fig. 3.11.11).
(4) The majority of such exceptional assemblages would necessarily have received extensive (e.g., monographic) treatment, but because considerable effort was made to include all major micropaleontological literature in the database, it is difficult to envision that a large number of papers reporting highly diverse Late Proterozoic assemblages would have been inadvertently overlooked.

In addition to the possible non-random omission of relevant data, it should be noted that several hundred reports of Proterozoic microfossils here regarded as questionable (Tables 22.1, 22.2) have purposefully not been included in the diversity analyses; because it has not been feasible to examine all of these particular specimens personally, it is possible that some fraction of these reports (listed as "dubiofossils" and "pseudofossils" in Chapter 22) may have been incorrectly evaluated.

Finally, this is a fast-moving field of science; it is thus conceivable that publications appearing after mid-1988 (and therefore not included in the database) may contain data that would alter the diversity patterns here reported.

With regard to *temporal coverage*, and despite the fact that between 1000 Ma and the beginning of the Phanerozoic there is a four- to five-fold increase in the number of microfossil assemblages analyzed (Fig. 11.3.1), it is probable that coverage of the database is more exhaustive for older Proterozoic assemblages (inclusion of which involved assessment of relatively fewer publications) than for those from the younger Proterozoic. For similar reasons, and because of the experience and predilections of the assemblers of the database (chiefly, C. V. Mendelson and J. W. S.), coverage seems likely to be somewhat more complete for benthic than for planktonic assemblages.

To avoid distortion of diversity trends by unsampled or poorly sampled portions of the Proterozoic fossil record, only those 50 Ma-long intervals containing ⩾ 20 occurrences of benthic or planktonic microfossils have been included in analyses of mean alpha, maximum alpha, and beta diversities. As is shown in Figure 11.3.7A (and listed in Tables 22.5 and 22.6), temporal coverage for these diversity measures is thus incomplete, for benthic species involving 18 of the 31 intervals (58%) from 2100 to 550 Ma (but including all intervals younger than 1100 Ma) with an average of 9.5 benthos-containing assemblages per interval, and for planktonic species involving 22 of the 31 intervals (71%) from 2100 to 550 Ma (but including all intervals younger than 1150 Ma) with an average of 10.8 plankton-containing assemblages per interval. Phanerozoic taxa have been largely omitted from this study, except for some Early Cambrian species (viz., taxa of benthic prokaryotes and acanthomorph acritarchs) and present-day "Lazarus taxa" (species that because they occur in the present biota, rather than reappearing in the fossil record following a mass extinction event, might more appropriately be referred to as "*Coelacanth* taxa"; pers. comm., J. J. Sepkoski, Jr.). This omission, however, is likely to have little affect on the diversity patterns here described: few Phanerozoic prokaryotes have been reported (Schopf 1974), and many of the Proterozoic eukaryotes here considered became extinct prior to the beginning of the Cambrian.

Geographic coverage of the Proterozoic portion of the database is summarized in Figure 11.3.2. Like the temporal coverage discussed above, geographic coverage reflects the distribution of the preserved rock record: coverage is greatest, both in terms of geographic extent and number of occurrences, in the younger Proterozoic, especially between 1000 and 550 Ma for which occurrences on all continents are represented.

Because some sampled formations contain only benthos, others contain only plankton, and others contain both of these components, the *lithologic (paleoenvironmental) coverage* of the database is perhaps best considered in terms of the distribution over time of "benthos-containing" and "plankton-containing formations" (as defined in Section 11.3.3D). As is shown in Figure 11.3.3, the ratio of benthos-containing to plankton-containing units analyzed in this study is more or less uniform, ranging from 0.6 to 1.2 (average = 0.9), 1850 to 600 Ma; during the latter portion of the Proterozoic (viz., 1300 to 600 Ma), the critical period of apparent change in biotic diversity discussed below (Section 11.3.4), this ratio is similarly uniform, ranging from 0.65 to 1.15 (average = 0.9).

The distribution of benthos and plankton also can be evaluated in terms of their reported occurrence in "offshore" shales/siltstones and "nearshore" carbonates/cherts (Fig. 11.3.3). With regard to differentiation of such units, however, it is important to note that the distinction here drawn is lithological, not necessarily environmental; for the vast majority of known Proterozoic microfossiliferous units, detailed sedimentological-paleoenvironmental data are not available by which to establish unequivocally proper use of such terms as "offshore" and "nearshore." Thus, although the majority of units here referred to as "'offshore' shales/siltstones" may indeed have been deposited in offshore marine settings, represented among such units may also be turbiditic, lagoonal, mudflat, and even playa or lacustrine facies. Similarly, although many of the stromatolitic units here referred to as "'nearshore' carbonates/cherts" were probably deposited in marine shallow-water settings, others seem certain to represent non-marine, especially hypersaline, environments (e.g., Southgate 1989), and some may represent sabkha deposits or such uncommonly preserved facies as microbe-containing hot spring sinter and freshwater carbonates. Use of quotation marks with the terms "offshore" and "nearshore" in this context is meant to emphasize these environmental uncertainties. The potential effect on diversity analyses of such uncertainties is addressed below and evaluated in more detail in Section 11.3.4D.

As shown in Figure 11.3.3, although shales/siltstones are over-represented among analyzed microfossiliferous units 700 Ma and younger and are somewhat under-represented among sampled formations of Early Proterozoic age, throughout much of the Proterozoic the ratio of analyzed benthos-containing "offshore" shales/siltstones to "nearshore" stromatolitic carbonates/cherts is more or less uniform (for intervals 1500 to 700 Ma: range = 0.5 to 2.5, average = 1.6, N = 118 formations; 1200 to 700 Ma: range = 1.5 to 2.3, average = 1.9) as is this ratio for sampled plankton-containing units (1800 to 700 Ma: range = 2.0 to 6.0, average = 3.8; N = 127 formations; 1200 to 700 Ma: range = 2.7 to 5.1, average = 3.7). As is also shown in Figure 11.3.3, between 1300 and 600 Ma, both "offshore" and "nearshore" sampled lithologies consistently exhibit a more or less uniform mix of benthos- and plankton-containing assemblages (a ratio over the intervals sampled of plankton- to benthos-containing shales/siltstones ranging from 1.1 to 2.1, with an average of 1.4, and of benthos- to plankton-containing stromatolitic carbonates/cherts ranging from 1.0 to 2.3, with an average of 1.6). Thus, lithologic coverage of the database appears to be more or less uniform throughout much of the Proterozoic.

The approximate uniformity of benthos- and plankton-related lithologic sampling of the units analyzed would not be inconsistent with a similar degree of uniformity in paleoenvironmental sampling; that is, with the suites of analyzed "offshore" and "nearshore" units each encompassing a more or less homogeneous mix of sampled paleoenvironments. Unfortunately, as noted above, more detailed paleoecological assessment of units included in the database is not feasible: sedimentological and detailed paleoenvironmental studies have been carried out on only a very few percent of reported Proterozoic microfossiliferous units (cf. Vidal and Knoll 1983). It is thus generally not possible to subdivide the analyzed units into discrete, ecologically relevant, habitat types (e.g., lower intertidal versus upper intertidal facies; marine stromatolitic versus hypersaline stromatolitic facies). Thus, at present, the influence of uneven environmental sampling on the diversity trends detected cannot be assessed directly. Indirect evidence, however, based on analyses of Late Proterozoic diversity trends (Section 11.3.4D), suggests that although such sampling problems may have influenced diversity patterns of assemblages from "nearshore" stromatolitic units, patterns based on as-

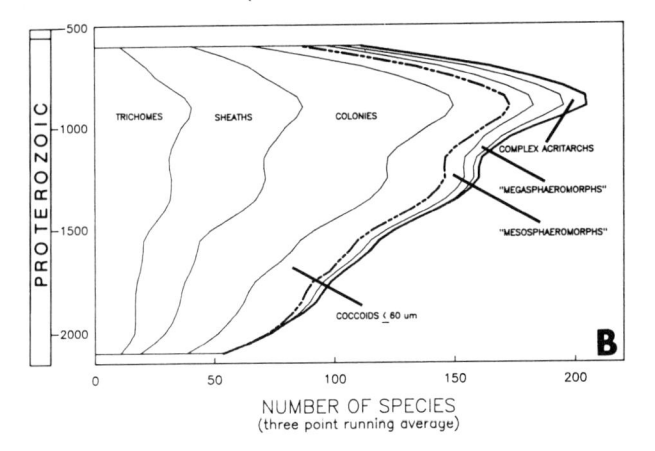

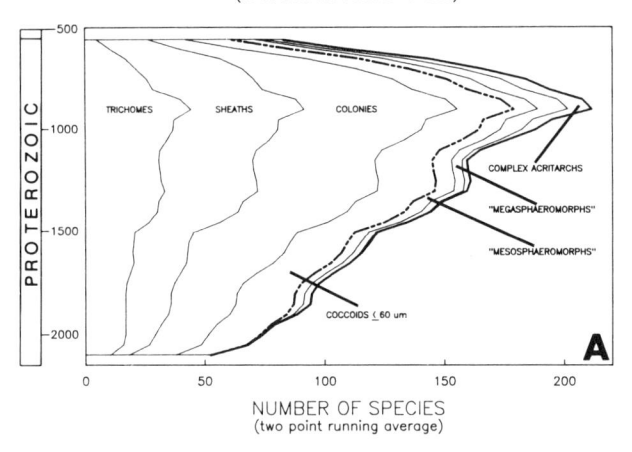

Figure 11.3.4. Observed species-level global biotic diversity of Proterozoic microfossils 2150 to 550 Ma (based on known Proterozoic distributions as reflected by first and last known Proterozoic occurrences as summarized in Fig. 25.4, but not including present-day Lazarus taxa). "Mesosphaeromorphs" (with maximum diameters ranging from 60 to 200 μm), "megasphaeromorphs" (≥200 μm in maximum diameter), and complex acritarchs (e.g., acanthomorphs) are grouped as "eukaryotes" (Table 25.1); these species, together with non-colonial coccoids ≤60 μm, are grouped as "plankton" (Table 25.1). (A) Number of species, plotted as a running average ±50 Ma of estimated formation ages. (B) Number of species, plotted as a running average ±75 Ma of estimated formation ages.

than vegetative planktonic unicells) have no doubt influenced the composition of the reported assemblages. And worker-related biases also may have influenced the range of fossils reported (e.g., it seems notable that until relatively recently, the occurrence of benthic filamentous cyanobacteria was rarely recorded in papers devoted to plankton-dominated shale microfloras; similarly, presumably overlooked because of their minute size non-cyanobacterial prokaryotes have only rarely been identified in Proterozoic strata). However, in view of the large number of fossil assemblages included in these analyses, the range of habitats thus sampled, and the fact that many Proterozoic taxa, like their modern counterparts, appear to have exhibited extensive (commonly cosmopolitan) distributions, the total mix of fossils included in the database seems likely to provide an adequate reflection of the global biota (and especially of its abundant components) over much of the Proterozoic.

11.3.3C Taxonomic Considerations

The Proterozoic micropaleontological literature is replete with redundancies (taxonomic synonymies) and inconsistencies that have resulted from past taxonomic practices (for discussion, see Chapter 25). In order to analyze diversity patterns potentially reflected by the known Proterozoic fossil record, it was therefore necessary first to construct a revised, internally consistent, species-level system of classification to include all known microfossils, regardless of their mode of preservation, technique of study, or the lithology from which they had been reported. The resulting classification is presented in Chapter 25. As documented in Section 5.4 and discussed in Chapter 25, this revised classification parallels that used for species of morphologically comparable modern microorganisms: all reported Proterozoic microfossils of the same or closely similar morphology, regardless of the binomials originally applied to them, have been grouped together into taxonomically informal fossil species; each of the several hundred taxa thus recognized exhibits a range of morphological variability similar to that characteristic of species of modern, living, analogues.

Any such comprehensive taxonomic revision, involving consideration of nearly 3000 microfossil occurrences, is fraught with difficulties, not the least of which derive from taxonomy-related deficiencies in the relevant literature. As noted in Section 11.2 and discussed more completely in Chapter 25: many published photomicrographs, even of holotype specimens, are of inadequate quality; reliable data regarding various preservable, taxonomically useful characters (e.g., ranges and patterns of distribution of cell sizes, wall and sheath thicknesses, and patterns of cell division) are lacking from many publications, especially the pre-1980 literature dealing with shale microfloras; other taxonomically useful characters (e.g., inferred life cycles) are rarely considered in the available literature or are not preservable in Proterozoic fossils (e.g., biochemical characters such as the various chlorophylls and accessory pigments); and changes in morphology due to preservational compression (Section 5.4) and diagenetic alteration commonly have been ignored or inadequately addressed. Nevertheless, and despite the obvious incompleteness of published data, the classification

semblages from "offshore" shales/siltstones may have been affected only minimally.

Finally, as might be expected, the *biologic coverage* of the database more or less parallels the temporal distribution of the facies sampled: benthic, prokaryote-dominated, fossil communities, occurring in "nearshore" stromatolitic facies, are particularly abundant among sampled Early Proterozoic units and are evidently relatively under-represented in the latest (≤700 Ma) Proterozoic intervals analyzed. In some assemblages, taphonomic effects and the preferential preservation of particular biotic components (e.g., of cyanobacterial sheaths rather than cellular trichomes, or of thick-walled cysts rather here used (Chapter 25) – like any meaningful system of paleontologic classification – necessarily has been based on, and constrained by, the known, documented, fossil record.

Bases for taxonomic differentiation of the species here recognized are specified in Chapter 25. Prior to construction of this classification, specimens of the great majority of these taxa were examined personally, as were microfossils, including many holotype and paratype specimens, from most of the known maximally diverse Proterozoic assemblages listed in Table 25.7 (viz., those from the Leningrad Blue Clays and the Lakhanda, Miroedikha, Zil'merdak, Satka, Bakal, Podinzer, Min'yar, Chichkan, Yudoma, and Sukhaya Tunguska Formations of the USSR; the Amelia Dolomite, Balbirini Dolomite, Skillogalee Dolomite, and the Bitter Springs and Barney Creek Formations of Australia; the Beck Spring Dolomite, the Gunflint Iron Formation, and the Kasegalik Formation of North America; and the Gaoyuzhuang Formation of China). Previously unpublished photomicrographs of many of these taxa are included in Chapter 24. Particularly diverse assemblages not personally examined include certain of those described from the USSR (viz., from the Kotujkan, Zigazino-Komarovsk, and Strel'nye Gory Formations), Scandinavia (from the Visingsö Group, Draken Conglomerate, and the Muhos Formation), China (from the Chuanlinggou and Doushantuo Formations), and India (from the Suket Shale). As is discussed in Chapter 25, morphometric data for the informal taxa here recognized have been derived from published descriptions augmented, where necessary, by measurement of individual specimens or of published photomicrographs. Biological implications of the classification here used (e.g., the resultant under- or overestimation of true biologic species) are addressed in Chapter 25.

Criteria used for recognition of Lazarus taxa (cf. "Coelacanth taxa"), known from the Proterozoic and present-day but evidently not reported from the intervening Phanerozoic, are also specified in Chapter 25. As was recognized early in the development of the field (Schopf 1968; Schopf and Blacic 1971), many Proterozoic microorganisms are "hypobradytelic" (Schopf 1987), exhibiting extreme morphological evolutionary conservatism, and most Proterozoic microfossil species are exceptionally long-ranging (Figure 25.4). It is thus not surprising that a relatively large proportion of such species are Lazarus taxa (Table 25.2), essentially indistinguishable in salient morphological characters from extant analogues.

Because of deficiencies in the published data, the large number of occurrences here assessed, and the degree of subjectivity inherent in the systematics of Proterozoic microfossils, it is a foregone conclusion that the informal classification used in these analyses will be open to criticism. Indeed, no claim is made that the revised classification presented in Chapter 25 is "perfect." Other workers might have approached the problem differently, used different taxonomic criteria, and identified a somewhat larger, or a somewhat smaller, suite of "recognized taxa." However, if it is granted that the object of such an exercise is that of defining Proterozoic taxa that are comparable in salient characteristics to modern, living, species, and if one is constrained by the documented fossil evidence as now available, it seems probable that any other comprehensive system of classification would differ in detail, rather than in substance, from that here employed. Despite its subjectivity, the classification of Proterozoic microfossils here used has the advantages of being comprehensive, biologically based, non-redundant, and internally consistent. It is the only such system yet devised; as such, it provides the best basis now available for assessment of patterns of Proterozoic biotic diversity.

11.3.3D Biologic and Habitat-Related Conventions

Of the 373 informal species of Proterozoic microfossils here recognized (Chapter 25), 351 species (94%) are included in the diversity analyses discussed below. Omitted from these analyses are 22 very rarely occurring taxa: one species of the distinctive, asymmetrically laminated, cyanobacterium *Polybessurus*, known from but two Proterozoic localities (code "POLY" in Tables 25.1 and 25.3); 13 species of morphologically "bizarre," primarily planktonic, prokaryotic(?) microproblematica (code "BZ"; for a discussion of these morphotypes, see Section 5.6); and eight species of non-acritarch, mostly benthic and filamentous, eukaryote-like morphotypes (code "EU"). Collectively, these exceptionally uncommon taxa comprise well less than 1% of all reported occurrences of Proterozoic microfossils.

For reasons discussed in detail in Section 5.4, the 259 benthic species included in these analyses (viz., non-septate or septate filaments; unordered, colonial, coccoidal or ellipsoidal cells; and ordered, colonial coccoidal cells; Table 25.1) are regarded as prokaryotes. "Benthos-containing formations" (Fig. 11.3.1A) are those reported to contain members of one or more of these categories of benthic, prokaryotic microfossils. It is assumed in these analyses that if any type of such benthos is known to occur in a particular unit, potential exists for all other categories of coeval benthos also to be represented. Available data seem consistent with this assumption: of the 164 benthos-containing formations 1150 to 530 Ma here analyzed, about one-quarter (42 = 26%) contain septate filaments, nearly half (78 = 48%) contain non-septate filaments, about three-quarters (129 = 79%) contain colonies, and more than half (85 = 52%) contain at least two of these three components. Relevant diversity data for benthos (e.g., Fig. 11.3.8) are therefore here presented as the number of benthic species known to occur per benthos-containing formation.

As is discussed in Sections 5.4 and 5.5 and in Chapter 25, the 92 planktonic species considered in these analyses include both "probable prokaryotes" (viz., 24 taxa of solitary coccoids $\leqslant 60\,\mu m$ in maximum diameter, code "S" in Tables 25.1 and 25.3) and "assured eukaryotes" (viz., 68 species of "mesosphaeromorphs," "megasphaeromorphs," and complex, non-sphaeromorph acritarchs, codes "MS," "MG," and "A," respectively). "Plankton-containing formations" (Fig. 11.3.1B) are those reported to contain members of one or more of these categories of planktonic microfossils. It is assumed in these analyses that if any type of such plankton is known to occur in a particular unit, potential exists for all other categories of coeval plankton also to be represented. Available data seem to support this assumption: of the 214 plankton-containing formations 1150 to 530 Ma here analyzed, a substantial majority (viz., 135 = 63%) contain both prokaryotic and eukaryotic plankton; decidedly fewer (60 = 28%) contain only prokaryotic plankton;

and a small minority (19 = 9%) contain only eukaryotic plankton. Relevant diversity data for plankton (e.g., Fig. 11.3.8) are therefore here presented as the number of planktonic species known to occur per plankton-containing formation.

11.3.3E Geochronologic Conventions

Because of limitations in techniques available for the dating of sedimentary rocks, ages of most of the Proterozoic microfossil assemblages included in these analyses are known only approximately. To obviate this problem, one that plagues Proterozoic paleobiologic studies in general, and to facilitate analyses of the large number of assemblages (and thus formations) considered, formations have been grouped according to their estimated ages and the resultant data analyzed in subgroups encompassing 50 Ma-long intervals. However, ages of many of the Proterozoic assemblages analyzed are less constrained by available geochronologic data than the ± 25 Ma implied by this convention, and assignment of an assemblage to a particular 50 Ma-long interval may therefore be open to question. Because of this uncertainty, diversity data in most of the analyses discussed below (Section 11.3.4) are presented as running averages that represent ± 50 Ma of estimated formation ages (viz., two point running means), ± 75 Ma of estimated formation ages (viz., three point running means), or $\pm 8\%$ to 10% of estimated formation ages. The variable effects of these forms of data presentation are compared graphically in Figures 11.3.5 and 11.3.7 and summarized in Table 11.3.2. The latter two forms of data presentation (viz., ± 75 Ma and $\pm 8\%$ to 10% of estimated formation ages), which yield closely similar plots for data 1150 to 550 Ma and comparable results throughout much of the remainder of the Proterozoic (Figs. 11.3.5 and 11.3.7), are regarded as reflecting most realistically current uncertainty in available geochronologic information. This convention of averaging data over consecutive 50 Ma-long intervals has the effect of obscuring details of the time-course, but not the times of maxima or minima, or the overall trends, of the diversity patterns detected.

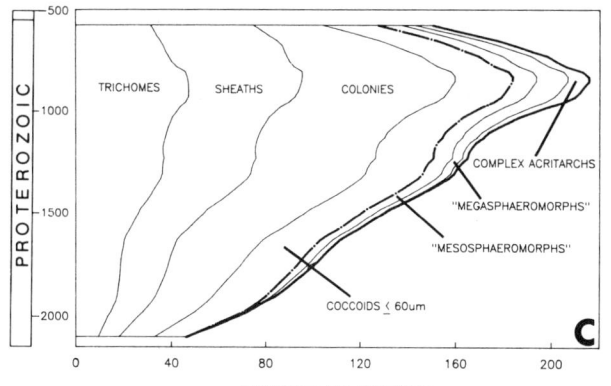

11.3.3F Measures of Diversity

Tables 25.5 and 25.6 summarize for each 50 Ma-long interval of the Proterozoic the number of reported microfossiliferous benthos- and plankton-containing formations ("FM"); number of species ("SP"), based on the revised informal classification of Chapter 25, reported to occur; number of known species-level taxonomic occurrences ("TX"); number of total reported taxonomic occurrences ("OCC"), which includes reported multiple occurrences (including those resulting from taxonomic synonymies) of particular species in single formations; and number of first and last reported species occurrences. Intervals shown in boldface in Tables 25.5 and 25.6 and indicated by stippling in Figure 11.3.7A, containing ≥ 20 occurrences of benthic or of planktonic species, are those used for analyses of alpha and beta diversities. Analyses of global diversity are based on first ("1sts") and last ("Lasts") known species occurrences as summarized in Tables 25.5 and 25.6 and in the range-charts shown in Figure 25.4. Data used to assess maximum alpha diversity are listed in Table 25.7.

Alpha diversity ("AD") is "within assemblage" taxonomic

Figure 11.3.5. Inferred species-level global biotic diversity of Proterozoic microfossils 2150 to 550 Ma (based on known Proterozoic distributions as reflected by first and last known Proterozoic occurrences plus inferred Phanerozoic distributions based on present-day Lazarus taxa as summarized in Table 25.2 and Fig. 25.4). "Mesosphaeromorphs" (with maximum diameters ranging from 60 to 200 μm), "megasphaeromorphs" (≥ 200 μm in maximum diameter), and complex acritarchs (e.g., acanthomorphs) are grouped as "eukaryotes" (Table 25.1); these species, together with non-colonial coccoids ≤ 60 μm, are grouped as "plankton" (Table 25.I). (A) Number of species, plotted as a running average ± 50 Ma of estimated formation ages. (B) Number of species, plotted as a running average ± 75 Ma of estimated formation ages. (C) Number of species, plotted as a running average ± 8 to 10% of estimated formation ages.

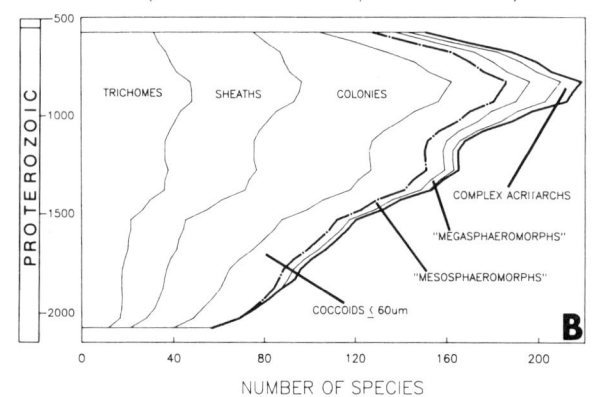

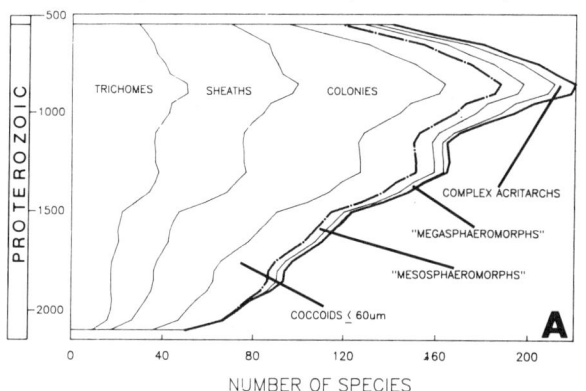

diversity, the richness of taxa known to occur in a particular assemblage or community, or at a single locality. Because benthic and planktonic species for each "FM" generally can be regarded as members of a "particular assemblage" or ecological community (and, with relatively few exceptions, have been reported from but a "single locality" of each formation), for benthic and planktonic species for each 50 Ma-long interval: *mean AD* = TX/FM, the mean species-level alpha diversity (or "average intra-assemblage species richness") of known fossiliferous formations within a particular 50 Ma-long interval. *Maximum AD*, the "maximum alpha diversity" for either benthos or plankton is the maximum number of species reported to occur in any single formation within a particular 50 Ma-long interval.

Beta diversity ("BD") is "between assemblage" taxonomic diversity, the amount of diversity gained when two or more samples or communities are combined, here measured over discrete (50 Ma-long) intervals (cf. Sepkoski 1988); in well-sampled sequences, "BD" is an index of rarely occurring (e.g., ecologically restricted) taxa. The simplest measure of beta diversity is the ratio of the total number of taxa in a pooled set of samples to the average number of taxa in each sample. Within each 50 Ma-long interval, "SP" is the total (pooled) number of known species. Thus, for benthic or planktonic species from each 50 Ma-long interval: BD = SP/(TX/FM) = SP/mean AD, the species-level beta diversity for a particular 50 Ma-long interval.

Global diversity ("GD"), the total known or inferred taxonomic diversity, is a measure based on putative originations ("1sts") and extinctions ("Lasts"). *Observed global diversity* is determined by "1sts" and "Lasts" as evidenced by the Proterozoic fossil record as now known. *Inferred global diversity* includes consideration of Lazarus taxa (species denoted by stars in the range-charts shown in Fig. 25.4) and thus reflects fewer putative extinctions than those inferred from a literal reading of the known fossil evidence.

Table 11.3.1 illustrates the calculation of six measures of diversity (viz., mean alpha diversity, maximum alpha diversity, mean alpha diversity excluding maximally diverse assemblages, mean alpha diversity excluding one-taxon assemblages, mean alpha diversity excluding maximum and one-taxon assemblages, and beta diversity) for a hypothetical 50 Ma-long interval containing five fossiliferous "formations" ("I" through "V") containing varying numbers of occurrences of six "species" ("A" through "F").

11.3.3G Evaluation of Detected Diversity Trends

As noted above (Section 11.3.1), the principal uncertainty regarding these analyses of patterns of Proterozoic biotic diversity is whether the data now available (i.e., available prior to mid-1988) are sufficiently representative of true diversity to yield meaningful generalizations. Although this issue cannot be resolved *a priori*, and a firm answer may therefore not be apparent for several decades (the cardinal test being whether

Table 11.3.1. *Calculation of alpha and beta diversities for a hypothetical 50 Ma-long interval*

Formations	Reported Occurrences of Species	Data Summary by Formation				Data for 50 Ma Interval			
		FM	SP	TX	OCC	FM	SP	TX	OCC
I	A, A, A, B, C, E, F	1	5	5	7	5	6	13	19
II	A, B, B, E, F	1	4	4	5				
III	B	1	1	1	1				
IV	A, A, A, A, D	1	2	2	5				
V	C	1	1	1	1				

Mean species-level alpha ("within assemblage") *diversity* for the 50 Ma-long interval:

mean AD = TX/FM = 13/5 = 2.6 species/fossiliferous formation

Maximum species-level alpha diversity for the 50 Ma-long interval:

maximum AD = 5 species (viz., Formation I, species A,B,C,E,F)

Mean species-level alpha diversity excluding the maximally diverse assemblage for the 50 Ma-long interval:

mean AD minus maximum assemblage = (TX − maxTX)/(FM − 1) = (13 − 5)/(5 − 1) = 8/4 = 2.0 species/fossiliferous formation

Mean species-level alpha diversity excluding one-taxon assemblages for the 50 Ma-long interval:

mean AD minus one-taxon assemblages = (TX − 2)/(FM − 2) = (13 − 2)/(5 − 2) = 11/3 = 3.7 species/fossiliferous formation

Mean species-level alpha diversity excluding the maximally diverse assemblage and one-taxon assemblages for the 50 Ma-long interval:

mean AD minus maximum and one-taxon assemblages = (TX − 7)/(FM − 3) = (13 − 7)/(5 − 3) = 6/2 = 3.0 species/fossiliferous formation

Species-level beta ("between assemblage") *diversity* for the 50 Ma-long interval:

BD = SP/mean AD = 6/2.6 = 2.3

currently detected trends are substantiated by additional discoveries and future analyses), a provisional basis for assessment of the detected trends can be provided by evaluation of at least the most obvious potential pitfalls. Among major questions to be asked are the following:

(1) Are available data both mutually consistent and mutually supportive? For example, do ecologically comparable components of the biota (e.g., all subgroups of plankton, or all subgroups of benthos) exhibit comparable diversity patterns?
(2) Are the detected diversity patterns a result of uneven sampling and/or investigation – do they reflect the influence(s) of monographic effects and/or of incomplete "grab-sampling," rather than of global biotic change? That is, have the observed trends been skewed due to the distorting influence(s) of a relatively few, exhaustively studied assemblages and/or a relatively large number of very incompletely studied assemblages, or do they reflect real changes in the total biota?
(3) Are the patterns detected a result of biased sampling of particular lithologies and/or environments? For example, do the biotas preserved in "offshore" shales/siltstones exhibit the same patterns of diversity as those in coeval "nearshore" stromatolitic carbonates/cherts?
(4) Are the patterns simply a result of the vagaries of local preservation – are the detected trends unique to particular locales, or are the same patterns repeated in widely separated stratigraphic sections?
(5) How "robust" are the detected patterns? What amount and quality of additional (currently "missing") evidence would be required to negate the observed trends?
(6) Are the diversity patterns consistent with independently inferred evolutionary trends? For example, do changes in diversity patterns coincide with major evolutionary events as inferred from the temporal distribution of various types of Proterozoic prokaryotes and eukaryotes?
(7) Are the observed trends consistent with other possibly correlative changes in the Proterozoic paleobiologic record such as variations in the carbon isotopic composition of sedimentary organic matter and, for benthic microfossils, in the abundance and diversity of stromatolites?

11.3.3H Additional Caveats

In addition to the foregoing, it should be recognized that for purposes of these analyses it had been assumed that:

(1) All benthic microfossils or planktonic microfossils reported from a particular formation can be regarded as members of single biotic "assemblages" (a term not necessarily synonymous with "communities," and a valid assumption, at least at present, inasmuch as microfossils have as yet been reported from only one or a very few closely spaced horizons in the vast majority of Proterozoic microfossiliferous units).
(2) All analyzed microfossiliferous Proterozoic formations can be treated as representing comparable, stratigraphically more or less equivalent, units (an assumption that may be valid on average, but one that is complicated by the tendency for Late Proterozoic formations to be stratigraphically more subdivided than their older counterparts).
(3) On average, comparable paleontological processing procedures and a similar degree of research effort have been applied to the fossiliferous units here considered (an assumption that may be appropriate, but one that in light of present data is virtually impossible to evaluate critically and is therefore here partially addressed by consideration of diversity trends based on data from which have been excluded maximally diverse assemblages as well as all units from which only a single taxon has been reported; see Section 11.3.4D).

11.3.3I Conclusions

As is noted at the outset of this discussion, emphasized by use of the term "tentative" in the title of this Section, and documented in the foregoing paragraphs, assessment of Proterozoic microfossil diversity based on the data currently available is a difficult, complex, multifaceted problem, fraught with uncertainty. Four principal deficiencies stand out, the need for (*i*) greatly increased geochronologic control on the ages of microfossiliferous Proterozoic strata; (*ii*) a thorough, formal, revision of Proterozoic microfossil taxonomy based on careful study of individual specimens and populations; (*iii*) a major increase in the number and quality of exhaustive, monographic, studies of Proterozoic microfossil assemblages and communities; and (*iv*) increased attention to the sedimentology and paleoecology of Proterozoic microfossil-bearing strata.

In some future "better world," several decades hence, after a much larger body of relevant data has been amassed, diversity studies of the type here attempted can be expected to become far more tractable. At present, however, this initial, tentative, analysis must necessarily be constrained by the actual evidence now at hand. The cardinal question is whether these currently limited data are sufficient to provide a coherent, meaningful, albeit a broad-brush and coarse-grained description of the overall pattern of Proterozoic biotic diversity, a description that will stand the test of time.

11.3.4 Analyses of Diversity

11.3.4A Global Diversity

As shown in Figures 11.2.1 and 11.3.1, for both benthos and plankton, the number of reported fossiliferous formations per analyzed 50 Ma-long interval prior to 1000 Ma is relatively low, approximately constant, and more or less paralleled by both number of species occurrences and of total reported occurrences. This parallelism indicates that prior to 1000 Ma, the number of known microfossil occurrences is a strong function of the number of formations sampled. Thus, the gradual increases during this period in observed (Fig. 11.3.4) and inferred (Fig. 11.3.5) species-level "global diversity" – i.e., in documented and extrapolated total biotic species richness – reflect the sequential addition of first-reported ("new") occurrences of (generally long-ranging) species as the total number of investigated formations increases. Between 1000 Ma and the beginning of the Phanerozoic, the number of known benthos- and plankton-containing formations increases by four- to more than five-fold (Fig. 11.3.1); many of these units contain exceptionally well-preserved microfossils (Chapter 24). Thus, observed Late Proterozoic declines in species occurrences (beginning at about 850 Ma; Fig. 11.3.1), total reported occurrences (Fig. 11.3.1), and (using a variety of plotting conventions to reflect variable degrees of age-related uncertainty) in both observed (Fig. 11.3.4) and inferred (Fig. 11.3.5) global biotic diversity, do not reflect a decrease in the number or quality of sampled geologic units. Similarly, correlative with the "pull of the Phanerozoic" (Section 11.3.3A), these evident declines have not been biased by such factors as extent of surface outcrop, volume of potentially fossiliferous rock, or degree of paleontological research effort, nor do they correlate with temporal variations in either geographic or lithologic sampling (see Section 11.3.3B and Figs. 11.3.2 and 11.3.3).

The Late Proterozoic declines in observed and inferred biotic global diversity (Figs. 11.3.4, 11.3.5) are paralleled by a comparable decrease in the observed species-level global diversity of planktonic eukaryotes (Fig. 11.3.6). Whether plotted as running averages (Figs. 11.3.4, 11.3.5) or simply as first and last known occurrences (Fig. 11.3.6), all of these declines are gradual and long-term, occurring over a several hundred Ma-long interval. As is shown in Figure 11.3.6, they thus contrast rather markedly with the short-term, sharp drop in latest Proterozoic eukaryotic plankton diversity inferred by Vidal and Knoll (1982), a difference evidently resulting from the much larger body of (post-1982, and especially Soviet) data included in the present analyses.

11.3.4B Mean alpha Diversity

Figure 11.3.7 summarizes patterns of Proterozoic species-level "mean alpha diversity" (the average number of species per assemblage) at various values of assumed uncertainty of estimated formation ages. It seems significant that the Late Proterozoic decline in global biotic diversity discussed above coincides with a major decrease in biotic mean alpha diversity (Fig. 11.3.7), a decrease occurring in all major components of the biota (Fig. 11.3.8) and in all planktonic subgroups (Fig. 11.3.9) that is at least roughly paralleled by a Late Proterozoic decline in both the abundance and the diversity of stromatolites (Fig. 11.3.10). As shown in Figure 11.3.11, the occurrence during the Late Proterozoic of a major decrease in biotic mean alpha diversity is inferred from analyses of a large number of formations and more than 1200 microfossil occurrences; as discussed below (Section 11.3.5), to negate this apparent decline would require future discovery of a nearly comparable body of currently "missing" data.

As discussed in Section 11.3.3A, because of the relative paucity of fossiliferous formations older than 1000 Ma, diversity data for the Early and Middle Proterozoic are less reliable than those available for the Late Proterozoic. Nevertheless, it is interesting to note that the data available suggest an inverse relationship between increasing planktonic eukaryotes and decreasing benthic prokaryotes 1500 to 1000 Ma (Fig. 11.3.8). If this pattern is substantiated by future investigations, and if total Proterozoic biomass was resource-limited (presumably by PO_4^{-3}), this inverse relationship may evidence parallel changes in mean alpha diversity and the prokaryote-eukaryote distribution of the Proterozoic global biomass.

11.3.4C Beta Diversity

Throughout much of the Proterozoic, "beta diversity" (a measure of between-assemblage species richness) strongly correlates with the number of formations sampled. Nevertheless, as is shown in Figure 11.3.12, near the close of the Proterozoic, the decline in mean species-level biotic alpha diversity noted above is paralleled by declines in the species-level beta diversity of various types of prokaryotic benthos (viz., benthic colonies, tubular sheaths, and cellular trichomes; Fig. 11.3.12A) and eukaryotic plankton (viz., "megasphaeromorphs"; Fig. 11.3.12B), suggesting that during this portion of the Late Proterozoic, relatively rare (e.g., ecologically restricted) species of fossil microorganisms may also have experienced significant decreases in diversity.

11.3.4D Maximum Alpha Diversity

This measure of diversity, the maximum number of benthic species and the maximum number of planktonic species reported for any of the assemblages within an analyzed 50 Ma-long interval (Table 25.7), is intended to assess the possible influence of monographic effects on patterns of mean alpha diversity. Because relatively few Proterozoic microfossil assemblages are currently available for analysis (on average, only about 10 assemblages per analyzed 50 Ma-long interval; Section 11.3.3B), there seems a strong possibility that analyses of patterns of average species diversity, based on comparison of data from sequential time intervals, may be overwhelmingly influenced by the relatively few, and perhaps temporally unevenly distributed assemblages that have been extensively (monographically) investigated.

Assessment of monographic effects. Comparison of patterns of maximum and mean alpha diversities provides a way to evaluate the influence of monographic effects. As shown in Figure 11.3.13, there is a lack of parallelism during the Late

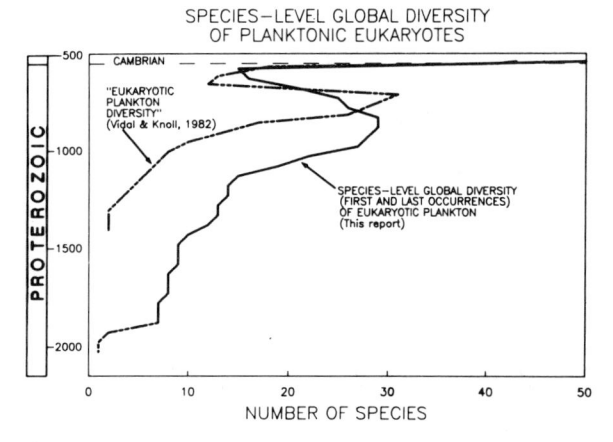

Figure 11.3.6. Comparison of Proterozoic "eukaryotic plankton diversity" as inferred by Vidal and Knoll (1982) and the species-level global diversity of Proterozoic eukaryotic plankton here reported (plotted as a cumulative curve based on the first and last known Proterozoic species occurrences specified in Fig. 25.4; not plotted as a running average).

Patterns of Proterozoic Microfossil Diversity

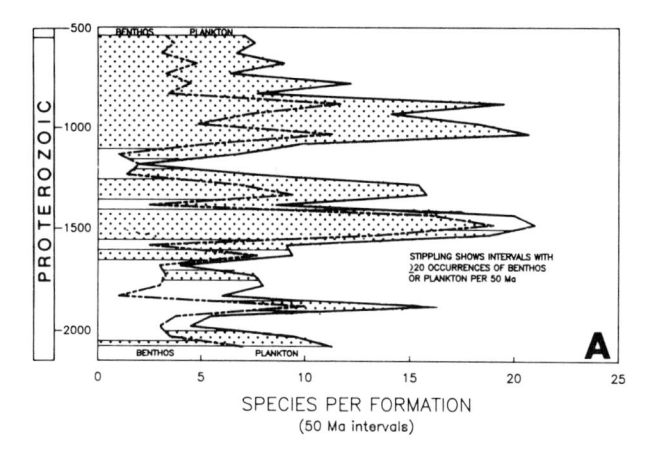

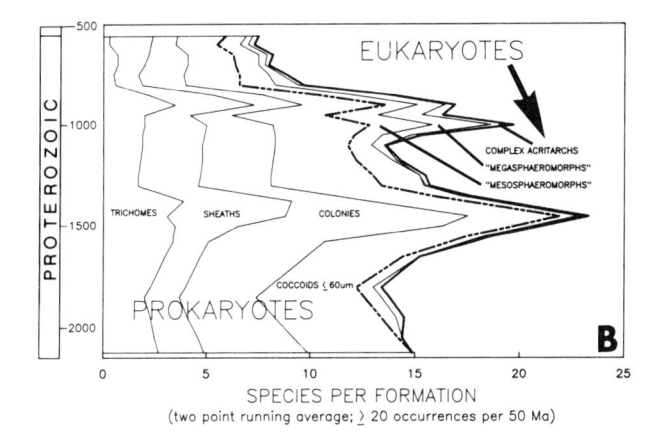

Proterozoic between these two measures of diversity for prokaryotes and (prokaryote-dominated) benthos, patterns that contrast strongly with the marked similarity of the mean and maximum alpha diversity curves for Late Proterozoic eukaryotes and plankton. In order to determine the source of this difference, a second analysis has been carried out, one based on the strategy of omitting from calculations of average diversity the maximally diverse assemblage known from each interval, and comparing the resulting modified mean alpha diversity pattern (viz., that for mean alpha diversity minus the maximally diverse assemblage for each interval) with the patterns of maximum alpha diversity and mean alpha diversity for all assemblages.

Comparison of maximum alpha diversity, mean alpha diversity, and mean alpha diversity excluding maximally diverse assemblages from each 50 Ma-long interval has been carried out for all analyzed formations (Figs. 11.3.13, 11.3.15, Table 11.3.2) as well as separately for assemblages from "offshore" shales/siltstones and from "nearshore" stromatolitic carbonates/cherts (Figs. 11.3.14, 11.3.16, Table 11.3.3). In addition, analyses have been carried out to determine the influence on the pattern of mean alpha diversity for assemblages from Late Proterozoic carbonates/cherts of the monographically studied Bitter Springs Formation of central Australia (Fig. 11.3.14), an exceptionally diverse microbiota (Table 25-7) and the only unquestionably non-marine stromatolitic assemblage now known from the Proterozoic fossil record (Southgate 1987, 1989).

Results obtained indicate that for eukaryotes and plankton, there is a marked comparability in the timing and magnitude of Late Proterozoic declines in maximum alpha diversity, mean alpha diversity and, for shales/siltstones, mean alpha diversity

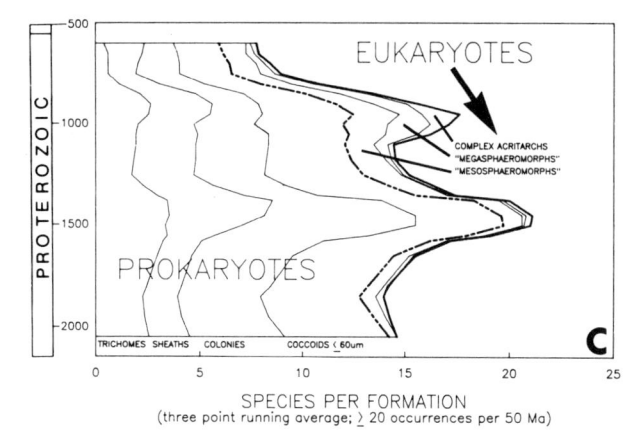

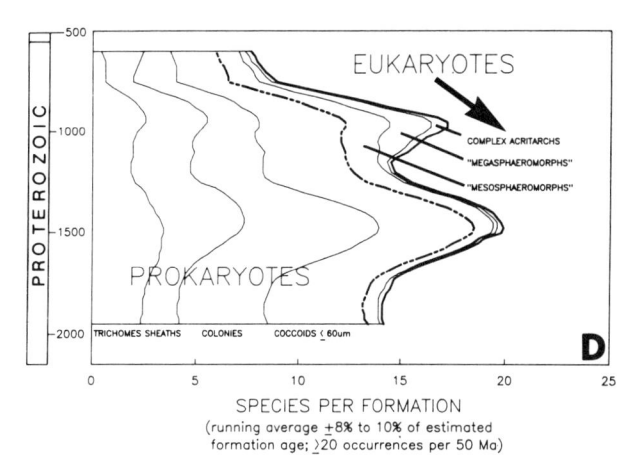

Figure 11.3.7. Observed mean species-level alpha biotic diversity for 50 Ma-long intervals, 2150 to 550 Ma (definitions of benthos and plankton, prokaryotes and eukaryotes, and plotted subgroups are given in Table 25.1; species included are listed in Table 25.3; data for analyzed intervals containing ≥20 occurrences of benthic or of planktonic species are summarized in Tables 25.5 and 25.6, respectively). (A) Mean number of species per formation for all 50 Ma-long intervals; stippling shows intervals included in analyses summarized in Figures 11.3.7B–D and 11.3.8–11.3.14 (for benthos, 58% of intervals from 2100 to 550 Ma, including all intervals younger than 1100 Ma, with an average of 9.5 benthos-containing assemblages per interval; for plankton, 71% of intervals from 2100 to 550, including all intervals younger than 1150 Ma, with an average of 10.8 plankton-containing assemblages per interval). (B) Mean number of species per formation, plotted as a running average ±50 Ma of estimated formation ages. (C) Mean number of species per formation, plotted as a running average ±75 Ma of estimated formation ages. (D) Mean number of species per formation, plotted as a running average ±8 to 10% of estimated formation ages.

540 Biotic Diversity and Rates of Evolution

Table 11.3.2. *Percent decrease during the Late Proterozoic in the species-level taxonomic diversity of prokaryotes, eukaryotes, benthos, plankton, and total biota*

Measures of Diversity (at various values of assumed uncertainty of estimated formation ages; Table 22.3)	Prokaryotes				Eukaryotes			
	Time (Ma) of Diversity		Duration of Decline	Percent Decrease	Time (Ma) of Diversity		Duration of Decline	Percent Decrease
	Max.	Min.			Max.	Min.		
Mean Alpha Diversity (all assemblages):								
(1) ±50 Ma	900	600	300 Ma	60%	1000	558	442 Ma	77%
(2) ±75 Ma	950	600	350 Ma	52%	1000	650	350 Ma	67%
(3) ±8 to 10%	975	600	375 Ma	52%	975	600	375 Ma	66%
Mean Alpha Diversity (minus maximum assemblages):								
(4) ±75 Ma	1050	600	450 Ma	57%	950	650	300 Ma	57%
Mean Alpha Diversity (minus one-taxon assemblages):								
(5) ±75 Ma	950	750	200 Ma	44%	1000	650	350 Ma	68%
Mean Alpha Diversity (minus maximum and one-taxon assemblages):								
(6) ±75 Ma	1050	650	400 Ma	47%	950	650	300 Ma	60%
Maximum Alpha Diversity:								
(7) ±50 Ma	900	558	342 Ma	52%	950	600	350 Ma	68%
(8) ±75 Ma	Variable				950	600	350 Ma	58%
Observed Global Diversity (not including Lazarus taxa):								
(9) ±50 Ma	900	558	342 Ma	66%	850	558	292 Ma	37%
(10) ±75 Ma	900	600	300 Ma	50%	800	600	200 Ma	30%
Inferred Global Diversity (including Lazarus taxa):								
(11) ±50 Ma	850	550	300 Ma	37%	900	550	350 Ma	32%
(12) ±75 Ma	825	575	250 Ma	32%	825	575	250 Ma	29%
(13) ±8 to 10%	825	575	250 Ma	30%	850	575	275 Ma	29%

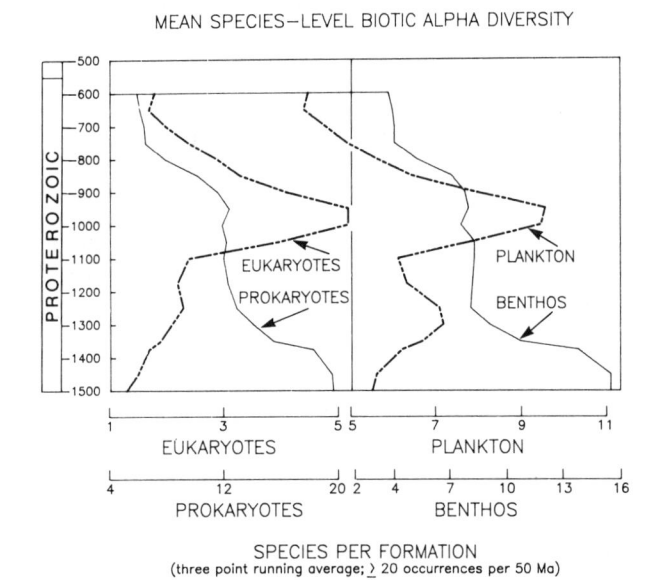

Figure 11.3.8. Observed mean species-level biotic alpha diversity of prokaryotes and eukaryotes, benthos and plankton, 1500 to 600 Ma, plotted as a running average ±75 Ma of estimated formation ages (definitions of prokaryotes and eukaryotes, benthos and plankton, are given in Table 25.1; species included are listed in Table 25.3; intervals analyzed are indicated in Fig. 11.3.7A).

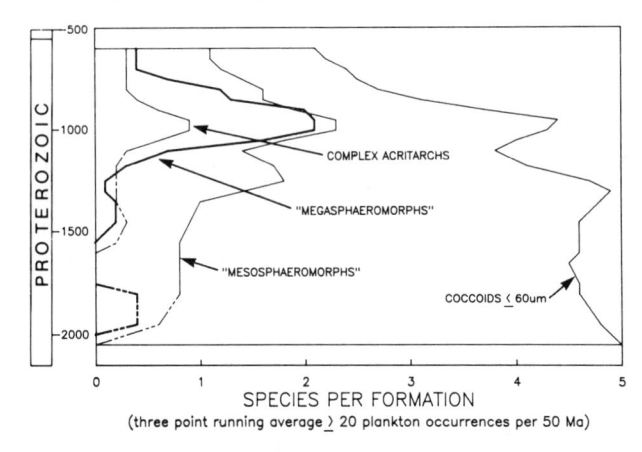

Figure 11.3.9. Observed mean species-level alpha diversity of various subgroups of Proterozoic plankton (coccoids ⩽60 μm, "mesosphaeromorphs," "megasphaeromorphs," and complex acritarchs), 2050 to 600 Ma, plotted as a running average ±75 Ma of estimated formation ages (plotted subgroups are defined in Figure 11.3.4; species included are listed in Table 25.3; intervals analyzed are indicated in Fig. 11.3.7A; dashed lines indicate questionable reported distributions). The numerically low values of species/formation here shown for subgroups of eukaryotic plankton reflect the fact that alpha diversity has been calculated relative to all plankton-containing formations, regardless of whether such units are known to contain eukaryotes (see Section 11.3.3D).

Patterns of Proterozoic Microfossil Diversity

Benthos				Plankton				Total Biota			
Time (Ma) of Diversity		Duration of Decline	Percent Decrease	Time (Ma) of Diversity		Duration of Decline	Percent Decrease	Time (Ma) of Diversity		Duration of Decline	Percent Decrease
Max.	Min.			Max.	Min.			Max.	Min.		
900	600	300 Ma	64%	1000	600	400 Ma	67%	1000	600	400 Ma	63%
950	600	350 Ma	54%	950	650	300 Ma	61%	950	600	350 Ma	56%
975	600	375 Ma	54%	975	600	375 Ma	58%	975	600	375 Ma	56%
1100	600	500 Ma	63%	950	650	300 Ma	46%	1050	650	400 Ma	59%
1050	600	450 Ma	46%	1000	650	350 Ma	60%	1000	650	350 Ma	46%
1100	600	500 Ma	48%	1000	650	350 Ma	48%	1050	650	400 Ma	48%
900	558	342 Ma	51%	950	600	350 Ma	66%	900	558	342 Ma	51%
Variable				950	650	300 Ma	56%	900	650	250 Ma	34%
900	558	342 Ma	73%	850	558	292 Ma	29%	900	558	342 Ma	61%
900	600	300 Ma	56%	850	600	250 Ma	20%	900	600	300 Ma	46%
850	550	300 Ma	42%	900	550	350 Ma	19%	850	550	300 Ma	36%
825	575	250 Ma	36%	825	575	250 Ma	17%	825	575	250 Ma	31%
825	575	250 Ma	35%	850	575	275 Ma	17%	850	575	275 Ma	30%

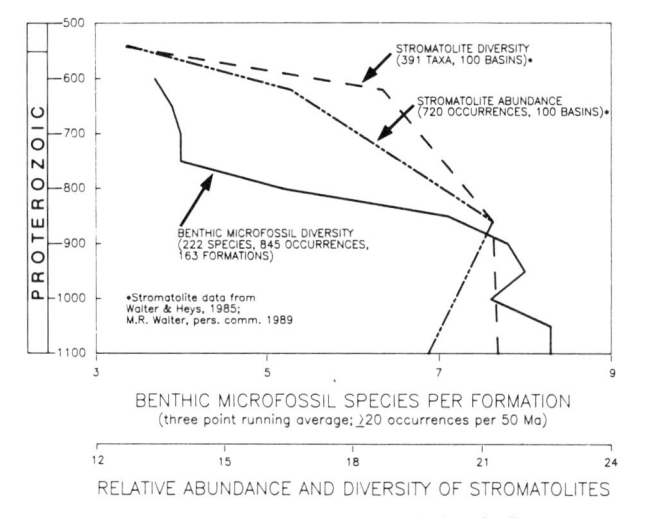

Figure 11.3.10. Comparison of the declines during the Late Proterozoic in the observed mean species-level alpha diversity of benthic microfossils (cf. Fig. 11.3.8; 1100 to 600 Ma, plotted as a running average ±75 Ma of estimated formation ages) and the relative abundance and diversity of stromatolites (1100 Ma to 540 Ma). Note that although the microfossil diversity plot is based on 11 data points, and the stromatolite plots are each based on only three such points ("Upper Riphean," "Vendian," and "Lower Cambrian"), all of the plots decline markedly, by 40% to >50%, between ~900 and ~600 Ma.

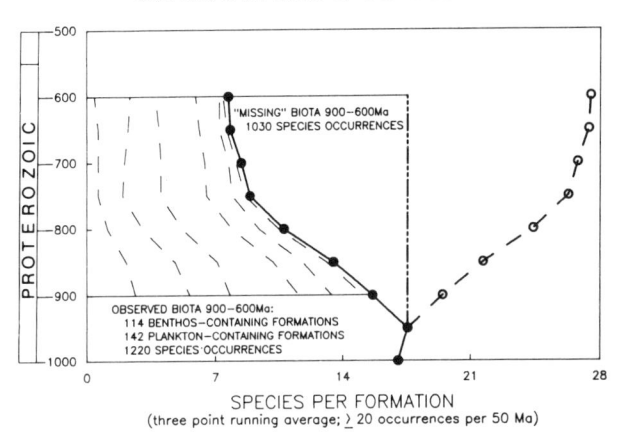

Figure 11.3.11. Graphical analysis of the "missing" Late Proterozoic biota, 900 to 600 Ma, plotted as a running average ±75 Ma of estimated formation ages. Documentation of the biotic decline from 900 to 600 Ma as measured by mean alpha diversity (cf. Fig. 11.3.7C) is based on 1220 reported species occurrences in 114 benthos- and 142 plankton-containing formations (solid line and filled circles); this apparent decline could be negated (i.e., uniform diversity during this period, graphed as the vertical dot-dot-dashed line, could be maintained) if an additional 1030 occurrences were to be discovered in these units, comprising a "missing fossil record" in which the initially overlooked components become gradually, increasingly, and ultimately exceptionally diverse (viz., exhibit the sequentially increasing mean alpha diversity plotted with open circles).

542 Biotic Diversity and Rates of Evolution

mean alpha diversity is clearly evident from this assessment, but it is notable that even if these highly diverse assemblages are omitted from the analyses, significant Late Proterozoic declines are nevertheless observed in mean alpha diversity excluding maximally diverse assemblages for all biotic components reported from shales/siltstones (viz., as summarized in Table 11.3.3, decreases ranging from a minimum of 40% for plankton, to 42% for both prokaryotes and eukaryotes, to a maximum of 58% for benthos). Thus, Late Proterozoic mean alpha diversity

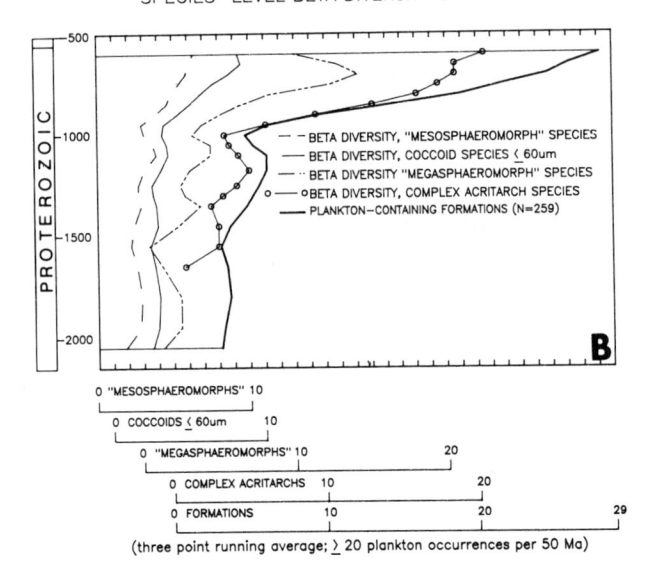

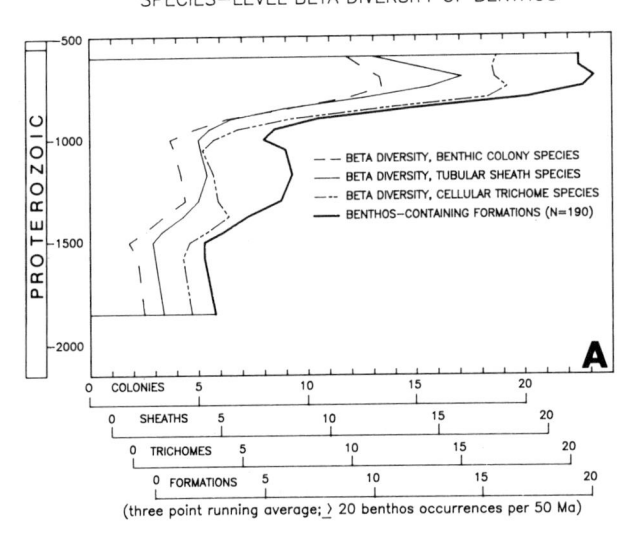

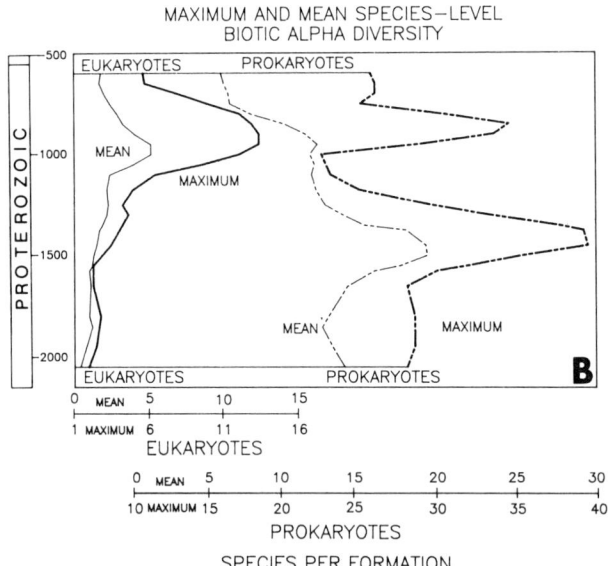

Figure 11.3.12. Temporal distribution of Proterozoic fossiliferous formations and species-level beta diversity, per analyzed 50 Ma-long interval, of subgroups of Proterozoic benthos and plankton, plotted as running averages ±75 Ma of estimated formation ages (species included are listed in Table 25.3; intervals analyzed are indicated in Fig. 11.3.7A). (A) Benthic colonies, tubular sheaths, and cellular trichomes (see Tables 25.3 and 25.5). (B) Planktonic "mesosphaeromorphs," coccoids ⩽60 μm, "megasphaeromorphs," and complex acritarchs (see Tables 25.3 and 25.6).

excluding maximally diverse assemblages or, for carbonates/cherts, mean alpha diversity excluding the Bitter Springs assemblage (Figs. 11.3.13, 11.3.14B, 11.3.14D); that comparable declines also occur in the prokaryotic and benthic components of assemblages reported from Late Proterozoic shales/siltstones (Fig. 11.3.14A, 11.3.14C); and that both the mean and maximum alpha diversity patterns of prokaryotes and benthos reported from carbonates/cherts have been markedly influenced by inclusion of data from the monographically studied Bitter Springs Formation (Fig. 11.3.14A, 11.3.14C).

The influence of maximally diverse assemblages on values of

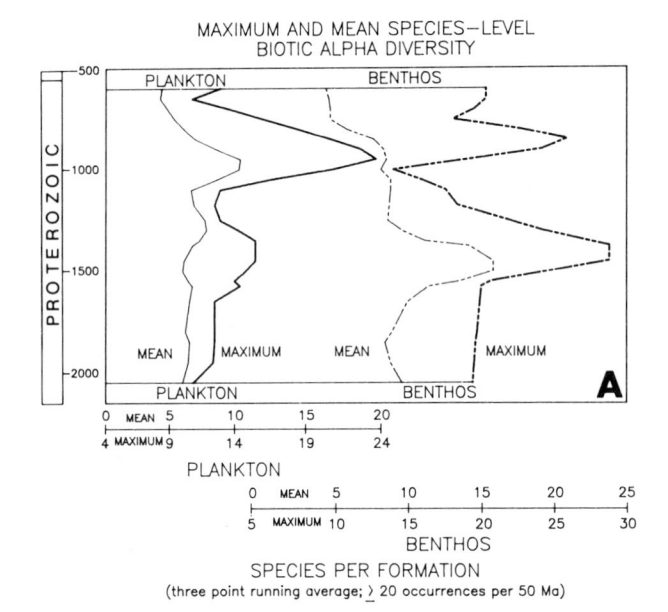

Figure 11.3.13. Comparison of maximum and mean species-level biotic alpha diversities of Proterozoic plankton and benthos, eukaryotes and prokaryotes, plotted as running averages ±75 Ma of estimated formation ages (definitions of plankton and benthos, eukaryotes and prokaryotes, are given in Table 25.1; species included are listed in Table 25.3; intervals analyzed are indicated in Fig. 11.3.7A). (A) Plankton and benthos: maximum (Table 25.7) and mean (Figs. 11.3.7C and 11.3.8) species-level alpha diversities. (B) Eukaryotes and prokaryotes: maximum (Table 25.7) and mean (Figs. 11.3.7C and 11.3.8) species-level alpha diversities.

Patterns of Proterozoic Microfossil Diversity

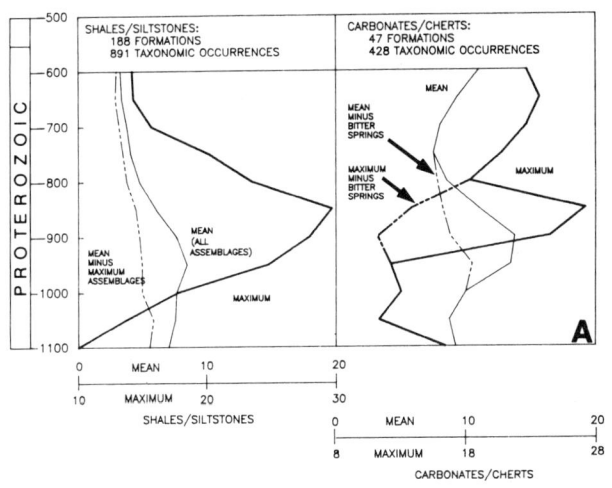

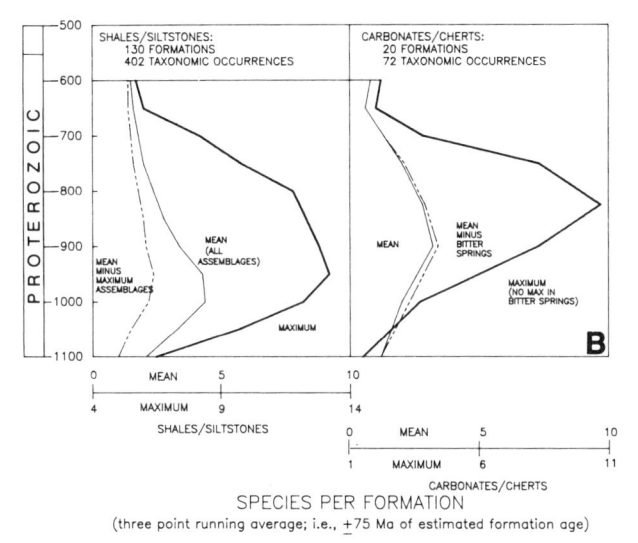

patterns for assemblages from shales/siltstones appear not to have been significantly influenced by monographic effects, whereas those for assemblages from carbonates/cherts appear to have been influenced by inclusion of data from a single, especially well-studied, Late Proterozoic unit.

Assessment of "grab-sampling" effects. In addition to the influence on patterns of mean alpha diversity of monographic effects, resulting from inclusion in the analyses of intensively studied assemblages, the possible influence of minimally studied "grab-sampled" assemblages (e.g., units reported to contain but a single fossil taxon) needs also to be assessed. Although some "one-taxon assemblages" presumably result from taphonomic effects and the vagaries of preservation, the majority seem likely to reflect incomplete sampling and/or inadequate study. Regardless of cause, however, such one-taxon assemblages

Figure 11.3.14. Comparison of maximum and mean species-level alpha diversities in shales/siltstones and stromatolitic carbonates/cherts of prokaryotes, eukaryotes, benthos, and plankton, 1100 to 600 Ma, plotted as running averages ±75 Ma of estimated formation ages (definitions of prokaryotes, eukaryotes, benthos, and plankton are given in Table 25.1; species included are listed in Table 25.3; maximum and minimum number of species/formation for each interval and percent decrease in diversity are summarized in Table 11.3.3). Plots of data for assemblages detected in shales/siltstones show the maximum diversity reported for each analyzed interval, the mean diversity for all analyzed assemblages, and the mean diversity excluding the maximally diverse assemblage for each interval. Plots of data for assemblages detected in carbonates/cherts show the maximum diversity reported for each analyzed interval, the mean diversity for all analyzed assemblages, and both the maximum and mean diversities excluding species reported from the monographically studied Bitter Springs Formation of central Australia. (A) Prokaryotes. (B) Eukaryotes. (C) Benthos. (D) Plankton.

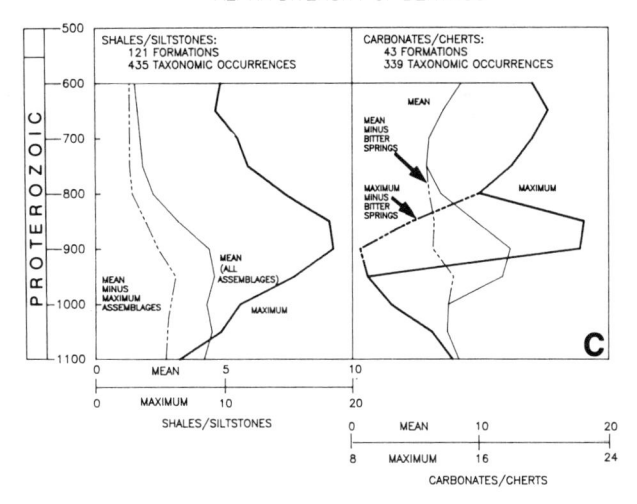

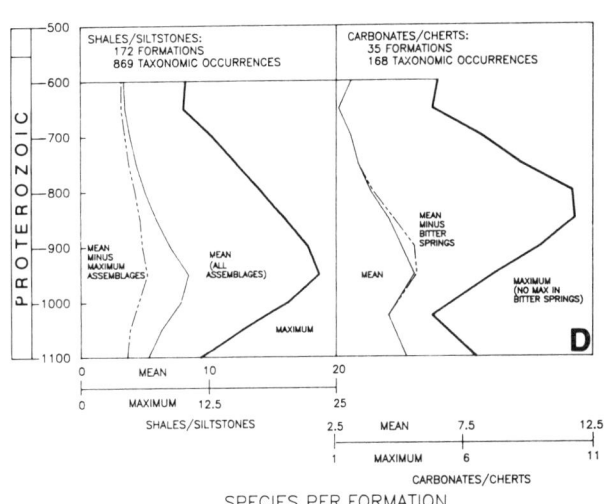

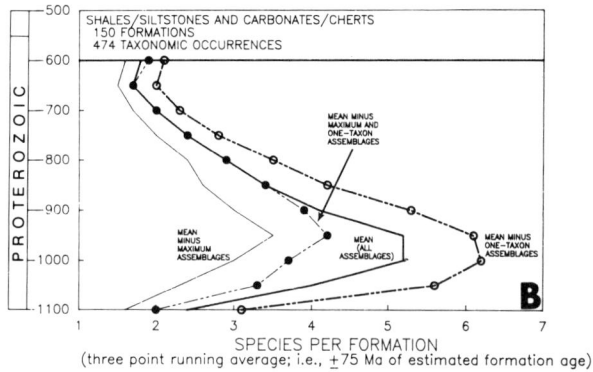

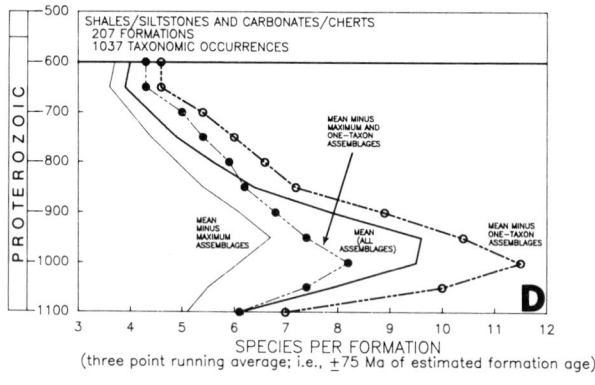

cannot reasonably be expected to reflect actual biotic diversity accurately. It seems important to determine, therefore, to what extent the inclusion of such assemblages in these analyses has affected the trends detected, an assessment of particular significance inasmuch as assemblages of this type are of rather common occurrence among reported microfossiliferous Proterozoic formations (viz, among analyzed units 1100 to 550 Ma, representing about 20% to more than 40% of prokaryote-, eukaryote-, benthos-, and plankton-containing shales/siltstones, and of eukaryote-, benthos-, and plankton-containing carbonates/cherts).

In order to assess the influence on Late Proterozoic diversity trends of inclusion of these minimally sampled and/or studied assemblages, patterns of mean alpha diversity excluding all one-taxon assemblages have been determined for data from all formations (Fig. 11.3.15, Table 11.3.2), and separately for assemblages from shales/siltstones and from stromatolitic carbonates/cherts (Fig. 11.3.16, Table 11.3.3). The picture that emerges from this analysis is much the same as that discussed above in the comparison of maximum alpha, mean alpha, and mean alpha excluding maximum diversities for Late Proterozoic assemblages: patterns for mean alpha diversity excluding all one-taxon assemblages show substantial declines for all biotic components using data from all formations (viz., 44% to 68%; Table 11.3.2); for all biotic components using data only from shales/siltstones (56% to 67%; Table 11.3.3); for plankton and eukaryotes reported from carbonates/cherts (43% and 88%, respectively); but show a high degree of variability for benthos

Figure 11.3.15. Comparison of species-level mean alpha diversities, mean alpha diversities excluding maximally diverse assemblages from each interval, mean alpha diversities excluding all one-taxon assemblages, and mean alpha diversities excluding both maximum and one-taxon assemblages from each interval, for prokaryotes, eukaryotes, benthos, and plankton from all sampled formations 1100 to 600 Ma, plotted as running averages ±75 Ma of estimated formation age (definitions of prokaryotes, eukaryotes, benthos, and plankton are given in Table 25.1; species included are listed in Table 25.3; duration and percent of decrease in diversity are summarized in Table 11.3.2). (A) Prokaryotes. (B) Eukaryotes. (C) Benthos. (D) Plankton.

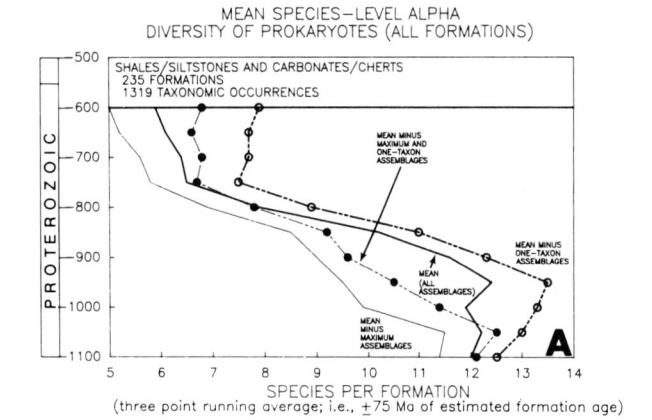

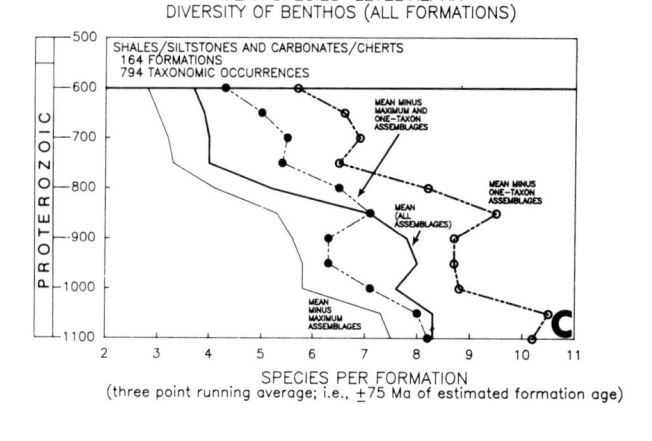

and prokaryotes reported from stromatolitic carbonates/cherts (Fig. 11.3.16A, 11.3.16C, Table 11.3.3).

Assessment of the combined influences of monographic and "grab-sampling" effects. Finally, the possibility exists that the combined effects of monographic and grab-sampling studies might together be responsible for influencing the trends detected – that is, although these two effects, considered separately, appear to have influenced only the Late Proterozoic diversity patterns of benthic prokaryotes from stromatolitic carbonates/cherts, in combination they may have influenced patterns of other components as well. To evaluate this possibility an additional analysis has been carried out, one based on the strategy of omitting from calculations of mean alpha diversity both the maximally diverse assemblage and all one-taxon assemblages known from each 50 Ma-long interval. This analysis thus deals with the "heart" of the database, that

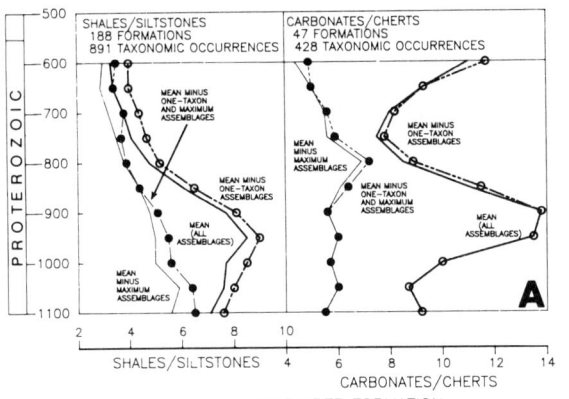

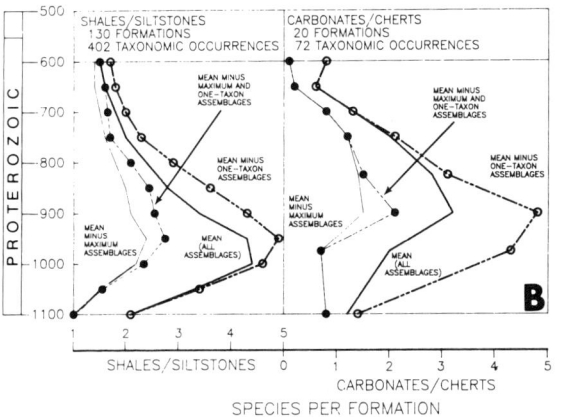

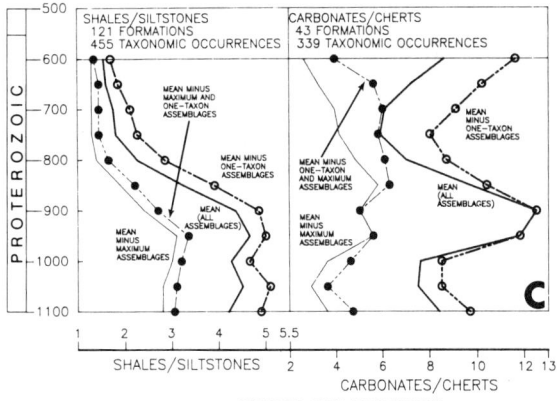

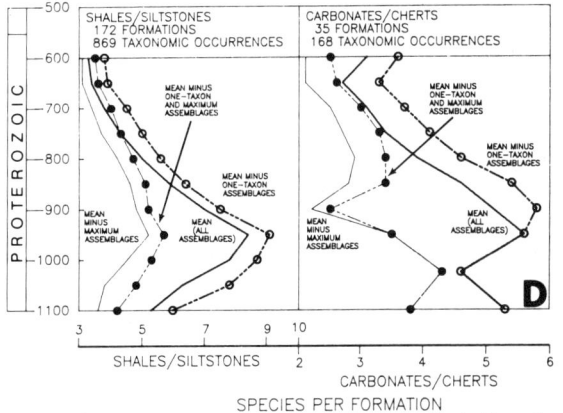

Patterns of Proterozoic Microfossil Diversity

Figure 11.3.16. Comparison of species-level mean alpha diversities, mean alpha diversities excluding maximally diverse assemblages from each interval, mean alpha diversities excluding all one-taxon assemblages, and mean alpha diversities excluding both maximum and one-taxon assemblages from each interval, in shales/siltstones and stromatolitic carbonates/cherts, for prokaryotes, eukaryotes, benthos, and plankton, 1100 to 600 Ma, plotted as running averages ±75 Ma of estimated formation age (definitions of prokaryotes, eukaryotes, benthos, and plankton are given in Table 25.1; species included are listed in Table 25.3; maximum and minimum number of species/formation for each interval and percent decrease in diversity are summarized in Table 11.3.3). (A) Prokaryotes. (B) Eukaryotes. (C) Benthos. (D) Plankton.

portion of the available data that seems least likely to have been influenced by uneven investigation: for Late Proterozoic taxonomic occurrences, about three-quarters of those for prokaryotes (76%) and plankton (76%) reported from shales/siltstones and for prokaryotes (72%) reported from carbonates/cherts; about half of those for eukaryotes (53%) and benthos (50%) reported from shales/siltstones and for benthos (51%) and plankton (49%) reported from carbonates/cherts; and about one-third of those for eukaryotes (35%) reported from carbonates/cherts. Results of this analysis, carried out both for assemblages from all formations and separately for assemblages from shales/siltstones and from carbonates/cherts, are presented graphically in Figures 11.3.15 and 11.3.16, tabulated in Tables 11.3.2 and 11.3.3, and are summarized below.

11.3.4E Summary Evaluation of the Effects of Uneven Sampling and/or Study

As they pertain to detected Late Proterozoic diversity patterns, comparison of results of the several types of analyses discussed above (viz., those of maximum alpha diversity, mean alpha diversity, and of mean alpha diversity from which have been excluded maximally diverse assemblages, one-taxon assemblages, and both maximum and one-taxon assemblages) can be summarized as follows:

(1) In assemblages from all formations, for eukaryotes (Figs. 11.3.13B, 11.3.15B) and plankton (Figs. 11.3.13A, 11.3.15D), there is a strong degree of parallelism in the patterns of (*i*) maximum alpha diversity, (*ii*) mean alpha diversity, (*iii*) mean alpha diversity excluding maximally diverse assemblages, (*iv*) mean alpha diversity excluding one-taxon assemblages, and (*v*) mean alpha diversity excluding both maximum and one-taxon assemblages.

(2) Comparable parallelism among the patterns of these five measures of diversity is exhibited in assemblages from shales/siltstones by eukaryotes (Figs. 11.3.14B, 11.3.16B), prokaryotes (Figs. 11.3.14A, 11.3.16A), plankton (Figs.

Table 11.3.3. *Percent decrease during the Late Proterozoic in the species-level taxonomic diversity of prokaryotes, eukaryotes, benthos, and plankton, and species per formation, in analyzed microfossil assemblages reported from shales/siltstones and carbonates/cherts*

Measures of Diversity	PROKARYOTES					EUKARYOTES				
	Number of Fms.	Time of Diversity (±75 Ma of estimated formation age)		Duration of Decline	Percent Decrease	Number of Fms.	Time of Diversity (±75 Ma of estimated formation age)		Duration of Decline	Percent Decrease
		Maximum (Sp./Fm.)	Minimum (Sp./Fm.)				Maximum (Sp./Fm.)	Minimum (Sp./Fm.)		
Shales/Siltstones:										
(14) Maximum Alpha Diversity:	11	850 Ma (29.7)	600 Ma (14.2)	250 Ma	52%	11	950 Ma (13.2)	600 Ma (5.7)	350 Ma	57%
(15) Mean Alpha Diversity (all assemblages):	188	950 Ma (8.5)	600 Ma (3.3)	350 Ma	61%	130	1000 Ma (4.4)	600 Ma (1.5)	400 Ma	66%
(16) Mean Alpha Diversity (minus maximum assemblages):	177	950 Ma (5.0)	650 Ma (2.9)	300 Ma	42%	119	950 Ma (2.4)	650 Ma (1.4)	300 Ma	42%
(17) Mean Alpha Diversity (minus one-taxon assemblages):	153	950 Ma (9.0)	600 Ma (4.0)	350 Ma	56%	80	950 Ma (4.9)	600 Ma (1.7)	350 Ma	65%
(18) Mean Alpha Diversity (minus maximum and one-taxon assemblages):	142	1100 Ma (6.5)	650 Ma (3.4)	450 Ma	48%	69	950 Ma (2.8)	600 Ma (1.5)	350 Ma	45%
Carbonates/Cherts:										
(19) Maximum Alpha Diversity	11	Variable				8	825 Ma (10.7)	650 Ma (2.0)	175 Ma	81%
(20) Maximum Alpha Diversity (minus Bitter Springs Formation):	11	Variable				8	825 Ma (10.7)	650 Ma (2.0)	175 Ma	81%
(21) Mean Alpha Diversity (all assemblages):	47	Variable				20	900 Ma (3.2)	650 Ma (0.6)	350 Ma	81%
(22) Mean Alpha Diversity (minus Bitter Springs Formation):	46	Approximately Constant (~9.5)				19	900 Ma (3.4)	650 Ma (0.6)	350 Ma	82%
(23) Mean Alpha Diversity (minus maximum assemblages):	36	Approximately Constant (~5.5)				12	900 Ma (1.5)	600 Ma (0.1)	300 Ma	93%
(24) Mean Alpha Diversity (minus one-taxon assemblages):	45	Variable				12	900 Ma (4.8)	650 Ma (0.6)	250 Ma	88%
(25) Mean Alpha Diversity (minus maximum and one-taxon assemblages):	34	Approximately Constant (~5.5)				7	900 Ma (2.1)	600 Ma (0.1)	300 Ma	95%

Patterns of Proterozoic Microfossil Diversity

BENTHOS					PLANKTON				
Number of Fms.	Time of Diversity (± 75 Ma of estimated formation age)		Duration of Decline	Percent Decrease	Number of Fms.	Time of Diversity (± 75 Ma of estimated formation age)		Duration of Decrease	Percent Decrease
	Maximum (Sp./Fm.)	Minimum (Sp./Fm.)				Maximum (Sp./Fm.)	Minimum (Sp./Fm.)		
11	900 Ma (18.5)	650 Ma (9.3)	250 Ma	50%	11	950 Ma (23.3)	650 Ma (10.0)	300 Ma	56%
121	950 Ma (4.6)	600 Ma (1.5)	350 Ma	67%	172	950 Ma (8.4)	600 Ma (3.3)	350 Ma	61%
110	950 Ma (3.1)	600 Ma (1.3)	350 Ma	58%	161	950 Ma (5.2)	600 Ma (3.1)	350 Ma	40%
71	1050 Ma (5.1)	600 Ma (1.7)	450 Ma	67%	142	950 Ma (9.1)	600 Ma (3.8)	350 Ma	58%
60	950 Ma (3.4)	600 Ma (1.3)	350 Ma	61%	131	950 Ma (5.7)	600 Ma (3.5)	350 Ma	39%
10	Variable				11	850 Ma (10.3)	600 Ma (5.0)	250 Ma	51%
10	Variable				10	850 Ma (10.3)	600 Ma (5.0)	250 Ma	51%
43	Variable				35	950 Ma (5.6)	650 Ma (2.7)	300 Ma	52%
42	Approximately Constant (~ 6.5)				34	950 Ma (5.7)	650 Ma (2.7)	300 Ma	53%
33	Approximately Constant (~ 4.0)				24	1025 Ma (4.3)	600 Ma (2.1)	425 Ma	51%
32	Variable				27	900 Ma (5.8)	650 Ma (3.3)	250 Ma	43%
22	Approximately Constant (~ 5.0)				17	1025 Ma (4.3)	600 Ma (2.5)	425 Ma	42%

11.3.14D, 11.3.16D), and benthos (Figs. 11.3.14C, 11.3.16C), and in assemblages from carbonates/cherts by eukaryotes (Figs. 11.3.14B, 11.3.16B) and, in part, by plankton (Figs. 11.3.14D, 11.3.16D).

(3) The degree of parallelism among the various measures of diversity noted in (1) and (2), above, can be interpreted as indicating that the patterns of diversity exhibited by all biotic components analyzed from shales/siltstones, and by eukaryotes and, to a lesser degree, plankton from carbonates/cherts, are robust and have been influenced minimally by monographic and/or grab-sampling effects.

(4) In assemblages from all formations, for prokaryotes (Figs. 11.3.13B, 11.3.15A) and benthos (Figs. 11.3.13A, 11.3.15C), the patterns of (*i*) maximum alpha diversity markedly deviate from those of (*ii*) mean alpha diversity, although the latter more or less parallel those of (*iii*) mean alpha diversity excluding maximally diverse assemblages, (*iv*) mean alpha diversity excluding one-taxon assemblages, and (*v*) mean alpha diversity excluding both maximum and one-taxon assemblages.

(5) The lack of parallelism between maximum and mean alpha diversities exhibited by prokaryotes and benthos for assemblages from all formations evidently reflects the influence of uneven sampling of the relatively small number of assemblages now known from Late Proterozoic carbonates/cherts and, most notably, inclusion of data from the monographically studied Bitter Springs Formation of central Australia (Figs. 11.3.14A, 11.3.14C).

(6) In summary, data for all five of these diversity measures for all components from shales/siltstones consistently yield patterns indicative of a gradual, long-term, Late Proterozoic decline in species-level biotic diversity. The consistency of these patterns, and the degree of parallelism exhibited by the patterns for all of the measures for each of the biotic components, suggest the absence of significant distortion due to uneven sampling. These characteristics may indicate that the analyzed data set contains a mix of sampled shale/siltstone environments (ranging from offshore marine facies to tubiditic, lagoonal, mudflat, and perhaps even playa and lacustrine settings) that on average is more or less homogeneous over time. In contrast, data for some groups of microfossils (viz., benthic prokaryotes) from carbonates/cherts yield divergent diversity patterns for certain of the measures analyzed. Evidence suggests that this divergence is due principally to monographic effects, the influence of which is rather strongly expressed because of the small total number of units analyzed, possibly augmented by uneven sampling of Late Proterozoic carbonate/chert environments (e.g., ranging from lower intertidal and higher intertidal marine carbonate facies to hypersaline, sabkha, and possibly hot springs and freshwater settings). Finally, although the pattern of gradual, long-term, Late Proterozoic diversity decline evidenced by microfossil assemblages from shales/siltstones (Figs. 3.11.14, 3.11.16) seems also to be reflected by diversity measures for microfossils from all formations (Fig. 11.3.15), in certain of these latter patterns (viz., those for prokaryotes and, especially, benthos) the strong signal from the shale/siltstone assemblages appears to have been somewhat diluted and distorted by the more variable diversity patterns of the prokaryote-dominated benthic assemblages occurring in stromatolitic carbonates/cherts (e.g., compare Figs. 11.3.15C and 11.3.16C).

11.3.4F Quantitative Assessment of Late Proterozoic Diversity Patterns

As is indicated by the data listed in Tables 11.3.2 and 11.3.3 (and summarized in part in the histograms shown in Fig. 11.3.17), all measures of diversity here considered, calculated at various values of uncertainty of estimated formation ages and determined for all microfossiliferous formations, for shales/siltstones, and for carbonates/cherts, exhibit significant declines during the Late Proterozoic for virtually all components of the biota. Percent declines in Late Proterozoic biotic diversity range from a low of ~30% (inferred global diversity), to 34% (maximum alpha diversity), 46% (observed global diversity), 46% (mean alpha diversity excluding one-taxon assemblages), 48% (mean alpha diversity excluding maximum and one-taxon assemblages), 56% (mean alpha diversity), to a high of 59% (mean alpha diversity excluding maximally diverse assemblages). As is summarized in Table 11.3.4, the apparent Late Proterozoic decline in species-level global diversity is similar in magnitude to (but differs significantly in duration from) declines in genus-level global diversity reported for major episodes of Phanerozoic extinction.

11.3.5 Evaluation of Detected Diversity Patterns

As noted above, all measures of diversity here considered for virtually all components of the biota exhibit significant declines between about 950 Ma and 575 Ma. However, as is also discussed above (Section 11.3.3), there are numerous defi-

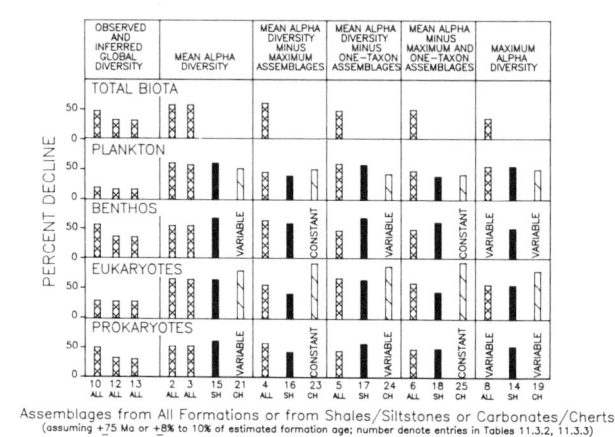

Figure 11.3.17. Histograms summarizing percent decrease during the Late Proterozoic of various measures of species-level taxonomic diversity in prokaryotes, eukaryotes, benthos, plankton, and total biota, in assemblages from all formations ("ALL"), shales/siltstones ("SH"), or from stromatolitic carbonates/cherts ("CH"), assuming uncertainties of ±75 Ma or ±8% to ±10% of estimated formation age; numbers along the abscissa denote entries in Tables 11.3.2 and 11.3.3.

Table 11.3.4. *Comparison of the Late Proterozoic decline in species-level global diversity with Phanerozoic episodes of major decline in genus-level global diversity*

Event	Age (Ma)	Percent Decline in Global Diversity
Middle Miocene extinction[a]	~14	10%
Late Eocene extinction[a]	~38	15%
End Cretaceous extinction[a]	~65	47%
Cenomanian extinction[a]	~94	26%
Aptian extinction[a]	~114	19%
Late Jurassic extinction[a]	~145	21%
Pliensbachian extinction[a]	~195	26%
Late Triassic extinction[a]	~210	47%
End Permian extinction[a]	~245	84%
Late Devonian extinction[b]	~365	50%
Late Ordovician extinction[b]	~440	57%
Late Proterozoic decline, Observed Value[c]	~900 to ~580	53%(±7%)
Late Proterozoic decline, Inferred Value[d]	~845 to ~560	33%(±3%)

[a] Genus-level, Phanerozoic marine animal fossils (Sepkoski 1989)
[b] Genus-level, Phanerozoic marine animal fossils (Sepkoski 1986)
[c] Species-level, total Proterozoic microfossils not including present-day Lazarus (cf. "*Coelacanth*") taxa, average value and range for assumed uncertainties of ±50 Ma and ±75 Ma of estimated formation ages (Table 11.3.2)
[d] Species-level, total Proterozoic microfossils plus Lazarus (cf. "*Coelacanth*") taxa, average value and range for assumed uncertainties of ±50 Ma, ±75 Ma, and ±8% to 10% of estimated formation ages (Table 11.3.2)

ciencies in the available data, and it is uncertain as to whether the quality and quantity of evidence currently available are sufficient to yield an accurate assessment of Proterozoic diversity patterns. The question posed in Section 11.3.3G provide at least a provisional basis for evaluation of the detected trends. Specifically:

(1) Are the available data mutually consistent? In general, yes, and particularly during the Late Proterozoic for which current data seem most reliable: (*i*) all subgroups of plankton exhibit similar diversity trends (Fig. 11.3.9); and (*ii*) all major components of the biota (prokaryotes and eukaryotes, benthos and plankton) exhibit comparable patterns of Late Proterozoic diversity decline (Fig. 11.3.15 and Tables 11.3.2, 11.3.3).

(2) Have the detected diversity patterns been significantly influenced by uneven (monographic or "grab-sample") sampling and/or study? As discussed in Section 11.3.4D and tested by comparison of mean alpha diversity, maximum alpha diversity, mean alpha diversity excluding maximally diverse assemblages, mean alpha diversity excluding one-taxon assemblages, and mean alpha diversity excluding both maximum and one-taxon assemblages: (*i*) Late Proterozoic diversity patterns for plankton and eukaryotes from all formations, and for benthos and prokaryotes from shales/siltstones, seem not to have been markedly distorted by such effects; whereas (*ii*) those for benthic prokaryotes from carbonates/cherts have evidently been so affected.

(3) Are the patterns detected a result of biased sampling of particular lithologies and/or environments? No, evidently not, at least during the Late Proterozoic and at the coarse-grained level of analysis permitted by available evidence: (*i*) as shown in Figure 11.3.3 (and discussed in Section 11.3.3B), during the Late Proterozoic (the critical period of apparent major diversity change) the mixes of sampled benthos- and plankton-containing assemblages and of benthos- and plankton-containing "offshore" shales/siltstones and "nearshore" carbonates/cherts are all more or less uniform, and the assemblages preserved in both lithologies consistently exhibit a more or less uniform mix of benthos and plankton; (*ii*) comparison of plots of the lithology-related data data shown in Figure 11.3.3 with the various diversity curves (Figs. 11.3.4 through 11.3.16) establishes that temporal variations in sampled lithologies (presumably a rough index of sampled paleoenvironments) do not correlate with changes in observed diversity patterns; (*iii*) data graphically summarized in Figures 11.3.14 and 11.3.16 (and tabulated in Table 11.3.3) indicate that for eukaryotes and, to a lesser degree, plankton, similar diversity patterns occur in coeval Late Proterozoic "offshore" shales/siltstones and "nearshore" stromatolitic carbonates/cherts; and (*iv*) data discussed in Section 11.3.4D suggest that on average the mix of shale/siltstone environments sampled over the Late Proterozoic may be more or less homogeneous whereas that represented by stromatolitic carbonates/cherts seems less likely to be so.

(4) Are the patterns a result of the vagaries of local preservation? No, evidently not: as evidenced by (*i*) comparison of fossil assemblages from varying levels and of markedly differing ages in relatively continuous local stratigraphic sequences (e.g., those at Jixian, China; South Australia; Bashkiria, USSR; and the Siberian Platform; Chapter 22), and by (*ii*) comparison of microfossils preserved in similarly aged sediments from geo-

graphically widely dispersed localities (Chapter 22, Fig. 25.4), changes in biotic diversity (including the sequential appearance during the Middle Proterozoic of phytoplankton of increasingly larger size, and the loss from the biota during the Late Proterozoic of relatively complex and of large-diameter acritarchs; Section 5.5.3D, Figs. 5.5.2, 5.5.6, 5.5.12, Table 5.5.2) appear to be global rather than local, isolated, occurrences.

(5) How "robust" are the detected patterns? Evidence for the occurrence of Late Proterozoic decline in biotic diversity seems strong: documentation of the decline from 900 Ma to 600 Ma as measured by mean alpha diversity is based on 1220 reported microfossil occurrences in 114 benthos- and 142 plankton-containing formations; as is illustrated in Figure 11.3.11, to negate this apparent decline by further study of these units would require that more than 1000 additional, initially "overlooked" occurrences be discovered and that this previously "missing fossil record" becomes gradually, increasingly, and ultimately exceptionally diverse over time.

How many additional maximally diverse assemblages would need be discovered to negate this apparent trend? As summarized in Table 25.7, the average diversity for the six maximally diverse plankton-containing units now known between 900 and 600 Ma (the Ushitsa Series, the Visingsö and Little Dal Groups, and the Miroedikha, Muhos, and Min'yar Formations) is 15.3 taxonomic occurrences per formation, and that for the six maximally diverse benthos-containing units from this period (the Draken Conglomerate, and the Bitter Springs, Doushantuo, Min'yar, Chichkan, and Kirgitej Formations) is 24.2 taxonomic occurrences per formation. The apparent Late Proterozoic diversity decline could thus be negated (i.e., uniform diversity 950 to 600 Ma could be maintained; Fig. 11.3.11) by future discovery of 59 additional exceptionally diverse assemblages (viz., 44 plankton- and 15 benthos-containing units) exhibiting an average level of diversity equivalent to that of the 12 maximally diverse assemblages now known from the period; the temporal distribution of these hypothetical assemblages would have to be such as to offset the apparent trend (viz., the assemblages would have to become increasingly diverse, and increasingly numerous over time) and concurrent discovery of any less diverse assemblages would, of course, necessitate discovery of an even larger number of such "maximally diverse assemblages."

(6) Are the diversity patterns consistent with inferred evolutionary trends? Yes, for those groups for which evolutionary trends can be inferred from the known fossil record (viz., eukaryotes and plankton): (i) the gradual appearance during the Middle Proterozoic of increasingly larger diameter phytoplankton (Figs. 5.5.2, 5.5.6, 5.5.12, 11.3.9) correlates with major increases in both mean and maximum alpha diversities (Figs. 11.3.13, 11.3.15); and (ii) the size-correlated sequential loss from the biota during the Late Proterozoic of large-diameter sphaeromorph and acanthomorph eukaryotic acritarchs (Figs. 5.5.12, 5.5.15) correlates with significant decreases in mean alpha, maximum alpha, and beta diversities (Figs. 11.3.12, 11.3.13, 11.3.15). Comparable evolutionary trends have not been detected for Proterozoic benthos (Section 5.4). Although a literal reading of available data (viz., Figs. 11.3.8, 11.3.10, 11.3.13, 11.3.15) suggests a winnowing of Late Proterozoic benthic communities and a resultant decrease in average species diversity, such data, at least for assemblages from stromatolitic carbonates/cherts, have evidently been affected by uneven sampling (Section 11.3.4D).

(7) Are the detected patterns consistent with other possibly correlative changes in the Proterozoic paleobiologic record? The Late Proterozoic carbon isotopic record has been interpreted as reflecting an extended period of enhanced organic burial (Knoll et al. 1986), with data from the latest Proterozoic interpreted as suggesting both minimal biomass and the occurrence of a mass extinction (Magaritz 1989); additional isotopic data consistent with these interpretations are presented in Chapter 3. Available carbon isotopic data thus seem consistent with the occurrence during the Late Proterozoic of a major decline in biotic diversity, as does a well-documented major decrease in the abundance and diversity of Late Proterozoic stromatolites (Fig. 11.3.10; see also Section 6.2, Fig. 6.2.3).

11.3.6 Interpretive Summary

Based on the foregoing analyses of Proterozoic biotic diversity, and on inferred major trends in the evolution of Proterozoic prokaryotes and micro-algae as discussed in Sections 5.4 and 5.5, the following is a broad-brush "best guess" summary of generalizations that seem consistent with currently available data.

11.3.6A Early Proterozoic (2500 to 1600 Ma) Microfossil Diversity

Assessment of diversity patterns detected for *Early Proterozoic prokaryotes* must be based on four principal considerations: (i) at present, relatively few fossiliferous formations are known from the Early Proterozoic (Chapter 22, Tables 25.5, 25.6); (ii) known Early Proterozoic species occurrences and total reported microfossil occurrences are strongly related to the number of formations sampled (Fig. 11.3.1); (iii) the size ranges of Early Proterozoic prokaryotic trichomes and tubular sheaths and the cell sizes and level of complexity of Early Proterozoic prokaryotic colonies do not differ significantly from those represented in assemblages known from the Middle and Late Proterozoic (Section 5.4, Figs. 5.4.26, 5.4.32); and (iv) all recognized species of planktonic prokaryotes are well represented in Early Proterozoic units (Fig. 25.4). Thus, although both the observed and inferred global diversities of prokaryotes increase dramatically during the Early Proterozoic (Figs. 11.3.4, 11.3.5), these increases, like variations during the Early Proterozoic in other measures of prokaryotic diversity (Figs. 11.3.7, 11.3.12, 11.3.13), presumably reflect the effects of incomplete sampling. Indeed, the limited relevant data available – particularly the size ranges, levels of complexity, patterns of colonial organization, and the taxonomic diversity exhibited by known Early Proterozoic prokaryotes (Section 5.4), together with the widespread occurrence of prokaryote-produced stromatolites in Early Proterozoic carbonate terranes (Section 6.2) – seem most plausibly interpreted as reflecting approximately the same level of species diversity as that known from the later Proterozoic.

In contrast with prokaryotes, the known record of *Early Proterozoic eukaryotes* suggests that although eukaryotic plankton were evidently extant as early as 1700 Ma or perhaps earlier (Section 5.5, Fig. 11.3.9), Early Proterozoic species diversity was relatively low, eukaryotes being represented solely by simple, unornamented, sphaeromorph acritarchs. The number of reported taxa of these predominantly (or perhaps entirely) "mesosphaeromorph" acritarchs increases toward the end of the Early Proterozoic (Fig. 25.4) whereas the number of sampled plankton-containing assemblages remains low and relatively constant (Fig. 11.3.1B). Thus, it appears likely that eukaryotic species diversity, as reflected by a variety of diversity measures (Figs. 11.3.4 through 11.3.7, 11.3.9, 11.3.12, 11.3.13), gradually increased toward the end of the Era.

11.3.6B Middle Proterozoic (1600 to 900 Ma) Microfossil Diversity

Throughout much of the Middle Proterozoic, the number of fossiliferous formations known per analyzed 50 Ma-long interval, like that of known species occurrences and of total microfossil occurrences, remains relatively low. The parallelism of these three factors (Fig. 11.3.1) indicates that they are probably interrelated, suggesting that were more assemblages to have been sampled, more species would have been reported. Thus, increases in both the observed and inferred global diversity of *Middle Proterozoic prokaryotes*, and the variations for prokaryotes in mean alpha (Fig. 11.3.7), maximum alpha (Fig. 11.3.13), and beta diversities (Fig. 11.3.12) during this period, seem most plausibly interpreted as reflecting incomplete sampling of a relatively diverse microbial biota that in terms of overall species composition may not have differed greatly from that of the remainder of the Proterozoic.

Unlike prokaryotes, however, the diversity of *Middle Proterozoic eukaryotes* evidently increased, first gradually, then markedly beginning between 1200 and 1100 Ma, as "megasphaeromorphs" and acanthomorph (spiny) acritarch became increasingly diverse and abundant (Section 5.5, Figs. 5.5.12, 11.3.6 through 11.3.9, 11.3.13, 11.3.15, 25-4). That this inferred Middle Proterozoic increase in eukaryotic diversity is not an artifact of sampling seems evidenced by its lack of correlation with the number of formations analyzed: the number of known plankton-containing formations begins to increase only after about 1000 Ma (Fig. 11.3.1B) whereas the number of reported eukaryotic taxa (Fig. 25.4) and both mean (Figs. 11.3.8, 11.3.9) and maximum alpha diversities (Fig. 11.3.13) begin to increase decidedly earlier. As is discussed in Section 5.5.2A, during the Middle Proterozoic planktonic eukaryotes also appear to have exhibited substantial evolutionary development, sequentially becoming both larger and increasingly complex (Figs. 5.5.2, 5.5.6, 5.5.12, 5.5.14, 5.5.15, 11.3.9), Thus, available data – numbers of species, species diversity, and inferred evolutionary trends – seem to indicate that the global diversity of planktonic eukaryotes increased decidedly during the Middle Proterozoic, reaching a maximum at about 950 Ma (Figs. 11.3.6, 11.3.15B).

11.3.6C Late Proterozoic (900 to 550 Ma) Microfossil Diversity

As is shown in Figures 11.2.1 and 11.3.1, between 1000 Ma and the beginning of the Phanerozoic, the number of known microfossiliferous formations increases dramatically, for both benthos (Fig. 11.3.1A) and plankton (Fig. 11.3.1B). During this same period, however, the number of benthos and plankton species occurrences and of total reported microfossil occurrences rises and then, beginning at about 850 Ma, systematically falls. The lack of correlation between these curves is substantial and seems significant, especially in comparison with the Early and Middle Proterozoic fossil record for which increases in sampled units yield correlative increases in number of reported fossil occurrences. Moreover, biotic global diversity, both observed (Fig. 11.3.4) and inferred (Fig. 11.3.5), declines significantly during this same period, as do such independent measures of biotic change as mean alpha diversity (Figs. 11.3.7, 11.3.8) and the beta diversities of various biotic components (Fig. 11.3.12). Assuming that the assemblages sampled are in fact representative of the overall Late Proterozoic biota, the conclusion seems inescapable that biotic diversity decreased gradually but markedly throughout much of Late Proterozoic time.

The patterns of diversity of *Late Proterozoic prokaryotes* (Fig. 11.3.8), particularly of plankton (Fig. 11.3.9), exhibit significant long-term declines. Observed declines in the diversity of prokaryotic benthos are less pronounced (Figs. 11.3.4, 11.3.5, 11.3.7), and although interpretation is complicated by the disparity between the mean and maximum alpha diversities of Late Proterozoic benthic prokaryotes reported from carbonates/cherts (Fig. 11.3.14), the occurrence of such declines seems consistent with the more or less coincident decline in the abundance of prokaryote-dominated stromatolites (Fig. 11.3.10). Global diversity of Late Proterozoic prokaryotes evidently decreased (Figs. 11.3.4, 11.3.5, 25-4), but not as markedly as that of eukaryotes. Although it seems clear that the true diversity of Late Proterozoic prokaryotes has as yet been inadequately sampled, available data seem plausibly interpreted as evincing a "winnowing" of the Late Proterozoic microbial flora that resulted in substantial decreases in the average number of species per assemblage and a correlative decrease in the abundance (and morphological diversity) of prokaryote-produced stromatolites, but in a less marked decrease in total species diversity.

With regard to the diversity patterns of *Late Proterozoic eukaryotes*, the picture seems clear: after reaching a maximum level of Proterozoic species diversity about 950 to 900 Ma (Figs. 11.3.6, 11.3.9, 11.3.15B, 11.3.16B), planktonic eukaryotes gradually but markedly declined in diversity until near the close of the Proterozoic. This long-term decline in planktonic eukaryotes is evidenced consistently by decreases during this period in global diversity (Figs. 11.3.4 through 11.3.6), mean alpha diversity (Figs. 11.3.7 through 11.3.9), maximum alpha diversity (Fig. 11.3.13), mean alpha diversity excluding maximally diverse assemblages (Fig. 11.3.15B), mean alpha diversity excluding one-taxon assemblages (Fig. 11.3.15B), mean alpha diversity excluding maximum and one-taxon assemblages (Fig. 11.3.15B), and the beta diversity of planktonic eukaryotic "megasphaeromorphs" (Fig. 11.3.12). Significantly, these decreases are evidently independent of the gross lithologies sampled (Fig. 11.3.16B), are long-term and gradual regardless of whether the data are plotted as running averages (Figs. 11.3.4, 11.3.5) or simply as first and last known occurrences (Fig.

11.3.6), and are exhibited by all subgroups of planktonic eukaryotes (Fig. 11.3.9). During this same period, large-diameter "megasphaeromorph" and acanthomorph acritarchs became extinct (Section 5.5.3, Figs. 5.5.6, 5.5.12, 5.5.15, Table 5.5.2) as did, near the close of the Proterozoic, numerous other types of relatively complex eukaryotic phytoplankton (Vidal and Knoll 1982). Thus, for eukaryotic plankton, the Late Proterozoic appears to have been a period of gradual but inexorable decline, a major, long-term decrease in species diversity from a peak of maximum development near the beginning of the Late Proterozoic to minimal diversity near the close of the Era.

11.3.7 Overview

The overall picture that emerges from this analysis is one of substantial prokaryote, but limited eukaryote, Early Proterozoic diversity; increasing Middle Proterozoic eukaryote diversity; an episode of maximum biotic diversity near the Middle to Late Proterozoic boundary; and a long-term, gradual decrease in prokaryote, but especially in eukaryote diversity, during the Late Proterozoic. The inferred Late Proterozoic decline in species-level biotic diversity is striking, both because of its similarity in magnitude to the major well-documented episodes of Phanerozoic extinction (Table 11.3.4) and because it differs so significantly from such Phanerozoic episodes in its gradual, exceptionally long-term character. Indeed, in duration and overall pattern, the decline resembles the Middle Cambrian to Late Permian decline in (family-level) diversity of the metazoan "Cambrian evolutionary fauna" (pers. comm. J. J. Sepkoski, Jr.; Fig. 2 of Sepkoski 1984). However, this Paleozoic decline correlates with coeval diversity increases in (and evidently resulted from supplanting of the Cambrian evolutionary fauna by) other evolutionary faunas, whereas prior to about 650 to 600 Ma and the first widespread appearance of Vendian metazoans and metaphytes, the Late Proterozoic biotic diversity decline does not correlate with the documented coeval rise of other biologic groups. Such groups seem certain to have existed but may well have left little trace – unmineralized, grazing, heterotrophic protists, for example, or perhaps millimetric, soft-bodied, early evolving metazoans; search for direct evidence of a correlative rise of such groups, and of the evolutionary radiation of the multicellular, presumably heterotrichous, micro-algal ancestors of Late Proterozoic metaphytes (taxa that may have supplanted Late Proterozoic prokaryotic benthos in some environments), seems a promising avenue for future research.

As is noted at the outset of this discussion, there are many deficiencies in the available evidence, and it is as yet uncertain whether data now available are of sufficient quality and quantity to justify firm acceptance of the generalizations here drawn. Indeed, as is explicitly indicated by the title of this Section, the foregoing discussion should be regarded as an "initial, tentative, analysis." However, if the overview here presented represents a more or less accurate, albeit coarse-grained description of the true pattern of Proterozoic biotic development, the Late Proterozoic biotic decline here inferred will warrent special explanation, presumably one involving (*i*) gradual, long-term change in the global environment (e.g., decreases in atmospheric P_{CO_2} resulting in a decline in photosynthetic efficiency, altered phytoplankton physiology, lowered global temperature, and increased environmental stress); (*ii*) competitive exclusion of prokaryote-dominated benthic autotrophic communities (resulting from the spread of multicellular micro-algae); (*iii*) diminution of autotrophic populations by newly evolved grazing heterotrophs (both protists and millimetric metazoans); or (*iv*) a combination of such causes.

Acknowledgments.

I thank Hans J. Hofmann, Andrew H. Knoll, Carl V. Mendelson, Simon Conway Morris, J. John Sepkoski, Jr., and Malcolm R. Walter for critical reviews of earlier versions of this discussion.

11.4

Proterozoic-Early Cambrian Diversification of Metazoans and Metaphytes

J. JOHN SEPKOSKI, JR.

11.4.1 Introduction

The radiation of metazoans in the latest Proterozoic and earliest Phanerozoic marks a fundamental change in the nature of the fossil record and the way we perceive macroevolution (Cloud 1948). Evolutionary rates among structurally complex multicellular animals, as indicated by preserved morphology, are much more rapid than evolutionary rates earlier (Section 11.3). Similarly, rates of diversification are much more rapid, with major increases in diversity occurring over a few tens of millions of years in contrast to the previous several tens to hundreds of million years.

In this Section, patterns of diversification and faunal change in the early metazoan fossil record are examined, and diversity patterns of some calcified photosynthesizers are considered briefly. In the following discussion, general patterns of metazoan diversity in the Vendian and Cambrian for taxonomic levels ranging from genus to class are first considered. Subsequently, patterns of faunal change among genera and how this relates to total diversity are analyzed. Finally, the hypothesis of exponential diversification across the Proterozoic-Cambrian boundary is examined, and a preliminary analysis of generic diversity among carbonaceous and calcareous "metaphytes" is presented.

This is the first time that an analysis of early metazoan diversification has been attempted at the taxonomic level of the genus. This has been made possible by the continuing exploration of the fossil record around the Proterozoic-Cambrian boundary and ongoing related studies of the systematics of early metazoans. Particularly significant efforts include (*i*) continued discovery and description of Ediacaran fossils throughout the world; (*ii*) ongoing study and publication of Proterozoic-Cambrian boundary sections and their sequences of body and trace fossils (see Cowie and Brasier 1989); (*iii*) increasing precision in correlation of these sections and in establishing a global sequence of evolutionary change; (*iv*) continued work on fossil systematics, especially the description of new taxa, elimination of taxonomic synonyms, and recognition of clades among early metazoans and metaphytes; and (*v*) new discovery and restudy of Cambrian *Lagerstätten*, providing fuller understanding of the total variety of body plans generated during the Proterozoic-Cambrian biotic explosion. These efforts are all very much ongoing, and new discoveries, syntheses, and insights are being made continuously. Thus, any discussion of the Proterozoic-Phanerozoic transition necessarily reflects only the current "state of the art." However, efforts have progressed sufficiently over the past decade so as to provide a taxonomic and stratigraphic precision that was not available even a few years ago. This new precision, coupled with the newly available data, necessitates revision of older ideas on early metazoan history; particular attention is therefore devoted below to updating and modifying the earlier interpretations of metazoan diversification by Sepkoski (1978, 1979, 1981).

11.4.2 Data

The data for diversity in the Vendian and Cambrian are based on literature compilations of the stratigraphic ranges of metazoan taxa. The databases used here are the following:

(1) Ediacaran taxa – the PPRG database for Vendian metazoans assembled by Runnegar and Fedonkin (Chapter 23), which includes most described genera and attempts informal synonymy of preservational variants as well as suppression of poorly known or preserved taxa. Genera are assembled into higher taxa that reflect the consensus of the PPRG "Metazoan Subgroup" (see Chapter 8).

(2) Lower Cambrian "small shelly fossils" – the previously unpublished database of Bengtson, which includes all described genera and stratigraphic occurrences and attempts to synonymize obvious variants (see Chapter 23). The classification of these genera into higher taxa (Chapter 26), compiled by Bengtson and Sepkoski, is conservative, attempting to use published names wherever possible.

(3) Trilobites and Middle and Upper Cambrian taxa – an updated version of Sepkoski's (1982) familial compendium (Section 26.1) and an unpublished generic compilation, both modified to be consistent with Bengtson's database. The stratigraphic and taxonomic data are not as precise or inclusive as Bengtson's but are adequate to provide a general overview of Middle and Late Cambrian diversity.

(4) Archaeocyathid genera – data from Debrenne and Rozanov (1983) for Regulares, and from Hill (1972) for Irregulares and other taxa. Both have been modified slightly from more recent literature, but this has not been exhaustive; thus, the archaeocyathid data constitute the least precise portion of the total data set.

Although these databases (excluding that for archaeocyathids) are the most complete that exist, the following limitations should be recognized: (*i*) The databases include a subjective element in the assessment of synonymies, certain of which will change with new discoveries and insight. (*ii*) There is considerable uncertainty in assignments of many genera to higher taxa, especially in the Ediacaran and Lower Cambrian. The approach taken in this analysis has been liberal, making ample use of *incertae sedis* and problematica (cf. Bengtson 1986) and not attempting to force all genera into well-established classes and phyla (see Chapters 23 and Section 26.1); however, no cladistic analysis of genealogical relationships has been attempted. This is important because in compiling numbers on diversity it is assumed that the higher taxa represent clades or paraclades (*sensu* Raup 1985) that contained at least one genus at all times between first and last fossil appearances. (*iii*) There is imprecision in the stratigraphic ages of some genera and higher taxa, especially in the Vendian and Lower Cambrian where global correlations are still somewhat tentative and subject to modification (see Brasier 1989).

The stratigraphic units employed in this analysis are similar to those used by Sepkoski (1979). The Ediacaran fauna is assumed to occupy a single stratigraphic unit, comparable to the Redkino of the Russian Platform. The earliest skeletal fossils (excluding *Cloudina*) are placed in a sub-Tommotian (*sensu stricto*) unit comparable to the *Anabarites-Protohertzina* Zone, which is assumed to include much of Meischucun member A and the Rovno and the lower Nemakit-Daldyn Horizons; this unit is treated as uppermost Vendian but may be assigned to the basal Cambrian when the Proterozoic-Cambrian boundary is formally defined. Lower Cambrian units used here follow the first three subdivisions of the Siberian Cambrian: Tommotian, Atdabanian, and Botomian. The Tommotian is subdivided into two units, with the lower being approximately equivalent to the Sunnagin Horizon. The Botomian is also subdivided into two subunits, subdivisions that are roughly comparable to the Kameshkov and Sanashtykgol Horizons. The Middle and Upper Cambrian are each divided into three units. The "stages" of the Middle Cambrian correspond roughly to the tripartite subdivision of the Australian Middle Cambrian, with the lower stage including the Siberian Tojonian (or "Elankian"). The units of the Upper Cambrian correspond approximately to the three traditional stages on the North American craton.

The precise ages of these units are rather uncertain, reflecting current debate about the age of the Proterozoic-Cambrian boundary and the dearth of reliable radiometric dates available from within the Cambrian (see Cowie and Harland 1989). The ages of the lower and upper boundaries of the Cambrian are assigned as 550 and 500 Ma, respectively. These ages are obviously subject to change, but such change will merely expand or compress apparent diversity patterns and should not alter overall trends. Ages of the units within the Cambrian have been estimated by rescaling the time scale of Sepkoski (1979) to the new beginning and ending ages of the System (Section 26.2). This rescaling leaves the estimated durations of the stages subequal, with the Botomian and upper Middle Cambrian being slightly longer, and the Upper Cambrian units slightly shorter, than the average. The subdivisions of the Tommotian and Botomian are treated as being equal in length.

The greatest uncertainty in ages is for the upper Vendian. The Ediacaran fauna is treated as being between 570 and 550 Ma, with the point for diversity plotted at 560 Ma. (All other diversity points are plotted at the ends of the units.) Note that changes in the estimated age of the Ediacaran fauna relative to the Proterozoic-Cambrian boundary could substantially alter the appearance of the earliest patterns of metazoan diversification. This is important in considering the question of exponential diversification (Section 11.4.5).

11.4.3 General Patterns of Early Metazoan Diversification

Among the various taxonomic levels of diversity that might be considered, genus-level diversity is of greatest interest because it is closest to species diversity, the level at which most evolution is presumed to occur. However, it is also important to examine the diversity of higher taxa for at least two reasons, one methodological and the other substantive: (*i*) the diversity of higher taxa should be less prone to sampling variance resulting from incomplete study of the fossil record (Raup 1972, 1975, 1979) because the occurrence of only one species or genus is sufficient to document the presence of a polytypic higher taxon; and (*ii*) higher taxa, especially orders, classes, and phyla, approximate the fundamental *baupläne* in the inimal kingdom and it is important to know whether the diversification of body plans differed from that of species (cf. Valentine 1969, 1980, 1986; Erwin et al. 1987).

Figure 11.4.1 illustrates patterns of diversity for genera, families, orders, and classes from the upper Vendian through the Cambrian. There is a striking similarity in the basic pattern of diversification at all of these taxonomic levels, despite some important differences in detail. The basic pattern is one of sigmoidal diversification, beginning with low diversity in the late Vendian, a rapid increase through the Tommotian and Atdabanian, and then a slow increase to a nearly constant level in the Botomian to Upper Cambrian. This basic pattern has important evolutionary implications:

(1) It indicates that the Proterozoic-Cambrian metazoan explosion was relatively rapid, spanning a period of perhaps 15 million years from the Tommotian to Atdabanian. This is short relative both to previous Proterozoic radiations and to subsequent Phanerozoic radiations (e.g., those of the Ordovician and Mesozoic). The only radiations of comparable length followed later mass extinctions, such as the end-Ordovician and end-Cretaceous events, although per-taxon rates of diversification were much lower at these times (cf. Sepkoski, 1984). Truly analogous events (in terms of rates of evolution) may be found only in very short-lived radiations in which taxa entered newly opened habitats, such as cardiaceans in the Pontian Sea or cichlids in large African lakes (see Stanley 1979).

Proterozoic-Early Cambrian Diversification   555

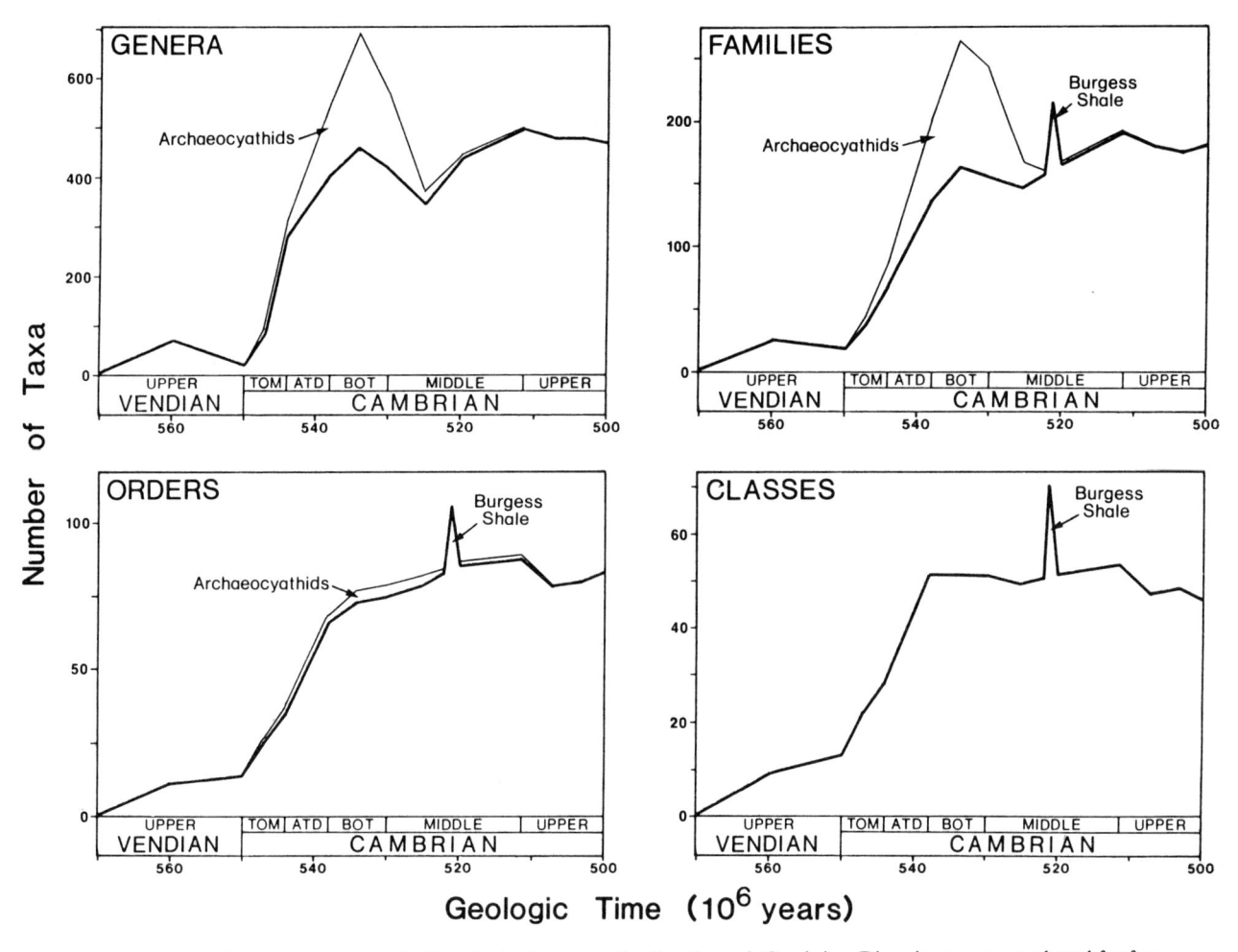

Figure 11.4.1. Patterns of metazoan taxonomic diversity in the upper Vendian through Cambrian. Diversity curves are plotted for four taxonomic levels: genera, families, orders, and classes. Except for archaeocyathids, the four curves are remarkably similar, exhibiting low diversity in the upper Vendian, rapidly rising diversity through the first half of the Lower Cambrian, and then relatively constant diversity to the end of the Cambrian. This indicates that the early Phanerozoic radiation of skeletal metazoans was indeed an explosive event, especially against the backdrop of the very slow diversity change among prokaryotes and eukaryotes in the preceding Proterozoic. Abbreviation are "TOM" = Tommotian; "ATD" = Atdabanian; "BOT" = Botomian.

(2) The similarity of pattern across taxonomic ranks suggests that there was little tendency in the earliest Phanerozoic for basic *baupläne* to diversify either earlier or more rapidly than species. Simpson (1953) and Valentine (1969, 1980) have argued that initially the metazoan biosphere should have filled by a proliferation of basic kinds of animals evolving in widely dispersed niches or adaptive zones in an uncrowded ecospace (or "empty ecological barrel," in the metaphor of Gould 1976), and that this proliferation should have been followed by diversification at lower taxonomic levels within successful *baupläne*. Although this scenario may be true for the Phanerozoic as a whole (Valentine 1969, 1973; Erwin et al. 1987), the data in Figure 11.4.1 indicate that it does not apply to the earliest metazoan radiation: genera and families expand at the same time as orders and classes, and all reach approximate "equilibrium" near the end of the Early Cambrian. Thus, it may have been that the early diversification of species controlled the appearance of *baupläne*, and not the reverse.

(3) The basic sigmoidal patterns in Figure 11.4.1 are consistent, at least to some extent, with the equilibrium, or "kinetic," model of taxonomic diversification developed by Sepkoski (1978, 1979, 1984) and tested with previously available data for families and orders (see also Raup et al. 1973; Carr and Kitchell 1980). This model predicts that early phases of radiations into ecologically vacant environments should be exponential (see also Stanley 1975, 1979) and should be followed by declining diversification resulting from decreased origination and increased extinction as the environment fills with species. If the environment or faunal constituents do not change radically, a dynamic equilibrium should be maintained, in which diversity fluctuates mildly about a constant level. (This interpretation is greatly oversimplified, as will be discussed in Section 11.4.4.)

Superimposed on the basic sigmoidal pattern shown in Figure 11.4.1 are a number of interesting details. In reverse chronologic sequence these include (*i*) the Middle Cambrian peak in families, orders, and classes, induced by the Burgess Shale fauna; (*ii*) the decline of genera and families around the Lower to Middle Cambrian boundary; (*iii*) the Lower Cambrian peak in genera and families produced by archaeocyathids; and (*iv*) the peak and subsequent decline in genera and families contributed by the Ediacaran fauna. Each of these is discussed below.

11.4.3A The Burgess Shale Fauna

The Burgess Shale is the greatest metazoan *Lagerstätten* of the Phanerozoic marine fossil record (Whittington 1985; Gould 1989), mostly because it provides a unique picture of the variety of soft-bodied and lightly sclerotized *baupläne* produced by the Proterozoic-Cambrian explosion. Many of the fossilized animals lack synapomorphies with other taxonomic groups and therefore must be treated as monotypic higher taxa (Conway Morris 1989a). This produces the sharp peak in Middle Cambrian diversity seen in Figure 11.4.1. This peak serves as a reminder that known fossil diversity samples only a portion of the total real diversity of any time interval. The important question is whether the known fossil sample is biased, providing a distorted picture of diversity that is without evolutionary meaning. The peaks in families, orders, and classes produced by the Burgess Shale fauna could suggest, for example, that proliferation of *baupläne* continued into the Middle Cambrian and that the appearance of an equilibrium is illusionary. However, soft-bodied Burgess Shale taxa recently have been discovered in the Atdabanian Chengjiang fauna of China (Hou 1987; Hou and Sun 1988). Description of this newly discovered fauna is still incomplete, but work to date suggests that many of the Burgess Shale taxa may have had substantial geologic ranges (Conway Morris 1989b) and that there is no reason to believe that the proliferation of higher taxa continued in any significant way past the sigmoidal diversification of the Early Cambrian.

11.4.3B Lower to Middle Cambrian Decline in Diversity

Genera and, to a lesser extent, families exhibit a substantial decline in diversity beginning in the upper Botomian and reaching a minimum in the lower Middle Cambrian. This decline is especially evident among archaeocyathids (see below) but also occurs among other metazoans, including trilobites (especially redlichiids) and hyoliths; a number of problematica also disappear at about this time (e.g., anabaritids and tommotiids). These disappearances suggest that the decline probably represents an extinction event near the end of the Early Cambrian, although some of the decline in diversity could reflect errors in correlation [e.g., uncertainty about how the Tojonian correlates with units outside Siberia; see Brasier (1989) for alternative correlations]. Palmer (1982) has conjectured that a wave of extinctions, the "olenellid biomere event," affected trilobites at about this time. Although this event has not yet been documented by rigorous biostratigraphic and taxonomic study, Palmer has suggested that it is comparable to the several biomere extinction events seen in the Upper Cambrian of North America, Australia, and China. (These latter extinctions do not appear in the diversity graphs of Figure 11.4.1 because of the long stratigraphic units used; each event was of very short duration and was followed by rapid re-radiations of trilobites over intervals of only a few zones). The long apparent duration of the Lower to Middle Cambrian event may reflect either the difficulties in correlation or the greater severity of the event. Significantly, diversity rebounded to approximately the same level as attained before, again suggesting that a quasistable equilibrium in diversity was attained by the late Early Cambrian. The same is evidently true for at least the first two of the Late Cambrian biomere events.

11.4.3C Archaeocyathids

The enigmatic archaeocyathids produce a major peak in generic and familial diversity over the Lower Cambrian that greatly perturbs the simple sigmoidal pattern shared by other metazoans at all taxonomic levels. This peak reflects the massive radiation of archaeocyathids during the late Early Cambrian that culminated in the occurrence of approximately 240 genera in the lower Botomian; this number is equal to half of the diversity of all other metazoans at that time (Figure 11.4.1). Archaeocyathids decline precipitously through the upper Botomian, and only about 24 genera are recorded from the early Middle Cambrian. Sepkoski (1979) suggested that this anomalous pattern of explosion and collapse might be because archaeocyathids were not animals but instead calcareous metaphytes (see also Öpik 1975; Nitecki and Debrenne 1979; Fisher and Nitecki 1982); this suggestion seems to have been rejected by most workers who accept the sponge-like (particularly sphinctozoan-like) architecture as indicating metazoan affinities (e.g., Pickett and Jell 1983; Debrenne and Vacelet 1984; Zuravleva and Nitecki 1985). Thus, there seems to be no way to negate the peak produced by archaeocyathids. It is true that many archaeocyathid genera are short lived (see Debrenne and Rozanov 1983) and that the units used to compile diversity may serve to combine non-overlapping genera (but not families); and, it is further possible that archaeocyathids, as highly variable bioherm-dwelling organisms, may be taxonomically oversplit (compare to the taxonomic problems affecting co-occurring calcareous "algae," discussed in Section 7.4). However, even radical lumping of genera and families would merely reduce the maximum in diversity and not eliminate the peak. Thus, it seems plausible to conclude that archaeocyathids were evolving under constraints different from those affecting other metazoans. Perhaps such constraints related to their ecology as bioherm-forming organisms that were concentrated in a habitat that, like Cretaceous rudists, admitted only limited diversity of other animals (e.g., James and Debrenne 1980).

11.4.3D Ediacaran Fauna

The description of the diversity curves as basically sigmoidal, with an early, low-diversity lag phase, assumes that members of the Ediacaran fauna were phylogenetically related to later animals and that the Early Cambrian diversification was a continuum that began with the Ediacaran (cf. Sepkoski 1978). There are several reasons, however, to question this assumption. First, recent reviews of Ediacaran fossils (e.g., Seilacher 1984, 1989; Fedonkin 1983, 1985; Chapter 23) have interpreted the great majority of genera as being unrelated to taxa in succeeding faunas. There is still considerable debate surrounding this interpretation (contrast Glaessner 1979, 1984), but it does seem that the peculiar "quilted" morphology of many "petal organisms" and the unusual concentric symmetries of many "medusoids" are unique to the Ediacaran. Thus, the peak in diversity among genera and families in Figure 11.4.1

may reflect an early radiation independent of the succeeding Early Cambrian diversification of "shelly" animals. Second, no soft-bodied Ediacaran animal has been definitively identified from Cambrian deposits (although see Conway Morris 1989a; McMenamin and McMenamin 1990), suggesting that a wave of extinctions may have terminated the Ediacaran (Seilacher 1984, 1989; McMenamin 1986; Sokolov and Fedonkin 1986). Alternative explanations for this disappearance involve hypothetical changes in preservation potential between the Vendian and Cambrian: Sokolov (1976b) suggested that the evolution of efficient scavengers greatly reduced the possibility that carcasses of soft-bodied organisms could survive long enough to be deeply buried; and Sepkoski (1979) conjectured that increasing bioturbation in the latest Vendian led to churning of sediments that previously encased Ediacaran-type fossils. While both of these explanations could be partially valid, the absence of Ediacaran-type animals from Cambrain *Lagerstätten* suggests that these metazoans did not persist in any abundance into the Cambrian. The Atdabanian Chengjiang fauna, in particular, occurs only about 15 million years after the Proterozoic-Cambrian boundary, and it includes a number of animals known from or closely related to the Burgess Shale fauna, indicating that the passage of time alone is not sufficient to account for the absence of Ediacaran animals.

There is other evidence, albeit imperfect, for some kind of extinction event. Data for "megasphaeromorph" and acanthomorph acritarchs, discussed in Sections 5.5 and 11.3 (and see especially Zang and Walter 1989), suggest that there may have been a wave of extinctions prior to the Proterozoic-Cambrian boundary, possibly terminating a mid-Vendian radiation. There are several taxa of trace fossils (e.g., *Bilinichnus, Harlaniella, Nenoxites*; see also Section 10.6C) that are confined to the Vendian (Crimes 1987, 1989) and which have comparatively short stratigraphic ranges; succeeding ichnotaxa of the "Nemakit-Daldyn" and Lower Cambrian have much longer ranges, suggesting they were not subject to mass extinction. These arguments are not compelling, but they do suggest that the early, Ediacaran diversity peak among metazoan genera and families is real and that the low diversity in the "Nemakit-Daldyn" is a consequence of the disappearance of these early metazoans.

The hypothesis presented here begs two obvious questions relating to Vendian diversification: how accurately do the diversity data reflect the magnitude of the Ediacaran radiation relative to that of the succeeding Cambrian, and what caused the collapse of Vendian diversity? There are several reasons to believe that the low diversity peak in the Vendian is probably real and not merely an artifact of the unusual preservation of Ediacaran animals. First, Cambrian *Lagerstätten* that preserve soft-bodied animals are as diverse, or even more so, than Ediacaran assemblages; the Burgess Shale, in particular, preserves approximately 120 genera (Conway Morris 1986), in contrast to the roughly 70 genera of Ediacaran animals known worldwide. Second, the diversity of ichnotaxa (Chapter 8) increases considerably into the Cambrian (Alpert 1977; Fedonkin 1978; Crimes 1987, 1989) as so does the depth and extent of bioturbation (Droser and Bottjer 1988, 1989); there is no reason to believe that the preservational potential of trace fossils changed in any significant way from the Vendian into the Cambrian; thus, their increase can be taken as an unbiased (although not entirely quantitative) indicator of increased faunal diversity after the Vendian.

The cause of the presumed extinction of Ediacaran animals is far from clear. However, the fact that this extinction seems to have affected a wide cross-section of the biota, additionally including planktonic eukaryotes (acritarchs) and some animals known only from trace fossils, implies that the cause was external, perhaps a climatic change associated with the regression near the end of the Vendian (cf. Stanley 1984b,c). However, biotic causes are not implausible. The large benthic soft-bodied animals and planktonic acritarchs of the late Vendian are replaced by a biota with smaller and more frequently armored individuals, including for the first time abundant calcified cyanobacteria and calcareous algae (Riding and Voronova 1982). These changes may reflect increased interaction among components of the ecosystem, including more intense herbivory and carnivory (e.g., Hutchinson 1961; Bengtson 1977a; Brasier 1979; McMenamin 1986; McMenamin and McMenamin 1990). Perhaps these pressures at the close of the Vendian induced a rapid but graded extinction of the dominant Ediacaran animals (Brasier 1979; McMenamin 1986; Seilacher 1989). Better stratigraphic resolution of the Ediacaran "extinctions" and further understanding of the functional morphology and ecology of Tommotian metazoans and protoctists may permit testing of this speculation.

11.4.4 Faunal Heterogeneity in the Vendian-Cambrian Diversification

Earlier discussions of Vendian to Cambrian diversification by Sepkoski (1979, 1981, 1984) treated the fauna of this interval as being homogeneous; that is, higher taxa were treated as if they shared a common evolutionary history without major differences in the times of their radiations and declines. Thus, most important Ediacaran and Cambrian classes were assembled into a "Cambrian evolutionary fauna" that dominated the early phase of metazoan diversification, as distinguished from the succeeding Paleozoic and Modern evolutionary faunas that radiated in the Ordovician and Mesozoic, respectively (see also Sepkoski 1990).

The limited temporal ranges of most Ediacaran and Tommotian taxa suggest that Sepkoski's earlier views may be greatly oversimplified. To investigate this possibility and to quantify the nature of early faunal change, a Q-mode factor analysis (Klovan and Imbrie 1971) was performed on the generic diversity data for the Vendian and Cambrian. This multivariate analysis was performed to summarize similarities in the faunal compositions of stratigraphic intervals in terms of a small number of uncorrelated "end members" that embody the major differences in temporal distributions of the taxa. This technique has been used by Flessa and Imbrie (1973) and by Sepkoski (1981) to investigate patterns of faunal change throughout the Phanerozoic and was the basis for recognizing the three "great evolutionary faunas" in familial diversity data.

Data used for the present analysis are the diversities of genera within 70 taxononic classes over 14 stratigraphic intervals from the upper Vendian to the top of the Cambrian. These intervals are the same as those used for plotting diversity in Figure 11.4.1 except that the Atdabanian has been subdivided into two units in order to gain more resolution on Early Cambrian faunal transitions. The total of 70 classes was chosen in part because of limitations of the available computer program. However, this number was sufficient to include all traditionally recognized classes as well as most class-level problematica that span more than one stratigraphic interval; only problematica known from single intervals or with ranges identical to others were excluded. Data for this analysis and selected output are listed in Sections 26.3 and 26.4.

The results of the factor analysis suggest that it would be useful to recognize at least three evolutionary faunas at the base of the metazoan diversification: an Ediacaran fauna, a Tommotian fauna, and a (now restricted) Cambrian fauna. This is not a straightforward result, however, because the two archaeocyathid classes again complicate the analysis. Without archaeocyathids, the analysis produces three strong factors that account for more than 94% of the data, contain the highest loadings on all stratigraphic intervals, and change little under varimax rotation. The smallest of these factors (7.6% of the data) receives a high loading from only the late Vendian Ediacaran interval; it is contributed principally by Petalonamae, Cyclozoa, and Scyphozoa (best considered a member of the "Modern" evolutionary fauna). The second factor (21 to 24% of the data) receives maximum loadings from the "Nemakit-Daldyn" and lower and upper Tommotian, as well as a substantial loading from the lower Atdabanian; it is contributed by orthothecimorph hyoliths, monoplacophorans, sabelliditids, and a variety of short-ranging problematica that originate in the Tommotian or early Atdabanian. The final factor (63 to 66% of the data) is the strongest and receives maximum loadings from the upper Atdabanian through Trempealeauan; classes contributing include trilobites, ostracods and other arthropods, inarticulate brachiopods, all enchinoderm classes, and a variety of other less diverse classes. This latter assemblage actually represents a mixture of members of the Cambrian (*sensu stricto*), Paleozoic, and Modern evolutionary faunas, listed in Table 11.4.1. As documented in Section 26.4, these factors exhibit regular rises and declines throughout the stratigraphic sequence, indicating that faunal change is regular instead of chaotic. The factors also exhibit increasing numbers of stratigraphic interval contributing high loadings, indicating that the rate of faunal turnover was initially high and slowed through the early metazoan diversification. Addition of more factors does not change this picture in any essential way: the "Nemakit-Daldyn" does receive its own factor in a varimax solution with four rotated factors, but this is probably best attributed to small sample (or low diversity) variation and not a separate fauna.

Inclusion of the two classes of archaeocyathids complicates the analysis, with these classes receiving their own factor (Section 26.4). Four factors become necessary to encompass more than 90% of the data and to receive high loadings on all stratigraphic intervals. Upon varimax rotation, the "Cambrian" factor of the previous analysis breaks into two, with one factor contributed mostly by archaeocyathids and receives loadings mostly from the lower Atdabanian through upper Botomian.

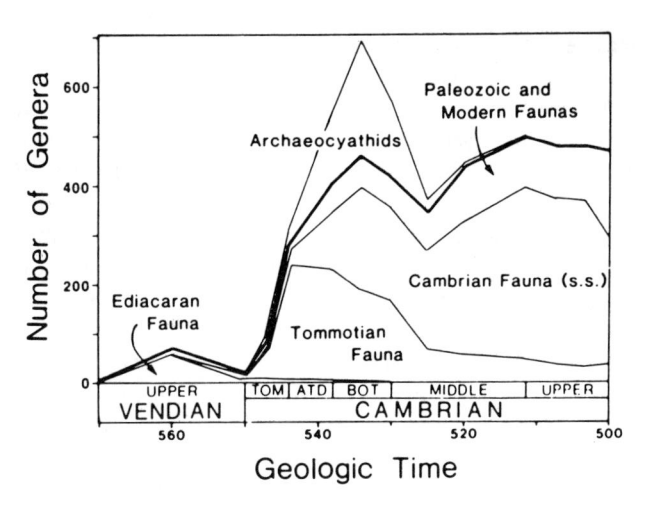

Figure 11.4.2. Diversity of metazoan genera in the upper Vendian through Cambrian, showing the contributions of the various evolutionary faunas, as defined here (with archaeocyathids unassigned). Each successive fauna, from Ediacaran to Tommotian to Cambrian and later faunas, has a higher level of diversity, slower rate of expansion, and longer span of dominance. The explosive radiation of skeletal metazoans in the earliest Cambrian resulted largely from expansion of the Tommotian fauna, which contains many short-lived taxa that did not survive the Early Cambrian.

This solution seems strongly determined by a single taxon and is thus less parsimonious than the three-factor solution. As in the analysis of overall diversity, it seems best to treat the Archaeocyatha as an outlier having a unique evolutionary history that is distinct from the common behavior of the other early metazoans.

The diversity history implied by the three-factor analysis above is illustrated in Figure 11.4.2, which shows how the Ediacaran, Tommotion, Cambrian, and later evolutionary faunas contribute to the total generic diversity in Figure 11.4.1. The diversity of each of these faunas was determined by summing the number of genera in each class assigned to the fauna, as listed in Table 11.4.1. The resulting pattern shows a regular sequence of expanding then declining faunas, much like that seen for the Phanerozoic as a whole (Sepkoski 1981, 1984, 1990). The Tommotian fauna, in particular, has the same form as seen among later evolutionary faunas: a rapid initial expansion producing an early maximum in diversity and followed by a long decline (here extending to the Recent with the few surviving monoplacophorans). Significantly, the more rapid initial expansion and lower maximum diversity relative to the Cambrian fauna (*sensu stricto*) fit neatly into the sequence of later evolutionary faunas, each of which has a slower diversification and higher "equilibrium" (Sepkoski 1984, 1990). The Tommotian fauna also appears to have suffered more than succeeding faunas at the Botomian extinction event, a pattern similar to that exhibited by younger Phanerozoic faunas during later mass extinctions. Data regarding the Ediacaran evolutionary fauna are too poorly constrained to permit any strong statement about its kinematics. However, the Ediacaran fauna does appear to have had a lower maximum diversity and a shorter span of dominance than the succeeding Tommotian fauna; it also may have had a longer phase of decline than of expansion if it is correct to include the triradiate, skeletal anabaritids among the Trilobozoa (cf. Fedonkin 1981; see also Chapter 8) or if some Burgess Shale taxa are indeed Ediacaran survivors (Conway Morris 1989a).

The patterns outlined above suggest that despite the rapidity of change during early metazoan diversification, the mode of faunal dominance did not differ substantively from that which occurred later in the Phanerozoic (excepting again the anomolous archaeocyathids). Thus, some important aspects of diversification and faunal replacement appear to be inherent to the metazoan condition, evidently having been established during the initial radiations of animals.

11.4.5 Exponential Diversification in the Proterozoic-Phanerozoic Transition

The simplest expectation of evolutionary expansion into an ecologically empty system is that diversification should proceed exponentially until the system begins to fill (Stanley 1975b, 1979; Sepkoski 1978, 1979). Thus, in the absence of severe environmental change, the diversity, D_t, at any time t should be approximately

$$D_t = D_o e^{rt} \qquad (11.4.1)$$

where D_o is the initial diversity at $t = 0$, and r is the per-taxon rate of diversification (i.e., the difference between the probabilities of cladogenesis and extinction).

Sepkoski (1978, 1979) plotted data for orders and families from the Vendian through Botomian on semi-logarithmic graphs and argued that the straight loci of points were consistent with exponential diversification over that interval. He thus concluded that the key to the Cambrian explosion of metazoans was not at the Proterozoic-Cambrian boundary but, rather, at the initial appearance of metazoans in the early Vendian. However, there are a number of problems with Sepkoski's data and analyses: (*i*) the time scale he employed, which is now very questionable; (*ii*) his separation of Ediacaran fossils into middle and upper Vendian components, which no longer can be distinguished (Glaessner 1984); (*iii*) his use of a taxonomy of Ediacaran fossils that extended many clades into the Phanerozoic; and (*iv*) his exclusion of many Lower Cambrian problematica from the ranks of higher taxa. There is also the problem, discussed above (Section 11.4.3), that the early metazoan radiation may have been heterogeneous, with a succession of taxonomic groups that radiated at differing rates. This suggests that the per-taxon rate of diversification was not constant through the Proterozoic-Phanerozoic transition but changed as different taxa rose to dominance.

It is of interest to determine to what degree the data now available on early metazoans still exhibit exponential patterns (assuming, of course, that the time scale employed is valid). Figure 11.4.3 illustrates semi-logarithmic graphs of the diversity data for all taxonomic levels shown in Figure 11.4.1. These graphs show that there is no long-term exponential diversification through the Vendian and Early Cambrian. The Ediacaran points for genera, families, and orders fall well above any trend that may be present in the earliest Cambrian. This is consistent with the interpretation that Ediacaran taxa represent a distinct radiation and a separate evolutionary fauna. It is also consistent with the hypothesis that the Ediacaran fauna was terminated by an extinction event.

Data for the Early Cambrian, however, do suggest limited exponential diversification among the earliest skeletal taxa, although the evidence is not strong (especially in view of the uncertainties in the time scale). For genera and families, three points line up in a straight locus: the "Nemakit-Daldyn," lower Tommotian, and upper Tommotian. Interestingly, this line extrapolates downward near to the estimated age of *Cloudina*, the oldest known skeletal metazoan, which is associated with Ediacaran fossils in the Nama Group of Namibia (Germs 1972). Thus, there may have been exponential diversification in the explosion of skeletal fossils of the Tommotian evolutionary faunas around the Proterozoic-Cambrian boundary, with a lag phase through the latest Vendian and an explosive phase through the Tommotian Stage. This exponential diversification was short lived, however, lasting only on the order of 15 million years. The data for genera and families suggest that by the Atdabanian the radiation was tapering, with diversity falling below the earlier exponential trend. This, of course, is the time when the Tommotian evolutionary fauna, as defined above, began to decline and the Cambrian fauna initiated its less rapid expansion. Orders also show this basic pattern but begin to drop below the apparent exponential trend by the upper Tommotian.

The semi-logarithmic graph for classes in Figure 11.4.3 is somewhat less easy to interpret, perhaps reflecting the smaller number of taxa included and the more arbitrary taxonomic definition. The easiest interpretation is that the pattern among classes, especially skeletalized classes, is similar to that of orders, exhibiting an exponential diversification from the appearance of *Cloudina* to the lower Tommotian. Alternatively, it is possible to interpret the data as showing more protracted exponential diversification from the late Vendian to the Atdabanian. The three points for total fossil diversity in the lower Tommotian, upper Tommotian, and Atdabanian fall along a straight line. This trends then extrapolates backward to about the position of the total Ediacaran fauna (although note the large potential error for time and taxonomy). Between, the "Nemakit-Daldyn" falls below the trend, although not substantially so. (No attempt has been made to extrapolate this trend back to the initial appearance of metazoan trace fossils because the timing of such appearance is too uncertain.) If this trend is correct, it defines two important aspects of the early metazoan radiation: (*i*) there is more temporal continuity in the exolution and diversification of *baupläne* than is seen among lower taxa, in which there is a succession of rapidly diversifying evolutionary faunas; and (*ii*) the diversification of *baupläne* is actually slower and more protracted than the diversification of lower taxa during the early metazoan radiation, extending over perhaps 50 million years. Both of these suggestions must be treated as mere speculations until better data are available for the systematics and ages of Vendian metazoan fossils.

11.4.6 Diversification of Metaphytes

It is important to know how similar the diversification of metaphytes is to that of metazoans during the Proterozoic-Phanerozoic transition. If temporal trends are congruent, this

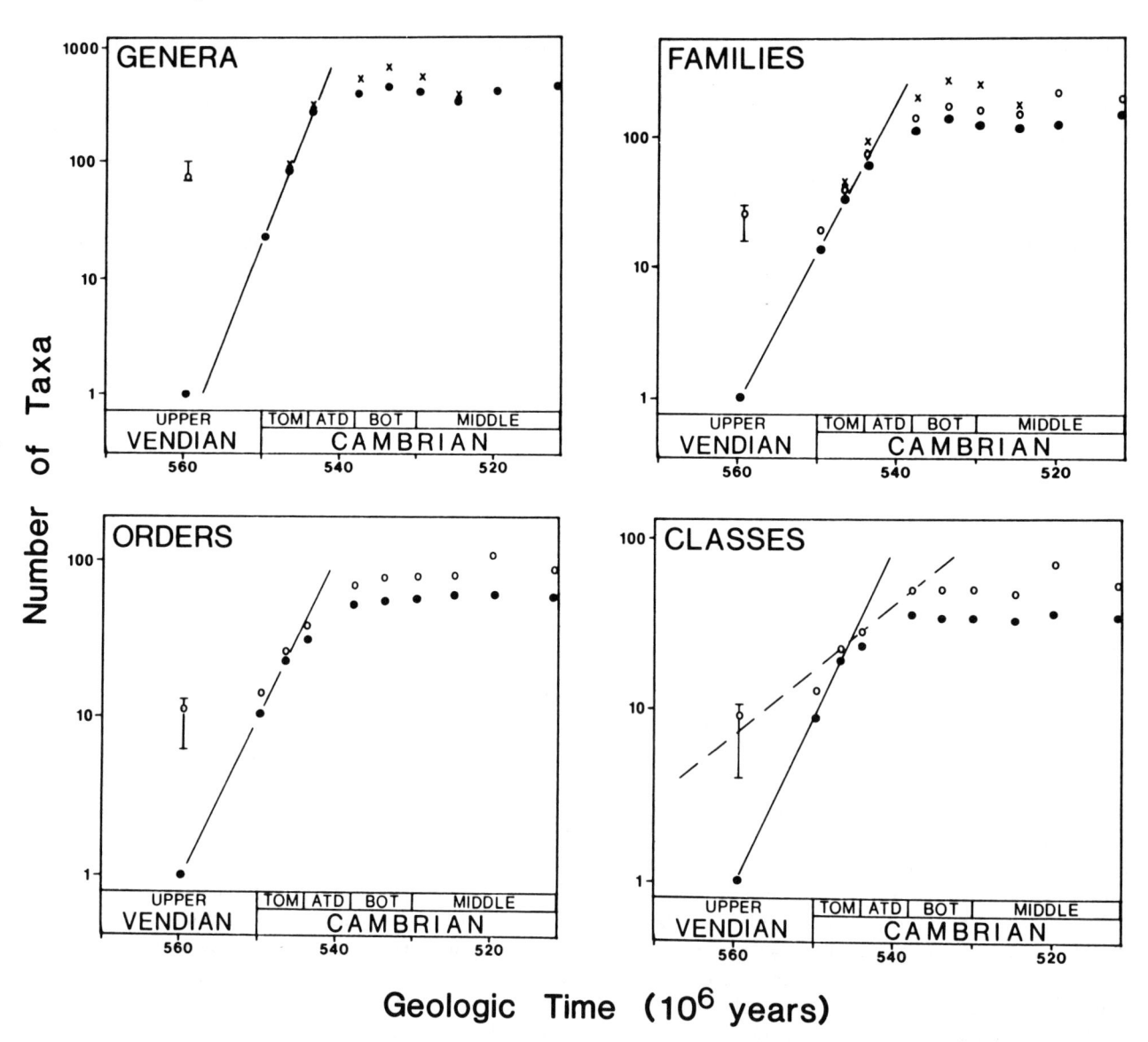

Figure 11.4.3. Semi-logarithmic graphs of diversity of metazoan genera, families, orders, and classes in the upper Vendian through Middle Cambrian, based on data in Figure 11.4.1. In each graph, solid dots indicate the diversity of skeletal taxa (excluding archaeocyathids); open dots the diversity of skeletal plus soft-bodied and lightly sclerotized taxa; and crosses the diversity of all taxa including archaeocyathids. Error bars on the open dot for the Edicaran fauna in the upper Vendian contrast very conservative estimates of number of taxa (e.g., Seilacher 1989) with more liberal estimates (e.g. Glaessner 1979d, 1984; Fedonkin 1983, 1985c). Straight line loci of points on semi-logarithmic graphs indicate exponential diversification at constant per-taxon rates. Solid lines were drawn through the three points for skeletal taxa in the "Nemakit-Daldyn" and Tommotian (two points in the case of classes). Note that these lines extrapolate backward to the first skeletal point, representing *Cloudina* in the upper Vendian Nama Group; these lines suggest that exponential diversification of skeletal taxa may have lasted for only about 15 million years around the Proterozoic-Cambrian boundary. The dashed line in the graph for classes was drawn through points for all taxa in the Tommotian and Atdabanian and might suggest a longer, roughly exponential radiation of *baupläne*.

might suggest that metazoans and metaphytes were responding to similar external changes in the environment. Alternatively, it might suggest that biotic interaction was important, with metaphytes evolving under increasing cropping pressure from metazoans (cf. McMenamin and McMenamin 1990).

Unfortunately, these questions can be addressed with only the coarsest generalizations at the current time. As discussed in Section 7.4, data on Vendian and Cambrian metaphytes, and particularly calcareous algae, are scanty and subject to considerable monographic noise. Few workers have studied these fossils, most come from limited stratigraphic intervals and geographic regions (e.g., portions of Siberia), and taxonomic work has been plagued by oversplitting and misidentification of pseudo- and dubiofossils. For macroscopic and/or calcareous fossils of undoubted biogenicity, it is often not clear whether they represent true metaphytes (i.e., eukaryotic algae) or calcified filaments or mat fragments of prokaryotes (see Sections 7.4 and 10.6). This situation makes it difficult at best to evaluate patterns of diversification.

Non-metazoan, carbonaceous megafossils range through most of the Proterozoic, predating the appearance of metazoan fossils by more than a billion years (see Sections 7.3 and 10.6). The approximate diversity of carbonaceous megafossils through the Proterozoic is shown in Figure 11.4.4A. This graph

Proterozoic-Early Cambrian Diversification   561

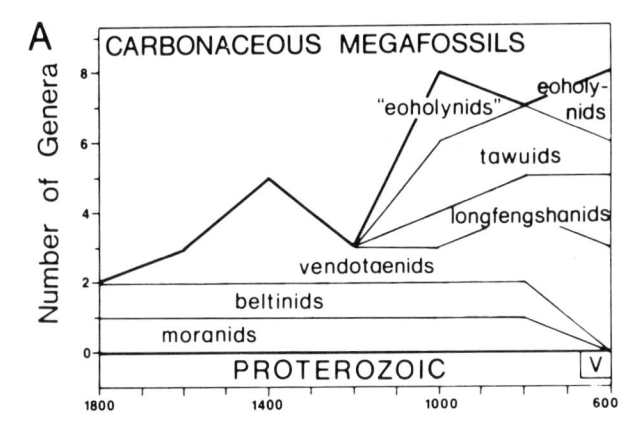

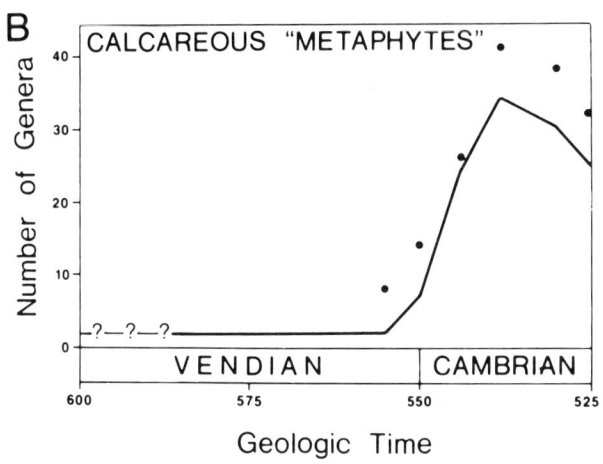

Figure 11.4.4. Preliminary generic diversity curves for potential metaphytes during the Proterozoic and the Proterozoic-Phanerozoic transition. (A) Diversity of carbonaceous megafossils (excluding chuarids) through the Proterozoic, based on the data and informal classification of Hofmann (1985b) and Section 10.6A and compiled at intervals of 200-myr duration. The data show that non-metazoan megafossils increased in diversity during the Middle Proterozoic. The significance of this diversification is unclear, however, since it is uncertain which, if any, of the form taxa of carbonaceous megafossils represent true, eukaryotic metaphytes. (B) Generic diversity of calcareous algae and calcified prokaryotes in late the Vendian to early Middle Cambrian. The solid curve represents a conservative estimate of generic diversity, whereas the unconnected points indicate total numbers of reported genera (excluding pseudofossils and dubiofossils). The data show that skeletal photosynthetic taxa diversified rapidly in the earliest Cambrian, much like skeletal animal taxa.

illustrates, at very coarse sampling intervals, both the total diversity of these megafossils and the contributions of the various form groups recognized by Hofman (1985b). (Chuarids are excluded since they are considered among sphaeromorph acritrachs in Section 11.3.) As is evident, carbonaceous megafossils display an irregular increase in diversity throughout the Middle and Upper Proterozoic. However, the significance of this diversification is far from clear since the taxonomic affinities of these fossils are largely unknown. Although some may be eukaryotic metaphytes (e.g., the later vendotaenids and eoholynids) or even eukaryotic heterotrophs (e.g., *Tawuia*), most, or even all, of the pre-Vendian carbonaceous megafossils might be simply forms of prokaryotic mats or colonies (see Section 7.3). Thus, the meaning of the increasing diversity of carbonaceous megafossils in the Proterozoic with respect to the later diversification of metaphytes is not clear. (True, non-skeletal metaphytes do occur in the pre-Vendian Proterozoic, however, as reported by Butterfield and Knoll 1989).

Calcareous "algae" (skeletalized cyanobacteria and eukaryotic algae) do not diversify until the Proterozoic-Phanerozoic transition, coincident with the radiation of metazoans. Figure 11.4.4B represents an attempt to graph the diversity of these calcareous taxa through the Vendian and Early Cambrian, based on data compiled by Mankiewicz in Chart 7.4.8 and Chapter 23. Only genera of well-documented fossils are plotted. The resulting pattern is basically congruent with that of metazoan diversity, although the uncertainties are large. There seems to be a very low diversity of calcified taxa through most of the Vendian (Section 7.4), a slight increase in the latest Vendian ("Nemaket-Daldyn"), and then rapid diversification in the Early Cambrian with most of the increase occurring in the Tommotian (cf. Riding 1982; Riding and Voronova 1982). By the Botomian, diversity may have leveled or even declined, although few reliable data are available for the Middle and Upper Cambrian.

The rapid increase in diversity in the Early Cambrian also parallels the increase in small, organic-walled, "ornamented" acritarchs, as documented in Sections 5.5 and 11.3. Thus, if diversity was responding to some external, environmental change, it would seem to be more complicated than simply something that made it easier for organisms to skeletonize (cf. Riding 1982). Alternatively, the change may have been biotic in nature, as suggested above: increasing cropping pressure from radiating herbivorous metazoans may have induced evolution of skeletal armor among a heterogeneous mixture of benthic cyanobacteria and algae (McMenamin and McMenamin 1990). In a similar vein, rapidly increasing suspension feeding among radiating metazoans may have selected for smaller, more ornamented planktonic photosynthetic eukaryotes that could float more efficiently. Further investigation of the nature and functional morphology of both metaphytes and metazoans in the early Phanerozoic is necessary for testing or modifying this admittedly speculative hypothesis.

Acknowledgments

I thank S. Bengtson for invaluable help in the attempt to systematize early matazoan families and also M. A. Fedonkin, C. Mankiewicz, and B. Runnegar for general discussions of aspects of early metazoan and metaphyte evolution. Research for this Section received partial support from NASA Grants NAG 2-282 and NAGW-1693.

11.5

Proterozoic and Earliest Phanerozoic Biotic Diversity: Unsolved Problems and Future Research Directions

J. JOHN SEPKOSKI, JR.   J. WILLIAM SCHOPF

Analyses of fossil diversity attempt to discover patterns in, and to interpret the macroevolutionary implications of, available paleontologic information. These endeavors are highly data dependent, as is well illustrated by the changing interpretations of the Proterozoic-Phanerozoic metazoan diversification that have emerged as more and better paleontologic data have accumulated. Similarly, it can be expected that the interpretations presented here of Proterozoic microorganism diversification are likely to be modified, perhaps markedly, as more data become available in future decades.

The first priority for future work on Proterozoic and earliest phanerozoic diversity patterns thus must be the standard efforts of paleontology: collecting, describing, and classifying – both of microfossils and, from the Proterozoic-Phanerozoic transition, of macroscopic body fossils and ichnofossils. These efforts will increase the sample size and enhance the accuracy of documented stratigraphic ranges available for diversity analyses. Work should also include increased attention to the paleoecology, taphonomy, and sedimentology of fossiliferous strata in order to aid in the assessment of sampling biases that can affect diversity patterns. Finally, there is urgent need for more accurate correlation and age determination of fossiliferous strata so that evolutionary rates and the timing of evolutionary events can be assessed more precisely throughout the Proterozoic and early Phanerozoic, and so that undesirable data manipulations, such as the use of moving averages, will not be necessary for compilation of Proterozoic diversity patterns. Obviously, these and related endeavors will be possible only with continued international communication and cooperation within the paleontologic and geologic communities.

All of the patterns and interpretations presented in this Chapter are tentative; all can be considered arenas for future research. Some of the more important problems that require attention including the following:

(1) The actual pattern of prokaryote diversity during the Early Proterozoic, in contrast with that inferred from the currently limited sample of investigated fossiliferous formations, needs to be determined. In addition to study of an increased number of microfossiliferous deposits, especially of Early Proterozoic clastic sediments, this might be greatly aided by comparative paleoecologic and taphonomic analyses of Early, Middle, and Late Proterozoic microfossil assemblages to assess in detail when and in what environments changes occurred in alpha diversity and in the taxonomic composition of preserved prokaryotic communities.

(2) The apparent decline in microfossil diversity during the Late Proterozoic needs to be tested further to ensure that this represents a real pattern. Such testing can be performed by comparative analyses such as those suggested above, and could be particularly useful if augmented by refined assessment of the classification, stratigraphic ranges, and modern analogues (viz., "Lazarus taxa") of the fossil prokaryotes and planktonic eukaryotes analyzed here.

(3) More precise data need to be acquired regarding the timing and nature of Middle and Late Proterozoic evolutionary events, especially (*i*) the origin of eukaryotic sexuality (the advent of which may have "triggered" the late Middle Proterozoic diversification of planktonic eukaryotes; Section 5.5); (*ii*) the apparent, sequential, size-related Late Proterozoic extinctions of large-diameter sphaeromorph and acanthomorph acritarchs (Section 5.5); (*iii*) the origin both of heterotrophic protists and of megascopic multicellular thallophytes, events that may have had significant impacts on patterns of Late Proterozoic biotic diversity; (*iv*) the extinction of numerous types of planktonic eukaryotes near the beginning of the Vendian; and (*v*) the possible occurrence of a subsequent extinction event very near the Proterozoic-Cambrian boundary.

(4) The early, presumably Late Proterozoic, history of metaphyte diversification needs to be determined. At present, due to inadequate sampling, taxonomic confusion, and uncertainty as to the nature and affinity of many Proterozoic carbonaceous megafossils, this history is very poorly known.

(5) The Vendian history of microfossils and of macroscopic body fossils and ichnofossils needs to be compiled with far greater precision so that it can be determined whether the flowering of the Ediacaran fauna reflects the advent of metazoans or simply the earliest manifest evidence of

metazoan diversification, what the history of diversity within the Ediacaran fauna actually is, and when and how components of the Ediacaran fauna became extinct.

(6) The pattern of diversification of the earliest skeletal metazoans (like that of early, mineralized, heterotrophic protists; Section 5.7) needs to be investigated further to determine whether it was in fact exponential (and possibly beginning with *Cloudina*); whether it reflects rates of diversification that were indeed higher than all later evolutionary radiations; and whether the early skeletal metazoans (here treated as the "Tommotian evolutionary fauna") actually represent a distinct phase of diversification, one separate from that represented by both earlier and later known metazoan faunas. In the relatively near future, such endeavors should become increasingly more feasible as a result of the continuing search for a global stratotype for the Proterozoic-Cambrian boundary and the related efforts in paleontologic exploration, stratigraphic correlation, isotopic and paleomagnetic characterization, and chronometric dating of boundary strata.

(7) The enigmatic archaeocyathids need additional study; in particular, their evolutionary history needs to be better defined, and it remains to be established why their pattern of diversification evidently differs so markedly from those of all other metazoan groups in the Cambrian.

(8) The apparent parallelism of metazoan diversity patterns across taxonomic levels from genus to class needs to be tested to determine if the appearances of *baupläne* really were tracking the diversification of species, as presumed here to be reflected among genera. This endeavor will be aided by further paleontologic exploration of Cambrian *Lagerstätten*, such as the Burgess Shale and similar deposits, to discover new organisms and to estend the known stratigraphic ranges of identified major groups; by more cladistic analyses of primitive metazoans; and by attempts to measure rates of morphologic as well as taxic evolution in the Early Cambrian.

11.6
Biotic Diversity During Proterozoic and Earliest Phanerozoic Time: Summary and Conclusions

J. JOHN SEPKOSKI, JR.　　J. WILLIAM SCHOPF

The following conclusions should be regarded as tentative; each is based on a limited suite of available data and all are thus in need of additional relevant evidence from the fossil record and of further testing, analyses, and subsequent modification.

(1) Because of limited sampling, the pattern of Early Proterozoic diversification of benthic and planktonic prokaryotes is as yet inadequately defined. Although the diversity of reported prokaryotic microfossils increases from about 2100 Ma (the beginning of the more or less continuous fossil record as now known) to 1000 Ma, this seems likely to be a result merely of a cumulative increase in the number of sampled formations; true diversity may have been essentially stable over this interval.

(2) Planktonic eukaryotes (viz., sphaeromorph acritarchs), evidently originating during the Early Proterozoic, exhibit a pattern of increasing diversity through the Middle Proterozoic which appears to be independent of the number of sampled fossiliferous formations; Proterozoic acritarchs attain an apparent maximum in diversity about 900 Ma.

(3) There is a major, long-term decline in the global and local diversities of planktonic eukaryotes, and apparently of some prokaryotic groups, through the Late Proterozoic, beginning about 900 to 800 Ma and evidently reaching a minimum in the Vendian.

(4) The "explosive" diversification of metazoans during the Proterozoic-Phanerozoic transition appears to have been a heterogeneous event, occurring in three more or less discrete phases: (*i*) a Vendian radiation and subsequent collapse of the Ediacaran fauna; (*ii*) a latest Vendian to earliest Cambrian radiation of the Tommotian fauna; and (*iii*) a later expansion of the Cambrian fauna, *sensu stricto*. Each of these evolutionary faunas is characterized in sequence by a higher level of diversity, by a longer history, and, perhaps, by a lower rate of diversification and general evolution.

(5) As documented by the limited fossil evidence now available, the known diversity of calcareous metaphytes and calcified prokaryotes (as well as that of mineralized protists; Section 5.7) appears to have been very low in the Vendian and to have increased markedly in the earliest Cambrian, evidently exhibiting a pattern paralleling that of the expansion of skeletal metazoans.

12
A Paleogeographic Model for Vendian and Cambrian Time

JOSEPH L. KIRSCHVINK

12.1

A Paleogeographic Model for Vendian and Cambrian Time

JOSEPH L. KIRSCHVINK

12.1.1 Introduction

In any attempt to construct a set of paleogeographic reconstructions, it is necessary to make sense out of a variety of sometimes conflicting sets of geological, paleontological, and geophysical constraints. Because the applicability of biostratigraphic correlation diminishes with increasing geologic age (Chapter 10), it is necessary that reconstructions of Late Proterozoic paleogeography (as well as those of the earliest Paleozoic) be anchored well within the Paleozoic, and only then be extended as far as feasible into the Precambrian. However, few reconstructions of Cambrian paleogeography have reconciled in a satisfactory manner both the pronounced lithologic and faunal variations that are known to occur with the often sparse paleomagnetic constraints (Jell 1974; Shergold 1988; Courjault-Radé 1987; 1988). Thus, it is necessary first to produce consistent tectonic and paleontological models for Cambrian time before attempting an extension into the Precambrian.

Temporal correlation is the most severe problem encountered in dealing with the Proterozoic time. Paleomagnetic data from rocks of this age are of highly variable quality, and are plagued with large uncertainties in the age of various components. After careful examination of the existing paleomagnetic data, we have chosen to limit our attempts at a detailed reconstruction to the Vendian, from the approximate time of the last glacial episode of the Neo-Proterozoic Cryogenian System (see Table 1.1.1) through to the Cambrian-Ordovician boundary. Several factors influence this decision, including the low number of well dated or stratigraphically constrained paleomagnetic determinations for Early and Middle Proterozoic time, and the general lack of intercontinental correlation between them. Precambrian paleomagnetic data have been used to support a wide variety of conflicting models of Precambrian plate motions, both for and against the existence of modern-style plate tectonics and even for an expanding earth (Piper 1982, 1983, 1986; Schmidt and Embleton 1981; Klootwijk 1980). In a recent review of the African Proterozoic paleomagnetic data, for example, Perrin et al. (1988) concluded that directional scatter in the published directions was so high, and the age constraints so lax, that any number of distinctly different apparent polar wander paths (APW paths) could be drawn through them.

Nevertheless, paleomagnetic data from Early and Middle Proterozoic time do seem to have an angular and temporal coherence that is difficult to explain without the presence of some sort of large supercontinent (Piper 1983, 1986; Schmidt and Embleton 1981). Furthermore, recent studies of thermal subsidence histories support the hypothesis that one or more large continental masses broke up shortly before the Precambrian-Cambrian boundary (Armin and Mayer 1983; Bond and Kominz 1984; Bond et al. 1984), with intermittent episodes of rifting which began somewhat earlier. These are appealing hypotheses because, if true, the presence of one or more amalgamations of continental masses would greatly simplify the problem of reconstructing their positions. However, our analysis of climatic and biostratigraphically constrained paleomagnetic information for Vendian and Cambrian time suggests that East and West Gondwana were separate and presumably did not join until Early or Middle Cambrian time, which is roughly coincident with the end of the pan-African metamorphic event along the eastern margin of Africa (Kröner 1980; Dixon and Golombek 1988). At the minimum, therefore, there were probably at least two large supercontinents in the Late Proterozoic.

In selecting the paleomagnetic poles used as constraints for our reconstructions for the Vendian and Cambrian, we have used almost exclusively data from paleontologically or stratigraphically constrained, unmetamorphosed sedimentary sequences, with virtually all such data derived from samples treated with progressive thermal demagnetization. Although such rocks are sometimes plagued by problems produced by secondary chemical overprints, their superior age control, relatively accurately determined paleo-horizontal orientation, and the numerous opportunities they provide for field tests (e.g., fold, reversal, conglomerate, etc.) to determine the age of the remanence, mark them as the best available for this purpose. We apologize to those readers whose preferred poles are not used here; however, we found that no coherent scheme is possible that incorporates all published results. Detailed discussions of the data used are given below for the four major

570   A Paleogeographic Model

reference continents (Australia, Siberia, North America, and Africa).

Finally, it also helps to use an additional assumption that extensive deposits of platform-type carbonates were deposited within an equatorial carbonate belt similar in character to that which has existed in post-Paleozoic time. Ziegler et al. (1984) argue that the major factor constraining the Cenozoic carbonate platforms to lie within about 33° of the equator is the need for carbonate-precipitating algae to have at least some direct sunlight each day; nearly total reflection of sunlight at the ocean surface will shut off light during the winter at higher latitudes, limiting algal growth. This model is compatible with a large body of Cambrian paleomagnetic data, and helps position plates (such as the Baltic Platform) for which there is as yet inadequate paleomagnetic control.

In our attempts to reconcile our reconstructions for Late Proterozoic and Cambrian time with those of other workers (e.g., Ziegler et al. 1979; Scotese 1984; Van der Voo 1988), we encountered a problem in the placement of the Siberian and Baltic Platforms relative to North America (Ripperdan and Kirschvink 1989a, b). The most parsimonious reconstruction that we could find to reunite Early Cambrian faunal provinces was to put these blocks to the *west* of North America, whereas in the most popular reconstructions for later Paleozoic times they are positioned off the *eastern* margin, closer to their present positions. Hence, our model implies oblique, left-lateral subduction during the post-Cambrian closure of the Iapetus Ocean between Baltica and the Appalachian margin of North America, and right-lateral motion of some sort between Siberia and the Cordilleran margin. Although other interpretations for the faunal distributions are certainly possible within the framework of the more conventional schemes, we felt this in particular might serve to stimulate discussion of the alternatives.

The following Sections will first provide an overview of our selection of paleomagnetic data used to partially constrain the orientation of the major continents, followed by a discussion concerning how smaller or less well-constrained fragments have been fit to these major fragments. The final discussion will concern the lithologic and faunal constraints.

12.1.2 Paleomagnetic Data

Of the four major continents (Australia, Siberia, North America, and Africa) that have more than one stratigraphically constrained paleomagnetic result for Vendian and Cambrian time, Australia and North America have the most and perhaps best data. Results for the Baltic Platform in this time interval are rather sparse (Pesonen et al. 1989), and many from Siberia are difficult to interpret. A summary of the data for each of these cratons follows below, and will focus primarily on results on two major time slices at the basal Vendian glaciation and the Precambrian-Cambrian boundary (taken as the time equivalent of the base of the Tommotian Stage of the Siberian Platform: Rozanov 1984; Cowie 1985). Compilations of paleontologically dated Cambrian data will be provided where needed for brevity. Figures 12.1 through 12.4 show our selected apparent polar wander (APW) paths for each of these cratons.

12.1.2A Australia

Australia has the best positional constraint for the basal Vendian glacial epoch of any continent, based on the paleomagnetic study of the Elatina Formation reported by Embleton and Williams (1986). The Elatina Formation con-

Australian Vendian-Cambrian APW Path

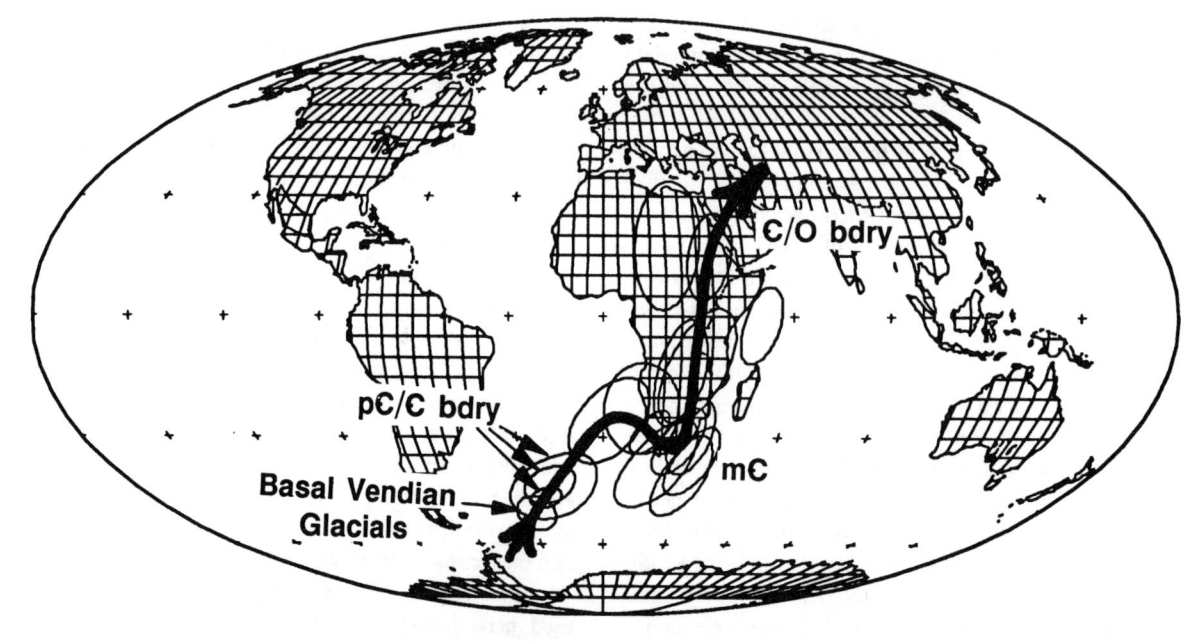

Figure 12.1 Vendian and Cambrian aparent polar wander (APW) path for Australia. The basal Vendian glacial pole is that from Embleton and Williams (1986), the Precambrian-Cambrian Boundary from Kirschvink (1978b), and the others from Klootwijk (1980).

A Paleogeographic Model 571

Siberian Platform Vendian-Cambrian APW Path

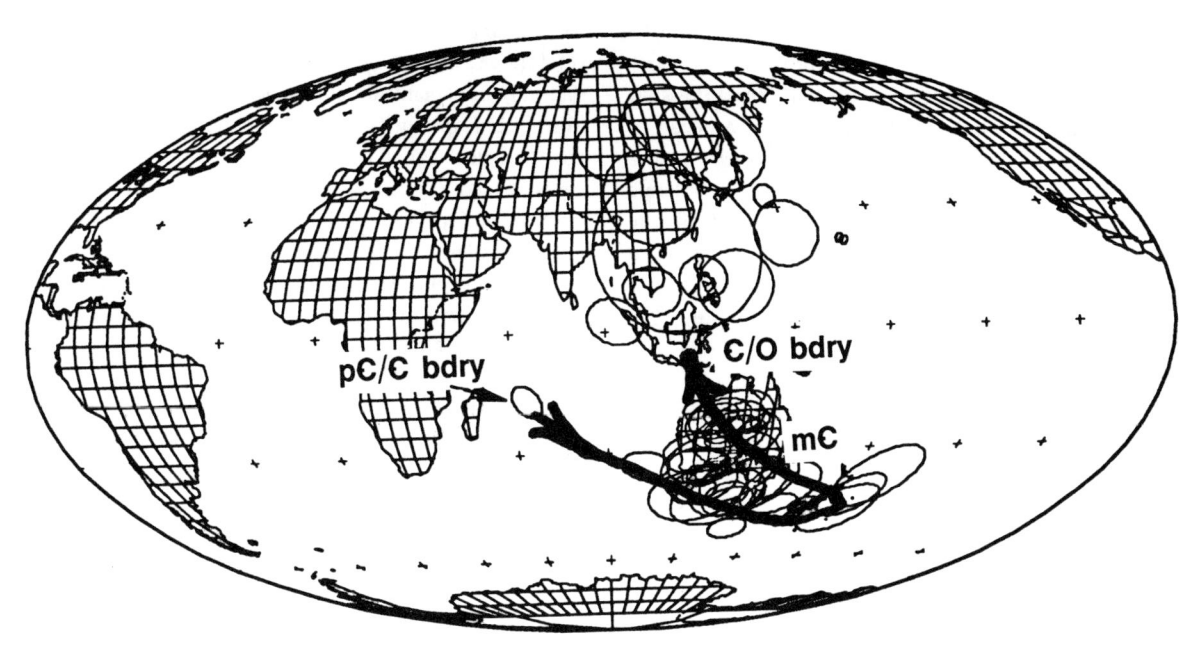

Figure 12.2. Vendian and Cambrian APW path for Siberia. The Precambrian-Cambrian pole is from Kirschvink and Rozanov (1984), and the others are taken from the Soviet compilations of Khramov (1982). Note that we have no good constraint for the Vendian of Siberia.

tains a series of rhythmic varves that formed on the outwash plain of the Marinoan glaciation, and is preserved in the Adelaide geosynclinal sequence of South Australia (Williams 1981). The magnetization in this unit is carried by detrital hematite with blocking temperatures in excess of 665°C, whereas studies of subgraphitic kerogen in the area reveal that the rocks have not been subjected to temperature-time equivalents more than 160°C for 300 Ma, well below that necessary to remagnetize the hematite (Embleton and Williams 1986). A recent study on the Elatina by Summer et al. (1987 and in prep.) also yielded a positive fold test, supporting the interpretation that the magnetization was gained early in its depositional history. Both of these studies imply that these widespread glacial deposits of the basal Vendian (or Edicaran) Period of Australia were deposited within a few degrees of the equator. The profound climatic implications of this are discussed elsewhere in this volume (Section 2.3) and by Chumikov and Elston (1989), and provide the rationale used in our oldest reconstructions, where we assume this uppermost Cryogenian (basal Vendian) glaciation was a globally synchronous event.

As shown in Figure 12.1, the pole position obtained from the Elatina Formation is indistinguishable from three others obtained from middle to late Vendian and earliest Cambrian rocks of the Amadeus Basin sequence in central Australia (Kirschvink 1978a, b). These magnetizations passed the fold, unconformity, and reversal tests, and the magnetic polarity stratigraphy could be traced for a substantial distance along the northern margin of the Amadeus Basin of central Australia. Thus, the Australian continent experienced no detectable polar wander for the entire Vendian period, and the Eocambrian quasi-static interval (a period of no apparent polar wander), first detected by Kirschvink (1978a, b), probably includes the entire Vendian.

Klootwijk (1980) published a major monograph on the paleomagnetism of Australia covering the remainder of Cambrian time, providing a large number of key, biostratigraphically dated pole positions which can be used to constrain the relative motions of Gondwana. In general, these data imply that this continent rotated 90° counter-clockwise (CCW) around a Euler pole located in the vicinity of Tasmania during Cambrian time. The rotation, however, was not continuous. Roughly half of the motion occurred during the Early Cambrian, with relatively little during the Middle and the first half of the Late Cambrian. The remaining 45° was confined entirely to the last half of the Late Cambrian (Klootwijk 1980). A new, magnetostratigraphically constrained pole position from the Cambrian-Ordovician boundary section at Black Mountain, Queensland, Australia (Ripperdan and Kirschvink, in prep.) confirms the earlier poles of this age as well as the magnitude of this rotation. Figure 12.1 shows the Vendian-Cambrian APW path for Australia.

12.1.2B Siberia

A key constraint on the position of the Siberian Platform during the Tommotian Stage of the Early Cambrian is provided by Kirschvink and Rozanov (1984). In addition to recognizing a correlatable and consistent magnetic polarity stratigraphy, their paleomagnetic data demonstrated that a narrow, 1,500-km-long belt of archaeocyathid bioherms was stretched out precisely along the equator during Tommotian time. Apparently, the density and diversity of these early reef-forming organisms were sensitive to the slight climatic maxima centered

572   A Paleogeographic Model

on the equator as is, for example, the diversity of modern tropical land vertebrates in Africa. Although the position of the reef-like complex remains stable during the Tommotian, it shifts towards the NE portion of the platform in the early Atdabanian. Kirschvink and Rozanov (1984) interpret this shift as marking the time at which the Siberian Platform began to move off the equator, as well as reflecting the onset of the Cambrian sea-level transgression. Unfortunately, few reliable paleomagnetic data are available for the Vendian of Siberia, and the result of Kirschvink and Rozanov (1984) and data from the compilation of Khramov (1982) imply a rotation of nearly 90° for Siberia during the Cambrian, as shown in Figure 12.2. The magnitude of this rotation, and the lack of reliable data within this period, introduce an ambiguity regarding how the APW path for Siberia should be connected between the Precambrian-Cambrian and Cambrian-Ordovician boundaries. This is complicated further by the use of a DC-demagnetization technique in much of the pre-1970 Soviet paleomagnetic data. However, it is clear from the extensive carbonates of this age that Siberia remained within the carbonate belt during both Vendian and Cambrian time, and faunal similarities in the Early and Middle Cambrian imply a close proximity to western Antarctica and Australia (Palmer and Gatehouse 1972). Thus, the most parsimonious position for Siberia is to fit its southeastern margin adjacent to the margins of western Antarctica and eastern Australia. In this orientation, the CCW rotation of Gondwana in the Early Cambrian will move Siberia northward (generating the proper relative motion for the archaeocyathid bioherms). Note that this orientation is *opposite* that used by Kirschvink and Rozanov (1984), and hence that their polarity interpretations should be switched.

12.1.2C North America (Laurentia)

Morris (1977) has conducted the most thorough paleomagnetic study of the uppermost Proterozoic glacial units yet done in North America. Using progressive alternating-field, thermal, and chemical demagnetization techniques on samples from the Rapitan Group of the Mackenzie Mountains of northwestern Canada, he discovered three magnetic components termed X, Y, and Z. Two of these (X and Z) imply a low-latitude of formation, while the Y component implies high paleolatitudes. Using a micro-fold test on units deformed by a glacial dropstone, Sumner et al. (1987) discovered that this Y component was a result of secondary chemical remagnetization, acquired sometime after emplacement of the dropstones. Hence, the X and Z components are the only possible primary directions, although there are no guarantees that either represents the original magnetization. For North America, however, there are several other paleomagnetic studies from radiometrically dated igneous bodies which yield similar poles (Fig. 12.3); hence, a position near the equator in the early Vendian is probable. Unfortunately, the Late Proterozoic glaciations are poorly dated and it is difficult to compare directions from the igneous bodies with those from the sedimentary record. For the purpose of these paleogeographic reconstructions, we have placed the Cordilleran margin of Laurentia adjacent to the northeastern margin of the Siberian Platform as suggested by Sears and Price (1978).

Between the time of the last Proterozoic glaciation and the early Middle Cambrian, there are very few stratigraphically constrained and reliable paleomagnetic directions for North America, particularly within the "Ediacaran" interval. Earlier results from the Late Proterozoic Desert Range, Nevada

North American Vendian-Cambrian APW Path

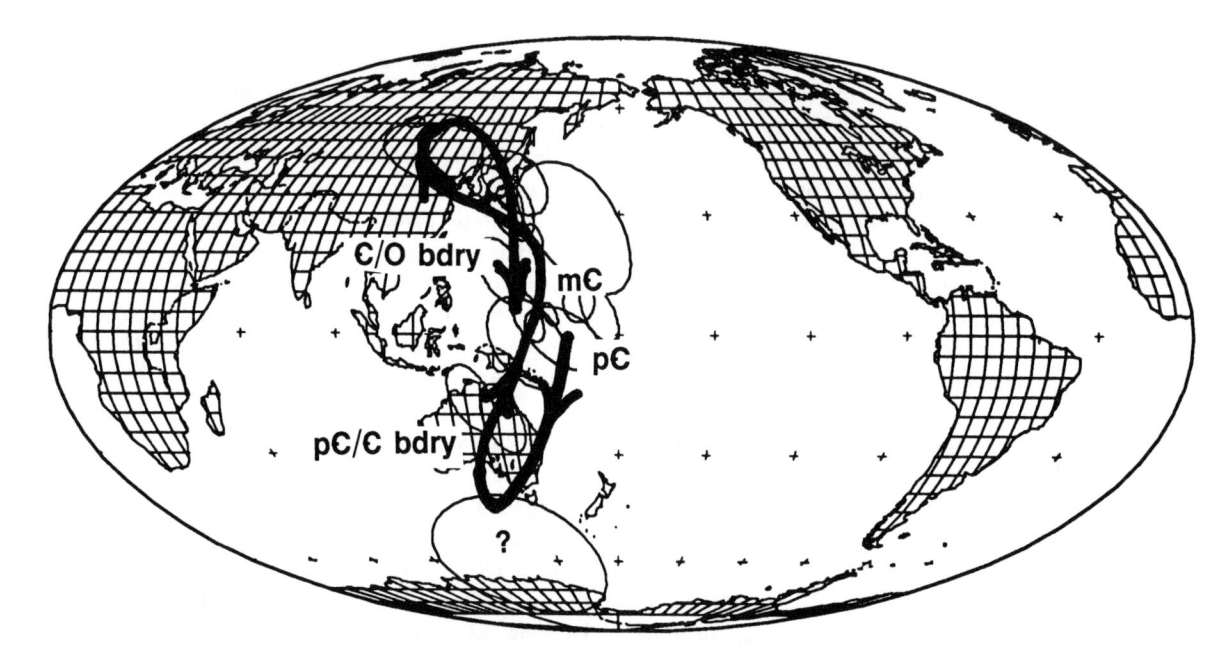

Figure 12.3. Vendian and Cambrian APW path for North America. Most of the poles are as used by Watts et al. (1980a, b), with the Tommotian result from Barr and Kirschvink (1983) which plots in Australia (both before and after correction for displacement on the Sonora-Mojave megashear). The querried direction is from Tanczyk et al. (1987), but is not biostratigraphically constrained in age.

African Vendian-Cambrian APW Path

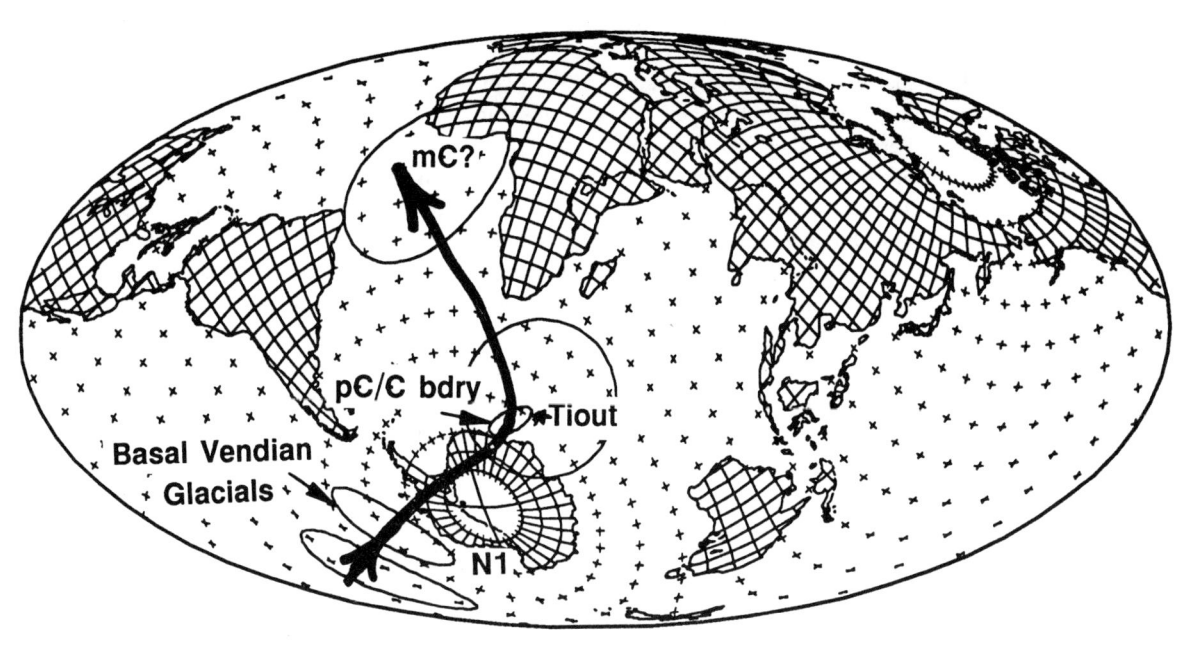

Figure 12.4. Vendian and Cambrian APW path for Africa. The basal Vendian galcial pole is the N1 direction of Kröner et al. 1980, and the mЄ? direction is their N3 overprint component. The Precambrian-Cambrian direction is that from Kirschvink (1980), supplemented with as yet unpublished data from additional samples from Morocco.

(Gillett and Van Alstine 1979; Van Alstine and Gillett 1979), were later found to be due to secondary magnetic components produced by thermal overprinting (Gillett and Van Alstine 1982), with the possible exception of a hematite-bearing unit of the (Ediacaran?) Rainstorm Member of the Johnnie Formation (from which we have recovered a few Ediacaran-like impressions in exposures at the Desert Range, Nevada). For the Tommotian, only one biostratigraphically constrained pole is available, from basaltic volcanics in the Puerto Blanco Formation near Caborca in northwestern Sonora, Mexico (Barr and Kirschvink 1983). McMenamin et al. (1983) described a fauna of small shelly fossils, similar in many respects to those of the Tommotian Stage in Siberia, from clastic units interbedded with the volcanics. The late Precambrian and early Paleozoic sequence at Caborca appears to be the lateral equivalent of the central Great Basin Sequence (Silver and Anderson 1974; Stewart et al. 1984) which has been displaced nearly 800 km in a left-lateral sense by the Sonora-Mojave megashear during Mesozoic time (Silver and Anderson 1974; Anderson and Schmidt 1983). The volcanics preserve a two-polarity characteristic component with a positive conglomerate test. Both before and after reconstruction for motion on the megashear (Anderson and Schmidt 1983), this paleomagnetic pole from Caborca is located in the vicinity of Australia as indicated on Figure 12.3. This implies that North America was in moderate southerly latitudes, with the Cordilleran margin to the north. A newly reported paleomagnetic pole, from intrusives dated at 540 ± 22 Ma (Rb/Sr) from the Sept-Iles layered mafic intrusion in Quebec, Canada (Tanczyk et al. 1987), also falls in this vicinity (although the age of the magnetizations in this unit are not constrained biostratigraphically).

Paleomagnetic data of reasonable quality are again available for North America from the early Middle Cambrian through the Cambrian-Ordovician boundary (Watts et al. 1980a, b). By early Middle Cambrian time, North America was again at low latitudes as it was in the late Precambrian. This similarity in position between Late Proterozoic and Middle Cambrian time makes it tempting to ignore otherwise anomalous paleomagnetic results, and to assume simply that the continent remained stably on the equator during Vendian time. However, within the Early Cambrian there is some biostratigraphic and lithostratigraphic evidence which supports this northward migration. Lithologically, the Cordilleran appears to have on the margin an equatorial carbonate belt, whereas the Appalachian margin does not. For example, the first archaeocyathids which appear in carbonates on the Cordilleran margin are biostratigraphically older than those which are first found in the first Cambrian carbonates of the Appalachian section in Labrador (Rowland and Gangloff 1988; F. Debrenne, written comm. 1988). Because the Caborca pole is the *only* biostratigraphically constrained result in this interval of time, and because it is consistent with other geological evidence, we use it here to constrain the position of North America.

Although a literal interpretation of the Middle and Late Cambrian paleomagnetic data shown in Figure 12.3 imply the presence of a hairpin loop in the Cambrian APW path for North America, this interpretation is also in dispute (Gillett 1982). Nevertheless, an equatorial position with the Cordilleran margin more or less to the north is reasonably consistent with the paleomagnetic data, whether or not this loop exists.

12.1.2D Africa

Perrin et al. (1988) have compiled an extensive list of published paleomagnetic pole positions for the Proterozoic and Cambrian of Africa which includes more than 50 entries. The majority of these, however, have such extraordinarily poor age constraints that it is difficult to place them in either the Proterozoic or Paleozoic, and Perrin et al. (1988) did not attempt to draw an APW path through them. For our reconstruction, however, it is necessary to make some selection of the data which is reasonably consistent with known geological and tectonic events. Figure 12.4 shows our best pick of these points for Vendian and Cambrian time. The N1 pole of Kröner et al. (1980), from the glaciogenic sediments of the Nama Group in Namibia, is perhaps the most reliable direction for the basal Vendian glaciation, and there is reasonable support from similar results from the Nosib Group (McWilliams and Kröner 1981).

The position of the Precambrian-Cambrian boundary on the African APW path has long been a mystery, principally because stratigraphic correlatives of the Tommotian fauna of small shelly fossils had not been located. Recent stratigraphic work in the Anti-Atlas Mountains of Morocco may have resolved this problem, however. A 3-km-thick sequence of Eocambrian shallow water carbonates is present in this area composed, in upward stratigraphic sequence, of the Calcaire Inferieur; the Série Lie de Vin; the Calcaire Supérieur; and the Série Schisto-Calcaire. Trilobites and archaeocyathids marking the Atdabanian Stage equivalents first appear in the middle horizons of the Calcaire Supérieur (Debrenne and Debrenne 1978; Sdzuy 1978), which led to the original suggestion that the Precambrian-Cambrian boundary lies near the top of the Série Lie de Vin. The late Brian Daily of the University of Adelaide, South Australia, however, claimed to have found tommotids near the base of the Série Lie de Vin (pers. comm. 1979), implying that horizons equivalent to the base of the Tommotian Stage of Siberia would lie somewhere in the uppermost Calcaire Inferieur. A comparison of the carbonate $\delta^{13}C$ pattern from Morocco of Tucker (1986) and with that from Siberia by Magaritz et al. (1986) and Magartiz (1989) is compatible with this interpretation, and suggests that the large isotopic excursion at the boundary level in Siberia correlates with a similar event about 100 m below the top of the Dolomie Inferieur of Morocco. Reddish carbonates within the Série Lie de Vin have been interpreted as forming in sabhka environments at the top of shallowing-upward cycles (Latham and Riding, in prep.), and preserve an extremely stable, two-polarity characteristic magnetization (Kirschvink 1980). More recent paleomagnetic work on these sediments reveals a polarity pattern similar to that of the Tommotian of Siberia after the inversion described above (Kirschvink, in prep.), arguing that the pole derived from this formation was acquired early after deposition. As shown in Figure 12.4, we use this pole as our best estimate for the basal Cambrian of Africa.

The subsequent Cambrian APW track for Africa is still poorly determined. However, basement cooling ages along the eastern African margin and in southern India imply that the assembly of the eastern and western halves of Gondwana was complete by ca. 550 Ma (Kröner 1980; and pers. comm. 1988), and it is therefore possible to use this tectonic constraint along with the more reliable Australian paleomagnetic data to constrain the path. The Nama overprint direction (N3 of Kröner et al. 1980) is in the correct vicinity as implied by these constraints (Fig. 12.4), and thus may have been acquired as a result of this final suturing event.

12.1.2E Baltic Platform and Fennoscandia

There are virtually no reliable, biostratigraphically dated paleomagnetic directions for the Baltic Platform from the base of the Vendian through to the Cambrian-Ordovician boundary, and this continent therefore poses more problems in interpreting its tectonic history than any other fragment (e.g., Pesonen et al. 1989). However, there are strong similarities between the Grenville-age basement in North America and Fennoscandia, so an initial placement of the Baltic Platform off the Appalachian margin of North America is reasonable. We have used the analysis of Gower and Owen (1984), which correlates the Sveconorwegian Orogenic Belt in southern Sweden with features surrounding the trans-Labrador batholith, to constrain the Precambrian position of the Baltic Platform relative to North America.

12.1.3 Tectonic Reconstructions

Figure 12.5 and Table 12.1 show the basic tectonic reconstructions and associated Euler poles used for producing the maps of Figures 12.6 through 12.11. On Figure 12.5, the continents and fragments are labeled using a two-letter code (AF, AU, etc.) which is translated in the caption for Table 12.1. Further details regarding these reconstructions for some of the major continental fragments are given below.

12.1.3A Gondwana

Several fundamentally different reconstructions for the paleocontinent of Gondwana have been proposed over the past 50 years (e.g., Du Toit 1937; Smith and Hallam 1970; Tarling 1972; Barron et al. 1978; Morgan 1981; Powell et al. 1980). Of these, there is usually a good consensus concerning which continental fragments belong in the separate assembles of the eastern and western portions, but sometimes a wide variance in how the eastern and western areas are reassembled and welded together finally. The western portion usually includes South America and the Falkland Plateau (Malvenas Platform), Africa and the Somolian and Arabian microplates, and the Florida basement, while the eastern area includes Australia, Antarctica, India, and Madagascar. These different reconstructions fall into two basic groups: those which fit Madagascar to Africa adjacent to Kenya (e.g., Du Toit 1937; Smith and Hallam 1970; Powell et al. 1980), and those which place it further south, offshore of Mozambique (Tarling 1972; Barron et al. 1978; Morgan 1981). Initial paleomagnetic evidence for the northerly position (McElhinny and Embleton 1976a) has been subsequently verified by the study of the marine magnetic lineations of the Indian Ocean (Powell et al. 1980), and versions of either the Smith and Hallam (1970) or Powell et al. (1980) reconstructions have been used commonly in paleomagnetic studies ever since (e.g., Embleton et al. 1980).

The basic reconstruction of Gondwana used in all of the

A Paleogeographic Model 575

Plate Identity Map

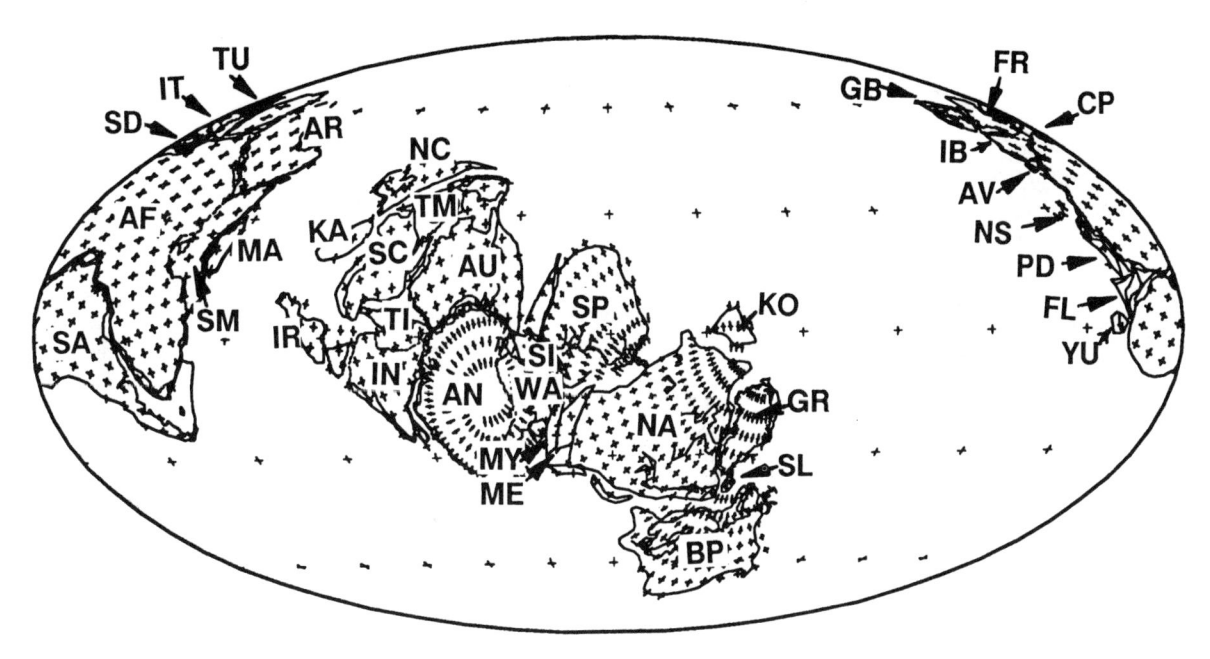

Figure 12.5. Tectonic plates used in these reconstructions. Two-letter abbreviations are as given in Table 12.1.

Cryogenian / Vendian Glaciations

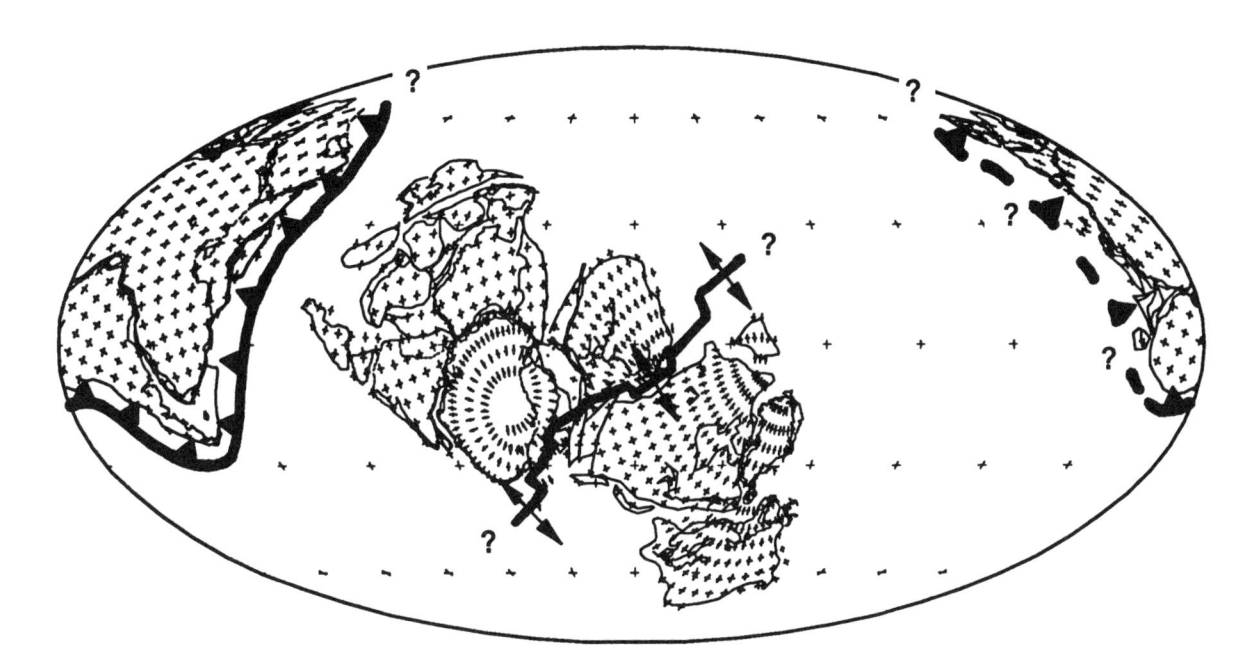

Figure 12.6. A cartoon for a possible basal Vendian glacial reconstruction and associated tectonic boundaries. In Figures 12.6 through 12.11, the possible locations of trenches are shown by the heavy black lines with tick marks on the upper plate; possible spreading centers are shown by heavy lines with divergent arrows; and possible locations for transform faults are given by heavy lines with opposing arrows. Note the extensive continental areas present in low latitudes implied by this reconstruction.

Mid-Vendian

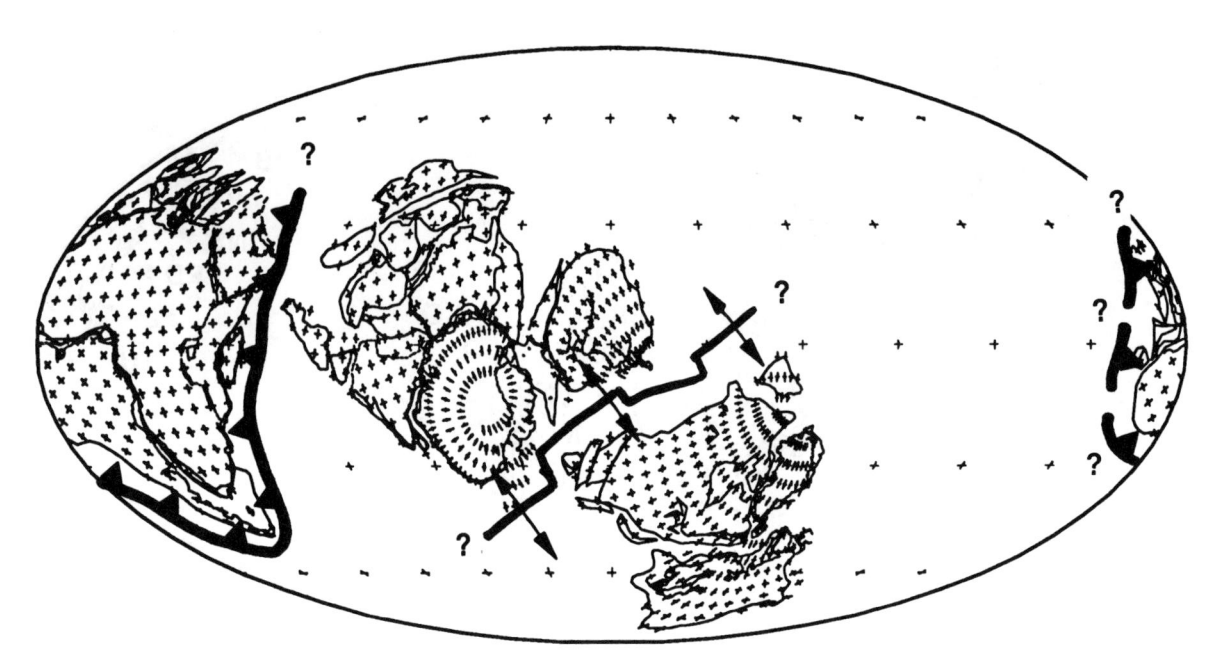

Figure 12.7. Mid-Vendian reconstruction and tectonic boundaries.

Precambrian / Cambrian Boundary

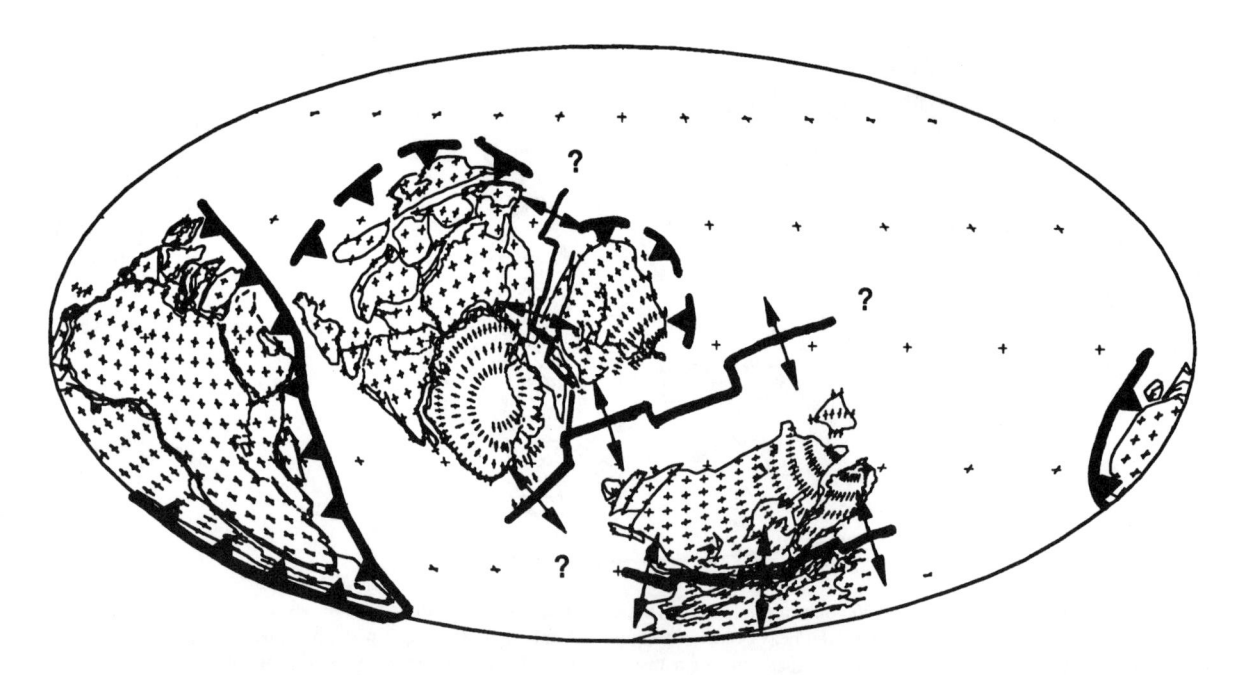

Figure 12.8. Precambrian-Cambrian boundary reconstruction and tectonic boundaries.

A Paleogeographic Model

Early / Middle Cambrian Boundary

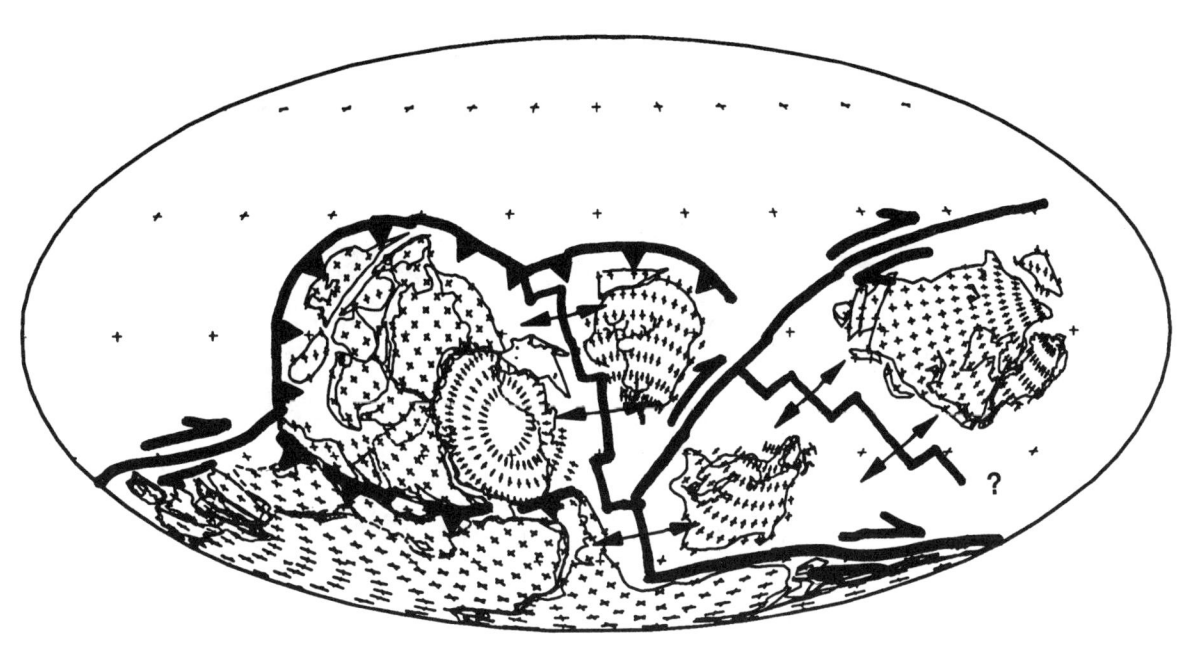

Figure 12.9. Early-Middle Cambrian boundary reconstruction and tectonic boundaries.

Middle / Upper Cambrian Boundary

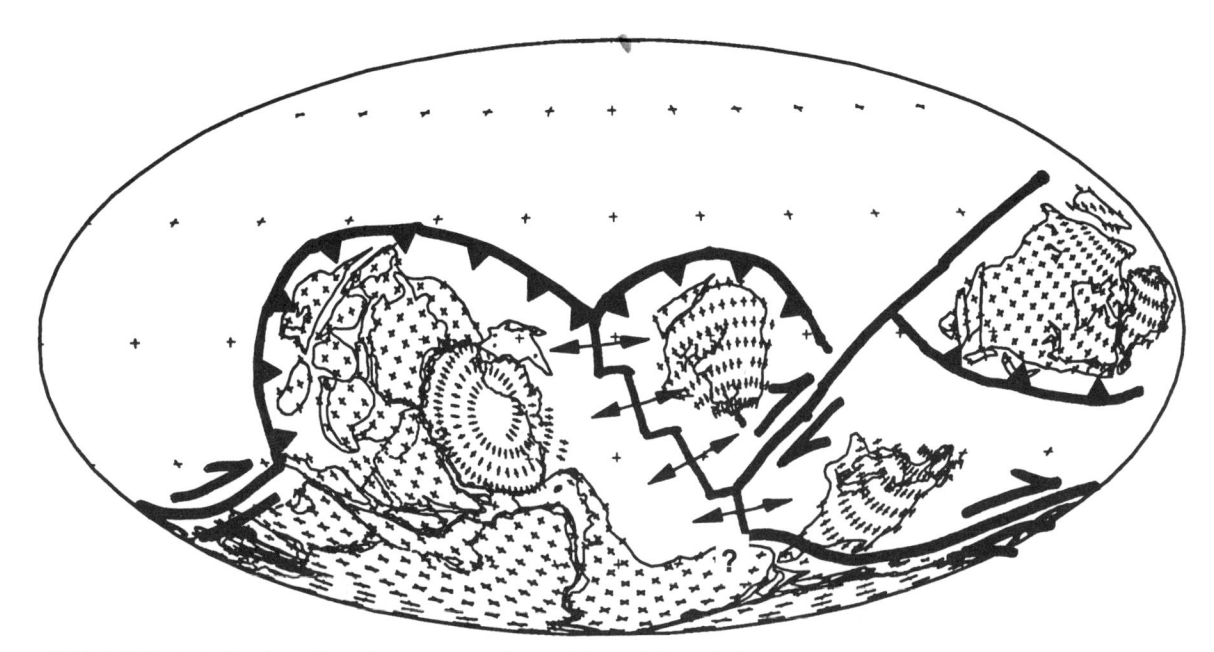

Figure 12.10. Middle-Late Cambrian boundary reconstruction and tectonic boundaries.

578 A Paleogeographic Model

Cambrian / Ordovician Boundary

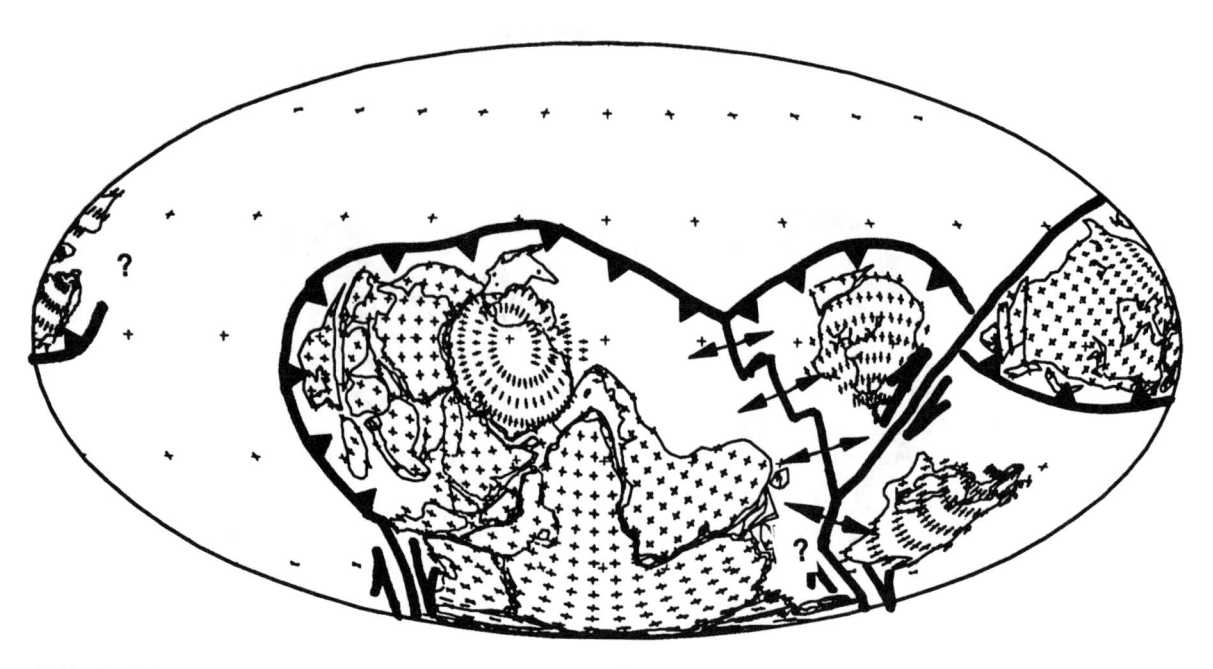

Figure 12.11. Cambrian-Ordovician boundary reconstruction and tectonic boundaries.

Table 12.1. *Absolute and relative Euler poles for the reconstructions for Vendian and Cambrian time. Euler poles give the right-handed (CCW) motion of the first plate relative to the second. AU = Australia; AN = E. Antarctica; WA = W. Antarctica; SI = South Island N.Z. + Campbell Plateau; IN = India; IR = Iran; TI = Tibet; TM = Tarim; NC = N. China; SC = S. China + Indochina + Burma; KA = Kazakhstan; AF = Africa; SA = S. America; MA = Madagascar; AR = Arabia; SM = Somalia; CP = Carpathians; PD = Piedmont Terrane; TU = Turkey; FL = Florida; NS = Nova Scotia; AV = Avalon Platform; IB = Iberia; FR = Central Europe; IT = Italy; SD = Sardinia; YU = Yukitan; GB = S. Great Britain; NA = North America; ME = Central America; MY = Northern Sonora Mexico; GR = Greenland; SL = Scotland; KO = Kolyma; BO = Bolivia; SJ = San Juan Argentina; BP = Baltic Platform; SP = Siberian Platform. We have obviously erred on the side of including too many microplates, and we specify Euler poles to two decimal places only to minimize cumulative errors for derived poles. Furthermore, the placement of many smaller terranes (such as the Avalon Platform) relative to larger continents, such as Africa and North America, is often poorly constrained. Euler poles given here for these reconstructions are not intended to be definitive; we present them here merely for completeness. Use them at your own risk! Relative longitudes are also unconstrained, except for time intervals in the Cambrian where an attempt is made to keep faunal provinces intact, but separated from each other*

Plate	Age	pLat.	pLong.	Angle	Reference
AU	€/O Bdry	43.74	−157.04	155.93	Klootwijk 1980
AU	M/L Camb	58.78	−168.90	154.18	Klootwijk 1980
AU	E/M Camb	63.51	−167.40	166.56	Klootwijk 1980
AU	P€/€ Bdry	67.40	157.67	179.99	Kirschvink 1978
AU	Cryo/Vend	67.40	157.67	179.99	Embleton and Williams 1986
AN-AU	Fit	−1.58	39.02	31.29	Lawver and Scotese 1987
WA-AN	Fit	62.27	21.84	13.27	Lawver and Scotese 1987
NI-AU	Fit	24.19	−19.91	44.61	Lawver and Scotese 1987
SI-WA	Fit	65.14	−52.00	62.38	Lawver and Scotese 1987
IN-AN	Fit	5.93	−158.43	94.53	Mod. from Morgan 1980
IR-IN	Fit	−31.37	−98.72	15.16	This paper
TI-AU	Fit	12.52	−149.61	64.20	This paper
TM-AU	Fit	21.42	111.17	164.81	This paper
NC-AU	Fit	22.27	120.53	179.33	This paper
SC-AU	Fit	5.76	111.02	162.82	This paper
KA-AU	Fit	11.46	−139.76	58.42	This paper
AF	€/O Bdry	26.71	−171.38	171.31	Derived from AU path
AF	M/L Camb	41.41	−179.24	177.57	Derived from AU path
AF	E/M Camb	−45.47	4.85	170.56	Kröner and McWilliams 1980
AF	P€/€ Bdry	−76.73	44.50	160.52	Kirschvink 1980
AF	Cryo/Vend	−75.95	−114.00	170.30	Kröner and McWilliams 1980

Table 12.1. (Continued)

Plate	Age	pLat.	pLong.	Angle	Reference
SA-AF	Fit	46.75	−32.65	56.40	Martin et al. 1981
MA-AF	Fit	0.17	−71.53	17.51	Mod. from Veevers et al. 1981
AR-AF	Fit	−36.58	−161.88	6.10	Morgan 1980
SM-AF	Fit	7.08	31.05	2.41	Morgan 1980
CP-AF	Fit	45.25	31.99	88.03	This paper
PD-AF	Fit	55.46	−30.29	96.68	This paper
TU-AF	Fit	−40.25	−153.11	89.47	This paper
FL-AF	Fit	43.36	−37.61	109.05	This paper
NS-AF	Fit	60.03	1.90	66.86	This paper
AV-AF	Fit	76.25	37.04	50.02	This paper
IB-AF	Fit	−37.35	176.23	38.33	This paper
FR-AF	Fit	50.78	118.08	5.74	This paper
IT-AF	Fit	−26.77	179.61	16.49	This paper
SD-AF	Fit	48.43	18.60	30.48	This paper
YU-AF	Fit	53.35	−35.38	81.03	This paper
GB-AF	Fit	50.78	118.08	5.74	This paper
NA	€/O Bdry	−56.52	−9.10	146.96	Watts et al. 1980a,b
NA	M/L Camb	−57.49	−16.60	159.55	Watts et al. 1980a,b
NA	E/M Camb	49.44	154.52	170.59	Elston and Bressler 1977
NA	P€/€ Bdry	33.37	101.44	152.28	Barr and Kirschvink 1983
NA	Cryo/Vend	34.92	90.07	121.28	Best guess
ME-NA	Fit	−41.71	88.68	55.85	Anderson and Schmidt 1983
MY-NA	Fit	−59.01	112.33	16.51	Anderson and Schmidt 1983
YA-NA	Fit	−53.23	102.65	15.00	Anderson and Schmidt 1983
GR-NA	Fit	−70.64	86.52	17.89	Morgan 1980
SL-NA	Fit	−43.83	−13.74	26.61	This paper
KO-NA	Fit	−43.50	37.78	32.77	This paper's guess
BO-NA	Fit	−16.06	136.34	90.60	This paper's guess
SJ-NA	Fit	−25.93	170.35	65.44	This paper's guess
BP-NA	Fit	−71.10	80.88	87.90	Est. from Gower and Owen 1984
SP-NA	Fit	64.70	−55.03	81.03	Est. from Sears and Price 1978
BP	€/O Bdry	25.74	−170.77	126.99	This paper's guess
BP	M/L Camb	18.69	176.17	114.77	This paper's guess
BP	E/M Camb	4.97	145.01	104.66	This paper's guess
BP	P€/€ Bdry	10.04	144.22	138.73	This paper's guess
BP	Cryo/Vend	0.78	131.32	117.07	Derived from BP-NA Fit
SP	€/O Bdry	30.37	96.08	169.66	ca. Khramov 1982
SP	M/L Camb	26.65	67.29	155.81	ca. Khramov 1982
SP	E/M Camb	17.31	66.19	130.82	ca. Khramov 1982
SP	P€/€ Bdry	37.32	40.02	153.73	Kirschvink and Rozanov 1984
SP	Cryo/Vend	37.32	40.02	153.73	Kirschvink and Rozanov 1984

base maps here presented incorporates features from several of these reconstructions. Most important are the Powell et al. arguments that the Falkland Plateau and the Mozambique Ridge are both shallow areas of thin continental crust; hence, the basement rocks of Antarctica are separated by several hundred km from those of the African mainland (unlike the Smith and Hallam reconstruction). Next, we use the Powell et al. placement of Madagascar against the *northern* margin of the west Indian shelf; this placement increases the area within the northern margin of Gondwana for use in reconstructing many of the Asian displaced terranes (discussed below). Third, several of the reconstructions assume that the island of Ceylon (Sri Lanka) has moved relative to the tip of India, yet we can find little geologic or paleomagnetic evidence to support this. W. J. Morgan's (1981) fit of India to Antarctica avoids this problem and is compatible with the fracture zone and magnetic anomaly patterns in the Indian Ocean (Powell et al. 1988).

The reconstructions (Figures 12.5 through 12.11) therefore use the basic Morgan (1981) fit of India to Antarctica, and the Powell et al. (1980) placement of Madagascar to India, to link the eastern and western halves of Gondwana, subsequent to their Early or Middle Cambrian suturing at the end of the Pan-African orogenic event. We have made minor adjustments in the fit of Madagascar against Africa, and India against Antarctica, so as to fit the Falkland Platform against the shelf of Antarctica (Euler poles for this are given in Table 12.1).

A third major chunk of this continent, here called "Northern Gondwana," presumably includes the scattered bits and pieces of Asia, such as Indochina, Burma, the Yangtzee Platform of South China, North China, Tarim, and slivers of Tibet. Although paleontological and lithological similarities in the early Paleozoic generally suggest that these fragments fit together somehow along the lithologically "broken" margins of western Australia, India, and northeastern Africa, the relative placement of the pieces has been hampered by inadequate paleomagnetic data and incomplete lithologic and biogeographic analyses. Data from both paleontological and paleomagnetic studies now seem to resolve the bulk of this problem. First, Burrett and Stait (1986) have recognized close similarities between the Cambrian and Ordovician faunas of Burma and the northwest margin of Australia, and argue that Burma should be positioned adjacent to the Canning Basin of Australia. They also note that the South China (Yangtzee) Platform also has similarities with Tibet; and, more recently, Apollonov et al. (1988; and pers. comm. 1988) notes a similar correspondence between Kazakhstan and South China. Several paleomagnetic attempts have also been made during the past ten years to resolve this question by constructing separate APW paths for each of these blocks (e.g., Lin and Fuller 1985; McElhinny et al. 1981), but large gaps in the mid-Paleozoic APW path still leave the *hemisphere* for the fragments unconstrained. However, this problem has been solved recently by comparing the magnetic polarity pattern across a small time interval around the Cambrian-Ordovician boundary on four continents (Ripperdan and Kirschvink 1989a, b; Ripperdan et al. in review). Results from North America reveal a distinctive normal polarity magnetochron during the *Cordyolotus proavis* conodont zone. A similar pattern with respect to the conodont zones is also present in North China, Kazakhstan, and South China, and hence the comparison uniquely removes the polarity ambiguity for the APW paths. The Cambrian-Ordovician paleolatitude and biogeographic constraints for these three blocks position them with the relative orientations shown on the reconstructions and in Table 12.1, and produce a remarkable convergence of the Early Cambrian pole positions between North China (Lin et al. 1985), South China (Wu et al. 1989), and Australia (Kirschvink 1978b).

A group of other minor continental fragments with Pan-African affinities also needs to be repositioned along the northwestern margin of Africa. These include terranes of the Appalachian margin of North America including Florida, the Piedmont, Nova Scotia, and the Avalon zone of Newfoundland (Opdyke et al. 1987; Venkatakrishnan and Culver 1988; O'Brien et al. 1983). We place the American Plate (including southern Great Britain, south-central Europe, and Iberia) adjacent to the northern African margin following the analyses of Perroud et al. (1984), Van der Voo (1988), and Scotese (1984).

12.1.4 Configuration and Breakup of Late Proterozoic Super(?) Continents

12.1.4A The Vendian World

Figure 12.6 shows our best guess at the early Vendian configuration of the Late Proterozoic continents, based partially on the paleomagnetic and lithologic constraints outlined above. To summarize, this reconstruction differs in several major respects from previously published reconstructions for the Late Proterozoic (e.g., Morel and Irving 1978; Piper 1986; Donovan 1987; and Bond et al. 1984). First, East and West Gondwana were probably separate entities until sometime in the Early or Middle Cambrian, based largely on the decreasing discrepancy in their relative orientations inferred from their basal Vendian glacial and basal Cambrian pole positions. This interpretation is also supported by the long APW path of Africa during this interval, as well as by the formation of extensive areas of new continental crust in Africa, particularly along the eastern margin (Bertrand and Caby 1978; Kröner 1980; Dixon and Golombek 1988; Venkatakrishnar and Culver 1988). Second, the Australasian fragments of Gondwana are reassembled on the western margin of Australia, greatly reducing the number of separately wandering fragments in the puzzle. Third, we fit Siberia adjacent to Australia and the basement of the Campbell Plateau, oriented according to the Tommotian pole from the Lena River. We assume that it, like Australia, had no motion during Vendian time, and use this fit principally to maintain its proximity to eastern Australia and Antarctica in the Early Cambrian (Palmer and Gatehouse 1972). Fourth, we fit the Baltic Platform to the Appalachian margin of North America following the analysis of Gower and Owen (1984) discussed above. The breakup between the Baltic Platform and Laurentia probably occurred near the end of the Vendian or in the Early Cambrian, based on an upper age constraint of 554 Ma for the rift-related Tibbit Hill Volcanics in Quebec and Vermont (Kumarapeli et al. 1989). Finally, the position of North America (Laurentia) is more problematic. For lack of better constraints, we fit the Cordilleran margin of North America adjacent to the northeast margin of Siberia, as was suggested by Sears and Price (1978) and used extensively in the reconstructions of Piper (1983, 1986). This is reasonably consistent with the lithologic and paleomagnetic constraints, although the timing of the rifting on the Cordilleran margin is not well constrained. In the tectonic model, however, it is necessary for Laurentia to move in a southerly direction during the late Vendian to accomodate the Caborca (and perhaps the Sept-Iles) paleomagnetic results. It is possible that this motion was initiated by a Late Proterozoic rifting event along the Cordilleran margin and some other craton (e.g., Armin and Mayer 1983; Bond et al. 1984), and from our previous analysis, Siberia, almost by default, happens to be in the correct orientation for this missing rift partner.

12.1.4B Cambrian Events

It appears that two major plate motions began approximately at the Precambrian-Cambrian boundary and continued until Middle Cambrian time. Paleomagnetic and lithologic evidence suggest that Laurentia reversed its southerly motion and returned to the equator by late Early Cambrian time (based on the Tapeats Sandstone pole of Elston and Bressler 1977), whereas Baltica does not reenter the carbonate belt until Early Ordovician time; this implies a Laurentia-Baltic rifting event involving the Appalachian margin. The consequent formation of the Iapetus Ocean may thus have been responsible for initiating the Cambrian sea-level transgression. Second, sometime in the Early Cambrian, Australia rotated nearly 45°CCW around an Euler pole located near Tasmania. We interpret this as happening more or less contemporaneously with the suturing event between the fragments of East and West Gondwana, resulting in the movement of northwest Africa into higher southerly latitudes and out of the carbonate belt (signaled by the transition from the Calcaire Supérieur to the

Série Schisto-Calcaire in the Anti-Atlas mountains of Morocco). Courjault-Radé (1987, 1988) also infers similar motion during Cambrian time for southern Europe, which was probably part of the northwest African promontory. The abrupt cessation of subduction may also be responsible for the extensive shallow epicontinental platforms along the eastern margin of Africa, such as the Faulkland Plateau.

During the Middle Cambrian and first half of the Late Cambrian, the CCW rotation of Australia was temporarily reduced (Klootwijk 1980). This time probably marks the onset of an east-dipping subduction zone and an island-arc along the Australasian margin of Gondwana (including Kazakhstan and perhaps North China). Presumably, the Siberian, Laurentian, and Baltic Plates would have been located due west of this convergent margin, and subduction may have helped all three plates to move eastward. Although this model implies left-lateral oblique subduction between Baltica and Laurentia in the early Paleozoic, there is no good constraint on the amount of separation between them, other than that the faunal provinces are distinct. It is not clear whether they could be as close as suggested by Mason (1988), however, because the deep-water trilobites of Siberia are similar to the shallow Baltic forms, yet both are distinct from those of North America (Jell 1974; Shergold 1988). Unfortunately, the APW paths for all three of these continents are constrained too poorly to resolve even the latitudinal issue at this time (e.g., Van der Voo 1988).

The interval from the mid-Late Cambrian to the Cambrian-Ordovician boundary contains another 45°CCW rotation of Gondwana in what may be a rather short period of time. Interestingly, the APW path for Laurentia is also consistent with roughly 45° of rotation during essentially the same time period, but in a CW sense rather than CCW (Watts et al. 1980a, b). Because the Gondwanan and Laurentian faunal provinces are quite distinct, the geography is such in this reconstruction that the two Euler poles for these rotations are nearly antipolar, and there is virtually no *relative* motion between these continents. Hence, *true polar wander*, which is a displacement between the earth's mantle and core resulting from the realignment of the principal moment of inertial axes of the earth, is a possible explanation for the synchrony and apparent rapidity of this shift. Thermal warming and absorption of a subducting slab stranded by the suture of East and West Gondwana, along with initiation of new subduction zones along the eastern margin of Australia and Antarctica, could provide such a switch in the earth's moment of inertia tensor.

Acknowlegments

Supported by the PPRG and NSF grants EAR-8721391 and PYI-8351370, and contributions from the Chevron Oil Field Research Company and the Arco Foundation. Contribution no. 4807 of the Division of Geological and Planetary Sciences of the California Institute of Technology.

13

Evolution of the Proterozoic Biosphere: Benchmarks, Tempo, and Mode

J. WILLIAM SCHOPF

13.1 Prologue

As originally envisioned, this concluding Chapter of Part 1 this monographic work was intended to be a multi-authored, collaborative effort prepared by participants in the Precambrian Paleobiology Research Group (PPRG) that would present a consensus view of the timing and nature of major events in the history of Proterozoic life. However, because of time constraints – at the time of the writing of this Chapter, it has already been more than two years since the end of the data-gathering stage of the project – this initial plan had to be abandoned. Moreover, even without such constraints, uncertainities in the available data are such that competing and, in some cases, mutually inconsistent interpretations can be reasonably entertained regarding a number of major aspects of Proterozoic history; it is probably too early in the development of the field for unanimity of opinion to be attained. This, of course, is not necessarily deleterious: the field is young, the data are limited, and important questions remain unanswered. At this stage in the development of the science, dogmatic adherence to one or another validly competing interpretation might well be a disservice, leading the field into a cul-de-sac by inhibiting fruitful lines of inquiry.

What follows, therefore, is an attempt to present a balanced discussion of the problems at hand. Embedded within any such discussion, however, are certain to be the assumptions, preferences, and biases of the author – for the science to progress, judgments must be made, commonly on the basis of limited evidence – and it is likely that not all will agree with the interpretations preferred here. The following synthesis is thus most appropriately regarded as a "best guess assessment" of the available relevant evidence. In large measure, it is based on data and discussions presented in foregoing Chapters of this work; it should be emphasized, however, that it is a *personal* assessment, not necessarily a PPRG consensus view.

13.2

Times of Origin and Earliest Evidence of Major Biologic Groups

The Archean and Proterozoic temporal distributions of major biologic groups inferred here are summarized in Figure 13.1. As is there indicated, and as is discussed below (for Archean distributions, an updated version of a similar discussion included in the previous PPRG collaborative volume; Schopf 1983a), varying degrees of uncertainty exist regarding the earliest dates of each of these inferred distributions.

13.2.1 Prokaryotes

13.2.1A Anaerobic bacteria

In addition to anaerobic photoautotrophs, discussed below (Section 13.2.1B), two broad categories of anaerobic bacteria need to be considered here: anaerobic chemoheterotrophs and anaerobic chemoautotrophs. Members of the former category, such as fermenting bacteria, obtain their cellular carbon, reducing power, and energy from assimilation and catabolism of exogenous organic matter. Although such microbes seem certain to have been extant since the Early Archean, direct evidence of their existence is sparse. Indirect lines of evidence include (*i*) the occurrence of "biogenic" sulfur-isotopic ratios, evidently indicating the presence of sulfate-reducing bacteria, earliest known with assurance from sediments 2.3 to 2.2 Ga old; (*ii*) the presence of fenestrate structures (originally gas-filled pockets) in stromatolites of the Late Archean (~2.8 Ga-old) Fortescue Group of Western Australia that may have been produced via microbial anaerobic decomposition of organic matter (Walter 1983); (*iii*) arguments based on the total organic carbon content of Archean sediments and on assumptions regarding the biological recycling of organic matter that suggest it is implausible for anaerobic chemoheterotrophs to have been absent from the earliest ecosystems (Schopf et al. 1983); and (*iv*) numerous biochemical and theoretical considerations apparently indicating the primitive nature of anaerobic chemoheterotrophic metabolism (e.g., Gest and Schopf 1983).

Similarly, and despite their presumed early evolution, direct evidence of anaerobic chemoautotrophs (microorganisms, such as methanogenic bacteria, that are capable of using CO_2 fixed by light-independent metabolic processes, as their immediate and major source of cellular carbon) is virtually lacking from the early fossil record. Relevant indirect evidence includes (*i*) anomalously light (^{13}C-poor) kerogens reported from several Late Archean and Early Proterozoic (~2.9 to 2.1 Ga-old) basins that have been plausibly interpreted as derived via fixation of methane produced by methanogenic bacteria (Hayes 1983); and (*ii*) studies of molecular phylogeny indicating that some extant anaerobic chemoautotrophs (e.g., archaebacterial methanogens) are members of very early-evolving lineages (Figs. 9.3.1, 9.3.2).

As outlined below, the known stromatolitic, microbial, and carbon-isotopic fossil records seem to establish firmly the existence of *photo*autotrophic prokaryotes (viz., photosynthetic bacteria and, possibly, O_2-producing cyanobacteria) as early as 3.5 to 3.4 Ga ago. Because such photoautotrophs presumably are evolutionary derivatives of anaerobic, non-photosynthetic prokaryotes, anaerobic bacteria are inferred to have been extant at least as early as ~3.5 Ga ago. Indeed, the occurrence of reduced carbon in the highly metamorphosed, ~3.8 Ga-old Isua Supracrustal Group of southwestern Greenland (Schidlowski et al. 1983) – if not a product of abiotic processes – may reflect the presence of such microbes even earlier in geologic time.

13.2.1B Photosynthetic bacteria

Included among this widespread biologic group are those prokaryotes capable of anoxygenic, but not oxygen-producing, photoautotrophy – microbes that during photosynthetic growth can use H_2, H_2S, or various organic substrates (but not H_2O) as a source of electrons for the light-driven reduction of CO_2 to produce cellular organic matter and yield cellularly usable chemical energy. As discussed elsewhere in this volume (Sections 6.5.2D, 6.9.2), such photosynthetic bacteria are important, indeed essentially ubiquitous, components of modern microbial mat-building communities. The occurrence of Archean stromatolites (Fig. 1.5.1), known earliest from sediments of the ~3.5 Ga-old Warrawoona Group of Western Australia, is thus strong evidence for the presence of photosynthetic bacteria (and is equally consistent with the existence of their presumed evolutionary derivatives, O_2-producing cyano-

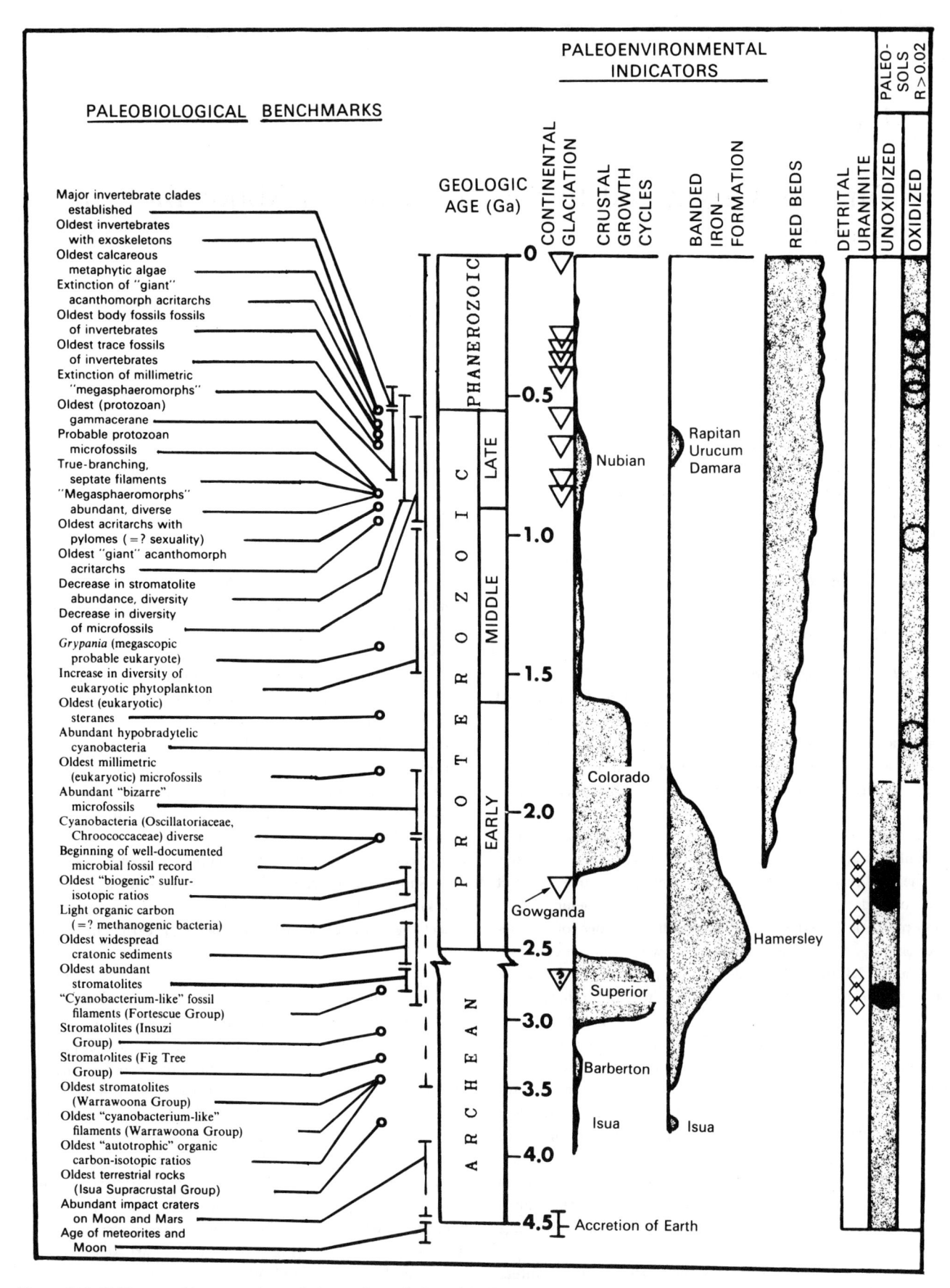

Figure 13.1. Evidence and interpretations relating to the evolution of the Proterozoic biosphere: paleobiological benchmarks; temporal distribution of paleoenvironmental indicators; inferred ranges of atmospheric CO_2, O_2, and organic carbon-isotopic ratios; and the inferred temporal distributions of major biologic groups.

Times of Origin and Earliest Evidence of Biologic Groups

INFERRED RANGES OF ATMOSPHERIC CO_2, O_2, AND ORGANIC CARBON–ISOTOPIC RATIOS

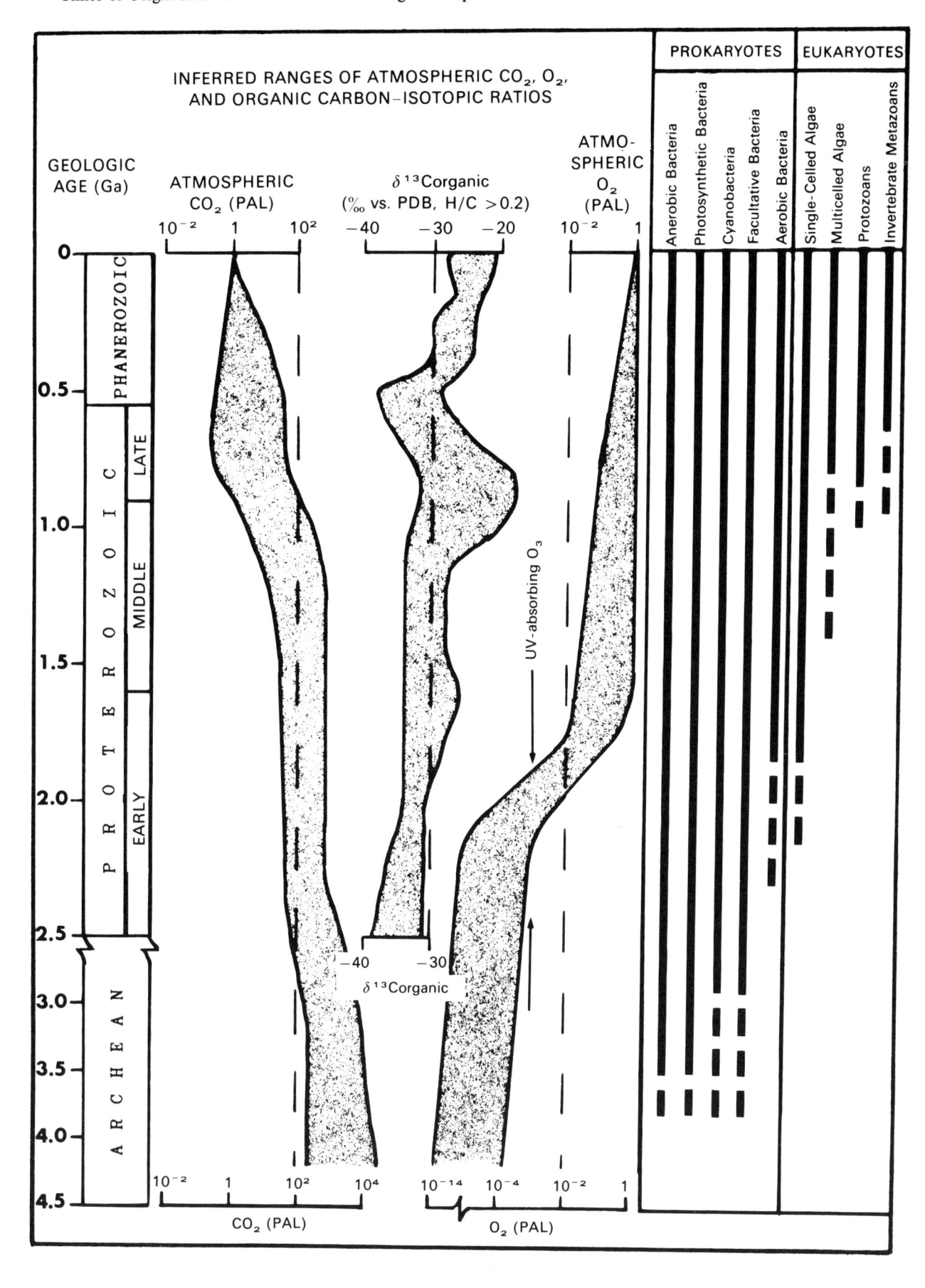

bacteria). Similarly, at least the narrower filamentous microfossils preserved in some Early Archean stromatolite-like laminae (e.g., those in cherty sediments of the ~ 3.4 Ga-old Onverwacht Group of South Africa; Fig. 1.5.2) are morphologically similar to extant photosynthetic bacteria (viz., *Chloroflexus*). And carbon-isotopic ratios in kerogens from both Warrawoona and Onverwacht Group sediments (Table 17.4) are consistent with photoautotrophic carbon fixation (e.g., by photosynthetic bacteria and/or cyanobacteria).

Thus, the conclusion seems clear: anoxygenic photoautotrophic bacteria had become established at least as early as ~ 3.5 Ga ago. Moreover, as suggested by the occurrence of graphitic carbonaceous matter in ~ 3.8 Ga-old sediments of the Isua Supracrustal Group, such organisms may have been extant even earlier (although the extreme metamorphism to which the Isua sediments have been subjected precludes clear-cut establishment of the presence of such microbes on the basis of either carbon isotopic or micropaleontological evidence).

13.2.1C Cyanobacteria

Together with prochlorophytes (an extant group of oxygenic photoautotrophs known from only two genera), cyanobacteria are chlorophyll *a*-containing prokaryotes that are capable of using light as their energy source, CO_2 as their immediate (and major) carbon source, and H_2O as the source of electrons necessary for the reduction of CO_2 to form cellular organic matter. Although some cyanobacteria are capable of temporarily performing anoxygenic photosynthesis (Section 6.5.1), the capability to perform oxygenic photosynthesis is a universal characteristic of the group; all cyanobacteria are thus oxygen-producing, oxygen-tolerant microorganisms.

As discussed elsewhere in this volume (Section 5.4), the relatively continuous fossil records of several cyanobacterial families extend well into the Early Proterozoic; oscillatoriaceans, chroococcaceans and, possibly, entophysalidaceans were evidently well established, exhibiting considerable morphological and taxonomic diversity, at least as early as 2.1 Ga ago. The earlier presence of these aerobic photoautotrophs is very likely: (*i*) sedimentological evidence – in particular, the widespread occurrence of major banded iron-formations prior to 2.1 Ga ago, requiring a sizable influx of (almost certainly biologically generated) free oxygen, and data from paleosols, implying the presence of atmospheric molecular oxygen 3.0 Ga ago (Section 4.4.2) – seems highly suggestive of their earlier existence, as does (*ii*) the well-documented Early Proterozoic morphological and taxonomic diversity of cyanobacteria; (*iii*) the presence of abundant, morphologically diverse stromatolites ~ 2.8 Ga ago; and (*iv*) the occurrence of "cyanobacterium-like" (viz., Oscillatoriaceae-like) filamentous microfossils in the ~ 2.4 Ga-old Gamohaan Formation of South Africa (Klein et al. 1987), the ~ 2.8 Ga-old Tumbiana Formation of Western Australia (Schopf and Walter 1983), and cherts of the ~ 3.4 Ga-old Apex Basalt of Western Australia (Figs. 1.5.4 through 1.5.6). Indeed (as is addressed at some length in Section 1.5), the paleoenvironmental, stromatolitic, microfossil, and carbon-isotopic data now available from units 3.5 to 3.3 Ga in age (as well as the occurrence of banded iron-formations and reduced carbon in these and sediments as old as

~ 3.8 Ga) are all consistent with the Early Archean existence of oxygen-producing cyanobacteria. As shown in Figure 13.1, cyanobacteria are inferred to have been extant during the Late Archean with their possible presence extending well into the Early Archean.

13.2.1D Facultative Bacteria

Included in this large, metabolically defined group are those prokaryotes that can use molecular oxygen as the terminal electron acceptor for intracellular energy conversion, thus carrying out aerobic respiration, but that in the absence of sufficient oxygen concentrations can survive and grow using alternative, anaerobic, metabolic processes. Because there are no known paleontological or geochemical criteria by which to distinguish unambiguously between anaerobes, aerobes, and the facultative microorganisms included in this group, direct evidence of facultative bacteria has yet to be recognized in the fossil record. However, molecular biological data suggest that such organisms evolved relatively early (Figs. 9.3.1 through 9.3.3). Moreover, they are ubiquitous components of extant oxygenic (Section 6.6.2) and intermittently oxygenic/anoxygenic microbial mats (Section 6.9.3), occupying the transitional, spatially and temporally fluctuating zone between aerobic and anaerobic conditions that occurs beneath the sediment-water interface within the uppermost few millimeters to centimeters of the sediment profile. To the extent that such varying conditions were globally widespread during the Archean and Early Proterozoic, and assuming that the distribution of such facultative microbes was not severely limited by other environmental factors (e.g., lack of nutrients or a high UV-flux), it seems reasonable to suppose that facultative prokaryotes would have diversified to occupy the available ecospace.

For reasons discussed elsewhere in this volume (Sections 1.4, 4.3, 4.6, 4.8), the advent of oxygen-producing photosynthesizers, even if such organisms were successful to the point of becoming the dominant components of the photic zone ecosystem, would not have resulted in an immediate transition from anoxic to oxygenic global environmental conditions. Rather, the molecular oxygen thus liberated would have been consumed, and ultimately sequestered in the rock record, by reaction with previously unoxidized substrates such as dissolved ferrous iron. Localized, oxygen-rich "oases" (Fischer 1965) may thus have been maintained, in and around sites of oxygen-production (e.g., cyanobacterium-dominated microbial mats), within an otherwise anoxic ocean system. As discussed in Section 4.6.2, it is even conceivable that oxidizing conditions, with dissolved O_2 concentrations approaching 10% of the current value, could have been maintained within large bodies of water while the atmosphere remained almost totally anoxic.

Global concentrations of atmospheric oxygen would have remained at low levels until the net rate of oxygen-production (in the presence of organisms capable of aerobic respiration, a rate stoichiometrically coupled to the rate of burial of organic carbon derived from O_2-producing photosynthesizers) exceeded the supply of dissolved reduced substrates from the deep ocean and of reduced gases evading to the atmosphere. But the rate of supply of all of these reactants no doubt varied temporally: photosynthetic O_2-production would have been locally diel as well as varying seasonally; the influx of reduced substrates from the deep ocean would have varied with seasonal upwelling and with episodic crustal movements and submarine volcanism; and the supply of reduced gases to the atmosphere would have been a function chiefly of temporally varying volcanic activity. Thus, local, and global, oxygen concentrations must have been temporally erratic. Under these conditions, the capability to use oxygen metabolically when it was available in sufficient concentrations, but to revert to less efficient anaerobic metabolism in the absence of such oxygen – i.e., to carry out facultative respiration – would have been of substantial selective advantage. Despite the absence of definitive fossil evidence, it therefore seems plausible to infer that facultative prokaryotes were probably abundant, widespread, and diverse throughout the period of transition from an essentially anoxic to an oxygenic environment (and that thereafter, they remained successful in local fluctuating settings). Moreover, it is possible, even probable, that such microorganisms played a major role in the evolving Archean to Early Proterozoic biosphere. Facultative bacteria are thus inferred here to have had a temporal distribution comparable to that of cyanobacteria.

13.2.1E Aerobic Bacteria

With the exception of what are anthropocentrically regarded as "marginal," "harsh," or "inhospitable" environments (e.g., anaerobic muds, metazoan entrails, sulfurous hot springs, and locales associated with deep-sea fumaroles), obligate aerobes – organisms wholly dependent on the use of molecular oxygen as a terminal electron acceptor for energy conversion – are abundant, widespread, indeed dominant components of virtually all extant ecosystems. However, prior to establishment of a stable, global, oxygenic environment, it is unlikely that this could have been true. The fundamental metabolic and biochemical capabilities exhibited by such oxygen-requiring microorganisms may well have evolved decidedly earlier (viz., within their presumed, facultative, evolutionary precursors), but it is difficult to envision how such obligately aerobic, and thus environmentally restricted, microbes could have become dispersed and been sustained on a global scale (e.g., as phytoplankton) were the oceanic photic zone to have been anaerobic, even intermittently and for short periods of time.

As is discussed elsewhere in this volume (Section 1.4, Chapter 4, and Sections 5.4, 5.5; see also Walker et al. 1983), analyses of (*i*) the geochemistry of paleosols, detrital uraninites, and banded iron-formations, of (*ii*) the temporal distributions of these sediment types as well as those of red beds and eukaryote-sized microfossils, and of (*iii*) theoretical models of the evolution of Proterozoic atmospheric chemistry, indicate that the transition to a stable oxygenic global environment, beginning to become evident during the early portion of the Early Proterozoic (~ 2.3 to 2.1 Ga ago), was completed by about 1.85 Ga ago. During this period, global O_2 levels appear to have increased by more than an order of magnitude, from $\leqslant 10^{-3}$ PAL (times the present atmospheric level) to values $\geqslant 10^{-2}$ PAL, never again to decrease to pre-aerobic levels (Fig. 13.1). As discussed below, the earliest firm records of millimetric (evidently eukaryotic and obligately aerobic) phytoplankton, ~ 1.85 Ga in age, and the even earlier reported occurrences (extending back to 2.1 to 2.2 Ga ago) of spheroidal planktonic microfossils in the 60 to 200 μm size range (larger than all extant prokaryotes, and thus similarly inferred here to be both eukaryotic and oxygen-requiring), are consistent with this interpretation. The microfossil-based aspects of this inference need to be tempered by acknowledgment of the existence of evidently primitive, mitochondrion-lacking (and therefore solely fermentative), modern non-photosynthetic protists. Nevertheless, the vast majority of modern eukaryotes, including all known eukaryotic phytoplankters, are oxygen-requiring in some portion of their life cycles or seem clearly to be evolutionary derivatives of obligate aerobes. The occurrence of microfossils interpretable as eukaryotic phytoplankton is thus regarded here as strong evidence of the presence of obligate aerobiosis; and obligately aerobic bacteria, which seem certain to have evolved either earlier than, or coincident with, the earliest "complete" eukaryote cells (see below), are thus inferred to have been extant at least as early as ~ 1.85 Ga ago.

13.2.2 Eukaryotes

13.2.2A Single-celled algae

Problems inherent in identifying early-evolving, single-celled eukaryotes in the fossil record are discussed in Sections 5.4 and 5.5. As is there indicated, analysis of the potentially preservable (and actually preserved) characters of modern prokaryotes and micro-algae suggest that cell size is the single, most useful, widely applicable criterion. Clearly, however, the occurrence of additional, more complex and relatively advanced characters would provide a firmer basis for identification. For example, eukaryotic affinities seem persuasively established for relatively large-celled filaments that exhibit true branching (a combination of typically thallophytic characters rarely exhibited by prokaryotes and known earliest from sediments ~ 850 Ma old) and by the occurrence of morphologically complex features unknown among prokaryotes (e.g., polygonal collars and pylomes, such as those exhibited by some Late Proterozoic vase-shaped fossils that are well represented in units ~ 850 Ma in age; or the distinctive, surficial, spiny processes characteristic of acanthomorph acritarchs and known earliest from ~ 950 Ma-old geologic units).

Similarly, because millimetric or larger size is exhibited among prokaryotes only by many-celled colonies, megascopic size is a generally useful (but not absolute) criterion for inference of eukaryotic affinity. For this reason, curved or coiled ribbon-like carbonaceous strands attributed to the genus *Grypania*, of well-established occurrence in ~ 1400 Ma-old sediments, are reasonably regarded as eukaryotes (although the nature of the cells composing such strands has not been established, and it is thus possible that they might represent filamentous colonies of *Nostoc*-like cyanobacteria). Other relatively large-sized "eukaryote-like" carbonaceous bodies are also known from units of this age or somewhat older, and the occurrence in sediments as old as ~ 1650 Ma of sterane hydrocarbons, hydrogenated geochemical derivatives of sterols and thus plausibly inferred to be of eukaryotic origin, suggests that "biochemically modern" eukaryotes were already extant during the Early Proterozoic.

There can be little doubt, however, that the earliest "complete eukaryotes" (viz., mitochondrion-, chloroplast-, and nucleus-containing walled cells capable of mitosis) were neither morphologically complex nor megascopic in size. Indeed, they almost certainly were simple, probably spheroidal, mitotic haploid unicells, perhaps morphologically similar to modern, small-celled, rhodophycean or chlorophycean micro-algae and thus essentially indistinguishable in cell size and shape from coexisting coccoid cyanobacteria. If so, and in the absence of other evidence by which to distinguish between fossils of prokaryotes and early-evolving eukaryotes (e.g., possibly the ultrastructure or the elemental composition of preserved cell walls), the actual time of origin of the eukaryotic cell may remain shrouded in mystery.

Thus, at present, the available fossil record can be expected to provide evidence only of the existence of "complete" eukaryotic *cells* (rather than the eukaryotic *genome*, the origin of which evidently occurred in anaerobic heterotrophs and predated, perhaps substantially, the endosymbiotic assembly of a "complete eukaryote"), and only after such eukaryotes had evolved sufficiently to become morphologically distinguishable from coexisting prokaryotes. For the reasons discussed in Section 5.5.2A, single-celled phytoplankters (sphaeromorph acritarchs) having a cell size larger than all extant prokaryotic unicells (viz., "mesosphaeromorphs," 60 to 200 μm in maximum diameter; and "megasphaeromorphs," 200 μm to more than 7.5 mm in maximum diameter), are in this sense regarded here as assuredly eukaryotic. Although a few scattered reports exist of "mesosphaeromorphs" from units as old as ~2250 Ma, the oldest well-documented occurrences of "meso-" and "megasphaeromorphs" are from ~1850 Ma-old sediments, the age inferred for the oldest clear-cut evidence of single-celled eukaryotic algae.

13.2.2B Multicelled algae

Grypania and similar taxa of relatively early-appearing Proterozoic ribbon-like carbonaceous films (Fig. 7.3.9) may well be remnants of multicellular thallophytes. As noted above, however, the nature of the cells originally making up such structures has yet to be demonstrated, and the possibility that they represent preserved colonial prokaryotes therefore cannot be excluded. Moreover, despite an increased search for such "possible megascopic algae" in recent years, reported occurrences remain few and temporally widely scattered, and strand-like remnants of aggregated, compressed sapropel can be easily confused with film-like fossils of this type. However, demonstrably cellular megascopic algae are well known from lowest Cambrian strata, at least two credible reports of cellular, Late Proterozoic, noncalcified megascopic (evidently rhodophycean) algae have recently appeared (e.g., Butterfield et al. 1990), and *Tawuia*, known from sediments ~800 Ma in age, seems convincingly thallophytic. Thus, multicelled algae are inferred here to have been extant at least as early as 800 Ma ago, with their possible occurrence extending back to ~1400 Ma.

13.2.2C Protozoans

As documented in Section 5.7, unquestionable skeletonized protozoans – those with agglutinated, calcareous, and siliceous tests or skeletal elements – are known earliest from Lower Cambrian strata. It has long been supposed, however, that early-evolving protozoans lacked mineralic tests, and so the absence of a known pre-Cambrian protozoan fossil record has been commonly assumed to reflect the "unpreservability" of unmineralized forms. Such an assumption may be unwarranted: the organic (dominantly peptidoglycan-containing) cell walls of prokaryotic bacteria and cyanobacteria, the originally polysaccharide or glycoprotein extracellular sheaths of such prokaryotes, as well as the (polysaccharide- or "sporopollenin"-containing) cell walls of eukaryotic acritarchs and megascopic thallophytes are demonstrably preservable. If so, why should this not also be true of the physically resilient organic ("proteinaceous") envelopes or pellicles that enclose protozoan cells, especially if such pellicles are relatively thick-walled as in cysts or "resting cells," and especially if they are preserved rapidly and by permineralization as in microfossiliferous Proterozoic cherts?

Significantly, new evidence suggests that exactly such preservation of unmineralized, Proterozoic, protozoan cysts may in fact occur. In particular, species of the heretofore enigmatic vase-shaped Proterozoic protist *Melanocyrillium* (Section 5.6.2B), well known from ~850 Ma-old sediments of the Chuar Group of northern Arizona, USA, appear to be indistinguishable in numerous salient characteristics (vesicle size, shape, surficial texture, and wall thickness, and apertural and opercular morphology) from the unmineralized organic-walled cysts of living species of several genera (e.g., *Cucurbitella, Nebela, Pontigulasia*) of testate amoebae (Sarcodina, Rhizopodea). Although as yet unpublished (J. W. Schopf and C. G. Ogden, in prep.), the evidence seems convincing. Moreover, this interpretation is supported by the occurrence in this same sequence of sediments of gammacerane, a geochemical derivative of the triterpenoid tetrahymenol that is characteristic of living foraminiferans, the ciliated protozoan *Tetrahymena*, and other animal protists, and that is thus regarded as a protozoan "biomarker." *Melanocyrillium* spp. (viz., certain of those species described from the Chuar Group; Chapter 24, Pl. 29) are therefore interpreted here to be the oldest known fossil protozoans. Although the Protozoa are therefore inferred to have been extant at least as early as ~850 Ma ago, their earlier history is as yet undocumented (and the possible thecamoebaean affinity of other reported occurrences of Late Proterozoic vase-shaped microfossils remains to be investigated).

13.2.2D Invertebrate Metazoans

During the past two decades, remarkable advances have been made in studies of the earliest fossil records of the Metazoa: numerous new occurrences of latest Proterozoic (Vendian) metazoan trace and body fossils have been discovered and reported, and enormous progress has been made toward documenting and deciphering the earliest (lowermost Cambrian) records of skeletal metazoans. As a result of this activity, new taxa have been described and new data and new syntheses have emerged; earlier tentative interpretations, proposed on the basis of far less evidence than that now available, seem largely to have been confirmed.

To a considerable extent, therefore, interpretations regarding the relative times of first occurrence of various important features of the early metazoan fossil record seem to have stabilized, a development reflecting either the probable correctness of the interpretations or the sustained intractability of particularly knotty, unsolved problems. And such problems do exist, especially with regard to the precise dating of fossiliferous Vendian units and Proterozoic/Cambrian "boundary strata." However, there seems consensus that where relevant stratigraphic relationships can be established, (i) all assuredly metazoan-containing fossiliferous strata, whether they contain trace or body fossils, overlie tillitic horizons marking the base of the Vendian; (ii) the earliest known metazoan trace fossils occur in sediments either contemporaneous with, or somewhat older than, those containing the oldest known metazoan body fossils; (iii) all members of this earliest (Ediacaran) fauna appear to have been essentially "soft-bodied" (i.e., in part exhibiting thickened coverings or carapaces but lacking shells or heavily mineralized skeletal elements); (iv) all known occurrences of the Ediacaran fauna are either in strata that underlie those containing the earliest skeletal fossils or occur in locations where other evidence indicates a latest Proterozoic age; (v) with the exception of *Cloudina*, a complex, tubular, evidently calcareous structure of Vendian age and of uncertain, but probably metazoan, affinity, all mineralized skeletal elements and skeletonized fossils occur in strata regarded as lowest Cambrian or younger; and (vi) virtually all major groups of potentially preservable marine invertebrates, including taxa exhibiting diverse life attitudes, feeding modes, and skeletal mineralogies, are known to have been represented in the Lower to Middle Cambrian fossil record.

On the basis of the foregoing, invertebrate metazoans are inferred here to have been extant at least as early as ~650 Ma ago with data from molecular biology (rRNA sequences) suggesting that this evidently monophyletic clade originated several tens to a few hundreds of millions of years earlier.

13.3

Tempo and Mode of Proterozoic Evolution

13.3.1 Introduction

As documented in this monograph, studies over the past two decades have provided much improved insight into many aspects of Proterozoic paleobiology. Broad quantitative constraints (Fig. 13.1) can now be placed on the Proterozoic histories both of increasing concentrations of atmospheric O_2 (Section 4.6) and of decreasing concentrations of atmospheric CO_2 (Section 4.7), as well as on the changing, and perhaps causally related carbon-isotopic compositions of well-preserved Proterozoic kerogens (Section 3.4). As alluded to above, particularly notable is the refined understanding of the timing and nature of the Early Proterozoic rise in ambient levels of metabolically important molecular oxygen. Similarly, studies of modern microbial mat-building communities analogous to those that produced Proterozoic stromatolites (Chapter 6), of the increasingly better known record of Proterozoic microfossils (Chapter 5 and Section 11.3), and of the early evolutionary history of the Metazoa (Chapters 7, 8 and Section 11.4) and interrelated Vendian-Cambrian paleogeography (Chapter 12) have all proven highly productive. And new data are also available regarding Archean and Proterozoic geology (Section 1.3, and Chapter 2), the rock record evidencing a cyclic development (Section 2.7) that provided the foundation, both literally and figuratively, for the evolution of the Proterozoic biosphere. These and many other aspects of the Proterozoic evolution of the planet, and of life, are ably addressed elsewhere in this work.

One additional set of questions deserves attention. Unlike the Phanerozoic history of life, the basic outlines of which have been known since early in the 19th century, understanding of the development of the Proterozoic biota has come into focus only in recent years. Combined with previous knowledge, what do these new data reveal about the totality of biotic evolution? Do the data now available provide some insight into why it seemingly took "so long" for advanced metaphytes and metazoans to appear on the planet? Were the tempos and modes of Phanerozoic and Proterozoic evolution basically similar, or might they have been markedly different? Based on the data summarized in Figure 13.1, these and related questions are addressed in the following portions of this Section.

13.3.2 Hypobradytely and the Evolution of Proterozoic Life

In his classic volume on the *Tempo and Mode in Evolution*, Simpson (1944) coined terms for "three decidedly different rate distributions" in evolution, inferred from morphological comparison of living and fossil taxa: *tachytelic*, or "fast" evolution; *horotelic*, the "standard" rate distribution, typical of the great majority of lineages; and *bradytelic*, "slow" evolution like that exhibited by various lingulid brachiopods, limulids (horseshoe crabs), crocodilians, coelacanth fish (e.g., *Latimeria chalumnae*), and other so-called living fossils, "groups that survive today and show relatively little change since the very remote time when they first appeared in the fossil record" (Simpson 1944, p. 125). In general concept, Simpson's bradytely closely approximates Ruedemann's (1918, 1922a, 1922b) "arrested evolution," both being based on comparisons of nearly identical modern and fossil taxa and both envisioning little or no morphological change over spans of geologic time as long as 100 million years or more.

Simpson and Ruedemann were each concerned almost exclusively with the evolutionary rates exhibited by megascopic, Phanerozoic, sexual eukaryotes, chiefly metazoans; at the times of their studies, data from the Precambrian fossil record were not yet available, and neither addressed the rates of evolution of asexual cyanobacteria (the fossil record of which is now far better known from the Proterozoic than from later geologic time). As documented in Section 5.4, however, cyanobacteria appear to have exhibited a remarkable lack of morphological change over enormous spans of earth history, a degree of arrested evolution (*i*) exhibited by several cyanobacterial families (viz., the Oscillatoriaceae, Chroococcaceae and, probably, the Entophysalidaceae) known from sediments at least as old as ~ 2.1 Ga (and possibly as old as ~ 3.5 Ga; Section 1.5); (*ii*) typical of numerous genera and a large number of species (viz., all but a very few of the 87 species of prokaryotic "Lazarus taxa" for which morphometric data permitting detailed comparison of Proterozoic and modern analogues are listed in Table 25.3); and (*iii*) evidenced by analysis of multiple morphologic characters in numerous taxa (e.g., Sections 5.4.4B, 5.4.7B). Recognized early in the development of Proterozoic micropaleontology and referred to as "morphological evolutionary conservatism" (Schopf 1968, 1970), the detailed similarities between Proterozoic and modern cyanobacterial species have been documented by numerous workers (to cite only a few: Schopf and Blacic 1971; Knoll et al. 1975; Golubić and Hoffman 1976; Golubić and Campbell 1979; Mendelson and Schopf 1982; Hoffman and Schopf 1983; Nyberg and Schopf 1984).

13.3.2A Hypobradytely: Definition and Caveats

Following the terminology of Simpson (1944), the term "hypobradytely" has been proposed (Schopf 1987) to refer to the exceptionally low rate of evolutionary change exhibited by cyanobacterial taxa, morphospecies that show little or no evident morphological change over many hundreds of millions of years and commonly over more than one or even two thousand million years. Hypobradytely (*hypo*, Gk., under, in English cognates commonly used to denote a low and usually the lowest position in a series, e.g., hyponitrous acid, hypoxanthine; *brady*, Gk., slow; thus, hypobradytely: the "lowest of slow rates of morphological evolution") is therefore simply an extension of Simpson's concept. Like tachytely, horotely, and bradytely, it is based solely on *morphological* comparison of living and fossil taxa (i.e., except as they can be related to morphology, use of the term is not intended necessarily to imply genomic, biochemical, or physiological identity); and it is applicable to cyanobacteria chiefly because the morphologic taxonomic descriptors traditionally used to define extant species (Desikachary 1959) are preservable in ancient sediments (Fig. 5.4.8).

Acceptance of the concept of cyanobacterial hypobradytely is not without potential pitfalls. In particular, it should be recognized that microscopic cyanobacteria exhibit few discernable morphologic taxonomic descriptors; were they as morphologically complex, for example, as many microscopic eukaryotes, or were their ultrastructure or biochemistry readily preservable, providing additional characters for comparison between modern and fossil taxa, some portion of their apparent hypobradytely might well disappear. Similarly, as illustrated by the fact that several of the Proterozoic Lazarus taxa listed in Table 25.3 are morphologically comparable to extant species *both* of cyanobacteria *and* bacteria, it is evident that a considerable degree of what might be termed "microbial mimicry" exists in the prokaryotic world; some "large" bacteria closely resemble small cyanobacteria, so some portion of the apparent hypobradytely of the group might reflect misidentification of the fossil taxa (a problem addressed at some length in Section 5.4).

Despite the foregoing caveats, it nevertheless seems impressive that although the Proterozoic microfossil record is very incompletely known – what passes now for the "encouraging progress" of the past two decades being certain to pale into near insignificance as data are increasingly amassed in future generations – the meager record that is available contains such a high proportion (as summarized in Table 25.2, 29% of the nearly 300 prokaryotic taxa here recognized) of Proterozoic microbial species that are comparable in morphological detail to specific, modern, microorganisms. Moreover, even based on this limited available record, a sizable majority of the fossil cyanobacterial species now known appear to have been extant over exceptionally long Proterozoic time ranges (Fig. 25.4), an observation firmly supporting their inferred hypobradytelic evolution (as well as indicating that such cyanobacteria cannot be applied to problems of biostratigraphic correlation).

13.2.2B To What Extent Might Morphological Similarity Between Fossil and Modern Taxa Reflect Biochemical Similarity?

Based on the foregoing, it seems reasonably clear that cyanobacteria (and possibly other prokaryotes as well) are, as defined here, hypobradytelic. The cardinal evolutionary question is the degree to which this evident morphological conservatism may parallel, and perhaps be a reflection of, a comparable evolutionary conservatism of genetics, biochemistry, and physiology. That is, did the arrested evolution of these microorganisms affect essentially all aspects of their biology, or might they have been subject to the so-called Volkswagen Syndrome (Schopf et al. 1983), a lack of change in external morphology masking significant changes in internal biochemical machinery?

With regard to genetics, little can be said aside from the observations that (*i*) morphological close comparability between modern and fossil taxa seems likely to reflect fundamental similarities in the genetic and developmental controls governing that morphology; (*ii*) direct fossil evidence seems to indicate that the modern mechanisms of prokaryotic (including cyanobacterial) cell division were already in place as early as ~3.4 Ga ago (Fig. 1.5.6); and (*iii*) virtually the entire range of morphological products of such cell division exhibited among modern cyanobacteria is also exhibited among Proterozoic taxa, whether considered in terms of the shape of the medial (e.g., discoidal, quadrate, barrel-shaped, elongate) or terminal cells (globose, conical, rounded, blunt-ended) of cyanobacterial trichomes, of the cell morphologies in non-filamentous colonies (coccoid or ellipsoid), or of the organization of cells that make up cyanobacterial colonial aggregates (e.g., occurring in uniseriate, unbranched trichomes; irregular colonies; and in tabular, decussate, cuboidal, rosette-like, and similarly highly ordered colonies).

Similar conclusions apply regarding the evolutionary conservatism of the biochemical and physiological characteristics of cyanobacteria: data from the fossil record are consistent with the proposition that these characters have not changed significantly over geologic time; no current evidence refutes (or even seriously challenges) the correctness of this proposition; but the proposition cannot be "proved" to be true by direct fossil evidence. (The situation may be somewhat similar to that of attempting to "prove" that Phanerozoic trilobites, or even dinosaurs, were obligately aerobic; there can be no doubt that they were, in fact, oxygen-requiring, but the relevant fossil evidence is necessarily indirect.)

Several lines of evidence suggest, however, that it is unlikely that either the biochemistry or the physiology of cyanobacteria has changed appreciably since at least ~2.1 Ga ago, the beginning of the relatively well-documented fossil record that links cyanobacterial fossils into an evolutionary continuum merging with the present. With respect to biochemistry, for example, the range of morphologies exhibited by Proterozoic cyanobacterial sheaths (viz., from thin, diffuse, and diaphanous to thick, robust, and multilamellated), essentially identical to that characteristic of the sheaths enclosing extant coccoid and filamentous cyanobacteria, is consistent with an original polysaccharide composition like that of the modern analogues, as is the commonly reported preferential preservation of such sheaths both in the Proterozoic fossil record and at depth in modern microbial mats. In terms of physiology, the occurrence of the isoprenoid hydrocarbons pristane and phytane (in part presumably chlorophyll-derived) in fossil cyanobacterium-containing stromatolitic Proterozoic sediments, the carbon-isotopic signatures of kerogens isolated from such units, and the laminar, mat-forming orientation of stromatolitic cyanobacterial microfossils are consistent with the occurrence of cyanobacterial photosynthesis, the oxygen thus liberated presumably being evidenced by the occurrence of oxidized Proterozoic sediments such as banded iron-formations and red beds. Similarly, where the paleoecology/paleobathymetry of fossiliferous units can be inferred confidently, all Proterozoic sediments containing non-transported fossil cyanobacteria appear to have been deposited within the photic zone, the fossils commonly occurring, like modern cyanobacteria, as components of stromatolitic biocoenoses. Moreover, whether at the familial or generic level, the compositions of Proterozoic and modern mat-building cyanobacterial communities are notably similar, both being dominated by oscillatoriaceans (e.g., *Oscillatoria, Lyngbya, Phormidium, Microcoleus*) with subordinant chroococcaceans (e.g., *Microcystis, Aphanocapsa, Gloeocapsa, Chroococcus*) and, in some settings, entophysalidaceans (e.g., *Entophysalis*) and pleurocapsaceans (e.g., *Pleurocapsa*). Finally, the sparse evidence now available from extract organic geochemistry (e.g., the Proterozoic distribution of hopanes, steranes, and other biomarkers; Section 3.3.2), although not definitive, is also consistent with the postulated absence of significant biochemical or physiological cyanobacterial evolution.

In short, therefore, and despite the fact that indirect evidence and "consistency arguments" like the foregoing are less than compelling, it seems a reasonable supposition that the hypobradytelic evolution of Proterozoic cyanobacteria was characteristic not only of their morphology, but of at least the most basic aspects of their genetics, biochemistry, and physiology as well. If so, what might this suggest about the nature of early evolution?

13.3.2C The "Normal Rules" of (Phanerozoic) Evolution

The keystone of biologic evolution, at least of all of Phanerozoic evolution, is change. Most Phanerozoic taxa have "species life-times" of a few to several million years; indeed, bradytelic species, so-called living fossils, are notable chiefly because they are rare. Yet the foregoing discussion leads to the conclusion that the dominant organisms on earth for the majority of earth history – prokaryotic, photoautotrophic, cyanobacteria – were characterized by a markedly different, hypobradytelic, rate of evolution. Perhaps Phanerozoic and Proterozoic organisms played the evolutionary "game" by different sets of rules. What might these rules have been?

With (*i*) the advent of the mitotic, haploid, eukaryotic cell; (*ii*) the development of meiosis and syngamy (viz., of eukaryotic sexuality) and a concomitant increase in genetic variability (inferred here to be reflected by the rise of eukaryotic phytoplankters ~1100 Ma ago; Section 5.5); (*iii*) the stepwise evolution to diploid dominant life cycles in eukaryotes (Schopf et al. 1973); and (*iv*) the subsequent rise to prominence of metaphytes and metazoans, the "normal rules" of Phanerozoic evolution gradually became established: specialization of the haploid phase of the life cycle for reproduction; specialization of the diploid phase for nutrient assimilation (via either heterotrophy or photosynthesis); and specialization of the dominant (usually diploid) phase to assure the dispersal and reproductive success of the subordinant (usually haploid) phase. The "normal rules" seem clear: with genetic variability being provided chiefly via the mechanisms of eukaryotic sexuality, a division of roles between, and specialization of, the two phases of the life cycle leading to reproductive success of environmentally well-adapted individuals.

13.3.2D Cyanobacteria: Unexcelled Ecologic "Generalists"

The "normal rules" of Phanerozoic evolution, however, do not apply to cyanobacteria: because they are strictly asexual, they do not have a haploid/diploid life cycle; and they are far from being "specialized." Indeed, cyanobacteria are the most successful (judged in terms of longevity) photoautotrophic generalists to have evolved in the history of the earth. Provided the necessities of photosynthesis (light, carbon dioxide, and water), they can survive, and grow, almost anywhere on the surface of the planet: deserts, tundra, forest-floor cover; within rocks as endoliths in the Dry Valleys of Antarctica; hot springs, snow fields, on the bottoms of permanently ice-covered lakes; freshwater, marine, extremely hypersaline settings; anoxic, microaerophilic, oxygenic environments; and low pH (~3.5) to exceptionally high pH (11 to perhaps 13) locales (for a more extensive tabulation, see Table 2 in Schopf 1974a). As a group, cyanobacteria are especially well adapted to low concentrations of ambient oxygen and to low light intensities. Some taxa are able to survive prolonged periods of total darkness (e.g., *Lyngbya* and *Microcoleus*, buried in anaerobic muds) and have even been recovered from the oceanic aphotic zone (viz., cf. *Nostoc planctonicum* at a depth of 1000 m in the Mediterranean Sea and the Indian Ocean; Bernard 1963). Many are highly resistant to desiccation. Virtually all have highly efficient DNA-repair mechanisms and thus tend to be notably resistant to UV-, X-, or gamma-irradiation. In addition to *Oscillatoria limnetica*, the classic example of a facultative photoautotroph, more than 20 other cyanobacterial strains are now known to be capable of facultative photosynthesis (viz., of using H_2S in place of H_2O as the source of electrons with which to reduce CO_2 photoautotrophically). And the ecologically advantageous ability to fix atmospheric nitrogen is widespread within the group, exhibited both by heterocystous (e.g., nostocacean) and by non-heterocystous taxa (e.g., *Trichodesmium* and *Gloeocapsa*).

In short, cyanobacteria are unexcelled ecologic generalists. In comparison with other extant aerobic photoautotrophs, they are neither the "fastest growers" nor the "most prolific reproducers." In many modern settings, they therefore are at a disadvantage in competition with (and can be heterotrophically consumed by) eukaryotes. Indeed, presumably as a result of their Proterozoic heritage, they are particularly well adapted to a eukaryote-free environment, and at present they thus tend to be particularly prevalent only in refugia that are relatively inhospitable to eukaryotes. Nevertheless, cyanobacteria are extremely difficult to exterminate, even when subjected to the most unusual conditions: *Nostoc commune* has been revived after 107 years of storage as a dried herbarium specimen (Davis 1972); *Oscillatoria princeps*, *O. subtilissima*, and *O. minima* have been reported to survive immersion in liquid helium at $-269°C$ for 7.5 hours (Davis 1972); several cyanobacterial species have been shown to survive nuclear test-site explosions within a distance of about 1 km from "ground zero," the radius of the zone from which surface soil was not completely stripped away by the nuclear blasts (Shields and Drouet 1962); and among these nuclear survivors was *Microcoleus vaginatus*, a species which also survived experimental exposure to 2,560 kr gamma-irradiation from a ^{60}Co source (Shields and Drouet 1962), a dosage more than two orders of magnitude greater than that lethal to eukaryotic micro-algae (Godward 1962). Such exceptional survivability, the ecologic versatility of the group, and the absence of the genetic variability provided in eukaryotes by sexual reproduction provide a plexus of convincing explanations for cyanobacterial hypobradytely – these hardy asexual generalists did not "need" to evolve more rapidly, and may have been inherently limited in their potential to do so.

13.3.2E Extinctions and the History of Life

In addition, the ecologic versatility and hardiness of cyanobacteria may explain what appears to be a salient difference between the evolutionary histories of Proterozoic and Phanerozoic life. Phanerozoic evolution is punctuated by numerous massive extinctions. In contrast, and with but one exception – an episode of extinction associated with the long-term "collapse" of the Late Proterozoic ecosystem (Section 11.3; Schopf 1989, 1991) that particularly affected large-celled, eukaryotic, phytoplankers (Sections 5.5 and 11.3; Vidal and Knoll 1982) – there is no evidence that such biotic calamities affected earlier, and very much longer, Precambrian biologic history. It should be stressed that relevant data for the Precambrian are minimal, and that only a single, exceptionally coarse-grained analysis of early biotic global diversity has as yet been attempted (Section 11.3). Nevertheless, at least as now known, this contrast between the Phanerozoic and earlier fossil records seems striking.

It has long been recognized that, regardless of postulated cause, the occurrence of a Phanerozoic mass extinction results in the opening of previously occupied ecospace, an event typically followed by the relatively rapid radiation of surviving lineages to exploit that space. As yet, prior to the very latest Proterozoic, there is no evidence for any such extinction-exploitation sequence in the Precambrian. If subsequent studies show that in fact none occurred (or if their effects, like those of the Late Proterozoic episode, are concentrated on eukaryotes rather than prokaryotes), their absence (or eukaryote-concentrated influence) may simply reflect the lack of susceptibility of cyanobacteria, the dominant components of the early biota, to extermination by the "normal" cause(s) of mass extinction. Certainly, the exceptionally long time-ranges of Proterozoic cyanobacterial species (Fig. 25.4) indicate that such extinction events, if they occurred during the Proterozoic, had only limited effect on cyanobacteria, and the hypobradytelic survival to the present of these hardy asexual generalists similarly indicates that they were little affected by the well-documented mass extinctions of the Phanerozoic. As Stanley (1984a, p. 280) has suggested, in some sense "living fossils... are simply champions at warding off extinction." If so, the "grand champions," over all of geologic time, must be the hypobradytelic cyanobacteria!

13.4

A Synoptic Comparison of Phanerozoic and Proterozoic Evolution

Based on the foregoing, there seems little doubt that a notable, indeed striking, contrast exists between the Phanerozoic and Proterozoic histories of life.

During the shorter, more recent, Phanerozoic Eon, the history of life was typified by the horotelic evolution of dominantly megascopic, sexual, aerobic, multicellular eukaryotes based on alternating life cycle phases specialized either for reproduction or for nutrient assimilation, with changes in the dominant (commonly diploid) phase chiefly resulting from modification of modular morphology:

(1) Evolution operated by "normal rules" (specialization of life cycle phases for organismal growth or for reproduction, with genetic variability provided largely by sexual recombination).

(2) Dominant organisms were megascopic, multicellular eukaryotes having aerobic metabolism and, in plants, oxygenic photosynthesis.

(3) The history of life was punctuated by recurrent mass extinctions, each followed by the evolutionary radiation of surviving lineages.

(4) The tempo of evolution was dominantly horotelic (but in some lineages tended to be tachytelic when it involved exploitation of newly available ecospace, a result predominantly either of an extinction event or of the development of new, successful, modular morphologies); species populations were small (relative to those of the Proterozoic); and species typically were relatively short-lived (having durations of a few to several million years).

(5) The mode of evolution was dominantly morphological and organismal, commonly involving an increase in complexity of muticellular "modules" (e.g., organs and tissues) via alterations of, and commonly addition to, previously established developmental pathways; modification of previously evolved modules to perform new functions was of particular evolutionary importance (e.g., in plants, development of megaphylls and cones by reduction and modification of branch systems, or of the conduplicate carpel from modified foliar organs; in animals, sequential modifications of limbs and jaw structures, or the development of avian feathers by modification of reptilian scales).

(6) Notable biochemical innovations were relatively rare, and were chiefly related to reproduction (pheromones in animals, floral volatiles and pigments in angiosperms) or the development of new organismal structures (e.g., collagenous connective tissue in metazoans, a Late Proterozoic innovation; bone in vertebrates; lignified xylem in land plants) rather than to metabolism or physiology.

(7) Overall, the history of Phanerozoic life was characterized by colonization of ecospace by autotrophs, followed by adaptation of heterotrophs to the autotroph-occupied environment (e.g., the origin of spore-producing vascular plants, tied to water for sexual reproduction, followed by development of semiterrestrial arthropods and vertebrate amphibians; the origin of pollen and seeds giving rise to fully terrestrial land plants, followed by the evolution of the amniotic egg in fully terrestrial vertebrates); as a result of this "sequential co-evolution," development of the Phanerozoic ecosystem was additive and expansive, increasingly complex layers being added to the previously existing ecologic structure as ecospace (e.g., the land surface) became progressively occupied and ecologically partitioned.

In contrast with the Phanerozoic history of life, that during the much longer Proterozoic Eon was typified by the hypobradytelic evolution of dominantly microscopic, asexual, metabolically diverse, unicellular or colonial prokaryotes, based chiefly on the intracellular development of new biochemical capabilities that in part were related to major environmental change:

(1) Evolution operated by "primitive rules" (asexual generalists were prevalent, both prokaryotes and early-evolving asexual eukaryotic phytoplankters, with genetic variability provided chiefly by mutations; relatively specialized, sexually reproducing, large-celled phytoplankton became abundant only during the Late Proterozoic, many of which became extinct toward the close of the Eon).

(2) Dominant organisms were microscopic, unicellular or colonial prokaryotes exhibiting a wide range of physiological (anaerobic, facultative, and aerobic) and metabolic capabilities (chemo- and photo- heterotrophy and autotrophy) and, in photoautotrophs, anoxygenic, facultative, and oxygenic photosynthesis.

(3) Unlike the Phanerozoic history of life, that of the Proter-

ozoic may not have been punctuated by recurrent mass extinctions; only one such episode, primarily affecting large-celled, eukaryotic phytoplankers rather than prokaryotic ecologic generalists, is now known to have occurred.

(4) Among Proterozoic prokaryotes and early-evolving, asexual, eukaryotic phytoplankton, the tempo of evolution was dominantly hypobradytelic (but with the development of meiosis and syngamy perhaps ~1100 Ma ago, was evidently more rapid among sexual eukaryotic phytoplankters); species populations were large (relative to those of the Phanerozoic); and species typically were exceptionally long-lived (commonly having durations of many hundreds of millions of years).

(5) The mode of evolution was dominantly biochemical and intracellular, involving (*i*) modifications of physiology (e.g., Early Proterozoic adaptations among facultative prokaryotes to increasing ambient oxygen concentrations, and development of the capabilities of dissimilatory sulfate reduction and obligate aerobiosis); (*ii*) development of biosynthetic capabilities related to metabolism (e.g., synthesis of photosynthetic accessory pigments in eukaryotic algae); and (*iii*) appearance of biochemical innovations resulting in formation of new cellular structures (e.g., synthesis of cell wall-constructing cellulosic polysaccharides; of intracellular membrane-constructing sterols and polyunsaturated fatty acids; and of structural proteins, such as those used in mitotic spindles or in the histone components of chromosomes); many of the new biosynthetic capabilities evolved by addition of molecular oxygen-requiring steps at the ends of previously established synthetic pathways.

(6) Modification of organismal morphology was notably rare among prokaryotes, having significant evolutionary impact only among Late Proterozoic protists, metaphytes, and metazoans.

(7) Like that of the Phanerozoic, the overall history of Proterozoic life was characterized by the sequential co-evolution of autotrophs and heterotrophs (e.g., the spread of oxygen-producing cyanobacterial photoautotrophs followed by the rise of facultative and obligately aerobic prokaryotic heterotrophs; the origin of planktonic, eukaryotic autotrophs, followed by the development of heterotrophic zooplankton; the origin of autotrophic thallophytes, followed by the development of heterotrophic metazoans); and like that of the Phanerozoic, development of the Proterozoic ecosystem was additive and expansive, increasingly complex layers being added to the previously existing ecologic structure as ecospace (e.g., the increasingly oxygenic, nearshore and oceanic photic zones) became progressively occupied and ecologically partitioned.

Acknowledgments

I thank Julie K. Bartley, Hans J. Hofmann, Cornelis Klein, Carol Mankiewicz, Carl V. Mendelson, Toby B. Moore, David M. Raup, Bruce N. Runnegar, and Jane Shen-Miller for helpful comments on an earlier version of this contribution.

PART II

14
Geographic and Geologic Data for PPRG Rock Samples
TOBY B. MOORE   J. WILLIAM SCHOPF

14.1

Introduction and Numerical Listing of Geologic Samples Included in the PPRG Collections

Table 14.1 lists all samples available for study during this project, ordered by an arbitrarily assigned PPRG Sample Number (e.g., 1001). For each sample or group of samples, the relevant stratigraphic unit (supergroup, group, formation, or member), tectonic unit, and regional geographic source are indicated.

For each PPRG sample, Table 14.2 lists relevant data regarding stratigraphic setting, geographic locality, lithology, and sample processing. The samples listed in Table 14.2 are grouped by continent; within each continent, samples are grouped by tectonic unit (ordered alphabetically); and within each tectonic unit, samples are ordered by PPRG Sample Number. Symbols enclosed in parentheses adjacent to the Sample Number indicate the following:

ts	Petrographic thin sections of this sample are available in the PPRG collections.
st	This sample is stromatolitic.
st*	The stromatolitic fabric of this sample is illustrated in the Figure indicated.
uf*	Microfossils from this sample are illustrated in the figure(s) or Chapter indicated.
TOC	The total organic carbon content of this sample is listed in this volume (see Section 3.4.1A and Table 17.1).
C_{org}	The organic carbon isotopic composition of this sample is listed in this volume (see Section 3.4.2B and Table 17.1).
C_{carb}	The carbonate carbon and oxygen isotopic compositions of this sample are listed in this volume (see Section 3.4.2A and Table 17.2).
S_{fide}	The sulfide sulfur isotopic composition and sulfur content of this sample are listed in this volume (see Section 3.5.1 and Table 17.6).
S_{fate}	The sulfate sulfur isotopic composition of this sample is listed in this volume (see Section 3.5.1 and Table 17.6).
ker	The elemental and carbon isotopic composition of kerogen isolated from this sample are listed in this volume (see Section 3.2 and Table 17.5).

Following these symbols is information enclosed in brackets pertaining to the collector and the field collection number of the sample. Collectors are identified by the following abbreviations:

A. V. N.	Albert V. Nyberg
B. P.	Bonnie Packer
C. M.	Carol Mankiewicz
C. V. M.	Carl V. Mendelson
D. C.	David Chapman
D. D.	David Des Marais
D. R. L.	Donald R. Lowe
D. W.	David Ward
E. K.	Earl Kauffman
G. J. R.	Greg J. Retallack
G. O. A.	G. O. Allard
H. J. H.	Hans J. Hofmann
H. S.	Harald Strauss
J. D.	J. Donald
J. H. L.	Jere H. Lipps
J. M. H.	John M. Hayes
J. W. H.	J. W. Halbich
J. W. S.	J. William Schopf
M. A. F.	Mikhail A. Fedonkin
M. R. W.	Malcolm R. Walter
M. S.	M. St-Ouge
N. B.	Nicolas Beukes
P. M.	Paul Myrow
R. J. H.	Robert J. Horodyski
R. I.	Raymond Ingersoll
S. B.	Stefan Bengtson
S. H.	Susan Hall
T. B. M.	Toby B. Moore
T. F.	Thomas Fairchild
Y. C.	Yehuda Cohen

Table 14.5 lists the estimated ages of the sampled geologic units included in the PPRG collections. For most units, the listed estimated ages have been extrapolated based on the stratigraphic position of the unit relative to known maximum and minimum ages of the relevant geologic sections, with

multiple units within single sections being ordered geochronologically. For other units, ages were estimated based on stratigraphic correlations with units of known (or relatively well estimated) age, on unpublished data, and/or personal knowledge of PPRG members acquainted with the stratigraphic sections and geologic units in question. As is indicated in Section 1.1, an age of 650 Ma has been assumed for the beginning of the Vendian, and an age of 550 Ma for the Proterozoic- (i.e., Vendian-) Cambrian "boundary."

Table 14.1. *Geologic samples included in the PPRG collections*

PPRG Sample Number	Geologic Unit	Tectonic Unit	Geographic Source
1001–1002	Siyeh Formation	Belt Basin	Montana, U.S.A.
1003–1010	Snowslip Formation	Belt Basin	Montana, U.S.A.
1011–1012	Shepard Formation	Belt Basin	Montana, U.S.A.
1013	Snowslip Formation	Belt Basin	Montana, U.S.A.
1014	Shepard Formation	Belt Basin	Montana, U.S.A.
1015–1048	Siyeh Formation	Belt Basin	Montana, U.S.A.
1049–1051	Altyn Formation	Belt Basin	Montana, U.S.A.
1052–1053	Siyeh Formation	Belt Basin	Montana, U.S.A.
1054–1061	Tanner Member	Colorado Plateau Province	Arizona, U.S.A.
1062–1077	Jupiter Member	Colorado Plateau Province	Arizona, U.S.A.
1078–1080	Tanner Member	Colorado Plateau Province	Arizona, U.S.A.
1081	Jupiter Member	Colorado Plateau Province	Arizona, U.S.A.
1082–1087	Awatubi Member	Colorado Plateau Province	Arizona, U.S.A.
1088–1123	Walcott Member	Colorado Plateau Province	Arizona, U.S.A.
1124–1125	Jupiter Member	Colorado Plateau Province	Arizona, U.S.A.
1126–1150	Carbon Canyon Member	Colorado Plateau Province	Arizona, U.S.A.
1151–1152	Duppa Member	Colorado Plateau Province	Arizona, U.S.A.
1153–1154	Carbon Canyon Member	Colorado Plateau Province	Arizona, U.S.A.
1155	Jupiter Member	Colorado Plateau Province	Arizona, U.S.A.
1156	Carbon Canyon Member	Colorado Plateau Province	Arizona, U.S.A.
1157–1161	Carbon Butte Member	Colorado Plateau Province	Arizona, U.S.A.
1162	Duppa Member	Colorado Plateau Province	Arizona, U.S.A.
1163	Awatubi Member	Colorado Plateau Province	Arizona, U.S.A.
1164–1210	Chapel Island Formation	Avalon Zone	southeastern Newfoundland, Canada
1211	Altyn Formation	Belt Basin	Montana, U.S.A.
1212–1218	Dismal Lakes Group	Coppermine Homocline	Northwest Territories, Canada
1219	River Wakefield Subgroup	Adelaide Geosyncline	South Australia
1220–1221	Auburn Dolomite	Adelaide Geosyncline	South Australia
1222	Rhynie Sandstone	Adelaide Geosyncline	South Australia
1223	Skillogalee Dolomite	Adelaide Geosyncline	South Australia
1224–1229	River Wakefield Subgroup	Adelaide Geosyncline	South Australia
1230–1233	Skillogalee Dolomite	Adelaide Geosyncline	South Australia
1234	Tapley Hill Formation	Adelaide Geosyncline	South Australia
1235	Beaumont Dolomite	Adelaide Geosyncline	South Australia
1236	Montacute Dolomite	Adelaide Geosyncline	South Australia
1237–1238	Tindelpina Shale Member	Adelaide Geosyncline	South Australia
1239–1242	Woocalla Dolomite Member	Adelaide Geosyncline	South Australia
1243–1247	Love's Creek Member	Amadeus Basin	Northern Territory, Australia
1248	Areyonga Formation	Amadeus Basin	Northern Territory, Australia
1249–1254	Love's Creek Member	Amadeus Basin	Northern Territory, Australia
1255	Areyonga Formation	Amadeus Basin	Northern Territory, Australia
1256	Bitter Springs Formation	Amadeus Basin	Northern Territory, Australia
1257–1259	Areyonga Formation	Amadeus Basin	Northern Territory, Australia
1260	Paradise Creek Formation	Lawn Hill Platform	Queensland, Australia
1261	Koolpin Formation	Limmen Geosyncline	Northern Territory, Australia
1262–1270	McMinn Formation	Limmen Geosyncline	Northern Territory, Australia
1271	Beck Spring Dolomite	Basin and Range Province	California, U.S.A.
1272	Windfall Formation	Basin and Range Province	Nevada, U.S.A.
1273–1278	Urquhart Shale	Mount Isa Inlier	Northern Territory, Australia

Table 14.1. *Continued.*

PPRG Sample Number	Geologic Unit	Tectonic Unit	Geographic Source
1279	McArthur Group	McArthur Basin	Northern Territory, Australia
1280	Auburn Dolomite	Adelaide Geosyncline	South Australia
1281–1282	River Wakefield Subgroup	Adelaide Geosyncline	South Australia
1283	Skillogalee Dolomite	Adelaide Geosyncline	South Australia
1284–1292	Gunflint Iron Formation	Port Arthur Homocline	southern Ontario, Canada
1293–1294	Keewatin Group	Port Arthur Homocline	southern Ontario, Canada
1295	Rove Formation	Port Arthur Homocline	southern Ontario, Canada
1296–1303	Gunflint Iron Formation	Port Arthur Homocline	southern Ontario, Canada
1304–1305	Rossport Formation	Port Arthur Homocline	southern Ontario, Canada
1306–1307	Jonesboro Limestone	Unaka Province	Tennessee, U.S.A.
1308–1309	Vempalle Formation	Cuddapah Basin	south-central India
1310	Guddadarangavanahalli Formation	Western Dharwar Craton	southwest India
1311	Love's Creek Member	Amadeus Basin	Northern Territory, Australia
1312	Bitter Springs Formation	Amadeus Basin	Northern Territory, Australia
1313–1314	Love's Creek Member	Amadeus Basin	Northern Territory, Australia
1315–1317	Bitter Springs Formation	Amadeus Basin	Northern Territory, Australia
1318	Pertatataka Formation	Amadeus Basin	Northern Territory, Australia
1319–1321	Love's Creek Member	Amadeus Basin	Northern Territory, Australia
1322	Areyonga Formation	Amadeus Basin	Northern Territory, Australia
1323–1324	Love's Creek Member	Amadeus Basin	Northern Territory, Australia
1325	Areyonga Formation	Amadeus Basin	Northern Territory, Australia
1326–1333	Love's Creek Member	Amadeus Basin	Northern Territory, Australia
1334–1349	Skillogalee Dolomite	Adelaide Geosyncline	South Australia
1350–1353	Love's Creek Member	Amadeus Basin	Northern Territory, Australia
1354–1356	Meentheena Carbonate Member	Hamersley Basin	Western Australia
1357–1358	Walcott Member	Colorado Plateau Province	Arizona, U.S.A.
1359	Carbon Canyon Member	Colorado Plateau Province	Arizona, U.S.A.
1360	Allamoore Formation	Van Horn Mobile Belt	Texas, U.S.A.
1361	Red Pine Shale	Uinta Uplift	Utah, U.S.A.
1362	Kona Dolomite	Lake Superior Basin	Michigan, U.S.A.
1363–1365	Tyler Formation	Lake Superior Basin	Michigan, U.S.A.
1366–1367	Nonesuch Shale	Lake Superior Basin	Michigan, U.S.A.
1368–1370	Pokegama Quartzite	Lake Superior Basin	Minnesota, U.S.A.
1371	Biwabik Iron Formation	Lake Superior Basin	Minnesota, U.S.A.
1372–1373	Soudan Iron Formation	Lake Superior Basin	Minnesota, U.S.A.
1374–1379	Gunflint Iron Formation	Port Arthur Homocline	southern Ontario, Canada
1380–1381	Gowganda Formation	Penokean Fold Belt	southern Ontario, Canada
1382	Union Island Group	Athapuscow Aulacogen	Northwest Territories, Canada
1383	Love's Creek Member	Amadeus Basin	Northern Territory, Australia
1384–1385	Theespruit Formation	Barberton Greenstone Belt	Transvaal, South Africa
1386–1389	Hooggenoeg Formation	Barberton Greenstone Belt	Transvaal, South Africa
1390–1397	Kromberg Formation	Barberton Greenstone Belt	Transvaal, South Africa
1398	Middle Marker Member	Barberton Greenstone Belt	Transvaal, South Africa
1399–1404	Fig Tree Group	Barberton Greenstone Belt	Transvaal, South Africa
1405	Kromberg Formation	Barberton Greenstone Belt	Transvaal, South Africa
1406–1407	Fig Tree Group	Barberton Greenstone Belt	Transvaal, South Africa
1408–1409	Insuzi Group	Wit Mfolozi Inlier	Natal, South Africa
1410	Carbon Leader Member	Kaapvaal Craton	Transvaal, South Africa
1411	Venterspost Formation	Kaapvaal Craton	Transvaal, South Africa
1412–1414	Gamohaan Formation	Kaapvaal Craton	Cape Province, South Africa
1415	Klipfonteinheuwel Formation	Kaapvaal Craton	Cape Province, South Africa
1416	Rietgat Formation	Kaapvaal Craton	Cape Province, South Africa
1417	Malmani Subgroup	Kaapvaal Craton	Cape Province, South Africa
1418	Kuruman Iron Formation	Kaapvaal Craton	Cape Province, South Africa
1419	Kameeldoorns Formation	Kaapvaal Craton	Transvaal, South Africa
1420–1422	Gaoyuzhuang Formation	North China Platform	northern China
1423	Zhangqu Formation	North China Platform	northern China

Table 14.1. *Continued.*

PPRG Sample Number	Geologic Unit	Tectonic Unit	Geographic Source
1424	Jiudingshan Formation	North China Platform	northern China
1425–1431	Wumishan Formation	North China Platform	northern China
1432	Gaoyuzhuang Formation	North China Platform	northern China
1433–1435	Wumishan Formation	North China Platform	northern China
1436	Gorge Creek Group	Pilbara Block	Western Australia
1437–1439	Towers Formation	Pilbara Block	Western Australia
1440–1441	Apex Basalt	Pilbara Block	Western Australia
1442–1443	Towers Formation	Pilbara Block	Western Australia
1444–1445	Cleaverville Formation	Pilbara Block	Western Australia
1446–1448	Towers Formation	Pilbara Block	Western Australia
1449	Mount Ada Basalt	Pilbara Block	Western Australia
1450	McPhee Formation	Pilbara Block	Western Australia
1451–1453	Towers Formation	Pilbara Block	Western Australia
1454	Hardy Sandstone	Hamersley Basin	Western Australia
1455–1457	Tumbiana Formation	Hamersley Basin	Western Australia
1458	Apex Basalt	Pilbara Block	Western Australia
1459–1460	Towers Formation	Pilbara Block	Western Australia
1461	Meentheena Carbonate Member	Hamersley Basin	Western Australia
1462	Tumbiana Formation	Hamersley Basin	Western Australia
1463	Kuruna Siltstone	Hamersley Basin	Western Australia
1464	Kylena Basalt	Hamersley Basin	Western Australia
1465–1466	Euro Basalt	Pilbara Block	Western Australia
1467	Wyman Formation	Pilbara Block	Western Australia
1468–1471	Meentheena Carbonate Member	Hamersley Basin	Western Australia
1472	Altyn Formation	Belt Basin	Montana, U.S.A.
1473–1474	Chichkan Formation	Malyj Karatau	southern Kazakhstan, U.S.S.R.
1475–1477	Chulaktau Formation	Malyj Karatau	southern Kazakhstan, U.S.S.R.
1478–1479	Min'yar Formation	Bashkirian Anticlinorium	central Bashkir, U.S.S.R.
1480–1483	Bianka Member	Bashkirian Anticlinorium	central Bashkir, U.S.S.R.
1484–1485	Katav Formation	Bashkirian Anticlinorium	central Bashkir, U.S.S.R.
1486	Revet Member	Bashkirian Anticlinorium	central Bashkir, U.S.S.R.
1487–1488	Tukan Member	Bashkirian Anticlinorium	central Bashkir, U.S.S.R.
1489	Bederysh Member	Bashkirian Anticlinorium	central Bashkir, U.S.S.R.
1490	Kataskin Member	Bashkirian Anticlinorium	central Bashkir, U.S.S.R.
1491	Ushakov Member	Bashkirian Anticlinorium	central Bashkir, U.S.S.R.
1492	Maloinzer Member	Bashkirian Anticlinorium	central Bashkir, U.S.S.R.
1493	Kuktur Member	Bashkirian Anticlinorium	central Bashkir, U.S.S.R.
1494	Estenilla Formation	Sierra de Guadalupe	south-central Spain
1495	Fuentes Formation	Sierra de Guadalupe	south-central Spain
1496	Pusa Formation	Sierra de Guadalupe	south-central Spain
1497–1498	Estenilla Formation	Sierra de Guadalupe	south-central Spain
1499–1501	Cijara Formation	Sierra de Guadalupe	south-central Spain
1502	Los Parrales Shale	Sierra de Guadalupe	south-central Spain
1503–1505	Almodovar del Rio Group	Sierra de Cordoba	southern Spain
1506	Sotillo Group	Sierra de Cordoba	southern Spain
1507–1508	Pedroche Formation	Sierra de Cordoba	southern Spain
1509–1528	Rocknest Formation	Wopmay Orogen	Northwest Territories, Canada
1529	Fontano Formation	Wopmay Orogen	Northwest Territories, Canada
1530–1531	Rocknest Formation	Wopmay Orogen	Northwest Territories, Canada
1532	Odjick Formation	Wopmay Orogen	Northwest Territories, Canada
1533–1587	Rocknest Formation	Wopmay Orogen	Northwest Territories, Canada
1588	Fontano Formation	Wopmay Orogen	Northwest Territories, Canada
1589–1595	Shezal Formation	MacKenzie Fold Belt	Northwest Territories, Canada
1596	Shezal Formation	MacKenzie Fold Belt	Northwest Territories, Canada
1597–1607	Shezal Formation	MacKenzie Fold Belt	Northwest Territories, Canada
1608–1610	Little Dal Group	MacKenzie Fold Belt	Yukon Territory, Canada
1611–1613	Twitya Formation	MacKenzie Fold Belt	Yukon Territory, Canada

Table 14.1. *Continued.*

PPRG Sample Number	Geologic Unit	Tectonic Unit	Geographic Source
1614–1615	Sheepbed Formation	MacKenzie Fold Belt	Yukon Territory, Canada
1616–1624	Keele Formation	MacKenzie Fold Belt	Yukon Territory, Canada
1625–1629	Shezal Formation	MacKenzie Fold Belt	Yukon Territory, Canada
1630–1632	Ekwi Supergroup	MacKenzie Fold Belt	Northwest Territories, Canada
1633	Whim Creek Group	Pilbara Block	Western Australia
1634	Ekwi Supergroup	MacKenzie Fold Belt	Northwest Territories, Canada
1635–1641	Little Dal Group	MacKenzie Fold Belt	Northwest Territories, Canada
1642–1650	Sheepbed Formation	MacKenzie Fold Belt	Northwest Territories, Canada
1651	Keele Formation	MacKenzie Fold Belt	Northwest Territories, Canada
1652–1661	Twitya Formation	MacKenzie Fold Belt	Northwest Territories, Canada
1662–1665	Redstone River Formation	MacKenzie Fold Belt	Northwest Territories, Canada
1666–1671	Coppercap Formation	MacKenzie Fold Belt	Northwest Territories, Canada
1672–1682	Little Dal Group	MacKenzie Fold Belt	Northwest Territories, Canada
1683–1707	Sayunei Formation	MacKenzie Fold Belt	Northwest Territories, Canada
1708–1739	Coppercap Formation	MacKenzie Fold Belt	Northwest Territories, Canada
1740–1750	Redstone River Formation	MacKenzie Fold Belt	Northwest Territories, Canada
1751	Little Dal Group	MacKenzie Fold Belt	Northwest Territories, Canada
1752–1753	Sayunei Formation	MacKenzie Fold Belt	Northwest Territories, Canada
1754–1757	Siyeh Formation	Belt Basin	Montana, U.S.A.
1758–1759	Bitter Springs Formation	Amadeus Basin	Northern Territory, Australia
1760–1762	Julie Member	Amadeus Basin	Northern Territory, Australia
1763	Bitter Springs Formation	Amadeus Basin	Northern Territory, Australia
1764–1765	Shezal Formation	MacKenzie Fold Belt	Northwest Territories, Canada
1766	Lookout Schist	Medicine Bow Range	Wyoming, U.S.A.
1767–1776	Shezal Formation	MacKenzie Fold Belt	Northwest Territories, Canada
1777–1787	Rapitan Group	MacKenzie Fold Belt	Northwest Territories, Canada
1788–1789	Twitya Formation	MacKenzie Fold Belt	Yukon Territory, Canada
1790	Shezal Formation	MacKenzie Fold Belt	Yukon Territory, Canada
1791	Franklin Mountain Formation	MacKenzie Fold Belt	Yukon Territory, Canada
1792–1798	Shezal Formation	MacKenzie Fold Belt	Yukon Territory, Canada
1799	Rapitan Group	MacKenzie Fold Belt	Northwest Territories, Canada
1800	Sayunei Formation	MacKenzie Fold Belt	Northwest Territories, Canada
1801	Shezal Formation	MacKenzie Fold Belt	Northwest Territories, Canada
1802–1804	Rapitan Group	MacKenzie Fold Belt	Northwest Territories, Canada
1805	Sayunei Formation	MacKenzie Fold Belt	Northwest Territories, Canada
1806	Shezal Formation	MacKenzie Fold Belt	Northwest Territories, Canada
1807–1809	Rapitan Group	MacKenzie Fold Belt	Northwest Territories, Canada
1810–1813	Rocknest Formation	Wopmay Orogen	Northwest Territories, Canada
1814–1816	Asiak Formation	Wopmay Orogen	Northwest Territories, Canada
1817–1818	Rocknest Formation	Wopmay Orogen	Northwest Territories, Canada
1819–1821	Odjick Formation	Wopmay Orogen	Northwest Territories, Canada
1822–1835	Sayunei Formation	MacKenzie Fold Belt	Northwest Territories, Canada
1836–1838	Rapitan Group	MacKenzie Fold Belt	Northwest Territories, Canada
1839	Little Dal Formation	MacKenzie Fold Belt	Northwest Territories, Canada
1840–1841	Redstone River Formation	MacKenzie Fold Belt	Northwest Territories, Canada
1842–1843	Sayunei Formation	MacKenzie Fold Belt	Northwest Territories, Canada
1844	Shezal Formation	MacKenzie Fold Belt	Northwest Territories, Canada
1845	Bar River Formation	Penokean Fold Belt	southern Ontario, Canada
1846	Gordon Lake Formation	Penokean Fold Belt	southern Ontario, Canada
1847	Lorrain Formation	Penokean Fold Belt	southern Ontario, Canada
1848–1852	Gowganda Formation	Penokean Fold Belt	southern Ontario, Canada
1853	Serpent Formation	Penokean Fold Belt	southern Ontario, Canada
1854	Espanola Formation	Penokean Fold Belt	southern Ontario, Canada
1855–1856	Bruce Formation	Penokean Fold Belt	southern Ontario, Canada
1857	Mississagi Formation	Penokean Fold Belt	southern Ontario, Canada
1858	Pecors Formation	Penokean Fold Belt	southern Ontario, Canada
1859	Ramsay Lake Formation	Penokean Fold Belt	southern Ontario, Canada

Table 14.1. Continued.

PPRG Sample Number	Geologic Unit	Tectonic Unit	Geographic Source
1860–1866	McKim Formation	Penokean Fold Belt	southern Ontario, Canada
1867–1869	Matinenda Formation	Penokean Fold Belt	southern Ontario, Canada
1870	Chamberlain Shale	Belt Basin	Montana, U.S.A.
1871	Newland Limestone	Belt Basin	Montana, U.S.A.
1872	Red Pine Shale	Uinta Uplift	Utah, U.S.A.
1873	Hornby Bay Group	Coppermine Homocline	Northwest Territories, Canada
1874	Torridon Group	Torridonian Succession	Scotland, Great Britain
1875	Torridon Group	Torridonian Succession	Scotland, Great Britain
1876–1889	Torridon Group	Torridonian Succession	Scotland, Great Britain
1890–1903	Dwaal Heuvel Quartzite	Kaapvaal Craton	Transvaal, South Africa
1904	Hekport Andesite	Kaapvaal Craton	Transvaal, South Africa
1905	Dwaal Heuvel Formation	Kaapvaal Craton	Transvaal, South Africa
1906–1939	Jeerinah Formation	Hamersley Basin	Western Australia
1940–1947	Kuruna Siltstone	Hamersley Basin	Western Australia
1948–1959	Nymerina Basalt	Hamersley Basin	Western Australia
1960–1969	Tumbiana Formation	Hamersley Basin	Western Australia
1970–1975	Jeerinah Formation	Hamersley Basin	Western Australia
1976–1994	Tumbiana Formation	Hamersley Basin	Western Australia
1995–2006	Apex Basalt	Pilbara Block	Western Australia
2007–2015	Towers Formation	Pilbara Block	Western Australia
2016–2018	Kylena Basalt	Hamersley Basin	Western Australia
2019–2048	Maddina Basalt	Hamersley Basin	Western Australia
2049–2121	Tumbiana Formation	Hamersley Basin	Western Australia
2123–2129	Gaoyuzhuang Formation	North China Platform	northern China
2130–2137	Changzhougou Formation	North China Platform	northern China
2138	Dahongyu Formation	North China Platform	northern China
2139–2141	Chuanlinggou Formation	North China Platform	northern China
2142–2143	Tuanshanzi Formation	North China Platform	northern China
2144	Chuanlinggou Formation	North China Platform	northern China
2145	Tuanshanzi Formation	North China Platform	northern China
2146–2148	Yangzhuang Formation	North China Platform	northern China
2149–2158	Wumishan Formation	North China Platform	northern China
2159–2160	Hongshuizhuang Formation	North China Platform	northern China
2161–2162	Tieling Formation	North China Platform	northern China
2163	Biegaizi Formation	North China Platform	northern China
2164	Erdaohe Formation	North China Platform	northern China
2165	Chengjiajiang Formation	North China Platform	northern China
2166–2167	Longjiayuan Formation	North China Platform	northern China
2168–2170	Xunjiansi Formation	North China Platform	northern China
2171–2172	Duguan Formation	North China Platform	southern China
2173	Fengjiawan Formation	North China Platform	northern China
2174–2187	Luoquan Formation	North China Platform	northern China
2188	Xingji Formation	North China Platform	northern China
2189	Jarrad Sandstone Member	Halls Creek Mobile Zone	Western Australia
2190	Ranford Formation	Halls Creek Mobile Zone	Western Australia
2191–2192	Jarrad Sandstone Member	Halls Creek Mobile Zone	Western Australia
2193–2194	Bungle Bungle Dolomite	Birrindudu Basin	Western Australia
2195	Moonlight Valley Tillite	Birrindudu Basin	Western Australia
2196–2197	Bungle Bungle Dolomite	Birrindudu Basin	Western Australia
2198	Yurabi Formation	Halls Creek Mobile Zone	Western Australia
2199–2202	Bungle Bungle Dolomite	Birrindudu Basin	Western Australia
2203–2212	Boonall Dolomite	Halls Creek Mobile Zone	Western Australia
2213–2214	Timperley Shale	Halls Creek Mobile Zone	Western Australia
2215	Mount Forster Sandstone	Halls Creek Mobile Zone	Western Australia
2216–2218	Timperley Shale	Birrindudu Basin	Western Australia
2219–2227	Egan Formation	Halls Creek Mobile Zone	Western Australia
2228	Timperley Shale	Birrindudu Basin	Western Australia

Table 14.1. *Continued.*

PPRG Sample Number	Geologic Unit	Tectonic Unit	Geographic Source
2229–2230	Egan Formation	Halls Creek Mobile Zone	Western Australia
2231	Ranford Formation	Halls Creek Mobile Zone	Western Australia
2232–2233	Bungle Bungle Dolomite	Birrindudu Basin	Western Australia
2234–2235	Elvire Formation	Birrindudu Basin	Western Australia
2236–2237	Timperley Shale	Birrindudu Basin	Western Australia
2238–2239	Ranford Formation	Halls Creek Mobile Zone	Western Australia
2240	Frank River Sandstone	Halls Creek Mobile Zone	Western Australia
2241–2244	Bungle Bungle Dolomite	Birrindudu Basin	Western Australia
2245–2248	Mount Forster Sandstone	Birrindudu Basin	Western Australia
2249–2252	Tertiary weathering zone	Kimberley Basin	Western Australia
2253	Yurabi Formation	Halls Creek Mobile Zone	Western Australia
2254	Egan Formation	Halls Creek Mobile Zone	Western Australia
2255–2258	Yurabi Formation	Halls Creek Mobile Zone	Western Australia
2259–2260	Egan Formation	Halls Creek Mobile Zone	Western Australia
2261	Yurabi Formation	Halls Creek Mobile Zone	Western Australia
2262–2269	Egan Formation	Halls Creek Mobile Zone	Western Australia
2270–2271	Red Rock Beds Formation	Halls Creek Mobile Zone	Western Australia
2272	Jarrad Sandstone Member	Halls Creek Mobile Zone	Western Australia
2273	Red Rock Beds Formation	Halls Creek Mobile Zone	Western Australia
2274	Louisa Downs Group	Halls Creek Mobile Zone	Western Australia
2275	Moonlight Valley Tillite	Halls Creek Mobile Zone	Western Australia
2276	Frank River Sandstone	Halls Creek Mobile Zone	Western Australia
2277	Jarrad Sandstone Member	Birrindudu Basin	Western Australia
2278	Mount Forster Sandstone	Birrindudu Basin	Western Australia
2279	Egan Formation	Halls Creek Mobile Zone	Western Australia
2280	Louisa Downs Group	Halls Creek Mobile Zone	Western Australia
2281	Olympio Formation	Halls Creek Mobile Zone	Western Australia
2282–2285	Bungle Bungle Dolomite	Birrindudu Basin	Western Australia
2286	Moonlight Valley Tillite	Birrindudu Basin	Western Australia
2287	Mount Forster Sandstone	Halls Creek Mobile Zone	Western Australia
2288	Timperley Shale	Halls Creek Mobile Zone	Western Australia
2289–2291	Egan Formation	Halls Creek Mobile Zone	Western Australia
2292–2293	Yurabi Formation	Halls Creek Mobile Zone	Western Australia
2294	Louisa Downs Group	Halls Creek Mobile Zone	Western Australia
2295	Yurabi Formation	Halls Creek Mobile Zone	Western Australia
2296	Fargoo Tillite	Halls Creek Mobile Zone	Western Australia
2297	Bungle Bungle Dolomite	Birrindudu Basin	Western Australia
2298–2305	Newland Limestone	Belt Basin	Montana, U.S.A.
2306	Chamberlain Shale	Belt Basin	Montana, U.S.A.
2307–2313	Newland Limestone	Belt Basin	Montana, U.S.A.
2314	Greyson Shale	Belt Basin	Montana, U.S.A.
2315–2320	Newland Limestone	Belt Basin	Montana, U.S.A.
2321	Greyson Shale	Belt Basin	Montana, U.S.A.
2322–2324	Newland Limestone	Belt Basin	Montana, U.S.A.
2325	Helena Formation	Belt Basin	Montana, U.S.A.
2326	Snowslip Formation	Belt Basin	Montana, U.S.A.
2327–2328	Prichard Formation	Belt Basin	Montana, U.S.A.
2329	Mount Shields Formation	Belt Basin	Montana, U.S.A.
2330–2331	Jarrad Sandstone Member	Halls Creek Mobile Zone	Western Australia
2332	Fargoo Tillite	Halls Creek Mobile Zone	Western Australia
2333	Moonlight Valley Tillite	Birrindudu Basin	Western Australia
2334–2336	Bungle Bungle Dolomite	Birrindudu Basin	Western Australia
2337	Mount Forster Sandstone	Birrindudu Basin	Western Australia
2338–2339	Timperley Shale	Birrindudu Basin	Western Australia
2340–2345	Egan Formation	Halls Creek Mobile Zone	Western Australia
2346–2347	Yurabi Formation	Halls Creek Mobile Zone	Western Australia
2348	Tean Formation	Kimberley Basin	Western Australia

Table 14.1. Continued.

PPRG Sample Number	Geologic Unit	Tectonic Unit	Geographic Source
2349	Not Known	Kimberley Basin	Western Australia
2350–2351	Not Known	Kimberley Basin	Western Australia
2352	Egan Formation	Kimberley Basin	Western Australia
2353–2355	Ranford Formation	Halls Creek Mobile Zone	Western Australia
2356	Rapitan Group	MacKenzie Fold Belt	Northwest Territories, Canada
2357	Louisa Downs Group	Halls Creek Mobile Zone	Western Australia
2358–2360	Dismal Lakes Group	Coppermine Homocline	Northwest Territories, Canada
2361	Kostov Formation	Russian Platform	western U.S.S.R.
2362	Blondeau Formation	Superior Province	northern Quebec, Canada
2363	Tamengo Formation	West-Central Brazil	Mata Grosso Do Sul, Brazil
2364	Paraopeba Formation	Sao Francisco Basin	Goias, Brazil
2365–2377	Rocknest Formation	Wopmay Orogen	Northwest Territories, Canada
2378–2385	Wyman Formation	Basin and Range Province	California, U.S.A.
2386–2394	Reed Dolomite	Basin and Range Province	California, U.S.A.
2395	Deep Spring Formation	Basin and Range Province	California, U.S.A.
2396–2404	Poleta Formation	Basin and Range Province	California, U.S.A.
2405	Deep Spring Formation	Basin and Range Province	California, U.S.A.
2406–2407	Montenegro Member	Basin and Range Province	California, U.S.A.
2408	Altyn Formation	Belt Basin	Montana, U.S.A.
2409–2410	Khatyspyt Formation	Olenyok Uplift	northern Yakutia, U.S.S.R.
2411	Not Applicable	Modern	Namibia, Africa
2412–2415	Biri Formation	Osen-roa Nappe Complex	southern Norway
2416–2421	Kona Dolomite	Marquette Range	Michigan, U.S.A.
2422–2428	Nonesuch Shale	Lake Superior Basin	Michigan, U.S.A.
2429–2432	Biwabik Iron Formation	Lake Superior Basin	Minnesota, U.S.A.
2433–2435	Soudan Iron Formation	Lake Superior Basin	Minnesota, U.S.A.
2436–2447	Gunflint Iron Formation	Port Arthur Homocline	southern Ontario, Canada
2448	Rove Formation	Port Arthur Homocline	southern Ontario, Canada
2449–2455	Gunflint Iron Formation	Port Arthur Homocline	southern Ontario, Canada
2456–2459	Rove Formation	Port Arthur Homocline	southern Ontario, Canada
2460–2462	Rossport Formation	Port Arthur Homocline	southern Ontario, Canada
2463–2495	Gunflint Iron Formation	Port Arthur Homocline	southern Ontario, Canada
2496	Gowganda Formation	Penokean Fold Belt	southern Ontario, Canada
2497	Espagnola Formation	Penokean Fold Belt	southern Ontario, Canada
2498	Carbon Leader Member	Kaapvaal Craton	Transvaal, South Africa
2499	North Leader Member	Kaapvaal Craton	Transvaal, South Africa
2500–2501	Carbon Leader Member	Kaapvaal Craton	Transvaal, South Africa
2502	North Leader Member	Kaapvaal Craton	Transvaal, South Africa
2503	Witwatersrand Supergroup	Kaapvaal Craton	Transvaal, South Africa
2504	Jeppestown Shale	Kaapvaal Craton	Transvaal, South Africa
2505	Government Subgroup	Kaapvaal Craton	Transvaal, South Africa
2506	Jeppestown Shale	Kaapvaal Craton	Transvaal, South Africa
2507	Venterspost Carbon Reef Member	Kaapvaal Craton	Transvaal, South Africa
2508	Kimberley Shale	Kaapvaal Craton	Transvaal, South Africa
2509	Venterspost Carbon Reef Member	Kaapvaal Craton	Transvaal, South Africa
2510–2511	Hotazel Formation	Kaapvaal Craton	Transvaal, South Africa
2512–2523	Gamohaan Formation	Kaapvaal Craton	Transvaal, South Africa
2524–2525	Rooinekke Formation	Kaapvaal Craton	Transvaal, South Africa
2526	Klipfonteinheuwel Formation	Kaapvaal Craton	Transvaal, South Africa
2527–2529	Ventersdorp Supergroup	Kaapvaal Craton	Cape Province, South Africa
2530–2535	Reivilo Formation	Kaapvaal Craton	Transvaal, South Africa
2536–2539	Kingston Peak Formation	Basin and Range Province	southwestern Nevada, U.S.A.
2540–2541	Anakeesta Formation	Blue Ridge Province	Tennessee, U.S.A.
2542	Pigeon Formation	Blue Ridge Province	Tennessee, U.S.A.
2543	Hekpoort Andesite	Kaapvaal Craton	Transvaal, South Africa
2544–2548	Theespruit Formation	Barberton Greenstone Belt	Transvaal, South Africa
2549–2563	Hooggenoeg Formation	Barberton Greenstone Belt	Transvaal, South Africa

Table 14.1. Continued.

PPRG Sample Number	Geologic Unit	Tectonic Unit	Geographic Source
2564	Middle Marker Member	Barberton Greenstone Belt	Transvaal, South Africa
2565–2566	Fig Tree Group	Barberton Greenstone Belt	Transvaal, South Africa
2567–2568	Msauli Chert Member	Barberton Greenstone Belt	Transvaal, South Africa
2569	Fig Tree Group	Barberton Greenstone Belt	Transvaal, South Africa
2570–2571	Kromberg Formation	Barberton Greenstone Belt	Transvaal, South Africa
2572	Msauli Chert Member	Barberton Greenstone Belt	Transvaal, South Africa
2573–2574	Fig Tree Group	Barberton Greenstone Belt	Transvaal, South Africa
2575–2576	Moodies Group	Barberton Greenstone Belt	Transvaal, South Africa
2577	Clutha Formation	Barberton Greenstone Belt	Transvaal, South Africa
2578–2579	Moodies Group	Barberton Greenstone Belt	Transvaal, South Africa
2580–2582	Pongola Supergroup	Barberton Greenstone Belt	Transvaal, South Africa
2583	Ecca Group	Karoo Basin	South Africa
2584–2588	Vryheid Formation	Karoo Basin	Transvaal, South Africa
2589–2596	Gorge Creek Group	Pilbara Block	Western Australia
2597–2601	Towers Formation	Pilbara Block	Western Australia
2602–2608	Apex Basalt	Pilbara Block	Western Australia
2609	Hardy Sandstone	Hamersley Basin	Western Australia
2610–2612	Apex Basalt	Hamersley Basin	Western Australia
2613–2617	Cleaverville Formation	Hamersley Basin	Western Australia
2618–2622	Towers Formation	Pilbara Block	Western Australia
2623–2628	Mount Ada Basalt	Pilbara Block	Western Australia
2629–2632	McPhee Formation	Pilbara Block	Western Australia
2633–2638	Towers Formation	Pilbara Block	Western Australia
2639	Rushall Slate	Pilbara Block	Western Australia
2640–2641	Whim Creek Group	Pilbara Block	Western Australia
2642	Louden Volcanics	Pilbara Block	Western Australia
2643	Hardy Sandstone	Hamersley Basin	Western Australia
2644	Apex Basalt	Pilbara Block	Western Australia
2645–2646	Towers Formation	Pilbara Block	Western Australia
2647–2651	Meentheena Carbonate Member	Hamersley Basin	Western Australia
2652	Meenthenna Carbonate Member	Hamersley Basin	Western Australia
2653–2659	Meentheena Carbonate Member	Hamersley Basin	Western Australia
2660	Mingah Tuff Member	Hamersley Basin	Western Australia
2661–2673	Meentheena Carbonate Member	Hamersley Basin	Western Australia
2674–2677	Kuruna Siltstone	Hamersley Basin	Western Australia
2678–2681	Kylena Basalt	Hamersley Basin	Western Australia
2682	Euro Basalt	Pilbara Block	Western Australia
2683–2684	Hardy Sandstone	Hamersley Basin	Western Australia
2685–2689	Lalla Rookh Sandstone	Pilbara Block	Western Australia
2690	Wyman Formation	Pilbara Block	Western Australia
2691	Lalla Rookh Sandstone	Pilbara Block	Western Australia
2692	Mosquito Formation	Pilbara Block	Western Australia
2693–2695	Hardy Sandstone	Pilbara Block	Western Australia
2696–2699	Warrie Member	Hamersley Basin	Western Australia
2700–2712	Tumbiana Formation	Hamersley Basin	Western Australia
2713–2714	Dales Gorge Member	Hamersley Basin	Western Australia
2715	Hardy Sandstone	Hamersley Basin	Western Australia
2716	Ashburton Formation	Gascoyne Province	Western Australia
2717	Whim Creek Group	Pilbara Block	Western Australia
2718	Euro Basalt	Pilbara Block	Western Australia
2719–2720	Apex Basalt	Pilbara Block	Western Australia
2721	McPhee Formation	Pilbara Block	Western Australia
2722	Ust-Pinega Formation	Russian Platform	Russia, U.S.S.R.
2723–2724	Khatyspyt Formation	Olenyok Uplift	northern Yakutia, U.S.S.R.
2725	Sukhaya Tunguska Formation	Khantay-Rybninsk Uplift	Siberia, U.S.S.R.
2726	Shorikha Formation	Khantay-Rybninsk Uplift	Siberia, U.S.S.R.
2727	Yudoma Formation	Uchur-Maya Block	Yakutia, U.S.S.R.

Table 14.1. *Continued.*

PPRG Sample Number	Geologic Unit	Tectonic Unit	Geographic Source
2728–2738	Visingsö Group	Lake Vättern Graben	southern Sweden
2739	Chambless Limestone	Basin and Range Province	California, U.S.A.
2740	Rencontre Formation	Avalon Zone	southeastern Newfoundland, Canada
2741	Lookout Schist	Medicine Bow Range	Wyoming, U.S.A.
2742	Rencontre Formation	Avalon Zone	southeastern Newfoundland, Canada
2743–2769	Chapel Island Formation	Avalon Zone	southeastern Newfoundland, Canada
2770	Brigus Formation	Avalon Zone	southeastern Newfoundland, Canada
2771	Random Formation	Avalon Zone	southeastern Newfoundland, Canada
2772–2775	Bonavista Formation	Avalon Zone	southeastern Newfoundland, Canada
2776–2778	Gamohaan Formation	Kaapvaal Craton	Transvaal, South Africa
2779–2781	Spartan Group	Cape Smith Belt	Quebec, Canada

Table 14.2. *Geographic and geologic data for rock samples included in the PPRG collections*

14.2.1 AFRICA

BARBERTON GREENSTONE BELT

Tectonic unit/location. Barberton Greenstone Belt, Transvaal, South Africa, and Swaziland.

Stratigraphy. Stratigraphic section, Barberton area, eastern Transvaal and northern Swaziland (Anhaeusser, 1981):

SWAZILAND SUPERGROUP
  MOODIES GROUP
    Baviaanskop Formation
    Joe's Luck Formation
    Clutha Formation
  FIG TREE GROUP
    Schoongezicht Formation
    Belvue Road Formation
    Sheba Formation
      Msauli Chert "Member"
  ONVERWACHT GROUP
    GELUK SUBGROUP
      Swartkoppie Formation
      Kromberg Formation
      Hooggenoeg Formation
        Middle Marker Member
    TJAKASTAD SUBGROUP
      Komati Formation
      Theespruit Formation
      Sandspruit Formation

PPRG samples/localities

1384 (ts, TOC, Corg) [J. W. S. field no. 1-6/17/84]. Recrystallized black chert from discontinuous outcrops from the middle of the Theespruit Formation; point "AT2" on Fig. 2 in Viljoen and Viljoen (1969), 20 km east-southeast of Badplaas, 260 km east of Johannesburg, Transvaal, South Africa.

1385 (ts, TOC, Corg, ker) [J. W. S. field no. 2-6/17/84]. Coarse-grained, laminated black chert from the middle to upper portion of the Theespruit Formation; 0.5 km north, locality cf. 1384.

1386 (ts, TOC, Corg, ker) [J. W. S. field no. 3-6/17/84]. Fine-grained, laminated, black and white chert from the lowest chert above "Middle Marker Member," immediately overlying pillow basalt, Hooggenoeg Formation; 600 m north-northeast of the letter "E" at north end of transect "D-E" of Viljoen and Viljoen (1969), on the west limb of the Onverwacht Anticline, 35 km east of Badplaas, 275 km east of Johannesburg, Transvaal, South Africa.

1387 (ts, TOC, Corg) [J. W. S. field no. 4-6/17/84]. Massive black chert, somewhat brecciated from a 14 m-thick horizon below the "middle chert horizon", above 1386, lower third of the Hooggenoeg Formation; 350 m north-northeast, locality cf. 1386.

Table 14.2. *Continued.*

1388 (ts, TOC, Corg) [J. W. S. field no. 5-6/17/84]. Black pyritic chert cf. 1387, but offset by a minor fault, from the lower third of the Hooggenoeg Formation; 500 m northeast of the west limb of the Onverwacht Anticline, 35 km east of Badplaas, 275 km east of Johannesburg, Transvaal, South Africa.

1389 (ts, TOC, Corg) [J. W. S. field no. 6-6/17/84]. Black chert from a 12 to 15 m-thick horizon associated with ash of the Hooggenoeg Formation; 1 km north-northeast of 1386, locality cf. 1388.

1390 (ts, TOC, Corg) [J. W. S. field no. 1-6/18/84]. Black chert from the uppermost chert horizon (1–2 m thick) of the Kromberg Formation; 35 m northwest of Silver Queen footbridge near M57 on Fig. 8 in Viljoen and Viljoen (1969), on the west side of the Komati River, 44 km east of Badplaas, 284 km east of Johannesburg, Transvaal, South Africa.

1391 (ts, TOC, Corg, ker) [J. W. S. field no. 1-6/18/84]. Black chert from the upper portion of the Kromberg Formation; locality cf. 1390.

1392 (ts, TOC, Corg, ker) [J. W. S. field no. 2-6/18/84]. Massive black chert from a 2 m-thick bed, upper portion of the Kromberg Formation; 65 m northwest, locality cf. 1390.

1393 (ts, TOC, Corg) [J. W. S. field no. 3-6/18/84]. Black chert from a 1 m-thick lens interbedded with volcanics of the Kromberg Formation; west side of Komati River, 2 km northwest (upstream), locality cf. 1390.

1394 (ts, st*, μf*, TOC, Corg) [J. W. S. field no. 4A-6/18/84]. Black and white banded chert from the "stratiform stromatolitic" facies containing microfossils (Fig. 1.5.2) reported by Walsh and Lowe (1985) from a major chert horizon at 1200 ft level in Viljoen and Viljoen (1969), lower third of the Kromberg Formation; 3 km northwest, locality cf. 1390.

1395 (ts, TOC, Corg) [J. W. S. field no. 4B-6/18/84]. Black chert directly below minor carbonate lens, 8 m stratigraphically above 1394, Kromberg Formation; along footpath, ca. 10 m southeast of 1394, locality cf. 1390.

1396 (ts, TOC, Corg, ker) [J. W. S. field no. 5-6/18/84]. Bedded black chert, 1 m thick from the lower sixth of the Kromberg Formation; footpath in Komati River Gorge, ca. 200 m northwest of 1394, Transvaal, South Africa; 26.03° S, 30.00° E.

1397 (ts, TOC, Corg) [J. W. S. field no. 4C-6/18/84]. Massive black chert from ca. 10 m stratigraphically above 1395 (same horizon), lower third of the Kromberg Formation; ca. 25 m southeast of 1394, Silver Queen footbridge near M57 on Fig. 8 in Viljoen and Viljoen (1969), on the west side of the Komati River, 44 km east of Badplaas, 284 km east of Johannesburg, Transvaal, South Africa.

1398 (ts, TOC, Corg) [J. W. S. field no. 6-6/18/84]. Massive banded black chert from a horizon ca. 1.5 m thick of the Middle Marker Member of the Hooggenoeg Formation; Fullerton Creek, 44 km southeast of Badplaas, Transvaal, South Africa.

1399 (ts, TOC, Corg) [J. W. S. field no. 1-6/19/84]. Black and white banded chert from a massive 50 m thick horizon, lower third of the Fig Tree Group; "Bruce's Hill" spherule locality, on north side of the Msauli River, ca. 300 m from Little Msauli Gorge on Granville Grove Farm, 19 km southwest of Barberton, Transvaal, South Africa; 25.91° S, 30.93° E.

1400 (ts, TOC, Corg, ker) [J. W. S. field no. 2-6/19/84]. Massive bedded black chert from the lower third of the Fig Tree Group; D. R. L. locality "SAF-186"; 100 m south of "mud pool structure" type locality, 200 m south of 1399, Transvaal, South Africa; 25.92° S, 30.93° E.

1401 (ts, TOC, Corg) [J. W. S. field no. 3-6/19/84]. Black chert from a dike within a 25 m-thick chert horizon from the basal horizon of the Fig Tree Group; Little Msauli Gorge of Msauli River on Granville Grove Farm, Transvaal, South Africa; 25.91° S, 30.93° E.

1402 (ts, TOC, Corg) [J. W. S. field no. 3-6/19/84]. Chert (same as 1401) from the basal horizon of the Fig Tree Group; locality cf. 1401.

1403 ts, st, TOC, Corg) [J. W. S. field no. 4-6/19/84]. Stromatolitic black chert, 10–20 cm thick from a mafic unit immediately above the Msauli Chert of the Fig Tree Group; "Greywacke Hill" locality, 1.2 km north-northeast of "Bruce's Hill Spherule" locality, Transvaal, South Africa; 25.91° S, 30.93° E.

1404 (ts, TOC, Corg) [J. W. S. field no. 5-6/19/84]. Chert pebbles in water-washed volcanic ash from the lowest horizon of the Fig Tree Group; near Eucalyptus Mill, 4.1 km west-northwest of 1399, 20.5 km southwest of Barberton, Transvaal, South Africa; 25.90° S, 30.05° E.

1405 (ts, TOC, Corg) [J. W. S. field no. 6-6/19/84]. Botryoidal black and white chert with gypsum molds, bed ca. 50–100 m thick from the lowest horizon of the Kromberg Formation; Geluk Beacon Hill, 2.7 km south of 1404, 3.8 km west-southwest of 1399, Transvaal, South Africa; 25.92° S, 30.88° E.

1406 (ts, TOC, Corg) [J. W. S. field no. 1A-6/20/84]. Well-sorted pebble conglomerate in scour channel from the lower portion of the Fig Tree Group; D. R. L. locality "SAF-16"; locality no. 6 in Lowe and Knauth (1977), Transvaal, South Africa; 25.89° S, 31.11° E.

1407 (ts, TOC, Corg) [J. W. S. field no. 1B-6/20/84]. Bedded and massive black chert, ca. 1 m thick from 10 m below 1406, 20 m below bedded barite, lower portion of the Fig Tree Group; locality cf. 1406.

2544 [J. M. H. field no. 840617-1a]. Black chert from the Theespruit Formation; Theespruit locality AT2, Transvaal, South Africa.

2545 (ts) [J. M. H. field no. 840617-1b]. Black chert from float of the Theespruit Formation; locality cf. 2544.

2546 [J. M. H. field no. 840617-1c]. Black chert from the upper third of the Theespruit Formation; downhill, locality cf. 2544.

2547 [J. M. H. field no. 840617-1d]. Black chert from the highest black chert of the Theespruit Formation; downhill, locality cf. 2544.

2548 [J. M. H. field no. 850617-1e]. Carbonate from the Theespruit Formation; collected next to a quartz vein, downhill, locality cf. 2544.

2549 [J. M. H. field no. 840617-4a]. Chert from the lowest chert of the Hooggenoeg Formation; footpath in Komati River Gorge, ca. 200 m northwest of 1394, Transvaal, South Africa; 26.03° S, 30.00° E.

Table 14.2. *Continued.*

2550 (ts) [J. M. H. field no. 840617-4b]. Black chert from the lower third of the Hooggenoeg Formation; locality cf. 2549.

2551 (ts) [J. M. H. field no. 840617-4c]. Black chert 70 m laterally away from 2550 of the Hooggenoeg Formation; locality cf. 2549.

2552 [J. M. H. field no. 840617-4d]. Black chert from the "Middle Chert" of the Hooggenoeg Formation; locality cf. 2549.

2553 [J. M. H. field no. 840618-1a]. Carbonate from the Hooggenoeg Formation; section started from footbridge, locality cf. 2549.

2554 (ts) [J. M. H. field no. 840618-1c]. Partially silicified shaley siltstone from the Hooggenoeg Formation; near top of second chert outcrop, locality cf. 2549.

2555 [J. M. H. field no. 840618-1d]. Carbonate from just below 2554, Hooggenoeg Formation; locality cf. 2549.

2556 (ts) [J. M. H. field no. 840618-1f]. Chert from just above pillow basalt of the Hooggenoeg Formation; locality cf. 2549.

2557 (ts) [J. M. H. field no. 840618-1g]. Chert from the Hooggenoeg Formation; near bottom of sediments at "Pflug locality"; locality cf. 2549.

2558 [J. M. H. field no. 840618-1h]. Chert-carbonate from the Hooggenoeg Formation; locality cf. 2549.

2559 [J. M. H. field no. 840618-1i]. Partially silicified shale from the Hooggenoeg Formation; locality cf. 2549.

2560 (ts) [J. M. H. field no. 840618-1j]. Carbonate from the highest evident carbonate band in section of the Hooggenoeg Formation; near 2559, locality cf. 2549.

2561 [J. M. H. field no. 840618-1k]. Banded carbonate from the Hooggenoeg Formation, 2 m below 2560; locality cf. 2549.

2562 [J. M. H. field no. 840618-1l]. Carbonate from the Hooggenoeg Formation, 3 m above 2560 and 1 m away from a massive tholeiite; locality cf. 2549.

2563 [J. M. H. field no. 840618-1m]. Silicified tuff, not a layered chert but entirely reworked sedimentary material, from the uppermost part of the Hooggenoeg Formation; 300 m past "Pflug-chert," locality cf. 2549.

2564 (ts) [J. M. H. field no. 840618-2a]. Black chert from the Middle Marker Member of the Hooggenoeg Formation; Fullerton Creek, Transvaal, South Africa.

2565 (ts) [J. M. H. field no. 840619-1a]. Black chert with two small extremely carbon-rich fragments from the lower portion of the Fig Tree Group; "Bruce's Hill" spherule locality, on north side of the Msauli River, ca. 300 m from Little Msauli Gorge on Granville Grove Farm, 19 km southwest of Barberton, Transvaal, South Africa; 25.91° S, 30.93° E.

2566 [J. M. H. field no. 840619-1b]. Chert from the basal part of the Fig Tree Group; D. R. L. "SAF-186" locality, 100 m south of "mud pool structure" type locality, Transvaal, South Africa.

2567 (ts) [J. M. H. field no. 840619-1c]. Dike from the Msauli Chert "Member" of the Sheba Formation; 1 m from J. W. S. Msauli chert sampling point, Little Msauli Gorge of Msauli River on Granville Grove Farm, Transvaal, South Africa; 25.91° S, 30.93° E.

2568 [J. M. H. field no. 840619-1d]. Rip-up breccia(?) from 4 m above the base of the Msauli Chert "Member"; close to 2567, locality cf. 2567.

2569 (ts) [J. M. H. field no. 840619-2a]. Black chert from just above Lowe's stromatolites of the Fig Tree Group; "Greywacke Hill" locality, 1.2 km north-northeast of "Bruce's Hill" spherule locality, Transvaal, South Africa; 25.91° S, 30.93° E.

2570 (ts) [J. M. H. field no. 840619-3a]. Chert from "Bucks Ridge Member" (informal), just above evaporite zone, Kromberg Formation; Silver Queen footbridge near M57 on Fig. 8 in Viljoen and Viljoen (1969), on the west side of the Komati River, 44 km east of Badplaas, 284 km east of Johannesburg, Transvaal, South Africa.

2571 (ts) [J. M. H. field no. 840619-3b]. Chert from "Bucks Ridge Member" (informal) of the Kromberg Formation; locality cf. 2570.

2572 [J. M. H. field no. 840619-4a]. Chert from the Msauli Chert "Member" of the Sheba Formation; Knoll and Barghoorn locality, Little Msauli Gorge of Msauli River on Granville Grove Farm, Transvaal, South Africa; 25.91° S, 30.93° E.

2573 [J. M. H. field no. 840620-1a]. Barite from the lower part of the Fig Tree Group; D. R. L. locality "SAF-16"; locality no. 6 in Lowe and Knauth (1977), Transvaal, South Africa; 25.89° S, 31.11° E.

2574 [J. M. H. field no. 840620-1b]. Black chert from the first black chert below barite of the Fig Tree Group; locality cf. 2573.

2575 [J. M. H. field no. 840620-2a]. Black chert or silicified mudstone(?) from the Moodies Group; Saddle Back Syncline (Eriksson sedimentology locality), Transvaal, South Africa.

2576 [J. M. H. field no. 840620-3a]. Chert pebbles from conglomerate from the Moodies Group; 300 m south of Sheba junction on the road from Barberton, Transvaal, South Africa.

2577 [J. M. H. field no. 840620-4a]. Fine-grained, laminated sandstone from the lower third of the Clutha Formation; Hollbrand's Pass, Transvaal, South Africa.

2578 [J. M. H. field no. 840620-5a]. Pebble from the basal conglomerate of the Moodies Group; Sheba Creek, Transvaal, South Africa.

2579 [J. M. H. field no. 840620-6a]. Siltstone from the Moodies Group; top of the Eureka Syncline, Transvaal, South Africa.

2580 [J. M. H. field no. 840621-1a]. Siltstone, possibly volcanogenic, contains magnetite, from the Pongola Supergroup; locality "R33," south of Amsterdam, 35 km north of Piet Retief, Pongola Poort section on the Usutu River, Transvaal, South Africa.

Table 14.2. *Continued.*

KAAPVAAL CRATON

Tectonic unit/location. Kaapvaal Craton, Transvaal and Cape Province, South Africa.

Stratigraphy. Composite stratigraphic section, Griqualand West, Central Rand, and Far West Rand areas, central and western Transvaal and northern Cape Province:

OLIFANTSHOEK GROUP/WATERBERG GROUP
    Bushveld Igneous Complex
TRANSVAAL SUPERGROUP
  PRETORIA GROUP/POSTMASBURG GROUP
    Magaliesberg Quartzite
    Daspoort Quartzite
    Dwaal Heuvel Quartzite/Hotazel Formation
    Hekpoort Andesite Formation/Ongeluk Lava
    Timeball Hill Quartzite
    Polo Ground Quartzite
  GHAAP GROUP/CHUNIESPOORT GROUP
    KOEGAS SUBGROUP
      Rooinekke Formation
    ASBESHEUWELS SUBGROUP
      Griquatown Iron Formation
      Kuruman Iron Formation
    CAMPBELLRAND SUBGROUP/MALMANI SUBGROUP
      Gamohaan Formation
      Klipfonteinheuwel Formation
      Reivilo Formation
    SCHMIDTSDRIF SUBGROUP
      Vryburg Formation
VENTERSDORP SUPERGROUP
  PNIEL GROUP
  PLATBERG GROUP
    Bothaville Formation
    Rietgat Formation
    Makawassi Quartz Porphyry
    Kameeldoorns Formation
    Venterspost Formation
      Venterspost Carbon Reef "Member"
WITWATERSRAND SUPERGROUP
  KLIPRIVIERSBERG GROUP
  CENTRAL RAND GROUP
    TURFFONTEIN SUBGROUP
    [UNNAMED UPPER ARCHEAN SUBGROUP]
      Booysens Shale
    JOHANNESBURG SUBGROUP
      Krugersdorp Formation
      Bird Conglomerate
      Luipaardsvlei Quartzite
      Livingstone Conglomerate
      Randfontein Quartzite
      Johnstone Conglomerate
      Langlaagte Quartzite
      Main Conglomerate
        Carbon Leader "Member"
        North Leader "Member"
      Maraisburg Quartzite
  WEST RAND GROUP
    Jeppestown Shale Formation
    Government Formation
  DOMINION GROUP
PONGOLA SUPERGROUP

Table 14.2. *Continued.*

PPRG samples/localities

1410 (ts, TOC, Corg, Sfide) [J. W. S. field no. 1-6/25/84]. Thucolite, 0.5-1.0 cm-thick beds from level no. 14, position 70, near the base of the Carbon Leader "Member" of the Main Conglomerate Formation; West Driefontein Mine 14-70, 5 km southeast of Carletonville, Transvaal, South Africa.

1411 (ts, TOC, Corg, Sfide) [J. W. S. field no. 1-6/26/84]. Pyritic quartz conglomerate from level 16, the middle third of the Venterspost Formation; Harvie Watt Shaft of the Libanon Gold Mine, near Carletonville, Transvaal, South Africa.

1412 (ts, st, TOC, Corg) [J. W. S. field no. 1-6/28/84]. Black vitreous chert, 5 mm- to 3 cm-thick bed within partially dolomitized carbonate stromatolite from 135 m below the top of Danielskuil G-3 stratigraphic section (Beukes, 1980), Gamohaan Formation; 7.5 km northeast of Danielskuil, 130 km northwest of Kimberley, Cape Province, South Africa.

1413 (ts, st, TOC, Corg) [J. W. S. field no. 2-6/28/84]. Black chert, from 5 mm-thick concentric layers within a 1–1.5 m-high *Conophyton* stromatolite, interbedded with carbonate from the top of a ridge, 60 m stratigraphically below top of Danielskuil G-3 section, Gamohaan Formation; locality cf. 1412.

1414 (ts, st, TOC, Corg, Sfide) [J. W. S. field no. 3-6/28/84]. Bedded black chert occurring as continuous beds 8 cm thick or as lenses 0.5–2 cm thick, interbedded with carbonate from 50 m stratigraphically below the top of Danielskuil G-3 section, Gamohaan Formation; locality cf. 1412.

1415 (ts, st, TOC, Corg) [J. W. S. field no. 5-6/28/84]. Stromatolitic black chert from 0.5 m-thick bed, interbedded with stromatolitic carbonate from near the base of a 40 m thick chert zone of the Klipfonteinheuwel Formation; at north side of road cut on the Griquatown-Kimberley road, 20 km east of Griquatown, Cape Province, South Africa.

1416 (ts, st, TOC, Corg) [J. W. S. field no. 1-6/29/84]. Stromatolitic calcareous chert from low discontinuous beds 0.5–1.5 m thick from the Rietgat Formation; Omdraii Vlei 93 farm in T'Kuip Hills area, ca. 65 km north of Britstown, Cape Province, South Africa; 30.03° S, 23.17° E.

1417 (ts, st, TOC, Corg) [J. W. S. field no. 2-6/30/84]. Black and gray chert forming margins of domical stromatolites 5–30 cm in diameter from 5 m above top of "Fall 4" of Truswell and Eriksson (1973), 10 m below the top of the Malmani Subgroup; at the south-side of Great Boetsap River (Hol River), 50 km northwest of Warrenton, Cape Province, South Africa.

1418 (ts, TOC, Corg) [J. W. S. field no. CN110]. Microbanded chert and siltstone from a core sample, DH CN110, Griqualand Exploration and Finance Co. (Kuruman), Kuruman Iron Formation; 10 km east of Hotazel and 47 km northwest of Kuruman, Cape Province, South Africa; 27.20° S, 23.07° E.

1419 (ts, TOC, Corg, Sfide) [J. W. S. field no. JWS-10/26/84]. Chert from a core sample, DH NVT1 at a depth of 3008 m, Kameeldoorns Formation; 15 km south of Bothaville at Nooitverwacht 248 farm, 210 km southwest of Johannesburg, Transvaal, South Africa.

1890 (ts) [G. J. R. field no. R341]. Schist from 10 cm below top of profile, at base of the Dwaal Heuvel Quartzite; roadcut 2.7 km west of Waterval Onder, Transvaal, South Africa.

1891 (ts) [G. J. R. field no. R342]. Schist from 30 cm below top of profile, Dwaal Heuvel Quartzite; locality cf. 1890.

1892 (ts) [G. J. R. field no. R343]. Schist from 50 cm below top of profile, Dwaal Heuvel Quartzite; locality cf. 1890.

1894 (ts) [G. J. R. field no. R344]. Schist from 70 cm below top of profile, Dwaal Heuvel Quartzite; locality cf. 1890.

1895 (ts) [G. J. R. field no. R345]. Schist from 110 cm below top of profile, Dwaal Heuvel Quartzite; locality cf. 1890.

1896 (ts) [G. J. R. field no. R346]. Schist from 130 cm below top of profile, Dwaal Heuvel Quartzite; locality cf. 1890.

1897 (ts) [G. J. R. field no. R347]. Schist from 150 cm below top of profile, Dwaal Heuvel Quartzite; locality cf. 1890.

1898 (ts) [G. J. R. field no. R348]. Schist from 200 cm below top of profile, Dwaal Heuvel Quartzite; locality cf. 1890.

1899 (ts) [G. J. R. field no. R349]. Schist from 280 cm below top of profile, Dwaal Heuvel Quartzite; locality cf. 1890.

1900 (ts) [G. J. R. field no. R350]. Schist from 310 cm below top of profile, Dwaal Heuvel Quartzite; locality cf. 1890.

1901 (ts) [G. J. R. field no. R351]. Schist from 330 cm below top of profile, Dwaal Heuvel Quartzite; locality cf. 1890.

1902 (ts) [G. J. R. field no. R352]. Schist from 360 cm below top of profile, Dwaal Heuvel Quartzite; locality cf. 1890.

1903 (ts) [G. J. R. field no. R353]. Schist from 420 cm below top of profile, Dwaal Heuvel Quartzite; locality cf. 1890.

1904 [G. J. R. field no. R356]. Andesitic basalt, paleosol from 560 cm below top of paleosol, Hekport Andesite; locality cf. 1890.

1905 [G. J. R. field no. R359]. Quartzite from the basal quartzite, 10 cm above clay paleosol, Dwaal Heuvel Formation; locality cf. 1890.

2498 (ts, TOC, Corg) [J. M. H. field no. 840625-1a/B638]. Thucolite from the Carbon Leader "Member" of the Main Conglomerate Formation; the 1470 raise, West Driefontein Mine 14–70, 5 km southeast of Carletonville, Transvaal, South Africa.

2499 (ts, TOC, Corg) [J. M. H. field no. 840625-1b/B639]. Thucolite from the North Leader "Member" of the Main Conglomerate Formation; the 1470 raise, locality cf. 2498.

2500 (Corg) [J. M. H. field no. 850625-1c/B640]. Fibrous thucolite from 100 m below 2498, Carbon Leader "Member" of the Main Conglomerate Formation; locality cf. 2498.

2501 (TOC, Corg) [J. M. H. field no. 840625-1d/B641]. Thucolite from below 2500, Carbon Leader "Member"; locality cf. 2498.

Table 14.2. *Continued.*

2502 (ts, TOC, Corg) [J. M. H. field no. 840625-1e/B642]. Thucolite from the north leader contact zone of the North Leader "Member"; locality cf. 2498.

2503 (ts, TOC, Corg) [J. M. H. field no. 840625-2a/B643]. Black shale from the Witwatersrand Supergroup; Connaught Farm, Transvaal, South Africa.

2504 (TOC, Corg) [J. M. H. field no. 840625-4a/B644]. Black shale from a core sample, depth of 39.2 m, from the middle part of the Jeppestown Shale; core 3SB1, West Driefontein Mine 14–70, 5 km southeast of Carletonville, Transvaal, South Africa.

2505 (TOC, Corg) [J. M. H. field no. 840625-4b/B645]. Black shale from a core sample, depth of 357 m, Government Subgroup; core 3SB1, locality cf. 2504.

2506 (ts, TOC, Corg) [J. M. H. field no. 840625-4c/B646]. Black shale from a core sample, depth of 155 m, Jeppestown Shale; core 3SB1, locality cf. 2504.

2507 (ts, TOC, Corg) [J. M. H. field no. 840626-1a/B647]. Thucolite from level 16, Venterspost Carbon Reef "Member" of the Venterspost Formation; Harvie Watt Shaft of the Libanon Gold Mine, near Carletonville, Transvaal, South Africa.

2508 (ts, TOC, Corg) [J. M. H. field no. 840626-1b/B648]. Shale from level 16, Kimberley Shale; locality cf. 2507.

2509 (ts) [J. M. H. field no. 840626-1d]. Thucolite with buckshot pyrite from level 16, Venterspost Carbon Reef "Member"; locality cf. 2507.

2510 (ts, Ccarb) [J. M. H. field no. 840627-2a/B649]. Manganese carbonate from the Hotazel Formation; Mamatwan manganese mine, Transvaal, South Africa.

2511 (ts, Ccarb) [J. M. H. field no. 840627-2b/B650]. Manganese carbonate from the Hotazel Formation; locality cf. 2510.

2512 (ts, st, TOC, Corg, ker) [J. M. H. field no. 840628-1a/B651]. Stromatolitic carbonate from the Gamohaan Formation; Danielskuil G3 section of Beukes (1980), Transvaal, South Africa.

2513 (ts, st, TOC, Corg, ker) [J. M. H. field no. 840628-1b/B652]. Well-laminated, stromatolitic carbonate from the Gamohaan Formation; locality cf. 2512.

2514 (TOC, Corg, ker) [J. M. H. field no. 840628-1c/B653]. Carbonate cemented tuff from the Gamohaan Formation; locality cf. 2512.

2515 (ts, st, TOC, Corg) [J. M. H. field no. 840628-1d/B654]. Stromatolitic chert domes from the Gamohaan Formation; locality cf. 2512.

2516 (ts, TOC, Corg) [J. M. H. field no. 840628-1e/B655]. Dolomite from the Gamohaan Formation; locality cf. 2512.

2517 (ts, st, TOC, Corg, ker) [J. M. H. field no. 840628-1f/B656]. Stromatolitic (*Conophyton*) sparry carbonate with contorted cryptalgal laminae from the Gamohaan Formation; locality cf. 2512.

2518 (ts, st, TOC, Corg, ker) [J. M. H. field no. 840628-1g/B657]. Sample cf. 2517 from the Gamohaan Formation; locality cf. 2512.

2519 (ts, st, TOC, Corg) [J. M. H. field no. 840628-1h/B658]. Center of a silicified stromatolite (*Conophyton*) from the Gamohaan Formation; locality cf. 2512.

2520 (ts, st, TOC, Corg) [J. M. H. field no. 840628-1i/B659]. Sample cf. 2519, but from the periphery of the sample, from the Gamohaan Formation; locality cf. 2512.

2521 (ts, st, TOC, Corg) [J. M. H. field no. 840628-1j/B660]. Stromatolitic carbonate from the Gamohaan Formation; locality cf. 2512.

2522 (ts, st, TOC, Corg) [J. M. H. field no. 840628-1k/B661]. Small domical carbonate stromatolites from the Gamohaan Formation; locality cf. 2512.

2523 (ts, TOC, Corg) [J. M. H. field no. 840628-1m/B662]. Black carbonate from the Gamohaan Formation; locality cf. 2512.

2524 (ts, TOC, Corg) [J. M. H. field no. 840628-2a/B663]. Carbonate from the Rooinekke Formation; Taiboschfontein farm, Transvaal, South Africa.

2525 (TOC, Corg) [J. M. H. field no. 840628-2a/B664]. Carbonate from the Rooinekke Formation; locality cf. 2524.

2526 (ts, st, TOC, Corg) [J. M. H. field no. 840628-3a/B665]. Stromatolitic black chert from the Klipfonteinheuwel Formation; on road from Kimberly to Griquatown, 20 km east of Griquatown, Transvaal, South Africa.

2527 (ts, TOC, Corg) [J. M. H. field no. 840629-1a/B666]. Silicified tuff from the Ventersdorp Supergroup; Omdraii Vlei 93 farm in T'Kuip Hills area, ca. 65 km north of Britstown, Cape Province, South Africa; 30.03° S, 23.17° E.

2528 (ts, TOC, Corg) [J. M. H. field no. 840629-2b/B667]. Massive, black carbonate from a core sample of the Ventersdorp Supergroup; locality cf. 2527.

2529 (ts, TOC, Corg) [J. M. H. field no. 840629-3a/B668]. Laminated, gray carbonate from a core sample of the Ventersdorp Supergroup; locality cf. 2527.

2530 (ts, st, Corg) [J. M. H. field no. 840630-2a/B669]. Dark black carbonate from a large domical stromatolite from section 37, Reivilo Formation; "Boetsap section," Transvaal, South Africa.

2531 (st) [J. M. H. field no. 840630-2b]. Dark gray carbonate with tufted mat on surface from the Reivilo Formation; locality cf. 2530.

2532 (TOC, Corg, ker) [J. M. H. field no. 840630-2c/B670]. Gray calcareous shale from the Reivilo Formation; above large sand ripples at the second waterfall of Truswell and Eriksson (1973), locality cf. 2530.

2533 (TOC, Corg, ker) [J. M. H. field no. 840630-2d/B671]. Dark gray shale from the Reivilo Formation; "waterfall no. 3" of Truswell and Eriksson (1973), locality cf. 2530.

2534 (TOC, Corg) [J. M. H. field no. 840630-2e/B672]. Gray shale from the Reivilo Formation; base of "waterfall no. 4" of Truswell and Eriksson (1973), locality cf. 2530.

2535 (TOC, Corg) [J. M. H. field no. 840630-2f/B673]. Black quartzite from recrystallized oolitic chert(?) from float of the Reivilo Formation; below "waterfall no. 3" of Truswell and Eriksson (1973), locality cf. 2530.

2543 (st) [J. M. H. field no. 840616-1a]. Massive carbonate from a stromatolitic sequence of the Hekpoort Andesite; Langspruit farm, Transvaal, South Africa.

2776 (Sfide) [J. W. H. field no. FIN 196-3]. Calcareous shale from 0.7 m below the top of the Gamohaan Formation; Finsch Diamond Mine, Transvaal, South Africa.

2777 (Sfide) [J. W. H. field no. FIN 208-2]. Carbonaceous shale from 9.6 m below the top of the Gamohaan Formation; locality cf. 2776.

2778 (Sfide) [J. W. H. field no. FIN 208-3]. Carbonaceous, dolomitic shale from 10.8 m below the top of the Gamohaan Formation; locality cf. 2776.

KAROO BASIN

Tectonic unit/location. Karoo Basin, Africa.

Stratigraphy.

    ECCA GROUP
        Vryheid Formation

PPRG samples/localities

2583 [J. M. H. field no. 840624-1a]. Siderite with some black micaceous siltstone from the Ecca Group; "Tunnel section," South Africa.

2584 [J. M. H. field no. 840624-1b]. Coal from the Vryheid Formation; near dike, "J. M. H. Karoo locality," Transvaal, South Africa.

2585 [J. M. H. field no. 840624-1c]. Coal from the Vryheid Formation; locality cf. 2584.

2586 [J. M. H. field no. 840624-1d]. Coal from the Vryheid Formation; locality cf. 2584.

2587 [J. M. H. field no. 840624-1e]. Coal from the Vryheid Formation; locality cf. 2584.

2588 [J. M. H. field no. 840624-1f]. Coal from the Vryheid Formation; locality cf. 2584.

MODERN SAMPLES

PPRG samples/localities

2411 (st) [Y. C. field collection]. Modern lithified microbial mat from the Doran Drraai Sabka, located near Namatoni in Etosha National Park, Namibia, Africa.

WIT MFOLOZI INLIER

Tectonic unit/location. Wit Mfolozi Inlier, Natal, South Africa.

Stratigraphy. Stratigraphic section, Wit Mfolozi River (Babanango area), northern Natal (von Brunn and Mason, 1977):

PONGOLA
    MOZAAN GROUP
    INSUZI GROUP

PPRG samples/localities

1408 (ts, st, TOC, Corg) [J. W. S. field no. 1-6/23/84]. Bulbose gray chert stromatolites (5–10 cm diameter) interbedded with carbonate from the Insuzi Group; 1.5 km south of the intersection of the White Umfolozi River and the Vryheid-Malmoth Road in the Wit-Mfolozi River valley, northern Natal Province, ca. 200 km north of Durban, Natal, South Africa.

1409 (ts, st, TOC, Corg) [J. W. S. field no. 2-6/23/84]. Gray chert stratiform and bulbose stromatolites capped with clastic carbonate exhibiting interference ripple marks from the Insuzi Group; on the west bank of the Wit-Mfolozi River, 100 m south, locality cf. 1408.

2581 [J. M. H. field no. 840623-1]. Calcareous siltstone or mudstone from the Pongola Supergroup; near new bridge, locality cf. 1408.

2582 [J. M. H. field no. 840623-1b]. Dolomite from the first stromatolite horizon of the Pongola Supergroup; locality cf. 1408.

14.2.2 ASIA

BASHKIRIA ANTICLINORIUM

Tectonic unit/location. Bashkiria Anticlinorium, Bashkir A. S. S. R., USSR.

Stratigraphy. Stratigraphic section, Ufa area, central Bashkir A. S. S. R. (Keller, 1975):

    ASHA GROUP
        Basa Formation
        Uryuk Formation

Geologic Samples

Table 14.2. *Continued.*

- **KARATAU GROUP**
  - Bakeevo Formation
  - Uk-Kudash Formation
  - Min'yar Formation
    - Upper Member
    - Bianka Member
    - Lower Member
  - Inzer Formation
  - Podinzer Formation/Sim Formation
  - Katav Formation
    - Upper Member
    - Middle Member
    - Lower Member
  - Zil'merdak Formation
    - Bederysh Member
    - Lemeza Member
    - Nugash Member
    - Biryan Member
- **YURMATA GROUP**
  - Avsyan Formation
    - Revet Member
    - Kuktur Member
    - Ushakov Member
    - Maloinzer Member
    - Kataskin Member
  - Zigazino-Komarovsk Formation
    - Tukan Member
    - Ambarka Member
    - Seregin Member
  - Zigal'ga Formation
  - Mashak Formation
- **BURZYAN GROUP**

PPRG samples/localities

1478 (ts, μf^*, TOC, Corg) [J. W. S. field no. R3MN2(ST)]. Microfossiliferous (Chapter 24) black chert occurring as lenses, nodules and more or less continuous beds 3–15 cm thick interbedded with light gray carbonates from the middle member of the Min'yar Formation; Zimlin River valley, 7 km northeast of the village of Bakeyvo, 150 km southeast of Ufa, central Bashkir, USSR; 53.82° N, 57.13° E.

1479 (ts, TOC, Corg) [J. W. S. field no. R3MN2-"1"]. Bedded black chert from the lowest chert horizon of the Bianka Member of the Min'yar Formation; locality cf. 1478.

1480 (ts, TOC, Corg) [J. W. S. field no. R3MN2-"2"]. Bedded black chert from the middle chert horizon of the Bianka Member; locality cf. 1478.

1481 (ts, TOC, Corg) [J. W. S. field no. R3MN2(RIV)]. Bedded black chert from the upper chert horizon of the Bianka Member; locality cf. 1478.

1482 (ts, TOC, Corg) [J. W. S. field no. R3MN2(BL)]. Bedded black chert from the upper chert horizon of the Bianka Member; locality cf. 1478.

1483 (ts, TOC, Corg) [J. W. S. field no. R3MN2-"3"]. Bedded black chert from the topmost chert horizon of the Bianka Member; locality cf. 1478.

1484 (ts, TOC, Corg) [J. W. S. field no. R3KT(MID)]. Chert-carbonate from the middle member of the Katav Formation; locality cf. 1478.

1485 (ts, st, TOC, Corg) [J. W. S. field no. R3KT(SM)]. Gray tabular stromatolitic, oncolitic (*Osagia*), biohermal limestone containing microphytolites from the upper "Simsk" member of the Katav Formation; locality cf. 1478.

1486 (ts, TOC, Corg) [J. W. S. field no. R2AV5]. Bedded, fine-grained black chert from the uppermost portion of the Revet Member of the Avsyan Formation; southern Urals, left bank of Bolshoy Avzyan River, upstream from Verkhny Avzyan village, central Bashkir, USSR.

1487 (ts, TOC, Corg) [J. W. S. field no. R2ZK3-C-1]. Dark gray to black, phyllitized shale from the uppermost portion of the Tukan Member of the Zigazino-Komarovsk Formation; locality cf. 1486.

1488 (ts, TOC, Corg) [J. W. S. field no. R2ZK3-C-2]. Dark gray to black, phyllitized shale from the uppermost portion of the Tukan Member; locality cf. 1486.

1489 (ts, TOC, Corg) [J. W. S. field no. R3ZLM4-C-3]. Brown siltstone and shale from the uppermost portion of the Bederysh Member of the Zil'merdak Formation; right bank of the Nugush River, near Biktashev, Ufa region, central Bashkir, USSR.

1490 (ts, TOC, Corg) [J. W. S. field no. R2AV1-C-9]. Shale interbedded with massive dolomitic carbonate from the lowermost portion of the Kataskin Member of the Avzyan Formation; southern Urals, left bank of Bolshoy Avzyan River, upstream from Verkhny Avzyan village, central Bashkir, USSR.

1491 (ts, TOC, Corg) [J. W. S. field no. R2AV3-C-15]. Gray shale interbedded with gray carbonate from the middle portion of the Ushakov Member of the Avzyan Formation; locality cf. 1490.

1492 (ts, TOC, Corg) [J. W. S. field no. R2AV2-C-12]. Medium gray carbonaceous shale from bed 12 of a measured section (Keller 1975) of the Maloinzer Member of the Avzyan Formation; locality cf. 1490.

1493 (ts, TOC, Corg) [J. W. S. field no. R2AV4-C-21]. Gray shale from bed 21 of a measured section (Keller 1975) of the Kuktur Member of the Avzyan Formation; locality cf. 1490.

CHITALDRUG SCHIST BELT

Tectonic unit/location. Chitaldrug Schist Belt, southern India.

Stratigraphy. Stratigraphic section, Dharwar region (Gowda and Sreenivasa 1969):

　Dharwar Group
　　Guddadarangayanhalli Formation

PPRG samples/localities

1310 (ts, st, TOC, Corg) [J. W. S. field no. 1 of 10/01/73]. Laminated, non-stromatolitic gray chert (0.5–2 cm thick) from the Guddadarangayanahalli Formation; vertical 15 m high outcrop, at the north side of unpaved cart road, 190 km northwest of Bangalore, southwest India.

CUDDAPAH BASIN

Tectonic unit/location. Cuddapah Basin, Andhra Pradesh, India.

Stratigraphy. Stratigraphic section, Kurnool area, western Andhra Pradesh (Meijerink et al. 1984);

　KURNOOL GROUP
　　Kundair Formation
　　Panem Formation
　　Banganapalle Formation
　NALLAMALAI GROUP
　　CUMBUM SUBGROUP
　　　Giddalur Formation
　　　Pullampet Formation
　　BAIRENKONDA SUBGROUP
　　　Dornala Formation
　　　Iswara Kuppam Formation
　CUDDAPAH GROUP
　　MOGAMURERU SUBGROUP
　　　Tadpatri Formation
　　　Pulivendla Formation
　　PAPAGHNI SUBGROUP
　　　Vempalle Formation
　　　Gulcheru Formation
　[UNNAMED UPPER ARCHEAN/LOWER PROTEROZOIC GROUP]
　　Closepet Granite

PPRG samples/localities.

1308 (ts, st, μf^*, TOC, Corg) [J. W. S. field no. 1. of 9/28/73]. Microfossiliferous (Chapter 24) black chert from carbonate *Collenia*-like stromatolites from the upper horizon of the Vempalle Formation; near the base of a 50 m sequence, 350 m west of main workings (as of 1973) of Bramanapalle Asbestos Mine, Andra Pradesh Mining Co.; ca. 240 km northwest of Madras, south-central India; 14.42°N, 78.20°E.

1309 (ts, TOC, Corg) [J. W. S. field no. 1 of 9/28/73]. Chert from the Vempalle Formation; locality cf. 1308.

MALYJ KARATAU

Tectonic unit/location. Malyj Karatau, southern Kazakhstan, USSR.

Stratigraphy. Stratigraphic section, Malyj Karatau region, Kazakhstan (Ogurtsova and Sergeev 1987).

　TAMDY SERIES
　　Shabakty Formation
　　Chulaktau Formation
　　Kyrshabakty Formation
　MALYJ KAROJ SERIES
　　Kurgan Formation
　　Chichkan Formation
　　Aktugaj Formation
　　Koksu Formation
　ZHANATUS SERIES

Geologic Samples

Table 14.2. Continued.

PPRG samples/localities.

1473 (ts, st, μf*, TOC, Corg) [J. W. S. field no. SOV-1054(1)-KC]. Microfossiliferous (Chapter 24) black chert stromatolites (*Conophyton gaubitza*) from the Chichkan Formation; Malyj Karatau Mountains, Ayusakan, southern Kazakhstan, USSR.

1474 (ts, st, TOC, Corg) [J. W. S. field no. SOV-1054(1)-KC]. Black chert conical stromatolite (*Conophyton*) from the Chichkan Formation; locality cf. 1473.

1475 (ts, TOC, Corg) [J. W. S. Field no. SOV-18-KC]. Bedded black chert from a core sample, from the upper part of the core, Chulaktau Formation; locality cf. 1473.

1476 (ts, TOC, Corg) [J. W. S. field no. SOV-19-KC]. Bedded black chert from a core sample, upper part of the drill core, Chulaktau Formation; locality cf. 1473.

1477 (ts, TOC, Corg) [J. W. S. field no. SOV-160-KC]. Bedded black chert from a core sample, upper part of the core, Chulaktau Formation; locality cf. 1473.

NORTH CHINA PLATFORM (EASTERN BORDER REGION)

Tectonic unit/location. North China Platform (eastern border region), Anhwei, People's Republic of China.

Stratigraphy. Stratigraphic section, Suxian area, northern Anhwei (S. Ouyang, pers. comm. to J. W. S. 1985):

SINIAN "SUPERGROUP"
  SUXIAN GROUP
    Gouhou Formation
    Jinshanzhai Formation
  XUHUAI GROUP
    Wanshan Formation
    Shijia Formation
    Weiji Formation
    Zhangqu Formation
    Jiudingshan Formation
    Niyuan Formation
    Zhaowei Formation
  HUAINAN GROUP

PPRG samples/localities.

1423 (ts, TOC, Corg) [J. W. S. field no. 2-5/12/81]. Bedded black and gray chert from the middle third of the Zhangqu Formation; outcrop on low hill 1.5 km east of Jaigon, 30 km north of Suxian, northern Anhui Province, northern China.

1424 (ts, TOC, Corg) [J. W. S. field no. 1-5/12/81]. Microfossiliferous bedded black and white chert from a continuous 0.5 m-thick bed interbedded with carbonate from the upper portion of the Jiudingshan Formation; locality cf. 1423.

NORTH CHINA PLATFORM (INTERIOR REGION)

Tectonic unit/location. North China Platform (interior region), Shaanxi, People's Republic of China.

Stratigraphy. Stratigraphic section, Luonan area, east-central Shaanxi (Qiu and Liu 1982):

[UNNAMED CAMBRIAN GROUP]
  Xingji Formation
SINIAN "GROUP" (SYSTEM)
  Luoquan Formation
QINGBAIKOU "GROUP" (SYSTEM)
  Dazhuang Formation
LUONAN GROUP
  Fengjiawan Formation
  Duguan Formation
  Xunjiansi Formation
  Longjiayuan Formation
GAOSHANHE GROUP
  Chenngjiajian Formation
  Erdaohe Formation
  Biegaizi Formation

PPRG samples/localities.

2163 (ts) [J. W. S. field no. 1-5/27/87]. Medium gray, quartzose sandstone from the lower third of the Biegaizi Formation; road cut, east side of Huanglongpu-Shimeng road, 100 m south of Huanlongpu, 4 km north of Chengjiajian village, 105 km east of Xian, Shaanxi (Shensi) Province, northern China (Fig. 14.3).

Table 14.2. *Continued.*

2164 (ts) [J. W. S. field no. 2-5/27/87]. Pink quartzite from the middle of the Erdaohe Formation; east side of roadcut, 2 km north of Chengjiajian Village, northern China (Fig. 14.3).

2165 (ts) [J. W. S. field no. 3-5/27/87]. White vitreous orthoquartzite from the uppermost beds, 1–2 m below top of the Chengjiajiang Formation; on footpath ca. 150 m east of road, at Shizhuang village, half the way between Chengjiajian and Yangsicheng villages, northern China (Fig. 14.3).

2166 (ts) [J. W. S. field no. 4-5/27/87]. Pisolitic red carbonate from ca. 2–3 m above 2165, Longjiayuan Formation; locality cf. 2165.

2167 (st) [J. W. S. field no. 5-05/27/87]. Red carbonate stromatolite (*Colonnella heishanensis*), from ca. 5–7 m above 2166, from the lower 10 m of the Longjiayuan Formation; locality cf. 2165.

2168 (ts, TOC, Corg) [J. W. S. field no. 6-5/27/87]. Gray dolostone from the middle third of the Xunjiansi Formation; ca. 1 km south of Yangsicheng village, northern China (Fig. 14.3).

2169 (ts, TOC, Corg) [J. W. S. field no. 7-5/27/87]. Gray chert lens, ca. 15 cm thick, interbedded with gray dolostone from the middle third of the Xunjiansi Formation; locality cf. 2168.

2170 (ts, TOC, Corg) [J. W. S. field no. 8-5/27/87]. Black chert and chert-carbonate from the middle third of the Xunjiansi Formation; locality cf. 2168.

2171 (ts, TOC, Corg) [J. W. S. field no. 9-5/27/87]. Massive, bedded gray dolostone from 35 m above the base of the Duguan Formation; east side of roadcut, 750 m north of Shangzhangwan village, northern China (Fig. 14.3).

2172 (ts) [J. W. S. field no. 10-5/27/87]. Slate from red and yellow slate unit, middle third of the Duguan Formation; locality cf. 2171.

2173 (ts, TOC, Corg) [J. W. S. field no. 11-5/27/87]. Gray dolostone, containing *Colonnella cormosa*, from the upper third of the Fengjiawan Formation; 250 m south, locality cf. 2171.

2174 [J. W. S. field no. 12-5/27/87]. Varves with dropstones from 50 cm below diamictite, 1 m above the lower glacial member, Luoquan Formation; locality cf. 2171.

2175 [J. W. S. field no. 13-5/27/87]. Diamictite from ca. 3 m above the base of the lower member of the Luoquan Formation; locality cf. 2171.

2176 (TOC, Corg, Sfide) [J. W. S. field no. 14-5/27/87]. Dark gray varves with dropstones, from 8–9 m above the base of the lower member of the Luoquan Formation; locality cf. 2171.

2177 [J. W. S. field no. 15-5/27/87]. Thin mass-flow diamictite unit within varves, from 8–9 m above the base of the lower member of the Luoquan Formation; locality cf. 2171.

2178 (ts, TOC, Corg, ker) [J. W. S. field no. 16-5/27/87]. Dark gray siltstone varves with fine dropstones, from 11–12 m above the base of the lower member of the Luoquan Formation; locality cf. 2171.

2179 (TOC, Corg) [J. W. S. field no. 17-5/27/87]. Gray shale from ca. 6.5 m above the contact with lower member of the Luoquan Formation; on shale slope east of road, locality cf. 2171.

2180 (TOC, Corg) [J. W. S. field no. 18-5/27/87]. Gray shale from ca. 9 m above the base of the upper member of the Luoquan Formation; locality cf. 2171.

2181 (TOC, Corg) [J. W. S. field no. 19-5/27/87]. Gray shale from 12 m above the base of the upper member of the Luoquan Formation; locality cf. 2171.

2182 (TOC, Corg) [J. W. S. field no. 20-5/27/87]. Gray shale from ca. 16 m above the base of the upper member of the Luoquan Formation; locality cf. 2171.

2183 (TOC, Corg) [J. W. S. field no. 21-5/27/87]. Gray shale from ca. 22 m above base of the upper member of the Luoquan Formation; locality cf. 2171.

2184 (TOC, Corg) [J. W. S. field no. 22-5/27/87]. Gray shale from ca. 27 m above base of the upper member of the Luoquan Formation; locality cf. 2171.

2185 (TOC, Corg, ker) [J. W. S. field no. 23-5/27/87]. Gray shale parting, ca. 2 cm thick, interbedded with blocky, sandy shale (massive) unit from ca. 32 m above the base of the upper member of the Luoquan Formation; on shale slope east of road, 0.5 km south of Shangzhangwan village, northern China (Fig. 14.3).

2186 (TOC, Corg, ker) [J. W. S. field no. 24-5/27/87]. Dark gray shale from the topmost sandy unit in the upper member, ca. 36 m above the base of Zz12 of the Luoquan Formation; locality cf. 2185.

2187 (TOC, Corg) [J. W. S. field no. 25-5/27/87]. Gray shale from ca. 40 m above the base of the upper member of the Luoquan Formation; on shale slope east of road, 750 m north of Shangzhangwan village, northern China (Fig. 14.3).

2188 [J. W. S. field no. 26-5/27/87]. Buff, sandy siltstone from ca. 3.5 m above 2187, Xingji Formation; locality cf. 2187.

NORTH CHINA PLATFORM (NORTHERN BORDER REGION)

Tectonic unit/location. North China Platform (northern border region), Tianjin Shi, People's Republic of China.

Stratigraphy. Stratigraphic section, Jixian area, northern Tianjin Shi (Tientsin Inst. Geol. Miner. Res. 1977):

Table 14.2. *Continued.*

QINGBAIKOU "GROUP" (SYSTEM)
    Jingeryu Formation/Luotuoling Formation
    Xiamaling Formation
JIXIAN "GROUP" (SYSTEM)
    Tieling Formation
    Hongshuizhuang Formation
    Wumishan Formation
    Yangzhuang Formation
CHANGCHENG "GROUP" (SYSTEM)
    Gaoyuzhuang Formation
    Dahongyu Formation
    Tuanshanzi Formation
    Chuanlinggou Formation
    Changzhougou Formation
QIANXI "GROUP" (SYSTEM)

PPRG samples/localities.

1420 (ts, st, TOC, Corg) [J. W. S. field no. 2-8/24/82]. Chert from silicified stromatolite (*Conophyton cylindricum*), weathering out of enclosing buff-colored dolomite, from the upper fourth of the Gaoyuzhuang Formation; north side of strike valley, ca. 1 km both to east and west of unpaved commune road adjacent to Sangshuan Elementary School, 12.5 km northeast of Jixian, northern China (Fig. 14.4).

1421 (ts, st, TOC, Corg) [J. W. S. field no. 1-8/24/82]. Stromatolitic bedded black chert, essentially flat-laminated, 10–30 cm (rarely 75 cm) thick, interbedded with carbonates, from the lower fourth of the Gaoyuzhuang Formation; south side of an east-trending creek bed, 50–60 m east of an unpaved commune road, ca. 1.2 km northeast of Sangshuan Elementary School, Yanshan Range, 14 km northeast of Jixian, northern China (Fig. 14.4).

1422 (ts, TOC, Corg) [J. W. S. field no. 1-4/28/81]. Bedded black chert from the lower fourth of the Gaoyuzhuang Formation; north side of strike valley, ca. 1 km both to east and west of unpaved commune road adjacent to Sangshuan Elementary School, 12.5 km northeast of Jixian, northern China (Fig. 14.4).

1425 (ts, TOC, Corg) [J. W. S. field no. 3-8/24/82]. Bedded black chert from 1 m above the base of the Wumishan Formation; 0.7 km east of Jixian-Huangyaguan Gate Road, 2–3 km south of road across river, on north side of small valley, northern China (Fig. 14.4).

1426 (ts, TOC, Corg) [J. W. S. field no. 2-8/25/82]. Bedded black chert with blocky fractures, from a 0.5–1.0 m-thick continuous bed from the lower portion of "member 2" of the Wumishan Formation; low ridge perpendicular to Jixian-Huangyaguan Gate Road, 100 m east of road, 1 km north of Ershilipu, northern China (Fig. 14.4).

1427 (ts, st, TOC, Corg) [J. W. S. field no. 1-8/26/82]. Stromatolitic black and white banded chert bed ca. 20 cm thick with nodular top, interbedded with buff to gray carbonate from "member 3" of the Wumishan Formation; roadcut on the west side of the Jixian-Huangyaguan Gate Road, ca. 0.5 km south of Ershilipu, ca. 9 km north of Jixian, northern China (Fig. 14.4).

1428 (ts, TOC, Corg) [J. W. S. field no. 3-8/26/82]. Banded black chert with a nodular upper surface, bed ca. 10 cm thick from "member 3" of the Wumishan Formation; north side of roadcut, ca. 1.7 km south of Ershilipu, ca. 8 km north of Jixian on the Jixian-Huangyaquan Gate Road, northern China (Fig. 14.4).

1429 (ts, st, TOC, Corg) [J. W. S. field no. 4-8/26/82]. Stromatolitic black and white banded chert (*Stratifera* with ca. 5 cm relief), bed ca. 0.75 m thick from "member 3" of the Wumishan Formation; roadcut, ca. 2 km south of Ershilipu, ca. 7.5 km north of Jixian on the Jixian-Huangyaguan Gate Road, northern China (Fig. 14.4).

1430 (ts, st, TOC, Corg) [J. W. S. field no. 5-8/26/82]. Stromatolitic black chert (*Stratifera*, and *Collenia* ca. 10 cm in diameter), bed ca. 4–15 cm thick from the upper portion of "member 3" of the Wumishan Formation; 1 km east of Hongshuizhuang, ca. 2.8 km south of Ershilipu, ca. 6.8 km north of Jixian on the Jixian-Huangyaguan Gate Road, northern China (Fig. 14.4).

1431 (ts, TOC, Corg) [J. W. S. field no. 6-8/26/82]. Finely bedded black chert, 3–8 cm thick, from "member 4" of the Wumishan Formation; roadcut on the north side of the Jixian-Huangyaguan Gate Road, 0.4 km east of Hongshuizhuan, 3.4 km south of Ershilipu, 6.2 km north of Jixian, northern China (Fig. 14.4).

1432 (ts, TOC, Corg) [J. W. S. field no. 2-8/24/82]. Black chert from the upper fourth of the Gaoyuzhuang Formation; north side of strike valley, ca. 1 km both to east and west of unpaved commune road adjacent to Sangshuan Elementary School, 12.5 km northeast of Jixian, northern China (Fig. 14.4).

1433 (ts, st, TOC, Corg) [J. W. S. field no. 1-8/25/82]. Stromatolitic chert, 5 m-thick bed containing microdigitate stromatolites (*Pseudogymnosolen mopangyuensis*) from the upper part of "member 1" (lowest) of the Wumishan Formation; on top of hill, 300 m east of Jixian-Huangyaguan Gate Road and 500 m south of where road turns west toward river, near Erlipu, northern China (Fig. 14.4).

1434 (ts, st, TOC, Corg) [J. W. S. field no. 81-001D]. Chert microdigitate stromatolite (*Pseudogymnosolen mopangyuensis*) from "member 1" (bed 6, outcrop A) of the Wumishan Formation; near Mopangyu Village, ca. 10 km northeast of Jixian China, outcrop A of Cao (1981), northern China (Fig. 14.4).

Table 14.2. *Continued.*

1435 (ts, st, TOC, Corg) [J. W. S. field no. 82-W2-F2]. Chert stromatolite (*Pseudogymnosolen mopangyuensis*) from "member 1" (bed 6, outcrop B) of the Wumishan Formation; locality cf. 1434.

2123 (ts, μf^*, TOC, Corg) [J. W. S. field no. 1-5/20/87]. Microfossiliferous (Chapter 24) black chert with carbonate from ca. 10 m above the base, lower third of the Gaoyuzhuang Formation; locality "A" of Schopf et al. 1984, northern China (Fig. 14.4).

2124 (ts, TOC, Corg) [J. W. S. field no. 2-5/20/87]. Black chert with carbonate from the lower third of the Gaoyuzhuang Formation; locality cf. 2123.

2125 (ts, TOC, Corg) [J. W. S. field no. 3-5/20/87]. Bedded black chert from the lower third of the Gaoyuzhuang Formation; locality cf. 2123.

2126 (ts, TOC, Corg, ker) [J. W. S. field no. 4-5/20/87]. Bedded black chert from the lower third of the Gaoyuzhuang Formation; locality cf. 2123.

2127 (ts) [J. W. S. field no. 5-5/20/87]. Laminated, stromatolitic black chert (*Conophyton cylindricum*) and carbonate matrix from the upper third of the Gaoyuzhuang Formation; west of road on slope above school house, near Sangshuan school (Jixian area), northern China (Fig. 14.4).

2128 (ts) [J. W. S. field no. 6-5/20/87]. Stromatolitic black chert (*Conophyton cylindricum*) from the upper third of the Gaoyuzhuang Formation; locality cf. 2127.

2129 (ts) [J. W. S. field no. 7-5/20/87]. Stromatolitic black chert (*Conophyton garganticum*) from the upper third of the Gaoyuzhuang Formation; locality cf. 2127.

2130 (ts) D. R. L. field no. DRL-1-052087]. Cross-stratified quartzite with laminations of magnetite from iron-formation from the basal 50 m of the Changzhougou Formation; protected reserve within the village of Changzhou at morth end of section, northern China.

2131 (ts) [D. R. L. field no. DRL-2-052087]. Conglomeratic sandstone from the basal 50 m of the Changzhougou Formation; locality cf. 2130.

2132 (ts) [D. R. L. field no. DRL-3-052087]. Very coarse-grained to granule-bearing quartzite from the middle part of the lower member, one fourth distance from the base of the Changzhougou Formation; locality cf. 2130.

2133 (ts) [D. R. L. field no. DRL-4-052087]. Granule-bearing quartzite from the upper part of the lower member of the Changzhougou Formation; along road marking protected section of Changcheng and Jixian Systems, 1–2 km south of Changzhou Village, northeast of Jixian, north of where the Great Wall crosses a stream, northern China.

2134 (ts) [D. R. L. field no. DRL-5-052087]. Cross-bedded quartzite with magnetite-rich laminations from the upper part of the lower member of the Changzhougou Formation; locality cf. 2133.

2135 (ts) [D. R. L. field no. DRL-6-052087]. Conglomeratic quartzite from the upper part of the lower member of the Changzhougou Formation; locality cf. 2133.

2136 (ts) [D. R. L. field no. DRL-7-052087]. White, medium-grained quartzite from the lower part of the upper member of the Changzhougou Formation; locality cf. 2133.

2137 [D. R. L. field no. DRL-8-052087]. White quartzite from the top of the upper member of the Changzhougou Formation; northern extent of section, ca. 1–2 km south of 2136, locality cf. 2133.

2138 (ts) [D. R. L. field no. DRL-9-052087]. White, vitreous, slightly weathered orthoquartzite from the basal quartzite of the Dahongyu Formation; protected section, ca. 1–2 km south of main road east of Xiaying Village, 1–1.5 km northwest of locality A in Schopf et al. (1984), northern China.

2139 (ker) [H. S. field no. HS-1 of 5/20/87]. Finely laminated, dark gray, fissile, micaceous shale from ca. 200 m below the top of the Chuanlinggou Formation; ca. 500 m north of Liuzhuangzi Village, northern China (Fig. 14.4).

2140 (TOC, Corg, ker) [H. S. field no. HS-2 of 5/20/87]. Micaceous shale from ca. 400 m above the base of the Chuanlinggou Formation; locality cf. 2139.

2141 (TOC, Corg) [H. S. field no. HS-3 of 5/20/87]. Thin, gray shale from ca. 50 m above the base of the Chuanlinggou Formation; ca. 150 m along the dirt road from the white stone which marks the contact, locality cf. 2139.

2142 (ts, TOC, Corg) [H. S. field no. HS-4 of 5/20/87]. Thinly laminated carbonaceous dolostone from the basal part of the Tuanshanzi Formation; along the road from Xiaying to Mashengiao, ca. 0.8 km east of dirt road leading to top of section, Tuanshanzi Village, northern China (Fig. 14.4).

2143 (ts, TOC, Corg, ker) [H. S. field no. HS-5 of 5/20/87]. Thinly bedded light gray dolomite with shale from the lower part of the Tuanshanzi Formation; locality cf. 2142.

2144 (TOC, Corg, ker) [H. S. field no. HS-6 of 5/20/87]. Dark gray shale from 6 m below the top of the Chuanlinggou Formation; locality cf. 2142.

2145 (ts, TOC, Corg) [H. S. field no. HS-7 of 5/20/87]. Gray carbonate from ca. 150 m below the top of the Tuanshanzi Formation; ca. 250 m on an unpaved road, south of road from Xiaying to Tuanshanzi, southwest of Tuanshanzi Village, northern China (Fig. 14.4).

2146 (ts, TOC, Corg) [J. W. S. field no. 1-5/21/87]. Black chert and carbonate from ca. 300 m above the bottom of the Yangzhuang Formation; ca. 40 m west of Jixian-Huangyaguan Gate Road, on a footpath, south side of east-west trending hill, ca. 3 km southwest of Xiaying Village, northern China (Fig. 14.4).

2147 (ts, TOC, Corg) [J. W. S. field no. 2-5/21/87]. Pisolitic chert in carbonate from the Yangzhuang Formation; locality cf. 2146.

2148 (ts, st, TOC, Corg) [J. W. S. field no. 3-5/21/87]. Black chert stringers in gray sparry dolomite from ca. 300 m above the bottom of the Yangzhuang Formation; locality cf. 2146.

Table 14.2. *Continued.*

2149 (ts, TOC, Corg) [J. W. S. field no. 4-5/21/87]. Chert stromatolite (*Pseudogymnosolen*) from ca. 350 m above the bottom of the Wumishan Formation; on footpath in valley, near Mopangyu Village, ca. 10 km northeast of Jixian China, outcrop A of Cao (1981), northern China (Fig. 14.4).

2150 (ts, TOC, Corg) [J. W. S. field no. 5-5/21/87]. Cherty pisolite in carbonate from ca. 300–400 m above the bottom of the Wumishan Formation; locality cf. 2149.

2151 (ts, TOC, Corg) [J. W. S. field no. 6-5/21/87]. Chert and carbonate, very similar to 2147 but has more carbonate, from ca. 300–400 m above the bottom of the Wumishan Formation; locality cf. 2149.

2152 (ts, TOC, Corg) [J. W. S. field no. 7-5/21/87]. Bedded black chert from a 0.5 m-thick bed interbedded with thick gray shale from "member 1" (lowest member) of the Wumishan Formation; footpath to east of Mopangyu, ca. 150 m west of 2149, locality cf. 2149.

2153 (ts, TOC, Corg) [J. W. S. field no. 8-5/21/87]. Gray shale from "member 1" (lowest member) of the Wumishan Formation; locality cf. 2149.

2154 (ts, TOC, Corg) [H. S. field no. 9-5/21/87]. Finely laminated dolomite from "member 2" of the Wumishan Formation; halfway up hill, locality cf. 2149.

2155 (TOC, Corg) [H. S. field no. 10-5/21/87]. Carbonate from the top of "member 2" of the Wumishan Formation; west side, Jixian road, west of Ershilipu Village, northern China (Fig. 14.4).

2156 (ts, TOC, Corg) [J. W. S. field no. 11-5/21/87]. Gypsum-arenite, possibly containing calcium sulfate, from the boundary between "member 2" and "member 3" of the Wumishan Formation; north side, locality cf. 2155.

2155 (ts, TOC, Corg) [J. W. S. field no. 12-5/21/87]. Carbonate from the middle to upper part of "member 3" of the Wumishan Formation; north side, locality cf. 2155.

2158 (TOC, Corg) [J. W. S. field no. 13-5/21/87]. Carbonate from the top of "member 4" (top of formation) of the Wumishan Formation; north side of Jixian road, just east of Hongshuizhuang village, northern China (Fig. 14.4).

2159 (TOC, Corg, ker) [J. W. S. field no. 14-5/21/87]. Dark gray shale from the basal part of the Hongshuizhuang Formation; north side of Jixian road, in road cut, locality cf. 2158.

2160 (ts, TOC, Corg, ker) [J. W. S. field no. 15-5/21/87]. Massive gray shale from ca. 30 m-thick bed at top of the Hongshuizhuang Formation; outcrop 100 m east of Jixian road on north side of dirt track leading to Tieling, at Shaolingzuh village, northern China (Fig. 14.4).

2161 (ts, st, TOC, Corg) [J. W. S. field no. 16-5/21/87]. Carbonate stromatolite (*Baicalia*) from the lower third of the Tieling Formation; east of Jixian road, uphill on dirt track to Shaolingzuh quarry, locality cf. 2160.

2162 (TOC, Corg) [J. W. S. field no. 17-5/21/87]. Gray shale from ca. 10 cm-thick zone interbedded with buff carbonate, ca. 15 m above the base of the Tieling Formation; locality cf. 2160.

OLENYOK UPLIFT

Tectonic unit/location. Olenyok Uplift

Stratigraphy.

    Khatyspyt Formation

PPRG samples/localities.

2409 (ts) [M. A. F. field no. 3995/601]. Chert lens in carbonate from the lower half of the Khatyspyt Formation; left bank of the Khorbusuonka River, 6 km downstream from the junction with the Anabyl River, northern Yakutia, USSR; 70.76°N, 123.7°E.

2410 (ts) [M. A. F. field no. 3995/600]. Chert lens in carbonate from the lower half of the Khatyspyt Formation; locality cf. 2409.

2723 (ts) [M. A. F. field no. K-1]. Dark thin organic films from the upper half of the Khatyspyt Formation; locality cf. 2409.

2724 (ts, TOC, Corg) [M. A. F. field no. K-2]. Thinly laminated silty carbonate with organic films on the bedding plane, from the lower half of the Khatyspyt Formation; locality cf. 2409.

RUSSIAN PLATFORM

Tectonic unit/location. Russian Platform.

Stratigraphy.

    Kostov Formation
    Ust-Pinega Formation

PPRG samples/localities.

2361 (ts) [G. J. R. field no. n/a]. Shale containing *Vendotaenia* from a core sample, depth of 398 m, Kostov Formation; Kranse Gory no. 6 drill hole, near Luga, 120 km east of Leningrad, western USSR.

2722 (ts) [M. A. F. field no. UP-361.5]. Gray-green silty claystone with fine laminations of pyrite, from a core sample, depth of 361.5 m, Ust-Pinega Formation; borehole Malinovka, 13 km southeast from Obozerskij (town), Arkhangelsk region, Russia, USSR; 63.47°N, 40.30°E.

Table 14.2. *Continued.*

14.2.3. AUSTRALIA

ADELAIDE GEOSYNCLINE

Tectonic unit/location. Adelaide Geosyncline, South Australia, Australia.

Stratigraphy. Composite stratigraphic section, Port Augusta and Adelaide areas, southern Flinders Ranges, and a portion of the Peake-Denison Range for the Burra Group, southeastern South Australia (Preiss 1987):

HAWKER GROUP
    Willkawillina Limestone/Ajax Limestone
    Parachilna Formation
    Uratanna Formation

HEYSEN SUPERGROUP
  WILPENNA GROUP
    POUND SUBGROUP
      Rawnsley Quartzite
        Ediacara Member
      Bonney Sandstone
    [UNNAMED UPPER PROTEROZOIC SUBGROUP]
      Wonoka Formation
      Bunyeroo Formation
      ABC Range Quartzite
      Brachina Formation
      Nuccaleena Formation
  UMBERTANA GROUP
    WILLOCHRA SUBGROUP
      Elatina Formation
      Wilmington Formation
      Angepena Formation
    FARINA SUBGROUP
     Brighton Limestone
      Tapley Hill Formation
        Woocalla Dolomite Member
        Tindelpina Shale Member Appila Tillite
  BURRA GROUP
    BELAIR SUBGROUP
      Kadunga Slate
      Gilbert Range Quartzite/Mitcham Range Quartzite
    UNNAMED SUBGROUP]
      Mintaro Shale/Glen Osmond Slate
      Saddleworth Formation
      Auburn Dolomite/Beaumont Dolomite
      Undalya Quartzite/Stonyfell Quartzite
    MUNDALLIO SUBGROUP
      Woolshed Flat Shale/Myrtle Springs Formation
      Skillogalee Dolomite/Montacute Dolomite
    EMEROO SUBGROUP
      Bungaree Formation
      River Wakefield Subgroup
        Blyth Dolomite "Member"
      Rhynie Sandstone
  CALLANNA GROUP
    CURDIMURKA SUBGROUP
    ARKAROOLA SUBGROUP
      Wooltana Volcanics equivalents

[UNNAMED MIDDLE PROTEROZOIC SUPERGROUP (SYSTEM)]
  [UNNAMED MIDDLE PROTEROZOIC GROUP]
    Roopena Volcanics

CARPENTARIAN "SUPERGROUP" (SYSTEM)
  [UNNAMED MIDDLE PROTEROZOIC GROUP]
    Moontana Porphyry
    Gawler Range Volcanics

Table 14.2. *Continued*.

PPRG samples/localities.

1219 (ts, Corg, ker) [J. W. S. field no. 1 of 5/7/68]. Black chert from the lower portion of the River Wakefield Subgroup; 100 m southeast of stream bed in a pasture, 8 km west of Rhynie, ca. 13 km west of Riverton, near B. P. Thomson ref. position no. 17 on photo no. 4359 (Mount Lofty Range), South Australia; Adelaide 1:250,000 map sheet; 34.00°S, 138.5°E.

1220 (ts, μf^*) [J. W. S. field no. 2 of 5/7/68]. Microfossiliferous (Chapter 24) minor black chert beds and nodules within a massive carbonaceous dolostone, from the upper third of the Auburn Dolomite; railroad cut 0.8 km west of Leasingham, 0.4 km north on railroad track from sealed road going west of Leasingham, South Australia; Burra 1:250,000 map sheet; 34.10°S, 138.6°E.

1221 (ts) [J. W. S. field no. 3 of 5/7/68]. Carbonaceous dolostone with minor interbedded black chert from the lower third of the Auburn Dolomite; low hills and ridges on south side of east-west secondary road (unpaved), ca. 1.6 km west-northwest of Wattle Hill, southwest of Spalding on Highway 32, South Australia; Burra 1:250,000 map sheet.

1222 (ts, TOC, Corg) [J. W. S. field no. 4 of 5/7/68]. Oolitic black chert from the upper third of the Rhynie Sandstone; stream bed, 11 km west-northwest of Yatina and 11 km southwest of Black Rock, 122.4 km south of Orroroo, South Australia; Orroroo 1:250,000 map sheet; 33.00°S, 138.5°E.

1223 (ts, TOC, Corg) [J. W. S. field no. 5 of 5/7/68]. Dark gray chert from the upper fourth of the Skillogalee Dolomite; low hill ca. 0.8 km west, locality cf. 1222.

1224 (ts, μf^*, TOC, ker) [J. W. S. field no. 6 of 5/8/68]. Microfossiliferous (Chapter 24) dark gray chert and carbonaceous dolostone from the bottom (Blyth Dolomite?) of the River Wakefield Subgroup; pasture ca. 180 m east of north-south paved road, 4.8 km south of Carrieton Railroad Station, 7.2 km southwest of Carrieton, South Australia; Orroroo 1:250,000 map sheet; 32.50°S, 138.5°F.

1225 (ts, TOC, Corg) [J. W. S. field no. 7 of 6/8/68]. Cherty, carbonaceous, laminated, biohermal(?) dolostone bed from the lower half of the River Wakefield Subgroup; "Carrieton Creek", ca. 0.8 km west, locality cf. 1224.

1226 (ts, TOC, Corg) [J. W. S. field no. 8A of 5/8/68]. Partly silicified and carbonaceous, weathered siltstone from the lower half of the River Wakefield Subgroup; ca. 18 m downstream and to the east of 1225, locality cf. 1224.

1227 (ts, TOC, Corg) [J. W. S. field no. 8B of 5/8/68]. Partly silicified and carbonaceous, weathered siltstone from 30 cm above 1226, lower half of the River Wakefield Subgroup; locality cf. 1224.

1228 (ts, TOC, Corg) [J. W. S. field no. 8C of 5/8/68]. Partly silicified and carbonaceous, weathered, siltstone from 30 cm above 1227, lower half of the River Wakefield Subgroup; locality cf. 1224.

1229 (ts, TOC, Corg) [J. W. S. field no. 9 of 5/8/68]. Bedded black chert from the upper third of the River Wakefield Subgroup; top of hill, ca. 1 km northwest of 1225, locality cf. 1224.

1230 (ts, μf^*, TOC, Corg) [J. W. S. field no. 10 of 5/8/68]. Microfossiliferous (Chapter 24) black chert from the upper third of the Skillogalee Dolomite; small ridge, just southeast of Mundallio Creek, ca. 15.5 km northeast of Port Augusta, South Australia; Port Augusta 1:250,000 map sheet; 32.50°S, 138.0°E.

1231 (ts, TOC, Dorg) [J. W. S. field no. 11 of 5/8/68]. Bedded chert within dark dolostone from immediately below the middle of the Skillogalee Dolomite; locality cf. 1230.

1232 (ts, μf^*, TOC, Corg) [J. W. S. field no. 12 of 5/9/68]. Microfossiliferous (Chapter 24) bedded black chert, within gray dolostone, from the middle third the Skillogalee Dolomite; at Depot Creek, 6.5 km east of Depot Creek rail siding, 14.5 km south-southeast of Wilkatanna, 34.5 km north-northwest of Port Augusta, South Australia; Port Augusta 1:250,000 map sheet; 32.20°S, 138.0°E.

1233 (ts, TOC, Corg) [J. W. S. field no. 13 of 5/9/68]. Bedded black chert occurring as nodules and thin 1–2 cm lenses interbedded with carbonate from the middle third(?) of the Skillogalee Dolomite; at low hills, ca. 100 m north of Crystal Brook, 5 km northeast of Crystal Brook rail siding, South Australia; Burra 1:250,000 map sheet; 33.30°S, 138.2°E.

1234 (ts, TOC, Corg) [J. W. S. field no. 1 of 5/11/68]. Carbonaceous, laminated slate from the lower half (Tindelpina Shale Mem.?) of the Tapley Hill Formation; road cut at side of Tapley's Hill, 11 km south-southwest of Adelaide City center, South Australia; 35.00°S, 138.5°E.

1235 (ts) [J. W. S. field no. 2 of 5/11/68]. Carbonaceous dolostone from the Beaumont Dolomite; on hillside in Brown Hill Reserve, ca 6.5 km south of Adelaide City center, South Australia.

1236 (ts) [J. W. S. field no. 3 of 5/11/68]. Carbonaceous dolostone with minor gray chert from the Montacute Dolomite; quarry on the south side of road between Castambul and Adelaide, ca. 9.5 km northeast of Adelaide City Center, South Australia; 35.00°S, 138.0°E.

1237 (ts, TOC, Corg) [J. W. S. field no. PC2-357ft]. Gray siltstone from a core sample, P. C. core #2 at a depth of 102 m, Tindelpina Shale Member of the Tapley Hill Formation; Paull Consolidated Copper Mine, ca. 6 km north of Leigh Creek and 234 km north of Port Augusta, South Australia.

1238 (ts, TOC, Corg, ker) [J. W. S. field no. PC2-441ft]. Gray siltstone from a core sample, P. C. core #2 at a depth of 133 m, Tindelpina Shale Member; locality cf. 1237.

1239 (ts, TOC, Corg, ker) [J. W. S. field no. WB-1136ft]. Silty, black chert, very finely laminated, from a core sample, Woomera Bore at a depth of 344 m, Woocalla Dolomite Member of the Tapley Hill Formation; Woomera Bore, ca. 75 km west of Lake Torrens and 175 km northwest of Port Augusta, South Australia; 31.20°S, 136.8°E.

1240 (ts, TOC, Corg) [J. W. S. field no. WB-1236ft]. Gray siltstone from a core sample, Woomera Bore at a depth of 374 m, Woocalla Dolomite Member; locality cf. 1239.

Table 14.2. *Continued.*

1241 (ts, TOC, Corg) [J. W. S. field no. WB-1335ft]. Black siltstone from a core sample, Woomera Bore at a depth of 404 m, Woocalla Dolomite Member; locality cf. 1239.

1242 (ts, TOC, Corg) [J. W. S. field no. WB-1430ft]. Dark gray siltstone from a core sample, Woomera Bore at a depth of 433 m, Woocalla Dolomite Member; locality cf. 1239.

1280 (ts, TOC, Corg) [J. W. S. field no. BGF-1]. Black chert from ca. 600 m below the Appila Tillite, Auburn Dolomite; along north-south paved road, 4.8 km south of Carrieton Railroad Station, 7.2 km southwest of Carrieton, South Australia; Ororoo 1:250,000 map sheet; 32.50°S, 138.5°E.

1281 (ts, μf^*, TOC, Corg) [J. W. S. field no. BGF-2]. Microfossiliferous (Chapter 24), laminated, gray dolostone with 2–9 cm-thick chert lenses, from ca. 1040 m below base of the Skillogalee Dolomite, lower portion (Blyth Dolomite?) of the River Wakefield Subgroup; 10.5 km northeast of Weira Hill and 50 km northeast of Carrieton (locality no. 23 of Forbes), South Australia; Ororoo 1:250,000 map sheet.

1282 (ts, TOC, Corg) [J. W. S. field no. BGF-3]. Black chert from ca. 300 m below the base of the Skillogalee Dolomite, River Wakefield Subgroup; 1 km north of locality 1281.

1283 (ts, st, μf^*, TOC, Corg) [J. W. S. field no. WOPF-BOOR2]. Microfossiliferous (Chapter 24), fine-grained, bedded, waxy, black chert, comprising domical stromatolites 0.5–1.0 m in diameter, from the Skillogalee Dolomite, 88 m below the base of the Myrtle Springs Fm.; 1.5 km east of Boorthanna-William Creek road, 75 km west of Lake Eyre North, 230 km northwest of Marree, and 450 km north-northwest of Port Augusta, Peake and Denison Ranges, South Australia; 28.67°S, 135.8°E.

1334 (ts, st, TOC, Corg) [J. W. S. field no. S1-8/21/73]. Stromatolitic black chert from the Skillogalee Dolomite; locality cf. 1283.

1335 (ts, st, TOC, Corg) [J. W. S. field no. 1-8/22/73]. Stromatolitic black chert from the Skillogalee Dolomite; locality cf. 1283.

1336 (ts, st, TOC, Corg) [J. W. S. field no. 2-8/22/73]. Stromatolitic black chert from the Skillogalee Dolomite; 100–200 m west, locality cf. 1283.

1337 (ts, st, TOC, Corg) [J. W. S. field no. 1-8/22/73]. Stromatolitic black chert from the Skillogalee Dolomite; locality cf. 1283.

1338 (ts, st, TOC, Corg) [J. W. S. field no. 3-8/22/73]. Stromatolitic black chert from the lower horizon of the Skillogalee Dolomite; locality cf. 1283.

1339 (ts, st, TOC) [J. W. S. field no. 3-8/22/73]. Stromatolitic black chert from the upper horizon (5 m above lower horizon) of the Skillogalee Dolomite; locality cf. 1283.

1340 (ts, st, TOC, Corg) [J. W. S. field no. 4-8/22/73]. Stromatolitic black chert from the Skillogalee Dolomite; locality cf. 1283.

1341 (ts, st, TOC, Corg) [J. W. S. field no. 5-8/22/73]. Stromatolitic black chert from the Skillogalee Dolomite; locality cf. 1283.

1342 (ts, st, TOC, Corg) [J. W. S. field no. 2-8/23/73]. Stromatolitic black chert from 35 m above the base of the Skillogalee Dolomite; locality cf. 1283.

1343 ts, st, TOC, Corg) [J. W. S. field no. 3-8/23/73]. Stromatolitic black chert from 820 m above the base of the Skillogalee Dolomite; locality cf. 1283.

1344 (ts, st, TOC, Corg) [J. W. S. field no. 4-8/23/73]. Stromatolitic black chert from the Skillogalee Dolomite; locality cf. 1283.

1345 (ts, st, TOC, Corg) [J. W. S. field no. 5-8/23/73]. Stromatolitic chert and carbonate from the Skillogalee Dolomite; locality cf. 1283.

1346 (ts, st, TOC, Corg) [J. W. S. field no. 6-8/23/73]. Stromatolitic black chert from the Skillogalee Dolomite; locality cf. 1283.

1347 (ts, st, TOC, Corg) [J. W. S. field no. 7-8/23/73]. Stromatolitic black chert from the Skillogalee Dolomite; locality cf. 1283.

1348 (ts, TOC, Corg) 8-8/23/73]. Chert from the Skillogalee Dolomite; locality cf. 1283.

1349 (ts, TOC, Corg) [J. W. S. field no. 9-8/23/73]. Chert from the Skillogalee Dolomite; locality cf. 1283.

AMADEUS BASIN

Tectonic unit/location. Amadeus Basin, Northern Territory, Australia.

Stratigraphy. Stratigraphic section, Alice Springs area, southern Northern Territory (Wells et al. 1967):

  PERTAOORRTA GROUP
    Arumbera Sandstone
  [UNNAMED UPPER PROTEROZOIC GROUP]
    Pertataka Formation
      Julie Member
      Waldo Pedlar Member
      Olympic Member
      Limbla Member
      Ringwood Member
      Cyclops Member
    Areyonga Formation
    Bitter Springs Formation
      Loves Creek Member
      Gillen Member
    Heavitree Quartzite

Table 14.2. Continued.

[UNNAMED MIDDLE PROTEROZOIC GROUP]
Arunta Igneous Complex

PPRG samples/localities.

1243 (ts, st, TOC) [J. W. S. field no. 15/14/68]. Black chert from the top of the Love's Creek Member of the Bitter Springs Formation; 46.4 km west-southwest of Alice Springs, Northern Territory, Australia.

1244 (ts, TOC, Corg) [J. W. S. field no. 16 of 5/14/68]. Weathered gray chert from the upper half of the Love's Creek Member; low hill at "Honeymoon Gap locality," 10.4 km southeast of Alice Springs, Northern Territory, Australia.

1245 (ts, μf^*, TOC, Corg) [J. W. S. field no. 17 of 5/15/68]. Microfossiliferous (Chapter 24) black chert from the uppermost unit of the Love's Creek Member; exposed on the south slope, 1.6 km north of Ross River Tourist Camp, Northern Territory, Australia.

1246 (ts, TOC) [J. W. S. field no. 17 of 5/15/68]. Black chert from the middle unit of the Love's Creek Member; locality cf. 1245.

1247 (ts, st) [J. W. S. field no. 17 of 5/15/68]. Gray chert and gray dolomite, stromatolitic, with irregular chert replacement, from the lowest third of the Love's Creek Member; locality cf. 1245.

1248 (ts) [J. W. S. field no. 17 of 5/15/68]. Chert from the middle unit of the Areyonga Formation; 64 km east-northeast of Alice Springs, Northern Territory, Australia; Alice Springs 1:250,000 map sheet.

1249 (ts) [J. W. S. field no. 18 of 5/15/68]. Chert from the middle third of the Love's Creek Member of the Bitter Springs Formation; 3.2 km east of Ross River Tourist Camp, Northern Territory, Australia.

1250 (ts, TOC, Dorg) [J. W. S. field no. 19 of 5/16/68]. Chert from the uppermost Love's Creek Member; 4.8 km east-southeast of Ross River Tourist Camp, Northern Territory, Australia.

1251 (ts, TOC, Corg) [J. W. S. field no. 20 of 5/17/86]. Chert from the upper third of the Love's Creek Member; on a low ridge to the southwest side of track ca. 3.2 km southeast of Shannon Bore, 14.4 km southeast of Ross River Tourist Camp, Northern Territory, Australia.

1252 (ts, TOC, Corg) [J. W. S. field no. 21 of 5/17/68]. Bedded black chert from the uppermost Love's Creek Member; ca. 6.4 km west of Julie Dam, 31.2 km southwest of Ross River Tourist Camp, Northern Territory, Australia.

1253 (ts, μf^*, TOC, Corg) [J. W. S. field no. 22 of 5/18/68]. Microfossiliferous (Chapter 24) bedded black chert from the middle of the Love's Creek Member; east side of north-south road at Ellery Gap, 1 km south of Ellery Creek Big Hole, 80 km west-southwest of Alice Springs, Northern Territory, Australia; Hermannsburg 1:250,000 map sheet.

1254 (ts, TOC, Corg) [J. W. S. field no. 23A of 5/19/68]. Black and gray chert from the upper third of the Love's Creek Member; 0.8 km north of Pioneer Creek, 12.8 km east-southeast of Glen Helen Tourist Camp, Northern Territory, Australia; Hermannsburg 1:250,000 map sheet.

1255 (ts, TOC, Corg) [J. W. S. field no. 23B of 5/19/68]. Chert from the middle third of the Areyonga Formation; locality cf. 1254.

1256 (ts, TOC, Corg) [J. W. S. field no. 24A of 5/19/68]. Fine-grained, waxy, black chert from the uppermost portion of the Bitter Springs Formation; ca. 0.8 km east of north-south road leading into Serpentine Gorge, 32 km southeast of Glen Helen Tourist Camp, 85 km west of Alice Springs, Northern Territory, Australia.

1257 (ts, TOC, Corg) [J. W. S. field no. 24B of 5/19/68]. Chert from the Areyonga Formation; locality cf. 1256.

1258 (ts, TOC, Corg) [J. W. S. field no. 25A of 5/19/68]. Black chert from the middle third of the Areyonga Formation; Macdonnell Ranges, 1 km north of Hugh River, 2 km south of Stuart Pass, 56 km west-southwest of Alice Springs, Northern Territory, Australia.

1259 (ts, TOC, Corg) [J. W. S. field no. 25B of 5/19/68]. Black chert from the middle third of the Areyonga Formation; occurring in low ridges and on flanks of gravel topped hills, locality cf. 1258.

1311 (ts, st, TOC, Corg) [J. W. S. field no. 1 of 9/6/73]. Carbonate stromatolite (*Gymnosolen*) from the Love's Creek Member of the Bitter Springs Formation; low ridge 1.2 km northwest of Ross River Tourist Camp, 64 km northeast of Alice Springs, Northern Territory, Australia.

1312 (ts, st, TOC, Corg) [J. W. S. field no. 1 of 9/6/73]. Silicified stromatolite (*Gymnosolen*) from the Bitter Springs Formation; locality cf. 1311.

1313 (ts, TOC, Corg) [J. W. S. field no. BS1-9/7/73]. Bedded chert, with 1-3 cm-thick beds, from 65 m below top of the Love's Creek Member of the Bitter Springs Formation; 0.6 km north of Ross River Tourist Camp, Northern Territory, Australia; Alice Springs 1:250,000 map sheet.

1314 (ts, TOC, Corg) [J. W. S. field no. BS2-9/7/73]. Bedded chert from 38 m beneath 1313, upper third of the Love's Creek Member; locality cf. 1313.

1315 (ts, st, TOC) [J. W. S. field no. BS3-9/7/73]. Stromatolitic chert from 54 m beneath 1314, Bitter Springs Formation; locality cf. 1313.

1316 (ts, TOC, Corg) [J. W. S. field no. BS4-9/7/73]. Bedded chert from the south-side of hill, 25 m beneath 1315, Bitter Springs Formation; locality cf. 1313.

1317 (ts, TOC, Corg) [J. W. S. field no. BS5-9/7/73]. Cherty nodular dolomite from the Gillen Member of the Bitter Springs Formation; locality cf. 1313.

1318 (ts, TOC, Corg) [J. W. S. field no. PERT1-9/8/73]. Chert from the Julie Member of the Pertataka Formation; from crest of hill, 400 m south of unpaved road, 64 km east-northeast of Alice Springs, Northern Territory, Australia; Alice Springs 1:250,000 map sheet.

1319 (ts, st, TOC, Corg) [J. W. S. field no. BS2-9/8/73]. Laminated gray chert (*Collenia*-like) from the Love's Creek Member of the Bitter Springs Formation; 50 m west of Allua Well, 48 km east of Alice Springs, Northern Territory, Australia.

Table 14.2. *Continued.*

1320 (ts, st, TOC, Corg) [J. W. S. field no. BS1-9/9/73]. Carbonate stromatolite from the uppermost portion of the Love's Creek Member; 2.4 km southwest of Jay Creek Native Settlement, 46 km west-southwest of Alice Springs, Northern Territory, Australia; Hermannsburg 1:250,000 map sheet.

1321 (ts, TOC) [J. W. S. field no. BS1-9/9/73]. Black chert in lenses and nodules from the uppermost portion of the Love's Creek Member; locality cf. 1320.

1322 (ts, TOC, Corg) [J. W. S. field no. ARY1-9/9/73]. Black chert, pebbles, cobbles and boulders in a conglomeratic bed, chert up to 0.5 m in diameter from the middle third of the Areyonga Formation; locality cf. 1320.

1323 (ts, st, TOC, Corg) [J. W. S. field no. BS2-9/9/73]. Gray dolostone with columnar stromatolites from the top of the Love's Creek Member; 0.8 km west of Jay Creek Mission, Northern Territory, Australia.

1324 (ts, st, TOC, Corg) [J. W. S. field no. BS2-9/9/73]. Carbonate stromatolite with flat laminations from the top of the Love's Creek Member; locality cf. 1323.

1325 (ts, TOC, Corg) [J. W. S. field no. ARY1-9/7/73]. Chert from the middle portion of the Areyonga Formation; 2.4 km southwest of Jay Creek Native Settlement, 46 km west-southwest of Alice Springs, Northern Territory, Australia; Hermannsburg 1:250,000 map sheet.

1326 (ts, TOC, Corg) [J. W. S. field no. BS1-9/10/73]. Bedded chert, 6–10 cm thick, from the Love's Creek Member of the Bitter Springs Formation; Katapata Gap, Gardener Range, ca. 192 km southwest of Alice Springs, Northern Territory, Australia.

1327 (ts, TOC, Corg) [J. W. S. field no. BS1-9/11/73]. Dark gray chert and dolomite from 83 m below the top of the Love's Creek Member; east side of north-south road at Ellery Gap, 1 km south of Ellery Creek Big Hole, 80 km west-southwest of Alice Springs, Northern Territory, Australia; Hermannsburg 1:250,000 map sheet.

1328 (ts, TOC, Corg) [J. W. S. field no. BS2-9/11/73]. Chert from a chert horizon ca. 1 m thick, 18 m below 1327, upper portion of the Love's Creek Member; locality cf. 1327.

1329 (ts, TOC, Corg) [J. W. S. field no. BS3-9/11/73]. Chert with interbedded carbonate from a blocky weathered horizon 1 m thick, 10 m below 1328, middle portion of the Love's Creek Member; locality cf. 1327.

1330 (ts, TOC, Corg) [J. W. S. field no. BS4-9/11/73]. Bedded carbonate and chert from 20 m below 1329, 131 m below top of section, middle portion of the Love's Creek Member; locality cf. 1327.

1331 (ts, TOC, Corg) [J. W. S. field no. BS5-9/11/73]. Bedded chert from a 1 m thick bed, 17 m below 1330, Love's Creek Member; locality cf. 1327.

1332 (ts, TOC, Corg) [J. W. S. field no. BS6-9/11/73]. Bedded carbonate and black chert from the upper portion of the Love's Creek Member; ca. 0.8 km east of north-south road leading into Serpentine Gorge, 32 km southeast of Glen Helen Tourist Camp, 85 km west of Alice Springs, Northern Territory, Australia.

1333 (ts, TOC, Corg) [J. W. S. field no. BS1-9/12/73]. Chert from a 4–8 cm-thick bed, ca. 35 m below top contact of the Love's Creek Member; 2.4 km east of north-south road leading to Ormiston Gorge, 115 km east of Alice Springs, Northern Territory, Australia.

1350 (ts, st, TOC, Corg) [J. W. S. field no. 1-9/6/73]. Columnar carbonate stromatolite from the Love's Creek Member; low ridge 1.2 km northwest of Ross River Tourist Camp, 64 km northeast of Alice Springs, Northern Territory, Australia.

1351 (ts, st, TOC, Corg) [J. W. S. field no. 1-9/6/73]. Columnar chert-carbonate stromatolite from the Love's Creek Member; locality cf. 1350.

1352 (ts, TOC, Corg) [J. W. S. field no. 2a-9/8/73]. Carbonate (cf. 1319) from the Love's Creek Member; 50 m west of Allua Well, 48 km east of Alice Springs, Northern Territory, Australia.

1353 (ts, st, TOC, Corg) [J. W. S. field no. 2b-9/8/73]. Carbonate stromatolite, part of a "mega-stromatolite" ca. 3 m in diameter, from a 10 cm cherty bed of the Love's Creek Member; locality cf. 1352.

1383 (ts, st, TOC, Corg) [J. W. S. field no. 15-5/14/68]. Carbonate stromatolite from near the base of the Love's Creek Member; 2.4 km southwest of Jay Creek Native Settlement, 46 km west-southwest of Alice Springs, Northern Territory, Australia; Hermannsburg 1:250,000 map sheet.

1758 (ts, st) [M. R. W. field no. 20.1b/5/86]. Carbonate stromatolite (*Inzeria intia*) from the Bitter Springs Formation; 64 km east-northeast of Alice Springs, Northern Territory, Australia; Alice Springs 1:250,000 map sheet.

1759 (ts) [M. R. W. field no. 20.2/5/86]. Carbonate with syneresis cracks from the Bitter Springs Formation; locality cf. 1758.

1760 (ts, TOC, Corg) [M. R. W. field no. 20.3/5/86-A]. Dolomite from the Julie Member of the Pertatataka Formation; locality cf. 1758.

1761 (ts, TOC, Corg) [M. R. W. field no. 20.3/5/86-B]. Dolomite from the Julie Member; locality cf. 1758.

1762 (ts, TOC, Corg) [M. R. W. field no. 20.3/5/86-C]. Dolomite from the Julie Member; locality cf. 1758.

1763 [M. R. W. field no. 21.1C/5/86]. Dolomite from the Bitter Springs Formation; 2.4 km southwest of Jay Creek Native Settlement, 46 km west-southwest of Alice Springs, Northern Territory, Australia; Hermannsburg 1:250,000 map sheet.

BIRRINDUDU BASIN

Tectonic unit/location. Birrindudu Basin, Western Australia and Northern Territory, Australia.

Stratigraphy. Stratigraphic section, Osmond Range area, northeastern Western Australia and northwestern Northern Territory (Thom 1975a).

Table 14.2. *Continued.*

[UNNAMED MIDDLE PROTEROZOIC GROUP]
    Helicopter Siltstone
    Wade Creek Sandstone
        Mount John Shale Member
    Bungle Bungle Dolomite
    Mount Parker Sandstone

PPRG samples/localities.

2193 (ts, TOC, Corg) [J. H. L. field no. A1-86]. Very dark chert from a cobble in a stream; Bungle Bungle Dolomite; 1 km south of Palm Spring, Western Australia.

2194 (ts, st, TOC, Corg) [M. R. W. field no. 860902.1]. Dark gray, 1 m-thick tabular biostrome of columnar branching stromatolite, interbedded in stratiform stromatolites, from the Bungle Bungle Dolomite; edge of small waterfall, locality cf. 2193.

2195 (ts, TOC) [M. R. W. field no. 860902.3]. Finely laminated (non-stromatolitic) dolomite from the cap dolomite of the Moonlight Valley Tillite; small rubbly outcrop, locality cf. 2193.

2196 (ts, TOC, Corg) [D. C. field no. 020986-1]. Light-colored chert from the Bungle Bungle Dolomite; west gorge, off main gorge, locality cf. 2193.

2197 (ts, TOC,) [D. C. field no. 020986-2]. Light-colored, coarse-grained chert from the Bungle Bungle Dolomite: west gorge, off main gorge, locality cf. 2193.

2199 (ts, TOC, Corg) [D. D. field no. 9286G]. Brecciated dolomite from ca. 8 m above 2198, Bungle Bungle Dolomite; side canyon, locality cf. 2193.

2200 (ts) [D. D. field no. 9286H]. Duplicate of 2199 from the Bungle Bungle Dolomite; side canyon, locality cf. 2193.

2201 (ts) [D. D. field no. 9286I]. Dolomitic breccia with large gray clasts cemented by carbonate which weathers to an orange color, from the Bungle Bungle Dolomite; small hill at exit to canyon, locality cf. 2193.

2202 [D. D. field no. 9286J]. Duplicate of 2201 from the Bungle Bungle Dolomite; small hill at exit to canyon, locality cf. 2193.

2216 (ts, TOC, Corg) [M. R. W. field no. 860903.2]. Medium green shale with 1–5 cm-thick interbeds of dark green to gray dolomite from the Timperley Shale; in creek bank, locality cf. 2193.

2217 (ts) [J. H. L. field no. A-7-86]. Green shale with interbeds of dark green to gray dolomite from the Timperley Shale; locality cf. 2193.

2218 (ts) [J. H. L. field no. A-8-86]. Medium green shale from the Timperley Shale; 800 m west of road crossing of Johnston River, adjacent to large burrow pit, locality cf. 2193.

2222 (ts) [D. D. field no. 9486D]. Medium grained carbonate from the base of a carbonate unit of the Egan Formation; roadside cut on old road, locality cf. 2193.

2228 (ts) [J. H. L. field no. A10–86]. Medium green shale from the Timperley Shale; locality cf. 2193.

2232 (ts) [J. H. L. field no. A2-86]. Chert from above the stratiform stromatolite layer of the Bungle Bungle Dolomite; locality cf. 2193.

2233 (ts, st) [J. H. L. field no. A3-86]. Stromatolitic chert from about 10 m stratigraphically above west side of canyon, Bungle Bungle Dolomite; locality cf. 2193.

2234 (ts) [J. H. L. field no. A-86]. Very fine-grained blue flaggy claystone from the Elvire Formation; locality cf. 2193.

2235 (ts) [J. H. L. field no. A5-86]. Greenish-gray mudstone from the Elvire Formation; locality cf. 2193.

2236 (J. H. L. field no. A6-86]. Gray-blue shale from 10–50 m above the base of the Timperley Shale; locality cf. 2193.

2237 (ts) [J. H. L. field no. A9-86]. Medium green shale from a washout in hillside of the Timperley Shale; locality cf. 2193.

2241 (ts, st) [D. R. L. field no. AUS-57-1]. Small dolomitic columnar stromatolite from the Bungle Bungle Dolomite; locality cf. 2193.

2242 (ts) [D. R. L. field no. AUS-57-2]. Chert from the Bungle Bungle Dolomite; locality cf. 2193.

2243 (ts) [D. R. L. field no. AUS-57-3]. Secondary(?) nodular radiating quartz masses in dolomite from the Bungle Bungle Dolomite; locality cf. 2193.

2244 (ts) [D. R. L. field no. AUS-57-4]. Silicified breccia within laminated black chert from the Bungle Bungle Dolomite; locality cf. 2193.

2245 (ts) [D. R. L. field no. AUS-57-5]. Black chert from the Mount Forster Sandstone; locality cf. 2193.

2246 (ts) [D. R. L. field no. AUS-57-6]. Quartz sandstone from the Mount Forster Sandstone; locality cf. 2193.

2247 (ts) [D. R. L. field no. AUS-57-7]. Granular chert conglomerate with a sandy matrix from the Mount Forster Sandstone; locality cf. 2193.

2248 (ts) [D. R. L. field no. AUS-57-8]. Chert pebble conglomerate from the Mount Forster Sandstone; locality cf. 2193.

2277 [R. I. field no. AUST-4]. Coarse grained weathered quartz sandstone from the Jarrad Sandstone Member of the Ranford Formation; locality cf. 2193.

2278 [R. I. field no. AUST-5]. Coarse grained chert-quartz sandstone from the Mount Forster Sandstone; locality cf. 2193.

2282 [H. J. H. field no. H-8609-2-1]. Black chert nodule in gray dolomite from the Bungle Bungle Dolomite; locality cf. 2193.

Table 14.2. *Continued.*

2283 [H. J. H. field no. H-8609-2-211.1m]. Black chert nodule from the Bungle Bungle Dolomite; locality cf. 2193.

2284 [H. J. H. field no. H-8609-2-2-38m]. Black chert nodule from the Bungle Bungle Dolomite; locality cf. 2193.

2285 [H. J. H. field no. H-8609-2-2-8m]. Black chert nodule from the Bungle Bungle Dolomite; locality cf. 2193.

2286 [H. J. H. field no. H-8609-2-3]. Loose clasts of chert weathered out of tillite from the Moonlight Valley Tillite; locality cf. 2193.

2297 [R. J. H. field no. PH-9-2-86-1]. Chert pebbles from a stream bed, Bungle Bungle Dolomite; locality cf. 2193.

2333 (ts) [D. W. field no. 9-2-86-A]. Black chert from the Moonlight Valley Tillite; locality cf. 2193.

2334 (ts) [D. D. field no. 9286B]. Black dolomitic chert from the Bungle Bungle Dolomite; locality cf. 2193.

2335 (ts, st) [D. D. field no. 9286E]. Stratiform dolomitic stromatolite with wavy laminations from the Bungle Bungle Dolomite; locality cf. 2193.

2336 [D. D. field no. 9286F]. Organic-rich dolomite, laminated, with biohermal shape and fenestrae, from the Bungle Bungle Dolomite; locality cf. 2193.

2337 [H. J. H. field no. H8609-3-1]. Sandstone with trace tossils(?) from the Mount Forster Sandstone; locality cf. 2193.

2338 [M. R. W. field no. 860903-3]. Olive-green, medium-grained shale from the Timperley Shale; locality cf. 2193.

2339 (ts) [M. R. W. field no. 860903-4]. Green, medium-grained shale from the Timperley Shale; locality cf. 2193.

GASCOYNE PROVINCE

Tectonic unit/location. Gascoyne Province, Western Australia.

Stratigraphy.

    Ashburton Formation

PPRG samples/localities.

2716 [J. M. H. field no. 820621-2a]. Gray schist from the Ashburton Formation; J. M. H. "Gascoyne" locality, Western Australia.

HAMERSLEY BASIN

Tectonic unit/location. Hamersley Basin, Western Australia, Australia.

Stratigraphy. Stratigraphic section, Mount Bruce area, northwestern Western Australia (Hickman 1978):

MOUNT BRUCE SUPERGROUP
  HAMERSLEY GROUP
      Boolgeeda Iron Formation
      Woongarra Volcanics
      Weeli Wolli Iron Formation
      Brockman Iron Formation
        Whaleback Shale Member
        Dales Gorge Member
      Mount McRae Shale
      Mount Sylvia Formation
      Wittenoom Dolomite/Carawine Dolomite
      Marra Mamba Iron Formation
  FORTESCUE GROUP
      Jeerinah Formation/Lewin Shale
        Roy Hill Member
        Warrie Member
        Woodiana Sandstone Member
      Maddina Basalt
      Kuruna Siltstone
      Nymerina Basalt
      Tumbiana Formation
        Meentheena Carbonate Member
        Mingah Tuff Member
      Kylena Basalt
      Hardy Sandstone
      Mount Roe Basalt

PPRG samples/localities.

1354 (ts, st, TOC, Corg) [J. W. S. field no. 1-6/19/82]. Black chert lenses interbedded with stromatolitic (*Colleniella*) carbonates from the Meentheena Carbonate Member of the Tumbiana Formation; top of hill, 100 m west of 211 km sign on Port Hedland-Newman railroad, 200 m west of railway maintenance road, Western Australia; Roy Hill 1:250,000 Geologic sheet no. SF-50-12; 22.05°S, 118.9°E.

Table 14.2. *Continued.*

1355 (ts, st, TOC, Corg) [J. W. S. field no. 2a-6/19/82]. Stromatolitic black chert and carbonate, 2 cm to 1 m in diameter *Colleniella*, cherts occur as lenses 1–5 mm thick, from the Meentheena Carbonate Member; small quarry 150 m east of maintenance road at km 211.3 on Port Hedland-Newman railroad, Western Australia; 22.05°S, 118.9°E.

1356 (ts, st, TOC, Corg) [J. W. S. field no. 2b-6/19/82]. Cherty carbonate stromatolitic domes from the Meentheena Carbonate Member; locality cf. 1355.

1454 (ts) [J. W. S. field no. 3-6/15/82]. Fine-grained, gray chert from the Hardy Sandstone; bed exposed at the side of a small creek, ca. 17 km west-southwest of Marble Bar, Western Australia; Marble Bar 1:100,000 map sheet 2588, grid reference 073561.

1455 (ts, st) [J. W. S. field no. 1-6/16/82]. Black chert from 0.5 m below bed 14, 8.5 m below the Nymerina Basalt, from the Tumbiana Formation; small gorge 1.6 km west of Nullagine River, 3 km northwest of Meentheena, 68 km south-southeast of Marble Bar, Western Australia; Mt. Edgar 1:100,000 map sheet 2955.

1456 (ts, st) [J. W. S. field no. 2-6/16/82]. Stromatolitic carbonate and black chert from 1.5 m below bed 11, 14.5 m below the Nymerina Basalt, from the Tumbiana Formation; locality cf. 1455.

1457 (ts) [J. W. S. field no. 5-6/16/82]. Carbonate and chert from cf. 1456, but 200 m east at the top of a low hill, from the Tumbiana Formation; locality cf. 1455.

1461 (ts, st) [J. W. S. field no. 6-6/16/82]. Chert stromatolites from discontinuous, thin (ca. 0.5 m thick), bed 19 of the Meentheena Carbonate Member of the Tumbiana Formation; 75 m southeast of mouth of small gorge, locality cf. 1455.

1462 (ts, st) [J. W. S. field no. 7-6/16/82]. Microstromatolitic carbonate from bed 11, 13 m below the Nymerina Basalt, from the Tumbiana Formation; locality cf. 1455.

1463 (ts, st) [J. W. S. field no. 8-6/16/82]. Stromatolitic carbonate and chert from above the pisolitic tuff unit of the Kuruna Siltstone; 100 m west of unpaved road, 300 m west Nullagine River, 22 km north-northeast of 1455; Yilgalong 1:100,000 map sheet 3055, grid reference 457645.

1464 (ts, st) [J. W. S. field no. 9-6/16/82]. Chert stromatolites from a bed ca. 0.8 m thick, immediately overlying a massive gray carbonate unit of the Kylena Basalt; immediately adjacent to unpaved road 3 km west of Nullagine River, Western Australia; map sheet 2955, grid reference 395556.

1468 (ts, st) [J. W. S. field no. 1-6/19/82]. Black chert lenses interbedded with stromatolitic (*Colleniella*) carbonates from the Meentheena Carbonate Member of the Tumbiana Formation; top of hill, 100 m west of 211 km sign on Port Hedland-Newman railroad, 200 m west of railway maintenance road, Western Australia; Roy Hill 1:250,000 Geologic sheet no. SF-50-12; 22.05°S, 118.9°E

1469 (ts, st) [J. W. S. field no. 2-6/19/82]. Stromatolitic black chert and carbonate from the Meentheena Carbonate Member; small quarry 150 m east of maintenance road at km 211.3 on Port Hedland-Newman railroad, Western Australia; 22.05°S, 118.9°E.

1470 (ts, st) [J. W. S. field no. 2-6/19/82]. Black chert comprising circular, flat, stromatolitic heads from the Meentheena Carbonate Member; locality cf. 1469.

1471 (ts, st) [J. W. S. field no. 3-6/19/82]. Black chert lenses capping and occurring between large, domical, stromatolites from the Meentheena Carbonate Member; top of cliff above, locality cf. 1469.

1906 [B. P. field no. 1-6/11/86-1]. Black chert from the Jeerinah Formation; "FVG1 locality," 150 km south of Marble Bar, Western Australia; 22.50°S, 119.5°E.

1907 (ts) [B. P. field no. 1-6/11/86-2]. Black chert with pyrite from the Jeerinah Formation; locality cf. 1906.

1908 (ts) [B. P. field no. 1-6/11/86-3]. Black shale from the Jeerinah Formation; locality cf. 1906.

1909 (ts) [B. P. field no. 1-6/11/86-4]. Black shale from the Jeerinah Formation; locality cf. 1906.

1910 (ts) [B. P. field no. 1-6/11/86-5]. Black shale from the Jeerinah Formation; locality cf. 1906.

1911 (ts) [B. P. field no. 1-6/11/86-6]. Black shale from the Jeerinah Formation; locality cf. 1906.

1912 (ts) [B. P. field no. 1-6/11/86-7]. Black shale and carbonate from the Jeerinah Formation; locality cf. 1906.

1913 (ts) [B. P. field no. 1-6/11/86-8]. Black shale from the Jeerinah Formation; locality cf. 1906.

1914 (ts) [B. P. field no. 1-6/11/86-9]. Black shale from the Jeerinah Formation; locality cf. 1906.

1915 (ts) [B. P. field no. 1-6/11/86-10]. Black shale from the Jeerinah Formation; locality cf. 1906.

1916 (ts) [B. P. field no. 1-6/11/86-11]. Black shale from the Jeerinah Formation; locality cf. 1906.

1917 (ts) [B. P. field no. 1-6/11/86-12]. Black carbonate from the Jeerinah Formation; locality cf. 1906.

1918 (ts) [B. P. field no. 1-6/11/86-13]. Black carbonate from the Jeerinah Formation; locality cf. 1906.

1919 (ts) [B. P. field no. 1-6/11/86-14]. Black shale from the Jeerinah Formation; locality cf. 1906.

1920 (ts) [B. P. field no. 1-6/11/86-15]. Black carbonate with pyrite from the Jeerinah Formation; locality cf. 1906.

1921 (ts) [B. P. field no. 1-6/11/86-16]. Black chert and carbonate from the Jeerinah Formation; locality cf. 1906.

1922 (ts) [B. P. field no. 1-6/11/86-17]. Black shale from the Jeerinah Formation; locality cf. 1906.

1923 (ts) [B. P. field no. 1-6/11/86-18]. Black carbonate with pyrite from the Jeerinah Formation; locality cf. 1906.

Table 14.2. *Continued.*

1924 (ts) [B. P. field no. 1-6/11/86-19]. Black carbonate with pyrite from the Jeerinah Formation; locality cf. 1906.

1925 (ts) B. P. Field no. 1-6/11/86-20]. Black carbonate from the Jeerinah Formation; locality cf. 1906.

1926 (ts) [B. P. field no. 1-6/11/86-21]. Black shale from the Jeerinah Formation; locality cf. 1906.

1927 (ts) [B. P. field no. 1-6/11/86-22]. Black shale from the Jeerinah Formation; locality cf. 1906.

1928 (ts) [B. P. field no. 1-6/11/86-23]. Black shale from the Jeerinah Formation; locality cf. 1906.

1929 (ts) [B. P. field no. 1-6/11/86-24]. Black shale from the Jeerinah Formation; locality cf. 1906.

1930 (ts) [B. P. field no. 1-6/11/86-25]. Black shale from the Jeerinah Formation; locality cf. 1906.

1931 (ts) [B. P. field no. 1-6/11/86-26]. Black shale from the Jeerinah Formation; locality cf. 1906.

1932 (ts, TOC, Corg, ker, Sfide) [B. P. field no. 1-6/11/86-27]. Black shale from the Jeerinah Formation; locality cf. 1906.

1933 (ts) [B. P. field no. 1-6/11/86-28]. Black shale from the Jeerinah Formation; locality cf. 1906.

1934 (ts) [B. P. field no. 1-6/11/86-29]. Black shale from the Jeerinah Formation; locality cf. 1906.

1935 (ts) [B. P. field no. 1-6/11/86-30]. Black shale from the Jeerinah Formation; locality cf. 1906.

1936 (ts) [B. P. field no. 1-6/11/86-31]. Calcite from the Jeerinah Formation; locality cf. 1906.

1937 (ts) [B. P. field no. 1-6/11/86-32]. Black carbonate from the Jeerinah Formation; locality cf. 1906.

1938 (ts) [B. P. field no. 1-6/11/86-33]. Black carbonate from the Jeerinah Formation; locality cf. 1906.

1939 (ts) [B. P. field no. 1-6/11/86-34]. Black carbonate from the Jeerinah Formation; locality cf. 1906.

1940 [B. P. field no. 1-6/12/86-1]. Gray carbonate from the Kuruna Siltstone; locality cf. 1906.

1941 [B. P. field no. 1-6/12/86-2]. Gray carbonate from the Kuruna Siltstone; locality cf. 1906.

1942 [B. P. field no. 1-6/12/86-3]. Gray carbonate from the Kuruna Siltstone; locality cf. 1906.

1943 [B. P. field no. 1-6/12/86-4]. Gray carbonate from the Kuruna Siltstone; locality cf. 1906.

1944 (TOC, Corg, Sfide) [B. P. field no. 1-6/12/86-5]. Gray carbonate from the Kuruna Siltstone; locality cf. 1906.

1945 [B. P. field no. 1-6/12/86-6]. Gray carbonate from the Kuruna Siltstone; locality cf. 1906.

1946 [B. P. field no. 1-6/12/86-7]. Gray carbonate from the Kuruna Siltstone; locality cf. 1906.

1947 [B. P. field no. 1-6/12/86-8]. Gray carbonate from the Kuruna Siltstone; locality cf. 1906.

1948 [B. P. field no. 1-6/12/86-9]. Gray carbonate from the Nymerina Basalt; locality cf. 1906.

1949 [B. P. field no. 1-6/12/86-10]. Gray carbonate from the Nymerina Basalt; locality cf. 1906.

1950 (TOC, Corg) [B. P. field no. 1-6/12/86-11]. Gray carbonate from the Nymerina Basalt; locality cf. 1906.

1951 [B. P. field no. 1-6/12/86-12]. Gray carbonate from the Nymerina Basalt; locality cf. 1906.

1952 [B. P. field no. 1-6/12/86-13]. Gray carbonate from the Nymerina Basalt; locality cf. 1906.

1953 [B. P. field no. 1-6/12/86-14]. Gray carbonate from the Nymerina Basalt; locality cf. 1906.

1954 [B. P. field no. 1-6/12/86-15]. Gray carbonate from the Nymerina Basalt; locality cf. 1906.

1955 [B. P. field no. 1-6/12/86-16]. Gray carbonate from the Nymerina Basalt; locality cf. 1906.

1956 [B. P. field no. 1-6/12/86-17]. Gray carbonate from the Nymerina Basalt; locality cf. 1906.

1957 [B. P. field no. 1-6/12/86-18]. Gray carbonate from the Nymerina Basalt; locality cf. 1906.

1958 [B. P. field no. 1-6/12/86-19]. Gray carbonate from the Nymerina Basalt; locality cf. 1906.

1959 [B. P. field no. 1-6/12/86-20]. Gray carbonate from the Nymerina Basalt; locality cf. 1906.

1960 (ts, TOC, Corg) [B. P. field no. 1-6/12/86-21]. Gray carbonate from the Tumbiana Formation; locality cf. 1906.

1961 (ts) [B. P. field no. 1-6/12/86-22]. Gray carbonate from the Tumbiana Formation; locality cf. 1906.

1962 (ts) [B. P. field no. 1-6/12/86-23]. Gray carbonate from the Tumbiana Formation; locality cf. 1906.

1963 (ts) [B. P. field no. 1-6/12/86-24]. Gray carbonate from the Tumbiana Formation; locality cf. 1906.

1964 (ts) [B. P. field no. 1-6/12/86-25]. Gray carbonate from the Tumbiana Formation; locality cf. 1906.

1965 (ts) [B. P. field no. 1-6/12/86-26]. Gray carbonate from the Tumbiana Formation; locality cf. 1906.

1966 (ts) [B. P. field no. 1-6/12/86-27]. Gray carbonate from the Tumbiana Formation; locality cf. 1906.

1967 (ts) [B. P. field no. 1-6/12/86-28]. Gray, oolitic carbonate from the Tumbiana Formation; locality cf. 1906.

1968 (ts) [B. P. field no. 1-6/12/86-29]. Gray carbonate from the Tumbiana Formation; locality cf. 1906.

Table 14.2. *Continued.*

1969 (ts) [B. P. field no. 1-6/12/86-30]. Gray carbonate from the Tumbiana Formation; locality cf. 1906.

1970 (ts) [B. P. field no. 1-6/13/86-1]. Gray carbonate from the Jeerinah Formation; "SGS1 locality," 175 km south-west of Marble Bar, Western Australia; 22.00°S, 118.0°E.

1971 (ts) [B. P. field no. 1-6/13/86-2]. Gray carbonate from the Jeerinah Formation; locality cf. 1970.

1972 (ts, TOC, Corg, ker, Sfide) [B. P. field no. 1-6/13/86-3]. Gray bedded siltstone from the Jeerinah Formation; locality cf. 1970.

1973 (ts) [B. P. field no. 1-6/13/86-4]. Siltstone from the Jeerinah Formation; locality cf. 1970.

1974 (ts) [B. P. field no. 1-6/13/86-5]. Gray siltstone from the Jeerinah Formation; locality cf. 1970.

1975 (ts) [B. P. field no. 1-6/13/86-6]. Gray siltstone from the Jeerinah Formation; locality cf. 1970.

1976 (ts) [B. P. field no. 1-6/13/86-7]. Gray siltstone from the Tumbiana Formation; locality cf. 1970.

1977 (ts) [B. P. field no. 1-6/13/86-8]. Gray siltstone from the Tumbiana Formation; locality cf. 1970.

1978 (ts) [B. P. field no. 1-6/13/86-9]. Gray siltstone from the Tumbiana Formation; locality cf. 1970.

1979 (ts) [B. P. field no. 1-6/13/86-10]. Gray siltstone from the Tumbiana Formation; locality cf. 1970.

1980 (ts) [B. P. field no. 1-6/13/86-11]. Gray siltstone from the Tumbiana Formation; locality cf. 1970.

1981 (ts) [B. P. field no. 1-6/13/86-12]. Gray siltstone from the Tumbiana Formation; locality cf. 1970.

1982 (ts) [B. P. field no. 1-6/13/86-13]. Gray siltstone from the Tumbiana Formation; locality cf. 1970.

1983 (ts) [B. P. field no. 1-6/13/86-14]. Gray siltstone from the Tumbiana Formation; locality cf. 1970.

1984 (ts) [B. P. field no. 1-6/13/86-15]. Gray siltstone from the Tumbiana Formation; locality cf. 1970.

1985 (ts, TOC, Corg) [B. P. field no. 1-6/13/86-16]. Gray siltstone from the Tumbiana Formation; locality cf. 1970.

1986 (ts) [B. P. field no. 1-6/13/86-17]. Gray siltstone from the Tumbiana Formation; locality cf. 1970.

1987 (ts) [B. P. field no. 1-6/13/86-18]. Gray siltstone from the Tumbiana Formation; locality cf. 1970.

1988 (ts) [B. P. field no. 1-6/13/86-19]. Gray siltstone from the Tumbiana Formation; locality cf. 1970.

1989 (ts) [B. P. field no. 1-6/13/86-20]. Gray siltstone from the Tumbiana Formation; locality cf. 1970.

1990 (ts) [B. P. field no. 1-6/13/86-21]. Gray siltstone from the Tumbiana Formation; locality cf. 1970.

1991 (ts) [B. P. field no. 1-6/13/86-22]. Gray siltstone from the Tumbiana Formation; locality cf. 1970.

1992 (ts) [B. P. field no. 1-6/13/86-23]. Gray siltstone from the Tumbiana Formation; locality cf. 1970.

1993 (ts) [B. P. field no. 1-6/13/86-24]. Gray siltstone from the Tumbiana Formation; locality cf. 1970.

1994 (ts) [B. P. field no. 1-6/13/86-25]. Gray siltstone from the Tumbiana Formation; locality cf. 1970.

2016 (ts) [B. P. field no. 1-6/22/86-1]. Gray lapilli tuff from the Kylena Basalt; Mt. Edgar, Western Australia.

2017 (ts, st, TOC, Corg) [B. P. field no. 1-6/22/86-2]. Stromatolitic carbonate from the Kylena Basalt; locality cf. 2016.

2018 (ts, st) [B. P. field no. 1-6/22/86-3]. Stromatolitic carbonate from the Kylena Basalt; locality cf. 2016.

2019 (ts) [B. P. field no. 1-7/5/86-1]. Chert breccia from the Maddina Basalt; B. P. "Tambrey V1" vent locality, Western Australia; Mount Billroth 1:100,000 map sheet 2454; 22.67°S, 117.5°E.

2020 (ts) [B. P. field no. 1-7/5/86-2]. Banded chert from the Maddina Basalt; locality cf. 2019.

2021 (ts) [B. P. field no. 1-7/5/86-3]. Gray laminated chert from the Maddina Basalt; locality cf. 2019.

2022 (ts, TOC, Corg) [B. P. field no. 1-7/5/86-4]. Laminated chert from the Maddina Basalt; locality cf. 2019.

2023 (ts) [B. P. field no. 1-7/5/86-5]. Gray laminated chert from the Maddina Basalt; locality cf. 2019.

2024 (ts) [B. P. field no. 1-7/5/86-6a+b]. Black and white chert from the Maddina Basalt; locality cf. 2019.

2025 (ts) [B. P. field no. 1-7/6/86-1]. Chert breccia from the Maddina Basalt; B. P. "Tambrey V2" vent locality, Western Australia; Mount Billroth 1:100,000 map sheet 2454; 22.67°S, 117.5°E.

2026 (ts) [B. P. field no. 1-7/6/86-2]. Chert breccia from the Maddina Basalt; locality cf. 2025.

2027 (ts) [B. P. field no. 1-7/6/86-3]. Chert from the Maddina Basalt; locality cf. 2025.

2028 (ts) [B. P. field no. 1-7/6/86-4]. Black and white chert from the Maddina Basalt; locality cf. 2025.

2029 (ts) [B. P. field no. 1-7/6/86-5]. Light gray laminated chert from the Maddina Basalt; locality cf. 2025.

2030 (ts) [B. P. field no. 1-7/7/86-1]. Black chert breccia from the Maddina Basalt; "Tambrey V3 vent locality," Western Australia; Mount Billroth 1:100,000 map sheet 2454; 22.67°S, 117.5°E.

2031 (ts) [B. P. field no. 1-7/7/86-2]. Chert breccia from the Maddina Basalt; locality cf. 2030.

Table 14.2. *Continued.*

2032 (ts) [B. P. field no. 1-7/7/86-3]. Chert breccia from the Maddina Basalt; locality cf. 2030.

2033 (ts) [B. P. field no. 1-7/7/86-4]. Black chert from the Maddina Basalt; locality cf. 2030.

2034 (ts) [B. P. field no. 1-7/7/86-5]. Black chert from the Maddina Basalt; locality cf. 2030.

2035 (ts) [B. P. field no. 1-7/7/86-6]. Black chert from the Maddina Basalt; locality cf. 2030.

2036 (ts) [B. P. field no. 1-7/7/86-7]. Black Quartzite from the Maddina Basalt; locality cf. 2030.

2037 (ts) [B. P. field no. 1-7/7/86-8]. Vesicular basalt from the Maddina Basalt; locality cf. 2030.

2038 (ts, st) [B. P. field no. 1-7/8/86-1]. Stromatolitic black chert from the Maddina Basalt; "Tambrey V4 vent locality," Western Australia; Mount Billroth 1:100,000 map sheet 2454; 22.67°S, 117.5°E.

2039 (ts, st) [B. P. field no. 1-7/8/86-2]. Stromatolitic black chert from the Maddina Basalt; locality cf. 2038.

2040 (ts) [B. P. field no. 1-7/8/86-3]. Laminated gray chert from the Maddina Basalt; locality cf. 2038.

2041 (ts, st, TOC, Corg) [B. P. field no. 1-7/8/86-4]. Stromatolitic black chert from the Maddina Basalt; locality cf. 2038.

2042 (ts, st) [B. P. field no. 1-7/8/86-5]. Stromatolitic black chert from the Maddina Basalt; locality cf. 2038.

2043 (ts, st) [B. P. field no. 1-7/8/86-6]. Stromatolitic black chert from the Maddina Basalt; locality cf. 2038.

2044 (ts, st) [B. P. field no. 1-7/8/86-7]. Stromatolitic black chert from the Maddina Basalt; locality cf. 2038.

2045 (ts, st) [B. P. field no. 1-7/8/86-8]. Stromatolitic black chert from the Maddina Basalt; locality cf. 2038.

2046 (ts) [B. P. field no. 1-7/8/86-9]. Laminated black chert from the Maddina Basalt; locality cf. 2038.

2047 (ts) [B. P. field no. 1-7/8/86-10]. Black chert breccia from the Maddina Basalt; locality cf. 2038.

2048 (ts) [B. P. field no. 1-7/9/86-1]. White chert with black inclusions from the Maddina Basalt; locality cf. 2038.

2049 (ts) [B. P. field no. 1-6/17/86-1]. Tuff from the Tumbiana Formation; small gorge 1.6 km west of Nullagine River, 3 km northwest of Meentheena, 68 km south-southeast of Marble Bar, Western Australia; Mt. Edgar 1:100,000 map sheet 2955.

2050 (ts, st, TOC, Corg) [B. P. field no. 1-6/17/86-2]. Stromatolitic gray limestone from the Tumbiana Formation; locality cf. 2049.

2051 (ts, st) [B. P. field no. 1-6/17/86-3]. Stromatolite from the Tumbiana Formation; locality cf. 2049.

2052 (ts, st) [B. P. field no. 1-6/17/86-4]. Stromatolite from the Tumbiana Formation; locality cf. 2049.

2053 (ts) [B. P. field no. 1-6/17/86-5]. Oolites(?) from the Tumbiana Formation; locality cf. 2049.

2054 (ts, st) [B. P. field no. 1-6/17/86-6]. Stromatolite from the Tumbiana Formation; locality cf. 2049.

2055 (ts, st) [B. P. field no. 1-6/17/86-7]. Stromatolite from the Tumbiana Formation; locality cf. 2049.

2056 (st) [B. P. field no. 1-6/17/86-8]. Stromatolite from the Tumbiana Formation; locality cf. 2049.

2057 (st) [B. P. field no. 1-6/17/86-9]. Stromatolite from the Tumbiana Formation; locality cf. 2049.

2058 (st) [B. P. field no. 1-6/17/86-10]. Stromatolite from the Tumbiana Formation; locality cf. 2049.

2059 (ts) [B. P. field no. 1-6/17/86-11]. Oolites(?) from the Tumbiana Formation; locality cf. 2049.

2060 (ts) [B. P. field no. 1-6/17/86-12]. Limestone from the Tumbiana Formation; locality cf. 2049.

2061 (ts, st) [B. P. field no. 1-6/17/86-13]. Stromatolite from the Tumbiana Formation; locality cf. 2049.

2062 (ts, st) [B. P. field no. 1-6/17/86-14]. Stromatolite from the Tumbiana Formation; locality cf. 2049.

2063 (ts) [B. P. field no. 1-6/17/86-15]. Oolites from the Tumbiana Formation; locality cf. 2049.

2064 (ts, st) [B. P. field no. 1-6/17/86-16]. Stromatolite from the Tumbiana Formation; locality cf. 2049.

2065 (ts, TOC, Corg, ker) [B. P. field no. 1-6/19/86-1]. Shale with sandy laminae and mudcracks from the Tumbiana Formation; Thornes Bluff, Western Australia.

2066 (ts, st) [B. P. field no. 1-6/19/86-2]. Dark gray dolomitic stromatolite from the Tumbiana Formation; locality cf. 2065.

2067 (ts) [B. P. field no. 1-6/19/86-3]. Calcite fenestrae from the Tumbiana Formation; locality cf. 2065.

2068 (ts, st) [B. P. field no. 1-6/19/86-4]. Gray dolomitic stromatolite from the Tumbiana Formation; locality cf. 2065.

2069 (ts) [B. P. field no. 1-6/19/86-5]. Oolite(?) sand from the Tumbiana Formation; locality cf. 2065.

2070 (ts, st) [B. P. field no. 1-6/19/86-6]. Gray dolomitic stromatolite from the Tumbiana Formation; locality cf. 2065.

2071 (ts, st) [B. P. field no. 1-6/19/86-7]. Dark gray limestone with black chert lenses from the Tumbiana Formation; locality cf. 2065.

2072 (ts, st) [B. P. field no. 1-6/19/86-8]. Gray dolomitic and cherty stromatolite from the Tumbiana Formation; locality cf. 2065.

2073 (ts, st) [B. P. field no. 1-6/19/86-9]. Dark gray, cherty, stromatolite from the Tumbiana Formation; locality cf. 2065.

2074 (ts) [B. P. field no. 1-6/19/86-10]. Oolite bed from the Tumbiana Formation; locality cf. 2065.

Table 14.2. *Continued.*

2075 [B. P. field no. 1-6/19/86-11]. Calcite from basalt from the Tumbiana Formation; locality cf. 2065.

2076 (ts) [B. P. field no. 1-6/26/86-1]. Carbonate sand from the Tumbiana Formation; Corkbark Springs, Western Australia.

2077 (ts) [B. P. field no. 1-6/26/86-2]. Carbonate sand from the Tumbiana Formation; locality cf. 2076.

2078 (ts) [B. P. field no. 1-6/26/86-3]. Volcanic lapilli from the Tumbiana Formation; locality cf. 2076.

2079 (ts, st, TOC, Corg) [B. P. field no. 1-6/26/86-4]. Dark gray dolomite, sparry calcite, stromatolite from the Tumbiana Formation; locality cf. 2076.

2080 (ts, st) [B. P. field no. 1-6/26/86-5]. Stromatolite from the Tumbiana Formation; locality cf. 2076.

2081 (ts, st) [B. P. field no. 1-6/26/86-6]. Stratiform stromatolite with pyrite from the Tumbiana Formation; locality cf. 2076.

2082 (st) [B. P. field no. 1-6/21/86-1]. Stromatolite from the Tumbiana Formation; Camel's Crossing locality, Western Australia.

2083 (ts, st) [B. P. field no. 1-6/21/86-2]. Stromatolite from the Tumbiana Formation; locality cf. 2082.

2084 (st) [B. P. field no. 1-6/21/86-3]. Stromatolite from the Tumbiana Formation; locality cf. 2082.

2085 (ts, st) [B. P. field no. 1-6/21/86-4]. Stromatolite from the Tumbiana Formation; locality cf. 2082.

2086 (ts, st, TOC, Corg) [B. P. field no. 1-6/21/86-5]. Stromatolite from the Tumbiana Formation; locality cf. 2082.

2087 [B. P. field no. 1-6/21/86-6]. Micrite from the Tumbiana Formation; locality cf. 2082.

2088 (ts, st) [B. P. field no. 1-6/26/86-7]. Stromatolite from the Tumbiana Formation; locality cf. 2082.

2089 (st) [B. P. field no. 1-6/24/86-1]. Stromatolite from the Tumbiana Formation; Cave Hill, Western Australia.

2090 (ts, st) [B. P. field no. 1-6/24/86-2]. Stromatolite from the Tumbiana Formation; locality cf. 2089.

2091 (st) [B. P. field no. 1-6/24/86-3]. Stromatolite from the Tumbiana Formation; locality cf. 2089.

2092 (ts, st, TOC) [B. P. field no. 1-6/24/86-4]. Stromatolite from the Tumbiana Formation; locality cf. 2089.

2093 (ts) [B. P. field no. 1-6/29/86-1]. Limestone with quartz sand from the Tumbiana Formation; small quarry 150 m east of maintenance road at km 211.3 on Port Hedland-Newman railroad, Western Australia; 22.05°S, 118.9°E.

2094 (ts) [B. P. field no. 1-6/29/86-2]. Quartz sand with calcite cement from the Tumbiana Formation; locality cf. 2093.

2095 (ts, st) [B. P. field no. 1-6/29/86-3]. Stromatolite from the Tumbiana Formation; locality cf. 2093.

2096 (ts, st) [B. P. field no. 1-6/29/86-4]. Stromatolite from the Tumbiana Formation; locality cf. 2093.

2097 (ts, st) [B. P. field no. 1-6/29/86-5]. Stromatolite from the Tumbiana Formation; locality cf. 2093.

2098 (ts, st) [B. P. field no. 1-6/29/86-6]. Stromatolite from the Tumbiana Formation; locality cf. 2093.

2099 (ts, st) [B. P. field no. 1-6/29/86-7]. Stromatolite from the Tumbiana Formation; locality cf. 2093.

2100 (ts, st, TOC, Corg) [B. P. field no. 1-6/29/86-8]. Stromatolite from the Tumbiana Formation; locality cf. 2093.

2101 (ts, st) [B. P. field no. 1-6/30/86-1]. Stromatolite from the Tumbiana Formation; locality cf. 2093.

2102 (ts, st) [B. P. field no. 1-6/30/86-2]. Stromatolite from the Tumbiana Formation; locality cf. 2093.

2103 (ts, st) [B. P. field no. 1-6/30/86-3]. Stromatolite from the Tumbiana Formation; locality cf. 2093.

2104 (ts) [B. P. field no. 1-7/1/86-1]. Limestone with fenestrae from the Tumbiana Formation; locality cf. 2093.

2105 (ts, st) [B. P. field no. 1-7/1/86-2]. Stromatolite from the Tumbiana Formation; locality cf. 2093.

2106 (ts, st) [B. P. field no. 1-7/1/86-3]. Stromatolite from the Tumbiana Formation; locality cf. 2093.

2107 (ts, st, TOC, Corg) [B. P. field no. 1-7/1/86-4]. Stromatolite from the Tumbiana Formation; locality cf. 2093.

2108 (ts, st) [B. P. field no. 1-7/1/86-5]. Stromatolite from the Tumbiana Formation; locality cf. 2093.

2109 (ts, st) [B. P. field no. 1-7/1/86-6]. Stromatolite from the Tumbiana Formation; locality cf. 2093.

2110 [B. P. field no. 1-7/1/86-7]. Gneiss from the Tumbiana Formation; locality cf. 2093.

2111 [B. P. field no. 1-7/1/86-8]. Sandstone from the Tumbiana Formation; locality cf. 2093.

2112 [B. P. field no. 1-7/1/86-9]. Limestone with quartz sandstone from the Tumbiana Formation; locality cf. 2093.

2113 (ts, st) [B. P. field no. 1-7/2/86-1]. Stromatolite from the Tumbiana Formation; locality cf. 2093.

2114 (st) [B. P. field no. 1-7/2/86-2]. Stromatolitic sandy limestone from the Tumbiana Formation; locality cf. 2093.

2115 (st) [B. P. field no. 1-7/2/86-3]. Stromatolite from the Tumbiana Formation; locality cf. 2093.

2116 [B. P. field no. 1-7/2/86-4]. Sandy limestone from the Tumbiana Formation; locality cf. 2093.

2117 (st) [B. P. field no. 1-7/2/86-5]. Stromatolite from the Tumbiana Formation; locality cf. 2093.

2118 (st) [B. P. field no. 1-7/2/86-6]. Stromatolite from the Tumbiana Formation; locality cf. 2093.

Table 14.2. *Continued.*

2119 (st, TOC, Corg) [B. P. field no. 1-7/2/86-7]. Dark gray, stromatolitic limestone from the Tumbiana Formation; locality cf. 2093.

2120 (st) [B. P. field no. 1-7/2/86-8]. Stromatolite from the Tumbiana Formation; locality cf. 2093.

2121 (st) [B. P. field no. 1-7/2/86-9]. Stromatolite from the Tumbiana Formation; locality cf. 2093.

2609 [J. M. H. field no. 820612-5a]. Very well-laminated shale from the Hardy Sandstone; Fortescue inlier near North Pole, Western Australia; North Shaw 1:100,000 map sheet 2755, grid reference 219622.

2610 [J. M. H. field no. 820612-5b]. Chert from the lower part of the Apex Basalt, 10 m from the contact with Fortescue Group; locality cf. 2609.

2611 [J. M. H. field no. 820612-4a]. Chert from the Apex Basalt; locality cf. 2609.

2612 [J. M. H. field no. 820612-4b]. Chert dike from the Apex Basalt; locality cf. 2609.

2613 [J. M. H. field no. 820613-1a]. Sandstone from the Cleaverville Formation; locality cf. 2609.

2614 [J. M. H. field no. 820613-1b]. Shaley mudstone from the Cleaverville Formation; locality cf. 2609.

2615 [J. M. H. field no. 820613-1c]. Shale from 100–200 m higher than 2614, middle third of the Cleaverville Formation; locality cf. 2609.

2616 [J. M. H. field no. 820613-1d]. Shale from 20 m higher than 2615, Cleaverville Formation; locality cf. 2609.

2617 [J. M. H. field no. 820613-1e]. Black shale from 10 m higher than 2616, Cleaverville Formation; locality cf. 2609.

2643 [J. M. H. field no. 820615-3a]. Shale from the Hardy Sandstone; bed exposed at the side of a small creek, ca. 17 km west-southwest of Marble Bar, Western Australia; Marble Bar 1:100,000 map sheet 2588, grid reference 073561.

2647 (ts) [J. M. H. field no. 820616-1a]. Thin bedded carbonate with shaley partings from the Meentheena Carbonate Member of the Tumbiana Formation; small gorge 1.6 km west of Nullagine River, 3 km northwest of Meentheena, 68 km south-southeast of Marble Bar, Western Australia; Mt. Edgar 1:100,000 map sheet 2955.

2648 (ts, st) [J. M. H. field no. 820616-1c]. Chert from central part of stromatolite domes from the Meentheena Carbonate Member; locality cf. 2647.

2649 (ts, st) [J. M. H. field no. 820616-1e]. Carbonate in 1 cm columnar stromatolite from the Meentheena Carbonate Member; locality cf. 2647.

2650 (ts) [J. M. H. field no. 820616-1h]. Chert from the Meentheena Carbonate Member; locality cf. 2647.

2651 (ts) [J. M. H. field no. 820616-1j]. Light to dark gray, fine-grained carbonate with wavy laminae from the Meentheena Carbonate Member; locality cf. 2647.

2652 (ts, st) [J. M. H. field no. 820616-1k]. Dark gray, fine-grained carbonate with 1 cm columnar stromatolites, from the Meentheena Carbonate Member; locality cf. 2647.

2653 [J. M. H. field no. 820616-1l]. Medium-grained, carbonate possibly tuffaceous(?) with small chert fragments and white carbonate (calcite?) filling, from the Meentheena Carbonate Member; locality cf. 2647.

2654 (ts, st) [J. M. H. field no. 820616-1b]. Black chert from a 1–2 cm band concordant with broad stromatolites of the Meentheena Carbonate Member; locality cf. 2647.

2655 (ts, st) [J. M. H. field no. 820616-1d]. Carbonate in broad stromatolitic domes from the Meentheena Carbonate Member; locality cf. 2647.

2656 [J. M. H. field no. 820616-1f]. Flat-laminated carbonate from below shale of the Meentheena Carbonate Member; locality cf. 2647.

2657 (st) [J. M. H. field no. 820616-1g]. Stromatolitic (columnar) carbonate from 40 cm above 2656 and 90 cm below 2647, Meentheena Carbonate Member; locality cf. 2647.

2658 [J. M. H. field no. 820616-1i]. Chert from the Meentheena Carbonate Member; locality cf. 2647.

2659 [J. M. H. field no. 820616-1m]. Light gray carbonate interbanded with black chert from the Meentheena Carbonate Member; locality cf. 2647.

2660 [J. M. H. field no. 820616-1n]. Tuff from the base of the Mingah Tuff Member of the Tumbiana Formation; locality cf. 2647.

2661 [J. M. H. field no. 820616-1o]. Dark carbonate with large calcite inclusions, from the Meentheena Carbonate Member of the Tumbiana Formation; locality cf. 2647.

2662 (ts) [J. M. H. field no. 820616-1p]. Very fine-grained siltstone(?) from the Meentheena Carbonate Member; locality cf. 2647.

2663 [J. M. H. field no. 820616-1q]. Dark gray, fine-grained carbonate with calcitic inclusions, from the Meentheena Carbonate Member; locality cf. 2647.

2664 [J. M. H. field no. 820616-1r]. Medium-gray, fine-grained chert with oxidation along fractures, significantly weathered, from the Meentheena Carbonate Member; locality cf. 2647.

2665 [J. M. H. field no. 820616-1s]. Stromatolitic carbonate from the Meentheena Carbonate Member; locality cf. 2647.

2666 [J. M. H. field no. 820616-1t]. Gray, massive carbonate from the Meentheena Carbonate Member; locality cf. 2647.

2667 [J. M. H. field no. 820616-1u]. Oolitic carbonate, from above 2666, of the Meentheena Carbonate Member; locality cf. 2647.

2668 (st) [J. M. H. field no. 820616-1v]. Stromatolitic carbonate with chert laminae, from the Meentheena Carbonate Member; locality cf. 2647.

2669 [J. M. H. field no. 820616-1w]. Carbonate-rich shale from the Meentheena Carbonate Member; locality cf. 2647.

Table 14.2. *Continued.*

2670 (st) [J. M. H. field no. 820616-1x]. Stromatolitic carbonate with chert laminae, from the Meentheena Carbonate Member; locality cf. 2647.

2671 [J. M. H. field no. 820616-1aa]. Medium gray shale from the Meentheena Carbonate Member; locality cf. 2647.

2672 (st) [J. M. H. field no. 820616-1y]. Medium gray carbonate bulbose stromatolites, with ca. 5 cm relief and lenses of dark gray chert, from the Meentheena Carbonate Member; locality cf. 2647.

2673 [J. M. H. field no. 820616-1z]. Dark gray, thin-bedded, flat-laminated (30–40 cm) chert-carbonate, from the Meentheena Carbonate Member; locality cf. 2647.

2674 [J. M. H. field no. 820616-2a]. Pisolitic carbonate from above 2675, Kuruna Siltstone; 100 m west of unpaved road, 300 m west Nullagine River, 22 km north-northeast of 1455, Western Australia; Yilgalong 1:100,000 map sheet 3055, grid reference 457645.

2675 [J. M. H. field no. 820616-2b]. Carbonate from a 3 cm-thick bed immediately overlying siltstone, Kuruna Siltstone; locality cf. 2674.

2676 [J. M. H. field no. 820616-2c]. Siltstone from the Kuruna Siltstone; locality cf. 2674.

2677 [J. M. H. field no. 820616-2d]. Cherty carbonate from above 2674, Kuruna Siltstone; locality cf. 2674.

2678 [J. M. H. field no. 820616-3a]. Dark gray siliceous carbonate from the Kylena Basalt; immediately adjacent to unpaved road 3 km west of Nullagine River, Western Australia; map sheet 2955, grid reference 395556.

2679 (st) [J. M. H. field no. 820616-3b]. Carbonate with irregular pustular stromatolitic laminae from below 2678, Kylena Basalt; locality cf. 2678.

2680 [J. M. H. field no. 820616-3c]. Chert fragment from 2679, from the Kylena Basalt; locality cf. 2678.

2681 [J. M. H. field no. 820616-4a]. Siliceous shale from the Kylena Basalt; side of road, near the Nullagine River, Western Australia; map sheet 2955, grid references 385540.

2683 [J. M. H. field no. 820617-2a]. Shale from the Hardy Sandstone; bed exposed at the side of a small creek, ca. 17 km west-southwest of Marble Bar, Western Australia; Marble Bar 1:100,000 map sheet 2588, grid reference 073561.

2684 [J. M. H. field no. 820617-2b]. Shale from the Hardy Sandstone; 2 m from 2683, locality cf. 2683.

2696 [J. M. H. field no. 820619-1a]. Shale from the Warrie Member of the Tumbiana Formation; "Warrie 1 locality," Western Australia; map sheet 2753, grid reference 086490.

2697 [J. M. H. field no. 820619-2a]. Black, tuffaceous, very fine-grained arenite, with some fractures, from a 750 cm-thick bed of the Warrie Member; "Warrie 2 locality," Western Australia; map sheet 2653, grid reference 060572.

2698 [J. M. H. field no. 820619-2b]. Black arenite, from 0.75 m below 2697, of the Warrie Member; locality cf. 2697.

2699 [J. M. H. field no. 820619-2c]. Black arenite from the Warrie Member; locality cf. 2697.

2700 (st) [J. M. H. field no. 820619-3a]. Chert from large domical stromatolites from the Tumbiana Formation; at top of cliff, small quarry 150 m east of maintenance road at km 211.3 on Port Hedland-Newman railroad, Western Australia; 22.05°S, 118.9°E.

2701 [J. M. H. field no. 820619-3b]. Flat-laminated chert-carbonate from the bed just below stromatolites in the section, Tumbiana Formation; locality cf. 2700.

2702 [J. M. H. field no. 820619-3c]. Chert-carbonate from below 2701, Tumbiana Formation; locality cf. 2700.

2703 [J. M. H. field no. 820619-3d]. Chert-carbonate from 2 m below 2702, Tumbiana Formation; locality cf. 2700.

2704 (st) [J. M. H. field no. 820619-3e]. Medium gray, flat-laminated cherty limestone, with stratiform and infrequent small domical stromatolites, from 2 m below 2703, Tumbiana Formation; locality cf. 2700.

2705 [J. M. H. field no. 820619-3f]. Light gray-brown, ripple-laminated limestone with thin cherty laminae, from 2 m below 2704, Tumbiana Formation; locality cf. 2700.

2706 (st) [J. M. H. field no. 820619-3g]. Dark gray cherty carbonate with domical stromatolites and *Alcheringa*, from a 30 cm-thick bed, 1.5 m below 2705, Tumbiana Formation; locality cf. 2700.

2707 [J. M. H. field no. 820619-3h]. Cherty carbonate from a 5–10 cm-thick bed of fenestrae limestone with calcite infillings, 4 m above bed 27, Tumbiana Formation; locality cf. 2700.

2708 [J. M. H. field no. 820619-3i]. Dark gray carbonate from a 40 cm-thick, brown weathering, coarse crystalline bed of the Tumbiana Formation; locality cf. 2700.

2709 [J. M. H. field no. 820619-3j]. Dark gray, non-stromatolitic, chert carbonate from a 10 cm-thick, deeply weathered, brown siltstone bed below 2708, Tumbiana Formation; locality cf. 2700.

2710 [J. M. H. field no. 820619-3k]. Dark blue-green chert, possibly tuffaceous, from a 2–5 cm-thick tabular bed, on the other side of the valley, between 2703 and 2704, Tumbiana Formation; locality cf. 2700.

2711 [J. M. H. field no. 820619-3l]. Dark green-gray chert from a 50 cm-thick siliceous bed with lenses of carbonate, ca. 20 m above unconformity, Tumbiana Formation; west side of valley, high on cliff above railroad tracks, locality cf. 2700.

2712 (ts) [J. M. H. field no. 820619-3m]. Black chert breccia in a calcareous siltstone matrix from a 2 cm-thick bed of the Tumbiana Formation; locality cf. 2700.

Table 14.2. *Continued.*

2713 [J. M. H. field no. 820620-1a]. Banded iron-formation with essentially pure bands of very coarsely crystalline siderite, from the Dales Gorge Member of the Brockman Iron Formation; Wittenoom Mine dump, Hamersley Exploration Camp, Wittenoom, Western Australia.

2714 [J. M. H. field no. 820620-2a]. Massive limestone from the bottom of the Dales Gorge Member; locality cf. 2713.

2715 [J. M. H. field no. 820621-1a]. Medium gray, fine-grained, thin-bedded, laminated shaley sandstone from a 1 m-thick bed of the Hardy Sandstone; west of Mount Tom Price, 1.5 km north of Rocklea Homestead, on the margin of Rocklea Dome, southern Hamersley Range, Western Australia.

KIMBERLEY BASIN (EAST KIMBERLEY)

Tectonic unit/location. Kimberley Basin (East Kimberley), Western Australia, Australia.

Stratigraphy. Stratigraphic section, Nicholson area, northeastern Western Australia (Thom 1975b):

ALBERT EDWARD GROUP
    Flat Rock Formation
    Nyuless Sandstone
    Timperley Shale
    Boonall Dolomite
    Elvire Formation
    Mount Forster Sandstone
LOUISA DOWNS GROUP
    Lubbock Formation
    Tean Formation
    McAlly Shale
    Yurabi Formation
    Egan Formation
DUERDIN GROUP
    Ranford Formation
        Johhny Cake Shale Member
        Jarrad Sandstone Member
    Moonlight Valley Tillite
    Frank River Sandstone
    Fargoo Tillite
KUNIANDI GROUP
    Mount Bertram Sandstone
    Wirara Formation
    Stein Formation
    Landrigan Tillite

PPRG samples/localities.

2224 (ts, st) [D. D. field no. 9686B]. Carbonate with wavy lamination (stromatolitic?) from the Egan Formation; half-way up hillside, Egan Range, Western Australia; 18.55°S, 127.1°E.

2225 (ts) [D. D. field no. 9686C]. Well indurated tillite with clasts from <1 mm to 5 mm, from 5 m below 2224, Egan Formation; locality cf. 2224.

2226 (ts) [D. D. field no. 9686D]. Duplicate of 2225, from ca. 5 m below 2224, Egan Formation; locality cf. 2224.

2227 (ts) [D. C. field no. 050986-1]. Dolomite from the Egan Formation; roadside cut 76 km from Hall's Creek, west side of cutting, on north face, locality cf. 2224.

2249 [D. R. L. field no. AUS-58-12]. Laterite from the Tertiary weathering zone; Cambrian(?) quartzite exposed on the rim of Wolf Creek crater, Western Australia; 19.17°S, 127.8°E.

2250 [D. R. L. field no. AUS-58-4]. Quartzite from the Tertiary weathering zone; locality cf. 2249.

2251 [D. R. L. field no. AUS-58-5]. Quartzite from the Tertiary weathering zone; locality cf. 2249.

2252 [D. R. L. field no. AUS-58-6]. Quartzite from the Tertiary weathering zone; locality cf. 2249.

2348 [M. R. W. field no. 860905-10]. Laminated, light-violet to grey, micaceous siltstone with problematic bedding plane structures, from the Tean Formation; roadcut on the highway near crossing of the Mary River, Western Australia; 18.71°S, 126.8°E.

2349 [D. R. L. field no. AUS-58-13]. Pisolitic, iron laterite nodules; Cambrian(?) quartzite exposed on the rim of Wolf Creek crater, Western Australia; 19.17°S, 127.8°E.

2350 [D. R. L. field no. AUS-58-8]. Massive iron laterite nodule; locality cf. 2349.

2351 [D. R. L. field no. AUS-58-14]. Massive iron laterite nodule; locality cf. 2349.

2352 (ts) [M. R. W. field no. 860906-1]. Dolomitic breccia from a 40 cm-thick bed near the middle of the Egan Formation; Egan Range, Western Australia; 18.55°S, 127.1°E.

Table 14.2. *Continued.*

KIMBERLEY BASIN (HALLS CREEK MOBILE ZONE)

Tectonic unit/location. Kimberley Basin (Halls Creek Mobile Zone), Western Australia, Australia.

Stratigraphy. Stratigraphic section, Halls Creek area, northeastern Western Australia (Thom 1975a):

    Red Rock Beds
  HALLS CREEK GROUP
    Olympio Formation
    Biscay Formation
    Saunders Creek Formation
    Ding Dong Downs Volcanics

PPRG samples/localities.

2189 [M. R. W. field no. 860901.2]. Thick-bedded, medium-grained, light red-brown sandstone with fluid escape structures and flute casts, from the Jarrad Sandstone Member of the Ranford Formation; Dixon Range, Western Australia; 17.08°S, 128.3°E.

2190 (ts) [M. R. W. field no. 860901.3]. Thick-bedded, medium-grained, light red-brown sandstone with fluid escape structures and flute casts, from the Ranford Formation; locality cf. 2189.

2191 (ts, TOC, Corg) [M. R. W. field no. 860901.4]. Thin-bedded, flat-laminated, medium pink dolomite interbedded in medium red-brown siltstone and shale, from the middle of a 7 m-thick sequence of the Jarrad Sandstone Member of the Ranford Formation; locality cf. 2189.

2192 (ts, TOC) [M. R. W. field no. 860901.5]. Thin-bedded, flat-laminated, medium pink dolomite, interbedded in medium red-brown siltstone and shale, from 30 cm above the base of the dolomite of the Jarrad Sandstone Member; locality cf. 2189.

2198 (ts, TOC) [R. J. H. field no. RH-9-5-86-1-#3]. Sandstone with dubiofossils from the Yurabi Formation; ca. 140 m east of and at the west end of a road cut in the Egan Range, Western Australia; 18.55°S, 127.1°E.

2203 (ts, TOC) [D. D. field no. 9386A]. Oolitic dolomite from 10 m above the base of the Boonall Dolomite; prominent ledge 8 m above creek bed, on the south side of gorge, Hall's Creek, Western Australia.

2204 (ts, TOC) [D. D. field no. 9386B]. Oolitic dolomite from 10 m above the base of the Boonall Dolomite; locality cf. 2203.

2205 (ts, st, TOC) [D. D. field no. 9386C]. Microbially laminated, micritic dolomite from 20–25 m above the base of the Boonall Dolomite; crest of slope on the south side of the gorge, locality cf. 2203.

2206 (ts, st) [D. D. field no. 9386D]. Microbially laminated, micritic dolomite from 20–25 m above the base of the Boonall Dolomite; locality cf. 2203.

2207 (ts) [D. D. field no. 9386E]. Conglomeratic dolomite, with angular carbonate clasts embedded in redish carbonate matrix, from 30 m above the base of the Boonall Dolomite; locality cf. 2203.

2208 (ts, st, TOC, Corg) [D. D. field no. 9386F]. Microbially laminated, micritic dolomite from 35 m above the base of the Boonall Dolomite; locality cf. 2203

2209 (ts, st) [D. D. field no. 9386G]. Microbially laminated, micritic dolomite from 35 m above the base of the Boonall Dolomite; locality cf. 2203.

2210 (ts) [D. D. field no. 9386H]. Dolomite with purple and gray mottling on surface, from 50 m above the base of the Boonall Dolomite; locality cf. 2203.

2211 (ts) [D. D. field no. 9386I]. Conglomeratic dolomite, with angular carbonate clasts embedded in the reddish carbonate matrix, from 30 m above the base of the Boonall Dolomite; locality cf. 2203.

2212 (ts) [D. D. field no. 9386J]. Light gray, micritic dolomite from above the base of the Boonall Dolomite; locality cf. 2203.

2213 (ts, TOC, Corg) [D. D. field no. 9386K]. Shaley carbonate from the Timperley Shale; 3 m high sandstone cliff on south side of road, east side of stream, locality cf. 2203.

2214 (ts) [D. D. field no. 9386L]. Shaley carbonate cf. 2213 from the Timperley Shale; locality cf. 2203.

2215 (ts) [M. R. W. field no. 860903.1]. Thin-bedded, flaggy, rippled, buff-colored, medium-grained sandstone, with sinuous incomplete desiccation cracks(?), from the Mount Forster Sandstone; on creek bank, locality cf. 2203.

2219 (ts) [D. D. field no. 9486A]. Massive micritic limestone from the Egan Formation; low limestone ridge in valley between two sandstone ridges, Louisa Downs, Western Australia; 18.73°S, 126.6°E.

2220 (ts) [D. D. field no. 9486B]. Massive, micritic limestone cf. 2219, from the Egan Formation; locality cf. 2219.

2221 (ts, TOC, Corg) [D. D. field no. 9486C]. Massive micritic carbonate that weathers to blue-gray, from 6 m above the base of a carbonate unit of the Egan Formation; low ridge (2 m high) in middle of creek bed, locality cf. 2219.

2223 (ts, st) [D. D. field no. 9686A]. Thin bedded, laminated, stromatolitic(?), pink, intraclastic carbonate from the Egan Formation; west end of road cut in the Egan Range, Western Australia; 18.55°S, 127.1°E.

2229 [J. H. L. field no. A11-86]. Massive micritic limestone from the Egan Formation; between two ridges west-southwest of fenceline, Louisa Downs, Western Australia; 18.73°S, 126.6°E.

Table 14.2. *Continued.*

2230 (ts) [J. H. L. field no. A12-86]. Massively bedded, red diamictite from the Egan Formation; west end of road cut in Egan Range, Western Australia; 18.55°S, 127.1°E.

2231 (ts) [J. H. L. field no. A13-86]. Red, siltstone, "zebra rock" containing jellyfish pseudofossils from the Ranford Formation; exposure near Lake Argyle, Western Australia.

2238 [D. R. L. field no. AUS-56-1]. Carbonate sandstone from the basal 5 cm of a turbidite, Ranford Formation; Dixon Range, Western Australia; 17.08°S, 128.3°E.

2239 (ts) [D. R. L. field no. AUS-56-2]. Micaceous siltstone from the upper part of a turbidite, Ranford Formation; locality cf. 2238.

2240 (ts) [D. R. L. field no. AUS-56-3]. Conglomeratic sandstone from a 2 m-thick unit of sand and conglomerate between the tillites of the Frank River Sandstone; locality cf. 2238.

2253 [D. R. L. field no. AUS-59-1]. Quartzite from the Yurabi Formation; "Yurabi" Locality, Western Australia; 18.58°S, 126.6°E.

2254 (ts) [D. R. L. field no. AUS-60-1]. Carbonate from the base of the Egan Formation; west end of road cut in Egan Range, Western Australia; 18.55°S, 127.1°E.

2255 (ts) [D. R. L. field no. AUS-60-10]. Medium-grained quartzose sandstone from 10 m above the base of the Yurabi Formation; locality cf. 2254.

2256 [D. R. L. field no. AUS-60-11]. Wave-rippled sandstone from 12 m above the base of the Yurabi Formation; locality cf. 2254.

2257 (ts) [D. R. L. field no. AUS-60-12]. Coarse-grained sandstone from 40 m above the base of the Yurabi Formation; locality cf. 2254.

2258 (ts) [D. R. L. field no. AUS-60-13]. Medium-grained sandstone from 40 m above the base of the Yurabi Formation; locality cf. 2254.

2259 (ts) [D. R. L. field no. AUS-60-14]. Red, clast-poor tillite from the Egan Formation; locality cf. 2254.

2260 (ts) [D. R. L. field no. AUS-60-15]. Fine-grained, detrital carbonate from near the top of the "cap dolomite" of the Egan Formation; locality cf. 2254.

2261 (ts) [D. R. L. field no. AUS-60-16]. Medium-grained, flat laminated, quartz sandstone from 1 m above the base of the Yurabi Formation; locality cf. 2254.

2262 (ts) [D. R. L. field no. AUS-60-2]. Carbonate from the basal portion of the Egan Formation; locality cf. 2254.

2263 (ts) [D. R. L. field no. AUS-60-3]. Dolomite from the basal portion of the Egan Formation; locality cf. 2254.

2264 [D. R. L. field no. AUS-60-4]. Coarse-grained, feldspathic sandstone from the basal unit of the Egan Formation; locality cf. 2254.

2265 [D. R. L. field no. AUS-60-5]. Sandy carbonate from the basal carbonate unit, below the diamictite, of the Egan Formation; locality cf. 2254.

2266 [D. R. L. field no. AUS-60-6]. Diamictite from the Egan Formation; locality cf. 2254.

2267 [D. R. L. field no. AUS 60-7]. Laminated, calcareous siltstone and sandstone from immediately above the diamictite of the Egan Formation; locality cf. 2254.

2268 [D. R. L. field no. AUS-60-8]. Laminated, calcareous siltstone and sandstone from the Egan Formation; locality cf. 2254.

2269 [D. R. L. field no. AUS-60-9]. Laminated calcareous siltstone and sandstone from immediately above the diamictite of the Egan Formation; locality cf. 2254.

2270 [D. R. L. field no. AUS-61-1]. Pebbly, feldspathic quartz sandstone from the Red Rock Beds Formation; stream cut, Halls Creek Fault Zone, northwest of Old Halls Creek, Western Australia.

2271 [D. R. L. field no. AUS-61-2]. Pebbly, feldspathic quartz sandstone from the Red Rock Beds Formation; locality cf. 2270.

2272 [R. I. field no. AUST-1]. Siliciclastic sandstone from the Jarrad Sandstone Member of the Ranford Formation; Dixon Range, Western Australia; 17.08°S, 128.3°E.

2273 [R. I. field no. AUST-10]. Pebbly, feldspathic quartz sandstone from the Red Rock Beds Formation; Halls Creek Fault Zone, northwest of Old Halls Creek, Western Australia.

2274 [R. I. field no. AUST-11]. Granitic sand derived from Proterozoic granite to the west, Louisa Downs Group; 40 km north of Halls Creek, Western Australia.

2275 [R. I. field no. AUST-2]. Carbonate sandstone from the Moonlight Valley Tillite; Dixon Range, Western Australia; 17.08°S, 128.3°E.

2276 [R. I. field no. AUST-3]. Mottled quartz sandstone from the Frank River Sandstone; locality cf. 2275.

2279 [R. I. field no. AUST-7]. Fine-grained graywacke(?) from the Egan Formation; Louisa Downs, Western Australia; 18.73°S, 126.6°E.

2280 [R. I. field no. AUST-8]. Quartz river sand derived from Proterozoic sediments and basement, from a recent river sand of the Louisa Downs Group; locality cf. 2279.

2281 [R. I. field no. AUST-9]. Medium-grained, gray, volcaniclastic(?) sandstone from the Olympio Formation; half-way between old Halls Creek and Palm Spring, Western Australia.

2287 [H. J. H. field no. H-8609-3-1]. Medium-grained, quartz sandstone, with two subcircular markings interpreted as dubiofossils (planolites?), from the Mount Forster Sandstone; Halls Creek, Western Australia.

Table 14.2. *Continued.*

2288 [H. J. H. field no. H8609-3-3]. Fine-grained, olive sandstone-siltstone from the Timperley Shale; locality cf. 2287.

2289 [H. J. H. field no. H8609-4-1]. Catagraphic gray limestone from the Egan Formation; in valley between two sandstone ridges, ca. 300 m southwest of fenceline, Louisa Downs site no. 2, Western Australia; 18.65°S, 126.5°E.

2290 [H. J. H. field no. H-8609-5-1]. Olive-weathering pink dolomite, from 30 cm above the top of the dolomite of the Egan Formation; west end of road cut in the Egan Range, Western Australia; 18.55°S, 127.1°E.

2291 [H. J. H. field no. H8609-5-2]. Pink-weathering dolomite from 3 m below the top of the dolomite of the Egan Formation; locality cf. 2290.

2292 [H. J. H. field no. H-8609-5-3]. Flaggy sandstone with circular markings, from the Yurabi Formation; locality cf. 2290.

2293 [H. J. H. field no. H8609-5-4]. Sandstone with dubiofossils from the Yurabi Formation; locality cf. 2290.

2294 [H. J. H. field no. H8609-5-5]. Diamictite from the Louisa Downs Group; locality cf. 2290.

2295 [H. J. H. field no. H8609-6-1]. Ripple marked quartz sandstone from the Yurabi Formation; from crest of sandstone ridge, north end of the Egan Range, Western Australia; 18.55°S, 127.1°E.

2296 (ts) [R. J. H. field no. RH-9-1-86-1]. Tillite from float, probably derived from of the Fargoo Tillite; Dixon Range, Western Australia; 17.08°S, 128.3°E.

2330 (ts) [D. D. field no. 9186A]. Cap Dolomite with vertical fractures, from the Jarrad Sandstone Member of the Ranford Formation; locality cf. 2296.

2331 [D. D. field no. 9186B]. Cap Dolomite from the Jarrad Sandstone Member; locality cf. 2296.

2332 (ts) [D. D. field no. 9186C]. Black chert from the Fargoo Tillite; locality cf. 2296.

2340 (ts) [M. R. W. field no. 860905-1]. Light pink and brown, poorly laminated, medium-bedded, vuggy carbonate from the Egan Formation; west end of road cut in Egan Range, Western Australia; 18.55°S, 127.1°E.

2341 (ts, st) [M. R. W. field no. 860905-2]. Medium-pink, thin-bedded, laminated, stromatolitic(?), intraclastic carbonate from the Egan Formation; locality cf. 2340.

2342 (ts) [M. R. W. field no. 860905-3]. Thick-bedded, poorly laminated (non-stromatolitic, with cross laminations), light-brown and pink intraclastic carbonate from the Egan Formation; locality cf. 2340.

2343 (ts, st) [M. R. W. field no. 860905-4]. Medium-pink, thin-bedded stratiform carbonate stromatolite with teepee structures from the Egan Formation; locality cf. 2340.

2344 (ts, st) [M. R. W. field no. 860905-5]. Light-pink, thin-bedded, stromatolitic(?) carbonate from the Egan Formation; locality cf. 2340.

2345 (ts, st) [M. R. W. field no. 860905-6]. Medium, stratiform and domical carbonate stromatolites from a bed ca. 3 m thick, Egan Formation; locality cf. 2340.

2346 (ts) [M. R. W. field no. 860905-7]. Light-brown, wavy-laminated carbonate from the Yurabi Formation; locality cf. 2340.

2347 [M. R. W. field no. 860905-8]. Thin-bedded, brown, medium-grained sandstone with cylindrical structures, from the Yurabi Formation; locality cf. 2340.

2353 [M. R. W. field no. 860907-1]. Medium red-brown, thin-bedded, flat-laminated siltstone "zebra rock," with jellyfish pseudofossils, from the Ranford Formation; exposure near Lake Argyle, Western Australia.

2354 [M. R. W. field no. 860907-2]. Siltstone cf. 2353, Ranford Formation; locality cf. 2353.

2355 [M. R. W. field no. 860907-3]. Siltstone cf. 2353, Ranford Formation; locality cf. 2353.

2357 [P. P. R. G. field no. N/A]. Iron laterite nodules from the Louisa Downs Group; Louisa Downs, Western Australia; 18.73°S, 126.6°E.

LAWN HILL PLATFORM

Tectonic unit/location. Lawn Hill Platform, Queensland and Northern Territory, Australia.

Stratigraphy. Stratigraphic section, Barkly Tableland area, northwestern Queensland and northeastern Northern Territory, Australia (Plumb et al. 1981; Blake 1987):

SOUTH NICOLSON GROUP
McNAMARA GROUP
    Lawn Hill Formation
    Termite Range Formation
    Riversleigh Siltstone
    Shady Bore Quartzite
    Lady Loretta Formation
    Esperanza Formation
    Paradise Creek Formation
    Gunpowder Creek Formation
    Torpedo Creek Quartzite

Table 14.2. *Continued.*

[UNNAMED LOWER PROTEROZOIC GROUP]
    Surprise Creek Formation/Carters Bore Rhyolite
    Kakadoon Granite/Wills Creek Granite

PPRG samples/localities.

1260 (ts, st, μf^*, TOC, Corg) [J. W. S. field no. 29 of 5/26/68]. Microfossiliferous (Chapter 24), stromatolitic, black chert from the upper third of the Paradise Creek Formation; hilltop 100 m east of unpaved road, 2.5 km north of Paradise Creek, 13 km southeast of Lady Agnes Mine, Queensland, Australia.

LIMMEN GEOSYNCLINE

Tectonic unit/location. Limmen Geosyncline, Northern Territory, Australia.

Stratigraphy.

    McMinn Formation
    Koolpin Formation

PPRG samples/localities.

1261 (ts, TOC, Corg, ker) [J. W. S. field no. 27 of 5/23/68]. Massive black chert interbedded with black shale from the upper fourth of the Koolpin Formation; 0.5 km northwest of Rockhole Creek Mine, on road to Rockhole Adit No. 3, 1 km east of El Sherana, 15 km east of Pine Creek, 190 km southeast of Darwin, Northern Territory, Australia.

1262 (ts, TOC, Corg) [J. W. S. field no. DDH-C1, 104′]. Bedded siltstone and shale from a core sample, DH C1 at a depth of 31.5 m, middle third of the McMinn Formation; Gum Creek area, complete core is located at the Roper Bar core shed, Northern Territory, Australia.

1263 (ts, TOC, Corg, ker) [J. W. S. field no. DDH-U32,135′]. Interlaminated siltstone and shale with mudcracks from a core sample, DH U32 at a depth of 41 m, McMinn Formation; Hodgson Downs area, locality cf. 1262.

1264 (ts, TOC, Corg, ker) [J. W. S. field no. DDH-W27,140′]. Interlaminated, dark-gray, micaceous siltstone and shale from a core sample, DH W27 at a depth of 40 m, middle third of the McMinn Formation; Hodgson Downs area, locality cf. 1262.

1265 (ts, TOC, Corg) [J. W. S. field no. DDH-M15,215′]. Interlaminated, dark-gray, micaceous siltstone and shale from a core sample, DH M15 at a depth of 65 m, middle third of the McMinn Formation; "M. Deposit" area, locality cf. 1262.

1266 (ts, TOC, Corg, ker) [J. W. S. field no. DDH-W27,142′]. Dark-gray micaceous siltstone from core sample, DH W27 at a depth of 40 m, middle third of the McMinn Formation; Hodgson Downs area, locality cf. 1262.

1267 (ts, TOC, Corg) [J. W. S. field no. DDH-U35,230′]. Dark-gray siltstone from a core sample, DH U35 at a depth of 70 m, middle third of the McMinn Formation; Hodgson Downs area, locality cf. 1262.

1268 (ts, TOC, Corg, ker) [J. W. S. field no. DDH-B7,40′]. Dark-gray siltstone with mudcracks from a core sample, DH B7 at a depth of 12 m, middle third of the McMinn Formation; Gum Creek area, locality cf. 1262.

1269 (ts, TOC, Corg) [J. W. S. field no. DDH-B7,45′]. Finely laminated, dark-gray micaceous siltstone from a core sample, DH B7 at a depth of 14 m, middle third of the McMinn Formation; Gum Creek area, locality cf. 1262.

1270 (ts, TOC, Corg, ker) [J. W. S. field no. DDH-M15,200′]. Dark-gray micaceous siltstone from a core sample, DH M15 at a depth of 60 m, middle third of the McMinn Formation; "M. Deposit" area, locality cf. 1262.

McARTHUR BASIN

Tectonic unit/location. McArthur Basin, Northern Territory, Australia.

Stratigraphy. Stratigraphic section, McArthur area, northeastern Northern Territory (Jackson et al. 1987):

ROPER GROUP
  MAIWOK SUBGROUP
    Chambers River Formation/(upper) Corbanbirini Formation
    McMinn Formation/(middle) Corbanbirini Formation
    Velkerri Formation/(lower) Corbanbirini Formation
  [UNNAMED MIDDLE PROTEROZOIC SUBGROUP]
    Bessie Creek Sandstone
    Corcoran Formation/Abner Sandstone
    Crawford Formation
    Mainoru Formation
    Limmen Sandstone
NATHAN GROUP
    Dungaminnie Formation
    Balbirini Dolomite
McARTHUR GROUP

BATTEN SUBGROUP
    Looking Glass Formation/Amos Formation
    Stretton Sandstone
    Yalco Formation
    Lynott Formation
UMBOLOOGA SUBGROUP
    Reward Dolomite
    Barney Creek Formation
        HYC Pyritic Shale Member/(upper) Cooley Dolomite Member
        W-Fold Shale Member/(lower) Cooley Dolomite Member
    Teena Dolomite
        Coxco Dolomite Member
        [Unnamed dololutitic-dolomitic Member]
    Emmerugga Dolomite
        Mitchell Yard Dolomite Member
        Mara Dolomite Member
    Myrtle Shale
    Leila Sandstone
    Tooganinie Formation
    Tatoola Sandstone
    Amelia Dolomite
    Mallapunyah Formation
    Masterton Sandstone
TAWALLAH GROUP

PPRG samples/localities.

1279 (ts, TOC, Corg) [J. W. S. field no. CROX-7/11/68]. Lead-zinc ore from the lower third of the McArthur Group; central trough of the McArthur Basin, Northern Territory, Australia.

MOUNT ISA INLIER

Tectonic unit/location. Mount Isa Inlier, Queensland, Australia.

Stratigraphy. Stratigraphic section, Mount Isa area, northwestern Queensland (Plumb et al. 1981; Blake 1987):

MOUNT ISA GROUP
    Magazine Shale
    Kennedy Siltstone
    Spear Siltstone
    Urquhart Shale
    Native Bee Siltstone
    Breakaway Shale
    Moodarra Siltstone
    Warrina Park Quartzite
[UNNAMED LOWER PROTEROZOIC GROUP]
    Surprise Creek Formation
    Kakadoon Granite/Wills Creek Granite

PPRG samples/localities.

1273 (ts, TOC, Corg) [J. W. S. field no. 30A of 5/27/68]. Black shale from level no. 13, 7300 N, footwall of no. 8 ore body, Urquhart Shale; Mount Isa Mine, Mount Isa, Northern Territory, Australia.

1274 (ts, TOC, Corg, Sfide) 30B of 5/27/68]. Dark gray shale containing pyrite and galena from level no. 13, 7000N, footwall of no. 5 ore body, Urquhart Shale; locality cf. 1273.

1275 (ts, TOC, Corg, Sfide) [J. W. S. field no. 30C of 5/27/68]. Shale from level no. 13, 7200N, footwall of no. 5 ore body, Urquhart Shale; locality cf. 1273.

1276 (ts, TOC, Corg) [J. W. S. field no. 30D of 5/27/68]. Dark gray siliceous shale from level no. 13, 6900N, footwall of no. 5 ore body, Urquhart Shale; locality cf. 1273.

1277 (ts, TOC, Corg, Sfide) [J. W. S. field no. 30E of 5/27/68]. Shale from level no. 13, 7200N, footwall of no. 500 ore body, Urquhart Shale; locality cf. 1273.

1278 (ts, TOC, Corg, ker) [J. W. S. field no. 30F of 5/27/68]. Dark gray shale containing pyrite and galena from level no. 13, 6500N, 2100E, Urquhart Shale; locality cf. 1273.

PILBARA BLOCK

Tectonic unit/location. Pilbara Block, Western Australia, Australia.

Stratigraphy. Stratigraphic section, Marble Bar area, northwestern Western Australia (Hickman 1983):

PILBARA SUPERGROUP
  [UNNAMED MIDDLE ARCHEAN GROUP]
    Negri Volcanics
    Louden Volcanics
  WHIM CREEK GROUP
    Rushall Slate
    Mons Cupri Volcanics
    Warambie Basalt
  GORGE CREEK GROUP
    [UNNAMED MIDDLE ARCHEAN SUBGROUP]
      Mosquito Creek Formation
      Lalla Rookh Sandstone
      Honeyeater Basalt
    SOANESVILLE SUBGROUP
      Cleaverville Formation
      Charteris Basalt
      Corboy Formation
  WARRAWOONA GROUP
    [UNNAMED MIDDLE ARCHEAN SUBGROUP]
      Wyman Formation
    SALGASH SUBGROUP
      Euro Basalt
      Panorama Formation
      Apex Basalt
      Towers Formation
    [UNNAMED LOWER ARCHEAN SUBGROUP]
      Duffer Formation
    TALGA TALGA SUBGROUP
      Mount Ada Basalt
      McPhee Formation
      North Star Basalt

PPRG samples/localities.

1436 (ts, TOC, Corg) [J. W. S. field no. 1-6/11/82]. Black and gray bedded chert from the upper Cleaverville Fm.(?) of the Gorge Creek Group; 800 km east of unpaved road, 5.5 km east of Shaw River, 9 km north of North Pole Mining Center, ca. 100 km southeast of Port Headland, Western Australia; North Shaw 1:100,000 map sheet 2755, grid reference 482738.

1437 (ts, TOC, Corg) [J. W. S. field no. 2-6/11/82]. Black and gray bedded chert from the Towers Formation; outcrops at tops and flanks of low hills, 6 km south-southeast of 1436, ca. 100 km southeast of Port Hedland, microfossil locality A of Awramik et al. (1983), Western Australia; North Shaw 1:100,000 map sheet 2755, grid reference 504638.5.

1438 (ts, TOC, Corg) [J. W. S. field no. 3-6/11/82]. Bedded, laminated, black chert from 1 m below the top of the outcrop, Towers Formation; south side of the gorge adjacent to unpaved road, 4.7 km southeast 1437 and 1 km south of Old Dresser Mineral Camp; locality B of Awramik et al. (1983), Western Australia; North Shaw 1:000,000 map sheet 2755, grid reference 536648.

1439 (ts, TOC, Corg) [J. W. S. field no. 3-6/11/82]. Black chert intrusive vein ca. 5 m from 1438, Towers Formation; locality cf. 1438.

1440 (ts, TOC, Corg) [J. W. S. field no. 1-6/12/82]. Fine-grained, gray and black chert from the Apex Basalt; small outcrop 5 m from an unpaved road, 600 m east of Miralga Creek, 4.5 km southeast of Old Dresser Mineral Camp, Western Australia; North Shaw 1:100,000 map sheet 2755, grid reference 575633.

1441 (ts, TOC, Corg) [J. W. S. field no. 2-6/12/82]. Black chert, thinly interbedded with carbonates, ca. 1 cm thick, from ca. 1000 m stratigraphically above 1440, Apex Basalt; east of creek bed, about 25 m above base of prominent basalt-capped hill and 20 m from unpaved road to Surprise Gold Mine, Western Australia; North Shaw 1:100,000 map sheet 2755, grid reference 572566.

1442 (ts, TOC, Corg) [J. W. S. field no. 3-6/12/82]. Bedded black chert from the same stratigraphic horizon as 1437, Towers Formation; near base of large hill, 9.6 km north of Panorama Ridge and 9.5 km south of North Pole Mining Center, Western Australia; North Shaw 1:100,000 map sheet 2755, grid reference 464553.

1443 (ts, Sfate) [J. W. S. field no. 4-6/12/82]. Bedded chert and barite from the Towers Formation; Old Dresser Mineral Camp, 7 km east of North Pole Mining Center, Western Australia; North Shaw 1:100,000 map sheet 2755, grid reference 537658.

Geologic Samples

Table 14.2. *Continued.*

1444 (ts, TOC, Corg) [J. W. S. field no. 1-6/13/82]. Banded black chert from a 8–10 cm-thick bed, interbedded with quartzitic shale, from near the base of the Cleaverville Formation; 500 m east of Six Mile Creek, 1.1 km south of Strelley Pool, Western Australia; North Shaw 1:100,000 map sheet 2755, grid reference 224638.

1445 (ts, TOC, Corg) [J. W. S. field no. 2-6/13/82]. Dark-gray, finely bedded shale from the middle third of the Cleaverville Formation; 3 to 15 m uphill from Six Mile Creek, 1.8 km south of Strelly Pool, Western Australia; North Shaw 1:100,000 map sheet 2755, grid reference 219622.

1446 (ts, TOC, Corg) [J. W. S. field no. 3-6/13/82]. Black and white banded chert from the uppermost of three chert horizons in the area, Towers Formation; ca. 20 m west of a minor unpaved road, 200 m west of Six Mile Creek, 800 m south of Strelley Pool, Western Australia; North Shaw 1:100,000 map sheet 2755, grid reference 218632.

1447 (ts, μf*, TOC, Corg) [J. W. S. field no. 4-6/13/82]. Microfossiliferous (Fig. 1.5.7), black and white banded chert with cross-bedding in 25–50 cm-thick pods and lenses, from the middle of three chert horizons in the area, Towers Formation; at top of minor ridge ca. 75 m from minor unpaved road, 350 m west of Six Mile Creek, 400 m south of Strelley Pool, locality 1 of Schopf and Packer (1987), Western Australia; North Shaw 1:100,000 map sheet 2755, grid reference 219636.

1448 (ts, TOC, Corg) [J. W. S. field no. 5-6/13/82]. Black and gray, bedded chert from the lowest of three chert horizons in the area, Towers Formation; outcrop forming platform at south end and west bank of Strelley Pool, Western Australia; North Shaw 1:100,000 map sheet 2755, grid reference 220639.

1449 (ts) [J. W. S. field no. 1-6/14/82]. Black chert and shale (silicified tuff?) from the top of the Mount Ada Basalt; at top of a low ridge, 35 m west of an unpaved road, 800 m west of the Talga River, 27 km north of Marble Bar, Western Australia; Coongan 1:100,000 map sheet 2856, grid reference 965797.

1450 (ts) [J. W. S. field no. 2-6/14/82]. Bedded black chert from the lower member ("banded chert member") of the McPhee Formation; at top of a hill southwest of valley of the McPhee Reward Mine area, Western Australia; Marble Bar 1:100,000 map sheet 2855, grid reference 931742.

1451 (ts) [J. W. S. field no. 3-6/14/82]. Black and gray chert from the Towers Formation; at top and flanks of 10 m-high hill at bank of Sandy Creek, Western Australia; Mt. Edgar 1:100,000 map sheet 2955, grid reference 950229.

1452 (ts) [J. W. S. field no. 3-6/14/82]. Silicified pseudofossil cf. "*Newlandia frondosa*" from the Towers Formation; locality cf. 1451.

1453 (ts, st) [J. W. S. field no. 1-6/15/82]. Stromatolitic(?), finely laminated, black chert from the lowest chert horizon at Strelley Pool, Towers Formation; at crest of 65 m high ridge, 13.7 km west-southwest of Strelley Pool, 7.2 km west-southwest of Strelley Gorge, Western Australia; North Shaw 1:100,000 map sheet 2755, grid reference 086613.

1458 (ts, μf*) [J. W. S. field no. 4-6/15/82]. Microfossiliferous (Figs. 1.5.4 through 1.5.6) gray to black, bedded chert unit concordant with volcanics which merge laterally into 20–30 m-thick brecciated chert unit, Apex Basalt; on hill just south of Chinaman Creek, 12 km west of Marble Bar, locality 2 of Schopf and Packer (1987), Western Australia; Marble Bar 1:100,000 map sheet 2855, grid reference 799558.

1459 (ts) [J. W. S. field no. 5-6/15/82]. Well-banded, dark- to light-gray, ferruginous chert, with bands 0.4-3 cm thick, from the Marble Bar chert "member" of the Towers Formation; near Marble Bar, west of Coongan road, Western Australia; Marble Bar 1:100,000 map sheet 2855, grid reference 808558.

1460 (ts) [J. W. S. field no. 6-6/15/82]. Bedded, black and white banded chert from Chinaman Pool Chert "member" of the Towers Formation; west side, Chinaman Pool, west of Marble Bar, Western Australia; Marble Bar 1:100,000 map sheet 2855, grid reference 814555.

1465 (ts) [J. W. S. field no. 1-6/17/82]. Black chert from a 8 m-thick massive chert unit of the Euro Basalt; at side of unpaved road, near Spinaway Creek, 4 km west of Great Northern Highway 95, Western Australia; Nullagine 1:250,000 Geologic sheet no. SF 51-5.

1466 (ts) [J. W. S. field no. 2-6/17/82]. Black chert from the same horizon as 1465, Euro Basalt; at side of unpaved road, Western Australia; Split Rock 1:100,000 map sheet 2854, grid reference 092998.

1467 (ts) [J. W. S. field no. 1-6/18/82]. Black and white banded chert from 10 m-thick chert dike in porphyritic rhyolite, from 10 m below the top of the Wyman Formation; in rolling hills, 23 km southwest of Copper Hills, Western Australia; Marble Bar 1:250,000 Geologic sheet no. SF 50-8.

1633 (ts) 2-6/15/82]. Chert from the Whim Creek Group; William Creek, Western Australia.

1995 (ts) [B. P. field no. 1-6/9/86-1]. Gray siltstone from the Apex Basalt; on hill just south of Chinaman Creek, 12 km west of Marble Bar, near (siltstone) or at (chert) locality 2 of Schopf and Packer (1987), Western Australia; cf. 1458; Marble Bar 1:100,000 map sheet 2855, grid reference 799558.

1996 (ts) [B. P. field no. 1-6/9/86-2]. Gray siltstone from the Apex Basalt; locality cf. 1995.

1997 (ts) [B. P. field no. 2-6/9/86-1]. Gray siltstone from the Apex Basalt; locality cf. 1995.

1998 (ts) [B. P. field no. 2-6/9/86-2]. Gray chert from the Apex Basalt; locality cf. 1995.

1999 (ts) [B. P. field no. 2-6/9/86-3]. Gray chert from the Apex Basalt; locality cf. 1995.

2000 (ts) [B. P. field no. 2-6/9/86-4]. Gray chert from the Apex Basalt; locality cf. 1995.

2001 (ts, μf*) [B. P. field no. 2-6/9/86-5]. Microfossiliferous (Figs. 1.5.4 through 1.5.6) gray chert from the Apex Basalt; locality cf. 1995.

2002 (ts) [B. P. field no. 2-6/9/86-6]. Gray chert from the Apex Basalt; locality cf. 1995.

2003 (ts, Corg) [B. P. field no. 3-6/9/86-1]. Gray chert from the Apex Basalt; locality cf. 1995.

Table 14.2. *Continued.*

2004 (ts) [B. P. field no. 3-6/9/86-2]. Gray chert from the Apex Basalt; locality cf. 1995.

2005 (ts) [B. P. field no. 3-6/9/86-3]. Gray chert from the Apex Basalt; locality cf. 1995.

2006 (ts, μf^*) [B. P. field no. 3-6/9/86-4]. Microfossiliferous (Figs. 1.5.4 through 1.5.6) gray chert from the Apex Basalt; locality cf. 1995.

2007 (ts, μf^*, Corg) [B. P. field no. 1-7/14/86-1]. Microfossiliferous (Fig. 1.5.7) gray chert from the Towers Formation; at top of minor ridge ca. 75 m from minor unpaved road, 350 m west of Six Mile Creek, 400 m south of Strelley Pool, locality 1 of Schopf and Packer (1987), Western Australia; cf. 1447; North Shaw 1:100,000 map sheet 2755, grid reference 219636.

2008 (ts) [B. P. field no. 1-7/14/86-2]. Gray chert from the Towers Formation; locality cf. 2007.

2009 (ts) [B. P. field no. 1-7/14/86-3]. Gray chert from the Towers Formation; locality cf. 2007.

2010 (ts) [B. P. field no. 1-7/14/86-4]. Gray chert from the Towers Formation; locality cf. 2007.

2011 (ts) [B. P. field no. 1-7/14/86-5]. Conglomerate with 1 cm-diameter chert clasts from the Towers Formation; locality cf. 2007.

2012 (ts) [B. P. field no. 1-7/14/86-6]. Carbonate from the Towers Formation; locality cf. 2007.

2013 (ts) [B. P. field no. 1-7/14/86-7]. Bedded chert from the Towers Formation; locality cf. 2007.

2014 (ts, TOC, Corg) [B. P. field no. 1-7/14/86-8]. Chert from the Towers Formation; locality cf. 2007.

2015 [B. P. field no. 1-7/14/86-9]. Chert from the Towers Formation; locality cf. 2007.

2589 [J. M. H. field no. 820611-1a]. Chert from the top of the Gorge Creek Group; from a small creek, 800 km east of unpaved road, 5.5 km east of Shaw River, 9 km north of North Pole Mining Center, ca. 100 km southeast of Port Hedland, Western Australia; North Shaw 1:100,000 map sheet 2755, grid reference 482738.

2590 [J. M. H. field no. 820611-1b]. Laminated chert (silicified shale?) from the Gorge Creek Group; near 2589, locality cf. 2589.

2591 [J. M. H. field no. 820611-1c]. Black chert from the Gorge Creek Group; 20 m from 2590, locality cf. 2589.

2592 [J. M. H. field no. 820611-2a]. Carbonate from the Gorge Creek Group; further along track to North Pole from "Lalla Rookh" site, Western Australia; North Shaw 1:100,000 map sheet 2755, grid reference 478726.

2593 [J. M. H. field no. 820611-2b]. Carbonate cemented siliciclastic with carbonate veins from the Gorge Creek Group; locality cf. 2592.

2594 [J. M. H. field no. 820611-2c]. Bedded carbonate from the Gorge Creek Group; locality cf. 2592.

2595 [J. M. H. field no. 820611-2d]. Carbonate vein from the Gorge Creek Group; right next to 2594, locality cf. 2592.

2596 [J. M. H. field no. 820611-2e]. Well-bedded carbonate from the lower part of the Gorge Creek Group; locality cf. 2592.

2597 [J. M. H. field no. 820611-3a]. Laminated, fine-grained chert from the Towers Formation; outcrops at tops and flanks of low hills, 6 km south-southeast of 1436, ca. 100 km southeast of Port Hedland, locality A of Awramik et al. (1983), Western Australia; cf. 1437; North Shaw 1:100,000 map sheet 2755, grid reference 504638.5.

2598 [J. M. H. field no. 820611-3b]. Coarse-grained chert from the Towers Formation; locality cf. 2597.

2599 [J. M. H. field no. 820611-3c]. Well-banded, dark chert from the Towers Formation; locality cf. 2597.

2600 [J. M. H. field no. 820611-3d]. Gray chert from the Towers Formation; locality cf. 2597.

2601 [J. M. H. field no. 820611-4b]. Bedded chert from the top of the sequence of the Towers Formation; south side of the gorge adjacent to unpaved road, 4.7 km southeast of 1437 and 1 km south of Old Dresser Mineral Camp; locality B of Awramik et al. (1983), Western Australia; cf. 1438; North Shaw 1:100,000 map sheet 2755, grid reference 536648.

2602 [J. M. H. field no. 820612-1a]. Green-gray chert with a clastic texture, 1–3 mm grains, some with concentric laminae, possibly a silicified lapilli tuff, from the base of the Apex Basalt; Old Dresser Mineral Camp, 7 km east of North Pole Mining Center, Western Australia; North Shaw 1:100,000 map sheet 2755, grid reference 537658.

2603 [J. M. H. field no. 820612-1b]. Black chert dike from the Apex Basalt; locality cf. 2602.

2604 [J. M. H. field no. 820612-2a]. Chert from ca. 500 m above the base of the Apex Basalt; locality cf. 2602.

2605 [J. M. H. field no. 820612-3a]. Bedded rippled sandstone and carbonate from ca. 500 m above 2604, Apex Basalt; 50 m from the Surprise Gold Mine, locality cf. 2602.

2602 [J. M. H. field no. 820612-3b]. Black carbonate with white quartz from the Apex Basalt; from top of cliff, locality cf. 2602.

2607 [J. M. H. field no. 820612-3d]. Banded chert from the Apex Basalt; 6 m below 2720, locality cf. 2602.

2608 [J. M. H. field no. 820612-3e]. Green chert, unbedded but stratiform from 0.5 m below 2607, Apex Basalt; locality cf. 2602.

2618 [J. M. H. field no. 820613-1f]. Coarse chert (silicified siltstone?) from the top chert of the Towers Formation; Fortescue inlier, near North Pole, Western Australia; North Shaw 1:100,000 map sheet 2755, grid reference 218632.

2619 [J. M. H. field no. 820613-1g]. Light-gray chert from the middle chert of the Towers Formation; 1.1 km south of Strelley Pool, Western Australia; North Shaw 1:100,000 map sheet 2755, grid reference 224638.

2620 [J. M. H. field no. 820613-1h]. Black chert from the Towers Formation; locality cf. 2619.

Table 14.2. Continued.

2621 [J. M. H. field no. 820613-1i]. Black chert with desiccation cracks from the Towers Formation; locality cf. 2619.

2622 [J. M. H. field no. 820613-1j]. Chert from the lowest chert horizon of the Towers Formation; locality cf. 2619.

2623 [J. M. H. field no. 820614-1a]. Shale from the Mount Ada Basalt; Shark Gully, Western Australia; Marble Bar 1:100,000 map sheet 2588, grid reference 709318.

2624 [J. M. H. field no. 820614-2a]. Shale interbedded with chert and volcanics from ca. 1000 m higher than the "Shark Gully locality" (2623) of the Mount Ada Basalt; at top of a low ridge, 35 m west of an unpaved road, 800 m west of the Talga River, 27 km north of Marble Bar, Western Australia; Coongan 1:100,000 map sheet 2856, grid reference 965797.

2625 [J. M. H. field no. 820614-2b]. Chert (may be a silicified tuff) from the Mount Ada Basalt; locality cf. 2624.

2626 [J. M. H. field no. 820614-2c]. Black chert from 10 m higher than 2625, Mount Ada Basalt; locality cf. 2624.

2627 [J. M. H. field no. 820614-1d]. Gray shale from the Mount Ada Basalt; locality cf. 2624.

2628 [J. M. H. field no. 820614-2e]. Chert from the Mount Ada Basalt; locality cf. 2624.

2629 [J. M. H. field no. 820614-3b]. Black chert from 5 m above 2721, McPhee Formation; at top of a hill southwest of valley of the McPhee Reward Mine area, Western Australia; Marble Bar 1:100,000 map sheet 2855, grid reference 931742.

2630 [J. M. H. field no. 820614-3c]. Carbonatic basalt from 10 m above 2629, McPhee Formation; locality cf. 2629.

2631 [J. M. H. field no. 820614-3d]. Carbonatic pillow basalt from the McPhee Formation; locality cf. 2629.

2632 [J. M. H. field no. 820614-3e]. Secondary carbonate from between pillows in pillow basalt from the McPhee Formation; locality cf. 2629.

2633 [J. M. H. field no. 820614-4a]. Gray chert pseudostromatolites from the Towers Formation; at top and flanks of 10 m high hill at bank of Sandy Creek, Western Australia; cf. 1452; Mt. Edgar 1:100,000 map sheet 2955, grid reference 950229.

2634 [J. M. H. field no. 820614-4b]. Secondary carbonate from the Towers Formation; locality cf. 2633.

2635 [J. M. H. field no. 820614-4c]. Black chert from the Towers Formation; locality cf. 2633.

2636 [J. M. H. field no. 820614-4d]. Rhythmically banded chert-carbonate from the Towers Formation; locality cf. 2633.

2637 (ts) [J. M. H. field no. 820614-4e]. Black chert from 20 m above 2635, Towers Formation; locality cf. 2633.

2638 (ts, st) [J. M. H. field no. 820615-1a]. Carbonaceous chert from a stratiform stromatolite from the lowest chert, 10 m above conical, columnar stromatolites, Towers Formation; at crest of 65 m-high ridge, 13.7 km west-southwest of Strelley Pool, 7.2 km west-southwest of Strelley Gorge, Western Australia; North Shaw 1:100,000 map sheet 2755, grid reference 086613.

2639 [J. M. H. field no. 820615-2a]. Shale from the top of the Rushall Slate; Mons Cupri Mine, Western Australia.

2640 [J. M. H. field no. 820615-2c]. Chert from 2 m above 2717, Whim Creek Group; locality cf. 2639.

2641 [J. M. H. field no. 820615-2d]. Shale from a core sample, DH CRD3, depths of 98.5–101.2/93.7–98.5 m, lower portion of the Whim Creek Group; locality cf. 2639.

2642 [J. M. H. field no. 820615-2e]. Black banded argillite from core sample DH NUD21, depth of 127 m, Louden Volcanics; locality cf. 2639.

2644 (ts) [J. M. H. field no. 820615-4a]. Black chert from the Apex Basalt; on hill just south of Chinaman Creek, 12 km west of Marble Bar, locality 2 of Schopf and Packer (1987), Western Australia; cf. 1458 and 1995; Marble Bar 1:100,000 map sheet 2855, grid reference 799558.

2645 (ts) [J. M. H. field no. 820615-4b]. Ferruginous, black chert from the Towers Formation; locality cf. 2644.

2646 (ts) [J. M. H. field no. 820615-4c]. Whitish-gray banded chert from the Towers Formation; Chinaman Pool, west of Marble Bar, Western Australia; Marble Bar 1:100,000 map sheet 2855, grid reference 814555.

2682 [J. M. H. field no. 820617-1a]. Massive, waxy, black and gray chert from a 8 m-thick bed of the Euro Basalt; at side of unpaved road, Western Australia; Split Rock 1:100,000 map sheet 2854, grid reference 092998.

2685 [J. M. H. field no. 820618-1a]. Dark brown-gray thin-bedded, fine-grained, feldspathic sandstone from the Lalla Rookh Sandstone; Budjan Creek, Western Australia; map sheet 2854, grid reference 962809.

2686 [J. M. H. field no. 820618-1b]. Dark gray, laminated, thin-bedded siltstone from the Lalla Rookh Sandstone; locality cf. 2685.

2687 (ts) [J. M. H. field no. 820618-1c]. Medium-gray, medium-grained, silty sandstone from the Lalla Rookh Sandstone; near Budjan Creek, Western Australia; map sheet 2854, grid reference 967807.

2688 [J. M. H. field no. 820618-1d]. Shale from the Lalla Rookh Sandstone; locality cf. 2687.

2689 [J. M. H. field no. 820618-1e]. Dark gray, laminated siltstone from the Lalla Rookh Sandstone; near Budjan Creek, Western Australia; map sheet 2854, grid reference 970812.

2690 (ts) [J. M. H. field no. 820618-1f]. Chert from a dike in felsic volcanics of the Wyman Formation; locality cf. 2689.

2691 [J. M. H. field no. 820618-1g]. Dark gray, thin-bedded, ripple-laminated siltstone from the basal part of the Lalla Rookh Sandstone; near Budjan Creek, Western Australia; map sheet 2854, grid reference 964813.

2692 [J. M. H. field no. 820618-2a]. Shale from the Mosquito Creek Formation; 0.5 km north, along road past Nullagine, Western Australia.

Table 14.2. *Continued.*

2693 [J. M. H. field no. 820618-3a]. Shale from the Hardy Sandstone; roadside 7.6 km north, locality cf. 2692.

2694 [J. M. H. field no. 820618-4a]. Shale from the Hardy Sandstone; north 2693, locality cf. 2692.

2695 [J. M. H. field no. 820618-4b]. Dark shale from the Hardy Sandstone; see 2693, locality cf. 2692.

2717 [J. M. H. field no. 820615-2b]. Chert from the Whim Creek Group; Mons Cupri Mine, Western Australia.

2718 (ts) [J. M. H. field no. 820617-3a]. Chert from the Euro Basalt; at side of unpaved road, Western Australia; Split Rock 1:100,000 map sheet 2854, grid reference 092998.

2719 [J. M. H. field no. 820612-4c]. Bedded black chert from the Apex Basalt; near base of large hill, 9.6 km north of Panorama Ridge and 9.5 km south of North Pole Mining Center, Western Australia; North Shaw 1:100,000 map sheet 2755, grid reference 464553.

2720 [J. M. H. field no. 820612-3c]. Fine-grained carbonate from an intermediate level in section, Apex Basalt; Old Dresser Mineral Camp, 7 km east of North Pole Mining Center, Western Australia; North Shaw 1:100,000 map sheet 2755, grid reference 537658.

2721 [J. M. H. field no. 820614-3a]. Black chert from the McPhee Formation; at top of a hill southwest of valley of the McPhee Reward Mine area, Western Australia; Marble Bar 1:100,000 map sheet 2855, grid reference 931742.

14.2.4 EUROPE

KHANTAY-RYBNINSK UPLIFT

Tectonic unit/location. Khantay-Rybninsk Uplift.

Stratigraphy.

    KIMAJ "COMPLEX"
        Turukhansk Formation
        Miroedikha Formation
        Shorikha Formation
    YAKUT "COMPLEX"
        Burovaya Formation
        Derevnya Formation
        Sukhaya Tunguska Formation
        Linok Formation
        Bezymyamyj Formation

PPRG samples/localities.

2725 (ts, st, μf^*, TOC, Corg) [J. W. S. field no. GIN 4324/98]. Microfossiliferous (Chapter 24), stromatolitic black chert from the upper half of the Sukhaya Tunguska Formation; locality B of Mendelson and Schopf (1982, p. 44, Fig. 1B), ca. 25 km northeast of Turukhansk, on the banks of the Nizhnyaya Tunguska River, Siberia, USSR; 65.83°N, 88.50°E.

2726 (ts, st, μf^*, TOC, Corg) [J. W. S. field no. GIN 4324/101]. Microfossiliferous (Chapter 24), stromatolitic black chert from the lower half of the Shorikha Formation; locality C of Mendelson and Schopf (1982), ca. 10 km east of Turukhansk, on the banks of the Nizhnyaya Tunguska River, Siberia, USSR; 65.75°N, 88.25°E.

LAKE VÄTTERN GRABEN

Tectonic unit/location. Lake Vättern Graben, southern Sweden.

Stratigraphy.

    Visingsö Group

PPRG samples/localities.

2728 (TOC, Corg) [C. V. M. field no. SV-8]. Medium green-gray shale containing *Chuaria* from the upper unit of the Visingsö Group; Omberg area between Alvarums udde and Mickelstorp near Lake Vättern (Vidal 1976, Fig. 1.4), southern Sweden; 58.50°N, 14.75°E.

2729 [C. V. M. field no. SV-9]. Cryptalgal(?) laminated dolostone containing vase-shaped microfossils from the ? upper unit of the Visingsö Group; Omberg area, along Storpissan (a stream), just north of Mickelstorp, near Lake Vättern (Vidal, 1976), southern Sweden; 58.50°N, 14.75°E.

2730 (ts) [C. V. M. field no. SV-12]. Phosphatized pebbles containing vase-shaped microfossils, from the upper unit of the Visingsö Group; beach of Lake Vättern, near mouth of Girabacken (Vidal 1976, Fig. 1.3), southern Sweden; 58.50°N, 14.75°E.

2731 [C. V. M. field no. SV-13]. Medium-gray, micaceous shale, containing acritarchs (including *Chuaria*), from the middle unit of the Visingsö Group Nas slott, Vidal's P2 locality (Vidal 1976, Fig. 1.1), southern Visingsö Island in Lake Vättern, southern Sweden; 58.50°N, 14.75°E.

2732 (TOC, Corg) [C. V. M. field no. SV-14]. Medium-gray, micaceous shale containing acritarchs, from the middle unit, Vidal's V70-182 (Fig. 3) of the Visingsö Group; Vidal's P3 locality (Vidal 1976, Fig. 1.1), locality cf. 2731.

2733 [C. V. M. field no. SV-24]. Medium-gray, micaceous shale from the middle unit, Vidal's V70-191 (Fig. 3) of the Visingsö Group; between Vidal's P3 and P4 localities (Vidal 1976, Fig. 1.1), locality cf. 2731.

Table 14.2. *Continued.*

2734 (TOC, Corg) [C. V. M. field no. SV-15]. Medium-gray, micaceous shale containing abundant *Chuaria*, from the middle unit, Vidal's V70-306/308 (Fig. 3) of the Visingsö Group; between Vidal's P3 and P4 localities (closer to P4), locality cf. 2731.

2735 [C. V. M. field no. SV-16]. Greenish-gray, micaceous, shaley sandstone containing acritarchs from the middle or ? upper unit, Vidal's V70-309/310 of the Visingsö Group; between Vidal's P3 and P4 localities, very close to P4 (Vidal 1976), locality cf. 2731.

2736 [C. V. M. field no. SV-17]. Dark-gray shale from the upper or ? middle unit, Vidal's V70-308 of the Visingsö Group; Vidal's P4 locality (Vidal 1976, Fig. 1.1), locality cf. 2731.

2737 [C. V. M. field no. SV-18]. Grayish-green, dolomitic shaley sandstone from the upper unit, Vidal's V70-322 (Fig. 3) of the Visingsö Group; Vidal's P6 locality (Vidal 1976, Fig. 1.1), locality cf. 2731.

2738 [C. V. M. field no. SV-22]. Medium gray, intraclastic dolostone, with an oxidized rind, from the upper unit of the Visingsö Group; Vidal's P7 locality, northern Visingsö Island in Lake Vättern, southern Sweden.

OSEN-RØA NAPPE COMPLEX

Tectonic unit/location. Osen-Røa Nappe Complex, Hedmark, Norway.

Stratigraphy. Stratigraphic section, Lillehammer-Lake Mjøsa area, southwestern Hedmark (Nystuen and Siedlecka 1988):

HEDMARK GROUP
    Vangsås Formation
    Ekre Formation
    Moelv Formation
    Ring Formation
    Biri Formation
    Biskopåsen Formation/Imsdalen Formation
    Brøttum Formation

PPRG samples/localities.

2412 (TOC, Corg, ker) [J. M. H. field no. V1/B589]. Organic-rich, laminated, limestone from the upper part of the Biri Formation; northern end of Lake Mjøsa, near the towns of Biri and Moelv, southern Norway.

2413 (TOC, Corg, ker) [J. M. H. field no. B2.1/B590]. Laminated, organic-rich limestone from the upper part of the Biri Formation; locality cf. 2412.

2414 (TOC, Corg, ker) [J. M. H. field no. B6.5/B591]. Laminated, organic-rich limestone from the upper part of the Biri Formation; locality cf. 2412.

2415 (TOC, Corg, ker) [J. M. H. field no. B20/B592]. Laminated, organic-rich limestone from the upper part of the Biri Formation; locality cf. 2412.

SIERRA DE CORDOBA

Tectonic unit/location.

Stratigraphy.

    [UNNAMED CAMBRIAN GROUP]
        Los Villares Formation
        Santo Domingo Formation
        Pedroche Formation
        Torrearboles Formation
    SOTILLO GROUP
    ALMODOVAR DEL RIO GROUP

PPRG samples/localities.

1503 (ts) [J. W. S. field no. 1-5/29/86]. Black-gray chert, laminated, deep-water facies of the Almodovar del Rio Group; west side of Almodovar del Rio-Casas de Villabillos Road, 2 km north of Castle at Almodovar Del Rio; 23 km southwest of center of Cordoba, southern Spain.

1504 (ts, Sfide) [J. W. S. field no. 1-5/30/86]. Pyritic black chert, non-stromatolitic, deep-water facies of the Almodovar del Rio Group; outcrop on west side of road, ca. 40 m west of Embalse de la Brena Reservoir, ca. 26 km southwest of center of Cordoba, southern Spain.

1505 (ts) [J. W. S. field no. 2-5/30/86]. Laminated black chert (silicified mudstone?), non-stromatolitic, deep-water facies of the Almodovar del Rio Group; outcrop on hilltop, ca. 25 m east of road, west of locality cf. 1504.

1506 (ts) [J. W. S. field no. 3-5/30/86]. Pyroclastic unit of the Sotillo Group; small quarry ca. 30 m west of unpaved road and 1 km west of farmhouse at Casas de Villalobillos, ca. 19 km west of the center of Cordoba, southern Spain.

1507 (ts) [J. W. S. field no. 4-5/30/86]. Greenish-gray shale from the basal portion of the Pedroche Formation; north side of Las Ninas-El Rosal Road near Cantera las Ermitas, ca. 6 km northwest of center of Cordoba, southern Spain.

1508 (ts) [J. W. S. field no. 1-5/31/86]. Siliceous carbonate from the lowermost horizon of the Pedroche Formation; west side of Arroyo del Pedroche, 30 m south of Highway no. 1, ca. 2 km north of center of Cordoba, southern Spain.

Table 14.2. *Continued.*

SIERRA DE GUADALUPE

Tectonic unit/location. Guadalupe area, Spain

Stratigraphy.

  [UNNAMED LOWER CAMBRIAN GROUP]
    Azorejo Sandstone Formation
    Pusa Formation
  [UNNAMED VENDIAN GROUP]
    Fuentes Formation
    Cijara Formation
    Estenilla Formation
    Monterrubio Formation

PPRG samples/localities.

1494 (ts, TOC, Corg) [J. W. S. field no. 1-5/27/86]. Finely laminated dolomitic mudstone from units 3, 4, 5 (undifferentiated) of the Estenilla Formation; ca. 3 km east of Ali on Guadalupe-Alia-Puerto de San Vicente Road, on a hill on the right side of road, at marker "56c" (painted white and on the outcrop), 13 km east of Guadalupe, south-central Spain.

1495 (ts, TOC, Corg) [J. W. S. field no. 2-5/27/86]. Mudstone matrix of olistostrome from 5 m above the top of El Membrillar Pond, from the Fuentes Formation; south side of El Membrillar Pond on Rio Estena, ca. 150 m west of Cijara-Gamonero Road, 4 km southeast of Gamonero, and 42 km east-southeast of Guadalupe, south-central Spain.

1496 (ts, TOC, Corg) [J. W. S. field no. 3-5/27/86]. Mudstone in turbidite sequence from the Pusa Formation; outcrop in roadcut on Cijara-Gamonero Road, ca. 7.5 km east-northeast of Cijara and ca. 5 km north-northwest of Gamonero, at Arroya de Valfernado, south-central Spain.

1497 (ts) [J. W. S. field no. 1A-5/28/86]. Pink, laminated (cryptalgal), recrystallized dolomite from 250 m northeast and 50 m below 1498, Estenilla Formation; outcrop from a steep carbonate ridge on northeast limb of anticline on the north side of Rio Tajo, Navalmoral de la Matu-Valdecanas de Tajo Road, 15 km southwest of Navalmoral de la Manta, south-central Spain.

1498 (ts) [J. W. S. field no. 1B-5/28/86]. Sandy dolomite from 250 m southwest and 50 m above 1497, Estenilla Formation; locality cf. 1497.

1499 (ts, TOC, Corg, Sfide) [J. W. S. field no. 2A-5/28/86]. Pyritic mudstone from the distal portion of a turbidite sequence, 10 m stratigraphically below 1500, Cijara Formation; outcrop on north side of Castanar de Ibor-Robledollano Road, ca. 4 km west of Castanar de Ibor, south-central Spain.

1500 (ts) [J. W. S. field no. 2B-5/28/86]. Organic-rich laminated mudstone, distal portion of turbidite sequence, 10 m stratigraphically above 1499, Cijara Formation; locality cf. 1499.

1501 (ts, Sfide) [J. W. S. field no. 3-5/28/86]. Pyritic, carbonaceous mudstone with dolomitic stringers, from distal portion of turbidite sequence, Cijara Formation; ca. 0.5 km east of locality 1499.

1502 (ts) [J. W. S. field no. 4-5/28/86]. Carbonaceous mudstone, shelf sediment with dolomitic and sand stringers from the Los Parrales Shale; ca. 0.5 km east of locality 1499.

TORRIDONIAN SUCCESSION

Tectonic unit/location: Torridonian Succession, Scotland, Great Britain.

Stratigraphy: Rayner 1981 (Fig. 10); Peat 1984; Williams 1968.

  TORRIDON GROUP
    Cailleach Head Formation
    Autbea Formation
    Applecross Formation
    Diabaig Formation
  STOER GROUP
  LEWISIAN BASEMENT

PPRG samples/localities.

1874 (ts, TOC, Corg) [G. J. R. field no. R317]. Conglomerate from 5 cm above paleosol on Lewisian biotite gneiss, basal unit of the Torridon Group; sea cliff, 2 km west-northwest of Sheigra, Scotland, Great Britain.

1875 (ts) [G. J. R. field no. R318]. Weathered biotite gneiss from 3 cm below top of profile of the Torridon Group; locality cf. 1874.

1876 (ts, TOC, Corg) [G. J. R. field no. R320]. Conglomerate from 20 cm below top of profile of the Torridon Group; locality cf. 1874.

1877 (ts, TOC, Corg) [G. J. R. field no. R321]. Conglomerate from 50 cm below top of profile of the Torridon Group; locality cf. 1874.

1878 (ts) [G. J. R. field no. R323]. Conglomerate from 100 cm below top of profile of the Torridon Group; locality cf. 1874.

1879 (ts) [G. J. R. field no. R324]. Conglomerate from 130 cm below top of profile of the Torridon Group; locality cf. 1874.

1880 (ts) [G. J. R. field no. R327]. Conglomerate from 190 cm below top of profile (within corestone) of the Torridon Group; locality cf. 1874.

Table 14.2. *Continued.*

1881 (ts) [G. J. R. field no. R328]. Conglomerate from 260 cm below top of profile of the Torridon Group; locality cf. 1874.

1882 (ts) [G. J. R. field no. R329]. Conglomerate from 420 cm below top of profile of the Torridon Group; locality cf. 1874.

1883 (ts) [G. J. R. field no. R330]. Conglomerate from 630 cm below top of profile of the Torridon Group; locality cf. 1874.

1884 (ts) [G. J. R. field no. R331]. Conglomerate from 15 cm above paleosol on Lewisian amphibolite, basal unit of the Torridon Group; locality cf. 1874.

1885 (ts) [G. J. R. field no. R332]. Weathered amphibolite from 10 cm below top of profile of the Torridon Group; locality cf. 1874.

1886 (ts) [G. J. R. field no. R334]. Conglomerate from 35 cm below top of profile of the Torridon Group; locality cf. 1874.

1887 (ts) [G. J. R. field no. R335]. Conglomerate from 100 cm below top of profile of the Torridon Group; locality cf. 1874.

1888 (ts) [G. J. R. field no. R337]. Conglomerate from 260 cm below top of profile of the Torridon Group; locality cf. 1874.

1889 (ts) [G. J. R. field no. R338]. Conglomerate from 320 cm below top of profile of the Torridon Group; locality cf. 1874.

UCHUR-MAYA BLOCK

Tectonic unit/location. Uchur-Maya Block, eastern Siberia, USSR (Mendelson and Schopf 1982)

Stratigraphy.

> Peszrotsvet Formation
> Yudoma Formation
> Kandyk Formation
> Lakhanda Formation
> Tsipanda Formation
> Malga Formation
> Maya Formation
> Uchur Formation
> [ARCHEAN BASEMENT]

PPRG samples/localities.

2727 (ts, μf^*, TOC, Corg) [J. W. S. field no. GIN B171]. Microfossiliferous (Chapter 24) black chert from the upper half of the Yudoma Formation; locality D of Mendelson and Schopf (1982, Fig. 1C), ca. 90 km east-northeast of Ust-Yudoma, on the banks of the Yudoma River, Yakutia, USSR; 59.38°N, 136.7°E.

14.2.5 NORTH AMERICA

ATHAPUSCOW AULACOGEN (EAST ARM FOLD BELT)

Tectonic unit/location. East Arm Thrust Belt (Athapuscow Aulacogen), Northern Territories (District of Mackenzie), Canada.

Stratigraphy. Stratigraphic section, Great Slave Lake area, south-central District of Mackenzie (Hoffman 1981):

GREAT SLAVE SUPERGROUP
    CHRISTIE BAY GROUP
    PETHEI GROUP
    KAHOCHELLA GROUP
    SOSAN GROUP
    UNION ISLAND GROUP

PPRG samples/localities.

1382 (ts, TOC, Corg, ker) [J. W. S. field no. JWS-3/77]. Black pyritic shale from the Union Island Group; Union Island, Northwest Territories, Canada.

AVALON ZONE

Tectonic unit/location. Avalon Zone, southeastern Newfoundland, Canada.

Stratigraphy.

> Brigus Formation
> Bonavista Formation
> Random Formation
> Chapel Island Formation

PPRG samples/localities.

1164 (TOC, Corg) [P. M. field no. FD 9.0]. Greenish-gray shale from the Chapel Island Formation; "Fortune Dump section," just west of "Fortune Head," southwest of the town of Fortune, southeastern Newfoundland, Canada.

1165 (ts, TOC, Corg) [P. M. field no. FD 17.8]. Greenish-gray shale from the Chapel Island Formation; locality cf. 1164.

Table 14.2. *Continued.*

1166 (ts, TOC) [P. M. field no. FD 17.6]. Greenish-gray shale from the Chapel Island Formation; locality cf. 1164.

1167 (ts, TOC, Corg) [P. M. field no. FD 21.5]. Greenish-gray shale from the Chapel Island Formation; locality cf. 1164.

1168 (ts, TOC, Corg) [P. M. field no. FD 26.8]. Greenish-gray shale from the Chapel Island Formation; locality cf. 1164.

1169 (ts, TOC, Corg) [P. M. field no. FD 27.3]. Greenish-gray shale from the Chapel Island Formation; locality cf. 1164.

1170 (ts, TOC, Corg) [P. M. field no. FD 67.7]. Greenish-gray shale from the Chapel Island Formation; locality cf. 1164.

1171 (ts, TOC, Corg) [P. M. field no. FD 89.5]. Greenish-gray shale from the Chapel Island Formation; locality cf. 1164.

1172 (ts, TOC, Corg) [P. M. field no. FD 171.4]. Greenish-gray shale from the Chapel Island Formation; locality cf. 1164.

1173 (ts, TOC, Corg) [P. M. field no. FD 190.7]. Greenish-gray shale from the Chapel Island Formation; locality cf. 1164.

1174 (TOC, Corg) [P. M. field no. FD 217.8]. Greenish-gray shale from the Chapel Island Formation; locality cf. 1164.

1175 (ts, TOC, Corg) [P. M. field no. FD 238.3]. Greenish-gray shale from the Chapel Island Formation; locality cf. 1164.

1176 (TOC, Corg) [P. M. field no. FD 259.8]. Greenish-gray shale from the Chapel Island Formation; locality cf. 1164.

1177 (ts, TOC, Corg) [P. M. field no. FD 280.15]. Greenish-gray shale from the Chapel Island Formation; locality cf. 1164.

1178 (ts, TOC, Corg) [P. M. field no. FD 299.5]. Greenish-gray shale from the Chapel Island Formation; locality cf. 1164.

1179 (ts, TOC, Corg) [P. M. field no. FD 318.85]. Greenish-gray shale from the Chapel Island Formation; locality cf. 1164.

1180 (ts, TOC, Corg) [P. M. field no. FD 339.85]. Greenish-gray shale from the Chapel Island Formation; locality cf. 1164.

1181 (ts, TOC, Corg) [P. M. field no. FD 357.9]. Greenish-gray shale from the Chapel Island Formation; locality cf. 1164.

1182 (ts, TOC, Corg) [P. M. field no. FD 377.3]. Greenish-gray shale from the Chapel Island Formation; locality cf. 1164.

1183 (ts, TOC) [P. M. field no. FD 398.6]. Greenish-gray shale from the Chapel Island Formation; locality cf. 1164.

1184 (ts, TOC) [P. M. field no. FD 417.9]. Greenish-gray shale from the Chapel Island Formation; locality cf. 1164.

1185 (ts, TOC, Corg) [P. M. field no. FD 438.3]. Greenish-gray shale from the Chapel Island Formation; locality cf. 1164.

1186 (ts, TOC, Corg, Sfide) [P. M. field no. DC 20]. Greenish-gray shale from the Chapel Island Formation; "Little Dantzic Cove section," at Pieduck Point, 18 km south of the city of Fortune via Route 220, southeastern Newfoundland, Canada; 46.96°N, 56.00°W.

1187 (ts, TOC, Corg) [P. M. field no. DC 30]. Greenish-gray shale from the Chapel Island Formation; locality cf. 1186.

1188 (ts, TOC, Corg) [P. M. field no. DC 40]. Greenish-gray shale from the Chapel Island Formation; locality cf. 1186.

1189 (ts, TOC, Corg) [P. M. field no. DC 50]. Greenish-gray shale from the Chapel Island Formation; locality cf. 1186.

1190 (ts, TOC, Corg) [P. M. field no. DC 60]. Greenish-gray shale from the Chapel Island Formation; locality cf. 1186.

1191 (ts, TOC) [P. M. field no. DC 69.9]. Greenish-gray shale from the Chapel Island Formation; locality cf. 1186.

1192 (ts, TOC, Corg) [P. M. field no. DC 80]. Greenish-gray shale from the Chapel Island Formation; locality cf. 1186.

1193 (ts, TOC) [P. M. field no. DC 89.9]. Greenish-gray shale from the Chapel Island Formation; locality cf. 1186.

1194 (ts, TOC, Corg) [P. M. field no. DC 100]. Greenish-gray shale from the Chapel Island Formation; locality cf. 1186.

1195 (ts, TOC, Corg) [P. M. field no. DC 110.2]. Greenish-gray shale from the Chapel Island Formation; locality cf. 1186.

1196 (ts, TOC, Corg) [P. M. field no. DC 120]. Greenish-gray shale from the Chapel Island Formation; locality cf. 1186.

1197 (ts, TOC) [P. M. field no. DC 129.8]. Greenish-gray shale from the Chapel Island Formation; locality cf. 1186.

1198 (ts, TOC, Corg) [P. M. field no. DC 139.8]. Greenish-gray shale from the Chapel Island Formation; locality cf. 1186.

1199 (ts, TOC, Corg) [P. M. field no. DC 150]. Greenish-gray shale from the Chapel Island Formation; locality cf. 1186.

1200 (ts, TOC, Corg) [P. M. field no. DC 160]. Greenish-gray shale from the Chapel Island Formation; locality cf. 1186.

1201 (ts, TOC, Corg) [P. M. field no. DC 170]. Greenish-gray shale from the Chapel Island Formation; locality cf. 1186.

1202 (ts, TOC, Corg) [P. M. field no. DC 180]. Greenish-gray shale from the Chapel Island Formation; locality cf. 1186.

1203 (ts, TOC, Corg) [P. M. field no. DC 229.7]. Greenish-gray shale from the Chapel Island Formation; locality cf. 1186.

1204 (ts, TOC, Corg) [P. M. field no. DC 241.0]. Greenish-gray shale from the Chapel Island Formation; locality cf. 1186.

1205 (ts, TOC, Corg) [P. M. field no. DC 250]. Greenish-gray shale from the Chapel Island Formation; locality cf. 1186.

1206 (ts, TOC, Corg) [P. M. field no. DC 260]. Greenish-gray shale from the Chapel Island Formation; locality cf. 1186.

1207 (ts, TOC, Corg) [P. M. field no. DC 270]. Greenish-gray shale from the Chapel Island Formation; locality cf. 1186.

1208 (ts, TOC) [P. M. field no. DC 280]. Greenish-gray shale from the Chapel Island Formation; locality cf. 1186.

1209 (ts, TOC, Corg) [P. M. field no. DC 289.7]. Greenish-gray shale from the Chapel Island Formation; locality cf. 1186.

1210 (ts, TOC, Corg) [P. M. field no. DC 299.6]. Greenish-gray shale from the Chapel Island Formation; locality cf. 1186.

Table 14.2. *Continued.*

2740 [S. B. field no. CAN 87-1-SB]. Conglomerate from the basal conglomerate of the Rencontre Formation; 200 m east of the parking spot, Grand Bank Head, 2 km northwest of Grand Bank, southeastern Newfoundland, Canada; 47.07°N, 55.81°W.

2742 [S. B. field no. CAN 87-2-SB]. Gray mudstone from the uppermost gray bed, 2 m below the top of the Rencontre Formation; west of parking spot, across baymouth bar, locality cf. 2740.

2743 [S. B. field no. CAN 87-3-SB]. Gray siltstone from 0.5 m below GBA-20 of the Chapel Island Formation; locality cf. 2740.

2744 [S. B. field no. CAN 87-4-SB]. Gray siltstone from GBB-11 of the Chapel Island Formation; several hundred meters southwest of 2743, locality cf. 2740.

2745 [S. B. field no. CAN 87-5-SB]. Calcareous nodule from 0.2 above GBB-40 of the Chapel Island Formation; southwest of 2744, locality cf. 2740.

2746 [S. B. field no. CAN 87-6-SB]. Gray mudstone from GBB-110 of the Chapel Island Formation; southwest of 2744, locality cf. 2740.

2747 [S. B. field no. CAN 87-7-SB]. Calcareous nodule from GBC-30 of the Chapel Island Formation; west of 2746, locality cf. 2740.

2748 [S. B. field no. CAN 87-8-SB]. Calcareous nodule from GBC-30 of the Chapel Island Formation; west of 2746, locality cf. 2740.

2749 [S. B. field no. CAN 87-9-SB]. Gray shale from level with *Harlaniella* and *Tyrasotaenia* of the Chapel Island Formation; locality cf. 2740.

2750 (Ccarb) [S. B. field no. CAN 87-10-SB]. Calcareous nodule from 3 m below top of "member 1" in guidebook of the Chapel Island Formation; 300 m west of the lighthouse, Fortune Head, 3.3 km west on gravel road, 2.7 km south of city of Fortune via Route 220, southeastern Newfoundland, Canada; 47.07°N, 55.81°W.

2751 (TOC, Corg) [S. B. field no. CAN 87-11-SB]. Gray siltstone from the top of "member 1" of the Chapel Island Formation; locality cf. 2750.

2752 (TOC, Corg) [S. B. field no. CAN 87-12-SB]. Gray mudstone from FH-29.5 of the Chapel Island Formation; 400 m west of the lighthouse, locality cf. 2750.

2753 (Ccarb) [S. B. field no. CAN 87-13-SB]. Calcareous nodule from FH30 of the Chapel Island Formation; locality cf. 2750.

2754 (Ccarb, TOC) [S. B. field no. CAN 87-14-SB]. Calcareous nodule from FH-34 of the Chapel Island Formation; locality cf. 2750.

2755 (Ccarb) [S. B. field no. CAN 87-15-SB]. Calcareous nodule from FH-66.6 of the Chapel Island Formation; locality cf. 2750.

2756 (TOC, Corg) [S. B. field no. CAN 87-16-SB]. Gray mudstone from FH-159.7 of the Chapel Island Formation; 400 m west of 2755, locality cf. 2750.

2757 [S. B. field no. CAN 87-17-SB]. Calcareous nodule from FH-162 of the Chapel Island Formation; same as 2756, locality cf. 2750.

2758 [S. B. field no. CAN 87-18-SB]. Shale from FH-167 of the Chapel Island Formation; same as 2756, locality cf. 2750.

2759 [S. B. field no. CAN 87-19-SB]. Gray shale from FH-250 of the Chapel Island Formation; same as 2756, locality cf. 2750.

2760 (Ccarb) [S. B. field no. CAN 87-20-SB]. Calcareous nodule from FH-259.1 of the Chapel Island Formation; same as 2756, locality cf. 2750.

2761 (Ccarb) [S. B. field no. CAN 87-21-SB]. Calcareous nodule from the Chapel Island Formation; 200 m west of 2760, locality cf. 2750.

2762 [S. B. field no. CAN 87-22-SB]. Gray mudstone from LDC 93 of the Chapel Island Formation; "Little Dantzic Cove section," at Pieduck Point, 18 km south of the city of Fortune via Route 220, southeastern Newfoundland, Canada; 46.96°N, 56.00°W.

2763 (Ccarb) [S. B. field no. CAN 87-23-SB]. Calcareous nodule from LDC 93 of the Chapel Island Formation; locality cf. 2762.

2764 (Ccarb) [S. B. field no. CAN 87-24-SB]. Calcareous nodule from LDC 104 of the Chapel Island Formation; locality cf. 2762.

2765 (Ccarb, TOC) [S. B. field no. CAN 87-25-SB]. Purple carbonate from the top of "member 3", LDC 135.0 of the Chapel Island Formation; locality cf. 2762.

2766 (Ccarb) [S. B. field no. CAN 87-26-SB]. Calcareous nodule from LDC 148.5 of the Chapel Island Formation; locality cf. 2762.

2767 (Ccarb) [S. B. field no. CAN 87-27-SB]. Limestone from the base of a 0.8 m-thick limestone, LDC 220 of the Chapel Island Formation; stop 4D, locality cf. 2762.

2768 (Ccarb) [S. B. field no. CAN 87-28-SB]. Limestone from the top of ca. 0.8 m-thick limestone, LDC-220.5 of the Chapel Island Formation; stop 4D, locality cf. 2762.

2769 [S. B. field no. CAN 87-29-SB]. Gray siltstone from "Member 5", LDC 300 of the Chapel Island Formation; stop 4E, locality cf. 2762.

2770 [S. B. field no. CAN 87-30-SB]. Purple limestone from 40 cm above the base of the Brigus Formation; stop 5, locality cf. 2762.

2771 [S. B. field no. CAN 87-31-SB]. Calcareous concretion from ca. 8 m below the top of the Random Formation; stop 6, north shore of Bull Arm fjord, on the beach in the village of Sunnyside, southeastern Newfoundland, Canada; 47.87°N, 53.92°W.

2772 [S. B. field no. CAN 87-32-SB]. Limestone from 3.5 m above the base of the Bonavista Formation; stop 6, locality cf. 2771.

2773 [S. B. field no. CAN 87-33-SB]. Limestone from ca. 35 m above the top of the Bonavista Formation; stop 6, locality cf. 2771.

2774 [S. B. field no. CAN 87-34-SB]. Calcareous nodule and adjacent shale from 83.5 m above the base of the Bonavista Formation; stop 6 (shore exposure), locality cf. 2771.

2775 [S. B. field no. CAN 87-35-SB]. Calcareous nodule from 103 m above the base of the Bonavista Formation; stop 6 (shore exposure), locality cf. 2771.

BASIN AND RANGE PROVINCE (EASTERN CALIFORNIA)

Tectonic unit/location. Basin and Range Province, Providence Mountains, eastern California, USA

Stratigraphy. Stratigraphic section, Providence Mountains, eastern California (Hazzard 1954; Stewart 1970):

[UNNAMED CAMBRIAN GROUP]
 Nopah Formation
 Bonanza King Formation
 Cadiz Formation
 Chambless Limestone
 Latham Shale
 Zabriskie Quartzite
 Wood Canyon Formation

PPRG samples/localities.

2739 (st) [C. V. M. field no. CL-1]. Dark gray, oncolitic limestone from the lower oncolite layer of the Chambless Limestone; southern Marble Mountains, ca. 8 km southeast of Chambless on US Route 66, San Bernardino County, California, USA; 34.52°N, 115.4°W.

BASIN AND RANGE PROVINCE (EASTERN CALIFORNIA AND SOUTHERN NEVADA)

Tectonic unit/location. Basin and Range Province, Death Valley and Kingston Range, USA.

Stratigraphy. Wright, 1954; Miller 1982; King 1976

 Wood Canyon Formation
 Stirling Quartzite
 Johnie Formation
 Noonday Dolomite
PAHRUMP GROUP
 Kingston Peak Formation
 Beck Spring Dolomite
 Crystal Spring Formation

PPRG samples/localities.

1271 (ts, st, μf*, TOC, Corg) [J. W. S. field no. 1 of 6/27/69]. Microfossiliferous (Chapter 24), cherty black dolostone, part of a domical stromatolite from the Beck Spring Dolomite; 35 km east of Tecopa in the northern Kingston Range, southeast of the Nopah Range, 16 km west of the California-Nevada border, California, USA.

2536 (TOC, Corg, ker) [J. M. H. field no. B6/B674]. Limestone from the Kingston Peak Formation; Death Valley National Park, southwestern Nevada, USA.

2537 (TOC, Corg, ker) [J. M. H. field no. B9/B675]. Limestone from the Kingston Peak Formation; locality cf. 2536.

2538 (TOC, Corg) [J. M. H. field no. B10/B676]. Limestone from the Kingston Peak Formation; locality cf. 2536.

2539 (TOC, Corg, ker) [J. M. H. field no. B22/B677]. Limestone from the Kingston Peak Formation; locality cf. 2536.

BASIN AND RANGE PROVINCE (UTAH)

Tectonic unit/location. Basin and Range Province, Eureka-House Range Region, western Utah, USA.

Stratigraphy. Palmer 1971.

 Windfall Formation
  Bullwhacker Member
  Catlin Member
 Dunderberg Formation
 Hamburg Formation

PPRG samples/localities.

1272 (ts, TOC, Corg) [J. W. S. field no. 9-26-67]. Interbedded black chert and massive limestone from the Catlin Member of the Windfall Formation; near the "Catlin Shaft" of the Croesus Mine in New York Canyon, Eureka Mining District, Eureka County, Nevada, USA.

BASIN AND RANGE PROVINCE (WESTERN REGION)

Tectonic unit/location. Basin and Range Province (western region), California and Nevada, USA.

Stratigraphy. Stratigraphic section, White Inyo Mountains area, eastern California and southwestern Nevada, USA (Stewart 1970; Signor and Onken 1987):

Table 14.2. *Continued.*

[UNNAMED CAMBRIAN GROUP]
   Harkless Formation
   Poleta Formation
   Campito Formation
      Montenegro Member
   Deep Spring Formation
   Reed Dolomite
   Wyman Formation

PPRG samples/localities.

2378 (ts, TOC, Corg) [C. V. M. field no. 871003-1]. Dark-gray, fissile, calcareous shale-siltstone from the Wyman Formation; Hines Ridge, California, USA; 37.11°N, 118.0°W.

2379 (ts) [C. V. M. field no. 871003-2]. Dark-gray, calcareous siltstone from stratigraphically below measured section, Wyman Formation; locality cf. 2378.

2380 (ts, TOC, Corg) [C. V. M. field no. 871003-3]. Dark-gray, micaceous siltstone-shale from 28.75 m above the base of the section, Wyman Formation; locality cf. 2378.

2381 (ts) [C. V. M. field no. 871003-4]. Dark-gray siltstone with calcareous fine sand streamers from 56.7 m above the base of the section, Wyman Formation; locality cf. 2378.

2382 (ts, TOC, Corg) [C. V. M. field no. 871003-5]. Dark-gray sandstone-siltstone from 79 m above the base of the section, Wyman Formation; locality cf. 2378.

2383 (ts) [C. V. M. field no. 871003-6]. Dark-gray siltstone with sandy light gray streamers from 105 m above the base of the section, Wyman Formation; locality cf. 2378.

2384 (ts, TOC, Corg) [C. V. M. field no. 871003-7]. Dark-gray siltstone from 127 m above the base of the section, Wyman Formation; locality cf. 2378.

2385 (ts) [C. V. M. field no. 871003-8]. Dark-gray sandy siltstone from 150.5 m above the base of the section, Wyman Formation; locality cf. 2378.

2386 (ts) [C. V. M. field no. 871003-9]. Medium-gray, medium crystalline, partially calcareous, massive dolomite, with 1–3 mm irregular vugs, from 175 m above the base of the section, Reed Dolomite; locality cf. 2378.

2387 (ts) [C. V. M. field no. 871003-10]. Light-gray, medium crystalline, dolomite containing concentrically laminated spheres 1–4 mm in diameter, from 198 m above the base of the section, Reed Dolomite; locality cf. 2378.

2388 (ts, Ccarb) [C. V. M. field no. 871003-11]. Light-gray, medium crystalline dolomite, from 254 m above the base of the section, Reed Dolomite; locality cf. 2378.

2389 (ts) [C. V. M. field no. 871003-12]. Light-gray, finely crystalline, dolomite from 302.5 m above the base of the section, Reed Dolomite; locality cf. 2378.

2390 (ts, Ccarb) [C. V. M. field no. 871003-13]. Light-gray, finely crystalline dolomite from 350.5 m above the base of the section, Reed Dolomite; locality cf. 2378.

2391 (ts, Ccarb) [C. V. M. field no. 871003-14]. White, finely crystalline, dolomite from 575.5 m above the base of the section, Reed Dolomite; locality cf. 2378.

2392 (ts) [C. V. M. field no. 871003-15]. White, very finely crystalline, dolomite from 601 m above the base of the section, Reed Dolomite; locality cf. 2378.

2393 (ts, Ccarb) [C. V. M. field no. 871003-16]. Very light-gray, finely crystalline dolomite from 625.5 m above the base of the section, Reed Dolomite; locality cf. 2378.

2394 (ts, Ccarb) [C. V. M. field no. 871003-17]. Light-gray, finely crystalline dolomite from 646.5 m above the base of the section, Reed Dolomite; locality cf. 2378.

2395 (ts) [C. V. M. field no. 871003-18]. Dark-gray siltstone from 681 m above the base of the section, Deep Spring Formation; locality cf. 2378.

2396 (ts) [C. M. field no. 871004-19]. Gray, oolitic limestone from the base of the oolitic unit above the archaeocyathid boundstone, from the lower part of the Poleta Formation; Rowland's Reef locality, T.6S R.41E, NW1/4 of the SW1/4 of Sec. 8, California, USA; Lida Quad 7.5 minute series; 37.43°N, 117.4°W.

2397 (ts) [C. M. field no. 871004-20]. Gray limestone, an archaeocyathid boundstone from the lower part of the Poleta Formation; locality cf. 2396.

2398 (ts) [C. M. field no. 871004-21]. Gray limestone, an archaeocyathid boundstone from the lower part of the Poleta Formation; locality cf. 2396.

2399 (ts) [C. M. field no. 871004-22]. Gray limestone, a calcareous algal and archaeocyathid boundstone from the lower half of the Poleta Formation; locality cf. 2396.

2400 (ts) [C. M. field no. 871004-23]. Oolitic limestone grainstone from the lower half of the Poleta Formation; locality cf. 2396.

2401 (ts) [C. M. field no. 871004-24]. Medium-gray, calcareous algal boundstone from the lower half of the Poleta Formation; locality cf. 2396.

Table 14.2. *Continued.*

2402 (ts) [C. M. field no. 871004-25]. Medium-gray, calcareous algal boundstone from the lower half of the Poleta Formation; locality cf. 2396.

2403 (ts) [C. M. field no. 871004-26]. Gray, calcareous wackestone from the lower half of the Poleta Formation; locality cf. 2396.

2404 [C. M. field no. 871004-27]. Gray, thrombolitic(?) limestone from the Poleta Formation; locality cf. 2396.

2405 (ts) [C. M. field no. 871005-28]. Light-brown limestone with *Wyattia* (calcareous tubes up to 2 m in outside diameter) from the lower part of the Deep Spring Formation; Mollie Gibson Canyon, ca. 3 mm north of the "o" in Mollie Gibson Mines on the Blanco Mtn. Quad. (type locality of *Wyattia*), California, USA; Blanco Mtn. quadrangle, T.7S, R.35E, sec. 16 NEI/4, NW1/4; 37.33°N, 118.1°W.

2406 (ts) [C. M. field no. 871005-29]. Gray archaeocyathid boundstone from the Montenegro Member of the Campito Formation; ridge on the north side of the drainage, "Montenegro bioherm locality," California, USA; Blanco Mtn. quadrangle, center of T.7S, R.35.E sec. 30; 37.33°N, 118.1°W.

2407 (ts) [C. M. field no. 871005-30]. Dark green-gray siltstone from the Montenegro Member of the Campito Formation; crest of White Inyo Range, ca. 48 km east of Big Pine, California, USA; 37.29°N, 118.1°W.

BELT BASIN

Tectonic unit/location. Belt Basin, Montana, USA.

Stratigraphy. Composite stratigraphic section, Glacier National Park, Little Belt Mountains, and Missoula-Saint Regis areas, northwestern and central Montana (Harrison 1972; Horodyski 1983):

BELT SUPERGROUP
    MISSOULA GROUP
      Mount Shields Formation
      Shepard Formation
      Snowslip Formation
    "PIEGAN GROUP"
      Siyeh Formation/Helena Formation
    RAVALLI GROUP
      (upper) Grinnell Argillite/Empire Formation
      (lower) Grinnell Argillite/Spokane Formation
      Appekunny Argillite/Greyson Shale
    "PRE-RAVELLI (GROUP)"
      Altyn Formation/Newland Limestone/(upper) Prichard Formation
      Chamberlain Shale/(middle) Prichard Formation
      Neihart Quartzite/(lower) Prichard Formation

PPRG samples/localities.

1001 (ts) [S. H. field no. SN-LC-1-(−10)]. Thinly bedded dolomite from 10 m below the top of the Siyeh Formation; traverse started from "Lunch Creek," 366 m north of Glacier Route 1, west side of Logan Pass, Glacier National Park, Montana, USA; 48.72°N, 113.6°W.

1002 (ts, st) [D. D. field no. D-71685-1]. Stromatolitic carbonate from 10 m below the top of the Siyeh Formation; locality cf. 1001.

1003 (ts) [S. H. field no. SN-LC-2-0]. Thinly bedded carbonate from the base of the Snowslip Formation; locality cf. 1001.

1004 (ts) [D. D. field no. D-71685-2]. Carbonate with pyrite from the basal part of the Snowslip Formation; locality cf. 1001.

1005 (ts) [S. H. field no. SN-LC-3-50]. Red argillite from the second red argillite within the basal part of the Snowslip Formation; locality cf. 1001.

1006 (st) [D. D. field no. D-71685-4]. Carbonate stromatolite (*Collenia*) with hematitic microfossils from the second red argillite, 67 m above the base of the Snowslip Formation; locality cf. 1001.

1007 (ts, st) [S. H. field no. SN-LC-4-95]. Stromatolitic carbonate from 95 m above the base of the Snowslip Formation; locality cf. 1001.

1008 (ts) [S. H. field no. SN-LC-5-125]. Dolomite from the top dolomite zone, 125 m above the base of the Snowslip Formation; locality cf. 1001.

1009 (ts) [S. H. field no. SN-LC-6-175]. Dolomite from 175 m above the base of the Snowslip Formation; locality cf. 1001.

1010 (ts, st) [S. H. field no. SN-LC-7-228]. Green argillite from the "green argillite stromatolite zone," 228 m above the base of the Snowslip Formation; locality cf. 1001.

1011 (ts) [S. H. field no. SH-LC-8-5]. Thinly bedded carbonate from 5 m above the base of the Shepard Formation; locality cf. 1001.

1012 (ts) [S. H. field no. SH-M-71685-1]. Chert from the Shepard Formation; locality cf. 1001.

1013 (ts, st) [D. D. field no. D-71685-5]. Stromatolitic carbonate from 228 m above the base of the Snowslip Formation; locality cf. 1001.

1014 (ts) [D. D. field no. D-71685-6]. Carbonate from the base of the Shepard Formation; locality cf. 1001.

1015 (ts) [S. H. field no. GTS-SY-1-10]. Carbonate from 10 m above the base of the Siyeh Formation; traverse on the south slope of the "Going-to-the-Sun Mountain," 8 km east of Logan pass on Glacier Route 1, Glacier National Park, Montana, USA; 48.70°N, 113.6°W.

1016 (ts) [S. H. field no. GTS-SY-2-30]. Carbonate from 30 m above the base of the Siyeh Formation; locality cf. 1015.

Table 14.2. *Continued.*

1017 (ts, st) [T. B. M. field no. 1A of 71785]. Carbonate stromatolite (*Baicalia*) from float, definitely of the Siyeh Formation; locality cf. 1015.

1018 (ts) [S. H. field no. GTS-SY-3-100]. Medium bedded carbonate from 100 m above the base of the Siyeh Formation; locality cf. 1015.

1019 (ts) [S. H. field no. GTS-SY-4-170]. Dolomite from 170 m above the base of the Siyeh Formation; locality cf. 1015.

1020 (ts) [S. H. field no. GTS-SY-5-210]. Carbonate from 210 m above the base of the Siyeh Formation; locality cf. 1015.

1021 (ts) [S. H. field no. GTS-SY-6-254]. Carbonate from 254 m above the base of the Siyeh Formation; locality cf. 1015.

1022 (ts) [S. H. field no. GTS-SY-7-295]. Dolomitic sandstone from 295 m above the base of the Siyeh Formation; locality cf. 1015.

1023 (ts) [S. H. field no. GTS-SY-8-320]. Dolomite from 320 m above the base of the Siyeh Formation; locality cf. 1015.

1024 (ts) [S. H. field no. GTS-SY-9-345]. Carbonate from 345 m above the base of the Siyeh Formation; locality cf. 1015.

1025 (ts, st) [D. D. field no. D-71785-3]. Stromatolitic black carbonate from 170 m above the base of the Siyeh Formation; locality cf. 1015.

1026 (ts, st) [D. D. field no. D-71785-4]. Stromatolitic chert from 175 m above the base of the Siyeh Formation; locality cf. 1015.

1027 (ts, st) [D. D. field no. D-71785-5]. Stromatolite black carbonate from 5 m above the base of the molartooth bed, 250 m above the base of the Siyeh Formation; locality cf. 1015.

1028 (ts, st) [D. D. field no. D-71785-6]. Stromatolitic carbonate from 320 m above the base of the Siyeh Formation; locality cf. 1015.

1029 (ts, st) [D. D. field no. D-71785-7]. Stromatolitic carbonate from 335 m above the base of the Siyeh Formation; locality cf. 1015.

1030 (ts) [S. H. field no. GTS-SY-10-373]. Oolitic carbonate from 373 m above the base of the Siyeh Formation; locality cf. 1015.

1031 (ts) [D. D. field no. D-71885-1]. Oolitic carbonate from 373 m above the base of the Siyeh Formation; locality cf. 1015.

1032 (ts) [S. H. field no. GTS-SY-11-455]. Dolomite from 455 m above the base and above a stromatolite unit of the Siyeh Formation; locality cf. 1015.

1033 (ts, st) [S. H. field no. GTS-SY-12-490]. Carbonate stromatolite from 490 m above the base of the Siyeh Formation; locality cf. 1015.

1034 (ts) [S. H. field no. GTS-SY-13-510]. Dolomite from 510 m above the base of the Siyeh Formation; locality cf. 1015.

1035 (ts) [S. H. field no. GTS-SY-14-545]. Carbonate with molartooth structure from 545 m above the base of the Siyeh Formation; locality cf. 1015.

1036 [D. D. field no. D-71885-2]. Halite casts from below the prominent diabase sill of the Siyeh Formation; locality cf. 1015.

1037 (ts, st) [D. D. field no. D-71885-3]. Stromatolite from 445 m above the base of the Siyeh Formation; locality cf. 1015.

1038 (ts, st) [D. D. field no. D-71885-4]. Stromatolitic carbonate from 525 m above the base of the Siyeh Formation; locality cf. 1015.

1039 (ts) [D. D. field no. D-71885-5]. Carbonate from 550 m above the base of the Siyeh Formation; locality cf. 1015.

1040 (ts) [D. D. field no. D-71985-1]. Carbonate from 25 m above the stromatolite zone and 595 m above the base of the Siyeh Formation; northeast slope of Piegan Mountain at "Piegan Pass"; trailhead is located 4 km east of Logan Pass along Glacier Route 1, Montana, USA; 48.72°N, 113.6°W.

1041 (ts) [S. H. field no. PP-SY-1-605]. Carbonate from 605 m above the base of the Siyeh Formation; locality cf. 1040.

1042 (ts, st) [D. D. field no. D-71985-2]. Domical carbonate stromatolite from 625 m above the base of the Siyeh Formation; locality cf. 1040.

1043 (ts) [S. H. field no. PP-SY-2-660]. Carbonate from 660 m above the base of the Siyeh Formation; locality cf. 1040.

1044 (ts) [S. H. field no. PP-SY-3-685]. Carbonate from 685 m above the base of the Siyeh Formation; locality cf. 1040.

1045 (ts) [D. D. field no. D-71985-3]. Carbonate from 690 m above the base of the Siyeh Formation; locality cf. 1040.

1046 (ts) [S. H. field no. PP-SY-4-725]. Carbonate from 725 m above the base of the Siyeh Formation; locality cf. 1040.

1047 (ts) [S. H. field no. PP-SY-5-775]. Calcitic pod from 775 m above the base of the Siyeh Formation; locality cf. 1040.

1048 (ts, st) [D. D. field no. D-71985-4]. Stromatolitic carbonate from 750 m above the base of the Siyeh Formation; locality cf. 1040.

1049 (ts) [S. H. field no. AF-AL-1-(−150)]. Carbonate from 150 m below the top of the Altyn Formation; Appekunny Falls, ca. 16 km west of Babb on Glacier Route 3, Glacier National Park, Montana, USA; 48.10°N, 113.1°W.

1050 (ts) [S. H. field no. AF-AL-2-(−30)]. Sandy dolomite with algal(?) laminations from 30 m below the top of the Altyn Formation; locality cf. 1049.

1051 (ts) [S. H. field no. AF-AL-3-(−80)]. Sandy dolomite from 80 m below the top of the Altyn Formation; locality cf. 1049.

1052 (ts, st) [T. B. M. field no. T-7/21/85-1]. Carbonate stromatolite (small *Conophyton*) from 8 m above the base of the "*Baicalia*"-*Conophyton* cycles of the Siyeh Formation; Swiftcurrent Glacier, 9.6 km west of Swiftcurrent Ranger Station on foot-trail (the last 800 m is cross-country); Station is at the west end of Glacier Route 3, Montana, USA; 48.78°N, 113.7°W.

1053 (ts, st) [T. B. M. field no. T-7/21/85-2]. Stromatolite (*Baicalia*) from the lower "*Baicalia*" unit, 5 m above the base of the "*Baicalia*"-*Conophyton* cycle of the Siyeh Formation; locality cf. 1052.

1211 (ts) [D. D. field no. D-72085-1]. Black chert and carbonate from the Altyn Formation; Appekunny Falls, ca. 16 km west of Babb on Glacier Route 3, Glacier National Park, Montana, USA; 48.10°N, 113.1°W.

Table 14.2. *Continued.*

1472 (ts, st, TOC, Corg) [R. J. H. field no. RHG-310]. Inclined, elongate, dolomitic stromatolite from the upper portion of the Altyn Formation; locality cf. 1211.

1754 (ts, st) [D. D. field no. D-71785-1]. Stromatolitic, gray carbonate from the Siyeh Formation; traverse on the south slope of the "Going-to-the-Sun Mountain"; 8 km east of Logan pass on Glacier Route 1, Glacier National Park, Montana, USA; 48.70°N, 113.6°W.

1755 (st) [D. D, field no. D-71785-2A]. Stromatolitic cherty carbonate from 100 m above the base of the Siyeh Formation; locality cf. 1754.

1756 [D. D. field no. D-71785-2B]. Cherty stromatolite from 100 m above the base of the Siyeh Formation; locality cf. 1754.

1757 [D. D. field no. D-71685-3]. Stromatolitic black carbonate from 170 m above the base of the Siyeh Formation; locality cf. 1754.

1870 (ts) [R. J. H. field no. CH-2]. Gray mudstone from ca. 100 m below the top of the Chamberlain Shale; 150 m west of Chamberlain Creek, along the National Forest Service road on the north side of Jefferson Creek, Little Belt Mountains, Montana, USA; U.T.M. zone 12, 5191620 mN., 524250 mE.

1871 [R. J. H. field collection]. Gray mudstone with rhythmic interbedding of dolomite from ca. 120 m above the base of the Newland Limestone; roadcut on US Highway 89, southeast of Neihart, Little Belt Mountains, Montana, USA; UTM (zone 12) 5191620 mN and 524250 mE.

2298 (ts, TOC, Corg) [H. S. field no. 870802-1]. Black shale with 3 mm siltstone bands from the Newland Limestone; outcrop located on east side of forest road no. 830 (Studhorse Creek Rd.), 300 m north of Donkey Gulch, northeast of White Sulphur Springs, 8 km east of US Highway 89, Montana, USA.

2299 (ts, TOC, Corg) [H. S. field no. 870802-2]. Black shale from the Newland Limestone; locality cf. 2298.

2300 (ts, TOC, Corg) [H. S. field no. 870802-3]. Dark gray-black carbonate from the Newland Limestone; locality cf. 2298.

2301 (ts, TOC, Corg) [H. S. field no. 870802-4]. Black limestone from an interbedded limestone-dolostone unit of the Newland Limestone; locality cf. 2298.

2302 (ts, TOC, Corg) [H. S. field no. 870802-5]. Cryptalgal(?) cherty dolomite with chert nodules from the second main carbonate horizon, upper portion of the Newland Limestone; south of Black Butte Mountain, Little Belt Mountains, Montana, USA.

2303 (ts, TOC, Corg) [H. S. field no. 870802-6]. Cryptalgal(?) dolostone with black chert nodules from the second main carbonate horizon, upper portion of the Newland Limestone; 2302.

2304 (ts) [H. S. field no. 870802-7]. Grayish-red barite from the upper third of the Newland Limestone; ca. 2.4 km along Sheep Creek Road off US Highway 89, locality cf. 2302.

2305 (ts, TOC, Corg) [H. S. field no. 870802-8]. Black carbonate from the Newland Limestone; outcrop located on the east side of forest road no. 830 (Studhorse Creek Rd.), 300 m north of Donkey Gulch, northeast of White Sulphur Springs, 8 km east of Highway 89, Montana, USA.

2306 (ts, TOC, Corg) [H. S. field no. 870803-9]. Fissile black shale from the Chamberlain Shale; ca. 1.6 km south of Jefferson Creek, Little Belt Mountains, Montana, USA.

2307 (ts, TOC, Corg, ker) [H. S. field no. 870803-10]. Dark-gray, micaceous shale from an interbedded dolostone-shale unit about 600 m above the base of the Newland Limestone; locality cf. 2306.

2308 (ts, TOC, Corg) [H. S. field no. 870803-11]. Black carbonate from an interbedded dolostone-shale unit of the Newland Limestone; locality cf. 2306.

2309 (ts, TOC, Corg) [H. S. field no. 870803-12]. Gray, dolomitic shale from the transition zone of JS-UTA of the Newland Limestone; roadcut along forest road, Miller Gulch, Little Belt Mountains, Montana, USA.

2310 (ts, TOC, Corg) [H. S. field no. 870804-13]. Gray shale from ca. 30 m below a pyrite-shale gossan transition of the Newland Limestone; locality cf. 2309.

2311 (ts, TOC, Corg, Sfide) [H. S. field no. 870804-14]. Dark-gray shale from the Newland Limestone; locality cf. 2309.

2312 (ts, TOC, Corg, Sfide) [H. S. field no. 870804-15]. Black shale from an interbedded limestone-shale unit, the uppermost main carbonate unit of the Newland Limestone; intersection of US Highway 89N and Newland Creek, Montana, USA.

2313 (ts, TOC, Corg) [H. S. field no. 870804-16]. Dark-gray shale from a carbonate-shale unit from the upper third of the Newland Limestone; locality cf. 2312.

2314 (ts, TOC, Corg, Sfide) [H. S. field no. 870804-17]. Greenish-gray shale from the Greyson Shale; along US Highway 89, north of White Sulfur Springs, Little Belt Mountains, Montana, USA.

2315 (ts, TOC, Corg, Sfide) [H. S. field no. 870805-18]. Dark-gray, slightly metamorphosed(?) shale from the Newland Limestone; north side of road, at mile post 14 along State Route 12 from White Sulphur Springs to Townsend, Deep Creek Valley, Little Belt Mountains, Montana, USA.

2316 (ts, TOC, Corg, Sfide) [H. S. field no. 870805-19]. Gray shale, possibly metamorphosed(?), from the Newland Limestone; locality cf. 2315.

2317 (ts, TOC, Corg) [H. S. field no. 870805-20]. Dark-gray, very compact, silicified shale from the Newland Limestone; roadcut on the north side of Duck Creek Road, Big Belt Mountains, Montana, USA.

2318 (ts, TOC, Corg, Sfide) [H. S. field no. 870805-21]. Dark-gray shale from the Newland Limestone; Trout Creek Road, along the upper Trout Creek, Big Belt Mountains, Montana, USA.

Table 14.2. *Continued.*

2319 (ts, TOC, Corg, ker) [H. S. field no. 870805-22]. Dark-gray shale from the uppermost carbonate-shale interbeds, Newland Limestone; locality cf. 2318.

2320 (ts, TOC, Corg) [H. S. field no. 870805-23]. Dark-gray limestone from the uppermost carbonate-shale interbeds, Newland Limestone; locality cf. 2318.

2321 (ts, TOC, Corg, ker) [H. S. field no. 870805-24]. Gray siltstone-shale from ca. 20 m above the base of the Greyson Shale; ca. 200 m west of 2320, locality cf. 2318.

2322 (ts, TOC, Corg) [H. S. field no. 870806-25]. Black dolomitic shale from the lower part of the Newland Limestone; roadcut on the northeast side of Horse Gulch Road, off Benton Gulch Road, Big Belt Mountains, Montana, USA.

2323 (ts, TOC, Corg, Sfide) [H. S. field no. 870806-26]. Gray shale with bedding planes speckled with euhedral pyrite (up to 1 cm per side), presumably late diagenetic, from the lower part of the Newland Limestone; half-way down Cement Gulch on a dirt road, Montana, USA.

2324 (ts, TOC, Corg) [H. S. field no. 870806-27]. Dark-gray shale from the lower part of the Newland Limestone; Avalanche Gulch, northeast of Townsend, Big Belt Mountains, Montana, USA.

2325 (ts, TOC, Corg) [H. S. field no. 870808-28]. Gray carbonate, metamorphosed(?), from the Helena Formation; ca. 800 m west of Georgetown Lake dam along Highway 10A on the right side of Road Stop 3 of the Belt Symposium II, Field Guide for trip no. 3, (Winston and Wallace 1983), Montana, USA; Georgetown Lake quadrangle.

2326 [H. S. field no. 870808-29]. Green argillite with mud cracks, from ca. 2 m above the base of the Snowslip Formation; 1.6 km west of Georgetown Lake Dam along US Highway 10A, stop 4 of the Belt Symposium II Field Guide for trip no. 3 (Winston and Wallace 1983), Montana, USA.

2327 (ts, TOC, Corg, Sfide) [H. S. field no. 870809-30]. Dark-gray, banded quartzite with some sulfide bands from "Member F" of the Prichard Formation; ca. 800 m south of Sigel Creek Road along State Route 135, 5 km south of the Lolo National Park entrance; stop 4 of the Belt Symposium II Field Guide to trip no. 4 (Cressman 1983), Montana, USA.

2328 (ts, TOC, Corg) [H. S. field no. 870809-31]. Slaty argillite from the Prichard Formation; ca. 4 km southwest of Siegel Creek road along State Route 135, stop 5 in the Belt Symposium II Field Guide for trip no. 4, Montana, USA.

2329 (ts, st, TOC, Corg) [H. S. field no. 870810-32]. Dolomite stromatolite from the Mount Shields Formation; ca. 14 km west of Libby, stop 9 in the Belt Symposium II Field Guide for trip no. 2 and 9, Montana, USA.

2408 [R. J. H. field no. AFI-30A]. Sandy dolarenite and muddy dololutite from the Altyn Formation; Appekunny Falls, ca. 16 km west of Babb on Glacier Route 3, Glacier National Park, Montana, USA; 48.10°N, 113.1°W.

BLUE RIDGE PROVINCE

Tectonic unit/location.

Stratigraphy.

    Anakeesta Formation
    Pigeon Formation

PPRG samples/localities.

2540 (ts) [H. S. field no. 860928-1]. Gray shale from the Anakeesta Formation; parking lot near Newfound Gap, Great Smoky Mountains National Park, Tennessee, USA.

2541 (ts) [H. S. field no. 860928-2]. Gray shale from the Anakeesta Formation; parking lot near Newfound Gap, locality cf. 2540.

2542 (ts) [H. S. field no. 860929-3]. Gray shale from the Pigeon Formation; near Pigeon Forge, locality cf. 2540.

CAPE SMITH BELT

Tectonic unit/location.

Stratigraphy.

    SPARTAN GROUP

PPRG samples/localities.

2779 [M. S. field no. B 289]. Graphitic pelite from the Spartan Group; 2 km east of the southern end of Lac Watts, Quebec, Canada; 62.00°N, 74.50°W.

2780 [M. S. field no. L 228]. Highly folded, graphitic pelite from the Spartan Group; ca. 5 km east of the southern end of Lac Watts, Quebec, Canada; 62.00°N, 74.50°W.

2781 [M. S. field no. L 229]. Graphitic pelite from the Spartan Group; ca. 5 km east of the southern end of Lac Watts, 1.5 km south of 2780, Quebec, Canada; 62.00°N, 74.50°W.

COLORADO PLATEAU PROVINCE

Tectonic unit/location. Colorado Plateau Province, Arizona, USA.

Stratigraphy. Stratigraphic section, Grand Canyon area, northwestern Arizona (Ford and Breed 1973):

GRAND CANYON SUPERGROUP
  CHUAR GROUP
    Sixtymile Formation
    Kwagunt Formation
      Walcott Member
      Awatubi Member
      Carbon Butte Member
    Galeros Formation
      Duppa Member
      Carbon Canyon Member
      Jupiter Member
      Tanner Member
  [UNNAMED UPPER PROTEROZOIC GROUP]
    Nankoweap Formation
  UNKAR GROUP
    Cardenas Lavas
    Dox Sandstone
    Shinumo Quartzite
    Hakatai Shale
    Bass Limestone

PPRG samples/localities.

1054 (ts) [T. B. M. field no. 91585-1]. Dolomite from the base of the Tanner Member of the Galeros Formation; east side of Basalt Canyon along Basalt Canyon Creek, 1 km northwest of Tanner Canyon Rapids, Grand Canyon National Park, Arizona, USA; 36.12°N, 111.8°W.

1055 (TOC, Corg, ker, Sfide) [T. B. M. field no. TM-91585-1]. Gray shale with fine laminations, 26 m above the base of the Tanner Member; locality cf. 1054.

1056 (ts) [T. B. M. field no. 1 of 9-15-85]. Red sandstone from 30 m above the base of the Tanner Member; locality cf. 1054.

1057 [T. B. M. field no. 2 of 9-15-85]. Gray shale from 49 m above the base of the Tanner Member; locality cf. 1054.

1058 (TOC, Corg) [T. B. M. field no. 3 of 9-15-85]. Gray shale with fine laminations from 53 m above the base of the Tanner Member; locality cf. 1054.

1059 (TOC, Corg, ker) [T. B. M. field no. 4 of 9-15-85]. Gray shale, partly weathered, from 70 m above the base of the Tanner Member; locality cf. 1054.

1060 (TOC, Corg) [T. B. M. field no. 5 of 9-15-85]. Gray shale, poorly laminated, from 88 m above the base of the Tanner Member; locality cf. 1054.

1061 (TOC, ker) [T. B. M. field no. 6 of 9-15-85]. Gray shale, poorly laminated, from 127 m above the base of the Tanner Member; locality cf. 1054.

1062 (TOC, Corg, ker) [T. B. M. field no. 7 of 9-15-85]. Gray shale, poorly laminated, from 134 m above the base of the Jupiter Member of the Galeros Formation; locality cf. 1054.

1063 (TOC, Corg, ker, Sfide) [T. B. M. field no. 8 of 9-15-85]. Gray micaceous shale from 223 m above the base of the Jupiter Member; locality cf. 1054.

1064 (ts, TOC, Corg) [T. B. M. field no. 9 of 9-15-85]. Carbonate from 180 m above the base of the Jupiter Member; locality cf. 1054.

1065 [T. B. M. field no. 10 of 9-15-85]. Red Shale from 181 m above the base of the Jupiter Member; locality cf. 1054.

1066 (ts) [T. B. M. field no. 11 of 9-15-85]. Carbonate from 175 m above the base of the Jupiter Member; locality cf. 1054.

1067 (ts, TOC, Corg, ker) [T. B. M. field no. 1 of 9-16-85]. Gray shale, finely laminated, from 21 m above the base of the Jupiter Member; locality cf. 1054.

1068 (ts) [T. B. M. field no. 2 of 9-16-85]. Volcanic tuff containing *Chuaria*, from 23 m above the base of the Jupiter Member; locality cf. 1054.

1069 (ts) [T. B. M. field no. 3 of 9-16-85]. Chert from the base of the Jupiter Member; locality cf. 1054.

1070 (ts, st, TOC) [T. B. M. field no. 4 of 9-16-85]. Stromatolitic carbonate from the base of the Jupiter Member; locality cf. 1054.

1071 (ts, st) [T. B. M. field no. 5 of 9-16-85]. Stromatolitic carbonate from the base of the Jupiter Member; locality cf. 1054.

1072 (ts) [T. B. M. field no. 6 of 9-16-85]. Cherty carbonate from the base of the Jupiter Member; locality cf. 1054.

1073 (ts) [T. B. M. field no. 7 of 9-16-85]. Chert from the base of the Jupiter Member; locality cf. 1054.

1074 (ts, st) [D. D. field no. 91685-1]. Stromatolite from the base of the Jupiter Member; locality cf. 1054.

Table 14.2. *Continued.*

1075 (ts, st) [D. D. field no. 91685-2]. Stromatolite from the base of the Jupiter Member; locality cf. 1054.

1076 (ts, st) [D. D. field no. 91685-3]. Cherty carbonate stromatolite from the base of the Jupiter Member; locality cf. 1054.

1077 (ts, st) [D. D. field no. 91685-4]. Stromatolite from the base of the Jupiter Member; locality cf. 1054.

1078 (ts) [B. P. field no. BP-9-16-85-1]. Gray sandstone from the base of the Tanner Member of the Galeros Formation; locality cf. 1054.

1079 (ts) [B. P. field no. BP-9-16-85-2]. Sandstone from the base of the Tanner Member; locality cf. 1054.

1080 (TOC, Corg, ker) [B. P. field no. BP-9-16-85-3]. Greenish-gray shale, poorly laminated, from 61 m above the base of the Tanner Member; locality cf. 1054.

1081 [B. P. field no. BP-9-16-85-4]. Shale from 107 m above the base of the Jupiter Member of the Galeros Formation; locality cf. 1054.

1082 (ts, st) [T. B. M. field no. 1 of 9-17-85]. Stromatolite from the base of the Awatubi Member of the Kwagunt Formation; south slope of Nankoweap Butte, 2.5 km southeast of Nankoweap Mesa, 5 km east of Kwagunt Rapids of the Colorado River, Arizona, USA; 36.25°N, 111.9°W.

1083 (TOC, Corg, ker, Sfide) [T. B. M. field no. 2 of 9-17-85]. Greenish-gray shale containing marcasite nodules from the first shale unit of the Awatubi Member; locality cf. 1082.

1084 [T. B. M. field no. 3 of 9-17-85]. Shale from 43 m above the base of the Awatubi Member; locality cf. 1082.

1085 (ts) [T. B. M. field no. 4 of 9-17-85]. Reddish-gray sandstone from 92 m above the base of the Awatubi Member; locality cf. 1082.

1086 [T. B. M. field no. 5 of 9-17-85]. Shale from 114 m above the base of the Awatubi Member; locality cf. 1082.

1087 (ts, TOC, Corg, ker, Sfide) [T. B. M. field no. 6 of 9-17-85]. Black shale from 18 m above the base of the Awatubi Member; locality cf. 1082.

1088 (ts, TOC, Corg, ker Sfide) [T. B. M. field no. 7 of 9-17-85]. Micaceous(?) black shale from 18 m above the base of the Walcott Member of the Kwagunt Formation; locality cf. 1082.

1089 (ts) [T. B. M. field no. 8 of 9-17-85]. Sandstone from the top of the Walcott Member; locality cf. 1082.

1090 (ts) [T. B. M. field no. 9 of 9-17-85]. Carbonate from 55 m above the base of the Walcott Member; locality cf. 1082.

1091 [D. D. field no. D-91785-2]. Carbonate from 55 m above the base of the Walcott Member; locality cf. 1082.

1092 (ts) [T. B. M. field no. 1 of 9-18-85]. Pisolitic chert from the uppermost part of the Walcott Member; locality cf. 1082.

1093 (ts, TOC, Corg) [T. B. M. field no. 2A of 9-18-85]. Pisolitic chert from the base of the Walcott Member; locality cf. 1082.

1094 (ts) [T. B. M. field no. 2B of 9-18-85]. Chert from the Walcott Member; locality cf. 1082.

1095 (ts, TOC, Corg, ker) [T. B. M. field no. 3 of 9-18-85]. Very fine-grained, gray sandstone with carbonate veins from the Walcott Member; locality cf. 1082.

1096 (ts, Sfide) [T. B. M. field no. 4A of 9-18-85]. Nodular chert within shale from the Walcott Member; locality cf. 1082.

1097 (ts, Corg) [T. B. M. field no. 4B of 9-18-85]. Gray shale with chert lenses from the Walcott Member; locality cf. 1082.

1098 (ts) [T. B. M. field no. 4C of 9-18-85]. Nodular chert within shale from the Walcott Member; locality cf. 1082.

1099 (ts) [T. B. M. field no. 5 of 9-18-85]. Black shale from 34 m above the base of the Walcott Member; locality cf. 1082.

1100 (ts) [T. B. M. field no. 6 of 9-18-85]. Black shale from 9 m above the base of the Walcott Member; locality cf. 1082.

1101 [T. B. M. field no. 7 of 9-18-85]. Black shale from 12 m above the base of the Walcott Member; locality cf. 1082.

1102 [T. B. M. field no. 8 of 9-18-85]. Gray shale from 17 m above the base of the Walcott Member; locality cf. 1082.

1103 (ts, Sfide) [T. B. M. field no. 9 of 9-18-85]. Black shale from 20 m above the base of the Walcott Member; locality cf. 1082.

1104 (ts) [T. B. M. field no. 10 of 9-18-85]. Shale from 5 m above the base of the Walcott Member; locality cf. 1082.

1105 [T. B. M. field no. 11 of 9-18-85]. Shale from 8 m above the base of the Walcott Member; locality cf. 1082.

1106 [T. B. M. field no. 12 of 9-18-85]. Black shale from 14 m above the base of the Walcott Member; locality cf. 1082.

1107 (ts) [T. B. M. field no. 13 of 9-18-85]. Pisolitic chert from the lower pisolite unit of the Walcott Member; locality cf. 1082.

1108 (ts) [T. B. M. field no. 14 of 9-18-85]. Black shale from the lower pisolite unit of the Walcott Member; locality cf. 1082.

1109 (ts) [D. D. field no. D-91885-1]. Pisolitic chert from the lower pisolite unit of the Walcott Member; locality cf. 1082.

1110 (ts) [D. D. field no. D-91885-2]. Chert from the lower pisolite unit of the Walcott Member; locality cf. 1082.

1111 (ts) [D. D. field no. D-91885-3]. Flakey dolomite from the basal dolomite unit of the Walcott Member; locality cf. 1082.

1112 (ts, TOC, Corg, ker) [D. D. field no. D-91885-6]. Dark gray, siliceous shale with chert lenses from 4 m above the base of the Walcott Member; locality cf. 1082.

1113 (ts) [R. J. H. field no. RJH91885-1]. Dark gray, finely laminated mudstone from 0.25 m above the base of the Walcott Member; locality cf. 1082.

1114 (ts) [R. J. H. field no. RJH91885-3]. Cherty pisolite with dolostone from 0.5 m above the base of the Walcott Member; locality cf. 1082.

1115 (ts) [R. J. H. field no. RJH91885-4]. Cherty pisolite with dolostone from 0.62 m above base of the Walcott Member; locality cf. 1082.

Table 14.2. *Continued.*

1116 (ts) [R. J. H. field no. RJH91885-5]. Dolostone with chert from 0.72 m above the base of the Walcott Member; locality cf. 1082.

1117 (ts) [R. J. H. field no. RJH91885-6]. Dolostone from 0.8 m above the base of the Walcott Member; locality cf. 1082.

1118 (ts) [R. J. H. field no. RJH91885-7]. Dolostone from 1.0 m above the base of the Walcott Member; locality cf. 1082.

1119 [R. J. H. field no. RJH91885-8]. Dolostone from 1.7 m above the base of the Walcott Member; locality cf. 1082.

1120 (ts) [R. J. H. field no. RJH91885-9]. Dolostone from 1.9 m above the base of the Walcott Member; locality cf. 1082.

1121 (ts) [R. J. H. field no. RJH91885-11]. Dolostone with chert laminae from 2.35 m above the base of the Walcott Member; locality cf. 1082.

1122 (ts) [R. J. H. field no. RJH91885-13]. Gypsum(?) nodule (50 cm long, 15 cm high) from 2.7 m above the base of the Walcott Member; locality cf. 1082.

1123 (ts, TOC, Corg, ker) [R. J. H. field no. RJH91885-14]. Carbonaceous sandstone from 2.7 m above the base of the Walcott Member; locality cf. 1082.

1124 (ts, TOC, Corg, Sfide) [T. B. M. field no. 1 of 9-19-85]. Siltstone from 15 m above the base of the Jupiter Member of the Galeros Formation; along west fork of Carbon Creek in Carbon Canyon, ca. 400 m upstream of east-west fork, Grand Canyon National Park, Arizona, USA; 36.17°N, 111.8°W.

1125 (ts, TOC, Corg, ker) [T. B. M. field no. 2 of 9-19-85]. Greenish-gray shale from 2 m above the base of the Jupiter Member; locality cf. 1124.

1126 (ts, TOC, ker) [T. B. M. field no. 4 of 9-19-85]. Brownish-gray siltstone from 6 m above the base of the Carbon Canyon Member of the Galeros Formation; locality cf. 1124.

1127 (ts) [T. B. M. field no. 5 of 9-19-85]. Sandstone from 5 m above the base of the Carbon Canyon Member; locality cf. 1124.

1128 (ts) [T. B. M. field no. 6 of 9-19-85]. Carbonate from 21 m above the base of the Carbon Canyon Member; locality cf. 1124.

1129 (ts) [T. B. M. field no. 7 of 9-19-85]. Sandstone from 34 m above the base of the Carbon Canyon Member; locality cf. 1124.

1130 (ts) [T. B. M. field no. 8 of 9-19-85]. Carbonate from 37 m above the base of the Carbon Canyon Member; locality cf. 1124.

1131 (ts) [T. B. M. field no. 9 of 9-19-85]. Carbonate from 41 m above the base of the Carbon Canyon Member; locality cf. 1124.

1132 (ts) [T. B. M. field no. 10 of 9-19-85]. Carbonate from 49 m above the base of the Carbon Canyon Member; locality cf. 1124.

1133 (ts) [T. B. M. field no. 11 of 9-19-85]. Carbonate from 56 m above the base of the Carbon Canyon Member; locality cf. 1124.

1134 (ts) [T. B. M. field no. 12 of 9-19-85]. Shale from 63 m above the base of the Carbon Canyon Member; locality cf. 1124.

1135 (ts) [T. B. M. field no. 13 of 9-19-85]. Carbonate from 76 m above the base of the Carbon Canyon Member; locality cf. 1124.

1136 (ts) [T. B. M. field no. 14 of 9-19-85]. Carbonate from 85 m above the base of the Carbon Canyon Member; locality cf. 1124.

1137 (ts) [T. B. M. field no. 15 of 9-19-85]. Carbonate from 24 m above the base of the Carbon Canyon Member; locality cf. 1124.

1138 (ts) [T. B. M. field no. 16 of 9-19-85]. Carbonate from 21 m above the base of the Carbon Canyon Member; locality cf. 1124.

1139 (ts) [T. B. M. field no. 17A of 9-19-85]. Carbonate from 15 m above the base of the Carbon Canyon Member; locality cf. 1124.

1140 (ts) [T. B. M. field no. 17B of 9-19-85]. Green shale with sandstone from 11 m above the base of the Carbon Canyon Member; locality cf. 1124.

1141 (ts, st) [T. B. M. field no. 18 of 9-19-85]. Stromatolite (*Baicalia*) from the base of the Carbon Canyon Member; locality cf. 1124.

1142 (ts) [T. B. M. field no. 19 of 9-19-85]. Carbonate from 9 m above the base of the Carbon Canyon Member; locality cf. 1124.

1143 (ts) [T. B. M. field no. 20 of 9-19-85]. Carbonate from 15 m above the base of the Carbon Member; locality cf. 1124.

1144 (ts) [T. B. M. field no. 21 of 9-19-85]. Carbonate from 23 m above the base of the Carbon Canyon Member; locality cf. 1124.

1145 (ts) [T. B. M. field no. 22 of 9-19-85]. Carbonate from 26 m above the base of the Carbon Canyon Member; locality cf. 1124.

1146 (ts) [T. B. M. field no. 23 of 9-19-85]. Carbonate from 29 m above the base of the Carbon Canyon Member; locality cf. 1124.

1147 (ts) [T. B. M. field no. 24 of 9-19-85]. Carbonate from 31 m above the base of the Carbon Canyon Member; locality cf. 1124.

1148 (ts) [T. B. M. field no. 25 of 9-19-85]. Carbonate from 40 m above the base of the Carbon Canyon Member; locality cf. 1124.

1149 [T. B. M. field no. 26 of 9-19-85]. Shale from 49 m above the base of the Carbon Canyon Member; locality cf. 1124.

1150 (ts) [T. B. M. field no. 27 of 9-19-85]. Carbonate from 61 m above the base of the Carbon Canyon Member; locality cf. 1124.

1151 [T. B. M. field no. 28 of 9-19-85]. Green shale from 8 m above the base of the Duppa Member of the Galeros Formation; locality cf. 1124.

1152 [T. B. M. field no. 29 of 9-19-85]. Green shale from 34 m above the base of the Duppa Member; locality cf. 1124.

1153 (TOC, Corg) [B. P. field no. BP1-9-19-85]. Shale from 12 m above the base of the Carbon Canyon Member of the Galeros Formation; locality cf. 1124.

1154 [B. P. field no. BP2-9-19-85]. Black shale from 8 m above the base of the Carbon Canyon Member; locality cf. 1124.

1155 (ts) [D. D. field no. D-91985-1]. Carbonate from 12 m above the base of the Jupiter Member of the Galeros Formation; locality cf. 1124.

Geologic Samples

Table 14.2. *Continued.*

1156 (ts, TOC, Corg) [D. D. field no. D-91985-2]. Light-gray dolomite with fine, dark-gray, silty laminae from 3 m above the base of the Carbon Canyon Member of the Galeros Formation; locality cf. 1124.

1157 [B. P. field no. BP1-9-20-85]. Shale from 6 m above the base of the Carbon Butte Member of the Kwagunt Formation; south slope of Carbon Butte, Grand Canyon National Park, Arizona, USA; 36.17°N, 111.8°W.

1158 [B. P. field no. BP2-9-20-85]. Red shale from 21 m above the base of the Carbon Butte Member; locality cf. 1157.

1159 [B. P. field no. BP-9-20-85]. Shale from 5 m above the base of the Carbon Butte Member; locality cf. 1157.

1160 (ts) [T. B. M. field no. 1 of 9-20-85]. Red sandstone from 5 m above the base of the Carbon Butte Member; locality cf. 1157.

1161 (ts) [T. B. M. field no. 2 of 9-20-85]. Sandstone from 14 m above the base of the Carbon Butte Member; locality cf. 1157.

1162 (ts) [D. D. field no. D-92085-1]. Carbonate from 24 m below a thick sandstone unit on the top of Carbon Butte, Duppa Member of the Galeros Formation; locality cf. 1157.

1163 (ts, st) [D. D. field no. D-92085-2]. Stromatolite from the base of the Awatubi Member of the Kwagunt Formation; locality cf. 1157.

1357 (ts, μf^*, TOC, Corg, ker) [J. W. S. field no. 1a-11/15/73]. Microfossiliferous (Chapter 24), pisolitic nodular chert, occurring in lenses, nodules are 3–10 cm thick from the upper of two pisolitic units, 200 m east of 1358 of the Walcott Member of the Kwagunt Formation; south slope of Nankoweap Butte, 2.5 km southeast of Nankoweap Mesa, 5 km east of Kwagunt Rapids of the Colorado River, Arizona, USA; 36.25°N, 111.9°W.

1358 (ts, TOC, Corg) [J. W. S. field no. 1b-11/15/73]. Pisolitic nodular chert from the upper of two pisolitic units, 22 m above the base of the Walcott Member; locality cf. 1357.

1359 (ts, st, μf^*, Corg) [J. W. S. field no. 2-11/16/73]. Microfossiliferous (Chapter 24), black chert, lenses and nodules interbedded with columnar and domical stromatolitic (*Baicalia*) carbonate from the middle third of the Carbon Canyon Member of the Galeros Formation; outcrop 2.4 km north of 1358, Arizona, USA.

COPPERMINE HOMOCLINE

Tectonic unit/location. Coppermine Homocline, Northwest Territories (District of Mackenzie), Canada.

Stratigraphy. Stratigraphic section, Coppermine area, northern District of Mackenzie (Baragar and Donaldson 1973):

RAE GROUP
COPPERMINE RIVER GROUP
DISMAL LAKES GROUP
HORNBY BAY GROUP

PPRG samples/localities.

1212 (ts, μf^*, TOC, Corg) [R. J. H. field no. A42a]. Microfossiliferous (Chapter 24), black and white, laminated chert from the Dismal Lakes Group; west of Dismal Lakes, Northwest Territories, Canada; 67.52°N, 117.7°W.

1213 (ts, TOC, Corg) [R. J. H. field no. A42f]. Chert from the Dismal Lakes Group; locality cf. 1212.

1214 (ts) [R. J. H. field no. A42g]. Chert from the Dismal Lakes Group; locality cf. 1212.

1215 (ts, TOC, Corg) [R. J. H. field no. A45c]. Chert from the Dismal Lakes Group; Dismal lakes, Northwest Territories, Canada; 67.52°N, 117.7°W.

1216 (ts, TOC, Corg) [R. J. H. field no. A45g]. Chert from the Dismal Lakes Group; locality cf. 1215.

1217 (ts, TOC, Corg, ker) [R. J. H. field no. A111a]. Black and white laminated chert from the Dismal Lakes Group; Dismal Lakes, Northwest Territories, Canada; 67.52°N, 117.6°W.

1218 (ts, TOC, Corg) [R. J. H. field no. A45h]. Chert from the Dismal Lakes Group; Dismal lakes, Northwest Territories, Canada; 67.52°N, 117.7°W.

1873 (st) [G. M. R. field no. GBO-86]. Chert in calcareous stromatolite from "Member E3, East River Fm." (informal name) of the Hornby Bay Group; collected from a Quaternary outwash channel incised into the East River Fm., north of Great Bear Lake and west of the west end of Dismal Lakes, Northwest Territories, Canada; Bebensee Lake map, National Topographic Map Series 860; 67.20°N, 118.0°W.

2358 (ts, μf^*) [R. J. H. field no. A-42b]. Microfossiliferous (Chapter 24) chert from the Dismal Lakes Group; west of Dismal Lakes, Northwest Territories, Canada; 67.52°N, 117.7°W.

2359 (ts) [R. J. H. field no. A42h]. Chert from the Dismal Lakes Group; locality cf. 2358.

2360 (ts) [R. J. H. field no. A-111i]. Black chert from the Dismal Lakes Group; Dismal Lakes, Northwest Territories, Canada; 67.52°N, 117.6°W.

LAKE SUPERIOR BASIN (NORTHERN PENINSULAR MICHIGAN)

Tectonic unit/location. Lake Superior Basin (Canadian Southern Province), Michigan, USA.

Stratigraphy. Stratigraphic section, Marquette-White Pine area, northern peninsular Michigan, USA (Cannon and Gair 1970):

Table 14.2. *Continued.*

KEWEENAWAN GROUP
    Freda Sandstone
    Nonesuch Shale
    Copper Harbor Conglomerate
    Portage Lake Lavas
MARQUETTE RANGE SUPERGROUP
  PAINT RIVER GROUP
  BARAGA GROUP
    Badwater Greenstone
    Tyler Formation/Michigamme Slate
    Amasa Formation
    Hemlock Formation
    Goodrich Quartzite
  MENOMINEE GROUP
  CHOCOLAY GROUP
    Wewe Slate
    Kona Dolomite/Randville Dolomite
    Mesnard Quartzite
    Enchantment Lake Formation

PPRG samples/localities.

1362 (ts, st, TOC, Corg) [J. W. S. field no. 1-6/8/83]. Carbonate stromatolite (*Collenia kona*) from the Kona Dolomite; south end of the Sands Quadrangle Quarry, 1.1 km east of Ragged Hills, 2 km south of Carp River Lake, 7.3 km northwest of Sands, and about 10 km southwest of Marquette, Michigan, USA; U.S.G.S. 1:24,000 Sands Quad., Michigan.

1363 (ts, μf^*, TOC, Corg, ker) [J. W. S. field no. 1-6/9/83]. Microfossiliferous (Chapter 24) granular black chert from the "chert-carbonate-iron facies," 1 m above 1364, upper chert of the Tyler Formation; on the west bank of the Black River, 50 m south of US Highway No. 2 bridge, Michigan, USA.

1364 (ts, TOC, Corg, ker) [J. W. S. field no. 2-6/9/83]. Chert from "lower chert horizon," 1 m below 1363, of the Tyler Formation; locality cf. 1363.

1365 (ts, TOC, Corg) [J. W. S. field no. 3-6/9/83]. Anthraxolite from carbonates (cf. 1363 facies) from the Tyler Formation; locality cf. 1363.

1366 (ts, TOC, Corg) [J. W. S. field no. 4-6/9/83]. Massive gray siltstone from the Parting Shale Member, lower third (bed 25) of the Nonesuch Shale; on the east bank of the Big Iron River at Iron River Falls, 1.3 km south of Lake Superior and 6.5 km north of White Pine Copper Mine, Ontonagon County, Michigan, USA.

1367 (ts, TOC, Corg) [J. W. S. field no. 5-6/9/83]. Chalcocite and native copper in gray shale from the Parting Shale Member, lower third (beds 21 and 22) of the Nonesuch Shale; White Pine Copper Mine, at White Pine, Ontonagon County, Michigan, USA.

1368 (ts, TOC, Corg) [J. W. S. field no. 1-6/11/83]. Quartzite from the Pokegama Quartzite; across the highway from the Holiday Inn, Eveleth, Minnesota, USA.

1369 (ts, TOC, Corg) [J. W. S. field no. 2-6/11/83]. Quartzite from the Pokegama Quartzite; locality cf. 1368.

1370 (ts, TOC, Corg) [J. W. S. field no. 3-6/11/83]. Quartzite from the Pokegama Quartzite; 1368.

1371 (ts, st, TOC, Corg) [J. W. S. field no. 4-6/11/83]. Chert from the Biwabik Iron Formation; Mary Ellen Mine in the city of Virginia, Minnesota, USA.

1372 (ts, TOC, Corg) [J. W. S. field no. 5-6/11/83]. Chert from the Tower Soudan Mine at 796 m depth at the mine shaft entrance, Soudan Iron Formation; Tower Soudan Mine, Minnesota, USA.

1373 (ts) [J. W. S. field no. 6-6/11/83]. Graphitic schist with chlorite from the Tower Soudan Mine, level 28 at spiral staircase, Soudan Iron Formation; locality cf. 1372.

2422 (ts, TOC, Corg) [J. M. H. field no. 830609-1a/B599]. Black, micaceous siltstone from the Upper Shale Member, at ca. bed 41–43 of the Nonesuch Shale; ca. 100 m south of parking lot, on the east bank of the Big Iron River at Iron River Falls, 1.3 km south of Lake Superior and 6.5 km north of White Pine Copper Mine, Ontonagon County, Michigan, USA.

2423 (ts, TOC, Corg) [J. M. H. field no. 830609-1b/B600]. Black siltstone from the Upper Shale Member of the Nonesuch Shale; 2 m below top of waterfall, locality cf. 2422.

2424 (TOC, Corg) [J. M. H. field no. 830609-1c/B601]. Black siltstone from the Upper Shale Member of the Nonesuch Shale; at the waterfall, locality cf. 2422.

2425 (TOC, Corg, ker) [J. M. H. field no. 830609-2a/B602]. Black shale with native copper from the lower transition and domino beds (21 and 23) of the Nonesuch Shale; White Pine Copper Mine, at White Pine, Ontonagon County, Michigan, USA.

2426 (ts, TOC, Corg) [J. M. H. field no. 830609-3a/B603]. Black chert with pyrite veins from the Nonesuch Shale; "Black River locality," Michigan, USA.

Table 14.2. *Continued.*

2427 (ts, TOC, Corg) [J. M. H. field no. 830609-3b/B604]. Brown carbonate from the Nonesuch Shale; locality cf. 2426.

2428 (ts, TOC, Corg) [J. M. H. field no. 830609-3d/B605]. Anthraxolite from the Nonesuch Shale; locality cf. 2426.

2429 (ts, TOC, Corg) [J. M. H. field no. 830610-1a/B606]. Banded iron-formation with pyrite nodules from the Biwabik Iron Formation; Mary Ellen Mine in the city of Virginia, Minnesota, USA.

2430 (ts, TOC, Corg) [J. M. H. field no. 830611-2a/B607]. Black chert from the Biwabik Iron Formation; locality cf. 2429.

2431 (ts, TOC, Corg) [J. M. H. field no. 830611-3a/B608]. Well-laminated chert and hematite from the upper cherty cycle of the Biwabik Iron Formation; locality cf. 2429.

2432 (ts, st, TOC, Corg) [J. M. H. field no. 830611-3b/B609]. *Collenia* stromatolites, closely associated with hematite from the Biwabik Iron Formation; locality cf. 2429.

2433 (ts, TOC, Corg) [J. M. H. field no. 830611-4a/B610]. Black chert from the Soudan Iron Formation; 27th level, near shaft of the Tower Soudan Mine, Minnesota, USA.

2434 (ts, TOC, Corg) [J. M. H. field no. 830611-4b/B611]. Graphitic, chlorite schist from the Soudan Iron Formation; shear zone near ore works on 27th level, locality cf. 2433.

2435 (ts, TOC, Corg) [J. M. H. field no. 830611-4c/B612]. Chert from the Soudan Iron Formation; locality cf. 2433.

MACKENZIE FOLD BELT

Tectonic unit/location. Mackenzie Fold Belt, Yukon Territory and Northwest Territories (District of Mackenzie), Canada.

Stratigraphy. Stratigraphic section, Mackenzie Mountains area, western District of Mackenzie (Young et al. 1979):

EKWI SUPERGROUP
   [UNNAMED UPPER PROTEROZOIC GROUP]
      Sheepbed Formation
   HAY CREEK GROUP
      Keele Formation
      Twitya Formation
   RAPITAN GROUP
      Shezal Formation
      Sayunei Formation
MACKENZIE MOUNTAINS SUPERGROUP
   [UNNAMED MIDDLE PROTEROZOIC GROUP]
      Coppercap Formation
      Redstone River Formation
   LITTLE DAL GROUP
   KATHERINE GROUP

PPRG samples/localities.

1589 [PPRG field no. 86-07-01-81]. Banded iron-formation (BIF), possibly including hematite and siltstone chips from the top of the BIF unit of the Shezal Formation; "Waterfall section," at Iron Creek and tributary to North Iron River, Northwest Territories, Canada; Yukon Territory 1:50,000 topographic sheet 106/6 (ed. 1); 65.24°N, 113.0°W.

1590 [PPRG field no. 86-07-01-82]. Red siltstone from stratigraphically immediately above 1589, Shezal Formation; locality cf. 1589.

1591 [PPRG field no. 86-07-01-83]. Hematite-rich sample with rounded grains or clasts from ca. 10 m below 1589, Rapitan iron-formation "member" of the Shezal Formation; locality cf. 1589.

1592 (ts) [PPRG field no. 86-07-01-84]. Very well-banded jasper and hematite from ca. 20 m below 1591, Rapitan iron-formation "member" of the Shezal Formation; locality cf. 1589.

1593 (ts) [PPRG field no. 86-07-01-95]. Nodular chert and hematite with jasper lenses from the Shezal Formation; locality cf. 1589.

1594 [PPRG field no. 86-07-01-86]. Banded hematite and jasper from the Shezal Formation; locality cf. 1589.

1595 [PPRG field no. 86-07-01-87]. Composite sample, with nodular top(?) and banded at bottom(?), from float collected by the creek, Shezal Formation; locality cf. 1589.

1596 [PPRG field no. 86-07-01-88]. Greenish-gray diamictite from the top of Iron Creek, far above the "Waterfall section," Shezal Formation; locality cf. 1589.

1597 [PPRG field no. 86-07-01-89]. Sandstone with jasper fragments that possibly represent reworked BIF in a sandstone (lithic) regime from further down-section from 1596, Shezal Formation; locality cf. 1589.

1598 [PPRG field no. 86-07-01-90]. Coarse- and fine-grained sandstone, possibly stratigraphically equivalent to 1589 and 1590, from the Shezal Formation; locality cf. 1589.

Table 14.2. Continued.

1599 (ts, TOC, Corg) [PPRG field no. 86-07-01-91]. Fine-grained sandstone from the Shezal Formation; locality cf. 1589.

1600 [PPRG field no. 86-07-01-92]. Jasper banded iron-formation from somewhere stratigraphically between 1591 and 1592, Shezal Formation; locality cf. 1589.

1601 [PPRG field no. 86-07-01-93]. Neptunian, black to dark-gray, sandstone dike from the Shezal Formation; locality cf. 1589.

1602 [PPRG field no. 86-07-01-94]. Nodular banded iron-formation from between two sandstone sills of the Shezal Formation; locality cf. 1589.

1603 86-07-01-95]. Iron-rich mudstone from just above mixtite band, Shezal Formation; locality cf. 1589.

1604 [PPRG field no. 86-07-01-96]. Banded iron-formation with carbonate nodules from the Shezal Formation; locality cf. 1589.

1605 [PPRG field no. 86-07-01-97]. Nodular banded iron-formation from the Shezal Formation; locality cf. 1589.

1606 [PPRG field no. 86-07-01-98]. Podded banded iron-formation from the Shezal Formation; locality cf. 1589.

1607 [PPRG field no. 86-07-01-99]. Nodular jasper from the Shezal Formation; locality cf. 1589.

1608 (ts, TOC, Corg) [PPRG field no. 86-07-02-100]. Thin bedded dark gray carbonate, non-stromatolitic but laminated, Little Dal Group; directly opposite the PPRG day 7 camp at the Snake River, Yukon Territory, Canada; Snake River 1:250,000 Geologic map 1529A; 65.37°N, 133.3°W.

1609 (ts, TOC, Corg, Sfide) [PPRG field no. 86-06-02-101]. Gray, fissile calcareous shale from further up in section than 1608, Little Dal Group; just east of, locality cf. 1608.

1610 (ts, TOC, Corg) [PPRG field no. 86-07-02-102]. Gray, laminated carbonate from immediately above 1609, Little Dal Group; directly east of, locality cf. 1608.

1611 (TOC, Corg) [PPRG field no. 86-07-02-103]. Medium green-gray shale from the Twitya Formation; ca. 16 km east of the Snake River, Yukon Territory, Canada; Snake River 1:250,000 Geologic map 1529A; 65.10°N, 132.8°W.

1612 (TOC, Corg) [PPRG field no. 86-07-02-104]. Medium green-gray shale from the upper third of the Twitya Formation; locality cf. 1611.

1613 (TOC, Corg, ker) [PPRG field no. 06-07-02-105]. Black shale from the middle of the Twitya Formation; locality cf. 1611.

1614 (TOC, Corg, Sfide) [PPRG field no. 86-07-02-106]. Laminated, fissile black shale from the lower part of the Sheepbed Formation; ca. 9 km east of the Snake River, Yukon Territory, Canada; Snake River 1:250,000 Geologic map 1529A; 65.07°N, 132.7°W.

1615 (Corg, Sfide) [PPRG field no. 86-07-02-107]. Laminated, fissile, black shale from the lowermost part of the Sheepbed Formation; locality cf. 1614.

1616 (ts, TOC, Corg) [PPRG field no. 86-07-02-108]. Wavy laminated, dark-gray carbonate from the uppermost part of the Keele Formation; locality cf. 1614.

1617 (ts, TOC, Corg) [PPRG field no. 86-07-02-109]. Wavy laminated, dark-gray carbonate from 3 m below 1616, Keele Formation; locality cf. 1614.

1618 (ts, TOC, Corg) [PPRG field no. 86-07-02-110]. Dark-gray micaceous, laminated, siltstone with red-brown mudstone and medium grained sandstone, from 5 m below 1617, Keele Formation; locality cf. 1614.

1619 (ts, TOC, Corg) [PPRG field no. 86-07-02-111]. Thinly bedded, flat-laminated, black carbonate from 5 m below 1618, Keele Formation; locality cf. 1614.

1620 (ts, TOC, Corg) [PPRG field no. 86-07-02-112]. Wavy laminated, medium bedded, black carbonate from 7 m below 1619, Keele Formation; locality cf. 1614.

1621 (ts, TOC, Corg) [PPRG field no. 86-07-02-113]. Wavy laminated, black carbonate from 7 m below 1620, Keele Formation; locality cf. 1614.

1622 (ts, TOC, Corg) [PPRG field no. 86-07-02-114]. Wavy laminated, black carbonate from 8 m below 1621, Keele Formation; locality cf. 1614.

1623 (ts, TOC, Corg) [PPRG field no. 86-07-02-115]. Wavy laminated, black carbonate from 10 m below 1622, Keele Formation; locality cf. 1614.

1624 (ts, TOC, Corg) [PPRG field no. 86-07-02-116]. Dark-gray bedded, poorly laminated carbonate with many stylolites, from very close to the base of the Keele Formation; locality cf. 1614.

1625 (ts) [PPRG field no. 86-07-03-117]. Diamictite (mixtite), maroon matrix with glacial (reworked) pebbles, Shezal Formation; ca. 30 km east of the Snake River, Yukon Territory, Canada; Snake River 1:250,000 Geologic map 1529A; 65.30°N, 132.7°W.

1626 (ts) [PPRG field no. 86-07-03-118]. Very coarse-grained to conglomeratic sandstone from the very top of the underlying banded iron-formation, Shezal Formation; locality cf. 1625.

1627 (ts) [PPRG field no. 86-07-03-119]. Banded and nodular iron-formation from the Shezal Formation; locality cf. 1625.

1628 (ts) [PPRG field no. 86-07-03-120]. Coarse-grained sandstone with jasper clasts from below an overlying sandstone, Shezal Formation; locality cf. 1625.

1629 (ts) [PPRG field no. 86-07-03-121]. Disc-rhythmite from banded iron-formation in a loose boulder in creek, Shezal Formation; "Waterfall section," at Iron Creek and tributary to North Iron River, Northwest Territories, Canada; Yukon Territory 1:50,000 topographic sheet 106/6 (ed. 1); 65.24°N, 113.0°W.

1630 (ts) [PPRG field no. 86-07-03-122]. Carbonate pisolites from a loose boulder in creek, Ekwi Supergroup; locality cf. 1629.

Table 14.2. *Continued.*

1631 (ts) [PPRG field no. 86-07-03-123]. Carbonate clasts in banded iron-formation from a loose boulder in creek, Ekwi Supergroup; locality cf. 1629.

1632 (ts) [PPRG field no. 86-07-03-124]. Intraclastic iron-formation from a loose boulder in creek, Ekwi Supergroup; locality cf. 1629.

1634 (ts) [PPRG field no. 86-07-03-125]. Hematite rhythmite in banded iron-formation from a loose boulder in creek, Ekwi Supergroup; locality cf. 1629.

1635 (ts, st, TOC) [PPRG field no. 86-07-05-126]. Black chert from a 10 cm-thick irregularly tabular bed interbedded in stromatolitic carbonate of the Little Dal Group; on hilltop just west of Lukas Creek, Keele River area, Northwest Territories, Canada; Wrigley Lake 1:250,000 Geologic map 1315A; 63.73°N, 127.3°W.

1636 (ts, st, TOC, Corg) [PPRG field no. 86-07-05-127]. Dolomitic, calcareous shale with pseudocolumnar stromatolites, from the Little Dal Group; locality cf. 1635.

1637 (ts, TOC, Corg) [PPRG field no. 86-07-05-128]. Black shale from float of the Little Dal Group; locality cf. 1635.

1638 (ts, TOC, Corg) [PPRG field no. 86-07-05-129]. Fissile, flat-laminated, gray carbonate from the lower portion of the Little Dal Group; locality cf. 1635.

1639 (ts, TOC, Corg) [PPRG field no. 86-07-05-130]. Fissile, thin-bedded, gray carbonate from above the intraclastic grainstone of 1638, Little Dal Group; locality cf. 1635.

1640 (ts, TOC, Corg) [PPRG field no. 86-07-05-131]. Flat-laminated, thin-bedded, gray carbonate from the lower part of the Little Dal Group; locality cf. 1635.

1641 (ts, TOC, Corg) [PPRG field no. 86-07-05-132]. Flat-laminated, bedded, gray carbonate with thin tabular gray chert, from the Little Dal Group; locality cf. 1635.

1642 (ts, TOC, Corg, ker, Sfide) [PPRG field no. 86-07-05-133]. Flat-laminated, black shale with numerous angular pits after pyrite, from 50 m above the base of the Sheepbed Formation; top of a plateau, 4 km north of North River, Keele River area, Northwest Territories, Canada; 63.62°N, 127.6°W.

1643 (ts, TOC, Corg, ker) [PPRG field no. 86-07-05-134]. Flat-laminated, black shale with numerous pits after pyrite, from 10 m above 1642, Sheepbed Formation; locality cf. 1642.

1644 (ts, TOC, Corg) [PPRG field no. 86-07-05-135]. Flat-laminated, black shale with numerous pits after pyrite, from 20 m above 1643, Sheepbed Formation; locality cf. 1642.

1645 (ts, TOC, Corg, ker, Sfide) [PPRG field no. 86-07-05-136]. Fissile black shale, with pyrite pits, from 25 m above 1644, Sheepbed Formation; locality cf. 1642.

1646 (ts, TOC, Corg, ker) [PPRG field no. 86-07-05-137]. Black shale from ca. 20 m above 1645, Sheepbed Formation; locality cf. 1642.

1647 (ts, TOC, Corg) [PPRG field no. 86-07-05-138]. Fissile black shale with pyrite pits, from ca. 20 m above 1646, Sheepbed Formation; locality cf. 1642.

1648 (ts, TOC, Corg) [PPRG field no. 86-07-05-139]. Flat-laminated black shale with numerous pits after pyrite, from ca. 10 m above 1647, Sheepbed Formation; locality cf. 1642.

1649 (ts) [PPRG field no. 86-07-05-140]. Fissile black shale with pyrite pits, from ca. 10 m above 1648, Sheepbed Formation; locality cf. 1642.

1650 (ts, TOC, Corg, ker) [PPRG field no. 86-07-05-141]. Fissile black shale with pyrite pits, from ca. 15 m above 1649, Sheepbed Formation; locality cf. 1642.

1651 (ts, TOC, Corg) [PPRG field no. 86-07-05-142]. Flat-laminated, thin bedded, light-brown carbonate, from ca. 15 m above 1650, Keele Formation; Keele River area, Northwest Territories, Canada; Wrigley Lake 1:250,000 Geologic map 1315A; 63.63°N, 127.4°W.

1652 (ts, TOC, Corg, Sfide) [PPRG field no. 86-07-05-143]. Fissile, flat-laminated, slightly micaceous black shale, from the base of the Twitya Formation; Keele River area, Northwest Territories, Canada; Wrigley Lake 1:250,000 Geologic Map 1315A; 63.72°N, 127.4°W.

1653 (ts, TOC, Corg, ker Sfide) [PPRG field no. 86-07-05-144]. Fissile, slightly micaceous, black shale with silty laminae, from ca. 20 m above 1652, Twitya Formation; locality cf. 1652.

1654 (ts, TOC, Corg, ker, Sfide) [PPRG field no. 86-07-05-145]. Fissile, slightly micaceous black shale, from ca. 20 m above 1653, Twitya Formation; locality cf. 1652.

1655 (ts, TOC, Corg, ker Sfide) [PPRG field no. 86-07-05-146]. Fissile, slightly micaceous black shale, from ca. 20 m above 1654, Twitya Formation; locality cf. 1652.

1656 (ts, TOC) [PPRG field no. 86-07-05-147]. Flat laminated gray carbonate, from ca. 25 m above 1655, Twitya Formation; locality cf. 1652.

1657 (ts, TOC, Corg) [PPRG field no. 86-07-05-148]. Fissile, micaceous gray shale, from ca. 20 m above 1656, Twitya Formation; locality cf. 1652.

1658 (ts, TOC, Corg) [PPRG field no. 86-07-05-149]. Fissile, micaceous gray shale, from ca. 15 m above 1657, Twitya Formation; locality cf. 1652.

1659 (ts, TOC, Corg) [PPRG field no. 86-07-05-150]. Fissile, micaceous, gray shale with some siltstone, from ca. 15 m above 1658, Twitya Formation; locality cf. 1652.

Table 14.2. *Continued.*

1660 (ts, TOC, Corg) [PPRG field no. 86-07-05-151]. Fissile, flat-laminated black shale, from ca. 15 m above 1659, Twitya Formation; locality cf. 1652.

1661 (ts, TOC, Corg) [PPRG field no. 86-07-05-152]. Fissile, flat-laminated black shale, from ca. 20 m above 1660, Twitya Formation; locality cf. 1652.

1662 (ts) [PPRG field no. 86-07-05-153]. Poorly laminated, thin bedded, white and pink gypsum from the middle of the Redstone River Formation; 1 km east of Keele River camp and airstrip, Northwest Territories, Canada; Wrigley Lake 1:250,000 Geologic map 1315A; 63.82°N, 127.8°W.

1663 (ts, Sfate) [PPRG field no. 86-07-05-154]. Silty gypsum with quartz granules from the Redstone River Formation; locality cf. 1662.

1664 (ts) [PPRG field no. 86-07-05-155]. Thin bedded, crudely laminated, medium red silty gypsum from the Redstone River Formation; locality cf. 1662.

1665 (ts, Sfate) [PPRG field no. 86-07-05-156]. White anhydrite(?) from the Redstone River Formation; locality cf. 1662.

1666 (ts, TOC, Corg) [PPRG field no. 86-07-05-157]. Laminated, light-gray carbonate with thin chert layers and lenses, from the lower part of the Coppercap Formation; locality cf. 1662.

1667 (ts, TOC, Corg) [PPRG field no. 86-07-05-158]. Gray carbonate with chert lenses from ca. 1.5 m above 1666, Coppercap Formation; locality cf. 1662.

1668 (ts, TOC, Corg) [PPRG field no. 86-07-05-159]. Laminated dark-gray carbonate from ca. 10 m above 1667, Coppercap Formation; locality cf. 1662.

1669 (ts, TOC, Corg, Sfide) [PPRG field no. 86-07-05-160]. Gray carbonate with copper veins ca. 20 cm thick, from the Coppercap Formation; locality cf. 1662.

1670 (ts, TOC, Corg) [PPRG field no. 86-07-05-161]. Dark-gray carbonate with thin shale layers from the Coppercap Formation; locality cf. 1662.

1671 (ts, TOC, Corg) [PPRG field no. 86-07-05-162]. Light-gray carbonate interbedded with thin shale from the Coppercap Formation; locality cf. 1662.

1672 (ts, TOC, Corg) [PPRG field no. 86-07-05-163]. Thin bedded, wavy laminated, light-gray carbonate with light- and dark-gray chert layers, from the upper part of the Little Dal Group; southwest of Lukas Creek, Keele River area, Northwest Territories, Canada; 63.77°N, 127.4°W.

1673 (ts, TOC, Corg) [PPRG field no. 86-07-05-164]. Wavy laminated, very fine-grained, black chert from 2 m below 1672, Little Dal Group; locality cf. 1672.

1674 (ts, TOC, Corg) [PPRG field no. 86-07-05-166]. Laminated dark-gray shaley carbonate from 3 m below 1673, Little Dal Group; locality cf. 1672.

1675 (ts, TOC, Corg) [PPRG field no. 86-07-05-166]. Thin bedded, laminated, gray carbonate from 3 m below 1674, Little Dal Group; locality cf. 1672.

1676 (ts) [PPRG field no. 86-07-05-167]. Gray cherty carbonate from 2 m below 1675, Little Dal Group; locality cf. 1672.

1677 (ts) [PPRG field no. 86-07-05-168]. Brown dolomite with desiccation cracks from ca. 10 m below 1676, Little Dal Group; locality cf. 1672.

1678 (ts) [PPRG field no. 86-07-05-169]. Thin bedded, dark red-brown shale with thin carbonate beds from ca. 5 m below a basalt bed, Little Dal Group; locality cf. 1672.

1679 (ts, st) [PPRG field no. 86-07-05-170]. Stromatolitic, light green-gray carbonate with pebbles of black chert, from ca. 20–30 m below a basalt bed, Little Dal Group; locality cf. 1672.

1680 (ts, st, TOC) [PPRG field no. 86-07-05-171]. Dark-gray carbonate, small domical stromatolites, from ca. 50 m below basalt bed, Little Dal Group; locality cf. 1672.

1681 (ts, TOC, Corg) [PPRG field no. 86-07-05-172]. Non-stromatolitic, laminated, nodular, gray carbonate from ca. 100 m stratigraphically below a basalt bed, Little Dal Group; locality cf. 1672.

1682 (ts, TOC, Corg) [PPRG field no. 86-07-05-173]. Thin bedded, black shale from ca. 150 m below a basalt bed, Little Dal Group; locality cf. 1672.

1683 (ts, TOC, Corg) [PPRG field no. 86-07-05-174]. Ferruginous shale from the Sayunei Formation; on hilltop just west of Lukas Creek, Keele River area, Northwest Territories, Canada; Wrigley Lake 1:250,000 Geologic map 1315A; 63.73°N, 127.3°W.

1684 (ts) [PPRG field no. 86-07-05-175]. Ferruginous sandstone from the Sayunei Formation; locality cf. 1683.

1685 (ts, TOC, Corg) [PPRG field no. 86-07-05-176]. Fissile ferruginous, slaty shale from above 1684, Sayunei Formation; locality cf. 1683.

1686 (ts, TOC) [PPRG field no. 86-07-05-177]. Finely laminated, ferruginous mudstone from the Sayunei Formation; locality cf. 1683.

1687 (ts) [PPRG field no. 86-07-05-178]. Ferruginous diamictite from the top of a mudstone sequence of the Sayunei Formation; locality cf. 1683.

1688 (ts) [PPRG field no. 86-07-05-179]. Graded sandstone from part of the lower sandstone and conglomerate unit of the Sayunei Formation; locality cf. 1683.

1689 (ts) [PPRG field no. 86-07-05-180]. Rippled carbonate (lacustrine?) from the lower shale unit of the Sayunei Formation; locality cf. 1683.

1690 (ts) [PPRG field no. 86-07-05-181]. Graded siltstone from the lower shale unit of the Sayunei Formation; locality cf. 1683.

Table 14.2. *Continued.*

1691 (ts) [PPRG field no. 86-07-05-182]. Mudchip sandstone from the lower shale unit of the Sayunei Formation; locality cf. 1683.

1692 (ts, TOC) [PPRG field no. 86-07-05-183]. Varved shale from the lower shale unit of the Sayunei Formation; locality cf. 1683.

1693 (ts, TOC) [PPRG field no. 86-07-05-184]. Very finely varved shale from the lower shale unit of the Sayunei Formation; locality cf. 1683.

1694 (ts) [PPRG field no. 86-07-05-185]. Varved shale with dropstones from the lower shale unit of the Sayunei Formation; locality cf. 1683.

1695 (ts) [PPRG field no. 86-07-05-186]. Cross-laminated quartz sandstone from the central sandstone unit of the Sayunei Formation; locality cf. 1683.

1696 (ts) [PPRG field no. 86-07-05-187]. Calcareous sandstone from the central sandstone unit of the Sayunei Formation; locality cf. 1683.

1697 (ts) [PPRG field no. 86-07-05-188]. Clastic textured limestone (marine?) from the upper shale unit of the Sayunei Formation; locality cf. 1683.

1698 (ts, TOC) [PPRG field no. 86-07-05-189]. Thickly laminated shale from the upper shale unit of the Sayunei Formation; locality cf. 1683.

1699 (ts, TOC) [PPRG field no. 86-07-05-190]. Thinly laminated shale from the upper shale unit of the Sayunei Formation; locality cf. 1683.

1700 (ts) [PPRG field no. 86-07-05-191]. Clastic textured limestone from the upper shale unit of the Sayunei Formation; locality cf. 1683.

1701 (ts) [PPRG field no. 86-07-05-192]. Clastic textured limestone from the upper shale unit of the Sayunei Formation; locality cf. 1683.

1702 (ts) [PPRG field no. 86-07-05-193]. Microbanded jasper from the iron-rich unit of the Sayunei Formation; locality cf. 1683.

1703 (ts) [PPRG field no. 86-07-05-194]. Limestone concretion from the iron-rich unit of the Sayunei Formation; locality cf. 1683.

1704 (ts) [PPRG field no. 86-07-05-195]. Iron-rich shale from the iron-rich unit of the Sayunei Formation; locality cf. 1683.

1705 (ts) [PPRG field no. 86-07-05-196]. Jasper pillow with syneresis cracks from the iron-rich unit of the Sayunei Formation; locality cf. 1683.

1706 (ts) [PPRG field no. 86-07-05-197]. Laminated, iron-rich limestone from the iron-rich unit of the Sayunei Formation; locality cf. 1683.

1707 (ts) [PPRG field no. 86-07-05-198]. Carbonate breccia or conglomerate (intraformational breccia) from the iron-rich unit of the Sayunei Formation; locality cf. 1683.

1708 (ts) 86-07-06-220]. Light-gray carbonate from a core sample, at a depth of 15.6 m, Coppercap Formation; Shell Canada Limited Diamond Drill Hole 3798F/12; core box located at the Shell Airstrip on the Keele River, Northwest Territories, Canada.

1709 (ts, TOC, Corg) [PPRG field no. 86-07-06-211]. Flat-laminated, dark-gray carbonate from a core sample, at a depth of 30.8 m, Coppercap Formation; locality cf. 1708.

1710 (ts) [PPRG field no. 86-07-06-222]. Dark-gray carbonate with some carbonate veins, from a core sample, at a depth of 45.8 m, Coppercap Formation; locality cf. 1708.

1711 (ts) [PPRG field no. 86-07-06-223]. Thinly bedded, dark-gray carbonate with one carbonate vein, from a core sample, at a depth of 59.8 m, Coppercap Formation; locality cf. 1708.

1712 (ts, TOC, Corg) [PPRG field no. 86-07-06-224]. Thinly bedded, laminated dark-gray carbonate from core sample, at a depth of 75.3 m, Coppercap Formation; locality cf. 1708.

1713 (ts) [PPRG field no. 86-07-06-225]. Dark-gray carbonate from a core sample, at a depth of 91.2 m, Coppercap Formation; locality cf. 1708.

1714 (ts) [PPRG field no. 86-07-06-226]. Dark-gray carbonate from a core sample, at a depth of 107.4 m, Coppercap Formation; locality cf. 1708.

1715 (ts) [PPRG field no. 86-07-06-227]. Dark-gray carbonate from a core sample, at a depth of 122 m, Coppercap Formation; locality cf. 1708.

1716 (ts, TOC, Corg) [PPRG field no. 86-07-06-228]. Dark-gray, laminated, carbonate with some carbonate veins, from a core sample at a depth of 137.3 m, Coppercap Formation; locality cf. 1708.

1717 (ts) [PPRG field no. 86-07-06-229]. Dark-gray carbonate from a core sample, at a depth of 90.9 m, Coppercap Formation; locality cf. 1708.

1718 (ts) [PPRG field no. 86-07-06-230]. Laminated, dark-gray carbonate from a core sample, at a depth of 169 m, Coppercap Formation; locality cf. 1708.

1719 (ts) [PPRG field no. 86-07-06-231]. Laminated, dark-gray carbonate from a core sample, at a depth of 179 m, Coppercap Formation; locality cf. 1708.

1720 (ts, TOC) [PPRG field no. 86-07-06-232]. Light-brown carbonate from a core sample, at a depth of 195.8 m, Coppercap Formation; locality cf. 1708.

1721 (ts) [PPRG field no. 86-07-06-233]. Thinly bedded, laminated, light-brown to gray carbonate from a core sample, at a depth of 202.2 m, Coppercap Formation; locality cf. 1708.

1722 (ts) [PPRG field no. 86-07-06-234]. Thinly bedded, laminated, light-brown to gray carbonate from a core sample, at a depth of 213.2 m, Coppercap Formation; locality cf. 1708.

1723 (ts) [PPRG field no. 86-07-06-235]. Laminated, light-brown to gray carbonate from a core sample, at a depth of 221.7 m, Coppercap Formation; locality cf. 1708.

1724 (ts) [PPRG field no. 86-07-06-236]. Laminated, dark-gray carbonate from core sample, at a depth of 232.7 m, Coppercap Formation; locality cf. 1708.

Table 14.2. *Continued.*

1725 (ts) [PPRG field no. 86-07-06-237]. Dark-gray carbonate with some carbonate veins from a core sample, at a depth of 244.6 m, Coppercap Formation; locality cf. 1708.

1726 (ts, TOC, Corg) [PPRG field no. 86-07-06-238]. Laminated, dark-gray carbonate from a core sample, at a depth of 261 m, Coppercap Formation; locality cf. 1708.

1727 (ts, TOC, Corg) [PPRG field no. 86-07-06-239]. Thinly bedded, laminated, dark-gray shale and light gray-brown carbonate from a core sample, at a depth of 272 m, Coppercap Formation; locality cf. 1708.

1728 (ts, TOC, Corg) [PPRG field no. 86-07-06-240]. Gray shale with gray-brown carbonate from a core sample, at a depth of 278.2 m, Coppercap Formation; locality cf. 1708.

1729 (ts) [PPRG field no. 86-07-06-241]. Thinly bedded, laminated gray shale from a core sample, at a depth of 290.1 m, Coppercap Formation; locality cf. 1708.

1730 (ts) [PPRG field no. 86-07-06-242]. Thin bedded, laminated, gray shale from a core sample, at a depth of 301.3 m, Coppercap Formation; locality cf. 1708.

1731 (ts) [PPRG field no. 86-07-06-243]. Laminated, dark-gray carbonate from a core sample, at a depth of 289.4 m, Coppercap Formation; locality cf. 1708.

1732 (ts) [PPRG field no. 86-07-06-244]. Laminated, dark-gray carbonate from a core sample, at a depth of 305 m, Coppercap Formation; locality cf. 1708.

1733 (ts) [PPRG field no. 86-07-06-245]. Dark-gray shaley carbonate from a core sample, at a depth of 320.3 m, Coppercap Formation; locality cf. 1708.

1734 (ts, TOC, Corg) [PPRG field no. 86-07-06-246]. Laminated, dark-gray shale and carbonate from a core sample, at a depth of 336.7 m, Coppercap Formation; locality cf. 1708.

1735 (ts, TOC, Corg) [PPRG field no. 86-07-06-247]. Dark-gray carbonate and shale with carbonate veins from a core sample, at a depth of 337.6 m, Coppercap Formation; locality cf. 1708.

1736 (ts, TOC, Corg) [PPRG field no. 86-07-06-248]. Gray shale and carbonate with carbonate veins from a core sample, at a depth of 341.0 m, Coppercap Formation; locality cf. 1708.

1737 (ts, TOC, Corg) [PPRG field no. 86-07-06-249]. Light-gray carbonate with sulfides from a core sample, at a depth of 353.8 m, Coppercap Formation; locality cf. 1708.

1738 (ts, TOC, Corg) [PPRG field no. 86-07-06-250]. Laminated dark-gray carbonate from a core sample, at a depth of 351.1 m, Coppercap Formation; locality cf. 1708.

1739 (ts, TOC) [PPRG field no. 86-07-06-251]. Light-gray carbonate from a core sample at a depth of 366.0 m, Coppercap Formation; locality cf. 1708.

1740 [PPRG field no. 86-07-06-252]. Gray silty gypsum from a core sample, at a depth of 147.5 m, Redstone River Formation; Shell Canada Limited Diamond Drill Hole 3698F/6, from the northeast slope of Fortress Mountain, Keele River area; core box located at the Shell Airstrip on the Keele River, Northwest Territories, Canada; Vanishing Ram Creek 1:50,000 topographic map 95M/13; 63.83°N, 127.9°W.

1741 [PPRG field no. 86-07-06-253]. Gray, silty gypsum from a core sample, at a depth of 151 m, Redstone Formation; locality cf. 1740.

1742 [PPRG field no. 86-07-06-254]. Gray, silty gypsum from a core sample, at a depth of 159 m, Redstone River Formation; locality cf. 1740.

1743 [PPRG field no. 86-07-06-255]. Gray, silty gypsum from a core sample, at a depth of 165 m, Redstone River Formation; locality cf. 1740.

1744 [PPRG field no. 86-07-06-256]. Gray, silty gypsum from a core sample, at a depth of 181 m, Redstone River Formation; locality cf. 1740.

1745 [PPRG field no. 86-07-06-275]. White gypsum and white-red siltstone from a core sample, at a depth of 186 m, Redstone River Formation; locality cf. 1740.

1746 [PPRG field no. 86-07-06-258]. White gypsum from a core sample, at a depth of 197.0 m, Redstone River Formation; locality cf. 1740.

1747 [PPRG field no. 86-07-06-259]. White gypsum from a core sample, at a depth of 203.5 m, Redstone River Formation; locality cf. 1740.

1748 [PPRG field no. 86-07-06-260]. Gray, silty gypsum from a core sample, at a depth of 232 m, Redstone River Formation; locality cf. 1740.

1749 [PPRG field no. 86-07-06-261]. Laminated, gray gypsum from a core sample, at a depth of 236 m, Redstone River Formation; locality cf. 1740.

1750 [PPRG field no. 86-07-06-262]. Nodular gypsum and carbonate from a core sample, at a depth of 215.5 m, Redstone River Formation; locality cf. 1740.

1751 (ts, TOC, Corg) [PPRG field no. 86-07-06-263]. Black shale with sulfide from a core sample, at a depth of 195.2 m, Little Dal Group; Shell Canada Limited Diamond Drill Hole 3798F/20; core box located at the Shell Airstrip on the Keele River, Northwest Territories, Canada.

1752 (ts) [PPRG field no. 86-07-06-264]. Magnetic shale from the top of the Sayunei Formation; 1 km east of Keele River camp and airstrip, Northwest Territories, Canada; Wrigley Lake 1:250,000 Geologic map 1315A; 63.82°N, 127.8°W.

1753 [PPRG field no. 86-07-06-265]. Magnetic shale from the Sayunei Formation; locality cf. 1752.

Table 14.2. *Continued.*

1764 [D. R. L. field no. YUK 1-1]. Clastic iron-formation, coarse- to very coarse-grained, from the upper part and overlying 30–60 m of sandstone, Shezal Formation; "Waterfall section," at confluence of Iron Creek and a tributary to the North Iron River, Northwest Territories, Canada; Yukon Territory 1:50,000 topographic sheet 106/6 (ed. 1); 65.24°N, 113.0°W.

1765 [D. R. L. field no. YUK 1-2]. Clastic iron-formation, coarse- to very coarse-grained, from the upper part and overlying 30–60 m of sandstone, Shezal Formation; locality cf. 1764.

1767 [D. R. L. field no. YUK 1-3]. Diamictite from ca. 45 m above iron-formation, Shezal Formation; locality cf. 1764.

1768 [D. R. L. field no. YUK 1-4]. Clastic intraformational iron-formation, coarse- to very-coarse grained sandstone from ca. 45 m above iron-formation, Shezal Formation; locality cf. 1764.

1769 [D. R. L. field no. YUK 1-5]. Varved siltstone-mudstone from float, but from the same general interval as samples 1764–1768, Shezal Formation; locality cf. 1764.

1770 [D. R. L. field no. YUK 1-7]. Sandstone, medium- to coarse-grained, terriginous from float, ca. 45 m above iron-formation, Shezal Formation; locality cf. 1764.

1771 [D. R. L. field no. YUK 1-8]. Sandstone, medium- to coarse-grained, terriginous from float, ca. 45 m above iron-formation, Shezal Formation; locality cf. 1764.

1772 [D. R. L. field no. YUK 1-9]. Conglomerate, granule to small pebbles of red mudstone, carbonate, and sedimentary debris, from float, Shezal Formation; locality cf. 1764.

1773 [D. R. L. field no. YUK 1-10]. Iron-formation with chert nodules from the Shezal Formation; locality cf. 1764.

1774 [D. R. L. field no. YUK 1-11]. Iron-formation with chert nodules from the Shezal Formation; locality cf. 1764.

1775 [D. R. L. field no. YUK 1-12]. Iron-formation with irregular, stringy chert patches from the Shezal Formation; locality cf. 1764.

1776 [D. R. L. field no. YUK 1-13]. Jasper with "pisolites" from iron-formation, Shezal Formation; locality cf. 1764.

1777 [D. R. L. field no. YUK 1-14]. Laminated jasper and hematite iron-formation from the Rapitan Group; locality cf. 1764.

1778 [D. R. L. field no. YUK 1-15]. Sandstone, medium- to coarse-grained, locally forming discordant sills as well as beds from interbeds within iron-formation, Rapitan Group; locality cf. 1764.

1779 [D. R. L. field no. YUK 1-16]. Diamictite interbedded with iron-formation, Rapitan Group; locality cf. 1764.

1780 [D. R. L. field no. YUK 1-17]. Laminated cherty iron-formation from float, Rapitan Group; locality cf. 1764.

1781 [D. R. L. field no. YUK 1-18]. Nodular cherty iron-formation from float, Rapitan Group; locality cf. 1764.

1782 [D. R. L. field no. YUK 1-19]. Laminated cherty iron-formation from float, Rapitan Group; locality cf. 1764.

1783 [D. R. L. field no. YUK 2-1]. Intraformational conglomerate composed of clasts of jasper iron-formation and iron-rich mudstone and carbonate from the Rapitan Group; east bank of Iron Creek, Northwest Territories, Canada; 65.30°N, 133.2°W.

1784 [D. R. L. field no. YUK 2-2]. Laminated cherty iron-formation from the Rapitan Group; locality cf. 1783.

1785 [D. R. L. field no. YUK 2-3]. Nodular carbonate-rich iron-formation from the Rapitan Group; locality cf. 1783.

1786 [D. R. L. field no. YUK 2-4]. Coarse-grained, clastic sandstone iron-formation from the Rapitan Group; locality cf. 1783.

1787 [D. R. L. field no. YUK 2-5]. Cherty iron-formation from float, Rapitan Group; locality cf. 1783.

1788 [D. R. L. field no. YUK 3-1]. Conglomerate with lithic-polymictic clasts of sedimentary rocks from the middle of the Twitya Formation; ca. 16 km east of the Snake River, Yukon Territory, Canada; Snake River 1:250,000 Geologic map 1529A; 65.10°N, 132.8°W.

1789 [D. R. L. field no. YUK 3-2]. Calcareous sandstone from the middle of the Twitya Formation; locality cf. 1788.

1790 [D. R. L. field no. YUK 3-3]. Diamictite from the top third of the Shezal Formation; locality cf. 1788.

1791 [D. R. L. field no. YUK 4-1]. Conglomerate made up of Rapitan Group clasts from the basal conglomerate of the Franklin Mountain Formation; ca. 30 km east of the Snake River, Yukon Territory, Canada; Snake River 1:250,000 Geologic map 1529A; 65.30°N, 132.7°W.

1792 [D. R. L. field no. YUK 4-2]. Coarse-grained sandstone from the upper fourth of the Shezal Formation; locality cf. 1791.

1793 [D. R. L. field no. YUK 4-3]. Diamictite from the upper fourth of the Shezal Formation; locality cf. 1791.

1794 [D. R. L. field no. YUK 4-4]. Diamictite with sandstone masses from the upper third of the Shezal Formation; locality cf. 1791.

1795 [D. R. L. field no. YUK 4-5]. Cross-laminated, fine-grained sandstone from the top third of the Shezal Formation; locality cf. 1791.

1796 [D. R. L. field no. YUK 4-6]. Sandy diamictite from the top half of the Shezal Formation; locality cf. 1791.

1797 [D. R. L. field no. YUK 4-7]. Diamictite from the middle to lower half of the Shezal Formation; locality cf. 1791.

1798 [D. R. L. field no. YUK 4-8]. Conglomerate bed in mudstone from the lower part of the Shezal Formation; locality cf. 1791.

1799 [D. R. L. field no. YUK 2-6]. Varved siltstone-sandstone couplets from float, Rapitan Group; east bank of Iron Creek, Northwest Territories, Canada; 65.30°N, 133.2°W.

Table 14.2. *Continued.*

1800 [D. R. L. field no. YUK 5-1]. Siltstone with conglomerate dropstones from a core sample, depth of 381.1 m from top of core, Sayunei Formation; Shell Canada Limited Diamond Drill Hole 3698F/41, Keele River area; core box located at the Shell Airstrip on the Keele River, Northwest Territories, Canada; Vanishing Ram Creek 1:50,000 topographic map 95M/13.

1801 [D. R. L. field no. YUK 5-2]. Siltstone and sandstone from a core sample, depth of 378 m from top of core, Shezal Formation; locality cf. 1800.

1802 [D. R. L. field no. YUK 5-3-A]. Stratified diamictite from a core sample, depth of 345.1 m from top of core, Rapitan Group; locality cf. 1800.

1803 [D. R. L. field no. YUK 5-3-B]. Layered mudstone and diamictite from a core sample, depth of 345.1 m from top of core, Rapitan Group; locality cf. 1800.

1804 [D. R. L. field no. YUK 5-4]. Diamictite from a core sample, depth of 126.9 m from top of core, Rapitan Group; locality cf. 1800.

1805 [D. R. L. field no. YUK 5-5]. Diamictite from a core sample, depth of 215.8 m from top of core, Sayunei Formation; locality cf. 1800.

1806 [D. R. L. field no. YUK 5-6]. Diamictite from a core sample, depth of 75.6 m from top of core, Shezal Formation; locality cf. 1800.

1807 [D. R. L. field no. YUK 5-7]. Diamictite with unusual iron-staining that appears to mark the Sayunei-Shezal contact, from a core sample, depth of 324.7 m from top of core, Rapitan Group; locality cf. 1800.

1808 [D. R. L. field no. YUK 5-8]. Conglomeratic sandstone from the basal part of the Rapitan Group; locality cf. 1800.

1809 [D. R. L. field no. YUK 5-9]. Layered mudstone and conglomerate sandstone from a core sample, depth of 346.2 m from top of core, Rapitan Group; locality cf. 1800.

1822 [D. R. L. field no. NWT 7-1]. Thin bedded, coarse- to fine-grained sandstone from 2 m above the base of the Sayunei Formation; Lukas Creek section, Keele River area, Northwest Territories, Canada; 63.71°N, 127.2°W.

1823 [D. R. L. field no. NWT 7-2]. Red-weathering, flaggy sandstone and conglomerate (shale clasts) from the Sayunei Formation; locality cf. 1822.

1824 [D. R. L. field no. NWT 7-3]. Thin bedded conglomerate from 10 m above the base of the Sayunei Formation; locality cf. 1822.

1825 [D. R. L. field no. NWT 7-4]. Coarse-grained sandstone graded bed, from 15 m above the base of the Sayunei Formation; locality cf. 1822.

1826 [D. R. L. field no. NWT 7-5]. Cross-laminated, calcareous sandstone, from 18 m above the base of the Sayunei Formation; locality cf. 1822.

1827 [D. R. L. field no. NWT 7-6]. Mudstone with graded units, from 22 m above the base of the Sayunei Formation; locality cf. 1822.

1828 [D. R. L. field no. NWT 7-7]. Varved mudstone with dropstones, from 27 m above the base of the Sayunei Formation; locality cf. 1822.

1829 [D. R. L. field no. NWT 7-8]. Varved mudstone-siltstone, from 30 m above the base of the Sayunei Formation; locality cf. 1822.

1830 [D. R. L. field no. NWT 7-10]. Varved mudstone, from 45 m above the base and within the lower part of the Sayunei Formation; locality cf. 1822.

1831 [D. R. L. field no. NWT 7-11]. Sandstone with graded bedding, from 55 m above the base and from within the middle of the Sayunei Formation; locality cf. 1822.

1832 [D. R. L. field no. NWT 7-12]. Coarse-grained sandstone from 80 m above the base and within the middle of the Sayunei Formation; locality cf. 1822.

1833 [D. R. L. field no. NWT 7-13]. Finely laminated red mudstone from the top of the Sayunei Formation; locality cf. 1822.

1834 [D. R. L. field no. NWT 7-14]. Cross-laminated sandstone from the top of the Sayunei Formation; locality cf. 1822.

1835 [D. R. L. field no. NWT 7-15]. Interbedded sandstone and mudstone from the top of the Sayunei Formation; locality cf. 1822.

1836 [D. R. L. field no. NWT 7-16]. Hematitic iron-formation from the Rapitan Group; locality cf. 1822.

1837 [D. R. L. field no. NWT 7-17]. Varved hematitic mudstone from the Rapitan Group; locality cf. 1822.

1838 [D. R. L. field no. NWT 7-18]. Hematitic mud-chip conglomerate from the top of the Rapitan Group; locality cf. 1822.

1839 [D. R. L. field no. NWT 8-1]. Sandstone and mudstone from the lower member of the Little Dal Formation; southwest of Lukas Creek, Keele River area, Northwest Territories, Canada; 63.77°N, 127.4°W.

1840 [D. R. L. field no. NWT 9-1]. Pink carbonate-rich conglomerate from the basal conglomerate of the Redstone River Formation; cliff above Redstone River, Keele River area, Northwest Territories, Canada; 63.79°N, 127.5°W.

1841 [D. R. L. field no. NWT 9-2]. Sandstone interbedded within basal conglomerate from the Redstone River Formation; locality cf. 1840.

1842 [D. R. L. field no. NWT 10-1]. Mudstone with dropstones from the topmost layers of the Sayunei Formation; 1 km east of Keele River camp and airstrip, Northwest Territories, Canada; Wrigley Lake 1:250,000 Geologic map 1315A; 63.82°N, 127.8°W.

1843 [D. R. L. field no. NWT 10-2]. Magnetite-rich varves from the topmost layers of the Sayunei Formation; locality cf. 1842.

1844 [D. R. L. field no. NWT 10-3]. Diamictite from the basal part of the Shezal Formation; locality cf. 1842.

2356 [D. R. L. field no. YUK 2-7]. Slightly varved siltstone from float, Rapitan Group; east bank of Iron Creek, Northwest Territories, Canada; 65.30°N, 133.2°W.

Table 14.2. *Continued.*

MARQUETTE RANGE

PPRG samples/localities.

2416 (ts, TOC, Corg) [J. M. H. field no. 830608-1a/B593]. Dark-gray chert from a debris pile of the Kona Dolomite; Kona Dolomite locality, Michigan, USA.

2417 (ts, TOC, Corg) [J. M. H. field no. 830608-1b/B594]. Black chert from a debris pile of the Kona Dolomite; locality cf. 2416.

2418 (ts, TOC, Corg) [J. M. H. field no. 830608-1c/B595]. Black chert from the Kona Dolomite; from the cliff face, locality cf. 2416.

2419 (ts, TOC, Corg) [J. M. H. field no. 830608-1d/B596]. Black carbonate from the Kona Dolomite; very top of quarry, locality cf. 2416.

2420 (ts, st, TOC, Corg) [J. M. H. field no. 830608-1e/B597]. Carbonate from huge columnar stromatolites from the Kona Dolomite; locality cf. 2416.

2421 (ts, TOC, Corg) [J. M. H. field no. 830608-1f/B598]. Black carbonate from the Kona Dolomite; locality cf. 2416.

MEDICINE BOW RANGE

Stratigraphy.

    Lookout Schist

PPRG samples/localities.

1766 [E. K. field no. LS-1]. Graphitic black schist, greenschist facies from the Lookout Schist; Medicine Bow Mountains, Wyoming, USA.

2741 [E. K. field no. LS-2]. Black graphitic schist, greenschist facies from the Lookout Schist; locality cf. 1766.

PENOKEAN FOLD BELT

Tectonic unit/location. Penokean Fold Belt (Canadian Southern Province), Ontario, Canada, and Michigan, USA.

Stratigraphy. Stratigraphic section, Blind River-Elliot Lake area, southern Ontario, Canada (Stockwell et al. 1970):

HURONIAN SUPERGROUP
  COBALT GROUP
    Bar River Formation
    Gordon Lake Formation
    Lorrain Formation
    Gowganda Formation
  QUIRKE LAKE GROUP
    Serpent Formation
    Espanola Formation
    Bruce Formation
  HOUGH LAKE GROUP
    Mississagi Formation
    Pecors Formation
    Ramsay Lake Formation
  ELLIOT LAKE GROUP
    McKim Formation
    Matinenda Formation

PPRG samples/localities.

1380 (ts, TOC) [J. W. S. field no. 1-6/18/83]. Glacial varves from the Gowganda Formation; 3.2 km east of junction with Highway 108 on Stanrock Road, southern Ontario, Canada.

1381 (ts, TOC) [J. W. S. field no. 3-6/16/83]. Tillite from the Gowganda Formation; "T. V. Hill," at roadcut along Highway 108, north of Elliot Lake, southern Ontario, Canada.

1845 (ts, TOC, Corg) [J. D. field no. 68-1/32]. Green-gray argillite from a core sample at a depth of 9.8 m, Bar River Formation; diamond drill hole 157-68-1, Claim no. SSM 97544, Township of Raimbault, cut during June, 1968 by Cdn. Johns-Manville Co. Ltd., southern Ontario, Canada.

1846 (ts, TOC, Corg) [J. D. field no. 68-1/1727]. Greenish graywacke with 7.5 cm beds of conglomerate from a core sample at a depth of 527 m, Gordon Lake Formation; locality cf. 1845.

1847 (ts, TOC, Corg) [J. D. field no. 150-4/192]. Quartzite from a core sample, at a depth of 58.6 m, Lorrain Formation; diamond drill hole 150-4, Claim no. S 138138, township of Bouck, cut during Sept. 1968 by Kerr-McGee, southern Ontario, Canada.

1848 (ts, TOC) [J. D. field no. 150-4/435-460]. Siltstone from a core sample, at a depth of 133–140 m, Gowganda Formation; locality cf. 1847.

1849 (ts, TOC, Corg) [J. D. field no. 150-4/460-485]. Siltstone from a core sample, at a depth of 140–148 m, Gowganda Formation; locality cf. 1847.

1850 (ts, TOC, Corg) [J. D. field no. 150-4/485-500]. Siltstone from a core sample, at a depth of 148–153 m, Gowganda Formation; locality cf. 1847.

Table 14.2. *Continued.*

1851 (ts, TOC) [J. D. field no. 150-4/500-517]. Siltstone from a core sample, at a depth of 153–158 m, Gowganda Formation; locality cf. 1847.

1852 (ts, TOC, Corg) [J. D. field no. 150-4/1115]. Conglomerate from a core sample, at a depth of 340 m, Gowganda Formation; locality cf. 1847.

1853 (ts, TOC, Corg) [J. D. field no. 150-4/2230]. Quartzite from a core sample, at a depth of 680 m, Serpent Formation; locality cf. 1847.

1854 (ts, TOC, Corg) [J. D. field no. 150-4/2985]. Siltstone from a core sample, at a depth of 910 m, Espanola Formation; locality cf. 1847.

1855 (ts, TOC, Corg) [J. D. field no. 150-4/3242]. Limestone from a core sample, at a depth of 989 m, Bruce Formation; locality cf. 1847.

1856 (ts, TOC, Corg) [J. D. field no. 150-4/3385]. Siltstone from a core sample, at a depth of 1032 m, Bruce Formation; locality cf. 1847.

1857 (ts, TOC, Corg) [J. D. field no. 150-4/3765]. Quartzite from a core sample, at a depth of 1148 m, Mississagi Formation; locality cf. 1847.

1858 (ts, TOC, Corg) [J. D. field no. 150-4/4820]. Siltstone from a core sample, at a depth of 1470 m, Pecors Formation; locality cf. 1847.

1859 (ts, TOC, Corg) [J. D. field no. 150-4/4960]. Conglomerate from a core sample, at a depth of 1513 m, Ramsay Lake Formation; locality cf. 1847.

1860 (ts, TOC, Corg) [J. D. field no. 150-4/5049-5054]. Siltstone from a core sample, at a depth of 1540–1542 m, McKim Formation; locality cf. 1847.

1861 (ts, TOC, Corg) [J. D. field no. 154-4/5185]. Quartzite from a core sample, at a depth of 1581 m, McKim Formation; locality cf. 1847.

1862 (ts) [J. D. field no. 150-4/5218-5225]. Siltstone and quartzite from a core sample, at a depth of 1592–1594 m, McKim Formation; locality cf. 1847.

1863 (ts, TOC, Corg) [J. D. field no. 150-4/5225-5250]. Quartzite from a core sample, at a depth of 1594–1601 m, McKim Formation; locality cf. 1847.

1864 (ts, TOC, Corg) [J. D. field no. 150-4/5250-5275]. Quartzite from a core sample, at a depth of 1601–1609 m, McKim Formation; locality cf. 1847.

1865 (ts, TOC, Corg) [J. D. field no. 150-4/5275-5300]. Siltstone from a core sample, at a depth of 1609–1617 m, McKim Formation; locality cf. 1847.

1866 (ts, TOC, Corg) [J. D. field no. 150-4/5300-5309]. Siltstone from a core sample, at a depth of 1617–1619 m, McKim Formation; locality cf. 1847.

1867 (ts, TOC, Corg) [J. D. field no. 150-4/5309-5334]. Quartzite from a core sample, at a depth of 1619–1627 m, Matinenda Formation; locality cf. 1847.

1868 (ts, TOC, Corg) [J. D. field no. 150-4/5345]. Quartzite from a core sample, at a depth of 1630 m, Matinenda Formation; locality cf. 1847.

1869 (ts, TOC, Corg) [J. D. field no. 150-4/5353-5359]. Siltstone from core sample 1633–1635 m, Matinenda Formation; locality cf. 1847.

2496 (ts, TOC, Corg) [J. D. field no. 830618-3a/B775]. Varved black shaley siltstone(?) from the Gowganda Formation; 3.2 km east of Highway 108 on Stanrock Road, southern Ontario, Canada.

2497 (ts, TOC, Corg) [J. D. field no. 830618-4a/B776]. Brown carbonate from the Espagnola Formation; Quirke Mine no. 2, southern Ontario, Canada.

PORT ARTHUR HOMOCLINE

Tectonic unit/location. Port Arthur Homocline (Canadian Southern Province), Ontario, Canada, and Minnesota, USA.

Stratigraphy. Stratigraphic section, Port Arthur-Schreiber area, southern Ontario, Canada (Stockwell et al. 1970):

SIBLEY GROUP
ANIMIKIE GROUP
    Rove Formation
    Gunflint Formation
KEEWATIN VOLCANICS

PPRG samples/localities.

1284 (ts, TOC, Corg) [J. W. S. field no. 1 of 6/26/68]. Black banded chert from the upper algal cycle of the Gunflint Iron Formation; "Frustration Bay locality," 1 km west of the mouth of the Blende River, along the north shore of Lake Superior west of the Sibley Peninsula, southern Ontario, Canada.

1285 (ts, TOC, Corg) [J. W. S. field no. 1 of 6/25/68]. Black chert from the upper algal cycle of the Gunflint Iron Formation; "Discovery Point locality," at point immediately west of "Frustration Bay", ca. 1 km west of the mouth of the Blende River, along the north shore of Lake Superior, west of the Sibley Peninsula, southern Ontario, Canada.

1286 (ts, TOC, Corg) [J. W. S. field no. 2 of 6/26/68]. Black chert from the lower algal cycle of the Gunflint Iron Formation; in river bed at bridge where Highway 588 crosses Whitefish River, ca. 1.75 km southwest of Nolalu; 54 km west of Thunder Bay, southern Ontario, Canada.

1287 (ts, st, TOC, Corg) [J. W. S. field no. 3 of 6/26/68]. Black chert from the lower algal cycle of the Gunflint Iron Formation; locality cf. 1286.

1288 (ts, TOC, Corg) [J. W. S. field no. 2 of 6/25/68]. Oolitic black chert from the lower algal cycle of the Gunflint Iron Formation; Pass Lake bridge, southern Ontario, Canada.

Table 14.2. *Continued.*

1289 (ts, µf*, TOC, Corg) [J. W. S. field no. 3 of 6/25/68]. Microfossiliferous (Chapter 24) black chert from the lower algal cycle of the Gunflint Iron Formation; "Schreiber Beach locality," 6.5 km west of Schreiber, along the north shore of Lake Superior, southern Ontario, Canada.

1290 (ts, TOC, Corg) [J. W. S. field no. 4 of 6/25/68]. Black chert from the lower algal cycle of the Gunflint Iron Formation; locality cf. 1289.

1291 (ts, TOC, Corg) [J. W. S. field no. 1 of 8/2/66]. Black chert from the lower algal cycle of the Gunflint Iron Formation; locality cf. 1289.

1292 (ts, TOC, Corg) [J. W. S. field no. 2 of 8/2/66]. Black chert from the lower algal cycle of the Gunflint Iron Formation; locality cf. 1289.

1293 (ts, TOC, Corg) [J. W. S. field no. 1 of 8/2/68]. Black chert from the Keewatin Group; Ontario Highway 17, 0.8 km northwest of Schreiber, southern Ontario, Canada.

1294 (ts, TOC, Corg) [J. W. S. field no. 2 of 8/2/68]. Black chert from the Keewatin Group; locality cf. 1293.

1295 (ts, TOC, Corg, ker) [J. W. S. field no. 5 of 6/25/68]. Black shale with kink folds from the Rove Formation; near railroad tracks, on the north shore of Pass Lake in Sibley Provincial Park, southern Ontario, Canada.

1296 (ts, st, TOC, Corg) [J. W. S. field no. 1 of 6/16/83]. Stromatolitic black chert from the Gunflint Iron Formation; on trail between the main site and the cobble beach, "Schreiber Beach locality," 6.5 km west of Schreiber, along the north shore of Lake Superior, southern Ontario, Canada.

1297 (ts, TOC, Corg) [J. W. S. field no. 1 of 6/14/83]. Algal chert from the lower algal cycle of the Gunflint Iron Formation; south side of Highway 11/17, at the junction with Highway 590, ca. 30 km west of Thunder Bay, southern Ontario, Canada.

1298 (ts, TOC, Corg) [J. W. S. field no. 2 of 6/14/83]. Algal chert from the Gunflint Iron Formation; northern-most "beach," 250 m upstream of Kakabeka Falls, Kakabeka Falls Provincial Park, ca. 25 km west of Thunder Bay, southern Ontario, Canada; 48.41°N, 89.62°W.

1299 (ts, TOC, Corg) [J. W. S. field no. 3 of 6/14/83]. Chert from the Gunflint Iron Formation; south side of Highway 11/17, at the junction with Highway 590, ca. 30 km west of Thunder Bay, southern Ontario, Canada.

1300 (ts, st, TOC, Corg) [J. W. S. field no. 4 of 6/14/83]. Black and red stromatolitic chert from the lower algal cycle of the Gunflint Iron Formation; in river bed at bridge where Highway 588 crosses Whitefish River, ca. 1.75 km southwest of Nolalu; 54 km west of Thunder Bay, southern Ontario, Canada.

1301 (ts, TOC, Corg) [J. W. S. field no. 1 of 6/15/83]. Black chert from the upper algal cycle of the Gunflint Iron Formation; "Frustration Bay locality," 1 km west of the mouth of the Blende River, along the north shore of Lake Superior west of the Sibley Peninsula, southern Ontario, Canada.

1302 (ts, st, TOC, Corg) [J. W. S. field no. 5 of 6/14/83]. Stromatolitic black chert from the lower algal cycle of the Gunflint Iron Formation; 100 m upstream (east), in river bed at bridge where Highway 588 crosses Whitefish River, ca. 1.75 km southwest of Nolalu; 54 km west of Thunder Bay, southern Ontario, Canada.

1303 (ts, TOC, Corg) [J. W. S. field no. 2 of 6/15/83]. Black chert from the upper algal cycle of the Gunflint Iron Formation; "Discovery Point" locality," at point immediately west of "Frustration Bay," ca. 1 km west of the mouth of the Blende River, along the north shore of Lake Superior, west of the Sibley Peninsula, southern Ontario, Canada.

1304 (ts, TOC, Corg) [J. W. S. field no. 3 of 6/15/83]. Black chert from the Rossport Formation; 13.3 miles east of Highway 11 and 17 junction, at Nipigon roadcut, southern Ontario, Canada.

1305 (ts, st, TOC, Corg) [J. W. S. field no. 4 of 6/15/83]. Stromatolitic cherty carbonate from the middle of the Rossport Formation; locality cf. 1304.

1374 (ts, st, TOC, Corg) [J. W. S. field no. 1-6/14/83]. Chert from the upper algal cycle of the Gunflint Iron Formation; "Mink Mountain locality," 60 km west of Thunder Bay, via Highway 11/17, southern Ontario, Canada.

1375 (ts, TOC, Corg, Sfide) [J. W. S. field no. 2-6/14/86]. Chert-carbonate "caprock" from above the argillite unit in the upper cycle of the Gunflint Iron Formation; Kakabeka Falls Provincial Park, ca. 25 km west of Thunder Bay, southern Ontario, Canada; 48.41°N, 89.62°W.

1376 (ts, TOC, Corg) [J. W. S. field no. 1-8/23/83]. Anthraxolite and calcite from the Gunflint Iron Formation; spillway section, 500 m upstream from the Ontario Hydro Station, access road is 0.5 m east of entrance to Kakabeka Falls Provincial Park, southern Ontario, Canada; 48.41°N, 89.62°W.

1377 (ts, TOC, Corg) [J. W. S. field no. 2-8/23/83]. Anthraxolite and calcite from the Gunflint Iron Formation; locality cf. 1376.

1378 (ts, TOC, Corg) [J. W. S. field no. 3-6/14/83]. Anthraxolite from the Gunflint Iron Formation; locality cf. 1376.

1379 (ts, st, TOC, Corg) [J. W. S. field no. 2-6/16/83]. Stromatolitic chert from the Gunflint Iron Formation; "Schreiber Beach locality," 6.5 km west of Schreiber, along the north shore of Lake Superior, southern Ontario, Canada.

2436 (ts, TOC, Corg, ker) [J. M. H. field no. 830614-1a/B613]. Black chert stringers in ferruginous carbonate from the Gunflint Iron Formation; Kakabeka Falls Provincial Park, ca. 25 km west of Thunder Bay, southern Ontario, Canada; 48.41°N, 89.62°W.

2437 (ts, TOC, Corg, ker) [J. M. H. field no. 830614-1b/B614]. Ferruginous dolomite overlying argillite from above the upper cherty "member" of the Gunflint Iron Formation; at main parking area near falls, locality cf. 2436.

2438 (ts, TOC, Corg) [J. M. H. field no. 830614-2a/B615]. Chert from the lower algal cycle of the Gunflint Iron Formation; 400 m above falls, near parking lot, locality cf. 2436.

2439 (ts, TOC, Corg) [J. M. H. field no. 830614-3a/B616]. Anthraxolite from veins from the Gunflint Iron Formation; north side of Highway 590, at junction with Highway 11/17, ca. 30 km west of Thunder Bay, southern Ontario, Canada.

Table 14.2. *Continued.*

2440 (ts, TOC, Corg) [J. M. H. field no. 830614-3b/B617]. Black, ferruginous slate from the Gunflint Iron Formation; locality cf. 2439.

2441 (ts, st, TOC, Corg) [J. M. H. field no. 830614-4a/B618]. Black chert with 1 cm-diameter columnar stromatolites from the Gunflint Iron Formation; in river bed at bridge where Highway 588 crosses Whitefish River, ca. 1.75 km southwest of Nolalu; 54 km west of Thunder Bay, southern Ontario, Canada.

2442 (ts, TOC, Corg) [J. M. H. field no. 830614-4b/B619]. Black chert from the Gunflint Iron Formation; float, 50 m downstream from, but locality cf. 2441.

2443 (ts, st) [J. M. H. field no. 830614-5a]. Red, stromatolitic chert from the Gunflint Iron Formation; locality cf. 2441.

2444 (ts, TOC, Corg, ker) [J. M. H. field no. 830615-1a/B620]. Gray chert with ankerite from the Gunflint Iron Formation; river bank, 150 m downstream from bridge, Current River Park, in the city of Thunder Bay, southern Ontario, Canada.

2445 (ts, TOC, Corg) [J. M. H. field no. 830615-1b/B621]. Chert from float of the Gunflint Iron Formation; locality cf. 2444.

2446 (ts, TOC, Corg) [J. M. H. field no. 830615-2a/B622]. Carbonate from the lower algal cycle of the Gunflint Iron Formation; south side of Highway 11/17, at the junction with Highway 590, ca. 30 km west of Thunder Bay, southern Ontario, Canada.

2447 (ts, TOC, Corg, Sfide) [J. M. H. field no. 830615-2b/B623]. Black pyritic chert with coarse carbonate spar from below the carbonate unit, lower cycle of the Gunflint Iron Formation; locality cf. 2446.

2448 (ts, TOC, Corg) [J. M. H. field no. 830615-2c/B624]. Shale from 2 m below large diabase sill, Rove Formation; locality cf. 2446.

2449 (ts, TOC, Corg) [J. M. H. field no. 830615-3a/B625]. Carbonate from the upper algal cycle of the Gunflint Iron Formation; "Frustration Bay locality," 1 km west of the mouth of the Blende River, along the north shore of Lake Superior west of the Sibley Peninsula, southern Ontario, Canada.

2450 (TOC, Corg) [J. M. H. field no. 830615-3a/B626]. Chert from the upper algal cycle of the Gunflint Iron Formation; locality cf. 2449.

2451 (ts, TOC, Corg) [J. M. H. field no. 830615-3b/B627]. Black chert from the upper algal cycle of the Gunflint Iron Formation; "Discovery Point locality," at point immediately west of "Frustration Bay," ca. 1 km west of the mouth of the Blende River, along the north shore of Lake Superior, west of the Sibley Peninsula, southern Ontario, Canada.

2452 (ts, TOC, Corg) [J. M. H. field no. 830615-3c/B628]. Chert from the upper algal cycle of the Gunflint Iron Formation; locality cf. 2451.

2453 (ts, TOC, Corg, ker) [J. M. H. field no. 830615-3d/B629]. Gray chert with anthraxolite veins from the upper algal cycle of the Gunflint Iron Formation; locality cf. 2451.

2454 (ts, TOC, Corg) [J. M. H. field no. 830615-3f/B630]. Chert from the upper algal cycle of the Gunflint Iron Formation; locality cf. 2451.

2455 (TOC, Corg) [J. M. H. field no. 830615-3f/B631]. Carbonate from the upper algal cycle of the Gunflint Iron Formation; locality cf. 2451.

2456 (ts, TOC, Corg) [J. M. H. field no. 830615-4a/B632]. Carbonate concretion in shale from the Rove Formation; near railroad tracks, on the north shore of Pass Lake in Sibley Provincial Park, southern Ontario, Canada.

2457 (ts, TOC, Corg) [J. M. H. field no 830615-5a/B633]. Carbonate from the Rove Formation; Rove Quarry on east side of Highway 587, Sibley Peninsula, southern Ontario, Canada.

2458 (TOC, Corg) [J. M. H. field no. 830615-5a/B634]. Shale from the Rove Formation; locality cf. 2457.

2459 (ts, TOC, Corg, ker) [J. M. H. field no. 830615-5b/B635]. Dark-gray carbonate fragment from huge concretion from the Rove Formation; locality cf. 2457.

2460 (ts, TOC, Corg) [J. M. H. field no. 830615-6a/B636]. Black banded carbonate from the Rossport Formation; east side of Kama Bay, north of Highway 17; east of Lake Nipigon, southern Ontario, Canada.

2461 (ts, TOC, Corg) [J. M. H. field no. 830615-6b/B637]. Dark banded carbonate from 1 m below igneous intrusion, Rossport Formation; locality cf. 2460.

2462 (ts) [J. M. H. field no. 830615-6c]. Carbonate from contact with igneous body, Rossport Formation; locality cf. 2460.

2463 [J. M. H. field no. 830616-1a]. Black chert from beach cobble, lower algal cycle of the Gunflint Iron Formation; "Schreiber Beach locality," 6.5 km west of Schreiber, along the north shore of Lake Superior, southern Ontario, Canada.

2464 (ts, TOC, Corg) [J. M. H. field no. 830616-1c/B746]. Stromatolitic black chert from the lower algal cycle of the Gunflint Iron Formation; locality cf. 2463.

2465 (ts, TOC, Corg) [J. M. H. field no. 830616-1d/B747]. Black chert with carbonate and pyrite veins from the lower algal cycle of the Gunflint Iron Formation; 200 m west, locality cf. 2463.

2466 (ts) [J. M. H. field no. 830616-1e]. Flat-laminated, gray chert with traces of carbonate, distinctly oolitic on top and bottom from the lower algal cycle of the Gunflint Iron Formation; 200 m west, locality cf. 2463.

2467 (ts, TOC, Corg) [J. M. H. field no. 830616-1f/B748]. Carbonate from chert-carbonate sample from the lower algal cycle of the Gunflint Iron Formation; 300 m from main site on "cobble beach," locality cf. 2463.

Table 14.2. *Continued.*

2468 (TOC, Corg) [J. M. H. field no. 830616-1f/B749]. Chert cf. 2467, lower part of the lower algal cycle of the Gunflint Iron Formation; 300 m west, locality cf. 2463.

2469 (TOC, Corg) [J. M. H. field no. 830616-1f/B750]. Carbonate from chert-carbonate sample, upper part of the lower algal cycle of the Gunflint Iron Formation; 300 m west, locality cf. 2463.

2470 (ts, TOC, Corg) [J. M. H. field no. 830616-1g/B751]. Dark-gray, laminated carbonate from above the lower algal cycle of the Gunflint Iron Formation; 200 m west, locality cf. 2463.

2471 (TOC, Corg) [J. M. H. field no. 830616-1h/B752]. Massive gray carbonate from above 2470, lower algal cycle of the Gunflint Iron Formation; 200 m west, locality cf. 2463.

2472 (ts) [J. M. H. field no. 830616-1i]. Cherty carbonate from above 2471, lower algal cycle of the Gunflint Iron Formation; 200 m west, locality cf. 2463.

2473 (ts) [J. M. H. field no. 830616-1j]. Cherty carbonate from above 2472, lower algal cycle of the Gunflint Iron Formation; 200 m west, locality cf. 2463.

2474 (ts, st, TOC, Corg) [J. M. H. field no. 830616-1l/B753]. Carbonate from top section of a large stromatolite from the lower algal cycle of the Gunflint Iron Formation; 200 m west, locality cf. 2463.

2475 (TOC, Corg) [J. M. H. field no. 830616-1l/B754]. "Anthraxolite-free chert," separated from anthraxolite-containing pieces, from the lower algal cycle of the Gunflint Iron Formation; 200 m west, locality cf. 2463.

2476 (ts, st, TOC, Corg) [J. M. H. field no. 830616-1n/B755]. Stromatolitic cherty carbonate from the lower algal cycle of the Gunflint Iron Formation; 200 m west, locality cf. 2463.

2477 (ts, st, TOC, Corg) [J. M. H. field no. 830616-1o/B756]. Stromatolitic cherty carbonate from the Gunflint Iron Formation; 200 m west, locality cf. 2463.

2478 (ts, st, TOC, Corg) [J. M. H. field no. 830616-1p/B757]. Chert with small domical stromatolites, irregularly laminated (no thimbles), from the Gunflint Iron Formation; 200 m west, locality cf. 2463.

2479 (TOC, Corg) [J. M. H. field no. 830616-1k/B758]. Anthraxolite from anthraxolite-chert sample from the lower algal cycle of the Gunflint Iron Formation; 200 m west, locality cf. 2463.

2480 (TOC, Corg) [J. M. H. field no. 830616-1n/B759]. Anthraxolite from the lower algal cycle of the Gunflint Iron Formation; 200 m west, locality cf. 2463.

2481 (TOC, Corg) [J. M. H. field no. 830616-1o/B760]. Anthraxolite from the lower algal cycle of the Gunflint Iron Formation; 200 m west, locality cf. 2463.

2482 (st, TOC, Corg) [J. M. H. field no. 820606-1a/B761]. Stromatolitic light gray chert from the lower algal cycle of the Gunflint Iron Formation; "Winston Point," 3 km west of the "Schreiber Beach locality," 9.5 km west of Schreiber, on the north shore of Lake Superior, southern Ontario, Canada.

2483 (st, TOC, Corg) [J. M. H. field no. 820608-1b/B762]. Stromatolitic chert from the lower algal cycle of the Gunflint Iron Formation; locality cf. 2482.

2484 (st, TOC, Corg) [J. M. H. field no. 820608-1c/B763]. Stromatolitic gray chert from stratigraphically above 2438, Gunflint Iron Formation; locality cf. 2482.

2485 (st, TOC, Corg) [J. M. H. field no. 820608-1d/B764]. Stromatolitic gray chert from the same level as 2483, but a different location laterally, Gunflint Iron Formation; locality cf. 2482.

2486 (st, TOC, Corg) [J. M. H. field no. 820608-1e/B765]. Stromatolitic gray chert from the Gunflint Iron Formation; locality cf. 2482.

2487 (st, TOC, Corg) [J. M. H. field no. 820608-1f/B766]. Stromatolitic gray chert from 9 cm above 2486, Gunflint Iron Formation; locality cf. 2482.

2488 (st, TOC, Corg) [J. M. H. field no. 820608-1g/B767]. Stromatolitic gray chert from the Gunflint Iron Formation; locality cf. 2482.

2489 (st, TOC, Corg) [J. M. H. field no. 820608-1h/B768]. Stromatolitic gray chert from the Gunflint Iron Formation; locality cf. 2482.

2490 (st, TOC, Corg) [J. M. H. field no. 820608-1i/B769]. Stromatolitic gray chert from the Gunflint Iron Formation; locality cf. 2482.

2491 (st, TOC, Corg) [J. M. H. field no. 820608-1j/B770]. Stromatolitic gray chert from the Gunflint Iron Formation; locality cf. 2482.

2492 (TOC, Corg) [J. M. H. field no. 820608-1k/B771]. Flat-laminated chert from the Gunflint Iron Formation; locality cf. 2482.

2493 (st, TOC, Corg) [J. M. H. field no. 820608-1l/B772]. Black chert from inner part of a domical stromatolite from the Gunflint Iron Formation; Kakabeka Falls Provincial Park, ca. 25 km west of Thunder Bay, southern Ontario, Canada; 48.41°N, 89.62°W.

2494 (st, TOC, Corg) [J. M. H. field no. 820608-1m/B773]. Chert from the outer part of a domical stromatolite cf. 2493, Gunflint Iron Formation; locality cf. 2493.

2495 (TOC, Corg) [J. M. H. field no. 820608-1n/B774]. Black chert from the upper algal cycle of the Gunflint Iron Formation; "Frustration Bay locality," 1 km west of the mouth of the Blende River, along the north shore of Lake Superior west of the Sibley Peninsula, southern Ontario, Canada.

Table 14.2. *Continued.*

SUPERIOR PROVINCE

Tectonic unit/location. Superior Province, northern Quebec, Canada

Stratigraphy.

    Blondeu Formation

PPRG samples/localities.

2362 [G. O. A. field collection]. Graphite-pyrite bearing argillite, low greenschist facies from the Blondeau Formation; Cummings Lake, northern Quebec, Canada.

UINTA UPLIFT

Tectonic unit/location. Uinta Uplift, Utah, USA.

Stratigraphy. King 1976.

  UINTA MOUNTAIN GROUP
    Red Pine Shale
    "Unnamed sediments"

PPRG samples/localities.

1361 (ts, TOC, Corg, ker) [J. W. S. field no. 1-8/13/77]. Gray, carbonaceous, fissile shale, containing *Chuaria* from 2 to 3 m above the base of the exposed section of clastic rocks of the Red Pine Shale; small southwest facing cliff on Yellowstone Creek, 2.3 km upstream from Swift Creek campground in the Uinta Mountains, 50 km north of Duchesne, Utah, USA; 40.60°N, 110.3°W.

1872 (TOC, Corg, ker) [R. J. H. field no. RH84-8-28-2]. Gray laminated micaceous siltstone from the Red Pine Shale; along Yellowstone Creek, 2.3 km upstream from National Park Service campground at the mouth of Swift Creek and 50 km north of Duchesne, southern Uinta Mountains, northern Utah, Utah, USA; 40.60°N, 110.3°W.

UNAKA PROVINCE

Tectonic unit/location. Unaka Province, northeastern Tennessee, USA.

Stratigraphy. Stratigraphic section (King and Ferguson 1960)

  KNOX GROUP
    Jonesboro Limestone
    Conococheague Limestone
  CONASAUGA GROUP
  [UNNAMED CAMBRIAN GROUP]
    Rome Formation
    Sandy Dolomite
  CHILHOWEE GROUP
  MOUNT ROGERS VOLCANIC GROUP

PPRG samples/localities.

1306 (ts, TOC, Corg) [J. W. S. field no. RCM-1/27/74-JWS]. Black chert from the Jonesboro Limestone near Cade's Cove in the Great Smokey Mountains National Park, ca. 30 km southeast of Gatlinburg, Tennessee, USA.

1307 (ts, TOC, Corg) [J. W. S. field no. RCM-1/27/74-JWS]. Chert from the Jonesboro Limestone; 35 to 45 m south of entrance to Gregory's cave, Tenessee, USA; Cade's Cove Quadrangle.

VAN HORN MOBILE BELT

Tectonic unit/location. Van Horn Mobile Belt, Texas, USA.

Stratigraphy. Stratigraphic section, Van Horn area, western Texas (King and Flawn 1953; King 1976):

  [UNNAMED MIDDLE PROTEROZOIC GROUP]
    Van Horn Sandstone
    Hazel Formation
    Allamoore Formation
    Carrizo Mountain Formation

PPRG samples/localities.

1360 (ts, st, TOC, Corg) [A. V. N. field no. AVN-1B-8/15/79]. Black chert in partially dolomitized stromatolitic limestones from the middle third(?) of the Allamoore Formation; 1 km north of Garren Ranch, 25 km northwest of Van Horn, Texas, USA; 31.13°N, 104.9°W.

Table 14.2. *Continued.*

WOPMAY OROGEN (FORELAND THRUST-FOLD BELT)

Tectonic unit/location. Wopmay Orogen (Foreland Thrust-Fold Belt), Northwest Territories (District of Mackenzie), Canada.

Stratigraphy. Stratigraphic section, Port Epworth area (on the south side of Coronation Gulf), northern District of Mackenzie (Hoffman 1981):

CORONATION SUPERGROUP
    [UNNAMED EARLY PROTEROZOIC GROUP]
        Takiyuak Formation
        Cowles Lake Formation
    RECLUSE GROUP
        Asiak Formation
        Fontano Formation
        Tree River Formation
    EPWORTH GROUP
        Rocknest Formation
        Odjick Formation
    AKAITCHO GROUP

PPRG samples/localities.

1509 (st) [PPRG field no. 86-06-26-1]. Small columnar branching stromatolites (ca. 2 cm in width), dolomitic carbonate with some small pyrite grains, from the base of the Rocknest Formation; stratigraphic section 17 of G.S.C. (Geological Survey of Canada) Open File Report 1278, "base camp section" of the PPRG 1986 N.W.T. field trip, Northwest Territories, Canada; 65.93°N, 114.2°W.

1510 (st) [PPRG field no. 86-06-26-2]. Stratiform carbonate stromatolite with microdigitate columns from the Lower Shale Member of the Rocknest Formation; stratigraphic section 13 of G.S.C. Open File Report 1278, "Autochton section locality," day 1 of PPRG 1986 N.W.T. field trip, Northwest Territories, Canada; 65.70°N, 113.4°W.

1511 (st, TOC, Corg) [PPRG field no. 86-06-26-3]. Gray dolomitic fenestrate domical stromatolites up to 1.5 m wide, from immediately below the *Conophyton* marker bed, Lower Shale Member of the Rocknest Formation; locality cf. 1510.

1512 (ts) [PPRG field no. 86-06-26-4]. Thin-bedded, flat-laminated, silty dolomite from the Lower Shale Member of the Rocknest Formation; locality cf. 1510.

1513 (ts, TOC, Corg) [PPRG field no. 86-06-26-5]. Thin-bedded, flat-laminated, silty dolomite with some cross-lamination, from the Lower Shale Member of the Rocknest Formation; locality cf. 1510.

1514 (st) [PPRG field no. 86-06-26-6]. Dolomite with minor chert lenses of domical and stratiform stromatolites with ribbon-like laminae, from the Lower Shale Member of the Rocknest Formation; locality cf. 1510.

1515 (ts) [PPRG field no. 86-06-26-7]. Intraclastic dolomitic grainstone from the Lower Shale Member of the Rocknest Formation; locality cf. 1510.

1516 (st, TOC, Corg) [PPRG field no. 86-06-26-8]. Dark-gray carbonate and chert domical and stratiform stromatolites from the Lower Shale Member of the Rocknest Formation; locality cf. 1510.

1517 (st) [PPRG field no. 86-06-26-9]. Carbonate and chert domical and stratiform stromatolites from the Lower Shale Member of the Rocknest Formation; locality cf. 1510.

1518 (st, TOC, Corg) [PPRG field no. 86-06-26-10]. Carbonate from the Lower Shale Member of the Rocknest Formation; locality cf. 1510.

1519 (ts) [PPRG field no. 86-06-26-11]. Carbonate intraclast grainstone from the bottom of the final cycle before the "intraclastic member," Lower Shale Member of the Rocknest Formation; locality cf. 1510.

1520 (st) [PPRG field no. 86-06-26-12]. Carbonate columnar stromatolites, 10 cm in diameter, forming small mounds up to 1 m-wide, from immediately above 1519, Lower Shale Member of the Rocknest Formation; locality cf. 1510.

1521 (st, TOC, Corg) [PPRG field no. 86-06-26-13]. Carbonate microdigitate "tufa" from the Lower Shale Member of the Rocknest Formation; locality cf. 1510.

1522 (st) [PPRG field no. 86-06-26-14]. Carbonate digitate "tufa" from the Lower Shale Member of the Rocknest Formation; locality cf. 1510.

1523 (st, TOC, Corg) [PPRG field no. 86-06-26-15]. Black chert in "tufa," inside microdigitate stromatolites, from the Lower Shale Member of the Rocknest Formation; locality cf. 1510.

1524 (st, TOC, Corg) [PPRG field no. 86-06-26-16]. Gray dolomitic bioherm, 3 m thick, with fenestrate fabric containing irregular patches of black chert, from the "Intraclastic Member" of the Rocknest Formation; locality cf. 1510.

1525 (st) [PPRG field no. 86-06-26-17]. Carbonate bioherm with columnar and pseudocolumnar stromatolites, exhibiting two types of internal fabric (fenestrate and thrombolitic), from the "Intraclastic Member" of the Rocknest Formation; locality cf. 1510.

1526 (st) [PPRG field no. 86-06-26-18]. Stromatolitic carbonate, columns ca. 10 cm in diameter with "smooth" fabric, from the "Intraclastic Member" of the Rocknest Formation; locality cf. 1510.

1527 (ts) [PPRG field no. 86-06-26-19]. Red mudstone from the Rocknest Formation; locality cf. 1510.

Table 14.2. *Continued.*

1528 (ts, TOC, Corg) [PPRG field no. 86-06-27-20]. Laminated pale-gray dolomite with some pyrite from "slope section", 208 m on section no. 22 of Grotzinger of the Rocknest Formation; section 22 of G.S.C. Open File Report 1278, day 2, stop 3 of the PPRG 1986 N.W.T. field trip, Northwest Territories, Canada; 66.03°N, 114.4°W.

1529 (ts, TOC, Corg) [PPRG field no. 86-06-27-21]. Dark gray, flat-laminated shale from the base of the Fontano Formation; locality cf. 1528.

1530 (ts) [PPRG field no. 86-06-27-22]. Dolomitized marine cement from the "reefal facies" of the Rocknest Formation; top of section 21 of G.S.C. Open File Report, day 2, stop 4 of the PPRG 1986 N.W.T. field trip, Northwest Territories, Canada; 66.03°N, 114.4°W.

1531 (st, TOC) [PPRG field no. 86-06-27-23]. Pale-gray, dolomitic, large domical and pseudocolumnar stromatolite, from reef core from the "reefal facies" of the Rocknest Formation; locality cf. 1530.

1532 (ts, TOC, Corg) [PPRG field no. 86-06-27-24]. Laminated dark gray shale with some pyrite, from the Odjick Formation; locality cf. 1530.

1533 (st) [PPRG field no. 86-06-28-25]. Laminated white dolomite columnar stromatolites, 5–20 cm in diameter, with steeply convex laminae and parallel branching, from the "reefal facies" of the Rocknest Formation; section 19 of G.S.C. Open File Report 1278, stop 5, day 3 of the PPRG 1986 N.W.T. field trip, Northwest Territories, Canada; 66.38°N, 114.3°W.

1534 (st) [PPRG field no. 86-06-28-26]. White dolomitic pseudocolumnar stromatolite with light gray to pink chert, from an irregular tabular bed of the "reefal facies" of the Rocknest Formation; locality cf. 1533.

1535 (st) [PPRG field no. 86-06-28-27]. White dolomite from large pseudocolumnar stromatolites from the "reefal facies," at base of 5 m-thick cycle of the Rocknest Formation; locality cf. 1533.

1536 (st, TOC) [PPRG field no. 86-06-28-28]. White dolomite from stratiform stomatolite, with numerous chert layers, from the "reefal facies" of the Rocknest Formation; locality cf. 1533.

1537 (ts) [PPRG field no. 86-06-28-29]. Teepee structure with internal cement, from the shoal part of the "reefal facies" of the Rocknest Formation; locality cf. 1533.

1538 (st, TOC) [PPRG field no. 86-06-28-30]. Chert pisolite from the "reefal facies" of the Rocknest Formation; locality cf. 1533.

1539 (st) [PPRG field no. 86-06-28-31]. Stratiform stromatolite with pink and green dolomite from the "reefal facies" of the Rocknest Formation; locality cf. 1533.

1540 (ts, st, TOC) [PPRG field no. 86-06-29-32]. Stratiform and low domical stromatolites with light gray dolomite, from the top of the "shelf interior" cycle 1 of the Rocknest Formation; interval of 230 m to 350 m on stratigraphic section 16 of G.S.C. Open File Report 1278; day 4, stop 5 of the PPRG 1986 N.W.T. field trip, Northwest Territories, Canada; 65.95°N, 114.1°W.

1541 (ts, st, TOC) [PPRG field no. 86-06-29-33]. Stratiform stromatolites, with light-gray dolomite, from the top of the "shelf interior" cycle 2 of the Rocknest Formation; locality cf. 1540.

1542 (ts, TOC) [PPRG field no. 86-06-29-34]. Red-gray, flat-laminated, shaley dolomite from the base of the "shelf interior" cycle 3 of the Rocknest Formation; locality cf. 1540.

1543 (ts, st) [PPRG field no. 86-06-29-35]. Stratiform to low domical stromatolites, with pale-gray dolomite, from the top of the "shelf interior" cycle 3 of the Rocknest Formation; locality cf. 1540.

1544 (ts) [PPRG field no. 86-06-29-36]. Intraclast packstone with light gray matrix, from the bottom of the "shelf interior" cycle 3 of the Rocknest Formation; locality cf. 1540.

1545 (ts, st) [PPRG field no. 86-06-29-37]. Stratiform stromatolite with carbonate, from the top of the "shelf interior" cycle 3 of the Rocknest Formation; locality cf. 1540.

1546 (ts, st, Ccarb) [PPRG field no. 86-06-29-38]. Stratiform stromatolite with light-gray dolomite, from the bottom of the "shelf interior" cycle 4 of the Rocknest Formation; locality cf. 1540.

1547 (ts, st, Ccarb) [PPRG field no. 86-06-29-39]. Stratiform stromatolites with light-gray dolomite, from the top of the "shelf interior" cycle 4 of the Rocknest Formation; locality cf. 1540.

1548 (ts) [PPRG field no. 86-06-29-40]. Light-gray, flat-laminated, dolomite from the bottom of the "self interior" cycle 5 of the Rocknest Formation; locality cf. 1540.

1549 (ts, st) [PPRG field no. 86-06-29-41]. Dolomite from a stratiform to low domical stromatolite, from the top of the "shelf interior" cycle 5 of the Rocknest Formation; locality cf. 1540.

1550 (ts) [PPRG field no. 86-06-29-42]. Light-gray, flat-laminated dolomite, with some stromatolitic layering, from the bottom of the "shelf interior" cycle 6 of the Rocknest Formation; locality cf. 1540.

1551 (ts, st) [PPRG field no. 86-06-29-43]. Light-gray dolomitic stratiform stromatolites, from the top of the "shelf interior" cycle 6 of the Rocknest Formation; locality cf. 1540.

1552 (ts, TOC) [PPRG field no. 86-06-29-44]. Medium-gray dolomite and siltstone, from the bottom of the "shelf interior" cycle 7 of the Rocknest Formation; locality cf. 1540.

1553 (ts, st, TOC) [PPRG field no. 86-06-29-45]. Light-gray dolomitic stratiform stromatolite, from the top of the "shelf interior" cycle 7 of the Rocknest Formation; locality cf. 1540.

Table 14.2. *Continued.*

1554 (ts, Ccarb) [PPRG field no. 86-06-29-46]. Shaley dolomite from the bottom of the "shelf interior" cycle 8 of the Rocknest Formation; locality cf. 1540.

1555 (ts, st, Ccarb) [PPRG field no. 86-06-29-47]. Light-gray dolomitic stratiform stromatolite from the top of the "shelf interior" cycle 8 of the Rocknest Formation; locality cf. 1540.

1556 (ts, Ccarb) [PPRG field no. 86-06-29-48]. Laminated dolomicrite from the bottom of the "shelf interior" cycle 9 of the Rocknest Formation; locality cf. 1540.

1557 (ts, st, Ccarb) [PPRG field no. 86-06-29-49]. Dolomitic stratiform and low domical stromatolite from the top of the "shelf interior" cycle 9 of the Rocknest Formation; locality cf. 1540.

1558 (ts, TOC) [PPRG field no. 86-06-29-50]. Gray, flat-laminated, shaley dolomite, from the bottom of the "shelf interior" cycle 10 of the Rocknest Formation; locality cf. 1540.

1559 (ts, st, TOC) [PPRG field no. 86-06-29-51]. Light-gray, dolomitic stratiform stromatolite, from the top of the "shelf interior" cycle 10 of the Rocknest Formation; locality cf. 1540.

1560 (ts) [PPRG field no. 86-06-29-52]. Medium-gray, flat-laminated, shaley dolomite, from the bottom of the "shelf interior" cycle 11 of the Rocknest Formation; locality cf. 1540.

1561 (ts, st) [PPRG field no. 86-06-29-53]. Light-gray, dolomitic, stratiform stromatolites, from the top of the "shelf interior" cycle 11 of the Rocknest Formation; locality cf. 1540.

1562 (ts) [PPRG field no. 86-06-29-54]. Light-gray dolomitic shale, from the bottom of the "shelf interior" cycle 12 of the Rocknest Formation; locality cf. 1540.

1563 (ts, st) [PPRG field no. 86-06-29-55]. Pale-gray dolomitic stratiform stromatolites, with some chert in irregular masses, lenses, and nodules, from the top of the "shelf interior" cycle 12 of the Rocknest Formation; locality cf. 1540.

1564 (ts, st) [PPRG field no. 86-06-29-56]. Light-gray dolomite from the base of the "shelf interior" cycle 13 of the Rocknest Formation; locality cf. 1540.

1565 (ts, st) [PPRG field no. 86-06-29-57]. Light-gray dolomite from the top of the "shelf interior" cycle 13 of the Rocknest Formation; locality cf. 1540.

1566 (ts, TOC) [PPRG field no. 86-06-29-58]. Light-gray shaley dolomite, with some pyrite, from the bottom of the "shelf interior" cycle 14 of the Rocknest Formation; locality cf. 1540.

1567 (ts, st, TOC) [PPRG field no. 86-06-29-59]. Light-gray dolomite from the top of the "shelf interior" cycle 14 of the Rocknest Formation; locality cf. 1540.

1568 (ts, st) [PPRG field no. 86-06-29-60]. Light-gray dolomite from the bottom of the "shelf interior" cycle 15 of the Rocknest Formation; locality cf. 1540.

1569 (ts, st) [PPRG field no. 86-06-29-61]. Dolomitic stratiform stromatolite, with some chert laminae, from the top of the "shelf interior" cycle 15 of the Rocknest Formation; locality cf. 1540.

1570 (ts, st) [PPRG field no. 86-06-29-62]. Light-gray dolomitic stratiform stromatolite from the bottom of the "shelf interior" cycle 16 of the Rocknest Formation; locality cf. 1540.

1571 (ts) [PPRG field no. 86-06-29-63]. Carbonate stratiform stromatolites from the top of the "shelf interior" cycle 16 of the Rocknest Formation; locality cf. 1540.

1572 (ts, st) [PPRG field no. 86-06-29-64]. Light-gray dolomite from the bottom of the "shelf interior" cycle 17 of the Rocknest Formation; locality cf. 1540.

1573 (ts, st) [PPRG field no. 86-06-29-65]. Light-gray dolomite from the top of the "shelf interior" cycle 17 of the Rocknest Formation; locality cf. 1540.

1574 (ts, TOC) [PPRG field no. 86-06-29-66]. Medium-gray, shaley dolomite, from the bottom of the "shelf interior" cycle 18 of the Rocknest Formation; locality cf. 1540.

1575 (ts, st, TOC) [PPRG field no. 86-06-29-67]. Light-gray dolomite, with some chert laminae, from the top of the "shelf interior" cycle 18 of the Rocknest Formation; locality cf. 1540.

1576 (ts) [PPRG field no. 86-06-29-68]. Medium-gray shaley dolomite from the bottom of the "shelf interior" cycle 19 of the Rocknest Formation; locality cf. 1540.

1577 (ts, st) [PPRG field no. 86-06-29-69]. Dolomitic, low domical stromatolite from the top of the "shelf interior" cycle 19 of the Rocknest Formation; locality cf. 1540.

1578 (ts) [PPRG field no. 86-06-29-70]. Medium-gray, shaley dolomite from the bottom of the "shelf interior" cycle 20 of the Rocknest Formation; locality cf. 1540.

1579 (ts, st) [PPRG field no. 86-06-29-71]. Dolomitic stratiform and low domical stromatolite, from the top of the "shelf interior" cycle 20 of the Rocknest Formation; locality cf. 1540.

Table 14.2. *Continued.*

1580 (ts, st, TOC, Corg, Ccarb) [PPRG field no. 86-06-29-72]. Medium-gray dolomite and white chert in domical stromatolites from the base of the lowermost "*Conophyton*" cycle of the Rocknest Formation; locality cf. 1540.

1581 (ts, st, TOC, Corg, Ccarb) [PPRG field no. 86-06-29-73]. Chertified stromatolites from the middle of the "*Conophyton*" cycle of the Rocknest Formation; locality cf. 1540.

1582 (ts, TOC, Corg, Ccarb) [PPRG field no. 86-06-29-74]. Tufa domes from the top of the "*Conophyton*" cycle of the Rocknest Formation; locality cf. 1540.

1583 (ts, st, TOC, Corg) [PPRG field no. 86-06-29-75]. Gray dolomite with some chert from the marker bed of the "*Conophyton*" cycle of the Rocknest Formation; locality cf. 1540.

1584 (ts, st, TOC, Corg) [PPRG field no. 86-06-29-76]. Gray dolomite, silicified stromatolite (*Conophyton*) from the middle of the "*Conophyton*" cycle of the Rocknest Formation; locality cf. 1540.

1585 (ts, st, TOC, Corg) [PPRG field no. 86-06-29-77]. Gray dolomite, silicified stromatolite (*Conophyton*) from the middle of the "*Conophyton*" cycle of the Rocknest Formation; locality cf. 1540.

1586 (ts, st, TOC, Corg) [PPRG field no. 86-06-29-78]. Dolomite, domical stromatolites from the top of the "*Conophyton*" cycle of the Rocknest Formation; locality cf. 1540.

1587 (ts, st, TOC, Corg) [PPRG field no. 86-06-29-79]. Black chert from the upper part of "*Conophyton*"-bearing domes of the Rocknest Formation; locality cf. 1540.

1588 (TOC, Corg, ker) [PPRG field no. 86-06-29-80]. Black weathered shale from the "Black Shale Member" of the Fontano Formation; locality cf. 1540.

1810 [D. R. L. field no. NWT 1-1]. Cross-bedded quartzite from the basal part of the Rocknest Formation; stratigraphic section 17 of G.S.C. Open File Report 1278, "base camp section" of the PPRG 1986 N.W.T. field trip, Northwest Territories, Canada; 65.93°N, 114.2°W.

1811 [D. R. L. field no. NWT 2-2]. Chert from the lower part of the Rocknest Formation; stratigraphic section 13 of GSC. Open File Report 1278, "Autochton section locality," day 1 of PPRG 1986 N.W.T field trip, Northwest Territories, Canada; 65.70°N, 113.4°W.

1812 [D. R. L. field no. NWT 3-1]. Metamorphosed chert from the "slope facies" of the Rocknest Formation; top of section 21 of G.S.C Open File Report, day 2, stop 4 of the PPRG 1986 N.W.T field trip, Northwest Territories, Canada; 66.03°N, 114.4°W.

1813 [D. R. L. field no. NWT 4-1]. Metamorphosed chert from the "reefal facies" of the Rocknest Formation; section 22 of G.S.C Open File Report 1278, day 2, stop 3 of the PPRG 1986 N.W.T field trip, Northwest Territories, Canada; 66.03°N, 114.4°W.

1814 [D. R. L. field no. NWT 4-3]. Sandstone-graywake from the basal 50 m of the Asiak Formation; locality cf. 1813.

1815 [D. R. L. field no. NWT 4-4]. Graywacke from the lower part of the Asiak Formation; locality cf. 1813.

1816 [D. R. L. field no. NWT 4-5]. Graywacke from the lower part of the Asiak Formation; locality cf. 1813.

1817 [D. R. L. field no. NWT 5-1]. Chert from the "pisolitic teepee facies" of the Rocknest Formation; section 19 of G.S.C. Open File Report 1278, stop 5, day 3 of the PPRG 1986 N.W.T. field trip, Northwest Territories, Canada; 66.38°N, 114.3°W.

1818 [D. R. L. field no. NWT 6-4]. Chert from the "cyclic shelf facies" of the Rocknest Formation; interval of 230 m to 350 m on stratigraphic section 16 of G.S.C. Open File Report 1278; day 4, stop 5 of the PPRG 1986 N.W.T. field trip, Northwest Territories, Canada; 65.95°N, 114.1°W.

1819 [D. R. L. field no. NWT 6-1]. Sandstone from the top of the Odjick Formation; locality cf. 1818.

1820 [D. R. L. field no. NWT 6-2]. Quartzite from the top of the Odjick Formation; locality cf. 1818.

1821 [D. R. L. field no. NWT 6-3]. Sandstone from the top of the Odjick Formation; locality cf. 1818.

2365 [D. R. L. field no. NWT-2-5]. Chert from the lower part of the Rocknest Formation; stratigraphic section 13 of G. S. C. Open File Report 1278, "Autochton section locality," day 1 of PPRG 1986 N.W.T. field trip, Northwest Territories, Canada; 65.70°N, 113.4°W.

2366 [D. R. L. field no. NWT 2-6]. Chert from the lower part of the Rocknest Formation; locality cf. 2365.

2367 [D. R. L. field no. NWT 3-2]. Chert from the "slope facies" of the Rocknest Formation; top of section 21 of G.S.C. Open File Report, day 2, stop 4 of the PPRG 1986 N.W.T. field trip, Northwest Territories, Canada; 66.03°N, 114.4°W.

2368 [D. R. L. field no. NWT 3-3]. Chert from the "slope facies" of the Rocknest Formation; locality cf. 2367.

2369 [D. R. L. field no. NWT 3-4]. Chert from the "slope facies" of the Rocknest Formation; locality cf. 2367.

2370 [D. R. L. field no. NWT 3-5]. Chert from the "slope facies" of the Rocknest Formation; locality cf. 2367.

2371 [D. R. L. field no. NWT 4-4]. Chert from the "reefal facies" of the Rocknest Formation; section 22 of G.S.C. Open File Report 1278, day 2, stop 3 of the PPRG 1986 N.W.T. field trip, Northwest Territories, Canada; 66.03°N, 114.4°W.

2372 [D. R. L. field no. NWT 5-2]. Chert from the "pisolitic teepee facies" of the Rocknest Formation; section 19 of G.S.C. Open File Report 1278, stop 5, day 3 of the PPRG 1986 N.W.T. field trip, Northwest Territories, Canada; 66.38°N, 114.3°W.

2373 [D. R. L. field no. NWT 5-3]. Chert from the "pisolitic teepee facies" of the Rocknest Formation; locality cf. 2372.

2374 [D. R. L. field no. NWT 5-5]. Chert from the "pisolitic teepee facies" of the Rocknest Formation; locality cf. 2372.

Geologic Samples

Table 14.2. *Continued.*

2375 [D. R. L. field no. NWT 6-5]. Chert from the "cyclic shelf facies" of the Rocknest Formation; interval of 230 m to 350 m on stratigraphic section 16 of G.S.C. Open File Report 1278; day 4, stop 5 of the PPRG 1986 NWT, field trip, Northwest Territories, Canada; 65.95°N, 114.1°W.

2376 [D. R. L. field no. NWT-6-6]. Recrystallized chert from the Rocknest Formation; locality cf. 2375.

2377 (st) [D. R. L. field no. NWT-6-7]. Recrystallized chert from a silicified "*Conophyton*" from the Rocknest Formation; locality cf. 2375.

14.2.6 SOUTH AMERICA

SÃO FRANCISCO BASIN

Stratigraphy.

    Paraopeba Formation

PPRG samples/localities.

2364 (ts) [T. F. field collection]. Black chert from the Paraopeba Formation; 9 km west of São Gabriel and ca. 80 km north-northeast of Brazilia, Goias, Brazil.

WEST-CENTRAL BRAZIL

Stratigraphy.

    Tamengo Formation

PPRG samples/localities.

2363 [T. F. field collection]. Carbonate from the Tamengo Formation; Ladario, (Brazil/Bolivia border), Mata Grosso Do Sul, Brazil.

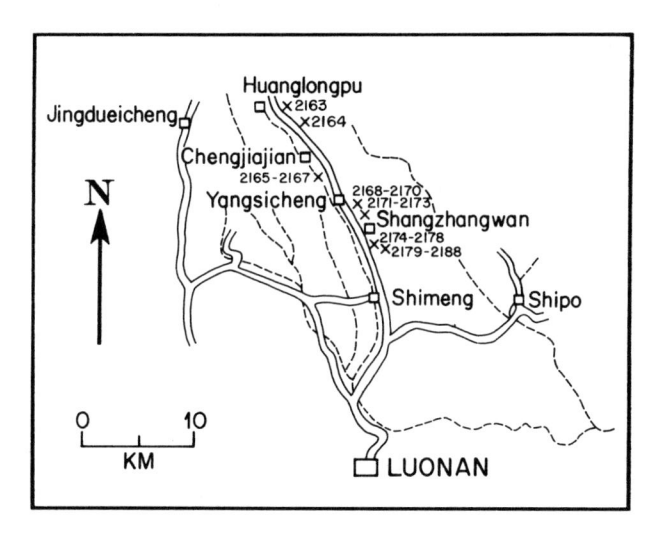

Figure 14.3. Geographic location of PPRG-P samples number 2163–2188 from the Luonan area, east-central Shaanxi, North China Platform.

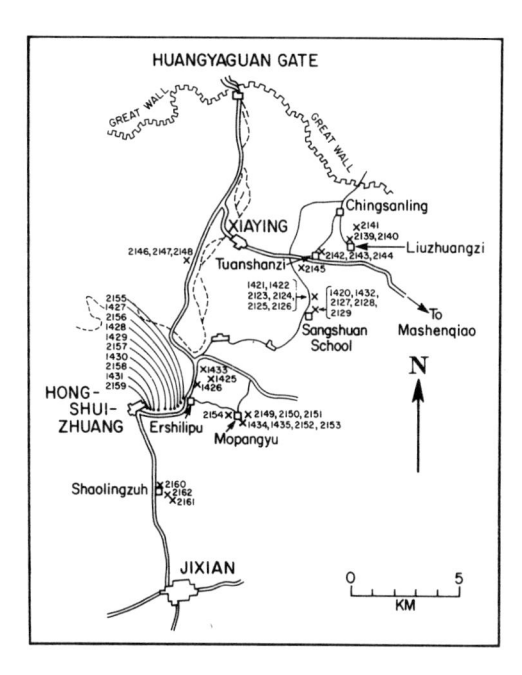

Figure 14.4. Geographic location of PPRG-P samples number 1420–1422, 1425–1435, 2123–2129, and 2139–2162 from the Jixian area, northern Tianjin Shi, North China Platform.

Table 14.5. *Estimated ages of geologic units represented in the PPRG collections*

GEOLOGIC UNIT	ESTIMATED AGE (Ma)	MAXIMUM AGE		MINIMUM AGE	
		AGE (Ma)	REFERENCES	AGE (Ma)	REFERENCES
Allamoore Fm	1050	1250	Denison 1980	1000	Denison 1980
Almodovar del Rio Grp	600				
Altyn Fm	1440	1700	Harrison 1984	1100	Elston 1984
Anakeesta Fm	600				
Apex Basalt	3400	3435	Richards et al. 1981	3330	Richards 1977
Areyonga Fm	760	897	Black et al. 1980	750	Wells et al. 1967
Asiak Fm	1900				
Auburn Dolomite	780	802	Preiss 1987	750	Preiss 1987
Awatubi Mem	862	1090	McKee and Noble 1976	830	Elston 1979
Bar River Fm	2250	2450	Fairbairn et al. 1969	2219	Corfu and Andrews 1986
Beaumont Dolomite	780	800	Preiss 1987	750	Preiss 1987
Beck Spring Dolomite	850	1400	Miller 1982	650	Labotka and Albee 1977
Bederysh Mem	1000	1100	Keller 1975	965	Keller 1975
Bianka Mem	740				
Biegaizi Fm	1400				
Biri Fm	700				
Bitter Springs Fm	850	897	Black et al. 1980	760	Wells et al. 1967
Biwabik Iron Fm	2090	2500	Faure and Kovach 1969	1880	Van Schmus, 1980
Blondeau Fm	2710			2500	
Bonavista Fm	545	584			
Boonall Dolomite	640	678	Plumb et al. 1981	639	Plumb et al. 1981
Brigus Fm	540	584			
Bruce Fm	2250	2450	Fairbairn et al. 1969	2219	Corfu and Andrews 1986
Bungle Bungle Dolomite	1364			1560	Plumb 1985
Carbon Butte Mem	870	1090	McKee and Nobel 1976	830	Elston 1979
Carbon Canyon Mem	900	1090	McKee and Nobel 1976	830	Elston 1979
Carbon Leader Mem	2710	2725		2643	
Chamberlain Shale	1450	1700	Harrison 1984	1100	Elston 1984
Chambless Limestone	540				
Changzhougou Fm	1950	1974	Chen et al. 1980; Schopf et al. 1984	1875	Chen et al. 1980; Schopf et al. 1984
Chapel Island Fm	560	608	Narboune et al. 1987	556	Narboune et al. 1987
Chengjiajiang Fm	1330				
Chichkan Fm	650				
Chuanlinggou Fm	1850	1875			
Chulaktau Fm	550				
Cijara Fm	610				
Cleaverville Fm	3300	3400	Hickman 1983	3000	Pidgeon 1984
Coppercap Fm	750	770			
Dohongyu Fm	1650	1678	Chen et al. 1980; Schopf et al. 1984	1621	Chen et al. 1980; Schopf et al. 1984

Table 14.5. Continued.

GEOLOGIC UNIT	ESTIMATED AGE (Ma)	MAXIMUM AGE		MINIMUM AGE	
		AGE (Ma)	REFERENCES	AGE (Ma)	REFERENCES
Dales Gorge Mem	2500				
Dismal Lakes Grp	1400	1663	Bowring and Ross 1985	1200	Bowring and Ross 1985
Duguan Fm	1200				
Duppa Mem	875	1090	Elston 1979	830	Elston 1979
Dwaal Heuvel Quartzite	2200	2224	Hamilton 1977	2050	Walraven et al. (In Press)
Egan Fm	640	605	Plumb et al. 1981	576	Plumb et al. 1981
Ekwi Supergrp	650	769	Park and Aitken 1986	570	Morris and Park 1981
Elvire Fm	639	678	Plumb et al. 1981	671	Plumb et al. 1981
Erdaohe Fm	1360				
Espanola Fm	2250	2450	Fairbairn et al. 1969	2219	Corfu and Andrews 1986
Estenilla Fm	620				
Euro Basalt Fm	3330	3340		3000	
Fargoo Tillite Fm	685			671	Plumb et al. 1981
Fengjiawan Fm	1190				
Fig Tree Grp	3350	3355	Martinez et al. 1984	3000	Tegtmeyer and Kroner 1987
Fontano Fm	1900				
Frank River Sandstone	680	680	Plumb et al. 1981	671	Plumb et al. 1981
Franklin Mountain Fm	515	530		500	
Fuentes Fm	620				
Gamohaan Fm	2450	2357	Beukes 1987	2239	
Gaoyuzhuang Fm	1425	1435			
Gordon Lake Fm	2259	2450	Fairbairn et al. 1969	2219	Corfu and Andrews 1986
Gorge Creek Grp	3200				
Government Subgrp	2710				
Gowganda Fm	2250	2450	Fairbairn et al. 1969	2219	Corfu and Andrews 1986
Greyson Shale	1420	1700	Harrison 1984	1100	Elston 1984
Guddadarangavanahalli Fm	1700	2400	Gowda and Sreenivasa 1969		
Gunflint Iron Fm	2090	2500	Faure and Kovach 1969	1880	Van Schmus 1980
Hardy Sandstone	2765	2770	Blake and McNaughton 1984	2760	Blake and McNaughton 1984
Hekpoort Andesite	2200	2224	Hamilton 1977	2050	Walraven et al. (In Press)
Helena Fm	1380	1430	Harrison 1984	1100	Elston 1984
Hongshuizhuang Fm	1250	1435		1185	
Hooggenoeg Fm	3450	3290	Sinha 1972	3200	Hurley et al. 1972
Hotazel Fm	2250				
Insuzi Grp	3000	3090	Burger and Coertze 1973	2870	Hunter 1974
Jarrad Sandstone	675	870	Plumb et al. 1981	639	Plumb et al. 1981

Table 14.5. Continued.

GEOLOGIC UNIT	ESTIMATED AGE (Ma)	MAXIMUM AGE		MINIMUM AGE	
		AGE (Ma)	REFERENCES	AGE (Ma)	REFERENCES
Jeerinah Fm	2650	2720	Blake and McNaughton 1984	2490	Blake and McNaughton 1984
Jeppestown Shale	2720				
Jiudingshan Fm	800	950		765	
Jonesboro Limestone	475				
Julie Mem	600	640		600	
Jupiter Mem	930	1090		830	
Kameeldoorns Fm	2705	2708		2699	
Kataskin Mem	1240				
Katav Fm	940				
Keele Fm	700		Eisbacher 1985		
Keewatin Grp	2500			2500	
Khatyspyt Fm	620				
Kimberley Shale	2750				
Kingston Peak Fm	850				
Klipfonteinheuwel Fm	2325	2357	Beukes 1987	2239	Beukes 1987
Kona Dolomite	2000	2500		1800	
Koolpin Fm	1885				
Kostov Fm	560				
Kromberg Fm	3300			3000	
Kuktur Mem	1240				
Kuruman Iron Fm	2350				
Kuruna Siltstone	2740	2760		2720	
Kylena Basalt	2768	2770	Blake and McNaughton 1984	2490	Blake and McNaughton 1984
Lalla Rookh Sandstone	3200				
Little Dal Grp	800	800		770	
Longjiayuan Fm	1300				
Lookout Schist	2000	2410		1650	
Lorrain Fm	2250	2450		2419	
Los Parrales Shale	550				
Louden Volcanics	3000				
Love's Creek Mem	850	897	Black et al. 1980	760	Wells et al. 1967
Luoquan Fm	700				
Maddina Basalt	2735	2760	Blake and McNaughton 1984	2720	Blake and McNaughton 1984
Malmani Subgrp	2400	2643		2145	
Maloinzer Mem	1240				
Matinenda Fm	2400	2450		2333	
McArthur Grp	1690	1820	Jackson et al. 1987	1620	Jackson et al. 1987
McKim Fm	2400	2450		2333	
McMinn Fm	1340	1690	Jackson et al. 1987	1280	Jackson et al. 1987
McPhee Fm	3525	3560	Blake and McNaughton 1984	3455	Blake and McNaughton 1984
Meentheena Carbonate Mem	2750	2760	Blake and McNaughton 1984	2720	Blake and McNaughton 1984
Middle Marker Mem	3450	3290		3200	
Min'yar Fm	740				

Table 14.5. Continued.

GEOLOGIC UNIT	ESTIMATED AGE (Ma)	MAXIMUM AGE AGE (Ma)	REFERENCES	MINIMUM AGE AGE (Ma)	REFERENCES
Mingah Tuff	2750	2760		2490	Blake and McNaughton 1984
Mississagi Fm	2250	2450	Fairbairn et al. 1969	2410	
Montacute Dolomite	770	800	Preiss 1987	750	Preiss 1987
Montenegro Mem	545				
Moonlight Valley Tillite	680	870	Plumb et al. 1981	671	Plumb et al. 1981
Mosquito Creek Fm	3200				
Mount Ada Basalt	3450				
Mount Forster Sandstone	640	671	Plumb et al. 1981	639	Plumb et al. 1981
Mount Shields Fm	1320	1430	Harrison 1984	1077	Elston 1979
Msauli Chert Mem	3375	3355	Stanistreet et al. 1981	3000	Viljoen and Viljoen 1969
Newland Limestone	1440	1700	Harrison 1984	1300	Elston 1984
Nonesuch Shale	1055	1023	Chandhuri and Faure 1967	1023	Chandhuri and Faure 1967
North Leader Mem	2710				
Nymerina Basalt	2700	2760	Blake and McNaughton 1984	2720	Blake and McNaughton 1984
Odjick Fm	1925				
Olympio Fm	1856			1854	
Paradise Creek Fm	1650	1670	Blake 1987	1620	Blake 1987
Paraopeba Fm	800	1000	Fairchild and Subacius 1986	600	Fairchild and Subacius 1986
Pecors Fm	2250	2450		2419	
Pedroche Fm	550				
Pertataka Fm	610	640			
Pigeon Fm	600				
Pokegama Quartzite	2090	2500	Faure and Kovach 1969	1880	Van Schmus 1980
Poleta Fm	540				Mount et al. 1983
Prichard Fm	1450	1700		1300	
Pusa Fm	620				
Ramsay Lake Fm	2250	2450		2419	
Random Fm	550	607			
Ranford Fm	671	870	Plumb et al. 1981	639	Plumb et al. 1981
Rapitan Grp	775		Eisbacher 1985		Park and Aitken 1986
Red Pine Shale	950	931		931	
Red Rock Beds	1800	1845		1760	
Redstone River Fm	770	770			
Reed Dolomite	570		Signor et al. 1987		
Reivilo Fm	2300				
Rencontre Fm	565				
Revet Mem	1260				
Rhynie Sandstone	785	800	Parkin 1969	750	Parkin 1969
Rietgat Fm	2650	2699		2357	
River Wakefield Subgrp	775	800	Preiss 1987	775	Ludbrook 1980
Rocknest Fm	1925	1930		1900	

Table 14.5. Continued.

GEOLOGIC UNIT	ESTIMATED AGE (Ma)	MAXIMUM AGE		MINIMUM AGE	
		AGE (Ma)	REFERENCES	AGE (Ma)	REFERENCES
Rooinekke Fm	2300				
Rossport Fm	1339	1536		1091	
Rove Fm	1900			1635	
Rushall Slate	3100				
Sayunei Fm	800		Yeo 1981		Yeo 1981
Serpent Fm	2250	2450		2419	
Sheepbed Fm	650		Eisbacher 1985		Eisbacher 1985
Shepard Fm	1350	1430	Harrison 1984	1100	Elston 1984
Shezal Fm	770		Yeo 1981		Eisbacher 1985
Shorikha Fm	900	827	Ivanovskaya et al. 1974	769	Ivanovskaya et al. 1974
Siyeh Fm	1380	1430	Harrison 1984	1100	Elston 1979
Skillogalee Dolomite	770	800	Preiss 1987	750	Preiss 1987
Snowslip Fm	1370	1430	Harrison 1984	1100	Elston 1979
Sotillo Grp	550				
Soudan Iron Fm	2650	2660		2500	
Spartan Grp	1900				
Sukhaya Tunguska Fm	1000	950	Ivanovskaya et al. 1974	850	Ivanovskaya et al. 1974
Tamengo Fm	570	623	Zaine and Fairchild 1985	547	Fairchild 1982
Tanner Mem	950	1090	Ford and Breed 1973	830	Ford and Breed 1973
Tapley Hill Fm	750	750	Preiss 1987	750	Preiss 1987
Tean Fm	630	576	Plumb et al. 1981		Plumb et al. 1981
Theespruit Fm	3550		Martinez et al. 1984	3510	Tegtmeyer and Kroner 1987
Tieling Fm	1185	1185		1185	
Timperley Shale	630	639	Plumb et al. 1981		Plumb et al. 1981
Tindelpina Shale Mem	750				
Torridon Grp	800	974		793	
Towers Fm	3435	3435	Blake and McNaughton 1984	3435	Blake and McNaughton 1984
Tuanshanzi Fm	1750	1875		1650	
Tukan Mem	1350				
Tumbiana Fm	2750	2768	Blake and McNaughton 1984	2490	Blake and McNaughton 1984
Twitya Fm	700		Eisbacher 1985		Eisbacher 1985
Tyler Fm	1950			1800	
Union Island Grp	1900	1928		1865	
Urquhart Shale	1670	1690	Blake 1987	1670	Blake 1987
Ushakov Mem	1240				Keller 1975
Ust-Pinega Fm	620				
Vempalle Fm	1700	2440	Crawford and Compston 1973		Crawford and Compston 1973
Ventersdorp Supergrp	2650				
Venterspost Fm	2670	2725		2643	
Venterspost Carbon Reef Mem	2670				
Visingsö Grp	775	1060	Magnasson 1960	663	Bonhomme and Welin 1984

Table 14.5. Continued.

GEOLOGIC UNIT	ESTIMATED AGE (Ma)	MAXIMUM AGE		MINIMUM AGE	
		AGE (Ma)	REFERENCES	AGE (Ma)	REFERENCES
Walcott Mem	850	1090	Ford and Breed 1973	830	Ford and Breed 1973
Warrie Mem	2650				
Whim Creek Grp	3000	3000	Blake and McNaughton 1984	3000	Blake and McNaughton 1984
Windfall Fm	510		Palmer 1965		Palmer 1971
Witwatersrand Supergrp	2710				
Woocalla Dolomite Mem	750	800		750	
Wumishan Fm	1325	1435		1185	
Wyman Fm (AUS)	550	3340	Pidgeon 1984 Richards et al. 1981	2920	Blake and McNaughton 1984; John et al. 1981
Wyman Fm (USA)	550				
Xingji Fm	560				
Xunjiansi Fm	1250				
Yangzhuang Fm	1350	1435		1185	
Yudoma Fm	600	675	Semikhatov et al. 1970	570	Semikhatov et al. 1970
Yurabi Fm	640	605	Plumb et al. 1981	576	Plumb et al. 1981
Zhangqu Fm	850	950		765	

15
Flow Chart and Processing Procedures for Rock Samples
HARALD STRAUSS DAVID J. DES MARAIS J. M. HAYES
TOBY B. MOORE J. WILLIAM SCHOPF

15.1

Summary of Processing Procedures

Investigation of a large number of samples during this project led to development of the following processing routine.

Curation of samples was performed at the University of California, Los Angeles. As appropriate, subsamples were subsequently distributed to PPRG members for analyses to be carried out at their home institutions. A flow chart outlining the various procedures involved is shown in Figure 15.2 and is summarized below.

The initial curation for every incoming rock sample consisted of assigning a PPRG Sample Number (e.g., "1001"). Pertinent geological information was compiled and entered into the databases "Inventory," "Site," and "Strat" (see Chapter 21).

For paleontological and mineralogical studies, petrographic thin sections were prepared of each sample: for microfossil studies, a 150 μm-thick "paleo"-section ("1001-1-A"); and for petrographic studies, either a standard 30 μm-thick section for non-carbonates ("1001-1-B") or a 5 to 15 μm-thick section for carbonates ("1001-1-C"). In addition, large-area thin sections (150 μm-thick, "1001-1-STROM") were prepared of selected stromatolitic samples.

Sample processing for geochemical and/or palynological studies was initiated by discarding any weathered surface or secondarily emplaced vein material and generating a mass of clean interior rock chips ≤1 cm in diameter ("1001-1-RC"). Chipping of small samples was performed using a geologic hammer; larger samples were chipped with a jawbone (i.e., "chipmunk") crusher. In order to remove any organic contaminants, the chips were etched in a 20% HF-10% HCl solution, then rinsed with large volumes of distilled water and dried in a drying oven at 75° C. This process was repeated until a 10 to 15% dry weight loss of the rock chips was attained. At this stage, ≤250 g of dry etched rock chips ("1001-1-ERC") were available for further processing.

Processing of samples selected for palynological studies involved maceration of the sample in HCl and HF; filter techniques were applied to concentrate the microfossiliferous residue and strew slides ("1001-1-SS") were prepared.

A representative suite of available rock samples was selected for geochemical investigation. Care was taken to consider all lithologies available from a sampled geologic unit. The analytical methods applied are described in detail in Chapters 16 and 18; a brief outline of the processing scheme is given below.

About 25 g of etched rock chips were pulverized in a shatterbox (ring or puck mill) to generate rock powder ("1001-1-RRP"). This powder was analyzed for its total organic carbon content (TOC, mgC/g) and organic carbon isotopic composition (del^{13}C‰ vs. PDB) as part of the "survey level" analyses discussed in Chapter 16.

Organic-rich samples (with TOC ≥ 5 mgC/g) were selected for further detailed organic geochemical studies of extract and kerogen fractions. The extract work involved an initial sample screening, using Rock–Eval to determine the maturity and type of organic matter present, followed by extraction of soluble components ("1001-1-EOM") with organic solvents. These extracts were analyzed by use of gas chromatography (GC) and combined gas chromatography-mass spectrometry (GC-MS). Kerogen studies consisted of isolating the pure kerogen fraction ("1001-1-KER") followed by analyses of the kerogen color, the elemental abundances of C, H, and N, and of the isotopic composition of organic carbon.

Bulk isotopic compositions of carbon and oxygen were determined for selected carbonate samples.

The abundance of sulfur in selected shales and sulfur-rich sediments was obtained through combustion in a LECO Analyzer. Samples containing appreciable amounts of sulfides or sulfates were prepared for sulfur isotopic measurements.

Throughout the entire sample processing procedure, care was taken to avoid any possible contamination. Blanks and standards were prepared to monitor the accuracy of the methods applied and the results obtained. The original samples as well as all subsamples were labelled as indicated on the flow chart (Fig. 15.1); unused portions of these samples are carefully stored for reference and future analyses in the PPRG rock storage facilities at the Center for the Study of Evolution and the Origin of Life (CSEOL), Geology Building, University of California, Los Angeles (J. W. Schopf, curator). These materials, as well as similarly processed samples listed in the previous PPRG collaborative work *Earth's Earliest Biosphere* (Schopf 1983a), are available for study by interested investigators.

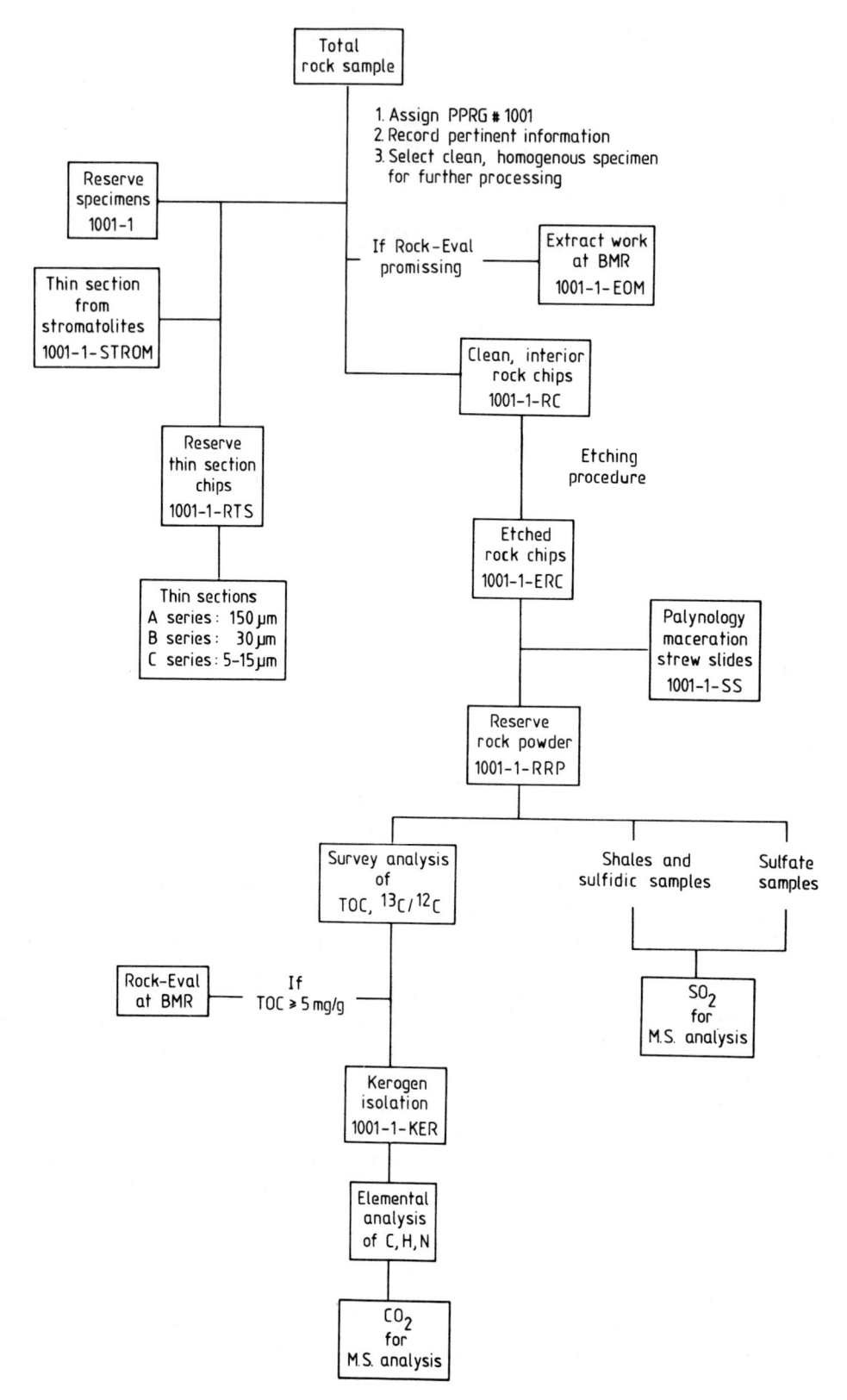

Flow Chart and Processing Procedures for Rock Samples

Figure 15.1. Flow chart summarizing the procedures of rock analysis.

16
Procedures of Whole Rock and Kerogen Analysis

HARALD STRAUSS   DAVID J. DES MARAIS   J. M. HAYES
IAN B. LAMBERT   ROGER E. SUMMONS

16.1

Carbon: Whole Rock Analysis

Samples selected for geochemical analyses were processed following the procedure outlined in Chapter 15 and summarized in Figure 15.1. In general, the methods applied were standard techniques that had been modified slightly in order to analyze efficiently a large number of samples. Great care was exercised to avoid any possible contamination, for example by extraneous organic matter.

16.1.1 Organic Carbon

A "survey level" examination, carried out on all samples, included determination of the total organic carbon content (TOC, mgC/g) and of its isotopic composition (del^{13}C‰ vs. PDB). A detailed description of this method has been given by Gelwicks et al. (pers. comm. to H. S.), an abbreviated version of which is outlined below.

Carbon dioxide produced for isotopic analysis was measured volumetrically and the total organic carbon content was calculated using the following equations:

$$n\mu m \cdot \frac{m}{10^6 \mu m} \cdot \frac{12\,g}{m} \cdot \frac{10^3\,mg}{g} = mg\,C \tag{16.1.1}$$

$$\frac{(mg\,C)(1000)}{mg_{sample}} = mg\,C/g\,(TOC) \tag{16.1.2}$$

Further processing for stable isotopic analysis involved four major steps:

(1) weighing of samples, followed by acidification to dissolve the carbonate fraction;
(2) combustion of samples to produce CO_2 for TOC and carbon-isotopic determination;
(3) purification and collection of CO_2, and calculation of TOC content; and
(4) isotopic analysis.

(1) Rock powder (50 to 300 mg) was weighted and transferred into a 9 mm-diameter quartz tube (one end of which had been sealed). Samples were treated at least twice with concentrated (37%) HCl to dissolve any carbonate minerals. They were then carefully washed with distilled water and dried overnight. A centrifuge was used to concentrate the HCl-resistant residue, and the supernatant was decanted. Neutrality was determined with standard pH paper.

(2) About 1 g of CuO (previously cleaned of any organic contaminants by heating at 550°C for several hours) was added to the dry samples, and the quartz tubes were attached to a vacuum line, evacuated, and sealed. Samples and CuO were mixed to homogeneity. All samples, standards, and blanks were placed in a muffle furnace and heated at 550°C for 1 h and then at 850°C for 3 h.

(3) After cooling to room temperature, each sample tube was attached to a vacuum line. Carbon dioxide was purified by cryogenic separation of H_2O and CO_2 by use of a pentane-liquid nitrogen slush. Incondensable gas was removed while the CO_2 remained frozen. Sufficient time was allowed for equilibration. Carbon dioxide was measured volumetrically (to permit calculation of TOC content) and then trapped in a 6 mm-diameter pyrex tube in preparation for mass spectrometric analysis.

(4) The CO_2 samples were transferred to a mass spectrometer and analyzed for their carbon-isotopic composition.

Analyses were carried out in batches of 24 (viz., 22 samples plus two blanks). All glass tubing used during these analyses had been previously cleaned of organic contaminants by heating to 550°C. Replicate analyses were performed in order to assure the accuracy of the analytical method (Table 16.1).

16.1.2 Rock–Eval Analyses

Selected samples (those with TOC $\geqslant$ 5 mgC/g) were subjected to Rock–Eval analysis to obtain information regarding their maturity and the type of organic matter present. Analyses were carried out on 100 mg aliquots of powdered rock using a GIRDEL IFP-Fina Mark 2 instrument. The critical parameters used to evaluate maturity were (i) T_{max} (°C), the maximum temperature to which the organic matter had been subjected; and (ii) S2(kg/tonne), a measure of the amount of kerogen preserved.

Table 16.1. *Replicate analyses of total organic carbon (TOC) content and del$^{13}C_{org}$ from whole rock samples.*

Sample Number	Lithology[a]	TOC (mg/g)	$\delta^{13}C$(‰)
2442	CH	0.97	−33.39
2442	CH	0.95	−33.58
1271	CH	0.39	−25.72
1271	CH	0.32	−25.61
2520	CARB	1.28	−33.01
2520	CARB	1.35	−33.19
1323	CARB	0.27	−29.57
1323	CARB	0.29	−29.61
1906	SH	34.28	−43.39
1906	SH	30.00	−43.25
1906	SH	33.77	−43.34
2426	SH	6.87	−32.44
2426	SH	6.99	−32.50
2426	SH	6.75	−32.38
2451	SH	29.32	−31.82
2451	SH	29.36	−31.74
2451	SH	29.75	−31.83

[a] CH = chert; CARB = carbonate; SH = shale

The following levels of maturity were distinguished:

< 420°C: data inconclusive
420° to 440°C: marginally mature;
440° to 460°C: mature
> 460°C: overmature.

Rock–Eval data were used to select samples for further processing by extraction of bitumen and detailed analysis of biomarkers (following the procedures outlined in Chapter 18).

16.1.3 Carbonate Carbon

Bulk carbon and oxygen isotopic compositions were measured for selected carbonate samples. Carbon dioxide was liberated via phosphorylization at 50°C for 48 h, using H_3PO_4 enriched with P_2O_5 to about 103% and having a density of > 1.9 (Wachter and Hayes 1985). Appropriate mineral and temperature correction factors were applied to the oxygen data (Table 16.2). Standard samples were analyzed to assure the accuracy of the results obtained (Table 16.3).

Table 16.2. *Fractionation factors for del^{18}O used for mineral and temperature corrections in carbonates.*

Mineral	50°C
Calcite	1.00925
Dolomite	1.010
Mixed	1.010

Table 16.3. *del^{13}C and del^{18}O values for carbonate standards.*

Standard	$\delta^{13}C$(‰) vs. PDB	$\delta^{18}O$(‰) vs. PDB
UCLA-Std.	−4.95	−19.98
	−4.98	−19.92
	−4.94	−19.72
	−4.97	−19.20
	−5.09	−19.34
NBS 20	−1.07	−4.42
	−1.06	−4.14[a]

[a] Coplen and Kendall (1982)

16.2

Sulfur: Whole Rock Analysis

16.2.1 Sulfide Sulfur

Selected shales and samples containing visible sulfide minerals were chosen for sulfur analysis. Total sulfur content was determined using a LECO Sulfur/Carbon Analyzer.

In preparation for isotopic analysis, some nodular or bedded pyrite was concentrated by microdrilling. Disseminated sulfides were prepared following the method described by Canfield et al. (1986). An initial HCl treatment was used to dissolve carbonates and monosulfides. Hydrogen sulfide, liberated by boiling of the sample in a chromium chloride solution for 2 h, was flushed by use of nitrogen into a 30% Zn-acetate trapping solution resulting in precipitation of ZnS. By use of 0.1 M $AgNO_3$, the ZnS was converted into Ag_2S which was then filtered, dried, and combusted on a vacuum line with CuO (Fritz et al. 1974) at 950°C. Pure mineral separates (viz., standards and some pyrite nodule samples) were combusted directly, without prior wet chemical preparation, at 1100°C. The SO_2 produced by combustion was purified, trapped in 6 mm-diameter pyrex tubes, and its fulfur-isotopic composition was measured by mass spectrometry. Results of replicate analyses of the sulfur-isotopic composition of a standard are presented in Table 16.4.

16.2.2 Sulfate Sulfur

Sulfates were reacted with "Thode-solution" (Thode et al. 1961). The resulting H_2S was flushed into a Zn-acetate trapping solution, the precipitated ZnS was converted in Ag_2S, and the sulfate sulfur-isotopic composition was analyzed as outlined above for sulfide sulfur. Results of replicate analyses of a sulfate sample are presented in Table 16.5.

Table 16.4. *Replicate analyses of del^{34}S versus CDT for sulfide standards.*

Standard	$\delta^{34}S$ (‰)
NBS 123	+17.15
	+17.34
	+17.11
	+17.09 ± 0.31 (IAEA)

Table 16.5. *Replicate analyses of del^{34}S versus CDT for sulfate samples.*

Sample Number	$\delta^{34}S$ (‰)
1663	+15.51
	+15.67

16.3 Notation and Precision of Isotopic Measurements

Stable isotope results obtained using the techniques outlined above are presented in Chapter 17. All values are reported in the usual delta notation. Carbon and oxygen data are given as per mil differences versus the PDB standard, and sulfur isotope results as per mil differences versus the Canyon Diablo troilite (CDT) standard. The precision determined from replicate measurements was usually better than 0.2‰ for carbon and oxygen, and better than 0.5‰ for sulfur.

16.4 Kerogen

On the basis of total organic carbon content (TOC > 1 mgC/g), lithology, facies, paleontological significance, and geologic age, a representative suite of samples was selected for kerogen analyses.

16.4.1 Kerogen Isolation

Kerogen was isolated from rock powder following a four-step procedure:

(1) Samples were treated with concentrated (37%) HCl for up to 12 h in order to remove carbonate minerals.
(2) Samples were next reacted with concentrated (48%) HF for up to 24 h in order to dissolve silicate minerals.
(3) The residue from the HF-treatment was reacted with concentrated HCl (for about 5 min, followed by thorough washing with hot distilled water) in order to remove neo-formed minerals (e.g., fluorites); shale residues, in particular, commonly contained alunite $[KAl_3(OH)_6(SO_4)_2]$, a likely breakdown product of aluminosilicates.
(4) Finally, samples containing a significant amount of pyrite ($\geq 50\%$ of the remaining "kerogen fraction") were boiled for 3 h in a $NaBH_4$ solution (2 g $NaBH_4$ in 150 ml distilled H_2O) in order to remove pyrite (Horvath and Jackson 1981).

All samples were carefully washed between, and after, acid treatments. All steps in the procedure were monitored by X-ray diffraction analysis in order to assure uniform purity of the final isolates. If impurities were detected (e.g., residual carbonate or silicate minerals, excess pyrite, or the occurrence of neo-formed minerals such as fluorites), the appropriate steps were repeated until a pure, low-ash kerogen fraction was obtained. The final isolates were freeze-dried, and then transferred into vials and stored in a desiccator.

16.4.2 Elemental Analysis of Carbon, Hydrogen, and Nitrogen

The elemental compositions of C, H, and N for the kerogen isolates were determined using a Carlo Erba Elemental Analyzer. Small amounts of pure kerogen were flash-combusted in an oxygen stream, and the abundances of C, H, and N were measured; on the basis of these, H/C and N/C ratios were calculated. Replicate analyses were made of an acetanilide standard and of kerogen samples encompassing a wide range of H/C ratios (Table 16.6). Carbon dioxide generated during elemental analyses was purified and collected on a vacuum line connected to the elemental analyzer and its carbon-isotopic composition was determined.

16.4.3 Kerogen Color

Kerogen color was determined relative to the "kerogen color index" (KCI), a modification developed by J. W. Schopf of the "index of alteration of microfossil colour (AMC)" presented by Rovnina (1981). Using standard palynological techniques, kerogen strew slides (cf. "1001-1-SS" of Fig. 15.1) were prepared by C. Mankiewicz and C. V. Mendelson; the KCI of the

Table 16.6. *Replicate analyses of elemental compositions (C, H, N) for kerogen samples.*

Sample Number	C(%)	H(%)	N(%)	H/C	N/C
1061	54.52	2.85	1.06	0.63	0.017
	55.53	2.78	1.07	0.60	0.017
	54.80	2.94	1.02	0.65	0.016
1087	60.21	4.51	2.47	0.90	0.035
	60.30	4.37	2.48	0.87	0.035
	56.79	4.22	2.30	0.89	0.035
1264	56.51	6.15	1.36	1.31	0.021
	58.62	6.49	1.51	1.33	0.022
	58.88	6.55	1.44	1.34	0.021
1268	75.39	2.25	1.06	0.36	0.012
	77.18	2.31	1.19	0.36	0.012
	73.31	2.17	1.02	0.36	0.012
1391	75.61	0.80	0.47	0.13	0.005
	77.77	0.84	0.48	0.13	0.005
	74.45	0.77	0.49	0.12	0.006
Acetanilide	70.59	6.70	10.29	1.14	0.125
Std. (n = 29)	±2.21	±0.45	±0.32	±0.06	±0.000

kerogen samples was determined by J. W. Schopf using a "blind technique" (viz., replicate, microscopic comparison of randomly ordered strew slides without knowledge of the geologic source of the samples or the aid of color-correlated geochemical data such as H/C ratios).

The elemental abundances, H/C and N/C ratios, carbon-isotopic compositions, and color of the kerogen samples analyzed during this project are tabulated in Table 17.5.

17

Abundances and Isotopic Compositions of Carbon and Sulfur Species in Whole Rock and Kerogen Samples

HARALD STRAUSS   TOBY B. MOORE

17.1
Introduction

In this Chapter are presented compilations of data regarding the abundance and isotopic composition of carbon and sulfur obtained for PPRG samples during the course of this project (Tables 17.1, 17.2, 17.5, 17.6), compilations of results of such analyses previously reported for Precambrian samples (Tables 17.3, 17.7, 17.9), and summary tables listing all available Precambrian results for the abundance and isotopic composition of carbon in whole rock samples (Table 17.4), for the abundance and isotopic composition of sulfur in sulfides and sulfates (Table 17.8), and for the elemental and carbon isotopic composition of kerogens (Table 17.10). Analytical procedures used to obtain the results here reported for PPRG samples are described in Chapters 15 and 16. Comparable compilations regarding the composition of organic matter extractable from Precambrian samples are presented in Chapter 19, the analytical procedures for which are discussed in Chapter 18.

In each of the ten tables presented in this Chapter, data are grouped by geologic unit, ordered geochronologically from youngest to oldest; within each unit, data for PPRG samples are ordered by PPRG Sample Number.

17.1.1 Explanation of symbols

Reference code. In Tables 17.3, 17.4, 17.7, 17.8, and 17.10, sources of previously published data are indicated by a ten-character reference code in which the first four characters are the first four letters in the surname of the first author of the publication; the following two characters are the last two numbers of the year of publication; and the final four characters encode the page number of the initial page of the publication. Thus, an article by "Smith and Jones" that was published in "1986" and that had as its initial page "p. 874" would be cited in these tables by the reference code: "SMIT86.874" (with the reference code "PPRG" referring to data obtained during the course of the present study).

Lithology code. In Tables 17.1–17.5, 17.9 and 17.10, lithologies are denoted as follows: ANH, anhydrite; ANK, ankerite; ANTHR, anthraxolite; BAR, barite; BIF, banded iron-formation; CARB, carbonate; CH, chert; CONGL, conglomerate; DOL, dolostone; GREYW, greywacke; H, schist; I, basalt; LS, limestone; N, gneiss; PYR, pyrite; Q, sandstone; QUAR, quartzite; SID, siderite; SILT, siltstone; SH, shale; UND, undetermined.

Mineral code. In Tables 17.6–17.8, minerals are denoted as follows: ANH, anhydrite; BAR, barite; CHALCOPYR, chalcopyrite; GAL, galena; GYP, gypsum; PO, pyrrhotite; PYR, pyrite; SFIDE, sulfide; SPH, sphene.

PPRG Sample Number suffixes. In Tables 17.1 and 17.2, the occurrence of a lower case "c," "k," and/or "x" immediately following the PPRG Sample Number indicates that the following analyses have been performed on this sample: c, organic carbon; k, kerogen; x, extractable organic matter.

N. In the three summary Tables (viz., Tables 17.4, 17.8, and 17.10), "N" refers to the number of analyses included in the calculation of the listed averaged value.

KCI. In Table 17.5, kerogen color is reported as the "kerogen color index" (KCI), a graded numerical scale developed by J. W. Schopf as a modification of the "Index of alteration of microfossil colour (AMC)" presented by Rovnina (1981). In the scale here used, the KCI numbers listed denote the following colors for thin films of kerogen observed microscopically in transmitted white light at 250X and 400X magnifications: 5.0, light brown; 5.5, brown; 6.0, dark brown; 6.5, very dark brown; 7.0, black (opaque); nd, not determined.

TOC, $delC_{org}$, and $del\,C_{carb}$. In Tables 17.1, 17.3, and 17.4, "TOC" refers to the total organic carbon content (mgC/g) of the sample. In Tables 17.1–17.5, 17.9, and 17.10, "del C_{org}" and "del C_{carb}" refer to the delta $^{13}C(‰)$ values of organic carbon and the delta $^{13}C(‰)$ values of carbonate carbon, respectively, relative to those of the PeeDee Formation belemnite (PDB) standard.

del O. In Table 17.2, "del O" refers to the delta $^{18}O(‰)$ values of carbonate oxygen relative to that of the PDB standard.

S, del S_{fide}, and del S_{fate}. In Table 17.6, "S" refers to the total sulfur content (mgS/g) of the sample. In Tables 17.6–17.8, "del S_{fide}" and "del S_{fate}" refer to the delta $^{34}S(‰)$ values of sulfide sulfur and the delta $^{34}S(‰)$ values of sulfate sulfur, respectively, relative to those of the Canyon Diablo troilite (CDT) standard.

Table 17.1. *New analyses of the abundance and isotopic composition of organic carbon in whole rock samples.*

PPRG Sample No.	Lithology	TOC (mg/g)	del C$_{org}$ (‰ vs. PDB)	PPRG Sample No.	Lithology	TOC (mg/g)	del C$_{org}$ (‰ vs. PDB)
Early Paleozoic				1201	SILT, SH	0.10	−25.0
Jonesboro Limestone (475 Ma)				1202	SILT, SH	0.10	−27.5
1306	CH	1.58	−28.2	1203	SILT, SH	0.11	−23.2
1307	CH	0.16	−29.3	1204	SILT, SH	0.10	−25.1
Windfall Formation (510 Ma)				1205	SILT, SH	0.06	−25.0
				1206	SILT, SH	0.06	−24.3
1272	CH, CARB	1.46	−27.7	1207	SILT, SH	0.06	−25.3
Late Proterozoic 550 to 500 Ma				1208	SILT, SH	0.05	
Chulaktau Formation (550 Ma)				1209	SILT, SH	0.09	−25.8
				1210	SILT, SH	0.07	−25.8
1475	CH	2.10	−36.2	2751	SILT, SH	0.19	−21.5
1476	CH	3.73	−36.0	2752	SILT, SH	0.13	−24.0
1477	CH	1.83	−35.8	2753c	SILT, SH	0.06	−31.1
Wyman Formation (550 Ma)				2754c	SILT, SH	<0.02	
				2756	SILT, SH	0.37	−23.2
2378	SILT, SH	0.93	−13.4	2763c	SILT, SH	0.07	−26.9
2380	SILT, SH	2.32	−12.9	2765c	SILT, SH	<0.02	
2382	SILT, SH	1.98	−13.1	*600 to 700 Ma*			
2384	SILT	1.56	−13.1	Julie Member (600 Ma)			
Chapel Island Formation (560 Ma)				1760	CARB	0.14	−24.3
1164	SILT, SH	0.12	−23.8	1761	CARB	0.18	−27.5
1165	SILT, SH	0.15	−24.7	1762	CARB	0.19	−28.3
1166	SILT, SH	0.04		Yudoma Formation (600 Ma)			
1167	SILT, SH	0.10	−26.4				
1168	SILT, SH	0.08	−26.2	2727	CH	1.44	−32.6
1169	SILT, SH	0.10	−24.8	Cijara Formation (610 Ma)			
1170	SILT, SH	0.12	−26.8				
1171	SILT, SH	0.08	−26.9	1499	SH	1.19	−30.0
1172	SILT, SH	0.19	−27.5	Pertatataka Formation (610 Ma)			
1173	SILT, SH	0.11	−27.9				
1174	SILT, SH	0.11	−27.0	1318	CH	0.11	−27.8
1175	SILT, SH	0.08	−27.6	Estenilla Formation (620 Ma)			
1176	SILT, SH	0.12	−27.4				
1177	SILT, SH	0.10	−25.1	1494	CARB	1.23	−27.1
1178	SILT, SH	0.12	−28.5	1498	CARB	0.80	−12.7
1179	SILT, SH	0.08	−25.8	Fuentes Formation (620 Ma)			
1180	SILT, SH	0.11	−25.3				
1181	SILT, SH	0.07	−26.4	1495	SH	0.78	−27.6
1182	SILT, SH	0.07	−27.2	Klatyspyt Formation (620 Ma)			
1183	SILT, SH	<0.02		2723 kx	SH	207.13	−36.4
1184	SILT, SH	0.04		2724	CARB	4.93	−37.1
1185	SILT, SH	0.10	−26.4	Pusa Formation (620 Ma)			
1186	SILT, SH	0.38	−31.7				
1187	SILT, SH	0.08	−27.3	1496	SH	1.62	−31.0
1188	SILT, SH	0.07	−28.8	Timperley Shale (630 Ma)			
1189	SILT, SH	0.14	−28.6				
1190	SILT, SH	0.11	−27.6	2213	CARB, SH	0.28	−25.7
1191	SILT, SH	<0.02		2216	SH, CARB	0.30	−26.0
1192	SILT, SH	0.14	−27.6	Boonall Dolomite (640 Ma)			
1193	SILT, SH	<0.02		2203	CARB	<0.02	
1194	SILT, SH	0.19	−29.1	2204 k	CARB	<0.02	
1195	SILT, SH	0.45	−32.2	2205	CARB	0.03	
1196	SILT, SH	0.08	−27.9	2208	CARB	0.09	−31.2
1197	SILT, SH	<0.02		Egan Formation (640 Ma)			
1198	SILT, SH	0.12	−28.5				
1199	SILT, SH	0.35	−17.9	2221	CARB	0.24	−29.5
1200	SILT, SH	0.10	−25.3				

Organic Carbon in Whole Rock Samples

Table 17.1. *Continued.*

PPRG Sample No.	Lithology	TOC (mg/g)	del C$_{org}$ (‰ vs. PDB)	PPRG Sample No.	Lithology	TOC (mg/g)	del C$_{org}$ (‰ vs. PDB)
Yurabi Formation (640 Ma)				Twitya Formation (700 Ma)			
2198	QUAR	<0.02		1611	SH	0.08	−26.8
Chichkan Formation (650 Ma)				1612	SH	0.08	−28.2
1473	CH	0.16	−29.0	1613 k	SH	2.14	−32.2
1474	CH	0.15	−28.9	1652	SH	2.40	−28.7
Sheepbed Formation (650 Ma)				1653 k	SH	2.13	−28.5
1614	SH	1.38	−26.4	1654 k	SH	1.35	−28.0
1615	SH	1.39	−26.1	1655 k	SH	1.44	−28.2
1642 k	SH, PYR	8.74	−30.0	1656	CARB	0.46	
1643 k	SH, PYR	6.41	−29.8	1657	SH	0.68	−29.0
1644	SH, PYR	6.06	−30.0	1658	SH	0.54	−28.9
1645 k	SH, PYR	10.23	−29.8	1659	SH	0.70	−29.3
1646 k	SH	9.18	−29.3	1660	SH	0.70	−29.0
1647	SH, PYR	7.27	−29.5	1661	SH	0.74	−27.9
1648	SH, PYR	3.07	−29.1	Bianka Member (740 Ma)			
1650 k	SH, PYR	1.90	−28.4	1480	CH	1.07	−29.4
Jarrad Sandstone Member (675 Ma)				1481	CH	0.71	−28.1
2191	CARB	0.07	−25.7	1482	CH	0.13	−28.6
2192	CARB	<0.02		1483	CH	0.71	−30.2
Moonlight Valley Tillite (680 Ma)				Min'yar Formation (740 Ma)			
2195	CARB	<0.02		1478	CH	1.44	−30.6
700 to 800 Ma				1479	CH	0.62	−29.7
Biri Formation (700 Ma)				Coppercap Formation (750 Ma)			
2412 k	CARB	23.00	−28.3	1666	CARB	0.08	−32.7
2413 k	CARB	16.00	−32.1	1667	CARB	0.11	−33.9
2414 k	CARB	20.00	−29.4	1668	CARB	0.84	−30.2
2415 k	CARB	34.00	−27.5	1669	CARB	0.08	−27.1
Keele Formation (700 Ma)				1670	CARB	1.04	−30.3
1616	CARB	0.27	−23.3	1671	CARB	3.28	−31.1
1617	CARB	0.21	−25.6	1709	CARB	0.37	−21.8
1618	CARB	0.19	−24.5	1712	CARB	0.37	−23.9
1619	CARB	1.35	−23.7	1716	CARB	0.85	−23.8
1620	CARB	0.87	−22.3	1720	CARB	<0.02	
1621	CARB	0.54	−24.2	1726	CARB	0.44	−24.9
1622	CARB	0.78	−22.2	1727	CARB, SH	14.48	−27.3
1623	CARB	0.67	−22.4	1728	CARB, SH	5.09	−27.2
1624	CARB	0.47	−23.5	1734	CARB, SH	3.98	−31.5
1651	CARB	0.06	−26.7	1735	CARB, SH	4.54	−31.3
Luoquan Formation (700 Ma)				1736	CARB, SH	5.06	−31.3
2176	SH	0.30	−18.9	1737	CARB	0.12	−28.9
2178 k	SH	3.68	−28.8	1738	CARB	0.08	−28.2
2179	SH	0.94	−29.2	1739	CARB	<0.02	
2180	SH	0.53	−27.9	Tapley Hill Formation (750 Ma)			
2181	SH	0.54	−27.0	1234	SH	2.61	−25.1
2182	SH	0.43	−26.2	Tindelpina Shale Member (750 Ma)			
2183	SH	0.57	−27.4	1237	SILT	3.27	−22.5
2184	SH	0.25	−26.7	1238 k	SILT	11.43	−23.5
2185 k	SH	3.85	−29.9	Woocalla Dolomite Member (750 Ma)			
2186 k	SH	1.78	−30.0	1239 k	SH	2.63	−29.3
2187	SH	0.92	−29.6	1240	SILT	3.40	−30.2
				1241	SILT	4.38	−30.8
				1242	SILT	5.06	−31.1

Table 17.1. Continued.

PPRG Sample No.	Lithology	TOC (mg/g)	del C$_{org}$ (‰ vs. PDB)	PPRG Sample No.	Lithology	TOC (mg/g)	del C$_{org}$ (‰ vs. PDB)
Areyonga Formation (760 Ma)				Auburn Dolomite (780 Ma)			
1248	CH	0.30		1280	CARB, CH	0.55	−19.3
1255	CH	0.08	−28.8	Rhynie Sandstone (785 Ma)			
1257	CH	0.15	−27.2	1222	CH	3.07	−17.6
1258	CH	0.10	−23.5	*800 to 900 Ma*			
1259	CH	0.08	−28.0	Jiudingshan Formation (800 Ma)			
1322	CH	0.09	−29.2	1424	CH	0.55	−23.6
1325	CH	0.09	−23.8	Little Dal Group (800 Ma)			
Shezal Formation (770 Ma)				1608	CARB	0.57	−24.2
1599	QUAR	0.12	−26.6	1609	SH, CARB	1.40	−26.1
Skillogalee Dolomite (770 Ma)				1610	CARB	0.38	−24.4
1223	CH	1.20	−19.8	1635	CH	0.11	
1230	CH	0.43	−21.8	1636	CARB	0.20	−26.2
1231	CH	0.53	−20.9	1637	SH	1.45	−25.0
1232	CH, CARB	0.82	−22.9	1638	CARB	1.08	−33.0
1233	CH, CARB	0.63	−20.9	1639	CARB	0.30	−31.7
1283	CH	0.34	−20.1	1640	CARB	0.10	−28.1
1334	CH	0.23	−18.1	1641	CARB	0.05	−28.1
1335	CH	0.76	−23.2	1672	CARB	0.12	−24.2
1336	CH	0.27	−18.6	1673	CH	0.69	−26.1
1337	CH	0.24	−19.6	1674	CARB, SH	5.05	−26.6
1338	CH	0.20	−20.4	1675	CARB	0.69	−24.4
1339	CH	0.41		1680	CARB	<0.02	
1340	CH	0.33	−22.8	1681	CARB	0.25	−20.2
1341	CH	0.56	−23.3	1682	SH	0.37	−24.8
1342	CH	0.48	−22.7	1751	SH	6.87	−25.1
1343	CH	0.08	−29.1	Sayunei Formation (800 Ma)			
1344	CH	0.27	−20.9	1683	SH	0.06	−16.4
1345	CARB, CH	0.39	−20.7	1685	SH	0.06	−17.0
1346	CH	0.52	−19.3	1686	SH	<0.02	
1347	CH	0.43	−16.6	1692	SH	<0.02	
1348	CH	0.46	−19.0	1693	SH	<0.02	
1349	CH	0.55	−20.7	1698	SH	0.03	
River Wakefield Subgroup (775 Ma)				1699	SH	0.06	
1219 k	CH	0.98	−11.0	Torridon Group (800 Ma)			
1224 k	CH, CARB	1.49		1874	CONGL	0.10	−27.0
1225	CH, CARB	0.34	−14.2	1876	CONGL	0.10	−25.6
1226	SH, CH	0.30	−19.0	1877	CONGL	0.04	−25.6
1227	SH, CH	0.31	−19.8	Beck Spring Dolomite (850 Ma)			
1228	SH, CH	0.33	−18.4	1271 k	CARB, CH	0.35	−25.7
1229	CH	0.69	−17.6	Bitter Springs Formation (850 Ma)			
1281	CARB, CH	0.27	−19.6	1256	CH	0.14	−26.7
1282	CARB, CH	0.38	−16.6	1312	CARB	0.17	−26.0
Visingsö Group (775 Ma)				1315	CH	0.21	
2728	SH	0.39	−27.5	1316	CH	0.16	−26.4
2729	CARB	0.46	−28.3	1317	CARB, CH	0.05	−24.6
2730	SH, SILT	5.38	−27.3	Kingston Peak Formation (850 Ma)			
2731	SH	1.26	−27.1	2536	CARB	15.80	−19.9
2732	SH	2.01	−28.7	2537	CARB	6.80	−21.5
2733	SH	0.86	−27.1	2538	CARB	14.90	−23.7
2734	SH	0.38	−26.8	2539 k	CARB	6.60	−24.2
2735	SH	0.50	−22.0				
2736	SH	2.11	−25.4				
2738	CARB	2.18	−27.9				

Table 17.1. Continued

PPRG Sample No.	Lithology	TOC (mg/g)	del C$_{org}$ (‰ vs. PDB)
Love's Creek Member (850 Ma)			
1243	CH	0.09	
1244	CH	0.45	−21.9
1245	CH	0.30	−21.9
1246	CH	0.20	
1250	CH	0.10	−28.8
1251	CH	0.15	−28.5
1252	CH	0.14	−24.9
1253	CH	0.24	−21.7
1254	CH	0.08	−28.1
1311	CARB	0.19	−30.2
1313	CH	0.09	−25.9
1314	CH	0.10	−25.4
1319	CH	0.11	−29.2
1320	CARB	0.18	−28.4
1321	CH	0.16	
1323 k	CARB	0.28	−29.6
1324	CARB	0.16	−29.8
1326	CH	0.06	−29.1
1327 k	CH	0.33	−24.0
1328	CH	0.11	−23.6
1329	CH, CARB	0.14	−26.8
1330	CARB, CH	0.11	−24.3
1331	CH	0.12	−25.9
1332	CARB, CH	0.10	−27.7
1333	CH	0.07	−24.9
1350	CARB	0.18	−25.5
1351	CH, CARB	0.16	−29.7
1352	CARB	0.13	−28.2
1353	CARB, CH	0.15	−29.7
1383	CARB	0.28	−25.0
Walcott Member (850 Ma)			
1088 k	SH	29.61	−27.7
1093	CH	0.52	−26.1
1095 kx	CARB	17.18	−26.5
1097 k	SH, CH	19.65	−26.5
1112 k	SH, CH	11.78	−26.4
1123 kx	CARB	18.21	−26.5
1357 k	CH	8.12	−27.0
1358	CH	0.80	−27.0
Zhangqu Formation (850 Ma)			
1423	CH	0.52	−29.4
Awatubi Member (862 Ma)			
1083 kx	SH, PYR	1.69	−32.0
1087 kx	SH	13.10	−27.0
Middle Proterozoic 900 to 1000 Ma			
Carbon Canyon Member (900 Ma)			
1153	SH	0.17	−24.4
1156	CARB	0.32	−29.1
1359	CH	0.96	−26.1
Shorikha Formation (900 Ma)			
2726	CH	0.38	−29.1
Jupiter Member (930 Ma)			
1062 k	SH	7.65	−25.5
1063 k	SH	7.56	−19.7
1064	CARB	0.09	−21.9
1067 kx	SH	2.94	−18.5
1070	CARB	0.04	
1124	SILT	11.44	−28.9
1125 k	SH	3.46	−30.4
Katav Formation (940 Ma)			
1484	CH, CARB	0.13	−29.1
1485	CARB	0.97	−28.5
Red Pine Shale (950 Ma)			
1361 kx	SH	2.20	−16.8
1872 k	SILT	1.48	−17.1
Tanner Member (950 Ma)			
1055 k	SH	4.80	−24.8
1058	SH	3.37	−23.2
1059 k	SH	2.24	−21.8
1060 k	SH	4.96	−19.5
1061 k	SH	1.89	−28.2
1080 k	SH	4.88	−20.7
1000 to 1200 Ma			
Bederysh Member (1000 Ma)			
1489	SILT, SH	1.31	−30.0
Sukhaya Tunguska Formation (1000 Ma)			
2725	CH	0.18	−26.5
Allamoore Formation (1050 Ma)			
1360 k	CH, CARB	0.44	−26.8
Nonesuch Shale (1055 Ma)			
1366	SILT	0.16	−26.9
1367	SH	2.15	−34.3
2422	SILT	0.18	−27.2
2423	SILT	0.09	−28.0
2424	SILT	0.22	−26.6
2425 k	SILT	3.67	−33.9
2426	SILT	6.93	−32.5
2427	CARB	1.08	−31.3
2428	ANTHR, PYR	14.69	−32.5
Tieling Formation (1185 Ma)			
2161	CARB	0.14	−24.5
2162	SH, CARB	0.97	−31.7
Fengjiawan Formation (1190 Ma)			
2173	CARB	0.08	−26.1
1200 to 1400 Ma			
Duguan Formation (1200 Ma)			
2171	CARB	0.09	−26.8
Kataskin Member (1240 Ma)			
1490	SH, CARB	0.87	−24.3

Table 17.1. *Continued*

PPRG Sample No.	Lithology	TOC (mg/g)	del C$_{org}$ (‰ vs. PDB)
Kuktur Member (1240 Ma)			
1493	SH	7.90	−21.6
Maloinzer Member (1240 Ma)			
1492	SH	4.33	−22.9
Ushakov Member (1240 Ma)			
1491	SH, CARB	0.86	−29.5
Hongshuizhuang Formation (1250 Ma)			
2159 k	SH	9.42	−23.6
2160 k	SH	7.25	−33.4
Xunjiansi Formation (1250 Ma)			
2168	CARB	0.15	−25.7
2169	CH, CARB	0.05	−28.3
2170	CH, CARB	0.05	−28.2
Revet Member (1260 Ma)			
1486	CH	0.43	−30.1
Mount Shields Formation (1320 Ma)			
2329	CARB	0.05	−26.3
Wumishan Formation (1325 Ma)			
1425	CH	0.56	−26.3
1426	CH	0.38	−26.6
1427	CH	0.13	−28.7
1428	CH	0.12	−29.6
1429	CH	0.39	−28.3
1430	CH	0.21	−28.6
1431	CH	0.23	−27.3
1433	CH	1.71	−25.2
1434	CH	0.50	−27.6
1435 k	CH	0.53	−30.4
2149	CH	0.31	−30.1
2150	CH, CARB	0.23	−29.1
2150	CH, CARB	0.23	−29.0
2151	CH, CARB	0.33	−30.2
2152	CH, CARB	0.38	−28.9
2153	SH	0.33	−28.8
2154	CARB	0.23	−27.4
2155	CARB	0.05	−28.0
2156	SILT	0.05	−28.0
2157	CARB	0.30	−28.5
2158	CARB	0.08	−29.2
Rossport Formation (1339 Ma)			
1304	CH	0.30	−27.3
1305	CH, CARB	0.10	−25.6
2460	CARB	0.85	−27.2
2461	CARB	0.44	−25.4
McMinn Formation (1340 Ma)			
1262	SILT	4.97	−31.7
1263 k	SILT	13.83	−33.0
1264 k	SILT	20.49	−33.5
1265	SILT	2.99	−33.2
1266 k	SILT	15.64	−32.9
1267	SILT	13.96	−32.9
1268 k	SILT	6.57	−32.9
1269	SILT	1.96	−32.2
1270 k	SILT	15.71	−32.8
Tukan Member (1350 Ma)			
1487	SH	1.69	−24.3
1488	SH	1.57	−25.6
Yangzhuang Formation (1350 Ma)			
2146	CH, CARB	0.29	−26.3
2146	CH, CARB	0.13	−27.4
2147	CH, CARB	0.16	−28.3
2148 k	CH	0.27	−28.0
Bungle Bungle Dolomite (1364 Ma)			
2193	CH	0.08	−17.9
2194	CARB	0.23	−24.0
2196	CH	0.08	−25.2
2197	CH	<0.02	
2199	CARB	0.08	−25.3
Helena Formation (1380 Ma)			
2325	CARB	0.08	−26.0
1400 to 1600 Ma			
Dismal Lakes Group (1400 Ma)			
1212	CH	1.35	−27.6
1213	CH	0.59	−28.3
1215	CH	0.32	−30.5
1216	CH	0.36	−27.2
1217 k	CH	1.16	−28.4
1218	CH	0.11	−29.1
Greyson Shale (1420 Ma)			
2314	SH	0.82	−31.1
2321 k	SILT, SH	1.63	−28.3
Gaoyuzhuang Formation (1425 Ma)			
1420	CH, CARB	0.35	−30.0
1421 k	CH, CARB	1.15	−32.0
1422	CH	1.30	−31.7
1432	CH	3.17	−31.3
2123	CH, CARB	0.44	−30.4
2124	CH, CARB	0.30	−29.9
2125	CH	1.62	−31.2
2126 k	CH	1.19	−31.2
Altyn Formation (1440 Ma)			
1472	CARB	0.07	−26.4
Newland Limestone (1440 Ma)			
2298	SH	0.50	−27.2
2299	SH	0.56	−27.2
2300	CARB	0.29	−24.0
2301	CARB	0.69	−28.2
2302	CARB, CH	0.46	−29.1

Table 17.1. Continued

PPRG Sample No.	Lithology	TOC (mg/g)	del C$_{org}$ (‰ vs. PDB)	PPRG Sample No.	Lithology	TOC (mg/g)	del C$_{org}$ (‰ vs. PDB)
2303	CARB, CH	0.27	−27.6	Koolpin Formation (1885 Ma)			
2305	CARB	0.21	−24.3	1261 k	CH	1.19	−30.6
2307 k	SH	1.37	−32.3	Fontano Formation (1900 Ma)			
2308	CARB	0.72	−31.0	1529	SH	4.04	−26.4
2309	SH, CARB	1.11	−28.4	1588 k	SH	9.06	−30.2
2310	SH	1.62	−30.5	Rove Formation (1900 Ma)			
2311	SH	0.86	−27.7	1295 k	SH	22.10	−31.6
2312	SH	0.31	−27.1	2448	SH	0.84	−27.6
2313	SH	1.03	−27.8	2456	CARB, SH	0.08	−25.6
2315	SH	0.68	−20.3	2457	CARB	0.51	−31.7
2316	SH	0.63	−20.1	2458 k	SH	23.17	−32.0
2317	SH	0.28	−15.8	2459 k	SH	14.60	−31.8
2318	SH	0.87	−24.7	Spartan Group (1900 Ma)			
2319 k	SH	2.18	−27.3	2779	SH	11.81	−30.5
2320	CARB	0.91	−26.0	2780	SH	3.54	−29.4
2322	SH	0.84	−24.7	2781	SH	4.04	−29.8
2323	SH, PYR	1.38	−13.6	Union Island Group (1900 Ma)			
2324	SH	0.80	−23.3	1382 k	SH	20.83	−39.4
Chamberlain Shale (1450 Ma)				Odjick Formation (1925 Ma)			
2306	SH	0.79	−32.0	1532	SH	0.22	−25.6
Prichard Formation (1450 Ma)				Rocknest Formation (1925 Ma)			
2327	QUAR	0.93	−23.1	1511 k	CARB	0.33	−24.3
2328	SH	0.30	−23.9	1513	CARB	0.09	−25.8
Early Proterozoic 1600 to 1800 Ma				1516 k	CARB	0.45	−22.7
Paradise Creek Formation (1650 Ma)				1518	CARB	0.65	−25.4
1260	CH	0.09	−28.8	1521	CARB	1.40	−24.7
Urquhart Shale (1670 Ma)				1523	CH	1.00	−24.7
1273	SH	0.27	−20.2	1524 k	CARB, CH	0.61	−23.9
1274	SH, PYR	4.00	−23.9	1528	CARB	0.06	−21.8
1275	SH	7.72	−28.5	1531	CARB	<0.02	
1276	SH	11.46	−28.3	1536	CARB, CH	<0.02	
1277	SH	15.30	−27.6	1538	CH	<0.02	
1278 k	SH, PYR	16.34	−27.9	1540	CARB	0.03	
McArthur Group (1690 Ma)				1541	CARB	0.03	
1279	SH	4.82	−32.6	1542	CARB	0.03	
Guddadarangavanahalli Formation (1700 Ma)				1552	CARB	<0.02	
1310	CH	0.08	−27.4	1553	CARB	<0.02	
Vempalle Formation (1700 Ma)				1558	CARB	<0.02	
1308	CH	0.05	−26.8	1559	CARB	<0.02	
1309	CH	0.12	−29.2	1566	CARB	<0.02	
Tuanshanzi Formation (1750 Ma)				1567	CARB	<0.02	
2142	CARB	0.57	−31.5	1574	CARB	<0.02	
2143 k	CARB	1.89	−30.5	1575	CARB	<0.02	
2145	CARB	0.13	−29.5	1580c	CARB	0.54	−12.6
1800 to 2000 Ma				1581 c	CH	0.53	−12.3
Chuanlinggou Formation (1850 Ma)				1582 ck	CARB, CH	0.94	−12.3
2139 k	SH	4.52		1583	CARB	0.33	−12.0
2140 k	SH	5.80	−33.1	1584	CARB	0.26	−14.2
2141	SH	0.76	−30.2	1585	CARB	0.25	−13.4
2144 k	SH	2.91	−33.0	1586	CARB	0.43	−16.9
				1587	CH	0.60	−20.4

Table 17.1. Continued.

PPRG Sample No.	Lithology	TOC (mg/g)	del C_{org} (‰ vs. PDB)
Tyler Formation (1950 Ma)			
1363 k	CH, SH	5.44	−31.2
1364 k	CH, SH	5.06	−31.5
1365	CARB	11.22	−32.3
2000 to 2250 Ma			
Kona Dolomite (2000 Ma)			
1362	CARB	0.06	−14.9
2416	CH	0.09	−24.1
2417	CH	0.03	−26.9
2418	CH	0.10	−23.0
2419	CARB	0.03	−25.9
2420	CARB	0.06	−27.7
2421	CARB	0.07	−23.7
Lookout Schist (2000 Ma)			
2741	SH	101.30	−20.6
Biwabik Iron Formation (2090 Ma)			
1371	CH	0.08	−30.3
2429	CH, PYR	0.13	−30.8
2430	CH	0.32	−32.3
2431	CH	0.12	−28.5
2432	CARB	0.14	−28.2
Gunflint Iron Formation (2090 Ma)			
1284	CH	2.16	−33.1
1285	CH	3.64	−32.9
1286	CH	0.56	−33.2
1287	CH	72.25	−34.3
1288	CH	1.14	−32.2
1289	CH	0.42	−33.5
1290	CH	0.18	−33.5
1291	CH	0.44	−33.3
1292	CH	0.85	−18.3
1296	CH	0.60	−33.3
1297	CH	1.38	−33.1
1298	CH	0.07	−27.3
1299	CH	0.78	−32.5
1300	CH	0.36	−32.7
1301	CH	0.16	−32.8
1302 k	CH	0.34	−34.0
1303	CH	0.28	−33.4
1374	CH	0.10	−30.6
1375	CH	5.35	−31.5
1376	CARB, ANTHR	103.72	−33.3
1377	CARB, ANTHR	390.36	−33.2
1378	CH, ANTHR	71.03	−33.2
1379	CH	0.81	−34.4
2436 k	CH, CARB	6.14	−31.9
2437 k	CARB	3.65	−31.7
2438	CH	0.14	−24.5
2439	ANTHR, CH	490.00	−26.5
2440	SH	16.06	−31.9
2441	CH	0.28	−33.4
2442	CH	0.96	−33.5
2444 k	CH	2.91	−33.1
2445	CH	0.17	−31.0
2446	CARB	0.73	−31.4
2447	CH, PYR	8.56	−32.8
2449	CARB	0.15	−31.4
2450	CH	0.26	−33.0
2451	CH	29.34	−31.8
2452	CH	0.16	−32.7
2453 k	CH, ANTHR	2.23	−33.1
2454	CH	0.19	−32.2
2455	CARB	0.18	−30.9
2464	CARB	0.81	−33.2
2465	CH, CARB	2.12	−33.0
2467	CH, CARB	3.25	−33.4
2468	CH	1.34	−33.5
2469	CARB	1.66	−33.2
2470	CARB	3.29	−34.0
2471	CARB	3.27	−34.4
2474	CH	14.26	−34.7
2475	CH	0.77	−33.6
2476	CH, CARB	50.18	−34.4
2477	CH, CARB	1.02	−33.6
2478	CH	0.78	−33.9
2479	ANTHR	78.86	−35.2
2480	ANTHR	63.57	−34.9
2481	ANTHR	62.36	−34.9
2482	CH	3.16	−33.5
2483	CH	2.19	−33.6
2484	CH	1.50	−33.4
2485	CH	3.06	−33.6
2486	CH	11.01	−33.9
2487	CH	2.13	−33.8
2488	CH	1.05	−33.0
2489	CH	2.77	−33.6
2490	CH	1.43	−33.4
2491	CH	0.97	−33.0
2492	CH	1.78	−32.6
2493	CH	0.64	−31.8
2494	CH	1.31	−32.4
2495	CH	0.61	−31.3
Pokegama Quartzite (2090 Ma)			
1368	QUAR	0.14	−32.7
1369	QUAR	0.22	−29.1
1370	QUAR	0.08	−30.8
2250 to 2500 Ma			
Bar River Formation (2250 Ma)			
1845	QUAR	0.07	−28.8
Bruce Formation (2250 Ma)			
1855	CARB	0.04	−21.4
1856	SILT	0.04	−28.3
Espanola Formation (2250 Ma)			
1854	SILT	0.17	−27.0
2497	CARB	0.16	−25.9

Table 17.1. Continued.

PPRG Sample No.	Lithology	TOC (mg/g)	del C$_{org}$ (‰ vs. PDB)
Gowganda Formation (2250 Ma)			
1380	SILT	0.04	
1381	SILT	0.06	
1848	SILT	0.02	
1849	SILT	0.04	−28.4
1850	SILT	0.04	−28.1
1851	SILT	0.02	
1852	CONGL	0.10	−28.4
2496	SILT	0.05	−26.2
Lorrain Formation (2250 Ma)			
1847	QUAR	0.06	−28.9
Mississagi Formation (2250 Ma)			
1857	QUAR	0.06	−29.0
Pecors Formation (2250 Ma)			
1858	SILT	0.17	−36.2
Ramsay Lake Formation (2250 Ma)			
1859	CONGL	0.18	−36.6
Serpent Formation (2250 Ma)			
1853	QUAR	0.07	−29.3
Gordon Lake Formation (2259 Ma)			
1846	QUAR	0.07	−26.7
Reivilo Formation (2300 Ma)			
2530	CARB	0.67	−31.2
2532 k	SH	4.38	−37.3
2533 k	SH	10.88	−37.3
2534	SH	2.56	−35.1
2535	SH	1.27	−38.1
Rooinekke Formation (2300 Ma)			
2524	CARB	0.03	−26.7
2525	CARB	0.07	−18.5
Klipfonteinheuwel Formation (2325 Ma)			
1415 k	CH, CARB	0.80	−38.5
2526	CH	0.86	−36.8
Kuruman Iron Formation (2350 Ma)			
1418	CH, SILT	0.38	−34.1
Malmani Subgroup (2400 Ma)			
1417	CH	0.22	−32.0
Matinenda Formation (2400 Ma)			
1867	QUAR	0.05	−29.0
1868	QUAR	0.09	−31.4
1869	SILT	0.03	−26.1
McKim Formation (2400 Ma)			
1860	SILT	0.06	−25.6
1861	QUAR	0.08	−28.7
1863	QUAR	0.10	−30.5
1864	QUAR	0.14	−35.1

PPRG Sample No.	Lithology	TOC (mg/g)	del C$_{org}$ (‰ vs. PDB)
1865	SILT	0.14	−25.2
1866	SILT	0.08	−29.5
Gamohaan Formation (2450 Ma)			
1412	CH	0.82	−34.5
1413	CH	0.54	−26.5
1414	CH	0.71	−36.2
2512 k	CARB	2.35	−32.2
2513 k	CARB	2.65	−32.8
2514 k	CARB	1.78	−34.8
2515	CH	0.52	−35.4
2516	CARB	0.52	−33.1
2517 k	CARB	1.69	−35.0
2518 k	CARB	2.02	−34.4
2519	CARB	0.76	−32.2
2520 k	CARB	1.32	−33.1
2521	CARB	1.68	−34.0
2522	CARB	0.45	−34.2
2523	CARB	1.39	−36.3
2776	SH	20.69	−40.5
2777	SH	21.13	−37.5
2778	SH	24.35	−39.3
Archean 2500 to 3000 Ma			
Keewatin Group (2500 Ma)			
1293	CH	0.50	−24.1
1294	CH	0.40	−15.3
Jeerinah Formation (2650 Ma)			
1932 k	SH	9.15	−48.6
1972 k	SILT	1.30	−42.9
Rietgat Formation (2650 Ma)			
1416	CH	0.69	−43.1
Soudan Iron Formation (2650 Ma)			
1372	CH	0.06	−29.5
1373	SH	0.13	
2433	CH	0.08	−24.6
2434	SH	0.23	−20.5
2435	CH	0.10	−28.7
Ventersdorp Supergroup (2650 Ma)			
2527	SILT	0.27	−36.5
2528	CARB	0.07	−27.8
2529	CARB	0.08	−30.9
Venterspost Formation (2670 Ma)			
1411	CONGL	0.09	−31.9
Venterspost Carbon Reef Member (2670 Ma)			
2507	SH	0.08	−23.7
Nymerina Basalt (2700 Ma)			
1950	CARB	0.86	−53.2
Kameeldoorns Formation (2705 Ma)			
1419	CH	0.38	−32.5

Table 17.1. *Continued.*

PPRG Sample No.	Lithology	TOC (mg/g)	del C$_{org}$ (‰ vs. PDB)
Carbon Leader Member (2710 Ma)			
1410	ANTHR	133.30	−37.0
2498	SILT	0.09	−25.2
2500	ANTHR	97.82	−38.8
2501	SILT	0.24	−30.6
Government Subgroup (2710 Ma)			
2505	SH	1.37	−36.7
North Leader Member (2710 Ma)			
2499	SILT	0.08	−24.3
2502	SILT	0.32	−32.4
Witwatersrand Supergroup (2710 Ma)			
2503	SH	0.05	−24.9
Jeppestown Shale (2720 Ma)			
2504	SH	2.77	−37.3
2506	SH	0.04	−25.3
Maddina Basalt (2735 Ma)			
2022	CH	0.18	−45.9
2041	CH	0.89	−46.7
Kuruna Siltstone (2740 Ma)			
1944	CARB	0.81	−55.8
Kimberley Shale (2750 Ma)			
2508	SH	0.08	−23.7
Meentheena Carbonate Member (2750 Ma)			
1354	CH, CARB	0.56	−50.7
1355 k	CH, CARB	0.46	−52.2
1356	CH	0.26	−47.8
Tumbiana Formation (2750 Ma)			
1960	CARB	0.05	−28.5
1985	SILT	0.63	−36.2
2050 k	CARB	0.43	−50.7
2065 k	SH	1.80	−60.9
2079	CARB, CH	0.43	−53.5
2086	CARB	0.18	−43.1
2092	CARB	0.05	
2100	CARB	0.49	−48.2
2107	CARB	0.45	−48.9
2119	CARB	0.48	−49.8
Kylena Basalt (2768 Ma)			
2017	CARB	0.38	−37.3
3000 to 3550 Ma			
Insuzi Group (3000 Ma)			
1408	CH	0.31	−17.8
1409	CARB	0.05	−28.1
Gorge Creek Group (3200 Ma)			
1436	CH	0.08	−29.8
Cleaverville Formation (3300 Ma)			
1444	CH	0.10	−26.3
1445	SH	0.56	−28.6
Kromberg Formation (3300 Ma)			
1390	CH	2.17	−30.9
1391 k	CH	3.86	−30.5
1392 k	CH	3.20	−26.2
1393	CH	0.63	−33.9
1394	CH	1.02	−30.4
1395	CH	1.90	−33.5
1396 k	CH	4.63	−32.9
1397	CH	0.73	−36.5
1405	CH	0.15	−28.7
Fig Tree Group (3350 Ma)			
1399	CH	0.16	−30.2
1400 k	CH	2.83	−28.5
1401	CH	0.20	−27.0
1402	CH	0.62	−28.5
1403 k	CH	0.51	−27.7
1404	CH	0.14	−26.9
1406	CH	0.46	−32.0
1407	CH	1.07	−35.4
Apex Basalt (3400 Ma)			
1440	CH	0.09	−30.0
1441	CH	0.13	−22.5
2003	CH	0.07	−28.6
Towers Formation (3435 Ma)			
1437	CH	0.22	−18.1
1438	CH	1.20	−36.9
1439	CH	0.08	−28.7
1442	CH	2.07	−31.7
1446	CH	0.17	−29.7
1447	CH	1.09	−33.2
1448	CH	0.82	−33.2
2007	CH	0.92	−33.6
2014	CH	0.16	−32.1
Hooggenoeg Formation (3450 Ma)			
1386 k	CH	3.25	−37.7
1387	CH	3.25	−32.4
1388	CH, PYR	2.45	−36.7
1389	CH	1.83	−38.6
Middle Marker Member (3450 Ma)			
1398	CH	2.31	−34.6
Theespruit Formation (3550 Ma)			
1384	CH	9.00	−16.3
1385 k	CH	3.30	−14.6

Table 17.2. *New analyses of the isotopic composition of carbonates*

PPRG Sample No.	Lithology	del C$_{carb}$ (‰ vs. PDB)	del O (‰ vs. PDB)	PPRG Sample No.	Lithology	del C$_{carb}$ (‰ vs. PDB)	del O (‰ vs. PDB)
Chapel Island Formation (560 Ma)				Rocknest Formation (1925 Ma)			
2750	LS	−14.9	−17.8	1546	DOL	0.8	−8.9
2753 c	LS	−22.3	−19.3	1547	DOL	1.0	−8.1
2754 c	LS	−21.9	−18.9	1554	DOL	0.9	−11.0
2755	LS	−23.1	−18.9	1555	DOL	1.1	−10.4
2760	LS	−21.7	−21.5	1556	DOL	1.0	−10.1
2761	LS	−20.6	−18.5	1557	DOL	1.0	−10.5
2763 c	LS	−15.9	−17.6	1580 c	DOL	1.6	−10.2
2764	LS	−13.9	−17.8	1581 c	DOL	1.7	−10.2
2765 c	LS	−3.1	−15.5	1582 ck	DOL	1.6	−9.8
2766	LS	0.4	−17.5				
2767	LS	−1.2	−11.2	Hotazel Formation (2250 Ma)			
2768	LS	0.6	−14.4	2510	LS	−7.5	−14.4
Reed Dolomite (570 Ma)				2511	LS	−11.1	−11.2
2388	DOL	1.0	−13.4	2511	DOL	−11.3	−12.0
2390	DOL	1.3	−11.7				
2391	DOL	2.6	−12.2	Dales Gorge Member (2500 Ma)			
2393	DOL	1.7	−11.7	487	UND	−5.1	−11.2
2394	DOL	1.8	−12.4	490	UND	−7.3	−12.7

Table 17.3. *Previously published analyses of the abundance and isotopic composition of carbon in whole rock samples.*

Lithology	TOC (mg/g)	del C$_{org}$ (‰ vs. PDB)	del C$_{carb}$ (‰ vs. PDB)	Reference Code	Lithology	TOC (mg/g)	del C$_{org}$ (‰ vs. PDB)	del C$_{carb}$ (‰ vs. PDB)	Reference Code
Early Cambrian					DOL			−1.3	MAGA86.258
Petstrostvet Formation (540 Ma)					DOL			−0.9	MAGA86.258
DOL			−1.9	MAGA86.258	DOL			−0.6	MAGA86.258
DOL			−1.4	MAGA86.258	DOL			−1.2	MAGA86.258
DOL			−1.7	MAGA86.258	DOL			0.1	MAGA86.258
DOL			−1.2	MAGA86.258	Tokammane Formation (540 Ma)				
DOL			−1.4	MAGA86.258	SH	0.60	−28.6		KNOL86.832
DOL			−1.1	MAGA86.258	DOL			−0.2	KNOL86.832
DOL			−1.1	MAGA86.258	DOL			−1.8	KNOL86.832
DOL			−1.0	MAGA86.258	DOL			0.4	KNOL86.832
DOL			0.1	MAGA86.258	SH			1.1	KNOL86.832
DOL			0.2	MAGA86.258					
DOL			0.2	MAGA86.258	*Late Proterozoic 550 to 600 Ma*				
DOL			0.1	MAGA86.258	Kirtonryggen Formation (550 Ma)				
DOL			−0.6	MAGA86.258					
DOL			−0.2	MAGA86.258	LS			−0.4	KNOL86.832
DOL			−0.2	MAGA86.258	LS			−0.4	KNOL86.832
DOL			0.6	MAGA86.258	Série Schisto Calcaire Formation (550 Ma)				
DOL			0.8	MAGA86.258					
DOL			1.4	MAGA86.258	SH			−0.3	TUCK86..48
DOL			1.3	MAGA86.258	SH			0.2	TUCK86..48
DOL			0.9	MAGA86.258	SH			−0.3	TUCK86..48
DOL			0.7	MAGA86.258	SH			−0.4	TUCK86..48
DOL			0.5	MAGA86.258	SH			−0.1	TUCK86..48
DOL			0.1	MAGA86.258	SH			−0.4	TUCK86..48
DOL			−0.3	MAGA86.258	SH			0.1	TUCK86..48
DOL			−0.9	MAGA86.258	SH			−0.4	TUCK86..48

Table 17.3. Continued.

Lithology	TOC (mg/g)	del C$_{org}$ (‰ vs. PDB)	del C$_{carb}$ (‰ vs. PDB)	Reference Code	Lithology	TOC (mg/g)	del C$_{org}$ (‰ vs. PDB)	del C$_{carb}$ (‰ vs. PDB)	Reference Code
Transition Beds (555 Ma)					LS			−4.9	TUCK86..48
LS			−0.2	TUCK86..48	LS			−1.1	TUCK86..48
LS			0.1	TUCK86..48	LS			−1.7	TUCK86..48
LS			−0.3	TUCK86..48	LS			−1.3	TUCK86..48
LS			−0.2	TUCK86..48	LS			−1.2	TUCK86..48
LS			0.7	TUCK86..48	LS			−1.8	TUCK86..48
LS			−0.6	TUCK86..48	LS			−0.4	TUCK86..48
LS			0.8	TUCK86..48	LS			−2.9	TUCK86..48
LS			−0.1	TUCK86..48	LS			−2.9	TUCK86..48
LS			0.3	TUCK86..48	LS			−2.4	TUCK86..48
Calcaire Supérieur Formation (560 Ma)					LS			−4.3	TUCK86..48
LS			−0.4	TUCK86..48	Dengying Formation (570 Ma)				
LS			−1.6	TUCK86..48	DOL		−26.0		LAMB87.140
LS			2.5	TUCK86..48	DOL		−31.7		LAMB87.140
LS			1.8	TUCK86..48	DOL			1.4	LAMB87.140
LS			2.4	TUCK86..48	DOL			0.8	LAMB87.140
LS			2.9	TUCK86..48	DOL			3.2	LAMB87.140
LS			2.5	TUCK86..48	DOL			2.3	LAMB87.140
LS			2.5	TUCK86..48	DOL			2.8	LAMB87.140
LS			1.7	TUCK86..48	DOL			2.8	LAMB87.140
LS			−0.3	TUCK86..48	DOL			0.8	LAMB87.140
"Cambrian" (560 Ma)					DOL			2.8	LAMB87.140
LS			−2.1	LAMB87.140	DOL			1.6	LAMB87.140
LS			−1.1	LAMB87.140	DOL			1.6	LAMB87.140
Schwarzrand "Series" (560 Ma)					DOL			3.9	LAMB87.140
LS			−1.4	SCHI75...1	DOL			0.9	LAMB87.140
DOL			−3.8	SCHI75...1	DOL			2.4	LAMB87.140
LS			−2.9	SCHI75...1	DOL			6.3	LAMB87.140
DOL			−3.1	SCHI75...1	DOL			4.1	LAMB87.140
LS			−4.0	SCHI75...1	Krol E Formation (570 Ma)				
LS			−2.8	SCHI75...1	DOL			2.0	AHAR87.699
DOL			−3.7	SCHI75...1	DOL			4.2	AHAR87.699
LS			−6.7	SCHI75...1	DOL			2.8	AHAR87.699
DOL			1.0	SCHI75...1	P€/€ Boundary Beds (570 Ma)				
LS			−0.6	SCHI75...1	LS			−0.5	LAMB87.140
LS			−0.6	SCHI75...1	Dolomie Inférieur Formation (580 Ma)				
LS			−0.2	SCHI75...1	DOL			−2.1	TUCK86..48
Tal Formation (560 Ma)					DOL			−1.7	TUCK86..48
SH	7.40	−33.0		BANE86.239	DOL			−1.0	TUCK86..48
SH	3.50	−33.6		BANE86.239	DOL			0.1	TUCK86..48
SH	8.10	−33.9		BANE86.239	DOL			1.3	TUCK86..48
SH	9.20	−34.4		BANE86.239	DOL			4.4	TUCK86..48
SH	21.90	−33.6		BANE86.239	DOL			5.0	TUCK86..48
SH	3.90	−31.4		BANE86.239	DOL			6.8	TUCK86..48
SH	20.10	−32.9		BANE86.239	DOL			−1.5	TUCK86..48
SH	8.90	−32.2		BANE86.239	DOL			−1.7	TUCK86..48
SH	10.10	−34.5		BANE86.239	Dracoisen Formation (580 Ma)				
DOL			0.7	AHAR87.699	SH	2.50	−25.6		KNOL86.832
DOL			0.3	AHAR87.699	SH	2.20	−25.5		KNOL86.832
Série Lie de Vin Formation (565 Ma)					SH	13.00	−32.2		KNOL86.832
LS			−3.0	TUCK86..48	SH	9.70	−34.0		KNOL86.832
LS			−3.0	TUCK86..48	SH	1.50	−33.3		KNOL86.832
LS			−4.0	TUCK86..48	DOL			10.0	KNOL86.832

Table 17.3. Continued.

Lithology	TOC (mg/g)	del C_{org} (‰ vs. PDB)	del C_{carb} (‰ vs. PDB)	Reference Code	Lithology	TOC (mg/g)	del C_{org} (‰ vs. PDB)	del C_{carb} (‰ vs. PDB)	Reference Code
SH			−3.9	KNOL86.832	DOL			−2.6	SCHI75...1
LS			−3.5	FAIR87.973	LS			4.8	EICH75.585
LS			−3.1	FAIR87.973	DOL			1.0	EICH75.585
LS			−2.9	FAIR87.973	Yudoma Formation (600 Ma)				
LS			10.0	FAIR87.973					
LS			10.9	FAIR87.973	DOL			0.6	MAGA86.258
LS			10.1	FAIR87.973	DOL			0.8	MAGA86.258
LS			5.1	FAIR87.973	DOL			1.0	MAGA86.258
LS			4.4	FAIR87.973	DOL			2.6	MAGA86.258
LS			10.2	FAIR87.973	DOL			2.7	MAGA86.258
LS			2.3	FAIR87.973	DOL			2.8	MAGA86.258
LS			2.4	FAIR87.973	DOL			3.4	MAGA86.258
Krol D Formation (580 Ma)					DOL			1.1	MAGA86.258
					DOL			0.6	MAGA86.258
DOL			3.5	AHAR87.699	DOL			−0.1	MAGA86.258
DOL			4.7	AHAR87.699	DOL			−0.3	MAGA86.258
DOL			6.0	AHAR87.699	DOL			0.3	MAGA86.258
DOL			3.5	AHAR87.699	DOL			0.8	MAGA86.258
DOL			3.2	AHAR87.699	DOL			1.4	MAGA86.258
DOL			3.2	AHAR87.699	DOL			0.3	MAGA86.258
DOL			−1.2	AHAR87.699	DOL			0.8	MAGA86.258
DOL			2.9	AHAR87.699	DOL			−0.8	MAGA86.258
DOL			1.6	AHAR87.699	DOL			−0.9	MAGA86.258
DOL			3.4	AHAR87.699	DOL			−1.7	MAGA86.258
DOL			0.8	AHAR87.699	DOL			−0.8	MAGA86.258
					DOL			−0.2	MAGA86.258
Krol C Formation (590 Ma)					DOL			−1.5	MAGA86.258
DOL			0.9	AHAR87.699	DOL			−1.5	MAGA86.258
DOL			3.4	AHAR87.699	DOL			−1.0	MAGA86.258
DOL			3.2	AHAR87.699	DOL			−2.0	MAGA86.258
DOL			3.2	AHAR87.699	DOL			−1.3	MAGA86.258
DOL			1.1	AHAR87.699	DOL			−3.0	MAGA86.258
DOL			0.6	AHAR87.699	DOL			−3.5	MAGA86.258
DOL			1.4	AHAR87.699	DOL			−4.6	MAGA86.258
LS			0.9	AHAR87.699	DOL			−3.5	MAGA86.258
LS			2.3	AHAR87.699	DOL			−4.4	MAGA86.258
					DOL			−4.6	MAGA86.258
600 to 700 Ma					Pertataka Formation (610 Ma)				
Huqf Formation (600 Ma)					DOL			4.8	VEIZ761387
OIL		−36.0		GRAN87..61	DOL			5.5	VEIZ761387
Krol Formation (600 Ma)					Canyon Formation (630 Ma)				
CARB	3.00	−27.7		BANE86.239	DOL	1.60	−30.7		KNOL86.832
CARB	3.10	−28.3		BANE86.239	SH	2.50	−20.1		KNOL86.832
Kuibis Formation (600 Ma)					DOL			−2.4	KNOL86.832
					SH			−3.0	KNOL86.832
LS			−10.9	EICH75.585	Spiral Creek Formation (630 Ma)				
DOL			−11.6	EICH75.585					
LS			−26.2	JACK78.335	DOL	3.00	−27.1		KNOL86.832
LS			4.8	SCHI75...1	DOL	0.60	−29.6		KNOL86.832
LS			3.1	SCHI75...1	DOL			−4.0	KNOL86.832
LS			1.5	SCHI75...1	DOL			−1.1	KNOL86.832
LS			1.4	SCHI75...1	Tillite Formation (630 Ma)				
LS			1.8	SCHI75...1					
LS			2.0	SCHI75...1	SH	1.10	−29.6		KNOL86.832
LS			1.7	SCHI75...1	SH	1.20	−29.3		KNOL86.832
DOL			−1.9	SCHI75...1	CARB	1.70	−25.5		KNOL86.832

Table 17.3. Continued.

Lithology	TOC (mg/g)	del C_{org} (‰ vs. PDB)	del C_{carb} (‰ vs. PDB)	Reference Code
SH			0.5	KNOL86.832
CARB			2.8	KNOL86.832
Egan Formation (640 Ma)				
DOL			−0.5	WILL79.377
Brachina Formation (650 Ma)				
SH		−27.9		MCKI74.591
SH		−18.8		MCKI74.591
SH		−14.7		MCKI74.591
SH		−23.8		MCKI74.591
Chichkan Formation (650 Ma)				
CH	0.76	−27.9		SCHO84.335
Doushantou Formation (650 Ma)				
SH		−38.1		LAMB87.140
SH		−28.0		LAMB87.140
SH		−28.0		LAMB87.140
SH		−24.5		LAMB87.140
SH		−28.8		LAMB87.140
SH		−29.5		LAMB87.140
SH		−26.2		LAMB87.140
SH		−26.7		LAMB87.140
SH		−27.0		LAMB87.140
SH		−28.4		LAMB87.140
CARB			3.0	LAMB87.140
CARB			−1.8	LAMB87.140
CARB			4.1	LAMB87.140
CARB			2.7	LAMB87.140
CARB			6.1	LAMB87.140
CARB			−0.5	LAMB87.140
CARB			6.6	LAMB87.140
CARB			6.7	LAMB87.140
CARB			6.8	LAMB87.140
CARB			3.5	LAMB87.140
CARB			4.2	LAMB87.140
CARB			7.1	LAMB87.140
CARB			5.6	LAMB87.140
CARB			3.1	LAMB87.140
CARB			4.6	LAMB87.140
CARB			3.7	LAMB87.140
Etina Limestone (650 Ma)				
LS			9.4	VEIZ761387
LS			9.5	VEIZ761387
Malmesbury Formation (650 Ma)				
LS		−25.5		EICH75.585
LS		−27.7		EICH75.585
LS		−24.1		EICH75.585
LS		−25.6		EICH75.585
LS		−21.2		EICH75.585
LS		−25.3		EICH75.585
LS		−20.7		EICH75.585
LS		−18.5		EICH75.585
LS		−23.8		EICH75.585
LS		−27.2		EICH75.585
LS			−0.9	SCHI75...1
LS			7.0	SCHI75...1
DOL			0.8	SCHI75...1
LS			2.6	SCHI75...1
LS			−3.0	SCHI75...1
LS			−0.4	SCHI75...1
LS			−0.9	SCHI75...1
DOL			−0.1	SCHI75...1
LS			8.5	SCHI75...1
LS			8.3	SCHI75...1
LS			0.8	SCHI75...1
LS			0.1	SCHI75...1
LS			0.3	SCHI75...1
LS			0.8	EICH75.585
LS			0.8	EICH75.585
Nuccaleena Formation (650 Ma)				
DOL			−2.0	VEIZ761387
DOL			−3.0	WILL79.377
DOL			−2.0	WILL79.377
Trezona Formation (650 Ma)				
LS			−7.7	VEIZ761387
LS			−7.5	VEIZ761387
LS			−7.4	VEIZ761387
Walsh Tillite (650 Ma)				
DOL			1.0	WILL79.377
DOL			−2.5	WILL79.377
Willochra Formation (650 Ma)				
LS			3.3	VEIZ761387
Wilsonbreen Formation (650 Ma)				
CARB	0.80	−29.8		KNOL86.832
CARB			4.0	KNOL86.832
LS			1.4	FAIR87.973
LS			2.7	FAIR87.973
LS			2.7	FAIR87.973
LS			4.4	FAIR87.973
LS			4.2	FAIR87.973
LS			−0.4	FAIR87.973
LS			4.0	FAIR87.973
LS			2.7	FAIR87.973
LS			−1.8	FAIR87.973
LS			1.0	FAIR87.973
LS			−1.3	FAIR87.973
Wonokaan Formation (650 Ma)				
LS			−7.2	VEIZ761387
Cango Formation (660 Ma)				
LS		−20.8		EICH75.585
LS		−21.6		EICH75.585
LS			−3.0	EICH75.585
LS			−0.4	EICH75.585

Table 17.3. Continued.

Lithology	TOC (mg/g)	del C_{org} (‰ vs. PDB)	del C_{carb} (‰ vs. PDB)	Reference Code
Barandium System (670 Ma)				
SH	10.10	−36.9		POUB86.225
SH	5.80	−35.5		POUB86.225
SH	5.00	−33.3		POUB86.225
SH	3.30	−33.0		POUB86.225
SH	2.50	−32.6		POUB86.225
SH	9.80	−32.2		POUB86.225
SH	2.20	−31.2		POUB86.225
SH	3.50	−30.5		POUB86.225
SH	2.90	−30.7		POUB86.225
SH	86.20	−35.9		POUB86.225
SH	32.00	−32.8		POUB86.225
SH	64.00	−32.2		POUB86.225
SH	68.00	−31.8		POUB86.225
Bhima Group (675 Ma)				
DOL			2.8	SATH87.147
DOL			3.8	SATH87.147
Umberatana Group (675 Ma)				
LS			2.1	SCHI75...1
LS			6.1	SCHI75...1
DOL			1.6	SCHI75...1
Bed 19 (680 Ma)				
SH	2.90	−31.0		KNOL86.832
LS	5.10	−26.2		KNOL86.832
LS	5.90	−27.2		KNOL86.832
LS	5.10	−26.5		KNOL86.832
LS	4.10	−17.0		KNOL86.832
LS	2.00	−26.6		KNOL86.832
LS	5.70	−25.4		KNOL86.832
LS			3.2	KNOL86.832
LS			2.9	KNOL86.832
LS			3.0	KNOL86.832
LS			6.6	KNOL86.832
LS			5.5	KNOL86.832
LS			7.2	KNOL86.832
Bed 19/20 (680 Ma)				
CARB	0.50	−30.0		KNOL86.832
CARB	1.20	−34.8		KNOL86.832
CARB			−4.2	KNOL86.832
LS			−6.7	KNOL86.832
CARB			−4.9	KNOL86.832
Blackaberget Formation (680 Ma)				
SILT	1.70	−34.6		KNOL86.832
SH	3.10	−31.0		KNOL86.832
SH	7.40	−27.7		KNOL86.832
SILT	4.90	−25.0		KNOL86.832
SILT	5.60	−26.0		KNOL86.832
SH	2.20	−31.6		KNOL86.832
SILT			−7.0	KNOL86.832
SH			3.6	KNOL86.832
SILT			5.3	KNOL86.832
SILT			5.6	KNOL86.832
SH			2.7	KNOL86.832
Datanpo Formation (680 Ma)				
SH		−33.5		LAMB87.140
SH		−16.3		LAMB87.140
CARB			−4.6	LAMB87.140
Klackberget Formation (680 Ma)				
DOL	2.00	−12.5		KNOL86.832
SH	0.70	−23.1		KNOL86.832
DOL			9.8	KNOL86.832
SH			−4.3	KNOL86.832
Moonlight Valley Tillite (680 Ma)				
DOL			−3.4	WILL79.377
DOL			−3.0	WILL79.377
DOL			−2.9	WILL79.377
DOL			−2.9	WILL79.377
DOL			−0.2	WILL79.377
DOL			1.5	WILL79.377
DOL			2.8	WILL79.377
Elbobreen Formation (690 Ma)				
SH	1.80	−28.7		KNOL86.832
SH	5.80	−30.0		KNOL86.832
DOL			3.7	KNOL86.832
SH			2.7	KNOL86.832
SH			2.1	KNOL86.832
LS			4.7	KNOL86.832
LS			5.7	FAIR87.973
LS			5.9	FAIR87.973
LS			5.8	FAIR87.973
LS			6.0	FAIR87.973
LS			6.0	FAIR87.973
LS			6.6	FAIR87.973
LS			4.3	FAIR87.973
LS			6.1	FAIR87.973
LS			5.1	FAIR87.973
LS			5.5	FAIR87.973
LS			3.0	FAIR87.973
LS			2.9	FAIR87.973
LS			1.0	FAIR87.973
LS			2.7	FAIR87.973
LS			4.4	FAIR87.973
LS			4.0	FAIR87.973
LS			3.7	FAIR87.973
LS			0.8	FAIR87.973
LS			4.1	FAIR87.973
LS			3.8	FAIR87.973
LS			1.4	FAIR87.973
LS			3.6	FAIR87.973
LS			3.6	FAIR87.973
LS			3.4	FAIR87.973
700 to 800 Ma				
Backlundtoppen Formation (700 Ma)				
CARB	0.40	−23.5		KNOL86.832
DOL	0.10	−24.1		KNOL86.832
DOL	0.90	−26.9		KNOL86.832

Table 17.3. Continued.

Lithology	TOC (mg/g)	del C$_{org}$ (‰ vs. PDB)	del C$_{carb}$ (‰ vs. PDB)	Reference Code
SH	3.60	−25.0		KNOL86.832
LS			3.9	KNOL86.832
LS			5.7	KNOL86.832
DOL			5.0	KNOL86.832
CARB			6.6	KNOL86.832
DOL			7.8	KNOL86.832
DOL			3.4	KNOL86.832
DOL			5.5	KNOL86.832
DOL			5.3	KNOL86.832
LS			6.5	FAIR87.973
LS			7.2	FAIR87.973
LS			6.0	FAIR87.973
Bed 17 (700 Ma)				
DOL	0.20	−23.1		KNOL86.832
DOL			11.0	KNOL86.832
LS			7.3	KNOL86.832
Bed 18 (700 Ma)				
DOL	0.40	−24.8		KNOL86.832
DOL	1.10	−33.1		KNOL86.832
LS	0.90	−23.9		KNOL86.832
LS	1.60	−23.9		KNOL86.832
DOL			4.8	KNOL86.832
DOL			−3.1	KNOL86.832
LS			7.2	KNOL86.832
LS			7.2	KNOL86.832
Biri Formation (700 Ma)				
LS	16.00			TUCK83.295
LS	20.00			TUCK83.295
LS	34.00			TUCK83.295
LS	23.00			TUCK83.295
LS			1.0	TUCK83.295
LS			1.2	TUCK83.295
LS			1.2	TUCK83.295
LS			1.7	TUCK83.295
LS			1.8	TUCK83.295
LS			1.8	TUCK83.295
LS			1.9	TUCK83.295
LS			1.9	TUCK83.295
LS			2.0	TUCK83.295
LS			2.2	TUCK83.295
LS			2.2	TUCK83.295
Brighton Limestone (700 Ma)				
LS			1.8	VEIZ761387
SILT			2.1	VEIZ761387
Conception Group (700 Ma)				
CH		−31.2		OEHL721246
Hector Formation (700 Ma)				
SH		−30.6		JACK78.335
SH		−30.4		JACK78.335
SH	2.00	−25.8		LEVE754706

Lithology	TOC (mg/g)	del C$_{org}$ (‰ vs. PDB)	del C$_{carb}$ (‰ vs. PDB)	Reference Code
Bed 14 (720 Ma)				
LS	0.90	−23.9		KNOL86.832
LS			4.4	KNOL86.832
Bed 15 (720 Ma)				
DOL	0.40	−23.8		KNOL86.832
DOL			5.6	KNOL86.832
Draken Formation (720 Ma)				
DOL	1.30	−20.9		KNOL86.832
SH	10.00	−24.4		KNOL86.832
DOL	0.90	−23.7		KNOL86.832
CARB	0.30	−22.2		KNOL86.832
DOL	0.20	−23.0		KNOL86.832
LS	0.30	−27.4		KNOL86.832
DOL			5.2	KNOL86.832
DOL			6.8	KNOL86.832
SH			2.9	KNOL86.832
SH			4.5	KNOL86.832
DOL			5.3	KNOL86.832
DOL			6.6	KNOL86.832
DOL			7.8	KNOL86.832
LS			6.6	KNOL86.832
LS			5.4	KNOL86.832
Ryssö Formation (720 Ma)				
DOL	1.20	−25.9		KNOL86.832
CARB	1.50	−24.7		KNOL86.832
DOL	0.80	−26.6		KNOL86.832
LS	1.20	−25.2		KNOL86.832
DOL	1.20	−25.2		KNOL86.832
DOL	1.10	−25.0		KNOL86.832
LS	0.90	−22.3		KNOL86.832
DOL	0.50	−17.3		KNOL86.832
DOL	1.20	−21.4		KNOL86.832
DOL	0.10	−25.2		KNOL86.832
DOL			4.4	KNOL86.832
CARB			6.1	KNOL86.832
DOL			5.2	KNOL86.832
LS			7.7	KNOL86.832
DOL			8.0	KNOL86.832
DOL			8.2	KNOL86.832
LS			5.1	KNOL86.832
DOL			4.6	KNOL86.832
DOL			3.0	KNOL86.832
DOL			4.3	KNOL86.832
DOL			5.8	KNOL86.832
Bed 12 (740 Ma)				
DOL	3.10	−28.2		KNOL86.832
DOL			1.7	KNOL86.832
Bed 13 (740 Ma)				
DOL	0.40	−23.9		KNOL86.832
SH	0.90	−24.2		KNOL86.832
SH	1.30	−23.1		KNOL86.832
DOL			4.7	KNOL86.832

Table 17.3. Continued.

Lithology	TOC (mg/g)	del C$_{org}$ (‰ vs. PDB)	del C$_{carb}$ (‰ vs. PDB)	Reference Code
Bjoranes Formation (740 Ma)				
SH		−25.6		JACK78.335
Damara Supergroup (750 Ma)				
DOL		−23.3		EICH75.585
DOL		−19.8		EICH75.585
LS		−20.4		EICH75.585
DOL			2.3	SCHI75...1
DOL			2.8	SCHI75...1
LS			9.9	SCHI75...1
DOL			6.2	SCHI75...1
DOL			5.5	SCHI75...1
DOL			3.9	SCHI75...1
DOL			−1.1	SCHI75...1
DOL			−1.0	SCHI75...1
LS			−0.8	SCHI75...1
DOL			2.5	SCHI75...1
LS			−2.1	SCHI75...1
DOL			−1.1	EICH75.585
DOL			2.8	EICH75.585
LS			9.9	EICH75.585
Diabaig Formation (750 Ma)				
SH		−30.1		JACK78.335
Eleonore Bay Group (750 Ma)				
DOL		−27.5		EICH75.585
LS		−25.0		EICH75.585
DOL		−22.4		EICH75.585
LS		−23.7		EICH75.585
DOL		−24.9		EICH75.585
LS		−23.9		EICH75.585
LS		−21.6		EICH75.585
DOL			5.2	SCHI75...1
DOL			5.6	SCHI75...1
LS			7.2	SCHI75...1
DOL			4.9	SCHI75...1
LS			5.2	SCHI75...1
CH			5.1	SCHI75...1
DOL			3.7	SCHI75...1
DOL			4.6	SCHI75...1
LS			5.4	SCHI75...1
LS			5.1	SCHI75...1
DOL			5.6	EICH75.585
LS			7.2	EICH75.585
DOL			4.9	EICH75.585
LS			5.2	EICH75.585
LS			3.7	EICH75.585
LS			5.4	EICH75.585
LS			5.1	EICH75.585
Landrigan Tillite (750 Ma)				
DOL			−5.4	WILL79.377
Sturtian (750 Ma)				
LS			2.2	VEIZ761387
LS			2.7	VEIZ761387
LS			3.8	VEIZ761387
Svanbergfjellet Formation (750 Ma)				
LS	0.80	−26.3		KNOL86.832
LS	3.40	−27.7		KNOL86.832
SH	2.80	−31.0		KNOL86.832
CARB	0.40	−27.4		KNOL86.832
LS			4.8	KNOL86.832
LS			4.6	KNOL86.832
DOL			2.4	KNOL86.832
DOL			−2.4	KNOL86.832
Tapley Hill Formation (750 Ma)				
SH		−22.9		MCKI74.591
SH		−23.4		MCKI74.591
SH		−24.2		MCKI74.591
SH		−17.5		MCKI74.591
SH		−24.2		MCKI75.345
SH		−18.8		MCKI75.345
SH		−22.9		MCKI75.345
SH		−23.4		MCKI75.345
SH		−23.8		MCKI75.345
SH		−14.7		MCKI75.345
SH		−17.5		MCKI75.345
LS			−0.2	SCHI75...1
DOL			−4.3	LAMB80...1
DOL			−3.2	LAMB80...1
DOL			−2.3	LAMB80...1
DOL			−1.9	LAMB80...1
LS			−1.6	LAMB80...1
LS			−1.4	LAMB80...1
LS			−1.4	LAMB80...1
LS			−1.2	LAMB80...1
LS			−1.0	LAMB80...1
LS			−1.0	LAMB80...1
LS			−0.8	LAMB80...1
LS			−0.6	LAMB80...1
DOL			−0.3	LAMB80...1
LS			0.4	LAMB80...1
DOL			1.1	LAMB80...1
LS			1.2	LAMB80...1
LS			2.6	LAMB80...1
LS			2.8	LAMB80...1
LS			3.8	LAMB80...1
LS			4.6	LAMB80...1
DOL			−3.8	LAMB84.266
DOL			−3.6	LAMB84.266
DOL			−3.5	LAMB84.266
DOL			−3.0	LAMB84.266
DOL			−2.9	LAMB84.266
DOL			−2.8	LAMB84.266
DOL			−2.8	LAMB84.266
DOL			−2.7	LAMB84.266
DOL			−2.5	LAMB84.266
DOL			−1.9	LAMB84.266
DOL			1.8	WILL79.377
DOL			−1.2	KNUT83.250
DOL			−0.7	KNUT83.250
DOL			−0.7	KNUT83.250
DOL			−0.2	KNUT83.250

Table 17.3. *Continued.*

Lithology	TOC (mg/g)	del C$_{org}$ (‰ vs. PDB)	del C$_{carb}$ (‰ vs. PDB)	Reference Code	Lithology	TOC (mg/g)	del C$_{org}$ (‰ vs. PDB)	del C$_{carb}$ (‰ vs. PDB)	Reference Code
DOL			0.3	KNUT83.250	Grusdievbreen Formation (770 Ma)				
DOL			−2.2	KNUT83.250					
DOL			−1.7	KNUT83.250	LS	0.50	−29.5		KNOL86.832
DOL			−1.7	KNUT83.250	LS	2.50	−25.4		KNOL86.832
DOL			−1.2	KNUT83.250	LS	0.90	−25.3		KNOL86.832
DOL			−1.2	KNUT83.250	LS			0.9	KNOL86.832
DOL			−1.2	KNUT83.250	CARB			7.0	KNOL86.832
DOL			−0.7	KNUT83.250	LS			6.6	KNOL86.832
DOL			−0.7	KNUT83.250	SH			5.1	KNOL86.832
DOL			−2.2	KNUT83.250	DOL			3.1	KNOL86.832
DOL			−1.7	KNUT83.250	LS			4.4	KNOL86.832
DOL			−1.7	KNUT83.250	Skillogalee Dolomite (770 Ma)				
DOL			−1.2	KNUT83.250					
DOL			−1.2	KNUT83.250	DOL		−25.0		EICH75.585
DOL			−1.2	KNUT83.250	CH		−21.0		OEHL721246
DOL			−0.7	KNUT83.250	CH		−23.0		OEHL721246
DOL			−0.2	KNUT83.250	CH		−25.2		OEHL721246
DOL			0.3	KNUT83.250	DOL		−10.8		MCKI74.591
DOL			0.3	KNUT83.250	DOL			−2.0	SCHI75...1
DOL			0.3	KNUT83.250	DOL			−0.3	SCHI75...1
DOL			1.8	KNUT83.250	DOL			−2.0	EICH75.585
DOL			1.8	KNUT83.250	DOL			−4.5	VEIZ761387
Tindelpina Shale Member (750 Ma)					DOL			3.3	VEIZ761387
					DOL			4.1	VEIZ761387
DOL			1.8	VEIZ761387	DOL			4.4	VEIZ761387
					DOL			5.0	VEIZ761387
Bed 10 (760 Ma)					Hunnberg Formation (775 Ma)				
DOL	0.50	−29.6		KNOL86.832	DOL	0.30	−28.8		KNOL86.832
DOL			−1.4	KNOL86.832	SH	1.10	−23.4		KNOL86.832
Bed 11 (760 Ma)					CARB	0.20	−27.1		KNOL86.832
					DOL	0.20	−28.8		KNOL86.832
SH	1.70	−26.6		KNOL86.832	SH	0.30	−24.8		KNOL86.832
CARB	2.90	−23.6		KNOL86.832	DOL			−1.9	KNOL86.832
LS	2.30	−25.6		KNOL86.832	SH			3.2	KNOL86.832
LS	0.50	−21.9		KNOL86.832	DOL			−2.1	KNOL86.832
CARB	0.30	−20.7		KNOL86.832	DOL			−1.1	KNOL86.832
SH			1.4	KNOL86.832	SH			3.7	KNOL86.832
CARB			5.6	KNOL86.832	DOL			2.6	KNOL86.832
LS			6.0	KNOL86.832	River Wakefield Subgroup (775 Ma)				
LS			4.9	KNOL86.832					
CARB			4.8	KNOL86.832	DOL			0.9	VEIZ761387
					DOL			2.9	VEIZ761387
Bed 9 (770 Ma)					DOL			3.3	VEIZ761387
					DOL			3.7	VEIZ761387
LS	0.40	−26.0		KNOL86.832	Bed 7 (780 Ma)				
LS	0.90	−23.3		KNOL86.832					
CARB	2.20	−23.8		KNOL86.832	LS	1.30	−23.7		KNOL86.832
LS	1.50	−23.0		KNOL86.832	LS			6.6	KNOL86.832
CARB	0.60	−26.3		KNOL86.832	Bed 8 (780 Ma)				
CARB	0.40	−23.0		KNOL86.832					
LS			4.6	KNOL86.832	LS	0.50	−22.4		KNOL86.832
LS			5.9	KNOL86.832	LS			4.3	KNOL86.832
CARB			5.7	KNOL86.832	Oxfordbreen Formation (780 Ma)				
LS			5.8	KNOL86.832					
CARB			7.7	KNOL86.832	SILT	0.80	−26.1		KNOL86.832
CARB			7.7	KNOL86.832	SH	3.30	−18.8		KNOL86.832
					LS	0.50	−23.7		KNOL86.832

Table 17.3. Continued.

Lithology	TOC (mg/g)	del C_{org} (‰ vs. PDB)	del C_{carb} (‰ vs. PDB)	Reference Code	Lithology	TOC (mg/g)	del C_{org} (‰ vs. PDB)	del C_{carb} (‰ vs. PDB)	Reference Code
SH			3.3	KNOL86.832	Unnamed Formation (830 Ma)				
LS			4.4	KNOL86.832	SH		−21.4		JACK78.335
Raudstop/Salodd Formation (780 Ma)					SH		−21.1		JACK78.335
SH	0.40	−21.8		KNOL86.832	SH		−21.8		JACK78.335
SH	0.50	−25.3		KNOL86.832	SH	34.00	−21.9		LEVE754706
SH			3.6	KNOL86.832	Kapp Lord Formation (840 Ma)				
SILT			3.4	KNOL86.832	LS	0.80	−23.0		KNOL86.832
800 to 900 Ma					SH	0.50	−22.5		KNOL86.832
Burra Group (800 Ma)					SH	1.50	−24.6		KNOL86.832
CH	1.30	−23.1		HAYE83..93	LS	0.50	−21.6		KNOL86.832
CH			4.6	SCHI83.149	CARB	0.10	−19.4		KNOL86.832
Glasgowbreen Formation (800 Ma)					SH	1.00	−23.1		KNOL86.832
SILT	2.10	−26.6		KNOL86.832	SH	0.20	−20.4		KNOL86.832
SH	1.20	−26.3		KNOL86.832	LS			5.1	KNOL86.832
SH	1.30	−22.9		KNOL86.832	LS			5.1	KNOL86.832
SH	1.50	−31.0		KNOL86.832	CARB			4.7	KNOL86.832
SH	1.10	−31.1		KNOL86.832	SH			5.4	KNOL86.832
SH	2.00	−33.1		KNOL86.832	Beck Spring Dolomite (850 Ma)				
SH	4.10	−29.8		KNOL86.832	CH		−25.0		OEHL721246
SH			−5.8	KNOL86.832	DOL		−15.2		JACK78.335
Norvik Formation (800 Ma)					DOL			3.8	TUCK82...7
SILT	0.40	−24.7		KNOL86.832	DOL			3.9	TUCK82...7
SILT	0.60	−27.6		KNOL86.832	DOL			4.0	TUCK82...7
Bambui Group (810 Ma)					DOL			4.1	TUCK82...7
CH		−30.6		FAIR86.323	DOL			4.1	TUCK82...7
Flora Formation (820 Ma)					DOL			4.1	TUCK82...7
SH	0.40	−31.9		KNOL86.832	DOL			4.2	TUCK82...7
SH	0.60	−26.2		KNOL86.832	DOL			4.3	TUCK82...7
Kingbreen Formation (830 Ma)					DOL			4.3	TUCK82...7
SH	0.70	−29.3		KNOL86.832	DOL			4.5	TUCK82...7
SH	1.20	−26.2		KNOL86.832	DOL			4.5	TUCK82...7
SH	1.70	−20.2		KNOL86.832	DOL			4.5	TUCK82...7
LS	1.00	−25.7		KNOL86.832	DOL			4.8	TUCK82...7
LS	0.80	−23.2		KNOL86.832	Bitter Springs Formation (850 Ma)				
CARB	0.80	−23.6		KNOL86.832	CH		−28.4		OEHL721246
CARB	0.30	−11.4		KNOL86.832	CH		−27.2		OEHL721246
SH	1.10	−28.1		KNOL86.832	DOL		−18.6		MCKI74.591
SH	1.10	−25.7		KNOL86.832	CH		−24.8		SMIT70.659
SH	1.80	−32.3		KNOL86.832	CH		−15.8		JACK78.335
LS			4.5	KNOL86.832	CH	0.50	−21.9		HAYE83..93
LS			5.0	KNOL86.832	CH	0.80	−22.4		HAYE83..93
CARB			5.0	KNOL86.832	CH			3.4	SCHI83.149
CARB			0.1	KNOL86.832	DOL			−3.8	VEIZ761387
Mineral Fork Tillite (830 Ma)					LS			−2.6	VEIZ761387
SILT		−27.5		JACK78.335	LS			−2.1	VEIZ761387
SILT		−26.2		JACK78.335	LS			−0.9	VEIZ761387
SILT		−24.4		JACK78.335	LS				VEIZ761387
SILT		−25.0		JACK78.335	LS			0.1	VEIZ761387
					DOL			1.4	VEIZ761387
					DOL			2.2	VEIZ761387
					DOL			2.2	VEIZ761387
					LS			3.1	VEIZ761387
					LS			3.7	VEIZ761387
					LS			4.1	VEIZ761387

Lithology	TOC (mg/g)	del C_{org} (‰ vs. PDB)	del C_{carb} (‰ vs. PDB)	Reference Code	Lithology	TOC (mg/g)	del C_{org} (‰ vs. PDB)	del C_{carb} (‰ vs. PDB)	Reference Code
LS			5.3	VEIZ761387	LS			0.7	SCHI75...1
LS			6.1	VEIZ761387	LS			1.3	SCHI75...1
Chattisgarh System (850 Ma)					LS			1.3	EICH75.585
DOL			2.5	SCHI75...1	LS			−0.9	EICH75.585
LS			4.4	SCHI75...1	LS			0.7	EICH75.585
LS			5.0	SCHI75...1	**Red Pine Shale (950 Ma)**				
DOL			4.4	SCHI75...1	SH	1.70	−17.0		HAYE83..93
LS			3.0	SCHI75...1	**Stoer Bay Formation (950 Ma')**				
LS			5.0	EICH75.585	SH		−29.1		JACK78.335
LS		−24.4		EICH75.585	*1000 to 1200 Ma*				
Kingston Peak Formation (850 Ma)					**Kundelungu "Series" (1000 Ma)**				
SILT		−19.5		JACK78.335	DOL			1.3	SCHI75...1
LS	3.50			TUCK86.818	**Kurnool System (1000 Ma)**				
LS	15.80	−19.9		TUCK86.818	LS		−25.0		EICH75.585
LS	6.80	−21.5		TUCK86.818	LS			3.4	EICH75.585
LS	14.90	−23.7		TUCK86.818	**Lower Kurnool System (1000 Ma)**				
LS	9.20			TUCK86.818	LS			3.4	SCHI75...1
LS	6.60	−24.2		TUCK86.818	**Lower Roan Group (1000 Ma)**				
LS	5.20			TUCK86.818	DOL			−1.9	SCHI75...1
LS	2.40			TUCK86.818	LS			−2.4	SCHI75...1
LS			1.1	TUCK86.818	DOL			4.6	SCHI75...1
LS			1.0	TUCK86.818	DOL			−1.5	SCHI75...1
LS			4.3	TUCK86.818	DOL			−1.0	SCHI75...1
LS			1.4	TUCK86.818	LS			−2.3	SCHI75...1
LS			2.8	TUCK86.818	DOL			1.3	SCHI75...1
LS			1.4	TUCK86.818	DOL			1.6	SCHI75...1
LS			1.8	TUCK86.818	DOL			1.4	SCHI75...1
Kwagunt Formation (850 Ma)					DOL			0.7	SCHI75...1
SH		−25.7		BLOE85.741	DOL			−18.5	SCHI75...1
Westmanbukta Formation (850 Ma)					CARB			−7.6	SCHI75...1
SH	1.40	−27.7		KNOL86.832	**Mwashia Group (1000 Ma)**				
SILT	13.30	−29.5		KNOL86.832	DOL			−3.5	SCHI75...1
Kortbreen Formation (860 Ma)					**Upper Kurnool System (1000 Ma)**				
SH	2.60	−25.0		KNOL86.832	LS			−0.6	SCHI75...1
SH	2.50	−24.4		KNOL86.832	**Upper Roan Group (1000 Ma)**				
SILT	2.80	−24.8		KNOL86.832	DOL			6.8	SCHI75...1
Persberget Formation (860 Ma)					**Allamoore Formation (1050 Ma)**				
SH	1.80	−21.6		KNOL86.832	CH	0.80	−31.3		HAYE83..93
SH	0.40	−21.7		KNOL86.832	**Oakway Formation (1050 Ma)**				
SH			1.1	KNOL86.832	DOL			−1.1	SCHI75...1
Middle Proterozoic 900 to 1000 Ma					**Nonesuch Shale (1055 Ma)**				
Taoudenni Basin (900 Ma)					SH		−27.1		HOER67.365
LS		−25.7		EICH75.585	SH		−33.1		JACK78.335
LS		−29.3		EICH75.585	SH		−33.0		JACK78.335
LS		−26.1		EICH75.585	SH		−33.4		JACK78.335
DOL			0.7	SCHI75...1	SH		−32.9		JACK78.335
DOL			−0.4	SCHI75...1					
LS			1.8	SCHI75...1					
LS			0.1	SCHI75...1					
LS			−0.9	SCHI75...1					
LS			−2.2	SCHI75...1					

Table 17.3. Continued.

Lithology	TOC (mg/g)	del C$_{org}$ (‰ vs. PDB)	del C$_{carb}$ (‰ vs. PDB)	Reference Code	Lithology	TOC (mg/g)	del C$_{org}$ (‰ vs. PDB)	del C$_{carb}$ (‰ vs. PDB)	Reference Code
SH		−33.5		JACK78.335	DOL			0.5	LAMB84.461
SH	2.70	−33.8		HAYE83..93	DOL			−1.1	LAMB84.461
SH	2.90			BURN72.895	DOL			2.9	LAMB84.461
SH	3.40			BURN72.895	DOL			−4.0	LAMB84.461
SH	1.50			BURN72.895	CARB			6.2	LAMB84.461
SH	1.50			BURN72.895	CARB			3.9	LAMB84.461
SH	1.00			BURN72.895	DOL			2.2	LAMB84.461
SH	8.90			BURN72.895	DOL			2.1	LAMB84.461
SH	4.00			BURN72.895	DOL			−0.3	LAMB84.461
SH	1.00			BURN72.895	CARB			3.0	LAMB84.461
SH	5.10			BURN72.895	DOL			4.1	LAMB84.461
SH	1.70			BURN72.895	DOL			4.4	LAMB84.461
SH	2.80			BURN72.895	CARB			1.7	LAMB84.461
SH	2.40			BURN72.895	CARB			2.5	LAMB84.461
SH	2.90			BURN72.895	CARB			3.0	LAMB84.461
SH	2.94			BURN72.895	CARB			3.3	LAMB84.461
SH	1.82			BURN72.895	CARB			3.5	LAMB84.461
SH	3.77			BURN72.895	CARB			3.5	LAMB84.461
SH	3.36			BURN72.895	DOL			0.8	LAMB84.461
SH	5.45			BURN72.895	CARB			2.0	LAMB84.461
SH	2.88			BURN72.895	CARB			2.0	LAMB84.461
SH	3.37			BURN72.895	CARB			−1.3	LAMB84.461
SH	3.48			BURN72.895	CARB			3.3	LAMB84.461
SH	3.40			BURN72.895	CARB			2.2	LAMB84.461
SH	3.99			BURN72.895	CARB			3.7	LAMB84.461
SH	2.81			BURN72.895	DOL			3.6	LAMB84.461
SH	3.26			BURN72.895	DOL			4.2	LAMB84.461
SH	3.31			BURN72.895	CARB			3.1	LAMB84.461
SH	3.80			BURN72.895	DOL			3.7	LAMB84.461
SH	2.62			BURN72.895	CARB			4.3	LAMB84.461
SH	3.38			BURN72.895	CARB			3.7	LAMB84.461
SH	3.64			BURN72.895	CARB			2.8	LAMB84.461
SH	6.90			BURN72.895	DOL			3.3	LAMB84.461
SH	7.52			BURN72.895	DOL			3.0	LAMB84.461
SH	4.40			BURN72.895	DOL			2.6	LAMB84.461
SH	5.06			BURN72.895	DOL			2.9	LAMB84.461
SH	1.61			BURN72.895	DOL			3.0	LAMB84.461
SH	3.11			BURN72.895	CARB			1.9	LAMB84.461
SH	5.71			BURN72.895	DOL			4.7	LAMB84.461
SH	8.44			BURN72.895	CARB			4.5	LAMB84.461
SH	2.51			BURN72.895	CARB			4.9	LAMB84.461
SH	7.20			BURN72.895	DOL			3.4	LAMB84.461
SH	9.95			BURN72.895	DOL			3.6	LAMB84.461
Bangemall Group (1100 Ma)					CARB			4.2	LAMB84.461
					CARB			3.8	LAMB84.461
DOL			2.3	VEIZ761387	CARB			4.1	LAMB84.461
DOL			2.2	VEIZ761387	CARB			4.9	LAMB84.461
DOL			−1.0	VEIZ761387	CARB			4.3	LAMB84.461
DOL			−0.1	VEIZ761387	DOL			2.8	LAMB84.461
DOL			2.7	VEIZ761387	DOL			4.7	LAMB84.461
DOL			0.4	VEIZ761387	CARB			3.6	LAMB84.461
DOL			−0.7	VEIZ761387	Mozambique Belt (1100 Ma)				
Eurelia Beds (1100 Ma)					LS			4.4	ARNE851553
DOL			−8.7	LAMB84.461	LS			2.0	ARNE851553
DOL			0.3	LAMB84.461	LS			1.2	ARNE851553
DOL			−1.2	LAMB84.461	LS			−0.3	ARNE851553
DOL			2.0	LAMB84.461	LS			−1.9	ANRE851553

Table 17.3. Continued.

Lithology	TOC (mg/g)	del C$_{org}$ (‰ vs. PDB)	del C$_{carb}$ (‰ vs. PDB)	Reference Code	Lithology	TOC (mg/g)	del C$_{org}$ (‰ vs. PDB)	del C$_{carb}$ (‰ vs. PDB)	Reference Code
LS			3.5	ARNE851553	Rossport Formation (1399 Ma)				
LS			−0.8	ARNE851553	DOL	0.13	−30.9		HAYE83..93
LS			1.6	ARNE851553	DOL	0.30	−28.4		HAYE83..93
LS			−0.6	ARNE851553	Sibley Group (1339 Ma)				
LS			−5.2	ARNE851553	DOL			−0.3	SCHI75...1
1200 to 1400 Ma					McMinn Formation (1340 Ma)				
Clarke "Series" (1200 Ma)					SH		−29.6		HOER67.365
LS			0.8	VEIZ761387	SH		−30.2		MCKI74.591
Mescal Limestone (1200 Ma)					SH	6.00	−32.9		HAYE83..93
DOL			3.7	BEEU85.737	SH	5.80	−32.9		HAYE83..93
DOL			2.8	BEEU85.737	SH	4.30	−32.6		HAYE83..93
DOL			3.1	BEEU85.737	SH	4.30	−32.7		HAYE83..93
DOL			3.4	BEEU85.737	SH	3.00	−33.0		HAYE83..93
DOL			3.0	BEEU85.737	SH	2.90	−32.3		HAYE83..93
DOL			3.2	BEEU85.737	SH	3.50	−32.1		HAYE83..93
DOL			2.7	BEEU85.737	SH	3.40	−31.9		HAYE83..93
DOL			2.8	BEEU85.737	SH	0.30	−29.2		HAYE83..93
DOL			3.5	BEEU85.737	SH	0.14	−28.4		HAYE83..93
DOL			2.3	BEEU85.737	SH	0.40	−29.5		HAYE83..93
DOL			3.6	BEEU85.737	SH	7.40	−31.0		HAYE83..93
DOL			3.0	BEEU85.737	SH	4.30	−32.6		HAYE83..93
DOL			3.2	BEEU85.737	SH	10.60	−32.5		HAYE83..93
DOL			2.7	BEEU85.737	SH	6.70	−32.9		HAYE83..93
DOL			2.8	BEEU85.737	SH	10.10	−32.3		HAYE83..93
DOL			3.7	BEEU85.737	SH	1.90	−31.9		HAYE83..93
Muhos Formation (1200 Ma)					SS	5.20	−33.0		HAYE83..93
SH		−27.6		HOER67.365	Belt Supergroup (1350 Ma)				
Vindhyan Supergroup (1200 Ma)					LS		−22.5		HOER62.190
SH		−33.1		KRIS86.119	LS			2.6	HOER62.190
SH		−33.4		KRIS86.119	Cuddapah System (1350 Ma)				
SH		−33.0		KRIS86.119	DOL		−12.7		EICH75.585
Spokane Formation (1250 Ma)					LS			−0.9	SCHI75...1
DOL			−3.8	LANG871334	DOL			−0.9	EICH75.585
DOL			−3.2	LANG871334	Lansen Creek Shale Member (1350 Ma)				
DOL			−3.8	LANG871334	SH	9.00	−33.4		POWE87...1
DOL			−3.7	LANG871334	Velkerri Formation (1350 Ma)				
DOL			−3.2	LANG871334	SH	72.00	−32.8		POWE87...1
DOL			−3.6	LANG871334	SH	47.00	−32.7		POWE87...1
DOL			−4.0	LANG871334	SH	64.00	−32.2		POWE87...1
DOL			−1.4	LANG871334	SH	35.20	−33.3		POWE87...1
DOL			−1.4	LANG871334	SH	38.60	−32.7		POWE87...1
DOL			−3.7	LANG871334	SH	20.20	−33.1		POWE87...1
Badami Group (1300 Ma)					SH	30.40	−33.0		POWE87...1
DOL			3.2	SATH87.147	Bungle Bungle Dolomite (1364 Ma)				
DOL			3.9	SATH87.147	CARB	0.40	−30.2		HAYE83..93
Dripping Springs Formation (1300 Ma)					CARB	0.20	−19.5		HAYE83..93
					CARB	0.20	−26.3		HAYE83..93
SH		−27.5		JACK78.335	CARB	0.09	−27.0		HAYE83..93
SH		−27.0		JACK78.335	CARB	0.07	−24.4		HAYE83..93
SH		−24.9		JACK78.335	SH	0.20	−27.8		HAYE83..93
					CARB	0.07	−22.9		HAYE83..93

Table 17.3. *Continued.*

Lithology	TOC (mg/g)	del C$_{org}$ (‰ vs. PDB)	del C$_{carb}$ (‰ vs. PDB)	Reference Code	Lithology	TOC (mg/g)	del C$_{org}$ (‰ vs. PDB)	del C$_{carb}$ (‰ vs. PDB)	Reference Code
CH	0.90	−25.9		HAYE83..93	SH	30.00	−31.6		POWE87...1
SH	1.80	−27.7		HAYE83..93	SH	46.80	−31.2		POWE87...1
SH	0.40	−29.9		HAYE83..93	SH	29.50	−30.1		POWE87...1
SH	2.60	−30.5		HAYE83..93	SH	45.00	−30.5		POWE87...1
SH	4.10	−32.4		HAYE83..93	SH	9.00	−29.9		POWE87...1
SH	0.10	−23.2		HAYE83..93	SH	10.40	−32.8		POWE87...1
SH	1.50	−32.7		HAYE83..93	SH	27.60	−32.8		POWE87...1
SH	26.60	−33.6		HAYE83..93	SH			−0.7	SMIT75.269
SH	11.50	−32.0		HAYE83..93	SH			−1.8	SMIT75.269
CARB	0.70	−28.8		HAYE83..93	LS			−1.6	SMIT75.269
CARB	0.40	−30.1		HAYE83..93	SILT			−1.3	SMIT75.269
DOL	0.50	−28.8		HAYE83..93	DOL			−3.8	SMIT75.269
DOL	3.60	−28.4		HAYE83..93	SH			−1.4	SMIT75.269
DOL			−0.9	SCHI83.149	SH			−1.2	SMIT75.269
CARB			−0.9	SCHI83.149	SH			−1.8	SMIT75.269
CARB			−0.8	SCHI83.149	DOL			−1.2	SMIT75.269
DOL			−0.5	SCHI83.149					
CH			−0.3	SCHI83.149	HYC Pyritic Shale Member (1500 Ma)				
DOL			−0.2	SCHI83.149	SH		−26.6		MCKI74.591
CARB			−1.0	SCHI83.149	CARB	5.90	−31.7		HAYE83..93
					DOL			−1.2	RYE.81...1
1400 to 1600 Ma					DOL			−1.1	RYE.81...1
Gaoyuzhuang Formation (1425 Ma)					DOL			−1.0	RYE.81...1
CH	1.54	−29.3		SCHO84.335	DOL			−1.0	RYE.81...1
CH	0.61	−26.5		SCHO84.335	DOL			−0.9	RYE.81...1
CH	0.61	−26.4		SCHO84.335	DOL			−0.8	RYE.81...1
CH	0.54	−27.6		SCHO84.335	DOL			−1.0	RYE.81...1
CH	0.61	−27.4		SCHO84.335	DOL			−1.0	RYE.81...1
					DOL			−1.2	RYE.81...1
Yalco Formation (1485 Ma)					DOL			−1.1	RYE.81...1
SH	13.60	−29.2		POWE87...1	DOL			−2.7	RYE.81...1
SH	53.70	−31.9		POWE87...1	DOL			−2.0	RYE.81...1
SH	39.90	−31.2		POWE87...1	DOL			−1.8	RYE.81...1
SH	31.50	−31.2		POWE87...1	DOL			−2.0	RYE.81...1
					DOL			−1.3	RYE.81...1
Barney Creek Formation (1500 Ma)					DOL			−0.7	RYE.81...1
SH	0.41	−30.0		SMIT75.269	DOL			−1.4	RYE.81...1
SH	0.26	−29.1		SMIT75.269	DOL			−1.1	RYE.81...1
SH	0.54	−31.0		SMIT75.269	DOL			−0.8	RYE.81...1
SH	1.20	−28.5		SMIT75.269	DOL			−1.0	RYE.81...1
SH	2.68	−28.6		SMIT75.269	DOL			−1.2	RYE.81...1
SH	2.88	−29.9		SMIT75.269	DOL			−1.0	RYE.81...1
SH	0.63	−28.6		SMIT75.269	DOL			−1.6	RYE.81...1
SILT	0.16	−29.6		SMIT75.269	LS			−1.6	RYE.81...1
SILT	0.47	−28.6		SMIT75.269	DOL			−1.8	RYE.81...1
SILT	0.16	−30.3		SMIT75.269	LS			−2.2	RYE.81...1
SILT	0.04	−28.9		SMIT75.269					
SILT	0.49	−27.5		SMIT75.269	Marimo Slate (1500 Ma)				
SILT	0.68	−29.5		SMIT75.269	SH		−25.8		MCKI74.591
DOL	0.45	−27.6		SMIT75.269	Uncompahgre Formation (1550 Ma)				
SH	8.20	−34.2		POWE87...1	SH	20.00	−29.3		BARK691403
SH	19.80	−32.9		POWE87...1	SH	22.00	−28.7		BARK691403
SH	27.50	−32.9		POWE87...1	SH	64.00	−29.6		BARK691403
SH	8.80	−32.2		POWE87...1	SH	5.30	−26.5		BARK691403
SH	7.10	−31.0		POWE87...1	SH	6.60	−29.1		BARK691403
SH	7.90	−34.2		POWE87...1	SH	1.50	−27.1		BARK691403
					SH	53.00	−30.9		BARK691403

Table 17.3. Continued.

Lithology	TOC (mg/g)	del C$_{org}$ (‰ vs. PDB)	del C$_{carb}$ (‰ vs. PDB)	Reference Code	Lithology	TOC (mg/g)	del C$_{org}$ (‰ vs. PDB)	del C$_{carb}$ (‰ vs. PDB)	Reference Code
SH	40.00	−29.5		BARK691403	Paradise Creek Formation (1650 Ma)				
SH	3.90	−26.0		BARK691403	CH		−29.3		SMIT70.659
SH	4.40	−25.5		BARK691403	CH		−28.9		JACK78.335
SH	1.40	−23.6		BARK691403	CH		−28.6		JACK78.335
SH	0.80	−24.3		BARK691403					
SH	4.00	−26.6		BARK691403	Urquhart Shale (1670 Ma)				
SH	4.20	−25.3		BARK691403	SH		−28.2		MCKI74.591
SH	2.80	−24.3		BARK691403	SH		−23.9		JACK78.335
SH	20.00	−30.4		BARK691403	SH	18.00	−22.2		LEVE754706
Einasleigh Metamorphics (1580 Ma)					SH	1.00	−25.1		SMIT78.369
LS			5.0	MCNA83.175	SH	11.80	−25.1		SMIT78.369
DOL			5.2	MCNA83.175	SH	3.00	−24.0		SMIT78.369
LS			4.8	MCNA83.175	SH	3.00	−24.3		SMIT78.369
DOL			5.1	MCNA83.175	SH	4.00	−22.9		SMIT78.369
LS			5.4	MCNA83.175	SH	8.00	−21.2		SMIT78.369
DOL			5.7	MCNA83.175	SH	11.00	−22.5		SMIT78.369
LS			4.5	MCNA83.175	SH	16.00	−23.9		SMIT78.369
DOL			6.1	MCNA83.175	SH	1.00	−21.4		SMIT78.369
Earley Proterozoic 1600 to 1800 Ma					SH		−24.0		SMIT78.369
Amelia Dolomite (1600 Ma)					SH		−23.9		SMIT78.369
					SH		−22.3		SMIT78.369
CH	0.70	−29.4		HAYE83..93	SH		−23.4		SMIT78.369
CARB	0.80	−29.1		HAYE83..93	SH	8.00	−22.9		SMIT78.369
CARB	0.90	−24.7		HAYE83..93	SH		−25.2		SMIT78.369
CARB	0.90	−24.4		HAYE83..93	SH	3.00	−24.9		SMIT78.369
CARB	1.70	−28.6		HAYE83..93	SH	2.00	−22.7		SMIT78.369
CARB	1.30	−27.8		HAYE83..93	SH		−23.5		SMIT78.369
DOL	0.10	−21.6		HAYE83..93	SH	2.00	−23.5		SMIT78.369
DOL			−0.1	SCHI83.149	SH	2.00	−23.5		SMIT78.369
CARB			−1.2	SCHI83.149	SH	2.00	−24.8		SMIT78.369
CARB			−1.1	SCHI83.149	SH		−26.1		SMIT78.369
Kaladgi Group (1600 Ma)					SH	2.00	−23.1		SMIT78.369
					SH		−24.4		SMIT78.369
DOL			0.6	SATH87.147	McArthur Group (1690 Ma)				
DOL			1.2	SATH87.147	DOL			−4.9	VEIZ761387
LS			1.1	SATH87.147	DOL			−0.8	VEIZ761387
LS			0.3	SATH87.147	DOL			−0.6	VEIZ761387
LS			1.5	SATH87.147	DOL			−0.5	VEIZ761387
LS			1.0	SATH87.147	DOL			−0.2	VEIZ761387
DOL			1.3	SATH87.147	Vermillion Limestone (1700 Ma)				
DOL			0.6	SATH87.147	CARB	0.30	−23.0		HAYE83..93
DOL			0.2	SATH87.147					
DOL			0.5	SATH87.147	*1800 to 2000 Ma*				
DOL			−3.6	SATH87.147	Amisk Group (1800 Ma)				
DOL			0.2	SATH87.147	SH	34.00	−28.4		STRA862653
DOL			0.2	SATH87.147	SH	51.00	−28.2		STRA862653
Tooganinie Formation (1600 Ma)					SH	17.00	−27.2		STRA862653
CH	0.40	−27.2		HAYE83..93	SH	32.00	−27.2		STRA862653
CH	0.20	−22.1		HAYE83..93	SH	20.00	−27.4		STRA862653
CARB			−2.4	SCHI83.149	SH	33.00	−28.3		STRA862653
DOL			−1.6	SCHI83.149	SH	19.00	−26.5		STRA862653
Mallapunyah Formation (1650 Ma)					SH	70.00	−28.8		STRA862653
CARB	3.90	−32.3		HAYE83..93	SH	27.00	−28.7		STRA862653
CARB			0.2	SCHI83.149	SH	17.00	−28.8		STRA862653

Carbon in Whole Rock Samples

Table 17.3. Continued.

Lithology	TOC (mg/g)	del C$_{org}$ (‰ vs. PDB)	del C$_{carb}$ (‰ vs. PDB)	Reference Code
SH	16.00	−28.4		STRA862653
SH	17.00	−28.1		STRA862653
SH	23.00	−27.9		STRA862653
SH	24.00	−28.5		STRA862653
SH	21.00	−29.4		STRA862653
SH	7.00	−27.0		STRA862653
SH	33.00	−28.0		STRA862653
SH	17.00	−28.2		STRA862653
SH	12.00	−27.3		STRA862653
SH	30.00	−29.0		STRA862653
SH	29.00	−29.8		STRA862653
SH	6.00	−25.2		STRA862653
SH	22.00	−17.9		STRA862653
SH	5.00	−25.4		STRA862653
SH	7.00	−23.8		STRA862653
SH	5.00	−23.5		STRA862653
SH	7.00	−22.7		STRA862653
SH	7.00	−22.4		STRA862653
SH	5.00	−23.1		STRA862653
SH	10.00	−23.0		STRA862653
SH	7.00	−23.5		STRA862653
SH	5.00	−23.0		STRA862653
SH	6.00	−22.0		STRA862653
SH	6.00	−20.7		STRA862653
SH	13.00	−28.6		STRA862653
SH	10.00	−28.0		STRA862653
SH	4.00	−27.9		STRA862653
SH	8.00	−27.8		STRA862653
SH	6.00	−27.1		STRA862653
SH	12.00	−28.5		STRA862653

Earaheedy Group (1800 Ma)

LS	0.70	−28.5		HAYE83..93
CARB	0.07	−15.1		HAYE83..93
LS	0.10	−21.1		HAYE83..93
CH	3.10	−32.8		HAYE83..93
CH	2.00	−29.0		HAYE83..93
CH	0.09	−10.6		HAYE83..93
CH	0.06	−11.2		HAYE83..93
CH	0.40	−32.2		HAYE83..93
LS			−0.9	SCHI83.149
LS			−1.8	SCHI83.149
CARB			−2.4	SCHI83.149

Frere Formation (1800 Ma)

DOL	0.08	−29.7		HAYE83..93
DOL	0.40	−25.0		HAYE83..93
DOL	0.20	−22.8		HAYE83..93
DOL	0.30	−26.8		HAYE83..93
DOL	0.13	−19.5		HAYE83..93
CARB	0.30	−24.8		HAYE83..93
CH	0.08	−20.0		HAYE83..93
CH	0.07	−11.8		HAYE83..93
CH	0.07	−20.2		HAYE83..93
CH	0.20	−14.1		HAYE83..93
DOL			−1.1	SCHI83.149

Lithology	TOC (mg/g)	del C$_{org}$ (‰ vs. PDB)	del C$_{carb}$ (‰ vs. PDB)	Reference Code
DOL			1.0	SCHI83.149
DOL			0.6	SCHI83.149

Jatulian Shungite Formation (1800 Ma)

SH		−32.0		LEVE754706

Kulele Creek Limestone (1800 Ma)

CARB	0.10	−20.0		HAYE83..93
CARB			−1.2	SCHI83.149

Windidda Formation (1800 Ma)

LS	0.20	−23.0		HAYE83..93
CH	0.20	−24.0		HAYE83..93

Yelma Formation (1800 Ma)

CARB	0.13	−23.0		HAYE83..93

Chelmsford Formation (1850 Ma)

SH	5.10	−31.6		HAYE83..93

Onwatin Formation (1850 Ma)

SH	40.00	−30.7		HAYE83..93
SH	36.00	−30.5		HAYE83..93
SH		−36.9		THOD62.565

Koolpin Formation (1885 Ma)

CH		−31.0		OEHL721246

Golden Dyke Formation (1900 Ma)

SH		−15.8		MCKI74.591

Pekanatui Point Formation (1900 Ma)

LS			0.4	VEIZ761387

Rove Formation (1900 Ma)

SH		−33.6		BARG77.425
SH		−34.8		JACK78.335
SH		−35.9		JACK78.335
SH	8.10	−32.4		HAYE83..93
SH	4.00	−32.0		LEVE754706

Umkondo Group (1900 Ma)

DOL			14.0	SCHI75...1

Utsingi Formation (1900 Ma)

LS			2.1	VEIZ761387

Epworth Group (1925 Ma)

DOL			0.2	SCHI75...1

Rocknest Formation (1925 Ma)

DOL			1.2	VEIZ761387
DOL			−0.6	VEIZ761387

Grythyttan Slate (1950 Ma)

SH		−30.3		EHLI80.145
SH		−27.3		EHLI80.145
SH		−28.2		EHLI80.145

Table 17.3. Continued.

Lithology	TOC (mg/g)	del C_{org} (‰ vs. PDB)	del C_{carb} (‰ vs. PDB)	Reference Code	Lithology	TOC (mg/g)	del C_{org} (‰ vs. PDB)	del C_{carb} (‰ vs. PDB)	Reference Code
2000 to 2250 Ma					LS			−9.3	PERR81..83
Barreiro Formation (2000 Ma)					LS			−9.6	PERR81..83
SH		−29.0		SCHI76.344	LS			−7.2	PERR81..83
SH		−24.6		SCHI76.344	LS			−7.4	PERR81..83
SH		−27.7		SCHI76.344	LS			−12.9	PERR81..83
SH		−27.7		SCHI76.344	LS			−12.4	PERR81..83
SH		−17.1		SCHI76.344	LS			−12.0	PERR81..83
SH		−24.3		SCHI76.344	LS			−13.9	PERR81..83
SH		−20.5		SCHI76.344	LS			−8.6	PERR81..83
CALC Zone (2000 Ma)					LS			−7.7	PERR81..83
CARB	2.10	−4.7		BANE86.239	LS			−6.8	PERR81..83
CARB	2.30	−5.9		BANE86.239	LS			−6.4	PERR81..83
Duck Creek Dolomite (2000 Ma)					LS			−7.7	PERR81..83
DOL	0.60	−23.6		HAYE83..93	LS			−9.0	PERR81..83
DOL	0.30	−19.7		HAYE83..93	LS			−13.6	PERR81..83
DOL	0.20	−19.8		HAYE83..93	LS			−10.7	PERR81..83
DOL	0.20	−24.8		HAYE83..93	LS			−10.3	PERR81..83
CH	0.10	−23.5		HAYE83..93	LS			−11.2	PERR81..83
CH	0.20	−25.1		HAYE83..93	LS			−9.3	PERR81..83
CH	0.20	−29.3		HAYE83..93	LS			−16.8	PERR81..83
CARB	0.20	−25.9		HAYE83..93	Menihek Formation (2000 Ma)				
CH	0.40	−22.2		HAYE83..93	SH		−31.1		LEVE754706
CH	0.20	−19.6		HAYE83..93	Randville Dolomite (2000 Ma)				
CARB	0.14	−24.6		HAYE83..93	DOL		−17.9		HOER62.190
CARB	0.14	−20.0		HAYE83..93	DOL			3.5	HOER62.190
CH	0.30	−22.7		HAYE83..93	Sabara Formation (2000 Ma)				
DOL			1.0	SCHI83.149	SH		−11.5		SCHI76.344
DOL			0.7	SCHI83.149	SH		−21.1		SCHI76.344
DOL			0.5	SCHI83.149	SH		−16.1		SCHI76.344
DOL			0.4	SCHI83.149	Taltheilei Formation (2000 Ma)				
CH			0.8	SCHI83.149	CARB	0.11	−22.8		HAYE83..93
LS			0.6	VEIZ761387	CARB			1.3	SCHI83.149
DOL			0.4	VEIZ761387	Karelian Dolomite (2050 Ma)				
DOL			−0.4	BECK72.577	DOL		−17.7		EICH75.585
LS			1.0	BECK72.577	DOL		−13.3		EICH75.585
DOL			−2.1	BECK72.577	DOL			8.1	SCHI75...1
DOL			−0.8	BECK72.577	DOL			3.1	SCHI75...1
Gibraltar Formation (2000 Ma)					DOL			4.1	SCHI75...1
CARB			−0.7	SCHI83.149	DOL			4.5	SCHI75...1
Hearne Formation (2000 Ma)					DOL			4.1	SCHI75...1
CARB	0.13	−25.1		HAYE83..93	DOL			8.6	SCHI75...1
CARB			1.6	SCHI83.149	DOL			3.5	SCHI75...1
Kahochella Group (2000 Ma)					DOL			5.2	SCHI75...1
CARB	0.09	−21.8		HAYE83..93	DOL			4.1	EICH75.585
CARB			−0.2	SCHI83.149	DOL			3.1	EICH75.585
Krivoj Rog Iron Formation (2000 Ma)					Biwabik Iron Formation (2090 Ma)				
BIF	0.07	−21.9		HAYE83..93	BIF		−34.5		PERR731110
LS			−10.8	PERR81..83	BIF		−33.6		PERR731110
LS			−9.4	PERR81..83	BIF		−33.6		PERR731110
LS			−9.1	PERR81..83	BIF		−30.8		PERR731110
LS			−0.8	PERR81..83	BIF		−33.2		PERR731110

Table 17.3. Continued.

Lithology	TOC (mg/g)	del C$_{org}$ (‰ vs. PDB)	del C$_{carb}$ (‰ vs. PDB)	Reference Code
Gunflint Iron Formation (2090 Ma)				
CH		−37.2		OEHL721246
CH		−34.1		OEHL721246
CH		−25.0		BARG77.425
CH		−29.1		BARG77.425
CH		−29.2		HEOR67..89
CH		−28.2		BARG77.425
CH		−37.0		BARG77.425
CH		−39.2		BARG77.425
SH		−34.0		BARG77.425
CH		−19.6		BARG77.425
CH		−15.7		BARG77.425
SH		−20.1		BARG77.425
CH		−30.7		HOER62.190
CH		−33.1		SMIT70.659
CH		−15.6		JACK78.335
CARB	9.40	−34.7		HAYE83..93
CH	0.40	−30.5		HAYE83..93
CH	0.60	−32.8		HAYE83..93
CH	0.20	−27.0		HAYE83..93
CH			−2.9	HOER62.190
Kasegalik Formation (2100 Ma)				
CH	0.13	−24.5		HAYE83..93
CH			−0.6	SCHI83.149
Lower Albanel Formation (2100 Ma)				
DOL	0.20	−22.8		HAYE83..93
DOL			6.4	SCHI83.149
DOL			6.3	SCHI83.149
Manitounuk Group (2100 Ma)				
DOL		−18.0		EICH75.585
DOL	0.14	−24.0		HAYE83..93
DOL	0.11	−21.1		HAYE83..93
DOL			−1.1	SCHI75...1
DOL			−2.2	SCHI75...1
DOL			−2.2	EICH75.585
McLeary Formation (2100 Ma)				
DOL	0.20	−25.6		HAYE83..93
CH	0.40	−25.3		HAYE83..93
CH	0.80	−26.4		HAYE83..93
CH	0.90	−22.4		HAYE83..93
CH			−0.4	SCHI83.149
Svecofennian (2100 Ma)				
LS			0.9	SCHI75...1
DOL			0.1	SCHI75...1
Tabooes Formation (2100 Ma)				
SH		−29.5		SCHI76.344
SH		−27.6		SCHI76.344
SH		−27.6		SCHI76.344
SH		−28.5		SCHI76.344
Upper Albanel Formation (2100 Ma)				
DOL	2.90	−31.4		HAYE83..93
DOL			0.9	SCHI83.149
Fecho do Funil Formation (2150 Ma)				
SH		−23.5		SCHI76.344
Gandarella Formation (2200 Ma)				
SH		−21.4		SCHI76.344
SH		−22.5		SCHI76.344
SH		−22.5		SCHI76.344
SH		−22.9		SCHI76.344
DOL			−1.0	SCHI76.344
DOL			−1.3	SCHI76.344
DOL			−3.2	SCHI76.344
DOL			0.1	SCHI76.344
DOL			−1.1	SCHI76.344
DOL			1.6	SCHI76.344

2250 to 2500 Ma

Lithology	TOC (mg/g)	del C$_{org}$ (‰ vs. PDB)	del C$_{carb}$ (‰ vs. PDB)	Reference Code
Espanola Formation (2250 Ma)				
LS			−3.4	SCHI75...1
Gowganda Formation (2250 Ma)				
SH		−27.4		JACK78.335
SH		−19.5		LEVE754706
Timeball Hill Formation (2250 Ma)				
DOL		−19.4		EICH75.585
DOL			0.4	EICH75.585
Asbestos Hills Formation (2300 Ma)				
BIF	0.20	−24.6		HAYE83..93
Caue Itabirito Formation (2300 Ma)				
SH		−26.0		SCHI76.344
SH		−24.2		SCHI76.344
SH		−20.3		SCHI76.344
SH		−25.1		SCHI76.344
SH		−25.4		SCHI76.344
SH		−21.4		SCHI76.344
DOL			−1.2	SCHI76.344
DOL			0.2	SCHI76.344
DOL			−2.4	SCHI76.344
Crocodile River Formation (2300 Ma)				
DOL	0.20	−29.0		HAYE83..93
Transvaal Supergroup (2300 Ma)				
CH		−28.0		OEHL721246
DOL		−18.5		EICH75.585
DOL		−24.8		EICH75.585
DOL		−23.2		EICH75.585
DOL		−27.0		EICH75.585
DOL		−26.7		EICH75.585
LS		−27.4		EICH75.585
DOL		−31.3		EICH75.585

Table 17.3. Continued.

Lithology	TOC (mg/g)	del C$_{org}$ (‰ vs. PDB)	del C$_{carb}$ (‰ vs. PDB)	Reference Code	Lithology	TOC (mg/g)	del C$_{org}$ (‰ vs. PDB)	del C$_{carb}$ (‰ vs. PDB)	Reference Code
DOL		−27.0		EICH75.585	LS			−2.8	SCHI75...1
LS		−39.9		EICH75.585	DOL			−0.6	SCHI75...1
LS		−29.5		EICH75.585	DOL			−1.9	SCHI75...1
LS		−31.5		EICH75.585	DOL			−1.0	SCHI75...1
DOL		−21.1		EICH75.585	DOL			0.3	SCHI75...1
DOL		−25.4		EICH75.585	DOL			−0.3	SCHI75...1
SH		−25.7		EICH75.585	DOL			−0.8	SCHI75...1
SH		−24.5		EICH75.585	DOL			−0.6	SCHI75...1
LS		−37.1		HOER67.365	DOL			−0.9	SCHI75...1
DOL		−28.9		HOER62.190	DOL			−1.3	SCHI75...1
DOL	0.14	−18.7		HAYE83..93	DOL			−1.0	SCHI75...1
DOL			−1.4	SCHI83.149	DOL			−1.3	SCHI75...1
LS			1.5	SCHI75...1	DOL			−0.5	SCHI75...1
DOL			0.4	SCHI75...1	DOL			−0.3	SCHI75...1
DOL			−0.9	SCHI75...1	DOL			−1.3	SCHI75...1
DOL			−0.9	SCHI75...1	DOL			−3.0	SCHI75...1
DOL			−1.0	SCHI75...1	DOL			−2.9	SCHI75...1
DOL			−0.5	SCHI75...1	DOL			−3.0	SCHI75...1
DOL			−0.9	SCHI75...1	DOL			−2.3	SCHI75...1
DOL			−0.8	SCHI75...1	DOL			−2.7	SCHI75...1
DOL			−0.2	SCHI75...1	DOL			−0.5	SCHI75...1
DOL			−1.2	SCHI75...1	DOL			−0.8	SCHI75...1
DOL			−1.1	SCHI75...1	DOL			−0.6	SCHI75...1
DOL			−0.7	SCHI75...1	DOL			−0.8	SCHI75...1
DOL			−1.1	SCHI75...1	DOL			−1.6	SCHI75...1
DOL			−0.3	SCHI75...1	DOL			−0.8	SCHI75...1
DOL			−0.3	SCHI75...1	DOL			−1.2	SCHI75...1
DOL			−0.7	SCHI75...1	DOL			−1.1	SCHI75...1
DOL			−0.9	SCHI75...1	LS			−2.2	SCHI75...1
DOL			0.3	SCHI75...1	LS			−1.7	SCHI75...1
LS			−0.8	SCHI75...1	DOL			−0.9	EICH75.585
DOL			0.2	SCHI75...1	DOL			−0.8	EICH75.585
LS			−1.1	SCHI75...1	DOL			−0.5	EICH75.585
LS			−0.7	SCHI75...1	DOL			0.2	EICH75.585
LS			−1.3	SCHI75...1	DOL			−1.0	EICH75.585
LS			−0.8	SCHI75...1	LS			0.3	EICH75.585
DOL			−0.2	SCHI75...1	DOL			−0.2	EICH75.585
DOL			−0.7	SCHI75...1	DOL			0.2	EICH75.585
LS			0.3	SCHI75...1	LS			−0.8	EICH75.585
LS			−0.7	SCHI75...1	LS			−1.1	EICH75.585
LS			−0.5	SCHI75...1	DOL			−0.7	EICH75.585
LS			6.6	SCHI75...1	DOL			−0.9	EICH75.585
DOL			0.7	SCHI75...1	DOL			1.9	HOER62.190
DOL			0.2	SCHI75...1					
DOL			−0.2	SCHI75...1	Aravalli Supergroup (2350 Ma)				
DOL			−0.6	SCHI75...1	CARB	1.60	−14.7		BANE86.239
DOL			−1.0	SCHI75...1	CARB	1.50	−15.9		BANE86.239
DOL			−0.5	SCHI75...1	CARB	1.90	−13.7		BANE86.239
DOL			−0.8	SCHI75...1	CARB	3.20	−14.1		BANE86.239
DOL			−1.3	SCHI75...1	CARB	2.80	−13.2		BANE86.239
DOL			−0.9	SCHI75...1	CARB	2.30	−14.5		BANE86.239
DOL			−1.2	SCHI75...1	CARB	1.30	−13.9		BANE86.239
DOL			−1.0	SCHI75...1	CARB	0.50			BANE86.239
DOL			−0.8	SCHI75...1	CARB	2.40	−21.8		BANE86.239
DOL			0.5	SCHI75...1	CARB	1.40			BANE86.239
LS			−2.3	SCHI75...1	CARB	4.50	−11.4		BANE86.239
CH			−2.1	SCHI75...1	CARB	0.90	−17.6		BANE86.239

Table 17.3. Continued.

Lithology	TOC (mg/g)	del C$_{org}$ (‰ vs. PDB)	del C$_{carb}$ (‰ vs. PDB)	Reference Code	Lithology	TOC (mg/g)	del C$_{org}$ (‰ vs. PDB)	del C$_{carb}$ (‰ vs. PDB)	Reference Code
Lomagundi Group (2350 Ma)					DOL			5.9	SCHI76.449
SH		−23.7		EICH75.585	DOL			6.6	SCHI76.449
SH		−24.1		EICH75.585	DOL			11.2	SCHI76.449
DOL			12.5	SCHI75...1	DOL			12.7	SCHI76.449
DOL			12.9	SCHI75...1	DOL			9.7	SCHI76.449
DOL			11.4	SCHI75...1	DOL			6.8	SCHI76.449
DOL			13.4	SCHI75...1	DOL			3.8	SCHI76.449
DOL			8.1	SCHI75...1	DOL			2.6	SCHI76.449
DOL			7.3	SCHI75...1	DOL			6.3	SCHI76.449
DOL			8.0	SCHI75...1	DOL			5.9	SCHI76.449
DOL			7.8	SCHI75...1	DOL			9.0	SCHI76.449
DOL			8.6	SCHI75...1	DOL			6.4	SCHI76.449
DOL			8.0	SCHI75...1	DOL			6.9	SCHI76.449
DOL			9.2	SCHI75...1	DOL			6.3	SCHI76.449
DOL			2.2	SCHI75...1	DOL			12.5	SCHI76.449
LS			−4.9	SCHI75...1	DOL			6.9	SCHI76.449
LS			−4.9	SCHI75...1	DOL			7.6	SCHI76.449
DOL			8.0	SCHI76.449	DOL			13.0	SCHI76.449
DOL			4.0	SCHI76.449	DOL			8.0	SCHI76.449
DOL			6.4	SCHI76.449	DOL			13.6	SCHI76.449
DOL			10.0	SCHI76.449	DOL			8.1	SCHI76.449
DOL			4.4	SCHI76.449	DOL			12.4	SCHI76.449
DOL			10.5	SCHI76.449	DOL			10.5	SCHI76.449
DOL			3.4	SCHI76.449	DOL			10.4	SCHI76.449
DOL			7.1	SCHI76.449	DOL			8.4	SCHI76.449
DOL			11.4	SCHI76.449	DOL			7.5	SCHI76.449
DOL			6.9	SCHI76.449	DOL			11.6	SCHI76.449
DOL			8.7	SCHI76.449	Batatal Formation (2400 Ma)				
DOL			7.7	SCHI76.449					
DOL			7.2	SCHI76.449	SH		−24.2		SCHI76.344
DOL			7.3	SCHI76.449	SH		−24.6		SCHI76.344
DOL			10.2	SCHI76.449	Malmani Subgroup (2400 Ma)				
DOL			2.7	SCHI76.449					
DOL			12.9	SCHI76.449	DOL	0.30	−28.7		HAYE83..93
DOL			11.4	SCHI76.449	DOL	0.80	−32.8		HAYE83..93
DOL			13.4	SCHI76.449	LS	0.30	−30.0		HAYE83..93
DOL			8.1	SCHI76.449	LS	0.40	−33.6		HAYE83..93
DOL			7.3	SCHI76.449	LS	1.20	−33.1		HAYE83..93
DOL			8.0	SCHI76.449	LS	1.30	−33.8		HAYE83..93
DOL			7.8	SCHI76.449	DOL	1.10	−34.4		HAYE83..93
DOL			8.6	SCHI76.449	CARB	0.70	−31.1		HAYE83..93
DOL			9.2	SCHI76.449	CARB	0.70	−33.1		HAYE83..93
DOL			4.0	SCHI76.449	CARB	1.10	−32.4		HAYE83..93
DOL			10.2	SCHI76.449	DOL	0.60	−34.2		HAYE83..93
DOL			10.8	SCHI76.449	DOL	0.90	−33.5		HAYE83..93
DOL			8.8	SCHI76.449	CH	0.20	−13.0		HAYE83..93
DOL			6.8	SCHI76.449	LS	0.90	−14.9		HAYE83..93
DOL			7.0	SCHI76.449	DOL	0.40	−30.4		HAYE83..93
DOL			8.3	SCHI76.449	DOL	0.80	−30.9		HAYE83..93
DOL			8.2	SCHI76.449	LS	0.80	−31.1		HAYE83..93
DOL			7.7	SCHI76.449	LS	0.80	−32.3		HAYE83..93
DOL			7.5	SCHI76.449	DOL	0.90	−34.4		HAYE83..93
DOL			5.8	SCHI76.449	LS	1.10	−34.9		HAYE83..93
DOL			9.8	SCHI76.449	LS	0.60	−31.2		HAYE83..93
DOL			5.9	SCHI76.449	CARB	1.80	−13.8		HAYE83..93
DOL			8.8	SCHI76.449	LS			−1.3	SCHI83.149
DOL			9.1	SCHI76.449	LS			−0.5	SCHI83.149

Table 17.3. Continued.

Lithology	TOC (mg/g)	del C$_{org}$ (‰ vs. PDB)	del C$_{carb}$ (‰ vs. PDB)	Reference Code	Lithology	TOC (mg/g)	del C$_{org}$ (‰ vs. PDB)	del C$_{carb}$ (‰ vs. PDB)	Reference Code
LS			−0.5	SCHI83.149	SID			−7.4	BECK72.577
LS			−0.7	SCHI83.149	ANK			−6.8	BECK72.577
CARB			−0.8	SCHI83.149	SID			−8.3	BECK72.577
CARB			−0.9	SCHI83.149	ANK			−6.5	BECK72.577
CARB			−0.1	SCHI83.149	SID			−7.8	BECK72.577
LS			−0.9	SCHI83.149	ANK			−9.0	BECK72.577
DOL			−0.6	SCHI83.149	SID			−9.8	BECK72.577
DOL			−1.0	SCHI83.149	ANK			−9.7	BECK72.577
LS			−0.8	SCHI83.149	ANK			−9.6	BECK72.577
LS			−0.8	SCHI83.149	ANK			−15.0	BECK72.577
CARB			−0.9	SCHI83.149	ANK			−11.4	BECK72.577
Gamohaan Formation (2450 Ma)					SID			−11.1	BECK72.577
					ANK			−12.9	BECK72.577
DOL	3.00	−36.1		KLEI87..81	ANK			−11.1	BECK72.577
CH	9.60	−37.4		KLEI87..81	SID			−12.5	BECK72.577
					ANK			−11.1	BECK72.577
Archean 2500 to 3000 Ma					SID			−12.6	BECK72.577
Carawine Dolomite (2500 Ma)					ANK			−10.8	BECK72.577
DOL	0.09	−17.3		HAYE83..93	SID			−11.0	BECK72.577
DOL	0.08	−13.6		HAYE83..93	ANK			−10.9	BECK72.577
DOL	0.10	−16.0		HAYE83..93	Keewatin Group (2500 Ma)				
SH	32.00	−32.7		HAYE83..93	CH		−23.2		OEHL721246
SH	41.00	−32.1		HAYE83..93	SH		−25.9		SCHO81.696
CH	0.20	−9.4		HAYE83..93	SH		−22.5		SCHO81.696
DOL	0.05	−21.7		HAYE83..93	SH		−26.9		SCHO81.696
CH			0.2	SCHI83.149	SH		−26.9		SCHO81.696
					SH		−24.6		SCHO81.696
Dales Gorge Member (2500 Ma)					SH		−21.0		SCHO81.696
BIF			−7.9	SCHI83.149	SH		−25.5		SCHO81.696
BIF			−9.3	SCHI83.149	SH		−35.5		SCHO81.696
BIF			−8.5	SCHI83.149	SH		−38.3		SCHO81.696
ANK			−9.7	BECK72.577	SH		−43.8		SCHO81.696
ANK			−9.7	BECK72.577	Keewatin Iron Formation (2500 Ma)				
ANK			−10.1	BECK72.577	CH	18.00	−27.1		Haye83..93
ANK			−9.9	BECK72.577	Keewatin, Abitibi Belt (2500 Ma)				
ANK			−9.7	BECK72.577					
ANK			−9.7	BECK72.577	SH	5.00	−36.5		STRA862653
ANK			−10.3	BECK72.577	SH	14.00	−40.3		STRA862653
ANK			−9.8	BECK72.577	SH	12.00	−40.6		STRA862653
ANK			−9.9	BECK72.577	SH	6.00	−41.0		STRA862653
ANK			−9.5	BECK72.577	SH	14.00	−43.2		STRA862653
ANK			−8.6	BECK72.577	SH	14.00	−43.7		STRA862653
SID			−9.3	BECK72.577	SH	74.00	−42.4		STRA862653
ANK			−9.0	BECK72.577	SH	125.00	−44.3		STRA862653
SID			−9.4	BECK72.577	SH	125.00	−44.6		STRA862653
ANK			−9.4	BECK72.577	SH	56.00	−44.6		STRA862653
SID			−9.3	BECK72.577	SH	65.00	−44.2		STRA862653
ANK			−9.1	BECK72.577	SH	49.00	−44.2		STRA862653
SID			−9.3	BECK72.577	SH	138.00	−44.5		STRA862653
ANK			−9.7	BECK72.577	SH	127.00	−44.6		STRA862653
ANK			−9.2	BECK72.577	SH	146.00	−44.8		STRA862653
ANK			−9.8	BECK72.577	SH	31.00	−42.9		STRA862653
ANK			−9.9	BECK72.577	SH	34.00	−41.6		STRA862653
ANK			−10.2	BECK72.577	SH	6.00	−41.7		STRA862653
ANK			−9.7	BECK72.577	SH	2.00	−37.6		STRA862653
ANK			−9.5	BECK72.577	SH	3.00	−33.6		STRA862653
ANK			−7.0	BECK72.577					

Table 17.3. Continued.

Lithology	TOC (mg/g)	del C_{org} (‰ vs. PDB)	del C_{carb} (‰ vs. PDB)	Reference Code	Lithology	TOC (mg/g)	del C_{org} (‰ vs. PDB)	del C_{carb} (‰ vs. PDB)	Reference Code
SH	5.00	−34.2		STRA862653	SH	62.00	−34.8		STRA862653
SH	93.00	−37.6		STRA862653	SH	47.00	−34.5		STRA862653
SH	42.00	−39.0		STRA862653	SH	51.00	−35.4		STRA862653
SH	146.00	−40.3		STRA862653	SH	103.00	−35.8		STRA862653
SH	70.00	−40.0		STRA862653	SH	8.00	−35.4		STRA862653
SH	86.00	−41.9		STRA862653	SH	46.00	−35.4		STRA862653
SH	16.00	−38.8		STRA862653	SH	52.00	−35.5		STRA862653
SH	76.00	−41.5		STRA862653	SH	11.00	−34.2		STRA862653
SH	118.00	−41.3		STRA862653	SH	15.00	−34.6		STRA862653
SH	36.00	−40.5		STRA862653	SH	20.00	−34.5		STRA862653
SH	60.00	−41.5		STRA862653	SH	14.00	−33.6		STRA862653
SH	21.00	−39.8		STRA862653	SH	78.00	−34.4		STRA862653
SH	80.00	−42.7		STRA862653	SH	80.00	−33.1		STRA862653
SH	54.00	−41.0		STRA862653	SH	66.00	−33.7		STRA862653
SH	129.00	−42.6		STRA862653	SH	28.00	−32.2		STRA862653
SH	4.00	−37.9		STRA862653	SH	58.00	−34.4		STRA862653
SH	5.00	−24.7		STRA862653	SH	29.00	−32.0		STRA862653
SH	68.00	−38.0		STRA862653	SH	47.00	−33.7		STRA862653
SH	57.00	−34.2		STRA862653	SH	57.00	−34.6		STRA862653
SH	104.00	−35.0		STRA862653	SH	41.00	−34.3		STRA862653
SH	80.00	−32.6		STRA862653	SH	64.00	−33.3		STRA862653
SH	19.00	−25.6		STRA862653	SH	27.00	−32.1		STRA862653
SH	52.00	−25.5		STRA862653	SH	3.00	−38.6		STRA862653
SH	1.00	−18.9		STRA862653	SH	7.00	−41.1		STRA862653
SH	2.00	−20.8		STRA862653	SH	81.00	−46.8		STRA862653
SH	3.00	−18.4		STRA862653	SH	3.00	−41.6		STRA862653
SH	28.00	−18.3		STRA862653	SH	32.00	−44.5		STRA862653
SH	7.00	−18.7		STRA862653	SH	54.00	−44.2		STRA862653
SH	20.00	−24.4		STRA862653	SH	15.00	−43.7		STRA862653
SH	124.00	−31.5		STRA862653	SH	7.00	−43.4		STRA862653
					SH	4.00	−40.4		STRA862653
Keewatin, Wabigoon Belt (2500 Ma)					SH	1.00	−33.2		STRA862653
SH	2.00	−21.7		STRA862653	SH	1.00	−24.4		STRA862653
SH	5.00	−20.5		STRA862653	SH	5.00	−28.8		STRA862653
SH	8.00	−20.1		STRA862653	SH	2.00	−23.7		STRA862653
SH	11.00	−20.0		STRA862653	SH	20.00	−27.1		STRA862653
SH	7.00	−22.1		STRA862653	SH	18.00	−27.9		STRA862653
SH	13.00	−21.2		STRA862653	SH	20.00	−28.8		STRA862653
SH	6.00	−23.1		STRA862653	SH	19.00	−29.6		STRA862653
SH	21.00	−21.3		STRA862653	SH	20.00	−29.0		STRA862653
SH	9.00	−22.1		STRA862653	SH	9.00	−29.3		STRA862653
SH	9.00	−21.5		STRA862653	SH	22.00	−29.2		STRA862653
SH	11.00	−21.3		STRA862653	SH	3.00	−30.7		STRA862653
SH	38.00	−21.3		STRA862653	SH	9.00	−29.2		STRA862653
SH	13.00	−20.9		STRA862653	SH	14.00	−29.1		STRA862653
SH	15.00	−21.3		STRA862653	SH	4.00	−28.5		STRA862653
SH	12.00	−21.8		STRA862653	SH	22.00	−30.1		STRA862653
SH	15.00	−21.2		STRA862653	SH	35.00	−27.8		STRA862653
SH	8.00	−21.4		STRA862653	SH	22.00	−33.0		STRA862653
SH	26.00	−20.8		STRA862653	SH	14.00	−31.2		STRA862653
SH	60.00	−21.3		STRA862653	SH	13.00	−33.0		STRA862653
SH	7.00	−20.8		STRA862653	SH	33.00	−33.3		STRA862653
SH	2.00	−27.0		STRA862653	SH	24.00	−33.1		STRA862653
SH	2.00	−31.0		STRA862653	SH	24.00	−32.9		STRA862653
SH	30.00	−33.4		STRA862653	SH	32.00	−32.8		STRA862653
SH	51.00	−34.6		STRA862653	SH	18.00	−32.6		STRA862653
SH	41.00	−33.9		STRA862653	SH	1.00	−23.6		STRA862653
SH	42.00	−34.6		STRA862653	SH	1.00	−24.4		STRA862653

Table 17.3. *Continued.*

Lithology	TOC (mg/g)	del C$_{org}$ (‰ vs. PDB)	del C$_{carb}$ (‰ vs. PDB)	Reference Code	Lithology	TOC (mg/g)	del C$_{org}$ (‰ vs. PDB)	del C$_{carb}$ (‰ vs. PDB)	Reference Code
SH	6.00	−15.4		STRA862653	SH	16.00	−20.6		STRA862653
SH	7.00	−15.5		STRA862653	SH	6.00	−21.8		STRA862653
SH	6.00	−15.4		STRA862653	Lewin Shale Formation (2500 Ma)				
SH	1.00	−21.3		STRA862653	SH	38.00	−43.1		HAYE83..93
SH	6.00	−16.2		STRA862653	Marra Mamba Iron Formation (2500 Ma)				
SH	1.00	−23.2		STRA862653					
SH	8.00	−24.8		STRA862653	BIF	0.30	−37.8		HAYE83..93
SH	13.00	−24.7		STRA862653	BIF	0.60	−39.8		HAYE83..93
SH	11.00	−24.6		STRA862653	BIF	14.00	−40.0		HAYE83..93
SH	12.00	−24.9		STRA862653	BIF	0.40	−34.2		HAYE83..93
SH	4.00	−24.7		STRA862653	BIF			−7.3	SCHI83.149
SH	4.00	−25.0		STRA862653	CARB			−2.8	BAUR85.270
SH	2.00	−25.7		STRA862653	CARB			−12.9	BAUR85.270
SH	5.00	−25.5		STRA862653	CARB			−5.3	BAUR85.270
SH	18.00	−24.2		STRA862653	CARB			−10.3	BAUR85.270
SH	26.00	−24.4		STRA862653	CARB			−8.3	BAUR85.270
SH	7.00	−24.2		STRA862653	CARB			−10.0	BAUR85.270
SH	7.00	−24.1		STRA862653	CARB			−10.5	BAUR85.270
SH	6.00	−18.9		STRA862653	CARB			−10.0	BAUR85.270
SH	8.00	−18.2		STRA862653	CARB			−9.1	BAUR85.270
SH	11.00	−18.2		STRA862653	CARB			−10.7	BAUR85.270
SH	12.00	−18.5		STRA862653	CARB			−11.1	BAUR85.270
SH	15.00	−18.1		STRA862653	CARB			−10.3	BAUR85.270
SH	45.00	−18.0		STRA862653	CARB			−10.4	BAUR85.270
SH	20.00	−19.8		STRA862653	CARB			−9.3	BAUR85.270
SH	4.00	−19.9		STRA862653	CARB			−10.1	BAUR85.270
SH	21.00	−17.6		STRA862653	CARB			−8.9	BAUR85.270
SH	4.00	−21.6		STRA862653	CARB			−11.5	BAUR85.270
SH	2.00	−21.5		STRA862653	CARB			−7.8	BAUR85.270
SH	5.00	−20.0		STRA862653	CARB			−7.8	BAUR85.270
SH	18.00	−22.7		STRA862653	CARB			−11.4	BAUR85.270
SH	20.00	−23.6		STRA862653	CARB			−13.1	BAUR85.270
SH	28.00	−23.6		STRA862653	CARB			−7.9	BAUR85.270
SH	17.00	−23.6		STRA862653	CARB			−8.6	BAUR85.270
SH	8.00	−23.5		STRA862653	CARB			−19.8	BAUR85.270
SH	2.00	−24.1		STRA862653	CARB			−8.4	BAUR85.270
SH	36.00	−19.4		STRA862653	CARB			−12.7	BAUR85.270
SH	23.00	−21.0		STRA862653	CARB			−7.2	BAUR85.270
SH	22.00	−22.5		STRA862653	CARB			−11.2	BAUR85.270
SH	45.00	−21.6		STRA862653					
SH	22.00	−21.3		STRA862653	Mt. McRae Shale (2500 Ma)				
SH	5.00	−19.0		STRA862653					
SH	6.00	−18.6		STRA862653	SH	37.00	−32.4		HAYE83..93
SH	18.00	−22.2		STRA862653	SH	65.00	−38.5		HAYE83..93
SH	28.00	−21.2		STRA862653	SH	2.30	−34.2		HAYE83..93
SH	11.00	−22.0		STRA862653	SH	45.00	−35.5		HAYE83..93
SH	66.00	−28.6		STRA862653	SH	18.00	−35.2		HAYE83..93
SH	24.00	−21.5		STRA862653	SH	41.00	−29.8		HAYE83..93
SH	3.00	−23.1		STRA862653	Mt. Sylvia Formation (2500 Ma)				
SH	13.00	−19.4		STRA862653					
SH	13.00	−21.0		STRA862653	BIF	0.90	−18.6		HAYE83..93
SH	42.00	−20.5		STRA862653	BIF			−12.0	SCHI83.149
SH	8.00	−19.7		STRA862653	CARB			−12.2	BAUR85.270
SH	23.00	−18.3		STRA862653	CARB			−13.3	BAUR85.270
SH	10.00	−19.9		STRA862653	CARB			−12.4	BAUR85.270
SH	9.00	−21.1		STRA862653	CARB			−12.6	BAUR85.270
SH	5.00	−22.0		STRA862653	CARB			−11.4	BAUR85.270
SH	9.00	−21.3		STRA862653	CARB			−13.9	BAUR85.270

Table 17.3. Continued.

Lithology	TOC (mg/g)	del C_{org} (‰ vs. PDB)	del C_{carb} (‰ vs. PDB)	Reference Code	Lithology	TOC (mg/g)	del C_{org} (‰ vs. PDB)	del C_{carb} (‰ vs. PDB)	Reference Code
CARB			−12.6	BAUR85.270	LS			−1.8	BECK72.577
CARB			−13.5	BAUR85.270	DOL			−1.8	BECK72.577
CARB			−7.6	BAUR85.270	LS			−6.3	BECK72.577
CARB			−12.7	BAUR85.270	LS			−0.8	BECK72.577
CARB			−13.5	BAUR85.270	LS			−0.1	BECK72.577
CARB			−20.7	BAUR85.270	LS			−1.6	BECK72.577
CARB			−16.8	BAUR85.270	LS			−1.2	BECK72.577
CARB			−22.0	BAUR85.270	LS			−0.3	BECK72.577
CARB			−9.4	BAUR85.270	LS			−1.1	BECK72.577
CARB			−5.8	BAUR85.270	LS			−0.8	BECK72.577
CARB			−14.1	BAUR85.270	DOL			0.1	BECK72.577
CARB			−7.7	BAUR85.270	DOL			0.4	BECK72.577
CARB			−21.1	BAUR85.270	Yellowknife Supergroup (2500 Ma)				
CARB			−9.1	BAUR85.270					
CARB			−12.5	BAUR85.270	CARB	0.30	−15.9		HAYE83..93
CARB			−10.2	BAUR85.270	CARB			1.8	SCHI83.149
CARB			−18.4	BAUR85.270	Brockman Iron Formation (2550 Ma)				
CARB			−14.2	BAUR85.270					
CARB			−11.2	BAUR85.270	CH		−33.4		OEHL721246
CARB			−11.4	BAUR85.270	CH		−29.3		OEHL721246
CARB			−12.6	BAUR85.270	SH	5.20	−31.4		HAYE83..93
CARB			−9.1	BAUR85.270	SH	14.00	−31.1		HAYE83..93
Pillingani Tuff (2500 Ma)					SH	15.00	−30.9		HAYE83..93
					SH	14.00	−33.2		HAYE83..93
CH			−1.6	VEIZ761387	SH	2.00	−31.1		HAYE83..93
Turee Creek Group (2500 Ma)					SH	1.60	−30.8		HAYE83..93
					SH	18.00	−33.2		HAYE83..93
CARB	0.10	−28.4		HAYE83..93	SH	0.11	−21.1		HAYE83..93
Wittenoom Dolomite (2500 Ma)					BIF	0.20	−22.3		HAYE83..93
					BIF	2.60	−21.4		HAYE83..93
CH		−29.0		OEHL721246	BIF	0.30	−18.0		HAYE83..93
CH		−31.2		OEHL721246	CARB			−8.5	BAUR85.270
CH		−30.2		OEHL721246	CARB			−12.4	BAUR85.270
DOL	0.20	−29.1		HAYE83..93	Belingwe Greenstone Belt (2600 Ma)				
DOL	7.30	−34.2		HAYE83..93					
DOL	21.00	−32.8		HAYE83..93	SH	1.69	−24.1		HAYE83..93
DOL	2.20	−43.0		HAYE83..93	Bulawayan Group (2600 Ma)				
DOL	3.60	−32.1		HAYE83..93					
SH	14.00	−36.0		HAYE83..93	CH		−33.5		OEHL721246
DOL			0.5	SCHI83.149	LS		−32.1		OEHL721246
DOL			−0.1	SCHI83.149	LS		−32.5		OEHL721246
DOL			0.4	SCHI83.149	CH		−31.8		OEHL721246
DOL			−0.1	SCHI83.149	LS		−26.9		EICH75.585
DOL			0.3	BAUR85.270	DOL		−30.2		EICH75.585
DOL			0.5	BAUR85.270	LS		−28.6		EICH75.585
DOL			0.6	BAUR85.270	LS		−31.3		EICH75.585
DOL			−0.1	VEIZ761387	LS		−30.6		EICH75.585
DOL			0.1	VEIZ761387	LS		−33.5		EICH75.585
DOL			−0.2	VEIZ761387	LS		−27.4		EICH75.585
LS			−4.9	BECK72.577	LS		−29.0		EICH75.585
LS			−4.7	BECK72.577	DOL		−15.9		EICH75.585
LS			−6.4	BECK72.577	SH		−39.3		EICH75.585
LS			−4.7	BECK72.577	LS		−28.1		HOER62.190
DOL			−5.7	BECK72.577	LS		−33.5		SCHO71.477
LS			−1.1	BECK72.577	LS		−32.5		SCHO71.477
LS			−1.8	BECK72.577	LS		−32.1		SCHO71.477
LS			−1.8	BECK72.577	LS		−31.8		SCHO71.477
DOL			−1.6	BECK72.577	LS		−30.5		SCHO71.477

Table 17.3. Continued.

Lithology	TOC (mg/g)	del C_{org} (‰ vs. PDB)	del C_{carb} (‰ vs. PDB)	Reference Code	Lithology	TOC (mg/g)	del C_{org} (‰ vs. PDB)	del C_{carb} (‰ vs. PDB)	Reference Code
CARB	1.00	−17.2		HAYE83..93	CH	0.90	−12.5		HAYE83..93
CARB	1.00	−30.6		HAYE83..93	SH	1.00	−27.1		HAYE83..93
CARB	1.80	−33.1		HAYE83..93	CH			−5.7	SCHI83.149
CARB			0.5	SCHI83.149	CH			−7.2	SCHI83.149
CH			−0.7	SCHI83.149	CH			−7.1	SCHI83.149
CH			−0.8	SCHI83.149					
CARB			0.8	SCHI83.149	Steep Rock Group (2600 Ma)				
CARB			1.2	SCHI83.149	LS			1.4	SCHI83.149
LS			0.6	SCHI75...1	CARB			1.1	SCHI83.149
LS			−0.9	SCHI75...1	CH			2.1	SCHI83.149
LS			−0.9	SCHI75...1	LS			2.0	VEIZ761387
LS			−2.5	SCHI75...1	LS			1.9	VEIZ761387
DOL			−3.3	SCHI75...1	LS	0.20	−25.0		HAYE83..93
LS			−3.2	SCHI75...1	LS	0.50	−26.4		HAYE83..93
LS			−0.5	SCHI75...1	LS	0.60	−22.1		HAYE83..93
LS			−0.2	SCHI75...1	CH	2.30	−30.7		HAYE83..93
LS			1.0	SCHI75...1	CH	0.40	−26.0		HAYE83..93
LS			0.2	SCHI75...1	SH	1.70	−26.9		HAYE83..93
LS			0.3	SCHI75...1					
LS			−0.2	SCHI75...1	Jeerinah Formation (2650 Ma)				
LS			−0.1	SCHI75...1	SH		−35.4		HOER67.365
LS			−0.5	SCHI75...1	CH	0.40	−33.0		HAYE83..93
LS			−0.8	SCHI75...1	SH	40.00	−41.2		HAYE83..93
LS			−0.2	SCHI75...1	SH	68.00	−41.3		HAYE83..93
LS			−0.4	SCHI75...1	SH	73.00	−41.4		HAYE83..93
LS			−0.6	SCHI75...1	SH	165.00	−40.5		HAYE83..93
LS			−0.1	SCHI75...1	SH	31.00	−42.0		HAYE83..93
LS			−0.1	SCHI75...1	SH	40.00	−42.4		HAYE83..93
LS			2.5	SCHI75...1	SH	12.00	−38.4		HAYE83..93
LS			0.4	SCHI75...1	CH	3.90	−27.9		HAYE83..93
LS			2.0	SCHI75...1					
LS			1.3	SCHI75...1	Rietgat Formation (2650 Ma)				
LS			2.0	SCHI75...1	CH	0.30	−34.1		HAYE83..93
LS			1.4	SCHI75...1	CH	1.00	−38.9		HAYE83..93
LS			1.3	SCHI75...1	CH	0.50	−35.0		HAYE83..93
LS			−5.4	SCHI75...1					
LS			−5.5	SCHI75...1	Shamvaian Group (2650 Ma)				
LS			0.6	EICH75.585	CARB			2.2	SCHI75...1
DOL			−3.3	EICH75.585	CARB			2.4	SCHI75...1
LS			1.0	EICH75.585	Soudan Iron Formation (2650 Ma)				
LS			−0.5	EICH75.585	SH		−33.7		HOER67.365
LS			0.2	EICH75.585					
LS			−0.2	EICH75.585	Ventersdorp Supergroup (2650 Ma)				
LS			0.3	EICH75.585	SH		−35.8		HOER67.365
LS			−0.2	EICH75.585	LS	0.40	−36.3		HAYE83..93
LS			−0.1	EICH75.585	LS	1.60	−40.6		HAYE83..93
LS			0.4	HOER62.190	LS	2.00	−41.6		HAYE83..93
LS			1.3	VEIZ761387	LS	1.00	−38.7		HAYE83..93
LS			1.3	VEIZ761387	LS	0.80	−36.7		HAYE83..93
Manjeri Formation (2600 Ma)					CARB	2.40	−39.0		HAYE83..93
CH	0.13	−28.7		HAYE83..93	LS	0.60	−31.6		HAYE83..93
CH	0.12	−29.4		HAYE83..93	LS	0.20	−28.4		HAYE83..93
CH	0.08	−13.0		HAYE83..93	LS	7.10	−39.6		HAYE83..93
CH	0.60	−8.8		HAYE83..93	CARB	0.90	−42.4		HAYE83..93
CH	0.60	−32.2		HAYE83..93	CARB	0.60	−41.8		HAYE83..93
CH	1.20	−8.2		HAYE83..93	LS	0.90	−36.7		HAYE83..93
CH	2.50	−29.6		HAYE83..93	LS	1.00	−37.0		HAYE83..93

Table 17.3. Continued.

Lithology	TOC (mg/g)	del C$_{org}$ (‰ vs. PDB)	del C$_{carb}$ (‰ vs. PDB)	Reference Code	Lithology	TOC (mg/g)	del C$_{org}$ (‰ vs. PDB)	del C$_{carb}$ (‰ vs. PDB)	Reference Code
LS	1.60	−37.2		HAYE83..93	BIF			−0.1	THOD83.337
LS	4.20	−39.8		HAYE83..93	BIF			0.1	THOD83.337
CARB	2.20	−41.1		HAYE83..93	BIF			−0.9	THOD83.337
CARB			−3.6	SCHI83.149	BIF			−0.2	THOD83.337
LS			−1.9	SCHI83.149	BIF			−5.8	THOD83.337
LS			−3.5	SCHI83.149	BIF			−7.1	THOD83.337
LS			−1.8	SCHI83.149	BIF			−3.3	THOD83.337
LS			−2.8	SCHI83.149	BIF			−3.8	THOD83.337
LS			−1.9	SCHI83.149	BIF			−2.4	THOD83.337
LS			−2.3	SCHI83.149	BIF			−1.3	THOD83.337
LS			−2.2	SCHI83.149	BIF			−2.5	THOD83.337
LS			−5.4	SCHI83.149	BIF			−2.9	THOD83.337
LS			−6.2	SCHI83.149	BIF			−1.1	THOD83.337
Witwatersrand Supergroup (2710 Ma)					BIF			0.5	THOD83.337
					BIF			1.9	THOD83.337
SH		−34.2		HOER67.365	BIF			1.0	THOD83.337
SH		−21.4		HOEF671096	BIF			−0.1	THOD83.337
SH		−21.9		HOEF671096	BIF			−0.2	THOD83.337
SH		−26.1		HOEF671096	BIF			1.2	THOD83.337
SH		−29.3		HOEF671096	BIF			0.1	THOD83.337
SH		−29.5		HOEF671096	BIF			1.1	THOD83.337
SH		−29.9		HOEF671096	BIF			2.3	THOD83.337
SH		−31.8		HOEF671096	BIF			−3.5	THOD83.337
					BIF			−7.6	THOD83.337
Cheshire Formation (2750 Ma)					BIF			−7.5	THOD83.337
LS		−30.0		ABEL85.357	BIF			−1.5	THOD83.337
LS		−30.6		ABEL85.357	BIF			−4.6	THOD83.337
LS		−31.2		ABEL85.357	Michipicoten Iron Formation (2750 Ma)				
LS		−30.5		ABEL85.357	BIF	19.00	−16.1		HAYE83..93
LS		−32.0		ABEL85.357	Tumbiana Formation (2750 Ma)				
LS		−31.2		ABEL85.357	LS	0.20	−45.8		HAYE83..93
LS		−32.8		ABEL85.357	CH	0.70	−49.9		HAYE83..93
LS		−31.0		ABEL85.357	CH	0.08	−45.4		HAYE83..93
LS		−31.3		ABEL85.357	CH	0.50	−50.3		HAYE83..93
LS		−31.2		ABEL85.357	CH	0.80	−50.6		HAYE83..93
LS		−29.0		ABEL85.357	CH	1.80	−51.7		HAYE83..93
LS		−27.2		ABEL85.357	CH	0.70	−51.6		HAYE83..93
Late Archean BIF, Ontario (2750 Ma)					CH	0.70	−51.9		HAYE83..93
BIF		−28.1		THOD83.337	CH	0.40	−47.6		HAYE83..93
BIF		−14.4		THOD83.337	CH	0.40	−46.9		HAYE83..93
BIF		−25.7		THOD83.337	CH	0.30	−44.6		HAYE83..93
BIF		−28.5		THOD83.337	CARB			−1.7	SCHI83.149
BIF		−28.2		THOD83.337	LS			−0.5	SCHI83.149
BIF		−33.5		THOD83.337	LS			−1.3	SCHI83.149
BIF		−23.7		THOD83.337	LS			−0.2	SCHI83.149
BIF		−24.3		THOD83.337	CH			0.5	SCHI83.149
BIF		−26.0		GOOD76.870	LS			0.2	SCHI83.149
BIF		−20.4		GOOD76.870	LS			0.8	SCHI83.149
BIF		−21.0		GOOD76.870	LS			0.7	SCHI83.149
BIF		−26.1		GOOD76.870	LS			0.7	SCHI83.149
BIF		−27.7		GOOD76.870	LS			0.6	SCHI83.149
BIF		−26.6		GOOD76.870	LS			−1.0	SCHI83.149
BIF			0.4	THOD83.337	Fortescue Group (2800 Ma)				
BIF			−1.3	THOD83.337					
BIF			−0.7	THOD83.337	CH		−28.4		OEHL721246
BIF			−1.1	THOD83.337	CH		−40.8		OEHL721246
BIF			0.1	THOD83.337					

Table 17.3. Continued.

Lithology	TOC (mg/g)	del C$_{org}$ (‰ vs. PDB)	del C$_{carb}$ (‰ vs. PDB)	Reference Code	Lithology	TOC (mg/g)	del C$_{org}$ (‰ vs. PDB)	del C$_{carb}$ (‰ vs. PDB)	Reference Code
Nova Lima Group (2800 Ma)					Kromberg Formation (3300 Ma)				
SH		−27.0		SCHI76.344	CH		−33.0		OEHL721246
SH		−25.8		SCHI76.344	CH		−27.5		OEHL721246
SH		−27.3		SCHI76.344	CH		−30.6		OEHL721246
SH		−27.2		SCHI76.344	CH		−26.1		OEHL721246
SH		−27.0		SCHI76.344	CH		−26.2		OEHL721246
SH		−20.4		SCHI76.344	CH		−26.1		JACK78.335
SH		−27.6		SCHI76.344	CH		−26.2		JACK78.335
SH		−26.6		SCHI76.344	CH	1.20	−32.4		HAYE83..93
SH		−22.2		SCHI76.344	CH	0.50	−27.1		HAYE83..93
SH		−25.2		SCHI76.344	CH	0.50	−31.5		HAYE83..93
Woman Lake Marble Formation (2800 Ma)					CH	2.60	−28.8		HAYE83..93
LS	0.90	−9.6		HAYE83..93	CH	1.90	−25.2		HAYE83..93
3000 to 3700 Ma					CH	6.00	−26.3		HAYE83..93
Insuzi Group (3000 Ma)					CH	1.10	−37.0		HAYE83..93
					SH	2.30	−26.1		HAYE83..93
DOL	0.06	−13.1		HAYE83..93	CH	2.00	−26.4		LEVE754706
CARB	0.50	−14.0		HAYE83..93	CARB			0.9	SCHI83.149
LS	0.20	−16.8		HAYE83..93	Sebakwian Group (3300 Ma)				
DOL	0.12	−16.6		HAYE83..93	LS		−14.8		EICH75.585
CARB	0.40	−18.3		HAYE83..93	LS			0.9	SCHI75...1
SH	0.30	−16.3		HAYE83..93	CARB			−3.5	SCHI75...1
SH	0.30	−24.4		HAYE83..93	CARB			−5.0	SCHI75...1
I	0.13	−18.4		HAYE83..93	LS			0.9	EICH75.585
DOL			1.8	SCHI83.149	Fig Tree Group (3350 Ma)				
DOL			1.4	SCHI83.149	CH		−28.7		OEHL721246
CARB			−0.1	SCHI83.149	CH		−28.0		OEHL721246
CARB			−0.6	SCHI83.149	CH		−26.8		EICH75.585
CARB			2.1	SCHI83.149	SH		−25.9		HOER67.365
DOL			1.8	SCHI83.149	SH		−28.6		JACK78.335
DOL			2.1	SCHI83.149	SH		−29.5		JACK78.335
LS			1.1	SCHI83.149	SH		−28.3		JACK78.335
DOL			2.9	SCHI83.149	CH	1.80	−26.9		HAYE83..93
Limpopo Belt (3000 Ma)					CH	2.20	−27.5		HAYE83..93
					CH	1.40	−31.5		HAYE83..93
DOL		−16.2		EICH75.585	CH	7.20	−26.1		HAYE83..93
Moodies Group (3000 Ma)					SH	3.00	−24.3		HAYE83..93
SH	1.00	−31.1		HAYE83..93	SH	62.00	−32.5		HAYE83..93
DOL			−3.4	SCHI75...1	SH	64.00	−32.8		HAYE83..93
Mozaan Group (3000 Ma)					SH	14.00	−30.4		HAYE83..93
					CH	1.40	−28.0		DUNG74.167
CH	0.20	−14.4		HAYE83..93	CH	0.37	−28.7		DUNG74.167
CH	0.20	−14.4		HAYE83..93	CH			1.5	SCHI75...1
SH	2.70	−26.6		HAYE83..93	DOL			0.3	SCHI75...1
SH	2.60	−25.6		HAYE83..93	Warrawoona Group (3400 Ma)				
Gorge Creek Group (3200 Ma)					CH	0.20	−29.8		HAYE83..93
					CH	0.80	−37.1		HAYE83..93
CH	4.80	−27.6		HAYE83..93	Towers Formation (3435 Ma)				
SH	40.00	−30.1		HAYE83..93	CH	0.70	−34.2		HAYE83..93
SH	43.00	−31.5		HAYE83..93	CH	0.50	−31.2		HAYE83..93
SH	39.00	−31.8		HAYE83..93	CH	0.50	−33.4		HAYE83..93
SH	40.00	−27.9		HAYE83..93	CH	0.50	−34.9		HAYE83..93

Table 17.3. Continued.

Lithology	TOC (mg/g)	del C$_{org}$ (‰ vs. PDB)	del C$_{carb}$ (‰ vs. PDB)	Reference Code	Lithology	TOC (mg/g)	del C$_{org}$ (‰ vs. PDB)	del C$_{carb}$ (‰ vs. PDB)	Reference Code
CH	0.11	−29.4		HAYE83..93	CH		−26.9		OEHL721246
CH	0.08	−25.5		HAYE83..93	CH		−31.4		OEHL721246
CH	0.05	−24.0		HAYE83..93	Theespruit Formation (3550 Ma)				
CH	0.20	−29.6		HAYE83..93					
CH	0.80	−33.0		HAYE83..93	CH		−19.5		OEHL721246
CH	1.40	−36.8		HAYE83..93	CH		−15.5		OEHL721246
CH	0.90	−34.7		HAYE83..93	CH		−15.1		OEHL721246
CH			2.0	SCHI83.149	CH	15.00	−15.2		HAYE83..93
CH			−0.8	SCHI83.149	SH	13.00	−15.2		HAYE83..93
Hooggenoeg Formation (3450 Ma)					I	0.40	−12.7		HAYE83..93
CH		−28.8		OEHL721246	Isua Supergroup (3700 Ma)				
CH		−28.4		OEHL721246	CARB		−22.2		SCHI79.189
CH		−32.1		OEHL721246	CARB		−21.6		SCHI79.189
CH		−32.9		OEHL721246	BIF		−9.2		SCHI79.189
CH	1.90	−31.3		HAYE83..93	BIF		−22.2		SCHI79.189
CH	3.30	−34.2		HAYE83..93	BIF		−14.4		SCHI79.189
CH	4.50	−26.8		HAYE83..93	CARB		−9.0		SCHI79.189
CARB			−0.2	SCHI83.149	LS		−24.9		SCHI79.189
Komati Formation (3450 Ma)					LS		−13.4		SCHI79.189
LS			−1.6	SCHI75...1	LS		−18.7		SCHI79.189
LS			−1.7	SCHI75...1	BIF		−16.8		SCHI79.189
LS			−2.3	SCHI75...1	BIF		−12.4		SCHI79.189
LS			0.5	SCHI75...1	CARB		−5.9		SCHI79.189
LS			−1.6	SCHI75...1	CARB		−8.3		SCHI79.189
Onverwacht Group (3450 Ma)					BIF		−10.3		PERR77.280
					BIF		−9.3		PERR77.280
CH	2.10	−31.4		DUNG74.167	BIF		−10.7		PERR77.280
CH	6.50			DUNG74.167	BIF		−12.5		PERR77.280
CH	2.00	−26.1		DUNG74.167	BIF		−16.3		PERR77.280
CH	7.40	−26.2		DUNG74.167	BIF		−15.3		PERR77.280
CH	1.00			DUNG74.167	BIF		−16.0		PERR77.280
CH	0.23			DUNG74.167	BIF	25.00	−9.8		HAYE83..93
CH	2.30	−33.0		DUNG74.167	BIF	0.06	−12.0		HAYE83..93
CH	3.60	−32.9		DUNG74.167	H	0.06	−14.3		HAYE83..93
CH	1.70	−28.4		DUNG74.167	H	0.06	−19.3		HAYE83..93
CH	11.50	−15.4		DUNG74.167	H	0.11	−28.2		HAYE83..93
CH	15.50	−14.3		DUNG74.167	N	0.70	−16.6		HAYE83..93
CARB			−1.8	SCHI75...1	N	22.00	−16.1		HAYE83..93
Swartkoppie Formation (3450 Ma)					N	0.10	−20.0		HAYE83..93
					N	0.09	−21.6		HAYE83..93
CH		−26.9		JACK78.335	N	0.06	−16.6		HAYE83..93
CH	0.30	−24.7		HAYE83..93	BIF			−4.7	SCHI83.149
CH	2.00	−23.4		HAYE83..93	BIF			−5.3	SCHI83.149
CH	0.80	−25.7		HAYE83..93	CARB			−4.5	PERR77.280
CH	1.60	−32.4		HAYE83..93	CARB			−3.3	PERR77.280
CH	16.00	−26.7		HAYE83..93	CARB			−4.8	PERR77.280
CH	1.60	−30.5		HAYE83..93	CARB			−0.3	PERR77.280
H	0.20	−21.3		HAYE83..93	CARB			−3.3	PERR77.280
CH	3.50	−27.6		LEVE754706	CARB			1.8	PERR77.280
CH		−24.9		OEHL721246	CARB			1.0	PERR77.280
					CARB			−4.1	PERR77.280

Table 17.4. *Summary of all results for abundance and isotopic composition of carbon in whole rock samples*

Lithology	TOC (mg/g)	del C$_{org}$ (‰ vs PDB)	del C$_{carb}$ (‰ vs PDB)	N	Reference Code
Early Cambrian					
Petstrostvet Formation (540 Ma)					
SH			−0.3	32	MAGA86.258
Tokammane Formation (540 Ma)					
SH		−28.6	1.1	1	KNOL86.823
			−0.1	3	KNOL86.832
Late Proterozoic					
Chulaktau Formation (550 Ma)					
CH	2.55	−36.0		3	PPRG
Kirtonryggen Formation (550 Ma)					
LS			−0.4	2	KNOL86.832
Série Schisto Calcaire Formation (550 Ma)					
SH			−0.2	8	TUCK86..48
Wyman Formation (550 Ma)					
SILT, SH	1.70	−13.1		4	PPRG
Transition Beds (555 Ma)					
LS			0.1	9	TUCK86..48
Calcaire Supérieur Formation (560 Ma)					
LS			1.4	10	TUCK86..48
"Cambrian" (560 Ma)					
LS			−1.6	2	LAMB87.140
Chapel Island Formation (560 Ma)					
SH	0.11	−26.5		47	PPRG
LS		−14.9		3	PPRG
LS		−21.9		5	PPRG
LS			−0.8	4	PPRG
Schwarzrand Series (560 Ma)					
CARB			−2.4	12	SCHI75...1
Tal Formation (560 Ma)					
SH	10.30	−33.3		9	BANE86.239
DOL			0.5	2	AHAR87.699
Série Lie de Vin Formation (565 Ma)					
LS			−2.5	14	TUCK86..48
Dengying Formation (570 Ma)					
DOL		−28.8		2	LAMB87.140
DOL			2.5	15	LAMB87.140
Krol E Formation (570 Ma)					
DOL			3.0	3	AHAR87.699
P∈/∈ Boundary Beds (570 Ma)					
LS			−0.5	1	LAMB87.140
Reed Dolomite (570 Ma)					
DOL			1.7	5	PPRG
Dolomie Inférieur Formation (580 Ma)					
DOL			1.0	10	TUCK86..48
Dracoisen Formation (580 Ma)					
SH	8.10	−33.2		3	KNOL86.823
SH	2.40	−25.6		2	KNOL86.823
CARB, SH			3.1	2	KNOL86.832
LS			4.2	11	FAIR87.973
Krol D Formation (580 Ma)					
DOL			2.9	11	AHAR86.699
Krol C Formation (590 Ma)					
CARB			1.9	9	AHAR86.699
Huqf Formation (600 Ma)					
(?)OIL		−36.0		1	GRAN87...6
Julie Member (600 Ma)					
CARB	0.17	−26.7		3	PPRG
Krol Formation (600 Ma)					
CARB	3.05	−28.0		2	BANE86.239
Kuibis Formation (600 Ma)					
CARB			−11.2	2	EICH75.585
CARB			−26.2	1	JACK78.335
CARB			1.3	9	SCHI75...1
CARB			2.9	2	EICH75.585
Yudoma Formation (600 Ma)					
DOL			−0.5	32	MAGA86.258
CH	1.44	−32.6		1	PPRG
Pertatataka Formation (610 Ma)					
CH	0.11	−27.8		1	PPRG
DOL			5.2	2	VEIZ761387
Khatyspyt Formation (620 Ma)					
SH	207.13	−36.4		1	PPRG
CARB	4.93		−37.1	1	PPRG
Late Proterozoic, Spain (625 Ma)					
CARB	1.05	−27.1		2	PPRG
SH	1.20	−29.7		3	PPRG
Canyon Formation (630 Ma)					
SH, CARB	2.05	−25.4		2	KNOL86.823
SH, CARB			−2.7	2	KNOL86.832
Spiral Creek Formation (630 Ma)					
CARB			−28.4	2	KNOL86.823
			−2.6	2	KNOL86.832
Tillite Formation (630 Ma)					
SH		−29.5		2	KNOL86.823
CARB			−25.5	1	KNOL86.823
SH, CARB			1.7	2	KNOL86.832

Table 17.4. Continued.

Lithology	TOC (mg/g)	del C$_{org}$ (‰ vs PDB)	del C$_{carb}$ (‰ vs PDB)	N	Reference Code	Lithology	TOC (mg/g)	del C$_{org}$ (‰ vs PDB)	del C$_{carb}$ (‰ vs PDB)	N	Reference Code
Timperley Shale (630 Ma)						**Barrandium System (670 Ma)**					
SH, CARB	0.29	−25.9		2	PPRG	SH	22.72	−33.0		13	POUB86.225
Boonall Dolomite (640 Ma)						**Bhima Group (675 Ma)**					
CARB	0.19	−31.2		4	PPRG	DOL			3.3	2	SATH87.147
Egan Formation (640 Ma)						**Jarrad Sandstone Member (675 Ma)**					
DOL			−0.5	1	WILL79.377	CARB	0.07	−25.7		2	PPRG
CARB	0.24	−29.5		1	PPRG	**Umberatana Group (675 Ma)**					
Yurabi Formation (640 Ma)						CARB			3.3	3	SCHI75...1
QUAR	0.02			1	PPRG	**Bed 19 (680 Ma)**					
Brachina Formation (650 Ma)						LS	4.65	−24.8		6	KNOL86.823
SH		−21.3		4	MCKI74.591	SH	2.90	−31.0		1	KNOL86.823
Chichkan Formation (650 Ma)						LS			4.7	6	KNOL86.832
CH	0.76	−27.9		1	SCHO84.335	**Bed 19/20 (680 Ma)**					
CH	0.16	−28.9		2	PPRG	CARB	0.85	−32.4		2	KNOL86.823
Doushantou Formation (650 Ma)						CARB			−5.3	3	KNOL86.832
SH		−28.5		10	LAMB87.140	**Blackaberget Formation (680 Ma)**					
CARB			4.1	16	LAMB87.140	SILT, SH	4.15	−29.3		6	KNOL86.823
Etina Limestone (650 Ma)						SILT, SH			2.0	5	KNOL86.832
LS			9.5	2	VEIZ761387	**Datanpo Formation (680 Ma)**					
Malmesbury Formation (650 Ma)						SH		−24.9		2	LAMB87.140
CARB		−24.0		10	EICH75.585	CARB			−4.6	1	LAMB87.140
CARB			1.5	13	SCHI75...1	**Klackaberget Formation (680 Ma)**					
CARB			0.8	2	EICH75.585	SH	0.70	−23.1		1	KNOL86.823
Nuccaleena Formation (650 Ma)						CARB	2.00	−12.5		1	KNOL86.823
DOL			−2.0	1	VEIZ761387	SH, CARB			2.8	2	KNOL86.832
DOL			−2.5	2	WILL79.377	**Moonlight Valley Tillite (680 Ma)**					
Sheepbed Formation (650 Ma)						DOL			−1.2	7	WILL79.377
SH	5.56	−28.8		10	PPRG	CARB	0.02			1	PPRG
Trezona Formation (650 Ma)						**Elbobreen Formation (690 Ma)**					
LS			−7.5	3	VEIZ761387	SH	3.80	−29.3		2	KNOL86.823
Walsh Tillite (650 Ma)						CARB, SH			3.3	4	KNOL86.832
DOL			−0.7	2	WILL79.377	CARB			4.1	24	FAIR87.973
Willochra Formation (650 Ma)						**Backlundtoppen Formation (700 Ma)**					
LS			3.3	1	VEIZ761387	CARB	0.46	−24.9		3	KNOL86.823
Wilsonbreen Formation (650 Ma)						SH	3.60	−25.0		1	KNOL86.823
CARB		−29.8		1	KNOL86.823	CARB			5.4	8	KNOL86.832
CARB			4.0	1	KNOL86.832	LS			6.6	3	FAIR87.973
CARB			1.8	11	FAIR87.973	**Bed 17 (700 Ma)**					
Wonokaan Formation (650 Ma)						CARB	0.20	−23.1		1	KNOL86.823
LS			−7.2	1	VEIZ761387	CARB			9.1	2	KNOL86.832
Cango Formation (660 Ma)						**Bed 18 (700 Ma)**					
CARB		−21.2		2	EICH75.585	CARB	1.00	−26.4		4	KNOL86.823
CARB			−1.7	2	EICH75.585	CARB			5.4	4	KNOL86.832
						Biri Formation (700 Ma)					
						LS	23.25	−28.6		4	TUCK83.295
						LS		−29.3		4	PPRG
						LS			1.7	11	TUCK83.295

Table 17.4. Continued.

Lithology	TOC (mg/g)	del C_{org} (‰ vs PDB)	del C_{carb} (‰ vs PDB)	N	Reference Code
Brighton Limestone (700 Ma)					
LS, SILT			2.0	2	VEIZ761387
Conception Group (700 Ma)					
CH		−31.2		1	OEHL721246
Hector Formation (700 Ma)					
SH		−30.5		2	JACK78.335
SH	2.00	−25.8		1	LEVE754706
Keele Formation (700 Ma)					
CARB	0.54	−23.8		10	PPRG
Luoquan Formation (700 Ma)					
SH	1.25	−27.4		11	PPRG
Twitya Formation (700 Ma)					
SH	1.03	−28.7		12	PPRG
CARB	0.46			1	PPRG
Bed 14 (720 Ma)					
LS	0.90	−23.9		1	KNOL86.823
LS			4.4	1	KNOL86.832
Bed 15 (720 Ma)					
DOL	0.40	−23.8		1	KNOL86.823
DOL			5.6	1	KNOL86.832
Draken Formation (720 Ma)					
CARB, SH	2.17	−23.6		6	KNOL86.823
CARB, SH			5.7	9	KNOL86.832
Ryssö Formation (720 Ma)					
CARB	0.97	−23.9		10	KNOL86.823
CARB			5.7	11	KNOL86.832
Bed 12 (740 Ma)					
DOL	3.10	−28.2		1	KNOL86.823
DOL			1.7	1	KNOL86.832
Bed 13 (740 Ma)					
SH	1.10	−23.7		2	KNOL86.823
DOL	0.40	−23.9		1	KNOL86.823
DOL			4.7	1	KNOL86.832
Bjoranes Formation (740 Ma)					
SH		−25.6		1	JACK78.335
Min'yar Formation (740 Ma)					
CH	0.78	−29.4		6	PPRG
Coppercap Formation (750 Ma)					
CARB	2.15	−28.6		19	PPRG
Damara Supergroup (750 Ma)					
CARB		−21.2		3	EICH75.585
CARB			2.6	11	SCHI75...1
CARB			5.8	3	EICH75.585
Diabaig Formation (750 Ma)					
SH		−30.1		1	JACK78.335
Eleonore Bay Formation (750 Ma)					
CARB		−24.1		7	EICH75.585
CARB			5.2	10	SCHI75...1
CARB			6.0	7	EICH75.585
Landrigan Tillite (750 Ma)					
DOL			−5.4	1	WILL79.377
Sturtian (750 Ma)					
LS			2.9	3	VEIZ761387
Svanbergfjellet Formation (750 Ma)					
CARB	1.53	−27.1		3	KNOL86.823
SH	2.80	−31.0		1	KNOL86.823
CARB			2.3	4	KNOL86.832
Tapley Hill Formation (750 Ma)					
SH		−22.0		4	MCKI74.591
SH		−20.8		7	MCKI75.345
SH	2.61	−25.1		1	PPRG
LS			−0.2	1	SCHI75...1
CARB			−0.2	20	LAMB80...1
DOL			1.8	1	WILL79.377
DOL			−0.8	26	KNUT83.250
DOL			−2.9	10	LAMB84.266
Tindelpina Shale Member (750 Ma)					
DOL			1.8	1	VEIZ761387
SILT	7.35	−23.0		2	PPRG
Woocalla Dolomite Member (750 Ma)					
SILT	3.70	−30.4		4	PPRG
Areyonga Formation (760 Ma)					
CH	0.13	−27.4		8	PPRG
Bed 10 (760 Ma)					
DOL	0.50	−29.6		1	KNOL86.823
DOL			−1.4	1	KNOL86.832
Bed 11 (760 Ma)					
CARB, SH	1.54	−23.7		5	KNOL86.823
CARB, SH			4.5	5	KNOL86.832
Bed 9 (770 Ma)					
CARB	1.00	−24.2		6	KNOL86.823
CARB			6.2	6	KNOL86.832
Grusdievbreen Formation (770 Ma)					
LS	1.30	−26.7		3	KNOL86.823
CARB			4.5	6	KNOL86.832
Skillogalee Dolomite (770 Ma)					
CARB		−25.0		1	EICH75.585
CH		−23.1		3	OEHL721246
DOLB		−10.8		1	MCKI74.591
CH	0.46	−20.1		22	PPRG

Table 17.4. Continued.

Lithology	TOC (mg/g)	del C$_{org}$ (‰ vs PDB)	del C$_{carb}$ (‰ vs PDB)	N	Reference Code
DOL			−1.1	2	SCHI75...1
DOL			−2.0	1	EICH75.585
DOL			2.5	5	VEIZ761387
Hunnberg Formation (775 Ma)					
CARB, SH	0.42	−26.6		5	KNOL86.823
CARB, SH			0.7	6	KNOL86.832
River Wakefield Subgroup (775 Ma)					
CH	0.88	14.3		4	PPRG
SILT	0.31	−19.1		3	PPRG
CARB	0.33	−18.1		2	PPRG
DOL			2.7	4	VEIZ761387
Shczal Formation (775 Ma)					
CARB	0.12	−26.6		1	PPRG
Visingsö Group (775 Ma)					
SH	1.61	−26.5		8	PPRG
CARB	1.32	−28.1		2	PPRG
Auburn Dolomite (780 Ma)					
CARB, CH	0.55	−19.3		1	PPRG
Bed 7 (780 Ma)					
LS	1.30	−23.7		1	KNOL86.823
LS			6.6	1	KNOL86.832
Bed 8 (780 Ma)					
LS	0.50	−22.4		1	KNOL86.823
LS			4.3	1	KNOL86.832
Oxfordbreen Formation (780 Ma)					
			2.9	2	KNOL86.832
SH		−22.5		2	KNOL86.823
Raudstop/Salodd Formation (780 Ma)					
CARB, SH	1.53	−23.5		3	KNOL86.823
			3.5	2	KNOL86.832
Rhynie Sandstone (785 Ma)					
CH	3.07	−17.6		1	PPRG
Burra Group (800 Ma)					
CH	1.30	−23.1		1	HAYE83..93
CH			4.6	1	SCHI83.149
Glasgowbreen Formation (800 Ma)					
SH	1.90	−28.7		7	KNOL86.823
SH			−5.8	1	KNOL86.832
Jiudingshan Formation (800 Ma)					
CH	0.55	−23.6		1	PPRG
Little Dal Group (800 Ma)					
CARB	0.34	−26.4		11	PPRG
SH	7.07	−25.5		5	PPRG
CH	0.40	−26.1		2	PPRG
Norvik Formation (800 Ma)					
SILT	0.50	−26.1		2	KNOL86.823
Sayunei Formation (800 Ma)					
SH, SILT	0.04	−16.7		7	PPRG
Torridon Group (800 Ma)					
CONGL	0.08	−26.1		3	PPRG
Bambui Group (810 Ma)					
CH		−30.6		1	FAIR86.323
Flora Formation (820 Ma)					
SH	0.50	−29.1		2	KNOL86.823
Kingbreen Formation (830 Ma)					
SH	1.27	−27.0		6	KNOL86.823
CARB	0.73	−21.0		4	KNOL86.823
CARB			3.6	4	KNOL86.832
Mineral Fork Tillite (830 Ma)					
SILT		−25.8		4	JACK78.335
Unnamed Formation (830 Ma)					
SH		−21.4		3	JACK78.335
SH	34.00	−21.9		1	LEVE754706
Kapp Lord Formation (840 Ma)					
CARB	0.47	−21.3		3	KNOL86.823
SH	0.80	−22.6		4	KNOL86.823
CARB, SH			5.1	4	KNOL86.832
Beck Spring Dolomite (850 Ma)					
CH		−25.0		1	OEHL721246
DOL		−15.2		1	JACK78.335
DOL			4.2	13	TUCK82...7
CARB, CH	0.35	−25.7		1	PPRG
Bitter Springs Formation (850 Ma)					
CH		−27.8		2	OEHL721246
DOL		−18.6		1	MCKI74.591
CH		−24.8		1	SMIT70.659
CH		−15.8		1	JACK78.335
CH	0.65	−22.1		2	HAYE83..93
CARB	0.18	−28.8		7	PPRG
CH	0.16	−25.8		29	PPRG
CH			3.4	1	SCHI83.149
CARB			1.3	14	VEIZ761387
Chattisgarh System (850 Ma)					
LS		−24.4		1	EICH75.585
CARB			3.9	5	SCHI75...1
LS			5.0	1	EICH75.585
Kingston Peak Formation (850 Ma)					
SILT		−19.5		1	JACK78.335
LS	8.05			4	TUCK86.818
LS		−22.3		4	PPRG
LS			2.0	7	TUCK87.818
Kwagunt Formation (850 Ma)					
SH		−25.7		1	BLOE85.741

Table 17.4. Continued.

Lithology	TOC (mg/g)	del C_{org} (‰ vs PDB)	del C_{carb} (‰ vs PDB)	N	Reference Code
Walcott Member (850 Ma)					
SH	20.35	−26.9		3	PPRG
CH	3.15	−26.7		3	PPRG
CARB	17.70		26.5	2	PPRG
Westmanbukta Formation (850 Ma)					
SH	7.35	−28.6		2	KNOL86.823
Zhangqu Formation (850 Ma)					
CH	0.52	−29.4		1	PPRG
Kortbreen Formation (860 Ma)					
SH	2.63	−24.7		3	KNOL86.823
Persberget Formation (860 Ma)					
SH	1.10	−21.6		2	KNOL86.823
SH			1.1	1	KNOL86.832
Awatubi Member (862 Ma)					
SH	7.40	−29.5		2	PPRG
Middle Proterozoic					
Carbon Canyon Member (900 Ma)					
CARB	0.32	−29.1		1	PPRG
SH	0.17	−24.4		2	PPRG
CH	0.96	−26.1		1	PPRG
Shorika Formation (900 Ma)					
CH	0.38	−29.1		1	PPRG
Taoudenni Basin (900 Ma)					
LS		−27.0		3	EICH75.585
CARB			0.1	8	SCHI75...1
CARB			0.6	3	EICH75.585
Jupiter Member (930 Ma)					
SH, CARB	4.74	−24.2		7	PPRG
Katav Formation (940 Ma)					
CARB	0.55	−28.8		2	PPRG
Red Pine Shale (950 Ma)					
SH	1.70	−17.0		1	HAYE83..93
SH	1.84	−17.0		2	PPRG
Stoer Bay Formation (950 Ma)					
SH		−29.1		1	JACK78.335
Tanner Member (950 Ma)					
SH	3.69	−22.0		6	PPRG
Bederysh Member (1000 Ma)					
SILT, SH	1.31	−30.0		1	PPRG
Kundelungu "Series" (1000 Ma)					
DOL			1.3	1	SCHI75...1
Kurnool System (1000 Ma)					
LS		−25.0		1	EICH75.585
LS			3.4	1	EICH75.585
Lower Kurnool System (1000 Ma)					
LS			3.4	1	SCHI75...1
Lower Roan Group (1000 Ma)					
CARB			−2.1	12	SCHI75...1
Mwashia Group (1000 Ma)					
DOL			−3.5	1	SCHI75...1
Sukhaya Tunguska Formation (1000 Ma)					
CH	0.18	−26.5		1	PPRG
Upper Kurnool System (1000 Ma)					
LS			−0.6	1	SCHI75...1
Upper Roan Group (1000 Ma)					
DOL			6.8	1	SCHI75...1
Allamoore Formation (1050 Ma)					
CH	0.80	−31.3		1	HAYE83..93
CH	0.44	−26.8		1	PPRG
Oakway Formation (1050 Ma)					
DOL			−1.0	1	SCHI75...1
Nonesuch Shale (1055 Ma)					
SH		−27.1		1	HOER67.365
SH		−33.2		5	JACK78.335
SH		−38.0		1	HAYE83..93
SH	3.24	−30.4		9	PPRG
SH	3.85			42	BURN72.895
Bangemall Group (1100 Ma)					
DOL			0.8	7	VEIZ761387
Eurelia Beds (1100 Ma)					
CARB			2.6	55	LAMB84.461
Mozambique Belt (1100 Ma)					
LS			0.4	10	ARNE851553
Tieling Formation (1185 Ma)					
CARB	0.14	−24.5		1	PPRG
SH, CARB	0.97	−31.7		1	PPRG
Fengjiawan Formation (1190 Ma)					
CARB	0.08	−26.1		1	PPRG
Clarke Series (1200 Ma)					
LS			0.8	1	VEIZ761387
Duguan Formation (1200 Ma)					
CARB	0.09	−26.8		1	PPRG
Mescal Limestone (1200 Ma)					
DOL			3.1	16	BEEU85.737
Muhos Formation (1200 Ma)					
SH		−27.6		1	HOER67.365

Table 17.4. *Continued.*

Lithology	TOC (mg/g)	del C$_{org}$ (‰ vs PDB)	del C$_{carb}$ (‰ vs PDB)	N	Reference Code
Vindhyan System (1200 Ma)					
SH		−33.2		3	KRIS86.119
Avsyan Formation (1240 Ma)					
SH	3.49	−23.9		4	PPRG
Hongshuizhuang Formation (1250 Ma)					
SH	8.34	−28.5		2	PPRG
Spokane Formation (1250 Ma)					
DOL			−3.2	10	LANG871334
Xunjiansi Formation (1250 Ma)					
CH, CARB	0.08	−27.4		3	PPRG
Revet Member (1260 Ma)					
CH	0.43	−30.1		1	PPRG
Badami Group (1300 Ma)					
DOL			3.6	2	SATH87.147
Dripping Springs Formation (1300 Ma)					
SH		−26.5		3	JACK78.335
Mount Shields Formation (1320 Ma)					
CARB	0.05	−26.3		1	PPRG
Wumishan Formation (1325 Ma)					
CH	0.42	−28.6		15	PPRG
SH	0.21	−28.4		2	PPRG
CARB	0.17	−28.3		4	PPRG
Rossport Formation (1339 Ma)					
DOL	0.22	−29.6		2	HAYE83..93
CARB, CH	0.42	−26.4		4	PPRG
Sibley Group (1339 Ma)					
DOL			−0.3	1	SCHI75...1
McMinn Formation (1340 Ma)					
SH		−29.6		1	HOER67.365
SH		−30.2		1	MCKI74.591
SH	4.46	−31.9		18	HAYE83..93
SILT	10.68	−32.8		9	PPRG
Belt Supergroup (1350 Ma)					
LS		−22.5		1	HOER62.190
LS			2.6	1	HOER62.190
Cuddapah System (1350 Ma)					
DOL		−12.7		1	EICH75.585
LS			−0.9	1	SCHI75...1
DOL			−0.9	1	EICH75.585
Lansen Creek Shale Member (1350 Ma)					
SH	9.00	−33.4		1	POWE87...1
Tukan Member (1350 Ma)					
SH	1.63	−25.0		2	PPRG
Velkerri Formation (1350 Ma)					
SH	43.91	−32.8		7	POWE87...1
Yangzhuang Formation (1350 Ma)					
CH, CARB	0.21	−27.5		4	PPRG
Bungle Bungle Dolomite (1364 Ma)					
CARB	0.62	−26.4		10	HAYE83..93
SH	5.42	−30.0		9	HAYE83..93
CARB	0.16	−24.7		2	PPRG
CH	0.08	−21.5		3	PPRG
CARB			−0.7	7	SCHI83.149
CH	0.90	−25.9		1	HAYE83..93
Helena Formation (1380 Ma)					
CARB	0.08	−26.0		1	PPRG
Dismal Lakes Group (1400 Ma)					
CH	0.77	−28.5		7	PPRG
Greyson Shale (1420 Ma)					
SH	1.22	−29.8		2	PPRG
Gaoyuzhuang Formation (1425 Ma)					
CH	1.30	−27.4		5	SCHO84.335
CH	1.19	−30.9		8	PPRG
Altyn Formation (1440 Ma)					
CARB	0.07	−26.4		1	PPRG
Newland Limestone (1440 Ma)					
SH	0.93	−24.7		15	PPRG
CARB	0.51	−27.3		7	PPRG
Chamberlain Shale (1450 Ma)					
SH	0.79	−32.0		1	PPRG
Prichard Formation (1450 Ma)					
QUAR	0.93	−23.1		1	PPRG
SH	0.30	−23.9		1	PPRG
Yalco Formation (1485 Ma)					
SH	34.68	−30.9		4	POWE87...1
Barney Creek Formation (1500 Ma)					
SH	0.79	−29.1		14	SMIT75.269
SH	21.35	−32.0		13	POWE87...1
SH	4.82	−32.6		1	PPRG
SH, CARB			−1.6	9	SMIT75.269
HYC Pyritic Shale Member (1500 Ma)					
SH		−26.6		1	MCKI74.591
SH	5.90	−31.7		1	HAYE83..93
DOL			−1.3	26	RYE81...1
Marimo Slate (1500 Ma)					
SH		−25.8		1	MCKI74.591
Uncompahgre Formation (1550 Ma)					
SH	15.87	−27.3		16	BARK691403

Table 17.4. Continued.

Lithology	TOC (mg/g)	del C_{org} (‰ vs PDB)	del C_{carb} (‰ vs PDB)	N	Reference Code
Einasleigh Metamorphics (1580 Ma)					
CARB			5.2	8	MCNA83.175
Early Proterozoic					
Amelia Dolomite (1600 Ma)					
CARB	0.95	−26.0		6	HAYE83..93
CARB			−0.8	3	SCHI83.149
CH	0.70	−29.4		1	HAYE83..93
Kaladgi Group (1600 Ma)					
CARB			0.3	16	SATH87.147
Tooganinie Formation (1600 Ma)					
CH		−24.6		2	HAYE83..93
CARB			−2.0	2	SCHI83.149
Mallapunyah Formation (1650 Ma)					
CARB	3.90		0.2	1	SCHI83.149
CARB		−32.3		1	HAYE83..93
Paradise Creek Formation (1650 Ma)					
CH		−29.3		1	SMIT70.659
CH		−28.7		2	JACK78.335
CH	0.09	−28.8		1	PPRG
Urquhart Shale (1670 Ma)					
SH		−28.2		1	MCKI74.591
SH		−23.9		1	JACK78.335
SH	18.00	−22.2		1	LEVE754706
SH	4.99	−23.7		24	SMIT78.369
SH	9.18	−26.1		6	PPRG
McArthur Group (1690 Ma)					
DOL			−1.2	6	VEIZ761387
Vempalle and G. R. Formations (1700 Ma)					
CH	0.08	−27.8		3	PPRG
Vermillion Limestone (1700 Ma)					
CARB	0.30	−23.0		1	HAYE83..93
Tuanshanzi Formation (1750 Ma)					
CARB	0.86	−30.5		3	PPRG
Amisk Group (1800 Ma)					
SH	16.78	−26.4		40	STRA862653
Earaheedy Group (1800 Ma)					
			−1.3	4	SCHI83.149
CH	1.13	−23.2		5	HAYE83..93
CARB	0.29	−21.6		3	HAYE83..93
Frere Formation (1800 Ma)					
CARB	0.24	−24.8		6	HAYE83..93
CH	0.11	−16.5		4	HAYE83..93
DOL			0.2	3	SCHI83.149
Jatulian Shungite Formation (1800 Ma)					
SH		−32.0		1	LEVE754706
Kulele Creek Limestone (1800 Ma)					
CARB	0.10	−20.0		1	HAYE83..93
CARB			−1.2	1	SCHI83.149
Windidda Formation (1800 Ma)					
LS	0.20	−23.0		1	HAYE83..93
CH	0.20	−24.0		1	HAYE83..93
Yelma Formation (1800 Ma)					
CARB	0.13	−23.0		1	HAYE83..93
Chelmsford Formation (1850 Ma)					
SH	5.10	−31.6		1	HAYE83..93
Chuanlinggou Formation (1850 Ma)					
SH	3.50	−32.1		4	PPRG
Onwatin Formation (1850 Ma)					
SH	38.00	−30.6		2	HAYE83..93
SH		−36.9		1	THOD62.565
Koolpin Formation (1885 Ma)					
CH		−31.0		1	OEHL721246
CH	1.13	−30.6		1	PPRG
Fontano Formation (1900 Ma)					
SH	6.55	−28.3		2	PPRG
Golden Dyke Formation (1900 Ma)					
SH		−15.8		1	MCKI74.591
Pekanatui Point Formation (1900 Ma)					
LS			0.4	1	VEIZ761387
Rove Formation (1900 Ma)					
SH		−33.6		1	BARG77.425
SH		−35.3		2	JACK78.335
SH	8.10	−32.4		1	HAYE83..93
SH	4.00	−32.0		1	LEVE754706
SH	15.18	−30.8		4	PPRG
CARB	0.51	−31.7		1	PPRG
Spartan Group (1900 Ma)					
SH	6.46	−29.9		3	PPRG
Umkondo Group (1900 Ma)					
DOL			14.0	1	SCHI75...1
Union Island Group (1900 Ma)					
SH	20.83	−39.4		1	PPRG
Utsingi Formation (1900 Ma)					
LS			2.1	1	VEIZ761387
Epworth Group (1925 Ma)					
DOL			0.2	1	SCHI75...1
Rocknest Formation (1925 Ma)					
CARB	0.22	−24.5		23	PPRG
CH	0.49	−14.3		7	PPRG

Table 17.4. Continued.

Lithology	TOC (mg/g)	del C_{org} (‰ vs PDB)	del C_{carb} (‰ vs PDB)	N	Reference Code
DOL			1.0	6	PPRG
DOL			1.6	3	PPRG
DOL			0.3	2	VEIZ761387
Grythyttan Slate (1950 Ma)					
SH		−28.6		3	EHLI80.145
Tyler Formation (1950 Ma)					
CH, CARB	7.24	−31.7		3	PPRG
Barreiro Formation (2000 Ma)					
SH		−24.4		7	SCHI76.344
Calc Zone (2000 Ma)					
CARB	2.20	−5.3		2	BANE86.239
Duck Creek Dolomite (2000 Ma)					
CARB	0.25	−22.6		7	HAYE83..93
CH	0.23	−23.7		6	HAYE83..93
CARB			0.7	5	SCHI83.149
CARB			0.5	2	VEIZ761387
CARB			−0.6	4	BECK72.577
Gibraltar Formation (2000 Ma)					
CARB			−0.7	1	SCHI83.149
Hearne Formation (2000 Ma)					
CARB	0.13	−25.1		1	HAYE83..93
CARB			1.6	1	SCHI83.149
Kahochella Group (2000 Ma)					
CARB	0.90	−21.8		1	HAYE83..93
CARB			−0.2	1	SCHI83.149
Kona Dolomite (2000 Ma)					
CARB	0.06	−23.1		4	PPRG
CH	0.07	−24.7		3	PPRG
Krivoj Rog Iron Formation (2000 Ma)					
BIF	0.70	−21.9		1	HAYE83..93
LS			−9.3	25	PERR81..83
Lookout Schist (2000 Ma)					
SH	101.30	−20.6		1	PPRG
Menihek Formation (2000 Ma)					
SH		−31.1		1	LEVE754706
Randville Dolomite (2000 Ma)					
DOL		−17.9		1	HOER62.190
DOL			3.5	1	HOER62.190
Sabara Formation (2000 Ma)					
SH		−16.2		3	SCHI76.344
Taltheilei Formation (2000 Ma)					
CARB	0.11	−22.8		1	HAYE83..93
CARB			1.3	1	SCHI83.149
Karelian Dolomite (2050 Ma)					
DOL		−15.5		2	EICH75.585
DOL			5.0	8	SCHI75...1
			3.6	2	EICH75.585
Biwabik Iron Formation (2090 Ma)					
BIF		−33.1		5	PERR731110
CH	0.16	−30.5		4	PPRG
CARB	0.14	−28.2		1	PPRG
Gunflint Iron Formation (2090 Ma)					
CH		−35.6		2	OEHL721246
CH	1.55	−32.6		57	PPRG
SH	16.06	−31.9		1	PPRG
CH		−29.2		1	HOER67..89
CARB	5.98	−32.7		10	PPRG
CH		−33.1		1	SMIT70.659
CH			−2.9	1	HOER62.190
CH	0.40	−30.1		3	HAYE83..93
CARB	9.40	−34.7		1	HAYE83..93
CH		−27.7		7	BARG77.425
CH		−30.7		1	HOER62.190
CH		−15.6		1	JACK78.335
SH		−27.1		2	BARG77.425
Pokegama Quartzite (2090 Ma)					
QUAR	0.15	−30.9		3	PPRG
Kasegalik Formation (2100 Ma)					
CH	0.13	−24.5		1	HAYE83..93
CH			−0.6	1	SCHI83.149
Lower Albanel Formation (2100 Ma)					
DOL	0.20	−22.8		1	HAYE83..93
DOL			6.4	2	SCHI83.149
Manitounuk Group (2100 Ma)					
DOL		−18.0		1	EICH75.585
DOL	0.13	−22.5		2	HAYE83..93
DOL			−1.6	2	SCHI75...1
DOL			−2.2	1	EICH75.585
McLeary Formation (2100 Ma)					
CH		−24.7		3	HAYE83..93
CARB		−25.0		1	HAYE83..93
CH			−0.4	1	SCHI83.149
Svecofennian (2100 Ma)					
CARB			0.5	2	SCHI75...1
Tabooes Formation (2100 Ma)					
SH		−28.3		4	SCHI76.344
Upper Albanel Formation (2100 Ma)					
DOL	2.90	−31.4		1	HAYE83..93
DOL			0.9	1	SCHI83.149
Fecho do Funil Formation (2150 Ma)					
SH		−23.5		1	SCHI76.344

Table 17.4. Continued.

Lithology	TOC (mg/g)	del C_{org} (‰ vs PDB)	del C_{carb} (‰ vs PDB)	N	Reference Code
Gandarella Formation (2200 Ma)					
SH		−22.3		4	SCHI76.344
DOL			−0.8	6	SCHI76.344
Bar River Formation (2250 Ma)					
QUAR	0.07	−28.8		1	PPRG
Bruce Formation (2250 Ma)					
CARB	0.04	−21.4		1	PPRG
SILT	0.04	−28.3		1	PPRG
Espanola Formation (2250 Ma)					
CARB	0.16	−25.9		1	PPRG
SILT	0.17	−27.0		1	PPRG
LS			−3.4	1	SCHI75…1
Gowganda Formation (2250 Ma)					
SH		−27.4		1	JACK78.335
SH		−19.5		1	LEVE754706
SILT	0.05	−27.8		8	PPRG
Hotazel Formation (2250 Ma)					
CARB			−10.0	3	PPRG
Lorrain Formation (2250 Ma)					
QUAR	0.06	−28.9		1	PPRG
Mississagi Formation (2250 Ma)					
QUAR	0.06	−29.0		1	PPRG
Pecors Formation (2250 Ma)					
SILT	0.17	−36.2		1	PPRG
Ramsay Lake Formation (2250 Ma)					
CONGL	0.18	−36.6		1	PPRG
Serpent Formation (2250 Ma)					
QUAR	0.07	−29.3		1	PPRG
Timeball Hill Formation (2250 Ma)					
DOL		−19.4		1	EICH75.585
DOL			0.4	1	EICH75.585
Gordon Lake Formation (2259 Ma)					
QUAR	0.07	−26.7		1	PPRG
Asbestos Hills Formation (2300 Ma)					
BIF	0.20	−24.6		1	HAYE83..93
Caue Itabirito Formation (2300 Ma)					
SH		−23.7		6	SCHI76.344
DOL			−1.1	3	SCHI76.344
Crocodile River Formation (2300 Ma)					
CARB	0.20	−29.0		1	HAYE83..93
Transvaal Supergroup (2300 Ma)					
CH		−28.0		1	OEHL721246
CARB		−26.9		15	EICH75.585
LS		−37.1		1	HOER67.365
DOL		−28.9		1	HOER62.190
DOL	0.14	−18.7		1	HAYE83..93
DOL			−1.4	1	SCHI83.149
CARB			−0.8	78	SCHI75…1
CARB			−0.5	13	EICH75.585
CARB			1.9	1	HOER62.190
CARB	0.26	−25.5		3	PPRG
SH	4.47	−37.0		4	PPRG
CH	0.52	−35.4		4	PPRG
Aravalli Supergroup (2350 Ma)					
CARB	2.02	−15.1		12	BANE86.239
Lomagundi Group (2350 Ma)					
CARB		−23.9		2	EICH75.585
CARB			7.1	14	SCHI75…1
DOL			8.2	67	SCHI76.344
Batatal Formation (2400 Ma)					
SH		−24.4		2	SCHI76.344
Malmani Subgroup (2400 Ma)					
CARB	0.84	−30.7		21	HAYE83..93
CH	0.20	−13.0		1	HAYE83..93
CARB			−0.8	13	SCHI83.149
Matinenda Formation (2400 Ma)					
QUAR	0.07	−30.2		2	PPRG
SILT	0.03	−26.1		1	PPRG
McKim Formation (2400 Ma)					
QUAR	0.11	−31.4		3	PPRG
SILT	0.09	−26.8		3	PPRG
Gamohaan Formation (2450 Ma)					
CH	9.60	−37.4		1	KLEI87..81
DOL	3.00	−36.1		1	KLEI87..81
CARB	1.51	−33.8		11	PPRG
SH	22.05	−39.1		3	PPRG
CH	0.65	−33.2		4	PPRG
Archean					
Carawine Dolomite (2500 Ma)					
CARB	0.08	−17.1		4	HAYE83..93
			0.2	1	SCHI83.149
SH	36.50	−32.4		2	HAYE83..93
CH	0.20	−9.4		1	HAYE83..93
Dales Gorge Member (2500 Ma)					
UND		−6.2		2	PPRG
BIF		−8.6		3	SCHI83.149
SID, ANK		−9.8		46	BECK72.577
Keewatin Group (2500 Ma)					
CH		−23.2		1	OEHL721246
SH		−29.1		10	SCHO81.696
CH	18.0	−27.1		1	HAYE83..93
CH	0.45	−19.7		2	PPRG

Table 17.4. Continued.

Lithology	TOC (mg/g)	del C$_{org}$ (‰ vs PDB)	del C$_{carb}$ (‰ vs PDB)	N	Reference Code
Keewatin, Abitibi Belt (2500 Ma)					
SH	53.04	−37.0		50	STRA862653
Keewatin, Wabigoon Belt (2500 Ma)					
SH	20.37	−26.6		144	STRA862653
Lewin Shale Formation (2500 Ma)					
SH	38.00	−43.1		1	HAYE83..93
Marra Mamba IF Formation (2500 Ma)					
BIF	3.83	−37.9		4	HAYE83..93
CARB			−9.9	28	BAUR85.270
Mt. McRae Shale (2500 Ma)					
SH	34.72	−34.3		6	HAYE83..93
Mt. Sylvia Formation (2500 Ma)					
BIF	0.90	−18.6		1	HAYE83..93
BIF			−12.0	1	SCHI83.149
CARB			−12.9	28	BAUR85.270
Pillingani Tuff (2500 Ma)					
CH			−1.6	1	VEIZ761387
Turee Creek Group (2500 Ma)					
CARB	0.10	−28.4		1	HAYE83..93
Wittenoom Dolomite (2500 Ma)					
CH		−30.1		3	OEHL721246
DOL	6.86	−34.2		5	HAYE83..93
SH	14.00	−36.0		1	HAYE83..93
DOL			0.2	4	SCHI83.149
DOL			0.5	3	BAUR85.270
CARB			−0.1	3	VEIZ761387
CARB			−2.3	21	BECK72.577
Yellowknife Supergroup (2500 Ma)					
CARB	0.30	−15.9		1	HAYE83..93
CARB			1.8	1	SCHI83.149
Brockman Iron Formation (2550 Ma)					
CH		−31.3		2	OEHL721246
SH	8.74	−30.3		8	HAYE83..93
BIF	1.03	−20.6		3	HAYE83..93
CARB			−10.5	2	BAUR85.270
Belingwe Greenstone Belt (2600 Ma)					
SH	1.69	−24.1		1	HAYE83..93
Bulawayan Group (2600 Ma)					
CH, CARB		−32.5		4	OEHL721246
CARB		−29.3		10	EICH75.585
LS		−28.1		1	HOER62.190
LS		−32.1		5	SCHO71.477
CARB	1.27	−27.0		3	HAYE83..93
CH, CARB			0.2	5	SCHI83.149
CARB			−0.4	30	SCHI75...1
CARB			−0.2	9	EICH75.585
LS			0.4	1	HOER62.190
LS			1.3	2	VEIZ761387
Manjeri Formation (2600 Ma)					
CH	0.77	−20.3		8	HAYE83..93
CH			−6.7	3	SCHI83.149
SH	1.00	−20.3		1	HAYE83..93
Steep Rock Group (2600 Ma)					
CARB, CH			1.5	3	SCHI83.149
LS			2.0	2	VEIZ761387
LS		−24.5		3	HAYE83..93
CH	1.35	−28.4		2	HAYE83..93
SH	1.70	−26.9		1	HAYE83..93
Jeerinah Formation (2650 Ma)					
SH		−35.4		1	HOER67.365
SH	64.14	−41.0		7	HAYE83..93
SILT	5.23	−45.8		2	PPRG
CH	2.15	−30.5		2	HAYE83..93
Rietgat Formation (2650 Ma)					
CH	0.60	−36.0		3	HAYE83..93
CH	0.69	−43.1		1	PPRG
Shamvaian Group (2650 Ma)					
CARB			2.3	2	SCHI75...1
Soudan Iron Formation (2650 Ma)					
SH		−33.7		1	HOER67.365
BIF	0.10	−29.5		2	PPRG
Ventersdorp Supergroup (2650 Ma)					
SH		−35.8		1	HOER67.365
CARB	1.72	−38.0		16	HAYE83..93
SILT	0.27	−36.5		1	PPRG
CARB			−3.2	10	SCHI83.149
Venterspost Formation (2670 Ma)					
CONGL	0.09	−31.9		1	PPRG
Venterspost Carbon Reef Member (2670 Ma)					
SH	0.08	−23.7		1	PPRG
Nymerina Basalt (2700 Ma)					
CARB	0.86	−53.2		1	PPRG
Kameeldoorns Formation (2705 Ma)					
CH	0.38	−32.5		1	PPRG
Witwatersrand Supergroup (2710 Ma)					
SH		−34.2		1	HOER67.365
SH		−27.1		7	HOEF671096
SH	0.62	−29.6		8	PPRG
Maddina Basalt (2735 Ma)					
CH	0.54	−46.3		2	PPRG
Kuruna Siltstone (2740 Ma)					
CARB	0.81	−55.8		1	PPRG
Cheshire Formation (2750 Ma)					
LS		−30.7		12	ABEL85.357

Table 17.4. Continued.

Lithology	TOC (mg/g)	del C$_{org}$ (‰ vs PDB)	del C$_{carb}$ (‰ vs PDB)	N	Reference Code	Lithology	TOC (mg/g)	del C$_{org}$ (‰ vs PDB)	del C$_{carb}$ (‰ vs PDB)	N	Reference Code
Kimberley Shale (2750 Ma)						CH	2.00	−26.4		1	LEVE754706
SH	0.08	−23.7		1	PPRG	CH	2.03	−31.5		9	PPRG
Late Archean BIF, Ontario (2750 Ma)						CARB			0.9	1	SCHI83.149
BIF		−25.8		6	THOD83.337	Sebakwian Group (3300 Ma)					
BIF		−24.6		6	GOOD76.870	LS		−14.8		1	EICH75.585
BIF			−1.5	33	THOD83.337	CARB			−2.5	3	SCHI75...1
Michipicoten Iron Formation (2750 Ma)						LS			0.9	1	EICH75.585
BIF	19.00	−16.1		1	HAYE83..93	Fig Tree Group (3350 Ma)					
Tumbiana Formation (2750 Ma)						CH		−28.3		2	OEHL721246
CH	0.60	−48.7		11	HAYE83..93	CH		−26.8		1	EICH75.585
CH	0.43	−51.1		4	PPRG	SH		−25.9		1	HOER67.365
SH	1.21	−48.6		2	PPRG	SH		−28.8		3	JACK78.335
CARB	0.30	−44.9		7	PPRG	CH	3.15	−28.0		4	HAYE83..93
CARB			−0.1	11	SCHI83.149	CH		−28.3		2	DUNG74.167
Kylena Basalt (2768 Ma)						CH	0.75	−24.5		8	PPRG
CARB	0.38	−37.3		1	PPRG	CH			1.5	1	SCHI75...1
Fortescue Group (2800 Ma)						DOL			0.3	1	SCHI75...1
CH		−34.6		2	OEHL721246	SH	35.75	−30.0		4	HAYE83..93
Nova Lima Group (2800 Ma)						Apex Basalt (3400 Ma)					
SH		−25.6		10	SCHI76.344	CH	0.10	−27.0		3	PPRG
Woman Lake Marble Formation (2800 Ma)						Warrawoona Group (3400 Ma)					
LS	0.90	−9.6		1	HAYE83..93	CH	0.50	−33.4		2	HAYE83..93
Insuzi Group (3000 Ma)						Towers Formation (3435 Ma)					
CARB	0.26	−15.8		3	HAYE83..93	CH	0.57	−30.9		10	HAYE83..93
CH	0.31	−17.8		1	PPRG	CH	0.75	−30.5		9	PPRG
CARB			1.4	9	SCHI83.149	CH			0.6	2	SCHI83.149
SH	0.30	−20.4		2	HAYE83.993	Hooggenoeg Formation (3450 Ma)					
CARB	0.05	−28.1		1	PPRG	CH		−30.5		4	OEHL721246
Limpopo Belt (3000 Ma)						CH	3.23	−32.7		3	HAYE83..93
DOL		−16.2		1	EICH75.585	CH	2.64	−36.0		5	PPRG
Moodies Group (3000 Ma)						CARB			−0.2	1	SCHI83.149
SH	1.00	−31.1		1	HAYE83..93	Komati Formation (3450 Ma)					
DOL			−3.4	1	SCHI75...1	LS		−1.3		5	SCHI75...1
Mozaan Group (3000 Ma)						Onverwacht Group (3450 Ma)					
CH	0.20	−14.4		2	HAYE83..93	CH		−26.0		8	DUNG74.167
SH	2.65	−26.1		2	HAYE83..93	CARB			−1.8	1	SCHI75...1
Gorge Creek Group (3200 Ma)						Swartkoppie Formation (3450 Ma)					
SH	40.50	−30.3		4	HAYE83..93	CH		−27.7		3	OEHL721246
CH	4.80	−27.6		1	HAYE83..93	CH		−26.9		1	JACK78.335
CH	0.08	−29.8		1	PPRG	CH	3.21	−26.4		7	HAYE83..93
Cleaverville Formation (3300 Ma)						CH		−27.6		1	LEVE754706
CH	0.10	−26.3		1	PPRG	Theespruit Formation (3550 Ma)					
SH	0.56	−28.6		1	PPRG	CH		−16.7		3	OEHL721246
Kromberg Formation (3300 Ma)						CH	14.00	−15.2		2	HAYE83..93
CH		−28.7		5	OEHL721246	CH	6.15	−15.5		2	PPRG
CH		−26.1		2	JACK78.335	Isua Supergroup (3700 Ma)					
CH	2.01	−29.3		8	HAYE83..93	BIF, CARB		−15.3		13	SCHI79.189
						BIF		−12.9		7	PERR77.280
						BIF		−17.4		10	HAYE83..93
						BIF			−5.0	2	SCHI83.149
						CARB			−2.2	8	PERR77.280

Kerogen Samples

Table 17.5. *New analyses of the elemental and carbon isotopic composition of kerogen samples.*

PPRG Sample No.	Lithology	C (%)	H (%)	N (%)	H/C	N/C	KCI	del C_{org} (‰ vs PDB)
Late Proterozoic								
Khatyspyt Formation (620 Ma)								
2723	SH	75.6	8.20	1.80	1.30	0.020	5.0	−36.2
Boonall Dolomite (640 Ma)								
2204	CARB	38.6	0.74	3.80	0.23	0.084	nd	−30.4
Sheepbed Formation (650 Ma)								
1642	SH	60.0	1.09	0.41	0.23	0.006	7.0	−29.9
1643	SH	58.7	1.08	0.42	0.22	0.006	7.0	−29.9
1645	SH	63.3	1.15	0.41	0.22	0.006	7.0	−29.7
1646	SH	63.8	1.23	0.43	0.23	0.006	7.0	−29.3
1650	SH	52.0	1.05	0.38	0.25	0.007	7.0	−28.5
Biri Formation (700 Ma)								
2412	CARB	20.2	1.15	0.66	0.69	0.028	nd	−30.5
2413	CARB	14.2	0.62	0.25	0.53	0.015	7.0	−32.4
2414	CARB	14.3	0.69	0.30	0.58	0.019	7.0	−32.0
2415	CARB	16.9	0.66	0.35	0.47	0.018	7.0	−31.4
Luoquan Formation (700 Ma)								
2178	SILT	39.4	0.84	0.35	0.26	0.008	7.0	−28.9
2185	SILT	27.3	0.70	0.39	0.31	0.013	7.0	−29.9
2186	SILT	24.5	0.73	0.34	0.36	0.012	7.0	−29.7
Twitya Formation (700 Ma)								
1613	SH	50.6	1.72	0.71	0.41	0.012	6.5	−32.1
1653	SH	25.0	0.84	0.39	0.41	0.014	7.0	−28.6
1654	SH	17.5	0.56	0.29	0.39	0.014	7.0	−28.1
1655	SH	21.4	0.97	0.39	0.55	0.016	7.0	−28.5
Tindelpina Shale Member (750 Ma)								
1238	SILT	45.0	0.66	0.39	0.18	0.008	7.0	−23.9
Woocalla Dolomite Member (750 Ma)								
1239	SILT	44.4	2.20	0.60	0.60	0.023	6.0	−29.3
River Wakefield Subgroup (775 Ma)								
1219	CH	38.1	2.57	1.08	0.82	0.025	6.0	−31.7
1224	CH	39.6	1.09	0.32	0.33	0.007	7.0	−16.4
Beck Spring Dolomite (850 Ma)								
1271	CARB	82.0	0.92	1.45	0.14	0.016	7.0	−26.5
Kingston Peak Formation (850 Ma)								
2539	CARB	84.7	1.91	1.10	0.27	0.011	nd	−26.0
Love's Creek Member (850 Ma)								
1323	CARB	52.8	1.55	1.80	0.35	0.030	nd	−26.1
1327	CH	52.0	1.70	1.95	0.39	0.032	nd	−29.7
Walcott Member (850 Ma)								
1088	SH	60.7	4.25	2.22	0.84	0.031	5.5	−27.8
1095	CARB	59.6	4.30	1.66	0.87	0.024	5.5	−26.5
1097	SH	53.4	4.06	1.76	0.91	0.028	6.0	−26.5
1112	SH	71.3	4.56	2.42	0.77	0.029	5.5	−26.3
1123	CARB	67.0	3.86	2.05	0.69	0.026	5.5	−26.5
1357	CH	50.0	3.31	1.48	0.79	0.025	6.0	−27.2

Table 17.5. *Continued.*

PPRG Sample No.	Lithology	C (%)	H (%)	N (%)	H/C	N/C	KCl	del C$_{org}$ (‰ vs PDB)
Awatubi Member (862 Ma)								
1083	SH	47.5	2.39	3.01	0.60	0.055	6.0	−32.2
1087	SH	62.1	4.10	2.72	0.79	0.038	5.5	−27.4
Middle Proterozoic								
Carbon Canyon Member (900 Ma)								
1126	SILT	66.4	4.11	2.00	0.74	0.026	5.5	−28.4
1141	CARB	70.0	3.70	1.90	0.64	0.024	nd	−26.3
Jupiter Member (930 Ma)								
1062	SH	57.1	2.37	1.79	0.50	0.027	5.5	−25.7
1063	SH	47.9	2.64	1.40	0.66	0.025	6.0	−20.6
1067	SH	40.6	1.67	1.11	0.49	0.023	5.5	−19.6
1125	SH	38.1	2.01	1.34	0.63	0.031	6.0	−30.0
Red Pine Shale (950 Ma)								
1361	SH	19.3	1.08	0.58	0.68	0.026	5.5	−17.7
1872	SH	32.6	1.69	0.84	0.62	0.023	6.5	−17.1
Tanner Member (950 Ma)								
1055	SH	43.1	1.83	1.02	0.51	0.020	5.5	−24.8
1059	SH	40.7	1.88	1.08	0.55	0.023	nd	−22.4
1060	SH	54.1	8.20	1.77	1.30	0.020	6.5	−36.2
1061	SH	55.2	2.52	1.04	0.55	0.017	5.5	−28.1
1080	SH	52.1	2.23	1.35	0.51	0.022	6.0	−20.9
Allamoore Formation (1050 Ma)								
1360	CH	65.8	1.65	1.10	0.31	0.015	7.0	−32.4
Nonesuch Shale (1055 Ma)								
2425	SILT	23.0	1.02	0.53	0.53	0.020	6.0	−33.5
Hongshuizhuang Formation (1250 Ma)								
2159	SH	64.0	3.75	1.42	0.71	0.019	6.0	−33.4
2160	SH	64.6	3.30	1.49	0.62	0.020	6.0	−33.0
Wumishan Formation (1325 Ma)								
1435	CH	72.4	3.20	2.70	0.54	0.031	6.0	−30.1
McMinn Formation (1340 Ma)								
1263	SILT	53.2	5.01	1.13	1.13	0.020	5.0	−32.1
1264	SILT	53.4	5.33	1.19	1.19	0.020	5.0	−33.2
1266	SILT	52.5	4.92	1.13	1.13	0.023	5.0	−32.7
1268	SILT	70.9	1.78	0.98	0.30	0.012	7.0	−32.4
1270	SILT	55.0	4.77	1.58	1.05	0.025	5.0	−32.5
Yangzhuang Formation (1350 Ma)								
2148	CH	72.1	3.00	2.20	0.49	0.026	nd	−28.4
Bungle Bungle Dolomite (1364 Ma)								
2234	CARB	29.7	0.83	1.50	0.34	0.043	nd	−22.6
2235	CARB	29.5	0.65	1.35	0.26	0.039	nd	−25.1
Dismal Lakes Group (1400 Ma)								
1217	CH	74.5	4.15	1.28	0.67	0.015	nd	−28.7
Greyson Shale (1420 Ma)								
2321	SILT	41.6	1.27	0.40	0.37	0.010	7.0	−28.3

Table 17.5. *Continued.*

PPRG Sample No.	Lithology	C (%)	H (%)	N (%)	H/C	N/C	KCI	del C$_{org}$ (‰ vs PDB)
Gaoyuzhuang Formation (1425 Ma)								
1421	CH	50.0	1.57	0.91	0.38	0.016	6.5	−31.4
2126	CH	60.7	2.65	0.85	0.52	0.012	6.5	−31.0
Newland Limestone (1440 Ma)								
2307	SH	55.5	1.94	0.90	0.42	0.014	6.5	−31.8
2319	SH	45.4	0.99	0.52	0.26	0.010	7.0	−27.5
Early Proterozoic								
Urquhart Shale (1670 Ma)								
1278	SH	54.1	0.49	0.27	0.11	0.004	7.0	−28.3
Tuanshanzi Formation (1750 Ma)								
2143	CARB	24.3	1.13	0.45	0.57	0.016	6.5	−31.4
Chuanlinggou Formation (1850 Ma)								
2139	SH	56.5	1.78	0.85	0.38	0.013	6.5	−31.3
2140	SH	67.5	1.86	0.84	0.33	0.011	7.0	−33.0
2144	SH	44.5	1.96	0.40	0.53	0.008	6.5	−32.7
Koolpin Formation (1885 Ma)								
1261	CH	83.9	1.21	0.17	0.17	0.007	7.0	−30.4
Fontano Formation (1900 Ma)								
1588	SH	61.2	0.69	0.28	0.14	0.004	7.0	−30.1
Rove Formation (1900 Ma)								
1295	SH	73.5	2.85	1.46	0.47	0.017	6.5	−31.6
2458	SH	64.4	2.25	1.30	0.42	0.018	6.5	−32.1
2459	SH	79.2	3.36	1.32	0.51	0.014	6.5	−31.6
Union Island Group (1900 Ma)								
1382	SH	79.6	1.53	1.06	0.23	0.012	7.0	−38.9
Rocknest Formation (1925 Ma)								
1511	CARB	64.2	0.96	1.10	0.18	0.015	7.0	−24.1
1516	CARB	67.0	2.50	0.99	0.45	0.013	7.0	−24.7
1524	CARB	76.1	0.70	1.08	0.12	0.012	7.0	−23.1
1524	CH	74.2	0.99	0.99	0.16	0.012	7.0	−24.2
1582	CARB	87.7	0.79	0.61	0.11	0.006	7.0	−12.6
Tyler Formation (1950 Ma)								
1363	SH	28.6	0.76	0.30	0.32	0.009	7.0	−31.8
1364	SH	55.2	1.19	0.53	0.26	0.008	7.0	−31.7
Gunflint Iron Formation (2090 Ma)								
1302	CH	22.6	0.60	0.39	0.32	0.015	6.5	−32.9
2436	CH	69.3	2.06	1.46	0.36	0.018	6.5	−31.7
2437	CARB	80.4	2.96	1.59	0.44	0.017	7.0	−31.7
2444	CH	76.1	3.41	0.43	0.54	0.005	6.5	−33.3
2453	CH	47.0	1.68	0.83	0.43	0.015	6.5	−33.1
Reivilo Formation (2300 Ma)								
2531	SH	71.7	0.77	0.67	0.13	0.008	nd	−34.5
2532	SH	74.5	2.14	0.46	0.35	0.005	7.0	−36.4
2533	SH	81.8	2.10	0.40	0.31	0.004	7.0	−36.7

Table 17.5. Continued.

PPRG Sample No.	Lithology	C (%)	H (%)	N (%)	H/C	N/C	KCl	del C_{org} (‰ vs PDB)
Klipfonteinheuwel Formation (2325 Ma)								
1415	CH	37.2	0.41	0.45	0.13	0.010	7.0	−37.9
Gamohaan Formation (2450 Ma)								
2512	CARB	75.7	1.71	0.60	0.27	0.007	7.0	−32.1
2513	CARB	73.0	1.47	0.63	0.25	0.008	7.0	−32.7
2514	CARB	78.4	1.71	0.78	0.26	0.009	7.0	−34.6
2517	CARB	86.0	1.98	0.95	0.28	0.010	7.0	−34.5
2518	CARB	68.9	1.49	0.74	0.26	0.009	7.0	−34.6
2520	SH	79.5	1.20	0.91	0.18	0.010	nd	−34.9
Archean								
Jeerinah Formation (2650 Ma)								
1932	SH	25.6	1.01	0.28	0.48	0.009	7.0	−45.1
1972	SH	31.7	0.93	0.20	0.36	0.006	7.0	−41.4
Meentheena Carbonate Member (2750 Ma)								
1355	CH	44.4	0.83	0.43	0.22	0.008	7.0	−51.8
Tumbiana Formation (2750 Ma)								
2050	CARB	28.7	0.68	0.41	0.28	0.012	7.0	−52.9
2065	SH	60.3	2.38	0.26	0.47	0.004	7.0	−58.7
2070	CARB	30.7	0.80	0.34	0.32	0.009	nd	−53.8
Kromberg Formation (3300 Ma)								
1391	CH	76.8	0.79	0.44	0.12	0.005	7.0	−30.3
1392	CH	88.8	1.06	0.54	0.14	0.005	7.0	−26.2
1396	CH	85.2	1.18	0.42	0.16	0.004	7.0	−32.6
Fig Tree Group (3350 Ma)								
1400	CH	73.2	1.03	0.63	0.18	0.007	7.0	−28.5
1403	CH	29.9	0.48	0.77	0.19	0.022	7.0	−28.6
Hooggenoeg Formation (3450 Ma)								
1386	CH	79.8	1.35	0.29	0.20	0.003	7.0	−36.7
Theespruit Formation (3550 Ma)								
1385	CH	88.3	0.45	0.21	0.06	0.002	7.0	−15.1

Table 17.6. *New analyses of sulfur abundance and isotopic composition of sulfides and sulfates.*

PPRG Sample No.	Mineral	S (mg/g)	del Sfide (‰ vs CDT)	PPRG Sample No.	Mineral	S (mg/g)	del Sfide (‰ vs CDT)
Late Proterozoic Sulfides				Sheepbed Formation (650 Ma)			
Chapel Island Formation (560 Ma)				1614	PYR	4.72	−8.7
1186	PYR	0.00	−18.0	1615	PYR	4.03	−12.6
				1642	PYR	0.61	9.7
				1643	PYR	0.30	
Almodovar del Rio Group (600 Ma)				1644	PYR	0.54	
1504	PYR	0.00	−2.1	1645	PYR	0.83	22.6
				1646	PYR	0.45	
Cijara Formation (610 Ma)				1647	PYR	0.29	
				1648	PYR	0.24	
1499	PYR	0.00	−2.9	1649	PYR	0.28	
1501	PYR	0.00	19.4	1650	PYR	0.30	

Table 17.6. Continued.

PPRG Sample No.	Mineral	S (mg/g)	del Sfide (‰ vs CDT)	PPRG Sample No.	Mineral	S (mg/g)	del Sfide (‰ vs CDT)
Luoquan Formation (700 Ma)				1060	PYR	0.34	
2176	PYR	2.89	23.0	1061	PYR	0.08	
2179	PYR	0.07		1080	PYR	0.15	
2180	PYR	0.04		Tieling Formation (1185 Ma)			
2181	PYR	0.05		2162	PYR	0.05	
2182	PYR	0.10		Hongshuizhuang Formation (1250 Ma)			
2183	PYR	0.15		2159	PYR	0.11	
2184	PYR	0.20		2160	PYR	0.12	
2185	PYR	0.07		Wumishan Formation (1325 Ma)			
2186	PYR	0.04		2153	PYR	0.20	
Twitya Formation (700 Ma)				Greyson Shale (1420 Ma)			
1611	PYR	0.24		2314	PYR	0.82	5.0
1612	PYR	0.14		2321	PYR	0.40	
1613	PYR	0.22		Newland Limestone (1440 Ma)			
1652	PYR	3.70	31.5	2298	PYR	0.08	
1653	PYR	3.16	26.2	2299	PYR	0.94	
1654	PYR	0.88	33.7	2307	PYR	0.24	
1655	PYR	0.78	30.3	2309	PYR	0.05	
1656	PYR	0.46		2310	PYR	0.33	
1657	PYR	0.04		2311	PYR	0.77	31.2
1658	PYR	0.12		2312	PYR	0.72	19.6
1659	PYR	0.13		2315	PYR	6.85	15.9
1660	PYR	0.13		2316	PYR	4.70	14.7
1661	PYR	0.09		2317	PYR	8.40	
Coppercap Formation (750 Ma)				2318	PYR	0.58	17.5
1669	SFIDE	0.00	6.3	2319	PYR	0.16	
Little Dal Group (800 Ma)				2322	PYR	0.09	
1609	PYR	1.94	−28.0	2323	PYR	14.04	22.3
Walcott Member (850 Ma)				2324	PYR	0.20	
1088	PYR	9.63	7.8	Chamberlain Shale (1450 Ma)			
1096	PYR	0.00	18.6	2306	PYR	0.09	
1103	PYR	0.00	8.8	Prichard Formation (1450 Ma)			
Awatubi Member (862 Ma)				2327	PYR	7.29	14.3
1083	PYR	0.12	−38.9	2328	PYR	0.22	
1087	PYR	1.53	1.4	*Early Proterozoic Sulfides*			
Middle Proterozoic Sulfides				Urquhart Shale (1670 Ma)			
Carbon Canyon Member (900 Ma)				1274	PYR	0.00	26.4
1126	PYR	1.96		1275	PYR	0.00	36.3
1153	PYR	0.24		1277	PYR	0.00	22.7
Jupiter Member (930 Ma)				Chuanlinggou Formation (1850 Ma)			
1062	PYR	0.15		2139	PYR	0.06	
1063	PYR	1.53	19.5	2140	PYR	0.07	
1067	PYR	1.91		2141	PYR	0.15	
1124	PYR	7.64	13.9	2144	PYR	0.08	
1125	PYR	0.27		Fontano Formation (1900 Ma)			
Tanner Member (950 Ma)				1588	PYR	0.53	
1055	PYR	0.69	15.7				
1058	PYR	0.20					
1059	PYR	0.19					

Table 17.6. Continued.

PPRG Sample No.	Mineral	S (mg/g)	del Sfide (‰ vs CDT)	PPRG Sample No.	Mineral	S (mg/g)	del Sfide (‰ vs CDT)
Gunflint Iron Formation (2090 Ma)				Venterspost Formation (2670 Ma)			
1375	PYR	0.00	7.6	1411	PYR	0.00	1.1
2447	PYR	0.00	5.6				
Gamohaan Formation (2450 Ma)				Kameeldoorns Formation (2705 Ma)			
1414	PYR	0.00	−2.0	1419	PYR	0.00	−0.9
2776	PYR	0.00	−0.2				
2777	PYR	0.00	1.8	Carbon Leader Member (2710 Ma)			
2778	PYR	0.00	1.6	1410	PYR	0.00	3.3
2778	PYR	0.00	−13.9				
Archean Sulfides				Kuruna Siltstone (2740 Ma)			
Dales Gorge Member (2500 Ma)				1944	PYR	0.75	0.3
480	PYR	0.00	6.0				
484	PYR	0.00	1.9	*Proterozoic and Archean Sulfates*			
485	PYR	0.00	−0.5	Redstone River Formation (770 Ma)			
486	PYR	0.00	−4.2	1663	ANH		15.6
487	PYR	0.00	−1.0	1665	ANH		19.5
490	PYR	0.00	3.1				
Jeerinah Formation (2650 Ma)				Towers Formation (3435 Ma)			
1932	PYR	6.14	9.4	1443	BAR		6.3
1972	PYR	1.48	6.4	1443	BAR		4.5

Table 17.7. *Previously published analyses of the isotopic composition of sulfides and sulfates.*

Mineral	del Sfide (‰ vs CDT)	del Sfate (‰ vs CDT)	Reference Code	Mineral	del Sfide (‰ vs CDT)	del Sfate (‰ vs CDT)	Reference Code
Late Proterozoic				Tapley Hill Formation (750 Ma)			
Ballachulish Slate (700 Ma)				SFIDE	−3.0		KNUT83.250
PYR	16.6		HALL87.305	SFIDE	2.5		KNUT83.250
PYR	15.6		HALL87.305	SFIDE	5.5		KNUT83.250
PYR	14.5		HALL87.305	SFIDE	7.5		KNUT83.250
PYR	16.1		HALL87.305	SFIDE	10.5		KNUT83.250
PYR	16.0		HALL87.305	SFIDE	12.5		KNUT83.250
PYR	13.0		HALL87.305	SFIDE	13.5		KNUT83.250
PYR	14.8		HALL87.305	SFIDE	17.5		KNUT83.250
PYR	13.6		HALL87.305	SFIDE	18.5		KNUT83.250
PO	15.8		HALL87.305	SFIDE	19.5		KNUT83.250
PO	15.0		HALL87.305	SFIDE	20.5		KNUT83.250
PO	13.3		HALL87.305	SFIDE	20.5		KNUT83.250
PO	12.8		HALL87.305	SFIDE	21.5		KNUT83.250
PYR	16.3		HALL87.305	SFIDE	21.5		KNUT83.250
PYR	15.8		HALL87.305	SFIDE	21.5		KNUT83.250
PO	15.7		HALL87.305	SFIDE	22.5		KNUT83.250
PO	14.2		HALL87.305	SFIDE	22.5		KNUT83.250
PO	13.5		HALL87.305	SFIDE	24.5		KNUT83.250
PYR	15.2		HALL87.305	SFIDE	26.5		KNUT83.250
PYR	14.3		HALL87.305	SFIDE	27.5		KNUT83.250
PYR	14.2		HALL87.305	SFIDE	29.5		KNUT83.250
				SFIDE	42.5		KNUT83.250
Shaler Group (700 Ma)				SFIDE	44.5		KNUT83.250
GYP		15.9	CLAY80.199	SFIDE	2.5		KNUT83.250

Table 17.7. *Continued.*

Mineral	del Sfide (‰ vs CDT)	del Sfate (‰ vs CDT)	Reference Code	Mineral	del Sfide (‰ vs CDT)	del Sfate (‰ vs CDT)	Reference Code
SFIDE	6.5		KNUT83.250	SFIDE	16.5		KNUT83.250
SFIDE	8.5		KNUT83.250	SFIDE	16.5		KNUT83.250
SFIDE	9.5		KNUT83.250	SFIDE	16.5		KNUT83.250
SFIDE	11.5		KNUT83.250	SFIDE	17.5		KNUT83.250
SFIDE	11.5		KNUT83.250	SFIDE	17.5		KNUT83.250
SFIDE	11.5		KNUT83.250	SFIDE	17.5		KNUT83.250
SFIDE	12.5		KNUT83.250	SFIDE	17.5		KNUT83.250
SFIDE	12.5		KNUT83.250	SFIDE	17.5		KNUT83.250
SFIDE	13.5		KNUT83.250	SFIDE	17.5		KNUT83.250
SFIDE	13.5		KNUT83.250	SFIDE	17.5		KNUT83.250
SFIDE	13.5		KNUT83.250	SFIDE	17.5		KNUT83.250
SFIDE	14.5		KNUT83.250	SFIDE	18.5		KNUT83.250
SFIDE	14.5		KNUT83.250	SFIDE	18.5		KNUT83.250
SFIDE	15.5		KNUT83.250	SFIDE	18.5		KNUT83.250
SFIDE	16.5		KNUT83.250	SFIDE	18.5		KNUT83.250
SFIDE	17.5		KNUT83.250	SFIDE	18.5		KNUT83.250
SFIDE	18.5		KNUT83.250	SFIDE	18.5		KNUT83.250
SFIDE	22.5		KNUT83.250	SFIDE	19.5		KNUT83.250
SFIDE	22.5		KNUT83.250	SFIDE	19.5		KNUT83.250
SFIDE	23.5		KNUT83.250	SFIDE	19.5		KNUT83.250
SFIDE	27.5		KNUT83.250	SFIDE	19.5		KNUT83.250
SFIDE	28.5		KNUT83.250	SFIDE	19.5		KNUT83.250
SFIDE	28.5		KNUT83.250	SFIDE	20.5		KNUT83.250
SFIDE	29.5		KNUT83.250	SFIDE	20.5		KNUT83.250
SFIDE	29.5		KNUT83.250	SFIDE	21.5		KNUT83.250
SFIDE	29.5		KNUT83.250	SFIDE	22.5		KNUT83.250
SFIDE	30.5		KNUT83.250	SFIDE	22.5		KNUT83.250
SFIDE	30.5		KNUT83.250	SFIDE	23.5		KNUT83.250
SFIDE	32.5		KNUT83.250	SFIDE	24.5		KNUT83.250
SFIDE	33.5		KNUT83.250	SFIDE	24.5		KNUT83.250
SFIDE	33.5		KNUT83.250	SFIDE	24.5		KNUT83.250
SFIDE	34.5		KNUT83.250	SFIDE	25.5		KNUT83.250
SFIDE	37.5		KNUT83.250	SFIDE	25.5		KNUT83.250
SFIDE	−0.5		KNUT83.250	SFIDE	26.5		KNUT83.250
SFIDE	2.5		KNUT83.250	SFIDE	26.5		KNUT83.250
SFIDE	3.5		KNUT83.250	SFIDE	26.5		KNUT83.250
SFIDE	5.5		KNUT83.250	SFIDE	27.5		KNUT83.250
SFIDE	7.5		KNUT83.250	SFIDE	29.5		KNUT83.250
SFIDE	7.5		KNUT83.250	SFIDE	29.5		KNUT83.250
SFIDE	8.5		KNUT83.250	SFIDE	29.5		KNUT83.250
SFIDE	10.5		KNUT83.250	SFIDE	30.5		KNUT83.250
SFIDE	10.5		KNUT83.250	SFIDE	30.5		KNUT83.250
SFIDE	11.5		KNUT83.250	SFIDE	31.5		KNUT83.250
SFIDE	12.5		KNUT83.250	SFIDE	32.5		KNUT83.250
SFIDE	12.5		KNUT83.250	SFIDE	32.5		KNUT83.250
SFIDE	13.5		KNUT83.250	SFIDE	34.5		KNUT83.250
SFIDE	14.5		KNUT83.250	SFIDE	36.5		KNUT83.250
SFIDE	15.5		KNUT83.250	SFIDE	37.5		KNUT83.250
SFIDE	15.5		KNUT83.250	SFIDE	38.5		KNUT83.250
SFIDE	15.5		KNUT83.250	SFIDE	38.5		KNUT83.250
SFIDE	15.5		KNUT83.250	SFIDE	39.5		KNUT83.250
SFIDE	15.5		KNUT83.250	SFIDE	45.5		KNUT83.250
SFIDE	15.5		KNUT83.250	PYR	−0.5		LAMB80...1
SFIDE	15.5		KNUT83.250	PYR	2.5		LAMB80...1
SFIDE	15.5		KNUT83.250	PYR	3.5		LAMB80...1
SFIDE	16.5		KNUT83.250	PYR	10.5		LAMB80...1
SFIDE	16.5		KNUT83.250	PYR	12.5		LAMB80...1
SFIDE	16.5		KNUT83.250	PYR	15.5		LAMB80...1

Table 17.7. *Continued*

Mineral	del Sfide (‰ vs CDT)	del Sfate (‰ vs CDT)	Reference Code	Mineral	del Sfide (‰ vs CDT)	del Sfate (‰ vs CDT)	Reference Code
PYR	17.5		LAMB80...1	CHALCOPYR	29.1		LAMB84.266
PYR	18.5		LAMB80...1	CHALCOPYR	29.6		LAMB84.266
PYR	19.5		LAMB80...1	CHALCOPYR	33.0		LAMB84.266
PYR	19.5		LAMB80...1	CHALCOPYR	9.0		LAMB84.266
PYR	22.5		LAMB80...1	CHALCOPYR	28.3		LAMB84.266
PYR	24.5		LAMB80...1	CHALCOPYR	35.5		LAMB84.266
PYR	29.5		LAMB80...1	CHALCOPYR	14.5		LAMB84.266
PYR	32.5		LAMB80...1	CHALCOPYR	22.6		LAMB84.266
PYR	34.5		LAMB80...1	CHALCOPYR	25.5		LAMB84.266
PYR	36.5		LAMB80...1	SFIDE	3.5		LAMB84.266
PYR	45.5		LAMB80...1	SFIDE	12.1		LAMB84.266
PO	8.5		LAMB80...1	SFIDE	13.9		LAMB84.266
PO	10.5		LAMB80...1	SFIDE	14.8		LAMB84.266
PO	12.5		LAMB80...1	SFIDE	15.3		LAMB84.266
PO	22.5		LAMB80...1	SFIDE	24.3		LAMB84.266
PO	24.5		LAMB80...1	SFIDE	31.0		LAMB84.266
PO	24.5		LAMB80...1	SFIDE	32.1		LAMB84.266
PO	25.5		LAMB80...1	SFIDE	44.4		LAMB84.266
PO	26.5		LAMB80...1	SFIDE	16.8		LAMB84.266
PO	26.5		LAMB80...1	SFIDE	19.8		LAMB84.266
PO	27.5		LAMB80...1	*Bitter Springs Formation (850 Ma)*			
PO	28.5		LAMB80...1				
PO	29.5		LAMB80...1	ANH		24.3	CLAY80.199
PO	30.5		LAMB80...1	GYP		23.4	CLAY80.199
PO	31.5		LAMB80...1	GYP		23.1	CLAY80.199
PO	37.5		LAMB80...1	GYP		16.4	CLAY80.199
PO	38.5		LAMB80...1	GYP		14.7	CLAY80.199
PO	38.5		LAMB80...1	GYP		15.1	CLAY80.199
PO	39.5		LAMB80...1	GYP		16.7	CLAY80.199
CHALCOPYR	9.5		LAMB80...1	GYP		17.2	CLAY80.199
CHALCOPYR	11.5		LAMB80...1	GYP		15.1	CLAY80.199
CHALCOPYR	16.5		LAMB80...1	GYP		16.4	CLAY80.199
CHALCOPYR	17.5		LAMB80...1	GYP		18.5	CLAY80.199
CHALCOPYR	18.5		LAMB80...1	GYP		18.0	CLAY80.199
CHALCOPYR	19.5		LAMB80...1	ANH		14.6	CLAY80.199
PYR	−3.9		LAMB84.266	ANH		17.4	SOLO71.259
PYR	2.5		LAMB84.266	ANH		17.5	SOLO71.259
PYR	14.8		LAMB84.266	ANH		20.1	SOLO71.259
PYR	18.6		LAMB84.266	ANH		16.6	SOLO71.259
PYR	19.2		LAMB84.266	ANH		11.6	SOLO71.259
PYR	19.2		LAMB84.266	ANH		19.0	SOLO71.259
PYR	22.4		LAMB84.266	ANH		16.6	SOLO71.259
PYR	29.0		LAMB84.266	ANH		19.7	SOLO71.259
PYR	30.5		LAMB84.266	ANH		25.2	SOLO71.259
PYR	31.4		LAMB84.266	*Middle Proterozoic*			
PYR	34.2		LAMB84.266				
PYR	41.8		LAMB84.266	*Bijaigarh Formation (950 Ma)*			
PYR	3.6		LAMB84.266	PYR	6.8		GUHA71.326
PYR	12.1		LAMB84.266	PYR	11.4		GUHA71.326
PYR	15.0		LAMB84.266	PYR	4.5		GUHA71.326
PYR	16.2		LAMB84.266	PYR	10.9		GUHA71.326
PYR	17.9		LAMB84.266	PYR	7.3		GUHA71.326
PYR	23.7		LAMB84.266	PYR	8.6		GUHA71.326
PYR	25.0		LAMB84.266	PYR	7.7		GUHA71.326
PYR	23.5		LAMB84.266	PYR	7.7		GUHA71.326
PYR	45.8		LAMB84.266	PYR	8.2		GUHA71.326
CHALCOPYR	10.0		LAMB84.266	PYR	10.0		GUHA71.326
CHALCOPYR	12.3		LAMB84.266	PYR	7.3		GUHA71.326

Table 17.7. *Continued.*

Mineral	del Sfide (‰ vs CDT)	del Sfate (‰ vs CDT)	Reference Code	Mineral	del Sfide (‰ vs CDT)	del Sfate (‰ vs CDT)	Reference Code
PYR	7.3		GUHA71.326	PYR	18.8		BROW73.362
PYR	11.4		GUHA71.326	SPH	14.4		BROW73.362
PYR	10.0		GUHA71.326	SPH	15.4		BROW73.362
PYR	10.5		GUHA71.326	SPH	14.0		BROW73.362
PYR	8.6		GUHA71.326	SPH	13.7		BROW73.362
PYR	8.6		GUHA71.326	SPH	16.3		BROW73.362
PYR	10.0		GUHA71.326	SPH	12.9		BROW73.362
PYR	8.6		GUHA71.326	SPH	14.4		BROW73.362
PYR	6.8		GUHA71.326	SPH	18.1		BROW73.362
PYR	9.5		GUHA71.326	SPH	15.2		BROW73.362
PYR	9.5		GUHA71.326	SPH	12.8		BROW73.362
PYR	8.2		GUHA71.326	SPH	14.5		BROW73.362
PYR	10.9		GUHA71.326	SPH	9.9		BROW73.362
PYR	6.3		GUHA71.326	SPH	14.9		BROW73.362
PYR	5.0		GUHA71.326	SPH	14.6		BROW73.362
PYR	8.6		GUHA71.326	SPH	14.1		BROW73.362
PYR	9.5		GUHA71.326	SPH	14.0		BROW73.362
PYR	8.2		GUHA71.326	SPH	14.5		BROW73.362
PYR	10.0		GUHA71.326	SPH	14.3		BROW73.362
PYR	9.1		GUHA71.326	SPH	15.2		BROW73.362
PYR	20.0		GUHA71.326	SPH	14.0		BROW73.362
PYR	15.1		GUHA71.326	PYR	−11.0		BUDD69.423
PYR	7.7		GUHA71.326	PYR	−10.4		BUDD69.423
PYR	19.2		GUHA71.326	PYR	−13.6		BUDD69.423
PYR	7.1		GUHA71.326	PYR	−13.2		BUDD69.423
				PYR	−8.8		BUDD69.423
Katanga System (1000 Ma)				PYR	4.9		BUDD69.423
ANH		13.0	DECH65.894	PYR	−15.3		BUDD69.423
ANH		16.2	DECH65.894	PYR	−12.2		BUDD69.423
ANH		15.9	DECH65.894	PYR	−9.8		BUDD69.423
ANH		17.6	DECH65.894	PYR	−11.4		BUDD69.423
ANH		16.1	DECH65.894	PYR	−11.4		BUDD69.423
ANH		19.3	DECH65.894	PYR	−8.7		BUDD69.423
GYP		15.6	DECH65.894	PYR	−9.9		BUDD69.423
ANH		21.0	DECH65.894	PYR	−11.6		BUDD69.423
ANH		17.0	DECH65.894	PYR	−11.2		BUDD69.423
ANH		14.2	DECH65.894	PYR	−9.2		BUDD69.423
				PYR	−7.4		BUDD69.423
Grenville Metasediments (1050 Ma)				PYR	−4.1		BUDD69.423
PYR	15.4		BROW73.362	PYR	−11.3		BUDD69.423
PYR	14.4		BROW73.362	PYR	−16.4		BUDD69.423
PYR	14.6		BROW73.362	PYR	−6.9		BUDD69.423
PYR	13.2		BROW73.362	PYR	−7.5		BUDD69.423
PYR	13.9		BROW73.362	PYR	−8.4		BUDD69.423
PYR	13.9		BROW73.362	PYR	−6.9		BUDD69.423
PYR	15.0		BROW73.362	PYR	−9.3		BUDD69.423
PYR	15.1		BROW73.362	PYR	−7.9		BUDD69.423
PYR	15.2		BROW73.362	PYR	−10.3		BUDD69.423
PYR	13.0		BROW73.362	ANH		22.4	BROW73.362
PYR	15.0		BROW73.362	ANH		21.9	BROW73.362
PYR	10.3		BROW73.362	ANH		28.6	BROW73.362
PYR	12.4		BROW73.362	ANH		26.1	BROW73.362
PYR	15.2		BROW73.362	ANH		14.5	BROW73.362
PYR	16.1		BROW73.362	ANH		16.2	BROW73.362
PYR	14.1		BROW73.362	ANH		20.8	BROW73.362
PYR	14.2		BROW73.362	ANH		19.7	BROW73.362
PYR	14.2		BROW73.362	ANH		20.6	BROW73.362
PYR	11.9		BROW73.362	ANH		22.0	BROW73.362

Table 17.7. Continued.

Mineral	del Sfide (‰ vs CDT)	del Sfate (‰ vs CDT)	Reference Code	Mineral	del Sfide (‰ vs CDT)	del Sfate (‰ vs CDT)	Reference Code
ANH		22.7	BROW73.362	PYR	11.5		BURN72.895
ANH		21.4	BROW73.362	PYR	21.3		BURN72.895
ANH		18.6	BROW73.362	PYR	−3.4		BURN72.895
				PYR	−6.8		BURN72.895
Nonesuch Shale (1055 Ma)				PYR	−1.8		BURN72.895
				PYR	2.9		BURN72.895
PYR	9.0		BURN72.895	PYR	9.2		BURN72.895
PYR	15.4		BURN72.895	PYR	−8.0		BURN72.895
PYR	29.2		BURN72.895	PYR	−6.6		BURN72.895
PYR	16.8		BURN72.895	PYR	15.4		BURN72.895
PYR	19.8		BURN72.895	PYR	17.9		BURN72.895
PYR	9.8		BURN72.895	PYR	−6.0		BURN72.895
PYR	15.2		BURN72.895	PYR	−1.7		BURN72.895
PYR	13.0		BURN72.895	PYR	−6.2		BURN72.895
PYR	8.4		BURN72.895	PYR	−6.8		BURN72.895
PYR	10.9		BURN72.895	PYR	20.2		BURN72.895
PYR	19.7		BURN72.895	PYR	5.9		BURN72.895
PYR	5.4		BURN72.895	PYR	−12.2		BURN72.895
PYR	−0.5		BURN72.895	PYR	−12.6		BURN72.895
PYR	31.3		BURN72.895	PYR	−4.8		BURN72.895
PYR	14.1		BURN72.895	PYR	−9.6		BURN72.895
PYR	8.0		BURN72.895	PYR	2.6		BURN72.895
PYR	−5.5		BURN72.895	PYR	2.0		BURN72.895
PYR	11.5		BURN72.895	PYR	17.9		BURN72.895
PYR	−10.0		BURN72.895	PYR	−7.7		BURN72.895
PYR	22.0		BURN72.895	PYR	7.7		BURN72.895
PYR	−5.5		BURN72.895	PYR	19.3		BURN72.895
PYR	17.8		BURN72.895	PYR	−12.8		BURN72.895
PYR	16.8		BURN72.895	PYR	7.4		BURN72.895
PYR	1.1		BURN72.895	PYR	−10.2		BURN72.895
PYR	12.5		BURN72.895	PYR	−15.8		BURN72.895
PYR	−6.6		BURN72.895	PYR	−8.9		BURN72.895
PYR	25.2		BURN72.895	Eurelia Beds (1100 Ma)			
PYR	2.4		BURN72.895				
PYR	9.2		BURN72.895	PYR	−16.4		LAMB84.461
PYR	16.8		BURN72.895	PYR	−13.4		LAMB84.461
PYR	−3.1		BURN72.895	PYR	−15.7		LAMB84.461
PYR	12.9		BURN72.895	PYR	−12.2		LAMB84.461
PYR	5.4		BURN72.895	PYR	−12.3		LAMB84.461
PYR	19.3		BURN72.895	PYR	−15.7		LAMB84.461
PYR	−1.4		BURN72.895	PYR	−17.3		LAMB84.461
PYR	−7.1		BURN72.895	PYR	−16.2		LAMB84.461
PYR	−1.4		BURN72.895	CHALCOPYR	−2.8		LAMB84.461
PYR	1.7		BURN72.895	CHALCOPYR	−4.8		LAMB84.461
PYR	−8.1		BURN72.895	CHALCOPYR	−9.7		LAMB84.461
PYR	0.4		BURN72.895	CHALCOPYR	15.8		LAMB84.461
PYR	9.1		BURN72.895	CHALCOPYR	−4.4		LAMB84.461
PYR	12.6		BURN72.895	PYR	−1.9		LAMB84.461
PYR	−9.6		BURN72.895	PYR	1.9		LAMB84.461
PYR	4.1		BURN72.895	PYR	0.1		LAMB84.461
PYR	−5.0		BURN72.895	CHALCOPYR	−8.7		LAMB84.461
PYR	3.6		BURN72.895	CHALCOPYR	2.4		LAMB84.461
PYR	−9.0		BURN72.895	PYR	5.0		LAMB84.461
PYR	0.7		BURN72.895	PYR	14.2		LAMB84.461
PYR	23.2		BURN72.895	CHALCOPYR	−0.3		LAMB84.461
PYR	−9.1		BURN72.895	PYR	7.2		LAMB84.461
PYR	−0.7		BURN72.895	PYR	6.0		LAMB84.461
PYR	−5.5		BURN72.895	PYR	6.5		LAMB84.461

Table 17.7. *Continued*

Mineral	del Sfide (‰ vs CDT)	del Sfate (‰ vs CDT)	Reference Code	Mineral	del Sfide (‰ vs CDT)	del Sfate (‰ vs CDT)	Reference Code
PYR	−10.4		LAMB84.461	SFIDE	−0.1		RYE.83.104
PYR	3.4		LAMB84.461	SFIDE	0.1		RYE.83.104
PYR	6.5		LAMB84.461	SFIDE	0.1		RYE.83.104
PYR	−5.3		LAMB84.461	SFIDE	0.1		RYE.83.104
CHALCOPYR	4.7		LAMB84.461	SFIDE	1.1		RYE.83.104
Gap Well Formation (1100 Ma)				SFIDE	1.1		RYE.83.104
				SFIDE	1.5		RYE.83.104
PYR	21.2		VOGT87.805	SFIDE	1.5		RYE.83.104
GAL	20.2		VOGT87.805	BAR		15.5	RYE.83.104
BAR		40.8	VOGT87.805	BAR		16.3	RYE.83.104
BAR		38.4	VOGT87.805	BAR		17.0	RYE.83.104
Roan Group (1155 Ma)				BAR		17.0	RYE.83.104
				BAR		17.5	RYE.83.104
ANH		21.0	CLAY80.199	BAR		17.5	RYE.83.104
ANH		20.9	CLAY80.199	BAR		17.9	RYE.83.104
ANH		20.5	CLAY80.199	BAR		18.5	RYE.83.104
ANH		19.9	CLAY80.199	BAR		19.0	RYE.83.104
ANH		15.8	CLAY80.199	BAR		19.0	RYE.83.104
ANH		17.1	CLAY80.199	BAR		19.8	RYE.83.104
ANH		16.7	CLAY80.199	Belt Supergroup (1350 Ma)			
ANH		17.2	CLAY80.199	SFIDE	1.4		MORT74.223
Spokane Formation (1250 Ma)				SFIDE	−7.3		MORT74.223
SFIDE	−16.6		LANG871334	SPH	11.6		MORT74.223
SFIDE	−16.8		LANG871334	CHALCOPYR	9.8		MORT74.223
SFIDE	−13.2		LANG871334	SFIDE	5.7		MORT74.223
SFIDE	−15.2		LANG871334	CHALCOPYR	6.2		MORT74.223
SFIDE	−15.7		LANG871334	SFIDE	12.5		MORT74.223
SFIDE	−14.1		LANG871334	SFIDE	8.3		MORT74.223
SFIDE	−7.0		LANG871334	CHALCOPYR	4.6		MORT74.223
SFIDE	−15.3		LANG871334	CHALCOPYR	10.5		MORT74.223
SFIDE	1.1		LANG871334	CHALCOPYR	4.8		MORT74.223
SFIDE	−15.3		LANG871334	SFIDE	−4.6		MORT74.223
SFIDE	−13.4		LANG871334	SFIDE	−4.6		MORT74.223
SFIDE	−17.0		LANG871334	SFIDE	0.6		MORT74.223
SFIDE	−8.6		LANG871334	SFIDE	0.6		MORT74.223
SFIDE	−7.5		LANG871334	PYR	10.9		MORT74.223
SFIDE	−9.1		LANG871334	Newland Limestone (1440 Ma)			
SFIDE	1.5		LANG871334	PYR	−14.0		STRAunpu
SFIDE	2.1		LANG871334	PYR	−5.2		STRAunpu
SFIDE	−14.7		LANG871334	PYR	−10.6		STRAunpu
SFIDE	−5.0		LANG871334	PYR	−10.6		STRAunpu
SFIDE	−2.4		LANG871334	PYR	−10.4		STRAunpu
SFIDE	1.6		LANG871334	PYR	−9.7		STRAunpu
SFIDE	−2.5		RYE.83.104	PYR	−7.1		STRAunpu
SFIDE	−2.2		RYE.83.104	PYR	−5.3		STRAunpu
SFIDE	−2.2		RYE.83.104	PYR	−4.1		STRAunpu
SFIDE	−1.9		RYE.83.104	PYR	−3.0		STRAunpu
SFIDE	−1.9		RYE.83.104	PYR	−5.0		STRAunpu
SFIDE	−1.9		RYE.83.104	PYR	−0.8		STRAunpu
SFIDE	−1.9		RYE.83.104	PYR	0.3		STRAunpu
SFIDE	−1.5		RYE.83.104	PYR	5.2		STRAunpu
SFIDE	−1.5		RYE.83.104	PYR	1.6		STRAunpu
SFIDE	−1.0		RYE.83.104	PYR	9.5		STRAunpu
SFIDE	−1.0		RYE.83.104	PYR	15.1		STRAunpu
SFIDE	−0.9		RYE.83.104	PYR	18.1		STRAunpu
SFIDE	−0.5		RYE.83.104	BAR		13.6	STRAunpu
SFIDE	−0.5		RYE.83.104				

Table 17.7. Continued.

Mineral	del Sfide (‰ vs CDT)	del Sfate (‰ vs CDT)	Reference Code	Mineral	del Sfide (‰ vs CDT)	del Sfate (‰ vs CDT)	Reference Code
BAR		14.4	STRAunpu	SFIDE	0.9		STAN66..16
BAR		18.3	STRAunpu	SFIDE	1.1		STAN66..16
Balbirini Dolomite (1482 Ma)				SFIDE	2.0		STAN66..16
				SFIDE	1.2		STAN66..16
CHALCOPYR	2.8		JACK88unpu	SFIDE	1.4		STAN66..16
Barney Creek Formation (1500 Ma)				SFIDE	1.0		STAN66..16
				SFIDE	0.5		STAN66..16
PYR	−2.2		SMIT75.269	SFIDE	1.6		STAN66..16
PYR	1.2		SMIT75.269	SFIDE	0.7		STAN66..16
PYR	5.6		SMIT75.269	SFIDE	0.6		STAN66..16
PYR	1.6		SMIT75.269	SFIDE	0.4		STAN66..16
PYR	−5.4		SMIT75.269	SFIDE	0.6		STAN66..16
PYR	18.2		SMIT75.269	SFIDE	0.1		STAN66..16
PYR	6.9		SMIT75.269	SFIDE	1.0		STAN66..16
PYR	12.9		SMIT75.269	SFIDE	0.2		STAN66..16
PYR	15.9		SMIT75.269	SFIDE	−2.0		STAN66..16
PYR	4.8		SMIT75.269	SFIDE	0.2		STAN66..16
PYR	21.2		SMIT75.269	SFIDE	1.2		STAN66..16
PYR	27.0		SMIT75.269	SFIDE	−2.2		STAN66..16
PYR	26.1		SMIT75.269	SFIDE	−0.7		STAN66..16
PYR	26.9		SMIT75.269	SFIDE	4.7		STAN66..16
PYR	23.4		SMIT75.269	SFIDE	2.4		STAN66..16
PYR	5.3		SMIT75.269	SFIDE	1.5		STAN66..16
PYR	13.8		SMIT75.269	SFIDE	1.0		STAN66..16
PYR	22.3		JACK88unpu	SFIDE	0.7		STAN66..16
PO	32.4		JACK88unpu	SFIDE	0.5		STAN66..16
PYR	20.4		JACK88unpu	SFIDE	0.2		STAN66..16
PYR	16.6		JACK88unpu	SFIDE	1.2		STAN66..16
PYR	8.3		JACK88unpu	SFIDE	0.4		STAN66..16
Early Proterozoic 1600 to 2000 Ma				SFIDE	0.5		STAN66..16
				SFIDE	−0.4		STAN66..16
Amelia Dolomite (1600 Ma)				SFIDE	−0.6		STAN66..16
PYR	7.2		JACK88unpu	SFIDE	0.8		STAN66..16
PYR	7.2		JACK88unpu	SFIDE	1.0		STAN66..16
PYR	5.4		JACK88unpu	SFIDE	0.9		STAN66..16
Broken Hill Formation (1600 Ma)				SFIDE	0.5		STAN66..16
				SFIDE	1.2		STAN66..16
SFIDE	1.0		STAN66..16	SFIDE	1.8		STAN66..16
SFIDE	0.1		STAN66..16	SFIDE	2.3		STAN66..16
SFIDE	−1.5		STAN66..16	SFIDE	0.9		STAN66..16
SFIDE	−0.1		STAN66..16	SFIDE	0.5		STAN66..16
SFIDE	1.1		STAN66..16	SFIDE	1.0		STAN66..16
SFIDE	0.7		STAN66..16	SFIDE	1.2		STAN66..16
SFIDE	2.2		STAN66..16	*Mount Isa Inlier (1600 Ma)*			
SFIDE	2.3		STAN66..16				
SFIDE	1.3		STAN66..16	GAL	10.6		CARR77.105
SFIDE	2.2		STAN66..16	GAL	13.5		CARR77.105
SFIDE	1.1		STAN66..16	GAL	9.9		CARR77.105
SFIDE	1.2		STAN66..16	GAL	16.0		CARR77.105
SFIDE	1.2		STAN66..16	GAL	11.5		CARR77.105
SFIDE	2.0		STAN66..16	GAL	11.6		CARR77.105
SFIDE	2.0		STAN66..16	GAL	14.0		CARR77.105
SFIDE	1.4		STAN66..16	SPH	12.9		CARR77.105
SFIDE	0.5		STAN66..16	SPH	17.3		CARR77.105
SFIDE	0.8		STAN66..16	SPH	12.4		CARR77.105
SFIDE	0.9		STAN66..16	SPH	18.7		CARR77.105
SFIDE	1.9		STAN66..16	SPH	16.0		CARR77.105
SFIDE	1.5		STAN66..16	SPH	16.3		CARR77.105

Table 17.7. Continued.

Mineral	del Sfide (‰ vs CDT)	del Sfate (‰ vs CDT)	Reference Code	Mineral	del Sfide (‰ vs CDT)	del Sfate (‰ vs CDT)	Reference Code
SPH	17.7		CARR77.105	GAL	0.1		SPRY87.109
SPH	15.4		CARR77.105	GAL	1.1		SPRY87.109
PYR	3.9		CARR77.105	GAL	0.5		SPRY87.109
PYR	6.8		CARR77.105	GAL	1.1		SPRY87.109
PYR	4.7		CARR77.105	GAL	−1.6		SPRY87.109
PYR	10.3		CARR77.105	GAL	0.1		SPRY87.109
PYR	11.8		CARR77.105	GAL	0.7		SPRY87.109
PYR	14.5		CARR77.105	GAL	1.7		SPRY87.109
PYR	9.5		CARR77.105	GAL	−0.8		SPRY87.109
PYR	5.4		CARR77.105	GAL	−0.3		SPRY87.109
PYR	13.5		CARR77.105	GAL	−0.1		SPRY87.109
PYR	17.8		CARR77.105	GAL	0.8		SPRY87.109
PYR	11.0		CARR77.105	GAL	2.2		SPRY87.109
PYR	14.9		CARR77.105	GAL	2.5		SPRY87.109
PYR	14.3		CARR77.105	GAL	2.1		SPRY87.109
PYR	11.3		CARR77.105	GAL	−0.6		SPRY87.109
PYR	16.3		CARR77.105	GAL	2.1		SPRY87.109
BAR		37.4	CARR77.105	GAL	−0.6		SPRY87.109
BAR		39.4	CARR77.105	GAL	2.5		SPRY87.109
BAR		39.7	CARR77.105	GAL	1.2		SPRY87.109
				GAL	0.2		SPRY87.109
Broken Hill Deposit (1650 Ma)				GAL	4.1		SPRY87.109
SPH	0.5		SPRY87.109	GAL	2.3		SPRY87.109
SPH	−0.6		SPRY87.109	CHALCOPYR	0.8		SPRY87.109
SPH	1.6		SPRY87.109	CHALCOPYR	0.7		SPRY87.109
SPH	1.6		SPRY87.109	CHALCOPYR	6.7		SPRY87.109
SPH	2.5		SPRY87.109	CHALCOPYR	3.3		SPRY87.109
SPH	1.9		SPRY87.109	CHALCOPYR	0.7		SPRY87.109
SPRY87.109	0.5		SPRY87.109	CHALCOPYR	−1.8		SPRY87.109
SPH	0.3		SPRY87.109	CHALCOPYR	0.5		SPRY87.109
SPH	2.1		SPRY87.109	CHALCOPYR	2.1		SPRY87.109
SPH	2.8		SPRY87.109	CHALCOPYR	0.3		SPRY87.109
SPH	2.1		SPRY87.109				
SPH	3.7		SPRY87.109	Mallapunyah Formation (1650 Ma)			
SPH	6.7		SPRY87.109	PYR	14.0		JACK88unpu
SPH	3.0		SPRY87.109	PYR	13.1		JACK88unpu
SPH	0.8		SPRY87.109	PYR	−34.8		JACK88unpu
SPH	3.1		SPRY87.109	PO	−34.8		JACK88unpu
SPH	2.7		SPRY87.109				
SPH	1.7		SPRY87.109	Willyama Complex (1650 Ma)			
PO	0.9		SPRY87.109	GAL	−0.8		BOTH75.308
PO	−0.6		SPRY87.109	SPH	−0.1		BOTH75.308
PO	1.4		SPRY87.109	GAL	−2.1		BOTH75.308
PO	2.1		SPRY87.109	SPH	−0.3		BOTH75.308
PO	2.1		SPRY87.109	GAL	0.5		BOTH75.308
PO	0.6		SPRY87.109	SPH	1.6		BOTH75.308
PO	−1.2		SPRY87.109	PO	1.5		BOTH75.308
PO	1.3		SPRY87.109	GAL	1.6		BOTH75.308
PO	1.3		SPRY87.109	SPH	2.0		BOTH75.308
PO	0.7		SPRY87.109	PO	2.2		BOTH75.308
PO	2.0		SPRY87.109	GAL	0.1		BOTH75.308
PO	2.9		SPRY87.109	SPH	0.9		BOTH75.308
PO	6.3		SPRY87.109	GAL	0.6		BOTH75.308
PO	−0.5		SPRY87.109	SPH	1.4		BOTH75.308
PO	2.8		SPRY87.109	GAL	1.1		BOTH75.308
GAL	1.0		SPRY87.109	SPH	1.7		BOTH75.308
GAL	−3.3		SPRY87.109	PO	1.4		BOTH75.308
GAL	−2.0		SPRY87.109	GAL	1.3		BOTH75.308

Table 17.7. Continued

Mineral	del Sfide (‰ vs CDT)	del Sfate (‰ vs CDT)	Reference Code	Mineral	del Sfide (‰ vs CDT)	del Sfate (‰ vs CDT)	Reference Code
SPH	2.4		BOTH75.308	CHALCOPYR	−0.6		BOTH75.308
GAL	0.5		BOTH75.308	GAL	−0.9		BOTH75.308
SPH	1.3		BOTH75.308	SPH	0.1		BOTH75.308
PO	1.2		BOTH75.308	SPH	−0.3		BOTH75.308
GAL	0.2		BOTH75.308	PYR	−0.3		BOTH75.308
SPH	1.1		BOTH75.308	PO	−0.1		BOTH75.308
PO	1.2		BOTH75.308	SPH	0.2		BOTH75.308
GAL	0.5		BOTH75.308	PYR	0.1		BOTH75.308
SPH	1.8		BOTH75.308	GAL	−0.1		BOTH75.308
GAL	0.3		BOTH75.308	SPH	0.2		BOTH75.308
SPH	1.5		BOTH75.308	GAL	−3.9		BOTH75.308
GAL	1.0		BOTH75.308	SPH	−1.6		BOTH75.308
SPH	1.7		BOTH75.308	GAL	−1.9		BOTH75.308
GAL	1.2		BOTH75.308	GAL	−1.6		BOTH75.308
SPH	1.5		BOTH75.308	GAL	−1.6		BOTH75.308
GAL	1.2		BOTH75.308	GAL	−1.8		BOTH75.308
SPH	1.9		BOTH75.308	GAL	−0.9		BOTH75.308
GAL	−3.1		BOTH75.308	GAL	−1.8		BOTH75.308
SPH	−1.4		BOTH75.308	GAL	16.5		BOTH75.308
GAL	−1.7		BOTH75.308	GAL	16.8		BOTH75.308
SPH	−1.0		BOTH75.308	GAL	2.6		BOTH75.308
SPH	1.3		BOTH75.308	GAL	1.9		BOTH75.308
GAL	−2.0		BOTH75.308	GAL	0.9		BOTH75.308
SPH	−1.1		BOTH75.308	GAL	1.6		BOTH75.308
GAL	−2.3		BOTH75.308	GAL	−1.3		BOTH75.308
SPH	−0.9		BOTH75.308	GAL	6.5		BOTH75.308
GAL	−2.2		BOTH75.308	GAL	4.0		BOTH75.308
SPH	−0.6		BOTH75.308	GAL	3.6		BOTH75.308
GAL	−3.5		BOTH75.308	GAL	1.8		BOTH75.308
SPH	−2.7		BOTH75.308	GAL	−4.9		BOTH75.308
GAL	−0.1		BOTH75.308	SPH	−2.8		BOTH75.308
GAL	0.9		BOTH75.308	GAL	−6.9		BOTH75.308
SPH	2.1		BOTH75.308	GAL	−3.8		BOTH75.308
PO	1.6		BOTH75.308	SPH	−2.0		BOTH75.308
GAL	1.6		BOTH75.308	CHALCOPYR	−2.9		BOTH75.308
SPH	3.0		BOTH75.308	GAL	−6.4		BOTH75.308
PO	1.8		BOTH75.308	GAL	−7.5		BOTH75.308
SPH	2.0		BOTH75.308	GAL	1.9		BOTH75.308
PO	0.5		BOTH75.308	GAL	2.7		BOTH75.308
GAL	0.2		BOTH75.308	GAL	−1.4		BOTH75.308
GAL	4.7		BOTH75.308	SPH	0.2		BOTH75.308
GAL	0.7		BOTH75.308				
GAL	0.7		BOTH75.308	Mt. Isa Group (1670 Ma)			
SPH	1.9		BOTH75.308				
GAL	−3.4		BOTH75.308	SFIDE	9.2		STAN66..16
GAL	−1.7		BOTH75.308	SFIDE	7.3		STAN66..16
GAL	−3.8		BOTH75.308	SFIDE	12.2		STAN66..16
GAL	−0.3		BOTH75.308	SFIDE	9.2		STAN66..16
SPH	1.1		BOTH75.308	SFIDE	7.4		STAN66..16
GAL	5.2		BOTH75.308	SFIDE	7.5		STAN66..16
GAL	4.2		BOTH75.308	SFIDE	9.6		STAN66..16
GAL	3.7		BOTH75.308	SFIDE	10.3		STAN66..16
GAL	5.4		BOTH75.308	SFIDE	12.2		STAN66..16
GAL	1.3		BOTH75.308	SFIDE	13.0		STAN66..16
GAL	4.2		BOTH75.308	SFIDE	13.1		STAN66..16
GAL	−0.7		BOTH75.308	SFIDE	14.5		STAN66..16
GAL	1.4		BOTH75.308	SFIDE	13.0		STAN66..16
SPH	−0.4		BOTH75.308	SFIDE	11.9		STAN66..16

Table 17.7. Continued.

Mineral	del Sfide (‰ vs CDT)	del Sfate (‰ vs CDT)	Reference Code	Mineral	del Sfide (‰ vs CDT)	del Sfate (‰ vs CDT)	Reference Code
Urquhart Shale (1670 Ma)				GAL	13.1		SMIT78.369
PO	9.5		SMIT78.369	GAL	11.2		SMIT78.369
PO	9.8		SMIT78.369	GAL	14.4		SMIT78.369
PO	9.6		SMIT78.369	GAL	14.8		SMIT78.369
PO	10.4		SMIT78.369	GAL	16.0		SMIT78.369
PO	17.5		SMIT78.369	GAL	16.6		SMIT78.369
PO	12.4		SMIT78.369	SPH	15.0		SMIT78.369
PO	10.8		SMIT78.369	SPH	15.7		SMIT78.369
PO	11.8		SMIT78.369	SPH	15.9		SMIT78.369
PO	11.1		SMIT78.369	SPH	16.4		SMIT78.369
PO	10.5		SMIT78.369	SPH	16.2		SMIT78.369
PO	11.6		SMIT78.369	SPH	16.0		SMIT78.369
GAL	5.0		SMIT78.369	SPH	17.2		SMIT78.369
GAL	10.3		SMIT78.369	SPH	18.4		SMIT78.369
GAL	16.4		SMIT78.369	SPH	18.2		SMIT78.369
GAL	9.2		SMIT78.369	SPH	18.6		SMIT78.369
SPH	12.2		SMIT78.369	PO	14.5		SMIT78.369
SPH	18.6		SMIT78.369	PO	13.6		SMIT78.369
SPH	11.0		SMIT78.369	PO	13.2		SMIT78.369
CHALCOPYR	11.7		SMIT78.369	PO	14.6		SMIT78.369
CHALCOPYR	11.9		SMIT78.369	PO	14.4		SMIT78.369
CHALCOPYR	10.8		SMIT78.369	PO	15.6		SMIT78.369
CHALCOPYR	12.0		SMIT78.369	PO	13.4		SMIT78.369
CHALCOPYR	11.8		SMIT78.369	PO	13.8		SMIT78.369
CHALCOPYR	11.4		SMIT78.369	PYR	15.9		SMIT78.369
CHALCOPYR	12.5		SMIT78.369	PYR	16.8		SMIT78.369
CHALCOPYR	11.6		SMIT78.369	PYR	18.1		SMIT78.369
CHALCOPYR	12.8		SMIT78.369	PYR	17.5		SMIT78.369
CHALCOPYR	12.5		SMIT78.369	PYR	19.1		SMIT78.369
CHALCOPYR	11.9		SMIT78.369	PYR	19.2		SMIT78.369
CHALCOPYR	13.6		SMIT78.369	PYR	18.0		SMIT78.369
PYR	10.8		SMIT78.369	PYR	17.0		SMIT78.369
PYR	11.1		SMIT78.369	PYR	21.6		SMIT78.369
PYR	14.4		SMIT78.369	PYR	22.8		SMIT78.369
PYR	21.0		SMIT78.369	PYR	18.8		SMIT78.369
PYR	12.9		SMIT78.369	PYR	13.4		SMIT78.369
PO	14.0		SMIT78.369	PYR	11.6		SMIT78.369
PO	14.3		SMIT78.369	PYR	17.7		SMIT78.369
PO	15.8		SMIT78.369	PYR	5.2		SMIT78.369
PO	15.1		SMIT78.369	PYR	8.1		SMIT78.369
CHALCOPYR	15.7		SMIT78.369	PYR	11.3		SMIT78.369
CHALCOPYR	14.6		SMIT78.369	PYR	15.4		SMIT78.369
CHALCOPYR	14.3		SMIT78.369	CHALCOPYR	16.9		SOLO65.737
CHALCOPYR	16.4		SMIT78.369	CHALCOPYR	13.0		SOLO65.737
CHALCOPYR	15.4		SMIT78.369	CHALCOPYR	16.1		SOLO65.737
CHALCOPYR	18.8		SMIT78.369	PYR	20.5		SOLO65.737
CHALCOPYR	17.5		SMIT78.369	PYR	21.0		SOLO65.737
PYR	15.4		SMIT78.369	PYR	8.4		SOLO65.737
PYR	19.4		SMIT78.369	PYR	13.7		SOLO65.737
PYR	18.6		SMIT78.369	PYR	15.8		SOLO65.737
PYR	19.2		SMIT78.369	PYR	18.7		SOLO65.737
PYR	19.0		SMIT78.369	PYR	18.8		SOLO65.737
PYR	20.0		SMIT78.369	PYR	27.6		SOLO65.737
PYR	20.3		SMIT78.369	PO	24.7		SOLO65.737
PYR	16.9		SMIT78.369	PO	8.8		SOLO65.737
GAL	11.5		SMIT78.369	PYR	21.0		SOLO65.737
GAL	10.6		SMIT78.369	PYR	18.2		SOLO65.737
GAL	13.4		SMIT78.369	PO	20.5		SOLO65.737

Table 17.7. Continued.

Mineral	del Sfide (‰ vs CDT)	del Sfate (‰ vs CDT)	Reference Code	Mineral	del Sfide (‰ vs CDT)	del Sfate (‰ vs CDT)	Reference Code
PYR	28.1		SOLO65.737	SPH	15.4		WALK83.214
PYR	13.4		SOLO65.737	SPH	14.5		WALK83.214
SPH	10.6		SOLO65.737	SPH	14.2		WALK83.214
GAL	4.0		SOLO65.737	SPH	14.5		WALK83.214
PO	12.5		SOLO65.737	SPH	5.2		WALK83.214
PYR	7.2		SOLO65.737	SPH	5.8		WALK83.214
PYR	11.1		SOLO65.737	SPH	7.0		WALK83.214
SPH	19.4		SOLO65.737	SPH	5.8		WALK83.214
SPH	23.0		SOLO65.737	SPH	2.9		WALK83.214
PYR	9.6		SOLO65.737	SPH	6.5		WALK83.214
PO	11.3		SOLO65.737	SPH	5.6		WALK83.214
PO	15.9		SOLO65.737	SPH	5.2		WALK83.214
SPH	19.0		SOLO65.737	SPH	7.3		WALK83.214
SPH	17.3		SOLO65.737	GAL	1.3		WALK83.214
SPH	18.2		SOLO65.737	GAL	3.1		WALK83.214
GAL	15.3		SOLO65.737	GAL	11.5		WALK83.214
SPH	16.6		SOLO65.737	GAL	2.0		WALK83.214
PYR	7.6		SOLO65.737	GAL	3.4		WALK83.214
PO	11.6		SOLO65.737	GAL	1.4		WALK83.214
PO	19.1		SOLO65.737	GAL	0.9		WALK83.214
GAL	8.5		SOLO65.737	GAL	1.0		WALK83.214
GAL	14.1		SOLO65.737	GAL	2.2		WALK83.214
GAL	13.6		SOLO65.737	PYR	6.8		RYE.81...1
GAL	0.2		SOLO65.737	PYR	10.1		RYE.81...1
PO	29.4		SOLO65.737	PYR	5.8		RYE.81...1
PYR	30.8		SOLO65.737	SPH	4.3		RYE.81...1
Wollogorang Formation (1670 Ma)				SPH	6.1		RYE.81...1
				SPH	5.5		RYE.81...1
PYR	−3.5		JACK88unpu	SPH	4.5		RYE.81...1
PYR	2.0		JACK88unpu	SPH	5.1		RYE.81...1
PYR	6.0		JACK88unpu	SPH	5.4		RYE.81...1
				SPH	4.9		RYE.81...1
McArthur Group (1690 Ma)				SPH	6.2		RYE.81...1
PYR	8.6		WALK83.214	SPH	4.8		RYE.81...1
PYR	18.4		WALK83.214	GAL	3.1		RYE.81...1
PYR	16.0		WALK83.214	GAL	2.8		RYE.81...1
PYR	13.4		WALK83.214	GAL	2.1		RYE.81...1
PYR	17.1		WALK83.214	GAL	3.8		RYE.81...1
PYR	17.6		WALK83.214	GAL	2.1		RYE.81...1
PYR	10.1		WALK83.214	GAL	0.6		RYE.81...1
PYR	21.0		WALK83.214	GAL	1.7		RYE.81...1
PYR	21.7		WALK83.214	GAL	1.3		RYE.81...1
PYR	15.4		WALK83.214	GAL	0.4		RYE.81...1
PYR	13.3		WALK83.214	GAL	2.8		RYE.81...1
PYR	16.1		WALK83.214	GAL	1.1		RYE.81...1
PYR	10.3		WALK83.214	GAL	5.7		SMIT73..10
PYR	5.2		WALK83.214	GAL	−0.5		SMIT73..10
PYR	3.6		WALK83.214	GAL	3.8		SMIT73..10
PYR	3.6		WALK83.214	GAL	2.5		SMIT73..10
PYR	−0.6		WALK83.214	GAL	−1.2		SMIT73..10
PYR	15.9		WALK83.214	GAL	0.3		SMIT73..10
PYR	16.6		WALK83.214	GAL	3.7		SMIT73..10
PYR	10.7		WALK83.214	GAL	−1.2		SMIT73..10
PYR	20.2		WALK83.214	GAL	0.6		SMIT73..10
PYR	19.2		WALK83.214	GAL	1.0		SMIT73..10
SPH	9.1		WALK83.214	GAL	1.9		SMIT73..10
SPH	6.5		WALK83.214	GAL	4.8		SMIT73..10
SPH	13.0		WALK83.214	GAL	1.7		SMIT73..10

Table 17.7. *Continued*

Mineral	del Sfide (‰ vs CDT)	del Sfate (‰ vs CDT)	Reference Code	Mineral	del Sfide (‰ vs CDT)	del Sfate (‰ vs CDT)	Reference Code
SPH	8.7		SMIT73..10	GAL	12.5		MUIR85.239
SPH	3.3		SMIT73..10	GAL	12.5		MUIR85.239
SPH	6.8		SMIT73..10	SPH	12.5		MUIR85.239
SPH	7.1		SMIT73..10	SPH	12.5		MUIR85.239
SPH	4.0		SMIT73..10	SPH	14.5		MUIR85.239
SPH	5.7		SMIT73..10	SPH	14.5		MUIR85.239
SPH	4.1		SMIT73..10	BAR		17.5	MUIR85.239
SPH	3.5		SMIT73..10	BAR		17.5	MUIR85.239
SPH	4.0		SMIT73..10	BAR		17.5	MUIR85.239
SPH	4.6		SMIT73..10	BAR		17.5	MUIR85.239
SPH	6.6		SMIT73..10	BAR		18.5	MUIR85.239
SPH	8.9		SMIT73..10	BAR		19.5	MUIR85.239
SPH	6.8		SMIT73..10	BAR		19.5	MUIR85.239
PYR	−3.9		SMIT73..10	BAR		19.5	MUIR85.239
PYR	9.3		SMIT73..10	BAR		19.5	MUIR85.239
PYR	−2.8		SMIT73..10	BAR		20.5	MUIR85.239
PYR	1.2		SMIT73..10	BAR		20.5	MUIR85.239
PYR	0.9		SMIT73..10	BAR		20.5	MUIR85.239
PYR	−0.1		SMIT73..10	BAR		24.5	MUIR85.239
PYR	7.9		SMIT73..10	Amisk Group (1800 Ma)			
PYR	5.3		SMIT73..10				
PYR	15.9		SMIT73..10	PYR	4.9		STRA862653
PYR	6.6		SMIT73..10	PYR	5.8		STRA862653
PYR	4.4		SMIT73..10	PYR	5.8		STRA862653
PYR	9.1		SMIT73..10	PYR	6.3		STRA862653
PYR	7.1		SMIT73..10	PYR	1.9		STRA862653
PYR	11.5		SMIT73..10	PYR	4.6		STRA862653
PYR	14.1		SMIT73..10	PYR	5.4		STRA862653
PYR	13.9		SMIT73..10	PYR	5.9		STRA862653
PYR	12.5		MUIR85.239	PYR	4.0		STRA862653
PYR	17.5		MUIR85.239	PYR	6.1		STRA862653
PYR	17.5		MUIR85.239	PYR	6.0		STRA862653
PYR	21.5		MUIR85.239	PYR	5.6		STRA862653
PYR	21.5		MUIR85.239	PYR	5.4		STRA862653
PYR	21.5		MUIR85.239	PYR	5.6		STRA862653
PYR	22.5		MUIR85.239	PYR	5.9		STRA862653
PYR	22.5		MUIR85.239	PYR	6.2		STRA862653
PYR	26.0		MUIR85.239	PYR	5.1		STRA862653
PYR	16.0		MUIR85.239	PYR	4.1		STRA862653
PYR	−7.0		MUIR85.239	PYR	5.4		STRA862653
PYR	−1.0		MUIR85.239	PYR	6.0		STRA862653
PYR	0.5		MUIR85.239	PYR	4.8		STRA862653
PYR	4.5		MUIR85.239	PYR	4.7		STRA862653
PYR	8.5		MUIR85.239	PYR	4.6		STRA862653
PYR	10.0		MUIR85.239	PYR	3.6		STRA862653
PYR	12.0		MUIR85.239	PYR	6.5		STRA862653
PYR	12.0		MUIR85.239	PYR	4.6		STRA862653
CHALCOPYR	2.5		MUIR85.239	PYR	6.5		STRA862653
CHALCOPYR	3.5		MUIR85.239	PYR	6.3		STRA862653
CHALCOPYR	10.0		MUIR85.239	PYR	5.6		STRA862653
GAL	5.0		MUIR85.239	PYR	6.1		STRA862653
GAL	6.5		MUIR85.239	PYR	6.8		STRA862653
GAL	6.5		MUIR85.239	PYR	6.6		STRA862653
GAL	8.0		MUIR85.239	PYR	6.5		STRA862653
GAL	11.5		MUIR85.239	PYR	7.6		STRA862653
GAL	12.5		MUIR85.239	PYR	6.1		STRA862653
GAL	10.5		MUIR85.239	PYR	6.6		STRA862653
GAL	10.5		MUIR85.239	PYR	5.7		STRA862653

Table 17.7. Continued.

Mineral	del Sfide (‰ vs CDT)	del Sfate (‰ vs CDT)	Reference Code	Mineral	del Sfide (‰ vs CDT)	del Sfate (‰ vs CDT)	Reference Code
PYR	2.2		STRA862653	PYR	−5.8		MAKE74...1
PYR	2.0		STRA862653	PYR	−8.5		MAKE74...1
PYR	2.4		STRA862653	PYR	−14.7		MAKE74...1
PYR	3.0		STRA862653	PYR	−9.8		MAKE74...1
PYR	2.4		STRA862653	PYR	−7.4		MAKE74...1
PYR	1.7		STRA862653	PYR	−7.9		MAKE74...1
PYR	3.4		STRA862653	PYR	−12.6		MAKE74...1
PYR	3.3		STRA862653	PYR	−13.1		MAKE74...1
PYR	1.7		STRA862653	PYR	−14.8		MAKE74...1
PYR	0.8		STRA862653	PYR	−13.6		MAKE74...1
PYR	3.2		STRA862653	PYR	−13.4		MAKE74...1
PYR	−4.7		STRA862653	PYR	−12.9		MAKE74...1
Chelmsford Formation (1850 Ma)				PYR	−13.1		MAKE74...1
				PYR	−10.2		MAKE74...1
CHALCOPYR	7.6		THOD62.565	PYR	−5.3		MAKE74...1
PYR	4.6		THOD62.565	PYR	−3.9		MAKE74...1
PYR	4.4		THOD62.565	PYR	−9.4		MAKE74...1
Onwatin Formation (1850 Ma)				PYR	−16.7		MAKE74...1
				PYR	−11.0		MAKE74...1
PYR	19.7		THOD62.565	PYR	−11.3		MAKE74...1
PYR	26.0		THOD62.565	PYR	−9.3		MAKE74...1
PO	24.1		THOD62.565	PYR	−15.7		MAKE74...1
PYR	20.1		THOD62.565	PYR	−17.3		MAKE74...1
PYR	15.7		THOD62.565	SFIDE	0.6		MAKE74...1
PYR	6.0		THOD62.565	SFIDE	−3.9		MAKE74...1
PYR	7.8		THOD62.565	SFIDE	−7.5		MAKE74...1
Outokumpu Ore Deposit (1850 Ma)				SFIDE	−6.6		MAKE74...1
				SFIDE	−4.4		MAKE74...1
PYR	−16.3		MAKE74...1	SFIDE	−3.2		MAKE74...1
PYR	−16.9		MAKE74...1	SFIDE	−2.8		MAKE74...1
PYR	−15.6		MAKE74...1	SFIDE	−1.4		MAKE74...1
PYR	−15.9		MAKE74...1	SFIDE	−3.6		MAKE74...1
PYR	−14.3		MAKE74...1	SFIDE	−0.4		MAKE74...1
PYR	−17.4		MAKE74...1	SFIDE	−1.8		MAKE74...1
PYR	−17.9		MAKE74...1	SFIDE	−2.3		MAKE74...1
PYR	−17.8		MAKE74...1	SFIDE	−1.1		MAKE74...1
PYR	−9.8		MAKE74...1	SFIDE	−2.0		MAKE74...1
PYR	1.9		MAKE74...1	SFIDE	−0.7		MAKE74...1
PYR	−19.2		MAKE74...1	SFIDE	−1.5		MAKE74...1
PYR	−17.7		MAKE74...1	SFIDE	−0.8		MAKE74...1
PYR	−19.1		MAKE74...1	SFIDE	−0.5		MAKE74...1
PYR	−18.9		MAKE74...1	SFIDE	−0.9		MAKE74...1
PYR	−5.4		MAKE74...1	SFIDE	−2.2		MAKE74...1
PYR	−11.4		MAKE74...1	SFIDE	0.3		MAKE74...1
PYR	−11.6		MAKE74...1	SFIDE	0.6		MAKE74...1
PYR	−11.4		MAKE74...1	SFIDE	−0.1		MAKE74...1
PYR	−18.6		MAKE74...1	SFIDE	0.6		MAKE74...1
PYR	−12.6		MAKE74...1	SFIDE	−0.2		MAKE74...1
PYR	−18.6		MAKE74...1	SFIDE	0.8		MAKE74...1
PYR	−14.6		MAKE74...1	SFIDE	−0.4		MAKE74...1
PYR	−12.3		MAKE74...1	SFIDE	0.4		MAKE74...1
PYR	−8.2		MAKE74...1	SFIDE	−1.5		MAKE74...1
PYR	−5.1		MAKE74...1	SFIDE	−1.2		MAKE74...1
PYR	−3.4		MAKE74...1	SFIDE	1.9		MAKE74...1
PYR	−2.1		MAKE74...1	SFIDE	2.9		MAKE74...1
PYR	−13.1		MAKE74...1	SFIDE	2.3		MAKE74...1
PYR	−1.4		MAKE74...1	SFIDE	5.8		MAKE74...1
PYR	−1.1		MAKE74...1	SFIDE	5.6		MAKE74...1

Table 17.7. Continued.

Mineral	del Sfide (‰ vs CDT)	del Sfate (‰ vs CDT)	Reference Code	Mineral	del Sfide (‰ vs CDT)	del Sfate (‰ vs CDT)	Reference Code
SFIDE	0.3		MAKE74...1	SFIDE	−0.6		MAKE74...1
SFIDE	0.2		MAKE74...1	SFIDE	−1.0		MAKE74...1
SFIDE	1.0		MAKE74...1	SFIDE	−0.4		MAKE74...1
SFIDE	0.7		MAKE74...1	SFIDE	1.2		MAKE74...1
SFIDE	−0.2		MAKE74...1	SFIDE	1.7		MAKE74...1
SFIDE	0.2		MAKE74...1	SFIDE	−0.6		MAKE74...1
SFIDE	0.9		MAKE74...1	SFIDE	−0.1		MAKE74...1
SFIDE	1.7		MAKE74...1	SFIDE	1.6		MAKE74...1
SFIDE	1.3		MAKE74...1	SFIDE	2.5		MAKE74...1
SFIDE	1.1		MAKE74...1	SFIDE	2.7		MAKE74...1
SFIDE	0.9		MAKE74...1	SFIDE	2.8		MAKE74...1
SFIDE	0.9		MAKE74...1	SFIDE	0.1		MAKE74...1
SFIDE	1.0		MAKE74...1	SFIDE	0.6		MAKE74...1
SFIDE	−0.9		MAKE74...1	SFIDE	1.6		MAKE74...1
SFIDE	0.5		MAKE74...1	SFIDE	3.2		MAKE74...1
SFIDE	1.1		MAKE74...1	SFIDE	1.2		MAKE74...1
SFIDE	−0.1		MAKE74...1	SFIDE	−0.3		MAKE74...1
SFIDE	−1.0		MAKE74...1	SFIDE	−1.7		MAKE74...1
SFIDE	0.3		MAKE74...1	SFIDE	−3.3		MAKE74...1
SFIDE	0.7		MAKE74...1	SFIDE	−4.5		MAKE74...1
SFIDE	1.0		MAKE74...1	SFIDE	−3.7		MAKE74...1
SFIDE	−1.1		MAKE74...1	SFIDE	2.4		MAKE74...1
SFIDE	1.2		MAKE74...1	SFIDE	1.0		MAKE74...1
SFIDE	1.3		MAKE74...1	SFIDE	3.5		MAKE74...1
SFIDE	1.0		MAKE74...1	SFIDE	3.5		MAKE74...1
SFIDE	0.7		MAKE74...1	SFIDE	5.0		MAKE74...1
SFIDE	0.1		MAKE74...1	SFIDE	3.0		MAKE74...1
SFIDE	−0.2		MAKE74...1	SFIDE	3.8		MAKE74...1
SFIDE	−1.4		MAKE74...1	SFIDE	1.2		MAKE74...1
SFIDE	−1.1		MAKE74...1	SFIDE	3.8		MAKE74...1
SFIDE	−1.8		MAKE74...1	SFIDE	0.8		MAKE74...1
SFIDE	−1.6		MAKE74...1	SFIDE	−0.2		MAKE74...1
SFIDE	−0.8		MAKE74...1	SFIDE	0.3		MAKE74...1
SFIDE	−1.6		MAKE74...1	SFIDE	−0.2		MAKE74...1
SFIDE	−0.9		MAKE74...1	SFIDE	−0.3		MAKE74...1
SFIDE	−1.1		MAKE74...1	SFIDE	−0.3		MAKE74...1
SFIDE	−2.0		MAKE74...1	SFIDE	1.6		MAKE74...1
SFIDE	−0.9		MAKE74...1	SFIDE	2.1		MAKE74...1
SFIDE	0.3		MAKE74...1	SFIDE	2.9		MAKE74...1
SFIDE	−0.9		MAKE74...1	SFIDE	3.6		MAKE74...1
SFIDE	0.4		MAKE74...1	SFIDE	1.2		MAKE74...1
SFIDE	−0.9		MAKE74...1	SFIDE	2.1		MAKE74...1
SFIDE	−1.7		MAKE74...1	SFIDE	0.7		MAKE74...1
SFIDE	−1.4		MAKE74...1	SFIDE	1.5		MAKE74...1
SFIDE	−1.6		MAKE74...1	SFIDE	0.8		MAKE74...1
SFIDE	0.3		MAKE74...1	SFIDE	1.6		MAKE74...1
SFIDE	−0.6		MAKE74...1	SFIDE	0.9		MAKE74...1
SFIDE	−0.1		MAKE74...1	SFIDE	0.5		MAKE74...1
SFIDE	−0.1		MAKE74...1	SFIDE	0.5		MAKE74...1
SFIDE	1.7		MAKE74...1	SFIDE	1.2		MAKE74...1
SFIDE	−0.8		MAKE74...1	SFIDE	−4.7		MAKE74...1
SFIDE	−2.0		MAKE74...1	SFIDE	−0.5		MAKE74...1
SFIDE	−0.7		MAKE74...1	SFIDE	−3.4		MAKE74...1
SFIDE	−0.3		MAKE74...1	SFIDE	−4.4		MAKE74...1
SFIDE	−0.4		MAKE74...1	SFIDE	−4.3		MAKE74...1
SFIDE	−0.9		MAKE74...1	SFIDE	−3.2		MAKE74...1
SFIDE	0.5		MAKE74...1	SFIDE	−4.4		MAKE74...1
SFIDE	−0.9		MAKE74...1	SFIDE	−4.0		MAKE74...1

Table 17.7. *Continued*

Mineral	del Sfide (‰ vs CDT)	del Sfate (‰ vs CDT)	Reference Code	Mineral	del Sfide (‰ vs CDT)	del Sfate (‰ vs CDT)	Reference Code
SFIDE	−4.0		MAKE74...1	BAR		8.4	RICK791060
SFIDE	−3.4		MAKE74...1	BAR		10.9	RICK791060
SFIDE	−3.6		MAKE74...1	BAR		11.4	RICK791060
SFIDE	−2.3		MAKE74...1	BAR		11.0	RICK791060
SFIDE	−3.5		MAKE74...1	BAR		4.9	GAVE60.510
SFIDE	−4.2		MAKE74...1	BAR		14.3	GAVE60.510
				BAR		1.5	GAVE60.510
Skellefte Field (1850 Ma)				BAR		13.0	GAVE60.510
PYR	−5.4		RICK791060	BAR		15.0	GAVE60.510
PYR	−6.1		RICK791060	Virginia Formation (1850 Ma)			
PYR	−3.9		RICK791060	SFIDE	5.6		RIPL87..87
PYR	−4.6		RICK791060	SFIDE	6.5		RIPL87..87
PYR	−1.8		RICK791060	SFIDE	7.1		RIPL87..87
PYR	−5.7		RICK791060	SFIDE	8.8		RIPL87..87
PYR	−7.5		RICK791060	SFIDE	8.0		RIPL87..87
PYR	−5.5		RICK791060	SFIDE	6.5		RIPL87..87
PYR	−3.0		RICK791060	Rove Formation (1900 Ma)			
PYR	−0.5		GAVE60.510	PYR	16.5		CAME80.181
PYR	−5.6		GAVE60.510	*2000 to 2250 Ma*			
PYR	−14.8		GAVE60.510	Animikie Group (2000 Ma)			
PYR	−1.7		GAVE60.510	PYR	1.3		BUDD69.423
PYR	−10.5		GAVE60.510	PYR	−0.3		BUDD69.423
PYR	−7.9		GAVE60.510	PYR	−0.4		BUDD69.423
PYR	−7.1		GAVE60.510	PYR	−2.1		BUDD69.423
PYR	−6.0		GAVE60.510	PYR	0.2		BUDD69.423
PYR	−10.5		GAVE60.510	PYR	2.4		BUDD69.423
PYR	−6.7		GAVE60.510	PYR	−2.7		BUDD69.423
PYR	−4.8		GAVE60.510	Kona Dolomite (2000 Ma)			
PYR	−6.0		GAVE60.510	ANH		13.3	HEMZ82.512
PYR	−11.8		GAVE60.510	Menihek Formation (2000 Ma)			
PYR	−7.6		GAVE60.510	PYR	11.0		CAME80.181
PYR	2.0		GAVE60.510	Temiscamie Iron Formation (2000 Ma)			
PYR	−12.3		GAVE60.510	PYR	15.8		CAME80.181
PYR	−2.3		GAVE60.510	Udokan "Series" (2000 Ma)			
PYR	3.3		GAVE60.510	CHALCOPYR	−22.6		CHUK711429
PYR	1.4		GAVE60.510	CHALCOPYR	−21.1		CHUK711429
PYR	0.4		GAVE60.510	SFIDE	−12.7		CHUK711429
PYR	1.4		GAVE60.510	SFIDE	−7.7		CHUK711429
PYR	1.4		GAVE60.510	Gunflint Iron Formation (2090 Ma)			
PYR	3.6		GAVE60.510	PYR	6.2		CAME80.181
PYR	1.4		GAVE60.510	Albanel Formation (2100 Ma)			
PYR	1.0		GAVE60.510	PYR	−0.3		CAME80.181
PYR	0.8		GAVE60.510	Attikamagen Formation (2100 Ma)			
PYR	−1.2		GAVE60.510	PYR	13.6		CAME80.181
PYR	−0.2		GAVE60.510	Daspoort Formation (2200 Ma)			
PYR	−1.3		GAVE60.510	SFIDE	8.5		CAME82.145
PYR	−2.3		GAVE60.510				
PYR	−1.3		GAVE60.510				
PYR	0.1		GAVE60.510				
PYR	0.3		GAVE60.510				
PYR	2.3		RICK791060				
PYR	2.3		RICK791060				
PYR	1.8		RICK791060				
PYR	0.3		RICK791060				
PYR	0.8		RICK791060				
PYR	1.8		RICK791060				
PYR	0.9		RICK791060				
PYR	1.3		RICK791060				

Table 17.7. *Continued.*

Mineral	del Sfide (‰ vs CDT)	del Sfate (‰ vs CDT)	Reference Code	Mineral	del Sfide (‰ vs CDT)	del Sfate (‰ vs CDT)	Reference Code
Silverton Formation (2200 Ma)				PYR	1.2		HATT83.549
SFIDE	−12.0		CAME82.145	PYR	1.9		HATT83.549
SFIDE	3.5		CAME82.145	PYR	2.4		HATT83.549
SFIDE	−14.5		CAME82.145	Pecors Formation (2250 Ma)			
SFIDE	−25.0		CAME82.145	PYR	−0.9		HATT83.549
SFIDE	−5.0		CAME82.145	PYR	1.9		HATT83.549
SFIDE	−5.5		CAME82.145	Ramsay Lake Formation (2250 Ma)			
Strubenkoop Formation (2200 Ma)				PYR	−1.0		HATT83.549
SFIDE	8.5		CAME82.145	PYR	3.6		HATT83.549
SFIDE	8.5		CAME82.145	Serpent Formation (2250 Ma)			
SFIDE	7.5		CAME82.145	PYR	−1.1		HATT83.549
SFIDE	10.0		CAME82.145	PYR	−0.1		HATT83.549
2250 to 2500 Ma				PYR	1.3		HATT83.549
Bruce Formation (2250 Ma)				Timeball Hill Formation (2250 Ma)			
PYR	0.8		HATT83.549	SFIDE	−13.5		CAME82.145
Espanola Formation (2250 Ma)				SFIDE	−14.5		CAME82.145
PYR	0.3		HATT83.549	SFIDE	−15.0		CAME82.145
PYR	0.5		HATT83.549	SFIDE	−16.0		CAME82.145
PYR	0.8		HATT83.549	SFIDE	−14.5		CAME82.145
Gowganda Formation (2250 Ma)				SFIDE	−16.0		CAME82.145
PYR	2.4		CAME80.181	SFIDE	−14.0		CAME82.145
PYR	−1.9		HATT83.549	SFIDE	−17.0		CAME82.145
PYR	−1.7		HATT83.549	SFIDE	−18.0		CAME82.145
PYR	−1.5		HATT83.549	SFIDE	−21.5		CAME82.145
PYR	−1.3		HATT83.549	SFIDE	−19.5		CAME82.145
PYR	−1.1		HATT83.549	SFIDE	−20.0		CAME82.145
PYR	−0.5		HATT83.549	SFIDE	−29.0		CAME82.145
PYR	0.6		HATT83.549	SFIDE	−31.0		CAME82.145
PYR	1.0		HATT83.549	Gordon Lake Formation (2259 Ma)			
PYR	1.1		HATT83.549	PYR	−10.6		HATT83.549
PYR	1.6		HATT83.549	PYR	−5.9		HATT83.549
PYR	1.9		HATT83.549	PYR	7.6		HATT83.549
PYR	2.3		HATT83.549	ANH		15.1	CAME83..54
PYR	4.5		HATT83.549	ANH		15.6	CAME83..54
PYR	5.0		HATT83.549	ANH		12.4	CAME83..54
Lorrain Formation (2250 Ma)				ANH		12.3	CAME83..54
PYR	−20.4		HATT83.549	ANH		13.3	CAME83..54
PYR	−19.0		HATT83.549	Penge Iron Formation (2300 Ma)			
PYR	−18.1		HATT83.549	SFIDE	−7.5		CAME82.145
PYR	−17.8		HATT83.549	SFIDE	−7.0		CAME82.145
PYR	−16.2		HATT83.549	SFIDE	−5.7		CAME82.145
PYR	−11.8		HATT83.549	SFIDE	−5.0		CAME82.145
PYR	−0.1		HATT83.549	SFIDE	−3.0		CAME82.145
PYR	1.0		HATT83.549	SFIDE	−5.0		CAME82.145
PYR	6.5		HATT83.549	SFIDE	−7.1		CAME82.145
PYR	15.3		HATT83.549	SFIDE	−7.0		CAME82.145
PYR	18.2		HATT83.549	SFIDE	−4.0		CAME82.145
Mississagi Formation (2250 Ma)				SFIDE	−3.5		CAME82.145
PYR	0.2		HATT83.549	Aravalli Supergroup (2350 Ma)			
PYR	0.5		HATT83.549	PO	6.5		DEB.86.313
PYR	0.6		HATT83.549	PO	9.0		DEB.86.313
PYR	1.1		HATT83.549				

Table 17.7. Continued.

Mineral	del Sfide (‰ vs CDT)	del Sfate (‰ vs CDT)	Reference Code	Mineral	del Sfide (‰ vs CDT)	del Sfate (‰ vs CDT)	Reference Code
PYR	8.5		DEB.86.313	PYR	−0.2		HATT83.323
PO	−2.4		DEB.86.313	PYR	1.1		HATT83.323
PO	−2.1		DEB.86.313	PYR	0.3		HATT83.323
PO	−2.0		DEB.86.313	PYR	1.4		HATT83.323
PYR	−4.7		DEB.86.313	PYR	0.4		HATT83.323
PYR	1.9		DEB.86.313	PYR	1.0		HATT83.323
PYR	−6.7		DEB.86.313	PYR	0.4		HATT83.323
PYR	2.0		DEB.86.313	PYR	1.0		HATT83.323
PYR	3.4		DEB.86.313	PYR	0.1		HATT83.323
PYR	5.6		DEB.86.313	PYR	0.8		HATT83.323
PYR	5.7		DEB.86.313	PYR	0.5		HATT83.323
PYR	6.8		DEB.86.313	PYR	−0.1		HATT83.323
PYR	3.5		DEB.86.313	PYR	0.7		HATT83.323
PYR	8.5		DEB.86.313	PYR	−0.1		HATT83.323
PYR	8.7		DEB.86.313	PYR	−0.4		HATT83.323
PO	−3.0		DEB.86.313	PYR	1.0		HATT83.323
PO	−5.5		DEB.86.313	PYR	1.0		HATT83.323
PO	0.9		DEB.86.313	PYR	0.3		HATT83.323
PO	−3.8		DEB.86.313	PYR	1.3		HATT83.323
PYR	−2.9		DEB.86.313	PYR	0.1		HATT83.323
Frood Formation (2400 Ma)				PYR	0.7		HATT83.323
				PYR	−0.3		HATT83.323
PYR	−0.1		THOD62.565	PYR	−0.3		HATT83.323
PO	0.2		THOD62.565	PYR	−0.1		HATT83.323
Malmani Subgroup (2400 Ma)				PYR	−0.5		HATT83.323
				PYR	0.9		HATT83.323
SFIDE	−3.5		CAME82.145	PYR	−0.2		HATT83.323
SFIDE	−7.0		CAME82.145	PYR	0.6		HATT83.323
SFIDE	−3.5		CAME82.145	PYR	−0.9		HATT83.323
SFIDE	−1.8		CAME82.145	PYR	−1.0		HATT83.323
SFIDE	−9.0		CAME82.145	PYR	−0.5		HATT83.323
SFIDE	−8.0		CAME82.145	PYR	0.8		HATT83.323
SFIDE	−1.0		CAME82.145	PYR	−0.4		HATT83.323
SFIDE	−4.0		CAME82.145	PYR	−0.7		HATT83.323
SFIDE	−2.4		CAME82.145	PYR	−0.8		HATT83.323
SFIDE	−3.5		CAME82.145	PYR	−0.5		HATT83.323
SFIDE	−5.0		CAME82.145	McKim Formation (2400 Ma)			
SFIDE	−9.5		CAME82.145	SFIDE	−0.3		THOD62.565
SFIDE	−3.5		CAME82.145	SFIDE	−0.5		THOD62.565
SFIDE	−7.0		CAME82.145	PYR	−0.3		HATT83.549
SFIDE	−3.5		CAME82.145	PYR	0.7		HATT83.549
SFIDE	−4.5		CAME82.145	PYR	0.8		HATT83.549
SFIDE	−2.6		CAME82.145	PYR	1.3		HATT83.549
SFIDE	−3.0		CAME82.145	PYR	3.8		HATT83.549
SFIDE	−3.0		CAME82.145	PYR	4.0		HATT83.549
SFIDE	13.0		CAME82.145	Snider Formation (2400 Ma)			
SFIDE	3.8		CAME82.145	SFIDE	−0.1		THOD62.565
SFIDE	4.0		CAME82.145	Black Reef Formation (2450 Ma)			
SFIDE	3.5		CAME82.145	SFIDE	4.8		CAME82.145
SFIDE	3.2		CAME82.145	SFIDE	3.5		CAME82.145
Matinenda Formation (2400 Ma)				*Archean 2500 to 3000 Ma*			
PYR	1.1		HATT83.323	Keewatin Group (2500 Ma)			
PYR	−0.1		HATT83.323	PYR	−2.7		SHEG78unpu
PYR	0.4		HATT83.323	PYR	−7.4		SHEG78unpu
PYR	−0.6		HATT83.323	PYR	−7.1		SHEG78unpu
PYR	−0.5		HATT83.323				
PYR	0.2		HATT83.323				
PYR	0.1		HATT83.323				

Table 17.7. Continued.

Mineral	del Sfide (‰ vs CDT)	del Sfate (‰ vs CDT)	Reference Code	Mineral	del Sfide (‰ vs CDT)	del Sfate (‰ vs CDT)	Reference Code
PYR	−2.1		SHEG78unpu	PYR	2.4		STRA862653
PYR	−0.5		SHEG78unpu	PYR	3.3		STRA862653
PYR	−6.6		SHEG78unpu	PYR	4.1		STRA862653
PYR	−4.6		SHEG78unpu	PYR	0.7		STRA862653
PYR	2.9		SHEG78unpu	PYR	2.1		STRA862653
PYR	1.0		SHEG78unpu	PYR	0.4		STRA862653
PYR	3.6		SHEG78unpu	PYR	1.3		STRA862653
PYR	3.7		SHEG78unpu	PYR	1.2		STRA862653
PYR	5.0		SHEG78unpu	PYR	0.1		STRA862653
PYR	2.1		SHEG78unpu	PYR	3.7		STRA862653
PYR	1.3		SHEG78unpu	PYR	2.9		STRA862653
PYR	−0.5		SHEG78unpu	PYR	4.1		STRA862653
PYR	4.0		SHEG78unpu	PYR	1.7		STRA862653
PYR	−6.1		SHEG78unpu	Keewatin, Wabigoon Belt (2500 Ma)			
PYR	−0.6		SHEG78unpu				
PYR	1.2		SHEG78unpu	PYR	4.2		STRA862653
PYR	−0.7		SHEG78unpu	PYR	3.7		STRA862653
PYR	−1.0		SHEG78unpu	PYR	5.0		STRA862653
PYR	−0.7		SHEG78unpu	PYR	3.5		STRA862653
PYR	1.4		SHEG78unpu	PYR	4.3		STRA862653
PYR	0.2		SHEG78unpu	PYR	3.7		STRA862653
PYR	0.5		SHEG78unpu	PYR	3.5		STRA862653
PYR	2.0		SHEG78unpu	PYR	3.9		STRA862653
PYR	1.1		SHEG78unpu	PYR	3.9		STRA862653
PYR	0.2		SHEG78unpu	PYR	4.0		STRA862653
PYR	−1.0		SHEG78unpu	PYR	5.7		STRA862653
PYR	−0.5		SHEG78unpu	PYR	3.6		STRA862653
PYR	−0.6		SHEG78unpu	PYR	3.7		STRA862653
PYR	−0.4		SHEG78unpu	PYR	3.4		STRA862653
PYR	−0.5		SHEG78unpu	PYR	2.0		STRA862653
PYR	0.1		SHEG78unpu	PYR	2.6		STRA862653
PYR	−1.1		SHEG78unpu	PYR	2.2		STRA862653
PYR	−1.7		SHEG78unpu	PYR	2.3		STRA862653
PYR	1.9		SHEG78unpu	PYR	2.0		STRA862653
PYR	1.5		SHEG78unpu	PYR	2.6		STRA862653
PYR	1.4		SHEG78unpu	PYR	2.2		STRA862653
PYR	−1.4		SHEG78unpu	PYR	2.2		STRA862653
PYR	−1.0		SHEG78unpu	PYR	2.3		STRA862653
PYR	−1.1		SHEG78unpu	PYR	2.9		STRA862653
PYR			SHEG78unpu	PYR	4.0		STRA862653
Keewatin, Abitibi Belt (2500 Ma)				PYR	4.3		STRA862653
				PYR	3.5		STRA862653
PYR	1.0		STRA862653	PYR	3.9		STRA862653
PYR	1.6		STRA862653	PYR	4.0		STRA862653
PYR	2.0		STRA862653	PYR	2.6		STRA862653
PYR	2.7		STRA862653	PYR	2.0		STRA862653
PYR	2.2		STRA862653	PYR	0.6		STRA862653
PYR	2.1		STRA862653	PYR	0.8		STRA862653
PYR	2.7		STRA862653	PYR	1.3		STRA862653
PYR	2.4		STRA862653	PYR	1.6		STRA862653
PYR	1.6		STRA862653	PYR	1.7		STRA862653
PYR	1.9		STRA862653	PYR	2.1		STRA862653
PYR	0.1		STRA862653	PYR	2.4		STRA862653
PYR	1.5		STRA862653	PYR	2.2		STRA862653
PYR	1.1		STRA862653	PYR	1.4		STRA862653
PYR	2.3		STRA862653	PYR	1.2		STRA862653
PYR	4.3		STRA862653	PYR	0.2		STRA862653
PYR	2.2		STRA862653	PYR	0.4		STRA862653
PYR	1.1		STRA862653	PYR	0.6		STRA862653

Table 17.7. Continued.

Mineral	del Sfide (‰ vs CDT)	del Sfate (‰ vs CDT)	Reference Code	Mineral	del Sfide (‰ vs CDT)	del Sfate (‰ vs CDT)	Reference Code
PYR	3.0		STRA862653	PYR	2.9		RIPL81.839
PYR	2.6		STRA862653	PYR	2.6		RIPL81.839
PYR	−3.1		STRA862653	PYR	3.3		RIPL81.839
PYR	−4.0		STRA862653	PYR	1.1		RIPL81.839
PYR	−5.7		STRA862653	PYR	1.2		RIPL81.839
PYR	−4.5		STRA862653	PYR	−1.4		RIPL81.839
PYR	−5.5		STRA862653	PYR	0.3		RIPL81.839
PYR	−4.5		STRA862653	PYR	−0.5		RIPL81.839
PYR	−3.9		STRA862653	PYR	−0.3		RIPL81.839
PYR	−0.2		STRA862653	PYR	0.3		RIPL81.839
PYR	−0.4		STRA862653	PYR	1.4		RIPL81.839
PYR	−0.1		STRA862653	PYR	3.7		RIPL81.839
PYR	−2.8		STRA862653	PYR	0.9		RIPL81.839
PYR	−0.7		STRA862653	PYR	−1.4		RIPL81.839
PYR	−0.9		STRA862653	PYR	−1.3		RIPL81.839
PYR	−1.9		STRA862653	PYR	1.7		RIPL81.839
PYR	−2.3		STRA862653	PYR	−0.5		RIPL81.839
PYR	−0.1		STRA862653	PYR	1.8		RIPL81.839
PYR	−0.1		STRA862653	PYR	−1.4		RIPL81.839
				PYR	2.7		RIPL81.839
"Lake Shore, Ontario" (2600 Ma)				PYR	1.8		RIPL81.839
BAR		12.0	HATT86..45	PYR	2.7		RIPL81.839
BAR		13.0	HATT86..45	PYR	1.7		RIPL81.839
BAR		14.0	HATT86..45	PYR	5.1		RIPL81.839
BAR		14.0	HATT86..45	PYR	2.4		RIPL81.839
				PYR	3.0		RIPL81.839
"Macassa, Ontario" (2600 Ma)				PYR	2.6		RIPL81.839
BAR		14.0	HATT86..45	PYR	2.9		RIPL81.839
				PYR	1.7		RIPL81.839
"McIntyre, Ontario" (2600 Ma)				PYR	1.1		RIPL81.839
ANH		8.0	HATT86..45	PYR	11.1		RIPL81.839
ANH		10.0	HATT86..45	PYR	6.0		RIPL81.839
				PYR	4.4		RIPL81.839
"Ross, Ontario" (2600 Ma)				PYR	8.0		RIPL81.839
ANH		8.0	HATT86..45	PYR	4.3		RIPL81.839
ANH		9.0	HATT86..45	PYR	5.3		RIPL81.839
ANH		10.0	HATT86..45	PYR	4.6		RIPL81.839
				PYR	5.2		RIPL81.839
Bulawayan/Shamvaian Group (2600 Ma)				PYR	1.5		RIPL81.839
PO	−2.1		FRIP79.205	PYR	−2.4		RIPL81.839
PYR	0.4		FRIP79.205				
PYR	0.4		FRIP79.205	Heron Bay Group (2700 Ma)			
PYR	3.3		FRIP79.205	BAR		2.0	HATT86..45
PO	3.2		FRIP79.205	BAR		8.0	HATT86..45
				BAR		10.0	HATT86..45
Deer Lake Complex (2600 Ma)				ANH		9.0	HATT86..45
PYR	1.5		RIPL81.839				
PYR	1.3		RIPL81.839	Hospital Hill Formation (2700 Ma)			
PYR	1.1		RIPL81.839	SFIDE	3.0		CAME82.145
PYR	2.7		RIPL81.839	SFIDE	2.0		CAME82.145
PYR	3.3		RIPL81.839				
PYR	3.6		RIPL81.839	Jacobina (2700 Ma)			
PYR	1.7		RIPL81.839	PYR	0.8		HATT83.323
PYR	4.7		RIPL81.839	PYR	−0.1		HATT83.323
PYR	2.9		RIPL81.839				
PYR	2.6		RIPL81.839	Montgomery Formation (2700 Ma)			
PYR	3.1		RIPL81.839	PYR	−1.3		HATT83.323
PYR	2.9		RIPL81.839	PYR	−1.7		HATT83.323

Table 17.7. Continued.

Mineral	del Sfide (‰ vs CDT)	del Sfate (‰ vs CDT)	Reference Code	Mineral	del Sfide (‰ vs CDT)	del Sfate (‰ vs CDT)	Reference Code
PYR	−2.0		HATT83.323	PYR	3.2		HOEF68.975
PYR	−1.5		HATT83.323	PYR	4.0		HOEF68.975
Yilgarn Block (2700 Ma)				PYR	3.9		HATT83.323
				PYR	4.0		HATT83.323
PYR	−1.4		DONN77.409	PYR	2.6		HATT83.323
PO	−1.9		DONN77.409	PYR	−0.5		HATT83.323
PYR	2.0		DONN77.409	PYR	4.0		HATT83.323
PO	0.2		DONN77.409	PYR	3.5		HATT83.323
PYR	0.2		DONN77.409	PYR	1.8		HATT83.323
PO	3.6		DONN77.409	PYR	2.4		HATT83.323
PYR	0.7		DONN77.409	PYR	1.7		HATT83.323
PYR	0.8		DONN77.409	PYR	1.9		HATT83.323
CHALCOPYR	1.3		DONN77.409	Jeppestown Shale (2720 Ma)			
PYR	−0.2		DONN77.409				
PO	−5.4		DONN77.409	SFIDE	1.4		CAME82.145
PYR	0.9		DONN77.409	SFIDE	1.5		CAME82.145
PYR	−2.0		DONN77.409	SFIDE	3.2		CAME82.145
PYR	0.1		DONN77.409	SFIDE	4.5		CAME82.145
PYR	−2.5		DONN77.409	SFIDE	2.5		CAME82.145
PYR	−7.6		DONN77.409	SFIDE	2.9		CAME82.145
PYR	−0.1		DONN77.409	SFIDE	2.5		CAME82.145
PO	3.1		DONN77.409				
SFIDE	5.2		DONN77.409	Late Archean BIF, Ontario (2750 Ma)			
PO	2.5		DONN77.409				
PO	0.9		DONN77.409	PO	−3.2		THOD83.337
PO	3.6		DONN77.409	PYR	−3.1		THOD83.337
PO	2.8		DONN77.409	PYR	−4.9		THOD83.337
PYR	2.3		DONN77.409	PO	−4.5		THOD83.337
PYR	2.7		DONN77.409	PYR	−3.9		THOD83.337
PO	2.3		DONN77.409	PO	−11.6		THOD83.337
PO	1.9		DONN77.409	PYR	−5.3		THOD83.337
PO	4.6		DONN77.409	PO	−5.4		THOD83.337
PO	2.4		DONN77.409	PYR	−4.3		THOD83.337
Government Formation (2710 Ma)				PO	−3.9		THOD83.337
				PYR	−4.6		THOD83.337
SFIDE	4.0		CAME82.145	PYR	5.0		THOD83.337
SFIDE	3.1		CAME82.145	PO	−4.2		THOD83.337
SFIDE	3.0		CAME82.145	PYR	−3.4		THOD83.337
SFIDE	3.5		CAME82.145	PYR	4.3		THOD83.337
SFIDE	3.0		CAME82.145	PO	4.2		THOD83.337
SFIDE	2.0		CAME82.145	PYR	5.8		THOD83.337
SFIDE	4.0		CAME82.145	PYR	−0.1		THOD83.337
				PO	−2.1		THOD83.337
Witwatersrand Supergroup (2710 Ma)				PYR	−2.8		THOD83.337
PYR	1.6		HOEF68.975	PO	2.7		THOD83.337
PYR	1.9		HOEF68.975	PYR	4.0		THOD83.337
PYR	1.9		HOEF68.975	PO	6.8		THOD83.337
PYR	2.0		HOEF68.975	PYR	7.8		THOD83.337
PYR	2.6		HOEF68.975	PYR	10.0		THOD83.337
PYR	2.6		HOEF68.975	PO	11.4		THOD83.337
PYR	2.7		HOEF68.975	PYR	13.2		THOD83.337
PYR	2.9		HOEF68.975	PYR	15.2		THOD83.337
PYR	3.3		HOEF68.975	PYR	19.0		THOD83.337
PYR	3.6		HOEF68.975	PO	−3.1		THOD83.337
PYR	1.0		HOEF68.975	PYR	−3.4		THOD83.337
PYR	1.8		HOEF68.975	PO	1.7		THOD83.337
PYR	1.9		HOEF68.975	PYR	2.7		THOD83.337
PYR	3.0		HOEF68.975	PYR	13.5		THOD83.337
PYR	3.1		HOEF68.975	PO	3.0		THOD83.337

Table 17.7. Continued.

Mineral	del Sfide (‰ vs CDT)	del Sfate (‰ vs CDT)	Reference Code	Mineral	del Sfide (‰ vs CDT)	del Sfate (‰ vs CDT)	Reference Code
PO	1.0		THOD83.337	SFIDE	3.4		GOOD76.870
PO	9.7		THOD83.337	SFIDE	−1.7		GOOD76.870
PO	14.4		THOD83.337	SFIDE	−10.5		GOOD76.870
PO	1.4		THOD83.337	SFIDE	−0.9		GOOD76.870
PO	−2.1		THOD83.337	SFIDE	−1.0		GOOD76.870
SFIDE	2.1		THOD83.337	SFIDE	−3.0		GOOD76.870
SFIDE	1.8		THOD83.337	SFIDE	2.7		GOOD76.870
SFIDE	4.8		THOD83.337	SFIDE	−0.5		GOOD76.870
SFIDE	9.1		THOD83.337	SFIDE	1.6		GOOD76.870
SFIDE	2.9		THOD83.337	SFIDE	2.4		GOOD76.870
SFIDE	13.6		THOD83.337	SFIDE	−4.5		GOOD76.870
SFIDE	0.8		THOD83.337	SFIDE	2.9		GOOD76.870
SFIDE	−1.8		THOD83.337	SFIDE	−1.1		GOOD76.870
SFIDE	−1.7		THOD83.337	SFIDE	5.4		GOOD76.870
SFIDE	−4.2		THOD83.337	SFIDE	3.2		GOOD76.870
SFIDE	1.5		THOD83.337	SFIDE	−2.3		GOOD76.870
SFIDE	−3.2		THOD83.337	SFIDE	−7.8		GOOD76.870
SFIDE	5.2		THOD83.337	SFIDE	3.8		GOOD76.870
SFIDE	1.1		THOD83.337	SFIDE	−2.8		GOOD76.870
SFIDE	1.4		THOD83.337	SFIDE	−4.9		GOOD76.870
SFIDE	1.8		THOD83.337	SFIDE	−2.9		GOOD76.870
SFIDE	−0.3		THOD83.337	SFIDE	−1.6		GOOD76.870
SFIDE	1.0		THOD83.337	SFIDE	−7.3		GOOD76.870
SFIDE	0.3		THOD83.337	SFIDE	1.0		GOOD76.870
SFIDE	0.4		THOD83.337	SFIDE	8.0		GOOD76.870
SFIDE	0.9		THOD83.337	SFIDE	1.6		GOOD76.870
SFIDE	0.8		THOD83.337	SFIDE	1.3		GOOD76.870
SFIDE	0.8		THOD83.337	SFIDE	2.0		GOOD76.870
SFIDE	1.9		THOD83.337	SFIDE	1.3		GOOD76.870
SFIDE	2.0		THOD83.337	SFIDE	8.3		GOOD76.870
SFIDE	1.3		THOD83.337	SFIDE	1.2		GOOD76.870
SFIDE	0.2		THOD83.337	SFIDE	1.6		GOOD76.870
SFIDE	0.1		THOD83.337	SFIDE	6.2		GOOD76.870
SFIDE	−0.2		THOD83.337	SFIDE	1.6		GOOD76.870
SFIDE	−0.4		THOD83.337	SFIDE	−1.0		GOOD76.870
SFIDE	−0.6		THOD83.337	SFIDE	−1.8		GOOD76.870
Michipicoten Iron Formation (2750 Ma)				SFIDE	3.1		GOOD76.870
				SFIDE	3.4		GOOD76.870
SFIDE	2.3		GOOD76.870	SFIDE	10.1		GOOD76.870
SFIDE	−3.8		GODD76.870	SFIDE	7.9		GOOD76.870
SFIDE	−4.4		GOOD76.870	SFIDE	6.8		GOOD76.870
SFIDE	−1.8		GOOD76.870	Uchi Greenstone Belt "Series" (2750 Ma)			
SFIDE	1.3		GOOD76.870				
SFIDE	2.0		GOOD76.870	PO	−1.6		SECC77.117
SFIDE	−2.4		GOOD76.870	PO	−0.1		SECC77.117
SFIDE	−7.5		GOOD76.870	PO	0.7		SECC77.117
SFIDE	−1.0		GOOD76.870	PO	3.3		SECC77.117
SFIDE	5.0		GOOD76.870	PYR	0.9		SECC77.117
SFIDE	−7.9		GOOD76.870	PYR	1.2		SECC77.117
SFIDE	−6.2		GOOD76.870	PYR	2.9		SECC77.117
SFIDE	−5.3		GOOD76.870	PYR	4.8		SECC77.117
SFIDE	−1.0		GOOD76.870	PO	−3.8		SECC77.117
SFIDE	−3.2		GOOD76.870	PO	−1.0		SECC77.117
SFIDE	−2.3		GOOD76.870	PO	0.4		SECC77.117
SFIDE	−6.3		GOOD76.870	PO	1.9		SECC77.117
SFIDE	6.3		GOOD76.870	PO	2.3		SECC77.117
SFIDE	−0.9		GOOD76.870	PO	3.0		SECC77.117
SFIDE	−0.9		GOOD76.870	PYR	−3.1		SECC77.117

Table 17.7. *Continued.*

Mineral	del Sfide (‰ vs CDT)	del Sfate (‰ vs CDT)	Reference Code	Mineral	del Sfide (‰ vs CDT)	del Sfate (‰ vs CDT)	Reference Code
PYR	−0.1		SECC77.117	PO	−3.1		SECC77.117
PYR	0.5		SECC77.117	PO	−3.1		SECC77.117
PYR	2.0		SECC77.117	PO	−1.9		SECC77.117
PYR	2.0		SECC77.117	PO	−1.3		SECC77.117
PO	−0.7		SECC77.117	PO	−0.8		SECC77.117
PO	−0.2		SECC77.117	PO	0.1		SECC77.117
PO	2.2		SECC77.117	PO	0.3		SECC77.117
PO	2.9		SECC77.117	PO	0.5		SECC77.117
PYR	−1.9		SECC77.117	PO	3.9		SECC77.117
PYR	2.0		SECC77.117	PYR	−9.9		SECC77.117
PO	−3.9		SECC77.117	PYR	−3.9		SECC77.117
PO	−3.5		SECC77.117	PYR	−0.1		SECC77.117
PO	−3.3		SECC77.117	PYR	0.9		SECC77.117
PO	−3.1		SECC77.117	PYR	1.1		SECC77.117
PO	−2.9		SECC77.117	PYR	3.7		SECC77.117
PO	−2.2		SECC77.117				
PO	−2.2		SECC77.117	Woman River Iron Formation (2750 Ma)			
PO	−2.1		SECC77.117	SFIDE	−1.7		GOOD76.870
PO	−1.1		SECC77.117	SFIDE	−2.6		GOOD76.870
PO	1.4		SECC77.117	SFIDE	−1.4		GOOD76.870
PO	2.3		SECC77.117	SFIDE	−3.1		GOOD76.870
PO	2.5		SECC77.117	SFIDE	−1.2		GOOD76.870
PYR	1.9		SECC77.117	SFIDE	−1.3		GOOD76.870
PO	−3.9		SECC77.117	SFIDE	−3.3		GOOD76.870
PO	−0.1		SECC77.117	SFIDE	−1.5		GOOD76.870
PO	0.1		SECC77.117	SFIDE	−6.9		GOOD76.870
PYR	−11.8		SECC77.117	SFIDE	−1.1		GOOD76.870
PYR	−9.8		SECC77.117	SFIDE	−2.8		GOOD76.870
PYR	−6.5		SECC77.117	SFIDE	3.3		GOOD76.870
PYR	0.1		SECC77.117	SFIDE	−2.2		GOOD76.870
PO	−8.9		SECC77.117	SFIDE	−1.8		GOOD76.870
PO	−8.7		SECC77.117	SFIDE	−3.6		GOOD76.870
PO	−7.0		SECC77.117	SFIDE	−3.9		GOOD76.870
PO	−6.5		SECC77.117	SFIDE	−4.9		GOOD76.870
PO	−5.9		SECC77.117	SFIDE	−3.9		GOOD76.870
PO	−5.1		SECC77.117	SFIDE	−5.3		GOOD76.870
PO	−4.9		SECC77.117	SFIDE	−4.3		GOOD76.870
PO	−3.8		SECC77.117	SFIDE	−0.9		GOOD76.870
PO	−3.8		SECC77.117	SFIDE	8.2		GOOD76.870
PO	−3.1		SECC77.117	SFIDE	−1.9		GOOD76.870
PO	−2.9		SECC77.117	SFIDE	−5.4		GOOD76.870
PO	−2.5		SECC77.117	SFIDE	6.5		GOOD76.870
PO	−1.9		SECC77.117	SFIDE	4.6		GOOD76.870
PO	−1.7		SECC77.117	SFIDE	4.8		GOOD76.870
PO	−0.9		SECC77.117	SFIDE	0.9		GOOD76.870
PO	−0.5		SECC77.117	SFIDE	−3.9		GOOD76.870
PO	−0.3		SECC77.117				
PYR	−8.3		SECC77.117	*3000 to 3700 Ma*			
PYR	−5.8		SECC77.117	Sebakwian Group (3300 Ma)			
PYR	−1.2		SECC77.117	SFIDE	3.2		FRIP79.205
PYR	0.1		SECC77.117	SFIDE	3.2		FRIP79.205
PO	−7.0		SECC77.117	PO	1.0		FRIP79.205
PO	−4.1		SECC77.117	SFIDE	1.8		FRIP79.205
PO	−3.8		SECC77.117	SFIDE	3.5		FRIP79.205
PO	−0.8		SECC77.117	PYR	−1.7		FRIP79.205
PYR	−6.7		SECC77.117	PYR	−3.9		FRIP79.205
PYR	−6.3		SECC77.117	PYR	0.1		FRIP79.205
PYR	0.1		SECC77.117				

Table 17.7. Continued.

Mineral	del Sfide (‰ vs CDT)	del Sfate (‰ vs CDT)	Reference Code	Mineral	del Sfide (‰ vs CDT)	del Sfate (‰ vs CDT)	Reference Code
PO	−0.9		FRIP79.205	SPH	−0.1		LAMB78.808
PYR	0.8		FRIP79.205	BAR		3.0	LAMB78.808
PYR	1.3		FRIP79.205	BAR		3.0	LAMB78.808
PYR	2.5		FRIP79.205	BAR		3.3	LAMB78.808
PYR	2.6		FRIP79.205	BAR		3.7	LAMB78.808
PYR	−2.1		FRIP79.205	BAR		3.7	LAMB78.808
PYR	−1.1		FRIP79.205	BAR		3.9	LAMB78.808
PYR	−3.2		FRIP79.205	BAR		3.9	LAMB78.808
PYR	2.7		FRIP79.205	BAR		3.9	LAMB78.808
PYR	−0.7		FRIP79.205	BAR		3.9	LAMB78.808
PYR	−0.9		FRIP79.205	BAR		4.1	LAMB78.808
PO	1.7		FRIP79.205	BAR		4.1	LAMB78.808
PYR	2.2		FRIP79.205	BAR		4.3	LAMB78.808
PYR	2.9		FRIP79.205	BAR		6.0	LAMB78.808
PO	−1.2		FRIP79.205				
PO	2.2		FRIP79.205	Onverwacht Group (3450 Ma)			
PO	0.1		FRIP79.205	SFIDE	−2.0		CAME82.145
Fig Tree Group (3350 Ma)				SFIDE	−0.5		CAME82.145
PYR	1.1		PERR711015	SFIDE	−2.0		CAME82.145
PYR	−0.1		PERR711015	SFIDE	−2.5		CAME82.145
PYR	1.6		PERR711015	SFIDE	1.0		CAME82.145
GAL	−0.9		AULT59.201	SFIDE	1.5		CAME82.145
SFIDE	−0.4		CAME82.145	BAR		3.3	LAMB78.808
SFIDE	2.4		CAME82.145	BAR		3.3	LAMB78.808
SFIDE	0.5		CAME82.145	BAR		3.3	LAMB78.808
SFIDE	4.0		CAME82.145	BAR		3.3	LAMB78.808
BAR		3.5	PERR711015	BAR		3.3	LAMB78.808
BAR		3.1	PERR711015	BAR		3.3	LAMB78.808
BAR		3.1	PERR711015	BAR		2.7	LAMB78.808
BAR		3.8	PERR711015	BAR		2.7	LAMB78.808
BAR		3.5	PERR711015	BAR		3.0	LAMB78.808
BAR		3.4	PERR711015	BAR		3.0	LAMB78.808
BAR		3.7	PERR711015	BAR		3.2	LAMB78.808
				BAR		3.2	LAMB78.808
Warrawoona Group (3400 Ma)				BAR		3.2	LAMB78.808
PYR	−3.1		LAMB78.808	BAR		3.5	LAMB78.808
PYR	−1.8		LAMB78.808	BAR		3.5	LAMB78.808
PYR	−1.5		LAMB78.808	BAR		3.5	LAMB78.808
PYR	−1.1		LAMB78.808	BAR		3.9	LAMB78.808
PYR	−1.1		LAMB78.808				
PYR	−1.1		LAMB78.808	Isua Supergroup (3700 Ma)			
PYR	−0.9		LAMB78.808	PYR	0.6		MONS79.405
PYR	−0.9		LAMB78.808	SFIDE	1.2		MONS79.405
PYR	−0.7		LAMB78.808	SFIDE	−0.4		MONS79.405
PYR	−0.7		LAMB78.808	SFIDE	2.0		MONS79.405
PYR	1.5		LAMB78.808	SFIDE	−0.4		MONS79.405
PYR	1.7		LAMB78.808	SFIDE	1.3		MONS79.405
PYR	2.8		LAMB78.808	SFIDE	1.8		MONS79.405
SPH	−4.4		LAMB78.808	SFIDE	−1.0		MONS79.405
SPH	−2.8		LAMB78.808	CHALCOPYR	0.6		MONS79.405
SPH	−2.8		LAMB78.808	CHALCOPYR	−0.3		MONS79.405
SPH	−2.6		LAMB78.808	CHALCOPYR	0.4		MONS79.405
SPH	−0.2		LAMB78.808	CHALCOPYR	0.1		MONS79.405

Sulfides and Sulfates

Table 17.8. *Summary of all results for isotopic composition of sulfides and sulfates.*

Mineral	del Sfide (‰ vs CDT)	del Sfate (‰ vs CDT)	N	Reference Code	Mineral	del Sfide (‰ vs CDT)	del Sfate (‰ vs CDT)	N	Reference Code
Late Proterozoic					*Middle Proterozoic*				
Chapel Island Formation (560 Ma)					Jupiter Member (930 Ma)				
PYR	−18.0		1	PPRG	PYR	16.7		2	PPRG
Almodovar del Rio Group (600 Ma)					Bijaigarh Formation (950 Ma)				
PYR	−2.1		1	PPRG	PYR	9.3		36	GUHA71.326
Cijara Formation (610 Ma)					Tanner Member (950 Ma)				
PYR	8.3		2	PPRG	PYR	15.7		1	PPRG
Bakoye Group (620 Ma)					Katanga System (1000 Ma)				
BAR		27.6	1	PPRG	SFIDE	15.1		10	DECH65.894
Sheepbed Formation (650 Ma)					SFIDE	5.1		9	DECH65.894
PYR	2.8		4	PPRG	SFIDE	2.5		8	DECH65.894
Ballachulish Slate (700 Ma)					SFIDE	−7.4		1	DECH65.894
PYR	15.1		13	HALL87.305	SFIDE	2.0		9	DECH65.894
PO	14.3		7	HALL87.305	GYP		15.6	1	DECH65.894
Luoquan Formation (700 Ma)					ANH		16.7	9	DECH65.894
PYR	23.0		1	PPRG	N. Rhodesian Copperbelt (1000 Ma)				
Shaler Group (700 Ma)					SFIDE	−3.6		22	DECH65.894
GYP		15.9	1	CLAY80.199	SFIDE	6.5		3	DECH65.894
					SFIDE	1.1		10	DECH65.894
Twitya Formation (700 Ma)					SFIDE	−2.3		14	DECH65.894
PYR	30.4		4	PPRG	SFIDE	6.7		9	DECH65.894
Coppercap Formation (750 Ma)					SFIDE	3.4		18	DECH65.894
SFIDE	6.3		1	PPRG	SFIDE	9.0		6	DECH65.894
Tapley Hill Formation (750 Ma)					SFIDE	5.2		13	DECH65.894
SFIDE	20.1		57	KNUT83.250	SFIDE	0.9		8	DECH65.894
SFIDE	20.4		77	KNUT83.250	SFIDE	8.2		6	DECH65.894
PYR	20.3		17	LAMB80...1	SFIDE	12.0		17	DECH65.894
PO	26.8		18	LAMB80...1	SFIDE	−13.0		11	DECH65.894
CHALCOPYR	15.5		6	LAMB80...1	SFIDE	1.5		26	DECH65.894
PYR	21.1		21	LAMB84.266	SFIDE	0.9		52	DECH65.894
CHALCOPYR	22.7		11	LAMB84.266	SFIDE	5.9		8	DECH65.894
SFIDE	20.7		11	LAMB84.266	SFIDE	7.6		8	DECH65.894
Redstone River Formation (770 Ma)					SFIDE	14.3		9	DECH65.894
ANH		20.2	3	PPRG	SFIDE	−5.6		9	DECH65.894
					SFIDE	−4.9		10	DECH65.894
Little Dal Group (800 Ma)					SFIDE	3.8		8	DECH65.894
PYR	−28.0		1	PPRG	SFIDE	3.8		11	DECH65.894
Bitter Springs Formation (850 Ma)					SFIDE	10.3		8	DECH65.894
ANH		19.5	2	CLAY80.199	SFIDE	3.9		7	DECH65.894
GYP		17.7	11	CLAY80.199	SFIDE	5.3		7	DECH65.894
ANH		18.2	9	SOLO71.259	SFIDE	9.6		8	DECH65.894
					SFIDE	5.6		2	DECH65.894
					SFIDE	9.3		7	DECH65.894
					SFIDE	12.4		2	DECH65.894
					SFIDE	−0.4		5	DECH65.894
Walcott Member (850 Ma)					SFIDE	12.8		6	DECH65.894
PYR	11.7		3	PPRG	SFIDE	7.0		6	DECH65.894
					SFIDE	3.9		11	DECH65.894
					Grenville Metasediments (1050 Ma)				
Awatubi Member (862 Ma)					PYR	14.3		20	BROW73.362
					SPH	14.4		20	BROW73.362
					PYR	−9.6		27	BUDD69.423
PYR	−18.8		2	PPRG	ANH		21.2	13	BROW73.362

Table 17.8. Continued.

Mineral	del Sfide (‰ vs CDT)	del Sfate (‰ vs CDT)	N	Reference Code	Mineral	del Sfide (‰ vs CDT)	del Sfate (‰ vs CDT)	N	Reference Code
Nonesuch Shale (1055 Ma)					Broken Hill Deposit (1650 Ma)				
PYR	4.5		86	BURN72.895	SFIDE	1.3		68	SPRY87.109
Eurelia Beds (1100 Ma)					Mallapunyah Formation (1650 Ma)				
PYR	−4.3		20	LAMB84.461	PYR	−2.6		3	JACK88unpu
CHALCOPYR	−0.9		9	LAMB84.461	PO	−34.8		1	JACK88unpu
Gap Well Formation (1100 Ma)					Willyama Complex (1650 Ma)				
PYR	21.2		1	VOGT87.805	GAL	0.4		71	BOTH75.308
GAL	20.2		1	VOGT87.805	SPH	0.5		36	BOTH75.308
BAR		39.6	2	VOGT87.805	PO	1.3		9	BOTH75.308
Roan Group (1155 Ma)					CHALCOPYR	−1.7		2	BOTH75.308
ANH		18.6	8	CLAY80.199	PYR	−0.1		2	BOTH75.308
Spokane Formation (1250 Ma)					Mt. Isa Group (1670 Ma)				
SFIDE	−9.5		21	LANG871334	SFIDE	10.7		14	STAN66..16
SFIDE	−0.7		22	RYE.83.104	Urquhart Shale (1670 Ma)				
BAR		17.7	11	RYE.83.104	PO	12.9		23	SMIT78.369
Belt Supergroup (1350 Ma)					GAL	12.5		13	SMIT78.369
SFIDE	1.4		9	MORT74.223	SPH	16.1		13	SMIT78.369
CHALCOPYR	7.2		5	MORT74.223	CHALCOPYR	13.5		19	SMIT78.369
SPH	11.6		1	MORT74.223	PYR	16.3		31	SMIT78.369
Greyson Shale (1420 Ma)					PYR	17.2		17	SOLO65.737
PYR	5.0		1	PPRG	PO	17.1		9	SOLO65.737
Newland Limestone (1440 Ma)					SPH	17.7		7	SOLO65.737
PYR	19.3		7	PPRG	GAL	9.3		6	SOLO65.737
BAR		11.8	1	PPRG	CHALCOPYR	15.3		3	SOLO65.737
PYR	−2.0		14	STRAunpu	PYR	28.5		3	PPRG
BAR		15.4	3	STRAunpu	Wollogorang Formation (1670 Ma)				
Prichard Formation (1450 Ma)					PYR	1.5		3	JACK88unpu
PYR	14.3		1	PPRG	McArthur Group (1690 Ma)				
Balbirini Dolomite (1482 Ma)					PYR	13.8		21	WALK83.214
CHALCOPYR	2.8		1	JACK88unpu	SPH	8.7		16	WALK83.214
Barney Creek Formation (1500 Ma)					GAL	3.0		9	WALK83.214
PYR	11.3		18	SMIT75.269	PYR	7.6		3	RYE.81...1
PYR	16.9		4	JACK88unpu	SPH	5.2		9	RYE.81...1
PO	32.4		1	JACK88unpu	GAL	2.0		11	RYE.81...1
Early Proterozoic					GAL	1.8		13	SMIT73..10
Amelia Dolomite (1600 Ma)					SPH	5.7		13	SMIT73..10
PYR	6.6		3	JACK88unpu	PYR	6.3		16	SMIT73..10
Broken Hill (1600 Ma)					PYR	13.3		18	MUIR85.239
SFIDE	0.9		63	STAN66..16	CHALCOPYR	5.3		3	MUIR85.239
Mount Isa Inlier (1600 Ma)					GAL	9.6		10	MUIR85.239
GAL	12.4		7	CARR77.105	SPH	13.5		4	MUIR85.239
SPH	15.8		8	CARR77.105	BAR		19.4	13	MUIR85.239
PYR	11.1		15	CARR77.105	Amisk Group (1800 Ma)				
BAR		38.8	3	CARR77.105	PYR	4.6		49	STRA862653
					Chelmsford Formation (1850 Ma)				
					CHALCOPYR	7.6		1	THOD62.565
					PYR	4.5		2	THOD62.565

Table 17.8. Continued.

Mineral	del Sfide (‰ vs CDT)	del Sfate (‰ vs CDT)	N	Reference Code	Mineral	del Sfide (‰ vs CDT)	del Sfate (‰ vs CDT)	N	Reference Code
Onwatin Formation (1850 Ma)					Strubenkoop Formation (2200 Ma)				
PO	24.1		1	THOD62.565	SFIDE	8.6		4	CAME82.145
PYR	15.9		6	THOD62.565	Bruce Formation (2250 Ma)				
Outokumpu Ore Deposit (1850 Ma)					PYR	0.8		1	HATT83.549
PYR	−3.1		212	MAKE74…1	Espanola Formation (2250 Ma)				
Skellefte Field (1850 Ma)					PYR	0.5		3	HATT83.549
PYR	−1.9		17	RICK791060	Gowganda Formation (2250 Ma)				
PYR	−3.1		34	GAVE60.510	PYR	2.4		1	CAME80.181
BAR		9.7	5	GAVE60.510	PYR	0.7		14	HATT83.549
BAR		10.4	4	RICK791060	Lorrain Formation (2250 Ma)				
Virginia Formation (1850 Ma)					PYR	−5.7		11	HATT83.549
SFIDE	7.1		6	RIPL87..87	Mississagi Formation (2250 Ma)				
Rove Formation (1900 Ma)					PYR	1.2		7	HATT83.549
PYR	16.5		1	CAME80.181	PYR	1.4		3	ROSC68…1
Ruth Shale (1900 Ma)					Pecors Formation (2250 Ma)				
PYR	5.4		7	CAME831069	PYR	0.5		2	HATT83.549
Animikie Group (2000 Ma)					Ramsay Lake Formation (2250 Ma)				
PYR	3.5		22	CAME831069	PYR	1.3		2	HATT83.549
PYR	−0.2		7	BUDD69.423	Serpent Formation (2250 Ma)				
Kona Dolomite (2000 Ma)					PYR	0.1		3	HATT83.549
ANH		13.3	1	HEMZ82.512	Timeball Hill Formation (2250 Ma)				
Menihek Formation (2000 Ma)					SFIDE	−18.5		14	CAME82.145
PYR	5.7		17	CAME831069	Gordon Lake Formation (2259 Ma)				
PYR	3.7		12	CAME831069	PYR	−3.0		3	HATT83.549
PYR	11.0		1	CAME80.181	ANH		13.7	5	CAME83..54
Temiscamie Iron Formation (2000 Ma)					Penge Iron Formation (2300 Ma)				
PYR	15.8		1	CAME80.181	PYR	−4.9		10	CAME831069
Udokan Series (2000 Ma)					SFIDE	−5.5		10	CAME82.145
CHALCOPYR	−21.8		2	CHUK711429	Aravalli Supergroup (2350 Ma)				
SFIDE	−10.2		2	CHUK711429	PYR	3.1		13	DEB.86.313
Gunflint Iron Formation (2090 Ma)					PO	−0.3		9	DEB.86.313
PYR	6.6		19	CAME831069	Huronian Supergroup (2350 Ma)				
PYR	6.2		1	CAME80.181	PYR	1.9		10	ROSC68…1
PYR	6.6		2	PPRG	PYR	2.7		1	ROSC68…1
Albanel Formation (2100 Ma)					SFIDE	5.6		7	ROSC68…1
PYR	−0.3		1	CAME80.181	Frood Formation (2400 Ma)				
Attikamagen Formation (2100 Ma)					PYR	−0.1		1	THOD62.565
PYR	13.6		1	CAME80.181	PO	0.1		1	THOD62.565
Daspoort Formation (2200 Ma)					Malmani Formation (2400 Ma)				
SFIDE	8.5		1	CAME82.145	SFIDE	−2.3		25	CAME82.145
Silverton Formation (2200 Ma)									
SFIDE	−9.7		6	CAME82.145					

Table 17.8. Continued.

Mineral	del Sfide (‰ vs CDT)	del Sfate (‰ vs CDT)	N	Reference Code	Mineral	del Sfide (‰ vs CDT)	del Sfate (‰ vs CDT)	N	Reference Code
Matinenda Formation (2400 Ma)					Kameeldoorns Formation (2705 Ma)				
PYR	0.2		50	HATT83.323	PYR	−0.9		1	PPRG
PYR	1.0		8	ROSC68...1	Carbon Leader Member (2710 Ma)				
PYR	2.9		4	ROSC68...1	PYR	3.3		1	PPRG
McKim Formation (2400 Ma)					Government Formation (2710 Ma)				
SFIDE	−0.4		2	THOD62.565	SFIDE	3.7		7	CAME82.145
PYR	1.7		6	HATT83.549	Witwatersrand Supergroup (2710 Ma)				
Snider Formation (2400 Ma)					PYR	2.5		17	HOEF68.975
SFIDE	−0.1		1	THOD62.565	PYR	2.3		11	HATT83.323
Black Reef Formation (2450 Ma)					Jeppestown Shale (2720 Ma)				
SFIDE	4.2		2	CAME82.145	SFIDE	2.3		8	CAME82.145
Gamohaan Formation (2450 Ma)					Kuruna Siltstone (2740 Ma)				
PYR	−2.5		5	PPRG	PYR	0.3		1	PPRG
Archean					Late Archean BIF, Ontario (2750 Ma)				
Dales Gorge Member (2500 Ma)					PO	0.8		21	THOD83.337
PYR	0.7		7	CAME831069	PYR	2.8		23	THOD83.337
PYR	0.9		6	PPRG	SFIDE	1.4		31	THOD83.337
Keewatin Group (2500 Ma)					Michipicoten Iron Formation (2750 Ma)				
PYR	−0.3		43	SHEG78unpu	PYR			61	GOOD76.870
Keewatin, Abitibi Belt (2500 Ma)					Uchi Greenstone Belt Series (2750 Ma)				
PYR	2.0		30	STRA862653	PO	−1.7		60	SECC77.117
Keewatin, Wabigoon Belt (2500 Ma)					PYR	−1.8		29	SECC77.117
PYR	1.3		64	STRA862653	Woman River Iron Formation (2750 Ma)				
Bulawayan/Shamvaian Group (2600 Ma)					PYR	−1.4		29	GOOD76.870
PO	0.6		2	FRIP79.205	Sebakwian Group (3300 Ma)				
PYR	1.4		3	FRIP79.205	SFIDE	2.9		4	FRIP79.205
Deer Lake Complex (2600 Ma)					PO	0.4		7	FRIP79.205
PYR	2.3		52	RIPL81.839	PYR	0.1		15	FRIP79.205
Jeerinah Formation (2650 Ma)					Fig Tree Group (3350 Ma)				
PYR	7.9		2	PPRG	PYR	0.9		3	PERR711015
Venterspost Carbon Reef Member (2670 Ma)					PYR	−0.9		1	AULT59.201
					SFIDE	1.6		4	CAME82.145
PYR	1.1		1	PPRG	BAR		3.4	7	PERR711015
Hospital Hill Formation (2700 Ma)					Warrawoona Group (3400 Ma)				
SFIDE	2.5		2	CAME82.145	PYR	−0.5		13	LAMB78.808
Jacobina (2700 Ma)					SPH	−2.1		6	LAMB78.808
PYR	0.5		2	HATT83.323	BAR		3.9	13	LAMB78.808
Montgomery Formation (2700 Ma)					Towers Formation (3435 Ma)				
PYR	−1.6		4	HATT83.323	BAR		5.4	2	PPRG
Yilgarn Block (2700 Ma)					Onverwacht Group (3450 Ma)				
PYR	−0.3		8	DONN77.409	SFIDE	−0.7		6	CAME82.145
PO	1.6		7	DONN77.409	BAR		3.3	17	LAMB78.808
					Isua Supergroup (3700 Ma)				
					SFIDE	0.5		13	MONS79.405

Kerogen Samples

Table 17.9. *Previously published analyses of elemental and carbon isotopic composition of kerogen samples.*

Reference code	Lithology	C (%)	H (%)	N (%)	H/C	N/C	del C$_{org}$ (‰ vs PDB)
Late Proterozoic							
Wonoka Formation (585 Ma)							
MCKI76.163	CARB	77.9	3.10	2.70	0.47	0.000	
Pertatataka Formation (610 Ma)							
MCKI80.187	SH	78.0	0.00	0.00	0.66	0.000	
Brachina Formation (650 Ma)							
MCKI74.591	SH	0.0	0.00	0.00	0.15	0.000	−27.9
MCKI74.591	SH	0.0	0.00	0.00	0.11	0.000	−18.8
MCKI74.591	SH	0.0	0.00	0.00	0.06	0.000	−14.7
Umberatana Group (675 Ma)							
MCKI76.163		81.2	3.40	0.90	0.49	0.000	
Tapley Hill Formation (750 Ma)							
MCKI75.345	SH	0.0	0.00	0.00	0.17	0.000	
MCKI75.345	SH	87.1	0.00	0.00	0.25	0.000	−24.2
MCKI75.345	SH	90.7	0.00	0.00	0.11	0.000	−18.8
MCKI75.345	SH	91.2	0.00	0.00	0.15	0.000	−22.9
MCKI75.345	SH	90.9	0.00	0.00	0.15	0.000	
MCKI75.345	SH	90.8	0.00	0.00	0.18	0.000	−23.4
MCKI75.345	SH	85.5	0.00	0.00	0.19	0.000	
MCKI75.345	SH	89.8	0.00	0.00	0.19	0.000	
MCKI75.345	SH	89.1	0.00	0.00	0.23	0.000	
MCKI75.345	SH	0.0	0.00	0.00	0.11	0.000	
MCKI75.345	SH	89.3	0.00	0.00	0.19	0.000	
MCKI75.345	SH	90.9	0.00	0.00	0.13	0.000	−23.8
MCKI75.345	SH	94.7	0.00	0.00	0.11	0.000	
MCKI75.345	SH	96.4	0.00	0.00	0.09	0.000	
MCKI75.345	SH	96.8	0.00	0.00	0.06	0.000	−14.7
MCKI75.345	SH	96.7	0.00	0.00	0.07	0.000	
MCKI75.345	SH	95.2	0.00	0.00	0.09	0.000	−17.5
MCKI75.345	SH	95.3	0.00	0.00	0.05	0.000	
MCKI75.345	SH	97.6	0.00	0.00	0.01	0.000	
MCKI75.345	SH	98.3	0.00	0.00	0.06	0.000	
MCKI75.345	SH	98.1	0.00	0.00	0.05	0.000	
MCKI75.345	SH	97.5	0.00	0.00	0.04	0.000	
MCKI75.345	SH	94.7	0.00	0.00	0.09	0.000	
MCKI75.345	SH	0.0	0.00	0.00	0.02	0.000	
MCKI76.163	CARB	84.0	1.70	0.90	0.24	0.000	
Skillogalee Dolomite (770 Ma)							
MCKI80.187	CH	92.2	0.00	0.00	0.19	0.000	
MCKI74.591	CARB	0.0	0.00	0.00	0.10	0.000	−10.8
DUNG72.699	CH	50.7	0.86	0.37	0.20	0.000	
MCKI76.163	CARB	90.9	0.80	0.50	0.10	0.000	
MCKI76.163	CARB	75.3	2.10	1.00	0.33	0.000	
MCKI76.163	CH	85.2	2.20	0.70	0.31	0.000	
MCKI76.163	CARB	82.2	2.00	0.70	0.27	0.000	
MCKI76.163	CARB	88.7	1.70	1.30	0.23	0.000	
Burra Group (800 Ma)							
HAYE83..93	CH	64.0	0.93	0.73	0.17	0.010	−23.2
Black Mudstone, Utah (830 Ma)							
LEVE754706	SH	0.0	0.00	0.00	0.10	0.000	−21.9

Table 17.9. Continued.

Reference code	Lithology	C (%)	H (%)	N (%)	H/C	N/C	del C$_{org}$ (‰ vs PDB)
Bitter Springs Formation (850 Ma)							
HAYE83..93	CH	97.0	2.73	2.85	0.34	0.025	
MCKI80.187	CARB	0.0	0.00	0.00	0.28	0.000	
MCKI80.187	CARB	0.0	0.00	0.00	0.46	0.000	
LEVE754706	CH	0.0	0.00	0.00	0.10	0.000	
MCKI74.591	CARB	0.0	0.00	0.00	0.73	0.000	−18.6
MCKI76.163	CARB	72.7	5.00	1.50	0.82	0.000	
Middle Proterozoic							
Nonesuch Shale (1055 Ma)							
LEVE754706	SH	0.0	0.00	0.00	0.50	0.000	
McMinn Formation (1340 Ma)							
HAYE83..93	SH	67.0	4.42	1.63	0.79	0.021	−32.4
HAYE83..93	SH	57.0	2.30	1.38	0.48	0.021	−32.0
HAYE83..93	SH	85.0	2.87	2.41	0.40	0.025	−32.1
MCKI80.187	SH	73.6	0.00	0.00	1.17	0.000	
MCKI80.187	SH	81.2	0.00	0.00	0.86	0.000	
MCKI80.187	SH	81.9	0.00	0.00	0.65	0.000	
MCKI80.197	SH	81.4	0.00	0.00	0.59	0.000	
MCKI80.187	SH	83.4	0.00	0.00	0.54	0.000	
MCKI80.187	SH	86.6	0.00	0.00	0.36	0.000	
MCKI74.591	SH	0.0	0.00	0.00	1.17	0.000	−30.2
Lansen Creek Shale Member (1350 Ma)							
POWE87...1	SH	55.1	5.70	0.25	1.24	0.000	
POWE87...1	SH	66.4	5.52	0.69	1.00	0.000	
POWE87...1	SH	80.8	5.99	1.11	0.89	0.000	
POWE87...1	SH	69.4	3.47	0.97	0.60	0.000	
Velkerri Formation (1350 Ma)							
POWE87...1	SH	86.6	7.26	0.74	1.01	0.000	
POWE87...1	SH	89.0	6.52	0.79	0.88	0.000	
POWE87...1	SH	87.8	6.28	0.97	0.86	0.000	
POWE87...1	SH	87.4	5.74	1.12	0.79	0.000	
POWE87...1	SH	89.1	5.65	1.12	0.76	0.000	
POWE87...1	SH	89.7	5.26	1.38	0.70	0.000	
POWE87...1	SH	88.6	4.81	0.87	0.65	0.000	
Bungle Bungle Dolomite (1364 Ma)							
HAYE83..93	CH	65.0	1.69	1.81	0.31	0.024	−30.0
HAYE83..93	SH	66.0	2.57	1.17	0.45	0.015	−33.5
HAYE83..93	CARB	85.0	2.73	0.96	0.38	0.010	−30.2
HAYE83..93	CARB	76.0	2.46	1.95	0.39	0.022	−30.7
HAYE83..93	CARB	89.0	3.19	1.77	0.43	0.017	−29.3
Balbirini Formation (1482 Ma)							
POWE87...1	CARB	87.0	7.19	0.51	0.99	0.000	
Yalco Formation (1485 Ma)							
POWE87...1	CARB	81.0	6.24	0.66	0.92	0.000	
POWE87...1	CARB	77.9	7.73	0.39	1.19	0.000	
POWE87...1	CARB	80.7	6.13	0.80	0.91	0.000	
POWE87...1	CARB	83.4	7.81	0.74	1.12	0.000	
POWE87...1	CARB	81.0	5.98	0.76	0.89	0.000	
POWE87...1	CARB	80.8	6.27	0.88	0.93	0.000	

Kerogen Samples

Table 17.9. Continued.

Reference code	Lithology	C (%)	H (%)	N (%)	H/C	N/C	del C$_{org}$ (‰ vs PDB)
Lynott-Don Formation (1490 Ma)							
POWE87...1	SH	78.6	6.17	1.11	0.94	0.000	
Barney Creek Formation (1500 Ma)							
POWE87...1	SH	73.6	10.19	4.17	1.66	0.000	
POWE87...1	SH	78.1	10.51	3.61	1.61	0.000	
POWE87...1	SH	78.9	9.78	4.71	1.49	0.000	
POWE87...1	SH	77.8	10.01	3.24	1.54	0.000	
POWE87...1	SH	73.6	9.28	2.21	1.51	0.000	
POWE87...1	SH	83.8	8.62	3.74	1.23	0.000	
POWE87...1	SH	78.3	7.48	3.05	1.15	0.000	
POWE87...1	SH	83.8	6.49	3.55	0.93	0.000	
POWE87...1	SH	77.0	7.12	3.14	1.11	0.000	
POWE87...1	SH	71.7	5.81	3.51	0.97	0.000	
POWE87...1	SH	87.7	6.68	3.89	0.91	0.000	
POWE87...1	SH	75.3	6.27	2.24	1.00	0.000	
POWE87...1	SH	82.2	5.99	1.29	0.87	0.000	
POWE87...1	SH	79.3	3.92	0.56	0.59	0.000	
POWE87...1	SH	52.7	3.41	0.88	0.78	0.000	
POWE87...1	SH	75.1	5.44	1.74	0.87	0.000	
POWE87...1	SH	83.7	4.34	0.95	0.62	0.000	
POWE87...1	SH	72.4	2.28	0.75	0.38	0.000	
POWE87...1	SH	81.3	2.57	2.67	0.38	0.000	
HYC Pyritic Shale Member (1500 Ma)							
HAYE83..93	SH	34.0	1.12	0.50	0.39	0.013	−38.9
MCKI74.591	SH	0.0	0.00	0.00	0.64	0.000	−26.6
Marimo Slate (1500 Ma)							
MCKI74.591	SH	0.0	0.00	0.00	0.05	0.000	−25.8
Early Proterozoic							
Tooganinie Formation (1600 Ma)							
MCKI76.163	CARB	77.2	4.20	1.80	1.80	0.65	0.000
Paradise Creek Formation (1650 Ma)							
LEVE754706	CH	0.0	0.00	0.00	0.20	0.000	
Urquhart Shale (1670 Ma)							
LEVE754706	SH	0.0	0.00	0.00	0.10	0.000	−22.2
MCKI74.591	SH	0.0	0.00	0.00	0.13	0.000	−28.2
Earaheedy Group (1800 Ma)							
HAYE83..93	CARB	97.0	2.18	0.54	0.27	0.005	−33.2
Golden Dyke Formation (1900 Ma)							
MCKI74.591	SH	0.0	0.00	0.00	0.04	0.000	−15.8
Rove Formation (1900 Ma)							
HAYE83..93	SH	61.0	2.10	0.83	0.41	0.012	−32.1
Duck Creek Dolomite (2000 Ma)							
HAYE83..93	CARB	76.0	0.92	1.02	0.14	0.011	−25.0
HAYE83..93	CH	40.0	1.08	0.23	0.32	0.005	−31.7
Gunflint Iron Formation (2090 Ma)							
HAYE83..93	CH	42.0	2.13	1.03	0.61	0.022	−33.5
LEVE754706	CH	0.0	0.00	0.00	0.10	0.000	

Table 17.9. Continued.

Reference code	Lithology	C (%)	H (%)	N (%)	H/C	N/C	del C$_{org}$ (‰ vs PDB)
McLeary Formation (2100 Ma)							
HAYE83..93	CH	80.0	1.71	1.19	0.26	0.013	−28.4
Upper Albanel Formation (2100 Ma)							
HAYE83..93	CARB	71.0	0.81	0.37	0.14	0.004	−29.8
Malmani Subgroup (2400 Ma)							
HAYE83..93	CARB	69.0	0.81	0.23	0.14	0.003	−35.0
HAYE83..93	CARB	52.0	0.86	0.40	0.21	0.007	−32.5
Archean							
Mt. McRae Shale (2500 Ma)							
HAYE83..93	SH	99.0	0.94	0.90	0.11	0.008	
HAYE83..93	SH	53.0	0.51	0.44	0.11	0.007	−35.3
Wittenoom Dolomite (2500 Ma)							
HAYE83..93	SH	52.0	0.59	0.33	0.14	0.005	−41.8
Bulawayan Group (2600 Ma)							
HAYE83..93	CARB	82.0	0.67	0.76	0.10	0.008	−31.2
Manjeri Formation (2600 Ma)							
HAYE83..93	CH	66.0	0.82	0.35	0.15	0.005	−33.0
Steeprock Group (2600 Ma)							
HAYE83..93	CH	76.0	1.09	0.77	0.17	0.009	−27.2
Jeerinah Formation (2650 Ma)							
HAYE83..93	SH	79.0	0.93	0.33	0.14	0.004	−41.0
Ventersdorp Supergroup (2650 Ma)							
HAYE83..93	CH	51.0	0.61	0.11	0.14	0.002	−38.5
Witwatersrand Supergroup (2710 Ma)							
ZUMB78.223	Q	0.0	0.00	0.00	1.25	0.000	
ZUMB78.223	Q	0.0	0.00	0.00	1.05	0.000	
ZUMB78.223	Q	0.0	0.00	0.00	1.08	0.000	
ZUMB78.223	Q	0.0	0.00	0.00	1.11	0.000	
ZUMB78.223	Q	0.0	0.00	0.00	0.83	0.000	
ZUMB78.223	Q	0.0	0.00	0.00	0.73	0.000	
ZUMB78.223	Q	0.0	0.00	0.00	0.67	0.000	
ZUMB78.223	Q	0.0	0.00	0.00	0.71	0.000	
ZUMB78.223	Q	0.0	0.00	0.00	0.71	0.000	
ZUMB78.223	Q	0.0	0.00	0.00	0.69	0.000	
ZUMB78.223	Q	0.0	0.00	0.00	0.67	0.000	
ZUMB78.223	Q	0.0	0.00	0.00	0.67	0.000	
ZUMB78.223	Q	0.0	0.00	0.00	0.69	0.000	
ZUMB78.223	Q	0.0	0.00	0.00	0.63	0.000	
ZUMB78.223	Q	0.0	0.00	0.00	0.59	0.000	
ZUMB78.223	Q	0.0	0.00	0.00	0.58	0.000	
ZUMB78.223	Q	0.0	0.00	0.00	0.57	0.000	
ZUMB78.223	Q	0.0	0.00	0.00	0.59	0.000	
ZUMB78.223	Q	0.0	0.00	0.00	0.54	0.000	
ZUMB78.223	Q	0.0	0.00	0.00	0.53	0.000	
ZUMB78.223	Q	0.0	0.00	0.00	0.53	0.000	
ZUMB78.223	Q	0.0	0.00	0.00	0.49	0.000	
ZUMB78.223	Q	0.0	0.00	0.00	0.48	0.000	

Kerogen Samples

Table 17.9. *Continued.*

Reference code	Lithology	C (%)	H (%)	N (%)	H/C	N/C	del C$_{org}$ (‰ vs PDB)
Tumbiana Formation (2750 Ma)							
HAYE83..93	CH	64.0	1.34	0.14	0.25	0.002	−51.2
Fortescue Group (2800 Ma)							
MCKI76.163	SH	73.8	2.70	0.60	0.43	0.000	
Mozaan Group (3000 Ma)							
HAYE83..93	SH	94.0	0.83	0.28	0.11	0.003	−27.3
Gorge Creek Group (3200 Ma)							
HAYE83..93	SH	74.0	0.53	0.00	0.09	0.000	−29.9
Kromberg Formation (3300 Ma)							
HAYE83..93	CH	82.0	0.88	0.30	0.13	0.003	−29.9
Fig Tree Group (3350 Ma)							
HAYE83..93	CH	80.0	1.07	0.61	0.16	0.006	−31.9
LEVE754706	CH	0.0	0.00	0.00	0.20	0.000	
LEVE754706	CH	0.0	0.00	0.00	0.10	0.000	−26.4
DUNG72.699	CH	69.4	0.90	0.40	0.16	0.000	
DUNG74.167	CH	0.0	0.00	0.00	0.41	0.000	−28.0
DUNG74.167	CH	0.0	0.00	0.00	0.49	0.000	−28.7
Towers Formation (3435 Ma)							
HAYE83..93	CH	64.0	1.58	0.35	0.30	0.005	−34.3
HAYE83..93	CH	77.0	1.06	0.60	0.16	0.007	−36.1
HAYE83..93	CH	50.0	1.27	0.14	0.30	0.002	−35.2
Hooggenoeg Formation (3450 Ma)							
HAYE83..93	CH	78.0	0.76	0.34	0.12	0.004	−31.6
MCKI74.591	CH	0.0	0.00	0.00	0.41	0.000	−26.0
MCKI74.591	CH	0.0	0.00	0.00	0.69	0.000	−27.9
MCKI74.591	CH	0.0	0.00	0.00	0.43	0.000	−26.3
MCKI74.591	CH	0.0	0.00	0.00	0.58	0.000	
Onverwacht Group (3450 Ma)							
DUNG72.699	CH	65.5	0.81	0.51	0.15	0.000	
DUNG74.167	CH	0.0	0.00	0.00	0.33	0.000	−31.4
DUNG74.167	CH	0.0	0.00	0.00	0.25	0.000	
DUNG74.167	CH	0.0	0.00	0.00	0.65	0.000	−26.2
DUNG74.167	CH	0.0	0.00	0.00	0.93	0.000	−26.2
DUNG74.167	CH	0.0	0.00	0.00	0.40	0.000	
DUNG74.167	CH	0.0	0.00	0.00	0.54	0.000	
DUNG74.167	CH	0.0	0.00	0.00	1.38	0.000	−33.0
DUNG74.167	CH	0.0	0.00	0.00	0.14	0.000	−32.9
DUNG74.167	CH	0.0	0.00	0.00	0.80	0.000	−28.4
DUNG74.167	CH	0.0	0.00	0.00	0.08	0.000	−15.8
DUNG74.167	CH	0.0	0.00	0.00	0.96	0.000	−14.3
Swartkoppie Formation (3450 Ma)							
HAYE83..93	CH	64.0	0.52	0.41	0.10	0.005	−32.0
HAYE83..93	CH	79.0	0.56	0.60	0.09	0.006	−26.6
LEVE754706	CH	0.0	0.00	0.00	0.30	0.000	−27.6
Theespruit Formation (3550 Ma)							
HAYE83..93	SH	98.0	0.16	0.06	0.02	0.001	−15.4
MCKI74.591	CH	0.0	0.00	0.00	0.09	0.000	
MCKI74.591	CH	0.0	0.00	0.00	0.04	0.000	−17.3

Table 17.9. Continued.

Reference code	Lithology	C (%)	H (%)	N (%)	H/C	N/C	del C_{org} (‰ vs PDB)
MCKI74.591	CH	0.0	0.00	0.00	0.04	0.000	−22.4
MCKI74.591	CH	0.0	0.00	0.00	0.03	0.000	−19.1
MCKI74.591	CH	0.0	0.00	0.00	0.05	0.000	−20.4
MCKI74.591	CH	0.0	0.00	0.00	0.05	0.000	−17.4
MCKI74.591	CH	0.0	0.00	0.00	0.04	0.000	−18.2
Isua Supergroup (3700 Ma)							
HAYE83..93	BIF	95.0	0.13	0.13	0.02	0.001	−10.0
HAYE83..93	H	57.0	0.06	0.06	0.01	0.001	−13.2

Table 17.10. *Summary of elemental and carbon isotopic composition of kerogen samples.*

Lithology	H/C	N/C	del C_{org} (‰ vs PDB)	N	Reference code
Late Proterozoic					
Wonoka Formation (585 Ma)					
CARB	0.47	0.000		1	MCKI76.163
Pertatataka Formation (610 Ma)					
SH	0.66	0.000		1	MCKI80.187
Khatyspyt Formation (620 Ma)					
SH	1.30	0.020	−36.2	1	PPRG
Boonall Dolomite (640 Ma)					
CARB	0.23	0.084	−30.4	1	PPRG
Brachina Formation (650 Ma)					
SH	0.11	0.000	−20.5	3	MCKI74.591
Sheepbed Formation (650 Ma)					
SH	0.23	0.006	−29.5	5	PPRG
Umberatana Group (675 Ma)					
CARB	0.49	0.000		1	MCKI76.163
Biri Formation (700 Ma)					
CARB	0.57	0.020	−31.6	4	PPRG
Luoquan Formation (700 Ma)					
SILT	0.31	0.011	−29.5	3	PPRG
Twitya Formation (700 Ma)					
SH	0.44	0.014	−29.3	4	PPRG
Tapley Hill Formation (750 Ma)					
CARB	0.24	0.000		1	MCKI76.163
SH	0.12	0.000	−20.8	24	MCKI75.345
Tindelpina Shale Member (750 Ma)					
SILT	0.18	0.008	−23.9	1	PPRG
Woocalla Dolomite Member (750 Ma)					
SILT	0.60	0.023	−29.3	1	PPRG
Skillogalee Dolomite (770 Ma)					
CARB	0.10	0.000	−10.8	1	MCKI74.591
CH	0.20	0.000		1	DUNG72.699
CH	0.31	0.000		1	MCKI76.163
CARB	0.23	0.000		4	MCKI80.187
River Wakefield Subgroup (775 Ma)					
CH	0.58	0.016	−24.1	2	PPRG
Burra Group (800 Ma)					
CH	0.17	0.010	−23.2	1	HAYE83..93
Black Mudstone, Utah (830 Ma)					
SH	0.10	0.000	−21.9	1	LEVE754706
Beck Spring Dolomite (850 Ma)					
CARB	0.14	0.016	−26.5	1	PPRG
Bitter Springs Formation (850 Ma)					
CARB	0.82	0.000		1	MCKI76.163
CH	0.10	0.000		1	LEVE754706
CH	0.34	0.025		1	HAYE83..93
CARB	0.73	0.000	−18.6	1	MCKI74.591
CARB	0.37	0.000		2	MCKI80.187
Kingston Peak Formation (850 Ma)					
CARB	0.27	0.011	−26.0	1	PPRG
Love's Creek Member (850 Ma)					
CARB	0.35	0.030	−26.1	1	PPRG
CH	0.39	0.032	−29.7	1	PPRG
Walcott Member (850 Ma)					
CH	0.79	0.025	−27.2	1	PPRG
SH	0.84	0.029	−26.9	3	PPRG
CARB	0.78	0.025	−26.5	2	PPRG
Awatubi Member (862 Ma)					
SH	0.70	0.047	−29.8	2	PPRG

Kerogen Samples

Table 17.10. *Continued.*

Lithology	H/C	N/C	del C$_{org}$ (‰ vs PDB)	N	Reference code
Middle Proterozoic					
Carbon Canyon Member (900 Ma)					
SILT	0.64	0.024	−26.3	1	PPRG
SILT	0.74	0.026	−28.4	1	PPRG
Jupiter Member (930 Ma)					
SH	0.57	0.027	−24.0	4	PPRG
Red Pine Shale (950 Ma)					
SH	0.65	0.025	−17.4	2	PPRG
Tanner Member (950 Ma)					
SH	0.68	0.020	−26.5	5	PPRG
Allamoore Formation (1050 Ma)					
CH	0.31	0.015	−32.4	1	PPRG
Nonesuch Shale (1055 Ma)					
SILT	0.53	0.020	−33.5	1	PPRG
SH	0.50	0.000		1	LEVE754706
Hongshuizhuang Formation (1250 Ma)					
SH	0.67	0.020	−33.2	2	PPRG
Wumishan Formation (1325 Ma)					
CH	0.54	0.031	−30.1	1	PPRG
McMinn Formation (1340 Ma)					
SH	0.56	0.022	−32.2	3	HAYE83..93
SH	1.17	0.000	−30.2	1	MCKI74.591
SH	0.70	0.000		6	MCKI80.187
SILT	0.96	0.020	−32.6	5	PPRG
Lansen Creek Shale Member (1350 Ma)					
SH	0.93	0.000		4	POWE87...1
Velkerri Formation (1350 Ma)					
SH	0.81	0.000		7	POWE87...1
Yangzhuang Formation (1350 Ma)					
CH	0.49	0.026	−28.4	1	PPRG
Bungle Bungle Dolomite (1364 Ma)					
CARB	0.40	0.016	−30.0	3	HAYE83..93
CARB	0.30	0.041	−23.9	2	PPRG
CH	0.31	0.024	−30.0	1	HAYE83..93
SH	0.45	0.015	−33.5	1	HAYE83..93
Dismal Lakes Group (1400 Ma)					
CH	0.67	0.015	−28.7	1	PPRG
Greyson Shale (1420 Ma)					
SILT	0.37	0.010	−28.3	1	PPRG
Gaoyuzhuang Formation (1425 Ma)					
CH	0.45	0.014	−31.2	2	PPRG
Newland Limestone (1440 Ma)					
SH	0.34	0.012	−29.7	2	PPRG
Balbirini Formation (1482 Ma)					
CARB	0.99	0.000		1	POWE87...1
Yalco Formation (1485 Ma)					
CARB	0.99	0.000		6	POWE87...1
Lynott-Donnegan Member (1490 Ma)					
SH	0.94	0.000		1	POWE87...1
Barney Creek Formation (1500 Ma)					
SH	1.03	0.000		19	POWE87...1
HYC Pyritic Shale Member (1500 Ma)					
SH	0.39	0.013	−38.9	1	HAYE83..93
SH	0.64	0.000	−26.6	1	MCKI74.591
Marimo Slate (1500 Ma)					
SH	0.05	0.000	−25.8	1	MCKI74.591
Early Proterozoic					
Tooganinie Formation (1600 Ma)					
CARB	0.65	0.000		1	MCKI76.163
Paradise Creek Formation (1650 Ma)					
CH	0.20	0.000		1	LEVE754706
Urquhart Shale (1670 Ma)					
SH	0.13	0.000	−28.2	1	MCKI74.591
SH	0.10	0.000	−22.2	1	LEVE754706
SH	0.11	0.004	−28.3	1	PPRG
Tuanshanzi Formation (1750 Ma)					
CARB	0.57	0.016	−31.4	1	PPRG
Earaheedy Group (1800 Ma)					
CARB	0.27	0.005	−33.2	1	HAYE83..93
Chuanlinggou Formation (1850 Ma)					
SH	0.41	0.011	−32.3	3	PPRG
Koolpin Formation (1885 Ma)					
CH	0.17	0.007	−30.4	1	PPRG
Fontano Formation (1900 Ma)					
SH	0.14	0.004	−30.1	1	PPRG
Golden Dyke Formation (1900 Ma)					
SH	0.04	0.000	−15.8	1	MCKI74.591
Rove Formation (1900 Ma)					
SH	0.47	0.016	−31.8	3	PPRG
SH	0.41	0.012	−32.1	1	HAYE83..93
Union Island Group (1900 Ma)					
SH	0.23	0.012	−38.9	1	PPRG
Rocknest Formation (1925 Ma)					
CARB	0.20	0.012	−21.7	5	PPRG

Table 17.10. Continued.

Lithology	H/C	N/C	del C_{org} (‰ vs PDB)	N	Reference code
Tyler Formation (1950 Ma)					
SH	0.29	0.009	−31.8	2	PPRG
Duck Creek Dolomite (2000 Ma)					
CARB	0.14	0.011	−25.0	1	HAYE83..93
CH	0.32	0.005	−31.7	1	HAYE83..93
Gunflint Iron Formation (2090 Ma)					
CH	0.41	0.013	−32.8	4	PPRG
CARB	0.44	0.017	−31.7	1	PPRG
CH	0.10	0.000		1	LEVE754706
CH	0.61	0.022	−33.5	1	HAYE83..93
Albanel Formation (2100 Ma)					
CARB	0.14	0.004	−29.8	1	HAYE83..93
McLeary Formation (2100 Ma)					
CH	0.26	0.013	−28.4	1	HAYE83..93
Reivilo Formation (2300 Ma)					
SH	0.26	0.006	−35.9	3	PPRG
Klipfonteinheuwel Formation (2325 Ma)					
CH	0.13	0.010	−37.9	1	PPRG
Malmani Subgroup (2400 Ma)					
CARB	0.18	0.005	−33.8	2	HAYE83..93
Gamonaan Formation (2450 Ma)					
SH	0.18	0.010	−34.9	1	PPRG
CARB	0.26	0.009	−33.7	5	PPRG
Archean					
Mt. McRae Shale (2500 Ma)					
SH	0.11	0.008	−35.3	2	HAYE83..93
Wittenoom Dolomite (2500 Ma)					
SH	0.14	0.005	−41.8	1	HAYE83..93
Bulawayan Group (2600 Ma)					
CARB	0.10	0.008	−31.2	1	HAYE83..93
Manjeri Formation (2600 Ma)					
CH	0.15	0.005	−33.0	1	HAYE83..93
Steeprock Group (2600 Ma)					
CH	0.17	0.009	−27.2	1	HAYE83..93
Jeerinah Formation (2650 Ma)					
SH	0.42	0.008	−43.3	2	PPRG
Jeerinah Formation (2650 Ma)					
SH	0.14	0.004	−41.0	1	HAYE83..93
Ventersdorp Supergroup (2650 Ma)					
CH	0.14	0.002	−38.5	1	HAYE83..93
Witwatersrand Supergroup (2710 Ma)					
Q	0.71	0.000		23	ZUMB78.223
Meentheena Carbonate Member (2750 Ma)					
CH	0.22	0.008	−51.8	1	PPRG
Tumbiana Formation (2750 Ma)					
CH	0.25	0.002	−51.2	1	HAYE83..93
SH	0.47	0.004	−58.7	1	PPRG
CARB	0.30	0.011	−53.4	2	PPRG
Fortescue Group (2800 Ma)					
SH	0.43	0.000		1	MCKI76.163
Mozaan Group (3000 Ma)					
SH	0.11	0.003	−27.3	1	HAYE83..93
Gorge Creek Group (3200 Ma)					
SH	0.09	0.000	−29.9	1	HAYE83..93
Kromberg Formation (3300 Ma)					
CH	0.14	0.005	−29.7	3	PPRG
CH	0.13	0.003	−29.9	1	HAYE83..93
Fig Tree Group (3350 Ma)					
CH	0.16	0.000		1	DUNG72.699
CH	0.45	0.000	−28.4	2	DUNG74.167
CH	0.16	0.006	−31.9	1	HAYE83..93
CH	0.19	0.015	−28.6	2	PPRG
CH	0.15	0.000	−26.4	2	LEVE754706
Towers Formation (3435 Ma)					
CH	0.25	0.005	−35.2	3	HAYE83..93
Hooggenoeg Formation (3450 Ma)					
CH	0.12	0.004	−31.6	1	HAYE83..93
CH	0.53	0.000	−26.7	3	MCKI74.591
CH	0.20	0.003	−36.7	1	PPRG
Onverwacht Group (3450 Ma)					
CH	0.15	0.000		1	DUNG72.699
CH	0.59	0.000	−26.0	11	DUNG74.167
Swartkoppie Formation (3450 Ma)					
CH	0.10	0.006	−29.3	2	HAYE83..93
CH	0.30	0.000	−27.6	1	LEVE754706
Theespruit Formation (3550 Ma)					
CH	0.06	0.002	−15.1	1	PPRG
SH	0.02	0.001	−15.4	1	HAYE83..93
CH	0.05	0.000	−19.1	7	MCKI74.591
Isua Supergroup (3700 Ma)					
BIF	0.02	0.001	−10.0	1	HAYE83..93
H	0.01	0.001	−13.2	1	HAYE83.93

18
Procedures for Analysis of Extractable Organic Matter
ROGER E. SUMMONS   HARALD STRAUSS

18.1
Sample Selection and Handling

As is discussed in Section 3.1.1, great care must be exercised to avoid contamination of extractable hydrocarbon samples by extraneous bitumens, particularly if the TOC (total organic carbon) content is low or if the amount of sample is small. In this study, sediment cores and outcrop samples were pre-rinsed with CH_2Cl_2 and dried to remove external contaminants. Where sample size permitted, external surfaces were removed by cutting with a diamond saw. Samples were then hammered to chips which were subsequently crushed to less than 200 mesh in a ring crusher. Rock powders were stored in clean glass containers. Lids were lined with pre-baked aluminium foil. All items used to handle the samples were scrupulously washed with hot water, and then distilled solvent, between each use.

TOC determination and Rock-Eval pyrolysis analysis were found to be particularly informative screening techniques. Samples with less than 0.2% TOC (i.e., <2mgC/g) were generally considered unsuitable for comprehensive hydrocarbon analysis because of the problems of contamination, although elemental and carbon isotopic analyses of their kerogens were parameters which could be reliably established at this low level of organic carbon. Samples with >0.2% TOC were usually assessed using Rock-Eval pyrolysis and the results interpreted using guidelines discussed by Espitalié et al. 1977. The relative proportions of bitumen (the Rock-Eval S1 peak in kg/tonne) and kerogen (S2 peak in kg/tonne), the pyrolysis temperature T_{max} (°C), and the overall appearance of the pyrograms all provided useful information. If the S2 peak was above 0.5 kg/tonne (preferably, above 1.0 kg/tonne), significant kerogen was present and a reliable indication of maturity could be established from the T_{max} value. However, care was exercised if a large S1 peak was present because this could have led to an apparent suppression of the pyrolysis temperature. Consistency of T_{max} values within the range of 420° to 470°C over a large suite of samples from a particular geologic unit was therefore considered a reliable indicator of the presence of marginally mature to mature kerogens with co-occurring bitumens, samples appropriate for hydrocarbon analysis. Data from isolated specimens or those from units having inadequate sampling density were regarded as less reliable. The occurrence of values of S1 ⩽ 30% of S2 together with values of T_{max} in the range of 440° to 460° were considered indicative of a sediment in the maturity phase for oil generation.

18.2

Extraction Procedures

18.2.1 Extraction and Chromatographic Separation of Bitumens

Distilled solvents were used in all phases of analyses. A Soxhlet apparatus was used for most extractions, although some small samples were extracted by sonication in a test tube. Prior to extraction of a sample, the Soxhlet apparatus and thimble were pre-extracted for 24 h; the rock powder was then extracted for 48 h (or until the extract was colorless), using a fresh batch of solvent (an azeotropic mixture of 87:13, $CHCl_3$:MeOH). Solvent was removed by use of a rotary evaporator and the bitumen ("EOM," the extractable organic matter) content was determined gravimetrically. A subsample (<50 mg) of each EOM sample, dissolved in <1 ml of 1:1 petroleum ether:CH_2Cl_2, was transferred to a Sep-Pak alumina N cartridge (Waters Associates) attached to a luer-lock syringe. An additional 6 ml of solvent, the eluate containing the saturated and aromatic hydrocarbons, was allowed to flow through the Sep-Pak. Asphaltenes and polar (nitrogen-, sulfur-, oxygen-containing, "NSO") compounds were then eluted with 6 ml of $CHCl_3$ followed by 4 ml of $CHCl_3$. Fraction weights were then determined, prior to the chromatographic separation of saturates and aromatics on a silica gel column. The silica gel (Merck 40, 70 to 230 mesh) was activated in a drying oven at 150°C overnight, and a bed (12 g) was prepared in a pre-rinsed column using petroleum ether as solvent. The sample (the first eluate from the Sep-Pak), dissolved in a minimum volume of petroleum ether (containing a trace of CH_2Cl_2, if necessary for complete dissolution), was added to the top of the column and then eluted with petroleum ether (40 ml) to yield the fraction containing saturated hydrocarbons. Aromatics were eluted using petroleum ether:CH_2Cl_2 (1:1, 50 ml), and polar compounds not trapped on the Sep-Pak were eluted with $CHCl_3$:MeOH (1:1, 40 ml). The various fractions were each reduced to 2 ml by rotary evaporation prior to their transfer into clean, pre-weighed sample vials for final, gentle drying under a stream of dry nitrogen (with special care being exercised at this point because exposure to excessive heat or a long drying time during final solvent removal can lead to significant loss of C_{12+} components).

Some sediment samples contain elemental sulfur which, in the above procedure, remains in the saturated hydrocarbon fraction. When the presence of sulfur was detected visually, the saturate fraction, in petroleum ether, was passed through a column of fresh elemental copper. Elemental copper was prepared using copper sulfate (45 g) in water (500 ml) acidified with HCl (20 ml, 10N) to which elemental zinc powder (15 g) was added in a water slurry. After the reaction, the solution was decanted and the copper exhaustively washed with water until neutral. Aliquots of the copper were stored in ice blocks until required. The column was pre-rinsed with acetone followed by petroleum ether, and eluted with four bed-volumes of petroleum ether. Complete removal of sulfur was achieved when the lower part of the column remained untarnished after passage of the hydrocarbon fraction.

18.2.1 Extraction and Fractionation of Signature Lipids from Extant Microbes and Modern Sediments

A reliable procedure for the isolation of intact complex lipids from the lyophilized biomass of extant microbes has been reported by Bligh and Dyer (1959) with modifications provided by Kates (1986). Total lipids may be further analyzed intact, or saponified and extracted with dilute base to give neutral lipid and fatty acid fractions. A convenient method used to fractionate a total (or neutral lipid) extract involved application (10 to 50 mg) to a preparative silica thin layer plate followed by plate development (twice) with CH_2Cl_2. Retention zones (with hydrocarbons traveling at the solvent front) for the major lipid classes (hydrocarbons, wax esters, alcohols, sterols, and polar components) were established using standards applied to a separate section at one edge of the plate and visualized by spraying with sulfuric acid (30%) and warming; the components were eluted from the silica in those zones using CH_2Cl_2 or CH_2Cl_2:MeOH.

18.3

Analyses of Extracts

18.3.1 Gas Chromatographic Analysis of Hydrocarbons and Other Lipids

Gas chromatography was carried out using Varian 3400, HP 5790, and HP 5890 instruments. The gas chromatographs were fitted with either 25 m or 50 m, 0.2 mm inner diameter, WCOT fused silica cross-linked methylsilicone capillary columns (HP Ultra-1), usually preceded by approximately 2 m of 0.3 mm inner diameter, deactivated but uncoated, fused silica as a pre-column and "retention gap." Data were collected and processed via an IBM compatible AT personal computer using DAPA acquisition software (DAPA Scientific Software P/L, Australia). Samples in hexane were injected on-column at 60°C and the oven was programmed at 3°, 4°, or 6°C/min. Hydrogen with a linear flow rate of 30 cm/sec was used as carrier gas. In the case of functionalized lipid components, such as sterols and fatty acids, pre-derivatization of the sample is preferable. Methods for these procedures have been reported (e.g., Dobson et al. 1988; Nichols and Johns 1986; Volkman et al. 1987, 1988).

18.3.2 Gas Chromatography-Mass Spectrometry

Gas chromatography–mass spectrometry (GC–MS) of saturated hydrocarbon fractions was carried out using a VG 7070E mass spectrometer equipped with an HP 5790 gas chromatograph and a VG 11-250 data system. In the full-scan mode, the mass spectrometer was scanned at a rate of 1.8 sec/decade over a mass range of 700 to 60 dalton with a 0.2 s delay. The source was operated at 240°C with 70 eV ionization potential. Some analyses employed selected ion monitoring (SIM) of 25 diagnostic ions using magnet peak-switching, or metastable reaction monitoring (MRM) of 25 diagnostic parent-daughter relationships, each complete cycle taking ca. 2.5 s. In the MRM mode (Warburton and Zumberge 1983; Summons and Powell 1987), unimolecular decompositions in the first field-free region were measured when a constant accelerating voltage (6 kV) was applied to the source; the magnet current and electrostatic analyzer were switched simultaneously under data system control with a sample time of 80 ms and dwell time of 20 ms. The gas chromatograph was usually equipped with a 50 m × 0.2 mm ID WCOT fused silica cross-linked methylsilicone capillary column (HP ULTRA-1). Samples, in hexane, were injected on-column at 60°C (S.G.E. OCI3 injector) and the oven was heated to 300°C at 3°C/min. Hydrogen with a linear flow of 30 cm/sec was used as a carrier gas. Analyses of some samples were repeated using more polar GC columns (S.G.E. BP-10 and BP-20) in order to examine the relative elution positions of the polycyclic terpenoids and to compare elution positions with those of authentic standards. Some standard oils and sediments (e.g., North Sea Kimmeridgian oil, Paris Basin Toarcian Shale, Wyoming Green River Shale) which have been extensively studied and reported in the literature were used to obtain relative retention data for standard sterane and triterpane hydrocarbons. Previously unreported compounds were assigned by mass spectral fragmentation data and, where possible, by unambiguous synthesis of standards and chromatographic comparisons (e.g., Hoffmann et al. 1987; Summons 1987; Summons and Powell 1987; Summons et al. 1987; Summons and Capon 1988).

18.4
Kerogen Analyses

18.4.1 Kerogen Hydrous Pyrolysis

A solvent-extracted subsample of kerogen concentrate (50 mg) and water (30 µl) was heated in an evacuated, sealed, Vycor tube at 330°C for 72 h. Liberated bitumen was obtained by thorough washing of the solid residue, and then fractionated and analyzed in the same way as the original rock extract. Pyrolysate yields were maturation-dependent and quite variable. Also, they did not account for gaseous and very volatile cracking products. However, in some cases (e.g., Summons et al. 1988b), hydrocarbon and cyclic biomarker composition was consistent with the distribution found in the co-occurring bitumen, bearing in mind that the kerogen being pyrolyzed had already yielded those hydrocarbons which were bound by weaker linkages. Pyrolysis and chemical degradation of kerogen concentrates are relatively under-exploited techniques and have the potential to yield valuable biomarker information when further developed.

18.4.2 Kerogen Pyrolysis-Gas Chromatography

Results of the flash pyrolysis of Proterozoic kerogens have been reported (Crick et al. 1988). In these experiments, a sample of kerogen concentrate (0.5 to 2 mg) was placed between glass wool plugs in a quartz tube. The sample was mixed with an aliquot of internal standard (1 µg of poly-α-methyl in 0.5 µl CH_2Cl_2) and pyrolyzed using a coil probe of a CDS Model 121 Pyroprobe. The probe was ballistically heated to 700°C and held for 1 min. The pyrolysis products were swept (using He carrier gas) through a modified split/splitless injector (SGE Unijector, split ratio 100:1) and onto a fused silica GC column (25 m, HP Ultra-1) at an oven temperature of −10°C. After 2 min the oven was programmed to 280°C at 4°C/min. Peak identifications were made by GC-MS and by retention time comparisons with authentic standards.

For calculation of pyrolysis yields, the following compounds was measured: total C_1 to C_5 fraction; C_7 to C_{30} n-alk-1-enes/n-alkanes; toluene; ethyl benzene; m + p + o − xylene; five C_3-benzenes; phenol; o + m + p − cresol; three C_2-phenols; thiophene; 2-methyl thiophene; and three C_2-thiophenes. Relative and absolute yields within and between compound classes are related to kerogen type, maturation, generative potential, and sulfur content.

18.5 Urea Adduction

In some experiments it is useful to remove n-alkanes prior to GS-MS analysis so that cyclic and isoprenoid biomarkers are effectively concentrated. This is usually achieved using adduction into a pre-dried 5 Å molecular seive in refluxing toluene, or by passage of the saturate fraction, in pentane solution, through a column of Silicalite. However, an alternative process of urea adduction allows the alkanes to be readily recovered for isotopic or other analyses and the method given here is an adaptation from one published earlier by Meinschein and Kenney (1957).

Urea adducts were formed by dissolving the saturated hydrocarbon fraction (10 mg in 2.5 ml of 25% v/v solution of methanol and toluene) and adding an equal volume of a saturated solution of urea in methanol. The mixture was refrigerated at 5°C for 24 h and filtered through a glass filter funnel. The crystalline product was washed twice with toluene, and the combined filtrate and washings was washed with water to remove excess urea. Evaporation of the solvent yielded the urea non-adduct fraction (containing branched and cyclic hydrocarbons). The crystalline adduct was then decomposed by addition of distilled water (2 to 5 ml); this solution was extracted twice with CH_2Cl_2 (5 ml aliquots) and the CH_2Cl_2 was evaporated to yield the n-alkanes (or n-alkanes and n-alkenes).

19
Composition of Extractable Organic Matter
ROGER E. SUMMONS   HARALD STRAUSS

19.1

Explanation of Tables Summarizing the Bitumen Content and Composition of Analyzed Samples

The data presented in this Chapter summarize the available, relevant, organic geochemical parameters for Proterozoic sediments which have, in the course of this project, been studied for their bitumen content and composition.

In Tables 19.1–19.4, "N/A" (not analyzed) indicates that an analysis was not carried out; and "NM" (not measured) indicates that a value for the parameter indicated was not determined, usually because some factor precluded its measurement. Two numbering systems are used here:

(1) Samples included in the PPRG sample collection which were subjected to initial Rock–Eval screening (Chapter 16) are identified by PPRG Sample Number only. Several of these samples, together with others that were acquired late in the project, were re-analyzed at the Bureau of Mineral Resources (BMR), Canberra, Australia, and are therefore denoted by dual PPRG and BMR Sample Numbers. Data regarding the geographic and geologic origin of PPRG samples are included in Chapter 14.

(2) Samples that were collected and analyzed specifically as part of BMR projects are denoted by BMR Sample Numbers only; information about the orgins of these samples is included in BMR publications and unpublished databases.

19.2 Discussion of the Data

In the initial screening exercise, all PPRG samples with a total organic carbon content in excess of 0.2% (i.e., with TOC $\geqslant$ 2.0 mgC/g) were analyzed by Rock–Eval pyrolysis. The vast majority of these samples either failed to show a significant S2 peak or had a T_{max} value exceeding 460°C, indicating advanced thermal alteration (see Section 18.1). The results for those PPRG samples which yielded measurable S2 peaks are presented in Table 19.1. From these results it became apparent that sediments from only two of the analyzed units, namely, (i) those from the Walcott Member of the Kwagunt Formation of the Chuar Group (Colorado Plateau Province, Arizona, USA), and (ii) those from the McMinn Formation of the Roper Group (McArthur Basin, Northern Territory, Australia), contained significant amounts of marginally mature to mature organic matter and were thus suitable for more detailed hydrocarbon analyses. Because the latter unit was included in a major project at the BMR (Jackson et al. 1986, 1988; Crick et al. 1988; Summons et al. 1988b), geochemical analyses of extracts from this unit were confined to BMR samples.

Having identified the Chuar Group as worthy of detailed

Table 19.1. *Rock–Eval data for PPRG samples*

Geologic Unit	PPRG Sample Number	TOC (mg/g)	T_{max} (°C)	S1 (Kg/tonne)	S2 (Kg/tonne)	Hydrogen Index	Maturity
Kwagunt Fm.:							
Awatubi Mbr.	1087	16.8	423	0.48	0.89	53	Mature
Walcott Mbr.	1095	11.3	447	0.37	1.81	160	Mature
Walcott Mbr.	1096	11.6	446	N/D	1.61	139	Mature
Walcott Mbr.	1123	12.0	470	0.11	1.05	88	Overmature
Walcott Mbr.	1357	8.6	447	0.22	0.79	92	Mature
Tapley Hill Fm.:							
Woocalla Dol. Mbr.	1241	0.5	485	0.09	0.19	33	Overmature
Woocalla Dol. Mbr.	1242	0.5	459	0.11	0.13	28	Overmature
McMinn Fm.	1262	5.0	401	0.01	0.20	40	Data Inconclusive
McMinn Fm.	1263	13.8	431	0.2	4.98	360	Marginally Mature
McMinn Fm.	1264	20.5	435	0.47	8.25	403	Marginally Mature
McMinn Fm.	1265	3.0	449	0.00	0.21	70	Mature
McMinn Fm.	1266	15.6	430	0.18	4.79	306	Marginally Mature
McMinn Fm.	1267	14.0	437	0.25	4.75	340	Marginally Mature
McMinn Fm.	1268	6.6	441	0.09	0.12	18	Marginally Mature
McMinn Fm.	1270	15.7	437	0.2	5.25	334	Marginally Mature
McArthur Grp.	1279	5.6	447	0.17	0.17	30	Mature
Gunflint Iron Fm.	1376	1.1	547	0.04	0.24	2	Overmature
Gunflint Iron Fm.	1377	2.5	546	0.02	0.58	2	Overmature
Biwabik Iron Fm.	2429	0.5	528	0.03	0.58	12	Overmature
Main Conglomerate:							
Carbon Leader Mbr.	1410	2.3	442	2.05	31.88	137	Mature
Carbon Leader Mbr.	2500	1.0	436	0.54	6.20	63	Mature

investigation, additional samples were obtained, including rock specimens from units other than the earlier analyzed Walcott Member of the Kwagunt Formation (Summons et al. 1988a). Detailed analyses were also carried out on three outcrop samples from the Khatyspyt Formation (Olenyok Uplift, northern Yakutia, USSR). Results of these studies are summarized in Table 19.2 in which Rock–Eval parameters are compared with bitumen extract data for specific samples. Gas chromatographic and GC-MS data for the saturated hydrocarbon fractions of these samples are presented and discussed in Section 3.3.2.

As is shown in Table 19.2, samples both from the Khatyspyt Formation and from the Chuar Group (viz., from the Walcott and Awatubi Members of the Kwagunt Formation) contain more than 10 mg/g of organic carbon and have a S2 peak that is large in comparison with S1. In comparison with samples from the Chuar Group, those from the Khatyspyt Formation have higher HI (hydrogen index) values and, generally, lower T_{max} values; these samples are marginally mature with respect to petroleum generation. Organic matter in the analyzed sediments from the Chuar Group is more thermally altered, as indicated the higher T_{max} values and the lower hydrogen indices. Organic matter in the Walcott and Awatubi Members is mature, and there is probably some suppression of T_{max} due to the high bitumen content of these sediments. Sediments underlying the Awatubi (e.g., those of the Carbon Canyon and Jupiter Members of the Galeros Formation) are likely to be overmature.

Gas chromatographic traces of extracts from samples of the Khatyspyt and Kwagunt Formations (Section 3.3.2) show surprisingly little evidence of the effects of alteration due to weathering despite the fact that all samples analyzed were obtained from surface outcrops. These GC traces exhibit high relative concentrations of acyclic isoprenoid hydrocarbons, low pristane:phytane ratios, and high sterane and triterpane contents, characteristics which are indicative of depositional conditions conducive to preservation of organic matter. Samples from the Galeros Formation of the Chuar Group contain lower amounts of organic matter and exhibit several characteristics, such as the relative absence of light alkanes and the presence of a large unresolved complex multiplet ("hump"), which seem likely to reflect outcrop weathering effects.

The data presented in Table 19.3 summarize the total organic carbon (TOC) and extractable organic matter (EOM) contents of Proterozoic samples analyzed during the course of several BMR studies into the nature and distribution of Proterozoic petroleums. Table 19.4 compares Rock–Eval parameters and extract compositions for selected individual samples that have been subject to the most extensive of these BMR analyses. Detailed discussion of the hydrocarbon distributions reflected by these data can be found in Section 3.3.2 and in the references there cited.

Acknowledgments

We are grateful to our PPRG colleagues, and particularly to M. A. Fedonkin, for providing samples of the Khatyspyt Formation, and to R. J. Horodyski and other participants in the PPRG field trip that resulted in collection of analyzed samples from the Chuar Group. With regard to other material included in Chapters 18 and 19, we also thank, for detailed discussion, support, and interpretation, C. J. Boreham, S. C. Brassell, P. J. Cook, M. J. Jackson, T. G. Powell, P. N. Southgate, and M. R. Walter; for access to samples and unpublished information, A. H. Knoll, L. M. Pratt, K. Swett, G. Vidal, and G. Weste; and for technical assistance and instrumental analyses, P. Fletcher, J. M. Hope, Z. Horvath, and Z. Roksandic.

Table 19.2. *Organic geochemical data for PPRG samples.*

Geologic Unit	PPRG Sample Number	BMR Sample Number	T max (°C)	S1 Kg/tonne	S2 Kg/tonne	TOC (mg/g)	HI	OI	EOM mg/g Org C	Extractable H/C'S SATS %	AROMS %	NSO %	Pr/Ph	Pr/nC_{17}
Khatyspyt Fm.														
Organic film	2723	4385K	445	0.50	50.10	100.8	496	24	91.8	31.50	20.10	31.50	1.72	0.08
Carbonate	2724	4496	434	0.18	1.33	4.8	277	29	22.5	29.10	9.70	52.30	1.38	0.32
Trace fossil		4495	441	0.29	7.56	13.0	581	11	87.5	22.40	4.10	70.70	0.96	0.38
Kwagunt Fm.														
Walcott Mbr.	1097	4418	N/A			19.7			110	36.60	15.90	47.60	1.01	0.78
Walcott Mbr.	1112	4419	N/A			11.8			110	53.00	9.00	37.80	1.04	0.52
Walcott Mbr.	1357	4422	N/A			8.6			100	36.70	3.30	60.00	1.37	0.25
Walcott Mbr.	1095	4216	447	0.37	1.81	11.3	160	34	108	35.50	5.70	38.80	1.11	0.63
Walcott Mbr.	1123	4217	470	0.11	1.05	12.0	88	72	42	60.60	9.60	18.30	1.16	0.64
Walcott Mbr.		4203	453	0.04	0.32	3.0	106	56	80.7	30.30	6.60	36.90	1.11	0.37
Walcott Mbr.		3239	446	0.27	1.61	11.6	139	10	105.6	43.60	10.00	30.60	1.11	0.81
Walcott Mbr.		3240	449	0.20	1.92	16.6	116	21	46.8	42.00	11.00	38.40	1.10	0.65
Awatubi Mbr.	1083	4214	N/A			1.9			38.7	30.00	3.00	60.00	0.67	0.44
Awatubi Mbr.	1087	4215	423	0.48	0.89	16.8	53	36	79	45.30	0.60	14.80	0.64	0.74
Awatubi Mbr.		4229	426	0.11	0.62	12.9	48	32	67.7	41.20	0.80	25.20	1.14	0.48
Galeros Fm.														
Carbon Canyon Mbr.	1126	4421	N/A			17.6				26.00	16.70	56.70	0.77	0.49
Jupiter Mbr.	1067	4218	436	0	0.02	2.9	6.8	3	11	30.00	0.00	70.00	0.54	0.72
Jupiter Mbr.	1125	4420	N/A			3.5				45.20	4.80	50.00	0.82	0.45

Discussion of the Data

Table 19.3. Extract data for BMR samples.

Stratigraphic Unit	Unit Type	Age (Ma)	TOC (%)	Total EOM (mg/g)	Saturates (mg/g)	Aromatics (mg/g)	Polar/Asph. (mg/g)	N
Rodda Beds	Fm.	610	0.130–0.80	0.03–0.23	0.006–0.08	0.004–0.06	0.020–0.10	9
Pertatataka	Fm.	610	0.110–0.860	0.02–0.38	0.007–0.23	0.001–0.02	0.005–0.13	11
Late Proterozoic, Svalbard		700	0.16–0.39	0.01–0.20	0.004–0.03	0.001–0.01	0.002–0.15	6
Visingsö	Grp.	775	0.14–0.26	0.08–0.16	0.020–0.04	0.010–0.02	0.030–0.10	3
Society Cliffs	Fm.	850	0.120–0.180	0.02–0.23	0.006–0.02	0.001–0.02	0.010–0.20	3
Bitter Springs	Fm.	850	0.30–0.45	0.04–1.47	0.010–0.28	0.010–0.09	0.008–1.15	19
Chuar	Grp.	850	0.30–1.66	0.18–0.76	0.070–0.39	0.020–0.09	0.090–0.30	3
Victor Bay	Fm.	1000	1.70–2.51	0.04–0.32	0.020–0.27	0.003–0.01	0.020–0.06	4
Nonesuch Shale	Fm.	1055	0.71	1.04	0.51	0.09	0.43	1
McMinn	Fm.	1340	1.06–2.96	0.09–2.29	0.030–0.69	0.040–0.52	0.010–1.41	7
Velkerri	Fm.	1350	0.67–8.68	1.93–30.0	0.57–10.38	0.290–3.52	0.93–19.02	25
Lansen Creek Shale	Mbr.	1350	0.420–4.18	0.21–4.19	0.020–1.28	0.010–0.78	0.180–2.12	9
Balbirini Dolomite	Fm.	1482	0.61	1.56	0.17	0.37	1.02	1
Yalco	Fm.	1485	0.840–5.37	0.45–3.24	0.010–0.50	0.040–1.30	0.340–1.44	6
Lynott Fm.:								
Donnegan	Mbr.	1490	1.21	0.69	0.18	0.11	0.41	1
Carabirini	Mbr.	1490	4.57	1.99	0.97	0.1	0.92	1
Reward Dolomite	Fm.	1490	3.65	1.92	0.63	0.27	1.02	1
Barney Creek	Fm.	1500	0.210–9.080	0.08–5.00	0.020–0.95	0.010–0.95	0.050–2.69	15

Table 19.4. Organic geochemical data for BMR samples.

WELL	SAMPLE Number	DEPTH (m)	Tmax (°C)	S1 (Kg/tonne)	S2 (Kg/tonne)	TOC (%)	HI	OI	EOM mg/g OrgC	Extractable H/carbons SATS %	AROMS %	NSO %	Pr/Ph	Pr/nC$_{17}$
Siberian Platform Oils														
V-Chonskaya	319	2500	N/A							56.7	25.4	13.9	0.78	0.40
Danilovskaya	320	1700	N/A							61.8	21.1	15.5	0.61	0.54
Duliominskaya	321	2600	N/A							65.7	16.4	15.7	0.98	0.52
Rodda Beds														
Marla No. 9	3501	256.30	N/A						NM	52.2	7.8	14.9	1.00	1.30
Marla No. 9	3581	202.95	439	0.23	0.86	0.69	124	0	30.70	31.3	18.0	46.7	1.05	0.34
Marla No. 9	3584	210.72	441	0.16	0.67	0.75	89	4	31.90	30.0	26.8	41.0	1.06	0.40
Marla No. 9	3587	234.65	443	0.28	0.92	0.75	123	0	22.80	41.4	15.0	41.4	1.12	0.37
Marla No. 9	3590	268.05	437	0.18	0.71	0.59	120	0	35.90	34.7	16.5	45.6	1.06	0.32
Ungoolya No. 1	3698	1208	426	0.06	0.68	0.26	262	50	67.50	13.2	4.4	80.9	0.59	0.70
Ungoolya No. 1	3561	1320	N/A			0.42			37.20	34.8	7.8	56.0	0.70	1.05
Ungoolya No. 1	3562	1323	N/A			0.30			41.50	35.4	7.5	53.2	0.58	1.00
Ungoolya No. 1	3567	1375	N/A			0.23			49.70	13.2	7.4	75.0	1.02	0.57
Ungoolya No. 1	3573	1394	N/A			0.13			NM	16.0	12.0	64.0	1.33	0.59
Ungoolya No. 1	3580	1526	N/A			0.11			68.00	21.4	14.9	56.2	0.79	0.33
Pertatataka Fm.														
Rodinga No. 4a	3780	128.3	367	0.02	0.04	0.18	22	0	N/A					
Rodinga No. 4a	3781	145.4	390	0.04	0.06	0.22	27	0	32.50	39.7	25.3	23.8	0.82	0.14
Rodinga No. 4a	3782	153.8	391	0.01	0.04	0.20	20	0	N/A					
Rodinga No. 4a	3783	169.5	343	0.03	0.03	0.19	16	0	N/A					
Mt. Charlotte No. 1	2459	1225.0	331	0.05	0.05	0.12	42	25	N/A					
Mt. Charlotte No. 1	2460	1535.0	304	0.05	0.01	0.22	5	300	N/A					
Ooraminna No. 1	3596	624.8	446	0.01	0.32	0.11	291	0	33.30	56.7	3.3	16.7	1.70	0.50
Ooraminna No. 1	3608	625.5	332	0.01	0.23	0.11	209	0	165.60	13.6	11.6	4.0	2.12	0.58
Ooraminna No. 1	3609	718.4	286	0.38	0.18	0.13	138	0	54.20	15.1	4.7	18.9	0.70	0.52
Ooraminna No. 1	3610	823.2	375	0.03	0.13	0.86	15	25	41.22	53.9	3.9	32.2	1.20	0.25

Table 19.4. Continued.

WELL	SAMPLE Number	DEPTH (m)	Tmax (°C)	S1 (Kg/tonne)	S2 (Kg/tonne)	TOC (%)	HI	OI	EOM mg/gOrgC	Extractable H/carbons SATS %	AROMS %	NSO %	Pr/Ph	Pr/nC$_{17}$
Visingsö Grp.														
Ooraminna No. 1	3368	o/crop	N/A						112.80	20.4	11.5	48.3	NM	NM
Ooraminna No. 1	4426	o/crop	435	0.13	0.65	0.52	125	N/A	116.00	26.1	11.0	38.0	1.96	0.61
Ooraminna No. 1	4427	o/crop	433	0.09	0.37	0.38	97	N/A	121.60	27.6	9.6	38.0	2.20	0.60
Ooraminna No. 1	4428	o/crop	433	0.12	0.49	0.46	106	N/A	129.60	30.8	11.1	42.9	2.18	0.63
Ooraminna No. 1	4429	o/crop	435	0.11	0.70	0.54	129	N/A	101.40	24.9	13.0	46.3	2.60	0.74
Ooraminna No. 1	4430	o/crop	440	0.18	1.12	0.67	167	N/A	118.50	14.4	11.0	29.4	2.10	0.64
Ooraminna No. 1	4431	o/crop	435	0.14	0.69	0.39	176	N/A	180.20	19.0	10.1	43.9	2.24	0.70
Ooraminna No. 1	4432	o/crop	440	0.26	1.90	0.97	195	N/A	100.10	24.0	11.6	59.5	2.53	0.89
Ooraminna No. 1	4433	o/crop	439	0.21	1.47	0.83	177	N/A	112.70	31.6	9.4	46.4	2.25	0.81
Ooraminna No. 1	4434	o/crop	437	0.23	2.24	0.96	233	N/A	92.60	24.0	11.6	44.7	2.24	0.71
Ooraminna No. 1	4435	o/crop	440	0.24	2.72	1.16	234	N/A	97.70	25.3	12.6	42.7	2.40	0.61
Ooraminna No. 1	4436	o/crop	437	0.11	0.77	0.71	108	N/A	94.70	17.6	14.3	47.5	2.65	0.94
Ooraminna No. 1	4437	o/crop	439	0.08	0.31	0.34	91	N/A	113.50	21.5	4.2	46.5	2.10	0.53
Ooraminna No. 1	4438	o/crop	437	0.10	0.39	0.45	87	N/A	128.80	26.5	4.7	39.9	1.50	0.41
Bitter Springs Fm.														
Alice Springs No. 3	3485	134.3	N/A			0.07			166.80	28.3	4.1	50.8	2.51	1.35
Alice Springs No. 3	3345	176.1	458	0.05	0.09	0.09	100	371	177.00	31.7	14.4	11.5	2.75	0.66
Ooraminna No. 1	2457	1417.0	245	0.07	0.00	0.06	0	83	N/A					
Ooraminna No. 1	2458	1690	221	0.03	0.00	0.10	0	400	N/A					
Mt. Charlotte No. 1	3339	1654	436	0.06	0.06	0.15	40	0	96.80	47.5	11.5	41.4	1.31	0.56
Mt. Charlotte No. 1	3340	1655	430	0.27	0.21	0.31	68	461	657.80	13.9	1.6	56.5	1.09	0.32
Bluebush No. 1	3595	2089	404	0.00	0.13	0.13	100	0	126.10	23.3	12.4	5.1	2.04	0.59
Mt. Winter No. 1	3592	2190	457	0.14	0.18	0.45	40	42	NM	7.2	1.2	0.0	1.98	0.52
Nonesuch Shale Fm.														
White Pine Region	3354		445	0.57	1.27	0.71	178	8	171.50	42.1	7.2	35.5	1.40	0.05
core samples	4552		437	0.23	10.62	1.96	117	4	18.60	16.1	9.3	49.1	2.32	0.51
core samples	4553		437	0.07	1.62	0.38	426	5	41.60	31.6	10.5	50.9	1.97	0.59
core samples	4554		436	0.10	0.83	0.43	193	41	47.70	34.0	9.4	54.7	1.74	0.75
core samples	4555		436	0.11	1.01	0.42	240	17	73.50	26.7	10.8	38.3	NM	NM
Nonesuch oils	306		N/A						NM	45.1	23.5	22.0	2.71	0.15
Nonesuch oils	366		N/A						NM	80.0	11.0	9.0	2.00	0.20
McMinn Fm.														
McMinn Fm. B8	3122	17.4	439	1.28	13.34	2.96	451	8	29.80	27.2	7.5	46.0	NM	NM
McMinn Fm. X28	3131	67.4	433	0.36	7.08	1.85	383	79	18.30	N/A	N/A	N/A	N/A	N/A
McMinn Fm. M13	3147	72.5	438	0.06	3.73	1.62	230	36	14.60	10.1	14.3	61.4	1.10	0.16
McMinn Fm. T19	3152	21.7	437	0.19	4.95	1.38	218	223	21.40	5.8	12.5	58.0	NM	NM
Velkerri Fm.														
Urapunga No. 4	3165	132.6	439	3.85	28.18	6.61	426	6	57.30	26.7	45.1	24.4	NM	NM
Urapunga No. 4	2467	155.6	443	4.88	23.73	5.28	449	5	78.00	39.7	44.9	13.2	NM	NM
Urapunga No. 4	2468	155.7	411	1.18	1.54	0.67	230	0	186.10	31.1	36.7	28.8	NM	NM
Urapunga No. 4	3180	216.9	445	3.26	11.25	3.88	290	35	82.20	39.3	24.4	34.1	1.00	<0.1
Urapunga No. 3	1522	39.9	441	3.03	30.20	7.20	419	2	30.50	48.9	16.6	34.3	NM	NM
Lansen Ck. Shale Mbr.														
82-1	1639	424.0	438	0.95	2.36	0.90	262	48	79.20	27.2	15.9	51.6	NM	NM
82-1	1640	427.8	439	1.73	3.15	0.91	346	121	204.90	42.4	20.8	35.3	NM	NM
Broadmere No. 1	2007	36.6	438	NM	12.18	2.98	408	17	19.90	33.8	22.1	43.8	NM	NM
Broadmere No. 1	2071	338.3	458	NM	6.29	3.44	183	8	26.40	48.2	15.4	34.1	NM	NM
Yalco Fm.														
Yalco 82-6	1169	125.7	434	1.49	31.55	5.37	588	10	33.50	13.2	34.2	37.9	1.70	0.24
Yalco 82-7	1172	186.3	440	0.76	14.28	3.99	358	0	9.90	14.5	31.1	46.8	NM	NM
Yalco 82-7	1173	187.2	437	0.45	8.77	3.15	278	108	9.70	3.1	33.0	46.5	NM	NM
Barney Ck. Fm.														
Barney Ck. GR7	2887	38.7	436	0.53	6.72	1.15	584	8	28.50	24.1	16.4	56.4	0.60	0.40
Barney Ck. GR7	2889	50.3	439	3.07	36.38	4.97	732	14	36.50	30.2	18.9	47.8	0.90	0.30
Barney Ck. GR7	2897	238.8	441	3.54	33.17	5.17	642	14	34.80	29.4	13.5	50.8	1.00	<0.1
Barney Ck. GR7	2911	601.5	445	1.08	2.90	1.25	232	43	110.30	42.8	14.0	39.8	NM	NM
Barney Ck. GR7	2919	866.0	446	1.71	4.48	2.81	154	22	49.80	43.5	22.4	40.1	NM	NM
Barney Ck. GR7	2926	49.2	433	1.03	17.16	3.11	552	30	23.20	21.1	16.3	59.6	NM	NM

Discussion of the Data

Table 19.4. *Continued.*

WELL	SAMPLE Number	DEPTH (m)	Tmax (°C)	S1 (Kg/tonne)	S2 (Kg/tonne)	TOC (%)	HI	OI	EOM mg/g Org C	Extractable H/carbons SATS %	AROMS %	NSO %	Pr/Ph	Pr/nC$_{17}$
Barney Ck. GR10	1808	80.1	435	1.83	17.82	2.48	719	33	N/A				1.00	0.30
Barney Ck. GR10	1809	88.4	438	2.81	22.12	3.00	737	48	87.00	33.4	18.8	47.3	1.00	0.30
Barney Ck. GR10	1810	143.3	440	2.75	36.85	4.68	787	39	16.30	26.4	25.0	50.0	1.00	0.30
Barney Ck. GR10	1831	360.0	439	0.12	0.28	0.31	90	548	92.70	32.0	14.0	51.9	1.00	0.00
Barney Ck. GR10	1835	481.4	446	0.75	2.62	1.04	252	131	74.80	25.9	14.6	57.2	1.00	<0.1
Barney Ck. GR10	1855	662.1	466	0.82	2.03	0.69	294	39	14.10	63.4	10.9	15.8	NM	NM
Barney Ck. GR11	1860	178.9	438	0.45	2.53	1.90	133	2	18.80	28.1	16.4	53.9	1.00	0.10
Barney Ck. GR11	1865	253.8	442	0.60	3.27	2.15	152	32	10.90	36.5	16.3	43.8	1.00	0.10
Barney Ck. GR11	1873	347.3	443	0.77	2.34	1.80	130	17	21.40	29.3	16.0	52.1	1.00	0.10

20
Modern Mat Building Microbial Communities: Methods of Investigation and Supporting Data

RICHARD W. CASTENHOLZ   ELISA D'AMELIO   JACK D. FARMER
BO BARKER JØRGENSEN   ANNA C. PALMISANO   BEVERLY K. PIERSON
DAVID M. WARD

20.1
Methods of Light Microscopy

The methods described below are those commonly used in the study of living microbial mats. Coverage here is not exhaustive; for some methods not here discussed only literature references are indicated. The material here presented (including appended tabulations of data: Tables 20.5–20.7) is relevant to the studies discussed in Chapter 6.

20.1.1 Preserved and Sectioned Material of Whole Mat (J. D. Farmer).

Syringe cores (30 cc) of mat are subsampled by slicing with a razor blade, and are fixed in a mixture of 3% glutaraldehyde and 1% formalin in pond water immediately following collection. Dehydrated samples are infiltrated with Spurr's Low Viscosity Embedding Medium (Spurr 1969; Polysciences, Inc.) under vacuum for several hours. A two-step graded series, beginning with a 50:50 mixture of resin and 100% ethanol for 2 hours, was found to enhance penetration. The hard cure schedule recommended by Polysciences is followed in order to obtain the hardest embedment for thin sections. Embedments are prepared in plastic "peel-a-way" molds (Polysciences, Inc.) and cured at 70°C for 8 hours.

Sections (20 to 30 μm thick, in order to facilitate comparisons with thin sections of fossil stromatolites) are prepared by hand grinding, following a modification of standard petrographic methods (Nye et al. 1972). Organisms in sections prepared from Spurr's resin can be stained using a 1% solution of toluidine blue in distilled water. A 1% solution of alizarine red in dilute HCl is an effective counterstain for detecting carbonate.

The principles of microscopy and applications of stains to bacteria have been reviewed elsewhere (Quesnel 1971; Norris and Swain 1971).

20.1.2 Living Material and Wet Mounts (B. K. Pierson).

One of the most informative ways to view microorganisms in mat samples is to make wet mounts. Wet mounts studied with bright field microscopy allow visualization of color which is a significant property of phototrophs. While phase contrast microscopy does not reveal color well, it permits visualization of considerable cellular detail such as sheaths, cross-walls, and sulfur inclusions. Motility (gliding, twitching, swimming) can be observed in live samples. For photography, live cells can be immobilized in agar (Pfennig and Wagener 1986). Because of the high density of many microorganisms in mats, however, organisms that may be significant ecologically but present in low densities are likely to be overlooked with light microscopy (Brock 1984).

Autofluorescence in wet mounts, dry mounts, or even permanent mounts (Crumpton 1987) can be used for identifying chlorophyll a (Chl a)- containing phototrophs in complex mat communities (Brock 1968). All Chl a-containing microorganisms fluoresce red or orange, while the bacteriochlorophyll (BChl)-containing anoxygenic phototrophs show no visible fluorescence. The infrared fluorescence of the Bchl-containing cells can be detected, however, with sensitized IR film, or the use of infrared image converters (Pierson and Howard 1972). Infrared-sensitive video cameras can also be used to detect infrared fluorescence. A usable system is a Zeiss epifluorescence microscope with a 200 W mercury lamp and a 510 dichroic mirror. The excitation beam is passed through a Calflex heat filter and Zeiss BG 38 and BG 12 filters. Objectives are 16X and 40X Zeiss Neofluar phase contrast dry objectives and a 63X phase contrast dry objective. The barrier system includes the 510 dichroic mirror and a 590 barrier filter for viewing Chl a red autofluorescence. An additional barrier filter consisting of a Wratten 88A gelatin filter (Kodak) which blocks all radiation below 720 nm is used to isolate Bchl infrared autofluorescence. The infrared fluorescence is detected with a Dage-MTI 67 M infrared-sensitive video camera and the image displayed on a high-resolution monochrome monitor (Lenco PMM-925). Using this system it is possible to readily distinguish between Chl-a and Bchl-containing microorganisms in fresh wet mounts of mat samples.

Autofluorescence of methanogens should be detectable in microbial mat ecosystems. The greenish-yellow fluorescence of F420 in methanogens can be seen using BG 3 and KP 425 filters in the excitation system and a K460 barrier filter (Doddema and Vogels 1978). The bluish-white fluorescence of F350 can be seen with an excitation system using the UG 1 filter and a K430 barrier filter (Doddema and Vogels 1978).

The enumeration of bacteria by use of acridine orange and other fluorescent techniques, the assessment of individual cell activities by use of light microscopy, and the determination of productivity by use of microscopy, microautoradiography, and immunolabelling in bacterial ecology have been reviewed recently (Newell et al. 1986).

20.2

Methods of Electron Microscopy

20.2.1 Stereoscopic Electron Microscopy (SEM) (B. K. Pierson)

The application of SEM in the study of microbial mats is useful in determining the relationships of microorganisms to each other and to inorganic and other debris. Most methods are similar (Jørgensen et al. 1983; Stal et al. 1986; Nicholson et al. 1987) and require fixation in glutaraldehyde (2 to 4% in natural water or buffer) followed by washing and dehydration in ethanol. The dehydrated samples are critical-point dried, gold or gold-palladium-coated, and analyzed with a scanning electron microscope.

20.2.2 Transmission Electron Microscopy (TEM): Special Techniques for Mats (E. D'Amelio)

The recommended procedure for small "cubicles" of mat (e.g., 8 mm wide by 4 mm long) is as follows:

(1) *Fixation*: immediately in 2.5% glutaraldehyde (e.g., Sigma, Grade 1, 25.0%) in sample water (filtered with a 0.22 μm membrane filter).
(2) *Rinse*: 0.2 M cacodylate buffer at pH 7.4, two rinses of 15 minutes each.
(3) *Post-fixation*: 2 hours in 1% osmium tetroxide in cacodylate buffer.
(4) *Rinse*: Same as step 2, above.
(5) *Dehydration*: in the following graded series:
   I. 50% ethanol in water, one change, 15 min.
   II. 70% ethanol in water, one change, 15 min.
   III. 90% ethanol in water, one change, 15 min.
   IV. 100% ethanol, three changes, 15 min each (using *fresh ethanol*).
   V. 100% ethanol/propylene oxide (50:50), one change, 15 min.
   VI. 100% propylene oxide, two changes, 15 min each.
(6) *Pre-embedding*: (*i*) Propylene oxide/epoxy resin (50:50), overnight (place the vials with the samples in a rotator). (*ii*) Propylene oxide/epoxy resin (20:80), two to three hours (in a rotator).
(7) *Embedding*: 100% epoxy resin, two to three hours (in a rotator); prepare epoxy resin (i.e., complete the formula).

After the complete epoxy resin is prepared, remove the samples from the vials and place them into embedding capsules. Place a few drops of the epoxy mix in the capsule, orient the sample so that the appropriate side is face down, and fill the capsule with epoxy mix. The next step is to place the embedded specimens in a 60°C oven for one to three days, after which the blocks will be ready to cut.

For staining ultrathin sections (60 to 90 nm thick), in order to locate chlorosomes or thylakoids, stain with *uranyl acetate* (2% in distilled water) for 20 to 30 minutes, followed by at least 10 minutes with *lead citrate* (at high pH) (Reynolds 1963). The sections should be *carefully rinsed* with *freshly boiled* (and cooled) distilled water after each stain.

20.3 Culture Methods

20.3.1 Filamentous Anoxygenic Phototrophs (B. K. Pierson).

Descriptions of methods are available for the isolation and culture of thermophilic *Chloroflexus aurantiacus* (Castenholz and Pierson 1981) and the *Chloroflexus* strains unable to grow chemotrophically (Giovannoni et al. 1987). Attempts have also been made to isolate and grow marine strains of *Chloroflexus*. One marine strain has been successfully enriched and maintained in mixed culture with two or three non-phototrophic bacteria (E. E. Mack, pers. comm.). Successful enrichments of *Chloroflexus* from hypersaline mats (Guerrero Negro, Baja California, Mexico) have been obtained but not beyond the level of secondary enrichment. Enrichments for these *Chloroflexus* types were only successful when 750 nm interference filters were used. Otherwise, cultures were rapidly dominated by purple sulfur bacteria (B. K. Pierson, unpubl.).

Heliothrix oregonensis from hot spring mats in Oregon has been grown in co-culture with *Isosphaera pallida* (Pierson et al. 1984), but axenic cultures were not established even after much effort.

20.3.2 Purple Bacteria and Green Sulfur Bacteria (R. W. Castenholz)

There are separate methods for sulfide-utilizing purple bacteria (Chromatiaceae) and for those (non-sulfur) types that grow primarily as photoheterotrophs (Rhodospirillaceae). Although many references exist, the simplest compilation for isolating and growing the purple non-sulfur bacteria is by Biebl and Pfennig (1981), for the purple sulfur bacteria and the green sulfur bacteria (Chlorobiaceae) by Pfennig and Trüper (1981), and for the genus *Ectothiorhodospira* by Trüper and Imhoff (1981). The cultivation of *Chlorobium* was further refined by Heda and Madigan (1986) and of anoxygenic bacteria in general by Malik (1983). Gibson et al. (1984) established cultivation methods for the marine *Chloroherpeton*. Gest et al. (1985) have utilized media free of combined nitrogen to enrich for phototrophic bacteria which are, in general, capable of nitrogen fixation. Methods for the cultivation of the more recently discovered "heliobacteria," such as *Heliobacterium* and *Heliobacillus*, are described by Beer-Romero and Gest (1987).

20.3.3 Cyanobacteria (R. W. Castenholz)

The isolation, culturing, and storage methods for cyanobacteria are described and reviewed by Rippka (1988), Rippka et al. (1981), Castenholz (1988a), and by Stevens (1988). Methods for more specialized groups of cyanobacteria are described in Waterbury and Stanier (1981), Waterbury and Willey (1988), and Ohki et al. (1986) [for marine forms]; Walsby (1981) [for gas vacuolate forms]; Castenholz (1981, 1988b) [for thermophilic forms]; Wolk (1988), and Mitsui and Cao [for nitrogen-fixing forms]; and Meeks (1988) [for symbiotic associations].

20.3.4 Other Conspicuous Bacteria of Microbial Mats (R. W. Castenholz).

Methods for the culturing of non-photosynthetic mat bacteria such as methanogens, sulfate- and sulfur-reducing bacteria, chemolithotrophs (e.g., sulfide-oxidizing bacteria), and for innumerable fermentative bacteria are described in the two volumes of *The Prokaryotes* (Starr et al. 1981; revised edition to appear in 1991) and in *Bergey's Manual of Systematic Bacteriology* (Holt 1984–1989).

20.4 Methods of Pigment Study

20.4.1 Organic Extracts of Chlorophyll a and Bacteriochlorophylls a, b, c, d, e, g (B. K. Pierson)

Many organic solvents can be used in the analysis of chlorophyll pigments, but solvents miscible with water such as acetone or methanol, or a mixture of the two, must be used to extract pigments from aqueous cells. There is wide variation in the efficiency of extraction of various chlorophylls in methanol and acetone (Bowles et al. 1985) but methanol appears to be an efficient solvent for quantitative extraction of the bacteriochlorophylls from photosynthetic bacteria (Oelze 1985) and for chlorophyll a from cyanobacteria (Holm-Hansen 1978). While some investigators prefer acetone or an acetone/methanol mix, 100% methanol offers some advantages. First, it is less volatile than acetone, resulting in more critical volume control for quantitative assays. Second, Bchl b, which is unstable in all organic solvents, is particularly unstable in acetone (Oelze 1985). Because all chlorophylls are unstable in organic solvents, extraction times should be short (preferably less than 30 minutes). All extracts should be kept dark and cold. The addition of 0.005 M sodium ascorbate can help prevent oxidations (Oelze 1985) and solvents can be buffered against acidification to prevent formation of pheophytins (Oelze 1985).

Our samples were extracted for 10 minutes in the dark and cold, and extraction was repeated until no further chlorophyll was removed. The quantitative samples were then combined and spectra recorded against 100% methanol using a CARY 2300 recording spectrophotometer (Varian Techtron, Mulgrave, Australia). The concentration of chlorophylls was determined using published molar or specific absorption coefficients for the far red absorbing band of each pigment in methanol Table 20.1). Because no absorption coefficient is available for Bchl b in methanol, we have used an estimated value of 66 mM^{-1}cm^{-1} based on the assumption that the ratio of molar absorption coefficients of Bchl a to Bchl b in diethyl ether is similar to the ratio in methanol. The value of 66 mM^{-1}cm^{-1} in methanol for Bchl b was calculated from the molar absorption coefficients of 96 mM^{-1}cm^{-1} for Bchl a in diethyl ether (Weigl 1953), 60 mM^{-1}cm^{-1} for Bchl a in methanol (Cohen-Bazine and Sistrom 1966), and 106 mM^{-1}cm^{-1} for Bchl b in diethyl ether (Steiner and Scheer as reported by Oelze 1985).

The biggest problem in the quantitative determination of chlorophylls in mat samples, as described above, is due to the fact that in mat samples with more than one type of pigment present, the absorption maxima of some of the pigments overlap making it difficult to determine how much each pigment contributes to the total absorption at a particular wavelength. In most mats the confusion occurs between Chl a (max = 663–665 nm in methanol) and Bchl c (max = 668–670 nm in methanol). In some mats, confusion can also arise due to overlap between Bchl.a (770 nm) and Bchl b (790 nm). Thin-layer chromatographic separation (Madigan and Brock 1976) can be used, but is too tedious and impractical when processing many small samples, and Bchl b is too labile to be quantitatively assayed in this way. Partitioning different chlorophylls between methanol and hexane (Stal et al. 1984) has limitations for accurate quantitative assay of samples because only Chl a separates entirely into the hexane phase while the other chlorophylls partition unevenly between the two phases. Likewise, intensive manipulations of Bchl b in organic solvents should always be avoided because of its instability. Estimates of the relative contribution of two chlorophylls to the same absorption band can be obtained by comparing the relative intensities of the well-separated far red absorption maxima *in vivo* and estimating the contribution made by each to the combined peaks in methanol, but this method leads to some inaccuracies. The best solution will be the use of HPLC (Korthals and Steenbergen 1985) or the determination of specific absorption coefficients for the chlorophylls *in vivo*.

20.4.2 Analysis of Lipophilic Pigments from Phototrophic Microbial Mats by High Performance Liquid Chromatography (A. C. Palmisano).

Lipophilic pigments from microbial mat communities can be separated, identified, and quantified by reverse phase-high performance liquid chromatography (HPLC). HPLC is a significant improvement over spectrophotometric analysis, in which overlapping absorbance spectra result in poor precision, and over thin layer chromatography (TLC), which is time-consuming and relatively insensitive. Using the following program (Palmisano et al. 1988), chlorophylls, bacteriochloro-

Table 20.1. *Absorption coefficients in organic solvents for cholorophylls commonly found in microbial mats*[a].

Pigment	Solvent	λ_{max}[nm]	Molar Absorption Coefficient [mM^{-1}cm^{-1}]	Specific Abs. Coeff. 1g^{-1}cm^{-1}	Reference
Chl a	methanol (90–100%)	665		75.0, 74.5	1, 2
	acetone (90%)			87.7	3
Chl b	acetone (90%)	646.8		51.4	3
Chl c_1	acetone (90%) (+1% pyridine)	630.6		44.8	4
Chl c_2	acetone (90%) (+1% pyridine)	630.9		40.4	4
Bchl a	methanol	770	60		4
	acetone/methanol [7:2, v/v]	767	76		5
Bchl b	methanol	790	66		6
Bchl c	methanol	670		86	7
	acetone (100%)	662.5		92.6	7
Bchl d	methanol	659		82.3	7
	acetone	654		98.0	7

[a]Data not available for Bchl e and g.

1. Lenz and Zeitschel (1968).
2. Seely and Jensen (1965).
3. Jeffrey and Humphrey (1975).
4. Jeffrey (1972).
5. Cohen-Bazire and Sistrom (1966).
6. Clayton (1966).
7. Unpublished, calculated by B. K. Pierson as explained in text.
8. Stanier and Smith (1960).

Table 20.2. *Absorption maxima (nm) in four solvents for carotenoids occurring in microbial mat communities.*

Carotenoid	Hexane	Light Petroleum	Acetone	Ethanol
anhydrorhodovibrin	451/480/516	451/480/512	460/485/495	—
caloxanthin	—	—	—	426/449/475
canthaxanthin	467	—	—	477
β-carotene	425/450/476	421/451/478	427/454/480	425/450/478
γ-carotene	431/462/494	437/461/491	—	447/477/510
chlorobactene	—	435/461/491	440/465/495	—
diadinoxanthin	450/474	—	—	425/445/477
diatoxanthin	450/479	430/451/482	—	428/452/479
echinenone	478/495	458	—	466
fucoxanthin	427/450/476	425/446/473	—	448
isorenieratene	443/465/493*	—	—	—
lycopene	447/471/501	446/472/505	448/474/505	443/472/502
myxoxanthophyll	—	—	450/478/510	448/471/503
nostoxanthin	—	—	—	426/448/475
okenone	—	(460)/484/516	(465)/487/518	—
rhodopin	443/470/504	—	448/474/506	455/470/501
rhodovibrin	—	455/483/516	460/488/522	—
spirilloxanthin	—	468/499/534	470/497/530	465/491/526
zeaxanthin	426/450/478	424/448/475	452/479	425/451/482

*in benzene
Reference: Foppen (1971).

phylls, their corresponding pheophytins, and a variety of carotenoids, can all be analyzed in a single 30 minute run.

Separations are performed on a Hewlett Packard 1084B chromatograph with a 250 mm × 4.6 mm ODS Hypersil column with 5 µm packing. A Whatman CSK guard column filled with Whatman ODS (Co:Pel) 30–38 µm packing is used to protect the analytical column. The ultraviolet-visible wavelength absorbance detector with a built-in scanning mode (250 to 550 nm) is set at 440 nm (550 nm reference). A non-linear gradient of 82% methanol, 13% acetonitrile, and 5% deionized water (Solvent A) containing tetrabutyl ammonium acetate as an ion-pairing agent is run against 100% acetone (Solvent B) in the following program: 0 to 10 min, linear increase from 5 to 28% B; 10 to 20 min, 28% B; 20 to 22 min, linear increase from 28 to 50% B; 22 to 25 min, linear increase from 50 to 60% B; 25 to 29 min, linear increase from 60–70% B; 29 to 30 min, decrease from 70 to 5% B; 30 to 35 min, reequilibrate at 5%% B.

Pigments are identified by their characteristic absorption maxima in organic solvents using ultraviolet-visible spectroscopy and by co-chromatography with standards obtained from Sigma Chemical Co., Hoffman-LaRoche Co., or isolated from pure cultures of cyanobacteria, diatoms, or *Chloroflexus aurantiacus* (Tables 20.1, 20.2). Pigments are quantified by comparing the integrated area beneath each peak to standard curves of individual pigments.

20.4.3 *"In Vivo"* Analysis of Pigments by Ultrasonic Disruption in Buffer (R. W. Castenholz).

The simplest method of assessing the total pigment complement of photosynthetic prokaryotes without chromatographic separation is by using ultrasonic disruptions giving essentially an "in vivo" absorption spectrum without the problems of scatter by high turbidity of whole cell suspensions. In addition, "in vivo" absorption spectra of mixed organisms allow the resolution of most of the types of chlorophylls and phycobilin pigments. Organic solvents do not extract phycobilins (which are commonly difficult to extract by any means). Organic solvents do not allow differentiation of some chlorophylls (see Section 20.4.1).

Ultrasonic disruption of cells in buffer is completed in three minutes with a Branson S-75 "sonifier" at 20,000 Hz in "TSM" buffer (per liter: 4.7 g NaCl, 1.2 g $MgSO_4 \cdot 7H_2O$, 2.4 g Tris, pH adjusted to 7.8). This is followed by centrifugation at about 2,800 × g to remove whole cells and cell wall debris (see Pierson et al. 1987; Sistrom and Clayton 1964).

20.5

Methods of Light Measurement

20.5.1 Definitions (B. B. Jørgensen).

The following light parameters are used to characterize the spectral radiation in microbial mats (cf. Jerlov 1976; Kirk 1983; Jørgensen, in press):

Radiant Flux (Φ) is the flow rate of radiant energy or quanta. Unit: Watts or quanta s^{-1}.

Radiance ($L_{(\theta,\phi)}$) is the radiant flux at a given point in space coming from a given direction. This is the light parameter measured by the fiber-optic microprobe. The radiance is expressed per unit solid angle, ω, and per unit area, A, at right angles to the direction of the flux. The direction is defined in spherical coordinates by its zenith angle, θ, and azimuth angle, ϕ:

$$L_{(\theta,\phi)} = d^2\Phi/dA \cdot d\omega. \tag{20.5.1}$$

The radiance has the units Watts m^{-2} steradian^{-1} or quanta m^{-2} s^{-1} steradian^{-1}.

Irradiance (E) is the radiant flux incident on a unit area of a surface. Most of the earlier studies of light penetration into mats and sediments have been based on large irradiance sensors over which mat slices of various thicknesses were placed. Irradiance measurements have the advantage of including both the collimated light and the forward (downward) scattered light. The downward irradiance, E_d, onto a horizontal surface is defined by integrating the radiance from all directions over the upper hemisphere (solid angle = 2π):

$$E_d = \int_{2\pi} L_{(\theta,\phi)} \cdot \cos\theta \cdot d\omega. \tag{20.5.2}$$

The irradiance has the units Watts m^{-2} or quanta m^{-2} s^{-1}.

Scalar irradiance (E_0) is the radiant flux coming from all directions. This parameter includes all collimated and scattered light and is therefore the most relevant measure of light available for a photosynthetic cell. A scalar irradiance sensor has not been developed in microscale, but E_0 can be calculated from radiance measurements made in relevant directions. The scalar irradiance is defined by integrating the radiance distribution at a point over all directions of the sphere (solid angle = 4):

$$E_o = \int_{4\pi} L_{(\theta,\phi)} \cdot d\omega. \tag{20.5.3}$$

The scalar irradiance has the units Watts m^{-2} or quanta m^{-2} s^{-1}.

20.5.2 Total Irradiance Within Mats (B. K. Pierson).

Total downward radiance over a unit area is measured at the surface of the mat and at depths within the mat by using a pyranometer sensor (LI-200SB, LI COR, Lincoln, Nebraska, USA) connected to the LI COR LI-185B radiometer. The pyranometer sensor is buried directly in the mat for measurements with depth. Alternatively, a core is mounted on the surface of the sensor and measurements made at various depths within the core (Pierson et al. 1987). Because the pigments within the various layers of mat strongly attenuate selected regions of the spectrum, the spectral composition of radiation at various depths in the mat is a more significant parameter to measure.

20.5.3 Spectral Distribution of Radiation Within Mats.

20.5.3A "Pierson Method" (B. K. Pierson).

Spectroradiometric measurements within mats can be made in the following two ways:

(1) A battery-operated field spectroradiometer (LI-1800, LI COR, Lincoln, Nebraska) is connected to an irradiance sensor (LI-1800-11 remote cosine receptor) via a quartz fiber optic cable (LI COR, LI-1800-10). The sensor is either covered with a core removed from the mat, and total downward spectral irradiance measured at different depths within the core, or the sensor is buried within the mat itself. Similar methods have been used to measure spectral irradiance in mats and soil (Tester and Morris 1987; Bliss and Smith 1985; Fenchel and Staarup 1971).

(2) The quartz fiber optic cable is connected to a tapered fiber optic tip 0.2 to 0.8 mm in diameter and approximately 3 to 4 cm in length pulled from a quartz fiber optic bundle (image conduit) that was originally 3.0 mm in diameter. The tips are filled to the end of the quartz fiber optic cable connected to the spectroradiometer. They are mounted in an aluminum housing that holds a core of mat in a darkened cup so that only the surface of the mat is exposed to natural sunlight. The tip is mounted to a micrometer

which can be used to advance it vertically through the mat, from the base to the surface, permitting measurements of the spectral distribution of relative intensity with depth intervals of about 250 μm. Measurements were recorded at 2 nm intervals from 400 to 1100 nm.

20.5.3.3B "Jørgensen Method" (B. B. Jørgensen).

A simple fiber-optic microprobe has been used to measure the spectral radiance distribution in microbial mats (Jørgensen and Des Marais 1986b; see Fig. 6.4.2). The probe consists of a single-stranded optic fiber of 80 μm diameter built into a supporting shaft. The fiber is tapered at the end and has a rounded sensing tip of 20 to 30 μm diameter. The spatial resolution is 50 μm. The microprobe is optically coupled to a low-noise photodiode detector attached directly to the end of the fiber. The detectability of "white" light is about 0.001 Watt m^{-2} or 0.01 μEinstein m^{-1}s^{-1}. The probe is attached to a micromanipulator with which it can be moved in three dimensions while the tip and mat are observed under a dissecting scope. The fiber measures directional light, i.e., radiance ($L_{(\theta,\phi)}$), with an acceptance half-angle of about 20°.

Measurements of downwelling light in microbial mats require that the fiber penetrates the mat from below. Small, undisturbed mat cores in plastic tubes are sealed at the bottom with an agar plug and positioned over a hole in a plate (Fig. 20.3). The mat surface is illuminated vertically by monochromatic light obtained by passing white light through a continuous interference filter. The spectrum is scanned at 10 nm intervals through the visible range. The fiber is advanced stepwise up through the mat at 0° light-fiber angle. At each depth and wavelength, the radiance is calculated as percent of the surface radiance into the mat.

Since the fiber-optic microprobe is a radiance sensor, it measures only a small part of the light field in the mat. When positioned vertically, it measures mostly the attenuation of the collimated light beam below the mat surface. In order to quantify the total flux of quanta which reach a cell at a given point within the mat, it is necessary to repeat the measurements in many directions. In practice, this can be done by pointing the fiber at five or more angles relative to the vertical light axis. For each angle the spectral radiance values are then multiplied by appropriate weighting factors (Vogelmann and Björn 1984;

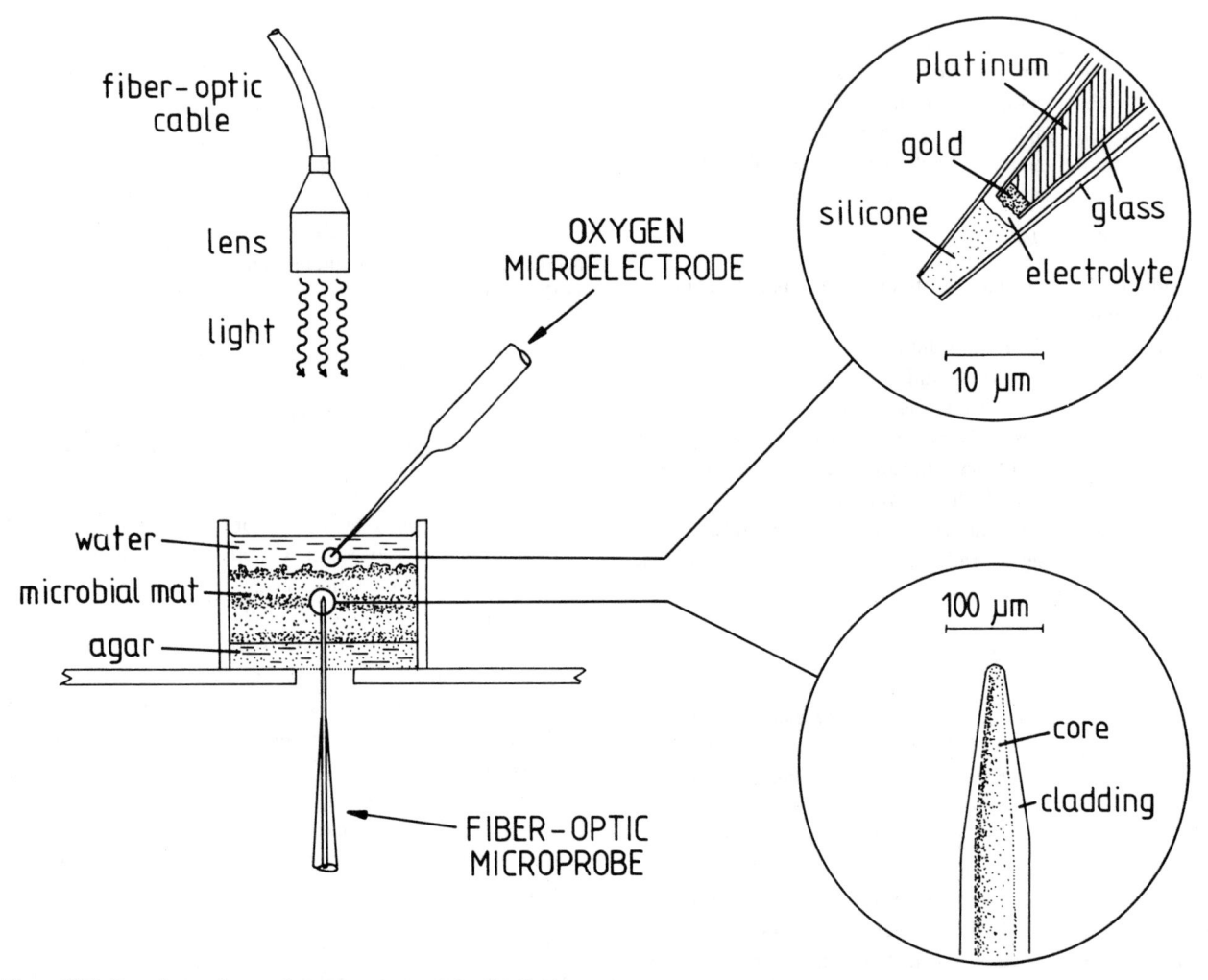

Figure 20.3. Experimental setup for measurement of O$_2$ distribution and photosynthesis with an oxygen microelectrode, and of light penetration with a fiber-optic microsensor. The oxygen microelectrode is coupled to a picoammeter and strip-chart recorder; the optical fiber of the light sensor is coupled to a photodiode and measuring circuit (see Jørgensen, in press).

Jørgensen and Des Marais 1988). A more ideal solution would be to use a microprobe with a sensing tip which inherently measured scalar irradiance. Such an isotropic fiber-optic sensor has so far been manufactured down to a size of 1 nm diameter (Star et. al. 1987). On the detector side, a much more useful instrument for laboratory measurements is obtained with a spectroradiometer instead of a simple diode. A sensitive optical multichannel analyzer has recently proven useful for mat studies.

20.5.4 Photoacoustic Methods (R. W. Castenholz).

Photoacoustic *in vivo* spectra of cyanobacteria and photosynthetic bacteria have been measured by Schubert et al. (1980).

20.6

Microelectrode Measurements

BO BARKER JØRGENSEN

The types of microsensors which have been applied in microbial mats include oxygen, sulfide, and pH microelectrodes. For details of their construction and experimental use see Revsbech and Jørgensen (1986), Jørgensen and Revsbech (in press), and references therein.

The applied oxygen microsensor is a small version of the conventional Clark electrode (Clark et al. 1953), scaled down to a sensing tip of about $5\,\mu m$ (Revsbech and Ward 1983) (Fig. 20.3). The internal cathode is polarized at -0.7 Volt and the electrode current is measured by a picoammeter. Due to the built-in reference electrode, the sensor is insensitive to the chemical composition of the external medium. The sulfide microsensor is a small Ag/Ag_2S electrode with a sensing tip of down to $20\,\mu m$ (Revsbech et al. 1983). Since it senses only the S^{2-} ion, its calibration is pH dependent and its sensitivity is limited at low or neutral pH where $\ll 10^{-6}$ of the total dissolved sulfide (H_2S, HS^-, S^{2-}) is present as S^{2-}. The pH microsensor is a small version of the glass pH electrode and has been used with tip diameters of 40 to $50\,\mu m$ (Thomas 1978). A mV-meter with high internal resistance is used to measure potentials of the pS^{2-} and pH electrodes.

New sensor types are currently being developed to measure other chemical species. A CO_2 microelectrode has been constructed based on a pH microelectrode in bicarbonate buffer behind a gas-permeable membrane. When used in phototrophic mats, this electrode has the disadvantage of a relatively slow response as well as insufficient sensitivity in the light due to the high pH and low CO_2 at the mat surface. Carbonate-sensitive liquid membrane electrodes have also been constructed (Herman and Rechnitz 1975), but they still have problems of interference from other common ions, which make them impractical for most ecological applications. Calcium-sensitive microelectrodes have important potential applications in the study of calcification and carbonate dissolution. A combined oxygen and nitrous oxide microsensor was recently developed based on similar principles to the Clark type oxygen microelectrode but with three cathodes enclosed within a sensing tip of $20\,\mu m$ (Revsbech et al. in press). When used experimentally in combination with acetylene addition, which inhibits the reduction of N_2O by denitrifying bacteria, denitrification can be quantified in mats at high resolution.

The microsensors are operated mechanically by micromanipulators attached to a heavy rack and can be moved in three dimensions at 0.1 to 0.01 mm precision. For laboratory studies of intact mat cores, the mats have been illuminated by a halogen lamp and the mat surface and electrode tip have been observed under a dissecting scope during measurements (Fig. 20.3). Measurements have also been made in the field under natural daylight. This has often precluded microscopic observation of the electrode tip, and its position relative to the mat surface has therefore been inferred from the measured microgradients (the use of a fiberscope may solve this problem). However, in some cases in the field, a small eyepiece telescope has been mounted on a stand and used to observe the electrode tip.

Transport of chemical species in mm scale at the surface of microbial mats is predominately driven by molecular diffusion. Based on the measured steady state microgradients and on information about diffusion coefficients of the molecules, it is possible to calculate chemical flux rates. This can be used to make simple estimates of metabolic rates of the whole microbial community, e.g., the O_2 uptake or release over the mat surface or flux within the mat (see Sections 6.4 and 6.6).

As a first approximation, it is assumed that the mat is laterally homogeneous and that only vertical gradients exist. The diffusion flux over a horizontal surface can then be calculated from Fick's first law of diffusion (Crank 1983):

$$J = -\rho \cdot D_s \cdot \delta C(z)/\delta z \; \text{mmol}\, O_2\, m^{-2} d^{-1}, \qquad (20.6.1)$$

where J is the flux of oxygen molecules through a unit area per unit time, ρ is the porosity of the mat, D_s is the apparent diffusion coefficient (Berner 1980) of oxygen in the mat at the given temperature, and $C(z)$ is the concentration of oxygen at depth z. The values of ρ and D_s can be determined by computer simulation of microelectrode data obtained from killed mat cores in which the ambient oxygen concentration in experimentally perturbed (N. P. Revsbech, in press). The concentration gradient is measured from the oxygen profile.

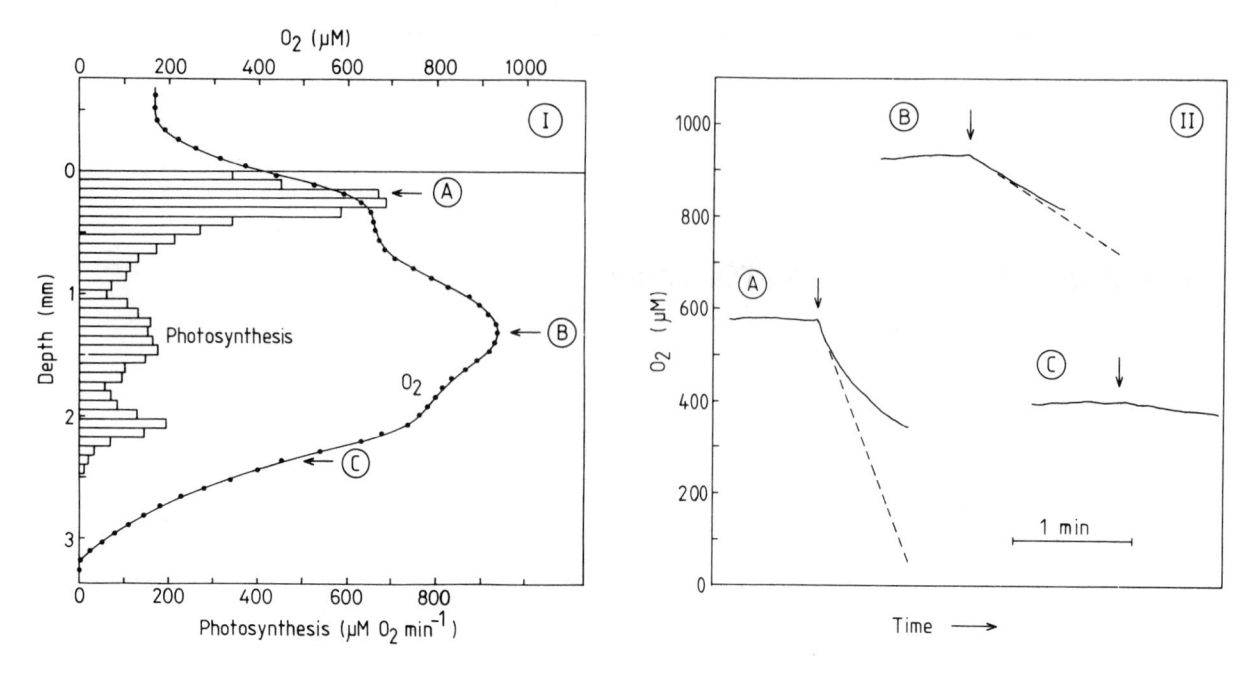

Figure 20.4. Experimental measurement of photosynthesis in a microbial mat (from a hypersaline pond at Guerrero Negro, Baja California Sur, Mexico) by the light-dark shift technique using oxygen microelectrodes. A steady state depth profile of O_2 was first measured in the light (curve plotted in I, left). The oxygen electrode was then postioned at the mat surface; the mat was shaded for several seconds during which the decrease in oxygen concentration was traced on a strip-chart recorder; the mat was then again illuminated until the oxygen concentration approached its former steady state value. The electrode was sequentially moved downward by 0.08 mm and the sequence was repeated, with rates of photosynthesis calculated at each depth from the slope of the plot of O_2 versus time. Three fragments of the strip-chart recording are shown (II, right) with the electrode tip positioned at depths A, B, and C, as indicated. Arrows indicate time of shading; only the intial slopes (indicated by the dashed lines in II) are valid measures of photosynthesis rates. The three depths of maximal photosynthesis recognized (histogram in I) correspond to dense layers of diatoms (uppermost maximum) and filamentous cyanobacteria (lower two maxima). (Data from Jørgensen et al. 1987.)

Metabolic rates can also be calculated from dynamic measurements during non-steady state based on Fick's second law of diffusion (Crank 1983). An experimental method for the determination of oxygenic photosynthesis is based on light-dark transients of oxygen distribution (Fig. 20.4; Revsbech et al. 1981; Revsbech and Jørgensen 1983). During steady state in the light, a stable oxygen distribution is reached where oxygen production by photosynthesis just balances oxygen removal by respiration, diffusion, etc., at each point. When light is suddenly removed, the oxygen cencentration will decrease at a rate which initially is equal to the former rate of photosynthesis. By this technique, photosynthesis can be measured over a few seconds with a spatial resolution of $\pm 50\,\mu m$ (see Section 6.5). By a more advanced computer simulation of dynamic oxygen data it is possible to calculate more precisely the photosynthesis distribution and also to calculate the distribution of aerobic respiration (Revsbech et al. 1986).

20.7
Meiofauna: Strategies for Field Studies
J. D. FARMER

In order to document the vertical distribution of meiofauna (viz., in microbial mats at Guerrero Negro, Baja California Sur, Mexico), fresh mat was cored to a depth of 2 cm, using modified 30 cc plastic syringes. Cores were extruded backwards in 1 mm increments, sliced with a sharp razor blade, and sub-cored with a small glass tube. Cores obtained had a surface area of 5 mm^2 (see Sections 6.3 and 6.10). The sub-cores were extruded into a glass syringe which was filled with native water, then forced through the tip of the syringe and repeatedly pumped back and forth until the mat was disaggregated. The preparation was placed on a coverslip, inverted on a depression slide, and examined at low power under a compound microscope. The entire slide was systematically traversed using the stage manipulator, and all meiofaunal species were identified and counted.

Spatial changes in meiofaunal species diversity and abundance were evaluated by sampling approximately every 0.5 km along a 6 km transect in Pond 4 (see Section 6.10). This transect was laid out so as to include the lower limit of mat distribution and the transition into the adjacent seagrass community. The transect was traversed by wading; samples were preserved immediately upon collection or used live. A small rubber raft functioned as a "floating lab" for sample preparation and transport. Alcohol dehydration and storage in 70% ethanol was the most convenient way to handle the large number of samples needed for microscopic examination. However, the hardening of materials made sample disaggregation more difficult, and most pigments were extracted by the higher alcohol concentrations.

20.8
Rate Measurements of Anaerobic Processes in Hot Spring Mats

DAVID M. WARD

Fermentation was studied by following accumulation of volatile and nonvolatile fatty acids and alcohols during anaerobic indubation using gas chromatographic procedures described in Anderson et al. (1987). This was possible because these fermentation products were not extensively metabolized under these conditions. The fate of fermentation products was studied by following [^{14}C]-labelled organic acids and alcohols into gaseous compounds (see below for cellular fractions as described by Anderson et al. 1987). Potential acetogenesis was studied by analyzing the sensitivity of the fate of [^{14}C]-labelled fermentation products in the presence of bromoethanesulfonic acid, an inhibitor of *methanogenic bacteria*, the dominant hydrogen consumers in low-sulfate hot springs (Anderson et al. 1987). Methanogenesis was measured by following accumulation of CH_4 in the headspace of samples incubated under dark anaerobic conditions as described by Sandbeck and Ward (1981). The conversion of the methane precursors to [^{14}C]-CO_2 or to [^{14}C]-CH_4, or of 2-[^{14}C]-acetate to [^{14}C]-CH_4 and/or [^{14}C]-CO_2, was measured by gas chromatography-gas proportion analysis, as described by Ward and Olson (1980). Sulfate reduction was measured by following conversion of [^{35}S]-SO_4^{2-} to [^{35}S]-H_2S as described by Ward and Olson (1980).

20.9

Chemical Analyses

BO BARKER JØRGENSEN

The following summarizes briefly a number of analytical techniques which have been applied to study the inorganic chemistry of microbial mats. Special emphasis is given to analytical techniques for sulfur compounds, but the list also includes inorganic nutrients, gases, and trace metals. Electrochemical measurements with microsensors are described above.

Dissolved inorganic ions (e.g., HS^-, SO_4^{2-}, NO_3^-, NO_2^-, NH_4^+, Ca^{2+}, and Mg^{2+}) have been analyzed in porewater samples obtained either by squeezing or by centrifugation of mat samples (the preferred technique depending on the ionic species and on the texture of the mat). Gases [e.g., H_2, CH_4, H_2S, CH_3SH, $(CH_3)_2S$] have been analyzed in the headspace of closed vials after equilibration with small mat samples. Heavy metals have been extracted by leaching with nitric acid while sorbed NH_4^+ was extracted with KCl.

Dissolved sulfide has been analyzed mostly by the sensitive, colorimetric methylene blue technique of Pachmayr (1960) or Cline (1969) (see Doemel and Brock 1976; Jørgensen and Cohen 1977; Stal et al. 1985). Sulfate analyses have been done by the gravimetric method (American Public Health Association 1971) or turbidimetric method (Tabatabai 1974) following barium precipitation (Jørgensen and Cohen 1977; Skyring et al. 1983). The more sensitive ion chromatographic methods would be preferable today at low sulfate concentrations. Elemental sulfur has been extracted by methanol or by carbon disulfide and the concentrations determined by UV spectroscopy (Jørgensen and Cohen 1977; Stal et al. 1985) or by colorimetry after cyanolysis (Doemel and Brock 1974; Troelsen and Jørgensen 1982). More complete suites of sulfur analytical techniques in microbial mats are summarized by Doemel and Brock (1974), Jørgensen and Cohen (1977), and Cohen et al. (1980).

Inorganic nutrients [NO_3^-, NO_2^-, NH_4^+, PO_4^{3-}, and $Si(OH)_4$] have been analyzed by standard colorimetric techniques (e.g., Grasshoff 1976; Parsons et al. 1988), either manually or by an autoanalyzer system (Lyons et al. 1984; Stal et al. 1985).

Gaseous species have been analyzed in cyanobacterial mats by gas chromatography with flame ionization and thermal conductivity detectors for CH_4 (Giani et al. 1984), thermal conductivity detectors for H_2 (Oremland 1983), and flame photometric detectors for S gases (Zinder et al. 1977).

Analyses of trace metals have been done by colorimetry (Stookey 1970) for iron, and by flame atomic absorption spectrophotometry for other metals such as cadmium, chromium, copper, lead, and zinc (Gaudette and Lyons 1984).

Table 20.5. *Summary of lipid and carotenoid biomarkers in modern microbial mats. (D. M. Ward).*

Mat	Phototrophic Micoorganisms[a]			Complex Lipids and Components			
	Photosynthetic bacteria	Cyanobacteria	Eukaryotic algae	intact polar lipids	glyco/phospho-lipid methano-lysis products	fatty acids	ether cleavage products
I. Anoxygenic Photosynthetic Bacterial Mats							
A. "New Pit" Spring							
Chloroflexus	—	—	—	BZ	—	6D
B. "Roland's Well"							
Chromatium							
Chloroflexus	—	—	—	—	—	—
C. "Travel Lodge" Stream							
Chlorobium	—	—	—	—	—	—
II. Thermal Cyanobacterial Mats							
A. Octopus Spring							
Chloroflexus	*Synechococcus*	—	GD	C_{16}, C_{18}, $C_{18:1}$ FAME > n,i,aiC_{15}-C_{17} FAME, br C_{20} diol, brC_{17} and nC_{18} mono-ethers > dialkyl ethers	—	tr phytanyl biphytanyl ethers
B. "Wiegert's Channel"							
Chloroflexus	*Synechococcus* *Phormidium*	—	—	—	—	—
C. Orakei Korako							
Chloroflexus	*Chlorogloeopsis*	—	—	—	—	—
III. Marine/Hypersaline Cyanobacterial							
A. Laguna Guerrero Negro shallow ponds							
—	*Microcoleus*	diatoms	—	—	—	—
salt pond 5							
Chloroflexus	*Microcoleus*	diatoms	—	—	—	—
salt pond 8							
Chloroflexus	*Microcoleus*	diatoms	—	—	—	—
B. Hamelin Pool							
Microcoleus mat							
	Microcoleus		—	—	—	—

Chemical Analyses

Free Lipids				Carotenoids[c]	Protokerogen	References
hydrocarbons	wax esters	alcohols	fatty acids			
nC_{29}-C_{33} di, trienes ($nC_{31:3}$)	n,n>i,i C_{29}-C_{36} (nC_{32})	n>i C_{15}-C_{19}(nC_{18}) tr sterols	nC_{14}-C_{18} (nC_{16})	—	—	1
nC_{16}-C_{25}, pr, ph	n,n>i,i C_{28}-C_{35}	n>iC_{14}-C_{19} (nC_{17}), ph	n>i C_{14}-C_{18}(nC_{16})	—	—	2
nC_{15}-C_{33}	n,:1 C_{28}-C_{32} (C_{30}, C_{32})	nC_{13}-C_{28}, ph, C_{30} hopanol	nC_{14}-C_{18}	—	—	2
ph:1, nC_{17}-C_{20}(nC_{17}), 7 MeC_{17}, tr hopenes, C_{27} C_{31} polyenes	n,n>i,i C_{29}-C_{37}	ph, n>i C_{15}-C_{20} (C_{17}), tr sterols, hopanols	nC_{16}, nC_{18}	—	—	1,3,4
nC_{15}-C_{19} (nC_{17}), suites of C_{16}-C_{19} monomethylalkanes	n,n>i,i C_{30}-C_{35} (C_{32}-C_{34}), tr C_{24}-C_{29}	ph, nC_{11}-C_{20}, tr sterols	n>i,ai C_{14}-C_{18} (C_{16}, C_{18})	—	—	2
C_{19}-C_{20} dimethylalkanes, C_{16}-C_{19} monomethylalkanes	n,n>i,i C_{30}-C_{36} (C_{32})	ph, nC_{14}-C_{19}, tr sterols	n>i,ai C_{13}-C_{20} (nC_{16})	—	—	2
nC_{17}, br$C_{17:1}$, ph:1, 7/8 MeC_{17}, hopene	—	>11 sterols	nC_{12}-C_{20} (nC_{16}), $C_{16:1}$	—	Pyr: n and :1 C_8-C_{20}, pr:1	1,5,6
nC_{14}-C_{19}(C_{16}-[b] C_{18}), C_{15}-C_{19} monomethylalkanes (7/8 MeC_{16}), iC_{20}	—	—	—	myx, βcar, γcar, zea, ech	—	6,7
nC_{16}-C_{19}, brC_{17}-C_{18} monomethylalkanes, ph:1, $C_{18:1}$, $C_{19:1}$, C_{30} hopene	—	—	—	myx, βcar, zea, γcar, ech	—	6,7
nC_{17}-C_{31}(nC_{17}), $nC_{17:1}$, ph:1, brC_{25} isoprenoid, $C_{30:6}$ hopene	—	—	—	βcar, myx, zea	—	7,8

Table 20.5. Continued.

Mat	Phototrophic Microorganisms[a]			Complex Lipids and Components			
	Photosynthetic bacteria	Cyanobacteria	Eukaryotic algae	intact polar lipids	glyco/phospho-lipid methano-lysis products	fatty acids	ether cleavage products
Lyngbya mat		Lyngbya Microcoleus		—	—	—	—
Entophysalis mat		Entophysalis		—	—	—	—
C. Laguna Mormona	Chloroflexus	Lyngbya Microcoleus	diatoms	—	—	—	—
D. Baffin Bay				—	—	—	—
E. Abu Dhabi	Chloroflexus Chromatium Thiocystis	Lyngbya Microcoleus		—	—	—	—
F. Solar Lake	Chloroflexus	Microcoleus Aphanothece	diatoms	—	—	Sap: $nC_{14}-C_{19}(nC_{16})$, $nC_{16:1}$, $nC_{18:1}$, tr tr aiC_{15}, $ai\,C_{17}$, iC_{16}, and $10\,MeC_{16}$ FAME	—
G. Gavish Sabkha		Microcoleus Oscillatoria		—	—	Sap: $nC_{14}-C_{18}$ (nC_{16}), i and ai $C_{15}-C_{19}$, $C_{18:1}$, $C_{18:2}$, cyC_{19}, tr $10\,MeC_{14}$ and $10\,MeC_{16}$	—
H. Hao Atoll		Phormidium Lyngbia Aphanocapsa		—	—	—	—
I. Roquetas de Mar				—	—	$H^{\pm}$-labile: nC_{16}, C_{14}-C_{18} hydroxy	—

Free Lipids						
hydrocarbons	wax esters	alcohols	fatty acids	Carotenoids[c]	Protokerogen	References
nC_{16}-C_{18} (nC_{17}), ph:1, $nC_{17:1}$, $nC_{20:2}$, $nC_{21:2}$, brC_{17}-C_{18} monomethylalkanes	—	—	—	βcar, ech, myx, zea, spir, can	—	7,8,9
nC_{17}, C_{30} hopene	—	—	—	βcar, can, zea	—	7,8
nC_{16}-C_{27} (nC_{17}), pr, ph, 7/8 MeC_{17}, C_{17}-C_{20} monoenes	—	ph, >15 sterols	nC_{14}-C_{29} (nC_{16}), $nC_{16:1}$, $nC_{18:1}$, i, aiC_{17}, cyC_{17}, cyC_{19}, ph, hopanoic acid	βcar, spir, tet, ech zea	Ox: nC_8-C_{27} FAME (C_{16}), C_{14}-C_{20}, isoprenoid FAME, brC_{17}-C_{18} FAME	10,11,12
nC_{15}-C_{19} (nC_{17}), 7/8 MeC_{17}, $C_{17:1}$, $C_{19:1}$	—	21 sterols	—	—	Pyr: n and :1 C_8-C_{19}, pr:1	5,13,14
nC_{17}-C_{20} (nC_{17}), pr, ph, 7/8 MeC_{17}	—	ph, hopanol, >15 sterols	nC_{14}-C_{18} (nC_{16}), $C_{16:1}$, $C_{18:1}$, $C_{19:1}$, i and ai C_{15}-C_{17}, ph, hopanoic acid	myx, rho, βcar derho, tor, ech, spir, zea	—	9,15
nC_{15}-C_{21} (nC_{17}), ph:1, $nC_{17:1}$, $nC_{18:1}$, 8 MeC_{16}, sq	—	ph, hopane tetrols, 26 sterols, nC_{14}-C_{18}	(see saponifiable)	βcar, ech, zea	SMEAH: phytol, nC_{16}, nC_{18}, $C_{18:1}$, and OH-C_{14}, C_{16} and C_{18}	16,17
nC_{12}-C_{19} (nC_{17}), 8 MeC_{16}, 6-9 MeC_{19}, 4-6 MeC_{17}, C_{17}-C_{18} dimethylalkanes, ph:1	—	ph, nC_{14}-C_{20} (nC_{16}), brC_{17}, ph:2, geranylgeraniol, 18 sterols, hopanols	(see saponifiable)	—	—	18,19,20
nC_{17}, $nC_{17:1}$, $nC_{17:2}$, $nC_{18:2}$, C_{29}-C_{31} mono, di-trienes, nC_{29}, nC_{31}, nC_{33}, sq, ph:1, hopenes, tr 4-6 MeC_{18} and 3/7 MeC_{17}	—	25 sterols, hopanol	—	—	—	21,22
—	—	—	—	—	—	23

Table 20.5. Continued.

Mat	Phototrophic Microorganisms[a]			Complex Lipids and Components			
	Photosynthetic bacteria	Cyanobacteria	Eukaryotic algae	intact polar lipids	glyco/phospho-lipid methano-lysis products	fatty acids	ether cleavage products
IV. Eucaryotic Alga-Dominated Mats							
A. Nymph Creek	—	—	*Cyanidium*	—	—	—	byphytanyls
B. Hamelin Pool colloform mat	+		diatoms	—	—	—	—
C. Lee Stocking Island colloform mat	+		diatoms	—	—	—	—
							prenoid, nC_{17}, 7/8 MeC_{17}

In general, the order in which compounds appear reflects their relative abundance.
n = normal; i = iso; ai = anteiso; :1 = monounsaturated; pr = pristane; ph = phytane (or phytol in alcohol fraction); ph:1 = phytene; ph:2 = phytadiene; tr = trace.
$xMeC_y$ = monomethylalkane of C number y and methylation at C number x.
Compounds in parentheses following a range of compounds is/are the major compound(s) within the range; absence of parentheses indicates lack of dominance of a compound(s).
— = no information available (most likely that no studies were performed); ND = not detected.
Degradative procedures: *Sap*: = saponification; *Pyr*: = pyrolysis; *Ox*: = oxidation with chromic acid; *SMEAH* = sodium bis-2-methoxy-ethoxy aluminum hydride (attacks ether, amide, and ester linkages); $H^{\pm}$-*labile* = HCl treatment of saponification residue.
BZ = personal communication from Bin Zeng to DMW; GD = personal communication from Gary, to DMW.
[a] For all lagoonal/hypersaline mats the dominant genus (general) reported in geochemical studies are given. More detailed microbiological analysis would probably reveal more representatives (e.g., see D'Amelio et al. in press for details of microflora of Solar Lake and Guerrero Negro Pond 5 mats).
[b] Results of analysis of hydrogenated sample.
[c] myx, myxoxanthophyll; βcar, β-carotene; γcar, γ-carotene; zea, zeaxanthin; ech, echinenone; can, canthaxanthin; spir, spirilloxanthin; tet, tetrahydrospirilloxanthin; rho, rhodopin; derho, 3.4-dehydrorhodopin; tor, torulene; fuc, fucoxanthin; diad, diadinoxanthin; diat, diatoxanthin.
References: 1. Ward et al. (1989b). 2. Shiea et al. (in prep. b). 3. Ward et al. (1985). 4. Dobson et al. (1988). 5. Philp (1980). 6. Palmisano et al. (1988). 7. Summons (unpublished). 8. Palmisano et al. (in prep.). 9. Watts et al. (1977). 10. Cordosa et al. (1976). 11. Philp and Calvin (1976). 12. Tibbets et al. (1978). 13. Winters et al. (1969). 14. Huang and Meinschein (1978). 15. Cardoso et al. (1978). 16. Boon et al. (1983). 17. Edmunds and Eglinton (1984). 18. De Leeuw et al. (1985). 19. Boon et al. (1985). 20. Ehrlich and Dor (1985). 21. Boudou et al. (1986a). 22. Boudou et al. (1986b). 23. Goossens et al. (1986).

Table 20.6. *Mid-chain branched monomethylalkane distribution in modern cyanobacteria and natural cyanobacterial samples (from Shiea et al., in prep.).*

Mid-chain branched monomethylalkanes (number indicates position of methylation)

Organism/Sample	methylpentadecanes					methylhexadecanes					methylheptadecanes				
	4	5	6	7	8	4	5	6	7	8	?	4	5	6	7	8	?
I. Cyanobacteria																	
Anabaena variabilis															+	+	
Anacystis nidulans[1]															+	+	
Anacystis cyanea[2]															+	+	
Calothrix sp.									+								+
Chlorogloea fritschii										+					+	+	
Chroococcus turgidus			+												+	+	

Chemical Analyses

Free Lipids						
hydrocarbons	wax esters	alcohols	fatty acids	Carotenoids[c]	Protokerogen	References
nC_{17}-C_{31} (nC_{17}), hopene	ND	ph, 14 sterols, tocopherols	nC_{14}-C_{18} (nC_{16}), $nC_{18:1}$	—	—	1
$brC_{25:1}$ isoprenoid, nC_{17}, $nC_{17:1}$, brC_{17}, C_{29} sterene	—	—	—	fuc, βcar, diad, diat	—	7,8
brC_{20} iso-	—	—	—	—	—	7

methyloctadecanes							methylnonadecanes					methyleicosanes			References
4	5	6	7	8	9	?	6	7	8	9	?	6	7	?	
															1
															2
															3
															4
															5
															3

Table 20.6. Continued.

Mid-chain branched monomethylalkanes (number indicates position of methylation)

Organism/Sample	methylpentadecanes					methylhexadecanes						methylheptadecanes				
	4	5	6	7	8	4	5	6	7	8	?	4	5	6	7	8	?
Lyngbya aestuarii															+	+	
Lyngbya lagerhaimii															+	+	
Lyngbya majuscula															+	+	
Microcystis aeruginosa										+							
Nostoc commune															+	+	
Nostoc muscorum[3]														+	+	+	
Nostoc sp.															+	+	
Phormidium luridum															+	+	
Spirulina platensis[4] Mao I										+					+	+	
Spirulina platensis[4] Mao II										+					+	+	
Spirulina sp. Mexico[4]									+						+	+	
Synechococcus bacillaris																	+
II. Natural Samples																	
A. Blooms																	
Gloeotrichia echinulata														+			
Microcystis aeruginosa															+	+	
Microcystis wesenbergii										+							
Oscillatoria agardhii															+	+	
B. Lichens																	
Cetraria nivalis															+	+	
Cetraria crispa															+	+	
Siphula ceratites															+	+	
C. Sediments																	
Rostherne Mere															+	+	
Esthwaite															+	+	
Great Pond, Virgin Is.															+	+	
B. Marine/lagoonal Cyanobacterial Mats																	
Abu Dhabi															+	+	
Baffin Bay															+	+	
Gavish Sabkha				+					+			+	+	+			
Hamelin Pool *Lyngbya* mat										+							+
Hao Atoll															+	+	
Laguna Guerrero Negro																	
shallow ponds															+	+	
salt pond 8[5]				+				+	+	+					+	+	
Laguna Mormona															+	+	
Solar Lake										+							
E. Hot Spring Cyanobacterial Mats																	
Octopus Spring															+		
"Wiegert's Channel"		+	+				+	+	+	+			+	+	+	+	
Orakei Korako	+	+	+	+	+		+						+	+	+	+	

No mid-chain branched alkanes were found in Anacystis montana (Gelpi, et al. 1970; Murray & Thompson 1977), *Agmenellum quadruplicatum*, *Coccochloris elabens*, *Oscillatoria williamsii*, or *Plectonema terebrans* (Winters et al. 1969), *Oscillatoria woronichinii* (Blumer et al. 1971), *Microcoleus chthonoplastes* (Winters et al. 1969; Summons, unpublished), or *Synechococcus* sp. (Goodloe & Light 1982); in blooms dominated by *Spirulina* sp. (Tulliez et al. 1975) or *Trichodesmium erythaeum* (Winters et al. 1969); in the lichen *Peltigera canina* (Gulz et al. 1978); or in Hamelin Pool *Microcoleus* and *Entophysalis* mats (Summons, unpublished).

[1]Possibly also found in *A. nidulans* by Oró, et al. (1967), but not found in *A. nidulans* by Gelpi et al. (1970) or Winters et al. (1969).
[2]Unialgal rather than pure culture.
[3]See also Han et al. (1968); brC_{17} also reported in *N. muscorum* by Winters et al. (1969).
[4]Reported to occur at carbon numbers between C_{17} and C_{28}. Mid-chain branched alkanes were not reported in *S. platensis* by Řezanka et al. (1982) or Gelpi et al. (1970).
[5]Sample was hydrogenated before analysis.

References: 1. Fehler and Light (1970). 2. Han and Calvin (1969). 3. Gelpi et al. (1970). 4. Paoletti et al. (1976a). 5. Han et al. (1968). 6. Winters et al. (1969). 7. Oehler (1976). 8. Jüttner (1976). 9. Han and Calvin (1970). 10. Blumer et al. (1971). 11. Cranwell (1976). 12. Brooks et al. (1976). 13. Gaskell et al. (1973). 14. Cardosa et al. (1978). 15. De Leeuw et al. (1985). 16. Summons (unpublished). 17. Boudou et al. (1986b). 18. Philp (1980). 19. Cardoso et al. (1976). 20. Boon et al. (1983). 21. Dobson et al. (1988). 22. Shiea et al. (in prep. a).

Chemical Analyses

	methyloctadecanes							methylnonadecanes					methyleicosanes			References
	4	5	6	7	8	9	?	6	7	8	9	?	6	7	?	
																3
																6
					+										+	7
																8
																4
																9
																3
																5
												+			+	
												+			+	4
												†			+	4
																10
																11
																12
																8
																11
																13
																13
																13
																12
																12
																12
																14
																6
							+	+	+	+	+		+	+		15
																16
	+		+	+												17
																18
				+						+						16
																19
																20
																21
			+	+		+										22
		+		+	+											22

Table 20.7. *Critical analysis of evidence for sterols in cyanobacteria and other prokaryotic microorganisms.* (D. M. Ward).

Organism	Culture purity	Medium purity	Analysis method	Sterol abundance (% dry wght)	sterol type	Radiolabelling evidence of biosynthesis	Reference
I. Cyanobacteria							
Anacystis nidulans	pure	defined	GCMS	not reported	common[4]	^{14}C-acetate incorporated into phytol, but not sterols	1
	—	—	—	—	—	^{14}CO$_2$, ^{14}C-acetate and ^{14}C-mevalonic acid incorporated into lipid fraction but not sterols	2
Anacystis variabilis	—	—	—	—	—	^{14}C-acetate and ^{14}C-mevalonic acid incorporated into lipid fraction but not sterols	2
Anabaena cylindrica	pure	defined	GLC+TLC/MS	0.1–0.2	common	<0.015% of added ^{14}C-acetate recovered in sterol column fraction after 21 h incubation	3
	pure	contained[5] seawater	GCMS	0.023	common[6]	none	4
Anabaena solitaria	pure	contained[5] seawater	GCMS	0.014	common[6]	none	4
Anabaena viquieri	pure	contained[5] seawater	GCMS	0.010	common[6]	none	4
Calothrix sp.	unialgal	"outdoor culture"	GLC+TLC/MS	not reported	common	none	5
Chlorogloea fritschii	?	?	?	0.01–0.13	common	none	6
Fremyella disposiphon	pure	defined	GCMS	not reported	common	none	1
Gleotrichia schinulata	natural bloom	—	—	—	—	no incorporation of digitonin ppt.	7
Microcystis aeruginosa	natural bloom	—	GCMS	not reported	common	no incorporation of ^{14}C-acetate into digitonin ppt.	7,8
Nostoc carneum	pure	contained[5] seawater	GCMS	0.030	common[6]	none	4
Nostoc commune	unialgal	"outdoor culture"	GLC+TLC/MS	not reported	common	none	5
Nostoc harveyana	pure	contained[5] seawater	GCMS	0.005	common[6]	none	4
Nostoc muscorum	?	?	—	—	—	^{14}C-acetate and ^{14}C-mevalonic acid incorporated into lipid fraction but not sterols	2
Phormidium luridum	pure	defined	GCMS	not reported	common	none	9
Prochloron sp.	natural symbiont collection	—	GC	not reported	common	none	10
Spirulina maxima	?	?	digitonin ppt. +TLC	not reported	common	none	11
Spirulina platensis							
Mao I	unialgal	"outdoor culture"	GLC+TLC/MS	not reported	common	none	5
Mao II	unialgal	"outdoor culture"	GLC+TLC/MS	not reported	common	none	5
	?	?	GCMS	not reported	common	none	12
Spirulina sp.	unialgal	"outdoor culture"	GLC+TLC/MS	not reported	common	none	5

Table 20.7. Continued.

Organism	Culture purity	Medium purity	Analysis method	Sterol abundance (% dry wght)	sterol type	Radiolabelling evidence of biosynthesis	Reference
II. Noncyanobacterial							
Azotobacter chroococcum	pure	defined	GCMS	0.01	common	none	13
Cellulomonas dehydrogens	pure	defined	GCMS	0.03–0.05	common	ca 0.02% of added ^{14}C-glucose incorporated into	14
Escherichia coli[1]	pure	defined	GCMS	0.0004	common	none	13
Methylobacterium organophilum	pure	defined	GCMS	0.00002	4,4 diMe	none	15
Methylococcus capsulatusa F. D.	pure	defined	GCMS/NMR	0.22	4 Me, 4,4 diMe, 8(13)	20% of added ^{3}H-squalene epoxide incorporated into sterol TLC spot within 4 h	16,17,18
Bath	pure	defined	GC	0.32	"	none	19
Micromonospora sp.	pure	defined	digitonin ppt.	0.001	—	none	20
Nannocystis exedens	pure	sterol-free	GCMS/NMR	0.4	common	5% of added ^{14}C-mevalonic acid incorporated into sterol HPLC peaks after 47 h	21
Staphylococcus aureus L-form[2]	pure	contaminated	GCMS	0.03–0.05	common	0.0025% of added ^{14}C-acetate incorporated into TLC sterol spot after growth to late log phase	22
Streptomyces olivaceous	pure	defined	GCMS	0.0035	common	none	13
Thiobacillus thioparas[3]	pure	defined	GCMS	not reported	common	none	23

[1] Not confirmed by Nes & Nes (1980).
[2] Parent strain did not contain sterols.
[3] Only one of two strains contained sterols.
[4] Sterols that are frequently observed in a variety of different organisms, lacking methylation at carbon number 4 or 14, with unsaturation at common positions (e.g., 5, 7, 22, 24), or with alkylation at C_{24}.
[5] Medium blanks, however, did not contain sterols.
[6] Traces of 4 Me sterols were reported, though dominant sterols were common.

References: 1. Reitz and Hamilton (1968). 2. Levin and Bloch (1964). 3. Teshima and Kanazawa (1972). 4. Kolhase and Pohl (1988). 5. Paoletti et al. (1976b). 6. Nichols (1973). 7. Lata (1974). 8. Nishimura and Koyama (1977). 9. De Souza and Nes (1968). 10. Perry et al. (1978). 11. Nadal (1971). 12. Forin et al. (1972). 13. Schubert et al. (1968). 14. Weeks and Fracesconi (1978). 15. Patt and Hanson (1978). 16. Bird et al. (1971). 17. Bouvier et al. (1976). 18. Rohmer et al. (1980). 19. Jahnke and Nichols (1986). 20. Fiertel and Klein (1959). 21. Kohl et al. (1983). 22. Hayami et al. (1979). 23. Christopher et al. (1980).

21

Construction and use of Geological, Geochemical, and Paleobiological Databases

J. M. HAYES   STEFAN BENGTSON   HANS J. HOFMANN
JERE H. LIPPS   DONALD R. LOWE   CAROL MANKIEWICZ
CARL V. MENDELSON   TOBY B. MOORE   BRUCE N. RUNNEGAR
HARALD STRAUSS

21.1

Sample Inventory and Curation

The earlier collaborative project of the PPRG (1979–1980; Schopf 1983a) used a great deal of paper. As analytical work neared completion, handwritten "scoreboards" and "hit lists" were compiled to be sure that work proceeded efficiently and that important samples were not missed. As tables of results were prepared, extensive bibliographies were developed relating to stratigraphic relationships and sedimentary ages. Participants in the project reworded the accumulating paper like so many burrowing animals. When, for example, a decision was reached about the age to be estimated for a particular rock unit, multiple tabular entries had to be changed. Much communication focused on keeping the records straight rather than on questions of interpretation.

The "personal-computer revolution" preceded the beginning of the current PPRG project. Many of the researchers involved had already developed computerized databases, and it was resolved that the power and flexibility of this technology would be applied to the sample-tracking and information-management problems of PPRG. Three problem areas were identified: (*i*) construction of unified bibliographic database that could be searched and which could be used for preparation of the reference list for the final publication; (*ii*) management of the sample inventory and laboratory work; and (*iii*) compilation of results and related information. Systems were eventually developed in all of these areas as described briefly below. In spite of efforts at coordination, the degree of integration initially hoped for was not achieved, principally because the databases were, in their organization a well as contents, the result of individual efforts. This is particularly true of the third area, which is the most complex and multidimensional. Even there, however, the use of related database structures represented an advance over the paper blizzard of the previous, 1980, venture.

With a single exception (noted below), the databases were maintained using computers based on 8088, 8086, or 80286 microprocessors and using the MS-DOS operating system. The database management program employed was "dBASE III plus" (Ashton-Tate, Torrance, California, USA). The specialized bibliographic database initially employed was "Sci-Mate," specifically, the subsection termed "The Manager" (Institute for Scientific Information, Philadelphia, Pennsylvania, USA); the final bibliography was prepared using "Papyrus" (Research Software Design, Portland, Oregon, USA).

Toby Moore constructed and maintained five databases which, together, provided information that facilitated maintenance and use of the sample collection. The inventory database, INVI, contains one record for each sample in the collection. Each record contains up to 377 characters, with specific fields devoted to sample number and storage location, the source of the sample, a description of the sample, and the stratigraphic location. Fields of special interest are SITE and UNIT. The first specifies, by use of an informal place name up to 25 characters in length, the site at which the sample was collected. More importantly, it refers to an entry in the SITE database, which contains one record for each site (typically, a single outcrop) at which samples were collected. Each record in SITE can contain up to 277 characters, with specific fields referring to political units, latitude and longitude, tectonic province, detailed description of location, and availability of related maps. Because the SITE field is common both to INVI and SITE, opening both databases simultaneously allows them to be linked so that an operator can obtain detailed site information without leaving the INVI database or, conversely, check availability of samples without leaving the SITE database. Use of two databases in this way is advantageous because detailed information (latitude, longitude, etc.) need only be entered once for each site, not repeated for each sample.

Stratigraphic and geochronometric information has been collated in the STRAT database, which contains one record for each rock unit, independent of whether the unit in question is a member, formation, group, or supergroup. Because each record specifies the identity of the next higher unit (e.g., the group within which a given formation occurs), the database provides a means of reconstructing stratigraphic relationships. The UNIT field is common to STRAT, INVI, and many other databases, thus providing a basis for multiple linkages. This feature was designed to circumvent a problem often encountered when collating data from literature reports. As with different authors, it was feared that different subgroups within the PPRG might use inconsistent ages when referring to the same rock unit. If the isotopic geochemists plotted the age of rock unit X at (for example) 1950 Ma while the organic geochemists and the paleontologists used 2200 and 1780 Ma, comparison and correlation of observations would be difficult. Features that would be noticed if consistent ages (whether accurate or not) were adopted could be obscured. The STRAT database contains fields for the best estimate of the age of each sedimentary unit, its *estimated* maximum and minimum ages, and the *dated* maximum and minimum ages (commonly of igneous units) from which the sedimentary ages were estimated. Additional fields provide space for references documenting the dated maximum and minimum ages and for the identity of the person constructing the estimates of the ages of the sedimentary unit. It was intended that ages would be associated with units only within STRAT. When ages were required within other databases, they were to be linked to STRAT via the UNIT field, with the assigned age then (*i*) being consistent across all databases and (*ii*) always representing the best available information. This approach proved not to be entirely workable because the growth of the STRAT database was outstripped by the growth of the results databases, particularly those collating results from non-PPRG investigations reported in the literature.

The THINMAIN database contains one record for each thin section in the collection. It is linked to STRAT through the UNIT field and contains information on the nature of each thin section and details of its production.

The INVCODES database can be linked to INVI by the sample-number field and to STRAT by the UNIT field. It contains one record for each PPRG sample. With each record, 26 logical fields summarize the information presented in Table 14.1.

21.2

Results and Correlative Information

21.2.1 Geochemistry and Lithology

Harald Strauss compiled 20 databases summarizing results of isotopic and elemental analyses of carbon (organic and carbonate carbon), sulfur (sulfides and sulfates), and kerogens in Proterozoic samples. These databases were used to produce the tables in Chapters 3 and 17. One set of databases dealt with results generated within the PPRG, another with data gleaned from the literature. Within each of these sets, databases of two types were maintained. In one type, a record was devoted to each analytical result. For example, if a paper reported elemental compositions of 12 different kerogens from the same stratigraphic unit, it resulted in 12 records in the KERL database (KERogen Literature). In the second type, results were averaged within each unit. For example, the 12 results postulated in the previous example would be collapsed to a single record in the KERLSUM database (KERogen Literature SUMmary). If an additional literature report dealt with the same stratigraphic unit, it would result in an additional record in that database (that is, results were not averaged across all literature reports of analyses of a given unit).

Don Lowe and coworkers compiled the LITH 1 database as a means of beginning an inventory of Proterozoic sediments. Each record contains 83 fields with a total capacity of 1004 characters. These fields summarize stratigraphic information including the nature of upper and lower contacts for the unit to which the record is devoted; the thickness, areal extent, metamorphic grade, and probable volume of the unit; the lithologic composition of the unit; paleobiologically relevant details; geological structures and related notes; and information on the depositional environment and tectonic setting.

21.2.2 Paleontology

Carl Mendelson and Bill Schopf compiled MICRO and MICOCCUR, two databases dealing with Proterozoic microfossils. The goals of this work were to provide an inventory of Proterozoic and Early Cambrian (2500 to 530 Ma) organic-walled microfossils (Table 22.3), to evaluate the authenticity and biological affinity of each reported object (Table 5.4.5, Chapter 22), to lay a foundation for future taxonomic revision (Chapter 25), and to provide a basis for evaluation of Proterozoic patterns of diversity in microscopic organisms (Section 11.3). The MICRO database contains most of the information about each reported object; MICOCCUR is devoted to the temporal, geographic, and stratigraphic occurrences of each object. The databases are linked via the field IND, a unique index number assigned to each object.

Each record in MICRO, which comprises 65 fields with a total capacity of 844 characters, covers a single microfossil or microfossil-like object. Most objects are identified formally by Linnean binomials; others have been given an informal designation (e.g., "long filament," "chroococcacean unicell"). Nearly all records make reference to a published illustration of the object. Those which have been included in spite of the absence of an illustration are deemed particularly important for resolving boundary problems or extinction events (e.g., the proposed Vendian extinction in acritarchs). Coverage is not exhaustive, but most reports dealing with microfossils having cyanobacterial or bacterial affinities have been included. The coverage of acritarchs is less complete; this is especially true for reports of Early Cambrian acritarchs. As of mid-August, 1988, the database included 3494 records.

Each reported object has been assigned to a general category denoted by the field CLASS (classification). Categories include nonfossil, pseudofossil, dubiofossil, bacterium, cyanobacterium, acritarch, eukaryote, and bizarre. A more detailed analysis is presented in the field INTERP (interpretation; e.g., "colonial chroococcacean cyanobacterium"). These evaluations are based on illustrations and descriptions and, to a lesser degree, on study of individual specimens. Also included is information on geographic and stratigraphic locations of the object and on its morphology. Records in MICOCCUR give geographic and stratigraphic locations for second and further occurrences of objects whose primary entries appear in MICRO. The microfossil databases were used to construct the tables in Chapters 22 and 25.

Two databases dealing with calcareous algae were compiled by Carol Mankiewicz with the objectives of providing an inventory of all described taxa of calcareous algae of Proterozoic through Early Cambrian (Pro-€l) age and of evaluating the relative quality or appropriateness of Pro-€l calcareous algal genera (Tables 23.2.1–23.2.5). The database GEN lists genera of Pro-Cl calcareous algae; ALGAE lists all species of Pro-Cl calcareous algae. There is only one record per taxon; occurrences are not listed separately. Genera were evaluated by grouping them into classes ranging from pseudofossils to well-documented calcified algae/cyanobacteria. The evaluation was subjective, commonly based on one or two photographs (or drawings) of the taxon. Thus, a "questionable" rating might reflect the quality of the illustration rather than the quality of the taxon. No taxonomic revisions were made, but the databases can provide a framework for revision of taxonomy.

Each record in GEN identifies the author of the genus and includes bibliographic data as well as a coded description of the morphology, the approximate number of described Pro-Cl species, the stratigraphic range, and the distribution in Proterozoic and Early Cambrian units. Each record in ALGAE includes similar information as well as details pertinent to the type specimen.

The "calcareous algae" comprise an artificial group including algae and cyanobacteria that precipitate $CaCO_3$ on or within the thallus while living. From authors' descriptions, it is not always clear whether taxa are calcareous. Distribution of taxa among various PPRG databases was handled on a case by case basis. In general, obvious stromatolites and oncolites, formally designated catagraphs, and radiocyathids were excluded from the calcareous algal databases. All genotypes of calcareous algae with Pro-Cl occurrences have been included. Consequently, some genotypes are not Pro-Cl, or are not calcareous. Also included are taxa originally described as calcareous algae, but now not considered to be algal; calcareous taxa originally described as nonalgal, but now considered to be algal; and taxa that may have been originally calcareous, but which have been diagenetically altered, or which might be ancestral uncalcified members of calcified taxa such as dasycladacean algae.

Two databases cataloguing reports of Ediacaran metazoans, trace fossils, dubiofossils, and pseudofossils, and of Late Proterozoic carbonaceous remains, are maintained by Hans Hofmann (see Tables 23.1 and 23.6).

A database cataloguing taxa of Late Proterozoic metazoans is maintained on a Macintosh computer by Bruce Runnegar (Tables 23.3.1 and 23.3.2). The software employed is dMacIII. A record is included for each named genus and species of Proterozoic soft-bodied metazoan (Ediacara fauna), trace fossils, dubiofossils, and pseudofossils. No entries are included for small shelly fossils or for carbonaceous films, and no attempt is made to list subsequent occurrences of previously described genera and species.

Three databases dealing with early skeletal fossils have been developed and are maintained by Stefan Bengtson (see Table 23.5). Together, these cover published taxa down to the species or subspecies level for skeletal fossils of Late Vendian to Early Cambrian age. "Skeletal" is taken to imply fossils having biomineralized structures (shells, tubes, spicules, sclerites, etc.), but skeleton-like structures of organic material or incorporating agglutinated sediment particles have also been included. Excluded from the database are calcareous algae, archaeocyathans, and trilobites.

The databases ESF, ESFFIG, and ESFLIT are linked by common fields. The records in ESF contain the basic data on taxonomy, distribution, affinities, and mineralogical composition. Records in ESFFIG list the places where the taxa have been figured or described in the literature subsequent to their original establishment. ESFLIT contains the literature references used for ESF and ESFFIG, and is compatible in format with the REFSYS database described below. Within ESF, specific fields refer to the original name applied to each taxon; the author and year of the taxon; information on type specimens, locality, and stratum; mineralogical composition of the skeleton; stratigraphic range; and distribution. ESFFIG designates the location of the cited figure and specifies the taxon in terms of the name given in the original reference and in terms of the catalogue numbers in ESF corresponding to (*i*) the original reference and (*ii*) the taxon identified by Bengtson.

21.3

Bibliographic System

The bibliographic system employed was that developed by John Hayes and Steve Studley at the Biogeochemical Laboratories, Indiana University. In this system, the database encoding information on references (books, papers, reports of all forms) exists in two forms: a "DBF" file constructed by and accessible via dBASE III+, and a "USR" (user) file accessible via Sci-Mate. This duality is advantageous because features of dBASE III make it easy to enter and edit bibliographic data and because specialized features of Sci-Mate allow (*i*) text searches 30 to 100 times more rapid than those available within dBASE and (*ii*) semi-automatic preparation of word-processor files in any format required. The latter feature can be very helpful in the preparation of manuscripts.

A menu-driven system has been developed to facilitate entry of bibliographic data (authors, titles, citations, etc.) and, by this process, an "update" database file can be accumulated. At intervals selected by the user, the update file can be processed to (*i*) add the new references to the master database (REFS) and (*ii*) prepare a text file that can be accepted as input by Sci-Mate. This processing step incorporates a number of operations of considerable importance. In the first, a "refcode" is derived for each reference in the update file. The refcode is meant to provide an automatically assignable, unique, and compact code for any given reference. Although it is the crude invention of earth scientists (as opposed to information specialists), very few problems have been experienced in its use, perhaps because the body of literature surveyed is not so large that author names are often duplicated. Each refcode is composed of ten characters: XXXXYYZZZZ, where XXXX are the first four characters of the first author's family name, YY the last two digits of the year of publication, and ZZZZ the four least significant digits of the beginning page number of the article (or the total number of pages in a book or report). If the author's family name contains less than four characters or if the page number is less than 1000, the place of any missing characters is taken by a point. For example, the refcode for the book resulting from the 1979–1980 PPRG project (J. W. Schopf, ed., *Earth's Earliest Biosphere*, Princeton, 1983, 543 pp.) would be Scho83.543. Because all reference citations in PPRG databases are in terms of refcodes, the REFCODE field allows REFS to be linked to all of these databases.

A second operation carried out during input processing is designed to avoid duplication of entries in REFS. Each refcode in the update database is compared to all refcodes already existing within the REFS database and, in case any duplication is found, both the old and the new references are printed for inspection. This step will, of course, also reveal any cases in which different references happen to lead to the same refcode.

Finally, various shorthand codes used during data entry are expanded. These codes are of two types. Names of commonly cited journals can be abbreviated during entry of citations (e.g., gca for *Geochimica et Cosmochimica Acta*), and *all* keywords are entered in terms of two-letter codes. The latter efficiency is possible because all references to be entered are accompanied by single-sheet forms on which appropriate keywords have been circled. On these forms, each keyword is associated with a two-letter code. Use of the keyword forms, which list only 353 possible keywords, avoids chaotic proliferation of individually chosen keywords. Location codes can be included among the keywords and, following the practice introduced by Hans Hofmann, the latter are simply the country codes specified by the International Standards Organization. The restricted range of keywords serves, in effect, as a prestructured index for the database. Its limitations are not troublesome because the Sci-Mate "manager" software can rapidly search *all* text fields (i.e., titles as well as keywords).

21.4 Summary

In retrospect, we have no clear formula for greater success. Problems encountered were of two types: (*i*) inadequate software- and database-management expertise within the group, and (*ii*) interdependent parts of the project sometimes proceeded at incompatible paces. The first problem *might* be solved by spending money to obtain high-level professional assistance, but the required financial commitment would likely be great enough that use of the funds for additional scientific manpower would ultimately be favored. The second problem must be inevitable. It can be "solved" by adoption of more flexible plans. For example, rather than requiring that all ages be drawn from a database that was completed only near the end of the project, we might have planned at the outset for what developed in practice: a system in which provisional ages could be assigned as required by individual investigators, then superseded by "best estimates" compiled by specialists.

Our computerized databases have served well; moreover, they provide a flexible basis for continuing work.

22

Proterozoic and Selected Early Cambrain Microfossils and Microfossil-Like Objects

CARL V. MENDELSON   J. WILLIAM SCHOPF

22.1 Introduction

The PPRG microfossil database (MICRO.DBF) contains information regarding approximately 3500 occurrences of Proterozoic and Early Cambrian microfossils and microfossil-like objects reported from about 470 stratigraphic units, the data included being based on analysis of 316 published papers, monographs, and books. This compilation forms the primary foundation for interpretations discussed in Sections 5.4, 5.5, 11.3, and Chapter 25. Tables 22.1–22.3, below, contain only a fraction of the information stored in the database; a brief description of the types of data stored (and of the database structure) is presented in Chapter 21.

22.1.1 Coverage

Table 22.3 includes an estimated 90 percent of all occurrences of authentic Proterozoic microfossils reported prior to August, 1988. Data from some relatively minor Soviet, Chinese, and Indian papers are not included, and a few significant Chinese monographs appeared too late to be covered. The Early Cambrian data are not comprehensive, but are perhaps representative; about 40% of the reported occurrences of Early Cambrian nonmineralized microfossils have been entered. The coverage in Tables 22.1 and 22.2 is a bit more eclectic. Because of time constraints, a number of papers containing descriptions of nonfossils, pseudofossils, and dubiofossils (see below for definitions) were not analyzed. The amassed data, although not complete, represent the largest compilation of this type now available for Proterozoic microfossils.

22.1.2 Content

Tables 22.1, 22.2, and 22.3 are based on the illustrated literature of Proterozoic and Early Cambrian microfossils and microfossil-like objects. We follow the terminology of Hofmann (1972; and revision of 1978) in allocating the described objects to the three tables. For the reader's convenience, Hofmann's (1987) definitions are reproduced below. [For more complete definitions and comments, consult Hofmann (1972, 1987) and Schopf (1983b).]

micrononfossils: contaminants and artifacts of preparation (Table 22.1).
micropseudofossils: structures resembling fossil organisms, but abiogenic (Table 22.1).
microdubiofossils: questionably biogenic remains (Table 22.2).
microfossils: morphologic remains of microorganisms (Table 22.3).

22.1.3 Caveats

The tables have been checked for accuracy but mistakes remain. For example, some authors (including us) have illustrated the same specimen in two or more papers. We have tried to eliminate these duplicate records, but are certain that we have missed some. We estimate that the actual number of entries in Table 22.3 is inflated by less than five percent.

We emphasize that we have *not* undertaken a formal taxonomic revision (see Chapter 25 for an *informal* revision); in addition, the tables should *not* be taken as guides to the proper nomenclature of Proterozoic and Early Cambrian microfossils. In most cases we present the taxon name as given by the authors, except that we have corrected obvious errors in orthography. We have also modified some specific epithets so that they agree in gender with their genus; such orthographic corrections are automatic under the International Code of Botanical Nomenclature (ICBN).

Note that a significant proportion of the names presented here are invalid or illegitimate, and that many transfers violate rules in the ICBN. For example, the very common Proterozoic genus *Protosphaeridium* Timofeev 1966 is illegitimate and must be rejected; it is a superfluous name for *Protoleiosphaeridium* Timofeev 1960. Unfortunately, few micropaleontologists studying the Proterozoic seem to be aware of this. Likewise, *Synsphaeridium* Timofeev 1966 is a superfluous name for *Protoleiosphaeridium*. These two cases, and others, have been reviewed by Loeblich and Tappan (1976); and they are just the tip of the iceberg. Clearly, a sorting out of this nomenclatural and taxonomic mess is a daunting task. Preliminary steps in this direction have been taken (Chapter 25; German et al. 1979).

22.1.4 Conventions

Entries in Tables 22.1–22.3 have the following format:

> Lithostratigraphic Unit
>
> Geographic Location
>
> Estimated Age

Name. (Reference Citation)
Interpretation (Initial).

In the case of an informal designation (e.g., "small spheroid"), the name is given within parentheses; in some cases we have added clarifying prose, in brackets. Formal taxonomic names are given in full: genus, specific epithet, taxon author(s), date of publication (but not necessarily of valid publication!). Question marks after the author or within the date indicate that we have been unable to ascertain a definite author or date of publication.

The interpretation is followed by one or two letters in parentheses, with or without quotation marks. The letters identify the person (M = Mendelson; S = Schopf) primarily responsible for analyzing the figured specimen and making the interpretation. Quotation marks surrounding the letter(s) indicate that the interpretation is the same as or similar to that given by the original author(s).

22.1.5 Abbreviations

22.1.5A Lithostratigraphic unit

Lithostratigraphic units are indicated as follows:

Sgrp.	Supergroup
Subgrp.	Subgroup
Grp.	Group
Fm.	Formation
Mbr.	Member

22.1.5B Geographic occurrence

The first two letters identify the continent:

AF	Africa
AN	Antarctica
AS	Asia
AU	Australasia
EA	Eurasia (e.g., localities in the Ural Mountains)
EU	Europe
SA	South America
NA	North America

The next group of three letters identify the country according to the standard ISO abbreviations to which we have added SIB (Siberia):

ATA	Antarctica
AUS	Australia
BEN	Benin
BRA	Brazil
BWA	Botswana
CAN	Canada
CHN	China
CSK	Czechoslovakia
DDR	East Germany
EGY	Egypt
ESH	Spanish Sahara
ESP	Spain
FIN	Finland
FRA	France
GAB	Gabon
GBR	Great Britain
GRL	Greenland
GUY	Guyana
HVO	Upper Volta
IND	India
MLI	Mali
MRT	Mauritania
NAM	Namibia
NOR	Norway
POL	Poland
PRT	Portugal
ROM	Romania
SAU	Saudi Arabia
SIB	Siberia
SJM	Svalbard and Jan Mayen Islands
SUN	Soviet Union
SWE	Sweden
USA	United States
ZAF	South Africa
ZAR	Zaire
ZMB	Zambia

Some entries have a final group of two letters; these identify the province or state for the following countries:

Australia:

NT	Northern Territory
QL	Queensland
SA	South Australia
WA	Western Australia

Canada:

AT	Alberta
BC	British Columbia
NF	Newfoundland
NW	Northwest Territories
ON	Ontario
QU	Québec
YK	Yukon Territory

India:

AP	Andhra Pradesh
HP	Himachal Pradesh
KA	Karnataka (= Mysore)
MP	Madhya Pradesh
UP	Uttar Pradesh

United States:

AZ	Arizona	MN	Minnesota
AK	Alaska	MT	Montana
CA	California	TX	Texas
ID	Idaho	UT	Utah
MI	Michigan	WI	Wisconsin

Table 22.1. *Proterozoic and Selected Early Cambrian Micrononfossils and Micropseudofossils*

Early Proterozoic
2550 to 2250 Ma:
Brockman Iron Fm.
AU AUS WA
(ca. 2550 Ma)

(Type "C" structure). (LaBerge 1967)
　Mineralic (hematite, stilpnomelane) colloid-derived spheroids (S).
(Type "D" structure). (LaBerge 1967)
　Mineralic (hematite, stilpnomelane) colloid-derived spheroids (S).
(fossil iron bacteria). (Karkhanis 1976)
　Modern filamentous contaminants (S).

Marra Mamba Iron Fm.
AU AUS WA
(ca. 2550 Ma)

(Type "C" structure). (LaBerge 1967)
　Mineralic (hematite, stilpnomelane) colloid-derived spheroids (S).
(Type "D" structure). (LaBerge 1967)
　Mineralic (hematite, stilpnomelane) colloid-derived spheroids (S).

Transvaal Sgrp., Malmani Dolomite
AF ZAF
(ca. 2400 Ma)

(coccoid forms). (Nagy 1978)
　Modern endoliths (S).
(filamentous forms). (Nagy 1978)
　Modern endoliths (S).
(oscillatoriacean and rivulariacean algae). (Cloud and Morrison 1979)
　Modern fungal hyphae ("S").
(round forms). (Nagy 1974)
　Modern endoliths (S).
Petraphera vivescenticula Nagy 1974. (Nagy 1978)
　Modern endoliths (S).
Petraphera vivescenticula Nagy 1974. (Nagy 1974)
　Modern filamentous endolith (S).
Petraphera vivescenticula Nagy 1974. (Cloud and Morrison 1979)
　Modern filamentous endolith ("S").

Transvaal Dolomite
AF ZAF
(ca. 2300 Ma)

(non-septate filaments). (MacGregor et al. 1974)
　Contaminant introduced during sample preparation (S).
(tapering filament with enlarged basal cell). (MacGregor et al. 1974)
　Contaminant introduced during sample preparation (S).
(tapering multicellular filament). (MacGregor et al. 1974)
　Contaminant introduced during sample preparation (S).

Ventersdorp Sgrp., Mogobane "Series"
AF BWA
(ca. 2300 Ma)

Botswanella-Konkretion. (Pflug and Strübel 1969)
　Nonbiogenic concretion (S).

Gowganda Fm.
NA CAN ON
(ca. 2250 Ma)

(actinomycete hyphae). (Jackson 1967)
　Contaminants and artifacts of TEM preparation (S).
(bacterial cells or actinomycete spores). (Jackson 1967)
　Artifact of TEM sample preparation (S).

Transvaal Sgrp.
AF ZAF
(ca. 2250 Ma)

(trichterförmige sporogene "Alge"). (Prashnowsky and Oberlies 1972)
　Contaminants and artifacts of TEM sample preparation (S).

2250 to 2000 Ma:
Roraima Fm.
SA GUY
(ca. 2200 Ma)

(fossil-like objects). (Bailey 1964)
　Mineralic pseudofossils (S).
(larger fossil-like objects). (Bailey 1964)
　Mineralic pseudofossils (S).

Belcher Grp.
NA CAN NW
(ca. 2150 Ma)

"*Palaeomicrocoleus*". (Hofmann and Jackson 1969)
　Trails of ambient pyrite grains ("S").

Ironwood Iron Fm.
NA USA MI
(ca. 2150 Ma)

(Type "B" structure [excluding *Eosphaera*]). (LaBerge 1967)
　Mineralic (hematite) colloid-derived spheroids (S)
　[also reported from other iron-formations].
(Type "G"structure). (LaBerge, 1996)
　Mineralic (hematite, greenalite, chlorite, stilpnomelane,
　minnesotaite, riebeckite, carbonate, magnetite, limonite),
　colloid-derived spheroids (S).

Francevillien
AF GAB
(ca. 2095 Ma)

(Algues Bothryococcacés). (Feys et al. 1966)
　Carbonaceous stain associated with botryoidal chalcedony (S).

Biwabik Iron Fm., lower cherty unit
NA USA MN
(ca. 2090 Ma)

(Type 'A" structure). (LaBerge 1967)
　Mineralic (greenalite, hematite) colloid-derived spheroids (S).

Biwabik Iron Fm., lower slaty Mbr
NA USA MN
(ca. 2090 Ma)

(spheroids). (Cloud and Licari 1968)
　Mineralic spherulites (S).

Gunflint Iron-Fm
NA CAN ON
(ca. 2090 Ma)

("filaments" [broad]). (Moorhouse and Beales 1962)
　Mineralic nonbiogenic "filaments" (S).
("spicules"). (Moorhouse and Beales 1962)
　Nonbiogenic mineralic needles (S).
(coccoid bacteria-like objects). (Schopf et al. 1965)
　Modern bacterial contaminants (S).
(colonial, actinomorphic aggregates). (Tyler and Barghoorn 1954)
　Botryoidal chalcedony (S).
(other [presumed organic] structures). (Moorhouse and Beales 1962)
　Nonbiogenic carbonate pseudomorphs (S).

Table 22.1 (Continued).

(rod-shaped bacterial cells). (Schopf et al. 1965)
  Modern bacterial contaminants (S).
(spherulitic structures and spherical bodies). (Moorhouse and Beales 1962)
  Nonbiogenic, inorganic spherulites and mineral grains (S).
Cumulosphaera lamellosa Edhorn 1973. (Edhorn 1973)
  Mineralic micropseudofossil (S).
Entosphaeroides amplus Barghoorn 1965. (Akiyama and Imoto 1975)
  Contaminants: modern fungal spores? (M S).

Gunflint Iron Fm.
NA CAN ON
(ca. 2090 Ma)
[also reported from other iron-formations]

(spherical to elliptical structures). (LaBerge 1973)
  Mineralic (hematite, siderite, pyrite, iron-silicate), colloid-derived spheroids (S).

Kipalu Iron Fm.
NA CAN NW
(ca. 2090 Ma)

(Type "A" structure). (LaBerge 1967)
  Mineralic (greenalite, hematite) colloid-derived spheroids (S).

Pokegama Quartzite
NA USA MN
(ca. 2090 Ma)

Palaeomicrocoleus gruneri Korde [Kordeh] 1965. (Vologdin and Kordeh 1965)
  Trails of ambient pyrite grains (S).

Temiscamie Iron Fm.
NA CAN QU
(ca. 2090 Ma)

(Type "A" structure). (LaBerge 1967)
  Mineralic (greenalite, hematite) colloid-derived spheroids (S).

Negaunee Iron Fm.
NA USA MI
(ca. 2050 Ma)

(hematite or limonite framboids). (Lougheed and Mancuso 1973)
  Hematite, probably pseudomorphic after pyrite ("S").
(possible fossils). (Mancuso et al. 1971)
  Planar, inorganic, nonbiogenic structures (recrystallized fracture fillings?) (S).

2000 to 1800 Ma:
Huronian Sgrp., Ogishke Conglomerate
NA CAN ON
(ca. 1950 Ma)

Inactis or *Microcoleus*. (Gruner 1923)
  Trails of ambient pyrite grains (S).

Transvaal Sgrp., Pretoria Grp.
AF ZAF
(ca. 1900 Ma)

(reticulate spheroids). (Cloud and Licari 1968)
  Mineralic spherulites (S).

Changcheng Grp., Chuanlinggou Fm.
AS CHN
(ca. 1850 Ma)

Glottimorpha asiatica Timofeev 1966. (Timofeev 1966)
  Angular sapropelic platelet (S).
Polyporata obsoleta Sin [Xing] and Liu 1973. (Xing and Liu 1973)
  Degraded, irregular sapropelic fragment (S).

Sortis Grp., Foselv Fm.
NA GRL
(ca. 1850 Ma)

(small short filaments). (Bondesen et al. 1967)
  Mineral grains or inclusions (S).
(type 3 microscopic sphere). (Bondesen et al. 1967)
  Bubble with adhering detritus produced during maceration (S).

Vallen Grp., Graensesø Fm.
NA GRL
(ca. 1850 Ma)

(large fragments and filaments). (Bondesen et al. 1967)
  Irregular organic debris (S).
(small irregular lumps). (Bondesen et al. 1967)
  Bedded organic debris (S).
Vallenia erlingii Raunsgaard Pedersen [Pedersen] 1967. (Bondesen et al. 1967)
  Recrystallized oöliths or spherulites (S).

Vallen Grp., Zig Zag Land Fm.
NA GRL
(ca. 1850 Ma)

(filaments). (Bondesen et al. 1967)
  Irregular organic debris (S).
(type 3 microscopic sphere). (Bondesen et al. 1967)
  Bubble with adhering detritus produced during maceration (S).

Vetlanda "Series"
EU SWE
(ca. 1850 Ma)

(spheroidal to sub-triangular bodies). (Vidal and Röshoff 1971)
  Modern contaminant (personal communication, G. Vidal 1987).
(spore-like bodies). (Vidal and Röshoff 1971)
  Modern contaminant (personal communication, G. Vidal 1987).
(unbranched filaments). (Vidal and Röshoff 1971)
  Modern contaminant (personal communication, G. Vidal 1987).

Jatulian
EU SUN
(ca. 1800 Ma)

Kareliana zonata Korde [Kordeh] 1965. (Vologdin and Kordeh 1965)
  Nonbiogenic mineralic aggregates (S).

Jatulian, marine Jatulian dolomites
EU FIN
(ca. 1800 Ma)

Hyypiana jatulica Tynni 1987. (Tynni and Sarapää 1987)
  Nonbiogenic (mineralic) crystallites and inclusions (S).

Minas "Series"
SA BRA
(ca. 1800 Ma)

(trichterförmige sporogene 'Algengebilde") (Prashnowsky and Oberlies 1972)
  Contaminants and artifacts of TEM sample preparation (S).

Table 22.1 (Continued).

1800 to 1600 Ma:
Bothnia Fm.
EU FIN
(ca. 1770 Ma)

Corycium enigmaticum Sederholm 1912. (Timofeev 1966)
  Figures do not permit interpretation (S M).

Mount Isa Shale
AU AUS QL
(ca. 1670 Ma)

(micro-fossils from solution of pyrite). (Love and Zimmerman 1961)
  Angular nonbiogenic bodies associated with pyrite (S).

Changcheng Grp., Dahongyu Fm.
AS CHN
(ca. 1650 Ma)

Polyporata sp. 2 of Sin [Xing] and Liu 1973. (Xing and Liu 1973)
  Irregular sapropelic fragments (S).

**Middle Proterozoic
1600 to 1200 Ma:**

Belt Sgrp., Wallace Fm.
NA USA ID
(ca. 1390 Ma)

(Delta-Typus). (Pflug 1968)
  Probably fungal or algal contaminant (S).
(Epsilon-Typus). (Pflug 1968)
  Contaminants or artifact of preparation (S).
(Gamma-Typus). (Pflug 1968)
  Mineralic (?) needles (S).
(Sigma-Typus). (Pflug 1968)
  Fungal contaminants (S).
Catinella polymorpha Pflug 1966. (Pflug 1966b)
  Modern fungal contaminant (S).
Fibularix Pflug 1965. (Pflug 1968)
  Fungal contaminants (S).
Fibularix porulosa Pflug 1966. (Pflug 1966b)
  Modern fungal contaminant (S).
Fibularix spinosa Pflug 1966. (Pflug 1966b)
  Modern fungal contaminant (S).
Fibularix verrucosa Pflug 1966. (Plug 1966b)
  Modern fungal contaminant (S).
cf. *Filamentella* Pflug 1965. (Pflug 1968)
  Fungal contaminants (S).
Millaria implexa Pflug 1966. (Pflug 1966b)
  Modern fungal contaminant (S).
cf. *Polycellaria* Pflug 1965. (Pflug 1968)
  Fungal contaminant (S).
Scintilla perforata Pflug 1966. (Pflug 1966b)
  Modern fungal contaminant (S).
Tormentella tubiformis Pflug 1966. (Pflug 1966b)
  Modern fungal contaminant (S).

Belt Sgrp., Missoula Grp., Striped Peak Fm.
NA USA ID
(ca. 1340 Ma)

Polycellaria bonnerensis Pflug 1965. (Pflug 1965a)
  Modern fungal contaminant (S).
Polycellaria levis Pflug 1965. (Pflug 1965a)
  Modern fungal contaminant (S).

Polycellaria longata Pflug 1965. (Pflug 1965a)
  Modern fungal contaminant (S).
Tricellaria delylensis Pflug 1965. (Pflug 1965a)
  Modern fungal contaminant (S).

Jixian Grp., Wumishan Fm.
AS CHN
(ca. 1325 Ma)

(fan-shaped crusts). (Cao 1984)
  Diagenetic mineralic fabric (pseudomorphic after evaporite minerals?) (S).
Cornutosphaera polycornuta Sin [Xing] 1973. (Xing and Liu 1973)
  Opaque, irregular sapropelic globule (S).
Lignum nematoideum Sin [Xing] 1973. (Xing and Liu 1973)
  Irregular, partly fibrous sapropelic (?) fragment (S).
Lignum punctulosum Sin [Xing] and Liu 1973. (Xing and Liu 1973)
  Irregular sapropelic fragment (S).
Lignum striatum Sin [Xing] and Liu 1973. (Xing and Liu 1973)
  Irregular sapropelic fragment (S).
Polyporata obsoleta Sin [Xing] and Liu 1973. (Xing and Liu 1973)
  Degraded, irregular sapropelic fragment (S).

BeltS grp., Missoula Sgrp., Libby Fm.
NA USA ID
(ca. 1300 Ma)

Fibularix funicula Pflug 1965. (Pflug 1966b)
  Modern fungal contaminant (S).
Fibularix funicula Pflug 1965. (Pflug 1965a)
  Moden fungal contaminant (S).
Montanella beltensis Pflug 1965. (Pflug 1965a)
  Modern fungal (?) contaminant (S).
Polycellaria indistincta Pflug 1965. (Pflug 1965a)
  Modern fungal (?) contaminant (S).

Jixian Grp., Hongshuizhuang Fm.
AS CHN
(ca. 1250 Ma)

Conodonta? sp. of Sin [Xing] and Liu 1973. (Xing and Liu 1973)
  Opaque fusiform sapropelic fragment (S).
Cornutosphaera polycornuta Sin [Xing] 1973. (Xing and Liu 1973)
  Opaque, irregular sapropelic globule (S).
Cornutosphaera sp. of Sin [Xing] and Liu 1973. (Xing and Liu 1973)
  Opaque sapropelic globule (S).
Glottimorpha ordinata Sin [Xing] 1973. (Xing and Liu 1973)
  Opaque organic (?) globule (S).
Glottimorpha sp. of Sin [Xing] and Liu 1973. (Xing and Liu 1973)
  Opaque sapropelic globule (S).
Lignum sp. of Sin [Xing] and Liu 1973. (Xing and Liu 1973)
  Irregular sapropelic fragment (S).
Pterospermopsimorpha? sp. 2 of Sin [Xing] and Liu 1973. (Xing and Liu 1973)
  Opaque irregular sapropelic globule (S).
Quadratimorpha florentis Sin [Xing] and Liu 1973. (Xing and Liu 1973)
  Fibrous degraded organic (?) particle (S).
Quadratimorpha sp. of Sin [Xing] and Liu 1973. (Xing and Liu 1973)
  Opaque sapropelic globule (S).
Triangumorpha striata Sin [Xing] and Liu 1973. (Xing and Liu 1973)
  Modern (fern?) spore contaminant with trilete mark (S).

Table 22.1 (Continued).

1200 to 1000 Ma:
Jixian Grp., Tieling Fm.
AS CHN
(ca. 1185 Ma)

Polyporata obsoleta Sin [Xing] and Liu 1973. (Xing and Liu 1973)
  Degraded, irregular sapropelic fragment (S).
Polyporata sp. 1 of Sin [Xing] and Liu 1973. (Xing and Liu 1973)
  Irregular sapropelic fragment (S).

Murandav Fm.
AS SIB
(ca. 1150 Ma)

Amurella pilosa Vologdin 1965. (Vologdin 1965)
  Globular mineralic aggregate (S).
Murandavia amurica Vologdin 1965. (Vologdin 1965)
  Globular mineralic aggregate (S).
Murandavia rotunda Vologdin 1965. (Vologdin 1965)
  Globular mineralic aggregate (S).

Nehlehgehr Fm.
AS SIB
(ca. 1150 Ma)

Costatosphaerina ramosa Lopuchin [Lopukhin] 1974. (Lopukhin 1974)
  Coagulated organic matter (M S).
Microtaenia granulosa Lopuchin [Lopukhin] 1974. (Lopukhin 1974)
  Filamentous artifact of sample preparation (S M).

Nonesuch Shale
NA USA MI
(ca. 1055 Ma)

(Type A). (Jost 1968)
  Modern contaminants (diatom frustules?) (S).
(Type B). (Jost 1968)
  Mineralic (siliceous?) pseudofossil or modern contaminant (S).

Siehtichan Fm.
AS SIB
(ca. 1050 Ma)

Costatosphaerina ramosa Lopuchin [Lopukhin] 1974. (Lopukhin 1974)
  Coagulated organic matter (M S).

Vindhyan Sgrp. (Lower)
AS IND
(ca. 1050 Ma)

(glauconite cast of organism). (Misra 1949)
  Broken glauconite grain (S).

Vindhyan Sgrp., Semri Grp., Suket Shale
AS IND
(ca. 1050 Ma)

(? Cyanophyceae). (Sitholey et al. 1953)
  Modern contaminants and artifacts of preparation (S).

Mbuji Mayi Sgrp. (= Bushimay Sgrp.)
AF ZAR
(ca. 1020 Ma)

(unbranched filament). (Maithy 1975)
  Elongate sapropel (S).
Kakabekia flabelliformis Maithy 1975. (Maithy 1975)
  Irregular mass of organic matter produced during maceration (S M).

Kakabekia rarea Maithy 1975. (Maithy 1975)
  Irregular mass of organic matter produced during maceration (S M).

Ehsehlehkh Fm.
AS SIB
(ca. 1000 Ma)

Costatosphaerina ramosa Lopuchin [Lopukhin] 1974. (Lopukhin 1974)
  Coagulated organic matter (M S).

1000 to 900 Ma:
Mwashya Grp.
AF ZMB
(ca. 975 Ma)

Fibularix mendelsohnii Binda 1977. (Binda 1977)
  Irregular pyritic and carbonaceous (?) detritus and (?) modern fungal contaminants (S).

"pre-Torridonian" pebbles in Torridon Grp., Applecross Fm.
EU GBR
(ca. 950 Ma)

(branching filaments). (Muir and Sutton 1970)
  Contaminant: fungal mycelia [Peat 1984, p. 68].
(irregularly branched chains of cells). (Muir and Sutton 1970)
  Contaminant: fungal conidiospores [Peat 1984, p. 68].
(large non-septate filament). (Muir and Sutton 1970)
  Contaminant: fungal hyphae [Peat 1984, p. 68].

Lakhanda Fm.
AS SIB
(ca. 950 Ma)

(piece of broad trichome [giant]). (Timofeev et al. 1976)
  Compressed sapropel (S).
Aimophyton varium Timofeev and Hermann [German] 1979. (Timofeev and German 1979)
  Globular and irregular sapropel (S). [See Chapter 24, Pl. 17, E_1–E_2.]
Majaphyton antiquaum Timofeev and Hermann [German] 1976. Timofeev et al. 1976)
  Irregularly coalesced organic debris (S). [See Chapter 24, Pl. 17, C.]
Majaphyton ceratum Timofeev and Hermann [German] 1976. (Timofeev et al. 1976)
  Irregularly coalesced organic debris (S).
Majaphyton cyatum Timofeev and Hermann [German] 1976. (Timofeev et al. 1976)
  Irregularly coalesced organic debris (S).
Majasphaeridium carpogenum Hermann [German] 1979. (German 1979)
  Coalesced globular sapropel (S). [See Chapter 24, Pl. 17, D.]
Mucorites rifeicus Herman [German] 1979. (German 1979)
  Irregularly aggregated filaments and globules on sapropel (S). [See Chapter 24, Pl. 16, C, E; Pl. 17, B.]
Mycosphaeroides aggregatus Hermann [German] 1979. (German 1979)
  Irregularly aggregated globules on sapropel (S). [See Chapter 24, Pl. 16, D.]
Mycosphaeroides caudatus Herman [German] 1979. (German 1979)
  Irregularly aggregated globules on sapropel (S). [See Chapter 24, Pl. 17, F.]

Table 22.1 (Continued).

Qingbaikou Grp., Xiamaling Fm.
AS CHN
(ca. 950 Ma)

Laminarites antiquissimus Eichwald 1854. (Xing and Liu 1973)
  Sapropelic fragment (S).
Laminarites sp. of Timofeev 1966. (Timofeev 1966)
  Irregular sapropelic platelet (S).

"Upper Precambrian of Guelb-er-Richât de l'Adrar"
AF MRT
(ca. 900 Ma)

Asterosphaeroides monodii Boureau 1976. (Boureau 1976)
  Nonbiogenic ööliths (S).
Asterosphaeroides richatensis Boureau 1973. (Boureau 1974)
  oöliths, micro-oncolites?, and spherulites (S).
Babetosphaera africana var. *africana* Boureau 1972. (Boureau 1974)
  Oöliths, micro-oncolites?, and spherulites (S).
Desmochitina grandis Boureau 1977. (Boureau 1977a)
  Nonbiogenic oölith (S).
Metallogenium personatum Perfil'ev 1961. (Boureau 1974)
  Mineralic actinomorphic aggregates (S).
Monodites princeps Boureau 1977. (Boureau 1977a)
  Nonbiogenic oölith (S).
Nostocites vesiculosa Maslov 1929. (Boureau 1977a)
  Recrystallized botyroidal spherulitic texture (S).
Polysphaerula globulosa Boureau 1975. (Boureau 1977a)
  Nonbiogenic oölith (S).

Kundelungu Grp. (Lower)
AF ZMB
(ca. 900 Ma)

Fibularix garlickii Binda 1977. (Binda 1977)
  Irregular detritus and spheroidal pyrite (S).
Fibularix mendelsohnii Binda 1977. (Binda 1977)
  Irregular pyritic and carbonaceous (?) detritus and (?) modern fungal contaminants (S).
Fibularix zambiana Binda 1977. (Binda 1977)
  Irregular nonbiogenic microstructures produced by alveolation during palynological maceration (S).

Telemark Suite, Bandak Grp.
EU NOR
(ca. 900 Ma)

Telemarkites enigmaticus Dons 1959. (Dons 1959)
  Nonbiogenic siliceous concretion (S).

Série de Nara
AF MLI
(ca. 900 Ma)

Asterosphaeroides darsii Boureau 1976. (Boureau 1976)
  Nonbiogenic pisolites and ööliths (S).

Late Proterozoic
900 to 800 Ma:
Bitter Springs Fm.
AU AUS NT
(ca. 850 Ma)

Palaeoanacystis vulgaris Schopf 1968. (Schopf 1968)
  Modern contaminant (fungal spores?) (S).

Miroedikha Fm.
AS SIB
(ca. 850 Ma)

Palaeocalothrix divaricatus Hermann [German] 1981. (German 1981b)
  Modern contaminant, possibly insect part or sclereid (S). [See Chapter 24, Pl. 26, F.]
Sibiriafilum tunicum Hermann [German] 1986. (German 1986)
  Modern contaminant, possibly insect part or sclereid (S). [See Chapter 24, Pl. 25, A; Pl. 26, D.]

Qingbaikou Grp., Jingeryu Fm.
AS CHN
(ca. 850 Ma)

Protoleiosphaeridium sp. of Sin [Xing] and Liu 1973. (Xing and Liu 1973)
  Modern (fungal spore?) contaminant (S).

Bambuí Grp., Paraopeba Fm.
SA BRA
(ca. 810 Ma)

(filaments). (Cloud and Morrison 1979)
  Modern fungal hyphae and spores ("S").

Kundelungu Grp.
AF ZMB
(ca. 800 Ma)

(algal forms). (Ashley 1937)
  Inorganic sinuous tubes produced by ambient pyrite (S).

800 to 700 Ma:
Drill-Hole Kabakovo-62, Bashkiria
EA SUN
(ca. 790 Ma)

Pellicularia tenera Jankauskas [Yankauskas] 1980. (Yankauskas 1980a)
  Compressed folded sapropel (S).
Plicatidium latum Jankauskas [Yankauskas] 1980. (Yankauskas 1980a)
  Compressed sapropel (S).

Kangjia [Kangchia] Grp.
AS CHN
(ca. 790 Ma)

Archaeofavosina cf. *simplex* Naumova 1960. (Xing and Liu 1973)
  Sapropelic fragment (M S).
Polyporata obsoleta Sin [Xing] and Liu 1973. (Xing and Liu 1973)
  Degraded, irregular sapropelic fragment (S).
Protoleiosphaeridium sp. of Sin [Xing] and Liu 1973. (Xing and Liu 1973)
  Modern (fungal spore?) contaminant (S).

Xihe Grp., Nanfen Fm.
AS CHN
(ca. 790 Ma)

Polyporata obsoleta Sin [Xing] and Liu 1973. (Xing and Liu 1973)
  Degraded, irregular sapropelic fragment (S).

"Infra-Cambrian Oölite"
AF MRT
(ca. 785 Ma)

Babetosphaera africana Boureau and Monod 1958. (Boureau 1958)
  Silicified oölitic structure (S).

Table 22.1 (Continued).

Visingsö Grp.
EU SWE
(ca. 775 Ma)

Laminarites sp. of Timofeev 1966. (Timofeev 1966)
    Irregular sapropelic platelet (S).

"chert" of Prague region
EU CSK
(ca. 760 Ma)

("form L" and spiny remains). (Konzalová 1972)
    Irregularly shaped opaque particulate organic matter (S).

"Late Precambrian" of Timan
EU SUN
(ca. 750 Ma)

Glottimorpha asiatica Timofeev 1966. (Timofeev 1966)
    Angular sapropelic platelet (S).
Laminarites sp. of Timofeev 1966. (Timofeev 1966)
    Irregular sapropelic platelet (S).

Damara Sgrp.
AF NAM
(ca. 750 Ma)

(trichterförmige sporogene "Algen", Gebilde). (Prashnowsky and Oberlies 1972)
    Contaminants and artifacts of TEM sample preparation (S).

Brioverian (Middle)
EU FRA
(ca. 740 Ma)

(sporomorphes globuleuses). (Roblot 1964b)
    Globular carbonaceous particles (S).

"Algonkian lydites"
EU CSK
(ca. 735 Ma)

cf. *Polycellaria* Pflug 1965. (Vavrdová 1966)
    Modern (probably fungal) contaminant (S).

"Algonkian shales"
EU CSK
(ca. 735 Ma)

(Acritarcha). (Pacltová 1970)
    Contaminants, including fungal hyphae, as suggested by author ("S").
Polycellaria Pflug 1965. (Pacltová 1970)
    Contaminants, as suggested by author ("S").

Gangolihat Dolomites, Chhera Mbr.
AS IND
(ca. 735 Ma)

Protoleiosphaeridium problematicum Venkatachala, Bhandari, Chaube and Rawat 1974. (Nautiyal 1980)
    Irregular clump of organic detritus (S M).

"Adelaidean"
AU AUS SA
(ca. 700 Ma)

Laminarites sp. of Timofeev 1966. (Timofeev 1966)
    Irregular sapropelic platelet (S).

700 to 600 Ma:
Vindhyan Sgrp., Bhander Grp., Sirbu Shale
AS IND
(ca. 690 Ma)

(? Cyanophyceae). (Sitholey et al. 1953)
    Modern contaminants and artifacts of preparation (S).
(fusiform bodies). (Sitholey et al. 1953)
    Modern contaminants (S).

Simla Slate Grp.
AS IND
(ca. 685 Ma)

Satpulispora minuta Nautiyal 1979. (Nautiyal 1979)
    Mineralic (hematite?) spherules (S).

Hedmark Grp., Biskopåsen Fm.
EU NOR
(ca. 650 Ma)

(problematic fossil type VIII). (Spjeldnæs 1968)
    Mineralic fibers, nonbiogenic (S).
(sporomorph type I). (Spjeldnaes 1968)
    Ellipsoidal nodule, nonbiogenic (S).
(sporomorph type II). (Spjeldnæs 1968)
    Mineralic fibers, nonbiogenic (S).
(sporomorph type I). (Spjeldnæs 1968).
    Phosphatic (?) nodule, nonbiogenic (S).
(sporomorph type IV). (Spjeldnæs 1968)
    Mineralic nodule, nonbiogenic (S).
(sporomorph type V). (Spjeldnæs 1968)
    Mineralic nodules, nonbiogenic (S).
(sporomorph type VI). (Spjeldnæs 1968)
    Mineralic spheroid, nonbiogenic (S).
(sporomorph type VII). (Spjeldnæs 1968)
    Nonbiogenic mineralic spherulites (S).

Muhos Fm.
EU FIN
(ca. 650 Ma)

Spirosaccus punctata Tynni and Uutela 1984. (Tynni and Uutela 1984)
    Irregular, spindle-shaped clumps of organic matter (S).

"Praha-Šárka cherts"
EU CSK
(ca. 625 Ma)

Bohemipora pragensis Pacltová 1977. (Pacltová 1977)
    Mineralic pseudofossils (S).

"Šárka Zone lydite"
EU CSK
(ca. 625 Ma)

Palaeocryptidium Deflandre 1955. (Pacltová 1972)
    Mineralic pseudofossil (S).

Mogilev Fm.
EU SUN
(ca. 620 Ma)

Laminarites sp. of Timofeev 1966. (Timofeev 1966)
    Irregular sapropelic platelet (S).

Table 22.1 (Continued).

"Nama System"
AF NAM
(ca. 610 Ma)

(trichterförmige sporogene "Algen"). (Prashnowsky and Oberlies 1972)
  Contaminants and artifacts of TEM sample preparation (S).

Yaryshev Fm.
EU SUN
(ca. 610 Ma)

Ljadovia perforata Assejeva [Aseeva] 1983. (Aseeva and Velikanov 1983)
  Organic film with pyritic perforations (S). [See Chapter 24, Pl. 52, A_1–A_3.]
Striatella coriacea Assejeva [Aseeva] 1983. (Aseeva and Velikanov 1983)
  Probably sapropel (S). [See Chapter 24, Pl. 51, H.]

"Vendian of Tien Shan"
AS SUN
(ca. 600 Ma)

Trachymarginata globulosa Lopukhin 1966. (Lopukhin 1966)
  Degraded mineralic material (M S).

"Vendian" of Moscow
EU SUN
(ca. 600 Ma)

Laminarites sp. of Timofeev 1966. (Timofeev 1966)
  Irregular sapropelic platelet (S).

"Vendian" of Moldavia
EU SUN
(ca. 600 Ma)

Laminarites sp. of Timofeev 1966. (Timofeev 1966)
  Irregular sapropelic platelet (S).

"Vendian" of Morsovo
EU SUN
(ca. 600 Ma)

Laminarites sp. of Timofeev 1966. (Timofeev 1966)
  Irregular sapropelic platelet (S).

Laminarites Clay of Leningrad
EU SUN
(ca. 600 Ma)

Laminarites sp. of Timofeev 1966. (Timofeev 1966)
  Irregular sapropelic platelet (S).

Roznichi Fm.
EU SUN
(ca. 600 Ma)

Striatella coriacea Assejeva [Aseeva] 1983. (Aseeva and Velikanov 1983)
  Probably sapropel (S).

Serebriya Fm.
EU SUN
(ca. 600 Ma)

Striatella coriacea Assejeva [Aseeva] 1983. (Aseeva and Velikanov 1983)
  Probably sapropel (S).

600 to 550 Ma
Valdaj [Valdai] "Series"
EU SUN
(ca. 590 Ma)

Laminarites sp. of Timofeev 1966. (Timofeev 1966)
  Irregular sapropelic platelet (S).

Redkino Fm.
EU SUN
(ca. 580 Ma)

Striatella coriacea Assejeva [Aseeva] 1983. (Aseeva and Velikanov 1983)
  Probably sapropel (S). [See Chapter 24, Pl. 55, D.]

Kotlin Fm.
EU SUN
(ca. 560 Ma)

Volyniella rara Paškevičienė [Pashkyavichene] 1980.
(Pashkyavichene 1980)
  Modern contaminant, possibly insect part (S). [See Chapter 24, Pl. 54, K_1–K_2.]

Early Cambrian:

Belaya Fm.
AS SIB
(ca. 540 Ma)

Polyporata nidia Pichova [Pykhova] 1966. (Pykhova 1967)
  Coagulated organic matter containing bubbles: artifact of preparation (M).
Polyporata nidia Pychova [Pykhova] 1966. (Pykhova 1966)
  Coagulated organic matter containing bubbles: artifact of preparation (M S).
Polyporata verrucosa Pychova [Pykhova] 1966. (Pykhova 1966)
  Coagulated organic matter containing bubbles: artifact of preparation (M S).
Uniporata nidia Pychova [Pykhova] 1966. (Pykhova 1966)
  Coagulated organic matter containing bubbles: artifact of preparation (M S).
Uniporata torosa Pychova [Pykhova] 1966. (Pykhova 1966)
  Coagulated organic matter containing bubbles: artifact of preparation (M S).

Moty Fm.
AS SIB
(ca. 540 Ma)

Asperatopsohosphaera cf. *magna* Shepeleva. (Pykhova 1967)
  Coagulated organic matter: artifact of preparation (M).
Asperatopsophosphaera cf. *medialis* Shepeleva.
(Pykhova 1967)
  Coagulated organic matter: artifact of preparation (M).

Vampire Fm.
NA CAN YK
(c. 540 M)

(Microdubiofossil Type C). (Hofmann 1984)
  Filamentous clump of carbonaceous matter ("H").

Table 22.2. *Proterozoic and Selected Early Cambrian Microdubiofossils*

Early Proterozoic
2350 to 2000 Ma:

"Svecofennian"
EU FIN
(ca. 2350 Ma)

Corycium enigmaticum Sederholm 1912. (Sederholm 1912, 1925)
  Graphitic saclike structure, possibly of cyanobacterial origin, cf. *Nostoc* balls? (S).

Belcher Grp.
NA CAN NW
(ca. 2150 Ma)

(Type 1). (Hofmann and Jackson 1969)
  Organic material that grew in place (H). Clumps of organic debris (S).
(grains of iron oxide). (Moore 1918)
  Not possible to evaluate from camera-lucida drawings (S).

Biwabik Iron Fm.
NA USA MN
(ca. 2090 Ma)

(algae). (Gruner 1922)
  Trails of ambient pyrite grains, at least in part (e.g., pl. 7C) (S).
(bacilli). (Gruner 1922)
  Possibly minute mineralic needles (S).
(iron bacteria). (Gruner 1922)
  Mineralic (iron oxide) filaments, possibly diagenetic (in chalcedony?) (S).

Gunflint Iron Fm.
NA CAN ON
(ca. 2090 Ma)

(algae). (Gruner 1922)
  Trails of ambient pyrite grains, at least in part (e.g., pl. 7C) (S).
(organisms [that] resemble Radiolaria). (Edhorn 1973)
  Mineralic objects of uncertain biogenicity (S).
(subspherical and rod-shaped bodies). (Cloud 1965)
  Finely divided organic detritus? (S).
Glenobotrydion aenigmatis Schopf 1968. (Edhorn 1973)
  Probably biogenic, cf. *Corymbococcus* (S).
Huroniospora microreticulata Barghoorn 1965. (Akiyama and Imoto 1975)
  Possibly degraded, reticulate *Huroniospora* (S).
Menneria levis Lopuchin [Lopukhin] 1971. (Lopukhin 1975a)
  No evidence of being indigenous to rock (S).
Palaeoanacystis irregularis Edhorn 1973. (Edhorn 1973)
  Hematitic, poorly preserved spheroids, possibly cf. *Corymbococcus* (S).
Sphaerophycus gigas Edhorn 1973. (Edhorn 1973)
  Hematitic, poorly preserved spheroids, possibly cf. *Corymbococcus* (S).

"unnamed shale" of High Falls, Québec
NA CAN QU
(ca. 2000 Ma)

(elliptical graphitic bodies). (Stinchcomb et al. 1965)
  Possibly sphaeromorph acritarch (e.g., *Chuaria*) or cyanobacterial "sea ball" or "lake ball" (e.g., *Nostoc* colony) (S); possibly nonbiogenic (M).

Attikamagen Fm.
NA CAN QU
(ca. 2000 Ma)

(elliptical graphitic bodies). (Stinchcomb et al. 1965)
  Possibly sphaeromorph acritarch (e.g., *Chuaria*) or cyanobacterial "sea ball" or "lake ball" (e.g., *Nostoc* colony) (S); possibly nonbiogenic (M).

2000 to 1800 Ma:

Marquette Range Sgrp., Baraga Grp.
NA USA MI
(ca. 1925 Ma)

(chains of spheroids). (Cloud and Morrison 1980)
  Microdubiofossil ("S").

Krivoj Rog "Series"
EU SUN
(ca. 1875 Ma)

Corycium? *oligomerum* Bielokrys [Belokrys] & Mordovez 1968. (Belokrys and Mordovets 1968)
  Deformed graphitic (?) sapropelic material? (S).

Changcheng Grp., Chuanlinggou Fm.
AS CHN
(ca. 1850 Ma)

Leiominuscula incrassata Sin [Xing] & Liu 1973. (Xing and Liu 1973)
  Figure does not permit interpretation (M S).
Pulvinomorpha mitis Timofeev 1966. (Timofeev 1966)
  Possibly sapropelic platelet (S).

Sortis Grp.
NA GRL
(ca. 1850 Ma)

(microscopic globules). (Pedersen and Lam 1970)
  Possibly poorly preserved sphaeromorph acritarchs (S M).

Vallen Grp., Grænsesø Fm
NA GRL
(ca. 1850 Ma)

(type 1 microscopic sphere). (Bondesen et al. 1967)
  Possible sphaeromorphs, based on two poorly preserved specimens (S).

Vallen Grp., Zig Zag Land Fm.
NA GRL
(ca. 1850 Ma)

(type 1 microscopic sphere). (Bondesen et al. 1967)
  Possible sphaeromorphs, based on two poorly preserved specimens (S).

Jatulian
EU SUN
(ca. 1800 Ma)

Jatuliana furcata Korde [Kordeh] 1965. (Vologdin and Kordeh 1965)
  Probably diagenetically produced nonbiogenic structure (S).
Menneria levis Lopuchin [Lopukhin] 1981. (Lopukhin 1975a)
  No evidence of being indigenous to rock (S).
Protosphaeridium sp. of Timofeev, German & Mikhajlova 1976. (Timofeev et al. 1976)
  Figure does not permit interpretation (M S).

Table 22.2 (Continued).

1800 to 1600 Ma:

Changcheng Grp., Tuanshanzi Fm.
AS CHN
(ca. 1750 Ma)

Pulvinomorpha angulata Timofeev 1966. (Timofeev 1966)
  Possibly sapropelic platelet (S).

Hornby Bay Grp.
NA CAN NW
(ca. 1732 Ma)

(problematic microstructure). (Horodyski et al. 1985)
  Microdubiofossil ("S").

McArthur Grp., Amelia Dolomite
AU AUS NT
(ca. 1600 Ma)

Gunflintia oehlerae Muir 1976. (Muir 1976)
  Figures do not permit interpretation (S).

Pechenga "Series" correlative (?)
EU SUN
(ca. 1600 Ma)

Leiosphaeridia bituminosa Timofeev 1966. (Timofeev et al. 1976)
  Figure does not permit interpretation (S).

Middle Proterozoic
1600 to 1400 Ma:

Belt Sgrp., Newland Limestone
NA USA MT
(ca. 1440 Ma)

(chains of cells from *Camasia spongiosa*). (Walcott 1914)
  Mineral grains or possible filamentous prokaryotes (S).

Changcheng Grp., Gaoyuzhuang Fm.
AS CHN
(ca. 1425 Ma)

Paleocalothrix xui Xu 1984. (Xu 1984a)
  Figure does not permit interpretation (S).

1400 to 1200 Ma:

Belt Sgrp., Wallace Fm.
NA USA ID
(ca. 1390 Ma)

(Foraminiferen). (Pflug 1965b)
  Nondescript elongate to globular mineralic (?) objects (S).
(Gruppe A). (Pflug 1964)
  Possibly contaminants (S).
(Gruppe B). (Pflug 1964)
  Possibly contaminants (S).
(Gruppe C). (Pflug 1964)
  Possibly contaminants (S).
(Gruppe D). (Pflug 1964)
  Possibly contaminants (S).
(Gruppe E). (Pflug 1964)
  Possibly contaminants (S).
(Gruppe F). (Pflug 1964)
  Possibly contaminants (S).
(Gruppe G). (Pflug 1964)
  Possibly contaminants (S).

Tottinsk Fm.
AS SIB
(ca. 1350 Ma)

Konderia elliptica Weiss [Vejs] 1983. (Vejs 1983)
  Possibly globular to elongate sapropel (S). [See Chapter 24, Pl. 5, G, H.]

Waterton Fm.
NA CAN AT
(ca.1350 Ma)

(small siliceous filaments). (J. Oehler 1976)
  Possibly silicified filamentous cyanobacterial? sheaths and trichomes ("S").

Zigazino-Komarovsk Fm.
EA SUN
(ca. 1350 Ma)

Satka colonialica Jankauskas [Yankauskas] 1979.
(Yankauskas 1979b)
  Possibly globular sapropel (S).

Belt Sgrp., Missoula Grp., Striped Peak Fm.
NA USA ID
(ca. 1340 Ma)

(Foraminiferen). (Pflug 1965b)
  Nondescript elongate to globular mineralic (?) objects (S).

Jixian Grp., Wumishan Fm.
AS CHN
(ca. 1325 Ma)

(spirochaetic thalli). (Cao 1984)
  Possibly aff. *Obruchevella*, but figure does not permit interpretation (S).
Quadratimorpha tenera Sin [Xing] & Liu 1973. (Xing and Liu 1973)
  Irregular globule, possibly degraded sphaeromorph acritarch (S).

Belt Sgrp., Hoadley Fm.
NA USA MT
(ca. 1300 Ma)

(Foraminiferen). (Pflug 1965b)
  Nondescript elongate to globular mineralic (?) objects (S).

Belt Sgrp., Missoula Grp., Libby Fm.
NA USA ID
(ca. 1300 Ma)

Filamentella plurima Pflug 1965. (Pflug 1965a)
  Figures do not permit interpretation (S).

Jixian Grp., Hongshuizhuang Fm.
AS CHN
(ca. 1250 Ma)

Leiofusa bicornuta Sin [Xing] & Liu 1973. (Xing and Liu 1973)
  Possibly preparational artifact (S).
Leiofusa digitata Sin [Xing] 1973. (Xing and Liu 1973)
  Opaque globular to elongate organic (?) particle (S).
Orygmatosphaeridium exile Sin [Xing] 1973. (Xing and Liu 1973)
  Figure does not permit interpretation (S).
Pterospermopsimorpha? sp. 1 of Sin [Xing] & Liu 1973. (Xing and Liu 1973)
  Figure does not permit interpretation (S).
Quadratimorpha ordinata Sin [Xing] & Liu 1973. (Xing and Liu 1973)
  Possible prismatomorph acritarch (S M).

Table 22.2 (Continued).

Quadratimorpha simplisis Sin [Xing] & Liu 1973. (Xing and Liu 1973)
  Possibly prismatomorph acritarch or inorganic (mineralic) particles (S).

Kokdzhot Fm.
AS SUN
(ca. 1200 Ma)

Menneria levis Lopuchin [Lopukhin] 1971. (Lopukhin 1971b)
  Contaminant? No evidence this is indigenous to rock (S).

1200 to 1000 Ma:

Jixian Grp., Tieling Fm.
AS CHN
(ca. 1185 Ma)

Taeniatum crassum Sin [Xing] & Liu 1973. (Xing and Liu 1973)
  Partly or wholly modern fibrous contaminants (S).

Roan Grp. (Lower)
AF ZMB
(ca. 1155 Ma)

(chains of cells). (Binda 1972)
  Possible prokaryotic filaments (S).

Nehlehgehr Fm.
AS SIB
(ca. 1150 Ma)

Angulisphaerina perfosoclathrata Lopuchin [Lopukhin] 1974. (Lopukhin 1974)
  Contaminant? Mineral grains? No evidence of being indigenous to rock (S).
Granomarginata typica Lopuchin [Lopukhin] 1974. (Lopukhin 1974)
  Contaminant? No evidence of being indigenous to rock (S).

"Riphean of South Urals"
EA SUN
(ca. 1125 Ma)

Leiopsophosphaera mugodjarica Lopuchin [Lopukhin]. (Lopukhin 1975a)
  No evidence of being indigenous to rock (S).

"sedimentary iron-formation" of Florence, Wisconsin
NA USA WI
(ca. 1100 Ma)

(Foraminiferen). (Pflug 1965y)
  Nondescript elongate to globular mineralic (?) objects (S).

Nonesuch Shale
NA USA MI
(ca. 1055 Ma)

(Type C). (Jost 1968)
  Modern contaminant? (S).

Aravalli Sgrp., Bijawar Grp.
AS IND UP
(ca. 1050 Ma)

Contrahofilum schopfii Nautiyal 1980. (Nautiyal 1983b)
  Figures do not permit interpretation (S).

Siehtichan Fm.
AS SIB
(ca. 1050 Ma)

Granomarginata typica Lopuchin [Lopukhin] 1974. (Lopukhin 1974)
  Contaminant? No evidence of being indigenous to rock (S).

Vindhyan Sgrp. (Lower)
AS IND
(ca. 1050 Ma)

(dasycladacean alga). (Misra 1949)
  Single specimen, not possible to evaluate from published figure (S).

Vindhyan Sgrp., Semri Grp.
AS IND
(ca. 1050 Ma)

Leisosphaeridia sarjeantii Nautiyal 1983. (Nautiyal 1983a)
  Figures do not permit interpretation (M S).

Vindhyan Sgrp., Semri Grp., Arangi Fm.
AS IND UP
(ca. 1050 Ma)

Contrahofilum schopfii Nautiyal 1980. (Nautiyal 1983b)
  Figures do not permit interpretation (S).
Leiofusa actinomorpha Maithy 1975. (Nautiyal 1983b)
  Figures do not permit interpretation (S M).

Vindhyan Sgrp., Semri Grp., Kheinjua Fm.
AS IND UP
(ca. 1050 Ma)

Pakria kheinjuaensis Nautiyal 1983. (Nautiyal 1983b)
  Figures do not permit interpretation (S).

Vindhyan Sgrp., Semri Grp., Patherwa Fm.
AS IND UP
(ca. 1050 Ma)

(Fungal filament Type B). (Nautiyal 1983b)
  Figures do not permit interpretation (S).
Paleonostochopsis vindhyanensis Nautiyal 1983. (Nautiyal 1983b)
  Figures do not permit interpretation (S).

Vindhyan Sgrp., Semri Grp., Rohtas Fm.
AS IND UP
(ca. 1050 Ma)

Contrahofilum minutum Nautiyal 1983. (Nautiyal 1983b)
  Figures do not permit interpretation (S).
Leiofusa actinomorpha Maithy 1975. (Nautiyal 1983b)
  Figures do not permit interpretation (S M).

Vindhyan Sgrp., Semri Grp., Suket Shale
AS IND MP
(ca. 1050 Ma)

(disc-like remains Type-5). (Maithy and Shukla 1977)
  Figure does not permit interpretation (S M).
Gloeocapsomorpha sp. of Maithy 1969. (Maithy 1969)
  Possibly colonial cyanobacterium, but figures do not permit interpretation (S).
cf. *Kakabekia* sp. of Maithy & Shukla 1977. (Maithy and Shukla 1977)
  Figures do not permit interpretation (S M).

"Riphean of northeast Prianabar'e"
AS SIB
(ca. 1040 Ma)

Menneria levis Lopuchin[Lopukhin] 1971. (Lopukhin 1975a)
  No evidence of being indigenous to rock (S).

Mbuji Mayi Sgrp. (= Bushimay Sgrp.)
AF ZAR
(ca. 1020 Ma)

Leiofusa actinomorpha Maithy 1975. (Maithy 1975)

Table 22.2 (Continued).

Figure does not permit interpretation. Possibly globular sapropel (S).
Sphaerophycus densus Maithy 1975. (Maithy 1975)
  Figures do not permit interpretation (S M).

Ehsehlehkh Fm.
AS SIB
(ca. 1000 Ma)

Angulisphaerina perfosoclathrata Lopuchin [Lopukhin] 1974. (Lopukhin 1974)
  Contaminant? Mineral grains? No evidence of being indigenous to rock (S).
Granomarginata typica Lopuchin [Lopukhin] 1974. (Lopukhin 1974)
  Contaminant? No evidence of being indigenous to rock (S).

1000 to 900 Ma:
Lakhanda Fm.
AS SIB
(ca. 950 Ma)

Annularia annulata Timofeev & Hermann [German] 1979. (Timofeev and German 1979)
  Possibly sapropel with adpressed sphaeromorph-like bodies (S). [See Chapter 24, Pl. 15, D_1-D_2.]
Eosaccharomyces ramosus Hermann [German] 1979. (German 1979)
  Possibly aggregated globules on sapropel (S). [See Chapter 24, Pl. 16, A_1-A_2.]
Ostiana microcystis Hermann [German] 1976. (Timofeev et al. 1976)
  Figures do not permit interpretation (S).
Ulophyton rifeicum Timofeev & Hermann [German] 1979. (Timofeev and German 1979)
  Possibly irregularly aggregated sapropelic strands (S). [See Chapter 24, Pl. 17, A.]

Qingbaikou Grp., Xiamaling Fm.
AS CHN
(ca. 950 Ma)

Microsphaera foveolata Sin [Xing] 1973. (Xing and Liu 1973)
  Possibly modern (unicellular alga?) contaminant (S).
Pseudochara sp. of Sin [Xing] & Liu 1973. (Xing and Liu 1973)
  Figure does not permit interpretation (S).
Trachysphaeridium chihsienense Liu & Sin [Xing] 1973. (Xing and Liu 1973)
  Globular to irregular bodies not interpretable from published figures (S).

Shtandin Fm.
EA SUN
(ca. 900 Ma)

Pellicularia tenera Jankauskas [Yankauskas] 1980. (German et al. 1989)
  Possibly elongate sapropel (S). [See Chapter 24, Pl. 21, A, D_1-D_2.]

Late Proterozoic
900 to 800 Ma:
Atar Grp., Atar Fm.
AF MRT
(ca. 890 Ma)

Synsphaeridium sp. of Amard 1986. (Amard 1986)
  Figure does not permit interpretation (M S).

Miroedikha Fm.
AS SIB
(ca. 850 Ma)

Caudiculophycus micronulatus Hermann [German] 1986. (German 1986)
  Possibly modern contaminant, insect part? or sclereid? (S). [See Chapter 24, Pl. 25, C.]
Cephalonyx sibiricus Weiss [Vejs] 1984. (Vejs 1984)
  Possibly modern contaminant, insect part? (S). [See Chapter 24, Pl. 25, E_1-E_2.]
Dictyosphaeridium vittaforme Timofeev 1969. (Timofeev 1969)
  Possibly tubular sheath of filamentous cyanobacterium (S).
Ostiana microcystis Hermann [German] 1976. (Timofeev et al. 1976)
  Figures do not permit interpretation (S). [See Chapter 24, Pl. 23, B, C_1-C_3.]

Pahrump Grp., Beck Spring Dolomite
NA USA CA
(ca. 850 Ma)

(spindle-shaped objects). (Licari 1978)
  Spindle-shaped objects of unknown composition (S).
Bullasphaera variegata Licari 1978. (Licari 1978)
  Spheroidal biogenic body with associated mineral needles or nonbiogenic spherules (S).
Protegocista Licari 1978. (Licari 1978)
  Spiny ellipsoid of unknown composition (S).

Delhi Sgrp., Ajabgarh Grp., Kushalgarh Fm.
AS IND
(ca. 818 Ma)

(solitary cell comparable with *Huroniospora*). (Mandal et al. 1984)
  Possible modern contaminant or solitary coccoid cyanobacterium (S M).
Animikiea septata Barghoorn 1965. (Mandal et al. 1984)
  Possible modern contaminant or tubular sheath of filamentous cyanobacterium (S M).
Ghoshia bifurcata Mandal & Maithy 1984. (Mandal et al. 1984)
  Possible modern contaminant or branched septate filamentous stigonematalean cyanobacterium (S M).
Gloeocapsomorpha karauliensis Maithy & Mandal 1983. (Mandal et al. 1984)
  Possible modern contaminant or colonial coccoid chroococcacean cyanobacterium (S M).
Gloeocapsomorpha prisca Zalessky [Zalesskij] 1916. (Mandal et al. 1984)
  Possible modern contaminant or colonial coccoid chroococcacean cyanobacterium (S M).
Myxococcoides compactus Mandal & Maithy 1984. (Mandal et al. 1984)
  Possible modern contaminant or colonial coccoid chroococcacean cyanobacterium (S M).
Myxococcoides inornata Schopf 1968. (Mandal et al. 1984)
  Possible modern contaminant or colonial coccoid chroococcacean cyanobacterium (S M).
Myxococcoides minor Schopf 1968. (Mandal et al. 1984)
  Possible modern contaminant or colonial coccoid chroococcacean cyanobacterium (S M).
Oscillatoriopsis obtusa Schopf 1968. (Mandal et al. 1984)

Table 22.2 (Continued).

Possible modern contaminant or ensheathed septate filamentous oscillatoriacean cyanobacterium (S M).

Palaeoanacystis vulgaris Schopf 1968. (Mandal et al. 1984)
Possible modern contaminant or colonial coccoid chroococcacean cyanobacterium (S M).

Palaeolyngbya baraudensis Mandal & Maithy 1984. (Mandal et al. 1984)
Possible modern contaminant or ensheathed septate filamentous oscillatoriacean cyanobacterium (S M).

Palaeolyngbya distincta Mandal & Maithy 1984. (Mandal et al. 1984)
Possible modern contaminant or ensheathed septate filamentous oscillatoriacean cyanobacterium (S M).

Palaeoscytonema indicum Mandal & Maithy 1984. (Mandal et al.1984)
Possible modern contaminant or septate ?false-branched filamentous ?scytonematacean cyanobacterium (S M).

Palaeoscytonema intermingla Mandal & Maithy 1984. (Mandal et al. 1984)
Possible modern contaminant or septate false-branched filamentous scytonematacean cyanobacterium (S M).

Palaeoscytonema misrae Mandal & Maithy 1984. (Mandal et al. 1984)
Possible modern contaminant or ensheathed septate filamentous ?oscillatoriacean cyanobacterium (S M).

Primorivularia robusta Mandal & Maithy 1984. (Mandal et al. 1984)
Possible modern contaminant or degraded filamentous cyanobacterium (S M).

Vesicophycus problematicus Mandal & Maithy 1984. (Mandal et al. 1984)
Figures do not permit interpretation (S).

Bambui Grp., Sete Lagoas Fm.
SA BRA
(ca. 810 Ma)

Bambuites erichsenii Sommer 1971. (Sommer 1971)
Possible acritarchs; figures do not permit interpretation (S).

800 to 700 Ma:
Kangjia [Kangchia] Grp.
AS CHN
(ca. 790 Ma)

Archaeofavosina? sp. of Sin [Xing] & Liu 1973. (Xing and Liu 1973)
Organic fragments not interpretable from published figures (S).

Protoleiosphaeridium bullatum Andreeva 1966. (Xing and Liu 1973)
Figure does not permit interpretation (S).

Pseudozonosphaera cf. *sinica* Sin [Xing] & Liu 1973. (Xing and Liu 1973)
Figures do not permit interpretation (M S).

Taeniatum crassum Sin [Xing] & Liu 1973. (Xing and Liu 1973)
Partly or wholly modern fibrous contaminants (S).

Taeniatum simplex Sin [Xing] 1962. (Xing and Liu 1973)
Possibly modern fibrous contaminant (S).

Trematosphaeridium sp. of Sin [Xing] & Liu 1973. (Xing and Liu 1973)
Opaque globule not interpretable from figure (S).

Qiaotou [Chiaotou] Fm.
AS CHN
(ca. 790 Ma)

Taeniatum crassum Sin [Xing] & Liu 1973. (Xing and Liu 1973)
Partly or wholly modern fibrous contaminants (S).

Xihe Grp., Nanfen Fm.
AS CHN
(ca. 790 Ma)

Archaeofavosina? sp. of Sin [Xing] & Liu 1973. (Xing and Liu 1973)
Organic fragments not interpretable from published figures (S).

Visingsö Grp.
EU SWE
(ca. 775 Ma)

(filament type 1). (Invert 1977)
Modern contaminant? (S).
(filament type 2). (Invert 1977)
Modern contaminant? (S).
(filament type 3). (Invert 1977)
Modern contaminant? (S).
(spheroidal structures). (Invert 1977)
Modern contaminant? (S).

Pterospermopsimorpha concentrica (Sin & Liu 1973) Vidal 1976. (Vidal 1976y)
Possible pteromorph acritarch (M S).

Chatkaragaj Fm.
AS SUN
(ca. 765 Ma)

Menneria foraminis Lopuchin [Lopukhin] 1971. (Lopukhin 1971a)
Contaminant? No evidence this is indigenous to rock (S).
Menneria foraminis Lopuchin [Lopukhin] 1971. (Lopukhin 1971b)
Contaminant? No evidence this is indigenous to rock (S).
Menneria levis Lopuchin [Lopukhin] 1971. (Lopukhin 1971b)
Contaminant? No evidence this is indigenous to rock (S).

Eleonore Bay Grp.
NA GRL
(ca. 750 Ma)

(endolithic microfossils). (Campbell 1982)
Possible microfossil; figures do not permit interpretation (S).

Brioverian
EU FRA
(ca. 740 Ma)

Hymenophacoïdes Roblot 1967. (Roblot 1967)
Aggregated carbonaceous spheroids, in part possibly fossil (S).

Brioverian "Schistes X"
EU FRA
(ca. 740 Ma)

(formes allongées et filamenteuses). (Fournier-Vinas and Debat 1970)
Poorly preserved. Possibly trichomes or sheaths of filamentous cyanobacteria (S).

Calcaires de Kakontwe
AF ZAR
(ca. 740 Ma)

(Sphaeromorphes). (Buffard and Willems 1982)
Probably irregular and globular sapropel (S).

"Algonkian lydites"
EU CSK
(ca. 735 Ma)

Palaeocryptidium cayeuxii Deflandre 1955. (Vavrdová 1966)
Possible sphaeromorph acritarch (S).

Table 22.2 (Continued).

Gangolihat Dolomites
AS IND
(ca. 735 Ma)

(Incertae sedis Type-1). (Nautiyal 1978a)
　Figure does not permit interpretation (S M).
(fungal remains). (Nautiyal 1978a)
　Figures do not permit interpretation (S M).
Baltisphaeridium gangolihatense Nautiyal 1978. (Nautiyal 1978a)
　Figures do not permit interpretation (M S).
Eomycetopsis (?) sp. A of Nautiyal 1978. (Nautiyal 1978a)
　Figures do not permit interpretation (S M).
Schismatosphaeridium kumauni Nautiyal 1978. (Nautiyal 1978a)
　Sphaeromorph acritarch? Figures do not permit interpretation (S M).

Gangolihat Dolomites, Chhera Mbr.
AS IND
(ca. 735 Ma)

Minutiafilum minutum Nautiyal 1980. (Nautiyal 1980)
　Possible modern contaminant (S).

"Upper Precambrian of Mali"
AF MLI
(ca. 725 Ma)

Menneria levis Lopuchin [Lopukhin] 1971. (Lopukhin 1973b)
　No evidence of being indigenous to rock (S).
Menneria roblotae Lopuchin [Lopukhin] 1971. (Lopukhin 1973b)
　No evidence of being indigenous to rock (S).

"Upper Precambrian of northwest Africa"
AF
(ca. 725 Ma)

Menneria levis Lopuchin [Lopukhin] 1971. (Lopukhin 1975a)
　No evidence of being indigenous to rock (S).

"late Precambrian of Tien Shan"
AS SUN
(ca. 725 Ma)

Menneria foraminis Lopuchin [Lopukhin] 1971. (Lopukhin 1975a)
　Irregular and curved organic (?) flakes. No evidence of being indigenous to rock (S).
Menneria roblotae Lopuchin [Lopukhin] 1971. (Lopukhin 1973b)
　No evidence of being indigenous to rock (S).
Menneria sp. of Lopukhin 1973. (Lopukhin 1973b)
　No evidence of being indigenous to rock (S).

Martyukhe Fm.
AS SIB
(ca. 725 Ma)

Azyrtalia fasciculata Vologdin & Drosdova [Drozdova] 1969. (Vologdin and Drozdova 1969c)
　Figures do not permit interpretation (S M).
Azyrtalia zonulata Vologdin & Drosdova [Drozdova] 1969. (Vologdin and Drozdova 1969c)
　Figures do not permit interpretation (S M).
Echaninia mucosa Vologdin & Drosdova [Drozdova] (1969. (Vologdin and Drozdova 1969b)
　Irregular globular clumps of kerogen (?) in recrystallized (?) matrix (S).
Murandavia granulosa Vologdin & Drosdova [Drozdova] 1969. (Vologdin and Drozdova 1969b)
　Irregular clump of kerogen (?) in recrystallized (?) matrix (S M).

Nelcanella annularia Vologdin & Drosdova [Drozdova] 1969. (Vologdin and Drozdova 1969c)
　Figures do not permit interpretation (S M).

Pachelma "Series"
[also reported from a variety of poorly dated Middle and Upper Riphean units in Tien Shan, Prianabar'e, and Mongolia]
EU SUN
(ca. 725 Ma)

Menneria levis Lopuchin [Lopukhin] 1971. (Lopukhin 1971a)
　Contaminant? No evidence this is indigenous to rock (S).

Pachelma "Series," Voron Fm.
[also reported from the Middle Riphean Kokdzhot Fm (Tien Shan) and the Vendian Kharayutehkh Fm (lower Lena River)]
EU SUN
(ca. 725 Ma)

Menneria roblotae Lopuchin [Lopukhin] 1971. (Lopukhin 1971a)
　Contaminant? No evidence of being indigenous to rock (S).

Uchur "Series," Gonam Fm.
AS SIB
(ca. 725 Ma)

Gonamophyton ovale Vologdin & Drosdova [Drozdova] 1964. (Vologdin and Drozdova 1964b)
　Probably diagenetically produced nonbiogenic structure (S).
Nelcanella radians Vologdin & Drosdova [Drozdova] 1964. (Vologdin and Drozdova 1964a)
　Possibly recrystallized (?) spherulite with radial fabric (S).
Nelcanella solaris Vologdin & Drosdova [Drozdova] 1964. (Vologdin and Drozdova 1964a)
　Possibly recrystallized (?) spherulite with radial fabric (S).
Nelcanella stellata Vologdin & Drosdova [Drozdova] 1964. (Vologdin and Drozdova 1964a)
　Possibly recrystallized (?) spherulite with radial fabric (S).

Windermere Sgrp., Miette Grp.
NA CAN BC
(ca. 700 Ma)

(pyrite framboids). (Javor and Mountjoy 1976)
　Framboidal pyrite, possibly biogenic ("S M").

700 to 600 Ma:

Vindhyan Sgrp., Bhander Grp., Sirbu Shale
AS IND
(ca. 690 Ma)

(fungal spores). (Sitholey et al. 1953)
　Figures do not permit interpretation (S).
(round bodies). (Sitholey et al. 1953)
　Possibly modern contaminants (S).

Amri unit
AS IND
(ca. 685 Ma)

(Incertae Sedis Type 1). (Nautiyal 1978b)
　Probably modern fungal hypha contaminant (S).
Granomarginata primitiva Salujha, Rehman & Arora 1971. (Nautiyal 1978b)
　Contaminant? Too well preserved to occur in schistose phyllite (S).
Myxococcoides minor Schopf 1968. (Nautiyal 1978b)
　Possibly sphaeromorph acritarch (S).
Myxococcoides minor Schopf 1968. (Nautiyal 1978b)

Table 22.2 (Continued).

Contaminant? Too well preserved and abundant to occur in schistose phyllite (S).
Protosphaeridium volkovae Maithy & Shukla 1977. (Nautiyal 1978b)
Contaminant? Too well preserved to occur in schistose phyllite (S).

Simla Slate Grp.
AS IND
(ca. 685 Ma)

Satpulispora major Nautiyal 1979. (Nautiyal 1979)
Possibly mineralic (hematite?) spherule (S).
Satpulispora microreticulata Nautiyal 1979. (Nautiyal 1979)
Possibly mineralic (hematite?) spherules (S).
Satpulispora psilata Nautiyal 1979. (Nautiyal 1979)
Possibly inorganic (hematite?) spherules (S).
Vavososphaeridium sp. A of Nautiyal 1982. (Nautiyal 1982c)
Possibly sphaeromorph acritarch (S).

Bhima "Series"
AS IND KA
(ca. 675 Ma)

(Incertae sedis Type 1). (Salujha et al. 1972)
Possibly sphaeromorph acritarch or coccoid solitary cyanobacterium (S).
(Incertae sedis Type 2). (Salujha et al. 1972)
Probably modern fungal contaminant (S).
(Incertae sedis Type 3). (Salujha et al. 1972)
Figures do not permit interpretation (S M).

Dalradian (Lower), Easdale Slate
EU GBR
(ca. 675 Ma)

Protarchaeosacculina atava Naumova 1960?. (Downie et al. 1971)
Figure does not permit interpretation (M S).

"Subcambrian tillites" of Tien Shan
AS SUN
(ca. 650 Ma)

Menneria foraminis Lopuchin [Lopukhin] 1971. (Lopukhin 1971b)
Contaminant? No evidence this is indigenous to rock (S).

Chichkan Fm.
AS SUN
(ca. 650 Ma)

Menneria foraminis Lopuchin [Lopukhin] 1971. (Lopukhin 1971a)
Contaminant? No evidence this is indigenous to rock (S).
Menneria foraminis Lopuchin [Lopukhin] 1971. (Lopukhin 1971b)
Contaminant? No evidence this is indigenous to rock (s).
Menneria granosa Lopuchin [Lopukhin] 1971. (Lopukhin 1971b)
Contaminant? No evidence this is indigenous to rock (S).
Menneria granosa Lopuchin [Lopukhin] 1971. (Lopukhin 1971a)
Contaminant? No evidence this is indigenous to rock (S).

Hailuoto Sequence (cf. Muhos Fm.)
EU FIN
(ca. 650 Ma)

? *Phycomycetes* sp. of Tynni & Donner 1980. (Tynni and Donner 1980)
Figure does not permit interpretation (S).
Podoliella prismatica Tynni & Donner 1980. (Tynni and Donner 1980)
Figure does not permit interpretation (S M).
Symplassosphaeridium sp. of Tynni & Donner 1980. (Tynni and Donner 1980)
Highly degraded sphaeromorph acritarch or sapropel (S M).

Muhos Fm.
EU FIN
(ca. 650 Ma)

(branched filament with cystoid protuberances). (Tynni and Uutela 1984)
Figure does not permit interpretation (S M).
(cylindrical form, ... resembling *Volyniella*). (Tynni and Uutela 1984)
Figures do not permit interpretation (S).
(giant filament). (Tynni and Uutela 1984)
Tubular sheath of filamentous cyanobacterium or elongate piece of sapropel (S).
? *Contortothrix* sp. of Tynni & Uutela 1984. (Tynni and Uutela 1984)
Figure does not permit interpretation (S M).
Eomicrhystridium sp. 1 of Tynni & Uutela 1984. (Tynni and Uutela 1984)
Possible solitary coccoid cyanobacterium (S).
Heliconema sp. of Tynni & Uutela 1984. (Tynni and Uutela 1984)
Figure does not permit interpretation (S M).
cf. *Polyedryxium* ? Deunff 1954. (Tynni and Uutela 1984)
Figure does not permit interpretation (S M).

Spilitic Grp., Lečice Mbr.
EU CSK
(ca. 635 Ma)

(spherical skeleton-like bodies: types A,B,C). (Konzalová 1973)
Figures do not permit interpretation (M S).

Post-Spilitic Grp.
EU CSK
(ca. 615 Ma)

cf. *Micrhystridium* sp. of Konzalová 1974. (Konzalová 1974a)
Poorly preserved, possible acritarch ("S").
Ooidium apertum Konzalová 1972. (Konzalová 1972)
Degraded ellipsoidal unicell? (S).
Ooidium apertum Konzalová 1972. (Konzalová 1974b)
Possibly degraded sphaeromorph acritarch (S).

Yaryshev Fm.
EU SUN
(ca. 610 Ma)

Taenitrichoides jaryschevicus Assejeva [Aseeva] 1983. (Aseeva and Velikanov 1983)
Possibly folded sapropel (S). [See Chapter 24, Pl. 51, K.]

?Murdama Grp.
AS SAU
(ca. 604 Ma)

cf. *Synsphaeridium* sp. of Binda & Bokhari 1980. (Binda and Bokhari 1980)
Irregular clump of organic matter (M S).

"Vendian of Tien Shan"
AS SUN
(ca. 600 Ma)

Menneria foraminis Lopuchin [Lopukhin] 1981. (Lopukhin 1973b)
No evidence of being indigenous to rock (S).

"Vendian of Ukrainian Carpathians"
EU SUN
(ca. 600 Ma)

Toromorpha sp. of Timofeev, German & Mikhajlova 1976. (Timofeev et al. 1976)
Figure does not permit interpretation (S M).

Table 22.2 (Continued).

Central Mount Stuart Fm.
AU AUS NT
(ca. 600 Ma)

cf. *Vendotaenia* sp. of Damassa & Knoll 1986. (Damassa and Knoll 1986)
  Possibly sapropelic fragments (S).

Derlo Fm.
EU SUN
(ca. 600 Ma)

Podolina angulata Hermann [German] 1976. (Timofeev et al. 1976)
  Possible polygonomorph or prismatomorph acritarch (S M).
Podolina echinata Hermann [German] 1976. (Timofeev et al. 1976)
  Possible polygonomorph or prismatomorph acritarch (S M).
Podolina minuta Hermann [German] 1976. (Timofeev et al. 1976)
  Figures do not permit interpretation (S).

Dzhetymtau Fm.
AS SUN
(ca. 600 Ma)

Menneria granosa Lopuchin [Lopukhin] 1971. (Lopukhin 1971a)
  Contaminant? No evidence this is indigenous to rock (S).

Kanilovka Fm. (=Zharnov Fm.)
EU SUN
(ca. 600 Ma)

Bicuspidata fusiformis Assejeva [Aseeva] 1982. (Aseeva 1982)
  Possibly sapropel or degraded tubular sheath of filamentous cyanobacterium (S). [See Chapter 24, Pl. 53, D.]
Menneria granosa Lopuchin [Lopukhin] 1971. (Lopukhin 1971a)
  Contaminant? No evidence this is indigenous to rock (S).
Studenicia bacotica Assejeva [Aseeva] 1982. (Aseeva 1982)
  Possibly modern contaminant, insect part or sclereid (S). [See Chapter 24, Pl. 53, F.]

Kyrshabaktin Fm.
AS SUN
(ca. 600 Ma)

Menneria granosa Lopuchin [Lopukhin] 1971. (Lopukhin 1971a)
  Contaminant? No evidence this is indigenous to rock (S).

Nama Grp., Kuibis Subgrp.
AF NAM
(ca. 600 Ma)

(*Comasphaeridium*-like microfossil). (Germs et al. 1986)
  Figure does not permit interpretation (S M)

Sinian (Upper), Doushantuo Fm.
AS CHN
(ca. 600 Ma)

(tripartite vesicles). (Awramik et al. 1985)
  Possibly degraded spiny acritarch, but figure does not permit interpretation (S).

Wilpena Grp., Tent Hill Fm., Arcoona Quartzite Mbr.
AU AUS SA
(ca. 600 Ma)

cf. *Vendotaenia* sp. of Damassa & Knoll 1986. (Damassa and Knoll 1986)
  Possibly sapropelic fragments (S).

600 to 550 Ma:
Redkino Fm.
EU SUN
(ca. 580 Ma)

Spumiosina rubiginosa (Andreeva 1966) Jankauskas & Medvedeva 1989. (German et al. 1989)
  Possibly globular sapropel (S). [See Chapter 24, Pl. 55, E_1–E_3.]
  "suite of bituminous rocks," Patom Highland

AS SIB
(ca. 555 Ma)

Archaeohystrichosphaeridium anatropum Kalin ? (Timofeev 1966)
  Figure does not permit interpretation (S).
Archaeohystrichosphaeridium vastum Kalin. ? (Timofeev 1966)
  Figure does not permit interpretation (SM).

Chulaktau Fm.
AS SUN
(ca. 550 Ma)

Archaeopsophosphaera adischevii Lopuchin [Lopukhin] 1969. (Lopukhin 1969)
  Contaminant? No evidence of being indigenous to rock (S).
Menneria adischevii (Lopukhin 1969) Lopuchin [Lopukhin] 1971. (Lopukhin 1971a)
  Contaminant? No evidence this is indigenous to rock (S).

Early Cambrian:
Belaya Fm.
AS SIB
(ca. 540 Ma)

Leiopsophosphaera rotunda Pichova [Pykhova] 1967. (Pykhova 1967)
  Figures do not permit interpretation (M S).

Moty Fm.
AS SIB
(ca. 540 M)

Spumosata nova Pichova [Pykhova] 1967. (Pykhova 1967)
  Figure does not permit interpretation (M S).

Shortor "Series"
AS SUN
(ca. 540 Ma)

Archaeopsophosphaera adischevii Lopuchin [Lopukhin] 1969. (Lopukhin 1969)
  Contaminant? No evidence of being indigenous to rock (S).
Latoporata naumovae Lopukhin 1966. (Lopukhin 1966)
  Figure does not permit interpretation (M S).
Menneria adischevii (Lopukhin 1969) Lopuchin [Lopukhin] 1971. (Lopukhin 1971a)
  Contaminant? No evidence this is indigenous to rock (S).
Mitroporata armata Lopukhin 1966. (Lopukhin 1966)
  Figure does not permit interpretation (M S).
Spumiosina cineraria Lopukhin1966. (Lopukhin 1966)
  Contaminant of preparation? (fern sporangium?) (S).
Trachymarginata rustosa Lopukhin 1966. (Lopukhin 1966).
  Figure does not permit interpretation (M S).

Table 22.2 (Continued).

Shortor "Series," ?Tamdy Fm. AS SUN (ca. 540 Ma)	Vampire Fm. NA CAN YK (ca. 540 Ma)

Spumiosina solida Lopukhin 1966. (Lopukhin 1966)
    Irregular sapropelic organic matter? (S).
Trachypsophosphaera exilis Lopukhin 1966. (Lopukhin 1966)
    Irregular sapropelic organic matter (?) (S).

    Tokammane Fm.
    EU SJM
    (ca. 540 Ma)

Ceratophyton vernicosum Kirjanov [Kir'yanov] 1979. (Knoll and Swett 1987)
    Possibly sapropel or degraded tubular sheath of filamentous prokaryote (S).

(Microdubiofossil Type B). (Hofmann 1984)
    Possibly diagenetically altered unicells ("S").

    Ushakovka Fm. (= Khuzhir Fm.)
    AS SIB
    (ca. 535 Ma)

Margominuscula regularis Naumova. (Pykhova 1966)
    Figure does not permit interpretation (M S).
Margominuscula tremata Naumov *in* Naumova and Pavlovskij 1961. (Pykhova 1966)
    Figures do not permit interpretation (M S).

Table 22.3. *Proterozoic and Selected Early Cambrian Microfossils*

Early Proterozoic
2450 to 2250 Ma:
Transvaal Sgrp., Ghaap Grp., Gamohaan Fm., Tsineng Mbr.
AF ZAF
(ca. 2450 Ma)

Siphonophycus transvaalense Beukes, Klein & Schopf 1987. (Klein et al. 1987)
    Tubular sheath of filamentous (oscillatoriacean) cyanobacterium ("S"). [See Chapter 24, Pl. 1, A–F.]

Transvaal Sgrp., Chuniespoort Grp., Monte Christo Fm.
AF ZAF
(ca. 2330 Ma)

(bacterial microfossils). (Lanier 1986)
    Ellipsoidal (rod-shaped) and coccoid bacteria ("S").
(coccoid microfossils). (Lanier 1986)
    Solitary, coccoid to ellipsoidal bacteria (S).
(unbranched segmented filaments). (Lanier 1986)
    Septate filamentous prokaryote ("S").

2250 to 2000 Ma:
Ikabijsk Fm.
AS SIB
(ca. 2200 Ma)

Polyedrosphaeridium bullatum Timofeev 1966. (Timofeev et al. 1976)
    Colonial coccoid (chroococcacean?) cyanobacterium (S).
Protosphaeridium densum Timofeev 1966. (Timofeev et al. 1976)
    Sphaeromorph acritarch ("S").
Stictosphaeridium pectinale Timofeev 1966. (Timofeev et al. 1976)
    Sphaeromorph acritarch ("S").

Belcher Grp.
NA CAN NW
(ca. 2150 Ma)

(Type 2). (Hofmann and Jackson 1969)
    Solitary coccoid cyanobacterium, cf. *Leptoteichos* Knoll et al. 1978 (S). [See Chapter 24, Pl. 1, H–J.]
(Type 3). (Hofmann and Jackson 1969)
    Possibly degraded colonial coccoid prokaryotes (S).
(Type 4). (Hofmann and Jackson 1969)
    Microfossil of unknown affinity (cf. ? *Eosphaera*) (S). [See Chapter 24, Pl. 1, K, L.]
cf. *Archaeotrichion* Schopf 1968. (Hofmann and Jackson 1969)
    Filamentous bacterium (S).
Biocatenoides sp. of Hofman & Jackson 1969. (Hofmann and Jackson 1969)
    Septate filamentous bacterium ("S"). [See Chapter 24, Pl. 1, O.]
Eomycetopsis sp. of Hofman & Jackson 1969. (Hofmann and Jackson 1969)
    Sheath of filamentous prokaryote (S). [See Chapter 24, Pl. 1, M, N.]
Huroniospora sp. of Hofmann & Jackson 1969. (Hofmann and Jackson 1969)
    Solitary coccoid cyanobacterium ("S"). [See Chapter 24, Pl. 1, G.]

Belcher Grp., Kasegalik Fm.
NA CAN NW
(ca. 2150 Ma)

(acritarchs). (Hofmann 1976)
    Spheroidal and ellipsoidal prokaryotic envelopes and unicells ("S").
(entophysalidacean colonies). (Hofmann 1975)
    Colonial entophysalidacean cyanobacteria ("S").
(nostocalean? filaments). Hofmann 1975)
    Degraded sheaths and trichomes of filamentous bacteria (M).
(sphaeromorph acritarchs). (Hofmann 1975)
    Solitary coccoid cyanobacteria, cf. *Huroniospora* (S).
Archaeotrichion sp. of Hofmann 1976. (Hofmann 1976)
    Sheath of filamentous prokaryote (cf. *Eomycetopsis filiformis*) ("S").
Biocatenoides sphaerula Schopf 1968. (Hofmann 1976)
    Septate filamentous bacterium (S).
Eoentophysalis belcherensis Hofmann 1976. (Hofmann 1976)
    Colonial, coccoid to ellipsoidal cyanobacterium ("S").
Eoentophysalis belcherensis Hofmann 1976. (Golubić and Hofmann 1976)
    Colonial coccoid cyanobacterium ("M").
Eomycetopsis filiformis Schopf 1968. (Hofmann 1976)
    Sheath of filamentous prokaryote ("S").
Eosynechococcus medius Hofmann 1976. (Hofmann 1976)
    Colonial ellipsoidal cyanobacterium ("S").
Eosynechococcus moorei Hofmann 1976. (Hofmann 1976)
    Colonial ellipsoidal bacterium ("S").

Table 22.3 (Continued).

Eosynechococcus moorei Hofmann 1976. (Golubić and Campbell 1979)
  Solitary and colonial ellipsoidal bacteria (M).
Globophycus sp. of Hofmann 1976. (Hofmann 1976)
  Sphaeromorph acritarch containing collapsed cytoplasmic remnants ("S").
Halythrix sp. of Hofmann 1976. (Hofmann 1976)
  Degraded filamentous prokaryote ("S").
Melasmatosphaera magna Hofmann 1976. (Hofmann 1976)
  Spheroidal cell-like bodies, possibly envelopes of colonial prokaryote (S).
Melasmatosphaera media Hofmann 1976. (Hofmann 1976)
  Plasmolyzed coccoid cyanobacteria ("S").
Melasmatosphaera parva Hofmann 1976. (Hofmann 1976)
  Plasmolyzed solitary coccoid cyanobacterium ("S").
Myxococcoides minor Schopf 1968. (Hofmann 1976)
  Colonial coccoid cyanobacterium ("S").
Palaeoanacystis vulgaris Schopf 1968. (Hofmann 1976)
  Colonial coccoid cyanobacterium ("S").
Pleurocapsa? sp. of Hofmann 1976. (Hofmann 1976)
  Ellipsoidal colonial bacterium (S).
Rhicnonema antiquum Hofmann 1976. (Hofmann 1976)
  Degraded filamentous ensheathed bacterium ("S").
Sphaerophycus parvum Schopf 1968. (Hofmann 1976)
  Colonial coccoid cyanobacterium ("S").

Belcher Grp., McLeary Fm.
NA CAN NW
(ca. 2150 Ma)

(acritarchs). (Hofmann 1976)
  Spheroidal and ellipsoidal prokaryotic envelopes and unicells ("S").
(entophysalidacean colonies). (Hofmann 1976)
  Colonial entophysalidacean cyanobacteria ("S").
Biocatenoides sphaerula Schopf 1968. (Hofmann 1976)
  Septate filamentous bacterium (S).
Caryosphaeroides sp. of Hofmann 1976. (Hofmann 1976)
  Plasmolyzed coccoid solitary cyanobacterium ("S").
Eoentophysalis belcherensis Hofmann 1976. (Golubić and Hofmann 1976)
  Colonial coccoid cyanobacterium ("M").
Eoentophysalis belcherensis Hofmann 1976. (Hofmann 1976)
  Colonial, coccoid to ellipsoidal cyanobacterium ("S").
Eomycetopsis filiformis Schopf 1968. (Hofmann 1976)
  Sheath of filamentous prokaryote ("S").
Eosynechococcus grandis Hofmann 1976. (Hofmann 1976)
  Ellipsoidal cell-like bodies, possibly envelopes of colonial prokaryote (S).
Eosynechococcus medius Hofmann 1976. (Hofmann 1976)
  Colonial ellipsoidal cyanobacterium ("S").
Eozygion minutum Schopf 1968. (Hofmann 1976)
  Ellipsoidal colonial cyanobacterium ("S").
Glenobotrydion majorinum Schopf & Blacic 1971. (Hofmann 1976)
  Sphaeromorph acritarch containing degraded internal cellular remnants (S).
Globophycus sp. of Hofmann 1976. (Hofmann 1976)
  Sphaeromorph acritarch containing collapsed cytoplasmic remnants ("S").
Kakabekia? sp. of Hofmann 1976. (Hofmann 1976)
  Degraded spheroidal or ellipsoidal cyanobacterium (S).
Melasmatosphaero magna Hofmann 1976. (Hofmann 1976)
  Spheroidal cell-like bodies, possibly envelopes of colonial prokaryote (S).
Melasmatosphaera media Hofmann 1976. (Hofmann 1976)
  Plasmolyzed coccoid cyanobacteria ("S").
Melasmatosphaera parva Hofmann 1976. (Hofmann 1976)
  Plasmolyzed solitary coccoid cyanobacterium ("S").
Myxococcoides inornata Schopf 1968. (Hofmann 1976)
  Colonial coccoid cyanobacterium ("S").
Myxococcoides minor Schopf 1968. (Hofmann 1976)
  Colonial coccoid cyanobacterium ("S").
Myxococcoides sp. of Hofmann 1976. (Hofmann 1976)
  Spheroidal, cell-like bodies, possibly envelopes of colonial prokaryote ("S").
Palaeoanacystis vulgaris Schopf 1968. (Hofmann 1976)
  Colonial coccoid cyanobacterium ("S").
Sphaerophycus parvum Schopf 1968. (Hofmann 1976)
  Colonial coccoid cyanobacterium ("S").
Zosterosphaera sp. of Hofmann 1976. (Hofmann 1976)
  Transected spheroid, probably envelope of a *Chroococcus*-like colonial cyanobacterium (S).

Karelia Complex
EA SUN
(ca. 2100 Ma)

Protosphaeridium sp. of Timofeev 1966. (Tomofeev 1966)
  Sphaeromorph acritarch ("S").
Synsphaeridium conglutinatum Timofeev 1966. (Timofeev 1966)
  Colonial coccoid chroococcacean? cyanobacterium (S).

Biwabik Iron Fm.
NA USA MN
(ca. 2090 Ma)

(Type "E" structure). (LaBerge 1967)
  Hematite-replaced, solitary coccoid cyanobacterium, cf. *Huroniospora* spp. ("S").
(Type "F" structure). (LaBerge 1967)
  Hematite-replaced filamentous bacterium, cf. *Gunflintia minuta* ("S").
Gunflintia Barghoorn 1965. (Cloud and Licari 1968)
  Septate filamentous prokaryote ("M").
Huroniospora Barghoorn 1965. (Cloud and Licari 1968)
  Solitary coccoid cyanobacterium ("M").

Gunflint Iron Fm.
NA CAN ON
(ca. 2090 Ma)

"*Schizothrix*" *atavia* Edhorn 1973. (Edhorn 1973)
  Septate filamentous bacterium, cf. *Gunflintia* (S).
("filaments" [narrow]). (Moorhouse and Beales 1962)
  Hematite-encrusted *Gunflintia*-like filamentous prokaryote (S).
(branching, nonseptate hyphae). (Tyler and Barghoorn 1954)
  Filamentous bacteria, probably *Gunflintia* (unbranched, not fungal) (S).
(dark spiral threads). (Cloud 1965)
  Degraded filamentous bacteria (=? *Gunflintia minuta*) ("S").
(eucaryotic cells). (Edhorn 1973)
  Solitary coccoid cyanobacterium, cf. *Huroniospora* (S).
(filaments suggestive of *Ankistrodesmus*). (Edhorn 1973)
  Iron-encrusted filamentous cyanobacterium (S).
(fine-textured ovoid body). (Cloud and Hagen 1965)
  Solitary ellipsoidal bacterium (S).
(hypothetical "Precambrian worm"). (Edhorn 1973)

Table 22.3 (*Continued*).

Tubular sheath of filamentous prokaryote, cf. *Animikiea* (S).
(large, apparently planktonic coccoid alga). (Knoll and Barghoorn 1976)
    Sphaeromorph acritarch (S).
(larger septate threads). (Cloud 1965)
    Septate filamentous cyanobacterium ("S").
(many-branched hyphae, bearing spores). (Tyler and Barghoorn 1954)
    Filamentous bacteria, probably *Gunflintia* (unbranched, not fungal) with *Huroniospora* (S).
(nonseptate or obscurely septate threads). (Cloud 1965)
    Degraded filamentous prokaryotes ("S").
(patterned globular bodies). (Cloud 1965)
    Assemblage of coccoid cyanobacteria ("S").
(radiate structures). (Cloud 1965)
    Actinomorphic colony(?) of radiating filaments (= *Eoastrion simplex*) (S).
(smaller threads). (Cloud and Hagen 1965)
    Sheath or trichome of filamentous bacterium (M S).
(thin septate threads). (Cloud 1965)
    Septate filamentous bacteria (= *Gunflintia minuta*) ("S").
(unicellular organisms). (Tyler and Barghoorn 1954)
    Probably = *Kakabekia umbellata* (S).
(unnamed sheath). (Awramik and Barghoorn 1977)
    Sheath of filamentous (oscillatoriacean?) cyanobacterium ("S").
(*Eosphaera*-like structures). (Kaźmierczak 1979)
    Concentrically layered, mineralic, nonbiogenic spheroids of colloidal origin (cf. Klein and Fink, 1976, fig. 13B) (S).
(*Micrasterias*-like organism). (Edhorn 1973)
    Probably iron-encrusted *Kakabekia* (S).
Anabaenidium barghoornii Edhorn 1973. (Edhorn 1973)
    Septate filamentous prokaryote, cf. *Gunflintia* spp. (S).
Animikiea septata Barghoorn 1965. (Barghoorn and Tyler 1965)
    Sheath of filamentous (oscillatoriacean?) cyanobacterium (S). [See Chapter 24, Pl. 2, L, M.]
Animikiea septata Barghoorn 1977. (Awramik and Barghoorn 1977)
    Sheath of filamentous (oscillatoriacean?) cyanobacterium ("S").
Archaeorestis magna Awramik & Barghoorn 1977. (Awramik and Barghoorn 1977)
    Prokaryote incertae sedis, possibly sheath or envelope of prokaryote (S).
Archaeorestis schreiberensis Barghoorn 1965. (Barghoorn and Tyler 1965)
    Prokaryote incertae sedis, possibly sheath or envelope of prokaryote (S).
Chlamydomonopsis primordialis Edhorn 1973. (Edhorn 1973)
    Sphaeromorph acritarch (S).
Corymbococcus hodgkissii Awramik & Barghoorn 1977. (Awramik and Barghoorn 1977)
    Colonial coccoid cyanobacterium ("S").
Entosphaeroides amplus Barghoorn 1965. (Barghoorn and Tyler 1965)
    Sheath of filamentous cyanobacterium (S). [See Chapter 24, Pl. 2, N.]
Eoastrion bifurcatum Barghoorn 1965. (Barghoorn and Tyler 1965)
    Prokaryote? incertae sedis (S). [See Chapter 24, Pl. 2, D.]
Eoastrion simplex Barghoorn 1965. (Barghoorn and Tyler 1965)
    Prokaryote? incertae sedis (S). [See Chapter 24, Pl. 2, C.]
Eomicrhystridium barghoornii Deflandre 1968. (Hofmann 1971a)
    Actinomorphic microfossil (= *Kakabekia* sp.) (S).
Eomicrhystridium barghoornii Deflandre 1968. (Deflandre 1968)
    Degraded sphaeromorph acritarch (S).

Eosphaera tyleri Barghoorn 1965. (Barghoorn and Tyler 1965)
    Colonial prokaryote? incertae sedis (S). [See Chapter 24, Pl. 2, A, B_1-B_2.]
Exochobrachium triangulum Awramik & Barghoorn 1977. (Awramik and Barghoorn 1977)
    Prokaryote incertae sedis, possibly degraded colonial coccoid prokaryote (S).
Frutexites microstroma Walter & Awramik 1979. (Walter and Awramik 1979)
    Filamentous bacterium (S).
Galaxiopsis melanocentra Awramik & Barghoorn 1977. (Awramik and Barghoorn 1977)
    Ellipsoidal, ensheathed cyanobacterium (S).
Gunflintia grandis Barghoorn 1965. (Barghoorn and Tyler 1965)
    Septate filamentous prokaryote (S). [See Chapter 24, Pl. H, I.]
Gunflintia minuta Barghoorn 1965. (Barghoorn and Tyler 1965)
    Septate filamentous bacterium (S). [See Chapter 24, Pl. 2, J, K.]
Gunflintia minuta Barghoorn 1965. (Cloud and Hagen 1965)
    Tubular sheath of filamentous bacterium ("S").
Huroniospora [spp.] of Strother & Tobin 1987. (Strother and Tobin 1987)
    Solitary coccoid to ellipsoidal prokaryotes (assemblage) (S).
Huroniospora macroreticulata Barghoorn 1965. (Cloud and Hagen 1965)
    Solitary coccoid cyanobacterium ("S").
Huroniospora macroreticulata Barghoorn 1965. (Barghoorn and Tyler 1965)
    Solitary coccoid prokaryote (assemblage) (S). [See Chapter 24, Pl. 2, G.]
Huroniospora microreticulata Barghoorn 1965. (Barghoorn and Tyler 1965)
    Solitary coccoid prokaryote (assemblage) (S). [See Chapter 24, Pl. 2, E.]
Huroniospora microreticulata Barghoorn 1965. (Cloud and Hagen 1965)
    Sphaeromorph acritarch (S).
Huroniospora miroreticulata Barghoorn 1965. (Darby 1974)
    Solitary coccoid cyanobacteria (S).
Huroniospora psilata Barghoorn 1965. (Barghoorn and Tyler 1965)
    Solitary coccoid prokaryote (assemblage) (S). [See Chapter 24, Pl. 2, F.]
Kakabekia umbellata Barghoorn 1965. (Barghoorn and Tyler 1965)
    Prokaryote? incertae sedis (S). [See Chapter 24, Pl. 2, O-R.]
Leptoteichos golubicii Knoll, Barghoorn & Awramik 1978. (Knoll et al. 1978)
    Colonial coccoid cyanobacterium (S).
Megalytrum diacenum Knoll, Barghoorn & Awramik 1978. (Knoll et al. 1978)
    Envelopes of solitary and paired cyanobacteria ("M").
Palaeoscytonema moorhousei Edhorn 1973. (Edhorn 1973)
    Septate filamentous bacterium, cf. *Gunflintia* (S).
Palaeospiralis canadensis Edhorn 1973. (Edhorn 1973)
    Septate filamentous cyanobacterium, cf. *Gunflintia* spp. (S).
Palaeospirulina arcuata Edhorn 1973. (Edhorn 1973)
    Septate filamentous prokaryote, cf. *Gunflintia* spp. (S).
Palaeospirulina minuta Edhorn 1973. (Edhorn 1973)
    Septate filamentous bacterium, cf. *Gunflintia* (S).
Primorivularia thunderbayensis Edhorn 1973. (Edhorn 1973)
    Poorly preserved septate filamentous bacterium, cf. *Gunflintia* spp. (S).
Thymos halis Awramik & Barghoorn 1977. (Awramik and Barghoorn

Table 22.3 (Continued).

1977)
  Colonial coccoid cyanobacterium (S).
Veryhachium? sp. of Hofmann 1971. (Hofmann 1971b)
  Microfossil of unknown affinities (S).
Xenothrix inconcreta Awramik & Barghoorn 1977. (Awramik and Barghoorn 1977)
  Prokaryote incertae sedis, possibly envelope of colonial prokaryote (S).

Pokegama Quartzite
NA USA MN
(ca. 2090 Ma)

Gunflintia minuta Barghoorn 1965. (Cloud and Licari 1972)
  Septate filamentous prokaryote (S).

Huto Grp.
AS CHN
(ca. 2075 Ma)

Favososphaeridium favosum Timofeev 1966. (Timofeev 1966)
  Sphaeromorph acritarch ("M S").

Ayan Fm.
AS SIB
(ca. 2000 Ma)

Favososphaeridium sp. of Timofeev, German & Mikhaljlova 1976. (Timofeev et al. 1976)
  Colonial coccoid cyanobacterium (S).
Stictosphaeridium sinapticuliferum Timofeev 1966. (Timofeev et al. 1976)
  Sphaeromorph acritarch ("S").

Duck Creek Dolomite
AU AUS WA
(ca. 2000 Ma)

(large coccoid alga). (Knoll and Barghoorn 1976)
  Sphaeromorph acritarch (S).
(small coccoid form). (Knoll and Barghoorn 1976)
  Assemblage of coccoid prokaryotes (S).
Eoastrion Barghoorn 1965. (Knoll and Barghoorn 1976)
  Prokaryote incertae sedis (S).
Eoastrion sp. of Knoll, Strother & Rossi 1988. (Knoll et al. 1988)
  Prokaryote incertae sedis (S).
Gunflintia Barghoorn 1965. (Knoll and Barghoorn 1976)
  Filamentous bacterium (S).
Gunflintia minuta Barghoorn 1965. (Knoll et al. 1988)
  Tubular sheath of filamentous prokaryote ("S").
Huroniospora [large sp.] of Knoll, Strother & Rossi 1988. (Knoll et al. 1988)
  Solitary coccoid cyanobacterium (S).
Huroniospora [small sp.] of Knoll, Strother & Rossi 1988. (Knoll et al. 1988)
  Solitary coccoid prokaryote ("S").
? *Leptoteichos golubicii* Knoll, Barghoorn & Awramik 1978. (Knoll et al. 1988)
  Sphaeromorph acritarch (S).
Oscillatoriopsis cuboides Knoll, Strother & Rossi 1988. (Knoll et al. 1988)
  Septate filamentous cyanobacterium ("S").
Oscillatoriopsis majuscula Knoll, Strother & Rossi 1988. (Knoll et al. 1988)
  Ensheathed septate filamentous oscillatoriacean cyanobacterium (S).

Siphonophycus [large] sp. of Knoll, Strother & Rossi 1988. (Knoll et al. 1988)
  Tubular sheath of filamentous cyanobacterium ("S").
Siphonophycus [small] sp. of Knoll, Strother & Rossi 1988. (Knoll et al. 1988)
  Tubular sheath of filamentous cyanobacterium (S).

Ladoga Fm.
EU SUN
(ca. 2000 Ma)

Leioligotriletum crassum (Naumova 1949) Timofeev 1960. (Timofeev 1960)
  Sphaeromorph acritarch ("S").
Leioligotriletum nitidum Timofeev 1960. (Timofeev 1960)
  Sphaeromorph acritarch ("S").
Protosphaeridium densum Timofeev 1966. (Timofeev 1969)
  Sphaeromorph acritarch ("S").
Protosphaeridium densum Timofeev 1966. (Timofeev et al. 1976)
  Sphaeromorph acritarch ("S").
Protosphaeridium flexuosum Timofeev 1966. (Timofeev 1966)
  Sphaeromorph acritarch ("S").
Protosphaeridium flexuosum Timofeev 1966. (Timofeev 1966)
  Sphaeromorph acritarch ("S").
Protosphaeridium parvulum Timofeev 1966. (Timofeev 1969)
  Sphaeromorph acritarch ("S").
Protosphaeridium rigidulum Timofeev 1966. (Timofeev et al. 1976)
  Sphaeromorph acritarch ("S").
Protosphaeridium rigidulum Timofeev 1966. (Timofeev 1969)
  Sphaeromorph acritarch ("S").
Protosphaeridium sp. of Timofeev 1966. (Timofeev 1966)
  Sphaeromorph acritarch ("S").
Stenozonoligotriletum sokolovii Timofeev 1960?. (Timofeev 1960)
  Sphaeromorph acritarch ("S").
Stenozonoligotriletum validum Timofeev 1960?. (Timofeev 1960)
  Sphaeromorph acritarch ("S").
Trachyoligotriletum laminaritum Timofeev 1960?. (Timofeev 1960)
  Sphaeromorph acritarch ("S").

2000 to 1800 Ma:

Butun Fm.
AS SIB
(ca. 1950 Ma)

Trematosphaeridium holtedahlii Timofeev 1966. (Timofeev et al. 1976)
  Sphaeromorph acritarch ("S").

Tyler Fm.
NA USA MI
(ca. 1950 Ma)

(20-μm diameter filament). (Cloud and Morrison 1980)
  Sheat of filamentous cyanobaterium (S).
cf. *Animikiea* sp. of Cloud & Morrison 1980. (Cloud and Morrison 1980)
  Tubular sheath of filamentous cyanobacterium ("S"). [See Chapter 24, Pl. 3, B.]
Gunflintia (smaller sp.) of Cloud & Morrison 1980. (Cloud and Morrison 1980)
  Filamentous prokaryote (S).
Gunflintia cf. *grandis* Barghoorn 1965. (Cloud and Morrison 1980)
  Septate filamentous cyanobacterium ("S"). [See Chapter 24, Pl. 3, A, C.]
cf. *Huroniospora* (larger spheroids) of Cloud & Morrison 1980. (Cloud and Morrison 1980)

Table 22.3 (*Continued*).

Sphaeromorph acritarchs (S).
cf. *Huroniospora* (smaller spheroids) Cloud and Morrison 1980. (Cloud and Morrison 1980)
    Sphaeromorph acritarchs (S).
Kakabekia cf. *umbellata* Barghoorn 1965. (Cloud and Morrison 1980)
    Prokaryotic (?) microfossil of uncertain affinity (S).
Metallogenium (larger) of Cloud and Morrison 1980. (Cloud and Morrison 1980)
    Species of *Eoastrion* (S).
Metallogenium (smaller) of Cloud and Morrison 1980. (Cloud and Morrison 1980)
    Species of *Eoastrion* (S).

Epworth Grp., Odjick Fm.
NA CAN NW
(ca. 1925 Ma)

Gunflintia sp. of Hofmann and Grotzinger 1985. (Hofmann and Grotzinger 1985)
    Septate filamentous prokaryote (S).
Huroniospora spp. of Hofmann and Grotzinger 1985. (Hofmann and Grotzinger 1985)
    Assemblage of coccoid prokaryotes ("S").
Siphonophycus spp. of Hofmann and Grotzinger 1985. (Hofmann and Grotzinger 1985)
    Sheath of filamentous cyanobacterium (S).

Epworth Grp., Rocknest Fm.
NA CAN NW
(ca. 1925 Ma)

(unnamed form A). (Hofmann and Grotzinger 1985)
    Sphaeromorph acritarch (S).
(unnamed form B). (Hofmann and Grotzinger 1985)
    Sphaeromorph acritarch (S).
Archaeonema sp. of Hofman and Grotzinger 1985. (Hofmann and Grozinger 1985)
    Septate filamentous, possibly eukaryotic, alga (S).
Archaeotrichion sp. of Hofmann and Grotzinger 1985. (Hofmann and Grotzinger 1985)
    Filamentous bacterium (M).
Biocatenoides incrustata J. Oehler 1977. (Hofmann and Grotzinger 1985)
    Septate filamentous bacterium (S).
Brevitrichoides sp. of Hofmann and Grotzinger 1985. (Hofmann and Grotzinger 1985)
    Elongate, rod-shaped, unicellular cyanobacterium (S).
Eomicrhystridium sp. of Hofmann and Grotzinger 1985. (Hofmann and Grotzinger 1985)
    Degraded and distorted sphaeromorph acritarch (S).
Eomycetopsis spp. of Hofmann and Grotzinger 1985. (Hofmann and Grotzinger 1985)
    Sheath of filamentous cyanobacterium (S).
Gunflintia? sp. of Hofmann and Grotzinger 1985. (Hofmann and Grotzinger 1985)
    Septate filamentous prokaryote (S).
Melasmatosphaera magna Hofmann 1976. (Hofmann and Grotzinger 1985)
    Ellipsoidal cyanobacterium (S).
Palaeoanacystis sp. of Hofmann and Grotzinger 1985. (Hofmann and Grotzinger 1985)
    Colonial coccoid cyanobacterium (S).
Siphonophycus spp. of Hofmann and Grotzinger 1985. (Hofmann and Grotzinger 1985)
    Sheath of filamentous cyanobacterium (S).
Sphaerophycus sp. of Hofmann and Grotzinger 1985. (Hofmann and Grotzinger 1985)
    Colonial coccoid cyanobacterium (S).

Onega Fm.
EU SUN
(ca. 1900 Ma)

Bothroligotriletum exasperatum Timofeev 1960?. (Timofeev 1960)
    Sphaeromorph acritarch ("S").
Mycteroligotriletum marmoratum Timofeev 1960?. (Timofeev 1960)
    Sphaeromorph acritarch ("S").
Protoleiosphaeridium conglutinatum Timofeev 1959. (Timofeev 1960)
    Colonial coccoid cyanobacterium (S).
Trachyoligotriletum asperatum (Naumova 1949) Timofeev 1960. (Timofeev 1960)
    Sphaeromorph acritarch ("S").

Sakukan Fm.
AS SIB
(ca. 1900 Ma)

Trachysphaeridium laminaritum Timofeev 1966. (Timofeev et al. 1976)
    Sphaeromorph acritarch ("S").

Sokoman Iron Fm.
NA CAN QU
(ca. 1900 Ma)

Animikiea septata Barghoorn 1965. (Knoll and Simonson 1981)
    Tubular sheath of filamentous cyanobacterium (M).
Gunflintia minuta Barghoorn 1965. (Knoll and Simonson 1981)
    Tubular sheath of filamentous bacterium (M).
Huroniospora Barghoorn 1965. (Knoll and Simonson 1981)
    Cell or envelope of coccoid cyanobacterium (M).

Sujsari Complex
EU SUN
(ca. 1900 Ma)

Protosphaeridium flexuosum Timofeev 1966. (Timofeev 1969)
    Sphaeromorph acritarch ("S").
Protosphaeridium sp. of Timofeev 1969. (Timofeev 1969)
    Sphaeromorph acritarch ("S").

Frere Fm.
AU AUS WA
(ca. 1875 Ma)

Animikiea sp. of Walter, Goode and Hall 1976. (Walter et al. 1976)
    Tubular sheath of filamentous cyanobacterium ("S").
Eoastrion simplex Barghoorn 1965. (Walter et al. 1976)
    Colonial radiating filamentous prokaryote (S).
Gunflintia minuta Barghoorn 1965. (Walter et al. 1976)
    Filamentous prokaryote (S).
Huroniospora spp. of Walter, Goode and Hall 1976. (Walter et al. 1976)
    Solitary coccoid cyanobacterium (S).
Kakabekia umbellata Barghoorn 1965. (Walter et al. 1976)
    Prokaryote incertae sedis (S).

Krivoj Rog "Series"
EU SUN
(ca. 1875 Ma)

Protosphaeridium acis Timofeev 1966. (Timofeev 1969)
    Sphaeromorph acritarch ("S").

Table 22.3 (Continued).

Protosphaeridium densum Timofeev 1966. (Timofeev 1966)
  Sphaeromorph acritarch ("S")
Protosphaeridium densum Timofeev 1966. (Timofeev 1969)
  Sphaeromorph acritarch ("S").
Protosphaeridium laccatum Timofeev 1966. (Timofeev 1966)
  Sphaeromorph acritarch ("S").
Protosphaeridium paleaceum Timofeev 1966. (Timofeev 1969)
  Sphaeromorph acritarch ("S").
Protosphaeridium rigidulum Timofeev 1966. (Timofeev 1969)
  Sphaeromorph acritarch ("S").
Protosphaeridium rigidulum Timofeev 1966. (Timofeev et al. 1976)
  Sphaeromorph acritarch ("S").
Protosphaeridium sp. of Timofeev 1966. (Timofeev 1966)
  Sphaeromorph acritarch ("S").
Stictosphaeridium sinapticuliferum Timofeev 1966. (Timofeev et al. 1976)
  Sphaeromorph acritarch ("S").
? *Stictosphaeridium sinapticuliferum* Timofeev 1969. (Timofeev 1969)
  Sphaeromorph acritarch ("S").

Changcheng Grp., Chuanlinggou Fm.
AS CHN
(ca. 1850 Ma)

Archaeotrichion sp. of Z. Zhang 1986. (Zhang 1986a)
  Flattened tubular sheath of filamentous bacterium (S).
Chuaria circularis Walcott 1899. (Zhang 1986a)
  Sphaeromorph acritarch ("S").
Chuaria sp. of Hofmann and Chen 1981. (Hofmann and Chen 1981)
  Sphaeromorph acritarch ("S").
Dictyosphaera macroreticulata Sin [Xing] and Liu 1973. (Xing and Liu 1973)
  Colonial coccoid (chroococcacean) cyanobacterium (S).
Dictyosphaera sinica Sin [Xing] and Liu 1973. (Xing and Liu 1973)
  Sphaeromorph acritarch ("S").
Eomycetopsis sp. of Z. Zhang 1986. (Zhang 1986a)
  Flattened tubular sheath of filamentous cyanobacterium ("S").
Gloeocapsomorpha hebeica Timofeev 1966. (Timofeev 1966)
  Colonial coccoid cyanobacterium (S).
Gloeocapsomorpha hebeica Timofeev 1966. (Timofeev 1969)
  Colonial coccoid cyanobacterium ("S").
Gloeocapsomorpha makrocysta Eisenack 1960. (Timofeev 1966)
  Colonial coccoid chroococcacean cyanobacterium ("S").
Gloeocapsomorpha prisca Zalessky [Zalesskij] 1917. (Timofeev 1966)
  Colonial coccoid chroococcacean? cyanobacterium ("S").
Gloeocapsomorpha sp. of Timofeev 1966. (Timofeev 1966)
  Colonial coccoid chroococcacean? cyanobacterium ("S").
Kildinosphaera sp. of Z. Zhang 1986. (Zhang 1986a)
  Sphaeromorph acritarch ("S").
Leiominuscula aff. *minuta* Naumova 1960. (Xing and Liu 1973)
  Solitary coccoid cyanobacterium (S).
Leiominuscula orientalis Sin [Xing] and Liu 1973. (Xing and Liu 1973)
  Colonial coccoid (chroococcacean?) cyanobacterium (S).
Leiominuscula pellucentis Sin [Xing] and Liu 1973. (Xing and Liu 1973)
  Solitary coccoid cyanobacterium (S).
Leiopsophaera apertus Schepeleva [Shepeleva] 1963. (Xing and Liu 1973)
  Sphaeromorph acritarch ("S").
Leiopsophaera aff. *effusus* Schepeleva [Shepeleva] 1963. (Xing and Liu 1973)
  Sphaeromorph acritarch ("S").
Leiopsophosphaera minor Schepeleva [Shepeleva] 1963. (Xing and Liu 1973)
  Sphaeromorph acritarch ("S").
Leiopsophosphaera pelucidus Schepeleva [Shepeleva] 1963. (Xing and Liu 1973)
  Sphaeromorph acritarch ("S").
Leiosphaeridia sp. of Z. Zhang 1986. (Zhang 1986a)
  Sphaeromorph acritarch ("S").
Margominuscula antiqua Naumova 1960. (Xing and Liu 1973)
  Colonial coccoid (chroococcacean?) cyanobacterium (S).
Margominuscula rugosa Naumova 1960. (Xing and Liu 1973)
  Solitary coccoid cyanobacterium (S).
Margominuscula aff. *tennela* Naumova 1961?. (Xing and Liu 1973)
  Solitary coccoid cyanobacterium (S).
Protosphaeridium acis Timofeev 1966. (Timofeev 1966)
  Sphaeromorph acritarch ("S").
Protosphaeridium densum Timofeev 1966. (Xing and Liu 1973)
  Sphaeromorph acritarch ("S").
Pterospermopsimorpha binata Timofeev 1966. (Timofeev 1966)
  Sphaeromorph acritarch containing degraded cytoplasmic remnants (M S).
Siphonophycus [larger] sp. of Z. Zhang 1986. (Zhang 1986a)
  Flattened tubular sheath of filamentous cyanobacterium ("S").
Siphonophycus [smaller] sp. of Z. Zhang 1986. (Zhang 1986a)
  Flattened tubular sheath of filamentous cyanobacterium ("S").
Stictosphaeridium inplexum Timofeev 1966. (Timofeev 1966)
  Sphaeromorph acritarch ("S").
Symplassosphaeridium sp. of Timofeev 1966. (Timofeev 1966)
  Colonial coccoid chroococcacean? cyanobacterium (S).
Trachysphaeridium hyalinum Sin [Xing] and Liu 1973. (Xing and Liu 1973)
  Sphaeromorph acritarch ("S").
Trachysphaeridium simplex Sin [Xing] 1962. (Xing and Liu 1973)
  Sphaeromorph acritarch ("S").

Vallen Grp., Zig Zag Land Fm.
NA GRL
(ca. 1850 Ma)

(type 2 microscopic sphere). (Bondesen et al. 1967)
  Sphaeromorph acritarch (S).

Besovets Fm.
EU SUN
(ca. 1800 Ma)

Gloeocapsomorpha sp. of Timofeev 1966. (Timofeev 1966)
  Colonial coccoid chroococcacean? cyanobacterium ("S").
Nucellosphaeridium minutum Timofeev 1966. (Timofeev 1966)
  Sphaeromorph acritarch containing cytoplasmic remnants (S M).
Protosphaeridium acis Timofeev 1966. (Timofeev 1966)
  Sphaeromorph acritarch ("S").
Protosphaeridium flexuosum Timofeev 1966. (Timofeev 1966)
  Sphaeromorph acritarch ("S").
Protosphaeridium laccatum Timofeev 1966. (Timofeev et al. 1976)
  Sphaeromorph acritarch ("S").
Protosphaeridium patelliforme Timofeev 1966. (Timofeev 1966)
  Sphaeromorph acritarch ("S").
Protosphaeridium scabridum Timofeev 1966. (Timofeev 1966)
  Sphaeromorph acritarch ("S").
Protosphaeridium scabridum Timofeev 1966. (Timofeev et al. 1976)
  Sphaeromorph acritarch ("S").

Table 22.3 (Continued).

Protosphaeridium sp. of Timofeev 1966. (Timofeev 1966)
  Sphaeromorph acritarch ("S").
Protosphaeridium tuberculiferum Timofeev 1966. (Timofeev 1969)
  Sphaeromorph acritarch ("S").

Jatulian
EU SUN
(ca. 1800 Ma)

Gloeocapsomorpha sp. of Timofeev 1969. (Timofeev 1969)
  Colonial coccoid cyanobacterium ("S").
Gloeocapsomorpha sp. of Timofeev, German and Mikhajlova 1976. (Timofeev et al. 1976)
  Colonial coccoid cyanobacterium ("S").
Protosphaeridium densum Timofeev 1966. (Timofeev 1969)
  Sphaeromorph acritarch ("S").
Protosphaeridium sp. of Timofeev 1969. (Timofeev 1969)
  Sphaeromorph acritarch ("S").
Turuchanica ternata Timofeev 1966. (Timofeev 1969)
  Flattened sphaeromorph? acritarch (S).

1800 to 1600 Ma:
Bothnia Fm.
EU FIN
(ca. 1770 Ma)

Favososphaeridium bothnicum Timofeev 1966. (Timofeev 1966)
  Colonial coccoid chroococcacean cyanobacterium (S).
Favososphaeridium sp. of Timofeev 1966. (Timofeev 1966)
  Colonial coccoid chroococcacean cyanobacterium (S).
Gloeocapsomorpha sp. of Timofeev 1966. (Timofeev 1966)
  Colonial coccoid chroococcacean? cyanobacterium ("S").
Leioligotriletum minutissimum (Naumova 1949) Timofeev 1960?. (Timofeev 1960)
  Sphaeromorph acritarch ("S").
Protosphaeridium acis Timofeev 1966. (Timofeev 1966)
  Sphaeromorph acritarch ("S").
Protosphaeridium densum Timofeev 1966. (Timofeev 1966)
  Sphaeromorph acritarch ("S").
Protosphaeridium laccatum Timofeev 1966. (Timofeev 1966)
  Sphaeromorph acritarch ("S").
Protosphaeridium patelliforme Timofeev 1966. (Timofeev 1966)
  Sphaeromorph acritarch ("S").
Protosphaeridium pusillum Timofeev 1966. (Timofeev 1966)
  Sphaeromorph acritarch ("S").
Trachyoligotriletum nevelense Timofeev 1957?. (Timofeev 1960)
  Sphaeromorph acritarch ("S").

Changcheng Grp., Tuanshanzi Fm.
AS CHN
(ca. 1750 Ma)

Chuaria sp. of Hofmann and Chen 1981. (Hofmann and Chen 1981)
  Sphaeromorph acritarch ("S").
Protosphaeridium patelliforme Timofeev 1966. (Timofeev 1966)
  Sphaeromorph acritarch ("S").
Tyrasotaenia cf. *podolica* Gnilovskaya 1971. (Hofmann and Chen 1981)
  Possibly sheath of filamentous cyanobacterium (S).

Hornby Bay Grp.
NA CAN NW
(ca. 1732 Ma)

(large spheroidal microfossil). (Horodyski et al. 1985)

  Sphaeromorph acritarch (S).
(narrow filaments). (Horodyski et al. 1985)
  Septate? filamentous bacterium (S).
(small spheroidal to ellipsoidal microfossils). (Horodyski et al. 1985)
  Solitary coccoid chroococcacean cyanobacterium ("S").
(smaller spheroids). (Horodyski et al. 1985)
  Solitary coccoid bacterium (S).
(tubular microfossil). (Horodyski et al. 1985)
  Tubular sheaths of filamentous cyanobacteria (S).

Brusno Fm.
EU SUN
(ca. 1700 Ma)

Orygmatosphaeridium sp. of Timofeev 1969. (Timofeev 1969)
  Sphaeromorph acritarch ("S").
Trematosphaeridium holtedahlii Timofeev 1966. (Timofeev 1969)
  Sphaeromorph acritarch ("S").
Trematosphaeridiym holtedahlii Timofeev 1966. (Timofeev et al. 1976)
  Sphaeromorph acritarch ("S").

Cuddapah Grp., Vempalle Fm.
AS IND AP
(ca. 1700 Ma)

(fine cellular filaments). (Schopf and Prasad 1978)
  Septate filamentous bacteria (S).
(less narrow cellular filaments). (Schopf and Prasad 1978)
  Septate filamentous prokaryotes ("S"). [See Chapter 24, Pl. 3, O.]
(organic "packets"). (Schopf and Prasad 1978)
  Envelope of colonial coccoid cyanobacteria ("S").
(spheroidal unicells). (Schopf and Prasad 1978)
  Assemblage of coccoid prokaryotes and sphaeromorph acritarchs ("S"). [See Chapter 24, Pl. 3, D–N.]
(tubular sheaths). (Schopf and Prasad 1978)
  Tubular sheath of filamentous (oscillatoriacean?) cyanobacteria ("S").

Ovruch "Series"
EU SUN
(ca. 1700 Ma)

Protosphaeridium densum Timofeev 1966. (Timofeev 1969)
  Sphaeromorph acritarch ("S").
Protosphaeridium densum Timofeev 1966. (Timofeev et al. 1976)
  Sphaeromorph acritarch ("S")
Protosphaeridium flexuosum Timofeev 1966. (Timofeev et al. 1976)
  Sphaeromorph acritarch ("S").
Protosphaeridium laccatum Timofeev 1966. (Timofeev 1969)
  Sphaeromorph acritarch ("S").
Protosphaeridium laccatum Timofeev 1966. (Timofeev et al. 1976)
  Sphaeromorph acritarch ("S").
Protosphaeridium paleaceum Timofeev 1966. (Timofeev 1969)
  Sphaeromorph acritarch ("S").
Protosphaeridium paleaceum Timofeev 1966. (Timofeev et al. 1976)
  Sphaeromorph acritarch ("S").
Protosphaeridium scabridum Timofeev 1966. (Timofeev et al. 1976)
  Sphaeromorph acritarch ("S").
Protosphaeridium tuberculiferum Timofeev 1966. (Timofeev et al. 1976)
  Sphaeromorph acritarch ("S").
Stictosphaeridium implexum Timofeev 1966. (Timofeev 1969)
  Sphaeromorph acritarch ("S").
Stictosphaeridium implexum Timofeev 1966. (Timofeev et al. 1976)

Table 22.3 (*Continued*).

Sphaeromorph acritarch ("S").
Stictosphaeridium sinapticuliferum Timofeev 1966. (Timofeev 1969)
  Sphaeromorph acritarch ("S").
Stictosphaeridium sinapticuliferum Timofeev 1966. (Timofeev et al. 1976)
  Sphaeromorph acritarch ("S").
Stictosphaeridium sinapticuliferum Timofeev 1966. (Timofeev 1966)
  Sphaeromorph acritarch ("S").
Stictosphaeridium sp. of Timofeev, German and Mikhajlova 1976. (Timofeev et al. 1976)
  Sphaeromorph acritarch ("S").
Stictosphaeridium sp. of Timofeev 1969. (Timofeev 1969)
  Colonial coccoid cyanobacterium (S).
Trematosphaeridium holtedahlii Timofeev 1966. (Timofeev et al. 1976)
  Sphaeromorph acritarch ("S").

Mount Isa Grp., Urquhart Shale
AU AUS QL
(ca. 1670 Ma)

Nanococcus vulgaris J. Oehler 1977. (Muir 1981)
  Colonial coccoid cyanobacterium ("S").
Sphaerophycus parvum Schopf 1968. (Muir 1981)
  Colonial coccoid cyanobacterium ("S")

Changcheng Grp., Dahongyu Fm.
AS CHN
(ca. 1650 Ma)

Eohyella campbellii Y. Zhang and Golubić 1987. (Zhang 1988)
  Colonial ellipsoidal cyanobacterium ("S").
Glenobotrydion granulosum Y. Zhang 1988. (Zhang 1988)
  Colonial coccoid bacterium (S).
Gloeodiniopsis sp. of Y. Zhang 1988. (Zhang 1988)
  Colonial coccoid chroococcacean cyanobacterium ("S").
Gloeothceopsis aggregata Y. Zhang 1988. (Zhang 1988)
  Colonial ellipsoidal bacterium (S).

Paradise Creek Fm.
AU AUS QL
(ca. 1650 Ma)

(Alpha Suite filaments). (Licari and Cloud 1972)
  Septate filamentous bacteria (S).
(Beta Suite filaments). (Licari and Cloud 1972)
  Septate filamentous prokaryote (S).
(Beta Suite unicells). (Licari and Cloud 1972)
  Colonial coccoid cyanobacteria ("S").
(Gamma Suite filaments). (Licari and Cloud 1972)
  Septate filamentous bacteria (S).
(Gamma Suite unicells). (Licari and Cloud 1972)
  Sphaeromorph acritarch (S).
Eucapsis? of Licari, Cloud and Smith 1969. (Licari et al. 1969)
  Cuboidal colonial coccoid cyanobacteria ("S"). [See Chapter 24, Pl. 4, A_1-A_3, B_1-B_2.]
Eucapsis? of Licari and Cloud 1972. (Licari and Cloud 1972)
  Cuboidal colonial coccoid prokaryote (S).

McArthur Grp., Amelia Dolomite
AU AUS
(ca. 1600 Ma)

(multicellular organism). (Croxford et al. 1973)
  Partially degraded, colonial cyanobacterium (S).
(tetrahedral tetrad). (Oehler et al. 1976)
  Sheath-enclosed colony of four coccoid cyanobacterial cells (S).

Ameliaphycus croxfordii Muir 1976. (Muir 1976)
  Colonial coccoid chroococcacean cyanobacterium (S).
Eomycetopsis filiformis Schopf 1968. (Muir 1976)
  Tubular sheath of filamentous cyanobacterium (S).
Eomycetopsis robusta Schopf 1968. (Muir 1976)
  Tubular sheaths of filamentous cyanobacteria (S).
Gunflintia cf. *minuta* Barghoorn 1965. (Muir 1976)
  Septate filamentous prokaryote (S).
Huroniospora microreticulata Barghoorn 1965. (Muir 1976)
  Solitary and colonial coccoid chroococcacean cyanobacterium (colonial, and so not *Huroniospora*) (S).
Huroniospora ornata Muir 1976. (Muir 1976)
  Diagenetically altered solitary or colonial (Chroococcaean?) cyanobacterium (colonial, so not *Huroniospora*) (S).
Huroniospora psilata Barghoorn 1965. (Muir 1976)
  Colonial coccoid chroococcacean cyanobacterium (colonial, and so not *Huroniospora*) (S).
Myxococcoides kingii Muir 1976. (Muir 1976)
  Colonial coccoid chroococcacean cyanobacterium (S).
Myxococcoides konzalovae Muir 1976. (Muir 1976)
  Colonial coccoid chroococcacean cyanobacterium (S).
Myxococcoides minor Schopf 1968. (Muir 1976)
  Colonial coccoid cyanobacterium (S).
Myxococcoides minuta Muir 1976. (Muir 1976)
  Colonial ellipsoidal ensheathed bacterium (S).
Myxococcoides reniformis Muir 1976. (Muir 1976)
  Colonial ellipsoidal chroococcacean cyanobacterium (S).
Myxomorpha janecekii Muir 1976. (Muir 1976)
  Colonial coccoid to ellipsoidal dermocarpacean-like cyanobacterium (S).
Palaeoanacystis plumbii Muir 1976. (Muir 1976)
  Colonial coccoid prokaryote (S).
Palaeoanacystis vulgaris Schopf 1968. (Muir 1974)
  Colonial coccoid prokaryote (S).
Palaeoanacystis vulgaris Schopf 1968. (Muir 1976)
  Colonial coccoid prokaryote (S).
Sphaerophycus parvum Schopf 1968. (Muir 1974)
  Solitary and paired coccoid prokaryote (S).
Sphaerophycus reticulatum Muir 1976. (Muir 1976)
  Colonial coccoid cyanobacterium (S).
Sphaerophycus tetragonale Muir 1976. (Muir 1976)
  Colonial coccoid and ellipsoidal bacterium (S).

McArthur Grp., Emmerugga Dolomite, Mara Dolomite Mbr.
AU AUS NT
(ca. 1600 Ma)

Bisacculoides grandis J. Oehler 1977. (Muir 1983)
  Colonial coccoid chroococcacean cyanobacterium (S).
Clonophycus elegans J. Oehler 1977. (Muir 1983)
  Colonial coccoid chroococcacean cyanobacterium (S).
Clonophycus laceyi Muir 1983. (Muir 1983)
  Colonial ensheathed coccoid cyanobacterium (S).
? *Clonophycus* sp. of Muir 1983. (Muir 1983)
  Spheroidal envelope of colonial coccoid prokaryote (S).
Eomycetopsis filiformis Schopf 1968. (Muir 1983)
  Tubular sheath of filamentous prokaryote (S).
Eomycetopsis robusta Schopf 1968. (Muir 1983)
  Tubular sheath of filamentous cyanobacterium (S).
Gunflintia minuta Barghoorn 1965. (Muir 1983)
  Septate filamentous bacterium (S).
Huroniospora microreticulata Barghoorn 1965. (Muir 1983)

Solitary chroococcacean cyanobacterium (S).
Huroniospora psilata Barghoorn 1965. (Muir 1983)
   Colonial coccoid chroococcacean cyanobacterium (colonial, and so not *Huroniospora*) (S).

Pechenga "Series"
EU SUN
(ca. 1600 Ma)

Protosphaeridium discum Timofeev 1976?. (Timofeev et al. 1976)
   Sphaeromorph acritarch ("S").
Protosphaeridium flexuosum Timofeev 1966. (Timofeev et al. 1976)
   Sphaeromorph acritarch ("S").
Protosphaeridium rigidulum Timofeev 1966. (Timofeev et al. 1976)
   Sphaeromorph acritarch ("S").
Protosphaeridium tuberculiferum Timofeev 1966. (Timofeev et al. 1976)
   Sphaeromorph acritarch ("S").
Stictosphaeridium implexum Timofeev 1966. (Timofeev et al. 1976)
   Sphaeromorph acritarch ("S").
Trematosphaeridium sp. of Timofeev, German and Mikhajlova 1976. (Timofeev et al. 1976)
   Sphaeromorph acritarch ("S").

Pechenga "Series" correlative (?)
EU SUN
(ca. 1600 Ma)

Gloeocapsomorpha priscata Timofeev 1973. (Timofeev et al. 1976)
   Colonial coccoid cyanobacterium ("S").

**Middle Proterozoic
1600 to 1400 Ma:**
Aj Fm.
EA SUN
(ca. 1590 Ma)

Brevitrichoides burzjanicus Jankauskas [Yankauskas] 1982. (Yankauskas 1982)
   Flattened, solitary, ellipsoidal cyanobacterium (S).
Kildinella hyperboreica Timofeev 1963. (Yankauskas 1982)
   Folded, originally spheroidal, sphaeromorph acritarch ("S").
Kildinella tschapomica Timofeev 1966. (Yankauskas 1982)
   Distorted, originally spheroidal, sphaeromorph acritarch ("S").
Leiominuscula minuta Naumova 1960. (Yankauskas 1982)
   Solitary coccoid cyanobacterium (S).
Symplassosphaeridium undosum Jankauskas [Yankauskas] 1979. (Yankauskas 1982)
   Sphaeromorph with botryoidal surface sculpture (or colony of cells?) ("S").

"Post-Jatulian deposits"
EU SUN
(ca. 1575 Ma)

Protosphaeridium acis Timofeev 1966. (Timofeev 1969)
   Sphaeromorph acritarch ("S").
Protosphaeridium laccatum Timofeev 1966. (Timofeev 1969)
   Sphaeromorph acritarch ("S").
Protosphaeridium scabridum Timofeev 1966. (Timofeev 1969)
   Sphaeromorph acritarch ("S").

Satka Fm.
EU SUN
(ca. 1550 Ma)

Brevitrichoides burzjanicus Jankauskas [Yankauskas] 1982. (Yankauskas 1982)
   Flattened, solitary, ellipsoidal cyanobacterium (S).
Eomarginata striata Jankauskas [Yankauskas] 1979. (Yankauskas 1982)
   Pterospermopsimorphid acritarch (S M).
Eomarginata striata Jankauskas [Yankauskas] 1979. (Yankauskas 1969a)
   Sphaeromorph acritarch ("S"). [See Chapter 24, Pl. 4, D.]
Gloeodiniopsis uralicus Kylov and Sergeev 1987.
   Colonial coccoid *Chroococcus*-like chroococcacean cyanobacterium ("S"). [See Chapter 24, Pl. 4, F_1-F_2.]
Kildinella hyperboreica Timofeev 1963. (Yankauskas 1982)
   Folded, originally spheroidal, sphaeromorph acritarch ("S").
Kildinella tschapomica Timofeev 1966. (Yankauskas 1982)
   Distorted, originally spheroidal, sphaeromorph acritarch ("S").
Leiominuscula minuta Naumova 1960. (Yankauskas 1982)
   Solitary coccoid cyanobacterium (S).
Micrhystridium sp. of Jankauskas [Yankauska] 1982.(Yankuskas 1982)
   Degraded sphaeromorph acritarch (S).
Palaeopleurocapsa kelleri Krylov and Sergeev 1987.
   Colonial coccoid ?pleurocapsalean cyanobacterium ("S"). [See Chapter 24, Pl. 4, E.]
Protosphaeridium flexuosum Timofeev 1966. (Yankauskas 1982)
   Thick-walled sphaeromorph acritarch ("S").
Satkafavosa Jankukas [Yankauskas] 1979. (Yankauskas 1982)
   Compressed colonial coccoid cyanobacterium (S).
Symplassosphaeridium undosum Jankauskas [Yankauskas] 1979. (Yankauskas 1982)
   Sphaeromorph with botryoidal surface sculpture (or colony of cells?) ("S").
Turuchanica ternata Timofeev 1966. (Yankauskas 1982)
   Thick-walled, originally spheroidal sphaeromorph acritarch; lobate due to compression ("S").

Bakal Fm.
EU SUN
(ca. 1500 Ma)

Brevitrichoides burzjanicus Jankauskas [Yankauskas] 1982. (Yankauskas 1982)
   Flattened, solitary, ellipsoidal cyanobacterium (S). [See Chapter 24, Pl.5, D.]
Eomarginata striata Jankauskas [Yankauskas] 1979. (Yankauskas 1982)
   Pterospermopsimorphid acritarch (S M).
Eomycetopsis psilata Maithy and Shukla 1977. (Yankauskas 1982)
   Tubular sheath of filamentous cyanobacterium ("S").
Kildinella hyperboreica Timofeev 1963. (Yankauskas 1982)
   Folded, originally spheroidal,phaeomorph acritrch (S').
Kildinella tschapomica Timofeev 1966. (Yankauskas 1982)
   Distorted, originally spheroidal, sphaeromorph acritarch ("S").
Leiominuscula minuta Naumova 1960. (Yankauskas 1982)
   Solitary coccoid cyanobacterium (S).
Margominuscula rugosa Naumova 1960. (Yankauskas in press)
   Solitary coccoid cyanobacterium (S). [See Chapter 24, Pl. 5, E.]
Margominuscula rugosa Naumova 1960. (Yankauskas 1982)
   Solitary cocoid to ellipsoidal cyanobacterium (S).
Micrhystridium sp. of Jankauskas [Yankauskas] 1982. (Yankauskas 1982)
   Degraded sphaeromorph acritarch (S).
Protosphaeridium flexuosum Timofeev 1966. (Yankauskas 1982)
   Thick-walled sphaeromorph acritarch ("S").

Table 22.3 (Continued).

Satka favosa Jankauskas [Yankauskas] 1979. (Yankauskas 1979a)
   Colonial coccoid cyanobacterium (S). [See Chapter 24, Pl. 5, A, C.]
Satka favosa Jankauskas [Yankauskas] 1979. (Yankauskas 1982)
   Compressed colonial coccoid cyanobacterium (S).
Symplassosphaeridium undosum Jankauskas [Yankauskas] 1979. (Yankauskas 1982)
   Sphaeromorph with botryoidal surface sculpture (or colony of cells?) ("S").
Symplassosphaeridium undosum Jankauskas [Yankauskas] 1979. (Yankauskas 1979a)
   Colonial coccoid cyanobacterium (S). [See Chapter 24, Pl. 5, B.]
Synsphaeridium spp. of Jankauskas [Yankauskas] 1982. (Yankauskas 1982)
   Colonial coccoid (?chroococcacean) cyanobacteria (S).
Trematosphaeridium holtedahlii Timofeev 1959. (Yankauskas 1982)
   Coarsely porate sphaeromorph acritarch ("S").
Turuchanica ternata Timofeev 1966. (Yankauskas 1982)
   Thick-walled, originally spheroidal sphaeromorph acritarch; lobate due to compression ("S").

McArthur Grp., Barney Creek Fm., Cooley Dolomite Mbr.
AU AUS NT
(ca. 1500 Ma)

(manganese-encrusted bacteria). (Muir 1978)
   Manganese-encrusted filamentous prokaryote ("S").

McArthur Grp., Barney Creek Fm., HYC Pyritic Shale Mbr.
AU AUS NT
(ca. 1500 Ma)

(biomorphic structures). (Hamilton and Muir 1974)
   Colonial coccoid cyanobacteria (M).
(filamentous alga [...similar to *Cyanonema*]). (J. Oehler 1977)
   Septate filamentous prokaryote (S).
(filamentous alga). (J. Oehler 1977)
   Septate filamentous (oscillatoriacean?) cyanobacterium (S).
(oscillatoriacean-type algal trichome). (J. Oehler 1977)
   Septate filamentous (oscillatoriacean) cyanobacterium (S).
(sheaths of filamentous algae). (J. Oehler 1977)
   Tubular sheath of filamentous (oscillatoriacean?) cyanobacterium (S).
(triad of cells). (J. Oehler 1977)
   Ensheathed colonial coccoid prokaryote (S).
(tubular structure). (J. Oehler 1977)
   Tubular sheath of filamentous cyanobacterium ("S").
Bigeminococcus mucidus Schopf and Blacic 1971. (J. Oehler 1977)
   Ensheathed colonial coccoid (chroococcacean) cyanobacterium ("S").
Biocatenoides incrustata J. Oehler 1977. (J. Oehler 1977)
   Septate filamentous bacterium ("S").
Biocatenoides pertenuis J. Oehler 1977. (J. Oehler 1977)
   Septate filamentous bacterium ("S").
Biocatenoides rhabdos Schopf 1968. (J. Oehler 1977)
   Septate filamentous bacterium ("S").
Bisacculoides grandis J. Oehler 1977. (J. Oehler 1977)
   Ensheathed colonial coccoid (chroococcacean) cyanobacterium ("S").
Bisacculoides tabeoviscus J. Oehler 1977. (J. Oehler 1977)
   Ensheathed colonial coccoid (chroococcacean) cyanobacterium ("S").
Bisacculoides vacua J. Oehler 1977. (J. Oehler 1977)
   Ensheathed colonial coccoid (chroococcacean) cyanobacterium ("S").
Clonophycus elegans J. Oehler 1977. (J. Oehler 1977)
   Ensheathed colonial coccoid (chroococcacean?) cyanobacterium (S).
Coleobacter primus J. Oehler 1977. (J. Oehler 1977)
   Ensheathed septate filamentous bacterium ("S").
Cyanonema inflatum J. Oehler 1977. (J. Oehler 1977)
   Septate filamentous prokaryote (S).
Cyanonema minor J. Oehler 1977. (J. Oehler 1977)
   Septate filamentous bacterium (S).
Eoastrion sp. of J. Oehler 1977. (J. Oehler 1977)
   Colonial branched filamentous bacterium ("S").
Ferrimonilis variabile J. Oehler 1977. (J. Oehler 1977)
   Pyritized filamentous bacterium ("S").
Globophycus minor J. Oehler 1977. (J. Oehler 1977)
   Ensheathed solitary coccoid cyanobacterium (S).
Gunflintia septata (Schopf 1968) J. Oehler 1977. (J. Oehler 1977)
   Septate filamentous bacterium (S).
Huroniospora sp. of J. Oehler 1977. (J. Oehler 1977)
   Sphaeromorph acritarch (S).
Myxococcoides kingii Muir 1976. (J. Oehler 1977)
   Colonial coccoid (chroococcacean) cyanobacterium ("S").
Nanococcus vulgaris J. Oehler 1977. (J. Oehler 1977)
   Colonial coccoid prokaryote (S).
Oscillatoriopsis schopfii J. Oehler 1977. (J. Oehler 1977)
   Septate filamentous oscillatoriacean cyanobacterium ("S").
Ramacia carpentariana J. Oehler 1977. (J. Oehler 1977)
   Branched filamentous bacterium ("S").
Sphaerophycus parvum Schopf 1968. (J. Oehler 1977)
   Coccoid colonial (chroococcacean) cyanobacterium ("S").

Omachtinsk Fm.
AS SIB
(ca. 1500 Ma)

Sivaglicania tadasii Weiss [Vejs] in press. (Vejs in press)
   Sphaeromorph acritarch with single ?horn (S). [See Chapter 24, Pl. 5, F.]

Nathan Grp., Balbirini Dolomite
AU AUS NT
(ca. 1483 Ma)

Balbiriniella praestans D. Oehler 1978. (Oehler 1978)
   Colonial coccoid (chroococcacean?) cyanobacterium (S).
Clonophycus biattina D. Oehler 1978. (Oehler 1978)
   Colonial coccoid (chroococcacean) cyanobacterium (S).
Clonophycus ostiolum D. Oehler 1978. (Oehler 1978)
   Colonial coccoid prokaryote (S).
Clonophycus refringens D. Oehler 1978. (Oehler 1978)
   Spheroidal envelope of spheroidal prokaryotic unicells (S).
Clonophycus sp. of D. Oehler 1978. (Oehler 1978)
   Globular to spheroidal envelopes of colonial prokaryotes (S).
Clonophycus vulgaris D. Oehler 1978. (Oehler 1978)
   Colonial coccoid (chroococcacean) cyanobacterium (S).
Eoentophysalis belcherensis Hofmann 1976. (Oehler 1978)
   Enheathed, colonial coccoid (entrophysalidacean) cyanobacterium ("S").
Eomycetopsis sp. of D. Oehler 1978. (Oehler 1978)
   Tubula sheath of filamentous prokaryote (S).
Myriasporella pyriformis D. Oehler 1978. (Oehler 1978)
   Colonial coccoid bacterium (S).

Table 22.3 (*Continued*).

Myxococcoides cracens D. Oehler 1978. (Oehler 1978)
  Colonial coccoid (chroococcacean) cyanobacterium ("S").
Myxococcoides minuta Muir 1976. (Oehler 1978)
  Colonial coccoid prokaryote (S).
Myxococcoides sp. of D. Oehler 1978. (Oehler 1978)
  Colonial coccoid (chroococcacean) cyanobacterium ("S").
Myxomorpha janecekii Muir 1976. (Oehler 1978)
  Colonial coccoid (dermocarpacean or pleurocapsacean) cyanobacterium ("S").
Nanococcus vulgaris J. Oehler 1977. (Oehler 1978)
  Colonial coccoid (chroococcacean) cyanobacterium ("S").
Palaeoanacystis plumbii Muir 1976. (Oehler 1978)
  Colonial coccoid prokaryote (S).
Palaeoanacystis vulgaris Schopf 1968. (Oehler 1978)
  Colonial coccoid (chroococcacean) cyanobacterium ("S").
Palaeolyngbya sp. of D. Oehler 1978. (Oehler 1978)
  Septate, filamentous (oscillatoriacean) cyanobacterium ("S").
Pilavia maculata D. Oehler 1978. (Oehler 1978)
  Solitay, coccoid to ellipsoidal cyanobacterium (S).
Siphonophycus sp. of D. Oehler 1978. (Oehler 1978)
  Tubular sheath of filamentous (oscillatoriacean) cyanobacterium ("S").
Sphaerophycus parvum Schopf 1968. (Oehler 1978)
  Colonial coccoid (chroococcacean) cyanobacterium ("S").
Sphaerophycus reticulatum Muir 1976. (Oehler 1978)
  Colonial coccoid prokaryote (S).
Tetraphycus acinulus D. Oehler 1978. (Oehler 1978)
  Colonial coccoid bacterium (S).
Tetraphycus diminutivus D. Oehler 1978. (Oehler 1978)
  Colonial coccoid bacterium (S).
Tetraphycus gregalis D. Oehler 1978. (Oehler 1978)
  Colonial coccoid prokaryote (S).
Tetraphycus major D. Oehler 1978. (Oehler 1979)
  Colonial coccoid (chroococcacean) cyanobacterium ("S").

Belt Sgrp., Chamberlain Shale
NA USA MT
(ca. 1450 Ma)

(larger filaments). (Horodyski 1980)
  Flattened tubular sheath of filamentous cyanobacteria ("S").
(other sphaeromorphs). (Horodyski 1980)
  Sphaeromorph acritarch ("S").
(polygonally segmented and bumpy surfaced sphaeromorphs). (Horodyski 1980)
  Sphaeromorph acritarch ("S").
Archaeotrichion/Siphonophycus spp. of Horodyski 1980. (Horodyski 1980)
  Flattened tubular sheath of filamentous cyanobacterium (S).
Kildinella sp. of Horodyski 1980. (Horodyski 1980)
  Sphaeromorph acritarch ("S").
Siphonophycus beltense Horodyski 1980. (Horodyski 1980)
  Flattened tubular sheath of filamentous cyanobacterium ("S").
Siphonophycus crassiusculum Horodyski 1980. (Horodyski 1980)
  Flattened tubular sheath of filamentous cyanobacterium ("S").

Belt Sgrp., Altyn Fm.
NA USA MT
(ca. 1440 Ma)

(tubular, nonseptate, branched filaments). (White 1974)
  Tubular sheaths of filamentous false-branching cyanobacteria ("S").

Belt Sgrp., Newland Limestone
NA USA MT
(ca. 1440 Ma)

Micrococcus sp. of Walcott 1915. (Walcott 1915)
  Probably coccoid and streptococcus-like bacteria (S).

Belt Sgrp., ?Newland Limestone
(carbonate facies of Chamberlain Shale?)
NA USA MT
(ca. 1440 Ma)

(larger filaments). (Horodyski 1980)
  Flattened tubular sheath of filamentous cyanobacteria ("S").
(other sphaeromorphs). (Horodyski 1980)
  Sphaeromorph acritarch ("S").
Archaeotrichion/Siphonophycus spp. of Horodyski 1980. (Horodyski 1980)
  Flattened tubular sheath of filamentous cyanobacterium (S).
Kildinella sp. of Horodyski 1980. (Horodyski 1980)
  Sphaeromorph acritarch ("S").
Siphonophycus beltense Horodyski 1980. (Horodyski 1980)
  Flattened tubular sheath of filamentous cyanobacterium (S).
Siphonophycus crassiusculum Horodyski 1980. (Horodyski 1980)
  Flattened tubular sheath of filamentous cyanobacterium ("S").

Zigal'ga Fm.
EA SUN
(ca. 1430 Ma)

Nucellosphaeridium minutum Timofeev 1966. (Timofeev 1969)
  Sphaeromorph acritarch ("S").
Protosphaeridium asaphum Timofeev 1966. (Timofeev 1969)
  Sphaeromorph acritarch ("S").

Changcheng Grp., Gaoyuzhuang Fm.
AS CHN
(ca. 1425 Ma)

(intermediate-diameter microbial sheaths). (Schopf et al. 1984)
  Tubular sheath of filamentous (oscillatoriacean?) cyanobacterium (S).
(large-diameter organic spheroids). (Schopf et al. 1984)
  Spheroidal envelope of colonial coccoid or ellipsoidal prokaryote (S).
(large-diameter organic tubules). (Schopf et al. 1984)
  Tubular lamellated sheath of filamentous (*Lyngbya*-like oscillatoriacean) cyanobacterium ("S"). [See Chapter 24, Pl. 6, B, E.]
(solitary and paired spheroidal unicells). (Schopf et al. 1984)
  Solitary coccoid (chroococcacean?) cyanobacteria (S).
(spheroidal, solitary or paired unicells). (Schopf et al. 1984)
  Solitary coccoid (chroococcacean?) cyanobacterium ("S"). [See Chapter 24, Pl. 8, A–C, I, K.]
(very fine filaments). (Schopf et al. 1984)
  Septate filamentous bacteria, cf. *Gunflintia* (S).
(*Eomycetopsis*-like forms). (Schopf et al. 1984)
  Tubular sheath of filamentous prokaryote ("S"). [See Chapter 24, Pl. 7, C, D.]
Anabaenidium sophoroides Xu 1984. (Xu 1984a)
  Septate filamentous bacterium (S).
Archaeoellipsoides grandis Horodyski and Donaldson 1980. (Zhang and Li 1985)
  Ellipsoidal envelope of colonial coccoid cyanobacterium (S).
Archaeonema longicellulare Schopf 1968. (Zang and Li 1985)
  Septate filamentous prokaryote (S).

Table 22.3 (Continued).

Asperatopsophosphaera umishanensis Sin [Xing] and Liu 1973. (Xing and Liu 1973)
    Sphaeromorph acritarch ("S").
Biocatenoides sp. of Xu 1984. (Xu 1984b)
    Septate filamentous bacterium ("S").
Biocatenoides sphaerula Schopf 1968. (Zhang and Li 1985)
    Septate filamentous bacterium (S).
Cephalophytarion constrictum Schopf and Blacic 1971. (Zhang and Li 1985)
    Septate filamentous prokaryote (S).
Cephalophytarion taenia Y. Zhang 1981. (Zhang 1981)
    Septate filamentous (oscillatoriacean?) cyanobacterium ("S").
Cephalophytarion? taenia Y. Zhang 1981. (Zhang and Li 1985)
    Septate filamentous oscillatoriacean cyanobacterium ("S").
Coniunctiophycus conglobatum Y. Zhang 1981. (Zhang 1981)
    Colonial coccoid bacterium (S).
Coniunctiophycus gaoyuzhuangense Y. Zhang 1981. (Zhang 1981)
    Colonial coccoid chroococcacean cyanobacterium ("S").
Cyanonema ligamen Y. Zhang 1981. (Zhang 1981)
    Septate filamentous bacterium (S).
Cyanonema ligamen Y. Zhang 1981. (Zhang and Li 1985)
    Septate filamentous prokaryote (S).
Eoaphanothece zhuiana Xu 1984. (Xu 1984a)
    Tubular sheath (or hormogone?) of filamentous prokaryote (S).
Eoentophysalis belcherensis Hofmann 1976. (Zhang 1981)
    Colonial coccoid (entophysalidacean?) cyanobacterium ("S").
Eomycetopsis filiformis Schopf 1968. (Zhang 1981)
    Tubular sheath of filamentous prokaryote (S).
Eomycetopsis robusta Schopf 1968. (Zhang and Li 1985)
    Tubular sheath of filamentous cyanobacterium ("S").
Eophormidium capitatum Xu 1984. (Xu 1984a)
    Septate filamentous bacterium (S).
Eophormidium liangii Xu 1984. (Xu 1984a)
    Septate filamentous prokaryote (S).
Eophormidium semicirculare Xu 1984. (Xu 1984a)
    Septate filamentous prokaryote (S).
Eosynechococcus? sp. of Y. Zhang 1981. (Zhang 1981)
    Colonial, coccoid to ellipsoidal chroococcacean cyanobacterium (S).
Glenobotrydion varioforme Y. Zhang 1981. (Zhang 1981)
    Solitary or colonial, coccoid chroococcacean cyanobacterium ("S").
Gloeodiniopsis hebeiensis Y. Zhang 1981. (Zhang 1981)
    Colonial coccoid prokaryote (S).
Gloeodiniopsis pangjapuensis Y. Zhang 1981. (Zhang 1981)
    Ensheathed solitary or colonial coccoid prokaryote (S).
Gunflintia minuta Barghoorn 1965. (Xu 1984a)
    Tubular sheath of filamentous bacterium (S).
Halythrix sp. of Y. Zhang 1981. (Zhang 1981)
    Septate filamentous bacterium (S).
Halythrix sp. of Z. Zhang and Li 1985. (Zhang and Li 1985)
    Septate filamentous prokaryote (S).
Heliconema australiense Schopf 1968. (Xu 1984b)
    Tubular sheath of filamentous prokaryote (S).
Microcystopsis yaoi Xu 1984. (Xu 1984a)
    Colonial coccoid bacterium (S).
Nanococcus vulgaris J. Oehler 1977. (Zhang 1981)
    Solitary and colonial coccoid to ellipsoidal prokaryote (S).
Nostocomorpha prisca Sin [Xing] and Liu 1978. (Xu 1984a)
    Septate filamentous prokaryote (S).
Oscillatoriopsis acuminata Xu 1984. (Xu 1984a)
    Septate filamentous prokaryote (S).

Oscillatoriopsis cf. *breviconvexa* Schopf and Blacic 1971. (Zhang and Li 1985)
    Septate filamentous oscillatoriacean cyanobacterium ("S").
Oscillatoriopsis disciformis Xu 1984. (Xu 1984a)
    Septate filamentous prokaryote (S).
Oscillatoriopsis glabra Xu 1984. (Xu 1984a)
    Septate filamentous prokaryote (S).
Oscillatoriopsis hemisphaerica Xu 1984. (Xu 1984a)
    Septate filamentous prokaryote (S).
Oscillatoriopsis maxima (Y. Zhang 1981) Z. Zhang 1985. (Zhang and Li 1985)
    Septate filamentous oscillatoriacean cyanobacterium ("S").
Oscillatoriopsis sp. I of Xu 1984. (Xu 1984b)
    Septate filamentous bacterium (S).
Oscillatoriopsis sp. II of Xu 1984. (Xu 1984b)
    Tubular sheath of filamentous bacterium (S).
Oscillatoriopsis tuberculata Xu 1984. (Xu 1984a)
    Tubular sheath of filamentous prokaryote (S).
Palaeoanacystis vulgaris Schopf 1968. (Zhang 1981)
    Colonial coccoid chroococcacean cyanobacterium ("S").l
Palaeolyngbya barghoorniana Schopf 1968. (Zhang 1981)
    Ensheathed septate filamentous oscillatoriacean cyanobacterium ("S").
Palaeolyngbya maxima Y. Zhang 1981. (Zhang 1981)
    Ensheathed, septate filamentous oscillatoriacean cyanobacterium ("S").
Paleoisocystis disporata Xu 1984. (Xu 1984a)
    Degraded tubular sheath of filamentous bacterium (S).
Paleoisocystis monosporata Xu 1984. (Xu 1984a)
    Tubular sheath of filamentous prokaryote (S).
Schizothropsis caudata Xu 1984. (Xu 1984a)
    Septate filamentous prokaryote (S).
Siphonophycus inornatum Y. Zhang 1981. (Zhang 1981)
    Tubular sheath of filamentous cyanobacterium (S).
Siphonophycus inornatum Y. Zhang 1981. (Zhang and Li 1985)
    Tubular sheath of filamentous cyanobacterium ("S").
Siphonophycus kestron Schopf 1968. (Xu 1984b)
    Tubular sheath of filamentous cyanobacterium ("S").
Siphonophycus sp. of Z. Zhang and Li 1985. (Zhang and Li 1985)
    Tubular sheath of filamentous cyanobacterium ("S").

Belt Sgrp., Appekunny Argillite
NA USA MT
(ca. 1400 Ma)

(sphaeromorph acritarch). (Horodyski 1981)
    Sphaeromorph acritarch ("S").

Dismal Lakes Grp.
NA CAN NW
(ca. 1400 Ma)

(colony of large spheroids). (Horodyski and Donaldson 1983)
    Colonial coccoid chroococcacean cyanobacterium ("S").
(filaments). (Horodyski and Donaldson 1983)
    Tubular sheaths of filamentous cyanobacteria ("S").
(laminated, spherical envelope). (Horodyski and Donaldson 1983)
    Solitary ensheathed coccoid chroococcacean cyanobacterium, cf. *Gloeodiniopsis* ("S").
(spindle-shaped microfossils). (Horodyski and Donaldson 1980)
    Degraded filamentous prokaryote (S).
(thin filaments). (Horodyski and Donaldson 1980)
    Degraded tubular sheaths of filamentous prokaryote (S).
Archaeoellipsoides grandis Horodyski and Donaldson 1980.

Table 22.3 (Continued).

(Horodyski and Donaldson 1980)
    Ellipsoidal sac-like envelopes of colonial coccoid prokaryote (S). [See Chapter 24, Pl. 9, I, K, L.]

Biocatenoides? sp. of Horodyski and Donaldson 1980. (Horodyski and Donaldson 1980)
    Septate filamentous bacterium (S).

Eoentophysalis dismallakesensis Horodyski and Donaldson 1980. (Horodyski and Donaldson 1980)
    Colonial ellipsoidal entophysalidacean cyanobacterium ("S"). [See Chapter 24, Pl. 9, E.]

Eoentophysalis cf. *dismallakesensis* Horodyski and Donaldson 1983. (Horodyski and Donaldson 1983)
    Colonial ellipsoidal entophysalidacean cyanobacterium ("S").

Eomicrocoleus crassus Horodyski and Donaldson 1980. (Horodyski and Donaldson 1980)
    Degraded filamentous (multitrichomic) bacterium (S).

Filiconstrictosus sp. of Horodyski and Donaldson 1980. (Horodyski and Donaldson 1980)
    Septate filamentous cyanobacterium ("S").

Myxococcoides grandis Horodyski and Donaldson 1980. (Horodyski and Donaldson 1980)
    Spheroidal envelope of colonial coccoid prokaryote (S). [See Chapter 24, Pl. 9, J.]

Myxococcoides? sp. of Horodyski and Donaldson 1980. (Horodyski and Donaldson 1980)
    Colonial coccoid chroococcacean cyanobacterium ("S").

Myxococcoides? sp. of Horodyski and Donaldson 1980. (Horodyski and Donaldson 1983)
    Colonial coccoid chroococcacean cyanobacterium ("S").

Oscillatoriopsis curta Horodyski and Donaldson 1980. (Horodyski and Donaldson 1980)
    Septate filamentous cyanobacterium ("S"). [See Chapter 24, Pl. 9, A.]

Oscillatoriopsis robusta Horodyski and Donaldson 1980. (Horodyski and Donaldson 1980)
    Septate filamentous oscillatoriacean cyanobacterium ("S").

Oscillatoriopsis sp. of Horodyski and Donaldson 1980. (Horodyski and Donaldson 1980)
    Septate filamentous oscillatoriacean cyanobacterium ("S").

Sphaerophycus medium Horodyski and Donaldson 1980. (Horodyski and Donaldson 1980)
    Colonial coccoid chroococcacean cyanobacterium ("S"). [See Chapter 24, Pl. 9, F, H.]

Sphaerophycus cf. *medium* Horodyski and Donaldson 1980. (Horodyski and Donaldson 1983)
    Colonial, coccoid and ellipsoidal chroococcacean cyanobacterium ("S").

Sphaerophycus parvum Schopf 1968. (Horodyski and Donaldson 1983)
    Colonial ellipsoidal bacterium (S).

Sphaerophycus parvum Schopf 1968. (Horodyski and Donaldson 1980)
    Colonial ellipsoidal bacterium (S). [See Chapter 24, Pl. 9, G.]

Sphaerophycus? sp. of Horodyski and Donaldson 1980. (Horodyski and Donaldson 1980)
    Colonial coccoid bacterium (S).

1400 to 1200 Ma:
Belt Sgrp., Missoula Grp., Snowslip Fm.
NA USA MT
(ca. 1370 Ma)

(larger tubular microfossil). (Horodyski 1975)
    Tubular sheath of filamentous cyanobacterium ("S").

(sinuous chainlike forms). (Horodyski 1975)
    Septate filamentous bacterium (S).

(smaller tubular microfossil). (Horodyski 1975)
    Tubular sheath of filamentous prokaryote (S).

Bungle Bungle Dolomite
AU AUS WA
(ca. 1364 Ma)

(colonial coccoids). (Diver 1974)
    Colonial coccoid cyanobacteria ("S").

(narrow filaments). (Diver 1974)
    Degraded(?) filamentous bacterium (=*Archaeotrichion*) (S).

(septate filaments). (Diver 1974)
    Oscillatoriacean or beggiatoacean prokaryote (S).

(spheroids with internal bodies). (Diver 1974)
    Paired and colonial coccoid cyanobacteria (=*Glenobotrydion*) (S)

(spheroids). (Diver 1974)
    Solitary and colonial coccoid cyanobacteria ("S").

(tubular filaments). (Diver 1974)
    Sheath of filamentous prokaryote (S).

Jixian Grp., Yangzhuang Fm.
AS CHN
(ca. 1350 Ma)

Pseudozonosphaera verrucosa Sin [Xing] and Liu 1973. (Xing and Liu 1973)
    Sphaeromorph acritarch ("S").

Zigazino-Komarovsk Fm.
EA SUN
(ca. 1350 Ma)

Kildinella hyperboreica Timofeev 1966. (Timofeev 1969)
    Sphaeromorph acritarch ("S").

Kildinella hyperboreica Timofeev 1963. (Yankauskas 1982)
    Folded, originally spheroidal, sphaeromorph acritarch ("S").

Kildinella lophostriata Jankauskas [Yankauskas] 1979. (Yankauskas 1979b]
    Sphaeromorph acritarch ("S").

Kildinella nordia Timofeev 1969. (Yankauskas 1982)
    Sphaeromorph acritarch containing globular, condensed cytoplasmic contents ("S").

Kildinella ripheica Timofeev 1969. (Yankauskas 1982)
    Wrinkled, originally smooth?, sphaeromorph acritarch (?cf. *Kildinella vesljanica*) (S).

Kildinella sinica Timofeev 1966. (Timofeev 1969)
    Sphaeromorph acritarch ("S").

Kildinella tschapomica Timofeev 1966. (Yankauskas 1982)
    Distorted, originally spheroidal, sphaeromorph acritarch ("S"),

Leiominuscula minuta Naumova 1960. (Yankauskas 1982)
    Solitary coccoid cyanobacterium (S).

Leiosphaeridia bicrura Jankauskas [Yankauskas] 1976. (Yankauskas 1982)
    Ruptured sphaeromorph acritarch ("S").

Margominuscula rugosa Naumova 1960. (Yankauskas 1982)
    Solitary coccoid to ellipsoidal cyanobacterium (S).

Protosphaeridium flexuosum Timofeev 1966. (Yankauskas 1982)
    Thick-walled sphaeromorph acritarch ("S").

Pterospermopsimorpha capsulata Jankauskas [Yankauskas] 1982. (Yankauskas 1982)

Table 22.3 (Continued).

Possible poorly preserved, diagenetically altered pteromorph acritarch (M S).

Satka elongata Jankauskas [Yankauskas] 1979. (Yankauskas 1979b)
  Compressed sphaeromorph acritarch (S).

Satka favosa Jankauskas [Yankauskas] 1979. (Yankauskas 1982)
  Compressed colonial coccoid cyanobacterium (S).

Symplassosphaeridium undosum Jankauskas [Yankauskas] 1979. (Yankauskas 1982)
  Sphaeromorph with botryoidal surface sculpture (or colony of cells?) ("S").

Synsphaeridium spp. of Jankauskas [Yankauskas] 1982. (Yankauskas 1982)
  Colonial coccoid (?chroococcacean) cyanobacteria (S).

Trematosphaeridium holtedahlii Timofeev 1959. (Yankauskas 1982)
  Coarsely porate sphaeromorph acritarch ("S").

Turuchanica ternata Timofeev 1966. (Yankauskas 1982)
  Thick-walled, originally spheroidal sphaeromorph acritarch; lobate due to compression ("S").

Roper Grp., McMinn Fm.
AU AUS NT
(ca. 1340 Ma)

(cellular filament type I). (Peat et al. 1978)
  Septate filamentous oscillatoriacean cyanobacterium (S).
(cellular filament type II). (Peat et al. 1978)
  Septate filamentous bacterium (S).
(colonies of closely-packed small cells). (Peat et al. 1978)
  Colonial coccoid (chroococcacean?) cyanobacterium (S).
(disphaeromorph acritarchs). (Peat et al. 1978)
  Degraded membranous sphaeromorph acritarchs (S M).
(giant [nonseptate] filament). (Peat et al. 1978)
  Tubular sheath of filamentous cyanobacterium (S).
(giant [tapered] filament). (Peat et al. 1978)
  Septate filamentous oscillatoriacean cyanobacterium (S).
(giant filament [with rounded terminal cell]). (Peat et al. 1978)
  Septate filamentous oscillatoriacean cyanobacterium (S).
(large sphaeromorphs). (Peat et al. 1978)
  Sphaeromorph acritarch ("S").
(sheets of cells embedded in mucilage). (Peat et al. 1978)
  Colonial coccoid (chroococcacean?) cyanobacterium (S).
(small spheres in various arrangements). (Peat et al. 1978)
  Colonial coccoid cyanobacterium (S).
(smooth spheres containing inclusions). (Peat et al. 1978)
  Sphaeromorph acritarch ("S").
(sphere with median split). (Peat et al. 1978)
  Sphaeromorph acritarch ("S").
(spheres having a granular surface). (Peat et al. 1978)
  Sphaeromorph acritarch ("S").
(spheres with crinkly surfaces). (Peat et al. 1978)
  Sphaeromorph acritarch ("S").
(type I reticulate sphere). (Peat et al. 1978)
  Sphaeromorph acritarch ("S").
(type II reticulate sphere). (Peat et al. 1978)
  Sphaeromorph acritarch ("S").

Billyakh Grp., Kotujkan Fm.
AS SIB
(ca. 1325 Ma)

Eogloeocapsa bella Golovenoc [Golovenok] and Belova 1984. (Yakshin 1986)
  Colonial coccoid chroococcacean cyanobacterium (S).

Eomicrocystis sp. of Yakshin 1986. (Yakshin 1986)
  Colonial coccoid cyanobacterium ("S").

Eomycetopsis sp. of Yakshin 1986. (Yakshin 1986)
  Septate filamentous prokaryote (S).

Eosynechococcus crassus Golovenoc [Golovenok] and Belova 1984. (Golovenok and Belova 1984)
  Ellipsoidal sac-like envelope of colonial prokaryote (S).

Eosynechococcus giganteus Golovenoc [Golovenok] and Belova 1984. (Yakshin 1986)
  Solitary ellipsoidal cyanobacterium incertae sedis or ellipsoidal envelope of colonial cyanobacterium (S M).

Eosynechococcus giganteus Golovenoc [Golovenok] and Belova 1984. (Golovenok and Belova 1984)
  Solitary ellipsoidal cyanobacterium incertae sedis or ellipsoidal envelope of colonial cyanobacterium (S M).

Eosynechococcus major Golovenoc [Golovenok] and Belova 1984. (Golovenok and Belova 1984)
  Solitary ellipsoidal chroococcacean cyanobacterium ("S").

Eosynechococcus major Golovenoc [Golovenok] and Belova 1984. (Yakshin 1986)
  Solitary ellipsoidal cyanobacterium (S).

Globophycus [sp.] of Yakshin 1986. (Yakshin 1986)
  Sphaeromorph acritarch ("S").

Gloeodiniopsis [sp.] of Yakshin 1986. (Yakshin 1986)
  Ensheathed solitary coccoid cyanobacterium (S).

Myxococcoides? [sp.] of Yakshin 1986. (Yakshin 1986)
  Colonial coccoid chroococcacean cyanobacterium (S).

Oscillatoriopsis sp. [broad] of Yakshin 1986. (Yakshin 1986)
  "Giant" septate filamentous oscillatoriacean cyanobacterium (S).

Oscillatoriopsis sp. [narrow] of Yakshin 1986. (Yakshin 1986)
  Septate filamentous oscillatoriacean cyanobacterium (S).

Palaeolyngbya sp. of Yakshin 1986. (Yakshin 1986)
  Septate filamentous oscillatoriacean cyanobacterium (S).

Phanerosphaerops [sp.] of Yakshin 1986. (Yakshin 1986)
  Sphaeromorph acritarch ("S").

Symplassosphaeridium [sp.] of Yakshin 1986. (Yakshin 1986)
  Colonial coccoid chroococcacean cyanobacterium (S),

Billyakh Grp., ?Kotujkan Fm.
AS SIB
(ca. 1325 Ma)

Eogloeocapsa bella Golovenoc [Golovenok] and Belova 1984. (Golovenok and Belova 1984)
  Colonial coccoid chroococcacean cyanobacterium ("S").

Eomicrocystis elegans Golovenoc [Golovenok] and Belova 1984. (Golovenok and Belova 1984)
  Colonial coccoid chroococcacean cyanobacterium ("S").

Eomicrocystis irregularis Golovenoc [Golovenok] and Belova 1984. (Golovenok and Belova 1984)
  Colonial coccoid chroococcacean cyanobacterium ("S").

Eomycetopsis sp. of Golovenok and Belova 1984. (Golovenok and Belova 1984)
  Tubular sheath of filamentous cyanobacterium (S).

Eosynechococcus elongatus Golovenoc [Golovenok] and Belova 1984. (Golovenok and Belova 1984)
  Hormogone-like fragment of tubular sheath of filamentous cyanobacterium (S).

Eosynechococcus grandis Hofmann 1976. (Golovenok and Belova 1984)
  Solitary ellipsoidal chroococcacean cyanobacterium ("S").

Tetraphycus amplus Golovenoc [Golovenok] and Belova 1984.

Table 22.3 (*Continued*).

(Golovenok and Belova 1984)
    Colonial coccoid chroococcacean cyanobacterium ("S").

Jixian Grp., Wumishan Fm.
AS CHN
(ca. 1325 Ma)

(cellular filaments). (Cao 1984)
    Septate filamentous prokaryote (S).
(colonial bodies). (Cao 1984)
    Colonial coccoid ?chroococcacean cyanobacteria ("S").
(tubular sheaths). (Cao 1984)
    Tubular sheath of filamentous cyanobacterium ("S").
(unicells). (Cao 1984)
    Assemblage of solitary coccoid prokaryotes ("S").
(unnamed form A). (Y. Zhang 1985)
    Spheroidal envelope of colonial (coccoid?) prokaryote (S).
(unnamed form B). (Y. Zhang 1985)
    Siphonophycus-like tubular sheath of filamentous cyanobacterium (S).
(unnamed form C). (Y. Zhang 1985)
    Septate filamentous prokaryote (S).
Archaeoellipsoides conjunctivus Y. Zhang 1985. (Y. Zhang 1985)
    Broad septate (ulotrichalean?) filament with elongate cells (S).
Archaeoellipsoides obesus Y. Zhang 1985. (Y. Zhang 1985)
    Ellipsoidal envelope of colonial (coccoid?) prokaryote (S).
Asperatopsophosphaera umishanensis var. *minor* Sin [Xing] and Liu 1973. (Xing and Liu 1973)
    Solitary coccoid cyanobacterium (S).
Asperatopsophosphaera umishanensis Sin [Xing] and Liu 1973. (Xing and Liu 1973)
    Sphaeromorph acritarch ("S").
Bactrophycus dolichum Y. Zhang 1985. (Y. Zhang 1985)
    Tubular sheath of filamentous prokaryote (S).
Bactrophycus oblongum Y. Zhang 1985. (Y. Zhang 1985)
    Tubular sheath of short segment (hormogone?) of filamentous (oscillatoriacean?) cyanobacterium (S).
Callosicoccus crauros Y. Zhang 1985. (Y. Zhang 1985)
    Spheroidal envelope of colonial (coccoid?) prokaryote (S).
Cephalophytarion taenia Y. Zhang 1981. (Y. Zhang 1985)
    Septate filamentous oscillatoriacean cyanobacterium ("S").
Clonophycus cf. *ostiolum* D. Oehler 1978. (Y. Zhang 1985)
    Colonial coccoid chroococcacean cyanobacterium (S).
Myxococcoides cf. *grandis* Horodyski and Donaldson 1980. (Y. Zhang 1985)
    Spheroidal envelope of colonial (coccoid?) prokaryote (S).
Myxococcoides sp. of Y. Zhang 1985. (Y. Zhang 1985)
    Spheroidal envelope of colonial (coccoid?) prokaryote (S).
Oscillatoriopsis sp. of Y. Zhang 1985. (Y. Zhang 1985)
    Septate filamentous oscillatoriacean cyanobacterium ("S").
Pseudozonosphaera verrucosa Sin [Xing] and Liu 1973. (Xing and Liu 1973)
    Sphaeromorph acritarch ("S").
Rhicnonema sp. of Y. Zhang 1985. (Y. Zhang 1985)
    Ensheathed septate filamentous bacterium (S).
Siphonophycus inornatum Y. Zhang 1981. (Y. Zhang 1985)
    Tubular sheath of filamentous (oscillatoriacean?) cyanobacterium (S).
Sphaerophycus medium Horodyski and Donaldson 1980. (Y. Zhang 1985)
    Colonial coccoid chroococcacean cyanobacterium ("S").
Synsphaeridium conglutinatum Timofeev 1966. (Xing and Liu 1973)
    Colonial, coccoid to ellipsoidal (chroococcacean?) cyanobacterium (S).

Jotnian
EU FIN
(ca. 1300 Ma)

Favososphaeridium bothnicum Timofeev 1966. (Timofeev 1966)
    Colonial coccoid chroococcacean cyanobacterium (S).
Leioligotriletum compactum Timofeev 1960?. (Timofeev 1960)
    Sphaeromorph acritarch ("S").
Leioligotriletum glumaceum Timofeev 1960?. (Timofeev 1960)
    Sphaeromorph acritarch ("S").
Protosphaeridium acis Timofeev 1966. (Timofeev 1966)
    Sphaeromorph acritarch ("S").
Protosphaeridium flexuosum Timofeev 1966. (Timofeev 1966)
    Sphaeromorph acritarch ("S").
Protosphaeridium laccatum Timofeev 1966. (Timofeev 1966)
    Sphaeromorph acritarch ("S").
Protosphaeridium paleaceum Timofeev 1966. (Timofeev 1966)
    Sphaeromorph acritarch ("S").
Protosphaeridium parvulum Timofeev 1966. (Timofeev 1966)
    Sphaeromorph acritarch ("S").
Protosphaeridium parvulum Timofeev 1966. (Timofeev 1966)
    Sphaeromorph acritarch ("S").
Protosphaeridium rigidulum Timofeev 1966. (Tomofeev 1966)
    Sphaeromorph acritarch ("S").
Protosphaeridium sp. of Timofeev 1966. (Timofeev 1966)
    Sphaeromorph acritarch ("S").
Protosphaeridium sp. of Timofeev 1966. (Timofeev 1966)
    Sphaeromorph acritarch ("S").
Stictosphaeridium sp. of Timofeev 1966. (Timofeev 1966)
    Sphaeromorph acritarch ("S").
Trachyoligotriletum minutum (Naumova 1949) Timofeev 1960?. (Timofeev 1960)
    Sphaeromorph acritarch ("S").
Trachyoligotriletum obsoletum (Naumova 1949) Timofeev 1960?. (Timofeev 1960)
    Sphaeromorph acritarch ("S").
Trachyoligotriletum planum Timofeev 1960?. (Timofeev 1960)
    Sphaeromorph acritarch ("S").
Trematosphaeridium sp. of Timofeev 1966. (Timofeev 1966)
    Sphaeromorph acritarch ("S").

Strel'nye Gory Fm.
AS SIB
(ca. 1300 Ma)

([intermediate-diameter sheath of] algal trichome). (Timofeev 1969)
    Flattened tubular sheath of filamentous cyanobacterium (S).
Dictyosphaeridium tungusum Timofeev 1969. (Timofeev 1969)
    Sphaeromorph acritarch ("S").
Kildinella hyperboreica Timofeev 1966. (Timofeev 1966)
    Sphaeromorph acritarch ("S").
Kildinella rifeica Timofeev 1969. (Timofeev 1969)
    Sphaeromorph acritarch ("S").
Kildinella sinica Timofeev 1966. (Timofeev 1969)
    Sphaeromorph acritarch ("S").
Kildinella sp. of Timofeev 1969. (Timofeev 1969)
    Colonial coccoid cyanobacterium (S).
Kildinella timanica Timofeev 1969. (Timofeev 1969)
    Sphaeromorph acritarch ("S").
Kildinella tschapomica Timofeev 1966. (Timofeev 1969)

Table 22.3 (*Continued*).

Sphaeromorph acritarch ("S").
Orygmatosphaeridium sp. of Timofeev 1969. (Timofeev 1969)
  Sphaeromorph acritarch ("S").
Protosphaeridium acis Timofeev 1966. (Timofeev 1969)
  Sphaeromorph acritarch ("S").
Protosphaeridium asaphum Timofeev 1966. (Timofeev 1969)
  Sphaeromorph acritarch ("S").
Protosphaeridium pusillum Timofeev 1966. (Timofeev 1969)
  Sphaeromorph acritarch ("S").
Protosphaeridium scabridum Timofeev 1966. (Timofeev 1969)
  Sphaeromorph acritarch ("S").
Protosphaeridium torulosum Timofeev 1966. (Timofeev 1969)
  Sphaeromorph acritarch ("S").
Stictosphaeridium implexum Timofeev 1966. (Timofeev 1966)
  Sphaeromorph acritarch ("S").
Stictosphaeridium pectinale Timofeev 1966. (Timofeev 1969)
  Sphaeromorph acritarch ("S").
Stictosphaeridium sibiricum Timofeev 1969. (Timofeev 1969)
  Sphaeromorph acritarch ("S").
Stictosphaeridium sinapticuliferum Timofeev 1966. (Timofeev 1969)
  Sphaeromorph acritarch ("S").
Stictosphaeridium tortulosum Timofeev 1966. (Timofeev 1969)
  Sphaeromorph acritarch ("S").
Synsphaeridium conglutinatum Timofeev 1966. (Timofeev 1969)
  Colonial coccoid cyanobacterium (S).
Trematosphaeridium holtedahlii Timofeev 1966. (Timofeev 1966)
  Sphaeromorph acritarch ("S").
Trematosphaeridium holtedahlii Timofeev 1966. (Timofeev 1969)
  Sphaeromorph acritarch ("S").
Trematosphaeridium sinuatum Timofeev 1966. (Timofeev 1966)
  Sphaeromorph acritarch ("S").
Trematosphaeridium sp. of Timofeev 1969. (Timofeev 1969)
  Sphaeromorph acritarch ("S").

"Sinian" of Jixian
AS CHN
(ca. 1250 Ma)

Acantholigotriletum? sp. of Timofeev 1966. (Timofeev 1966)
  Possible acanthomorph acritarch (S).
Favososphaeridium favosum Timofeev 1966. (Timofeev 1966)
  Sphaeromorph acritarch ("M S")
Gloeocapsomorpha makrocysta Eisenack 1960. (Timofeev 1966)
  Colonial coccoid chroococcacean cyanobacterium ("S").
Gloeocapsomorpha prisca Zalessky [Zalesskij] 1917. (Timofeev 1966)
  Colonial coccoid chroococcacean? cyanobacterium ("S").
Protosphaeridium acis Timofeev 1966. (Timofeev 1966)
  Sphaeromorph acritarch ("S").
Protosphaeridium densum Timofeev 1966. (Timofeev 1966)
  Sphaeromorph acritarch ("S").
Protosphaeridium patelliforme Timofeev 1966. (Timofeev 1966)
  Sphaeromorph acritarch ("S").
Protosphaeridium pusillum Timofeev 1966. (Timofeev 1966)
  Sphaeromorph acritarch ("S").
Protosphaeridium sp. of Timofeev 1966. (Timofeev 1966)
  Sphaeromorph acritarch ("S").
Pterospermopsimorpha pileiformis Timofeev 1966. (Timofeev 1966)
  Sphaeromorph acritarch containing degraded cytoplasmic remnants (M S).

Jixian Grp., Hongshuizhuang Fm.
AS CHN
(ca. 1250 Ma)

Gloeocapsomorpha makrocysta Eisenack 19??. (Tomofeev 1966)

Colonial coccoid chroococcacean cyanobacterium ("S").
Gloeocapsomorpha prisca Zalessky [Zalesskij] 1917. (Timofeev 1966)
  Colonial coccoid chroococcacean? cyanobacterium ("S").
Nucellosphaeridium zonale Sin [Xing] and Liu 1973. (Xing and Liu 1973)
  Multilamellated sheath-enclosed coccoid chroococcacean cyanobacterium (S).
Polynucella biconcentrica Sin [Xing] and Liu 1973. (Xing and Liu 1973)
  Multilamellate sheath-enclosed colonial coccoid (chroococcacean) cyanobacterium (S).
Protoleiosphaeridium solidum Liu and Sin [Xing] 1973. (Xing and Liu 1973)
  Sphaeromorph acritarch ("S").
Protosphaeridium asaphum Timofeev 1966. (Timofeev 1966)
  Sphaeromorph acritarch ("S").
Pterospermopsis concentricus Sin [Xing] and Liu 1973. (Xing and Liu 1973)
  Multilamellated sheath-enclosed solitary coccoid chroococcacean cyanobacterium (S).
Pterospermopsis oculatus Sin [Xing] and Liu 1973. (Xing and Liu 1973)
  Multilamellated sheath-enclosed solitary coccoid (chroococcacean?) cyanobacterium (S).
Synsphaeridium conglutinatum Timofeev 1966. (Timofeev 1966)
  Colonial coccoid chroococcacean? cyanobacterium (S).

Malgin Fm.
AS SIB
(ca. 1250 Ma)

Kildinella hyperboreica Timofeev 1966. (Timofeev 1969)
  Sphaeromorph acritarch ("S").
Kildinella hyperboreica Timofeev 1966. (Timofeev 1966)
  Sphaeromorph acritarch ("S").
Kildinella sinica Timofeev 1966. (Timofeev 1969)
  Sphaeromorph acritarch ("S").
Orygmatosphaeridium sp. of Timofeev 1966. (Timofeev 1966)
  Sphaeromorph acritarch ("S").
Polyedrosphaeridium bullatum Timofeev 1966. (Timofeev 1966)
  Colonial coccoid chroococcacean? cyanobacterium (S).
Protosphaeridium acis Timofeev 1966. (Timofeev 1966)
  Sphaeromorph acritarch ("S").
Protosphaeridium acis Timofeev 1966. (Timofeev 1966)
  Sphaeromorph acritarch ("S").
Protosphaeridium laccatum Timofeev 1966. (Timofeev 1966)
  Sphaeromorph acritarch ("S").
Protosphaeridium laccatum Timofeev 1966. (Timofeev 1969)
  Sphaeromorph acritarch ("S").
Protosphaeridium scabridum Timofeev 1966. (Timofeev 1969)
  Sphaeromorph acritarch ("S").
Protosphaeridium sp. of Timofeev 1966. (Timofeev 1966)
  Sphaeromorph acritarch ("S").
Protosphaeridium tuberculiferum Timofeev 1966. (Timofeev 1966)
  Sphaeromorph acritarch ("S").
Protosphaeridium tuberculiferum Timofeev 1966. (Timofeev 1969)
  Sphaeromorph acritarch ("S").
Stenozonoligotriletum validum Timofeev 1960?. (Timofeev 1966)
  Sphaeromorph acritarch ("S").
Stictosphaeridium tortulosum Timofeev 1966. (Timofeev 1966)
  Sphaeromorph acritarch ("S").

Table 22.3 (Continued).

Trematosphaeridium sp. of Timofeev 1966. (Timofeev 1966)
  Sphaeromorph acritarch ("S").
Turuchanica alara Rudavskaja [Rudavskaya] 1964?. (Timofeev 1969)
  Compressed sphaeromorph? acritarch (S).

Avzyan Fm.
EA SUN
(ca. 1240 Ma)

Leiominuscula minuta Naumova 1960. (Yankauskas 1982)
  Solitary coccoid cyanobacterium (S).

Verkhnij Avzyan Fm.
EA SUN
(ca. 1230 Ma)

Kildinella hyperboreica Timofeev 1966. (Timofeev 1969)
  Sphaeromorph acritarch ("S").
Kildinella sinica Timofeev 1966. (Timofeev 1969)
  Sphaeromorph acritarch ("S").
Protosphaeridium tuberculiferum Timofeev 1966. (Timofeev 1966)
  Sphaeromorph acritarch ("S").
Trematosphaeridkum sp. of Timofeev 1966. (Timofeev 1966)
  Sphaeromorph acritarch ("S").
Turuchanica alara Rudavskaja [Rudavskaya] 1964?. (Timofeev 1969)
  Compressed sphaeromorph? acritarch (S).
Turuchanica ternata Timofeev 1966. (Timofeev 1969)
  Flattened sphaeromorph? acritarch (S).

Goloustnaya Fm.
AS SIB
(ca. 1200 Ma)

Favosophaeridium sp. of Timofeev 1966. (Timofeev 1966)
  Colonial coccoid chroococcacean cyanobacterium ("S").
Gloeocapsomorpha prisca Zalessky [Zalesskij] 1917. (Timofeev 1966)
  Colonial coccoid chroococcacean? cyanobacterium ("S").
Protosphaeridium densum Timofeev 1966. (Timofeev 1966)
  Sphaeromorph acritarch ("S").
Protosphaeridium flexuosum Timofeev 1966. (Timofeev 1966)
  Sphaeromorph acritarch ("S").
Protosphaeridium gibberosum Timofeev 1966. (Timofeev 1966)
  Sphaeromorph acritarch ("S").
Protosphaeridium laccatum Timofeev 1966. (Timofeev 1969)
  Sphaeromorph acritarch ("S").
Protosphaeridium papyraceum Timofeev 1966. (Timofeev 1966)
  Sphaeromorph acritarch ("S").
Trematosphaeridium holtedahlii Timofeev 1966. (Timofeev 1966)
  Sphaeromorph acritarch ("S").
Turuchanica alara Rudavskaja [Rudavskaya] 1964?. (Timofeev 1966)
  Compressed sphaeromorph? acritarch (S).

Il'yushkana Fm.
AS SIB
(ca. 1200 Ma)

Dictyosphaeridium eniseicum Timofeev 1969. (Timofeev 1969)
  Sphaeromorph acritarch ("S").
Favososphaeridium cf. *bothnicum* Timofeev 1966. (Timofeev 1969)
  Sphaeromorph acritarch ("S").
Favososphaeridium bothnicum Timofeev 1966. (Timofeev 1966)
  Sphaeromorph acritarch ("S").
Gloeocapsomorpha sp. of Timofeev 1966. (Timofeev 1966)
  Colonial coccoid chroococcacean? cyanobacterium ("S").
Kildinella? *hyperboreica* Timofeev 1966. (Timofeev 1966)
  Sphaeromorph acritarch ("S").
Kildinella sinica Timofeev 1966. (Timofeev 1969)
  Sphaeromorph acritarch ("S").
Kildinella sinica Timofeev 1966. (Timofeev 1966)
  Sphaeromorph acritarch ("S").
Kildinella sp. of Timofeev 1969. (Timofeev 1969)
  Colonial coccoid cyanobacterium (S).
Kildinella tschapomica Timofeev 1966. (Timofeev 1969)
  Sphaeromorph acritarch ("S").
Protosphaeridium crispum Timofeev 1969?. (Timofeev 1969)
  Sphaeromorph acritarch ("S").
Protosphaeridium parvulum Timofeev 1966. (Timofeev 1969)
  Sphaeromorph acritarch ("S").
Pterospermopsimorpha insolita Timofeev 1969. (Timofeev 1969)
  Pteromorph acritarch ("S").
Stictosphaeridium pectinale Timofeev 1966. (Timofeev 1969)
  Sphaeromorph acritarch ("S").
Stictosphaeridium sinapticuliferum Timofeev 1966. (Timofeev 1969)
  Sphaeromorph acritarch ("S").
Stictosphaeridium tortulosum Timofeev 1966. (Timofeev 1969)
  Sphaeromorph acritarch ("S").
Synsphaeridium conglutinatum Timofeev 1966. (Timofeev 1969)
  Colonial coccoid cyanobacterium (S).
Trematosphaeridium sp. of Timofeev 1969. (Timofeev 1969)
  Sphaeromorph acritarch ("S").
Zonosphaeridium crassum Timofeev 1969. (Timofeev 1969)
  Sphaeromorph acritarch ("S").

Kil'din "Series"
EU SUN
(ca. 1200 Ma)

Favososphaeridium favosum Timofeev 1966. (Timofeev 1969)
  Sphaeromorph acritarch ("S").
Favososphaeridium sp. of Timofeev 1969. (Timofeev 1969)
  Colonial coccoid cyanobacterium (S).
Kildinella hyperboreica Timofeev 1966. (Timofeev 1969)
  Sphaeromorph acritarch ("S").
Kildinella hyperboreica Timofeev 1966. (Timofeev 1966)
  Sphaeromorph acritarch ("S").
Kildinella nordia Timofeev 1969. (Timofeev 1969)
  Sphaeromorph acritarch ("S").
Kildinella sinica Timofeev 1966. (Timofeev 1966)
  Sphaeromorph acritarch ("S").
Kildinella sinica Timofeev 1966. (Timofeev 1969)
  Sphaeromorph acritarch ("S").
Kildinella sp. of Timofeev 1966. (Timofeev 1966)
  Sphaeromorph acritarch ("S").
Kildinella sp. of Timofeev 1969. (Timofeev 1969)
  Colonial coccoid cyanobacterium (S).
Kildinella timanica Timofeev 1969. (Timofeev 1969)
  Sphaeromorph acritarch ("S").
Protosphaeridium densum Timofeev 1966. (Timofeev 1969)
  Sphaeromorph acritarch ("S").
Protosphaeridium gibberosum Timofeev 1966. (Timofeev 1966)
  Sphaeromorph acritarch ("S").
Pterospermopsimorpha annulare Timofeev 1969. (Timofeev 1969)
  Pteromorph acritarch ("S").
Stictosphaeridium implexum Timofeev 1966. (Timofeev 1969)
  Sphaeromorph acritarch ("S").
Stictosphaeridium pectinale Timofeev 1966. (Timofeev 1969)
  Sphaeromorph acritarch ("S").

Table 22.3 (Continued).

Stictosphaeridium sp. of Timofeev 1969. (Timofeev 1969)
  Colonial coccoid cyanobacterium (S).
Turuchanica ternata Timofeev 1966. (Timofeev 1969)
  Flattened sphaeromorph? acritarch (S).

Tsipanda Fm.
AS SIB
(ca. 1200 Ma)

Kildinella hyperboreica Timofeev 1966. (Timofeev 1969)
  Sphaeromorph acritarch ("S").
Protosphaeridium flexuosum Timofeev 1966. (Timofeev 1966)
  Sphaeromorph acritarch ("S").
Protosphaeridium paleaceum Timofeev 1966. (Timofeev 1966)
  Sphaeromorph acritarch ("S").
Protosphaeridium papyraceum Timofeev 1966. (Timofeev 1966)
  Sphaeromorph acritarch ("S").
Protosphaeridium parvulum Timofeev 1966. (Timofeev 1969)
  Sphaeromorph acritarch ("S").
Protosphaeridium patelliforme Timofeev 1966. (Timofeev 1966)
  Sphaeromorph acritarch ("S").
Protosphaeridium? patelliforme Timofeev 1966. (Timofeev 1969)
  Sphaeromorph acritarch ("S").
Protosphaeridium patelliforme Timofeev 1966. (Timofeev 1969)
  Sphaeromorph acritarch ("S").
Protosphaeridium tuberculiferum Timofeev 1966. (Timofeev 1969)
  Sphaeromorph acritarch ("S").
Stictosphaeridium sinapticuliferum Timofeev 1966. (Timofeev 1966)
  Sphaeromorph acritarch ("S").
Turuchanica ternata Timofeev 1966. (Timofeev 1969)
  Flattened sphaeromorph? acritarch (S).

Vtorokamensk Fm.
AS SIB
(ca. 1200 Ma)

Leiosphaeridia bituminosa Timofeev 1966. (Timofeev 1969)
  Sphaeromorph acritarch ("S").
Protosphaeridium pusillum Timofeev 1966. (Timofeev 1969)
  Sphaeromorph acritarch ("S").
Stictosphaeridium implexum Timofeev 1966. (Timofeev 1969)
  Sphaeromorph acritarch ("S").
Synsphaeridium sorediforme Timofeev 1969?. (Timofeev 1969)
  Colonial coccoid cyanobacterium (S).

1200 to 1000 Ma:
Jixian Grp., Tieling Fm.
AS CHN
(ca. 1185 Ma)

Gloeocapsomorpha makrocysta Eisenack 1960. (Timofeev 1966)
  Colonial coccoid chroococcacean cyanobacterium ("S").
Protosphaeridium densum Timofeev 1966. (Timofeev 1966)
  Sphaeromorph acritarch ("S").
Stenozonoligotriletum validum Timofeev 1960?. (Timofeev 1966)
  Sphaeromorph acritarch ("S").

Roan Grp. (Lower)
AF ZMB
(ca. 1155 Ma)

(clusters of spherical bodies). (Binda 1972)
  Colonial coccoid cyanobacteria (S).
(indistinct ovoidal bodies). (Binda 1972)
  Ellipsoidal microfossil (S).

Torridonian (Lower), Stoer Fm.
EU GBR
(ca. 1115 Ma)

Favososphaeridium variabilis Cloud and Germs 1971. (Cloud and Germs 1971)
  Sphaeromorph acritarch ("M").
Protosphaeridium cf. *acis* Timofeev 1966. (Cloud and Germs 1971)
  Sphaeromorph acritarch ("M").
Protosphaeridium cf. *parvulum* Timofeev 1966. (Cloud and Germs 1971)
  Sphaeromorph acritarch ("M").

Kachergat Fm.
AS SIB
(ca. 1100 Ma)

Kildinella hyperboreica Timofeev 1966. (Timofeev 1969)
  Sphaeromorph acritarch ("S").
Kildinella sinica Timofeev 1966. (Timofeev 1969)
  Sphaeromorph acritarch ("S").
Kildinella? sp. of Timofeev 1969. (Timofeev 1969)
  Sphaeromorph acritarch ("S").
Lopholigotriletum? spathaeforme Timofeev 1959?. (Timofeev 1966)
  Sphaeromorph acritarch ("S").
Orygmatosphaeridium sp. of Timofeev 1969. (Timofeev 1969)
  Sphaeromorph acritarch ("S").
Protosphaeridium densum Timofeev 1966. (Timofeev 1966)
  Sphaeromorph acritarch ("S").
Protosphaeridium rigidulum Timofeev 1966. (Timofeev 1966)
  Sphaeromorph acritarch ("S").
Protosphaeridium tuberculiferum Timofeev 1966. (Timofeev 1969)
  Sphaeromorph acritarch ("S").
Stictosphaeridium implexum Timofeev 1966. (Timofeev 1966)
  Sphaeromorph acritarch ("S").
Stictosphaeridium pectinale Timofeev 1966. (Timofeev 1969)
  Sphaeromorph acritarch ("S").
Trematosphaeridium sp. of Timofeev 1969. (Timofeev 1969)
  Sphaeromorph acritarch ("S").
Turuchanica ternata Timofeev 1966. (Timofeev 1966)
  Flattened, compressed, (sphaeromorph?) acritarch (S).
Turuchanica ternata Timofeev 1966. (Timofeev 1969)
  Flattened sphaeromorph? acritarch (S).

Allamoore Fm.
NA USA TX
(ca. 1050 Ma)

(broad tubular sheaths). (Nyberg and Schopf 1981)
  Sheath of filamentous (oscillatoriacean?) cyanobacterium ("S")
(intermediate-diameter tubular sheaths). (Nyberg and Schopf 1981)
  Sheath of filamentous (oscillatoriacean?) cyanobacterium ("S").
(large unicells). (Nyberg and Schopf 1981)
  Assemblage of sphaeromorph acritarchs (S).
(narrow cellular filaments). (Nyberg and Schopf 1981)
  Septate filamentous bacteria, cf. *Biocatenoides* or cf. *Gunflintia* ("S").
(narrow tubular sheaths). (Nyberg and Schopf 1981)
  Tubular sheaths of filamentous cyanobacteria (S).
(sheath-enclosed colonies of unicells). (Nyberg and Schopf 1981)
  Colonial coccoid cyanobacterium ("S").
(solitary unicells). (Nyberg and Schopf 1981)
  Assemblage of solitary coccoid prokaryotes ("S").

Table 22.3 (Continued).

Aravalli Sgrp., Bijawar Grp.
AS IND UP
(ca. 1050 Ma)

(Fungal filament Type A). (Nautiyal 1983b)
  Septate filamentous prokaryote (S).
(Fungal spore Type A). (Nautiyal 1983b)
  Colonial coccoid cyanobacterium (S).
Eomycetopsis filiformis Schopf 1968. (Nautiyal 1983b)
  Tubular sheath of filamentous bacterium (S).
Granomarginata regia Salujha, Rehman and Arora 1972. (Nautiyal 1983b)
  Sphaeromorph acritarch ("S").
Myxococcoides inornata Schopf 1968. (Nautiyal 1983b)
  Colonial coccoid cyanobacterium ("S").
Palaeoanacystis vulgaris Schopf 1968. (Nautiyal 1983b)
  Colonial coccoid cyanobacterium ("S").
Palaeopleurocapsa wopfneri Knoll, Barghoorn and Golubić 1975. (Nautiyal 1983b)
  Colonial coccoid cyanobacterium ("S").

Uluntuj Fm.
AS SIB
(ca. 1050 Ma)

(large ghost filament). (Pierce and Cloud 1979)
  Ensheathed filamentous (oscillatoriacean?) cyanobacterium, and sheaths of same (S).
Gloeocapsomorpha makrocysta Eisenack 19??. (Timofeev 1966)
  Colonial coccoid chroococcacean cyanobacterium ("S").
Lopholigotriletum? spathaeforme Timofeev 1959?. (Timofeev 1966)
  Sphaeromorph acritarch ("S").
Protosphaeridium densum Timofeev 1966. (Timofeev 1969)
  Sphaeromorph acritarch ("S").
Protosphaeridium flexuosum Timofeev 1966. (Timofeev 1969)
  Sphaeromorph acritarch ("S").
Protosphaeridium ? laccatum Timofeev 1966. (Timofeev 1969)
  Sphaeromorph acritarch ("S").
Protosphaeridium tuberculiferum Timofeev 1966. (Timofeev 1969)
  Sphaeromorph acritarch ("S").
Stictosphaeridium pectinale Timofeev 1966. (Timofeev 1966)
  Sphaeromorph acritarch ("S").
Stictosphaeridium sinapticuliferum Timofeev 1966. (Timofeev 1966)
  Sphaeromorph acritarch ("S").
Trachysphaeridium sp. of Timofeev 1969. (Timofeev 1969)
  Sphaeromorph acritarch ("S").
Trematosphaeridium sinuatum Timofeev 1966. (Timofeev 1966)
  Sphaeromorph acritarch ("S").

Vindhyan Sgrp., Semri Grp.
AS IND
(ca. 1050 Ma)

Leiosphaeridia microgranulosa Nautiyal 1983. (Nautiyal 1983a)
  Sphaeromorph acritarch ("S").
Leiosphaeridia porcellanitensis Nautiyal 1983. (Nautiyal 1983a)
  Sphaeromorph acritarch ("S").
Satpulispora major Nautiyal 1979. (Nautiyal 1983a)
  Solitary coccoid (chroococcacean?) cyanobacterium ("S").
Satpulispora microreticulata Nautiyal 1979. (Nautiyal 1983a)
  Solitary coccoid (chroococcacean?) cyanobacterium ("S").
Satpulispora minuta Nautiyal 1979. (Nautiyal 1983a)
  Solitary coccoid (chroococcacean?) cyanobacterium ("S").
Satpulispora psilata Nautiyal 1979. (Nautiyal 1983a)
  Colonial coccoid chroococcacean cyanobacterium ("S").

Vindhyan Sgrp., Semri Grp., Arangi Fm.
AS IND UP
(ca. 1050 Ma)

Baltisphaeridium sp. of Maithy 1975. (Nautiyal 1983b)
  Possible acanthomorph acritarch, but figures do not permit a confident interpretation (M S).
Eomycetopsis filiformis Schopf 1968. (Nautiyal 1983b)
  Tubular sheath of filamentous bacterium (S).
Glenobotrydion majorinum Schopf and Blacic 1971. (Nautiyal 1983b)
  Colonial coccoid (chroococcacean) cyanobacterium (S).
Granomarginata vetula Salujha, Rehman and Arora 1972. (Nautiyal 1983b)
  Sphaeromorph acritarch ("S").
Kildinella minuta Maithy and Shukla 1977. (Nautiyal 1983b)
  Sphaeromorph acritarch ("S").
Leiosphaeridia densum (Maithy 1975) Nautiyal 1983. (Nautiyal 1983b)
  Solitary and colonial coccoid cyanobacteria (S M).
Leiosphaeridia sp. A of Nautiyal 1983. (Nautiyal 1983b)
  Sphaeromorph acritarch ("S").
Palaeoanacystis suketensis Maithy and Shukla 1977. (Nautiyal 1983b)
  Colonial coccoid (chroococcacean) cyanobacterium ("S").
Palaeopleurocapsa wopfneri Knoll, Barghoorn and Golubić 1975. (Nautiyal 1983b)
  Colonial coccoid cyanobacterium ("S").

Vindhyan Sgrp., Semri Grp., Kajrahat Fm.
AS IND UP
(ca. 1050 Ma)

Aphanocapsaopsis sitholeyi Maithy and Shukla 1977. (Nautiyal 1983b)
  Colonial coccoid cyanobacterium ("S").
Baltisphaeridium sp. of Maithy 1975. (Nautiyal 1983b)
  Possible acanthomorph acritarch, but figures do not permit a confident interpretation (M S).

Vindhyan Sgrp., Semri Grp., Kheinjua Fm.
AS IND
(ca. 1050 Ma)

Eoentophysalis belcherensis Hofmann 1976. (McMenamin et al. 1983)
  Colonial coccoid (entophysalidacean?) cyanobacterium ("S").
Eoentophysalis magna McMenamin, Kumar and Awramik 1983. (McMenamin et al. 1983)
  Colonial coccoid (entophysalidacean?) cyanobacterium ("S").
Eomycetopsis psilata Maithy and Shukla 1977. (Nautiyal 1983b)
  Tubular sheath of filamentous prokaryote (S).
Eomycetopsis septata Maithy 1975. (Nautiyal 1983b)
  Tubular sheath of filamentous prokaryote (S).
Eomycetopsis? siberiensis Lo 1980. (McMenamin et al. 1983)
  Tubular sheath of filamentous cyanobacterium (S).
Eosynechococcus isolatus McMenamin, Kumar and Awramik 1983. (McMenamin et al. 1983)
  Colonial ellipsoidal bacterium (S).
Glenobotrydion aenigmatis Schopf 1968. (McMenamin et al. 1983)
  Colonial (pseudofilamentous) coccoid chroococcacean cyanobacterium ("S").
Gunflintia cf. *minuta* Barghoorn 1965. (McMenamin et al. 1983)
  Tubular sheath of filamentous bacterium ("S").
Kheinjuasphaera vulgaris McMenamin, Kumar and Awramik 1983. (McMenamin et al. 1983)

Table 22.3 (Continued).

Solitary coccoid cyanobacterial cells (and/or possibly sheaths of spherical colonies) (S).
aff. *Kildinella* Timofeev 1963. (Nautiyal 1983b)
Sphaeromorph acritarch ("S").
Myxococcoides minor Schopf 1968. (McMenamin et al. 1983)
Colonial coccoid chroococcacean cyanobacterium ("S").
Palaeoanacystis vulgaris Schopf 1968. (Nautiyal 1983b)
Colonial coccoid cyanobacterium ("S").
Tetraphycus congregatus McMenamin, Kumar and Awramik 1983. (McMenamin et al. 1983)
Colonial ellipsoidal (pleurocapsalean?) cyanobacterium ("S").

Vindhyan Sgrp., Semri Grp., Patherwa Fm.
AS IND UP
(ca. 1050 Ma)

(Fungal spore Type A). (Nautiyal 1983b)
Colonial coccoid cyanobacterium (S).
Eomycetopsis filiformis Schopf 1968. (Nautiyal 1983b)
Tubular sheath of filamentous bacterium (S).
Globophycus sp. A of Nautiyal 1983. (Nautiyal 1983b)
Solitary coccoid cyanobacterium (S).
Granomarginata regia Salujha, Rehman and Arora 1972. (Nautiyal 1983b)
Sphaeromorph acritarch ("S").
Lophosphaeridium vetulum Salujha, Rehman and Arora 1971. (Nautiyal 1983b)
Sphaeromorph acritarch ("S").
Palaeoanacystis vulgaris Schopf 1968. (Nautiyal 1983b)
Colonial coccoid cyanobacterium ("S").
Palaeopleurocapsa wopfneri Knoll, Barghoorn and Golubić 1975. (Nautiyal 1983b)
Colonial coccoid cyanobacterium ("S").

Vindhyan Sgrp., Semri Grp., Rohtas Fm.
AS IND UP
(ca. 1050 Ma)

(Fungal spore Type A). (Nautiyal 1983b)
Colonial coccoid cyanobacterium (S).
Eozygion grande Schopf and Blacic 1971. (Nautiyal 1983b)
Colonial coccoid chroococcacean cyanobacterium ("S").
Granomarginata regia Salujha, Rehman and Arora 1972. (Nautiyal 1983b)
Sphaeromorph acritarch ("S").
Palaeopleurocapsa wopfneri Knoll, Barghoorn and Golubić 1975. (Nautiyal 1983b)
Colonial coccoid cyanobacterium ("S").

Vindhyan Sgrp., Semri Grp., Suket Shale
AS IND MP
(ca. 1050 Ma)

(disc-like remains Type-1). (Maithy and Shukla 1977)
Sphaeromorph acritarch, cf. *Chuaria* sp. (S).
(disc-like remains Type-2). (Maithy and Shukla 1977)
Sphaeromorph acritarch, cf. *Chuaria* sp. (S).
(disc-like remains Type-3). (Maithy and Shukla 1977)
Sphaeromorph acritarch, cf. *Chuaria* sp. (S).
(disc-like remains Type-4). (Maithy and Shukla 1977)
Sphaeromorph acritarch, cf. *Chuaria* sp. (S).
(disc-like remains Type-6). (Maithy and Shukla 1977)
Sphaeromorph acritarch, cf. *Chuaria* sp. (S).
Aphanocapsaopsis ramapuraensis Maithy and Shukla 1977. (Maithy and Shukla 1977)
Colonial coccoid (chroococcacean?) cyanobacterium ("S").
Aphanocapsaopsis sitholeyi Maithy and Shukla 1977. (Maithy and Shukla 1977)
Colonial coccoid (chroococcacean) cyanobacterium ("S").
Archaeofavosina reticulata Maithy and Shukla 1977. (Maithy and Shukla 1977)
Sphaeromorph acritarch ("S").
Eomycetopsis pflugii Maithy and Shukla 1977. (Maithy and Shukla 1977)
Tubular sheath of filamentous cyanobacterium (S).
Eomycetopsis psilata Maithy and Shukla 1977. (Maithy and Shukla 1977)
Tubular sheath of filamentous prokaryote (S).
Eomycetopsis reticulata Maithy and Shukla 1977. (Maithy and Shukla 1977)
Tubular sheath of filamentous cyanobacterium (S).
Granomarginata rotata Maithy and Shukla 1977. (Maithy and Shukla 1977)
Sphaeromorph acritarch ("S").
Kildinella minuta Maithy and Shukla 1977. (Maithy and Shukla 1977)
Sphaeromorph acritarch ("S").
Kildinella suketensis Maithy and Shukla 1977. (Maithy and Shukla 1977)
Sphaeromorph acritarch ("S").
Leiosphaeridia sp. of Maithy 1969. (Maithy 1969)
Sphaeromorph acritarch ("S").
Myxococcoides globosa Maithy and Shukla 1977. (Maithy and Shukla 1977)
Colonial coccoid (chroococcacean?) cyanobacterium ("S").
Myxococcoides magnus Maithy and Shukla 1977. (Maithy and Shukla 1977)
Colonial coccoid chroococcacean cyanobacterium ("S")
Myxococcoides ramapuraensis Maithy and Shukla 1977. (Maithy and Shukla 1977)
Colonial coccoid (chroococcacean?) cyanobacterium ("S").
Nucellosphaeridium minimum Maithy and Shukla 1977. (Maithy and Shukla 1977)
Sphaeromorph acritarch containing degraded cellular remnants (S M).
Nucellosphaeridium zonatum Maithy and Shukla 1977. (Maithy and Shukla 1977)
Sphaeromorph acritarch containing degraded cellular remnants (S M).
Orygmatosphaeridium plicatum Maithy and Shukla 1977. (Maithy and Shukla 1977)
Sphaeromorph acritarch ("S").
Oscillatoriopsis psilata Maithy and Shukla 1977. (Maithy and Shukla 1977)
Septate filamentous (oscillatoriacean) cyanobacterium ("S").
Palaeoanacystis punctatus Maithy and Shukla 1977. (Maithy and Shukla 1977)
Colonial coccoid (chroococcacean?) cyanobacterium ("S").
Palaeoanacystis reticulatus Maithy and Shukla 1977. (Maithy and Shukla 1977)
Colonial coccoid (chroococcacean) cyanobacterium ("S").
Palaeoanacystis suketensis Maithy and Shukla 1977. (Maithy and Shukla 1977)
Colonial coccoid (chroococcacean?) cyanobacterium ("S").
Palaeoanacystis verrucosus Maithy and Shukla 1977. (Maithy and Shukla 1977)

Table 22.3 (Continued).

Colonial coccoid (chroococcacean) cyanobacterium ("S").
Palaeoscytonema srivastavae Maithy and Shukla 1977. (Maithy and Shukla 1977)
    Septate filamentous (oscillatoriacean) cyanobacterium ("S").
Protoleiosphaeridium densum Maithy 1975. (Maithy and Shukla 1977)
    Sphaeromorph acritarch ("S").
Protoleiosphaeridium sp. of Maithy 1969. (Maithy 1969)
    Sphaeromorph acritarch ("S").
Protosphaeridium volkovae Maithy and Shukla 1977. (Maithy and Shukla 1977)
    Sphaeromorph acritarch ("S").
Retisphaeridium vindhyanense Maithy 1969. (Maithy 1969)
    Degraded sphaeromorph acritarch (M S).
Symplassosphaeridium bulbosum Maithy and Shukla 1977. (Maithy and Shukla 1977)
    Colonial coccoid (chroococcacean) cyanobacterium (S).
Tasmanites sp. of Maithy 1969. (Maithy 1969).
    Sphaeromorph acritarch ("S").
Tasmanites vindhyanensis Maithy and Shukla 1977. (Maithy and Shukla 1977)
    Sphaeromorph acritarch ("S").
Vavososphaeridium vindhyanense (Maithy 1969) Maithy and Shukla 1977. (Maithy and Shukla 1977)
    Sphaeromorph acritarch ("S").
Zonosphaeridium punctatum Maithy and Shukla 1977. (Maithy and Shukla 1977)
    Sphaeromorph acritarch ("S").

Kadali Grp., Valyukhta Fm.
AS SIB
(ca. 1025 Ma)

(broad, unbranched trichomes). (Schopf et al. 1977)
    Septate, sheath-enclosed oscillatoriacean cyanobacterium, cf. *Symploca* or *Phormidium* ("S").
(tubular, unbranched filaments). (Schopf et al. 1977)
    Tubular sheaths of filamentous (oscillatoriacean) cyanobacteria ("S").
(unbranched narrow filaments). (Schopf et al. 1977)
    Septate filamentous prokaryote, cf. *Gunflintia* ("S").

Mbuji Mayi Sgrp. (=Bushimay Sgrp.)
AF ZAR
(ca. 1020 Ma)

Archaeofavosina naumovae Maithy 1975. (Maithy 1975)
    Sphaeromorph acritarch ("S").
Archaeofavosina sinuta Maithy 1975. (Maithy 1975)
    Sphaeromorph acritarch ("S").
Baltisphaeridium sp. of Maithy 1975. (Maithy 1975)
    Possible acanthomorph acritarch (S M).
Chlorogloeaopsis zairensis Maithy 1975. (Maithy 1975)
    Colonial coccoid (entophysalidacean?) cyanobacterium ("S").
Entosphaeroides bilinearis Maithy 1975. (Maithy 1975)
    Colonial coccoid (chroococcacean?) cyanobacterium ("S").
Entosphaeroides irregularis Maithy 1975. (Maithy 1975)
    Colonial coccoid mucilage-embedded (chroococcacean?) cyanobacterium ("S").
Eomycetopsis cylindrica Maithy 1975. (Maithy 1975)
    Tubular sheath of filamentous cyanobacterium ("S").
Eomycetopsis rugosa Maithy 1975. (Maithy 1975)
    Tubular sheath of filamentous cyanobacterium (S).
Eomycetopsis septata Maithy 1975. (Maithy 1975)
    Tubular sheath of filamentous (oscillatoriacean?) cyanobacterium (S).
Glenobotrydion kanshiensis Maithy 1975. (Maithy 1975)
    Colonial (pseudofilamentous) coccoid (chroococcacean?) cyanobacterium (S).
Glenobotrydion tetragonalum Maithy 1975. (Maithy 1975)
    Colonial (pseudofilamentous) coccoid (chroococcacean?) cyanobacterium (S).
Granomarginata minuta Maithy 1975. (Maithy 1975)
    Sphaeromorph acritarch ("S").
Gunflintia barghoornii Maithy 1975. (Maithy 1975)
    Septate filamentous eukaryotic alga (S).
Gunflintia magna Maithy 1975. (Maithy 1975)
    Septate filamentous (oscillatoriacean?) cyanobacterium ("S").
Kildinella timofeevii Maithy 1975. (Maithy 1975)
    Sphaeromorph acritarch ("S").
Leiosphaeridia kanshiensis Maithy 1975. (Maithy 1975)
    Sphaeromorph acritarch ("S").
Lophosphaeridium granulatum Maithy 1975. (Maithy 1975)
    Sphaeromorph acritarch ("S").
Myxococcoides congoensis Maithy 1975. (Maithy 1975)
    Colonial coccoid (chroococcacean?) cyanobacterium ("S").
Myxococcoides verrucosa Maithy 1975. (Maithy 1975)
    Colonial coccoid chroococcacean cyanobacterium ("S").
Nucellosphaeridium magnum Maithy 1975. (Maithy 1975)
    Sphaeromorph acritarch containing degraded cellular remnants (S M).
Nucellosphaeridium triangulatum Maithy 1975. (Maithy 1975)
    Sphaeromorph acritarch containing degraded cellular remnants (S M).
Nucellosphaeridium zonatum Maithy 1975. (Maithy 1975)
    Sphaeromorph acritarch containing degraded cellular remnants (S M).
Orygmatosphaeridium trizonatum Maithy 1975. (Maithy 1975)
    Sphaeromorph acritarch ("S").
Orygmatosphaeridium vulgarum Maithy 1975. (Maithy 1975)
    Sphaeromorph acritarch ("S").
Palaeoanacystis psilata Maithy 1975. (Maithy 1975)
    Colonial coccoid chroococcacean cyanobacterium ("S").
Palaeomicrocystis schopfii Maithy 1975. (Maithy 1975)
    Colonial coccoid (chroococcacean?) cyanobacterium ("S").
Protoleiosphaeridium densum Maithy 1975. (Maithy 1975)
    Sphaeromorph acritarch ("S").
Protoleiosphaeridium laevigatum Maithy 1975. (Maithy 1975)
    Sphaeromorph acritarch ("S").
Siphonophycus punctatus Maithy 1975. (Maithy 1975)
    Tubular sheath of filamentous (oscillatoriacean?) cyanobacterium ("S").
Symplassosphaeridium bushimayense Maithy 1975. (Maithy 1975)
    Sphaeromorph acritarch ("S").
Tasmanites sp. of Maithy 1975. (Maithy 1975)
    Sphaeromorph acritarch ("S").
Trematosphaeridium zairense Maithy 1975. (Maithy 1975)
    Sphaeromorph acritarch ("S").
Vavososphaeridium bharadwajii Salujha, Rehman and Rawat 1971. (Maithy 1975)
    Sphaeromorph acritarch ("S").
Vavososphaeridium densum Maithy 1975. (Maithy 1975)
    Sphaeromorph acritarch ("S").
Zonosphaeridium densum Maithy 1975. (Maithy 1975)
    Sphaeromorph acritarch ("S").

Table 22.3 (*Continued*).

Zonosphaeridium foveolatum Maithy 1975. (Maithy 1975)
  Sphaeromorph acritarch ("S").

Bobrovka Fm.
EU SUN
(ca. 1000 Ma)

Protosphaeridium torulosum Timofeev 1966. (Timofeev 1966)
  Sphaeromorph acritarch ("S").

Sukhaya Tunguska Fm.
AS SIB
(ca. 1000 Ma)

("larger chroococcacean cyanobacteria"). (Mendelson and Schopf 1982)
  Solitary and colonial (chroococcacean) cyanobacteria ("M S"). [See Chapter 24, Pl. 10, J.]
("smaller chroococcacean cyanobacteria"). (Mendelson and Schopf 1982)
  Solitary and ?colonial (chroococcacean) cyanobacteria ("M S"). [See Chapter 24, Pl. 10, N.]
(colonial unicells). (Schopf et al. 1977)
  Colonial, coccoid and ellipsoid (chroococcacean or entophysalidacean) cyanobacteria ("S").
(degraded cellular filaments). (Mendelson and Schopf 1982)
  Degraded trichome of filamentous prokaryote ("M S"). [See Chapter 24, Pl. 10, G.]
(filamentous microfossils). (Schopf et al. 1977)
  Tubular sheaths of filamentous cyanobacterium ("S").
(large-diameter "oscillatoriacean" sheaths). (Mendelson and Schopf 1982)
  Sheaths of filamentous cyanobacteria ("M S").
(long, thin filament). (Mendelson and Schopf 1982)
  Sheath or trichome of filamentous bacterium ("M S").
(small rod-shaped cells). (Schopf et al. 1977)
  Solitary and chainlike colonies of rod-shaped bacteria ("S").
(solitary algal unicells). (Schopf et al. 1977)
  Solitary coccoid (chroococcacean) cyanobacterium, cf. *Chroococcus* or *Gloeocapsa* ("S").
(tubular sheath with internal "trichome"). (Mendelson and Schopf 1982)
  Ensheathed filamentous prokaryote (M S). [See Chapter 24, Pl. 10, D.]
(*Chroococcus*-like morphotype). (Mendelson and Schopf 1982)
  Colonial, coccoid (chroococcacean) cyanobacterium, aff. *Gloeodiniopsis lamellosa* Schopf ("M S"). [See Chapter 24, Pl. 10, L.]
(*Globophycus*-like morphotype). (Mendelson and Schopf 1982)
  Solitary or ?colonial coccoid chroococcacean cyanobacterium ("M S")
Eoentophysalis arcata Mendelson and Schopf 1982. (Mendelson and Schopf 1982)
  Colonial (entophysalidacean?) cyanobacterium ("M S"). [See Frontispiece and Chapter 24, Pl. 10, E.]
Eomycetopsis spp. of Mendelson and Schopf 1982. (Mendelson and Schopf 1982)
  Sheaths of filamentous cyanobacteria ("M S"). [See Chapter 24, Pl. 10, I, M.]
Eosynechococcus medius Hofmann 1976. (Mendelson and Schopf 1982)
  Ellipsoidal, possibly ensheathed, solitary and ?colonial cyanobacteria ("M S"). [See Chapter 24, Pl. 10, O.]
Gloeodiniopsis lamellosa Schopf 1968. (Mendelson and Schopf 1982)
  Solitary, ensheathed coccoid (chroococcacean) cyanobacterium ("M S"). [See Chapter 24, Pl. 10, K, P.]
Kildinella hyperboreica Timofeev 1966. (Timofeev 1966)
  Sphaeromorph acritarch ("S").
Kildinella sinica Timofeev 1966. (Timofeev 1969)
  Sphaeromorph acritarch ("S").
Kildinella tschapomica Timofeev 1966. (Timofeev 1969)
  Sphaeromorph acritarch ("S").
Leiosphaeridia bituminosa Timofeev 1966. (Timofeev 1966)
  Sphaeromorph acritarch ("S").
Leiosphaeridia sp. of Timofeev 1969. (Timofeev 1969)
  Sphaeromorph acritarch ("S").
Oscillatoriopsis media Mendelson and Schopf 1982. (Mendelson and Schopf 1982)
  Septate filamentous oscillatoriacean cyanobacterium ("M S"). [See Chapter 24, Pl. 10, F, H.]
Protosphaeridium densum Timofeev 1966. (Timofeev 1969)
  Sphaeromorph acritarch ("S").
Protosphaeridium pusillum Timofeev 1966. (Timofeev 1969)
  Sphaeromorph acritarch ("S").
Protosphaeridium scabridum Timofeev 1966. (Timofeev 1966)
  Sphaeromorph acritarch ("S").
Protosphaeridium torulosum Timofeev 1966. (Timofeev 1969)
  Sphaeromorph acritarch ("S").
Protosphaeridium tuberculiferum Timofeev 1966. (Timofeev 1966)
  Sphaeromorph acritarch ("S").
Stictosphaeridium implexum Timofeev 1966. (Timofeev 1969)
  Sphaeromorph acritarch ("S").
Synsphaeridium sorediforme Timofeev 1969?. (Timofeev 1969)
  Colonial coccoid cyanobacterium (S).
Trematosphaeridium holtedahlii Timofeev 1966. (Timofeev 1966)
  Sphaeromorph acritarch ("S").
Turuchanica alara Rudavskaja [Rudavskaya] 1964?. (Timofeev 1966)
  Compressed sphaeromorph? acritarch (S).
Turuchanica ternata Timofeev 1966. (Timofeev 1966)
  Flattened, compressed, (sphaeromorph?) acritarch (S).

Zil'merdak Fm.
EA SUN
(ca. 1000 Ma)

Kildinella hyperboreica Timofeev 1963. (Yandauskas 1982)
  Folded, originally spheroidal, sphaeromorph acritarch ("S").
Kildinella ripheica Timofeev 1969. (Yankauskas 1982)
  Wrinkled, originally smooth?, sphaeromorph acritarch (?cf. *Kildinella vesljanica*) (S).
Kildinella vesljanica Timofeev 1963. (Yankauskas 1982)
  Wrinkled to coarsely rugulate, originally smooth?, sphaeromorph acritarch ("S").

Zil'merdak Fm., Bederysh Mbr.
EA SUN
(ca. 1000 Ma)

Arctacellularia ellipsoidea Hermann [German] 1976. (Yankauskas 1982)
  Pseudofilamentous (uniseriate) colonial coccoid cyanobacterium (?cf. *Glenobotrydion aenigmatis*) (S).
Arctacellularia sp. 1 of Jankauskas [Yankauskas] 1982. (Yankauskas 1982)
  Solitary (disarticulated) trichomic cell of filamentous or pseudofilamentous (colonial) cyanobacterium, or solitary cylindrical unicell incertae sedis (S).
Arctacellularia sp. 2 of Jankauskas [Yankauskas] 1982. (Yankauskas 1982)

Table 22.3 (Continued).

Solitary (disarticulated) trichomic cell of filamentous or pseudofilamentous (colonial) cyanobacterium, or solitary cylindrical unicell incertae sedis (viz., pl. 37, fig. 1, 3, 8; pl. 38, fig. 2). Pl. 37, fig. 14 = tubular sheath of filamentous cyanobacterium; pl. 37, fig. 7 = *Brevitrichoides bashkiricus* Jankauskas 1980 (S).

Arctacellularia sp. 3 of Jankauskas [Yankauskas] 1982. (Yankauskas 1982)
  Solitary or paired (disarticulated) trichomic cells of filamentous or pseudofilamentous (colonial) cyanobacterium, or solitary or paired cylindrical cells incertae sedis (S).

Brevitrichoides bashkiricus Jankauskas [Yankauskas] 1980. (Yankauskas 1982)
  Tubular sheath (or trichome without preserved lateral cell walls) of filamentous (oscillatoriacean) cyanobacterium (S).

Brevitrichoides bashkiricus Jankauskas [Yankauskas] 1980. (Yankauskas, 1980a)
  Tubular sheath (or trichome without preserved lateral cell walls) of filamentous (oscillatoriacean) cyanobacterium (S). [See Chapter 24, Pl. 10, B.]

Brevitrichoides karatavicus Jankauskas [Yankauskas] 1980. (Yankauskas 1982)
  Solitary, ellipsoidal cyanobacterial cell or sheath; preservational variant of *Glenobotrydion solutum* Jankauskas 1980 (S).

Brevitrichoides karatavicus Jankauskas [Yankauskas] 1980. (Yankauskas, 1980a)
  Solitary, ellipsoidal cyanobacterial cell or sheath; preservational variant of *Glenobotrydion solutum* Jankauskas 1980 (S). [See Chapter 24, Pl. 10, A.]

Calyptothrix alternata Jankauskas [Yankauskas] 1980. (Yankauskas 1982)
  Septate filamentous (oscillatoriacean) cyanobacterium ("S").

Calyptothrix geminata Jankauskas [Yankauskas] 1980. (Yankauskas 1982)
  Septate filamentous (oscillatoriacean) cyanobacterium ("S").

Cephalophytarion sp. of Jankauskas [Yankauskas] 1982. (Yankauskas 1982)
  Septate filamentous cyanobacterium (cf. Dismal Lakes taxon) (S).

Chuaria aff. *circularis* Walcott 1899. (Yankauskas 1982)
  Large, thick-walled sphaeromorph acritarch ("S").

Entosphaeroides aff. *irregularis* Maithy 1975. (Yankauskas 1982)
  Pseudofilamentous (biseriate) colonial coccoid cyanobacterium (?cf. *Glenobotrydion aenigmatis* Schopf and ?cf. *Arctacellularia ellipsoidea* Hermann) (S).

Entosphaeroides sp. of Jankauskas [Yankauskas] 1982. (Yankauskas 1982)
  Pseudofilamentous (bi- or triseriate) colonial coccoid cyanobacterium (?cf. *Glenobotrydion aenigmatis* Schopf and ?cf. *Arctacellularia ellipsoidea* Hermann) (S).

Eomycetopsis psilata Maithy and Shukla 1977. (Yankauskas 1982)
  Tubular sheath of filamentous cyanobacterium ("S").

Eomycetopsis rimata Jankauskas [Yankauskas] 1980. (Yankauskas 1982)
  Flattened sheath of filamentous cyanobacterium ("S").

Eomycetopsis rimata Jankauskas [Yankauskas] 1980. (Yankauskas 1980a)
  Flattened tubular sheath of filamentous cyanobacterium ("S").

Glenobotrydion solutum Jankauskas [Yankauskas] 1980. (Yankauskas 1982)
  Solitary, ellipsoidal cyanobacterial cell or sheath; preservational variant of *Brevitrichoides karativicus* Jankauskas (S). [See Chapter 24, Pl. 10, C.]

Kildinella kulgunica (Jankauskas 1980) Jankauskas 1982 (Yankauskas 1982)
  Sphaeromorph acritarch with large circular pylome ("S").

Kildinella nordia Timofeev 1969. (Yankauskas 1982)
  Sphaeromorph acritarch containing globular, condensed cytoplasmic contents ("S").

Kildinella tschapomica Timofeev 1966. (Yankauskas 1982)
  Distorted, originally spheroidal, sphaeromorph acritarch ("S").

Leiominuscula minuta Naumova 1960. (Yankauskas 1982)
  Solitary coccoid cyanobacterium (S).

Leiothrichoides typicus Hermann [German] 1974. (Yankauskas 1982)
  Tubular sheath of filamentous cyanobacterium ("S").

Oscillatoriopsis sp. 3 of Jankauskas [Yankauskas] 1982. (Yankauskas 1982)
  Septate filamentous prokaryote (S).

Partitiofilum? sp. of Jankauskas [Yankauskas] 1982. (Yankauskas 1982)
  Septate filamentous (oscillatoriacean) cyanobacterium ("S").

Polythrichoides lineatus Hermann [German] 1974. (Yankauskas 1982)
  Ensheathed multitrichomic filamentous (aff. *Microcoleus*) prokaryote ("S").

Pterospermopsimorpha capsulata Jankauskas [Yankauskas] 1982. (Yankauskas 1982)
  Possible poorly preserved, diagenetically altered pteromorph acritarch (M S).

Siphonophycus costatus Jankauskas [Yankauskas] 1980. (Yankauskas 1980a)
  Tubular sheath of filamentous cyanobacterium ("S").

Siphonophycus costatus Jankauskas [Yankauskas] 1980. (Yankauskas 1982)
  Tubular sheath of filamentous cyanobacterium ("S").

Synsphaeridium spp. of Jankauskas [Yankauskas] 1982. (Yankauskas 1982)
  Colonial coccoid (?chroococcacean) cyanobacteria (S).

Tortunema eniseica Hermann [German] 1976. (Yankauskas, 1982)
  Flattened, intertwined, degraded, monotrichomic, ensheathed filamentous prokaryote (S).

Tortunema sibirica Hermann [German] 1976. (Yankauskas 1982)
  Tubular sheath of filamentous cyanobacterium ("S").

Trematosphaeridium holtedahlii Timofeev 1959. (Yankauskas 1982)
  Coarsely porate sphaeromorph acritarch ("S").

1000 to 900 Ma:

Mwashya Grp.
AF ZMB
(ca. 975 Ma)

(spherical organic microfossil). (Binda 1972)
  Solitary and colonial coccoid cyanobacteria (S).

Chuar Grp., Galeros Fm., Tanner Mbr.
NA USA AZ
(ca. 950 Ma)

Leiosphaeridia asperata (Naumova 1950) Lindgren 1982. (Vidal and Ford 1985)
  Sphaeromorph acritarch ("S").

Derevnya Fm.
AS SIB
(ca. 950 Ma)

Partitiofilum tungusum Mikhailova [Mikhajlova] 1989. (German et al. 1989)

Table 22.3 (Continued).

Septate filamentous oscillatoriacean cyanobacterium ("S"). [See Chapter 24, Pl. 5, I_1-I_2.]
Turuchanica alara Rudavskaja [Rudavskaya] 1964?. (Timofeev 1966)
Compressed sphaeromorph? acritarch (S).

Lakhanda Fm.
AS SIB
(ca. 950 Ma)

"*Phycomycetes*" Timofeev, German and Mikhajlova 1976. (Timofeev et al. 1976)
Sphaeromorph acritarch with associated "tail-like" sapropelic debris (S).
"*Phycomycetes*" Timofeev 1969. (Timofeev 1969)
Sphaeromorph acritarch with coalesced "tail-like" associated strand (S).
([narrow tubular sheath of] algal trichome). (Timofeev 1969)
Flattened tubular sheath of filamentous cyanobacterium (S).
(piece of trichome [narrow]). (Timofeev et al. 1976)
Flattened tubular sheath of filamentous cyanobacterium ("S").
Aimia delicata Hermann [German] 1979. (German 1979)
Colonial coccoid cyanobacterium with prominent colonial envelope (S). [See Chapter 24, Pl. 16, B_1-B_2.]
Eomycetopsis sp. of German in press. (German, in press)
Flattened tubular sheath of filamentous bacterium (S). [See Chapter 24, Pl. 13, A_1-A_2.]
Jacutianema solubila Timofeev and Hermann [German] 1979. (Timofeev and German 1979)
Flattened tubular sheath of filamentous cyanobacterium ("S"). [See Chapter 24, Pl. 13, B.]
Kildinella hyperboreica Timofeev 1966. (Timofeev 1966)
Sphaeromorph acritarch ("S").
Kildinells hyperboreica Timofeev 1966. (Timofeev 1966)
Sphaeromorph acritarch ("S").
Kildinella hyperboreica Timofeev 1966. (Timofeev 1969)
Sphaeromorph acritarch ("S").
Kildinella jacutica Timofeev 1966. (Timofeev et al. 1976)
Sphaeromorph acritarch ("S").
Kildinella jacutica Timofeev 1966. (Timofeev 1966) Sphaeromorph acritarch ("S").
Kildinella jacutica Timofeev 1966. (Timofeev 1969)
Sphaeromorph acritarch ("S").
Kildinella sinica Timofeev 1966. (Timofeev 1969)
Sphaeromorph acritarch ("S").
Kildinella sp. of German in press. (German, in press)
Sphaeromorph acritarch ("S"). [See Chapter 24, Pl. 14, B.]
Kildinella vesljanica Timofeev 1969. (Timofeev et al. 1976)
Sphaeromorph acritarch ("S").
Lakhandinia prolata Timofeev and Hermann [German] 1979. (Timofeev and German 1979)
Probably globular sapropel or sac-like sheath of colonial coccoid cyanobacterium (S). [See Chapter 24, Pl. 15, A_1-A_2.]
Leiothrichoides tenuitunicatus Hermann [German] 1981. (German 1981a)
Ensheathed septate filamentous prokaryote (S). [See Chapter 24, Pl. 11, B_1-B_3.]
Lomentunella vaginata Hermann [German] 1981. (German 1981a)
Ensheathed septate filamentous oscillatoriacean cyanobacterium (S). [See Chapter 24, Pl. 15, C.]
Nucellohystrichosphaera megalea Timofeev 1976. (Timofeev et al. 1976)
Acanthomorph acritarch ("S"). [See Chapter 24, Pl. 14, D_1-D_2.]
Nucellosphaeridium bellum Timofeev 1969. (Timofeev et al. 1976)
Sphaeromorph acritarch ("S").
Nucellosphaeridium bellum Timofeev 1969. (Timofeev 1969)
Sphaeromorph acritarch ("S").
Nucellosphaeridium deminatum Timofeev 1969. (Timofeev et al. 1976)
Sphaeromorph acritarch containing degraded cytoplasmic contents (S).
Nucellosphaeridium deminatum Timofeev 1969. (Timofeev 1969)
Sphaeromorph acritarch ("S").
Nucellosphaeridium sp. of Timofeev, German and Mikhajlova 1976. (Timofeev et al. 1976)
Sphaeromorph acritarch with degraded cytoplasmic contents (S).
Octoedryxium sp. of Timofeev, German and Mikhajlova 1976. (Timofeev et al. 1976)
Prismatomorph acritarch ("S M")
Palaeolyngbya helva Hermann [German] 1981. (German 1981a)
Ensheathed septate filamentous oscillatoriacean cyanobacterium ("S"). [See Chapter 24, Pl. 11, A_1-A_2.]
Palaeovaucheria clavata Hermann [German] 1981. (German 1981a)
Branched septate *Vaucheria*-like, possibly eukaryotic, filaments ("S"). [See Chapter 24, Pl. 11, C; Pl. 12, A_1-A_5; Pl. 13, D_1-D_2.]
Protosphaeridium densum Timofeev 1966. (Timofeev 1969)
Sphaeromorph acritarch ("S").
Protosphaeridium gibberosum Timofeev 1966. (Timofeev 1969)
Sphaeromorph acritarch ("S").
Protosphaeridium laccatum Timofeev 1966. (Timofeev 1966)
Sphaeromorph acritarch ("S").
Protosphaeridium patelliforme Timofeev 1966. (Timofeev 1966)
Sphaeromorph acritarch ("S").
Protosphaeridium rigidulum Timofeev 1966. (Timofeev 1969)
Sphaeromorph acritarch ("S").
Pterospermopsimorpha pileiformis Timofeev 1966. (Timofeev 1966)
Sphaeromorph acritarch containing degraded cytoplasmic remnants (M S).
Pterospermopsimorpha sp. of German in press. (German in press)
Flattened sphaeromorph acritarch (S). [See Chapter 24, Pl. 15, E_1-E_2.]
Rugosoopsis tenuis Timofeev and Hermann [German] 1979. (Timofeev and German 1979)
Flattened tubular sheath of filamentous cyanobacterium ("S"). [See Chapter 24, Pl. 13, C.]
Stictosphaeridium pectinale Timofeev 1966. (Timofeev 1969)
Sphaeromorph acritarch ("S").
Stictosphaeridium sinapticuliferum Timofeev 1966. (Timofeev 1966)
Sphaeromorph acritarch ("S").
Stictosphaeridium sinapticuliferum Timofeev 1966. (Timofeev 1969)
Sphaeromorph acritarch ("S").
Synsphaeridium conglutinatum Timofeev 1966. (Timofeev 1969)
Colonial coccoid cyanobacterium (S).
Synsphaeridium sorediforme Timofeev 1969?. (Timofeev 1966)
Colonial coccoid chroococcacean cyanobacterium (S).
Tetrasphaera antiqua Timofeev and Hermann [German] 1979. (Timofeev and German 1979)
Colonial ensheathed sphaeromorph acritarch ("S"). [See Chapter 24, Pl. 14, C, F_1-F_2.]
Trachyhystrichosphaera aimika Hermann [German] 1976. (Timofeev et al. 1976)
Acanthomorph acritarch ("S"). [See Chapter 24, Pl. 14, E_1-E_2.]
Trachysphaeridium lachandinum Timofeev 1969. (Timofeev 1969)
Sphaeromorph acritarch ("S").

Table 22.3 (*Continued*).

Trachysphaeridium lachandinum Timofeev 1969. (Timofeev et al. 1976)
  Sphaeromorph acritarch ("S").
Trachysphaeridium laminaritum Timofeev 1966. (German, in press)
  Sphaeromorph acritarch ("S"). [See Chapter 24, Pl. 14, A.]
Trachysphaeridium maicum Timofeev 1969. (Timofeev 1969)
  Sphaeromorph acritarch ("S").
Trachysphaeridium maicum Timofeev 1969. (Timofeev et al. 1976)
  Sphaeromorph acritarch ("S").
Trachysphaeridium sp. of Timofeev 1969. (Timofeev 1969)
  Sphaeromorph acritarch ("S").
Trachysphaeridium sp. of Timofeev, German and Mikhajlova 1976. (Timofeev et al. 1976)
  Sphaeromorph acritarch ("S").
Trematosphaeridium sp. of Timofeev 1966. (Timofeev 1966)
  Sphaeromorph acritarch ("S").
Turuchanica ternata Timofeev 1966. (Timofeev 1966)
  Flattened, compressed, (sphaeromorph?) acritarch (S).
Turuchanica ternata Timofeev 1966. (German, in press)
  Flattened sphaeromorph acritarch ("S"). [See Chapter 24, Pl. 15, B.]
Turuchanica ternata Timofeev 1966. (Timofeev 1969)
  Flattened sphaeromorph? acritarch (S).

Qingbaikou Grp., Xiamaling Fm.
AS CHN
(ca. 950 Ma)

Acantholigotriletum primigenum (Naumova 1949) Timofeev 1959. (Timofeev 1966)
  Acanthomorph acritarch ("S").
Asperatopsophosphaera bavlensis Schepeleva [Shepeleva] 1963. (Xing and Liu 1973)
  Sphaeromorph acritarch ("S").
Chuaria circularis Walcott 1899. (Du and Tian 1985)
  Sphaeromorph acritarch ("S").
Leiopsophosphaera apertus Schepeleva [Shepeleva] 1963. (Xing and Liu 1973)
  Sphaeromorph acritarch ("S").
Leiopsophosphaera minor Schepeleva [Shepeleva] 1963. (Xing and Liu 1973)
  Sphaeromorph acritarch ("S").
Microconcentrica induplicata Liu and Sin [Xing] 1973. (Xing and Liu 1973)
  Colonial coccoid (chroococcacean?) cyanobacterium [pl. 10, fig. 7 not interpretable] ("S").
Protoleiosphaeridium aff. *faveolatum* Timofeev 1959. (Xing and Liu 1973)
  Sphaeromorph acritarch ("S").
Protoleiosphaeridium infriatum Andreeva 1966. (Xing and Liu 1973)
  Sphaeromorph acritarch ("S").
Protoleiosphaeridium solidum Liu and Sin [Xing] 1973. (Xing and Liu 1973)
  Sphaeromorph acritarch ("S").
Protosphaeridium acis Timofeev 1966. (Timofeev 1966)
  Sphaeromorph acritarch ("S").
Protosphaeridium asaphum Timofeev 1966. (Timofeev 1966)
  Sphaeromorph acritarch ("S").
Protosphaeridium densum Timofeev 1966. (Xing and Liu 1973)
  Sphaeromorph acritarch ("S").
Pseudozonosphaera sinica Sin [Xing] and Liu 1973. (Xing and Liu 1973)
  Colonial, coccoid cyanobacterium (pl. 6, fig. 13 and pl. 12, fig. 10) (S).
Pterospermopsis oculatus Sin [Xing] and Liu 1973. (Xing and Liu 1973)
  Multilamellated sheath-enclosed solitary coccoid (chroococcacean?) cyanobacterium (S).
Shouhsienia longa Xing 1979. (Du and Tian 1985)
  Sphaeromorph acritarch ("S").
Souhsienia shouhsienensis Xing 1979. (Du and Tian 1985)
  Sphaeromorph acritarch ("S").
Stictosphaeridium implexum Timofeev 1966. (Timofeev 1966)
  Sphaeromorph acritarch ("S").
Stictosphaeridium sinapticuliferum Timofeev 1966. (Timofeev 1966)
  Sphaeromorph acritarch ("S").
Synsphaeridium conglutinatum Timofeev 1966. (Xing and Liu 1973)
  Colonial, coccoid to ellipsoidal (chroococcacean?) cyanobacterium (S).
Trachysphaeridium cultum (Andreeva 1966) Sin [Xing] 1973. (Xing and Liu 1973)
  Sphaeromorph acritarch ("S").
Trachysphaeridium hyalinum Sin [Xing] and Liu 1973. (Xing and Liu 1973)
  Sphaeromorph acritarch ("S").
Trachysphaeridium aff. *laminaritum* Timofeev 1966. (Xing and Liu 1973)
  Degraded sphaeromorph acritarch ("S").
Trachysphaeridium laminaritum Timofeev 1966. (Timofeev 1966)
  Sphaeromorph acritarch ("S").
Trachysphaeridium minor Liu and Sin [Xing] 1973. (Xing and Liu 1973)
  Colonial coccoid cyanobacterium (S).
Trachysphaeridium stipticum Sin [Xing] 1973. (Xing and Liu 1973)
  Degraded sphaeromorph acritarch ("S").
Turuchanica alara Rudavskaja [Rudavskaya] 1964? (Timofeev 1966)
  Compressed sphaeromorph? acritarch (S).

Uinta Mountain Grp., Mount Watson Fm.
NA USA UT
(ca. 950 Ma)

Leiosphaeridia asperata (Naumova 1950) Lindgren 1982. (Vidal and Ford 1985)
  Sphaeromorph acritarch ("S").
Trachysphaeridium laufeldii Vidal 1976. (Vidal and Ford 1985)
  Short-spined acanthomorph acritarch ("S M").

Uinta Mountain Grp., Red Pine Shale
NA USA UT
(ca. 950 Ma)

(cluster of spherical cells). (Vidal and Ford 1985)
  Colonial coccoid (chroococcacean?) cyanobacterium (S).
Chuaria circularis Walcott 1899. (Hofmann 1977)
  Sphaeromorph acritarch ("S").
Kildinosphaera chagrinata Vidal 1983. (Vidal and Ford 1985)
  Sphaeromorph acritarch ("S").
Leiosphaeridia asperata (Naumova 1950) Lindgren 1982. (Vidal and Ford 1985)
  Sphaeromorph acritarch ("S").
cf. *Stictosphaeridium* sp. of Vidal and Ford 1985. (Vidal and Ford 1985)
  Colonial coccoid (chroococcacean?) cyanobacterium (S).
Tasmanites rifejicus Jankauskas [Yankauskas] 1978. (Vidal and Ford 1985)
  Sphaeromorph acritarch ("S").
Trachysphaeridium laminaritum Timofeev 1966. (Vidal and Ford 1985)

Table 22.3 (Continued).

Sphaeromorph acritarch ("S").
Trachysphaeridium laufeldii Vidal 1976. (Vidal and Ford 1985)
 Short-spined acanthomorph acritarch ("S M").
Trachysphaeridium sp. A of Vidal and Ford 1985. (Vidal and Ford 1985)
 Sphaeromorph acritarch ("S").

Katav Fm.
EA SUN
(ca. 940 Ma)

Kildinella hyperboreica Timofeev 1966. (Timofeev 1969)
 Sphaeromorph acritarch ("S").
Kildinella sinica Timofeev 1966. (Timofeev 1966)
 Sphaeromorph acritarch ("S").
Kildinella sp. of Timofeev 1969. (Timofeev 1969)
 Colonial coccoid cyanobacterium (S).
Protosphaeridium flexuosum Timofeev 1966. (Timofeev 1969)
 Sphaeromorph acritarch ("S").
Protosphaeridium parvulum Timofeev 1966. (Timofeev 1969)
 Sphaeromorph acritarch ("S").
Protosphaeridium scabridum Timofeev 1966. (Timofeev 1969)
 Sphaeromorph acritarch ("S").
Protosphaeridium tuberculiferum Timofeev 1966. (Timofeev 1969)
 Sphaeromorph acritarch ("S").
Protosphaeridium tuberculiferum Timofeev 1966. (Timofeev 1966)
 Sphaeromorph acritarch ("S").
Stictosphaeridium pectinale Timofeev 1966. (Timofeev 1969)
 Sphaeromorph acritarch ("S").
Turuchanica ternata Timofeev 1966. (Timofeev 1969)
 Flattened sphaeromorph? acritarch (S).

Chuar Grp., Galeros Fm.
NA USA AZ
(ca. 930 Ma)

Kildinosphaera chagrinata Vidal 1983. (Vidal and Ford 1985)
 Sphaeromorph acritarch ("S").
Kildinosphaera lophostriata (Jankauskas 1979) Vidal 1983. (Vidal and Ford 1985)
 Sphaeromorph acritarch ("S").

Chuar Grp., Galeros Fm., Jupiter Mbr.
NA USA AZ
(ca. 930 Ma)

Leiosphaeridia asperata (Naumova 1950) Lindgren 1982. (Vidal and Ford 1985)
 Sphaeromorph acritarch ("S").
Leiosphaeridia sp. A of Vidal and Ford 1985. (Vidal and Ford 1985)
 Sphaeromorph acritarch ("S").
Tasmanites rifejicus Jankauskas [Yankauskas] 1978. (Vidal and Ford 1985)
 Sphaeromorph acritarch ("S").

Burovaya Fm.
AS SIB
(ca. 925 Ma)

Kildinella hyperboreica Timofeev 1966. (Timofeev 1966)
 Sphaeromorph acritarch ("S").
Kildinella hyperboreica Timofeev 1966. (Timofeev 1969)
 Sphaeromorph acritarch ("S").
Kildinella jacutica Timofeev 1966. (Timofeev 1969)
 Sphaeromorph acritarch ("S").
Kildinella sinica Timofeev 1966. (Timofeev 1966)
 Sphaeromorph acritarch ("S").
Kildinella sinica Timofeev 1966. (Timofeev 1969)
 Sphaeromorph acritarch ("S").
Protosphaeridium laccatum Timofeev 1966. (Timofeev 1969)
 Sphaeromorph acritarch ("S").
Protosphaeridium parvulum Timofeev 1966. (Timofeev 1966)
 Sphaeromorph acritarch ("S").
Protosphaeridium tuberculiferum Timofeev 1966. (Timofeev 1966)
 Sphaeromorph acritarch ("S").
Protosphaeridium vermium Timofeev 1969. (Timofeev 1969)
 Sphaeromorph acritarch ("S").
Stictosphaeridium pectinale Timofeev 1966. (Timofeev 1969)
 Sphaeromorph acritarch ("S").
Trematosphaeridium holtedahlii Timofeev 1966. (Timofeev 1966)
 Sphaeromorph acritarch ("S").

Podinzer Fm. (= Sim Fm.)
EU SUN
(ca. 925 Ma)

Calyptothrix alternata Jankauskas [Yankauskas] 1980. (Yankauskas 1982)
 Septate filamentous (oscillatoriacean) cyanobacterium ("S").
Calyptothrix geminata Jankauskas [Yankauskas] 1980. (Yankauskas 1980a)
 Septate filamentous (oscillatoriacean) cyanobacterium ("S").
Caudiculophycus acuminatus Schopf and Blacic 1971. (Yankauskas 1982)
 Septate filamentous (rivulariacean?) cyanobacterium (S).
Chuaria aff. *circularis* Walcott 1899. (Yankauskas 1982)
 Large, thick-walled sphaeromorph acritarch ("S").
Eomycetopsis psilata Maithy and Shukla 1977. (Yankauskas 1982)
 Tubular sheath of filamentous cyanobacterium ("S").
Eomycetopsis rimata Jankauskas [Yankauskas] 1980. (Yankauskas 1982)
 Flattened sheath of filamentous cyanobacterium ("S").
Eomycetopsis rimata Jankauskas [Yankauskas] 1980. (Yankauskas 1980a)
 Flattened tubular sheath of filamentous cyanobacterium ("S").
Eomycetopsis aff. *rugosa* Maithy 1975. (Yankauskas 1982)
 Tubular sheath of filamentous cyanobacterium ("S").
Heliconema uralense Jankauskas [Yankauskas] 1980. (Yankauskas 1982)
 Tubular sheath of filamentous bacterium (S).
Heliconema uralense Jankauskas [Yankauskas] 1980. (Yankauskas 1980a)
 Coiled filamentous bacterium (S). [See Chapter 24, Pl. 19, C.]
Kildinella hyperboreica Timofeev 1963. (Yankauskas 1982)
 Folded, originally spheroidal, sphaeromorph acritarch ("S").
Kildinella kulgunica (Jankauskas 1980) Jankauskas 1982. (Yankauskas 1982)
 Sphaeromorph acritarch with large circular pylome ("S").
Kildinella nordia Timofeev 1969. (Yankauskas 1982)
 Sphaeromorph acritarch containing globular, condensed cytoplasmic contents ("S").
Kildinella ripheica Timofeev 1969. (Yankauskas 1982)
 Wrinkled, originally smooth?, sphaeromorph acritarch (?cf. *Kildinella vesljanica*) (S).
Kildinella tschapomica Timofeev 1966. (Yankauskas 1982)
 Distorted, originally spheroidal, sphaeromorph acritarch ("S").
Kildinella vesljanica Timofeev 1963. (Yankauskas 1982)
 Wrinkled to coarsely rugulate, originally smooth?, sphaeromorph

Table 22.3 (Continued).

acritarch ("S").
Leiofusidium dubium (Jankauskas 1980) Jankauskas [Yankauskas] 1982. (Yankauskas 1982)
  Unusual sphaeromorph acritarch having horns (S M).
Leiominuscula minuta Naumova 1960. (Yankauskas 1982)
  Solitary coccoid cyanobacterium (S).
Leiosphaeridia kulgunica Jankauskas [Yankauskas] 1980. (Yankauskas 1980y)
  Compressed sphaeromorph acritarch ("S"). [See Chapter 24, Pl. 19, G.]
Oscillatoriopsis sp. 1 of Jankauskas [Yankauskas] 1982. (Yankauskas 1982)
  Septate filament (or crenulated sheath?) of filamentous (oscillatoriacean?) cyanobacterium (S).
Oscillatoriopsis sp. 2 of Jankauskas [Yankauskas] 1982. (Yankauskas 1982)
  Septate filamentous (oscillatoriacean) cyanobacterium (S).
Oscillatoriopsis sp. 3 of Jankauskas [Yankauskas] 1982. (Yankauskas 1982)
  Septate filamentous prokaryote (S).
Palaeolyngbya minor Schopf and Blacic 1971. (Yankauskas 1982)
  Septate filamentous (oscillatoriacean) cyanobacterium ("S").
Polythrichoides lineatus Hermann [German] 1974. (Yankauskas 1982)
  Ensheathed multitrichomic filamentous (aff. *Microcoleus*) prokaryote ("S").
Protosphaeridium flexuosum Timofeev 1966. (Yankauskas 1982)
  Thick-walled sphaeromorph acritarch ("S").
Pterospermella (?) *simica* (Jankauskas 1980) Jankauskas [Yankauskas] 1982. (Yankauskas 1982)
  Possible degraded pteromorph acritarch (S M).
Pterospermella simica Jankauskas [Yankauskas] in press (Yankauskas, in press)
  Possibly poorly preserved pteromorph acritarch (S M).
Pterospermopsimorpha capsulata Jankauskas [Yankauskas] 1982. (Yankauskas 1982)
  Possible poorly preserved, diagenetically altered pteromorph acritarch (M S). [See Chapter 24, Pl. 19, B.]
Pterospermopsis dubius Jankauskas [Yankauskas] 1980. (Yankauskas 1980y)
  Possibly degraded sphaeromorph acritarch (S M). [See Chapter 24, Pl. 19, D.]
Pterospermopsis simicus Jankauskas [Yankauskas] 1980. (Yankauskas 1980y)
  Possibly poorly preserved pteromorph acritarch (S M). [See Chapter 24, Pl. 19, E, F.]
Stictosphaeridium implexum Timofeev 1966. (Yankauskas 1982)
  Wrinkled to rugulate, originally smooth?, sphaeromorph acritarch (?cf. *Kildinella vesljanica*) ("S").
Synsphaeridium spp. of Jankauskas [Yankauskas] 1982. (Yankauskas 1982)
  Colonial coccoid (?chroococcacean) cyanobacteria (S).
Tasmanites refejicus Jankauskas [Yankauskas] 1978. (Yankauskas 1978)
  Sphaeromorph acritarch ("S"). [See Chapter 24, Pl. 19, A_1–A_2.]
Tasmanites ripheicus Jankauskas [Yankauskas] 1978. (Yankauskas 1982)
  Finely porate sphaeromorph acritarch (S).
Tortunema sibirica Hermann [German] 1976. (Yankauskas 1982)
  Tubular sheath of filamentous cyanobacterium ("S").
Trematosphaeridium holtedahlii Timofeev 1959. (Yankauskas 1982)
  Coarsely porate sphaeromorph acritarch ("S").
Turuchanica ternata Timofeev 1966. (Yankauskas 1982)
  Thick-walled, originally spheroidal, sphaeromorph acritarch; lobate due to compression ("S").

Chuar Grp.
NA USA AZ
(ca. 900 Ma)

(small filament). (Pierce and Cloud 1979)
  Sheath-enclosed filamentous bacterium (S).

Deoban Limestone
AS IND
(ca. 900 Ma)

Animikiea septata Barghoorn 1965. (Shukla et al. 1987)
  Tubular sheath of filamentous cyanobacterium ("S")
Archaeotrichion sp. of Shukla, Tewari and Yadav 1987. (Shukla et al. 1987)
  Flattened tubular sheath of filamentous cyanobacterium (S).
Biocatenoides sp. of Shukla, Tewari and Yadav 1987. (Shukla et al. 1987)
  Septate filamentous prokaryote ("S").
Cyanonema sp. of Shukla, Tewari and Yadav 1987. (Shukla et al. 1987)
  Septate filamentous prokaryote (S).
Eomycetopsis robusta Schopf 1968. (Shukla et al. 1987)
  Flattened tubular sheath of filamentous cyanobacterium ("S").
Glenobotrydion aenigmatis Schopf 1968. (Shukla et al. 1987)
  Colonial coccoid ?chroococcacean cyanobacterium (S).
Glenobotrydion majorinum Schopf and Blacic 1971. (Shukla et al. 1987)
  Sphaeromorph acritarch (S).
Globophycus sp. of Shukla, Tewari and Yadav 1987. (Shukla et al. 1987)
  Sphaeromorph acritarch ("S").
Gunflintia minuta Barghoorn 1965. (Shukla et al. 1987)
  Tubular sheath of filamentous cyanobacterium ("S").
Kildinosphaera sp. of Shukla, Tewari and Yadav 1987. (Shukla et al. 1987)
  Sphaeromorph acritarch ("S").
Melasmatosphaera media Hofmann 1976. (Shukla et al. 1987)
  Sphaeromorph acritarch ("S").
Myxococcoides minor Schopf 1968. (Shukla et al. 1987)
  Colonial coccoid chroococcacean cyanobacterium ("S").
Oscillatoriopsis media Mendelson and Schopf 1982. (Shukla et al. 1987)
  Septate filamentous oscillatoriacean cyanobacterium ("S").
Siphonophycus kestron Schopf 1968. (Shukla et al. 1987)
  Tubular sheath of filamentous cyanobacterium (S).
Sphaerophycus parvum Schopf 1968. (Shukla et al. 1987)
  Colonial coccoid chroococcacean cyanobacterium ("S").

Kundelungu Grp. (Lower)
AF ZMB
(ca. 900 Ma)

(Sphaeromorph D). (Binda 1977)
  Sphaeromorph acritarch ("S").

Polese "Series"
EU SUN
(ca. 900 Ma)

Favososphaeridium sp. of Timofeev 1966. (Timofeev 1966)

Table 22.3 (Continued).

Colonial coccoid chroococcacean cyanobacterium (S).
Gloeocapsomorpha prisca Zalessky [Zalesskij] 1917. (Timofeev 1966)
　Colonial coccoid chroococcacean? cyanobacterium ("S").
Stictosphaeridium implexum Timofeev 1966. (Timofeev 1966)
　Sphaeromorph acritarch ("S").

Shorikha Fm.
AS SIB
(ca. 900 Ma)

("undifferentiated spheroids" [Shorikha Fm]). (Mendelson and Schopf 1982)
　Solitary and colonial (chroococcacean?) cyanobacteria ("MS"). [See Chapter 24, Pl. 19, H–K.]
(filamentous microfossils). (Schopf et al. 1977)
　Tubular sheaths of filamentous prokaryote, cf. *Eomycetopsis* (S).
　(solitary algal unicells). (Schopf et al. 1977)
　Sphaeromorph acritarch ("S").
Eoentophysalis sp. of Sergeev 1984. (Sergeev 1984)
　Colonial coccoid chroococcalean cyanobacterium (S).
Eomycetopsis robusta Schopf 1968. (Mendelson and Schopf 1982)
　Tubular sheath of filamentous prokaryote (M S). [See Chapter 24, Pl. 19, L'L$_2$.]
Eomycetopsis robusta Schopf 1968. (Sergeev 1984)
　Tubular sheath of filamentnus prokaryote (S).
Eosynechococcus grandis Hofmann 1976. (Sergeev 1984)
　Solitary ellipsoidal chroococcacean cyanobacterium (S).
Kildinella sinica Timofeev 1966. (Timofeev 1969)
　Sphaeromorph acritarch ("S").
Palaeopleurocapsa aff. *wopfneri* Knoll, Barghoorn and Golubić 1975. (Sergeev 1984)
　Colonial coccoid cyannbacterium (S).
Protosphaeridium densum Timofeev 1966. (Timofeev 1969)
　Sphaeromorph acritarch ("S").
Protosphaeridium flexuosum Timofeev 1966. (Timofeev 1969)
　Sphaeromorph acritarch ("S").
Protosphaeridium tuberculiferum Timofeev 1966. (Timofeev 1969)
　Sphaeromorph acritarch ("S").

Shtandin Fm.
EA SUN
(ca. 900 Ma)

(unnamed filament). (Yankauskas, in press)
　Flattened septate filamentous oscillatoriacean cyanobacterium ("S"). [See Chapter 24, Pl. 20, D$_1$–D$_2$.]
Arctacellularia sp. of Yankauskas in press. (Yankauskas, in press)
　Coccoid, possibly pseudofilamentous cyanobacterium (S). [See Chapter 24, Pl. 21, C.]
Tortunema sp. of Yankauskas in press. (Yankauskas, in press)
　Septate filamentous oscillatoriacean cyanobacterium ("S") [See Chapter 24, Pl. 20, B$_1$–B$_2$.]
Volyniella glomerata Jankauskas [Yankauskas] 1980. (Yankauskas, in press)
　Flattened tubular sheath of filamentous bacterium (S). [See Chapter 24, Pl. 21, F, G.]

**Late Proterozoic
900 to 800 Ma:**
Atar Grp., Atar Fm.
AF MRT
(ca. 980 Ma)

(fragment énigmatique). (Amard 1986)
　Tubular sheath of filamentous cyanobacterium (S).
Brevitrichoides bashkiricus Jankauskas [Yankauskas] 1982. (Amard 1986)
　Tubular sheath of filamentous cyanobacterium (S).
Chuaria circularis Walcott 1899. (Amard 1986)
　Sphaeromorph acritarch ("S").
Eomycetopsis robusta Schopf 1968. (Amard 1986)
　Tubular sheath of filamentous cyanobacterium (S).
Kildinosphaera chagrinata Vidal 1983. (Amard 1986)
　Sphaeromorph acritarch ("S").
Kildinosphaera verrucata Vidal 1983. (Amard 1986)
　Sphaeromorph acritarch ("S").
Leiosphaeridia asperata (Naumova 1950) Lindgren 1982. (Amard 1986)
　Sphaeromorph acritarch ("S").
Leiosphaeridia sp. of Amard 1986. (Amard 1986)
　Sphaeromorph acritarch ("S").
Palaeolyngbya cf. *minor* Schopf and Blacic 1971. (Amard 1986)
　Septate filamentous (oscillatoriacean?) cyanobacterium (S).
Protosphaeridium cf. *flexuosum* Timofeev 1966. (Amard 1986)
　Sphaeromorph acritarch ("S").
Protosphaeridium cf. *gibberosum* Timofeev 1966. (Amard 1986)
　Sphaeromorph acritarch ("S").
Protospaeridium cf. *laccatum* Timofeev 1966. (Amard 1986)
　Sphaeromorph acritarch ("S").
Protosphaeridium cf. *parvulum* Timofeev 1966. (Amard 1986)
　Sphaeromorph acritarch ("S").
Protosphaeridium sp. of Amard 1986. (Amard 1986)
　Sphaeromorph acritarch ("S").
Pterospermopsimorpha sp. of Amard 1986. (Amard 1986)
　Highly degraded sphaeromorph acritarch ("S M").
Stictosphaeridium cf. *sinapticuliferum* Timofeev 1966. (Amard 1986)
　Sphaeromorph acritarch ("S").
Trachysphaeridium levis (Lopukhin 1971) Vidal 1974. (Amard 1986)
　Sphaeromorph acritarch ("S").
Trachysphaeridium sp. of Amard 1986. (Amard 1986)
　Sphaeromorph acritarch ("S").
Trematosphaeridium holtedahlii Timofeev 1966. (Amard 1986)
　Sphaeromorph acritarch ("S").

Vindhyan Sgrp.
AS IND
(ca. 890 Ma)

Fermoria Chapman 1934. (Sahni and Shrivastava 1954)
　Sphaeromorph acritarch, aff. *Chuaria* (S).
Fermoria minima Chapman 1934. (Sahni 1936)
　Sphaeromorph acritarch, cf. *Chuaria* (S).
Krishnania acuminata Sahni and Shrivastava 1954. (Sahni and Shrivastava 1954)
　Distorted, compressed sphaeromorph acritarch, cf. *Chuaria*? (S).

Torridon Grp., Aultbea Fm.
EU GBR
(ca. 880 Ma)

(nematomorph cryptarchs). (Zhang et al. 1981)
　Tubular sheath of filamentous cyanobacterium (S).
(sphaeromorph cryptarchs). (Zhang et al. 1981)
　Sphaeromorph acritarch ("S").
(unnamed larger filaments). (Zhang 1982a)
　Degraded, flattened tubular sheaths of filamentous cyanobacteria (S).

Table 22.3 (Continued).

Eomycetopsis crassiusculum (Horodyski 1980) Z. Zhang 1982. (Zhang 1982a)
　Flattened tubular sheath of filamentous prokaryote (S).
Siphonophycus beltense Horodyski 1980. (Zhang 1982a)
　Flattened tubular sheath of filamentous cyanobacterium (S).
Siphonophycus sp. of Z. Zhang 1982. (Zhang 1982a)
　Degraded flattened tubular sheath of filamentous cyanobacterium (S).
Torridoniphycus lepidus Z. Zhang 1982. (Zhang 1982a)
　Ensheathed colonial coccoid (pleurocapsalean?) cyanobacterium ("S").

Inzer Fm.
EA SUN
(ca. 875 Ma)

Kildinella hyperboreica Timofeev 1963. (Yankauskas 1982)
　Folded, originally spheroidal, sphaeromorph acritarch (S).
Protosphaeridium sp. of Timofeev 1966. (Timofeev 1966)
　Sphaeromorph acritarch ("S").

Qingbaikou Grp., Changlongshan Fm.
AS CHN
(ca. 875 Ma)

Chuaria annularis Zheng 1981. (Du and Tian 1985)
　Sphaeromorph acritarch ("S").
Chuaria circularis Walcott 1899. (Du and Tian 1985)
　Sphaeromorph acritarch ("S").
Chuaria multirugosa Du and Tian 1985. (Du and Tian 1985)
　Sphaeromorph acritarch ("S").
Ovidiscina bagongshanica Zheng 1981. (Du and Tian 1985)
　Sphaeromorph acritarch ("S").
Ovidiscina bagongshanica Zheng 1981. (Du and Tian 1985)
　Sphaeromorph acritarch ("S").
Ovidiscina longa Du and Tian 1985. (Du and Tian 1985)
　Sphaeromorph acritarch ("S").
Shouhsienia longa Xing 1979. (Du and Tian 1985)
　Sphaeromorph acritarch ("S").
Shouhsienia multirugosa Du and Tian 1985. (Du and Tian 1985)
　Sphaeromorph acritarch ("S").
Shouhsienia shouhsienensis Xing 1979. (Du and Tian 1985)
　Sphaeromorph acritarch ("S").

Chuar Grp., Kwagunt Fm., Awatubi Mbr.
NA USA AZ
(ca. 863 Ma)

(filamentous microfossils). (Vidal and Ford 1985)
　Tubular (flattened) sheath of filamentous prokaryote (S).
Chuaria circularis Walcott 1899. (Ford and Breed 1973)
　Sphaeromorph acritarch ("S").
Chuaria circularis Walcott 1899. (Vidal and Ford 1985)
　Sphaeromorph acritarch ("S").
Cymatiosphaeroides cf. *kullingii* Knoll 1984. (Vidal and Ford 1985)
　Degraded membranous acanthomorph acritarch (S).
Kildinosphaera chagrinata Vidal 1983. (Vidal and Ford 1985)
　Sphaeromorph acritarch ("S").
Kildinosphaera verrucata Vidal 1983. (Vidal and Ford 1985)
　Sphaeromorph acritarch ("S").
Leiosphaeridia sp. A of Vidal and Ford 1985. (Vidal and Ford 1985)
　Sphaeromorph acritarch ("S").
Trachysphaeridium laminaritum Timofeev 1966. (Vidal and Ford 1985)
　Sphaeromorph acritarch ("S").
Trachysphaeridium laufeldii Vidal 1976. (Vidal and Ford 1985)
　Short-spined acanthomorph acritarch ("S M").

Bitter Springs Fm.
AU AUS NT
(ca. 850 Ma)

(large septate filaments). (Barghoorn and Schopf 1965)
　Septate filamentous prokaryotes (M).
(nonseptate filaments). (Barghoorn and Schopf 1965)
　Sheaths of filamentous cyanobacteria ("S").
(small septate filaments). (Barghoorn and Schopf 1965)
　Septate filamentous bacterium (S).
(spheroidal bodies). (Barghoorn and Schopf 1965)
　Solitary and colonial coccoid cyanobacteria (S).
Anabaenidium johnsonii Schopf 1968. (Schopf 1968)
　Septate filamentous prokaryote, possibly nostocacean (S).
Archaeonema longicellulare Schopf 1968. (Schopf 1968)
　Septate filamentous prokaryote (S).
Archaeotrichion contortum Schopf 1968. (Schopf 1968)
　Filamentous bacterium or tubular bacterial sheath ("S").
Bigeminococcus lamellosus Schopf and Blacic 1971. (Schopf and Blacic 1971)
　Colonial ellipsoidal (chroococcacean) cyanobacterium ("S"). [See Chapter 24, Pl. 33, A_1-A_2.]
Bigeminococcus mucidus Schopf and Blacic 1971. (Schopf and Blacic 1971)
　Colonial coccoid (chroococcacean) cyanobacterium ("S").
Biocatenoides rhabdos Schopf 1968. (Schopf 1968)
　Colonial (filamentous) rod-shaped bacterium ("S").
Biocatenoides sphaerula Schopf 1968. (Schopf 1968)
　Colonial (filamentous) coccoid bacterium ("S").
Calyptothrix annulata Schopf 1968. (Schopf 1968)
　Septate filamentous prokaryote (S).
Caryosphaeroides pristina Schopf 1968. (Schopf 1968)
　Plasmolyzed coccoid alga (chlorophycean?) or cyanobacterium (chroococcacean?) (S). [See Chapter 24, Pl. 33, G.]
Caryosphaeroides tetras Schopf 1968. (Schopf 1968)
　Plasmolyzed coccoid alga (chlorophycean?) or cyanobacterium (chroococcacean?) (S).
Caudiculophycus acuminatus Schopf and Blacic 1971. (Schopf and Blacic 1971)
　Septate filamentous cyanobacterium, possibly rivulariacean ("S"). [See Chapter 24, Pl. 30, O.]
Caudiculophycus rivularioides Schopf 1968. (Schopf 1968)
　Septate filamentous prokaryote, possibly rivulariacean (S). [See Chapter 24, Pl. 30, M, N.]
Cephalophytarion constrictum Schopf and Blacic 1971. (Schopf and Blacic 1971)
　Septate filamentous (oscillatoriacean) cyanobacterium ("S").
Cephalophytarion delicatulum Schopf and Blacic 1971. (Schopf and Blacic 1971)
　Septate filamentous prokaryote (S).
Cephalophytarion grande Schopf 1968. (Schopf 1968)
　Septate filamentous prokaryote (S). [See Chapter 24, Pl. 32, A, B.]
Cephalophxtarion grande Schopf 1968. (Schopf and Blacic 1971)
　Septate filamentous prokaryote (S).
Cephalophytarion laticellulosum Schopf and Blacic 1971. (Schopf and Blacic 1971)
　Septate filamentous (oscillatoriacean) cyanobacterium ("S"). [See Chapter 24, Pl. 31, A, F, L.]
Cephalophytarion minutum Schopf 1968. (Schopf 1968)

Table 22.3 (Continued).

Septate filamentous bacterium (S). [See Chapter 24, Pl. 31, I.]
Cephalophytarion variabile Schopf and Blacic 1971. (Schopf and Blacic 1971)
  Septate filamentous prokaryote (S). [See Chapter 24, Pl. 32, D.]
Contortothrix vermiformis Schopf 1968. (Schopf 1968)
  Septate filamentous prokaryote (S).
Cyanonema attenuatum Schopf 1968. (Schopf 1968)
  Septate filamentous bacterium (S).
Eoentophysalis cumulus Knoll and Golubić 1979. (Knoll and Golubić, 1979)
  Colonial ellipsoidal bacterium (S).
Eomycetopsis filiformis Schopf 1968. (Schopf 1968)
  Tubular sheath of filamentous prokaryote (S).
Eomycetopsis robusta Schopf 1968. (Knoll and Golubić, 1979)
  Tubular sheath of filamentous prokaryote (S).
Eomycetopsis robusta Schopf 1968. (Schopf 1968)
  Tubular sheath of filamentous prokaryote (S). [See Chapter 24, Pl. 30, S.]
Eosynechococcus amadeus Knoll and Golubić 1979. (Knoll and Golubić 1979)
  Colonial ellipsoidal bacterium ("S").
Eotetrahedrion princeps Schopf and Blacic 1971. (Schopf and Blqcic 1971)
  Colonial coccoid (chroococcacean) cyanobacterium (S). [See Chapter 24, Pl. 33, D_1–D_2.]
Eozygion grande Schopf and Blacic 1971. (Schopf and Blacic 1971)
  Colonial ellipsoidal (chroococcacean) cyanobacterium ("S"). [See Chapter 24, Pl. 33, C_1–C_3.]
Eozygion minutum Schopf and Blacic 1971. (Schopf and Blacic 1971)
  Colonial ellipsoidal (chroococcacean) cyanobacterium ("S"). [See Chapter 24, Pl. 33, H, J.]
Filiconstrictosus diminutus Schopf and Blacic 1971. (Schopf and Blacic 1971)
  Septate filamentous cyanobacterium ("S"). [See Chapter 24, Pl. 30, J; Pl. 31, H.]
Filiconstrictosus majusculus Schopf and Blacic 1971. (Schopf and Blacic 1971)
  Septate filamentous (oscillatoriacean) cyanobacterium ("S"). [See Chapter 24, Pl. 30, R.]
Glenobotrydion aenigmatis Schopf 1968. (Schopf 1968)
  Colonial coccoid alga (chlorophycean?) or cyanobacterium (chroococcacean?) with condensed protoplast (S). [See Chapter 24, Pl. 32, E; Pl. 33, B.]
Glenobotrydion aenigmatis Schopf 1968. (D. Oehler 1977)
  Colonial coccoid ?cyanobacterium (S).
Glenobotrydion aenigmatis Schopf 1968. (D. Oehler 1976)
  Colonial coccoid ?cyanobacterium (S).
Glenobotrydion majorinum Schopf and Blacic 1971. (Schopf and Blacic 1971)
  Solitary or colonial coccoid (chroococcacean) cyanobacterium (S).
Glenobotrydion majorinum Schopf and Blacic 1971. (D. Oehler 1976)
  Colonial coccoid ?cyanobacterium (S).
Globophycus rugosum Schopf 1968. (Schopf 1968)
  Solitary sheath-enclosed coccoid (chroococcacean) cyanobacterium (S). [See Chapter 24, Pl. 33, E.]
Gloeodiniopsis gregaria Knoll and Golubić 1979. (Knoll and Golubić, 1979)
  Colonial coccoid chroococcacean cyanobacterium ("S").
Gloeodiniopsis lamellosa Schopf 1968. (Schopf and Blacic 1971)
  Solitary sheath-enclosed ellipsoidal (chroococcacean) cyanobacterium, cf. *Chroococcus* or cf. *Gloeocapsa* ("S"). [See Chapter 24, Pl. 32, J.]
Gloeodiniopsis lamellosa Schopf 1968. (Schopf 1968)
  Solitary sheath-enclosed ellipsoidal (chroococcacean) cyanobacterium, cf. *Chroococcus* or *Gloeocapsa* (S).
Gloeodiniopsis lamellosa Schopf 1968. (Knoll and Golubić 1979)
  Colonial coccoid chroococcacean cyanobacterium ("S").
Halythrix nodosa Schopf 1968. (Schopf 1968)
  Septate filamentous cyanobacterium. Spool-shape of cells similar to *Oscillatoria princeps* ("S"). [See Chapter 24, Pl. 31, K.]
Heliconema australiense Schopf 1968. (Schopf 1968)
  Helical filamentous prokaryote, aff. *Spirulina* or *Arthrospira*? (S).
Heliconema funiculum Schopf and Blacic 1971. (Schopf and Blacic 1971)
  Helical filamentous cyanobacterium, aff. *Spirulina* or *Arthrospira*? ("S"). [See Chapter 24, Pl. 31, B.]
Myxococcoides inornata Schopf 1968. (Schopf 1968)
  Colonial coccoid (chroococcacean) cyanobacterium ("S").
Myxococcoides minor Schopf 1968. (D. Oehler 1976)
  Colonial coccoid chroococcacean cyanobacterium ("S").
Myxococcoides minor Schopf 1968. (D. Oehler 1977)
  Colonial coccoid cyanobacterium ("S").
Myxococcoides minor Schopf 1968. (Schopf 1968)
  Colonial coccoid (chroococcacean) cyanobacterium ("S"). [See Chapter 24, Pl. 32, H, I.]
Myxococcoides reticulata Schopf 1968. (Schopf 1968)
  Colonial coccoid (chroococcacean) cyanobacterium ("S").
Obconicophycus amadeus Schopf and Blacic 1971. (Schopf and Blacic 1971)
  Septate filamentous (oscillatoriacean) cyanobacterium ("S"). [See Chapter 24, Pl. 31, E.]
Oscillatoriopsis breviconvexa Schopf and Blacic 1971. (Schopf and Blacic 1971)
  Septate filamentous (oscillatoriacean) cyanobacterium ("S"). [See Chapter 24, Pl. 31, C.]
Oscillatoriopsis obtusa Schopf 1968. (Schopf 1968)
  Septate filamentous cyanobacterium ("S"). [See Chapter 24, Pl. 31, G.]
Palaeoanacystis vulgaris Schopf 1968. (D. Oehler 1976)
  Colonial coccoid chroococcacean cyanobacterium ("S").
Palaeolyngbya barghoorniana Schopf 1968. (Schopf 1968)
  Septate filamentous (oscillatoriacean) cyanobacterium ("S"). [See Chapter 24, Pl. 30, Q; Pl. 32, C.]
Palaeolyngbya cf. *barghoorniana* Schopf 1968. (D. Ohler 1976)
  Septate filamentous oscillatoriacean cyanobacterium ("S").
Palaeolyngbya minor Schopf and Blacic 1971. (Schopf and Blacic 1971)
  Septate filamentous (oscillatoriacean) cyanobacterium ("S"). [See Chapter 24, Pl. 30, K.]
Partitiofilum gongyloides Schopf and Blacic 1971. (Schopf and Blacic 1971)
  Septate filamentous cyanobacterium ("S"). [See Chapter 24, Pl. 30, P.]
Phanerosphaerops capitaneus Schopf and Blacic 1971. (Schopf and Blacic 1971)
  Solitary coccoid (chroococcacean) cyanobacterium ("S").
Siphonophycus kestron Schopf 1968. (Schopf 1968)
  Tubular sheath of filamentous (oscillatoriacean?) cyanobacterium ("S"). [See Chapter 24, Pl. 31, J.]
Sphaerophycus parvum Schopf 1968. (Schopf 1968)
  Solitary and paired coccoid (chroococcacean) cyanobacterium ("S"). [See Chapter 24, Pl. 33, F.]

Table 22.3 (*Continued*).

Tenuofilum septatum Schopf 1968. (Schopf 1968)
  Septate filamentous bacterium, cf. *Gunflintia* (S).
Veteronostocale amoenum Schopf and Blacic 1971. (Schopf and Blacic 1971)
  Septate filamentous prokaryote, possibly nostocacean (S). [See Chapter 24, Pl. 30, L.]
Zosterosphaera tripunctata Schopf 1968. (Schopf 1968)
  Sheath of colonial coccoid prokaryote (S).

Bitter Springs Fm., Gillen Mbr.
AU AUS NT
(ca. 850 Ma)

(interwoven tubules). (Oehler et al. 1979)
  Tubular sheath of filamentous cyanobacterium ("S").
(larger unicells). (Oehler et al. 1979)
  Colonial coccoid chroococcacean cyanobacteria ("S").
(smaller unicells). (Oehler et al. 1979)
  Colonial coccoid ?entophysalidacean cyanobacteria ("S").

Chuar Grp., Kwagunt Fm.
NA USA AZ
(ca. 850 Ma)

(flask-shaped chitinozoan). (Bloeser et al. 1977)
  Flask-shaped microfossil incertae sedis (S).
(solid filament). (Horodyski and Bloeser 1983)
  Tubular sheath of filamentous cyanobacterium (S).
(tear-shaped chitinozoan). (Bloeser et al. 1977)
  Tear-shaped microfossil incertae sedis (S).
Kildinosphaera lophostriata (Jankauskas 1979) Vidal 1983. (Vidal and Ford 1985)
  Sphaeromorph acritarch ("S").

Chuar Grp., Kwagunt Fm., Walcott Mbr.
NA USA AZ
(ca. 850 Ma)

(robust, tubular, filamentous thallophytes). (Schopf et al. 1973)
  Tubular sheath of filamentous prokaryote ("S"). [See Chapter 24, Pl. 30, B, H, I.]
(spheroidal unicells). (Schopf et al. 1973)
  Solitary and colonial coccoid (chroococcacean?) cyanobacteria ("S"). [See Chapter 24, Pl. 29, I, J; Pl. 30, G.]
Chuaria circularis Walcott 1899. (Walcott 1899)
  Sphaeromorph acritarch (S).
Melanocyrillium fimbriatum Bloeser 1985. (Bloeser 1985)
  Microfossil incertae sedis ("S"). [See Chapter 24, Pl. 29, A, E_1-E_2, G_1-G_2.]
Melanocyrillium hexodiadema Bloeser 1985. (Bloeser 1985)
  Microfossil incertae sedis ("S"). [See Chapter 24, Pl. 29, C, D, F, H_1-H_4.]
Melanocyrillium horodyskii Bloeser 1985. (Bloeser 1985)
  Microfossil incertae sedis ("S").
Satka colonialica Jankauskas (Yankauskas) 1979. (Vidal and Ford 1985)
  Colonial coccoid cyanobacterium (S).
Vandalosphaeridium walcottii Vidal and Ford 1985. (Vidal and Ford 1985)
  Degraded membranous acanthomorph acritarch ("S M").

Miroedikha Fm.
AS SIB
(ca. 850 Ma)

"*Phycomycetes*" Timofeev, German and Mikhajlova 1976. (Timofeev et al. 1976)
  Sphaeromorph acritarch with associated "tail-like" sapropelic debris (S).
((?) trichome with pseudovacuoles). (Timofeev et al. 1976)
  Septate filamentous (oscillatoriacean?) cyanobacterium ("S").
([broad tubular sheath of] algal trichome). (Timofeev 1969)
  Flattened tubular sheath of filamentous cyanobacterium (S).
([intermediate-diameter sheath of] algal trichome). (Timofeev 1969)
  Flattened tubular sheath of filamentous cyanobacterium (S).
(algal trichome [with barrel-shaped cells]). (Timofeev 1969)
  Septate filamentous (nostocalean) cyanobacteria ("S").
(algal trichome [with quadrate cells]). (Timofeev 1969)
  Septate filamentous oscillatoriacean cyanobacterium ("S").
(sheath of trichome [broad]). (Timofeev et al. 1976)
  Flattened tubular sheath of filamentous cyanobacterium ("S").
(sheath of trichome [intermediate diameter]). (Timofeev et al. 1976)
  Flattened tubular sheath of filamentous cyanobacterium ("S").
Arctacellularia doliiformis Hermann [German] 1976. (Timofeev et al. 1976)
  Trichome and trichomic cells of filamentous cyanobacterium (S M). [See Chapter 24, Pl. 26. 26, $B_1 - B_2$.]
Arctacellularia ellipsoidea Hermann [German] 1976. (Timofeev et al. 1976)
  Ensheathed septate filamentous oscillatoriacean cyanobacterium (S). [See Chapter 24, Pl. 24, A_1-A_2.]
Arctacellularia ellipsoidea Hermann [German] 1976. (Timofeev et al. 1976)
  Pseudofilamentous colonies of spheroidal to ellipsoidal cyanobacteria (S). [See Chapter 24, Pl. 24, A_1-A_2.]
Arthrosiphon cornutus Weiss [Vejs] 1984. (Vejs 1984)
  Septate filamentous oscillatoriacean cyanobacterium ("S"). [See Chapter 24, Pl. 22, E.]
Arthrosiphon typicus Weiss [Vejs] 1984. (Vejs 1984)
  Septate filamentous prokaryote (S). [See Chapter 24, Pl. 22, F.]
Calyptothrix perfecta Weiss [Vejs] 1984. (Vejs 1984)
  Septate filamentous oscillatoriacean cyanobacterium ("S"). [See Chapter 24, Pl. 22, D.]
Cephalophytarion turukhanicum Weiss [Vejs] 1984. (Vejs 1984)
  Septate filamentous oscillatoriacean cyanobacterium ("S"). [See Chapter 24, Pl. 25, D_1-D_2; Pl. 26, C.]
Dictyosphaeridium eniseicum Timofeev 1969. (Timofeev 1969)
  Sphaeromorph acritarch ("S").
Filiconstrictosus diminutus Schopf 1968. (Timofeev et al. 1976)
  Septate filamentous oscillatoriacean cyanobacterium ("S").
Filiconstrictosus eniseicum Weiss [Vejs] 1984. (Vejs 1984)
  Septate filamentous oscillatoriacean cyanobacterium ("S"). [See Chapter 24, Pl. 25, B.]
Heliconema turukhania Hermann [German] 1981. (German 1981b)
  Filamentous, *Spirulina*-like prokaryote (S). [See Frontispiece, and Chapter 24, Pl. 22, H.]
Kildinella hyperboreica Timofeev 1966. (Timofeev 1969)
  Sphaeromorph acritarch ("S").
Kildinella hyperboreica Timofeev 1966. (Timofeev 1966)
  Sphaeromorph acritarch ("S").
Kildinella jacutica Timofeev 1966. (Timofeev 1969)
  Sphaeromorph acritarch ("S").
Kildinella jacutica Timofeev 1966. (Timofeev 1966)
  Sphaeromorph acritarch ("S").
Kildinella magna Timofeev 1969. (Timofeev 1969)
  Sphaeromorph acritarch ("S").
Kildinella miroedichia Timofeev 1969. (Timofeev 1969)
  Sphaeromorph acritarch ("S").

Table 22.3 *(Continued).*

Kildinella sinica Timofeev 1966. (Timofeev 1966)
  Sphaeromorph acritarch ("S").
Kildinella sinica Timofeev 1966. (Timofeev 1969)
  Sphaeromorph acritarch ("S").
Kildinella sinica Timofeev 1966. (Timofeev 1969)
  Sphaeromorph acritarch ("S").
Kildinella sp. of Timofeev 1969. (Timofeev 1969)
  Colonial coccoid cyanobacterium (S).
Kildinella sp. of Timofeev, German and Mikhajlova 1976. (Timofeev et al. 1976)
  Sphaeromorph acritarch ("S").
Kildinella tschapomica Timofeev 1966. (Timofeev 1969)
  Sphaeromorph acritarch ("S").
Leiosphaeridia asperata (Naumova 1950) Lindgren 1982. (Lindgren 1982)
  Colonial coccoid cyanobacteria and sphaeromorph acritarchs (S).
Leiosphaeridia bituminosa Timofeev 1966. (Timofeev 1969)
  Sphaeromorph acritarch ("S").
Leiothrichoides maculatus Hermann [German] 1986. (German 1986)
  Tubular sheath of filamentous cyanobacterium containing degraded cellular remnants (S). [See Chapter 24, Pl. 28, B.]
Leiothrichoides typicus Hermann [German] 1974. (German 1974)
  Flattened tubular sheath of filamentous cyanobacterium ("S").
Leiothrichoides typicus Hermann [German] 1974. (Timofeev et al. 1976)
  Flattened tubular sheath of filamentous cyanobacterium (S).
Leiothrichoides typicus Hermann [German] 1974. (Timofeev and German 1979)
  Tubular sheath of filamentous cyanobacterium ("S"). [See Chapter 24, Pl. 27, B_1-B_4.]
Macroptycha biplicata Timofeev 1976. (Timofeev et al. 1976)
  Sphaeromorph acritarch ("S"). [See Chapter 24, Pl. 24, F.]
Macroptycha multiplicata Timofeev 1976. (Timofeev et al. 1976)
  Sphaeromorph acritarch ("S").
Macroptycha triplicata Timofeev 1976. (Timofeev et al. 1976)
  Sphaeromorph acritarch ("S").
Macroptycha uniplicata Timofeev 1973. (Timofeev et al. 1976)
  Sphaeromorph acritarch ("S"). [See Chapter 24, Pl. 24, E.]
Microconcentrica sp. of Timofeev, German and Mikhajlova 1976. (Timofeev et al. 1976)
  Colonial coccoid cyanobacterium (S).
Nucellosphaeridium bellum Timofeev 1969. (Timofeev 1969)
  Sphaeromorph acritarch ("S").
Nucellosphaeridium deminatum Timofeev 1969. (Timofeev 1969)
  Sphaeromorph acritarch ("S").
Nucellosphaeridium deminatum Timofeev 1969. (Timofeev et al. 1976)
  Sphaeromorph acritarch containing degraded cytoplasmic contents (S).
Nucellosphaeridium sp. of Timofeev, German & Mikhajlova 1976. (Timofeev et al. 1976)
  Sphaeromorph acritarch with degraded cytoplasmic contents (S).
Oscillatoriopsis bacilaris Hermann [German] 1981. (German 1981b)
  Flattened tubular sheaths (of hormogones?) of oscillatoriacean cyanobacterium ("S"). [See Chapter 24, Pl. 28, D.]
Oscillatoriopsis sp. of Timofeev, German & Mikhajlova 1976. (Timofeev et al. 1976)
  Septate filamentous oscillatoriacean cyanobacterium ("S").
Oscillatorites wernadskii [intermediate diameter] Schepelewa [Shepeleva] 1960. (Timofeev 1966)
  Septate filamentous oscillatoriacean cyanobacterium ("S").

Oscillatorites wernadskii Schepelewa [Shepeleva] 1960. (Timofeev 1966)
  Septate filamentous oscillatoriacean cyanobacterium ("S").
Palaeolyngbya catenata Hermann [German] 1974. (Timofeev et al. 1976)
  Ensheathed septate filamentous oscillatoriacean cyanobacterium ("S").
Palaeolyngbya catenata Hermann [German] 1974. (German 1974)
  Septate filamentous oscillatoriacean cyanobacterium ("S").
Palaeolyngbya sphaerocephala Hermann [German] and Pylina 1986. (German 1986)
  Ensheathed septate filamentous prokaryote (S). [See Chapter 24, Pl. 22, G_1-G_2; Pl. 28, A_1-A_2.]
Polysphaeroides contextus Hermann [German] 1976. (Timofeev et al. 1976)
  Septate filamentous oscillatoriacean cyanobacterium ("S"). [See Chapter 24, Pl. 24, B_1-B_2.]
Polysphaeroides filliformis Hermann [German] 1976. (Timofeev et al. 1976)
  Septate filamentous oscillatoriacean cyanobacterium ("S"). [See Chapter 24, Pl. 23, A_1-A_2.]
Polythrichoides lineatus Hermann [German] 1974. (German 1974)
  Flattened tubular sheath of filamentous prokaryote ("S"). [See Chapter 24, Pl. 27, A_1-A_2.]
Primorivularia dissimilara Hermann [German] 1986. (German 1986)
  Septate filamentous prokaryote (S). [See Chapter 24, Pl. 22, C.]
Protosphaeridium paleaceum Timofeev 1966. (Timofeev 1969)
  Sphaeromorph acritarch ("S").
Protosphaeridium tuberculiferum Timofeev 1966. (Timofeev 1969)
  Sphaeromorph acritarch ("S").
Protosphaeridium vermium Timofeev 1969. (Timofeev 1969)
  Sphaeromorph acritarch ("S").
Pterospermopsimorpha insolita Timofeev 1969. (German in press)
  Sphaeromorph acritarch containing degraded globular internal remnants (S M). [See Chapter 24, Pl. 24, D.]
Pterospermopsimorpha pileiformis Timofeev 1966. (Timofeev 1969)
  Sphaeromorph acritarch containing coalesced cytoplasmic remnants (M S).
Pterospermopsimorpha sp. of Timofeev, German and Mikhajlova 1976. (Timofeev et al. 1976)
  Sphaeromorph acritarch containing coalesced cytoplasmic remnants (M S).
Scaphita eniseica Timofeev 1976. (Timofeev et al. 1976)
  Sphaeromorph acritarch ("S").
Stictosphaeridium pectinale Timofeev 1966. (Timofeev 1969)
  Sphaeromorph acritarch ("S").
Stictosphaeridium sinapticuliferum Timofeev 1966. (Timofeev 1969)
  Sphaeromorph acritarch ("S").
Stictosphaeridium tortulosum Timofeev 1966. (Timofeev 1969)
  Sphaeromorph acritarch ("S").
Stictosphaeridium? tortulosum Timofeev 1966. (Timofeev 1969)
  Sphaeromorph acritarch ("S").
Symplassosphaeridium sp. of Timofeev 1969. (Timofeev 1969)
  Colonial coccoid chroococcacean cyanobacterium (S).
Synsphaeridium sp. of Timofeev 1969. (Timofeev 1969)
  Colonial coccoid cyanobacterium (S).
Tortunema eniseica Hermann [German] 1976. (Timofeev et al. 1976)
  Flattened tubular sheath of filamentous cyanobacterium (S). [See Chapter 24, Pl. 26, A_1-A_2.]
Tortunema sibirica Hermann [German] 1976. (Timofeev et al. 1976)
  Septate filamentous oscillatoriacean cyanobacterium ("S"). [See Chapter 24, Pl. 28, E, F_1-F_2.]

Table 22.3 (Continued).

Trachysphaeridium lachandinum Timofeev 1969. (Timofeev 1969)
Sphaeromorph acritarch ("S").
Trachysphaeridium laminaritum Timofeev 1966. (Timofeev 1969)
Sphaeromorph acritarch ("S").
Trachysphaeridium maicum Timofeev 1969. (Timofeev 1969)
Sphaeromorph acritarch ("S").
Trachysphaeridium sp. of Timofeev 1969. (Timofeev 1969)
Sphaeromorph acritarch ("S").
Trachysphaeridium sp. of Timofeev, German and Mikhajlova 1976. (Timofeev et al. 1976)
Sphaeromorph acritarch ("S").
Trachythrichoides ovalis Hermann [German] 1976. (Timofeev et al. 1976)
Septate filamentous eukaryotic alga (S). [See Chapter 24, Pl. 26, E; Pl. 28, C.]
Trachythrichoides ovalis [large] Hermann [German] 1976. (Timofeev et al. 1976)
Septate filamentous eukaryotic alga (S). [See Chapter 24, Pl. 26, E; pl. 28, C.]
Turuchanica ternata Timofeev 1966. (Timofeev 1966)
Flattened, compressed, sphaeromorph? acritarch (S).
Turuchanica ternata Timofeev 1966. (Timofeev 1969)
Flattened sphaeromorph? acritarch (S).
Zonosphaeridium sp. of Timofeev, German and Mikhajlova 1976. (Timofeev et al. 1976)
Sphaeromorph acritarch ("S").

Pahrump Grp., Beck Spring Dolomite
NA USA CA
(ca. 850 Ma)

(chains of hematite crystals). (Gutstadt and Schopf 1969)
Filamentous cyanobacterium ("S").
(type "a" [filaments]). (Cloud et al. 1969)
Septate filamentous prokaryotes (S).
(type "b" [unicells]). (Cloud et al. 1969)
Solitary and colonial coccoid cyanobacteria (S).
(type "c" [unicells]). (Cloud et al. 1969)
Solitary and colonial coccoid cyanobacteria (S M).
(type "d" [unicells]). (Cloud et al. 1969)
Colonial coccoid cyanobacteria (S).
(type "e" [unicells]). (Cloud et al. 1969)
Solitary coccoid cyanobacteria associated with mineralic needles (S).
(type "f" [unicells]). (Cloud et al. 1969)
Solitary coccoid cyanobacteria associated with mineralic needles (S).
Abundacapsa impages Licari 1978. (Licari 1978)
Colonial coccoid cyanobacterium ("S").
Beckspringia communis Licari 1978. (Licari 1978)
Septate filamentous prokaryote (S). [See Chapter 24, Pl. 21, H, I.]
Conglobocella troxelii Licari 1978. (Licari 1978)
Colonial coccoid cyanobacterium (S).
Latisphaera wrightii Licari 1978. (Licari 1978)
Algal unicell (?) or sheath of colonial coccoid cyanobacterium (S).
Maculosphaera kingstonensis Licari 1978. (Licari 1978)
Solitary and colonial coccoid cyanobacterium (S).
Palaeosiphonella cloudii Licari 1978. (Licari 1978)
Possibly branching filamentous eukaryotic alga ("S"), or sheath of filamentous false-branching cyanobacterium (S). [See Chapter 24, Pl. 22, A, B.]

Pahrump Grp., Kingston Peak Fm.
NA USA CA
(ca. 850 Ma)

(elongate spheroidal forms). (Pierce and Cloud 1979)
Solitary and colonial coccoid cyanobacteria and sphaeromorph acritarchs (S).
(ghost filaments). (Pierce and Cloud 1979)
Sheaths of filamentous (oscillatoriacean?) cyanobacteria ("S").
cf. *Conglobocella* Licari 1978. (Pierce and Cloud 1979)
Colonial coccoid cyanobacterium ("S").
Girvanella Nicholson and Etheridge 1878. (Pierce and Cloud 1979)
Sheath of filamentous (oscillatoriacean?) cyanobacterium, cf. *Siphonophycus* (not *Girvanella*) (S).

Qingbaikou Grp., Jingeryu Fm.
AS CHN
(ca. 850 Ma)

Acantholigotriletum primigenum (Naumova 1949) Timofeev 1959. (Timofeev 1966)
Acanthomorph acritarch ("S").
Gloeocapsomorpha prisca Zalessky [Zalesskij] 1917. (Timofeev 1966)
Colonial coccoid chroococcacean? cyanobacterium ("S").
Orygmatosphaeridium rubiginosum Andreeva 1966. (Xing and Liu 1973)
Sphaeromorph acritarch ("S").
Protosphaeridium acis Timofeev 1966. (Timofeev 1966)
Sphaeromorph acritarch ("S").
Protosphaeridium gibberosum Timofeev 1966. (Timofeev 1966)
Sphaeromorph acritarch ("S").
Protosphaeridium ostiolatum Rudavskaja [Rudavskaja] 19??. (Timofeev 1966)
Sphaeromorph acritarch ("S").
Protosphaeridium patelliforme Timofeev 1966. (Timofeev 1966)
Sphaeromorph acritarch ("S").
Protosphaeridium sp. of Timofeev 1966. (Timofeev 1966)
Sphaeromorph acritarch ("S").
Trematosphaeridium holtedahlii Timofeev 1966. (Timofeev 1966)
Sphaeromorph acritarch ("S").
Trematosphaeridium holtedahlii Timofeev 1966. (Xing and Liu 1973)
Sphaeromorph acritarch ("S").
Trematosphaeridium sp. of Timofeev 1966. (Timofeev 1966)
Sphaeromorph acritarch ("S").

Veteranen Grp., Glasgowbreen Fm.
EU SJM
(ca. 850 Ma)

(filamentous microfossils). (Knoll and Swett 1985)
Tubular sheaths of filamentous prokaryotes (S).
(small coccoidal unicells). (Knoll and Swett 1985)
Solitary and colonial coccoid cyanobacteria and sphaeromorph acritarchs (M S).
Bavlinella faveolata Schepeleva [Shepeleva] 1962. (Knoll and Swett 1985)
Colonial coccoid bacterium (S).
Chuaria? circularis Walcott 1899. (Knoll and Swett 1985)
Sphaeromorph acritarch ("S").
Eosynechococcus sp. of Knoll and Swett 1985. (Knoll and Swett 1985)
Solitary ellipsoidal bacterium (S).
Favososphaeridium sp. of Knoll and Swett 1985. (Knoll and Swett 1985)
Sphaeromorph acritarch ("S").

Table 22.3 (Continued).

Kildinosphaera chagrinata Vidal 1983. (Knoll and Swett 1985)
  Sphaeromorph acritarch ("S").
Kildinosphaera granulata Vidal 1983. (Knoll and Swett 1985)
  Sphaeromorph acritarch ("S").
Leiosphaeridia asperata (Naumova 1950) Lindgren 1982. (Knoll and Swett 1985)
  Sphaeromorph acritarch ("S").
Satka colonialica Jankauskas [Yankauskas] 1979. (Knoll and Swett 1985)
  Colonial ellipsoidal cyanobacterium (S).
Synsphaeridium sp. of Knoll and Swett 1985. (Knoll and Swett 1985)
  Colonial coccoid cyanobacterium (S).
Tasmanites rifejieus Jankauskas [Yankauskas] 1978. (Knoll and Swett 1985)
  Sphaeromorph acritarch (S).

Veteranen Grp., Kingbreen Fm.
EU SJM
(ca. 850 Ma)

(filamentous microfossils). (Knoll and Swett 1985)
  Tubular sheaths of filamentous prokaryotes (S).
(small coccoidal unicells). (Knoll and Swett 1985)
  Solitary and colonial coccoid cyanobacteria and sphaeromorph acritarchs (M S).
Bavlinella faveolata Schepeleva [Shepeleva] 1962. (Knoll and Swett 1985)
  Colonial coccoid bacterium (S).
Chuaria? circularis Walcott 1899. (Knoll and Swett 1985)
  Sphaeromorph acritarch ("S").
Eosynechococcus sp. of Knoll and Swett 1985. (Knoll and Swett 1985)
  Solitary ellipsoidal bacterium (S).
Kildinosphaera chagrinata Vidal 1983. (Knoll and Swett 1985)
  Sphaeromorph acritarch ("S").
Leiosphaeridia asperata (Naumova 1950) Lindgren 1982. (Knoll and Swett 1985)
  Sphaeromorph acritarch ("S").
Synsphaeridium sp. of Knoll and Swett 1985. (Knoll and Swett 1985)
  Colonial coccoid cyanobacterium (S).
Tasmanites rifejicus Jankauskas [Yankauskas] 1978. (Knoll and Swett 1985)
  Sphaeromorph acritarch (S).

Veteranen Grp., Kortbreen Fm.
EU SJM
(ca. 850 Ma)

(filamentous microfossils). (Knoll and Swett 1985)
  Tubular sheaths of filamentous prokaryotes (S).
(small coccoidal unicells). (Knoll and Swett 1985)
  Solitary and colonial coccoid cyanobacteria and sphaeromorph acritarchs (M S).
Chuaria? circularis Walcott 1899. (Knoll and Swett 1985)
  Sphaeromorph acritarch ("S").
Kildinosphaera chagrinata Vidal 1983. (Knoll and Swett 1985)
  Sphaeromorph acritarch ("S").
Leiosphaeridia asperata (Naumova 1950) Lindgren 1982. (Knoll and Swett 1985)
  Sphaeromorph acritarch ("S").

Veteranen Grp., Oxfordbreen Fm.
EU SJM
(ca. 850 Ma)

(filamentous microfossils). (Knoll and Swett 1985)
  Tubular sheaths of filamentous prokaryotes (S).
(small coccoidal unicells). (Knoll and Swett 1985)
  Solitary and colonial coccoid cyanobacteria and sphaeromorph acritarchs (M S).
Chuaria? circularis Walcott 1899. (Knoll and Swett 1985)
  Sphaeromorph acritarch ("S").
Kildinosphaera chagrinata Vidal 1983. (Knoll and Swett 1985)
  Sphaeromorph acritarch ("S").
Kildinosphaera granulata Vidal 1983. (Knoll and Swett 1985)
  Sphaeromorph acritarch ("S").
Leiosphaeridia asperata (Naumova 1950) Lindgren 1982. (Knoll and Swett 1985)
  Sphaeromorph acritarch ("S").
Leiosphaeridia asperata (Naumova 1950) Lindgren 1982. (Knoll and Swett 1985)
  Sphaeromorph acritarch ("S").
Synsphaeridium sp. of Knoll and Swett 1985. (Knoll and Swett 1985)
  Colonial coccoid cyanobacterium (S).

Delhi Sgrp., Ajabgarh Grp., Kushalgarh Fm.
AS IND
(ca. 818 Ma)

Palaeolyngbya elongata Mandal and Maithy 1984. (Mandal et al. 1984)
  Possibly modern contaminant or ensheathed septate filamentous oscillatoriacean cyanobacterium (S M).

Bambuí Grp.
SA BRA
(ca. 810 Ma)

(sac-like bodies). (Fairchild and Subacius 1986)
  Cyanobacterial/bacterial envelope (M).
(spheroids and ellipsoids). (Fairchild and Subacius 1986)
  Solitary coccoid cyanobacteria (M).

Bambuí Grp. (Lower)
SA BRA
(ca. 810 Ma)

(small-celled colonies). (Fairchild et al. 1980)
  Colonial coccoid and ellipsoidal cyanobacteria (M).
(unbranched filaments). (Fairchild et al. 1980)
  Filamentous prokaryotes (M).

Vadsø Grp., Klubbnes Fm.
EU NOR
(ca. 810 Ma)

Chuaria circularis Walcott 1899. (Vidal 1981b)
  Sphaeromorph acritarch ("S").
Kildinella hyperboreica Timofeev 1966. (Vidal 1981b)
  Sphaeromorph acritarch ("S").
Kildinella sinica Timofeev 1966. (Vidal 1981b)
  Sphaeromorph acritarch ("S").
Kildinella sp. A of Vidal 1981. (Vidal 1981b)
  Sphaeromorph acritarch ("S").
Kildinella sp. B of Vidal 1981. (Vidal 1981b)
  Sphaeromorph acritarch ("S").
cf. *Stictosphaeridium* sp. of Vidal 1981. (Vidal 1981b)

Table 22.3 (Continued).

Colonial coccoid cyanobacterium (fig. 19A–F) (S).
Tasmanites rifejicus Jankauskas [Yankauskas] 1978. (Vidal and Siedlecka 1983)
  Sphaeromorph acritarch ("S").

Akberdin Fm.
EA SUN
(ca. 800 Ma)

Valeria lophostriata (Jankauskas 1979) Jankauskas 1982. (Yankauskas, in press)
  Sphaeromorph acritarch ("S"). [See Chapter 24, Pl. 42, A_1-A_2, D_1-D_2.]

Bodajbo Subgrp.
AS SIB
(ca. 800 Ma)

Gloeocapsomorpha priscata Timofeev 1973. (Timofeev et al. 1976)
  Colonial coccoid cyanobacterium ("S").
Kildinella hyperboreica Timofeev 1966. (Timofeev 1969)
  Sphaeromorph acritarch ("S").
Kildinella vesljanica Timofeev 1969. (Timofeev et al. 1976)
  Sphaeromorph acritarch ("S").
Ocridosphaeridium sp. of Timofeev, German and Mikhajlova 1976. (Timofeev et al. 1976)
  Sphaeromorph acritarch ("S").
Orygmatosphaeridium sp. of Timofeev, German and Mikhajlova 1976. (Timofeev et al. 1976)
  Colonial coccoid (chroococcacean?) cyanobacterium (S).
Protosphaeridium acis Timofeev 1966. (Timofeev et al. 1976)
  Sphaeromorph acritarch ("S").
Protosphaeridium densum Timofeev 1966. (Timofeev 1969)
  Sphaeromorph acritarch ("S").
Protosphaeridium discum Timofeev 19??. (Timofeev et al. 1976)
  Sphaeromorph acritarch ("S").
Protosphaeridium flexuosum Timofeev 1966. (Timofeev et al. 1976)
  Sphaeromorph acritarch ("S").
Protosphaeridium rigidulum Timofeev 1966. (Timofeev 1969)
  Sphaeromorph acritarch ("S").
Protosphaeridium scabridum Timofeev 1966. (Timofeev et al. 1976)
  Sphaeromorph acritarch ("S").
Protosphaeridium torulosum Timofeev 1966. (Timofeev et al. 1976)
  Sphaeromorph acritarch ("S").
Protosphaeridium tuberculiferum Timofeev 1966. (Timofeev et al. 1976)
  Sphaeromorph acritarch ("S").
Protosphaeridium vermium Timofeev 1969. (Timofeev et al. 1976)
  Sphaeromorph acritarch ("S").
Stictosphaeridium implexum Timofeev 1966. (Timofeev 1969)
  Sphaeromorph acritarch ("S").

Huainan Grp., Liulaobei Fm.
AS CHN
(ca. 800 Ma)

Chuaria circularis Walcott 1899. (Sun 1987)
  Sphaeromorph acritarch (S).

Hyperborean Fm.
EU SUN
(ca. 800 Ma)

Protosphaeridium parvulum Timofeev 1966. (Timofeev 1966)
  Sphaeromorph acritarch ("S").
Protosphaeridium pusillum Timofeev 1966. (Timofeev 1966)
  Sphaeromorph acritarch ("S").
Trematosphaeridium holtedahlii Timofeev 1966. (Timofeev 1966)
  Sphaeromorph acritarch ("S").

Jiudingshan Fm.
AS CHN
(ca. 800 Ma)

Animikiea cf. *septata* [nonseptate] Barghoorn 1965. (Liu et al. 1984)
  Tubular sheath of filamentous cyanobacterium ("S").
Animikiea cf. *septata* [septate] Barghoorn 1965. (Liu et al. 1984)
  Septate filamentous oscillatoriacean cyanobacterium ("S").
Astercapsoides borealis Liu 1982. (Liu et al. 1984)
  Sphaeromorph acritarch (S).
Bigeminococcus lamellosus Schopf and Blacic 1971. (Liu et al. 1984)
  Colonial coccoid chroococcacean cyanobacterium ("S").
Eoentophysalis belcherensis Hofmann 1976. (Liu et al. 1984)
  Colonial coccoid ?entophysalidacean cyanobacterium ("S").
Eomycetopsis robusta Schopf 1968. (Liu et al. 1984)
  Tubular sheath of filamentous cyanobacterium ("S").
Eoplectonema minimum Liu 1984. (Liu et al. 1984)
  Tubular (unbranched) sheath of filamentous cyanobacterium (S).
Eopleurocapsa cf. *sinica* Liu 1982. (Liu et al. 1984)
  Colonial coccoid chroococcacean cyanobacterium (S).
Gloeocapsoides media Liu 1982. (Liu et al. 1984)
  Colonial coccoid chroococcacean cyanobacterium ("S").
Leiopsophosphaera pelucidus Schepeleva [Shepeleva] 1963. (Liu et al 1984)
  Sphaeromorph acritarch (S).
Melasmatosphaera magna Hofmann 1976. (Liu et al. 1984)
  Sphaeromorph acritarch (S).
Obruchevella condensata Liu 1984. (Liu et al. 1984)
  Spirally coiled *Spirulina*- or *Arthrorpira*-like prokaryote (S).
Protosphaeridium densum [colonial] Timofeev 1966. (Liu et al. 1984)
  Colonial coccoid chroococcalean cyanobacterium (S).
Protosphaeridium densum [solitary] Timofeev 1966. (Liu et al. 1984)
  Sphaeromorph acritarch (S).
Rhicnonema crassivaginatum Zhu 1982. (Liu et al. 1984)
  Degraded ensheathed septate filamentous prokaryote (S).
Siphonophycus lamellosum Liu 1984. (Liu et al. 1984)
  Tubular sheath of filamentous cyanobacterium ("S").
Tetraphycus hebeiense Liu 1982. (Liu et al. 1984)
  Colonial coccoid chroococcacean cyanobacterium (S).

Kirgitej Fm.
AS SIB
(ca. 800 Ma)

Angaronema septatum Golovenoc [Golovenok] and Belova 1985. (Golovenok and Belova 1985)
  Ensheathed septate filamentous cyanobacterium ("S").
Archaeotrichion contortum Schopf 1968. (Golovenok and Belova 1985)
  Filamentous prokaryote (S).
Beckspringia sp. of Golovenok and Belova 1985. (Golovenok and Belova 1985)
  Septate filamentous cyanobacterium (S).
Biocatenoides ferrata Golovenoc [Golovenok] and Belova 1985. (Golovenok and Belova 1985)
  Septate filamentous prokaryote ("S").
Distichococcus minutus Golovenoc [Golovenok] and Belova 1985. (Golovenok and Belova 1985)
  Planar, colonial coccoid cyanobacterium ("S").
Eomycetopsis lata Golovenoc [Golovenok] and Belova 1985.

Table 22.3 (Continued).

(Golovenok and Belova 1985)
  Tubular sheath of filamentous cyanobacterium ("S").
Eomycetopsis riberiensis Lo 1980. (Golovenok and Belova 1985)
  Tubular sheath of filamentous cyanobacterium ("S").
Eosynechococcus grandis Hofmann 1976. (Golovenok and Belova 1985)
  Solitary, ellipsoidal chroococcacean cyanobacterium (S).
Eucapsomorpha rara Golovenoc [Golovenok] and Belova 1985. (Golovenok and Belova 1985)
  Cuboidal colonial coccoid chroococcacean cyanobacterium ("S").
Glenobotrydion compressus Golovenoc [Golovenok] and Belova 1985. (Golovenok and Belova 1985)
  Colonial coccoid chroococcacean cyanobacterium (S).
Huroniospora rimosa Golovenoc [Golovenok] and Belova 1985. (Golovenok and Belova 1985)
  Sphaeromorph acritarch (S).
Myxococcoides inornata Schopf 1968. (Golovenok and Belova 1985)
  Colonial coccoid cyanobacterium (S).
Shuntaria evidens Golovenoc [Golovenok] and Belova 1985. (Golovenok and Belova 1985)
  Ensheathed septate filamentous cyanobacterium ("S").

Kundelungu Grp.
AF ZMB
(ca. 800 Ma)

(spherical organic microfossil). (Binda 1972)
  Solitary and colonial coccoid cyanobacteria (S).

Little Dal Grp.
NA CAN NW
(ca. 800 Ma)

Archaeotrichion sp. of Hofmann and Aitken 1979. (Hofmann and Aitken 1979)
  Filamentous bacterium ("S").
Chuaria circularis Walcott 1899. (Hofmann and Aitken 1979)
  Sphaeromorph acritarch ("S").
Kildinella spp. of Hofmann and Aitken 1979. (Hofmann and Aitken 1979)
  Sphaeromorph acritarch ("S").
Nucellosphaeridium spp. of Hofmann and Aitken 1979. (Hofmann and Aitken 1979)
  Sphaeromorph acritarch ("S").
Siphonophycus spp. of Hofmann and Aitken 1979. (Hofmann and Aitken 1979)
  Sheaths of filamentous cyanobacteria ("S").
Taeniatum spp. of Hofmann and Aitken 1979. (Hofmann and Aitken 1979)
  Flattened sheath of filamentous cyanobacterium ("S").
Trachysphaeridium spp. of Hofmann and Aitken 1979. (Hofmann and Aitken 1979)
  Sphaeromorph acritarch ("S").

Lopatinskij Fm.
AS SIB
(ca. 800 Ma)

Cyanothrixoides mirabilis Golovenoc [Golovenok] and Belova 1985. (Golovenok and Belova 1985)
  Pseudofilamentous colonial coccoid ?chroococcacean cyanobacterium (S).
Myxococcoides inornata Schopf 1968. (Golovenok and Belova 1985)
  Colonial coccoid cyanobacterium (S).
Myxococcoides minor Schopf 1968. (Golovenok and Belova 1985)
  Colonial coccoid chroococcacean cyanobacterium (S).

Malka Fm.
AS SIB
(ca. 800 Ma)

Protosphaeridium patelliforme Timofeev 1966. (Timofeev 1966)
  Sphaeromorph acritarch ("S").

Nel'kan Fm.
AS SIB
(ca. 800 Ma)

Dictyosphaeridium tungusum Timofeev 1969. (Timofeev 1969)
  Sphaeromorph acritarch ("S").
Kildinella hyperboreica Timofeev 1966. (Timofeev 1969)
  Sphaeromorph acritarch ("S").
Protosphaeridium densum Timofeev 1966. (Timofeev 1969)
  Sphaeromorph acritarch ("S").
Protosphaeridium rigidulum Timofeev 1966. (Timofeev 1969)
  Sphaeromorph acritarch ("S").

Onguren Fm.
AS SIB
(ca. 800 Ma)

Protosphaeridium acis Timofeev 1966. (Timofeev 1966)
  Sphaeromorph acritarch ("S").
Protosphaeridium densum Timofeev 1966. (Timofeev 1966)
  Sphaeromorph acritarch ("S").
Protosphaeridium densum Timofeev 1966. (Timofeev 1969)
  Sphaeromorph acritarch ("S").
Protosphaeridium papyraceum Timofeev 1966. (Timofeev 1966)
  Sphaeromorph acritarch ("S").
Protosphaeridium rigidulum Timofeev 1966. (Timofeev 1969)
  Sphaeromorph acritarch ("S").
Protosphaeridium scabridum Timofeev 1966. (Timofeev 1969)
  Sphaeromorph acritarch ("S").
Trachyoligotriletum incrassatum (Naumova 1949) Timofeev 1959. (Timofeev 1960)
  Sphaeromorph acritarch ("S").

Paun Fm.
EU SUN
(ca. 800 Ma)

Protosphaeridium asaphum Timofeev 1966. (Timofeev 1966)
  Sphaeromorph acritarch ("S").

800 to 700 Ma:
Drill-Hole Kabakovo-62, Bashkiria
EA SUN
(ca. 790 Ma)

Arctacellularia ellipsoidea Hermann [German] 1976. (Yankauskas 1982)
  Pseudofilamentous (uniseriate) colonial coccoid cyanobacterium (?cf. *Glenobotrydium aenigmatis*) (S).
Arctacellularia sp. 2 of Jankauskas [Yankauskas] 1982. (Yankauskas 1982)
  Solitary (disarticulated) trichomic cell of filamentous or pseudofilamentous (colonial) cyanobacterium, or solitary cylindrical unicell incertae sedis (viz., pl. 37, Fig. 1, 3, 8; pl. 38, Fig. 2). Pl. 37, Fig. 14 = tubular sheath of filamentous cyanobacterium; pl. 37, Fig. 7 = *Brevitrichoides bashkiricus* Jankauskas 1980 (S).
Brevitrichoides bashkiricus Jankauskas [Yankauskas] 1980.

(Yankauskas 1980a)
  Tubular sheath (or trichome without preserved lateral cell walls) of filamentous (oscillatoriacean) cyanobacterium (S).
Brevitrichoides bashkiricus Jankauskas [Yankauskas] 1980. (Yankauskas 1982)
  Tubular sheath (or trichome without preserved lateral cell walls) of filamentous (oscillatoriacean) cyanobacteria (S).
Calyptothrix alternata Jankauskas [Yankauskas] 1980. (Yankauskas 1980a)
  Septate filamentous prokaryote (S).
Calyptothrix geminata Jankauskas [Yankauskas] 1980. (Yankauskas 1980a)
  Septate filamentous (oscillatoriacean) cyanobacterium ("S").
Eomycetopsis psilata Maithy and Shukla 1977. (Yankauskas 1982)
  Tubular sheath of filamentous cyanobacterium ("S").
Eomycetopsis rimata Jankauskas [Yankauskas] 1980. (Yankauskas 1982)
  Flattened sheath of filamentous cyanobacterium ("S").
Eomycetopsis rimata Jankauskas [Yankauskas] 1980. (Yankauskas 1980a)
  Flattened tubular sheath of filamentous cyanobacterium ("S").
Eomycetopsis aff. *rugosa* Maithy 1975. (Yankauskas 1982)
  Tubular sheath of filamentous cyanobacterium ("S").
Kildinella hyperboreica Timofeev 1963. (Yankauskas 1982)
  Folded, originally spheroidal, sphaeromorph acritarch ("S").
Kildinella nordia Timofeev 1969. (Yankauskas 1982)
  Sphaeromorph acritarch containing globular, condensed cytoplasmic contents ("S").
Leiothrichoides typicus Hermann [German] 1974. (Yankauskas 1982)
  Tubular sheath of filamentous cyanobacterium ("S").
Oscillatoriopsis sp. 1 of Jankauskas [Yankauskas] 1982. (Yankauskas 1982)
  Septate filament (or crenulated sheath?) of filamentous (oscillatoriacean?) cyanobacterium (S).
Palaeolyngbya minor Schopf and Blacic 1971. (Yankauskas 1982)
  Septate filamentous (oscillatoriacean) cyanobacterium ("S").
Partitiofilum aff. *gongyloides* Schopf and Blacic 1971. (Yankauskas 1982)
  Septate filamentous (oscillatoriacean) cyanobacterium (S).
Partitiofilum? sp. of Jankauskas [Yankauskas] 1982. (Yankauskas 1982)
  Septate filamentous (oscillatoriacean) cyanobacterium ("S").
Plicatidium sp. of Janiauskas [Yankauskas] 1982. (Yankauskas 1982)
  Flattened tubular sheath of filamentous cyanobacterium ("S").
Protosphaeridium flexuosum Timofeev 1966. (Yankauskas 1982)
  Thick-walled sphaeromorph acritarch ("S").
Pterospermopsimorpha capsulata Jankauskas [Yankauskas] 1982. (Yankauskas 1982)
  Possible poorly preserved, diagenetically altered pteromorph acritarch (M S).
Siphonophycus costatus Jankauskas [Yankauskas] 1980. (Yankauskas 1982)
  Tubular sheath of filamentous cyanobacterium ("S").
Siphonophycus costatus Jankauskas [Yankauskas] 1980. (Yankauskas 1980a)
  Tubular sheath of filamentous cyanobacterium ("S").
Symplassosphaeridium undosum Jankauskas [Yankauskas] 1979. (Yankauskas 1982)
  Sphaeromorph with botryoidal surface sculpture (or colony of cells?) ("S").
Tortunema eniseica Hermann [German] 1976. (Yankauskas 1982)
  Flattened, intertwined, degraded, monotrichomic, ensheathed filamentous prokaryote (S).
Tortunema sibirica Hermann [German] 1976. (Yankauskas 1982)
  Tubular sheath of filamentous cyanobacterium ("S").
Trachyhystrichosphaera aimika Hermann [German] 1976. (Yankauskas 1982)
  Large acanthomorph acritarch having short processes ("S M").
Valeria lophostriata (Jankauskas 1979) Jankauskas 1982 (Yankauskas 1982)
  Folded, originally spheroidal, striate sphaeromorph acritarch (S).
Volyniella glomerata Jankauskas [Yankauskas] 1980. (Yankauskas 1980a)
  Coiled tubular sheath of filamentous bacterium (S).

Kangjia [Kangchia] Grp.
AS CHN
(ca. 790 Ma)

Leiopsophosphaera apertus [Shepeleva] 1963. (Xing and Liu 1973)
  Sphaeromorph acritarch ("S").
Leiopsophosphaera minor Schepeleva [Shepeleva] 1963. (Xing and Liu 1973)
  Sphaeromorph acritarch ("S").
Leiopsophosphaera pelucidus Schepeleva [Shepeleva] 1963. (Xing and Liu 1973)
  Sphaeromorph acritarch ("S").
Margominuscula rugosa Naumova 1960. (Xing and Liu 1973)
  Solitary coccoid cyanobacterium (S).
Microconcentrica aff. *induplicata* Liu and Sin [Xing] 1973. (Xing and Liu 1973)
  Sphaeromorph acritarch ("S").
Microconcentrica induplicata Liu and Sin [Xing] 1973. (Xing and Liu 1973)
  Colonial coccoid (chroococcacean?) cyanobacterium [pl. 10, Fig. 7 not interpretable] (S).
Protoleiosphaeridium infriatum Andreeva 1966. (Xing and Liu 1973)
  Sphaeromorph acritarch ("S").
Protoleiosphaeridium pusillum Sin [Xing] 1962. (Xing and Liu 1973)
  Sphaeromorph acritarch ("S").
Protosphaeridium densum Timofeev 1966. (Xing and Liu 1973)
  Sphaeromorph acritarch ("S").
Synsphaeridium conglutinatum Timofeev 1966. (Xing and Liu 1973)
  Colonial, coccoid to ellipsoidal (chroococcacean?) cyanobacterium (S).
Trachysphaeridium cultum (Andreeva 1966) Sin [Xing] 1973. (Xing and Liu 1973)
  Sphaeromorph acritarch ("S").
Trachysphaeridium hyalinum Sin [Xing] and Liu 1973. (Xing and Liu 1973)
  Sphaeromorph acritarch ("S").
Trachysphaeridium incrassatum Sin [Xing] 1973. (Xing and Liu 1973)
  Sphaeromorph acritarch ("S").
Trachysphaeridium aff. *laminaritum* Timofeev 1966. (Xing and Liu 1973)
  Degraded sphaeromorph acritarch ("S").
Trachysphaeridium planum Sin [Xing] 1973. (Xing and Liu 1973)
  Sphaeromorph acritarch ("S").
Trachysphaeridium rugosum Sin [Xing] 1973. (Xing and Liu 1973)
  Sphaeromorph acritarch ("S").
Trachysphaeridium simplex Sin [Xing] 1962. (Xing and Liu 1973)
  Sphaeromorph acritarch ("S").

Table 22.3 (Continued).

Zonosphaeridium minutum Sin [Xing] 1973. (Xing and Liu 1973)
  Sphaeromorph acritarch ("S").

Longmyndian Sgrp., Stretton "Series"
EU GBR
(ca. 790 Ma)

(broad ribbon [without transverse striae]). (Peat 1984)
  Probably degraded tubular sheath of filamentous cyanobacterium (S).
(broad ribbon with transverse striae). (Peat 1984)
  Septate filamentous oscillatoriacean cyanobacterium (S).
(carbonaceous discs). (Peat 1984)
  Sphaeromorph acritarch ("S").
(interwoven nematomorph cryptarchs). (Peat 1984)
  Probably degraded tubular sheaths of filamentous cyanobacteria (S).

Qiaotou [Chiaotou] Fm.
AS CHN
(ca. 790 Ma)

Leiopsophosphaera apertus Schepeleva [Shepeleva] 1963. (Xing and Liu 1973)
  Sphaeromorph acritarch ("S").
Leiopsophosphaera aff. *effusus* Schepeleva [Shepeleva] 1963. (Xing and Liu 1973)
  Sphaeromorph acritarch ("S").
Leiopsophosphaera minor Schepeleva [Shepeleva] 1963. (Xing and Liu 1973)
  Sphaeromorph acritarch ("S").
Leiopsophosphaera pelucidus Schepeleva [Shepeleva] 1963. (Xing and Liu 1973)
  Sphaeromorph acritarch ("S").
Margominuscula rugosa Naumova 1960. (Xing and Liu 1973)
  Solitary coccoid cyanobacterium (S).
Protoleiosphaeridium infriatum Andreeva 1966. (Xing and Liu 1973)
  Sphaeromorph acritarch ("S").
Protosphaeridium densum Timofeev 1966. (Xing and Liu 1973)
  Sphaeromorph acritarch ("S").
Pseudozonosphaera sinica Sin [Xing] and Liu 1973. (Xing and Liu 1973)
  Colonial, coccoid cyanobacterium [pl. 6, Fig. 13 and pl. 12, Fig. 10] (S).
Synsphaeridium conglutinatum Timofeev 1966. (Xing and Liu 1973)
  Colonial, coccoid to ellipsoidal (chroococcacean?) cyanobacterium (S).
Trachysphaeridium aff. *chihrienense* Liu and Sin [Xing] 1973. (Xing and Liu 1973)
  Sphaeromorph acritarch ("S").
Trachysphaeridium cultum (Andreeva 1966) Sin [Xing] 1973. (Xing and Liu 1973)
  Sphaeromorph acritarch ("S").
Trachysphaeridium hyalinum Sin [Xing] and Liu 1973. (Xing and Liu 1973)
  Sphaeromorph acritarch ("S").
Trachysphaeridium incrassatum Sin [Xing] 1973. (Xing and Liu 1973)
  Sphaeromorph acritarch ("S").
Trachysphaeridium planum Sin [Xing] 1973. (Xing and Liu 1973)
  Sphaeromorph acritarch ("S").
Trachysphaeridium simplex Sin [Xing] 1962. (Xing and Liu 1973)
  Sphaeromorph acritarch ("S").
Zonosphaeridium minutum Sin [Xing] 1973. (Xing and Liu 1973)
  Sphaeromorph acritarch ("S").
Zonosphaeridium sp. of Sin [Xing] and Liu 1973. (Xing and Liu 1973)
  Sphaeromorph acritarch ("S").

Xihe Grp., Nanfen Fm.
AS CHN
(ca. 790 Ma)

Chuaria circularis Walcott 1899. (Sun 1987)
  Sphaeromorph acritarch (S).
Leiopsophosphaera apertus Schepeleva [Shepeleva] 1963. (Xing and Liu 1973)
  Sphaeromorph acritarch ("S").
Trachysphaeridium hyalinum Sin [Xing] and Liu 1973. (Xing and Liu 1973)
  Sphaeromorph acritarch ("S").
Trachysphaeridium incrassatum Sin [Xing] 1973. (Xing and Liu 1973)
  Sphaeromorph acritarch ("S").
Trachysphaeridium simplex Sin [Xing] 1962. (Xing and Liu 1973)
  Sphaeromorph acritarch ("S").

Brioverian (Lower and Middle)
EU FRA
(ca. 785 Ma)

(forme A). (Chauvel and Mansuy 1981)
  Colonial coccoid chroococcacean cyanobacteria (S).
(forme B). (Chauvel and Mansuy 1981)
  Colonial coccoid bacterium (S).
(forme C). (Chauvel and Mansuy 1981)
  Colonial coccoid prokaryote (S).

Vadsø Grp., Andersby Fm.
EU NOR
(ca. 785 Ma)

Chuaria circularis Walcott 1899. (Vidal 1981b)
  Sphaeromorph acritarch ("S").
Kildinella hyperboreica Timofeev 1966. (Vidal 1981b)
  Sphaeromorph acritarch ("S").
Kildinella sinica Timofeev 1966. (Vidal 1981b)
  Sphaeromorph acritarch ("S").
Kildinella sp. A of Vidal 1981. (Vidal 1981b)
  Sphaeromorph acritarch ("S").
cf. *Stictosphaeridium* sp. of Vidal 1981. (Vidal 1981b)
  Colonial coccoid cyanobacterium (Fig. 19A–F) (S).
cf. *Stictosphaeridium* sp. of Vidal 1981. (Vidal 1981b)
  Highly degraded sphaeromorph acritarch containing degraded internal remnants (Fig. 11I) (S).

Vadsø Grp., ?Andersby Fm.
EU NOR
(ca. 785 Ma)

Kildinella sp. B of Vidal 1981. (Vidal 1981b)
  Sphaeromorph acritarch ("S").

Vadsø Grp., Golneselv Fm.
EU NOR
(ca. 785 Ma)

Chuaria circularis Walcott 1899. (Vidal 1981b)
  Sphaeromorph acritarch ("S").
Kildinella hyperboreica Timofeev 1966. (Vidal 1981b)
  Sphaeromorph acritarch ("S").

Table 22.3 (Continued).

Kildinella sinica Timofeev 1966. (Vidal 1981b)
   Sphaeromorph acritarch ("S").

"Barrandian flysch facies"
EU CSK
(ca. 775 Ma)

Chabiosphaera bohemica Drábek 1972. (Drabek 1972)
   Colonial coccoid cyanobacterium ("S").

Albinia Fm.
AU AUS NT
(ca. 775 Ma)

Eomycetopsis sp. of Walter and Cloud 1983. (Walter and Cloud 1983)
   Tubular sheath of filamentous cyanobacterium (S).
Myxococcoides? inornata Schopf 1968. (Walter and Cloud 1983)
   Colonial coccoid (chroococcacean) cyanobacterium (S).l

Hunnberg Fm.
EU SJM
(ca. 775 Ma)

(pterospermopsimorphid form A). (Knoll 1984)
   Sphaeromorph acritarch ("S").
(pterospermopsimorphid form B). (Knoll 1984)
   Sphaeromorph acritarch ("S").
(unnamed form A). (Knoll 1984)
   Preservationally altered (scalloped) sphaeromorph acritarch (S).
(unnamed form B). (Knoll 1984)
   Sphaeromorph acritarch ("S").
(unnamed form C). (Knoll 1984)
   Envelope of colonial coccoid prokaryote ("S").
(unnamed form D). (Knoll 1984)
   Sphaeromorph acritarch (S).
(unnamed form E). (Knoll 1984)
   Sphaeromorph acritarch (S).
Chuaria circularis Walcott 1899. (Knoll 1984)
   Sphaeromorph acritarch ("S").
Cymatiosphaeroides kullingii Knoll 1984. (Knoll 1984)
   Spheroidal acritarch bearing short spines, which support outer membrane ("M S").
Eomycetopsis robusta Schopf 1968. (Knoll 1984)
   Tubular sheath of filamentous prokaryote (S).
Glenobotrydion aenigmatis Schopf 1968. (Knoll 1984)
   Colonial coccoid (chroococcacean) cyanobacterium (S).
Glenobotrydion sp. of Knoll 1984. (Knoll 1984)
   Colonial coccoid chroococcacean cyanobacterium (S).
Kildinella hyperboreica Timofeev 1966. (Knoll 1984)
   Sphaeromorph acritarch ("S").
Kildinella cf. *jacutica* Timofeev 1966. (Knoll 1984)
   Sphaeromorph acritarch ("S").
Leptoteichos? sp. of Knoll 1984. (Knoll 1984)
   Colonial coccoid chroococcacean cyanobacterium (S).
Myxococcoides cantabrigiensis Knoll 1982. (Knoll 1984)
   Colonial coccoid chroococcacean cyanobacterium (S).
Myxococcoides inornata Schopf 1968. (Knoll 1984)
   Colonial coccoid chroococcacean cyanobacterium (S).
Myxococcoides sp. A of Knoll 1982. (Knoll 1984)
   Colonial coccoid chroococcacean cyanobacterium (S).
Myxococcoides sp. C of Knoll 1982. (Knoll 1984)
   Colonial coccoid chroococcacean cyanobacterium (S).
Myxococcoides sp. D of Knoll 1984. (Knoll 1984)
   Sphaeromorph acritarch (S).
Oscillatoriopsis media Mendelson and Schopf 1982. (Knoll 1984)
   Septate filamentous (oscillatoriacean) cyanobacterium ("S").
Phanerosphaerops capitaneus Schopf and Blacic 1971. (Knoll 1984)
   Sphaeromorph acritarch ("S").
Protosphaeridium cf. *flexuosum* Timofeev 1966. (Knoll 1984)
   Sphaeromorph acritarch ("S").
Siphonophycus sp. of Knoll 1984. (Knoll 1984)
   Tubular sheath of filamentous cyanobacterium (S).
Stictosphaeridium sinapticuliferum Timofeev 1966. (Knoll 1984)
   Possibly sheath of coccoid prokaryote (S).
cf. *Stictosphaeridium* sp. of Knoll 1984. (Knoll 1984)
   Possibly sheath of coccoid prokaryote (S).
Trachyhystrichosphaera vidalii Knoll 1984. (Knoll 1984)
   Membranous acanthomorph acritarch (S).
Trachysphaeridium levis (Lopukhin 1971) Vidal 1974. (Knoll 1984)
   Sphaeromorph acritarch ("S").
Trachysphaeridium sp. A of Knoll 1984. (Knoll 1984)
   Sphaeromorph acritarch ("S").
Trachysphaeridium sp. B of Knoll 1984. (Knoll 1984)
   Sphaeromorph acritarch ("S").
Trachysphaeridium timofeevii Vidal 1976. (Knoll 1984)
   Sphaeromorph acritarch ("S").
Trematosphaeridium holtedahlii Timofeev 1966. (Knoll 1984)
   Sphaeromorph acritarch ("S").

River Wakefield Subgrp., Blyth Dolomite
AU AUS SA
(ca. 775 Ma)

(spheroidal unicells). (Schopf 1977)
   Colonial coccoid cyanobacterium (S).

Visingsö Grp.
EU SWE
(ca. 775 Ma)

(enigmatic sphaeromorphs). (Vidal 1976b)
   Sphaeromorph acritarchs, and solitary and colonial cyanobacteria (S).
(sphaeromorphs of unknown affinity). (Vidal 1974)
   Sphaeromorph acritarch ("S").
(vase-shaped microfossils). (Knoll and Vidal 1980)
   Microfossils incertae sedis ("S").
Bavlinella faveolata Shepeleva 1962. (Vidal 1976b)
   Colonial coccoid bacterium (S).
Chuaria circularis Walcott 1899. (Vidal 1976b)
   Sphaeromorph acritarch ("S").
Chuaria wimanii Brotzen 1943. (Timofeev 1966)
   Sphaeromorph acritarch ("S").
Favososphaeridium favosum Timofeev 1966. (Vidal 1976b)
   Sphaeromorph acritarch ("S").
Kildinella hyperboreica Timofeev 1966. (Vidal 1976b)
   Sphaeromorph acritarch ("S").
Kildinella magna Timofeev 1969. (Timofeev 1969)
   Sphaeromorph acritarch ("S").
Kildinella cf. *sinica* Timofeev 1966. (Vidal 1976b)
   Sphaeromorph acritarch ("S").
Kildinella sp. of Vidal 1974. (Vidal 1974)
   Colonial coccoid cyanobacterium (S).
Octoedryxium truncatum Rudavskaja [Rudavskaya] 1973. (Vidal 1976b)
   Prismatomorph acritarch ("S").

Table 22.3 (Continued).

Peteinosphaeridium reticulatum Vidal 1976. (Vidal 1976b)
  Membranous herkomorph/acanthomorph acritarch ("S M").
Protosphaeridium acis Timofeev 1966. (Timofeev 1966)
  Sphaeromorph acritarch ("S").
Protosphaeridium densum Timofeev 1966. (Timofeev 1966)
  Sphaeromorph acritarch ("S").
Protosphaeridium cf. *flexuosum* Timofeev 1966. (Vidal 1976b)
  Sphaeromorph acritarch ("S").
Protosphaeridium flexuosum Timofeev 1966. (Timofeev 1966)
  Sphaeromorph acritarch ("S").
Protosphaeridium laccatum Timofeev 1966. (Vidal 1976b)
  Sphaeromorph acritarch ("S").
Protosphaeridium papyraceum Timofeev 1966. (Vidal 1976b)
  Sphaeromorph acritarch ("S").
Protosphaeridium patelliforme Timofeev 1966. (Timofeev 1966)
  Sphaeromorph acritarch ("S").
Protosphaeridium rigidulum Timofeev 1966. (Timofeev 1966)
  Sphaeromorph acritarch ("S").
Protosphaeridium wimanii Timofeev 1960?. (Timofeev 1966)
  Sphaeromorph acritarch ("S").
Pterospermopsimorpha? densicoronata Vidal 1976. (Vidal 1976b)
  Sphaeromorph acritarch ("S").
Stictosphaeridium cf. *sinapticuliferum* Timofeev 1966. (Vidal 1976b)
  Sphaeromorph acritarch ("S").
cf. *Stictosphaeridium* sp. of Vidal 1976. (Vidal 1976b)
  Sphaeromorph acritarch ("S").
Stictosphaeridium verrucatum Vidal 1976. (Vidal 1976b)
  Sphaeromorph acritarch ("S").
Synsphaeridium sp. of Vidal 1976. (Vidal 1976b)
  Colonial coccoid cyanobacterium (S).
Trachysphaeridium apertum Vidal 1976. (Vidal 1976b)
  Sphaeromorph acritarch ("S").
Trachysphaeridium laminaritum Timofeev 1966. (Vidal 1976b)
  Sphaeromorph acritarch ("S").
Trachysphaeridium laufeldii Vidal 1976. (Vidal 1976b)
  Sphaeromorph acritarch ("S M").
Trachysphaeridium levis (Lopukhin 1971) Vidal 1974. (Vidal 1976b)
  Sphaeromorph acritarch ("S").
Trachysphaeridium timofeevii Vidal 1976. (Vidal 1976b)
  Sphaeromorph acritarch (specimens $>40\,\mu m$, Fig. 22D–F) (S).
Trachysphaeridium timofeevii Vidal 1976. (Vidal 1976b)
  Colonial coccoid cyanobacteria (specimens $<25\,\mu m$: Fig. 22G–K), cf. *Glenobotrydion aenigmatis* Schopf 1968 (S).
Trachysphaeridium vetterni Timofeev 1969. (Timofeev 1969)
  Sphaeromorph acritarch ("S").
Trematosphaeridium holtedahlii Timofeev 1966. (Vidal 1976b)
  Sphaeromorph acritarch ("S").

Visingsö Grp., lower sandstone formation
EU SWE
(ca. 775 Ma)

Chuaria circularis Walcott 1899. (Vidal 1974)
  Sphaeromorph acritarch ("S").
Kildinella hyperboreica Timofeev 1966. (Vidal 1974)
  Sphaeromorph acritarch (S).
Kildinella sinica Timofeev 1966. (Vidal 1974)
  Sphaeromorph acritarch ("M").
Kildinella cf. *vesljanica* Timofeev 1963. (Vidal 1974)
  Sphaeromorph acritarch ("S").
Protosphaeridium laccatum Timofeev 1966. (Vidal 1974)
  Sphaeromorph acritarch ("S").

Trachysphaeridium laminaritum Timofeev 1966. (Vidal 1974)
  Sphaeromorph acritarch ("S").
Trachysphaeridium levis (Lopukhin 1971) Vidal 1974. (Vidal 1974)
  Sphaeromorph acritarch ("S").
Trachysphaeridium sp. of Vidal 1974. (Vidal 1974)
  Sphaeromorph acritarch ("S").
Trachysphaeridium sp. A of Vidal 1974. (Vidal 1974)
  Sphaeromorph acritarch ("S").

Skillogalee Dolomite
AU AUS SA
(ca. 770 Ma)

"*Polybessurus*" Schopf 1977. (Schopf 1977)
  Colonial mucilage-secreting column-forming cyanobacterium (Chamaesiphonales?) (S).
(ascus-like microfossil). (Schopf and Barghoorn 1969)
  Probably colonial, ellipsoidal, relatively large-celled cyanobacterium (S).
(cellular filament). (Schopf 1977)
  Septate filamentous eukaryotic alga (S). [See Chapter 24, Pl. 40, A_1–A_2.]
(cellular filaments). (Schopf and Fairchild 1973)
  Diverse septate filamentous prokaryotes ("S"). [See Chapter 24, Pl. 40, B.]
(colonial unicells: rounded aggregations). (Schopf and Fairchild 1973)
  Colonial ellipsoidal cyanobacterium, cf. *Myxosarcina* ("S"). [See Chapter 24, Pl. 39, C_1–C_2.]
(colonial unicells: tabular groupings). (Schopf and Fairchild 1973)
  Colonial coccoid cyanobacterium, cf. *Merismopedia* ("S"). [See Chapter 24, Pl. 41, A.]
(nonseptate filaments). (Schopf and Fairchild 1973)
  Tubular sheaths of filamentous (oscillatoriacean?) cyanobacteria ("S").
(solitary unicells). (Schopf and Fairchild 1973)
  Diverse solitary coccoid (chroococcacean?) cyanobacteria and sphaeromorph acritarchs (S).
(unicells containing granular bodies). (Schopf and Fairchild 1973)
  Diverse solitary or colonial coccoid (chroococcacean?, entophysalidacean?) cyanobacteria (S).
Archaeonema longicellulare Schopf 1968. (Schopf and Barghoorn 1969)
  Septate filamentous prokaryote (S).
Myxococcoides Schopf 1968. (Schopf and Barghoorn 1969)
  Solitary or colonial coccoid (chroococcacean?) cyanobacterium ("S").
Myxococcoides muricata Schopf and Barghoorn 1969. (Schopf and Barghoorn 1969)
  Sphaeromorph acritarch (S).
Palaeopleurocapsa wopfneri Knoll, Barghoorn and Golubić 1975. (Knoll et al. 1975)
  Colonial pleurocapsacean cyanobacterium ("M").
Palaeosiphonella Schopf 1977. (Schopf 1977)
  Possibly tubular sheath of false-branching filamentous cyanobacterium (S). [See Chapter 24, Pl. 39, A_1–A_3.]

"greywacke" of Bohemia
EU CSK
(ca. 760 Ma)

(filamentous alga-like remains ("form S"]). (Konzalová 1972)
  Tubular sheaths of filamentous cyanobacteria ("S").

Table 22.3 (*Continued*).

(spherical bodies gathered in colonies). (Konzalová 1972)
  Solitary and colonial coccoid cyanobacteria (S).

"Late Precambrian" of the Azov region
EU SUN
(ca. 750 Ma)

Gloeocapsomorpha makrocysta Eisenack 1960. (Timofeev 1966)
  Colonial coccoid chroococcacean cyanobacterium ("S").
Protosphaeridium acis Timofeev 1966. (Timofeev 1966)
  Sphaeromorph acritarch ("S").
Protosphaeridium asaphum Timofeev 1966. (Timofeev 1966)
  Sphaeromorph acritarch ("S").
Protosphaeridium densum Timofeev 1966. (Timofeev 1966)
  Sphaeromorph acritarch ("S").
Protosphaeridium patelliforme Timofeev 1966. (Timofeev 1966)
  Sphaeromorph acritarch ("S").
Protosphaeridium sp. of Timofeev 1966. (Timofeev 1966)
  Sphaeromorph acritarch ("S").
Protosphaeridium tuberculiferum Timofeev 1966. (Timofeev 1966)
  Sphaeromorph acritarch ("S").

"Late Precambrian" of the Holy Cross Mountains
EU POL
(ca. 750 Ma)

Gloeocapsomorpha makrocysta Eisenack 1960. (Timofeev 1966)
  Colonial coccoid chroococcacean cyanobacterium ("S").
Protosphaeridium densum Timofeev 1966. (Timofeev 1966)
  Sphaeromorph acritarch ("S").
Protosphaeridium gibberosum Timofeev 1966. (Timofeev 1966)
  Sphaeromorph acritarch ("S").
Protosphaeridium tuberculiferum Timofeev 1966. (Timofeev 1966)
  Sphaeromorph acritarch ("S").
Pterospermopsimorpha sp. of Timofeev 1966. (Timofeev 1966)
  Sphaeromorph acritarch containing coalesced cytoplasmic remnants (S M).
Stictosphaeridium implexum Timofeev 1966. (Timofeev 1966)
  Sphaeromorph acritarch ("S").

"Late Precambrian" of Dobruja
EU ROM
(ca. 750 Ma)

Gloeocapsomorpha prisca Zalessky (Zalesskij] 1917. (Timofeev 1966)
  Colonial coccoid chroococcacean? cyanobacterium ("S").
Kildinella hyperboreica Timofeev 1966. (Timofeev 1966)
  Sphaeromorph acritarch ("S").
Lopholigotriletum crispum Timofeev 1959. (Timofeev 1966)
  Sphaeromorph acritarch ("S").
Stictosphaeridium sinapticuliferum Timofeev 1966. (Timofeev 1966)
  Sphaeromorph acritarch ("S").

"Late Precambrian" of Timan
EU SUN
(ca. 750 Ma)

Gloeocapsomorpha prisca Zalessky [Zalesskij] 1917. (Timofeev 1966)
  Colonial coccoid chroococcacean? cyanobacterium ("S").
Kildinella hyperboreica Timofeev 1966. (Timofeev 1966)
  Sphaeromorph acritarch ("S").
Kildinella sp. of Timofeev 1966. (Timofeev 1966)
  Sphaeromorph acritarch ("S").
Lopholigotriletum? *spathaeforme* Timofeev 1959. (Timofeev 1966)
  Sphaeromorph acritarch ("S").
Protosphaeridium densum Timofeev 1966. (Timofeev 1966)
  Sphaeromorph acritarch ("S").
Protosphaeridium laccatum Timofeev 1966. (Timofeev 1966)
  Sphaeromorph acritarch ("S").
Stictosphaeridium implexum Timofeev 1966. (Timofeev 1966)
  Sphaeromorph acritarch ("S").
Stictosphaeridium sinapticuliferum Timofeev 1966. (Timofeev 1966)
  Sphaeromorph acritarch ("S").
Tasmanites sp. of Timofeev 1966. (Timofeev 1966)
  Sphaeromorph acritarch ("S").
Turuchanica alara Rudavskaja [Rudavskaya] 1964?. (Timofeev 1966)
  Compressed sphaeromorph? acritarch (S).

"Late Precambrian" of Carpathia
EU SUN
(ca. 750 Ma)

Gloeocapsomorpha prisca Zalessky [Zalesskij] 1917. (Timofeev 1966)
  Colonial coccoid chroococcacean? cyanobacterium ("S").
Protosphaeridium densum Timofeev 1966. (Timofeev 1966)
  Sphaeromorph acritarch ("S").
Stictosphaeridium implexum Timofeev 1966. (Timofeev 1966)
  Sphaeromorph acritarch ("S").
Tasmanites sp. of Timofeev 1966. (Timofeev 1966)
  Sphaeromorph acritarch ("S").
Trematosphaeridium holtedahlii Timofeev 1966. (Timofeev 1966)
  Sphaeromorph acritarch ("S").
Turuchanica alara Rudavskaja [Rudavskaya] 1964?. (Timofeev 1966)
  Compressed sphaeromorph? acritarch (S)

"Late Precambrian" of the Rybachij Peninsula
EU SUN
(ca. 750 Ma)

Protosphaeridium densum Timofeev 1966. (Timofeev 1966)
  Sphaeromorph acritarch ("S").
Protosphaeridium tuberculiferum Timofeev 1966. (Timofeev 1966)
  Sphaeromorph acritarch ("S").

"Late Precambrian" of Trans-Carpathia
EU SUN
(ca. 750 Ma)

Protosphaeridium laccatum Timofeev 1966. (Timofeev 1966)
  Sphaeromorph acritarch ("S").

Akademikerbreen Grp., Draken Conglomerate
EU SJM
(ca. 750 Ma)

(Unnamed Form A). (Knoll 1982b)
  Incertae sedis, possibly distorted tubular sheath of filamentous prokaryote (S).
(Unnamed Form B). (Knoll 1982b)
  Poorly preserved, sheath-encompassed, colonial coccoid (chroococcacean?) cyanobacterium (S).
(Unnamed Form C). (Knoll 1982b)
  Sphaeromorph acritarch ("S").
(Unnamed Form D). (Knoll 1982b)
  Sphaeromorph acritarch containing degraded internal elmnants (S).
(Unnamed Form E). (Knoll 1982b)
  Solitary coccoid (chroococcacean) cyanobacterium (S).
(filamentous microfossils). (Swett and Knoll 1985)
  Tubular sheaths of filamentous cyanobacteria ("S").
(vase-shaped protistan microfossil). (Swett and Knoll 1985)
  Vase-shaped acritarch (S).

Table 22.3 (*Continued*).

Eomycetopsis robusta Schopf 1968. (Knoll 1982b)
 Tubular sheath of filamentous prokaryote (S).
Eosynechococcus brevis Knoll 1982. (Knoll 1982b)
 Solitary or paired, ellipsoidal bacterium (S).
Eosynechococcus depressus Knoll 1982. (Knoll 1982b)
 Solitary ellipsoidal cyanobacterium ("S").
Eosynechococcus medius Hofmann 1976. (Knoll 1982b)
 Solitary or paired, ellipsoidal chroococcacean cyanobacterium ("S").
Eosynechococcus sp. of Knoll 1982. (Knoll 1982b)
 Solitary or paired, ellipsoidal chroococcacean cyanobacterium ("S").
Gloeodiniopsis gregaria Knoll and Golubić 1979. (Knoll 1982b)
 Colonial coccoid chroococcacean cyanobacterium ("S").
Gloeodiniopsis mikros Knoll 1982. (Knoll 1982b)
 Solitary or colonial coccoid chroococcacean cyanobacterium ("S").
cf. *Gloeodiniopsis* sp. of Knoll 1982. (Knoll 1982b)
 Solitary or paired coccoid chroococcacean cyanobacterium ("S").
Myxococcoides cantabrigiensis Knoll 1982. (Knoll 1982b)
 Solitary coccoid chroococcacean cyanobacterium (S).
Myxococcoides minor Schopf 1968. (Knoll 1982b)
 Colonial coccoid chroococcacean cyanobacterium (S).
Myxococcoides ovata Knoll 1982. (Knoll 1982b)
 Colonial coccoid chroococcacean cyanobacterium (S).
Myxococcoides sp. A of Knoll 1982. (Knoll 1982b)
 Colonial coccoid chroococcacean cyanobacterium (S).
Myxococcoides sp. B of Knoll 1982. (Knoll 1982b)
 Colonial coccoid chroococcacean cyanobacterium (S).
Myxococcoides sp. C of Knoll 1982. (Knoll 1982b)
 Sphaeromorph acritarch (S).
Salome svalbardensis Knoll 1982. (Knoll 1982b)
 Ensheathed septate (oscillatoriacean) cyanobacterium ("S").
Siphonophycus inornatum Y. Zhang 1981. (Knoll 1982b)
 Tubular sheath of filamentous cyanobacterium ("S").
Sphaerophycus medium Horodyski and Donaldson 1982. (Knoll 1982b)
 Paired coccoid chroococcacean cyanobacterium ("S").
Sphaerophycus parvum Schopf 1968. (Knoll 1982b)
 Solitary or paired coccoid prokaryote (S).
Sphaerophycus wilsonii Knoll 1982. (Knoll 1982b)
 Colonial coccoid chroococcacean cyanobacterium ("S").
Synodophycus euthemos Knoll 1982. (Knoll 1982b)
 Rosette-shaped colonial coccoid cyanobacterium (S)
Tenuofilum septatum Schopf 1968. (Knoll 1982b)
 Tubular sheath of filamentous bacterium (S).
Tetraphycus diminutivus D. Oehler 1978. (Knoll 1982b)
 Colonial coccoid bacterium (S).

"Ancient Series" of Caucasia
EA SUN
(ca. 750 Ma)

Gloeocapsomorpha prisca Zalessky [Zalesskij] 1917?. (Timofeev 1966)
 Colonial coccoid chroococcacean? cyanobacterium ("S").
Protosphaeridium densum Timofeev 1966. (Timofeev 1966)
 Sphaeromorph acritarch ("S").
Protosphaeridium parvulum Timofeev 1966. (Timofeev 1966)
 Sphaeromorph acritarch ("S").
Protosphaeridium pusillum Timofeev 1966. (Timofeev 1966)
 Sphaeromorph acritarch ("S").

Chapoma Fm.
EU SUN
(ca. 750 Ma)

Fasososphaeridium sp. of Timofeev 1969. (Timofeev 1969)
 Colonial coccoid cyanobacterium (S).
Kildinella exsculpta Timofeev 1969. (Timofeev 1969)
 Sphaeromorph acritarch ("S").
Kildinella hyperboreica Timofeev 1966. (Timofeev 1966)
 Sphaeromorph acritarch ("S").
Kildinella hyperboreica Timofeev 1966. (Timofeev 1969)
 Sphaeromorph acritarch ("S").
Kildinella jacutica Timofeev 1966. (Timofeev 1966)
 Sphaeromorph acritarch ("S").
Kildinella jacutica Timofeev 1966. (Timofeev 1969)
 Sphaeromorph acritarch ("S").
Kildinella sinica Timofeev 1966. (Timofeev 1966)
 Sphaeromorph acritarch ("S").
Kildinella sinica Timofeev 1966. (Timofeev 1969)
 Sphaeromorph acritarch ("S").
Kildinella sp. of Timofeev 1969. (Timofeev 1969)
 Colonial coccoid cyanobacterium (S).
Kildinella timanica Timofeev 1969. (Timofeev 1969)
 Sphaeromorph acritarch ("S").
Kildinella tschapomica Timofeev 1966. (Timofeev 1966)
 Sphaeromorph acritarch ("S").
Kildinella tschapomica Timofeev 1966. (Timofeev 1969)
 Sphaeromorph acritarch ("S").
Leiosphaeridia sp. of Timofeev 1969. (Timofeev 1969)
 Sphaeromorph acritarch ("S").
Protosphaeridium densum Timofeev 1966. (Timofeev 1969)
 Sphaeromorph acritarch ("S").
Stictosphaeridium implexum Timofeev 1966. (Timofeev 1969)
 Sphaeromorph acritarch ("S").
Stictosphaeridium implexum (Timofeev 1966)
 Sphaeromorph acritarch ("S").
Stictosphaeridium sinapticuliferum Timofeev 1966. (Timofeev 1969)
 Sphaeromorph acritarch ("S").
Stictosphaeridium sinapticuliferum Timofeev 1966. (Timofeev 1966)
 Sphaeromorph acritarch ("S").
Trachysphaeridium laminaritum Timofeev 1966. (Timofeev 1969)
 Sphaeromorph acritarch ("S").

Chitkanda Fm.
AS SIB
(ca. 750 Ma)

Protosphaeridium discum Timofeev 19??. (Timofeev et al. 1976)
 Sphaeromorph acritarch ("S").
Stictosphaeridium implexum Timofeev 1966. (Timofeev et al. 1976)
 Sphaeromorph acritarch ("S").

Dashka Fm.
AS SIB
(ca. 750 Ma)

Calyptothrix obsoletus Mikhailova [Mikhajlova] 1986. (Mikhajlova 1986)
 Septate filamentous (oscillatoriacean) cyanobacterium ("S"). [See Chapter 24, Pl. 42, F.]
Cephalophytarion piliformis Mikhailova [Mikhajlova] 1986.

(Mikhajlova 1986)
  Septate filamentous (oscillatoriacean) cyanobacterium ("S"). [See Chapter 24, Pl. 43, A.]
Germinosphaera bispinosa Mikhailova [Mikhajlova] 1986. (Mikhajlova 1986)
  Possibly reproductive body (sporangium?) with attached vegetative filament (S). [See Chapter 24, Pl. 42, E.]
Germinosphaera unispinosa Mikhailova [Mikhajlova] 1986. (Mikhajlova 1986)
  Degraded sphaeromorph acritarch or reproductive body (sporangium?) with attached vegetative filament (S). [See Chapter 24, Pl. 42, I.]
Nucellosphaeridium spumosum Mikhailova [Mikhajlova] 1986. (Mikhajlova 1986)
  Sphaeromorph acritarch ("S"). [See Chapter 24, Pl. 42, H.]
Palaeoaphanizomenon scabratus Mikhailova [Mikhajlova] 1986. (Mikhajlova 1986)
  Septate filamentous (oscillatoriacean?) cyanobacterium ("S"). [See Chapter 24, Pl. 42, G.]
Polysphaeroides nuclearis Mikhailova [Mikhajlova] 1986. (Mikhajlova 1986)
  Colonial coccoid (chroococcacean?) cyanobacterium (S). [See Chapter 24, Pl. 43, B.]

Ehjno [Eino] Fm.
EU SUN
(ca. 750 Ma)

Kildinella hyperboreica Timofeev 1966. (Timofeev 1969)
  Sphaeromorph acritarch ("S").
Kildinella sinica Timofeev 1966. (Timofeev 1969)
  Sphaeromorph acritarch ("S").
Kildinella sp. of Timofeev 1969. (Timofeev 1969)
  Colonial coccoid cyanobacterium (S).
Stictosphaeridium pectinale Timofeev 1966. (Timofeev 1969)
  Sphaeromorph acritarch ("S").
Stictosphaeridium sinapticuliferum Timofeev 1966. (Timofeev 1969)
  Sphaeromorph acritarch ("S").
Trematosphaeridium holtedahlii Timofeev 1966. (Timofeev 1969)
  Sphaeromorph acritarch ("S").

Eleonore Bay Grp.
NA GRL
(ca. 750 Ma)

("chitinozoan-like" microfossil). (Vidal 1979)
  Microfossil incertae sedis (S).
Chuaria circularis Walcott 1899. (Vidal 1979)
  Sphaeromorph acritarch ("S").
Kildinella hyperboreica Timofeev 1966. (Vidal 1979)
  Sphaeromorph acritarch ("S").
Kildinella sinica Timofeev 1966. (Vidal 1979)
  Sphaeromorph acritarch ("S").
Kildinella sp. of Vidal 1979. (Vidal 1979)
  Sphaeromorph acritarch ("S").
Stictosphaeridium cf. *sinapticuliferum* Timofeev 1966. (Vidal 1976a)
  Sphaeromorph acritarch ("S").
cf. *Stictosphaeridium* sp. of Vidal 1976. (Vidal 1979)
  Colonial coccoid cyanobacterium (S).
cf. *Stictosphaeridium* sp. of Vidal 1976. (Vidal 1976a)
  Sphaeromorph acritarch ("S").

Trachysphaeridium laufeldii Vidal 1976. (Vidal 1976a)
  Short-spined acanthomorph acritarch ("S M").

Eleonore Bay Grp., Limestone/Dolomite "Series"
NA GRL
(ca. 750 Ma)

(microbial endoliths [cf. *Hyella gigas*]). (Knoll et al. 1986)
  Pleurocapsalean (Hyellaceae) cyanobacterium ("S").
Cunicularius halleri Green, Knoll and Swett 1988. (Green et al. 1988)
  Tubular sheath of endolithic filamentous prokaryote (S).
Eoentophysalis dismallakesensis Horodyski and Donaldson 1980. (Green et al. 1988)
  Colonial ellipsoidal entophysalidacean cyanobacterium ("S").
Eohyella dichotoma Green, Knoll and Swett 1988. (Green et al. 1988)
  Pseudofilamentous colonial endolithic coccoid pleurocapsalean cyanobacterium ("S").
Eohyella endoatracta Green, Knoll and Swett 1988. (Green et al. 1988)
  Pseudofilamentous colonial endolithic coccoid pleurocapsalean cyanobacterium ("S").
Eohyella rectoclada Green, Knoll and Swett 1988. (Green et al. 1988)
  Pseudofilamentous colonial endolithic coccoid pleurocapsalean cyanobacterium ("S").
Eomycetopsis robusta Schopf 1968. (Green et al. 1988)
  Tubular sheath of filamentous cyanobacterium ("S").
Glenobotrydion sp. of Green, Knoll and Swett 1988. (Green et al. 1988)
  Pseudofilamentous colonial coccoid chroococcacean? cyanobacterium (S).
Graviglomus incrustus Green, Knoll and Swett 1988. (Green et al. 1988)
  Colonial endolithic ellipsoidal cyanobacterium (S).
Melanocyrillium spp. of Green, Knoll and Swett 1988. (Green et al. 1988)
  Microfossils incertae sedis ("S").
Myxococcoides cantabrigiensis Knoll 1982. (Green et al. 1988)
  Colonial coccoid chroococcacean cyanobacterium (S).
Myxococcoides sp. C of Knoll 1982. (Green et al. 1988)
  Colonial coccoid chroococcacean cyanobacterium (S).
Palaeopleurocapsa sp. of Green, Knoll and Swett 1988. (Green et al. 1988)
  Colonial ellipsoidal pleurocapsalean cyanobacterium ("S").
Parenchymodiscus endolithicus Green, Knoll and Swett 1988. (Green et al. 1988)
  Colonial ellipsoidal to coccoid endolithic ?chroococcacean cyanobacterium (S).
Perulagranum obovatum Green, Knoll and Swett 1988. (Green et al. 1988)
  Colonial coccoid chroococcalean or pleurocapsalean cyanobacterium (S).
Perulagranum? sp. of Green, Knoll and Swett 1988. (Green et al. 1988)
  Colonial coccoid ?chroococcalean cyanobacterium (S).
Polybessurus bipartitus Fairchild ex Green, Knoll, Golubić and Swett 1987. (Green et al. 1987)
  Dermocarpacean cyanobacterium ("S").
Thylacocausticus globorum Green, Knoll and Swett 1988. (Green et al. 1988)
  Colonial coccoid endolithic chroococcalean (?dermocarpacean) cyanobacterium (S).

Kan'onsk Fm.
AS SIB
(ca. 750 Ma)

Halythrix leningradica Schenfil [Shenfil'] 1983. (Shenfil', 1983)
  Septate filamentous oscillatoriacean cyanobacterium ("S").
Oscillatoriopsis taimirica Schenfil [Schenfil'] 1983. (Shenfil', 1983)
  Septate filamentous oscillatoriacean cyanobacterium ("S").

Kulindinsk Fm.
AS SIB
(ca. 750 Ma)

Favososphaeridium bothnicum Timofeev 1966. (Timofeev 1966)
  Colonial coccoid chroococcacean cyanobacterium (S).
Protosphaeridium sp. of Timofeev 1966. (Timofeev 1966)
  Sphaeromorph acritarch ("S").

Ryssö Fm
EU SJM
(ca. 750 Ma)

(multilamellated sheath). (Knoll and Calder 1983)
  Tubular multilamellated sheath of filamentous cyanobacterium ("S").
(vase-shaped microfossils). (Knoll and Calder 1983)
  Vase-shaped acritarchs (S).
Chuaria circularis Walcott 1899. (Knoll and Calder 1983)
  Sphaeromorph acritarch ("S").
Coniunctiophycus sp. of Knoll and Calder 1983. (Knoll and Calder 1983)
  Colonial ellipsoidal bacterium (S).
Eomycetopsis robusta Schopf 1968. (Knoll and Calder 1983)
  Tubular sheath of filamentous prokaryote (S).
Glenobotrydion aenigmatis Schopf 1968. (Knoll and Calder 1983)
  Colonial coccoid (chroococcacean) cyanobacterium (S).
Kildinella hyperboreica Timofeev 1966. (Knoll and Calder 1983)
  Sphaeromorph acritarch ("S").
Kildinella sinica Timofeev 1966. (Knoll and Calder 1983)
  Sphaeromorph acritarch ("S").
Myxococcoides spp. of Knoll and Calder 1983. (Knoll and Calder 1983)
  Diverse solitary and colonial coccoid (chroococcacean?) cyanobacteria (S).
Phanerosphaerops capitaneus Schopf and Blacic 1971. (Knoll and Calder 1983)
  Sphaeromorph acritarch ("S").
Pterospermopsimorpha sp. of Knoll and Calder 1983. (Knoll and Calder 1983)
  Double-walled sphaeromorph (or disphaeromorph?) acritarch ("S").
Scissilisphaera regularis Knoll and Calder 1983. (Knoll and Calder 1983)
  Colonial coccoid pleurocapsalean cyanobacterium, cf. *Chroococcidiopsis* ("S").
Siphonophycus kestron Schopf 1968. (Knoll and Calder 1983)
  Tubular sheath of filamentous cyanobacterium (S).
cf. *Stictosphaeridium* sp. of Knoll and Calder 1983. (Knoll and Calder 1983)
  Sphaeromorph acritarch ("S").
Tenuofilum septatum Schopf 1968. (Knoll and Calder 1983)
  Tubular sheath of filamentous bacterium (S).
Trachyhystrichosphaera vidalii Knoll and Calder 1983. (Knoll and Calder 1983)
  Acanthomorph acritarch ("S").
Trachysphaeridium laufeldii Vidal 1976. (Knoll and Calder 1983)
  Short-spined acanthomorph acritarch ("S M").
Trachysphaeridium levis (Lopukhin 1971) Vidal 1974. (Knoll and Calder 1983)
  Sphaeromorph acritarch ("S").
Trachysphaeridium sp. A of Knoll 1984 [as "Knoll 1983"]. (Knoll and Calder 1983)
  Sphaeromorph acritarch ("S").
Trachysphaeridium sp. B of Knoll 1984 [as "Knoll 1983"]. (Knoll and Calder 1983)
  Sphaeromorph acritarch ("S").

Serebryanka Fm.
AS SIB
(ca. 750 Ma)

Protosphaeridium acis Timofeev 1966. (Timofeev 1966)
  Sphaeromorph acritarch ("S").
Protosphaeridium papyraceum Timofeev 1966. (Timofeev 1966)
  Sphaeromorph acritarch ("S").

Terskij Fm.
EU SUN
(ca. 750 Ma)

Gloeocapsomorpha sp. of Timofeev 1966. (Timofeev 1966)
  Colonial coccoid chroococcacean? cyanobacterium ("S").
Kildinella hyperboreica Timofeev 1966. (Timofeev 1969)
  Sphaeromorph acritarch ("S").
Kildinella sinica Timofeev 1966. (Timofeev 1966)
  Sphaeromorph acritarch ("S").
Protosphaeridium acis Timofeev 1966. (Timofeev 1969)
  Sphaeromorph acritarch ("S").
Protosphaeridium densum Timofeev 1966. (Timofeev 1966)
  Sphaeromorph acritarch ("S").
Protosphaeridium rigidulum Timofeev 1966. (Timofeev 1969)
  Sphaeromorph acritarch ("S").
Protosphaeridium scabridum Timofeev 1966. (Timofeev 1969)
  Sphaeromorph acritarch ("S").
Protosphaeridium torulosum Timofeev 1966. (Timofeev 1966)
  Sphaeromorph acritarch ("S").

Veslyana Fm.
EU SUN
(ca. 750 Ma)

(algal trichome [with disc-shaped cells]). (Timofeev 1969)
  Septate filamentous oscillatoriacean cyanobacterium (S).
Favososphaeridium favosum Timofeev 1966. (Timofeev 1969)
  Sphaeromorph acritarch ("S").
Kildinella hyperboreica Timofeev 1966. (Timofeev 1969)
  Sphaeromorph acritarch ("S").
Kildinella sinica Timofeev 1966. (Timofeev 1969)
  Sphaeromorph acritarch ("S").
Kildinella sp. of Timofeev 1969. (Timofeev 1969)
  Colonial coccoid cyanobacterium (S).
Kildinella timanica Timofeev 1969. (Timofeev 1969)
  Sphaeromorph acritarch ("S").
Kildinella tschapomica Timofeev 1966. (Timofeev 1969)
  Sphaeromorph acritarch ("S").
Kildinella vesljanica Timofeev 1969. (Timofeev 1969)
  Sphaeromorph acritarch ("S").
Protosphaeridium acis Timofeev 1966. (Timofeev 1969)

Table 22.3 (Continued).

Sphaeromorph acritarch ("S").
Protosphaeridium densum Timofeev 1966. (Timofeev 1969)
Sphaeromorph acritarch ("S").
Protosphaeridium vermium Timofeev 1969. (Timofeev 1969)
Sphaeromorph acritarch ("S").
Pterospermopsimorpha insolita Timofeev 1969. (Timofeev 1969)
Pteromorph acritarch ("S").
Pterospermopsimorpha sp. of Timofeev 1969. (Timofeev 1969)
Degraded sphaeromorph containing coalesced cytoplasmic remnants (M S).
Stictosphaeridium pectinale Timofeev 1966. (Timofeev 1969)
Sphaeromorph acritarch ("S").
Synsphaeridium conglutinatum Timofeev 1966. (Timofeev 1969)
Colonial coccoid cyanobacterium (S).
Turuchanica ternata Timofeev 1966. (Timofeev 1969)
Flattened sphaeromorph? acritarch (S).

Zubovsk Fm.
EU SUN
(ca. 750 Ma)

Kildinella hyperboreica Timofeev 1966. (Timofeev 1969)
Sphaeromorph acritarch ("S").
Kildinella sinica Timofeev 1966. (Timofeev 1969)
Sphaeromorph acritarch ("S").

Brioverian
EU FRA
(ca. 740 Ma)

(spherical cell-aggregates). (Mansuy and Vidal 1983)
Colonial coccoid bacteria (S).
(sporomorphes). (Roblot 1964a)
Sphaeromorph acritarch ("S").
(type I). (Chauvel and Schopf 1978)
Colonial and solitary cyanobacteria (S). [See Chapter 24, Pl. 44, L. N–P.]
(type II). (Chauvel and Schopf 1978)
Colonial coccoid prokaryote (S). [See Chapter 24, Pl. 44, K, M.]
(type III). (Chauvel and Schopf 1978)
Colonial coccoid cyanobacteria (S).
(type IV). (Chauvel and Schopf 1978)
Colonial coccoid (chroococcacean?) cyanobacterium ("S"). [See Chapter 24, Pl. 44, F–H, J_1–J_2.]
Eomicrhystridium aremoricanum Deflandre 1968. (Deflandre 1968)
Distorted, degraded solitary coccoid cyanobacterium (S).
Palaeocryptidium cayeuxii Deflandre 1955. (Deflandre 1965)
Solitary, coccoid cyanobacterium (S).

Brioverian "Schistes X"
EU FRA
(ca. 740 Ma)

(formes elliptiques). (Fournier-Vinas and Debat 1970)
Poorly preserved sphaeromorph acritarchs ("S").
(formes sphériques à subsphériques). (Fournier-Vinas and Debat 1970)
Poorly preserved sphaeromorph acritarchs ("S").

Brioverian (Middle)
EU FRA
(ca. 740 Ma)

(sporomorphes discoïdales). (Roblot 1964b)
Sphaeromorph acritarch ("S").

Min'yar Fm.
EU SUN
(ca. 740 Ma)

(colonial unicells). (Schopf et al. 1977)
Colonial coccoid (chroococcacean) cyanobacteria ("S").
(intermediate-diameter oscillatoriacean sheaths). (Nyberg and Schopf 1984)
Tubular sheath of filamentous oscillatoriacean cyanobacterium ("S"). [See Chapter 24, Pl. 46, F, G.]
(narrow tubular sheaths...). (Nyberg and Schopf 1984)
Tubular sheaths of filamentous prokaryote, cf. *Eomycetopsis* (S).
(solitary unicells). (Schopf et al. 1977)
Lamellated, sheath-enclosed solitary coccoid (chroococcacean) cyanobacterium, cf. *Gloeocapsa* ("S").
(tubular, unbranched, organic filaments). (Schopf et al. 1977)
Tubular sheaths of filamentous cyanobacteria ("S").
(undifferentiated unicells). (Nyberg and Schopf 1984)
Sphaeromorph acritarch (S).
Biocatenoides? sp. of Nyberg and Schopf 1984. (Nyberg and Schopf 1984)
Septate filamentous bacterium (S).
Caudiculophycus? sp. of Nyberg and Schopf 1984. (Nyberg and Schopf 1984)
Septate filamentous prokaryote (S).
Entosphaeroides? sp. of Nyberg and Schopf 1984. (Nyberg and Schopf 1984)
Tubular sheath of filamentous cyanobacterium containing degraded cellular remnants ("S").
Eoaphanocapsa oparinii Nyberg and Schopf 1984. (Nyberg and Schopf 1984)
Colonial, ellipsoidal chroococcacean cyanobacterium ("S"). [See Chapter 24, Pl. 47, E_1–E_2, F?.]
Eomycetopsis robusta Schopf 1968. (Nyberg and Schopf 1984)
Tubular sheath of filamentous prokaryote (S). [See Chapter 24, Pl. 46, A, B.]
Eosynechococcus amadeus Knoll and Golubić 1979. (Nyberg and Schopf 1984)
Colonial ellipsoidal bacterium (S).
Glenobotrydion majorinum Schopf and Blacic 1971. (Nyberg and Schopf 1984)
Probably coccoid cyanobacterium with degraded protoplasmic remnants, but possibly eukaryotic (chlorphyte or rhodophyte) alga (S). [See Chapter 24, Pl. 47, A–D, G.]
Gloeodiniopsis grandis Sergeev and Krylov 1986. (Sergeev and Krylov 1986)
Colonial Gloeocapsa- and *Chroococcus*-like chroococcacean cyanobacterium ("S"). [See Chapter 24, Pl. 45, B.]
Gloeodiniopsis lamellosa Schopf 1968. (Nyberg and Schopf 1984)
Solitary or paired, coccoid chroococcacean cyanobacterium ("S"). [See Chapter 24, Pl. 45, E, H, I.]
Gloeodiniopsis lamellosa Schopf 1968. (Sergeev and Krylov 1986)
Gloeocapsa-like chroococcacean cyanobacterium ("S").
Gloeodiniopsis magna Nyberg and Schopf 1984. (Nyberg and Schopf 1984)
Solitary, coccoid, ensheathed chroococcacean cyanobacterium ("S"). [See Chapter 24, Pl. 45, D, F, G.]
Kildinella hyperboreica Timofeev 1966. (Timofeev 1969)
Sphaeromorph acritarch ("S").
Kildinella hyperboreica Timofeev 1966. (Timofeev 1966)
Sphaeromorph acritarch ("S").

Table 22.3 (Continued).

Kildinella hyperboreica Timofeev 1963. (Yankauskas 1982)
  Folded, originally spheroidal, sphaeromorph acritarch ("S").
Palaeolyngbya? sp. of Nyberg and Schopf 1984. (Nyberg and Schopf 1984)
  Ensheathed septate filamentous oscillatoriacean cyanobacterium ("S").
Palaeopleurocapsa kamaelgensis Sergeev and Krylov 1986. (Sergeev and Krylov 1986)
  Colonial coccoid chroococcacean cyanobacterium ("S"). [See Chapter 24, Pl. 45, C.]
Protosphaeridium tuberculiferum Timofeev 1966. (Timofeev 1969)
  Sphaeromorph acritarch ("S").
Ramivaginalis uralensis Nyberg and Schopf 1984. (Nyberg and Schopf 1984)
  Sheath of filamentous, false-branching(?), cyanobacterium ("S"). [See Chapter 24, Pl. 46, E.]
Rhicnonema antiquum Hofmann 1976. (Nyberg and Schopf 1984)
  Tubular sheath of filamentous cyanobacterium containing degraded cellular remnants ("S").
Siphonophycus capitaneum Nyberg and Schopf 1984. (Nyberg and Schopf 1984)
  Tubular sheath of filamentous oscillatoriacean cyanobacterium ("S"). [See Chapter 24, Pl. 46, C.]
Sphaerophycus medium Horodyski and Donaldson 1980. (Nyberg and Schopf 1984)
  Colonial, ellipsoidal chroococcacean cyanobacterium ("S").
Trematosphaeridium sp. of Timofeev 1966. (Timofeev 1966)
  Sphaeromorph acritarch ("S").

Olkha Fm.
AS SIB
(ca. 738 Ma)

(branched, septate filaments). (Schopf et al. 1977)
  Branched septate filament, possibly fungal or green algal ("S"). [See Chapter 24, Pl. 45, A.]
(unbranched cellular trichomes). (Schopf et al. 1977)
  Sheath-enclosed septate oscillatoriacean cyanobacterium, cf. *Oscillatoria* and *Lyngbya* ("S").
(unbranched, tubular filaments). (Schopf et al. 1977)
  Tubular sheath of filamentous oscillatoriacean cyanobacterium, cf. *Lyngbya* ("S").

"Algonkian lydites"
EU CSK
(ca. 735 Ma)

Huroniospora macroreticulata Barghoorn 1965. (Vavrdová 1966)
  Sphaeromorph acritarch (S).
Nevidia multicellaria Vavrdová 1966. (Vavrdová 1966)
  Colonial coccoid prokaryote (S).
Nevidia sphaerocellaria Vavrdová 1966. (Vavrdová 1966)
  Colonial coccoid prokaryote (S).

Gangolihat Dolomites
AS IND
(ca. 735 Ma)

Eomycetopsis filiformis Schopf 1968. (Nautiyal 1978a)
  Tubular sheath of filamentous bacterium (S).
Eozygion minutum Schopf and Blacic 1971. (Nautiyal 1978a)
  Colonial ellipsoidal chroococcacean cyanobacterium ("S").
Gunflintia grandis Barghoorn 1965. (Nautiyal 1978a)
  Tubular sheath of filamentous prokaryote (S).
Gunflintia minuta Barghoorn 1965. (Nautiyal 1978a)
  Septate filamentous bacterium (S).
Huroniospora microreticulata Barghoorn 1965. (Nautiyal 1978a)
  Sphaeromorph acritarch (S).
Huroniospora psilata Barghoorn 1965. (Nautiyal 1978a)
  Solitary coccoid chroococcacean cyanobacterium (S).
Myxococcoides indicus Venkatachala, Bhandari, Chaube and Rawat 1974. (Nautiyal 1978a)
  Colonial coccoid chroococcacean cyanobacterium ("S").

Gangolihat Dolomites, Chhera Mbr.
AS IND
(ca. 735 Ma)

Archaeorestis minuta Nautiyal 1980. (Nautiyal 1980)
  Tubular sheath of filamentous bacterium. ("Branches" not evident in figure.) (S).
Contrahofilum schopfii Nautiyal 1980. (Nautiyal 1980)
  Tubular sheath of filamentous cyanobacterium (S).
Myxococcoides bansensis Nautiyal 1980. (Nautiyal 1980)
  Colonial coccoid chroococcacean cyanobacterium ("S").
Siphonophycus indicum Nautiyal 1980. (Nautiyal 1980)
  Tubular sheath of filamentous cyanobacterium (S).
Trachysphaeridium decorum Venkatachala, Bhandari, Chaube and Rawat 1974. (Nautiyal 1980)
  Sphaeromorph acritarch ("pylome" may be diagenetic artifact) ("S").

Gangolihat Dolomites, Hiunpani Mbr.
AS IND
(ca. 735 Ma)

Eomycetopsis (?) *schopfii* Nautiyal 1980. (Nautiyal 1980)
  Branched septate filaments, possibly fungal hyphae ("S").
Eosynechococcus minutus Nautiyal 1980. (Nautiyal 1980)
  Ensheathed colonial coccoid prokaryote (not solitary and not ellipsoidal, so not *Eosynechococcus*) (S).
Glenobotrydion majorinum Schopf and Blacic 1971. (Nautiyal 1980)
  Sphaeromorph acritarch (S).
Palaeonostoc barghoornii Nautiyal 1980. (Nautiyal 1980)
  Septate filamentous cyanobacterium (S).
Trachysphaeridium sp. A of Nautiyal 1980. (Nautiyal 1980)
  Sphaeromorph acritarch ("S").

Penganga Fm.
AS IND
(ca. 735 Ma)

Orygmatosphaeridium plicatum Maithy and Shukla 1977. (Maithy 1980)
  Sphaeromorph acritarch ("S").

Zona de Ossa-Morena, Série Negra
EU PRT
(ca. 735 Ma)

Eomicrhystridium? sp. of Gonçalves and Palacios 1984. (Gonçalves and Palacios 1984)
  Degraded sphaeromorph acritarch (S M).

Barents Sea Grp.
EU NOR
(ca. 730 Ma)

(sphaeromorphs sp. indet.). (Vidal and Siedlecka 1983)
  Colonial and isolated coccoid cyanobacteria and sphaeromorph acritarchs (S).
Kildinosphaera chagrinata Vidal 1983. (Vidal and Siedlecka 1983)
  Sphaeromorph acritarch ("S").

Table 22.3 *(Continued).*

Protosphaeridium sp. of Vidal and Siedlecka 1983. (Vidal and Siedlecka 1983)
  Solitary coccoid cyanobacterium (S).
Synsphaeridium sp. of Vidal and Siedlecka 1983. (Vidal and Siedlecka 1983)
  Colonial coccoid cyanobacterium (S).
Taeniatum sp. of Vidal and Siedlecka 1983. (Vidal and Siedlecka 1983)
  Flattened tubular sheath of filamentous prokaryote (S).

Barents Sea Grp., Båsnæring Fm.
EU NOR
(ca. 730 Ma)

Leiosphaeridia asperata (Naumova 1950) Lindgren 1982. (Vidal and Siedlecka 1983)
  Sphaeromorph acritarch ("S").

Barents Sea Grp., Båtsfjord Fm.
EU NOR
(ca. 730 Ma)

Kildinosphaera granulata Vidal 1983. (Vidal and Siedlecka 1983)
  Sphaeromorph acritarch ("S").
Kildinosphaera lophostriata (Jankauskas 1979) Vidal 1983. (Vidal and Siedlecka 1983)
  Sphaeromorph acritarch ("S").
Kildinosphaera verrucata Vidal 1983. (Vidal and Siedlecka 1983)
  Sphaeromorph acritarch ("S").
Leiosphaeridia asperata (Naumova 1950) Lindgren 1982. (Vidal and Siedlecka 1983)
  Sphaeromorph acritarch ("S").
Octoedryxium truncatum Rudavskaya 1973. (Vidal and Siedlecka 1983)
  Prismatomorph acritarch ("S").
Podolina minuta Hermann [German] 1976. (Vidal and Siedlecka 1983)
  Microfossil incertae sedis, possibly polygonomorph acritarch ("S").

Barents Sea Grp., Kongsfjord Fm.
EU NOR
(ca. 730 Ma)

(enigmatic sphaeromorphic acritarchs). (Vidal and Siedlecka 1983)
  Sphaeromorph acritarch containing degraded cellular contents (S M).
Bavlinella faveolata Shepeleva 1962. (Vidal and Siedlecka 1983)
  Spheroidal colony of coccoid bacteria (S).

Beda Fm. or Samra Fm.
AF EGY
(ca. 725 Ma)

Sinaiicoccus avnimelechii Shimron and Horowitz 1972. (Shimron and Horowitz 1972)
  Sphaeromorph acritarch (S).
Sinaiicoccus delicatus Shimron and Horowitz 1972. (Shimron and Horowitz 1972)
  Sphaeromorph acritarch (S).
Sinaiicoccus minutus Shimron and Horowitz 1972. (Shimron and Horowitz 1972)
  Sphaeromorph acritarch (S).
Spirillinema bentorii Shimron and Horowitz 1972. (Shimron and Horowitz 1972)
  Coiled filamentous bacterium (S).

Martyukhe Fm.
AS SIB
(ca. 725 Ma)

Murandavia aff. *rotunda* Vologdin 1965. (Vologdin and Drozdova 1969b)
  Colonial coccoid (chroococcacean?) cyanobacterium ("S").
Vesicophyton punctatum Vologdin and Drosdova [Drozdova] 1969. (Vologdin and Drozdova 1969a)
  Colonial coccoid (chroococcacean) cyanobacterium ("S").

"Adelaidean"
AU AUS SA
(ca. 700 Ma)

Protosphaeridium densum Timofeev 1966. (Timofeev 1966)
  Sphaeromorph acritarch ("S").

"unnamed black mudstone" of the Uinta Mountains
NA USA UT
(ca. 700 Ma)

Sphaerocongregus variabilis Moorman 1974. (Cloud et al. 1975)
  Colonial coccoid prokaryote (S).

Hedmark Grp.
EU NOR
(ca. 700 Ma)

Kildinella hyperboreica Timofeev 1966. (Timofeev 1966)
  Sphaeromorph acritarch ("S").
Kildinella sinica Timofeev 1966. (Timofeev 1966)
  Sphaeromorph acritarch ("S").
Trematosphaeridium holtedahlii Timofeev 1966. (Timofeev 1966)
  Sphaeromorph acritarch ("S").

Hedmark Grp., Biri Fm.
EU NOR
(ca. 700 Ma)

(Form A). (Manum 1967)
  Colonial coccoid prokaryote, cf. *Bavlinella* (S).
(Form B). (Manum 1967)
  Solitary and colonial coccoid cyanobacteria (S).

Jacadigo Grp., Santa Cruz Fm.
SA BRA
(ca. 700 Ma)

(filamentous microfossils). (Fairchild, Barbour, and Haralyi 1978)
  Tubular sheath of filamentous cyanobacterium ("S").

Jacadigo Grp., Urucum Fm.
SA BRA
(ca. 700 Ma)

(filamentous microfossils). (Fairchild, Barbour, and Haralyi 1978)
  Tubular sheath of filamentous cyanobacterium ("S").

Kandyk Fm.
AS SIB
(ca. 700 Ma)

Cymatiosphaera? sp. of Timofeev 1966. (Timofeev 1966)
  Sphaeromorph acritarch having degraded cellular contents (S).
Dictyosphaeridium sp. of Timofeev 1969. (Timofeev 1969)
  Sphaeromorph acritarch ("S").
Gloeocapsomorpha hebeica Timofeev 1966. (Timofeev 1969)
  Colonial coccoid cyanobacterium ("S").
Kildinella hyperboreica Timofeev 1966. (Timofeev 1969)

Table 22.3 (Continued).

Sphaeromorph acritarch ("S").
Kildinella hyperboreica Timofeev 1966. (Timofeev 1966)
   Sphaeromorph acritarch ("S").
Kildinella sinica Timofeev 1966. (Timofeev 1966)
   Sphaeromorph acritarch ("S").
Kildinella sinica Timofeev 1966. (Timofeev 1969)
   Sphaeromorph acritarch ("S").
Protosphaeridium densum Timofeev 1966. (Timofeev 1966)
   Sphaeromorph acritarch ("S").
Protosphaeridium paleaceum Timofeev 1966. (Timofeev 1966)
   Sphaeromorph acritarch ("S").
Protosphaeridium pusillum Timofeev 1966. (Timofeev 1966)
   Sphaeromorph acritarch ("S").
Protosphaeridium rigidulum Timofeev 1966. (Timofeev 1969)
   Sphaeromorph acritarch ("S").
Protosphaeridium tuberculiferum Timofeev 1966. (Timofeev 1966)
   Sphaeromorph acritarch ("S").
Trematosphaeridium sinuatum Timofeev 1966. (Timofeev 1966)
   Sphaeromorph acritarch ("S").

Mineral Fork Fm.
NA USA UT
(ca. 700 Ma)

(simple unornamented unicells). (Knoll et al. 1981)
   Diverse, solitary and colonial, coccoid cyanobacteria (S).
Bavlinella faveolata Shepeleva 1962. (Knoll et al. 1981)
   Colonial coccoid bacterium (S).

Tindir Grp.
NA USA AK
(ca. 700 Ma)

Sphaerocongregus variabilis Moorman 1974. (Cloud et al. 1975)
   Colonial coccoid prokaryote (S).

Tindir Grp., dolomitic sandstone and shale?
NA USA AK
(ca. 700 Ma)

(type "A"). (Allison and Moorman 1973)
   Colonial coccoid bacteria, with cells replaced by pyrite ("M").
(type "B"). (Allison and Moorman 1973)
   Colonial coccoid prokaryote, with cells replaced by pyrite ("M").
(type "C"). (Allison and Moorman 1973)
   Solitary or poorly preserved colonial coccoid cyanobacteria ("M").

Vadsø Grp., Ekkerøy Fm.
EU NOR
(ca. 700 Ma)

Chuaria circularis Walcott 1899. (Vidal 1981b)
   Sphaeromorph acritarch ("S").
Kildinella hyperboreica Timofeev 1966. (Vidal 1981b)
   Sphaeromorph acritarch ("S").
Kildinella sp. B of Vidal 1981. (Vidal 1981b)
   Sphaeromorph acritarch ("S").
Pterospermopsimorpha mogilevica Timofeev 1973. (Vidal 1981b)
   Degraded sphaeromorph acritarch containing internal cellular remnants (S M).
cf. *Stictosphaeridium* sp. of Vidal 1981. (Vidal 1981b)
   Colonial coccoid cyanobacterium (Fig. 19A–F) (S).
Vandalosphaeridium varangeri Vidal 1981. (Vidal 1981b)
   Degraded, membranous, possibly acanthomorph acritarch ("S M")

Windermere Sgrp., Miette Grp., Hector Fm.
NA CAN AT
(ca. 700 Ma)

Chuaria cf. *circularis* Walcott 1899. (Gussow 1973)
   Sphaeromorph acritarch ("S").
Sphaerocongregus variabilis Moorman 1974. (Cloud et al. 1975)
   Colonial coccoid prokaryote (S). [See Chapter 24, Pl. 48, A, B.]
Sphaerocongregus variabilis Moorman 1974. (Moorman 1974)
   Colonial coccoid prokaryote (M).

700 to 600 Ma:
Rechka Fm.
AS SIB
(ca. 690 Ma)

(broad tubular sheath of] algal trichome). (Timofeev 1969)
   Flattened tubular sheath of filamentous cyanobacterium (S).
Leiosphaeridia sp. of Timofeev 1969. (Timofeev 1969)
   Sphaeromorph acritarch ("S").
Protosphaeridium densum Timofeev 1966. (Timofeev 1969)
   Sphaeromorph acritarch ("S").
Protosphaeridium flexuosum Timofeev 1966. (Timofeev 1969)
   Sphaeromorph acritarch ("S").
Protosphaeridium sp. of Timofeev 1969. (Timofeev 1969)
   Sphaeromorph acritarch ("S").
Protosphaeridium tuberculiferum Timofeev 1966. (Timofeev 1969)
   Sphaeromorph acritarch ("S").
Synsphaeridium sorediforme Timofeev 1969?. (Timofeev 1969)
   Colonial coccoid cyanobacterium (S).

Vindhyan Sgrp., Bhander Grp., Sirbu Shale
AS IND
(ca. 690 Ma)

(disc-like forms). (Sitholey et al. 1953)
   Sphaeromorph acritarchs (S).
(filamentous body). (Sitholey et al. 1953)
   Septate filamentous cyanobacterium (S).

Thule Grp., Narssârssuk Fm., Aorfêrneq Dolomite Mbr.
NA GRL
(ca. 688 Ja)

(chroococcoid unicell type A). (Strother et al. 1983)
   Colonial coccoid chroococcacean cyanobacterium ("S").
(chroococcoid unicell type B). (Strother et al. 1983)
   Solitary ellipsoidal chroococcacean cyanobacterium ("S").
(spheroid type A). (Strother et al. 1983)
   Degraded, solitary coccoid cyanobacterium (S).
(spheroid type B). (Strother et al. 1983)
   Spheroidal to globular envelope of colonial, coccoid or ellipsoidal, prokaryote (S).
Avictuspirulina minuta Strother, Knoll and Barghoorn 1983. (Strother et al. 1983)
   Tubular helical bacterium (S).
Coleogleba auctifica Strother, Knoll and Barghoorn 1983. (Strother et al. 1983)
   Colonial coccoid prokaryote, possibly aff. *Microcystis* (S).
Eoentophysalis cf. *belcherensis* Hofmann 1976. (Strother et al. 1983)
   Colonial, coccoid to ellipsoidal bacterium (S).
Eomycetopsis robusta Schopf 1968. (Strother et al. 1983)
   Tubular sheath of filamentous cyanobacterium ("S").
Eosynechococcus amadeus Knoll and Golubić 1983. (Strother et al. 1983)

Colonial ellipsoidal bacterium (M S).
Eosynechococcus cf. *amadeus* Knoll and Golubić 1979. (Strother et al. 1983)
Colonial ellipsoidal bacterium (S).
Eosynechococcus thuleënsis Strother, Kholl and Barghoorn 1983. (Strother et al. 1983)
Solitary ellipsoidal chroococcacean cyanobacterium ("S").
Gloeodiniopsis cf. *lamellosa* Schopf 1968. (Strother et al. 1983)
Solitary coccoid chroococcacean cyanobacterium ("S").
Gyalosphaera cf. *fluitans* Strother, Knoll and Barghoorn 1983. (Strother et al. 1983)
Colonial coccoid bacterium (S).
Gyalosphaera fluitans Strother, Knoll and Barghoorn 1983. (Strother et al. 1983)
Colonial coccoid cyanobacterium ("S").
Myxococcoides sp. of Strother, Knoll and Barghoorn 1983. (Strother et al. 1983)
Sphaeromorph acritarch (S).
Oscillatoriopsis variabilis Strother, Knoll and Barghoorn 1983. (Strother et al. 1983)
Septate filamentous oscillatoriacean cyanobacterium ("S").
Siphonophycus sp. of Strother, Knoll and Barghoorn 1983. (Strother et al. 1983)
Tubular sheath of filamentous cyanobacterium (S).
Tetraphycus sp. of Strother, Knoll and Barghoorn 1983. (Strother et al. 1983)
Colonial coccoid chroococcacean cyanobacterium ("S").

Amri unit
AS IND
(ca. 685 Ma)

Eomycetopsis septata Maithy 1975. (Nautiyal 1978b)
Tubular sheath of filamentous cyanobacterium (S).

Simla Slate Grp.
AS IND HP
(ca. 685 Ma)

Granomarginata dhalii Nautiyal 1982. (Nautiyal 1982c)
Sphaeromorph acritarch ("S").
Granomarginata simlaensis Nautiyal 1982. (Nautiyal 1982c)
Sphaeromorph acritarch ("S").

Priozersk Fm.
EU SUN
(ca. 680 Ma)

Favososphaeridium sp. of Timofeev 1966. (Timofeev 1966)
Colonial coccoid chroococcacean cyanobacterium (S).
Kildinella hyperboreica Timofeev 1966. (Timofeev 1966)
Sphaeromorph acritarch ("S").
Protosphaeridium acis Timofeev 1966. (Timofeev 1966)
Sphaeromorph acritarch ("S").
Protosphaeridium asaphum Timofeev 1966. (Timofeev 1966)
Sphaeromorph acritarch ("S").
Protosphaeridium flexuosum Timofeev 1966. (Timofeev 1966)
Sphaeromorph acritarch ("S").
Protosphaeridium papyraceum Timofeev 1966. (Timofeev 1966)
Sphaeromorph acritarch ("S").
Protosphaeridium parvulum Timofeev 1966. (Timofeev 1966)
Sphaeromorph acritarch ("S").
Protosphaeridium pusillum Timofeev 1966. (Timofeev 1966)
Sphaeromorph acritarch ("S").
Protosphaeridium torulosum Timofeev 1966. (Timofeev 1966)
Sphaeromorph acritarch ("S").
Synsphaeridium conglutinatum Timofeev 1966. (Timofeev 1966)
Colonial coccoid chroococcacean? cyanobacterium (S).

Bavly "Series"
EU SUN
(ca. 675 Ma)

Bavlinella faveolata Schepelewa [Shepeleva] 1962. (Timofeev 1966)
Colonial coccoid prokaryote (S).

Bhima "Series"
AS IND KA
(ca. 675 Ma)

(Dasycladaceous alga). (Salujha et al. 1972)
Sphaeromorph acritarch (S).
(Discoidal body). (Salujha et al. 1972)
Sphaeromorph acritarch (S).
Archaeofavosina compta Salujha, Rehman and Arora 1972. (Salujha et al. 1972)
Sphaeromorph acritarch ("S").
Chuaria circularis Walcott 1899. (Suresh and Raju 1983)
Sphaeromorph acritarch ("S").
Granomarginata exquisita Salujha, Rehman and Arora 1972. (Salujha et al. 1972)
Sphaeromorph acritarch ("S").
Granomarginata primitiva Salujha, Rehman and Arora 1971. (Salujha et al. 1972)
Sphaeromorph acritarch ("S").
Granomarginata sp. of Salujha, Rehman and Arora 1972. (Salujha et al. 1972)
Sphaeromorph acritarch ("S").
Lophosphaeridium bellus Salujha, Rehman and Arora 1972. (Salujha et al. 1972)
Sphaeromorph acritarch ("S").
Lophosphaeridium sp. of Salujha, Rehman and Arora 1972. (Salujha et al. 1972)
Sphaeromorph acritarch ("S").
Ooidium sp. of Salujha, Rehman and Arora 1972. (Salujha et al. 1972)
Sphaeromorph acritarch (M).
cf. *Tasmanites* sp. of Salujha, Rehman and Arora 1972. (Salujha et al. 1972)
Sphaeromorph acritarch ("S").
Trematosphaeridium bhimaii Salujha, Rehman and Arora 1972. (Salujha et al. 1972)
Sphaeromorph acritarch ("S").
Trematosphaeridium sp. of Salujha, Rehman and Arora 1972. (Salujha et al. 1972)
Sphaeromorph acritarch ("S").
Vavososphaeridium reticulatum Salujha, Rehman and Arora 1972. (Salujha et al. 1972)
Sphaeromorph acritarch ("S").
Vavososphaeridium sp. of Salujha, Rehman and Arora 1972. (Salujha et al. 1972)
Sphaeromorph acritarch ("S").

Dalradian (Lower), Easdale Slate
EU GBR
(ca. 675 Ma)

(sphaeromorph acritarchs). (Downie et al. 1971)
Sphaeromorph acritarch ("S").

Table 22.3 (*Continued*).

Dalradian (Lower), Islay Quartzite "Series"
EU GBR
(ca. 675 Ma)

(possible acanthomorph acritarch). (Downie et al. 1971)
  Possible poorly preserved acanthomorph acritarch ("S M").
(Sphaeromorph acritarchs). (Downie et al. 1971)
  Sphaeromorph acritarch ("S").
?*Leiopsophaera convexiplicata* Naumova 1960?. (Downie et al. 1971)
  Sphaeromorph acritarch ("S").
?*Leiopsophaera microrugosa* Naumova 1960?. (Downie et al. 1971)
  Sphaeromorph acritarch ("S").

Izluchinsk Fm.
AS SIB
(ca. 675 Ma)

([intermediate-diam. sheath of] algal trichome). (Timofeev 1969)
  Flattened tubular sheath of filamentous cyanobacterium (S).
Kildinella hyperboreica Timofeev 1966. (Timofeev 1969)
  Sphaeromorph acritarch ("S").
Protosphaeridium flexuosum Timofeev 1966. (Timofeev 1969)
  Sphaeromorph acritarch ("S").
Protosphaeridium papyraceum Timofeev 1966. (Timofeev 1969)
  Sphaeromorph acritarch ("S").
Protosphaeridium tuberculiferum Timofeev 1966. (Timofeev 1969)
  Sphaeromorph acritarch ("S").
Protosphaeridium tuberculiferum Timofeev 1966. (Timofeev 1969)
  Sphaeromorph acritarch ("S").
Stictosphaeridium pectinale Timofeev 1966. (Timofeev 1969)
  Sphaeromorph acritarch ("S").
Stictosphaeridium sinapticuliferum Timofeev 1966. (Timofeev 1969)
  Sphaeromorph acritarch ("S").
Trematosphaeridium holtedahlii Timofeev 1966. (Timofeev 1969)
  Sphaeromorph acritarch ("S").

Kamovsk Fm.
AS SIB
(ca. 675 Ma)

Cucumiforma vanavaria Mikhailova [Mikhajlova] 1986. (Mikhajlova 1986)
  Sphaeromorph acritarch ("S"). [See Frontispiece of Part 1, and Chapter 24, Pl. 48, E_1-E_2.]
Gloeocapsomorpha sp. of Timofeev 1966. (Timofeev 1966)
  Colonial coccoid chroococcacean? cyanobacterium ("S").
Satka squamifera Pjatiletov [Pyatiletov] 1980. (Pyatiletov 1980)
  Colonial coccoid chroococcacean cyanobacterium (S). [See Chapter 24, Pl. 48, F_1-F_2.]

Tanafjord Grp., Stangenes Fm.
EU NOR
(ca. 675 Ma)

Kildinella hyperboreica Timofeev 1966. (Vidal 1981b)
  Sphaeromorph acritarch ("S").
Vandalosphaeridium varangeri Vidal 1981. (Vidal 1981b)
  Degraded, membranous, possibly acanthomorph acritarch ("S M").

Uk Fm. (= Kudash Fm.)
EA SUN
(ca. 675 Ma)

Arctacellularia ellipsoidea Hermann [German] 1976. (Yankauskas 1982)
  Pseudofilamentous (uniseriate) colonial coccoid cyanobacterium (?cf. *Glenobotrydion aenigmatis*) (S).

Arctacellularia sp. 3 of Jankauskas [Yankauskas] 1982. (Yankauskas 1982)
  Solitary or paired (disarticulated) trichomic cells of filamentous or pseudofilamentous (colonial) cyanobacterium, or solitary or paired cylindrical cells incertae sedis (S).
Brevitrichoides bashkiricus Jankauskas [Yankauskas] 1980. (Yankauskas 1982)
  Tubular sheath (or trichome without preserved lateral cell walls) of filamentous (oscillatoriacean) cyanobacterium (S).
Calyptothrix alternata Jankauskas [Yankauskas] 1980. (Yankauskas 1980a)
  Septate filamentous prokaryote (S).
Eomycetopsis psilata Maithy and Shukla 1977. (Yankauskas 1982)
  Tubular sheath of filamentous cyanobacterium ("S").
Eomycetopsis rimata Jankauskas [Yankauskas] 1980. (Yankauskas 1982)
  Flattened sheath of filamentous cyanobacterium ("S").
Eomycetopsis rimata Jankauskas [Yankauskas] 1980. (Yankauskas 1980a)
  Flattened tubular sheath of filamentous cyanobacterium ("S").
Eomycetopsis aff. *rugosa* Maithy 1975. (Yankauskas 1982)
  Tubular sheath of filamentous cyanobacterium ("S").
Kildinella hyperboreica Timofeev 1963. (Yankauskas 1982)
  Folded, originally spheroidal, sphaeromorph acritarch ("S").
Kildinella nordia Timofeev 1969. (Yankauskas 1982)
  Sphaeromorph acritarch containing globular, condensed cytoplasmic contents ("S").
Kildinella tschapomica Timofeev 1966. (Yankauskas 1982)
  Distorted, originally spheroidal, sphaeromorph acritarch ("S").
Leiothrichoides typicus Hermann [German] 1974. (Yankauskas 1982)
  Tubular sheath of filamentous cyanobacterium ("S").
Palaeolyngbya zilimica Jankauskas [Yankauskas] 1980. (Yankauskas 1982)
  Septate hormogone of filamentous (oscillatoriacean) cyanobacterium ("S").
Palaeolyngbya zilimica Jankauskas [Yankauskas] 1980. (Yankauskas 1980a)
  Septate hormogone of filamentous (oscillatoriacean) cyanobacterium ("S").
Palaeolyngbya zilimica Jankauskas [Yankauskas] in press (Yankauskas in press)
  Hormogonia-like septate filamentous oscillatoriacean cyanobacterium ("S"). [See Chapter 24, Pl. 48, C.]
Polythrichoides lineatus Hermann [German] 1974. (Yankauskas 1982)
  Ensheathed multitrichomic filamentous (aff. *Microcoleus*) prokaryote ("S").
Siphonophycus costatus Jankauskas [Yankauskas] 1980. (Yankauskas 1982)
  Tubular sheath of filamentous cyanobacterium ("S"). [See Chapter 24, Pl. 48, D.]
Synsphaeridium spp. of Jankauskas [Yankauskas] 1982. (Yankauskas 1982)
  Colonial coccoid (?chroococcacean) cyanobacteria (S).
Tortunema sibirica Hermann [German] 1976. (Yankauskas 1982)
  Tubular sheath of filamentous cyanobacterium ("S").

Bakeevo Fm.
EA SUN
(ca. 670 Ma)

Kildinella hyperboreica Timofeev 1963. (Yankauskas 1982)
  Folded, originally spheroidal, sphaeromorph acritarch ("S").

Table 22.3 (Continued).

Kildinella nordia Timofeev 1969. (Yankauskas 1982)
Sphaeromorph acritarch containing globular, condensed cytoplasmic contents ("S").
Turuchanica ternata Timofeev 1966. (Yankauskas 1982)
Thick-walled, originally spheroidal sphaeromorph acritarch; lobate due to compression ("S").

Durnoj Mys Fm.
AS SIB
(ca. 670 Ma)

Protosphaeridium tuberculiferum Timofeev 1966. (Timofeev 1969)
Sphaeromorph acritarch ("S").

Nemchany Fm.
AS SIB
(ca. 670 Ma)

Kildinella hyperboreica Timofeev 1966. (Timofeev 1966)
Sphaeromorph acritarch ("S").
Polyedrosphaeridium bullatum Timofeev 1966. (Timofeev 1966)
Colonial coccoid chroococcacean? cyanobacterium (S).
Protosphaeridium papyraceum Timofeev 1966. (Timofeev 1966)
Sphaeromorph acritarch ("S").
Stictosphaeridium implexum Timofeev 1966. (Timofeev 1966)
Sphaeromorph acritarch ("S").
Turuchanica alara Rudavskaja [Rudavskaya] 1964?. (Timofeev 1966)
Compressed sphaeromorph? acritarch (S).

Tsyp-Navolok shales
EU SUN
(ca. 670 Ma)

Kildinella hyperboreica Timofeev 1966. (Timofeev 1969)
Sphaeromorph acritarch ("S").
Kildinella sinica Timofeev 1966. (Timofeev 1969)
Sphaeromorph acritarch ("S").
Protosphaeridium densum Timofeev 1966. (Timofeev 1969)
Sphaeromorph acritarch ("S").
Stictosphaeridium sp. of Timofeev 1969. (Timofeev 1969)
Colonial coccoid cyanobacterium (S).
Trematosphaeridium holtedahlii Timofeev 1966. (Timofeev 1969)
Sphaeromorph acritarch ("S").
Turuchanica ternata Timofeev 1966. (Timofeev 1969)
Flattened sphaeromorph? acritarch (S).

Volokov "Series"
EU SUN
(ca. 670 Ma)

Kildinella hyperboreica Timofeev 1966. (Timofeev 1969)
Sphaeromorph acritarch ("S").
Kildinella hyperboreica Timofeev 1966. (Timofeev 1966)
Sphaeromorph acritarch ("S").
Kildinella sinica Timofeev 1966. (Timofeev 1969)
Sphaeromorph acritarch ("S").
Protosphaeridium laccatum Timofeev 1966. (Timofeev 1969)
Sphaeromorph acritarch ("S").
Protosphaeridium scabridum Timofeev 1966. (Timofeev 1969)
Sphaeromorph acritarch ("S").
Protosphaeridium tuberculiferum Timofeev 1966. (Timofeev 1966)
Sphaeromorph acritarch ("S").
Stictosphaeridium implexum Timofeev 1966. (Timofeev 1969)
Sphaeromorph acritarch ("S").
Stictosphaeridium pectinale Timofeev 1966. (Timofeev 1969)
Sphaeromorph acritarch ("S").

Stictosphaeridium sinapticuliferum Timofeev 1966. (Timofeev 1966)
Sphaeromorph acritarch ("S").
Turuchanica ternata Timofeev 1966. (Timofeev 1969)
Flattened sphaeromorph? acritarch (S).

Pendkari Grp., Pendjari Fm.
AF HVO BEN
(ca. 660 Ma)

Chuaria circularis Walcott 1899. (Amard and Affaton 1984)
Sphaeromorph acritarch ("S").

Cambridge Argillite
NA USA MA
(ca. 650 Ma)

(filaments). (Lenk et al. 1982)
Septate filamentous cyanobacterium (S).
(simple spherical cells). (Lenk et al. 1982)
Solitary and colonial coccoid cyanobacteria (S).
Bavlinella cf. *faveolata* Shepeleva 1962. (Lenk et al. 1982)
Colonial coccoid bacterium (S).

Chichkan Fm.
AS SUN
(ca. 650 Ma)

(broad oscillatoriacean sheaths). (Schopf and Sovietov 1976)
Lamellated tubular sheath of filamentous oscillatoriacean cyanobacterium, cf. *Lyngbya* ("S"). [See Chapter 24, Pl. 49, A_1–A_4, E.]
(broad oscillatoriacean trichomes). (Schopf and Sovietov 1976)
Septate filamentous oscillatoriacean, cf. *Lyngbya* or cf. *Oscillatoria* ("S"). [See Chapter 24, Pl. 49, B, D.]
(broad tubular sheaths). (Schopf et al. 1977)
Lamellated tubular sheaths of filamentous oscillatoriacean cyanobacteria, cf. *Lyngbya* ("S"). [See Chapter 24, Pl. 49, A_1–A_4, E.]
(colonial unicells). (Schopf and Sovietov 1976)
Colonial coccoid (chroococcacean) cyanobacteria ("S"). [See Chapter 24, Pl. 51, A.]
(narrow tubular filaments). (Schopf and Sovietov 1976)
Tubular sheaths of filamentous cyanobacteria ("S").
(solitary algal unicells). (Schopf and Sovietov 1976)
Sphaeromorph acritarchs (S). [See Chapter 24, Pl. 51, B–D.]
(solitary algal unicells). (Schopf et al. 1977)
Diverse solitary coccoid (chroococcacean) cyanobacteria and possibly (chlorophycean and/or rhodophycean) algae ("S"). [See Chapter 24, Pl. 51, B–D.]
Cyanonema disjuncta Ogurtsova and Sergeev 1987. (Ogurtsova and Serveev 1987)
Septate filamentous cyanobacterium ("S"). [See Chapter 24, Pl. 49, F, G_1–G_2.]
Eoentophysalis dismallakesensis Horodyski and Donaldson 1980. (Ogurtsova and Serveev 1987)
Colonial coccoid chroococcalean cyanobacterium ("S").
Globophycus rugosum Schopf 1968. (Ogurtsova and Serveev 1987)
Sphaeromorph acritarch (S).
Gloeodiniopsis dilutus Ogurtsova and Sergeev 1987. (Ogurtsova and Serveev 1987)
Colonial coccoid chroococcacean cyanobacterium ("S").
Obruchevella sp. of Ogurtsova and Sergeev 1987. (Ogurtsova and Serveev 1987)
Spirally coiled *Spirulina*- or *Arthrospira*-like prokaryote (S).

Table 22.3 (Continued).

Oscillatoriopsis media Mendelson and Schopf 1982. (Ogurtsova and Serveev 1987)
　Septate filamentous oscillatoriacean cyanobacterium ("S").
Palaeopleurocapsa fusiforma Ogurtsova and Sergeev 1987. (Ogurtsova and Serveev 1987)
　Colonial coccoid chroococcalean cyanobacterium ("S"). [See Chapter 24, Pl. 50, B_1–B_2.]
Palaeopleurocapsa reniforma Ogurtsova and Sergeev 1987. (Ogurtsova and Serveev 1987)
　Colonial coccoid chroococcalean cyanobacterium ("S"). [See Chapter 24, Pl. 50, C_1–C_2.]
Tetraphycus bistratosus Ogurtsova and Sergeev 1987. (Ogurtsova and Serveev 1987)
　Colonial coccoid chroococcacean cyanobacterium ("S"). [See Chapter 24, Pl. 50, A_1–A_2.]
Veteronostocale copiosus Ogurtsova and Sergeev 1987. (Ogurtsova and Serveev 1987)
　Septate filamentous oscillatoriacean cyanobacterium ("S").

Churochnaya Fm.
EA SUN
(ca. 650 Ma)

Kildinella hyperboreica Timofeev 1966. (Timofeev 1966)
　Sphaeromorph acritarch ("S").
Protosphaeridium acis Timofeev 1966. (Timofeev 1966)
　Sphaeromorph acritarch ("S").
Protosphaeridium paleaceum Timofeev 1966. (Timofeev 1966)
　Sphaeromorph acritarch ("S").
Protosphaeridium patelliforme Timofeev 1966. (Timofeev 1966)
　Sphaeromorph acritarch ("S").
Protosphaeridium sp. of Timofeev 1966. (Timofeev 1966)
　Sphaeromorph acritarch ("S").
Stenozonoligotriletum sokolovii Timofeev 1959. (Timofeev 1966)
　Sphaeromorph acritarch ("S").

Hailuoto Sequence (cf. Muhos Fm.)
EU FIN
(ca. 650 Ma)

(algal cells resembling *Bavlinella faveolata*.) (Tynni and Donner 1980)
　Colonial coccoid bacteria (S).
Anabaenidium hailuotoense Tynni and Donner 1980. (Tynni and Donner 1980)
　Tubular sheath of filamentous cyanobacterium (S).
Caudiculophycus curvata Tynni and Donner 1980. (Tynni and Donner 1980)
　Flattened tubular (partially degraded) sheath of filamentous cyanobacterium (S).
Chuaria circularis Walcott 1899. (Tynni and Donner 1980)
　Sphaeromorph acritarch ("S").
Cymatiosphaera precambrica Tynni and Donner 1980. (Tynni and Donner 1980)
　Degraded sphaeromorph or possible herkomorph acritarch (S M).
Cymatiosphaera sp. of Tynni and Donner 1980. (Tynni and Donner 1980)
　Herkomorph acritarch or diagenetically altered sphaeromorph acritarch (M).
Favososphaeridium sp. of Tynni and Donner 1980. (Tynni and Donner 1980)
　Sphaeromorph acritarch ("S").

Gloeocapsomorpha sp. of Tynni and Donner 1980. (Tynni and Donner 1980)
　Colonial coccoid (chroococcacean?) cyanobacterium ("S").
Granomarginata sp. ? of Tynni and Donner 1980. (Tynni and Donner 1980)
　Sphaeromorph acritarch ("S").
Kildinella cf. *sinica* Timofeev 1966. (Tynni and Donner 1980)
　Sphaeromorph acritarch ("S").
Leiosphaeridia sp. of Tynni and Donner 1980. (Tynni and Donner 1980)
　Sphaeromorph acritarch ("S").
Oscillatoriopsis bothnica Tynni and Donner 1980. (Tynni and Donner 1980)
　Septate filamentous oscillatoriacean cyanobacterium ("S").
Oscillatoriopsis constricta Tynni and Donner 1980. (Tynni and Donner 1980)
　Septate filamentous oscillatoriacean cyanobacterium ("S").
Oscillatoriopsis magna Tynni and Donner 1980. (Tynni and Donner 1980)
　Septate filamentous oscillatoriacean cyanobacterium ("S").
Palaeolyngbya lata Tynni and Donner 1980. (Tynni and Donner 1980)
　Septate ensheathed filamentous oscillatoriacean cyanobacterium ("S").
Synsphaeridium sp. of Tynni and Donner 1980. (Tynni and Donner 1980)
　Colonial coccoid cyanobacterium (S).
Tortunema bothnica Tynni and Donner 1980. (Tynni and Donner 1980)
　Tubular sheath of filamentous cyanobacterium ("S").
Trachyhystrichosphaera bothnica Tynni and Donner 1980. (Tynni and Donner 1980)
　Possible poorly preserved acanthomorph acritarch (S M).
? *Trachyhystrichosphaera* sp. of Tynni and Donner 1980. (Tynni and Donner 1980)
　Poorly preserved sphaeromorph acritarch (M S).
Trachysphaeridium levis (Lopukhin 1971) Vidal 1974. (Tynni and Donner 1980)
　Sphaeromorph acritarch ("S").
Trachysphaeridium sp. of Tynni and Donner 1980. (Tynni and Donner 1980)
　Sphaeromorph acritarch ("S").
Volyniella cylindrica Tynni and Donner 1980. (Tynni and Donner 1980)
　Filamentous oscillatoriacean cyanobacterium ("S").

Hedmark Grp., Biskopåsen Fm.
EU NOR
(ca. 650 Ma)

(Form A). (Manum 1967)
　Colonial coccoid prokaryote, cf. *Bavlinella* (S).
(Form B). (Manum 1967)
　Solitary and colonial coccoid cyanobacterium (S).
(Form C). (Manum 1967)
　Colonial coccoid cyanobacterium (S).
(Form D). (Manum 1967)
　Sphaeromorph acritarch (S).
(Form E). (Manum 1967)
　Sphaeromorph acritarch (S).
(Form F). (Manum 1967)
　Ellipsoidal (originally coccoid?) solitary cyanobacterium (S).

Table 22.3 (*Continued*).

Muhos Fm.
EU FIN
(ca. 650 Ma)

(algal colonies). (Tynni and Siivola 1966)
  Colonial coccoid (chroococcacean?) cyanobacterium ("S").
(cylindrical form with ringlike distensions). (Tynni and Uutela 1984)
  Tubular sheath of filamentous prokaryote (S).
(large sphaeromorph forms). (Tynni and Uutela 1984)
  Sphaeromorph acritarch ("S").
(sheathlike tube). (Tynni and Uutela 1984)
  Tubular sheath of filamentous cyanobacterium (S).
Eomicrhystridium sp. 2 of Tynni and Uutela 1984. (Tynni and Uutela 1984)
  Solitary coccoid cyanobacterium (S).
Eomycetopsis sp. of Tynni and Uutela 1984. (Tynni and Uutela 1984)
  Degraded collapsed sheaths of filamentous prokaryote (S).
Eosphaera sp. of Tynni and Uutela 1984. (Tynni and Uutela 1984)
  Sphaeromorph acritarch (not *Eosphaera*) (S).
Eosynechococcus moorei Hofmann 1976. (Tynni and Uutela 1984)
  Colonial ellipsoidal chroococcacean cyanobacterium ("S").
Favososphaeridium sp. type 1 of Tynni and Uutela 1984. (Tynni and Uutela 1984)
  Colonial coccoid (chroococcacean?) cyanobacterium (S).
Favososphaeridium sp. type 2 of Tynni and Uutela 1984. (Tynni and Uutela 1984)
  Solitary and possibly colonial coccoid cyanobacteria (S).
Floritheca muhosensis Tynni and Uutela 1984. (Tynni and Uutela 1984)
  Colonial ellipsoidal bacterium (S).
Gloeocapsomorpha sp. of Tynni and Uutela 1984. (Tynni and Uutela 1984)
  Colonial coccoid (chroococcacean?) cyanobacterium ("S").
Granomarginata sp. of Tynni and Uutela 1984. (Tynni and Uutela 1984)
  Sphaeromorph acritarch ("S").
Leiosphaeridia spp. of Tynni and Uutela 1984. (Tynni and Uutela 1984)
  Sphaeromorph acritarch ("S").
Leiovalia sp. of Tynni and Uutela 1984. (Tynni and Uutela 1984)
  Ellipsoidal envelope of colonial coccoid prokaryote (S).
Lunulidia nana Tynni and Uutela 1984. (Tynni and Uutela 1984)
  Solitary ellipsoidal cyanobacterium (S).
Microvalia spinosa Tynni and Uutela 1984. (Tynni and Uutela 1984)
  Solitary ellipsoidal cyanobacterium (S).
Muhosspora reticulata Tynni 1966. (Tynni and Siivola 1966)
  Sphaeromorph acritarch ("S").
Nucellosphaeridium sp. of Tynni and Uutela 1984. (Tynni and Uutela 1984)
  Sphaeromorph acritarch containing degraded cellular remnants ("S").
cf. *Octoedryxium*? Rudavskaja [Rudavskaya] 1973. (Tynni and Uutela 1984)
  Folded, sphaeromorph acritarch ("S").
Palaeoanacystis sp. of Tynni and Uutela 1984. (Tynni and Uutela 1984)
  Colonial coccoid (chroococcacean) cyanobacterium ("S").
Palaeopleurocapsa sp. of Tynni and Uutela 1984. (Tynni and Uutela 1984)
  Colonial coccoid prokaryote (S).
Pterospermella simica (Jankauskas 1980) Jankauskas 1982. (Tynni and Uutela 1984)
  Sphaeromorph acritarch (very poorly preserved) ("S").
Pterospermopsimorpha ornata Tynni and Uutela 1984. (Tynni and Uutela 1984)
  Sphaeromorph acritarch ("S").
Siphonophycus sp. of Tynni and Uutela 1984. (Tynni and Uutela 1984)
  Tubular sheath of filamentous cyanobacterium ("S").
Sphaerophycus aff. *parvum* Schopf 1968. (Tynni and Uutela 1984)
  Colonial coccoid prokaryote (S).
Symplassosphaeridium parvum Tynni 1978. (Tynni and Uutela 1984)
  Colonial coccoid (chroococcacean?) cyanobacterium (S).
Symplassosphaeridium sp. of Tynni and Uutela 1984. (Tynni and Uutela 1984)
  Sphaeromorph acritarch (?) (S).
Synsphaeridium spp. of Tynni and Uutela 1984. (Tynni and Uutela 1984)
  Colonial coccoid (chroococcacean) cyanobacterium (S).
Trachysphaeridium laminaritum Timofeev 1966. (Tynni and Uutela 1984)
  Sphaeromorph acritarch ("S").
Trachysphaeridium levis (Lopukhin 1971) Vidal 1974. (Tynni and Uutela 1984)
  Sphaeromorph acritarch ("S").
Trachysphaeridium sp. of Tynni and Uutela 1984. (Tynni and Uutela 1984)
  Sphaeromorph acritarch ("S").
Turuchanica aff. *kulgunica* (Jankauskas 1982) Tynni and Uutela 1984?. (Tynni and Uutela 1984)
  Sphaeromorph acritarch ("S").
Turuchanica maculata Tynni and Uutela 1984. (Tynni and Uutela 1984)
  Sphaeromorph acritarch ("S").

Tanafjord Grp., Dakkovarre Fm.
EU NOR
(ca. 650 Ma)

Spiromorpha sp. of Vidal 1981. (Vidal 1981b)
  Coiled tubular sheath of filamentous bacterium (S).
cf. *Stictosphaeridium* sp. of Vidal 1981. (Vidal 1981b)
  Colonial coccoid cyanobacterium (Fig. 19A–F) (S).

Ostrog "Series"
EU SUN
(ca. 640 Ma)

Gloeocapsomorpha prisca Zalessky [Zalesskij] 1917. (Timofeev 1966)
  Colonial coccoid chroococcacean? cyanobacterium ("S").
Lophosphaeridium sp. of Timofeev 1966. (Timofeev 1966)
  Sphaeromorph acritarch ("S").
Protosphaeridium densum Timofeev 1966. (Timofeev 1966)
  Sphaeromorph acritarch ("S").
Symplassosphaeridium incrustatum Timofeev 1959. (Timofeev 1966)
  Colonial coccoid chroococcacean? cyanobacterium (S).
Synsphaeridium conglutinatum Timofeev 1966. (Timofeev 1966)
  Colonial coccoid chroococcacean? cyanobacterium (S).
Turuchanica alara Rudavskaja [Rudavskaya] 1964?. (Timofeev 1966)
  Compressed sphaeromorph? acritarch (S).
Tyloligotriletum? *asper* Timofeev 1959. (Timofeev 1966)
  Sphaeromorph acritarch ("S").

Table 22.3 (Continued).

Barma Fm.
EU SUN
(ca. 630 Ma)

Protosphaeridium densum Timofeev 1966. (Timofeev 1966)
  Sphaeromorph acritarch ("S").

Brioverian (Upper)
EU FRA
(ca. 625 Ma)

(forme D). (Chauvel and Mansuy 1981)
  Colonial coccoid chroococcacean cyanobacterium (S).
(forme E). (Chauvel and Mansuy 1981)
  Sphaeromorph acritarch (S).
(forme G). (Chauvel and Mansuy 1981)
  Colonial coccoid cyanobacterium (S). [See Chapter 24, Pl. 44; I.]
Eosphaera aff. *tyleri* Barghoorn 1965. (Chauvel and Mansuy 1981)
  Degraded colonial coccoid cyanobacterium (S).

Nibel' Fm.
EU SUN
(ca. 625 Ma)

Leiosphaeridia eisenackia Timofeev 1959. (Timofeev 1966)
  Sphaeromorph acritarch ("S").
Protosphaeridium acis Timofeev 1966. (Timofeev 1966)
  Sphaeromorph acritarch ("S").
Protosphaeridium flexuosum Timofeev 1966. (Timofeev 1966)
  Sphaeromorph acritarch ("S").
Protosphaeridium tuberculiferum Timofeev 1966. (Timofeev 1966)
  Sphaeromorph acritarch ("S").

Zigan Fm.
EA SUN
(ca. 625 Ma)

Omalophyma gracilia Golub 1979. (Yankauskas 1982)
  Degraded carbonized tubular sheath of filamentous cyanobacterium (S).

Gdov Beds
EU SUN
(ca. 620 Ma)

Trachyoligotriletum incrassatum (Naumova 1949) Timofeev 1959. (Timofeev 1966)
  Sphaeromorph acritarch ("S").
Trachysphaeridium attenuatum Timofeev 1959. (Timofeev 1966)
  Sphaeromorph acritarch ("S").

Mogilev Fm.
EU SUN
(ca. 620 Ma)

Circumiella mogilevica Aseeva 1974. (Aseeva 1974)
  Tubular sheath of filamentous cyanobacterium (S).
Gloeocapsomorpha prisca Zalessky [Zalesskij] 1917. (Timofeev 1966)
  Colonial coccoid chroococcacean? cyanobacterium ("S").
Polycavita bullata (Andreeva 1966) Assejeva [Aseeva] 1982. (Aseeva 1982)
  Colonial coccoid cyanobacterium (S). [See Chapter 24, Pl. 51, F.]
Protosphaeridium acis Timofeev 1966. (Timofeev 1966)
  Sphaeromorph acritarch ("S").
Protosphaeridium densum Timofeev 1966. (Timofeev 1966)
  Sphaeromorph acritarch ("S").
Protosphaeridium densum Timofeev 1966. (Timofeev 1966)
  Sphaeromorph acritarch ("S").

Stictosphaeridium implexum Timofeev 1966. (Timofeev 1966)
  Sphaeromorph acritarch ("S").

Sediol' Fm.
EU SUN
(ca. 620 Ma)

Bavlinella faveolata Schepelewa [Schepeleva] 1963. (Timofeev 1969)
  Colonial coccoid prokaryote (S).
Bavlinella faveolata Schepelewa [Schepeleva] 1962. (Timofeev 1966)
  Colonial coccoid prokaryote (S).

Staraya Rechka Fm.
AS SIB
(ca. 620 Ma)

Gloeocapsomorpha sp. of Timofeev 1966. (Timofeev 1966)
  Colonial coccoid chroococcacean? cyanobacterium ("S").
Kildinella sp. of Timofeev 1966. (Timofeev 1966)
  Sphaeromorph acritarch ("S").
Leiosphaeridia eisenackia Timofeev 1959. (Timofeev 1966)
  Sphaeromorph acritarch ("S").
Protosphaeridium asaphum Timofeev 1966. (Timofeev 1966)
  Sphaeromorph acritarch ("S").
Protosphaeridium flexuosum Timofeev 1966. (Timofeev 1966)
  Sphaeromorph acritarch ("S").
Protosphaeridium gibberosum Timofeev 1966. (Timofeev 1966)
  Sphaeromorph acritarch ("S").
Protosphaeridium laccatum Timofeev 1966. (Timofeev 1966)
  Sphaeromorph acritarch ("S").
Protosphaeridium paleaceum Timofeev 1966. (Timofeev 1966)
  Sphaeromorph acritarch ("S").
Protosphaeridium pusillum Timofeev 1966. (Timofeev 1966)
  Sphaeromorph acritarch ("S").
Protosphaeridium rigidulum Timofeev 1966. (Timofeev 1966)
  Sphaeromorph acritarch ("S").
Protosphaeridium sp. of Timofeev 1966. (Timofeev 1966)
  Sphaeromorph acritarch ("S").
Protosphaeridium torulosum Timofeev 1966. (Timofeev 1966)
  Sphaeromorph acritarch ("S").
Stictosphaeridium sinapticuliferum Timofeev 1966. (Timofeev 1966)
  Sphaeromorph acritarch ("S").

Suirovo Fm.
EA SUN
(ca. 620 Ma)

Retiforma tolparica Mikhailova [Mikhajlova] 1987. (Mikhajlova and Podkovyrov 1987)
  Sphaeromorph acritarch ("S"). [See Chapter 24, Pl. 51, G.]

Zherbinsk Fm.
AS SIB
(ca. 620 Ma)

Archaeohystrichosphaeridium ignotum Timofeev 19??. (Timofeev 1966)
  Possible sphaeromorph acritarch having coarse surface ornament (cf. *Lophosphaeridium*) (M S).

Portfjeld Fm.
NA GRL
(ca. 615 Ma)

(thread-formed microfossil). (Pedersen 1970)
  Tubular sheath of filamentous cyanobacterium, cf. *Eomycetopsis* ("S").

Table 22.3 (Continued).

Post-Spilitic Grp.
EU CSK
(ca. 615 Ma)

(algal colony—cf. Botryococcaceae). (Konzalová 1973)
    Pila-like colonial alga (S).
Baltisphaeridium bohemicum Konzalová 1972. (Konzalová 1972)
    Highly degraded sphaeromorph acritarch (S M).
Baltisphaeridium bohemicum Konzalová 1972. (Konzalová 1974b)
    Degraded sphaeromorph (not acanthomorph) acritarch (S).
cf. *Cymatiogalea* Deunff 1961. (Konzalová 1974b)
    Poorly preserved sphaeromorph acritarch (S).
cf. *Cymatiogalea* sp. of Konzalová 1974. (Konzalová 1974a)
    Degraded sphaeromorph acritarch (M S)
Favososphaera conglobata Burmann 1972. (Konzalová 1974a)
    Colonial coccoid bacterium (S).
Favososphaera conglobata Burmann 1972. (Konzalová 1974b)
    Colonial coccoid cyanobacterium (S).
Favososphaera sola Burmann 1972. (Konzalová 1974a)
    Colonial coccoid bacterium (S).
Granomarginata cf. *squamacea* Volkova 1968. (Konzalová 1974b)
    Sphaeromorph acritarch ("S").
Granomarginata cf. *squamacea* Volkova 1968. (Konzalová 1974a)
    Sphaeromorph acritarch ("S").
aff. *Leiosphaeridia* sp. of Konzalová 1974. (Konzalová 1974a)
    Sphaeromorph acritarch ("S").
cf. *Leiosphaeridia* sp. of Konzalová 1974. (Konzalová 1974b)
    Sphaeromorph acritarch ("S").
aff. *Lophosphaeridium* sp. of Konzalová 1974. (Konzalová 1974b)
    Sphaeromorph acritarch ("S").
Micrhystridium nannacanthum Deflandre 1945. (Konzalová 1974a)
    Degraded solitary coccoid cyanobacterium (M S).
Micrhystridium spp. of Konzalová 1974. (Konzalová 1974b)
    Degraded sphaeromorph (not acanthomorph) acritarch (S).

Polarisbreen Grp., Wilsonbreen Fm.
EU SJM
(ca. 610 Ma)

(small leiosphaerids). (Knoll and Swett 1987)
    Solitary coccoid cyanobacteria (S).
Bavlinella faveolata Schepeleva [Shepeleva] 1963. (Knoll and Swett 1987)
    Colonial coccoid bacterium (S).
Protosphaeridium sp. of Knoll and Swett 1987. (Knoll and Swett 1987)
    Sphaeromorph acritarch ("S").

Ushitsa "Series"
EU SUN
(ca. 610 Ma)

Gloeocapsomorpha prisca Zalessky [Zalesskij] 1917. (Timofeev 1966)
    Colonial coccoid chroococcacean? cyanobacterium ("S").
Leioligotriletum crassum (Naumova 1949) Timofeev 1960. (Timofeev 1966)
    Sphaeromorph acritarch ("S").
Leiosphaeridia eisenackia Timofeev 1959. (Timofeev 1966)
    Sphaeromorph acritarch ("S").
Lopholigotriletum crispum Timofeev 1959. (Timofeev 1966)
    Sphaeromorph acritarch ("S").
Oscillatorites wernadskii Schepelewa [Shepeleva] 1960. (Timofeev 1966)
    Septate filamentous oscillatoriacean cyanobacterium ("S").
Protosphaeridium densum Timofeev 1966. (Timofeev 1966)
    Sphaeromorph acritarch ("S").
Protosphaeridium laccatum Timofeev 1966. (Timofeev 1966)
    Sphaeromorph acritarch ("S").
Protosphaeridium sp. of Timofeev 1966. (Timofeev 1966)
    Sphaeromorph acritarch ("S").
Protosphaeridium tuberculiferum Timofeev 1966. (Timofeev 1966)
    Sphaeromorph acritarch ("S").
Stenozonoligotriletum validum Timofeev 1958. (Timofeev 1966)
    Sphaeromorph acritarch ("S").
Stictosphaeridium implexum Timofeev 1966. (Timofeev 1966)
    Sphaeromorph acritarch ("S").
Stictosphaeridium sinapticuliferum Timofeev 1966. (Timofeev 1966)
    Sphaeromorph acritarch ("S").
Symplassosphaeridium incrustatum Timofeev 1959. (Timofeev 1966)
    Colonial coccoid chroococcacean? cyanobacterium (S).
Turuchanica alara Rudavskaja [Rudavakaya] 1964?. (Tomofeev 1966)
    Compressed sphaeromorph? acritarch (S).
Turuchanica alara Rudavskaja [Rudavskaya] 1964?· (Timofeev 1966)
    Compressed sphaeromorph? acritarch (S).

Ushitsa "Series," Min'kovets Horizon
EU SUN
(ca. 610 Ma)

Oscillatorites sp. of Timofeev 1966. (Timofeev 1966)
    Septate filamentous oscillatoriacean cyanobacterium ("S").

Ushitsa "Series," Yaryshev Fm.
EU SUN
(ca. 610 Ma)

Redkinia fedonkinispis Asseeva in press. (Asseeva in press)
    Possibly parts of invertebrate (?arthropod) carapace (S). [See Chapter 24, Pl. 52, B, C, D_1-D_2.]
Volyniella valdaica Schepeleva ex Aseeva 1974. (Aseeva 1974)
    Coiled tubular sheath of filamentous cyanobacterium (S). [See Chapter 24, Pl. 51, J.]

?Murdama Grp.
AS SAU
(ca. 604 Ma)

(chitinozoanlike microfossils). (Binda and Bokhari 1980)
    Tear-shaped microfossil ("S") = *Melanocyrillium* sp. (S).

"Vendian (?)" of the White Sea
EU SUN
(ca. 600 Ma)

Leiosphaeridia bituminosa Timofeev 1966. (Timofeev 1966)
    Sphaeromorph acritarch ("S").
Protosphaeridium densum Timofeev 1966. (Timofeev 1966)
    Sphaeromorph acritarch ("S").
Trachysphaeridium sp. of Timofeev 1966. (Timofeev 1966)
    Sphaeromorph acritarch ("S").

"Vendian" of Moldavia
EU SUN
(ca. 600 Ma)

Protosphaeridium acis Timofeev 1966. (Timofeev 1966)
    Sphaeromorph acritarch ("S").
Trachysphaeridium attenuatum Timofeev 1959. (Timofeev 1966)
    Sphaeromorph acritarch ("S").
Trachysphaeridium laminaritum Timofeev 1966. (Timofeev 1966)
    Sphaeromorph acritarch ("S").

Table 22.3 (Continued).

Trachysphaeridium uspenskyi Timofeev 1959. (Timofeev 1966)
   Sphaeromorph acritarch ("S").

"Vendian of Tien Shan"
AS SUN
(ca. 600 Ma)

Trachymarginata speciosa Lopukhin 1966. (Lopukhin 1966)
   Sphaeromorph acritarch ("S").

"Vendian of Ukrainian Carpathians"
EU SUN
(ca. 600 Ma)

Cephalophytarion sp. of Timofeev, German and Mikhajlova 1976. (Timofeev et al. 1976)
   Septate filamentous cyanobacterium ("S").
Ethmosphaeridium sp. of Timofeev, German and Mikhajlova 1976. (Timofeev et al. 1976)
   Sphaeromorph acritarch ("S").
Kildinella hyperboreica Timofeev 1966. (Timofeev et al. 1976)
   Sphaeromorph acritarch ("S").
Macroptycha biplicata Timofeev 1976. (Timofeev et al. 1976)
   Sphaeromorph acritarch ("S").
Nucellosphaeridium cf. *minutum* Timofeev 1966. (Timofeev et al. 1976)
   Sphaeromorph acritarch ("S").
Oscillatorites sp. of Timofeev, German and Mikhajlova 1976. (Timofeev et al. 1976)
   Septate filamentous oscillatoriacean cyanobacterium ("S").
Podoliella irregulare Timofeev 1973. (Timofeev et al. 1976)
   Sphaeromorph acritarch ("S").
Podoliella regulare Timofeev 1973. (Timofeev et al. 1976)
   Colonial coccoid cyanobacterium (S).
Protosphaeridium flexuosum Timofeev 1966. (Timofeev et al. 1976)
   Sphaeromorph acritarch ("S").
Protosphaeridium tuberculiferum Timofeev 1966. (Timofeev et al. 1976)
   Sphaeromorph acritarch ("S").
Symplassosphaeridium sp. of Timofeev, German and Mikhajlova 1976. (Timofeev et al. 1976)
   Colonial coccoid (chroococcacean?) cyanobacterium (S).
Synsphaeridium sorediforme Timofeev 1969?. (Timofeev et al. 1976)
   Colonial coccoid (chroococcacean?) cyanobacterium (S).
Trachysphaeridium laminaritum Timofeev 1966. (Timofeev et al. 1976)
   Sphaeromorph acritarch ("S").

"Vendian" of Moscow
EU SUN
(ca. 600 Ma)

Orygmatosphaeridium sp. of Timofeev 1966. (Timofeev 1966)
   Sphaeromorph acritarch ("S").
Trachysphaeridium attenuatum Timofeev 1959. (Timofeev 1966)
   Sphaeromorph acritarch ("S").

"Vendian" of Arkhangel'sk
EU SUN
(ca. 600 Ma)

Polyedrosphaeridium bullatum Timofeev 1966. (Timofeev 1966)
   Colonial coccoid chroococcacean? cyanobacterium (S).

"Vendian" of Leningrad
EU SUN
(ca. 600 Ma)

Stenozonoligotriletum sokolovii Timofeev 1959. (Timofeev 1966)
   Sphaeromorph acritarch ("S").

Central Mount Stuart Fm.
AU AUS NT
(ca. 600 Ma)

(non-septate tubular filaments). (Damassa and Knoll 1986)
   Tubular sheath of filamentous cyanobacterium ("S").
Leiosphaeridia spp. of Damassa and Knoll 1986. (Damassa and Knoll 1986)
   Sphaeromorph acritarch ("S").

Derlo Fm.
EU SUN
(ca. 600 Ma)

Octoedryxium simmetricum Timofeev 1973. (Timofeev 1973)
   Prismatomorph acritarch ("S"). [See Chapter 24, Pl. 53, H.]
Octoedryxium truncatum Rudavskaja [Rudavskaya] 1973. (German in press)
   Prismatomorph acritarch ("S"). [See Chapter 24, Pl. 53, G_1–G_3, I.]

Grant Bluff Fm.
AU AUS NT
(ca. 600 Ma)

Leiosphaeridia spp. of Damassa and Knoll 1986. (Damassa and Knoll 1986)
   Sphaeromorph acritarch ("S").

Kairovo Fm. (Drill-Hole Sergeevskaya-800) (=?Redkino Fm.)
EU SUN
(ca. 600 Ma)

Baltisphaeridium perrarum Jankauskas [Yankauskas] 1980. (Yankauskas 1980c)
   Acanthomorph acritarch ("S"). [See Chapter 24, Pl. 55, B.]
Leiosphaeridia incrassatula Jankauskas [Yankauskas] 1980. (Yankauskas 1980c)
   Sphaeromorph acritarch with coalesced cytoplasmic contents ("S"). [See Chapter 24, Pl. 55, C.]
Satka granulosa Jankauskas [Yankauskas] 1980. (Yankauskas 1980c)
   Flattened sphaeromorph acritarch ("S"). [See Chapter 24, Pl. 55, A_1–A_2.]

Kanilovka Fm. (=Zharnov Fm.)
EU SUN
(ca. 600 Ma)

Flagellis tenuis Assejeva [Aseeva] 1982. (Aseeva 1982)
   Twisted tubular sheath of filamentous cyanobacterium (S). [See Chapter 24, Pl. 53, E_1–E_2.]
Podoliella irregulare Timofeev 1973. (German in press)
   Sphaeromorph acritarch ("S"). [See Chapter 24, Pl. 53, B, C.]
Volyniella canilovica Aseeva 1974. (Aseeva 1974)
   Tubular sheath of filamentous cyanobacterium (S). [See Chapter 24, Pl. 53, A.]

Laminarites Beds of Moscow
EU SUN
(ca. 600 Ma)

Protosphaeridium torulosum Timofeev 1966. (Timofeev 1966)

Laminarites Clay of Leningrad
EU SUN
(ca. 600 Ma)

Orygmatosphaeridium sp. of Timofeev 1966. (Timofeev 1966)

Sphaeromorph acritarch ("S").
Protosphaeridium laccatum Timofeev 1966. (Timofeev 1966)
Sphaeromorph acritarch ("S").

Laminarites Horizon of Morsovo
EU SUN
(ca. 600 Ma)

Leiosphaeridia bituminosa Timofeev 1966. (Timofeev 1966)
Sphaeromorph acritarch ("S").

Laminarites Horizon of Vologda
EU SUN
(ca. 600 Ma)

Synsphaeridium conglutinatum Timofeev 1966. (Timofeev 1966)
Colonial coccoid chroococcacean? cyanobacterium (S).

Nama Grp., Kuibis Subgrp.
AF NAM
(ca. 600 Ma)

Bavlinella faveolata Shepeleva 1963. (Germs et al. 1986)
Colonial coccoid prokaryote (S).

Nama Grp., Schwarzrand Subgrp.
AF NAM
(ca. 600 Ma)

(filamentous sheaths). (Germs et al. 1986)
Flattened tubular sheath of filamentous cyanobacterium ("S").
Bavlinella faveolata Shepeleva 1963. (Germs et al. 1986)
Colonial coccoid prokaryote (S).
Chuaria circularis Walcott 1899. (Germs et al. 1986)
Sphaeromorph acritarch ("S").
Leiosphaeridia spp. of Germs, Knoll and Vidal 1986. (Germs et al. 1986)
Sphaeromorph acritarch ("S").
Vendotaenia sp. of Germs, Knoll and Vidal 1986. (Germs et al. 1986)
Possibly multitrichomic sheath of filamentous cyanobacteria ("S").

Olistostroma del Membrillar
EU ESP
(ca. 600 Ma)

(tipo A). (Palacios 1983)
Colonial coccoid prokaryote (S).
(tipo B). (Palacios 1983)
Solitary coccoid ensheathed chroococcacean cyanobacterium (S).
Bavlinella faveolata Shepeleva 1963. (Palacios 1983)
Colonial coccoid bacterium (S).
Trachysphaeridium? laufeldii Vidal 1976. (Palacios 1983)
Sphaeromorph acritarch ("S").

Onon Fm.
AS SIB
(ca. 600 Ma)

Acanthodiacrodium sp. of Timofeev 1966. (Timofeev 1966)
Acanthomorph acritarch ("S").
Favososphaeridium sp. of Timofeev 1966. (Timofeev 1966)
Colonial coccoid chroococcacean cyanobacterium (S).
Protosphaeridium pusillum Timofeev 1966. (Timofeev 1966)
Sphaeromorph acritarch ("S").
Protosphaeridium sp. of Timofeev 1966. (Timofeev 1966)
Sphaeromorph acritarch ("S").

Roznichi Fm. (=?Yaryshev Fm.)
EU SUN
(ca. 600 Ma)

Tubulosa corrugata Assejeva [Aseeva] 1982. (Aseeva 1982)
Tubular sheath of filamentous cyanobacterium (S). [See Chapter 24, Pl. 51, I.]

Sinian (Upper), Doushantuo Fm.
AS CHN
(ca. 600 Ma)

(backfilled burrows). (Awramik et al. 1985)
Possibly *Polybessurus*-like cyanobacterium (S).
(ensheathed trichome). (Awramik et al. 1985)
Ensheathed septate filamentous oscillatoriacean (*Lyngbya*-like) cyanobacterium ("S").
(partitioned tubes). (Awramik et al. 1985)
Tubular sheath of pseudoseptate filamentous cyanobacterium, possibly aff. *Polychlamydum insigne* (S).
(possible multitrichomous bundles). (Awramik et al. 1985)
Aggregated tubular sheaths of filamentous cyanobacteria (S).
(unnamed double-walled spheroids). (Z. Zhang 1985)
Ensheathed solitary coccoid chroococcacean cyanobacterium ("S").
(unnamed filaments). (Zhang 1982b)
Tubular sheaths of filamentous oscillatoriacean cyanobacteria containing possible trichome remnants ("M").
(unnamed larger spheroids). (Z. Zhang 1985)
Spheroidal envelope of colonial coccoid cyanobacterium (cf. *Myxococcoides* sp.) ("S").
Aphetospora euthenia Lo 1980. (Z. Zhang 1985)
Colonial coccoid chroococcacean cyanobacterium ("S").
Baltisphaeridium [sp.] of Awramik et al. 1985. (Awramik et al. 1985)
Acanthomorph acritarch ("S").
Comasphaeridium magnum Z. Zhang 1984. (Zhang 1984a)
Acanthomorph acritarch ("S").
Doushantuonema peatii Z. Zhang 1981. (Zhang 1982b)
Septate filamentous oscillatoriacean cyanobacterium ("S").
Eomycetopsis robusta Schopf 1968. (Zhang 1982b)
Tubular sheath of filamentous prokaryote (S).
Gunflintia cf. *minuta* Barghoorn 1965. (Zhang 1982b)
Septate filamentous bacterium (S).
Huroniospora spp. of Z. Zhang 1985. (Z. Zhang 1985)
Assemblage of coccoid prokaryotes (S).
Myxococcoides inornata Schopf 1968. (Z. Zhang 1985)
Colonial coccoid chroococcacean cyanobacterium ("S").
Myxococcoides sp. of Z. Zhang 1985. (Z. Zhang 1985)
Ensheathed, colonial coccoid chroococcacean cyanobacteriuj ("S").
Nannococcus vulgaris J. Oehler 1977. (Z. Zhang 1985)
Colonial coccoid chroococcacean cyanobacterium ("S").
Obruchevella minor Z. Zhang 1984. (Zhang 1984c)
Spirally coiled, tubular *Spirulina*- or *Arthrospira*-like cyanobacterium ("S").
Oscillatoriopsis maxima (Y. Zhang 1981) Z. Zhang 1985. (Zhang 1986b)
Septate filamentous oscillatoriacean cyanobacterium ("S").
Oscillatoriopsis sp. of Z. Zhang 1984. (Zhang 1986b)
Septate filamentous oscillatoriacean cyanobacterium ("S").
Palaeolyngbya sp. of Z. Zhang 1982. (Zhang 1982b)
Septate filamentous oscillatoriacean cyanobacterium ("S").
Paratetraphycus giganteus Z. Zhang 1985. (Z. Zhang 1985)
Colonial (planar tetrahedral) coccoid chroococcacean (cf. *Gloeocapsa*) cyanobacterium ("S").

Microfossils

Table 22.3 *(Continued)*.

Rhicnonema antiquum Hofmann 1976. Zhang 1982b)
  Septate filamentous prokaryote (S).
Salome hubeiensis Z. Zhang 1986. (Zhang 1986b)
  Ensheathed filamentous ?oscillatoriacean cyanobacterium ("S").
Salome svalbardensis Knoll 1982. (Zhang 1986b)
  Lamellated tubular sheath of filamentous cyanobacterium ("S").
Siphonophycus inornatum Y. Zhang 1981. (Zhang 1986b)
  Tubular sheath of filamentous cyanobacterium (S).
Siphonophycus sinense Z. Zhang 1986. (Zhang 1986b)
  Tubular sheath of filamentous cyanobacterium ("S").
Siphonophycus sp. of Z. Zhang 1984. (Zhang 1984b)
  Tubular sheath of filamentous cyanobacterium ("S").
Siphonophycus sp. of Z. Zhang 1982. (Zhang 1982b)
  Tubular sheath of filamentous cyanobacterium ("S").

  Tanafjord Grp., Dakkovarre Fm.
  EU NOR
  (ca. 600 Ma)

Chuaria circularis Walcott 1899. (Vidal 1981b)
  Sphaeromorph acritarch ("S").
Kildinella hyperboreica Timofeev 1966. (Vidal 1981b)
  Sphaeromorph acritarch ("S").
Kildinella sinica Timofeev 1966. (Vidal 1981b)
  Sphaeromorph acritarch ("S").
Pterospermopsimorpha mogilevica Timofeev 1973. (Vidal 1981b)
  Degraded sphaeromorph acritarch containing internal cellular remnants (S M).
Vandalosphaeridium varangeri Vidal 1981. (Vidal 1981b)
  Degraded, membranous, possibly acanthomorph acritarch ("S M").

  Tanafjord Grp., Grasdal Fm.
  EU NOR
  (ca. 600 Ma)

Chuaria circularis Walcott 1899. (Vidal 1981b)
  Sphaeromorph acritarch ("S").
Vandalosphaeridium varangeri Vidal 1981. (Vidal 1981b)
  Degraded, membranous, possibly acanthomorph acritarch ("S M").

  Tillite Grp.
  NA GRL
  (ca. 600 Ma)

Bavlinella faveolata Shepeleva 1962. (Vidal 1976a)
  Colonial coccoid bacterium (S).
Bavlinella faveolata Shepeleva 1962. (Vidal 1979)
  Colonial coccoid bacterium (S).
Chuaria circularis Walcott 1899. (Vidal 1979)
  Sphaeromorph acritarch ("S").
Kildinella sp. of Vidal 1979. (Vidal 1979)
  Sphaeromorph acritarch ("S").
Leiosphaeridia sp. of Vidal 1979. (Vidal 1979)
  Sphaeromorph acritarch ("S").
Octoedryxium truncatum Rudavskaya 1973. (Vidal 1979)
  Prismatomorph acritarch ("S").
cf. *Stictosphaeridium* sp. of Vidal 1976. (Vidal 1979)
  Colonial coccoid cyanobacterium (S).
Trachysphaeridium timofeevii Vidal 1976. (Vidal 1976a)
  Sphaeromorph acritarch ("S").

  Vestertana Grp., Lower Tillite Fm.
  EU NOR
  (ca. 600 Ma)

Chuaria circularis Walcott 1899. (Vidal 1981b)
  Sphaeromorph acritarch ("S").
Kildinella hyperboreica Timofeev 1966. (Vidal 1981b)
  Sphaeromorph acritarch ("S").
Kildinella sinica Timofeev 1966. (Vidal 1981b)
  Sphaeromorph acritarch ("S").
cf. *Stictosphaeridium* sp. of Vidal 1981. (Vidal 1981b)
  Colonial coccoid cyanobacterium (Fig. 19A–F) (S).

  Vestertana Grp., Mortensnes Tillite Fm.
  EU NOR
  (ca. 600 Ma)

Kildinella hyperboreica Timofeev 1966. (Vidal 1981b)
  Sphaeromorph acritarch ("S").
cf. *Stictosphaeridium* sp. of Vidal 1981. (Vidal 1981b)
  Colonial coccoid cyanobacterium (Fig. 19A–F) (S).

  Vestertana Grp., Nyborg Fm.
  EU NOR
  (ca. 600 Ma)

Chuaria circularis Walcott 1899. (Vidal 1981b)
  Sphaeromorph acritarch ("S").
Kildinella hyperboreica Timofeev 1966. (Vidal 1981b)
  Sphaeromorph acritarch ("S").

  Wilpena Grp., Tent Hill Fm., Arcoona Quartzite Mbr.
  AU AUS SA
  (ca. 600 Ma)

Chuaria cf. *circularis* Walcott 1899. (Damassa and Knoll 1986)
  Sphaeromorph acritarch ("S").
Leiosphaeridia spp. of Damassa and Knoll 1986. (Damassa and Knoll 1986)
  Sphaeromorph acritarch ("S").

  Yudoma Fm.
  AS SIB
  (ca. 600 Ma)

("undifferentiated spheroids" [Yudoma Fm.]). (Mendelson and Schopf 1982)
  Solitary (chroococcacean?) cyanobacteria ("M S"). [See Chzpter 24, Pl. 54, A–C.]
(branched tubes with ellipsoidal inclusions). (Lo 1980)
  Incertae sedis. Resemble fungal hyphae ("S").
(spheroids in linear arrangement). (Lo 1980)
  Colonial coccoid (chroococcacean?) cyanobacterium ("S").
(spheroids in triads). (Lo 1980)
  Colonial coccoid chroococcacean cyanobacterium ("S").
(*Diplococcus*-shaped microstructures). (Lo 1980)
  Colonial coccoid chroococcacean cyanobacterium ("S").
Aphetospora euthenia Lo 1980. (Lo 1980)
  Solitary coccoid chroococcacean cyanobacterium ("S").
Brachypleganon khandanum Lo 1980. (Lo 1980)
  Solitary ellipsoidal (chroococcacean?) cyanobacterium ("S").
Caryosphaeroides? sp. of Lo 1980. (Lo 1980)
  Colonial coccoid chroococcacean cyanobacterium ("S").
Eoentophysalis yudomatica Lo 1980. (Lo 1980)
  Colonial ellipsoidal entophysalidacean cyanobacterium ("S").
Eomycetopsis? campylomitus Lo 1980. (Lo 1980)
  Tubular sheath of filamentous bacterium (S).
Eomycetopsis? siberiensis Lo 1980. (Lo 1980)
  Tubular sheath of filamentous cyanobacterium (S). [See Chapter 24, Pl. 54, F.]
Eosphaera? sp. of Lo 1980. (Lo 1980)

Table 22.3 (Continued).

Incompletely preserved ensheathed colonial coccoid cyanobacterium (not *Eosphaera*) (S).
Eozygion sp. of Lo 1980. (Lo 1980)
Colonial coccoid chroococcacean cyanobacterium ("S").
Euryaulidion cylindratum Lo 1980. (Lo 1980)
Tubular sheath of filamentous cyanobacterium (S).
Huroniospora spp. of Mendelson and Schopf 1982. (Mendelson and Schopf 1982)
Solitary, coccoid (chroococcacean?) cyanobacterium ("M S"). [See Chapter 24, Pl. 54, D, E.]
Kildinella jacutica Timofeev 1966. (Timofeev 1966)
Sphaeromorph acritarch ("S").
Micrhystridium sp. of Lo 1980. (Lo 1980)
Acanthomorph acritarch ("S").
Myxococcoides staphylidion Lo 1980. (Lo 1980)
Colonial coccoid chroococcacean cyanobacterium ("S"). [See Chapter 24, Pl. 54, G.]
Protosphaeridium sp. of Timofeev 1966. (Timofeev 1966)
Sphaeromorph acritarch ("S").
Tetraphycus conjunctum Lo 1980. (Lo 1980)
Colonial coccoid chroococcacean cyanobacterium ("S"). [See Chapter 24, Pl. 54, H.]
Trachysphaeridium attenuatum Timofeev 1959. (Timofeev 1966)
Sphaeromorph acritarch ("S").

Zin'kov Fm.
EU SUN
(ca. 600 Ma)

Leiosphaeridia undulata Timofeev 1973. (Timofeev 1973)
Sphaeromorph acritarch ("S"). [See Chapter 24, Pl. 53, J.]

600 to 550 Ma:
Valdaj [Valdai] "Series"
EU SUN
(ca. 590 Ma)

Polyedrosphaeridium bullatum Timofeev 1966. (Timofeev 1966)
Colonial coccoid chroococcacean? cyanobacterium (S).
Protosphaeridium acis Timofeev 1966. (Timofeev 1966)
Sphaeromorph acritarch ("S").
Protosphaeridium asaphum Timofeev 1966. (Timofeev 1966)
Sphaeromorph acritarch ("S").
Protosphaeridium flexuosum Timofeev 1966. (Timofeev 1966)
Sphaeromorph acritarch ("S").
Protosphaeridium laccatum Timofeev 1966. (Timofeev 1966)
Sphaeromorph acritarch ("S").
Stictosphaeridium sinapticuliferum Timofeev 1966. (Timofeev 1966)
Sphaeromorph acritarch ("S").
Symplassosphaeridium incrustatum Timofeev 1959. (Timofeev 1966)
Colonial coccoid chroococcacean? cyanobacterium (S).
Trachysphaeridium attenuatum Timofeev 1959. (Timofeev 1966)
Sphaeromorph acritarch ("S").

Conception Grp., Drook Fm.
NA CAN NF
(ca. 585 Ma)

Taeniatum sp. of Hofmann, Hill & King 1979. (Hofmann et al. 1979)
Sheath of filamentous cyanobacterium ("S").

Conception Grp., Mall Bay Fm.
NA CAN NF
(ca. 585 Ma)

Taeniatum sp. of Hofmann, Hill & King 1979. (Hofmann et al. 1979)
Sheath of filamentous cyanobacterium ("S").

Conncecting Point Grp.
NA CAN NF
(ca. 583 Ma)

Eomicrhystridium? sp. of Hofmann, Hill & King 1979. (Hofmann et al. 1979)
Deformed sphaeromorph acritarch (S).
Taeniatum sp. of Hofmann, Hill & King 1979. (Hofmann et al. 1979)
Sheath of filamentous cyanobacterium ("S").
Trachysphaeridium sp. of Hofmann, Hill & King 1979. (Hofmann et al. 1979)
Sphaeromorph acritarch ("S").
Trematosphaeridium holtedahlii Timofeev 1966. (Hofmann et al. 1979)
Sphaeromorph acritarch ("S").

Manykaj Fm.
AS SIB
(ca. 580 Ma)

Kildinella hyperboreica Timofeev 1966. (Timofeev 1966)
Sphaeromorph acritarch ("S").
Protosphaeridium acis Timofeev 1966. (Timofeev 1966)
Sphaeromorph acritarch ("S").
Protosphaeridium torulosum Timofeev 1966. (Timofeev 1966)
Sphaeromorph acritarch ("S").

Aga Fm.
AS SIB
(ca. 575 Ma)

Favososphaeridium favosum Timofeev 1966. (Timofeev 1966)
Sphaeromorph acritarch ("M S").
Protosphaeridium rigidulum Timofeev 1966. (Timofeev 1966)
Sphaeromorph acritarch ("S").

Map unit 11
NA CAN YK
(ca. 575 Ma)

Eomycetopsis spp. of Hofmann 1984. (Hofmann 1984)
Tubular sheath of filamentous prokaryote ("S").

Polarisbreen Grp., Dracoisen Fm.
EU SJM
(ca. 575 Ma)

(filamentous microfossils). (Knoll and Swett 1987)
Flattened tubular sheath of filamentous cyanobacterium ("S"). Fig. 8.19 not interpretable (S).
(small leiosphaerids). (Knoll and Swett 1987)
Solitary coccoid cyanobacteria (S).
Bavlinella faveolata Schepeleva [Shepeleva] 1963. (Knoll and Swett 1987)
Colonial coccoid bacterium (S).
Leiosphaeridia spp. of Knoll and Swett 1987. (Knoll and Swett 1987)
Sphaeromorph acritarch ("S").
Micrhystridium tornatum Volkova 1968. (Knoll and Swett 1987)
Acanthomorph acritarch (very short-spined) ("S").
Protosphaeridium sp. of Knoll and Swett 1987. (Knoll and Swett 1987)
Sphaeromorph acritarch ("S").

Table 22.3 (Continued).

Unnamed siltstone unit 1
NA CAN YK
(ca. 575 Ma)

Bavlinella sp. of Hofmann 1984. (Hofmann 1984)
 Colonial coccoid bacterium (S).
Eomicrhystridium? sp. of Hofmann 1984. (Hofmann 1984)
 Degraded sphaeromorph acritarch (S).
Eomycetopsis spp. of Hofmann 1984. (Hofmann 1984)
 Tubular sheath of filamentous prokaryote ("S").
Leiosphaeridia spp. of Hofmann 1984. (Hofmann 1984)
 Sphaeromorph acritarch ("S").
Nostocomorpha sp. of Hofmann 1984. (Hofmann 1984)
 Iron oxide-replaced filamentous (nostocalean) cyanobacterium (S).
Nucellosphaeridium sp. of Hofmann 1984. (Hofmann 1984)
 Degraded sphaeromorph acritarch ("S").
Paleamorpha? sp. of Hofmann 1984. (Hofmann 1984)
 Flattened tubular sheath of filamentous cyanobacterium (S).
Siphonophycus spp. of Hofmann 1984. (Hofmann 1984)
 Tubular sheaths of filamentous cyanobacteria (S).

Unnamed siltstone unit 2
NA CAN YK
(ca. 575 Ma)

Bavlinella sp. of Hofmann 1984. (Hofmann 1984)
 Colonial coccoid bacterium (S).
Eomycetopsis spp. of Hofmann 1984. (Hofmann 1984)
 Tubular sheath of filamentous prokaryote ("S").
Nostocomorpha sp. of Hofmann 1984. (Hofmann 1984)
 Iron oxide-replaced filamentous (nostocalean) cyanobacterium (S).

"Eocambrian" of Finnmark
EU NOR
(ca. 570 Ma)

Leioligotriletum cf. *crassum* (Naumova 1949) Timofeev 1959. (Timofeev 1963)
 Sphaeromorph acritarch ("S").
Monotrematum sp. of Timofeev 1963. (Timofeev 1963)
 Sphaeromorph acritarch ("S").
Symplassosphaeridium incrustatum Timofeev 1959. (Timofeev 1966)
 Colonial coccoid chroococcacean? cyanobacterium (S).
Trachyoligotriletum cf. *incrassatum* (Naumova 1949) Timofeev 1959. (Timofeev 1963)
 Sphaeromorph acritarch ("S").
Trematosphaeridium sp. of Timofeev 1963. (Timofeev 1963)
 Sphaeromorph acritarch ("S").

"Eocambrian" of Prague
EU CSK
(ca. 570 Ma)

Protosphaeridium acis Timofeev 1966. (Timofeev 1966)
 Sphaeromorph acritarch ("S").
Protosphaeridium patelliforme Timofeev 1966. (Timofeev 1966)
 Sphaeromorph acritarch ("S").
Protosphaeridium rigidulum Timofeev 1966. (Timofeev 1966)
 Sphaeromorph acritarch ("S").
Stenozonoligotriletum sokolovii Timofeev 1959. (Timofeev 1966)
 Sphaeromorph acritarch ("S").

Dengying Fm.
AS CHN
(ca. 570 Ma)

Clonophycus J. Oehler 1977. (Wang 1981)
 Colonial coccoid cyanobacteria ("S").

Nigrit "Series"
AF ESH
(ca. 570 Ma)

Kildinella hyperboreica Timofeev 1966. (Timofeev 1966)
 Sphaeromorph acritarch ("S").
Protosphaeridium densum Timofeev 1966. (Timofeev 1966)
 Sphaeromorph acritarch ("S").

Vestertana Grp., Stappogiedde Fm.
EU NOR
(ca. 570 Ma)

Vendotaenia cf. *antiqua* Gnilovskaya 1971. (Vidal 1981b)
 Possibly flattened carbonized cyanobacterial sheath (S).

Lausitzer Grauwackenformation
EU DDR
(ca. 565 Ma)

(fadenförmige Reste [filamentous remains]). (Burmann 1966)
 Degraded sheaths of filamentous cyanobacteria (S).

Kotlin Fm.
EU SUN
(ca. 560 Ma)

Ambiguaspora parvula Volkova 1976. (Volkova 1976)
 Probably meiotically produced algal (chlorophycean? or rhodophycean?) spores ("S").
Bavlinella faveolata Schepeleva [Shepeleva] 1962. (German et al. 1989)
 Colonial coccoid bacterium (S). [See Chapter 24, Pl. 54, J_1–J_3.]

Hodgewater Grp., Halls Town Fm.
NA CAN NF
(ca. 558 Ma)

Taeniatum sp. of Hofmann, Hill and King 1979. (Hofmann et al. 1979)
 Sheath of filamentous cyanobacterium ("S").
Trachysphaeridium sp. of Hofmann, Hill and King 1979. (Hofmann et al. 1979)
 Sphaeromorph acritarch ("S").

Hodgewater Grp., Snows Pond Fm.
NA CAN NF
(ca. 558 Ma)

Taeniatum sp. of Hofmann, Hill and King 1979. (Hofmann et al. 1979)
 Sheath of filamentous cyanobacterium ("S").
Trachysphaeridium sp. of Hofmann, Hill and King 1979. (Hofmann et al. 1979)
 Sphaeromorph acritarch ("S").

Musgravetown Grp.
NA CAN NF
(ca. 558 Ma)

Taeniatum sp. of Hofmann, Hill and King 1979. (Hofmann et al. 1979)
 Sheath of filamentous cyanobacterium ("S").

Musgravetown Grp., Trinny Cove Fm.
NA CAN NF
(ca. 558 Ma)

Trachysphaeridium sp. of Hofmann, Hill and King 1979. (Hofmann et al. 1979)

Table 22.3 (Continued).

Sphaeromorph acritarch ("S").

Signal Hill Grp., Cappahayden Fm.
NA CAN NF
(ca. 558 Ma)

Taeniatum sp. of Hofmann, Hill and King 1979. (Hofmann et al. 1979)
Sheath of filamentous cyanobacterium ("S").
Trachysphaeridium sp. of Hofmann, Hill and King 1979. (Hofmann et al. 1979)
Sphaeromorph acritarch ("S").

Signal Hill Grp., Gibbett Hill Fm.
NA CAN NF
(ca. 558 Ma)

Trachysphaeridium sp. of Hofmann, Hill and King 1979. (Hofmann et al. 1979)
Sphaeromorph acritarch ("S").

St. John's Grp., Fermeuse Fm.
NA CAN NF
(ca. 558 Ma)

Taeniatum sp. of Hofmann, Hill and King 1979. (Hofmann et al. 1979)
Sheath of filamentous cyanobacterium ("S").
Trachysphaeridium sp. of Hofmann, Hill and King 1979. (Hofmann et al. 1979)
Sphaeromorph acritarch ("S").

St. John's Grp., Renews Head Fm.
NA CAN NF
(ca. 558 Ma)

Trachysphaeridium sp. of Hofmann, Hill and King 1979. (Hofmann et al. 1979)
Sphaeromorph acritarch ("S").

"Late Precambrian" of Antarctica
AN ATA
(ca. 555 Ma)

Protosphaeridium densum Timofeev 1966. (Timofeev 1966)
Sphaeromorph acritarch ("S").

"Late Precambrian/Cambrian" of Antarctica
AN ATA
(ca. 555 Ma)

Acanthodiacrodium sp. of Timofeev 1966. (Timofeev 1966)
Acanthomorph acritarch ("S").
Protosphaeridium acis Timofeev 1966. (Timofeev 1966)
Sphaeromorph acritarch ("S").

"Late Precambrian/Cambrian," Patom Highland
AS SIB
(ca. 555 Ma)

Kildinella hyperboreica Timofeev 1966. (Timofeev 1966)
Sphaeromorph acritarch ("S").
Protosphaeridium pusillum Timofeev 1966. (Timofeev 1966)
Sphaeromorph acritarch ("S").
Protosphaeridium tuberculiferum Timofeev 1966. (Timofeev 1966)
Sphaeromorph acritarch ("S").
Synsphaeridium conglutinatum Timofeev 1966. (Timofeev 1966)
Colonial coccoid chroococcacean? cyanobacterium (S).
Trematosphaeridium sinuatum Timofeev 1966. (Timofeev 1966)
Sphaeromorph acritarch ("S").

"suite of bituminous rocks," Patom Highland
AS SIB
(ca. 555 Ma)

Archaeohystrichosphaeridium acer Timofeev 1959. (Timofeev 1966)
Acanthomorph acritarch ("S").
Archaeohystrichosphaeridium glebosum Kalin.? (Timofeev 1966)
Sphaeromorph acritarch ("S").
Archaeohystrichosphaeridium sp. of Timofeev 1966. (Timofeev 1966)
Acanthomorph acritarch ("S").

Kada Fm.
AS SIB
(ca. 555 Ma)

Protosphaeridium tuberculiferum Timofeev 1966. (Timofeev 1966)
Sphaeromorph acritarch ("S").

Nerchinskij Zavod Fm.
AS SIB
(ca. 555 Ma)

Gloeocapsomorpha sp. of Timofeev 1966. (Timofeev 1966)
Colonial coccoid chroococcacean? cyanobacterium ("S").
Protosphaeridium laccatum Timofeev 1966. (Timofeev 1966)
Sphaeromorph acritarch ("S").
Protosphaeridium paleaceum Timofeev 1966. (Timofeev 1966)
Sphaeromorph acritarch ("S").

Adeyton Grp., Random Fm.
NA CAN NF
(ca. 550 Ma)

Gunflintia brueckneri Nautiyal 1982. (Nautiyal 1982b)
Flattened tubular sheath of filamentous (oscillatoriacean?) cyanobacterium (nonseptate, broad, so not *Gunflintia*) (S).
Heliconema randomense Nautiyal 1982. (Nautiyal 1982b)
Flattened tubular sheath of filamentous (oscillatoriacean?) cyanobacterium (not spirally coiled, so not *Heliconema*) (S)
Siphonophycus hughesii Nautiyal 1982. (Nautiyal 1982b)
Flattened tubular sheath of filamentous (oscillatoriacean?) cyanobacterium ("S").
Siphonophycus kestron Schopf 1968. (Nautiyal 1976)
Flattened tubular sheath of filamentous (oscillatoriacean?) cyanobacterium ("S").
Taeniatum sp. of Hofmann, Hill and King 1979. (Hofmann et al. 1979)
Sheath of filamentous cyanobacterium ("S").

Blue Clays of Estonia
EU SUN
(ca. 550 Ma)

Archaeohystrichosphaeridium acanthaceum Timofeev 1959. (Timofeev 1966)
Acanthomorph acritarch ("S").
Archaeohystrichosphaeridium cellulare Timofeev 1959. (Timofeev 1966)
Acanthomorph or herkomorph acritarch (S M).
Kildinella hyperboreica Timofeev 1966. (Timofeev 1966)
Sphaeromorph acritarch ("S").
Leiosphaeridia ochroleuca Timofeev 19??. (Timofeev 1966)
Sphaeromorph acritarch ("S").
Lopholigotriletum crispum Timofeev 1959. (Timofeev 1966)
Sphaeromorph acritarch ("S").
Protosphaeridium asaphum Timofeev 1966. (Timofeev 1966)

Table 22.3 (Continued).

Sphaeromorph acritarch ("S").
Protosphaeridium densum Timofeev 1966. (Timofeev 1966)
Sphaeromorph acritarch ("S").
Stenozonoligotriletum sokolovii Timofeev 1959. (Timofeev 1966)
Sphaeromorph acritarch ("S").
Trachysphaeridium patellare Timofeev 1959. (Timofeev 1966)
Sphaeromorph acritarch ("S").

Blue Clays of Vologda
EU SUN
(ca. 550 Ma)

Archaeohystrichosphaeridium dorofeevii Timofeev 1959. (Timofeev 1966)
Acanthomorph acritarch ("S").
Archaeohystrichosphaeridium innominatum Timofeev 1959. (Timofeev 1966)
Acanthomorph acritarch ("S").
Archaeohystrichosphaeridium stipiforme Timofeev 1959. (Timofeev 1966)
Acanthomorph acritarch ("S").

Blue Clays of the White Sea
EU SUN
(ca. 550 Ma)

Kildinella sinica Timofeev 1966. (Timofeev 1966)
Sphaeromorph acritarch ("S").
Polyedrosphaeridium bullatum Timofeev 1966. (Timofeev 1966)
Colonial coccoid chroococcacean? cyanobacterium (S).
Protosphaeridium asaphum Timofeev 1966. (Timofeev 1966)
Sphaeromorph acritarch ("S").
Protosphaeridium densum Timofeev 1966. (Timofeev 1966)
Sphaeromorph acritarch ("S").
Protosphaeridium flexuosum Timofeev 1966. (Timofeev 1966)
Sphaeromorph acritarch ("S").
Protosphaeridium patelliforme Timofeev 1966. (Timofeev 1966)
Sphaeromorph acritarch ("S").
Protosphaeridium tuberculiferum Timofeev 1966. (Timofeev 1966)
Sphaeromorph acritarch ("S").
Stictosphaeridium sinapticuliferum Timofeev 1966. (Timofeev 1966)
Sphaeromorph acritarch ("S").
Stictosphaeridium tortulosum Timofeev 1966. (Timofeev 1966)
Sphaeromorph acritarch ("S").
Trachysphaeridium laminaritum Timofeev 1966. (Timofeev 1966)
Sphaeromorph acritarch ("S").

Blue Clays of Leningrad
EU SUN
(ca. 550 Ma)

Lophorytidodiacrodium tosnaense Timofeev 1959. (Timofeev 1966)
Distorted, folded sphaeromorph acritarch (S).
Nodularites maslovii Schepelewa [Shepeleva] 1960. (Timofeev 1966)
Septate filamentous prokaryote (S).
Oscillatorites wernadskii Shepeleva 1960. (Shepeleva 1960)
Septate filamentous cyanobacterium ("S").
Protosphaeridium densum Timofeev 1966. (Timofeev 1966)
Sphaeromorph acritarch ("S").
Trachysphaeridium attenuatum Timofeev 1966. (Timofeev 1966)
Sphaeromorph acritarch ("S").
Zonooidium guttiforme Timofeev 1957. (Timofeev 1966)
Sphaeromorph acritarch ("S").
Zonooidium mirabile Timofeev 1957. (Timofeev 1966)
Sphaeromorph acritarch ("S").

Kanilovka Fm. (ca. 600 Ma), Kotlin Fm. (ca. 560 Ma), and Lower Cambrian (Tommotian, Atdabanian, Lenian)
EU SUN POL
(ca. 545 Ma)

Leiosphaeridia div. sp. of Volkova et al. 1979. (Volkova et al. 1979)
Sphaeromorph acritarch ("S").

Kotlin Fm. (ca. 560 Ma), and
Lower Cambrian (Tommotian, Atdabanian, Lenian)
EU SUN POL
(ca. 545 Ma)

Micrhystridium tornatum Volkova 1968. (Volkova et al. 1979)
Acanthomorph acritarch (very short-spined) ("S").

Early Cambrian:

Lower Cambrian (Tommotian)
EU SUN POL
(ca. 545 Ma)

Ceratophyton vernicosum Kirjanov [Kir'yanov] 1979. (Volkova et al. 1979)
Conical bodies, possibly of animal (rather than plant) origin (S).
Dictyotidium birvetense Paškevičienė [Pashkyavichene] 1979. (Volkova et al. 1979)
Herkomorph acritarch ("S").
Leiosphaeridia dehisca Paškevičienė [Pashkyavichene] 1979. (Volkova et al. 1979)
Sphaeromorph acritarch ("S").
Leiosphaeridia pylomifera Paškevičienė [Pashkyavichene] 1979. (Volkova et al. 1979)
Sphaeromorph acritarch ("S").
Leiosphaeridia sp. 1 of Volkova et al. 1979. (Volkova et al. 1979)
Sphaeromorph acritarch ("S").
Retisphaeridium densum Paškevičienė [Pashkyavichene] 1979. (Volkova et al. 1979)
Folded sphaeromorph acritarch (S).
Teophipolia lacerata Kirjanov [Kir'yanov] 1979. (Volkova et al. 1979)
Netromorph acritarch (S M).

Meishucun Fm.
AS CHN
(ca. 545 Ma)

(coiled filament). (Wang 1981)
Sheath of filamentous cyanobacterium (S).
(degraded trichomes). (Wang 1981)
Septate filamentous cyanobacterium (S).
(filaments). (Wang 1981)
Septate filamentous cyanobacteria ("S").
(fragment of cellular filament). (Wang 1981)
Septate filamentous oscillatoriacean cyanobacterium ("S").
(non-ensheathed colonies). (Wang 1981)
Colonial coccoid cyanobacterium ("S").
Baltisphaeridium. (Wang 1981)
Possible poorly preserved acanthomorph acritarch (S M).
Circulinema jinningense F. Wang 1981. (Wang 1981)
Septate filamentous cyanobacterium ("S").
Cyanonema inflatum J. Oehler 1977. (Wang 1981)
Septate filamentous cyanobacterium ("S").
Eomycetopsis robusta Schopf 1968. (Wang 1981)
Sheath of filamentous cyanobacterium ("S").

Table 22.3 (Continued).

Myxococcoides grandis Horodyski and Donaldson 1980. (Wang 1981)
    Sphaeromorph acritarch (S).
Myxococcoides kingii Muir 1976. (Wang 1981)
    Colonial coccoid cyanobacterium ("S").
Obruchevella parva Reitlinger [Rejtlinger]. (Wang 1981)
    Tubular sheath of filamentous cyanobacterium ("S").
Protosphaeridium Timofeev 1966. (Wang 1981)
    Solitary coccoid cyanobacterium (S).

Niutitang Fm.
AS CHN
(ca. 545 Ma)

Paracymatiosphaera annularis F. Wang 1985. (Wang 1985)
    Pterospermopsimorphid acritarch (S M).

Yangjiaping Fm.
AS CHN
(ca. 545 Ma)

Micrhystridium ampliatum F. Wang 1985. (Wang 1985)
    Acanthomorph acritarch ("S").
Paracymatiosphaera hunnanensis F. Wang 1985. (Wang 1985)
    Pterospermopsimorphid acritarch (S).
Paracymatiosphaera irregularis F. Wang 1985. (Wang 1985)
    Pterospermopsimorphid acritarch (S M).
Paracymatiosphaera regularis F. Wang 1985. (Wang 1985)
    Pterospermopsimorphic acritarch (S M).
Protosolenopora distincta F. Wang 1985. (Wang 1985)
    Colonial ellipsoidal cyanobacterium (S).
Radiophycus yangjiapingensis F. Wang 1985. (Wang 1985)
    Acanthomorph acritarch (S).

Lower Cambrian (Tommotian, Atdabanian)
EU SUN POL
(ca. 543 Ma)

Granomarginata prima Naumova 1960. (Volkova et al. 1979)
    Sphaeromorph acritarch ("S").
Leiomarginata simplex Naumova 1960. (Volkova et al. 1979)
    Sphaeromorph acritarch ("S").
Tasmanites tenellus Volkova 1968. (Volkova et al. 1979)
    Sphaeromorph acritarch (S).

"Lower Cambrian" of Utai
AS CHN
(ca. 540 Ma)

Gloeocapsomorpha prisca Zalessky [Zalesskij] 1917. (Timofeev 1966)
    Colonial coccoid chroococcacean? cyanobacterium ("S").
Nucellosphaeridium sp. of Timofeev 1966. (Timofeev 1966)
    Sphaeromorph acritarch ("S").
Protosphaeridium densum Timofeev 1966. (Timofeev 1966)
    Sphaeromorph acritarch ("S").
Protosphaeridium patelliforme Timofeev 1966. (Timofeev 1966)
    Sphaeromorph acritarch ("S").
Protosphaeridium sp. of Timofeev 1966. (Timofeev 1966)
    Sphaeromorph acritarch ("S").
Symplassosphaeridium tumidulum Timofeev 1959. (Timofeev 1966)
    Colonial coccoid chroococcacean cyanobacterium (S).

"Lower Cambrian of Tien Shan"
AS SUN
(ca. 540 Ma)

Margoporata conflata Lopukhin 1966. (Lopukhin 1966)
    Sphaeromorph acritarch ("S").

"Lower Cambrian" of the Holy Cross Mountains
EU POL
(ca. 540 Ma)

Archaeohystrichosphaeridium acanthaceum Timofeev 1959. (Timofeev 1966)
    Acanthomorph acritarch ("S").
Kildinella sinica Timofeev 1966. (Timofeev 1966)
    Sphaeromorph acritarch ("S").
Lopholigotriletum grumosum Timofeev 1959. (Timofeev 1966)
    Sphaeromorph acritarch ("S").
Stictosphaeridium implexum Timofeev 1966. (Timofeev 1966)
    Sphaeromorph acritarch ("S").
Stictosphaeridium sinapticuliferum Timofeev 1966. (Timofeev 1966)
    Sphaeromorph acritarch ("S").

"Lower Cambrian" of Vologda
AS SIB
(ca. 540 Ma)

Archaeohystrichosphaeridium vologdaense Timofeev 1959. (Timofeev 1966)
    Acanthomorph acritarch ("S").

"Lower Cambrian" of Sayan
AS SIB
(ca. 540 Ma)

Lopholigotriletum grumosum Timofeev 1959. (Timofeev 1966)
    Sphaeromorph acritarch ("S").

"Lower Cambrian" of Yakutia
AS SIB
(ca. 540 Ma)

Protosphaeridium densum Timofeev 1966. (Timofeev 1966)
    Sphaeromorph acritarch ("S").

"Lower Cambrian" of Georgia
EU SUN
(ca. 540 Ma)

Protosphaeridium tuberculiferum Timofeev 1966. (Timofeev 1966)
    Sphaeromorph acritarch ("S").
Symplassosphaeridium tumidulum Timofeev 1959. (Timofeev 1966)
    Colonial coccoid chroococcacean cyanobacterium (S).

"1aβ"
EU NOR
(ca. 540 Ma)

Liepaina? sp. of Vidal 1981. (Vidal 1981c)
    Acanthomorph acritarch ("S").

"1bα"–"1bβ"
EU NOR
(ca. 540 Ma)

Comasphaeridium sp. of Vidal 1981. (Vidal 1981c)
    Spheroidal microfossil incertae sedis (S).
Stellinium? sp. of Vidal 1981. (Vidal 1981c)
    Acanthomorph acritarch ("S").

Angara Fm.
AS SIB
(ca. 540 Ma)

Nucellosphaeridium minutum Timofeev 1966. (Timofeev 1966)
    Sphaeromorph acritarch containing cytoplasmic remnants (S M).

Table 22.3 (Continued).

Bastion Fm.
NA GRL
(ca. 540 Ma)

Leiosphaeridia sp. of Vidal 1979. (Vidal 1979)
Sphaeromorph acritarch ("S").

Belaya Fm.
AS SIB
(ca. 540 Ma)

Archaeohystrichosphaeridium cellulare Timofeev 1959. (Timofeev 1966)
Acanthomorph or herkomorph acritarch (S M).
Archaeohystrichosphaeridium janischewskyi Timofeev 1959. (Timofeev 1966)
Acanthomorph or herkomorph acritarch (S M).
Archaeohystrichosphaeridium operculatum Timofeev 1959. (Timofeev 1966)
Acanthomorph acritarch ("S").
Favososphaeridium sp. of Timofeev 1966. (Timofeev 1966)
Colonial coccoid chroococcacean cyanobacterium (S).
Polyedrosphaeridium sp. of Timofeev 1966. (Timofeev 1966)
Sphaeromorph acritarch displaying (diagenetically altered?) surface sculpture (M S).
Psophosphaera obscura Pichova [Pykhova] 1967. (Pykhova 1967)
Sphaeromorph acritarch (S).
Psophosphaera obscura Pychova [Pykhova] 1966. (Pykhova 1966)
Sphaeromorph acritarch (S).
Psophosphaera selebrosa Pychova [Pykhova] 1966. (Pykhova 1966)
Sphaeromorph acritarch (S).

Dalradian (Upper), Tayvallich Limestone
EU GBR
(ca. 540 Ma)

(acanthomorph acritarch). (Downie et al. 1971)
Possible acanthomorph acritarch ("S").
(sphaeromorph). (Downie et al. 1971)
Sphaeromorph acritarch ("S").
Polyporata nidia Pychova [Pykhova] 1966. (Downie et al. 1971)
Sphaeromorph acritarch ("S").
Polyporata verrucosa Pychova [Pykhova] 1966. (Downie et al. 1971)
Sphaeromorph acritarch ("S").
Uniporata nidia Pychova [Pykhova] 1966. (Downie et al. 1971)
Sphaeromorph acritarch ("S").

Ella Island Fm.
NA GRL
(ca. 540 Ma)

Baltisphaeridium compressum Volkova 1968. (Vidal 1979)
Acanthomorph acritarch ("S").
Baltisphaeridium orbiculare Volkova 1968. (Vidal 1979)
Acanthomorph acritarch ("S").
Baltisphaeridium sp. of Vidal 1979. (Vidal 1979)
Acanthomorph acritarch ("S").
Baltisphaeridium? *strigosum* Yankauskas 1976. (Vidal 1979)
Acanthomorph acritarch ("S").
Leiosphaeridia sp. of Vidal 1979. (Vidal 1979)
Sphaeromorph acritarch ("S").
Tasmanites variabilis Volkova 1968. (Vidal 1979)
Sphaeromorph acritarch ("S").

Eophyton Horizon of Estonia
EU SUN
(ca. 540 Ma)

Archaeohystrichosphaeridium acanthaceum Timofeev 1959. (Timofeev 1966)
Acanthomorph acritarch ("S").
Leiosphaeridia ochroleuca Timofeev 19??. (Timofeev 1966)
Sphaeromorph acritarch ("S").
Lopholigotriletum? *spathaeforme* Timofeev 1959. (Timofeev 1966)
Sphaeromorph acritarch ("S").
Polyedrosphaeridium bullatum Timofeev 1966. (Timofeev 1966)
Colonial coccoid chroococcacean? cyanobacterium (S).
Protosphaeridium laccatum Timofeev 1966. (Timofeev 1966)
Sphaeromorph acritarch ("S").
Tylosphaeridium tallinicum Timofeev 1966. (Timofeev 1966)
Sphaeromorph acritarch ("S").

Kacha Fm.
AS SIB
(ca. 540 Ma)

Protosphaeridium parvulum Timofeev 1966. (Timofeev 1966)
Sphaeromorph acritarch ("S").
Protosphaeridium torulosum Timofeev 1966. (Timofeev 1966)
Sphaeromorph acritarch ("S").

Lingulid Sandstone
EU SWE
(ca. 540 Ma)

Micrhystridium sp. of Vidal 1981. (Vidal 1981c)
Acanthomorph acritarch ("S").

Lower Cambrian (Tommotian, Atdabanian, Lenian)
EU SUN POL
(ca. 540 Ma)

Granomarginata squamacea Volkova 1968. (Volkova et al. 1979)
Sphaeromorph acritarch ("S").
Leiosphaeridia bicrura Jaukauskas [Yankauskas] 1976. (Volkova et al. 1979)
Sphaeromorph acritarch ("S").

Moty Fm.
AS SIB
(ca. 540 Ma)

Archaeodiscina nova Pichova [Pykhova] 1967?. (Pykhova 1967)
Sphaeromorph acritarch ("S").
Archaeohystrichosphaeridium acanthaceum Timofeev 1959. (Timofeev 1966)
Acanthomorph acritarch ("S").
Archaeopsophosphaera cf. *asperata* Naumova. (Pykhova 1966)
Sphaeromorph acritarch ("S").
Asperatopsophosphaera partialis Shepeleva 1963. (Pykhova 1967)
Sphaeromorph acritarch? (M).
Brochopsophosphaera simplex Pychova [Pykhova] 1966. (Pykhova 1966)
Sphaeromorph acritarch ("S").
Gloeocapsomorpha makrocysta Eisenack 1960. (Timofeev 1966)
Colonial coccoid chroococcacean cyanobacterium ("S").
Gloeocapsomorpha prisca Zalessky [Zalesskij] 1917. (Timofeev 1966)
Colonial coccoid chroococcacean? cyanobacterium ("S").
Leiopsophosphaera effusa Schepeleva [Shepeleva] 1963. (Pykhova 1966)

Sphaeromorph acritarch (S).
Octoedryxium truncatum Rudavskaja [Rudavskaya] 1973.
(Rudavskaya *in* Myatlyuk et al. 1973)
  Prismatomorph acritarch ("S"). [See Chapter 24, Pl. 56, C_1-C_3.]
Paracrassosphaera dedalea Rudavskaja [Rudavskaya] 1979.
(Fajzulina and Treshchetenkova 1979)
  Pterospermopsimorphid acritarch (S M). [See Chapter 24, Pl. 56, A_1-A_2.]
Polyedryxium neftelenicum Rudavskaja [Rudavskaya] 1971.
(Rudavskaya 1971)
  Sphaeromorph acritarch with coalesced cytoplasmic contents (S). [See Chapter 24, Pl. 56, B.]
Spumosata minor Pichova [Pykhova] 1967. (Pykhova 1967)
  Sphaeromorph acritarch ("S").
Spumosata minor Pychova [Pykhova] 1966. (Pykhova 1966)
  Sphaeromorph acritarch ("S").
Spumosata nova Pychova [Pykhova] 1966. (Pykhova 1966)
  Sphaeromorph acritarch? (S).
Spumosata simplex Naumova?. (Pykhova 1967)
  Sphaeromorph acritarch (S).
Spumosata simplex Naumova?. (Pykhova 1966)
  Sphaeromorph acritarch (M).
Stictosphaeridium tortulosum Timofeev 1966. (Timofeev 1966)
  Sphaeromorph acritarch ("S").
Trematosphaeridium holtedahlii Timofeev 1966. (Timofeev 1966)
  Sphaeromorph acritarch ("S").

Okhonojsk Fm.
AS SIB
(ca. 540 Ma)

Acantholigotriletum primigenum (Naumova 1949) Timofeev 1959. (Timofeev 1966)
  Acanthomorph acritarch ("S").
Kildinella hyperboreica Timofeev 1966. (Timofeev 1966)
  Sphaeromorph acritarch ("S").
Protosphaeridium densum Timofeev 1966. (Timofeev 1966)
  Sphaeromorph acritarch ("S").
Protosphaeridium patelliforme Timofeev 1966. (Timofeev 1966)
  Sphaeromorph acritarch ("S").
Pterospermopsimorpha sp. of Timofeev 1966. (Timofeev 1966)
  Sphaeromorph acritarch containing coalesced cytoplasmic remnants (S M)
Trachyoligotriletum gyratum Timofeev 1959. (Timofeev 1966)
  Sphaeromorph acritarch ("S").

Olekma Fm.
AS SIB
(ca. 540 Ma)

Symplassosphaeridium subcoalitum Timofeev 1959. (Timofeev 1966)
  Colonial coccoid chroococcacean? cyanobacterium (S).

Oselochnaya Fm.
AS SIB
(ca. 540 Ma)

Cymatiosphaera lavrovii Rudavskaja [Rudavskaya] 1964?. (Timofeev 1966)
  Acanthomorph acritarch within "capsule"? (M S).
Polyedrosphaeridium bullatum Timofeev 1966. (Timofeev 1966)
  Colonial coccoid chroococcacean? cyanobacterium (S).

Pestrotsvet Fm.
AS SIB
(ca. 540 Ma)

Leiofusa sp. of Timofeev 1966. (Timofeev 1966)
  Sphaeromorph acritarch. Not a convincing netromorph (S M).
Lophosphaeridium rarum Timofeev 1959. (Timofeev 1966)
  Sphaeromorph acritarch ("S").
Polyedrosphaeridium bullatum Timofeev 1966. (Timofeev 1966)
  Colonial coccoid chroococcacean? cyanobacterium (S).
Symplassosphaeridium subcoalitum Timofeev 1959. (Timofeev 1966)
  Colonial coccoid chroococcacean? cyanobacterium (S).
Trachysphaeridium laminaritum Timofeev 1966. (Timofeev 1966)
  Sphaeromorph acritarch ("S").

Shortor "Series"
AS SUN
(ca. 540 Ma)

Angulisphaerina korolevii Lopukhin 1966. (Lopukhin 1966)
  Sphaeromorph acritarch ("S").
Constrictosphaerina alaica Lopukhin 1966. (Lopukhin 1966)
  Sphaeromorph acritarch ("S").
Costatosphaerina septata Lopukhin 1966. (Lopukhin 1966)
  Sphaeromorph acritarch ("S").
Dictyopsophosphaera perrara Lopukhin 1966. (Lopukhin 1966)
  Sphaeromorph acritarch ("S").
Gyratosphaerina aspera Lopukhin 1966. (Lopukhin 1966)
  Sphaeromorph acritarch ("S").
Vesiculosphaerina singularis Lopukhin 1966. (Lopukhin 1966)
  Sphaeromorph acritarch ("S").

Shortor "Series,"? Tamdy Fm.
AS SUN
(ca. 540 Ma)

Angulisphaerina keminica Lopukhin 1966. (Lopukhin 1966)
  Sphaeromorph acritarch ("S").
Trachymarginata aberrantis Lopukhin 1966. (Lopukhin 1966)
  Sphaeromorph acritarch ("S").
Uniporata striata Lopukhin 1966. (Lopukhin 1966)
  Sphaeromorph acritarch ("S").

Tokammane Fm.
EU SJM
(ca. 540 Ma)

(filamentous microfossils). (Knoll and Swett 1987)
  Flattened tubular sheath of filamentous cyanobacterium ("S"). Fig. 8.19 not interpretable (S).
(small leiospaerids). (Knoll and Swett 1987)
  Solitary coccoid cyanobacteria (S).
Archaeodiscina umbonulata Volkova 1968. (Knoll and Swett 1987)
  Sphaeromorph acritarch ("S").
Baltisphaeridium cerinum Volkova 1968. (Knoll and Swett 1987)
  Acanthomorph acritarch ("S").
Baltisphaeridium implicatum Fridrichsone [Fridrikhsone] 1971. (Knoll and Swett 1987)
  Acanthomorph acritarch ("S").
Bavlinella faveolata Schepeleva [Shepeleva] 1963. (Knoll and Swett 1987)
  Colonial coccoid bacterium (S).
Celtiberium sp. A of Knoll and Swett 1987. (Knoll and Swett 1987)
  Distinctive sphaeromorph acritarch with horns ("S").
Celtiberium sp. B of Knoll and Swett 1987. (Knoll and Swett 1987)

Table 22.3 (Continued).

Acanthomorph acritarch ("S").
Comasphaeridium strigosum (Jankauskas 1976) Downie 1982. (Knoll and Swett 1987)
  Acanthomorph acritarch ("S").
Cymatiosphaera sp. of Knoll and Swett 1987. (Knoll and Swett 1987)
  Herkomorph acritarch ("S M").
Estiastra minima Volkova 1969. (Knoll and Swett 1987)
  Polygonomorph acritarch ("S M").
Evittia cf. *irregulare* Downie 1982. (Knoll and Swett 1987)
  Acanthomorph acritarch ("S").
Goniosphaeridium primarium (Jankauskas 1979) Downie 1982. (Knoll and Swett 1987)
  Acanthomorph acritarch ("S").
Granomarginata prima Naumova 1960. (Knoll and Swett 1987)
  Sphaeromorph (or pteromorph?) acritarch ("S M").
Granomarginata squamacea Volkova 1968. (Knoll and Swett 1987)
  Sphaeromorph (or pteromorph?) acritarch ("S M").
Leiosphaeridia spp. of Knoll and Swett 1987. (Knoll and Swett 1987)
  Sphaeromorph acritarch ("S").
Lophosphaeridium truncatum Volkova 1969. (Knoll and Swett 1987)
  Sphaeromorph acritarch ("S M").
Micrhystridium dissimilare Volkova 1969. (Knoll and Swett 1987)
  Acanthomorph acritarch ("S").
Micrhystridium lanatum Volkova 1969. (Knoll and Swett 1987)
  Acanthomorph acritarch ("S").
Micrhystridium cf. *minutum* Downie 1982. (Knoll and Swett 1987)
  Acanthomorph acritarch ("S").
Micrhystridium sp. A of Knoll and Swett 1987. (Knoll and Swett 1987)
  Acanthomorph acritarch ("S").
Micrhystridium tornatum Volkova 1968. (Knoll and Swett 1987)
  Acanthomorph acritarch (very short-spined) ("S").
Protosphaeridium sp. of Knoll and Swett 1987. (Knoll and Swett 1987)
  Sphaeromorph acritarch ("S").
Pterospermella cf. *solida* (Volkova 1968) Volkova 1979. (Knoll and Swett 1987)
  Pteromorph acritarch (S).
Skiagia ciliosa (Volkova 1969) Downie 1982. (Knoll and Swett 1987)
  Acanthomorph acritarch ("S").
Skiagia compressa (Volkova 1968) Downie 1982. (Knoll and Swett 1987)
  Acanthomorph acritarch ("S").
Skiagia orbiculare (Volkova 1968) Downie 1982. (Knoll and Swett 1987)
  Acanthomorph acritarch ("S").
Skiagia ornata (Volkova 1968) Downie 1982. (Knoll and Swett 1987)
  Acanthomorph acritarch ("S").
Skiagia scottica Downie 1982. (Knoll and Swett 1987)
  Acanthomorph acritarch ("S").
Skiagia sp. A of Knoll & Swett 1987. (Knoll and swett 1987)
  Acanthomorph acritarch ("S").
Tasmanites bobrowskii Ważyńska 1967. (Knoll and Swett 1987)
  Sphaeromorph acritarch ("S").
Tasmanites tenellus Volkova 1968. (Knoll and Swett 1987)
  Sphaeromorph acritarch ("S").
Tasmanites volkovae Kirjanov [Kir'yanov] 1974. (Knoll and Swett 1987)
  Sphaeromorph acritarch ("S").
Trachysphaeridium timofeevii Vidal 1976. (Knoll and Swett 1987)
  Sphaeromorph acritarch ("S").

Ushakovka Fm. (Lower)
AS SIB
(ca. 540 Ma)

Favososphaeridium scandicum Timofeev 1966. (Timofeev 1966)
  Colonial coccoid chroococcacean cyanobacterium (S).
Protosphaeridium densum Timofeev 1966. (Timofeev 1966)
  Sphaeromorph acritarch ("S").

Usol'e Fm.
AS SIB
(ca. 540 Ma)

Archaeohystrichosphaeridium dasyacanthum Timofeev 1959. (Timofeev 1966)
  Acanthomorph acritarch ("S").
Protosphaeridium acis Timofeev 1966. (Timofeev 1966)
  Sphaeromorph acritarch ("S").
Protosphaeridium densum Timofeev 1966. (Timofeev 1966)
  Sphaeromorph acritarch ("S").
Protosphaeridium patelliforme Timofeev 1966. (Timofeev 1966)
  Sphaeromorph acritarch ("S").
Stictosphaeridium implexum Timofeev 1966. (Timofeev 1966)
  Sphaeromorph acritarch ("S").

Vampire Fm.
NA CAN YK
(ca. 540 Ma)

(Microdubiofossil Type A). (Hofmann 1984)
  Incompletely preserved colonial coccoid prokaryote (S).
Archaeodiscina? sp. of Hofmann 1984. (Hofmann 1984)
  Compressed sphaeromorph acritarch ("S").
Archaeorestis? sp. of Hofmann 1984. (Hofmann 1984)
  Probably branched sheath of false branching nostocalean cyanobacterium (S).
Archaeotrichion sp. of Hpfmann 1984. (Hofmann 1984)
  Sheath or trichome of filamentous bacterium (S).
Bavlinella sp. of Hofmann 1984. (Hofmann 1984)
  Colonial coccoid bacterium (S).
Eomycetopsis spp. of Hofmann 1984. (Hofmann 1984)
  Tubular sheath of filamentous prokaryote ("S").
Leiosphaeridia spp of Hofmann 1984. (Hofmann 1984)
  Sphaeromorph acritarch ("S").
Nostocomorpha sp. of Hofmann 1984. (Hofmann 1984)
  Iron oxide-replaced filamentous (nostocalean) cyanobacterium (S).
Oscilatoriopsis psilata Maithy and Shukla 1977. (Hofmann 1984)
  Septate filamentous (oscillatoriacean) cyanobacterium (S).
Siphonophycus spp. of Hofmann 1984. (Hofmann 1984)
  Tubular sheaths of filamentous cyanobacteria (S).

Vampire Fm. (basal?)
NA CAN YK
(ca. 540 Ma)

Paleamorpha? sp. of Hofmann 1984. (Hofmann 1984)
  Flattened tubular sheath of filamentous cyanobacterium (S).

Lower Cambrian (Atdabanian)
EU SUN POL
(ca. 538 Ma)

Aranidium aff. *pycnacanthum* Jankauskas [Yankauskas] 1975. (Volkova et al. 1979)
  Acanthomorph acritarch (S).

Table 22.3 (*Continued*).

Archaeodiscina? *bicostata* Volkova 1979. (Volkova et al. 1979)
  Sphaeromorph acritarch with internal degraded cytoplasmic remnants (S).
Baltisphaeridium brachyspinosum Kirjanov [Kir'yanov] 1974. (Volkova et al. 1979)
  Acanthomorph acritarch ("S").
Baltisphaeridium cerinum Volkova 1968. (Volkova et al. 1979)
  Acanthomorph acritarch ("S").
Baltisphaeridium dubium Volkova 1968. (Volkova et al. 1979)
  Short-spined acanthomorph acritarch ("S").
Baltisphaeridium papillosum (Timofeev 1959) Volkova 1968. (Volkova et al. 1979)
  Acanthomorph acritarch ("S").
Baltisphaeridium pilosiusculum Jankauskas [Yankauskas] 1979. (Volkova et al. 1979)
  Acanthomorph acritarch ("S").
Baltisphaeridium primarium Jankauskas [Yankauskas] 1979. (Volkova et al. 1979)
  Acanthomorph acritarch ("S").
Baltisphaeridium sp. 2 of Volkova et al. 1979. (Volkova et al. 1979)
  Acanthomorph acritarch ("S").
Cymatiosphaera cristata Jankauskas [Yankauskas] 1976. (Volkova et al. 1979)
  Herkomorph acritarch ("S").
Cymatiosphaera favosa Jankauskas [Yankauskas] 1976. (Volkova et al. 1979)
  Herkomorph acritarch ("S").
Cymatiosphaera? *membranacea* Kirjanov [Kir'yanov] 1974. (Volkova et al. 1979)
  Pteromorph acritarch (S).
Cymatiosphaera minuta Jankauskas [Yankauskas] 1979. (Volkova et al. 1979)
  Possible herkomorph acritarch (M S).
Cymatiosphaera nerisica Jankauskas [Yankauskas] 1976. (Volkova et al. 1979)
  Herkomorph acritarch ("S").
Dominopolia lata Kirjanov [Kir'yanov] 1974. (Volkova et al. 1979)
  Netromorph acritarch having paired filamentous appendages ("S").
Dominopolia longispinosa Kirjanov [Kir'yanov] 1974. (Volkova et al. 1979)
  Netromorph acritarch having filamentous appendage(s) ("S").
Leiosphaeridia subgranulata Kirjanov [Kir'yanov] 1974. (Volkova et al. 1979)
  Sphaeromorph acritarch ("S").
Lophosphaeridium tentativum Volkova 1968. (Volkova et al. 1979)
  Sphaeromorph acritarch ("S").
Micrhystridium brevicornum Jankauskas [Yankauskas] 1976. (Volkova et al. 1979)
  Acanthomorph acritarch ("S").
Micrhystridium pallidum Volkova 1968. (Volkova et al. 1979)
  Acanthomorph acritarch ("S").
Pseudotasmanites parvus Kirjanov [Kir'yanov] 1974. (Volkova et al. 1979)
  Sphaeromorph acritarch (S).
Pterospermella vitalis Jankauskas [Yankauskas] 1979. (Volkova et al. 1979)
  Possible poorly preserved pteromorph acritarch (M S).
Pterospermopsimorpha wolynica Kirjanov [Kir'yanov] 1974. (Volkova et al. 1979)
  Sphaeromorph acritarch with degraded cytoplasmic contents (S).
Tasmanites piritaensis Posti and Jankauskas [Yankauskas] 1976. (Volkova et al. 1979)
  Sphaeromorph acritarch (S).

Altacha Fm.
AS SIB
(ca. 535 Ma)

Gloeocapsomorpha prisca Zalessky [Zalesskij] 1917. (Timofeev 1966)
  Colonial coccoid chroococcacean? cyanobacterium ("S").
Protosphaeridium patelliforme Timofeev 1966. (Timofeev 1966)
  Sphaeromorph acritarch ("S").
Protosphaeridium tuberculiferum Timofeev 1966. (Timofeev 1966)
  Sphaeromorph acritarch ("S").

Chara Fm.
AS SIB
(ca. 535 Ma)

Pterospermopsimorpha sp. of Timofeev 1966. (Timofeev 1966)
  Sphaeromorph acritarch containing coalesced cytoplasmic remnants (S M).
Tasmanites sp. of Timofeev 1966. (Timofeev 1966)
  Sphaeromorph acritarch ("S").

Lower Cambrian (Atdabanian, Lenian)
EU SUN POL
(ca. 535 Ma)

Alliumella baltica Vanderflit [Fanderflit] 1971. (Volkova et al. 1979)
  Sphaeromorph acritarch with filamentous appendage (S).
Archaeodiscina umbonulata Volkova et al. 1979)
  Sphaeromorph acritarch containing condensed organic material(?) (S M).
Baltisphaeridium acerosum Jankauskas [Yankauskas] and Posti 1976. (Volkova et al. 1979)
  Short-spined acanthomorph acritarch ("S").
Baltisphaeridium ciliosum Volkova 1969. (Volkova et al. 1979)
  Acanthomorph acritarch ("S").
Baltisphaeridium compressum Volkova 1968. (Volkova et al. 1979)
  Acanthomorph acritarch ("S").
Baltisphaeridium implicatum Fridrichsone [Fridrikhsone] 1971. (Volkova et al. 1979)
  Acanthomorph acritarch ("S").
Baltisphaeridium insigne (Fridrichsone 1971) Volkova 1974. (Volkova et al. 1979)
  Acanthomorph acritarch ("S").
Baltisphaeridium orbiculare Volkova 1968. (Volkova et al. 1979)
  Acanthomorph acritarch ("S").
Baltisphaeridium ornatum Volkova 1968. (Volkova et al. 1979)
  Acanthomorph acritarch ("S").
Baltisphaeridium? *strigosum* Jankauskas [Yankauskas] 1976. (Volkova et al. 1979)
  Acanthomorph acritarch ("S").
Baltisphaeridium varium Volkova 1969. (Volkova et al. 1979)
  Acanthomorph acritarch ("S").
Cymatiosphaera capsulara Jankauskas [Yankauskas] 1976. (Volkova et al. 1979)
  Herkomorph acritarch ("S").
Cymatiosphaera div. sp. of Volkova et al. 1979. (Volkova et al. 1979)
  Herkomorph acritarch ("S").
Cymatiosphaera postii Jankauskas [Yankauskas] 1979. (Volkova et al. 1979)
  Herkomorph acritarch ("S").
Dictyotidium priscum Kirjanov [Kir'yanov] and Volkova 1979. (Volkova et al. 1979)

Table 22.3 (*Continued*).

Herkomorph acritarch ("S").
Estiastra minima Volkova 1969. (Volkova et al. 1979)
  Polygonomorph acritarch ("S").
Leiovalia tenera Kirjanov [Kir'yanov] 1974. (Volkova et al. 1979)
  Netromorph acritarch ("S M").
Lophosphaeridium truncatum Volkova 1969. (Volkova et al. 1979)
  Sphaeromorph acritarch ("S").
Micrhystridium dissimilare Volkova 1969. (Volkova et al. 1979)
  Acanthomorph acritarch ("S").
Micrhystridium lanatum Volkova 1969. (Volkova et al. 1979)
  Acanthomorph acritarch ("S").
Micrhystridium lubomlense Kirjanov [Kir'yanov] 1974. (Volkova et al. 1979)
  Acanthomorph acritarch ("S").
Micrhystridium obscurum Volkova 1969. (Volkova et al. 1979)
  Acanthomorph acritarch ("S").
Micrhystridium radzynicum Volkova 1979. (Volkova et al. 1979)
  Acanthomorph acritarch ("S").
Micrhystridium spinosum Volkova 1969. (Volkova et al. 1979)
  Acanthomorph acritarch ("S").
Micrhystridium villosum Kirjanov [Kir'yanov] 1974. (Volkova et al. 1979)
  Acanthomorph acritarch ("S").
Multiplicisphaeridium dendroideum (Jankauskas 1976) Jankauskas and Kirjanov [Kir'yanov] 1979. (Volkova et al. 1979)
  Acanthomorph acritarch ("S").
Multiplicisphaeridium vilnense (Jankauskas 1976) Jankauskas [Yankauskas] 1979. (Volkova et al. 1979)
  Acanthomorph acritarch ("S").
Ovulum lanceolatum Jankauskas [Yankauskas] 1975. (Volkova et al. 1979)
  Ellipsoidal sphaeromorph acritarch (S).
Ovulum saccatum Jankauskas [Yankauskas] 1975. (Volkova et al. 1979)
  Sphaeromorph acritarch (S).
Pterospermella solida (Volkova 1969) Volkova 1979. (Volkova et al. 1979)
  Pteromorph acritarch ("S").
Tasmanites bobrowskii Ważyńska 1967. (Volkova et al. 1979)
  Sphaeromorph acritarch (S).
Tasmanites volkovae Kirjanov [Kir'yanov] 1974. (Volkova et al. 1979)
  Sphaeromorph acritarch (S).

Ushakovka Fm. (= Khuzhir Fm.)
AS SIB
(ca. 535 Ma)

Archaeodiscina minor Pychova [Pykhova] 1966. (Pykhova 1966)
  Solitary coccoid cyanobacterium (S).
Archaeodiscina prima Pychova [Pykhova] 1966. (Pykhova 1966)
  Solitary coccoid cyanobacterium (S).
Archaeosacculina atava Pychova [Pykhova] 1966. (Pykhova 1966)
  Solitary coccoid cyanobacterium (S).
Archaeosacculina salebrosa Pychova [Pykhova] 1966. (Pykhova 1966)
  Sphaeromorph acritarch (very poorly preserved) (M).
Archaeosacculina torosa Pychova [Pykhova] 1966. (Pykhova 1966)
  Sphaeromorph acritarch ("S").
Kildinella hyperboreica Timofeev 1966. (Timofeev 1966)
  Sphaeromorph acritarch ("S").
Leiominuscula minuta Naumova 1960. (Pykhova 1966)
  Sphaeromorph acritarch ("S").
Lophominuscula rugosa Naumova in Naumova and Pavlovskij 1961. (Pykhova 1966)
  Sphaeromorph acritarch ("S").
Margominuscula antiqua Naumova 1960. (Pykhova 1966)
  Sphaeromorph acritarch (S).
Margominuscula antiqua Naumova 1960. (Pykhova 1967)
  Solitary ellipsoidal cyanobacterium (S).
Margominuscula prima Pychova [Pychova] 1966. (Pykhova 1966)
  Solitary coccoid cyanobacterium (S).
Margominuscula prima Pichova [Pykhova] 1966. (Pykhova 1967)
  Solitary coccoid cyanobacterium (S).
Margominuscula rugosa Naumova 1960. (Pykhova 1967)
  Sphaeromorph acritarch ("S").
Margominuscula sp. of Pykhova 1967. (Pykhova 1967)
  Sphaeromorph acritarch ("S").
Margominuscula tremata Naumova in Naumova and Pavlovskij 1961. (Pykhova 1967)
  Solitary coccoid cyanobacterium (S).
Polyedrosphaeridium sp. of Timofeev 1966. (Timofeev 1966)
  Sphaeromorph acritarch displaying (diagenetically altered?) surface sculpture (M S).

Lower Cambrian (Lenian)
EU SUN POL
(ca. 533 Ma)

Aranidium sparsum Volkova 1979. (Volkova et al. 1979)
  Acanthomorph acritarch (S).
Cymatiosphaera sp. 1 of Volkova et al. 1979. (Volkova et al. 1979)
  Folded sphaeromorph acritarch (S).
Deunffia dentifera Volkova 1969. (Volkova et al. 1979)
  Netromorph acritarch having filamentous appendage ("S M").
Micrhystridium notatum Volkova 1969. (Volkova et al. 1979)
  Acanthomorph acritarch ("S").
Micrhystridium oligum Jankauskas [Yankauskas] 1976. (Volkova et al. 1979)
  Acanthomorph acritarch ("S").
Pterospermella vitrea (Volkova 1974) Volkova 1979. (Volkova et al. 1979)
  Possible degraded pteromorph acritarch (S M).
Synsphaeridium switjasium Kirjanov [Kir'yanov] 1974. (Volkova et al. 1979)
  Sphaeromorph acritarch ("S").

Bystraya Fm.
AS SIB
(ca. 530 Ma)

Protosphaeridium tuberculiferum Timofeev 1966. (Timofeev 1966)
  Sphaeromorph acritarch ("S").

Ushakovka Fm. (Upper)
EU SUN
(ca. 530 Ma)

Archaeohystrichosphaeridium cuneidentatum Timofeev 1959. (Timofeev 1966)
  Polygonomorph or acanthomorph acritarch (S M).
Archaeohystrichosphaeridium vologdaense Timofeev 1959. (Timofeev 1966)
  Acanthomorph acritarch ("S").
Stictosphaeridium pectinale Timofeev 1966. (Timofeev 1966)
  Sphaeromorph acritarch ("S").

23

Described Taxa of Proterozoic and Selected Earliest Cambrian Carbonaceous Remains, Trace and Body Fossils

KENNETH M. TOWE STEFAN BEGTSON MIKHAIL A. FEDONKIN
HANS J. HOFMANN CAROL MANKIEWICZ BRUCE N. RUNNEGAR

Introduction

KENNETH M. TOWE

Included in this Chapter are tabulations of data on which are based discussions of Proterozoic and earliest Cambrian carbonaceous remains, trace and body fossils (Chapter 7); the Late Proterozoic–Early Cambrian evolution of metaphytes and metazoans (Chapter 8); and the Proterozoic–Early Cambrian diversification of metazoans and metaphytes (Section 11.4). Specifically, tabulations (including an evaluation of the nature and origin of the taxa and objects listed) are presented below for the following six categories of megascopic remains:

(1) Proterozoic and selected Cambrian megascopic *carbonaceous films* (Table 23.1);

(2) described genera and species of Proterozoic and Early Cambrian *calcareous algae* (Tables 23.2.1–23.2.5);
(3) taxonomically described, as well as figured but not formally described, Ediacaran (Table 23.3.1) and other Proterozoic (Table 23.3.2) *metazoan body fossils*;
(4) taxonomically described, as well as figured but not formally described, Proterozoic (Vendian) *metazoan trace fossils* (Table 23.4);
(5) genera of Late Proterozoic–Early Cambrian *skeletal fossils* (Table 23.5); and
(6) Proterozoic and selected Cambrian megascopic *dubiofossils* and *pseudofossils* (Table 23.6).

23.1

Proterozoic and Selected Cambrian Megascopic Carbonaceous Films

HANS J. HOFMANN

In Table 23.1 are listed, alphabetically and by year of publication, reported Proterozoic and selected Cambrian megascopic films. The conventions noted below have been used in Table 23.1.

Type code: ∗∗ = holotype of the type species of the indicated genus; ∗ = holotype of a species other than the type species of the indicated genus.

Reference code: the first four characters of the code are the first four letters in the surname of the principal author of the referenced publication (characters 1–4); the next two characters are the last two numbers of the year of publication (characters 5 and 6); the final four characters denote the page number of the first page in the referenced publication (characters 7–10). Thus, a publication by "Smith and Jones" published in "1987" and beginning on "page 93" would here be referenced as "Smit87..93."

Age code: X = Early Proterozoic (2500 to 1600 Ma); Y = Middle Proterozoic (1600 to 900 Ma); Z = Late Proterozoic, including Vendian (900 to 550 Ma); C = Cambrian.

See Section 7.3 for a discussion of the content of Table 23.1.

Described Taxa of Proterozoic and Selected Earliest Cambrian Megafossils

Table 23.1. *Proterozoic and selected Cambrian megascopic carbonaceous films*

Number	Taxon	Reference Code and Page	Continent and Country	Locality
1	(Algal dust)	Misr50..88 p. 89, Fig. 2.	AS IN	58 Banjari, Shahabad Dist.
2	(Bag-shaped body)	Chap35.109 p. 118, Pl. 2, Fig. 2.	AS IN	57 Neemuch
3	(Branched filament)	Walt76.872 p. 878, pl. 2, Fig. 11.	NA US	8 No locality given
4	(Carbonaceous disc)	Misr50..88 p. 88, Fig. 1.	AS IN	58 Banjari, Shahabad Dist.
5	(Carbonaceous films)	Horo80.649 p. 661, text–Fig. 10A–E.	NA US	7 SE of Neihart, US Hwy. 89
6	(Coaly seam)	Math82.125 p. 130.	AS IN	94 S. Mirzapur Dist.
7	(Disc-shaped fossils)	Dutt75.149 p. 151–152, Pl. 1, Figs. 1–3.	AS IN	56 Ferozpur Jhirka
8	(Discoid bodies)	King72.313 p. 69.	AS IN	59 No locality given
9	(Discs)	Misr52..46 p. 47, Figs. 1–4, 6–7.	AS IN	57 Rampura
10	(Eoholynides)	Ishc86..91 p. 154, Pl. 1, Fig. 1.	EU SU	40 Bernashevka, Vinnitsa obl.
11	(Fucoids)	Stoc64..29 p. 14, Pl. 1, Fig. 2.	AS IR	42 Chapoghlu
12	(Graphitic compression)	Tyle571293 p. 1296, Pl. 1, Figs. 1–3, 5,6.	NA US	12 6.5 mi N, 1 mi E of Iron River
13	(Graphitic compression)	Stin65..75 p. 75–76, Fig. 1 f,g.	NA US	12 6.5 mi N, 1 mi E of Iron River
14	(Graphitic compression)	Stin65..75 p. 75–76, Fig. 1 a,b	NA CA	13 High Falls, Swampy Bay River
15	(Graphitic compression)	Stin65..75 p. 75–76, Fig. 1 c–e.	NA CA	13 Schefferville, 1.6 km N airport
16	(Kidney shaped form)	Misr57..54 p. 55, Pl. 7, Fig. 5.	AS IN	57 Rampura
17	(Kidney-shaped form)	Misr52..46 p. 47, Fig. 5.	AS IN	57 Rampura
18	(Large sphaeromorph)	Horo80.649 p. 657, text–Fig. 9.	NA US	6 SE of Neihart, US Hwy. 89
19	(Minute organic bodies)	Holl09..70 p. 66.	AS IN	57
20	(Ovoid films)	Aitk73...1 p. 150–151.	NA CA	66 Cap Mountain, AC-541
21	(Plant impressions)	Glae63.113 p. 117, Pl. 2, Fig. 2.	AF NA	99 Farm Ibenstein 55, Rehoboth Dist.
22	(Problematica)	Asse63.503 p. 502–503, Fig. 2.	AS IR	42 Gasir, in Karaj valley
23	(Rounded intraclasts)	Horo80.649 p. 662, text–Fig. 10 F.	NA US	6 SE of Neihart, US Hwy. 89
24	(Sabelliditidae-like)	Chen82.339 p. 339–340, Fig. 1.	AS CN	91
25	(Small black disks)	Wima94.109 p. 109–113, Pl. 5, Figs. 1–5.	EU SE	24 Mullskredena, W of Omberg
26	(Small discinoid shell)	Walc83.437 p. 195.	NA US	11 Grand Canyon, Nankoweap Butte
27	(Spiral impression)	Beer19.139 p. 139, Pl. 30, Figs. 1–2.	AS IN	58 Saraidanr near Rohtas
28	(Thin layers of carbon)	Low03.DD1 p. 16DD.	NA CA	14 E side of Cotter Id.
29	(Trace fossil)	Du85...5 p. 9, Pl. 1, Fig. 28.	AS CN	84 Jixian (Sangshuan?)
30	?*Chuaria circularis*	Vida76..19 p. 6, 17.	NA GL	21 Kong Oscars Fjord
31	?*Chuaria circularis*	Vida76..19 p. 8, 16.	NA GL	19 Kong Oscars Fjord
32	? *Chuaria circularis*	Vida76..19 p. 15.	NA GL	20 Kong Oscars Fjord
33	? *Chuaria circularis*	Vida79..40 p. 13, 16, 36.	NA GL	18 Kong Oscars Fjord
34	? *Chuaria circularis*	Knol85.451 p. 457, Pl. 51, Figs. 1, 2.	EU SJ	71 Faksevagen
35	? *Vendotaenia antiqua*	Zhan86..67 p. 84, 86, Pl. 4, Fig. 4.	AS CN	90 Mt. Gaojiashan, Beiwan, Hujiaba
36	*Aataenia reticularis*	**Gnil76..10 p. 11–12, Pl. 1, Fig. 6.	EU SU	40 Aa well, Moscow Basin
37	*Aataenia reticularis*	Gnil79..39 p. 41, Pl. 46, Fig. 6.	EU SU	40 Aa well
38	*Aataenia reticularis*	Gnil83..46 p. 49, Pl. 46, Fig. 6.	EU SU	40 Aa well, Moscow Basin 138–140 m
39	*Aataenia* sp.	Soko76.126 p. 139, Fig. b.	EU SU	40 Aa well, Moscow Basin
40	*Anhuiella sinensis*	Xing85.182 p. 189, Pl. 39, Fig. 5.	AS CN	53
41	*Anhuiella sinensis*	Chen86.221 p. 231, Pl. 2, Fig. 11	AS CN	53 Mt. Sidingshan, Huainan
42	*Bagongshanella striolata*	Chen86.221 p. 231, Pl. 2, Fig. 9.	AS CN	52 Mt. Bagongshan, Huainan
43	*Beltanelloides sorichevae*	Soko72.114 Pl. 4, Figs. 1–7.	EU SU	38 Various localities
44	*Beltanelloides sorichevae*	Soko72..48 p. 53, (unnumbered fig.).	EU SU	38 Mezen Riv., Leshukonskoe well
45	*Beltanelloides sorichevae*	Soko73.204 p. 213, Fig. 5, no. 4.	EU SU	38 L. Onega Riv., well
46	*Beltanelloides sorichevae*	Soko73.204 p. 213, Fig. 5, no. 2.	EU SU	38 Loino well
47	*Beltanelloides sorichevae*	Soko73.204 p. 213, Fig. 5, no. 1.	EU SU	38 Nenoxa well
48	*Beltanelloides sorichevae*	Soko76.126 p. 138, Fig. a.	EU SU	38 Kirs well

Latitute	Longitude	Geologic Unit	Age	Original Author's Interpretation	Hofmann Interpretation
25.2	80.9	Semri Gp., Rohtas Ls.	YZ	Algal spores?	
24.4	74.9	Semri Gp., Suket Sh.	YZ	Egg mass	Chuaria?
		Belt Supgp., Greyson Sh.	Y	Algae	
25.2	80.9	Semri Gp., Rohtas Ls.	YZ	Plant remains	
46.9	−110.7	Belt Supgp., Newland Fm.	Y	Fragments of mats	
25.0	82.5	Semri Gp., Arangi Fm.	YZ	Carbonaceous matter	
27.8	76.9	Delhi Supgp., Alwar Qte.	Y	Megafossil	
		Kurnool Supgp., Jammalamadugu Gp., Owk Sh.	Z	Fossil	Chuaria?
24.5	75.5	Semri Gp., Suket Sh.	YZ	Inorganic	Chuaria
		Mogilev-Podolosky Gp., Yaryshev Fm.	Z	Noncalcareous alga	Vendotaenid
36.2	48.9	Chapoghlu Sh.	Z	Fossils	
46.3	−88.7	Michigamme Sh.	X	Nostoc-like colonies	cf. Morania?
46.3	−88.7	Michigamme Sh.	X	Possibly biogenic	cf. Morania?
56.1	−68.3	Kaniapiskau Supgp., Knob Lake Gp., Hautes Chutes Fm.	X	Possibly biogenic	cf. Morania?
54.8	−66.8	Kaniapiskau Supgp., Knob Lake Gp., Le Fer Fm. (Attik.?)	X	Possibly biogenic	cf. Morania?
24.5	75.5	Semri Gp., Suket Sh.	YZ	Inorganic	short Tawuia
24.5	75.5	Semri Gp., Suket Sh.	YZ	Inorganic	short Tawuia
46.9	−110.7	Belt Supgp., Chamberlain Sh.	Y	Sphaeromorph	
		Semri Gp., Suket Sh.	YZ	Incertae sedis	Chuaria
63.4	−123.2	Lone Land Fm.	Z	"almost certainly organic"	cf. Morania?
−22.8	16.5	pre-Nama quartzite	Z	Probably algae	Inorganic?
36.0	51.3	Chapoghlu Sh.	Z	Problematic structures	Chuaria or Beltanelliformis
46.9	−110.7	Belt Supgp., Chamberlain Sh.	Y	Mat fragments	
		Gaojiatun Fm. [cf. Changlingzi Fm.]	Z	Sabelliditid	Sinosabelliditid
58.3	14.7	Visingsö Fm.	YZ	No interpretation	Chuaria circularis
36.3	−111.9	Chuar Gp., Kwagunt Fm.	YZ	Possible brachipod	
24.6	84.0	Semri Gp., Rohtas Ls.	YZ	Worm or burrow	Body fossil?
57.8	−77.1	Manitounuk Supgp., Nastapoka Gp.	X	Lowly organized plant life	cf. Morania?
40.2	117.4	Changcheng Sys., Gaoyuzhuang Fm.	Y	Trace fossil	Grypania
72.6	−25.1	Eleonore Bay Gp., Limestone-Dolomite Ser.	Z	Acritarch	Sphaeromorph acritarch
72.5	−25.0	Eleonore Bay Gp., Quartzite Ser.	Z	Acritarch	Sphaeromorph acritarch
72.4	−24.5	Eleonore Bay Gp., Multicolored Ser.	Z	Acritarch	Sphaeromorph acritarch
72.5	−25.4	Eleonore Bay Gp., U. Argillaceous-Arenaceous Ser.	Z	Acritarch	
79.5	17.5	Veteranen Gp. (in 4 formations)	Z	Acritarcha	Acritarcha
33.0	106.5	Dengying Fm., Gaojiashan Mem.	Z	Algae	
		Povarovo Gp., Lyubim Fm.	Z	Metaphyta	Vendotaenid
		Povarovo Gp., Lyubim Fm.	Z	Metaphyta	Vendotaenid
		Povarovo Gp., Lyubim Fm.	Z	Vendotaenides	Vendotaenid
		Valdai Ser.	Z	Metaphyta	Vendotaenid
		Jiuliqiao Fm.	Z	Pogonophora, Saarindae	Sabelliditid
32.6	116.8	Jiuliqiao Fm.	Z	Worm	
32.6	116.8	Huainan Gp., Liulaobei Fm.	Z	Tawuid	
		Redkino Fm.	Z	Fossil	
64.5	45.5	Redkino Ser.	Z	Fossil	
59.4	52.2	Redkino Ser.	Z	Fossil	
59.5	52.3	Redkino Ser., Kairovo Fm.	Z	Fossil	
64.3	39.5	Redkino Ser.	Z	Fossil	
59.4	52.2	Redkino Ser., Kairovo Fm.	Z	Macrophytoplankton	

Described Taxa of Proterozoic and Selected Earliest Cambrian Megafossils

Table 23.1. (Continued).

Number	Taxon	Reference Code and Page	Continent and Country	Locality
49	*Beltanelloides sorichevae*	Bras79.379 p. 381, Fig. 2b–c.	EU ES	23 Rio Uso, near Fuentes
50	*Beltanelloides sorichevae*	Chou80..85 p. 158.	EU ES	23 Rio Uso, near Fuentes
51	*Beltanelloides sorichevae major*	Soko73.204 p. 213, Fig. 5, no. 5.	EU SU	38 L. Onega Riv., well
52	*Beltanelloides sorichevae minor*	Soko73.204 p. 213, Fig. 5, no. 7.	EU SU	38 L. Onega Riv., well
53	*Beltanelloides sorichevae minor*	Soko73.204 p. 213, Fig. 5, no. 3.	EU SU	38 Mezen Riv., Leshukonskoe well
54	*Beltanelloides sorichevae minor*	Soko73.204 p. 213, Fig. 5, no. 6.	EU SU	38 L. Onega Riv., well
55	*Beltanelloides sorichevae*	Pere79.197 p. 198.	EU ES	23 Rio Uso, near Fuentes
56	*Beltina*	Fent43..83 p. 84.	NA US	5 Many Glacier
57	*Beltina*	Aitk81..47 p. 48, 56, 60, 65.	NA CA	2 Mackenzie Mts. 3
58	*Beltina* cf. *danai*	Ishc86..91 p. 156, Pl. 2, Figs. 1–2.	EU SU	40 Serebriya well, Vinnitza reg.
59	*Beltina* cf. *danaii*	Fent371873 p. 1949, Pl. 2, Fig. 4.	NA US	5 Many Glacier
60	*Beltina danai*	Walc99.199 p. 237, Pls. 25, 26. Pl. 27, Figs. 2–6.	NA US	8 Sawmill and Deep Creek Canyons
61	*Beltina danai*	Will02.305 p. 317.	NA US	5 Many Glacier, E of hotel
62	*Beltina danai*	Walc11..17 p. 21, Pl. 7, Figs. 2, 2a, 3.?	NA CA	5 Cameronian Mtn., Oil City
63	*Beltina danai*	Walc11..17 p. 21, Pl. 7, Fig. 4.	NA US	5 Swift Current Ck., Canyon Ck.
64	*Beltina danai*	Daly12.528 p. 65, 183.	NA CA	5 Cameronian Mtn., Oil City
65	*Beltina danai*	Hofm79.150 p. 162–163, Fig. 17 B.	NA CA	1 Mackenzie Mts. 1
66	*Beltina danai*	Hofm85.331 p. 344–346, Pl. 39, Figs. 5–8.	NA CA	1 Mackenzie Mts. 2
67	*Beltina danai*	Horo86.640 p. 640	NA US	8
68	*Beltina danai*?	Fent31.670 p. 686.	NA US	5 Many Glacier, E of hotel
69	*Beltina*?	Walc11..17 p. 21.	NA CA	4 Purcell Mts., Meachem Ck.
70	*Beltina*?	Hofm83.321 p. 348–349, Photo 14-10A–O.	AU WA	60 Tributary to Osmond Ck.
71	*Bipatinella cervicalis*	Chen86.221 p. 231. Pl. 2, Fig. 6.	AS CN	52 Mt. Bagongshan, Huainan
72	*Calyptrina* aff. *striata*	Xing85.182 p. 189, Pl. 39, Fig. 1.	AS CN	88 N Anhui
73	*Caudina cauda*	**Gnil79..39 p. 43, Pl. 47, Figs. 1–4.	EU SU	38 Dorogobuzh well
74	*Caudina cauda*	Gnil79.611 p. 616, Pl. 2, Figs. 2–3.	EU SU	38 Dorogobuzh well
75	*Caudina cauda*	Gnil83..46 p. 51, Pl. 47, Figs. 1–4.	EU SU	38 Dorogobuzh well, 813 m
76	*Chuaria*	Whit28.389 p. 389.	NA US	68 Grand Canyon
77	*Chuaria*	Aitk78.481 p. 483.	NA US	1 Mackenzie Mts.
78	*Chuaria*	Nybe80.299 p. 299	NA US	9 NE Utah, no locality given
79	*Chuaria*	Tapp801028 p. 818	NA US	11 Grand Canyon
80	*Chuaria*	Aitk81..47 p. 48, 56, 60, 65.	NA CA	2 Mackenzie Mts. 2
81	*Chuaria*	Du85.183 p. 186, 189, Pl. 1, Figs. 10–11 (partim).	AS CN	45 Longfengshan, Xinglong
82	*Chuaria*	Stra86.765 p. 765	NA US	9 Uinta Mts., Yellowstone Ck.
83	*Chuaria*	Vida87.345 p. 346.	AS CN	52 Mt. Bagongshan, Huainan
84	*Chuaria*	Sun87.349 p. 351.	AS CN	53 Huainan
85	*Chuaria annularis*	Yang80.231 p. 253, Pl. 17, Figs. 3–4.	AS CN	52 Mt. Bagongshan
86	*Chuaria annularis*	*Zhen80..49 p. 59–60, Pl. 1, Figs. 3–4.	AS CN	52 Mt. Bagongshan, Huainan
87	*Chuaria annularis*	Du85...5 p. 7, 9, Pl. 1, Figs. 7–9.	AS CN	45 Longfengshan
88	*Chuaria annularis*	Chen86.221 p. 231. Pl. 1, Figs. 2, 3, 12a	AS CN	52 Mt. Bagongshan, Huainan
89	*Chuaria* cf. *C. circularis*	Guss731108 p. 1109–1111, Text-Fig. 2A–D.	NA CA	3 E base of Storm Mtn.
90	*Chuaria* cf. *circularis*	Du82...1 p. 2–3, Pl., Fig. 4.	AS CN	45 Longfengshan, Huailai
91	*Chuaris* cf. *circularis*	Dama86.417 p. 423, Fig. 5K.	AU SA	102 Stuart Shelf: AMOCO well SCYWIA
92	*Chuaria circularis*	**Walc99.199 p. 234–235, Pl. 27, Figs. 12–13. 12–13.	NA US	11 Grand Canyon, Nankoweap Butte
93	*Chuaria circularis*	Ford69.114 p. 117, 119–120, Fig. 2.	NA US	11 Grand Canyon, Nankoweap Butte

Latitute	Longitude	Geologic Unit	Age	Original Author's Interpretation	Hofmann Interpretation
39.7	−4.7	Pusa Sh.	Z	Body Fossil	
39.7	−4.7	Pusa Sh.	Z	Fossil	
59.4	52.2	Redkino Ser.	Z	Fossil	
59.4	52.2	Redkino Ser., Kairovo Fm.	Z	Fossil	
64.5	45.5	Redkino Ser.	Z	Fossil	
59.4	52.2	Redkino Ser.	Z	Fossil	
39.7	−4.7	Pusa Sh.	Z	Fossil	
48.8	−113.7	Belt Supgp., Altyn Fm.	Y	Brown alga	Carbonaceous film
64.1	−128.5	Little Dal Gp., Rusty Shale	Z	Fossil	
48.5	27.8	Mogilev-Podolsky Gp., Yaryshev Fm.	Z	Noncalcareous alga	Beltinid
48.8	−113.7	Belt Supgp., Altyn Fm.	Y	Thallophytes, brown algae?	
		Belt Supgp., Greyson Sh. (and Newland Ls.?)	Y	Merostomata	Carbonaceous films
48.8	−113.7	Belt Supgp., Altyn Fm. [originally: Siyeh Fm.]	Y	Crustacean remains, Merostomata	Carbonaceous films or clasts
49.1	−114.0	Belt (Purcell) Supgp., Altyn Fm.	Y	Merostomata	Carbonaceous film or clast
48.8	−113.7	Belt Supgp., Altyn Fm.	Y	Merostomata	Carbonaceous film or clast
49.1	−114.0	Belt (Purcell) Supgp., Altyn Fm.	Y	Crustacean	
64.8	−129.9	Little Dal Gp.	Z	Algal thalli or mats, scums	Carbonaceous films
64.6	−128.8	Little Dal Gp.	Z	Algae?	Carbonaceous films
		Lower Belt Spgp.	Y	Fragments of microbial mats	
48.8	−113.7	Belt Supgp., Altyn Fm.	Y	Algae	Carbonaceous film fragments
49.9	−116.4	Belt (Purcell) Supgp., Aldridge Fm.	Y	Crustacean	Carbonaceous film or fragment
−17.2	128.4	Bungle Bungle Dol.	X	Dubiofossils	Carbonaceous films
32.6	116.8	Huainan Gp., Liulaobei Fm.	Z	Carbonaceous fossil	cf. short *Tawuia*
		Jinshanzai Fm.	ZC	Pogonophora, Saarinidae	Sabelliditid
54.9	33.3	Redkino Ser., Borodino Fm.	Z	Vendotaenid	
54.9	33.3	Redkino Fm.	Z	Metaphyta	
54.9	33.3	Redkino Ser., Borodino Fm.	Z	Vendotaenid	
		Unkar Gp.	Y	Alga	
		Little Dal Gp.	Z	Fossil	
		Uinta Mountain Gp., Red Pine Sh.	YZ	Eucaryotic phytoplankton	
		Chuar Gp.	YZ	Prasinophyta, Pyramimonadales	Megascopic acritarch
64.1	−128.5	Little Dal Gp., Rusty Shale	Z	Fossil	
40.4	117.5	Qingbaikou Sys., Changlongshan Fm.	Z	Algae	
40.6	−110.4	Uinta Mountain Gp., Red Pine Sh.	YZ	Form taxon of several species	
32.6	116.8	Huainan Gp., Liulaobei Fm.	Z	Acritarch	
		Liulaobei Fm., Jiuliqiao Fm.	Z	Fossil	
32.6	116.8	Huainan Gp., Liulaobei Fm.	Z	Carbonaceous fossil	
32.6	116.8	Huainan Gp., Liulaobei Fm.	Z	Chuaridae	
40.4	115.5	Qingbaikou Sys., Changlongshan Fm.	Z	Algae	
32.6	116.8	Huainan Gp., Liulaobei Fm.	Z	Chuarid	
52.2	−116.0	Miette Gp., Hector Fm.	Z	Fossil	cf. *Beltanelliformis*
40.4	115.5	Qingbaikou Sys., Changlongshan Fm.	Z	Chuaridae	
−30.0	137.0	Tent Hill Fm., Arcoona Qte. Mem.	Z	Thick-walled sphaeromorph	
36.3	−111.9	Chuar Gp., Kwagunt Fm.	YZ	Fossil	Megascopic acritarch
36.3	−111.9	Chuar Gp.	YZ	Problematicum	

Table 23.1. (Continued).

Number	Taxon	Reference Code and Page	Continent and Country	Locality
94	*Chuaria circularis*	Ford72..11 p. 11–17, Pl. 1, Figs. 1–6 Pl. 2, Fig. 4.	NA US	11 Grand Canyon, Nankoweap Butte
95	*Chuaria circularis*	Ford72..11 p. 13, Pl. 2, Figs. 2–3.	EU SE	24 L. Vättern area
96	*Chuaria circularis*	Ford73.535 p. 539–546, Pl. 62, Figs. 2–6	EU SE	24
97	*Chuaria circularis*	Ford73.535 p. 546.	NA US	10
98	*Chuaria circularis*	Ford73.535 p. 539–546, Pl. 61, Figs. 1–7, Pl. 62, Fig. 1, Pl. 63, Fig. 4.	NA US	11 Grand Canyon, Nankoweap Butte
99	*Chuaria circularis*	Ford73.535 p. 539–546, Pl. 63, Fig. 3	AU NT	61 Mt. Skinner area, C. Australia
100	*Chuaria circularis*	Ford73.535 p. 539–546, Pl. 63, Figs. 1–2.	AS IR	42 Chapoghlu
101	*Chuaria circularis*	Hofm77...1 p. 1–11, Figs. 2–4.	NA US	9 Uinta Mts., Yellowstone Ck.
102	*Chuaria circularis*	Jux77...1 p. 9–12, Pl. 5.	NA US	11 Grand Canyon, Nankoweap Butte
103	*Chuaria circularis*	Hofm79.150 p. 157, Figs. 13 I, K–M, 14.	NA CA	1 Mackenzie Mts. 1
104	*Chuaria circularis*	Ghar79..91 p. 91–92, Pl. 7, Figs. 1–2 Pl. 8, Figs. 3–4.	AS IN	57 Rampura
105	*Chuaria circularis*	Pere79.197 p. 198.	EU ES	23 Rio Uso, near Fuentes
106	*Chuaria circularis*	Vida79..40 p. 11, 16	NA GL	22 Kong Oscars Fjord
107	*Chuaria circularis*	Vida79..40 p. 13, 16	NA GL	19 Kong Oscars Fjord
108	*Chuaria circularis*	Vida79..40 p. 13, 16	NA GL	20 Kong Oscars Fjord
109	*Chuaria circularis*	Vida79..40 p. 12, 16, Pl. 4, Fig. a.	NA GL	21 Kong Oscars Fjord
110	*Chuaria circularis*	Gowd80.156 p. 156.	AS IN	93 No locality given
111	*Chuaria circularis*	Zhen80..49 p. 58–59, Pl. 1, Figs. 1–2.	AS CN	52 Mt. Bagongshan, Huainan
112	*Chuaria circularis*	Yang80.231 p. 253, Pl. 17, Figs. 1–2.	AS CN	52 Mt. Bagongshan
113	*Chuaria circularis*	Du80.341 p. 353, Pl. 25, Fig. 3.	AS CN	45 Longfengshan, Huailai
114	*Chuaria circularis*	Chou80..85 p. 158.	EU ES	23 Rio Uso, near Fuentes
115	*Chuaria circularis*	Tynn80..27 p. 12, 22.	EU FI	75 Hailuoto
116	*Chuaria circularis*	Knol81..55 p. 55	EU SJ	25 Murchisonfjorden
117	*Chuaria circularis*	Knol81..55 p. 55	EU SJ	29 Murchisonfjorden
118	*Chuaria circularis*	Vida81..53 p. 9.	EU NO	37 E Finnmark, Austertana
119	*Chuaria circularis*	Vida81..53 p. 15.	EU NO	33 E Finnmark, Store Ekkerøy
120	*Chuaria circularis*	Vida81..53 p. 17.	EU NO	31 E Finnmark, Andersby
121	*Chuaria circularis*	Vida81..53 p. 13.	EU NO	34 E Finnmark, Vagge
122	*Chuaria circularis*	Vida81..53 p. 11.	EU NO	35 E Finnmark, Grasdalen
123	*Chuaria circularis*	Vida81..53 p. 10.	EU NO	36 E Finnmark, Veinesfjorden
124	*Chuaria circularis*	Vida81..53 p. 17.	EU NO	32 E Finnmark, Vadsø
125	*Chuaria circularis*	Vida81..53 p. 18, Fig. 11 J–K.	EU NO	30 E Finnmark, Klubbnasen
126	*Chuaria circularis*	Duan82..57 p. 58–61, Figs. 3 G–J.	AS CN	49 Guanjiatun, Fuxian
127	*Chuaria circularis*	Duan82..57 p. 58–61, Figs. 5 A–J, O–P.	AS CN	52 Mt. Bagongshan, Huainan
128	*Chuaria circularis*	Knol82.269 p. 273, Pl. 2, Fig. 1.	EU SJ	29 Murchisonfjorden
129	*Chuaris circularis*	Knol82.269 p. 273, 275, Pl. 4, Fig. 13.	EU SJ	25 Murchisonfjorden
130	*Chuaria circularis*	Math82.125 p. 128–129, Fig. 3A.	AS IN	57 Rampura
131	*Chuaria circularis*	Duan82..57 p. 58–61, Fig. 3 F	AS CN	49 Qinggouzi, Hunjiang City
132	*Chuaria circularis*	Du82...1 p. 2, Pl., Figs. 1–3.	AS CN	45 Longfengshan, Huailai
133	*Chuaria circularis*	Duan82..57 p. 58–61, Figs. 3 A–E, 4	AS CN	48 Jixian, W. Jingeryu Village
134	*Chuaria circularis*	Knol82.269 p. 273	EU SJ	28 Murchisonfjorden
135	*Chuaria circularis*	Knol82.269 p. 275, Pl. 3, Fig. 12.	EU SJ	26 Murchisonfjorden
136	*Chuaria circularis*	Sure83..79 p. 79–84, Figs. 2–3.	AS IN	93 Chitapur Taluk, Gulbarga Dist.
137	*Chuaria circularis*	Math83.363 p. 364, Fig. 1 A.	AS IN	57 Rampura
138	*Chuaria circularis*	Vida83..45 p. 52.	EU NO	73 E Finnmark, Batsfjord (and oth.)

Latitute	Longitude	Geologic Unit	Age	Original Author's Interpretation	Hofmann Interpretation
36.3	−111.9	Chuar Gp.	YZ	Problematicum	
58.3	14.7	Visingsö Fm.	YZ	Problematicum	
58.3	14.7	Visingsö Fm.	YZ	Large acritarch-like organism	
		Chuar Gp., Galeros Fm.	YZ		
36.3	−111.9	Chuar Gp., Kwagunt Fm.	YZ	Large acritarch-like organism	
−22.2	134.3	Central Mt. Stuart beds	Z	*Chuaria*-like fossil	cf. *Beltanelliformis*
36.2	48.9	Chapoghlu Sh.	Z	Large acritarch-like organism	
40.6	−110.4	Uinta Mountain Gp., Red Pine Sh.	YZ	Planktonic problematicum	
36.3	−111.9	Chuar Gp., Kwagunt Fm., Awatubi Mem.	YZ	Cysts	
64.8	−129.9	Little Dal Gp.	Z	Fragments of thalli or mats	Fragments of thalli or mats
24.5	75.5	Semri Gp., Suket Sh.	YZ	No affinities given	
39.7	−4.7	Pusa Sh.	Z	Carbonaceous fossil	
72.6	−25.1	Tillite Gp., L. Tillite & Inter-Tillite beds	Z	Acritarch	
72.6	−25.6	Eleonore Bay Gp., Quartzite Ser.	Z	Acritarch	
72.6	−25.4	Eleonore Bay Gp., Multicolored Ser.	Z	Acritarch	
72.6	−25.1	Eleonore Bay Gp., Limestone-Dolomite Ser.	Z	Acritarch	
		Bhima Gp.	Z	No affinities given	
32.6	116.8	Huainan Gp., Liulaobei Fm.	Z	Chuaridae	
32.6	116.8	Huainan Gp., Liulaobei Fm.	Z		
40.4	115.5	Qingbaikou Sys., Changlongshan Fm.	Z		
39.7	−4.7	Pusa Sh.	Z	Fossil	
60.0	24.8	Hailuoto Fm.	YZ	Acritarch	Sphaeromorph acritarch
80.3	18.7	Franklinsundet Gp., Westmanbukta Fm.	Z	Acritarch	
79.8	18.5	Roaldtoppen Gp., Ryssö Fm.	Z	Acritarch	
70.4	28.5	Vestertana Gp., Nyborg Fm.	Z	Acritarch	Sphaeromorph acritarch
70.1	30.1	Vadsø Gp., Ekkerøy Fm.	Z	Acritarch	Sphaeromorph acritarch
70.1	29.1	Vadsø Gp., Andersby Fm.	Z	Acritarch	Sphaeromorph acritarch
70.5	28.5	Tanafjord Gp., Dakkovarre Fm.	Z	Acritarch	Sphaeromorph acritarch
70.7	28.5	Tanafjord Gp., Grasdal Fm.	Z	Acritarch	Sphaeromorph acritarch
70.1	28.7	Vestertana Gp., L. Tillite (Smalfjord Fm.)	Z	Acritarch	Sphaeromorph acritarch
70.1	29.9	Vadsø Gp., Golneselv Fm.	Z	Acritarch	Sphaeromorph acritarch
70.1	29.2	Vadsø Gp., Klubbnes Fm.	Z	Acritarch	Sphaeromorph acritarch
39.7	121.9	Xihe Gp., Nanfen Fm.	Z	Chuariaceae	
32.6	116.8	Huainan Gp., Liulaobei Fm.	Z	Chuariaceae	
79.8	18.5	Roaldtoppen Gp., Ryssö Fm.	Z	Acritarch	
80.3	18.7	Franklinsundet Gp., Westmanbukta Fm.	Z	Acritarch	
24.5	75.5	Semri Gp., Suket Sh.	YZ	Large leiosphaerid acritarch	
41.8	126.2	Xihe Gp., Nanfen Fm.	Z	Chuariaceae	
40.4	115.5	Qingbaikou Sys., Changlongshan Fm.	Z	Chuaridae	
40.0	117.4	Qingbaikou Sys., Jingeryu Fm.	Z	Chuariaceae	
79.9	18.6	Roaldtoppen Gp., Hunnberg Fm.	Z	Acritarch	
80.0	19.0	Franklinsundet Gp., Kapp Lord Fm.	Z	Acritarch	
17.2	77.1	Bhima Sh., Gangurthi sh.	Z	Algae	
24.5	75.5	Semri Gp., Suket Sh.	YZ	Sphaeromorph acritarch	
70.6	29.8	Barents Sea Gp., Båtsfjord Fm., Annijokka Mem.	Z	Acritarch	Sphaeromorph acritarch

Table 23.1. (Continued).

Number	Taxon	Reference Code and Page	Continent and Country	Locality
139	*Chuaria circularis*	Vida83..45 p. 49.	EU NO	72 E Finnmark, Langbunes
140	*Chuaria circularis*	Amar841405 p. 1406, Pl. Fig. 3a.	AF MR	97 Atar
141	*Chuaria circularis*	Wang84.271 p. 278, Pl. 2, Figs. 6, 7a, 8b, 9	AS CN	53 Mt. Sidingshan, Huainan
142	*Chuaria circularis*	Amar84.975 p. 975–980, Pl. Figs. 3a–e.	AF HV	98 Pendjari Riv.
143	*Chuaria circularis*	Dawe85..22 p. 24.	NA GL	15 Thule
144	*Chuaria circularis*	Du85...5 p. 7, 9, Pl. 1, Fig. 2.	AS CN	45 Jixian
145	*Chuaria circularis*	Du85...5 p. 7, 9, Pl. 1, Figs. 1.	AS CN	45 Longfengshan
146	*Chuaria circularis*	Vida85.349 p. 383	NA US	11 Grand Canyon, Nankoweap Butte
147	*Chuaria circularis*	Vida85.349 p. 355–359	NA US	11 Grand Canyon, Nankoweap Butte
148	*Chuaria circularis*	Du85...5 p. 7, 9, Pl. 1, Figs. 3, 5.	AS CN	49 Fuxian
149	*Chuaria circularis*	Dawe85..22 p. 24.	NA GL	16 Thule, 100 km NW of settlement
150	*Chuaria circularis*	Hofm85.331 p. 342–343, pl. 35 (2, 4–6, 8) Pl. 36 (6, 12) Pl. 37 (1) Pl. 39 (2) text–Fig. 4.	NA CA	1 Mackenzie Mts 1
151	*Chuaria circularis*	Vida85.349 p. 355–359, Fig. 3A	NA US	10 Grand Canyon
152	*Chuaria circularis*	Vida85.349 p. 383	NA US	10 Grand Canyon, Technicolor Cliff
153	*Chuaria circularis*	Vida85.349 p. 355–359	NA US	9 Setting Road, 16 km E Kamas
154	*Chuaria circularis*	Hofm85.331 p. 342–343	NA CA	1 Mackenzie Mts. 5
155	*Chuaria circularis*	Hofm85.331 p. 342–343, Pl. 38 (4) text–Figs. 3, 5.	NA CA	1 Mackenzie Mts. 4
156	*Chuaria circularis*	Hofm85..20 p. 26, Fig. 4 A,B.	NA CA	1 Mackenzie Mts. 1
157	*Chuaria circularis*	Vida85.349 p. 382	NA US	10 Grand Canyon, Basalt Canyon
158	*Chuaria circularis*	Vida85.349 p. 383	NA US	11 Grand Canyon
159	*Chuaria circularis*	Duan85..68 p. 70, Pl. 16, Fig. 1.	AS CN	52
160	*Chuaria circularis*	Duan85..68 p. 70, Pl. 16, Fig. 2.	AS CN	48
161	*Chuaria circularis*	Pyat86.129 p. 144, 153, Pl. 1, Fig. 12.	AS SU	79 Vanavara well 2, 2830 m
162	*Chuaria circularis*	Amar86..69 p. 74–77, Pl. 1, Figs. 1, 6.	AF MR	97 Atar
163	*Chuaria circularis*	Germ86..45 p. 53–54, Fig. 4f	AF NA	101 Tses well
164	*Chuaria circularis*	Chen86.221 p. 231. Pl. 1, Figs. 1, 4, 12b, 14b.	AS CN	52 Mt. Bagongshan, Huainan
165	*Chuaria circularis*	Sun87.109 p. 128	NA CA	1 Mackenzie Mts.
166	*Chuaria circularis*	Sun87.109 p. 115–123, Pl. 1, Figs. 1–2, Pl. 2, Figs. 1–8, Pl. 4, Figs. 1, 2.	AS CN	52 Huainan
167	*Chuaria circularis*	Sun87.109 p. 115–123, Pl. 1, Figs. 3–8	AS CN	49 Fuxian
168	*Chuaria circularis*	Sun87.109 p. 112	AS CN	54 Changshan
169	*Chuaria circularis*	Sun87.109 p. 112	AS CN	50 W. Honan
170	*Chuaria circularis*	Sun87.109 p. 112, 113.	AS CN	85 Fuxian
171	*Chuaria circularis*	Veys88..47 p. 51, 52.	AS SU	Uchur-Maya region, Yudoma Riv.
172	*Chuaria circularis*	Veys88..47 p. 51, 52.	AS SU	Uchur-Maya region
173	*Chuaria circularis*	Veys88..47 p. 52, 55.	AS SU	Turukhansk region
174	*Chuaria circularis*	Veys88..47 p. 50, 52.	AS SU	Uchur-Maya region, Omya Riv.
175	*Chuaria circularis*	Veys88..47 p. 52, 56.	AS SU	Turukhansk region
176	*Chuaria circularis*	Veys88..47 p. 51, 52.	AS SU	Uchur-Maya region, Bol. Kandyk
177	*Chuaria circularis*	Veys88..47 p. 52, 56.	AS SU	Turukhansk region
178	*Chuaria circularis*	Veys88..47 p. 51, 52.	AS SU	Uchur-Maya region, Belaya Riv.
179	*Chuaria circularis*	Veys88..47 p. 52, 56.	AS SU	Turukhansk region

Latitute	Longitude	Geologic Unit	Age	Original Author's Interpretation	Hofmann Interpretation
70.3	30.7	Barents Sea Gp., Båsnaering Fm., Naeringselva Mem.	Z	Acritarch	Sphaeromorph acritarch
20.5	−13.1	Atar Fm., Unit 3	YZ	Acritarch	
32.6	116.8	Jiuliqiao Fm.	Z	Acritarch	
11.6	1.4	Pendjari Fm.	Z	Acritarch	Acritarch
77.5	−69.2	Thule Gp., Wolstenholme Fm.	YZ	Acritarch	
		Qingbaikou Sys., Changlongshan Fm.	Z	Algae	
40.4	115.5	Qingbaikou Sys., Changlongshan Fm.	Z	Algae	
36.3	−111.9	Chuar Gp., Kwagunt Fm., Awatubi Mem.	YZ	Acritarch	
36.3	−111.9	Chuar Gp., Kwagunt Fm.	YZ	Acritarch	
39.7	121.9	Xihe Gp., Nanfen Fm.	Z	Algae	
78.0	−71.7	Thule Gp., Dundas Fm.	YZ	Acritarch	
64.8	−129.9	Little Dal Gp., lower part	Z	*Incertae sedis*, Algae?	
		Chuar Gp., Galeros Fm.	YZ	Acritarch	
		Chuar Gp., Galeros Fm., Jupiter Mem.	YZ	Acritarch	
40.6	−111.1	Uinta Mountain Gp., Red Pine Sh.	YZ	Acritarch	
62.6	−126.6	Little Dal Gp., lower part	Z	*Incertae sedis*, Algae?	
63.8	−127.6	Little Dal Gp.	Z	*Incertae sedis*, Algae?	
64.8	−129.9	Little Dal Gp., lower part	Z	Sphaeromorph	
36.1	−111.9	Chuar Gp., Galeros Fm., Tanner Mem.	YZ	Acritarch	
		Chuar Gp., Kwagunt Fm., Walcott Mem.	YZ	Acritarch	
		Huainan Gp., Liulaobei Fm.	Z	Chuariaceae	
		Qingbaikou Sys., Jingeryu Fm.	Z	Chuariaceae	
60.4	102.2	Kamo Fm.	Z		Large *Leiosphaeridia*
20.5	−13.1	Atar Fm., Unit 3	YZ		
−25.8	18.1	Kuibis and Schwarzrand Fms.	Z	Thick-walled sphaeromorph	
32.6	116.8	Huainan Gp., Liulaobei Fm.	Z	Chuarid	
		Little Dal Gp.	Z	*Nostoc*-like balls	
		Huainan Gp., Liulaobei Fm.	Z	Algae, *Nostoc*-like colony	
39.7	121.9	Xihe Gp., Nanfen Fm.	Z	*Nostoc*-like colonies	
30.2	117.5	Tumen Gp., Xingxing Fm.	Z	*Nostoc*-like colonies	
34.5	111.5	Puyu Fm.	Z	*Nostoc*-like colonies	
39.8	121.8	Wuhangshan Gp., Changlingzi Fm.	Z	*Nostoc*-like colonies	
59.6	135.9	Ui Gp., Ust'kyrbinsk Fm.	YZ	Acritarch	
58.9	128.2	Yudoma Gp., Ust'yudoma Fm.	Z	Acritarch	
60.9	88.5	Bezymenskaya Fm.	YZ	Acritarch	
58.5	133.3	Tottinskaya Fm.	YZ	Acritarch	
60.8	88.1	Dereviniskaya Fm.	YZ	Acritarch	
59.0	135.2	Ui Gp., Kandyk Fm.	YZ	Acritarch	
60.8	88.1	Dereviniskaya Fm.	YZ	Acritarch	
61.6	137.5	Lakhanda Gp., Neryuen Fm.	YZ	Acritarch	
60.6	88.1	Miroedikha Fm.	YZ	Acritarch	

Table 23.1. (Continued).

Number	Taxon	Reference Code and Page	Continent and Country	Locality
180	*Chuaria circularis?*	Coop82.629 p. 631.	AN	63 Molar Massif
181	*Chuaria circularis?*	Coop82.629 p. 631.	AN	62 Mt. McCarthy
182	*Chuaria circularis?*	Coop82.629 p. 632.	AN	64 Houliston Glacier
183	*Chuaria fermorei*	*Math83.363 p. 364, Fig. 1B.	AS IN	57 Rampura
184	*Chuaria minima*	Mait84.146 p. 146–150, Pl. 1, Figs. 1–10.	AS IN	57 Rampura (Rampura)
185	*Chuaria multirugosa*	Du85...5 p. 7, 9, Pl. 1, Figs. 10, 11.	AS CN	45 Longfengshan
186	*Chuaria multirugosa*	*Duan85..68 p. 70, Pl. 16, Fig. 3.	AS CN	48
187	*Chuaria nerjuenica*	Veys88..47 p. 52, 56.	AS SU	Turukhansk region
188	*Chuaria nerjuenica*	Veys88..47 p. 51, 52.	AS SU	Uchur-Maya region, Belaya Riv.
189	*Chuaria olavarriensis*	*Bald83..73 p. 78–79, Pl. 1, Figs. k, 3, 10.	SA AR	69 Olavarria
190	*Chuaria* sp.	Hofm81.443 p. 446, Fig. 3 C–F.	AS CN	47 Liuzhuangzi, NNE of Jixian
191	*Chuaria* sp.	Hofm81.443 p. 446, Fig. 3 G.	AS CN	46 Liuzhuangzi, NNE of Jixian
192	*Chuaria* sp.	Du85...5 p. 7, 9, Pl. 1, Fig. 6.	AS CN	47 Jixian
193	*Chuaria* sp. cf. *circularis*	Clou75.131 p. 150.	NA CA	3 E base of Storm Mtn.
194	*Chuaria wimani*	*Brot41.245 p. 258–259.	EU SE	24 Gränna
195	*Chuaria wimani*	Regn55.546 p. 546, 555.	EU SE	24 L. Vättern area
196	*Chuaria wimani*	Timo60..28 p. 29, 30, 40.	EU SE	24 Mullskredena, W of Omberg
197	*Chuaria wimani*	Eise66..52 p. 52–54, Figs. 1–2.	EU SE	24 L. Vättern area
198	*Chuaria?*	Debr81..53 p. 30, Photo 7.	EU IT	70 Sardinia, Sarrabus, Tuviois
199	*Chuaria?*	Knol82.269 p. 274, Pl. 4, Fig. 7.	EU SJ	27 Murchisonfjorden
200	*Chuaria?*	Hofm83.321 p. 338, 349, Photo 14-10 M, N, P–S.	AU WA	60 Tributary to Osmond Ck.
201	*Chuaria?* sp.	Zhan86..67 p. 84, 86, Pl. 4, Figs. 7, 8.	AS CN	90 Mt. Gaojiashan, Beiwan, Hujiaba
202	*Conicina obtusa*	Chen86.221 p. 231, Pl. 1, Fig. 14a	AS CN	52 Mt. Bagongshan, Huainan
203	*Corycium enigmaticum*	Rank48.389 p. 392–414, Pls. 1–3, text–Fig. 2.	EU FI	74 Tähtinen, Aitolahti
204	*Corycium enigmaticum*	Ohls61.377 p. 384–389	EU FI	74 Tähtinen, Aitolahti
205	*Corycium? oligomerum*	*Belo68.196 p. 196–199, Figs. 1–2.	EU SU	76 Krivoi Rog, Frunze mine hole
206	*Daltaenia mackenziensis*	**Hofm85.331 p. 346–348, Pl. 39, Figs. 1–3 text–Fig. 6.	NA CA	1 Mackenzie Mts. 1
207	*Ellipsophysa axicula*	**Zhen80..49 p. 60, Pl. 1, Figs. 5–6.	AS CN	52 Mt. Bagongshan, Huainan
208	*Ellipsophysa axicula*	Yang80.231 p. 253, Pl. 17, Figs. 5, 6.	AS CN	52 Mt. Bagongshan
209	*Ellipsophysa proceriaxis*	*Zhen80..49 p. 60–61. Pl. 1, Figs. 7, 13, Pl. 2, Figs. 11, 15.	AS CN	52 Mt. Bagongshan, Huainan
210	*Ellipsophysa proceriaxis*	Yang80.231 p. 254, Pl. 17, Figs. 7, 11, 13, 15.	AS CN	52 Mt. Bagongshan
211	*Enteromophites siniansis*	**Zhu 84.558 p. 559–560, Pl. 1, Figs. 1–2.	AS CN	89 Yangtze Gorge
212	*Eoholynia mosquensis*	**Gnil75.258 p. 260–261, Fig. 1/5–9.	EU SU	40 Soligalich-1 well
213	*Eoholynia mosquensis*	Gnil76..10 p. 12, 295, Pl. 1, Figs. 4–5.	EU SU	38 Soligalich-1 well
214	*Eoholynia mosquensis*	Gnil79..39 p. 42, Pl. 45, Figs. 1–8.	EU SU	38 Soligalich-1 well
215	*Eoholynia mosquensis*	Gnil79.611 p. 614, Pl. 1, Figs. 6–7.	EU SU	38 Soligalich-1 well
216	*Eoholynia mosquensis*	Gnil83..46 p. 49, Pl. 45, Figs. 1–8.	EU SU	38 Soligalich-1 well, 1983 m
217	*Eurycyphus altilis*	*Fu86..76 p. 80–81, Pl. 1, Figs. 1, 2, 5, 8, 11, 13.	AS CN	52 Mt. Bagongshan, Huainan
218	*Eurycyphus lycotropus*	**Fu86..76 p. 79–80, Pl. 1, Figs. 9, 10, 12	AS CN	52 Mt. Bagongshan, Huainan
219	*Fasciculella bagungshanensis*	**Duan85..68 p. 77, Pl. 17, Fig. 14.	AS CN	52
220	*Fengyangella doedica*	Chen86.221 p. 231. Pl. 1, Fig. 11, Pl. 2, Figs. 4, 5	AS CN	52 Mt. Bagongshan, Huainan
221	*Fermoria*	Sahn56..25 p. 25–26.	AS IN	57 Neemuch
222	*Fermoria*	Ford72..11 p. 15, Pl. 2, Fig. 1.	AS IR	42 Chapoghlu
223	*Fermoria capsella*	*Chap35.109 p. 117, Pl. 2, Figs. 3–4.	AS IN	57 Rampura

Latitute	Longitude	Geologic Unit	Age	Original Author's Interpretation	Hofmann Interpretation
−71.5	163.5	Bowers Supgp., Molar Fm.	ZC	Acritarch fragments	
−72.6	166.2	Robertson Bay Gp.	ZC	Acritarch fragments	
−72.0	164.7	Clasts in Houliston Glacier beds (Cambrian)	ZC	Acritarch fragments	
24.5	75.5	Semri Gp., Suket Sh.	YZ	Sphaeromorph acritarch	
24.5	75.5	Semri Gp., Suket Sh.	YZ	Cryptarch	
40.4	115.5	Qingbaikou Sys., Changlongshan Fm.	Z	Algae	
		Qingbaikou Sys., Jingeryu Fm.	Z	Chuariaceae	
60.6	88.1	Miroedikha Fm.	YZ	Acritarch	
61.6	137.5	Lakhanda Gp., Neryuen Fm.	YZ	Acritarch	
−36.9	−60.2	Sierra Bayas Fm. (?=La Tinta Fm.)	Z	Fossil	
40.2	117.5	Changcheng Sys., Tuanshanzi Fm.	X	Megascopic compressions	Carbonaceous compression
40.2	117.5	Changcheng Sys., Chuanlinggou Fm.	X	Megascopic compression	
40.2	117.5	Changcheng Sys., Tuanshanzi Fm.	X		
52.2	−116.0	Miette Gp., Hector Fm.	Z	Planktonic algal spheres	cf. *Beltanelliformis*
58.0	14.5	Visingsö Fm.	YZ	Chitinous foraminifera	
58.3	14.7	Visingsö Fm.	YZ	Acritarch	
58.3	14.7	Visingsö Fm.	YZ	Megaspore	
58.3	14.7	Visingsö Fm.	YZ	Chitinous foraminifera	
39.4	9.3	San Vito Fm.	ZC	Dubious fossil	Dubiofossil
80.1	18.7	Celsiusberget Gp., Flora Fm.	Z	Acritarch	Sphaeromorph acritarch
−17.2	128.4	Bungle Bungle Dol.	X	Algae or colonial microbes	Algae or colonial microbes
33.0	106.5	Dengying Fm., Gaojiashan Mem.	Z	Algae	
32.6	116.8	Huainan Gp., Liulaobei Fm.	Z	Chuarid	cf. *Tawuia*
61.7	23.7	Bothnian phyllites	W	Biogenic carbon, real fossil	Carbon biogenic, diffusion
61.7	23.7	Bothnian phyllites	W	Lake balls	Diffusion phenomenon
47.9	33.4	Krivoi Rog 'ser.'	X	Lower algae	Biogenic carbon around grains
64.8	−129.9	Little Dal Gp.	Z	*Incertae sedis*, Algae?	
32.6	116.8	Huainan Gp., Liulaobei Fm.	Z	Chuaridae	*Tawuia*
32.6	116.8	Huainan Gp., Liulaobei Fm.	Z	Fossil	*Tawuia*
32.6	116.8	Huainan Gp., Liulaobei Fm.	Z	Chuaridae	*Tawuia*
32.6	116.8	Huainan Gp., Liulaobei Fm.	Z	Fossil	*Tawuia*
30.8	111.1	Doushantuo Fm.	Z	Green alga	
59.0	42.2	Redkino Ser., Nelidovo, Borodino, Petrovskaya Fms.	Z	Vendotaenides, algae	
59.0	42.2	Redkino Fm.	Z	Metaphyta	
59.0	42.2	Redkino Ser., Nelidovo, Borodino Fms.	Z	Metaphyta	
59.0	42.2	Redkino Fm.	Z	Metaphyta	
59.0	42.2	Redkino Ser., Borodino Fm.	Z	Vendotaenides	
32.6	116.8	Huainan Gp., Liulaobei Fm.	Z	Cyphomegacritarch	*Tawuia*?
32.6	116.8	Huainan Gp., Liulaobei Fm.	Z	Cyphomegacritarch	*Tawuia*?
		Huainan Gp., Liulaobei Fm.	Z	Megascopic algae	Vendotaenid
32.6	116.8	Huainan Gp., Liulaobei Fm.	Z	Tawuid	cf. *Tawuia*
24.4	74.9	Semri Gp., Suket Sh.	YZ	Problematicum	*Chuaria*
36.2	48.9	Chapoghlu Sh.	Z	Problematicum	*Chuaria*
24.5	75.5	Semri Gp., Suket Sh.	YZ	Brachiopoda, Atremata?	*Chuaria*

Table 23.1. (Continued).

Number	Taxon	Reference Code and Page	Continent and Country	Locality
224	*Fermoria granulosa*	*Chap35.109 p. 116, Pl. 1, Figs. 2, 4, Pl. 2, Fig. 5.	AS IN	57 Rampura
225	*Fermoria minima*	**Chap35.109 p. 114–116, Pl. 1, Figs. 1, 3.	AS IN	57 Neemuch
226	*Fermoria minima*	Sahn36.458 p. 465–467, Pl. 43, Figs. 1–4.	AS IN	57 Neemuch
227	*Fermoria minima*	Sahn77.289 p. 289–299, Pls. 1–2, Pl. 3, Figs. 1–2.	AS IN	57 Rampura
228	*Fermoria* spp.	Misr57..54 p. 55, Pl. 7, Figs. 1, ?2–3.	AS IN	57 Rampura
229	*Fermoria* types 1–6	Mait77.176 p. 183, Pl. 5, Figs. 35–40.	AS IN	57 Rampura (Ramapura)
230	*Fermoria* with filament	Sahn77.289 p. 293, 298, Pl. 2, Figs. 3–4, Pl. 3, Fig. 1.	AS IN	57 Rampura
231	*Fermoria?*	Raju63.306 p. 306–307, Fig. 1.	AS IN	59 Ankireddipalle
232	*Fermoria?* sp.	Stoc64..29 p. 14, Pl. 1, Figs. 3–5.	AS IR	42 Chapoghlu
233	*Fermoria?* sp.	Stoc64..29 p. 14	AS IR	42 Gasir, in Karaj valley
234	*Fermoria?* sp.	Stoc64..29 p. 14	AS IR	42 Barut-Aghaji
235	*Fusosquamula vlasovi*	**Asee76..40 p. 58, Pl. 20, Figs. 1–2.	EU SU	41 Volynia well 350 Grabov 65 m
236	*Fusosquamula vlasovi*	Asee83..96 p. 97, Pl. 1, Fig. 6, Pl. 7, Figs. 3–5.	EU SU	41 Volynia well 350 Grabov 65 m
237	*Glossophyton foliformis*	Du85...5 p. 7, 9, Pl. 1, Fig. 23.	AS CN	45 Longfengshan
238	*Glossophyton foliformis*	*Duan85..68 p. 73, Pl. 16, Fig. 15.	AS CN	48 Huailai
239	*Glossophyton hailaiensis*	Du85...5 p. 7, 9, Pl. 1, Fig. 22.	AS CN	45 Longfengshan
240	*Glossophyton huailaiensis*	**Duan85..68 p. 72–73, Pl. 17, Fig. 7.	AS CN	48 Huailai
241	*Glossophyton mucronatus*	Du85...5 p. 7, 9, Pl. 1, Fig. 24.	AS CN	45 Longfengshan
242	*Glossophyton?* *mucronatus*	*Duan85..68 p. 73, Pl. 17, Fig. 5.	AS CN	48 Huailai
243	*Grypania spiralis*	*Walt76.872 p. 877–878, Pl. 2, Figs. 4–10.	NA US	8 Mouth of Deep Creek Canyon
244	*Grypania spiralis*	Hofm85.331 p. 348–349, Pl. 39, Fig. 4.	NA CA	1 Mackenzie Mts. 1
245	*Grypania spiralis*	Horo86.640 p. 640.	NA US	8
246	*Helminthoidichnites meeki*	*Walc99.199 p. 236.	NA US	8 Mouth of Deep Creek Canyon
247	*Helminthoidichnites?* *meeki*	Walt76.872 p. 878, Pl. 2, Fig. 12.	NA US	8 Mouth of Deep Creek Canyon
248	*Helminthoidichnites?* *meeki*	Horo86.640 p. 640.	NA US	8 Mouth of Deep Creek Canyon
249	*Helminthoidichnites?* *neihartensis*	Walc99.199 p. 236.	NA US	8 Sawmill Canyon, above Neihart
250	*Helminthoidichnites?* sp.	Soko72..48 p. 50, (Fig. 1).	AS SU	80 Kuk-Yurt well, 404–414 m
251	*Helminthoidichnites?* *spiralis*	Walc99.199 p. 236.	NA US	8 Mouth of Deep Creek Canyon
252	*Huainanella cylindrica*	**Wang82...9 p. 12, Pl. 1, Figs. 5, 9.	AS CN	52 Mt. Bagongshan, Huainan
253	*Huainanella cylindrica*	Wang84.271 p. 278, Pl. 1, Fig. 6.	AS CN	52 Mt. Bagongshan, Huainan
254	*Huainanella striata*	Chen86.221 p. 231. Pl. 2, Fig. 13.	AS CN	53 Mt. Sidingshan, Huainan
255	*Huainania comma*	Wang84.271 p. 278, Pl. 2, Figs. 1, 5, 7b.	AS CN	52 Mt. Bagongshan, Huainan
256	*Huaiyuanella* aff. *marginata*	*Xing85.182 p. 191, Pl. 39, Fig. 7.	AS CN	87 Shandong-Xuhai region
257	*Huaiyuanella baiguashanensis*	Xing85.182 p. 189–190, Pl. 40, Figs. 1–3.	AS CN	53
258	*Huaiyuanella baiuashanensis*	**Xing84.151 p. 151–152, Pl. 1, Figs. 1–4.	AS CN	53 Mt. Baiguashan, Huaiyuan
259	*Huaiyuanella marginata*	*Xing85.182 p. 190–191, Pl. 39, Fig. 6.	AS CN	53
260	*Huaiyuanella minuta*	*Xing84.151 p. 152, Pl. 1, Figs. 5–9.	AS CN	53 Huaiyuan
261	*Huaiyuanella minuta*	Xing85.182 p. 190, Pl. 40, Figs. 4–7.	AS CN	53
262	*Huaiyuanella striata*	*Xing84.151 p. 152, Pl. 1, Figs. 10–11.	AS CN	53 Huaiyuan
263	*Huaiyuanella striata*	Xing85.182 p. 190, Pl. 39, Figs. 8–9.	AS CN	53
264	*Kanilovia insolita*	**Ishc83.181 p. 202–203, Pls. 17, 18.	EU SU	40 Chernovitskaya Ob., Kuleshovka
265	*Kanilovia insolita*	Ishc86..91 p. 156, Pl. 1, Fig. 6.	EU SU	40 Bakota well, Khmelnitzkiy reg.
266	*Kanilovia insolita*	Ishc86..91 p. 156, Pl. 1, Fig. 5.	EU SU	40 Kuleshevka, Chernovitza region
267	*Kildinella jacutica*	Timo69.146 p. 11, Pl. 2, Fig. 3.	AS SU	43 Turukhansk, Miroedikha Riv.
268	*Kildinella magna*	*Timo69.146 p. 14, Pl. 6, Figs. 4, 5.	EU SE	24 L. Vättern area
269	*Kildinella magna*	Timo69.146 p. 14, Pl. 28, Fig. 3.	AS SU	43 Turukhansk, Miroedikha Riv.

Latitute	Longitude	Geologic Unit	Age	Original Author's Interpretation	Hofmann Interpretation
24.5	75.5	Semri Gp., Suket Sh.	YZ	Brachiopoda, Atremata?	*Chuaria*
24.4	74.9	Semri Gp., Suket Sh.	YZ	Brachiopoda, Atremata?	*Chuaria*
24.4	74.9	Semri Gp., Suket Sh.	YZ	Fossil	*Chuaria*
24.5	75.5	Semri Gp., Suket Sh.	YZ	Fossil	*Chuaria*
24.5	75.5	Semri Gp., Suket Sh.	YZ	Inorganic	*Chuaria*
24.5	75.5	Semri Gp., Suket Sh.	YZ	Fossil	*Chuaria*
24.5	75.5	Semri Gp., Suket Sh.	YZ	Fossil	*Tawuia* (Hofmann 1985, p. 334).
15.0	75.0	Kurnool Supgp., Jammalamadugu Gp., Owk Sh.	Z	Affinities not given	*Chuaria*
36.2	48.9	Chapoghlu Sh.	Z	Fossil	*Chuaria*
36.0	51.3	Chapoghlu Sh.	Z	Fossils	cf. *Vendotaenia*
36.5	48.9	Chapoghlu Sh.	Z	Fossils	cf. *Vendotaenia*
50.8	26.0	Kanilov Fm.	Z	Vendotaenid	
50.8	26.0	Kanilov Fm.	Z	Macrophytofossils	
40.4	115.5	Qingbaikou Sys., Changlongshan Fm.	Z	Algae	short *Tawuia* or *Pumilibaxa*
		Qingbaikou Sys., Jingeryu Fm.	Z	Megascopic algae	short *Tawuia* or *Pumilibaxa*
40.4	115.5	Qingbaikou Sys., Changlongshan Fm.	Z	Megascopic algae	short *Tawuia* or *Pumilibaxa*
		Qingbaikou Sys., Jingeryu Fm.	Z	Megascopic algae	short *Tawuia* or *Pumilibaxa*
40.4	115.5	Qingbaikou Sys., Changlongshan Fm.	Z	Megascopic algae	short *Tawuia* or *Pumilibaxa*
		Qingbaikou Sys., Jingeryu Fm.	Z	Megascopic algae	short *Tawuia* or *Pumilibaxa*
46.3	−111.2	Belt Supgp., Greyson Sh.	Y	Algae	
64.8	−129.9	Little Dal Gp.	Z	*Incertae sedis*, Algae?	
		Lower Belt Spgp.	Y	Nonbiologic, fracture fill	
46.3	−111.2	Belt Supgp., Greyson Sh. (and Newland Ls.?)	Y	Annelid trail	Carbonaceous flm
46.3	−111.2	Belt Supgp., Greyson Sh.	Y	Algae	
		Lower Belt Spgp.	Y	Nonbiologic, fracture surface	
46.9	−110.7	Belt Supgp., Greyson Sh.	Y	Mollusc or crustacean trail	Carbonaceous film
52.1	104.1	Olkha Fm. (Olkhin) (U. Riphean)	Z		*Sangshuania, Spiroichnus*
46.3	−111.2	Belt Supgp., Greyson Sh.	Y	Annelid trail	Carbonaceous film
32.6	116.8	Huainan Gp., Liulaobei Fm.	Z	Annelida, Arenicolidae	
32.6	116.8	Huainan Gp., Liulaobei Fm.	Z	Annelid	
32.6	116.8	Jiuliqiao Fm.	Z	Worm	*Sinosabellidites*
32.6	116.8	Huainan Gp., Liulaobei Fm.	Z	Acritarch	cf. short *Tawuia* or *Shouhsienia*
		Shijia Fm.	Z	Pogonophora, Huaiyuanellidae	Sabelliditid
		Jiuliqiao Fm.	Z	Pogonophora, Huaiyuanellidae	Sabelliditid
32.7	117.2	Jiuliqiao Fm.	Z	Vermes, Huaiyuanellidae	
		Jiuliqiao Fm.	Z	Pogonophora, Huaiyuanellidae	Sabelliditid
32.7	117.2	Jiuliqiao Fm.	Z	Vermes, Huaiyuanellidae	
		Jiuliqiao Fm.	Z	Pogonophora, Huaiyuanellidae	Sabelliditid
32.7	117.2	Jiuliqiao Fm.	Z	Vermes, Huaiyuanellidae	
		Jiuliqiao Fm.	Z	Pogonophora, Huaiyuanellidae	Sabelliditid
		Kanilov Gp., Zharnovka Fm.	Z	Chuariamorphida	Vendotaenid
		Kanikov Gp., Studenitza Fm.	Z	Noncalcareous alga	Vendotaenid
		Kanilov Gp., Zharnovka Fm.	Z	Noncalcareous alga	Vendotaenid
66.0	88.0	Mirodeikha Fm.	YZ	Megasphaeromorphid	?=*Chuaria circularis*
58.3	14.7	Visingsö Fm.	YZ	Megasphaeromorphid	*Chuaria circularis*
66.0	88.0	Miroedikha Fm.	YZ	Megasphaeromorphid	?=*Chuaria circularis*

970 Described Taxa of Proterozoic and Selected Earliest Cambrian Megafossils

Table 23.1. (Continued).

Number	Taxon	Reference Code and Page	Continent and Country	Locality
270	*Kildenella magna*	Timo70. 157 p. 158, Pl. 1 A–B.	EU SE	24 L. Vättern area
271	*Krishnania acuminata*	Misr57..54 p. 55, Pl. 7, Fig. 4.	AS IN	57 Neemuch
272	*Krishnania acuminata*	Sahn77. 289 p. 291, 293, Fig. 1a, Pl. 3, Fig. 3.	AS IN	57 Rampura
273	*Lakhandinia prolata*	**Timo79. 137 p. 140–141, Pl. 25, Figs. 5, 6.	AS SU	44 Maya Riv.
274	*Lanceoforma striata*	**Walt76. 872 p. 877, Pl. 2, Figs. 2–3.	NA US	8 Mouth of Deep Creek Canyon?
275	*Lanceoforma striata*	Horo86. 640 p. 640	NA US	8
276	*Linguiformis loeris*	Chen86. 221 p. 231, Pl. 1, Fig. 13.	AS CN	52 Mt. Bagongshan, Huainan
277	*Lingulella, Obolella*	Powe76. 218 p. 79.	NA US	11 Grand Canyon, Kwagunt Valley
278	*Liulaobeia mesacosta*	Wang84. 271 p. 278, Pl. 2, Fig. 2.	AS CN	52 Mt. Bagongshan, Huainan
279	*Ljadlovites reticulatus*	**Ishc83. 181 p. 201, Pl. 12, Fig. 2.	EU SU	40 Vinnitza Obl.
280	*Longfengshania elongata*	Du85... 5 p. 7, 13, Pl. 2, Figs. 11, 12.	AS CN	45 Longfengshan
281	*Longfengshania elongata*	Du85. 183 p. 187, Pl. 1, Figs. 5, 6.	AS CN	45 Longfengshan
282	*Longfengshania elongata*	*Duan85..68 p. 75–76, Pl. 17, Figs. 1–2, 6.	AS CN	48
283	*Longfengshania gemmiforma*	Du85... 5 p. 7, 13, Pl. 2, Fig. 10.	AS CN	45 Longfengshan
284	*Longfengshania longepetiolata*	*Du85. 183 p. 187, Pl. 2, Figs. 1–4.	AS CN	45 Longfengshan
285	*Longfengshania longipetiolata*	Du85... 5 p. 7, 13, Pl. 2, Figs. 14, 17–18.	AS CN	45 Longfengshan, Xinglong
286	*Longfengshania longipetiolata*	Du85... 5 p. 7, 13, Pl. 2, Fig. 16.	AS CN	83 Longfengshan, Xinglong
287	*Longfenghania ovalis*	Du85... 5 p. 7, 13, Pl. 2, Figs. 6–9, 19.	AS CN	45 Longfengshan, Xinglong
288	*Longfengshania ovalis*	Du85. 183 p. 187, Pl. 1, Figs. 2–4, 7, 8.	AS CN	45 Longfengshan
289	*Longfengshania ovalis*	*Duan85..68 p. 75, Pl. 17, Figs. 3, 8–9.	AS CN	48
290	*Longfengshania* sp.	Du80. 341 p. 354, Pl. 25, Figs. 7–8.	AS CN	45 Longfengshan, Huailai
291	*Longfengshania spheria*	Du85... 5 p. 7, 13, Pl. 2, Fig. 15.	AS CN	45 Longfengshan
292	*Longfengshania stipitata*	**Du82... 1 p. 3, Pl., Figs. 11–15.	AS CN	45 Longfengshan, Huailai
293	*Longfengshania stipitata*	Hofm85. 331 p. 343–344, Pl. 38, Fig. 4, text–Fig. 5.	NA CA	2 Mackenzie Mts. 4
294	*Longfengshania stipitata*	Du85... 5 p. 7, 13, Pl. 2, Figs. 1–5	AS CN	45 Longfengshan
295	*Longfengshania stipitata*	Du85. 183 p. 187, Pl. 1, Fig. 1, Pl. 2, Fig. 2.	AS CN	45 Longfengshan, Xinglong
296	*Longfengshanis stipitata*	Duan85..68 p. 74–75, Pl. 16, Figs. 8, 10, 12–16.	AS CN	48
297	*Loriforma closta*	Chen86. 221 p. 231. Pl. 2, Fig. 8.	AS CN	52 Mt. Bagongshan, Huainan
298	*Mezenia kossovoyi*	**Soko76. 126 p. 138, Fig. b.	EU SU	38 Mezen Riv., Leshukonskoe well
299	*Misraea psilata*	*Mait86. 223 p. 225, Pl. 1, Fig. 7.	AS IN	58 Chopan, Dala-Chopan roadcut
300	*Misraea vindhyanensis*	**Mait86. 223 p. 224–225, Pl. 1, Figs. 1–4, 6, text–Fig. 1 A–C.	AS IN	95 Chopan, Dala-Chopan roadcut
301	*Misraea vindhyanensis*	Mait86. 223 p. 224–225, Pl. 1, Fig. 5.	AS IN	96 Chopan
302	*Morania*	Aitk81..47 p. 48, 56, 60, 65.	NA CA	2
303	*Morania antiqua*	*Fent371873 p. 1949–1950, Pl. 2, Fig. 5.	NA US	5 Many Glacier, Roes Creek
304	*Morania antiqua*	Fent43..83 p. 84.	NA US	5 Many Glacier
305	*Morania antiqua*	Math83. 363 p. 364, Fig. 1 C.	AS IN	57 Rampura
306	*Morania* spp.	Yang80. 231 p. 254, Pl. 17, Fig. 10.	AS CN	52 Mt. Bagongshan
307	*Morania* spp.	Zhen80..49 p. 65, Pl. 2, Fig. 14.	AS CN	52 Mt. Bagongshan, Huainan
308	*Morania?* *antiqua*	Hofm79. 150 p. 160, 162, Figs. 13 J, 17 A.	NA CA	1 Mackenzie Mts. 1
309	*Morania?* *antiqua*	Hofm85. 331 p. 344.	NA CA	1 Mackenzie Mts. 2
310	*Morania?* *antiqua*	Chen86. 221 p. 231. Pl. 1, Fig. 12 c.	AS CN	52 Mt. Bagongshan, Huainan

Latitute	Longitude	Geologic Unit	Age	Original Author's Interpretation	Hofmann Interpretation
58.3	14.7	Visingsö Fm.	YZ	Megasphaeromorphida	*Chuaria*
24.4	74.9	Semri Gp., Suket Sh.	YZ	Inorganic	cf. *Longfengshania*
24.5	75.5	Semri Gp., Suket Sh.	YZ		cf. *Longfengshania*
59.0	135.0	Lakhanda Fm.	YZ	Microfossil	Small *Tawuia*?
		Belt Supgp., Greyson Sh.	Y	Algae	
		Lower Belt Spgp.	Y	Fragments of microbial mats	
32.6	116.8	Huainan Gp., Liulaobei Fm.	Z	Chuarid	cf. *Tawuia*
		Chuar Gp., Kwagunt Fm.	YZ	Brachiopod	*Chuaria*
32.6	116.8	Huainan Gp., Liulaobei Fm.	Z	Acritarch	cf. *Tawuia*
		Mogilev-Podolsky Gp., Yaryshev Fm.	Z	Chuariamorphida	
40.4	115.5	Qingbaikou Sys., Changlongshan Fm.	Z	Algae	
40.4	115.5	Qingbaikou Sys., Changlongshan Fm.	Z	Algae	
		Qingbaikou Sys., Jingeryu Fm.	Z	Longfengshanides	
40.4	115.5	Qingbaikou Sys., Changlongshan Fm.	Z	Algae	
40.4	115.5	Qingbaikou Sys., Changlongshan Fm.	Z	Algae	
40.4	117.5	Qingbaikou Sys., Changlongshan Fm.	Z	Algae	
40.4	117.5	Changlongshou Fm.	Z	Algae	
40.4	117.5	Qingbaikou Sys., Changlongshan Fm.	Z	Algae	
40.4	115.5	Qingbaikou Sys., Changlongshan Fm.	Z	Algae	
		Qingbaikou Sys., Jingeryu Fm.	Z	Longfengshanides	
40.4	115.5	Qingbaikou Sys., Changlongshan Fm.	Z	Algae	
40.4	115.5	Qingbaikou Sys., Changlongshan Fm.	Z	Algae	
40.4	115.5	Qingbaikou Sys., Changlongshan Fm.	Z	Chuaridae	
63.8	−127.6	Little Dal Gp., Rusty Shale	Z	*Incertae sedis*, Algae?	
40.4	115.5	Qingbaikou Sys., Changlongshan Fm.	Z	Algae	
40.4	117.5	Qingbaikou Sys., Changlongshan Fm.	Z	Algae	
		Qingbaikou Sys., Jingeryu Fm.	Z	Longfengshanides	
32.6	116.8	Huainan Gp., Liulaobei Fm.	Z	Tawuid	
64.5	45.5	Redkino Ser.	Z	Makroplankton, Chuariamorph	cf. *Tawuia*
24.5	83.0	Semri Gp., Rohtas Ls.	YZ	Miscellaneous fossil	Uninterpretable
24.5	83.0	Semri Gp., Chopan Porcellanite	YZ	Miscellaneous fossil	Uninterpretable
24.5	83.0	Semri Gp., Kheinjua Sh.	YZ	Miscellaneous fossil	Uninterpretable
		Little Dal Gp., Rusty Shale	Z		
48.8	−113.7	Belt Supgp., Altyn Fm.	Y	Microbial colonies	
48.8	−113.7	Belt Supgp., Altyn Fm.	Y	*Nostoc*-like form	Nostocalean colony?
24.5	75.5	Semri Gp., Suket Sh.	YZ	Sphaeromorph acritarch	short *Tawuia*?
32.6	116.8	Huainan Gp., Liulaobei Fm.	Z		
32.6	116.8	Huainan Gp., Liulaobei Fm.	Z	Cyanobacterial aggregates	
64.8	−129.9	Little Dal Gp.	Z	*Nostoc*-like colonies	
64.6	−128.8	Little Dal Gp.	Z	Algae? or procaryotic colonies	Prokaryotic colonies
32.6	116.8	Huainan Gp., Liulaobei Fm.	Z	Carbonaceous fossil	

Table 23.1. (Continued).

Number	Taxon	Reference Code and Page	Continent and Country	Locality
311	*Nephroformia liulaobeiensis*	**Zhen80..49 p. 62–63, Pl. 1, Fig. 31.	AS CN	52 Mt. Bagongshan, Huainan
312	*Nephroformia liulaobeiensis*	Chen86.221 p. 231. Pl. 1, Fig. 8.	AS CN	52 Mt. Bagonshan, Huainan
313	*Nephroformia liulaopeiensis*	Yang80.231 p. 254, Pl. 17, Fig. 31.	AS CN	52 Mt. Bagongshan
314	*Orbisiana simplex*	**Soko76.126 p. 138, Fig. v.	EU SU	38 Moscow syneclise
315	*Ovidiscina bagongshanica*	**Zhen80..49 p. 61, Pl. 1, Fig. 8.	AS CN	52 Mt. Bagongshan, Huainan
316	*Ovidiscina bagongshanica*	Du85...5 p. 7, 9, Pl. 1, Fig. 18	AS CN	45 Longfengshan
317	*Ovidiscina bagongshanica*	Du85...5 p. 7, 9, Pl. 1, Fig. 19.	AS CN	82 Zhaojiashan Mtn.
318	*Ovidiscina bagongshanica*	Chen86.221 p. 231 Pl. 1, Fig. 9	AS CN	52 Mt. Bagongshan, Huainan
319	*Ovidiscina longa*	Du85...5 p. 7, 9, Pl. 1, Fig. 20.	AS CN	45 Longfengshan
320	*Ovidiscina pakungshan(n)i(c)a*	Yang80.231 p. 254, Pl. 17, Fig. 8.	AS CN	52 Mt. Bagongshan
321	*Palaeolina evenkiana*	Xing85.182 p. 188–189, Pl. 39, Figs. 3–4.	AS CN	88 N Anhui
322	*Paleolina tortuosa*	Wang82...9 p. 13. Pl. 2, Figs. 1, 5.	AS CN	53 Mt. Baiguashan, Huaiyuan
323	*Paleolina tortuosa*	Wang84.271 p. 278, Pl. 1, Fig. 10.	AS CN	53 Mt. Baiguashan, Huaiyuan
324	*Paleolina tortuosa*	Chen86.221 p. 231. Pl. 2, Fig. 12.	AS CN	53 Mt. Sidingshan, Huainan
325	*Paleorhyncus anhuiensis*	**Wang82...9 p. 13, 19. Pl. 1, Fig. 3.	AS CN	53 Mt. Baiguashan, Huaiyuan
326	*Paleorhyncus anhuiensis*	Wang84.271 p. 278, Pl. 1, Fig. 9.	AS CN	53 Mt. Baiguashan, Huaiyuan
327	*Paralongfengshania sicyoides*	Du85...5 p. 7, 13, Pl. 2, Fig. 13.	AS CN	45 Longfengshan
328	*Paralongfengshania sicyoides*	**Duan85..68 p. 76, Pl. 17, Fig. 10.	AS CN	48
329	*Paraenicola huaiyuanensis*	Wang84.271 p. 278, Pl. 1, Figs. 1–3.	AS CN	53 Mt. Baiguashan, Huaiyuan
330	*Pararenicola huaiyuanensis*	**Wang82...9 p. 11, Pl. 1, Figs. 1, 2, 4, 6, 7.	AS CN	53 Mt. Baiguashan, Huaiyuan
331	*Pararenicola huaiyuanensis*	Sun86.377 p. 390–394. Figs. 4:4–6, 7:1–12.	AS CN	53 Mt. Baiguashan, Huainan
332	*Phascolites symmetricus*	Du85...5 p. 7, 9, Pl. 1, Fig. 25.	AS CN	45 Longfengshan
333	*Phascolites symmetricus*	**Duan85..68 p. 73–74, Pl. 17, Fig. 4.	AS CN	48
334	*Pilitella composita*	**Asee76..40 p. 57–58, Pl. 20. Figs. 3–5.	EU SU	40 Naslavcha, rt. bank of Dnjestr
335	*Pilitella composita*	Asee83..96 p. 96, Pl. 2, Figs. 5–6.	EU SU	40 Naslavcha, rt. bank of Dnjestr
336	*Pilitella composita*	Asee86.148 Pl. Figs. 7–11.	EU SU	40 Naslavcha, rt. bank of Dnjestr
337	*Primoflagella speciosa*	Gnil79.611 p. 614, Pl. 1, Figs. 8–9.	EU SU	40 Drissa well
338	*Primoflagella speciosa*	Gnil83..46 p. 51–52, Pl. 47, Figs. 5–6 text–Fig. 1.	EU SU	40 Drissa (Verkhnedvinsk), 537 m
339	*Primoflagella speciosa*	Asee83..96 p. 98, Pl. 5.	EU SU	40
340	*Primophlagella speciosa*	**Gnil79..39 p. 44, Pl. 47, Figs. 5–6 text–Fig. 1.	EU SU	40 Drissa well
341	*Protarenicola baiguashanensis*	**Wang82...9 p. 11, Pl. 2, Fig. 3.	AS CN	53 Mt. Baiguashan, Huaiyuan
342	*Protarenicola baiguashanensis*	Sun86.377 p. 394–396. Fig. 4:7.	AS CN	53 Mt. Baiguashan, Huaiyuan
343	*Proterotainia montana*	**Walt76.872 p. 874, Pl. 1, Fig. 1a, b.	NA US	8 Sawmill Canyon, above Neihart
344	*Proterotainia montana*	Horo86.640 p. 640	NA US	8
345	*Proterotainia neihartensis*	*Walt76.872 p. 874, 877, Pl. 1, Figs. 2–4, Pl. 2, Fig. 1.	NA US	8 Sawmill Canyon, above Neihart
346	*Proterotainia neihartensis*	Horo86.640 p. 640	NA US	8
347	*Protoarenicola baiguashanensis*	Wang84.271 p. 278, Pl. 1, Fig. 8.	AS CN	53 Mt. Baiguashan, Huaiyuan
348	*Protobolella jonesi*	**Chap35.109 p. 117–118, Pl. 1, Figs. 5–6 Pl. 2, Fig. 1.	AS IN	57 Rampura
349	*Protobolella minima*	Rowe71..71 p. 72–73, Pl. 1, Figs. 1–3.	AS IN	57 Neemuch
350	*Pumilibaxa* cf. *huaiheiana*	Du85...5 p. 7, 9, Pl. 1, Fig. 21.	AS CN	45 Longfengshan
351	*Pumilibaxa huaiheiana*	**Zhen80..49 p. 61–62, Pl. 1, Figs. 9, 12, 32.	AS CN	52 Mt. Bagongshan, Huainan
352	*Pumilibaxa huaiheiana*	*Duan85..68 p. 72, Pl. 16, Fig. 9.	AS CN	52
353	*Pumilibaxa huaiheiana*	Chen86.221 p. 231, Pl. 1, Fig. 10	AS CN	52 Mt. Bagongshan, Huainan
354	*Pumilibaxa huaihoiana*	Yang80.231 p. 254, Pl. 17, Figs. 9, 12, 32.	AS CN	52 Mt. Bagongshan
355	*Radicula podolicina*	Chen86.221 p. 231, Pl. 1, Fig. 27	AS CN	52 Mt. Bagongshan, Huainan
356	*Radicula?*	Yang80.231 p. 254, Pl. 17, Fig. 33.	AS CN	52 Mt. Bagongshan
357	*Radicula?*	Zhen80..49 p. 65, Pl. 2, Fig. 33.	AS CN	52 Mt. Bagongshan, Huainan
358	*Ruedemannella minuta*	*Wang82...9 p. 12, Pl. 1, Fig. 8, Pl. 2, Fig. 4.	AS CN	53 Mt. Baiguashan, Huaiyuan

Latitute	Longitude	Geologic Unit	Age	Original Author's Interpretation	Hofmann Interpretation
32.6	116.8	Huainan Gp., Liulaobei Fm.	Z	Chuaridae	short *Tawuia*
32.6	116.8	Huainan Gp., Liulaobei Fm.	Z	Chuarid	short *Tawuia*
32.6	116.8	Huainan Gp., Liulaobei Fm.	Z		short *Tawuia*
		Redkino Ser.	Z	Macrophytoplankton	
32.6	116.8	Huainan Gp., Liulaobei Fm.	Z	Chuaridae	deformed *Chuaria*
40.4	115.5	Qingbaikou Sys., Changlongshan Fm.	Z	Algae	deformed *Chuaria*
40.3	115.4	Qingbaikou Sys., Xiamaling Fm.	YZ	Algae	*Chuaria*
32.6	116.8	Huainan Gp., Liulaobei Fm.	Z	Chuarid	deformed *Chuaria*
40.4	115.5	Qingbaikou Sys., Changlongshan Fm.	Z	Algae	deformed *Chuaria*
32.6	116.8	Huainan Gp., Liulaobei Fm.	Z		deformed *Chuaria*
		Jinshanzai Fm.	ZC	Pogonophora, Sabelliditidae	Sabelliditid
32.7	117.2	Jiuliqiao Fm.	Z	Pogonophora, Sabelliditidae	
32.7	117.2	Jiuliqiao Fm.	Z	Pogonophora	
32.6	116.8	Jiuliqiao Fm.	Z	Worm	cf. *Sinosabellidites*
32.7	117.2	Jiuliqiao Fm.	Z	Annelida, Class uncertain	
32.7	117.2	Jiuliqiao Fm.	Z	Annelid	
40.4	115.5	Qingbaikou Sys., Changlongshan Fm.	Z	Algae	
		Qingbaikou Sys., Jingeryu Fm.	Z	Megascopic algae	Tawuid
32.7	117.2	Jiuliqiao Fm.	Z	Annelid	
32.7	117.2	Jiuliqiao Fm.	Z	Annelida, Arenicolidae	
32.6	116.8	Jiuliqiao Fm.	Z	Worm-like animal	
40.4	115.5	Qingbaikou Sys., Changlongshan Fm.	Z	Algae	two overlapping Shouhsienias?
		Qingbaikou Sys., Jingeryu Fm.	Z	Megascopic algae	two overlapping Shouhsienias?
48.4	27.7	Mogilev-Podolsky Gp., Nagoryany Fm.	Z	Vendotaenid	
48.4	27.7	Mogilev-Podolsky Gp., Nagoryany Fm.	Z	Macrophytofossils	
48.4	27.7	Mogilev-Podolsky Gp., Nagoryany Fm.	Z	Noncalcareous alga	Vendotaenid
55.8	28.0	Kotlin Fm.	Z	Metaphyta	
55.8	28.0	Kotlin Fm.	Z	Vendotaenides	
		Kotlin Fm.	Z	Macrophytofossils	
55.8	28.0	Povarovo Gp., Makaryevo Fm.	Z	Vendotaenides	
32.7	117.2	Jiuliqiao Fm.	Z	Annelida, Arenicolidae	
32.7	117.2	Jiuliqiao Fm.	Z	Worm-like animals	
46.9	−110.7	Belt Supgp., Greyson Sh.	Y	Algae	
		Lower Belt Spgp.	Y	Nonbiologic	
46.9	−110.7	Belt Supgp., Greyson Sh.	Y	Algae	
		Lower Belt Spgp.	Y	Nonbiologic	
32.7	117.2	Jiuliqiao Fm.	Z	Annelid	
24.5	75.5	Semri Gp., Suket Sh.	YZ	Brachiopoda, Atremata?	*Chuaria*
24.4	74.9	Semri Gp., Suket Sh.	YZ	Not brachiopods	*Chuaria*
40.4	115.5	Qingbaikou Sys., Changlongshan Fm.	Z	Algae	short *Tawuia*?
32.6	116.8	Huainan Gp., Liulaobei Fm.	Z	Chuaridae	short *Tawuia*?
		Huainan Gp., Liulaobei Fm.	Z	Megascopic algae	short *Tawuia*?
32.6	116.8	Huainan Gp., Liulaobei Fm.	Z	Chuarid	short *Tawuia*?
32.6	116.8	Huainan Gp., Liulaobei Fm.	Z	Megascopic algae	short *Tawuia*?
32.6	116.8	Huainan Gp., Liulaobei Fm.	Z	Carbonaceous fossil	cf. *Beltina*
32.6	116.8	Huainan Gp., Liulaobei Fm.	Z	Fossil	cf. *Beltina*
32.6	116.8	Huainan Gp., Liulaobei Fm.	Z	Fossil	cf. *Beltina*
32.7	117.2	Jiuliqiao Fm.	Z	Annelida, Chloraemidae?	

Table 23.1. (Continued).

Number	Taxon	Reference Code and Page	Continent and Country	Locality
359	*Ruedemannella minuta*	Wang84.271 p. 278, Pl. 1, Figs. 4–5.	AS CN	53 Mt. Baiguashan, Huaiyuan
360	*Sabellidites cambriensis*	Xing85.182 p. 188, Pl. 40, Fig. 9.	AS CN	85
361	*Sabellidites* sp.	Wang82...9 p. 14, Pl. 2, Fig. 2.	AS CN	53 Mt. Baiguashan, Huaiyuan
362	*Sabellidites* sp.	Wang84.271 p. 272, 274, Pl. 2, Figs. 15–17	AS CN	53 Mt. Baiguashan, Huaiyuan
363	*Sabellidites?* sp.	Yang80.231 p. 254, Pl. 17, Fig. 30.	AS CN	53 Mt. Baiguashan
364	*Sabellidites?* sp.	Chen86.221 p. 231, Pl. 2, Fig. 10	AS CN	53 Mt. Sidingshan, Huainan
365	*Sangshuania linearis*	*Du86.115 p. 117, Pl. 1, Figs. 6–9, 12–14.	AS CN	84 Sangshuan, 15 km NNE Jixian
366	*Sangshuania sangshuanensis*	**Du86.115 p. 117, Pl. 1, Figs. 1–5, 10, 11.	AS CN	84 Sangshuan, 15 km NNE Jixian
367	*Sarmenta capitata*	Gnil79.611 p. 616, Pl. 2, Fig. 1.	EU SU	40 Vorobyevo well, 1029 m
368	*Sarmenta capitula*	**Gnil79..39 p. 42–43, Pl. 47, Fig. 7.	EU SU	40 Vorobyevo well, 1029 m
369	*Sarmenta capitula*	Gnil83..46 p. 50–51, Pl. 47, Fig. 7.	EU SU	40 Vorobyevo well, 1029–1033 m
370	*Shouhsienia longa*	Du82...1 p. 3, Pl., Fig. 9.	AS CN	45 Longfengshan, Huailai
371	*Shouhsienia longa*	Wang84.271 p. 278, Pl. 2, Fig. 4, 8a	AS CN	52 Mt. Bagongshan, Huainan
372	*Shouhsienia longa*	Du85...5 p. 7, 9, Pl. 1, Figs. 14, 15.	AS CN	45 Longfengshan, Xinglong
373	*Shouhsienia magna*	*Duan85..68 p. 71–72, Pl. 16, Figs. 2, 13.	AS CN	52
374	*Shouhsienia multirugosa*	Du85...5 p. 7, 9, Pl. 1, Figs. 16, 17.	AS CN	45 Longfengshan
375	*Shouhsienia shouhsienensis*	Du80.341 p. 353, Pl. 25, Figs. 4–6.	AS CN	45 Longfengshan, Huailai
376	*Shouhsienia shouhsienensis*	Du82...1 p. 3, Pl., Figs. 6–8, 10.	AS CN	45 Longfengshan, Huailai
377	*Shouhsienia shouhsienensis*	Xing84..34 p. 153, Pl. 21, Figs. 11–12.	AS CN	92 Wangjiawan
378	*Shouhsienia shouhsienensis*	Wang84.271 p. 278, Pl. 2, Fig. 3	AS CN	52 Mt. Bagongshan, Huainan
379	*Shouhsienia shouhsienensis*	Du85...5 p. 7, 9, Pl. 1, Figs. 12, 13.	AS CN	45 Longfengshan
380	*Shouhsienia shouhsienensis*	Duan85..68 p. 71, Pl. 16, Figs. 10–11.	AS CN	86
381	*Shouhsienia shouhsienensis*	Duan85..68 p. 71, Pl. 16, Fig. 6–7.	AS CN	52
382	*Shouhsienia shouhsienensis*	Du85...5 p. 7, 9, Pl. 1, Figs. 12, 13.	AS CN	82 Longfengshan
383	*Shouhsienia shouhsienensis*	Chen86.221 p. 231, Pl. 1, Figs. 5, 6, 7	AS CN	52 Mt. Bagongshan, Huainan
384	*Shouhsienia?* sp.	Xing84..34 p. 34, 115, Pl. 21, Fig. 13	AS CN	92 Meishucun
385	*Sicyus anacanthus*	Chen86.221 p. 231, Pl. 2, Fig. 7	AS CN	52 Mt. Bagongshan, Huainan
386	*Sinosabellidites*	Sun87.349 p. 350.	AS CN	52 Huainan
387	*Sinosabellidites huainanensis*	Yang80.231 p. 254, Pl. 17, Figs. 14, 16–23.	AS CN	52 Mt. Bagongshan
388	*Sinosabellidites huainanensis*	**Zhen80..49 p. 63, Pl. 2, Figs. 14, 16–23.	AS CN	52 Mt. Bagongshan, Huainan
389	*Sinosabellidites huainanensis*	Chen86.221 p. 231, Pl. 2, Figs. 1a, b, 2a–c	AS CN	52 Mt. Bagongshan, Huainan
390	*Sinosabellidites huainanensis*	Sun86.377 p. 385–390. Fig. 4:1, 2	AS CN	52 Mt. Bagongshan, Huainan
391	*Sinotaenia liulaobeiensis*	Chen86.221 p. 231, Pl. 1, Fig. 30	AS CN	52 Mt. Bagongshan, Huainan
392	*Stenocyphus subtilis*	**Fu86..76 p. 81, Pl. 1, Figs. 3, 4, 6, 7.	AS CN	52
393	*Tasmanites vindhyanensis*	Mait77.176 p. 182, Pl. 4, Figs. 32, 33.	AS IN	57 Rampura (Ramapura)
394	*Tawuia*	Aitk81..47 p. 48, 56, 60, 65.	NA CA	2
395	*Tawuia*	Vida87.345 p. 346.	AS CN	52 Mt. Bagongshan, Huainan
396	*Tawuia*	Sun87.349 p. 351.	AS CN	53 Huainan
397	*Tawuia* cf. *dalensis*	Yang80.231 p. 254, Pl. 17, Fig. 29.	AS CN	52 Mt. Bagongshan
398	*Tawuia* cf. *dalensis*	Zhen80..49 p. 64, Pl. 1, Fig. 29.	AS CN	52 Mt. Bagongshan, Huainan
399	*Tawuia* cf. *dalensis*	Xing84..34 p. 153, Pl. 21, Fig. 8.	AS CN	92 Wangjiawan
400	*Tawuia dalensis*	**Hofm79.150 p. 157–160, Figs. 13, 15, 16.	NA CA	1 Mackenzie Mts. 1
401	*Tawuia dalensis*	Knol81..55 p. 55	EU SJ	26 Murchisonfjorden
402	*Tawuia dalensis*	Knol82.269 p. 275, Pl. 3, Figs. 13, 14, text-Fig. 3.	EU SJ	26 Murchisonfjorden
403	*Tawuia dalensis*	Duan82..57 p. 63, Fig. 5K–N.	AS CN	52 Mt. Bagongshan, Huainan
404	*Tawuia dalensis*	Mait84.146 p. 213, Pl. 1, Fig. 4.	AS IN	57 Rampura (Ramapura)
405	*Tawuia dalensis*	Wang84.271 p. 278, Pl. 2, Figs. 11–13	AS CN	53 Mt. Baiguashan, Huaiyuan
406	*Tawuia dalensis*	Hofm85..20 p. 26, Figs. 4 A, B.	NA CA	1 Mackenzie Mts. 1
407	*Tawuia dalensis*	Hofm85.331 p. 334–342, Pl. 36, Fig. 1, text-Fig. 3.	NA CA	1 Mackenzie Mts. 3

Latitute	Longitude	Geologic Unit	Age	Original Author's Interpretation	Hofmann Interpretation
32.7	117.2	Jiuliqiao Fm.	Z	Annelida	
		Changlingzi Fm.	Z	Pogonophora, Sabelliditidae	Sabelliditid
32.7	117.2	Jiuliqiao Fm.	Z	Pogonophora, Sabelliditidae	
32.7	117.2	Jiuliqiao Fm.	Z	Pogonophora	
32.7	117.2	Jiuliqiao Fm.	Z	Worm	
32.6	116.8	Jiuliqiao Fm.	Z	Worm	
40.2	117.4	Changcheng Sys., Gaoyuzhuang Fm.	Y	Algae	*Grypania*
40.2	117.4	Changcheng Sys., Gaoyuzhuang Fm., 5th mem.	Y	Primitive algae	*Grypania*
54.5	37.2	Kotlin Fm.	Z	Metaphyta	
54.5	37.2	Povarovo Gp., Makaryevo Fm.	Z	Vendotaenides	
54.5	37.2	Povarovo Gp., Makaryevo Fm.	Z	Vendotaenides	
40.4	115.5	Qingbaikou Sys., Changlongshan Fm.	Z	Algae	short *Tawuia*
32.6	116.8	Huainan Gp., Liulaobei Fm.	Z	Acritarch	short *Tawuia*
40.4	117.5	Qingbaikou Sys., Changlongshan Fm.	Z	Algae	short *Tawuia*
		Huainan Gp., Liulaobei Fm.	Z	Megascopic algae	Tawuid or chuarid
40.4	115.5	Qingbaikou Sys., Changlongshan Fm.	Z	Algae	short *Tawuia*
40.4	115.5	Qingbaikou Sys., Changlongshan Fm.	Z	Algae	short *Tawuia*?
40.4	115.5	Qingbaikou Sys., Changlongshan Fm.	Z	Algae	short *Tawuia*?
24.6	102.7	Yuhucun Fm., Jiucheng Mem.	ZC	Mega-algae	short *Tawuia*?
32.6	116.8	Huainan Gp., Liulaobei Fm.	Z	Acritarch	short *Tawuia*?
40.4	115.5	Qingbaikou Sys., Changlongshan Fm.	Z	Algae	short *Tawuia*?
		Diaoyutai Fm.	Z	Megascopic algae	short *Tawuia*?
		Huainan Gp., Liulaobei Fm.	Z	Megascopic algae	short *Tawuia*?
40.4	115.5	Qingbaikou Sys., Xiamaling Fm.	YZ	Algae	short *Tawuia*?
32.6	116.8	Huainan Gp., Liulaobei Fm.	Z	Chuarid	short *Tawuia*?
24.7	102.5	Yuhucun Fm., Jiucheng Mem.	ZC	Mega-algae	
32.6	116.8	Huainan Gp., Liulaobei Fm.	Z	Tawuid	cf. short *Tawuia*
		Huainan Gp., Liulaobei Fm.	Z	Algal or metazoan, worm-like	
32.6	116.8	Huainan Gp., Liulaobei Fm.	Z	Worm-like fossil	
32.6	116.8	Huainan Gp., Liulaobei Fm.	Z	Sabelliditidae	
32.6	116.8	Huainan Gp., Liulaobei Fm.	Z	Worm	
32.6	116.8	Huainan Gp., Liulaobei Fm.	Z	Worm-like fossil	
32.6	116.8	Huainan Gp., Liulaobei Fm.	Z	Vendotaenid	*Grypania*
		Huainan Gp., Liulaobei Fm.	Z	Cyphomegacritarch	cf. *Tawuia*
24.5	75.5	Semri Gp., Suket Sh.	YZ	Fossil	
		Little Dal Gp., Rusty Shale	Z	Carbonaceous fossil	
32.6	116.8	Huainan Gp., Liulaobei Fm.	Z	Fossil	
		Liulaobei Fm., Jiuliqiao Fm.	Z	*Nostoc*-like	
32.6	116.8	Huainan Gp., Liulaobei Fm.	Z	Algae	
32.6	116.8	Huainan Gp., Liulaobei Fm.	Z	Vendotaenides?	
24.6	102.7	Yuhucun Fm., Jiucheng Mem.	ZC	Mega-algae	
64.8	−129.9	Little Dal Gp.	Z	Algae or metazoans	
80.0	19.0	Franklinsundet Gp., Kapp Lord Fm.	Z	Macrofossil	
80.0	19.0	Franklinsundet Gp., Kapp Lord Fm.	Z	Macrofossil	
32.6	116.8	Huainan Gp., Liulaobei Fm.	Z	Chuariaceae	
24.5	75.5	Semri Gp., Suket Sh.	YZ	Not given	*Tawuia*
32.7	117.2	Jiuliqiao Fm.	Z	Phaeophyta	
64.8	−129.9	Little Dal Gp., lower part	Z	*Incertae sedis*, Algae?	
64.1	−128.5	Little Dal Gp.	Z	*Incertae sedis*, Algae?	

976 Described Taxa of Proterozoic and Selected Earliest Cambrian Megafossils

Table 23.1. (Continued).

Number	Taxon	Reference Code and Page	Continent and Country	Locality
408	*Tawuia dalensis*	Hofm85.331 p. 334–342, Pl. 35:1–3, Pl. 36:3, 7–11, Pl. 37:1,2,4–7, Pl. 38:1, text-Figs. 3, 4.	NA CA	1 Mackenzie Mts. 1
409	*Tawuia dalensis*	Du85...5 p. 7, 9, Pl. 1, Fig. 26.	AS CN	45 Longfengshan
410	*Tawuia dalensis*	Hofm85.331 p. 334–342, text-Fig. 3	NA CA	1 Mackenzie Mts. 5
411	*Tawuia dalensis*	Hofm85.331 p. 334–342, Pl. 36, Fig. 2, Pl. 38, Figs. 1, 2, text-Fig. 3.	NA CA	2 Mackenzie Mts. 4
412	*Tawuia dalensis*	Duan85..68 p. 71, Pl. 16, Fig. 14.	AS CN	52
413	*Tawuia dalensis*	Sun87.109 p. 124–126, 128	NA CA	1 Mackenzie Mts.
414	*Tawuia dalensis*	Sun87.109 p. 123–126, Pl. 4, Figs. 1–9.	AS CN	53 Huainan
415	*Tawuia fusiformis*	*Xing84..34 p. 34, 115, Pl. 21, Fig. 7.	AS CN	92 Wangjiawan
416	*Tawuia hippocrepica*	Chen86.221 p. 231, Pl. 1, Fig. 24–26	AS CN	52 Mt. Bagongshan, Huainan
417	*Tawuia rampuraensis*	*Math83.363 p. 364, Fig. 1E.	AS IN	57 Rampura
418	*Tawuia sinensis*	*Duan82..57 p. 63–64, Figs. 3K–Q, 5P in part, 6.	AS CN	52 Mt. Bagongshan, Huainan
419	*Tawuia sinensis*	Wang84.271 p. 278, Pl. 2, Fig. 10	AS CN	52 Mt. Bagongshan, Huainan
420	*Tawuia sinensis*	Duan85..68 p. 71, Pl. 16, Fig. 5.	AS CN	52
421	*Tawuia sinensis*	Sun86.377 p. 387, Fig. 4:3	AS CN	52 Mt. Bagongshan, Huainan
422	*Tawuia sinensis*	Chen86.221 p. 231, Pl. 1, Figs. 15–21, Pl. 2, Fig. 3	AS CN	52 Mt. Bagongshan, Huainan
423	*Tawuia* spp.	Yang80.231 p. 402, Pl. 17, Figs. 24–27.	AS CN	52 Mt. Bagongshan
424	*Tawuia* spp.	Zhen80..49 p. 64–65, Pl. 2, Figs. 24–27.	AS CN	52 Mt. Bagongshan, Huainan
425	*Tawuia striatia*	*Zhen80..49 p. 63–64, Pl. 2, Fig. 28.	AS CN	52 Mt. Bagongshan, Huainan
426	*Tawuia striatia*	Chen86.221 p. 231, Pl. 1, Fig. 22, 23	AS CN	53 Mt. Sidingshan, Huainan
427	*Tawuia striatis*	*Yang80.231 p. 254, Pl. 17, Fig. 28.	AS CN	52 Mt. Bagongshan
428	*Tawuia suketensis*	*Math82.125 p. 128–129, Fig. 3B.	AS IN	57 Rampura
429	*Tawuia suketensis*	*Math83.363 p. 364, Fig. 1D.	AS IN	57 Rampura
430	*Trachysphaeridium lachandinum*	*Timo69.146 p. 20, Pl. 6, Fig. 1.	AS SU	44 Maya Riv.
431	*Trachysphaeridium lachandinum*	Timo69.146 p. 20–21, Pl. 28, Fig. 2.	AS SU	43 Turukhansk, Miroedikha Riv.
432	*Trachysphaeridium* sp.	Timo70.157 p. 158, Pl. 1E	AS SU	44 Maya Riv.
433	*Trachysphaeridium vetterni*	Timo69.146 p. 21, Pl. 6, Fig. 3.	EU SE	24 L. Vättern area
434	*Tyrasotaenia* cf. *podolica*	Xing78.109 p. 125, Pl. 9, Figs. 9–10, Pl. 10, Figs. 1–2.	AS CN	55 Yangtze Gorge
435	*Tyrasotaenia* cf. *podolica*	Xing84..34 p. 153, Pl. 21, Fig. 10.	AS CN	92 Wangjiawan
436	*Tyrasotaenia* cf. *podolica*	Duan85..68 p. 77, Pl. 17, Figs. 11–12.	AS CN	55
437	*Tyrasotaenia* cf. *podolica*	Xing85.108 p. 127.	AS CN	55 Yangtze Gorge, Nantuo-Shipai
438	*Tyrasotaenia filiforma*	Wang84.271 p. 278, Pl. 2, Figs. 15–17	AS CN	52 Mt. Bagongshan, Huainan
439	*Tyrasotaenia filiforma*	Chen86.221 p. 231, Pl. 1, Fig. 28	AS CN	52 Mt. Bagongshan, Huainan
440	*Tyrasotaenia podolica*	**Gnil71.101 p. 106–107, Pl. 11, Figs. 1–5.	EU SU	41 Kitaygorod, Ternava Riv.
441	*Tyrasotaenia podolica*	Gnil79..39 p. 41, Pl. 44, Figs. 1,2,4–7.	EU SU	40 Bolotino
442	*Tyrasotaenia podolica*	Gnil79.611 p. 614, Pl. 1, Fig. 5.	EU PL	40 Kaplonosy well
443	*Tyrasotaenia podolica*	Gnil83..46 p. 48, Pl. 44, Figs. 1–7.	EU SU	40 Bolotino-1 well, 135 m
444	*Tyrasotaenia* sp.	Soko76.126 p. 139, Fig. 6.	EU SU	40
445	*Tyrasotaenia* sp.	Wang84.271 p. 278, Pl. 2, Fig. 14	AS CN	52 Mt. Bagongshan, Huainan
446	*Tyrasotaenia* sp.	Hofm85.331 p. 349	NA CA	1 Mackenzie Mts. 2
447	*Tyrasotaenia* sp.	Hofm85.331 p. 349, Pl. 35, Fig. 10, Pl. 39, Figs. 9–11.	NA CA	1 Mackenzie Mts. 1
448	*Tyrasotaenia* sp.	Chen86.221 p. 231, Pl. 1, Fig. 29	AS CN	52 Mt. Bagongshan, Huainan
449	*Tyrasotaenia* sp.	Narb871277 p. 1284	NA CA	67 Burin Peninsula, Fortune
450	*Tyrasotaenia* sp. cf. *podolica*	Hofm81.443 p. 446, Fig. 3A–B.	AS CN	47 Liuzhuangzi, NNE of Jixian
451	*Tyrasotaenia tungusica*	*Gnil79..39 p. 41, Pl. 44, Fig. 6 in part.	AS SU	79 Sukhaya Tunguska Riv.
452	*Tyrasotaenia tungusica*	Gnil79..39 p. 41, Pl. 44, Fig. 6 in part.	EU PL	40 Kaplonosy well
453	*Tyrasotaenia tungusica*	Gnil83..46 p. 48–49, Pl. 44, Fig. 6 in part.	AS CN	47 Liuzhuangzi, NNE of Jixian
454	*Tyrasotaenia tungusica*	Gnil83..46 p. 48–49.	AS SU	79 Sukhaya Tunguska Riv.
455	*Tyrasotaenia tungusica*	Gnil83..46 p. 48–49, Pl. 44, Figs. 4 in part.	EU SU	40 Bolotino-1 well, 465 m
456	*Tyrasotaenia?* sp.	Xing78.109 p. 125, Pl. 10, Fig. 3	AS CN	55 Yangtze Gorge

Latitute	Longitude	Geologic Unit	Age	Original Author's Interpretation	Hofmann Interpretation
64.8	−129.9	Little Dal Gp., lower part	Z	*Incertae sedis*, Algae?	
40.4	115.5	Qingbaikou Sys., Changlongshan Fm.	Z	Algae	
62.6	−126.6	Little Dal Gp.	Z	*Incertae sedis*, Algae?	
63.8	−127.6	Little Dal Gp., Rusty Shale	Z	*Incertae sedis*, Algae?	
		Huainan Gp., Liulaobei Fm.	Z	Megascopic algae	Tawuid
		Little Dal Gp.	Z	*Nostoc*-like colonies	
		Liulaobei Fm., Jiuliqiao Fm.	Z	*Nostoc*-like colony	
24.6	102.7	Yuhucun Fm., Jiucheng Mem.	ZC	Mega-algae	
32.6	116.8	Huainan Gp., Liulaobei Fm.	Z	Tawuid	curved *Tawuia dalensis*
24.5	75.5	Semri Gp., Suket Sh.	YZ	Algae	
32.6	116.8	Huainan Gp., Liulaobei Fm.	Z	Chuariaceae	
32.6	116.8	Huainan Gp., Liulaobei Fm.	Z	Phaeophyta	
		Huainan Gp., Liulaobei Fm.	Z	Megascopic algae	Tawuid
32.6	116.8	Huainan Gp., Liulaobei Fm.	Z	Cyanobacterial colonies	
32.6	116.8	Huainan Gp., Liulaobei Fm.	Z	Tawuid	*Tawuia dalensis*
32.6	116.8	Huainan Gp., Liulaobei Fm.	Z	Macrofossil	
32.6	116.8	Huainan Gp., Liulaobei Fm.	Z	Vendotaenides	
32.6	116.8	Huainan Gp., Liulaobei Fm.	Z	Vendotaenides	
32.6	116.8	Jiuliqiao Fm.	Z	Tawuid	curved *Tawuia dalensis*
32.6	116.8	Huainan Gp., Liulaobei Fm.	Z	Algae	
24.5	75.5	Semri Gp., Suket Sh.	YZ	Macrofossil	
24.5	75.5	Semri Gp., Suket Sh.	YZ	Algae	
59.0	135.0	Lakhanda Fm.	YZ	Megasphaeromorphid	? = *Chuaria circularis*
66.0	88.0	Miroedikha Fm.	YZ	Megasphaeromorphid	? = *Chuaria circularis*
59.0	135.0	Lakhanda Fm.	YZ	Megasphaeromorphida	
58.3	14.7	Visingsö Fm.	YZ	Megasphaeromorphid	*Chuaria circularis*
30.6	111.1	Dengying Fm.	Z	Vendotaenid	
24.6	102.7	Yuhucun Fm., Jiucheng Mem.	ZC	Mega-algae	
		Dengying Fm.	Z	Vendotaenid	Vendotaenid
30.6	111.1	Dengying Fm., Shibantan Mem.	Z	Algae	
32.6	116.8	Huainan Gp., Liulaobei Fm.	Z	Vendotaenid	
32.6	116.8	Huainan Gp., Liulaobei Fm.	Z	Vendotaenid	
48.6	26.8	Kanilov Fm.	Z	Vendotaenides, brown algae?	
47.7	27.3	Povarovo Gp., Makaryevo Fm.	Z	Vendotaenides	
51.6	23.3	Kotlin Fm.	Z	Metaphyta	
47.7	27.3	Ferapontyevo Fm.	Z	Vendotaenides	
		Vendian	Z	Metaphyta	
32.6	116.8	Huainan Gp., Liulaobei Fm.	Z	Vendotaenid	
64.6	−128.8	Little Dal Gp.	Z	Algae?	
64.8	−129.9	Little Dal Gp.	Z	Algae?	
32.6	116.8	Huainan Gp., Liulaobei Fm.	Z	Vendotaenid	
47.5	−55.8	Chapel Island Fm., Mem. 1	Z		
40.2	117.5	Changcheng Sys., Tuanshanzi Fm.	X	Megascopic compressions	
65.2	88.5	Platonov Fm., Nemakit-Daldyn hor.	C	Vendotaenides	
51.6	23.3	Lublin Fm.	Z	Vendotaenides	
40.2	117.5	Changcheng Sys., Tuanshanzi Fm.	X	Vendotaenides	
65.2	88.5	Platonov Fm., Nemakit-Daldyn hor.	C	Vendotaenid	
47.7	27.3	Ferapontyevo Fm.	Z	Vendotaenides	
30.6	111.1	Dengying Fm.	Z	Vendotaenid	

Table 23.1. (Continued).

Number	Taxon	Reference Code and Page	Continent and Country	Locality
457	*Tyrasotaenia*? sp.	Hofm85.331 p. 349–350, Pl. 35, Figs. 4, 8, Pl. 36, Fig. 12, Pl. 37, Fig. 8.	NA CA	1 Mackenzie Mts. 1
458	*Tyrasotaenia*? sp.	Du85...5 p. 7, 9, Pl. 1, Fig. 27.	AS CN	46 Jixian
459	*Vendotaenia antiqua*	*Gnil71.101 p. 105–106, Pl. 11, Figs. 6–8.	EU SU	39 Leningrad subway
460	*Vendotaenia antiqua*	Gnil75.258 p. 258, Fig. 1/3–4.	EU SU	40 Rovno well
461	*Vendotaenia antiqua*	Soko76.126 p. 139, Fig. a.	EU SU	40
462	*Vendotaenia antiqua*	Asee76..40 p. 56–57, Pl. 19, Figs. 1–3.	EU SU	41 Volynia well 350 Grabov 22-115
463	*Vendotaenia antiqua*	Gnil76..10 p. 11, 295, Pl. 1, Figs. 1–3.	EU SU	40 Rovno well
464	*Vendotaenia antiqua*	Gnil79..39 p. 40, Pls. 39–43.	EU SU	40 Many localities
465	*Vendotaenia antiqua*	Gnil79.611 p. 614, Pl. 1, Figs. 1–4.	EU SU	40 Rovno well
466	*Vendotaenia antiqua*	Gnil83..46 p. 48, Pls. 39–43.	EU SU	40 Many localities in E Europe
467	*Vendotaenia antiqua*	Asee83..96 p. 96, Pl. 1, Figs. 1–5, Pl. 2, Figs. 1–4, Pl. 3, Fig. 1.	EU SU	40 Kishinev well 924–925 m
468	*Vendotaenia antiqua*	Gnil84..58 p. 61–64, Pl. 7, Figs. 1–8	EU SU	40 Various localities
469	*Vendotaenia antiqua*	Ishc86..91 p. 154, Pl. 1, Fig. 4.	EU SU	40 Kitaygorod, Khmelnitzkiy reg.
470	*Vendotaenia antiqua*	Asee86.148 Pl. Figs. 12–17	EU SU	41 Kishinev-1 well, 924–925 m
471	*Vendotaenia* sp.	Gnil75.258 p. 258, Fig. 1/1–2.	EU SU	40 Toropets well, 842.3–846.3 m
472	*Vendotaenia* sp.	Asee76..40 p. 57, Pl. 19, Figs. 4–5.	EU SU	40 Volynia
473	*Vendotaenia* sp.	Xing78.109 p. 125, Pl. 10, Fig. 4	AS CN	55 Yangtze Gorge
474	*Vendotaenia* sp.	Du80.341 p. 353, Pl. 25, Figs. 1–2.	AS CN	45 Longfengshan, Huailai
475	*Vendotaenia* sp.	Asee83..96 p. 96, Pl. 3, Figs. 4, 6.	EU SU	40 Kishinev well?
476	*Vendotaenia* sp.	Xing84..34 p. 153, Pl. 21, Fig. 9.	AS CN	92 Wangjiawan
477	*Vendotaenia* sp.	Xing85.108 p. 115, 127, Pl. 6, Figs. 10–12.	AS CN	55 Yangtze Gorge
478	*Vendotaenia* sp.	Duan85..68 p. 76, Pl. 17, Fig. 13.	AS CN	55
479	*Vendotaenia* sp.	Germ86..45 p. 55–57, Fig. 5a	AF NA	100 Nutupsdrift well
480	*Vendotaenia* sp.	Germ86..45 p. 53, 55–57, Fig. 5c	AF NA	101 Tses well
481	*Vendotaenia* sp.	Sun86.361 p. 372, Fig. 5	AS CN	55 Yangtze Gorge, Shibantan
482	*Vendotaenia* sp. nov.	Soko75.112 p. 113, 240, Pl. 2, Figs. 1–3	AS SU	81 Khidusa Riv., Lake Baikal area
483	*Vendotaenia*? sp.	Du82...1 p. 4, Pl. Fig. 16.	AS CN	45 Longfengshan, Huailai
484	*Vendotaenia*? sp.	Narb87.647 p. 671, text–Fig. 10g	NA CA	65 Wernecke Mts.
485	*Vindhyanella jonesi*	*Sahn36.458 p. 467.	AS IN	57
486	*Vindhyania jonesi*	**Math82.125 p. 129, Fig. 3C.	AS IN	57 Rampura
487	*Vindhyania jonesi*	**Math83.363 p. 364, Fig. 1F.	AS IN	57 Rampura
488	cf. *Vendotaenia* sp.	Dama86.417 p. 423, Fig. 6E–G, I.	AU SA	102 Stuart Shelf: AMOCO well SCYWIA

Latitute	Longitude	Geologic Unit	Age	Original Author's Interpretation	Hofmann Interpretation
64.8	−129.9	Little Dal Gp.	Z	Filament or trace fossil	
40.2	117.5	Changcheng Sys., Chuanlinggou Fm.	X	Algae	
59.9	30.4	Kotlin Fm. Kanilov Fm.	Z	Vendotaenides, Brown algae?	
50.6	26.2	Kotlin Gp., Seliger Fm.	Z	Vendotaenides, algae	
		Valdai Ser.	Z	Metaphyta	
50.8	26.0	Kanilov Fm. Kotlin Fm.	Z	Vendotaenid	
50.6	26.2	Povarovo Gp., Lyubim Fm.	Z	Metaphyta	
		Many Vendian fms. on Russian Platform	Z	Vendotaenides	
50.6	26.2	Kotlin Fm.	Z	Metaphyta	
		Many formations (Nagoryani, Kaushany, Sokoletzkoye, etc.)	Z	Vendotaenides	
47.0	28.8	Sokoletzkaya Fm.	ZC	Macrophytofossil	
		Kotlin Fm.	Z	Vendotaenid algae	
		Kanilov Gp., Studenitza Fm.	Z	Noncalcareous alga	Vendotaenid
47.0	28.8	Kanilov Fm.	Z	Noncalcareous alga	Vendotaenid
55.6	31.5	Kotlin Gp., Seliger Fm.	Z	Vendotaenides, algae	
		Gorynskoy Fm.	Z	Vendotaenid	
30.6	111.1	Dengying Fm.	Z	Vendotaenid	
40.4	115.5	Qingbaikou Sys., Changlongshan Fm.	Z		
		Sokoletzkaya Fm.	ZC	Macrophytofossil	
24.6	102.7	Yuhucun Fm., Jiucheng Mem.	ZC	Mega-algae	
30.6	111.1	Dengying Fm., Shibantan Mem.	Z	Vendotaenid algae	
		Dengying Fm.	Z	Vendotaenid	Vendotaenid
−24.7	16.9	Schwarzrand Subgp., Nudaus Fm., Bingerbreek Mem.	ZC	Problematic systematic position	
−25.8	18.1	Kuibis and Schwarzrand Fms.	Z	Problematic systematic position	
30.8	111.2	Dengying Fm., Shibantan Mem.	Z	Fossil	
52.0	105.0	Moty Fm., U. part	Z	Vendotaenid	Vendotaenid
40.4	115.5	Qingbaikou Sys., Changlongshan Fm.	Z	Vendotaenides	Vendotaenid
64.5	−132.9	Windermere Supgp., Siltstone 2	Z	Vendotaenid	
		Semri Gp., Suket Sh.	YZ	Brachiopod	
24.5	75.5	Semri Gp., Suket Sh.	YZ	Macrofossil	cf. *Tawuia*
24.5	75.5	Semri Gp., Suket Sh.	YZ	Algae	cf. *Tawuia*
−30.0	137.0	Tent Hill Fm., Arcoona Qte. Mem.	Z	Vendotaenid	

23.2

Proterozoic and Early Cambrian Calcareous Algae

CAROL MANKIEWICZ

In Tables 23.2.1–23.2.5 are listed, alphabetically by taxon, described Proterozoic and Early Cambrian calcareous algae. Table 23.2.1 is an annotated listing of genera described as Proterozoic and/or Early Cambrian calcareous algae; Table 23.2.2 is a species-level listing of described taxa that are here regarded to be bona fide fossils; Table 23.2.3 lists reported species that are not evaluated here; Table 23.2.4 lists described species that are interpreted here to be dubiofossils; and Table 23.2.5 lists described species that are interpreted here to be pseudofossils.

Table 3.2.1 lists all(?) genera that have been described as Proterozoic to Early Cambrian calcareous algae. It does not include stromatolites, radiocyathans, or catagraphs. A literature reference is included in parentheses following the taxon name only if the author(s) of the taxon differ from the author(s) of the publication. An asterisk (*) denotes the type species of the genus, listed in parentheses. The known range of the genus is listed in brackets. Morphological terms are after Riding and Voronova (1985).

Each genus is subjectively rated 0, 1, 2, or 3. Those genera lacking Proterozoic or Early Cambrian representatives received a rating of '0'. A rating of '1' was given to those genera recognized by "splitters"; '3' to those recognized by "lumpers"; and '2' to those genera which I believe are distinct, potentially recognizable taxa. Reasons for the ratings are coded as follows: 1 = pseudofossil; 2 = no definitive Early Cambrian or older representative; 3 = probably no Early Cambrian or older representative; 4 = junior synonym of other recognized taxon or transferred to other taxon; 5 = original calcareous nature questionable; 6 = poor illustration/copy that could not be interpreted; 7 = limited occurrence; and 8 = recognition by only one person.

Additional information is provided in the remarks.

Table 23.2.1. *Proterozoic and Early Cambrian genera of calcareous algae.*

Acanthina Korde 1973 (*A. multiformis*).
 [Early Cambrian] Morphology: botryoidal. Rating: 1. Reason(s): 5, 7, 8?.
 Remarks: Radial fabric and "variable shape" suggest possible alteration of oölith or of void-filling cement.

Actinophycus Tsao & Zhao 1974 (Cao and Zhao 1974) (*Varicamanicosiphonia quadricella*).
 [Sinian] Morphology: see remarks. Rating: 0. Reason(s): 1.
 Remarks: Pseudofossils/nonbiogenic stromatolites? (Hofmann and Jackson 1987).

Actinophycus Korde 1954 (*A. orbrutschevi*).
 [Sinian-Cambrian] Morphology: see remarks. Rating: 0. Reason(s): 1, see remarks.
 Remarks: Originally described as stromatolith, but several Chinese authors classify as a rhodophyte.

Acus Tsao & Zhao 1974 (Cao and Zhao 1974) (*A. platypluteus*).
 [Sinian] Morphology: see remarks. Rating: 0. Reason(s): 1.
 Remarks: Looks like radially fibrous cement. Hofmann and Jackson (1987) interpret similar taxa of the Manicosiphoniaceae of Cao and Zhao (1978) as nonbiological.

Amgaina Korde 1973 (*A. compacta*).
 [Middle Cambrian] Morphology: tubiform. Rating: 0. Reason(s): 2, 5, 6, 7, 8.
 Remarks: Photograph of specimen resembles a digitate stromatolite.

Amganella Reitlinger 1959 (*Proaulopora glabra*).
 [Yudomian?-Middle Cambrian] Morphology: tubiform. Rating: 0. Reason(s) 4.
 Remarks: Reitlinger erected *Amganella* to differentiate the obviously segmented *Proaulopora rarissima* Vologdin from the simple smooth tube illustrated by Krasnopeeva for *P. glabra*. Range only valid if *Amganella* = *Proaulopora*.

Angulocellularia Vologdin 1962 (*A. anisotoma*).
 [Early Cambrian] Morphology: dendritic. Rating: 2. Reason(s): see remarks.
 Remarks: Interpretive drawings of Vologdin do not look at all like photographs (see Riding and Voronova 1982 for discussion). See remarks for *Angusticellularia*.

Angusticellularia Vologdin 1962. (*A. anisotoma*).
 [Early Cambrian] Morphology: dendritic. Rating: 0. Reason(s): 4, see remarks.

Table 23.2.1 (Continued).

Remarks: = *Angulocellularia* Vologdin 1962 (Vologdin 1962b); same illustrations used for both. Vologdin (1962a) sent to printers in February, 1962; 1962b sent in July, but publication date unknown, and therefore can not determine priority of name.

Anomas Vologdin 1932 (**A*. ovisimilis*).
[Early Cambrian] Morphology: spherical. Rating: 0. Reason(s): 1?, 6, 7, 8.
Remarks: Coated grains?

Antiquus Butin 1959 (**A*. cusarandicus*).
[Middle Proterozoic] Morphology: see remarks. Rating: 1. Reason(s): 4?, 5, 6, 7.
Remarks: Poor photocopy. Pl. 2, Fig. 7 looks a bit like *Razumovskya*.

Azyrtalia Vologdin 1969 (**A*. zonulata*).
[Late Proterozoic-Early Cambrian] Morphology: spherical. Rating: 1. Reason(s): 6, see remarks.
Remarks: In Drozdova (1980), illustration looks like it could be altered? fibrous isopachous cement around nucleus of undetermined affinity; Vologdin's original illustrations are not easily evaluated. Glaessner (1979, p. A110) listed as doubtful taxa.

Bajanophyton Drosdova 1980 (**B*. mucosum*).
[Early Cambrian] Morphology: dentric. Rating: 2. Reason(s): 8, see remarks.
Remarks: Diagenetic taxon?

Batinevia Korde 1966 (**B*. ramosa*).
[Early Cambrian-Ordovician?] Morphology: tubiform. Rating: 2.

Bestjachica Kolosov 1970 (**B*. rara*).
[Late Proterozoic] Morphology: botryoidal. Rating: 1. Reason(s): 1?, 5, 6, 7.
Remarks: Possibly a form of *Renalcis*?

Bija Vologdin 1932 (**B*. sibirica*).
[Early Cambrian] Morphology: tubiform. Rating: 1. Reason(s): see remarks.
Remarks: *Originally described as alcyonarian. *Bija* not always easy to distinguish from *Solenopora*, *Rothpletzella*, and *Hedstroemia*. *Bija*: irregular arrangement of cells; *Solenopora*: regular arrangement; *Hedstroemia*: obvious branching.

Bogutschanophycus Korde 1954 (**B*. mariae*).
[Late Cambrian?-Ordovician] Morphology: tubiform. Rating: 0. Reason(s): 2, 4.
Remarks: = *Nuia* Maslov 1954. See comments under *Nuia* regarding classification.

Botomaella Korde 1958 (**B*. zelenovi*).
[Early Cambrian] Morphology: tubiform. Rating: 3.
Remarks: Riding and Voronova (1985) put it in continuous series: *Botomaella* (discrete branching tubes in fan form) > *Hedstroemia* > *Solenopora* (fused, predominantly nonbranching tubes).

Botominella Reitlinger 1959 (**B*. lineata*).
[Early Cambrian] Morphology: tubiform. Rating: 2. Reason(s): see remarks.
Remarks: Difficult to differentiate from *Subtiflora* Maslov (cellular filaments) and *Batinevia* Korde (branching bundles). Luchinina (1975): *Subtiflora* = senior synonym of *Botominella*. Riding and Voronova (1985), separated all 3 genera, but didn't in 1984.

Burchalaella Kolosov 1970 (**B*. pulchella*).
[Late Proterozoic] Morphology: botryoidal. Rating: 1. Reason(s): 1?, 5, 6, 7.

Cambriocodium Jiang 1982 (**C*. capilloodes*).
[Sinian?-Early Cambrian?] Morphology: tubiform. Rating: 2. Reason(s): 5, 6, 7, 8.

Remarks: Can't evaluate. Looks like a tangled mass of filaments.

Cambrina Korde 1973 (**C*. fruticulosa*).
[Early Cambrian] Morphology: dendritic. Rating: 1. Reason(s): 5, 6, 7, 8.

Cambroporella Korde 1950 (**C*. tuvensis*).
[middle Early Cambrian] Morphology: cup. Rating: 2. Reason(s): 6, 7, 8.
Remarks: Kordeh's reconstruction looks like a modern dasyclad, but the actual photographs are difficult to interpret.

Canadiophycus Voronova and Drosdova 1987 (**C*. fibrosus*).
[Early Cambrian] Morphology: tubiform. Rating: 1. Reason(s): 7.

Cavifera Reitlinger 1948 (**C*. concinna*).
[Early Cambrian] Morphology: tubiform. Rating: 1. Reason(s): 4?, 7.
Remarks: As originally described, probably a fragment of *Obruchevella*, *Syniella*, or *Girvanella*, etc.

Chabakovia Vologdin 1939 (**C*. ramosa*).
[Early Cambrian-Devonian] Morphology: dendritic. Rating: 3. Reason(s): see remarks.
Remarks: See Riding and Wray (1972) for discussion of distinctiveness of *Chabakovia*.

Charaussaia Vologdin 1940 (**C*. camptotaenia*).
[late Early Cambrian] Morphology: ribbon. Rating: 1. Reason(s): 7, 8.

Chomustachia Korde 1973 (**Chabakovia tuberosa*).
[Early Cambrian] Morphology: dendritic. Rating: 1.
Remarks: From photos difficult to differentiate from *Chabakovia* and *Parachabakovia*. According to Kordeh, *Chabakovia* is dendritic, *Parachabakovia* rarely branches, and *Chomustachia* shows a more zoned appearance. These characters, however, are not obvious.

Confervites Brongniart 1828 (**C*. thoreaeformis*).
[Cambrian?-Paleozoic?] Morphology: dendritic. Rating: 0. Reason(s): 3, 4.
Remarks: Early Cambrian example is *Confervites primordialis* Bornemann which many workers say = *Epiphyton*.

Corbularia Vologdin 1962 (**C*. conglutinata*).
[Early Cambrian] Morphology: botryoidal. Rating: 1. Reason(s): 7.
Remarks: Illustrations look like it could be altered *Renalcis* or just a clotted, nonbiogenic fabric. Brasier (1977) illustrates also, but within an oncolith.

Cornutula Korde 1973 (**C*. kaltatica*).
[Early Cambrian] Morphology: see remarks. Rating: 1. Reason(s): 5, 7, 8.
Remarks: Illustrations resemble altered coids.

Dasycirriphycus Vologdin 1962 (**D*. frutuculosus*).
[Early Cambrian] Morphology: botryoidal. Rating: 1. Reason(s): 5, 6, 7, 8.

Edelsteina Vologdin 1934 (*?).
[Early Cambrian] Morphology: tubiform. Rating: 1. Reason(s): 6, 8.
Remarks: Never found original reference.

Eoepiphyton Butin 1959 (**E*. jalgamicum*).
[Middle Proterozoic] Morphology: dendritic. Rating: 1. Reason(s): 4, 5, 6, 7.
Remarks: Possibly synonymous with *Epiphyton*, but much older that any other occurrences of that genus. Not stated explicitly that is is calcareous. Glaessner (1979, p. A109) listed with microscopic algae).

Epiphyton Bornemann 1886 (**E*. flabellatum*).
[late Vendian-Devonian?] Morphology: dendritic. Rating: 3.

Table 23.2.1 (*Continued*).

Reason(s): 3.
Remarks: Predominantly a ubiquitous Lower-Middle Cambrian genus. Gowda (1970) reported it from the Archean of southern India, but did not illustrate it.

Epiphytonoides Korde 1973 (**E. sanashtykgolicus*).
[Early Cambrian-Middle Cambrian] Morphology: dendritic. Rating: 2. Reason(s): see remarks.
Remarks: A diagenetic form of other dendritic form?

Erbina Korde 1973 (**E. aristata*).
[Early Cambrian] Morphology: dendritic. Rating: 1. Reason(s): 5, 6, 7, 8.

Filaria Korde 1973 (**F. seriata*).
[Early Cambrian] Morphology: dendritic. Rating: 1. Reason(s): 4?, 8, see remarks.
Remarks: Less well-preserved *Epiphyton*? Not obvious that all species are calcified, but *F. calcarata* is according to Kordeh.

Fistulella Korde 1973 (**F. decipiens*).
[Early Cambrian] Morphology: tubiform. Rating: 1. Reason(s): 4?, 7, 8.
Remarks: Riding and Voronova (1985, p. 63) say that *F. decipiens* belongs in the *Botomaella* group and *F. sanashtykgolica* belongs in the *Hedstroemia* group; agreed. Danielli (1981) said = *Girvanella*.

Flabellina Korde 1986 (**F. multiformis*).
[Early Cambrian] Morphology: dendritic. Rating: 1. Reason(s): 1?, 5, 7, 8.
Remarks: Cement fabrics?

Foliaceria Vologdin 1962 (**F. polymorpha*).
[Early Cambrian] Morphology: see remarks. Rating: 1. Reason(s): 5, 6, 7, 8, see remarks.
Remarks: From drawing, looks like *Chabakovia*, but photographs are not interpretable.

Foninia Korde 1973 (**F. fasciculata*).
[Late Proterozoic] Morphology: spherical. Rating: 1. Reason(s): 1?, 7, 8.
Remarks: Probably an altered coid (see Riding and Voronova (1985). Glaessner (1979, p. A111) treated as problematic microfossils.

Gemma Luchinina 1982 (Zhuravleva et al. 1982) (**G. inclusa*).
[late Vendian-Early Cambrian] Morphology: botryoidal. Rating: 3. Reason(s): 7.
Remarks: Has not been widely illustrated, but is a distinct, recognizable taxa due to the peloidal wall.

Girvanella Nicholson and Etheridge 1878 (**G. problematica*).
[late Middle Proterozoic-Recent] Morphology: tubiform. Rating: 3.
Remarks: Wood (1957) reevaluated topotype Ordovician material containing *G. problematica* and amended original description of Nicholson and Etheridge (1878).

Globuloella Korde 1958 (**G. botomensis*).
[Early Cambrian] Morphology: spherical. Rating: 1. Reason(s): 5.
Remarks: Altered oolith?

Globulus Voronova and Drosdova 1987 (**G. gregalis*).
[Early Cambrian] Morphology: botryoidal. Rating: 1. Reason(s): 7, 8, see remarks.
Remarks: Diagenetically altered *Renalcis*?

Glomovertella Reitlinger 1948 (**G. firma*).
[Early Cambrian] Morphology: tubiform. Rating: 1. Reason(s): 4?, 7.
Remarks: Probably a fragment of *Obruchevella*, *Syniella*, etc., at least as originally described.

Gordonophyton Korde 1973 (**G. distinctum*).

[Early Cambrian] Morphology: dendritic. Rating: 2. Reason(s): see remarks.
Remarks: Similar to *Chabakovia* but with more disc-like "chambers".

Hedstroemia Rothpletz 1913 (**H. halimedoidea*).
[late Vendian?-Early Carboniferous] Morphology: tubiform. Rating: 3. Reason(s): 3, see remarks.
Remarks: Only Cambrian or earlier occurrence was reported by Bertrand-Sarfati (1979) from Mali, but questioned the generic assignment. If include the 2 species of *Rothpletzella* and *Protoortonella*, then range extends into the Nemakit-Daldyn Horizon.

Honanella Vologdin 1958 (**H. densa*).
[Early Cambrian] Morphology: tubiform. Rating: 1. Reason(s): 4, 5, 6, 7, 8.
Remarks: Poor reproduction of plates made evaluation impossible.

Jatuliana Korde 1965 (**J. furcata*).
[Middle Proterozoic] Morphology: spherical. Rating: 0. Reason(s): 1?, 5, 7, 8.
Remarks: No mention of calcification, but looks mineralized. Does not look biogenic—cement fabric? Glaessner (1979, p. A111) listed as doubtful taxa.

Kadvoya Korde 1973 (**K. mirabilis*).
[Early Cambrian] Morphology: dendritic. Rating: 1. Reason(s): 6, 7, 8.
Remarks: May be a cement fabric.

Kareliana Krode 1965 (**K. zonata*).
[Middle Proterozoic] Morphology: botryoidal. Rating: 0. Reason(s): 1?, 5, 7, 8.
Remarks: No mention of calcification.

Kenella Korde 1973 (**K. ornata*).
[Early Cambrian] Morphology: tubiform. Rating: 1. Reason(s): 4?, 7, 8.
Remarks: Danielli says = *Girvanella*. Kordeh would differentiate them on the basis of distinct, vertically oriented thallus that branches.

Ketemella Pjanovskaya 1974 (**K. lenaica*).
[late Early Cambrian] Morphology: tubiform. Rating: 1. Reason(s): 1?, 5, 7, 8.
Remarks: Possibly an altered archaeocyathid?

Kordephyton Radugin and Stepanova 1964 (**Epiphyton crinitum*).
[Early Cambrian-Middle Cambrian] Morphology: dendritic. Rating: 1.

Korilophyton Voronova 1976 (**Epiphyton inopinatum*).
[Late Vendian-Early Cambrian] Morphology: dendritic. Rating: 1. Reason(s): see remarks.
Remarks: Voronova says differs from *Epiphyton* in having irregular branching. Branches are very short. Riding and Voronova (1984) placed *K. angustum* in *Angulocellularia* thus reducing the number of Cambrian species to 1.

Kundatia Korde 1973 (**K. composita*).
[Early Cambrian] Morphology: ?. Rating: 1. Reason(s): 5, 7, 8.

Kyzassia Korde 1973 (**K. formosa*).
[Early Cambrian] Morphology: dendritic. Rating: 1. Reason(s): 4?, 5, 7, 8.

Lenaella Korde 1959 (**L. reticulata*).
[Early Cambrian] Morphology: cup. Rating: 0. Reason(s): see remarks.
Remarks: Cribricyathan? Kordeh originally said they resemble coelenterates (hydrozoans) or sponges, but also referred to as a possible alga.

Table 23.2.1 (Continued).

Mackenziephycus Voronova and Drosdova 1987 (**M. medullaris*).
[Early Cambrian] Morphology: tubiform. Rating: 1. Reason(s): 7, 8.
Remarks: Similar to *Batinevia* and *Botominella* but can be distinguished by complicated structure of filaments (Voronova and Drosdova 1987).

Majaella Vologdin and Maslov 1960 (**M. verkhojanica*).
[Yudomian] Morphology: plate/sheet. Rating: 0. Reason(s): see remarks.
Remarks: Vologdin and Maslov said *Majaella* and *Suvorovella* are similar to archaeocyaths and even to receptaculitids. They did consider them to be metazoans, but others (e.g., Missarzhevskij and Rozanov 1968) have listed as algae, thus, warranting inclusion here.

Manicosiphonia Tsao and Zhao 1978 (Cao and Zhao 1978) (**M. bambusa*).
[Sinian] Morphology: see remarks. Rating: 0. Reason(s): 1.
Remarks: Pseudofossils/nonbiogenic stromatolites? (Hofmann and Jackson 1987).

Marenita Korde 1973 (**M. kundatica*).
[Vendian-Early Cambrian] Morphology: spherical. Rating: 1. Reason(s): 5, 7, 8.
Remarks: Riding and Voronova (1985) suggested that *Marenita* might be an ooid or microfabric of ooids. Agreed. Glaessner (1979, A111) treated them as problematic fossils.

Mawsonella Chapman 1927 (**M. wooltanensis*).
[Early Cambrian?] Morphology: see remarks. Rating: 0. Reason(s): 1, 6, 7, 8.
Remarks: Glaessner (1979, p. A112) stated that it is "now considered as intraformational carbonate breccia."

Microcodium Glück 1912 (**M. elegans*).
[Permian?-Recent] Morphology: botryoidal. Rating: 0. Reason(s): 2.
Remarks: Voronova (1976) said that *M. laxus* Voronova 1969 (only Paleozoic species) = altered *Renalcis*.

Multisiphonia Tsao and Liang 1974 (Cao and Liang 1974) (**M. nanshanensis*).
[Sinian] Morphology: see remarks. Rating: 0. Reason(s): 1.
Remarks: Looks like radially fibrous-cement fabric. Hofmann and Jackson (1987) concluded that similar taxa of the Manicosiphoniaceae are nonbiologic.

Murandavia Vologdin 1965 (**M. amurica*).
[Middle Proterozoic] Morphology: botryoidal. Rating: 1. Reason(s): 1?, 8.
Remarks: Glaessner (1979, p. A111) listed as doubtful taxa.

Nanamanicosiphonia Tsao and Zhao 1978 (Cao and Zhao 1978) (**N. minuta*).
[Sinian] Morphology: see remarks. Rating: 0. Reason(s): 1.
Remarks: Pseudofossils/nonbiogenic stromatolites? (Hofmann and Jackson 1987).

Nephelostroma Dangeard and Doré 1958 (**N. lecomtei*).
[Cambrian] Morphology: botryoidal. Rating: 0. Reason(s): 4, 5, 7, 8.
Remarks: Probably = *Renalcis*.

Nicholsonia Korde 1973 (**N. glomerata*).
[Late Proterozoic?-Early Cambrian] Morphology: tubiform. Rating: 1. Reason(s): see remarks.
Remarks: Danielli 1979 said = *Girvanella*. Kordeh says *Nicholsonia* has protosporangia, but it is not obvious from photographs.

Nubecularities Maslov 1937 (**N. polymorphus*).
[Middle Cambrian] Morphology: botryoidal. Rating: 0. Reason(s): 3, see remarks.
Remarks: *N. polymorphus* = *Renalcis polymorphus* according to Rejtlinger (1959) and many others; agreed. According to Riding and Voronova (1985), all other *Nubecularities* are probably catagraphs.

Nuia Maslov 1954 (**N. sibirica*).
[Ordovician] Morphology: tubiform. Rating: 0. Reason(s): 2, 4.
Remarks: Has been classified as cyanobacteria and as codiacean algae (see Wray 1977); general form similar to *Microcodium* and primitive green algae, but nothing diagnostic is preserved.

Obruchevella Reitlinger 1948 (**O. delicata*).
[Late Proterozoic-Devonian] Morphology: tubiform. Rating: 3.

Ortonella Garwood 1914 (**O. furcata*).
[Silurian-Early Carboniferous] Morphology: tubiform. Rating: 0. Reason(s): 3.

Palaeogirvanella Krasnopeeva 1937 (**P. erbiensis*).
[Late Proterozoic?] Morphology: tubiform. Rating: 1. Reason(s): 5, 6, 7, 8.
Remarks: Illustrated as a nondescript drawing—can't evaluate. Only *P. erbiensis* is described, but in Pl. 1, Kordeh mentions *P. sajanica* n. sp.

Palaeomicrocystis Korde 1955 (**P. cambrica*).
[Early Cambrian] Morphology: spherical. Rating: 0. Reason(s): 5, see remarks.
Remarks: Can't really tell what it is—nothing diagnostic. Not sure if belongs in this database. I do not include all species—several reported from China and other parts of Siberia (e.g., Kolosov 1975; Yin et al. 1980).

Paleonites Maslov 1956 (**P. jacutii*).
[Early Cambrian] Morphology: tubiform. Rating: 0. Reason(s): 4, 6.
Remarks: Original drawing looks like *Proaulopora*, but photo of specimen looks like *Tubophyllum*. See discussion of these two genera.

Panomninella Kolosov 1966 (**P. ornata*).
[late Vendian-Early Cambrian] Morphology: spherical. Rating: 1. Reason(s): 4?.
Remarks: Kolosov transferred *Microcodium laxus* Voronova to *Panomninella*. But Voronova later (1976) said *M. laxus* = diagenetically altered *Renalcis*. Are all *Panomninella* diagenetic taxa? Similar to *Acanthina* and *Utchurella*?

Papillomembrana Spjeldnaes 1963 (**P. compta*).
[Ediacaran] Morphology: see remarks. Rating: 0. Reason(s): 5, 7, 8, see remarks.
Remarks: All specimens are compressed. Spjeldnaes said could have been spherical or cylindrical. Compression may suggest not originally calcified. Also, type of preservation is not characteristic of dasyclads (typically = molds). Glaessner (1979): megascopic algae.

Parachabakovia Korde 1973 (**P. dura*).
[Early Cambrian] Morphology: dendritic. Rating: 1. Reason(s): 4?, 7, 8.
Remarks: Difficult to differentiate among *Chabakovia*, *Chomustachia*, and *Parachabakovia*.

Parasolenopora Tsao and Zhao 1974 (Cao and Zhao 1974) (**P. subradiata*).
[late Sinian] Morphology: see remarks. Rating: 0. Reason(s): 1.
Remarks: Looks like radially fibrous-cement fabric. Hofmann and Jackson (1987) conclude that similar taxa of the Manicosiphoniaceae are nonbiologic.

Phacelofimbria Tsao and Zhao 1974 (Cao and Zhao 1974) (**P. emeishanensis*).

Table 23.2.1 (*Continued*).

[late Sinian] Morphology: see remarks. Rating: 0. Reason(s): 1.
Remarks: Looks like radially fibrous-cement fabric. Hofmann and Jackson (1987) concluded that similar taxa of the Manicosiphoniaceae are nonbiologic.

Pinnulina Korde 1986 (**P. cambrica*).
[Early Cambrian] Morphology: dendritic. Rating: 1. Reason(s): 1?, 5, 7, 8.
Remarks: Classification: Kordeh said either red or brown algae. Can't tell much from photograph—could be a cement fabric.

Potentillina Korde 1973 (**P. campanulata*).
[Early Cambrian] Morphology: tubiform. Rating: 1. Reason(s): 4, 7, 8, see remarks.
Remarks: Altered *Chabakovia*?

Praesolenopora Tsao and Zhao 1974 (Cao and Zhao 1974) (**P. flabella*).
[late Sinian] Morphology: see remarks. Rating: 0. Reason(s): 1.
Remarks: Looks like radially fibrous-cement fabric. Hofmann and Jackson (1987) concluded that similar taxa of the Manicosiphoniaceae are nonbiologic.

Proaulopora Vologdin 1934 (**P. rarissima*).
[Early Cambrian-Middle Cambrian+] Morphology: tubiform. Rating: 3.
Remarks: Taxonomic confusion! When described?, by whom?, relation to *Amganella* and *Tubophyllum*?, etc. Some have it credited to Vologdin (1937), but it's listed as n. gen. in Vologdin (1962). Possibly Vologdin 1934—see Krasnopeeva (1937). Similar to modern genus *Calothrix*?

Protoortonella Luchinina 1985 (Voronova and Luchinina 1985) (**Rothpletzella igarcaensis*).
[late Vendian-Early Cambrian] Morphology: tubiform. Rating: 1. Reason(s): 4?.
Remarks: Luchinina transferred both Cambrian species of *Rothpletzella* to *Protoortonella*. But Riding and Voronova (1984) said that the 2 species of *Rothpletzella* belonged in *Hedstroemia*.

Protorivularia Butin 1959 (**P. onega*).
[Middle Proterozoic] Morphology: spherical. Rating: 0. Reason(s): 6, 7, 8, see remarks.
Remarks: An oncolith? Glaessner (1979, p. A109) listed with microscopic algae.

Protosolenopora Wang 1985 (**P. distincta*).
[late Sinian] Morphology: tubiform. Rating: 0. Reason(s): 5, 7, 8.
Remarks: Cell-wall is organic, not calcareous. It may be algal, but can't tell for sure from the illustration.

Protuberantia Vologdin 1962 (**P. vesicularis*).
[Early Cambrian] Morphology: see remarks. Rating: 0. Reason(s): 5, 6, 7, 8, see remarks.
Remarks: Micrite-coated grains?

Pseudoacus Tsao and Zhao 1974 (Cao and Zhao 1974) (**P. renalis*).
[late Sinian] Morphology: see remarks. Rating: 0. Reason(s): 1.
Remarks: Looks like radially fibrous-cement fabric. Hofmann and Jackson (1987) conclude similar taxa of Manicosiphoniaceae are nonbiologic.

Pustularia Vologdin 1955 (**P. taeniata*).
[Middle Proterozoic?-Late Proterozoic] Morphology: ribbon. Rating: 1. Reason(s): 1?, 6, 7, 8.
Remarks: Glaessner (1979, p. A111) listed as doubtful taxa. Agreed.

Razumovskya Vologdin 1939 (**R. uralica*).
[Sinian-Middle Cambrian] Morphology: tubiform. Rating: 3.
Remarks: Vologdin (1939) described as n. gen., *R. uralica* as type. In 1937, Krasnopeeva referred to *R.* cf. *alta* Vologdin, so genus was described earlier; unable to obtain the earlier publication. But *R. alta* = n. sp. in Vologdin (1962).

Renalcis Vologdin 1932 (**R. granosus*).
[late Vendian-Devonian] Morphology: botryoidal. Rating: 3. Reason(s): see remarks.
Remarks: "represents the remains of irregularly stacked, rubbery, gelatinous colonies of non calcareous, chroococcalean algae with marked pigmentation gradients" (Hofmann 1975).

Rothpletzella Wood 1948 (**Sphaerocodium gotlandica*).
[Ordovician?-Triassic?] Morphology: tubiform. Rating: 0. Reason(s): 2, 4, 7.
Remarks: Riding and Voronova (1984) put both Cambrian species in *Hedstroemia*; Luchinina (in Voronova and Luchinina 1985) put both in *Protoortonella*. In either case, there are no Cambrian species of this genus.

Sajania Vologdin 1962 (**S. frondosa*).
[Early Cambrian-Middle Cambrian] Morphology: dendritic. Rating: 1. Reason(s): see remarks.
Remarks: According to Vologdin, differs from *Epiphyton* "by having threadlike thallus and irregular branching" (per Johnson 1966). Kordeh's interpretation of *Sajania*, however, looks more like *Epiphyton*.

Serligia Korde 1973 (**S. fragilis*).
[Early Cambrian] Morphology: dendritic. Rating: 2. Reason(s): 1?, 8.
Remarks: No one else ever mentions, except Kobluk (1979): *S.* cf. *fragilis*. Has distinct form.

Shujana Korde 1979 (**S. shulgini*).
[Middle Proterozoic] Morphology: tubiform. Rating: 1. Reason(s): 1?, 7, 8.
Remarks: Presumably calcareous. Title of article is "rock-building algae..." Diagenetic?

Sinocapsa Vologdin 1958 (**S. honanica*).
[Early Cambrian] Morphology: tubiform. Rating: 1. Reason(s): 5, 6, 7, 8.
Remarks: Illustration difficult to interpret.

Siphonia Tsao and Zhao 1974 (Cao and Zhao 1974) (**S. herbacea*).
[late Sinian] Morphology: see remarks. Rating: 0. Reason(s): 1.
Remarks: Looks like radially fibrous-cement fabric. Hofmann and Jackson (1987) concluded that similar taxa of the Manicosiphoniaceae are nonbiologic.

Siphonophycus Schopf 1968 (**S. kestron*).
[Middle Proterozoic-?] Morphology: tubiform. Rating: 1. Reason(s): 5, 7, 8.
Remarks: Only one "calcareous" occurrence (Klein et al. 1987). Calcification is weak—single crystals may have adhered to sheath. No evidence of cyanobacteria promoting calcification.

Solenopora Dybowski 1879 (**S. spongioïdes*).
[Vendian?-Paleozoic] Morphology: tubiform. Rating: 2. Reason(s): 3?, 6, see remarks.
Remarks: Originally described as chaetetid. Neither of the Proterozoic/Early Cambrian occurrences are convincing.

Sporinula Korde 1973 (**S. palmata*).
[Early Cambrian] Morphology: dendritic. Rating: 1. Reason(s): 7, 8.
Remarks: Altered form of other dendritic alga?

Subtiflora Maslov 1956 (**S. delicata*).
[Early Cambrian-Carboniferous?] Morphology: tubiform. Rating: 2. Reason(s): see remarks.
Remarks: Many place *Subtiflora* as senior synonym of *Botominella*. The difference, as originally described, is that *Subtiflora* has

Table 23.2.1 (Continued).

preserved cells within filaments; *Botominella* does not.

Suvorovella Vologdin and Maslov 1960 (*S. aldanica*).
   [late Vendian] Morphology: plate/sheet. Rating: 0. Reason(s): see remarks.
   Remarks: Vologdin and Maslov said that *Suvorovella* and *Majaella* have similarities to archaeocyatids and receptaculitids. They considered them to be metazoans, but others (e.g., Missarzhevskij and Rozanov 1968) have listed them with the algae—thus, inclusion here.

Syniella Reitlinger 1948 (*S. invensuta*).
   [Early Cambrian] Morphology: tubiform. Rating: 0. Reason(s): 4, 7, 8, see remarks.
   Remarks: *Syniella* is probably a preservational variation of *Obruchevella*.

Taninia Korde 1973 (*T. tomentosa*).
   [Early Cambrian] Morphology: dendritic. Rating: 1. Reason(s): 4?, 7, 8, see remarks.
   Remarks: Possibly synonymous with *Chabakovia*. Kordeh (1973) separated *Taninia*, *Tomentula*, and *Sporinula* from the *Chabakovia* family because all, according to her, had a basal attachment structure and presence of reproductive structures; these aren't obvious!

Tannuolaia Vologdin 1967 (*T. fonini*).
   [Early Cambrian] Morphology: tuberous. Rating: 0. Reason(s): 7, 8, see remarks.
   Remarks: Good fossil, but possibly a strange archaeocyath?

Tarthinia Drosdova 1975 (*T. rotunda*).
   [Early Cambrian] Morphology: botryoidal. Rating: 2. Reason(s): 8, see remarks.
   Remarks: Fibrous wall distinguishes from *Renalcis*. Fibrous nature is not blatant in the illustration. Riding and Voronova (1984) transferred *Renalcis gelatinosus* Korde to *Tarthinia*.

Templuma Zhang 1979 (*T. sinica*).
   [Late Proterozoic] Morphology: tubiform. Rating: 2. Reason(s): 5, 6, 7, 8.
   Remarks: Can't evaluate from photograph. Does have general dasycladacean shape. It's small: about 0.2 mm × 1.2 mm.

Tersia Vologdin 1931 (*T. filiforma*).
   [Early Cambrian] Morphology: see remarks. Rating: 0. Reason(s): 6, see remarks.
   Remarks: An archaeocyath (as listed in treatise and even by Vologdin himself by year 1937).

Thaumatophycus Korde 1950 (*T. furcatus*).
   [Late Cambrian] Morphology: tubiform. Rating: 0. Reason(s): 2, 6, 7, 8.

Timanella Vologdin 1966 (*T. gigas*).
   [Sinian?] Morphology: tuberous. Rating: 0. Reason(s): 5, 7, 8.
   Remarks: Non-calcified. Very large (about 6 × 15 cm). Resembles a cross section through a palm? or related? plant. Glaessner (1979, p. A112) listed with megascopic algae.

Tomentula Korde 1973 (*T. villosa*).
   [Early Cambrian] Morphology: dendritic. Rating: 1. Reason(s): 4?, 8, see remarks.
   Remarks: Similar to *Epiphyton*? Kordeh differentiates on basis of felted structure and position of protosporangia.

Tubercularia Vologdin 1962 (*T. latiuscula*).
   [Early Cambrian] Morphology: see remarks. Rating: 0. Reason(s): 5, 6, 7, 8, see remarks.
   Remarks: Stromatolite fabric?

Tubomorphophyton Korde 1973 (*Epiphyton benignum*).
   [Early Cambrian-Devonian?] Morphology: dendritic. Rating: 3.

Tubophyllum Krasnopeeva 1955 (*T. victori*).
   [Early Cambrian] Morphology: tubiform. Rating: 0. Reason(s): 4, 7, 8.
   Remarks: Krasnopeeva used the same illustration (drawn) of *T. victori* that she used earlier for *Proaulopora glabra*, with no discussion of why.

Unbellula Korde 1973 (*U. minuta*).
   [Early Cambrian] Morphology: dendritic. Rating: 1. Reason(s): 6, 7, 8.
   Remarks: Could be disseminated organic matter or possibly related somehow to *Epiphyton*.

Uranovia Korde 1958 (*U. granosa*).
   [Proterozoic?-Early Cambrian] Morphology: botryoidal. Rating: 0. Reason(s): 5, 6, 7, 8.
   Remarks: Doubtful that it's calcareous.

Utchurella Kolosov 1970 (*U. explicata*).
   [Proterozoic] Morphology: spherical. Rating: 1. Reason(s): 1?, 5, 6, 7.
   Remarks: Similar to *Acanthina* and *Panomninella*?

Varicamanicosiphonia Tsao and Zhao 1978 (Cao and Zhao 1978) (*Actinophycus quadricella*).
   [late Sinian] Morphology: see remarks. Rating: 0. Reason(s): 1.
   Remarks: Hofmann and Jackson (1987) interpret as nonbiologic: diagenetic radial fibrous fabric.

Vologdinella Korde 1957 (*V. fragile*).
   [Early Cambrian] Morphology: tubiform. Rating: 0. Reason(s): 4, 7, 8.
   Remarks: Kordeh (1973) later classified under rhodophytes and differentiates *Vologdinella* from *Proaulopora* by thallus structure. Difficult to tell the two genera apart; see for example Voronova (1976).

Vologdinia Korde 1973 (*V. verticillata*).
   [Early Cambrian] Morphology: dendritic. Rating: 1. Reason(s): 7, 8, see remarks.
   Remarks: Highly altered *Sajania*?, *Epiphyton*?, or something related. Kordeh differentiates it from *Sajania* and *Cambrina* by having cells of various sizes and shapes (diagenetic?) and "complex structure of reproductive organs."

Wetheredella Wood 1948 (*W. silurica*).
   [Early Cambrian-Devonian?] Morphology: tubiform. Rating: 3.
   Remarks: Chuvashov et al. (1987) said *Wetheredella* ranges from Silurian to Devonian, but Kobluk (1979) illustrated it from Lower Cambrian of southern Labrador.

Yentaiia Vologdin 1958 (*Y. liaoyangensis*).
   [late Early Cambrian] Morphology: tubiform. Rating: 1. Reason(s): 4, 5, 6, 7, 8.
   Remarks: Could not obtain original publication. On basis of photocopy, it does look like a filamentous alga/cyanobacteria, possibly in the *Girvanella* or *Subtiflora* groups.

Zaganolomia Drosdova 1980 (*Z. buralica*).
   [Early Cambrian] Morphology: tubiform. Rating: 1. Reason(s): 1?, 7, 8.
   Remarks: Cement fabrics? (see James and Klappa 1983, Figs. 8 and 9.)

Table 23.2.2 lists species of Proterozoic and Early Cambrian calcareous algae that are interpreted here to be fossils. A plus (+) symbol denotes the possible Proterozoic occurrence of the taxon indicated. For the most part, only figured and described taxa are included in Table 23.2.2; multiple occurrences are not included. Coverage of taxa in open nomenclature is incomplete. A literature reference is included in parentheses after the taxon only if the author(s) of the taxon differ from the author(s) of the publication. Stratigraphic information is included if available in the original reference; otherwise, limited locality information is provided. "Remarks" includes additional taxonomic information.

Table 23.2.2 *Species of bona fide Proterozoic and/or Early Cambrian calcareous algae.*

+(dasyclad) Misra 1949.
    Vindhyan (lower) of India.
+(dasyclad) Lyubtsov 1962.
    Kola Peninsula, Soviet Union.
    Remarks: Absolute age date of 1720 to 1780 Ma, but presence of good dasyclads, nautiloids, and corals? suggest a more recent age, perhaps Ordovician as indicated on the published stratigraphic column.
+(encrusting alga) Grant and Knoll 1987.
    Urusis Fm. of Namibia.
    Remarks: Not yet published.
(*Renalcis*-like alga) James and Kobluk 1978.
    Forteau Fm. (lower) of Canada.
(*Renalcis*-like alga) Kobuluk and James 1979.
    Forteau Fm. (lower) of Canada.
Amganella glabra (Krasnopeeva 1937) Reitlinger 1959.
    Remarks: Reitlinger (1959) created genus for "smooth" *Proaulopora* to clear up Krasnopeeva's taxa. She used same drawing for *P. glabra* in 1937 and for *Tubophyllum victori* in 1955. Most consider *Amganella* a junior synonym of *Proaaulopora*.
Angulocellularia mansurkaensis Vologdin 1962.
    Manzurska Fm. of Siberia.
Bajanophyton egiingolicum Voronova and Drosdova 1983.
    Lenian Stage of western Mongolia.
Bajanophyton mucosum Drosdova 1980.
    Tommotian Stage of Siberia.
Batinevia bayankolica Korde 1973.
    Bagrad Horizon (upper) of Siberia.
    Remarks: Danielli (1981) said = *Girvanella*.
Batinevia nodosa Stepanova 1974.
    Sakharovsk Fm. of Siberia.
Batinevia ramosa Korde 1966.
    Elansk Stage, Obruchev Horizon of Siberia.
    Remarks: Danielli (1981) said = *Girvanella*.
Bija canadensis Voronova and Drosdova 1987.
    Mackenzie Mountains, Canada.
Bija grandis Korde 1973.
    Sanashtygol Horizon of Siberia.
Bija sibirica Vologdin 1932.
    Altai Sayan, Siberia.
    Remarks: Originally described as an alcyonarian.
Bija cf. *sibirica* Kobluk and James 1979.
    Forteau Fm. (lower) of Canada.
Botomaella aequalis Voronova and Drosdova 1987.
    Mackenzie Mountains, Canada.
Botomaella anabarica Voronova 1969.
    Atdabanian Stage of Siberia.
Botomaella crassa Voronova and Drosdova 1987.
    Mackenzie Mountains, Canada.
Botomaella dubia Voronova and Drosdova 1987.
    Mackenzie Mountains, Canada.
Botomaella mitis Voronova 1969.
    Atdabanian Stage of Siberia.
Botomaella sibirica Voronova 1976.
    Tommotian Stage of Siberia.
Botomaella zelenovi Korde 1958.
    Atdabanian Stage, Kameshki Horizon of Siberia.
Botominella lineata Reitlinger 1959.
    Pestrotsvet Fm. of Siberia.
    Remarks: Luchinina (1975) referred to *Subtiflora lineata*. Danielli (1981) said = *Girvanella*.
Botominella aff. *lineata* Reitlinger 1959.
    Botoma River, Siberia.
Botominella lineata var. *elanskensis* Reitlinger 1959.
    Elanka Fm. of Siberia.
Cambrina composita Korde 1973.
    Bagrad Horizon of Siberia.
Cambrina fruticulosa Korde 1973.
    Bagrad Horizon of Siberia.
Cambroporella tuvensis Korde 1950.
    Tuva, Siberia.
Canadiophycus fibrosus Voronova and Drosdova 1987.
    Mackenzie Mountains, Canada.
Cavifera concinna Reitlinger 1948.
    Kutorgina Fm. of Siberia.
Cavifera cf. *concinna* Reitlinger 1948. (Kobluk 1985)
    Shady Dolomite of the United States.
Chabakovia [sp.] Cherchi and Schroeder 1984.
    Archaeocyath Limestone of Italy.
Chabakovia sp. Drobkova 1979.
    Usol'e Fm. (lower) of Siberia.
Chabakovia cavitata Vologdin 1962.
    Yangud Fm. (lower) of Siberia.
    Remarks: Described as new species in Vologdin (1962) also.
Chabakovia chabakoviformis (Voronova 1973) Drosdova 1980.
    Atdabanian Stage, Kameshki Horizon of western Mongolia.
    Remarks: = *Renalcis chabakoviformis*.
Chabakovia flabellata Korde 1973.
    Bagrad Horizon of Siberia.
Chabakovia fungiformis Korde 1973.
    Bagrad Horizon of Soviet Union.
Chabakovia monstrata Korde 1960.
    Bazaikhe Horizon of Siberia.
Chabakovia nana Korde 1973.
    Bagrad Horizon of Siberia.
Chabakovia nodosa Korde 1960.
    Bazaikhe Horizon of Siberia.
Chabakovia ramosa Vologdin 1939.
    Limestone exposure 188b of Soviet Union.
    Remarks: Riding and Wray (1972) accept drawn specimen as holotype.

Chabakovia subglobosa Luchinina 1975. (Belyaeva et al. 1975)
  Gerbikan Horizon of Siberia.
Chabakovia tuberosa Korde 1961.
  Elansk Stage of Siberia.
Charaussaia camptotaenia Vologdin 1940.
  Khara-Usu Lake, western Mongolia.
Chomustachia diadroma Korde 1973.
  Bagrad Horizon of Siberia.
Chomustachia tuberosa (Korde 1961) Korde 1973.
  Remarks: = *Chabakovia tuberosa*.
Confervites primordialis Bornemann 1886.
  Marbles of Italy.
  Remarks: = *Epiphyton*.
Epiphyton sp. Buggish and Webers 1982.
  Whiteout Conglomerate of Antarctica.
Epiphyton [sp.] Read and Pfeil 1983.
  Shady Dolomite of the United States.
  Remarks: Flat-lying "*Epiphyton*" might be referred to *Kordephyton* by some.
Epiphyton [sp.] Cherchi and Schroeder 1984.
  Archaeocyath Limestone of Italy.
Epiphyton [sp.] Pfeil and Read 1980.
  Shady Dolomite of the United States.
Epiphyton sp. Riding and Voronova 1985.
  Atdabanian Stage of Siberia.
  Remarks: List as *Epiphyton* sp. (cf. *Tubomorphophyton* sp.).
Epiphyton [sp.] Zamarreño and Debrenne 1977.
  Córdoba, Spain.
Epiphyton [sp.] Rowland 1978.
  Poleta Fm. of the United States.
Epiphyton sp. Reitlinger 1959.
  Pestrotsvet Fm. of Siberia.
Epiphyton [sp.] Selg 1986.
  Nebida Fm. of Italy.
Epiphyton [sp.] Kobluk 1985.
  Shady Dolomite of the United States.
Epiphyton [sp.] Kobluk and James 1979.
  Forteau Fm. (lower) of Canada.
Epiphyton [sp.] Nikolaeva, Borodaev., Peroz, and Belob. 1987.
  Pestrotsvet Fm. of Siberia.
Epiphyton absimilis Voronova 1969.
  Tommotian Stage of Siberia.
  Remarks: = *E. cristatum* Korde 1961 per Voronova (1976).
Epiphyton achoricum Gudymovich 1966.
  Ungut Fm. (upper) of Siberia.
Epiphyton amplificatum Korde 1960.
  Bol'shaya Erba Horizon (lower) of Siberia.
Epiphyton anguinum Korde 1960.
  Bol'shaya Erba Horizon of Siberia.
Epiphyton benignum Korde 1960.
  Bol'shaya Erba Horizon of Siberia.
  Remarks: = *Tubomorphophyton benignum* per Kordeh (1973). No holotype assigned in 1969; Figure 3, Pl. 31 was designated as holotype in Kordeh (1961).
Epiphyton bifidum Korde 1961.
  Atdabanian Horizon of Siberia.
Epiphyton bisporanginum Korde 1973.
  Bagrad Horizon-Sanashtygol Horizon of Siberia.
Epiphyton botomense Korde 1955.
  Atdabanian Horizon of Siberia.
  Remarks: = *Tubomorphophyton botomense* per Kordeh (1973).

Epiphyton cf. *botomense* Reitlinger 1959.
  Pestrotsvet Fm. (up) of Siberia.
Epiphyton breviramosum Korde 1973.
  Atdabanian Horizon of Siberia.
Epiphyton carptum Korde 1961.
  Atdabanian Horizon of Siberia.
Epiphyton celsum Korde 1960.
  Bol'shaya Erba Horizon of Siberia.
Epiphyton complexum forma *semenica* Gudymovich 1967.
  Ungut Fm. (lower) of Siberia.
Epiphyton complexum forma *ungutica* Gudymovich 1967.
  Ungut Fm. of Siberia.
Epiphyton confractum Korde 1961.
  Atdabanian Stage of Siberia.
Epiphyton crassum Korde 1961.
  Atdabanian Horizon of Siberia.
Epiphyton crebrum Drosdova 1980.
  Lenian Stage, Sanashtygol Horizon of western Mongolia.
Epiphyton crinitum Korde 1955.
  Ust'Botoma Fm (lower) of Siberia.
  Remarks: = *Kordephyton crinitum* per Radugin and Stepanova (1964). In Kordeh (1960), this species was drawn (no photograph) and listed as new species.
Epiphyton crispum Korde 1960.
  Bol'shaya Erba Horizon of Siberia.
  Remarks: = *Kordephyton crispum* per Kordeh (1973).
Epiphyton cristatum Korde 1961.
  Atdabanian Horizon of Siberia.
  Remarks: = *Tubomorphophyton cristatum* per Kordeh (1973).
+*Epiphyton cudi* Stepanova 1964. (Radugin and Stepanova, 1964)
  E Sayan (SW part), Siberia.
Epiphyton curvatum Korde 1973.
  Sanashtygol Horizon of Siberia.
+*Epiphyton decumanum* Gudymovich 1966.
  Ungut Fm. (lower) of Siberia.
Epiphyton decumanum forma *anastasica* Gudymovich 1967.
  Ungut Fm. (lower) of Siberia.
Epiphyton decumanum forma *kolbaica* Gudymovich 1967.
  Ungut Fm. of Siberia.
Epiphyton decumanum forma *zherzhulica* Gudymovich 1967.
  Anastas'ino Fm. (upper) of Siberia.
Epiphyton dembovi Korde 1960.
  Bazaikhe Horizon of Siberia.
  Remarks: = *Gordonophyton dembovi* per Kordeh (1973).
Epiphyton durum Korde 1961.
  Pestrotsvet Fm. of Siberia.
  Remarks: = *Gordonophyton durum* per Drozdova (1980).
Epiphyton elegans Korde 1973.
  Bagrad Horizon of Siberia.
Epiphyton evolutum Korde 1960.
  Bol'shaya Erba Horizon of Siberia.
Epiphyton ezhimicum Korde 1973.
  Sanashtygol Horizon of Siberia.
Epiphyton falcifruticosum Korde 1973.
  Bagrad Horizon of Siberia.
Epiphyton fasciatum Gudymovich 1966.
  Ungut Fm. (upper) of Siberia.
Epiphyton fibratum Korde non Krasnopeeva 1960.
  Kuznetsk Alatau, Siberia.
Epiphyton flabellatum Bornemann 1886.
  Limestone of Sardinia, Italy.

Table 23.2.2 (Continued).

Epiphyton frequens Drosdova 1980.
   Lenian Stage, Sanashtygol Horizon of western Mongolia.
Epiphyton frondosum Korde 1961.
   Tankha Fm. of Siberia.
Epiphyton fruticosum Vologdin 1939.
   Limestone exposure 438d of Soviet Union.
Epiphyton furcatum Korde 1960.
   Podobruchev Horizon of Siberia.
Epiphyton geniculatum Voronova 1969.
   Tommotian Stage of Siberia.
Epiphyton gigam Korde 1973.
   Sanashtygol Horizon of Siberia.
Epiphyton grande Gordon 1920.
   Weddell Sea dredgings of Antarctica.
   Remarks: = *Gordonophyton grandis* per Kordeh (1973).
+ *Epiphyton improcerum* Gudymovich 1966.
   Anastas'ino Fm. (upper) of Siberia.
Epiphyton induratum Korde 1961.
   Atdabanian Horizon of Siberia.
Epiphyton inexpectatum Korde 1960.
   Podobruchev Horizon of Siberia.
   Remarks: = *Gordonophyton inexpectatum* per Kordeh (1973).
Epiphyton inobservabile Korde 1961.
   Atdabanian Horizon of Siberia.
+ *Epiphyton inopinatus* Voronova 1969.
   Nemakit Daldyn Horizon of Siberia.
   Remarks: = *Korilophyton inopinatum* per Voronova (1976).
Epiphyton intergerinum Gudymovich 1967.
   Anastas'ino Fm. (upper) of Siberia.
Epiphyton ? jacutii Maslov 1937.
   Lena River, Siberia.
   Remarks: = *Paleonites jacutii* per Maslov (1956).
   = *Proaulopora glabra*.
Epiphyton kiyanicum Korde 1973.
   Bagrad Horizon of Siberia.
Epiphyton longum Korde 1955.
   Amga Fm. (lower) of Siberia.
+ *Epiphyton manaense* Gudymovich 1966.
   Ungut Fm. (lower) of Siberia.
Epiphyton manaense forma *giganta* Gudymovich 1967.
   Ungut Fm. (lower) of Siberia.
Epiphyton mirabile Korde 1961.
   Atdabanian Horizon of Siberia.
Epiphyton naturale Korde 1960.
   Bazaikhe Horizon of Siberia.
Epiphyton novum Korde 1961.
   Atdabanian Horizon of Siberia.
Epiphyton nubilum Korde 1961.
   Atdabanian Horizon (lower) of Siberia.
   Remarks: = *Tubomorphophyton nubilum* per Kordeh (1973).
Epiphyton ordonatum Korde 1973.
   Sanashtygol Horizon of Siberia.
Epiphyton ornatum Korde 1961.
   Amga Fm. of Siberia.
   Remarks: = *Epiphytonoides ornatus* Korde 1973 ex Drosdova 1980.
Epiphyton parapusillum Korde 1973.
   Sanashtygol Horizon of Siberia.
Epiphyton pencillatum Korde 1973.
   Bagrad Horizon of Siberia.
Epiphyton cf. *pencillatum* Kobluk 1981.
   Poleta Fm. of the United States.
   Remarks: Listed under specific name in text, but under generic name only in figure captions.
+ *Epiphyton periodicum* Radugin 1964.
   E Sayan (NW part), Siberia.
Epiphyton plumosum Korde 1955.
   Atdabanian Horizon of Siberia.
Epiphyton pretiosum Korde 1961.
   Atdabanian Horizon of Siberia.
Epiphyton procerum Gudymovich 1966.
   Ungut Fm. (upper) of Siberia.
Epiphyton pseudoflexuosum Korde 1961.
   Atdabanian Horizon of Siberia.
Epiphyton pusillum Korde 1961.
   Atdabanian Horizon of Siberia.
Epiphyton racemosum Korde 1961.
   Atdabanian Horizon of Siberia.
Epiphyton ramosum Drosdova 1980.
   Tommotian Stage, Bazaikhe Horizon of western Mongolia.
Epiphyton rectum Korde 1960.
   Podobruchev Horizon of Siberia.
Epiphyton rosulare Korde 1973.
   Obruchev Horizon of Siberia.
Epiphyton saturum Drosdova 1980.
   Tommotian Stage, Bazaikhe Horizon of western Mongolia.
Epiphyton scapulum Korde 1961.
   Atdabanian Horizon (lower) of Siberia.
Epiphyton scoparium Korde 1960.
   Atdabanian Stage, Kameshki Horizon of Siberia.
Epiphyton simplex Korde 1960.
   Sanashtygol Horizon of Siberia.
   Remarks: = *Gordonophyton simplex* per Kordeh (1973).
Epiphyton spissum Korde 1961.
   Tankha Fm. of Siberia.
Epiphyton subfruticosum Voronova 1969.
   Tommotian Stage of Siberia.
Epiphyton suvorovae Korde 1961.
   Elanka Horizon-Amga Horizon of Siberia.
Epiphyton tenue Vologdin 1932.
   Altai Sayan, Siberia.
Epiphyton tuberculosum Korde 1961.
   Atdabanian Horizon (lower) of Siberia.
Epiphyton ulinicum Drosdova 1980.
   Atdabanian Stage, Kameshki Horizon of western Mongolia.
Epiphyton varium Korde 1961.
   Bol'shaya Erba Horizon of Siberia.
Epiphyton vulgare Korde 1961.
   Atdabanian Horizon of Siberia.
Epiphyton zhuravlevae Korde 1960.
   Bol'shaya Erba Horizon of Siberia.
Epiphytonoides affinis Korde 1973.
   Bagrad Horizon of Siberia.
Epiphytonoides fasciculatus (Chapman 1916) Korde 1973.
   Remarks: = *Epiphyton fasciculatum*.
Epiphytonoides nurmogoicus Drosdova 1980.
   Atdabanian Stage, Kameshki Horizon of western Mongolia.
Epiphytonoides ornatus (Korde 1961) Korde 1973 ex Drosdova 1980.
   Remarks: = *Epiphyton ornatum*.
Epiphytonoides roselatus Korde 1973.
   Bagrad Horizon of Siberia.
Epiphytonoides sanashtykgolicus Korde 1973.
   Sanashtygol Horizon of Siberia.

Table 23.2.2 *(Continued)*.

Epiphytonoides shevelicus Korde 1973.
  Sanashtygol Horizon of Siberia.
Epiphytonoides tenuiramosus Korde 1973.
  Bagrad Horizon of Siberia.
Filaria sp. Kolosov 1977.
  Yuedej Fm. of Siberia.
Filaria calcarata Korde 1973.
  Sanashtygol Horizon of Siberia.
Filaria seriata Korde 1973.
  Bagrad Horizon of Siberia.
Filaria sporifera Korde 1973.
  Bagrad Horizon of Siberia.
Fistulella sanashtykogolica Korde 1973.
  Sanashtygol Horizon of Siberia.
  Remarks: = *Girvanella* per Danielli (1981). = *Hedstroemia* per Riding and Voronova (1984).
Fistulella decipiens Korde 1973.
  Sanashtygol Horizon of Siberia.
  Remarks: Danielli (1981) said = *Girvanella*.
+*Gemma inclusa* Luchinina 1982. (Zhuravleva et al. 1982)
  Nemakit Daldyn Horizon of Siberia.
Gemma maculosa Voronova and Drosdova 1982.
  Tommotian Stage (lower) of western Mongolia.
Girvanella sp. Luchinina 1975.
  Atdabanian Stage (lower) of Siberia.
+*Girvanella* [sp.] Pierce and Cloud 1979.
  Pahrump Grp. of the United States.
Girvanella sp. 1 Cao 1986.
  Shanxi Province (N part), China.
Girvanella sp. 2 Cao 1986.
  Shanxi Province (N part), China.
+*Girvanella* sp. Yin 1980.
  Sichuan, China.
+*Girvanella* sp. Chen et al. 1983.
  China.
  Remarks: Looks like unwound *Obruchevella*.
Girvanella [sp.] Read and Pfeil 1983.
  Shady Dolomite of the United States.
  Remarks: Much of *Girvanella* might be referred by some to *Razumovskya*. Read and Pfeil described *Girvanella* boundstones as consisting of subparallel platy crusts with vertical to radiating filaments; a morphology characteristic of *Razumovskya*.
Girvanella ? [sp.] Pfeil and Read 1980.
  Shady Dolomite of the United States.
  Remarks: Possibly = *Razumovskya*.
Girvanella [sp.] Selg 1986.
  Nebida Fm. of Italy.
+*Girvanella* [sp.] Li and Yang 1984.
  Doushantuo Fm. of China.
Girvanella [sp.] Dangeard and Doré 1958.
  Manche, Carteret, France.
  Remarks: Pl. 46, Figure 1 might be referred to *Razumovskya* by some.
Girvanella [sp.] Kobluk 1985.
  Shady Dolomite of the United States.
  Remarks: Might be referred by some to *Razumovskya*.
Girvanella [sp.] Kobluk and James 1979.
  Forteau Fm. (lower) of Canada.
Girvanella [sp.] Nikolaeva et al. 1987.
  Lena River, Siberia.
Girvanella [sp.] Nikolaeva et al. 1987.
  Atdabanian Stage of Siberia.
Girvanella sp. Pak and Terleev 1986.
  Kureninsk Fm. of Siberia.
  Remarks: Section had been described as Riphean. Finding of Cambrian algae puts this age assignment into question.
Girvanella antiqua Maslov non Dawson 1937.
  Sinyaya River, Siberia.
  Remarks: = *Girvanella sibirica* per Maslov (1956).
Girvanella iyangensis Vologdin 1958.
  Henan Province, China.
Girvanella manchurica Yabe and Ozaki 1930.
  Limestone, Liaoning Province, China.
Girvanella mexicana Johnson 1952.
  Buelna Limestone (lower) of Mexico.
+*Girvanella problematica* Nicholson and Etheridge 1878.
  Craighead Limestone of the United Kingdom.
Girvanella sibirica Maslov 1956.
  Tuva, Siberia.
  Remarks: Maslov places *G. antiqua* Maslov 1937 in synonymy, but may have assigned a new type, but did not figure it; type number/locality does not match that of Maslov 1937.
Girvanella sinensis Yabe 1912.
  Limestone, Hubei Province, China.
  Remarks: Originally thought to be Carboniferous or Ordovician, but later referred to Early Cambrian (Yabe and Ozaki 1930).
Girvanella staminae Garwood 1931.
  Main algal series of the United Kingdom.
Globulus gregalis Voronova and Drosdova 1987.
  Mackenzie Mountains, Canada.
  Remarks: May be diagenetic alteration of *Renalcis*.
Glomovertella [sp.] Pyatiletov et al. 1981.
  Atdabanian Stage of Siberia.
Glomovertella firma Reitlinger 1948.
  Kutorgina Fm. of Siberia.
Gordonophyton axillare Korde 1973.
  Sanashtygol Horizon of Siberia.
Gordonophyton demboi (Korde 1960) Korde 1973.
  Remarks: = *Epiphyton dembovi*.
Gordonophyton distinctum Korde 1973.
  Bagrad Horizon of Siberia.
Gordonophyton durum (Korde 1961) Korde 1973 ex Drosdova 1980.
  Remarks: = *Epiphyton durum*.
Gordonophyton grandis (Gordon 1920) Korde 1973.
  Remarks: = *Epiphyton grande*.
Gordonophyton inexpectatum (Korde 1960) Korde 1973.
  Remarks: = *Epiphyton inexpectatum*.
Gordonophyton nodosum Drosdova 1980.
  Tommotian Stage, Bazaikhe Horizon of western Mongolia.
Gordonophyton parvulum Voronova and Drosdova 1983.
  Lenian Stage of western Mongolia.
Gordonophyton simplex (Korde 1960) Korde 1973.
  Remarks: = *Epiphyton simplex*.
+*Hedstroemia* ? [sp.] Bertrand-Sarfati 1976.
  Sarnyere Fm. of Mali.
Hedstroemia sp. Riding and Voronova 1985.
  Botomian of western Mongolia.
Hedstroemia flabellata (Voronova 1976) Voronova and Drosdova 1987.
  Remarks: = *Rothpletzella flabellata*.
Hedstroemia series Voronova and Drosdova 1987.
  Mackenzie Mountains, Canada.

Table 23.2.2 (Continued).

Honanella densa Vologdin 1958.
    Henan Province, China.
Kadvoya mirabilis Korde 1973.
    Sanashtygol Horizon of Siberia.
Kenella ornata Korde 1973.
    Bagrad Horizon of Siberia.
    Remarks: Danielli (1981) said = *Girvanella*.
Ketemella lenaica Pjanovskaya 1974.
    Keteme Fm. of Siberia.
Kordephyton sp. Drobkova 1979.
    Usol'e Fm. (lower) of Siberia.
Kordephyton conglutinatum Korde 1973.
    Bagrad Horizon of Siberia.
Kordephyton crinitum (Korde 1955) Radugin and Stepanova 1964.
    Remarks: = *Epiphyton crinitum*.
Kordephyton crispum (Korde 1960) Korde 1973.
    Remarks: = *Epiphyton crispum*.
Korilophyton angustum Voronova 1976.
    Kessyuse Fm. of Siberia.
    Remarks: = *Angulocellularia* per Riding and Voronova (1984).
+*Korilophyton inopinatum* (Voronova 1969) Voronova 1976.
    Remarks: = *Epiphyton inopinatum*.
Kundatia composita Korde 1973.
    Bagrad Horizon of Siberia.
Kyzassia elegans Korde 1973.
    Obruchev Horizon of Siberia.
Kyzassia formosa Korde 1973.
    Sanashtygol Horizon of Siberia.
Lenaella longa Korde 1959.
    Atdabanian Horizon of Siberia.
Lenaella reticulata Korde 1959.
    Atdabanian Horizon (upper) of Siberia.
Mackenziephycus medullaris Voronova and Drosdova 1987.
    Mackenzie Mountains, Canada.
+*Majaella* sp. Vologdin and Maslov 1960.
    Yudoma Fm. (lower) of Siberia.
+*Majaella verkhojanica* Vologdin and Maslov 1960.
    Yudoma Fm. (lower) of Siberia.
Nephelostroma lecomtei Dangeard and Doré 1958.
    Carteret, France.
Nicholsonia composita Korde 1973.
    Sanashtygol Horizon of Siberia.
    Remarks: Danielli (1981) said = *Girvanella*.
Nicholsonia glomerata Korde 1973.
    Bagrad Horizon (upper) of Siberia.
    Remarks: Danielli (1981) said = *Girvanella*.
Nicholsonia grandis Korde 1973.
    Sanashtygol Horizon of Siberia.
    Remarks: Danielli (1981) said = *Girvanella*.
+*Nicholsonia involuntans* Korde 1973.
    Enisej Fm. of Siberia.
    Remarks: Danielli (1981) said = *Girvanella*.
Nubecularites polymorphus Maslov 1937.
    Nizhnie Horizon of Siberia.
    Remarks: = *Renalcis polymorphus* per Reitlinger (1959).
Nuia sibirica Maslov 1954.
    Ust'Kut Fm. of Siberia.
Obruchevella sp. Wang, Zhang and Guo 1983.
    Meishucun Fm. (lower) of China.
    Remarks: Calcified?
Obruchevella spp. Cloud et al. 1979.
    Tindir Grp of Alaska.
    Remarks: Calcified? Kline (1977) mentioned the occurrence of these in an abstract, but had never illustrated any specimens.
+*Obruchevella* sp. Kolosov 1977.
    Tinnov Fm. (upper) of Siberia.
+*Obruchevella blandita* Shenfil' 1980.
    Seryj Klyuch of Siberia.
    Remarks: Calcified?
+*Obruchevella condensata* Liu 1984.
    Jiudingshan Fm. (upper) of China.
    Remarks: Calcified?
Obruchevella delicata Reitlinger 1948.
    Kutorgina Fm. of Siberia.
Obruchevella delicata var. *elongata* Reitlinger 1948.
    Kutorgina Fm. (upper) of Siberia.
+*Obruchevella delicata* forma *semeikini* Shenfil' 1983.
    Zabit Fm. of Siberia.
    Remarks: Calcified?
+*Obruchevella ditissimus* Shipitzyn and Yakshin 1981.
    (Yakshin and Luchinina 1981).
    Martyukhinsk Fm. of Siberia.
    Remarks: Calcified? Similar to *Syniella* Reitlinger 1948, which in turn may be a variation of *Obruchevella*.
Obruchevella meishucunensis Song 1984.
    Yuhucun Fm. of China.
    Remarks: Calcified?
+*Obruchevella minor* Zhang 1984.
    Toushantou Fm. of China.
    Remarks: Calcified?
+*Obruchevella parva* Reitlinger 1959.
    Tinnov Fm. of Siberia.
Obruchevella parvissima Song 1984.
    Yuhucun Fm. of China.
    Remarks: Calcified?
+*Obruchevella pusilla* Golovenok and Belova 1983.
    Valyukhta Fm. of Siberia.
    Remarks: Calcified?
Obruchevella sibirica Reitlinger 1959.
    Siberian Platform (N part).
+*Obruchevella tungusica* Pyatiletov 1986.
    Vanavarsk Fm. of Siberia.
Ortonella furcata Garwood 1914.
    Westmorland of the United Kingdom.
Paleonites jacutii (Maslov 1937) Maslov 1956.
    Remarks: = *Epiphyton* (?) *jacutii*.
+*Panomninella laxa* (Voronova 1969) Kolosov 1977.
    Remarks: = *Microcodium laxus*.
+*Panomninella ornata* Kolosov 1966.
    Porokhtakh Fm. (lower) of Siberia.
Panomninella petrosa Drosdova 1980.
    Atdabanian Stage, Kameshki Horizon of western Mongolia.
+*Papillomembrana compta* Spjeldnaes 1963.
    Hedmark Grp. of Norway.
    Remarks: See Spjeldnaes (1967) for better photographs of same specimens.
Parachabakovia dura Korde 1973.
    Sanashtygol Horizon of Siberia.
Potentillina campanulata Korde 1973.
    Sanashtygol Horizon of Siberia.
Potentillina monstrata Korde 1973.
    Bagrad Horizon of Siberia.

Table 23.2.2 (Continued).

Remarks: Kordeh transferred *Chabakovia monstrata* to *Potentillina* but assigned a new holotype. In description, she does not say it is a new species, but does in the figure caption.

Proaulopora [sp.] Nikolaeva et al. 1987.
  Pestrotsvet Fm. of Siberia.
Proaulopora composita Korde 1973.
  Sanashtygol Horizon of Siberia.
Proaulopora crassa Korde 1973.
  Sanashtygol Horizon of Siberia.
Proaulopora extincta Korde 1973.
  Sanashtygol Horizon of Siberia.
Proaulopora flexuosa Korde 1973.
  Bagrad Horizon of Siberia.
Proaulopora longa Korde 1973.
  Bagrad Horizon of Siberia.
Proaulopora microspora Korde 1973.
  Sanashtygol Horizon of Siberia.
Proaulopora sajanica Korde 1960.
  Chesnokovaya Horizon of Siberia.
+*Protoortonella flabellata* (Voronova 1976) Luchinina 1985.
  (Voronova and Luchinina 1985)
  Remarks: = *Rothpletzella flabellata*. Two misspellings: *plabellata* in text and *Habellata* in figure caption.
Protoortonella igarcaensis (Voronova 1976) Luchinina 1985.
  Remarks: = *Rothpletzella igarcaensis*.
+*Protosolenopora distincta* Wang Fuxing 1985.
  Yangjiaping Fm. of China.
Razumovskia ethmoidale Drosdova 1980.
  Atdabanian Stage, Kameshki Horizon of western Mongolia.
Razumovskia fibrosa Drosdova 1980.
  Atdabanian Stage, Kameshki Horizon of western Mongolia.
Razumovskia grandis Korde 1961.
  Atdabanian Horizon of Siberia.
Razumovskia hispida Korde 1973.
  Lenian Stage, Sanashtygol Horizon of Siberia.
Razumovskia kiyanica Korde 1973.
  Bagrad Horizon of Siberia.
Razumovskia lata Drosdova 1980.
  Atdabanian Stage, Kameshki Horizon of western Mongolia.
Razumovskia multispora Korde 1973.
  Bagrad Horizon of Siberia.
Razumovskia seriata Korde 1973.
  Bagrad Horizon of Siberia.
Razumovskya uralica Vologdin 1939.
  Limestone exposure 4-M of Soviet Union.
  Remarks: In original publication, *Razumovskya* was spelled with a "y," but in all subsequent publications, it is spelled with an "i".
Renalcis [sp.] Read and Pfeil 1983.
  Shady Dolomite of the United States.
Renalcis [sp.] James and Kobluk 1978.
  Forteau Fm. of Canada.
Renalcis [sp.] Cherchi and Schroeder 1984.
  Archaeocyath Limestone of Italy.
Renalcis [sp.] Pfeil and Read 1980.
  Shady Dolomite of the United States.
Renalcis [sp.] Rowland 1978.
  Poleta Fm. of the United States.
  Remarks: Due to diagenesis, Rowland classified some renalcids as "probable *Renalcis*."
Renalcis sp. Reitlinger 1959.
  Olekma Fm. of Siberia.
Renalcis [sp.] Selg 1986.
  Nebida Fm. of Italy.
Renalcis [sp.] Kobluk 1981.
  Poleta Fm. of the United States.
Renalcis [sp.] Kobluk 1985.
  Shady Dolomite of the United States.
Renalcis [sp.] Kobluk and James 1979.
  Forteau Fm. (lower) of Canada.
Renalcis [sp.] Nikolaeva et al., 1987.
  Atdabanian Stage of Siberia.
Renalcis chabakoviformis Voronova 1973.
  Atdabanian Stage of Siberia.
  Remarks: = *Chabakovia chabakoviformis* per Drozdova (1980).
Renalcis cibus Vologdin 1939.
  Ural Mountains, Soviet Union.
Renalcis compositus Korde 1973.
  Atdabanian Stage, Bagrad Horizon of Siberia.
Renalcis conchaeformis Titorenko 1974.
  Usol'e Fm. of Siberia.
Renalcis densum Titorenko 1974.
  Usol'e Fm. of Siberia.
Renalcis elegans Titorenko 1974.
  Usol'e Fm. of Siberia.
Renalcis erbinatus Korde 1973.
  Sanashtygol Horizon of Siberia.
Renalcis gelatinosus Korde 1961.
  Atdabanian Horizon (upper) of Siberia.
  Remarks: = *Tarthinia* per Riding and Voronova (1984).
Renalcis granosus Vologdin 1932.
  Altai Sayan, Siberia.
Renalcis granosus var. *feguratus* Titorenko 1966. (Lysova et al. 1966)
  Siberia.
  Remarks: Figured but not described.
Renalcis granosus var. *plenus* Titorenko 1966. (Lysova et al. 1966)
  Beloj Fm. of Siberia.
  Remarks: Figured but not described.
Renalcis granulatus Korde 1973.
  Atdabanian Stage, Bagrad Horizon of Siberia.
Renalcis jacuticus Korde 1955.
  Tommotian Stage, Kenyade Horizon of Siberia.
  Remarks: Not figured in Kordeh (1960), but she refers to a holotype number (PIN 984/456). In Kordeh (1961), where species is figured, holotype designation is different (PIN 1298/169).
Renalcis aff. *jacuticus* Reitlinger 1959.
  Pestrotsvet Fm. (upper) of Siberia.
Renalcis ex gr. *jacuticus* Reitlinger 1959.
  Pestrotsvet Fm. (upper) of Siberia.
Renalcis lenaicum Titorenko 1974.
  Usol'e Fm. of Siberia.
Renalcis levis Vologdin 1940.
  Sehr' Range, western Mongolia.
Renalcis minutus Voronova and Drosdova 1983.
  Lenian Stage of western Mongolia.
Renalcis nodularis Voronova and Drosdova 1987.
  Mackenzie Mountains, Canada.
Renalcis novum Voronova 1976.
  Lenian Stage of Siberia.
Renalcis pectunculus Korde 1961.
  Tolbachan Horizon of Siberia.
+*Renalcis* ? *polymorphus* (Maslov 1937) Reitlinger 1959.
  Remarks: = *Nubecularites polymorphus*.

Table 23.2.2 (Continued).

Renalcis ? aff. *polymorphus* Reitlinger 1959.
Lena Fm. of Siberia.
Renalcis rotundus (Drosdova 1975) Voronova 1982.
Remarks: = *Tarthinia rotunda*.
Renalcis seriata Korde 1955.
Kurorgina Horizon of Siberia.
Renalcis tuberculatus Korde 1973.
Bagrad Horizon of Siberia.
Rothpletzella flabellata Voronova 1976.
Tommotian Stage of Siberia.
Remarks: = *Hedstroemia flabellata* per Riding and Voronova (1984). = *Protoortonella flabellata* per Luchinina (1985).
Rothpletzella igarcaensis Voronova 1976.
Lenian Stage of Siberia.
Remarks: = *Hedstroemia igarcaensis* per Riding and Voronova (1984). = *Protoortonella igarcaensis* per Luchinina (1985).
Sajania fasciculata Korde 1973.
Bagrad Horizon of Siberia.
Sajania frondosa Vologdin 1962.
Torgashino Fm. of Siberia.
Sajania pennata Drosdova 1980.
Atdabanian Stage, Kameshki Horizon of western Mongolia.
Serligia fragilis Korde 1973.
Bagrad Horizon (upper) of Siberia.
Serligia gracilis Korde 1973.
Sanashtygol Horizon of Siberia.
Serligia cf. *gracilis* Kobluk and James 1979.
Forteau Fm. (lower) of Canada.
Sinocapsa honanica Vologdin 1958.
Henan Province, China.
+*Siphonophycus kestron* Schopf 1968.
Bitter Springs Fm. of Australia.
+*Siphonophycus transvaalense* Beukes et al. 1987. (Klein et al. 1987)
Gamohaan Fm. of South Africa.
Remarks: Encrusted with minute randomly distributed carbonate needles or outlined on their inner and outer surfaces by minute hematitic granules.
Solenopora ? Priestly and David 1912.
Limestone breccia of Antarctica.
Solenopora sp. Riding and Voronova 1985.
Botomian of western Mongolia.
Solenopora compacta (Billings 1865) Rothpletz 1908.
Trenton Limestone of Canada.
Remarks: = *Stromatopora compacta*. Can't evaluate transfer; could not find an illustration of the original in Billings.
Sporinula palmata Korde 1973.
Bagrad Horizon of Siberia.
Subtiflora delicata Maslov 1956.
Khanrkhan Mtn., Tuva, Siberia.
Subtiflora mazasia Stepanova 1974.
Obruchev Horizon of Siberia.

+*Suvorovella aldanica* Vologdin and Maslov 1960.
Yudoma Fm. (lower) of Siberia.
Syniella invenusta Reitlinger 1948.
Kutorgina Fm. of Siberia.
Taninia tomentosa Korde 1973.
Bagrad Horizon of Siberia.
Tannuolaia fonini Vologdin 1967.
Shiveligsk Fm. (middle) of Siberia.
Tersia sp. Vologdin 1932.
Limestone of Altai Sayan, Siberia.
Remarks: An archaeocyath.
Tersia filiforma Vologdin 1931.
Kuznetsk Alatau, Siberia.
Remarks: An archaeocyath.
+*Timanella gigas* Vologdin 1966.
Timan (W part), Soviet Union.
Tomentula interrupta Korde 1973.
Bagrad Horizon of Siberia.
Tomentula villosa Korde 1973.
Bagrad Horizon of Siberia.
Tubomorphophyton benignum (Korde 1960) Korde 1973.
Remarks: = *Epiphyton benignum*.
Tubomorphophyton botomense (Korde 1955) Korde 1973.
Remarks: = *Epiphyton botomense*.
Tubomorphophyton cristatum (Korde 1961) Korde 1973.
Remarks: = *Epiphyton cristatum*.
Tubomorphophyton latum Drosdova 1980.
Atdabanian Stage, Kameshki Horizon of western Mongolia.
Tubomorphophyton limpidum Drosdova 1980.
Lenian Stage, Sanashtygol Horizon of western Mongolia.
Tubomorphophyton nubilum (Korde 1961) Korde 1973.
Remarks: = *Epiphyton nubilum*.
Tubomorphophyton pseudofruticosum Drosdova 1980.
Tommotian Stage of western Mongolia.
Unbellula minuta Korde 1973.
Bagrad Horizon of Siberia.
Vologdinella fragile Korde 1957.
Atdabanian Horizon of Siberia.
Remarks: = *Proaulopora glabra*.
Vologdinella grandis Korde 1973.
Sanashtygol Horizon of Siberia.
Vologdinella pulchra Korde 1973.
Obruchev Horizon of Siberia.
Vologdinia verticillata Korde 1973.
Bagrad Horizon (upper) of Siberia.
Wetheredella [sp.] Kobluk and James 1979.
Forteau Fm. (lower) of Canada.
Remarks: Authors describe as foraminifer, but Loeblich and Tappan refer to the calcareous algae.
Yentaiia liaoyangensis Vologdin 1958.
Shijiac = Shihchiao Fm. of China.

Table 23.2.3 lists reported species of Proterozoic and Early Cambrian calcareous algae that are not evaluated here. These reports have not been interpreted due to (*i*) an inability to obtain the original publication; (*ii*) poor quality of photocopies of the original publication; (*iii*) poor quality of originally published photographs of the reported taxa; or (*iv*) an absence of published photographs (in older publications in particular) of the reported taxa. See the introduction to Table 23.2.2, above, for explanation of entries.

Table 23.2.3 *Species of Proterozoic and/or Early Cambrian calcareous algae not here evaluated.*

+(algae fossil) Zhao et al. 1984. (Zhou et al. 1984)
  Doushantuo Fm. of China.
+(dasycladaceous algae) Salujha et al. 1972.
  Bhima Series of India.
  Remarks: Can't evaluate from photograph. From macerations. Thus, don't know if calcified.
(small algal? spheres) Kobluk 1985.
  Shady Dolomite of the United States.
  Remarks: Difficult to evaluate simple sphere. Kobluk suggests possible affinity with *Pseudoanthus cambricum* Korde 1973 or *Mucilina fossilis* Korde 1973, but his specimens are smaller, thicker walled, and do not occur in clusters.
+ *Actinophycus obrutschevi* Korde 1954.
  Krasnoyarsk region, Siberia.
  Remarks: Kordeh originally described as stromatolite, but many Chinese workers classify as rhodophyte.
Angulocellularia anisotoma Vologdin 1962.
  Yangud Fm. (lower) of Siberia.
Angusticellularia anisotoma Vologdin 1962.
  Yangud Fm. (lower) of Siberia.
  Remarks: Don't know if *Angusticellularia anisotoma* or *Angulocellularia anisotoma* Vologdin (1962b) has priority.
Anomas ovisimilis Vologdin 1932.
  Limestone of Altai Sayan, Siberia.
+ *Antiquus cusarandicus* Butin 1959.
  Karelia (S part), Soviet Union.
+ *Azyrtalia fasciculata* Vologdin and Drosdova 1969.
  Martyukhinsk Fm. of Siberia.
+ *Azyrtalia globosa* Kolosov 1975.
  Chenchinsk Fm. of Siberia.
+ *Azyrtalia zonulata* Vologdin and Drosdova 1969.
  Martyukhinsk Fm. of Siberia.
+ *Bestjachica rara* Kolosov 1970.
  Khopychsk Fm. of Siberia.
Botominella ? sp. Drobkova 1979.
  Usol'e Fm. (lower) of Siberia.
+ *Burchalaella parvula* Kolosov 1970.
  Tsipanda Fm. (low) of Siberia.
+ *Burchalaella pulchella* Kolosov 1970.
  Tsipanda Fm. (low) of Siberia.
+ *Cambricodium capilloodes* Jiang 1982.
  Yunnan (E part), China.
  Remarks: Also known as *C. capilloides*.
Corbularia sp. Brasier 1977.
  Eilean Dubh Fm. of the United Kingdom.
Edelsteinia cylindrica Vologdin 1934.
  W Sayan, Siberia.
  Remarks: Unable to locate reference.
Edelsteinia mongolica Vologdin 1940.
  Burgastaj River, Mongolia.
+ *Eoepiphyton jalgamicum* Butin 1959.
  Karelia (S part), Soviet Union.
+ *Eoepiphyton jatulicum* Butin 1959.
  Karelia (S part), Soviet Union.
+ *Epiphyton* [sp.] Shukla 1984.
  Deoban Fm. of India.
+ *Epiphyton* [sp.] Cahen et al. 1946.
  Bushimale System of Zaire.
  Remarks: Looks more like *Frutexites*. There is no description of the fossil, so don't know if calcareous or ferruginous.
Epiphyton [sp.] Edhorn 1977.
  Bonavista Fm. of Canada.
+ *Epiphyton baicalicum* Vologdin 1962.
  Uluntuj Fm. (upper) of Siberia.
+ *Epiphyton buguldeicum* Vologdin 19??.
  E Sayan, Siberia.
  Remarks: Mentioned in Vologdin (1960), but not able to obtain publication in which it was described and figured.
Epiphyton condensum Vologdin 1962.
  Yangud Fm. (lower) of Siberia.
+ *Epiphyton corporatum* 19??.
  E Sayan, Siberia.
  Remarks: Mentioned as new species in Gudymovich (1967), but not able to obtain publication in which it was described and figured.
Epiphyton fasciculatum Chapman 1916.
  Limestone, 85° S latitude of Antarctica.
  Remarks: = *Epiphytonoides fasciculatus* per Kordeh (1973).
Epiphyton fibratus Krasnopeeva 1937.
  Khakassia, Siberia.
Epiphyton aff. *fruticosum* Vologdin 1962.
  Yangud Fm. (middle of lower) of Siberia.
Epiphyton cf. *inopinatus* Drobkova 1979.
  Usol'e Fm. (lower) of Siberia.
Epiphyton neodensum Vologdin 1962.
  Yangud Fm. (middle of lower) of Siberia.
Erbina aristata Korde 1973.
  Kuznetsk Alatau, Siberia.
Filaria mira Korde 1973.
  Bagrad Horizon of Siberia.
Foliaceria polymorpha Vologdin 1962.
  Yangud Fm. (lower) of Siberia.
Girvanella [sp.] Edhorn and Anderson 1977.
  Bonavista Fm. of Canada.
  Remarks: Can't evaulate from photograph; possibly cement fabric.
Girvanella [sp.] Dangeard 1957.
  Manche, Carteret, France.
+ *Girvanella* [sp.] Edhorn 1978.
  Mwashya Grp. of Zaire.
Girvanella antiqua Dawson 1896.
  Sillery Fm. of Canada.
  Remarks: No photograph.
+ *Girvanella antiquoformis* Vologdin 1962.
  Nizhnej Tungusk River of Siberia.
Girvanella ? *ocellatus* (Seely 1885) Walcott 1890.
  Nevada, the United States.
  Remarks: Can't evaluate from drawing of oncolite. Seely (1885) originally described as a calcareous sponge called *Strephochetus ocellatus* from the Ordovician (Chazy) of Vermont.
+ *Girvanella recta* 19??.
  Anastas'ino Fm. of Siberia.
  Remarks: Mentioned as new species in Gudymovich (1967), but not able to obtain publication in which it was described and figured.
+ *Girvanella roberti* Hacquaert 1943.
  Mine Series of Zaire.
Globuloella botomensis Korde 1958.
  Atdabanian Horizon (upper) of Siberia.
Globuloella incompacta Korde 1958.
  Atdabanian Horizon (upper) of Siberia.
+ *Globuloella notabila* Kolosov 1975.
  Porokhtakh Fm. of Siberia.

Table 23.2.3 (*Continued*).

Hedstroemia borealis Luchinina 19??.
  Kessyusinsk Fm. of Siberia.
  Remarks: Have not seen original publication.
+*Jatuliana furcata* Vologdin 1965.
  Karelia, Olenij Island, Soviet Union.
+*Kareliana zonata* Korde 1965. (Vologdin and Kordeh 1965)
  Karelia, Olenij Island, Soviet Union.
+*Korilophyton* sp. Kolosov 1977.
  Manykajkaya Fm. of Siberia.
+*Microcodium laxus* Voronova 1969.
  Nemakit Daldyn Horizon of Siberia.
  Remarks: = *Panomninella laxa* per Kolosov (1977).
+*Palaeogirvanella erbiensis* Krasnopeeva 1937.
  Khakassia, Siberia.
+*Panomninella copiosa* Kolosov 1975.
  Tinnov Fm. (lower) of Siberia.
+*Panomninella floribunda* Kolosov 1975.
  Tokko Fm. of Siberia.
+*Panomninella lecta* Kolosov 1975.
  Chenchinsk Fm. (lower) of Siberia.
+*Panomninella rotunda* Kolosov 1975.
  Chenchinsk Fm. (lower) of Siberia.
+*Panomninella silicea* Kolosov 1984.
  Yukandinsk Fm. of Siberia.
+*Proaulopora* ? sp. Yin 1980.
  Sichuan, China.
+*Proaulopora* sp. 2 Kolosov 1975.
  Tinnov Fm. (lower) of Siberia.
+*Proaulopora* sp. 3 Kolosov 1975.
  Porokhtakh Fm. of Siberia.
+*Proaulopora* sp. 1 Kolosov 1975.
  Tinnov Fm. (middle) of Siberia.
Proaulopora glabra Krasnopeeva 1937.
  Kuznetsk Alatau, Siberia.
  Remarks: = *Amganella glabra* per Reitlinger (1959).
Proaulopora rarissima Vologdin 1934.
  Remarks: Never found original description. Kordeh (1960, p. 258) refers to Vologdin (1933) as original description; Krasnopeeva (1937) refers to it as Vologdin (1934).
+*Protorivularia onega* Butin 1959.
  Karelia (S part), Soviet Union.
Protuberantia vesicularis Vologdin 1962.
  Yangud Fm. (lower) of Siberia.
+*Razumovskia* ? sp. Xing and Liu 1978.
  Toushantou Fm. of China.
Razumovskia sp. Drobkova 1979.
  Usol'e Fm. (lower) of Siberia.
Razumovskia alta Vologdin 19??.
  Remarks: Never found original reference. See remarks under genus.

Razumovskia cf. *alta* Krasnopeeva 1937.
  Chesnokovaya Mountains of Siberia.
+*Razumovskia gracilis* Xing and Liu 1978.
  Toushantou Fm. of China.
+*Renalcis* ? Yin 1980.
  Sichuan, China.
Renalcis sp. Kolosov 1984.
  Karakatty Fm. of Soviet Union.
+*Renalcis* [sp.] Shukla 1984.
  Deoban Fm. of India.
Renalcis sp. Brasier 1977.
  Eilean Dubh Fm. of the United Kingdom.
Renalcis halisiteformis Krasnopeeva 1937.
  Chesnokovaya Mountains, Siberia.
Renalcis pseudoradiatus Titorenko 1966. (Lysova et al. 1966)
  Bulajskaya Fm. of Siberia.
Renalcis simplex Drobkova 1981. (Akul'cheva et al. 1981)
  Moty Fm. of Siberia.
Renalcis textularites Titorenko 1966. (Lysova et al. 1966)
  Beloj Fm. of Siberia.
+*Sajania* [sp.] Shukla 1984.
  Deoban Fm. of India.
Serligia sp. Kolosov 1977.
  Yuedej Fm. of Siberia.
+*Solenopora* sp. Chen and Liu 1986.
  phosphorites of China.
+*Solenopora* [sp.] Zhao 1986.
  Doushantuo Fm. of China.
  Remarks: Can't evaluate from photograph—has some resemblance to solenoporid structure and cell size is about right.
Solenopora tjanshanica Vologdin 1955.
  Tyan-shan (N part), Soviet Union.
Tarthinia rotunda Drosdova 1975.
  Lenian Stage, Kameshki Horizon of western Mongolia.
  Remarks: = *Renalcis rotundus* per Voronova (1982).
Tarthinia zachirica Drosdova 1975.
  Lenian Stage, Sanashtygol Horizon of western Mongolia.
+*Templuma sinica* Zhang 1979.
  Wumishan Fm. of China.
Tubercularia latiuscula Vologdin 1962.
  Yangud Fm. (middle of lower) of Siberia.
Tubophyllum victori Krasnopeeva 1955.
  Kuznetsk Alatau, Siberia.
  Remarks: = *Amganella glabra* per Reitlinger (1959).
+*Uranovia granosa* Korde 1958.
  Kuznetsk Alatau, Siberia.
+*Uranovia multa* Korde 1958.
  Kuznetsk Alatau, Siberia.
+*Utchurella explicata* Kolosov 1970.
  Omakhtinsk Fm. of Siberia.

Table 23.2.4 lists species of Proterozoic and Early Cambrian calcareous algae that are interpreted here to be dubiofossils. See introduction to Table 23.2.2, above, for explanation of entries.

Table 23.2.5 lists species of Proterozoic and Early Cambrian calcareous algae interpreted here to be pseudofossils. See introduction to Table 23.2.2, above, for explanation of entries.

Table 23.2.4. *Described species of Proterozoic and/or Early Cambrian calcareous algae interpreted to be dubiofossils.*

+(Dasycladacea thallus) Rao and Mohan 1954.
  Dogra Slates of India.
  Remarks: In phyllitic rocks.
+(algue microscopique) Cahen et al. 1946.
  Schisto-calcarie Grp of Zaire.
+(dasyclad) Rao 1943.
  Cuddupah Limestone (lower) of India.
  Remarks: An oölith? Refer to Rao (1944) in which he makes less of a case for a dasycladacean affinity and says only possible alga.
+(solenoporid precursor) Cassedanne 1965.
  Bambui Series of Brazil.
  Remarks: Some photos look like crossbedding. Lithology is marble.
Acanthina multiformis Korde 1973.
  Atdabanian Stage, Bagrad Horizon of Siberia.
  Remarks: Calcareous? Other species are not included in the database.
Azyrtalia telmenica Drosdova 1980.
  Atdabanian Stage, Kameshki Horizon of western Mongolia.
Corbularia conglutinata Vologdin 1962.
  Lenian Stage, Ketema Horizon of Siberia.
Cornutula kaltatica Korde 1973.
  Bagrad Horizon of Siberia.
Cornutula kyzassica Korde 1973.
  Sanashtygol Horizon of Siberia.
Dasycirriphycus fruticulosus Vologdin 1962.
  Burajskij Horizon of Siberia.
  Remarks: Calcareous? Illustration similar to some *Renalcis* and *Tarthinia*.
Epiphyton sp. Sedlak 1980.
  Quartzites of Poland.
  Remarks: Not algal. May be a metaphyte or pseudofossil—can't evaluate from photograph.
Epiphyton [sp.] Edhorn and Anderson 1977.
  Bonavista Fm. of Canada.
  Remarks: Stromatolite microstructure?
Erbina sp. Sedlak 1980.
  Quartzites of Poland.
  Remarks: Not algal. May be a metaphyte or pseudofossil—can't evaluate from photograph.
Flabellina multiformis Korde 1986.
  Dzhide Fm. of Siberia.
+*Foninia fasciculata* Korde 1973.
  Enisej Fm. of Siberia.
Globuloella pellucida Drosdova 1980.
  Tommotian Stage, Bazaikhe Horizon of western Mongolia.
+*Kareliana ukrainica* Snezhko 1986.
  Gleevatsk Fm. of Soviet Union.
Marenita bayankolica Korde 1973.
  Bagrad Horizon of Siberia.
+*Marenita kundatica* Korde 1973.
  Nikitinsk Horizon of Siberia.
+*Murandavia amurica* Vologdin 1965.
  Far East of Siberia.
+*Murandavia granulosa* Vologdin and Drosdova 1969.
  Martyukhinsk Fm. of Siberia.
+*Murandavia magna* Vologdin 1967.
  Shchekur'insk Fm. of Soviet Union.
+*Murandavia marginata* Vologdin 1967.
  Murandavsk Fm. of Siberia.
+*Murandavia rotunda* Vologdin 1965.
  Far East of Siberia.
+*Murandavia* aff. *rotunda* Vologdin 1969.
  Martyukhinsk Fm. of Siberia.
Palaeomicrocystis cambrica Korde 1955.
  Ketema Horizon of Siberia.
Pinnulina cambrica Korde 1986.
  Sanashtygol Horizon of Siberia.
+*Pustularia taeniata* Vologdin 1955.
  Dzhur Fm. (upper) of Siberia.
  Remarks: The reconstruction looks like an alga, but no distinctive features are visible in the photograph.
+*Pustularia vetusta* Vologdin 1962.
  Aladin Fm., Enisej System of Siberia.
+*Shujana praefulgida* Snezhko 1986.
  Gleevatsk Fm. of Soviet Union.
+*Shujana shulgini* Korde 1979.
  Karelia, Soviet Union.
+*Solenopora* sp. Vologdin 1962.
  Uluntuj Fm. (upper) of Siberia.
  Remarks: Photographs are not convincing—may be radially fibrous cement.

Table 23.2.5. *Described species of Proterozoic and/or Early Cambrian calcareous algae interpreted to be pseudofossils.*

+(dasyclad alga?) Pantin 1955.
  Dalradian hornfels of the United Kingdom.
  Remarks: Inorganic colloidal accretion (Downie 1971).
+*Actinophycus divaricatus* Yin 1980.
  Sichuan, China.
  Remarks: Diagenetic fabric?
+*Actinophycus liangshanensis* Tsao and Zhao 1978. (Cao and Zhao 1978)
  Dengying Fm. of China.
  Remarks: Diagenetic fabric?
+*Actinophycus nanjiangensis* Tsao and Zhao 1974. (Cao and Zhao 1974)
  Dengying? Fm. of China.
  Remarks: Diagenetic fabric?
+*Actinophycus quadricella* Tsao and Zhao 1974. (Cao and Zhao 1974)
  Dengying Fm. of China.
  Remarks: Diagenetic fabric.
+*Acus concentricus* Tsao and Zhao 1974. (Cao and Zhao 1974)
  Dengying? Fm. of China.

Table 23.2.5 (Continued).

Remarks: Diagenetic fabric.
+ *Acus fasciatus* Tsao and Zhao 1974. (Cao and Zhao 1974)
    Dengying? Fm. of China.
    Remarks: Diagenetic fabric.
+ *Acus muricatus* Tsao and Zhao 1974. (Cao and Zhao 1974)
    Dengying? Fm. of China.
    Remarks: Diagenetic fabric.
+ *Acus platypluteus* Tsao and Zhao 1974. (Cao and Zhao 1974)
    Dengying? Fm. of China.
    Remarks: Diagenetic fabric.
+ *Manicosiphonia bambusa* Tsao and Zhao 1978. (Cao and Zhao 1978)
    Dengying Fm. of China.
    Remarks: Diagenetic fabric.
+ *Manicosiphonia conica* Tsao and Zhao 1978. (Cao and Zhao 1978)
    Dengying Fm. of China.
    Remarks: Diagenetic fabric.
+ *Manicosiphonia conserta* Tsao and Zhao 1978. (Cao and Zhao 1978)
    Dengying Fm. of China.
    Remarks: Diagenetic fabric. Also known as *M. concerta*.
+ *Manicosiphonia fissilis* Tsao and Zhao 1978. (Cao and Zhao 1978)
    Dengying Fm. of China.
    Remarks: Diagenetic fabric.
+ *Manicosiphonia furcata* Tsao and Zhao 1978. (Cao and Zhao 1978)
    Dengying Fm. of China.
    Remarks: Diagenetic fabric.
+ *Manicosiphonia hanyuanensis* Tsao and Zhao 1978. (Cao and Zhao 1978)
    Dengying Fm. of China.
    Remarks: Diagenetic fabric.
Mawsonella wooltanensis Chapman 1927.
    Above Proterozoic tillites of Australia.
    Remarks: Intraformational breccia (Glaessner 1979).
+ *Multisiphonia hemicirculis* Tsao and Liang 1974. (Cao and Liang 1974)
    China
    Remarks: Diagenetic fabric.
+ *Multisiphonia nanshanensis* Tsao and Liang 1974. (Cao and Liang 1974)
    China.
    Remarks: Diagenetic fabric.
+ *Nanamanicosiphonia lepradosa* Tsao and Zhao 1978. (Cao and Zhao 1978)
    Dengying Fm. of China.
    Remarks: Diagenetic fabric.
+ *Nanamanicosiphonia liangshanensis* Tsao and Zhao 1978. (Cao and Zhao 1978)
    Dengying Fm. of China.
    Remarks: Diagenetic fabric.
+ *Nanamanicosiphonia minuta* Tsao and Zhao 1978. (Cao and Zhao 1978)
    Dengying Fm. of China.
    Remarks: Diagenetic fabric.
+ *Nanamanicosiphonia ninglangensis* Tsao and Zhao 1978. (Cao and Zhao 1978)
    Dengying Fm. of China.
    Remarks: Diagenetic fabric.
+ *Nanamanicosiphonia yunnanensis* Tsao and Zhao 1978. (Cao and Zhao 1978)
    Dengying Fm. of China.
    Remarks: Diagenetic fabric.
+ *Parasolenopora irregularis* Tsao and Zhao 1974. (Cao and Zhao 1974)
    Dengying? Fm. of China.
    Remarks: Diagenetic fabric.
+ *Parasolenopora subradiata* Tsao and Zhao 1974. (Cao and Zhao 1974)
    Dengying? Fm. of China.
    Remarks: Diagenetic fabric.
+ *Phacelofimbria emeishanensis* Tsao and Zhao 1974. (Cao and Zhao 1974)
    Dengying? Fm. of China.
    Remarks: Diagenetic fabric.
+ *Phacelofimbria minor* Tsao and Liang 1974. (Cao and Liang 1974)
    China.
    Remarks: Diagenetic fabric.
+ *Praesolenopora fascicularis* Tsao and Zhao 1974. (Cao and Zhao 1974)
    Dengying? Fm. of China.
    Remarks: Diagenetic fabric.
+ *Praesolenopora flabella* Tsao and Zhao 1974. (Cao and Zhao 1974)
    Dengying? Fm. of China.
    Remarks: Diagenetic fabric.
+ *Praesolenopora formosa* Tsao and Liang 1974. (Cao and Liang 1974)
    China.
    Remarks: Diagenetic fabric.
+ *Praesolenopora furcata* Tsao and Liang 1974. (Cao and Liang 1974)
    China.
    Remarks: Diagenetic fabric.
+ *Praesolenopora hanyuanensis* Tsao and zhao 1978. (Cao and Zhao 1978)
    Dengying Fm. of China.
    Remarks: Diagenetic fabric.
+ *Praesolenopora liaoningensis* Tsao and Liang 1974. (Cao and Liang 1974)
    China.
    Remarks: Diagenetic fabric.
+ *Praesolenopora magniflabella* Tsao and Zhao 1974. (Cao and Zhao 1974)
    Dengying? Fm. of China.
    Remarks: Diagenetic fabric.
+ *Pseudoacus renalis* Tsao and Zhao 1974. (Cao and Zhao 1974)
    Dengying? Fm. of China.
    Remarks: Diagenetic fabric.
+ *Siphonia* sp. Yin 1980.
    Sichuan China.
    Remarks: Diagenetic fabric.
+ *Siphonia columella* Tsao and Zhao 1974. (Cao and Zhao 1974)
    Dengying? Fm. of China.
    Remarks: Diagenetic fabric.
+ *Siphonia decussa* Tsao and Zhao 1974. (Cao and Zhao 1974)
    Dengying? Fm. of China.
    Remarks: Diagenetic fabric.
+ *Siphonia florisglobosa* Tsao and Zhao 1974. (Cao and Zhao 1974)
    Dengying? Fm. of China.
    Remarks: Diagenetic fabric.
+ *Siphonia herbacea* Tsao and Zhao 1974. (Cao and Zhao 1974)

Table 23.2.5 (Continued).

Dengying? Fm. of China.
Remarks: Diagenetic fabric.
+*Siphonia songlinensis* Tsao and Zhao 1974. (Cao and Zhao 1974)
Dengying? Fm. of China.
Remarks: Diagenetic fabric.
+*Varicamanicosiphonia quadricella* (Tsao and Zhao 1974) Tsao and Zhao 1978.
Remarks: = *Actinophycus quadricella*. Diagenetic fabric.
+*Varicamanicosiphonia segmenta* Tsao and Zhao 1978. (Cao and Zhao 1978)
Dengying Fm. of China.
Zaganolomia buralica Drosdova 1980.
Atdabanian Stage, Kameshki Horizon of western Mongolia.
Remarks: Cement fabric. See remarks under genus.

23.3

Proterozoic Fossils of Soft-Bodied Metazoans (Ediacara Faunas)

BRUCE N. RUNNEGAR

In Tables 23.3.1 and 23.3.2 are listed structures described or reported as Proterozoic (Vendian) metazoan body fossils. Interpretations based on these tabulations are presented in Section 7.5.

Table 23.3.1 lists, alphabetically by genus and species (or informal name), the named taxa and some unnamed structures that comprise the Ediacara Fauna. Type species of genera are identified by an asterisk. Each "good" taxon/form is given a rating of one, two, or three stars which correspond, respectively, to its low, medium, or high quality as judged from the number of specimens known, the morphological complexity of the fossil, and its reported geographic distribution. As indicated, other listed "taxa" are interpreted to be junior synonyms, fossils *incertae sedis* (with unique specimens denoted by an asterisk), or pseudofossils. Note that many "genera" are monospecific and that many "species" are based on unique specimens.

Table 23.3.2 lists, alphabetically by genus and species, fossils and pseudofossils that have been interpreted as body fossils of Proterozoic metazoans other than the Ediacara Fauna. The tabulation includes annulated, organic-walled tubes; calcareous tubes; and the enigmatic "denticulate" fossil *Redkinia*. Type species of genera are identified by an asterisk. Each "good" taxon/form is given a rating of one, two, or three stars which correspond, respectively, to its low, medium, or high quality as judged from the number of specimens known, the morphological complexity of the fossil, and its geographic distribution. As indicated, other "taxa" are interpreted to be junior synonyms, fossils *incertae sedis*, or pseudofossils.

Table 23.3.1. *Formally described species (and figured but not formally described structures) of Vendian soft-bodied metazoans that comprise the Ediacara faunas.*

Taxon	Author	Date	Locality	Lat	Lon
Affinovendia arctosa	Sokolov	1984	Olenyok Uplift	71.2°N	113.6°E
Aksumbensis aksumbensis	Borovikov and Kraskov	1977	Kazakhstan		
*Albumares brunsae**	Fedonkin	1976	Onega Peninsula	65.7°N	39.9°E
*Anabylia improvisa**	Vodanyuk	1989	Olenyok Uplift	71.2°N	113.6°E
*Anafesta stankovskii**	Fedonkin	1984	Winter Shore	65.5°N	39.7°E
*Arborea arborea**	(Glaessner)	1959	Ediacara	30.8°S	138.2°E
*Archaeichnium haughtoni**	Glaessner	1963	Groendoorn	27.5°S	18.3°E
*Archangelia valdaica**	Fedonkin	1979	Summer Shore	65.5°N	39.7°E
*Arkarua adami**	Gehling	1987	Chace Range	31.7°S	138.7°E
*Armillifera parva**	Fedonkin	1980	Winter Shore	65.5°N	39.7°E
*Arumberia banksi**	Glaessner and Walter	1975	Alice Springs	23.7°S	133.7°E
Aspidella costata	Vodanyuk	1989	Olenyok Uplift	71.2°N	113.6°E
Aspidella hatyspytia	Vodanyuk	1989	Olenyok Uplift	71.2°N	113.6°E
*Aspidella terranovica**	Billings	1872	Newfoundland	47.5°N	52.5°W
*Ausia fenestrata**	Hahn and Pflug	1985	Plateau	26.6°S	16.5°E
*Baikalina sessilis**	Sokolov	1972	Lake Baikal	52.5°N	105.4°E
*Beltanella gilesi**	Sprigg	1947	Ediacara	30.8°S	138.2°E
Beltanella podolica	Zaika-Novatsky	1965	Dniester River	48.4°N	26.1°E
*Beltanelliformis brunsae**	Menner	1974	Loino Borehole	59.5°N	52.3°E
*Beltanelloides sorichevae**	Sokolov	1972	Loino Borehole	59.5°N	52.3°E
*Bomakiella kelleri**	Fedonkin	1985	Syuzma River	64.7°N	39.7°E
*Bonata septata**	Fedonkin	1980	Winter Shore	65.5°N	39.7°E
*Brachina delicata**	Wade	1972	Brachina Gorge	31.3°S	138.6°E
*Bronicella podolica**	(Zaika-Novatsky)	1965	Dniester River	48.4°N	26.1°E
Brooksella canyonensis	Bassler	1941	Arizona	36.3°N	111.8°W
*Buchholzbrunnichnus kroeneri**	Germs	1973	Buchholzbrunn	26.7°S	17.1°E
*Bunyerichnus dalgarnoi**	Glaessner	1969	Bunyeroo Gorge	31.4°S	138.5°E
Bush-like form	Misra	1969	Newfoundland	46.7°N	53.0°W
Charnia grandis	(Glaessner and Wade)	1966	Ediacara	30.8°S	138.2°E
*Charnia masoni**	Ford	1958	Charnwood Forest	52.7°N	1.1°W
Charnia sibirica	(Sokolov)	1972	Olenyok Uplift	71.2°N	113.6°E
Charniodiscus arboreus	(Glaessner)	1959	Ediacara	30.8°S	138.2°E
*Charniodiscus concentricus**	Ford	1958	Charnwood Forest	52.7°N	1.1°W
Charniodiscus longa	(Glaessner and Wade)	1966	Ediacara	30.8°S	138.2°E
Charniodiscus oppositus	Jenkins and Gehling	1978	Ediacara	30.8°S	138.2°E
Charniodiscus planus	Sokolov	1972	Dniester River	48.4°N	26.1°E
*Chondroplon bilobatum**	Wade	1971	Ediacara	30.8°S	138.2°E
Circular structures	Plummer	1980	Flinders Ranges	31.6°S	138.7°E
*Conomedusites lobatus**	Glaessner and Wade	1966	Ediacara	30.8°S	138.2°E
*Cyclomedusa davidi**	Sprigg	1947	Ediacara	30.8°S	138.2°E
Cyclomedusa delicata	Fedonkin	1981	Syuzma River	64.7°N	39.7°E
Cyclomedus gigantea	Sprigg	1947	Ediacara	30.8°S	138.2°E
Cyclomedusa gracilis	Xing and Liu	1979	Liaoning	40°N	122°E
Cyclomedusa minus	Xing and Liu	1979	Liaoning	40°N	122°E
Cyclomedusa minuta	Fedonkin	1979	Syuzma River	64.7°N	39.7°E
Cyclomedusa plana	Glaessner and Wade	1966	Ediacara	30.8°S	138.2°E
Cyclomedusa radiata	Sprigg	1949	Ediacara	30.8°S	138.2°E
Cyclomedusa serebrina	Palij	1969	Dniester River	48.4°N	26.1°E
Cyclomedusa simplicus	Xing and Liu	1979	Liaoning	40°N	122°E
Dickinsonia brachina	Wade	1972	Brachina Gorge	31.3°S	138.6°E
*Dickinsonia costata**	Sprigg	1947	Ediacara	30.8°S	138.2°E
Dickinsonia elongata	Glaessner and Wade	1966	Ediacara	30.8°S	138.2°E
Dickinsonia lissa	Wade	1972	Ediacara	30.8°S	138.2°E
Dickinsonia minima	Sprigg	1949	Ediacara	30.8°S	138.2°E
Dickinsonia spriggi	Harrington and Moore	1955	Ediacara	30.8°S	138.2°E
Dickinsonia tenuis	Glaessner and Wade	1966	Ediacara	30.8°S	138.2°E
Disk-like structure	Webby	1970	Broken Hill	31.2°S	141.8°E
*Ediacaria flindersi**	Sprigg	1947	Ediacara	30.8°S	138.2°E
*Elasenia aseevae**	Fedonkin	1983	Dniester River	48.4°N	26.1°E
*Eoporpita medusa**	Wade	1972	Ediacara	30.8°S	138.2°E
*Erniaster apertus**	Pflug	1972	Aar or Plateau	26.6°S	16.5°E
Erniaster patellus	Pflug	1972	Aar or Plateau	26.6°S	16.5°E
Ernietta aarensis	Pflug	1972	Aar	26.6°S	16.5°E
*Ernietta plateauensis**	Pflug	1966	Plateau	26.6°S	16.5°E

Formation	Group	Interpretation	References
Nemakit-Daldyn		*incertae sedis**	Sokolov 1984.
Aksumbinsk		pseudofossils?	Borovikov and Kraskov 1977.
Ust'-Pinega	Valdai	★★★	Keller and Fedonkin 1976.
Khatyspyt		*incertae sedis*	Vodanyuk 1989.
Ust'-Pinega	Valdai	★★★	Fedonkin 1984.
Rawnsley	Wilpena	*Charniodiscus arboreus*	Glaessner and Daily 1959; Glaessner and Wade 1966.
Nasep	Nama	★★	Glaessner 1963.
Ust'-Pinega	Valdai	*incertae sedis**	Palij, Posti and Fedonkin 1979.
Rawnsley	Wilpena	★★	Gehling 1987.
Ust'-Pinega	Valdai	*incertae sedis**	Fedonkin 1980a.
Arumbera II	Pertaoorrta	pseudofossil	Glaessner and Walter 1975; Jenkins et al. 1981.
Khatyspyt		*Aspidella hatyspytia*?	Vodanyuk 1989.
Khatyspyt		★	Vodanyuk 1989.
	St John's	pseudofossil	Billings 1872; Hofmann 1971.
Dabis	Nama	★	Hahn and Pflug 1985.
Kurtun		*Pteridinium* or *Ernietta*?	Sokolov 1972, 1973.
Rawnsley	Wilpena	*incertae sedis**	Sprigg 1947, 1949.
Yaryshev		*Bronicella podolica*	Zaika-Novatsky 1965, 1968; Palij 1976; Palij et al. 1979.
	Redkino	★★	Keller et al. 1974; Narbonne and Hofmann 1987.
	Redkino	*Beltanelliformis brunsae*	Fedonkin 1972; Narbonne and Hofmann 1987.
Ust'-Pinega	Valdai	*incertae sedis**	Fedonkin 1985a.
Ust'-Pinega	Valdai	★★	Fedonkin 1980a.
Rawnsley	Wilpena	★	Wade 1972.
Yaryshev		*incertae sedis*	Zaika-Novatsky 1965, 1968; Palij 1976; Palij et al. 1979.
	Nankoweap	pseudofossil	Bassler 1941; Fürsich and Bromley 1985; Runnegar and Stait 1990.
Dabis	Nama	*incertae sedis*	Germs 1973b.
Brachina	Wilpena	*incertae sedis*	Glaessner 1969; Cloud and Glaessner 1982; Glaessner 1984.
Mistaken Pt	Conception	★★	Misra 1969; Anderson 1978; Anderson and Conway Morris 1982.
Rawnsley	Wilpena	★	Glaessner and Wade 1966; Germs 1972a, 1973a.
Woodhouse	Maplewood	★★★	Ford 1958; Jenkins 1985.
Khatysput		★	Sokolov 1972, 1973; Glaessner 1984.
Rawnsley	Wilpena	★★	Glaessner and Daily 1959; Jenkins and Gehling 1978.
Woodhouse	Maplewood	★★★	Ford 1958; Jenkins and Gehling 1978.
Rawnsley	Wilpena	★★★	Glaessner and Wade 1966; Sun 1986b.
Rawnsley	Wilpena	★	Jenkins and Gehling 1978.
Yaryshev		*incertae sedis**	Sokolov 1972, 1973.
Rawnsley	Wilpena	*incertae sedis**	Wade 1971; Hofmann 1988.
Moorillah	Wilpena	pseudofossils	Plummer 1980.
Rawnsley	Wilpena	*incertae sedis*	Glaessner and Wade 1966; Glaessner 1971; Oliver 1984.
Rawnsley	Wilpena	holotype poorly preserved	Sprigg 1947, 1949; Jenkins 1984a; Sun 1986b.
Ust'-Pinega	Valdai	inorganic?	Fedonkin 1981.
Rawnsley	Wilpena	*incertae sedis**	Sprigg 1947.
Changlingzi	Sinian	pseudofossil	Xing and Liu 1979; Sun 1986a.
Changlingzi	Sinian	pseudofossil	Xing and Liu 1979; Sun 1986a.
Ust'-Pinega	Valdai	*incertae sedis**	Palij, Posti and Fedonkin 1979.
Rawnsley	Wilpena	*incertae sedis*	Glaessner and Wade 1966.
Rawnsley	Wilpena	*Tateana inflata*	Sprigg 1949; Jenkins 1984a.
Yaryshev		*Pollukia serebrina*	Palij, 1969; Palij, Posti and Fedonkin 1979.
Changlingzi	Sinian	pseudofossil	Xing and Liu 1979; Sun 1986a.
Rawnsley	Wilpena	★★	Wade 1972b.
Rawnsley	Wilpena	★★★	Sprigg 1947; Wade 1972; Runnegar 1982b.
Rawnsley	Wilpena	★★	Glaessner and Wade 1966; Wade 1972; Runnegar 1990a.
Rawnsley	Wilpena	★★	Wade 1972.
Rawnsley	Wilpena	★★	Sprigg 1949; Wade 1972.
Rawnsley	Wilpena	*Dickinsonia costata*	Harrington and Moore 1955; Glaessner and Wade 1966.
Rawnsley	Wilpena	★★	Glaessner and Wade 1966; Wade 1972.
Fowlers Gap	Torrowangee	pseudofossil	Webby 1970a.
Rawnsley	Wilpena	★★★	Sprigg 1947, 1949.
Mogilev		*incertae sedis*	Veikanov, Aseeva and Fedonkin 1983.
Rawnsley	Wilpena	★	Wade 1972; Fedonkin 1985a.
Dabis	Nama	*Ernietta plateauensis*	Pflug 1972; Jenkins et al. 1981.
Dabis	Nama	*Ernietta plateauensis*	Pflug 1972; Jenkins et al. 1981.
Dabis	Nama	*Ernietta plateauensis*	Pflug 1972; Jenkins et al. 1981.
Dabis	Nama	★★★	Pflug 1966, 1972; Jenkins et al. 1981.

Table 23.3.1. (Continued).

Taxon	Author	Date	Locality	Lat	Lon
Ernietta tsachanabis	Pflug	1972	Aar or Plateau	26.6°S	16.5°E
Erniobaris baroides*	Pflug	1972	Aar or Plateau	26.6°S	16.5°E
Erniobaris epistula	Pflug	1972	Aar or Plateau	26.6°S	16.5°E
Erniobaris gula	Pflug	1972	Aar or Plateau	26.6°S	16.5°E
Erniobaris parietalis	Pflug	1972	Aar or Plateau	26.6°S	16.5°E
Erniobeta forensis	Pflug	1972	Aar or Plateau	26.6°S	16.5°E
Erniobeta scapulosa*	Pflug	1972	Aar or Plateau	26.6°S	16.5°E
Erniocarpus carpoides*	Pflug	1972	Aar or Plateau	26.6°S	16.5°E
Erniocarpus sermo	Pflug	1972	Aar or Plateau	26.6°S	16.5°E
Erniocentris centriformis*	Pflug	1972	Aar or Plateau	26.6°S	16.5°E
Erniocoris orbiformis*	Pflug	1972	Aar or Plateau	26.6°S	16.5°E
Erniodiscus clipeus*	Pflug	1972	Aar or Plateau	26.6°S	16.5°E
Erniodiscus rutilis	Pflug	1972	Aar or Plateau	26.6°S	16.5°E
Erniofossa prognatha*	Pflug	1972	Aar or Plateau	26.6°S	16.5°E
Erniograndis paraglossa	Pflug	1972	Aar or Plateau	26.6°S	16.5°E
Erniograndis sandalix*	Pflug	1972	Aar or Plateau	26.6°S	16.5°E
Ernionorma abyssoides*	Pflug	1972	Aar or Plateau	26.6°S	16.5°E
Ernionorma clausula	Pflug	1972	Aar or Plateau	26.6°S	16.5°E
Ernionorma corrector	Pflug	1972	Aar or Plateau	26.6°S	16.5°E
Ernionorma peltis	Pflug	1972	Aar or Plateau	26.6°S	16.5°E
Ernionorma rector	Pflug	1972	Aar or Plateau	26.6°S	16.5°E
Ernionorma tribunalis	Pflug	1972	Aar or Plateau	26.6°S	16.5°E
Erniopelta scrupula*	Pflug	1972	Aar or Plateau	26.6°S	16.5°E
Erniotaxis segmentrix*	Pflug	1972	Aar or Plateau	26.6°S	16.5°E
Evmiaksia aksionovi*	Fedonkin	1984	Winter Shore	65.5°N	39.7°E
Fossil jellyfish	Sisodiya	1982	Madhya Pradesh	24°N	80°E
Frond-like fossils	Dyson	1985	Mt Remarkable	30.4°S	130.0°E
Glaessneria imperfecta*	Gureev	1987	Dniester River	48.4°N	26.1°E
Glaessnerina grandis*	(Glaessner and Wade)	1966	Ediacara	30.8°S	138.2°E
Hagenetta aarensis*	Hahn and Pflug	1988	Aar or Plateau	26.6°S	16.5°E
Hallidaya brueri*	Wade	1969	Mt Skinner	22.2°S	134.3°E
Hiemalora pleiomorphus	Vodanyuk	1989	Olenyok Uplift	71.2°N	113.6°E
Hiemalora stellaris*	Fedonkin	1982	Winter Shore	65.5°N	39.7°E
Ichnusa cocozzi*	Debrenne and Naud	1981	Sardinia	39.4°N	9.3°E
Inaria karli*	Gehling	1988	Chace Range	31.7°S	138.7°E
Inkrylovia lata*	Fedonkin	1979	Summer Shore	65.5°N	39.7°E
Irridinitus multiradiatus*	Fedonkin	1983	Dniester River	48.4°N	26.1°E
Jixiella capistratus*	Liu	1981	Heilongjiang	45°N	132°E
Kaisalia levis	Gureev	1987	Dniester River	48.4°N	26.1°E
Kaisalia mensae*	Fedonkin	1984	Winter Shore	65.5°N	39.7°E
Khatyspytia grandis*	Fedonkin	1985	Olenyok Uplift	71.2°N	113.6°E
Kimberella quadrata*	(Glaessner and Wade)	1966	Ediacara	30.8°S	138.2°E
Kimberia quadrata*	Glaessner and Wade	1966	Ediacara	30.8°S	138.2°E
Kuibisia glabra*	Hahn and Pflug	1985	Plateau	26.6°S	16.5°E
Kullingia concentrica*	Glaessner	1979	Norway	61°N	11°E
Liaoningia fuxianensis	Xing and Liu	1979	Liaoning	40°N	122°E
Lomosovis malus*	Fedonkin	1983	Dniester River	48.4°N	26.1°E
Lorenzinites rarus*	Glaessner and Wade	1966	Ediacara	30.8°S	138.2°E
Madigania annulata*	Sprigg	1949	Ediacara	30.8°S	138.2°E
Majaella verkhojanica*	Vologdin and Maslov	1960	Maya River		
Marywadea ovata*	(Glaessner and Wade)	1966	Ediacara	30.8°S	138.2°E
Mashania angusta	Liu	1981	Heilongjiang	45°N	132°E
Mashania annulata	Liu	1981	Heilongjiang	45°N	132°E
Mashania deformata	Liu	1981	Heilongjiang	45°N	132°E
Mashania longshanensis*	Liu	1981	Heilongjiang	45°N	132°E
Mawsonites randallensis	Sun	1986	Ediacara	30.8°S	138.2°E
Mawsonites spriggi	Glaessner and Wade	1966	Ediacara	30.8°S	138.2°E
Medusina asteroides	Sprigg	1949	Ediacara	30.8°S	138.2°E
Medusina filamentus	Sprigg	1949	Ediacara	30.8°S	138.2°E
Medusina mawsoni	Sprigg	1949	Ediacara	30.8°S	138.2°E
Medusinites asteroides*	(Sprigg)	1949	Ediacara	30.8°S	138.2°E
Medusinites paliji	Gureev	1987	Dniester River	48.4°N	26.1°E
Medusinites patellaris	Sokolov	1972	Dniester River	48.4°N	26.1°E
Medusinites simplex	Xing and Liu	1979	Liaoning	40°N	122°E

Formation	Group	Interpretation	References
Dabis	Nama	*Ernietta plateauensis*	Pflug 1972; Jenkins et al. 1981.
Dabis	Nama	*Ernietta plateauensis*	Pflug 1972; Jenkins et al. 1981.
Dabis	Nama	*Ernietta plateauensis*	Pflug 1972; Jenkins et al. 1981.
Dabis	Nama	*Ernietta plateauensis*	Pflug 1972; Jenkins et al. 1981.
Dabis	Nama	*Ernietta plateauensis*	Pflug 1972; Jenkins et al. 1981.
Dabis	Nama	*Ernietta plateauensis*	Pflug 1972; Jenkins et al. 1981.
Dabis	Nama	*Ernietta plateauensis*	Pflug 1972; Jenkins et al. 1981.
Dabis	Nama	*Ernietta plateauensis*	Pflug 1972; Jenkins et al. 1981.
Dabis	Nama	*Ernietta plateauensis*	Pflug 1972; Jenkins et al. 1981.
Dabis	Nama	*Ernietta plateauensis*	Pflug 1972; Jenkins et al. 1981.
Dabis	Nama	*Ernietta plateauensis*	Pflug 1972; Jenkins et al. 1981.
Dabis	Nama	*Ernietta plateauensis*	Pflug 1972; Jenkins et al. 1981.
Dabis	Nama	*Ernietta plateauensis*	Pflug 1972; Jenkins et al. 1981.
Dabis	Nama	*Ernietta plateauensis*	Pflug 1972; Jenkins et al. 1981.
Dabis	Nama	*Ernietta plateauensis*	Pflug 1972; Jenkins et al. 1981.
Dabis	Nama	*Ernietta plateauensis*	Pflug 1972; Jenkins et al. 1981.
Dabis	Nama	*Ernietta plateauensis*	Pflug 1972; Jenkins et al. 1981.
Dabis	Nama	*Ernietta plateauensis*	Pflug 1972; Jenkins et al. 1981.
Dabis	Nama	*Ernietta plateauensis*	Pflug 1972; Jenkins et al. 1981.
Dabis	Nama	*Ernietta plateauensis*	Pflug 1972; Jenkins et al. 1981.
Dabis	Nama	*Ernietta plateauensis*	Pflug 1972; Jenkins et al. 1981.
Dabis	Nama	*Ernietta plateauensis*	Pflug 1972; Jenkins et al. 1981.
Ust'-Pinega	Valdai	*incertae sedis**	Fedonkin 1984.
Nimbahera	Vindhyan	*incerate sedis*	Sisodiya 1982.
Elatina	Umberatana	pseudofossil; ice tracks	Dyson 1985; Jenkins 1986.
Yaryshev		*incertae sedis*	Gureev 1987.
Rawnsley	Wilpena	*Charnia grandis*	Glaessner and Wade 1966; Germs 1972a, 1973a.
Dabis	Nama	*Beltanelliformis brunsae*	Hahn and Pflug 1988.
Central Mt Stuart		★	Wade 1969.
Khatyspyt		★★	Vodanyuk 1989.
Ust'-Pinega	Valdai	★	Fedonkin 1982.
San Vito		pseudofossil?	Debrenne and Naud 1981.
Rawnsley	Wilpena	★★★	Gehling 1988.
Ust'-Pinega	Valdai	*Pteridinium latum*	Palij, Posti and Fedonkin 1979.
Mogilev		*incertae sedis*	Veikanov, Aseeva and Fedonkin 1983.
Liumao	Mashan	pseudofossil; patterned cone	Liu 1981; Boyd and Ore 1963.
Mogilev		*incertae sedis*	Gureev 1987.
Ust'-Pinega	Valdai	*incertae sedis**	Fedonkin 1984.
Khatyspyt		*incertae sedis*	Fedonkin 1985.
Rawnsley	Wilpena	*incertae sedis*	Glaessner and Wade 1966; Wade 1972; Jenkins 1984.
Rawnsley	Wilpena	*Kimberella quadrata*	Glaessner and Wade 1966.
Dabis	Nama	*Ernietta plateauensis*	Hahn and Pflug 1985.
	Dividal	*incertae sedis*	Føyn and Glaessner 1979.
Changlingzi	Sinian	pseudofossil	Xing and Liu 1979; Sun 1986a.
Mogilev		tissue fragments	Veikanov, Aseeva and Fedonkin 1983.
Rawnsley	Wilpena	★	Glaessner and Wade 1966.
Rawnsley	Wilpena	*Spriggia annulata*	Sprigg 1949; Southcott 1958.
Yudoma		*incertae sedis**	Vologdin and Maslov 1960; Glaessner 1979a.
Rawnsley	Wilpena	*Spriggina floundersi*	Glaessner 1976b.
Liumao	Mashan	pseudofossil; patterned cone	Liu 1981; Boyd and Ore 1963.
Liumao	Mashan	pseudofossil; patterned cone	Liu 1981; Boyd and Ore 1963.
Liumao	Mashan	pseudofossil; patterned cone	Liu 1981; Boyd and Ore 1963.
Liumao	Mashan	pseudofossil; patterned cone	Liu 1981; Boyd and Ore 1963.
Rawnsley	Wilpena	*incertae sedis*	Sun 1986c.
Rawnsley	Wilpena	*incertae sedis*	Glaessner and Wade 1966; Sun 1986c.
Rawnsley	Wilpena	*Medusinites asteroides*	Sprigg 1949.
Rawnsley	Wilpena	*incertae sedis**	Sprigg 1949.
Rawnsley	Wilpena	*Medusinites asteroides*	Sprigg 1949.
Rawnsley	Wilpena	*incertae sedis*	Sprigg 1949; Glaessner and Wade 1966.
Mogilev		*incertae sedis*	Gureev 1987.
Mogilev		*Planomedusites patellaris*	Sokolov 1972, 1973.
Changlingzi	Sinian	pseudofossil	Xing and Liu 1979; Sun 1986a.

Table 23.3.1. (Continued).

Taxon	Author	Date	Locality	Lat	Lon
Medusinites sokolovi	Gureev	1985	Ukraine	48°N	26°E
Mialsemia semichatovi*	Fedonkin	1985	Winter Shore	65.5°N	39.7°E
Nadalina yukonensis*	Narbonne and Hofmann	1987	Yukon	64.6°N	133°W
Namalia villiersiensis*	Germs	1968	Buchholzbrunn	26.7°S	17.2°E
Nasepia altae*	Germs	1972	Arimas	27.6°S	16.9°E
Nemiana simplex*	Palij	1976	Dniester River	48.4°N	26.1°E
Nimbia dniesteri	Fedonkin	1983	Dniester River	48.4°N	26.1°E
Nimbia occlusa*	Fedonkin	1980	Onega Peninsula	62.7°N	39.9°E
Nimbia paula	Gureev	1985	Ukraine	48.4°N	26.1°E
Onega stepanovi*	Fedonkin	1976	Syuzma River	64.7°N	39.7°E
Onegia nenoxa*	(Keller)	1974	Syuzma River	64.7°N	39.7°E
Orthogonium parallelum	Gürich	1930	Kuibis	26.5°S	16.7°E
Ovatoscutum concentricum*	Glaessner and Wade	1966	Ediacara	30.8°S	138.2°E
Palaeoplatoda segmentata*	Fedonkin	1979	Syuzma River	64.7°N	39.7°E
Paliella patelliformis*	Fedonkin	1980	Winter Shore	65.5°N	39.7°E
Papillionata eyrei*	Sprigg	1947	Ediacara	30.8°S	138.2°E
Paracharnia dengyingensis	Sun	1986	Yangtze Gorge	30.8N	111.1E
Paramedusium africanum*	Gürich	1930	Groendoorn	27.5°S	18.3°E
Parvancorina minchami*	Glaessner	1958	Ediacara	30.8°S	138.2°E
Pectinate form	Misra	1969	Newfoundland	46.7°N	53.0°W
Persimedusites chahgazensis*	Hahn and Pflug	1980	Iran	32°N	55°E
Petalostroma kuibis*	Pflug	1973	Aar	26.6°S	16.5°E
Phyllozoon hanseni*	Jenkins and Gehling	1978	Devil's Peak	32.5°S	138.0°E
Pinegia stellaris*	Fedonkin	1980	Winter Shore	65.5°N	39.7°E
Planomedusites grandis*	Sokolov	1972	Dniester River	48.4°N	26.1°E
Planomedusites patellaris	(Sokolov)	1972	Dniester River	48.4°N	26.1°E
Platypholinia pholiata*	Fedonkin	1985	Winter Shore	65.5°N	39.7°E
Podolimirus mirus*	Fedonkin	1983	Dniester River	48.4°N	26.1°E
Pollukia serebrina*	(Palij)	1969	Dniester River	48.4°N	26.1°E
Pollukia shulgae	Gureev	1987	Ternopol	49.5°N	26°E
Pomoria corolliformis*	Fedonkin	1980	Winter Shore	65.5°N	39.7°E
Praecambridium sigillum*	Glaessner and Wade	1966	Ediacara	30.8°S	138.2°E
Protodipleurosoma rugulosum	Fedonkin	1980	Winter Shore	65.5°N	39.7°E
Protodipleurosoma wardi*	Sprigg	1949	Ediacara	30.8°S	138.2°E
Protoechiurus edmondsi*	Glaessner	1979	Plateau	26.6°S	16.5°E
Pseudorhizostomites howchini*	Sprigg	1949	Ediacara	30.8°S	138.2°E
Pseudorhopilema chapmani*	Sprigg	1949	Ediacara	30.8°S	138.2°E
Pseudovendia charnwoodensis*	Boynton and Ford	1979	Charnwood Forest	52.7°N	1.1°W
Pteridinium carolinaense	(St Jean)	1973	North Carolina	35.4°N	80.3°W
Pteridinium latum	(Fedonkin)	1979	Summer Shore	65.5°N	39.7°E
Pteridinium nenoxa	Keller	1974	Syuzma River	64.7°N	39.7°E
Pteridinium simplex*	Gürich	1933	Kuibis	26.5°S	16.7°E
Pteridium simplex*	Gürich	1930	Kuibis	26.5°S	16.7°E
Ramellina pennata*	Fedonkin	1980	Winter Shore	65.5°N	39.7°E
Rangea arborea	Glaessner	1959	Ediacara	30.8°S	138.2°E
Rangea brevior	Gürich	1930	Kubis	26.5°S	16.7°E
Rangea grandis	Glaessner and Wade	1966	Ediacara	30.8°S	138.2°E
Rangea longa	Glaessner and Wade	1966	Ediacara	30.8°S	138.2°E
Rangea schneiderhoehni*	Gürich	1929	Kuibis	26.5°S	16.7°E
Rangea sibirica	Sokolov	1972	Olenyok Uplift	71.2°N	113.6°E
Rugoconites enigmaticus*	Glaessner and Wade	1966	Ediacara	30.8°S	138.2°E
Rugoconites tenuirugosus	Wade	1972	Bunyeroo Gorge	31.4°S	138.5°E
Sajanella arshanica	Vologdin	1966	Irkutsk area	52°N	104°E
Sekwia excentrica*	Hofmann	1981	Mackenzie Mts	63.4°N	128.4°W
Sekwia kaptarenkoe	Gureev	1987	Dniester River	48.4°N	26.1°E
Skinnera brooksi*	Wade	1969	Mt Skinner	22.2°S	134.3°E
Spindle-shaped form	Misra	1969	Newfoundland	46.7°N	53.0°W
Spriggia annulata*	(Sprigg)	1949	Ediacara	30.8°S	138.2°E
Spriggia wadea	Sun	1986	Ediacara	30.8°S	138.2°E
Spriggina borealis	Fedonkin	1979	Syuzma River	64.7°N	39.7°E
Spriggina floundersi*	Glaessner	1958	Ediacara	30.8°S	138.2°E
Spriggina ovata	Glaessner and Wade	1966	Ediacara	30.8°S	138.2°E
Star-shaped form	Anderson	1978	Newfoundland	46.7°N	53.0°W
Staurinidia crucicula*	Fedonkin	1985	Winter Shore	65.5°N	39.7°E
Suvorovella aldanica*	Vologdin and Maslov	1960	Maya River		

Formation	Group	Interpretation	References
Zharnovki		*Vendella sokolovi*	Gureev 1985.
Ust'-Pinega	Valdai	incertae sedis*	Fedonkin 1985a.
Siltstone unit 1	Windermere	incerate sedis*	Narbonne and Hofmann 1987.
Dabis	Nama	*Ernietta plateauensis*	Germs 1968, 1972a; Jenkins et al. 1981.
Nasep	Nama	incertae sedis	Germs 1972a, 1973a.
Mogilev		*Beltanelliformis brunsae*	Palij 1976.
Mogilev		incertae sedis	Veikanov, Aseeva and Fedonkin 1983.
Ust'-Pinega	Valdai	incertae sedis	Fedonkin 1980a.
Nagoriany		incertae sedis*	Gureev 1985.
Ust'-Pinega	Valdai	★★★	Keller and Fedonkin 1976.
Ust'-Pinega	Valdai	*Pteridinium carolinaense*	Keller et al. 1974; Sokolov 1976.
Dabis	Nama	lost in Second World War	Gürich 1930, 1933; Häntzschel 1975.
Rawnsley	Wilpena	★★★	Glaessner and Wade 1966; Wade 1971.
Ust'-Pinega	Valdai	*Dickinsonia?*	Palij, Posti and Fedonkin 1979.
Ust'-Pinega	Valdai	incertae sedis	Fedonkin 1980a.
Rawnsley	Wilpena	*Dickinsonia costata*	Sprigg 1947, 1949.
Dengying		★★	Sun 1986b.
Nasep?	Nama	lost in Second World War	Gürich 1930, 1933; Germs 1972a.
Rawnsley	Wilpena	★★★	Glaessner 1958, 1980.
Mistaken Pt	Conception	★★	Misra 1969; Anderson 1978; Anderson and Conway Morris 1982.
Esfordi		incertae sedis	Hahn and Pflug 1980.
Dabis	Nama	incertae sedis*	Pflug 1973.
Rawnsley	Wilpena	★★★	Jenkins and Gehling 1978.
Ust'-Pinega	Valdai	*Hiemalora stellaris*	Fedonkin 1980a.
Mogilev		incertae sedis*	Sokolov 1972, 1973.
Mogilev		incertae sedis*	Sokolov 1972, 1973; Gureev 1987.
Ust'-Pinega	Valdai	incertae sedis*	Fedonkin 1985a.
Mogilev		incertae sedis*	Veikanov, Aseeva and Fedonkin 1983.
Yaryshev		incertae sedis	Gureev 1987.
Zburch		Cambrian	Gureev 1987.
Ust'-Pinega	Valdai	incertae sedis	Fedonkin 1980a.
Rawnsley	Wilpena	*Spriggina floundersi?*	Glaessner and Wade 1966, 1971; Birket-Smith 1981a.
Ust'-Pinega	Valdai	incertae sedis	Fedonkin 1980a.
Rawnsley	Wilpena	incertae sedis*	Sprigg 1949.
Dabris	Nama	dubiofossil	Glaessner 1979b.
Rawnsley	Wilpena	gas escape structure	Sprigg 1949; Glaessner and Wade 1966; Wade 1968.
Rawnsley	Wilpena	*Pseudorhizostomites howchini*	Sprigg 1949; Glaessner and Wade 1966.
Woodhouse	Mapleton	incertae sedis*	Boynton and Ford 1979.
McManus	Albemarle	★★★	St Jean 1973; Gibson et al. 1984.
Ust'-Pinega	Valdai	★★	Palij, Posti and Fedonkin 1979.
Ust'-Pinega	Valdai	*Pteridinium carolinaense*	Keller et al. 1974; Fedonkin 1981; Gibson et al. 1984.
Dabis	Nama	★★★	Gürich 1933; Richter 1955; Pflug 1970a.
Dabis	Nama	*Pteridinium simplex*	Gürich 1930.
Ust'-Pinega	Valdai	incertae sedis*	Fedonkin 1980a.
Rawnsley	Wilpena	*Charniodiscus arboreus*	Glaessner and Daily 1959; Jenkins and Gehling 1978.
Dabis	Nama	*Rangea schneiderhoehni*	Gürich 1930, 1933; Jenkins 1985.
Rawnsley	Wilpena	*Charnia grandis*	Glaessner and Wade 1966.
Rawnsley	Wilpena	*Charniodiscus longa*	Glaessner and Wade 1966.
Dabis	Nama	★★★	Gürich 1929; Pflug 1970b; Germs 1973a; Jenkins 1985.
Khatysput		*Charnia sibirica*	Sokolov 1972.
Rawnsley	Wilpena	incertae sedis	Glaessner and Wade 1966.
Rawnsley	Wilpena	incertae sedis	Wade 1972.
Karagas		deformed algal mat	Vologdin 1966.
Blueflower	Windermere	★	Hofmann 1981.
Mogilev		incertae sedis	Gureev 1987.
Central Mt Stuart		★★	Wade 1969.
Mistaken Pt	Conception	★★★	Misra 1969; Anderson 1978; Anderson and Conway Morris 1982.
Rawnsley	Wilpena	★★★	Southcott 1958; Jenkins 1984a; Sun 1986b.
Rawnsley	Wilpena	incertae sedis	Sun 1986a.
Ust'-Pinega	Valdai	incertae sedis*	Palij, Posti and Fedonkin 1979.
Rawnsley	Wilpena	★★★	Glaessner 1958; Birket-Smith 1981a–b.
Rawnsley	Wilpena	*Spriggina floundersi*	Glaessner and Wade 1966.
Mistaken Pt	Conception	★	Anderson 1978; Anderson and Conway Morris 1982.
Ust'-Pinega	Valdai	incertae sedis	Fedonkin 1985a.
Yudoma		incertae sedis*	Vologdin and Maslov 1960; Sokolov 1973; Glaessner 1979a.

Table 23.3.1. (Continued).

Taxon	Author	Date	Locality	Lat	Lon
Suzmites tenuis	Fedonkin	1981	Syuzma River	64.7°N	39.7°E
*Suzmites volutatus**	Fedonkin	1976	Syuzma River	64.7°N	39.7°E
*Tateana inflata**	Sprigg	1949	Ediacara	30.8°S	138.2°E
Tirasiana cocarda	Bekker	1985	Perm	54.3°N	58.1°E
Tirasiana concentralis	Bekker	1985	Perm	54.3°N	58.1°E
Tirasiana coniformis	Palij	1976	Moldavia	46°N	29°E
*Tirasiana disciformis**	Palij	1976	Moldavia	46°N	29°E
*Tribrachidium heraldicum**	Glaessner	1959	Ediacara	30.8°S	138.2°E
*Vaizitsinia sophia**	Sokolov and Fedonkin	1983	Winter Shore	65.5°N	39.7°E
*Valdainia plumosa**	Fedonkin	1983	Dniester River	48.4°N	26.1°E
*Vaveliksia velikanovi**	Fedonkin	1983	Dniester River	48.4°N	26.1°E
*Velancorina martina**	Pflug	1966	Aar	26.6°S	16.5°E
Vendella haelenicae	Gureev	1987	Dniester River	48.4°N	26.1°E
Vendella larini	Gureev	1987	Ukraine	48°N	26°E
*Vendella sokolovi**	(Gureev)	1985	Dniester River	48.4°N	26.1°E
*Vendia sokolovi**	Keller	1969	Yarensk Borehole		
*Vendomia menneri**	Keller	1976	Syuzma River	64.7°N	39.7°E
*Veprina undosa**	Fedonkin	1980	Winter Shore	65.5°N	39.7°E
*Vermiforma antiqua**	Cloud	1976	North Carolina	36.2°N	78.8°W
*Vladmissa missarzhevskii**	Fedonkin	1985	Winter Shore	65.5°N	39.7°E
*Wigwamiella enigmatica**	Runnegar	1990	Mt Scott Range		
*Zolotytsia biserialis**	Fedonkin	1981	Winter Shore	65.5°N	39.7°E

Table 23.3.2. *Formally described taxa (excluding members of the Ediacara faunas) of fossils and pseudofossils that have been interpreted as the remains of Proterozoic metazoans.*

Taxon	Author	Date	Locality	Lat	Lon
*Acuticloudina borrelloi**	(Yochelson and Herrera)	1974	Argentina	31.4°S	68.5°W
Brabbinthes churkini	Allison	1975	Alaska	62°N	140°W
Brooksella canyonensis	Bassler	1941	Arizona	36.3°N	111.8°W
*Buchholzbrunnichnus kroeneri**	Germs	1973	Buchholzbrunn	26.7°S	17.1°E
*Buschmania roeringi**	Kaever and Richter	1976	Grünental	23°S	17°E
Calyptrina partita	Sokolov	1965	Anabar Uplift		
Cloudina borrelloi	Yochelson and Herrera	1974	Argentina	31.4°S	68.5°W
*Cloudina hartmanae**	Germs; Glaessner	1976	Driedoornvlakte	23.8°S	16.6°E
*Cloudina hartmannae**	Germs	1972	Driedoornvlakte	23.8°S	16.6°E
Cloudina lucianoi	(Beurlen and Sommer)	1957	Mato Grosso	19.1°S	57.7°W
Cloudina riemkeae	Germs	1972	Driedoornvlakte	23.8°S	16.6°E
Cloudina waldei	Hahn and Pflug	1985	Mato Grosso	19.1°S	57.7°W
*Corumbella werneri**	Hahn et al.	1982	Mato Grosso	19.1°S	57.7°W
Gdowia assatkini	Yanashevskiy				
Huainanella cylindrica	Wang	1982	Anhui	32.6°N	116.7°E
Palaeolina sp.	Fedonkin	1985	Arkhangelsk bore	65°N	41°E
Paleolina evenkiana	Sokolov	1965	Siberia	66°N	88°E
Paleolina tortuosa	Wang	1982	Anhui	32.6°N	116.7°E
Paleorhynchus anhuiensis	Wang	1982	Anhui	32.6°N	116.7°E
Pararenicola huaiyuanensis	Wang	1982	Anhui	32.6°N	116.7°E
Protoarenicola baiguashanensis	Wang	1982	Anhui	32.6°N	116.7°E
Redkinia spinosa	Sokolov	1976	Nepeitsino bore		
Ruedemannella minuta	Wang	1982	Anhui	32.6°N	116.7°E
Saarina sp.	Runnegar	1990	Arkhangelsk bore	65°N	41°E
Sinosabellidites huainanensis	Zheng	1980	Anhui	32.6°N	116.7°E
Tyrkanispongia tenua	Vologdin and Drozdova	1970	Urchur River		

Formation	Group	Interpretation	References
Ust'-Pinega	Valdai	pseudofossil	Fedonkin 1988.
Ust'-Pinega	Valdai	pseudofossil	Fedonkin 1976.
Rawnsley	Wilpena	★★★	Sprigg 1949; Jenkins 1984a; Sun 1986a.
Chernakamenka		*incertae sedis**	Bekker 1985.
Chernakamenka		*incertae sedis**	Bekker 1985.
Mogilev		*incertae sedis**	Palij 1976.
Mogilev		★★	Palij 1976; Fedonkin 1981.
Rawnsley	Wilpena	★★★	Glaessner and Daily 1959; Fedonkin 1981.
Ust'-Pinega	Valdai		
Mogilev		*Pteridinium latum*?	Veikanov, Aseeva and Fedonkin 1983.
Mogilev		*incertae sedis**	Veikanov, Aseeva and Fedonkin 1983.
Dabis	Nama	*incertae sedis**	Pflug 1966.
Yaryshev		*incertae sedis*	Gureev 1987.
Zharnovki		*incertae sedis*	Gureev 1987.
Mogilev		*incertae sedis*	Gureev 1985, 1987.
Ust'-Pinega	Valdai	only specimen	Keller 1969.
Ust'-Pinega	Valdai	only specimen	Keller and Fedonkin 1976.
Ust'-Pinega	Valdai	*incertae sedis*	Fedonkin 1980a.
Hyco		trace fossil?	Cloud et al. 1976.
Ust'-Pinega	Valdai	*incertae sedis**	Fedonkin 1985a.
Rawnsley	Wilpena	★	Runnegar 1991b.
Ust'-Pinega	Valdai	*incertae sedis**	Fedonkin 1981.

Formation	Group	Interpretation	References
Cámbrico inferior		*Cloudina borrelloi*	Yochelson and Herrera 1974; Hahn and Pflug 1985.
	Tindir	hexactinellid sponge spicule	Allison 1975; Cloud, Wright and Glover 1976.
	Nankoweap	pseudofossil	Bassler 1941; Glaessner 1969; Fürsich and Bromley 1985.
Dabis	Nama	*incertae sedis*	Germs 1973b.
Buschmann	Nama	pseudofossils; crystal aggregates	Kaever and Richter 1976; Debrenne and Lafuste 1979.
Nemakit-Daldyn		Cambrian	Sokolov 1965, 1967, 1968.
Cámbrico inferior		*Cloudina*?	Yochelson and Herrera 1974; Hahn and Pflug 1985.
Zaris	Nama	★★★	Germs 1972b; Glaessner 1976a, 1984.
Zaris	Nama	*Cloudina hartmanae*	Germs 1972b; Glaessner 1976a, 1984.
Tamengo	Corumbá	★	Beurlen and Sommer 1957; Hahn and Pflug 1985; Zaine and Fairchild 1987.
Zaris	Nama	*Cloudina hartmanae*	Germs 1972b.
Tamengo	Corumbá	*Cloudina luicanoi*	Hahn and Pflug 1985; Zaine and Fairchild 1987.
Tamengo	Corumbá	★★	Hahn et al. 1982; Zaine and Fairchild 1987.
Liulaobei	Huainan	*Sinosabellidites huainanensis*	Wang 1982; Sun, Wang and Zhou 1986.
Ust'-Pinega	Valdai	*Paleolina sp.*	Sokolov and Ivanovsky 1985.
Platonovka		★★★	Sokolov 1965.
Jiuliqiao	Feishui	*Pararenicola huaiyuanensis*	Wang 1982; Sun, Wang and Zhou 1986.
Jiuliqiao	Feishui	*Pararenicola huaiyuanensis*	Wang 1982; Sun, Wang and Zhou 1986.
Jiuliqiao	Feishui	★★★	Wang 1982; Sun, Wang and Zhou 1986.
Jiuliqiao	Feishui	*incertae sedis*	Wang 1982; Sun, Wang and Zhou 1986.
Redkino		★★★	Sokolov 1976, 1985.
Jiuliqiao	Feishui	*Pararenicola huaiyuanensis*	Wang 1982; Sun, Wang and Zhou 1986.
Ust'-Pinega	Valdai	★★★	Sokolov and Ivanovsky 1985.
Liulaobei	Huainan	★★★	Zheng 1974; Sun, Wang and Zhou 1986.
Gonam		volcanic glass shards	Vologdin and Drozdova 1970; Glaessner 1979a.

23.4

Proterozoic Metazoan Trace Fossils

BRUCE N. RUNNEGAR

In Table 23.4.1 are listed structures described or reported as Proterozoic (Vendian) metazoan trace fossils. Interpretations based on these tabulations are presented in Section 7.6.

Table 23.4.1 lists, alphabetically by genus and species or informal name, described taxa and some unnamed structures that have been reported from Vendian and/or earliest Cambrian strata. Type species of genera are identified by an asterisk.

Particularly significant taxa are given a rating of either two or three stars which correspond, respectively, to their medium or high quality as judged from the complexity of the trace fossil and the evidence for its Precambrian age. As indicated, other listed forms are interpreted variously as junior synonyms, younger than Precambrian, dubiofossils, or pseudofossils.

Table 23.4.1. *Structures formally described or reported as Proterozoic (Vendian) metazoan trace fossils.*

Taxon	Author	Date	Type	Locality
Annelid crawling traces	Tynni and Hokkanen	1982	horizontal burrow	Finland
Arenicolites sp.	Crimes and Jiang	1986	vertical burrow	Yunnan
Asteriacites sp.	Crimes and Jiang	1986	resting trace?	Yunnan
Aulichnites sp.	Fedonkin	1985	horizontal trail	White Sea
Bedding-plane markings	Horodyski	1982	dubiofossil	Montana
*Beltanelliformis brunsae**	Menner; Narbonne and Hofmann	1987	vertical burrows?	Wernecke Mtns
Bergaueria sp.	Fedonkin	1985	horizontal trail	Podolia
Bergaueria sp.	Crimes and Germs	1982	dwelling structure	Kuibis
Bergaueria? tumulus	Elphinstone and Walter	1990	pseudofossil	Georgina Basin
*Bilinichnus simplex**	Fedonkin and Palij	1979	pseudofossil	White Sea
bilobated trails	Webby	1970	horizontal	New South Wales
Brooksella canyonensis	Bassler	1941	pseudofossil	Arizona
*Buchholzbrunnichnus kroeneri**	Germs	1973	surface trail?	Buchholzbrunn
*Bunyerichnus dalgarnoi**	Glaessner	1969	dubiofossil	Flinders Ranges
Cochlichnus serpens	Webby	1970	horizontal burrow	New South Wales
Cochlichnus serpens	Webby; Aitken	1989	horizontal burrow	Mackenzie Mtns
Cochlichnus sp.	Palij, Posti and Fedonkin	1979	horizontal burrow	Podolia
Cochlichnus sp.	Palij, Posti and Fedonkin	1979	horizontal burrow?	Podolia
Cochlichnus sp.	Cope	1983	surface trail	Wales
Curvolithus aequus	Elphinstone and Walter	1990	horizontal burrow	Amadeus Basin
Curvolithus sp.	Elphinstone and Walter	1990	horizontal burrow	Amadeus Basin
Curvolithus? sp.	Fedonkin	1977	horizontal trace	Poland
Cylindichnus sp.	Glaessner	1969	body fossil	Ediacara
*Didymaulichnus lyelli**	(Roualt); Elphinstone and Walter	1850	horizontal burrow	Amadeus Basin
Didymaulichnus meanderiformis	Fedonkin	1985	horizontal trace	Olenek Uplift
Didymaulichnus miettensis	Young	1972	horizontal trace	British Columbia
Didymaulichnus miettensis	Young; Elphinstone and Walter	1990	horizontal trace	Amadeus Basin
Didymaulichnus miettensis	Young; Fritz and Crimes	1985	horizontal trace	British Columbia
Didymaulichnus tirasensis	Palij	1974	horizontal burrow	Podolia
Diplichnites sp.	Elphinstone and Walter	1990	surface trail	Amadeus Basin
Diplocraterion parallelum	Torell	1870	vertical burrow	Sweden
Diplocraterion parallelum	Torell; Daily	1976	vertical burrow	Flinders Ranges
Fossil burrows	Faul	1949	pseudofossils?	Michigan
Gordia antiqua	Runnegar	1990	horizontal trail	Ediacara
Gordia arcuata	Ksiazkiewicz	1977	horizontal trace	Poland
Gordia arcuata	Ksiazkiewicz; Fritz and Crimes	1985	horizontal trace	British Columbia
Gordia arcuata	Ksiazkiewicz; Narbonne et al.	1987	horizontal trace	Newfoundland
Gordia arcuata	Ksiazkiewicz; Elphinstone and Walter	1990	surface trail	Amadeus Basin
Gordia arcuata?	Ksiazkiewicz; Gibson	1989	horizontal burrow	North Carolina
Gordia marina	Emmons; Narbonne and Hofmann	1987	horizontal burrow	Wernecke Mtns
Gordia sp.	Hofmann	1981	horizontal burrow	Mackenzie Mtns
Gordia sp.	Elphinstone and Walter	1990	horizontal burrow	Amadeus Basin
Gordia sp.	Fedonkin	1985	horizontal burrow	Olenek Uplift
Gordia sp.	Crimes and Anderson	1985	horizontal burrow	Burin Peninsula
*Hagenetta aarensis**	Hahn and Pflug	1988	vertical burrow?	Namibia
*Harlaniella podolica**	Sokolov	1972	horizontal burrow	Podolia
*Harlaniella podolica**	Sokolov; Narbonne et al.	1987	horizontal burrow	Newfoundland
Helminthoida sp.	Fedonkin	1985	horizontal burrow	White Sea
*Helminthoidichnites tenuis**	Fitch	1850	narrow surface trail	New York
*Helminthoidichnites tenuis**	Fitch; Hofmann and Patel	1989	surface trail	New Brunswick
Helminthopsis sp.	Alpert	1977	horizontal burrow	California
Helminthopsis sp.	Fritz and Crimes	1985	horizontal burrow	British Columbia
Helminthopsis? sp.	Gibson	1989	horizontal trace	North Carolina
Hormosiroidea arumbera	Elphinstone and Walter	1990	series of burrows?	Amadeus Basin
Hormosiroidea canadensis	Crimes and Anderson	1985	series of burrows?	Newfoundland
*Intrites punctatus**	Fedonkin	1980	unknown	White Sea
Large sinuous trails	Glaessner	1985	horizontal tube?	Ediacara
Margaritichnus linearis	Fedonkin	1976	one row of pellets	White Sea
Meandering feeding trail	Cope	1983	horizontal trace	Wales
Meandering grazing patterns	Glaessner	1969	horizontal burrow	Ediacara
*Medvezhichnus pudicum**	Fedonkin	1985	horizontal burrow	White Sea
*Monocraterion tentaculatum**	Torell	1870	vertical burrow	Sweden
Monocraterion sp.	Hofmann and Patel	1989	vertical burrow	New Brunswick
Monomorphichnus bilinearis	Crimes; Elphinstone and Walter	1990	surface marking	Amadeus Basin
Monomorphichnus lineatus	Crimes et al.; Elphinstone and Walter	1990	surface marking	Amadeus Basin

Formation	Group	Comments	References
Lauhanvuoren		Cambrian?	Tynni and Hokkanen 1982.
Zhongyicun	Dengying	Cambrian	Crimes and Jiang 1986.
Zhongyicun	Dengying	Cambrian	Crimes and Jiang 1986.
Ust'-Pinega	Valdai	*Sellaulichnus meishucunensis*	Fedonkin 1985a.
Appekunny	Belt	dubiofossils	Horodyski 1982.
Siltstone units 1–2	Windermere	★★★	Keller et al. 1974; Narbonne and Hofmann 1981.
Mogilev		*Beltanelliformis brunsae*	Fedonkin 1985a.
Zaris	Nama	Vendian?	Crimes and Germs 1982.
Central Mt Stuart		pseudofossils	Walter, Elphinstone and Heys 1990.
Ust'-Pinega	Valdai	pseudofossil	Palij, Posti and Fedonkin 1979; Fedonkin 1981, 1985.
Lintiss Vale	Farnell	*Didymaulichnus lyelli*	Webby 1970b.
	Nankoweap	pseudofossil	Fürsich and Bromley 1985; Runnegar and Stait 1990.
Dabis	Nama	*incertae sedis*	Germs 1973b.
Brachina	Wilpena	may be a medusoid body fossil	Glaessner 1969; Cloud and Glaessner 1981; Glaessner 1984.
Lintiss Vale	Farnell	Cambrian?	Webby 1970b.
Blueflower	Windermere	Cambrian?	Aitken 1989.
Khmelnitski		Namakit-Daldyn	Palij, Posti and Fedonkin 1979.
Mogilev		Vendian	Palij, Posti and Fedonkin 1979.
unnamed		*Helminthoidichnites tenuis*	Cope 1983.
Arumbera 3 and 4		Cambrian	Walter Elphinstone and Heys 1990.
Arumbera 3		Cambrian	Walter, Elphinstone and Heys 1990.
Tommotian		*Curvolithus aequus*	Fedonkin 1977; Walter, Elphinstone and Heys 1990
Rawnsley	Wilpena	casts of twisted tissue	Glaessner 1969.
Arumbera 3 and 4		Cambrian	Walter, Elphinstone and Heys 1990.
Tommotian		*Taphrhelminthopsis dailyi?*	Fedonkin 1985a.
	Miette	★★★	Young 1972; Hofmann, Mountjoy and Teitz 1985.
Arumbera 3 and 4		Cambrian	Glaessner 1969; Walter, Elphinstone and Heys 1990.
Stelkuz		Vendian?	Fritz and Crimes 1985.
Khmelnitskiy		Nemakit-Daldyn	Palij 1974b; Palij, Posti and Fedonkin 1979; Fedonkin 1985.
Arumbera 3		Cambrian	Glaessner 1969; Walter, Elphinstone and Heys 1990.
Mickwitzia		★★★	Torell 1870; Häntzschel 1975.
Parachilna		Cambrian	Daily 1976.
Ajibik		pseudofossils?	Faul 1949, 1950.
Rawnsley	Wilpena	with Ediacara fauna	Runnegar 1990b.
Oligicene		★★	Ksiazkiewicz 1977.
Stelkuz		Vendian?	Fritz and Crimes 1985.
Chapel Island 1		Vendian?	Narbonne, Myrow, Landing and Anderson 1987.
Arumbera 3 and 4		Cambrian	Walter, Elphinstone and Heys 1990.
McManus	Albermarle	with Ediacara fuana	Gibson 1989.
Silstone unit 1	Windermere	with Ediacara fuana	Narbonne and Hofmann 1987.
Blueflower	Windermere	float below Ediacara fuana	Hofmann 1981.
Arumbera 3 and 4		*Helminthoidichnites tenuis*	Walter, Elphinstone and Heys 1990.
Nemakit-Daldyn		Cambrian	Fedonkin 1985a.
Chapel Island 1		*Helminthoidichnites tenuis*	Crimes and Anderson 1985.
Dabis	Nama	Vendian	Hahn and Pflug 1988.
Komarovo	Studientsa	Vendian	Sokolov 1972, 1973; Palij 1976.
Chapel Island 1		Cambrian?	Bengtson and Fletcher 1983; Narbonne et al. 1987.
Ust'Pinega	Valdai	Vendian	Fedonkin 1985a.
Ordovician		★★★	Fitch 1950; Hofmann and Patel 1989.
Ratcliffe Brook	Etcheminian	Cambrian	Hofmann and Patel 1989.
Campito		Cambrian	Alpert 1977.
Stelkuz		Vendian?	Fritz and Crimes 1985.
McManus	Albermarle	with Ediacara fuana	Gibson 1989.
Arumbera 3		Cambrian	Walter, Elphinstone and Heys 1990.
Chapel Island 2		Cambrian?	Crimes and Anderson 1985.
Ust'-Pinega	Valdai	Vendian	Fedonkin 1980b, 1985a.
Rawnsley	Wilpena	cast of ?mucilagenous tube	Glaessner 1969.
Ust'-Pinega	Valdai	*Neonereites uniserialis*	Fedonkin 1976.
unnamed		*Yelovichnus gracilis*	Cope 1983.
Rawnsley	Wilpena	*Yelovichnus gracilis*	Glaessner 1969.
Ust'-Pinega?	Valdai	Vendian	Fedonkin 1985a.
Early Cambrian		★★★	Torell 1870; Boyd 1966.
Ratcliffe Brook	Etcheminian	Cambrian	Hofmann and Patel 1989.
Arumbera 3 and 4		Cambrian	Walter, Elphinstone and Heys 1990.
Arumbera 3		Cambrian	Walter, Elphinstone and Heys 1990.

Described Taxa of Proterozoic and Selected Earliest Cambrian Megafossils

Table 23.4.1. (Continued).

Taxon	Author	Date	Type	Locality
Monomorphichnus lineatus	Crimes et al. Elphinstone and Walter	1990	surface marking	Amadeus Basin
*Nenoxites curvus**	Fedonkin	1976	horizontal trace	White Sea
*Neonereites biserialis**	Seilacher	1960	two rows of pellets	Germany
*Neonereites biserialis**	Seilacher; Fedonkin	1976	two rows of pellets	White Sea
*Neonereites biserialis**	Seilacher; Crimes and Jiang	1986	two rows of pellets	Yunnan
*Neonereites biserialis**	Seilacher; Gibson	1989	horizontal burrow	North Carolina
Neonereites renarius	Fedonkin	1980	row of pellets?	White Sea
Neonereites uniserialis	Seilacher; Fedonkin	1976	one row of pellets	White Sea
Neonereites uniserialis	Seilacher; Crimes and Jiang	1986	one row of pellets	Yunnan
Neonereites uniserialis	Seilacher; Gibson	1989	horizontal trace	North Carolina
Nereites sp.	Elphinstone and Walter	1990	horizontal trace	Amadeus Basin
Oldest metazoan trace fossils?	Kauffman and Steidtmann	1981	pseudofossils	Wyoming
*Olenichnus irregularis**	Fedonkin	1985	pseudofossil	Olenek Uplift
*Palaeopascichnus delicatus**	Palij	1976	horizontal trace	Podolia
*Palaeopascichnus delicatus**	Palij; Narbonne et al.	1987	horizontal trace	Newfoundland
Palaeopascichnus sinuosus	Fedonkin	1981	horizontal trace	White Sea
Palaeophycus alternatus	Pemberton and Frey; Elphinstone and Walter	1990	horizontal burrow	Amadeus Basin
*Palaeophycus tubularis**	Hall; Hofmann and Patel	1989	horizontal burrow	New Brunswick
*Palaeophycus tubularis**	Hall; Elphinstone and Walter	1990	horizontal burrow	Amadeus Basin
Phycodes pedum	Seilacher	1955	horizontal trace	Salt Range
Phycodes pedum	Seilacher; Glaessner	1990	horizontal trace	Amadeus Basin
Phycodes pedum	Seilacher; Banks	1970	horizontal trace	Norway
Phycodes pedum	Seilacher; Fedonkin	1985	horizontal trace	Podolia
Phycodes pedum	Seilacher; Narbonne et al.	1987	horizontal trace	Newfoundland
Phycodes sp.	Fedonkin	1985	horizontal burrow	Olenek Uplift
*Plagiogmus arcuatus**	Roedel	1926	horizontal burrow	Sweden
*Plagiogmus arcuatus**	Roedel: Elphinstone and Walter	1990	horizontal burrow	Amadeus Basin
*Planispiralichnus grandis**	Fedonkin	1985	inorganic?	Olenek Uplift
Planolites ballandus	Webby	1970	horizontal burrow	New South Wales
Planolites ballandus	Webby; Elphinstone and Walter	1990	horizontal burrow	Georgina Basin
*Planolites beverleyensis**	Narbonne et al.	1987	horizontal burrow	Newfoundland
*Planolites beverleyensis**	(Billings); Gibson	1989	horizontal trace	North Carolina
Planolites montanus	Richter; Narbonne and Hofmann	1987	horizontal burrow	Wernecke Mtns
Planolites montanus	Richter; Narbonne et al.	1987	horizontal burrow	Newfoundland
Planolites montanus	Richter; Gibson	1989	horizontal burrow	North Carolina
Planolites sp.	Alpert	1976	horizontal burrow	California
Planolites sp.	Fedonkin	1985	horizontal burrow	Olenek Uplift
Planolites sp.	Hofmann and Patel	1989	horizontal burrow	New Brunswick
Planolites sp.	Aitken	1989	horizontal burrow	Mackenzie Mtns
Precambrian fossil burrows	Faul	1950	pseudofossils?	Michigan
Precambrian trace fossils	Squire	1973	unresolved	Jersey
Problematical structure	Rayner	1957	tool mark?	Yorkshire
*Protospiralichnus circularis**	Fedonkin	1985	dubiofossil	Olenek Uplift
*Psammichnites gigas**	Torell; Hofmann and Patel	1989	crawling trails	New Brunswick
Roll- und Rieselmarken	Glaessner	1963	pseudofossil	Namibia
Rugoinfractus ovruchensis	Palij	1974	horizontal burrow	Ukraine
Scolicia sp.	Aitken	1989	horizontal trail	Mackenzie Mtns
*Sellaulichnus mcishucunensis**	Jiang	1982	horizontal trail	Yunnan
*Sellaulichnus meishucunensis**	Jiang	1990	horizontal trail	White Sea
Shallow branching burrows	Cope	1983	horizontal trail	Wales
Sinuous trails	Webby	1970	horizontal trail	New South Wales
Skolithos declinatus	Fedonkin	1985	inclined burrow	White Sea
Skolithos ramosus	Elphinstone and Walter	1990	vertical burrow	Amadeus Basin
Skolithos? sp.	Germs	1972	pseudofossil	Namibia
Skolithos sp.	Fritz and Crimes	1985	vertical burrow	British Columbia
Subphyllochorda? sp.	Hofmann and Patel	1989	horizontal burrow	New Brunswick
Surface trails	Glaessner	1969	surface trails	Ediacara
*Suzmites volutatus**	Fedonkin	1976	dubiofossil	White Sea
Syringomorpha nilssoni?	Nathorst; Gibson	1989	horizontal burrow	North Carolina
Taphrhelminthoida dailyi	Hofmann and Patel	1989	horizontal trail	New Brunswick
Taphrhelminthopsis circularis	Crimes	1977	horizontal trace	Spain
Taphrhelminthopsis circularis	Crimes; Fritz and Crimes	1985	horizontal trace	British Columbia
Three-ridged trails	Germs	1972	horizontal trace	Arimas
*Torrowangea rosei**	Webby	1970	horizontal burrow	New South Wales

Formation	Group	Comments	References
Arumbera 3		Cambrian	Walter, Elphinstone and Heys 1990.
Ust'-Pinega	Valdai	Vendian	Fedonkin 1976, 1977, 1981, 1985.
Jurassic		★★★	Seilacher 1960; Häntzschel 1975.
Ust'-Pinega	Valdai	Vendian	Fedonkin 1976, 1977, 1981, 1985a.
Zhongyicun	Dengying	Cambrian	Crimes and Jiang 1986.
McManus	Albermarle	with Ediacara fauna	Gibson 1989.
Ust'-Pinega	Valdai	Vendian	Fedonkin 1980b, 1985a.
Ust'-Pinega	Valdai	Vendian	Fedonkin 1976, 1977, 1981, 1985a.
Zhongyicun	Dengying	Cambrian	Crimes and Jiang 1986.
McManus	Albermarle	with Ediacara fauna?	Gibson 1989.
Arumbera 3		Cambrian	Walter, Elphinstone and Heys 1990.
Medicine Peak	Libby Creek	pseudofossils; metamorphic	Kauffman and Steidtmann 1981; Runnegar and Stait 1990.
Nemakit-Daldyn		pseudofossil	Fedonkin 1985a.
Mogilev		Vendian	Palij 1976; Fedonkin 1977, 1981, 1985a.
Chapel Island 1		above Ediacara fauna	Narbonne, Myrow, Landing and Anderson 1987.
Ust'-Pinega		Vendian	Fedonkin 1981; 1985a.
Arumbera 3 and 4		Cambrian	Walter, Elphinstone and Heys 1990.
Ratcliffe Brook	Etcheminian	Cambrian	Hofmann and Patel 1989.
Arumbera 3 and 4		Cambrian	Walter, Elphinstone and Heys 1990.
Cambrian		★★★	Seilacher 1955; Häntzschel 1975.
Arumbera 3 and 4		Cambrian	Glaessner 1969; Walter, Elphinstone and Heys 1990.
lower Breivik		Cambrian?	Banks 1970
Khmelnitski		Nemakit-Daldyn	Fedonkin 1985a.
Chapel Island 2		Cambrian	Narbonne, Myrow, Landing and Anderson 1987.
Nemakit-Daldyn		*Palaeophycus?*	Fedonkin 1985a.
Kalmarsund		★★★	Roedel 1926; Jaeger and Martinsson 1980.
Arumbera 3 and 4		Cambrian	Glaessner 1969; Walter, Elphinstone and Heys 1990.
Nemakit-Daldyn		inorganic or body fossil?	Fedonkin 1985a.
Lintiss Vale	Farnell	Cambrian?	Webby 1970b.
Elkera		Vendian	Glaessner 1984; Walter, Elphinstone and Heys 1990.
Chapel Island 1		Vendian?	Narbonne, Myrow, Landing and Anderson 1987.
McManus	Albermarle	with Ediacara fauna?	Gibson 1989.
Siltstone unit 1	Windermere	with Ediacara fauna	Narbonne and Hofmann 1987.
Chapel Island 1		Vendian?	Narbonne, Myrow, Landing and Anderson 1987.
McManus	Albermarle	with Ediacara fauna?	Gibson 1989.
Wyman		*Gordia marina?*	Alpert 1975, 1975; Fedonkin and Lipps, in press.
Nemakit-Daldyn		Cambrian	Fedonkin 1985a.
Ratcliffe Brook	Etcheminian	Cambrian	Hofmann and Patel 1989.
Blueflower	Windermere	Vendian	Aitken 1989.
Ajibik		pseudofossils?	Faul 1950.
Briverian		unresolved	Squire 1973.
Ingletonian		Cambrian?	Rayner 1957; Kelling and Whitaker 1970; O'Nions et al. 1973.
Nemakit-Daldyn		origin unclear	Fedonkin 1985a.
Ratcliffe Brook	Etcheminian	Cambrian	Hofmann and Patel 1989.
Schwarzkalk	Nama	pseudofossil	Glaessner 1963.
Ovruch	Riphean	probably Paleozoic	Palij 1974a.
Ingta	Windermere	Cambrian?	Aitken 1989.
Zhongyicun	Dengying	Cambrian	Jiang, Luo and Zhang 1982; Crimes and Jiang 1986.
Ust'-Pinega	Valdai	★★★	Jiang, Luo and Zhang 1982.
unnamed		with Ediacara fauna	Cope 1983.
Fowlers Gap	Farnell	Vendian?	Webby 1970b.
Ust'-Pinega	Valdai	above Ediacara fauna	Fedonkin 1985a.
Arumbera 3 and 4		Cambrian	Walter, Elphinstone and Heys 1990.
Zaris		pseudofossil	Germs 1972c.
Stelkuz		Vendian?	Fritz and Crimes 1985.
Ratcliffe Brook	Etcheminian	Cambrian	Hofmann and Patel 1989.
Rawnsley	Wilpena	*Helminthoidichnites tenuis*	Glaessner 1969; Walter, Elphinstone and Heys 1990.
Ust'-Pinega	Valdai	dubifossil	Fedonkin 1976, 1977.
McManus	Albermarle	*Vimenites bacillaris?*	Gibson 1989.
Ratcliffe Brook	Etcheminian	Cambrian	Hofmann and Patel 1989.
Cayetano		Cambrian	Crimes 1976.
Stelkuz		Vendian?	Fritz and Crimes 1985.
Nasep	Nama	*Curvolithus sp.*	Germs 1972c.
Lintiss Vale	Farnell	★★	Webby 1970b.

Table 23.4.1. (Continued).

Taxon	Author	Date	Type	Locality
*Torrowangea rosei**	Webby; Elphinstone and Walter	1990	horizontal burrow	Amadeus Basin
Torrowangea sp.	Hofmann	1981	horizontal burrow	Mackenzie Mtns
Torrowangea sp.	Narbonne et al.	1987	horizontal burrow	Newfoundland
Torrowangea? sp.	Gibson	1989	horizontal burrow	North Carolina
Traces of mud-eater	Sokolov	1973	horizontal burrow	Podolia
Transversely segmented trail	Webby	1970	horizontal trace	New South Wales
*Treptichnus bifurcus**	Miller; Fedonkin	1985	horizontal burrow	Podolia
Treptichnus triplex	Palij	1976	horizontal trace	Podolia
*Vendichnus vendicus**	Fedonkin	1979	unknown	White Sea
Vendovermites	Sokolov and Fedonkin	1984	unknown	
*Vermiforma antiqua**	Cloud	1976	horizontal burrow?	North Carolina
*Vimenites bacillaris**	Fedonkin	1980	horizontal trace	White Sea
World's oldest animal traces	Clemmy	1976	modern burrows	Zambia
Worm trail	Glaessner	1985	horizontal trail	Ediacara
*Yelovichnus gracilis**	Fedonkin	1985	grazing structure	White Sea

Formation	Group	Comments	References
Arumbera 3		Cambrian	Walter, Elphinstone and Heys 1990.
Blueflower	Windermere	Vendian	Hofmann 1981; Aitken 1989.
Chapel Island 1		above Ediacara fauna	Narbonne, Myrow, Landing and Anderson 1987.
McManus	Albermarle	with Ediacara fauna?	Gibson 1989.
Yaryshev		*Torrowangea sp.*	Sokolov 1973.
Lintiss Vale	Farnell	*Palaepascichnus delicatus?*	Webby 1970b.
Khmelnitski		Cambrian	Fedonkin 1985a.
Khmelnitski		Cambrian	Palij 1976; Palij, Posti and Fedonkin 1979; Fedonkin 1985a.
Ust'-Pinega	Valdai	Vendian	Palij, Posti and Fedonkin 1979; Fedonkin 1981.
		nomen nudum?	Sokolov and Fedonkin 1984.
		Taphrhelminthopsis circularis?	Cloud et al. 1976.
Ust'-Pinega	Valdai	★★	Fedonkin 1980b, 1985a.
Ore horizon	Roan	modern termite burrows	Clemmy 1976; Cloud et al. 1980.
Rawnsley	Wilpena	*Helminthoidichnites tenuis**	Glaessner 1969.
Ust'-Pinega	Valdai	★★	Fedonkin 1985a.

23.5

Proterozoic and Earliest Cambrian Skeletal Metazoans

STEFAN BENGTSON

In Table 23.5 are listed all published genera of Late Proterozoic-Early Cambrian skeletal fossils and metazoan body fossils, exclusive of calcareous algae (Section 23.2) and Ediacaran fossils (Sections 23.3, 23.4), used for analyses presented in Sections 7.7, Chapter 8, and Section 11.4. They have been separated into higher taxonomical categories to the phylum level (see Sections 7.7 and 8.5 regarding the definition of phyla), but the taxonomy presented herein is simplified, because the usage of higher-level taxa is frequently unstable, and with the exception of the phylum level they play no part in the analyses.

The genera have been culled from the database on early skeletal fossils compiled by Bengtson, complemented by Sepkoski's generic database (in particular for arthropods and —for archaeocyathans—the compilations of Debrenne and Rozanov (1983) and Hill (1972). This is not a complete list of known occurrences of organism groups, as it does not include forms that have not been described and named.

Names denoted by asterisks are deemed not valid in the sense of the International Code of Zoological Nomenclature. Available synonyms and some *nomina nuda* have been selectively included. The ranges within brackets are where feasible expressed in Siberian or Chinese stages (Tomm–Tommotian; Atda–Atdabanian; Boto–Botomian; Tojo–Tojonian; Meish–Meishucunian; Qiong–Qiongzhusian; Cang–Canglangpuan; Vend–Vendian; N-D–Nemakit-Daldyn; Camb–Cambrian; L–Lower; M–Middle; U–Upper; cf. also Fig. 7.5.2).

Table 23.5. *Genera of Late Proterozoic-Early Cambrian skeletal fossils and metazoan body fossils, exclusive of Ediacaran fossils and calcareous algae.*

PORIFERANS
PHYLUM PORIFERA GRANT 1872
    Class Calcarea Bowerbank 1864
        Dodecaactinella Reif 1968 [U Atda-Silurian]
        Phobetractinia Reif 1968 [U L Camb-L Carboniferous]
       *Polyactinella Mostler 1985 (= *Dodecaactinella* Reif 1968)
       *Sardospongia Mostler 1985 (= *Dodecaactinella* Reif 1968)
        Zangerlispongia Rigby and Nitecki 1975 [L Camb?-Carboniferous]
        Zigzagella Jiang and Huang 1986 [M Camb?]
    Order Heteractinida Hinde 1888
        Family Eiffeliidae Rigby 1986
       * *Actinoites* Duan 1984 (= *Eiffelia* Walcott 1920)
        Eiffelia Walcott 1920 [U Meish?-M Camb]
       * *Lenastella* Missarzhevsky 1981 in Missarzhevsky and Mambetov, 1981 (= *Eiffelia* Walcott 1920)
       * *Niphadus* Duan 1984 (= *Eiffelia* Walcott 1920)
    Class Demospongea Sollas 1875
        Choia Walcott 1920 [Boto?-Ordovician]
        Hazelia Walcott 1920 [U L Camb-M Camb]
        Heterostella Fedorov 1987 *in* Shabanov et al., 1987 [U Atda]
        Inflexiostella Fedorov 1987 *in* Shabanov et al. 1987 [U Atda]
        Lenica Goryanskij 1977 [Boto]
        Leptomitus Walcott 1886 [U L Camb-M Camb]
        Paraleptomitella Chen et al. 1989 *in* Chen Junyuan et al. 1989 [L Qiong]
        Protohyalostelia Chapman 1940 [Atda?]
        Taraxaculum Bengtson 1990 *in* Bengtson et al. 1990 [U Atda]
        Vauxia Walcott 1920 [L Qiong-M M Camb]
    ?Class Demospongea Sollas 1875
        Karatubulus Missarzhevsky 1981 *in* Missarzhevsky and Mambetov, 1981 [U Atda]
    Class Hexaxtinellida Sollas 1887
        Azyrtalia Nazarov 1973 [U Atda?]
      *Brabbinthes Allison 1975 (= Hexactinellida gen. indet.)
        Calcihexactina Sdzuy 1969 [M Camb]
        Hunanospongia Qian & Ding 1988 *in* Ding and Qian 1988 [Meish]
        Protospongia Salter 1864 [L Tomm-Ordovician]
    'Sphinctozoans'
        Jawonya Kruse 1987 [U Atda?-Ordian]
        Wagima Kruse 1987 [Ordian]
?PHYLUM PORIFERA GRANT 1872
        Drosdovia Sayutina 1980 [Atda-Boto]
        Eocoryne Matthew 1886 [Boto-M Camb]

Table 23.5 (Continued).

  Edelsteinia Vologdin 1940 [Atda-Boto]
  Khasaktia Sayutina 1980 [L Atda-L Boto]
  Korovinella Radugin 1960 *in* Khalfina, 1960 [Boto]
  Rackovskia Vologdin 1940 [Atda-Boto]
  Vittia Sayutina 1980 [L Atda]

ARCHAEOCYATHANS
PHYLUM ARCHAEOCYATHA BORNEMANN 1884
[The list has been compiled from Debrenne and Rozanov (1983) and Hill 1972).]

 Class Irregulares Vologdin 1937
  Order Archaeocyathida Okulitch 1935
   Abakanicyathus [Boto]
   Acanthopyrgus [L Camb]
   Agastrocyathus [Atda]
   Anthomorpha [U Atda-L Tojo]
   Archaeocyathus [Boto-Tojo]
   Archaeofungia [U Atda-L Boto]
   Archaeopharetra [U Atda-L Boto]
   Archaeosycon [U Tomm-U Tojo]
   Ardrossocyathus [L Camb]
   Batenevia [U Tojo]
   Beltanacyathus [L Camb]
   Beticocyathus [L Camb]
   Bicyathus [U Tomm-Boto]
   Bottonaecyathus [Boto]
   Cambrocyathellus [M Tomm-Atda]
   Cambrocyathus [L Camb]
   Cambronanus [Boto]
   Chouberticyathus [U Atda-Boto]
   Claruscoscinus [Boto-U Tojo]
   Claruscyathus [Boto-U Tojo]
   Copleicyathus [U Atda-L Boto]
   Dendrocyathus [L Camb]
   Dictyocoscinus [U Atda-L Boto]
   Dictyocyathus [M Tomm-Boto]
   Fenestrocyathus [U Atda-L Boto]
   Flindersicoscinus [U Atda-L Boto]
   Fragilicyathus [Boto]
   Hupecyathus [Atda]
   Kazakhstanicyathus [Tojo]
   Metacoscinus [U Atda-L Boto]
   Metafungia [U Atda-L Boto]
   Metaldetes [U Atda-L Boto]
   Metethmophyllum Atda]
   Okulitchicyathus [L Tomm-L Atda]
   Palmericyathellus [U Atda-L Boto]
   Paracoscinus [U Atda-L Boto]
   Paranacyathus [M Tomm-Boto]
   Pinacocyathus [U Atda-L Boto]
   Prismocyathus [Boto]
   Protocyclocyathus [L Camb]
   Protopharetra [M Tomm-Tojo]
   Pseudosyringocnema [Atda]
   Pycnoidocyathus [Atda-Tojo]
   Shiveligocyathus [U Boto]
   Sigmofungia [U Atda-L Boto]
   Sphinctocyathus [M Tomm-Atda]
   Spirillicyathus [Boto]
   Syringocnema [U Atda-Boto]
   Syringocoscinus [L Camb]
   Syringsella [U Tojo]

  Tabellaecyathus [Boto]
  Tabulacyathus [U Atda-Boto]
  Tolliccyathus [Boto]
  Tubocyathus [L Camb]
  Volvacyathus [Atda]
  Voznesenskicyathus [U Boto]
 Order Thasassocyathida Vologdin 1962
  Bacatocyathus [Tomm-Boto]
  Thalassocyathus [Boto]

 Class Regulares Vologdin 1937
  Order Ajacicyathida Bedford and Bedford 1939
   Afiacyathus [M Atda-U Boto]
   Ajacicyathellus [M Atda-U Boto]
   Ajacicyathus [U Atda-U Boto]
   Alconeracyathus [M Atda-M Boto]
   Aldanocyathus [L Tomm-U Boto]
   Ambistapis [L Boto-U Boto]
   Angaricyathus [U Atda-M Tojo]
   Annulocyathella [L Boto-U Boto]
   Annulocyathus [L Boto-U Boto]
   Annulofungia [M Boto]
   Aporosocyathus [L Boto-U Boto]
   Arthurocyathus [L Atda-L Boto]
   Baikalocyathus [L Atda-U Boto]
   Botomocyathus [L Boto-M Boto]
   Cadniacyathus [L Boto-U Boto]
   Carinacyathus [L Atda-U Atda]
   Carpicyathus [L Boto-M Boto]
   Chakassicyathus [M Atda-U Boto]
   Chankacyathus [L Boto-U Boto]
   Clathricyathus [L Boto-U Boto]
   Compositocyathus [L Atda-U Boto]
   Conannulofungia [M Atda-L Boto]
   Cordilleracyathus [L Boto-U Boto]
   Cricopectinus [L Boto-U Boto]
   Cyathocricus [L Boto-U Boto]
   Cyclocyathella [M Atda-U Boto]
   Dailycyathus [L Boto-U Boto]
   Degeletiicyathus [M Atda-U Boto]
   Degeletticyathellus [M Atda-U Boto]
   Denaecyathus [L Boto-U Boto]
   Densocyathus [L Boto-U Boto]
   Didymocyathus [L Boto-U Boto]
   Diplocyathellus [L Boto-U Boto]
   Dupliporocyathus [M Atda-U Atda]
   Eladicyathus [M Atda-U Atda]
   Erbocyathus [M Boto-U Tojo]
   Ethmocyathus [L Boto-U Boto]
   Ethmopectinus [L Bono-U Boto]
   Ethmophyllum [U Atda-U Boto]
   Fallocyathus [M Atda-M Boto]
   Fansycyathus [U Atda-L Boto]
   Flexannulus [L Boto-U Boto]
   Formosocyathus [M Atda-U Boto]
   Frinalicyathus [M Atda-U Boto]
   Gagarinicyathus [M Atda-M Boto]
   Geocyathus [L Atda-L Boto]
   Glaessnericyathus [L Boto-U Boto]
   Gloriosocyathus [L Boto]
   Gnaltacyathus [L Boto-U Boto]
   Gordonicyathellus [U Atda]

Table 23.5 (Continued).

Gordonicyathus [M Atda-U Boto]
Gordonifungia [M Atda-L Boto]
Gorskinocyathus [L Boto-U Boto]
Gumbycyathus [L Boto-U Boto]
Halysicyathus [M Boto]
Heckericyathus [L Atda-U Atda]
Hupecyathellus [U Atda-L Boto]
Hyptocyathus [M Boto]
Ichnusocyathus [L Boto-U Boto]
Inacyathella [M Boto-U Boto]
Inessocyathus [M Atda-U Boto]
Irinaecyathus [L Boto-U Tojo]
Isiticyathus [U Atda-L Boto]
Jakutocarinus [L Atda-L Tojo]
Jangudacyathus [U Atda-L Boto]
Japhanicyathus [U Atda-L Boto]
Kandatocyathus [M Atda-U Boto]
Kellericyathus [M Atda-U Atda]
Kijacyathus [M Atda]
Kisasaecyathus [M Boto-U Boto]
Kordecyathus [L Boto-U Boto]
Kotuyicyathellus [M Atda-U Atda]
Kotuyicyathus [M Tomm-M Boto]
Krasnopeevacyathus [M Boto]
Ladaecyathus [L Boto-U Boto]
Lebediscyathus [L Boto-U Boto]
Lenocyathus [M Atda-U Atda]
Leptosocyathus [L Atda-U Boto]
Loculicyathellus [M Boto]
Loculicyathus [L Atda-U Boto]
Mackenziecyathus [U Atda-U Boto]
Mattajacyathus [L Boto]
Nalivkinicyathus [M Atda-U Atda]
Neoloculicyathus [L Atda-U Boto]
Nochoriocyathellus [M Atda]
Nochoriocyathus [L Tomm-U Boto]
Nuchacyathus [L Boto-U Boto]
Olgaecyathus [L Boto-U Boto]
Orbiasterocyathus [U Atda-L Boto]
Orbicyathellus [M Atda-U Atda]
Orbicyathus [U Tomm-M Boto]
Palmericyathus [U Atda-U Boto]
Pectenocyathus [M Boto]
Peregrinicyathus [L Boto-U Boto]
Piamaecyathellus [L Boto-U Boto]
Piamaecyathus [L Boto-U Boto]
Plicocyathus [M Boto-U Boto]
Porocyathellus [M Boto-U Boto]
Porocyathus [M Atda-U Boto]
Prethmophyllum [L Boto-U Boto]
Pretiosocyathus [M Atda-U Boto]
Raropectinus [M Atda]
Rasetticyathus [L Atda-U Boto]
Rectannulus [M Boto-U Boto]
Rewardocyathus [U Atda-M Boto]
Ringifungia [M Atda-L Boto]
Robertocyathus [M Atda-U Boto]
Robustocyathellus [M Atda-U Boto]
Robustocyathus [M Tomm-U Boto]
Rossocyathella [U Atda-U Boto]
Rotundocyathus [U Tomm-U Boto]
Russocyathus [M Atda-U Atda]
Sagacyathus [L Boto-U Boto]
Sajanacyathus [M Boto]
Sanarkocyathus [M Boto-U Boto]
Schiderticyathellus [M Tojo-U Tojo]
Sibirecyathus [U Tomm-U Boto]
Sichotecyathus [M Atda-U Boto]
Sigmocyathus [L Boto-U Boto]
Squamosocyathus [M Atda-M Boto]
Stapicyathus [L Boto-U Boto]
Svetlanocyathus [M Boto-U Boto]
Taylorcyathus [L Atda-U Boto]
Tegerocyathus [M Boto-U Tojo]
Tennericyathus [L Atda-M Atda]
Tercyathus [L Boto-U Boto]
Terraecyathus [U Atda-U Boto]
Tersicyathus [M Atda-U Boto]
Thalamocyathus [M Atda-U Boto]
Thalamopectinus [L Boto-U Boto]
Torosocyathus [M Atda-U Boto]
Trininaecyathus [L Boto-M Boto]
Tumulifungia [M Atda-U Boto]
Tumulocyathellus [M Atda-U Boto]
Tumulocyathus [U Tomm-U Boto]
Urcyathella [L Boto-U Boto]
Urcyathus [M Atda-U Boto]
Vologdinocyathellus [M Atda-U Boto]
Vologdinocyathus [M Atda-U Tojo]
Voroninicyathus [M Atda-U Boto]
Wrighticyathus [M Boto-U Boto]
Yudjaicyathus [U Atda]
Yukonocyathus [U Atda-U Boto]
Zonacyathus [M Atda-M Tojo]
Order Coscinocyathida Zhuravleva 1955
Agyrekocyathus [U Atda-U Boto]
Alataucyathus [M Atda-M Boto]
Anaptychocyathus [L Boto-M Boto]
Asterocyathus [L Boto-U Boto]
Asterotumulus [M Atda-M Boto]
Axiculifungia [M Atda-U Atda]
Bractocyathus [L Boto-U Boto]
Calyptocoscinus [L Boto-U Boto]
Chengkoucyathus [L Boto-L Tojo]
Clathricoscinus [U Atda-U Boto]
Coscinocyathellus [L Boto-U Boto]
Coscinobyathus [M Tomm-U Boto]
Coscinoptycta [L Boto-U Boto]
Erugatocyathus [M Atda-U Boto]
Ethmocoscinus [M Boto-U Boto]
Flexicyathus [L Boto-L Tojo]
Geniculicyathus [M Atda-U Atda]
Ijinicyathus [M Atda-U Atda]
Jebiletticoscinus [L Boto-U Boto]
Kazyrycyathus [L Boto-U Boto]
Lanicyathus [L Boto-U Boto]
Lunulacyathus [M Boto]
Membranacyathus [M Atda-U Atda]
Mennericyathus [M Atda-U Boto]
Mootwingeecyathus [L Boto-U Boto]
Mrassucyathus [M Atda-U Atda]
Mucchatocyathus [L Boto]

Table 23.5 (Continued).

Orbicoscinus [L Boto-M Boto]
Orienticyathus [L Boto-M Boto]
Pluralicoscinus [L Boto-U Boto]
Polycoscinus [L Boto-U Boto]
Polystillicidocyathus [L Boto-U Boto]
Porocoscinus [L Boto-U Boto]
Retecoscinus [M Tomm-M Boto]
Retetumulus [M Boto-U Boto]
Rozanovicoscinus [L Boto-U Boto]
Rozanovicyathus [L Boto-M Boto]
Rudanalus [M Boto]
Salairocyathus [L Boto-U Boto]
Schumnycyathus [L Boto]
Sigmocoscinus [M Boto]
Silviacoscinus [L Boto-U Boto]
Somphocyathus [L Boto-M Boto]
Statanulocyathus [L Boto-U Boto]
Stillicidocyathus [L Boto-U Boto]
Tatijanaecyathus [L Boto]
Tegerocoscinus [L Boto-U Boto]
Tomocyathus [M Atda-L Boto]
Tubicoscinus [L Boto-U Boto]
Tumulocoscinus [M Atda-L Boto]
Veronicacyathus [U Atda-M Boto]
Xestecyathus [L Boto-U Boto]
Zonacoscinus [L Boto-U Boto]
Order Monocyathida Okulitch 1935
Alphacyathus [M Boto]
Aptocyathella [L Boto-U Boto]
Aptocyathus [L Boto-U Boto]
Archaeolynthus [L Tomm-U Boto]
Batschikicyathus [U Atda]
Butakovicyathus [U Atda-L Boto]
Capsolynthus [U Atda-M Boto]
Capsulocyathus [L Atda-U Boto]
Chabakovicyathus [M Atda-U Boto]
Cordobicyathus [M Atda-U Atda]
Cryptoporocyathus [L Tomm-M Boto]
Debrennecyathus [L Boto-U Boto]
Dokidocyathella [M Atda-U Boto]
Dokidocyathus [L Tomm-U Boto]
Favilynthus [M Boto]
Fransuasaecyathus [L Atda-U Boto]
Galinaecyathus [U Atda-U Boto]
Gerbikanaecyathus [M Boto-U Boto]
Globosocyathus [M Atda-U Boto]
Incurvocyathus [L Boto-U Boto]
Kaltatocyathus [L Atda-U Boto]
Kazakovicyathus [L Boto-U Boto]
Kidrjassocyathus [M Atda-U Boto]
Kyarocyathus [L Boto-U Boto]
Melkanicyathus [U Boto]
Mirandocyanthus [M Boto]
Papillocyathus [M Atda-U Atda]
Phymatocyathus [M Tojo-U Tojo]
Propriolynthus [M Tomm-U Boto]
Rhabdocyathella [L Boto-L Tojo]
Sajanolynthus [L Boto-U Boto]
Sekwicyathus [U Atda-U Boto]
Soanicyathus [L Boto-U Boto]
Subtilocyathus [L Boto-U Boto]

Tubericyathus [M Tojo-U Tojo]
Tumuliolynthus [M Tomm-U Boto]
Tumuloglobosus [U Boto]
Uralocyathella [L Boto-U Boto]
Zhuravlevaecyathus [M Boto]
?PHYLUM ARCHAEOCYATHA
Palaeoconularia Chudinova 1959 [L Camb]

RADIOCYATHANS
PHYLUM NOT ASSIGNED
  Class Radiocyatha Debrenne et al. 1970
    Blastasteria Debrenne et al. 1971 [U Atda]
    Girphanovella Zhuravleva 1967 in Zhuravleva et al. 1967 [U Atda?-Boto]
    Gonamispongia Korshunov 1968 [Tomm]
    Heterocyathus Bedford and Bedford 1934 (= *Radiocyathus* Okulitch 1937)
    Kuraya Romanenko 1968 [Boto]
    Radiocyathus Okulitch 1937 [U Atda]
    Uranosphaera Bedford and Bedford 1934 [U Atda]

CNIDARIANS
PHYLUM CNIDARIA HATSCHEK 1888
  Class Hydrozoa Owen 1843
    Order Siphonophorida Eschscholtz 1829
      Family Porpitidae Brandt 1835
        Velumbrella Stasińska 1960 [U L Camb]
      Family Scenellidae Wenz 1938
        Scenella Billings 1872 [L Camb]
      Family not assigned
        Rotadiscus Sun and Hou 1987 [L Qiong]
  Class Hydroconozoa Kordeh 1963
    Order Hydryconida Kordeh 1963
      Dasyconus Kordeh 1963 [L Camb]
      Gastroconus Kordeh 1963 [L Camb]
      Hydroconus Kordeh 1963 [L Camb]
      Tabulaconus Handfield 1969 [Boto]
      Tuvaeconus Kordeh 1963 [L Camb]
?PHYLUM CNIDARIA HATSCHEK 1888
    Cambrotrypa [L Camb-U M Camb]
    Coelenteratella Kordeh 1959 [L M Camb]
    Lenaella Kordeh 1959 [Atda]
    Protolyellia Torell 1870 [L Atda-M Camb]
    Rosellatana Kobluk [L Boto]
    Spatangopsis Torell 1870 [L Atda]
  ?Class Anthozoa Ehrenberg 1834
    ?Order Octocorallia Haeckel 1866
      Microcoryne Bengtson 1990 in Bengtson et al. 1990 [U Atda]

PRIAPULIDS
PHYLUM PRIAPULIDA
    Cricocosmia Hou and Sun 1988 [L Qiong]
    Maotianshania Sun and Hou 1987 [L Qiong]
    Ottoia Walcott 1911 [L Qiong-M M Camb]
    Selkirkia Walcott 1911 [Boto-M M Camb]

ANNELIDS
PHYLUM ANNELIDA LAMARCK 1809
    Myoscolex Glaessner 1979 [U L Camb-M M Camb]
    Palaeoscolex Whittard 1953 [L Qiong-Pridolian]
    Vetustovermis Glaessner 1979 [U L Camb]

TRILOBOZOANS
PHYLUM TRILOBOZOA FEDONKIN 1985
    Family Anabaritidae Missarzhevsky 1974

Table 23.5 (Continued).

Aculeochrea Val'kov and Sysoev 1970 [N-D]
Anabaritellus Missarzhevsky 1974 [L Tomm-U Atda?]
Anabarites Missarzhevsky 1969 in Voronova and Missarzhevsky, 1969 [N-D-U Atda?]
* *Angustiochrea* Val'kov and Sysoev 1970 (= *Anabarites* Missarzhevsky 1969)
Cambrotubulus Missarzhevsky 1969 in Rozanov et al. 1969 [L Tomm-U Tomm]
Gastreochrea Val'kov 1982 [L Tomm]
* *Jakutiochrea* Val'kov and Sysoev 1970 (= *Anabarites* Missarzhevsky 1969)
Kotuites Missarzhevsky 1989 [N-D]
* *Kotyikanites* Bokova 1985 (= *Anabarites* Missarzhevsky 1969)
Lobiochrea Val'kov and Sysoev 1970 [N-D]
Longiochrea Val'kov and Sysoev 1970 [N-D]
Mariochrea Val'kov 1982 [L Tomm]
* *Selindeochrea* Val'kov 1982 (= *Anabarites* Missarzhevsky 1969)
Tiksitheca Missarzhevsky 1969 in Rozanov et al. 1969 [L Tomm]
* *Udzhaites* Vasil'eva 1986 (= *Anabarites* Missarzhevsky 1969)

?PHYLUM TRILOBOZOA
  ?Family Anabaritidae Missarzhevsky 1974
    Eogloborilus Qian 1977 [L Meish]
    Lophotheca Qian 1977 [L Meish]
    Paragloborilus Qian 1977 [Meish]
    Salanytheca Missarzhevsky 1981 [N-D-Tomm]

COLEOLIDS
PHYLUM, CLASS AND ORDER NOT ASSIGNED
  Family Coleolidae Fisher 1962
    Coleoloides Walcott 1890 [Tomm-L Camb]
    Coleolus Hall 1879 [L Tomm-Carboniferous]
    * *Glauderia* Poulsen 1967 (= *Coleoloides* Walcott 1890)

CRIBRICYATHANS
PHYLUM CRIBRICYATHA VOLOGDIN 1964
  Class Cribricyathea Vologdin 1961
    Order Conoidocyathida Vologdin 1964
      Azyricyathus Vologdin 1964 [Tojo]
      Conoidocyathus Vologdin 1964 [Boto]
      Pubericyathus Vologdin 1964 [Boto]
    Order Cribricyathida Vologdin 1964
      Apocyathus Vologdin 1964 [Boto]
      Capillicyathus Vologdin 1964 [Boto]
      Cribricyathus Vologdin 1964 [Boto-Tojo]
      Dolichocyathus Vologdin 1964 [Tojo]
      Gracilicyathus Vologdin and Jankauskas 1968 in Vologdin and Yankauskas, 1968 [Atda]
      Lagenicyathus Vologdin 1964 [Boto]
      Lomatiocyathus Vologdin 1964 [Tojo]
      Longicyathus Vologdin 1964 [Boto]
      Peripteratocyathus Vologdin 1964 [Tojo]
      Pyxidocyathus Vologdin 1964 [Boto]
      Radicicyathus Vologdin 1964 [Tojo]
      Rarocyathus Vologdin and Jankauskas 1968 in Vologdin and Yankauskas 1968 [Atda]
      Sunicyathus Vologdin 1964 [Boto]
      Thecocyathus Vologdin 1964 [Boto]
      Tortocyathus Vologdin and Jankauskas 1968 in Vologdin and Yankauskas 1968 [Atda-Boto]
      Turricyathus Vologdin 1964 [Boto]
    Order Vologdinophyllacea Radugin 1964
      Achorocyathus Jankauskas 1965 in Jankauska 1965 [Atda]
      Akademiophyllum Radugin 1964 [Atda]
      Cardiophyllum Radugin 1964 [Atda]
      Crispus Jankauskas 1965 in Jankauska 1965 [Atda]
      Dubius Jankauskas 1969 [Atda]
      Erphyllum Radugin 1964 [Atda]
      Lacerathus Jankauskas 1965 in Jankauska 1965 [Atda]
      Leibaella Jankauskas 1964 [Atda]
      Longaevus Jankauskas 1965 in Jankauska 1965 [Atda]
      Pterocyathus Jankauskas 1965 in Jankauska 1965 [Atda]
      Ramifer Jankauskas 1965 in Jankauska 1965 [Atda]
      Topolinocyathus Jankauskas 1965 in Jankauska 1965 [Atda]
      Vologdinophyllum Radugin 1962 [Atda]
  Class, order and family not assigned
    Abicyathus Jankauskas 1972 [Tojo]
    Archaeophyllum Simon 1939 [U Atda]
    Butovia Vologdin 1931 [U Atda]
    Manacyathus Jankauskas 1969 [Atda]
    Striatocyathus Vologdin and Jankauskas 1968 in Vologdin and Yankauskas 1968 [Atda-Boto]

SABELLIDITIDS
PHYLUM AND CLASS NOT ASSIGNED
  Order Sabelliditida Sokolov 1965
    Calyptrina Sokolov 1965 [U Vend]
    Paleolina Sokolov 1965 [U Vend-L Camb]
    Parasabellidites Sokolov 1967 [L L Camb]
    Saarina Sokolov 1965 [N-D]
    Sabellidites Yanishevskij [N-D-L L Camb]
    Sokolovina Kir'yanov 1968 [Tomm]
  ?Order Sabelliditida Sokolov 1965
    Anhuiella Yan and Xing 1984 in Xing Yusheng 1985 [U Vend]
    * *Huainanella* Zheng 1982 (= *Sinosabellidites* Zheng 1980)
    Huaiyuanella Xing et al. in Xing Yusheng 1985 [U Vend]
    * *Paleorhychus* Wang 1982 (= *Pararenicola* Wang 1982)
    Pararenicola Wang 1982 [U Vend?]
    Protoarenicola Wang 1982 [U Vend?]
    Sinosabellidites Zheng 1980 [U Vend?]

AGMATANS
PHYLUM AGMATA YOCHELSON 1977
  * *Campitius* Firby and Durham 1974 (= *Volborthella* Schmidt 1888)
  Salterella Billings 1861 [Atda-U L Camb]
  Volborthella Schmidt 1888 [Atda-U L Camb]

HYOLITHELMINTHS
PHYLUM AND CLASS NOT ASSIGNED
  Order Hyolithelminthes Fisher 1962
    Byronia Matthew 1899 [U Atda-Permian]
    Hyolithellus Billings 1871 [N-D-M Camb]
    Koksuja Missarzhevsky 1981 in Missarzhevsky and Mambetov 1981 [U Atda]
    Pseudorthotheca Cobbold 1935 [L Camb]
    Rushtonia Cobbold and Pocock 1934 [Tomm-Boto?]
    Torellella Holm 1893 [L Tomm]
    Torellelloides Meshkova 1969 [M Tomm?]

PAIUTIIDS
PHYLUM AND CLASS NOT ASSIGNED
  Order Paiutiida Tynan 1983
    Family Paiutitubulitidae Tynan 1983

Cambrotubulites Tynan 1983 (= *Paiutitubulites* Tynan 1983)
  Paiutitubulites Tynan 1983 [U Atda]
CONULARIIDS
PHYLUM AND CLASS NOT ASSIGNED
  Order Conulariida Miller and Gurley 1896
    Family Carinachitidae He 1987
      Aciconularia He 1987 [Meish]
      Carinachites Qian 1977 [Meish]
      Mabianoconullus He 1984 *in* Xing et al. 1984 [Meish]
     *Paraconularoides He 1984 (= *Quadrosiphogonuchites* Chen 1982)
     *Paranabarites Jiang 1982 *in* Luo et al. 1982 (= *Quadrosiphogonuchites* Chen 1982)
      Quadrosiphogonuchites Chen 1982 [L Meish-M Meish]
    Family Conulariellidae Kiderlen 1937
      Arthrochites Chen 1982 [L Meish-M Meish]
     *Barbitositheca Qian and Jiang 1982 *in* Luo et al. 1982 (= *Arthrochites* Chen 1982)
      Hexangulaconularia He 1984 *in* Xing et al. 1984 [Meish]
      Lagenaconularia He 1984 *in* Xing et al. 1984 [Meish]
     *Sacciella He 1984 (= *Arthrochites* Chen 1982)
MOBERGELLANS
PHYLUM, CLASS, ORDER AND FAMILY NOT ASSIGNED
  Aktugaia Missarzhevsky 1976 [U Atda]
  *Brastadella Missarzhevsky 1989 (= *Mobergella* Hedström 1923)
  Discinella Hall 1872 [U L Camb]
  Mobergella Hedström 1923 [U Tomm-U Atda]
  Thorslundella Nyers 1984 [U Atda]
ARTHROPODS
PHYLUM ARTHROPODA SIEBOLD AND STANNIUS 1845
SUBPHYLUM CRUSTACEA
Class Ostracoda Latreille 1806
  Order Archaeocopida Sylvester-Bradley 1961
    Acanthomeridion Huo 1956 *in* Hou Xianguang et al. 1989 [L Qiong]
    Aluta Matthew 1896 [U L Camb]
    Antihipponicharion Huo and Shu 1985 [L Qiong]
    Auriculatella Tan 1980 *in* Yin Jicheng et al. 1980 [Qiong?]
    Bajiella Jiang 1982 *in* Luo et al. 1982 [Qiong]
    Beyrichona Matthew 1886 [L Camb]
    Bradoria Matthew 1899 [U Atda-Camb U]
   *Bradorona Matthew 1903 (= *Bradoria* Matthew 1899)
    Cambria Netskaya and Ivanova 1956 [L Camb]
    Changshabaella Huo and Shu 1982 [Qiong?]
    Combinivalvula Hou 1987 [L Qiong]
    Comptaluta Öpik 1968 [Ordian]
    Dabashanella Huo et al. 1983 *in* Huo et al. 1983 [Qiong?]
    Dahaiella Jiang 1982 *in* Luo et al., 1982 [Qiong?]
    Dielymella Ulrich and Bassler 1931 [Atda?-M L Camb]
    Epactridion Bengtson 1990 *in* Bengtson et al. 1990 [U Atda]
   *Escasona Matthew 1902 (= *Beyrichona* Matthew 1886)
    Guangyuanella Chang 1974 [Atda]
    Hanchiangella Huo 1956 [Qiong]
    Hanchungella Huo 1956 [Qiong]
    Hipponicharion Matthew 1886 [U Atda-U L Camb]
    Houlongdongella Li 1975 [Cang]
    Indiana Matthew 1902 [U Atda]
   *Indianites Ulrich and Bassler 1931 (= *Indiana* Matthew 1902)
    Indoto Öpik 1968 [Boto-Ordian]
    Konicekion Snajdr 1975 [U Atda-M Camb]
    Kunmingella Huo 1956 [Qiong]
    Kunyangella Huo 1965 [Qiong]
    Leshanella Li 1975 [Qiong?]
    Liangshanella Huo 1956 [Qiong]
    Luella Huo 1965 [L Camb]
    Malongella Chang 1974 [Atda]
    Meishucunella Jiang 1982 *in* Luo et al. 1982 [Qiong]
    Monasterium Fleming 1973 [Cang-M M Camb]
    Mononotella Ulrich and Bassler 1931 [U L Camb?]
    Nanchengella Huo 1956 [Qiong]
    Nanchengella jinningensis Hou 1987 [Qiong]
    Neokunmingella Chang 1974 [Boto]
    Ophiosema Öpik 1968 [Ordian]
    Ovaluta Zhang 1987 [Cang]
    Parakunmingella Chang 1974 [Atda]
    Paraphaseolella Tong 1987 [L Camb]
    Phaseolella Zhang 1987 [Cang]
    Polyphyma Groom 1902 [L Camb]
    Pseudokunmingella Huo and Shu 1982 [Qiong?]
    Sellula Wiman 1903 [U Atda]
    Shensiella Huo 1956 [Qiong]
    Songlinella Yin 1978 [Qiong?]
    Sunella Huo 1965 [L Camb]
    Tropidiana Öpik 1968 [Ordian]
    Tsunyiella Chang 1974 [Boto]
    Walcottella Ulrich and Bassler 1931 [M Camb]
    Wuchiapingella Huo 1956 [Qiong]
    Wutingella Chang 1974 [Atda]
    Yaoyingella Chang 1974 [Atda]
    Yeshanella Lin 1987 [U L Camb]
    Zepaera Fleming 1973 [Qiong?-L M Camb]
    Zhenpingella Li 1975 [L Qiong-M Cang]
    Zhijinella Yin 1978 [Qiong?]
    Zhongbaoella Huo and Shu 1982 [Qiong?]
  Order Palaeocopida Henningsmoen 1953
    Family Tetradellidae Swartz 1936
      Ushkarella Koneva 1978 [Boto]
Class Malacostraca Latreille 1806
  Order Archaeostraca
    Silesicaris [L Camb]
  Order Canadaspidida
    Perspicaris Briggs 1977 [Qiong-M M Camb]
  Order Hymenostraca
    Hymenocaris [Boto-Arenigian]
SUBPHYLUM CHELICERATA
Class Merostomata Dana 1852
  Eolimulus Bergström 1968 [L L Camb]
  *Paleomerus Størmer 1956 (= *Strabops* Beecher 1901)
  Strabops Beecher 1901 [Atda-Camb U]
SUBPHYLUM TRILOBITA WALCH 1771
[The list has been compiled from J. J. Sepkoski's database on Phanerozoic genera]
  Order Agnostida Kobayashi 1935
    Acidiscus [U L Camb]
    Archaeagnostus [U L Camb]
    Bathydiscus [U L Camb]
    Calodiscus [Atda-U L Camb]
    Chelediscus [U L Camb]
    Dicerodiscus [Boto]

Table 23.5 (Continued).

    Dipharus [Camb]
    Eoagnostus [U L Camb]
    Eodiscus [U L Camb-U M Camb]
    Guizhoudiscus [U Atda-Boto]
    Hebediscus [L Atda-Boto]
    Hupeidiscus [U Atda-Boto]
    Lenadiscus [Boto]
    Leptochilodiscus [U L Camb]
    Mallagnostus [U L Camb]
    Microdiscus [U L Camb]
    Neocobboldia [Boto-L M Camb]
    Neopagetina [Boto-U M Camb]
    Pagetides [U Atda-Tojo]
    Pagetiellus [L Atda-L M Camb]
    Serrodiscus [U Atda-Tojo]
    Shivelicus [U L Camb]
    Shizudiscus [U Atda]
    Sinodiscus [U Atda-U Camb]
    Szechuanaspis [U Atda-Boto]
    Triangulaspis [U Atda-Boto]
    Triangullina [L Atda]
    Tsunyidiscus [U Atda]
    Weymouthia [L Atda-U Atda]
    Yukonia [L Camb]
Order Corynexochida Kobayashi 1935
    Abakania [U L Camb]
    Arthricocephalus [Tojo]
    Atdabanella [Boto]
    Babakovia [U L Camb-M Camb?]
    Balangia [Tojo]
    Bathyuriscellus [Boto]
    Bonnaria [U Atda-Boto]
    Bonnaspis [Boto-M Camb]
    Bonnia [Boto-M Camb]
    Bonniella [U L Camb]
    Bonniopsis [U L Camb]
    Changaspis [Tojo]
    Cheiruroides [Tojo-L M Camb]
    Chondrinouvina [Boto]
    Compsocephalus [L Atda]
    Dolichometopsis [U L Camb-L M Camb]
    Duotingia [Atda]
    Erbiella [Boto]
    Erbiopsidella [U L Camb]
    Erbiopsis [U L Camb]
    Fordaspis [U Atda-U L Camb]
    Granularia [Boto-M Camb]
    Hunnanocephalus [U L Camb]
    Inuoyina [Boto-L M Camb]
    Jakutus [Boto-L M Camb]
    Kootenia [U Atda-U M Camb]
    Labradoria [Boto]
    Lancastria [Boto-Tojo]
    Lenaspis [Tojo]
    Micmaccopsis [Boto-Tojo]
    Ogygopsis [Tojo-M M Camb]
    Olenoides [Tojo-U Camb]
    Oryctocephalus [Boto-U M Camb]
    Poliella [Tojo-M M Camb]
    Poliellaspis [U L Camb]
    Poliellina [Boto]

    Proerbia [Tojo-L M Camb]
    Prokootenia [L Camb]
    Protypus [U L Camb]
    Prozacanthoides [Boto-L M Camb]
    Pseudolancastria [Camb]
    Rondocephalus [U L Camb]
    Shabaella [U L Camb]
    Sinijanella [Boto]
    Strettonia [U L Camb]
    Tabatopygellina [U L Camb?]
    Tarynaspis [Boto]
    Uktaspis [U Atda]
    Zacanthopsis [Tojo-U M Camb]
Order not assigned
    Acimetopus [U L Camb]
    Aguaraya [U L Camb?]
    Analox [U L Camb]
    Arthricocephalites [Tojo]
    Avalonia [L Camb]
    Bolboparia [U L Camb]
    Botomella [Atda]
    Chakasskia [U L Camb?-M M Camb]
    Chengkouia [U L Camb]
    Feilongshania [Tojo]
    Gelasene [U Atda]
    Giordanella [U L Camb?]
    Goldfieldia [U L Camb]
    Jangudaspis [U L Camb?]
    Judaiella [U L Camb?]
    Keeleaspis [U Atda?]
    Kuanyangia [L Camb]
    Kueichowia [L Camb]
    Laticephalus [Boto]
    Menghinella [L Camb]
    Metisaspina [U L Camb?]
    Miranella [Boto]
    Nehannisaspis [U Atda?]
    Otekmaspis [U L Camb]
    Paragraulos [U L Camb]
    Polliaxis [Tojo]
    Sailycaspis [Boto]
    Sekwiaspis [U Atda]
    Tuyunaspis [Tojo]
    Variopelta [U L Camb]
    Xiuqiella [U L Camb]
    Yukonides [U Atda?]
Order Odontopleuriida Whittington 1959
    Eodontopleura [Tojo]
Order Ptychopariida Swinnerton 1915
    Agraulos [U L Camb-U M Camb]
    Altitudella [Boto]
    Antagmus [U L Camb-L M Camb]
    Atops [U Atda]
    Bicella [Tojo]
    Binodaspis [Boto-L M Camb]
    Chondrastaulina [U L Camb]
    Crassifimbra [U L Camb]
    Eoptychoparia [U L Camb-M M Camb]
    Kolbinella [Boto]
    Luxella [Tojo]
    Oncchocephalus [Tojo-L M Camb]

Table 23.5 (Continued).

Pachyaspis [Tojo-M M Camb]
Periomma [Tojo]
Periomella [L Camb]
Piaziella [Tojo-L M Camb]
Poulsenia [Tojo-L M Camb]
Proliostracus [Tojo-L M Camb]
Pseudaptos [U Atda]
Ptychoparopsis [Camb]
Rimouskia [U Atka?]
Sanaschtykgolia [U L Camb]
Solenopleurella [Tojo-U M Camb]
Syspacephalus [Tojo-L M Camb]
Xiangquianaspis [U L Camb-L M Camb]
Xilingxia [Tojo]
Yuehsienszella [U L Camb-L M Camb]
Order Redlichiida Richter 1933
Abadiella [Atda]
Alanisia [L Camb]
Alataurus [U L Camb?]
Aldonaia [Boto-L M Camb]
Anadoxides [Atda]
Angusteva [L Camb]
Antatlasia [L Camb]
Archaeaspis [L Atda]
Archaeops [L Camb]
Argunaspis [U L Camb?]
Asiatella [U L Camb?]
Bathynotus [U L Camb-L M Camb]
Bathyuriscellus [U L Camb]
Belliceps [U L Camb?]
Bergeroniaspis [Boto-Tojo]
Bergeroniellus [Boto-L M Camb]
Biceratops [U L Camb]
Bidjinella [U L Camb?]
Bigotina [L Atda-U L Camb?]
Bigotinops [L Camb]
Blayacina [U Atda-Boto]
Bondonella [L Camb]
Bradyfallotaspis [U Atda]
Breviredlichia [U L Camb]
Bristolia [U L Camb]
Bulaiaspis [U L Camb]
Callavia [L Atda-Boto]
Chaoaspis [Atda]
Choubertella [Atda]
Clariondia [L Camb]
Collyrolenus [L Camb]
Conoredlichia [U L Camb]
Daguinaspis [Atda]
Despujolsia [L Atda]
Dolerolenus [Boto?]
Drepanopyge [Boto]
Drepanuriodes [Boto]
Elganellus [U L Camb?]
Ellipsocephalus [L Atda-L M Camb]
Ellipsostrenua [L Camb]
Elliptocephala [U Atda-Boto]
Enammocephalus [U L Camb]
Eops [L Camb]
Eoredlichia [U Atda]
Esmeraldina [U L Camb]
Estrangia [Camb]
Fallotaspis [L Atda]
Ferralsia [U Atda-Tojo]
Fremontella [Boto]
Gigantopygus [U L Camb]
Gigoutella [U L Camb?]
Guangyuanaspis [Atda]
Habrocephalus [U L Camb?]
Hamatolenus [L Camb]
Hicksia [U L Camb]
Hindermeyeria [L Camb]
Hoffetella [Tojo-L M Camb]
Holmia [L Atda-U Atda]
Holmiella [U Atda]
Hsuaspis [U Atda-Boto]
Hupeia [U L Camb?]
Ichangia [U L Camb?]
Judomia [U Atda-Boto]
Judomiella [Boto]
Kadyella [U L Camb]
Kameschkoviella [U L Camb?]
Kijanella [U L Camb?]
Kingaspis [U L Camb-M M Camb]
Kjerulfia [L Atda]
Krolina [U L Camb?]
Kuanyangia [U Atda]
Kueichowia [Boto]
Latiredlichia [L Camb]
Laudonia [Boto]
Longduia [Boto]
Longianda [Boto]
Luaspis [L Camb]
Lunolenus [Camb]
Lusatiops [U L Camb]
Malungia [Atda-U L Camb]
Mayiella [Boto]
Megapalaeolenus [Tojo]
Mesetaia [L Camb]
Mesodema [L Camb]
Metadoxides [U Atda-U L Camb]
Metaredlichioides [Atda-U L Camb]
Micmacca [Atda-L M Camb]
Minusella [U L Camb?]
Mundocephalina [U L Camb?]
Myopsolenus [L Camb]
Neltneria [L Camb]
Neoredlichia [U L Camb]
Nevadella [L Atda-Boto]
Nevadia [U Atda]
Olekmaspis [U L Camb]
Olenellus [U Atda-Tojo]
Ouijjana [L Camb]
Pachyredlichia [Atda]
Palaeolenella [U L Camb?]
Palaeolenides [Atda]
Palaeolentus [Tojo]
Palaeolenus [Tojo]
Paokannia [U L Camb]
Parabadiella [L Atda]
Parafallotaspis [L Atda]
Paramalungia [U L Camb]

Table 23.5 (Continued).

Pararedlichia [Atda]
Paratermierella [L Camb]
Paratungusella [U L Camb?]
Pareops [L Camb]
Peachella [U L Camb]
Planaspis [U L Camb?]
Poletaevella [U L Camb]
Profallotaspis [L Atda]
Proichangia [U Atda]
Protagraulos [U L Camb]
Protolenella [U L Camb]
Protolenus [U Atda-M M Camb]
Pruvostina [L Camb]
Pruvostinoides [L Camb-M Camb]
Pseudoichangia [Tojo]
Pseudokadyella [U L Camb?]
Pseudolenus [L Camb]
Pseudoredlichia [U L Camb]
Pseudoresserops [L Atda]
Pseudosaukianda [Camb]
Pteroredlichia [U L Camb-L M Camb]
Qiaotingaspis [Boto]
Redlichaspis [L Camb]
Redlichia [Atda?-L M Camb]
Redlichina [Boto-L M Camb?]
Redlichops [L Camb]
Resimopsis [U L Camb?]
Resserops [U L Camb]
Rinconia [L Camb]
Sajanaspis [U L Camb]
Saukianda [Boto]
Saukiandiops [L Camb]
Schmidtiellus [U Atda-Boto]
Shatania [Boto]
Shifangia [U L Camb]
Shipaiella [U L Camb]
Sibiriaspis [U L Camb?]
Sichuanolenus [Tojo]
Sinolenus [L Camb]
Sinskia [Atda-Boto]
Strenuaeva [U L Camb]
Strenuella [L Atda-U L Camb?]
Syndianella [U L Camb]
Terechtaspis [U L Camb?]
Termieraspis [U L Camb?]
Termierella [U L Camb]
Thoralaspis [U L Camb]
Tungusella [U L Camb]
Tuvanella [U L Camb?]
Ushbaspis [Atda]
Validaspis [Boto]
Wangzishia [U Atda]
Wanneria [Boto-Tojo]
Wutingaspis [U Atda]
Yiliangella [Boto]
Yiliangellina [U L Camb]
Yinites [Boto]
Yunnanaspidella [Boto]
Yunnanaspis [Boto]
Yunnanocephalus [U Atda]
Zhenbaspis [Atda]

Order and family not assigned
   Gdowia Yanishevskij 1950 [U Tomm?]
   Naraoia Walcott 1912 [Qiong-M Camb]
Suphylum, class, order and family not assigned
   Alalcomenaeus Simonetta 1970 [L Qiong-M M Camb]
   Anomalocaris Whiteaves 1892 [L L Camb-M M Camb]
   Branchiocaris Briggs 1976 [L Qiong-M M Camb]
   Cassubia Lendzion 1977 [U Tomm]
   Chiella Huo 1965 [L Camb]
   *Cymbia Jiang 1982 *in* Luo et al. 1982 (= *Isoxys* Walcott 1891)
   Dioxycaris [L Camb-L M Camb]
   Emeiella Li 1975 [Qiong]
   Fuxianhuia Hou 1987 [L Qiong]
   Gaoqiaoella Li 1975 [Qiong]
   Isoxys Walcott 1891 [L Qiong-U M Camb]
   Jianfengia Hou 1987 [L Qiong]
   Kuamaia Hou 1987 [L Qiong]
   Leanchoilia Walcott 1931 [U L Camb-M M Camb]
   *Livia Lendzion 1975 (= *Liwia* Dzik and Lendzion 1988)
   Liwia Dzik and Lendzion 1988 [L Atda]
   Nothozoe Barrande 1872 [L Camb-M Ordovician]
   *Peytoia Walcott 1911 (= *Anomalocaris* Whiteaves 1892)
   *Pomerania Lendzio 1975 (= *Cassubia* Lendzion 1977)
   Protocaris [Boto-U M Camb]
   Retifacies Hou et al. 1989 *in* Hou Xianguang et al. 1989 [L Qiong]
   Rhombicalvaria Hou 1987 [L Qiong]
   Serracaris Briggs 1978 [L Camb]
   Tuzoia Walcott 1912 [Atda-M Camb]
   Vetulicola Hou 1987 [L Qiong]
?PHYLUM ARTHROPODA SIEBOLD AND STANNIUS 1845
   Motina Galperova 1981 *in* Akul'cheva et al. 1981 [L Camb]
   Parapunctella Jiang 1982 *in* Luo et al. 1982 [M Meish]
   Yangtzedonta Yü 1985 [M Meish]
COELOSCLERITOPHORANS
PHYLUM NOT ASSIGNED
   Class Coeloscleritophora Bengtson and Missarzhevsky 1981
     Order Chancelloriida Walcott 1920
       Family Chancelloriidae Walcott 1920
         Aldania Vasil'eva 1985 *non* Moore 1896 (Lepidoptera) [L Tomm]
         Aldanospina Missarzhevsky 1989 [L Tomm]
         Allonnia Doré and Reid 1965 [Atda-M Camb]
         Archiasterella Sdzuy 1969 [Atdy-M Camb?]
         Archicladium Qian and Xiao 1984 [Meish]
         Chancelloria Walcott 1920 [L Tomm-Camb U]
         *Dimidia Jiang 1982 *in* Luo et al. 1982 (= *Allonnia* Doré and Reid 1965)
         Elkanospina Missarzhevsky 1989 [L Tomm]
         Eremactis Bengtson and Conway Morris 1990 *in* Bengtson et al. 1990 [U Atda]
         *Fangxianites Duan 1984 (= *Chancelloria* Walcott 1920)
         Ginospina Missarzhevsky 1989 [L Tomm]
         *Onychia Jiang 1982 *in* Luo et al. 1982 (= *Allonnia* Doré and Reid 1965)
         *Polycladium Qian and Xiao 1984 (= *Chancelloria* Walcott 1920)
         *Sissospina Missarzhevsky 1989 (= *Chancelloria* Walcott 1920)

*Stellaria Vasil'eva 1985, non Schmidt 1832; Nardo 1834; Bonaparte 1838 (= Ginospina Missarzhevsky 1989)
*Tuserospina Missarzhevsky 1989 (= Archiasterella Sdzuy 1969)
?Order Chancelloriida Walcott 1920
    Platyspinites Vasil'eva 1985 [M Tomm]
Order Sachitida He 1980
    Family Halkieriidae Poulsen 1967
        *Acrosquama Qian and Xiao 1984 (= Halkieria Poulsen 1967)
        *Acuminachites Qian and Yin 1984 (= Halkieria Poulsen 1967)
        *Adversella Jiang 1982 in Luo et al. 1982 (? = Halkieria Poulsen 1967)
        *Dactyosachites He 1981 (= Halkieria Poulsen 1967)
        Halkieria Poulsen 1967 [L Tomm-U Atda]
        *Microsachites He 1981 (= Halkieria Poulsen 1967)
        Sinosachites He 1980 in Yin Jicheng et al. 1980 [Meish]
        Thambetolepis Jell 1981 [U Atda?]
    Family Sachitidae Meshkova 1969
        Hippopharangites Bengtson 1990 in Bengtson et al. 1990 [U Atda]
        Sachites Meshkova 1969 [M Tomm?-U Atda]
    Family Siphogonuchitidae Qian 1977
        Drepanochites Qian and Jiang 1984 in Qian, 1984 [M Meish]
        *Lepochites Zhong (Chen) 1977 [nomen nudum] (= Lopochites Qian 1977)
        Lomasulcachites Qiang and Jiang 1982 in Luo et al. 1982 [M Meish]
        *Lomasulcavichites Jiang 1980 [nomen nudum] (= Lomasulcachites Qiang and Jiang 1982)
        Lopochites Qian 1977 [M Meish]
        *Palaeosulcachites Qian 1977 (= Siphogonuchites Qian 1977)
        Siphogonuchites Qian 1977 [Meish]
        Tianzhushania Qian et al. 1979 [Meish]
        *Trapezochites Qian and Jiang 1982 in Luo et al. 1982 (= Siphogonuchites Qian 1977)
    ?Family Siphogonuchitidae Qian 1977
        Dabashanites Chen 1979 [Meish]
    Family not assigned
        Diplospinella Vasil'eva 1988 in Vasil'eva and Sayutina 1988 [M Tomm]
        Monospinites Sayutina 1988 in Vasil'eva and Sayutina 1988 [Boto]
        Rosella Vasil'eva 1988 non Cossman 1925 (mollusc) in Vasil'eva and Sayutina 1988 [L Tomm]

PARACARINACHITIDS
PHYLUM, CLASS AND ORDER NOT ASSIGNED
    Family Paracarinachitidae Qian 1984
        *Luyanhaochiton Yu 1984 (= Paracarinachites Qian and Jiang 1982)
        Paracarinachites Qian and Jiang 1982 in Luo et al. 1982 [M Meish]
        *Protopterygotheca Zhong (Chen) 1977 [nomen nudum] (= Protopterygotheca Chen 1979)
        Protopterygotheca Chen 1979 in Qian et al. 1979 [M Meish]
        *Yangtzechiton Yu 1984 (= Paracarinachites Qian and Jiang 1982)
    ?Family Paracarinachitidae Qian 1984
        Scoponodus Jiang 1982 in Luo et al. 1982 [M Meish]

CAMBROCLAVES
PHYLUM, CLASS AND ORDER NOT ASSIGNED
    Family Zhijinitidae Qian 1978
        Cambroclavus Mambetov 1979 in Mambetov and Repina 1979 [Atda]
        Deiradoclavus Conway Morris and Chen 1990 [U Meish]
        Deltaclavus Conway Morris and Chen 1990 [U Meish]
        *Isoclavus Qian and Zhang 1983 (= Cambroclavus Mambetov 1979)
        Pseudoclavus Mambetov 1979 in Mambetov and Repina 1979 [U Atda]
        *Sinoclavus Duan 1984 (= Cambroclavus Mambetov 1979)
        *Sugaites Qian and Xiao 1984 (= Cambroclavus Mambetov 1979)
        *Tanbaoites Duan 1984 (= Cambroclavus Mambetov 1979)
        Zhijinites Qian 1979 [M Meish-Atda?]

MOLLUSCS
PHYLUM MOLLUSCA CUVIER 1797
    Class Bivalvia Linnaeus 1758
        Order Modiomorphoida Newell 1969
            Family Fordillidae
                *Buluniella Ermak 1986 (= Fordilla Barrande 1881)
                Fordilla Barrande 1881 [Tomm-Acta]
        Order Nuculoida Dall 1889
            Family Praenuculidae McAlester 1969
                *Oryzoconcha He and Pei 1985 (= Pojetaia Jell 1980)
                Pojetaia Jell 1980 [Atda]
    Class Rostroconchia Pojeta et al. 1972
        Order Ribeirioida Kobayashi 1933
            Family Ribeiriidae Kobayashi 1933
                *Heraultia Cobbold 1935 (= Watsonella Grabau 1900)
                *Heraultipegma Poejta and Runnegar 1976 (= Watsonella Grabau 1900)]
                Watsonella Grabau 1900 [Tomm]
    Class Gastropoda Cuvier 1797
        Order Archaeogastropoda Thiele 1925
            Family Aldanellidae Linsley and Kier 1984
                Aldanella Vostokova 1962 [N-D-U Tomm?]
                Barskovia Golubev 1976 [N-D-L Atda]
                Nomgoliella Missarzhevsky 1981 [N-D-Tomm]
                Paraaldanella Golubev 1976 [L Tomm]
                *Philoxenella Vostokova 1962 (= Aldanella Vostokova 1962)
            Family onychochilidae Koken 1925
                Beshtashella Missarzhevsky 1981 in Missarzhevsky and Mambetov, 1981 [U Atda]
                Yuwenia Runnegar 1981 [U Atda]
            Family Coreospiridae Knight 1947
                Coreospira Saito 1936 [U L Camb]
    Class Monoplacophora Wenz in Knight 1952
        Order Bellerophontida Ulrich and Scofield 1897
            Protowenella Runnegar and Jell 1976 [U Atda-M Camb]
        Order Cyrtonellida Horný 1963
            Anabarella Vostokova 1962 [N-D-Atda?]
            Archaeospira Yü 1979 [M Meish]
            *Auriculaspira Zhou and Xiao 1984 (= Pelagiella Matthew 1895)
            Bemella Missarzhevsky 1969 in Rozanov et al. 1969 [N-D-M Tomm]
            Eotebenna Runnegar and Jell 1976 [L Boto-M Camb]
            *Ginella Missarzhevsky 1969 in Rozanov et al. 1969

Table 23.5 (Continued).

(= *Ilsanella* Missarzhevsky 1981)
Hamusella Val'kov 1987 [Tomm]
Helcionella Grabau and Shimer 1909 [L Tomm-L Camb]
Hubeispira Yü 1981 [M Meish]
Igorella Missarzhevsky 1969 in Rozanov et al. 1969 [N-D-M Camb]
Ilsanella Missarzhevsky 1981 [Tomm-Boto]
Khairkhania Missarzhevsky 1981 [Tomm-Boto]
Latouchella Cobbold 1921 [N-D-M Camb]
Leptostega Geyer 1986 [U Atda-M Camb]
Mackinnonia Runnegar 1990 in Bengtson et al. 1990 [U Atda]
Oelandia Westergård 1936 (= *Latouchella* Cobbold 1921)
Oelandiella Vostokova 1962 (= *Latouchella* Cobbold 1921)
Parailsanella Zhegallo 1987 in Voronova et al. 1987 [Meish-U Atda]
Pararaconus Runnegar 1990 in Bengtson et al. 1990 [U Atda]
Pelagiella Matthew 1895 [Atda-Camb U]
Proecceylipterus Kobayashi 1939 (= *Pelagiella* Matthew 1895)
Randomia Matthew 1899 [L Atda-M Camb]
Stenotheca Hicks 1872 [Atda?]
Tannuella Missarzhevsky 1969 in Rozanov et al. 1969 [Atda-Atda?]
Yangtzeconus Yü 1979 [M Meish]
Yangtzespira Yü 1979 (= *Archaeospira* Yü 1979)
Yochelcionella Runnegar and Pojeta 1974 [U Atda-M L Camb]
Yunnanospira Jiang 1980 (= *Archaeospira* Yü 1979)
Order Tryblidiida Lemche 1957
Kalbyella Berg-Madsen and Peel 1978 [U Atda-M M Camb]
Proplina Kobayashi 1933 [Atda?-Camb U]
Class Stenothecoida Yochelson 1968
Family Cambridiidae Horný 1957
Bagenovia Radugin 1937 [L Camb]
Bagenoviella Aksarina 1968 [Boto?-Tojo]
Cambridium Horný 1957 [U L Camb]
Dignus Pel'man 1985 [Boto]
Kaschkadakia Aksarina 1968 [Atda?-Tojo]
Katunioides Aksarina 1978 in Aksarina and Pel'man, 1978 [Boto?]
Sargaella Aksarina 1978 in Aksarina and Pel'man 1978 [Atsa?]
Serioides Pel'man 1985 non Curtis 1844 (Coleoptera) [Boto]
Stenothecella Aksarina 1978 in Aksarina and Pel'man 1978 [Boto?]
Stenothecoides Resser 1938 [L Camb-M Camb]
Sulcocarina Aksarina 1968 [U L Camb]
Class, order, and family not assigned
Absidaticonus Yue 1984 in Xing et al. 1984 [M Meish]
Acutirostriconus Yue 1984 in Xing et al. 1984 [M Meish]
Algomella Val'kov and Karlova 1984 [N-D]
Anconochilus Knight 1947 [M U Camb]
Anhuiconus Zhou and Xiao 1984 [Boto?]
Asperoconus Yü 1979 [M Meish]
Bucania Hall 1847 [Camb U?-Silurian]
Cambroconus Yü 1981 [M Meish]
Ceratoconus Chen and Zhang 1980 [M Meish]
Eocyrtolites Yu 1986 [M Cang-Cang]

Eosoconus Yü 1979 [M Meish]
Gibbaspira He 1984 [M Meish]
Gutticonus Yue 1984 in Xiang et al. 1984 [M Meish]
Hampilina Kobayashi 1958 [Camb]
Huanglingella Chen et al. 1981 [M Meish]
Huangshandongoconus Yü 1979 [M Meish]
Hujiagouella Chen and Zhang 1980 [M Meish]
Igorellina Missarzhevsky 1989 [M Tomm]
Isitella Missarzhevsky 1989 (= *Isitiella* Val'kov and Karlova 1984)
Isitiella Val'kov and Karlova 1984 [L Tomm-U Tomm]
Kistasella Missarzhevsky 1989 [L Boto]
Latirostratus Yü 1979 [M Meish]
Maidipingoconus Yü 1979 [M Meish]
Makarakia Aksarina 1968 [Tojo]
Mastakhella Missarzhevsky 1989 [M Tomm]
Mellopegma Runnegar and Jell 1976 [Boto?-M Camb]
Michniakia Missarzhevsky 1966 in Rozanov and Missarzhevsky 1966 [U Atda]
Mirabella Barskova 1988 [Atda]
Nothamusium [L Camb]
Obtusoconus Yü 1979 [M Meish]
Palaeacmaea Hall and Whitfield 1872 [Camb]
Parmorphorella Matthew 1886 [U L Camb]
Planuspira Jiang 1980 in Jiang 1980 [M Meish]
Porcaconus Qian and Xiao 1984 [Meish]
Prosinuites Poulsen 1967 [L Atda]
Salanyella Missarzhevsky 1981 [N-D-Tomm]
Scenellopsis Resser 1938 [L Camb-M Camb]
Scutatestomaconus Chen and Zhang 1980 [M Meish]
Shabaktiella Missarzhevsky 1981 in Missarzhevsky and Mambetov 1981 [Tomm]
Sichuanospira He 1984 [M Meish]
Songlingella Chen et al. 1981 [M Meish]
Tannuspira Missarzhevsky 1989 [Boto]
Tichkaella [Boto-L M Camb]
Tuberoconus Zhou and Xiao 1984 in Xing et al. 1984 [M Meish]
Tuoraconus Missarzhevsky 1989 [M Tomm]
Uncinaspira He 1984 [M Meish]
Xianfengella He and Yang 1982 [M Meish]
Xilingziaconus Chen and Zhang 1980 [M Meish]
?PHYLUM MOLLUSCA
Class, order and family not assigned
Actinoconus Yü 1979 [M Meish]
Aculopileus Kerber 1988 [U Atda]
Aegides Jiang 1980 [M Meish]
Cambroscutum Kerber 1988 [U Atda]
Cambrospira Yü 1979 [M Meish]
Canopoconus Jiang 1982 in Luo et al. 1982 (= *Maikhanella* Zhegallo 1982)
Cassidina Jiang 1980 (= *Maikhanella* Zhegallo 1982)
Centriconus Yü 1979 [M Meish]
Chengjiangoconus He and Yang 1982 (= *Ocruranus* Liu 1979)
Codonoconus Chen and Zhang 1980 [M Meish]
Cremnodinotus Liu 1987 (= *Eohalobia*? Jiang 1982)
Crestoconus Jiang 1980 [M Meish]
Cycloconchoides Zhang 1980 [Qiong?]
Dengyingoconus Chen and Xiong 1984 in Xing et al. 1984 [Meish]

Table 23.5 (Continued).

*Diandongoconus He and Yang 1982 (= Emarginoconus Yü 1979)
Dolichomocelypha Liu 1987 [M Meish]
Dysnoetopla Liu 1987 [M Meish]
Ebianella He and Lin 1980 in Yin Jicheng et al. 1980 [Meish]
Emarginoconus Yü 1979 [M Meish]
*Emeiconus He and Yang 1982 (= Purella Missarzhevsky 1974)
Eohalobia Jiang 1982 in Luo et al. 1982 [M Meish]
Granoconus Yü 1979 [M Meish]
Hamatoconus Chen and Xiong 1984 in Xing et al. 1984 [Meish]
Heosomocelypha Liu 1987 [M Meish]
Hubeinella Zhang 1980 [Qiong?]
*Jakobina Kerber 1988 (= Purella Missarzhevsky 1974)
*Jinkenites Yu 1988 (= Maikhanella Zhegallo 1982)
Kuanchuanella Yue 1984 in Xing et al. 1984 [Meish]
Laticonus Yü 1979 [M Meish]
Lepidites Zhong [Chen] 1977 in Zhong Hua 1977 [L Camb]
Liantuoconus Yü 1979, nomen dubium [M Meish]
Ligyrokala Liu 1987 [M Meish]
*Liorichita Liu 1987 (= Ocruranus Liu 1979)
*Mackenziella Zhegallo 1987 in Voronova et al. 1987 (= Ocruranus Liu 1979)
Maikhanella Zhegallo 1982 in Voronin et al. 1982 [N-D-Tomm]
Maishucunconus Jiang 1980 [M Meish]
*Meishucunchiton Yu 1987 (= Eohalobia Jiang 1982)
Merismoconcha Yü 1979 [M Meish]
Minymerisma Yü 1984 [Meish]
Ocruranus Liu 1979 [M Meish]
Omalenlina Liu 1987 [M Meish]
Paraceratoconus Zhou and Xiang 1984 [Boto?]
Paraformichella Qian and Zhang 1983 [U Atda?]
Parascenella Chen and Xiong 1984 in Xing et al. 1984 [Meish]
Pileconus Jiang 1980 [M Meish]
Postacanthella Yue 1984 in Xing et al. 1984 [Meish]
*Postestephaconus Jiang 1980 (= Ocruranus Liu 1979)
Praelamellodonta Zhang 1980 [Qiong?]
Protoconus Yü 1979 [M Meish]
Pseudopollicina Vostokova 1962 [M Camb]
Purella Missarzhevsky 1974 [N-D-L Tomm]
*Ramenta Jiang 1982 in Luo et al. 1982 (= Maikhanella Zhegallo 1982)
Rostroconus Jiang 1980 [M Meish]
Rozanoviella Missarzhevsky 1981 [Tomm]
*Runnegarochiton Yu 1987 (= Ocruranus Liu 1979)
Sacciconus Jiang 1980 [M Meish]
Scamboscamna Liu 1987 [M Meish]
Securiconus Jiang 1980 [M Meish]
Sinuconus Yü 1979 [M Meish]
Spatuloconus Yü 1979 [M Meish]
*Stephaconus Jiang 1980 [nomen nudum] (= Ocruranus Liu 1979)
*Stoliconus Jiang 1980 (= Ocruranus Liu 1979)
Tianzhushanospira Yü 1979 [M Meish]
Truncatoconus Yü 1979 [M Meish]
Xiadongoconus Yü 1979 [M Meish]

Xianfengoconcha Zhang 1980 [Qiong?]
Yangtzemerisma Yü 1984 [Meish]
Yunnanopleura Yu 1987 [M Meish]
Zeugites Qian et al. 1979, nomen dubium [Meish]

HYOLITHS
PHYLUM NOT ASSIGNED

Class Hyolitha Marek 1963
Order Hyolithida Matthew 1899
Aimitus Sysoev 1966 [Tomm]
Altaicornus Sysoev 1970 [U Atda?]
Ambrolinevitus Sysoev 1958 [L Camb]
Angusticornus Sysoev 1968 [U Atda-L Boto]
Atdabanithes Meshkova 1974 [U Tomm-U Atda]
Brevilabiatus Sysoev 1965 [Atda]
Burithes Missarzhevsky 1969 in Rozanov et al. 1969 [M Tomm-Atda]
Carinolithes Sysoev 1958 [L Camb]
Crestijachites Sysoev 1968 [U Tomm-L Atda]
Diplotheca Matthew 1885 [Boto]
Doliutus Missarzhevsky and Sysoev 1969 in Rozanov et al. 1969 [U Tomm-Atda]
Dorsojugatus Sysoev 1968 [L Atda]
Dorsolinevitus Sysoev 1958 [L Camb]
Erraticornus Sysoev 1973 [L Boto]
Galicornus Val'kov 1975 [U [U Atda-Boto?]
Goniocornus Sysoev 1970 [U Atda?]
Helenia Walcott 1890 [L Camb]
Hyolithes Eichwald 1840 [L Camb?-Ordovician?]
Hyptiotheca Bengtson 1990 in Bengtson et al. 1990 [U Atda]
Insignicornus Sysoev 1973 [L Boto]
Jacuticornus [U Tomm]
Ketemecornus Sysoev 1974 [U Atda-L M Camb]
Korilithes Missarzhevsky 1969 in Rozanov et al. 1969 [M Tomm]
Kuonamkicornus Val'kov 1975 [U Atda-Boto]
Laticornus Mambetov 1981 in Missarzhevsky and Mambetov 1981 [U Atda]
Lenalituus Missarzhevsky 1981 [Atda]
Linevitus Sysoev 1958 [L Camb]
Meitanovitus Qian 1978 [Meish?]
Microcornus Mambetov 1972 [M Tomm-U Atda]
Nelegerocornus Meshkova 1974 [U Atda-L Boto]
Nitoricornus Sysoev [L Boto]
Notabilitus Sysoev 1968 [U Tomm]
Oblisicornus Sysoev 1968 [U Tomm]
Parakorilithes He and Pei 1984 in He Tinggui et al. 1984 [Cang]
Parkula Bengtson 1990 in Bengtson et al. 1990 [U Atda]
Planotheca Meshkova 1974 [L Boto]
Quadrotheca Sysoev 1958 [L Camb-Camb U?]
Quinquelithes Sysoev 1958 [L Camb]
Rarissimetus Sysoev 1968 [U Tomm-L Atda]
Sulcavitus Sysoev 1958 [L Camb]
Trapezovitus Sysoev 1958 [L Camb]
Tulenicornus Val'kov 1970 [U L Camb-LM Camb]
Tuojdachites Missarzhevsky 1969 in Rozanov et al. 1969 [M Tomm]
Yacutolituus Missarzhevsky 1974 [M Tomm]
Yankongovitus Qian 1978 [Meish?]

Table 23.5 (Continued).

Class Orthothecida Marek 1966
  Order Camerothecida Sysove 1957
    Camerotheca Matthew 1885 [Boto]
  Order Circothecida Sysoev 1968
    Argatheca Missarzhevsky 1989 [L Tomm]
    Asijatheca Mambetov 1981 in Missarzhevsky and Mambetov 1981 [U Atda]
    Circotheca Sysoev 1958 [L Tomm-Camb U?]
    *Coleolella Missarzhevsky 1969 in Rozanov et al. 1969 (? = *Spinulitheca* Sysoev 1958)
    Conotheca Missarzhevsky 1969 in Rozanov et al. 1969 [L Tomm-Atda]
    Costatheca Missarzhevsky 1969 in Rozanov et al. 1969 [Atda-U Atda]
    Crossbitheca Missarzhevsky 1974 [N-D-L Tomm]
    Dabanitheca Missarzhevsky 1989 [U Tomm]
    Ensitheca Val'kov 1982 [L Tomm]
    Extentitheca Sysoev 1972 [U Tomm]
    Khetatheca Missarzhevsky 1989 [N-D]
    Kotuyitheca Missarzhevsky 1974 [N-D-L Tomm]
    Kugdatheca Missarzhevsky 1969 in Rozanov et al. 1969 [L Tomm]
    Kunyangotheca Qian 1978 [Meish]
    Ladatheca Sysoev 1968 [N-D-U Tomm]
    Laratheca Missarzhevsky 1969 in Rozanov et al. 1969 [L Tomm]
    Leibotheca Qian 1978 [Meish?]
    *Simplotubus Singh and Shukla 1981 (= *Conotheca* Missarzhevsky 1969)
    Spinulitheca Sysoev 1968 [L Tomm]
    Stimulitheca Duan 1984 [U Meish?]
    Tchuranitheca Meshkova 1974 non Sysoev 1968 [U Tomm-L Atda]
    Tchuranitheca Sysoev 1968 [Tomm]
    Turcutheca Missarzhevsky 1969 in Rozanov et al. 1969 [L Tomm-U Tomm]
    Virgatotheca Meshkova 1974 [M Tomm]
  Order Globorilida Sysoev 1957
    Cobboldiella Kerber 1988 [U Atda]
    Globorilus Sysoev 1958 [L Camb-M Camb]
    Globoritubulus Qian and Zhang 1985 [L Camb]
    Neogloborilus Qian and Zhang 1983 [U Atda?]
    Sulcagloborilus Jiang 1982 in Luo et al. 1982 [Meish?]
    Xiadongtubulus Qian and Zhang 1985 [L Comb]
  Order Orthothecida Marek 1966
    Adyshevitheca Mambetov 1979 in Mambetov and Repina 1979 [U Atda-L Boto]
    Aldanotheca Meshkova 1974 [L Tomm-M Tomm]
    Allatheca Missarzhevsky 1969 in Rozanov et al. 1969 [L Tomm-Ordovician]
    Ancheilotheca Qian 1978 [Meish?]
    Antiquatheca Missarzhevsky 1974 [M Tomm]
    Carinitheca Sysoev 1968 [Atda]
    Contitheca Sysoev 1972 [U L Camb?-M Camb]
    Curtitheca Sysoev 1968 [Tomm]
    Ebianotheca He 1980 in Yin Jicheng et al. 1980 [Meish]
    Egdetheca Missarzhevsky 1969 in Rozanov et al. 1969 [L Tomm-M Tomm]
    Eonovitatus Sysoev 1968 [Tomm-Atda]
    Exilitheca Sysoev 1968 [L Tomm-U Tomm]
    Gracilitheca Sysoev 1968 [U Tomm-M Camb]
    Holmitheca Sysoev 1968 [L Boto]
    Isititheca Sysoev 1968 [U Tomm]
    Lenatheca Missarzhevsky 1969 in Rozanov et al. 1969 [U Tomm-Ordivician]
    Lentitheca Sysoev 1958 [L Camb]
    Majatheca Missarzhevsky 1969 in Rozanov et al. 1969 [U Tomm]
    Malykanotheca Meshkova 1974 [L Atda]
    Micatheca Sysoev 1972 [L Boto]
    Minitheca Meshkova 1969 [Atda]
    Mooritheca Val'kov 1975 [U Tomm-Boto]
    Nikatheca Val'kov 1975 [L Tomm]
    Novitatus Sysoev 1968 [Atda-L Boto]
    Obliquatheca Sysoev 1968 [Boto?]
    Orthotheca Novak 1886 [L Tomm-Devonian M]
    Plicitheca Sysoev 1968 [U Atda-U Boto]
    Renitheca Sysoev 1968 [L Boto]
    Semielliptotheca Sysoev 1958 [L Camb]
    Sokolovitheca Sysoev 1972 [L Boto-U Boto]
    Tcharatheca Sysoev 1972 [Boto?]
    Tetratheca Sysoev 1968 [U Tomm-L Boto]
    Trapezotheca Sysoev 1958 [M Tomm-Camb U?]
  Order and family not assigned
    Calcitheca Galperova 1981 in Akul'cheva et al. 1981 [L Camb]
    Loculitheca Sysoev 1968 [L Tomm-U Tomm]
    Ovalitheca Sysoev 1968 [Tomm-Atda]
    Uniformitheca Sysoev 1968 [U Tomm-L Atda]
Class, order and family not assigned
  Borealicornus [Atda-L Boto]
  Decoritheca Sysoev 1968 [L Camb-M U Camb]
  Firmicornus [U Atda-L Boto]
  Oxytus [U Tomm-L Atda]
  Persicitheca Duan 1984 [U Meish?]
?PHYLUM HYOLITHA MAREK 1963
  *Actinotheca Xiao and Zhou 1984 [Atda-Boto?]
  Arcitheca Duan 1984 (= *Actinotheca* Xiao and Zhou 1984)
  Bucanotheca Qian and Jiang 1982 in Luo et al. 1982 [M Meish]
  *Cupitheca Qian and Xiao 1984 [nomen nudum] (= *Actinotheca* Xiao and Zhou 1984)
  *Cupitheca Duan 1984 (= *Actinotheca* Xiao and Zhou 1984)
  *Emeitheca Duan 1984 (= *Actinotheca* Xiao and Zhou 1984)
  *Ensitheca Duan 1984 (= *Actinotheca* Xiao and Zhou 1984)
  Parahyolithes Poulsen 1967 [U L Camb]
  *Varitheca Duan 1984 (= *Actinotheca* Xiao and Zhou 1984)
MICRODICTYON
PHYLUM, CLASS, ORDER AND FAMILY NOT ASSIGNED
  *Eoconcharium Hao and Shu 1987 (= *Microdictyon* Bengtson et al. 1981)
  *Eoncharium Shu and Chen 1988 non Hao and Shu 1987 (= *Microdictyon* Bengtson et al. 1981)
  Fusuconcharium Hao and Shu 1987 [L Camb]
  Microdictyon Bengtson et al. 1981 in Missarzhevsky and Mambetov 1981 [U Atda-L M Camb]
  Quadratapora Hao and Shu 1987 [L Camb]
TOMMOTIIDS
PHYLUM, CLASS AND ORDER NOT ASSIGNED
  Family Kelanellidae Missarzhevsky and Grigor'eva 1981
    *Bengtsonia Missarzhevsky and Grigor'eva 1981 (= *Kelanella* Missarzhevsky 1966)

Table 23.5 (Continued).

Bercutia Missarzhevsky 1981 *in* Missarzhevsky and Mambetov 1981 [Tomm?]
Geresia Missarzhevsky 1981 *in* Missarzhevsky and Mambetov 1981 [U Tomm]
Kelanella Missarzhevsky 1966 *in* Rozanov and Missarzhevsky 1966 [Atda]
Sonella Missarzhevsky and Grigor'eva 1981 [L M Camb]
*Tesella Missarzhevsky and Grigor'eva 1981 (= *Sonella* Missarzhevsky and Grigor'eva 1981)
Family Lapworthellidae
Lapworthella Cobbold 1921 [M Tomm-M Camb]
?Family Lapworthellidae
Stenothecopsis Cobbold 1935 [Atda?]
Family Sunnaginiidae Landing 1984
Eccentrotheca Landing et al. 1980 [Atda]
Sunnaginia Missarzhevsky 1969 *in* Rozanov et al. 1969 [L Tomm-Atda]
Family Tannuolinidae Fonin and Smirnova 1967
Micrina Laurie 1986 [Atda?]
Tannuolina Fonin and Smirnova 1967 [Boto]
Family Tommotiidae Bengtson 1970
*Camena Missarzhevsky 1966 *in* Rozanov and Missarzhevsky 1966 (= *Camenella* Missarzhevsky 1966)
Camenella Missarzhevsky 1966 *in* Rozanov and Missarzhevsky 1966 [L Tomm-L Atda?]
*Tommotia Missarzhevsky 1970 (= *Camenella* Missarzhevsky 1966)
Family not assigned
Dailyatia Bischoff 1977 [U Atda]
Kennardia Laurie 1986 [U Atda]
Kulparina Conway Morris and Bengtson 1990 *in* Bengtson et al. 1990 [U Atda]
Lugoviella Grigor'eva 1983 *in* Grigor'eva et al. 1983 [L Boto]
Ninella Missarzhevsky 1981 *in* Missarzhevsky and Mambetov 1981 [U Atda]
Paterimitra Laurie 1986 [U Atda]
Porcauricula Qian and Bengtson 1989 *in* Qian Yi and Bengtson 1989 [M Meish]
*Yunnanotheca Jiang 1980 [*nomen nudum*] (= *Porcauricula* Qian and Bengtson 1989)

BRACHIOPODS
PHYLUM BRACHIOPODA
  Class Articulata
    Order Orthida Schuchert and Cooper 1932
      Billingsella Hall and Clarke 1892 *in* Hall and Clarke 1892 [Boto-Arenigian]
      Eoconcha Cooper 1951 [L Camb]
      Glyptoria Cooper 1976 [Boto?]
      Israelaria Cooper 1976 [Boto?]
      Kotujella Andreeva 1962 [L Camb]
      Kundatella Aksarina 1978 *in* Aksarina and Pel'man 1978 [Atda?]
      Leioria Cooper 1976 [Boto?]
      Matutella Cooper 1951 [Tojo-M Camb]
      Narynella Andreeva 1987 [Boto-M Camb]
      Nisusia Walcott 1905 [L Atda-M Camb]
      Psiloria Cooper 1976 [U Atda-Boto]
      Tcharella Andreeva 1987 [Tojo?-M Camb]
    Order and family not assigned
      Bojarinovia Aksarina 1978 *in* Aksarina and Pel'man 1978 [Boto?]
  Class Inarticulata
    Order Acrotretida Kuhn 1949
      Acrothele Linnarsson 1876 [U Tomm-U M Camb]
      Acrothyra Matthew 1901 [L Boto-U M Camb]
      Acrotreta Kutorga 1848 [Atda-L Llanvirnian]
      Botsfordia Matthew 1891 [Atda-M Camb]
      Diandongia Rong 1974 *in* Xu et al. 1974 [L Camb]
      Dysoristus Bell 1941 *in* Lochman and Duncan 1944 [Tojo-M Camb]
      Edreja Koneva 1979 [U L Camb]
      Glyptias Walcott 1901 [L Camb]
      Hadrotreta Rowell 1966 [Boto-M Camb?]
      Homotreta Bell 1941 [U Boto-M Camb]
      Linnarssonia Walcott 1885 [L Boto-U M Camb]
      Prototreta Bell 1941 [Boto-M Camb]
      Schizopholis Waagen 1885 [Boto]
      Spinulothele Rowell 1977 [L Boto]
    Order Kutorginida Kuhn 1949
      Agyrekia Koneva 1979 [Boto]
      Haupiria MacKinnon 1983 [L Camb-U M Camb]
      Kutorgina Billings 1861 [Atda-U M Camb]
      Trematosia Cooper 1976 [Boto?]
    Order Lingulida Waagen 1885
      Lingulella Salter 1866 [U Tomm?-U Ashgillian]
      Obolopsis Saito 1936 [U L Camb]
      Obolus Eichwald 1829 [U Atda?-Llanvirnian]
      Palaeoschmidtites Koneva 1979 [Boto]
      Westonia Walcott 1901 [Boto-Caradocian]
    Order Obolellida Rowell 1968
      Alisina Rowell 1962 [U L Camb?]
      Bicia Walcott 1901 [Boto]
      Ivshinella Koneva 1979 [Boto]
      Magnicanalis Rowell 1962 [L Camb]
      Monoconvexa Pel'man 1977 [Atda]
      Nochoroiella Pel'man 1983 *in* Grigor'eva et al. 1983 [U Tomm-L Atda]
      Obolella Billings 1861 [L Atda-Boto]
      Sibiria Goryanskij 1977 *in* Pel'man 1977 [L Atda-Boto]
      Trematobolus Matthew 1893 [U Atda-M Camb]
      Yorkia Walcott 1897 [Atda-Boto]
    Order Paterinida Rowell 1965
      Askepasma Laurie 1986 [Atda?]
      Dictyonina Cooper 1942 [L Camb-L M Camb]
      *Iphidea Billings 1872 (= *Paterina* Beecher 1891)
      Mickwitzia Schmidt 1888 [U Tomm-M Camb]
      Micromitra Meek 1873 [U Tomm-M U Camb]
      Paterina Beecher 1891 [U Tomm?-U M Camb]
      Walcottina Cobbold 1921 [Atsa?]
    Order and family not assigned
      Aldanotreta Pel'man 1977 [L Tomm-U Tomm]
      Cryptotreta Pel'man 1977 [U Tomm-U Atda]
      Khasagtina Ushatinskaya 1987 [Tomm-Atda]
      Salanygolina Ushatinskaya 1987 [Boto]
  Class, order and family not assigned
    Curticia Walcott 1905 [Boto-L U Camb]
    Eothele Rowell 1980 [U Boto]
    Heliomedusa Sun and Hou 1987 [L Qiong]
    Neobolus Waagen 1885 [L Camb]

Table 23.5 (Continued).

    Pegmatreta Bell 1941 [Tojo-L M Camb]
    Pompeckium [L Boto]
    Quebecia Walcott 1905 [L Camb]
    Rustella Walcott 1905 [Atda-Boto]
    Swantonia Walcott 1905 [L Boto]
?PHYLUM BRACHIOPODA
  Class, order and family not assigned
    Acidotocarena Liu 1979 [M Meish]
    Apistoconcha Conway Morris 1990 in Bengtson et al. 1990 [U Atda]
    Aroonia Bengtson 1990 in Bengtson et al. 1990 [U Atda]
    Artimyctella Liu 1979 [M Meish]
    Ernogia Jiang 1982 in Luo et al. 1982 [M Meish]
    Hanshuiella Yue 1984 in Xing et al. 1984 [Meish]
    Lathamella Liu 1979 [M Meish]
    Plicatolingula Liu 1979 [M Meish]
    Protobolus Liu 1979 [M Meish]
    Psamathopalass Liu 1979 [M Meish]
    Scambocris Liu 1979 [M Meish]
    Surindia Galimova 1988 [L Camb]
    Tianzhushanella Liu 1979 [M Meish]
    *Yuanjiapingella Yue 1984 in Xing et al. 1984 (= *Hanshuiella* Yue 1984)
ECHINODERMS
PHYLUM ECHINODERMATA
  Class Camptostromatoidea
    Order Camptostromatidae
      Camptostroma Ruedemann 1933 [U Boto]
  Class Edriosteroidea
    Order Stromatocystitida
      Edriodiscus Jell et al. 1985 in Jell et al. 1985 [Ordian]
      Stromatocystites Pompeckj 1896 [Boto-M Camb]
  Class Eocrinoidea
    Gogia Walcott 1917 [Boto?-M Camb]
  Class Helicoplacoidea Durham and Caster 1963
    Helicoplacus Durham and Caster 1963 [U Atda?]
    Polyplacus Durham 1967 [U Atda?]
    Waucobella Durham 1967 [U Atda?]
  Class not assigned
    Order Lepidocystoidea
      Kinzercystis Sprinkle 1973 [U Boto]
      Lepidocystis Foerste 1938 [U Boto]
CHAETOGNATHS
PHYLUM CHAETOGNATHA
  Class and order not assigned
    Family Amphigeisinidae Miller 1979
      Amphigeisina Bengtson 1976 [L Camb-M Camb]
      *Emeidus Chen 1982 (= *Protohertzina* Missarzhevsky 1973)
      *Hastina Yang and He 1984 (= *Protohertzina* Missarzhevsky 1973)
      *Hastina Yang and He 1983 in Yang et al. 1983 [*nomen nudum*] (= *Protohertzina* Missarzhevsky 1973)
      Protohertzina Missarzhevsky 1973 [N-D-Atda]
OTHER TOOTH-SHAPED FOSSILS
    Cyrtochites Qian 1984 [M Meish]
    *Ganloudina He 1984 in Xing et al. 1984 [Meish]
    Ganloudina Yang and He 1983 in Yang et al. 1983 [*nomen nudum*] (= *Ganloudina* He 1984)
    Indocera Singh and Shukla 1981 [L Tomm]
    Jiangshanodus Yue 1989 in Yue Zhao et al. 1989 [U Qiong]
    Kijacus Missarzhevsky 1981 in Missarzhevsky and Mambetov 1981 [U Tomm]
    Leguminella He 1984 [M Meish]
    Maldeotaia Singh and Shukla 1981 [L Tomm?]
    Mongolodus Missarzhevsky 1977 [L Boto]
    Quadrorites Qian and Ding 1988 in Ding and Qian 1988 [Meish]
    Rhombocorniculum Walliser 1958 [L Atda-U Atda]
    Spinocera Singh and Shukla 1981 [L Tomm]
    Yunnanodus Wang and Jiang 1980 in Jiang 1980 [M Meish]
UTAHPHOSPHIDS
PHYLUM, CLASS AND ORDER NOT ASSIGNED
    Family Utahphosphidae Wrona 1987
      Hadimopanella Gedik 1977 [U Atda-Camb U?]
      *Lenargyrion Bengtson 1977 (= *Hadimopanella* Gedik 1977)
PROBLEMATIC FOSSILS NOT REFERRED TO HIGHER TAXA OR GROUP
  Sclerites
    Acanthocassis He and Xie 1989 in He and Xie 1989 [M Meish]
    Acanthosphaera Duan 1984 [U Meish?]
    Acidocharacus Qin and Ding 1988 [M Meish]
    Amoebinella He and Xie 1989 [M Meish]
    Aurisella Qian and Xiao 1984 [M Meish]
    Bioistodina He and Pei 1984 in He Tinggui et al. 1984 [Cang]
    Brushenodus Jiang 1982 in Luo et al. 1982 [M Meish]
    Cowiella Hinz 1987 in Hinz 1987 [Atda?]
    Emeithella Qian 1977 [Meish]
    Fomitchella Missarzhevsky 1969 in Rozanov et al. 1969 [L Tomm-M Tomm]
    Henaniodus He and Pei 1984 in He Tinggui et al. 1984 [Cang]
    Huizenodus He and Xie 1989 [M Meish]
    Kaiyangites Qian and Yin 1984 [U Meish]
    Koksodus Missarzhevsky 1981 in Missarzhevsky and Mambetov 1981 [Tomm]
    Korilacus Missarzhevsky 1989 [N-D]
    *Lunachites Qian and Yin 1984 (= *Solenotia* Qian and Yin 1984)
    Miriella Jiang and Huang 1986 [M Camb?]
    Mongolitubulus Missarzhevsky 1977 [L Boto-L M Camb]
    Ningqiangsclerites Yue 1984 in Xing et al. 1984 [Meish]
    Paracanthodus Chen 1982 [Meish]
    Parazhijinites Qiang and Yin 1984 [Qiong]
    Parazhijinites Qian and Yin 1984 in Xing et al. 1984 [Qiong]
    *Phyllochites Qian and Yin 1984 (= *Solenotia* Qian and Yin 1984)
    Phyllochiton Duan 1984 [U Meish?]
    Primaconulariella He 1987 [Meish]
    Rhabdochites He 1984 in Xing et al. 1984 [Meish-Qiong]
    Rushtonites Hinz 1987 [Atda?]
    Salanacus Grigor'eva 1982 in Voronin et al. 1982 [Tomm]
    Solenotia Qian and Yin 1984 [Meish]
    Stefania Grigor'eva 1982 in Voronin et al. 1982 [Boto?]
    Tchangsichiton Yu 1987 [M Meish]
    Tumulduria Missarzhevsky 1969 in Rozanov et al. 1969 [L Tomm]

Wushichites Qian and Xiao 1984 [U Meish?]
Yuliunia He and Xie 1989 [M Meish]
?Sclerites
    Dentachites He and Lin 1980 *in* Yin Jicheng et al. 1980 [Meish]
    Flabetheca Qian and Yin 1984 [Meish]
    Humboldtochaeta Gedik 1989 [L Camb]
    Mabiania He 1984 *in* Xing et al. 1984 [Meish]
    Mongoliacus Missarzhevsky 197 [L Boto]
    Pyrgites Yue 1984 *in* Xing et al. 1984 [Meish]
    Punctatus He 1980 *in* Yin Jicheng et al. 1980 [Meish]
    Redkinia Sokolov 1977 [U Vend]
    Stoibostrombus Conway Morris and Bengtson 1990 *in* Bengtson et al. 1990 [U Atda]
    Triplicatella Conway Morris 1990 *in* Bengtson et al. 1990 [U Atda]
    Yangtzesclerites Chen et al. 1981 [Meish]
Shells
    Aldanolina Pel'man 1976 [M Tomm]
    Archaeopetasus Conway Morris and Bengtson et al. 1990 [U Atda]
    Ardrossania Runnegar 1990 *in* Bengtson et al. 1990 [U Atda]
    Aviculocephaloconus Chen and Zhang 1980 [M Meish]
    Cambrocassis Missarzhevsky 1977 [U Atda]
    Gonamella Val'kov and Karlova 1984 [L Tomm]
    Stictoconus Qian and Bengtson 1989 *in* Qian Yi and Bengtson 1989 [M Meish]
?Shells
    Archaeotremaria Yü 1979 [Meish]
    Resegia Missarzhevsky 1981 *in* Missarzhevsky and Mambetov 1981 [U Atda]
Tubular fossils
    Acanthoclava Li 1984 [L Camb]
    Acuticloudina Hahn and Pflug 1985 [U L Camb]
    Acutitheca Yue 1984 *in* Xing et al. 1984 [Meish]
    Cerabonusoides Qian 1978 [Meish?]
    Cloudina Germs 1972 [U Vend-Camb?]
    Corumbella Hahn et al. 1982 [Vend]
    Heterosculpotheca Jiang 1982 *in* Luo et al. 1982 [Meish?]
    Lidaconus Onken and Signor [Boto]
    Miratheca Yue 1984 *in* Xing et al. 1984 [Meish]
    Nevadatubulus Signor et al. 1987 *in* Signor et al. 1987 [Tomm?]
    Onuphionella Kir'yanov 1968 [Tomm]
    Platysolenites Eichwald 1860 [L L Camb]
    Plinthoconion Landing 1988 [Tomm-Atda]
    Quadrochites Qian et al. 1979 [Meish]
    Rugatotheca He 1980 *in* Yin Jicheng et al. 1980 [Meish]
    Sagittitheca Yue 1984 *in* Xing et al. 1984 [Meish]
    Salopiella Cobbold 1921 *in* Cobbold 1919 [U Atda?]
    Scissotheca Yue 1984 *in* Xing et al. 1984 [Meish]
    Sinotubulites Chen et al. 1981 [U Vend-Tomm?]
    *Spirosolenites Føyn and Glaessner 1979 (= *Platysolenites* Eichwald 1860)
    Tommotitubulus Fedorov 1986 [L Tomm]
    Tubulella Howell 1949 [Boto-M Camb]
    Wyattia Taylor 1966 (L L Camb]
    *Yanischevskyites Sokolov 1965 (= *Platysolenites* Eichwald 1860)

Spherical fossils
    Aetholicopalla Conway Morris 1990 *in* Bengtson et al. 1990 [U Atda]
    Ambarchaeooides Qian et al. 1979 [Meish]
    Archaeooides Qian 1977 [Meish-Atda?]
    *Gaparella Missarzhevsky 1981 *in* Missarzhevsky and Mambetov 1981 (= *Archaeooides* Qian 1977)
    Jaraktina Galperova 1981 *in* Akul'cheva et al. 1981 [L Camb]
    *Lithapium Nazarov 1973 (= *Paleoxiphosphera* Nazarov 1973)
    Markuelia Val'kov 1983 [N-D-L Tomm]
    Megasphaera Chen and Liu 1986 [U Vend]
    Meghystrichosphaeridium Chen and Liu 1986 [U Vend]
    Nephrooides Qian 1977, nomen dubium [Meish]
    Olivooides Qian 1977, nomen dubium [Meish]
    Paleocenosphaera Nazarov 1973 [U Atda?-Camb U?]
    Paleoxiphosphaera Nazarov 1973 [U Atda?]
    *Paramobergella Zhong (Chen) 1977 [*nomen nudum*] (? = *Archaeooides* Qian 1977)
    *Protosphaerites Chen 1982 (= *Archaeooides*? Qian 1977)
    Pseudooides Qian 1977, nomen dubium [Meish]
    *Tianshandiscus Qian and Xiao 1984 (= *Archaeooides* Qian 1977)
    Ulcundia Nazarov 1974 [L Camb-Ordivician U]
Flask-shaped fossils
    Cambrothyra Qian and Zhang 1983 [Meish-Qiong]
    *Globifructus Geng and Zhang 1987 (= *Cambrothyra* Qian and Zhang 1983)
    *Mirabichitina Yang and He 1983 *in* Yang et al. 1983 [*nomen nudum*] (= *Cambrothyra* Qian and Zhang 1983)
    *Mirabifolliculus Yang and He 1984 (= *Cambrothyra* Qian and Zhang 1983)
    *Nanjiangochitina Yang and He 1983 *in* Yang et al. 1983 [*nomen nudum*] (= *Cambrothyra* Qian and Zhang 1983)
    **Nanjiangofolliculus* Yang and He 1984 (= *Cambrothyra* Qian and Zhang 1983)
    Pollofructus Geng and Zhang 1987 [Qiong]
Siliceous scale-like fossils
    Altarmilla Allison and Hilgert 1986 [L Camb?]
    Aqualisquama Allison and Hilgert 1986 [L Camb?]
    Archeoxybaphon Allison and Hilgert 1986 [L Camb?]
    Bicorniculum Allison and Hilgert 1986 [L Camb?]
    Characodictyon Allison and Hilgert 1986 [L Camb?]
    Chilodictyon Allison and Hilgert 1986 [L Camb?]
    Confinisquama Allison and Hilgert 1986 [L Camb?]
    Hyaloxybaphon Allison and Hilgert 1986 [L Camb?]
    Invaginatibalteus Allison and Hilgert 1986 [L Camb?]
    Paleocrassilimbus Allison and Hilgert 1986 [L Camb?]
    Paleohexadictyon Allison and Hilgert 1986 [L Camb?]
    Paleomegasquama Allison and Hilgert 1986 [L Camb?]
    Paterisquama Allison and Hilgert 1986 [L Camb?]
    Patinisquama Allison and Hilgert 1986 [L Camb?]
    Petasisquama Allison and Hilgert 1986 [L Camb?]
    Radiocerniculum Allison and Hilgert 1986 [L Camb?]
    Spinicerniculum Allison and Hilgert 1986 [L Camb?]
Calcified Cyanobacteria
    Endoconchia Runnegar 1990 *in* Bengtson et al. 1990 [U Atda]
    *Jiangispirellus Peel 1988 (= *Spirellus* Jiang 1982)

Table 23.5 (Continued).

Spirellus Jiang 1982 *in* Luo et al. 1982 [L Meish-M Meish]	*Micropylepora* Jiang 1982 *in* Luo et al. 1982 [M Meish]
Other fossils	*Poratites* Jiang 1980 [M Meish]
Aksuglobulus Qian and Xiao 1984 [Meish]	*Radiaxialia* Pyanovskaya 1985 [U M Camb-L U Camb]
Anzalia [L Camb]	*Stellostomites* Sun and Hou 1987 (= *Eldonia* Walcott 1911)
Bija Vologdin 1932 [Boto]	*Tuvinia* Krasnopeeva 1972 [Boto]
Carubacgutes [Tomm]	*Yunnanomedusa* Sun and Hou 1987 (= *Eldonia* Walcott 1911)
Dinomischus Conway Morris 1977 [U Atda-M M Camb]	*Xenusion* Pompeckj 1927 [L Camb?]
Eldonia Walcott 1911 [U Atda-M M Camb]	Dubiofossils and pseudofossils
Emmensaspis [Boto]	*Beltania* Bedford and Bedford 1936 [U Atda]
Eoescharopora Jiang 1982 *in* Luo et al. 1982 [M Meish]	*Eocucumaria* Qian and Ding 1988 *in* Ding and Qian 1988 [Meish]
Facivermis Hou and Chen 1989 [L Qiong]	
Labyrinthus Kobluk 1969 [Boto?]	
Margaretia Walcott 1931 [Boto-M M Camb]	*Huangshandongella* Qian et al. 1979 [Meish]

23.6

Proterozoic and Selected Cambrian Megascopic Dubiofossils and Pseudofossils

HANS J. HOFMANN

In Table 23.6 are listed, alphabetically and by year of publication, reported Proterozoic and selected Cambrian megascopic dubiofossils and pseudofossils. See Section 7.8 for a discussion of the contents of Table 23.6.

The following conventions have been used in Table 23.6:

Type code: ** = holotype of the type species of the indicated genus; * = holotype of a species other than the type species of the indicated genus.

Reference code: the first four characters of the code are the first four letters in the surname of the principal author of the referenced publication (characters 1–4); the next two characters are the last two numbers of the year of publication (characters 5 and 6); the final four characters denote the page number of the first page of the referenced publication (characters 7–10). Thus, a publication by "Smith and Jones" published in "1987" and beginning on "page 93" would be referenced in Table 23.6 as "Smit87..93."

Age code: X = Early Proterozoic (2500 to 1600 Ma); Y = Middle Proterozoic (1600 to 900 Ma); Z = Late Proterozoic, including Vendian (900 to 550 Ma); C = Cambrian.

Lithology code: CA = carbonate; MS = mudstone; SH = shale; SS = sandstone; CH = chert; IF = iron-formation; IG = igneous.

Class code: D = dubiofossil; P = pseudofossil.

Described Taxa of Proterozoic and Selected Earliest Cambrian Megafossils

Table 23.6. *Proterozoic and selected Cambrian megascopic dubiofossils and pseudofossils.*

Number	Taxon	Reference Code	Continent and Country	Locality	Latitude	Longitude
1	"*Arborea*"	Boyn78.291	EU GB	Charnwood Forest, Loc. witheld	0.0	0.0
2	"*Archaeocyathus*"	Guri30.637	AF NA	Groendoorn?	−28.5	18.7
3	(Algae or colloidal bodies)	Fent39..89	NA CA	Mystery Id., Echo Bay	66.1	−118.0
4	(Algal-like forms)	Thom60..65	NA CA	Errington No. 2 mine	46.5	−81.3
5	(Annelid burrows)	Fent371873	NA US	Dawson Pass	48.5	−113.5
6	(Apparently cylindrical body)	Ross59.125	NA US	Swiftcurrent Falls	48.8	−113.7
7	(*Arthrophycus*-like markings)	Aitk73.178	NA CA	Tributary to Mountain Riv.	64.6	−129.6
8	(*Arthrophycus*-like markings)	Aitk73.178	NA CA	Deca Creek	64.2	−128.4
9	(Bioturbation structures)	Knig73.156	NA CA	No locality given	58.9	−63.2
10	(Bioturbation structures)	Morg75..42	NA CA		58.9	−63.2
11	(Bivalve-like structures)	Smit68.639	NA US	Grand Canyon, Bass Trail	36.2	−112.4
13	(Burrow-like tube filling)	Song86.831	AS CN	Ming Tombs	40.3	116.3
14	(Burrows)	Daws66.608	NA CA	Madoc	44.5	−77.5
15	(Burrows)	Faul49..72	NA US	5 km WNW Ishpeming, Ropes Mine	46.5	−87.7
16	(Burrows)	Faul50.102	NA US	5 km WNW Ishpeming, Ropes Mine	46.5	−87.7
17	(Burrows)	Gure81...5	EU SU	Zharnovki Cr., 2km from Dnestr	0.0	0.0
18	(Burrows)	Sing85.422	AS IN	Bhasawar	27.0	77.0
19	(Burrows?)	Aitk73.178	NA CA	Tawu Range, Artic Red Riv.	65.2	−131.0
20	(Corrugated vermiform structures)	Youn69.795	NA CA	Flack Lake	46.6	−82.7
21	(Crawling traces)	Sing85.422	AS IN	Bairat	27.5	76.2
22	(Crawling trails?)	Song86.831	AS CN	Ming Tombs	40.3	116.3
23	(*Cyclomedusa*-like pseudofossils)	Sun 86.325	AS CN	Wuhangshan-Paoya, Fuxian	39.6	121.9
24	(Dimpled mound structure)	Stew84..36	NA MX	Caborca, Cerros de la Ciénega	30.2	−112.1
25	(Disk-like structure)	Webb70.191	AU NS	Carnies Tank, Surts Meadows	−31.2	141.7
27	(Dubiofossil A)	Narb87.647	NA CA	Wernecke Mts. C	64.7	−132.3
28	(Dubiofossil B)	Narb87.647	NA CA	Wernecket Mts. B	64.7	−132.9
29	(Dubiofossil C)	Narb87.647	NA CA	Wernecke Mts. D	64.5	−133.0
30	(Flexuous annelid burrows)	Fent371873	NA US	Boulder Pass	48.7	−113.6
31	(Fossil-like objects)	Elst72...1	NA US	Sierra Ancha Mts.	34.1	−111.2
32	(Fucoidal markings)	Gabr73.153	NA CA	10 km SE of Coates L.	62.6	−126.5
33	(Horizontal tubes resembling *Planolites*)	Tay186...8	AF ZA	Thabazimbi?	−24.6	27.4
34	(Incertae sedis)	Fedo85..70	EU SU	Zimniy Bereg	65.5	39.7
35	(Jellyfish impressions)	Alf 59..60	NA US	Grand Canyon, Kaibab Trail	36.1	−112.1
36	(Lined subvertical tube)	Tayl86...8	AF ZA	Lydenburg?	−25.0	30.8

Geologic Unit	Age	Lithology	Class	Original Author's Interpretation and other interpretations	Hofmann Interpretation and Remarks
Woodhouse Beds	Z	MS	D	Problematica. Inorganic (Ford 1980, p. 82)	Dubiofossil cf. *Arborea arborea*
Kuibis Fm.	Z		D	Archaeocyathid. Trace fossil (Glaessner)	Trace Fossil? specimen reported by Haughton = *Archaeichnium* Glaessner
Labine Gp., [Echo Bay Fm.]	X		P	Algae or Colloidal Bodies	Pseudofossil. Chemical
Whitewater Gp., Onwatin Fm. [Vermillion Mem.]	X	CA	P	Algal. Fissure fillings (Hofmann 1971, p. 34–35.)	Vein Fillings
Siyeh Fm.	Y	CA	D	Annelid Burrows	Dubiofossils. Mudcrack fillings?
Altyn Fm.	Y	CA	P	Organic. Compared with spine of trilobite or chitinous brachiopod (J. H. Johnson in Ross (1959, p. 20)	Ooid
Proterozoic Unit H-5 [Little Dal Gp.]	Y	SS	D	Trace Fossil	Mudcrack Filling?
Katherine Gp.	Y	SS	D	Trace Fossil	Mudcrack Filling?
Ramah Gp., Reddick Bight Fm.	X	SS	D	Trace Fossil	Probably Mudcrack Filling
Ramah Gp., Reddick Bight Fm.	X		D	Trace Fossil	Probably Mudcrack Filling. Refers to Knight (1973, p. 157)
Unkar Gp., Bass Ls. or Hakatai Sh.	Y	MS	D	Fossils	Dubiofossils. Insufficient information
Changcheng Sys., Changzhougou Fm.	X	SH	D	Metazoan Burrow	Dubiofossil
Grenville Supgp.	Y?	SS?	D	Annelid Burrows	Dubiofossils. Material inadeuate. = *Sabellarites* Dawson 1890, p. 608.
Ajibik Qte.	X	SS	D	Burrows. Mudcrack fillings (Schindewolf 1956, p. 460)	Dubiofossils
Ajibik Qte.	X	SS	D	Burrows. Mudcrack fillings (Schindewolf 1956, p. 460)	Dubiofossils
Kanilov Fm.	Z	MS	D	Vertical Burrows, Cones + Pits	Dubiofossil, Water Escape Pits cf. *Chomatichnus*, resemblance to *Arumberia* mounds with pits
Delhi Supgp., Alwar Qte.	XY	SS	D	Trace Fossils	Dubiofossils Mudcracks?, but also looks like *Arenicolites* bottoms
Tsezotene Fm., unit 3	Y	SS	D	Trace Fossil	Probably Mudcrack Filling
Gordon Lake Fm. (? Bar River Fm.)	X	SS	P	Subaqueous Crack Fillings	Inorganic. Mobilized shrinkage crack fillings
Delhi Supgp., Alwar Qte.	XY	SS	P	Trace Fossils	Mudcrack Fillings. cf. *Manchuriophycus*
Changcheng Sys., Changzhougou Fm.	X	SS	D	Metazoan Trace Fossil	Dubiofossil. Epirelief with parallel ridges along furrow
Wuhangshan Gp., Changlingzi Fm.	Z	SS	D	Gas Pits	Dubiofossils. Not *Cyclomedusa*, gas pits
Clemente Fm., Unit 6	Z?	MS	D	Metazoan Origin Likely	Dubiofossil. Fluid escape structure?
Torrowangee Gp., Fowlers Gap Beds	Z	MS	P	Dubiofossil	Inorganic. Medusoid-like structures [probably concretionary]
Windermere Supgp., Siltstone 1	Z	SS	D	Dubiofossil	Dubiofossil. Possibly a distorted *Cyclomedusa*
Windermere Supgp., Siltstone 1	Z	SS	D	Dubiofossil	Dubiofossil. Possible spheroidal or sac-shaped organism
Windermere Supgp., Siltstone 2	Z	SS	D	Dubiofossil	Dubiofossil. Possible juvenile forms of *Beltanelliformis* or *Bergaueria*
Miller Peak Fm., Kintla Mem.	Y	CA	D	Annelid Burrows	Dubiofossil
Troy Qte. (1.2 Ga)	Y	SS	P	Arthropods. Inorganic (Glaessner 1973a, Chowns 1973, Teichert 1973, Lindström 1973, Häntschel 1975)	Curled Mudcracks. cf. *Caragassia*?
Tigonankweine Fm. [Katherine Gp.]	Y	SS	D	Undetermined Structures	Possibly Mudcrack Fillings
Transvall Spgp., Black Reef Qte.	X	SS	D	Possible Trace Fossils	Dubiofossils. cf. dewatering structures of Rousell (1984) in Chelmsford Fm.
Ust-Pinega Fm.	Z		D	Undetermined	Burrow?
Unkar Gp., Bass Ls. or Hakatai Sh.	Y	MS	D	Medusoids. Raindrop splash marks (Cloud 1968, p. 22, 53), algal colonies (Glaessner (1962, 1969, p. 375–376)	Dubiofossils
Transvaal Spgp., Malmani Dol.	X	CA	D	Possible Trace Fossils	Dubiofossils. Compared with *Muensteria*

Table 23.6. (Continued).

Number	Taxon	Reference Code	Continent and Country	Locality	Latitude	Longitude
37	(Linguloid brachiopod)	Bart71.293	NA US	Washington Co. (no loc. given)	36.7	−81.7
38	(Medusa-like markings)	Van 37.314	NA US	Grand Canyon, Basalt Cliffs	36.1	−111.9
39	(Medusoid imprints)	Houz79.379	AF MA	Irherm, S slope of Fou Woudid	30.0	−8.5
40	(Organic (?) structure)	Reza63..33	NA US	Swiftcurrent Falls	48.8	−113.7
41	(Plant impressions)	Glae63.113	AF NA	Farm Ibenstein 55, Rehoboth	−22.5	16.5
42	(Polyp-like form)	Haug64.257	AF ZA	Hartheespoortdam	−25.8	27.8
43	(Possible biogenic structure)	Rous84.211	NA CA	Hwy. 144	46.6	−81.3
44	(Possible biogenic structures)	Rous72..79	NA CA	Hwy. 144	46.6	−81.3
45	(Possible biogenic structures)	Morg75..42	NA CA	4 km W of Mount Dalhousie	59.0	−63.3
46	(Possible burrows)	Toom83.345	NA US	Fusselman Canyon, Rd. 375	31.9	−106.5
47	(Possible fossil burrows)	Chap75.477	NA US	Haywood Co.	35.7	−83.1
48	(Possible metazoans)	Frar63.461	NA CA	Desbarats	46.4	−83.9
49	(Possible organic structures)	Youn67.565	NA CA	Flack Lake	46.6	−82.7
50	(Possible organic structures)	Hard86.684	NA CA	Flack Lake	46.6	−82.7
51	(Possible trace fossil)	Stew84..36	NA MX	Caborca, Cerros de la Ciénega	30.2	−112.1
52	(Probable burrows)	Poul73.292	NA CA	N Dogtooth Mts.	51.4	−117.4
53	(Probable trace fossils)	McMe83.227	NA MX	Caborca, Cerros de la Ciénega	30.2	−112.1
54	(Problematic markings)	Ding85.115	AS CN	Yangtze Gorge	32.8	111.1
55	(Problematic structures)	Fent371873	NA US	Glacier Nat. Park	0.0	0.0
56	(Protomedusoids)	Zhan81..217	AS CN	Xupu, Anhua	27.9	110.6
57	(Pseudo-fucoids)	Math83.111	AS IN	Nilgarh, Raisen district	23.0	78.0
58	(Rootlike burrow)	Glae63.113	AF ZA	Kleine Kloof Farm, Calvinia	−31.2	19.2
59	(*Scolithus* tubes?)	Hunt66.162	NA US	S Panamint Rge., Galena Canyon	36.0	−116.9
60	(Sediment-filled tubes)	Kauf81.923	NA US	Medicine Bow Mts.	41.3	−106.3
61	(*Skolithos*-like tubes)	Hoff68..93	NA CA	N shore Charlton Bay	62.7	−109.1
62	(Small bilobate traces)	Seil56.155	NA US	Grand Canyon, Kaibab Trail	36.1	−112.1
63	(Spiculae of sponges)	Walc93.639	NA US	Grand Canyon	0.0	0.0
64	(Spicules)	Daws66.608	NA CA	Madoc?	44.5	−77.5
65	(Spongelike nodules)	Alf 59..60	NA US	Grand Canyon, Kaibab Trail	36.1	−112.1
66	(Strings of beads)	Grey87..36	AU WA	Bangemall Basin, E of Newman	−23.5	120.9
67	(Strings of flattened beads)	Horo82.882	NA US	S side of Appekunny Mountain	48.8	−113.7
68	(Stromatolites?)	Youn67.565	NA CA	Washagami Lake	46.7	−80.5
69	(Structure resembling Nautiloid)	Wads84.208	NA US	Copper Harbor or Eagle River	47.5	−87.9

Geologic Unit	Age	Lithology	Class	Original Author's Interpretation and other interpretatons	Hofmann Interpretation and Remarks
Mount Rogers Volcanic Gp. (570 feet below top)	ZC	MS	D	Phosphatic Brachiopod Valve	Insufficient Data for Interpr. 18 × 7 mm, phosphatic, 86 growth lines
Nankoweap Fm.	Y	SS	D	Medusoid	Dubiofossil = Brooksella canyonensis Bassler 1941
Adoudounian, Basal Ser.	Z	MS		Medusoid	Dubiofossil. Compacted carbonate nodules?
Altyn Fm.	Y	CA	P	Fossil	Ooid? = diagenetically modified ooid
pre-Nama quartzite	YZ	SS	D	Plants	Dubiofossils. Crystal casts?
Transvaal Supgp., Dolomite Gp.	X	CH	D	Animal with Symbiotic Algae. Concretion (Cloud 1968, p. 53)	Dubiofossil (Chemogenic)
Whitewater Gp., Chelmsford Fm.	X	SS	P	Organic. Dewatering structure (Cloud 1988)	Dewatering Structure
Whitewater Gp., Chelmsford Fm.	X	SS	P	Possible Trace Fossil. Dewatering structure (Cloud 1988)	Dewatering Structure
Ramah Gp., Rowsell Harbour Fm.	X	SS	D	Questionably Biogenic	Dubiofossil. Insufficient data
Castner Marble (basic intrusion within)	Y	IG	P	Possible Trace Fossils	Pseudofossils. Weathering pattern in diabase
Snowbird Gp., Longarm Qte.	Z	SS	D	Trace Fossils	Insufficient Data for Interpr. 5–25 mm wide, 2.08 m long, branching = ? secondary, chatter
Huronian Supgp., Lorrain Fm.	X	SS	P	Possible Worm Tubes. Injection or load phenomena in contraction cracks (Cloud 1968, p. 54 Hofmann 1971, p. 36–39)	Pseudofossils = Rhysonetron. Compacted sinuous mudcrack fillings
Bar River Fm.	X	SS	P	Casts of Vermiform Organism	Mudcrack Fillings = Rhysonetron
Bar River Fm.	X	SS	P	Possibly Organic	Pseudofossils. Authors present inconclusive microprobe analysis
Clemente Fm., Unit 6	Z?	MS	D	Possible Trace Fossils	Dubiofossil. Possible spreite, may be diffusion structure
Windermere Supgp., Horsethief Creek Gp., Subunits 4, 7	Z	CA	D	Probable Trace Fossils	Dubiofossils. No illustrations or descriptions
Clemente Fm.	Z	MS	D	Probable Trace Fossils	Dubiofossil. Diffusion banding?
Dengying Fm.	Z		D	Questionable Fossil	Dubiofossil
Shepard Fm.	Y	MS	D	Annelid Castings?	Dubiofossils. Possibly compacted crack filling
Liuchapo Fm. (Dengying Fm.)	Z		D	Protomedusae	Dubiofossils. Resemble rosetted crystal structures
Sirbu Sh.		SS	D	Trace Fossils	Dubiofossils. See Vredenburg 1908, [markings not interpretable]
Nama Gp., Kuibis Fm., Schwarzkalk Mem.	Z		D	Burrows	Dubiofossils
Noonday Dol.	Z	CA	D	Trace Fossils	Dubiofossils Usgs pp 494–A
Libby Creek Gp., Medicine Peak Qte. (2410–2000 Ma)	X	SS	D	Possible Burrows	Dubiofossils
Sosan Gp., Akaitcho River Fm.	X	SS	P	Possible Trace Fossils. Inorganic (Hofmann 1971, p. 39–40)	Pseudofossils. Sedimentary lamination pass across structures
Unkar Gp., Bass Ls. or Hakatai Sh.	Y	MS	D	Trace Fossils	Dubiofossils
Unkar Gp.	Y	CA	D	Sponge Spicules	Dubiofossils
Grenville Supgp.	Y?		D	Sponges	Dubiofossils. Insufficient data
Unkar Gp., Bass Ls. or Hakatai Sh.	Y	MS	D	Possible Sponge. Silica nodule (Cloud 1968, p. 57)	Dubiofossil
Manganese Gp. (ca. 1.1 Ga)	Y		DB	Possible Megascopic Algae	Fossil, Possibly Algal. cf. beads described by Horodyski (1982) from Appekunny Fm.
Appekunny Fm.	Y	MS	DB	Dubiofossils. Possible megascopic algae (Grey and Williams 1987)	Dubiofossils. cf. Neonereites, Hormosiroidea, algae, microbial colonies
Huronian Supgp., Gowganda Fm.	X	MS	P	Stromatolites. Diffusion structures (Hofmann 1971)	Pseudofossils = Kempia huronense
Keweenawan	Y	IG	D	Nautiloid	Dubiofossil. Uninterpretable, no specimens available

Table 23.6. (*Continued*).

Number	Taxon	Reference Code	Continent and Country	Locality	Latitude	Longitude
70	(Supposed fossil)	Loga63.983	NA CA	Grand Calumet, near Bryson	45.7	−76.6
71	(Trace fossils)	Dawe76.249	NA GL	Narssaq, N of Olrik Fjord	77.2	−68.5
72	(Trace fossils, bedding plane markings)	Gres96.622	NA US	Chapin Mine, Iron Mountain	45.9	−88.1
73	(Trace fossils, bedding plane markings)	Gres97.527	NA US	Chapin Mine, Iron Mountain	45.9	−88.1
74	(Traces of pelecypods)	Fent371873	NA US	Dawson Pass	48.5	−113.5
75	(Triact spicules)	Dunn64.195	AU NT	Roper River	−14.8	134.7
76	(Tube-shaped structures)	Pede66..40	NA GL	Zig Zag Land	61.4	−45.0
77	(U-shaped tubes)	Tayl86...8	AF ZA	Lydenburg?	−25.0	30.8
78	(Various problematic remains)	Whit29.392	NA US	Grand Canyon, Bright Angel Ck.	36.1	−112.1
79	(Vermiform markings)	Alf59..60	NA US	Grand Canyon, Kaibab Trail	36.1	−112.1
80	(Vermiform structures)	Dona871273	NA CA	Flack Lake	46.6	−82.7
81	(Vermiform structures)	Dawe75..38	NA GL	Narssaq, N of Olrik Fjord	77.2	−68.5
82	(Worm borings)	Ford58.211	EU GB	Bradgate Park, Deer Park Spin.	52.6	−1.3
83	(Worm casts and moulds)	Visw72.422	AS IN	Hirenandi near Gokak	16.1	74.9
84	(Worm trace)	Glae63.113	AF NA	Chamaites 113, S of Seeheim	−27.1	17.7
85	(Worm trails)	Knig77..31	NA CA	SW of Reddick Bight	58.9	−63.2
86	*Acanthichus*	Das84.251	AS IN	Bankuiyan, Rewa Dist.	24.6	81.2
87	*Allatheca* sp.	Mait84.212	AS IN	Rampura (Ramapura)	24.5	75.5
88	*Amanlisia simplex*	**Lebe91.200	EU FR	Amanlis, Montfort	48.0	−1.9
89	*Antholithina rosacea*	**Chou51...9	AF MA	Taghdout	0.0	0.0
90	*Archaeophyton Newberryanum*	Brit88..89	NA US	Highlands of Sussex Co.	41.2	−74.6
91	*Archaeophyton Newberryanum*	**Brit88.123	NA US	Highlands of Sussex Co.	41.2	−74.6
92	*Archaeoprotospongia* sp.	Cao83	AS CN	Lientuo	0.0	0.0
93	*Archaeospherina*	*Daws75.239	NA CA	Côte St. Pierre	45.8	−75.1
94	*Archaeospongia radiata*	Vasi68.106	AS SU	Kuznetz Alatau, Tamalyk	53.7	89.7
95	*Archaeospongia radiata*	Vasi68.106	AS SU	Kuznetz Alatau, Tamalyk	53.7	89.7
96	*Archaeospongia* sp.	Vasi68.106	AS SU	Kuznetz Alatau, Komunar	54.5	89.4
97	*Archaeoxylon Krasseri*	**Krau24..31	EU CS	Bohemia, Mratim well, 228 m	0.0	0.0
98	*Arenicolites spiralis*	Bill72.478	NA CA	Near St. John's	47.6	−52.7
99	*Armelia Barrandei*	**Lebe91.200	EU FR	Corps-Nuds, St. Armel	48.0	−1.6
100	*Arumberia*	Blan82.445	NA CA	No locality given	47.6	−52.7
101	*Arumberia*	Blan82.445	NA CA	No locality given	47.3	−53.8
102	*Arumberia*	Blan84.625	NA CA	No locality given	47.6	−52.7
103	*Arumberia banksi*	**Glae75..59	AU NT	22 km SW of Alice Springs	−23.8	133.7

Proterozoic and Selected Cambrian Megascopic Dubiofossils and Pseudofossils

Geologic Unit	Age	Lithology	Class	Original Author's Interpretation and other interpretatons	Hofmann Interpretation and Remarks
Grenville Supgp.	Y?	CA	P	Fossil?. Inorganic (Hofmann 1971)	Pseudofossil. Metamorphic. = *Eozoon canadense* (first illustrated specimen)
Thule Gp., Wolstenholme Fm.	Y	SS	P	Trace Fossils	Syneresis Crack Fillings. Refers to Dawes and Bromley (1975, p. 39)
Menominee Iron Fm.?	X	IF	D	Trace Fossils	Dubiofossils. No specimens. Found as fragments at ore terminal at Erie, PA
Menominee Iron Fm.?	X	IF	D	Trace Fossils	Dubiofossils. No specimens available
Siyeh Fm.	Y	CA	D	Pelecypod Burrows	Dubiofossils
McArthur Gp.?	X	CH	P	Sponge Spicules. Possibly volcanic shards (Cloud 1968, p. 56)	Volcanic Shards
Vallen Gp., Zigzagland Fm.	X	CA	D	Undetermined Affinities	Dubiofossil. cf. Camasiidae
Transvaal Spgp., Malmani Dol.	X	CA	D	Possible Trace Fossils	Dubiofossils. Compared with *Arenicolites*
Unkar Gp., Bass Ls. or Hakatai Sh.	Y	MS	D	Problematic Markings	Dubiofossils. Insufficient information
Unkar Gp., Bass Ls. or Hakatai Sh.	Y	MS	D	Fossil	Dubiofossil
Bar River Fm.	X	SS	P	Deformed Microbial Mat. Compacted crack filling (Hofmann 1971, p. 37)	Compacted Crack Filling = *Rhysonetron*
Thule Gp., Wolstenholme Fm.	Y	SS	P	Burrows	Syneresis Crack Fillings
Brand Ser., Charnian	Z	SS	D	Worm Boring. Inorganic (Watts 1947, p. 104) (Friedman 1950, p. 441)	Dubiofossil
Cuddapah Supgp., Badami Gp.	Y	SS	P	Worm Casts and Moulds	Mudcrack Fillings. cf. *Manchuriophycus*
Nama Gp., Fish River beds	ZC	SS	D	Spiral Worm Trace	Dubiofossil. Curved mudcrack filling?
Ramah Gp., Reddick Bight Fm., unit 20	X	SS	D	Trace Fossil	Probably Mudcrack Filling. Insufficient information
Bhander Ls.			D	Trace Fossils	Dubiofossil. Insufficient data for interpretation
Semri Gp., Suket Sh.	Y	SH	D	Body Fossils of Uncertain Aff.	Dubiofossil. Small rounded grains
Rennes Sh.	Z	MS	I	Alga. Crawling trace (Seilacher 1956, p. 167) uncharacteristic form (Häntzschel 1965, p. 8)	Trace Fossil {OD, M}
Taghdout Ls.	YZ	CA	P	Calcareous Alga. Inorganic, ooid grains with envelope of iron hydroxide (Schindewolf 1956 p. 468)	Allochem, Ooid
Precambrian	Y?	CA	DC	Algae, Plant Material. Inorganic (Häntzschel 1965, p. 9)	Dubiofossil, Chemofossil. Black graphite films, up to 0.5 mm thick, 3 mm wide
Precambrian	Y?	CA	DC	Algae, Plant Material. Inorganic (Häntzschel 1965, p. 9)	Dubiofossil, Chemofossil. Metamorphic graphite, cf. slickenfiber faults
Doushantuo Fm.	ZC		D	Sponge	Dubiofossil. Citing Tang, Zhang and Jiang 1978
Grenville Supgp.	Y?	CA	P	Foraminifera like *Globigerina*. Inorganic (Hofmann 1971)	Pseudofossil. No species designated. Serpentine grains in marble
Tyurim Fm.	Y	CA	D	Fossils	Dubiofossils. Insufficient data, cf. Camasiids
Tyurim Fm.	Y	CA	D	Fossils	Dubiofossils. Insufficient data, cf. Camasiids
Tyurim Fm.	Y	CA	D	Fossils	Dubiofossils. Insufficient data, cf. Camasiids
Pribram Fm.	YZ	MS	DC	Pteridophyte?	Organic Carbon? n. nud.
St. John's Gp.	Z	SS	D	Trace Fossil. Dubiofossil (Hofmann 1971)	Trace Fossil or Grypania. No specimens available
Rennes Sh.	Z		D	Echinoderm (Cystoid). Problematic body fossil (Häntzschel 1965, p. 10 Seilacher in Häntzschel 1975, p. W180)	Dubiofossil. Uninterpretable
Signal Hill Gp., Gibbett Hill Fm.	Z	SS	D	Benthic Colonial Organism	Dubiofossil. Flow marks
Musgravetown Gp.	Z		D	Benthic Colonial Organism	Dubiofossil. Flow marks
Signal Hill Gp., Gibbett Hill Fm.	Z	SS	D	Problematic Colonial Organism	Dubiofossil. Flow marks
Arumbera Qte.	ZC	SS	D	Coelenterata. Pseudofossil (Brasier (1979) (Jenkins et al. 1981)	Dubiofossil. Flow marks

Table 23.6. (Continued).

Number	Taxon	Reference Code	Continent and Country	Locality	Latitude	Longitude
104	*Arumberia banksi*	Bekk80.480	EU SU	Mouth of Koyva Riv.	58.3	58.2
105	*Arumberia banksi*	Liu 81..71	AS CN	Jixi	45.1	130.5
106	*Arumberia banksi*	Wang84.136	AS CN	Jixi	45.1	130.5
107	*Aspidella terranovica*	** Bill72.478	NA CA	St. John's area	47.6	−52.7
108	*Asteriradiatus karauliensis*	** Math82.127	AS IN	Panna Dist.	24.7	80.2
109	*Asterosoma*	Glae69.369	NA US	Grand Canyon	36.1	−111.9
110	*Beaumontia eckersleyi*	** Davi28.191	AU SA	Goldsack Qy., Beaumont	−35.0	138.7
111	*Beltanelliformis* (?)	Acen86.367	SA AR	Sierra de la Ovejerîa	−27.5	−68.0
112	*Bergaueria*?	Lina87.211	EU ES	Coria, Alagón Riv.	40.0	−6.5
113	*Bergaueria*? sp.	Hofm79.150	NA CA	Mackenzie Mts. 2	64.6	−128.8
114	*Bergaueria*? sp.	Hofm85.331	NA CA	Mackenzie Mts. 2	64.6	−128.8
115	*Bestricophyton*	Das 84.251	AS IN	Bankuiyan, Rewa Dist.	24.6	81.2
116	*Bohaimedusa fuxianensis*	** Chan80.266	AS CN	Fuxian	39.6	121.9
117	*Bostricophyton bankuiyanensis*	* Verm68.557	AS IN	Bankuiyan	24.0	81.2
118	*Brooksella canyonensis*	* Bass41.519	NA US	Grand Canyon, Basalt Cliffs	36.1	−111.9
119	*Brooksella canyonensis*	Kauf83.608	NA US	Grand Canyon	36.1	−111.9
120	*Butunia enigmatica*	** Kell821187	AS SU	Butun Riv., Kalar Riv. area	56.0	118.0
121	*Camasia fruticulata*	* Sosn81..73	AS SU	Batenev Ridge, Karysh Creek	54.2	90.2
122	*Camasia spongiosa*	** Walc14..77	NA US	Forks of Birch Creek (400c)	46.6	−111.1
123	*Caragassia karassevi*	** Volo651426	AS SU	Arshan, Iya Riv.	53.9	99.8
124	*Carelozoon jatulicum*	** Metz24..86	EU SU	Suojärvi (formerly E. Finland)	62.3	32.3
125	*Caryschia cyathiformis*	** Sosn81..73	AS SU	Batenev Ridge	54.2	90.2
126	*Caryschia magna*	* Sosn81..73	AS SU	Batenev Ridge	54.2	90.2
127	*Chomatichnus loevcichnus*	** Gure84...5	EU SU	Zharnovki Cr., 2 km from Dnestr	0.0	0.0
128	*Chordoichnus latouchei*	** Math83.111	AS IN	Barui, near Osian	26.8	72.9
129	*Chuaria*?	Debr81..23	EU IT	Sardinia, Sarrabus, Tuviois	39.4	9.3
130	*Coleolella billingsi*	Mait84.212	AS IN	Ramapura (Ramapura)	24.5	75.5
131	*Collinsia mississagiense*	** Bain27.281	NA CA	Vernon Twp., 22 km N Espanola	46.5	−81.8
132	*Conophyllum minor*	Vasi68.106	AS SU	Gornoy Shoria, Mrassu, Saga	53.0	89.0
133	*Copperia tubiformis*	** Walc14..77	NA US	Forks of Birch Creek (400c)	46.6	−111.1
134	*Ctenichnites*	Matt90.145	NA CA	Thunder Bay area	48.4	−89.3
135	*Cyathospongia? Eozoica*	* Matt90..42	NA CA	Drury Cove, 6.5 km NE St. John	45.3	−66.0
136	*Cyclomedusa* aff. *davidi*	Xing79.167	AS CN	Wuhangshan-Paoya, Fuxian	39.6	121.9
137	*Cyclomedusa annulata*	Xing79.167	AS CN	Wuhangshan-Paoya, Fuxian	39.6	121.9
138	*Cyclomedusa* cf. *davidi*	Xing79.167	AS CN	Wuhangshan-Paoya, Fuxian	39.6	121.9

Geologic Unit	Age	Lithology	Class	Original Author's Interpretation and other interpretatons	Hofmann Interpretation and Remarks
Ust'-Sylvitsa Fm.	Z	SS	D	Soft-Bodied Animal	Dubiofossil. Flow marks
Mashan Gp., Liumao Fm., Shichang Mem.	Z?	CA	D	Coelenterata	Dubiofossil. Karren weathering?
Mashan Gp., Zhongsanyang Fm.	Z?	CA	D	Cnidaria	Dubiofossil. Karren weathering?
St. John's Gp.	Z	MS	D	Fossils (cf. Chiton, Patella). Compaction and spall marks (Cloud 1968, p. 55), focussed rupture surface (Hofmann 1971, p. 14–17)	Dubiofossil. Cnidarian?
Rewah Gp., Karauli Qte.		SS	D	Trace Fossil	Dubiofossil. n. nud. Radiating burrows?
Nankowap Gp.	Z	SS	D	Trace Fossil	Dubiofossil = *Brooksella*
Burra Gp., Beaumont Dol.	YZ	CA	P	Eurypterid. Inorganic (Glaessner in Häntzschel 1965, p. 15), mudflakes (Cloud 1968)	Inorganic. Non *Beaumontia* Edw. and Haime 1851 nec Deslongchamps, 1856
Puncoviscana and Suncho Fms.	ZC		D	Coelenterate Impressions	Burrows?
Schist-Graywacke Complex, Conglomerate Beds	Z	MS	D	Trace Fossil	Dubiofossil. cf. *Nemiana* or *Beltanelloides*, "*Chuaria*" from central AU
Little Dal Gp.	Z	SS	P	Trace Fossil?	Inorganic. Dewatering craters
Little Dal Gp.	Z	SS	D	Dubiofossils	Probably Inorganic. cf. '*Zoophycos*' Rodriguez and Gutschick
Bhander Ls.	Z	SS	D	Trace Fossils	Dubiofossil, Insufficient Data. n. null. for *Bostricophyton*
Changlingzi Fm.?	Z	SS	D	Coelenterate. Dewatering pits (Sun 1986a, p. 349)	Dubiofossil. n. nud.
Bhander Ls.	Z	CA	D	Repichnia of Worm or Arthropod.	Dubiofossil
Nankoweap Fm.	Y	SS	D	Medusoid. Trace fossil (Fürsich and Bromley 1985)	Dubiofossil: Compaction = *Asterosoma*? Glaessner 1969, and Crimes and Germs 1982
Nankoweap Fm.	Z	SS	D	Trace Fossil with Spreite	Dubiofossil. Compaction?
Butun Fm., Udokan Gp.	X	MS	D	Dubious Organism	Dubiofossil. cf. *Arumberia*
Polundennaya Gp., Tyurin Fm.	Y	CA	D	Camasiida	Dubiofossil. Insufficient data, cf. *Camasia*
Newland Fm.	Y	CA	D	Algae. Inorganic segregation (Fenton and Fenton 1936, p. 615), diagenetic (Häntzschel 1975, p. W171)	Dubiofossil. Contains "chains of blue-green algae"
Karagas Fm.	YZ	SS	P	Merostomoidea	Pseudofossils, Mudflakes
Jatulian	X	CA	PS	Coelenterate, Algae. Possible concretions (Seilacher 1956, p. 158, Cloud 1968, p. 54)	Matrix between Stromatolites
Poludennaya Gp., Tyurim Fm.	Y	CA	D	Camasiida	Dubiofossil. Insufficient data, cf. *Camasia*
Poludennaya Gp., Tyurim Fm.	Y	CA	D	Camasiida	Dubiofossil. Insufficient data, cf. *Camasia*
Zharnov Fm., Kanilov Gp.	Z	MS	D	Trace Fossil	Dubiofossil, Water Escape Pits = Burrows of Gureev 1981, p. 6–7.
Marwar Gp., Jodhpur Ss.		SS	D	Trace Fossil	Dubiofossil. n.g. n.sp. see Vredenburg 1908, Pl. 34
San Vito Fm.	ZC	MS	D	Dubious Fossil	Dubiofossil. Not *Chuaria*
Semri Gp., Suket Sh.	Y	SH	D	Body Fossil of Uncertain Aff.	Dubiofossil. Small doughnut-like grain
Huronian Supgp., Mississagi Fm.	X	SS	P	Stromatoporoid?. Chemical (Hofmann 1971, p. 29)	Pseudofossil
Poludennaya Fm.	Y	CA	D	Fossils	Dubiofossils. Insufficient data, cf. Camasiids
Newland Fm.	Y	CA	D	Algae. = *Greysonia*, inorganic segregation (Fenton and Fenton 1936, p. 614) diagenetic (Häntzschel 1975)	Dubiofossil. Parallel to bedding
Animikie Gp.	X	SS	D	Trace Fossil. Sole (tool and flute) markings (Hofmann 1971, p. 20–21)	Dubiofossil
Green Head Gp.	Y?	SS	P	Sponge. Not sponge (Rauff 1893, p. 59), probably crystals (Cloud 1968, p. 57)	Pseudofossil. No specimens available
Nanguanling Fm. [Changlingzi Fm.]	Z	SS	D	Coelenterata. Inorganic (Sun 1968a, p. 349)	Dubiofossils
Nanguanling Fm. [Changlingzi Fm.]	Z	SS	D	Coelenterata. Inorganic (Sun 1968a, p. 349)	Dubiofossils
Nanguanling Fm. [Changlingzi Fm.]	Z	SS	D	Coelenterata. Inorganic (Sun 1986a, p. 349)	Dubiofossils

Table 23.6. (Continued).

Number	Taxon	Reference Code	Continent and Country	Locality	Latitude	Longitude
139	*Cyclomedusa gracilis*	*Xing79.167	AS CN	Wuhangshan-Paoya, Fuxian	39.6	121.9
140	*Cyclomedusa minus*	*Xing79.167	AS CN	Wuhangshan-Paoya, Fuxian	39.6	121.9
141	*Cyclomedusa minus*	Chan80.266	AS CN	Fuxian	39.6	121.9
142	*Cyclomedusa simplicis*	*Xing79.167	AS CN	Wuhangshan-Paoya, Fuxian	39.6	121.9
143	*Cyclomedusa sp.*	Chen84..51	AS CN	Fuxian	39.6	121.9
144	*Cylindrocraterion heroni*	**Ghar77.205	AS IN	Rawatbhata	25.0	75.6
145	*Dactyloidites canyonensis*	Furs85.199	NA US	Grand Canyon, Basalt Cliffs	36.1	−111.9
146	*Eophyton*	Hant75W269			0.0	0.0
147	*Eosfroma prima*	Vasi68.106	AS SU	Gornoy Shoria, Mrassu, Saga	53.0	89.0
148	*Eosomedusa sp.*	Chen84..51	AS CN	Liaotung Penisnula	0.0	0.0
149	*Eospicula cayeuxi*	**de L55..21	EU FR	Brittany	48.5	−2.5
150	*Eozoon Canadense*	**Daws64.218	NA CA	Côte St. Pierre	45.8	−75.1
151	*Eozoon Canadense*	Edwa70.226	NA US	Thurman, Warren Co., NY	43.5	−73.9
152	*Eozoon canadense*	Loga66...1	NA CA	North Burgess Twp., S of Perth	44.9	−76.3
153	*Eozoon canadense*	Daws67.257	NA CA	Millbridge, Tudor Twp. (loose)	44.7	−77.6
154	*Eozoon canadense*	Merr99.189	NA US	Thurman, Warren Co., NY	43.5	−73.9
155	*Eozoon canadense*	Merr99.363	NA US	Thurman, Warren Co., NY	43.5	−73.9
156	*Eozoon canadense*	Mill14.151	NA CA	Madoc, Huntington Twp.	44.5	−77.5
157	*Exoculatus arschaniensis*	**Dodi661169	AS SU	Iya Riv., above Arshan	53.9	99.8
158	*Exoculatus selgineikensis*	*Dodi661169	AS SU	Iya Riv., above Arshan	53.9	99.8
159	*Gakarusia addisoni*	**Haug64.257	AF ZA	Gakarusa, near Daniels Kuil	−27.9	23.6
160	*Gallatinia pertexa*	**Walc14..77	NA US	N of Barlow Bridge (400j)	45.9	−117.3
161	*Gallatinia pretexa* [sic]	Ghar77.205	AS IN	Rawatbhata	25.0	75.6
162	*Gallatinia scalariformis*	*Ghar77.205	AS IN	Rawatbhata	25.0	75.6
163	*Glaessnerina sp.*	Liu 81..71	AS CN	Jixi	45.1	130.5
164	*Greysonia basaltica*	**Walc14..77	NA US	Forks of Birch Creek	46.6	−111.1
165	*Halichondrites graphitiferus*	*Matt90..42	NA CA	Reversing Falls, Saint John	45.3	−66.1
166	*Halichondrites graphitiferus*	Mill871913	NA CA	Reversing Falls, Saint John	45.3	−66.1
167	*Ichnusa cocozzi*	**Debr81..23	EU IT	Sarrabus, 2 km W of Tuviois	39.4	9.3
168	*Ichnusina*		EU IT	Sarrabus, 2 km W of Tuviois	39.4	9.3
169	*Ikeyia tumida*	**Volo651426	AS SU	Ikey village	54.2	100.1
170	*Iyaia sayanica*	**Volo651426	AS SU	Arshan, Iya Riv.	53.9	99.8
171	*Jixiella capistratus*	**Liu 81..71	AS CN	Jixi	45.1	130.5
172	*Jixiella capistratus*	Wang84.136	AS CN	Jixi	45.1	130.5
173	*Jussenia*	Vasi68.106	AS SU	Gornoy Shoria, Mrassu, Saga	53.0	89.0
174	*Jussenia cf. edelsteini*	Vasi68.106	AS SU	Kuznetz Alatau, Tamalyk	53.7	89.7

Proterozoic and Selected Cambrian Megascopic Dubiofossils and Pseudofossils 1045

Geologic Unit	Age	Lithology	Class	Original Author's Interpretation and other interpretatons	Hofmann Interpretation and Remarks
Nanguanling Fm. [Changlingzi Fm.]	Z	SS	D	Coelenterata. Inorganic (Sun 1986a, p. 349)	Dubiofossils
Nanguanling Fm. [Changlingzi Fm.]	Z	SS	D	Coelenterata. Inorganic (Sun 1986a, p. 349)	Dubiofossils
Changlingzi Fm.?	Z	SS	D	Colonial Algae?.	Dubiofossil
Nanguanling Fm. [Changlingzi Fm.]	Z	SS	D	Coelenterata. Inorganic (Sun 1968a, p. 349)	Dubiofossils
Wuhangshan Gp., Nanguanling Fm.	Z	SS	D	Dubiofossils	Dubiofossils
Kaimur Gp.	Z	SS	D	Dwelling Burrow	Dubiofossil. cf. *Skolithos*, *Monocraterion*
Nankoweap Fm.	Y	SS	D	Trace Fossil with Spreite	Dubiofossil = *Brooksella canyonensis* Bassler
			P	Drag Marks	Drag Marks
Tyurim Fm.	Y	CA	D	Fossils	Dubiofossils. Insufficient data, cf. Camasiids
Wuhangshan Gp., Nanguanling Fm.		SS	D	Dubiofossil	Dubiofossil
Lamballe	Z	CH	P	Sponge Spicules. Inorganic (Rauff 1896, Schindewolf 1956), crystals? (Cloud 1968, p. 56)	Inorganic = structures described by Cayeux 1985
Grenville Supgp.	Y?	CA	P	Foraminifera. Metamorphic banding of serpentine and dolomite (Hofmann 1971)	Pseudofossil. Metamorphic banding
Precambrian	Y?	CA	P	Fossils?	Pseudofossil. Metamorphic banding
Grenville Supgp.	Y?	CA	D	Fossil. Metamorphic banding (Hofmann 1971)	Dubiofossil. Possibly metamorphosed stromatolite
Grenville Supgp.	Y?	CA	P	Foraminifera. Rhythmic calcite veins in limestone (Hofmann 1971, p. 9)	Pseudofossil
Precambrian	Y?	CA	P	Fossils?.	Pseudofossils
Precambrian	Y?	CA	P	Fossils?	Pseudofossils
Grenville Supgp.	Y?	CA	D	Fossil	Dubiofossil. Possibly metamorphosed stromatolite
Karagas Fm. Horizon II	YZ	SS	D	Merostomata, Eurypterida	Dubiofossil, Mudcrack Curls
Karagas Fm. Horizon II	YZ	SS	D	Merostomata, Eurypterida	Dubiofossil, Mudcrack Curls
Transvaal Supgp., Pretoria Gp., L. Griquatown Fm.	X	CH	P	Medusoid. Concretions (Cloud 1968, p. 53), trace fossil or medusa (Häntzschel 1975, p. W147)	Dubiofossil, Chemogenic?
Newland Fm.	Y	CA	P	Algae. Septarian nodule (Raymond 1935, Schindewolf 1956, Häntzschel 175, p. W175)	Pseudofossil. Septarian nodule
Kaimur Gp.	Z	SS	P	Pseudofossil	Pseudofossil. Classified as pseudofossil, but given Linnéan name
Kaimur Gp.	Z	SS	P	Pseudofossil	Pseudofossil. Classified by auth. as pseudofossil, but given Linnéan name
Mashan Gp., Liumao Fm.	Z?	CA	D	Pennatulacean	Dubiofossil. Karren weathering?
Newland Fm.	Y	CA	D	Algae. Shrinkage cracks (Raymond 1935), inorganic segregation (Fenton and Fenton 1936, p. 614)	Dubiofossil. Chemical
Green Head Gp.	Y?	SH	P	Sponge Spicules. Crystallographic markings on graphite (Rauff 1893), probably crystals (Cloud 1968, p. 57)	Inorganic. Accidental scratch marks (Hofmann 1971, p. 22)
Green Head Gp.	Y?	SH	P	Inorganic	Inorganic. Holotype rediscovered. Scratch marks
San Vito Fm.	ZC	MS	D	Medusae	Dubiofossils. n. inval. (=) *Ichnusina* in corrigendum, also n. inval.)
San Vito Fm.	ZC	MS	D	Medusae	Dubiofossils = *Ichnusa* Debrenne and Naud 1981a
Karagas Fm.	YZ	SS	D	Merostomooidea	Dubiofossils, Mudflake Curls
Karagas Fm.	YZ	SS	D	Merostomooidea, Prochelicerata	Dubiofossils, Mudflake Curls
Mashan Gp., Liumao Fm.	Z?	CA	D	Cnidaria	Dubiofossil. cf. *Arumberia*
Mashan Gp., Zhongsanyang Fm.	Z?	CA	D	Cnidaria	Dubiofossils. cf. *Arumberia*
Tyurim Fm.	Y	CA	D	Fossils	Dubiofossils. Insufficient data, cf. Camasiids
Tyurim Fm.	Y	CA	D	Fossils	Dubiofossils. Insufficient data, cf. Camasiids

1046 Described Taxa of Proterozoic and Selected Earliest Cambrian Megafossils

Table 23.6. (*Continued*).

Number	Taxon	Reference Code	Continent and Country	Locality	Latitude	Longitude
175	*Jussenia edelsteini*	Vasi68.106	AS SU	Batenev Ridge, Son Riv.	54.5	89.4
176	*Jussenia edelsteini*	Vasi68.106	AS SU	Kuznetz Alatau, Sarala	55.0	89.8
177	*Kaimuria chambalensis*	** Ghar77.205	AS IN	Rawatbhata	25.0	75.6
178	*Kempia huronense*	** Bain27.281	NA CA	Vernon Twp. (and Porter Twp.)	46.5	−81.8
179	*Kempia huronense*	Hofm71.146	NA CA	6.5 km NE Iron Bridge	46.3	−83.2
180	*Kinneyia simulans*	** Walc14..77	NA US	Forks of Birch Creek	46.6	−111.1
181	*Liaoningia* cf. *fuxianensis*	Xing79.167	AS CN	Wuhangshan-Paoya, Fuxian	39.6	121.9
182	*Liaoningia fuxianensis*	** Xing79.167	AS CN	Wuhangshan-Paoya, Fuxian	39.6	121.9
183	*Liaoningia* (?) sp.	Xing79.167	AS CN	Wuhangshan-Paoya, Fuxian	39.6	121.9
184	*Lingula calumet*	* Winc85.196	NA US	Pipestone	44.0	−96.3
185	*Lingulella montana*	* Fent36.609	NA US	Little Birch Creek	46.6	−111.1
186	Macro-spherical structures	Bond67..41	NA GL	1.6 km S of Vallen L.	61.4	−45.0
187	*Manchuriophycus inexpectans*	* Endo33..43	AS CN	Nanshan Mtn. near Chiaotou	41.2	123.7
188	*Manchuriophycus sawadai*	* Yabe39.205	AS CN	Ryozyun [? = Dailien]	39.0	127.7
189	*Manchuriophycus sawadai*	Ghar77.205	AS IN	Rawatbhata? Rampura?	25.0	75.6
190	*Manchuriophycus yamamotoi*	** Endo33..43	AS CN	Nanshan Mtn. near Chiaotau	41.2	123.7
191	*Mashania angusta*	* Liu81..71	AS CN	Jixi	45.1	130.5
192	*Mashania angusta*	Wang84.136	AS CN	Jixi	45.1	130.5
193	*Mashania annulata*	* Liu81..71	AS CN	Jixi	45.1	130.5
194	*Mashania annulata*	Wang84.136	AS CN	Jixi	45.1	130.5
195	*Mashania deformata*	* Liu81..71	AS CN	Jixi	45.1	130.5
196	*Mashania deformata*	Wang84.136	AS CN	Jixi	45.1	130.5
197	*Mashania longshanensis*	** Liu81..71	AS CN	Jixi	45.1	130.5
198	*Mashania longshanensis*	Wang84.136	AS CN	Jixi	45.1	130.5
199	*Mashania minuta*	* Liu83...1	AS CN	Jixi	45.1	130.5
200	*Mashania minuta*	Wang84.136	AS CN	Jixi	45.1	130.5
201	*Mashania sinensis*	* Liu83...1	AS CN	Jixi	45.1	130.5
202	*Mashania sinensis*	Wang84.136	AS CN	Jixi	45.1	130.5
203	*Mashania* sp.	Wang84.136	AS CN	Jixi	45.1	130.5
204	*Mawsonella wooltanensis*	** Chap27.123	AU SA	14 km W Wooltana Head Station	−30.5	139.4
205	*Medusichnites* form gamma	Matt91.123	NA CA	Thunder Bay area	48.4	−89.3
206	*Medusinites simplex*	* Xing79.167	AS CN	Wuhangshan-Paoya, Fuxian	39.6	121.9
207	Medusoids?	Edge64.235	AU WA	1.6 km N Hamersley Homestead	−22.3	117.7
208	*Megagrapton regulare*	* Ghar77.205	AS IN	Rampura	24.5	75.5
209	*Misracyathus vindhicanus*	Volo59..79	AS IN		0.0	0.0
210	*Muniaichnus* sp.	* Kuma78.144	AS IN	Mirzapur	20.5	82.5
211	*Neantia deformata*	* Lebe87.776	EU FR	Montfort	48.0	−2.0

Geologic Unit	Age	Lithology	Class	Original Author's Interpretation and other interpretatons	Hofmann Interpretation and Remarks
Poludennaya Fm.	Y	CA	D	Fossils	Dubiofossil. Insufficient data, cf. Camasiids
Poludennaya Fm.	Y	CA	D	Fossils	Dubiofossils. Insufficient data, cf. Camasiids
Kaimur Gp.	Z	SS	D	Grazing Trail	Dubiofossil. Meandering trail?
Huronian Supgp., Mississagi Fm.	X	SS	P	Stromatoporoid?. Diffusion banding (Hofmann 1971, p. 27–29)	Pseudofossil. Diffusion banding
Huronian Supgp., Gowganda Fm.	X	MS	P	Diffusion Banding	Pseudofossil. Diffusion banding
Newland Fm.	Y	CA	P	Algae. Inorganic segregation (Raymond 1935, p. 376, Fenton and Fenton 1936, p. 615)	Pseudofossil. Diagenetic ripples (Schindewolf 1956)
Nanguanling Fm. [Changlingzi Fm.]	Z	SS	D	Coelenterata. Inorganic (Sun 1986a, p. 349)	Dubiofossil. Dewatering marks?
Nanguanling Fm. [Changlingzi Fm.]	Z	SS	D	Coelenterata. Inorganic (Sun 1986, p. 349)	Dubiofossil. Dewatering marks?
Nanguanling Fm. [Changlingzi Fm.]	Z	SS	D	Coelenterata. Inorganic (Sun 1986a, p. 349)	Dubiofossils. Dewatering marks
Sioux Fm.	Y	MS	D	Brachiopod. Inorganic (Darby 1972, p. 267)	Inorganic? Insufficient data
Newland Fm.	Y	CA	D	Obolid Brachiopod. Stromatolite (Glaessner 1962, Cloud 1968, p. 52), slippage structures (Rowell 1971)	Dubiofossil. Unrecognizable structure (Häntzschel 1975, p. W186)
Vallen Gp., Zigzagland Fm.	X	CA	D	Possibly Organic	Dubiofossil. Uninterpretable
Nanshan Fm.	Z	SS	P	Fillings of Stems of Algae. Shrinkage cracks (Häntzschel 1949, 1975, p. W176)	Pseudofossil. Mudcrack fillings
Precambrian		SS	P	Algae. Shrinkage cracks (Häntzschel 1949, 1975, p. W176)	Pseudofossil. Mudcrack fillings
Kaimur Gp.	Z	SS	P	Pseudofossil	Mudcrack Fillings. Classified as pseudofossil, but given Linnéan name
Nanshan Fm.	Z	SS	P	Algae. Shrinkage crack (Häntzschel 1949, 1975, p. W176)	Pseudofossils. Mudcrack fillings
Mashan Gp., Liumao Fm.	Z?		D	Coelenterata	Dubiofossil. Resemblance to *Pteridinium*, *Arumberia*
Mashan Gp., Zhongsanyang Fm.	Z?		D	Cnidaria	Dubiofossil. cf. *Arumberia*
Mashan Gp., Liumao Fm.	Z?		D	Coelenterata	Dubiofossil. cf. *Arumberia*, *Pteridinium*
Mashan Gp., Zhongsanyang Fm.	Z?		D	Cnidaria	Dubiofossil. cf. *Arumberia*
Mashan Gp., Liumao Fm.	Z?		D	Coelenterata	Dubiofossil. cf. *Arumberia*, *Pteridinium*
Mashan Gp., Zhongsanyang Fm.	Z?		D	Cnidaria	Dubiofossil. cf. *Arumberia*
Mashan Gp., Liumao Fm.	Z?		D	Coelenterata	Dubiofossil. cf. *Arumberia*, *Pteridinium*
Mashan Gp., Zhongsanyang Fm.	Z?		D	Cnidaria	Dubiofossil. cf. *Arumberia*
Mashan Gp.	Z?		D	Coelenterata	Dubiofossil
Mashan Gp., Zhongsanyang Fm.	Z?		D	Cnidaria	Dubiofossil. cf. *Arumberia*
Mashan Gp.	Z?		D	Coelenterata, Pennatulacean	Dubiofossil
Mashan Gp., Zhongsanyang Fm.	Z?		D	Cnidaria	Dubiofossils. cf. *Arumberia*
Mashan Gp., Zhongsanyang Fm.	Z?		D	Cnidaria	Dubiofossil. cf. *Arumberia*
	Z	CA	D	Green Alga. Intraformational carbonate breccia (Glaessner 1979, p. A112)	Dubiofossil
Animikie Gp.	X	SS	P	Drag Marks of Medusoid. Sole marks (Hofmann 1971, p. 18–20)	Pseudofossils. Shear drag marks on sole of beds
Nanguanling Fm. [Changlingzi Fm.]	Z	SS	P	Coelenterata. Inorganic (Sun 1968a, p. 349)	Pseudofossil. Fluid escape mark
Hamersley Gp., Brockman Iron Fm.	X	CH IF	P	Coelenterata	Pseudofossil. cf. "medusoids" of Houzay
Kaimur Gp.	Z	SS	D	Trace Fossil	Dubiofossil. Mudcrack fillings?
			D	Archaeocyathid. Not archaeocyathid (Zhuravleva 1960b, Finks and Hill 1967, p. 342, Hill 1972, p. E142)	Dubiofossil, Inadequate Data. Reference not seen
Semri Gp., Kheinjua Fm.	Y	SS	D	Trace Fossil	Dubiofossil. cf. *Cochlichnus* n. nud. (no species designated)
Rennes Sh.	Z		P	Sponge. Inorganic (Häntzschel 1965, p. 59), ripple marks (Cloud 1968, p. 52)	Inorganic, Runnel Marks. nom. invalid.: non Recluz 1843 (Moll.) fide Häntzschel 1965

1048 Described Taxa of Proterozoic and Selected Earliest Cambrian Megafossils

Table 23.6. (Continued).

Number	Taxon	Reference Code	Continent and Country	Locality	Latitude	Longitude
212	Neantia reticulata	*Lebe87.776	EU FR	Montfort	48.0	−2.0
213	Neantia rhedonensis	**Lebe87.776	EU FR	Montfort	48.0	−2.0
214	Neantia rhedonensis	Lebe91.200	EU FR	Montfort	48.0	−2.0
215	Neantia verrucosa	*Lebe87.776	EU FR	Montfort	48.0	−2.0
216	Newlandia	Evan33106A	NA CA	Twelvemile Creek?	51.2	−117.0
217	Newlandia cf. frondosa	Vasi68.106	AS SU	Kuznetz Alatau, Sarala	55.0	89.8
218	Newlandia cf. frondosa	Vasi68.106	AS SU	Kuznetz Alatau, Tamalyk	53.7	89.7
219	Newlandia concentrica	*Walc14..77	NA US	Forks of Birch Creek	46.6	−111.1
220	Newlandia concentrica	Vasi68.106	AS SU	Gornoy Shoria, Mrassu, Saga	53.0	89.0
221	Newlandia concentrica	Vasi68.106	AS SU	Batenev Ridge, Son Riv.	54.5	89.4
222	Newlandia concentrica	Vasi68.106	AS SU	Kuznetz Alatau, Tamalyk	53.7	89.7
223	Newlandia cristata	Vasi68.106	AS SU	Kuznetz Alatau, Komunar	54.5	89.4
224	Newlandia cristata	Vasi68.106	AS SU	Kuznetz Alatau, Sarala	55.0	89.8
225	Newlandia frondosa	**Walc14..77	NA US	Forks of Birch Creek (400c)	46.6	−111.1
226	Newlandia frondosa	Vasi68.106	AS SU	Kuznetz Alatau, Sarala	55.0	89.8
227	Newlandia lamellosa	*Walc14..77	NA US	Forks of Birch Creek (400c)	46.6	−111.1
228	Newlandia lamellosa	Vasi68.106	AS SU	Kuznetz Alatau, Sarala	55.0	89.8
229	Newlandia major	*Walc14..77	NA US	Forks of Birch Creek	46.6	−111.1
230	Newlandia obrutchevi	Vasi68.106	AS SU	Kuznetz Alatau, Sarala	55.0	89.8
231	Newlandia prava	Vasi68.106	AS SU	Kuznetz Alatau, Sarala	55.0	89.8
232	Newlandia sarcinula	*Fent31.670	NA US	Spot Mtn.	48.5	−113.4
233	Newlandia sarcinula	Fent371873	NA US	Spot Mtn.	48.5	−113.4
234	Newlandia tchurakovi	Vasi68.106	AS SU	Kuznetz Alatau, Sarala	55.0	89.8
235	Newlandia usovi	Vasi68.106	AS SU	Kuznetz Alatau, Sarala	55.0	89.8
236	Oldhamia	Murr68...1	NA CA	St. John's area	47.7	−52.7
237	Olivooides papillatus	*Zhan86..67	AS CN	Mt. Gaojiashan	33.0	106.5
238	Oniscoichnus	Das 84.251	AS IN	Bankuiyan, Rewa Dist.	24.6	81.2
239	Ovolites primus	Vasi68.106	AS SU	Gornoy Shoria, Mrassu, Saga	53.0	89.0
240	Palaeocuniculichnites osangustus	**Ghar77.205	AS IN	Rampura	24.5	75.5
241	Palaeoplatoda? segmentata	Liu 83...1	AS CN	Jixi	45.1	130.5
242	Palaeotrochis	Wils63..94	NA US	Orange Co. 1 mi NW Cross Roads	36.0	−79.1

Geologic Unit	Age	Lithology	Class	Original Author's Interpretation and other interpretatons	Hofmann Interpretation and Remarks
Rennes Sh.	Z		P	Sponge. Inorganic (Häntzschel 1965, p. 59)	Inorganic, Runnel Marks. nom. invalid.: non Recluz 1843 (Moll.) fide Häntzschel 1965
Rennes Sh.	Z		P	Sponge. Inorganic (Häntzschel 1965, p. 59), ripple marks (Cloud 1968, p. 57)	Inorganic, Runnel Marks. nom. invalid.: non Recluz 1843 (Moll.) fide Häntzschel 1965
Rennes Sh.	Z		P	Sponge. Inorganic (Häntzschel 1965, p. 59)	Inorganic, Runnel Marks. nom. invalid.: non Recluz 1843 (Moll.) fide Häntzschel 1965
Rennes Sh.	Z		P	Sponge. Inorganic (Häntzschel 1965, p. 59)	Inorganic, Runnel Marks. nom. invalid.: non Recluz 1843 (Moll.) fide Häntzschel 1965
Windermere Supgp., Horsethief Creek Gp.	Z	CA?	D	Fossil	Dubiofossil. Not described, no specimens available
Tyurim Fm.	Y	CA	D	Fossils	Dubiofossils. Insufficient data, cf. Camasiids
Tyurim Fm.	Y	CA	D	Fossils	Dubiofossils. Insufficient data, cf. Camasiids
Newland Fm.	Y	CA	D	Algae. (cf. Blue-Green Algae) Pseudo-alga-or alga (Fenton and Fenton 1936, p. 615)	Dubiofossil = *Newlandia frondosa*: Fenton Fenton 1936, p. 615
Poludennaya Fm.	Y	CA	D	Fossils	Dubiofossils. Insufficient data, cf. Camasiids
Poludennaya Fm.	Y	CA	D	Fossils	Dubiofossils. Insufficient data, cf. Camasiids
Tyurim Fm.	Y	CA	D	Fossils	Dubiofossils. Insufficient data, cf. Camasiids
Tyurim Fm.	Y	CA	D	Fossils	Dubiofossils. Insufficient data, cf. Camasiids
Tyurim Fm.	Y	CA	D	Fossils	Dubiofossils. Insufficient data, cf. Camasiids
Newland Fm.	Y	CA	D	Algae. Pseudo-alga or alga (Fenton and Fenton 1936, p. 615)	Dubiofossils. Diffusion structures
Pouldennaya Fm.	Y	CA	D	Fossils	Dubiofossils. Insufficient data, cf. Camasiids
Newland Fm.	Y	CA	D	Algae	Dubiofossil. Diffusion structures
Poludennaya Fm.	Y	CA	D	Fossils	Dubiofossils. Insufficient data, cf. Camasiids
Newland Fm.	Y	CA	D	Algae. Pseudo-algae or alga (Fenton and Fenton 1936, p. 615)	Dubiofossil. Diffusion structures
Poludennaya Fm.	Y	CA	D	Fossils	Dubiofossils. Insufficient data, cf. Camasiids
Poludennaya Fm.	Y	CA	D	Fossils	Dubiofossils. Insufficient data, cf. Camasiids
Altyn Fm.	Y	CA	D	Algae (Stromatolites)	Dubiofossils. Diffusion banding
Altyn Fm., Hellroaring Mem.	Y	CA	D	Calcareous Algae	Dubiofossils. Diffusion banding
Poludennaya Fm.	Y	CA	D	Fossils	Dubiofossils. Insufficient data, cf. Camasiids
Poludennaya Fm.	Y	CA	D	Fossils	Dubiofossils. Insufficient data, cf. Camasiids
Conception Gp.	Z		D	Fossil (Weston 1895, p. 139). Concretion (Hofmann 1971, p. 14)	Dubiofossil. No specimens available.? = *Aspidella*
Dengying Fm., Gaojiashan Mem.	Z		D	"Globomorpha"	Dubiofossil
Bhander Ls.			D	Trace Fossils	Dubiofossil. Insufficient data for interpretation
Poludennaya Fm.	Y	CA	D	Fossils	Dubiofossils. Insufficient data, cf. Camasiids
Kaimur Gp.	Z	SS	D	Dwelling Burrow	Dubiofossil. cf. *Skolithos*
Mashan Gp.	Z?		D	Platyhelminthes	Dubiofossil
Precambrian	Z		P	Spherulites	Pseudofossils

Table 23.6. (Continued).

Number	Taxon	Reference Code	Continent and Country	Locality	Latitude	Longitude
243	*Palaeotrochis major*	* Emmo56.389	NA US	Troy, Montgomery Co.	35.4	−79.9
244	*Palaeotrochis minor*	** Emmo56.389	NA US	Troy, Montgomery Co.	35.4	−79.9
245	*Paradoxides barberi*	* Winc85.196	NA US	Pipestone	44.0	−96.3
246	*Planolites corrugatus*	* Walc99.199	NA US	Sawmill Canyon	46.9	−110.7
247	*Planolites superbus*	* Walc99.199	NA US	Sawmill Canyon	46.9	−110.7
248	*Protadelaidea browni*	* Davi36.122	AU SA	Teatree Gully	−34.8	138.7
249	*Protadelaidea howchini*	** Davi36.122	AU SA	Teatree Gully	−34.8	138.7
250	*Protoniobia wadea*	** Spri49..72	AU WA	Mt. John, Osmond Rge.	−17.2	128.7
251	*Protovirgularia*	Das 84.251	AS IN	Bankuiyan, Rewa Dist.	24.6	81.2
252	*Pseudojussenia parva*	Vasi68.106	AS SU	Kuznetz Alatau, Tamalyk	53.7	89.7
253	*Pseudojussenia parva*	Vasi68.106	AS SU	Kuznetz Alatau, Komunar	54.5	89.4
254	*Puratanichnus bijawarensis*	** Math86.249	AG IN	Bijawar	24.6	79.5
255	*Ramapuraea vindhyanensis*	** Mait84.212	AS IN	Rampura (Ramapura)	24.5	75.5
256	*Reynella howchini*	** Davi28.191	AU SA	Reynella, 27 km S of Adelaide	−35.0	138.5
257	*Rhysonetron byei*	* Hofm67.500	NA CA	Desbarats	46.4	−83.9
258	*Rhysonetron lahtii*	** Hofm67.500	NA CA	Flack Lake	46.6	−82.7
259	*Rhysonetron ramapuraensis*	* Ghar77.205	AS IN	Rampura (Ramapura)	24.5	75.5
260	*Roualtia rewaensis*	* Verm68.557	AS IN	Bankuiyan	24.0	81.2
261	*Rugoinfractus ovruchensis*	** Pali74..34	EU SU	Perchotravneve, Ovruch	51.3	28.7
262	*Sabellarites*	* Daws90.595	NA CA	Madoc	44.5	−77.5
263	*Sajanella arshanica*	Volo66.434	AS SU	Mt. Pykhtun, Arshan, Iya Riv.	53.9	99.8
264	*Saralinskia boulinnikovi*	Vasi68.106	AS SU	Kuznetz Alatau, Sarala	55.0	89.8
265	*Saralinskia glomeria*	Vasi68.106	AS SU	Kuznetz Alatau, Sarala	55.0	89.8
266	*Saralinskia multiangulata*	Vasi68.106	AS SU	Kuznetz Alatau, Sarala	55.0	89.8
267	*Saralinskia multiangulata*	Vasi68.106	AS SU	Batenev Ridge, Son Riv.	54.5	89.4
268	*Saralinskia plana*	Vasi68.106	AS SU	Batenev Ridge, Son Riv.	54.5	89.4
269	*Saralinskia radiata*	Vasi68.106	AS SU	Kuznetz Alatau, Sarala	55.0	89.8
270	*Saralinskia radiata*	Vasi68.106	AS SU	Batenev Ridge, Son Riv.	54.5	89.4
271	*Saralinskia ramosa*	Vasi68.106	AS SU	Kuznetz Alatau, Sarala	55.0	89.8
272	*Saralinskia serrata*	Vasi68.106	AS SU	Kuznetz Alatau, Sarala	55.0	89.8
273	*Saralinskia sp.*	Vasi68.106	AS SU	Kuznetz Alatau, Tamalyk	53.7	89.7
274	*Saralinskia stellata*	Vasi68.106	AS SU	Kuznetz Alatau, Sarala	55.0	89.8

Proterozoic and Selected Cambrian Megascopic Dubiofossils and Pseudofossils

Geologic Unit	Age	Lithology	Class	Original Author's Interpretation and other interpretatons	Hofmann Interpretation and Remarks
Precambrian	Z?		P	Coral. Concretion (Hall 1857). Cone-in-cone structure (Marsh 1868)	Inorganic. No type species designated. In volcanic rock
Precambrian	Z?		P	Coral. Concretion (Hall 1857). Cone-in-cone structure (Marsh 1868)	Inorganic. In volcanics. For Cenozoic *Palaeotrochis*, see Williams (1899)
Sioux Fm. (1470 ± 50 Rb/Sr)	Y	MS	P	Trilobite. Current or load structure (Darby 1972, p. 267)	Inorganic
Greyson Sh.	Y	MS	P	Worm Burrow. Algal? (Cloud 1968, p. 55)	Pseudofossil. Sand injection structure
Greyson Sh.	Y	MS	P	Annelid Burrow. Alagl? (Cloud 1968, p. 55)	Pseudofossil, Linear Fold. Soft-sediment deformation
Teatree Gully Qte.	YZ	SS	D	Giant Arthropods. Mud flakes (Häntzschel 1975, p. W177–178). Possibly pyritized plant tissue (Glaessner 1959)	Mudchips
Teatree Gully Qte.	YZ	SS	D	Giant Arthropods. Mud flakes (Häntzschel 1975, p. W177–178). Possibly pyritized plant tissue (Glaessner 1959)	Mudchips
"Lower Cambrian flags, Mount John"	Z	SS	D	Hydroid, Suborder Anthomedusae. Concretion (Harrington and Moore 1949, p. F159, Cloud 1968, p. 54)	Dubiofossil. Concretion?
Bhander Ls.			D	Trace Fossils	Dubiofossil. Insufficient data for interpretation
Tyurim Fm.	Y	CA	D	Fossils	Dubiofossils. Insufficient data, cf. Camasiids
Tyurim Fm.	Y	CA	D	Fossils	Dubiofossils. Insufficient data, cf. Camasiids
Bijawar Gp., Amronia Qte.	X	SS	D	Annelid Trails	Dubiofossils. Mudcrack fillings?, cf. *Manchuriophycus* and *Rhysonetron*
Semri Gp., Suket Sh.	Y	SH	D	Jellyfish	Dubiofossil. Crystal rosettes?
Brighton Ls.	Z		P	Crustacean. Inorganic (Glaessner 1959b, p. 525. Häntzschel 1975, p. W178). Mudflakes (Cloud 1968)	Mudflakes
Huronian Supgp., Lorrain Fm.	X	SS	P	Questionable Worm Tubes. Compacted mudcrack fillings (Hofmann 1971)	Pseudofossils. Corrugated mudcrack fillings
Huronian Supgp., Bar River Fm.	X	SS	P	Possible Worm Tubes. Corrugated mudcrack fillings (Hofmann 1971)	Pseudofossil. Corrugated mudcrack fillings
Kaimur Gp.	Z	SS	P	Pseudofossil	Mudcrack Fillings. Classified as pseudofossil, but given Linnéan name
Bhander Ls.	Z		D	Trace Fossil, Repichnia	Dubiofossil. Insufficient data
Ovruch Ser. (Riphean)	YZC	SS	D	Trace Fossil	Age Questionable?
Grenville Supgp., Hastings Gp.	Y?	SS?	D	Annelid Burrows	Dubiofossil. Material inadequate. = Burrows of Dawson 1866, p. 608.
Karagas Fm., L pt.	YZ	SS	D	Protomedusa, Brooksellidae.	Dubiofossil
Tyurim Fm.	Y	CA	D	Fossils	Dubiofossils. Insufficient data, cf. Camasiids
Tyurim Fm.	Y	CA	D	Fossils	Dubiofossils. Insufficient data, cf. Camasiids
Tyurim Fm.	Y	CA	D	Fossils	Dubiofossils. Insufficient data, cf. Camasiids
Poludennaya Fm.	Y	CA	D	Fossils	Dubiofossils. Insufficient data, cf. Camasiids
Poludennaya Fm.	Y	CA	D	Fossils	Dubiofossils. Insufficient data, cf. Camasiids
Poludennaya Fm.	Y	CA	D	Fossils	Dubiofossils. Insufficient data, cf. Camasiids
Tyurim Fm.	Y	CA	D	Fossils	Dubiofossils. Insufficient data, cf. Camasiids
Tyurim Fm.	Y	CA	D	Fossils	Dubiofossils. Insufficient data, cf. Camasiids
Tyurim Fm.	Y	CA	D	Fossils	Dubiofossils. Insufficient data, cf. Camasiids
Tyurim Fm.	Y	CA	D	Fossils	Dubiofossils. Insufficient data, cf. Camasiids
Poludennaya Fm.	Y	CA	D	Fossils	Dubiofossils. Insufficient data, cf. Camasiids

Table 23.6. (Continued).

Number	Taxon	Reference Code	Continent and Country	Locality	Latitude	Longitude
275	Sidneyia groenlandica	*Clea35.463	NA GL	Ymer Id., 1 mi N Blomsterbukta	73.9	−25.4
276	Skolithos? miaoheensis	Ding85.115	AS CN	Yangtze Gorge	32.8	111.1
277	Sonskia prava	Vasi68.106	AS SU	Batenev Ridge, Son Riv.	54.5	89.4
278	Spiroscolex	Hant75.269	NA CA	Near St. John's	47.6	−52.7
279	Spongoides grandis	Vasi68.106	AS SU	Gornoy Shoria, Mrassu, Saga	53.0	89.0
280	Squamodictyon	Durh78..21		Siberia and Australia	0.0	0.0
281	Suzmites? sp.	Hofm81.303	NA CA	Sekwi Brook	63.4	−128.4
282	Taonichnites	*Matt90.123	NA CA	Thunder Bay area	48.4	−89.3
283	Tasmanadia dassii	*Verm68.557	AS IN	Bankuiyan	24.0	81.2
284	Telemarkites enigmaticus	**Dons59.249	EU NO	Haugli, 10 km WNW of Kviteseid	59.5	8.4
285	Templuma sinica	**Zhan79..87	AS CN	Jixian Co.	0.0	0.0
286	Tigillites bohmei	*Bron64...4	NA US	Giacona's Camp	33.5	−110.9
287	Tricuspidatia trigonata	**Sosn81..73	AS SU	Batenev Ridge, Karysh Creek	54.2	90.2
288	Tridia koptevi	**Sosn84.128	AS SU	Batenev Ridge, Karysh Creek	54.2	90.2
289	Tridia salebrosa	*Sosn84.128	AS SU	Karysh Creek, Chalgystag	54.2	90.2
290	Tubiphyton taghdoutensis	**Chou51...9	AF MA	Taghdout near Tazenakht	0.0	0.0
291	Vallenia erlingi	Pede66..40	NA GL	Graenseland	61.5	−48.0
292	Vallenia erlingi	**Bond67..41	NA GL	Graenseland	61.5	−48.0
293	Vendovermites	Soko65..78	EU SU	Rybachiy Peninsula	69.7	32.5
294	Volodia annulata	**Sosn80.145	AS SU	Batenev Ridge, Karysh Creek	54.2	90.2
295	Vologdinia concentrica	Vasi68.106	AS SU	Gornoy Shoria, Mrassu, Saga	53.0	89.0
296	Vologdinia major	Vasi68.106	AS SU	Gornoy Shoria, Mrassu, Saga	53.0	89.0
297	Vologdinia shorica	Vasi68.106	AS SU	Gornoy Shoria, Mrassu, Saga	53.0	89.0

Geologic Unit	Age	Lithology	Class	Original Author's Interpretation and other interpretatons	Hofmann Interpretation and Remarks
Ymer Fm. (Eleonore Bay Gp., Multicolored Ser.?)	ZC	SH	D	Arachnida, Xenopoda. Damaged ripple marks (Eha 1953, p. 15–16), pseudofossil (Häntzschel 1975, p. W179)	Dubiofossil. Originally considered M Cam.
Dengying Fm.	Z		D	Domichnia	Questionably Organic. Intersecting straight structures
Poludennaya Fm.	Y	CA	D	Fossils	Dubiofossils. Insufficient data, cf. Camasiids
St. John's Gp.	Z		D	Unrecognizable Genus	Dubiofossil = *Arenicolites spiralis* Dawson (1897, p. 53, 54, Fig. 13)
Poludennaya Fm.	Y	CA	D	Fossils	Dubiofossils. Insufficient data, cf. Camasiids
Not given	Z		D	Trace Fossil	Dubiofossil. Insufficient data, citation only
Blueflower Fm. (Map-unit 10b)	Z	SS	D	Dubiofossil	Dubiofossil
Animikie Gp.	X	SS	D	Drag Marks of Medusoid. Sole marks (Hofmann 1971, p. 18–20)	Dubiofossil = *Medusichnites* form gamma (Matthew 1891)
Bhander Ls.	Z		D	Trace Fossil, Pascichnia	Dubiofossil, Inadequate Illus
Bandak Gp., Telemark Fm.	YZ	SS	DC	Sponges or Algal Concretions. Pseudofossil cf. *Botswanella* (Häntzschel 1975, p. W179). Algal (Pflug and Strübel 1969, p. 149).	Dubiofossil, Chemogenic? Structures contain colonial microbes (Häntzschel 1975)
Wumishan Fm.	Y	CH	BH	Chlorophyceae, Dasycladaceae. Dubiofossil (Zhang Z. 1980, p. 341)	Fossil cf. *Papillomembrana* Spjeldnaes, cf. *Paracharnia* Sun 1986
Apache Gp., Mescal Ls.	Y	CA	D	Vertical Burrows	Dubiofossils. Uninterpretable
Poludennaya Gp., Aramon Fm.	Y	CA	D	Camasiida	Dubiofossil. Insufficient data, cf. *Camasia*
Poludennaya Gp., Aramon Fm.	Y	CA	D	Camasiida, Animals	Dubiofossils, cf. *Camasia*. Insufficient data, Tridiidae Sosnovskaya n. fam.
Poludennaya Gp., Aramon Fm.	Y	CA	D	Camasiida, Animals	Dubiofossils, cf. *Camasia*. Insufficient data, Tridiidae Sosnovskaya n. fam.
	Z	CA	P	Algae. Tectonic-metamorphic structures (Schindewolf 1956, p. 468–469)	Catagraphs. Intraclasts
Vallen Gp., Graenseso Fm.	X	CA	D	Undetermined Affinities Microproblematicum (Häntzschel 1975, p. W167)	Dubiofossils, Replaced Ooids?
Vallen Gp., Graenseso Fm.	X	CA	D	Undetermined Affinities Microproblematicum (Häntzschel 1975, p. W167)	Dubiofossils, Replaced Ooids?
Vendian, shales with Laminarites	Z	MS	D	Vermiform Organism	Dubiofossil. N. nud., no species, description, illustration, or data
Poludennaya Gp., Tyurim Fm.	Y	CA	D	Tubular Fossil (Camasiida)	Dubiofossil, cf. *Camasia*. Insufficient data
Poludennaya Fm.	Y	CA	D	Fossils	Dubiofossils. Insufficient data, cf. Camasiids
Tyurim Fm.	Y	CA	D	Fossils	Dubiofossils. Insufficient data, cf. Camasiids
Poludennaya Fm.	Y	CA	D	Fossils	Dubiofossils. Insufficient data, cf. Camasiids

24

Atlas of Representative Proterozoic Microfossils

J. WILLIAM SCHOPF

24.1 Introduction

24.1.1 Introduction

On the following plates are illustrated microfossils from 48 geologic units (representing about 10% of the units listed in Table 22.3). Although coverage is by no means exhaustive, the fossils illustrated are representative of the diversity and level of organization exhibited by particularly well-preserved Proterozoic assemblages. In selecting the fossils illustrated, emphasis has been placed on inclusion of holotype specimens, particularly holotypes of taxa described from the Soviet Union, relatively few of which have been previously illustrated in the non-Russian language literature (and none of which has been previously illustrated using interference contrast optics). As summarized below, emphasis has also been placed on illustration of assemblages of Middle and Late Proterozoic age; this atlas therefore complements, but does not duplicate, illustrations of Early Proterozoic communities included in Hofmann and Schopf (1983). The atlas contains more than 300 new illustrations, including photomicrographs of specimens not previously figured from the Dismal Lakes Group, River Wakefield Subgroup, Skillogalee Dolomite, the Gunflint Iron Formation, and the Gaoyuzhuang, Galeros, Kwagunt, and Myrtle Springs Formations.

24.1.2 Acknowledgments

Soviet type specimens were provided for study by E. Aseeva, T. N. Hermann, T. Jankauskas, N. Mikhailova, V. Sergeev, and A. Weiss; photomicrography of these specimens was carried out by C. V. Mendelson, J. W. Schopf, and G. Vidal; this work, conducted at the Palaeontological Institute in Moscow, was facilitated through the courtesy of M. A. Fedonkin and B. S. Sokolov. Studies of new material from the Gaoyuzhuang Formation were carried out in collaboration with R. Mikawa (University of Chicago); studies of new specimens from the River Wakefield Subgroup and the Skillogalee Dolomite were carried out by or in collaboration with T. R. Fairchild (Universidade de São Paulo, Brazil). Photographic negatives showing specimens previously illustrated from the Tyler, Paradise Creek, Beck Spring, and Yudoma Formations were kindly provided by P. Cloud (University of California, Santa Barbara); these and all other photos in the atlas were printed by R. Mantonya (University of California, Los Angeles).

24.2

Illustrated Type Specimens

For Soviet type specimens, figure descriptions include the identification number of the specimen-containing slide and the location of the specimen in England Finder Slide coordinates. The final entry (in brackets) for many figure descriptions lists a code indicating the taxonomic category to which the fossil has been referred in the informal revised classification presented in Chapter 25. In addition to numerous paratypes and topotypes, 164 "name-carrying" specimens (viz., holotypes, lectotypes, and neotypes), listed below, are illustrated from the following geologic units:

Age (Ma)	Geologic Unit/ Illustrated Holotypes	Country	Plates and Figures
2450	Gamohaan Fm.	S. Africa	Pl. 1, A–F
	Siphonophycus transvaalense Beukes, Klein and Schopf		
2150	Kasegalik Fm.	Canada	Pl. 1, G–O; see also Hofmann and Schopf (1983), Photo 14-2, A–L, N, P, Q
[2150	McLeary Fm.	Canada	Hofmann and Schopf (1983), Photo 14-2, M, O, R–Z]
2090	Gunflint Iron Fm.	Canada	Pl. 2, A–R; see also Hofmann and Schopf (1983), Photo 14-1, A–V
	Animikiea septata Barghoorn		
	Entosphaeroides amplus Barghoorn		
	Eosphaera tyleri Barghoorn		
	Gunflintia grandis Barghoorn		
	Gunflinta minuta Barghoorn		
	Huroniospora macroreticulata Barghoorn		
	Huroniospora microreticulata Barghoorn		
	Huroniospora psilata Barghoorn		
	Kakabekia umbellata Barghoorn		
[2000	Duck Creek Dolomite	Australia	Hofmann and Schopf (1983), Photos 14-3, A–J and 14-4, A–E]
1950	Tyler Fm.	USA	Pl. 3, A–C
[1890	Windidda Fm.	Australia	Hofmann and Schopf (1983), Photo 14-5, A–C]
[1875	Frere Fm.	Australia	Hofmann and Schopf (1983), Photo 14-5, D–G]
1700	Vempalle Fm.	India	Pl. 3, D–O
1650	Paradise Creek Fm.	Australia	Pl. 4, A–C; see also Hofmann and Schopf (1983), Photo 14-11, A–C
[1600	Amelia Dolomite	Australia	Hofmann and Schopf (1983), Photo 14-6, A–O]
1550	Satka Fm.	USSR	Pl. 4, D–F
	Eomarginata striata Jankauskas		
	Gloeodiniopsis uralicus Krylov and Sergeev		
	Palaeopleurocapsa kelleri Krylov and Sergeev		
1500	Bakal Fm.	USSR	Pl. 5, A–E
	Brevitrichoides burzjanicus Jankauskas		
	Margominuscula rugosa Naumova		
	Satka favosa Jankauskas		
	Symplassosphaeridium undosum Jankauskas		
1500	Omachtinsk Fm.	USSR	Pl. 5, F
	Sivaglicania tadasii Weiss		
[1500	Barney Creek Fm.	Australia	Hofmann and Schopf (1983) Photo 14-7, A–Z]
[1482	Balbirini Dolomite	Australia	Hofmann and Schopf (1983) Photo 14-8, A–W]
1425	Gaoyuzhuang Fm.	China	Pl. 6, A–E; Pl. 7, A–G; Pl. 8, A–M
1400	Dismal Lakes Grp.	Canada	Pl. 9, A–L
	Archaeoellipsoides grandis Horodyski and Donaldson		
	Myxococcoides grandis Horodyski and Donaldson		

Age (Ma)	Geologic Unit/ Illustrated Holotypes	Country	Plates and Figures
[1364	Bungle Bungle Dolomite	Australia	Hofmann and Schopf (1983) Photos 14-9, A–Q and 14-10, A–S]
1350	Tottinsk Fm.	USSR	Pl. 5, G, H
	Konderia elliptica Weiss		
1000	Zil'merdak Fm.	USSR	Pl. 10, A–C
	Brevitrichoides bashkiricus Jankauskas		
	Brevitrichoides karatavicus Jankauskas		
	Glenobotrydion solutum Jankauskas		
1000	Sukhaya Tunguska Fm.	USSR	Pl. 10, D–P; see also Frontispiece
	Eoentophysalis arcata Mendelson and Schopf		
	Oscillatoriopsis media Mendelson and Schopf		
950	Derevnya Fm.	USSR	Pl. 5, I
	Partitiofilum tungusum Mikhailova		
950	Lakhanda Fm.	USSR	Pl. 11, A–C; Pl. 12, A; Pl. 13, A–D; Pl. 14, A–F, Pl. 15, A–E; Pl. 16, A–E; Pl. 17, A–F;
	Aimia delicata Hermann		
	Aimophyton varium Timofeev and Hermann		
	Annularia annulata Timofeev and Hermann		
	Eosaccharomyces ramosus Hermann		
	Jacutianema solubila Timofeev and Hermann		
	Lakhandinia prolata Timofeev and Hermann		
	Leiothrichoides tenuitunicatus Hermann		
	Lomentunella vaginata Hermann		
	Majaphyton antiquaum Timofeev and Hermann		
	Majasphaeridium carpogenum Hermann		
	Mucorites rifeicus Hermann		
	Mycosphaeroides aggregatus Hermann		
	Mycosphaeroides caudatus Hermann		
	Nucellohystrichosphaera megalea Timofeev		
	Palaeolyngbya helva Hermann		
	Palaeovaucheria clavata Hermann		
	Rugosoopsis tenuis Timofeev and Hermann		
	Tetrasphaera antiqua Timofeev and Hermann		
	Trachyhystrichosphaera aimika Hermann		
	Trachysphaeridium laminaritum Timofeev		
	Turuchanica ternata (Timofeev) Jankauskas		
	Ulophyton rifeicum Timofeev and Hermann		
930	Galeros Fm.	USA	Pl. 18, A–M
925	Sim (Podinzer) Fm.	USSR	Pl. 19, A–G
	Heliconema uralense Jankauskas		
	Leiosphaeridia kulgunica Jankauskas		
	Pterospermopsimorpha capsulata Jankauskas		
	Pterospermopsis dubius Jankauskas		
	Pterospermopsis simicus Jankauskas		
	Tasmanites rifejicus Jankauskas		
900	Shorikha Fm.	USSR	Pl. 19, H–L
900	Shtandin Fm.	USSR	Pl. 19, M; Pl. 20, A–F; Pl. 21, A–G

Age (Ma)	Geologic Unit/ Illustrated Holotypes	Country	Plates and Figures
	Calyptothrix alternata Jankauskas		
	Calyptothrix geminata Jankauskas		
	Eomycetopsis rimata Jankauskas		
	Pellicularia tenera Jankauskas		
	Plicatidium latum Jankauskas		
	Siphonophycus costatus Jankauskas		
	Volyniella glomerata Jankauskas		
850	Beck Spring Fm.	USA	Pl. 21, H, I; Pl. 22, A, B
	Beckspringia communis Licari		
	Palaeosiphonella cloudii Licari		
850	Miroedikha Fm.	USSR	Pl. 22, C–H; Pl. 23, A–C; Pl. 24, A–F; Pl. 25, A–E; Pl. 26, A, B; Pl. 27, A–F; Pl. 28, A–F;
	Arctacellularia doliiformis Hermann		
	Arctacellularia ellipsoidea Hermann		
	Arthrosiphon cornutus Weiss		
	Arthrosiphon typicus Weiss		
	Calyptothrix perfecta Weiss		
	Caudiculophycus micronulatus Hermann		
	Cephalonyx sibiricus Weiss		
	Cephalophytarion turukhanicum Weiss		
	Dictyosphaeridium tungusum Timofeev		
	Filiconstrictosus eniseicum Weiss		
	Heliconema turukhania Hermann		
	Leiothrichoides maculatus Hermann		
	Macroptycha biplicata Timofeev		
	Macroptycha uniplicata Timofeev		
	Oscillatoriopsis bacilaris Hermann		
	Ostiana microcystis Hermann		
	Palaeocalothrix divaricatus Hermann		
	Palaeolyngbya sphaerocephala Hermann		
	Polysphaeroides filliformis Hermann		
	Polysphaeroides contextus Hermann		
	Primorivularia dissimilara Hermann		
	Pterospermopsimorpha insolita Timofeev		
	Sibiriafilum tunicum Hermann		
	Tortunema eniseica Hermann		
	Tortunema sibirica Hermann		
	Trachythrichoides ovalis Hermann		
850	Kwagunt Fm.	USA	Pl. 29, A–J; Pl. 30, A–I
	Melanocyrillium hexodiadema Bloeser		
850	Bitter Springs Fm.	Australia	Pl. 30, J–S; Pl. 31, A–L; Pl. 32, A–J; Pl. 33, A–L
	Bigeminococcus lamellosus Schopf and Blacic		
	Caryosphaeroides pristina Schopf		
	Caudiculophycus acuminatus Schopf and Blacic		
	Caudiculophycus rivularioides Schopf		
	Cephalophytarion grande Schopf		
	Cephalophytarion laticellulosum Schopf and Blacic		
	Cephalophytarion minutum Schopf		
	Cephalophytarion variabile Schopf and Blacic		

Illustrated Type Specimens

Age (Ma)	Geologic Unit/ Illustrated Holotypes	Country	Plates and Figures
	Eotetrahedrion princeps Schopf and Blacic		
	Eozygion grande Schopf and Blacic		
	Eozygion minutum Schopf and Blacic		
	Filiconstrictosus diminutus Schopf and Blacic		
	Filiconstrictosus majusculus Schopf and Blacic		
	Globophycus rugosum Schopf		
	Halythrix nodosa Schopf		
	Heliconema funiculum Schopf and Blacic		
	Myxococcoides minor Schopf		
	Obconicophycus amadeus Schopf and Blacic		
	Oscillatoriopsis breviconvexa Schopf and Blacic		
	Oscillatoriopsis obtusa Schopf		
	Palaeolyngbya barghoorniana Schopf		
	Palaeolyngbya minor Schopf and Blacic		
	Partitiofilum gongyloides Schopf and Blacic		
	Sphaerophycus parvum Schopf		
	Veteronostocale amoenum Schopf and Blacic		
800	Akberdin Fm.	USSR	Pl. 42, A–D
	Kildinella lophostriata Jankauskas		
	Satka colonialica Jankauskas		
	Satka elongata Jankauskas		
780	Auburn Dolomite	Australia	Pl. 44, A–E
775	River Wakefield Subgroup	Australia	Pl. 34, A–M; Pl. 35, A–E; Pl. 36, A–G; Pl. 37, A–I
770	Skillogalee Dolomite	Australia	Pl. 37, J; Pl. 38, A–H; Pl. 39, A–C; Pl. 40, A–H; Pl. 41, A–L
750	Dashka Fm.	USSR	Pl. 42, E–I; Pl. 43, A, B
	Calyptothrix obsoletus Mikhailova		
	Cephalophytarion piliformis Mikhailova		
	Germinosphaera bispinosa Mikhailova		
	Germinosphaera unispinosa Mikhailova		
	Nucellosphaeridium spumosum Mikhailova		
	Palaeoaphanizomanon scabratus Mikhailova		
	Polysphaeroides nuclearis Mikhailova		
750	Myrtle Springs Fm.	Australia	Pl. 43, C–I
740	Brioverian	France	Pl. 44, F–P
740	Min'yar Fm.	USSR	Pl. 45, B–I; Pl. 46, A–G; Pl. 47, A–G
	Eoaphanocapsa oparinii Nyberg and Schopf		
	Gloeodiniopsis grandis Sergeev and Krylov		
	Gloeodiniopsis magna Nyberg and Schopf		
	Palaeopleurocapsa kamaelgensis Sergeev and Krylov		
	Ramivaginalis uralensis Nyberg and Schopf		
	Siphonophycus capitaneum Nyberg and Schopf		
738	Olkha Fm.	USSR	Pl. 45, A
700	Hector Fm.	Canada	Pl. 48, A, B
675	Uk Fm.	USSR	Pl. 48, C, D
	Palaeolyngbya zilimica Jankauskas		
675	Kamovsk Fm.	USSR	Pl. 48, E, F; see also Frontispiece
	Cucumiforma vanavaria Mikhailova		
650	Chichkan Fm.	USSR	Pl. 49, A–G; Pl. 50, A–C; Pl. 51, A–D
	Palaeopleurocapsa fusiforma Ogurtsova and Sergeev		
	Palaeopleurocapsa reniforma Ogurtsova and Sergeev		
	Tetraphycus bistratosus Ogurtsova and Sergeev		
640	Chartorysk Fm.	USSR	Pl. 48, G
620	Mogilev Fm.	USSR	Pl. 51, E, F
	Circumiella mogilevica Aseeva		
	Polycavita bullata (Andreeva) Assejeva		
620	Suirovo Fm.	USSR	Pl. 51, G
	Retiforma tolparica Mikhailova		
610	Yaryshev Fm.	USSR	Pl. 51, H–K; Pl. 52, A–D
	Ljadovia perforata Assejeva		
	Redkinia fedonkinispis Aseeva		
	Striatella coriacea Assejeva		
	Taenitrichoides jaryschevicus Assejeva		
	Tubulosa corrugata Assejeva		
	Volyniella valdaica (Shepeleva) Aseeva		
600	Kanilovka Fm.	USSR	Pl. 53, A–F
	Bicuspidata fusiformis Assejeva		
	Flagellis tenuis Assejeva		
	Podoliella irregulare Timofeev		
	Studenicia bacotica Assejeva		
	Volyniella canilovica Aseeva		
600	Derlo Fm.	USSR	Pl. 53, G–I
	Octoedryxium simmetricum Timofeev		
600	Zin'kov Fm.	USSR	Pl. 53, J
	Leiosphaeridia undulata Timofeev		
600	Yudoma Fm.	USSR	Pl. 54, A–H
	?*Eomycetopsis siberiensis* Lo		
	Myxococcoides staphylidion Lo		
	Tetraphycus conjunctum Lo		
580	Redkino Fm.	USSR	Pl. 55, A–E
	Baltisphaeridium perrarum Jankauskas		
	Leiosphaeridia incrassatula Jankauskas		
	Satka granulosa Jankauskas		
	Spumosina rubiginosa (Naumova) Jankauskas		
560	Kotlin Fm.	USSR	Pl. 54, I–K
	Bavlinella faveolata Shepeleva		
	Volyniella rara Paskeviciene		
540	Moty Fm.	USSR	Pl. 56, A–C
	Paracrassosphaera dedalea Rudavskaya		
	Polyedryxium neftelenicum Rudavskaya		
	Octoedryxium truncatum Rudavskaya		

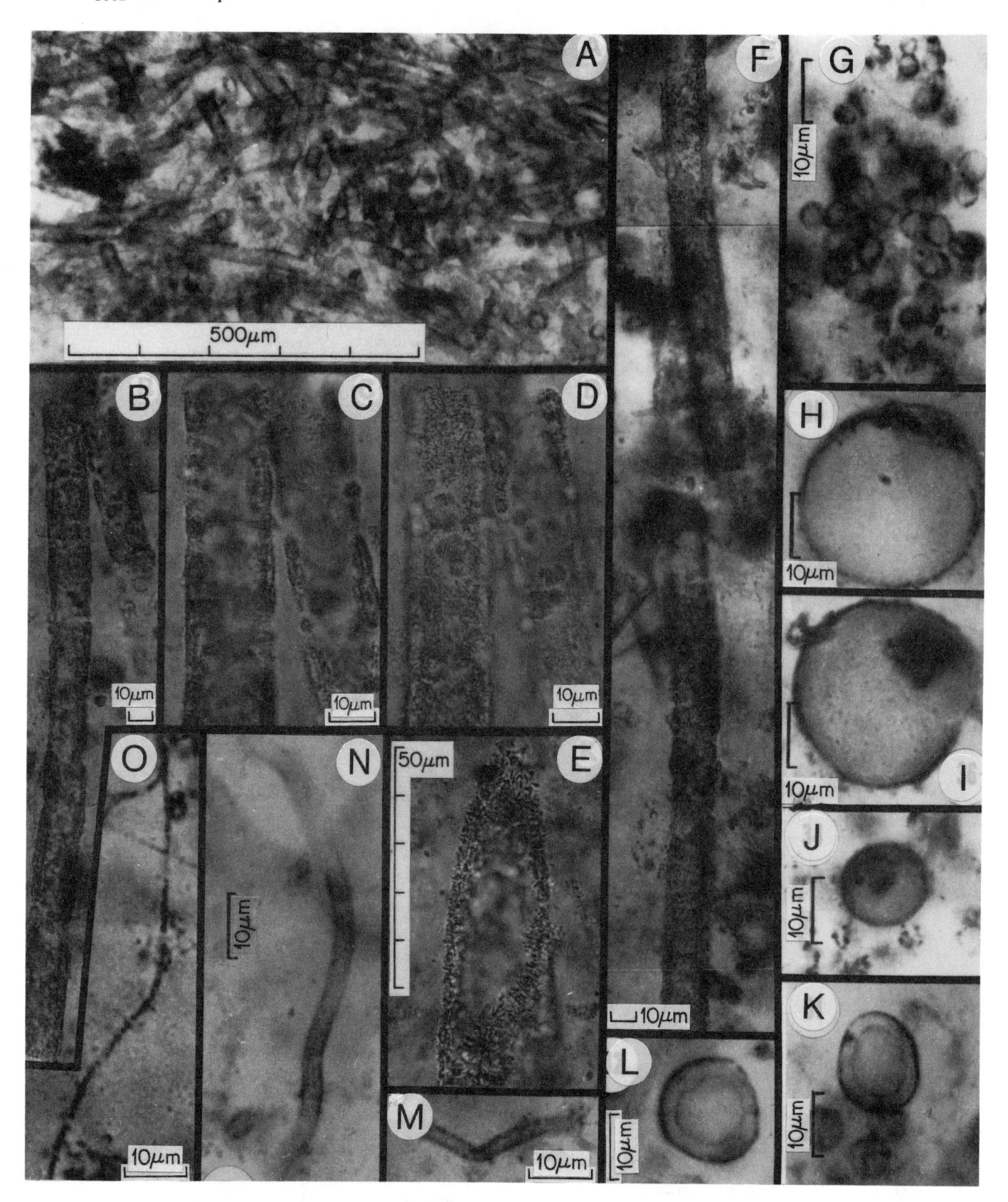

Plate 1. A–F, Gamohaan Formation, Griqualand West, northern Cape Province, South Africa (2450 Ma). [A–D, F, petrographic thin sections, transmitted light; E, polished microprobe section, plane polarized light.] A, E, F, *Siphonophycus transvaalense* Beukes, Klein & Schopf 1987, PARATYPES [TU50]; B, C, D, *Siphonophycus transvaalense* Beukes, Klein & Schopf 1987, HOLOTYPE [TU50]. G–O, Kasegalik Formation, Hudson Bay, North West Territories, Canada (2150 Ma). [G–O, petrographic thin sections, transmitted light.] G, *Huroniospora* sp. [S2]; H, I, J, "Type 2 microfossil"; K, L, "Type 4 microfossil" [BZ4]; M, N, *Eomycetopsis* sp.; O, *Biocatenoides* sp. [3N1] (Hofmann and Jackson 1969).

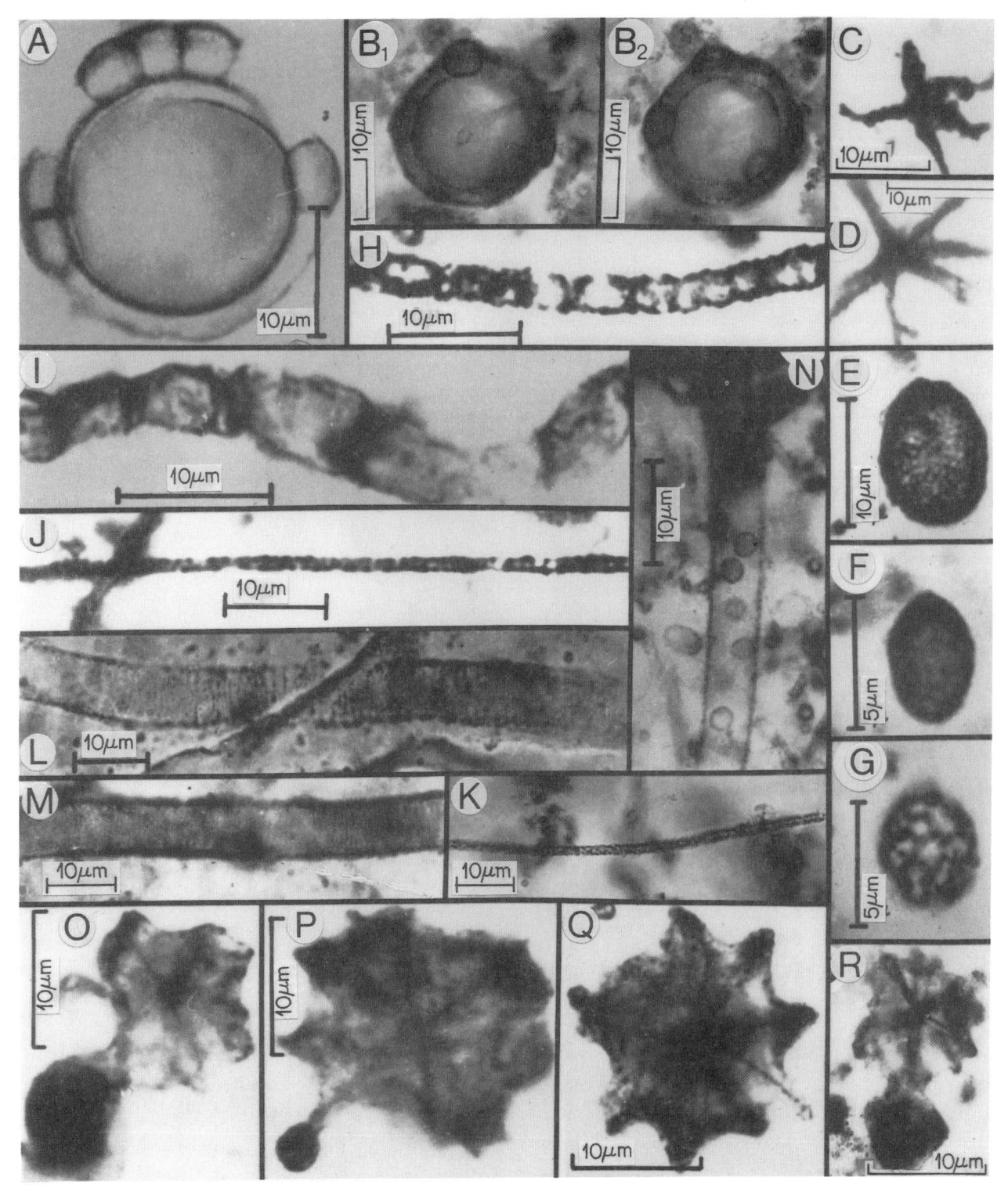

Plate 2. A–R, Gunflint Iron Formation, southern Ontario, Canada (2090 Ma; localities cf. PPRG 1289 and 2463). [A–R, petrographic thin sections, transmitted light; B, K, specimens not previously illustrated; B_1, B_2 show the same specimen at two focal depths.] A, *Eosphaera tyleri* Barghoorn 1965, HOLOTYPE [BZ4]; B_1, B_2, *Eosphaera tyleri* Barghoorn 1965, TOPOTYPE [BZ4]; C, *Eoastrion simplex* Barghoorn 1965, PARATYPE [BZ3]; D, *Eoastrion bifurcatum* Barghoorn 1965, PARATYPE [BZ3]; E, *Huroniospora microreticulata* Barghoorn 1965, HOLOTYPE; F, *Huroniospora psilata* Barghoorn 1965, HOLOTYPE; G, *Huroniospora macroreticulata* Barghoorn 1965, HOLOTYPE; H, *Gunflintia grandis* Barghoorn 1965, PARATYPE [3N27]; I, *Gunflintia grandis* Barghoorn 1965, HOLOTYPE [3N27]; J, *Gunflintia minuta* Barghoorn 1965, HOLOTYPE [3N7]; K, *Gunflintia minuta* Barghoorn 1965, TOPOTYPE [3N7]; L, *Animikiea septata* Barghoorn 1965, HOLOTYPE [TU37]; M, *Animikiea septata* Barghoorn 1965, PARATYPE [3N37]; N, *Entosphaeroides amplus* Barghoorn 1965, HOLOTYPE [TU29]; O, *Kakabekia umbellata* Barghoorn 1965, PARATYPE [BZ8]; P, *Kakabekia umbellata* Barghoorn 1965, HOLOTYPE [BZ8]; Q, *Kakabekia umbellata* Barghoorn 1965, PARATYPE [BZ8]; R, *Kakabekia umbellata* Barghoorn 1965, PARATYPE [BZ8] (Barghoorn and Tyler 1965).

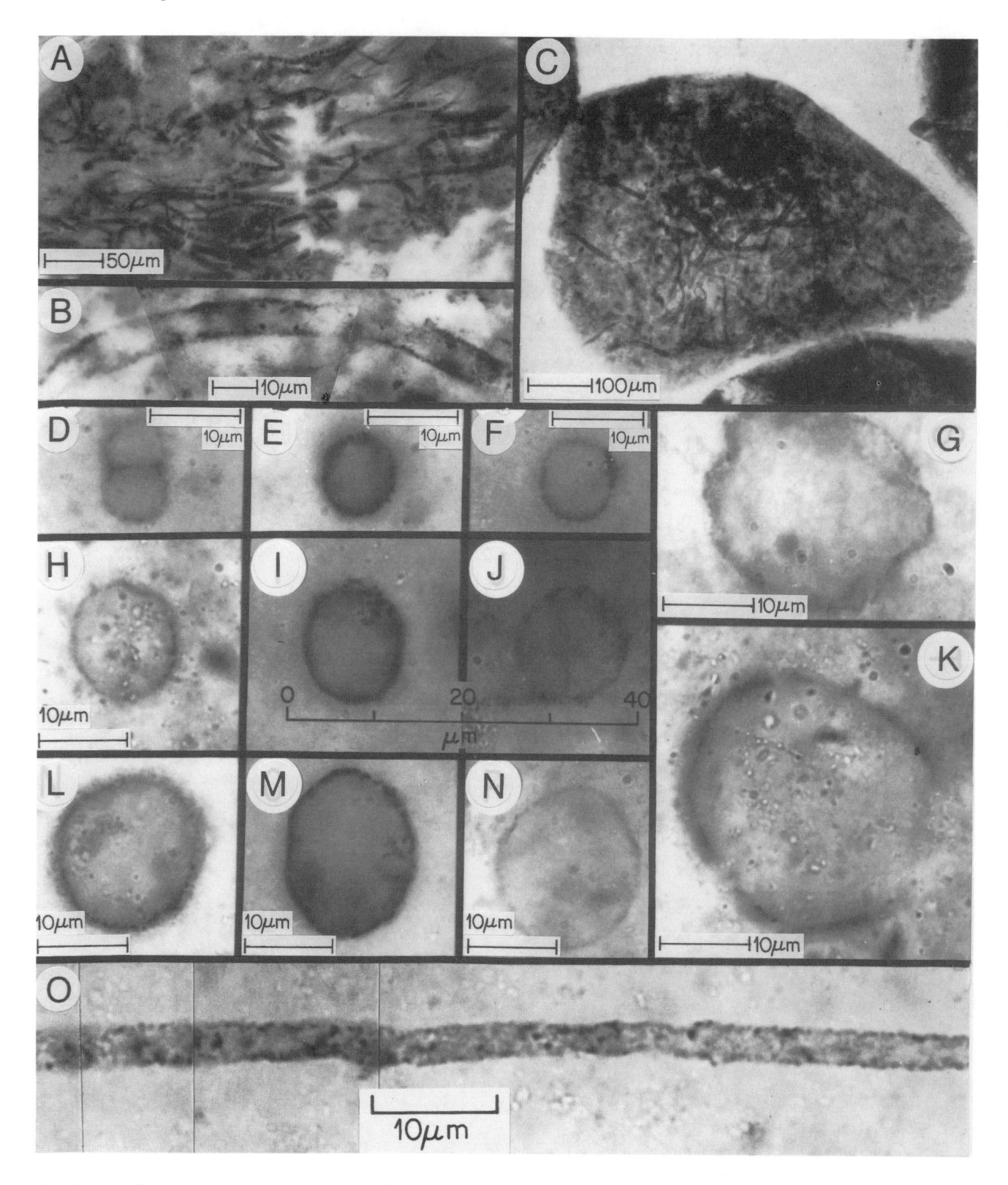

Plate 3. A–C, Tyler Formation, northern peninsular Michigan, USA (1950 Ma; locality cf. PPRG 1363). [A–C, petrographic thin sections, transmitted light; B, composite photomicrograph.] A, cf. *Gunflintia grandis* Barghoorn 1965 [3N35]; B, cf. *Animikiea* sp. [TU40]; C, cf. *Gunflintia grandis* Barghoorn 1965 [3N35] (Cloud and Morrison 1980). D–O, Vempalle Formation, Andhra Pradesh, south-central India (1700 Ma; locality cf. PPRG 1308). [D–O, petrographic thin sections, transmitted light; O, composite photomicrograph.] D–N, unnamed spheroidal to ellipsoidal unicells; O, unnamed cellular filament [3N11] (Schopf and Prasad 1978).

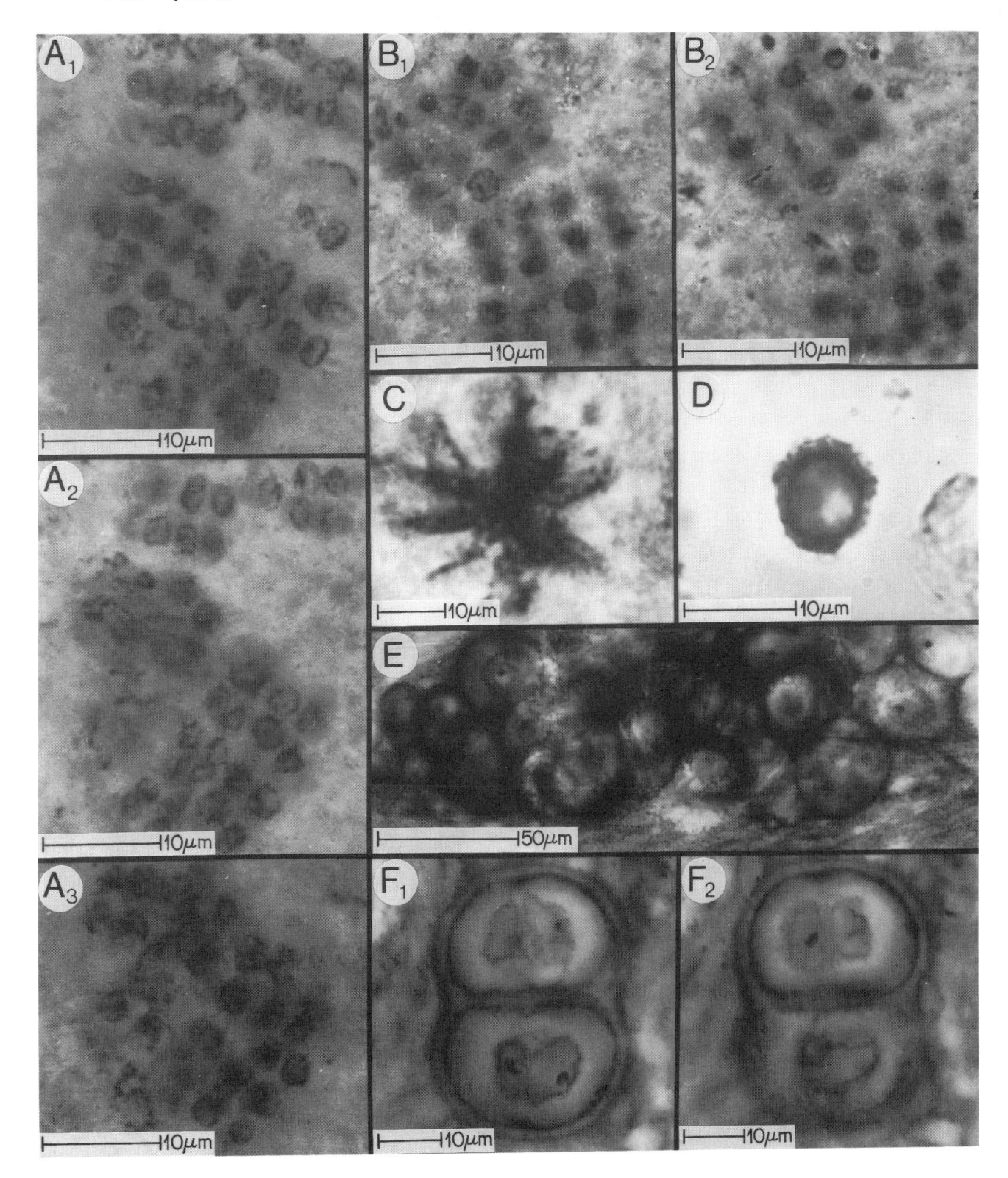

Plate 4. A–C, Paradise Creek Formation, Queensland, Australia (1650 Ma; locality cf. PPRG 1260). [A–C, petrographic thin sections, transmitted light; A_1, A_3 and B_1, B_2 show different colonies at three and two focal depths, respectively]. A_2–A_3, B_1, B_2, ?*Eucapsis* [CCU1] (Licari et al. 1969); C, *Metallogenium* sp. = *Eoastrion* sp. [BZ3] (Cloud 1976). D–F, Satka Formation, southern Ural Mountains, Bashkiria, USSR (1550 Ma). [D, acid resistant residue, interference contrast; E, F, petrographic thin sections, transmitted light; F_1, F_2 show the same specimen at two focal depths.] D, *Eomarginata striata* Jankauskas (=Yankauskas) 1979, HOLOTYPE (slide 16-2133-990/2, England slide P16/4, slide label left) [A1]; E, *Palaeopleurocapsa kelleri* Krylov & Sergeev 1987, HOLOTYPE (slide 10-80, England slide K9, slide label left) [PLEU4]; F_1, F_2, *Gloeodiniopsis uralicus* Krylov & Sergeev 1987, HOLOTYPE (slide 130-82, England slide N23, slide label left) [G6].

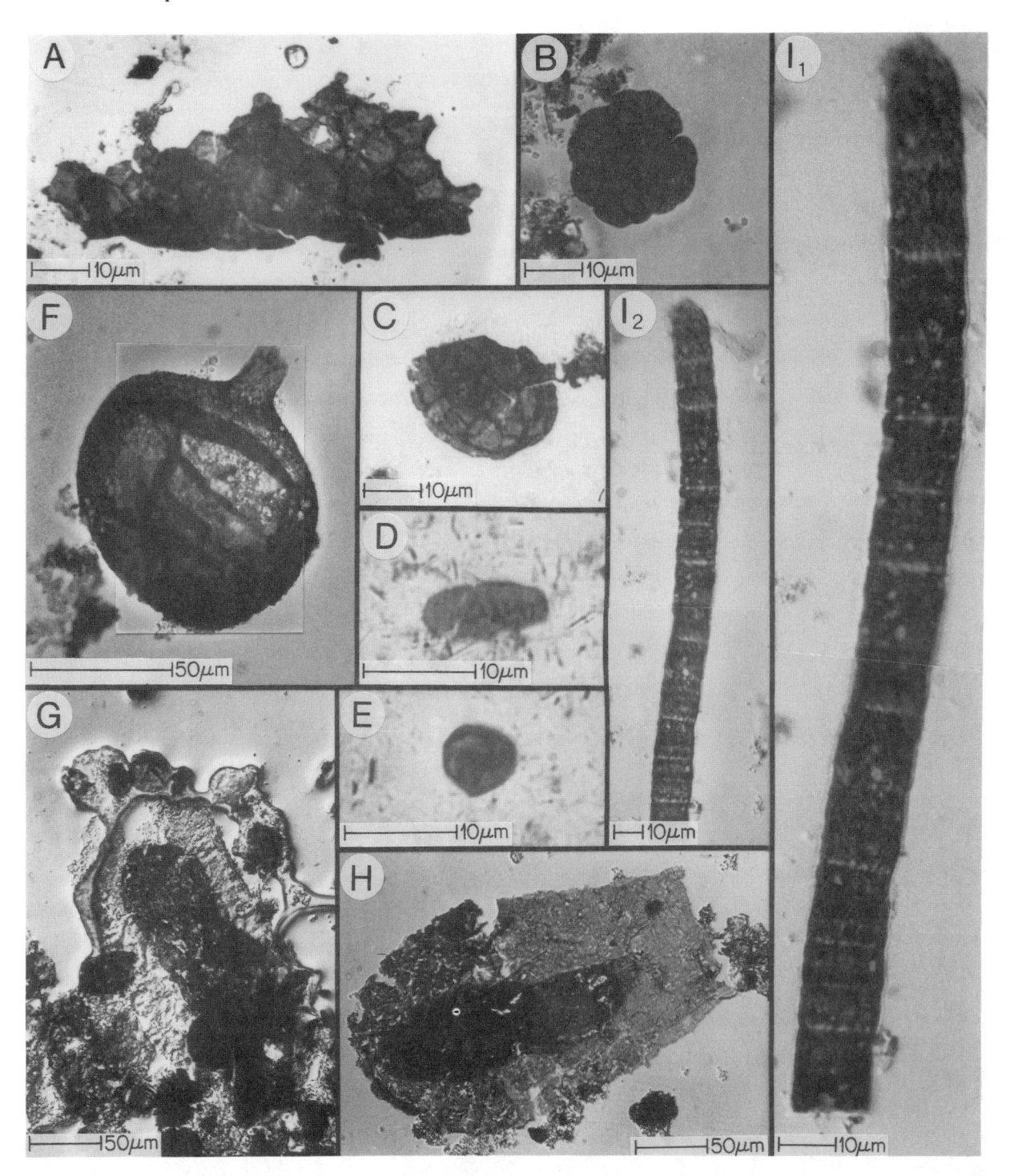

Plate 5. A–E, Bakal Formation, southern Ural Mountains, Bashkiria, USSR (1500 Ma). [A–E, acid resistant residues; A, C, D, E, interference contrast; B, transmitted light.] A, *Satka favosa* Jankauskas (=Yankauskas) 1979, PARATYPE (slide 16-1815-635, England slide 011/4, slide label left) [CRCF3]; B, *Symplassosphaeridium undosum* Jankauskas (=Yankauskas) 1979, HOLOTYPE (slide 16-1815-230, England slide N25/4, slide label left); C, *Satka favosa* Jankauskas (=Yankauskas) 1979, HOLOTYPE (slide 16-1815-635, England slide S36, slide label left) [CRCF3]; D, *Brevitrichoides burzjanicus* Jankauskas (=Yankauskas) 1982, HOLOTYPE (slide 16-1815-222/2, England slide L29, slide label left); E, *Margominuscula rugosa* Naumova 1960, NEOTYPE (slide 16-1815-532/9, England slide K32/4, slide label left) (Yankauskas, in press) [S2]. F, Omachtinsk Formation, Uchur-Maya region, Siberia, USSR (1500 Ma). [F, acid resistant residue, transmitted light, composite photomicrograph.] F, *Sivaglicania tadasii* Weiss (=Vejs) in press, HOLOTYPE (slide 80-32 = slide 80/30-6, England slide X35, read slide label from right). G, H, Tottinsk Formation, Uchur-Maya region, Siberia, USSR (1350 Ma). [G, H, acid resistant residues, interference contrast.] G, *Konderia elliptica* Weiss (=Vejs) 1983, HOLOTYPE (slide 43/24 = slide 43/24-42, England slide Z52, read slide label from right); H, *Konderia elliptica* Weiss (=Vejs) 1983, PARATYPE (slide 43/24 = slide 43/24-71, England slide P50/3, read slide label from right). I, Derevnya Formation, Turukhansk region, Siberia, USSR (950 Ma). [I_1, I_2, acid resistant residue, interference contrast, composite photomicrographs showing the same specimen at two magnifications.] I_1, I_2, *Partitiofilum tungusum* Mikhailova (=Mikhajlova) (in German 1989) HOLOTYPE (slide 1821/4, England slide K33/1, read slide label from right) [3N58].

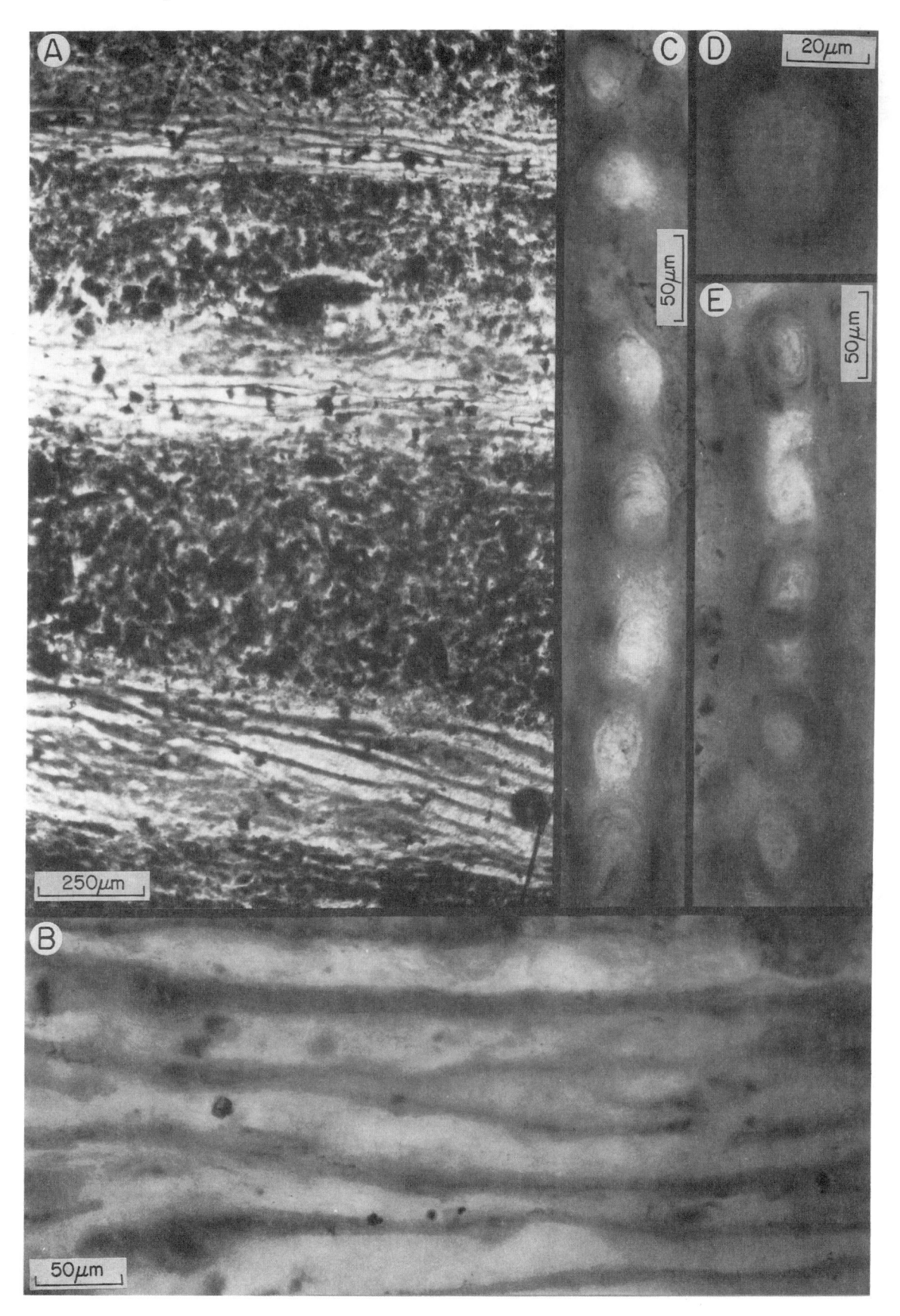

Plate 6. A–E, Gaoyuzhuang Formation, Yanshan Range, Jixian region, northern China (1425 Ma; locality cf. PPRG 2123). [A–E, petrographic thin sections, transmitted light; A, C, D, specimens not previously illustrated.] A, thin section cut perpendicular to bedding illustrating alternation of "sheath mats" with layers composed of finely divided organic detritus; B–E unnamed multilamellated *Lyngbya*-like sheaths, in longitudinal (B) and transverse sections (C–E), comprising the "sheath mats" shown in part A (B, E, Schopf et al. 1984) [TUL4].

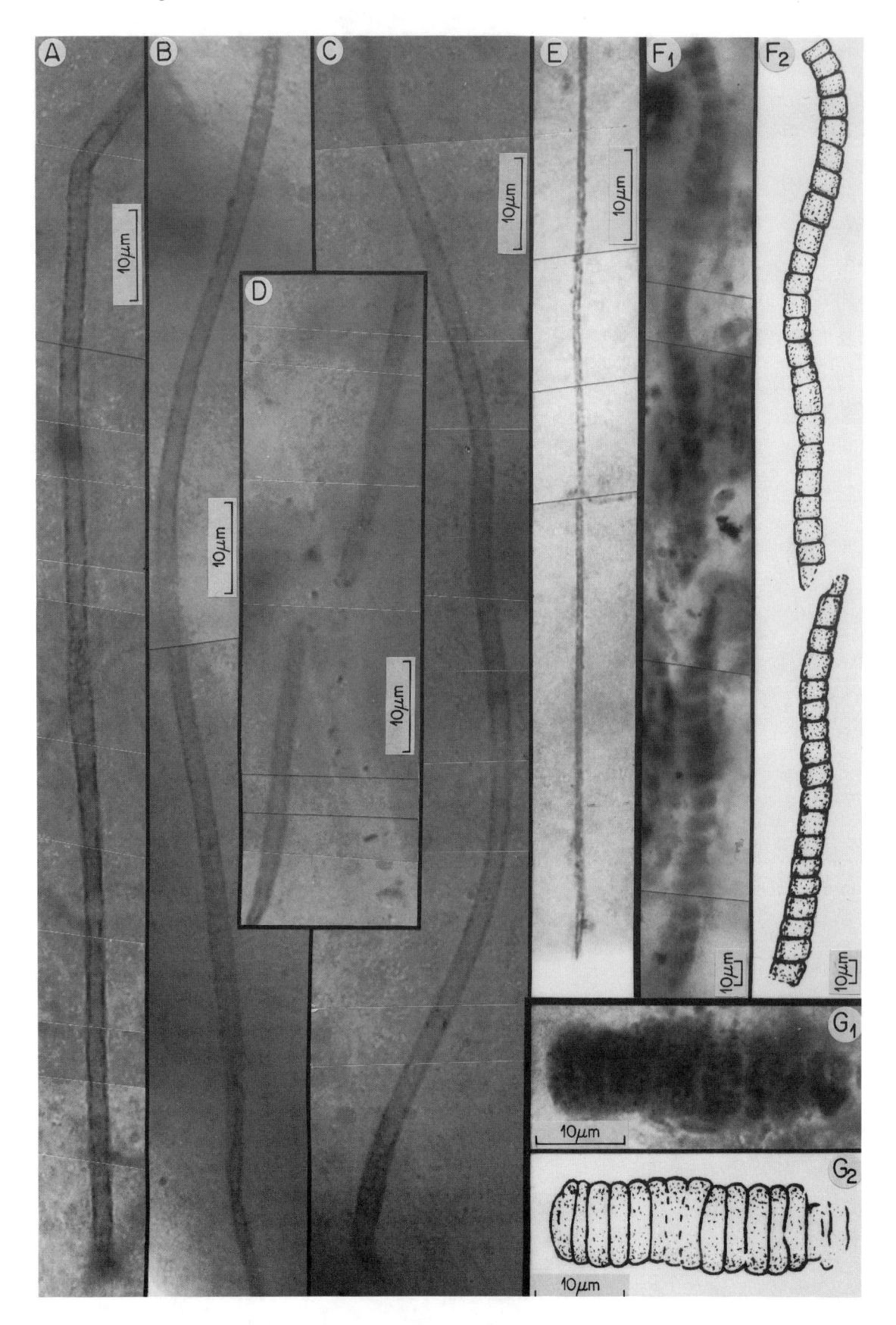

Plate 7. A–G, Gaoyuzhuang Formation, Yanshan Range, Jixian region, northern China (1425 Ma; locality cf. PPRG 2123). [A–G, petrographic thin sections, transmitted light; A–F, composite photomicrographs; A, B, E–G, specimens not previously illustrated; F_2 and G_2 are interpretive drawings based on tracings of F_1 and G_1, respectively.] A–D, *Eomycetopsis* sp. (C, D, Schopf et al. 1984) [TU4]; E, *Gunflintia* sp. with well-defined quadrate cells; F_1, F_2, unnamed cellular trichome; G_1, G_2, *Oscillatoriopsis* sp.

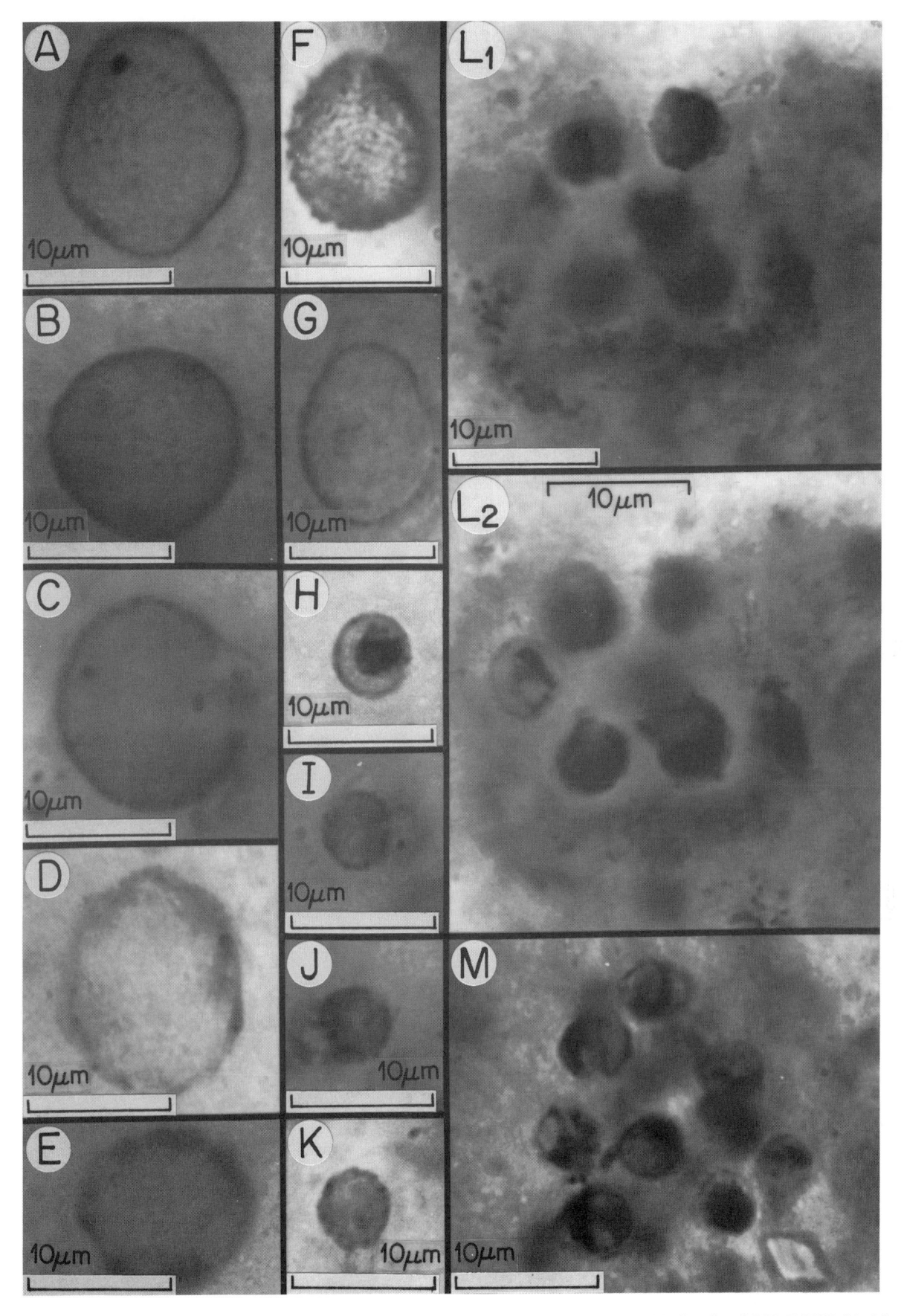

Plate 8. A–M, Gaoyuzhuang Formation, Yanshan Range, Jixian region, northern China (1425 Ma; locality cf. PPRG 2123). [A–M, petrographic thin sections, transmitted light; D–H, J, L, M, specimens not previously illustrated; L_1, L_2 show two focal depths of the same colony.] A–K, unnamed spheroidal to ellipsoidal unicells (A–C, I, K, Schopf et al. 1984); L, M, unnamed colonial coccoid cells.

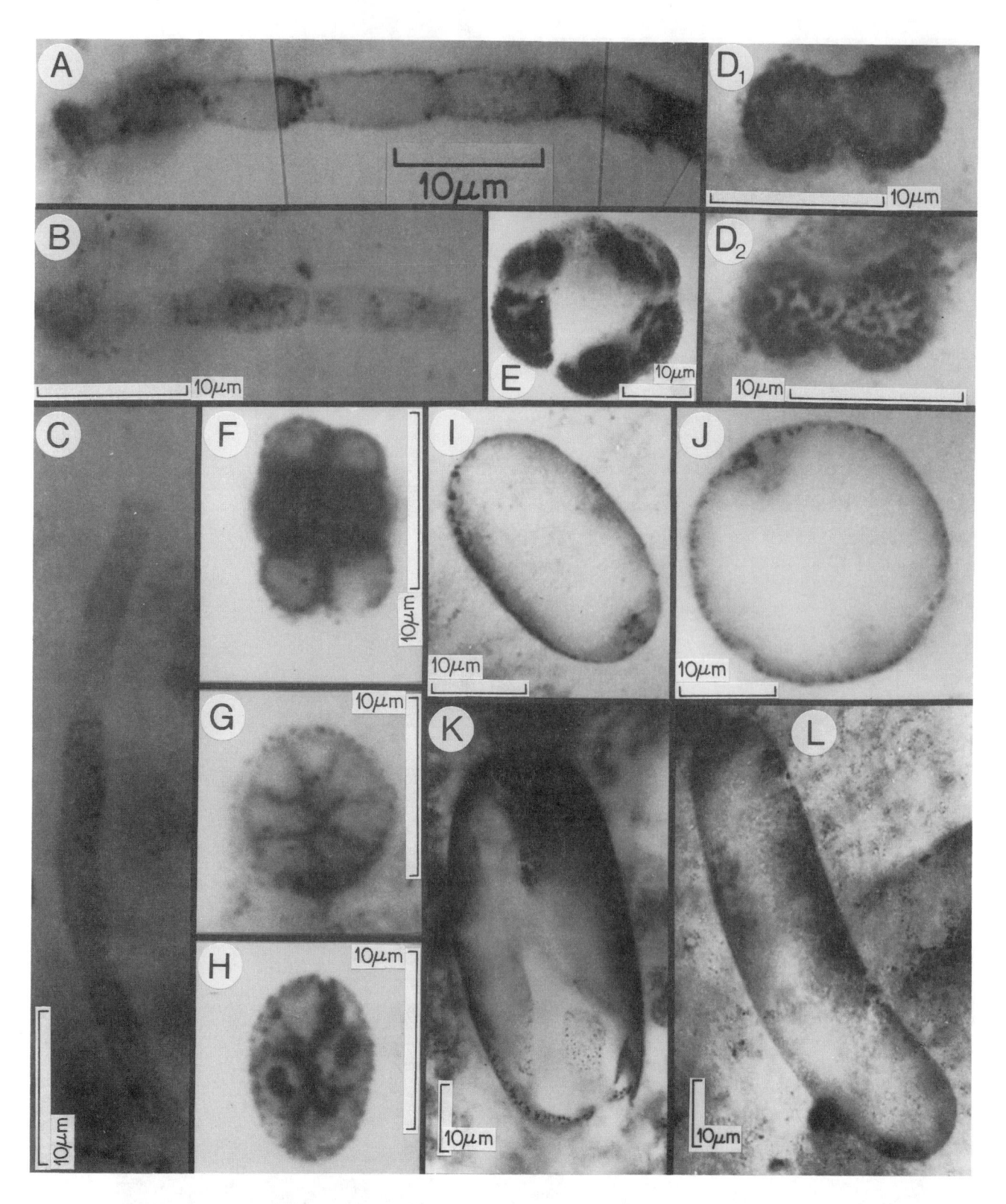

Plate 9. A–L, Dismal Lakes Group, District of Mackenzie, North West Territories, Canada (1400 Ma; locality cf. PPRG 1212 and PPRG 2358). [A–L, petrographic thin sections, transmitted light; A, composite photomicrograph; B–D, specimens not previously illustrated; D$_1$, D$_2$ show the same specimen at two focal depths.] A, *Oscillatoriopsis curta* Horodyski & Donaldson 1980, PARATYPE [3N37]; B, *Oscillatoriopsis ?curta* Horodyski & Donaldson 1980, TOPOTYPE [3N37]; C, *Oscillatoriopsis* sp.; D$_1$, D$_2$, unnamed paired cells; E, *Eoentophysalis dismallakesensis* Horodyski & Donaldson 1980, PARATYPE [EIC11]; F, *Sphaerophycus ?medium* Horodyski & Donaldson 1980, PARATYPE [EIC8]; G, *Sphaerophycus ?parvum* Schopf 1968 (Horodyski and Donaldson 1980) [EIC6]; H, *Sphaerophycus ?medium* Horodyski & Donaldson 1980, PARATYPE [EIC8]; I, *Archaeoellipsoides grandis* Horodyski & Donaldson 1980, PARATYPE; J, *Myxococcoides grandis* Horodyski & Donaldson 1980, HOLOTYPE; K, *Archaeoellipsoides grandis* Horodyski & Donaldson 1980, PARATYPE; L, *Archaeoellipsoides grandis* Horodyski & Donaldson 1980, HOLOTYPE.

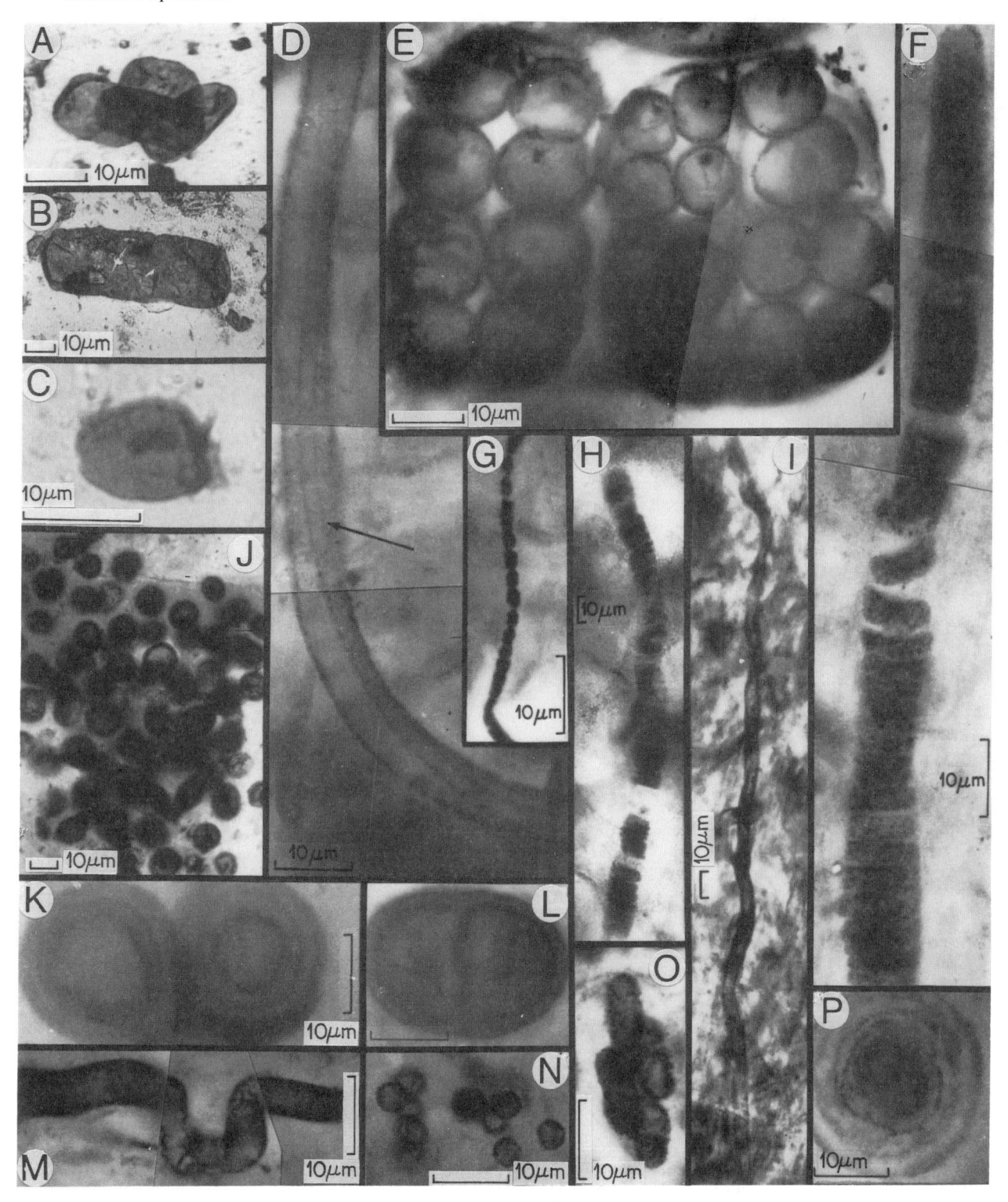

Plate 10. A–C, Zil'merdak Formation, southern Ural Mountains, Bashkiria, USSR (1000 Ma). [A–C, acid resistant residues; A, transmitted light; B, C, interference contrast.] A, *Brevitrichoides karatavicus* Jankauskas (= Yankauskas) 1980, HOLOTYPE (slide 16-50/3, England slide U10, slide label left); B, *Brevitrichoides bashkiricus* Jankauskas (= Yankauskas) 1980, HOLOTYPE (slide 16-50/7, England slide R37, slide label left); C, *Glenobotrydion solutum* Jankauskas (= Yankauskas) 1980, LECTOTYPE (slide 16-50/4, England slide L15/1-3, slide label left). D–P, Sukhaya Tunguska Formation, Turukhansk region, Siberia, USSR (1000 Ma; locality cf. PPRG 2725). [D–P, petrographic thin sections, transmitted light; D, E, F, M, composite photomicrographs.] D, "tubular sheath with internal trichome" (at arrow) [3S5]; E, *Eoentophysalis arcata* Mendelson & Schopf 1982, HOLOTYPE (see also Frontispiece) [CCU4]; F, *Oscillatoriopsis media* Mendelson & Schopf 1982, HOLOTYPE [3N56]; G, "degraded cellular filament"; H, *Oscillatoriopsis media* Mendelson & Schopf 1982, PARATYPE [3N56]; I, *Eomycetopsis* sp. [TU7]; J, "larger chroococcacean cyanobacteria"; K, P, *Gloeodiniopsis lamellosa* Schopf 1968 [G4]; L, "*Chroococcus*-like morphotype" [CHR9]; M, *Ecomycetopsis* sp.; N, "smaller chroococcacean cyanobacteria" [CIR8]; O, *Eosynechococcus medius* Hofmann 1976 [EIC5] (Mendelson and Schopf 1982).

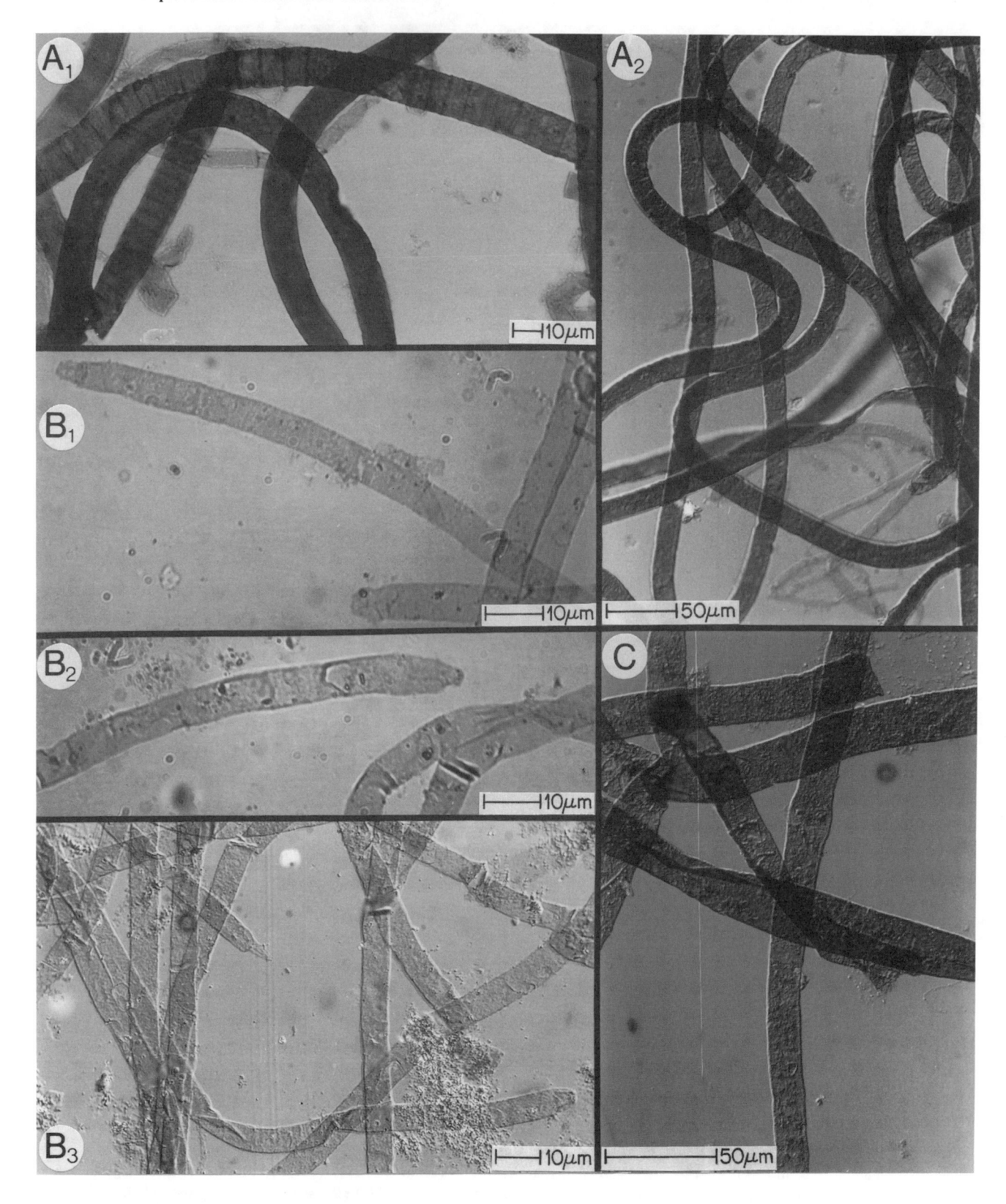

1072 Atlas of Representative Proterozoic Microfossils

Plate 11. A–C, Lakhanda Formation, Khabarovsk region, Siberia, USSR (950 Ma). [A–C, acid resistant residues; A_1, B_1, B_2, transmitted light; A_2, B_3, C, interference contrast; A_1, A_2 and B_1–B_3 each show portions of specimens at differing magnifications.] A_1, A_2, *Palaeolyngbya helva* Hermann (=German) 1981, HOLOTYPE (slide 27/6, England slide V38/2, read label from right) [3S13]; B_1–B_3, *Leiothrichoides tenuitunicatus* Hermann (=German) 1981, HOLOTYPE (slide 5/4, England slide L34, read label from right) [3N28]; C, *Palaeovaucheria clavata* Hermann (=German) 1981, PARATYPE (slide 27/10, England slide T33/2, read label from right) [EU5].

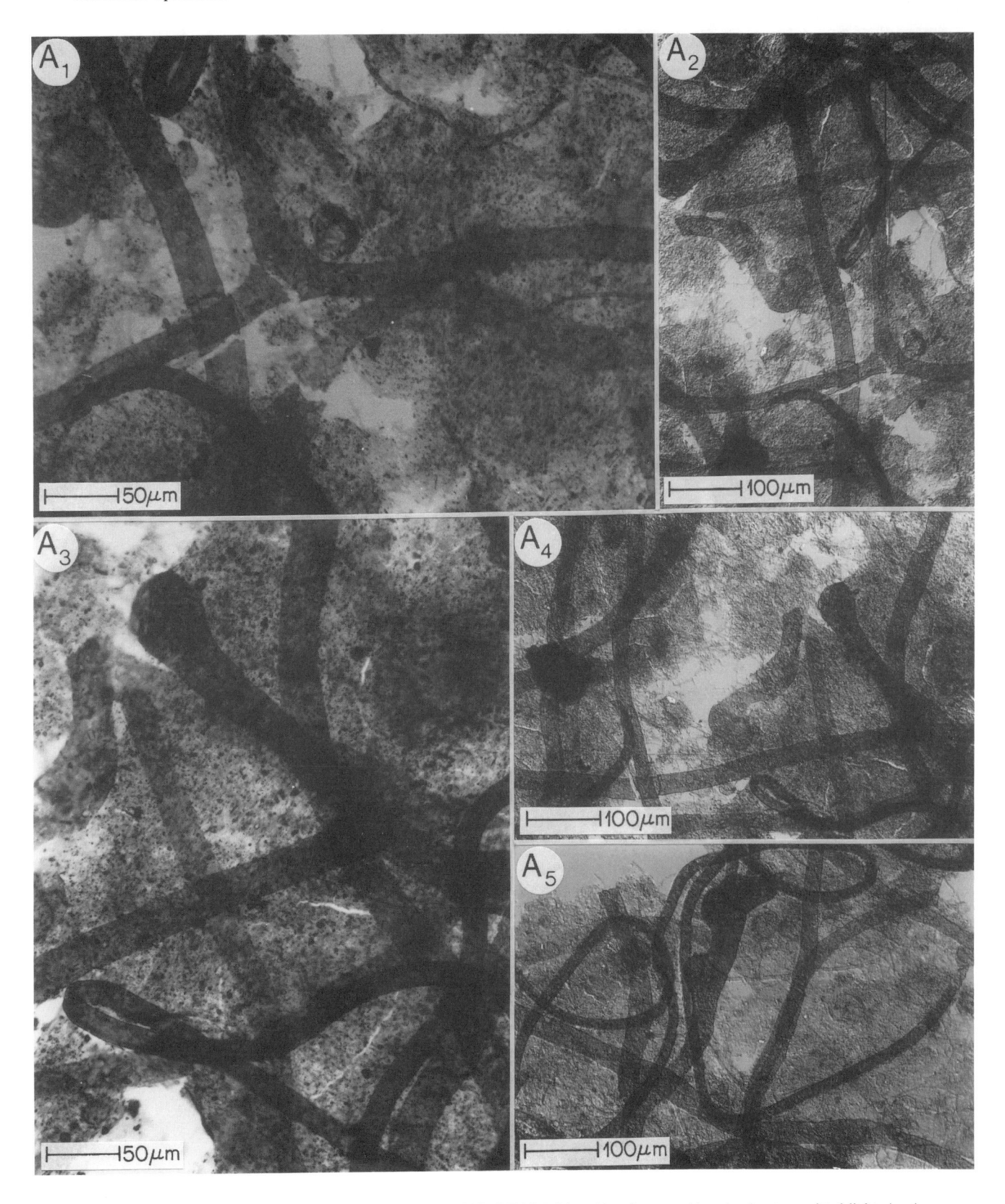

Plate 12. A, Lakhanda Formation, Khabarovsk region, Siberia, USSR (950 Ma). [A, acid resistant residue; A_1, A_3, transmitted light; A_2, A_4, A_5, interference contrast; A_1–A_5 show portions of the same specimen at differing magnifications.] A_1–A_5, *Palaeovaucheria clavata* Hermann (=German) 1981, HOLOTYPE (slide 28/4, England slide S38, read label from right) [EU5].

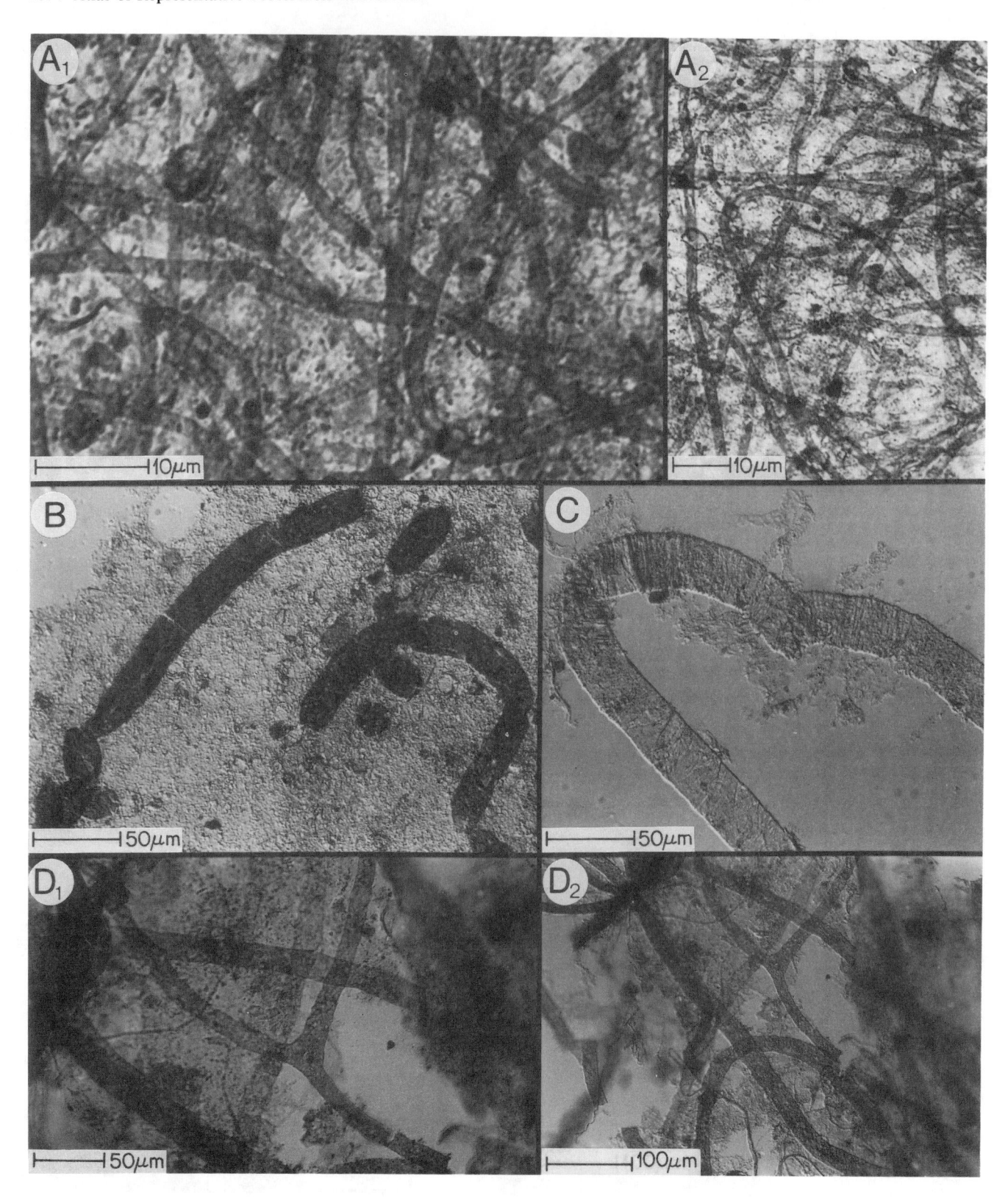

Plate 13. A–D, Lakhanda Formation, Khabarovsk region, Siberia, USSR (950 Ma). [A–D, acid resistant residues; A–C, D_2, interference contrast; D_1, transmitted light; A_1, A_2 and D_1, D_2 each show specimens at two magnifications.] A_1, A_2, *Eomycetopsis* sp. (German, in press; slide 27-5; England slide 046, read label from right) [TU3]; B, *Jacutianema solubila* Timofeev & Hermann (= German) 1979, HOLOTYPE (slide 22/3 = slide 22/3-76/2, England slide H42, read label from right); C, *Rugosoopsis tenuis* Timofeev & Hermann (= German) 1979, HOLOTYPE (slide 1-22 = slide 1-22/1-77/1, England slide Z28/3, read label from right) [TU53]; D_1, D_2, *Palaeovaucheria clavata* Hermann (= German) 1981, PARATYPE (slide 28/2, England slide K43/4, read label from right) [EU5].

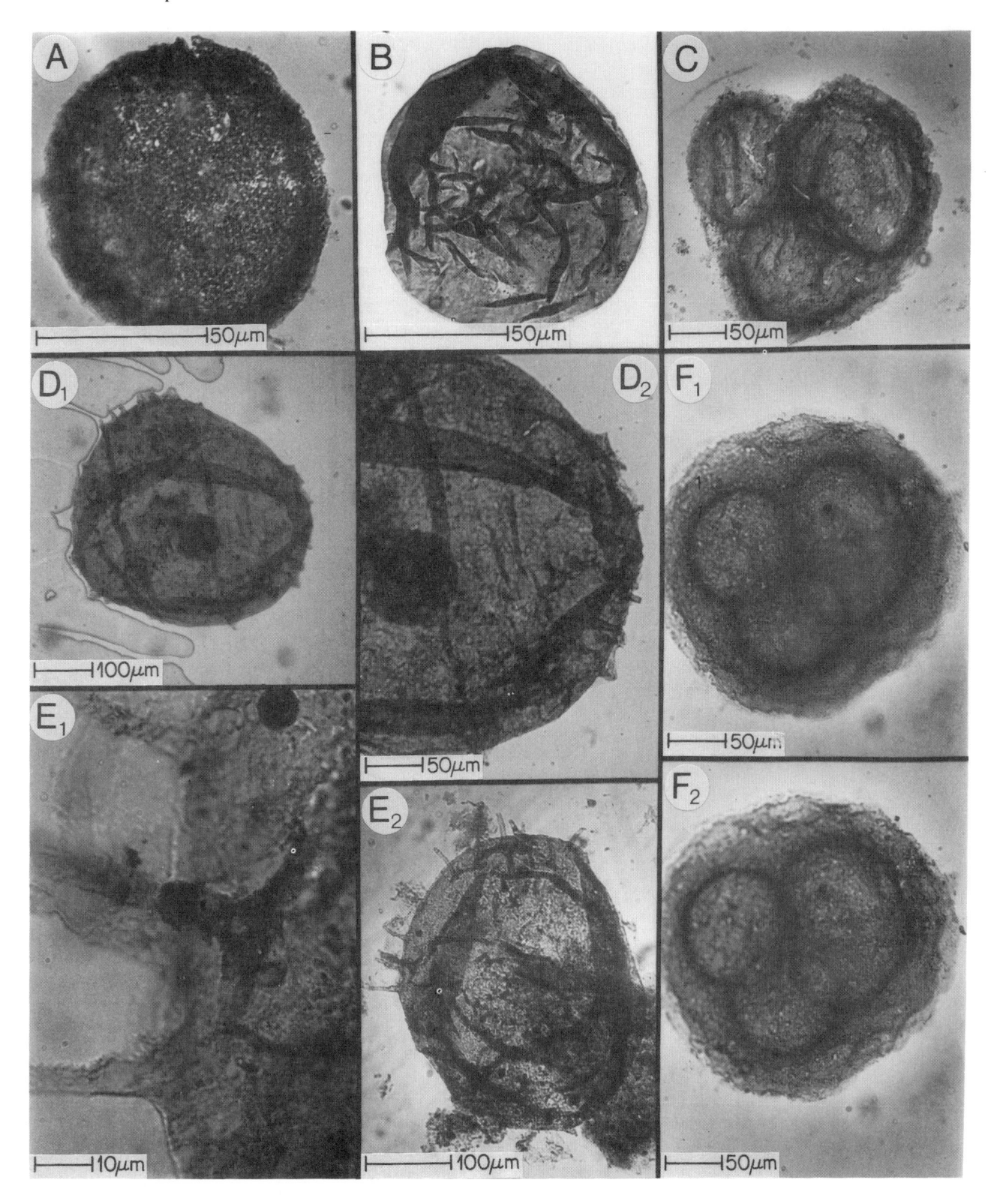

Plate 14. A–F, Lakhanda Formation, Khabarovsk region, Siberia, USSR (950 Ma). [A–F, acid resistant residues; A–C, E, F, interference contrast; D_1, D_2, transmitted light; D_1, D_2, E_1, E_2, and F_1, F_2 each show single specimens at differing focal depths and/or magnifications.] A, *Trachysphaeridium laminaritum* Timofeev in press, HOLOTYPE (slide 26/1, England slide H43/4, read label from right); B, *Kildinella* sp. (= *Leiosphaeridia* sp. = *Chuaria jacutia* Timofeev in press) (slide 26/4, England slide G37/1, read label from right) [MS3]; C, *Tetrasphaera antiqua* Timofeev & Hermann (= German) 1979, HOLOTYPE (slide 6/1-77/2, England slide P40/1, read label from right); D_1, D_2, *Nucellohystrichosphaera megalea* Timofeev 1976 (in Timofeev and German 1976), HOLOTYPE (slide 10/1, England slide R19/4, read label from right) [A10]; E_1, E_2, *Trachystrichosphaera aimika* Hermann (= German) 1976 (in Timofeev and German, 1976), HOLOTYPE (slide 23a/4-11.XI.77, England slide M35/4, read label from right) [A11]; F_1, F_2, *Tetrasphaera antiqua* Timofeev & Hermann (= German) 1979, PARATYPE (slide 13/12, England slide R36/1, read label from right).

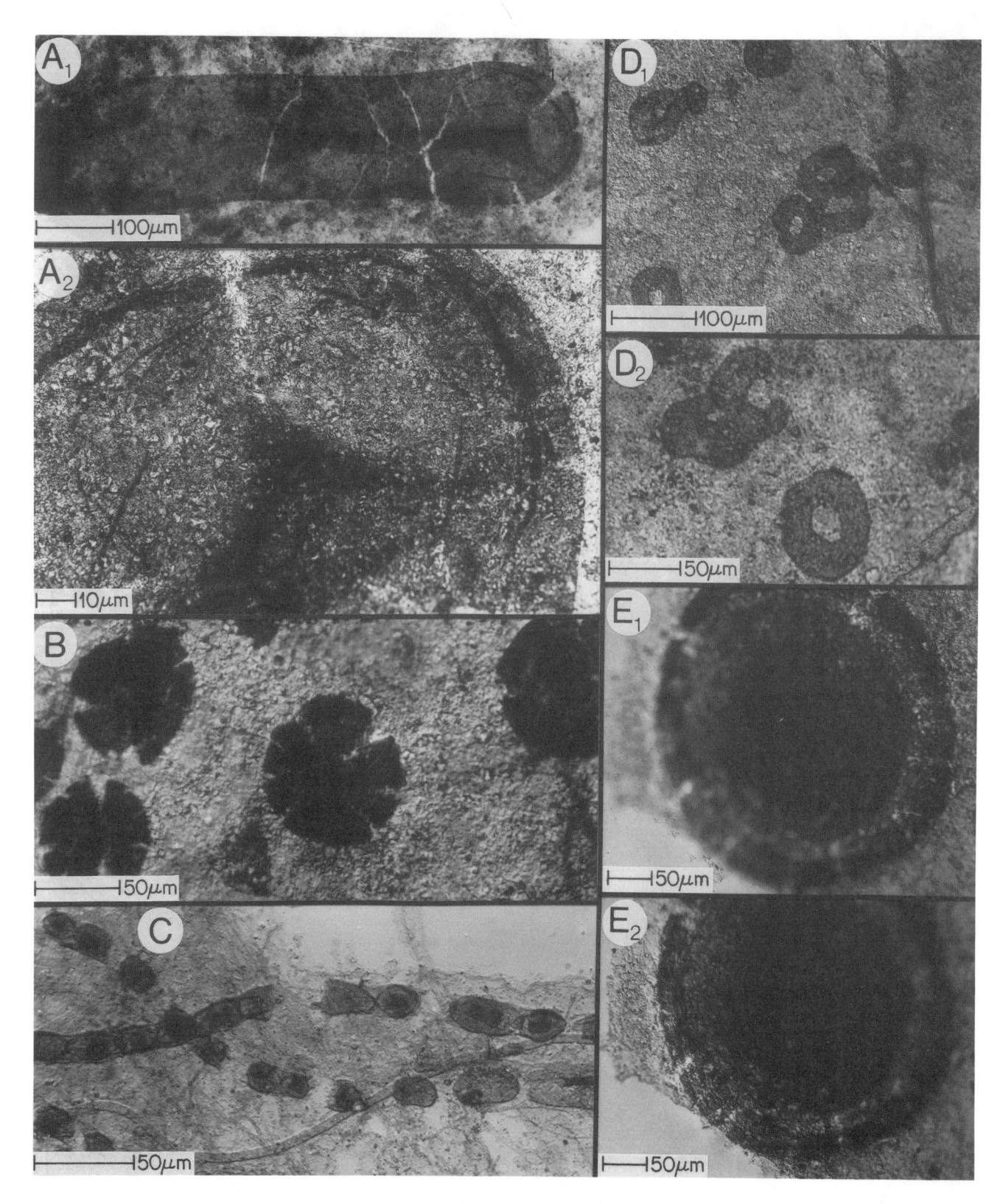

Plate 15. A–E, Lakhanda Formation, Khabarovsk region, Siberia, USSR (950 Ma). [A–E, acid resistant residues; A_1, D_2, transmitted light; A_2, B, C, D_1, E, interference contrast; A_1, A_2, D_1, D_2, and E_1, E_2 each show single specimens at differing focal depths and/or magnifications.] A_1, A_2, *Lakhandinia prolata* Timofeev & Hermann (=German) 1979, HOLOTYPE (slide 22/6-75/2, England slide R34/3, read label from right); B, *Turuchanica ternata* (Timofeev) Jankauskas (=Yankauskas) comb. nov. in press, HOLOTYPE (slide 29/2-8.II.77, England slide S32, read label from right) [S23]; C, *Lomentunella vaginata* Hermann (=German) 1981, HOLOTYPE (slide 27/1, England slide 038/1, read label from right); D_1, D_2, *Annularia annulata* Timofeev & Hermann (=German) 1979, HOLOTYPE (slide 19/76-3, England slide B39/2, read label from right); E_1, E_2, *Pterospermopsimorpha* sp. (German, in press) (slide 22/13-18.IX.75, England slide H41/1, read label from right).

Illustrated Specimens 1077

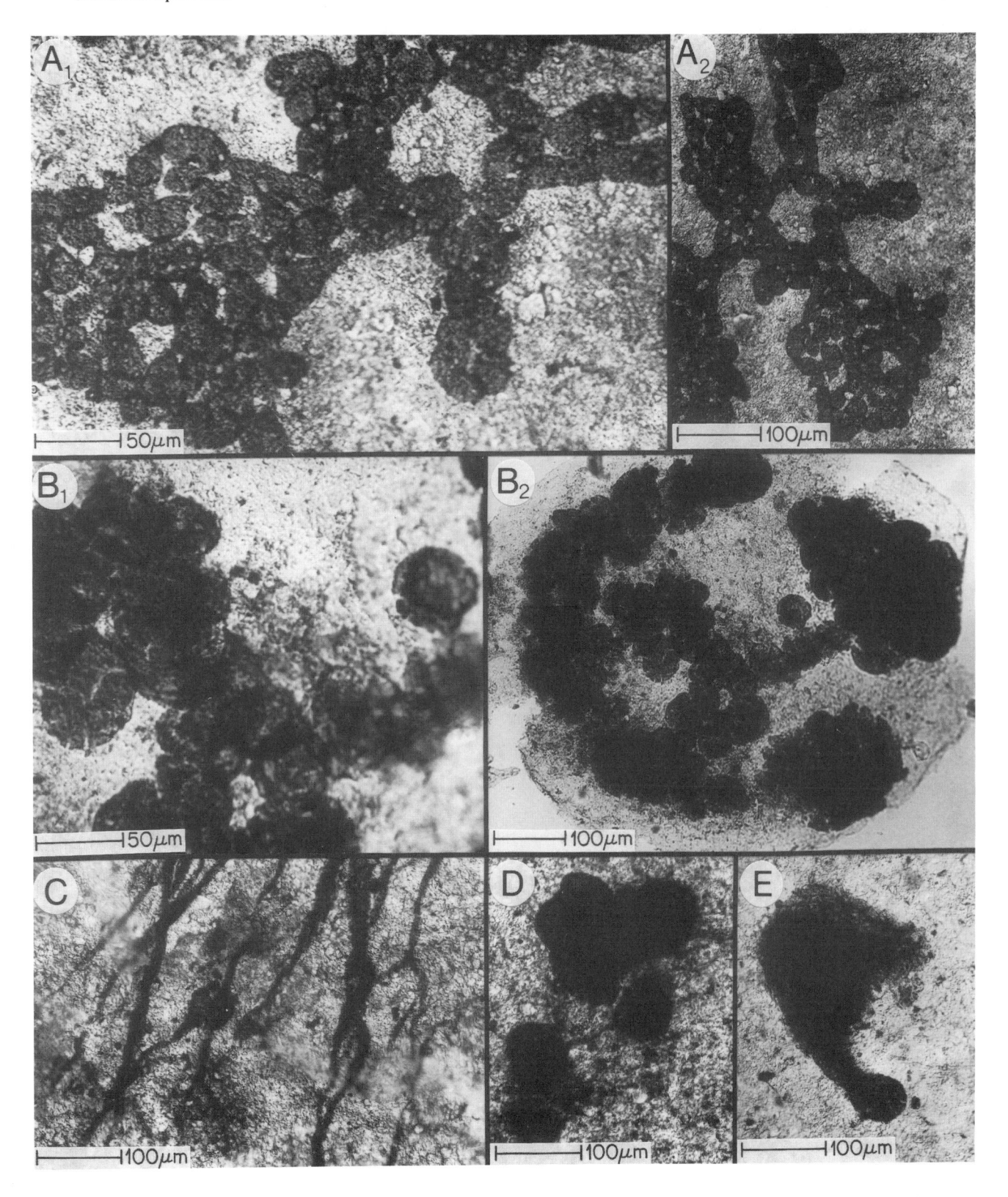

Plate 16. A–E, Lakhanda Formation, Khabarovsk region, Siberia, USSR (950 Ma). [A–E, acid resistant residues, interference contrast; A_1, A_2 and B_1, B_2 each show single specimens at two magnifications.] A_1, A_2, *Eosaccharomyces ramosus* Hermann (=German) in press, NEOTYPE (slide 29/4, England slide S33, read label from right); B_1, B_2, *Aimia delicata* Hermann (=German) 1979, HOLOTYPE (slide 22/2-75/2, England slide R28/3, read label from right); C, *Mucorites rifeicus* Hermann (=German) 1979, PARATYPE (slide 22/1-14.II.76, England slide M47/4, read label from right); D, *Mycosphaeroides aggregatus* Hermann (=German) 1979, HOLOTYPE, (slide 22/24-75/3, England slide T32, read label from right); E, *Mucorites rifeicus* Hermann (=German) 1979, PARATYPE (=*Flabellaforma compacta* Hermann 1979 = "zygospores") (slide 22/1-12.09.75, England slide T34/1, read label from right).

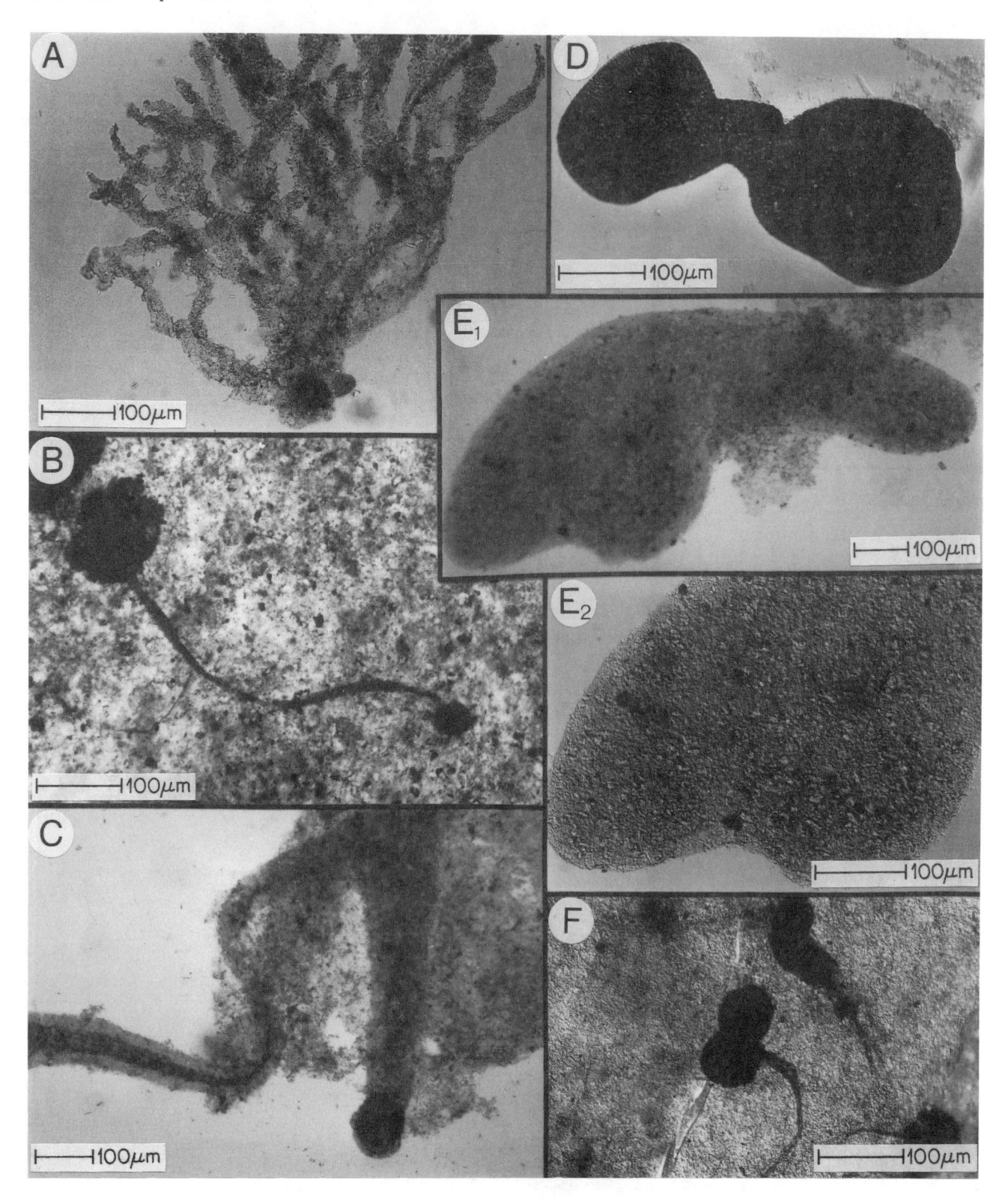

1078 Atlas of Representative Proterozoic Microfossils

Plate 17. A–F, Lakhanda Formation, Khabarovsk region, Siberia, USSR (950 Ma). [A–F, acid resistant residues; A, D, E_2, F, interference contrast; B, C, E_1, transmitted light; E_1, E_2 show the same specimen at two magnifications.] A, *Ulophyton rifeicum* Timofeev & Hermann (=German) 1979, HOLOTYPE (slide 22/2-76/1, England slide H32, read label from right); B, *Mucorites rifeicus* Hermann (=German) 1979, HOLOTYPE, (slide 22/24-75/3, England slide V30, read label from right); C, *Majaphyton antiquaum* Timofeev & Hermann (=German) 1976, HOLOTYPE (slide 22/2-25.II.75, England slide M41/4, read label from right); D, *Majasphaeridium carpogenum* Hermann (=German) 1979, HOLOTYPE (slide 22/19-75/4, England slide M40/1, read label from right); E_1, E_2, *Aimophyton varium* Timofeev & Hermann (=German) 1979, HOLOTYPE (slide 22/22-75/1, England slide S33, read label from right); F, *Mycosphaeroides caudatus* Hermann (=German) 1979, HOLOTYPE (slide 19/1 = slide 19/1-76/4, England slide K31/4, read label from right).

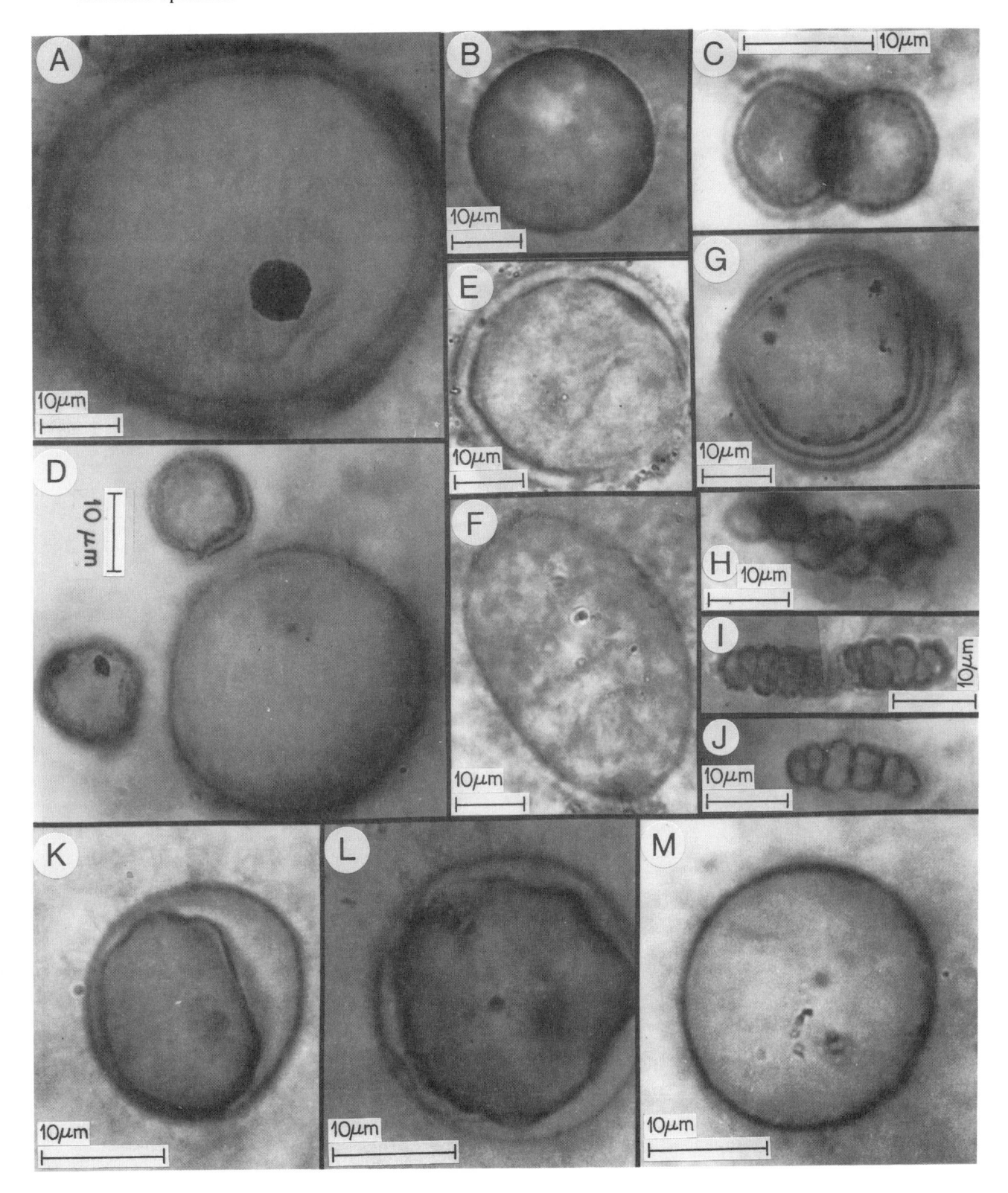

Plate 18. A–M, Galeros Formation, Grand Canyon National Park, northwestern Arizona, USA (930 Ma; locality cf. PPRG 1359). [A–M, petrographic thin sections, transmitted light; I, composite photomicrograph; A–C, E, F, H–K, M, specimens not previously illustrated.] A–G, K–M, unnamed spheroidal to ellipsoidal unicells (D, G, L, Schopf 1975); H–J, unnamed colonial coccoid cells.

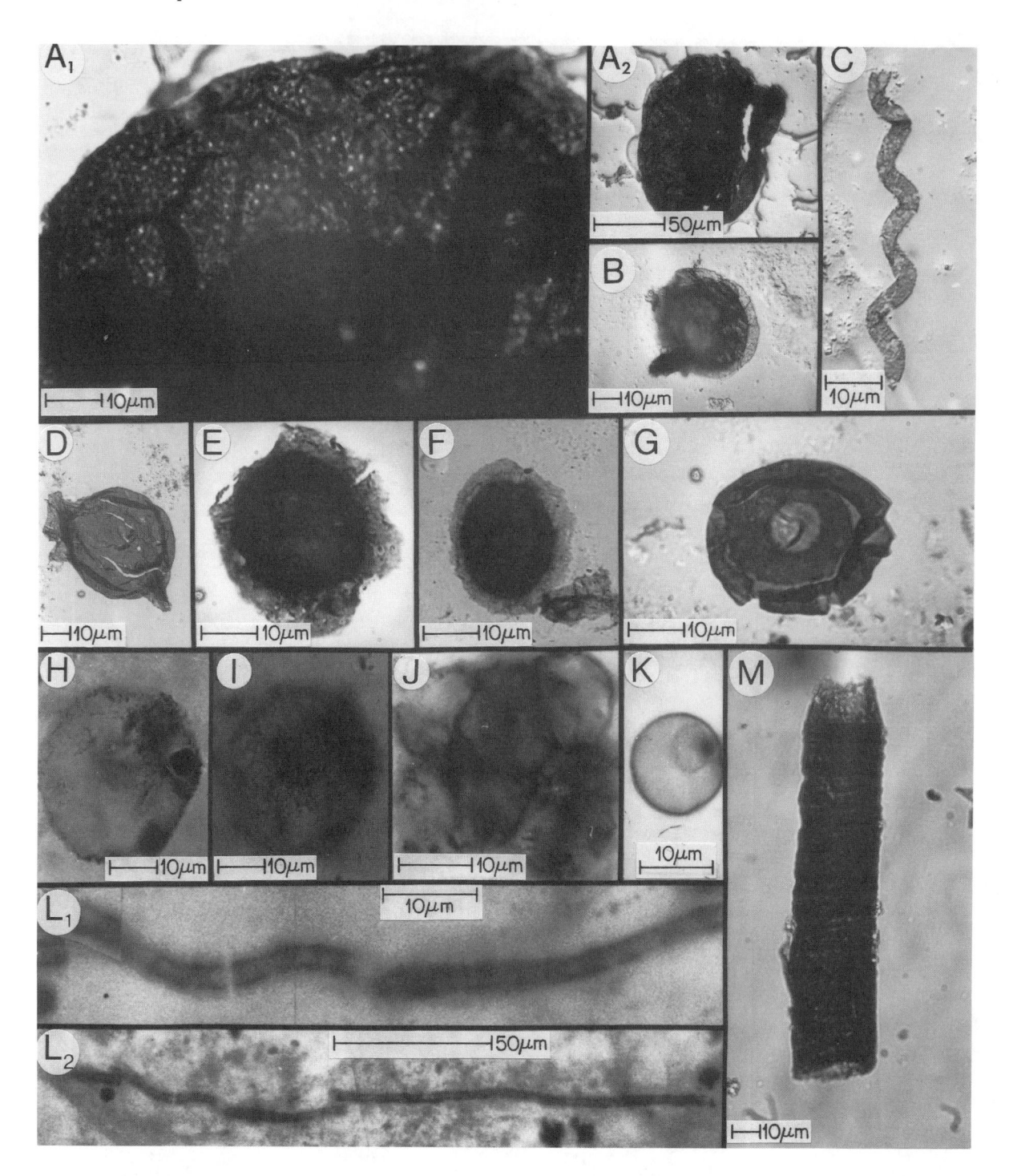

Plate 19. A–G, Sim (=Podinzer) Formation, southern Ural Mountains, Bashkiria, USSR (925 Ma). [A–G, acid resistant residues; A, transmitted light; B–G, interference contrast; A_1, A_2 show the same specimen at two magnifications.] A_1, A_2, *Tasmanites rifejicus* Jankauskas (=Yankauskas) 1978, HOLOTYPE (slide 16-3-25/7-6, England slide J16, slide label left) [MS10]; B, *Pterospermopsimorpha capsulata* Jankauskas (=Yankauskas) 1982, HOLOTYPE (slide 16-25/7-1, England slide V32/1, slide label left) [A2]; C, *Heliconema uralense* Jankauskas (=Yankauskas) 1980, HOLOTYPE (slide 16-25-9/1, England slide V8/4, slide label left) [TUC2]; D, *Pterospermopsis dubius* Jankauskas (=Yankauskas) 1980, HOLOTYPE (=*Leiofusidium dubium* Jankauskas) (slide 16-25/1-5, England slide G10, slide label left) [A19]; E, *Pterospermopsis simicus* Jankauskas (=Yankauskas) 1980, HOLOTYPE (=*Pterospermella simica* Jankauskas) (slide 16-25/7-7, England slide P5, slide label left) [A18]; F, *Pterospermopsis simicus* Jankauskas (=Yankauskas) 1980, PARATYPE (=*Pterospermella simica* Jankauskas) (slide 16-25/13-2, England slide U34/1, slide label left) [A18]; G, *Leiosphaeridia kulgunica* Jankauskas (=Yankauskas) 1980, HOLOTYPE (slide 16-25/8-5, England slide C9, slide label left). H–L, Shorikha Formation, Turukhansk region, Siberia, USSR (900 Ma; locality cf. PPRG 2726). [H–L, petrographic thin sections, transmitted light; L_1, L_2, composite photomicrographs showing the same specimen at two magnifications.] H–K, "undifferentiated spheroids"; L_1, L_2, *Eomycetopsis robusta* Schopf 1968 emend. Knoll & Golubic 1979 (Mendelson and Schopf 1982). M, Shtandin Formation, forelands of the Ural Mountains, Bashkiria, USSR (900 Ma). [M, acid resistant residue, transmitted light.] M, *Siphonophycus costatus* Jankauskas (=Yankauskas) 1980, HOLOTYPE (slide 16-62-3526/20, England slide Q11, slide label left).

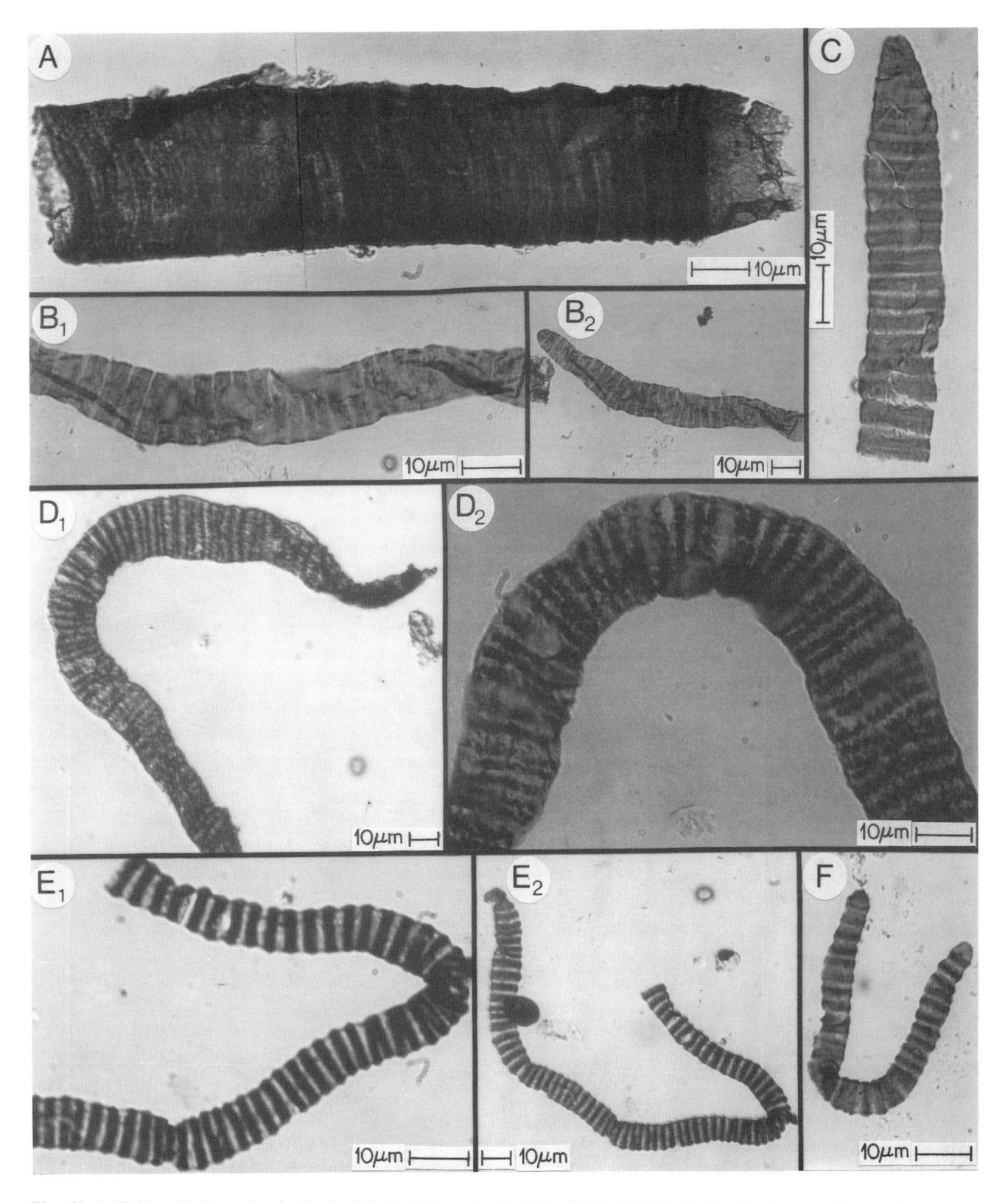

Plate 20. A–F, Shtandin Formation, forelands of the Ural Mountains, Bashkiria, USSR (900 Ma). [A–F, acid resistant residues; A, B$_1$, C, D$_1$, E$_2$, F, interference contrast; B$_2$, D$_2$, E$_1$, transmitted light; A, composite photomicrograph; B$_1$, B$_2$, D$_1$, D$_2$ and E$_1$, E$_2$ each show single specimens at two magnifications.] A, *Siphonophycus costatus* Jankauskas (=Yankauskas) 1980, HOLOTYPE (slide 16-62-3526/20, England slide Q11, slide label left) [TU47]; B$_1$, B$_2$, *Tortunema* sp. (Yankauskas, in press) (slide 16-62-3526/6, England slide Q11, slide label left) [3N44]; C, *Calyptothrix geminata* Jankaukas (=Yankauskas) 1980, HOLOTYPE (slide 16-62-3526/6, England slide V12/3, slide label left) [3N52]; D$_1$, D$_2$, unnamed trichome (Yankauskas, in press) (slide 16-62-3526/18, England slide U23/4, slide label left) [3N61]; E$_1$, E$_2$, *Calyptothrix alternata* Jankauskas (=Yankauskas) 1980, PARATYPE (Yankauskas 1982) (slide 16-62-3526/9, England slide R8/4, slide label left) [3N36]; F, *Calyptothrix alternata* Jankauskas (=Yankauskas) 1980, HOLOTYPE (slide 16-62-3526/18, England slide Q15/3, slide label left) [3N36].

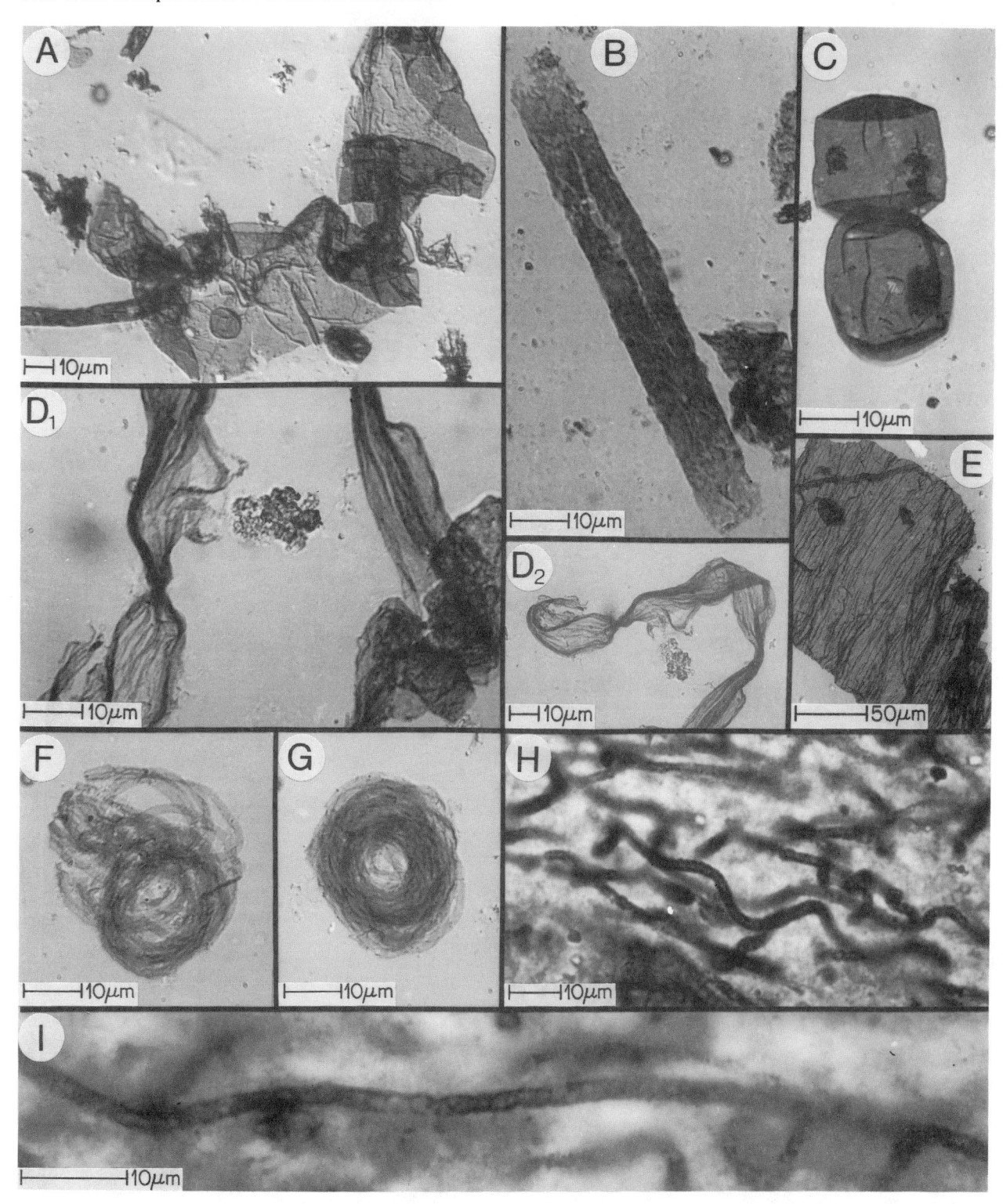

Plate 21. A–G, Shtandin Formation, forelands of the Ural Mountains, Bashkiria, USSR (900 Ma). [A–G, acid resistant residues; A–D$_1$, E–G, interference contrast; D$_2$, transmitted light; D$_1$, D$_2$ show the same specimen at two magnifications.] A, *Pellicularia tenera* Jankauskas (=Yankauskas) 1980, PARATYPE (Yankauskas, in press) (slide 16-62-3526/19, England slide M25/3, slide label left); B, *Eomycetopsis rimata* Jankauskas (=Yankauskas) 1980, HOLOTYPE (slide 16-62-3526/6, England slide D10, slide label left) [TU15]; C, *Arctacellularia* sp. (Yankauskas, in press) (slide 3575-78 = slide 16-4-3625/10, England slide G12/4, slide label left); D$_1$, D$_2$, *Pellicularia tenera* Jankauskas (=Yankauskas) 1980 (in German 1989), HOLOTYPE (slide 16-62-3526/24, England slide S18/2, slide label left); E, *Plicatidium latum* Jankauskas (=Yankauskas) 1980, HOLOTYPE (slide 16-4-3526/10, England slide R22/1, slide label left); F, *Volyniella glomerata* Jankauskas (=Yankauskas) 1980, HOLOTYPE (slide 16-62-3526/11, England slide T25/2, slide label left) [TU2]; G, *Volyniella glomerata* Jankauskas (=Yankauskas) 1980, PARATYPE (Yankauskas, in press) (slide 16-62-3526/12, England slide E8/3, slide label left) [TU2]. H, I, Beck Spring Dolomite, southeastern California, USA (850 Ma; locality cf. PPRG 1271). [H, I, petrographic thin sections, transmitted light.] H, *Beckspringia communis* Licari 1978, PARATYPE [3N14]; I, *Beckspringia communis* Licari 1978, HOLOTYPE [3N14].

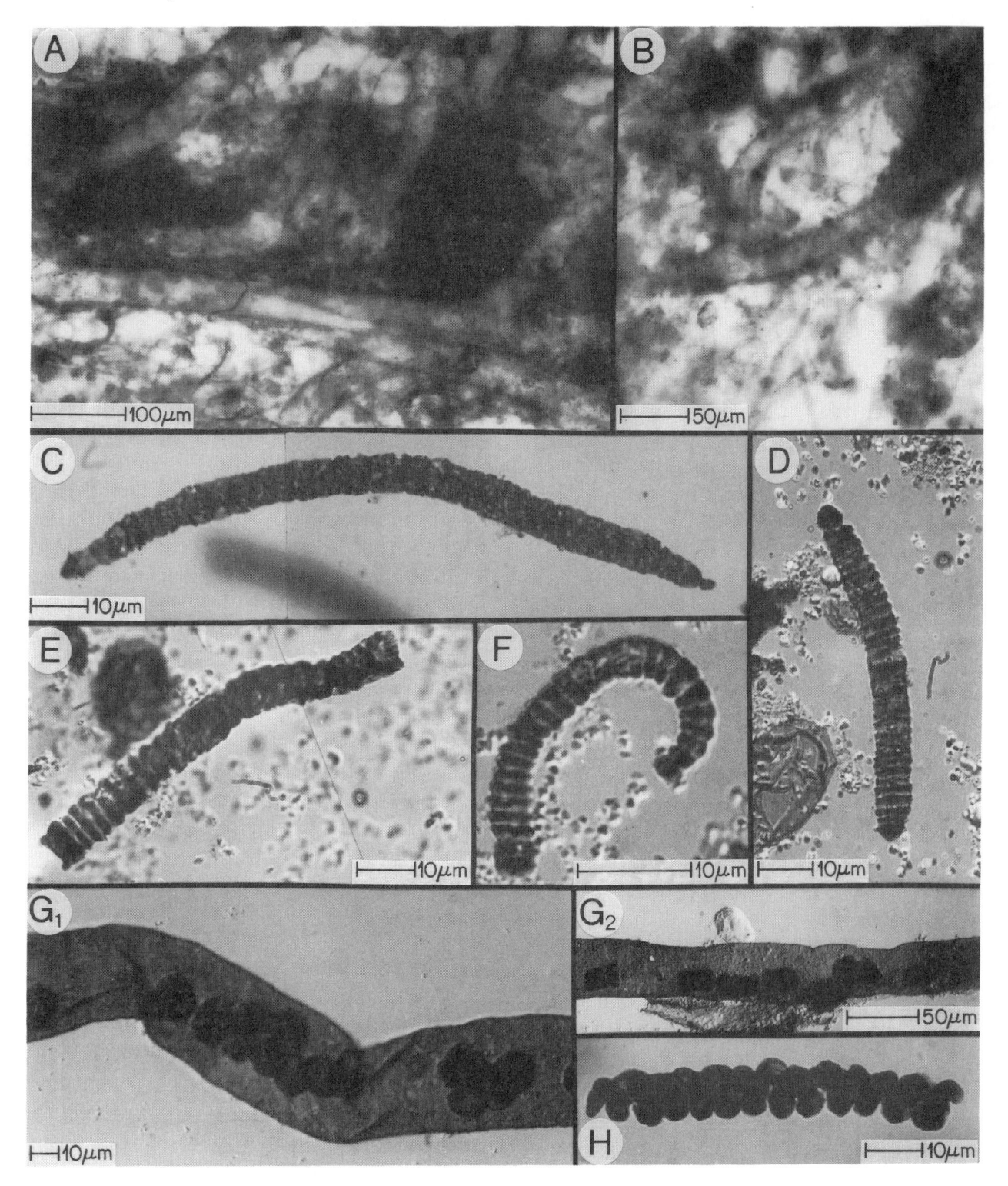

Plate 22. A, B, Beck Spring Dolomite, southeastern California, USA (850 Ma; locality cf. PPRG 1271). [A, B, petrographic thin sections, transmitted light.] A, *Palaeosiphonella cloudii* Licari 1978, HOLOTYPE [TB5]; B, *Palaeosiphonella cloudii* Licari 1978, PARATYPE [TB5]. C–H, Miroedikha Formation, Turukhansk region, Siberia, USSR (850 Ma). [C–H, acid resistant residues; C, H, transmitted light; D–G, interference contrast; C, E, composite photomicrographs; G_1, G_2 show the same specimen at two magnifications.] C, *Primorivularia dissimilara* Hermann (=German) 1986, HOLOTYPE (slide 49a/3, England slide R37/2, read label from right) [3N23]; D, *Calyptothrix perfecta* Weiss (=Vejs) 1984, HOLOTYPE (slide 2678/4-2(5)II = slide 2678/4-22(5), England slide N38, read label from right) [3N37]; E, *Arthrosiphon cornutus* Weiss (=Vejs) 1984, HOLOTYPE (slide 2678/4-2(5)II = slide 2678-428, England slide N67/2, read label from right, i.e., bottom) [3N36]; F, *Arthrosiphon typicus* Weiss (=Vejs) 1984, HOLOTYPE (slide 2678/4-2(5)II = slide 2678-425, England slide J35, read label from right, i.e., bottom) [3N21]; G_1, G_2, *Palaeolyngbya sphaerocephala* Hermann (=German) 1986, HOLOTYPE (slide 964/5a(8), England slide J41/3, read label from right) [3S8]; H, *Heliconema turukhania* Hermann (=German) 1981, HOLOTYPE (slide 49a/3, England slide M41/1, read label from right) [TUC3].

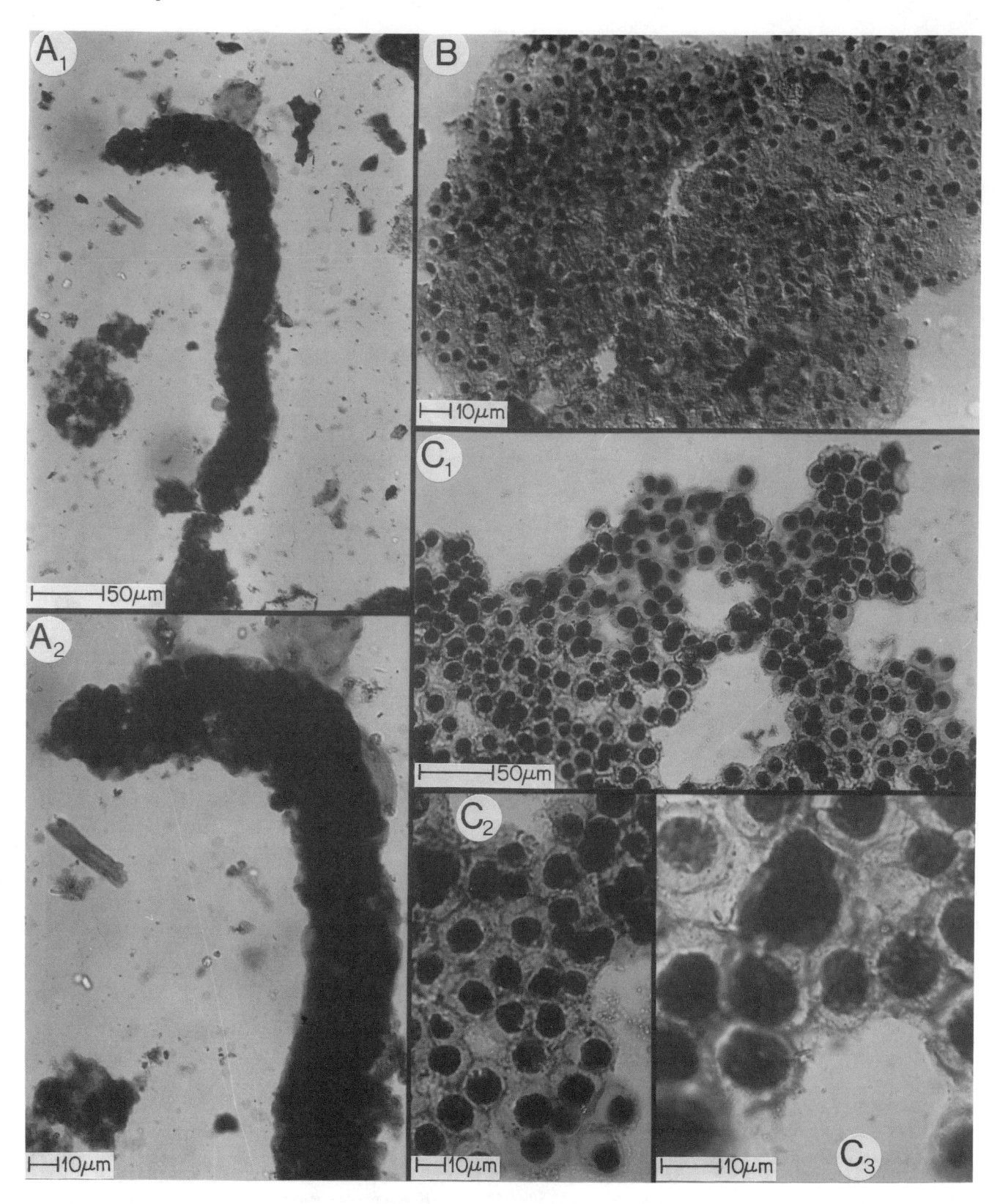

1084 Atlas of Representative Proterozoic Microfossils

Plate 23. A–C, Miroedikha Formation, Turukhansk region, Siberia, USSR (850 Ma). [A–C, acid resistant residues; A, C$_1$, C$_2$, transmitted light; B, C$_3$, interference contrast; A$_1$, A$_2$, and C$_1$–C$_3$ each show single specimens at differing magnifications.] A$_1$, A$_2$, *Polysphaeroides filliformis* Hermann (=German) 1976 (in Timofeev and German, 1976), HOLOTYPE (slide 504/6, England slide V44, read label from right); B, *Ostiana microcystis* Hermann (=German) 1976 (in Timofeev and German, 1976), HOLOTYPE (slide 49b/3, England slide L36, read label from right); C$_1$–C$_3$, *Ostiana microcystis* Hermann (=German) 1976 (in Timofeev and German, 1976), LECTOTYPE (German, in press) (slide 28-6, England slide H40, read label from right).

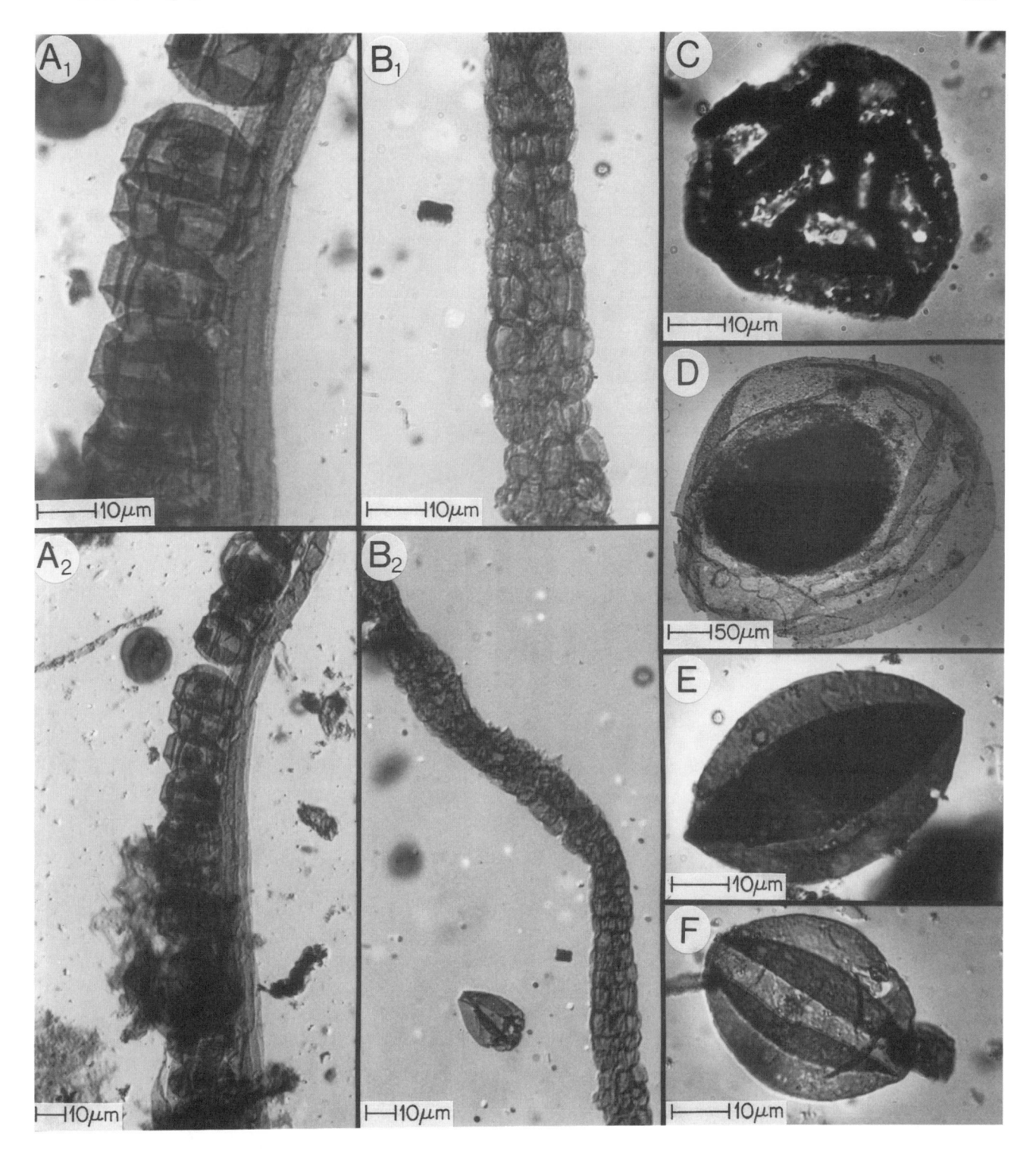

Plate 24. A–F, Miroedikha Formation, Turukhansk region, Siberia, USSR (850 Ma). [A–F, acid resistant residues; A_1 transmitted light; A_2–F, interference contrast; A_1, A_2, B_1, B_2 each show single specimens at two magnifications.] A_1, A_2, *Arctacellularia ellipsoidea* Hermann (=German) 1976 (in Timofeev and German, 1976), HOLOTYPE (slide 49/24-XI.72, England slide T45/4, read label from right) [CFL4]; B_1, B_2, *Polysphaeroides contextus* Hermann (=German) 1976 (in Timofeev and German 1976), HOLOTYPE (slide 49a/3, England slide X45, read label from right); C, *Dictyosphaeridium tungusum* Timofeev 1969, HOLOTYPE (slide 2/3, England slide R42, read label from right) [S16]; D, *Pterospermopsimorpha insolita* Timofeev 1969, LECTOTYPE (German in press) (slide 16/42, England slide J41, read label from right) [MG4]; E, *Macroptycha uniplicata* Timofeev 1976 (in Timofeev and German, 1976), HOLOTYPE (slide 49/4, England slide S41/2, read label from right); F, *Macroptycha biplicata* Timofeev 1976 (in Timofeev and German, 1976), HOLOTYPE (slide 49/4, England slide 041/2, read label from right).

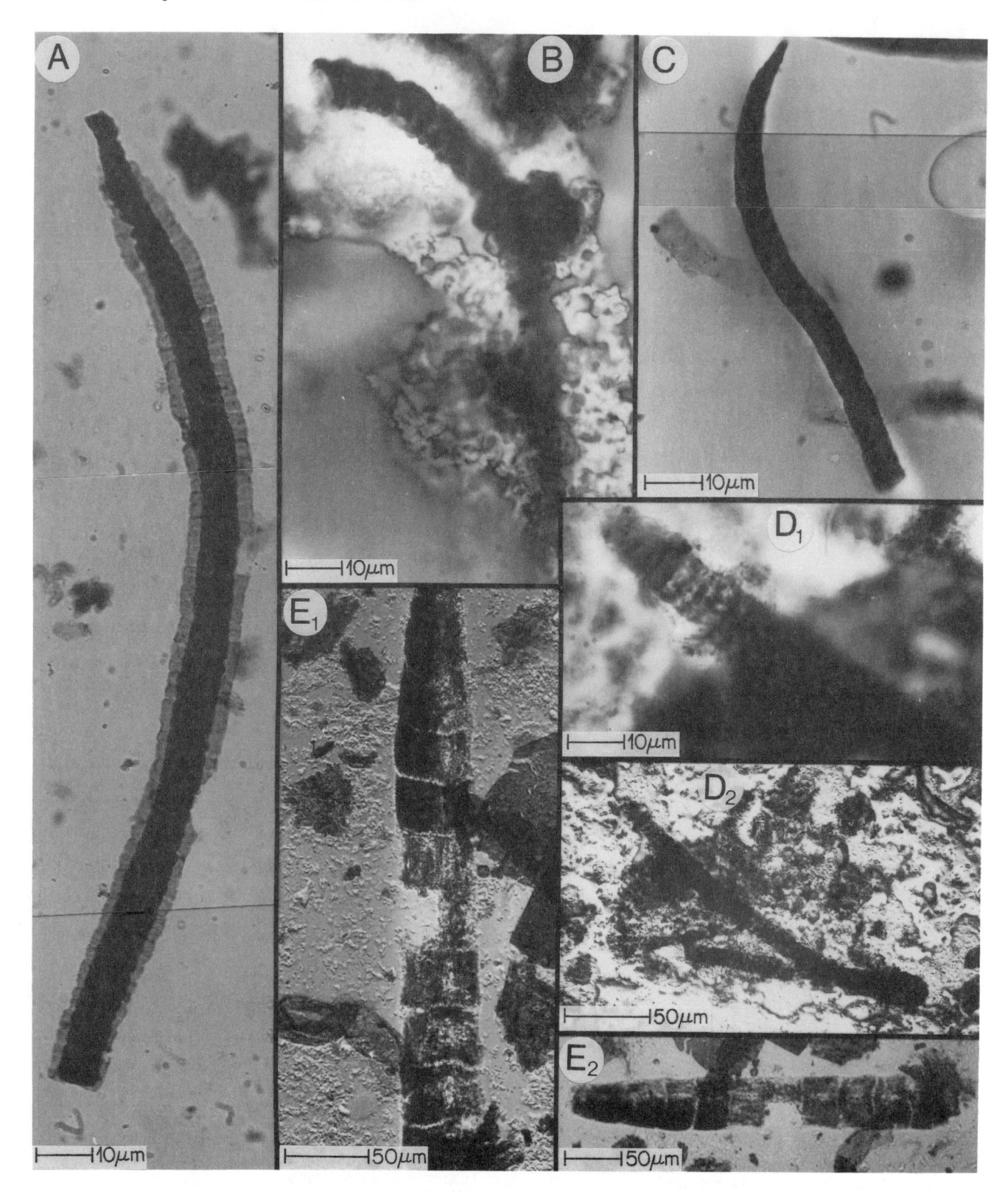

Plate 25. A–E, Miroedikha Formation, Turukhansk region, Siberia, USSR (850). [A–E, acid resistant residues; A–C, D_1, E_2, transmitted light; D_2, E_1, interference contrast; A, C, composite photomicrographs; D_1, D_2 and E_1, E_2 each show single specimens at two magnifications.] A, *Sibiriafilum tunicum* Hermann (=German) 1986, PARATYPE (slide 49a/5, England slide L31, read label from right); B, *Filiconstrictosus eniseicum* Weiss (=Vejs) 1984, HOLOTYPE (slide 2678/4-2(5)VI = slide 2678-426VI, England slide N23, read label from right, i.e., bottom); C, *Caudiculophycus micronulatus* Hermann (=German) 1986, HOLOTYPE (slide 49a/5, England slide K35, read label from right); D_1, D_2, *Cephalophytarion turukhanicum* Weiss (=Vejs) 1984, HOLOTYPE (slide 2678/4-2(5)VI = slide 2678-421, England slide N17/3, read label from right, i.e., bottom); E_1, E_2, *Cephalonyx sibiricus* Weiss (=Vejs) 1984, HOLOTYPE (slide 2678/4-2(5)VI = slide 2678-424, England slide J13/2, read label from right, i.e., bottom).

Illustrated Specimens 1087

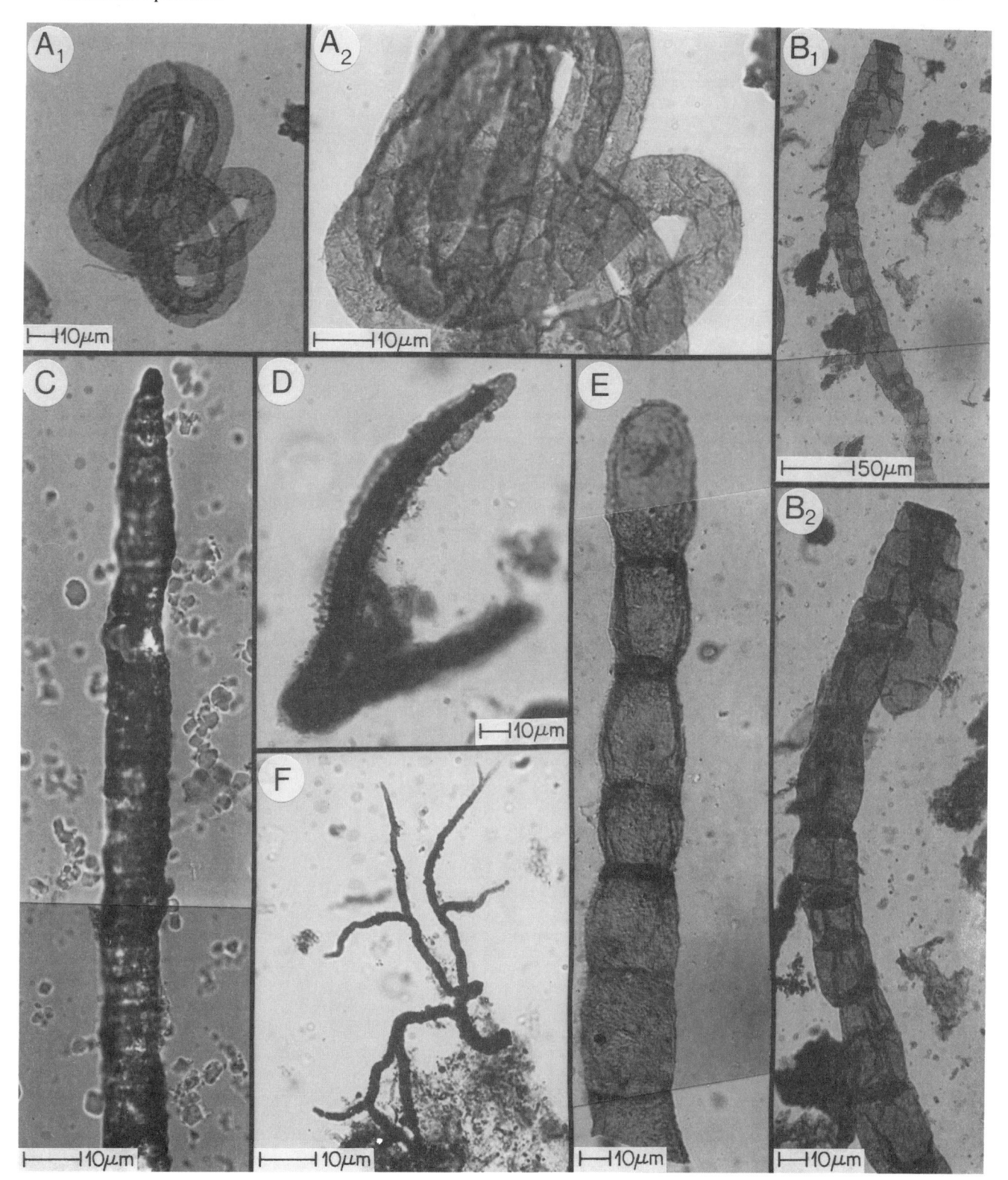

Plate 26. A–F, Miroedikha Formation, Turukhansk region, Siberia, USSR (850 Ma). [A–F, acid resistant residues; A_1, B–D, F, transmitted light; A_2, E, interference contrast; B_1, C, E, composite photomicrographs; A_1, A_2 B_1, B_2 show single specimens at two magnifications.] A_1, A_2, *Tortunema eniseica* Hermann (=German) 1976 (in Timofeev and German, 1976), HOLOTYPE (slide 49/47, England slide N47/1, read label from right) [TU31]; B_1, B_2, *Arctacellularia doliiformis* Hermann (=German) 1976 (in Timofeev and German 1976), HOLOTYPE (slide 49/45-XI.72, England slide P44, read label from right); C, *Cephalophytarion turukhanicum* Weiss (=Vejs) 1984, PARATYPE (slide 2678/4-2(5)V = slide 2678-415, England slide T44, read label from right); D, *Sibiriafilum tunicum* Hermann (=German) 1986, HOLOTYPE (slide 49a/5, England slide L31, read label from right); E, *Trachythrichoides ovalis* Hermann (=German) 1976 (in Timofeev and German 1976), HOLOTYPE (slide 49/5, England slide V33/3, read label from right) [EU8]; F, *Palaeocalothrix divaricatus* Hermann (=German) 1981, HOLOTYPE (slide 49/35, England slide U41/2, read label from right).

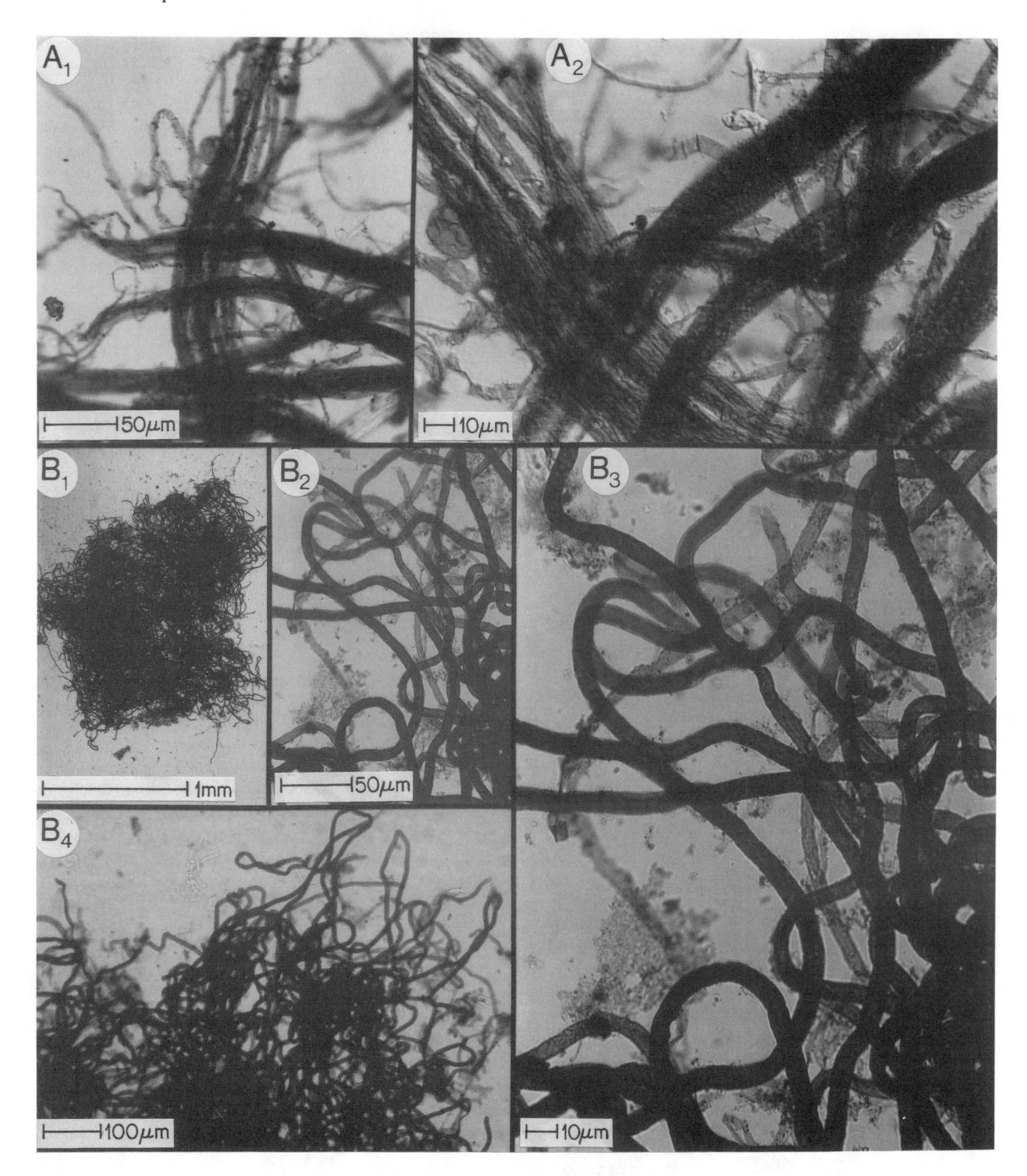

Plate 27. A, B, Miroedikha Formation, Turukhansk region, Siberia, USSR (850 Ma). [A, B, acid resistant residues; A_1, B_1–B_4, transmitted light; A_2, interference contrast; A_1, A_2 and B_1, B_4 show single specimens at differing magnifications.] A_1, A_2, *Polythrichoides lineatus* Hermann (=German) 1974, PARATYPE (German, in press) (slide 49/29, England slide L34, read label from right); B_1–B_4. *Leiothrichoides typicus* (Hermann 1974) emend. Hermann (=German) 1979, PARATYPE (slide 49/20T, England slide K54, read label from right) [TU18].

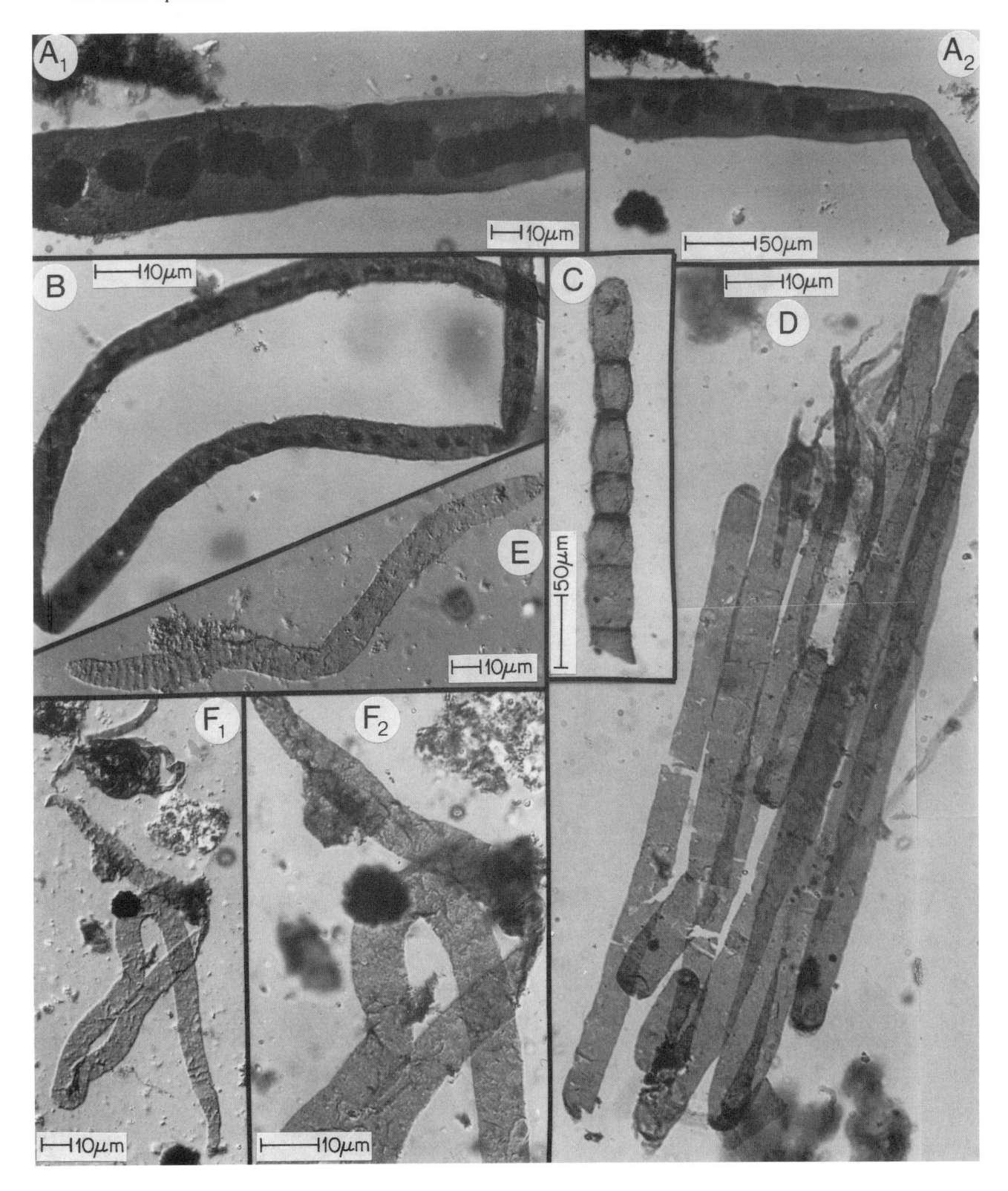

Plate 28. A–F, Miroedikha Formation, Turukhansk region, Siberia, USSR (850 Ma). [A–F, acid resistant residues; A–C, E, F, interference contrast; D, transmitted light; D, composite photomicrograph; A_1, A_2, F_1, F_2 show single specimens at two magnifications.] A_1, A_2, *Palaeolyngbya sphaerocephala* Hermann (=German) & Pylina 1986, PARATYPE (slide 964/5a(4), England slide P39, read label from right) [3S8]; B, *Leiothrichoides maculatus* Hermann (=German) 1986, HOLOTYPE (slide 49/6 = slide 49/6-1167, England slide G46/1, read label from right); C, *Trachythrichoides ovalis* Hermann (=German) 1976 (in Timofeev and German 1976), HOLOTYPE (slide 49/5, England slide V33/3, read label from right) [EU8]; D, *Oscillatoriopsis bacilaris* Hermann (=German) 1981, HOLOTYPE (slide 49/35, England slide L39/1, read label from right); E, *Tortunema sibirica* Hermann (=German) 1976 (in Timofeev and German 1976), HOLOTYPE (slide 504b/3, England slide N42, read label from right); F_1, F_2, *Tortunema sibirica* Hermann (=German) 1976 (in Timofeev and German 1976), PARATYPE (slide 49/47b, England slide P41, read label from right).

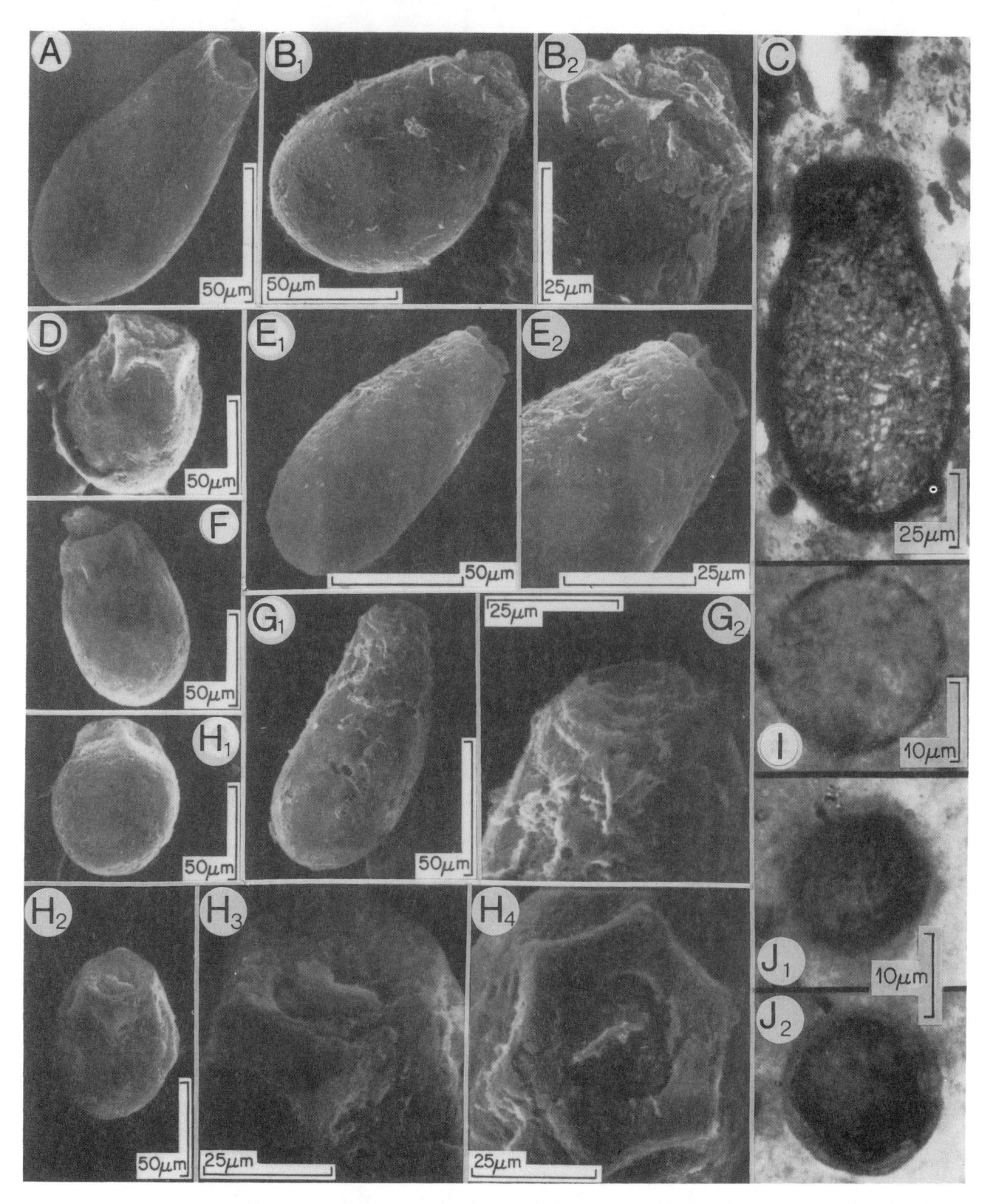

1090 Atlas of Representative Proterozoic Microfossils

Plate 29. A–J, Kwagunt Formation, Grand Canyon National Park, northwestern Arizona, USA (850 Ma; locality cf. PPRG 1357). [A, B, D–H, acid resistant residues, scanning electron micrographs; C, I, J, petrographic thin sections, transmitted light photomicrographs; B_1, B_2, E_1, E_2, G_1, G_2, H_1–H_4, J_1, J_2 each show single specimens at differing magnifications and/or orientations.] A, *Melanocyrillium fimbriatum* Bloeser 1985, PARATYPE [A23]; B_1, B_2, *Melanocyrillium* sp. (Bloeser et al. 1977); C, *Melanocyrillium hexodiadema* Bloeser 1985, PARATYPE [A21]; D, *Melanocyrillium hexodiadema* Bloeser 1985, PARATYPE [A21]; E_1, E_2, *Melanocyrillium fimbriatum* Bloeser 1985, PARATYPE [A23]; F, *Melanocyrillium hexodiadema* Bloeser 1985, PARATYPE [A21]; G_1, G_2, *Melanocyrillium fimbriatum* Bloeser 1985, PARATYPE [A23]; H_1–H_4, *Melanocyrillium hexodiadema* Bloeser 1985, HOLOTYPE [A21]; I, J, unnamed spheroidal unicells (Schopf et al. 1973).

Illustrated Specimens 1091

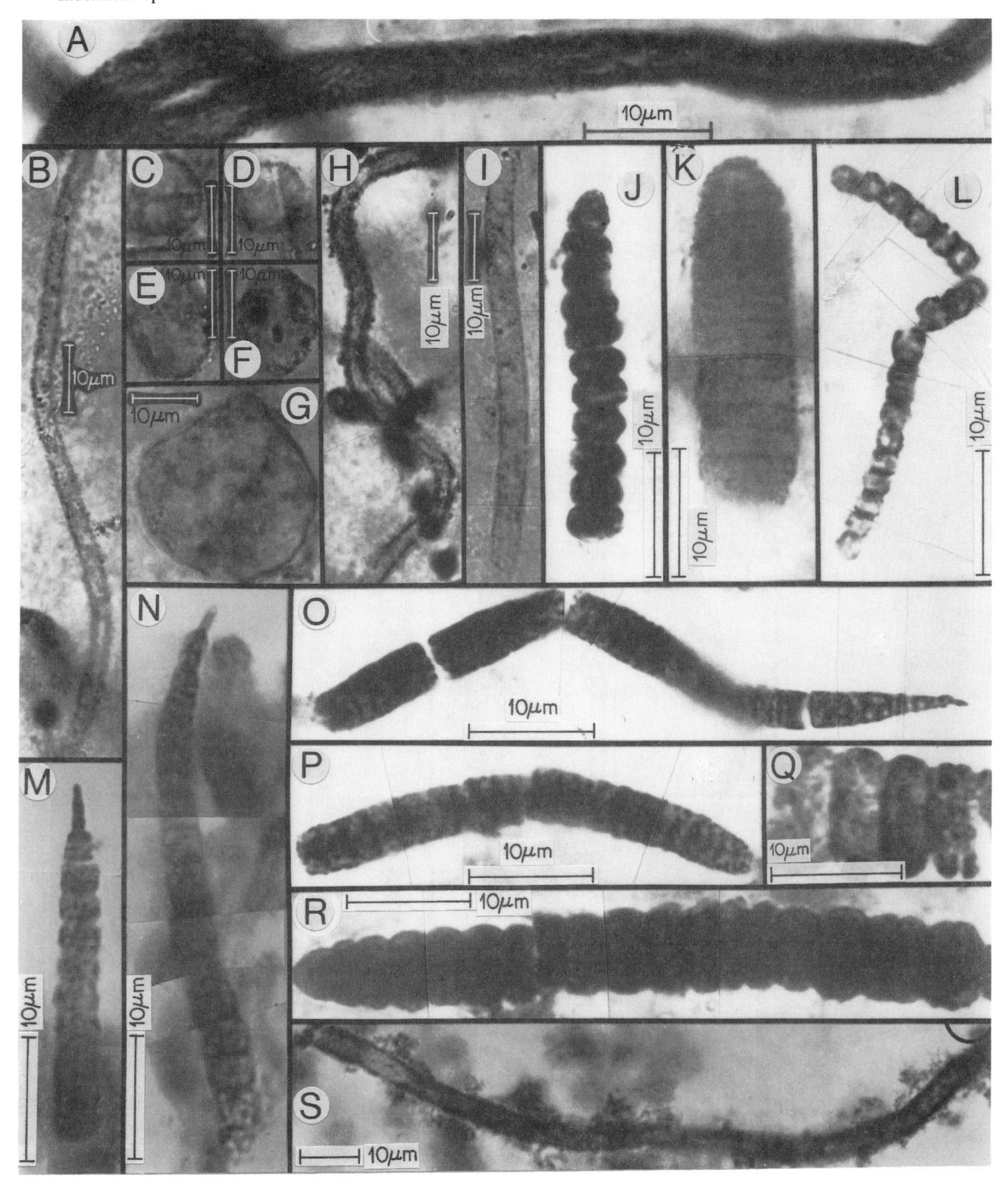

Plate 30. A–I, Kwagunt Formation, Grand Canyon National Park, northwestern Arizona, USA (850 Ma; locality cf. PPRG 1357). [A–I, petrographic thin sections, transmitted light; H, I, composite photomicrographs; A, C–F, specimens not previously illustrated.] A, B, H, I, *Eomycetopis* sp. (B, H, I, Schopf et al. 1973); C–G, unnamed spheroidal unicells (G, Schopf et al. 1973). J–S, Bitter Springs Formation, Alice Springs region, Northern Territory, Australia (850 Ma; localities cf. PPRG 1245 and 1253). [J–R, petrographic thin sections; S, acid resistant residue; J–S, transmitted light; J–P, R, composite photomicrographs.] J, *Filiconstrictosus diminutus* Schopf & Blacic 1971, HOLOTYPE [3N29]; K, *Paleolyngbya minor* Schopf & Blacic 1971, HOLOTYPE [3N55]; L, *Veteronostocale amoenum* Schopf & Blacic 1971, HOLOTYPE [3N22]; M, *Caudiculophycus rivularioides* Schopf 1968, PARATYPE [3N32]; N, *Caudiculophycus rivularioides* Schopf 1968, HOLOTYPE [3N32]; O, *Caudiculophycus acuminatus* Schopf & Blacic 1971, HOLOTYPE [3N32]; P, *Partitiofilum gongyloides* Schopf & Blacic 1971, HOLOTYPE [3N36]; Q, *Palaeolyngbya barghoorniana* Schopf 1968, PARATYPE [3N56]; R, *Filiconstrictosus majusculus* Schopf & Blacic 1971, HOLOTYPE [3N44]; S, *Eomycetopsis robusta* Schopf 1968, PARATYPE [TU16].

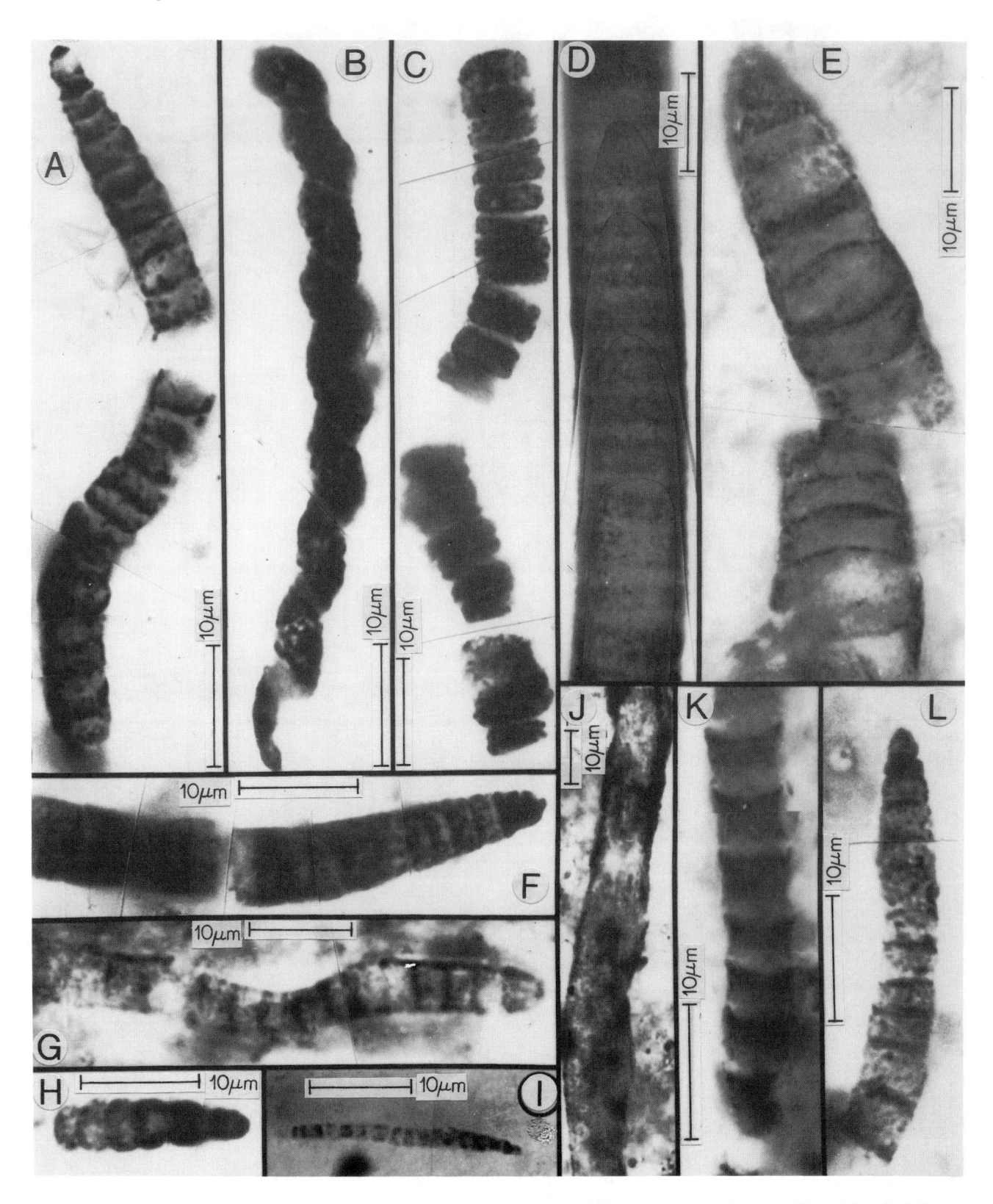

Plate 31. A–L, Bitter Springs Formation, Alice Springs region, Northern Territory, Australia (850 Ma; localities cf. PPRG 1245 and 1253). [A–L, petrographic thin sections, transmitted light; A–I, K, L, composite photomicrographs.] A, *Cephalophytarion laticellulosum* Schopf & Blacic 1971, HOLOTYPE [3N43]; B, *Heliconema funiculum* Schopf & Blacic 1971, HOLOTYPE [TUC4]; C, *Oscillatoriopsis breviconvexa* Schopf & Blacic 1971, HOLOTYPE [3N44]; D, unnamed oscillatoriacean trichome (Schopf 1974); E, *Obconicophycus amadeus* Schopf & Blacic 1971, HOLOTYPE [3N60]; F, *Cephalophytarion laticellulosum* Schopf & Blacic, PARATYPE [3N43]; G, *Oscillatoriopsis obtusa* Schopf 1968, HOLOTYPE [3N33]; H, *Filiconstrictosus diminutus* Schopf & Blacic 1971, PARATYPE [3N29]; I, *Cephalophytarion minutum* Schopf 1968, HOLOTYPE [3N11]; J, *Siphonophycus kestron* Schopf 1968, PARATYPE [TU42]; K, *Halythrix nodosa* Schopf 1968, HOLOTYPE [3N41]; L, *Cephalophytarion laticellulosum* Schopf & Blacic 1971, PARATYPE [3N43].

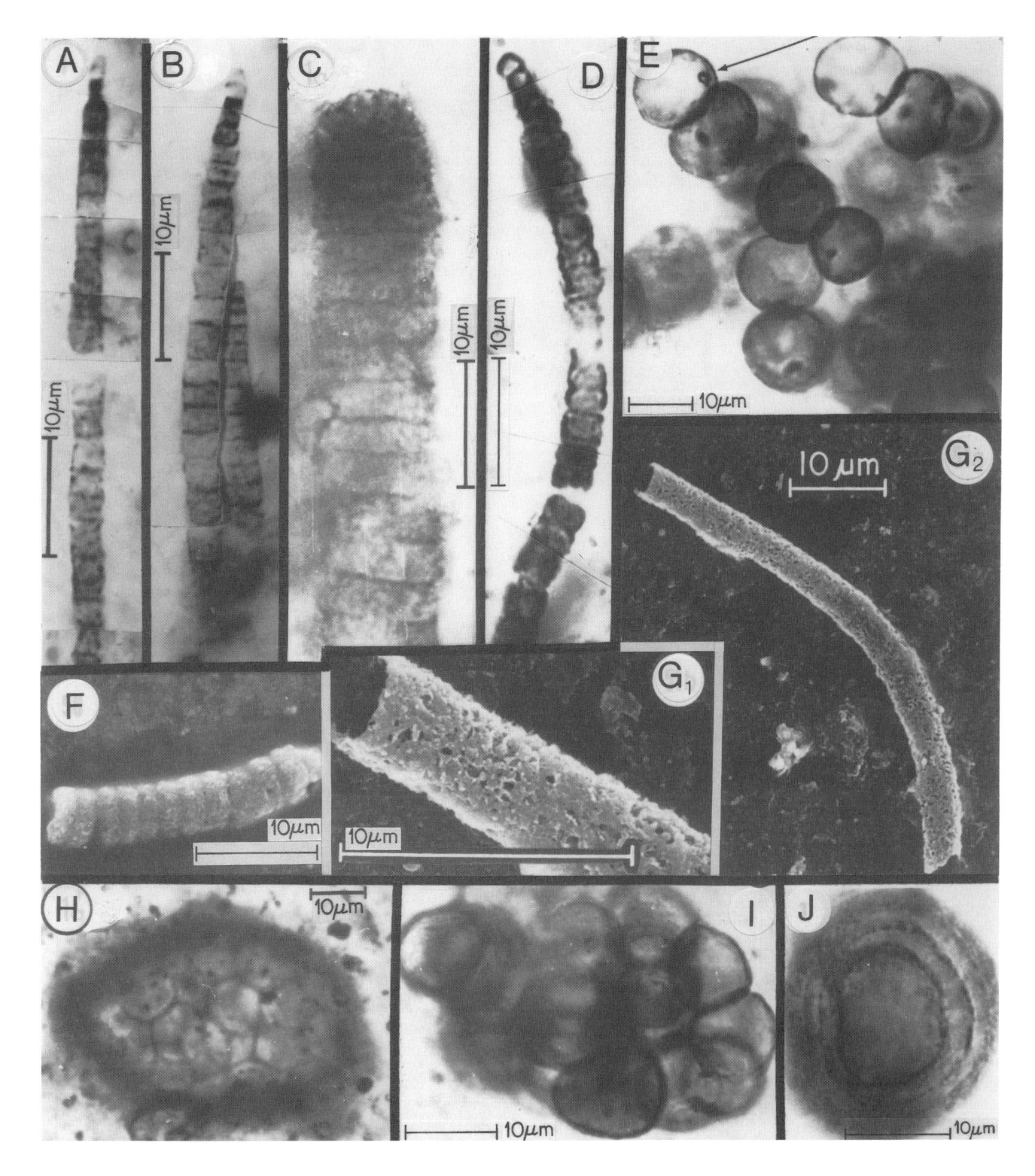

Plate 32. A–J, Bitter Springs Formation, Alice Springs region, Northern Territory, Australia (850 Ma; localities cf. PPRG 1245 and 1253). [A–E, H–J, petrographic thin sections, transmitted light photomicrographs; F, G, acid resistant residues, scanning electron micrographs; A, B, D, composite photomicrographs; G_1, G_2 show a single specimen at two magnifications.] A, *Cephalophytarion grande* Schopf 1968, PARATYPE [3N25]; B, *Cephalophytarion grande* Schopf 1968, HOLOTYPE [3N25]; C, *Palaeolyngbya barghoorniana* Schopf 1968, HOLOTYPE [3N56]; D, *Cephalophytarion variabile* Schopf & Blacic 1971, HOLOTYPE [3N25]; E, *Glenobotrydion aenigmatis* Schopf 1968, PARATYPE, showing characteristic circular "spot-like" body (arrow); F, unnamed oscillatoriacean trichome (Schopf 1972); G_1, G_2, *Eomycetopsis robusta* Schopf 1968 (Schopf 1972), TOPOTYPE; H, *Myxococcoides minor* Schopf 1968, HOLOTYPE [CIR9]; I, *Myxococcoides minor* Schopf 1968, PARATYPE [CIR9]; J, *Gloeodiniopsis lamellosa* Schopf 1968 (Schopf and Blacic 1971), TOPOTYPE [G3].

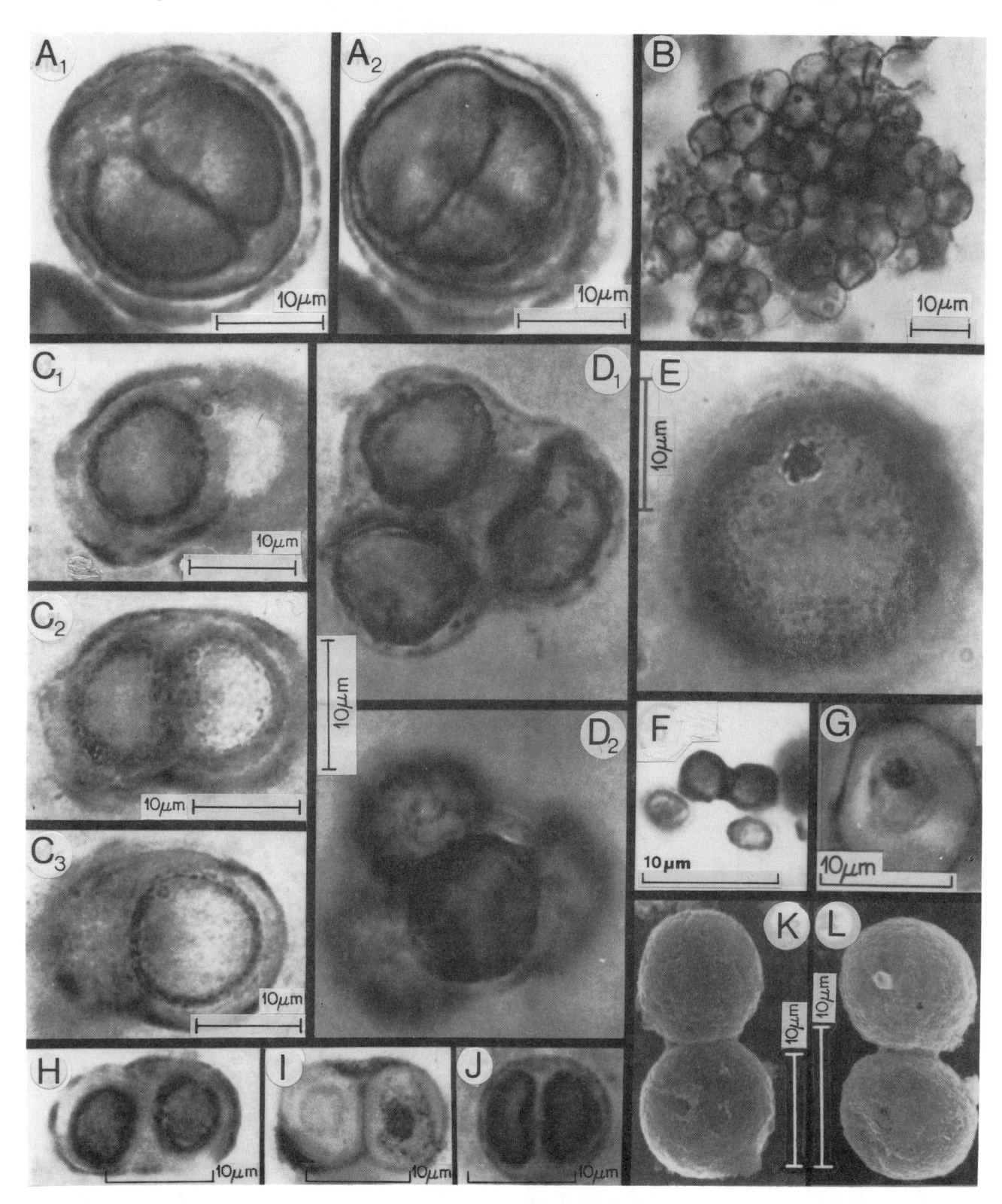

Plate 33. A–L, Bitter Springs Formation, Alice Springs region, Northern Territory, Australia (850 Ma; localities cf. PPRG 1245 and 1253). [A, C–J, petrographic thin sections; B, K, L, acid resistant residues; A–J transmitted light photomicrographs; K, L, scanning electron micrographs; A_1, A_2, C_1–C_3, D_1, D_2 show single specimens at differing focal depths.] A_1, A_2, *Bigeminococcus lamellosus* Schopf & Blacic 1971, HOLOTYPE [CHR6]; B, *Glenobotrydion aenigmatis* Schopf 1968, PARATYPE; C_1–C_3, *Eozygion grande* Schopf & Blacic 1971, HOLOTYPE [CHR6]; D_1, D_2, *Eotetrahedrion princeps* Schopf & Blacic 1971, HOLOTYPE [CHR7]; E, *Globophycus rugosum* Schopf 1968, HOLOTYPE; F, *Sphaerophycus parvum* Schopf 1968, HOLOTYPE; G, *Caryosphaeroides pristina* Schopf 1968, HOLOTYPE [S15]; H, *Eozygion minutum* Schopf & Blacic 1971, HOLOTYPE [CHR5]; I, *Caryosphaeroides tetras* Schopf 1968 (Schopf and Blacic 1971), TOPOTYPE [CHR5]; J, *Eozygion minutum* Schopf & Blacic 1971, PARATYPE [CHR5]; K, L, unnamed paired coccoid cyanobacteria (Schopf 1972).

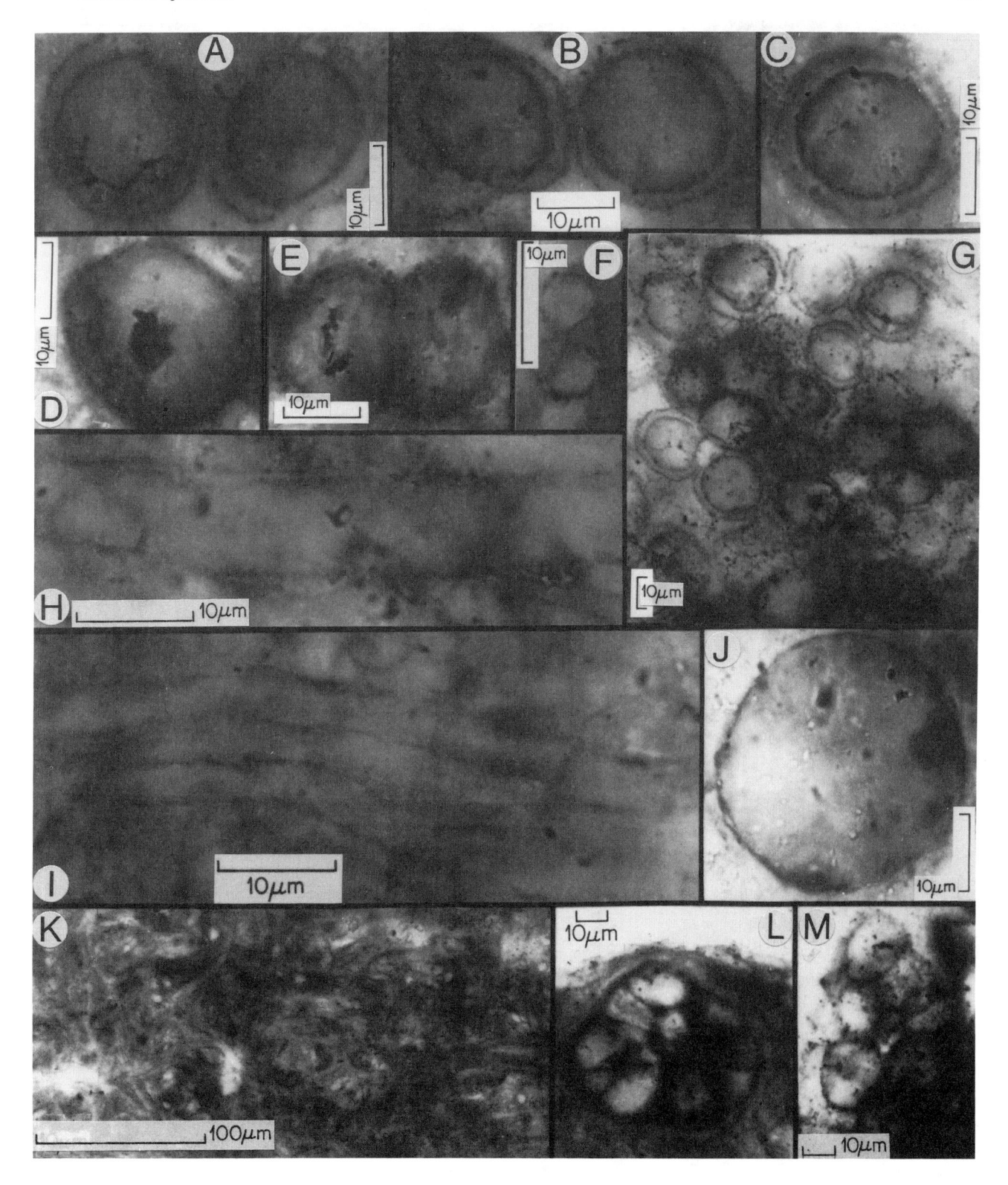

Plate 34. A–M, River Wakefield Subgroup, Port Augusta region, South Australia, Australia (775 Ma; locality cf. PPRG 1281). [A–M, petrographic thin sections, transmitted light; A–C, M, specimens not previously illustrated.] A–C, G, unnamed sheath-enclosed colonial unicells (G; Schopf 1975); D, E, J, unnamed spheroidal unicells (Schopf 1975); F, H, I, unnamed tubular microbial sheaths shown in transverse (F) and longitudinal (H, I) sections (Schopf 1975); K, unnamed tubular sheaths comprising an entangled "sheath mat" (Schopf 1975); L, M, unnamed colonial coccoid cells (L, Schopf 1975).

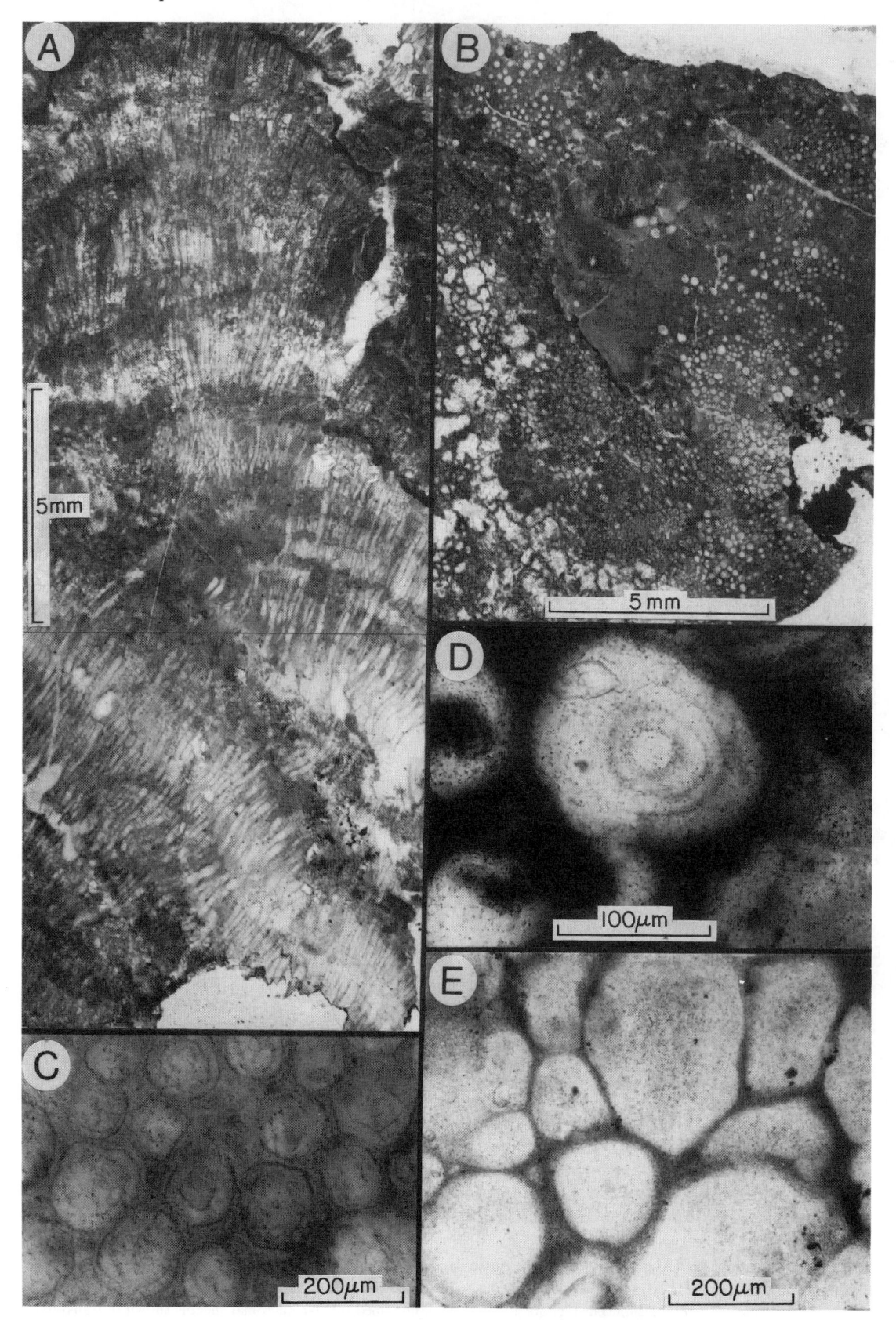

Plate 35. A–E, River Wakefield Subgroup, Port Augusta region, South Australia, Australia (775 Ma; locality cf. PPRG 1281). [A–E, petrographic thin sections, transmitted light; A, composite photomicrograph; B–E, specimens not previously illustrated.] A–E, *Polybessurus* cf. *bipartitus* Fairchild ex Green, Knoll, Golubic & Swett 1987, colonial, predominantly vertically oriented, asymmetrically laminated stalks secreted by pleurocapsacean or chamaesiphonacean cyanobacteria, shown in longitudinal (A, Schopf 1977) and transverse sections (B–E) and containing distinct, fine, laminae (C, D) [POLY1].

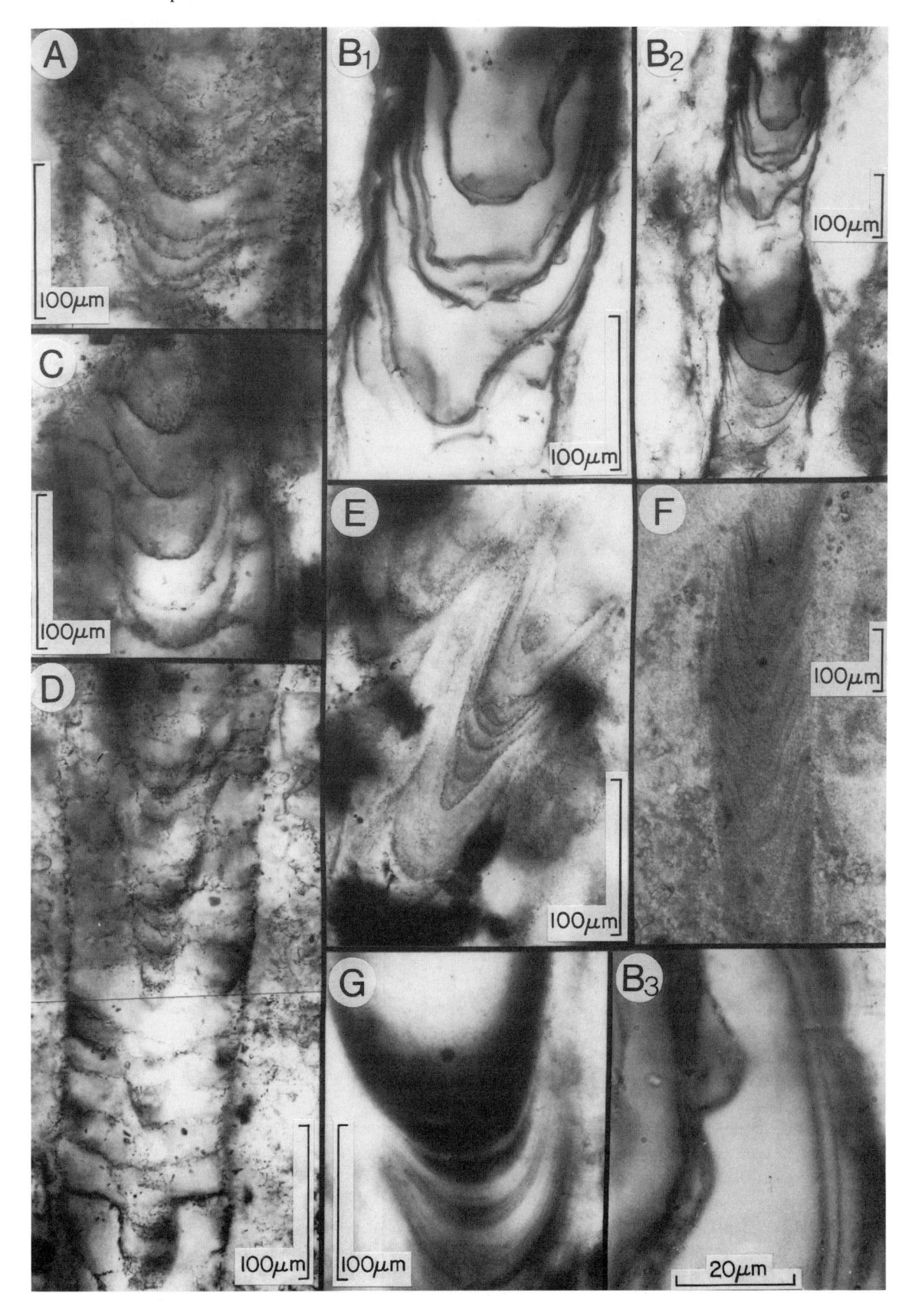

Plate 36. A–G, River Wakefield Subgroup, Port Augusta region, South Australia, Australia (775 Ma; localities cf. PPRG 1224 and 1281). [A–G, petrographic thin sections, transmitted light; D, composite photomicrograph; A–C, E–G, specimens not previously illustrated; B_1–B_3 show a single specimen at differing magnifications.] A–G, *Polybessurus* cf. *bipartitus* Fairchild ex Green, Knoll, Golubic & Swett 1987, longitudinal sections of colonial, vertically oriented, asymmetrically laminated stalks secreted by pleurocapsacean or chamaesiphonacean cyanobacteria, illustrating the observed range of preserved laminar shape, texture, thickness, and spacing (D, Schopf 1977) [POLY1].

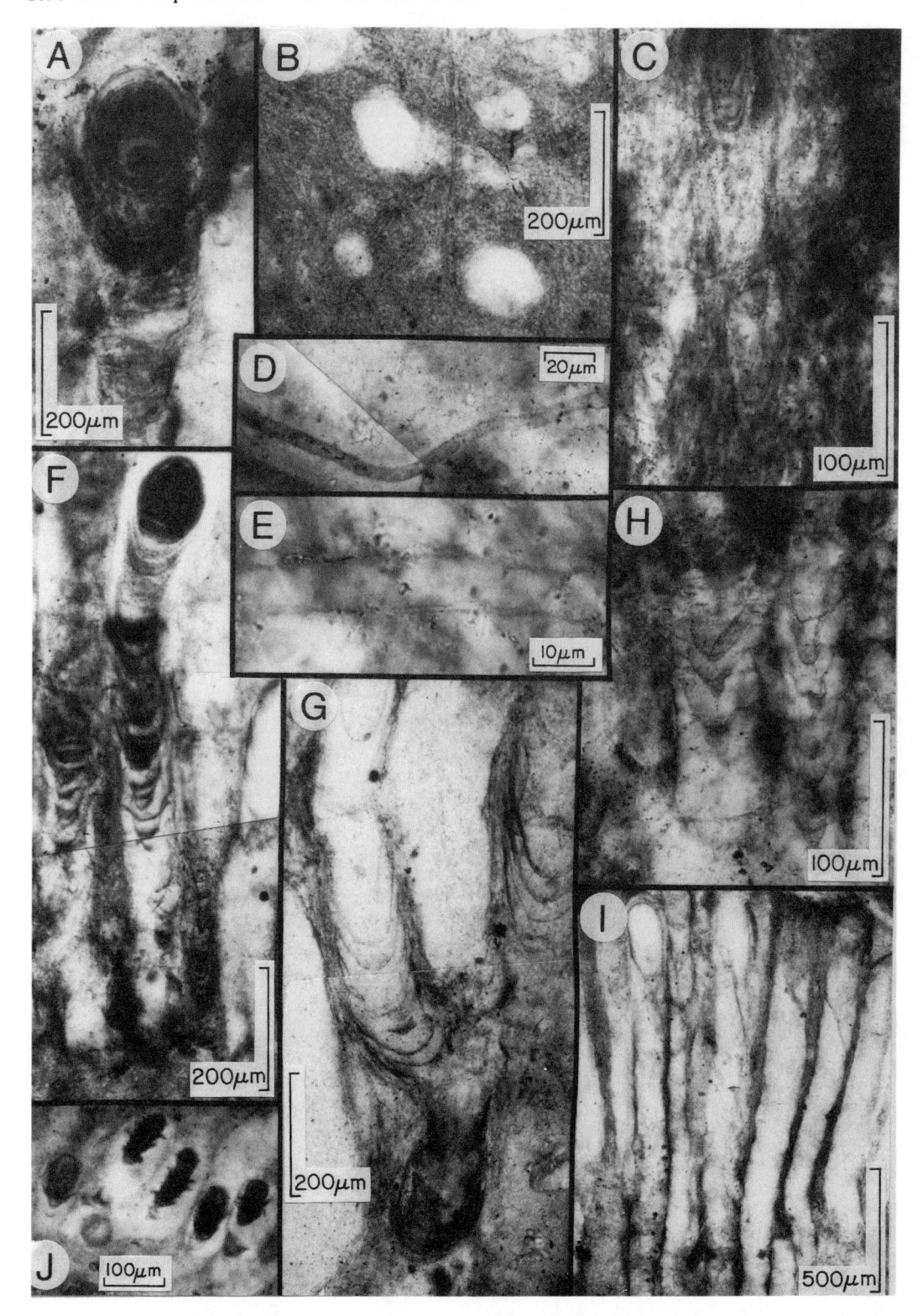

Plate 37. A–I, River Wakefield Subgroup, Port Augusta region, Flinders Ranges, southern South Australia, Australia (775 Ma; localites cf. PPRG 1224 and 1281). [A–I, petrographic thin sections, transmitted light; D, F, G, composite photomicrographs; A–G, I, specimens not previously illustrated.] A, C, F–I, *Polybessurus* cf. *bipartitus* Fairchild ex Green, Knoll, Golubic & Swett 1987, longitudinal sections of colonial, vertically oriented, asymmetrically laminated (A, F–H) stalks secreted by large, ellipsoidal, unicellular (A, F) pleurocapsacean or chamaesiphonacean cyanobacteria (H, Schopf 1977) [POLY1]; B, transverse sections of *Polybessurus* sp. stalks surrounded by interwoven, broad to narrow, tubular cyanobacterial sheaths (also shown surrounding longitudinal sections of *Polybessurus* sp. stalks in C, above); D, E, tubular cyanobacterial sheaths (*Siphonophycus* sp.) of the type intermeshed with the *Polybessurus* sp. stalks shown in B and C, above. J, Skillogalee Dolomite, William Creek region, Peake and Denison Ranges, central South Australia, Australia (770 Ma; locality cf. PPRG 1334). [J, petrographic thin section, transmitted light.] J, *Polybessurus* cf. *bipartitus* Fairchild ex Green, Knoll, Golubic & Swett 1987, ellipsoidal unicells with asymmetrically laminated sheaths (Fairchild 1975) [POLY1].

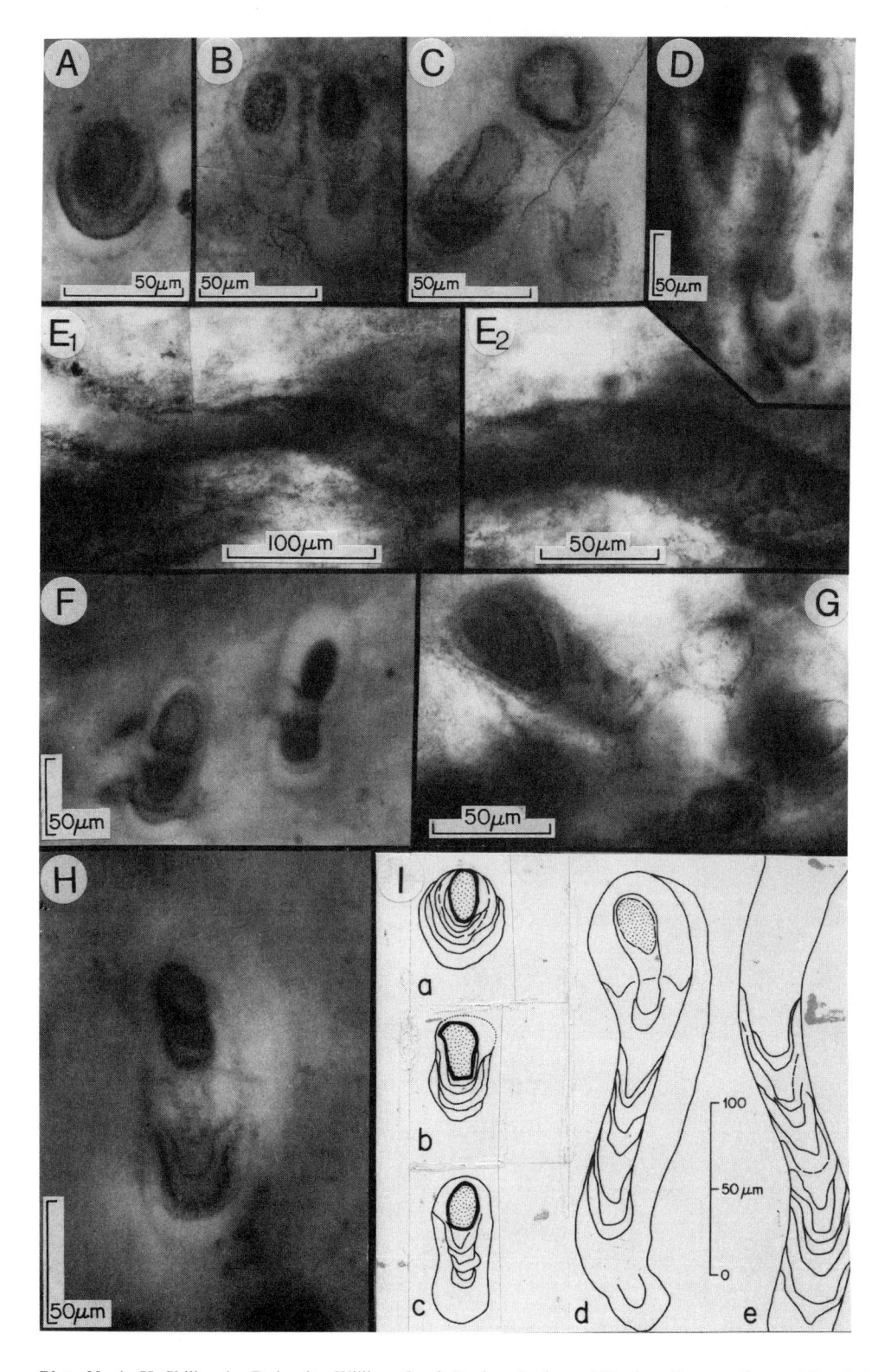

Plate 38. A–H, Skillogalee Dolomite, William Creek Region, Peake and Denison Ranges of central South Australia, and Port Augusta region, Flinders Ranges of southern South Australia, Australia (770 Ma; localities cf. PPRG 1230, 1232, 1283, and 1334). [A–H, petrographic thin sections, transmitted light, specimens not previously illustrated; B, C, E_1, composite photomicrographs; E_1, E_2 show different portions of a single specimen.] A–C, F–H, *Polybessurus* cf. *bipartitus* Fairchild ex Green, Knoll, Golubic & Swett 1987, ellipsoidal unicells with asymmetrically laminated sheaths (A, C, F, H, Fairchild 1975) [POLY1]; D, *Polybessurus* cf. *bipartitus* Fairchild ex Green, Knoll, Golubic & Swett 1987, elongate ellipsoidal cell at upper terminus of cylindrical, asymmetrically laminated stalk [POLY1]; E_1, E_2, *Polybessurus* cf. *bipartitus* Fairchild ex Green, Knoll, Golubic & Swett 1987, longitudinal sections of asymmetrically laminated cylindrical stalk [POLY1]. I, River Wakefield Subgroup and Skillogalee Dolomite, South Australia, Australia (775 Ma and 770 Ma, respectively; localities cf. PPRG 1224, 1230, 1232, 1281, 1283, and 1334.] I, generalized reconstructed ontogenetic sequence showing production of cylindrical laminated stalk (d, e) by ellipsoidal, asymmetrically ensheathed *Polybessurus* unicells (stippled, a–d).

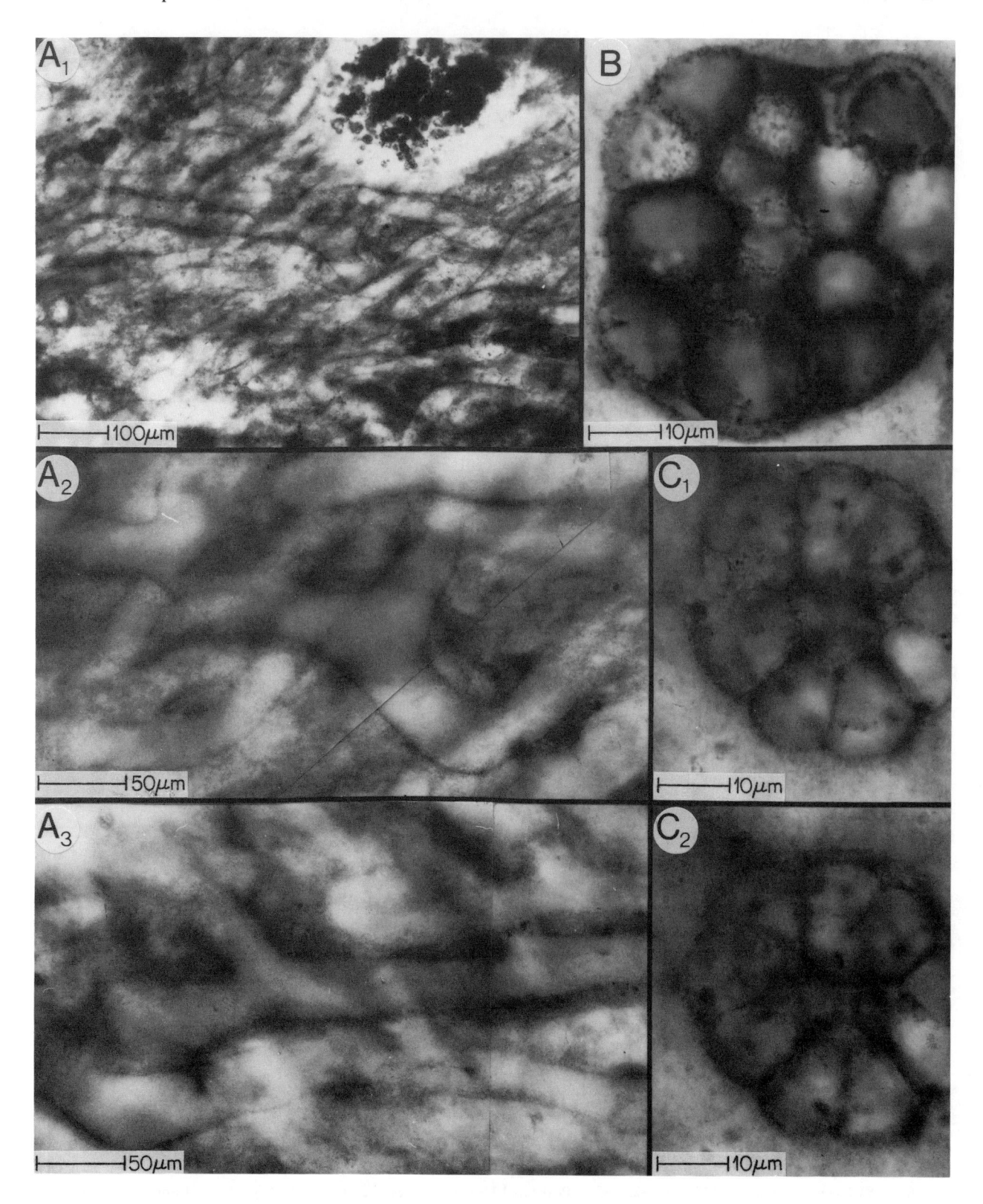

Plate 39. A–C, Skillogalee Dolomite, William Creek region, Peake and Denison Ranges, central South Australia, Australia (770 Ma; localities cf. PPRG 1283, 1334). [A–C, petrographic thin sections, transmitted light; A_2, A_3, composite photomicrographs; B, specimen not previously illustrated; A_1–A_3 C_1, C_2 show specimens at differing focal depths.] A_1–A_3, *Palaeosiphonella* sp. (Schopf 1977) [TB5]; B, unnamed sheath-enclosed cyanobacterial colony; C_1, C_2, rosette-like cyanobacterial colony cf. modern *Myxosarcina*, Pleurocapsaceae (Schopf and Fairchild 1973) [MYX1].

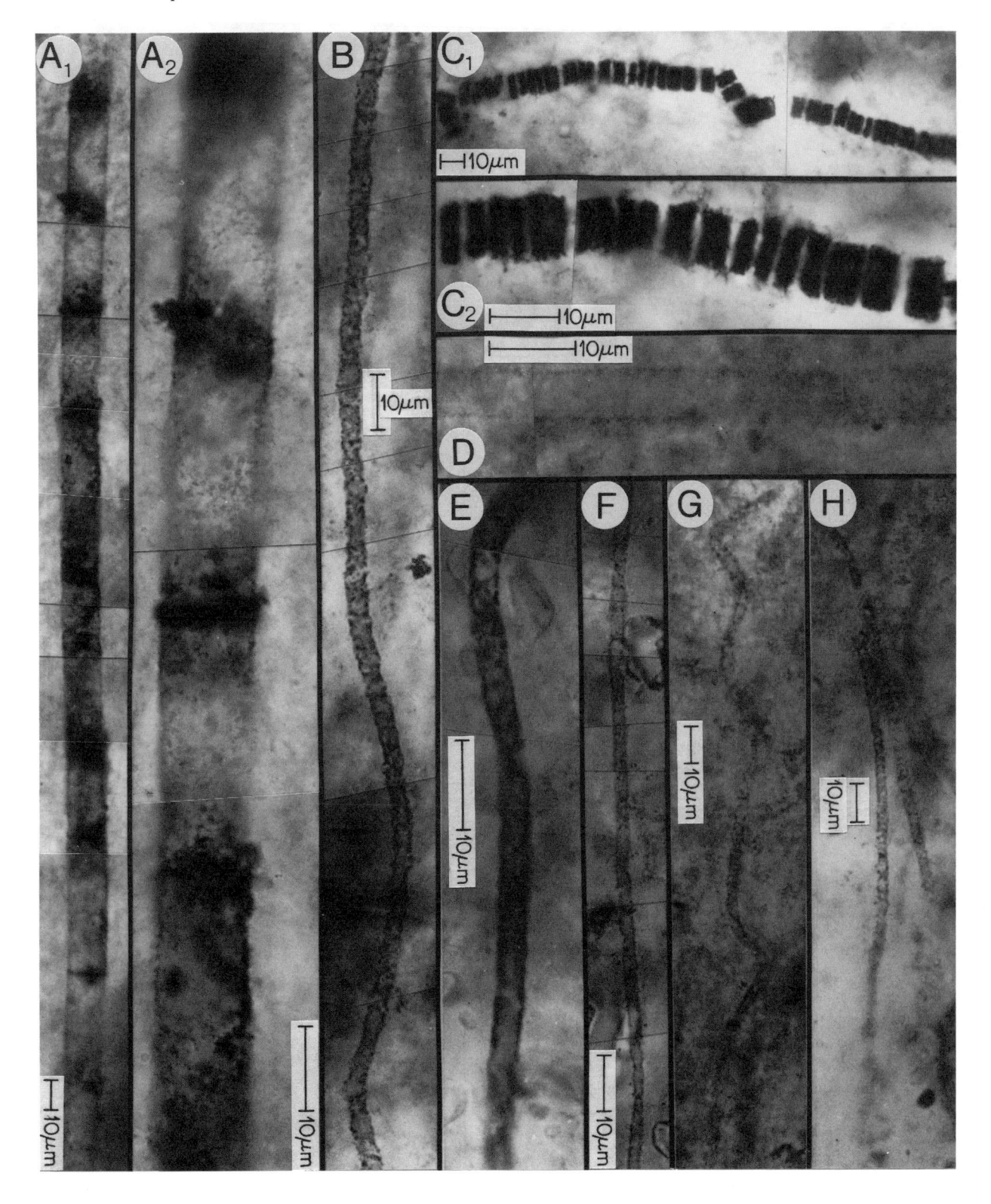

Plate 40. A–H, Skillogalee Dolomite, William Creek region, Peake and Denison Ranges, central South Australia, Australia (770 Ma; localities cf. PPRG 1283, 1334). [A–H, petrographic thin sections, transmitted light; A–F, composite photomicrographs; C, E, G, H, specimens not previously illustrated; A_1, A_2, C_1, C_2 each show single specimens at two magnifications.] A_1, A_2, unnamed chlorophycean, possibly ulothricacean, broad filament with elongate cells (Schopf 1977) [EU6]; B, G, H, unnamed trichomes with poorly preserved quadrate to elongate cells (B, Schopf and Fairchild 1973); C_1, C_2 unnamed oscillatoriacean trichome with disc-shaped cells (Fairchild 1975); D, *Siphonophycus* sp. (Schopf and Fairchild 1973); E, *Eomycetopsis* sp.; F, unnamed narrow cellular microbial trichome (Schopf and Fairchild 1973).

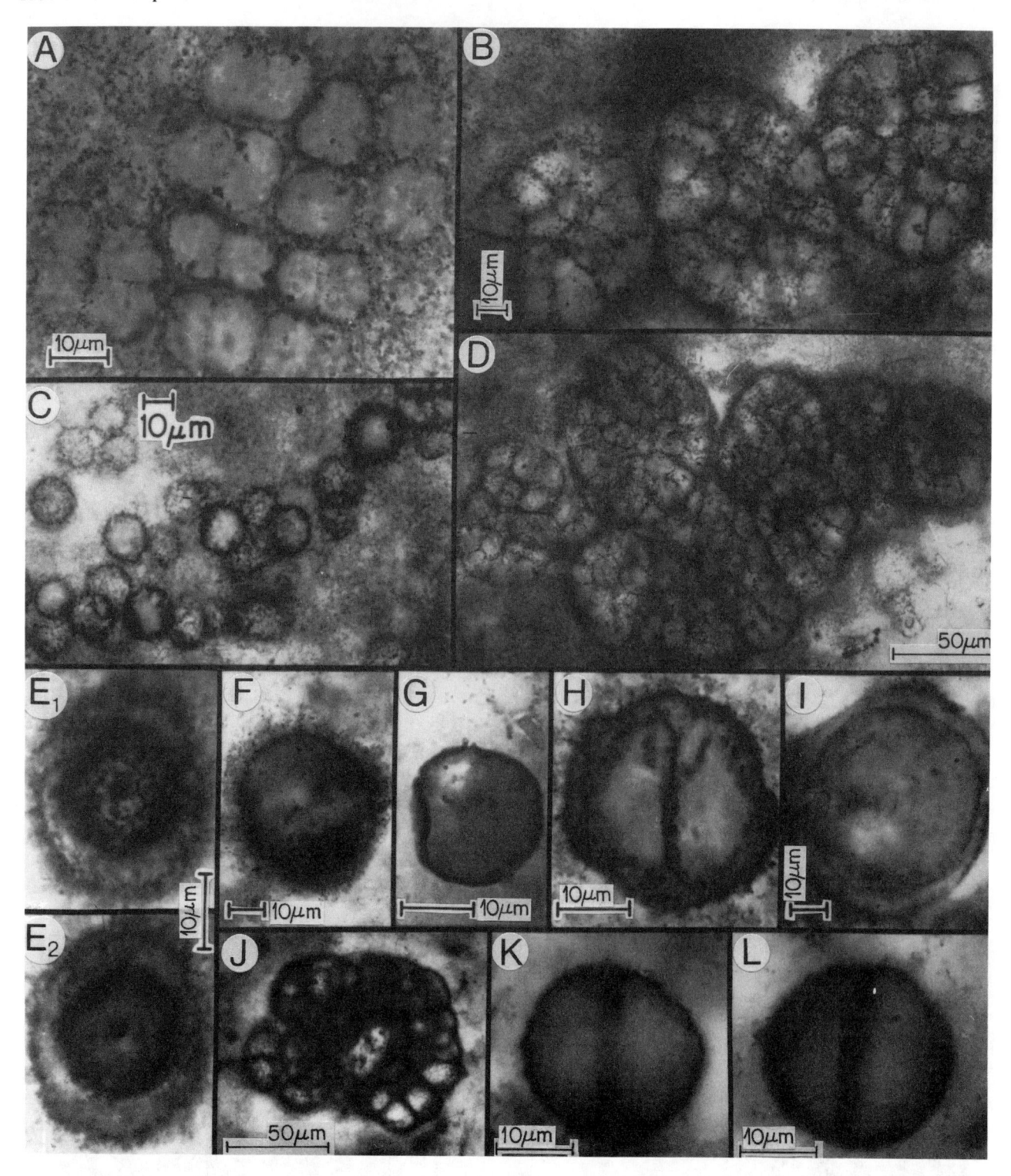

Plate 41. A–L, Skillogalee Dolomite, William Creek region, Peake and Denison Ranges, central South Australia, Australia (770 Ma; localities cf. PPRG 1283, 1334). [A–L, petrographic thin sections, transmitted light; B–E, J, specimens not previously illustrated; E_1, E_2 show a single specimen at two focal depths.] A, unnamed tabular, cyanobacterial colony cf. modern *Merismopedia*, Chroococcaceae (Schopf and Fairchild 1973) [CPL2]; B, D, J, unnamed rosette-like cyanobacterial colony cf. modern *Myxosarcina*, Pleurocapsaceae (D, Fairchild 1975) [MYX1]; C, unnamed coccoid unicells (Fairchild 1975); E_1, E_2, F, I, unnamed sheath-enclosed unicells (E, Fairchild 1975; F, I, Schopf and Fairchild 1973); G, unnamed unicell (Schopf and Fairchild 1973); H, K, L, unnamed, sheath-enclosed, paired unicells (H, Schopf and Fairchild 1973; K, L, Schopf 1977).

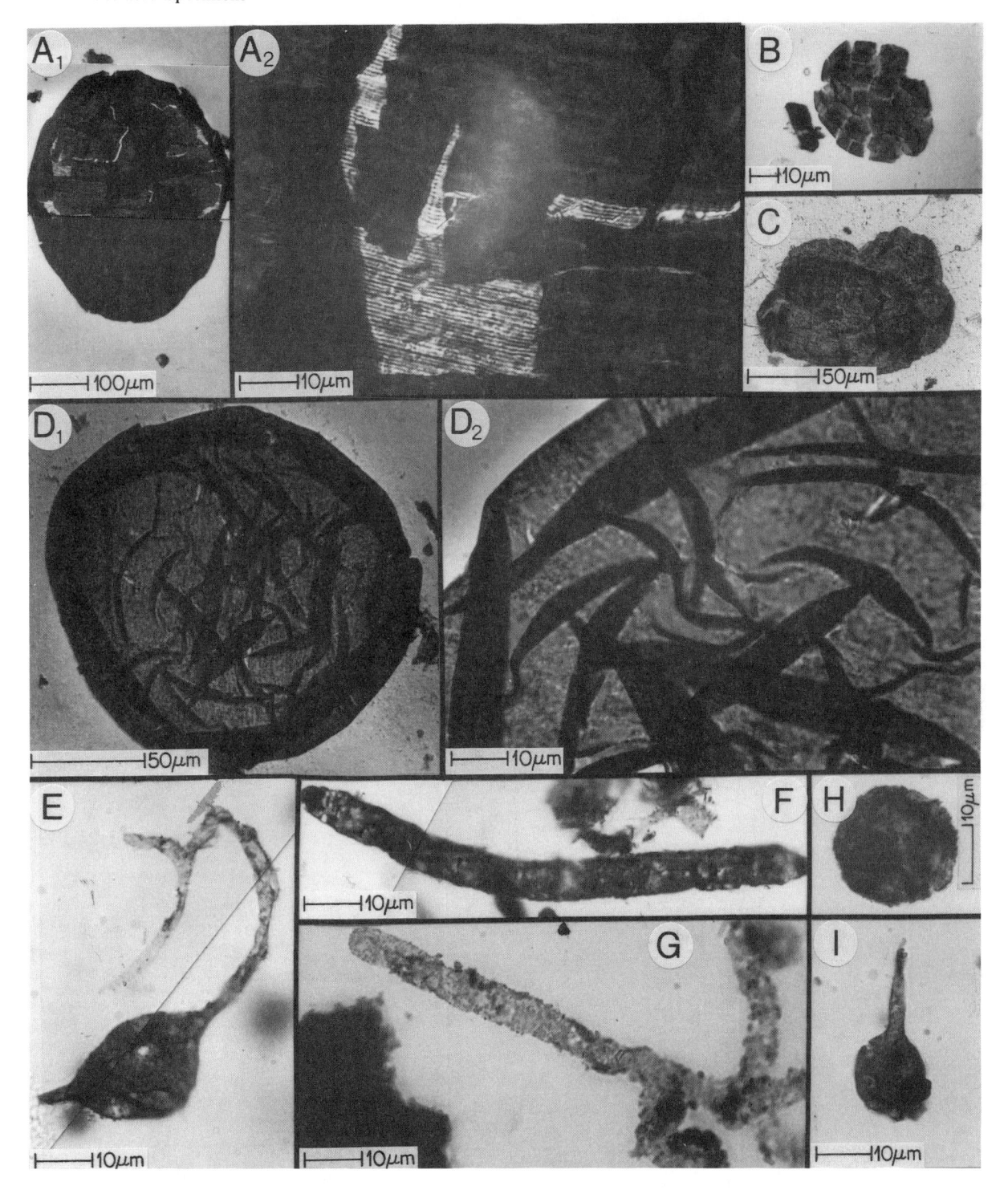

Plate 42. A–D, Akberdin Formation, forelands of the Ural Mountains, Bashkiria, USSR (800 Ma). [A–D, acid resistant residues; A_1, $B-D_1$, interference contrast; A_2, D_2, transmitted light; A_1, composite photomicrograph.] A_1, A_2, *Valeria lophostriata* Jankauskas (=Yankauskas) 1979, PARATYPE (Yankauskas, in press) (slide 16-62-4762/19, England slide L20/1, label left) [MG6]; B, *Satka elongata* Jankauskas (=Yankauskas) 1979, HOLOTYPE (slide 16-62-4762/9, England slide 011/1, label left); C, *Satka colonialica* Jankauskas (=Yankauskas) 1979, HOLOTYPE (slide 16-62-4762/22, England slide Q34/1, label left); D_1, D_2, *Kildinella lophostriata* Jankauskas (=Yankauskas) 1979, HOLOTYPE (=*Valeria lophostriata* Jankauskas) (slide 16-62-4762/16, England slide 019/1, label left) [MG6]. E–I, Dashka Formation, Krasnoyarsk region, Siberia, USSR (750 Ma). [E–I, acid resistant residues; E, F, H, I, interference contrast; G, transmitted light; E, F, composite photomicrographs.] E, *Germinosphaera bispinosa* Mikhailova (=Mikhajlova) 1986, HOLOTYPE (slide 882/2, England slide R49/1, label right) [BZ6]; F, *Calyptothrix obsoletus* Mikhailova (=Mikhajlova) 1986, HOLOTYPE (slide 882/3, England slide F34, read label from right) [3N34]; G, *Palaeoaphanizomenon scabratus* Mikhailova (=Mikhajlova) 1986, HOLOTYPE (slide 885/4, England slide K50/2-1, read label from right) [3N27]; H, *Nucellosphaeridium spumosum* Mikhailova (=Mikhajlova) 1986, HOLOTYPE (slide 885/3, England slide P38/2, read label from right) [S11]; I, *Germinosphaera unispinosa* Mikhailova (=Mikhajlova) 1986, HOLOTYPE (slide 882/1, England slide R42, label right) [BZ7].

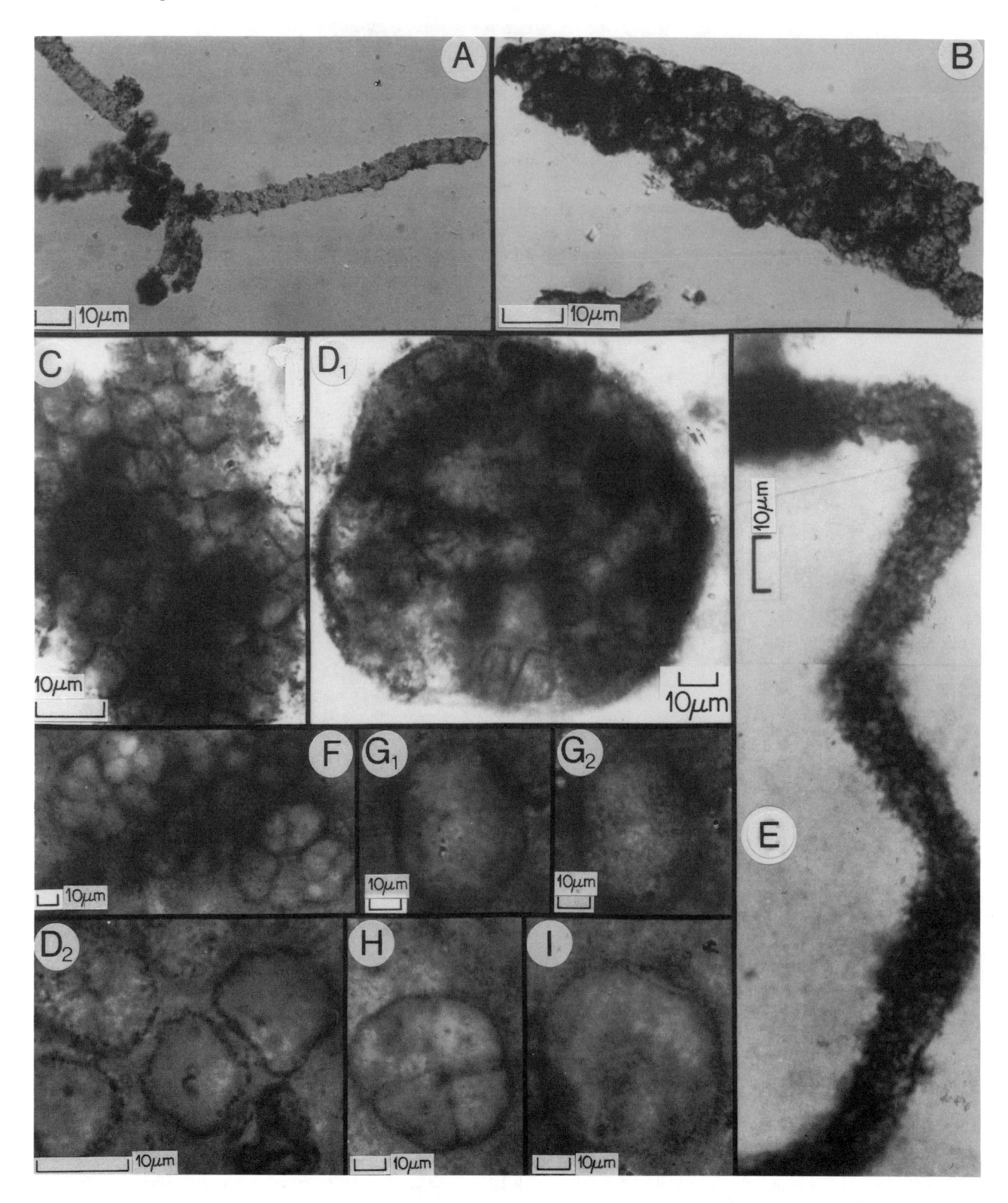

1104 Atlas of Representative Proterozoic Microfossils

Plate 43. A, B, Dashka Formation, Krasnoyarsk region, Siberia, USSR (750 Ma). [A, B, acid resistant residues, interference contrast.] A, *Cephalophytarion piliformis* Mikhailova (=Mikhajlova) 1986, HOLOTYPE (slide 885/4, England slide K44/1, read label from right) [3N26]; B, *Polysphaeroides nuclearis* Mikhailova (=Mikhajlova) 1986, HOLOTYPE (slide 878/1, England slide W29, read label from right) [CIR9]. C–I, Myrtle Springs Formation, William Creek region, Peake and Denison Ranges, central South Australia (750 Ma; locality cf., but stratigraphically 100 m above, PPRG 1283). [C–I, petrographic thin sections, transmitted light; C, F–I, specimens not previously illustrated; D_1, D_2, G_1, G_2 each show single specimens at two focal depths.] C, D, F, H, unnamed colonial unicells (D, Schopf 1975); E, unnamed sheath-like filament (Schopf 1975); G, I, unnamed ellipsoidal to spheroidal unicells.

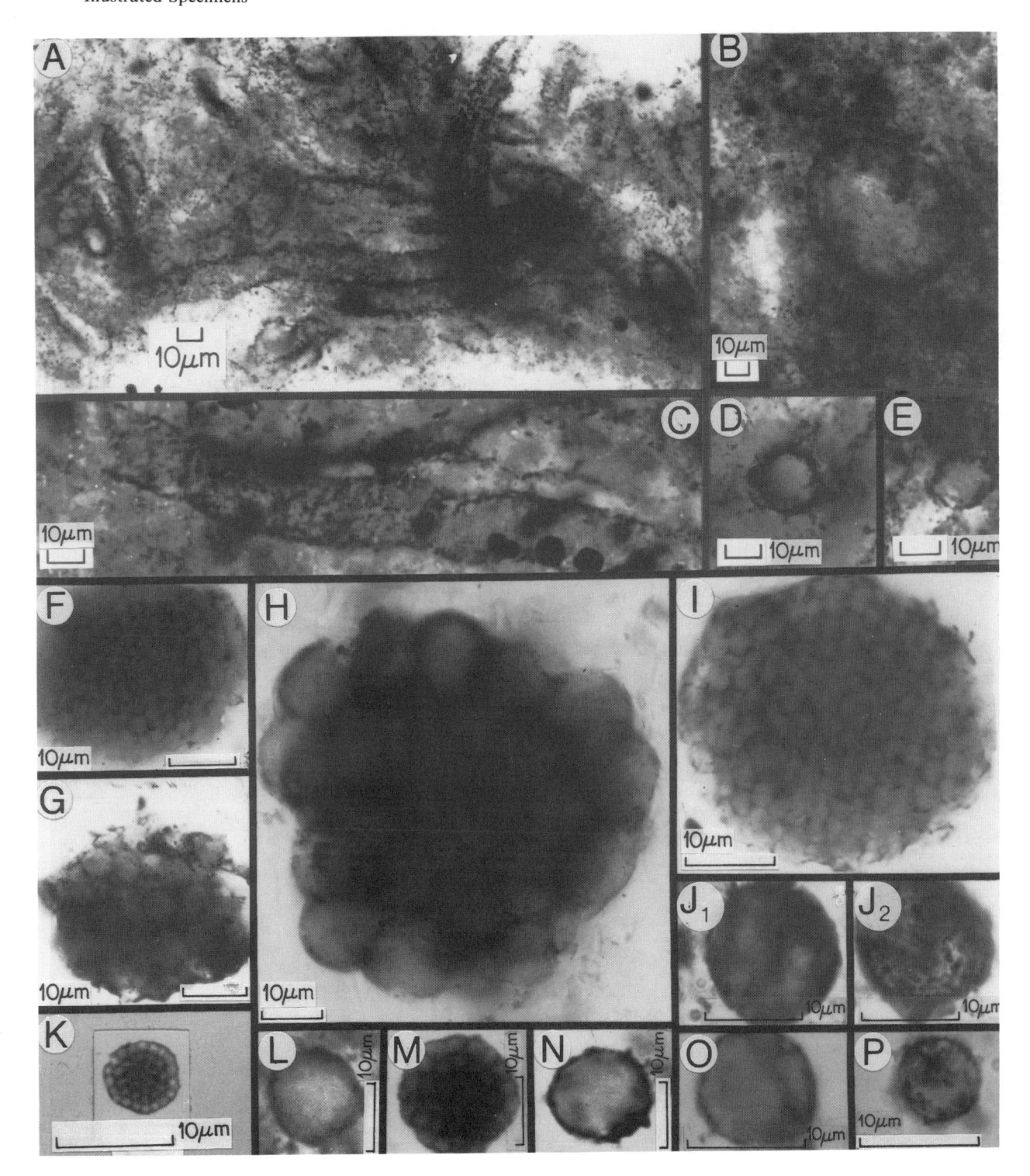

Plate 44. A–E, Auburn Dolomite, Port Augusta region, southern South Australia, Australia (780 Ma; locality cf. PPRG 1220). [A–E, petrographic thin sections, transmitted light]. A, C–E, unnamed tubular cyanobacterial sheaths in longitudinal (A, C) and transverse (D, E) sections (Schopf 1975); B, unnamed ellipsoidal unicell (Schopf 1975). F–P, Brioverian cherts and shales, Brittany and Normandy, France (740 Ma). [F–N, petrographic thin sections; J, N–P, acid resistant residues; F–N, transmitted light; J_1, J_2 show a single specimen at two focal depths.] F–H, J_1, J_2, "Type IV" microfossil (Chauvel and Schopf 1978); I, "Type G" microfossil (Chauvel and Mansuy 1981); K, M, "Type II" microfossil (Chauvel and Schopf 1978) [CSF2]; L, N–P, "Type I" microfossil (Chauvel and Schopf 1978).

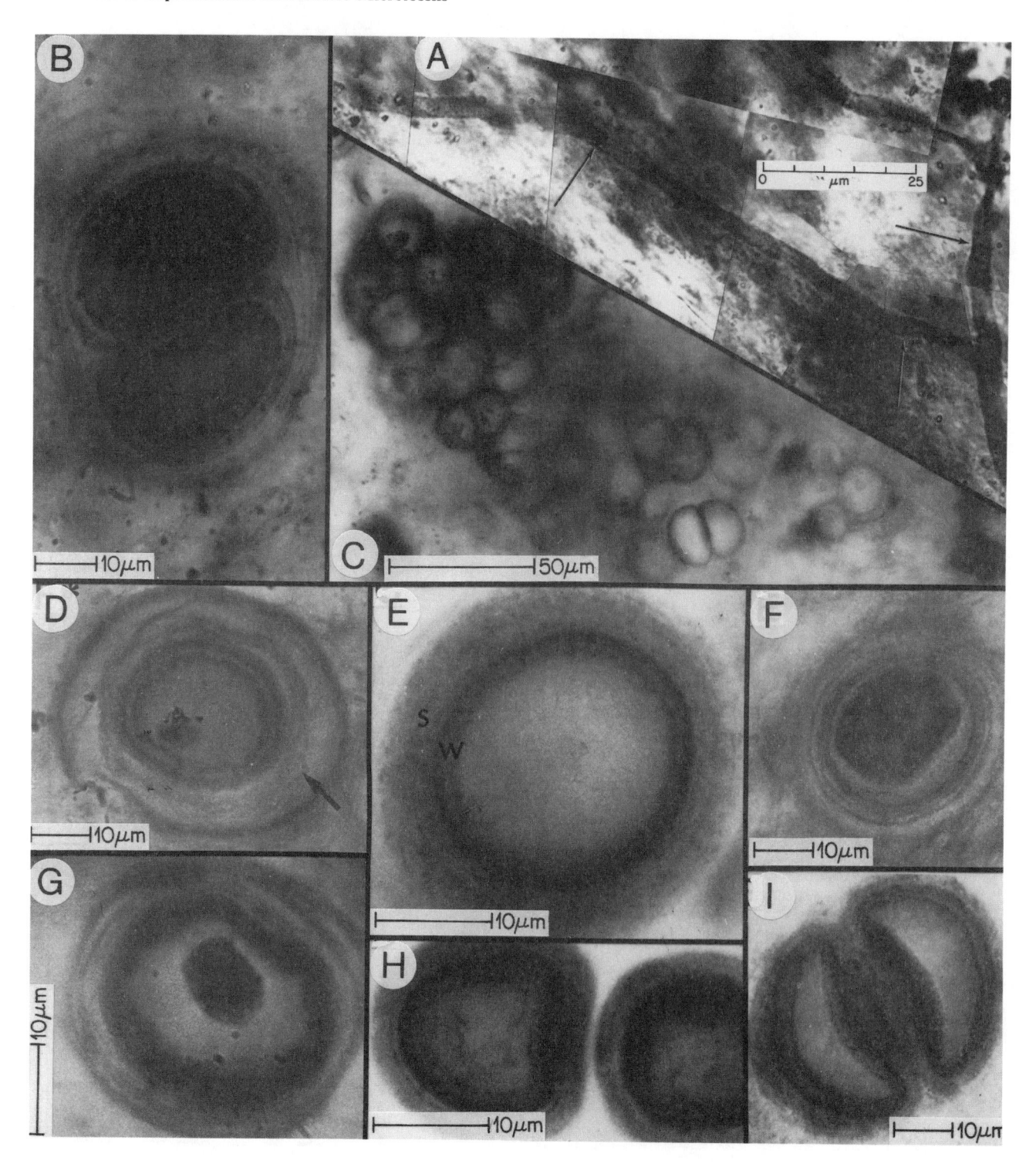

Plate 45. A, Olkha Formation, Irkutsk region, Siberia, USSR (738 Ma). [A, petrographic thin section, transmitted light, composite photomicrograph.] A, unnamed branched, septate (at arrows) filamentous microfossil (Schopf et al. 1977) [EU4]. B–I, Min'yar Formation, Ufa region, southern Ural Mountains, Bashkiria, USSR (740 Ma; D–I, locality cf. PPRG 1478). [B–I, petrographic thin sections, transmitted light.] B, *Gloeodiniopsis grandis* Sergeev & Krylov 1986, HOLOTYPE (slide 71-81, England slide K27/1, label left) [G5]; C, *Palaeopleurocapsa kamaelgensis* Sergeev & Krylov 1986, HOLOTYPE (slide 71-81, England slide K28/1, label left) [PLEU4]; D, *Gloeodiniopsis magna* Nyberg & Schopf 1984, HOLOTYPE, showing wrinkled sheath laminae (at arrows) [G5]; E, H, I, *Gloeodiniopsis lamellosa* Schopf 1968 emend. Knoll & Golubic 1979 showing (E) faintly lamellated hyaline sheath (s) and well-defined cell wall (w) (Nyberg and Schopf 1984) [G2]; F, G, *Gloeodiniopsis magna* Nyberg & Schopf 1984, PARATYPES [G5].

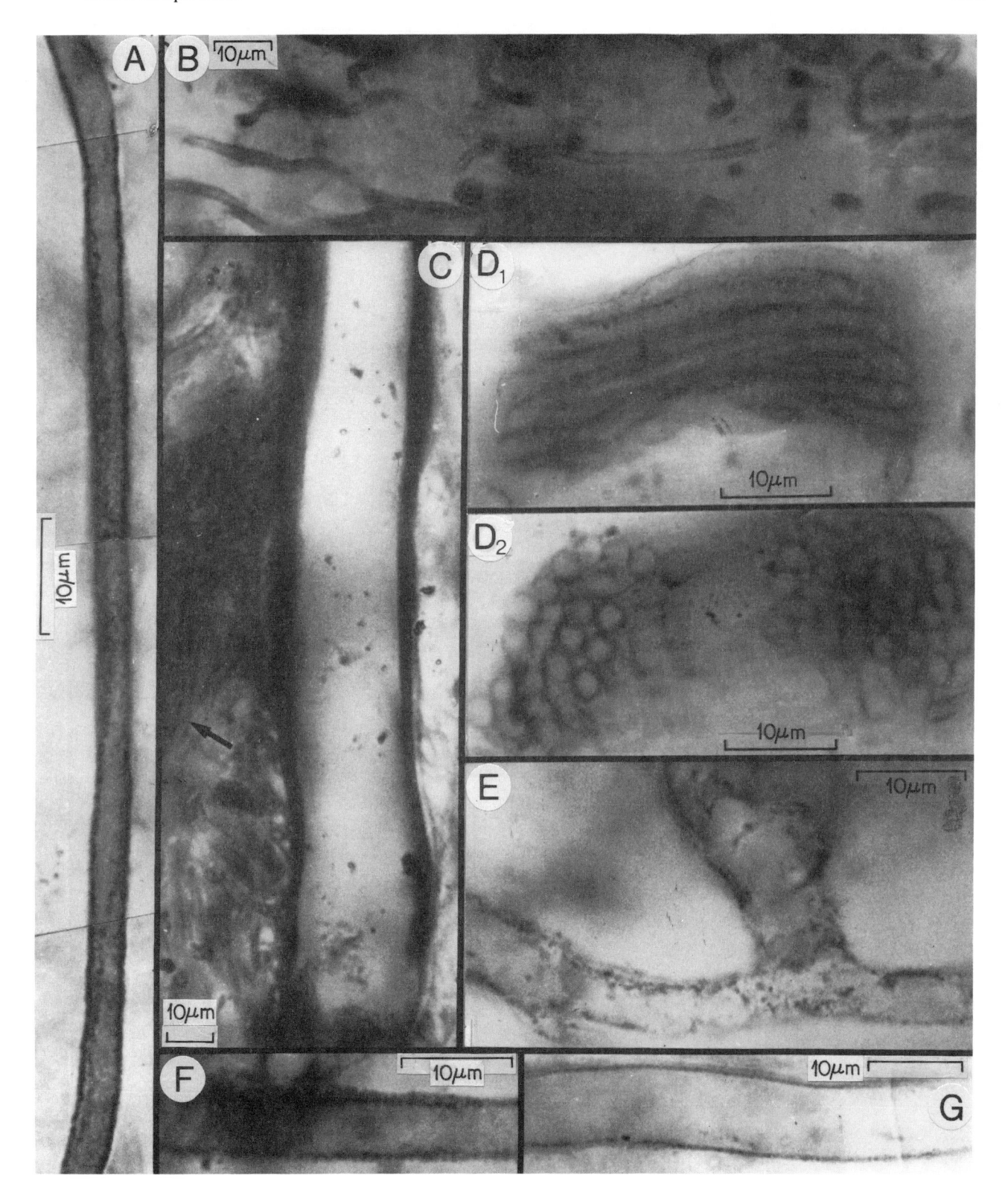

Plate 46. A–G, Min'yar Formation, Ufa region, southern Ural Mountains, Bashkiria, USSR (740 Ma; locality cf. PPRG 1478). [A–G, petrographic thin sections, transmitted light; A, composite photomicrograph; D_1, D_2 show the same specimens at two focal depths.] A, B, *Eomycetopsis robusta* Schopf 1968 emend. Knoll & Golubic 1979 [TU9]; C, *Siphonophycus capitaneum* Nyberg & Schopf 1984, HOLOTYPE (note *Eomycetopis* sp. at arrow) [TU55]; D_1, D_2, *Eomycetopsis* sp. [TU10]; E, *Ramivaginalis uralensis* Nyberg & Schopf 1984, HOLOTYPE [TB3]; F, G, "intermediate-diameter oscillatoriacean sheaths" (Nyberg and Schopf 1984).

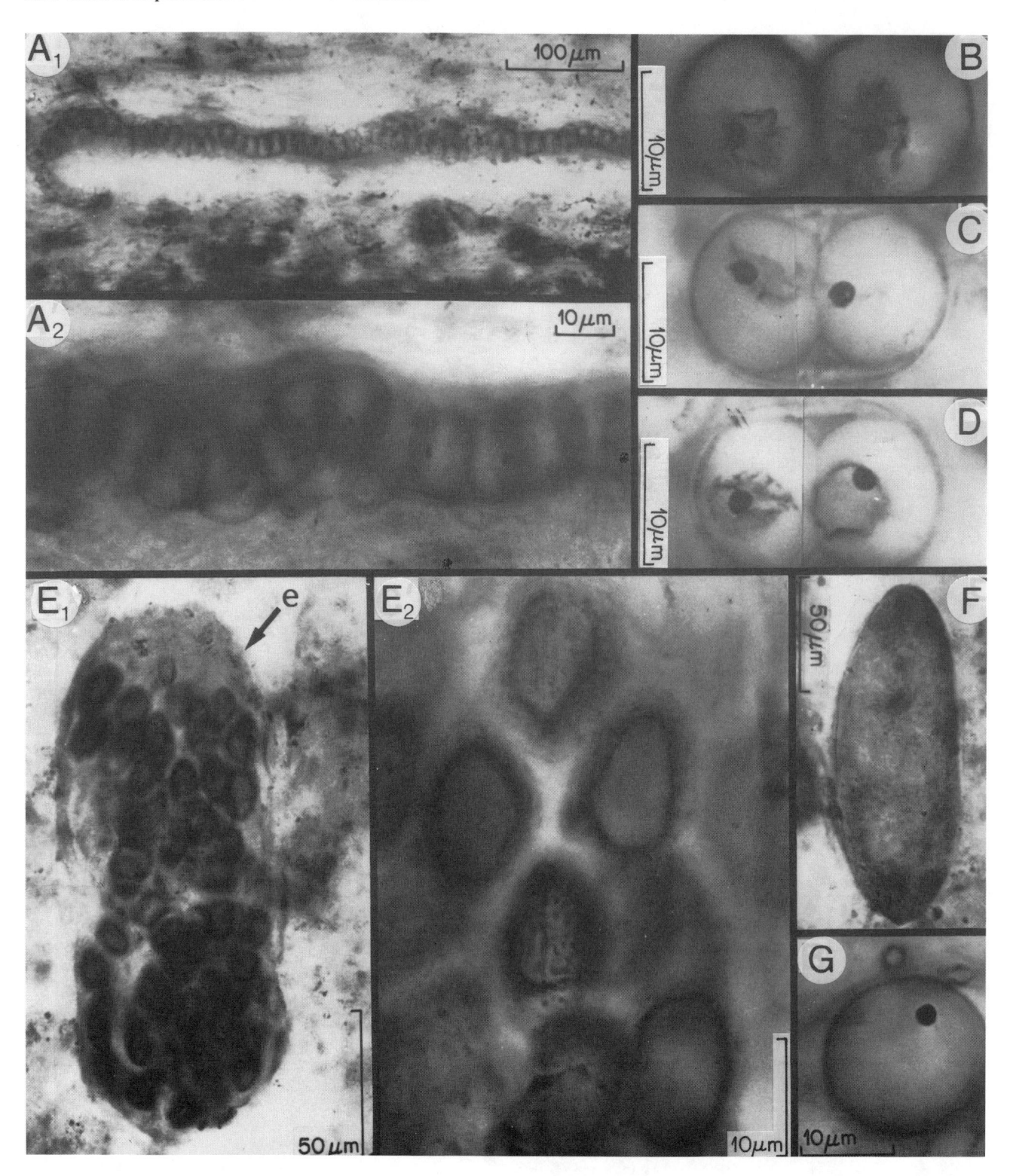

1108 Atlas of Representative Proterozoic Microfossils

Plate 47. A–G, Min'yar Formation, Ufa region, southern Ural Mountains, Bashkiria, USSR (740 Ma; locality cf. PPRG 1478). [A–G, petrographic thin sections, transmitted light; A_1, A_2, E_1, E_2 each show specimens at two magnifications.] A–D, G, *Glenobotrydion majorinum* Schopf & Blacic 1971 emend. Nyberg & Schopf 1984; E_1, E_2, *Eoaphanocapsa oparinii* Nyberg & Schopf 1984, HOLOTYPE, showing a distinct outer boundary of the colonial envelope (e); F, ellipsoidal envelope, possibly of *Eoaphanocapsa oparinii* Nyberg & Schopf 1984 (Nyberg and Schopf 1984).

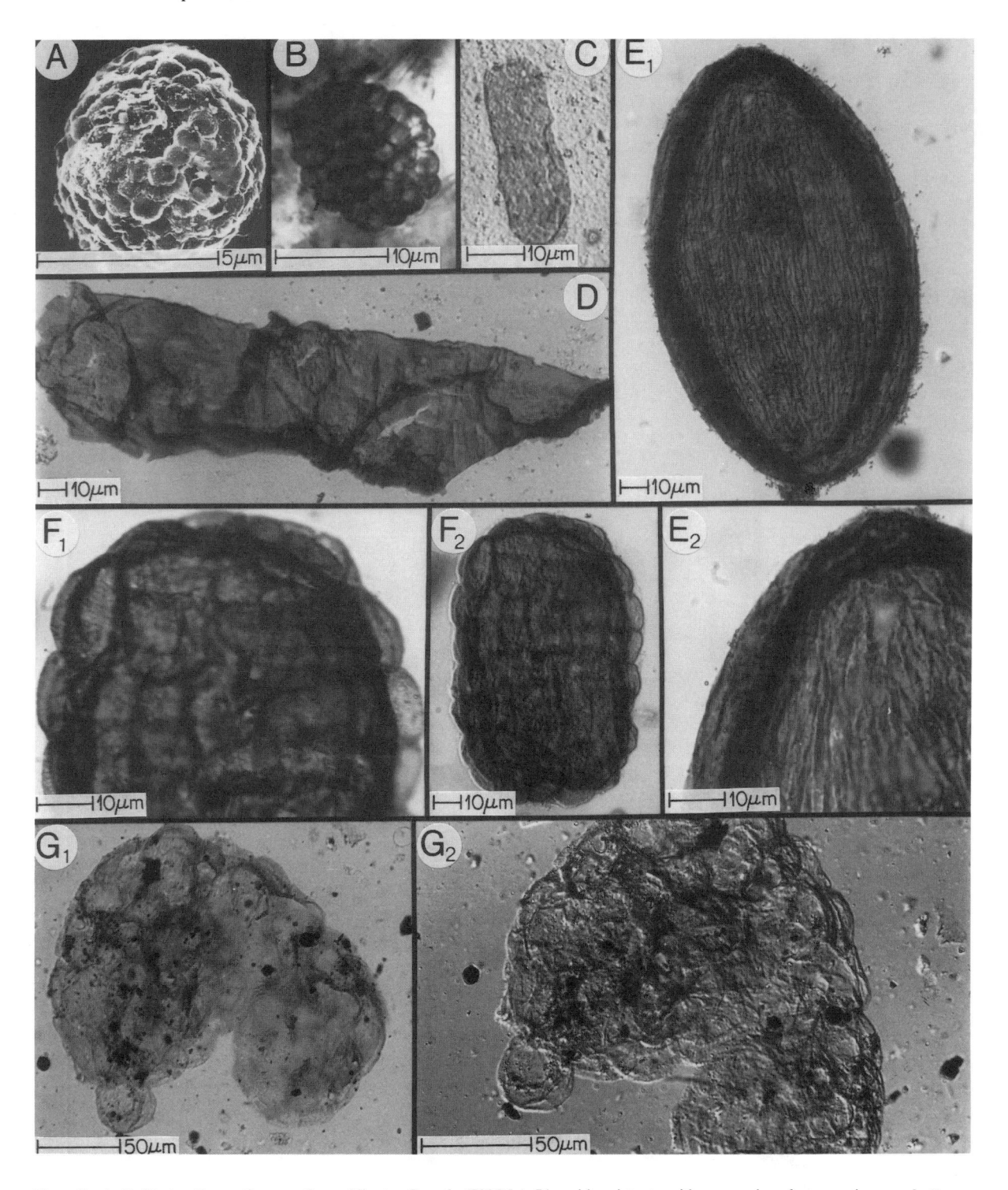

Plate 48. A, B, Hector Formation, southern Alberta, Canada (700 Ma). [A, acid resistant residue, scanning electron micrograph; B, petrographic thin section, transmitted light.] A, B, *Sphaerocongregus variabilis* Moorman 1974, PARATYPES (Cloud et al. 1975) [CSF2]. C, D, Uk Formation, Ufa region, southern Ural Mountains, Bashkiria, USSR (675 Ma). [C, D, acid resistant residues, interference contrast.] C, *Palaeolyngbya zilimica* Jankauskas (=Yankauskas) 1980, NEOTYPE (Yankauskas, in press) (slide 16-15/4-1, England slide P30, label left); D, *Siphonophycus costatus* Jankauskas (=Yankauskas) 1980 (slide 16-14/2-2, England slide K20/2, label left). E, F, Kamovsk Formation, Krasnoyarsk region, Siberia, USSR (675 Ma). [E, F, acid resistant residues; E_1, E_2, F_2, interference contrast; F_1, transmitted light; E_1, E_2 and F_1, F_2 each show single specimens at two magnifications.] E_1, E_2, *Cucumiforma vanavaria* Mikhailova (=Mikhajlova) 1986, HOLOTYPE (see also Frontispiece) (slide 1178/1, England slide S59, label right); F_1, F_2, *Satka squamifera* Pyatiletov 1980 (Mikhajlova, in press) (slide 1183/1, England slide J63/4, read label from right) [CRCF4]. G, Chartorysk Formation, Volyn region, Ukraine, USSR (640 Ma). [G, acid resistant residue; G_1, transmitted light; G_2, interference contrast; G_1, G_2 show the same specimen at two magnifications.] G_1, G_2, *Polycavita bullata* (Andreeva 1966) Assejeva (=Aseeva) 1982, PARATYPE (=*Polycavita frillata* Aseeva in press) (slide 249 CK 350, England slide F56/2, label right).

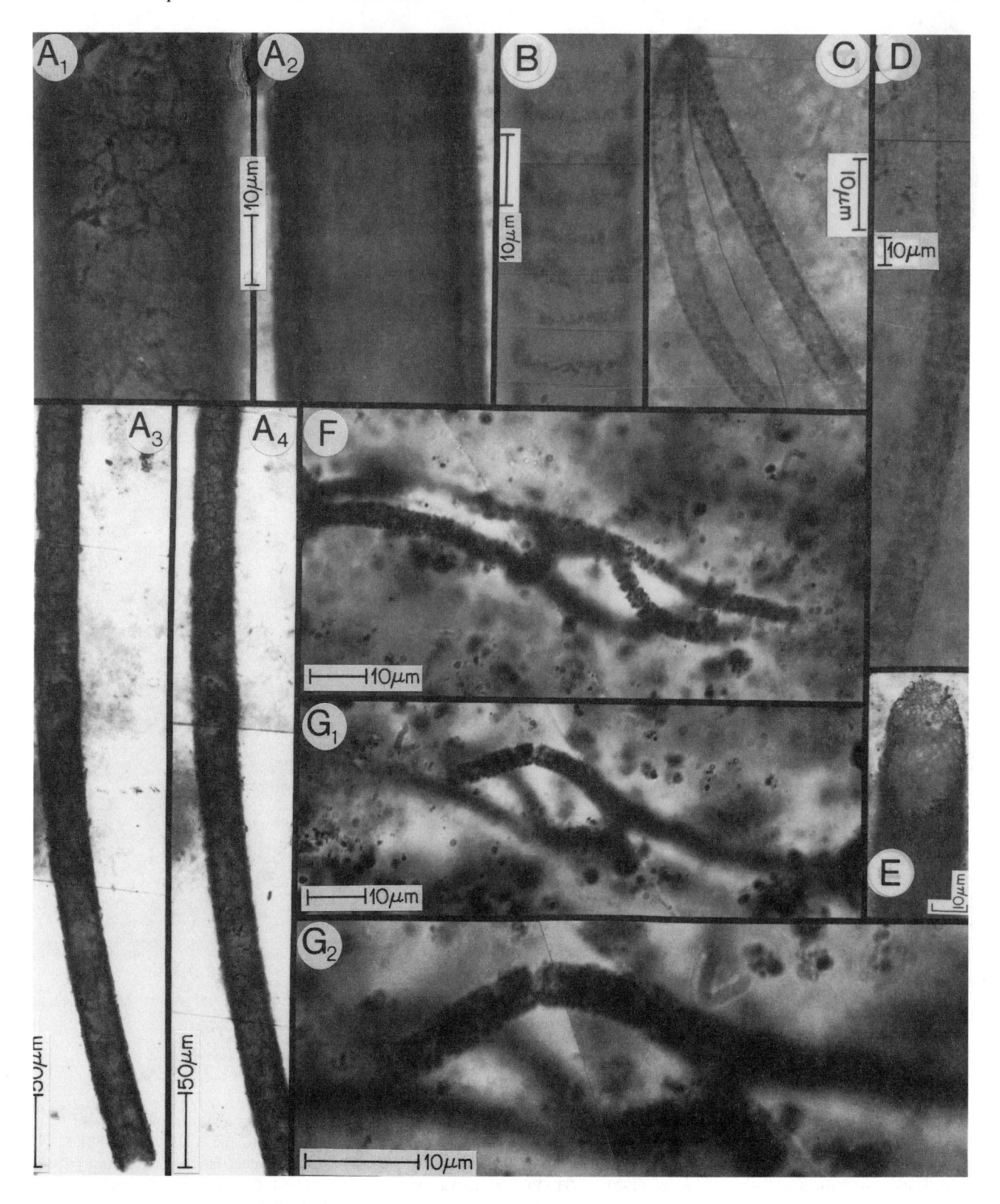

1110 Atlas of Representative Proterozoic Microfossils

Plate 49. A–G, Chichkan Formation, Zhanatas region, southern Kazakhstan, USSR (650 Ma; A–E, locality cf. PPRG 1473). [A–G, petrographic thin sections; A_3, A_4, B–D, G_2, composite photomicrographs; A_1–A_4, G_1, G_2 each show different magnifications and focal depths of single specimens.] A_1–A_4, E, unnamed broad oscillatoriacean tubular sheaths shown in longitudinal (A_1–A_4) and oblique sections (E) and in both medial (A_2, A_3) and surfical views (A_1, A_4) (Schopf and Sovietov 1976) [TUL5]; B, D, unnamed oscillatoriacean cellular trichomes (Schopf and Sovietov 1976) [3N64]; C, *Eomycetopsis* sp. (Schopf and Sovietov 1976) [TU17]; F, G_1, G_2, *Cyanonema disjuncta* Ogurtsova & Sergeev 1987, PARATYPES (slide 284-84, England slide N35/2, label left).

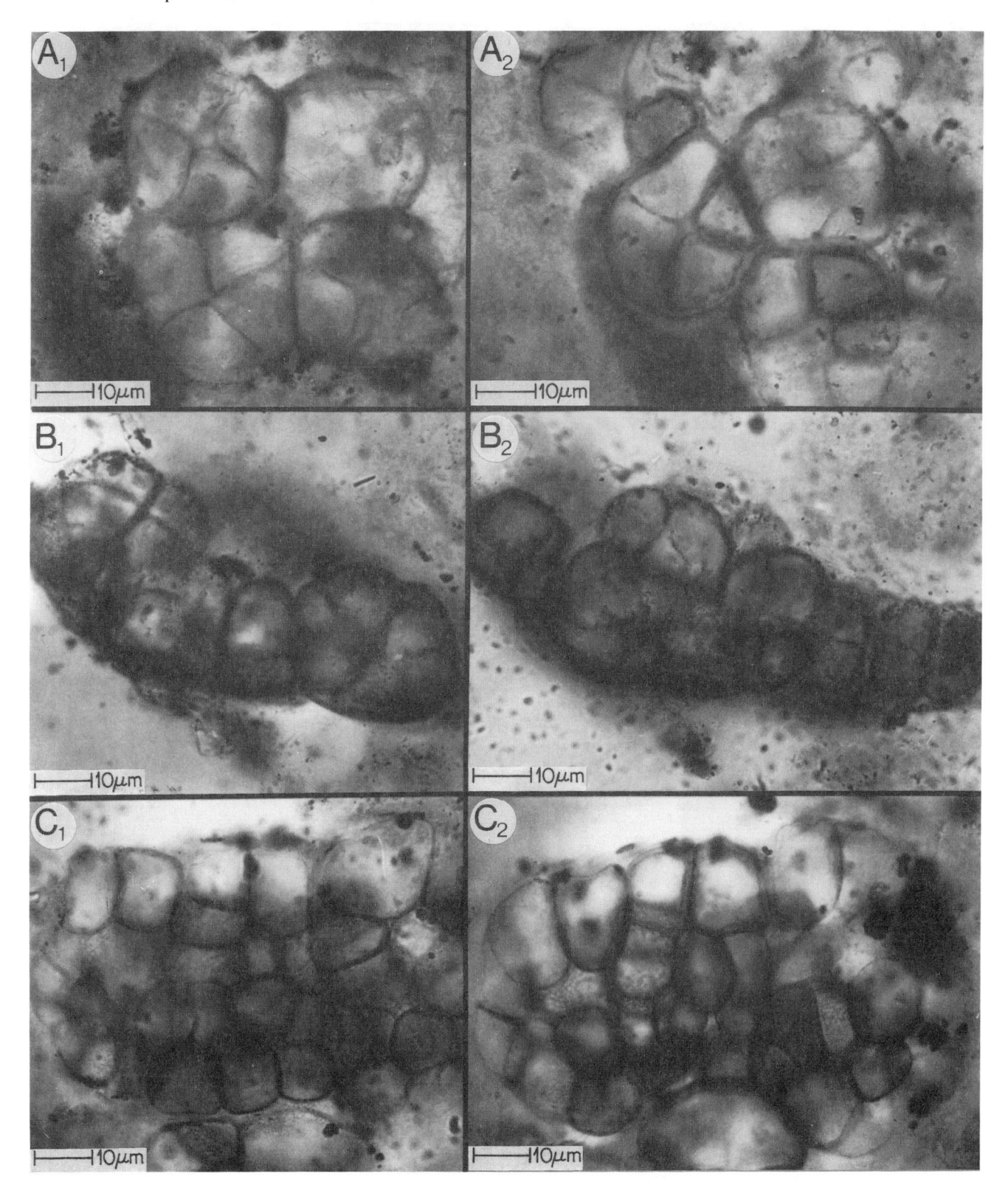

Plate 50. A–C, Chichkan Formation, Zhanatas region, southern Kazakhstan, USSR (650 Ma). [A–C, petrographic thin sections, transmitted light; A_1, A_2, B_1, B_2, C_1, C_2 each show single specimens or groups of specimens at two focal depths.] A_1, A_2, *Tetraphycus bistratosus* Ogurtsova & Sergeev 1987, HOLOTYPE (slide 288-84, England slide U30/1, label left) [TET6]; B_1, B_2, *Palaeopleurocapsa fusiforma* Ogurtsova & Sergeev 1987, HOLOTYPE (slide 315-84, England slide E38, label left) [PLEU3]; C_1, C_2, *Palaeopleurocapsa reniforma* Ogurtsova & Sergeev 1987, HOLOTYPE (slide 294-84, England slide T34/4, label left) [PLEU3].

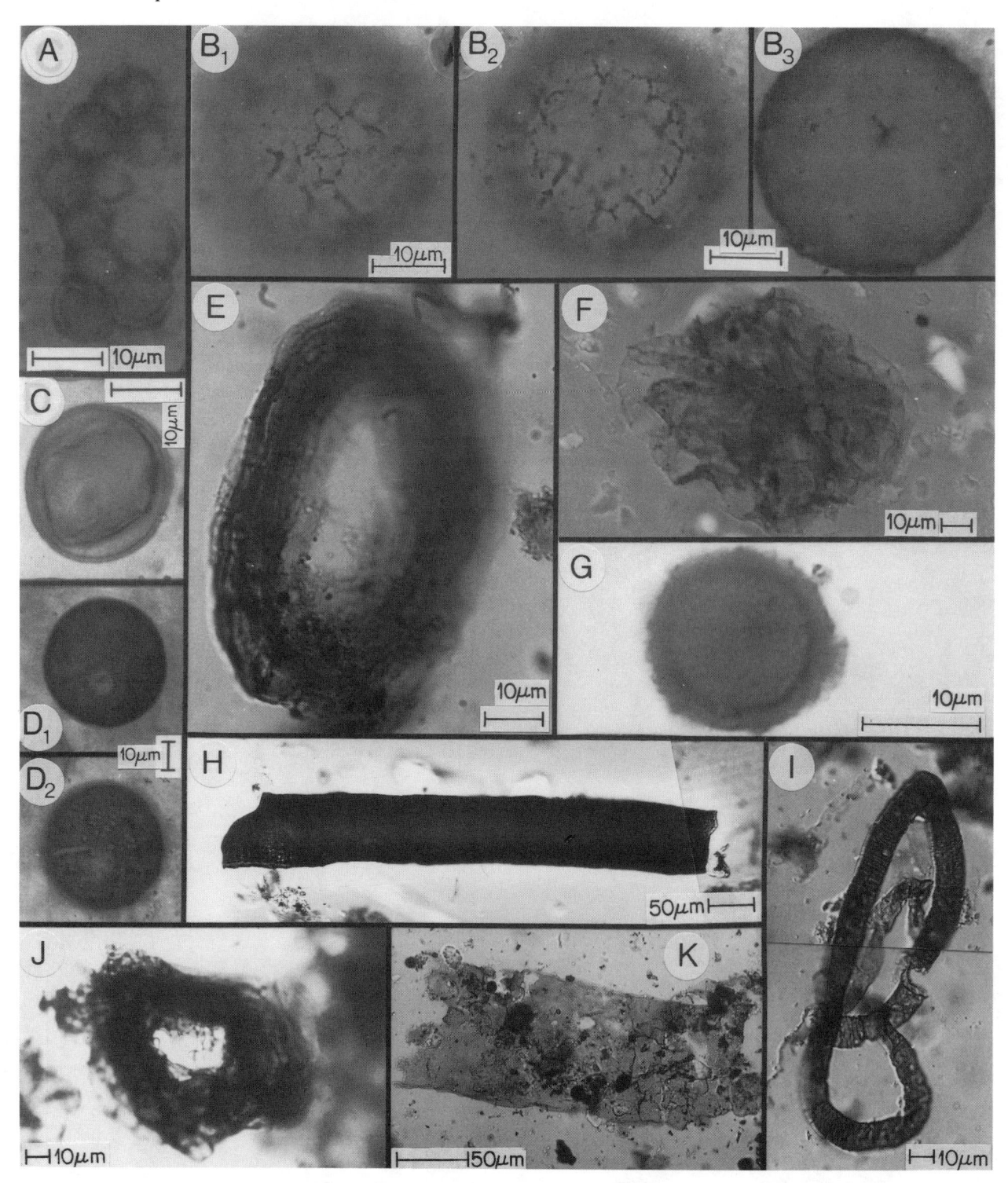

Figure 51. A–D, Chichkan Formation, Zhanatas region, southern Kazakhstan, USSR (650 Ma; locality cf. PPRG 1473). [A–D, petrographic thin sections, transmitted light; B_1–B_3 and D_1, D_2 show single specimens at differing focal depths.] A, unnamed, sheath-enclosed, colonial unicells; B–D, unnamed solitary unicells (Schopf & Sovietov 1976). E, F, Mogilev Formation, Podolia, Ukraine, USSR (620 Ma). [E, F, acid resistant residues; transmitted light.] E, *Circumiella mogilevica* Aseeva 1974, LECTOTYPE (Aseeva, in press) (slide 2269, England slide U29, label right); F, *Polycavita bullata* (Andreeva 1966) Assejeva (= Aseeva) 1982, NEOTYPE (= *Polycavita frillata* Aseeva in press) (slide 1/2, England slide N58/3, label right). G, Suirovo Formation, southern Ural Mountains, Bashkiria, USSR (620 Ma). [G, acid resistant residue, interference contrast.] G, *Retiforma tolparica* Mikhailova (= Mikhajlova) 1987 (in Mikhajlova and Podkovyrov 1987), HOLOTYPE (slide 3137/2, England slide L40/2, read label from right). H–K, Yaryshev Formation, Podolia, Ukraine, USSR (610 Ma). [H–K, acid resistant residues; H, I, interference contrast; J, K, transmitted light; H, I, composite photomicrographs.] H, *Striatella coriacea* Assejeva (= Aseeva) 1983, HOLOTYPE (slide 1796/2, England slide W40/3, label right); I, *Tubulosa corrugata* Assejeva (= Aseeva) 1982, HOLOTYPE (slide 1428/1); J, *Volyniella valdaica* (Shepeleva 1967) Aseeva 1974, LECTOTOPE (Aseeva, in press) (slide 2614, England slide J57, label right); K, *Taenitrichoides jaryschevicus* Assejeva (= Aseeva) 1983, HOLOTYPE (slide 1796/1, England slide P30/2, label right).

Illustrated Specimens 1113

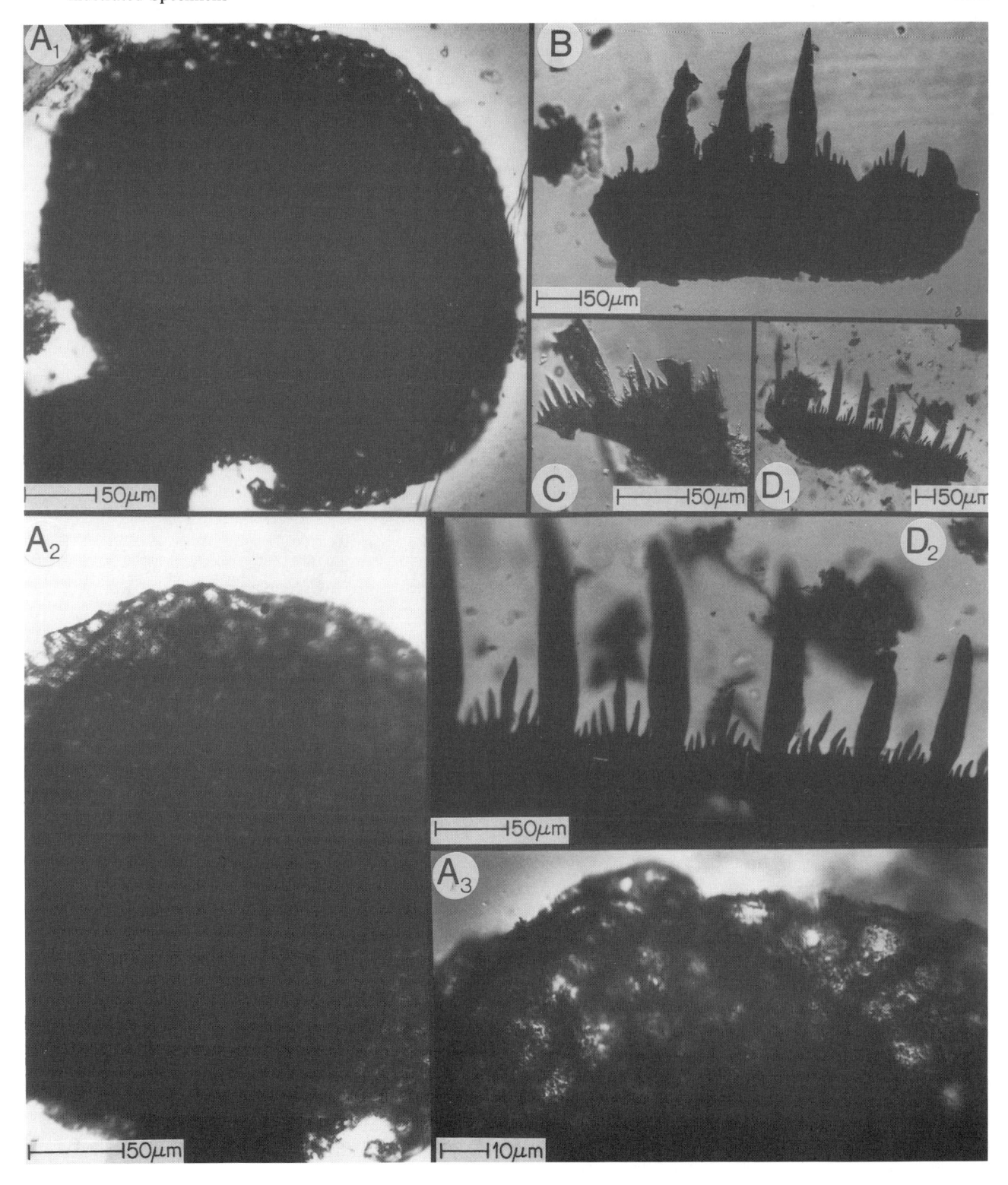

Plate 52. A–D, Yaryshev Formation, Podolia, Ukraine, USSR (610 Ma). [A–D, acid resistant residues; A$_1$, B, D$_2$, transmitted light; A$_2$, A$_3$, C, D$_1$, interference contrast; A$_1$–A$_3$ and D$_1$, D$_2$ each show single specimens at differing magnifications.] A$_1$–A$_3$, *Ljadovia perforata* Assejeva (=Aseeva) 1983, HOLOTYPE (slide 1796/3, England slide U41/2, label right); B, *Redkinia fedonkinispis* Aseeva in press, PARATYPE (slide 1876/1, England slide S59/4, label right); C, *Redkinia fedonkinispis* Aseeva in press, PARATYPE (slide 2009, England slide L62/1, label right); D$_1$, D$_2$, *Redkinia fedonkinispis* Aseeva in press, HOLOTYPE (slide 1656/1, England slide L59/1, text to right).

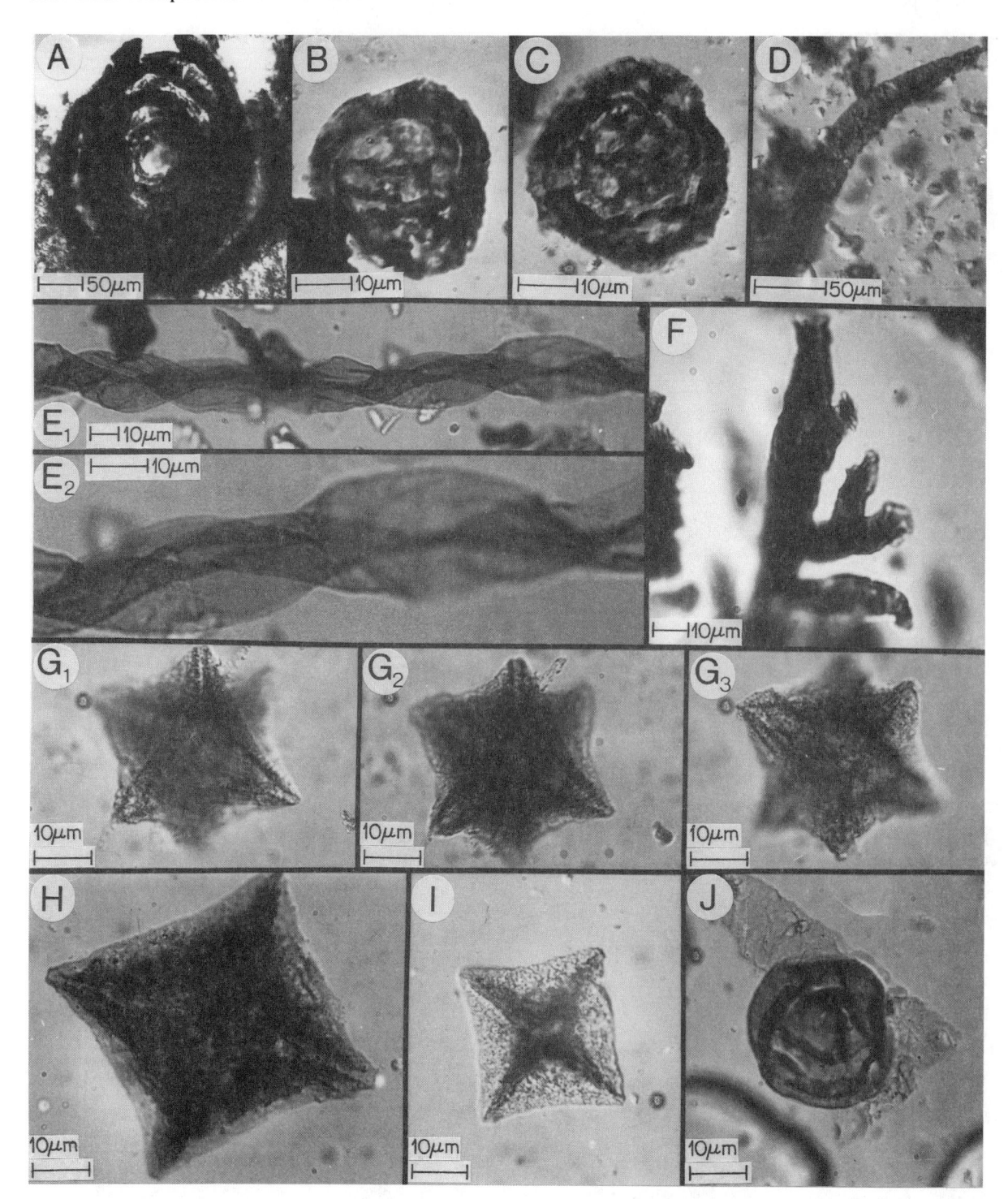

Plate 53. A–F, Kanilovka Formation, Podolia, Ukraine, USSR (600 Ma). [A–F, acid resistant residues; A–D, interference contrast; E, F, transmitted light; E_1, E_2 show a single specimen at two magnifications.] A, *Volyniella canilovica* Aseeva 1974, LECTOTYPE (= *Cocheatina canilovica* Aseeva, in press) (slide 1813, England slide N48, label right); B, *Podoliella irregulare* Timofeev 1973, HOLOTYPE (slide 744/1, England slide X47/3, read label from right) [S21]; C, *Podoliella irregulare* Timofeev 1973, PARATYPE (slide 744/1, England slide U29/3, read label from right) [S21]; D, *Bicuspidata fusiformis* Assejeva (= Aseeva) 1982, HOLOTYPE (slide 1598/1, England slide K68/4, label right); E_1, E_2, *Flagellis tenuis* Assejeva (= Aseeva) 1982, HOLOTYPE (slide 1483/1/22, England slide K68/1, label right); F, *Studenicia bacotica* Assejeva (= Aseeva) 1982, HOLOTYPE (slide 68/4 = slide 1428/1, England slide U42, label right). G–I, Derlo Formation, Podolia, Ukraine, USSR (600 Ma). [G–I, acid resistant residues, interference contrast; G_1–G_3 show a single specimen at three focal depths.] G_1–G_3, *Octoedryxium truncatum* (Rudavskaya) Vidal 1976, PARATYPE (German, in press) (slide 679/3, England slide V35/1, read label from right) [A15]; H, *Octoedryxium simmetricum* Timofeev 1973, HOLOTYPE (slide 679/1, England slide P44/1, read label from right) [A16]; I, *Octoedryxium truncatum* (Rudavskaya) Vidal 1976, PARATYPE (German, in press) (slide 679/3, England slide 041, read label from right) [A15]. J, Zin'kov Formation, Podolia, Ukraine, USSR (600 Ma). [J, acid resistant residue, interference contrast.] J, *Leiosphaeridia undulata* Timofeev 1973, HOLOTYPE (slide 1080/3, England slide T26/2, read label from right).

Illustrated Specimens   1115

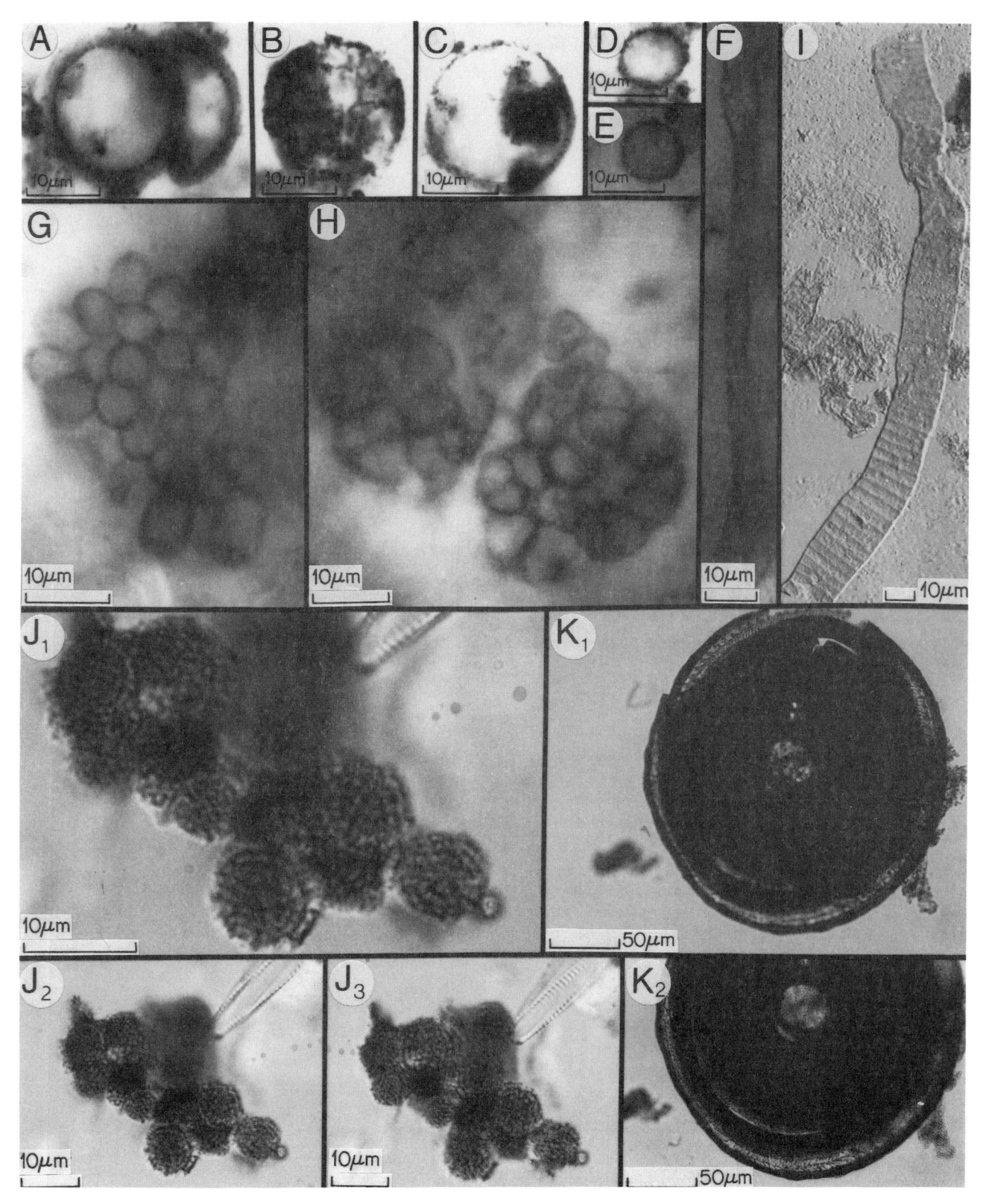

Plate 54. A–H, Yudoma Formation, Yudoma-Maya region, Siberia, USSR (600 Ma; A–E, locality cf. PPRG 2727). [A–H, petrographic thin sections, transmitted light.] A–C, "undifferentiated spheroids" (Mendelson and Schopf 1982) [S17]; D, E, *Huroniospora* spp. (Mendelson and Schopf 1982) [S6]; F, ?*Eomycetopsis siberiensis* Lo 1980, HOLOTYPE; G, *Myxococcoides staphylidion* Lo 1980, HOLOTYPE [CIR6]; H, *Tetraphycus conjunctum* Lo 1980, HOLOTYPE [TET3]. I–K, Kotlin Formation, East European Platform, Latvia and Belorussia, USSR (560 Ma). [I–K, acid resistant residues; I, J, interference contrast; K, transmitted light; J_1–J_3, K_1, K_2 each show single specimens at differing magnifications and/or focal depths.] I, *Oscillatorites wernadskii* Shepeleva 1960 (Yankauskas, in press) (slide 15-101-697/15, England slide Q4, scribed label left); J_1–J_3, *Bavlinella faveolata* Shepeleva 1962, LECTOTYPE (German et al. 1989) (slide Shepeleva 16/1893, England slide S34/4, white label left) [CSF1]; K_1, K_2, *Volyniella rara* Paskeviciene 1980, HOLOTYPE (slide N15-86-3, England slide R37/4, scribed label left).

1116 Atlas of Representative Proterozoic Microfossils

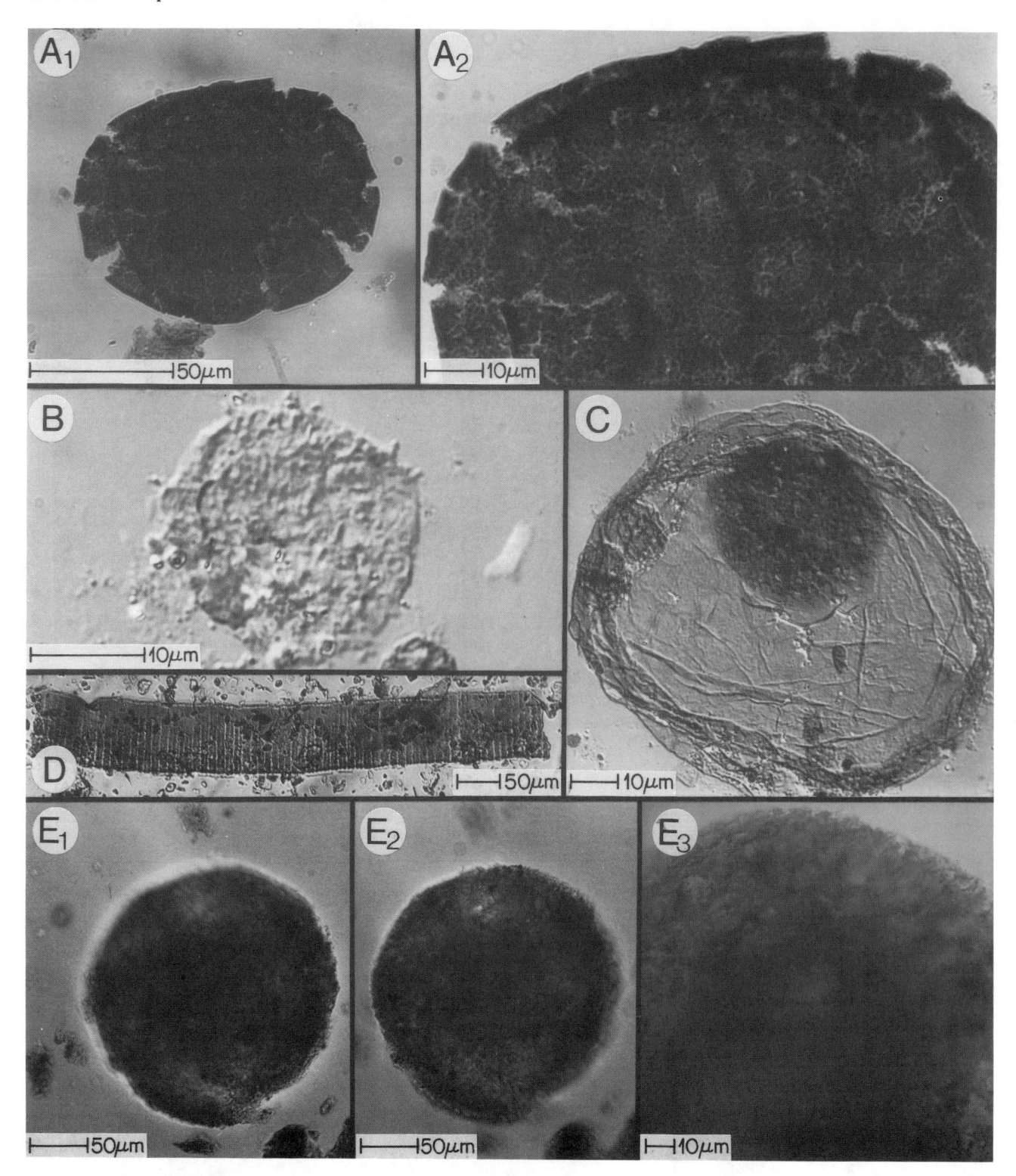

Plate 55. A–E, Redkino Formation, forelands of the Ural Mountains, Bashkiria, USSR (580 Ma). [A–E, acid resistant residues; A, D, E_3, transmitted light; B, C, E_1, E_2, interference contrast; A_1, A_2 E_1–E_3 each show single specimens at differing focal depths and/or magnification.] A_1, A_2, *Satka granulosa* Jankauskas (=Yankauskas) 1980, HOLOTYPE (slide 16-800-2942/2, England slide M25, label left); B, *Baltisphaeridium perrarum* Jankauskas (=Yankauskas) 1980, HOLOTYPE (slide 16-800-2942-1, England slide C20/1, label left) [A9]; C, *Leiosphaeridia incrassatula* Jankauskas (=Yankauskas) 1980, HOLOTYPE (slide 16-800-2942/8, England slide K16/3, label left); D, *Striatella coriacea* Assejeva (=Aseeva) 1983, PARATYPE (slide 536, England slide J59/2, label right); E_1–E_3, *Spumosina rubiginosa* (Andreeva) emend. Jankauskas (=Yankauskas) & Medvedeva (in German 1989), HOLOTYPE (slide Shepeleva N44/17265, England slide H32/4, white label left).

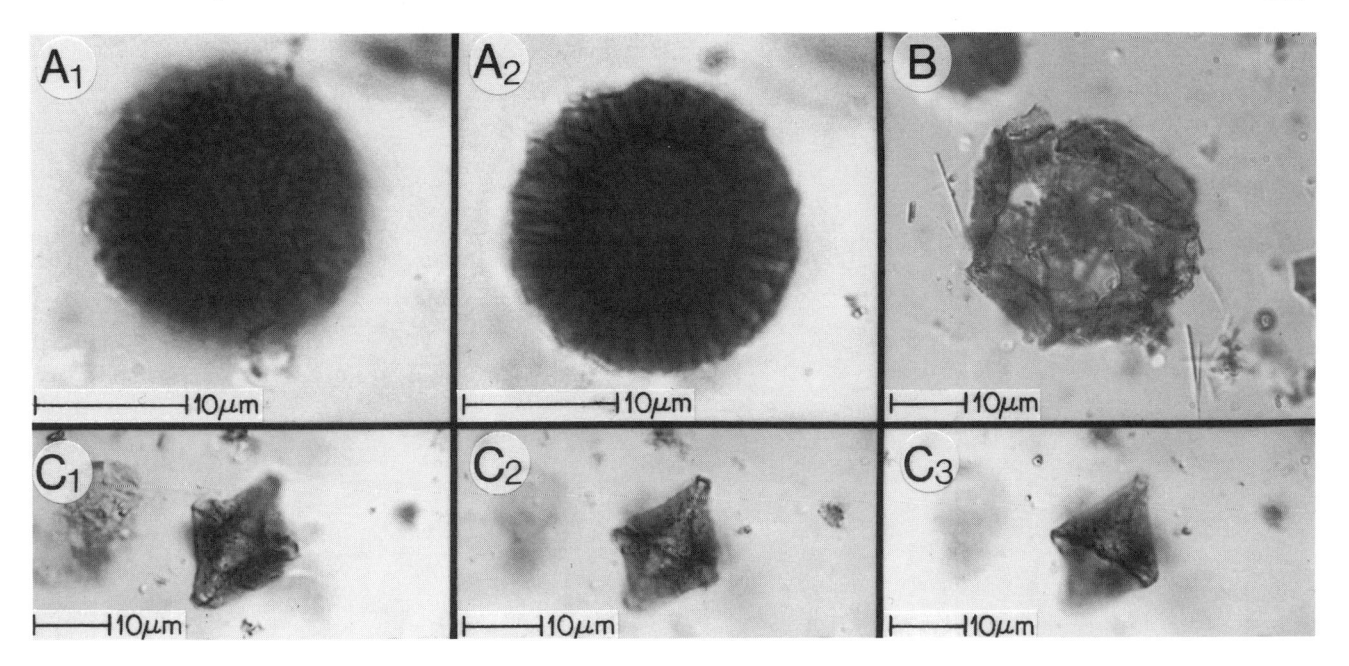

Plate 56. A–C, Moty Formation, Irkutsk region, Siberia, USSR (Lower Cambrian, 540 Ma). [A–C, acid resistant residues, interference contrast; A_1, A_2 C_1–C_3 each show single specimens at differing focal depths.] A_1, A_2, *Paracrassosphaera dedalea* Rudavskaya 1979 (in Fajzulina and Treshchetenkova 1979), HOLOTYPE (slide 604/1, England slide F20, read label from right) [A43]; B, *Polyedryxium neftelenicum* Rudavskaya 1971, HOLOTYPE (slide 116/1/6, England slide M40/3, read label from right); C_1–C_3 *Octoedryxium truncatum* (Rudavskaya) Vidal 1976, HOLOTYPE (slide 93/1-4, England slide M25/2, read label from right) [A15].

25

Informal Revised Classification of Proterozoic Microfossils

J. WILLIAM SCHOPF

25.1 Introduction

Although micropaleontological study of the Proterozoic was already underway nearly a century ago (Walcott 1899), significant progress has been of surprisingly recent vintage. Indeed, of the 2800 occurrences of authentic microfossils now known from sediments of Proterozoic age (Table 22.3), more than 85% have been reported during the past two decades (Section 5.2). Reported chiefly from shales (62% of occurrences) and stromatolitic cherts (38%), planktonic (1580 occurrences in 248 formations) and benthic taxa (1219 occurrences in 208 formations) are now well known from units worldwide (Sections 5.4, 5.5; Chapter 22).

Unfortunately, however, meaningful interpretation of these newly available data is difficult, largely because of the lithology-related differing systems of classification and taxonomy of such microfossils that are currently in use. In particular, the majority of microfossils preserved as essentially two-dimensional compressions in Proterozoic shales, extracted from their surrounding matrix by palynological techniques and studied in acid-resistant residues, have been classified as "acritarchs"—organic-walled microfossils, commonly spheroidal, of uncertain systematic position—and have therefore been grouped into morphological categories by use of an "artificial" (non-biologic based) system of classification. In contrast, because of their three-dimensional preservation and morphological comparability to extant bacteria, cyanobacteria, and eukaryotic microalgae, permineralized microfossils detected in Proterozoic cherts and studied in petrographic thin sections have generally been grouped into the same categories as those used for classification of living microorganisms. Thus, because of differences in preservation, methods of study, and systematics, morphologically similar fossils occurring in shales and cherts have rarely been assigned to the same taxon. Moreover, because virtually all such Proterozoic studies, whether of assemblages from shales or from cherts, have involved detection and description of microfossils from but a single type of lithology, comparison of shale and chert microfloras and systematic assessment of the alteration of original morphology occurring during preservation have rarely been attempted. Finally, for numerous reasons—chiefly, the relative youth and rapid growth of the field and the corresponding absence of an accepted worldwide compendium of relevant data (e.g., a Proterozoic equivalent of the almost exclusively Phanerozoic *Treatise on Invertebrate Paleontology*)—virtually identical fossils detected by different workers in different locales have commonly been assigned different binomials; perhaps more than in any other area of paleontology, taxonomic synonymies abound in the Proterozoic microfossil literature.

In order to bring into some degree of coherence the large body of micropaleontological data now available from the Proterozoic: (*i*) an extensive database, estimated to include perhaps 90% of all Proterozoic microfossils reported prior to mid-1988, was amassed from the published literature (Chapter 22); (*ii*) using the approach discussed in Section 5.4, appropriate morphometric data for shale microfloras were corrected for the effects of preservational compression (see Table 5.4.4); and (*iii*) as is discussed in the following paragraphs, a biologically based informal system of classification was established to encompass Proterozoic microfossils from both shales and cherts.

25.2

Informal Revised Classification

As listed in Table 25.3, 373 Proterozoic informal "species," grouped and subgrouped based on morphology and inferred biological affinity, are here recognized. The groups and subgroups used in this classification are summarized in Table 25.1; Table 25.2 summarizes the distribution within these groups of fossil species having morphologically closely similar modern analogues (i.e., "Lazarus taxa," or perhaps more properly, "Coelacanth taxa"; see Section 11.3). Geochronologic range charts, indicating the first, last, and intermediate-age Proterozoic taxonomic occurrences of these 373 taxa, are presented in Figure 25.4; photomicrographs of many of these species are included in Chapter 24.

Forms classified in Table 25.3 include only those entities here regarded (Table 22.3) as bona fide fossils; forms regarded as nonfossils (artifacts, contaminants, and pseudofossils; Table 22.1) or as dubiofossils (Table 22.2) are not included. Taxa listed in Table 25.3 should be regarded as "taxonomically informal species"; synonymies have not been formally specified, and although indistinguishable or closely similar forms have been grouped together, regardless of previously applied binomials, species descriptions have not been formally emended. Effort has been made to refer to each species by use of the applicable binomial having taxonomic priority; however, for those taxa for which the earliest described forms were inadequately defined or illustrated, a later published binomial has been used. Undescribed taxa are referred to by use of informal nomenclature (e.g., "*Archaeotrichion*" sp. #1; "*Siphonophycus*" sp. #9). As feasible (i.e., for all taxa for which biological affinities can be inferred with a reasonable degree of confidence; see Section 5.4.3, Table 5.4.5), the species recognized have been defined based on comparison with appropriate extant microorganisms; that is, effort has been made to derive a biologically based system of classification, one in which fossil species are similar in concept to living, modern, analogues.

25.2.1 Non-Septate Filaments

In order to provide a biologic basis for taxonomic differentiation of Proterozoic non-septate filaments, morphometric data (ranges of diameter) were compiled for the following categories of living microorganisms: the helical oscillatoriacean cyanobacterium *Spirulina* (14 species and varieties; Desikachary 1959); helical "thread cell" bacteria greater than 10 μm in length and greater than 0.5 μm in diameter (26 taxa; Buchanan and Gibbons 1974); the tubular sheaths of monotrichomic (33 taxa) and polytrichomic (five taxa) filamentous bacteria and of non-septate, non-helical "thread cell" bacteria (26 taxa; Buchanan and Gibbons 1974; Laskin and Lechevalier 1977; Clayton and Sistrom 1978; Starr et al. 1981); the unbranched (197 taxa) and branched (three taxa) tubular unlamellated (157 taxa) and lamellated (43 taxa) sheaths of monotrichomic oscillatoriaceans (Desikachary 1959); and the unbranched (two taxa) and branched (28 taxa), unlamellated (16 taxa) and lamellated (14 taxa) tubular sheaths of polytrichomic oscillatoriaceans (Desikachary 1959). These data are tabulated in Table 5.4.8 and are summarized graphically in Figures 5.4.23–5.4.25.

Comparable morphometric data (ranges of minimum and maximum diameters) were compiled for 447 occurrences of tubular non-septate microfossils reported from 104 Proterozoic formations having the temporal distribution summarized in Table 25.5. Data used in this compilation (summarized graphically in Figures 5.4.6, 5.4.7, and 5.4.26–5.4.28) were derived either from direct measurement of the specimens illustrated in Chapter 24 or from measurement of previously published illustrations and published size data (including those in taxonomic descriptions), corrected, where appropriate, for the effects of preservational compression (Section 5.4.2). Tabulated occurrences of non-septate filamentous microfossils were subdivided into four morphological categories: (*i*) unlamellated unbranched tubular; (*ii*) lamellated unbranched tubular; (*iii*) *Spirulina*-like (viz., spirally coiled); and (*iv*) branched tubular. Morphometric data for members of each category were compared with those compiled for living species of filamentous bacteria and cyanobacteria and the categories were subdivided into the 68 informal species of unbranched (TU, TUL, and TUC) and branched (TB) tubular Proterozoic taxa listed in Table 25.3. Each of the non-septate filamentous taxa there listed exhibits a range of morphological variability similar to or somewhat broader than comparable living species; that is, this revised informal classification is biologically based with a slight bias toward the "lumping" rather than the "splitting" of taxa.

25.2.2 Septate Unbranched Filaments

To provide a biologic basis for taxonomic differentiation of Proterozoic septate unbranched filaments, morphometric data (range of cell diameter, range of cell length, medial cell shape, and terminal cell shape) were compiled for 83 taxa of free-living modern septate filamentous bacteria (Buchanan and Gibbons 1974; Laskin and Lechevalier 1974; Clayton and Sistrom 1978; Starr et al. 1981; Jannasch 1984); for 247 species and varieties of living, naked (non-sheath-enclosed), non-heterocystous nostocalean (chiefly oscillatoriacean) cyanobac-

Table 25.1. *Summary of informal revised classification of Proterozoic microfossils*

Category	Code	Number of Species
PROKARYOTIC FILAMENTS		
Non-septate Filaments		
Unlamellated Tubular Sheaths	TU	55
Lamellated Tubular Sheaths	TUL	5
Spirulina-like Morphotypes	TUC	5
Branched Tubular Morphotypes	TB	3
Septate Filaments		
Naked Cellular Trichomes	3N	69
Sheath-enclosed Cellular Trichomes	3S	24
	Subtotal:	161
PROKARYOTIC COCCOIDAL OR EILIPSOIDAL CELLS		
Solitary Coccoidal Cells		
Solitary Coccoid Cells $\leq 60\,\mu m$	S	24
Unordered Colonial Coccoidal Cells		
Irregular Colonies of Coccoid Cells	CIR	17
Colonial Coccoids with Lamellated Sheaths	CLI	3
Spheroidal Colonies of Coccoid Cells	CSF	9
Irregular or Spheroidal Colonial Coccoids	CRCF	7
Pseudofilamentous Colonies of Coccoids	CFL	8
Unordered Colonial Ellipsoidal Cells		
Irregular Colonies of Ellipsoidal Cells	EIC	20
Ordered Colonial Coccoidal Cells		
Gloeocapsa-like Morphotypes	G	6
Chroococcus-like Morphotypes	CHR	11
Myxosarcina-like Morphotypes	MYX	1
Pleurocapsa-like Morphotypes	PLEU	4
Colonial Planner Tetrads of Coccoid Cells	TET	6
Planar (Non-tetrad) Colonial Coccoids	CPL	2
Cuboidal Colonies of Coccoid Cells	CCU	4
Miscellaneous		
Polybessurus sp.	POLY	1
Bizarre Morphotypes	BZ	13
	Subtotal:	136
EUKARYOTES		
Sphaeromorph Acritarchs		
"Mesosphaeromorphs" 60–200 μm	MS	10
"Megasphaeromorphs" $\geq 200\,\mu m$	MG	15
Non-sphaeromorph Acritarchs		
Complex (Non-sphaeromorph) acritarchs	A	43
Filaments and Other Morphotypes		
Non-acritarch Eukaryote-like Morphotypes	EU	8
	Subtotal:	76

Table 25.2. *Phylogenetic and habitat-related distribution of Lazarus species of Proterozoic microfossils*

Category of Microfossil	Code	Microfossil Species (no.)	Lazarus Species (no.)	Lazarus Taxa (%)
PROKARYOTES:				
Non-septate Filaments	TU, TUL, TUC, TB	68	30	44
Septate Filaments	3N, 3S	93	26	28
Solitary Coccoidal Cells $\leq 60\,\mu m$	S	24	17	71
Unordered Colonial Coccoidal Cells	CIR, CLI, CSF, CRCF, CFL	44	4	9
Unordered Colonial Ellipsoidal Cells	EIC	20	4	20
Ordered Colonial Coccoidal Cells	G, CHR, MYX, PLEU, TET, CPL, CCU	34	6	18
Miscellaneous	POLY, BZ	14	0	0
TOTAL PROKARYOTES:		297	87	29
EUKARYOTES:				
Sphaeromorph Acritarchs	MS, MG	25	4	16
Non-sphaeromorph Acritarchs	A	43	0	0
Filaments and Other Morphotypes	EU	8	0	0
TOTAL EUKARYOTES:		76	4	5
BENTHOS:				
Prokaryotic Filaments	TU, TUL, TUC, TB, 3N, 3S	161	56	35
Prokaryotic Colonies (and Miscellaneous)	CIR, CLI, CSF, CRCF, CFL, EIC, G, CHR, MYX, PLEU, TET, CPL, CCU, POLY, BZ	112	14	13
Eukaryotic Filaments and Other Morphotypes	EU	8	0	0
TOTAL BENTHOS:		281	70	25
PLANKTON:				
Prokaryotic Solitary Coccoidal Cells $\leq 60\,\mu m$	S	24	17	71
Eukaryotic Sphaeromorph Acritarchs	MS, MG	25	4	16
Eukaryotic Non-sphaeromorph Acritarchs	A	43	0	0
TOTAL PLANKTON:		92	21	23

Table 25.3. *Informal revised classification of Proterozoic microfossils*

	Range of Diameter (μm)		Number of Taxonomic Occurrences
	Minimum	Maximum	

PROKARYOTIC FILAMENTS
TU—UNLAMELLATED UNBRANCHED TUBULAR SHEATHS

"*Archaeotrichion*" sp. #1

| *TU 1 | 0.2 | 0.3 | 1 |

Modern bacterial (Cytophagaceae) analogue:

Cytophga johnsonae Stanier [Buchanan and Gibbons 1974, p. 102]

| | 0.2 | 0.4 | |

Archaeotrichion contortum Schopf 1968

| TU 2 | 0.3–0.5 | 0.7–1.2 | 6 |

[see Chapter 24, Pl. 21, F, G]

"*Archaeotrichion*" sp. 2

| *TU 3 | 0.5–0.8 | 0.8–1.5 | 6 |

[see Chapter 24, Pl. 13, A]

Modern bacterial (Cytophagaceae) analogue:

Cytophaga krzemieniewskae Stanier [Buchanan and Gibbons 1974, p. 102]

| | 0.5 | 1.5 | |

Modern cyanobacterial (Oscillatoriaceae) analogue:

Phormidium bigranulatum Gardner [Desikachary 1959, p. 261]

| | 0.8 | 1.4 | |

Eomycetopsis campylomitus Lo 1980

| *TU 4 | 0.8–1.0 | 1.0–2.2 | 7 |

[see Chapter 24, Pl. 7, A–D]

Modern cyanobacterial (Oscillatoriaceae) analogue:

Lyngbya limnetica Lemmermann [Desikachary 1959, p. 294]

| | 1.0 | 2.0 | |

"*Eomycetopsis*" sp. #1

| TU 5 | 1.0 | 2.4–2.5 | 2 |

"*Eomycetopis*" sp. #2

| *TU 6 | 1.2 | 1.6 | 3 |

Modern cyanobacterial (Oscillatoriaceae) analogue:

Lyngbya chaetomorphae Iyengar and Desikachary [Desikachary 1959, p. 281]

| | 1.3 | 1.5 | |

"*Eomycetopis*" sp. #3

| *TU 7 | 1.5 | 1.5–2.0 | 3 |

[see Chapter 24, Pl. 10, I]

Modern cyanobacterial (Oscillatoriaceae) analogues:

Lyngbya epiphytica Hieronymus [Desikachary 1959, p. 284]

| | 1.5 | 2.0 | |

Lyngbya bipunctata Lemmermann [Desikachary 1959, p. 290]

| | 1.5 | 2.0 | |

Lyngbya perelegans Lemmermann [Desikachary 1959, p. 309]

| | 1.5 | 2.0 | |

"*Eomycetopsis*" sp #4

| TU 8 | 1.6–2.0 | 2.1–3.5 | 8 |

Siphonophycus crassiusculum Horodyski 1980

| TU 9 | 1.9–2.0 | 3.8–4.5 | 20 |

[see Chapter 24, Pl. 46, A, B]

"*Eomycetopsis*" sp. #5

| TU 10 | 2.0 | 5.0 | 3 |

[see Chapter 24, Pl. 46, D]

Eomycetopsis filiformis Schopf 1968

| *TU 11 | 2.1–2.3 | 2.3–3.4 | 4 |

Modern cyanobacterial (Oscillatoriaceae) analogue:

Lyngbya kuetzingii Schmidle [Desikachary 1959, p. 282]

| | 2.0 | 3.5 | |

Eomycetopis cylindrica Maithy 1975

| *TU 12 | 2.4–2.5 | 3.8 | 2 |

Modern cyanobacterial (Oscillatoriaceae) analogue:

Lyngbya loriae Forti [Desikachary 1959, p. 321]

| | 2.5 | 3.5 | |

Eomycetopsis septata Maithy 1975

| TU 13 | 2.4–2.5 | 4.5–5.0 | 4 |

"*Eomycetopsis*" sp. #6

| *TU 14 | 2.5 | 2.5–3.2 | 4 |

Modern cyanobacterial (Oscillatoriaceae) analogue:

Lyngbya digueti Gomont [Desikachary 1959, p. 310]

| | 2.5 | 3.0 | |

Eomycetopsis rimata Jankauskas (=Yankauskas) 1980

| TU 15 | 2.5 | 6.4 | 4 |

[see Chapter 24, Pl. 21, B]

Eomycetopis robusta Schopf 1968

| *TU 16 | 2.8–3.0 | 4.0–4.2 | 5 |

[see Chapter 24, Pl. 30, S]

Modern cyanobacterial (Oscillatoriaceae) analogue:

Phormidium purpurascens (Kützing) Gomont [Desikachary 1959, p. 262]

| | 3.0 | 4.0 | |

Table 25.3. (Continued).

	Range of Diameter (μm)		Number of Taxonomic Occurrences
	Minimum	Maximum	
Eomycetopsis siberiensis Lo 1980			
TU 17	3.0	6.0–7.0	2
[see Chapter 24, Pl. 49, C]			
"*Eomycetopis*" sp. #7			
∗TU 18	3.0–3.2	4.5–5.0	5
[see Chapter 24, Pl. 27, B]			
Modern cyanobacterial (Oscillatoriaceae) analogue:			
Lyngbya mesotricha Skuja [Desikachary 1959, p. 282]			
	3.0	5.0	
"*Eomycetopis*" sp. #8			
∗TU 19	3.2	3.2	2
Modern cyanobacterial (Oscillatoriaceae) analogue:			
Lyngbya polysiphoniae Fremy [Desikachary 1959, p. 287]			
	3.0	3.3	
Volyniella canilovica Aseeva 1974			
TU 20	3.2	6.4	5
Eomycetopsis rugosa Maithy 1975			
TU 21	3.2	7.6	1
Contrahofilum schopfii Nautiyal 1980			
TU 22	3.8	4.5–5.0	6
Siphonophycus hughesii Nautiyal 1982			
TU 23	3.8	6.1–6.4	4
Leiothrichoides typicus Hermann (=German) 1974			
∗TU 24	3.8–3.9	8.9–9.2	6
Modern cyanobacterial (Oscillatoriaceae) analogue:			
Lyngbya cryptovaginata Schkorbatow [Desikachary 1959, p. 297]			
	4.0	9.0	
"*Eomycetopsis*" sp. #9			
TU 25	4.0	4.0	1
Siphonophycus inornatum Y. Zhang 1981			
TU 26	4.0	7.1–8.0	5
Leiothrichoides maculatus Hermann (=German) 1986			
TU 27	4.2–4.4	9.6–10.4	2
"*Siphonophycus*" sp. #1			
∗TU 28	4.4–4.5	5.1–5.8	3
Modern cyanobacterial (Oscillatoriaceae) analogue:			
Lyngbya borgerti Lemmermann [Desikachary 1959, p. 293–294]			
	4.5	6.0	
Eomycetopsis pflugii Maithy and Shukla 1977			
∗TU 29	5.0	5.7–6.0	3
[see Chapter 24, Pl. 2, N]			

	Range of Diameter (μm)		Number of Taxonomic Occurrences
	Minimum	Maximum	
Modern cyanobacterial (Oscillatoriaceae) analogue:			
Lyngbya spiralis Geitler [Desikachary 1959, p. 289]			
	5.0	6.0	
"*Siphonophycus*" sp. #2			
TU 30	5.0	7.0–7.6	2
Siphonophycus beltense Horodyski 1980			
TU 31	5.0–6.0	8.0–9.0	7
[see Chapter 24, Pl. 26, A]			
Omalophyma gracilia Jankauskas (=Yankauskas) 1982			
TU 32	5.0–5.7	9.6–10.0	2
"*Siphonophycus*" sp. #3			
∗TU 33	5.5–5.7	6.5–6.7	2
Modern cyanobacterial (Oscillatoriaceae) analogue:			
Phormidium subincrustatum Fritsch and Rich [Desikachary 1959, p. 267]			
	5.3	6.5	
"*Siphonophycus*" sp. #4			
TU 34	5.6–6.0	12.0	3
Flagellis tenuis Assejeva (=Aseeva) 1982			
∗TU 35	6.4–7.0	7.6–8.6	4
Modern cyanobacterial (Oscillatoriaceae) analogue:			
Lyngbya rubida Fremy [Desikachary 1959, p. 298–299]			
	6.0	8.0	
Tubulosa corrugata Assejeva (=Aseeva) 1982			
∗TU 36	6.4	9.6–10.8	2
Modern cyanobacterial (Oscillatoriaceae) analogue:			
Phormidium retzii (Agardh) Gomont f. *major* Parukutty [Desikachary 1959, p. 268]			
	6.0	10.5	
Animikiea septata Barghoorn 1965			
TU 37	7.0–7.6	10.0–11.0	4
[see Chapter 24, Pl. 2, L, M]			
Jacutianema solubila Timofeev and Hermann (=German) 1979			
∗TU 38	7.0–7.6	12.2–12.7	2
Modern cyanobacterial (Oscillatoriaceae) analogue:			
Symploca muscorum (Agardh) Gomont [Desikachary 1959, p. 337]			
	7.0	12.0	
"*Siphonophycus*" sp. #5			
∗TU 39	7.6	7.6	1
Modern cyanobacterial (Oscillatoriaceae) analogue:			
Phormidium mucosum Gardner [Desikachary 1959, p. 265]			
	7.2	7.8	

Table 25.3. (Continued).

	Range of Diameter (μm)		Number of Taxonomic Occurrences		Range of Diameter (μm)		Number of Taxonomic Occurrences
	Minimum	Maximum			Minimum	Maximum	
"*Siphonophycus*" sp. #6				*Siphonophycus punctatus* Maithy 1975			
∗TU 40	8.0–8.9	9.6–10.0	4	∗TU 55	24.0–25.5	31.8–33.0	3
[see Chapter 24, Pl. 3, B]				[see Chapter 24, Pl. 46, C]			
Modern cyanobacterial (Oscillatoriaceae) analogue:				Modern cyanobacterial (Oscillatoriaceae) analogue:			
Lyngbya calcifera Bruhl and Biwas [Desikachary 1959, p. 301]				*Lyngbya magnifica* Gardner [Desikachary 1959, p. 320]			
	8.0	10.0			24.0	33.0	
Eomycetopsis lata Golovenoc and Belova 1985				"*Siphonophycus*" sp. #10			
TU 41	8.0–8.3	12.0–12.4	3	∗TU 56	32.0	45.0	1
Siphonophycus kestron Schopf 1968				Modern cyanobacterial (Oscillatoriaceae) analogue:			
TU 42	8.0–9.0	15.0–18.0	4	*Microcoleus lacustris* (Rabehnorst) Farlow f. *minor* Desikachary [1959, p. 345–346]			
[see Chapter 24, Pl. 31, J]					30.0	45.0	
"*Siphonophycus*" sp. #7				"*Siphonophycus*" sp. #11			
∗TU 44	9.6–10.2	15.3	2	∗TU 57	40.0	100.0	1
Modern cyanobacterial (Oscillatoriaceae) analogue:				Modern cyanobacterial (Oscillatoriaceae) analogue (unbranched portion of sheath):			
Lyngbya martensiana Meneghini ex Gomont [Desikachary 1959, p. 318]				*Hydrocoleum heterotrichum* Kützing em. Gomont [Desikachary 1959, p. 348]			
	9.5	16.0			45.0	105.0	
Brevithrichoides bashkiricus Jankauskas (=Yankauskas) 1982				"*Siphonophycus*" sp. #12			
TU 45	9.6	19.1	3	TU 58	75.3	78.2	1
"*Siphonophycus*" sp. #8				**TUL—LAMELLATED UNBRANCHED TUBULAR SHEATHS**			
TU 46	10.8–12.1	13.4–16.6	5	Unnamed "hypothetical lamellated tubular sheath taxon" needed to complete size range reported for an undifferentiated assemblage			
Siphonophycus costatus Jankauskas (=Yankauskas) 1980				TUL 1	6.0	14.0	1
∗TU 47	12.0–14.1	20.0–22.3	8	Unnamed "hypothetical lamellated tubular sheath taxon" needed to complete size range reported for an undifferentiated assemblage			
[see Chapter 24, Pl. 20, A]				TUL 2	18.0	30.0	1
Modern cyanobacterial (Oscillatoriaceae) analogue:				"*Siphonophycus*" sp. #13			
Microcoleus subtorulosus (Breb.) Gomont [Desikachary 1959, p. 345]				TUL 3	22.0	36.0	1
	13.2	20.0		"*Siphonophycus*" sp. #14			
"*Siphonophycus*" sp. #9				TUL 4	30.0	34.0–40.0	3
TU 49	13.0–14.0	17.5–17.8	3	[see Chapter 24, Pl. 6, B–E]			
Siphonophycus transvaalense Beukes, Klein and Schopf 1987				*Siphonophycus lamellosum* Liu 1983			
TU 50	13.2–15.0	26.0–27.0	2	∗TUL 5	13.0	17.5	2
[see Chapter 24, Pl. 1, A–F]				[see Chapter 24, Pl. 49, A, E]			
Rugosoopsis sp. #1				Modern cyanobacterial (Oscillatoriaceae) analogue:			
TU 51	15.4–16.0	17.9	2	*Lyngbya semiplena* (C. Ag.) J. Ag. ex Gomont [Desikachary 1959, p. 315]			
Rugosoopsis sp. #2					12.0	18.0	
TU 52	18.0	19.0	2				
Rugosoopsis tenuis Timofeev and Hermann (=German) 1979							
TU 53	19.1	23.9–28.7	3				
[see Chapter 24, Pl. 13, C]							

Table 25.3. (Continued).

	Range of Diameter (μm)		Number of Taxonomic Occurrences		Range of Diameter (μm)		Number of Taxonomic Occurrences
	Minimum	Maximum			Minimum	Maximum	
TUC—*Spirulina*-LIKE MORPHOTYPES				*Heliconema funiculum* Schopf and Blacic 1971			
Avictuspirulina minuta Strother, Knoll and Barghoorn 1983				∗TUC 4	3.0–4.0	3.8–6.0	5
∗TUC 1	0.7	0.8	1	[see Chapter 24, Pl. 31, B]			
Modern cyanobacterial (Oscillatoriaceae) analogue:				Modern cyanobacterial (Oscillatoriaceae) analogues:			
Spirulina laxissima West [Desikachary 1959, p. 196]				*Arthrospira khannae* Drouet and Strickland [Desikachary 1959, p. 189–190]			
	0.7	0.8			3.0	5.0	
Spirillinema bentorii Shimron and Horowitz 1972				*Arthrospira platensis* (Nordst.) Gomont Var. *tenuis* (Rao) Desikachary [1959, p. 190–191]			
∗TUC 2	1.3–1.8	1.6–2.0	3		4.0	6.0	
[see Chapter 24, Pl. 19, C]				*Arthrospira jenneri* Stizenberger ex Gomont [Desikachary 1959, p. 192]			
Modern bacterial (Spirillaceae) analogue:					4.0	6.0	
Spirillum volutans Ehrenberg [Buchanan and Gibbons 1974, p. 198]				*Volyniella cylindrica* Tynni and Donner 1980			
	1.4	1.7		TUC 5	6.4	9.6	1
Modern cyanobacterial (Oscillatoriaceae) analogue:				**TB-BRANCHED TUBULAR MORPHOTYPES**			
Spirulina meneghiniana Zanardini ex Gomont [Desikachary 1959, p. 195]				*Ramivaginalis uralensis* Nyberg and Schopf 1984			
	1.0	2.0		TB 3	4.0	9.0	1
Heliconema australiense Schopf 1968				[see Chapter 24, Pl. 46, E]			
∗TUC 3	2.2–2.5	2.8–3.5	3	"*Ramivaginalis*" sp. #1			
[see Chapter 24, Pl. 22, H]				TB 4	14.0	18.0	1
Modern cyanobacterial (Oscillatoriaceae) analogue:				*Palaeosiphonella cloudii* Licari 1978			
Arthrospira spirulinoides Ghose f. *tenuis* (Singh) Desikachary [1959, p. 189]				TB 5	14.0–24.0	38.0–50.0	2
	2.0	3.3		[see Chapter 24, Pl. 22, A, B; Pl. 39, A]			

	Cell Diameter Range (μm)	Cell Length Range (μm)	Medial Cell Shape	Terminal Cell Shape	Number of Taxonomic Occurrences
3N—NAKED CELLULAR TRICHOMES					
Biocatenoides sphaerula Schopf 1968					
3N 1	0.1/0.3–0.3/0.5	0.1/0.3–0.3/0.5	Q		6
[see Chapter 24, Pl. 1, O]					
Biocatenoides incrustata J. Oehler 1977					
∗3N 2	0.2/0.6–0.8/1.0	0.3/0.6–0.8/1.0	Q		7
Modern bacterial (Beggiatoaceae) analogue:					
Beggiatoa minima Winogradsky [Buchanan and Gibbons 1974, p. 114]					
	≤1.0	1.0	Q	Trd	
Biocatenoides rhabdos Schopf 1968					
3N 3	0.3/0.8–0.4/1.5	0.8/1.5–1.4/1.7	Q, L, Q/L		8

1130 Informal Revised Classification

Table 25.3 (Continued).

	Cell Diameter Range (μm)	Cell Length Range (μm)	Medial Cell Shape	Terminal Cell Shape	Number of Taxonomic Occurrences
"*Gunflintia*" sp. #1					
3N 4	0.4/1.0–2.8/3.3	1.0/1.5–3.0	Q, Q/L		3
Ferrimonilis variabile J. Oehler 1977					
3N 5	0.4–3.4	1.0–8.0	L		1
Eomicrocoleus crassus Horodyski 1980					
3N 6	0.5–1.0	0.2–0.4	D		1
Gunflintia minuta Barghoorn 1965					
3N 7	0.5–2.0	0.8/1.0–2.5	Q, Q/L		2
[see Chapter 24, Pl. 2, J, K]					
"*Gunflintia*" sp. #2					
3N 8	0.5/0.8–1.7/2.0	1.4/1.5–3.5	Q, L		2
"*Gunflintia*" sp. #3					
3N 9	0.6–2.4/2.6	0.9/1.0–2.0/2.3	Q		2
Gunflintia septata (Schopf) J. Oehler 1977					
3N 10	0.8/1.1–0.9/1.5	0.5/1.0–0.9/1.2	Q		5
Cephalophytarion minutum Schopf 1968					
3N 11	0.8/0.9–1.4/1.5	1.0/1.3–2.0/2.1	Q	Tgl	2
[see Chapter 24, Pl. 3, 0; Pl. 31, I]					
"*Gunflintia*" sp. #4					
3N 12	1.0/1.2–1.4/2.0	1.0/1.5–1.5/2.5	Q	Trd	6
Cyanonema minor J. Oehler 1977					
3N 13	1.0/1.6–1.5/2.6	1.4/1.6–2.2/2.9	Q		5
Beckspringia communis Licari 1978					
3N 14	1.0/1.2–3.0/3.4	1.5/5.0–2.0/5.0	Q, Q/L		4
[see Chapter 24, Pl. 21, H, I]					
"*Cyanonema*" sp. #1					
3N 15	1.1–1.5	0.6–3.2	Q, D, L		1
"*Cyanonema*" sp. #2					
*3N 16	1.0–1.2	4.0–6.0	L		1
Modern cyanobacterial (Oscillatoriaceae) analogue:					
Oscillatoria trichoides Szafer [Desikachary 1959, p. 228]					
	1.0–1.5	4.0–5.0	L	Trd	
Anabaenidium barghoornii Edhorn 1973					
*3N 17	1.2/1.3–1.7/2.0	2.0/2.1–2.5	Q		2
Modern cyanobacterial (Oscillatoriaceae) analogue:					
Oscillatoria pseudogeminata Schmid [Desikachary 1959, p. 228]					
	1.3–2.2	2.0–2.6	Q	Trd	
Cyanonema attenuata Schopf 1968					
3N 18	1.2/1.3–2.2/2.4	1.9/2.5–4.5/4.8	L, Q/L	Tgl	2

Table 25.3 (*Continued*).

	Cell Diameter Range (μm)	Cell Length Range (μm)	Medial Cell Shape	Terminal Cell Shape	Number of Taxonomic Occurrences
Calyptothrix annulata Schopf 1968					
3N 19	1.5/1.8–2.4/3.0	0.6/0.8–1.0/1.3	D	Tta	2
Cyanonema ligamen Z. Zhang 1985					
*3N 20	1.5/1.8–2.0/2.8	1.7/2.5–3.3/4.0	Q	Tta	5
Modern cyanobacterial (Oscillatoriaceae) analogue:					
Oscillatoria animalis Agardh ex Gomont f. *tenuior* Stockmeyer [Desikachary 1959, p. 240]					
	1.8–2.5	2.0–4.8	Q	Tta	
Contortothrix vermiformis Schopf 1968					
*3N 21	1.7/2.3–2.3/2.7	0.8/1.8–1.2/2.5	Q, Q/D	Trd	6
[see Chapter 24, Pl. 22, F]					
Modern bacterial (*Achroonema*: gliding bacterium; *Thiothrix*: Leucothricaceae) analogues:					
Achroonema inaequale Skuja [Buchanan and Gibbons 1974, p. 124]					
	2.0–2.7	1.5–3.0	Q	Trd	
Thiothrix nivea (Rabenhorst) Winogradsky [Buchanan and Gibbons 1974, p. 119]					
	1.5–3.0	1.5–3.0	Q	Trd	
Veteronostocale amoenum Schopf 1968					
3N 22	1.8/2.0–2.5/3.5	1.8/2.5–2.6/3.0	Q	Trd	3
[see Chapter 24, Pl. 30, L]					
Oscillatoriopsis acuminata Xu 1984					
*3N 23	1.9/2.0–3.0/3.2	1.5/2.0–3.0	Q	Tta	2
[see Chapter 24, Pl. 22, C]					
Modern cyanobacterial (Oscillatoriaceae) analogue:					
Oscillatoria schultzii Lemmermann [Desikachary 1959, p. 232]					
	2.5–3.5	1.5–4.0	Q	Tta	
Cyanonema inflatum J. Oehler 1977					
3N 24	2.0/2.1–2.9/3.6	2.1/3.0–3.5/5.4	Q		4
Cephalophytarion grande Schopf 1968					
3N 25	2.0/2.7–2.8/4.7	1.6/2.0–2.7/6.7	Q	Tgl	5
[see Chapter 24, Pl. 32, A, B, D]					
Cephalophytarion piliformis Mikhailova (=Mikajlova) 1986					
*3N 26	2.3/2.8–3.0/4.5	2.0/3.5–3.5/5.0	Q	Tgl	5
[see Chapter 24, Pl. 43, A]					
Modern cyanobacterial (Oscillatoriaceae) analogue:					
Oscillatoria amoena (Kützing) Gomont [Desikachary 1959, p. 230]					
	2.5–5.0	2.5–4.2	Q	Tgl	
Gunflintia grandis Barghoorn 1965					
3N 27	2.3/2.5–3.8/5.5	5.0–7.0/8.0	L, Q/L	Trd	3
[see Chapter 24, Pl. 2, H, I; Pl. 42, G]					
"*Cephalophytarion*" sp. #1					
3N 28	2.5/3.2–3.5/4.4	4.2/6.0–7.0/8.0	Q, L, Q/L	Tta	3
[see Chapter 24, Pl. 11, B]					

Table 25.3 (*Continued*).

	Cell Diameter Range (μm)	Cell Length Range (μm)	Medial Cell Shape	Terminal Cell Shape	Number of Taxonomic Occurrences
Filiconstrictosus diminutus Schopf and Blacic 1971					
3N 29	2.5/2.9–3.2/4.7	1.3/1.8–2.5/3.0	Q	Trd	4
[see Chapter 24, Pl. 30, J; Pl. 31, H]					
Archaeonema longicellulare (Schopf) Schopf and Barghoorn 1969					
3N 30	2.1/2.8–3.3/4.4	2.6/6.6–10.2/11.0	L, Q/L		3
"*Calyptothrix*" sp. #1					
3N 31	2.8–3.3	1.7–4.0	Q	Tta	1
Caudiculophycus rivularioides Schopf 1968					
3N 32	3.0/3.3–4.0/4.2	2.0–3.1/3.3	Q	Thr	2
[see Chapter 24, Pl. 30, M–O]					
Oscillatoriopsis obtusa Schopf 1968					
3N 33	3.0/3.8–4.0/6.0	2.0/4.0–3.0/5.0	Q	Trd	12
[see Chapter 24, Pl. 31, G]					
"*Oscillatoriopsis*" sp. #1					
*3N 34	3.0/4.0–3.6/6.1	3.3/6.0–5.4/8.0	Q	Trd	10
[see Chapter 24, Pl. 42, F]					
Modern cyanobacterial (Oscillatoriaceae) analogue:					
Oscillatoria chlorina Kützing ex Gomont [Desikachary 1959, p. 215]					
	3.5–6.0	3.7–8.0	Q	Trd	
"*Oscillatoriopsis*" sp. #2					
3N 35	3.0–8.0	1.5–2.5	D		1
[see Chapter 24, Pl. 3, A, C]					
Partitiofilum gongyloides Schopf and Blacic 1971					
*3N 36	3.7/5.4–4.7/6.1	1.0/1.8–1.8/3.0	D, Q/D	Trd	7
[see Chapter 24, Pl. 20, E, F; Pl. 22, E; Pl. 30, P]					
Modern cyanobacterial (Oscillatoriaceae) analogue:					
Oscillatoria grunowiana Gomont [Desikachary 1959, p. 216]					
	3.5–6.0	1.4–4.0	D, Q/D	Trd	
Oscillatoriopsis curta Horodyski 1980					
3N 37	2.5/3.8–5.0	0.5/0.6–1.0	D		2
[see Chapter 24, Pl. 9, A, B; Pl. 22, D]					
Tortunema sibirica Timofeev and Hermann (=German) 1976					
3N 38	3.8–5.7	1.6–2.3	D	Tta	1
Polysphaeroides contextus Timofeev and Hermann (=German) 1976					
*3N 40	3.8/5.1–6.1/6.4	2.5/3.1–5.5/6.5	Q		2
Modern cyanobacterial (Oscillatoriaceae) analogue:					
Oscillatoria terebriformis Agardh ex Gomont [Desikachary 1959, p. 217]					
	4.0–6.5	2.5–6.0	Q	Trd	
Halythrix nodosa Schopf 1968					
*3N 41	4.0/4.3–4.7/5.0	3.0/4.1–4.0/4.3	Q	Tta	2
[see Chapter 24, Pl. 31, K]					

Informal Revised Classification 1133

Table 25.3 (Continued).

	Cell Diameter Range (μm)	Cell Length Range (μm)	Medial Cell Shape	Terminal Cell Shape	Number of Taxonomic Occurrences
Modern cyanobacterial (Oscillatoriaceae) analogue:					
Oscillatoria acuta Bruhl and Biwas orth. mut. Geitler [Desikachary 1959, p. 240–241]					
	4.0–6.0	3.0–4.0	Q	Tta	
"*Oscillatoriopsis*" sp. #3					
3N 42	4.4–9.6	1.5–2.0	D		2
Cephalophytarion constrictum Schopf and Blacic 1971					
*3N 43	4.7–5.3/6.3	2.0/2.2–4.0/4.3	Q/D	Tgl	2
[see Chapter 24, Pl. 31, A, F, L]					
Modern cyanobacterial (Oscillatoriaceae) analogue:					
Oscillatoria rubescens D. C. ex Gomont [Desikachary 1959, p. 235]					
	4.8–6.4	1.2–4.0	D, Q/D	Tgl	
Oscillatoriopsis breviconvexa Schopf and Blacic 1971					
*3N 44	5.0/6.5–6.4/8.3	2.0/3.0–2.7/4.5	Q, D, Q/D	Trd	8
[see Chapter 24, Pl. 20, B; Pl. 30, R; Pl. 31, C]					
Modern cyanobacterial (Oscillatoriaceae) analogue:					
Oscillatoria tenuis Agardh ex Gomont [Desikachary 1959, p. 223]					
	5.0–9.0	2.6–5.0	D, Q/D	Trd	
"*Oscillatoriopsis*" sp. #4					
3N 45	5.0–6.0/8.5	3.5/5.0–5.0/9.0	Q	Trd	6
"*Oscillatoriopsis*" sp. #5					
3N 46	5.0–14.0	3.0–4.5	Q/D		1
Filiconstrictosus eniseicum Weiss (= Vejs) 1984					
3N 47	5.4–6.1	8.0–10.0	Q		1
Gunflintia magna Maithy 1975					
3N 48	5.7–6.4/7.0	10.0/12.0–14.0/17.0	L, Q/L	Trd	2
"*Caudiculophycus*" sp. #1					
3N 49	6.1–7.9	9.5–14.0	Q	Thr	1
"*Calyptothrix*" sp. #2					
*3N 50	6.4–7.6	1.6–2.4	D	Tta	1
Modern cyanobacterial (Oscillatoriaceae) analogue:					
Oscillatoria salina Biwas f. *major* Desikachary [1959, p. 239]					
	6.6–7.9	1.5–2.0	D	Tta	
"*Oscillatoriopsis*" sp. #6					
*3N 51	6.4/7.6–7.6/9.6	3.9/6.0–6.0/8.0	Q	Trd	5
Modern cyanobacterial (Oscillatoriaceae) analogue:					
Oscillatoria nigra Vaucher [Desikachary 1959, p. 223]					
	6.0–10.0	2.8–8.5	Q, Q/D	Trd	
Calyptothrix geminata Jankauskas (= Yankauskas) 1980					
3N 52	7.0/8.3–8.9/9.6	2.0/3.0–4.0/5.0	D, Q/D	Tta	4
[see Chapter 24, Pl. 20, C]					

1134 Informal Revised Classification

Table 25.3 (*Continued*).

	Cell Diameter Range (μm)	Cell Length Range (μm)	Medial Cell Shape	Terminal Cell Shape	Number of Taxonomic Occurrences
"*Cephalophytarion*" sp. #3					
3N 53	7.4–14.7	5.3–9.3	Q	Tgl	1
Oscillatoriopsis bothnica Tynni and Donner 1980					
*3N 54	7.5/8.3–8.9/11.0	3.0/5.0–6.5/7.0	Q, Q/D	Trd	3
Modern cyanobacterial (Oscillatoriaceae) analogue:					
Oscillatoria chalybea (Mertens) Gomont [Desikachary 1959, p. 219]					
	8.0–13.0	3.6–8.0	Q, Q/D	Trd	
"*Oscillatoriopsis*" sp. #7					
3N 55	7.9–8.1	1.5–2.1	D		1
[see Chapter 24, Pl. 30, K]					
Oscillatoriopsis media Mendelson and Schopf 1982					
*3N 56	8.0/8.9–10.2/12.1	2.0/3.0–3.1/5.5	D	Trd	5
[see Chapter 24, Pl. 10, F, H; Pl. 30, Q; Pl. 32, C]					
Modern cyanobacterial (Oscillatoriaceae) analogue:					
Oscillatoria chalybea (Mertens) Gomont [Desikachary 1959, p. 219]					
	6.6–13.0	2.0–6.6	D, D/Q	Trd	
Oscillatorites wernadskii Shepeleva 1960					
3N 57	8.0/8.4–11.5/19.0	6.0/6.6–8.0/8.2	Q, Q/D	Trd	2
Partitiofilum tungusum Mikhailova (=Mikhajlova) in press					
3N 58	8.3–8.9	13.5–15.0	Q/D	Trd	1
[see Chapter 24, Pl. 5, I]					
"*Oscillatoriopsis*" sp. #8					
3N 59	8.3–9.6	7.0–12.0	Q		1
Obconicophycus amadeus Schopf and Blacic 1971					
3N 60	10.0/11.5–12.0/14.0	4.0–4.7/7.0	D	Trd	3
[see Chapter 24, Pl. 31, E]					
Oscillatoriopsis taimirica Shenfil' 1983					
*3N 61	10.5/12.4–11.5/16.6	2.0/3.0–3.5/4.0	D		6
[see Chapter 24, Pl. 20, D]					
Modern cyanobacterial (Oscillatoriaceae) analogue:					
Oscillatoria curviceps Agardh ex Gomont [Desikachary 1959, p. 209]					
	10.0–17.0	2.0–5.0	D	Trd	
Halythrix leningradica Shenfil' 1983					
3N 62	11.0/14.0–13.0/15.0	7.5/10.0–12.0/12.5	Q	Tta	2
"*Oscillatoriopsis*" sp. #9					
3N 63	11.3–25.4	3.3–7.6	D		1
Oscillatoriopsis variabilis Strother, Knoll and Barghoorn 1983					
*3N 64	14.0/15.0–15.9/19.0	4.0/7.0–6.5/9.0	D, Q/D	Trd	3
[see Chapter 24, Pl. 49, B, D]					

Table 25.3 (*Continued*).

	Cell Diameter Range (μm)	Cell Length Range (μm)	Medial Cell Shape	Terminal Cell Shape	Number of Taxonomic Occurrences
Modern cyanobacterial (Oscillatoriaceae) analogue:					
Oscillatoria antillarum Kutzing [Desikachary 1959, p. 242]					
	15.0–18.0	5.0–9.0	D, Q/D	Trd	
Oscillatoriopsis robusta Horodyski 1980					
*3N 65	17.0/20.4–19.5/24.0	4.2/6.5–7.2/9.0	D	Trd	2
Modern cyanobacterial (Oscillatoriaceae) analogue:					
Oscillatoria miniata (Zanardini) Hauck ex Gomont [Desikachary 1959, p. 202]					
	16.0–24.0	4.0–11.0	D	Trd	
"*Oscillatoriopsis*" sp. #10					
3N 66	20.6–22.4	9.5–10.3	D	Tta	1
Oscillatoriopsis maxima (Y. Zhang) Y. Zhang 1986					
3N 67	28.0–38.0	6.0–8.0	D		1
"*Oscillatoriopsis*" sp. #11					
3N 69	37.0–45.0	5.5–10.0	D	Trd	1
Oscillatoriopsis magna Tynni and Donner 1980					
3N 70	45.9–79.6	18.0–35.0	D		1
"*Oscillatoriopsis*" sp. #12					
3N 71	75.0/76.4–76.4/80.0	16.0/23.0–20.0/26.0	D	Trd	2

	Cell Diameter Range (μm)	Cell Length Range (μm)	Sheath Diameter Range (μm)	Medial Cell Shape	Terminal Cell Shape	Number of Taxonomic Occurrences
3S—SHEATH-ENCLOSED CELLULAR TRICHOMES						
Coleobacter primus J. Oehler 1977						
3S 1	0.5–0.5	0.5–1.0	0.7–0.7	Q/L		1
Eophormidium capitatum Xu 1984						
3S 2	1.0–1.5	1.0–1.5	1.7–2.2	Q	Tgl	1
"*Eophormidium*" sp. #1						
3S 3	1.1–2.1	0.9–1.3	3.0–4.3	Q		2
Rhicnonema crassivaginatum Zhu 1982						
3S 4	1.5–3.0	3.0–6.0	11.0–12.5	L		1
Tortunema eniseica Jankauskas (= Yankauskas) 1982						
3S 5	2.0–3.2	2.0–4.5	4.5–7.0	Q		3
[see Chapter 24, Pl. 10, D]						
Eophormidium liangii Xu 1984						
*3S 6	2.0–2.6	1.1–2.2	3.0–3.6	Q, Q/D		2
Modern cyanobacterial (Oscillatoriaceae) analogue:						
Phormidium bohneri Schmidle [Desikachary 1959, p. 262–262]						
	2.2–2.8	1.6–2.8	3.0–3.7	Q	Trd	

Table 25.3 (Continued).

	Cell Diameter Range (μm)	Cell Length Range (μm)	Sheath Diameter Range (μm)	Medial Cell Shape	Terminal Cell Shape	Number of Taxonomic Occurrences
"*Palaeolyngbya*" sp. #1						
*3S 7	3.0–9.0	4.0–10.0	4.5–10.5	Q		1
Modern cyanobacterial (Oscillatoriaceae) analogue:						
Phormidium favosum (Bory) Gomont [Desikachary 1959, p. 275]						
	4.0–9.0	3.0–7.0	4.5–9.5	Q	Trd	
Palaeoscytonema srivastavae Maithy 1977						
3S 8	3.2–5.0	1.0–4.0	10.1–12.9	Q, Q/D	Trd	3
[see Chapter 24, Pl. 22, G; Pl. 28, A]						
Angaronema septatum Golovenoc and Belova 1985						
3S 9	4.0–5.0	7.0–10.0	16.0–17.0	Q/L		1
Palaeolyngbya elongata Mandal and Maithy 1984						
*3S 10	4.1–4.7	5.0–5.5	6.6–7.2	Q		1
Modern cyanobacterial (Oscillatoriaceae) analogue:						
Phormidium incrustatum (Nageli) Gomont [Desikachary 1959, p. 269]						
	4.0–5.0	3.5–5.2	6.0–7.0	Q	Tta	
Palaeolyngbya catenata Hermann (=German) 1976						
*3S 11	6.4–8.3	1.0–4.5	7.1–10.6	D, Q/D		2
Modern cyanobacterial (Oscillatoriaceae) analogue:						
Phormidium uncinatum (Agardh) Gomont [Desikachary 1959, p. 276]						
	6.0–9.0	2.0–6.0	8.0–11.0	D, Q/D	Trd	
Lomentunella vaginata Hermann (=German) 1981						
3S 12	6.4–9.6	8.0–15.0	7.6–10.8	Q		1
Palaeolyngbya helva Hermann (=German) 1981						
3S 13	7.0–8.9	6.0–8.0	8.6–10.5	Q	Trd	1
[see Chapter 24, Pl. 11, A]						
"*Palaeolyngbya*" sp. #2						
3S 14	10.8–15.3	18.0–22.0	12.8–17.2	Q	Trd	1
"*Palaeolyngbya*" sp. #3						
3S 15	12.5–15.5	2.5–3.8	27.0–29.5	D		1
Palaeolyngbya lata Tynni and Donner 1980						
*3S 16	14.5–33.0	2.5–8.0	18.9–37.4	D	Trd	3
Modern cyanobacterial (Oscillatoriaceae) analogue:						
Lyngbya sordida (Zanardini) Gomont [Desikachary 1959, p. 285]						
	14.0–31.0	4.0–10.0	17.0–37.4	D	Trd	
"*Palaeolyngbya*" sp. #4						
3S 17	26.0–27.0	9.0–12.0	27.0–28.0	D		1
Circulinema jinningense Wang 1981						
3S 18	24.0–33.0	3.0–8.0	26.0–43.0	D		2
"*Palaeolyngbya*" sp. #5						
3S 19	36.0–42.0	13.0–27.0	52.0–58.0	Q/D		1

Table 25.3 (Continued).

	Cell Diameter Range (μm)	Cell Length Range (μm)	Sheath Diameter Range (μm)	Medial Cell Shape	Terminal Cell Shape	Number of Taxonomic Occurrences
Oscillatoriopsis majuscula Knoll, Strother and Rossi 1988						
3S 20	60.0–74.0	6.0–11.0	72.0–86.0	D		1
Salome svalbardensis Knoll 1982						
3S 21	6.0–15.0	1.8–2.2	31.0–40.0 (lamellated)	D	Trd	1
"*Salome*" sp. #1						
*3S 22	8.2–11.5	2.5–5.0	13.2–16.5 (lamellated)	D		1
Modern cyanobacterial (Oscillatoriaceae) analogue:						
Lyngbya stagnina Kutzing [Desikachary 1959, p. 317]						
	9.5–12.0	2.0–4.0	11.0–16.0 (lamellated)	D	Trd	
Polythrichoides lineatus Jankauskas (= Yankauskas) 1982						
3S 23	1.6–2.5	3.0–6.0	2.5–7.6	L		3
Salome hubeiensis Z. Zhang 1986						
3S 24	14.0–29.0	5.0–7.0	62.0–80.0 (lamellated)	D		1

	Range of Diameter (μm)		Divisional Dispersion Index (DDI)	Number of Taxonomic Occurrences
	Minimum	Maximum		

PROKARYOTIC COCCOIDAL OR ELLIPSOIDAL CELLS

S—SOLITARY COCCOID CELLS (MAXIMUM DIAMETER ≤60 μm)

	Minimum	Maximum	DDI	Occurrences
"*Huroniospora*" sp. #1				
*S 1	0.2	2.0	6	1
Modern bacterial (Peptococcaceae) analogue:				
Ruminococcus albus Hungate [Buchanan and Gibbons 1974, p. 527]				
	0.8	2.0	4	
"*Huroniospora*" sp. #2				
*S 2	2.0	7.0	6	1
[see Chapter 24, Pl. 1, G; Pl. 5, E]				
Modern cyanobacterial (Chroococcaceae) analogues (non-colonial, single cells):				
Microcystis viridis (Braun) Lemmerman [Desikachary 1959, p. 87]				
	3.0	7.0	4	
Microcystis aeruginosa Kützing [Desikachary 1959, p. 93]				
	3.0	7.0	4	
Microcystis flos-aquae (Wittr.) Kirchner [Desikachary 1959, p. 94]				
	3.0	7.0	4	
Microcystis pseudofilamentosa Crow [Desikachary 1959, p. 94–95]				
	3.0	7.0	4	
Pilavia maculata J. H. Oehler 1978				
*S 3	2.7–4.0	4.0–5.0	3	3

1138 Informal Revised Classification

Table 25.3 (*Continued*)

	Range of Diameter (μm)		Divisional Dispersion Index (DDI)	Number of Taxonomic Occurrences
	Minimum	Maximum		

Modern cyanobacterial (Chroococcaceae) analogue (non-colonial, single cells):

Microcystis ramosa Bharadwaja [Desikachary 1959, p. 95]

| | 3.0 | 5.0 | 3 | |

Modern eukaryotic (Chlorellaceae) analogue:

Chlorella pyrenoidsa Chick [West and Fritsch 1927, p. 119]

| | 3.0 | 5.0 | 3 | |

"*Huroniospora*" sp. #3

| *S 4 | 3.8–5.0 | 7.0–8.0 | 5 | 5 |

Modern cyanobacterial (Chroococcaceae) analogue (non-colonial, single cells:

Microcystis scripta (Richter) Lemmermann [Desikachary 1959, p. 91, 93]

| | 4.5 | 8.0 | 3 | |

"*Huroniospora*" sp. #4

| *S 5 | 4.0–5.0 | 10.0–13.0 | 5 | 5 |

Modern cyanobacterial (Entophysalidaceae) analogue (non-colonial, single cells):

Chlorogloea fritschii Mitra [Desikachary 1959, p. 163]

| | 4.0 | 12.0 | 5 | |

"*Protosphaeridium*" sp. #1

| *S 6 | 5.0 | 15.0–20.0 | 6 | 2 |

[see Chapter 24, Pl. 54, D, E]

Modern sulfur metabolizing bacterial analogue:

Thiovulum majus Hinze [Buchanan and Gibbons 1974, p. 463]

| | 5.0 | 25.0 | 7 | |

Modern eukaryotic (Chlorococcaceae) analogue:

Chlorococcum humicolum (Nägeli) Rabenhorst [West and Fritsch 1927, p. 105]

| | 5.0 | 20.0 | 6 | |

Melasmatosphaera parva Hofmann 1976

| *S 7 | 5.0–8.0 | 6.1–9.0 | 3 | 10 |

Modern cyanobacterial (Chroococcaceae) analogues (non-colonial, single cells):

Microcystis robusta (Clark) Nygaard [Desikachary 1959, p. 85]

| | 6.0 | 9.0 | 2 | |

Aphanocapsa roeseana de Bary [Desikachary 1959, p. 131, 132]

| | 5.0 | 8.0 | 2 | |

Modern eukaryotic (Chlorellaceae) analogue:

Chlorella vulgaris Beijernck [West and Fritsch 1927, p. 118]

| | 5.0 | 10.0 | 3 | |

Leiominuscula pellucentis Sin (=Xing) and Liu 1973

| S 8 | 6.0–7.0 | 10.0–14.0 | 4 | 4 |

"*Protosphaeridium*" sp. #2

| S 9 | 7.0–9.1 | 19.0–21.0 | 5 | 4 |

Myxococcoides muricata Schopf and Barghoorn 1969

| *S 10 | 7.9–10.0 | 14.5–17.6 | 3 | 7 |

Table 25.3 (*Continued*)

	Range of Diameter (μm)		Divisional Dispersion Index (DDI)	Number of Taxonomic Occurrences
	Minimum	Maximum		

Modern eukaryotic (Palmellaceae) analogue:

Urococcus foslieanus Hansgrig [Taylor 1957, p. 37]

| | 8.0 | 18.0 | 4 | |

Protosphaeridium rigidulum Timofeev 1966

| S 11 | 8.0–10.0 | 24.0–30.0 | 6 | 3 |

[see Chapter 24, Pl. 42, H]

Asperatopsophosphaera umishanensis Sin (=Xing) and Liu 1973

| *S 12 | 8.0–12.0 | 9.5–12.0 | 2 | 11 |

Modern cyanobacterial (Chroococcaceae) analogue (non-colonial, single cells):

Aphanocapsa crassa Ghose [Desikachary 1959, p. 136]

| | 7.0 | 12.0 | 3 | |

Modern eukaryotic (Palmellaceae) analogue:

Gloeocystis gigas (Kützing) Langerheim [Prescott 1962, p. 84]

| | 9.0 | 12.0 | 2 | |

Protosphaeridium patelliforme Timofeev 1966

| *S 13 | 10.0–14.0 | 30.0–40.0 | 6 | 7 |

Modern eukaryotic (Oocystaceae, Chlorococcaceae) analogues:

Palmellococcus marinus Collins [Taylor 1957, p. 43]

| | 10.0 | 40.0 | 6 | |

Chlorococcum infusionum (Schrank) Meneghini [Archibald and Bold 1970, p. 21]

| | 9.0 | 40.0 | 7 | |

Protosphaeridium scabridum Timofeev 1966

| *S 14 | 10.0–24.0 | 17.0–24.5 | 4 | 34 |

Modern eukaryotic (Chlorococcaceae) analogues:

Chlorococcum endozoicum Collins [Taylor 1957, p. 39]

| | 10.0 | 25.0 | 4 | |

Chlorococcum aerenosum Archibald and Bold [1970, p. 22]

| | 10.0 | 25.0 | 4 | |

Sinaiicoccus minutus Shimron and Horowitz 1972

| *S 15 | 11.0–16.0 | 13.0–16.5 | 2 | 15 |

[see Chapter 24, Pl. 33, G]

Modern eukaryotic (Chlorococcaceae) analogue:

Chlorococcum infusorium Beijernck [West and Fritsch 1927, p. 106]

| | 10.0 | 15.0 | 2 | |

Protosphaeridium paleaceum Timofeev 1966

| S 16 | 10.0–25.0 | 40.0–49.0 | 6 | 28 |

[see Chapter 24, Pl. 24, C]

Protosphaeridium parvulum Timofeev 1966

| *S 17 | 13.0–30.0 | 25.0–30.0 | 4 | 35 |

[see Chapter 24, Pl. 54, A–C]

1140 Informal Revised Classification

Table 25.3 (*Continued*)

	Range of Diameter (μm)		Divisional Dispersion Index (DDI)	Number of Taxonomic Occurrences
	Minimum	Maximum		
Modern eukaryotic (Chlorococcaceae) analogue:				
Spongiochloris excentrica Starr [Bischoff and Bold 1963, p. 29]				
	12.0	30.0	4	
Protosphaeridium tuberculiferum Timofeev 1966				
*S 18	14.0–21.0	52.0–60.0	6	9
Modern eukaryotic (Chlorococcaceae) analogue:				
Neochloris pseudostigmatica Bischoff and Bold [1963, p. 33]				
	15.0	65.0	7	
Stictosphaeridium verrucatum Vidal 1976				
*S 19	15.0–18.8	33.0–38.0	4	11
Modern eukaryotic (Chlorococcaceae) analogue:				
Chlorococcum diploibionticoideum Chantanachat and Bold [1962, p. 23]				
	14.0	36.0	4	
Trachysphaeridium hyalinum Sin (=Xing) and Liu 1973				
*S 20	20.0–23.4	30.7–37.0	3	11
Modern eukaryotic (Oocystaceae) analogue:				
Bracteacoccus grandis Bischoff and Bold [1963, p. 36]				
	20.0	35.0	3	
Trematosphaeridium sinuatum Timofeev 1966				
S 21	25.0–39.0	31.0–39.0	2	29
[see Chapter 24, Pl. 53, B, C]				
Protosphaeridium papyraceum Timofeev 1966				
S 22	25.0–34.0	50.0–60.0	5	16
Stictosphaeridium tortulosum Timofeev 1966				
S 23	29.0–36.0	40.0–49.0	3	15
[see Chapter 24, Pl. 15, B]				
Kildinella rifeica Timofeev 1969				
*S 24	38.0–49.0	40.0–49.0	2	15
Modern eukaryotic (Chlorococcaceae) analogue:				
Spongiochloris incrassata Chantanachat and Bold [1962, p. 28]				
	38.0	54.0	2	

CIR—IRREGULAR COLONIES OF COCCOID CELLS

"*Myxococcoides*" sp. #1				
*CIR 1	0.2–0.8	1.0–1.2	6	3
Modern bacterial (Methanobacteriaceae, Micrococcaceae) analogues:				
Methanococcus vannielii Stadtman and Barker [Buchanan and Gibbons 1974, p. 477]				
	0.5	1.0	3	
Staphylococcus saprophyticus (Fairbrother) Shaw, Stitt and Cowan [Buchanan and Gibbons 1974, p. 488]				
	0.5	1.5	5	

Table 25.3 (Continued)

	Range of Diameter (μm)		Divisional Dispersion Index (DDI)	Number of Taxonomic Occurrences
	Minimum	Maximum		
Modern cyanobacterial (Chroococcaceae) analogue:				
Aphanocapsa delicatissima West and West [Desikachary 1959, p. 133]				
	0.5	0.8	2	
Myxococcoides minuta Muir 1976				
CIR 2	0.7–1.5	3.0–4.0	6	7
Myriasporella pyriformis D. Z. Oehler 1978				
CIR 3	0.9–2.0	1.8–2.7	3	6
Myxomorpha janecekii Muir 1976				
CIR 4	1.1–1.5	6.0–6.2	6	2
Nannococcus vulgaris J. H. Oehler 1977				
CIR 5	1.3–1.6	4.5–4.6	4	3
Myxococcoides staphylidion Lo 1980				
CIR 6	1.6–2.0	11.5–12.1	6	2
[see Chapter 24, Pl. 54, G]				
Dictyosphaera macroreticulata Sin (=Xing) and Liu 1973				
CIR 7	2.0–3.0	4.0–6.7	4	15
Microconcentrica induplicata Liu and Sin (=Xing) 1973				
CIR 8	2.8–6.0	7.7–8.5	4	8
[see Chapter 24, Pl. 10, N]				
Myxococcoides minor Schopf 1968				
CIR 9	3.0–10.0	9.5–12.8	5	18
[see Chapter 24, Pl. 32, H, I; Pl. 42, B]				
Palaeoanacystis vulgaris Schopf 1968				
*CIR 10	3.1–5.4	5.0–7.5	3	15
Modern cyanobacterial (Chroococcaceae) analogues:				
Microcystis viridis (Braun) Lemmerman [Desikachary 1959, p. 87]				
	3.0	7.0	4	
Microcystis aeruginosa Kützing [Desikachary 1959, p. 93]				
	3.0	7.0	4	
Symplassosphaeridium incrustatum Timofeev 1966				
CIR 11	4.0–8.0	15.0–25.0	6	17
Gloeocapsomorpha makrocysta Timofeev 1966				
CIR 12	4.0–9.0	30.0–34.0	6	6
Myxococcoides reticulata Schopf 1968				
CIR 13	8.0–16.2	13.0–21.0	3	18
Protoleiosphaeridium conglutinatum Timofeev 1960				
CIR 14	10.0–20.0	19.0–25.6	3	21
Polyedrosphaeridium bullatum Timofeev 1966				
CIR 15	13.0–17.0	29.0–40.0	4	9

1142 Informal Revised Classification

Table 25.3 (Continued)

	Range of Diameter (μm)		Divisional Dispersion Index (DDI)	Number of Taxonomic Occurrences
	Minimum	Maximum		
Latisphaera wrightii Licari 1978				
CIR 16	15.0–40.0	54.0–62.0	4	3
Myxococcoides congoensis Maithy 1975				
CIR 17	20.0–30.0	28.0–40.0	3	6

	Range of Cell Size (μm)	Range of Sheath Thickness (μm)	Divisional Dispersion Index (DDI)	Number of Taxonomic Occurrences
CLI—COLONIAL COCCOIDS WITH LAMELLATED SHEATHS				
"*Gloeodiniopsis*" sp. #3				
CLI 1	3.0–5.0	1.0–2.0	2	1
Gloeodiniopsis gregaria Knoll and Golubić 1979				
CLI 2	4.3–8.5	0.5–2.5	2	2
Vesicophyton punctatum Vologdin 1969				
CLI 3	15.0–60.0	7.0–16.0	4	1

	Range of Diameter (μm)		Divisional Dispersion Index (DDI)	Number of Taxonomic Occurrences
	Minimum	Maximum		
CSF—SPHEROIDAL COLONIES OF COCCOID CELLS				
Bavlinella faveolata (Shepeleva) Vidal 1976				
CSF 1	0.3–1.0	0.6–1.3	5	9
[see Chapter 24, Pl. 54, J]				
Sphaerocongregus variablis Moorman 1974				
CSF 2	0.5	2.0–3.0	6	2
[see Chapter 24, Pl. 44, K, M; Pl. 48, A, B]				
Favosphaera conglobata Konzalova 1974				
CSF 3	1.0–2.0	2.0–3.6	4	7
Sphaerophycus reticulatum Muir 1976				
CSF 4	2.0–4.0	4.0–7.5	4	9
Clonophycus biattina J. H. Oehler 1977				
CSF 5	2.7–4.0	8.0–13.3	5	6
Favosphaeridium bothnicum Timofeev 1966				
CSF 6	6.0–8.3	13.0–16.0	3	6
Favosphaeridium scandium Timofeev 1966				
*CSF 7	6.0–10.0	8.0–11.0	2	7
Modern cyanobacterial (Chroococcaceae) analogue:				
Aphanocapsa littoralis Hansgrig Var. *macrococca* Hansgrig [Desikachary 1959, p. 131]				
	6.0	11.0	3	
"*Favosphaeridium*" sp. #1				
CSF 8	12.0–13.0	18.0–18.7	1	2

Informal Revised Classification

Table 25.3 (*Continued*)

	Range of Diameter (μm)		Divisional Dispersion Index (DDI)	Number of Taxonomic Occurrences
	Minimum	Maximum		
"*Favosphaeridium*" sp. #2				
CSF 9	17.0–18.0	20.0–22.0	2	2
CRCF—IRREGULAR OR SPHEROIDAL COLONIAL COCCOIDS				
Nevidia multicellularia Vavrdova 1966				
CRCF 1	0.8–2.0	2.0–3.6	5	5
Myxococcoides cracens D. Z. Oehler 1978				
CRCF 2	2.0–4.5	5.7–7.5	4	10
Symplassosphaeridium tumidulum Timofeev 1966				
*CRCF 3	3.6–6.0	8.5–10.0	4	5
[see Chapter 24, Pl. 5, A, C]				
Modern cyanobacterial (Chroococcaceae) analogue:				
Aphanocapsa littoralis Hansgrig Var. *macrococca* Hansgrig [Desikachary 1959, p. 131]				
	4.5	10.0	4	
Symplassosphaeridium subcoalitum Timofeev 1966				
CRCF 4	5.0–10.0	12.0–20.0	4	8
[see Chapter 24, Pl. 48, F]				
Satka colonialica Jankauskas (=Yankauskas) 1979				
CRCF 5	6.0–8.0	28.0–32.0	5	2
Scissilisphaera regularis Knoll and Calder 1983				
CRCF 6	11.0	45.0	6	1
Myxococcoides globosa Maithy and Shukla 1977				
CRCF 7	15.0–20.0	20.0–30.0	3	3

	Range of Cell Size (μm)	Divisional Dispersion Index (DDI)	Number of Taxonomic Occurrences
CFL—PSEUDOFILAMENTOUS COLONIES OF COCCOIDS			
Glenobotrydion aenigmatis Schopf 1968			
CFL 1	3.9–12.8	4	6
Eohyella endoatracta Green, Knoll and Swett 1988			
CFL 2	4.0–19.0 × 7.5–42.0	5	1
Arctacellularia ellipsoidea Hermann (=German) 1976			
CFL 4	6.0–20.0 × 9.0–35.0	6	5
[see Chapter 24, Pl. 24, A]			
Eohyella rectoclada Green, Knoll and Swett 1988			
CFL 5	6.5–21.0	4	2
Entosphaeroides irregularis Maithy 1975			
CFL 6	9.0–18.0	2	3
Entosphaeroides bilinearis Maithy 1975			
CFL 7	16.0–24.0	2	2

Table 25.3 (Continued).

	Range of Cell Size (μm)	Divisional Dispersion Index (DDI)	Number of Taxonomic Occurrences
Glenobotrydion tetragonalum Maithy 1975			
CFL 8	19.0–43.0 × 30.0–58.0	4	2
Arctacellularia doliiformis Hermann (=German) 1976			
CFL 9	20.0–23.0 × 24.0–35.0	2	2

	Range of Width (μm)	Range of Length (μm)	Number of Taxonomic Occurrences

EIC—IRREGULAR COLONIES OF ELLIPSOIDAL CELLS

Eosynechococcus amadeus Knoll and Golubić 1979

| *EIC 1 | 0.7–2.0 | × 1.7–6.0 | 4 |

Modern cyanobacterial (Chroococcaceae) analogue:

Aphanothece saxicola Nageli [Desikachary 1959, p. 138]

| | 1.0–2.0 | × 2.0–6.0 | |

Brachypelganon khandanum Lo 1980

| EIC 2 | 0.8–2.7 | × 6.0–20.3 | 1 |

Eosynechococcus moorei Hofmann 1976

| EIC 3 | 1.0–3.5 | × 2.0–9.0 | 3 |

Eosynechococcus brevis Knoll 1982

| EIC 4 | 1.0–4.5 | × 1.5–6.0 | 3 |

Eosynechococcus isolatus McMenamin, Kumar and Awramik 1983

| EIC 5 | 1.0–6.8 | × 1.7–11.0 | 3 |

[see Chapter 24, Pl. 10, 0]

Eoentophysalis cumulus Knoll and Golubić 1979

| EIC 6 | 1.5–4.0 | × 2.0–5.0 | 7 |

[see Chapter 24, Pl. 9, G]

Eosynechococcus depressus Knoll 1982

| EIC 7 | 1.5–4.0 | × 5.0–10.0 | 3 |

Eosynechococcus medius Hofmann 1976

| *EIC 8 | 2.5–6.0 | × 3.0–7.5 | 6 |

[see Chapter 24, Pl. 9, F, H]

Modern cyanobacterial (Chroococcaceae) analogue:

Aphanothece pallida (Kützing) Rabenhorst f. minor Dixit [Desikachary 1959, p. 141]

| | 2.5–4.0 | × 3.6–6.5 | |

Eosynechococcus thuleensis Strother, Knoll and Barghoorn 1983

| EIC 9 | 3.0–4.6 | × 5.0–25.0 | 1 |

Brevitrichoides burzjanicus Jankauskas (=Yankauskas) 1982

| *EIC 10 | 3.0–6.0 | × 5.0–13.0 | 4 |

Modern cyanobacterial (Chroococcaceae) analogues:

Aphanothece bullosa (Meneghini) Rabenhorst [Desikachary 1959, p. 142]

| | 3.0–5.0 | × 5.0–12.5 | |

Aphanothece bullosa (Meneghini) Rabenhorst [Desikachary 1959, p. 142]

| | 3.8–5.4 | × 6.0–13.2 | |

Eoentophysalis dismallakensis Horodyski and Donaldson 1980

| EIC 11 | 3.0–10.0 | × 4.0–13.0 | 1 |

[see Chapter 24, Pl. 9, E]

Eoentophysalis yudomatica Lo 1980

| EIC 12 | 4.0–21.0 | × 6.0–32.0 | 2 |

"*Eosynechococcus*" sp. #1

| *EIC 13 | 8.0–15.0 | × 10.0–25.0 | 1 |

Modern sulfur metabolizing bacterial analogue:

Macromonas mobilis (Lauterborn) Utermohl and Koppe [Buchanan and Gibbons 1974, p. 463]

| | 6.0–10.0 | × 14.0–30.0 | |

Modern cyanobacterial (Chroococcaceae) analogue:

Synechococcus aeruginosus Nägeli [Desikachary 1959, p. 143]

| | 5.0–16.0 | × 12.0–30.0 | |

Eosynechococcus grandis Hofmann 1976

| EIC 14 | 5.0–9.0 | × 11.0–24.0 | 5 |

Margominuscula antiqua Pichova 1966

| EIC 15 | 8.0–12.0 | × 12.0–14.0 | 3 |

"*Eosynechococcus*" sp. #2

| EIC 16 | 9.5–11.0 | × 32.0–38.0 | 1 |

Eosynechococcus major Golovenok and Belova 1984

| EIC 17 | 10.0–20.0 | × 24.0–50.0 | 1 |

Eosynechococcus giganteus Golovenok and Belova 1984

| EIC 18 | 12.0–30.0 | × 50.0–100.0 | 2 |

Graviglomus incrustus Green, Knoll and Swett 1988

| EIC 19 | 16.0–52.0 | × 21.0–61.0 | 1 |

"*Eosynechococcus*" sp. #3

| EIC 20 | 4.0–8.0 | × 35.0–40.0 | 1 |

Table 25.3 (*Continued*)

	Range of Cell Diameter (μm)	Range of Sheath Thickness (μm)	Divisional Dispersion Index (DDI)	Number of Taxonomic Occurrences
G—*Gloeocapsa*-LIKE MORPHOTYPES				
Gloeodiniopsis mikros Knoll 1982				
∗G 1	3.0–7.0	1.6–8.0	3	4
Modern cyanobacterial (Chroococcaceae) analogue:				
Gloeocapsa repestris Kützing [Desikachary 1959, p. 117]				
	3.0–6.0	3.0–7.0	1	
"*Gloeodiniopsis*" sp. #1				
G 2	5.0–21.5	3.0–6.0	5	2
[see Chapter 24, Pl. 45, E, H, I]				
Gloeodiniopsis lamellosa Schopf 1968				
G 3	8.0–14.0	1.5–10.0	2	3
[see Chapter 24, Pl. 32, J]				
"*Gloeodiniopsis*" sp. #2				
G 4	9.0–24.0	1.5–12.0	3	3
[see Chapter 24, Pl. 10, K, P]				
Gloeodiniopsis magna Nyberg and Schopf 1984				
G 5	13.6–35.0	2.0–14.1	3	2
[see Chapter 24, Pl. 45, B, D, F, G]				
Gloeodiniopsis uralicus Krylov and Sergeev 1987				
G 6	18.0–22.0	1.0–5.0	1	2
[see Chapter 24, Pl. 4, F]				

	Range of Cell Size (μm)	Range of Sheath Thickness (μm)	Divisional Dispersion Index (DDI)	Number of Taxonomic Occurrences
CHR—*Chroococcus*-LIKE MORPHOTYPES				
Glenobotrydion granulosum Y. Zhang 1988				
CHR 1	2.6–6.5	0.7–2.0	4	1
Gloeodiniopsis pangjapuensis Y. Zhang 1981				
∗CHR 2	0.8–3.0	0.2–2.0	6	5
Modern cyanobacterial (Chroococcaceae) analogue:				
Gloeocapsa punctata Nägeli [Desikachary 1959, p. 115]				
	0.7–2.8	1.5–2.0	6	
Sphaerophycus parvum Schopf 1968				
CHR 3	1.0–4.5	0.2–1.3	6	6
Sphaerophycus medium Horodyski and Donaldson 1980				
∗CHR 4	2.0–7.0	0.3–1.0	6	3
Modern cyanobacterial (Chroococcaceae) analogue:				
Chroococcus cohaerens (Breb.) Nägeli [Desikachary 1959, p. 111]				
	2.0–7.0	0.3–1.0	6	

1146 Informal Revised Classification

Table 25.3. (Continued).

	Range of Cell Size (μm)	Range of Sheath Thickness (μm)	Divisional Dispersion Index (DDI)	Number of Taxonomic Occurrences
Eozygion minutum Schopf and Blacic 1971				
*CHR 5	4.0–10.6	0.5–3.0	4	5
[see Chapter 24, Pl. 33, H–J]				
Modern cyanobacterial (Chroococcaceae) analogue:				
Chroococcus minutus (Kützing) Nageli [Desikachary 1959, p. 103]				
	4.0–10.0	1.0–2.5	4	
Bigeminococcus lamellosus Schopf and Blacic 1971				
CHR 6	6.7–18.7	2.0–8.0	4	2
[see Chapter 24, Pl. 33, A, C]				
Eotetrahedrion princeps Schopf and Blacic 1971				
CHR 7	7.3–14.8	0.3–4.0	3	7
[see Chapter 24, Pl. 33, D]				
Myxococcoides cantabrigiensis Knoll 1982				
CHR 8	7.0–21.8	0.3–1.8	4	2
Bisacculoides grandis J. H. Oehler 1977				
CHR 9	10.1–24.0	0.7–5.0	4	9
[see Chapter 24, Pl. 10, L]				
Globophycus rugosum Schopf 1968				
*CHR 10	15.7–22.0	1.0–5.0	2	4
Modern cyanobacterial (Chroococcaceae) analogue:				
Chroococcus tenax (Kirchner) Hieronymus [Desikachary 1959, p. 103]				
	16.0–21.0	2.0–2.5	2	
Gloeodiniopsis dilutus Ogurtsova and Sergeev 1987				
CHR 11	19.4–50.0	2.0–5.0	4	2

	Range of Cell Size (μm)	Range of Sheath Thickness (μm)	Divisional Dispersion Index (DDI)	Number of Taxonomic Occurrences
MYX—*Myxosarcina*-LIKE MORPHOTYPE				
unnamed *Myxosarcina*-like colony (Schopf and Fairchild 1973)				
MYX 1	8.0–12.0 × 9.0–14.5	1.5–2.5	2	1
[see Chapter 24, Pl. 39, C; Pl. 41, B, D, J]				

	Range of Cell Size (μm)	Divisional Dispersion Index (DDI)	Number of Taxonomic Occurrences
PLEU—*Pleurocapsa*-LIKE MORPHOTYPES			
Palaeopleurocapsa wopfneri Knoll, Barghoorn and Golubić 1975			
PLEU 1	4.0–26.0	6	2
"*Palaeopleurocapsa*" sp. #1			
PLEU 2	5.0–10.0	2	2

Table 25.3 (Continued).

	Range of Cell Size (μm)	Divisional Dispersion Index (DDI)	Number of Taxonomic Occurrences
Palaeopleurocapsa fusiforma Ogurtsova and Sergeev 1987			
PLEU 3	8.0–21.0	3	2
[see Chapter 24, Pl. 50, B, C]			
Palaeopleurocapsa kamaelgensis Sergeev and Krylov 1986			
PLEU 4	14.0–35.0	3	2
[see Chapter 24, Pl. 4, E; Pl. 45, C]			

	Range of Cell Size (μm)	Range of Sheath Thickness (μm)	Divisional Dispersion Index (DDI)	Number of Taxonomic Occurrences
TET—COLONIAL PLANAR TETRADS OF COCCOID CELLS				
Tetraphycus acinulus D. Z. Oehler 1978				
TET 1	0.4–1.9	0.0–3.0	5	3
Sphaerophycus tetragonalis Muir 1976				
TET 2	1.1–4.1	0.0–3.0	4	2
Tetraphycus major D. Z. Oehler 1978				
TET 3	2.0–6.5	0.5–3.0	4	4
[see Chapter 24, Pl. 54, H]				
Tetraphycus congregatus McMenamin, Kumar and Awramik 1983				
TET 4	3.4–9.1	0.2–1.0	3	2
Conglobocella troxelii Licari 1978				
TET 5	4.0–15.0	0.0–4.0	4	4
Paratetraphycus giganteus Z. Zhang 1985				
TET 6	7.0–22.0	0.0–3.0	4	3
[see Chapter 24, Pl. 50, A]				

	Range of Cell Size (μm)	Divisional Dispersion Index (DDI)	Number of Taxonomic Occurrences
CPL—PLANAR (NON-TETRAD) COLONIAL COCCOIDS			
Distichococcus minutus Golovenoc and Belova 1985			
∗CPL 1	4.0–6.0	2	1
Modern cyanobacterial (Chroococcaceae) analogue:			
Merismopedia glauca (Ehrenberg) Nägeli [Desikachary 1959, p. 155–156]			
	3.0–6.0	3	
Unnamed *Merismopedia*-like colony (Schopf and Fairchild 1973)			
CPL 2	8.0–12.0	2	1
[see Chapter 24, Pl. 41, A]			

Table 25.3 (Continued).

	Range of Cell Size (μm)	Range of Sheath Thickness (μm)	Divisional Dispersion Index (DDI)	Number of Taxonomic Occurrences
CCU—CUBOIDAL COLONIES OF COCCOID CELLS				
?Eucapsis sp.				
CCU 1	2.0–3.0	0.0	2	2
[see Chapter 24, Pl. 4, A, B]				
Eucapsomorpha rara Golovenoc and Belova 1985				
CCU 2	2.0–6.6	0.3–2.5	4	3
"*Eucapsomorpha*" sp. #1				
CCU 3	3.0–11.0 × 5.0–12.0	0.5–1.0	4	1
Eoentophysalis arcata Mendelson and Schopf 1982				
CCU 4	6.0–20.0	2.0–4.0	4	1
[see Chapter 24, Pl. 10, E]				

	Range of Cell Dimensions (μm)	Divisional Dispersion Index (DDI)	Number of Taxonomic Occurrences
POLY—*Polybessurus*, ASSYMETRICALLY LAMELLATED ELLIPSOIDS			
Polybessurus bipartitus Fairchild ex Green, Knoll, Golubić and Swett 1987			
POLY 1	19.5–58.1 × 25.8–83.9	5	2
[see Chapter 24, Pl. 35, A–E; Pl. 36, A–G; Pl. 37, A, C. F–I, J; Pl. 38, A–H]			

	Dimensions	Number of Taxonomic Occurrences
BZ—BIZARRE MORPHOTYPES		
BZ 1	*Archaeorestis schreiberensis* Barghoorn 1965 (= *A. magna* Awramik and Barghoorn 1977)	2
BZ 2	*Ceratophyton vernicosum* Kirjanov 1979	1
BZ 3	*Eoastrion simplex* Barghoorn 1965 (= *E. bifurcatum* Barghoorn 1965 = "*Metallogenium*")	5
[see Chapter 24, Pl. 2, C, D; Pl. 4, C]		
BZ 4	*Eosphaera tyleri* Barghoorn 1965	2
[see Chapter 24, Pl. 1, K, L; Pl. 2, A, B]		
BZ 5	*Exochobrachium triangulum* Awramik and Barghoorn 1977	1
BZ 6	*Germinosphaera bispinosa* Mikhailova (= Mikhajlova) 1986	1
[see Chapter 24, Pl. 42, E]		
BZ 7	*Germinosphaera unispinosa* Mikhailova (= Mikhajlova) 1986	1
[see Chapter 24, Pl. 42, I]		

Informal Revised Classification 1149

Table 25.3 (*Continued*)

	Dimensions	Number of Taxonomic Occurrences
BZ 8 [see Chapter 24, Pl. 2, O–R]	*Kakabekia umbellata* Barghoorn 1965	4
BZ 9 [see Chapter 24, Pl. 52, B–D]	*Redkinia fedonkinispis* Aseeva in press	1
BZ 10	*Xenothrix inconcreta* Awramik and Barghoorn 1977	1
BZ 11	*Veryhachium*? Hofmann 1971	1
BZ 12	Unnamed organic tube with globose end (Knoll, 1982)	1
BZ 13	Unnamed branched tube with internal ellipsoids Lo, 1980)	1

	Range of Diameter (μm)		Divisional Dispersion Index (DDI)	Number of Taxonomic Occurrences
	Minimum	Maximum		

EUKARYOTES

MS—"MESOSPHAEROMORPHS" (MAXIMUM DIAMETER 60–200 μm)

Kildinella hyperboreica Timofeev 1960				
MS 1	20.0–35.0	62.0–78.0	6	15
Stictosphaeridium implexum Timofeev 1966				
*MS 2	26.0–45.0	80.0–103.0	6	17
Modern eukaryotic (Chlorococcaceae) analogue:				
Spongiochloris spongiosa (Vischer) Starr [Chantanachat and Bold 1962, p. 28]				
	30.0	100.0	6	
Kildinella nordia Timofeev 1969				
MS 3	36.0–62.0	50.0–62.5	3	26
[see Chapter 24, Pl. 14, B]				
Leiosphaeridia bituminosa Timofeev 1966				
MS 4	40.0–75.0	64.0–78.0	3	18
Vavosphaeridium densum Maithy 1975				
MS 5	40.0–75.0	125.0–140.0	6	9
Tasmanites volkovae Kirjanov 1974				
MS 6	42.0–80.0	52.0–82.0	3	4
Tylosphaeridium tallinicum Timofeev 1966				
MS 7	50.0–98.0	80.0–98.0	3	20
Kildinella timofeevii Maithy 1975				
*MS 8	60.0–80.0	100.0–110.0	3	6
Modern eukaryotic (Oocystaceae) analogue:				
Eremosphaera gigas (Archer) Fott and Kalina [Smith and Bold 1966, p. 21]				
	60.0	110.0	3	
Kildinella jacutica Timofeev 1966				
*MS 9	60.0–110.0	108–186.0	5	16

Table 25.3 (*Continued*).

	Range of Diameter (μm)		Divisional Dispersion Index (DDI)	Number of Taxonomic Occurrences
	Minimum	Maximum		

Modern eukaryotic (Chlorococcaceae) analogue:

Planktosphaeria maxima Bischoff and Bold [1963, p. 39]

| | 62.0 | 200.0 | 5 | |

Tasmanites bobrowskii Wazynska 1967

| MS 10 | 60.0–140.0 | 120.0–160.0 | 5 | 5 |

[see Chapter 24, Pl. 19, A]

MG—"MEGASPHAEROMORPHS" (MAXIMUM DIAMETER ≥ 200 μm)

Retisphaeridium vindhyanense Maithy 1969

| *MG 1 | 75.0–104.0 | 200.0–230.0 | 5 | 6 |

Modern eukaryotic (Oocystaceae) analogue:

Eremosphaera viridus de Bary [Smith and Bold 1966, p. 21]

| | 80.0 | 200.0 | 4 | |

Tasmanites variabilis Volkova 1968

| MG 2 | 90.0–175.0 | 184.0–296.0 | 6 | 3 |

Nucellosphaeridium sp. #1

| MG 3 | 80.0–114.0 | 319.0–430.0 | 5 | 4 |

Leiosphaeridia ochroleuca Timofeev 1966

| MG 4 | 130.0–260.0 | 140.0–260.0 | 3 | 13 |

[see Chapter 24, Pl. 24, D]

Nucellosphaeridium magnum Maithy 1975

| MG 5 | 150.0–210.0 | 300.0–365.0 | 4 | 4 |

Nucellosphaeridium deminatum Timofeev 1969

| MG 6 | 174.0–500.0 | 500.0–600.0 | 5 | 6 |

[see Chapter 24, Pl. 42, A, D]

Tasmanites tenellus Volkova 1968

| MG 7 | 200.0–320.0 | 400.0–550.0 | 5 | 2 |

Nucellosphaeridium bellum Timofeev 1969

| MG 8 | 220.0–305.0 | 290.0–420.0 | 3 | 6 |

Trachysphaeridium maicum Timofeev 1969

| MG 9 | 290.0–320.0 | 710.0–820.0 | 5 | 3 |

Tasmanites vindhyanensis Maithy and Shukla 1977

| MG 10 | 500.0 | 1000.0–1250.0 | 4 | 2 |

Kildinella magna Timofeev 1969

| MG 11 | 563.0–690.0 | 840.0–1780.0 | 5 | 2 |

Trachysphaeridium vetterni Timofeev 1969

| MG 12 | 1060.0–3100.0 | 1350.0–4000.0 | 6 | 7 |

Chuaria circularis Walcott 1899

| MG 13 | 2000.0–2900.0 | 5000.0–6800.0 | 5 | 3 |

Table 25.3 (Continued).

	Range of Diameter (μm)		Divisional Dispersion Index (DDI)	Number of Taxonomic Occurrences
	Minimum	Maximum		
Chuaria multirugosa Du 1985				
MG 14	4000.0–5000.0	6300.0–7500.0	3	2
Unnamed "hypothetical megasphaeromorph taxon" needed to complete size range reported for an undifferentiated assemblage				
MG 15	7500.0	10000.0	2	1

	Number of Taxonomic Occurrences		Number of Taxonomic Occurrences

A—COMPLEX (NON-SPHAEROMORPH) ACRITARCHS

Eomarginata striata Jankauskas (=Yankauskas) 1979

A-1   2

[see Chapter 24, Pl. 4, D]

Pterospermopsimorpha capsulata Jankauskas (=Yankauskas) 1982

A-2   4

[see Chapter 24, Pl. 19, B]

Pterospermopsimorpha annulare Timofeev 1969

A-3   1

Pterospermopsimorpha insolita Timofeev 1969

A-4   2

Pterospermopsimorpha sp.

A-5   1

Acantholigotriletum sp.

A-6   1

Acantholigotriletum primigenum Timofeev 1966

A-7   2

Baltisphaeridium sp.

A-8   3

Baltisphaeridium perrarum Jankauskas (=Yankauskas) 1980

A-9   1

[see Chapter 24, Pl. 55, B]

Nucellohystrichosphaera megalea Timofeev 1976 (in Timofeev and German 1976)

A-10   1

[see Chapter 24, Pl. 14, D]

Trachyhystrichosphaera aimika Hermann (=German) 1976 (in Timofeev and German 1976)

A-11   2

[see Chapter 24, Pl. 14, E]

Trachyhystrichosphaera vidalii Knoll 1983

A-12   1

Trachyhystrichosphaera bothnica Tynni and Donner 1980

A-13   1

Octoedryxium sp.

A-14   1

Octoedryxium truncatum (Rudavskaya) Vidal 1976

A-15   4

[see Chapter 24, Pl. 53, G, I; Pl. 56, C]

Octoedryxium simmetricum Timofeev 1973

A-16   1

[see Chapter 24, Pl. 53, H]

Trachysphaeridium laufeldii Vidal 1976

A-17   3

Pterospermopsis simicus Jankauskas (=Yankauskas) 1980 (=*Pterospermella simica* Jankauskas 1982)

A-18   1

[see Chapter 24, Pl. 19, E, F]

Pterospermopsis dubius Jankauskas (=Yankauskas) 1980

A-19   1

[see Chapter 24, Pl. 19, D]

Unnamed tear- and flask-shaped microfossils (Bloeser et al. 1977; Vidal 1979; Binda and Bokhari 1980; Knoll and Vidal 1980; Knoll 1982; Swett and Knoll 1985; Green et al. 1988)

A-20   5

Melanocyrillium hexodiadema Bloeser 1985

A-21   1

[see Chapter 24, Pl. 29, C, D, F, H]

Melanocyrillium horodyskii Bloeser 1985

A-22   1

Melanocyrillium fimbriatum Bloeser 1985

A-23   1

[see Chapter 24, Pl. 29, A, E, G]

Cymatiospheroides kullingii Knoll 1983

A-24   2

Table 25.3 (Continued).

	Number of Taxonomic Occurrences		Dimensions	Number of Taxonomic Occurrences
Vandalosphaeridium walcottii Vidal and Ford 1985			EU—NON-ACRITARCH EUKARYOTE-LIKE MORPHOTYPES	
A-25	1	EU 1	Unnamed Botryococcaceae-like colony (Konzalova 1973)	
Vandalosphaeridium varangeri Vidal 1981			Cell Diameter (μm) / 2.5–3.0	1
A-26	3			
Peteinosphaeridium reticulatum Vidal 1976		EU 2	*Ambiguaspora parvula* Volkova 1976 (spore-like bodies with trilete scars)	
A-27	1		Cell Diameter (μm) / 3.0–9.0	1
Podolina minuta Hermann (=German) 1976				
A-28	1	EU 3	*Eomycetopsis schopfii* Nautiyal 1980 (branched, hypha-like septate filament)	
Unnamed possible acanthomorph acritarch (Downie et al. 1971)			Cell Width (μm) Cell Length (μm)	
A-29	1		3.0–4.8 8.4–12.0	1
Cymatiosphaera precambrica Tynni and Donner 1980				
A-30	1	EU 4	Unnamed branched septate filament compared with "fungi or chlorophycean algae" (Schopf et al. 1977)	
Spumosata nova Pychova (=Pykhova) 1966			Cell Width (μm) Cell Length (μm)	
A-31	1		2.0–4.0 32.0–59.0	1
Acanthodiacrodium sp.		[see Chapter 24, Pl. 45, A]		
A-32	2			
Archaeohystrichosphaeridium acer Timofeev 1966		EU 5	*Palaeovaucheria clavata* Hermann (=German) 1981 (Vaucheriaceae-like filaments)	
A-33	1		Filament Width (μm) / 18.0–50.0	1
Archaeohystrichosphaeridium sp.				
A-34	1	[see Chapter 24, Pl. 11, C; Pl. 12, A; Pl. 13, D]		
Micrhystridium tornatum Volkova 1968		EU 6	*Gunflintia barghoornii* Maithy 1975 (unbranched septate filament, possibly Ulothricacean)	
A-35	1		Cell Width (μm) Cell Length (μm)	
Micrhystridium sp.			9.0–14.0 33.0–60.0	2
A-36	1	[see Chapter 24, Pl. 40, A]		
Archaeohystrichosphaeridium cellulare Timofeev 1966		EU 7	*Archaeoellipsoides conjunctivus* Y. Zhang 1985 (unbranched septate filament)	
A-37	1		Cell Width (μm) Cell Length (μm)	
Archaeohystrichosphaeridium stipiforme Timofeev 1966			16.8–18.0 29.4–39.0	2
A-38	1			
Archaeohystrichosphaeridium innominatum Timofeev 1966		EU 8	*Trachythrichoides ovalis* Hermann (=German) 1976 (unbranched septate filament)	
A-39	1		Cell Width (μm) Cell Length (μm)	
Archaeohystrichosphaeridium dorofeevii Timofeev 1966			16.0–23.0 30.0–50.0	1
A-40	1	[see Chapter 24, Pl. 26, E; Pl. 28, C]		
Archaeohystrichosphaeridium acanthaceum Timofeev 1966				
A-41	1			
Asperatopsophosphaera partialis Shepeleva 1963				
A-42	1			
Paracrassosphaera dedalea Rudavskaya 1979 (in Fajzulina and Treshchetenkova 1979)				
A-43	1			
[see Chapter 24, Pl. 56, A]				

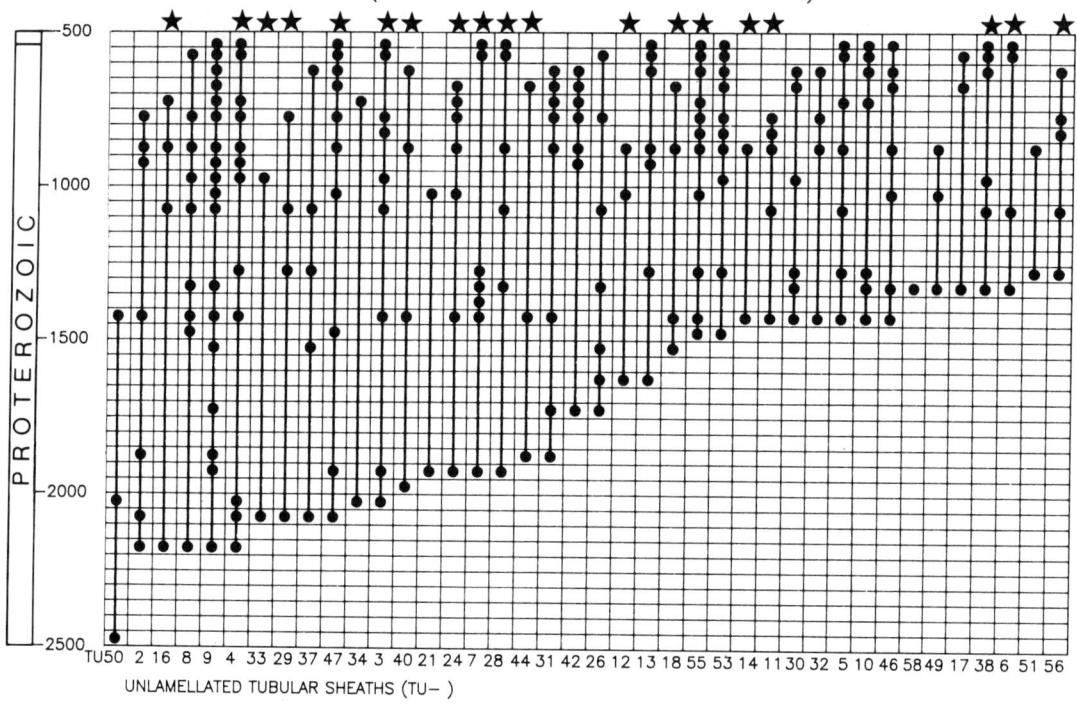

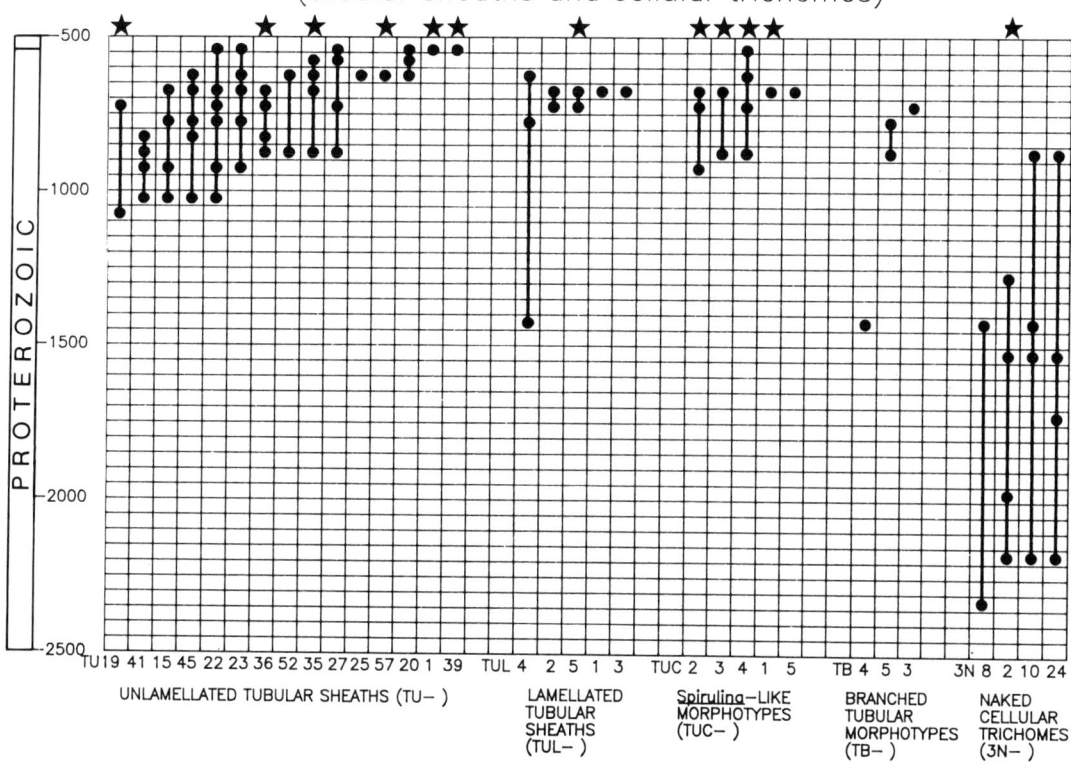

Figure 25.4. Geochronologic range charts for species of Proterozoic microfossils listed in Table 25.3 (solid circles indicate that one *or more* occurrences of the species have been reported from the indicated 50 Ma-long interval; stars at the tops of columns denote Lazarus taxa).

1154 Informal Revised Classification

RANGE CHART – PROKARYOTIC FILAMENTS
(naked cellular trichomes)

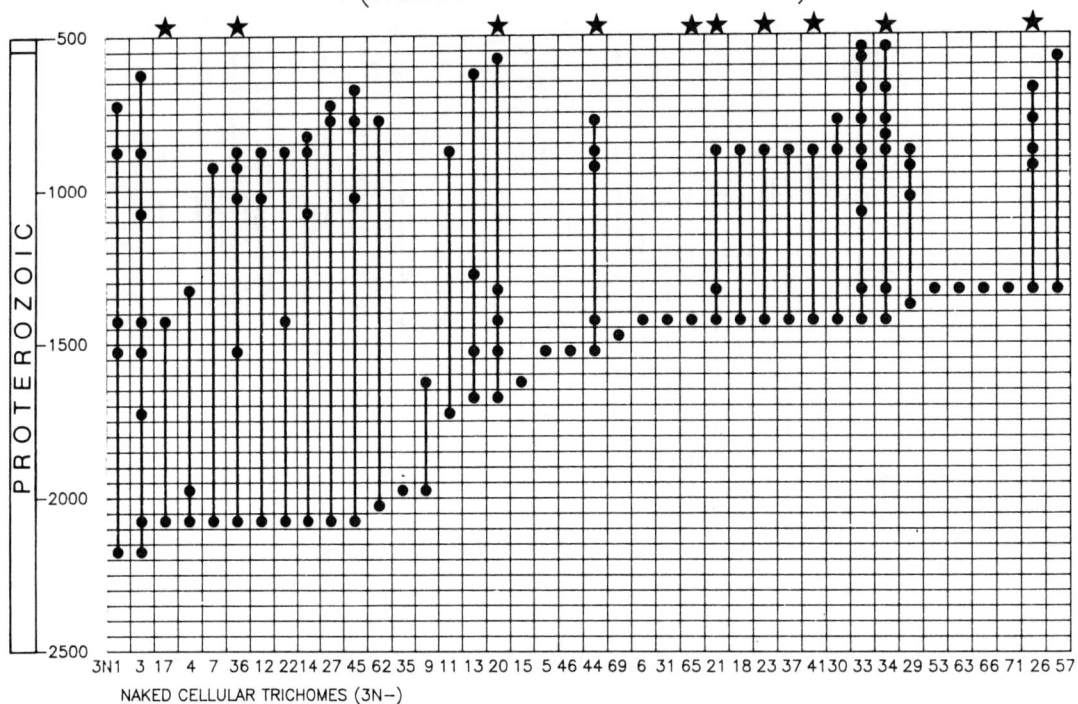

NAKED CELLULAR TRICHOMES (3N–)

RANGE CHART – PROKARYOTIC FILAMENTS
(naked and sheath–enclosed cellular trichomes)

NAKED CELLULAR TRICHOMES (3N–)   SHEATH–ENCLOSED CELLULAR TRICHOMES (3S–)

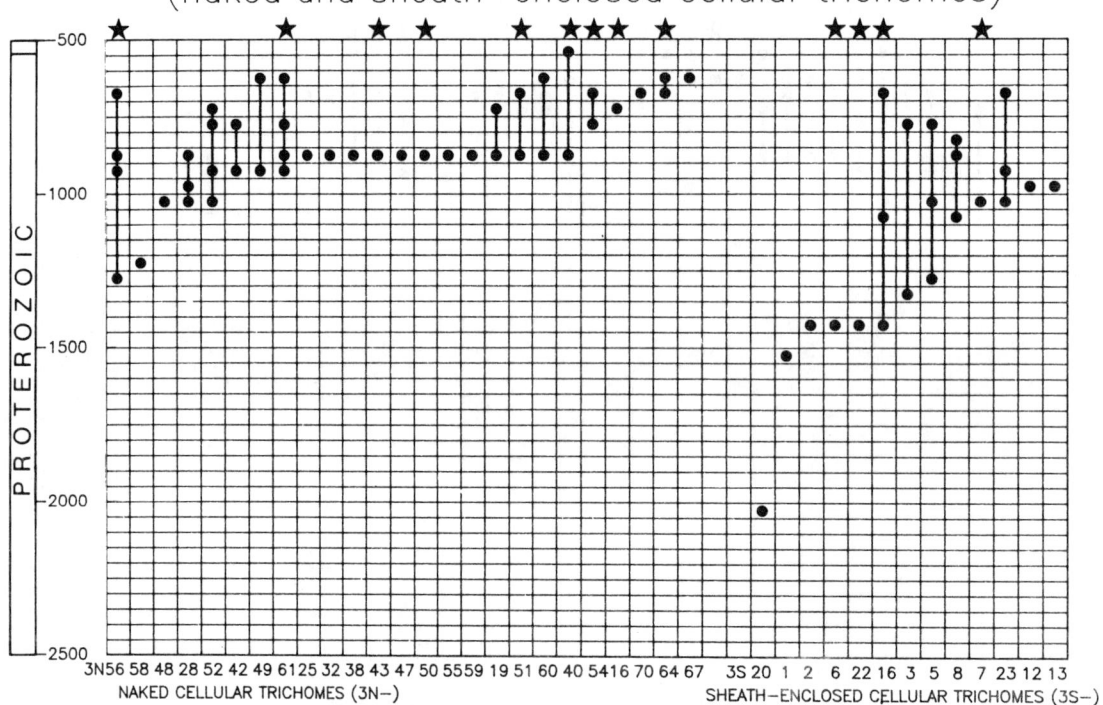

RANGE CHART — PROKARYOTIC FILAMENTS AND SOLITARY UNICELLS
(sheath-enclosed cellular trichomes and solitary coccoids)

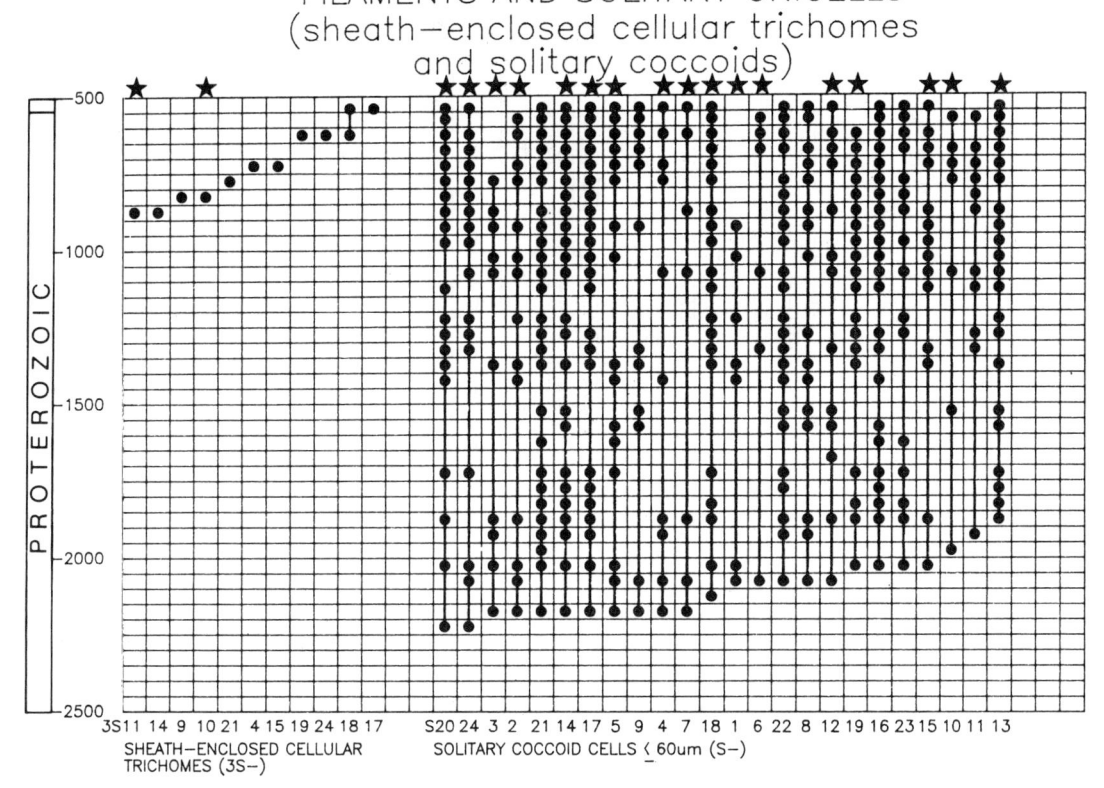

SHEATH-ENCLOSED CELLULAR TRICHOMES (3S-)

SOLITARY COCCOID CELLS ≤ 60um (S-)

RANGE CHART — PROKARYOTIC COLONIES
(irregular or spheroidal colonial coccoids)

IRREGULAR COLONIES OF COCCOID CELLS (CIR-)

COLONIAL COCCOIDS WITH LAMELLATED SHEATHS (CLI-)

SPHEROIDAL COLONIES OF COCCOID CELLS (CSF-)

IRREGULAR OR SPHEROIDAL COLONIAL COCCOIDS (CRCF-)

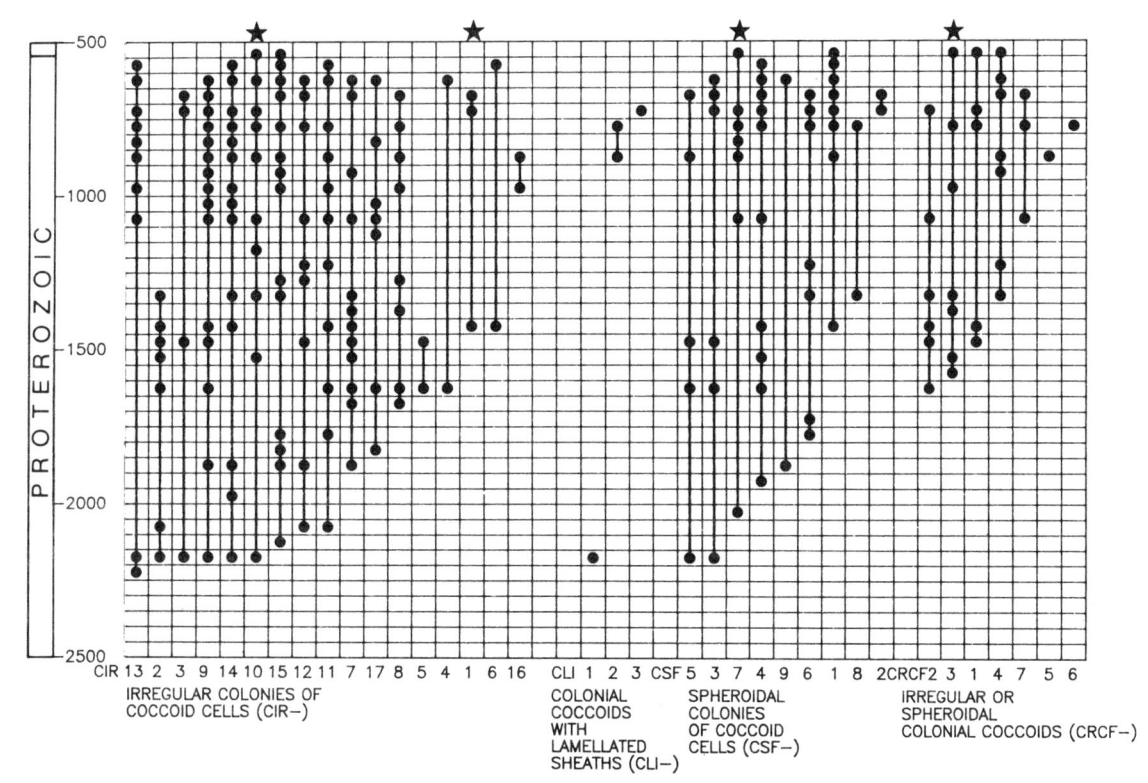

RANGE CHART — PROKARYOTIC COLONIES
(pseudofilamentous colonies, irregular colonies of ellipsoids, and Gloeocapsa-like morphotypes)

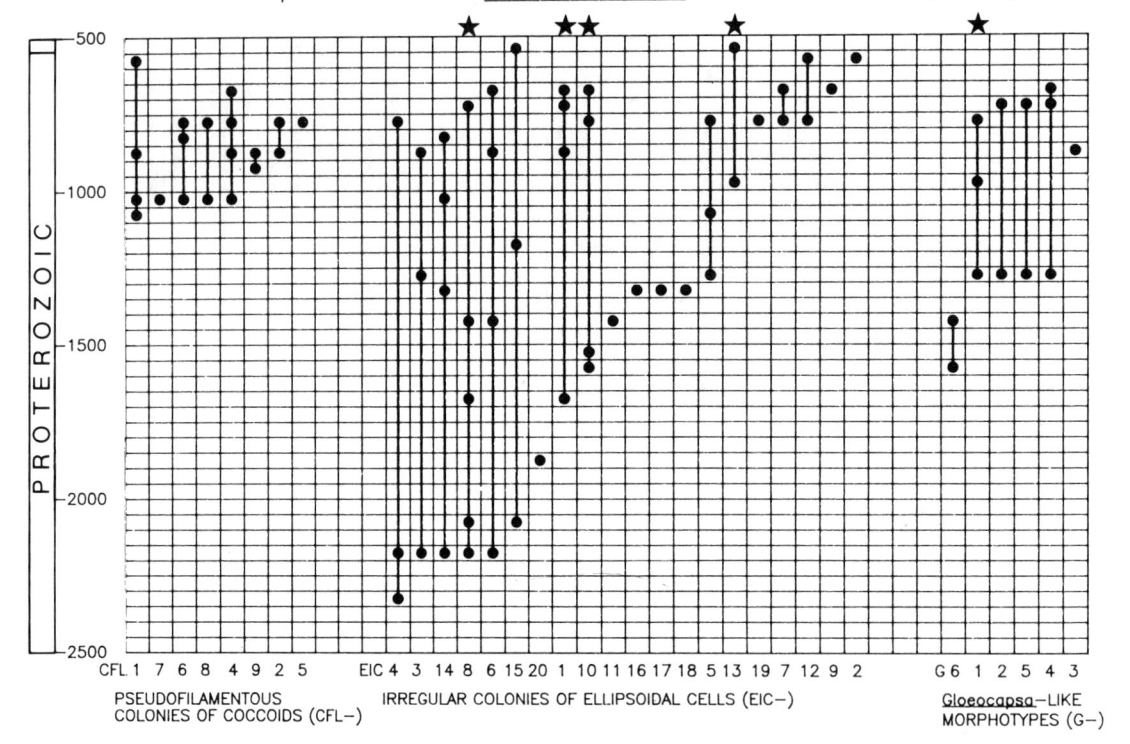

RANGE CHART — ORDERED PROKARYOTIC COLONIES
(Chroococcus-, Myxosarcina-, and Pleurocapsa-like morphotypes and planar and cuboidal colonies)

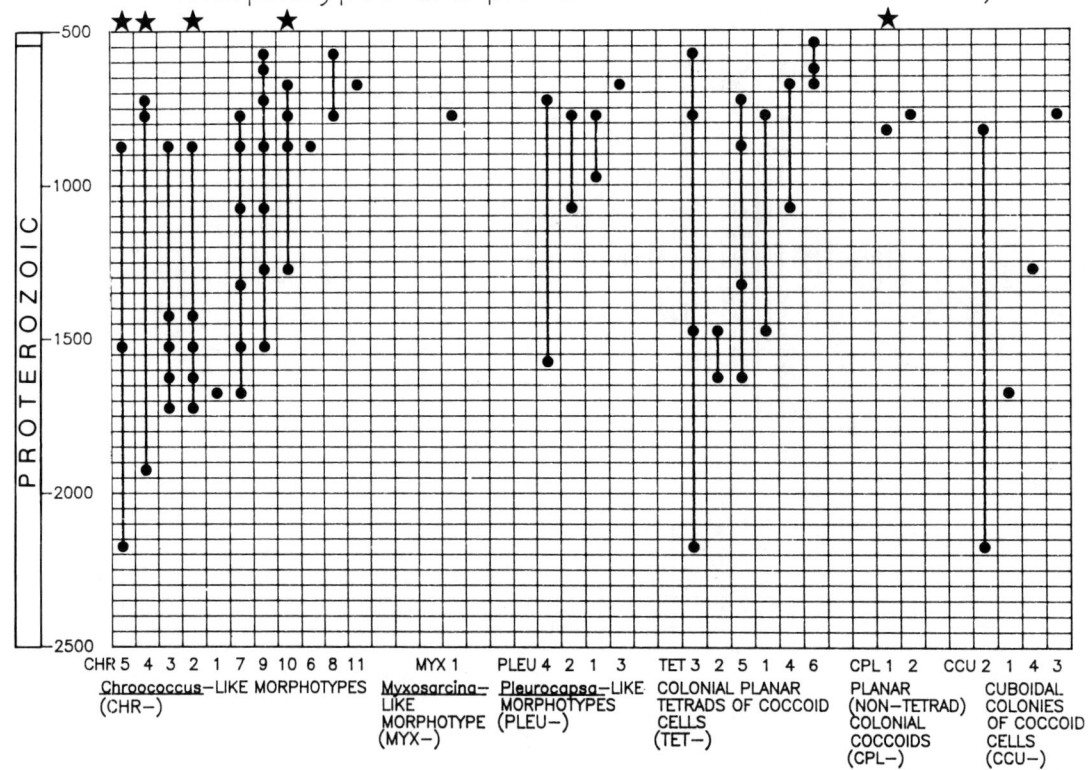

RANGE CHART — PROKARYOTES AND EUKARYOTES
(Polybessurus, bizarre morphotypes, and sphaeromorph acritarchs)

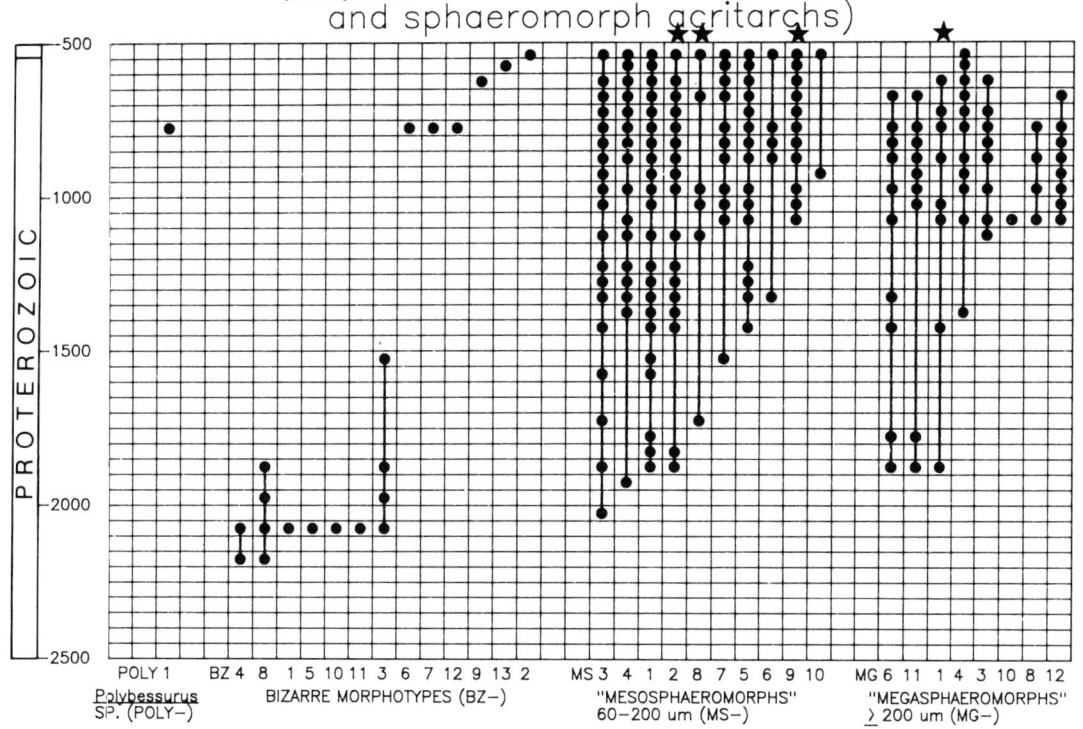

RANGE CHART — EUKARYOTES
("megasphaeromorphs" and complex acritarchs)

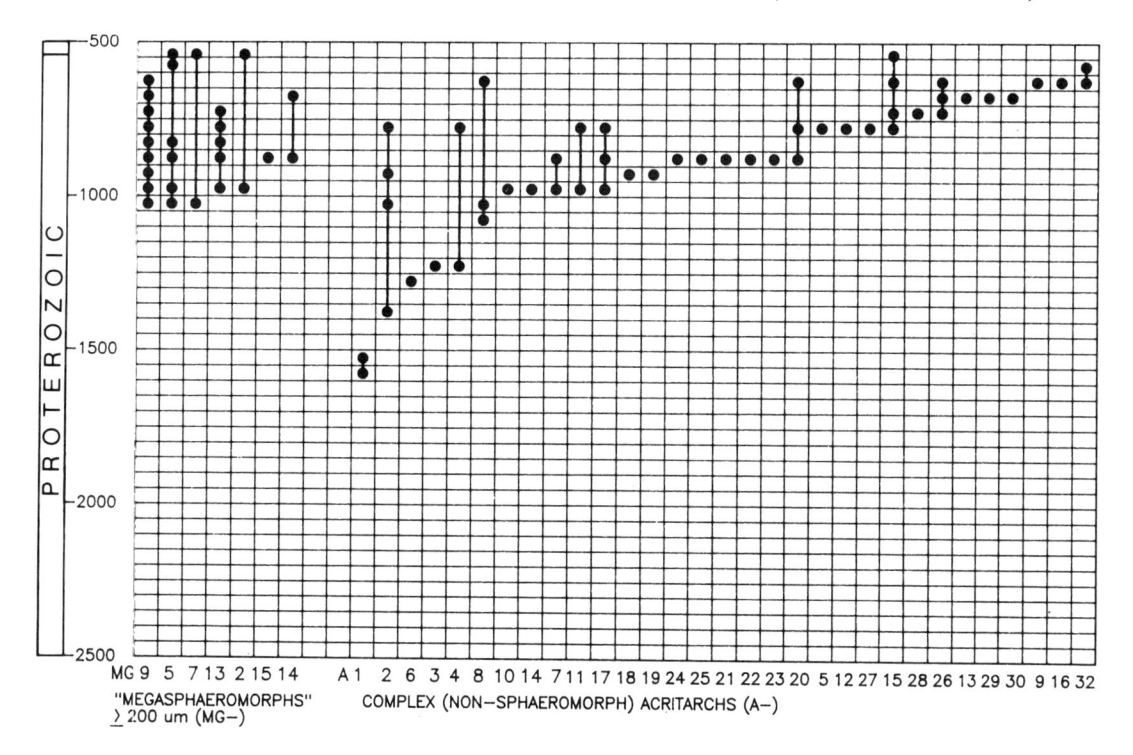

RANGE CHART – EUKARYOTES
(complex acritarchs and filamentous and other non-acritarch eukaryotes)

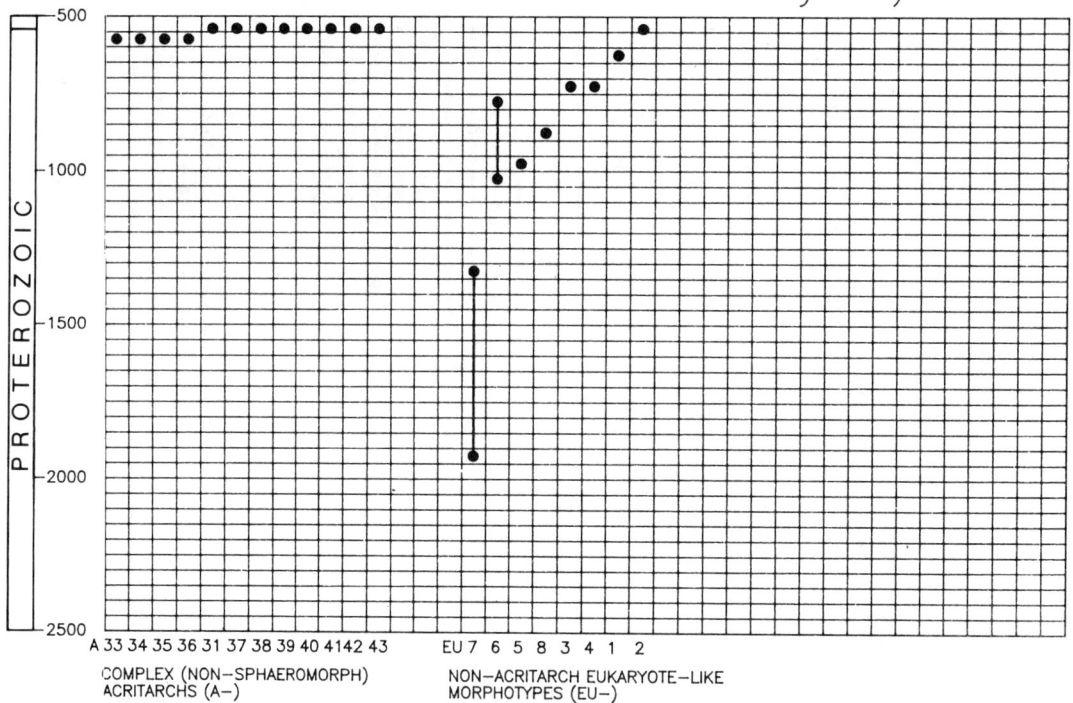

teria (Desikachary 1959); and for an additional 200 species and varieties of sheath-enclosed oscillatoriaceans, taxa for which data regarding sheath thickness and diameter were also compiled (Desikachary 1959). These data are tabulated in Table 5.4.9 and summarized graphically in Figures 5.4.29–5.4.31.

Comparable morophometric data (viz., ranges of cell diameter and length, medial cell shape and, where applicable, terminal cell shape and sheath thickness and diameter) were compiled for 251 occurrences of septate unbranched fossil filaments, both naked and sheath-enclosed, reported from 69 Proterozoic formations having the temporal distribution summarized in Table 25.5. Data used in this compilation (summarized graphically in Figures 5.4.32–5.4.34) were derived either from direct measurement of the specimens illustrated in Chapter 24 or from measurement of previously published illustrations and published size data (including those in taxonomic descriptions), corrected, where appropriate, for the effects of preservational compression (Section 5.4.2). Tabulated occurrences of septate filamentous microfossils were subdivided into two morphological categories: (*i*) naked cellular trichomes; and (*ii*) sheath-enclosed cellular trichomes. Morphometric data for members of each category were compared with those compiled for living species of septate filamentous bacteria and cyanobacteria and the categories were subdivided into the 93 informal species of naked (3N) and sheath-enclosed (3S) cellular filamentous taxa listed in Table 25.3.

In Table 25.3, the following letter codes are used to denote medial cell shape: "L" = "*long*" cells, with lengths ⩾ two times cell diameters; "D" = "*discoidal*" cells, with diameters ⩾ two times cell lengths; "Q" = "*quadrate*" cells having a shape between L and D; "Q/D" and "Q/L" = cell shapes intermediate between quadrate and discoidal, and quadrate and long, respectively. Terminal cell shapes are denoted in this listing by the following letter codes: "Trd" = *rounded*; "Tta" = *tapered and/or conical*; "Tgl" = *globose and/or capitate*; "Thr" = attenuated into a terminal *hair*. Number of reported taxonomic occurrences, reference to illustrations in this volume, relevant morphometric data, and data regarding modern morphological analogues of "Lazarus taxa" (see Section 25.3, below) are listed in Table 25.3 for each fossil species here recognized. For example, with regard to filaments here referred to informal species "3N 36" (p. 1132):

(1) Seven occurrences of naked septate filaments meeting the criteria used to define this informal species have been reported from Proterozoic strata.

(2) This taxon was originally described as *Partitiofilum gongyloides* Schopf and Blacic 1971.

(3) Medial cells of these filaments are discoidal or intermediate between quadrate and discoidal.

(4) Cell diameters range from a minimum of between 3.7 and 5.4 μm to a maximum of between 4.7 and 6.1 μm, and cell lengths range from a minimum of between 1.0 and 1.8 μm to a maximum of between 1.8 and 3.0 μm.

(5) Where reported, terminal cells are rounded.

(6) Filaments referred to this informal species are illustrated in Chapter 24, Pl. 20, E, F; Pl. 22, E; and Pl. 30, P.

(7) In the salient morphological features noted, this fossil

taxon is closely comparable to naked cellular trichomes of the modern oscillatoriacean cyanobacterium *Oscillatoria grunowiana* Gomont as described by Desikachary (1959, p. 216).

Each of the informal species of fossil septate filaments listed in Table 25.3 exhibits a range of morphological variability similar in salient respects to that of comparable living prokaryotic species.

25.2.3 Solitary Coccoidal Cells and Sphaeromorph Acritarchs

In order to provide a basis for taxonomic differentiation of Proterozoic coccoid unicells and sphaeromorph acritarchs, morphometric data (ranges of cell diameter) and divisional dispersion indices (Schopf 1976) were compiled for the following categories of living microorganisms: 138 species and varieties of free-living coccoid bacteria (Buchanan and Gibbons 1974; Laskin and Lechevalier 1977; Starr et al. 1981); 121 species and varieties of coccoid chroococcalean (viz., 114 chroococcacean and seven entophysalidacean taxa) cyanobacteria (Desikachary 1959); and the coccoid vegetative cells of 234 species and varieties (viz. 226 chlorophycean and eight rhodophycean taxa) of eukaryotic micro-algae (West and Fritsch 1927; Prescott 1954, 1962; Taylor 1957; Smith and Bold 1966; Bourrelly 1970; La Rivers 1978). Morphometric data for these categories of modern coccoid microorganisms are tabulated in Table 5.4.6 and summarized graphically in Figures 5.4.10 – 5.4.13, 5.4.16, and 5.4.17.

For species populations of vegetative unicells of modern spheroidal microorganisms, the Divisional Dispersion Index (DDI)—an index designed to interrelate endpoints of a population size range—has been defined as "the least number of sequential vegetative divisions required to mathematically 'reduce' the largest cell of a population to the smallest cell of that population" (Schopf 1976, p. 31); the DDI of a species is a genetically determined trait characteristic of that taxon in exactly the same sense as is the size range of the taxon. Previous studies have shown that, regardless of the cell size of the species studied, DDIs measured for numerous species of cyanobacteria and green algae cluster in the range of 2 to 4 (Schopf 1976). Analyses of the nearly 500 species of coccoid microorganisms noted above confirm this result and establish that it is applicable also to coccoid bacteria. Specifically, whether eukaryote or prokaryote, and regardless of the median cell size of the species population investigated, the DDIs of the vegetative cells of coccoid taxa cluster in the range of 2 to 4, are commonly about 3 (average DDI = 3.28, N = 473 species and varieties), and rarely range as high as 8 with the great majority (ca. 94%) having DDIs of 6 or less.

Morphometric data of the same type as that compiled for modern coccoid unicells (viz., ranges of cell diameter) were assembled for 1661 occurrences of spheroidal non-colonial microfossils reported from 272 Proterozoic formations. With the exceptions noted below, data used in this compilation (summarized graphically in Figures 5.4.15–5.4.17, 5.5.1–5.5.4, 5.5.5–5.5.7, and 5.5.12) were derived either from direct measurement of the specimens illustrated in Chapter 24, from measurement of previously published photomicrographs, or from published size data (including those in taxonomic descriptions). As appropriate (Section 5.4.2), size data were corrected for effects of preservational compression and radial cracking.

Unlike size data published for most chert microfloras (which commonly include histograms, scatter diagrams, and other relatively detailed descriptions of patterns of size distribution), many publications on shale microfloras, especially those published prior to the past decade, include few data regarding measured size distributions. This problem is especially acute for the large number of taxa described prior to 1980 by B. V. Timofeev and his colleagues in several particularly important and extensive monographs (Timofeev 1960, 1966, 1969, 1973; Timofeev and German 1976). Therefore, to avoid uncertainties arising from inclusion of unsubstantiated size data for specimens reported in these monographic works, rather than relying on data provided in texts or taxonomic descriptions, only morphometric data based on direct measurement of microfossils (e.g., of the type specimens re-photographed for this volume and illustrated in Chapter 24) or of published photomicrographs have been here used.

Tabulated morphometric data were used to group together similar sized fossil spheroids. Because of variable states of preservation and the poor quality of many published photomicrographs, it did not prove possible to use wall thickness as a secondary taxonomic character; nevertheless, where distinctive wall structure has been reported (e.g., taxa inferred by their authors to have porate walls and therefore referred to *Tasmanites* spp.), vesicles of similar size were grouped together (although question remains as to whether such features are in fact original or diagenetic; see Section 5.5 and Figure 5.5.6). In sum, a total of 49 informal species was recognized, each with a DDI of 6 or less.

To aid in interpretation of the biologic affinities of these microfossils, taxa were grouped into three morphological categories: (*i*) 24 species of solitary coccoid microfossils ⩽ 60 μm in maximum diameter (S), represented in the Proterozoic record by 1051 occurrences reported from 259 formations; (*ii*) ten species of "mesosphaeromorph" acritarchs 60 to 200 μm in maximum diameter (MS), represented by 439 occurrences reported from 163 formations; and (*iii*) 15 species of "megasphaeromorph" acritarchs ⩾ 200 μm in maximum diameter (MG), represented by 171 occurrences reported from 69 formations; the temporal distribution of these occurrences is summarized in Table 25.6. Although each of these species has a "biologic size range"—each having a DDI of 6 or less—the general lack of useful secondary taxonomic characters has necessitated substantial "lumping." Moreover, some of these coccoids, particularly the larger sphaeromorphs, probably represent cysts (for which applicability of the DDI has not been demonstrated) rather than vegetative cells. In view of these considerations, it seems probable that the 49 taxa recognized underestimates, perhaps markedly, the actual species-level diversity of known Proterozoic coccoids and sphaeromorphs.

As indicated in Table 5.4.6, virtually all living species of coccoid prokaryotes (viz., 100% of bacteria and 95% of cyanobacteria) are less than 25 μm in maximum diameter; the largest extant coccoid cyanobacterium is about 58 μm in diameter. Thus, although some members of the smallest morphological

Table 25.5. *Temporal distribution of Proterozoic benthic microfossils*

AGE (Ma)	CELLULAR TRICHOMES						TUBULAR SHEATHS					
	FM	SP	TX	OCC	1st	Last	FM	SP	TX	OCC	1st	Last
2499–2450	—	—	—	—	—	—	1	1	1	1	1	0
2449–2400	—	—	—	—	—	—	—	—	—	—	—	—
2399–2350	—	—	—	—	—	—	—	—	—	—	—	—
2349–2300	1	1	1	1	1	0	—	—	—	—	—	—
2299–2250	—	—	—	—	—	—	—	—	—	—	—	—
2249–2200	—	—	—	—	—	—	—	—	—	—	—	—
2199–2150	3	6	6	8	5	0	2	5	7	7	5	0
2149–2100	—	—	—	—	—	—	—	—	—	—	—	—
2099–2050	3	10	10	13	9	0	1	6	6	9	4	0
2049–2000	1	2	2	2	2	1	1	4	4	4	2	0
1999–1950	1	1	1	1	1	1	1	1	2	2	1	0
1949–1900	2	3	3	3	1	0	3	7	8	9	4	0
1899–1850	—	—	—	—	—	—	1	4	4	4	2	0
1849–1800	—	—	—	—	—	—	—	—	—	—	—	—
1799–1750	—	—	—	—	—	—	—	—	—	—	—	—
1749–1700	2	3	3	3	1	0	2	4	4	4	2	0
1699–1650	1	2	2	3	2	0	—	—	—	—	—	—
1649–1600	2	2	2	2	1	2	2	3	4	4	2	0
1599–1550	—	—	—	—	—	—	—	—	—	—	—	—
1549–1500	1	12	12	12	3	3	2	4	4	4	1	0
1499–1450	1	1	1	1	1	1	1	4	4	4	1	0
1449–1400	2	23	26	32	15	8	5	22	37	41	10	2
1399–1350	1	1	1	1	1	0	1	1	1	1	0	0
1349–1300	3	12	13	13	8	5	4	13	15	16	4	0
1299–1250	2	4	4	4	2	1	2	12	14	14	3	1
1249–1200	1	1	1	1	1	1	—	—	—	—	—	—
1199–1150	—	—	—	—	—	—	—	—	—	—	—	—
1149–1100	—	—	—	—	—	—	—	—	—	—	—	—
1099–1050	4	5	5	5	1	0	6	14	17	17	2	0
1049–1000	3	10	10	10	5	2	3	13	13	14	3	1
999–950	1	3	3	3	2	2	2	8	8	9	0	1
949–900	3	12	13	13	3	1	3	10	11	12	2	0
899–850	4	40	44	53	14	24	10	34	57	67	7	4
849–800	2	5	5	5	2	3	3	7	10	10	0	1
799–750	8	15	16	17	2	7	8	22	37	38	0	4
749–700	3	7	7	7	3	7	8	18	21	24	3	4
699–650	6	11	15	16	2	8	8	23	24	25	4	14
649–600	3	10	10	11	4	10	9	24	32	33	3	11
599–550	2	3	2	2	0	1	14	19	51	51	0	4
549–530	3	5	8	9	1	5	4	21	23	23	2	21
TOTALS:	69		226	251	93	93	104		419	447	68	68

category of non-colonial fossil coccoids—the 24 "S" species with a maximum diameter $\leqslant 60\,\mu m$—may actually be eukaryotes, all are within the size range of modern prokaryotic unicells. Because of their relatively larger size, however, fossil taxa including vesicles greater than $60\,\mu m$ in maximum diameter, the 25 species included in the "MS" and "MG" categories, seem reasonably regarded as eukaryotes. Although one modern chlorophycean species with unicells as large as $800\,\mu m$ in diameter has been reported (Prescott 1962), virtually all species (ca. 99%) of living planktonic eukaryotic micro-algae are less than $200\,\mu m$ in maximum diameter (Table 5.4.8) as are the vast majority of known Phanerozoic sphaeromorphic taxa (Tappan 1980). Thus, "mesosphaeromorphs" included in the intermediate-size category—the ten "MS" species with a maximum diameter of between 60 and $200\,\mu m$—are all within the size range tyoical of large-diameter modern and Phanerozcic eukaryotic planktonic algae; such forms were evidently also relatively abundant during the Middle and Late Proterozoic (viz., represented in the known fossil record by 395 occurrences, 1400 to 550 Ma; Figure 5.2.2). Finally, members of the largest-diameter category—the 15 "MG" species with a maximum diameter greater than $200\,\mu m$—are commonly a few to several millimeters in diameter and more than an order of magnitude larger than typical Phanerozoic (Tappan 1980) or modern planktonic sphaeromorphic algae; these "megasphaeromorphic" eukaryotes constitute a characteristic Late Pro-

COLONIES OF COCCOIDAL OR ELLIPSOIDAL CELLS						TOTAL BENTHOS					
FM	SP	TX	OCC	1st	Last	FM	SP	TX	OCC	1st	Last
—	—	—	—	—	—	1	1	1	1	1	0
—	—	—	—	—	—	0	0	0	0	0	0
—	—	—	—	—	—	0	0	0	0	0	0
1	1	1	1	1	0	2	2	2	2	2	0
—	—	—	—	—	—	0	0	0	0	0	0
1	1	1	1	1	0	1	1	1	1	1	0
3	17	25	26	16	1	3	28	38	41	26	1
1	1	1	1	1	0	1	1	1	1	1	0
1	5	5	5	3	0	3	21	21	27	16	0
1	1	1	1	1	0	2	7	7	7	5	1
—	—	—	—	—	—	1	2	3	3	2	1
2	3	4	4	2	0	4	13	15	16	7	0
1	6	6	8	3	1	1	10	10	12	5	1
2	2	2	2	1	0	2	2	2	2	1	0
1	3	3	3	1	0	1	3	3	3	1	0
2	3	3	3	1	0	3	10	10	10	4	0
3	7	7	9	5	2	3	9	9	12	7	2
3	16	17	24	6	0	3	21	23	30	9	2
2	4	5	5	4	1	2	4	5	5	4	1
2	11	11	15	1	0	2	27	27	31	5	3
1	14	14	19	2	2	1	19	19	24	14	3
2	17	17	19	4	1	5	62	80	92	29	11
2	3	3	3	0	0	2	5	5	5	1	0
5	16	19	22	4	4	5	41	47	51	16	9
4	12	16	20	6	1	5	28	34	38	11	3
4	5	6	7	1	0	5	6	7	8	2	1
1	2	2	2	0	0	1	2	2	2	0	0
1	1	1	1	0	0	1	1	1	1	0	0
10	18	31	36	3	0	10	37	53	58	6	0
3	9	11	17	5	1	3	32	34	41	13	4
5	11	13	14	3	0	5	22	24	26	5	3
3	5	5	6	1	0	4	27	29	31	6	1
10	28	38	50	4	8	12	102	139	170	25	36
6	9	12	13	1	3	8	21	27	28	3	7
25	43	74	79	10	19	28	80	127	134	12	30
15	24	34	39	3	8	19	49	62	70	9	19
13	29	38	45	3	18	16	63	77	86	9	40
17	17	31	34	0	7	23	51	73	78	7	28
9	14	19	19	1	11	20	36	72	72	1	16
13	11	19	19	0	10	15	37	50	51	3	36
175		495	572	98	98	223	1140		1270	259	259

terozoic assemblage (represented by 119 occurrences, 1100 to 530 Ma; Figure 5.5.6), many taxa of which became extinct prior to the beginning of the Phanerozoic (Figure 5.5.12).

25.2.4 Colonial Coccoidal and Ellipsoidal Cells

In addition to the morphometric and DDI data noted above for 138 taxa of free-living coccoid bacteria and 121 taxa of coccoid chroococcaleans, size data (minimum and maximum cell lengths and widths) were compiled for 45 species and varieties of free-living ellipsoidal bacteria (Buchanan and Gibbons 1974; Starr et al. 1981) and for 47 species and varieties of ellipsoidal chroococcalean cyanobacteria (Desikachary 1959). These data are tabulated in Table 5.4.7 and summarized graphically in Figures 5.4.18–5.4.20. As applicable, data were also compiled both for coccoid and for ellipsoidal taxa regarding colony shape (viz., irregular, spheroidal, pseudofilamentous, tabular, cuboidal), intra-colony cellular organization (unordered, rosette-like, ordered in plate-like sheets, three-dimensionally ordered in ranks and files), and thickness and nature of colony-encompassing and of individual cell-encompassing sheaths (diffuse, well defined, variable, lamellated or unlamellated).

Comparable morphometric data were compiled for 572 occurrences of fossil colonial coccoidal and ellipsoidal cells reported from 175 Proterozoic formations having the temporal distribution summarized in Table 25.5. Data used in this

1162 Informal Revised Classification

Table 25.6. *Temporal distribution of Proterozoic planktonic microfossils*

	"Prokaryotes"						"Eukaryotes"					
	SOLITARY COCCOIDS <60 μm diameter						"MESOSPHAEROMORPHS" 60–200 μm diameter					
AGE (Ma)	FM	SP	TX	OCC	1st	Last	FM	SP	TX	OCC	1st	Last
2499–2450	—	—	—	—	—	—	—	—	—	—	—	—
2449–2400	—	—	—	—	—	—	—	—	—	—	—	—
2399–2350	—	—	—	—	—	—	—	—	—	—	—	—
2349–2300	—	—	—	—	—	—	—	—	—	—	—	—
2299–2250	—	—	—	—	—	—	—	—	—	—	—	—
2249–2200	1	2	2	2	2	0	—	—	—	—	—	—
2199–2150	3	9	17	20	9	0	—	—	—	—	—	—
2149–2100	1	1	1	1	1	0	—	—	—	—	—	—
2099–2050	3	12	13	26	5	0	—	—	—	—	—	—
2049–2000	2	15	17	20	4	0	1	1	1	1	1	0
1999–1950	2	2	3	3	1	0	—	—	—	—	—	—
1949–1900	5	6	9	10	1	0	1	1	1	1	1	0
1899–1850	4	17	18	22	1	0	2	3	4	8	2	0
1849–1800	2	8	8	11	0	0	2	2	2	2	0	0
1799–1750	2	6	7	7	0	0	1	1	1	1	0	0
1749–1700	4	12	15	20	0	0	2	2	2	2	1	0
1699–1650	1	1	1	1	0	0	—	—	—	—	—	—
1649–1600	3	4	5	8	0	0	—	—	—	—	—	—
1599–1550	3	10	16	16	0	0	3	2	3	3	0	0
1549–1500	2	13	13	17	0	0	2	2	2	2	1	0
1499–1450	1	2	2	2	0	0	—	—	—	—	—	—
1449–1400	4	8	10	10	0	0	2	4	4	4	1	0
1399–1350	4	15	18	20	0	0	2	2	4	4	0	0
1349–1300	6	13	28	40	0	0	3	6	9	12	1	0
1299–1250	4	13	23	30	0	0	4	5	10	15	0	0
1249–1200	8	12	32	48	0	0	7	5	14	20	0	0
1199–1150	—	—	—	—	—	—	—	—	—	—	—	—
1149–1100	4	10	16	23	0	0	3	5	6	12	0	0
1099–1050	9	19	26	30	0	0	2	4	5	9	1	0
1049–1000	3	12	14	19	0	0	2	6	6	13	0	0
999–950	3	13	18	25	0	0	4	8	16	28	0	0
949–900	5	16	27	36	0	1	3	4	6	9	1	0
899–850	13	19	53	65	0	0	11	8	28	37	0	0
849–800	9	10	22	29	0	0	7	8	16	17	0	0
799–750	27	18	89	124	0	1	19	8	55	83	0	0
749–700	20	18	43	53	0	0	8	6	16	20	0	0
699–650	19	18	58	82	0	0	16	8	30	40	0	0
649–600	27	20	61	74	0	1	20	7	37	39	0	0
599–550	22	16	48	50	0	4	18	5	21	23	0	0
549–530	33	17	82	107	0	17	18	10	30	34	0	10
TOTALS:	259		815	1051	24	24	163		329	439	10	10

compilation (summarized graphically in Figures 5.4.18, 5.4.21, 5.4.22) were derived either from direct measurement of the specimens illustrated in Chapter 24 or from measurement of previously published illustrations and published size data (including those in taxonomic descriptions). Tabulated occurrences were subdivided into the morphological categories listed in Table 25.1 and morphologically similar forms were grouped together into 44 informal species of unordered colonial coccoids (viz., categories CIR, CLI, CSF, CRCF, CFL), 34 species of ordered colonial coccoids (viz., categories G, CHR, MYX, PLEU, TET, CPL, CCU), and 20 species of unordered colonial ellipsoids (category EIC). All of the coccoid colonial taxa thus recognized have DDIs of 6 or less; all colonial taxa, both coccoidal and ellipsoidal, exhibit a range of morphological variability comparable to that of living analogues.

25.2.5 Miscellaneous Prokaryotes

One species of *Polybesserus* has been recognized, a distinctive, asymmetrically lamellated and commonly stalked and colonial ellipsoidal cyanobacterium (Table 25.3, POLY-1), as have 13 species of "bizarre morphotypes" (BZ). With the exceptions of *Eoastrion* and *Kakabekia* (BZ-3 and BZ-8, re-

Informal Revised Classification

"Eukaryotes"

"MEGASPHAEROMORPHS" 200–10,000 μm diam.						COMPLEX ACRITARCHS						TOTAL PLANKTON					
FM	SP	TX	OCC	1st	Last	FM	SP	TX	OCC	1st	Last	FM	SP	TX	OCC	1st	Last
—	—	—	—	—	—	—	—	—	—	—	—	0	0	0	0	0	0
—	—	—	—	—	—	—	—	—	—	—	—	0	0	0	0	0	0
—	—	—	—	—	—	—	—	—	—	—	—	0	0	0	0	0	0
—	—	—	—	—	—	—	—	—	—	—	—	0	0	0	0	0	0
—	—	—	—	—	—	—	—	—	—	—	—	0	0	0	0	0	0
—	—	—	—	—	—	—	—	—	—	—	—	1	2	2	2	2	0
—	—	—	—	—	—	—	—	—	—	—	—	3	9	17	20	9	0
—	—	—	—	—	—	—	—	—	—	—	—	1	1	1	1	1	0
—	—	—	—	—	—	—	—	—	—	—	—	3	12	13	26	5	0
—	—	—	—	—	—	—	—	—	—	—	—	3	16	18	21	5	0
—	—	—	—	—	—	—	—	—	—	—	—	2	2	3	3	1	0
—	—	—	—	—	—	—	—	—	—	—	—	6	7	10	11	2	0
1	3	3	3	3	0	—	—	—	—	—	—	4	23	25	33	6	0
—	—	—	—	—	—	—	—	—	—	—	—	2	10	10	13	0	0
1	2	2	2	0	0	—	—	—	—	—	—	2	9	10	10	0	0
—	—	—	—	—	—	—	—	—	—	—	—	4	14	17	22	1	0
—	—	—	—	—	—	—	—	—	—	—	—	1	1	1	1	0	0
—	—	—	—	—	—	—	—	—	—	—	—	3	4	5	8	0	0
—	—	—	—	—	—	1	1	1	1	1	0	3	13	20	20	1	0
—	—	—	—	—	—	1	1	1	1	0	1	3	15	16	20	1	1
—	—	—	—	—	—	—	—	—	—	—	—	1	2	2	2	0	0
1	2	2	2	0	0	—	—	—	—	—	—	4	14	16	16	1	0
1	1	1	1	1	0	1	1	1	1	1	0	4	19	24	26	2	0
1	1	1	1	0	0	—	—	—	—	—	—	6	20	38	53	1	0
—	—	—	—	—	—	1	1	1	1	1	1	4	19	34	46	1	1
—	—	—	—	—	—	2	2	2	2	2	1	8	19	48	70	2	1
—	—	—	—	—	—	—	—	—	—	—	—	0	0	0	0	0	0
1	1	1	1	1	0	—	—	—	—	—	—	4	16	23	36	1	0
2	7	7	11	2	1	2	1	2	2	1	0	9	31	40	52	4	1
2	5	6	6	1	0	2	2	2	2	0	0	3	25	28	40	1	0
3	9	14	19	4	0	4	5	6	6	5	2	4	35	54	78	9	2
2	4	4	0	0	0	1	3	3	5	2	2	6	27	40	54	3	3
9	12	28	39	3	1	2	8	8	9	6	3	15	47	116	150	9	4
4	7	14	15	0	0	—	—	—	—	—	—	12	25	52	61	0	0
9	9	25	25	0	1	7	9	14	15	4	7	27	44	183	247	4	9
4	5	5	5	0	1	2	3	3	3	2	1	21	32	67	81	2	2
4	6	8	8	0	4	3	4	4	4	3	3	23	36	100	134	3	7
9	4	14	14	0	3	8	7	10	10	3	8	34	38	122	137	3	12
8	2	8	8	0	0	4	5	5	5	4	5	22	28	82	86	4	9
7	4	7	7	0	4	2	9	10	10	8	9	34	40	129	158	8	40
69		149	171	15	15	43		73	77	43	43	282		1366	1738	92	92

spectively), members of this latter category are represented in the known Proterozoic record by but one or two occurrences; all taxa are morphologically distinctive and are not readily assignable to modern biologic families or classes; many (but not BZ-9) are evidently of bacterial affinity, in part possibly members of now extinct microbial groups (see Section 5.6).

25.2.6 Miscellaneous Eukaryotes

Included here (Table 25.3) are 43 species of "complex" (i.e., non-sphaeromorph, mostly acanthomorph) acritarchs and eight species of non-acritarch (chiefly filamentous) possible eukaryotes. In all cases, specific names listed for members of the former grouping (A-1–A-43) are those given in the publications reporting their occurrence; other than to distinguish between questionable and acceptable reports of such microfossils (Chapter 22), and in contrast to all of the other informal species included in this revised classification, no attempt has been made here to bring together morphologically similar forms.

Members of the final grouping (EU-1–EU-8) have been brought together because of their possible, or probable, eukaryotic affinities. The minute cell size of one of these species (viz., EU-1) suggests that it may actually be prokaryotic. Small

1164 Informal Revised Classification

Table 25.7. *Maximally diverse Proterozoic benthic and planktonic microfossil assemblages*

Age (Ma)	Geologic Units Benthos-Containing	Geologic Units Plankton-Containing	Benthos Cellular Trichomes TX	Benthos Cellular Trichomes OCC	Benthos Tubular Sheaths TX	Benthos Tubular Sheaths OCC	Benthos Colonial Coccoids or Ellipsoids TX	Benthos Colonial Coccoids or Ellipsoids OCC
2499–2450	Gamohaan Fm.	—	0	0	1	1	0	0
2449–2400	—	—	—	—	—	—	—	—
2399–2350	—	—	—	—	—	—	—	—
2349–2300	Monte Cristo Fm.	—	1	1	0	0	1	1
2299–2250	—	—	—	—	—	—	—	—
2249–2200	Ikabijsk Fm.	Ikabijsk Fm.	0	0	0	0	1	1
2199–2150	Kasegalik Fm.	Kasegalik Fm.	4	6	5	5	13	14
2149–2100	Karelia Complex	Karelia Complex	0	0	0	0	1	1
2099–2050	Gunflint Fm.	Gunflint Fm.	8	11	6	9	5	5
2049–2000	Duck Creek Dolomite	Ladoga Fm.	2	2	4	4	0	0
1999–1950	Tyler Fm.	Tyler Fm.	1	1	2	2	0	0
1949–1900	Rocknest Fm.	Onega Fm.	2	2	2	2	3	3
1899–1850	Chuanlinggou Fm.	Chuanlinggou Fm.	0	0	4	4	6	8
1849–1800	Besovets Fm.	Besovets Fm.	0	0	0	0	1	1
1799–1750	Bothnia Fm.	Bothnia Fm.	0	0	0	0	3	3
1749–1700	Hornby Bay Grp.	Ovruch Series	1	1	1	1	2	2
1699–1650	Paradise Creek Fm.	Paradise Creek Fm.	2	3	0	0	2	3
1649–1600	Amelia Dolomite	Penchenga Series	1	1	2	2	13	19
1599–1550	Satka Fm.	Satka Fm.	0	0	0	0	4	4
1549–1500	Barney Creek Fm.	Bakal Fm.	12	12	3	3	9	11
1499–1450	Balbirini Dolomite	Balbirini Dolomite	1	1	4	4	14	19
1449–1400	Gaoyuzhuang Fm.	Gaoyuzhuang Fm.	20	26	16	20	10	10
1399–1350	Bungle Bungle Dolomite	Zigazino-Komarovsk Fm.	1	1	1	1	2	2
1349–1300	Kotujkan Fm.	Strel'nye Gory Fm.	4	4	3	3	12	14
1299–1250	Sukhaya Tunguska Fm.	Sukhaya Tunguska Fm.	3	3	12	12	9	10
1249–1200	Kil'din Series	Il'yushkina Fm.	0	0	0	0	2	2
1199–1150	Roan Grp.	—	0	0	0	0	2	2
1149–1100	Tieling Fm.	Kachergat Fm.	0	0	0	0	1	1
1099–1050	Kheinjua Fm.	Suket Shale	0	0	6	6	6	7
1049–1000	Zil'merdak Fm.	Zil'merdak Fm.	7	7	6	7	4	8
999–950	Lakhanda Fm.	Lakhanda Fm.	3	3	5	5	3	3
949–900	Podinzer Fm.	Podinzer Fm.	8	8	5	5	0	0
899–850	Bitter Springs Fm.	Miroedikha Fm.	23	30	9	11	16	27
849–800	Kirgitej Fm.	Little Dal Grp.	4	4	3	3	4	5
799–750	Draken Conglomerate	Visingsö Grp.	1	1	6	6	18	19
749–700	Min'yar Fm.	Min'yar Fm.	3	3	6	8	8	11
699–650	Chichkan Fm.	Muhos Fm.	4	4	5	5	4	5
649–600	Doushantuo Fm.	Ushitsa Series	7	7	19	20	5	5
599–550	Yudoma Fm.	Yudoma Fm.	0	0	6	6	10	10
549–530	Vampire Fm.	Blue Clays	2	2	10	10	2	2

Total Benthos		Plankton									
		Solitary Coccoids <60 μm		"Meso-sphaero-morphs"		"Mega-sphaero-morphs"		Complex Acritarchs		Total Plankton	
TX	OCC	TX	OCC	TX	OCC	TX	OCC	TX	OCC	TX	OCC
1	1	—	—	—	—	—	—	—	—	—	—
—	—	—	—	—	—	—	—	—	—	—	—
—	—	—	—	—	—	—	—	—	—	—	—
2	2	—	—	—	—	—	—	—	—	—	—
—	—	—	—	—	—	—	—	—	—	—	—
1	1	2	2	0	0	0	0	0	0	2	2
22	25	7	7	0	0	0	0	0	0	7	7
1	1	1	1	0	0	0	0	0	0	1	1
19	25	10	23	0	0	0	0	0	0	10	23
6	6	10	12	0	0	0	0	0	0	10	12
3	3	2	2	0	0	0	0	0	0	2	2
7	7	2	3	0	0	0	0	0	0	2	3
10	14	9	12	3	5	3	3	0	0	15	20
1	1	6	8	1	1	0	0	0	0	7	9
3	3	6	6	1	1	0	0	0	0	7	7
4	4	10	15	1	1	0	0	0	0	11	16
4	6	1	1	0	0	0	0	0	0	1	1
16	22	3	6	0	0	0	0	0	0	3	6
4	4	9	9	1	1	0	0	1	1	11	11
24	26	12	16	1	1	0	0	1	1	14	18
19	24	2	2	0	0	0	0	0	0	2	2
46	56	6	6	0	0	0	0	0	0	6	6
4	4	14	16	3	3	1	1	1	1	19	21
19	21	6	15	5	8	0	0	0	0	11	23
24	25	10	12	4	7	0	0	0	0	14	19
2	2	7	11	4	4	1	1	0	0	12	16
2	2	—	—	—	—	—	—	—	—	—	—
1	1	7	10	3	6	0	0	0	0	10	16
12	13	3	3	4	8	6	10	0	0	13	21
17	22	10	12	2	2	3	3	1	1	16	18
11	11	8	13	8	16	8	13	3	3	27	45
13	13	14	18	3	4	3	3	3	5	23	30
48	68	9	15	8	15	7	13	0	0	24	43
11	12	2	2	3	4	7	8	0	0	12	14
25	26	10	15	7	19	5	10	3	3	25	47
17	22	5	6	3	4	0	0	0	0	8	10
13	14	5	10	5	7	1	1	2	2	13	20
31	32	7	9	3	3	0	0	0	0	10	12
16	16	4	5	2	2	0	0	1	1	7	8
14	14	6	10	4	6	1	1	5	5	16	22

spore-like bodies referred to a second species (viz. EU-2), kindly provided for study by N. A. Volkova, exhibit clearly defined trilete scars and, thus, are evidently of eukaryotic, meiotic, origin; nevertheless, they have been detected only in drill cores and their precise geologic age is therefore somewhat open to question. All other taxa in this category are filamentous, assigned to this grouping on the basis of their typically eukaryotic (and non-prokaryotic) cell width and/or length.

25.3

Temporal Distribution of Benthic and Planktonic Proterozoic Microfossils

Geochronologic range charts, indicating first, last, and intermediate-age Proterozoic taxonomic occurrences of the informal taxa listed in Table 25.3 are presented in Figure 25.4. Within these range charts, the location of a solid circle indicates that one or more taxonomic occurrences of the species denoted by the code at the bottom of the figure (see Table 25.3) has been reported from the indicated 50-Ma long interval. Solid stars at the top of taxon columns denote "Lazarus taxa," Proterozoic species having morphologically indistinguishable or closely similar modern species-level analogues (as listed in Table 25.3, with the relevant codes for the fossil taxa there denoted by asterisks) but which evidently are as yet unreported from intervening Phanerozoic strata. Data presented in Figure 25.4 are those here used (see Section 11.3) for analysis of global diversity. As discussed below, however, earliest and latest known occurrences indicated in Figure 25.4 may not necessarily represent "originations" and "extinctions" of taxa.

The known temporal distribution per 50 Ma-long interval of benthic and planktonic microfossils occurring in the Proterozoic formations listed in Chapter 22 are summarized in Tables 25.5 and 25.6, respectively. In each of these tables, entries listed in **boldface** are those for 50 Ma-long intervals containing ⩾ 20 reported occurrences of benthic or planktonic species, intervals here used (see Section 11.3) for analysis of mean alpha and beta diversity. Data regarding the maximally diverse benthic and planktonic assemblages known for each 50 Ma-long interval, used to assess maximum alpha diversity (Section 11.3), are presented in Table 25.7.

The following definitions apply to the entries in Tables 25.5, 25.6, and/or 25.7:

"*FM*" ("FOSSILIFEROUS FORMATIONS"), the number of formations in which one or more species of the indicated category of microfossil has been reported to occur within the indicated 50 Ma-long interval.

"*SP*" ("REPORTED SPECIES"), the number of informal species here recognized (Table 25.3) of the indicated category of microfossil that have been reported to occur within the indicated 50 Ma-long interval.

"*TX*" ("SPECIES-LEVEL TAXONOMIC OCCURRENCES"), the number of times (not including multiple occurrences in single formations) that each species of the indicated category of microfossil has been reported to occur within the indicated 50 Ma-long interval.

"*OCC*" ("REPORTED OCCURRENCES"), the number of times, including multiple occurrences in single formations, that each species of the indicated category of microfossil has been reported to occur within the indicated 50 Ma-long interval. "OCC" minus "TX" for a particular 50 Ma-long interval is thus the number of reported "redundant" or "multiple" occurrences, i.e., the number of times that species have been reported more than once to occur (including redundant occurrences resulting from taxonomic synonymies) in particular formations of the indicated 50 Ma-long interval.

"*1st*" ("FIRSTS"), the number of species of the indicated category of microfossil that have their earliest known Proterozoic occurrence in the indicated 50 Ma-long interval; "1sts" are thus putative "originations." It should be noted, however, that solitary coccoids, colonies of coccoid or ellipsoidal cells, trichomes, and tubular sheaths are known from the pre-Proterozoic (Archean) fossil record (Section 1.5); listed "1sts" for species in these categories should therefore not be interpreted necessarily as "originations." In contrast, "1sts" listed for "mesosphaeromorphs," "megasphaeromorphs," and for complex acritarchs, categories of microfossils unknown earlier than 2050 Ma, are plausibly interpretable as "originations."

"*Last*" ("LASTS"), the number of species of the indicated category of microfossil that have their latest known Proterozoic/Early Cambrian occurrence in the indicated interval; "Lasts" are thus putative "extinctions." However, for solitary coccoids, "mesosphaeromorphs," trichomes, tubular sheaths, and for colonies of coccoidal and ellipsoidal cells, "Lasts" do not necessarily represent "extinctions"—a substantial portion of these species (as listed in Table 25.2: 29% of prokaryotes and 5% of eukaryotes, 25% of benthic species and 23% of planktonic species) are "Lazarus taxa," forms that are similar or virtually identical in morphology to living taxa but which evidently have yet to be reported from the intervening Phanerozoic. Moreover, because Phanerozoic units younger than Early Cambrian have not been analyzed quantitatively in this study, "Lasts" for non-Lazarus taxa listed as occurring in the 550–530 Ma (Lower Cambrian) interval, the youngest interval here included, should not be regarded necessarily as "extinctions." However, "Lasts" listed for all "megasphaeromorph" species $\geqslant 600\,\mu$m in diameter, for many species of complex acritarchs, and for other distinctive and/or relatively abundant Proterozoic taxa that are known neither from the Phanerozoic fossil record nor the modern biota, appear plausibly interpretable as "extinctions."

26
Models for Vendian-Cambrian Biotic Diversity and for Proterozoic Atmospheric and Ocean Chemistry

J. JOHN SEPKOSKI, JR.   JAMES F. KASTING

26.1

Taxonomy and Stratigraphic Ranges of Animal Families in the Vendian and Cambrian: Data and Analytical Results for Section 11.4

J. JOHN SEPKOSKI, JR.

In Table 26.1, formally defined families and genera assessed to be of familial status are listed under a consensus classification to order, class, and phylum. Informal taxa are used for many problematica, and this classification may change considerably with new discoveries and insights into phylogenetic relationships. Problematical taxa followed in parentheses by an asterisk were treated as equivalent in rank to orders and classes in the analyses of diversity patterns in Section 11.4.

Stratigraphic ranges listed in Table 26.1 represent "best guesses," especially for the Lower Cambrian, and may change as global correlation becomes more accurate. The units employed represent an attempt to standardize stratigraphy to the Siberian sequence for the Lower Cambrian and to the North American cratonic sequence for the Upper Cambrian. Abbreviations for stratigraphic units are as follows: "V" = Vendian; "Cm" = Cambrian; "Edia" = Ediacaran (upper Vendian); "N-Da" = Nemakit-Daldyn (uppermost Vendian); "Tomm" = Tommotian; "Atda" = Atdabanian; "Boto" = Botomian; "1Mid" = lower Middle Cambrian (as defined in Sepkoski 1979); "mMid" = middle Middle Cambrian; "uMid" = upper Middle Cambrian; "Dres" = Dresbachian; "Fran" = Franconian; "Trep" = Trempealeauan; "l" = lower; "u" = upper. Other stratigraphic abbreviations are as defined in Sepkoski (1982).

Table 26.2 summarizes the time scale used in Section 11.4 in plotting diversity patterns for Vendian-Cambrian metazoans and metaphytes.

Data used in Section 11.4 for Q-mode factor analysis of faunal heterogeneity in the Vendian-Cambrian radiation are listed in Table 26.3. Tables 26.4.1–26.4.4 summarize results of the Q-mode factor analysis of Venidan-Cambrian animal genera within 68 classes (excluding archaeocyathids), whereas Tables 26.4.5–26.4.8 summarize the results for genera within 70 classes (including archaeocyathids).

Table 26.1 *Taxonomy and stratigraphic ranges of animal families in the Vendian and Cambrian.*

PROTOZOA		Leptomidae	Cm(Atda)–Cm(uMid)
Cl. RHIZOPODEA		Piraniidae	Cm(mMid)–S
Or. FORAMINIFERIDA		Takakkawiidae	Cm(mMid)
Ammodiscidae	Cm(Boto)–R	Wapkiidae	Cm(mMid)
Bathysiphonidae			
(+ Platysiphonidae)	Cm(Tomm-l)–R	Or. LITHISTIDA	
Maylisoriidae	Cm(u)–S	Anthaspidellidae	Cm(lMid)–P(Tatr)
		?Vauxiidae	Cm(lMid)–Cm(uMid)
Cl. RADIOLARIA			
Or. SPUMELLARIA		Or. INCERTAE SEDIS	
Entactiniidae	Cm(Atda)–C(Serp)	Lenica	Cm(Boto)
		?Protohyalostelia	Cm(Atda)
PORIFERA			
Cl. DEMOSPONGIA		?Or. STROMATOPOROIDEA	
Or. MONAXONIDA		?Khasaktidae	Cm(Atda)–Cm(Boto-l)
Choiidae	Cm(Boto)–O		
Corralioidae	Cm(mMid)	Cl. CALCAREA	
Halichondritidae	Cm(mMid)	Or. HETERACTINIDA	
Hamptoniidae	Cm(mMid)–Cm(uMid)	Eiffeliidae	Cm(Atda)–C(Mosc)
Hazeliidae	Cm(Boto)–Cm(mMid)	Or. SPHINCTOZOA	
		Sebargasiidae	Cm(lMid)–Tr(Nori)

Table 26.1 (Continued).

Or. INCERTAE SEDIS		Fallocyathidae	Cm(Atda)–Cm(Boto-l)
?"phobetractinids"	Cm(Atda)–O (l)?	Fansycyathidae	Cm(Atda)–Cm(Boto-l)
Cl. HEXACTINELLIDA		Formosocyathidae	Cm(Atda)–Cm(Boto-u)
Or. RETICULOSA		Geocyathidae	Cm(Atda)–Cm(Boto-l)
Dierespongiidae	Cm(uMid)–O (Cara)	Gloriosocyathidae	Cm(Atda)–Cm(Boto-u)
Hintzespongiidae	Cm(mMid)–Cm(uMid)	Gnaltacyathidae	Cm(Boto)
Hydnodictyidae	Cm(uMid)–O (Ashg)	Hupecyathellidae	Cm(Atda)–Cm(Boto-l)
Multivasculatidae	Cm(Fran)–Cm(Trep)	Hyptocyathidae	Cm(Boto)
Protospongiidae	Cm(Tomm-l)–D (u)	Irinaecyathidae	Cm(Atda)–Cm(lMid)
Teganiidae	Cm(uMid)–P (Leon)	Kaltatocyathidae	Cm(Tomm-u)–Cm(Boto-u)
		Kasyricyathidae	Cm(Boto-l)–Cm(Boto-u)
Or. INCERTAE SEDIS		Kidrjasocyathidae	Cm(Atda)–Cm(Boto-u)
Konyriidae	Cm(Dres)–O (Lide)	Kijacyathidae	Cm(Atda)–Cm(Boto-u)
Azyrtalia	Cm(Atda)	Kolbicyathidae	Cm(Atda)
		Kordecyathidae	Cm(Boto-l)–Cm(Boto-u)
ARCHAEOCYATHA		Lenocyathidae	Cm(Atda)–Cm(Boto-l)
Cl. REGULARES		Mrassucyathidae	Cm(Atda)
Or. MONOCYATHIDA		Nochoroicyathidae	Cm(Tomm-l)–Cm(Boto-u)
Capsolynthidae	Cm(Atda)–Cm(Boto)	Peregrinicyathidae	Cm(Boto-l)–Cm(Boto-u)
Capsulocyathidae	Cm(Tomm-u)–Cm(Boto-u)	Piamaecyathidae	Cm(Boto-l)–Cm(Boto-u)
Cryptoporocyathidae	Cm(Tomm-l)–Cm(Boto-l)	Polycoscinidae	Cm(Atda)–Cm(Boto-u)
Ethmolynthidae	Cm(Atda)–Cm(Boto)	Porocoscinidae	Cm(Boto-l)–Cm(Boto-u)
Fransuasaecyathidae	Cm(Tomm-u)–Cm(Boto-u)	Porocyathidae	Cm(Atda)–Cm(Boto-u)
Globosocyathidae	Cm(Atda)–Cm(Boto-u)	Pretiosocyathidae	Cm(Atda)–Cm(Boto-u)
Monocyathidae	Cm(Tomm-l)–Cm(Boto-u)	Robertocyathidae	Cm(Atda)–Cm(Boto-u)
Propriolynthidae	Cm(Atda)–Cm(Boto)	Robustocyathidae	Cm(Tomm-l)–Cm(Boto-u)
Rhabdocyathellidae	Cm(Atda)–Cm(mMid)	Rozanovicyathidae	Cm(Boto-l)
Tumuliolynthidae	Cm(Tomm-u)–Cm(Boto-u)	Sajanocyathidae	Cm(Boto-l)–Cm(lMid)
Uralocyathellidae	Cm(Boto-l)–Cm(Boto-u)	Sanarkocyathidae	Cm(Boto-l)–Cm(Boto-u)
Uralocyathidae	Cm(Tomm-u)–Cm(Boto-u)	Schidertycyathidae	Cm(Boto)–Cm(lMid)
		Sigmocoscinidae	Cm(Boto-u)
Or. PUTAPACYATHIDA		Sigmocyathidae	Cm(Boto-l)
Aptocyathidae	Cm(Atda)–Cm(Boto-u)	Soanicyathidae	Cm(Atda)–Cm(Boto-u)
Gerbicanicyathidae	Cm(Boto-u)	Squamosocyathidae	Cm(Atda)–Cm(Boto-l)
Putapacyathidae	Cm(Atda)	Stillicidocyathidae	Cm(Boto-l)–Cm(Boto-u)
		Tannuolacyathidae	Cm(Boto-l)
Or. AJACICYATHIDA		Tegerocyathidae	Cm(Boto-u)–Cm(lMid)
Acanthinocyathidae	Cm(Boto-l)	Tennericyathidae	Cm(Atda)–Cm(Boto-l)
Agyrekocyathidae	Cm(Boto-l)–Cm(Boto-u)	Tercyathidae	Cm(Boto-l)–Cm(Boto-u)
Ajacicyathidae	Cm(Tomm-l)–Cm(Boto-u)	Tumulifungiidae	Cm(Atda)–Cm(Boto-u)
Alataucyathidae	Cm(Atda)–Cm(Boto-u)	Tumulocyathidae	Cm(Tomm-u)–Cm(Boto-u)
Anaptyctocyathidae	Cm(Boto-l)–Cm(Boto-u)	Vologdinocyathidae	Cm(Atda)–Cm(lMid)
Annulocyathidae	Cm(Boto-l)–Cm(Boto-u)	Wrighticyathidae	Cm(Boto-u)
Baikalopectinidae	Cm(Atda)	Cl. IRREGULARES	
Bosceculcyathidae	Cm(Boto)–Cm(lMid)	Or. THALASSOCYATHIDA	
Botomocyathidae	Cm(Boto-l)–Cm(Boto-u)	(=Rhizacyathida)	
Bronchocyathidae	Cm(Boto-l)	Bacatocyathidae	Cm(Tomm-u)–Cm(mMid)
Calyptocoscinidae	Cm(Atda)–Cm(Boto-u)		
Carinacyathidae	Cm(Atda)–Cm(Boto-u)	Or. ARCHAEOCYATHIDA	
Chankacyathidae	Cm(Boto)	Anthomorphidae	Cm(Atda)–Cm(lMid)
Clathricoscinidae	Cm(Atda)–Cm(Boto-u)	Archaeocyathidae	Cm(Boto-l)–Cm(Dres)
Compositocyathidae	Cm(Atda)–Cm(Boto-u)	Archaeofungiidae	Cm(Atda)–Cm(Boto-l)
Coscinocyathellidae	Cm(Boto-l)–Cm(Boto-u)	Archaeopharetridae	Cm(Boto-l)
Coscinocyathidae	Cm(Tomm-u)–Cm(Boto-u)	Archaeosyconidae	Cm(Tomm-l)–Cm(lMid)
Crassicoscinidae	Cm(Boto)	Ardossacyathidae	Cm(Atda)
Cyclocyathellidae	Cm(Atda)–Cm(Boto-u)	Beltanacyathidae	Cm(Atda)
Dokidocyathidae	Cm(Tomm-l)–Cm(Boto-u)	Bicyathidae	Cm(Tomm-u)–Cm(Boto)
Erbocyathidae	Cm(Atda)–Cm(lMid)	Claruscoscinidae	Cm(Boto-l)–Cm(lMid)
Ethmocoscinidae	Cm(Atda)–Cm(Boto-u)	Copleicyathidae	Cm(Boto-l)
Ethmocyathidae	Cm(Boto)	Dictyocoscinidae	Cm(Boto-l)
Ethmopectinidae	Cm(Boto-l)–Cm(Boto-u)	Dictyocyathidae	Cm(Tomm-u)–Cm(Boto)
Ethmophyllidae	Cm(Atda)–Cm(lMid)		

Table 26.1 (*Continued*).

Flindersicyathidae	Cm(Boto-l)–Cm(lMid)	Or. UNCERTAIN	
Graphoscyphiidae	Cm(Boto)	*Ovatoscutum*	V(Edia)
Hawkeicyathidae	Cm(Atda)	*Spriggia*	V(Edia)
Jugalicyathidae	Cm(Atda)	Cl. SCYPHOZOA	
Metacoscinidae	Cm(Boto-l)–Cm(lMid)	Or. INCERTAE SEDIS	
Metacyathidae	Cm(Tomm-u)–Cm(lMid)	Conchopeltidae	V(Edia)–O(Cara)
Prismocyathidae	Cm(Boto)	?Corumbellidae	V(Edia)
Protocyclocyathidae	Cm(Boto)	Tirasianidae	V(Edia)
Protopharetridae	Cm(Tomm-u)–Cm(lMid)	Cl. INCERTAE SEDIS	
Pycnoidocoscinidae	Cm(l)	Bonatiidae	V(Edia)
Sigmofungiidae	Cm(Boto-l)	Hiemaloriidae	V(Edia)
Spirillicyathidae	Cm(Atda)	Kimberellidae	V(Edia)
Tabellaecyathidae	Cm(Boto)–Cm(lMid)	?Pomoriidae	V(Edia)
Tabulacyathidae	Cm(Atda)–Cm(Boto)	Staurinidae	V(Edia)
Or. SYRINGOCNEMIDIDA		*Cambromedusa*	Cm(uMid)
Pseudosyringocnemidae	Cm(Boto)	Cl. HYDROCONOZOA	
Syringocnemididae	Cm(Boto)	Gastroconidae	Cm(Boto)
Syringocoscinidae	Cm(Boto)	Hydroconidae	Cm(Atda)–Cm(Boto)
Cl. INCERTAE SEDIS		?Tabulaconidae	Cm(Boto)–Cm(lMid)
Or. KAZAKHSTANICYATHIDA		incertae sedis	Cm(Boto)
Kazakhstanicyathidae	Cm(Boto)–Cm(lMid)	Cl. ANTHOZOA	
Or. ARCHAEOPHYLLIDA		Or. ALCYONIDA	
Archaeophyllidae	Cm(Atda)	?*Margaretia*	Cm(Boto)
Or. INCERTAE SEDIS		Or. TABULATA	
Acanthopyrgidae	Cm(Boto)	?Lipoporidae	Cm(lMid)–Cm(mMid)
?Korovinellidae	Cm(Boto-u)	Or. RUGOSA	
?Mattewcyathidae	Cm(m)	?Cothoniidae	Cm(lMid)
TRILOBOZOA		Or. ACTINARIA	
Albumaresidae	V(Edia)	*Mackenzia*	Cm(mMid)
?Anabaritidae	V(N-Da)–Cm(Boto-u)	Or. UNCERTAIN	
Tribrachidium	V(Edia)	*Beltanelliformis*	V(Edia)–V(N-Da)
CNIDARIA		Cl. INCERTAE SEDIS	
?Cl. PETALONOMAE		?*Rosellatona*	Cm(Boto-l)
Or. ERNIETTAMORPHA		CTENOPHORA	
?Bomakellidae	V(Edia)	?*Fasciculus*	Cm(mMid)
?Dickinsonidae	V(Edia)	PRIAPULIDA	
Erniettidae	V(Edia)	Ancalagonidae	Cm(mMid)
Pteridiniidae	V(Edia)	Fieldiidae	Cm(mMid)
Or. RANGEOMORPHA		Miskoiidae	Cm(mMid)
Charniidae	V(Edia)	Ottoiidae	Cm(mMid)–Cm(uMid)
Rangeidae	V(Edia)	Selkirkiidae	Cm(Boto-u)–Cm(uMid)
Cl. CYCLOZOA		*Cricocosmia*	Cm(Atda)
Cyclomedusidae	V(Edia)–V(N-Da)?	*Lecythioscopa*	Cm(mMid)
?Medusinitidae	V(Edia)	*Maotianshania*	Cm(Atda)
Cl. HYDROZOA		*Scolecofura*	Cm(mMid)
Or. TRACHYLINIDA		PROBLEMATICA	
?*Velumbrella*	Cm(Boto)	Cl. COELOSCLERITOPHORA	
Or. HYDROIDA		Or. SACHITIDA	
Aeguoreidae	Cm(lMid)–R	Halkieriidae	Cm(Tomm-l)–Cm(Atda)
Or. SIPHONOPHORIDA		Sachitidae	Cm(Tomm-u)–Cm(Atda)
Porpitidae	Cm(Atda)–R	Siphogonuchitidae	Cm(Tomm-l)–Cm(Boto)
?Palaeacmaeidae	Cm(Trep)–O(Trem)	Wiwaxiidae	Cm(mMid)
?Scenellidae	Cm(Tomm-u)–D(Sieg)	Or. "chancelloriida"	
?*Eoporpita*	V(Edia)	Chancelloriidae	Cm(Tomm-l)–Cm(Fran)

Cl. CRIBRICYATHIDA
   Achorocyathidae   Cm(Atda)
   Akademiophyllidae   Cm(Atda)
   Capillicyathidae   Cm(Boto)
   Cloudinidae   V(Edia)–Cm(Boto-l)?
   Conoidocyathidae   Cm(Boto)–Cm(lMid)
   Cribricyathidae   Cm(Boto)–Cm(lMid)
   Leibaellidae   Cm(Atda)
   Manacyathidae   Cm(Atda)
   Pyxidocyathidae   Cm(Boto)–Cm(lMid)
   Vologdinophyllidae   Cm(Atda)

Cl. STENOTHECOIDA
   Cambridiidae   Cm(Atda)–Cm(uMid)

Cl. TOMMOTIIDA
Or. MITROSAGOPHORA
   Tannuolinidae   Cm(Boto-l)
   Tommotiidae   Cm(Tomm-l)–Cm(lMid)

Or. unnamed
   Kelanellidae   Cm(Atda)–Cm(mMid)
   Lapworthellidae   Cm(Tomm-l)–Cm(mMid)
   Sunnaginidae   Cm(Tomm-l)–Cm(Boto-l)

Cl. UNCERTAIN
Or. "cambroclava" (*)
   Zhijinitidae   Cm(Tomm-u)–Cm(Boto-l)

Or. COLEOLIDA (*)
   Coleolidae   V(N-Da)–C(Tour)

Or. CONULARIIDA (*)
   Carinachitidae   V(N-Da)–Cm(Tomm-u)
   Conulariellidae   Cm(Tomm-l)–O (Aren)
   Conulariidae   Cm(u)–Tr(Nori)

Or. HYOLITHELMINTHES (*)
   Hyolithellidae   V(N-Da)–Cm(mMid)
   Torellellidae   V(N-Da)–P(Sakm)
   ?Byronia   Cm(mMid)–O(Cara)

Or. PAIUTIIDA (*)
   Paiutitubulitidae   Cm(Atda)

Or. SABELLIDITIDA (*)
   Saarinidae   V(N-Da)–Cm(Tomm-u)
   Sabelliditidae   V(N-Da)–Cm(lMid)

Or. VOLBORTHELLIDA (= Agmata) (*)
   Salterellidae   Cm(Boto-l)–Cm(lMid)
   Volborthellidae   Cm(Atda)–Cm(Boto-l)
   Vologdinellidae   Cm(uMid)

Or. INCERTAE SEDIS
   Amiskwiidae (*)   Cm(mMid)
   Anomalocaridae (*)   Cm(Atda)–Cm(uMid)
   Dinomischidae (*)   Cm(Atda)–Cm(mMid)
   Hallucigeniidae (*)   Cm(mMid)
   Lenargyrionidae (*)   Cm(Atda)–O(Aren)
   Microdictyonidae (*)   Cm(Atda)–Cm(lMid)
   Odontogriphidae (*)   Cm(mMid)
   Opabiniidae (*)   Cm(mMid)
   Paracarinachitidae (*)   Cm(Tomm-u)–Cm(Atda)
   Group (*Mongolitubulus*
     + *Rushtonites*) (*)   Cm(Atda)–Cm(Boto-u)
   Banffia (*)   Cm(mMid)
   Cowiella   Cm(Atda)

   Cyrtochites   Cm(Tomm-u)
   Eldonia (*)   Cm(Atda)–Cm(uMid)
   Fomitchella (*)   Cm(Tomm-l)–Cm(Atda)
   Nectocaris (*)   Cm(mMid)
   Oesia (*)   Cm(mMid)
   Paleobotryllus   Cm(Trep)
   Pollingeria   Cm(mMid)
   Portalia   Cm(mMid)
   Pyrgites   Cm(Tomm)
   Redkinia (*)   V(Edia)
   Redoubtia (*)   Cm(mMid)
   Rhombocorniculum (*)   Cm(Tomm-u)–Cm(Boto-u)
   Sinotubulites   V (N-Da)–Cm(Tomm-l)
   Stefania   Cm(Boto)
   Tubulella (*)   Cm(Boto-u)–Cm(uMid)
   Tumulduria   Cm(Tomm-l)
   Westgardia   Cm(Boto-l)
   Worthenella (*)   Cm(mMid)
   Yunnanodus   Cm(Tomm-u)

MOLLUSCA

Cl. POLYPLACOPHORA
Or. PALAEOLORICATA
   Mattheviidae   Cm(Fran)–O(Trem)
   Praecanthochitonidae   Cm(Trep)–O(Trem)

Cl. MONOPLACOPHORA
Or. CYRTONELLIDA
   Helcionellidae   Cm(Tomm-l)–O(Aren)
   Hypseloconidae   Cm(Dres)–O (Llde)
   Shelbyoceridae   Cm(uMid)–O(Trem)
   Stenothecidae   Cm(Tomm-u)–Cm(uMid)
   Yochelcionellidae   Cm(Atda)–Cm(uMid)
   ?Group (*Canopoconus*,
     Purella, Ramentia)   V(N-Da)–Cm(Atda)
   ?*Maikhanella*   Cm(Tomm)
   ?*Ocruranus*   Cm(Tomm-u)
   ?*Xianfengella*   Cm(Tomm-u)

Or. TRYBLIDIIDA
   Tryblidiidae   Cm(lMid)–R

Or. BELLEROPHONTIDA
   Bellerophontidae   Cm(Atda)–Tr(Indu)
   ?Multifariidae   Cm(Atda)–O(Aren)
   Sinuitidae   Cm(uMid)–Tr(Olen)

Or. PELAGIELLIDA
   Pelagiellidae   Cm(Atda)–O(Trem)

Or. TUARANGIIDA
   Tuarangiidae   Cm(uMid)–Cm(Dres)

Cl. GASTROPODA
Or. ARCHAEOGASTROPODA
   Aldanellidae   Cm(Tomm-l)–Cm(Atda)
   Clisospiridae   Cm(Dres)–D(Give)
   Eotomariidae   Cm(Trep)–J(Plie)
   Macluritidae   Cm(Dres)–O(Ashg)
   Onychochilidae   Cm(Atda)–D(Sieg)
   Sinuopeidae   Cm(Trep)–P(Tatr)

Cl. ROSTROCONCHIA
Or. RIBEIROIDA
   Ischyriniidae   Cm(Trep)–O(Ashg)

Table 26.1 (*Continued*).

Ribeiriidae	Cm(Tomm-l)–O(Ashg)	Sulcavitidae	Cm(Tomm-u)–S(Ldov)
Technophoridae	Cm(Dres)–S(Ldov)	Trapezovitidae	Cm(Tomm-u)–Cm(Boto-u)

Or. CONOCARDIOIDA
   Eopteriidae   Cm(Dres)–O(Ashg)

Cl. BIVALVIA
Or. NUCULOIDA
   Praenuculidae
     (+ *Pojetaia*)   Cm(Atda)–D(l)

Or. MODIOMORPHOIDA
   ?Fordillidae   Cm(Atda)–O(Cara)

Cl. CEPHALOPODA
Or. ELLESMEROCERIDA
   Acaroceratidae   Cm(Trep)–Cm(Trep)
   Ellesmeroceratidae   Cm(Trep)–O(Aren)
   Huaiheceratidae   Cm(Trep)–Cm(Trep)
   Xiaoshanoceratidae   Cm(Trep)

Or. PLECTONOCERIDA
   Balkoceratidae   Cm(Trep)
   Plectonoceratidae   Cm(Fran)–Cm(Trep)

Or. PROTACTINOCERIDA
   Protactinoceratidae   Cm(Trep)–Cm(Trep)

Or. YANHECERIDA
   Yanheceratidae   Cm(Trep)

Or. ENDOCERIDA
   Proterocameroceratidae   Cm(Trep)–O(Llde)

HYOLITHA
Cl. ORTHOTHECIMORPHA
Or. CIRCOTHECIDA
   Circothecidae   V(N-Da)–O(Cara)
   Spinulithecidae   V(N-Da)–Cm(Atda)

Or. EXILITHECIDA
   Exilithecidae   Cm(Tomm-l)–Cm(Boto-u)
   Gracilithecidae   Cm(Tomm-u)–Cm(mMid)
   Obliquathecidae   Cm(Tomm-u)–Cm(Boto)

Or. GLOBORILIDA
   Globorilidae   Cm(Tomm-u)–Cm(lMid)

Or. ORTHOTHECIDA
   Allathecidae   Cm(Tomm-l)–Cm(Boto)
   Isitithecidae   Cm(Tomm-u)–Cm(Boto-u)
   Novitatidae   Cm(Atda)–Cm(Boto-l)
   Orthothecidae   V(N-Da)–D(m)
   Tchuranothecidae   Cm(Tomm-l)–Cm(Atda)
   Tetrathecidae   V(N-Da)–O(l)

Cl. "hyolithomorpha"
Or. HYOLITHIDA
   Aimitidae   Cm(Tomm-u)–Cm(Atda)
   Altaicornidae   Cm(Boto-u)
   Amydaicornidae   Cm(Atda)–Cm(mMid)
   Angusticornidae   Cm(Atda)–Cm(lMid)
   Crestjahitidae   Cm(Tomm-u)–Cm(Boto-l)
   Dorsojugatidae   Cm(Tomm-u)–Cm(Boto-l)
   Galicornidae   Cm(Atda)–Cm(Boto-l)
   Hyolithidae   Cm(Tomm-u)–P(Guad)
   Nelegerocornidae   Cm(Tomm-u)–Cm(Boto-l)
   Notabilitidae   Cm(Tomm-u)–Cm(Boto-l)
   Pauxillitidae   Cm(Boto)–O(Ashg)

Cl. INCERTAE SEDIS
Or. CAMEROTHECIDA
   Camerothecidae   Cm(Boto)
   Diplothcidae   Cm(Boto)

ANNELIDA
Cl. POLYCHAETA
Or. SPIOMORPHA
   Spionidae   Cm(Boto)?–R

Or. DRILOMORPHA
   Opheliidae   Cm(Boto)?–R

Or. TEREBELLOMORPHA
   Terebellidae   Cm(mMid)–R

Or. INCERTAE SEDIS
   Burgessochaetidae   Cm(mMid)
   Canadiidae   Cm(mMid)–Cm(uMid)
   Insilicoryphidae   Cm(mMid)
   Peronochaetidae   Cm(mMid)
   Stephenoscolecidae   Cm(mMid)

Cl. PALAEOSCOLECIDA
   Palaeoscolecidae   Cm(Atda)–S(Prid)

ARTHROPODA
?Cl. PARATRILOBITA
   Sprigginidae   V(Edia)

Cl. MARRELOMORPHA
Or. MIMETASTERIDA   Cm(mMid)–D(Emsi)

Cl. INCERTAE SEDIS
   Sanctacaris   Cm(mMid)

Cl. MEROSTOMATA
Or. LEANCHOILIIDA (*)
   Leanchoiliidae   Cm(lMid)–Cm(mMid)
   ?*Burgessia*   Cm(mMid)

Or. unnamed (*)
   Yohoiidae   Cm(Atda)–Cm(mMid)
   ?*Actaeus*   Cm(mMid)
   ?*Sarotrocercus*   Cm(mMid)

Or. INCERTAE SEDIS
   ?Eolimulidae   Cm(Atda)

Cl. TRILOBITA
Or. AGNOSTIDA
   Agnostidae   Cm(Boto)–O(Ashg)
   Clavagnostidae   Cm(uMid)–Cm(Dres)
   Condylopygidae   Cm(Boto-l)–Cm(uMid)
   Diplagnostidae   Cm(mMid)–O(Trem)
   Discagnostidae   Cm(Dres)
   Eodiscidae   Cm(Atda)–Cm(uMid)
   Pagetiidae   Cm(Atda)–Cm(mMid)
   Phalacromidae   Cm(uMid)–Cm(Dres)
   Trinodidae   Cm(Dres)–O(Ashg)

Or. REDLICHIIDA
   Abadiellidae   Cm(Atda)–Cm(lMid)
   Archaeaspididae   Cm(Atda)
   Bathynotidae   Cm(Boto)–Cm(lMid)
   Chengkouiidae   Cm(Boto)

Table 26.1 (*Continued*).

Daguinaspididae	Cm(Atda)	Harpididae	Cm(Fran)–S(Ldov)
Despujolsiidae	Cm(Atda)	Housiidae	Cm(Dres)–Cm(Fran)
Dolerolenidae	Cm(Atda)–Cm(Boto)	Hungaiidae	Cm(Trep)–O(Ashg)
?Ellipsocephalidae	Cm(Atda)–Cm(mMid)	Idahoiidae	Cm(Dres)–Cm(Trep)
Emuellidae	Cm(lMid)	Illaenuridae	Cm(Fran)–Cm(Trep)
Gigantopygidae	Cm(Boto)	Kainellidae	Cm(Fran)–O(m)
Hicksiidae	Cm(Boto)	Kaolishaniidae	Cm(Dres)–Cm(Trep)
Kueichowiidae	Cm(Boto)	Kingstoniidae	Cm(uMid)–O(Trem)
Longduiidae	Cm(Boto)	Komaspididae	Cm(uMid)–O(Cara)
Mayiellidae	Cm(Boto-u)	Lecanopygidae	Cm(Fran)–O(Aren)
Neoredlichiidae	Cm(Atda)–Cm(Boto)	Leiostegiidae	Cm(uMid)–O(Aren)
Olenellidae	Cm(Atda)–Cm(mMid)	Liostracinidae	Cm(uMid)–Cm(Trep)
Paradoxididae	Cm(Atda)–Cm(uMid)	Loganellidae	Cm(Fran)–Cm(Trep)
Protolenidae	Cm(Atda)–Cm(mMid)	Lonchocephalidae	Cm(uMid)–Cm(Trep)
Redlichiidae	Cm(Atda)–Cm(mMid)	Marjumiidae	Cm(mMid)–O(Trem)
Saukiandidae	Cm(Boto)	Menomoniidae	Cm(uMid)–Cm(Fran)
Yinitidae	Cm(Atda)–Cm(Boto)	Missisquoiidae	Cm(Fran)–O(Trem)
Yunnanocephalidae	Cm(Atda)	Nepeidae	Cm(uMid)–Cm(Dres)
Or. CORYNEXOCHIA		Nileidae	Cm(Trep)–O(Ashg)
Balangiidae	Cm(Boto)–Cm(lMid)	Norwoodiidae	Cm(uMid)–O(Trem)
Corynexochidae	Cm(Boto-l)–Cm(Fran)	Olenidae	Cm(Dres)–O(Ashg)
Dinesidae	Cm(Boto)–Cm(uMid)	Pagodiidae	Cm(uMid)–Cm(Trep)
Dolichometopidae	Cm(Boto-l)–Cm(Dres)	Papyriaspididae	Cm(lMid)–Cm(Trep)
Dorypygidae	Cm(Atda)–Cm(Trep)	Parabolinoididae	Cm(Dres)–Cm(Fran)
?Edelsteinaspidae	Cm(Boto-l)–Cm(mMid)	Placosematidae	Cm(uMid)–Cm(Dres)
?Granulariidae	Cm(Boto-l)–Cm(uMid)	Plectriferidae	Cm(uMid)–Cm(Dres)
Jakutidae	Cm(Boto-l)–Cm(lMid)	Plethopeltidae	Cm(uMid)–O(Trem)
Namanoiidae	Cm(lMid)–Cm(uMid)	Polycyrtaspididae	Cm(Dres)
Ogygopsidae	Cm(Boto-u)–Cm(mMid)	Pterocephalidae	Cm(Dres)–Cm(Trep)
Oryctocephalidae	Cm(Atda)–Cm(uMid)	Ptychaspididae	Cm(Fran)–O(Trem)
Zacanthoididae	Cm(Boto-l)–Cm(uMid)	Ptychopariidae	Cm(Boto-l)–O(Trem)
Or. PTYCHOPARIIDA		Rasettaspiidae	Cm(Fran)
Agraulidae	Cm(lMid)–Cm(Dres)	Raymodinidae	Cm(uMid)–Cm(Trep)
Alokistocaridae	Cm(Boto)–Cm(Fran)	Remopleurididae	Cm(uMid)–O(Ashg)
Andrarinidae	Cm(mMid)–Cm(Fran)	Rhyssometopidae	Cm(uMid)–Cm(Dres)
Annamitiidae	Cm(uMid)	Saukiidae	Cm(Dres)–Cm(Trep)
Anomocaridae	Cm(Boto-l)–O(Trem)	Shirakiellidae	Cm(Fran)
Asaphidae	Cm(uMid)–O(Ashg)	Shumardiidae	Cm(Fran)–O(Ashg)
Asaphiscidae	Cm(lMid)–O(Trem)	Solenopleuridae	Cm(Boto-u)–O(Trem)
Aulacodigmatidae	Cm(uMid)–Cm(Dres)	Tricrepicephalidae	Cm(uMid)–Cm(Dres)
Auritamidae	Cm(uMid)–Cm(Dres)	Tsinaniidae	Cm(Fran)–Cm(Trep)
Avoninidae	Cm(uMid)–Cm(Fran)	Or. PHACOPIDA	
Bestjubellidae	Cm(Fran)	Cheiruridae	Cm(Fran)–D(Give)
Bolaspididae	Cm(mMid)–Cm(uMid)	Or. ODONTOPLEURIDA	
Burlingiidae	Cm(mMid)–Cm(Fran)	Eoacidaspididae	Cm(uMid)–O(Trem)
Cattillicephalidae	Cm(uMid)–O(Trem)	Eodontopleuridae	Cm(Boto-u)
Ceratopygidae	Cm(uMid)–O(Trem)	Odontopleuridae	Cm(uMid)–D(Fras)
Cheilocephalidae	Cm(Dres)–Cm(Fran)	Or. NEKTASPIDA	
Conocoryphidae	Cm(Atda)–O(Trem)	Naraoidae	Cm(Atda)–Cm(uMid)
Cooseliidae	Cm(uMid)–Cm(Dres)	Or. INCERTAE SEDIS	
Crepicephalidae	Cm(lMid)–Cm(Dres)	Amgawpidae	Cm(lMid)–Cm(uMid)
Damesellidae	Cm(uMid)–Cm(Dres)	Hanburiidae	Cm(mMid)
Diceratocephalidae	Cm(Dres)	Tegopeltidae	Cm(mMid)
Dikelocephalidae	Cm(Dres)–Cm(Trep)	Cl. MEROSTOMOIDEA	
Dokimocephalidae	Cm(Dres)–O(Trem)	Or. LIMULAVIDA	
Elviniidae	Cm(Dres)–O(Trem)	Sidneyiidae	Cm(mMid)–Cm(uMid)
Emmrichellidae	Cm(Boto)–O(Trem)	Or. EMERALDELLIDA	
Entomaspididae	Cm(Trep)–O(Trem)	Emerallidae	Cm(mMid)
Erixaniidae	Cm(Dres)		
Eurekiidae	Cm(Fran)–O(Trem)		

Table 26.1 (Continued).

Or. INCERTAE SEDIS
    Paleomeridae   Cm(Atda)
    Strabopidae   Cm(Fran)–O(Ashg)
    Fuxianhuia   Cm(Atda)
    Helmetia   Cm(mMid)
    Kuamaia   Cm(Atda)
    Rhombicalvaria   Cm(Atda)

Cl. INCERTAE SEDIS
Or. SKARACARIDA
    Skara   Cm(Dres)

Or. ORSTENOCARIDA
    Bredocaris   Cm(Fran)

Or. INCERTAE SEDIS
    Martinssonia   Cm(Dres)
    Oelandocaris   Cm(Dres)
    Rehbachiella   Cm(Dres)
    Walossekia   Cm(Dres)

Or. WAPTIIDA
    Waptiidae   Cm(mMid)
    ?*Plenocaris*   Cm(mMid)

Cl. CEPHALOCARIDA
    ?*Dala*   Cm(Dres)

Cl. OSTRACODA
Or. ARCHAEOCOPIDA
    (= Bradoriida + Phosphatocopida)

    Alutidae   Cm(Atda)–Cm(lMid)
    Beyrichonidae   Cm(Atda)–O(Trem)
    Bradoriidae   Cm(Atda)–Cm(Fran)
    Comptalutidae   Cm(lMid)–Cm(uMid)
    Falitidae   Cm(Dres)–Cm(Trep)
    Hesslandonidae   Cm(Dres)
    Indianidae   Cm(Atda)–Cm(mMid)
    Monasteriidae   Cm(mMid)
    Oepikalutidae   Cm(mMid)–Cm(uMid)
    Svealutidae   Cm(mMid)–Cm(Dres)
    Vestrogothiidae   Cm(lMid)–Cm(Dres)

Or. LEPERDITICOPIDA
    ?*Ushkarella*   Cm(lMid)

Or. PALAEOCOPIDA
    Tetradellidae   Cm(Boto)?–S(Ldov)

Cl. CIRRIPEDIA
Or. THORACICA
    ?*Priscansermarinus*   Cm(mMid)

Cl. MALACOSTRACA
Or. CANADASPIDIDA
    Canadaspididae   Cm(mMid)
    ?Odaraidae   Cm(Atda)–Cm(mMid)
    ?Perspicarididae   Cm(Atda)–Cm(uMid)

Or. HYMENOSTRACA
    Hymenocarididae   Cm(Boto)–O(Aren)

Cl. ONYCHOPHORA
    ?*Aysheaia*   Cm(mMid)–Cm(uMid)
    ?*Xenusion*   Cm(Tomm-u)

Cl. MYRIAPODA
    Cambropodus   Cm(uMid)

Cl. INCERTAE SEDIS
Or. AGLASPIDA
    Aglaspididae   Cm(uMid)–Cm(Trep)

Or. MARRIOCARIDA (∗)   Cm(mMid)–O(Ashg)

Or. INCERTAE SEDIS
    Mollisoniidae (∗)   Cm(mMid)–Cm(uMid)
    Protocarididae (∗)   Cm(Atda)–Cm(uMid)
    Tuzoiidae   Cm(Atda)–Cm(uMid)
    ?*Branchiocaris*   Cm(mMid)–Cm(uMid)
    Carnarvonia   Cm(mMid)
    Combinivalvula   Cm(Atda)
    Dioxycaris   Cm(Boto)–Cm(uMid)
    Fieldia   Cm(mMid)
    Gdowia   Cm(Tomm-u)
    Hurdia   Cm(mMid)
    Molaria   Cm(mMid)
    Isoxys   Cm(Atda)–Cm(mMid)
    Nothozoe   Cm(Boto)–O(Llde)
    Pahvantia   Cm(uMid)
    ?*Pomerania*   Cm(Tomm-u)
    Proboscicaris   Cm(mMid)–Cm(uMid)
    Pseudoarctolepis   Cm(mMid)–Cm(uMid)
    Saccocaris   Cm(u)–O(l)
    Serracaris   Cm(Boto)
    Shafferia   Cm(mMid)
    Skania   Cm(mMid)
    Vetulicola   Cm(Atda)

BRACHIOPODA

Cl. INARTICULATA
Or. LINGULIDA
    Elkaniidae   Cm(Fran)–O(Aren)
    Obolidae   Cm(Tomm-u)–O(Ashg)
    Zhanatellidae   Cm(Boto)–Cm(u)

Or. ACROTRETIDA
    Acrothelidae   Cm(Tomm-u)–O(Trem)
    Acrotretidae   Cm(Tomm-l)–D(Emsi)
    Botsfordiidae   Cm(Atda)–Cm(mMid)
    Curticiidae   Cm(Boto)–Cm(Fran)
    Siphonotretidae   Cm(uMid)–O(Cara)

Or. OBOLELLIDA
    Cryptotretidae   Cm(Tomm-l)–Cm(Atda)
    Obolellidae   Cm(Atda)–Cm(uMid)

Or. PATERINIDA
    Paterinidae   Cm(Tomm-l)–O(Cara)

Or. KUTORGINIDA
    Kutorginidae   Cm(Tomm-u)–Cm(mMid)
    Yorkiidae   Cm(Atda)–Cm(Boto)

Cl. ARTICULATA
Or. ORTHIDA
    Billingsellidae   Cm(Boto)–O(Aren)
    Eoorthidae   Cm(Boto)–O(Aren)
    Finkelnburgiidae   Cm(Fran)–O(Llvi)
    Leioriidae   Cm(Boto)
    Nisusiidae   Cm(Atda)–Cm(uMid)
    Orthidae   Cm(m)–P(Tatr)
    Protorthidae   Cm(Atda)–Cm(uMid)

Table 26.1 (*Continued*).

Or. PENTAMERIDA		Cl. CAMPTOSTROMATOIDEA	
Eostrophiidae	Cm(mMid)	Camptostromatidae	Cm(Boto-u)
Huenellidae	Cm(Dres)–O(l)	Cl. EDRIOASTEROIDEA	
Tetralobulidae	Cm(Trep)–O(Aren)	Or. STROMATOCYSTITIDA	
CHAETOGNATHA		Stromatocystitidae	Cm(Boto-u)–Cm(mMid)
?Or. "protoconodontida"		Or. EDRIOASTERIDA	
Amphigeisinidae	V(N-Da)–Cm(Dres)	Edrioasteridae	Cm(mMid)–O(Cara)
ECHINODERMATA		Totiglobidae	Cm(mMid)–Cm(u)
Cl. CTENOCYSTOIDEA	Cm(mMid)–Cm(uMid)	Or. ISOROPHIDA	
Cl. STYLOPHORA		Cambrasteridae	Cm(mMid)
Or. CORNUTA		Cyclocystoididae	Cm(lMid)–D(Fras)
Ceratocystidae	Cm(mMid)–Cm(Fran)	HEMICHORDATA	
Cothurnocystidae	Cm(lMid)–O(Ashg)	Cl. GRAPTOLITHINA	
Phyllocystidae	Cm(uMid)–O(Llvi)	Or. DENDROIDEA	
Cl. HOMOSTELEA		Acanthograptidae	Cm(Trep)–D(Sieg)
Or. CINCTA		Dendrograptidae	Cm(uMid)–C(Serp)
Gyrocystidae	Cm(mMid)	Or. TUBOIDEA	
Trochocystitidae	Cm(mMid)–Cm(uMid)	Idiotubidae	Cm(u)–S(Ludl)
Cl. HOMOIOSTELEA		Or. DITHECOIDEA	
Or. SOLUTA		Bulmanidendridae	Cm(uMid)
Dendrocystitidae	Cm(Boto-l)–D(Gedi)	?Chuanograptidae	Cm(mMid)–D(Gedi)
Minervaecystidae	Cm(Trep)–O(Trem)	Dithecodendridae	Cm(uMid)–Cm(u)
Cl. HELICOPLACOIDEA		Siberiograptidae	Cm(u)–S(Ldov)
Helicoplacidae	Cm(Atda)–Cm(Boto-l)	Or. ARCHAEODENDRIDA	
Cl. EOCRINOIDEA		Archaeodendridae	Cm(uMid)
Or. IMBRICATA		Cl. PTEROBRANCHIA	
(= Lepidocystoidea)		Or. RHABDOPLEURIDA	
Lepidocystidae	Cm(Boto-u)	*Rhabdotubus*	Cm(mMid)–Cm(uMid)
Or. "number 1"		Cl. ENTEROPNEUSTA	
Eocrinidae	Cm(Atda)–Cm(uMid)	?incertae sedis	Cm(mMid)
Lichenoididae	Cm(uMid)	CHORDATA	
Or. "number 2"		Cl. ACRANIA (Cephalochordata)	
Group		*Pikaia*	Cm(mMid)
(*Cambrocrinus* + *Eocystites*)	Cm(Boto)–Cm(Dres)	?Cl. CONODONTA	
Group (*Eustypocystis*		Or. PARACONODONTIDA	
+ *Nolichuckia* + *Pareocrinus*)	Cm(uMid)	Furnishinidae	Cm(mMid)–O(Trem)
Or. INCERTAE SEDIS		Westergaardodinidae	Cm(mMid)–O(Llde)
Trachelocrinidae	Cm(mMid)–Cm(Fran)	Or. CONODONTOPHORIDA	
Cymbionites	Cm(lMid)	Clavohamulidae	Cm(Trep)–O(Aren)
Peridionites	Cm(lMid)	Cordylododontidae	Cm(Trep)–O(Llvi)
Cl. CRINOIDEA		Fryxellodontidae	Cm(Trep)–O(Trem)
Or. ECHMATOCRINIDA		Oneotodontidae	Cm(mMid)–O(Cara)
Echmatocrinidae	Cm(mMid)	Proconodontidae	Cm(Dres)–O(Trem)
		Teridontidae	Cm(Fran)–O(Trem)

Taxonomy and Stratigraphic Ranges of Animal Families

Table 26.2. *Time scale employed in plotting Vendian-Cambrian diversity of metazoans and metaphytes.*

Series	"Stage"	Ma	Series	"Stage"	Ma
Upper Cambrian:	Trempealeauan	—500	Lower Cambrian:	Botomian	
		—503			—538
	Franconian			Atdabanian	
		—507			—544
	Dresbachian			Tommotian	
		—511.5			—550
Middle Cambrian:	upper Middle		Vendian:	Nemakit-Daldyn	
		—520			—555
	middle Middle			Ediacaran	
		—525			—570
	lower Middle				
		—530			

Table 26.3. *Data used in the Q-mode factor analysis of faunal heterogeneity in the Vendian-Cambrian metazoan radiation. The Table lists number of genera within classes and class-level problematica for 14 Vendian to Upper Cambrian stratigraphic units (dashes indicate zero genera). Abbreviations are "Edi" = Ediacaran; "NDa" = Nemakit Daldyn; "lT" = lower Tommotian; "uT" = upper Tommotian; "lA" = lower Atdabanian; "uA" = upper Atdabanian; "lB" = lower Botomian; "uB" = upper Botomian; "lM" = lower Middle Cambrian; "mM" = middle Middle Cambrian; "uM" = upper Middle Cambrian; "Dr" = Dresbachian; "Fr" = Franconian; "Tr" = Trempealeauan.*

	Edi	NDa	lT	uT	lA	uA	lB	uB	lM	mM	uM	Dr	Fr	Tr
Rhizopodea	—	—	1	1	1	1	1	1	1	2	4	4	4	5
Radiolaria	—	—	—	—	—	2	2	2	2	2	3	3	3	3
Demospongia	—	—	—	—	3	8	8	8	6	16	7	5	6	7
Calcarea	—	—	—	—	—	2	2	3	5	4	3	3	3	3
Hexactinellida	—	—	1	2	1	2	1	2	2	6	9	8	6	6
Regulares	—	—	3	14	62	110	198	123	11	1	—	—	—	—
Irregulares	—	—	1	9	15	32	39	27	12	2	1	1	—	—
Trilobozoa	4	3	8	5	2	2	1	1	—	—	—	—	—	—
Petalonamae	22	1	—	—	—	—	—	—	—	—	—	—	—	—
Cyclozoa	6	1	—	—	—	—	—	—	—	—	—	—	—	—
Hydrozoa	3	1	1	2	2	3	3	3	5	5	6	6	6	6
Scyphozoa	4	1	1	1	1	1	1	1	1	1	1	1	1	1
Hydroconozoa	—	—	—	—	—	1	5	5	1	—	—	—	—	—
Anthozoa	1	1	1	1	1	1	1	2	4	4	4	4	4	4
Ctenophora	—	—	—	—	—	—	—	—	—	1	1	1	1	1
Priapulida	—	—	—	—	2	1	1	1	7	2	1	1	1	1
Polyplacophora	—	—	—	—	—	—	—	—	—	—	—	—	1	3
Monoplacophora	—	3	13	87	21	30	26	26	27	24	24	16	12	20
Gastropoda	—	—	1	1	1	2	1	1	1	1	1	4	5	10
Rostroconchia	—	—	1	3	2	1	1	1	1	1	1	6	4	10
Bivalvia	—	—	—	1	2	2	2	2	2	2	2	2	2	2
Cephalopoda	—	—	—	—	—	—	—	—	—	—	—	—	1	42
Orthothecimorpha	—	3	23	42	28	37	27	21	14	11	10	10	9	9
hyolithomorphs	—	—	—	18	12	25	22	9	8	11	7	7	7	5
Polychaeta	—	—	—	—	—	—	4	3	8	4	3	3	3	3
Palaeoscolecida	—	—	—	—	1	1	1	1	1	1	1	1	1	1
Coeloscleritophora	—	1	6	14	6	9	4	3	3	2	1	1	1	—
Cribricyathida	1	1	1	1	1	11	5	3	3	—	—	—	—	—
Stenothecoidea	—	—	—	—	1	3	3	2	2	1	1	—	—	—
Tommotiida	—	—	5	6	6	9	7	2	2	2	—	—	—	—
Cambroclaves	—	—	—	1	1	2	1	—	—	—	—	—	—	—
Coeloloida	—	—	3	3	3	3	3	1	1	1	1	1	1	1
Conulariida	—	2	3	7	1	1	1	1	1	2	2	2	2	3
Hyolithelminthes	—	1	3	6	5	7	6	5	3	3	3	3	3	4
Sabelliditida	—	4	3	3	2	2	1	1	1	—	—	—	—	—
Volborthellida	—	—	—	—	1	1	2	1	1	1	1	—	—	—

1180 Models

Table 26.3 (Continued).

	Edi	NDa	lT	uT	lA	uA	lB	uB	lM	mM	uM	Dr	Fr	Tr
Anomalocaridae	—	—	—	—	1	1	1	1	1	1	1	—	—	—
Dinomischidae	—	—	—	—	—	1	1	1	1	1	—	—	—	—
Mongolitubulus + *Rustonites*	—	—	—	—	—	1	1	1	—	—	—	—	—	—
Fomitchella	—	—	1	1	—	—	—	—	—	—	—	—	—	—
Lenargyrionidae	—	—	—	—	—	1	1	1	1	1	1	1	1	1
Microdictyon	—	—	—	—	—	1	1	1	1	—	—	—	—	—
Rhombocorniculum	—	—	—	1	1	1	1	1	—	—	—	—	—	—
Paratrilobita	3	—	—	—	—	—	—	—	—	—	—	—	—	—
Merostomata	—	—	—	1	1	1	1	1	1	1	2	2	4	9
Trilobita	—	—	—	—	20	82	195	177	156	171	278	290	302	221
Merostomoidea	—	—	—	—	1	4	1	1	1	4	2	1	1	1
Marrelomorpha	—	—	—	—	—	—	—	—	—	1	1	1	1	1
Cephalocarida	—	—	—	—	—	—	—	—	—	—	—	1	1	1
Ostracoda	—	—	—	—	4	27	12	12	20	17	13	13	4	3
Cirripedia	—	—	—	—	—	—	—	—	—	1	1	1	1	1
Malacostraca	—	—	—	—	—	1	2	2	2	3	2	1	1	1
Odaraidae + *Branchiocaris*	—	—	—	—	—	1	1	1	1	2	—	—	—	—
Onychophora	—	—	—	1	1	1	1	1	1	1	1	1	1	1
Myriapoda	—	—	—	—	—	—	—	—	—	—	1	1	1	1
Yohoidae + *Actaeus* + *Alalcomenaeus*	—	—	—	—	—	2	2	2	2	3	—	—	—	—
Inarticulata	—	—	2	22	18	16	35	27	26	25	29	26	20	13
Articulata	—	—	—	—	2	4	12	12	15	16	10	8	12	11
protoconodonts	—	1	3	4	2	3	3	2	2	2	1	1	—	—
Ctenocystoidea	—	—	—	—	—	—	—	—	—	1	2	—	—	—
Stylophora	—	—	—	—	—	—	—	—	—	1	3	3	3	2
Homostelea	—	—	—	—	—	—	—	—	—	2	2	—	—	—
Homoiostelea	—	—	—	—	—	—	1	1	1	1	1	1	1	2
Helicoplacoidea	—	—	—	—	—	1	2	—	—	—	—	—	—	—
Eocrinoidea	—	—	—	—	1	1	1	4	4	5	10	3	3	2
Crinoidea	—	—	—	—	—	—	—	—	1	1	1	1	1	1
Camptostromatoidea	—	—	—	—	—	—	1	—	—	—	—	—	—	—
Edrioasteroidea	—	—	—	—	—	—	1	4	5	3	3	2	2	—
Graptolithina	—	—	—	—	—	—	—	—	—	—	6	6	11	8
Conodonta	—	—	—	—	—	—	—	—	—	5	8	9	10	16

Tables 26.4.1–26.4.4 Summary of results of the Q-mode factor analysis of Vendian-Cambrian animal genera within 68 classes (excluding archaeocyathids). Italicized numbers indicate maximum loadings or scores for each stage or taxon, respectively.

Table 26.4.1. *Loadings on first three principal factors.*

stratigraphic intervals	factors			stratigraphic intervals	factors		
	1	2	3		1	2	3
Trempealeauan	*0.9540*	−0.2126	0.0349	l Atdabanian	*0.7040*	0.6283	−0.1414
Franconian	*0.9608*	−0.2525	0.0407	u Tommotian	0.3098	*0.8458*	−0.1708
Dresbachian	*0.9674*	−0.2358	0.0354	l Tommotian	0.2785	*0.9020*	−0.0815
u Middle	*0.9730*	−0.2148	0.0299	Nemakit-Daldyn	0.2084	*0.8013*	0.2622
m Middle	*0.9840*	−0.1485	0.0128	Ediacaran	0.0169	0.1521	*0.9649*
l Middle	*0.9897*	−0.1103	0.0068				
u Botomian	*0.9920*	−0.1119	0.0073				
l Botomian	*0.9926*	−0.0933	−0.0012	Eigenvalues:	65.9%	20.8%	7.6%
u Atdabanian	*0.9352*	0.2316	−0.0608	Cumulative eigenvalues:	65.9%	86.7%	94.3%

Table 26.4.2. *Scores for first three principal factors.*

	taxonomic class	factors				taxonomic class	factors		
		1	2	3			1	2	3
1	Rhizopodea	*0.013*	0.012	−0.005	35	Anomalocaridae	*0.005*	0.004	−0.003
2	Radiolaria	*0.011*	−0.004	0.001	36	Dinomischidae	*0.003*	0.000	0.000
3	Demospongia	*0.045*	0.004	−0.008	37	Lenargyrionidae	*0.005*	−0.001	0.000
4	Calcarea	*0.015*	−0.005	0.001	38	*Mongol + Rushton*	*0.002*	0.000	−0.001
5	Hexactinellida	*0.023*	0.011	−0.006	39	*Formitchella*	0.003	*0.018*	−0.007
6	Trilobozoa	0.025	*0.227*	0.215	40	*Microdictyon*	*0.003*	0.000	0.000
7	Petalonamae	0.005	0.085	*0.867*	41	*Rhombocorniculum*	0.004	*0.008*	−0.005
8	Cyclozoa	0.003	0.050	*0.260*	42	Paratrilobita	*0.000*	0.007	*0.114*
9	Hydrozoa	0.030	0.060	*0.137*	43	Merostomata	*0.012*	0.002	−0.003
10	Scyphozoa	0.011	0.062	*0.178*	44	Trilobita	*0.933*	−0.310	0.053
11	Hydroconoza	*0.007*	−0.001	0.000	45	Merostomoidea	*0.012*	0.004	−0.004
12	Anthozoa	0.020	0.050	*0.066*	46	Marrelomorpha	*0.002*	−0.001	0.001
13	Ctenophora	*0.002*	−0.001	0.001	47	Cephalocarida	*0.001*	−0.001	0.000
14	Priapulida	*0.010*	−0.002	0.000	48	Ostracoda	*0.081*	0.016	−0.020
15	Polyplacophora	*0.002*	−0.001	0.001	49	Cirripedia	*0.002*	−0.001	0.001
16	Monoplacophora	0.202	*0.565*	−0.132	50	Malacostraca	*0.008*	−0.003	0.000
17	Gastropoda	*0.015*	0.012	−0.005	51	Odaraidae + Branch	*0.004*	0.000	0.000
18	Rostroconchia	*0.017*	*0.022*	−0.010	52	Onychophora	*0.007*	0.006	−0.004
19	Bivalvia	*0.013*	0.009	−0.007	53	Myriapoda	*0.002*	−0.001	0.000
20	Cephalopoda	*0.019*	−0.014	0.006	54	Yohoiidae + Act	*0.007*	−0.001	−0.001
21	Orthothecimorpha	0.183	*0.597*	−0.117	55	Inarticulata	*0.153*	0.127	−0.084
22	hyolithomorphs	0.088	*0.106*	−0.070	56	Articulata	*0.055*	−0.012	−0.001
23	Polychaeta	*0.014*	−0.008	0.002	57	protoconodonts	0.019	*0.088*	0.012
24	Palaeoscolecida	*0.005*	−0.001	0.000	58	Ctenocystoidea	*0.001*	−0.001	0.000
25	Coeloscleritophora	0.040	*0.169*	−0.025	59	Stylophora	*0.005*	−0.003	0.001
26	Cribricyathida	0.023	*0.063*	0.058	60	Homostelea	*0.002*	−0.001	0.000
27	Stenothecoidea	*0.010*	0.005	−0.004	61	Homoiostelea	*0.004*	−0.002	0.001
28	Tommotiida	0.033	*0.100*	−0.044	62	Helicoplacoidea	*0.002*	0.000	−0.001
29	cambroclaves	0.004	0.009	−0.005	63	Eocrinoidea	*0.018*	−0.003	−0.001
30	Coleolida	0.017	0.053	−0.022	64	Crinoidea	*0.002*	−0.001	0.001
31	Conulariida	0.024	*0.154*	0.030	65	Camptostromatoidea	*0.001*	0.000	0.000
32	Hyolithelminthes	0.038	*0.106*	0.000	66	Edrioasteroidea	*0.010*	−0.005	0.002
33	Sabelliditida	0.023	*0.197*	0.113	67	Graptolithina	*0.012*	−0.009	0.004
34	Volborthellida	*0.006*	0.004	−0.003	68	Conodonta	*0.020*	−0.014	0.006

Table 26.4.3. *Loadings on varimax rotated factors.*

stratigraphic intervals	factors			stratigraphic intervals	factors		
	1	2	3		1	2	3
Trempealeauan	*0.9771*	0.0417	0.0126	u Atdabanian	*0.8409*	0.4725	−0.0397
Franconian	*0.9941*	0.0046	0.0145	l Atdabanian	0.5124	*0.8007*	−0.0812
Dresbachian	*0.9960*	0.0229	0.0108	u Tommotian	0.0746	*0.9094*	−0.0888
u Middle	*0.9958*	0.0450	0.0074	l Tommotian	0.0322	*0.9470*	0.0055
m Middle	*0.9888*	0.1132	−0.0032	Nemakit-Daldyn	0.0000	*0.8000*	0.3380
l Middle	*0.9841*	0.1519	−0.0055	Ediacaran	0.0027	0.0609	*0.9750*
u Botomian	*0.9868*	0.1509	−0.0053	Eigenvalues:	62.8%	23.7%	7.7%
l Botomian	*0.9824*	0.1697	−0.0119	Cumulative eigenvalues:	62.8%	86.5%	94.2%

1182 Models

Table 26.4.4. *Scores for rotated factors.*

	taxonomic class	factors				taxonomic class	factors		
		1	2	3			1	2	3
1	Rhizopodea	0.010	*0.016*	−0.004	35	Anomalocaridae	0.004	*0.006*	−0.003
2	Radiolaria	*0.012*	−0.001	0.000	36	Dinomiscidae	*0.003*	0.001	0.000
3	Demospongia	*0.042*	0.016	−0.008	37	Lenargyrionidae	*0.005*	0.000	0.000
4	Calcarea	*0.016*	−0.001	0.000	38	Mongol + Rushton	*0.002*	0.001	0.000
5	Hexactinellida	*0.019*	0.017	−0.005	39	Fomitchella	−0.002	*0.018*	−0.005
6	Trilobozoa	−0.029	0.205	*0.236*	40	Microdictyon	*0.003*	0.001	0.000
7	Petalonamae	0.006	0.002	*0.871*	41	Rhombocorniculum	0.002	*0.009*	−0.004
8	Cyclozoa	−0.003	0.025	*0.264*	42	Paratrilobita	0.002	−0.004	*0.114*
9	Hydrozoa	0.017	0.053	*0.142*	43	Merostomata	*0.011*	0.005	−0.002
10	Scyphozoa	−0.001	0.046	*0.183*	44	Trilobita	*0.983*	−0.059	0.021
11	Hydroconozoa	*0.007*	0.001	0.000	45	Merostomoidea	*0.010*	0.008	−0.004
12	Anthozoa	0.008	0.047	*0.070*	46	Marrelomorpha	*0.002*	−0.001	0.000
13	Ctenophora	*0.002*	−0.001	0.000	47	Cephalocarida	*0.001*	−0.001	0.000
14	Priapulida	*0.010*	0.000	0.000	48	Ostracoda	*0.074*	0.038	−0.018
15	Polyplacophora	*0.002*	−0.001	0.000	49	Cirripedia	*0.002*	−0.001	0.000
16	Monoplacophora	0.045	*0.607*	−0.077	50	Malacostraca	*0.009*	0.000	0.000
17	Gastropoda	0.011	*0.016*	−0.004	51	Odaraidae + Branch	*0.004*	0.001	0.000
18	Rostroconchia	0.010	*0.026*	−0.008	52	Onychophora	0.005	*0.008*	−0.004
19	Bivalvia	0.010	*0.013*	−0.006	53	Myriapoda	*0.002*	−0.001	0.000
20	Cephalopoda	*0.022*	−0.009	0.005	54	Yohoiidae + *Act*	*0.007*	0.001	−0.001
21	Orthothecimorpha	0.019	*0.633*	−0.059	55	Inarticulata	0.112	*0.170*	−0.072
22	hyolithomorphs	0.055	*0.131*	−0.060	56	Articulata	*0.057*	0.003	−0.002
23	Polychaeta	*0.016*	−0.004	0.002	57	protoconodonts	−0.004	*0.089*	0.021
24	Palaeoscolecida	*0.005*	0.000	0.000	58	Ctenocystoidea	*0.002*	0.000	0.000
25	Coeloscleritophora	−0.006	*0.176*	−0.009	59	Stylophora	*0.005*	−0.002	0.001
26	Cribricyathida	0.007	0.061	*0.064*	60	Homostelea	*0.002*	−0.001	0.000
27	Stenothecoidea	*0.008*	0.008	−0.004	61	Homoiostelea	*0.005*	−0.001	0.001
28	Tommotiida	0.004	*0.109*	−0.034	62	Helicoplacophora	*0.002*	0.001	−0.001
29	cambroclaves	0.002	*0.010*	−0.005	63	Eocrinoidea	*0.018*	0.002	−0.001
30	Coleolida	0.002	*0.058*	−0.017	64	Crinoidea	*0.002*	−0.001	0.000
31	Conulariida	−0.016	*0.151*	0.044	65	Camptostromatoidea	*0.001*	0.000	0.000
32	Hyolithelminthes	0.009	*0.111*	0.010	66	Edrioasteroidea	*0.011*	−0.003	0.001
33	Sabellidita	−0.026	*0.185*	0.131	67	Graptolithina	*0.014*	−0.006	0.003
34	Volborthellida	0.005	*0.006*	−0.003	68	Conodonta	*0.023*	−0.009	0.004

Tables 26.4.5–25.4.8 Summary of results of the Q-mode factor analysis of Vendian-Cambrian animal genera within 70 classes (including archaeocyathids). Italicized numbers indicate maximum loadings or scores for each stage or taxon, respectively.

Table 26.4.5. *Loadings on first four principal factors.*

stratigraphic intervals	factors				Stratigraphic intervals	factors			
	1	2	3	4		1	2	3	4
Trempealeauan	*0.9196*	−0.2831	0.1961	−0.0667	l Atdabanian	*0.6241*	0.6109	−0.4599	0.1306
Franconian	*0.9264*	−0.3194	0.1796	−0.0508	u Tommotian	0.3134	*0.7978*	0.1518	−0.2499
Dresbachian	*0.9321*	−0.3043	0.1798	−0.0549	l Tommotian	0.2709	*0.8395*	0.2601	−0.2162
u Middle	*0.9369*	−0.2855	0.1846	−0.0626	Nemakit-Daldyn	0.1842	*0.7212*	0.5297	−0.0052
m Middle	*0.9467*	−0.2228	0.1837	−0.0757	Ediacaran	0.0162	0.1406	0.4392	*0.8749*
l Middle	*0.9684*	−0.1580	0.1363	−0.0608	Eigenvalues:	58.7%	20.3%	9.7%	7.0%
u Botomian	*0.9505*	0.0560	−0.2633	0.1209	Cumulative eigenvalues:	58.7%	79.0%	88.7%	95.7%
l Botomian	*0.8819*	0.1418	−0.3967	0.1784					
u Atdabanian	*0.8142*	0.3675	−0.4039	0.1458					

Table 26.4.6. *Scores for first four principal factors.*

	taxonomic class	factors					Taxonomic class	factors			
		1	2	3	4			1	2	3	4
1	Rhizopodea	*0.013*	0.009	0.010	−0.011	36	Volborthellida	0.005	0.003	−*0.007*	0.003
2	Radiolaria	*0.011*	−0.004	0.001	0.000	37	Anomalocaridae	0.004	0.003	−*0.006*	0.003
3	Demospongia	*0.040*	−0.002	−0.015	0.007	38	Dinomiscidae	*0.003*	0.000	−0.002	0.001
4	Calcarea	*0.015*	−0.006	0.004	−0.001	39	Lenargyrionidae	*0.005*	−0.001	0.000	0.000
5	Hexactinellida	*0.023*	0.006	0.018	−0.017	40	Mongol + Rushton	0.002	0.001	−*0.004*	0.002
6	Regulares	0.284	0.367	−*0.742*	0.341	41	Fomitchella	0.002	*0.015*	0.003	−0.008
7	Irregulares	0.078	0.104	−*0.163*	0.060	42	Microdictyon	0.002	0.001	−*0.003*	0.002
8	Trilobozoa	0.024	0.210	*0.253*	0.082	43	Rhombocorniculum	0.003	0.006	−*0.007*	0.001
9	Petalonamae	0.005	0.079	0.350	*0.818*	44	Paratrilobita	0.000	0.006	0.041	*0.112*
10	Cyclozoa	0.003	0.046	0.134	*0.223*	45	Merostomata	*0.012*	−0.001	0.004	−0.004
11	Hydrozoa	0.028	0.051	*0.100*	0.099	46	Trilobita	0.902	−0.369	0.167	−0.035
12	Scyphozoa	0.010	0.056	0.109	*0.141*	47	Merostomoidea	*0.010*	0.002	−0.008	0.003
13	Hydroconozoa	0.006	0.002	−*0.011*	0.006	48	Marrelomorpha	0.002	−0.002	*0.003*	−0.001
14	Anthozoa	0.020	0.043	*0.078*	0.024	49	Cephalocarida	0.001	−0.001	*0.002*	−0.001
15	Ctenophora	0.002	−0.002	*0.003*	−0.001	50	Ostracoda	*0.070*	0.010	−0.051	0.025
16	Priapulida	*0.009*	−0.003	0.003	−0.001	51	Cirripedia	0.002	−0.002	*0.003*	−0.001
17	Polyplacophora	*0.002*	−0.002	0.002	−0.001	52	Malacostraca	*0.008*	−0.003	0.000	0.000
18	Monoplacophora	0.182	*0.504*	0.208	−0.251	53	Odaraidae + Branch	*0.004*	0.000	−0.002	0.001
19	Gastropoda	*0.015*	0.008	0.010	−0.011	54	Onychophora	*0.006*	0.004	−0.004	0.000
20	Rostronchia	*0.016*	0.015	0.010	−0.015	55	Myriapoda	*0.002*	−0.002	0.002	−0.001
21	Bivalvia	*0.011*	0.006	−0.008	0.002	56	Yohoiidae + Act	*0.006*	0.000	−0.004	0.002
22	Cephalopoda	0.021	−0.019	*0.027*	−0.013	57	Inarticulata	*0.135*	0.096	−0.053	−0.025
23	Orthothecimorpha	0.154	*0.530*	0.149	−0.181	58	Articulata	*0.053*	−0.016	0.003	0.000
24	hyolithomorphs	0.070	*0.088*	−0.084	0.006	59	protoconodonts	0.017	*0.081*	0.060	−0.025
25	Polychaeta	*0.015*	−0.009	0.011	−0.005	60	Ctenocystoidea	0.001	−0.001	*0.002*	−0.001
26	Palaeoscolecida	*0.005*	−0.001	0.000	0.000	61	Stylophora	0.005	−0.005	*0.006*	−0.003
27	Coeloscleritophora	0.034	*0.152*	0.061	−0.059	62	Homostelea	*0.002*	−0.002	0.002	−0.001
28	Cribricyathida	0.018	*0.060*	0.042	0.043	63	Homoiostelea	*0.004*	−0.002	0.002	−0.001
29	Stenothecoidea	0.007	0.004	−*0.012*	0.006	64	Helicoplacophora	0.001	0.001	−*0.004*	0.002
30	Tommotiida	0.026	*0.089*	−0.010	−0.029	65	Eocrinoidea	*0.017*	−0.006	0.004	−0.002
31	cambroclaves	0.003	0.007	−*0.008*	0.002	66	Crinoidea	0.002	−0.002	*0.003*	−0.001
32	Coleolida	0.014	*0.047*	0.003	−0.021	67	Camptostromatoidea	*0.001*	0.000	−0.001	*0.001*
33	Conulariida	0.024	*0.142*	0.141	−0.064	68	Edrioasteroidea	*0.011*	−0.007	0.010	−0.005
34	Hyolithelminthes	0.032	*0.093*	0.043	−0.021	69	Graptolithina	0.013	−0.012	*0.015*	−0.007
35	Sabelliditida	0.021	0.180	*0.218*	−0.026	70	Conodonta	0.021	−0.019	*0.026*	−0.012

Table 26.4.7. *Loadings on varimax rotated factors.*

stratigraphic intervals	factors				stratigraphic intervals	factors			
	1	2	3	4		1	2	3	4
Trempealeauan	*0.9653*	0.1902	0.0261	0.0037	l Atdabanian	0.1220	*0.9189*	0.3632	−0.0083
Franconian	*0.9780*	0.1957	−0.0147	0.0090	u Tommotian	0.0329	0.2900	*0.8512*	−0.1021
Dresbachian	*0.9775*	0.2032	0.0000	0.0065	l Tommotian	0.0123	0.2161	*0.9193*	−0.0254
u Middle	*0.9767*	0.2074	0.0206	0.0029	Nemakit-Daldyn	0.0493	−0.0105	*0.8731*	0.2642
m Middle	*0.9616*	0.2338	0.0774	−0.0044	Ediacaran	−0.0002	−0.0147	0.0824	*0.9856*
l Middle	*0.9371*	0.3082	0.1089	−0.0042	Eigenvalues:	46.7%	23.4%	18.1%	7.5%
u Botomian	0.6836	*0.7212*	0.0503	0.0203	Cumulative Eigenvalues:	46.7%	70.1%	88.2%	95.7%
l Botomian	0.5406	*0.8325*	0.0331	0.0260					
u Atdabanian	0.3959	*0.8829*	0.2145	0.0104					

Table 26.4.8. *Scores on rotated factors.*

	taxonomic class	factors					taxonomic class	factors			
		1	2	3	4			1	2	3	4
1	Rhizopodea	0.012	−0.001	*0.017*	−0.005	36	Volborthellida	0.000	*0.010*	−0.001	0.001
2	Radiolaria	*0.011*	0.003	−0.001	0.000	37	Anomalocaridae	0.000	*0.008*	−0.001	0.000
3	Demospongia	0.029	*0.032*	−0.004	0.001	38	Dinomiscidae	0.001	*0.004*	−0.001	0.000
4	Calcarea	*0.016*	0.002	0.000	0.000	39	Lenargyrionidae	*0.004*	0.002	−0.001	0.000
5	Hexactinellida	*0.025*	−0.004	0.021	−0.007	40	Mongol + Rushton	−0.001	*0.005*	−0.001	0.001
6	Regulares	−0.190	*0.915*	−0.083	0.049	41	Fomitchella	−0.002	0.003	*0.017*	−0.005
7	Irregulares	−0.036	*0.214*	0.008	0.000	42	Microdictyon	0.000	*0.004*	−0.001	0.000
8	Trilobozoa	0.012	−0.067	*0.272*	0.192	43	Rhombocorniculum	−0.003	*0.009*	0.002	−0.001
9	Petalonamae	−0.008	0.004	0.004	*0.894*	44	Paratrilobita	−0.002	0.004	−0.007	*0.119*
10	Cyclozoa	0.003	−0.016	0.040	*0.260*	45	Merostomata	*0.012*	0.002	0.004	−0.002
11	Hydrozoa	0.025	−0.011	0.066	*0.135*	46	Trilobita	*0.971*	0.178	−0.070	0.014
12	Scyphozoa	0.006	−0.013	0.060	*0.176*	47	Merostomoidea	0.004	*0.012*	−0.001	0.000
13	Hydroconozoa	0.000	*0.013*	−0.004	0.002	48	Marrelomorpha	*0.004*	−0.002	0.000	0.000
14	Anthozoa	0.023	−0.023	*0.068*	0.057	49	Cephalocarida	*0.002*	−0.001	0.000	0.000
15	Ctenophora	*0.004*	−0.002	0.000	0.000	50	Ostracoda	0.036	*0.082*	−0.010	0.004
16	Priapulida	*0.010*	0.001	0.000	0.000	51	Cirripedia	*0.004*	−0.002	0.000	0.000
17	Polyplacophora	*0.003*	−0.002	0.000	0.000	52	Malacostraca	*0.008*	0.003	−0.001	0.000
18	Monoplacophora	0.056	0.068	*0.612*	−0.105	53	Odaraidae + *Branch*	0.002	*0.003*	−0.001	0.000
19	Gastropoda	0.014	0.001	*0.017*	−0.005	54	Onychophora	0.002	*0.007*	0.003	−0.001
20	Rostronconchia	0.013	0.003	*0.024*	−0.008	55	Myriapoda	*0.003*	−0.001	0.000	0.000
21	Bivalvia	0.004	*0.014*	0.002	−0.001	56	Yohoiidae + *Act*	0.004	*0.007*	−0.001	0.000
22	Cephalopoda	*0.036*	−0.020	0.004	−0.002	57	Inarticulata	0.062	*0.136*	0.085	−0.035
23	Orthothecimorpha	−0.007	0.126	*0.583*	−0.063	58	Articulata	*0.052*	0.018	−0.003	0.000
24	hyolithomorphs	−0.004	*0.132*	0.045	−0.020	59	protoconodonts	0.006	−0.011	*0.104*	0.007
25	Polychaeta	*0.020*	−0.005	0.001	−0.001	60	Ctenocystoidea	*0.002*	−0.001	0.000	0.000
26	Palaeoscolecida	*0.004*	0.002	−0.001	0.000	61	Stylophora	*0.009*	−0.004	0.001	0.000
27	Coeloscleritomorpha	−0.004	0.016	*0.176*	−0.017	62	Homostelea	*0.003*	−0.002	0.001	0.000
28	Cribricyathida	0.000	0.014	*0.061*	*0.061*	63	Homoiostelea	*0.005*	−0.001	0.000	0.000
29	Stenothecoidea	0.000	*0.016*	−0.002	0.001	64	Helicoplacophora	−0.001	*0.005*	−0.001	0.001
30	Tommotiida	−0.013	0.047	*0.081*	−0.023	65	Eocrinoidea	*0.019*	0.003	0.000	−0.001
31	cambroclaves	−0.003	*0.011*	0.002	−0.001	66	Crinoidea	*0.004*	−0.002	0.000	0.000
32	Coleolida	−0.003	0.017	*0.048*	−0.014	67	Camptostromatoidea	0.000	*0.001*	0.000	0.000
33	Conulariida	0.019	−0.052	*0.204*	0.009	68	Edrioasteroidea	*0.016*	−0.006	0.002	−0.001
34	Hyolithelminthes	0.008	0.015	*0.108*	0.006	69	Graptolithina	*0.021*	−0.011	0.001	−0.001
35	Sabellidita	0.022	−0.084	*0.260*	0.077	70	Conodonta	*0.036*	−0.019	0.003	−0.002

26.2

Models Relating to Proterozoic Atmospheric and Ocean Chemistry

JAMES F. KASTING

26.2.1 Box Models Relating to the Rise in Atmospheric Oxygen (Section 4.6).

A simple, but nonetheless useful way of modeling the atmosphere/ocean system is to divide it into three "boxes" representing the atmosphere, the surface ocean, and the deep ocean. The surface ocean includes the uppermost 75 m or so of water that is stirred rapidly by the action of the winds. The mixing time of the present surface ocean is approximately one month. By comparison, the turnover time for the modern deep ocean, as estimated from radioactive tracer studies, is on the order of 1000 years. The relatively long mixing time for the deep ocean makes it possible to establish steep chemical gradients between surface water and deep water. As demonstrated below, these gradients have significance for understanding the rise of atmospheric oxygen.

Chemical gradients can also exist between the atmosphere and the surface ocean. Transfer of a gas between the atmosphere and ocean can be viewed as being limited by the rate of diffusion of the gas through a thin stagnant film at the ocean surface (Broecker and Peng 1982). The maximum transfer rate of the gas is given by the "piston velocity"

$$v_{pis} = D/z_{film}, \qquad (26.2.1)$$

where D is the diffusion coefficient of the gas in seawater and z_{film} is the thickness of the stagnant film. In practice, z_{film} is determined empirically by studying the oceanic uptake of carbon-14. A typical value for the modern ocean is about 4×10^{-3} cm (Broecker and Peng 1982). When combined with an aqueous diffusion coefficient of 1.9×10^{-5} cm^2 s^{-1} for O_2 at 15°C, this yields a piston velocity of 4.8×10^{-3} cm/s, or roughly 4 m/day. This relatively slow rate of exchange allows for significant disequilibrium between the atmosphere and the surface ocean. This disequilibrium is also relevant to understanding the rise of atmospheric oxygen.

In terms of the three-box model, the oxidation state of the atmosphere/ocean system should have progressed through three stages (Fig. 26.5). In Stage I, which Walker et al. (1983) have termed "reducing," all three reservoirs are devoid of free oxygen, except perhaps in localized, high-productivity areas of the surface ocean (represented by the dashed box). The redox balance of the deep ocean is dominated by the presence of dissolved ferrous iron. The Stage I atmosphere contains appreciable concentrations (10^{-5} to 10^{-3}, by volume) of volcanically-generated hydrogen (Walker 1977; Kasting et al. 1984). Atmospheric O_2 concentrations during Stage I can be calculated from atmospheric photochemical models. Typical ground-level O_2 concentrations are on the order of 10^{-14} PAL (times the present atmospheric level). This type of weakly reducing atmosphere should have been present prior to the advent of oxygenic photosynthesis. However, a Stage I atmosphere could have persisted long after this evolutionary invention, provided that a substantial fraction of the oxygen produced by photosynthesis was consumed by reaction with volcanic gases or with dissolved ferrous iron.

Parts of the surface ocean where photosynthetic activity was high could have become local "oxygen oases" in this otherwise anoxic Stage I environment. One can estimate the maximum concentration of O_2 that could have built up in the water column by equating its production rate with the rate at which it was lost to the atmosphere. (Once in the atmosphere, free O_2 would have been rapidly consumed by reaction with reduced

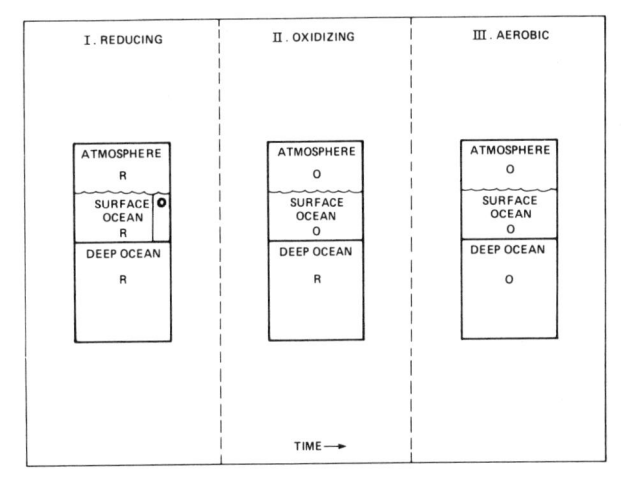

Figure 26.5. "Three-box model" of the atmosphere-ocean system relating to the rise of atmospheric oxygen.

gases present during Stage I.) Today, the rate of primary production by phytoplankton in regions of high productivity is about $1 \text{ g C m}^{-2} \text{ day}^{-1}$, which is equivalent to an oxygen production rate (P) of 0.083 moles $\text{m}^{-2} \text{ day}^{-1}$ (Riley and Chester 1971, p. 264). More primitive prokaryotic phytoplankton may not have been able to maintain this same level of photosynthetic activity, but they were probably not greatly inferior in their capabilities. Because the O_2 concentration directly above the ocean surface was effectively zero, the rate at which oxygen would have evaded to the atmosphere is given by $v_{\text{pis}}[O_2]_s$, where v_{pis} is the piston velocity defined above, and $[O_2]_s$ is the concentration of dissolved oxygen in surface water. If one ignores other loss processes for oxygen, $[O_2]_s$ would be given by

$$[O_2]_s = P/v_{\text{pis}} = 2.1 \times 10^{-5} \text{ moles } l^{-1}. \qquad (26.2.2)$$

This is a substantial dissolved oxygen concentration. One can illustrate this by calculating the effective atmospheric O_2 partial pressure with which it would be in equilibrium at $15°C$, given a Henry's Law coefficient α of 1.3×10^{-3} moles l^{-1} atm^{-1}. The "effective" pO_2 is 0.016 atm, or 0.08 PAL.

The actual concentration of dissolved oxygen in even a very productive ocean basin would probably have been lower than the value derived here because some oxygen should have reacted with upwelled ferrous iron and some may have been transported laterally to less productive regions of the surface ocean. The rate of oxygen consumption by ferrous iron can be estimated by analogy with processes in the modern ocean. Maximum upwelling rates in the open ocean are about 66 m yr^{-1}, or 0.2 m day^{-1} (Broecker and Peng 1982, p. 429). Upwelling rates in some coastal zones might be higher than this but are not likely to exceed 1 m day^{-1}. If the ferrous iron concentration in Archean deep water was 3 parts per million by weight (Holland 1984), the rate of O_2 consumption should therefore have been less than 0.14 moles $\text{m}^{-2} \text{ day}^{-1}$, or about 1/6 of the primary production rate estimated above. Hence, upwelling of ferrous iron should have reduced dissolved O_2 concentrations by only a modest amount. Lateral transport rates are more difficult to estimate since they depend on both the velocity of surface currents and on the geometry of the particular ocean basin. Basins with large horizontal extent and weak surface currents would have been most capable of maintaining high dissolved O_2 concentrations. Small basins with strong surface currents would have had substantially lower O_2 concentrations.

This calculation is instructive, despite its limitations, because it indicates that localized areas of the surface ocean could have been highly oxidizing even though the atmosphere was anoxic. Thus, oxidized chemical precipitates such as banded iron-formations should not be considered as evidence for an O_2-rich atmosphere.

At some time during the Precambrian, the rate at which photosynthetically generated oxygen entered the atmosphere must have exceeded the rate of supply of reduced gases. The environment would then have entered Stage II, which Walker et al. (1983) have termed "oxidizing." Both the atmosphere and the surface ocean should by this time have contained significant levels of free O_2. In contrast, a large influx of ferrous iron from the mid-ocean ridges (Veizer et al. 1982; Veizer 1983), combined with lower primary productivity in the surface oceans (Section 4.6), would have kept the deep oceans reduced. Reducing conditions must have persisted in the deep oceans until around 1.85 Ga, as evidenced by the continued deposition of massive banded iron-formations until about that time (Sections 4.2 and 4.3).

As shown by Kasting (1987), the three-box model can also be used to derive an approximate upper limit on the O_2 concentration in the Stage II atmosphere. Over time scales in excess of about one month, the atmosphere and surface ocean would have stayed in approximate equilibrium and can therefore be treated as a single "surface" reservoir. The net rate at which oxygen was supplied to this reservoir would have been equivalent to the rate (B_{org}) at which organic carbon was buried in sediments. (Organic matter that was decomposed anaerobically in the deep ocean could not have contributed to oxygen production because the gaseous by-products of this process, CH_4 and H_2, would have percolated back to the surface and reacted with O_2.) The approximate constancy of the organic carbon content of shales throughout geologic time, along with the relatively small variation in the carbon isotopic composition of marine carbonates, indicates that B_{org} has always been close to its present value of about 10^{13} moles yr^{-1} (Holland 1978).

Oxygen would have been lost from the surface reservoir in several different ways, including transfer to the deep ocean (where it would have reacted with ferrous iron), reaction with ferrous iron upwelled from the deep ocean, and oxidative weathering on the continents. The rate at which oxygen was consumed by the latter two processes is difficult to estimate; thus, it is not possible to calculate the atmospheric O_2 level directly. One can, however, obtain an upper limit on P_{O_2} by considering only the loss of oxygen to the deep ocean. If the rate of deep ocean circulation was the same as today, then the rate at which oxygen was downwelled would have been given by $k[O_2]_s$, where k ($= 1.4 \times 10^{21}$ l/10^3 yr) represents the present rate of deep ocean mixing and $[O_2]_s$ is, as above, the dissolved O_2 concentration in surface water. The upper limit on P_{O_2} under these conditions is

$$P_{O_2} \leqslant B_{\text{org}}/k\alpha = 6 \times 10^{-3} \text{ bar} = 0.03 \text{ PAL}. \qquad (26.2.3)$$

This upper limit would obviously change if the rate of organic carbon burial or the rate of deep ocean mixing were different in the past.

The atmosphere evidently left Stage II and entered Stage III around 1.85 Ga ago, as indicated by the disappearance of the banded iron-formations. Indeed, the fact that such deposition essentially ceased can be used to estimate a lower limit on atmospheric P_{O_2} at this time. As discussed in Section 4.8, the rate at which reduced materials (mostly iron and sulfide) are presently emanating from the mid-ocean ridges is equivalent to an oxygen sink (S_{MOR}) of about 4×10^{11} moles O_2 yr^{-1}. The reductant flux at 1.85 Ga should have been higher than this by at least a factor of two. In order for the deep oceans to have become oxygenated, the rate of downwelling of O_2, $k[O_2]_s$, must have exceeded this influx of reductants. Thus, the lower limit on P_{O_2} at 1.85 Ga is

$$P_{O_2} \geq S_{MOR}/k\alpha = 4 \times 10^{-4} \text{ bar} = 0.002 \text{ PAL}.$$

As in the previous example, this limit would change if ocean mixing rates were different than have been assumed.

After this time the utility of the three-box model becomes greatly diminished. Additional changes in atmospheric P_{O_2} should have occurred during Stage III, but they would have depended on factors more complex than those considered here. More complicated models (e.g., Lasaga et al. 1985; Kump and Garrels 1986) may ultimately provide some insight regarding this final stage of atmospheric oxygen evolution, but the quantitative predictions of such models are presently unreliable.

26.2.2 One-Dimensional Climate Modeling of Past CO_2 Concentrations (Section 4.7).

The climate model used to derive the estimates of Proterozoic CO_2 levels given in Section 4.7 is similar to the model described by Kasting and Ackerman (1986). This model includes absorption coefficients for H_2O and CO_2 derived from the Air Force Geophysical Laboratories (AFGL) tape. The coefficients span 55 separate intervals in the thermal infrared and 38 intervals in the visible and near infrared. These absorption coefficients are entirely independent of those used by Kasting (1987) in his earlier study of Proterozoic CO_2 levels and climate. The fact that the CO_2 levels found here agree reasonably well with those predicted by the earlier study provides some evidence that the radiative transfer calculations were done correctly. The current model is also in good agreement with the results of Kuhn and Kasting (1983) and Kiehl and Dickinson (1987), who have performed similar calculations of the greenhouse effect resulting from very high CO_2 levels.

Most of the other modeling assumptions are the same as in Kasting (1987). Clouds were excluded from the model, but their effect was accounted for implicitly by adopting an enhanced surface albedo. (The model value is 0.22, as compared to an observed value of ~ 0.05 to 0.1 for the modern earth's surface.) This allows the model to reproduce the present mean surface temperature (15°C) at the present intensity of solar insolation. The positive feedback effect of water vapor on the global climate was simulated by fixing the tropospheric relative humidity at current values (Manabe and Wetherald 1967). The moist adiabatic lapse rate was used to parameterize the effects of convection. The background atmosphere was assumed to consist of 0.77 bar of N_2. Variable amounts of CO_2, O_2, and O_3 were assumed, as described in Section 4.7. The inclusion of oxygen and ozone results in a substantial amount of warming that was not taken into account in the earlier Kasting (1987) study. In the present atmosphere, O_2 and O_3 together warm the surface by some 6°C. Roughly three degrees of this comes from the absorption of solar radiation in the Chappuis bands of O_3 and in the 0.76 μm band of O_2, one degree from absorption of infrared energy by the 9.6 μm band of O_3, and the remainder from pressure broadening of H_2O and CO_2 lines by O_2.

The climate calculations presented here are, of course, subject to many uncertainties. The largest of these is the effect of possible changes in cloud cover and cloud properties. Clouds currently obscure about 40% of the earth's surface and reflect perhaps 20% of the incoming solar radiation. (The earth's albedo as a whole is about 0.3.) The approach taken here is equivalent to assuming zero cloud feedback; that is, the effect of clouds on the radiation budget is assumed to remain the same at all surface temperatures. Although one strongly suspects that this is not true, there is currently no reliable way of assessing what the actual feedback might be. This uncertainty is somewhat mitigated by the fact that the Proterozoic glacial climates simulated in this study are not altogether different from the present climate. Although one typically thinks of the modern epoch as representing an interglacial period, the existence of massive polar caps makes it a "glacial climate" in the sense considered here.

27

Glossary of Technical Terms

J. WILLIAM SCHOPF CORNELIS KLEIN

Abiotic. Pertaining to substances or objects that are of nonbiologic origin; used especially in reference to organic matter produced via chemical reactions in the absence of living systems.

Acetic acid. An organic compound, the two-carbon carboxylic acid CH_3COOH.

Acetogenesis. Formation of acetic acid (e.g., from alcohol by bacterial fermentation).

Acetyl-coenzyme A (acetyl-co A). An organic complex involved in enzymatic acetyl transfer reactions, $CH_3COSCoA$.

Acritarch. A "phytoplankton-like" microfossil of unknown or uncertain biologic relationships, unicellular or apparently unicellular, organic-walled, commonly spheroidal, smooth (e.g., sphaeromorphs) or spiny (e.g., acanthomorphs); "Acritarcha" is an artificial group composed of such microfossils.

Actinolite. A mineral of the amphibole group, $Ca_2(Mg, Fe)_5Si_8O_{22}(OH)_2$.

Adenosine triphosphate (ATP). An organic compound, adenosine 5'-triphosphoric acid, $C_{10}H_{16}N_5O_{13}P_3$, a coenzyme involved in the transfer of phosphate bond energy.

Adiabatic. In thermodynamics, pertaining to a relationship of pressure and volume when a gas or fluid is compressed or expanded without either giving or receiving heat.

Aerobe. An organism able to live in the presence of free (uncombined) molecular oxygen; aerobes are termed "obligate" if they can exist and grow only in the presence of O_2.

Aerotaxis. The locomotory movement of an organism in response to an environmental gradient of molecular oxygen, either toward a higher oxygen concentration ("positive aerotaxis") or away from a higher oxygen concentration ("negative aerotaxis").

Akinete. Specialized reproductive cells (resting spores) occurring singly or in uniseriate groups, commonly adjacent to heterocysts, in various types of (predominantly filamentous) cyanobacteria.

Albedo. The ratio of the amount of electromagnetic energy reflected by a surface to the amount of energy incident upon it.

Albite. A mineral of the feldspar group, $NaAlSi_3O_8$.

Albite-epidote-amphibolite grade or facies. Metamorphic rocks formed under intermediate pressures (3 to 7 kb) and temperatures (250° to 450°C), conditions intermediate between those of the greenschist and amphibolite facies; also referred to as the "epidote-amphibolite facies" or the "quartz-albite-epidote-almadine subfacies."

Alcohol. Any of various organic compounds analogous to ethyl alcohol, hydroxyl (OH-containing) derivatives of hydrocarbons.

Aldehyde. Any of various organic compounds characterized by the group CHO.

Algae. Photosynthetic, eukaryotic, nonvascular (thallophytic) plants, unicellular or multicellular, commonly aquatic; "seaweeds" and their freshwater equivalents.

Alkali. Any basic substance such as a hydroxide or carbonate of an alkali metal (e.g., sodium or potassium).

Alkalic. A material rich in the alkali metals (e.g., sodium, potassium); pertaining to an igneous rock that contains more alkali metals than is considered average for the group of rocks to which it belongs.

Alkane. Any of a series of saturated hydrocarbons having the empirical formula $C_nH_{(2n+2)}$, such as methane, CH_4.

Alkene. Any of a series of hydrocarbons with one C=C bond (one site of unsaturation) having the empirical formula C_nH_{2n}, such as ethylene, $H_2C=CH_2$.

Allochthonous. Formed or occurring elsewhere than where found, of foreign origin; opposite of autochthonous.

Alluvium. A general term for unconsolidated detrital sediment (clay, silt, sand, etc.) deposited during relatively recent geologic time by running water, especially as a result of flood.

Amino acid. Any of various organic carboxylic (COOH-containing) acids containing the amino (NH_2) group such as the alpha-amino acids that are the chief components of proteins (e.g., glycine, $NH_2C(H_2)COOH$).

Amphibolite grade or facies. Metamorphic rocks formed under moderate to high pressure (3 to 8 kb) and temperatures (450° to 700°C).

Anaerobe. An organism able to live in the absence of free

(uncombined) molecular oxygen; anaerobes are termed "obligate" if they can exist and grow only in the absence of O_2.

Anatexis. Melting of preexisting rock, commonly resulting in formation of magma.

Andalusite. A mineral, Al_2SiO_5, trimorphous with kyanite and sillimanite.

Andesine. A mineral of the plagioclase feldspar group common in andesite and diorite.

Andesite. Igneous, extrusive (volcanic) rock, composed chiefly of acid plagioclase (especially andesine), one or more mafic minerals (e.g., biotite, hornblende, pyroxene), and some quartz; the extrusive equivalent of diorite.

Anhydrite. A mineral, anhydrous calcium sulfate, $CaSO_4$ (see *Gypsum*).

Ankerite. A mineral, $Ca(Fe, Mg, Mn)(CO_3)_2$.

Anorogenic. Not orogenic; unrelated to tectonic disturbance.

Anorogenic granite. An anorogenic plutonic rock composed mainly of quartz and alkali feldspar.

Anorthosite. Igneous, plutonic rock, composed almost entirely of calcic plagioclase feldspar (usually labradorite).

Anoxic. Pertaining to the absence of uncombined molecular oxygen.

Anticline. A geologic structure, a fold convex upward whose core contains the stratigraphically older rocks.

Antiform. An anticline-like geologic structure in which the stratigraphic sequence is unknown.

Apatite. A calcium phosphatic mineral with the general formula $Ca_5(PO_4, CO_3)_3(F, OH, Cl)$.

Apparent polar wander (APW) path. An interpretive path, inferred from the thermoremanent magnetization of rocks, showing temporal variation in the position of the magnetic north pole relative to tectonic plates.

Aragonite. A mineral, $CaCO_3$, a metastable form of calcite.

Arc. Island arc.

Archaebacteria. Members of the Archaebacteriae, a kingdom proposed by Woese and Fox (1977) to include methanogenic, extremely halophilic, and some thermoacidophilic bacteria, taxa regarded by them as being separable from all other prokaryotes (viz., the Eubacteriae) based on the chemistry of their cell walls, membranes, transfer RNAs and RNA polymerase subunits.

Archean. The earliest Eon of earth history, extending from the time of formation of the earth (~4.5 Ga ago) to the beginning of the Proterozoic Eon (2.5 Ga ago); commonly regarded as composed of four subdivisions: the Hadean (4.5 to 3.9 Ga ago), the Early Archean (3.9 to 3.3 Ga ago), the Middle Archean (3.3 to 2.9 Ga ago), and the Late Archean (2.9 to 2.5 Ga ago); Precambrian earth history is composed of the Archean and the subsequent Proterozoic Eon (2.5 to 0.57 Ga ago).

Argillite. Clastic sedimentary rock, derived from, but more indurated than, either a mudstone or a shale.

Arkose. Feldspar-rich sandstone, typically coarse-grained and reddish; usually derived from disintegration of granitic rocks.

Aromatic hydrocarbon. Any of various cyclic organic compounds composed of carbon and hydrogen and characterized by the presence of at least one benzene ring.

Asteroid. Any of the small "planetoid" bodies in orbit about the Sun.

Asthenosphere. The structurally weak shell of the earth below the lithosphere; essentially equivalent to the upper mantle.

Atdabanian. The third oldest of three Stages of the Lower Cambrian (immediately younger than the Tommotian and immediately older than the Botomian) as defined in Siberia, USSR (see Fig. 7.5.2).

Augite. A mineral, the clinopyroxene $(Ca, Na)(Mg, Fe^{2+}, Al)(Si, Al)_2O_6$.

Aulocogen. A geologic structure, a fault-bounded intracratonal trough.

Authigenesis. The process by which minerals form in place within an enclosing sediment or sedimentary rock during or after deposition.

Autochthonous. Formed or occurring in the place where found; opposite of allochthonous.

Autotroph. An organism that uses CO_2, present in the environment or generated from some other source (e.g., HCO_3^-), as the immediate and major source of cellular carbon (e.g., methanogenic bacteria, photosynthetic bacteria, cyanobacteria, eukaryotic phytoplankton, metaphytes).

Back arc basin. Small marine basin, marginal to an oceanic basin, floored by rocks that are much younger than those of the adjacent oceanic basin.

Bacteria. As used here, prokaryotic organisms other than cyanobacteria (viz., archaebacteria, prochlorophytes, and most eubacteria).

Banded iron-formation (BIF). Chemical sedimentary rock, typically thin bedded and/or finely laminated, commonly containing ≥15% iron and layers of chert; following Western usage (e.g., Brandt et al. 1972), "iron-formation" is here hyphenated when the term is used in a lithologic sense, but is capitalized and non-hyphenated when used in a stratigraphic sense as in "Marra Mamba Iron Formation"; see also *Iron-formation, Jaspilite.*

Barite. A mineral, $BaSO_4$.

Basalt. Igneous, extrusive (volcanic) rock, composed mainly of calcic plagioclase feldspar (usually labradorite) and clinopyroxene; extrusive equivalent of gabbro.

Basement. Complex of rocks that underlies an area; e.g., the rocks of a basement terrane.

Batholith. A mass of plutonic igneous rocks ≥100 km² in surface exposure and composed predominantly of granodiorite and quartz monzonite.

Benthos. Subaqueous bottom-dwelling organisms.

Benzene. An organic compound, the aromatic (cyclic) hydrocarbon C_6H_6.

BIF. See *Banded iron-formation.*

Biocoenose. A community of organisms that live closely together and that form a natural ecologic unit.

Biogenetic element. A general term for the principal chemical elements of living systems, C, H, O, N, (S and P).

Biogenic. Substances or objects of biological origin; formed by the activity of organisms.

Bioherm. A moundlike mass of rock built by sedentary organisms (commonly corals, calcified algae, etc.); cf. *Biostrome.*

Biomarker. An organic compound or molecular fossil containing specific structures of identifiable biosynthetic origin or *derivation and thus of known or plausibly inferred biological source.*

Biomass. The total amount (in weight or volume) of living organisms in a particular area or environment.

Biostratigraphy. Differentiation of rock units on the basis of the description and study of the fossils that they contain.

Biostrome. A laterally extensive or broadly lenticular commonly "reef-like" rock mass built by and composed mainly of the remains of sedentary organisms; cf. *Bioherm.*

Biosynthetic pathway. Any of numerous series of enzyme-mediated reactions by which biochemical products are synthesized in organisms.

Biota. All of the organisms of a particular area or time.

Bioturbation. The disruption, churning, and/or stirring of an unlithified sediment by organisms.

Bitumen. A generic term for naturally occurring substances of variable color, hardness, and volatility, composed principally of a mixture of hydrocarbons (e.g., petroleums, asphalts, natural mineral waxes).

Black smoker. A vent occurring along the volcanically active parts of mid-ocean ridge systems; the superheated (300° to 400°C) water discharged at high flow rates from these vents is less dense than seawater, rising above the vents to form large dark plumes rich in many elements including sulfur.

Blocking temperature. The narrow range of temperatures over which the relaxation time for an assemblage of cooling magnetized mineral phases increases from less than 10^3 seconds to more than 10^2 years.

Blue-green algae. See *Cyanobacteria.*

Blueschist grade or facies. Metamorphic rocks, blue in color due to the presence of sodic amphibole, glaucophane, or crossite, formed under high pressures (>5 kb) and moderate temperatures (300° to 400°C); also referred to as the "glaucophane schist facies."

Botomian. The youngest of four Stages (immediately younger than the Atdabanian) of the Lower Cambrian as recognized in Siberia, USSR (see Fig. 7.5.2).

Breccia. Sedimentary, igneous, or tectonic rock, clastic and coarse-grained, composed of large (>2 mm), angular, broken rock fragments cemented together in a finer-grained matrix; consolidated equivalent to rubble.

Burial metamorphism. A type of low-grade regional metamorphism (at 200° to 450°C) affecting sediments and interlayered volcanic rocks in a geosyncline without the influence of orogenesis or magmatic intrusion.

C3 metabolism. A metabolic pathway exhibited by autotrophs in which the three-carbon compound phosphoglyceric acid is the first identified product of carbon fixation.

C4 metabolism. A metabolic pathway exhibited by autotrophs in which the four-carbon compound oxalacetic acid is the first identified product of carbon fixation.

Calc-alkaline. A term applied to an igneous rock containing plagioclase feldspar.

Calcite. A mineral, $CaCO_3$, trimorphic with aragonite and vaterite.

Calvin cycle. Ribulose bis-phosphate-utilizing biosynthetic cycle of carbon fixation exhibited by C3 autotrophs.

Cambrian. The earliest geologic Period of the Paleozoic Era (and of the Phanerozoic Eon) of earth history.

CAM plant. An autotroph exhibiting "crassulacean acid metabolism," a specialized pathway of carbon fixation typical especially of plants of the Crassulaceae.

Carbohydrate. Any of various organic compounds consisting of a chain of carbon atoms in which hydrogen and oxygen are attached in a 2:1 ratio as in cellulose, $(C_6H_{10}O_5)_n$.

Carbonaceous. Pertaining to an object or substance that contains or is composed of organic matter.

Carbonate. A mineral, characterized by CO_3^{2-} (e.g., calcite, $CaCO_3$), or a rock consisting chiefly of carbonate minerals (e.g., limestone, dolostone, or carbonatite).

Carbonatite. An igneous rock of carbonate composition.

Carbon fixation. Biological assimilation of inorganic carbon resulting in its conversion from a soluble or exchangeable form (CO_2, HCO_3^-) to a relatively insoluble form (viz., an organic compound).

Carbonyl. The functional group —C=O of many organic compounds.

Carboxylase. An enzyme that catalyzes decarboxylation or carboxylation.

Carboxylic acid. Any of various organic acids containing one or more carboxylic (—COOH) groups.

Carotenoid. Any of various usually yellow to red, biologically widely distributed pigments characterized chemically by a long aliphatic polyene chain composed of isoprene subunits.

Catagenesis. The pre-metamorphic stage of alteration of sedimentary organic matter immediately between diagenesis and metagenesis, occurring between 50° and 150°C.

Catalase. Any of a group of protein complexes with hematin groups that catalyzes the decomposition of hydrogen peroxide into water and oxygen.

Ce (cesium) anomaly. In geologic materials, a departure from the concentration of Ce expected on the basis of a reference rock standard (e.g., chondritic meteorites).

Cenozoic. Most recent Era of the Phanerozoic Eon of earth history.

Chalcedony. A mineral, crytocrystalline quartz, SiO_2, commonly microscopically fibrous.

Chasmolith. Any of the various microorganisms growing in the crevices of a rock.

Chemical fossil. Broadly, any direct chemical evidence (geochemical compound, isotope effect, etc.) indicative of pre-existent life; in a more restricted sense, pertaining to organic molecular fossils containing specific structures of identifiable biosynthetic origin or derivation (in this sense, cf. *Biomarker*).

Chemical sediment. Sediment or sedimentary rock composed primarily of material formed by precipitation from solution or colloidal suspension, as by evaporation, that typically exhibits a crystalline texture (e.g., gypsum, halite, and many cherts and limestones).

Chemocline. The boundary between the circulating and non-circulating water masses or layers of a large body of water (e.g., a lake or ocean); in such bodies, a zone of rapid chemical change with depth.

Chemotroph. An organism that uses inorganic (chemoautotrophs) or organic substances (chemoheterotrophs), rather than light (as in phototrophs), as energy sources.

1192 Glossary of Technical Terms

Chert. Chemical sedimentary rock (or a product of secondary replacement), consisting mainly of microcrystalline or cryptocrystalline quartz and lesser amounts of chalcedony.

Chlorite. Minerals of a group having the general formula $(Mg, Fe^{2+}, Fe^{3+})_6 AlSi_3 O_{10}(OH)_8$.

Chloroplast. Chlorophyll-containing, membrane-bound organelle of plants and photosynthetic protists; the site of photosynthesis in eukaryotic cells.

Cholestane. An organic compound, a hydrogenated geochemical derivative of cholesterol.

Cholesterol. A steroid alcohol, $C_{27}H_{45}OH$, biosynthesized from squalene via a series of O_2-requiring steps; an important component of eukaryotic membranes.

Chondrite. A stony meteorite characterized by the presence of chondrules, spheroidal granules usually about 1 mm in diameter that consist chiefly of olivine and/or orthopyroxene.

Citric acid cycle. The energy-yielding, molecular oxygen-requiring metabolic cycle of aerobic respiration.

Clast. An individual constituent, grain, or fragment of a sedimentary rock, produced by mechanical weathering.

Clastic. A term pertaining to a rock or sediment composed principally of broken fragments derived from preexisting rocks that have been transported individually from their place of origin; sandstones and shales are commonly occurring clastic rocks.

Clinopyroxene. A mineral group composed of pyroxenes of the monoclinic crystal system and sometimes containing considerable calcium (see *Orthopyroxene*).

Coenocytic. A term pertaining to a filamentous organism that is tubular, lacking transverse walls to separate protoplasts into a series of cells; in eukaryotes (especially thallophytes) the term is used commonly in reference to a multinucleate cell.

Conglomerate. Clastic, sedimentary, coarse-grained rock, composed of rounded fragments larger than 2 mm in diameter cemented in a finer-grained matrix; consolidated equivalent of gravel.

Conglomerate test. A field-based test used to investigate the age of stability of magnetization in individual rounded fragments (e.g., cobbles) of a conglomerate.

Continental crust. The type of earth's crust, composed of sial above and sima below, that underlies the continents and continental shelves and is commonly about 35 km deep with depths under mountain ranges as much as 60 km.

Continental shelf. That part of the continental margin between the shoreline and the continental slope.

Convection cell. In plate tectonics, a pattern of mass movement of mantle material in which the central area is uprising and the outer area is downflowing, due to heat variations.

Convergent plates. Two tectonic plates characterized by movement toward each other; in collision, such plates produce a "convergent plate boundary."

Core of the earth. The spherical central zone of the earth's interior, below the Gutenberg Discontinuity at a depth of 2,900 km to the center of the planet (at a depth of 6,371 km), divided into two zones, the liquid outer core (2,900 to 5,080 km deep) and the solid inner core (5,080 km to the center of the earth).

Crater. An approximately circular depression formed by meteorite impact; also, a basin-like, rimmed structure usually at the summit of a volcanic cone.

Craton. The extensive, central, tectonically stable region of a continent composed of shields and platforms.

Critical point. In a system of one component, the temperature and pressure at which a liquid and its vapor become identical in all properties.

Crossbedding. A sedimentary structure with an internal arrangement of the layers of a stratified rock characterized by the laminae being inclined in concave forms by changing currents of air (as in dune sandstones) or water (e.g., in stream, channel, or delta deposits).

Crust of the earth. The outermost shell of the earth (representing less than 0.1% of the earth's total volume) that lies above the Mohorovičić Discontinuity (i.e., extending to the depth of 30 to 50 km beneath most continents and about 10 to 12 km beneath most oceans).

Cryogenian. Recently proposed geologic Period, extending from 850 to 650 Ma ago, that includes the Late Proterozoic glacial epochs (see Table 1.1.1).

Cumulate. Pertaining to an igneous rock formed by the accumulation of crystals that due to gravity settled out from the magma.

Curie temperature. The temperature of transition from ferromagnetism to paramagnetism above which thermal agitation prevents spontaneous magnetic ordering.

Cyanobacteria. Prokaryotic, bacterium-like microorganisms containing phycocyanin and/or phycoerythrin, chlorophyll *a* (but not chlorophyll *b*), and capable of aerobic (oxygen-producing) photosynthesis; numerous strains are also capable temporarily of anaerobic (non-oxygen-producing) photosynthesis; also referred to as blue-green algae, cyanophytes, and myxophytes (see *Prochlorophyte*).

Cyanophytes. Cyanobacteria.

Cyclohexane. A cyclic hydrocarbon, C_6H_{12}.

Cytochrome. Any of a group of iron-containing hydrogen or electron carriers involved in cellular metabolism.

Dacite. A fine-grained, extrusive (volcanic) rock having the same general chemical composition as andesite.

Decollement. A type of geologic structure produced by the detachment of strata during folding and overthrusting resulting in independent styles of deformation in the rocks above and below.

Dehydrogenase. Any of the various enzymes capable of mediating the biochemical removal of hydrogen from an appropriate substrate.

Demagnetization. The stepwise removal of natural remanent magnetization from a geologic sample by subjecting it, for example, to a particular peak alternating field or temperature.

Dendrite. A mineralic object, occurring either as a surficial deposit or an inclusion, that has crystallized in a branching pattern.

Detritus. A collective term for fragmental mineral and rock material (e.g., sand, silt, clay) produced by mechanical erosion.

Glossary of Technical Terms

Diabase. Igneous, intrusive rock, composed chiefly of labradorite and pyroxene and exhibiting ophitic texture.

Diagenesis. Broadly, all of the chemical, physical, and biological changes undergone by a sediment and its component minerals, fossils, etc., after its initial deposition and during and after its lithification exclusive of subsequent weathering or metamorphism; in organic geochemistry, the low temperature ($<50°C$), in part microbially mediated, earliest stage of alteration of organic matter in sediments preceding catagenesis, metagenesis, and metamorphism.

Diamictite. A non-sorted or poorly sorted clastic rock, produced by unspecified sedimentary processes, that contains a wide variety of particle sizes (e.g., a tillite, pebbly mudstone, or turbiditic deposit).

Dike. A tabular intrusion of igneous rock that cuts across the planar structure of the surrounding rock (see *Sill*).

Dinoflagellate. A one-celled, microscopic, chiefly marine, usually solitary, eukaryotic phytoplankter, characterized by one transverse flagellum and one posterior flagellum; the organic wall (theca) varies from simple and smooth to sculptured and subdivided into plates and grooves; some produce a spiny resting stage (cyst) that differs markedly in morphology from the theca.

Diorite. Igneous, intrusive rock, composed chiefly of acid plagioclase feldspars (oligoclase and andesine), amphiboles (especially hornblende), pyroxene, and sometimes small amounts of quartz; intrusive equivalent of andesite.

Dip. The angle that a structural surface (e.g., a bedding or fault plane) makes with the horizontal, measured perpendicular to the strike of the structure.

Divergent plates. Two tectonic plates characterized by movement away from each other.

Dolomicrite. A sedimentary rock consisting of clay-sized dolomite crystals, interpreted as lithified dolomite mud.

Dolomite. A mineral, $CaMg(CO_3)_2$, or sedimentary rock (also referred to as "dolostone") consisting chiefly of the mineral dolomite.

Dolostone. A sedimentary rock consisting chiefly of the mineral dolomite.

Downwelling. Sinking, subsidence; in geology, the downwarping of a large area of the earth's crust relative to its surrounding parts.

Dropstone. A clast that has fallen from a floating body (e.g., glacial ice) and been incorporated into a sedimentary unit.

Dubiofossil. A megascopic or microscopic fossil-life object or structure (or chemical component) of possible but unestablished biogenicity; structures (or chemical components) are assigned to this category temporarily, pending availability of additional information that would permit them to be classed either as bona fide fossils or as assuredly nonfossil.

Dubiomicrofossil. See *Microdubiofossil*.

Dunite. Igneous, plutonic rock, a peridotite composed almost entirely of olivine.

Duricrust. A general term for a hard soil crust formed in semiarid environments due to evaporation of vadose waters and resultant precipitation of calcareous material (calcrete), siliceous material (silcrete), phosphatic material (phoscrete), and the like.

Eclogite. A granular rock composed essentially of garnet and sodium-rich clinopyroxene (omphacite).

Eclogite grade or facies. Metamorphic rocks formed under high pressures (7.5 to >10 kb) and high temperatures (generally, 600° to 700°C).

Ediacarian System. A proposed name for rocks of the latest Precambrian, including and stratigraphically equivalent to those containing megascopic soft-bodied invertebrate fossils of the Ediacaran Fauna of South Australia; included within, or regarded as equivalent to, the Vendian and/or Eocambrian.

Embden-Meyerhof-Parnas (E-M-P) Pathway. The energy-yielding, anaerobic, metabolic pathway resulting in formations of lactic acid and adenosine triphosphate from the catabolism of glucose; also referred to as glycolysis.

Endogenic. A term meaning derived from within; pertaining to geologic processes or their resultant features that originate within the earth (e.g., volcanism, extrusive igneous rocks, etc.); opposite of exogenic.

Endolith. Any of the various microorganisms growing within the pore spaces of a rock or lithified soil crust.

Ensialic. A term pertaining to material, usually sedimentary, accumulating on a sialic (e.g., continental) crust.

Ensimatic. A term pertaining to material, usually effusive (volcanic), accumulating on a simatic (e.g., oceanic) crust.

Eolian. Pertaining to a geologic process or its resultant features caused by wind (e.g., loess, dunes, and similar wind-deposited or wind-eroded materials); equivalent to aeolian.

Epeiric sea. A sea on the continental shelf or within a continent; also referred to as an epicontinental sea.

Epiclastic rock. A sedimentary rock the fragments in which are derived by weathering or erosion.

Epidote. A mineral, $Ca_2(Al, Fe)_3Si_3O_{12}(OH)$.

Epigenetic. A term pertaining to a sedimentary mineral, texture, or structure formed after the deposition of the enclosing sediment.

Ester. Any of the various organic compounds characterized by a —C(O)O— linkage.

Ether. Any of various organic compounds characterized by a —C—O—C— linkage.

Eu (europium) anomaly. In geologic materials, a departure from the concentration of Eu expected on the basis of a reference rock standard (e.g., chondritic meteorites).

Eugeosyncline. A geosyncline in which volcanism is associated with clastic sedimentation; the portion of an orthogeosyncline located away from the craton; see *Miogeosyncline*.

Eukaryotes. Unicellular or multicellular organisms (viz., protists, fungi, plants, and animals) typically characterized by nucleus-, mitochondrion-, and (in plants and some protists) chloroplast-containing cells that are capable of mitotic cell division.

Euxinic. A term pertaining to an aqueous environment of restricted circulation with anaerobic bottom waters (e.g., a fjord or silled basin) or to the material deposited in such an environment (e.g., organic- and hydrogen sulfide-rich muds).

Evaporite. Chemical sedimentary rock, composed principally *of minerals (e.g., gypsum, anhydrite, rock salt, etc.) precipitated as a result of evaporation of an aqueous saline solution.

Exergonic. A term referring to a biochemical reaction that results in liberation of energy.

Exogenic. A term meaning derived from without; pertaining to geologic processes or their resultant features that originate externally to the earth (e.g., weathering, erosion, clastic rocks, etc.); opposite of endogenic.

Exosphere. The upper region of an atmosphere where some particles achieve the velocity necessary for their escape from the earth's gravitational field.

Fabric. In sedimentology, the distribution and orientation of elements (particles, cement, etc.) of which a sedimentary rock is composed; with regard to stromatolites, the microstructure, laminar shape, and related microscopic details of a stromatolitic sediment.

Facies. The sum of all primary lithologic and paleontologic characteristics exhibited by a sedimentary rock and from which its origin and environment of formation may be inferred (see *Metamorphic Facies*).

Facultative. Having the capability to live under different conditions (e.g., facultative bacteria capable of living under either aerobic or anaerobic conditions).

Fault. A surface or zone of rock fracture along which there has been displacement.

Feldspar. A mineral group composed of compounds of the general formula $(K, Na, Ca, Ba, Rb, Sr, Fe)Al(Al, Si)_3O_8$.

Feldspathic. A term pertaining to a rock or mineral aggregate containing feldspars.

Felsic. A general term for light-colored igneous rocks consisting largely of feldspar and silica (see *Mafic*).

Femicrite. Iron-rich micrite, a lithified, chemically precipitated iron-rich carbonate mud with grains $\leq 4 \mu m$ in size.

Fenestra. In sedimentology, a shrinkage pore or open space structure in a rock that may be completely or partly filled by secondarily introduced sediment and cement.

Fermentation. Any of a variety of energy-yielding biologic processes in which energy derived from metabolism or catabolism is used for generation of ATP by substrate-level phosphorylation (e.g., glycolytic fermentation of glucose to lactic acid).

Ferredoxins. Any of a group of non-heme iron-containing proteins functioning as cellular electron carriers in photosynthesis, nitrogen fixation, and other biological oxidation-reduction reactions.

Ferruginous. Pertaining to or containing iron.

Filament. In microbiology, a collective term referring to the cylindrical external sheath and the internal cellular trichome of a filamentous prokaryotic microorganism.

Fischer-Tropsch Reaction. The chemical synthesis of hydrocarbons, aliphatic alcohols, aldehydes, and ketones by the catalytic hydrogenation of carbon monoxide using enriched synthesis gas from the passage of steam over heated coke.

Fluvial deposit. A sedimentary deposit consisting of material transported and laid down by a stream or river.

Flysch. A marine sedimentary facies characterized by a thick sequence of poorly fossiliferous, thinly bedded, graded deposits (chiefly of calcareous shales and muds) rhythmically interbedded with conglomerates, coarse-grained sandstones, and graywackes; the term is commonly applied to turbidites.

Fold. A bend of rock strata or bedding planes resulting from tectonic deformation.

Fold test. In paleomagnetic studies, a field-based test to assess the age stability of magnetization based on determinations over a range of orientations in folded strata or specimens.

Foreland. A stable area marginal to an orogenic belt toward which the rocks of the belt were thrust or folded.

Formaldehyde. An organic compound, the simplest aldehyde, HCHO.

Formic acid. An organic compound, the simplest monocarboxylic acid, HCOOH.

Fossil. Any direct physical (viz., morphological fossil) or chemical (viz., chemical fossil) remains, object, structure, trace, imprint, etc., indicative of preexistent life).

Fractional crystallization. Crystallization in which the early-formed crystals do not equilibrate with the liquid from which they grew, resulting in a series of residual liquids increasingly concentrated in particular elements.

Fumarole. A vent, usually volcanic or associated with volcanism, from which gases and fluids are emitted.

Ga. Giga anna, 10^9 years; also, 10^9 years ago.

Gabbro. Igneous, intrusive rock, composed chiefly of basic plagioclase (commonly labradorite or bytownite) and clinopyroxene; intrusive equivalent to basalt.

Galena. A mineral, PbS.

Garnet. A mineral group composed of compounds with the general formula
$(Ca, Mg, Fe^{2+}, Mn^{2+})_3(Al, Fe^{3+}, Mn^{3+}, Cr)_2(SiO_4)_3$.

Geosyncline. A mobile, geographically extensive, elongate or basin-like downwarping of earth's crust where thick sequences (thousands of meters) of sedimentary and volcanic rocks accumulate.

Geothermal gradient. The increase in temperature within the earth's crust occurring with increasing depth, approximately $25°C/km$.

Glaucophane. A mineral of the amphibole group, $Na_2(Mg, Fe^{2+})_3Al_2Si_8O_{22}(OH)_2$.

Glaucophane schist grade or facies. Blueschist facies.

Gliding mobility. A type of biologic locomotion, slow, smooth to jerky, not involving flagella, pseudopodia, or similar structures, typical especially of the sheath-enclosed trichomes of filamentous prokaryotes.

Glucose. An organic compound, the six-carbon (hexose) sugar $C_6H_{12}O_6$.

Glycolysis. Anaerobic fermentation of sugars via the Embden-Meyerhof-Parnas pathway to yield lactic acid and energy.

Gneiss. Metamorphic rock, foliated and banded as a result of regional deformation.

Goethite. A mineral, alpha-FeO(OH).

Gondwana. A supercontinent of the Paleozoic and early Mesozoic, composed of India and the present-day southern hemisphere continents, derived from the splitting of Pangea; equivalent to Gondwanaland.

Glossary of Technical Terms

Grainstone. Sedimentary, clastic rock, essentially mud-free and composed of grain-supported carbonate particles.

Granite. Igneous, plutonic rock with quartz and feldspar as principal constituents; intrusive equivalent of rhyolite.

Granitoid. A term implying tentative identification as "granite-like" of a plutonic rock composed mainly of quartz and alkali feldspar.

Granodiorite. Igneous, plutonic rock, intermediate in composition between quartz diorite and quartz monzonite.

Granule. A rock fragment, 2 to 4 mm in size, larger than very coarse sand and smaller than a pebble.

Granulite grade or facies. Metamorphic gneissic rocks formed at low to high pressures (3 to 12 kb) and high temperatures ($>650°C$).

Graphite. A mineral, crystalline carbon, dimorphous with diamond.

Graywacke. Clastic sedimentary rock, a firmly indurated coarse-grained sandstone consisting of poorly sorted, angular to subangular grains of quartz, feldspar, and abundant rock and mineral fragments.

Greenalite. A mineral, $(Fe^{2+}, Fe^{3+})_{5-6}Si_4O_{10}(OH)_8$.

Greenhouse effect. A warming of the earth's surface and lower atmosphere resulting from a process involving selective transmission of short wave solar radiation by the atmosphere, absorbtion of this radiation by the earth's surface, and its reradiation as infrared which is absorbed and partly reradiated back to the surface by atmospheric "greenhouse gases" such as CO_2, CH_4, and water vapor.

Greenschist grade or facies. Metamorphic schistose rocks formed under low to moderate pressures (3 to 8 kb) and low to moderate temperatures (250° to 450°C).

Greenstone belt. A folded, structurally distinct, generally elongate region containing abundant, dark-green, altered mafic to ultramafic igneous rocks.

Green sulfur bacteria. Anaerobic photosynthetic prokaryotes of the Chlorobiaceae.

Gypsum. A mineral, hydrous calcium sulfate, $CaSO_4 2H_2O$ (see *Anhydrite*).

Halite. A mineral, rock salt, NaCl.

Halophile. An organism that is well adapted to a high salinity (including hypersaline) environment.

Heliotropic. A term pertaining to growth toward the light in which sunlight is the orienting stimulus.

Hematite. A mineral, ferric oxide, alpha-Fe_2O_3.

Heterocyst. A specialized, thick-walled cell, the site of nitrogen fixation, occurring in members of several families of filamentous cyanobacteria.

Heterotrichous. A term pertaining to organisms, chiefly algae and one order of cyanobacteria, composed of two or more structurally distinct types of trichomes or filaments.

Heterotroph. An organism (prokaryote, protist, fungus, or animal) that uses organic carbon compounds as sources of cellular carbon.

Hexose. Any of various six-carbon sugars such as glucose.

Hopanoid. Any of various members of a subgroup of organic pentacyclic triterpanes.

Hornblende. A mineral, the commonest member of the ampibole group, $(Ca, Na)_2(Mg, Fe^{2+})_4(Al, Fe^{3+}, Ti)(Al, Si)_8 O_{22}(O, OH)_2$.

Humic acid. A general term for dark-colored acidic organic matter extractable by alkali from soils, low rank coals, and decayed plant material that is insoluble in acids and organic solvents.

Huronian. A subdivision of the Early Proterozoic of the Canadian Shield.

Hydrocarbon. Any of the various members of a class of organic compounds composed only of hydrogen and carbon.

Hydrogenase. Any of the various enzymes capable of mediating biochemical production of hydrogen from an appropriate substrate.

Hydrothermal. Of or pertaining to hot water, or to the action of hot water or the products of this action, such as mineral deposits precipitated from hot aqueous solutions.

Hypersthene. A mineral of the orthopyroxene group $(Mg, Fe)SiO_3$.

Hypobradytely. An exceptionally slow rate of morphological evolution typical of many species of cyanobacteria and, possibly, other prokaryotes.

Igneous. A term pertaining to a rock or mineral that has solidified from molten or partly molten material (e.g., from a magma).

Ilmenite. A mineral, $FeTiO_3$.

Immature sediment. A clastic sediment that has been differentiated from its parent rock by processes acting over a short time and/or at low intensity such that it is characterized by relatively unstable minerals and poorly sorted and angular grains (see *Mature sediment*).

Index fossil. Morphologically distinctive, abundant, and widespread fossil taxon restricted to, and characteristic of, a defined stratigraphic range and thus used in biostratigraphic correlation; equivalent to guide fossil.

Intraclast. A component of a sedimentary rock, usually of a limestone, that represents a torn-up and reworked fragment of a penecontemporaneous sediment.

Intracratonic basin. A sedimentary basin occurring on a craton.

Iron-formation. Banded iron-formation.

Iron-formation (carbonate-rich). A banded iron-formation mainly composed of siderite, with lesser ankerite, as well as chert and iron-oxide minerals.

Iron-formation (oxide-rich). A banded iron-formation mainly composed of iron-oxide minerals such as magnetite and/or hematite, as well as chert.

Iron-formation (silicate-rich). A banded iron-formation mainly composed of silicate minerals such as greenalite, stilpnomelane, or minnesotaite (or their metamorphic reaction products), as well as chert and iron-oxide minerals.

Island Arc. A curved chain of islands, chiefly volcanic, rising from the deep sea floor near to a continent; see *Arc*.

Isocline. A geologic structure, a fold with limbs so compressed that they have the same dip; characteristic of strong regional deformation.

Isoprenoid hydrocarbon. Any of various branched organic compounds composed of one or more isoprene (C_5H_8) subunits.

Isotope. One of two or more, radioactively stable or unstable, atomic species of a chemical element, viz., species of an element having the same number of protons in the nucleus but having a different number of neutrons.

Jasper. A variety of chert commonly occurring in banded iron-formations and containing iron-oxide impurities giving it a characteristic red color.

Jaspilite. Oxide-rich banded iron-formation.

Kaolinite. A mineral of the kaolin clay group, $Al_2Si_2O_5(OH)_4$.

Kerogen. Particulate, geochemically altered, macromolecular organic matter lacking regular chemical structure, insoluble in organic solvents and mineral acids, present in sedimentary rocks.

Ketone. Any of various organic compounds, such as acetone, that contain a carbonyl (—C=O) group attached to two carbon atoms.

K-feldspar. A general term for the three different structural types (microcline, orthoclase, and sanidine) of potassium feldspar.

Kimberlite. Igneous, plutonic rock, a porphyritic alkalic peridotite containing abundant phenocrysts of olivine and phlogopite.

Komatiite. Igneous, extrusive rock, picritic (i.e., olivine- and pyroxene-rich) and of ultramafic composition.

Kurnahorite. A mineral, $CaMn(CO_3)_2$.

Kyanite. A mineral, Al_2SiO_5, trimorphous with andalusite and sillimanite.

Labradorite. A mineral, the plagioclase feldspar with the approximate composition $(NaAlSi_3O_8)(CaAl_2Si_2O_8)$.

Lactic acid. An organic compound, the three carbon monocarboxylic acid $CH_3CHOHCOOH$.

Lacustrine. Pertaining to, produced by, or formed in a lake.

Lahar. A mudflow (or similar mass move) composed chiefly of volcaniclastic materials on the flank of a volcano.

Large ion lithophile (LIL). Any of various silicate-forming chemical elements of relatively high atomic number.

Lawsonite. A mineral, $CaAl_2(Si_2O_7)(OH)_2(H_2O)$.

Lawsonite-albite grade or facies. Metamorphic rocks of the glaucophane schist (blueschist) facies formed at high pressures (6 to 7.5 kb) and moderate temperatures (250° to 400°C).

Lazarus taxa. As used here, species known from the Proterozoic and the present-day, but evidently not reported from the intervening Phanerozoic; in this sense, equivalent to "coelacanth taxa."

Leaching. In geology, the dissolution and selective removal of soluble constituents from a rock or ore body by the action of percolating groundwater (including hydrothermal solutions).

Lenian. A Stage of the Early and Middle Cambrian, equivalent to the Botomian plus the Tojonian (Elankian) of Siberia, recognized in portions of Eastern Europe and the adjoining USSR (see Fig. 7.5.2).

Lherzolite. Igneous, plutonic rock, a peridotite composed chiefly of olivine, orthopyroxene, and clinopyroxene.

Lignin. A general term for a group of polyphenolic, aromatic organic compounds that occur with cellulose in the woody portions of vascular plants.

Limestone. Clastic or chemical sedimentary rock composed mainly of calcium carbonate minerals.

Limonite. A mineral group composed of naturally occurring hydrous ferric oxides and hydroxides.

Lipid. Any of various saponifiable oxygenated fats or fatty acid-containing substances such as waxes, exclusive of hydrocarbons, that generally are soluble in organic solvents.

Lithic. A term pertaining to a medium-grained sedimentary rock, or to a pyroclastic deposit, containing abundant fragments of previously formed rocks.

Lithification. The conversion of unconsolidated sediment into a coherent, solid rock, involving such processes as cementation, compaction, desiccation, and the like.

Lithofacies. A lateral, mappable, subdivision of a stratigraphic unit, distinguished from adjacent subdivisions on the basis of its lithology (including mineralogical and petrographic characters and those paleontologic characters that influence the appearance, composition, or texture of the rock).

Lithosphere. The solid portion of the earth as compared with the hydrosphere and atmosphere; also, the crust and the upper portion of the upper mantle of the earth.

Lithostratigraphy. The portion of stratigraphy that deals with the lithology of strata and with their organization into units based on lithological characteristics.

Ma. Mega anna, 10^6 years; also, 10^6 years ago.

Maceration. The act or process of disintegrating sedimentary rocks (such as shale, chert, or coal) by various chemical and physical techniques in order to extract and concentrate acid-resistant kerogen and carbonaceous and other microfossils from them.

Mafic. A general term for dark-colored igneous rocks consisting largely of magnesian and ferric (i.e., ferromagnesian), basic, minerals (see *Felsic*).

Magma. Molten or partly molten rock material.

Magnetite. A mineral, ferro-ferric oxide, $(Fe, Mg)Fe_2O_4$.

Magnetochron. A time period during which orientation of the earth's magnetic poles is constant.

Mantle of the earth. The shell of the earth below the crust and above the core, divided into three zones: the upper mantle (extending from the Mohorovičić Discontinuity to a depth of about 250 to 400 km), probably peridotite, eclogite, and pyrolite in composition; the transition zone (extending from the depth of about 300 to 900 km); and the lower mantle (900 to 2,900 km deep), composed probably of material of pyrolite-like composition.

Mature sediment. A clastic sediment that has been differentiated from its parent rock by processes acting over a long time and/or at high intensity such that it is characterized by stable minerals (e.g., quartz) and well-sorted but subangular to angular grains (see *Immature sediment*).

Marble. A metamorphic rock, commonly a metamorphosed limestone or dolostone, consisting predominantly of fine- to coarse-grained recrystallized calcite and/or dolomite.

Mare (lunar). One of several dark, low-lying, relatively level and smooth large areas on the surface of the moon, floored by mafic and ultramafic volcanic rock.

Megafossil. A morphological fossil large enough to be studied without use of a microscope (generally >0.2 mm); also referred to as a macrofossil.

Meiosis. The type of cell (or nuclear) division resulting in formation of four daughter cells ("spores" or their equivalents), each containing a copy of one-half of the chromosomes of the parent cell; meiosis and syngamy (fusion of gametes) are fundamental aspects of eukaryotic sexual reproduction.

Mesozoic. An Era of the Phanerozoic Eon of earth history intermediate between the Paleozoic and Cenozoic Eras.

Meta-. In igneous or sedimentary petrology, a prefix indicating that the rock type has been metamorphosed as in "metabasalt," "metaquartzite," or "metapelite."

Metagenesis. The pre-metamorphic stage of alteration of sedimentary organic matter intermediate between catagenesis and greenschist facies metamorphism and occurring at about 150° to 250°C.

Metamorphic grade or facies. All the rocks (or any chemical composition and mineralogy) that have reached chemical equilibrium within the limits of a certain pressure-temperature range defined by the stability of specific index minerals (see *Amphibolite facies, Blueschist facies, Eclogite facies, Granulite facies, Greenschist facies, Prehnite-pumpellyite facies,* and *Zeolite facies*).

Metamorphic rock. Rock derived as a result of mineralogical, chemical, and structural changes in preexisting rocks in response to marked changes in temperature, pressure, and chemical environment at depth in the earth's crust.

Metamorphism. The mineralogical and structural adjustment of rocks to physical and chemical conditions that have been imposed at depth in the earth's crust and that differ from the conditions under which the rocks originated.

Metaphyte. Eukaryotic, multicellular, usually megascopic plant, whether vascular (i.e., tracheophytic "land plants") or non-vascular (e.g., "seaweeds").

Metasediment. Metamorphosed sedimentary rock, as in metaconglomerate, metashale, metachert, and metaquartzite.

Metazoan. Eukaryotic, multicellular, usually megascopic animal.

Methanogen. Bacterium capable of producing methane (CH_4) as a metabolic byproduct of the reduction of carbon dioxide.

Methanol. An organic compound, methyl alcohol, CH_3OH.

Methylotroph. Bacterium capable of metabolically oxidizing methane; extant methylotrophs are predominantly (but not exclusively) aerobes.

Mica. A mineral group composed of phyllosilicates (i.e., silicates with a sheet-like substructure) with the general formula
$(K, Na, Ca)(Mg, Fe, Li, Al)_{2-3}(Al, Si)_4O_{10}(OH, F)_2$.

Micrite. A descriptive term for the crystalline matrix of limestones or dolostones that consist of lithified, chemically precipitated carbonate muds with crystals $\leq 4 \mu m$.

Micro-banded. A term pertaining to rocks that display microscopically visible banding on the 1.0 to 0.1 mm scale.

Microbial mat. As used here, an unlithified accretionary organosedimentary structure, commonly laminated and megascopic, produced as a result of the growth and metabolic activities of (and usually due to the attendant trapping, binding, and/or precipitation of mineralic material by) benthic, mat-building communities of mucilage-secreting microorganisms, principally filamentous photoautotrophic prokaryotes such as cyanobacteria; the unlithified equivalent of a stromatolite.

Microbiota. A localized group of microscopic organisms that comprise a biocoenose, used especially in reference to communities of microorganisms or microfossils that occur within a microbial mat, stromatolite, or a particular stromatolitic horizon.

Microdubiofossil. A mineralic or carbonaceous microfossil-like object or structure of possible but unestablished biogenicity; a morphological dubiofossil bearing resemblance to a fossil microscopic organism or other microfossil; equivalent to dubiomicrofossil.

Microfossil. A morphological fossil too small to be studied without use of a microscope (generally, < 0.2 mm), either the remains of a microscopic organism or of a larger organism (e.g., spores and pollen of higher plants).

Micropseudofossil. A naturally occurring, non-biogenic, mineralic or carbonaceous microfossil-like object or structure; a morphological pseudofossil bearing resemblance to a microscopic organism or other microfossil; equivalent to pseudomicrofossil.

Migmatite. Rock, composed of igneous or igneous-looking and/or metamorphic materials that are megascopically distinguishable.

Milankovitch cycle. A solar system cycle, named after Milutin M. Milankovitch, based on orbital perturbations which change the caloric half-year radiation in different regions of the earth.

Minnesotaite. A mineral, $(Fe, Mg)_3Si_4O_{10}(OH)_2$.

Miogeosyncline. A geosyncline in which volcanism is not associated with sedimentation; the portion of an orthogeosyncline located near the craton (see *Eugeosyncline*).

Mitochondrion. An organelle of eukaryotic cells, the site of aerobic respiration.

Mitosis. The type of cell (or nuclear) division resulting in formation of two daughter cells, the chromosomes of each of which are exact copies of those of the parent cell; in unicellular eukaryotic organisms, a type of asexual reproduction.

Mobile belt. A long, relatively narrow crustal region of tectonic activity having the potential to develop into an "orogenic belt," a folded and deformed linear region giving rise to a "mountain belt" via post-organic processes.

Mohorovičić Discontinuity. The boundary surface (or seismic-velocity discontinuity) that separates the earth's crust from the subjacent mantle; its depth varies from about 5 to 15 km beneath the ocean floor to about 30 to 50 km below the continents (up to 70 km beneath some mountain ranges).

Monzonite. Igenous, plutonic rock containing little or no quartz, composed chiefly of orthoclase and plagioclase feldspar and augite.

MORB. An abbreviation referring to the typical chemical composition of Mid-ocean ridge basalts.

Morphological fossil. Any direct morphological evidence (remains, trace, imprint, etc.) indicative of preexistent life.

Mudstone. An indurated mud having the texture and composition of shale but lacking its fine lamination or fissility.

Nappe. Sheet-like, allochthonous geologic unit introduced into an area via thrust faulting and/or recumbent folding.

Nebula. Interstellar cloud of gas and dust.

Nemakit-Daldyn. The earliest Stage of the Lower Cambrian, older than the Tommotian Stage, as defined in Siberia, USSR (see Fig. 7.5.2).

Nitrogenase. An enzyme complex present in some prokaryotes involved in the fixation and reduction of N_2 to yield NH_3 or similarly reduced nitrogen compounds.

Nonfossil. A structure or object (or organic geochemical or biochemical component) that may be mistaken for a morphological (or a chemical) fossil but that is not a true fossil (e.g., non-biogenic organic or inorganic artifacts of sample preparation; mineralic or carbonaceous pseudofossils; and modern biogenic contaminants, both morphological and chemical).

Nontronite. An iron-rich clay mineral, $Na_{0.33}Fe_2^{3+}(Si, Al)_4O_{10} \cdot nH_2O$.

Nucleic acid. Any of the various organic acids (such as DNA, deoxyribonucleic acid, or RNA, ribonucleic acid) composed of linked nucleotides, each made up of a five-carbon sugar (deoxyribose or ribose), a phosphate, and a nitrogenous organic base (e.g., adenine, guanine, thymine, uracil, or cytosine).

Oceanic crust. That type of earth's crust that underlies ocean basins; it is equivalent to sima, composed of rocks that are rich in silica and magnesia.

Olivine. A mineral, especially common in ultramafic igneous rocks, $(Mg, Fe)_2SiO_4$.

Oncolite. An accretionary organosedimentary structure, viz., an unattached stromatolite, usually small (<10 cm) and more or less spheroidal, with encapsulating, concentric or overlapping laminae.

Oolite. Small, spheroidal accretionary granules, 0.25 to 2 mm in diameter, commonly products of inorganic precipitation, occurring in sedimentary rocks; also referred to as ooids or ooliths (in which case "oolite" refers to the rock made up of ooids or ooliths); small-size pisolites (pisoliths).

Opal. A mineral, amorphous hydrous silica, $SiO_2 \cdot nH_2O$.

Open space structure. A small-scale sedimentary structure, commonly lenticular, usually in chert or carbonate rock, produced by the partial or complete mineralic infilling of an originally gas-filled pocket or void.

Ophiolite. A general term for a group of mafic and ultramafic igneous rocks, ranging from basalt to gabbro and peridotite, the origin of which is associated with an early phase in the development of a geosyncline.

Ophitic texture. A term pertaining to the halocrystalline texture of an igneous rock, especially diabase, in which lath-shaped plagioclase crystals are partially or completely included in pyroxene crystals (typically of augite).

Organelle. A membrane-bound, intracellular structure or body of a eukaryotic cell having a specific function (e.g., mitochondria, nuclei, chloroplasts).

Organic. A term pertaining to a compound, structure, or substance containing carbon, and usually hydrogen, oxygen and/or nitrogen, of the type characteristic of, but not limited to, biologic systems (e.g., abiotically produced organic matter).

Organic preservation. A term pertaining to morphological fossils preserving some portion of the original (but geochemically altered) organic matter, as in permineralization.

Orogenic belt. A linear or arcuate region that has been subjected to folding and other deformation during an orogenic cycle.

Orogeny. The process of the formation of mountains.

Ortho-. In sedimentary petrology, a prefix that indicates the primary origin of a sedimentary rock as in "orthoquartzite" as distinguished from "metaquartzite," or in "orthoconglomerate" as distinguished from "metaconglomerate."

Orthochemical. Pertaining to an essentially normal precipitate formed by direct chemical action within a depositional basin or within a sediment.

Orthopyroxene. A mineral group composed of pyroxenes of the orthorhombic crystal system and usually containing no calcium and little aluminum (see *Clinopyroxene*).

Orthoquartzite. Clastic sedimentary rock composed almost entirely of quartz sand; a "pure quartz sandstone."

Outgassing. The removal of occluded gases, usually by heating, such as the process involving the release of gases and water vapor from molten rocks during volcanism.

Overthrust. A geologic structure, a low-angle thrust fault of large scale.

Oxalic acid. An organic compound, the simplest dicarboxylic acid, HOOCCOOH.

Oxic. A term pertaining to the presence of free (uncombined) molecular oxygen.

Oxidase. Any of various enzymes that catalyze oxidations, especially one capable of interacting directly with molecular oxygen.

Oxidation-reduction reaction. A chemical reaction in which one or more electrons are transferred from one atom or molecule to another.

Oxide-rich iron-formation. See *iron-formation*.

Oxidizing. A term pertaining to a process (or environment) in which a chemical element or ion is oxidized via the loss of electrons (i.e., changed from a lower to a higher positive valence); an oxidizing environment need not necessarily be aerobic (i.e., capable of supporting aerobiosis).

Oxygenic. A term pertaining to the presence of uncombined molecular oxygen; equivalent to oxic.

Paleomagnetism. Pertaining to the ancient geomagnetic field of the earth, recorded as permanent magnetizations in geologic materials.

Paleosol. A buried, commonly lithified, soil horizon of the geologic past.

Paleozoic. Earliest Era of the Phanerozoic Eon.

Palimpsest stromatolitic microstructure. Microstructure in a stromatolitic sediment in which the distribution of kerogen, iron oxide, pyrite, or some other pigmenting material can be interpreted as indicating the former occurrence of microbial remains.

Palynology. A subdiscipline of botany concerned chiefly with the study of the pollen and spores of higher plants, whether living or fossil.

Pangea. Pre-Mesozoic supercontinent, the precursor of Gondwana and Laurasia.

Pegmatite. Igneous, plutonic rock, exceptionally coarse-grained, commonly having the composition of granite.

Pelagic. A term pertaining to the open ocean environment or to the organisms inhabiting that environment.

Pelite. Clastic sedimentary rock composed of very fine-grained detrius (viz., of clay- or mud-size particles).

Pentose. Any of various five-carbon sugars.

Peptide. Any of various amides that are derived from two or more amino acids by combination of the amino group of one acid and the carboxylic group of another, thus characterized by the peptide linkage, the bivalent group HN—C(O)R.

Peridotite. Igneous, plutonic, coarse-grained rock composed chiefly of olivine with or without other mafic minerals.

Permineralization. A process of fossilization whereby the original hard parts of an animal, the mineralic test of a protist, or the physically resiliant organic structures (especially, the cell walls) of a plant, fungus, protist, or prokaryote, have mineral material (commonly silica or calcium carbonate) deposited in pore spaces, cell lumina, intramicellar spaces, etc.; cf. petrification in which original (but geochemically altered) organic matter is preserved.

Petrographic thin section. A slice of rock or mineral mechanically ground to a thickness of approximately 30 μm (for petrologic study) or 150 μm (for micropaleontologic study of translucent chert), mounted on a glass microscope slide, for study by optical microscopy.

Phanerozoic. The more recent Eon of earth history, composed of the Paleozoic (550 to 230 Ma ago), Mesozoic (230 to 62 Ma ago), and the Cenozoic Eras (62 Ma ago to the present); the Phanerozoic Eon and the preceeding Proterozoic (2500 to 550 Ma ago) and Archean Eons (4500 to 2500 Ma ago) together comprise all of earth history.

Phase diagram. A graph showing the boundaries of the fields of stability, usually in terms of pressure and temperature, of the various phases of a chemical system.

Phenocrysts. Relatively large, conspicuous crystals in a porphyritic rock.

Phlogopite. A mineral of the mica group, $K(Mg, Fe)_3AlSi_3O_{10}(OH, F)_2$.

Phosphoenolpyruvate (PEP). An organic compound, a phosphorylated reactive intermediate in glycolysis, $CH_2=COPO_3^{2-}COO^-$.

3-phosphoglycerate. An organic compound, a phosphorylated reactive intermediate in glycolysis, $CH_2OPO_3^{2-}CHOHCOO^-$.

Phosphorite. A sedimentary rock composed principally of phosphate minerals; broadly, a rock unit containing sufficient quantities of phosphate minerals to be of economic interest.

Photic zone. That part of an aqueous body where there is sufficient light penetration to support biological photosynthesis (of variable depth to about 200 m, but commonly about 50 m).

Photoautotroph. An organism (photosynthetic bacterium, cyanobacterium, photosynthetic protist, plant) that can use light as its energy source and CO_2 as its immediate (and major) source of cellular carbon.

Photoheterotroph. An organism that can use light as its energy source and organic carbon compounds as the sources of cellular carbon.

Photolysis. Chemical decomposition by the action of radiant energy (e.g., the splitting of H_2O into H— and OH— by ultraviolet light).

Photophosphorylation. In biologic systems, the light-driven conversion of adenosine diphosphate (ADP) to adenosine triphosphate (ATP).

Photorespiration. In biologic systems, the light-driven oxidation of organic compounds resulting in generation of carbon dioxide.

Photosynthetic bacteria. Prokaryotic microorganisms capable of anaerobic, but not oxygen-producing, photosynthesis.

Phototaxis. Locomotory movement of an organism (e.g., via gliding motility) in response to an environmental gradient of light, either toward a higher light concentration ("positive phototaxis") or away from a higher light concentration ("negative phototaxis").

Phototroph. An organism, either a photoautotroph or a photoheterotroph, that can use light as its energy source.

Phylogeny. The lineal evolutionary relationships among a particular group of organisms.

Phytane. An isoprenoid hydrocarbon, $C_{20}H_{42}$, a hydrogenated geochemical derivative of the phytyl alcohol moiety of chlorophyll.

Pillow basalt. Igneous, extrusive volcanic rock, solidified in a subaqueous environment to form discontinuous pillow-shaped masses generally 30 to 60 cm across.

Pisolite. Small, spheroidal, accretionary granules, 2 to 10 mm in diameter, commonly products of inorganic precipitation, occurring in sedimentary rocks; also termed pisoliths (in which case "pisolite" refers to the rock made up of pisoliths); pea-size oolites (ooliths).

Placer. A surficial mineral deposit (e.g., of a heavy metal such as gold) formed by mechanical concentration of mineral particles from weathered debris, commonly of alluvial origin.

Plagioclase. A mineral group composed of feldspars having the general formula $(Na, Ca)Al(Si, Al)Si_2O_8$.

Planetesmal. Small (1 to 10 km), rocky "meteorite-like" bodies, the accretion of which leads to the formation of protoplanets.

Plankton. Aquatic, floating (or weakly swimming) pelagic organisms.

Plate tectonics. Global tectonics based on an earth model characterized by several to many thick rigid plates (i.e., blocks composed of both continental and oceanic crust and upper mantle) each of which "floats" on a viscous *underlayer within the mantle and which moves slowly across the global surface, propelled via convection cells within the mantle, commonly with centers of sea-floor spreading at the rear of the plate and a subduction zone at its leading edge.

Platform. That of a continent covered by flat-lying or gentle tilted, mainly sedimentary strata; together with shields, platforms are components of cratons.

Pluton. A deep-seated, igneous intrusive rock mass.

Plutonism. A general term for the phenomena associated with formation of intrusive igneous rock masses.

Polar wander path. See *Apparent polar wander (APW) path.*

Porphyrin. Any of various metal-free organic compounds derived from pyrrole-containing (C_4H_5N-containing) compounds, especially from chlorophyll or hemoglobin.

Porphyry. An igneous rock on any composition that contains phenocrysts in a fine-grains groundmass.

Porphyry copper deposit. A large body of rock, typically a porphyry, having copper mineralization.

Potash. Potassium carbonate, K_2CO_3; also, a term used loosely for potassium oxide, potassium hydroxide, or potassium.

Precambrian. The earlier seven-eighths of earth history, composed of the Archean (4.5 to 2.5 Ga ago) and the Proterozoic Eons (2.5 to 0.55 Ga ago); the Precambrian and the subsequent Phanerozoic Eon (0.55 Ga ago to the present) together comprise all of the earth history.

Prehnite. A mineral, $Ca_2Al_2Si_3O_{10}(OH)_2$.

Prehnite-pumpellyite grade or facies. Metamorphic rocks formed under conditions of very low grade metamorphism; see *Zeolite facies.*

Pristane. An isoprenoid hydrocarbon, $C_{19}H_{40}$, that commonly co-occurs with phytane.

Proalgae. A term proposed by Van Valen and Maiorana (1980) to include prokaryotes capable of oxygen-producing photosynthesis (viz., cyanobacteria and prochlorophytes).

Prochlorophytes. Members of the Prochlorophyta, a taxonomic division proposed by Lewin (1976) to include prokaryotic, bacterium-like microorganisms containing chlorophylls *a* and *b*, lacking phycocyanin, phycoerythrin, and other bilin pigments, and capable of aerobic (oxygen-producing) photosynthesis; the two genera of known prochlorophytes occur either as obligate symbionts in tropical ascidians (*Prochloron*) or as free-living plankton (*Prochlorothrix*; see *Cyanobacteria* and *Proalgae*).

Progradation. The building outward toward the sea of a coastline by nearshore deposition of river-borne sediment or by accumulation of wave-deposited beach material.

Prokaryotes. Microbial microorganisms (viz., bacteria, cyanobacteria, archaebacteria, and prochlorophytes) characterized by cells that lack membrane-bound nuclei, mitochondria, chloroplasts, and similar organelles and that reproduce by non-mitotic and non-meiotic processes.

Protein. An organic compound, any of numerous polymers composed of amino acids.

Proterozoic. An Eon of earth history, as used here composed of three subdivisions: the Early Proterozoic (2.5 to 1.6 Ga ago), the Middle Proterozoic (1.6 to 0.9 Ga ago), and the Late Proterozoic (0.9 to 0.55 Ga ago, including the Vendian, 0.65 to 0.55 Ga ago); Precambrian earth history is composed of the Proterozoic and the preceeding Archean Eon (4.5 to 2.5 Ga ago).

Protist. Any of a variety of "animal-like" protozoans, "plant-like" phytoplankton, and similar unicellular eukaryotes.

Protolith. The unmetamorphosed rock from which, via metamorphism, a metamorphic rock was formed.

Pseudofossil. A naturally occurring, non-biogenic, mineralic or carbonaecous object or structure (viz., morphological pseudofossil) or chemical component (viz., chemical pseudofossil) that bears resemblance to, and may be mistaken for, a morphological or chemical fossil.

Pseudomicrofossil. See *Micropseudofossil.*

Pseudomorph. A secondary mineral developed by alteration, substitution, or incrustation of a primary mineral and having the superficial crystal form of the primary mineral.

Purine. Organic compounds with the empirical formula $C_5H_4N_4$, nitrogen-containing bases such as adenine or guanine that are components of nucleotides and of nucleic acids.

Purple sulfur bacteria. Anaerobic photosynthetic bacteria of the Chromatiaceae.

Pycnocline. A density gradient, especially a vertical gradient making a sharp change as in a layer of oceanic water that is characterized by a rapid change of density with depth.

Pylome. A pore-like opening, commonly circular, in the wall of an acritarch, probably functioning in excystment.

Pyrimidine. Organic compounds with the general formula $C_4H_4N_2$, nitrogen-containing bases such as thymine, cytosine, and uracil that are components of nucleotides and of nucleic acids.

Pyrite. A mineral, FeS_2.

Pyrobitumen. Dark-colored, solid, fairly hard, nonvolative substances composed of geochemically altered, largely insoluble, hydrocarbon complexes.

Pyroclastic. A term pertaining to clastic rock material formed by a volcanic explosion.

Pyrolite. The proposed composition of mantle material, consisting of one part basalt and three parts dunite.

Pyrolysis. Chemical breakdown brought about by heating.

Pyroxene. A mineral group composed of compounds with the general formula $(Ca, Na, Mg, Fe^{2+})(Mg, Fe^{3+}, Al)Si_2O_6$; see *Clinopyroxene* and *Orthopyroxene.*

Pyrrhotite. A mineral, $Fe_{1-x}S$ (having a defect lattice in which some of the ferrous ions are lacking).

Pyruvic acid. An organic compound, the three-carbon carboxylic acid $CH_3COCOOH$, the precursor of lactic acid in fermentation.

Quartz. A mineral, crystalline silica, SiO_2 (see *Opal* and *Chalcedony*).

Quartzite. Clastic, sedimentary rock, sandstone consisting chiefly of cemented quartz grains.

Rapakivi texture. A texture occurring in igneous and metamorphic rocks in which rounded crystals of K-feldspar, a few centimeters in size and enclosed in a fine-grained matrix of quartz and other minerals, are surrounded by a rim of Na-feldspar.

Glossary of Technical Terms

Rare earth element (REE). Any of a group of trivalent metallic elements with atomic numbers 57 (lanthanum) to 71 (lutetium), inclusive.

Recumbent fold. Geologic structure, an overturned fold with a nearly horizontal axial surface.

Red bed. Clastic, sedimentary rock, chiefly sandstone, siltstone, and shale, red or reddish-brown in color due to the presence of ferric oxide minerals (mainly hematite) usually coating individual grains.

Reducing. A term pertaining to a process (or environment) in which the chemical element or ion is reduced via the gain of electrons (i.e., changed from a higher to a lower positive valence); opposite of oxidizing.

Refractory. An element or substance resistant to a particular treatment and/or process.

Regression. The withdrawal of the sea from land areas and the consequent evidence of such withdrawal (e.g., enlargement of the area of deltaic deposition).

Remanence. The permanent magnetization in any geologic material.

Respiration. Biological, light-independent, electron transport processes in which electrons flow from inorganic or organic compounds to molecular oxygen (aerobic respiration) or to some other terminal electron acceptor (anaerobic respiration).

Reversal test. A laboratory determination establishing the presence of dual (including opposite) polarity magnetizations in a suite of rocks, the presence of which indicates that magnetization was acquired over at least one reversal of geomagnetic polarity.

Rheology. The study of the deformation and flow of matter.

Rhyolite. Igneous, extrusive rock, generally porphyritic with phenocrysts of quartz and feldspar and exhibiting flow structure; extrusive equivalent to granite.

Ribulose 1,5-bis-phosphate (RuBP) carboxylase/oxygenase. Enzyme of the Calvin cycle catalyzing the carboxylation (with CO_2) and cleavage of ribulose 1,5-diphosphate to yield two molecules of 3-phosphoglycerate; equivalent to Rubisco.

Riebeckite. A mineral of the amphibole group, $Na_2(Fe, Mg)_5Si_8O_{22}(OH)_2$.

Ripidolite. A mineral of the chlorite group, $(Mg, Fe^{2+})_9Al_6Si_5O_{20}(OH)_{16}$.

Ripple mark. A sedimentary structure with an undulatory surface formed at the interface between a fluid (e.g., wind or water currents) and incoherent sedimentary material (e.g., snow particles or loose sand).

Rubisco. See *Ribulose 1,5-bis-phosphate (RuBP) carboxylase/oxygenase.*

Sabhka. A supratidal environment of sedimentation, formed under arid conditions on restricted coastal plains immediately above normal high-tide level.

Saccharolytic. Pertaining to the breakdown of sugars.

Sandstone. A medium-grained clastic sedimentary rock composed of abundant rounded or angular sand sized fragments set in a fine-grained matrix or united by a cementing material.

Schist. Metamorphic rock, highly foliated as a result of regional deformation.

Secular. A term pertaining to a process or event persisting for an indefinitely (e.g., geologically) long period of time.

Sericite. A fine-grained micaceous mineral, usually muscovite, $KAl_2Si_3O_{10}(OH)_2$.

Serpentine. A mineral group composed of compounds having the general formula $(Mg, Fe)_3Si_2O_5(OH)_4$.

Serpentinization. The process of hydrothermal alteration by which magnesium-rich silicate minerals (olivine, pyroxene, etc.) are converted into minerals of the serpentine group.

Sexual reproduction. In eukaryotes, a type of biological reproduction involving meiosis and syngamy (fusion of gametes) and most commonly a regularized alteration of haploid and diploid phases in the life cycle; analogous (but not evolutionary homologous) to non-meiotic, parasexual processes of reproduction that occur in some prokaryotes.

Shale. Clastic, sedimentary rock, fine-grained and finely layered; a thinly laminated claystone, siltstone, or mudstone.

Shard. A vitric fragment in pyroclastic rocks, usually produced by disintegration of pumice during or after an eruption.

Sheath. In prokaryotic microorganisms, an extracellular, generally polysaccharide, mucilaginous investment surrounding individual cells or colonies of cells; in filamentous prokaryotes, a cylindrical, hollow, mucilaginous organic tube that encompasses the cellular trichome (a sheath and a trichome being collectively referred to as a "filament").

Shield. A large area of exposed continental basement rock in a craton, generally of Precambrian age, surrounded by sediment-covered platforms.

Sial. A petrologic term for the granitic, silica- and alumina-rich upper layer of the earth's crust characteristic of continental masses (see *Sima*).

Siderite. A mineral, $FeCO_3$.

Siderite-rich iron-formation. See *iron-formation.*

Siliciclastic. Made up of broken fragments of silicate materials.

Sill. A tabular intrusion of igneous rock that parallels the planar structure of the surrounding rock (see *Dike*).

Sillimanite. A mineral, Al_2SiO_5, trimorphous with kyanite and andalusite.

Siltstone. An indurated silt having the texture and composition of shale but lacking fine lamination or fissility.

Sima. A petrologic term for the basaltic, silica- and magnesium-rich, oceanic crust and the lower portion of the earth's continental curst (see *Sial*).

Slate. A compact, fine-grained metamorphic rock that possesses slaty cleavage and hence can be split into slabs and thin plates.

Solidus. On a temperature-composition diagram, the locus of points in a chemical system above which solid and liquid are in equilibrium and below which the system is completely solid.

Sparry calcite. Coarse-grained crystalline calcite.

Spinel. A mineral, $MgAl_2O_4$, or members of a mineral group having similar general composition.

Spreading center. In plate tectonics, a linear, principally submarine ridge system (for example, the mid-Atlantic ridge) at which new oceanic crust is added, and from which adjacent tectonic plates diverge (at a rate of about 1 to 10 cm per year) as a result of a convection upwelling of magma.

Squalene. An acyclic hydrocarbon, $C_{30}H_{50}$, a biosynthetic precursor of sterols.

Staurolite. A mineral, $(Fe, Mg)_2Al_9Si_4O_{23}(OH)$.

Sterol. Any of various polycyclic organic compounds such as cholesterol, containing a molecular skeleton of four fused carbon rings.

Stilpnomelane. A mineral, approximately of the composition $K(Fe, Mg, Al)_3Si_4O_{10}(OH)_2 \cdot nH_2O$.

Stochastic. A term pertaining to a process in which the dependent variable is random and the outcome at any instant cannot be predicted with certainty.

Stratigraphy. The arrangement (or study of the arrangement) of rock strata, especially with regard to the geographic position and chronologic order of the sequence.

Strike. The direction or trend that a structural surface (e.g., a bedding or fault plane) takes as it intersects the horizontal.

Strike-slip. In a geologic fault, the component of the movement that is parallel to the strike of the fault.

Stromatolite. As used here, a lithified accretionary organosedimentary structure, commonly laminated, megascopic, and calcareous, produced as a result of the growth and metabollic activities of (and usually due to the attendant trapping, binding, and/or precipitation of mineralic material by) benthic, mat-building communities of mucilage-secreting microorganisms, principally filamentous photoautotrophic prokaryotes such as cyanobacteria; the lithified equivalent of a microbial mat; stromatolites can be stratiform (with flat-lying, continuous laminae), columnar (with the dimension in the direction of accretion greater than at least one of the transverse dimensions), or spheroidal (viz., oncolites); a "thrombolite" is an unlaminated stromatolite characterized by a megascopic clotted fabric.

Stylolite. A thin seam usually occurring in more or less homogeneous carbonate rocks produced via pressure solution in which the insoluble constituents of the rock (e.g., clay, kerogen, etc.) form an irregular, interlocking layer of mutually interpenetrating projections.

Subduction zone. In plate tectonics, an elongate region along which a crustal block (i.e., tectonic plate) descends relative to an adjacent crustal block.

Subsolidus. A chemical system below its melting point in which reactions may occur in the solid state.

Sulfate reduction. The energy yielding bacterial process by which SO_4^{2-} is reduced to form H_2S.

Supracrustal. A term pertaining to rocks that overlie a complex of basement rocks.

Synclinorium. A composite geologic structure of regional extent composed of a series of synclinal folds the cores of which contain the stratigraphically younger rocks of the sequence.

Syngamy. The fusion of gametes (sex cells) occurring during eukaryotic sexual reproduction.

Syngenetic. A term pertaining to structures (e.g., ripple marks, microfossils) or substances (e.g., organic compounds, minerals) of primary origin, formed and deposited comtemporaneously with the deposition of the surrounding sediment.

Talc. A mineral, $Mg_3Si_4O_{10}(OH)_2$.

Taxon. In taxonomy, a unit of any rank (e.g., a particular species, genus, family, or class) or the formal name applied to that unit.

Tectonic. A term pertaining to the forces involved in, or the resulting structures or features produced by, diastrophism, orogeny, etc. (i.e., "tectonism").

Terpenoid. Any of numerous hydrocarbons having the empirical formula $(C_5H_8)_n$.

Terrestrial planets. The inner planets of the solar system, similar to the earth in terms of size, mean density, and rocky composition (viz., Mercury, Venus, Earth, Moon and Mars).

Thallophyte. A eukaryotic nonvascular plant (viz., an alga) or a fungus composed of a "thallus," a multicellular or coenocytic body without differentiation into true roots, leaves, or tracheid-containing stems; although similarly nonvascular, bryophytes (mosses and liverworts) exhibit a greater degree of cellular differentiation than do thallophytes.

Thermal overprinting. A postdepositional heating event affecting igneous, sedimentary, or metamorphic rocks.

Thermoacidophole. Bacteria (chiefly archaebacteria) that are well adapted to low pH, high temperature environments.

Thiophene. A heterocyclic organic liquid, C_4H_4S.

Tholeiite. Igneous, extrusive rock, a member of a major subgroup of alkali basalts.

Tholeitic basalt. A tholeiite.

Thrust fault. A fault with a dip of 45° or less, the overlying side of which (the "hanging wall") appears to have moved upward relative to the other wall.

Thucholite. Sedimentary, organic-rich rock, a brittle, jet-black mixture of solidified hydrocarbons and uraninite, with some sulfide minerals, occurring especially in Early Proterozoic gold-bearing conglomerates.

Tillite. Clastic, sedimentary rock, lithified glacial till (unsorted and unstratified glacial debris).

Tommotian. The second oldest Stage of the Lower Cambrian, younger than the Namakit-Daldyn and immediately older than the Atdabanian, as defined in Siberia, USSR (see Table 7.5.2).

Tonalite. Igneous, intrusive rock, essentially identical in composition to quartz diorite.

Trichome. In filamentous prokaryotic microorganisms, the threadlike, usually many-celled strand that is encompassed commonly by a tubular sheath to form a filament.

Troilite. A mineral, FeS, present in small amounts in many meteorites.

Trondhjemite. A light-colored plutonic rock composed of sodic plagioclase, quartz, biotite, and containing little or no alkali feldspar.

Tuff. Clastic, sedimentary rock, a compacted pyroclastic deposit of volcanic ash and dust.

Turbidite. Clastic, sedimentary rock characterized by graded bedding, moderate sorting, and well-developed primary structures, and inferred to have been deposited from a turbidity current.

Unconformity. A substantial break or gap in the geologic record where a rock unit is overlain by another that is not next in the stratigraphic succession, such as an interuption in the continuity of a depositional sequence of sedimentary rocks.

Unconformity test. Determination of the consistency of magnetization, in both polarity and direction, across an unconformity.

Ultramafic. A term pertaining to an igneous rock composed mainly of mafic (basic) minerals.

Uniseriate. Organized into a single row, like beads on a string.

Upwelling. The upward movement of fluids, e.g., of molten magma or of a cold, subsurface oceanic water mass toward the surface.

Uraninite. A mineral approximately of the composition UO_2.

Urea. An organic compound, H_2NCONH_2.

Varve. A sedimentary bed or lamina, or sequence of laminae, deposited in a body of still water within a one year period (e.g., a thin pair of graded glaciolacustrine layers seasonally deposited in a glacial lake).

Vendian. A latest Proterozoic time period, as used here extending from 650 to 550 Ma ago (see Fig. 7.5.2).

Vitrinite. An oxygen-rich micropetrological unit of coals and other carbonaceous sediments, characteristic of vitrain (a macroscopically visible band, especially in bituminous coal, characterized by brilliant vitreous luster) and composed of humic material associated with peat formation.

Volcaniclastic. A term pertaining to a clastic rock containing volcanic material.

Volcanogenic. A term pertaining to a volcanic origin.

Wackestone. Clastic, sedimentary rock, a mud-supported carbonate containing more than 10% of grains (i.e., particles $> 20\ \mu m$).

Wilson cycle (Wilson-type) plate tectonics. The plate tectonic regime characteristic of the Phanerozoic Eon, as described by J. T. Wilson, involving large convection cells, rifting and dispersal of large continental fragments (tectonic plates), formation of paired metamorphic belts, mountain building at plate margins, and related features.

Xenocryst. A crystal in an igneous rock that resembles a phenocryst but is foreign to the body in which it occurs.

Xenolith. A foreign inclusion in an igneous rock.

Xenotopic. A term pertaining to the fabric of a crystalline sedimentary rock where the majority of the constituent crystals are anhedral (i.e., lacking well-defined crystal faces).

Zeolite grade or facies. Metamorphic rocks formed in the zone of transition from diagenesis to metamorphism, under low pressures (2 to 3 kb) and temperatures (200° to 300°C).

Zircon. A mineral, $ZrSiO_4$.

References Cited

Abaimova, G. P. 1978. Anabaritidy—drevnejshie iskopaemye s karbonatnym skeletom [Anabaritids—ancient fossils with carbonate skeleton]. *Trudy Sibirskogo Nanchno-Isskdovatel'skogo Instituta Geologii, Geofiziki i Mineral'nogo Syr'ya* 260: 77–83.

Abelson, P. H. 1959. Geochemistry of organic substances. In: P. H. Abelson (Ed.), *Researches in Geochemistry* (John Wiley & Sons: New York), pp. 79–103.

Achenbach-Richter, L., Gupta, R., Zillig, W. and Woese, C. R. 1988. Were the original eubacteria thermophiles? *Systematics and Applied Microbiology* 9: 34–39.

Achenbach-Richter, P., Gupta, R., Zillig, W. and Woese, C. R. 1988. Rooting the archaebacterial tree: The pivotal role of *Thermococcus celer* in archaebacterial evolution. *Systematics and Applied Microbiology* 10: 231–240.

Adhikary, S. P., Weckesser, J., Jurgens, U. J., Golecki, J. R. and Borowiak, D. 1986. Isolation and chemical characterization of the sheath from the cyanobacterium *Chroococcus minutus* SAG B.41.79. *Journal of General Microbiology* 132: 2595–2599.

Aharon, P., Schidlowski, M. and Singh, I. B. 1987. Chronostratigraphic markers in the end-Precambrian carbon isotope record of the Lesser Himalaya. *Nature* 327: 699–702.

Åhman, E. and Martinsson, A. 1965. Fossiliferous Lower Cambrian at äspelund on the Skäggenäs Peninsula. *Geologiska Föreningens i Stockholm Förhandlingar* 87(1): 139–151.

Aitken, J. D. 1989. Uppermost Proterozoic formations in central Mackenzie Mountains, Northwest Territories. *Geological Survey of Canada, Bulletin* 368: 1–26.

Aizenshtat, Z., Stoler, A., Cohen, Y. and Nielsen, H. 1983. The geochemical sulphur enrichment of recent organic matter by polysulfides in the Solar Lake. In: M. Bjorøy, C. Albercht, K. Cornford, K. de Groot, G. Eglinton, E. Galimov, D. Leythaeuser, R. Pelet, J. Rullkotter and G. Speers (Eds.), *Advances in Organic Geochemistry, 1981* (John Wiley & Sons: New York), pp. 279–288.

Aizenshtat, Z., Lipiner, G. and Cohen, Y. 1984. Biogeochemistry of carbon and sulfur cycle in the microbial mats of the Solar Lake (Sinai). In: Y. Cohen, R. W. Castenholz and H. O. Halvorson (Eds.), *Microbial Mats: Stromatolites* (Alan R. Liss, Inc.: New York), pp. 281–312.

Aksarina, N. A. 1968. Probivalvia—novyoj klass drevnejshikh mollyuskov [Probivalvia—a new class of ancient molluscs]. *Novye dannye po geologii i poleznym iskopaemym zapadnoj Sibiri* 3: 77–89.

Aksarina, N. A. and Pel'man, Y. L. 1978. Kemabrijskie brakhiopody i dvustvorchatye mollyuski Sibiri [Cambrian brachiopods and bivalved molluscs from Siberia]. *Trudy Institut geologii i geofiziki Sibirskoe Otdolenie, Akademiya Nauk SSSR* 362: 1–180.

Akul'cheva, Z. A., Galperova, E. M., Drobkova, E. L., Lysova, L. A., Titorenko, T. N., Treshchetenkova, A. A. and Fajzulina, Z. K. 1981. Motskie otlozheniya i ikh analogi v Irkutskom amfiteatre [Moty deposits and their analogues in the Irkutsk amphiteatre]. In: N. P. Meshkova and I. Nikolaeva V (Eds.), *Pogranichnye Otlozheniya Dokembriya i Kembriya Sibirskoj Platformy (Biostratigrafiya, Paleontologiya, Usloviya Obrazovaniya)* [*Boundary Deposits of the Precambrian and Cambrian of the Siberian Platform (Biostratigraphy, Paleontology, Conditions of Formation)*]. *Trudy Inst. Geol. Geofiz. Sibirsk. Otd. Akad. Nauk SSSR* 475 (Nauka: Novosibirsk), pp. 65–139.

Alf, R. M. 1959. Possible fossils from the early Proterozoic Bass Formation, Grand Canyon, Arizona. *Plateau* 31: 60–63.

Allaart, J. A. 1976. The pre-3760 m.y. old supracrustal rocks of the Isua area, central West Greenland, and the associated occurence of quartz-banded ironstone. In: B. F. Windley (Ed.), *The Early History of the Earth* (John Wiley & Sons: London), pp. 177–189.

Allègre, C. J., Staudacher, T., Sarda, P. and Kurz, M. 1983. Constraints on evolution of earth's mantle from rare gas systematics. *Nature* 303: 762–766.

Allen, P. M. and Jackson, A. A. 1978. Bryn-teg Borehole, North Wales. *Geological Survey of Great Britain Bulletin* 61: 1–48.

Allison, C. W. 1975. Primitive fossil flatworm from Alaska: new evidence bearing on ancestry of the Metazoa. *Geology* 3(11): 649–652.

Allison, C. W. and Hilgert, J. W. 1986. Scale microfossils from the Early Cambrian of Northwest Canada. *Journal of Paleontology* 60(5): 973–1015.

Al-Marjeby, A. and Nash, D. 1986. A summary of the geology and oil habitat of the Eastern Flank Hydrocarbon Province of South Oman. *Marine and Petroleum Geology* 3: 306–314.

Almeida, F. F. M. de and Hasui, Y. 1984. *O pre-cambriano do Brasil* (Edgard Blucher: São Paulo, Brazil), 378 pp.

Alperin, M. J. and Reeburgh, W. S. 1985. Inhibition experiments on anaerobic methane oxidation. *Applied and Environmental Microbiology* 50: 940–945.

Alpert, S. P. 1973. *Bergaueria* Prantl (Cambrian and Ordovician), a probable actinian trace fossil. *Journal of Paleontology* 47: 919–924.

Alpert, S. P. 1975. *Planolites* and *Skolithos* from the upper Precambrian–Lower Cambrian, White–Inyo Mountains, California. *Journal of Paleontology* 49: 508–521.

Alpert, S. P. 1977. Trace fossils and the basal Cambrian Boundary. In: T. P. Crimes and J. C. Harper (Eds.), *Trace Fossils*, vol. 2 (Seel House Press: Liverpool), pp. 1–8.

Alpert, S. P. and Moore, J. N. 1975. Lower Cambrian trace fossil evidence for predation of trilobites. *Lethaia* 8: 223–230.

Alsharhan, A. S. and Kendall, C. G. 1986. Precambrian to Jurassic rocks of the Arabian Gulf and adjacent areas: Their facies, depositional setting and hydrocarbon habitat. *American Association of Petroleum Geologists Bulletin* 14: 977–1002.

Alvarez, W. 1986. Toward a theory of impact crises. *EOS* 67(35): 649–655.

Amati, B. B., Goldschmidt-Clermont, M., Wallace, C. J. A. and

Rochaix, J.-D. 1988. cDNA and deduced amino acid sequences of cytochrome c from *Chlamydomonas reinhardtii*: unexpected functional and phylogenetic implications. *Journal of Molecular Evolution* 28: 151–160.

American Public Health Association. 1971. *Standard Methods for the Examination of Water and Wastewater*, 13th ed. (American Public Health Assoc.; American Water Works Assoc.; Water Pollution Control Federation: Washington, D.C.), 1193 pp.

Anders, D. E. and Robinson, W. E. 1971. Cycloalkane constituents of the bitumen from Green River Shale. *Geochimica et Cosmochimica Acta* 35: 661–678.

Anderson, K. L., Tayne, T. A. and Ward, D. M. 1987. Formation and fate of fermentation products in hot spring cyanobacterial mats. *Applied and Environmental Microbiology* 53: 2343–2352.

Anderson, M. M. 1978. Ediacaran fauna. In: D. N. Lapeded (Ed.), *Yearbook of Science and Technology* (McGraw Hill Book Company: New York), pp. 146–149.

Anderson, M. M. and Conway Morris, S. 1982. A review, with descriptions of four unusual forms, of the soft-bodied fauna of the Conception and St. John's Groups (late-Precambrian), Avalon Peninsula, Newfoundland. *Third North American Paleontological Convention, Proceedings* 1: 1–8.

Anderson, M. M. and Misra, S. B. 1968. Fossils found in the Precambrian Conception Group of south-eastern Newfoundland. *Nature* 220: 680–681.

Anderson, T. H. and Schmidt, V. A. 1983. The evolution of Middle America and the Gulf of Mexico-Caribbean Sea region during Mesozoic time. *Geological Society of America Bulletin* 94: 941–966.

Andreeva, C. W. 1962. Nekotorye kembrisjskie brakhiopody Sibiri i Srednej Azii [Some Cambrian brachiopods from Siberia and Middle Asia]. *Paleontologicheskij Zhurnal* 1962: 87–96.

Andreeva, O. N. 1987. Kembrijskie zamkovye brakhiopody [Cambrian articulate brachiopods]. *Paleontologicheskij Zhurnal* 1987: 31–40.

Angelucci, A. 1970. Su alcune strutture sedimentaire nella formazione delle arenarie del Cambriano inferiore dell'Iglesiente (Sardegna Sud-Occidenale). *Rediconti del Seminario della Facolta di Scienze della Univesità di Cagliari* 40: 1–27.

Anhaeusser, C. R. 1973. The evolution of the early Precambrian crust of southern Africa. *Philosophical Transactions of the Royal Society of London A* 273: 359–388.

Anhaeusser, C. R. (Ed.). 1981. *Barberton Excursion Guidebook. Archean Geology of the Barberton Mountain Land. South Africa Geodynamics Project* (Geological Society of South Africa: Johannesburg), 78 pp.

Appel, P. W. U. 1988. On an Sn-W-bearing iron-formation in the Archean Malene supracrustals, West Greenland. *Precambrian Research* 39: 131–137.

Appollonov, M. K., Chugeava, S., V, Dubinina, S., V and Azemchuzhnikov, V. G. 1988. Batyrbay Section, South Kazakhstan, USSR—Potential stratotype for the Cambrian–Ordovician Boundary. *Geological Magazine* 125: 445–449.

Aquino, N. F. R., Restle, A., Connan, J., Albrecht, P. and Ourisson, G. 1982. Novel tricyclic terpanes (C_{19}, C_{20}) in sediments and petroleums. *Tetrahedron Letters* 23: 2027–2030.

Arai, M. N. and McGugan, A. 1968. A problematical coelenterate (?) from the Lower Cambrian, near Moraine Lake, Banff Area, Alberta. *Journal of Paleontology* 42: 205–209.

Arai, M. N. and McGugan, A. 1969. A problematical Cambrian coelenterate (?). *Journal of Paleontology* 43: 93–94.

Archibald, P. A. and Bold, H. C. 1970. *The Genus* Chlorococcum *Meneghini. Phycological Studies*, XI No. 7015, (University of Texas: Austin), 114 pp.

Aref'ev, O. A., Zabrodina, M. N., Makushina, V. M. and Petrov, A. A. 1980. Relict tetra- and pentacyclic hydrocarbons in ancient oils of the Siberian Platform. *Izvestiya Akademiya Nauk SSSR, seriia Geologicheskaya* 3: 135–140 (in Russian).

Armin, R. A. and Mayer, L. 1983. Subsidence analysis of the Cordilleran miogeocline: Implications for timing of Late Proterozoic rifting and amount of extension. *Geology* 11: 702–705.

Armitage, J. P. 1988. Tactic responses in photosynthetic bacteria. *Canadian Journal of Microbiology* 34: 475–481.

Armstrong, N. V., Hunter, D. R. and Wilson, A. H. 1982. Stratigraphy and petrology of the Archean Nsuze Group, northern Natal and southeastern Transvaal, South Africa. *Precambrian Research* 19: 75–107.

Armstrong, R. L. 1981. Radiogenic isotopes: the case for crustal recycling on a near-steady-state no-continental-growth earth. *Philosophical Transactions of the Royal Society of London A* 301: 443–472.

Arthur, M. A., Anderson, T. F., Kaplan, I. R., Veizer, J. and Land, L. S. (Eds.). 1983. *Stable Isotopes in Sedimentary Geology* (Society of Economic Paleontologists and Mineralogists: Tulsa), 432 pp.

Aseeva, E. A. 1974. O spirale- i kol'tsevidnykh obrazovaniyakh v verkhnedokembrijskikh otlozheniyakh Podolii [About spiral- and ring-like structures in the Upper Precambrian deposits of Podolia]. *Paleontologicheskii Sbornik*: 95–98.

Ashley, B. A. 1937. Fossil algae from the Kundelungu Series of northern Rhodesia. *Journal of Geology* 45: 332–335.

Ashlock, P. D. 1971. Monophyly and associated terms. *Systematic Zoology* 20: 63–69.

Aspler, L. B. and Donaldson, J. A. 1985. The Nonacho Basin (early Proterozoic), Northwest Territories, Canada; Sedimentation and deformation in a strike-slip setting. In: K. T. Biddle and N. Christie-Blick (Eds.), *Strike-slip Deformation, Basin Formation and Sedimentation. Society of Economic Paleontologists and Mineralogists Special Publication* 37: 193–210.

Atkinson, A. W., Jr., Gunning, B. E. S. and John, P. C. L. 1972. Sporopollenin in the cell wall of *Chlorella* and other algae: ultrastructure, chemistry, and incorporation of ^{14}C-acetate, studied in synchronous cultures. *Planta* 107: 1–32.

Awramik, S. and Margulis, L. 1974. Definition of stromatolite. *Stromatolite Newsletter* 2: 1–5.

Awramik, S. M. 1971. Precambrian columnar stromatolite diversity. *Science* 174: 825–827.

Awramik, S. M. 1981. The pre-Phanerozoic biosphere—three billion years of crises and opportunities. In: M. H. Nitecki (Ed.), *Biotic Crises in Ecological and Evolutionary Time* (Academic Press: New York), pp. 83–102.

Awramik, S. M. 1986. The Precambrian-Cambrian boundary and geochemical perturbations. *Nature* 319: 696.

Awramik, S. M. and Barghoorn, E. S. 1977. The Gunflint microbiota. *Precambrian Research* 5: 121–142.

Awramik, S. M. and Riding, R. 1988. Role of algal eukaryotes in subtidal columnar stromatolite formation. *Proceedings of the National Academy of Science USA* 85: 1327–1329.

Awramik, S. M. and Semikhatov, M. A. 1979. The relationship between morphology, microstructure, and microbiota in three vertically intergrading stromatolites from the Gunflint Iron Formation. *Canadian Journal of Earth Sciences* 16: 484–495.

Awramik, S. M. and Vanyo, J. P. 1986. Heliotropism in modern stromatolites. *Science* 231: 1279–1281.

Awramik, S. M., Golubić, S. and Barghoorn, E. S. 1972. Blue-green algal cell degradation and its implication for the fossil record. *Geological Society of America Abstracts with Programs* 4(7): 438.

Awramik, S. M., Margulis, L. and Barghoorn, E. S. 1976. Evolutionary processes in the formation of stromatolites. In: M. R. Walter (Ed.), *Stromatolites* (Elsevier: Amsterdam), pp. 149–162.

Awramik, S. M., Schopf, J. W. and Walter, M. R. 1983. Filamentous fossil bacteria from the Archean of Western Australia. *Precambrian Research* 20: 357–374.

Awramik, S. M., McMenamin, D. S., Yin, C., Zhao, Z., Ding, Q. and Zhang, S. 1985. Prokaryotic and eukaryotic microfossils from a Proterozoic/Phanerozoic transition in China. *Nature* 315: 655–658.

Awramik, S. M., Schopf, J. W. and Walter, M. R. 1988. Carbonaceous filaments from North Pole, Western Australia: are they fossil bacteria in Archean stromatolites? A discussion. *Precambrian*

References

Research 39: 303–309.

Ax, P. 1989. Basic phylogenetic systematization of the Metazoa. In: B. Fernholm, K. Bremer and H. Jörnvall (Eds.), *The Hierarchy of Life* (Elsevier: Amsterdam), pp. 229–246.

Axelrod, D. I. 1958. Early Cambrian marine fauna. *Science* 128: 7–9.

Azbel, I. Y. and Tolstikhin, I. N. 1990. Geodynamics, magmatism, and degassing of the earth. *Geochimica et Cosmochimica Acta* 54: 139–154.

Azmi, R. J. 1983. Microfauna and age of the Lower Tal phosphorite of Mussoorie Syncline, Garhwal Lesser Himalaya, India. *Himalayan Geology* 11: 373–409.

Azmi, R. J. and Pancholi, V. P. 1983. Early Cambrian (Tommotion) conodonts and other shelly microfauna from the upper Krol of the Mussoorie Syncline, Garhwal Lesser Himalaya with remarks on the Precambrian-Cambrian Boundary. *Himalayan Geology* 11: 360–372.

Azmi, R. J., Joshi, M. N. and Juyal, K. P. 1981. Discovery of the Cambro-Ordovician conodonts from the Mussoorie Tal Phosphorite: its significance in correlation of the Lesser Himalaya. *Contemporary Geoscience Research—Himalaya* 1: 245–250.

Babcock, L. E. and Feldmann, R. M. 1986. The phylum Conulariida. In: A. Hoffman and M. H. Nitecki (Eds.), *Problematic Fossil Taxa. Oxford Monographs on Geology and Geophysics*, 5 (Oxford University Press: Oxford), pp. 135–147.

Babcock, L. E. and Robison, R. A. 1989. Preferences of Paleozoic predators. *Nature* 337: 695–696.

Badger, M. R. and Andrews, T. J. 1982. Photosynthesis and inorganic carbon usage by the marine cyanobacterium *Synechococcus* sp. *Plant Physiology* 70: 517–523.

Badger, M. R. and Andrews, T. J. 1987. Co-evolution of rubisco and CO_2-concentrating mechanisms. *Progress in Photosynthesis Research* 3: 601–609.

Baer, A. J. 1983. Proterozoic orogenies and crustal evolution. In: L. G. Medaris Jr., C. W. Byers, D. M. Mickelson and W. C. Shanks (Eds.), *Proterozoic Geology: Selected Papers from an International Proterozoic Symposium. Geological Society of America Memoir* 161: 47–58.

Bahareen, S. and Vishniac, H. S. 1984. 25S ribosomal RHA homologies of basidiomycetous yeasts: taxonomic and phylogenetic implications. *Canadian Journal of Microbiology* 30: 613–621.

Baker, A. J. and Fallick, A. E. 1989a. Evidence from Lewisian limestones for isotopically heavy carbon in two-thousand-million-year-old sea water. *Nature* 337: 352–354.

Baker, A. J. and Fallick, A. E. 1989b. Heavy carbon in two-billion-year-old marbles from Lofoten-Vesterålen, Norway: Implications for the Precambrian carbon cycle. *Geochimica et Cosmochimica Acta* 53: 1111–1115.

Baltscheffsky, H. 1986. Light and organic pyrophosphate as possible key compounds in the development of the earliest bioenergetic systems. *Origins of Life* 16: 377–378.

Bambach, R. K. 1977. Species richness in marine benthic habitats through the Phanerozoic. *Paleobiology* 3: 152–167.

Banks, N. L. 1970. Trace fossils from the late Precambrian and Lower Cambrian of Finnmark, Norway. In: T. P. Crimes and J. C. Harper (Eds.), *Trace Fossils* (Seel House Press: Liverpool), pp. 19–34.

Baragar, W. R. A. and Scoates, F. J. 1981. The Circum-Superior belt: a Proterozoic plate margin? In: A. Kröner (Ed.), *Precambrian Plate Tectonics* (Elsevier: Amsterdam), pp. 295–330.

Barghoorn, E. S. 1957. Origin of life. *Treatise on Marine Ecology and Paleoecology 2. Geological Society of America Memoir* 67: 75–85.

Barghoorn, E. S. and Schopf, J. W. 1965. Microorganisms from the Late Precambrian of central Australia. *Science* 150: 337–339.

Barghoorn, E. S. and Schopf, J. W. 1966. Microorganisms three billion years old from the Precambrian of South Africa. *Science* 152: 758–763.

Barghoorn, E. S. and Tyler, S. A. 1965. Microorganisms from the Gunflint chert. *Science* 147: 563–577.

Barghoorn, E. S., Meinschein, W. G. and Schopf, J. W. 1965. Paleobiology of a Precambrian shale. *Science* 148: 461–472.

Barghoorn, E. S., Knoll, A. H., Dembicki Jr, H. and Meinschein, W. G. 1977. Variation in stable carbon isotopes in organic matter from the Gunflint Iron Formation. *Geochimica et Cosmochimica Acta* 41: 425–430.

Barley, M. E. 1987. The Archean Whim Creek Belt, an ensialic fault-bounded basin in the Pilbara Block, Australia. *Precambrian Research* 37: 199–215.

Barley, M. E., Dunlop, J. S. R., Glover, J. E. and Groves, D. I. 1979. Sedimentary evidence for an Archean shallow-water volcanic-sedimentary facies, eastern Pilbara Block, Western Australia. *Earth and Planetary Science Letters* 43: 74–84.

Barley, M. E., Sylvester, G. C. and Groves, D. I. 1984. Archean calc-alkaline volcanism in the Pilbara Block, Western Australia. *Precambrian Research* 24: 285–319.

Barnes, R. D. 1985. Current perspectives on the origins and relationships of lower invertebrates. In: S. Conway Morris, J. D. George, R. Gibson and H. M. Platt (Eds.), *The Origin and Relationships of Lower Invertebrates. Systematics Association Special Volume*, 30 (Clarendon Press: Oxford), pp. 360–367.

Baroin, A., Perasso, R., Qu, L.-H., Brugerolle, G., Bachelleria, J.-P. and Adoutte, A. 1988. Partial phylogeny of the unicellular eukaryotes based on rapid sequencing of a portion of 28S ribosomal RNA. *Proceedings of the National Academy of Science USA* 85: 3474–3478.

Bar-Or, Y. and Shilo, M. 1987. Characterization of macromolecular flocculants produced by *Phormidium* sp. strain J-1 and by *Anabaenopsis circularis* PCC 6720. *Applied and Environmental Microbiology* 53: 2226–2230.

Bar-Or, Y. and Shilo, M. 1988. The role of cell-bound flocculants in coflocculation of benthic cyanobacteria with clay particles. *FEMS Microbiology Ecology* 53: 169–174.

Baross, J. A. and Hoffman, S. E. 1985. Submarine hydrothermal vents and associated gradient environments as sites for the origin and evolution of life. *Origins of Life* 15: 327–345.

Barovich, K. M., Patchett, P. J., Peterman, Z. E. and Sims, P. K. 1989. Nd isotopes and the origin of 1.9–1.7 Ga Penokean continental crust of the Lake Superior region. *Geological Society of America Bulletin* 101: 333–338.

Barr, T. D. and Kirschvink, J. L. 1983. The paleoposition of North America in the Early Paleozoic: New data from the Caborca sequence in Sonora, Mexico. *EOS* 64: 689–690.

Barrande, J. 1872. *Système Silurien du centre de la Bohême*, vol. 1 (Chez l'auteur et e'diteur: Paris, Prague), 647 pp.

Barrande, J. 1881. *Système Silurien du centre de la Bohême*, vol. 6 (Chez l'auteur et e'diteur: Paris, Prague), 342 pp.

Barron, E. J., Harrison, C. G. A. and Hay, W. W. 1978. A revised reconstruction of the southern continents. *EOS* 59: 436–449.

Barskova, M. I. 1988. Novye mollyuski iz nizhnekembrijskikh otlozhenij Prikolymskogo podnyatiya [New molluscs from Lower Cambrian deposits of the Kolyma Uplift]. *Paleontologicheskij Zhurnal* 1988(1): 101–105.

Barthel, K. W. 1978. *Solnhofen. Ein Blick in die Erdgeschichte* (Ott Verlag Thun: Basel), 369 pp.

Barton, J. M., Robb, L. J., Anhaeusser, C. R. and van Nierop, D. A. 1983. Geochronologic and Sr-isotopic studies of certain units in the Barberton granite-greenstone terrane, South Africa. In: C. R. Anhaeusser (Ed.), *Contributions to the Geology of the Barberton Mountain Land. Geological Society of South Africa Special Publication* 9: 63–72.

Bassler, R. S. 1941. A supposed jellyfish from the Precambrian of the Grand Canyon. *Proceedings of the United States National Museum* 89: 519–522.

Bassoullet, J.-P., Bernier, P., Deloffre, R., Genot, P., Jaffrenzo, M. and Vachard, D. 1979. Essai de classification des Dasycladales en tribus [Attempt to classify Dasycladales in tribes]. *Centre de Recherche Exploration-Production Elf-Aquitaine, Bulletin* 3: 429–442.

Bateson, M. M. and Ward, D. M. 1988. Photoexcretion and fate of glycolate in a hot spring cyanobacterial mat. *Applied and Environ-*

mental Microbiology 54: 1738–1743.

Bateson, M. M., Wiegel, J. and Ward, D. M. 1989. Comparative analysis of 16S ribosomal RNA sequences of thermophilic fermentative bacteria isolated from hot spring cyanobacterial mats. *Systematics and Applied Microbiology* 12: 1–7.

Bateson, M. M., Thibault, K. J. and Ward, D. M. 1990. Comparative analysis of 16S ribosomal RNA sequences of *Thermus* species. *Systematics and Applied Microbiology* 13: 8–13.

Bathurst, R. G. C. 1967. Subtidal gelatinous mat, sand stabilizer and food, Great Bahama Bank. *Geology* 75: 736–738.

Bauchop, T. 1979. Rumen anaerobic fungi of cattle and sheep. *Applied and Environmental Microbiology* 38(1): 148–158.

Bauer, J. E., Haddad, R. I., Des Marais, D. J. and Hguyen, M. 1988. Stable carbon isotope measurements of dissolved organic carbon in pore waters. *EOS* 69: 1130.

Bauld, J. 1981a. Geobiological role of cyanobacterial mats in sedimentary environments: production and preservation of organic matter. *Bureau of Mineral Resources Journal of Australian Geology and Geophysics* 6: 307–317.

Bauld, J. 1981b. Occurrence of benthic microbial mats in saline lakes. *Hydrobiologia* 81: 87–111.

Bauld, J. 1984. Microbial Mats in Marginal Marine Environments: Shark Bay, Western Australia, and Spencer Gulf, South Australia. In: Y. Cohen, R. W. Castenholz and H. O. Halvorson (Eds.), *Microbial Mats: Stromatolites* (Alan R. Liss, Inc.: New York), pp. 39–58.

Bauld, J. 1986. Benthic microbial communities of Australian saline lakes. In: P. De Deckker and W. D. Williams (Eds.), *Limnology In Australia* (CSIRO and Dr. W. Junk Publishers: Melbourne and Dordrecht), pp. 95–111.

Bauld, J. 1987. (Photo)heterotrophic activity in benthic microbial communities. *International Symposium on Environmental Biogeochemistry* 8: 48.

Bauld, J. and Brock, T. D. 1973. Ecological studies of *Chloroflexus*, a gliding photosynthetic bacterium. *Archives of Microbiology* 92: 267–284.

Bauld, J. and Brock, T. D. 1974. Algal excretion and bacterial assimilation in hot spring algal mats. *Phycologia* 10: 101–106.

Bauld, J., Chambers, L. A. and Skyring, G. W. 1979. Primary productivity, sulfate reduction, and sulfur isotope fractionation in algal mats and sediments of Hamlin Pool, Shark Bay, Western Australia. *Journal of Marine and Freshwater Research* 30: 753–764.

Bauld, J., Burne, R. V., Chambers, L. A., Ferguson, J. and Skyring, G. W. 1980. Sedimentological and geobiological studies of intertidal cyanobacterial mats in north-eastern Spencer Gulf, S. A. In: P. A. Trudinger, M. R. Walter and B. J. Ralph (Eds.), *Biogeochemistry of Ancient and Modern Environments* (Australian Academy of Science: Canberra), pp. 157–166.

Baur, M. E., Hayes, J. M., Studley, S. A. and Walter, M. R. 1985. Millimeter-scale variations of stable isotope abundances in carbonates from banded iron-formations in the Hamersley Group of Western Australia. *Economic Geology* 80: 270–282.

Baverstock, P. R., Ilana, S., Christy, P. E., Robinson, B. S. and Johnson, A. M. 1989. srRNA evolution and phylogenetic relationships of the genus *Naegleria* (Protista: Rhizopoda). *Molecular Biology and Evolution* 6: 243–257.

Beadle, L. C. 1974. *The Inland Waters of Tropical Africa: An Introduction to Tropical Limnology* (Longman Group Ltd.: London), 356 pp.

Beardsmore, T. J., Newbery, S. P. and Laing, W. P. 1988. The Maronan Supergroup: an inferred early volcanosedimentary rift sequence in the Mount Isa Inlier, and its implications for ensialic rifting in the Middle Proterozoic of northwest Queensland. *Precambrian Research* 40/41: 487–507.

Beaumont, C. 1981. Foreland basins. *Geophysical Journal of the Royal Astronomical Society* 65: 291–329.

Bebout, B. M., Paerl, H. W., Crocker, K. M. and Prufert, L. E. 1987. Diel interactions of oxygenic photosynthesis and N_2 fixation (acetylene reduction) in a marine microbial mat community. *Applied and Environmental Microbiology* 53: 2353–2362.

Becchstadt, T., Boni, M. and Selg, M. 1985. The lower Cambrian of SW-Sardinia: From a clastic tidal shelf to an isolated carbonate platform. *Facies* 12: 113–140.

Becker, R. H. and Clayton, R. N. 1972. Carbon isotopic evidence for the origin of a banded iron-formation in Western Australia. *Geochemica et Cosmochimica Acta* 36: 577–595.

Bedford, L. E. and Bedford, W. R. 1934. New species of Archaeocyathinae. *Memoirs of the Kyancutta Museum* 1: 1–7.

Bedford, R. and Bedford, J. 1936. Further notes on Cyathospongia (Archaeocyathi) and other organisms from the Lower Cambrian of Beltana, South Australia. *Memoirs of the Kyancutta Museum* 3: 21–26.

Beech, E. M. and Chadwick, B. 1980. The Malene Supracrustal gneisses of northwest Buksefjorden: their origin and significance in the Archean crustal evolution of southern West Greenland. *Precambrian Research* 11: 329–355.

Beecher, C. E. 1891. Development of the Brachiopoda. Pt. 1. Introduction. *American Journal of Science* 41: 343–357.

Beecher, C. E. 1901. Discovery of eurypterid remains in the Cambrian of Missouri. *American Journal of Science* 12: 364–366.

Beer, E. J. 1919. Note on a spiral impression on Lower Vindhyan limestone. *Geological Survey of India, Records* 50: 139.

Beer-Romero, P. and Gest, H. 1987. *Heliobacillus mobilis*, a peritrichously flagellated anoxyphototroph containing bacteriochlorophyll g. *FEMS Microbiology Letters* 41: 109–114.

Behr, H. J., Ahrendt, H., Porada, H., Rohrs, J. and Weber, K. 1983. Upper Proterozoic playa and sabkha deposits in the Damara Orogen, SWA/Namibia. In: R. Miller (Ed.), *Evolution of the Damara Orogen of South West Africa/Namibia*. Geological Society of South Africa Special Publication 11: 1–20.

Behrens, E. W. and Frishman, S. A. 1971. Stable carbon isotopes in blue-green algal mats. *Geology* 79: 94–100.

Bekker, Y. R. 1985. Metazoa iz venda Urala [Vendian Metazoa of the Urals]. In: B. S. Sokolov and A. B. Ivanovsky (Eds.), *Vendskaya Sistema 1* (Nauka: Moscow), pp. 107–112.

Beklemishev, W. N. 1969. *Principles of Comparative Anatomy of Invertebrates. 1. Promorphology* (Oliver and Boyd: Edinburgh), 449 pp.

Belkin, S. and Jannasch, H. W. 1988. Microbial mats at deep-sea hydrothermal vents: New observations. Chapter 2. In: Y. Cohen and E. Rosenberg (Eds.), *Microbial Mats: Physiological Ecology of Benthic Microbial Communities* (American Society for Microbiology: Washington, D.C.), pp. 16—21.

Belkin, S. and Padan, E. 1978a. Hydrogen metabolism in the facultative anoxygenic cyanobacteria (blue-green algae) *Oscillatoria limnetica* and *Aphanothece halophytica*. *Archives of Microbiology* 116: 109–111.

Belkin, S. and Padan, E. 1978b. Sulfide-dependent hydrogen evolution in the cyanobacterium *Oscillatoria limnetica*. *FEBS Letters* 94: 291–294.

Belkin, S. and Padan, E. 1983. Low redox potential promotes sulphide- and light-dependent hydrogen evolution in *Oscillatoria limnetica*. *General Microbiology* 129: 3091–3098.

Bell, W. C. 1938. *Prototreta*, a new genus of brachiopod from the Middle Cambrian of Montana. *Michigan Academy of Science, Arts and Letters* Paper 23: 403–408.

Bell, W. C. 1941. Cambrian Brachiopoda from Montana. *Journal of Paleontology* 15: 193–255.

Belly, R. T., Tansey, M. R. and Brock, T. D. 1973. Algal excretion of ^{14}C-labeled compounds and microbial interactions in *Cyanidium caldarium* mats. *Journal of Phycology* 9: 123–127.

Belaeva, G., V, Luchinina, V. A., Nazarov, V., V, Repina, L. N. and Sobolev, L. P. 1975. Kembrijskaya fauna i flora khrebta Dzhagdy (Dal'nij Vostok) [Cambrian Fauna and Flora from the Dzhagdy Range (Far East)]. *Trudy Inst. Geol. Geofiz. Sibirsk. Otd. Akad. Nauk SSSR*, 226 (Nauka: Novosibirsk), 208 pp.

Ben-Amotz, A. and Avron, M. 1983. On the factors which determine massive β-carotene accumulation in the halotolerant *Dunaliella*

References

bardawil. *Plant Physiology* 72: 593–597.

Ben-Bassat, A. and Zeikus, J. G. 1981. *Themobacteroides acetoethylicus* gen. nov. and spec. nov., a new chemoorganotrophic, anaerobic, thermophilic bacterium. *Archives of Microbiology* 128: 365–370.

Bendix-Almgren, S. E. and Peel, J. S. 1988. *Handimopanella* from the Lower Cambrian of North Greenland: structure and affinities. *Bulletin of the Geological Society of Denmark* 37: 83–103.

Bengtson, S. 1968. The problematic genus *Mobergella* from the Lower Cambrian of the Baltic area. *Lethaia* 1(4): 325–351.

Bengtson, S. 1970. The Lower Cambrian fossil *Tommotia*. *Lethaia* 3(4): 363–392.

Bengtson, S. 1976. The structure of some Middle Cambrian conodonts, and the early evolution of conodont structure and function. *Lethaia* 9(2): 185–206.

Bengtson, S. 1977a. Aspects of problematic fossils in the early Palaeozoic. *Acta Universitatis Upsaliensis, Abstracts of Uppsala dissertations from the Faculty of Science* 415: 1–71.

Bengtson, S. 1977b. Early Cambrian button-shaped phosphatic microfossils from the Siberian Platform. *Palaeontology* 20(4): 751–762.

Bengtson, S. 1983. The early history of the Conodonta. *Fossils and Strata* 15: 5–19.

Bengtson, S. 1985. Taxonomy of disarticulated fossils. *Journal of Paleontology* 59: 1350–1358.

Bengtson, S. 1986a. A new Mongolian species of the Lower Cambrian genus *Camenella* and the problem of scleritome-based taxonomy of the Tommotiida. *Paläontologische Zeitschrift* 60(1–2): 45–55.

Bengtson, S. 1986b. The problem of the Problematica. In: A. Hoffman and M. H. Nitecki (Eds.), *Problematic Fossil Taxa. Oxford Monographs on Geology and Geophysics*, 5 (Oxford University Press: New York), pp. 3–11.

Bengtson, S. 1986c. Siliceous microfossils from the Upper Cambrian of Queensland. *Alcheringa* 10: 195–216.

Bengtson, S. and Conway Morris, S. 1984. A comparative study of Lower Cambrian *Halkieria* and Middle Cambrian *Wiwaxia*. *Lethaia* 17(4): 307–329.

Bengtson, S. and Fletcher, T. P. 1983. The oldest sequence of skeletal fossils in the Lower Cambrian of southeastern Newfoundland. *Canadian Journal of Earth Sciences* 20(4): 525–536.

Bengtson, S. and Missarzhevsky, V. V. 1981. Coeloscleritophora—a major group of enigmatic Cambrian metazoans. In: M. E. Taylor (Ed.), *Short Papers for the Second International Symposium on the Cambrian System 1981. U.S. Geological Survey Open File Report* 81-743: 19–21.

Bengtson, S. and Urbanek, A. 1986. *Rhabdotubulus*, a Middle Cambrian rhabdopleurid hemichordate. *Lethaia* 19: 293–308.

Bengtson, S., Matthew, S. C. and Missarzhevsky, V. V. 1986. The Cambrian netlike fossil *Microdictyon*. In: A. Hoffman and M. H. Nitecki (Eds.), *Problematic Fossil Taxa. Oxford Monographs on Geology and Geophysics*, 5 (Oxford University Press: New York), pp. 97–115.

Bengtson, S., Conway Morris, S., Cooper, B. J., Jell, P. A. and Runnegar, B. N. 1990. Early Cambrian fossils from South Australia. *Memoirs of the Association of Australasian Palaeontologists* 9: 1–364.

Benlow, A. and Meadows, A. J. 1977. The formation of the atmospheres of the terrestrial planets by impact. *Astrophysics and Space Science* 46: 293–300.

Benton, M. J. 1985. Patterns in the diversification of Mesozoic non-marine tetrapods and problems in historical diversity analysis. *Palaeontological Association. Special Papers in Palaeontology* 33: 185–202.

Benus, A. P. 1989. Sedimentological context of a deep-water Ediacaran fauna (Mistaken Point Formation, Avalon Zone, eastern Newfoundland). *New York State Museum and Geological Survey Bulletin* 463: 8–9.

Berg-Madsen, V. and Peel, J. S. 1978. Middle Cambrian monoplacophorans from Bornholm and Australia, and the systematic position of the bellerophontiform molluscs. *Lethaia* 11(2): 113–125.

Bergquist, P. R. 1985. Poriferan relationships. In: S. Conway Morris, J. D. George, R. Gibson and H. M. Platt (Eds.), *The Origins and Relationships of Lower Invertebrates. The Systematics Association Special Volume 28* (Clarendon Press: Oxford), pp. 14–27.

Bergström, J. 1968. *Eolimulus*, a Lower Cambrian xiphosurid from Sweden. *Geologiska Föreningens i Stockholm Förhandlingar* 90(4): 489–503.

Bergström, J. 1973. Organization, life, and systematics of trilobites. *Fossils and Strata* 2: 1–69.

Bergström, J. 1986. Metazoan evolution–a new model. *Zoologica Scripta* 15: 189–200.

Bergström, J. 1989. The origin of animal phyla and the new phylum Procoelomata. *Lethaia* 22: 259–269.

Berkaloff, C., Casadevall, E., Largeau, C., Metzger, P., Peracca, S. and Virlett, J. 1983. The resistant biopolymer of the walls of the hydrocarbon-rich alga *Botryococcus braunii*. *Phytochemistry* 22: 389–397.

Berkner, L., V and Marshall, L. C. 1964a. The history of oxygenic concentration in the earth's atmosphere. *Discussions of the Faraday Society* 37: 122–141.

Berkner, L., V and Marshall, L. C. 1964b. The history of oxygenic concentration in the earth's atmosphere. In: C. J. Brancuzio and G. W. Cameron (Eds.), *The Origin and Evolution of Atmospheres and Oceans* (John Wiley and Sons: New York), pp. 102–126.

Berkner, L., V and Marshall, L. C. 1965a. History of major atmospheric components. *Proceedings of the National Academy of Science USA* 53: 1215–1225.

Berkner, L., V and Marshall, L. C. 1965b. On the origin and rise of oxygen concentration in the earth's atmosphere. *Journal of Atmospheric Science* 22: 225–261.

Berkner, L. V. and Marshall, L. C. 1967. The rise of oxygen in the earth's atmosphere with notes on the martian atmosphere. *Advances in Geophysics* 12: 309–331.

Bernard, F. 1963. Density of flagellates and Myxophyceae in the heterotrophic layers related to environment. In: C. Oppenheimer (Ed.), *Symposium on Marine Microbiology* (Charles C. Thomas Publishing Co.: Springfield), pp. 215–228.

Bernasconi, A. 1987. The major Precambrian terranes of eastern South America: A study of their regional and chronological evolution. *Precambrian Research* 37: 107–124.

Berner, R. A. 1980. *Early Diagenesis, a Theoretical Approach* (Princeton University Press: Princeton), 241 pp.

Berner, R. A. 1984. Sedimentary pyrite formation: an update. *Geochimica et Cosmochimica Acta* 48: 605–615.

Berner, R. A. 1989. Biogeochemical cycles of carbon and sulfur and their effects on atmospheric oxygen over Phanerozoic time. *Palaeogeography, Palaeoclimatology, Palaeoecology* 75: 97–122.

Berner, R. A. and Canfield, D. E. 1989. A model for atmospheric oxygen over Phanerozoic time. *American Journal of Science* 289: 333–361.

Berner, R. A., Lasaga, A. C. and Garrels, R. M. 1983. The carbonate-silicate geochemical cycle and its effect on the atmospheric carbon dioxide over the past 100 million years. *American Journal of Science* 283: 641–683.

Berry, W. B. N. and Wilde, P. 1978. Progressive ventilation of the oceans—an explanation for the distribution of Lower Paleozoic black shales. *American Journal of Science* 278: 257–275.

Bertrand, J. M. L. and Caby, R. 1978. Geodynamic evolution of the Pan-Africa Orogenic Belt: A new interpretation of the Hoggar Shield (Algerian Sahara). *Geologische Rundschau* 67(2): 357–388.

Bertrand-Sarfati, J. 1972. Stromatolites Columnaires du Precabrien Superieur du Sahara Nord—Occidental [Columnar stromatolites of the Upper Precambrian from North-Western Sahara]. *Inventaire, Morphologie et Microstructure des Laminations, Correlations Stratigraphiques. Centre de Recherches sur les zones arides, series geologie* 14: 1–245.

Bertrand-Sarfati, J. 1979. Une algue inhabituelle verte, rouge ou bleue dans une formation dolomitique présumée d'age Précambrien supérieur [An unusual green, red or blue-green alga in a dolomitic formation presumed of late Precambrian age]. *Centre de*

Recherche Exploration-Production Elf-Aquitaine, Bulletin 3: 453–461.

Bertrand-Sarfati, J. and Walter, M. R. 1981. Stromatolite biostratigraphy. *Precambrian Research* 15: 353–371.

Beukes, N. J. 1983. Palaeoenvironmental setting of iron-formation in the depositional basin of the Transvaal Supergroup, South Africa. In: A. F. Trendall and R. C. Morris (Eds.), *Iron-formation, Facts and Problems* (Elsevier: Amsterdam), pp. 131–210.

Beukes, N. J. 1984. Sedimentology of the Kuruman and Griquatown iron-formations, Transvaal Supergroup, Griqualand West, South Africa. *Precambrian Research* 24: 47–84.

Beukes, N. J. 1985. *Final Report on the Geology of the Sishen Iron Ore and the Kalahari Manganese Deposits* (C.S.I.R., C.S.P., Open file report: Pretoria), 49 pp.

Beukes, N. J. 1987. Facies relations, depositional environments and diagenesis in a major early Proterozoic stromatolitic carbonate platform to basinal sequence, Campbellrand Subgroup, Transvaal Supergroup, southern Africa. *Sedimentary Geology* 54: 1–46.

Beukes, N. J. and Klein, C. 1990. Geochemistry and sedimentology of a facies transition—from microbanded to granular iron-formation—in the early Proterozoic Transvaal Supergroup, South Africa. *Precambrian Research* 47: 99–139.

Beukes, N. J. and Lowe, D. R. 1989. Environmental control on diverse stromatolite morphologies in the 3000 Myr-old Pongola Supergroup, South Africa. *Sedimentology* 36: 383–397.

Beukes, N. J., Klein, C., Kaufman, A. J. and Hayes, J. M. 1990. Carbonate petrography, kerogen distribution and carbon and oxygen isotope variations in an early Proterozoic transition from limestone to iron-formation deposition, Transvaal Supergroup, South Africa. *Economic Geology* 85: 663–690.

Beurlen, K. and Sommer, F. W. 1957. Oservações estratigráficas e paleontolólogicas sobre o calcário Corumbá. *Boletim Divisão Geologia e Mineralogia, Rio de Janeiro* 168: 1–35.

Beveridge, T. J. 1981. Ultrastructure, chemistry and function of the bacterial wall. *International Review of Cytology* 72: 229–317.

Beveridge, T. S., Stewart, M., Doyle, R. J. and Sprott, G. D. 1985. Unusual stability of the *Methanospirillum hungatei* sheath. *Journal of Bacteriology* 162: 728–737.

Beyer, P., Falk, H. and Kleinig, H. 1983. Particulate fractions from *Chlorflexus aurantiacus* and distribution of lipids and polyprenoid forming activities. *Archives of Microbiology* 134: 60–63.

Beyers, C. W. 1976. Bioturbation and the origin of the metazoans: evidence from the Belt Supergroup, Montana. *Geology* 4: 565–567.

Bhatt, D. K., Mamgain, V. D., Misra, R. S. and Srivastava, J. P. 1983. Shelly microfossils of Tommotian age (Lower Cambrian) from the Chert-Phosphorite Member of Lower Tal Formation, Maldeota, Dehra Dun District, Uttar Pradesh. *Geophytology* 13(1): 116–123.

Bhatt, D. K., Mamgain, V. D. and Misra, R. S. 1985. Small shelly fossils of early Cambrian (Tommotian) age from Chert-Phosphorite Member, Tal Formation, Mussoorie Syncline, Lesser Himalaya, India, and their chronostratigraphic evaluation. *Journal of the Palaeontological Society of India* 30: 92–102.

Bhattacharya, D. and Druehl, L. D. 1988. Phylogenetic comparison of the small subunit-ribosomal DNA sequence of *Costaria costata* (Phaeophyta) with those of other algae, vascular plants and oomycetes. *Journal of Phycology* 24: 539–543.

Bhattacharya, D., Elwood, H. J., Goff, L. J. and Sogin, M. L. 1990. Phylogeny of *Gracilaria lemaneiformis* (Rhodophyta) based on sequence analysis of its small subunit ribosomal RNA coding region. *Journal of Phycology* 26: 181–186.

Bickford, M. E. 1988. The accretion of Proterozoic crust in Colorado: Igneous, sedimentary, deformational, and metamorphic history. In: W. G. Ernst (Ed.), *Metamorphic and Crustal Evolution of the Western United States* (Prentice-Hall: Englewood Cliffs, N. J.), pp. 411–430.

Bickle, M. J. 1978. Heat loss from the earth: A constraint on Archean tectonics from the relation between geothermal gradients and the rate of plate production. *Earth and Planetary Science Letters* 40: 301–315.

Biebl, H. and Pfennig, N. 1979. Anaerobic CO_2 uptake of phototrophic bacteria. A review. *Arch. Hydrobiol. Beih. Ergebn. Limnol.* 12: 48–58.

Biebl, H. and Pfennig, N. 1981. Isolation of members of the family Rhodospirillaceae. In: M. P. Starr, H. Stolp, H. G. Trüper, A. Balows and H. G. Schelegel (Eds.), *The Prokaryotes: A Handbook on Habitat. Isolation and Identification* (Springer-Verlag: Berlin), pp. 268–273.

Billings, E. 1861. On some new or little known species of Lower Silurian fossils from the Potsdam Group (Primordial Zone). *Rep. Geol. Vermont* 2: 942–960.

Billings, E. 1865. *Palaeozoic Fossils. Volume 1. Silurian Rocks* (Geological Survey of Canada, Dawson Brothers: Montrial), 426 pp.

Billings, E. 1871. On some new species of Palaeozoic fossils. *Canadian Naturalist* 6(2): 213–223, 240.

Billings, E. 1872a. Fossils in the Huronian rocks. *Canadian Naturalist* 6: 478.

Billings, E. 1872b. On some fossils from the primordial rocks of Newfoundland. *Canadian Naturalist* 6(4): 465–479.

Binda, P. L. and Bokhari, M. M. 1980. Chitinozoanlike microfossils in a late Precambrian dolostone from Saudi Arabia. *Geology* 8: 70–71.

Birch, F. 1965. Speculations on the earth's thermal history. *Geological Society of America Bulletin* 76: 133–154.

Bird, C. W., Lynch, J. M., Pirt, S. J., Reid, W. W., Brooks, C. J. W. and Middleditch, B. S. 1971. Steroids and squalene in *Methylococcus capsulatus* grown on methane. *Nature* 230: 473–474.

Birket-Smith, S. J. R. 1981a. Is *Praecambridium* a juvenile *Spriggina*? *Zoologische Jahrbucher. Abteilung fuer Anatomie und Ontogenie der Tiere* 105: 233–235.

Birket-Smith, S. J. R. 1981b. A reconstruction of the Precambrian *Spriggina*. *Zoologische Jahrbucher. Abteilung fuer Anatomie und Ontogenie der Tiere* 105: 237–258.

Bischoff, G. C. O. 1976. *Dailyatia*, a new genus of the Tommotiidae from Cambrian strata of SE Australia (Crustacea, Cirripedia). *Senckenbergiana lethaea* 57: 1–33.

Bisseret, P., Zundel, M. and Rohmer, M. 1985. Prokaryotic triterpenoids. 2. 2β-methylhopanoids from *Methylobacterium organophilum* and *Nostoc muscorum*, a new series of prokaryotic triterpenoids. *European Journal of Biochemistry* 150: 29–34.

Bitton, G. and Freihofer, V. 1978. Influence of extracellular polysaccharides on the toxicity of copper and cadmium towards *Klebsiella aerogenes*. *Microbial Ecology* 4: 119–125.

Black, L. P., Shaw, R. D. and Offe, L. A. 1980. The age of the Stuart Dyke Swarm and its bearing on the onset of late Precambrian sedimentation in central Australia. *Journal of the Geological Society of Australia* 27: 151–155.

Black, L. P., James, P. R. and Harley, S. L. 1983. The geochronology, structure and metamorphism of early Archean rocks at Fyfe Hills, Enderby Land, Antarctica. *Precambrian Research* 21: 197–222.

Blackburn, C. E. 1980. Towards a mobilistic tectonic model for part of the Archean of northwestern Ontario. *Geoscience Canada* 7: 64–72.

Blackburn, C. E., Bond, W. D., Breaks, F. W., Davis, D. W., Edwards, G. R., Poulsen, K. H., Trowell, N. F. and Wood, J. 1985. Evolution of Archean volcanic-sedimentary sequences of the western Wabigoon Subprovince and its margins. In: L. D. Ayres, P. C. Thurston, K. C. Card and W. Weber (Eds.), *Evolution of Archean Supracrustal Sequences. Geological Association of Canada Special Paper* 28: 90–116.

Blackburn, T. H. 1983. The microbial nitrogen cycle. In: W. E. Krumbein (Ed.), *Microbial Geochemistry* (Blackwell Scientific: Oxford), pp. 63–89.

Blackmore, S. and Barnes, S. H. 1987. Embryophyte spore walls: origin, development, and homologies. *Cladistics* 3: 185–195.

Blake, D. H. 1987. *Geology of the Mount Isa Inlier and Environs, Queensland and Northern Territory. Department of Resources and Energy, Bureau of Mineral Resources, Geology and Geophysics, Bulletin* 225 (Australian Government Publishing Service: Canberra), 82 pp.

References

Blake, T. S. and McNaughton 1984. A geochronological framework for the Pilbara region. *University of Western Australia Geology Department and University Extension Publication* 9: 1–22.

Blaxland, A. E. 1974. Geochemistry and geochronology of chemical weathering, Butler Hill Granite, Missouri. *Geochimica et Cosmochimica Acta* 38: 843–852.

Bligh, E. G. and Dyer, W. J. 1959. A rapid method of total lipid extraction and purification. *Canadian Journal of Biochemistry and Physiology* 37: 911–917.

Bliss, D. and Smith, H. 1985. Penetration of light into soil and its role in the control of seed germination. *Plant, Cell and Environment* 8: 475–483.

Bloch, K. 1983. Sterol structure and membrane function. *CRC Critical Reviews in Biochemistry* 14: 47–92.

Bloeser, B. 1979. *Melanocyrillium*-new acritarch genus from Kwagunt Formation (Late Precambrian), Chuar Group, Grand Canyon Supergroup, Arizona. *American Association of Petroleum Geologists Bulletin* 63: 420–421.

Bloeser, B. 1980. Structurally complex microfossils from shales of the Late Precambrian Kwagunt Formation (Walcott Member, Chuar Group) of the eastern Grand Canyon, Arizona. M.S. thesis, University of California, Los Angeles, 188 pp.

Bloeser, B. 1985. *Melanocyrillium*, a new genus of structurally complex late Proterozoic microfossils from the Kwagunt Formation (Chuar Group), Grand Canyon, Arizona. *Journal of Paleontology* 59: 741–765.

Bloeser, B., Schopf, J. W., Horodyski, R. J. and Breed, W. J. 1977. Chitinozoans from the Late Precambrian Chuar Group of the Grand Canyon, Arizona. *Science* 195: 676–679.

Blumer, M., Guillard, R. R. L. and Chase, T. 1971. Hydrocarbons of marine phytoplankton. *Marine Biology* 8: 183–189.

Boak, J. L. and Dymek, R. F. 1982. Metamorphism of the ca. 3800 Ma Supracrustal rocks at Isua, West Greenland: Implications for early Archean crustal evolution. *Earth and Planetary Science Letters* 59: 155–176.

Boardman, S. J. 1986. Early Proterozoic bimodal volcanic rocks in central Colorado, U.S.A., Part I: Petrography, stratigraphy, and depositional history. *Precambrian Research* 34: 1–36.

Boardman, S. J. and Condie, K. C. 1986. Early Proterozoic bimodal volcanic rocks in central Colorado U.S.A., Part II: geochemistry, petrogenesis, and tectonic setting. *Precambrian Research* 34: 37–68.

Bockelie, T. and Fortey, R. A. 1976. An early Ordivician vertebrate. *Nature* 260: 36–38.

Bokova, A. R. 1985. Drevnejshij kompleks organizmov kembriya zapaknogo Prianabar'ya [The oldest complex of organisms in the Cambrian of the western Anabar region]. In: V. Khomentovskij, A. A. Terleev and S. S. Bragin (Eds.), *Stratigrafiya Pozknego Dokembriya i Rannego Kembriya Paleozoya Sibiri*. Vend i r. Institut Geol. Geof. Sibirsk. Otd. Akad. SSSR (Nauka: Novosibirsk), pp. 13–28.

Bond, G. C. and Kominz, M. A. 1984. Construction of tectonic subsidence curves for the early Paleozoic miogeocline, southern Canadian Rocky Mountains: implications for subsidence mechanisms, age of breakup, and crustal thinning. *Geological Society of America Bulletin* 95: 155–173.

Bond, G. C., Nickerson, P. A. and Kominz M. A. 1984. Breakup of a supercontinent between 625 Ma and 555 Ma: new evidence and implications for continental histories. *Earth and Planetary Science Letters* 70: 325–345.

Bond, G. C., Kominz, M. A., Grotzinger, J. P. and Steckler, M. S. 1989. Role of thermal subsidence, flexure, and eustasy in the evolution of early Paleozoic passive-margin carbonate platforms. In: P. Crevello, J. L. Wilson, R. Sarg and J. F. Read (Eds.), *Controls on Carbonate Platform and Basin Development*. Society of Economic Paleontologists and Mineralogists Special Publication 44: 39–62.

Bonen, L. and Doolittle, W. F. 1975. On the prokaryotic nature of red algal chloroplasts. *Proceedings of the National Academy of Science USA* 72: 2310–2314.

Bonen, L. and Doolittle, W. F. 1976. Partial sequences of 16S rRNA and the phylogeny of blue-green algae and chloroplasts. *Nature* 261: 669–673.

Bonhomme, M. G. and Welin, E. 1984. Rb-Sr and K-Ar isotopic data on shale and silstone from the Visingsö Group, Lake Vättern Basin, Sweden. *Geologiska Föreningens i Stockholm Förhandlingar* 105: 363–366.

Bonner, J. T. 1965. *Size and Cycle: An Essay on the Structure of Biology* (Princeton University Press: Princeton, NJ), 219 pp.

Bonner, J. T. 1974. *On Development. The Biology of Form* (Harvard University Press: Cambridge, Mass.), 282 pp.

Bonnot-Courtois, C. 1981. Distribution des terres rares dans les depôts hydrothermaux de la zone FAMOUS et des Galapagos-comparison avec les sediments metalliferes. *Marine Geology* 39: 1–14.

Boon, J. J. and De Leeuw, J. W. 1987. Organic geochemical aspects of cyanobacterial mats. In: F. Fay and C. Von Baalen (Eds.), *The Cyanobacteria* (Elsevier: Amsterdam), pp. 471–492.

Boon, J. J., Hines, H. H., Burlingame, A. L., Klok, J., Rijpstra, W. I. C., Leeuw, J. W. de, Edmunds, K. E. and Eglinton, G. 1983. Organic geochemical studies of Solar Lake laminated cyanobacterial mats. In: M. Bjorøy, K. Albercht, K. Cornford, K. de Groot, G. Eglinton, E. Galimov, D. Leythaeuser, R. Pelet, J. Rullkotter and G. Speers (Eds.), *Advances in Organic Geochemistry* 1981 (John Wiley & Sons: New York), pp. 207–227.

Boon, J. J., de Leeuw, J. W. and Krumbein, W. E. 1985. Biogeochemistry of Gavish Sabkha sediments. II. Pyrolysis mass spectrometry of the laminated microbial mat in the permanently water-covered zone before and after the desert sheetflood of 1979. In: G. M. Friedman and W. E. Krumbein (Eds.), *Hypersaline Ecosystems: The Gavish Sabkha* (Springer-Verlag: Heidelberg), pp. 368–380.

Borass, M. E. 1984. Predator-mediated algal evolution in a chemostat culture. *American Geophysics Union, Transactions* 64: 1102.

von der Borch, C. C. 1976. Stratigraphy of stromatolite occurrences in carbonate lakes of the Coorong Lagoon area, South Australia. In: M. R. Walter (Ed.), *Stromatolites: Developments in Sedimentology*, vol. 20 (Elsevier: Amsterdam), pp. 413–420.

Boreham, C. J., Crick, I. H. and Powell, T. G. 1988a. Alternative calibration of the Methylphenanthrene Index against vitrinite reflectance: Application to maturity measurements on oils and sediments. *Organic Geochemistry* 12: 289–294.

Boreham, C. J., Powell, T. G. and Hutton, A. C. 1988b. Chemical and petrographic characterisation of the Australian Tertiary, Duaringa oil shale deposit. *Fuel* 67: 1369–1377.

Boreham, C. J., Fookes, C. J. R., Popp B. N. and Hayes, J. M. 1989. Origins of etioporphyrins in sediments, evidence from stable isotopes. *Geochimica et Cosmochimica Acta* 53: 2451–2455.

Borg, G. 1988. The Koras-Sinclair-Ghanzi Rift in southern Africa. Volcanism, sedimentation, age relationships and geophysical signature of a late middle Proterozoic rift system. *Precambrian Research* 38: 75–90.

Bornemann, J. G. 1986. Die Versteinerungen des cambrischen Schichtensystems der Insel Sarinien [The fossils of the Cambrian System of the island of Sardinia]. *Nova Acta Kaiser. Leopold. Carolin. Deutsch. Akad. Naturforscher (Halle)* 51: 1–148.

Borovikov, L. I. 1976. Pervaya nakhodka isokopayemkh ostatkov *Dickinsonia* v nizhnekembriyskikh otlozheniyakh na territorii SSSR [First find of fossil *Dickinsonia* in Lower Cambrian sediments in the USSR]. *Doklady Akademii Nauk SSSR* 231: 1182–1184.

Borovikov, L. I. and Kraskov, L. N. 1977. Pervaya nakhodka iskopaemykh ostatkov v Aksumbinskoy svite khrebta bolshoy Karatau (yuzhnyy Kazakhstan) [First find of fossils from the Aksumbinsk Suite of the Karatau Ridge (southern Kazakhstan)]. *Izvestiya Akademii Nauk Kazakhskoy SSR, seriya geologicheskaya* 3: 52–57.

Borowitzka, M. A. 1986. Physiology and biochemistry of calcification in the Chlorophyceae. In: B. S. Leadbeater and R. Riding (Eds.), *Biomineralization in Lower Plants and Animals* (Oxford University

Press: New York), pp. 107–124.
Borowska, Z. and Mauzerall, D. 1988. Photoreduction of carbon dioxide by aqueous ferrous ion: An alternative to the strongly reducing atmosphere for the chemical origin of life. *Proceedings of the National Academy of Science USA* 85: 6577–6580.
Bosence, D. W. J. 1973. Recent serpulid reefs, Connemara, Eire. *Nature* 242: 40–41.
Botha, B. J., V (Ed.). 1983. *Namaqualand Metamorphic Complex. Geological Society of South Africa Special Publication*, 10 (Geological Society of South Africa: Johannesburg), 198 pp.
Boudou, J. P., Trichet, J., Robinson, N. and Brassell, S. C. 1986a. Profile of aliphatic hydrocarbons in a recent Polynesian microbial mat. *International Journal of Environmental Analytical Chemistry* 26: 137–155.
Boudou, J. P., Trichet, J., Rocinson, N. and Brassell, S. C. 1986b. Lipid composition of a recent Polynesian microbial mat sequence. *Organic Geochemistry* 10: 705–709.
Boudreau, B. P. and Guinasso, N. L. 1982. The influence of a diffuse sublayer on accretion, dissolution, and diagenesis at the sea floor. In: K. A. Fanning and F. T. Manheim (Eds.), *The Dynamic Environment of the Ocean Floor* (Lexington Books: Lexington, MA.), pp. 115–145.
Bourrelly, P. 1970. *Les Algues d'eau Douce: Algues Bleues et Rouges*, vol. 3 (Éditions N. Boubée & Cie: Paris), 512 pp.
Bouvier, P. M., Rohmer, P., Beneveniste, P. and Ourisson, G. 1976. $\Delta^{8(14)}$-Steroids in the bacterium *Methylococcus capsulatus*. *The Biochemical Journal* 159: 267–271.
Bovee, E. C. 1981. Distribution and forms of siliceous structures among Protozoa. In: T. L. Simpson and B. E. Volcani (Eds.), *Silicon and Siliceous Structures in Biological Systems* (Springer-Verlag: New York), pp. 233–279.
Bowers, T. S., Campbell, A. C., Measures, C. I., Spivack, A. J., Khadem, M. and Edmond, J. M. 1988. Chemical controls on the composition of vent fluids at 13–11°N and 21°N, East Pacific Rise. *Journal of Geophysical Research* 93: 4522–4536.
Bowles, N. D., Paerl, H. W. and Tucker, J. 1985. Effective solvents and extraction periods employed in photoplankton carotenoid and chlorophyll determinations. *Canadian Journal of Fisheries and Aquatic Science* 42: 1127–1131.
Bowring, S. A. 1989. Remnants of the earth's oldest continental crust. *Geological Society of America Abstracts with Programs* 21: A22.
Bowring, S. A. and Grotzinger, J. P. 1989. Implications of new U-Pb dating and stratigraphic correlations for current tectonic models for Wopmay Orogen and Thelon Tectonic Zone. *Program with Abstracts. Geological Association of Canada/Mineralogical Association of Canada, Annual Joint Meeting, Abstracts* 14: 74.
Bowring, S. A. and Ross, G. M. 1985. Geochronology of the Narakay Volcanic Complex: implications for the age of the Coppermine Homocline and Mackenzie igneous events. *Canadian Journal of Earth Sciences* 22: 774–781.
Bowring, S. A. and Van Schnus, W. R. 1984. U-Pb zircon constraints on evolution of Wopmay Orogen, N.W.T. *Program with Abstracts. Geological Association of Canada/Mineralogical Association of Canada, Annual Joint Meeting, Abstracts* 9: 47.
Bowring, S. A., Williams, I. S. and Compston, W. 1989. 3.96 Ga gneisses from the Slave Province, Northwest Territories, Canada. *Geology* 17: 971–975.
Boyd, D. W. 1966. Lamination deformed by burrowers in Flathead Sandstone (Middle Cambrian) of central Wyoming. *University of Wyoming Contributions to Geology* 5: 45–53.
Boyd, D. W. 1974. Wyoming specimens of the trace fossil *Bergaueria*. *University of Wyoming Contributions to Geology* 13: 11–15.
Boyd, D. W. and Ore, H. T. 1963. Patterned cones in Permo-Triassic redbeds of Wyoming and adjacent areas. *Journal of Sedimentary Petrology* 33: 438–451.
Boyd, F. R., Gurney, J. J. and Richardson, S. H. 1985. Evidence for a 150–200 km thick Archean lithosphere from diamond inclusion thermobarometry. *Nature* 315: 387–389.
Boynton, H. E. and Ford, T. D. 1979. *Pseudovendia charnwoodensis*—a new Precambrian arthropod from Charnwood Forest, Leicestershire. *Mercian Geologist* 7: 175–177.
Bralower, T. J. and Thierstein, H. R. 1984. Low productivity and slow deep-water circulation in mid-Cretaceous oceans. *Geology* 12: 614–618.
Branagan, D. F. 1976. Jelly fish trails. *Journal of Sedimentary Petrology* 46: 240–242.
Brannon, D. K. and Caldwell, D. E. 1986. Ecology and metabolism of *Thermothrix thiopara*. *Advances in Applied Microbiology* 31: 233–270.
Brasier, M. D. 1977. An early Cambrian chert biota and its implications. *Nature* 268: 719–720.
Brasier, M. D. 1979. The Cambrian radiation event. In: M. R. House (Ed.), *The Origin of Major Invertebrate Groups* (Academic Press: London), pp. 103–159.
Brasier, M. D. 1982. Sea-level changes, facies changes and the late Precambrian-early Cambrian evolutionary explosion. *Precambrian Research* 17: 105–123.
Brasier, M. D. 1984. Microfossils and small shelly fossils from the Lower Cambrian *Hyolithes* Limestone at Nuneaton, English Midlands. *Geological Magazine* 121(3): 229–253.
Brasier, M. D. 1985. Evolutionary and geological events across the Precambrian–Cambrian boundary. *Geology Today* 1985: 141–146.
Brasier, M. D. 1986. The succession of small shelly fossils (especially conoidal microfossils) from English Precambrian-Cambrian boundary beds. *Geology Magazine* 123: 237–256.
Brasier, M. D. 1989. Towards a biostratigraphy of the earliest skeletal biotas. In: J. W. Cowie and M. D. Brasier (Eds.), *The Precambrian-Cambrian Boundary* (Oxford Science Publishers, Clarendon Press: Oxford), pp. 117–165.
Brasier, M. D. In Press. In: J. H. Lipps and P. W. Signor (Eds.), *Origin and Early Evolution of Metazoa* (Pergamon Press: New York).
Brasier, M. D. and Cowie, J. W. 1989. Other areas: North-west Canada; California, Nevada, and Mexico; Morocco, Spain, and France. In: J. W. Cowie and M. D. Brasier (Eds.), *The Precambrian-Cambrian Boundary* (Oxford Science Publishers, Clarendon Press: Oxford), p. 213.
Brasier, M. D. and Hewitt, R. A. 1979. Environmental setting of fossiliferous rocks from the uppermost Proterozoic–lower Cambrian of central England. *Palaeogeography, Palaeoclimatology, Palaeoecology* 27: 35–57.
Brasier, M. D. and Singh, P. 1987. Microfossils and Precambrian-Cambrian boundary stratigraphy at Maldeota, Lesser Himalayas. *Geological Magazine* 124(4): 323–345.
Brassell, S. C. and Eglinton, G. 1981. Biogeochemical significance of a novel C27 stanol. *Nature* 290: 579–582.
Brassell, S. C., Wardroper, A. M. K., Thomson, I. D., Maxwell, J. R. and Eglinton, G. 1981. Specific acyclic isoprenoids as biological markers of methanogenic bacteria in marine sediments. *Nature* 290: 693–696.
Brassell, S. C., Lewis, C. A., De Leeuw, J. W., De Lange, F. and Sinninghe Damste, J. S. 1986. Isoprenoids thiophenes: novel diagenetic products in sediments. *Nature* 320: 160–162.
Brassell, S. C., Eglinton, G. and Howell, V. J. 1987. Palaeoenvironmental assessment of marine organic-rich sediments using molecular organic geochemistry. In: J. Brooks and A. J. Fleet (Eds.), *Marine Petroleum Source Rocks. Geological Society Special Publication*, 26 (Blackwell: Oxford), pp. 79–98.
Braterman, P. S., Cairns-Smith, A. G. and Sloper, R. W. 1983. Photooxidation of hydrated Fe^{+2}: The significance for banded iron formations. *Nature* 303: 163–164.
Breiteneder, H., Seiser, C., Löfelhardt, W., Michalowski, C. and Bohnert, H. J. 1988. Physical map and protein gene map of cyanelle DNA from the second known isolate of *Cyanophora paradoxa* (Kies Straub). *Current Genetics* 13: 199–206.
Breitkopf, J. H. 1988. Iron formations related to mafic volcanism and ensialic rifting in the southern margin zone of the Damara Orogen, Namibia. *Precambrian Research* 38: 111–130.
Brewer, A. M., Dunster, J. N., Gatehouse, C. G., Henry, R. L. and

Weste, G. 1987. A revision of the stratigraphy of the Eastern Officer Basin. *Geological Society of South Australia Quarterly Geological Notes* 102: 1–15.

Bridgwater, D., Keto, L., McGregor, V. R. and Myers, J. S. 1976. Archean gneiss complex of Greenland. In: A. Escher and W. S. Watts (Eds.), *Geology of Greenland* (Geological Survey of Greenland: Copenhagen), pp. 18–75.

Briggs, D. E. 1977. Bivalved arthropods from the Cambrian Burgess Shale of British Columbia. *Palaeontology* 20(3): 67–72.

Briggs, D. E. G. 1976. The arthropod *Branchiocaris* n. gen., Middle Cambrian, Burgess Shale Columbia. *Bulletin of the Geological Survey of Canada* 264: 1–29.

Briggs, D. E. G. 1978. A new trilobite-like arthropod from the Lower Cambrian Kinzers Formation, Pennsylvania. *Journal of Paleontology* 52(1): 132–140.

Briggs, D. E. G. 1979. *Anomalocaris*, the largest known Cambrian arthropod. *Palaeontology* 22(3): 631–664.

Briggs, D. E. G. and Fortey, R. A. 1989. The early radiation and relationships of the major arthropod groups. *Science* 246: 241–243.

Briggs, D. E. G. and Mount, J. D. 1982. The occurrence of the giant arthropod *Anomalocaris* in the Lower Cambrian of southern California, and the overall distribution of the genus. *Journal of Paleontology* 56(5): 1112–1118.

Brinkman, R. T. 1969. Dissociation of water vapor and evolution of oxygen in the terrestrial atmosphere. *Journal of Geophysical Research* 74: 5355–5368.

Brock, M. L., Wiegert, R. G. and Brock, T. D. 1969. Feeding by *Paracoenia* and *Ephydra* (Diptera: Ephydridae) on the microorganisms of hot springs. *Ecology* 50(2): 192–200.

Brock, T. D. 1968. Taxonomic confusion concerning certain filamentous blue-green algae. *Journal of Phycology* 4: 178–179.

Brock, T. D. 1973. Lower pH limit for the existence of blue-green algae: evolutionary and ecological implications. *Science* 179: 480–483.

Brock, T. D. 1976. Halophilic blue-green algae. *Archives of Microbiology* 107: 109–111.

Brock T. D. 1978. *Thermophilic Microorganisms and Life at High Temperatures* (Springer-Verlag: New York), 465 pp.

Brock, T. D. 1984. How sensitive is the light microscope for observations on microorganisms in natural habitats? *Microbial Ecology* 10: 297–300.

Brock, T. D. 1985. Life at high temperatures. *Science* 230: 132–138.

Brock, T. D. and Brock, M. L. 1971. Microbiological studies of thermal habitats of the Central Volcanic Region, North Island, New Zealand. *New Zealand Journal of Marine and Freshwater Research* 5: 233–258.

Brock, T. D. and Freeze, H. 1969. *Thermus aquaticus* gen. n. and sp. n., a non-sporulating extreme thermophile. *Journal of Bacteriology* 98: 289–297.

Broda, E. 1975. *The Evolution of Bioenergetic Processes* (Pergamon Press: New York), 220 pp.

Broecker, W. S. 1970. A boundary condition on the evolution of atmospheric oxygen. *Journal of Geophysical Research* 75(18): 3553–3557.

Broecker, W. S. and Peng, T. H. 1982. *Tracers in the Sea* (Lamont-Doherty Geological Observatory: Palisades, N.Y.), 690 pp.

Bro Larson, E. 1936. Biologische Studien über die tunnelgrabenden Käfer auf Skallingen. *Videnskabelige Meddelelser fra Dansk Naturhistorish Forening, Krobenhavn* 100: 1–231.

Brongniart, A. T. 1928. *Histoire des Végétaux Fossiles 1* [*History of Plant Fossils 1*] (A Asher & Co.: Amsterdam), 488 pp.

Brooke, C. and Riding, R. 1988. A new look at the Solenoporaceae. *4th International Symposium on Fossil Algae, Cardiff, Great Britain*: 7.

Brooks, J., Grant, P. R., Muir, M. D., Van Gijzel, P. and Shaw, P. (Eds.). 1971. *Sporopollenin* (Academic Press: London), 718 pp.

Brooks, P. W., Eglinton, G., Gaskell, S. J., McHugh, D. J., Maxwell, J. R. and Philp, R. P. 1976. Lipids of recent sediments, Part I: straight-chain hydrocarbons and carboxylic acids of some temperate lacustrine and subtropical lagoonal/tidal flat sediments. *Chemical Geology* 18: 21–38.

Brooks, W. K. 1894. The origin of the oldest fossils and the discovery of the bottom of the ocean. *Journal of Geology* 2: 455–479.

Brown, A. C. 1978. Stratiform copper deposits—evidence for their post-sedimentary origin. *Minerals Science and Engineering* 10: 172–181.

Brune, D. C., Nozawa, T. and Blankenship, R. E. 1987. Antenna organization in green photosynthetic bacteria. 1. Oligomeric bacteriochlorophyll c in *Chloroflexus aurantiacus* chlorosomes. *Biochemistry* 26: 8644–8652.

Buchanan, R. E. and Gibbons, M. E. 1974. *Bergey's Manual of Determinative Bacteriology*, 8th ed. (Williams & Wilkins Co.: Baltimore), 1268 pp.

Buekes, N. J. 1980. Lithofacies and stratigraphy of the Kuruman and Griquatown iron-formations, northern Cape Province, South Africa. *Transactions of the Geological Society of South Africa* 83: 69–86.

Buggisch, W. and Flügel, E. 1988. The Precambrian/Cambrian boundary in the Anti-Atlas (Morocco). Discussion and new results. In: V. H. Jacobshagen (Ed.), *The Atlas System of Morocco. Lecture Notes in Earth Sciences 15* (Springer-Verlag: Berlin), pp. 81–90.

Buggisch, W. and Webers, G. F. 1982. Zur Fazies der Karbonatgesteine in den Ellsworth Mountains (Paläozoikum, Westantarktis) [Facies of carbonate rocks in the Ellsworth Mountains (Paleozoic, West Antarctica)]. *Facies* 7: 199–228.

Buick, R. 1984. Carbonaceous filaments from North Pole, Western Australia: are they fossil bacteria in Archean stromatolites? *Precambrian Research* 24: 157–172.

Buick, R. 1988. Carbonaceous filaments from North Pole, Western Australia: are they fossil bacteria in Archean stromatolites? A reply. *Precambrian Research* 39: 311–317.

Buick, R. and Dunlop, J. S. R. 1987. Early Archean evaporitic sediments from the Warrawoona Group, North Pole, Western Australia. *Geological Society of America Abstracts with Programs* 19: 604.

Buick, R., Dunlop, J. S. R. and Groves, D. I. 1981. Stromatolite recognition in ancient rocks: an appraisal of irregularly laminated structures in an early Archean chert-barite unit from North Pole, Western Australia. *Alcheringa* 5: 161–181.

Burger, A. J. and Coertze, F. J. 1973. Radiometric age measurements on rocks from southern Africa to the end of 1971. *Bulletin of the Geological Survey of South Africa* 58: 46–58.

Burger-Wiersma, T., Veenhuis, M., Korthals, H. J., van der Wiel, C. C. M. and Muir, L. R. 1986. A new prokaryote containing chlorophylls a and b. *Nature* 320: 262–264.

Burke, K. and Dewey, J. F. 1973. An outline of Precambrian plate development. In: D. H. Tarling and S. K. Runcorn (Eds.), *Implications of Continental Drift to the Earth Sciences* (Academic Press: London), pp. 1035–1045.

Burke, K., Dewey, J. F. and Kidd, W. S. F. 1976. Precambrian paleomagnetic results compatible with the Wilson Cycle. *Tectonophysics* 33: 287–299.

Burke, K., Kidd, W. S. F. and Kusky, T. 1985a. Is the Ventersdorp rift system of southern Africa related to a continental collision between the Kaapvaal and Zimbabwe Cratons at 2.64 Ga ago? *Tectonophysics* 115: 1–24.

Burke, K., Kidd, W. S. F. and Kusky, T. M. 1985b. The Pongola structure of southeastern Africa: the world's oldest preserved rift? *Journal of Geodynamics* 2: 35–49.

Burkholder, P. R., Repak, A. and Sibert, J. 1965. Studies on some Long Island Sound littoral communities of microorganisms and their primary productivity. *Bulletin of the Torrey Botanical Club* 92: 378–402.

Burlingame, A. L., Haug, P., Belski, T. and Calvin, M. 1965. Occurrence of biogenic steranes and pentacyclic triterpanes in the Eocene shale (52 million years) and in the early Precambrian shale (2.7 billion years): a preliminary report. *Proceedings of the National Academy of Science USA* 54: 1406–1412.

Burnap, R. L. and Trench, R. K. 1989. The biogenesis of the cyanelle of *Cyanophora paradoxa*. III. In vitro synthesis of cyanellar polypeptides using separated cytoplasmic and cyanellar RNA. *Proceedings of the Royal Society of London, Series B* 238: 89–102.

Burne, R. V. and Moore, L. S. 1987. Microbialites: organosedimentary deposits of benthic microbial communities. *Palaios* 2: 241–254.

Burnett, D. S. 1975. Lunar science: the Apollo legacy. *Reviews of Geophysics and Space Physics* 13: 13–34.

Burnett, C. and Stait, B. 1986. China and southeast Asia as part of the Tethyan margin of Cambro-Ordovician Gondwanaland. In: K. G. McKenzie (Ed.), *Shallow Tethyhs*, vol. 2 (A. A. Balkema: Rotterdam), pp. 65–77.

Burwood, R., Drodz, R. J., Halpern, H. I. and Sedivy, R. A. 1988. Carbon isotopic variations of kerogen pyrolysates. *Organic Geochemistry* 12: 195–205.

Buss, L. W. 1987. *The Evolution of Individuality* (Princeton University Press: Princeton, NJ), 201 pp.

Butin, R., V 1959. Iskopaemye Cyanophyceae v proterozojskikh karbonatnykh otlozheniyakh Yuzhnoj Karelii [Fossil Cyanophyceae from Proterozoic carbonate deposits of South Karelia]. *Izvestiya Karel'skogo Kol'skogo Filialov Akademii Nauk SSSR:* 47–51.

Butin, R., V 1966. Iskopaemye vodorosli proterozoya Karelii [Fossil algae from the Proterozoic of Karelia]. In: *Ostaki Organizmov i Problematika Proterozojskikh Obrazovanij Karelii* (Karel'skow Knizhnow Izd.: Petrozabodsk), pp. 34–63.

Butterfield, N. J. and Knoll, A. H. 1989. Metaphytes and multicellularity in the Proterozoic: Examples from Svalbard and arctic Canada. *Geological Society of America Abstracts with Programs* 21(6): A146.

Butterfield, N. J. and Rainbird, R. H. 1988. The paleobiology of two Proterozoic shales. *Geological Society of America Abstracts with Programs* 20: 103.

Butterfield, N. J., Knoll, A. H. and Swett, K. 1988. Exceptional preservation of fossils in an Upper Proterozoic shale. *Nature* 334: 424–427.

Butterfield, N. J., Knoll, A. H. and Swett, K. 1990. A bangiophyte red alga from the Proterozoic of arctic Canada. *Science* 250: 104–107.

Button, A. 1979. Early Proterozoic weathering profile on the 2200 m.y. old Hekpoort Basalt, Pretoria Group, South Africa: preliminary results. *Economic Geology Research Unit. University of Witwatersrand Information Circular* 133.

Button, A., Brock, T. D., Cook, P. J., Eugster, H. P., Goodwin, A. M., James, H. L., Margulis, L., Nealson, K. H., Nriagu, J. O., Trendall, A. F. and Walter, M. R. 1982. Sedimentary iron deposits, evaporites and phosphorites; State of the art report. In: H. D. Holland and M. Schidlowski (Eds.), *Mineral Deposits and the Evolution of the Biosphere* (Springer-Verlag: New York), pp. 259–273.

Byerly, G. R., Lowe, D. R. and Walsh, M. M. 1986. Stromatolites from the 3,300–3,500-Myr Swaziland Supergroup, Barberton Mountain Land, South Africa. *Nature* 319: 489–491.

Byers, C. W. 1982. Geologic significance of marine biogenic sedimentary structures. In: P. L. McCall and M. J. S. Tevesz (Eds.), *Animal-Sediment Relations* (Plenum Press: New York), pp. 221–256.

Bykova, E., V 1961. *Foraminfery Karadoka vostochnogo Kazakhstana [Caradocian Foraminifers from Eastern Kazakhstan]*. Inst. Geol. Nauk, Akad. Nauk Kazakhskoj SSR (Akademiya Nauk Kazakhstoj SSR: Alma-Ata), 120 pp.

Cabioch, J. and Giraud, G. 1986. Structural aspects of biomineralization in the coralline algae (calcified Rhodophyceae). In: B. S. Leadbeater and R. Riding (Eds.), *Biomineralization in Lower Plants and Animals* (Oxford University Press: New York), pp. 141–156.

Caby, R., Bertrand, J. M. L. and Black, R. 1981. Pan-African ocean closure and continental collision in the Hoggar-Iforas segment, central Sahara. In: A. Kröner (Ed.), *Precambrian Plate Tectonics* (Elsevier: Amsterdam), pp. 407–434.

Cadée, G. C. and Hegeman, J. 1974. Primary production of the benthic microflora living on tidal flats in the Dutch Wadden Sea. *Netherlands Journal of Sea Research* 8: 260–291.

Cahen, L., Jamotte, A., Lepersonne, J. and Mortlemans, G. 1946. Aperçu sur la question des algues des séries calcaires anciennes du Congo belge et essai de corrélation. Présentation d'échantillons. *Soc. Belge de Géologie, de Paléontologie et d'hydrologie* 55: 164–192.

Cahen, L., Snelling, N. J., Delhal, J. and Vail, J. R. 1984. *The Geochronology and Evolution of Africa* (Clarendon Press: Oxford), 512 pp.

Cairns-Smith, A. G. 1978. Precambrian solution photochemistry, inverse segregation and banded iron formations. *Nature* 276: 807–808.

Cairns-Smith, A. G. 1982. *Genetic Takeover and the Mineral Origins of Life* (Cambridge University Press: Cambridge), 477 pp.

Caldwell, D. E., Caldwell, S. J. and Laycock, J. P. 1976. Thermothrix thioparus gen. et sp. nov., a facultative chemolithotroph living at neutral pH and high temperature. *Canadian Journal of Microbiology* 10: 1509–1517.

Cameron, A. G. W. 1978. Physics of the primitive solar disk. *Moon Planets* 18: 5–40.

Cameron, E. M. 1982. Sulphate and sulphate reduction in early Precambrian oceans. *Nature* 296: 145–148.

Campbell, A. C., Bowers, T. S., Measures, C. I., Falkner, K. K., Khadem, M. and Edmond, J. M. 1988. A time series of vent fluid compositions from 21°N, East Pacific Rise (1979, 1981, 1985), and the Guaymas Basin, Gulf of California (1982, 1985). *Journal of Geophysical Research* 93: 4537–4549.

Campbell, S. E. 1979. Soil stabilization by a prokaryotic desert crust: implications for Precambrian land biota. *Origins of Life* 9: 335–348.

Canfield, D. E., Raiswell, R., Westrich, J. T., Reaves, C. M. and Berner, R. A. 1986. The use of chromium reduction in the analysis of reduced inorganic sulfur in sediments and shales. *Chemical Geology* 54: 149–155.

Canning, E. U. 1988. Nuclear division and chromosome cycles in microsporida. *Biosystems* 21: 333–340.

Cannon, R. T. 1965. Age of the transition in the Precambrian atmosphere. *Nature* 205: 586.

Cannon, W. F. and Gairn, J. E. 1970. A revision of stratigraphic nomenclature for middle Precambrian rocks in northern Michigan. *Geological Society of America Bulletin* 81: 2843–2846.

Cánovas, J. L., Ornston, L. N. and Stanier, R. Y. 1967. Evolutionary significance of metabolic control systems. *Science* 156: 1695–1699.

Cao, R. 1986. On the role of algae in the origin of iron-bearing stromatolites. *Acta Micropalaeontologica Sinica* 3(2): 185–192.

Cao, R. and Liang, Y. 1974. [On the classification and correlation of the Sinian System in China, based on a study of algae and stromatolites]. *Memoirs of the Nanjing Institute of Geology and Palaeontology:* 1–26.

Cao, R. and Zhao, W. 1974. [Sinian Ancient Algae]. In: Nanjing Institute of Geology and Palaeontology (Ed.), *A Handbook of the Stratigraphy and Paleontology in Southwest China* (Science Press: Beijing), pp. 66–72.

Cao, R. and Zhao, W. 1978a. [The algal flora of the Tongyong Formation (Upper Sinian System) in southwestern China]. *Memoirs of the Nanjing Institute of Geology and Palaeontology:* 1–30.

Cao, R. and Zhao, W. 1978b. Manicosiphoniaceae, a new family of fossil algae from the Sinian System of SW China with reference to its systematic position. *Acta Palaeontologica Sinica* 17(1): 29–40.

Capo, R. C. 1984. Petrology and geochemistry of a Cambrian paleosol developed on Precambrian granite, Llano Uplift, Texas. M.A. thesis, University of Texas, Austin.

Card, K. D. 1986. Tectonic setting and evolution of Late Archean greenstone belts of Superior Province, Canada. In: M. J. deWit and L. D. Ashwal (Eds.), *Workshop on Tectonic Evolution of Greenstone Belts. Lunar and Planetary Institute Tecnical Report* 86-10: 74–76.

Cardoso, J., Brooks, P. W., Eglinton, G., Goodfellow, R., Maxwell, J. R. and Philp, R. P. 1976. Lipids of recently deposited algal mats at Laguna Mormona, Baja California. In: J. O. Nriagu (Ed.), *Environmental Biogeochemistry*, vol. 1 (Ann Arbor Science Publ. Inc.: Ann Arbor), pp. 149–174.

Cardoso, J. N., Watts, C. D., Maxwell, J. R., Goodfellow, R., Eglinton, G. and Golubić, S. 1978. A biogeochemical study of the Abu Dhabi algal mats: a simplified ecosystem. *Chemical Geology* 70: 273–291.

References

Caron, D. A., Pick, F. R. and Lean, D. R. S. 1985. Chroococcoid cyanobacteria in Lake Ontario: vertical and seasonal distributions during 1982. *Journal of Phycology* 21: 171–175.

Carozzi, A., V 1962. Observations on algal biostromes in the Great Salt Lake, Utah. *Journal of Geology* 70: 246–252.

Carpenter, R. C. 1985. Relationships between primary production and irradiance in coral reef algal communities. *Limnology and Oceanography* 30: 784–793.

Carr, T. R. and Kitchell, J. A. 1980. Dynamics of Taxonomic diversity. *Paleobiology* 6: 427–443.

Cassedanne, J. 1965. Decouverte d'Algue dans le Calcaire de Bambuí (Etat de Minas Gerais, Brésil) [Discovery of algae from the Bambuí Limestone (State of Minas Gerais, Brazil)]. *Anais de Academia Brasileira de Ciências* 37: 79–81.

Castenholz, R. W. 1961. The effect of grazing of marine littoral diatom populations. *Ecology* 42: 783–794.

Castenholz, R. W. 1968. The behavior of *Oscillatoria terebriformis* in hot springs. *Journal of Phycology* 4: 132–139.

Castenholz, R. W. 1969a. Thermophilic blue-green algae and the thermal environment. *Bacteriological Reviews* 33: 476–504.

Castenholz, R. W. 1969b. The thermophilic cyanophytes of Iceland and the upper temperature limit. *Journal of Phycology* 5: 360–368.

Castenholz, R. W. 1973a. Ecology of blue-green algae in hot springs. In: N. G. Carr and B. A. Whitton (Eds.), *The Biology of Blue-Green Algae* (Blackwell: Oxford), pp. 379–414.

Castenholz, R. W. 1973b. The possible photosynthetic use of sulfide by the filamentous phototrophic bacteria of hot springs. *Limnology and Oceanography* 18: 863–876.

Castenholz, R. W. 1976. The effect of sulfide on the blue-green algae of hot springs. I. New Zealand and Iceland. *Journal of Phycology* 12: 54–68.

Castenholz, R. W. 1977. The effect of sulfide on the blue-green algae of hot springs. II. Yellowstone National Park. *Microbial Ecology* 3: 79–105.

Castenholz, R. W. 1979. Evolution and ecology of thermophilic microorganisms. In: M. Shilo (Ed.), *Strategies of Microbial Life in Extreme Environments*. Dahlem Konferenzen, 13 (Verlag Chemie: Weinheim), pp. 373–392.

Castenholz, R. W. 1981. Isolation and cultivation of thermophilic cyanobacteria. In: M. P. Starr, H. Stolp, H. G. Trüper, A. Balows and H. G. Schelegel (Eds.), *The Prokaryotes: A Handbook on Habitats. Isolation and Identification* (Springer-Verlag: Berlin), pp. 236–246.

Castenholz, R. W. 1982. Motility and taxis. In: N. G. Carr and B. A. Whitton (Eds.), *The Biology of Cyanobacteria* (University of California Press: Berkeley), pp. 413–440.

Castenholz, R. W. 1984. Composition of hot spring microbial mats: a summary. In: Y. Cohen, R. W. Castenholz and H. O. Halvorson (Eds.), *Microbial Mats: Stromatolites* (Alan R. Liss: New York), pp. 101–119.

Castenholz, R. W. 1988a. Culturing methods for cyanobacteria. In: L. Packer and A. N. Glazer (Eds.), *Cyanobacteria, Vol. 167 Methods in Enzymology* (Academic Press: London, New York), pp. 68–93.

Castenholz, R. W. 1988b. The green sulfur and non-sulfur bacteria of hot springs. In: J. M. Olson, J. G. Ormerod, J. Amesz, E. Stackebrandt and H. G. Trüper (Eds.), *Green Photosynthetic Bacteria* (Plenum Press: New York), pp. 243–255.

Castenholz, R. W. 1988c. Thermophilic cyanobacteria: special problems. In: L. Packer and A. N. Glazer (Eds.), *Cyanobacteria, Vol. 167 Methods in Enzymology* (Academic Press: London, New York), pp. 96–100.

Castenholz, R. W. and Pierson, B. K. 1981. Isolation of members of the family Chloroflexaceae. In: M. P. Starr, H. Stolp, H. G. Trüper, A. Balows and H. G. Schelegel (Eds.), *The Prokaryotes: A Handbook on Habitats. Isolation and Identification* (Springer-Verlag: Berlin), pp. 290–298.

Castenholz, R. W. and Utkilen, H. C. 1984. Physiology of sulfide tolerance in a thermophilic *Oscillatoria*. *Archives of Microbiology* 138: 299–305.

Castenholz, R. W., Bauld, J. and Jørgensen, B. B. In Press—a. Anoxygenic microbial mats of hot springs: thermophilic *Chlorobium* sp. *FEMS Microbiology Ecology*.

Castenholz, R. W., Jørgensen, B. B., D'Amelio, E. and Bauld, J. In Press—b. The versatility of *Oscillatoria boryana* (cyanobacterium) in sulfide-rich microbial mats in New Zealand. *FEMS Microbiology Ecology*.

Caster, K. E. 1957. Problematica. *Geological Society of America Memoir* 67: 1025–1032.

Cattolico, R. A. 1986. Chloroplast evolution in algae and land plants. *Trends in Ecology and Evolution* 1: 64–67.

Caumette, P., Baulaigue, R. and Matheron, R. 1988. Characterization of *Chromatium salexigens* sp. nov., a halophilic *Chromatiaceae* isolated from Mediterranean salinas. *Systematics and Applied Microbiology* 10: 284–292.

Cavalier-Smith, T. 1981. The origin and early evolution of the eukaryotic cell. *Symposium of the Society of General Microbiology* 32: 33–84.

Cavalier-Smith, T. 1982. Origins of plastids. *Biological Journal of the Linnean Society* 71: 289–306.

Cavalier-Smith, T. 1986. The kingdom Chromista: Origin and systematics. *Progress in Phycological Research* 4: 319–358.

Cavalier-Smith, T. 1987a. Eukaryotes with no mitochondria. *Nature* 326: 332–333.

Cavalier-Smith, T. 1987b. The origin of cells: A symbiosis between genes, catalysts, and membranes. *Cold Spring Harbor Symposia on Quantitative Biology LII*: 805–824.

Cavalier-Smith, T. 1987c. The origin of eukaryote and archaebacterial cells. In: J. L. Lee and J. F. Frederick (Eds.), *Endocytobiology III*. *Annals of the New York Academy of Sciences* 503: 17–54.

Cavalier-Smith, T. 1987d. The simultaneous symbiotic origin of mitochondria, chloroplasts, and microbodies. *Annals of the New York Academy of Sciences* 503: 55–71.

Cavalier-Smith, T. 1989a. Archaebacteria and Archezoa. *Nature* 339: 100–101.

Cavalier-Smith, T. 1989b. Making a real discovery. *Nature* 342: 870.

Cayeux, M. L. 1894a. Les preuves de l'existence d'organismes dans le terrains precambrien: Premiere note sur les radiolares precambriens. *Bulletin de la Societe Géologique de France* 22: 197–228.

Cayeux, M. L. 1894b. Sur la presence des restes de foraminiferes dans les terrains precambriens de Bretagne. *Société géologique du Nord. Annales* 22: 116–119.

Cayeux, M. L. 1895. De l'existence de nombreux debris de spongiares dans le Precambrien de Bretagne. *Société géologique du Nord. Annales* 23: 52–65.

Cech, T. R. 1985. Self-splicing RNA: Implications for evolution. *International Review of Cytology* 93: 3–22.

Cech, T. R., Zaug, A. J. and Grabowski, P. J. 1981. In vitro splicing of the ribosomal RNA precursor of tetrahymena: Involvement of a guanosine nucleotide in the excision of the intervening sequence. *Cell* 27: 487–496.

Cedergren, R., Gray, M. W., Abel and Sankoff, D. 1988. The evolutionary relationships among known life forms. *Journal of Molecular Evolution* 28: 98–112.

Chalansonnet, S., Largeau, C., Casadevall, E., Berkaloff, C., Peniguel, G. and Couderc, R. 1988. Cyanobacterial resistant biopolymers. Chemical implications of the properties of *Schizothrix* sp. resistant material. In: L. Mattavelli and L. Novelli (Eds.), *Advances in Organic Geochemistry 1987* (Pergamon Press: Oxford), pp. 1003–1010.

Chamberlain, W. M. and Marland, G. 1977. Precambrian evolution in a stratified global sea. *Nature* 265: 135–136.

Chambers, L. A. and Trudinger, P. A. 1978. Microbiological fractionation of stable sulfur isotopes: a review and critique. *Geomicrobiology Journal* 1: 249–293.

Chameides, W. L. and Walker, J. C. G. 1981. Rates of fixation by lightning of carbon and nitrogen in possible primitive atmospheres. *Origins of Life* 11: 291–302.

Chandhuri, S. and Faure, G. 1967. Geochronology of the Keweenawan rock, White Pine, Michigan. *Economic Geology* 62: 1011–1033.

Chang, S. 1979. Cosmic connections with carbonaceous meteorites, interstellar molecules and the origin of life. In: M. Neugebar, D. K. Yeomans and J. C. Brandt (Eds.), *Space Missions to Comets. NASA Conference Publication*, 2089 (NASA Scientific and Technical Information Branch: Washington D.C.), pp. 59–111.

Chang, S. 1988. Planetary environments and the conditions of life. *Philosophical Transactions of the Royal Society of London A* 325: 601–610.

Chang, S., Des Marais, D., Mack, R., Miller, S. L. and Strathearn, G. E. 1983. Prebiotic organic syntheses and the origin of life. Chapter 4. In: J. W. Schopf (Ed.), *Earth's Earliest Biosphere: Its Origin and Evolution* (Princeton University Press: Princeton), pp. 53–92.

Chang, S.-B. R. and Kirschvink, J. L. 1989. Magnetofossils, the magnetization of sediments, and the evolution of magnetite biomineralization. *Annual Review of Earth and Planetary Sciences* 1989: 169–195.

Chang Wentang 1974. *Handbook of Stratigraphy and Palaeontology of Southwestern China* (Science Press: Beijing), 454 pp.

Chapman, D. J. and Schopf, J. W. 1983. Biological and biochemical effects of the development of an aerobic environment. Chapter 13. In: J. W. Schopf (Ed.), *Earth's Earliest Biosphere: Its Origin and Evolution* (Princeton University Press: Princeton), pp. 302–320.

Chapman, D. J. and Trench, R. K. 1982. Prochlorophyceae: Introduction and bibliography. In: J. R. Rosowski and B. C. Parker (Eds.), *Selected Papers in Phycology II* (Phycological Society of America: Lawrence, Kansas), pp. 656–658.

Chapman, F. 1908. On the relationship of the genus *Girvanella*, and its occurrence in the Silurian limestones of Victoria. *Report of the Australian Association for the Advancement of Science (1907-1908)*: 377–386.

Chapman, F. 1916. Report on a probable calcareous alga from the Cambrian limestone breccia found in Antarctica at 85°S. *Rept. Scientific Invest. Geology* 2: 81–84.

Chapman, F. 1927. On a new genus of calcareous algae, from the Lower Cambrian (?), west of Wooltana, South Australia. *Transactions of the Royal Society of South Australia* 51: 123–125.

Chapman, F. 1940. On a new genus of sponges from the Cambrian of the Flinders Range, South Australia. *Transactions of the Royal Society of South Australia* 64(1): 101–108.

Chappe, B., Michaelis, W. and Albrecht, P. 1980. Molecular fossils of archaebacteria as selective degradation products of kerogen. In: A. G. Douglas and J. R. Maxwell (Eds.), *Advances in Organic Geochemistry* (Pergamon Press: Oxford), pp. 265–274.

Chappe, B., Albrecht, P. and Michaelis, W. 1982. Polar lipids of archaebacteria in sediments and petroleums. *Science* 217: 65–66.

Chappe, B., Albrecht, P. and Michaelis, W. 1983. Archaebacterial molecular markers in sediments and petroleums. *Terra Cognita* 3: 216.

Chauhan, D. S. 1979. Phosphate-bearing stromatolites of the Precambrian Aravalli phosphorite deposits of the Udaipur region, their environmental significance and genesis of phosphorite. *Precambrian Research* 8: 95–126.

Chauvel, J. J. and Dimroth, E. 1974. Facies types and depositional environment of the Sokoman Iron Formation, Central Labrador Trough, Quebec, Canada. *Journal of Sedimentary Petrology* 44: 299–327.

Chen, J., Zhang, H., Zhu, S., Zhao, Z. and Wang, Z. 1980. Research on the Sinian Suberathem of Jixian, Tianjin. In: Tianjin Institute of Geology and Mineral Resources (Ed.), *Research on Precambrian Geology, Sinian Suberathem of China* (Tianjin Science and Technology Press: Tianjin), pp. 56–114.

Chen, J.-Y. 1988. Precambrian metazoans of the Huai River drainage area (Anhui, E. China): their taphonomic and ecological evidence. *Senckenbergiana Lethaea* 69: 189–215.

Chen, J.-Y., Hou, X.-G. and Lu, H.-Z. 1989a. [Early Cambrian netted scale-bearing worm-like sea animal]. *Acta Palaeontologica Sinica* 28: 1–16.

Chen, J.-Y., Hou, X.-G. and Lu, H.-Z. 1989b. [Lower Cambrian Leptomitids (Demospongea), Chengjiang, Yunnan]. *Acta Palaeontologica Sinica* 28: 17–31.

Chen, M. and Cao, R. 1966. [Note on a new species of ancient fossil algae from the Sinian Dengying Formation of Eastern Yunnan]. *Scientia Geologica Sinica* 2: 185–188.

Chen, M. and Liu, K. 1986. The geological significance of newly discovered microfossils from the upper Sinian (Doushantuo age) phosphorites. *Scientia Geologica Sinica* 1: 46–53.

Chen, M. and Zheng, W. 1986. On the pre-Ediacaran Huainan biota. *Scientia Geologica Sinica* 7(3): 221–231 (In Chinese, with English abstract).

Chen, M.-G. 1979. [Some skeletal fossils from the phosphatic sequence, early Lower Cambrian, south China]. *Scientia Geologica Sinica* 4: 187–189.

Chen, M.-G. 1982. [The new knowledge of the fossil assemblages from Maidiping section, Emei County, Sichuan with reference to the Sinian-Cambrian boundary]. *Scientia Geologica Sinica* 1982: 253–262.

Chen, M.-G. and Liu, K.-W. 1986. The geological significance of newly discovered microfossils from the Upper Sinian (Doushantuo age) phosphorites. *Scientia Geologica Sinica* 1986(1): 46–53.

Chen, M.-G., Chen, Y.-Y. and Qian, Y. 1981a. [Some tubular fossils from Sinian-Lower Cambrian boundary sequences, Yangtze Gorge]. *Bulletin of the Tianjin Institute of Geology and Mineral Resources* 34: 117–124.

Chen, M.-G., Chen, Y.-Y. and Zhang, S.-S. 1981b. The small shell fossil assemblage in the limestone of the uppermost part of Dengying Formation at Songlingpo, Yichang. *Chikyu Kagaku [Earth Science], Journal of the Association for Geological Collaboration in Japan* 1(14): 32–41.

Chen, M.-W., Anne, J., Volkaert, G., Huysmans, E., Vandenberghe, A. and De Wachter, R. 1984. Nucleotide sequence of the 5S rRNAs of seven molds and a yeast and their use in studying ascomycete phylogeny. *Nucleic Acids Research* 12: 4881–4892.

Chen, Y., Zhang, S., Liu, G., Xiong, X., Chen, P., et al. 1983. [The Sinian-Cambrian boundary in eastern Yunnan]. *The Sinnian-Cambrian Boundary of China. Bulletin of the Institute of Geology, Chinese Academy of Geological Sciences* 10: 14–35 (IGCP Project No. 29).

Chen, Y.-Y. and Zhang, S.-S. 1980. *Chikyu Kagaku [Earth Science], Journal of the Association for Geological Collaboration in Japan* 26: 190–197.

Cherchi, A. and Schroeder, R. 1985a. Intergrowth of the Early Cambrian algae *Epiphyton* Bornemann and *Renalcis* Vologdin from SW Sardinia. *Bollettino della Società Paleontologica Italiana* 23(2): 141–147.

Cherchi, A. and Schroeder, R. 1985b. Middle Cambrian foraminifera and other microfossils from SW Sardinia. *Bollettino della Società Paleontologica Italiana* 23(2): 149–160.

Chevé, S. R. and Machado, N. 1988. Reinvestigation of the Castignon Lake carbonatite complex, Labrador Trough, New Quebec. *Geological Association of Canada Program with Abstracts* 13: 20.

Christie-Blick, N. and Biddle, K. T. 1985. Deformation and basin formation along strike slip faults. In: K. T. Biddle and N. Christie-Blick (Eds.), *Strike-slip Deformation, Basin, and Sedimentation. Society of Economic Paleontologists and Mineralogists Special Publication* 37: 1–34.

Christopher, R. K., Duffield, A. M. and Ralph, B. J. 1980. Identification of some neutral lipids of *Thiobacillus thioparus* using gas chromatography-chemical ionization mass spectrometry. *Australian Journal of Biological Science* 33: 737–741.

Chudinova, I. I. 1959. O nakhodke konulyarii v nizhnem kembrii zapadnykh Sayan [On the discovery of conularians in the Lower Cambrian of western Sayan]. *Paleontologicheskij Zhurnal* 1959(2): 53–55.

Chumakov, N. M. and Elston, D. P. 1989. The paradox of Late Proterozoic glaciations at low latitudes. *Episodes* 12: 115–120.

Chumakov, N. M. and Semikhatov, M. A. 1981. Riphean and Vendian of the USSR. *Precambrian Research* 15(3–4): 229–253.

Chuvashov, B. I. and Riding, R. 1984. Principal floras of Paleozoic marine calcareous algae. *Palaeontology* 27: 487–500.

Chuvashov, B. I., Yuferev, O. V. and Luchinina, V. A. 1985. Vodorosli srednego i verkhnego Devona zapadnoj Sibiri i Urala [Algae of the

References

Middle and Upper Devonian of western Siberia and the Urals]. In: V. N. Dubatolov (Ed.), *Biostratigrafiya Paleozoya Zapanoj Sibiri* [*Paleozoic Biostratigraphy of Western Siberia*]. *Trudy Inst. Geol. Geofiz. Sibirsk. Otd. Akad. Nauk SSSR* 619 (Nauka: Novosibirsk), pp. 72–98.

Chuvashov, B. I., Luchinina, V. A., Shujskij, V. P., Shajkin, I. M., Berchenko, O. I., Ishchenko, A., Saltovskaya, V. D. and Shirsova, D. I. 1987. Iskopaemye izvestkovye vodorosli [Fossil calcareous algae]. *Trudy Institut Geologii i Geofiziki Sibirskoe Otdolenie, Akademiya Nauk SSSR* 674: 1–225.

Ciferri, O. 1983. *Spirulina*, the edible micoorganism. *Microbiological Reviews* 47: 551–578.

Cisne, J. L. 1974. Trilobites and the origin of arthropods. *Science* 186: 13–18.

Clark, C. G. and Cross, G. A. M. 1988. Small-subunit ribosomal RNA sequence from *Naegleria gruberi* supports the polyphyletic origin of amoebas. *Molecular Biology and Evolution* 5: 512–518.

Clark, L. C., Wolf, R., Granger, D. and Taylor, A. 1953. Continuous recording of blood oxygen tension by polarography. *Journal of Applied Physiology* 6: 189–193.

Clark, R. B. 1964. *Dynamics in Metazoan Evolution. The Origin of the Coelom and Segments* (Clarendon Press: Oxford), 313 pp.

Clark, R. B. 1979. Radiation of the Metazoa. In: M. R. House (Ed.), *The Origin of Major Invertebrate Groups* (Academic Press: London), pp. 55–102.

Clark, E. de C. and Teichert, C. 1946. Algal structures in a Western Australia salt lake. *American Journal of Science* 244: 271–276.

Clauer, N., Caby, R., Jeanette, D. and Trompette, R. 1982. Geochronology of sedimentary and metasedimentary Precambrian rocks of the West African Craton. *Precambrian Research* 18: 53–71.

Claypool, G. E., Holser, W. T., Kaplan, I. R., Sakai, H. and Zak, I. 1980. The age curves of sulfur and oxygen isotopes in marine sulfate and their mutual interpretation. *Chemical Geology* 28: 199–260.

Clayton, R. K. 1966. Spectroscopic analysis of bacteriochlorophylls *in vitro* and *in vivo*. *Photochemistry and Photobiology* 5: 669–677.

Clayton, R. K. and Sistrom, W. R. (Eds.). 1978. *The Photosynthetic Bacteria* (Plenum: New York), 946 pp.

Clemmey, H. 1976. World's oldest animal traces. *Nature* 261: 576–578.

Clemmey, H. and Badham, N. 1982. Oxygen in the Precambrian atmosphere: an evaluation of the geological evidence. *Geology* 10: 141–146.

Cline, J. D. 1969. Spectrophotometric determination of hydrogen sulfide in natural waters. *Limnology and Oceanography* 14: 454–458.

Cloud, P. 1942. Notes on stromatolites. *American Journal of Science* 240: 363–379.

Cloud, P. 1948. Some problems and patterns of evolution exemplified by fossil invertebrates. *Evolution* 2: 322–350.

Cloud, P. 1965. Significance of the Gunflint (Precambrian) microfloro. *Science* 148: 27–35.

Cloud, P. 1968. Atmospheric and hydrospheric evolution on the primitive earth. *Science* 160: 729–736.

Cloud, P. 1972. A working model of the primitive earth. *American Journal of Science* 272: 537–548.

Cloud, P. 1973. Paleoecological significance of the banded iron-formation. *Economic Geology* 68: 1135–1143.

Cloud, P. 1974. Evolution of ecosystems. *American Scientist* 62: 54–66.

Cloud, P. 1976a. Beginnings of biospheric evolution and their biogeochemical consequences. *Paleobiology* 2: 351–387.

Cloud, P. 1976b. Major features of crustal evolution. *Geological Society of South Africa Special Publication* 79(annexe): 1–32.

Cloud, P. 1980. Early biogeochemical systems. In: P. A. Trudinger, M. R. Walter and B. J. Ralph (Eds.), *Biogeochemistry of Ancient and Modern Environments* (Australian Academy of Science: Canberra), pp. 7–21.

Cloud, P. 1983. Early biogeologic history: The emergence of a paradigm. Chapter 1. In: J. W. Schopf (Ed.), *Earth's Earliest Biosphere: Its Origin and Evolution* (Princeton University Press: Princeton), pp. 14–31.

Cloud, P. 1985. Vestiges of a beginning. *Geological Society of America Centennial Special* 1: 151–156.

Cloud, P. 1986. Reflections on the beginnings of metazoan evolution. *Precambrian Research* 31(4): 405–408.

Cloud, P. 1987. Trends, transitions, and events in cryptozoic history and their calibration: apropos recommendations by the Subcommission on Precambrian Stratigraphy. *Precambrian Research* 37: 257–264.

Cloud, P. and Abelson, P. H. 1961. Woodring conference on major biologic innovations and the geologic record. *Proceedings of the National Academy of Science USA* 47: 1705–1712.

Cloud, P. and Hagen, H. 1965. Electron microscopy of the Gunflint microflora: preliminary results. *Proceedings of the National Academy of Science USA* 54: 1–8.

Cloud, P. and Morrison, K. 1980. New microbial fossils from 2 Gyr old rocks in northern Michigan. *Geomicrobiology Journal* 2: 161–178.

Cloud, P., Awramik, S. M., Morrison, K. and Hadley, D. G. 1979. Earliest Phanerozoic or latest Proterozoic fossils from the Arabian Shield. *Precambrian Research* 10: 73–93.

Cloud, P., Gustafson, L. B. and Watson, J. A. L. 1980. The works of living social insects as pseudofossils and the age of the oldest known Metazoa. *Science* 210: 1013–1015.

Cloud, P. E. 1960. Gas as a sedimentary and diagenetic agent. *American Journal of Science* 258A: 35–45.

Cloud, P. E. 1968. Pre-Metazoan evolution and the origins of the Metazoa. In: E. T. Drake (Ed.), *Evolution and Environment* (Yale University Press: New Haven), pp. 1–72.

Cloud, P. E. 1983. Are the Medicine Peak Quartzite dubiofossils fluid-evasion tracks? *Geology* 11: 618–619.

Cloud, P. E. and Dardenne, M. 1973. Proterozoic Age of the Bambuí Group in Brazil. *Geological Society of America Bulletin* 84: 1673–1676.

Cloud, P. E. and Glaessner, M. F. 1982. The Ediacarian Period and System. Metazoa inherit the earth. *Science* 217: 783–792.

Cloud, P. E. and Nelson, C. A. 1966. *Pteridinium* and the Precambrian-Cambrian boundary. *Science* 154: 766–770.

Cloud, P. E., Wright, J. and Glover, L. 1976. Traces of animal life from 620-million-year-old rocks in North Carolina. *American Scientist* 64: 396–406.

Cloud, P. E., Jr. and Semikhatov, M. A. 1969. Proterozoic stromatolite zonation. *American Journal of Science* 267: 1017–1061.

Cobbold, E. S. 1919. Cambrian Hyolithidae, etc., from Hartshill in the Nuneaton District, Warwickshire. *Geological Magazine* 56: 149–158.

Cobbold, E. S. 1921. The Cambrian horizons of Comley (Shropshire) and their Brachiopoda, Pteropoda, Gasteropoda, etc. *Quarterly Journal of the Geological Society of London* 76: 325–386.

Cobbold, E. S. 1935. Lower Cambrian fauna from Herault, France. *Annals and Magazine of Natural History* 16: 25–48.

Cobbold, E. S. and Popcock, R. W. 1934. The Cambrian area of Rushton, Shropshire. *Philosophical Transactions of the Royal Society of London B* 223: 305–409.

Cohen, Y. 1984a. Oxygenic photosynthesis, anoxygenic photosynthesis and sulfate reduction in cyanobacterial mats. In: M. J. Klug and C. A. Reddy (Eds.), *Current Perspectives in Microbial Ecology* (American Society for Microbiology: Washington, D.C.), pp. 435–441.

Cohen, Y. 1984b. The Solar Lake cyanobacterial mats: strategies of photosynthetic life under sulfide. In: Y. Cohen, R. W. Castenholz and H. O. Halvorson (Eds.), *Microbial Mats: Stromatolites* (Alan R. Liss, Inc: New York), pp. 133–148.

Cohen, Y. and Rosenberg, E. (Eds.). 1989. *Microbial Mats: Physiological Ecology of Benthic Microbial Communities* (American Society for Microbiology: Washington, D.C.), 494 pp.

Cohen, Y., Padan, E. and Shilo, M. 1975. Facultative anoxygenic photosynthesis in the cyanobacterium *Oscillatoria limnetica*. *Journal of Bacteriology* 123: 855–861.

Cohen, Y., Aizenshtat, Z., Stoler, A. and Jørgensen, B. B. 1980. The microbial geochemistry of Solar Lake, Sinai. In: P. A. Trudinger and M. R. Walter (Eds.), *Biogeochemistry of Ancient and Modern*

Environments (Australian Academy of Science: Canberra), pp. 167–171.
Cohen, Y., Castenholz, R. W. and Halvorson, H. O. (Eds.). 1984. *Microbial Mats: Stromatolites* (Alan R. Liss: New York), 498 pp.
Cohen, Y., Jørgensen, B. B., Revsbech, N. P. and Poplawski, R. 1986. Adaptation to hydrogen sulfide of oxygenic and anoxygenic photosynthesis among cyanobacteria. *Applied and Environmental Microbiology* 51(2): 398–407.
Cohen-Bazire, G. and Bryant, D. A. 1982. Phycobilisomes: composition and structure. In: N. G. Carr and B. A. Whitton (Eds.), *The Biology of Cyanobacteria* (University of California Press: Berkeley), pp. 143–190.
Cohen-Bazire, G. and Sistrom, W. R. 1966. The prokaryotic photosynthetic apparatus. In: L. P. Vernon and G. R. Seely (Eds.), *The Chlorophylls* (Academic Press: New York), pp. 313–341.
Colbath, G. K. 1983. Fossil prasinophycean phycomata (Chlorophyta) from the Silurian Bainbridge Formation, Missouri, U.S.A. *Phycologia* 22: 249–265.
Coleman, A. W. 1985. Cyanophyte and cyanelle DNA: A search for the origins of plastids. *Journal of Phycology* 21: 371–379.
Coleman, D. C., Anderson, R., Cole, C., Elliott, E. T., Woods, L. and Campion, M. K. 1978. Trophic interactions in soils as they affect energy and nutrient dynamics. IV. Flows of metabolic and biomass carbon. *Microbial Ecology* 4: 373–380.
Collins, N. C., Mitchell, R. and Wiegert, R. G. 1976. Functional analysis of a thermal spring ecosystem, with an evaluation of the role of consumers. *Ecology* 57(6): 1221–1232.
Compston, W. and Pidgeon, R. T. 1986. Jack Hills, evidence of more very old detrital zircons in Western Australia. *Nature* 321: 766–769.
Compston, W., Williams, I. S., Campbell, I. H. and Gresham, J. 1985. Zircon xenocrysts from the Kambalda volcanics: age constraints and direct evidence of older continental crust below the Kambalda-Norseman greenstones. *Earth and Planetary Science Letters* 76: 299–311.
Compston, W., Williams, I. S., Jenkins, R. J. F., Gostin, V. A. and Haines, P. W. 1987. Zircon age evidence for the late Precambrian Acraman ejecta blanket. *Australian Journal of Earth Sciences* 34: 435–445.
Condie, K. C. 1980. Origin and early development of the earth's crust. *Precambrian Research* 11: 183–197.
Condie, K. C. and Allen, P. 1984. Origin of Archaean charnockites from southern India. In: A. Kröner, G. N. Hanson and A. M. Goodwin (Eds.), *Archaean Geochemistry* (Springer-Verlag: Berlin), pp. 182–203.
Condie, K. C. and Martell, C. 1983. Early Proterozoic metasediments from north-central Colorado: Metamorphism, provenance, and tectonic setting. *Geological Society of America Bulletin* 94: 1215–1224.
Condie, K. C. and Moore, J. C., Jr. 1977. Geochemistry of Proterozoic volcanic rocks from the Grenville Province, eastern Ontario. In: W. R. A. Baragar, L. C. Coleman and J. M. Hall (Eds.), *Volcanic Regimes in Canada. Geological Association of Canada Special Paper* 83-1B: 243–252.
Condie, K. C. and Shadel, C. A. 1984. An Early Proterozoic volcanic arc succession in southeastern Wyoming. *Canadian Journal of Earth Sciences* 21: 415–427.
Coniglio, M. and James, N. P. 1985. Calcified algae as sediment contributors to Early Paleozoic limestones: evidence from deep-water sediments of the Cow Head Group, western Newfoundland. *Journal of Sedimentary Petrology* 55(5): 746–754.
Connan, J. and Restle, A. 1984. La biodegradation des hydrocarbures dans les reservoirs. *Centre de Recherche Exploration-Production Elf-Aquitaine, Bulletin* 8: 291–302.
Constantz, B. R. 1986. Coral skeleton construction: a physiochemically dominated process. *Palaios* 1: 152–157.
Conway Morris, S. 1977a. Fossil priapulid worms. *Palaeontological Association. Special Papers in Palaeontology* 20: 1–95.
Conway Morris, S. 1977b. A new entoproct-like organism from the Burgess Shale of British Columbia. *Palaeontology* 20: 833–845.
Conway Morris, S. 1979. Middle Cambrian polychaetes from the Burgess Shale of British Columbia. *Philosophical Transactions of the Royal Society of London* B285: 227–274.
Conway Morris, S. 1985a. The Ediacaran biota and early metazoan evolution. *Geological Magazine* 122(1): 77–81.
Conway Morris, S. 1985b. The Middle Cambrian Metazoan *Wiwaxia corrugata* (Matthew) from the Burgess Shale and *Ogygopsis* Shale, British Columbia, Canada. *Philosophical Transactions of the Royal Society of London* B307: 507–586.
Conway Morris, S. 1985c. Non-skeletalized lower invertebrate fossils: a review. In: S. Conway Morris, J. D. George, R. Gibson and H. M. Platt (Eds.), *The Origins and Relationships of Lower Invertebrates. The Systematics Association Special Volume* 28 (Clarendon Press: Oxford), pp. 343–359.
Conway Morris, S. 1986. The community structure of the Middle Cambrian Phyllopod Bed (Burgess Shale). *Palaeontology* 29(3): 423–467.
Conway Morris, S. 1987. The search for the Precambrian-Cambrian boundary. *American Scientist* 75: 157–167.
Conway Morris, S. 1989a. Burgess Shale faunas and the Cambrian explosion. *Science* 246: 339–346.
Conway-Morris, S. 1989b. Early Metazoans. *Science Progress* 73: 81–99.
Conway Morris, S. 1989c. The persistence of Burgess Shale-type faunas: implication for the evolution of deeper water faunas. *Transactions of the Royal Society of Edinburgh* 80: 271–284.
Conway Morris, S. 1989d. Radiometric dating of the Precambrian-Cambrian boundary in the Avalon Zone. *New York State Museum and Geological Survey Bulletin* 463: 53–58.
Conway Morris, S. 1989e. South-eastern Newfoundland and adjacent areas (Avalon Zone). In: J. W. Cowie and M. D. Brasier (Eds.), *The Precambrian-Cambrian Boundary* (Oxford Science Publishers, Clarendon Press: Oxford), pp. 7–39.
Conway Morris, S. and Bengtson, S. 1986. The Precambrian-Cambrian boundary and geochemical perturbations. *Nature* 319: 696–697.
Conway Morris, S. and Chen Menge 1990. *Blastulospongia polytreta* n. sp., an enigmatic organism from the Lower Cambrian of Hubei, China. *Journal of Paleontology* 64: 26–30.
Conway Morris, S. and Chen Menge In Prep. Anabaritids from the Lower Cambrian of South China.
Conway Morris, S. and Chen Menge In Press. Cambroclaves and paracarinachitids, early skeletal problematica from the Lower Cambrian of South China. *Palaeontology*.
Conway Morris, S. and Fritz, W. H. 1980. Shelly microfossils near the Precambrian-Cambrian boundary, Mackenzie Mountains, northwestern Canada. *Nature* 286(577): 381–384.
Conway Morris, S. and Jenkins, R. J. F. 1985. Healed injuries in Early Cambrian trilobites from South Australia. *Alcheringa* 9: 167–177.
Conway Morris, S. and Robison, R. A. 1982. The enigmatic medusoid *Peytoia* and a comparison of some Cambrian biotas. *Journal of Paleontology* 56: 116–122.
Conway Morris, S., George, J. D., Gibson, R. and Platt, H. M. (Eds.). 1985. *The Origins and Relationships of Lower Invertebrates. Systematics Association Special Volume* 28 (Clarendon Press: Oxford), 397 pp.
Conway Morris, S., Peel, J. S., Higgins, A. K., Soper, N. J. and Davis, N. C. 1987. A Burgess shale-like fauna from the Lower Cambrian of North Greenland. *Science* 326: 181–183.
Cook, P. J. and McElhinny, M. W. 1979. A reevaluation of the spatial and temporal distribution of sedimentary phosphate deposits in light of plate tectonics. *Economic Geology* 74: 315–330.
Cook, P. J. and Shergold, J. H. 1984. Phosphorus, phosphorites and skeletal evolution at the Precambrian-Cambrian boundary. *Nature* 308: 231–236.
Cook, P. J. and Shergold, J. H. 1986. Proterozoic and Cambrian phosphorites—nature and origin. In: P. J. Cook and J. H. Shergold (Eds.), *Phosphate Deposits of the World. 1. Proterozoic and Cambrian Phosphorites* (Cambridge University Press: Cambridge), pp. 369–386.
Cooper, G. A. 1942. New genera of North America brachiopods.

References

Washington Academy of Science Journal 32: 228–235.

Cooper, G. A. 1951. New brachiopods from the Lower Cambrian of Virginia. *Washington Academy of Science Journal* 41: 4–8.

Cooper, G. A. 1976. Lower Cambrian brachiopods from the Rift Valley (Israel and Jordan). *Journal of Paleontology* 50: 269–289.

Cooper, J. A., James, P. R. and Rutland, R. W. R. 1982. Isotopic dating and structural relationships of granitoids and greenstones in the East Pilbara, Western Australia. *Precambrian Research* 18: 199–236.

Cope, J. W. 1977. An Ediacara-type fauna from South Wales. *Nature* 268: 624.

Cope, J. W. 1983. Precambrian fossils of the Carmarthen area, Dyfed. *Nature in Wales* 1(2): 11–16.

Coplen, T. B. and Kendall, C. 1982. Preparation and Stable Isotope Determination of NBS-16 and NBS-17 Carbon Dioxide Reference Samples. *Analytical Chemistry* 54: 2611–2612.

Corfu, F. and Andrews, A. J. 1986. A U-Pb age for mineralized Nipissing diabase, Gowganda, Ontario. *Canadian Journal of Earth Sciences* 23: 107–109.

Corliss, J. B., Lyle, M., Dymond, J. and Krane, K. 1978. The chemistry of hydrothermal mounds near the Galapagos Rift. *Earth and Planetary Science Letters* 40: 12–24.

Corliss, J. B., Baross, J. A. and Hoffman, S. E. 1981. An hypothesis concerning the relationship between submarine hot springs and the origin of life on earth. *Suppl. C4. Oceanologica Acta* 4: 59–69.

Costerton, J. W., Irvin, R. T. and Cheng, K. J. 1981. The bacterial glycocalyx in nature and disease. *Annual Review of Microbiology* 35: 229–324.

Courjault-Radé, P. 1987. Evolution tectono-sédimentaire du cambrien Supérieur et des couches de transition avec l'Ordovicien inférieur (Trémadoc) dans la nappe du Minervios (versant sud de la Montagne Noire, France). *Comptes Rendus de l'Academie des sciences Paris, Serie D* 305(Série II): 293–296.

Courjault-Radé, P. 1988. Analyse sédimentologique de la formation de l'Orbiel ("alternances gréso-calcaires" auct., Cambrien inférieur). Evolution tectono-sédimentaire et climatique (versant sud de la montagne Noire, Massif central, France). *Bulletin de la Societe Géologique de France* IV(6): 1003–1013.

Cowie, J. 1967. Life in Pre-Cambrian and early Cambrian time. In: *The Fossil Record, a Symposium with Documentation* (Geological Society of London: London), pp. 17–35.

Cowie, J. W. 1985. Continuing work on the Precambrian-Cambrian boundary. *Episodes* 8(2): 93–97.

Cowie, J. W. and Brasier, M. D. (Eds.). 1989. *The Precambrian-Cambrian Boundary* (Clarendon Press: Oxford), 213 pp.

Cowie, J. W. and Harland, W. B. 1989. Chronometry. In: J. W. Cowie and M. D. Brasier (Eds.), *The Precambrian-Cambrian Boundary. Oxford monographs on geology and geophysics*, 12 (Clarendon Press: Oxford), pp. 186–198.

Cowie, J. W. and Johnson, M. R. W. 1985. Late Precambrian and Cambrian geological time scale. *The Chronology of the Geological Records. Geological Society of London Memoir* 10: 47–65.

Cowie, J. W., Ziegler, W. and Remane, J. 1989. Stratigraphic commission accelerates progress, 1984–1989. *Episodes* 12: 79–80.

Crank, J. 1975. *The Mathematics of Diffusion* (Clarendon Press: Oxford), 414 pp.

Cranwell, P. A. 1976. Decomposition of aquatic biota and sediment formation: lipid components of two blue-green algal species and of detritus resulting from microbial attack. *Freshwater Biology* 6: 481–488.

Crawford, A. R. and Compston, W. 1973. The age of the Cuddapah and Kurnool Systems, southern India. *Journal of the Geological Society of Australia* 19: 453–464.

Crawford, A. R. and Daily, B. 1971. Probable non-synchroneity of late Precambrian glaciations. *Nature* 230: 111–112.

Crick, I. H., Boreham, C. J., Cook, A. C. and Powell, T. G. 1988. Petroleum geology and geochemistry of the Middle Proterozoic McArthur Basin, North Australia. II. Assessment of source rocks. *American Association of Petroleum Geologists Bulletin* 72: 1495–1514.

Crimes, T. P. 1974. Colonization of the early ocean floor. *Nature* 248: 328–330.

Crimes, T. P. 1982. Trace fossils from the Nama Group (Precambrian-Cambrian) of Southwest Africa (Namibia). *Journal of Paleontology* 56: 890–907.

Crimes, T. P. 1987. Trace fossils and fossils and correlation of late Precambrian and early Cambrian Strata. *Geological Magazine* 124(2): 97–119.

Crimes, T. P. 1989. Trace fossils. In: J. W. Cowie and M. D. Brasier (Eds.), *The Precambrian-Cambrian Boundary* (Clarendon Press: Oxford), pp. 166–185.

Crimes, T. P. and Anderson, M. M. 1985. Trace fossils from Late Precambrian-Early Cambrian strata of southeastern Newfoundland (Canada): temporal and environmental implications. *Journal of Paleontology* 59(2): 310–343.

Crimes, T. P. and Jiang Zhiwen 1986. Trace fossils from the Precambrian-Cambrian boundary candidate at Meishucun, Jinning, Yunnan, China. *Geological Magazine* 123: 641–649.

Crimes, T. P., Legg, I., Marcos, A. and Arboleya, M. 1976. Late Precambrian-Lower Cambrian trace fossils from Spain. In T. P. Crimes and J. C. Harper (Eds.), *Trace Fossils 2* (Seel House Press: Liverpool), pp. 91–138.

Crowell, J. C. 1974. Origin of late Cenozoic basins in southern California. In: W. R. Dickinson (Ed.), *Tectonics and Sedimentation. Society of Economic Paleontologists and Mineralogists Special Publication* 22: 190–204.

Crowell, J. C. 1983. Ice ages recorded on Gondwanan continents. *Transactions of the Geological Society of South Africa* 86: 237–262.

Crowley, T. J. 1983. The geologic record of climatic change. *Reviews of Geophysics and Space Physics* 21: 828–877.

Crumpton, W. G. 1987. A simple and reliable method for making permanent mounts of phytoplankton for light and fluorescence microscopy. *Limnology and Oceanography* 32: 1154–1158.

Cuffey, R. J. 1988. Incomplete thin-lamina mudcracks and reported "annelid burrows" in the Proterozoic Siyeh (Helena) Limestone on Dawson Pass. *Geological Society of America Abstracts with Programs* 20(7): A227.

Culver, S. J., Buzas, M. A. and Collins, L. S. 1987. On the value of taxonomic standardization in evolutionary studies. *Paleobiology* 13: 169–176.

Cyr, T. D., Payzant, J. D., Montgomery, D. S. and Strausz, O. P. 1986. A homologous series of novel hopane sulfides in petroleum. *Organic Geochemistry* 9: 139–143.

Daily, B. 1972. The base of the Cambrian and the first Cambrian faunas. *University of Adelaide, Centre for Precambrian Research, Special Paper* 1: 13–41.

Daily, B. 1973. Discovery and significance of basal Cambrian Uratanna Formation, Mt. Scott Range, Flinders Ranges, South Australia. *Search* 4: 202–205.

Daily, B. 1976. The base of the Cambrian in Australia. *International Geological Congress* 25(3): 857 (Abstract).

Dale, B. 1983. Dinoflagellate resting cysts: "benthic plankton". In: G. A. Fryxell (Ed.), *Survival Strategies of the Algae* (Cambridge University Press: Cambridge), pp. 69–136.

Dalrymple, D. W. 1965. Calcium carbonate deposition associated with blue-green algal mats, Baffin Bay, Texas. *Publications of the Institute for Marine Science, Austin* 10: 187–200.

Daly, R. A. 1907. The limeless oceans of Precambrian time. *American Journal of Science* 23: 93–115.

D'Amelio, E., Cohen, Y. and Des Marais, D. J. 1989. Comparative functional ultrastructure of two hypersaline submerged cyanobacterial mats: Guerrero Negro, Baja California Sur, Mexico, and Solar Lake, Sinai, Egypt. Chapter 9. In: Y. Cohen and E. Rosenberg (Eds.), *Microbial Mats: Physiological Ecology of Benthic Microbial Communities* (American Society for Microbiology: Washington, D.C.), pp. 97–113.

D'Amelio, E. D., Cohen, Y. and Des Marais, D. J. 1987. Association of a new type of gliding, filamentous, purple phototrophic bacterium inside bundles of *Microcoleus chthonoplastes* in hypersaline cyanobacterial mats. *Archives of Microbiology* 147: 213–220.

Dangeard, L. 1957. Algues microscopiques a structure conservee dans le Cambrien de Carteret, Manche [Microscopic algal structures preserved in the Cambrian of Carteret, Manche]. *Bulletin de la Société linnéenne de Normandie, Série 9* 8: 54–55 (1954–1955).

Dangeard, L. and Doré, F. 1958. Observations nouvelles sur les Algues et les stromatolithes du Cambrien de Carteret (Manche) [New observations on Cambrian algae and stromatolites of Carteret (Manche)]. *Bulletin de la Societe Géologique de France* 7(6th series): 1069–1075.

Danielli, H. M. C. 1981. The fossil alga *Girvanella* Nicholson & Etheridge. *Bulletin of the British Museum (Natural History), Geological Series* 35(2): 79–107.

Danner, W. R. 1955. Some fossil worm tubes of western Washington. *Rocks and Minerals*: 451–458.

Darwin, C. 1859. *On the Origin of Species by Means of Natural Selection* (John Murray: London), 490 pp.

Darwin, C. 1895. *The Origin of Species. 6th edition, 1902* (John Murray: London), 703 pp.

Dasgupta, A., Ayanoglu, E. and Djerassi, C. 1984. Phospholipid studies of marine organisms: New branched fatty acids from *Stronylophora durissima*. *Lipids* 19: 768–776.

Davey, A. 1983. Effects of abiotic factors on nitrogen fixation by blue-green algae in Antarctica. *Polar Biology* 2: 95–100.

Davey, A. and Marchant, H. J. 1983. Seasonal variation in nitrogen fixation by *Nostoc commune* Vaucher at the Vestfold Hills, Antarctica. *Phycologia* 22: 377–385.

Davidson, C. F. and Cosgrove, M. E. 1955. On the impersistence of uraninite as a detrital mineral. *Geological Survey of Great Britain Bulletin* 10: 74–80.

Davies, G. R. 1970. Algal-laminated sediments, Gladstone Embayment, Shark Bay, Western Australia. *American Society of Petroleum Geologists Memoir* 13: 169–205.

Davis, C. C. 1966. Notes on the ecology and reproduction of *Trichocorixa reticulata* in a Jamaican salt-water pool. *Ecology* 46: 850–852.

Davis, D. W., Corfu, F. and Krogh, T. E. 1986. High precision U-Pb geochronology and implications for the tectonic evolution of the Superior Province. In: M. J. deWitt and L. D. Ashwal (Eds.), *Workshop on Tectonic Evolution of Greenstone Belts. Lunar and Planetary Institute Technical Report* 86-10: 77–79.

Davis, J. S. 1972. Survival records in the algae, and the survival role of certain algal pigments, fat, and mucilagenous substances. *The Biologist* 54: 52–93.

Dawes, C. J. 1981. *Marine Botany* (John Wiley & Sons: New York), 628 pp.

Dawes, E. A. 1986. *Microbial Energetics* (Blackie & Sons: Glasgow), 187 pp.

Dawson, J. W. 1866. On supposed burrows of worms in the Laurentian rocks of Canada. *Quarterly Journal of the Geological Society of London* 22: 608–609.

Dawson, J. W. 1875. *The Dawn of Life* (Hodder and Stoughton: London), 239 pp.

Dawson, J. W. 1890. On burrows and tracks of invertebrate animals in Paleozoic rocks and other markings. *Quarterly Journal of the Geological Society of London* 46: 595–618.

Dawson, W. 1896. Note on Cryptzoon and other ancient fossils. *The Canadian Record of Science* 7(4): 202–219.

Dayhoff, M.-O. and Schwartz, R. M. 1981. Evidence on the origin of eukaryotic mitochondria from protein and nucleic acid sequence. *Annals of the New York Academy of Sciences* 361: 92–103.

Deamer, D. 1986. Role of amphiphilic compounds in the evolution of membrane structure on the early earth. *Origins of Life* 17: 3–26.

Dean, W. E. and Eggleston, J. R. 1975. Comparative anatomy of marine and freshwater algal reefs, Bermuda and central New York. *Geological Society of America Bulletin* 86: 665–676.

Debrenne, F. 1975. Formations organogènes du Cambrien inférieur du Maroc [Organic structures from the Cambrian of Morocco]. In: B. S. Sokolov (Ed.), *Drevnie Cnidaria [Ancient Cnidaria]* (Nauka, Sibirskoe Otdelenie: Novosibirsk), pp. 19–24.

Debrenne, F. and Courjault-Radé, P. 1986. Découverte de faunules d'Archéocyathes dans l'Est des monts de Lacaune, flanc nord de la Montagne Noire. Implications biostratigraphiques [Discovery of Archaeocyathan fauna in the east of monts de Lacaune, northern flank of Montagne Noire]. *Bulletin de la Societe Géologique de France* 2(2): 285–292.

Debrenne, F. and Debrenne, M. 1978. Archaeocyatid fauna of the lowest fossiliferous levels of Tiout (Lower Cambrian, Southern Morocco). *Geological Magazine* 115: 101–119.

Debrenne, F. and Lafuste, J. G. 1979. *Buschmannia roeringi* (Kaever & Richter 1976) a so-called archaecyatha, and the problem of the Precambrian or Cambrian age of the Nama System (S. W. Africa). *Geological Magazine* 116: 143–144.

Debrenne, F. and Naud, G. 1981. Méduse et traces fossiles supposées précambriennes dans la formation de San Vito, Sarrabus, Sud-Est de la Sardaigne. *Bulletin de la Societe Géologique de France* 23: 23–31.

Debrenne, F. and Rozanov, A. 1983. Paleogeographic and stratigraphic distribution of regular Archaeocyatha (Lower Cambrian fossils). *Geobios* 16: 727–736.

Debrenne, F. and Vacelet, J. 1984. Archaeocyatha: is the sponge model consistent with their structural organization? *Palaeontographica Americana* 54: 358–369.

Debrenne, F., Termier, H. and Termier, G. 1970. Radiocyatha. Une nouvelle classe d'organismes primitifs du Cambrien inférieur. *Bulletin de la Societe Géologique de France* 12: 120–125.

Debrenne, F., Termier, H. and Termier, G. 1971. Sur de nouveaux représentans de la classe des *Radiocyatha*. Essai sur l'évolution des Métazoaires primitifs. *Bulletin de la Societe Géologique de France* 13(7 ser): 439–444.

Debrenne, F., Rozanov, A. Y. and Webers, G. F. 1984. Upper Cambrian Archaeocyatha from Antarctica. *Geological Magazine* 121: 291–299.

Decho, A. W. and Castenholz, R. W. 1986. Spatial patterns and feeding of meiobenthic harpacticoid copepods in relation to resident microbial flora. *Hydrobiologia* 131: 87–96.

Defarge, C., Trichet, J. and Siu, P. 1985. First data on the biogeochemistry of kopara deposits from Rangiroa Atoll. *Proceedings of the Fifth International Coral Reef Congress, Tahiti* 3: 365–370.

Degens, E. T. 1969. Biogeochemistry of stable carbon isotopes. In: G. Eglinton and M. T. J. Murphy (Eds.), *Organic Geochemistry* (Springer-Verlag: New York), pp. 304–328.

Degens, E. T. 1984. Warum vekalken organismen? *Acta Univeristatis Carolinae Geologica* 2: 109–121.

Degens, E. T., Kamierczak, J. and Ittekott, V. 1985. Cellular response to Ca^{2+} stress and its geological implications. *Acta Palaeontologica Polonica* 30: 115–135.

Deines, P. 1980. The isotopic composition of reduced organic carbon. In: P. Fritz and J. C. Fontes (Eds.), *Handbook of Environmental Isotope Geochemistry* (Elsevier: Amsterdam), pp. 329–406.

de Laeter, J. R., Libby, W. G. and Trendall, A. F. 1981. The older Precambrian geochronology of Western Australia. In: J. E. Glover and D. I. Groves (Eds.), *Archaean Geology. Geological Society of Australia Special Publication*, 7 (Geological Society of Australia: Perth), pp. 145–158.

de Leeuw, J. W., Sinnighe Damste, J. S., Klok, J., Schenck, P. A. and Boon, J. J. 1985. Biogeochemistry of Gavish Sabkha sediments. I. Studies on neutral reducing sugars and lipid moieties by gas chromatography-mass spectrometry. In: G. M. Friedman and W. E. Krumbein (Eds.), *Hypersaline Ecosystems: The Gavish Sabkha* (Springer-Verlag: Berlin), pp. 350–367.

Delihas, N. and Fox, G. E. 1987. Origins of the plant chloroplasts and mitochondria based on comparisons of 5S ribosomal RNAs. *Annals of the New York Academy of Sciences* 503: 92–102.

De Luca, P. and Moretti, A. 1983. Floridosides in *Cyanidium caldarium*, *Cyanidioschyzon merolae*, and *Galdieria sulphuraria* (Rhodophyta, Cyanidiophyceae). *Journal of Phycology* 19: 368–369.

Delwiche, C. F., Graham, L. E. and Thomson, N. 1989. Lignin-like compounds and sporopollenin in *Coleochaete*, an algal model for land plant ancestry. *Science* 245: 399–401.

Demoulin, V. 1974. The origin of the Ascomycetes and Basidomycetes. The case for a red algal ancestry. *Botanical Review* 40: 315–345.

Demoulin, V. 1985. The red algal-higher fungi phylogenetic link: the last ten years. *Biosystems* 18: 347–356.

Den Hartog, C. 1970. *The Seagrasses of the World* (North-Holland: Amsterdam), 275 pp.

Denison, R. E. 1980. Pre-Bliss (pre-Cambrian) rocks in the Van Horn region, Trans-Pecos, Texas. In: P. W. Dickerson and J. M. Hoffer (Eds.), *31st Field Conference, Trans-Pecos Region* (New Mexico Geological Society: Socorro, New Mexico), pp. 155–158.

DePaolo, D. J. 1981. Nd isotopic studies: some new perspectives on earth structure and evolution. *Transactions of the American Geophysics Union* 62: 137–140.

Derenne, S., Largeau, C., Casadevall, E. and Connan, J. 1988. Comparison of torbanites of various origins and evolutionary stages. Bacterial contribution to their formation. Cause of the lack of botryococcane in bitumens. *Organic Geochemistry* 12: 43–59.

De Rosa, M., Gambacorta, A., Minale, L. and Bullock, J. D. 1971. Bacterial triterpanes. *Chemical Communications*: 619–620.

De Rosa, M., Gambacorta, A. and Gliozzi, A. 1986. Structure, biosynthesis and physiochemical properties of archaebacterial lipids. *Microbiological Reviews* 50: 70–80.

Derrick, G. M. 1982. A Proterozoic rift zone at Mount Isa, Queensland, and implications for mineralisation. *Journal of Australian Geology and Geophysics* 7: 81–92.

Derry, L. A. and Jacobsen, S. B. 1988. The Nd and Sr isotopic evolution of Proterozoic seawater. *Geophysics Research Letters* 15: 397–400.

Derstler, K. 1981. Morphological diversity of Early Cambrian echinoderms. *Short papers for the Second International Symposium on the Cambrian System 1981. U.S. Geological Survey Open File Report* 81-743: 71–75.

Desikachary, T. V. 1959. *Cyanophyta* (Indian Council of Agricultural Research: New Delhi), 686 pp.

Des Marais, D. J. and Moore, J. G. 1984. Carbon and its isotopes in mid-oceanic basaltic glasses. *Earth and Planetary Science Letters* 69: 43–57.

Des Marais, D. J., Cohen, Y., Nguyen, H., Cheatham, M., Cheatham, T. and Munoz, E. 1989. Carbon isotopic trends in the hypersaline ponds and microbial mats at Guerrero Negro, Baja California Sur, Mexico: Implications for Precambrian stromatolites. In: Y. Cohen and E. Rosenberg (Eds.), *Microbial Mats: Physiological Ecology of Benthic Microbial Communities* (American Society for Microbiology: Washington), pp. 191–203.

De Souza, N. J. and New, W. R. 1968. Sterols: isolation from a blue-green alga. *Science* 162: 363–365.

Deuser, W. G. 1970. Carbon-13 in Black Sea waters and implications for the origin of hydrogen sulfide. *Science* 168: 1575–1577.

De Wachter, R., Huysmans, E. and Vandenberghe, E. 1985. 5S ribosomal RNA as a tool for studying evolution. In: K. H. Schleifer and E. Stackenbrandt (Eds.), *Evolution of Prokaryotes* (Academic Press: London), pp. 114–141.

Dewey, J. F. and Burke, K. C. A. 1973. Tibetan, Variscan, and Precambrian basement reactivation: products of continental collision. *Journal of Geology* 81: 683–692.

de Wit, R. and van Gemerden, H. 1987a. Chemolithotrophic growth of a phototrophic sulfur bacterium, *Thiocapsa roseopersicina* (FEC 00112). *FEMS Microbiology Ecology* 45: 117–126.

de Wit, R. and van Gemerden, H. 1987b. Oxidation of sulfide to thiosulfate by *Microcoleus chthonoplastes*. *FEMS Microbiology Ecology* 45: 7–13.

de Wit, R., van Boekel, W. H. M. and van Gemerden, H. 1988. Growth of the cyanobacterium *Microcoleus chthonoplastes* on sulfide. *FEMS Microbiology Ecology* 53: 203–209.

Deynoux, M. and Trompette, R. 1976. Late Precambrian mixtites: glacial and/or non-glacial? A discussion dealing expecially with the mixtites of West Africa. *American Journal of Science* 276: 1302–1315.

Dickas, A. B. 1986. Comparative Precambrian stratigraphy and structure along the Mid-Continent Rift. *American Association of Petroleum Geologists Bulletin* 70: 225–238,

Dickerson, R. E. 1971. The structure of cytochrome *c* and the rates of molecular evolution. *Journal of Molecular Evolution* 1: 26–45.

Dickin, A. P. and McNutt, R. H. 1989. Nd model age mapping of the southeast margin of the Archean foreland in the Grenville Province of Ontario. *Geology* 17: 299–302.

Dickinson, W. R. and Suczek, C. A. 1979. Plate tectonics and sandstone compositions. *American Association of Petroleum Geologists Bulletin* 63: 2164–2182.

Didyk, B. M., Simoneit, B. R. T., Brassell, S. C. and Eglinton, G. 1978. Organic geochemical indicators of palaeoenvironment and conditions of sedimentation. *Nature* 272: 216–222.

Dill, R. F., Shinn, E. A., Jones, A. T., Kelly, K. and Steinen, R. P. 1986. Giant subtidal stromatolites forming in normal salinity waters. *Nature* 324: 55–58.

Dimroth, E. and Chauvel, J. J. 1973. Petrography of the Sokoman Iron Formation in part of the central Labrador Trough. *Geological Society of America Bulletin* 84: 111–134.

Dimroth, E. and Lichtblau, A. P. 1978. Oxygen in the Archean ocean: Comparison of ferric oxide crusts on Archean and Cainozoic pillow basalts. *Neues Jahrbuch für Mineralogie, Abhandlungen* 133: 1–22.

Dimroth, K. and Kimberley, M. M. 1976. Precambrian atmospheric oxygen: evidence in the sedimentary distributions of carbon, sulfur, uranium, and iron. *Canadian Journal of Earth Sciences* 13(9): 1161–1185.

Ding Weiming and Qian Yi 1988. Late Sinian to Early Cambrian small shelly fossils from Yangjiaping, Shimen, Hunan. *Acta Micropalaeontologica Sinica* 5: 39–55.

Dixon, T. H. and Golombek, M. P. 1988. Late Precambrian crustal accretion rates in northeast Africa and Arabia. *Geology* 16: 991–994.

Djerassi, C. 1981. Recent studies in the marine sterol field. *Pure and Applied Chemistry* 53: 873–890.

Dobson, G., Ward, D. M., Robinson, N. and Eglinton, G. 1988. Biogeochemistry of hot spring environments: Extractable lipids of a cyanobacterial mat. *Chemical Geology* 68: 155–179.

Dobzhansky, T., Ayala, F. J., Ledyard Stebbins, G. and Valentine, J. W. 1977. *Evolution* (W. H. Freeman & Co.: San Francisco), 572 pp.

Doddema, H. J. and Vogels, G. D. 1978. Improved identification of methanogenic bacteria by fluorescence microscopy. *Applied and Environmental Microbiology* 36: 752–754.

Dodge, J. D. 1971. A dinoflagellate with both a mesocaryotic and a eukaryotic nucleus 1. Fine structure. *Protoplasma* 73: 145–157.

Dodge, J. D. 1983. A re-examination of the relationship between unicellular host and eucaryotic endosymbiont with special reference to *Glenodinium foliaceum* Dinophyceae. In: H. E. A. Schenk and W. Schwemmler (Eds.), *Endocytobiology II* (W. de Gruyter: Berlin), pp. 1015–1026.

Doemel, W. N. and Brock, T. D. 1974. Bacterial stromatolites: origin of laminations. *Science* 184: 1083–1085.

Doemel, W. N. and Brock, T. D. 1976. Vertical distribution of sulfur species in benthic algal mats. *Limnology and Oceanography* 21: 237–244.

Doemel, W. N. and Brock, T. D. 1977. Structure, growth and decomposition of laminated algal-bacterial mats in alkaline hot springs. *Applied and Environmental Microbiology* 34: 433–452.

Donaldson, J. A. 1963. Stromatolites in the Denault Formation, Marion Lake, Coast of Labrador, Newfoundland. *Geological Survey of Canada, Bulletin* 102: 1–33.

Donaldson, J. A. 1976. Paleoecoloby of *Conophyton* and associated stromatolites in the Precambrian Dismal Lakes and Rae Groups, Canada. In: M. R. Walter (Ed.), *Stromatolites: Developments in Sedimentology*, vol. 20 (Elsevier: Amsterdam), pp. 523–534.

Donaldson, J. A. and Taylor, A. H. 1972. Conical-columnar stromatolites and subtidal environment. *American Association of Petroleum Geologists Bulletin* 56(abs): 614.

Donovan, S. K. 1987. The fit of the continents in the late Precambrian. *Nature* 327: 139–141.

Doolittle, R. F., Anderson, K. L. and Feng, D.-F. 1989. Estimating the prokaryote-eukaryote divergence time from protein sequence. In: B. Fernholm, K. Bremer and H. Jörnwall (Eds.), *The Hierarchy of Life* (Elsevier: Amsterdam), p. 73.

Doolittle, W. F. and Bonen, L. 1981. Molecular sequence data indicating an endosymbiotic origin for plastids. *Annals of the New York Academy of Sciences* 361: 248–258.

Doré, F. and Reid, R. E. 1965. *Allonnia tripodophora* nov. gen., nov. sp., nouvelle Eponge du Cambrien inferieur de Carteret (Manche). *Société géologique de France. Compte Rendu sommaire des séances* 1965(1): 20–21.

Dorr, J. V. N., Jr. 1973a. Iron-formation and associated manganese in Brazil. In: *Genesis of Precambrian Iron and Manganese Deposits. Earth Sciences*, 9 (Unesco: Paris), pp. 105–113.

Dorr, J. V. N., Jr. 1973b. Iron-formation in South America. *Economic Geology* 68: 1005–1022.

Dott, R. H., Jr. 1983. The Proterozoic red quartzite enigma in the north-central United States: Resolved by plate collision? In: L. G. Medaris (Ed.), *Early Proterozoic Geology of the Great Lakes Region. Geological Society of America Memoir* 160: 129–141.

Doubilet, D. and Kohl, L. 1980. British Columbia's cold emerald sea. *National Geographic* 157: 526–551.

Dowling, N. J. E., Widdel, F. and White, D. C. 1986. Phospholipid ester-linked fatty acid biomarkers of acetate oxidising sulfate reducers and other sulfide forming bacteria. *Journal of General Microbiology* 132: 1815–1825.

Downie, C. 1982. Lower Cambrian acritarchs from Scotland, Norway, Greenland, and Canada. *Transactions of the Royal Society of Edinburgh: Earth Sciences* 72: 257–285.

Downie, C., Evitt, W. R. and Sarjeant, W. A. S. 1963. Dinoflagellates, hystrichospheres, and the classification of the acritarchs. *Stanford University Publications. Geological Sciences* 7(3): 1–16.

Dravis, J. J. 1983. Hardened subtidal stromatolites, Bahamas. *Science* 219: 385–386.

Drever, J. I. 1974. Geochemical model for the origin of Precambrian banded iron-formations. *Geological Society of America Bulletin* 85: 1099–1106.

Drobkova, E. L. 1979. Mikrofitolity i vodorosli nizhnepaleozojskikh otlozhenij Sredne-Botuobinskogo rajona [Microphytoliths and algae from the lower Paleozoic deposits of the North-Botuobisk area]. In: B. S. Sokolov (Ed.), *Paleontologiya Dokembriya i Rannego Kembriya* (Nauka, Leningradskoe Otdelenie: Leningrad), pp. 94–96.

Droser, M. L. and Bottjer, D. J. 1988. Trends in depth and extent of bioturbation in Cambrian carbonate marine environments, western United States. *Geology* 16: 233–236.

Droser, M. L. and Bottjer, D. J. 1989. Ordovician increase in extent and depth of bioturbation: implications for understanding early Paleozoic ecospace utilization. *Geology* 17: 850–852.

Drozdova, N. A. 1975a. Vodorosli iz otlozhenij nizhnego kembriya Zapadnoj Mongolii [Algae in the Lower Cambrian deposits of Western Mongolia]. *Iskopaemaya Fauna i Flora Mongolii, Trudy Sovmestnaia Sovetsko-Mongol'skaia Paleontologicheskaia Ekspeditsiia* 2: 300–302.

Drozdova, N. A. 1975b. Vodorosli nizhnego kembriya gory Tsakhir (zapadnaya Mongoliya) [Lower Cambrian algae from the Tsakhir Mountains (western Mongolia)]. In: N. N. Kramarenko (Ed.), *Iskopaemaya Fauna i Flora Mongolii* [*Fossil Fauna and Flora of Mongolia*], Trudy Sovmest. Sovet.-Mongo, vol. 2 (Nauka: Moscow), pp. 303–305.

Drozdova, N. A. 1980. Vodorosli v organogennykh postrojkakh nizhnego kembriya Zapadnoj Mongolii [Algae in Lower Cambrian organic mounds of West Mongolia]. *Trudy Sovmestnaia Sovetsko-Mongol'skaia Paleontologicheskaia Ekhspeditsiia* 10: 1–140.

Du, R., Tian, L. and Li, H. 1986. Discovery of megafossils in the Gaoyuzhuang Formation of the Changchengian System, Jixian. *Acta Geologica Sinica* 1986(2): 115–120 (In Chinese, with English abstract).

Duan, C., Xing, Y., Du, R., Yin, Y. and Liu, G. 1985. Macroscopic fossil algae. In: Y. Xing and et al. (Eds.), *Late Precambrian Palaeontology of China. Ministry of Geology and Mineral Resources, Geological Memoirs*, Series 2, No. 2 (Geological Publishing House: Beijing), pp. 68–77. (In Chinese).

Duan, C.-H. 1982. Late Precambrian algal megafossils *Chuaria* and *Tawuia* in some areas of China. *Alcheringa* 6: 57–68.

Duan Chenghua 1984. [Small shelly fossils from the Lower Cambrian Xihapoing Formation in the Shennonghia District, Hubei Province—hyoliths and fossil skeletons of unknown affinities]. *Bulletin of the Tianjin Institute of Geology and Mineral Resources* 7: 143–188.

Dunlop, J. S. R., Muir, M. D., Milne, V. A. and Groves, D. I. 1978. A new microfossil assemblage from the Archean of Western Australia. *Nature* 274: 676–678.

Durand, B. (Ed.). 1980. *Kerogen, Insoluble Organic Matter from Sedimentary Rocks* (Editions Technip: Paris), 519 pp.

Durham, J. W. 1967. Notes on the Helicoplacoidea and early echinoderms. *Journal of Paleontology* 41(1): 97–102.

Durham, J. W. 1971. The fossil record and the origin of the Deuterostomata. *Proceedings of the North American Paleontological Convention* Pt. H: 1104–1132.

Durham, J. W. 1974. Systematic position of *Eldonia ludwigi* Walcott. *Journal of Paleontology* 48: 750–755.

Durham, J. W. 1978a. A Lower Cambrian eocrinoid. *Journal of Paleontology* 52(1): 195–199.

Durham, J. W. 1978b. The probable Metazoan biota of the Precambrian as indicated by the subsequent record. *Annual Review of Earth and Planetary Sciences* 6: 21–42.

Durham, J. W. and Caster, K. E. 1963. Helicoplacoidea, a new class of echinoderms. *Science* 140: 820–822.

Du Toit, A. L. 1937. *Our Wandering Continents* (Oliver and Boyd: Edinburgh), 366 pp.

Dybowski, W. 1879. Die Chaetetiden der ostbaltischen Silur-Formation [The chaetids of the east Baltic Silurian Formation]. *Verh. Russisch-Kaiser. Mineral. Gesell. Zapiski Imperator. Sanktpetersburgs. Mineral. Obshch.*, Ser. 2 14: 1–134.

Dymek, R. F. and Klein, C. 1988. Chemistry, petrology and origin of banded iron-formation lithologies from the 3800 Ma Isua supracrustal belt, West Greenland. *Precambrian Research* 39: 247–302.

Dyson, I. A. 1985. Frond-like fossils from the base of the late Precambrian Wilpena Group, South Australia. *Nature* 318: 283–285.

Dzik, J. 1986. Turrilepadida and other Machaeridia. In: A. Hoffman and M. H. Nitecki (Eds.), *Problematic Fossil Taxa. Oxford Monographs on Geology and Geophysics*, 5 (Oxford University Press: New York), pp. 116–134.

Dzik, J. and Lendzion, K. 1988. The oldest arthropods of the East European Platform. *Lethaia* 21(1): 29–38.

Easton, R. M. 1981. Stratigraphy of the Akaitcho Group and the development of an Early Proterozoic continental margin, Wopmay orogen, Northwest Territories. In: F. H. A. Campbell (Ed.), *Proterozoic Basins of Canada. Geological Survey of Canada, Paper* 81-10: 79–96.

Easton, R. M. 1986. Geochronology of the Grenville Province. In: J. M. Moore, A. Davidson and A. J. Baer (Eds.), *The Grenville Province. Geological Association of Canada Special Paper* 31: 127–174.

Edhorn, A.-S. 1977. Early Cambrian algae croppers. *Canadian Journal of Earth Sciences* 14(5): 1014–1020.

Edhorn, A.-S. 1978. Microorganisms preserved in cherty stromatolitic dolomite of the Mwasha Group from N'Guba, Mulungwishi, and Shituru localities, Shaba, Zaire, Central Africa. In: C. Ponnamperuma (Ed.), *Comparative Planetology* (Academic Press: New York), pp. 257–268.

Edhorn, A.-S. 1979. *Girvanella* in the "Button Algae" Horizon of the Forteau Formation (Lower Cambrian), Western Newfoundland. *Centre de Recherche Exploration-Production Elf-Aquitaine, Bulletin* 3(2): 557–567.

Edhorn, A.-S. and Anderson, M. M. 1977. Algal remains in the Lower Cambrian Bonavista Formation, Conception Bay, southeastern

Newfoundland. In: E. Flügel (Ed.), *Fossil Algae*, vol. 12 (Springer-Verlag: Berlin), pp. 113–123.

Edman, J. C., Kovacs, J. A., Masur, H., Santi, D., V, Elwood, H. J. and Sogin, M. L. 1988. Ribosomal RNA sequences show *Pneumocystis carinii* to be closely related to the yeasts. *Nature* 334: 519–522.

Edman, J. C., Kovacs, J. A., Masur, H., Santi, D., V, Elwood, H. J. and Sogin, M. L. 1989. Ribosomal RNA genes of *Pneumocystis carinii*. *Journal of Protozoology* 36: 18s–20s.

Edmunds, K. L. H. and Eglinton, G. 1984. Microbial lipids and carotenoids and their early diagenesis in the Solar Lake laminated microbial mat sequence. *MBL Lectures in Biology* 3: 343–389.

Edwards, L. E. 1987. Dinoflagellates. In: T. W. Broadhead (Ed.), *Fossil Prokaryotes and Protists: Notes for a Short Course. Studies in Geology 18* (University of Tennessee Department of Geological Sciences: Knoxville), pp. 34–61.

Eganov, E. A., Sovetov, Y. K. and Yanshin, A. L. 1986. Proterozoic and Cambrian phosphorite-deposits: Karatau, southern Kazakhstan, USSR. In: P. J. Cook and J. H. Shergold (Eds.), *Phosphate Deposits of the World. 1. Proterozoic and Cambrian Phosphorites* (Cambridge University Press: Cambridge), pp. 175–189.

Eglinton, G. and Calvin, M. 1967. Chemical fossils. *Scientific American* 216: 32–43.

Eglinton, G., Scott, P. M., Belsky, T., Burlingame, A. L. and Calvin, M. 1964. Hydrocarbons of biological origin from a one-billion-year-old sediment. *Science* 145: 263–264.

Ehrlich, A. and Dor, I. 1985. Photosynthetic microorganisms of the Gavish Sabkha. In: M. Friedman and W. E. Krumbein (Eds.), *Hypersaline Ecosystems: the Gavish Sabkha* (Springer-Verlag: Berlin), pp. 296–321.

Eichwald, E. 1854. *The Paleontology of Russia. The Ancient Period*, 245 pp.

Eichwald, E. von 1829. *Zoologia specialis, quam expositis animalibus tum vivis, tum fossilibus* (Vilnius), 314 pp.

Eichwald, E. von 1840. Ueber das silurische Schichtensystem in Esthland. *Zeitschrift Nat. Heilkunde Medchirur. K. Akad. St. Petersburg* 1–2.

Eichwald, E. von 1860. *Lethaea rossica ou paléontologie de la Russie* (E. Schweizerbart: Stuttgart), 681 pp.

Eisbacher, G. H. 1985. Late Proterozoic rifting, glacial sedimentation, and sedimentary cycles in the light of Windermere deposition, western Canada. *Palaeogeography, Palaeoclimatology, Palaeoecology* 51: 231–254.

Ekweozor, C. M. and Strausz, O. P. 1983. Tricyclic terpanes in the Athabasca Oil Sands: their geochemistry. In: M. Bjorøy, C. Albercht, K. Cornford, K. de Groot, G. Eglinton, E. Galimov, D. Leythaeuser, R. Pelet, J. Rullkoter and G. Speers (Eds.), *Advances in Organic Geochemistry 1981* (Wiley Heyden Ltd.: Chichester), pp. 746–766.

Elderfield, H. and Greaves, M. J. 1982. The rare earth elements in seawater. *Nature* 296: 214–219.

Elsasser, W. M. 1963. Early history of the earth. In: J. Geiss and E. Goldbert (Eds.), *Earth Science and Meteorites* (North Holland: Amsterdam), pp. 1–30.

Elston, D. D. 1979. Late Precambrian Sixtymile Formation and orogony at the top of the Grand Canyon Supergroup, Northern Arizona. *U.S. Geological Survey Professional Paper*: 1–20.

Elston, D. P. 1984. Magnetostratigraphy of the Belt Supergroup—A synopsis. In: S. W. Hobbs (Ed.), *Montana Bureau of Mines and Geology Special Publication* 90: 88–90.

Elston, D. P. and Bressler, S. 1977. Paleomagnetic poles and polarity zonation from Cambrian and Devonian strata of Arizona. *Earth and Planetary Science Letters* 36: 423–433.

Elwood, H. J., Olsen, G. J. and Sogin, M. L. 1985. The small-subunit ribosomal gene sequences from the hypotrichous ciliates *Oxytrichia nova* and *Stylonychia postulata*. *Molecular Biology and Evolution* 2: 399–410.

Embleton, B. J. J. and Williams, G. E. 1986. Low palaeoaltitude of deposition for Late Precambrian periglacial varvites from South Australia: Implications for palaeoclimatology. *Earth and Planetary Science Letters* 79: 419–430.

Embleton, B. J. J., Veevers, J. J., Johnson, B. D. and Powell, C. McA. 1980. Palaeomagnetic comparison of a new fit of east and west Gondwanaland with the Smith and Hallam fit. *Tectonophysics* 61: 381–390.

Ensminger, A., van Dorsselaer, A., Spyckerelle, C. and Albrecht, P. 1974. Pentacyclic triterpanes of the hopane type as ubiquitous geochemical markers: origin and significance. In: B. Tissot and F. Bienner (Eds.), *Advances in Organic Geochemistry, 1973* (Editions Technip: Paris), pp. 245–260.

Erdmann, V. A., Wolters, J., Peiler, T., Digweed, M., Specht, T. and Ulbrich, N. 1987. Evolution of organisms and organelles as studied by comparative computer and biochemical analyses of ribosomal 5S RNA structure. *Annals of the New York Academy of Sciences* 504: 103–124.

Eriksson, K. A. 1977. Tidal deposits from the Archaean Moodies Group, Barberton Mountain Land, South Africa. *Sedimentary Geology* 18: 257–281.

Eriksson, K. A. 1978. Alluvial and destructive beach facies from the Archaean Moodies Group, Barberton Mountain Land, South Africa and Swaziland. In: A. D. Miall (Ed.), *Fluvial Sedimentology*. *Canadian Society of Petroleum Geologists Memoirs* 5: 287–311.

Eriksson, K. A. 1979. Marginal marine depositional processes from the Archaean Moodies Group, Barberton Mountain Land, South Africa: Evidence and significance. *Precambrian Research* 8: 153–182.

Ermak, V. V. 1986. Rannekembrijskie fordillidy (Bivalvia) severa Sibirskoj platformy [Early Cambrian fordillids (Bivalvia) of the Siberian Platform]. In: I. T. Zhuravleva (Ed.), *Biostratigrafiya i Paleontologiya Kembriya Severnoj Azii. Trudy Inst. Geol. Geofiz. Sibirsk Otd. Akad. Nauk SSSR 669*, vol. 669 (Nauka: Novosibirsk), pp. 183–188.

Ernst, W. G. 1983. The early earth and the Archean rock record. Chapter 3. In: J. W. Schopf (Ed.), *Earth's Earliest Biosphere: Its Origin and Evolution* (Princeton University Press: Princeton), pp. 41–52.

Ernst, W. G. In Press. Evolution of the lithosphere and inferred increasing size of mantle convection cells over geologic time. In: L. L. Perchuk (Ed.), *Progress in Metamorphic and Magmatic Petrology* (Cambridge University Press: Cambridge).

Erwin, D. H. 1989. Molecular clocks, molecular phylogenies and the origin of phyla. *Lethaia* 22: 251–257.

Erwin, D. H. and Valentine, J. W. 1984. "Hopeful monsters," transposons, and metazoan radiation. *Proceedings of the National Academy of Science USA* 81: 5482–5483.

Erwin, D. H., Valentine, J. W. and Sepkoski, J. J., Jr. 1987. A comparative study of diversification events: the early Paleozoic versus the Mesozoic. *Evolution* 41(6): 1177–1186.

Espitalié, J., Laporte, J., Madec, M., Marquis, F., Leplat, P., Paulet, J. and Boutefeu, A. 1977. Méthode rapide de caractérisation des roches mères, de leur potentiel pétrolier et de leur degré d'évolution. *Revue de l'institut français du pétrole* 32: 23–42.

Esteve, I., Debón, R., Massana, R. and Mir, J. 1988. Isolation and characterization of purple phototrophic bacteria from two laminated microbial communities. *VI International Symposium of Photosynthetic Prokaryotes*. Noordwijkerhout, The Netherlands: 178 (Abstract).

Ethier, V. G., Campbell, F. A., Both, R. A. and Krouse, H. R. 1976. Geological setting of the Sullivan orebody and estimates of temperatures and pressures of metamorphism. *Economic Geology* 71: 1570–1588.

Ethridge, M. A., Rutland, R. W. R. and Wyborn, L. A. I. 1987. Orogenesis and tectonic processes in the Early to Middle Proterozoic of northern Australia. In: A. Kröner (Ed.), *Proterozoic Lithospheric Evolution. American Geophysical Union Geodynamics Series* 17: 131–147.

Eugster, H. P. and Chou, I.-M. 1973. The depositional environments of Precambrian banded iron formations. *Economic Geology* 68: 1144–1169.

Eugster, H. P. and Jones, B. R. 1968. Gels composed of sodium-aluminium-silicate, Lake Magadi, Kenya. *Science* 161: 160–163.

Evans, J. S. 1912. The sudden appearance of the Cambrian fauna. *International Geological Congress* 11(1): 543–546.

Evitt, W. R. 1963. A discussion and proposals concerning fossil dinofagellates, hystrichospheres and acritarchs. *Proceedings of the National Academy of Science USA* 49: 158–164.

Evitt, W. R. 1985. *Sporopollenin Dinoflagellate Cysts: Their Morphology and Interpretation* (American Association of Stratigraphic Palynologists Foundation: Dallas), 333 pp.

Evrard, J. L., Kuntz, M. and Weil, J. L. 1990. The nucleotide sequence of five ribosomal protein genes from the cyanelles of *Cyanophora paradoxa*: Implications concerning the phylogenetic relationship between cyanelles and chloroplasts. *Journal of Molecular Evolution* 30: 16–25.

Ewers, W. E. 1983. Chemical factors in the deposition and diagenesis of banded iron-formation. In: A. F. Trendall and R. C. Morris (Eds.), *Iron-Formations: Facts and Problems* (Elsevier: Amsterdam), pp. 491–512.

Ewers, W. E. and Morris, R. C. 1981. Studies of the Dales Gorge Member of the Brockman Iron Formation, Western Australia. *Economic Geology* 76: 1929–1953.

Ewitz, C. E. 1933. Einige neue Fossilfunde in der Visingsöformation. *Geologiska Föreningens i Stockholm Förhandlingar* 55: 506–518.

Faggart, B. E., Jr. and Basu, A. R. 1986. Origin of the Sudbury Complex by meteoritic impact: Neodymium isotopic evidence. *Science* 230: 436–439.

Fairbairn, H. W., Hurley, P. M., Carol, K. D. and Knight, C. J. 1969. Correlation of radiometric ages of Nipissing diabase and Huronian metasediments with Proterozoic orogenic events in Ontario. *Canadian Journal of Earth Sciences* 6: 489–497.

Fairchild, I. J. and Spiro, B. 1987. Petrological and isotopic implications of some contrasting Late Precambrian carbonates, NE Spitsbergen. *Sedimentology* 34: 973–989.

Fairchild, T. R. 1982. Evidence of life in the Proterozoic of Brazil: its practical use and potential theoretical significance. In: R. Weber (Ed.), *Joint Meeting of IGCP-Projects 157 and 160, Development and Interactions of Precambrian Lithosphere, Biosphere, and Atmosphere, Abstracts* (Instituto de Geologia, Universidad Nacional Autonoma de Mexico: Mexico City), p. 13. (Abstract).

Fairchild, T. R. and Subacius, S. M. R. 1986. Microfossils associated with silicified *Stratifera undata* Komar 1966 from the Late Proterozoic Bambuí Group, south-central Brazil. *Precambrian Research* 33: 323–339.

Fairchild, T. R., Barbour, A. P. and Haralyi, N. L. E. 1978. Microfossils in the "Eopaleozoic" Jacadigo Group at Urucum, Mato Grosso, southwest Brazil. *Boletim IG, Instituto de Geociências, Universidade de Sao Paulo* 9: 74–79.

Fajzulina, Z. K. and Treshchetetenkova, A. A. 1979. Rastitel'nye mikrofossilii nizhnepaleozojskikh otlozhenij (analogov motskoj i usol'skoj svit) Botuobinskogo podnyatiya [Plant microfossils of lower Paleozoic deposits (analogues of the Moty and Usol'e Formations) Botuoba uplift]. In: B. S. Sokolov (Ed.), *Paleontologiya Dokembriya i Rannego Kembriya* [*Paleontology of the Precambrian and Early Cambrian*] (Nauka: Leningrad), pp. 163–165.

Fajzulina, Z. K., Lysova, L. A. and Treshchetenkova, A. A. 1973. Mikrofossilii iz nizhnekembrijskikh otlozhenij Irkutskogo amfiteratra [Microfossils from the Lower Cambrian deposits of the Irkutsk amphitheatre]. In: T. F. Vozzhennikova and B. V. Timofeev (Eds.), *Mikrofossilii Drevnejshikh Otlozhenij* [*Microfossils of the Oldest Deposits*] (Nauka: Moscos), pp. 25–28.

Fajzulina, Z. K., Lysova, L. A., Treshchetenkova, A. A., Galperova, E. M. and Drobkova, E. L. 1982. Biostratigrafiya pozdnedokembrijskikh i rannekembrijskikh otlozhenij Nepsko-Botuobinskogo rajona [Biostratigraphy of the Late Precambrian and Early Cambrian deposits of the Nepsko-Botuoba region]. *Izvestiya Akademiya Nauk SSSR, seriia Geologicheskaya* 1982(2): 13–26.

Fajzulina, Z. K., Stanevich, A. M. and Treshchetenkova, A. A. 1984. Rastitel'nye mikrofossilii kholodninskof, oldakitskoj i tukalomijskoj svit severnogo Pribajkal'ya [Plant microfossils from the Kholodnaya, Oldakit and Tukamij Formations of northern Pribaikal]. In: V. Khomentovskij (Ed.), *Stratigrafiya Pozdnego Dokembriya i Rannego Paleozoya. Srednyaya Sibir'* [*Stratigraphy of the Late Precambrian and Early Paleozoic, Central Siberia*]. *Inst. Geol. Geofiz. Sibirsk. Otd. Akad. Nauk SSSR* (Nauka: Novosibirsk), pp. 80–93.

Fanning, C. M., Flint, R. B., Parker, A. J., Ludwig, K. R. and Blissett, A. H. 1988. Refined Proterozoic evolution of the Gawler Craton, South Australia, through U-Pb zircon geochronology. *Precambrian Research* 40/41: 363–386.

Farmer, J. D. and Richardson, L. L. 1988. Origin of microfabric in laminated microbial mats: Implications for interpreting stromatolites. *EOS* 69: 1131.

Farquhar, G. D. and Richards, R. A. 1984. Isotopic composition of plant carbon correlates with water-use efficiency of wheat genotypes. *Australian Journal of Plant Physiology* 11: 539–552.

Farquhar, G. D., O'Leary, M. H. and Berry, J. A. 1982. On the relationship between carbon isotope discrimination and the intercellular carbon dioxide concentration in leaves. *Australian Journal of Plant Physiology* 9: 539–552.

Farquhar, G. D., Ehleringer, J. R. and Hubick, K. T. 1989. Carbon Isotope discrimination and photosynthesis. *Annual Review of Plant Physiology and Plant Molecular Biology* 40: 503–537.

Farris, J. S. 1972. Estimating phylogenetic trees from distance matrices. *American Naturalist* 106: 645–668.

Fattom, A. and Shilo, M. 1985. Production of emulcyan by *Phormidium* J-1: its activity and function. *FEMS Microbiology Ecology* 31: 3–9.

Faul, H. 1949. Fossil burrows from the Pre-Cambrian Ajibik Quartzite of Michigan. *Nature* 164: 32.

Faul, H. 1950. Fossil burrows from the Precambrian Ajibik Quartzite of Michigan. *Journal of Paleontology* 24: 102–106.

Fawley, M. W., Morton, S. J., Stewart, K. D. and Mattox, K. R. 1987. Evidence for a common evolutionary origin of light-harvesting fucoxanthin chlorophyll a/c-protein complexes of *Pavlova gyrans* (Prymnesiophyceae) and *Phaeodactylum tricornutum* (Bacillariophyceae). *Journal of Phycology* 23: 377–381.

Feakes, C. R., Zbinden, E. A. and Holland, H. D. 1989. Paleosols at Arisaig, Nova Scotia and the evolution of the atmosphere. *supplement. Catena* 16: 207–232.

Fedonkin, M. A. 1976. Sledy mnogokletochnykh iz Valdajskoj serii [Traces of multicellular animals from the Valdaj Series]. *Izvestiya Akademiya Nauk SSSR, seriia Geologicheskaya* 1976(4): 129–132.

Fedonkin, M. A. 1977. Precambrian-Cambrian ichnocoenoses of the east European Platform. In: T. P. Crimes and T. C. Harper (Eds.), *Trace Fossils 2* (Seel House Press: Liverpool), pp. 183–194.

Fedonkin, M. A. 1978a. Ancient trace fossils and the behavioral evolution of mud-eaters. *Paleontological Journal* 12: 241–246.

Fedonkin, M. A. 1978b. Drevneyshiye iskopayemyye sledy i puti evolyutsii povedeniya gruntoyedov [Ancient trace fossils and the behavioural evolution of mud-eaters]. *Paleontologicheskij Zhurnal* 1978(2): 106–112.

Fedonkin, M. A. 1980a. [Early stages of evolution of Metazoa on the basis of paleontological data]. *Zhurnal Obshchei Biologii* 41: 226–233 (In Russian).

Fedonkin, M. A. 1980b. Iskopaemye sledy dokembriyskikh Metazoa [Fossil traces of Precambrian Metazoa]. *Izvestiya Akademiya Nauk SSSR, seriia Geologicheskaya* 1980(1): 39–46.

Fedonkin, M. A. 1980c. Novye predstaviteli dokembriyskikh kishechnopolostnykh na severe Russkoy platformy [New examples of Precambrian coelenterates from the northern Russian platform]. *Paleontologicheskij Zhurnal* 1980(2): 7–15.

Fedonkin, M. A. 1981. Belomorskaya biota venda [The Vendian White Sea biota]. *Akademiia Nauk SSSR, Trudy* 342: 1–100.

Fedonkin, M. A. 1982. Novoe rodovoe nazvanie dokembriyskikh kishechnopolostnykh [A new generic name for a Precambrian coelenterate]. *Paleontologicheskij Zhurnal* 1982(2): 137.

References

Fedonkin, M. A. 1983. *Organicheskii Mir Venda* (Nauka: Moscow), 126 pp.

Fedonkin, M. A. 1984. Promofologiya vendckikh Radialia [Promorphology of the Vendian Radialia]. In: *Stratigrafiya i Paleontologiya Drevneyshego Fanerozoya* (Nauka: Moscow), pp. 30–58.

Fedonkin, M. A. 1985a. Besskeletnaya fauna venda: promorfologicheskij analiz [The non-skeletal fauna of the Vendian: a promorphological analysis]. In: B. S. Sokolov and A. B. Ivanovskij (Eds.), *Vendskaya Sistema 1* (Nauka: Moscow), pp. 10–69.

Fedonkin, M. A. 1985b. Paleochronology of the Vendian Metazoa. In: B. S. Sokolov and A. B. Ivanovskiy (Eds.), *Vendskaya Sistema*, vol. 1 (Nauka: Moscow), pp. 112–117.

Fedonkin, M. A. 1985c. Precambrian metazoans: the problems of preservation, systematics and evolution. *Philosophical Transactions of the Royal Society of London B* 311: 27–45.

Fedonkin, M. A. 1985d. Sistematicheskoe opisanie vendskikh Metazoa [Systematic descriptions of the Vendian Metazoa]. In: B. S. Sokolov and A. B. Ivanovsky (Eds.), *Vendskaya Sistema 1* (Nauka: Moscow), pp. 70–106.

Fedonkin, M. A. 1987. Besskeletnaya fauna venda i ee mesto v evolyutsii Metazoa [The non-skeletal fauna of the Vendian and its place in the evolution of the Metazoa]. *Trudy Paleontologicheskogo Instityuta, Akademiya Nauk SSSR* 226: 1–174.

Fedonkin, M. A. and Lipps, J. In Press. Wyman trace fossils. *Geology*.

Fedorov, A. B. 1986. Novye trubchatye problematiki iz stratotipa tommotskogo yarusa [New tubular problematica from the stratotype of the Tommotian Stage]. *Paleontologicheskij Zhurnal* 1986(3): 110–112.

Fehler, S. W. G. and Light, R. J. 1970. Biosynthesis of hydrocarbons in *Anabaena variabilis*. Incorporation of (methyl-^{14}C)- and (methyl-$^{2}H_3$)- methionine into 7- and 8- methylheptadecanes. *Biochemistry* 9: 418–422.

Fehler, S. W. G. and Light, R. J. 1972. Biosynthesis of methylheptadecanes in *Anabaena variabilis*. In vitro incorporation of S- (methyl-^{14}C) adenosymethionine. *Biochemistry* 11: 2411–2416.

Felsenstein, J. 1979. Alternative methods of phylogenetic inference and their inter-relationship. *Systematic Zoology* 28: 49–62.

Fenchel, T. 1977. The significance of bactivorous protozoa in the microbial community of detrital particles. In: J. Cairns Jr. (Ed.), *Aquatic Microbial Communities* (Garland Publishing: New York), pp. 529–544.

Fenchel, T. 1987. *Ecology of Protozoa. The Biology of Free-living Phagotrophic Protists* (Science Tech. Publishers: Madison, Wisc.), 197 pp.

Fenchel, T. M. and Jørgensen, B. B. 1977. Detritus food chains of aquatic ecosystems. The role of bacteria. In: M. Alexander (Ed.), *Advances in Microbial Ecology. Vol. 1* (Plenum Press: New York), pp. 1–58.

Fenchel, T. M. and Straarup, B. J. 1971. Vertical distribution of photosynthetic pigments and the penetration of light in marine sediments. *Oikos* 22: 172–182.

Fenton, C. L. and Fenton, M. A. 1931. Algae and algal beds in the Belt Series of Glacier National Park. *Journal of Geology* 39: 670–686.

Fenton, C. L. and Fenton, M. A. 1936. Walcott's "Pre-Cambrian Algonkian algal flora" and associated animals. *Geological Society of America Bulletin* 47: 609–620.

Fenton, C. L. and Fenton, M. A. 1937. Belt Series of the north: Stratigraphy, sedimentation, paleontology. *Geological Society of America Bulletin* 48: 1873–1970.

Ferezou, J. P., Devys, M., Allais, J. P. and Barbier, M. 1974. Sur le stérol à 26 atomes de carbone de l'algue rouge *Rhodymenia palmata*. *Phytochemistry* 13: 593–598.

Ferguson, J. and Goleby, A. (Eds.). 1980. *Uranium in the Pine Creek Geosyncline* (International Atomic Energy Agency: Vienna), 760 pp.

Field, K. G., Olsen, G. J., Lane, D. L., Giovannono, S. J., Ghiselin, M. T., Raff, E. C., Pace, N. R. and Raff, R. A. 1988. Molecular phylogeny of the animal kingdom. *Science* 239: 748–753.

Fiertel, A. and Klein, H. P. 1959. On sterols in bacteria. *Journal of Bacteriology* 78: 738–739.

Firby, J. B. and Durham, J. W. 1974. Molluscan radula from earliest Cambrian. *Journal of Paleontology* 48(6): 1109–1119.

Firby-Durham, J. B. 1977. *Platysolenites* iz nizhnego kembriya kalifornii [*Platysolenites* in the Lower Cambrian of California]. *Izvestiya Akademiya Nauk SSSR, seriia Geologicheskaya* 1977(9): 146–149.

Fischer, A. G. 1965. Fossils, early life, and atmospheric history. *Proceedings of the National Academy of Science USA* 53: 1205–1215.

Fisher, D. C. and Nitecki, M. H. 1982. Problems in the analysis of receptaculitid affinities. *Third North American Paleontological Convention, Proceedings* 1: 181–186.

Fisher, D. W. 1957. Lithology, paleoecology and paleontology of the Vernon Shale (Late Silurian) in the type area. *New York State Museum and Science Service Bulletin* 364: 1–31.

Fisher, D. W. 1962. Small conoidal shells of uncertain affinities. In: R. C. Moore (Ed.), *Treatise on Invertebrate Paleontology, Part W, Miscellanea* (Geological Society of America and University of Kansas: Lawrence), pp. W98–W143.

Fisher, W. K. and MacGintie, G. E. 1928. The natural history of an echiuroid worm. *Annals and Magazine of Natural History* 1 (Series 10): 204–213.

Fitch, A. 1850. A historical, topographical and agricultural survey of the County of Washington. Part 2–5. *New York Agricultural Society Transactions* 9: 753–944.

Fitch, W. M. 1971. Toward defining the course of evolution: Minimum change for a specific tree topology. *Systematic Zoology* 20: 406–416.

Fitch, W. M. 1976. An evaluation of molecular evolutionary clocks. In: F. J. Ayala (Ed.), *Molecular Evolution* (Sinauer Associates: Sunderland, Mass.), pp. 160–178.

Fitch, W. M. and Langley, C. H. 1976. Protein evolution and the molecular clock. *Federation Proceedings* 35: 2092–2097.

Fitch, W. M. and Smith, T. F. 1983. Optical sequence alignments. *Proceedings of the National Academy of Science USA* 80: 1382–1386.

Flaser, F. M. and Birch, F. 1973. Energetics of core formation: a correction. *Journal of Geophysical Research* 78: 6101–6103.

Fleischer, V. D., Garlick, W. G. and Haldane, R. 1976. Geology of the Zambian copper belt. In: K. H. Wolf (Ed.), *Handbook of Stratabound and Stratiform Ore Deposits*, vol. 6 (Elsevier: Amsterdam), pp. 223–352.

Fleming, P. J. G. 1973. Bradoriids from the *Xystridura* Zone of the Georgina Basin, Queensland. *Geological Survey of Queensland Publication* 356: 1–9.

Flessa, K. and Imbrie, J. 1973. Evolutionary pulsations: evidence from Phanerozoic diversity patterns. In: D. H. Tarling and S. K. Runcorn (Eds.), *Implications of Continental Drift to the Earth Sciences*, vol. 1 (Academic Press: London), pp. 247–285.

Fletcher, I. R., Rosman, K. J. R., Williams, I. R., Hickman, A. H. and Baxter, J. L. 1984. Sm-Nd geochronology of greenstone belts in the Yilgarn Block, Western Australia. *Precambrian Research* 26: 333–361.

Foerste, A. F. 1938. Echinodermata. *Lower Cambrian Olenellus Zone of the Appalachians. Geological Society of America Bulletin* 49: 212–213.

Fogg, G. E. and Stewart, W. D. P. 1968. In situ determinations of biological nitrogen fixation in Antarctica. *British Antarctic Survey Bulletin* 15: 39–46.

Fonin, V. D. and Smirnova, T. N. 1967. Novaya gruppa problematicheskih rannekembrijskikh organizmov i nekotorye metody ikh preparirovaniya [a new group of problematic Early Cambrian organisms and some methods of preparing them]. *Paleontologicheskij Zhurnal* 1967(2): 15–27.

Foppen, F. H. 1971. Table for identification of carotenoid pigments. *Chromatography Review* 14: 133–298.

Ford, T. D. 1958. Pre-Cambrian fossils from Charnwood Forest. *Proceedings of the Yorkshire Geological Society* 31: 211–217.

Ford, T. D. 1963. The Pre-Cambrian fossils of Charnwood Forest.

Transactions of the Leicester Library and Philosophical Society 57: 57–62.

Ford, T. D. 1979. Precambrian fossils and the origin of the Phanerozoic phyla. In: M. R. House (Ed.), *The Origin of Major Invertebrate Groups* (Academic Press: London), pp. 7–21.

Ford, T. D. 1980. The Ediacaran fossils of Charnwood Forest, Leicestershire. *Proceedings of the Geologists Association* 91: 81–83.

Ford, T. D. and Breed, W. J. 1973. Late Precambrian Chuar Group, Grand Canyon, Arizona. *Geological Society of America Bulletin* 84: 1243–1260.

Fordyce, R. E. 1980. Trace fossils from Ohika Formation (Pororari Group, Lower Cretaceous), lower Buller Gorge, Buller, New Zealand. *New Zealand Journal of Geology and Geophysics* 23: 121–124.

Forin, M.-C., Maume, B. and Baron, C. 1972. Sur les stérols et alcohols triterpéniques d'une Cynaphycée: *Spirulina platensis* Gietler. *Comptes Rendus de l'Academie des sciences Paris, Serie D* 274: 133–136.

Fortey, R. A. and Jefferies, R. P. S. 1982. Fossils and phylogeny—a comprehensive approach. In: K. A. Joysey and A. E. Friday (Eds.), *Problems of Phylogenetic Reconstruction. Systematics Association Special Volume*, 21 (Academic Press: London; New York), pp. 197–234.

Foster, H. J., Biemann, K., Haigh, W., Tattrie, N. H. and Colvin, J. R. 1973. The structure of novel C35 triterpanes from *Acetobacter xylinum. The Biochemical Journal* 135: 133–143.

Fowler, M. G. and Douglas, A. G. 1984. Distribution and structure of hydrocarbons in four organic-rich Ordovician rocks. *Organic Geochemistry* 6: 105–114.

Fowler, M. G. and Douglas, A. G. 1987. Saturated hydrocarbon biomarkers in oils of Late Precambrian age from Eastern Siberia. *Organic Geochemistry* 11: 201–203.

Fowler, M. G., Abolins, P. and Douglas, A. G. 1986. Monocyclic alkanes in Ordovician organic matter. In: D. Leytthhaeuser and J. Kullkötter (Eds.), *Advances in Organic Geochemistry 1985. Organic Geochemistry* 10: 815–823.

Fox, G., Stackebrandt, E., Hepspell, R. B., Gibson, J., Maniloff, J., Dyer, T. A., Wolfe, R. S., Balch, W. E., Tanner, R. S., Magrum, L. J., Zablen, L. B., Blakemore, R., Gupta, R., Bonen, L., Lewis, B. J., Stahl, D. A., Luehrsen, K. R., Chen, K. N. and Woese, C. R. 1980. The phylogeny of prokaryotes. *Science* 209: 457–463.

Fox, G. E. and Stackebrandt, E. 1987. The application of 16S rRNA cataloguing and 5S rRNA sequencing in bacterial systematics. *Methods of Microbiology* 19: 406–458.

Fox, L. A. B. 1975. An adaptive model for the origin of invertebrate skeletons. M. S. thesis, University of California, Davis, 44 pp.

Fox, S. W. and Dose, K. 1972. *Molecular Evolution and the Origin of Life* (Freeman: San Francisco), 347 pp.

Føyn, S. and Glaessner, M. F. 1979. *Platysolenites*, other animal fossils, and the Precambrian-Cambrian transition in Norway. *Norsk Geologisk Tiddsskrift* 59: 25–46.

Frakes, L. A. 1979. *Climate Throughout Geologic Time* (Elsevier: New York), 310 pp.

François, L. M. 1986. Extensive deposition of banded iron formations was possible without photosynthesis. *Nature* 320: 352–354.

Frarey, M. J. and McLaren, D. J. 1963. Possible metazoans from the early Proterozoic of the Canadian shield. *Nature* 200: 461–462.

Freeman, K. H., Hayes, J. M., Trendal, J. M. and Albrecht, P. 1990. Evidence from carbon isotope measurements for diverse origins of sedimentary hydrocarbons. *Nature* 343: 254–256.

Fridrikhsone, A. I. 1971. Akritarkhi *Baltisphaeridium* i gistrokhosfery(?) iz kembrijskikh otlozhenij Latvii [The acritarch *Baltisphaeridium* and hystrichospheres(?) from Cambrian deposits of Latvia]. In: A. A. Grigyalis (Ed.), *Paleontologiya i Stratigrafiya Pribaltiki i Belorussii, Sbornik 3* [*Palentology and Stratigraphy of the Baltic Region and Byelorussia, Number 3*] (Mintis: Vilnius), pp. 5–22.

Friedman, A. L. and Alberte, R. S. 1987. Phylogenetic distribution of the major diatom light-harvesting pigment-protein determined by immunological methods. *Journal of Phycology* 23: 427–431.

Friedman, E. I. 1982. Endolithic microorganisms in the Antarctic cold desert. *Science* 215: 1045–1053.

Friedman, E. I. and Ocpampo-Friedmann, R. 1984. Endolithic microorganisms in extreme dry environments: analysis of a litho-biontic microbial habitat. In: M. J. Klug and C. A. Reddy (Eds.), *Current Perspectives in Microbial Ecology* (ASM: Washington, D.C.), pp. 177–185.

Friedman, G. M. and Krumbein, W. E. (Eds.). 1985. *Hypersaline Ecosystems: The Gavish Sabkha* (Springer-Verlag: New York), 484 pp.

Fripp, R. E. P. 1983. The Precambrian geology of the area the Sand River near Messina, Central Zone, Limpopo Mobile Belt. *Geological Society of South Africa Special Publication* 8: 89–102.

Fritz, P., Drimmie, R. J. and Nowicki, V. K. 1974. Preparation of sulfur dioxide for mass spectrometer analyses by cumbustion of sulfides with copper oxide. *Analytical Chemistry* 46: 164–166.

Fritz, W. H. and Crimes, T. P. 1985. Lithology, trace fossils, and correlation of Precambrian-Cambrian boundary beds, Cassiar Mountains, north-central British Columbia. *Geological Survey of Canada, Paper* 83-13: 1–24.

Fritz-Sheridan, R. P. 1987. Nitrogen fixation on a tropical volcano, La Soufriere. II. Nitrogen fixation by *Scytonema* sp. and *Stereocaulon virgatum* Ach. during colonization of phreatic material. *Biotropica* 19: 297–300.

Froelich, P. N., Bender, M. L., Luedtke, N. A., Heath, G. R. and DeVries, T. 1982. The marine phosphorus cycle. *American Journal of Science* 282: 474–511.

Froude, D. O., Ireland, T. R., Kinny, P. D., Williams, I. S., Compston, W., Williams, I. R. and Myers, J. S. 1983. Ion microprobe identification of 4100 to 4200 Ma-old terrestrial zircons. *Nature* 304: 616–618.

Fry, B. D. 1986. Sources of carbon and sulfur nutrition for consumers in three meromictic lakes of New York state. *Limnology and Oceanography* 31: 79–88.

Fry, J. C. 1982. Interactions between bacteria and benthic invertebrates. In: D. B. Nedwell and C. M. Brown (Eds.), *Sediment Microbiology. Special Publications of the Society for General Microbiology*, Vol. 7 (Academic Press: London), pp. 171–201.

Fry, W. G. 1970. The sponge as a population: a biometric approach. In: W. G. Fry (Ed.), *The Biology of the Porifera. Symposium of the Zoological Society of London* 25: 135–162.

Fryer, B. J. 1972. Age determinations in the Circum-Ungava Geosyncline and the evolution of Precambrian banded iron-formations. *Canadian Journal of Earth Sciences* 9: 652–663.

Fryer, B. J. 1973. Trace element geochemistry of the Sokoman Iron Formation. *Canadian Journal of Earth Sciences* 14: 1598–1610.

Fryer, B. J. 1983. Rare earth elements in iron-formation. In: A. F. Trendall and R. C. Morris (Eds.), *Iron-Formations: Facts and Problem* (Elsevier: Amsterdam), pp. 345–359.

Fryer, B. J., Fyfe, W. S. and Kerrich, R. 1979. Archean volcanogenic oceans. *Chemical Geology* 24: 25–33.

Fu, J. 1986. Cyphomegacritarchs and their mathematical simulation from the Liulaobei Formation, Bagong Mountain, Shouxian County, Anhui Province. *Journal of Northwest University* [*China*] 16(1): 76–88 (In Chinese, with English abstract).

Fulco, A. J. 1983. Fatty acid metabolism in bacteria. *Progress in Lipid Research* 22: 133–160.

Fürsich, F. T. and Bromley, R. G. 1985. Behavioural interpretation of a rosetted spreite trace fossil: *Dactyloidites ottoi* (Geinitz). *Lethaia* 18: 199–207.

Fyfe, W. S. 1978. The evolution of the earth's crust: modern plate tectonics to ancient hot spot tectonics? *Chemical Evolution* 23: 89–114.

Gale, G. H. 1983. Proterozoic exhalative massive sulfide deposits. *Geological Society of America Memoir* 161: 191–207.

Galimov, E. M. 1980. C13/C12 in kerogen. In: B. Durand (Ed.), *Kerogen, Insoluble Organic Matter from Sedimentary Rocks* (Editions Technip: Paris), pp. 271–299.

Galimova, V. S. 1988. Nakhodka problematiki *Surindia surindaensis*

References

gen. et sp. nov. v motskoj svite Irkutskogo amfiteatra [Discovery of the problematicum *Surindia surindaensis* gen. et sp. nov. in the Moty Formation of the Irkutsk Amphitheatre]. In: I. T. Zhuravleva and L. N. Repina (Eds.), *Kembrij Sibiri i Srednej Azii. Inst. Geol. Geofiz. Sibirsk. Otd. Akad. Nauk SSSR*, 720 (Nauka: Novosibirsk), pp. 185–190.

Gallardo, V. A. 1977. Large benthic microbial communities in sulphide biota under Peru-Chile subsurface countercurrent. *Nature* 268: 331–332.

Games, L. M. and Hayes, J. M. 1976. On the mechanisms of CO_2 and CH_4 production in natural anaerobic environments. In: J. O. Nriagu (Ed.), *Environmental Biogeochemistry* (Ann Arbor Science: Ann Arbor), pp. 51–73.

Garcia, D., Parot, P., Vermeglio, A. and Madigan, M. T. 1986. The light-harvesting complexes of a thermophilic purple sulfur photosynthetic bacterium *Chromatium tepidum*. *Biochimica et Biophysica Acta* 850: 390–395.

Gargas, E. 1970. Measurements of primary production, dark fixation and vertical distribution of the microbenthic algae in the Øresund. *Ophelia* 8: 231–253.

Garlick, S., Oren, A. and Padan, E. 1977. Occurrence of facultative anoxygenic photosynthesis among filamentous and unicellular cyanobacteria. *Journal of Bacteriology* 129: 623–629.

Garrels, R. M. 1987. A model for the deposition of the microbanded Precambrian iron formations. *American Journal of Science* 287: 81–106.

Garrels, R. M. and Mackenzie, F. T. 1971. *Evolution of Sedimentary Rocks* (W. W. Norton: New York), 397 pp.

Garrett, P. 1970a. Deposit feeders limit development of stromatolites. *Bulletin of the American Association of Petroleum Geologists* 54: 848.

Garrett, P. 1970b. Phanerozoic stromatolites: Non-competitive ecologic restriction by grazing and burrowing animals. *Science* 169: 171–173.

Garwood, E. J. 1914. Some new rock-building organisms from the Lower Carboniferous beds of Westmorland. *Geological Magazine* 6(1): 265–271.

Garwood, E. J. 1931. The Tuedian Beds of Northern Cumberland and Roxburghshire east of the Liddle Water. *Quarterly Journal of the Geological Society of London* 87: 97–159.

Gaskell, S. J., Eglinton, G. and Bruun, T. 1973. Hydrocarbon constituents of three species of Norwegian lichen: *Cetraria nivalis, C. crispa, Siphula ceratites*. *Phytochemistry* 12: 1174–1176.

Gaudette, H. E. and Lyons, W. B. 1984. Trace metal concentrations in modern marine sabkha sediments. In: Y. Cohen, R. W. Castenholz and H. O. Halvorson (Eds.), *Microbial Mats: Stromatolites* (Alan R. Liss, Inc.: New York), pp. 425–434.

Gay, A. L. and Grandstaff, D. E. 1980. Chemistry and mineralogy of Precambrian paleosols at Elliot Lake, Ontario, Canada. *Precambrian Research* 12: 349–373.

Gebelein, C. D. 1969. Distribution, morphology, and accretion rate of recent subtidal algal stromatolites. *Journal of Sedimentary Petrology* 39: 49–69.

Gebelein, C. D. 1974. Biological control of stromatolite microstructure: implications for Precambrian time stratigraphy. *American Journal of Science* 274(6): 575–598.

Gebelein, C. D. 1976a. The Effects of the Physical, Chemical and Biological Evolution of the Earth. In: M. R. Walter (Ed.), *Stromatolites: Developments in Sedimentology*, vol. 20 (Elsevier: Amsterdam), pp. 499–516.

Gebelein, C. D. 1976b. Open marine subtidal and intertidal stromatolites (Florida, The Bahamas and Bermuda). In: M. R. Walter (Ed.), *Stromatolites: Developments in Sedimentology*, vol. 20 (Elsevier: Amsterdam), pp. 381–388.

Gebelein, C. D. 1977. *Dynamics of Recent Carbonate Sedimentology and Ecology* (Brill: Leiden), 120 pp.

Gedik, I. 1977. Orta Toroslar'da konodont biyostratigrafisi [Conodont biostratigraphy in the Middle Taurus]. *Türkiye Jeoloji Kurumu Bülteni* 20: 35–48.

Gedik, I. 1981. *Hadimopanella* Gedik, 1977 nin stratigrafik daglilmi ve mikroyapisi konusunda bazi gözlemler [Some remarks on microstructure and stratigraphic range of *Hadimopanella* Gedik, 1977]. *Karadeniz Teknik Üniversitesi Yer Blimleri Dergisi, Jeoloji* 1: 159–163.

Gedik, I. 1989. Bati Toroslar Kambriyen'inde Hadimopanellid biyostratigrafisi: Kambriyen'de yeni bir biyostratigrafik zonlama [Hadimopanellid biostratigraphy in the Cambrian of the Western Taurids: A new biostratigraphic tool in the subdivision of the Cambrian System]. *Türkiye Jeoloji Kurumu Bülteni* 32: 65–78.

Gee, R. D., Baxter, J. L., Wilde, S. A. and Williams, I. R. 1981. Crustal development in the Archean Yilgarn Block, Western Australia. In: J. E. Glover and D. I. Groves (Eds.), *Archean Geology. Geological Society of Australia Special Publication*, 7 (Geological Society of Australia: Perth), pp. 43–56.

Gee, R. D., Myers, J. S. and Trendall, A. F. 1986. Relation between Archean high-grade gneiss and granite-greenstone terrain in Western Australia. *Precambrian Research* 33: 87–102.

Gehling, J. G. 1986. Algal binding of siliciclastic sediments: a mechanism in the preservation of Ediacaran fossils. *12th International Sedimentological Congress Abstracts*: 117.

Gehling, J. G. 1987. Earliest known echinoderm—a new Ediacaran fossil from the Pound Subgroup of South Australia. *Alcheringa* 11: 337–345.

Gehling, J. G. 1988. A cnidarian of actinian-grade from the Ediacaran Pound Subgroup, South Australia. *Alcheringa* 12: 299–314.

Gehring, W. J. 1985. The molecular basis of development. In: W. J. Gehring (Ed.), *Readings from Scientific American: The Molecules of Life* (Freeman: New York), pp. 107–118.

Geider, R. J. and Gunter, P. A. 1989. Evidence for the presence of Phycoerythrin in *Dinophysis norvegica*, a pink dinoflagellate. *British Phycological Journal* 24: 195–198.

Gelpi, E., Schneider, H., Mann, J. and Oro, J. 1970. Hydrocarbons of geochemical significance in microscopic algae. *Phytochemistry* 9: 603–612.

Geng Lianyu; Zhang Shiben 1987. Early Cambrian problematic fossils from Fangxian, Hubei, China. In: A. S. Nanjing Institute of Geology and Paleontology (Ed.), *Stratigraphy and Palaeontology of Systematic Boundaries in China–PC–C boundary 1* (Nanjing University Publishing House: Nanjing), pp. 523–534.

George, D. G., Hunt, L. T. and Dayhoff, M. O. 1983. Sequence evidence for the symbiotic origins of chloroplasts and mitochondria. In: H. E. A. Schenk and W. Schwemmler (Eds.), *Endocytobiology II* (W. de Gruyter: Berlin), pp. 845–862.

Gerdes, G. and Krumbein, W. E. 1984. Animal communities in recent potential stromatolites of hypersaline origin. In: Y. Cohen, R. W. Castenholz and H. O. Halvorson (Eds.), *Microbial Mats: Stromatolites* (Alan R. Liss: New York), pp. 59–83.

Gerdes, G. and Krumbein, W. E. 1987. *Biolaminated Deposits*. In: S. Bhattacharji, G. M. Friedman, H. J. Neugebauer and A. Seilacher (Eds.), *Lecture Notes in Earth Sciences*, 9 (Springer-Verlag: Berlin), 183 pp.

Gerdes, G., Holtkamp, E. M. and Krumbein, W. E. 1985a. Salinity and water activity related zonation of microbial communities and potential stromatolites of the Gavish Sabkha. In: G. M. Friedman and W. E. Krumbein (Eds.), *Hypersaline Ecosystems–The Gavish Sabkha* (Springer-Verlag: Heidelberg), pp. 238–266.

Gerdes, G., Spira, J. and Dimentman, C. 1985b. The Fauna of the Gavish Sabkha and the Solar Lake—a comparative study. Chapter 15. In: G. M. Friedman and W. E. Krumbein (Eds.), *Hypersaline Ecosystems: The Gavish Sabkha* (Springer-Verlag: New York), pp. 322–345.

German, T. N. 1981. Nitchatye mikroorganizmy lakhandinskoj svity reki Mai [Filamentous microorganisms from the Lankhanda Formation on the Maya River]. *Paleontologicheskij Zhurnal* 1981(2): 126–131.

German, T. N. 1985. Nitchatye vodorosli venda [Vendian filamentous algae]. In: B. S. Sokolov (Ed.), *Vendskaya Sistema 1* (Nauka: Moscow), pp. 146–153.

German, T. N., Mikhajlova, N. S., Yankauskas, T., V, et al. 1989. Sistematicheskoe opisanie mikrofossilij [Systematic description of microfossils]. In: T. Yankauskas V (Ed.), *Mikrofossilii Dokembriya SSSR [Precambrian Microfossils of the USSR]* (Nauka: Leningrad), pp. 34–151.

Germs, G. B. 1972. New shelly fossils from the Nama Group, South West Africa. *American Journal of Science* 272: 752–761.

Germs, G. J. B. 1968. Discovery of a new fossil in the Nama System, South West Africa. *Nature* 219: 53–54.

Germs, G. J. B. 1972a. New shelly fossils from Nama Group, South West Africa. *American Journal of Science* 272: 752–761.

Germs, G. J. B. 1972b. The stratigraphy and paleontology of the lower Nama Group, South West Africa. *University of Cape Town, Department of Geology, Chamber of Mines Precambrian Research Unit Bulletin* 12: 1–250.

Germs, G. J. B. 1972c. Trace fossils from the Nama Group, South-west Africa. *Journal of Paleontology* 46: 864–870.

Germs, G. J. B. 1973a. Possible new spriggind worm and a new trace fossil from the Nama Group, South West Africa. *Geology* 1: 69–70.

Germs, G. J. B. 1973b. A reinterpretation of *Rangea schneiderhoehni* and the discovery of a related new fossil from the Nama Group, South West Africa. *Lethaia* 6: 1–10.

Germs, G. J. B. 1983. Implications of a sedimentary facies and depositional environmental analysis of the Nama Group in South West Africa/Namibia. In: R. Miller (Ed.), *Evolution of the Damara Orogen of South West Africa/Namibia. Geological Society of South Africa Special Publication* 11: 89–114.

Gest, H. and Schopf, J. W. 1983. Biochemical evolution of anaerobic energy conversion: the transition from fermentation to anoxygenic photosynthesis. Chapter 6. In: J. W. Schopf (Ed.), *Earth's Earliest Biosphere: Its Origin and Evolution* (Princeton University Press: Princeton), pp. 135–148.

Gest, H., Favinger, J. and Madigan, M. T. 1985. Exploitation of N_2-fixation capacity for enrichment of anoxygenic photosynthetic bacteria in ecological Solar Lake (Sinai). *FEMS Microbiology Ecology* 31: 317–322.

Geyer, G. 1986. Mittelkabrische Mollusken aus Marokko und Spanien. *Senckenbergiana Lethaea* 67: 55–118.

G-Farrow, C. E. and Mossman, D. J. 1988. Geology of Precambrian paleosols at the base of the Huronian supergroup, Elliot Lake, Ontario, Canada. *Precambrian Research* 42: 107–139.

Ghiorse, W. C. 1984. Biology of iron-and manganese-depositing bacteria. *Annual Review of Microbiology* 38: 515–550.

Ghiselin, M. T. 1988. The origin of molluscs in the light of molecular evidence. *Oxford Surveys in Evolutionary Biology* 5: 66–95.

Ghiselin, M. T. 1989. Summary of our present knowledge of metazoan phylogeny. In: B. Fernholm, K. Bremer and H. Jörnvall (Eds.), *The Hierarchy of Life* (Elsevier: Amsterdam), pp. 261–272.

Giani, D., Giani, L., Cohen, Y. and Krumbein, W. E. 1984. Methanogenesis in the hypersaline Solar Lake (Sinai). *FEMS Microbiology Letters* 25: 219–224.

Giani, D., Seeler, J., Giani, L. and Krumbein, W. E. 1989. Microbial mats and physicochemistry in a saltern in the Bretagne (France) and in a laboratory scale saltern model. *FEMS Microbiology Ecology* 62: 151–162.

Gibbs, A. K. and Barron, C. 1983. The Guiana Shield reviewed. *Episodes* 2: 7–14.

Gibbs, A. K. and Olszewski, W. J., Jr. 1982. Zircon U-Pb ages of Guyana greenstone-gneiss terrane. *Precambrian Research* 17: 199–214.

Gibbs, A. K., Montgomery, C. W., O'Day, P. A. and Erslev, E. A. 1986. The Archean-Proterozoic transition: Evidence from the geochemistry of metasedimentary rocks of Guyana and Montana. *Geochimica et Cosmochimica Acta* 50: 2125–2141.

Gibbs, S. P. 1978. The chloroplasts of *Euglena* may have evolved from symbiotic green algae. *Canadian Journal of Botany* 56: 2883–2889.

Gibbs, S. P. 1981a. The chloroplast endoplasmic reticulum: structure, function and evolutionary significance. *International Review of Cytology* 72: 49–99.

Gibbs, S. P. 1981b. The chloroplasts of some algal groups may have evolved from endosymbiotic eukaryotic algae. *Annals of the New York Academy of Sciences* 361: 193–207.

Gibson, G. G. 1989. Trace fossils from late Precambrian Carolina Slate Belt, south-central North Carolina. *Journal of Paleontology* 63: 1–10.

Gibson, G. G., Teeter, S. A. and Fedonkin, M. A. 1984. Ediacaran fossils from the Carolina slate belt, Stanly County, North Carolina. *Geology* 12: 387–390.

Gibson, J. 1981. Movement of acetate across the cytoplasmic membrane of the unicellular cyanobacteria *Synechococcus* and *Aphanocapsa*. *Archives of Microbiology* 130: 175–179.

Gibson, J., Pfennig, N. and Waterbury, J. B. 1984. *Chloroherpeton thalassium* gen. nov. et spec. nov., a non-filamentous, flexing and gliding green sulfur bacterium. *Archives of Microbiology* 138: 96–101.

Gillett, S. L. 1982. Paleomagnetism of the Late Cambrian *Crepicephalus-Aphelaspis* trilobite zone boundary in North America—Divergent poles from isochronous strata. *Earth and Planetary Science Letters* 58: 383–394.

Gillett, S. L. and Van Alstine, D. R. 1979. Paleomagnetism of Lower and Middle Cambrian sedimentary rocks from the Desert Range, Nevada. *Journal of Geophysical Research* 84: 4475–4489.

Gillett, S. L. and Van Alstine, D. R. 1982. Remagnetization and tectonic rotation of upper Precambrian and lower Paleozoic strata from the Desert Range, southern Nevada. *Journal of Geophysical Research* 87: 10,929–10,953.

Giovannoni, S. J., Revsbech, N. P., Ward, D. M. and Castenholz, R. W. 1987a. Obligately phototrophic *Chloroflexus*: primary production in anaerobic hot spring microbial mats. *Archives of Microbiology* 147: 80–87.

Giovannoni, S. J., Schabtach, E. and Castenholz, R. W. 1987b. *Isosphaera pallida* gen. nov. and sp. nov., a gliding budding eubacterium from hot springs. *Archives of Microbiology* 147: 276–284.

Giovannoni, S. J., Turner, S., Olsen, G. J., Barns, S., Lane, D. J. and Pace, N. R. 1988. Evolutionary relationships among cyanobacteria and green chloroplasts. *Journal of Bacteriology* 170(8): 3584–3592.

Given, R. K. and Lohmann, K. C. 1986. Isotopic evidence for the early meteoric diagenesis of the reef facies, Permian Reef Complex of West Texas and New Mexico. *Journal of Sedimentary Petrology* 56: 183–193.

Glaessner, M. F. 1958a. New fossils from the base of the Cambrian in South Australia. *Transactions of the Royal Society of South Australia* 81: 185–188.

Glaessner, M. F. 1958b. The oldest fossil faunas of South Australia. *Geologische Rundschau* 47(2): 522–531.

Glaessner, M. F. 1962. Pre-Cambrian fossils. *Biological Reviews* 37: 467–494.

Glaessner, M. F. 1963. Zur Kenntnis der Nama-Fossilien Südwest-Afrikas. *Annalen des Naturhistorischen Museums in Wien* 66: 113–120.

Glaessner, M. F. 1966. Precambrian evolution. *Earth-Science Reviews* 1: 29–50.

Glaessner, M. F. 1969. Trace fossils from the Precambrian and basal Cambrian. *Lethaia* 2: 369–393.

Glaessner, M. F. 1971. The genus *Conomedusites* Glaessner & Wade and the diversification of the Cnidaria. *Paläontologische Zeitschrift* 43: 7–17.

Glaessner, M. F. 1972a. Precambrian Palaeozoology. In: J. B. Jones and B. McGowran (Eds.), *Stratigraphic Problems of the Later Precambrian and Early Cambrian. University of Adelaide, Center Precambrian Research, Special Paper*, 1 (University of Adelaide: Adelaide), pp. 43–52.

Glaessner, M. F. 1972b. Preface. In: M. R. Walter (Ed.), *Stromatolites and the Biostratigraphy of the Australian Precambrian and Cambrian. Palaeontological Association. Special Papers in Palaeon-

References

tology 11: v–vi.

Glaessner, M. F. 1976a. Early Phanerozoic annelid worms and their geological and biological significance. *Journal of the Geological Society of London* 132: 259–275.

Glaessner, M. F. 1976b. A new genus of Late Precambrian polychaete worms from South Australia. *Transactions of the Royal Society of South Australia* 100: 169–170.

Glaessner, M. F. 1978. The oldest foraminifera. *Australia Bureau of Mineral Resources, Geology and Geophysics Bulletin* 192: 61–65.

Glaessner, M. F. 1979a. Biogeography and biostratigraphy. Precambrian. In: R. A. Robinson and C. Teichert (Eds.), *Treatise on Invertebrate Paleontology, Part A, Introduction* (Geological Society of America: Boulder), pp. 79–118.

Glaessner, M. F. 1979b. An echiuroid worm from the late Precambrian. *Lethaia* 12: 121–124.

Glaessner, M. F. 1979c. Lower Cambrian Crustacea and annelid worms from Kangaroo Island, South Australia. *Alcheringa* 3(1): 21–31.

Glaessner, M. F. 1980a. *Parvancorina*–an arthropod from the Late Precambrian (Ediacarian) of South Australia. *Annalen des Naturhistorischen Museums in Wien* 83: 83–90.

Glaessner, M. F. 1980b. Pseudofossils from the Precambrian, including '*Buschmannia*' and '*Praesolenopora*'. *Geological Magazine* 117(2): 199–200.

Glaessner, M. F. 1983. The emergence of Metazoa in the early history of life. *Precambrian Research* 20: 427–441.

Glaessner, M. F. 1984. *The Dawn of Animal Life. A Biohistorical Study* (Cambridge University Press: Cambridge), 244 pp.

Glaessner, M. F. and Daily, B. 1959. The geology and late Precambrian fauna of the Ediacara fossil reserve. *Records of the South Australian Museum* 13: 369–401.

Glaessner, M. F. and Wade, M. 1966. The late Precambrian fossils from Ediacara, South Australia. *Palaeontology* 9: 599–628.

Glaessner, M. F. and Wade, M. 1971. *Praecambridium*–a primitive arthropod. *Lethaia* 4: 71–77.

Glaessner, M. F. and Walter, M. R. 1975. New Precambrian fossils from the Arumbera Sandstone, Northern Territory, Australia. *Alcheringa* 1: 59–69.

Glagolev, A. N. 1984. *Motility and Taxis in Prokaryotes. Soviet Scientific Reviews, Suppl. Ser. Physiochemical Biology*, Vol. 3 (Harwood Acad. Publ.: Chur, Switzerland), 279 pp.

Glazer, A. N. 1983. Comparative biochemistry of photosynthetic light-harvesting systems. *Annual Review of Biochemistry* 52: 125–127.

Glazer, A. N. and Apell, G. S. 1977. A common evolutionary origin for the biliproteins of cyanobacteria, rhodophyta and cryptophyta. *FEMS Microbiology Letters* 1: 113–116.

Glickson, A. 1986. The oldest age determined for Pilbara greenstones. *Bureau Mineral Resources Research Newsletter* 4: 7.

Glikson, A. Y. 1984. Significance of Early Archean mafic-ultramafic xenolith patterns. In: A. Kröner, G. N. Hanson and A. C. Goodwin (Eds.), *Archaean Geochemistry* (Springer-Verlag: Berlin), pp. 262–282.

Gloe, A. and Risch, N. 1978. Bacteriochlorophyll c_s, a new bacteriochlorophyll from *Chloroflexus aurantiacus*. *Archives of Microbiology* 118: 153–156.

Gloe, A., Pfennig, N., Brockmann, H. and Trowitzsch, W. 1975. A new bacteriochlorophyll from brown-colored Chlorobiaceae. *Archives of Microbiology* 102: 103–108.

Glück, H. 1912. Eine neue gesteinbildende Siphonee (Codiacee) aus dem marinen Tertiär von Süddeutschland [A new rock-building siphonous alga (codiacean) from the Tertiary marine sediments of southern Germany]. *Mitteilungen der grosserzoglich Badischen geolischen Landesanstalt* 7: 1–24.

Gnilovskaya, M. B. 1971a. Drevneyshie vodnye rasteniya venda Russkoy platformy (pozdniy dokembriy) [The oldest aquatic plants of the Vendian of the Russian Platform]. *Paleontological Journal* 1971(3): 372–378.

Gnilovskaya, M. B. 1971b. Drevneyshie vodnye rasteniya venda Russkoy platformy (pozdniy dokembriy) [The oldest aquatic plants of the Vendian of the Russian Platform]. *Paleontologicheskij Zhurnal* 1971(3): 101–107.

Gnilovskaya, M. B. 1979. The Vendian Metaphyta. *Centre de Recherche Exploration-Production Elf-Aquitaine, Bulletin* 3(2): 611–618.

Gnilovskaya, M. B. 1985. Vendotaenids—Vendian Metaphyta. In: B. S. Sokolov and A. B. Ivanovskiy (Eds.), *Vendskaya Sistema*, vol. 1 (Nauka: Moscow), pp. 117–125.

Gnilovskaya, M. B. and Kolesnikov, C. H. M. 1988. The Vendotaenian flora: the most ancient tissue algae on the earth (morphology, histochemistry, system). In: *Indo-Soviet Symposium on Stromatolites and Stromatolitic Deposits* (Wadia Institute of Himalayan Geology: Dehra Dun), pp. 3–7. (Abstracts).

Gnilovskaya, M. B., Ishchenko, A. A., Kolesnikov, C. H. M., Korenchuk, L. B. and Udal'Nov, A. P. 1988. *Vendotenidy Vostochno-Evropeiskoi Platformy [Vendotaenids of the East European Platform]* (Nauka: Leningrad), 143 pp.

Godward, M. B. E. 1962. Invisible radiations. In: R. A. Lewin (Ed.), *Physiology and Biochemistry of Algae* (Academic Press: New York), pp. 551–556.

Goetz, P. A. 1980. Depositional environment of the Sherridon Group and related mineral deposits near Sherridon, Manitoba. Unpub. Ph.D thesis, Carleton University, Ottawa, Ontario.

Gogarten, J. P., Kibak, H., Dittrich, P., Taiz, L., Bowman, E. M., Bowman, B. J., Manolson, M. F., Poole, R. J., Date, T., Oshima, T., Konishi, J., Denda, K. and Yoshida, M. 1989. Evolution of the vacuolar H^+-ATPase: Implications for the origin of eukaryotes. *Proceedings of the National Academy of Science USA* 86: 6661–6665.

Goldich, S. S. 1968. Geochronology in the Lake Superior region. *Canadian Journal of Earth Sciences* 5: 715–727.

Goldich, S. S. and Wooden, J. L. 1980. Geochemistry of the Archean rocks in the Morton and Granite Falls areas, southern Minnesota. *Precambrian Research* 11: 267–296.

Goldman, C. R., Mason, D. T. and Wood, B. J. B. 1972. Comparative study of limnology of two small lakes on Ross Island, Antarctica. *Antarctic Research Series* 20: 1–50.

Goldring, R. 1965. Sediments into rocks. *New Scientist* 26: 863–865.

Goldring, R. and Curnow, C. N. 1967. The stratigraphy and facies of the late Precambrian at Ediacara, South Australia. *Journal of the Geological Society of Australia* 14: 195–214.

Goldring, R. and Seilacher, A. 1971. Limulid undertracks and their sedimentological implications. *Neues Jahrbuch für Geologie und Paläontologie, Abhandlungen* 137: 422–442.

Gole, M. J. 1981. Archean banded iron-formations, Yilgarn Block, Western Australia. *Economic Geology* 76: 1954–1975.

Gole, M. J. and Klein, C. 1981. Banded iron-formations through much of Precambrian time. *Journal of Geology* 89: 169–183.

Golovenok, V. K. 1985. Stromatolites and microphytolites in Precambrian stratigraphy: expectations and reality. *Soviet Geology*: 78–83.

Golovenok, V. K. and Belova, M. Y. 1983. Nakhodki *Obruchevella* v rifee Patomskogo nagor'ya i v vende Yuzhnogo Kazakhstana [*Obruchevella* from the Riphean of the Patom Highland and the Vendian of Southern Kazakhstan]. *Doklady Akademii Nauk SSSR* 272(6): 1462–1465.

Golubev, S. N. 1976. Ontogeneticheskie izmeneiya i ehvolutsionnye tendentsii rannekembrijskikh spiral'nykh gastropod Pelagiellacea [Ontogenetic changes and evolutionary trends in the early Cambrian spiral gastropods Pelagiellacea]. *Paleontologicheskij Zhurnal* 1976(2): 34–40.

Golubić, S. 1967. Algenvegetation der Felsen. In: H.-J. Elster and W. Ohle (Eds.), *Die Binnengewässer*, vol. XXIII (E. Schweizerbartsche: Stuttgart), pp. 1–183.

Golubić, S. 1973. The relationship between blue-green algae and carbonate deposits. In: N. G. Carr and B. A. Whitton (Eds.), *The Biology of Blue-Green Algae* (Blackwell: Oxford), pp. 434–472.

Golubić, S. 1976. Organisms that build stromatolites. In: M. R. Walter

(Ed.), *Stromatolites: Developments in Sedimentology*, vol. 20 (Elsevier: Amsterdam), pp. 113–126.

Golubić, S. 1985. Microbial mats and modern stromatolites in Shark Bay, Western Australia. In: D. E. Caldwell, J. A. Brierley and C. L. Brierley (Eds.), *Planetary Ecology* (Van Nostrand Reinhold: New York), pp. 3–16.

Golubić, S. and Barghoorn, E. S. 1977. Interpretation of microbial fossils with special reference to the Precambrian. In: E. Flügel (Ed.), *Fossil Algae, Recent Results and Developments* (Springer-Verlag: Berlin), pp. 1–14.

Golubić, S. and Focke, J. W. 1978. *Phormidium hendersonii* Howe: identity and significance of a modern stromatolite building microorganism. *Journal of Sedimentary Petrology* 48: 751–764.

Golubić, S. and Hofmann, H. J. 1976. Comparison of Holocene and mid-Precambrian Entophysalidacea (Cyanophyta) in stromatolitic algal mats: cell division and degradation. *Journal of Paleontology* 50: 1074–1082.

Golyshev, S. E., Padalko, N. L. and Pechenkin, S. A. 1981. Fractionation of stable oxygen and carbon isotopes in carbonate systems. *Geochemistry International* 18: 85–99.

Gomes, N. A. de N. C. 1985. Modern stromatolites in a karst structure from the Malmani Subgroup, Transvaal Sequence, South Africa. *Transactions of the Geological Society of South Africa* 88: 1–9.

Gooday, A. J. 1988. The genus *Bathysiphon* (Protista, Foraminiferida) in the north-east Atlantic: A neotype for *B. filiformis* G.O. & M. Sars, 1872 and the description of a new species. *Journal of Natural History* 22: 95–105.

Goode, A. D. T. 1981. Proterozoic geology of Western Australia. In: D. R. Hunter (Ed.), *Precambrian of the Southern Hemisphere* (Elsevier: Amsterdam), pp. 105–204.

Goode, A. D. T., Hall, W. D. M. and Bunting, J. A. 1983. The Naberru basin of Western Australia. In: A. F. Trendall and R. C. Morris (Eds.), *Iron-Formations: Facts and Problems* (Elsevier: Amsterdam), pp. 295–323.

Goodloe, R. S. and Light, R. J. 1982. Structure and composition of hydrocarbons and fatty acids from a marine blue-green alga, *Synechococcus* sp. *Biochimica et Biophysica Acta* 710: 485–492.

Goodwin, A. M. 1956. Facies relations in the Gunflint Iron Formation. *Economic Geology* 51: 565–595.

Goodwin, A. M., Monster, J. and Thode, H. G. 1976. Carbon and sulfur isotope abundances in Archean iron-formations and early Precambrian life. *Economic Geology* 71: 870–891.

Goodwin, N. S., Mann, A. L. and Patience, R. L. 1988. Structure and significance of C_{30} 4-methylsteranes in lacustrine shales and oils. *Organic Geochemistry* 12: 495–506.

Goodwin, T. H. 1980. *The Biochemistry of Carotenoids* (Chapman and Hall: London), 377 pp.

Goody, R. and Walker, J. C. G. 1972. *Atmospheres* (Prentice-Hall, Inc.: Englewood Cliffs, N.J.), 150 pp.

Goossens, H., de Leeuw, J. W., Schenck, P. A. and Brassell, S. C. 1984. Tocopherols as likely precursors of pristane in ancient sediments and crude oils. *Nature* 312: 440–442.

Goossens, H., Rijpstra, I. C., Duren, R. R., de Leeuw, J. W. and Schenk, P. A. 1986. Bacterial contribution to sedimentary organic matter: a comparative study of lipid moieties in bacteria and recent sediments. *Organic Geochemistry* 10: 683–696.

Gorbatschev, R. 1985. Precambrian basement of the Scandinavian Caledonides. In: D. G. Gee and B. A. Sturt (Eds.), *The Caledonide Orogen–Scandinavia and Related Areas* (John Wiley & Sons: New York), pp. 197–212.

Gorbatschev, R. and Gaal, G. 1987. The Precambrian history of the Baltic Shield. In: A. Kröner (Ed.), *Proterozoic Lithospheric Evolution*. *American Geophysical Union Geodynamics Series* 17: 149–159.

Gorden, R. W. and Wiegert, R. G. 1977. Bacterial types and interactions in a thermal blue-green algae-Ephydrid fly ecosystem. Chapter 7. In: J. Cairns Jr. (Ed.), *Aquatic Microbial Communities* (Garland Publishing Co.: New York), pp. 205–241.

Gordon, W. T. 1920. Scottish National Antarctic Expedition, 1902–1904: Cambrian organic remains from a dredging in the Weddell Sea. *Transactions of the Royal Society of Edinburgh* 52: 681–714.

Gorin, G. E., Racz, L. G. and Walter, M. R. 1982. Late Precambrian-Cambrian sediments of Huqf Group, Sultanate of Oman. *American Association of Petroleum Geologists Bulletin* 66: 2609–2627.

Gorjansky, W. J. and Popov, L. Y. 1986. On the origin and systematic position of the calcareous-shelled inarticulate brachiopods. *Lethaia* 19: 233–240.

Gorlenko, V. M. and Yurkov, V., V 1988. Ecology of phototrophic microorganisms of Bolshereschensky Thermal Springs. *Poster presented at VI International Symposium on Photosynthetic Prokaryotes, Noordwijkerhout, The Netherlands, 8-13 August, 1988*: 179 (Abstract).

Gorlenko, V. M., Kompantseva, E. I. and Puchkova, N. N. 1985. Influence of temperature on prevalence of phototrophic bacteria in hot springs. *Mikrobiologiya* 54: 848–853.

Goryanskij, V. Y. 1977. Pervaya nakhodka ostatkov gubki v nizhnem kembrij vostochnoj Sibiri [The first find of sponge remains in the Lower Cambrian of Siberia]. *Ezhegodnik Vsesoyuznogo Paleontologicheskogo Obshchestva* 20: 274–276.

Gottschalk, G. 1986. *Bacterial Metabolism, Second Edition* (Springer-Verlag: Berlin), 359 pp.

Gottschalk, M. and Blanz, P. A. 1984. Highly conserved 5S ribosomal RNA sequences in four rust fungi and atypical 5S rRNA secondary structure in *Microstroma juglandis*. *Nucleic Acids Research* 12: 3951–3958.

Gottschalk, M. and Blanz, P. A. 1985. Untersuchungen an 5S ribosomalen Ribonukleinsäuren als Beitrag zur Klärung von Systematik und Phylogenie der Basidiomyceten. *Zeitschrift für Mykologie* 51: 205–243.

Gough, D. O. 1981. Solar interior structure and luminosity variations. *Solar Physics* 74: 21–34.

Gould, S. J. 1976. The interpretation of diagrams. *Natural History* 85(7): 18–28.

Gould, S. J. 1977. *Ever Since Darwin* (Norton: New York). 285 pp.

Gould, S. J. 1983. Nature's great era of experiments. *Natural History* 92(7): 12–21.

Gould, S. J. 1988. Trends as changes in variance: a new slant on progress and directionality in evolution. *Journal of Paleontology* 62: 319–329.

Gould, S. J. 1989. *Wonderful Life. The Burgess Shale and the Nature of History* (Norton: New York), 347 pp.

Gouy, M. and Li, W.-H. 1989a. Molecular phylogeny of the kingdoms Animalia, Plantae and Fungi. *Molecular Biology and Evolution* 6: 109–122.

Gouy, M. and Li, W.-H. 1989b. Phylogenetic analysis based on rRNA sequences support the archaebacterial rather than the eocyte tree. *Nature* 339: 145–147.

Govindjee and Braun, B. Z. 1974. Light absorption, emission and photosynthesis. In: W. D. P. Stewart (Ed.), *Algal Physiology and Biochemistry* (University of California Press: Berkeley), pp. 346–390.

Gowda, S. S. and Sreenivasa, T. N. 1969. Microfossils from the Archean Complex of Mysore. *Journal of the Geological Society of India* 10: 201–208.

Gower, C. F. and Owen, V. 1984. Pre-Grenvillian and Grenvillian lithotectonic regions in eastern Labrador—Correlations with the Sveconorwegian Orogenic Belt in Sweden. *Canadian Journal of Earth Sciences* 21: 678–693.

Grabau, A. W. 1900. Paleontology of the Cambrian Terranes of the Boston Basin. *Occasional papers of the Boston Society of Natural History* 4: 601–694.

Grabau, A. W. and Shimer, H. W. 1909. *North American Index Fossils. Invertebrates* (Seiler & Co.: New York), 853 pp.

Grambling, J. A., Williams, M. L. and Mawer, C. K. 1988. Proterozoic tectonic assembly of New Mexico. *Geology* 16: 724–727.

Grandstaff, D. E. 1976. A kinetic study of the dissolution of uraninite. *Economic Geology* 71: 1493–1506.

Grandstaff, D. E. 1980. Origin of uraniferous conglomerates at Elliot

References

Lake, Canada and Witwatersrand, South Africa: Implications for oxygen in the Precambrian Atmosphere. *Precambrian Research* 13: 1–26.

Grandstaff, D. E., Edelmann, M. J., Forster, R. W., Zbinden, E. and Kimberley, M. M. 1986. Chemistry and mineralogy of Precambrian paleosols at the base of the Dominion and Pongola Groups (Transvaal, South Africa). *Precambrian Research* 32: 97–131.

Grant, S. W. F., Knoll, A. H. and Germs, G. J. B. 1987. Metaphyte biomineralization in the uppermost Proterozoic Nama Group, Namibia. *Geological Society of America Abstracts with Programs* 19: 681.

Grantham, P. J. 1986a. The occurrence of unusual C_{27} and C_{29} sterane predominances in two types of Oman crude oil. *Organic Geochemistry* 9: 1–10.

Grantham, P. J. 1986b. Sterane isomerisation and moretane/hoptane ratios in crude oils derived from Tertiary source rocks. *Organic Geochemistry* 9: 293–304.

Grantham, P. J., Posthuma, J. and De Groot, K. 1980. Variations and significance of the C_{27} and C_{28} triterpane content of a North Sea core and various North Sea crude oils. In: J. R. Maxwell and A. G. Douglas (Eds.), *Advances in Organic Geochemistry 1979* (Pergamon Press: Oxford), pp. 29–38.

Grantham, P. J., Lijmbach, G. W. M., Posthuma, J., Hughes, C. M. W. and Willink, R. J. 1988. Origin of crude oils in Oman. *Journal of Petroleum Geology* 11: 61–79.

Grasshoff, K. (Ed.). 1976. *Methods of Seawater Analysis* (Verlag Chemie: Weinheim), 419 pp.

Gray, M. W. 1983. The bacterial ancestry of plastids and mitochondria. *Bioscience* 33: 693–699.

Gray, M. W. 1988. Organelle origins and ribosomal RNA. *Biochemistry and Cell Biology* 66: 325–348.

Gray, M. W. 1989. The evolutionary origin of organelles. *Trends in Genetics* 5: 294–299.

Gray, M. W. and Boer, P. H. 1988. Organization and expression of algal (*Chlamydomonas reinhardtii*) mitochondrial DNA. *Philosophical Transactions of the Royal Society of London* B 319: 135–147.

Gray, M. W. and Doolittle, W. F. 1982. Has the endosymbiont hypothesis been proven? *Microbiological Reviews* 46: 1–42.

Gray, M. W., Sankoff, D. and Cedergen, R. J. 1984. On the evolutionary descent of organisms and organelles: a global phylogeny based on a highly conserved structural core in small subunit ribosomal RNA. *Nucleic Acids Research* 12: 5837–5852.

Gray, M. W., Cedegren, R., Abel, Y. and Sankoff, D. 1989. On the evolutionary origin of the plant mitochondrion and its genome *Proceedings of the National Academy of Science USA* 86: 2267–2271.

Gregory, S., V 1983. Plant-herbivore interactions in stream systems. In: J. R. Barnes and G. W. Minshall (Eds.), *Stream Ecology: Application and Testing of General Ecological Theory* (Plenum Press: New York), pp. 157–189.

Grey, K. 1982. Aspects of Proterozoic stromatolite biostratigraphy in Western Australia. *Precambrian Research* 18(4): 347–365.

Grey, K. 1984. Biostratigraphic Studies of Stromatolites from the Proterozoic Earaheedy Group Nabberu Basin, Western Australia. *Geological Survey of Western Australia Bulletin* 130: 1–123.

Grey, K. 1987. *Acaciella* cf. *Australica*, a Late Proterozoic stromatolite from the Savory Group, Robertson 1:250,000 sheet area, W. A. Geol. Surv. Western Australia. *Palaeontology Report* 9(1987): 13.

Grey, K. 1989. Handbook on Stromatolites. *Stromatolite Newsletter*.

Grey, K. and Thorne, A. M. 1985. Biostratigraphic significance of stromatolites in upward shallowing sequences of the early Proterozoic Duck Creek Dolomite, Western Australia. *Precambrian Research* 29: 183–206.

Grey, K. and Williams, I. R. 1987. Possible megascopic algae from the Middle Proterozoic Manganese Group, Bangemall Basin, Western Australia. In: S. Beadle (Ed.), *Abstracts, 4th International Symposium on Fossil Algae. Friends of the Algae Newsletter* 8: 36–37.

Grey, K. and Williams, I. R. 1990. Problematic bedding-plane markings from the Middle Proterozoic Manganese Subgroup, Bangemall Basin, Western Australia. *Precambrian Research* 46: 307–327.

Grey, K., Moore, L. S., Burne, R., V, Pierson, B. K. and Bauld, J. 1990. Lake Thetis, Western Australia: an example of saline lake sedimentation dominated by benthic microbial processes. *Australian Journal of Marine and Freshwater Research* 41: 275–300.

Grigor'eva, N. V. and Zhegallo, E. A. 1979. K issledovanii mikrostruktur nekotorykh tommotskikh iskopaemykh [To the study of the microstructure of some Tommotian fossils]. *Paleontologicheskij Zhurnal* 1979(2): 142–144.

Grigor'eva, N., V, Mel'nikova, L. M. and Pel'man, Y. L. 1983. Briiopody, ostrakody, (bradoriidy) i problematika iz stratitypicheskogo rayona yarusov nizhnego kembriya [Brachiopods, ostracodes (bradoriids) and a problematicum from the stratotype region of the Lower Cambrian stages]. *Paleontologicheskij Zhurnal* 1983: 54–58.

Gromet, L. P., Dymek, R. F., Haskin, L. A. and Korotev, R. L. 1984. The "North American Shale Composite": its composition, major and trace element characteristics. *Geochimica et Cosmochimica Acta* 48: 2469–2482.

Groom, T. T. 1902. On *Polyphyma*, a new genus belonging to the Leperditiidae from the Cambrian shales of Malvern. *Quarterly Journal of the Geological Society of London* 58: 83–88.

Gross, G. A. and Zajac, I. S. 1983. Iron-formation in fold belts marginal to the Ungava Craton. In: A. F. Trendall and R. C. Morris (Eds.), *Iron-Formations: Facts and Problems* (Elsevier: Amsterdam), pp. 253–294.

Grotzinger, J. P. 1986a. Cyclicity and paleoenvironmental dynamics, Rocknest platform, northwest Canada. *Geological Society of America Bulletin* 97: 1208–1231.

Grotzinger, J. P. 1986b. Evolution of early Proterozoic passive-margin carbonate platform, Rocknest Formation, Wopmay Orogen, Northwest Territories, Canada. *Journal of Sedimentary Petrology* 56(6): 831–847.

Grotzinger, J. P. 1989. Introduction to Precambrian reefs. In: H. Geldsetzer, N. P. James and G. Tebbutt (Eds.), *Reefs, Canada and Adjacent Areas. Canadian Society of Petroleum Geologists Memoirs* 13: 9–12.

Grotzinger, J. P. In Press. Facies and evolution of Precambrian carbonate depositional systems: emergence of the modern platform archetype. In: P. Crevello, J. F. Read, R. Sarg and J. Wison (Eds.), *Controls on Carbonate Platform and Basin Development. Society of Economic Paleontologists and Mineralogists Special Publication* 44.

Grotzinger, J. P. and Gall, Q. 1986. Preliminary investigations of Early Proterozoic Western River and Burnside River Formations: Evidence for foredeep origin of Kilohigok Basin, N.W.T., Canada. *Current Research, Part A. Geological Survey of Canada, Paper* 86-1A: 95–106.

Grotzinger, J. P. and McCormick, D. S. 1988. Flexure of the lithosphere and the evolution of Kilohigok Basin (1.9 Ga), northwest Canadian Shield. In: K. Kleinsphehn and C. Paola (Eds.), *Perspectives in Basin Analysis* (Springer-Verlag: Berlin), pp. 405–430.

Grotzinger, J. P. and Read, J. F. 1983. Evidence for primary aragonite precipitation, Lower Proterozoic (1.9 Ga) dolomite, Wopmay Orogen, Northwest Canada. *Geology* 11: 710–713.

Grotzinger, J. P., McCormick, D. S. and Pelechaty, S. M. 1987. Progress report on the stratigraphy, sedimentology, and significance of the Kimerot and Bear Creek Groups, Kilohigok Basin, District of Mackenzie. *Current Research, Part A. Geological Survey of Canada, Paper* 87-1A: 219–238.

Grotzinger, J. P., Gamba, C., Pelechaty, S. M. and McCormick, D. S. 1988. Stratigraphy of a 1.9 Ga foreland basin shelf-to-slope transition: Bear Creek Group, Tinney Hills area of Kilohigok Basin, District of Mackenzie. *Current Research, Part C. Geological Survey of Canada, Paper* 88-1C: 313–320.

Groves, D. I. and Batt, W. D. 1984. Spatial and temporal variations of Archean metallogenic associations in terms of evolution of granitoid-greenstone terrains with particular emphasis on the

Western Australian Shield. In: A. Kröner, G. N. Hansen and A. M. Goodwin (Eds.), *Archaean Geochemistry* (Springer-Verlag: Berlin), pp. 73–98.

Gruau, G., Martin, H., Leveque, B., Capdevilla, R. and Marot, A. 1985. Rb-Sr and Sm-Nd geochronology of lower Proterozoic granite-greenstone terrains in French Guiana, South America. *Precambrian Research* 30: 63–80.

Gruner, J. W. 1923. Algae, believed to be Archean. *Journal of Geology* 31: 146–148.

Gruner, J. W. 1924. Discovery of life in the Archean. *Journal of Geology* 33: 151–152.

Gudymovich, S. S. 1966. Ov ehpifitonakh Anastas'inskoj i Ungutskoj svit pozdnego dokembriya (?)—nizhnego kembriya severo-zapadnoj chasti Vostochnogo Sayana [On *Epiphyton* from the Anastas'inc and Ungut Formations of Late Precambrian (?)—Lower Cambrian age in the northwest part of eastern Sayan]. *Izvestiya Politekhnicheskogo Instituta [Tomsk]* 151: 109–115.

Gudymovich, S. S. 1967. Izvestkovye vodorsli Anastas'inskoj i Ungutskoj svit pozdnego dokembriya (?)-nizhnego kembriya severo-zapadnoj chasti Vostochnogo Sayana [Calcaeous algae from the Anastasyinskaya and Ungutskaya suites of upper Precambrian (?)—Lower Cambrian age in the northwest of East Sayan. In: T. F. Vozzhennikova (Ed.), *Iskopaemye Vodorosli SSSR [Fossil Algae of the USSR]* (Nauka: Moscow), pp. 134–138.

Gulz, P., Faxel, P., Fiege, G. B., Schmitz, B. and Egge, H. 1978. Die Kohlenwasserstoffe der blauaber-flechte *Peltigera canina* (L.) WILLD. *Zeitschrift für Pflanzenphysiologie* 89: 159–167.

Gunatilaka, A. 1975. Some aspects of the biology and sedimentology of laminated algae mats from Mannar Lagoon, Northwest Ceylon. *Sedimentary Geology* 14: 275–300.

Gunderson, J. H., McCutchan, T. F. and Sogin, M. L. 1986. Sequence of the small subunit ribosomal RNA gene expressed in the bloodstream stages of *Plasmodium berghei*: evolutionary implications. *Journal of Protozoology* 33: 525–529.

Gunderson, J. H., Elwood, H., Ingold, A., Kindle, K. and Sogin, M. L. 1987. Phylogenetic relationships between chlorophytes, chrysophytes and oomycetes. *Proceedings of the National Academy of Science USA* 84: 5823–5827.

Gureev, Y. A. 1985. Vendiata—primitivnye dokembriyskie Radialia [Vendiata—primitive Precambrian Radialia]. *Akademiya Nauk SSSR, Sibirskoe Otdelenie, Trudy Institut Geologicheskikh Nauk* 632: 93–103.

Gureev, Y. A. 1987. Morfologicheskiy analiz i systematika Vendiat [Morphological analysis and systematics of the Vendiata]. *Akademiya Nauk Ukrainskoy SSSR, Institut Geologicheskikh Nauk* 87-15: 1–54 (Preprint).

Gürich, G. 1906. Les Spongiostromides du Viséen de la province de Namur. *Musée Royal d'Histoire Naturelle de Belgique Memoir* 3: 55.

Gürich, G. 1929. Die ältesten Fossilien Südafrikas. *Zeitschrift für Praktische Geologie* 37: 85–86.

Gürich, G. 1930a. Die bislang ältesten Spuren von Organismen in Südafrika. *International Geological Congress* 15(2): 670–680.

Gürich, G. 1930b. Über den Kuibis-Quartzit in Südwestafrika. *Zeitschrift der Deutschen Geologischen Gesellschaft* 82: 637.

Gürich, G. 1933. Die Kuibis-Fossilien der Nama-Formation von Südwestafrika. *Paläontologische Zeitschrift* 15: 137–154.

Guss, J. M. and Freeman, H. C. 1983. Structure of oxidized Poplar plastocyanin at 1.6 Å resolution. *Journal of Molecular Biology* 169: 521–563.

Gustafson, L. B. and Williams, N. 1981. Sediment-hostend stratiform deposits of copper, lead, and zinc. In: B. J. Skinner (Ed.), *Economic Geology 75th Anniversary Volume* (Economic Geology Publishing Company: El Paso), pp. 139–178.

Gutschick, R. C. and Rodriguez, J. 1990. By-the-wind-sailors from a Late Devonian foreshore environment in western Montana. *Journal of Paleontology* 64: 31–39.

Haack, T. K. and McFeters, G. A. 1982. Microbial dynamics of an epilithic mat in a high alpine stream. *Applied and Environmental Microbiology* 43: 702–707.

Hacquaert, A. 1943. Over het Voorkomen van *Girvanella* in een oolithisch Gesteente van de Serie van Mwashya uit Katanga [On the occurrence of *Girvanella* in an oolitic rock from the Mwashya Series of Katanga]. *Natuurwetenschappelijk Tijdschrift voor Nederlandsch Indiee* 25(2–4): 33–38.

Häder, D.-P. 1987a. Photomovement. In: P. Fay and C. Van Baalen (Eds.), *The Cyanobacteria* (Elsevier: Amsterdam), pp. 325–345.

Häder, D.-P. 1987b. Photosensory behavior in prokaryotes. *Microbiological Reviews* 51: 1–21.

Häder, D.-P. and Tevini, M. 1987. *General Photobiology* (Pergamon Press: Oxford, New York), 323 pp.

Hadzi, J. 1963. *The Evolution of the Metazoa* (Pergamon Press: Oxford), 499 pp.

Hahn, G. and Pflug, H. D. 1980. Ein neuer Medusen-Fund aus dem Jung-Präkambrium von Zentral-Iran. *Senckenbergiana Lethaea* 60: 449–460.

Hahn, G. and Pflug, H. D. 1985a. Die Cloudinidae n. fam., Kalk-Röhren aus dem Vendium und Unter-Cambrium. *Senckenbergiana Lethaea* 65: 413–431.

Hahn, G. and Pflug, H. D. 1985b. Polypenartige Organismen aus dem Jung-Präkambrium (Nama-Gruppe) von Namibia. *Geologica et Palaeontologica* 19: 1–8.

Hahn, G. and Pflug, H. D. 1988. Zweischalige Organismen aus dem Jung-Präkambrium (Vendium) von Namibia (SE-Afrika). *Geologica et Palaeontologica* 22: 1–19.

Hahn, G., Hahn, R., Leonardos, O. H., Pflug, H. D. and Walde, D. H. G. 1982. Körperlich erhaltene Scyphozoen-Reste aus dem Jungpräkambrium Brasiliens. *Geologica et Palaeontologica* 16: 1–18.

Haldane, J. B. S. 1928. The origin of life. *Rationalist Annual* 148: 3–10.

Hall, J. 1847. Palaeontology. *Geological Survey of New York* 1: 1–338.

Hall, J. 1872. On some new or imperfectly known forms among the Brachiopoda. *Annual Report of the New York State Museum of Natural History* 23: 244–247.

Hall, J. 1879. Containing descriptions of the Gasteropoda, Pteropoda and Cephalopoda of the Upper Helderberg, Hamilton, Portage and Chemung groups. In: *Palaeontology of New York* (New York State Geological Survey: New York), pp. 1–492.

Hall, J. and Clarke, J. M. 1892. An introduction to the study of the genera of Paleozoic Brachiopoda. *New York Geological Survey* 8: 1–367.

Hall, J. and Whitfield, R. P. 1872. Notice of two new species of fossil shells from the Potsdam sandstone of New York. *Annual Report of the New York State Museum of Natural History* 23: 241–242.

Hall, J. L., Ramanis, Z. and Luck, D. J. L. 1989. Basal body/centriolar DNA: Molecular genetics studies in *Chlamydomonas*. *Cell* 59: 121–132.

Hallberg, J. A. 1986. Archean basin development and crustal extension in the northeastern Yilgarn Block, Western Australia. *Precambrian Research* 31: 133–156.

Halley, R. B. 1976. Textural variation within Great Salt Lake algal mounds. In: M. R. Walter (Ed.), *Stromatolites: Developments in Sedimentology*, vol. 20 (Elsevier: Amsterdam), pp. 435–445.

Halpern, Y. 1988. The effects of lithology, diagenesis, and low-grade metamorphism on the ultrastructure and surface sculpture of acritarchs from the Late Proterozoic Chuar Group, Grand Canyon, Arizona. M.S. thesis, Tulane University (Department of Geology).

Hamar, G. 1967. *Platysolenites antiquissimus* Eicw. (Vermes) from the lower Cambrian of Norway. *Norges Geologiske Undersoekelse* 249: 88–95.

Hamdi, B., Brasier, M. D. and Zhiwen, J. 1989. Earliest skeletal fossils from Precambrian-Cambrian boundary strata, Elburz Mountains, Iran. *Geological Magazine* 126: 283–289.

Hamilton, J. 1977. Sr. isotope and trace element studies of the Great Dyke and Bushveld Mafic Phase and their relation to early Proterozoic magma genesis in southern Africa. *Journal of Petrology* 18: 24–52.

Hamilton, W. A. 1987. Biofilms: microbial interactions and metabolic

References

activities. In: M. Fletcher, T. R. G. Gray and J. G. Jones (Eds.), *Ecology of Microbial Communities. Society of General Microbiology Symposium*, 41 (Cambridge University Press: Cambridge), pp. 361–385.

Han, J. and Calvin, M. 1969a. Hydrocarbon distribution of algae and bacteria, and microbiological activity in sediments. *Proceedings of the National Academy of Science USA* 64: 436–443.

Han, J. and Calvin, M. 1969b. Occurrences of fatty acids and alophatic hydrocarbons in a 3-4 billion-year-old sediment. *Nature* 224: 576–577.

Han, J. and Calvin, M. 1970. Branched alkanes from blue-green algae. *Chemical Communications* 22: 1490–1491.

Han, J., McCarthy, E. D., Calvin, M. and Benn, M. H. 1968. Hydrocarbon constituents of the blue-green algae *Nostoc muscorum*, *Anacystis nidulans*, *Phormidium luridum* and *Chlorogloea fritschii*. *Journal of the Chemical Society (C)*: 2785–2790.

Han, J., Chan, H. W.-S. and Calvin, M. 1969. Biosynthesis of alkanes in *Nostoc muscorum*. *Journal of the American Chemical Society* 91: 5156–5159.

Handfield, R. C. 1969. Early Cambrian coral-like fossils from the Northern Cordillera of Western Canada. *Canadian Journal of Earth Sciences* 6(4): 782–785.

Hanert, H. H. 1981. The genus *Gallionella*. In: M. P. Starr, H. Stolp, H. G. Trüper, A. Ballows and H. G. Schlegel (Eds.), *The Prokaryotes*, Vol 1 (Springer-Verlag: Berlin), pp. 509–515.

Hannsmann, P. 1988. Ultrastructural localisation of RNA in Cryptomonads. *Protoplasma* 146: 81–88.

Hannsmann, P., Falk, H., Scheer, U. and Sitte, P. 1986. Ultrastructural localization of DNA in two *Cryptomonas* species by use of a monoclonal DNA antibody. *European Journal of Cell Biology* 42: 152–160.

Hanson, E. C., Newton, R. C. and Janardhan, A. S. 1984. Pressures, temperatures and metamorphic fluids across an unbroken amphibolite facies to granulite facies transition in southern Karnataka, India. In: A. Kröner, G. N. Hansen and A. M. Goodwin (Eds.), *Archean Geochemistry* (Springer-Verlag: Berlin), pp. 161–181.

Hanson, R. E., Wilson, T. J., Brueckner, H. K., Onstott, T. C., Wardlaw, M. S., Johns, C. C. and Hardcastle, K. C. 1988a. Reconnaissance geochronology, tectonothermal evolution, and regional significance of the Middle Proterozoic Choma-Kalomo block, southern Zambia. *Precambrian Research* 42: 39–61.

Hanson, R. E., Wilson, T. J. and Wardlaw, M. S. 1988b. Deformed batholiths in the Pan-African Zambesi belt, Zambia. *Geology* 16: 1134–1137.

Häntzschel, W. 1949. Zur Deutung von *Manchuriophycus* ENDO und ähnlichen Problematika. *Mitteilungen des Geologischen Staatsinstituts, Hamburg* 19: 77–84.

Häntzschel, W. 1962. Trace fossils and problematica. In: R. C. Moore (Ed.), *Treatise on Invertebrate Paleontology, Part W* (Geological Society of America and University of Kansas Press: Lawrence), pp. W177–W245.

Häntzschel, W. 1965. *Vestigia invertebratorum et problematica*. *Fossilium Catalogus, I: Animalia, pars 108*. F. Westphal (Ed.), (Uitgeverij Dr. E. Junk's: Gravenhage), 142 pp.

Häntzschel, W. 1975. *Treatise on Invertebrate Paleontology. Supplement 1. Trace Fossils and Problematica, 2nd ed. Part W. Miscellanea*, (Geological Society of America and University of Kansas: Lawrence), 269 pp.

Hao, Y.-C. and Shu, D.-G. 1987. [The oldest known well-preserved Phaeodaaria (Radiolaria) from southern Shaanxi]. *Geoscience [Chinese]* 1: 301–310.

Hardie, L. A. 1967. The gypsum anhydrite equilibrium at one atmosphere pressure. *American Mineralogist* 52: 171–200.

Hargraves, R. B. 1976. Precambrian geologic history. *Science* 193: 363–371.

Hargraves, R. B. 1981. Precambrian tectonic style: a liberal uniformatarian interpretation. In: A. Kröner (Ed.), *Precambrian Plate Tectonics* (Elsevier: Amsterdam), pp. 21–56.

Harland, B. W. 1964. Evidence of late Precambrian glaciation and its significance. In: A. E. M. Nairn (Ed.), *Problems in Palaeoclimatology* (Interscience: London), pp. 119–149.

Harland, W. B. 1964. Critical evidence for a great infra-Cambrian glaciation. *Geologische Rundschau* 54: 45–61.

Harland, W. B. 1983. The Proterozoic glacial record. In: L. G. Medaris Jr., C. W. Byers, D. M. Mickleson and W. C. Shanks (Eds.), *Proterozoic Geology: Selected Papers from an International Proterozoic Symposium*. Geological Society of America Memoir 161: 279–288.

Harrington, H. J. and Moore, R. C. 1955. Fossil jellyfish from Kansas Pennsylvanian rocks and elsewhere. *State Geological Survey of Kansas Bulletin* 114: 153–164.

Harrington, H. J. and Moore, R. C. 1956. Protomedusae. In: R. C. Moore (Ed.), *Treatise on Invertebrate Paleontology. Part F. Coelenterata* (Geological Society of America and University of Kansas Press: Lawrence), pp. 21–23.

Harris, C. W. and Glover, L. 1988. The regional extent of the ca. 600 Ma Virgilina deformation: implications for stratigraphic correlation in the Carolina terrane. *Geological Society of America Bulletin* 100: 200–217.

Harrison, J. E. 1972. Precambrian Belt Basin of northwestern United States: Its geometry, sedimentation, and copper occurrences. *Geological Society of America Bulletin* 83: 1215–1240.

Harrison, J. E. 1984. Session on geochronology and geophysics. *Montana Bureau of Mines and Geology Special Publication* 90: 98–100.

Harrison, J. E. and Peterman, Z. E. 1980. North American Commission on Stratigraphic Nomenclature Note 52—a preliminary proposal for a chronometric time scale for the Precambrian of the United States and Mexico. *Geological Society of America Bulletin* 91(Part 1): 377–380.

Hart, D. D. 1985. Grazing insects mediate algal interactions in a stream benthic community. *Oikos* 44: 40–46.

Hart, T. J. 1986. A petrographic and geochemical study of the 2.2 b.y. Hekpoort paleosol at the Daspoort Tunnel, Pretoria, in the Republic of South Africa. B.A. thesis, Harvard University.

Hartman, H. 1984. The evolution of photosynthesis and microbial mats: a speculation on the banded iron-formations. In: Y. Cohen, R. W. Castenholz and H. O. Halvorson (Eds.), *Microbial Mats: Stromatolites* (Alan Liss: New York), pp. 449–453.

Hartman, W. D. 1981. Form and distribution of silica in sponges. In: T. L. Simpson and B. E. Volcani (Eds.), *Silicon and Siliceous Structures in Biological Systems* (Springer-Verlag: New York), pp. 453–493.

Hartmann, W. K., Phillips, R. J. and Taylor, G. J. 1986. *Origin of the Moon* (Lunar and Planetary Institute: Houston), 781 pp.

Hasegawa, M., Iida, Y., Yano, T., Takaiwa, F. and Iwabuchi, M. 1985. Phylogenetic relationships among eukaryotic kingdoms inferred from ribosomal RNA sequences. *Journal of Molecular Evolution* 22: 32–38.

Hatai, K. and Hayasaka, S. 1961. Tongallen from the Atsumi Penninsula, Aichi Prefecture, Japan. *Japanese Journal of Geology and Geography* 32: 5–8.

Haug, P. and Curry, D. J. 1974. Isoprenoids in a Costa Rican seep oil. *Geochimica et Cosmochimica Acta* 33: 601–619.

Hawley, J. E. 1926. An evaluation of evidence of life in the Archean. *Journal of Geology* 34: 441–461.

Hayami, M., Okabe, A., Sasai, K., Hayashi, H. and Kanemasa, Y. 1979. Presence and synthesis of cholesterol in stable staphyococcal L-forms. *Journal of Bacteriology* 140: 859–863.

Hayes, J. M. 1983a. Geochemical evidence bearing on the origin of aerobiosis, a speculative hypothesis. Chapter 12. In: J. W. Schopf (Ed.), *Earth's Earliest Biosphere: Its Origin and Evolution* (Princeton University Press: Princeton), pp. 291–301.

Hayes, J. M. 1983b. Practice and principles of isotopic measurements in organic geochemistry. In: W. G. Meinschein (Ed.), *Organic Geochemistry of Contemporaneous and Ancient Sediments* (Great Lakes Section, Society of Economic Paleontologists and Mineralogists:

Bloomington, Indiana), pp. 1–31.

Hayes, J. M., Kaplan, I. R. and Wedeking, K. W. 1983. Precambrian organic geochemistry, preservation of the record. Chapter 5. In: J. W. Schopf (Ed.), *Earth's Earliest Biosphere: Its Origin and Evolution* (Princeton University Press: Princeton), pp. 93–134.

Hayes, J. M., Takigiku, R., Ocampo, R., Calloth, H. J. and Albrecht, P. 1987. Isotopic compositions and probable origins of organic molecules in the Eocene Messel Shale. *Nature* 329: 48–51.

Hayes, J. M., Popp, B. N., Takigiku, R. and Johnson, M. W. 1989. An isotopic study of biogeochemical relationships between carbonates and organic carbon in the Greenhorn Formation. *Geochimica et Cosmochimica Acta* 53: 2961–2972.

Haynes, D. W. and Bloom, M. S. 1987. Stratiform copper deposits hosted by low energy sediments: III. Aspects of metal transport. *Economic Geology* 82: 635–648.

Hazzard, J. C. 1954. Rocks and Structure of the northern Province Mountains, San Bernadino County, California. In: R. H. Jahns (Ed.), *Geology of Southern California. California Division of Mines and Geology Bulletin* 170(4): 27–35.

He, T.-G. 1981. [Lower Cambrian (Meishucunian) sachiatids and their stratigraphic significance]. *Chengdu Dizhi Xueyuan Xiebao* 1981: 84–90.

He, T.-G. 1984. [Discovery of *Lapworthella bella* assemblage from Lower Cambrian Meishucun Stage in Niuniuzhai, Leibo County, Sichuan Province]. *Professional Papers in Stratigraphy and Paleontology* 13: 23–24.

He, T.-G. 1987. Early Cambrian conulariids from the Yangtze Platform and their early evolution. *Journal of Chengdu College of Geology* 14(2): 7–18.

He, T.-G. and Pei, F. 1985. The discovery of bivalves from the Lower Cambrian Xinji Formation in Fangcheng County, Henan Province. *Journal of Chengdu College of Geology* 1985(1): 61–66.

He, T.-G. and Xie, Y.-S. 1989. [Some problematic small shelly fossils from the Meishucunian of the Lower Cambrian in the western Yangtze Region]. *Acta Micropalaeontologica Sinica* 6: 111–127.

He, T.-G. and Yang, X.-H. 1982. Lower Cambrian Meishucun Stage of the western Yangtze stratigraphic region and its small shelly fossils. *Bulletin of the Chengdu Institute of Geological and Mineral Research* 1982(3): 69–95.

He, T.-G., Pei, F. and Fu, G.-H. 1984. [Some small shelly fossils from the Lower Cambrian Xinji Formation in Fangcheng County, Henan Province]. *Acta Palaeontologica Sinica* 23: 350–357.

Heda, G. D. and Madigan, M. T. 1986. Aspects of nitrogen fixation in *Chlorobium*. *Archives of Microbiology* 143: 330–336.

Hedström, H. 1923. On "*Discinella holsti* Mbg." and *Scapha antiquissima* (Markl.) of the division Patellacea. *Sveriges Geologiska Undersökning* 313(Series C): 1–13.

Hedström, H. 1930. *Mobergella* versus *Discinella*; *Paterella* versus *Scapha* & *Archaeophiala* (some questions on nomenclature), *Sveriges Geologiska Undersökning* 362(Series C): 1–8.

Henderson, J. B. 1975. Archean stromatolites in the northern Slave Province, Northwest Territories, Canada. *Canadian Journal of Earth Sciences* 12: 1619–1630.

Henderson-Sellers, A. 1979. Clouds and the long term stability of the earth's atmosphere and climate. *Nature* 279: 786–788.

Hendricks, L., Goris, A., Neefs, J.-M., Vande Peer, Y., Hennebert, G. and de Wachter, R. 1989. The nucleotide sequence of the small subunit RNA of the yeast *Candida albicans* and the evolutionary position of the fungi along the eukaryotes. *Systematics and Applied Microbiology* 12: 223–229.

Hennig, W. 1966. *Phylogenetic Systematics* (University of Illinois Press: Urbana, Ill.), 263 pp.

Henrichs, S. M. and Farrington, J. W. 1984. Peru upwelling region sediments near 15°S. 1. Remineralization and accumulation of organic matter. *Limnology and Oceanography* 29: 1–19.

Henry, G., Stanistreet, I. G. and Maiden, K. J. 1986. Preliminary results of a sedimentological study of the Chuos Formation in the central zone of the Damara orogen: Evidence for mass flow processes and glacial activity. *Communications Geological Survey of South West Africa/Namibia* 2: 75–92.

Hensel, H., Zwicke, P., Fabry, S., Lang, J. and Palm, P. 1989. Sequence comparison of glyceraldehyde-3-phosphate dehydrogenases from the three urkingoms: evolutionary implication. *Canadian Journal of Microbiology* 35: 81–85.

Herman, H. B. and Rechnitz, G. A. 1975. Preparation and properties of a carbonate ion-selective membrane electrode. *Analytica Chimica Acta* 76: 155–164.

Herzog, M. and Maroteaux, L. 1986. Dinoflagellate 17S rRNA sequence inferred from the gene sequence: Evolutionary implications. *Proceedings of the National Academy of Science USA* 83: 8644–8648.

Hewitt, R. A. 1980. Microstructural contrasts between some sedimentary francolites. *Journal of the Geological Society* 137(6): 661–667.

Hibberd, D. J. 1977. Ultrastructure of the cryptomonad endosymbiont of the red-water ciliate *Mesodinium rubrum*. *Journal of the Marine Biology Association of the UK* 57: 45–61.

Hibberd, D. J. and Norris, R. E. 1984. Cytology and ultrastructure of *Chlorarachnion reptans* (Chlorarachniophyta division nova, Chlorarachniophyceae classis nova). *Journal of Phycology* 20: 310–330.

Hickman, A. H. 1978. *Nullagine, Western Australia, 1:250,000 Geological Series—Explanatory Notes (Sheet SF 51-5 International Index)* (Geological Survey of Western Australia: Perth), 22 pp.

Hickman, A. H. 1983. Geology of the Pilbara Block and its environs. *Bulletin of the Geological Survey of Western Australia* 127: 1–267.

Hickman, A. H. and Lipple, S. L. 1978. *Marble Bar Western Australia, 1:250,000 Geological Series—Explanatory Notes* (Geological Survey of Western Australia: Perth), 24 pp.

Hicks, H. 1872. Undescribed fossils from the Menevian group. *Quarterly Journal of the Geological Society of London* 28: 173–184.

Hildebrand, R. S., Hoffman, P. F. and Bowring, S. A. 1987. Tectonomagmatic evolution of the 1.9-Ga Great Bear magmatic zone, Wopmay orogen, northwestern Canada. *Journal of Volcanological Geothermal Research* 32: 99–118.

Hill, D. 1972. Archaeocyatha. In: C. Teichert (Ed.), *Treatise on Invertebrate Paleontology, Part E*, vol. 1 (Geological Society of America and University of Kansas Press: Lawrence, Kansas), pp. E1–E158.

Hill, W. R. and Knight, A. W. 1987. Experimental analysis of the grazing interaction between a mayfly and stream algae. *Ecology* 68: 1955–1965.

Hill, W. R. and Knight, A. W. 1988. Concurrent grazing effects of two stream insects on periphyton. *Limnology and Oceanography* 33(1): 15–26.

Hiller, R. G., Larkum, A. W. D. and Wrench, P. M. 1988. Chlorophyll proteins of the prymnesiophyte *Pavlova lutherii* (Droop) Comb. nov.: identification of the major light-harvesting complex. *Biochimica et Biophysica Acta* 932: 223–231.

Hills, I. R. and Whitehead, E., V 1966. Triterpanes in optically active petroleum distillates. *Nature* 209: 977–979.

Hinz, I. 1987. The Lower Cambrian microfauna of Comley and Rushton, Shropshire/England. *Palaeontographica* A 198(1–3): 41–100.

Hiscott, R. N., James, N. P. and Pemberton, S. G. 1984. Sedimentology and ichnology of the Lower Cambrian Bradore Formation, coastal Labrador: Fluvial to shallow-marine transgressive sequence. *Bulletin of Canadian Petroleum Geologists* 32: 11–26.

Hoefs, J. 1987. *Stable Isotope Geochemistry*, 3rd ed. (Springer-Verlag: Berlin), 241 pp.

Hoeg, O. A. 1932. Ordivician algae from the Trondheim area. *Norske Videnskaps Akademii i Oslo Skrifter I. Matematisk-Naturvidenskapelig Klasse* 1932: 63–96.

Hoering, T. C. 1961. The stable isotopes of carbon in the carbonate and reduced carbon of Precambrian sediments. *Carnegie Institution of Washington Yearbook* 61: 190–191.

Hoering, T. C. 1967a. Criteria for suitable rocks in Precambrian organic geochemistry. *Carnegie Institution of Washington Yearbook* 65: 365–372.

Hoering, T. C. 1967b. The organic geochemistry of Precambrian rocks.

References

In: P. H. Abelson (Ed.), *Researches in Geochemistry* (John Wiley & Sons: New York), pp. 89–111.

Hoering, T. C. 1976. Molecular fossils from the Precambrian Nonesuch Shale. *Carnegie Institute of Washington Yearbook* 75: 806–813.

Hoering, T. C. 1981. Monomethyl acyclic hydrocarbons in petroleum and rock extracts. *Carnegie Institute of Washington Yearbook* 80: 389–393.

Hoering, T. C. and Navale, V. 1987. A search for molecular fossils in the kerogen of Precambrian sedimentary rocks. *Precambrian Research* 34: 247–267.

Hoffman, A. and Nitecki, M. H. (Eds.). 1986. *Problematic Fossil Taxa. Oxford Monographs on Geology and Geophysics*, 5 (Oxford University Press: New York), 267 pp.

Hoffman, P. 1967. Algal stromatolites: use in stratigraphic correlation and paleocurrent determination. *Science* 157: 1043–1045.

Hoffman, P. F. 1973. Evolution of an early Proterozoic continental margin: the Coronation geosyncline and associated aulacogens of the northwestern Canadian Shield. *Philosophical Transactions of the Royal Society of London A* 273: 547–581.

Hoffman, P. F. 1974. Shallow and deep-water stromatolites in lower Proterozoic platform-to-basin facies change, Great Slave Lake, Canada. *American Association of Petroleum Geologists Bulletin* 58: 856–867.

Hoffman, P. F. 1975. Shoaling-upward shale-to-dolomite cycles in the Rocknest Formation, Northwest Territories. In: R. N. Ginsburg and G. de Klein V (Eds.), *Tidal Deposits* (Springer-Verlag: New York), pp. 257–265.

Hoffman, P. F. 1976. Environmental diversity of middle Precambrian stromatolites. In: M. R. Walter (Ed.), *Stromatolites: Developments in Sedimentology*, vol. 20 (Elsevier: Amsterdam), pp. 599–612.

Hoffman, P. F. 1980. Wopmay orogen: A Wilson cycle of Early Proterozoic age in the northwest of the Canadian Shield. In: D. W. Strangway (Ed.), *The Continental Crust and its Mineral Deposits. Geological Association of Canada Special Paper* 20: 523–549.

Hoffman, P. F. 1981. Revision of stratigraphic nomenclature, foreland thrust-fold belt of Wopmay Orogen, District of Mackenzie. *Current Research, Part A. Geological Survey of Canada, Paper* 81-1A: 247–250.

Hoffman, P. F. 1986. Crustal accretion in a 2.7–2.5 Ga "granite-greenstone" terrane, Slave Province, NWT: A prograding trench-arc system? In: M. J. deWit and L. D. Ashwal (Eds.), *Workshop on Tectonic Evolution of Greenstone Belts. Lunar and Planetary Institute Technical Report* 86-10: 120.

Hoffman, P. F. 1987. Early Proterozoic foredeeps, foredeep magmatism, and Superior-type iron-formations of the Canadian Shield. In: A. Kröner (Ed.), *Proterozoic Lithospheric Evolution. American Geophysical Union Geodynamics Series* 17: 85–98.

Hoffman, P. F. 1988. United plates of America, the birth of a craton: Early Proterozoic assembly and growth of Laurentia. *Annual Review of Earth and Planetary Sciences* 16: 543–603.

Hoffman, P. F. 1989a. Pethei Reff Complex (1.9 Ga), Great Slave Lake, N.W.T. In: H. Geldsetzer, N. P. James and G. Tebbutt (Eds.), *Reefs, Canada and Adjacent Areas. Canadian Society of Petroleum Geologists Memoirs* 13.

Hoffman, P. F. 1989b. Speculations on Laurentia's first gigayear (2.0–1.0 Ga). *Geology* 17: 135–138.

Hoffman, P. F. and Bowring, S. A. 1984. A short-lived 1.9 Ga continental margin and its destruction, Wopmay Orogen, northwest Canada. *Geology* 12: 68–72.

Hoffmann, C. F., Foster, C. B., Powell, T. G. and Summons, R. E. 1987. Hydrocarbon biomarkers from Ordovician sediments and the fossil alga *Gloeocapsomorpha prisca* Zalessky 1917. *Geochimica et Cosmochimica Acta* 51: 2681–2697.

Hofmann, H. J. 1969a. Attributes of stromatolites. *Geological Survey of Canada, Paper* 69-39: v–58.

Hofmann, H. J. 1969b. Stromatolites from the Proterozoic Animikie and Sibley Groups, Ontario. *Geological Survey of Canada, Paper* 68-69: 1–77.

Hofmann, H. J. 1971a. Polygonomorph acritarch from the Gunflint Formation (Precambrian), Ontario. *Journal of Paleontology* 45: 522–524.

Hofmann, H. J. 1971b. Precambrian fossils, pseudofossils and problematica in Canada. *Geological Survey of Canada, Bulletin* 189: 1–146.

Hofmann, H. J. 1972. Precambrian remains in Canada: Fossils, dubiofossils, and pseudofossils. In: A. M. Goodwin and H. R. Wynne-Edwards (Eds.), *Proceedings of the 24th International Geological Congress, Section 1* (Hapell's Press Co-operative: Gardenvale, Quebec), pp. 20–30.

Hofmann, H. J. 1973. Stromatolites: characteristics and utility. *Earth-Science Reviews* 9: 339–373.

Hofmann, H. J. 1975. Stratiform Precambrian stromatolites, Belcher Islands, Canada: relations between silicified microfossils and microstructure. *American Journal of Science* 275: 1121–1132.

Hofmann, H. J. 1976. Precambrian microflora, Belcher Islands, Canada: significance and systematics. *Journal of Paleontology* 50: 1040–1073.

Hofmann, H. J. 1981a. First record of a Late Proterozoic faunal assemblage in the North American Cordillera. *Lethaia* 14: 303–310.

Hofmann, H. J. 1981b. Precambrian fossils in Canada—the 1970's in retrospect. In: F. H. A. Campbell (Ed.), *Proterozoic Basins in Canada. Geological Survey of Canada, Paper* 81-10: 419–443.

Hofmann, H. J. 1982. J. W. Dawson and 19th century Precambrian paleontology. *Third North American Paleontological Convention, Proceedings* 1: 243–249.

Hofmann, H. J. 1985a. The mid-Proterozoic Little Dal macrobiota, Mackenzie Mountains, north-west Canada. *Palaeontology* 28: 331–354.

Hofmann, H. J. 1985b. Precambrian carbonaceous megafossils. In: D. F. Toomey and M. H. Nitecki (Eds.), *Paleoalgology: Contemporary Research and Applications* (Springer-Verlag: Berlin), pp. 20–33.

Hofmann, H. J. 1987. Precambrian biostratigraphy. *Geoscience Canada* 14(3): 134–154.

Hofmann, H. J. 1988. An alternative interpretation of the Ediacaran (Precambrian) chondrophore *Chondroplon* Wade. *Alcheringa* 12: 315–318.

Hofmann, H. J. and Aitken, J. D. 1979. Precambrian biota from the Little Dal Group, Mackenzie Mountains, northwest Canada. *Canadian Journal of Earth Sciences* 16: 150–166.

Hofmann, H. J. and Chen Jinbiao 1981. Carbonaceous megafossils from the Precambrian (1800 Ma) near Jixian, northern China. *Canadian Journal of Earth Sciences* 18: 443–447.

Hofmann, H. J. and Grotzinger, J. P. 1985. Shelf-facies microbiotas from the Odjick and Rocknest Formations (Epworth Group; 1.89 Ga), northwestern Canada. *Canadian Journal of Earth Sciences* 22: 1781–1792.

Hofmann, H. J. and Jackson, G. D. 1969. Precambrian (Aphebian) microfossils from Belcher Islands, Hudson Bay. *Canadian Journal of Earth Sciences* 6: 1137–1144.

Hofmann, H. J. and Jackson, G. D. 1987. Proterozoic ministromatolites with radial-fibrous fabric. *Sedimentology* 34: 963–971.

Hofmann, H. J. and Schopf, J. W. 1983. Early Proterozoic microfossils. Chapter 14. In: J. W. Schopf (Ed.), *Earth's Earliest Biosphere: Its Origin and Evolution* (Princeton University Press: Princeton), pp. 321–360.

Hofmann, H. J., Hill, J. and King, A. F. 1979. Late Precambrian microfossils, southeastern Newfoundland. *Current Research Part B. Geological Survey of Canada, Paper* 79-1B: 83–98.

Hofmann, H. J., Pearson, D. A. B. and Wilson, B. H. 1980. Stromatolites and fenestral fabric in Early Proterozoic Huronian Supergroup, Ontario. *Canadian Journal of Earth Sciences* 17: 1351–1357.

Hofmann, H. J., Fritz, W. H. and Narbonne, G. M. 1983. Ediacaran (Precambrian) fossils from the Wernecke Mountains, northwestern Canada. *Science* 221: 455–457.

Hofmann, H. J., Mountjoy, E. W. and Teitz, M. W. 1985a. Ediacaran fossils from the Miette Group, Rocky Mountains, British Columbia, Canada. *Geology* 13: 819–821.

Hofmann, H. J., Thurston, P. C. and Wallace, H. 1985b. Archean stromatolites from Uchi Greenstone Belt, northwestern Ontario. In: D. Ayres, P. C. Thuston, K. D. Card and W. Weber (Eds.), *Evolution of Archean Supracrustal Sequences. Geological Association of Canada Special Paper* 28: 125–132.

Holland, H. D. 1962. Model of the evolution of the earth's atmosphere. In: A. E. J. Engle, H. L. James and B. F. Leonard (Eds.), *Petrologic Studies: A Volume to Honor A. F. Buddington* (Geological Society of America: New York), pp. 447–477.

Holland, H. D. 1973a. Ocean water, nutrients, and atmospheric oxygen. In: *Proceedings of a Symposium on Hydrogeochemistry and Biochemistry* (Clarke: Washington, D.C.), pp. 66–81.

Holland, H. D. 1973b. The oceans: a possible source of iron in ironformations. *Economic Geology* 68: 1169–1172.

Holland, H. D. 1978. *The Chemistry of the Atmosphere and Oceans* (John Wiley & Sons: New York), 351 pp.

Holland, H. D. 1984. *The Chemical Evolution of the Atmosphere and Oceans* (Princeton University Press: Princeton, N.J.), 582 pp.

Holland, H. D. 1989. Volcanic gases and isotopic record of carbon and sulfur in sedimentary rocks. *International Geological Congress* 28(2): 66 (Abstract).

Holland, H. D. and Beukes, N. J. In Press. A paleoweathering profile from Griqualand West, South Africa: Evidence for a dramatic rise in atmospheric oxygen between 2.2 and 1.8 b.y.b.p. *American Journal of Science*.

Holland, H. D. and Freakes, C. R. 1989. Paleosols and their relevance to Precambrian atmospheric composition: a discussion. *Journal of Geology* 97: 761–762.

Holland, H. D. and Zbinden, E. A. 1988. Paleosols and the evolution of the atmosphere. Part I. In: A. Lerman and M. Meybeck (Eds.), *Physical and Chemical Weathering in Geochemical Cycles* (Reidel: Dordrecht), pp. 61–82.

Holland, H. D., Feakes, C. R. and Zbinden, E. N. 1989. The Flin Flon paleosol and the composition of the atmosphere 1.8 b.y.b.p. *American Journal of Science* 289: 362–389.

Holm, G. 1893. Sveriges kambrisk-siluriska Hyolithidae och Conulariidae. *Sveriges Geologiska Undersökning* 112(Ser. C): 1–172.

Holmer, L. E. 1989. Middle Ordovician Phosphatic inarticulate brachiopods from Västergötland and Dalarna, Sweden. *Fossils and Strata* 26: 1–172.

Holm-Hansen, O. 1978. Chl *a* determination: improvements in methodology. *Oikos* 30: 438–447.

Holo, H. and Sirevåg, R. 1986. Autotrophic growth and CO_2 fixation of *Chloroflexus auranticus*. *Archives of Microbiology* 145: 173–180.

Holser, W. T. 1984. Gradual and abrupt shifts in ocean chemistry during Phanerozoic time. In: H. D. Holland and A. F. Trendall (Eds.), *Patterns of Change in Earth Evolution* (Springer-Verlag: Berlin), pp. 123–143.

Holser, W. T., Schidlowski, M., Mackenzie, F. T. and Maynard, J. B. 1988. Geochemical cycles of carbon and sulfur. In: C. B. Gregor, R. M. Garrels, F. T. MacKenzie and J. B. Maynard (Eds.), *Chemical Cycles in the Evolution of the Earth* (John Wiley & Sons: New York), pp. 105–174.

Holt, J. G. (Ed.), In Press. *Bergey's Manual of Systematic Bacteriology*, vol 1–4 (Williams and Wilkins: Baltimore, London).

Hommeril, P. and Rioult, M. 1962. Phénomènes d'erosion et de sédimentation marines entre Sainte-Honorine-des Perthes et Port-en-Bessin (Calvados). Role des *Rhodothamniella floridula* dans la retenue des sédiments fins. *Cahiers de Oceanographie* 14: 25–45.

Hommeril, P. and Rioult, M. 1965. Étude de la fixation des sédiment meubles par deux algues marines: *Microcoleus chthonoplastes* Thur. *Marine Geology* 3: 131–155.

Hori, H. and Osawa, S. 1986. Evolutionary change in 5S rRNA secondary structure and a phylogenic tree of 352 5S rRNA species. *Biosystems* 19: 163–172.

Hori, H. and Osawa, S. 1987. Origin and evolution of organisms as deduced from 5S ribosomal RNA sequences. *Molecular Biology and Evolution* 4(5): 445–472.

Hori, H., Itoh, T. and Osawa, S. 1982. The phylogenic structure of the Metabacteria. *Zentralblatt für Bakteriologie, Mikrobiologie und Hygiene 1. Abteilung Originale C* 3: 18–30.

Hori, H., Lim, B. and Osawa, S. 1985. Evolution of green plants as deduced from 5S rRNA sequences. *Proceedings of the National Academy of Science USA* 82: 820–823.

Horny, R. 1957. [Problematic molluscs (?Amphineura) from the Lower Cambrian of South and East Siberia (USSR)]. *Sbornik Geologickych Ved. Paleontologie* 23: 397–413.

Horodyski, R. J. 1976. Stromatolites of the upper Siyeh Limestone (Middle Proterozoic), Belt Supergroup, Glacier National Park, Montana. *Precambrian Research* 3: 517–536.

Horodyski, R. J. 1977a. Environmental influences on columnar stromatolite branching patterns: examples from the Middle Proterozoic Belt Supergroup, Glacier National Park, Montana. *Journal of Paleontology* 51: 661–671.

Horodyski, R. J. 1977b. *Lyngbya* mats at Laguna Mormona, Baja California, Mexico: Comparison with Proterozoic stromatolites. *Journal of Sedimentary Petrology* 47(3): 1305–1320.

Horodyski, R. J. 1980. Middle Proterozoic shale-facies microbiota from the lower Belt Supergroup, Little Belt Mountains, Montana. *Journal of Paleontology* 54: 649–663.

Horodyski, R. J. 1981. Pseudomicrofossils and altered microfossils from a Middle Proterozoic shale, Belt Supergroup, Montana. *Precambrian Research* 16: 143–154.

Horodyski, R. J. 1982. Problematic bedding-plane markings from the middle Proterozoic Appekunny Argillite, Belt Supergroup, northwestern Montana. *Journal of Paleontology* 56: 882–889.

Horodyski, R. J. 1983. Sedimentary geology and stromatolites of the middle Proterozoic Belt Supergroup, Glacier National Park, Montana. *Precambrian Research* 20: 391–425.

Horodyski, R. J. 1987. A new occurrence of the vase-shaped fossil *Melanocyrillium* and new data on this relatively complex Late Precambrian fossil. *Geological Society of America Abstracts with Programs* 19(7): 707.

Horodyski, R. J. and Bloeser, B. 1983. Possible eukaryotic algal filaments from the Late Proterozoic Chuar Group Grand Canyon, Arizona. *Journal of Paleontology* 57: 321–326.

Horodyski, R. J. and Donaldson, J. A. 1980. Microfossils from the Middle Proterozoic Dismal Lakes Group, Arctic Canada. *Precambrian Research* 11: 125–159.

Horodyski, R. J. and Donaldson, J. A. 1983. Distribution and significance of microfossils in cherts of the Middle Proterozoic Dismal Lakes Group, District of Mackenzie, Northwest Territories, Canada. *Journal of Paleontology* 57: 271–288.

Horodyski, R. J. and Vonder Haar, S. 1975. Recent calcareous stromatolites from Laguna Mormona, Mexico. *Journal of Sedimentary Petrology* 45: 894–906.

Horodyski, R. J., Bloeser, B. and Vonder Haar, S. 1977. Laminated algal mats from a coastal lagoon, Laguna Mormona, Baja California, Mexico. *Journal of Sedimentary Petrology* 47(2): 680–696.

Horvath, Z. and Jackson, K. S. 1981. *Procedure for the Isolation of Kerogen from Sedimentary Rocks Report* 1981/82, (Bureau of Mineral Resources: Canberra), 6 pp.

Hou, X. 1987. Oldest Cambrian bradoriids from eastern Yunnan. In: *Stratigraphy and Palaeontology of Systemic Boundaries in China— PC-C Boundary*, vol. 1 (Nanjing University Publishing House: Nanjing), pp. 537–545.

Hou, X.-G. 1978a. Early Cambrian large bivalved arthropods from Chengjiang, eastern Yunnan. *Acta Palaeontologica Sinica* 26(3): 286–298.

Hou, X.-G. 1987b. Three new large arthropods from Lower Cambrian, Chengjiang, eastern Yunnan. *Acta Palaeontologica Sinica* 26(3): 272–285.

Hou, X.-G. 1987c. Two new arthropods from Lower Cambrian, Chengjiang, eastern Yunnan. *Acta Palaeontologica Sinica* 26(3): 236–256.

Hou, X.-G. and Chen, J.-Y. 1989. [Early Cambrian tenatacled worm-like animals (*Facivermis* gen. nov.) from Chengjiang, Yunnan].

References

Acta Palaeontologica Sinica 28: 32–41.
Hou, X.-G. and Sun, W.-G. 1988. Discovery of Chengjiang fauna at Meishucun, Jinning, Yunnan. *Acta Palaeontologica Sinica* 27: 1–12.
Hou, X.-G., Chen, J.-Y. and Ju, H.-Z. 1989. [Early Cambrian new arthropods from Chenjiang, Yunnan]. *Acta Palaeontologica Sinica* 28: 42–57.
Howard-Williams, C., Vincent, C. L., Broady, P. A. and Vincent, W. F. 1986. Antarctic stream ecosystems: variability in environmental properties and algal community structure. *Internationale Revue der Gesamten Hydrobiologie* 71: 511–544.
Howell, B. F. 1949. New hydrozoan and brachiopod and new genus of worms from the Ordovician Schenectady Formation of New York. *Bulletin of the Wagner Free Institute of Science* 24: 1–10.
Howell, B. F. 1962. Worms. In: R. C. Moore (Ed.), *Treatise on Invertebrate Paleontology, Part W, Miscellanea* (Geological Society of America and University of Kansas Press: Lawrence), pp. W144–W177.
Howsley, K. and Pearson, H. W. 1979. pH dependent sulfide toxicity to oxygenic photosynthesis in cyanobacteria. *FEMS Microbiology Letters* 6: 287–292.
Hoyle, B. and Beveridge, T. J. 1984. Metal binding by the peptidoglycan sacculus of *Escherichia coli* K-12. *Canadian Journal of Microbiology* 30: 204–221.
Hsü, K. J., Oberhänsli, H., Gao, J. Y., Shu, S., Haihong, C. and Krähenbühl, U. 1985. "Strangelove ocean" before the Cambrian explosion. *Nature* 316: 809–811.
Hu, H., Yu, M. and Zhang, X. 1980. Discovery of phycobilin in *Gymnodinium cyaneum* Hu. sp. nov. and its phylogenetic significance. *Kexue Tongbao* 25: 882–884.
Huang, W. and Meinschein, W. G. 1978. Sterols in sediments from Baffin Bay, Texas. *Geochimica et Cosmochimica Acta* 42: 1391–1396.
Hubregtze, J. J. M. W. 1980. The Archean Pikwitonei granulite domain and its position at the margin of the northwestern Superior Province (Central Manitoba). *Manitoba Geological Survey Paper* GP80-3: 1–16.
Hughes, C. M. W. 1988. Stratigraphy and rock unit nomenclature in the oil producing area of interior Oman. *Journal of Petroleum Geology* 11: 5–60.
Hunt, J. M. 1979. *Petroleum Geochemistry and Geology* (W. H. Freeman & Co.: San Francisco), 595 pp.
Hunt, L. T., George, D. G. and Barker, W. C. 1985. The prokaryote-eukaryote interface. *Biosystems* 18: 223–241.
Hunter, D. R. 1974. Crustal Development in the Kaapvaal Craton, II. The Proterozoic. *Precambrian Research* 1: 295–326.
Huo, S.-C. 1956. [Brief notes of Lower Cambrian Archaeostraca from Shensi and Yunnan]. *Acta Palaeontologica Sinica* 4: 425–445.
Huo, S.-C. 1965. Additional notes on Lower Cambrian Archaeostraca from Shensi and Yunnan. *Acta Palaeontologica Sinica* 13(2): 291–303.
Huo, S.-C. and Shu, D.-G. 1982. [Notes on Lower Cambrian Bradoriida (Crustacea) from western Sichan and southern Shaanxi]. *Acta Palaeontologica Sinica* 13: 291–303.
Huo, S.-C. and Shu, D.-G. 1985. [Cambrian Bradoriida of South China]: 1–251.
Hurd, D. C. 1972. Factors affecting solution rate of biogenic opal in sea water. *Earth and Planetary Science Letters* 15: 411–417.
Hurd, D. C. 1973. Interactions of biogenic opal, sediment and seawater in the Central Equatorial Pacific. *Geochimica et Cosmochimica Acta* 37: 2257–2282.
Hurley, P. M. and Rand, J. R. 1969. Pre-drift continental nuclei. *Science* 164: 1229–1242.
Hurley, P. M., Pinson Jr., W. H., Nagy, B. and Teska, T. M. 1972. Ancient age of the middle marker horizon, Onverwacht Group, Swaziland sequence, South Africa. *Earth and Planetary Science Letters* 14: 360–366.
Hurst, J. M. and Hewitt, R. A. 1977. Tubular Problematica from the type Caradoc (Ordovician) of England. *Neues Jahrbuch für Geologie und Paläontologie, Abhandlungen* 153: 147–169.
Hutchinson, G. E. 1961. The biologist poses some problems. In: M. Sears (Ed.), *Oceanography. American Association for the Advancement of Science, Publication* 67: 85–94.
Huysmans, E., Dams, E., Vandenberghe, A. and De Wachter, R. 1983. The nucleotide sequences of the 5S rRNAs of four mushrooms and their use in studying the phylogenetic position of basidomycetes among the eukaryotes. *Nucleic Acids Research* 11: 2871–2880.
Hyde, R. S. 1980. Sedimentary facies in the Archean Timiskaming Group and their tectonic implications, Abitibi greenstone belt, northeastern Ontario, Canada. *Precambrian Research* 12: 161–195.
Hynes, A. J. and Francis, D. M. 1982. Komatiitic basalts of the Cape Smith foldbelt, New Quebec, Canada. In: N. T. Arndt and E. G. Nisbet (Eds.), *Komatiites* (Allen and Unwin: New York), pp. 159–170.
Iams, W. I. and Stevens, R. K. 1988. Radiolaria and other siliceous microfossils of the Cow Head Group (Upper Cambrian-Middle Ordovician) of Western Newfoundland. *Geologica et Palaeontologica* 22: 192–193.
Il'yasova, Z. K. and Lysova, L. A. 1959. Spory nizhnekembrijskikh otlozhenij yuzhnoj chasti Sibirskoj platformy [Spores of the Lower Cambrian deposits of the southern part of the Siberian Platform]. In: E. G. Pershina (Ed.), *Geologiya i Neftegazonosnost' Vostochnoj Sibiri [Geology and Oil-and-Gas potential(?) of Eastern Siberia]* (Gostoptekhizdat: Moscow), pp. 304–311.
Imbus, S. W., Engel, M. H., Elmore, R. D. and Zumberge, J. E. 1988. The origin, distribution and hydrocarbon generation potential of the organic-rich facies in the Nonesuch Formation, central North American rift system: A regional study. In: L. Mattavelli and L. Novelli (Eds.), *Advances in Organic Geochemistry 1987* (Pergamon Press: Oxford), pp. 207–219.
Imhoff, H. F., Sahl, H. G., Soliman, G. S. H. and Trüper, H. G. 1979. The Wadi Natrun: chemical composition and microbial mass developments in alkaline brines of eutrophic desert lakes. *Geomicrobiology Journal* 1: 219–234.
Immega, I. P. and Klein, C. 1976. Mineralogy and petrology of some metamorphic Precambrian iron-formations in southwestern Montana. *American Mineralogist* 61: 1117–1144.
Ingersoll, R., V 1988. Tectonics of sedimentary basins. *Geological Society of America Bulletin* 100: 1704–1719.
Inoue, T. and Orgel, L. E. 1983. A non-enzymatic RNA polymerase model. *Science* 219: 859–862.
Ittekkot, V. 1988. Global trends in the nature of organic matter in river suspensions. *Nature* 332: 436–438.
Ivanovskaya, A. V., Kazanskij, Y. P. and Timofeev, B. V. 1974. Raspredeleniye mikrofitofossilij v razlichnykh litologofatsial'nykh Zonakh rifeya Vostochnoj Sibiri. *Akademiya Nauk SSSR, Trudy* 81: 99–102.
Ivanovskaya, A., V and Timofeev, B., V 1971. Zavisimost' mezhdu solenost'yu i raspredeleniem fitoplanktona [Relationship between salinity and the distribution of phytoplankton]. *Geologiya i Geofizika* 1971(8): 113–117.
Jaag, O. 1945. Untersuchungen über die Vegetation un Biologie der Algen des nackten Gesteins in den Alpen, im Jura und im schweizerischen Mittelland. *Beiträge zur Kryptogamenflora der Schweiz* 9(3): 1–560.
Jackson, K. S., McKirdy, D. M. and Deckelman 1984. Hydrocarbon generation in the Amadeus Basin. *Australian Petroleum Exploration Association Journal* 24: 42–65.
Jackson, M. J., Powell, T. G., Summons, R. E. and Sweet, I. P. 1986. Hydrocarbon shows and petroleum source rocks in sediments as old as 1.7×10^9 years. *Nature* 322: 727–729.
Jackson, M. J., Muir, M. D. and Plumb, K. A. 1987. *Geology of the Southern McArthur Basin, Northern Territory* Bulletin 220, (Bureau of Mineral Resources, Geology and Geophysics: Canberra), 173 pp.
Jackson, M. J., Sweet, I. P. and Powell, T. G. 1988. Petroleum geology and geochemistry of the Middle Proterozoic, McArthur Basin,

Northern Australia 1: Petroleum potential. *Australian Petroleum Exploration Association Journal* 29: 283–302.

Jackson, M. P. A., Eriksson, K. A. and Harris, C. W. 1987. Early Archean foredeep sedimentation related to crustal shortening: a reinterpretation of the Barberton sequence, southern Africa. *Tectonophysics* 136: 197–221.

Jackson, N. J. 1980. Correlations of Late Proterozoic stratigraphies, NE Africa and Arabia: summary of an IGCP Project Report. *Journal of the Geological Society of London* 137: 629–634.

Jackson, T. J., Ramaley, R. F. and Meinschein, W. G. 1973. *Thermomicrobium*, a new genus of extremely thermophilic bacteria. *International Journal of Systematic Bacteriology* 23: 28–36.

Jacobsen, S. B. 1984. Isotopic constraints on the development of the early crust. *Transactions of the American Geophysics Union* 65: 550.

Jacobsen, S. B. and Dymek, R. F. 1988. Nd and Sr isotope systematics of clastic metasediments from Isua, West Greenland: Identification of pre-3.8 Ga differentiated crustal components. *Journal of Geophysical Research* 93: 338–354.

Jacobsen, S. B. and Pimentel-Klose, M. R. 1988. A Nd isotopic study of the Hamersley and Michipicoten banded iron-formations: the source of REE and Fe in Archaean oceans. *Earth and Planetary Science Letters* 87: 29–44.

Jacobsen, S. B. and Wasserburg, G. J. 1981. Transport models for crust and mantle evolution. *Tectonophysics* 75: 163–179.

Jager, H. and Martinsson, A. 1980. The Early Cambrian trace tossil *Plagiogmus* in its type area. *Geologiska Föreningens i Stockholm Förhandlingar* 102: 117–126.

Jägersten, G. 1968. *Livscykelns Evolution hos Metazoa* (Läromedelsförlagen: Stockholm), 295 pp.

Jahnke, L. L. and Nichols, P. D. 1986. Methyl sterol and cyclopropane fatty acid composition of *Methylococcus capsulatus* grown at low oxygen tensions. *Journal of Bacteriology* 167: 238–242.

James, H. L. 1954. Sedimentary facies of iron-formation. *Economic Geology* 49: 235–293.

James, H. L. 1983. Distribution of banded iron-formations in space and time. In: A. F. Trendall and R. C. Morris (Eds.), *Iron Formations: Facts and Problems* (Elsevier: Amsterdam), pp. 471–490.

James, H. L. and Trendall, A. F. 1982. Banded iron-formation: distribution in time and paleoenvironmental significance. In: H. D. Holland and M. Schidlowski (Eds.), *Mineral Deposits and the Evolution of the Biosphere* (Springer-Verlag: New York), pp. 199–218.

James, N. P. 1981. Megablocks of calcified algae in the Cow Head Breccia, western Newfoundland: vestiges of a Cambro-Ordovician platform margin. *Geological Society of America Bulletin* 92: 799–811.

James, N. P. 1984. Shallowing-upward Sequences in Carbonates. In: R. G. Walker (Ed.), *Facies Models, Second Edition. Geoscience Canada Reprint Series*, 1 (Geological Association of Canada Publications: Toronto), pp. 213–228.

James, N. P. and Debrenne, F. 1980. Lower Cambrian bioherms: pioneer reefs of the Phanerozoic. *Acta Palaeontologica Polonica* 25: 655–668.

James, N. P. and Gravestock, D. I. 1986. Lower Cambrian carbonate shelf and shelf margin buildups, South Australia. *12th International Sedimentological Congress Abstracts*: 154.

James, N. P. and Klappa, C. F. 1983. Petrogenesis of Early Cambrian reef limestones, Labrador, Canada. *Journal of Sedimentary Petrology* 53(4): 1051–1096.

James, N. P. and Kobluk, D. R. 1978. Lower Cambrian patch reefs and associated sediments: southern Labrador, Canada. *Sedimentology* 25: 1–35.

Jankauskas, T. 1964. O nekotorykh problematicheskikh organicheskikh ostatkakh iz nizhnego kembriya Vostochnogo Sayana [On some problematic remains from the Lower Cambrian of Western Sayan]. In: M. A. Usova (Ed.), *Materialy po Geologii i Poleznym Iskopaemym so Dnya Rozhdeniya* (Gosudarstvennyj Univeritet: Tomsk). pp. 56–59.

Jankauskas, T. 1965. Pterotsiadty—novyj otryad kribitsiat [Ptercyathids—a new order of cribicyathans]. *Doklady Akademii Nauk SSSR* 162: 438–440.

Jankauskas, T. V. 1969. Pterotsiatidy nizhnego kembriya Krasnoyarskogo kryazha (vostochnyj Sayan) [Lower Cambrian pterocyathids from the Krasnoyarsk Hills (eastern Sayan)]. In: I. T. Zhuravleva (Ed.), *Biostratigrafiya i Paleontologiya Nezhnego Kembriya Sibiri i Dal'nego Vostoka* (Nauka: Moscow), pp. 114–157.

Jankauskas, T. V. 1972. Kribitsiaty nizhnego kembriya Sibiri [Lower Cambrian cribricyathans from Siberia]. In: I. T. Zhuravleva (Ed.), *Problemy Biostratigraffi i Paleontologii Nizhnego Kembriya Sibiri* (Nauka: Moscow), pp. 161–183.

Jankauskas, T. V. 1973. Opyt izychenii kribritsiat kembriya SSSR [Experiences from the study of Cambrian cribricyathans of the USSR]. In: I. T. Zhuraleva (Ed.), *Problemy Paleontologii i Biostratigrafii Nizhnego Kembriya Sibiri i Dal'nego Vostoka Trudy Inst. Geol. Geofiz. Sibirsk. Otd. Akad. Nauk SSSR* 49 (Nauka: Novosibirsk), pp. 45–52.

Janks, M. J., Matherly, R. M. and Koster van Groos, A. F. 1985. Free fatty acids of the Early Proterozoic Animikie Group, Minnesota and Ontario. *Organic Geochemistry* 8: 215–219.

Jannasch, H. W. 1984a. Chemosynthetic microbial mats of deep-sea hydrothermal vents. In: Y. Cohen, R. W. Castenholz and H. O. Halvorson (Eds.), *Microbial Mats: Stromatolites* (Alan R. Liss, Inc.: New York), pp. 121–131.

Jannasch, H. W. 1984b. Microbial processes at deep sea hydrothermal vents. In: R. A. Rona (Ed.), *Hydrothermal Processes at Sea Floor Spreading Centers* (Plenum: New York), pp. 677–709.

Jannasch, J. W. and Wirsen, C. O. 1981. Morphological survey of microbial mats near deep-sea thermal vents. *Applied and Environmental Microbiology* 41: 528–538.

Javor, B. 1983a. Nutrients and ecology of the Western Salt and Exportadora de Sal Saltern brines. In: B. C. Schreiber and H. L. Harner (Eds.), *Sixth International Symposium on Salt* (Salt Institute: Alexandria, VA), pp. 195–205.

Javor, B. 1983b. Planktonic standing crop and nutrients in a saltern ecosystem. *Limnology and Oceanography* 28: 153–159.

Javor, B. J. and Castenholz, R. W. 1981. Laminated microbial mats, Laguna Guerrero Negro, Mexico. *Geomicrobiology Journal* 2(3): 237–273.

Javor, B. J. and Castenholz, R. W. 1984a. Invertebrate grazers of microbial mats, Laguna Guerrero Negro, Mexico. In: Y. Cohen, R. W. Castenholz and H. O. Halvorson (Eds.), *Microbial Mats: Stromatolites* (Alan R. Liss, Inc.: New York), pp. 85–94.

Javor, B. J. and Castenholz, R. W. 1984b. Productivity studies of microbial mats, Laguna Guerrero Negro, Mexico. In: Y. Cohen, R. W. Castenholz and H. O. Halvorson (Eds.), *Microbial Mats: Stromatolites* (Alan R. Liss, Inc.: New York), pp. 149–170.

Jefferies, R. P. S. 1986. *The Ancestry of the Vertebrates* (Cambridge University Press: Cambridge), 376 pp.

Jeffrey, S. W. 1972. Preparation and some properties of crystalline chlorophyll c_1 and c_2 from marine algae. *Biochimica et Biophysica Acta* 279: 15–33.

Jeffrey, S. W. and Humphrey, G. F. 1975. New spectrophotometric equations for determining chlorophylls a, b, c_1 and c_2 in higher plants, algae and natural phytoplankton. *Biochem. Physiol. Pflazen* 167: 191–194.

Jeffrey, S. W. and Vesk, M. 1976. Further evidence for a membrane-bound endosymbiont within the dinoflagellate *Peridinium foliaceum*. *Journal of Phycology* 12: 450–455.

Jell, J. S. 1984. Cambrian cnidarians with mineralized skeleton. In: W. A. Oliver Jr., W. J. Sando, S. D. Cairns, A. G. Coates and et al. (Eds.), *Recent Advances in the Paleobiology and Geology of the Cnidaria*. *Palaeontographica Americana* 54: 105–109.

Jell, P., Burrett, C. F. and Banks, M. R. 1985. Cambrian and Ordovician echinoderms from eastern Australia. *Alcheringa* 9(3): 183–208.

Jell, P. A. 1974. Faunal provinces and possible planetary reconstruction of the middle Cambrian. *Journal of Geology* 82: 319–350.

Jell, P. A. 1979. *Plumulites* and the machaeridian problem. *Alcheringa* 3: 253–259.

References

Jell, P. A. 1980. Earliest known pelecypod on earth—a new Early Cambrian genus from South Australia. *Alcheringa* 4: 233–239.

Jell, P. A. 1981. *Thambetolepis delicata* gen. et sp. nov., an enigmatic fossil from the Early Cambrian of South Australia. *Alcheringa* 5: 85–93.

Jenkins, R. J. F. 1981. The concept of an 'Ediacaran Period' and its stratigraphic significance in Australia. *Transactions of the Royal Society of South Australia* 105: 179–194.

Jenkins, R. J. F. 1984a. Ediacaran events: boundary relationships and correlation of key sections, especially in 'Armorica'. *Geological Magazine* 121: 635–643.

Jenkins, R. J. F. 1984b. Interpreting the oldest fossil cnidarians. *Palaeontographica Americana* 54: 95–104.

Jenkins, R. J. F. 1985. The enigmatic Ediacaran (late Precambrian) genus *Rangea* and related forms. *Paleobiology* 11(3): 336–355.

Jenkins, R. J. F. 1986. Are enigmatic markings in Adelaidean of Flinders Ranges fossil ice-tracks? *Nature* 323: 472.

Jenkins, R. J. F. 1989. The "supposed terminal Precambrian extinction event" in relation to the Cnidaria. *Memoirs of the Association of Australasian Palaeontologists* 8: 307–317.

Jenkins, R. J. F. and Gehling, J. G. 1978. A review of the frond-like fossils of the Ediacara assemblage. *Records of the South Australian Museum* 17: 347–359.

Jenkins, R. J. F., Plummer, P. S. and Moriarty, K. C. 1981. Late Precambrian pseudofossils from the Flinders Ranges, South Australia. *Transactions of the Royal Society of South Australia* 105: 67–83.

Jenkins, R. J. F., Ford, C. H. and Gehling, J. G. 1983. The Ediacara Member of the Rawnsley Quartzite: the context of the Ediacara assemblage (late Precambrian, Flinders Ranges). *Journal of the Geological Society of Australia* 30: 101–119.

Jensen, L. S. 1985. Stratigraphy and petrogenesis of Archean metavolcanic sequences, southwestern Abitibi subprovince, Ontario. In: L. D. Ayres, P. C. Thurston, K. D. Card and W. Weber (Eds.), *Evolution of Archean Supracrustal Sequences*. Geological Association of Canada Special Paper 28: 65–88.

Jensen, S. 1990. Predation by early Cambrian trilobites on infaunal worms—evidence from the Swedish Mickwitzia Sandstone. *Lethaia* 23: 29–42.

Jerlov, N. 1976. *Marine Optics*, 2nd ed. (Elsevier: Amsterdam, New York), 231 pp.

Jiang, Z. 1982. [*Thallophyta*]. In: *The Sinian-Cambrian boundary in Eastern Yunnan, China*, pp. 215–216.

Jiang, Z. 1980a. The Meishucun Stage and fauna of the Jinning County, Yunnan. *Bulletin of the Chinese Academy of Geological Sciences, Series I* 2(1): 75–92.

Jiang, Z. 1980b. Monoplacophorans and gastropods of the Meishucun Stage from the Meishucun section, Yunnan. *Acta Geologica Sinica* 1980(2): 112–123.

Jiang, Z.-W., Luo, H.-L. and Zhang, S.-H. 1982. Trace fossils of the Meishucun Stage (lowermost Cambrian) from the Meishucan Section in China. *Geological Review* 28: 7–13 (In Chinese).

Jiang Zhiwen 1985. Evolution of shelly fossils and the end of the late Precambrian. *Precambrian Research* 29: 45–52.

Johannes, R. E., Alberts, J., D'Elia, C., Kinzie, R. A., Pomeroy, L. R., Sottile, W., Wiebe, W., Marsh, J. A., Jr., Helfrich, P., Maragos, J., Meyer, J., Smith, S., Crabtree, D., Roth, A., McCloskey, L. R., Betzer, S., Marshall, N., Pilson, M. E. Q., Telek, G., Clutter, R. I., DuPaul, W. D., Webb, K. L. and Wells, J. R., Jr. 1972. The metabolism of some coral reef communities: a team study of nutrient and energy flux at Eniwetok. *Bioscience* 22: 541–543.

Johansen, T., Johansen, S. and Haugli, F. B. 1988. Nucleotide sequence of the *Physarum polycephalum* small subunit ribosomal RNA as inferred from the gene sequence: secondary structure and evolutionary implications. *Current Genetics* 14: 265–273.

John, B., Glikson, A. V., Peucat, J. J. and Hickman, A. H. 1981. REE geochemistry and isotopic data of Archean silicic volcanics and granitoids from the Pilbara Block, Western Australia: implications for the early crustal evolution. *Geochimica et Cosmochimica Acta* 45: 1633–1652.

John, P. 1987. *Paracoccus* as a free living mitochondrion. *Annals of the New York Academy of Sciences* 503: 140–150.

John, P. and Whatley, F. R. 1975. *Paracoccus denitrificans* and the evolutionary origin of the mitochondrion. *Nature* 254: 495–498.

John, P. and Whatley, F. R. 1977. *Paracoccus denitrificans* Davis (*Micrococcus denitrificans* Beijerinck) as a mitochondrion. *Advances in Botanical Research* 4: 51–115.

Johns, R. B., Belsky, T., McCarthy, E. D., Burlingame, A. L., Haug, P., Schnoes, H. K., Richter, W. J. and Calvin, M. 1966. The organic geochemistry of ancient sediments. *Geochimica et Cosmochimica Acta* 30: 1191–1222.

Johnson, J. H. 1952. *Girvanella*. Cambrian stratigraphy and paleontology near Caborca, northwestern Sonora, Mexico. *Smithsonian Miscellaneous Collections* 119: 24–46.

Johnson, J. H. 1966. A review of the Cambrian algae. *Colorado School Mines Quart.* 61(1): 162.

Johnson, P. W. and Sieburth, J. McN. 1979. Chroococooid cyanobacteria in the sea: a ubiquitous biomass. *Limnology and Oceanography* 24: 928–935.

Johnson, R. G. 1964. The community approach to paleoecology. In: J. Imbrie and N. Newell (Eds.), *Approaches to Paleoecology* (John Wiley & Sons: New York), pp. 107–134.

Johnson, T. C. 1976. Controls of the preservation of biogenic opal in seawater. *Science* 192: 887–890.

Jones, W. J., Nagle, D. P., Jr. and Whitman, W. B. 1987. Methanogens and the diversity of archaebacteria. *Microbiological Reviews* 51: 135–177.

Jordan, D. B. and Ogren, W. L. 1981. Species variation in the specificity of ribulose bisphosphate carboxylase/oxygenase. *Nature* 291: 513–515.

Jordan, T. E. 1981. Thrust loads and foreland basin evolution, Cretaceous, western United States. *American Association of Petroleum Geologists Bulletin* 65: 2506–2520.

Jørgensen, B. and Revsbech, N. P. 1983. Colorless sulfur bacteria, *Beggiatoa* spp. and *Thiovulum* spp., in O_2 and H_2S microgradients. *Applied and Environmental Microbiology* 45(4): 1261–1270.

Jørgensen, B. B. 1977. The sulfur cycle of a coastal marine sediment (Limfjorden, Denmark). *Limnology and Oceanography* 22: 814–832.

Jørgensen, B. B. 1980. Mineralization and the bacterial cycling of carbon, nitrogen and sulfur in marine sediments. In: D. C. Ellwood, J. N. Hedger, M. J. Latham, J. M. Lynch and J. H. Slater (Eds.), *Contemporary Microbiology Ecology* (Academic Press: London), pp. 239–251.

Jørgensen, B. B. 1982a. Ecology of the bacteria of the sulfur cycle with special reference to anoxic-oxic interface environment. *Philosophical Transactions of the Royal Society of London B* 298: 543–561.

Jørgensen, B. B. 1982b. Mineralization of organic matter in the sea bed—the role of sulphate reduction. *Nature* 296: 643–645.

Jørgensen, B. B. 1987. Ecology of the sulphur cycle: oxidative pathways in sediments. In: J. A. Cole and S. Ferguson (Eds.), *The Nitrogen and Sulfur Cycles* (Cambridge University Press: Cambridge), pp. 31–63.

Jørgensen, B. B. 1988. Ecology of the sulphur cycle: oxidative pathways in sediments. *Symposium of the Society of General Microbiology* 42: 31–63.

Jørgensen, B. B. 1989. Light penetration, absorption, and action spectra in cyanobacterial mats. In: Y. Cohen and E. Rosenberg (Eds.), *Microbial Mats: Physiological Ecology of Benthic Microbial Communities* (American Society for Microbiology: Washington, D.C.), pp. 123–137.

Jørgensen, B. B. In Press—a. Biogeochemistry of autotrophic bacteria. In: H. G. Schelegel and B. Bowien (Eds.), *Biology of Autotrophic Bacteria* (Science Tech Publishers: Madison).

Jørgensen, B. B. In Press—b. The diffusive boundary layer of sediments: oxygen microgradients over a microbial mat. *Limnology and Oceanography*.

Jørgensen, B. B. and Cohen, Y. 1977. Solar Lake (Sinai). 5. The sulfur cycle of the benthic cyanobacterial mats. *Limnology and Oceanography* 22: 657–666.

Jørgensen, B. B. and Cohen, Y. 1987. Photosynthetic potential and light-dependent oxygen consumption in a benthic cyanobacterial mat. *Applied and Environmental Microbiology* 54: 176–182.

Jørgensen, B. B. and Des Marais, D. J. 1986a. Competition for sulfide among colorless and purple sulfur bacteria in cyanobacterial mats. *FEMS Microbiology Ecology* 38: 179–186.

Jørgensen, B. B. and Des Marais, D. J. 1986b. A simple fiber-optic microprobe for high resolution light measurements: Application in marine sediment. *Limnology and Oceanography* 31: 1376–1383.

Jørgensen, B. B. and Des Marais, D. J. 1988. Optical properties of benthic photosynthetic communities: fiber-optic studies of cyanobacterial mats. *Limnology and Oceanography* 33: 99–113.

Jørgensen, B. B. and Nelson, D. C. 1988. Bacterial zonation, photosynthesis and spectral light distribution in hot spring microbial mats of Iceland. *Microbial Ecology* 16: 133–147.

Jørgensen, B. B. and Revsbech, N. P. 1985. Diffusive boundary layers and the oxygen uptake of sediments and detritus. *Limnology and Oceanography* 30(1): 111–122.

Jørgensen, B. B. and Revsbech, N. P. 1988. Microsensors. In: L. Packer and A. N. Glazer (Eds.), *Methods in Enzymology*, vol. 167 (Academic Press: San Diego), pp. 639–659.

Jørgensen, B. B., Revsbech, N. P. and Cohen, Y. 1983. Photosynthesis and structure of benthic microbial mats: microelectrode and SEM studies of four cyanobacterial communities. *Limnology and Oceanography* 28: 1075–1093.

Jørgensen, B. B., Cohen, Y. and Revsbech, N. P. 1986. Transition from anoxygenic to oxygenic photosynthesis in a *Microcoleus chthonoplastes* cyanobacterial mat. *Applied and Environmental Microbiology* 51(2): 408–417.

Jørgensen, B. B., Cohen, Y. and Des Marais, D. J. 1987. Photosynthetic action spectra and adaptation to spectral light distribution in a benthic cyanobacterial mat. *Applied and Environmental Microbiology* 53: 879–886.

Jørgensen, B. B., Cohen, Y. and Revsbech, N. P. 1988. Photosynthetic potential and light-dependant oxygen consumption in a benthic cyanobacterial mat. *Applied and Environmental Microbiology* 54: 176–182.

Joset-Espardellier, F., Astier, C., Evans, E. H. and Carr, N. G. 1978. Cyanobacteria grown under photoautotrophic, photoheterotrophic, and heterotrophic regimes: sugar metabolism and carbon dioxide fixation. *FEMS Microbiology Letters* 4: 261–264.

Jukes, T. H. and Cantor, C. R. 1969. Evolution of protein molecules. In: H. N. Munro (Ed.), *Mammalian Protein Metabolism III* (Academic Press: New York), pp. 22–132.

Jupe, E. R., Chapman, R. L. and Zimmer, E. 1988. Nuclear ribosomal RNA genes and algal phylogeny—the *chlamydomonas* example. *Biosystems* 21: 223–230.

Jürgens, U. J. and Weckesser, J. 1985. The fine structure and chemical composition of the cell wall sheath layers of cyanobacteria. *Annales de l'Institut Pasteur. Microbiologie* 136A: 41–44.

Jüttner, F. 1976. β-cyclociral and alkanes in *Microcystis* (cyanophyceae). *Zeitschrift für Naturforschung* 31: 491–495.

Jux, U. 1977. Uber die Wandstrukturen sphaeromorpher Acritarchen: *Tasmanites* Newton, *Tapajonites* Sommer & Van Boekel, *Chuaria* Walcott. *Paleontographica Abteilung B* 160: 1–16.

Kadouri, A., Derenne, S., Largeau, C., Casadevall, E. and Berkaloff, C. 1988. Resistant biopolymer in the outer walls of *Botryococcus braunii*, B race. *Phytochemistry* 27: 551–557.

Kaever, M. and Richter, P. 1976. *Buschmannia roeringi* n.gen., n.sp. (Archaeocyatha) aus der Nama-Gruppe Südwestafrikas. *Paläontologische Zeitschrift* 50: 27–33.

Kalkowsky, E. 1908. Oölith and stromatolith im Norddeutschen Bundsandstein. *Zeitschrift der Deutschen Geologischen Gesellschaft* 60: 68–125.

Kandler, O. and König, H. 1985. Cell envelopes of archaebacteria. In: C. R. Woese and R. S. Wolf (Eds.), *Archaebacteria. The Bacteria: A Treatise on Structure and Function*, 8 (Academic Press: Orlando), pp. 413–457.

Kaneda, T. 1977. Fatty acids of the genus *Bacillus*: an example of branched chain preference. *Bacterial Review* 41: 391–418.

Karhu, J. and Epstein, S. 1966. The implication of the oxygen isotope records in coexisting cherts and phosphates. *Geochimica et Cosmochimica Acta* 50: 1745–1756.

Karl, D. M. 1987. Bacterial production at deep-sea hydrothermal vents and cold seeps: evidence for chemosynthetic primary production. *Symposium of the Society of General Microbiology* 41: 319–360.

Karlstrom, K. E. and Bowring, S. A. 1988. Early Proterozoic assembly of tectonostratigraphic terranes in southwestern North America. *Journal of Geology* 96: 561–576.

Karlstrom, K. E. and Houston, R. S. 1984. The Cheyenne Belt: analysis of a Proterozoic suture in southern Wyoming. *Precambrian Research* 25: 415–446.

Kasting, J. F. 1982. Stability of ammonia in the primitive terrestrial atmosphere. *Journal of Geophysical Research* 87: 3091–3098.

Kasting, J. F. 1987. Theoretical constraints on oxygen and carbon dioxide concentrations in the Precambrian atmosphere. *Precambrian Research* 34: 205–229.

Kasting, J. F. In Press. Bolide impacts and the oxidation state of carbon in the earth's early atmosphere. *Origins of Life*.

Kasting, J. F. and Ackerman, T. P. 1986. Climatic consequences of very high CO_2 levels in earth's early atmosphere. *Science* 234: 1383–1385.

Kasting, J. F., Zahnle, K. J. and Walker, J. C. G. 1983. Photochemistry of methane in the earth's early atmosphere. *Precambrian Research* 20: 121–148.

Kasting, J. F., Pollack, J. B. and Crisp, D. 1984. Effects of high CO_2 levels on surface temperature and atmospheric oxidation state on the early earth. *Journal of Atmospheric Chemistry* 1: 403–408.

Kasting, J. F., Holland, H. D. and Pinto, J. P. 1985. Oxidant abundances in rainwater and the evolution of atmospheric oxygen. *Journal of Geophysical Research* 90: 10,497–10,510.

Kasting, J. F., Toon, O. B. and Pollack, J. B. 1988. How climate evolved on the terrestrial planets. *Scientific American* 256: 90–97.

Kasting, J. F., Zahnle, K. J., Pinto, J. P. and Young, A. T. 1989. Sulfur, ultraviolet radiation, and the early evolution of life. *Origins of Life* 19: 95–108.

Kates, M. 1978. The phytanyl ether-linked polar lipids and isoprenoid neutral lipids of extremely halophilic bacteria. *Progress in Chemistry of Fats and other Lipids* 15: 301–342.

Kates. M. 1986. *Techniques in Lipidology: Isolation, Analysis and Identification of Lipids*, vol. 3 (Elsevier: Amsterdam), 464 pp.

Katz, B. J. and Elrod, L. W. 1983. Organic geochemistry of DSDP 467, offshore California, Middle Miocene to Lower Pliocene strata. *Geochimica et Cosmochimica Acta* 47: 267–275.

Kauffman, E. G. and Fürsich, F. T. 1983. *Brooksella canyonensis*: a billion year old complex metazoan trace fossil from the Grand Canyon. *Geological Society of America Abstracts with Programs* 15(6): 608.

Kauffman, E. G. and Steidtmann, J. R. 1981. Are these the oldest metazoan trace fossils? *Journal of Paleontology* 55: 923–947.

Kauffman, E. G. and Steidtmann, J. R. 1983. No, they are still dubiofossils! *Geology* 11: 619–621.

Kauffman, S. A. 1989. Cambrian explosion and Permian quiescence: implications of rugged fitness landscapes. *Evolutionary Ecology* 3: 274-281.

Kaufman, A. J., Hayes, J. M. and Klein, C. 1990. Mineralogic control of isotopic compositions in microbanded carbonates and cherts from a banded iron formation. *Geochimica et Cosmochimica Acta* 54: 3461–3473.

Kaufman, A. J., Hayes, J. M., Knoll, A. H. and Germs, G. J. B. In Press—b. Isotopic compositions of carbonates and organic carbon from upper Proterozoic successions in Namibia: stratigraphic variation and the effect of diagenesis and metamorphism. *Precambrian Research*.

Kaufman, A. J., Hayes, J. M., Knoll, A. H. and Germs, J. G. B. In Press—c. Secular variation of stable carbon isotope ratios in the Upper Proterozoic Otavi and Nama groups, Namibia/South West Africa. *Precambrian Research*.

References

Kaula, W. M. 1979. Thermal evolution of earth and moon growing by planetesimal impacts. *Journal of Geophysical Research* 84: 999–1008.

Kazansky, V. I. and Moralev, V. M. 1981. Archean geology and metallogeny of the Aldan Shield. In: J. E. Glover and D. I. Groves (Eds.), *Archean Geology. Geological Society of Australia Special Publication* 7: 111–120.

Kaz'mierczak, J., Ittekott, V. and Degens, E. T. 1985. Biocalcification through time: Environmental challenge and cellular response. *Paläontologische Zeitschrift* 59: 15–33.

Keen, J. N., Pappin, D. J. C. and Evans, L. 1988. Amino acid sequence analysis of the small subunit of rigulose biphosphate carboxylase from *Fucus* (Phaeophyceae). *Journal of Phycology* 24: 324–327.

Keller, B. M. 1969. Otpechatok neizvestnogo zhivotnogo iz Valdayskoy serii Russkoi platformy. In: A. Y. Rozanov and et al. (Eds.), *Tommotskiy Yarus i Problema Nizhney Granitsy Kembriya*, vol. 206 (Trudy geologicheskiy Institut, Akademiya Nauk SSSR: Moscow), pp. 175–176.

Keller, B. M. 1975. *Guidebook of the Excursion in Bashkiria* (UNESCO-IGCP: Moscow), 47 pp.

Keller, B. M. 1979. Precambrian stratigraphic scale of the USSR. *Geological Magazine* 116: 419–429.

Keller, B. M. and Fedonkin, M. A. 1976. Novyye nakhodki okamenelostey v Valdayskoy serii Dokembriya po r. Syuz'me. *Izvestiya Akademii Nauk SSSR, seriia Geologicheskaya* 1976(3): 38–44 (Republished in 1977 as: New organic fossil finds in the Precambrian Valday series along the Syuz'ma River. *International Geological Review* 19: 924–930).

Keller, B. M. and Rozanov, A. Y. (Eds.). 1979. *Paleontologiya Verkhnedokembriyskikh i Kembriyskikh Otlozheniy Vostochno-Evropeyskoy Platformy* [Upper Precambrian and Cambrian Paleontology of East-European Platform] (Nauka: Moscow), 212 pp.

Keller, B. M., Menner, V. V., Stepanov, V. A. and Chumakov, N. M. 1974. Novye nakhodki Metazoa v vendomii Russkoy platformy [New discoveries of Metazoa in the Vendian of the Russian Platform]. *Izvestiya Akademii Nauk SSSR, serii Geologicheskaya* 1974(12): 130–134.

Keller, B. M., ed. 1982. Stratotyp rifeya [Stratotype of Riphean]. *Transactions, Nauk SSSR, Geological Institute* 368: 175.

Kelling, G. and Whitaker, J. H. McD. 1970. Tool marks made by ribbed orthoconic nautiloids. *Journal of Geology* 78: 371–374.

Kempe, S. and Degens, E. T. 1985. An early soda ocean? *Chemical Geology* 53: 95–108.

Kennard, J. M. and James, N. P. 1986. Thrombolites and stromatolites: two distinct types of microbial structures. *Palaios* 1: 492–503.

Kenyon, D. H. and Steinman, G. 1969. *Biochemical Predestination* (McGraw-Hill: San Francisco), 301 pp.

Kepkay, P. E., Cooke, R. C. and Novitsky, J. A. 1979. Microbial autotrophy: a primary source of organic carbon in marine sediments. *Science* 204: 68–69.

Kerans, C. 1982. Sedimentology and stratigraphy of the Dismal Lakes Group. Ph.D. thesis, Carlton University, Ottawa, 404 pp.

Kerans, C. and Donaldson, J. A. 1989. Deeper water conical stromatolite reef, Sulky Formation, Middle Proterozoic, N.W.T. In: H. Gedsetzer, N. P. James and G. Tebbutt (Eds.), *Reefs, Canada and Adjacent Areas. Canadian Society of Petroleum Geologists Memoirs* 13.

Kerber, M. 1988. Mikrofossilien aus unterkambrischen Gesteinen der Montagne Noire, Frankreich. *Palaeontographica* A202: 127–203.

Kerby, N. W. and Raven, J. A. 1985. Transport and fixation of inorganic carbon by marine algae. *Advances in Botanical Research* 11: 71–123.

Khain, V. E. 1985. *Geology of the USSR* (Gebruder Borntraeger: Berlin), 272 pp.

Khalfina, V. K. 1960. Stromatoporoidei iz kembrijskikh otlozhenij Sibiri [Stromatoporoids from Cambrian deposits of Siberia]. *Materialy po paleontologiya i stratigrafiya zapadnoj Sibiri. Trudy Sibirskogo Nanchno-Isskdovatel'skogo Instituta Geologii, Geofiziki i Mineral'nogo Syr'ya* 8: 79–83.

Kharmov, A. N. 1982. *Paleomagnetology* (Nedra Press: Leningrad), 312 pp.

Khomentovskij, V. V. and Karlova, G. A. 1986. O nizhnej granitse pestrotsvetnoj svity v bassejne r. Aldan [On the lower boundary of the Pestrovet Formation in the River Aldan basin]. In: V. V. Khomentovskij (Ed.), *Pozdnij Dokembrij i Rannij Paleozoj Sibiri. Sibirskaya Platforma i Vneshnyaya* (Nauka: Novosibirsk), pp. 3–22.

Khomentovskij, V. V., Shenfil, B. Y., Yakshin, M. S. and Butakov, E. P. 1972. Opornye razrezy otlozheniy verkhnego dokenbriya i nizhnego kembriya Sibirskoy platformy [Base sections of Upper Precambrian and Lower Cambrian deposits of the Siberian Platform]. *Transactions of the Akademiya Nauk SSSR, Siberian Branch, Institute of Geology and Geophysics* 141: 1–356.

Kiderlen, H. 1937. Die Conularien, über Bau und Leben der ersten Scyphozoa. *Neues Jahrbuch für Mineralogie B, Beil. Band* 77: 113–169.

Kiehl, J. T. and Dickinson, R. E. 1987. A study of the radiative effects of enhanced atmospheric CO_2 and CH_4 on early earth surface temperatures. *Journal of Geophysical Research* 92: 2991–2998.

King, G. M. 1988. Methanogenesis from methylated amines in a hypersaline algal mat. *Applied and Environmental Microbiology* 54: 130–136.

King, P. B. 1964. *Geology of the central Great Smoky Mountains, Tennessee. United States Geological Survey Professional Paper*, 349-C (United States Geological Survey: Washington, D.C.), 148 pp.

King, P. B. 1976. Precambrian geology of the United States: an explanatory text to accompany the geological map of the United States. *U.S. Geological Survey Professional Paper* 902: 1–85.

King, P. B. and Ferguson, H. W. 1960. Geology of Northeasternmost Tennessee. *U.S. Geological Survey Professional Paper* 311: 1–136.

King, P. B. and Flawn, P. T. 1953. Geology and mineral deposits of the Precambrian rocks of the Van Horn area, Texas. *Texas University, Bureau of Economic Geology, Publication* 5301: 218.

King, W. 1872. On the Kadapah and Karnul Formations in the Madras Presidency. *Memoirs of the Geological Survey of India* 8: 1–313.

Kinny, P. D., Williams, I. S., Froude, D. O., Ireland, T. R. and Compston, W. 1988. Early Archean zircon ages from othogneisses and anorthosites at Mount Narryer, Western Australia. *Precambrian Research* 35: 325–341.

Kinsman, D. J. J. and Park, R. K. 1976. Algal belt and coastal sabkha evolution, Trucial Coast, Persian Gulf. In: M. R. Walter (Ed.), *Stromatolites: Developments in Sedimentology*, vol. 20 (Elsevier: Amsterdam), pp. 421–433.

Kirk, J. T. O. 1983. *Light and Photosynthesis in Aquatic Ecosystems* (Cambridge University Press: Cambridge), 401 pp.

Kirschvink, J. L. 1978a. The Precambrian-Cambrian boundary problem: Magnetostratigraphy of the Amadeus Basin, Central Australia. *Geological Magazine* 115: 139–150.

Kirschvink, J. L. 1978b. The Precambrian-cambrian boundary problem: Paleomagnetic directions from the Amadeus Basin, Central Australia. *Earth and Planetary Science Letters* 40: 91–100.

Kirschvink, J. L. 1980. The least-squares line and plane and the analysis of paleomagnetic data: examples from Siberia and Morocco. *Geophysical Journal of the Royal Astronomical Society* 62: 699–718.

Kirschvink, J. L. and Rozanov, A. Y. 1984. Magnetostratigraphy of lower Cambrian strata from the Siberian Platform: a palaeomagnetic pole and a preliminary polarity time-scale. *Geological Magazine* 121(3): 189–203.

Kir'yanov, V. 1968. Paleontologicheskie ostatki i stratigrafiya otlozhenij Baltijskoj serii Volyno-Podolii [Palaeontological remains and stratigraphy of the deposits of the Baltic Series of Volyno-Podolia]. In: V. S. Krandievskij, T. A. Ishchenko and V. Kir'yanov (Eds.), *Paleontologiya i Stratigrafiya Nizhnego Paleozoya Volyno-Podolii* (Naukova Dumka: Kiev), pp. 5–25.

Kir'yanov, V., V 1974. Novye akritarkhi iz kembrijskikh otlozhenij Volyni [New acritarchs from the Cambrian deposits of Volhynia]. *Paleontologicheskij Zhurnal* 1974(2): 117–129.

Kite, G. C. and Dodge, J. D. 1985. Structural organization of plastid

DNA in two anomalously pigmented dinoflagellates. *Journal of Phycology* 21: 50–56.

Kite, G. C. and Dodge, J. D. 1988. Cell and chloroplast ultrastructure in *Gyrodinium aureoleum* and *Gymnodinium galatheanum*. Two marine dinoflagellates containing an unusual carotenoid. *Sarsia* 73: 131–138.

Kite, G. C., Rothschild, L. T. and Dodge, J. D. 1988. Nuclear and plastid DNAs from the binucliate dinoflagellates *Glenodinium (Peridinium) foliaceum* and *Peridinium balticum*. *Biosystems* 21: 151–163.

Kivic, P. A. and Waine, P. H. 1984. An evaluation of a possible phylogenetic relationship between the Euglenophyta and Kinetoplastida. *Origins of Life* 13: 269–288.

Klappa, C. F. 1978. Biolithogenesis of *Microcodium*; elucidation. *Sedimentology* 25: 489–522.

Klein, C. 1974. Greenalite, stilpnomelane, minnesotaite, crocidolite and carbonates in a very low-grade metamorphic Precambrian iron-formation. *Mineralogist* 12: 475–498.

Klein, C. 1978. Regional metamorphism of Proterozoic iron-formation, Labrador Trough, Canada. *American Mineralogist* 63: 898–912.

Klein, C. 1983. Diagenesis and metamorphism of Precambrian banded iron-formations. In: A. F. Trendall and R. C. Morris (Eds.), *Iron Formations: Facts and Problems* (Elsevier: Amsterdam), pp. 417–471.

Klein, C. and Beukes, N. J. 1989. Geochemistry and sedimentology of a facies transition from limestone to iron-formation deposition in the early Proterozoic Transvaal Supergroup, South Africa. *Economic Geology* 84: 1733–1774.

Klein, C. and Bricker, O. P. 1977. Some aspects of the sedimentary and diagenetic environment of Proterozoic banded iron formation. *Economic Geology* 72: 1457–1470.

Klein, C. and Gole, M. J. 1981. Mineralogy and petrology of parts of the Marra Mamba iron-formation, Hamersley Basin, Western Australia. *American Mineralogist* 66: 507–525.

Klein, C., Beukes, N. J. and Schopf, J. W. 1987. Filamentous microfossils in the Early Proterozoic Transvaal Supergroup: Their morphology, significance, and paleoenvironmental setting. *Precambrian Research* 36: 81–94.

Klein, C., Jr. and Fink, R. P. 1976. Petrology of the Sokoman Iron Formation in the Howells River Area, at the western edge of the Labrador Trough. *Economic Geology* 71: 453–487.

Klerkx, J., Liegeois, J. P., Lavreau, J. and Claessens, W. 1987. Crustal evolution of the northern Kibaran belt, eastern and central Africa. In: A. Kröner (Ed.), *Proterozoic Lithospheric Evolution. American Geophysical Union Geodynamics Series* 17: 217–234.

Kline, G. 1977. Earliest Cambrian (Tommotian) age of the upper Tindir Group, east-central Alaska. *Geological Society of America Abstracts with Programs* 9(4): 448.

Klok, J., Cox, H. C., Baas, M., Schuyl, P. J. W., de Leeuw, J. W. and Schenck, P. A. 1984. Carbohydrates in recent marine sediments—I. Origin and significance of deoxy- and O-methyl-monosaccharides. *Organic Geochemistry* 7: 73–84.

Klomp, U. C. 1986. The chemical structure of a pronounced series of iso-alkanes in South Oman crudes. In: D. Leythaeuser and J. Rullkötter (Eds.), *Advances in Organic Geochemistry 1985* (Pergamon Press: Oxford), pp. 807–814.

Klootwijk, C. T. 1980. Early Palaeozoic paleomagnetism in Australia. *Tectonophysics* 64: 249–332.

Klovan, J. E. and Imbrie, J. 1971. An algorithm and FORTRAN-IV program for large-scale Q-mode factor analysis and calculation of factor scores. *Mathematical Geology* 3: 61–77.

Knauth, L. P. and Epstein, S. 1976. Hydrogen and oxygen isotope ratios in nodular and bedded cherts. *Geochimica et Cosmochimica Acta* 40: 1095–1108.

Knight, I. and Morgan, W. C. 1981. The Aphebian Ramah Group, Northern Labrador. In: F. H. A. Campbell (Ed.), *Proterozoic Basins of Canada. Geological Survey of Canada, Paper* 81-10: 313–330.

Knight, J. B. 1947. Some new Cambrian bellerophont gastropods. *Smithsonian Miscellaneous Collections* 106: 1–11.

Knoll, A. H. 1977. Paleomicrobiology. In: A. Laskin and H. A. Lechevalier (Eds.), *CRC Handbook of Microbiology. Volume I*, Bacteria, 2nd ed. (CRC Press: Cleveland), pp. 9–29.

Knoll, A. H. 1978. Did emerging continents trigger metazoan evolution? *Nature* 276: 701–703.

Knoll, A. H. 1982a. Microfossil-based biostratigraphy of the Precambrian Hecla Hoek sequence, Nordaustlandet, Svalbard. *Geological Magazine* 119: 269–279.

Knoll, A. H. 1982b. Microfossils from the Late Precambrian Draken Conglomerate, Ny Feiesland, Svalbard. *Journal of Paleontology* 56: 755–790.

Knoll, A. H. 1983. Biological interactions and Precambrian eukaryotes. In: M. J. S. Tevesz and P. L. McCall (Eds.), *Biological Interactions in Recent and Fossil Benthic Communities* (Plenum: New York), pp. 251–282.

Knoll, A. H. 1984. Microbiotas of the Late Precambrian Hunnberg Formation, Nordaustlandet, Svalbard. *Journal of Paleontology* 58: 131–162.

Knoll, A. H. 1985a. The distribution and evolution of microbial life in the late Proterozoic era. *Annual Review of Microbiology* 39: 391–417.

Knoll, A. H. 1985b. Exceptional preservation of photosynthetic organisms in silicified carbonates and silicified peats. *Philosophical Transactions of the Royal Society of London B* 311: 111–122.

Knoll, A. H. 1985c. A paleobiological perspective on sabkhas. In: G. M. Friedman and W. E. Krumbein (Eds.), *Ecological Studies: Hypersaline Ecosystems*, vol. 53 (Springer-Verlag: New York), pp. 407–425.

Knoll, A. H. 1987. Why did the Proterozoic Eon end? *Geological Society of America Abstracts with Programs* 19: 730.

Knoll, A. H. 1989. The paleomicrobiological information in Proterozoic rocks. In: Y. Cohen and E. Rosenberg (Eds.), *Microbial Mats: Physiological Ecology of Benthic Microbial Communities* (American Society for Microbiology: Washington, D.C.), pp. 469–484.

Knoll, A. H. In Press. Biological and biogeochemical preludes in the Ediacaran radiation. In: J. H. Lipps and P. W. Signor (Eds.), *Origins and Early Evolutionary History of the Metazoa* (Plenum: New York).

Knoll, A. H. and Barghoorn, E. S. 1975. Precambrian eukaryotic organisms: a reassessment of the evidence. *Science* 190: 52–54.

Knoll, A. H. and Barghoorn, E. S. 1976. A Gunflint-type microbiota from the Duck Creek Dolomite, Western Australia. *Origins of Life* 7: 417–423.

Knoll, A. H. and Barghoorn, E. S. 1977. Archean microfossils showing cell division from the Swaziland System of South Africa. *Science* 198: 396–398.

Knoll, A. H. and Bauld, J. 1989. The evolution of ecological tolerance in prokaryotes. *Transactions of the Royal Society of Edinburgh: Earth Sciences* 80: 209–223.

Knoll, A. H. and Butterfield, N. J. 1989. New window on Proterozoic life. *Nature* 337: 602–603.

Knoll, A. H. and Calder, S. 1983. Microbiotas of the Late Precambrian Ryssö Formation, Nordaustlandet, Svalbard. *Palaeontology* 26: 467–496.

Knoll, A. H. and Golubić, S. 1979. Anatomy and taphonomy of a Precambrian algal stromatolite. *Precambrian Research* 10: 115–151.

Knoll, A. H. and Ohta, Y. 1988. Microfossils in metasediments from Prins Karls Forland, western Svalbard. *Polar Research* 6: 59–67.

Knoll, A. H. and Simonson, B. 1981. Early Proterozoic microfossils and penecontemporaneous quartz-cementation in the Sokoman Iron Formation, Canada. *Science* 211: 478–480.

Knoll, A. H. and Swett, K. 1985. Micropaleontology of the Late Proterozoic Veteranen Group, Spitsbergen. *Palaeontology* 28: 451–473.

Knoll, A. H. and Swett, K. 1987. Micropaleontology across the Precambrian-Cambrian boundary in Spitsbergen. *Journal of*

Paleontology 61: 898–926.
Knoll, A. H. and Vidal, G. 1980. Late Proterozoic vase-shaped microfossils from the Visingsö Beds, Sweden. *Geologiska Föreningens i Stockholm Förhandlingar* 102: 207–211.
Knoll, A. H., Barghoorn, E. S. and Awramik, S. M. 1978. New microorganisms from the Aphebian Gunflint Iron Formation, Ontario. *Journal of Paleontology* 52: 976–992.
Knoll, A. H., Hayes, J. M., Kaufman, A. J., Swett, K. and Lambert, I. B. 1986. Secular variation in carbon isotope ratios from Upper Proterozoic successions of Svalbard and East Greenland. *Nature* 321: 832–838.
Knoll, A. H., Strother, P. K. and Rossi, S. 1988. Distribution and diagenesis of microfossils from the Lower Proterozoic Duck Creek Dolomite, Western Australia. *Precambrian Research* 38: 257–279.
Knoll, A. H., Swett, K. and Burkhardt, E. 1989. Paleoenvironmental distribution of microfossils and stromatolites in the Upper Proterozoic Backlundtoppen Formation, Spitsbergen. *Journal of Paleontology* 63: 129–145.
Knudsen, E., Jantzen, E., Bryn, K., Ormerod, J. G. and Sirevåg, R. 1982. Quantitative and structural characteristics of lipids in *Chlorobium* and *Chloroflexus*. *Archives of Microbiology* 132: 149–154.
Kobayashi, T. 1933. Upper Cambrian of the Wuhutsui Basin, Liaotung, with special reference to the limit of the Chaumitian (or Upper Cambrian) of eastern Asia, and its subdivision. *Japanese Journal of Geology and Geography* 11: 55–155.
Kobayashi, T. 1939. Restudy of Lorenz's *Raphistoma bröggeri* from Shantung with a note on *Pelagiella*. In: *Jubilee Publication in Commemoration of Professor Hisakatou Yake's 60th Birthday* (Tohoku Imperial University: Tohoku), pp. 283–288.
Kobayashi, T. 1958. On some Cambrian gastropods from Korea. *Japanese Journal of Geology and Geophysics* 29: 112–118.
Kobluk, D. R. 1979. A new and unusual skeletal organism from the Lower Cambrian of Labrador. *Canadian Journal of Earth Sciences* 16(10): 2040–2045.
Kobluk, D. R. 1981. Earliest cavity-dwelling organisms (coelobionts), Lower Cambrian Poleta Formation, Nevada. *Canadian Journal of Earth Sciences* 18(4): 669–679.
Kobluk, D. R. 1984. A new compound skeletal organism from the Rosella Formation (Lower Cambrian), Atan Group, Cassiar Mountains, British Columbia. *Journal of Paleontology* 58: 703–708.
Kobluk, D. R. 1985. Biota preserved within cavities in Cambrian *Epiphyton* mounds, upper Shady Dolomite, southwestern Virginia. *Journal of Paleontology* 59(5): 1158–1172.
Kobluk, D. R. and James, N. P. 1979. Cavity-dwelling organisms in Lower Cambrian patch reefs from southern Labrador. *Lethaia* 12: 193–218.
Kohl, W., Gloe, A. and Reichenbach, H. 1983. Steroids from the myxobacterium *Nannocystis exedens*. *Journal of General Microbiology* 129: 1629–1635.
Kohlhase, M. and Pohl, P. 1988. Saturated and unsaturated sterols of nitrogen-fixing blue-green algae (cyanobacteria). *Phytochemistry* 27: 1735–1740.
Kolosov, P. N. 1966. Novye vidy dokembrijskikh vodoroslej bassejna reki Olekmy [New species of Precambrian algae from the Olekma River basin]. *Doklady Akademii Nauk SSSR* 171(4): 978–980 (1967, Doklady Earth Science Sections 171: 235–237).
Kolosov, P. N. 1970. Organicheskie ostaki verknego dokembriya Yuga Yakutii [Organic remains from the upper Precambrian of southern Yakutia]. In: A. K. Bobrov (Ed.), *Stratigrafiya i Paleontologiya Proterozoya i Kembriya Vostoka Sibirskoj Plaformy* (Yakutskow Knizhoe Izdatel'stvo: Yakutsk), pp. 57–70.
Kolosov, P. N. 1975. *Stratigrafiya Verkhnego Dokembriya Yuga Yakutii* [*Upper Precambrian Stratigraphy of Southern Yakutia*] (Nauka: Novosibirsk), 156 pp.
Kolosov, P. N. 1977. *Drevnie Neftegazonosnye Tolshchi Yugo-Vostoka Sibirskoj Platformy* [*Ancient Oil-and Gas-Bearing Deposits of the Southeast Siberian Platform*] (Nauka, Sibirskoe Otdelenie: Novosibirsk), 92 pp.

Kolosov, P. N. 1979. On time of appearance of Cyanophyta, widely distributed in the Cambrian. *Centre de Recherche Exploration-Production Elf-Aquitaine, Bulletin* 3(2): 665–667.
Kolosov, P. N. 1982. *Verkhnedokembrijskie paleoal'goloicheski ostaki Sibirskoj platformy* [*Upper Precambrian paleo-algal remains from the Siberian Platform*] (Nauka: Moscow), 96 pp.
Kolosov, P. N. 1983. Nitchatye mikrofitofossilii v Kursovskoj svite venda Yakutii [Filamentous microphytofossils in the Vendian Kursovskaya Suite in Yakutia]. *Doklady Akademii Nauk SSSR* 269(4): 944–946 (1984, Doklady Earth Science Sections 269: 184–187.
Kolosov, P. N. 1984. *Pozdnedokembrijskie Microorganizmy Vostoka Sibirskoj Platformy* [*Late Precambrian Microorganisms of the East Siberian Platform*] (Yakutskij filial, Akademiya Nauk SSSR: Moscow), 84 pp.
Komar, V. A., Raaben, M. E. and Semikhatov, M. A. 1965. Conophytons in the Riphean of the USSR and their stratigraphic importance. *Trudy geologicheskii institut, Akademiya Nauk SSSR* 131: 1–72 (In Russian).
Kominz, M. A. 1986. Geophysical Modeling Studies. Ph.D thesis, Columbia University, 218 pp.
Koneva, S. P. 1978. O pervoj nakhodke ostrakod v nozhnem kembrii Kazakhstana [On the first find of ostracodes in the Lower Cambrian of Kazakhstan]. *Paleontologicheskij Zhurnal* 1978: 150–152.
Koneva, S. P. 1979. *Stenotekoidy i Bezzamkovye Brakhiopody Nizhnego i Nizov Srednego Kembriya Tsent* (Nauka: Alma-Ata), 123 pp.
Köppel, V. 1980. Lead isotope studies of stratiform ore deposits of the Namaqualand, NW Cape Province, South Africa, and their implications on the age of the Bushmanland Sequence. In: J. D. Ridge (Ed.), *Proceedings of the 5th IAGOD Symposium* (E. Schweizerbart'sche Verlag: Stuttgart), pp. 195–207.
Kordeh, K. B. 1950a. Dasycladaceae iz kembriya Tuvy [Dasycladaceae from the Cambrian of Tuva]. *Doklady Akademii Nauk SSSR* 73(2): 371–374.
Kordeh, K. B. 1950b. Mikroskopicheskaya struktura nasloenij stromatolitov i tipy sokhrannosti iskopaemykh Cyanophyceae [Microscopic structure of layered stromatolites and types of preservation of fossil Cyanophyeae]. *Doklady Akademii Nauk SSSR* 71(6): 1109–1112.
Kordeh, K. B. 1954. Kembrijskie vodorsli iz okrestnostej s. Boguchany na r. Angare [Cambrian algae from the region of Boguchany village on the Angara River]. *Voprosy Geologii Azii* 1: 531–555.
Kordeh, K. B. 1955. Vodorsli iz kembrijskikh otlozhenij pek Leny, Botomy i Amgi [Algae from the Cambrian deposits of the Lena, Botom, and Amga Rivers]. *Trudy Paleontologicheskogo Instityuta, Akademiya Nauk SSSR* 56: 79–92.
Kordeh, K. B. 1957. Novye spedstaviteli sifonnikovykh vodoroslej [New specimens of siphonaceous algae]. *Materialy Osnovam Paleontol.* [*Moscow*] 1: 67–75.
Kordeh, K. B. 1958. O neskol'kikh vidakh iskopaemykh sinezelenykh vodoroslej [On some species of fossil blue-green algae]. *Materialy Osnovam Paleontol.* [*Moscow*] 2: 113–118.
Kordeh, K. B. 1959. Problematicheskie ostatki iz kembrijskikh otlozhenij yugo-vostoka Sibirskoj platformy [Problematic fossils from Cambrian deposits of the southeast of the Siberian Platform]. *Doklady Akademii Nauk SSSR* 125(3): 625–627 (1960, *Doklady Earth Science Sections*, 125: 358–360.
Kordeh, K. B. 1960. Vodorosli. Cyanophyta [Algae. Cyanophyta]. In: L. L. Khalfina (Ed.), *Biostratigrafiya Paleozoya Sayano-Altajskoj Gornoj oblasti. Trudy Sibirskogo Nanchno-Isskdovatel'skogo Instituta Geologii, Geofiziki i Mineral'nogo Syr'ya* 19: 256–274.
Kordeh, K. B. 1961. Vodorosli kembriya Yugo-Vostoka Sibirskoj platformy [Algae of the Cambrian Period on the southeastern part of the Siberian Platform]. *Trudy Paleontologicheskogo Instityuta, Akademiya Nauk SSSR* 89: 1–148 (translation of pp. 14–40, Associated Technical Services, Inc., East Orange, N.J.).
Kordeh, K. B. 1963. Hydroconozoa—novyj klass kishechnopolostnykh zhivotnykh [Hydrochronozoa—a new class of

coelenterate animals]. *Paleontologicheskij Zhurnal* 1963(2): 20–25.

Kordeh, K. B. 1966. Novye materialy k sistematike i ehvolyutsii krasnykh vodoroslej rannego Paleozoya [Fresh data on the taxonomy and evolution of early Paleozoic rhodophyceans]. *Doklady Akademii Nauk SSSR* 166(6): 1440–1442.

Kordeh, K. B. 1967. Geologicheskaya istoriya drevnikh vodoroslej i ikh stratigraficheskie kompleksy [The geological history of the earliest algae and their stratigraphical assemblages]. In: T. F. Vozzhennikova, Z. I. Glezer and A. P. Zhuze (Eds.), *Iskopaemye Vodorosli SSSR* [*Fossil Algae of the USSR*] (Nauka: Moscow), pp. 5–11. (1969, translated by G. K. Beedle, National Lending Library for Science and Technology, Boston Spa, Yorkshire, Eng., 9–14).

Kordeh, K. B. 1973. *Vodorosli Kembriya* [*Cambrian Algae*]. *Trudy Paleontol. Inst. Akad. Nauk SSSR 139* (Nauka: Moscow), 351 pp.

Kordeh, K. B. 1986. Novye vidy vodoroslej rannego paleozoya Vostochnogo Sayana i Zabajkal'ya [New species of algae from the early Paleozoic of eastern Sayan and Zabajkal]. In: B. S. Sokolov (Ed.), *Aktual'nye Voprosy Sovremennoj Paleoal'gologii* [*Current Questions in Contemporary Paleoalgology*] (Naukova Dumka: Kiev), pp. 105–109.

Kordeh, K. B., Maslov, V. P. and Krylov, I. N. 1963. Tip Cyanophyta (Schizophyceae). Sinezelenye vodorosli. In: Y. A. Orlov (Ed.), Osnovy Paleontologii [Fundamental Paleontology] (Akademiya Nauk SSSR: Moscow), pp. 29–54.

Kordeh, K. B., Sergeenko, V. N. and Shul'gin, V. N. 1979. Porodoobrazuyushchie vodorosli iz vesovetskoj serii Karelii [Rock-building algae from the Besovetskoj series of Karelia]. In: B. S. Sokolov (Ed.), *Paleontologiya Dokembriya i Rannego Kembriya* (Nauka, Leningradskoe Otdelenie: Leningrad), pp. 109–111.

Korolev, V. G. and Ogurtsova, R. N. 1982. Korrelyatsiya pogranichnykh otlozhenij vendanizhnego kembriya Talaso-Karatauskoj zony (khr. Malyj Karatau) s opornymi razrezami Vostochno-Evropejskoj i Sibriskoj platform [Correlation of Vendian-Lower Cambrian boundary deposits in the Talas-Karatau zone (Malyj Karatau Range) with the reference sections of the East European and Siberian Platforms]. *Izvestiya Akademiya Nauk SSSR, seriia Geologicheskaya* 1982(6): 27–36.

Korotev, R. L. 1987. National Bureau of Standard Coal Flyash (SRM 1633A) as a multi-element standard for instrumental neutron activation analysis. *Journal of Radioanalytical and Nuclear Chemistry* 110: 159–177.

Korshunov, V. I. 1968. Gonamispongia—novyj rod gubok semejstva Chancelloriidae [*Gonamispongia*–a new sponge genus of the family Chancelloriidae]. *Paleontologicheskij Zhurnal* 1968(3): 127–129.

Korthals, H. J. and Steenbergen, C. L. M. 1985. Separation and quantification of pigments from natural phototrophic microbial populations. *FEMS Microbiology Ecology* 31: 177–185.

Koshevoj, V., V 1987. Nekotorioe kolokolovidioe infuzorii iz dokembriiskikh osadochnikh porod chexoslovakii. *Izvestiya Bi'sshix Uchebni'x Zabedenii, Geologiya i Razbedka* 2: 20–24.

Kössel, H., Edwards, K., Fretzsche, E., Koch, W. and Schwarz, Z. 1983. Phylogenetic significance of nucleotide sequence analysis. In: U. Jensen and D. E. Fairbrothers (Eds.), *Proteins and Nucleic Acids in Plant Systematics* (Springer-Verlag: Berlin), pp. 36–57.

Koyabashi, T. 1930. On the occurrence of *Girvanella manchurica* Yabe & Ozaki in northern Korea. *Chishitsugaku Zasshi* [*Journal of the Geological Society of Tokyo*] 37: 464.

Kralik, M. 1982. Rb-Sr age determinations on Precambrian carbonate rocks of the Carpentarian McArthur Basin, Northern Territories, Australia. *Precambrian Research* 18: 157–170.

Krasnopeeva, P. S. 1937. *Vodorosli i Arkheotsiaty Drevnejskikh Tolshch Potekhinskogo Plansheta Khakassii* [*Algae and Archaeocyathids of the Oldest Formations of the Potekeen District in Khakassia*] (Materialy Geol. Krasnoyarskogo Kraya, Rec. Geol. Krasnojarsky Region), 51 pp.

Krasnopeeva, P. S. 1955. Vodorosli [Algae]. In: L. L. Khalfina (Ed.), *Atlas Rukobodyashchikh Form Iskopaemykh Fauny i Flory Zapadnoj Sibiri* (Gosudarstvennoe Nauch.-Tekh. Iz.: Moscow), pp. 145–148.

Krasnopeeva, P. S. 1972. Nekotorye novye okamenelosti kembriya zapadnoj Sibiri. [Some new Cambrian fossils from western Siberia]. In: I. T. Zhuravleva (Ed.), *Problemy Biostratigrafii i Paleontologii Nizhnego Kembriya Sibiri* (Nauka: Moskva), pp. 144–146.

Kröner, A. 1977. Precambrian mobile belts of southern and eastern Africa—Ancient sutures or sites of ensialic mobility? A case for crustal evolution towards plate tectonics. *Tectonophysics* 40: 101–135.

Kröner, A. 1978. The Namaqua mobile belt within the framework of Precambrian crustal evolution in southern Africa. In: W. J. Vorwoerd (Ed.), *Mineralisation in Metamorphic Terranes. Geological Society of South Africa Special Publication* 4: 181–188.

Kröner, A. 1980. Pan African Crustal Evolution. *Episodes* 1980(2): 3–8.

Kröner, A. 1981a. Late Precambrian diamictites of South Africa and Namibia. In: M. J. Hambrey and W. B. Harland (Eds.), *Earth's Pre-Pleistocene Glacial Record* (Cambridge University Press: Cambridge), pp. 167–177.

Kröner, A. 1981b. Precambrian plate tectonics. In: A. Kröner (Ed.), *Precambrian Plate Tectonics* (Elsevier: Amsterdam), pp. 57–90.

Kröner, A. 1983. Proterozoic mobile belts compatible with the plate tectonic concept. In: L. G. Medaris, C. W. Byers, D. M. Mickleson and W. C. Shanks (Eds.), *Proterozoic Geology: Selected Papers from an International Proterozoic Symposium. Geological Society of America Memoir* 161: 59–74.

Kröner, A. 1984. Late Precambrian plate tectonics and orogeny: A need to redefine the term Pan-African. In: J. Klerkx and J. Michot (Eds.), *Geologie Africaine: Volume en hommage à L. Cahen* (Musee royal de l'Afrique centrale: Tervuren), pp. 23–28.

Kröner, A. 1985. Ophiolites and the evolution of tectonic boundaries in the Late Proterozoic Arabian-Nubian shield of northeast Africa and Arabia. *Precambrian Research* 27: 277–300.

Kröner, A. (Ed.), 1987. *Proterozoic Lithospheric Evolution. American Geophysical Union Geodynamics Series*, 17 (American Geophysical Union: Washington), 273 pp.

Kröner, A., McWilliams, M. O., Germs, G. J. B., Reid, A. B. and Schalk, K. E. L. 1980. Paleomagnetism of Late Precambrian to early Paleozoic mixtite-bearing formations in Namibia (Southwest Africa): The Nama Group and Blaubeker Formation. *American Journal of Science* 280: 942–968.

Kröner, A., Greiling, R., Reischmann, T., Hussein, I. M., Stern, R. J., Durr, S., Kruger, J. and Zimmer, M. 1987. Pan-African crustal evolution in the Nubian segment of northeast Africa. In: A. Kröner (Ed.), *Proterozoic Lithospheric Evolution. American Geophysical Union Geodynamics Series* 17: 235–258.

Krumbein, W. E. 1983. Stromatolites—the challenge of a term in space and time. *Precambrian Research* 20: 493–531.

Krumbein, W. E. and Cohen, Y. 1977. Primary production, mat formation and lithification: contribution of oxygenic and facultative anoxygenic cyanobacteria. In: E. Flügel (Ed.), *Fossil Algae* (Springer-Verlag: New York), pp. 37–56.

Krumbein, W. E., Cohen, Y. and Shilo, M. 1977. Solar Lake (Sinai). 4. Stromatolitic cyanobacterial mats. *Limnology and Oceanography* 22: 635–656.

Krumbein, W. E., Buchholz, H., Franke, P., Giani, D., Giele, C. and Wonneberger, K. 1979. O_2 and H_2S coexistence in stromatolites: a model for the origin of mineralogical lamination on stromatolites and banded iron formations. *Naturwissenschaften* 66: 381–389.

Kruse, P. 1987. Further Australian Cambrian sphinctozoans. *Geological Magazine* 124(6): 543–553.

Krylov, I. N. 1976. Approaches to the Classification of Stromatolites. In: M. R. Walter (Ed.), *Stromatolites: Developments in Sedimentology*, vol. 20 (Elsevier: Amsterdam), pp. 31–44.

Krylov, I. N. and Semikhatov, M. A. 1976. Table of time-ranges of the principal groups of Precambrian Stromatolites. In: M. R. Walter (Ed.), *Stromatolites: Developments in Sedimentology*, vol. 20 (Elsevier: Amsterdam), Appendix, pp. 693–694.

References

Ksiazkiewicz, M. 1977. Trace fossils in the flysch of the Polish Carpathians. *Palaeontologica Polonica* 36: 1–208.

Kuhn, W. R. and Atreya, S. K. 1979. Ammonia photolysis and the greenhouse effect in the primordial atmosphere of the earth. *Icarus* 37: 207–213.

Kuhn, W. R. and Kasting, J. F. 1983. The effects of increased CO_2 concentrations on surface temperature of the early earth. *Nature* 301: 53–55.

Kumar, G., Bhatt, D. K. and Raina, B. K. 1987. Skeletal microfauna of Meishucunian and Qiongzhusian (Precambrina-Cambrian boundary) age from the Ganga Valley, Lesser Himalaya, India. *Geological Magazine* 124(2): 167–171.

Kumarapeli, S., Pintson, H. and Dunning, G. R. 1989. Age of the Tibbit Hill formation and its implications on the timing of Iapetan rifting. *Geological Association of Canada Mineralogical Association of Canada Annual Meeting, Program with Abstracts* V(14): A125.

Kumazaki, T., Hori, H. and Osawa, S. 1983. Phylogeny of protozoa deduced from 5S rRNA sequences. *Journal of Molecular Evolution* 19: 411–419.

Kump, L. R. 1988. Terrestrial feedback in atmospheric oxygen regulation by fire and phosphorus. *Nature* 335: 152–154.

Kump, L. R. and Garrels, R. M. 1986. Modeling atmospheric O_2 in the global sedimentary redox cycle. *American Journal of Science* 286: 337–360.

Küntzel, H. and Köchel, H. G. 1981. Evolution of rRNA and origin of mitochondria. *Nature* 293: 751–755.

Küntzel, H., Heidrich, M. and Piechulla, B. 1981. Phylogenetic tree derived from bacterial, cytosol and organelle 5S rRNA sequences. *Nucleic Acids Research* 9: 1451–1456.

Kusky, T. M. 1989. Accretion of the Archean Slave province. *Geology* 17: 63–67.

Kutorga, S. S. 1848. Über die Brachiopoden-Familie der Siphonotretaceae. *Russisch Kaiserl. Min. Gesell., Verhandlugen:* 250–286.

Kvale, E. P., Archer, A. W. and Johnson, H. R. 1989. Daily, monthly, and yearly tidal cycles within laminated siltstones of the Mansfield Formation (Pennsylvanian) of Indiana. *Geology* 17: 365–368.

Kvenvolden, K. A. and Hodgson, G. W. 1969. Evidence of porphyrins in Early Precambrian Swaziland System sediments. *Geochimica et Cosmochimica Acta* 33: 1195–1202.

Kvenvolden, K. A., Peterson, E. and Pollock, G. E. 1969. Optical configurations of amino acids in Precambrian Fig Tree chert. *Nature* 221: 141–143.

Kwok, S., White, T. J. and Taylor, J. W. 1986. Evolutionary relationships between fungi, red algae, and other simple eukaryotes inferred from total DNA hybridization to a cloned basidiomycete ribosomal DNA. *Experimental Mycology* 10: 196–204.

LaBarbara, M. 1978. Precambrian geological history and the origin of the Metazoa. *Nature* 273: 22–25.

LaBerge, G. L. 1966a. Altered pyroclastic rocks in iron-formation in the Hamersley Range, Western Australia. *Economic Geology* 61: 147–161.

LaBerge, G. L. 1966b. Altered pyroclastic rocks in South African iron-formation. *Economic Geology* 61: 572–581.

Labotka, T. C. and Albee, A. C. 1977. Late Precambrian depositional environment of the Pahrump Group, Panamint Mountains, California. *California Division of Mines and Geology Special Report* 129: 93–100.

Lake, J. A. 1983. Ribosome evolution: The structural bases of protein synthesis in Archaebacteria, Eubacteria, and Eukaryotes. *Progress in Nucleic Acid Research and Molecular Biology* 30: 163–192.

Lake, J. A. 1987a. Prokaryotes and archebacteria are not monophyletic: rate invariant analyses of rRNA genes indicate that eukaryotes and eocytes form a monophyletic taxon. *Cold Spring Harbor Symposia on Quantitative Biology* 52: 839–846.

Lake, J. A. 1987b. A rate-independent technique for analysis of nucleic acid sequences: evolutionary parsimony. *Molecular Biology and Evolution* 4(2): 167–191.

Lake, J. A. 1988. Origin of the eukaryotic nucleus determined by rate-invariant analysis of rRNA sequences. *Nature* 331: 184–186.

Lake, J. A. 1989a. Origin of the eukaryotic nucleus determined by rate-invariant analyses of ribosomal RNA genes. In: B. Fernholm, K. Bremer and H. Jörnwall (Eds.), *The Hierarchy of Life* (Elsevier: Amsterdam), pp. 87–99.

Lake, J. A. 1989b. Origin of the eukaryotic nucleus: eukaryotes and eocytes are genotypically related. *Canadian Journal of Microbiology* 35: 108–109.

Lake, J. A. 1990. Origin of the Metazoa. *Proceedings of the National Academy of Science USA* 87: 763–766.

Lake, J. A., Henderson, E., Oakes, M. and Clarke, M. W. 1984. Eocytes: a new ribosome structure indicates a kingdom with a close relationship to eukaryotes. *Proceedings of the National Academy of Science USA* 81: 3786–3790.

Lamb, S. H. 1984. Structures on the eastern margin of the Archean Barberton greenstone belt, northwest Swaziland. In: A. Kröner and R. Greiling (Eds.), *Precambrian Tectonics Illustrated* (E. Schweizerbart'sche Verlagsbuchhandlung: Stuttgart, Germany), pp. 19–40.

Lambert, I. B. 1983. The major stratiform lead-zinc deposits of the Proterozoic. *Geological Society of America Memoir* 161: 209–226.

Lambert, I. B. and Donnelly, T. H. In Press—a. Global oxidation and a supercontinent in the Proterozoic: evidence from stable isotopic trends. In: M. Schidlowski (Ed.), *Early Organic Evolution: Implications for Mineral and Energy Resources* (Springer-Verlag: Berlin).

Lambert, I. B. and Donnelly, T. H. In Press—b. The paleoenvironmental significance of trends in sulfur isotope compositions in the Precambrian: a critical review. In: H. K. Herbert (Ed.), *Stable Isotope and Fluid Processes in Mineralization. Geological Society of Australia Special Publication.*

Lambert, I. B. and Groves, D. J. 1981. Early earth evolution and metallogeny. In: K. H. Wolf (Ed.), *Handbook of Stratabound and Stratiform Ore Deposits*, vol. 8 (Elsevier: Amsterdam), pp. 339–447.

Lambert, I. B., Donnelly, T. H., Dunlop, J. S. R. and Groves, D. I. 1978. Stable isotopic compositions of early Archean sulphate deposits of probable evaporitic and volcanogenic origins. *Nature* 276: 808–811.

Lambert, I. B., Knutson, J., Donnelly, T. H. and Etminan, H. 1985. The diverse styles of sediment-hosted copper deposits in Australia. In: G. H. Friedrich (Ed.), *Geology and Metallogeny of Copper Deposits* (Springer-Verlag: Berlin), pp. 540–558.

Lambert, I. B., Walter, M. R., Zang, W., Lu, S. and Ma, G. 1987. Paleoenvironment and carbon isotope stratigraphy of Upper Proterozoic carbonates of the Yangtze Platform. *Nature* 325: 140–142.

Lambert, R. St. J. 1976. Archean thermal regimes, crustal and upper mantle temperatures, and a progressive evolutionary model for the earth. In: B. F. Windley (Ed.), *The Early History of the Earth* (John Wiley & Sons: New York), pp. 363–373.

Lamberti, G. A. and Moore, J. W. 1984. Aquatic insects as primary consumers. In: V. H. Resh and D. M. Rosenberg (Eds.), *The Ecology of Aquatic Insects* (Praeger Publ.: New York), pp. 164–195.

Landing, E. 1984. Skeleton of lapworthellids and the suprageneric classification of tommotiids (Early and Middle Cambrian phosphatic problematica). *Journal of Paleontology* 58(6): 1380–1398.

Landing, E. 1988. Lower Cambrian of eastern Massachusetts: stratigraphy and small shelly fossils. *Journal of Paleontology* 62: 661–695.

Landing, E. 1989. The Plancentian Series: Appearance of the oldest skeletalized faunas in southeastern Newfoundland. *Journal of Paleontology* 63: 739–769.

Landing, E., Nowlan, G. S. and Fletcher, T. P. 1980. A microfauna associated with Early Cambrian trilobites of the *Callavia* Zone, northern Antigonish Highlands, Nova Scotia. *Canadian Journal of Earth Sciences* 17(3): 400–418.

Lange, M. A. and Ahrens, T. J. 1982. The evolution of an impact-generated atmosphere. *Icarus* 51: 96–120.

Langford, F. F. 1983. Proterozoic uranium deposits and the Precambrian atmosphere. In: L. G. Medaris Jr., C. W. Byers, D. M.

Mickelson and W. C. Shanks (Eds.), *Proterozoic Geology: Selected Papers from an International Proterozoic Symposium.* Geological Society of America Memoir 161: 237–244.

Langford, F. F. and Morin, J. A. 1976. The development of the Superior Province of northwestern Ontario by merging island arcs. *American Journal of Science* 276: 1023–1034.

Langworthy, T. A. 1985. Lipids of Archaebacteria. In: C. R. Woese and R. S. Wolfe (Eds.), *The Bacteria. A Treatise on Structure and Function*, vol. 8 (Academic Press: New York), pp. 459–497.

Lanier, W. P. 1986. Approximate growth rates of Early Proterozoic microstromatolites as deduced by biomass productivity. *Palaios* 1: 525–542.

Lanier, W. P. 1989. Interstitial and peloid microfossils from the 2.0 Ga Gunflint Formation: Implications for the paleoecology of the Gunflint stromatolites. *Precambrian Research* 45: 291–318.

Largeau, C., Casadevall, E., Kadouri, A. and Metzger, P. 1984. Formation of *Botryococcus*-derived kerogens. Comparative study of immature torbanites and of the extant alga *Botryococcus braunii*. *Organic Geochemistry* 6: 327–332.

Largeau, C., Derenne, S., Casadevall, E., Kadouri, A. and Sellier, N. 1986. Pyrolysis of immature torbanite and of the resistant biopolymer (PRB A) isolated from extant alga *Botryococcus braunii*. Mechanism of formation and structure of torbanite. *Organic Geochemistry* 10: 1023–1032.

La Rivers, I. 1978. *Algae of the Western Great Basin* (University of Nevada Publishing: Las Vegas), 390 pp.

Larkin, J. M. and Strohl, W. R. 1983. *Beggiatoa, Thiotrix* and *Thioploca*. *Annual Review of Microbiology* 37: 341–367.

Larter, S. and Senftle, J. T. 1985. Improved kerogen typing for petroleum source rock analysis. *Nature* 318: 277–280.

Lasaga, A. C., Berner, R. A. and Garrels, R. M. 1985. An improved geochemical model of atmospheric CO_2 fluctuations over the past 100 million years. In: E. T. Sundquist and W. S. Broecker (Eds.), *The Carbon cycle and Atmospheric CO_2: Natural Variations Archean to Present. Geophysical Monograph*, 32 (American Geophysics Union: Washington, D.C.), pp. 397–411.

Laskin, A. I. and Lechevalier, H. A. (Eds.), 1977. *CRC Handbook of Microbiology*, 2nd ed., vol. 1 (CRC Press: Boca Raton), 757 pp.

Lata, G. F. 1974. An investigation of the possible occurrence of sterols in blue-green algae. *Proceedings of the Iowa Academy of Science* 81: 89–90.

Latham, A. and Riding, R. 1990. The Precambrian-Cambrian Boundary in Morocco: Fossil evidence calibrates isotope curve. *Nature* 344: 752–754.

Laughan, A. and Scott, M. P. 1984. Sequence of a *Drosophila* segmentation gene: protein structure homology with DNA-binding proteins. *Nature* 310: 25–31.

Laureillard, J., Largeau, C. and Casadevall, E. 1988. Oleic acid in the biosynthesis of the resistant biopolymers of *Botryococcus braunii*. *Phytochemistry* 27: 2095–2098.

Laurie, J. R. 1986. Phosphatic fauna of the Early Cambrian Todd River Dolomite, Amadeus Basin, central Australia. *Alcheringa* 10(3–4): 431–454.

Lawver, L. A. and Scotese, C. R. 1987. A revised reconstruction of Gondwana. In: G. D. McKenzie (Ed.), *Gondwana Six: Structure, Tectonics & Geophysics. Geophysical Monograph*, 40 (American Geophysics Union: Washington, DC), pp. 17–23.

Lazarek, S. 1982. Structure and function of a cyanophytan mat community in an acidified lake. *Canadian Journal of Botany* 60: 2235–2240.

Leadbeater, B. S. C. and Riding, R. (Eds.), 1986. *Biomineralization in Lower Plants and Animals. The Systematics Association Special Volume 30* (Clarendon Press: Oxford), 401 pp.

Lebesconte, P. 1887. Constitution générale du Massif breton comparée á celle du Finistère. ser. 3. *Société géologique de France, Bulletin* 14(3): 776–820.

Lebesconte, P. 1891. Les poudingues rouges de Montfort. *Revue des Sciences Naturelles de l'Ouest, 1891* fasc. 3: 200–207.

Lee, W. H. K. 1970. On the global variations of terrestrial heat flow. *Physics of the Earth and Planetary Interiors* 2: 332–341.

Leffers, H., Kjems, J., Ostergaard, L., Larsen, J. and Garrett, R. A. 1987. Evolutionary relationships amongst Archaebacteria. A comparative study of 23S ribosomal RNAs of a sulphur-dependent extreme thermophile, an extreme halophile and a thermophilic methanogen. *Journal of Molecular Biology* 195: 43–61.

LeGallais, C. J. and Lavoie, S. 1982. Basin evolution of the lower Proterozoic Kaniapiskau Supergroup, central Labrador miogeocline (trough), Quebec. *Bulletin of Canadian Petroleum Geologists* 30: 150–156.

Legendre, L., Demers, S., Yentsch, C. M. and Yentsch, C. S. 1983. The ^{14}C method: Patterns of dark CO_2 fixation and DCMU correction to replace the dark bottle. *Limnology and Oceanography* 28: 996–1003.

Lenaers, G., Maroteaux, L., Michot, B. and Herzog, M. 1989. Dinoflagellates in evolution. A molecular phylogenetic approach analysis of large subunit ribosomal RNA. *Journal of Molecular Evolution* 29: 40–51.

Lendzion, K. 1975. Fauna of the *Mobergella* Zone in the Polish Lower Cambrian. *Kwartalnik Geologiczny* 19: 237–242.

Lendzion, K. 1977. *Cassubia*–a new generic name for *Pomerania* Lendzion, 1975. *Kwartalnik Geologiczny* 21: 211.

Lenn, R. C. 1966. Primary Productivity of Drakesbad Hot Springs. M.A. thesis, University of California, Davis.

Lenz, J. and Zeitzschell, B. 1968. Zur Bestimmung des Extinktionskoeffizienten für Chlorophyll *a* in Methanol. *Kieler Meeres Forschungen* 24: 41–50.

Lepp, H. 1966. Chemical composition of the Biwabik Iron Formation, Minnesota. *Economic Geology* 61: 243–250.

Lesher, C. M. 1978. Mineralogy and petrology of the Sokoman iron formation near Ardua Lake, Quebec. *Canadian Journal of Earth Sciences* 15: 480–500.

Leventhal, J., Suess, S. E. and Cloud, P. 1975. Nonprevalence of biochemical fossils in kerogen from pre-Phanerozoic sediments. *Proceedings of the National Academy of Science USA* 72: 4706–4710.

Leventhal, J. S. 1987. Carbon and sulfur relationships in Devonian shales from the Appalachian basin as an indicator of environment of deposition. *American Journal of Science* 287: 33–49.

Levin, E. Y. and Bloch, K. 1964. Absence of sterols in blue-green algae. *Science* 202: 90–91.

Levine, J. S., Boughner, R. E. and Smith, K. A. 1980. Ozone, ultraviolet flux, and the temperature of the paleoatmosphere. *Origins of Life* 10: 199–213.

Lewan, M. D. 1986. Stable carbon isotopes of amorphous kerogens from Phanerozoic sedimentary rocks. *Geochimica et Cosmochimica Acta* 50: 1583–1591.

Lewan, M. D., Bjorøy, M. and Dolcater, D. L. 1986. Effects of thermal maturation on steroid hydrocarbons as determined by hydrous pyrolysis of Phosphoria Retort Shale. *Geochimica et Cosmochimica Acta* 50: 1977–1987.

Lewin, R. 1984. Alien beings here on earth. *Science* 223: 39.

Lewin, R. A. 1984. *Prochloron*–a status report. *Phycologia* 23: 203–208.

Lewin, R. A. and Cheng, L. (Eds.), 1989. *Prochloron: A Microbial Enigma* (Chapman and Hall: New York), 129 pp.

Lewis, H. P. 1942. On *Girvanella* in the "Shumardia Limestone" of Lévis, Quebec. *Annals and Magazine of Natural History* 9 (Series 11): 49–55.

Lewis, S. M., Norris, J. N. and Searles, R. B. 1987. The regulation of morphological plasticity in tropical reef algae by herbivory. *Ecology* 68: 636–641.

Li, W.-H., Tonimura, M. and Sharp, P. H. 1987. An evaluation of the molecular clock using mammalian DNA sequences. *Journal of Molecular Evolution* 25: 330–342.

Li, Y. and Yang, X. 1984. [On the geologic feature and minergenetic mechanism of Jing-Xiang phosphate deposit, Hubei Province]. *International Field Workshop and Seminar on Phosphorite*, 5th, 1982, K'un ming shis, China 2: 285–306.

Li, Y.-W. 1975. [Cambrian ostracodes and their new knowledge from Nichuuan, Yunnan, and Shaanxi]. In: A. G. S. Editorial Committee (Ed.), *Stratigraphy, Palaeontology* (Geology Press:

References

Beijing), pp. 37–72.
Li, Z.-P. 1984. The discovery and its significance of small shelly fossils in the Hexi area, Xixiang, Shaanxi. *Geology of Shannxi* 2(1): 73–77.
Liang, Y. and Cao, R. 1976. *Biostratigraphic Significance of Stromatolites and Red Algae from the Sinian Subera of China*, 15 pp.
Liang, Y. and Cao, R. 1979. [Late Precambrian stromatolites and red algae]. In: *Collection of Essays on International Geological Correlation 2. Stratigraphy & Paleontology* (Geological Publishing House: Beijing), pp. 47–54.
Liang, Y., Zhu, S., Zhang, L., Cao, R., Gao, Z. and Bu, D. 1985. Stromatolite Assemblages of the Late Precambrian in China. *Precambrian Research* 29(1-3): 15–32.
Licari, G. R. 1978. Biogeography of the late Pre-Phanerozoic Beck Spring Dolomite of Eastern California. *Journal of Paleontology* 52(4): 767–792.
Likens, G. E., Borman, F. H. and Johnson, N. M. 1981. Interactions between major biogeochemical cycles in terrestrial ecosystems. In: G. E. Likens (Ed.), *Some Perspectives of the Major Biogeochemical Cycles* (John Wiley & Sons: New York), pp. 93–122.
Lin, J.-L., Fuller, M. and Zhang, W.-Y. 1985. Paleogeography of the North and South China Blocks during the Cambrian. *Journal of Geodynamics* 2: 91–114.
Lin, Tianrui 1987. Early Cambrian bradoriids from Yeshan, Luhe District, Jiangsu. *Acta Palaeontologica Sinica* 26(1): 84–85.
Liñan, E., Moreno-Eiris, E., Perejón, A. and Schmitt, M. 1981. Fossils from the basal levels of the Pedroche Formation, Lower Cambrian (Sierra Morena, Cordoba, Spain). *Bol. Real Soc. Española Hist. Natural (Geológica)* 79: 277–286.
Lindgren, S. 1981. Remarks on the taxonomy, botanical affinities, and distribution of leiospheres. *Stockholm Contributions in Geology* 38: 1–20.
Lindholm, T., Lindroos, P. and Mörk, A.-C. 1988. Ultrastructure of the photosynthetic ciliate *Mesodinium rubin*. *Biosystems* 21: 141–149.
Lindsay, J. F. 1987a. Late Proterozoic evaporites in the Amadeus Basin, central Australia, and their role in basin tectonics. *Geological Society of America Bulletin* 99: 852–865.
Lindsay, J. F. 1987b. Sequence stratigraphy and depositional controls in late Proterozoic-Early Cambrian sediments of Amadeus Basin, central Australia. *American Association of Petroleum Geologists Bulletin* 71: 1387–1403.
Linnarsson, J. G. O. 1876. On the Brachiopoda of the *Paradoxides* Bed of Sweden. *Bih. Kongl. Svensk Vet. Akad. Förhandlung* 1875: 1–34.
Lipps, J. H. 1981. What, if anything, is micropaleontology? *Paleobiology* 7: 167–199.
Lipps, J. H. 1987. Cambrian foraminifera. *Geological Society of America Abstracts with Programs* 19: 747.
Lipps, J. H. and Sylvester, A. G. 1968. The enigmatic Cambrian fossil *Volborthella* and its occurrence in California. *Journal of Paleontology* 42(2): 329–336.
Lipscomb, D. L. 1985. The eukaryotic kingdoms. *Cladistics* 1: 127–140.
Liu, D. 1987. Brachiopods and tommotiids near Precambrian-Cambrian boundary in SW China. In: *Stratigraphy and Palaeontology of Systemic Boundaries in China—PC-C Boundary*, vol. 1 (Nanjing University Publishing House: Nanjing), pp. 345–414.
Liu, D. 1979. Earliest Cambrian brachiopods from southwest China. *Acta Palaeontologica Sinica* 18(5): 505–512.
Liu, X.-L. 1981. Metazoa fossils from Mashan Group near Jixi, Heilongjiang. *Bulletin of the Chinese Academy of Geological Sciences* 3: 71–83.
Liu, X.-X., Liu, Z.-L., Zhang, L. and Xu, X.-S. 1984. [A study of Late Precambrian microfossil algal community from Suining County, Jiangsu Province]. *Acta Micropalaeontologica Sinica* 1(2): 171–182 (In Chinese, with English abstract).
Ljungdahl, L. J. 1986. The autotrophic pathway of acetate synthesis in acetogenic bacteria. *Annual Review of Microbiology* 40: 415–450.
Lo, S. C. 1980. Microbial fossils from the Lower Yudoma Suite, earliest Phanerozoic, eastern Siberia. *Precambrian Research* 13: 109–166.
Lobach-Zhuchenk, S. B., Levchenko, O. A., Chekulaev, V. P. and Krylov, I. N. 1986. Geological evolution of the Karelian granite-greenstone terrain. *Precambrian Research* 33: 45–65.
Lochman, C. and Duncan, D. 1944. Early Upper Cambrian faunas of central Montana. *Geological Society of America Special Paper* 54: 1–181.
Loeblich, A. R., Jr. and Tappan, H. 1969. Acritarch excystment and surface ultrastructure with descriptions of some Ordovician taxa. *Revista Española de Micropaleontología* 1: 45–57.
Loeblich, A. R., Jr. and Tappan, H. 1976. Some new and revised organic-walled phytoplankton microfossil genera. *Journal of Paleontology* 50: 301–308.
Loeblich, A. R., Jr. and Tappan, H. 1987. *Foraminiferal Genera and Their Classification* (Van Nostrand Reinhold Co.: New York), 970 pp.
Logan, B. W., Hoffman, P. and Gebelein, C. D. 1974a. Algal mats, cryptalgal fabrics, and structures, Hamelin Pool, Western Australia. *American Society of Petroleum Geologists Memoir* 22: 140–194.
Logan, B. W., Read, J. F., Hagan, G. M., Hoffman, P. F., Brown, R. G., Woods, P. J. and Gebelein, C. D. 1974b. Evolution and Diagenesis of Quarternary Carbonate Sequences, Shark Bay, Western Australia. *American Association of Petroleum Geologists Memoir* 22: 1–358.
Loiseaux-de Goer, S., Markowicz, Y., Dalmon, J. and Audren, H. 1988. Physical maps of the two circular plastid DNA molecules of the brown alga *Pylaiella littoralis* (L.) Kjellm. *Current Genetics* 14: 155–162.
Longhi, J. 1978. Pyroxene stability and the composition of the lunar magma ocean. In: R. B. Merrill (Ed.), *Proceedings of the Ninth Lunar and Planetary Sciences Conference* (Pergamon Press: New York), pp. 285–306.
Lorenzen, S. 1985. Phylogenetic aspects of pseudocoelomate evolution. In: S. Conway Morris, J. D. George, R. Gibson and H. M. Platt (Eds.), *The Origin and Relationships of Lower Invertebrates. Systematics Association Special Volume*, 28 (Clarendon Press: Oxford), pp. 210–223.
Love, F. G., Simmons, G. M., Jr., Parker, B. C., Wharton, R. A., Jr. and Seaburg, K. G. 1983. Modern Conophyton-like microbial mats discovered in Lake Vanda, Antarctica. *Geomicrobiology Journal* 3: 33–48.
Lowe, D. R. 1980a. Archean sedimentations. *Annual Review of Earth and Planetary Sciences* 8: 145–167.
Lowe, D. R. 1980b. Stromatolites 3,400-Myr old from the Archean of Western Australia. *Nature* 284: 441–443.
Lowe, D. R. 1982. Comparative sedimentology of the principal volcanic sequences of Archean greenstone belts in South Africa, Western Australia, and Canada: implications for crustal evolution. *Precambrian Research* 17: 1–29.
Lowe, D. R. 1983. Restricted shallow-water sedimentation of early Archean stromatolitic and evaporitic strata of the Strelley Pool Chert, Pilbara Block, Western Australia. *Precambrian Research* 19: 239–283.
Lowe, D. R. and Byerly, G. R. In Press. Stratigraphy of the west-central part of the Barberton Greenstone Belt, South Africa. In: D. R. Lowe and G. R. Byerly (Eds.), *Geologic Evolution of the Barberton Greenstone Belt. Geological Society of America Memoir*.
Lowe, D. R. and Knauth, L. P. 1977. Sedimentology of the Onverwacht Group (3.4 billion years), Transvaal, South Africa, and its bearing on the characteristics and evolution of the early earth. *Journal of Geology* 85: 699–723.
Lowe, D. R., Byerly, G. R., Asaro, F. and Kyte, F. J. 1989. Geological and geochemical record of 3400-million-year-old terrestrial meteorite impacts. *Science* 245: 959–962.
Lowenstam, H. and Weiner, S. 1983. Mineralization by organisms and the evolution of biomineralization. In: P. Westbroek and E. W. De Jong (Eds.), *Biomineralization and Biological Metal Accumulation* (Reidel: Dordrecht), pp. 191–203.
Lowenstam, H. A. 1950. Niagaran reefs of the Great Lakes Area. *Geology* 58: 430–487.
Lowenstam, H. A. and Margulis, L. 1980a. Calcium regulation and the appearance of calcareous skeletons in the fossil record. In: M. Omori and N. Watabe (Eds.), *Mechanisms of Biomineralization in*

Animals and Plants (Tokai University Press: Tokyo), pp. 289–300.

Lowenstam, H. A. and Margulis, L. 1980b. Evolutionary prerequisites for early Phanerozoic calcareous skeletons. *Biosystems* 12: 27–41.

Lowenstam, H. A. and Weiner, S. 1985. Transformation of amorphous calcium phosphate to crystalline dahllite in the radular teeth of chitons. *Science* 227: 51–53.

Lowenstam, H. A. and Weiner, S. 1989. *On Biomineralization* (Oxford University Press: New York), 324 pp.

Lubchenco, J. and Cubit, J. D. 1980. Heteromorphic life histories of certain marine algae as adaptations to variations in herbivory. *Ecology* 61: 676–687.

Luchinina, V. A. 1969a. On pervoj nakhodke trubchatykh vodoroslej v Yuedejskoj svite Yakutii [On the first discovery of tubular algae from the Yuedej Formation of Yakutia]. In: I. T. Zhuravleva (Ed.), *Biostratigrafiya i Paleontologiya Nizhnego Kembiya Sibiri i Al'nego Vostoka* (Nauka: Moscow), pp. 182–183.

Luchinina, V. A. 1969b. *Renalcis polymorphus* Maslov iz Yudomskogo kompleksa p. Sukharikhi [*Renalcis polymorphus* Maslov from the Yudomian complex on the Sukharika River]. In: I. T. Zhuravleva (Ed.), *Biostratigrafiya i Paleontologiya Nizhnego Kembriya Sibiri i Dal'nego Vostoka* (Nauka: Moscow), pp. 184–185.

Luchinina, V. A. 1971. K sistematike poda *Proaulopora* Vologdin [On the systematics of the genus *Proaulpora* Vologdin]. In: T. F. Vozzhennikova (Ed.), *Vodorosli Paleozoya i Mezozoya Sibiri [Algae of the Paleozoic and Mesozoic of Siberia]* (Nauka: Moscow), pp. 5–8.

Luchinina, V. A. 1972. Kembrijskie izvestkovistye vodorsli podov *Subtiflora* Maslov i *Batinevia* Korde [Cambrian calcareous algae of the genera *Subtiflora* Maslov and *Batinevia* Korde]. In: I. T. Zhuravleva (Ed.), *Problemy Biostratigrafii i Paleontologii Nizhnego Kembriya Sibiri* (Nauka: Moscow), pp. 217–221.

Luchinina, V. A. 1975. *Paleoal'gologicheskaya Kharakteristika Rannego Kembriya Sibirskoj Platformy [Paleo-Algal Characteristics of the Early Cambrian of the Southeastern Siberian Platform]* (Nauka, Siberskoe Otdelenie: Novosibirsk), 100 pp.

Luchinina, V. A. 1985. Vodoroslevye postrokojki rannego paleozoya Severa Sibirskoj Platformy [Algal buildups of the early Paleozoic of the north Siberian Platform] Sreda i zhizn' v geologicheskom proshlom. *Trudy Institut Geologii i Geofiziki Sibirskoe Otdolenie, Akademiya Nauk SSSR* 628: 45–50.

Luchinina, V. A. 1986. Izvestkovye vodorosli v kembrijskikh organogennykh postrojkakh Manskogo progiba [Calcareous algae from Cambrian biogenic structures in the Mansk trough]. In: I. T. Zhuravleva (Ed.), *Biostratigrafiya i Paleontologiya Kembriya Severnoj Azii. Trudy Inst. Geol. Geofiz. Sibirsk Otd. Akad. Nauk SSSR 669* (Nauka: Novosibirsk), pp. 77–85.

Luchinina, V. A. and Voronova, L. G. 1983. Izvestkovye vodorosli [Calcareous algae]. In: B. S. Sokolov and I. T. Zhuravleva (Eds.), *Yarusnoe Raschlenenie Nizhnego Kembriya Sibiri. Atlas Okamenelostej. Trudy Inst. Geol. Geofiz. Sibirsk. Otd. Akad. Nauk SSSR 558* (Nauka: Moscow), pp. 170–177.

Ludbrook, N. H. 1980. *A Guide to the Geology and Mineral Resources of South Australia* (Department of Mines and Energy of South Australia: Adelaide), 230 pp.

Ludden, J., Hubert, C. and Gariépy, C. 1986. The tectonic evolution of the Abitibi greenstone belt of Canada. *Geology Magazine* 123: 153–166.

Ludwig, M. and Gibbs, S. P. 1985. DNA is present in the nucleomorph of cryptomonads: further evidence that the chloroplast evolved from a eukaryotic symbiont. *Protoplasma* 127: 9–20.

Ludwig, M. and Gibbs, S. P. 1987. Are the nucleomorphs of cryptomonads and *Chlorarachnion* the vestigal nuclei of eukaryotic endosymbionts? *Annals of the New York Academy of Sciences* 503: 198–212.

Ludwig, M. and Gibbs, S. P. 1989. Evidence that the nucleomorphs of *Chlorarachnion reptans* (Chlorarachniophyceae) are vestigial nuclei: morphology, division and DNA-DAPI fluorescence. *Journal of Phycology* 25: 385–394.

Luo, H., Jiang, Z., Wu, X., Song, X. and Ouyang, L. 1982. [*The Sinian-Cambrian Boundary in Eastern Yunnan, China*] (People's Republic of China), 265 pp.

Luo, H.-L., Jiang, Z.-W., Xu, Z.-J., Song, X.-L. and Xue, X.-F. 1980. [On the Sinian-Cambrian boundary of Meischuun and Wangjiawan, Jinning County, Yunnan]. *Acta Geologica Sinica* 1980(2): 95–111.

Luo, H.-L., Jiang, Z.-W., Wu, X., Song, H., Ouyang, L., Xing, Y. and Liu, G. 1984. *Sinian-Cambrian Boundary Stratotype Section at Meishucun, Jinning, Yunnan, China* (People's Publishing House: Yunnan), 154 pp.

Luo, Q. 1984. [A preliminary study on the uppermost Sinian-lowermost Cambrian age microfossils from Qingzhen-Zhijin County in Guizhou]. In: *The Upper Precambrian and Sinian-Cambrian Boundary in Guizhou* (People's Publishing House of Guizhou: Guizhou), pp. 107–116.

Luo, Q., Wang, F. and Wang, Y. 1982. [Uppermost Sinian-lowermost Cambrian age microfossils from Qingzhen-Zhijin County, Guizhou Province]. *Bulletin of the Tianjin Institute of Geology and Mineral Resources*: 23–41.

Luyendyk, B. P. and Hornafius, J. S. 1987. Neogene crustal rotations, fault lip, and basin development in southern California. In: R. Ingersoll V and W. G. Ernst (Eds.), *Cenozoic Basin Development of Coastal California. Rubey*, Volume VI (Prentice-Hall: Englewood Cliffs, N.J.), pp. 259–283.

Lynn, D. H. and Sogin, M. L. 1988. Assessment of phylogenetic relationships among ciliated protists using partial ribosomal RNA sequences derived from reverse transcripts. *Biosystems* 21: 249–254.

Lyons, W. B., Gaudette, H. E. and Gustafson, N. C. 1982. Dissolved organic carbon in pore waters from a hypersaline environment. *Organic Geochemistry* 3: 133–135.

Lyons, W. B., Hines, M. E. and Gaudette, H. E. 1984a. Major and minor element pore water geochemistry of modern marine sabkhas: the influence of cyanobacterial mats. In: Y. Cohen, R. W. Castenholz and H. O. Halvorson (Eds.), *Microbial Mats: Stromatolites* (Alan R. Liss, Inc.: New York), pp. 411–423.

Lyons, W. B., Long, D. T., Hines, M. E., Gaudette, H. E. and Armstrong, P. B. 1984b. Calcification of cyanobacterial mats in Solar Lake, Sinai. *Geology* 12: 623–626.

Lysova, L. A., Galimova, B. S., Titorenko, T. N. and Fajzulina, Z. K. 1966. Paleontologicheskaya kharakteristika nizhnekembrijskikh otlozhenij vskrytykh Markovskoj opornoj skvazhinoj [Paleontological characteristics of Lower Cambrian deposits discovered in the Markovsk borehole]. In: *Geologiya i Gazoneftenosnost' Vostochnoj Sibiri* (Nedra: Moscow), pp. 345–356.

Lyubtsov, V. V. 1962. Organicheskiye ostatki drevnejshikh osadochno-metamorficheskikh tolshch Kolskogo poluostrova [Organic remains in most ancient sedimentary and metamorphic sequences of the Kola Peninsula]. *Izvestiya Akademii Nauk SSSR, seriia Geologicheskaya* 1962(10): 69–73 (1964, International Geology Review 6 (8): 1408–1412).

Macdonald, A. J. 1987. The platinum group element deposits: classification and genesis. *Geoscience Canada* 14: 155–166.

Mack, E. E. and Pierson, B. K. 1988. Preliminary characterization of a temperate marine member of the Chloroflexaceae. In: J. M. Olson, J. G. Ormerod, J. Amesz, E. Stackebrandt and H. G. Trüper (Eds.), *Green Photosynthetic Bacteria* (Plenum Press: New York), pp. 237–241.

MacKay, R. M., Salgado, D., Bonen, L., Stackebrandt, E. and Doolittle, W. F. 1982. The 5S ribosomal RNAs of *Paracoccus denitrificans* and *Prochloron*. *Nucleic Acids Research* 10: 2963–2970.

Mackenzie, A. S. and McKenzie, D. 1983. Isomerisation and aromatisation of hydrocarbons in sedimentary basins formed by extension. *Geological Magazine* 120: 417–470.

Mackenzie, A. S., Brassell, S. C., Eglinton, G. and Maxwell, J. R. 1982. Chemical fossils: The geological fate of steroids. *Science* 217: 491–504.

Mackenzie, S. A., Patience, R. L., Maxwell, J. R., Vandenbrouke, M. and Durand, B. 1980. Molecular parameters of maturation in the Toarcian shales, Paris Basin, France—I. Changes in the con-

References

figuration of acyclic isoprenoid alkanes, steranes and triterpanes. *Geochimica et Cosmochimica Acta* 44: 1709–1721.

Mackie, G. O. 1963. Siphonophores, bud colonies, and superorganisms. In: E. C. Dougherty et al. (Eds.). *The Lower Metazoa, Comparative Biology and Phylogeny* (University of California Press: Berkeley), pp. 329–337.

MacKinnon, D. I. 1983. A late middle cambrian orthide-kutorginide brachiopod from northwest Nelson, New Zealand. *New Zealand Journal of Geology and Geophysics* 26: 97–102.

Madigan, M. T. 1986. *Chromatium tepidum* sp. nov., a thermophilic photosynthetic bacterium of the family Chromatiaceae. *International Journal of Systematic Bacteriology* 36: 222–227.

Madigan, M. T. and Brock, T. D. 1976. Quantitative estimation of bacteriochlorophyll *c* in the presence of chlorophyll *a* in aquatic environments. *Limnology and Oceanography* 21: 462–467.

Madigan, M. T., Takigiku, U., Lee, R. G., Gest, H. and Hayes, J. M. 1989. Carbon isotope fractionation by thermophilic phototrophic sulfur bacteria: evidence for autotrophic growth in natural populations. *Applied and Environmental Microbiology* 55: 639–644.

Märss, T. 1988. Early Palaeozoic hadimopanellids of Estonia and Kirgizia (USSR). *Eesti NSV Teaduste Akadeemia Toimetised* 37(1): 10–17.

Magaritz, M. 1989. $\delta^{13}C$ minima follow extinction events: a clue to faunal radiation. *Geology* 17: 337–340.

Magaritz, M., Holser, W. T. and Kirschvink, J. L. 1986. Carbon-isotope events across the Precambrian/Cambrian boundary on the Siberian Platform. *Nature* 320: 258–259.

Magnasson, N. H. 1960. Age determinations of Swedish Precambrian rocks. *Geologiska Föreningens i Stockholm Förhandlingar* 82: 407–431.

Maher, K. A. and Stevenson, D. J. 1988. Impact frustration of the origin of life. *Nature* 331: 612–614.

Maier, S. and Gallardo, V. A. 1984a. Nutritional characteristics of two marine thioplocas determined by autoradiography. *Archives of Microbiology* 139: 218–220.

Maier, S. and Gallardo, V. A. 1984b. *Thioploca araucae* sp. nov. and *Thioploca chieae* sp. nov. *International Journal of Systematic Bacteriology* 34: 414–418.

Maithy, P. K. 1973. Micro-organisms from the Bushimay System (Late Precambrian) of Kanshi, Zaire. *The Palaeobotanist* 22(2): 133–149.

Maithy, P. K. and Shukla, M. 1977. Microbiota from the Suket Shales, Ramapura, Vindhyan System (Late Pre-Cambrian), Madhya Pradesh. *The Palaeobotanist* 23: 176–188.

Majoran, S. 1987. Structural investigations of octocoral sclerites. *Zoologica Scripta* 16: 277–287.

Maki, Y. 1986. Factors on habitat preference *in situ* of sulfur-turfs growing in hot springs effluents: dissolved oxygen and current velocities. *Journal of General and Applied Microbiology* 32: 203–213.

Maki, Y. 1987. Biological oxidation of sulfide and elemental sulfur by the A-type sulfur-turf growing in hot spring effluents. *Journal of General and Applied Microbiology* 22: 123–134.

Makushina, V. M., Aref'ev, O. A., Zabrodina, M. N. and Petrov, A. A. 1978. New relic alkanes of petroleums. *Neftekhimiya* 18: 847–854. (in Russian).

Malik, K. A. 1983. A modified method for the cultivation of phototrophic bacteria. *Journal of Microbiological Methods* 1: 343–352.

Malin, G. and Walsby, A. E. 1985. Chemotaxis of a cyanobacterium in concentration gradients of carbon disxide, bicarbonate and oxygen. *Journal of General Microbiology* 131: 2643–2652.

Malinky, J. M. 1983. New taxa in the Lower Cambrian of Ny Friesland, Spitsbergen, and their biostratigraphic significance. *Geological Society of America Abstracts with Programs* 15: 634.

Mambetov, A. M. 1972. Novyj rod khiolitov iz nizhnego kembriya Malogo Karatau (severo-zapadnyj Tian'-Shan) [A new genus of hyoliths from the Lower Cambrian of Malyj Karatau (north-western Tien-Shan)]. *Paleontologicheskij Zhurnal* 1972(2): 140–142.

Mambetov, A. M. and Repina, L. N. 1979. Nizhnij kembrij Talasskogo Ala-Too i ego korrelyatsiya s razrezami Malogo Karatau i Sibirskoj Platformy [The Lower Cambrian of Talasskij Ala-Too and its correlation with the sections of Malyj Karatau and the Siberian Platform]. *Trudy Institut Geologii i Geofiziki Sibirskoe Otdolenie, Akademiya Nauk SSSR* 406: 98–158.

Manabe, S. and Wetherald, R. T. 1967. Thermal equilibrium of the atmosphere with a given distribution of relative humidity. *Journal of Atmospheric Science* 24: 241–259.

Mannella, C. A., Frank, J. and Delihas, N. 1985. Interrelatedness of 5S RNA sequences investigated by correspondence analysis. *Journal of Molecular Evolution* 24: 228–235.

Manske, C. L. and Chapman, D. J. 1987. Nonuniformity of nucleotide substitution rates in molecular evolution: Computer simulation and analysis of 5S ribosomal RNA sequences. *Journal of Molecular Evolution* 26: 226–251.

Marek, L. 1963a. The Class Hyolitha in the Caradoc of Bohemia. *Sbornik Geologickych Ved. Paleontologie* 9: 51–113.

Marek, L. 1963b. New knowledge on the morphology of *Hyolithes*. *Sbornik Geologickych Ved. Paleontologie* 9: 53–113.

Marek, L. and Yochelson, E. L. 1964. Paleozoic mollusk: *Hyolithes*. *Science* 146: 1674–1675.

Marek, L. and Yochelson, E. L. 1976. Aspects of the Biology of *Hyolitha* (Mollusca). *Lethaia* 9: 65–82.

Margulis, L. 1970. *Origin of Eukaryotic Cells* (Yale University Press: New Haven), 349 pp.

Margulis, L. and Obar, R. 1985. *Heliobacterium* and the origin of chrysoplasts. *Biosystems* 17: 317–325.

Margulis, L., Walker, J. C. G. and Rambler, M. 1976. Reassessment of roles of oxygen and ultraviolet light in Precambrian evolution. *Nature* 264: 620–624.

Margulis, L., Barghoorn, E. S., Ashendorf, D., Banerjee, S., Chase, D., Francis, S., Giovannoni, S. and Stolz, J. 1980. The microbial community in the layered sediments at Laguna Figueroa, Baja California, Mexico: does it have Precambrian analogues? *Precambrian Research* 11: 93–123.

Markowicz, Y., Loiseaux-de Goer, S. and Mache, R. 1988a. Presence of a 16S rRNA pseudogene in the bi-molecular plastid genome of the primitive brown alga *Pykaiella littoralis*. Evolutionary implications. *Current Genetics* 14: 599–608.

Markowicz, Y., Mache, R. and Loiseaux-de Goer, S. 1988b. Sequence of the plastid rDNA spacer region of the brown alga *Pylaiella littoralis* (L.) Kjellm. Evolutionary significance. *Plant Molecular Biology* 10: 465–469.

Maront, S. 1987. Uncomformity-type uranium deposits. *Geoscience Canada* 14: 219–229.

Marshall, H. G., Walker, J. C. G. and Kuhn, W. R. 1988. Long-term climate change and the geochemical cycle of carbon. *Journal of Geophysical Research* 93: 791–802.

Martin, A., Nisbet, E. G. and Bickle, M. J. 1980. Archean stromatolites of the Belingwe Greenstone Belt, Zimbabwe (Rhodesia). *Precambrian Research* 13: 337–362.

Martin, A. K., Hartnady, C. J. H. and Goodlad, S. W. 1981. A revised fit of South America and south central Africa. *Earth and Planetary Science Letters* 54: 293–305.

Martin, H. and Porada, H. 1977. The intracratonic branch of the Damara Orogen in South West Africa. I. Discussion of geodynamic models. *Precambrian Research* 5: 311–338.

Martin, J. M. and Meybeck, M. 1979. Elemental mass-balance of material carried by major world rivers. *Marine Chemistry* 7: 173–206.

Martin, J. P., Ervin, J. O. and Richards, S. J. 1972. Decomposition and binding action in soil of some mannose-containing microbial polysaccharides and their Fe, Al, Zn and Cu complexes. *Soil Science* 113: 322–327.

Martinez, M. L., York, D., Hall, C. M. and Hanes, J. A. 1984. Oldest reliable 40Ar/39Ar ages for terrestrial rocks: Barberton Mountain Komatiites. *Nature* 307: 352–354.

Maslov, V. P. 1973a. Nizhne-Paleozojskie porodoobrazuyuschie vodorosli Vostochnoj Sibiri [Lower-Paleozoic rock-building algae of East Siberia]. *Problemy Paleontologii* 2-3: 249-325.

Maslov, V. P. 1937b. O pasprostranenii carbonatnikh vodoroslej v Vostochnoj Sibiri [On the distribution of calcareous algae in East Siberia]. *Problemy Paleontologii* 2-3: 327-348.

Maslov, V. P. 1949. Vodorosl' *Girvalella*, ee ekologiya i stratigraficheskoe znachenie [The alga *Girvanella*, its ecology and stratigraphic significance]. *Byulleten' Moskovskogo Obshchestva Ispytatelej Prirody, Otdel Geologicheskij* 24(2): 89-100.

Maslov, V. P. 1954. O nizhnem Silure Vostochnoj Sibiri [On the Lower Silurian of east Siberia]. *Voprosy Geologii Azii* 1: 525-530.

Maslov, V. P. 1956. *Iskopaemye Izvestkolye Vodorosli SSSR* [*Fossil Calcareous Algae of the USSR*] (Akademiya Nauk SSSR: Moscow), 303 pp.

Mason, R. 1973. The Limpopo mobile belt—southern Africa. *Philosophical Transactions of the Royal Society of London A* 273: 463-485.

Mason, R. 1988. Did the Iapetus Ocean really exist? *Geology* 16: 823-826.

Mason, T. R. and Von Brunn, V. 1977. 3-Gyr-old stromatolites from South Africa. *Nature* 266: 47-49.

Matheke, G. E. M. and Horner, R. 1974. Primary productivity of the benthic microalgae in the Chukchi Sea near Barrow, Alaska. *Journal of the Fisheries Research Board of Canada* 31: 1779-1986.

Mathur, S. M. 1983. A reappraisal of trace fossils described by Vrendenburg (1908) and Beer (1919) in rocks of the Vindhyan Supergroup. *Records of the Geological Survey of India* 113: 111-113.

Mathur, V. K. and Shanker, R. 1989. First record of Ediacaran fossils from the Krol Formation of Naini Tal Syncline. *Journal of the Geological Society of India* 34: 245-254.

Matsubara, H. and Hase, T. 1983. Phylogenetic consideration of ferredoxin sequences in plants, particularly algae. In: U. Jensen and D. E. Fairbrothers (Eds.), *Proteins and Nucleic Acids in Plant Systematics* (Springer-Verlag: Berlin), pp. 168-181.

Matsui, T. and Abe, Y. 1986. Evolution of an impact-induced atmosphere and magma ocean on the accreting earth. *Nature* 319: 303-305.

Matthew, G. F. 1885a. A new genus of Cambrian pteropods. *Canadian Rec. Science* 1: 149-152.

Matthew, G. F. 1885b. Notice of a new genus of pteropods from the St. John Group. *American Journal of Science* 30: 293-294.

Matthew, G. F. 1886a. Illustrations of the fauna of the St. John Group continued. No. 3. Descriptions of new genera and species, (including a description of a new species of *Solenopleura* by J. F. Whiteaves). *Transactions of the Royal Society of Canada* 3: 29-84.

Matthew, G. F. 1886b. On the Cambrian Faunas of Cape Breton and Newfoundland. *Transactions of the Royal Society of Canada* 4(4): 147-157.

Matthew, G. F. 1889. On Cambrian organisms in Acadia. *Transactions of the Royal Society of Canada* 7(4): 135-162.

Matthew, G. F. 1890. On the existence of organisms in the Pre-Cambrian rocks. *Bulletin of the Natural History Society of New Brunswick* 2(9): 28-33.

Matthew, G. F. 1891. Illustrations of the fauna of the St. John Group No. 5. *Royal Society of Canada, Proceedings and Transactions* 8(Ser. 1): 123-166.

Matthew, G. F. 1893. *Trematobolus*. *The Canadian Record of Science* 5: 276-279.

Matthew, G. F. 1895. The *Protolenus* fauna. *Transactions of the New York Academy of Science* 14: 101-153.

Matthew, G. F. 1896. Faunas of the *Paradoxides* Beds in eastern North America. No. 1. *Transactions of the New York Academy of Science* 15: 192-247.

Matthew, G. F. 1899a. The Etcheminian fauna of Smith Sound, Newfoundland. *Transactions of the Royal Society of Canada* 7(Ser. 2, 1899-1900): 97-119.

Matthew, G. F. 1899b. Preliminary notes on the Etcheminian fauna of Cape Breton. *Bulletin of the Natural Histroy Society of New Brunswick* 18: 198-208.

Matthew, G. F. 1899c. Studies on Cambrian faunas No. 3—Upper Cambrian fauna of Mount Stephen, British Columbia—the trilobites and worms. *Transactions of the Royal Society of Canada* 7(Ser. 2): 39-66.

Matthew, G. F. 1901. *Acrothyra*. A new genus of Etcheminian brachiopods. *Bulletin of the Natural History Society of New Brunswick* 19: 303-304.

Matthew, G. F. 1902. Ostracoda of the basal Cambrian rocks on Cape Breton. *Canadian Rec. Science* 8: 437-470.

Matthew, G. F. 1903. *Report on the Cambrian rocks of Cape Breton* (Canada Geological Survey: Ottawa), 246 pp.

Matthews, S. C. 1973. Lapworthellids from the Lower Cambrian *Strenuella* Limestone at Comley, Shropshire. *Palaeontology* 16(1): 139-148.

Mattox, K. R. and Stewart, K. D. 1984. Classification of the green algae: a concept based on comparative cytology. In: D. E. G. Irvine and D. M. John (Eds.), *Systematics of the Green Algae* (Academic Press: London), pp. 29-72.

Mawson, D. 1925. Evidence and indications of algal contributions in the Cambrian and Pre-Cambrian limestones of South Australia. *Transactions of the Royal Society of South Australia* 49: 186-190.

Maxwell, E. S., Liu, J. and Shivley, J. M. 1986. Nucleotide sequences of *Cyanophora paradoxa* cellular and cyanelle-associated 5S ribosomal RNAs: The cyanelle as a potential intermediate in plastid evolution. *Journal of Molecular Evolution* 23: 300-304.

Mazur, P. 1980. Limits to life at low temperatures and at reduced water contents and water activities. *Origins of Life* 10: 137-159.

Mazzullo, S. J. and Friedman, G. M. 1977. Competitive algal colonization of peritidal flats in a schizohaline environment: the Lower Ordovician of New York. *Journal of Sedimentary Petrology* 47: 398-410.

McAlester, A. L. 1962. Mode of preservation in early Paleozoic pelecypods and its morphologic and ecologic significance. *Journal of Paleontology* 36: 69-73.

McAuliffe, J. R. 1984. Resource depression by a stream herbivore: Effects on distributions and abundances of other grazers. *Oikos* 42: 327-333.

McBride, O. W. and Harrington, W. F. 1967. *Ascaris* cuticle collagen: on the disulfide cross-linkages and the molecular properties of the subunits. *Biochemistry* 6: 1484-1498.

McCabe, B. 1985. The Dynamics of ^{13}C in Several New Zealand Lakes. Ph.D. thesis, University of Waikato, New Zealand, 278 pp.

McCarroll, R., Olsen, G., Stahl, Y. D., Woese, C. R. and Sogin, M. L. 1983. Nucleotide sequence of the *Dictyosterium discoideum* small sub-unit ribosomal ribonucleic acid inferred from the gene sequence: Evolutionary implications. *Biochemistry* 22: 5858-5868.

McCulloch, M. T. 1987. Sm-Nd constraints on the evolution of Precambrian crust in the Australian continent. In: A. Kröner (Ed.), *Proterozoic Lithospheric Evolution. American Geophysical Union Geodynamics Series* 17: 115-130.

McElhinny, M. W. and Embleton, B. J. J. 1976a. The palaeoposition of Madagascar: remanence and magnetic properties of Late Palaeozoic sediments. *Earth and Planetary Science Letters* 31: 101-112.

McElhinny, M. W. and Embleton, B. J. J. 1976b. Precambrian and Early Cambrian palaeomagnetism in Australia. *Philosophical Transactions of the Royal Society of London A* 280: 417-431.

McElhinny, M. W., Embleton, B. J. J., Ma, X. H. and Zhang, Z. K. 1981. Fragmentation of Asia in the Permian. *Nature* 293: 212-216.

McGinnis, W., Garber, R. L., Wirz, J., Kuroiwa, A. and Gehring, W. T. 1984a. A homologous protein-coding sequence in *Drosophila* homeotic genes and its conservation in other metazoans. *Cell* 37: 403-408.

McGinnis, W., Levine, M. S., Hafen, E., Kuroiwa, A. and Gehring, W. J. 1984b. A conserved DNA sequence in homeotic genes of *Drosophila* antennapedia and bithorax complexes. *Nature* 308: 428-433.

McKay, C. P. and Friedman, E. I. 1985. The cryptoendolithic microbial environment in the Antarctic cold desert: temperature variations in nature. *Polar Biology* 4: 19-25.

References

McKee, E. H. and Noble, D. C. 1976. Age of the Cardenas Lavas, Grand Canyon, Arizona. *Geological Society of America Bulletin* 87: 1188–1190;.

McKenzie, D. 1978. Some remarks on the development of sedimentary basins. *Earth and Planetary Science Letters* 340: 25–32.

McKenzie, D. and Weiss, N. 1980. The thermal history of the earth. In: D. W. Strangway (Ed.), *The Continental Crust and its Mineral Deposits. Geological Association of Canada Special Paper* 20: 575–590.

McKerracher, L. and Gibbs, S. P. 1982. Cell and nucleomorph division in the alga *Cryptomonas. Canadian Journal of Botany* 69: 2440–2452.

McKinley, W. R. and Wetzel, R. G. 1979. Photolithotrophy, photoheterotrophy, and chemoheterotrophy: patterns of resource utilization on an annual and a diurnal basis within a pelagic microbial community. *Microbial Ecology* 5: 1–15.

McKirdy, D. M. 1974. Organic Geochemistry in Precambrian Research. *Precambrian Research* 1: 75–137.

McKirdy, D. M. 1976. Biochemical markers in stromatolites. In: M. R. Walter (Ed.), *Stromatolites: Developments in Sedimentology*, vol. 20 (Elsevier: Amsterdam), pp. 163–191.

McKirdy, D. M. and Hahn, J. H. 1982. The composition of kerogen and hydrocarbons in Precambrian rocks. In: H. D. Holland and M. Schidlowski (Eds.), *Mineral Deposits and the Evolution of the Biosphere, Dahlem Konferenzen* (Springer-Verlag: Berlin), pp. 123–154.

McKirdy, D. M., Aldridge, A. K. and Ypma, P. J. M. 1983. A geochemical comparison of some crude oils from Pre-Ordovician carbonate rocks. In: M. Bjorøy, C. Albercht, K. Cornford, K. de Groot, G. Eglinton, E. Galimov, D. Leythaeuser, R. Pelet, J. Rullkotter and G. Speers (Eds.), *Advances in Organic Geochemistry 1981* (John Wiley & Sons: New York), pp. 99–107.

McKirdy, D. M., Kantsler, A. S., Emmett, J. K. and Aldridge, A. K. 1984. Hydrocarbon genesis and organic facies in Cambrian carbonates of the eastern Officer Basin, South Australia. In: J. G. Palacas (Ed.), *Petroleum Geochemistry and Source Rock Potential of Carbonate Rocks. American Association of Petroleum Geologists Studies in Geology* 18: 13–32.

McKnight, D. M., Kimball, B. A. and Bencala, K. E. 1988. Iron photoreduction and oxidation in an acidic mountain stream. *Science* 240: 637–640.

McLennan, S. M., Taylor, S. R. and McGregor, V. R. 1984. Geochemistry of Archean metasedimentary rocks from West Greenland. *Geochimica et Cosmochimica Acta* 48: 1–13.

McMenamin, M. A. S. 1986. The garden of Ediacara. *Palaios* 1: 178–182.

McMenamin, M. A. S. 1987. The emergence of animals. *Scientific American* 256(4): 94–102.

McMenamin, M. A. S. and McMenamin, D. L. S. 1989. *The Emergence of Animals. The Cambrian Breakthrough* (Columbia University Press: New York), 217 pp.

McMenamin, M. A. S., Awramik, S. M. and Stewart, J. H. 1983. Precambrian-Cambrian transition problem in Western North America: Part II. Early Cambrian skeletonized fauna and associated fossils from Sonora, Mexico. *Geology* 11: 227–230.

McNamara, K. J. 1982. Heterochrony and phylogenetic trends. *Paleobiology* 8: 130–142.

McWilliams, M. O. and Kröner, A. 1981. Paleomagnetic and tectonic evolution of the pan-African Damara belt, Southern Africa. *Journal of Geophysical Research* 86: 5147–5162.

Medlin, L., Elwood, H. J., Stickel, S. and Sogin, M. L. 1988. The characterization of enzymatically amplified 16S-like rRNA-coding regions. *Gene* 71: 491–499.

Meek, F. B. 1873. Preliminary palaeontological report. *Annual Report of the U.S. Geological Survey of Montana, Idaho, Wyoming and Utah; Being a Report of the Exploration for the Year 1872*: 429–518.

Meeks, J. C. 1988. Symbiotic associations. In: L. Packer and A. N. Glazer (Eds.), *Cyanobacteria, Vol. 167 Method in Enzymology* (Academic Press: London, New York), pp. 113–121.

Meijerink, A. M. J., Rao, D. P. and Rupke, J. 1984. Stratigraphic and structural development of the Precambrian Cuddapah basin, S. E. India. *Precambrian Research* 26: 57–104.

Meinschein, W. G. and Kenny, G. S. 1957. Analysis of a chromotographic fraction of organic extracts from soils. *Analytical Chemistry* 29: 1153–1161.

Mendelson, C. V. and Schopf, J. W. 1982. Proterozoic microfossils from the Sukhaya Tunguska, Shorikha, and Yudoma Formations of the Siberian Platform, USSR. *Journal of Paleontology* 56: 42–83.

Meshkova, N. P. 1969a. K voprosu o paleontologicheskoj kharakteristike nizhnekembrijskikh otlozhenij Sibirskoh platformy [To the question of the palaeontological character of the Lower Cambrian deposits of the Siberian Platform]. In: I. T. Zhuravleva (Ed.), *Biostratigrafiya i Paleontologiya Nizhnego Kembriya Sibiri i Dal'nego Vostoka* (Nauka: Moscow.), pp. 158–174.

Meshkova, N. P. 1969b. Novye predstaviteli otryada Hyolithida v nizhnem kembrii Sibirskoj platformy [New representatives of the order Hyolithida in the Lower Cambrian of the Siberian Platform]. In: I. T. Zhuravleva (Ed.), *Biostratigrafiya i Paleontologiya Nizhnego Kembriya Sibiri i Dal'nego Vostoka* (Nauka: Moscow), pp. 175–179.

Meshkova, N. P. 1974. Khiolity nizhnego kembriya Sibirskoj platformy [Lower Cambrian hyoliths of the Siberian platform]. *Trudy Institut Geologii i Geofiziki Sibirskoe Otdolenie, Akademiya Nauk SSSR* 97: 1–110.

Metz, R. 1987. Sinusoidal trail formed by a recent biting midge (Family Ceratopogonidae): trace fossil implications. *Journal of Paleontology* 61: 312–314.

Meyer, C. 1981. Ore forming processes in geologic history. In: B. J. Skinner (Ed.), *Economic Geology 75th Anniversary Volume* (Economic Geology Publishing Company: El Paso, Texas), pp. 6–41.

Meyer, C. 1988. Ore deposits as guides to geologic history of the earth. *Annual Review of Earth and Planetary Sciences* 16: 147–171.

Meyerhoff, A. A. 1980. Geology and petroleum fields in Proterozoic and lower Cambrian strata, Lena-Tunguska Petroleum Province, Eastern Siberia. In: M. T. Halbouty (Ed.), *Giant Oil and Gas Fields of the Decade 1968—1978. American Association of Petroleum Geologists Memoir* 30: 225–252.

Michaelis, W. and Albrecht, P. 1979. Molecular Fossils of archaebacteria in kerogen. *Naturwissenschaften* 66: 420–422.

Michard, A., Albarede, R., Michard, G., Minster, J. F. and Charlou, J. L. 1983. Rare-earth elements and uranium in high-temperature solutions from East Pacific Rise hydrothermal vent field (13N). *Nature* 303: 795–797.

Michener, H. D. and Elliot, R. P. 1964. Minimum growth temperatures for food-poisoning, fecal-indicator, and psychrophilic microorganisms. *Advances in Food Research* 13: 349–396.

Mikhajlova, N. S. 1986. Novye nakhodki mikrofitofossilij iz otlozhenij verkhnego rifeya Krasnoyarskogo kraya [New occurrences of microphytofossils from the Upper Riphean of the Krasnoyarsk region]. In: B. S. Sokolov (Ed.), *Aktual'nye Voprosy Sovremennoj Paleoal'gologii [Current Questions in Contemporary Paleoalgology]* (Naukova Dumka: Kiev), pp. 31–37.

Mikhajlova, N. S. and Podkovyrov, V. N. 1987. Mikkrofitologicheskaya kharakteristika pogranichnykh gorizontov rifeya i venda Yuzhnogo Urala [Microphytological characteristics of the Riphean and Vendian boundary horizons in the southern Urals]. *Izvestiya Akademiya Nauk SSSR, seriia Geologicheskaya* 1987(9): 75–83.

Miller, J. M. G. 1982. Kingston Peak Formation in the southern Panamint Range: a glacial interpretation. In: J. P. Cooper, B. W. Troxel and L. A. Wright (Eds.), *Geology of Selected Areas in the San Bernardino Mountains, Western Mojave Desert, and Southern Great Basin, California. Geological Society of America Cordilleran Section Meeting Guidebook*, Field Trip 9 (Death Valley Publishing Co.: Shoshone), pp. 155–164.

Miller, M. M. and Hardwood, D. S. 1989. Paleozoic and early Mesozoic paleogeographic relations between the Klamath Mountains, northern Sierra Nevada, and western North America. *Geology* 17: 369–372.

Miller, R. F. 1987. On the inorganic character of *Halichondrites graphitiferus* Matthew, a supposed sponge from the Precambrian of Saint John, New Brunswick. *Canadian Journal of Earth Sciences* 24(9): 1913–1915.

Miller, S. L. 1953. A production of amino acids under possible primitive earth conditions. *Science* 117: 528–529.

Miller, S. L. 1955. Production of some organic compounds under possible earth conditions. *Journal of the American Chemical Society* 77: 2351–2361.

Miller, S. L. and Orgel, L. E. 1974. *The Origins of Life on Earth* (Prentice Hall: Englewood Cliffs, N.J.), 299 pp.

Miller, S. L. and Schlesinger, G. 1984. Carbon and energy yields in prebiotic syntheses using atmospheres containing CH_4, CO, and CO_2. *Origins of Life* 14: 83–89.

Miller, S. L. and Urey, H. C. 1959. Organic compound synthesis on the primitive earth. *Science* 130: 245–251.

Miller, W., III 1988a. Giant agglutinated foraminiferids from Franciscan turbidites at Redwood Creek, northwestern California, with the description of a new species of *Bathysiphon*. *Tulane Studies in Geology and Paleontology* 21: 81–84.

Miller, W., III 1988b. Giant *Bathysiphon* (Foraminiferida) from Cretaceous turbidites, northern California. *Lethaia* 21: 363–374.

Milliman, J. D. and Meade, R. H. 1983. World-wide delivery of river sediments to the oceans. *Journal of Geology* 91: 1–21.

Minami, Y., Sugimura, Y., Wakabayashi, S., Wada K., Takahashi, Y. and Matsubara, H. 1985. Isolation, properties and amino acid sequence of a ferredoxin from a multinucleate, unicellular green algae *Bryopsis maxima*. *Physiologie Végétale* 23: 669–678.

Mishler, B. D. and Churchill, S. P. 1985. Transition to a land flora: phylogenetic relationships of the green algae and bryophytes. *Cladistics* 1: 305–328.

Misra, R. C. 1949. On organic remains from the Vindhyans (Precambrian). *Current Science* 18: 439.

Misra, S. B. 1969. Late Precambrian (?) fossils from Southeastern Newfoundland. *Geological Society of America Bulletin* 80: 2133–2140.

Missarzhevskij, V. V. 1966. Pervye nakhodki *Lapworthella* v nizhnem kembrii Sibirskoj platformy [The first finds of *Lapworthella* in the Lower Cambrian of the Siberian Platform]. *Paleontologicheskij Zhurnal* 1966(2): 13–18.

Missarzhevskij, V. V. 1970. Novoe rodovoe nazvanie *Tommotia* Missarzhevsky nom. nov [The new generic name *Tommotia* Missarzhevsky nom. nov.]. *Paleontologicheskij Zhurnal* 1970(4): 100.

Missarzhevskij, V. V. 1973. Kondonontoobraznye organizmy iz pogranichnykh sloev kembriya i dokembriya Sibirskoj platformy i Kazakhstana [Conodont-shaped organisms from the Precambrian-Cambrian boundary beds of the Siberian Platform and Kazakhstan]. In: I. T. Zhuraleva (Ed.), *Problemy Paleontologii i Biostratigrafii Nizhnego Kembriya Sibiri i Dal'nego vostoka. Trudy Inst. Geol. Geofiz. Sibirsk. Otd. Akad. Nauk SSSR 49* (Nauka: Moscow), pp. 53–57.

Missarzhevskij, V. V. 1974. Novye dannye o drevnejshikh okamenelostyakh rannego kembriya Sibirskoj Platformy [New data on the oldest Lower Cambrian fossils of the Siberian Platform]. [In: I. T. Zhuravleva and A. Y. Rozanov (Eds.), *Biostratigrafiya i Paleontologiya Nizhnego Kembriya Evropy i Severnoj Azii* (Nauka: Moscow), pp. 179–189.

Missarzhevskij, V. V. 1977. Konodonty (?) i fosfatnye problematiki kembrii Mongolii i Sibiri [Conodonts (?) and phosphatic problematica from the Cambrian of Mongolia and Siberia]. In: L. P. Tatarinov (Ed.), *Bespozvonochnye Paleozoya Mongolii, Sovmestnaya Sovetsko-Mongol'skaya Paleontologicheskaya Ehkspedi* (Nauka: Moscow), pp. 10–19.

Missarzhevskij, V. V. 1981a. Novoe rodovoe nazvanie dlya gastropod. [A new generic name for gastropods]. *Paleontologicheskij Zhurnal* 1981: 123.

Missarzhevskij, V. V. 1981b. Rannekembrijskie khiolity i gastropody Mongolii [Early Cambrian hyoliths and brachiopods of Mongolia]. *Paleontologicheskij Zhurhnal* 1981(1): 21–28.

Missarzhevskij, V. V. 1989. Drevnejshie skeletnye okamenelosti i stratigrafiya pogranichnykh tolshch dokembriya i kembriya [The oldest skeletal fossils and stratigraphy of the Precambrian-Cambrian boundary beds]. *Trudy Institut Geologii i Geofiziki Sibirskoe Otdolenie, Akademiya Nauk SSSR* 443: 1–237.

Missarzhevskij, V. V. and Grigor'eva, N. V. 1981. Novye predstaviteli otryada Tommotiida [New representatives of the order Tommotiida]. *Paleontologicheskij Zhurnal* 1981(4): 91–97.

Missarzhevskij, V. V. and Mambetov, A. M. 1981. Stratigrafiya i fauna pogranichnykh sloev kembriya i dokembriya Malogo Karatau [Stratigraphy and fauna of the Precambrian-Cambrian boundary beds in Malyj Karatau]. *Trudy geologicheskii institut, Akademiya Nauk SSSR* 326: 1–92.

Missarzhevskij, V. V. 1982. Raschlenenie i korrelyatsiya pogranichnykh tolshch dokembriya i kembriya po nelotorym drevnejshim gruppam skeletnoh okamenemostej [Subdivision and correlation of Precambrian-Cambrian boundary beds using some groups of early skeletal fossils]. *Byulleten' Moskovskogo Obshchestva Ispytatelej Prirody, Otdel Geologicheskij* 57(5): 52–57.

Missarzhevskij, V. V. 1983. Stratigrafiya drevnejshikh tolshch fanerozoya Anabarskogo massiva [Stratigraphy of the oldest Phanerozoic beds of the Anabar Massif]. *Sovetskaya geologiya* 1983(9): 62–73.

Mitsui, A. and Cao, S. 1988. Isolation and culture of the marine nitrogen-fixing unicellular cyanobacteria, *Synechococcus*. In: L. Packer and A. N. Glazer (Eds.), *Cyanobacteria, Vol. 167 Method in Enzymology* (Academic Press: London, New York), pp. 105–113.

Mizutani, H. and Wada, E. 1982. Effect of high atmospheric CO_2 concentration on $\delta^{13}C$ of algae. *Origins of Life* 12: 377–390.

Møller, M. M., Nielsen, L. P. and Jørgensen, B. B. 1985. Oxygen responses and mat formation by *Beggiatoa* spp. *Applied and Environmental Microbiology* 50(2): 373–382.

Moberg, J. C. 1892. Om en nyupptäckt fauna i block af kambrisk sandsten, insamlade af Dr. N. O. Holst. *Geologiska Föreningens i Stockholm Förhandlingar* 14(2): 103–120.

Moczydłowska, M. 1980. *Acritarcha* z osadów kambru wiercenia Okuniew IG 1 [*Acritarcha* from the Cambrian of the Borehole Okuniew IG 1]. *Kwartalnik Geologiczny* 24: 461–487. (in Polish with English abstract).

Moczydłowska, M. 1988a. New Lower Cambrian acritarchs from Poland. *Review of Palaeobotany and Palynology* 54: 1–10.

Moczydłowska, M. 1988b. Thermal alteration of the organic matter around the Precambrian-Cambrian transition in the Lublin Slope of the East European Platform in Poland. *Geologiska Föreningens i Stockholm Förhandlingar* 110: 351–361.

Moczydłowska, M. and Vidal, G. 1986. Lower Cambrian acritarch zonation in southern Scandinavia and southeastern Poland. *Geologiska Föreningens i Stockholm Förhandlingar* 108: 201–223.

Moczydłowska, M. and Vidal, G. 1988a. Early Cambrian acritarchs from Scandinavia and Poland. *Palynology* 12: 1–10.

Moczydłowska, M. and Vidal, G. 1988b. How old is the Tommotian? *Geology* 16(2): 166–168.

Moeller, P. and Danielson, A. 1988. Significance of Eu anomalies in banded iron-formation. *Geological Society of America Abstracts with Programs* 20: 381.

Moldowan, J. M. and Seifert, W. K. 1979. Head-to-head linked isoprenoid hydrocarbons in petroleum. *Science* 204: 169–171.

Moldowan, J. M., Seifert, W. K. and Gallegos, E. J. 1983. Identification of an extended series of tricyclic terpanes in petroleum. *Geochimica et Cosmochimica Acta* 47: 1531–1534.

Moldowan, J. M., Seifert, W. K. and Gallegos, E. J. 1985. Relationship between petroleum composition and depositional environment of petroleum source rocks. *American Association of Petroleum Geologists Bulletin* 69: 1255–1268.

Monson, K. D. and Hayes, J. M. 1980. Biosynthetic control of the

References

natural abundance of carbon 13 at specific positions within fatty acids in *Escherichia coli*. *The Journal of Biological Chemistry* 255: 11435–11441.

Monty, C. L., V 1965. Recent algal stromatolites in the Windward Lagoon, Andros Island, Bahamas. *Annales de la Societe geologique de Belgique* 88: 269–276.

Monty, C. L., V 1967. Distribution and structure of Recent stromatolitic algal mats, eastern Andros Island, Bahamas. *Société Géologie Belgique. Annales* 90: 55–100.

Monty, C. L., V 1976. The origin and development of cryptalgal fabrics. In: M. R. Walter (Ed.), *Stromatolites: Developments in Sedimentology*, vol. 20 (Elsevier: Amsterdam), pp. 193–249.

Monty, C. L., V 1979. Monospecific stromatolites from the Great Barrier Reef Tract and their paleontological significance. *Annales de la Societe geologique de Belgique* 101: 163–171.

Monty, C. L., V and Hardie, L. A. 1976. The geological significance of the freshwater blue-green algal calcareous marsh. In: M. R. Walter (Ed.), *Stromatolites: Developments in Sedimentology*, vol. 20 (Elsevier: Amsterdam), pp. 447–477.

Moorbath, S. 1975. Evolution of Precambrian crust from strontium isotopic evidence. *Nature* 254: 395–399.

Moorbath, S. and Windley, B. 1981. The origin and evolution of the earth's continental crust. *Philosophical Transactions of the Royal Society of London A* 301: 183–487.

Moorbath, S., O'Nions, R. K., Pankhurst, J. R., Gale, N. H. and McGregor, V. R. 1972. Further Rb-Sr age determinations on the Early Precambrian rocks of the Godthaab district, West Greenland. *Nature* 240: 78–82.

Moore, E. S. 1918. The iron formation on Belcher Island, Hudson Bay, with reference to its origin and its associated algal limestones. *Journal of Geology* 26: 412–438.

Moore, J. M., Davidson, A. and Baer, A. J. 1986. The Grenville Province. *Geological Association of Canada Special Paper* 31: 1–358.

Moore, L. S. 1987. Water chemistry of the coastal saline lakes of the Clifton-Preston Lakeland System, South-Western Australia, and its influence of stromalolite formation. *Australian Journal of Marine and Freshwater Research* 38: 647–660.

Moore, L. S., Knott, B. and Stanley, N. F. 1983. The stromatolites of Lake Clifton, Western Australia. *Search* 14: 309–314.

Moore, R. C. and Harrington, H. J. 1956. Conulata. In: R. C. Moore (Ed.), *Treatise on Invertebrate Paleontology. F. Coelenterata* (Geological Society of America and University of Kansas Press: Lawrence), pp. F54–F66.

Morden, C. W. and Golden, S. S. 1989. *psbA* genes indicate a common ancestry of prochlorophytes and chloroplasts. *Nature* 337: 382–385.

Morel, P. and Irving, E. 1978. Tentative paleocontinental maps for the early Phanerozoic and Proterozoic. *Journal of Geology* 86: 535–561.

Morey, G. B. 1983. Animikie Basin, Lake Superior Region, U.S.A. In: A. F. Trendall and R. C. Morris (Eds.), *Iron Formations: Facts and Problems* (Elsevier: Amsterdam), pp. 13–67.

Morgan, W. J. 1981. Hotspot tracks and the opening of the Atlantic and Indian Oceans. In: C. Emiliani (Ed.), *The Sea*, vol. 7 (John Wiley & Sons: New York), pp. 443–487.

Morowitz, H. J., Heinz, B. and Deamer, D. W. 1988. The chemical logic of a minum protocell. *Origins of Life* 18: 281–288.

Morrall, S. and Greenwood, A. D. 1982. Ultrastructure of nucleomorph division in species of cryptophyceae and its evolutionary implications. *Journal of Cell Science* 54: 311–328.

Morrill, L. C. and Loeblich, A. R., III 1981. The dinoflagellate pellicular wall layer and its occurrence in the Division Pyrrhophyta. *Journal of Phycology* 17: 315–323.

Morris, R. C. and Horwitz, R. C. 1983. The origin of the iron-formation-rich Hamersley Group of Western Australia—deposition on a platform. *Precambrian Research* 21: 273–297.

Morris, R. C. and Trendall, A. F. 1988. A model for the deposition of the microbanded Precambrian iron-formations. *American Journal of Science* 288: 664–669.

Morris, W. A. 1977. Palaeolatitude of glaciogenic upper Precambrian Rapitan Group and the use of tillites as chronostratigraphic marker horizons. *Geology* 5: 85–88.

Mostler, H. 1985. Neue heteractinide Spongien (*Calcispongea*) aus dem Unter-und Mittelkambrium Südwestsardiniens. *Berichte Naturwissenschaftlich-Medizinischen Vereins in Innsbruck* 72: 7–32.

Mostler, H. 1986. Beitrag zur stratigraphischen Verbreitung und phylogenetischen Stellung der Amphidiscophora und Hexasterophora (Hexactinellida, Porifera). *Mitteilungen der österreichischen Geologischen Gesellschaft* 78: 319–359.

Mottl, M. J. 1983. Metabasalts, axial hot springs and the structure of hydrothermal systems at mid-ocean ridges. *Geological Society of America Bulletin* 94: 161–180.

Mottl, M. J. and Holland, H. D. 1978. Chemical exchange during hydrothermal alteration of basalt by seawater. I. Experimental results for major and minor components of seawater. *Geochimica et Cosmochimica Acta* 42: 1103–1115.

Mount, J. D. 1980. Characteristics of Early Cambrian faunas from eastern San Bernadino County, California. *Southern California Paleontological Society Special Publication* 2: 19–29.

Mount, J. F. 1989a. Ediacaran-Cambrian transition in South Australia: reevaluation of boundary unconformities and apparent stratigraphic separation between Ediacaran and Cambrian faunas. *International Geological Congress* 28(2): 470–471 (Abstract).

Mount, J. F. 1989b. Re-evaluation of unconformities separating the "Ediacaran" and Cambrian systems, South Australia. *Palaios* 4: 366–373.

Mount, J. F. and Signor, P. W. 1985. Early Cambrian innovation in shallow subtidal environments: Paleoenvironments of Early Cambrian shelly fossils. *Geology* 13: 730–733.

Mount, J. F. and Signor, P. W. 1988. Environmental stratigraphy of the Proterozoic-Cambrian transition and the record of the metazoan radiation event in western North America. In: E. Landing and G. Narbonne (Eds.), *Trace fossils, Small Shelly Fossils, and the Precambrian-Cambrian Boundary. New York State Museum and Geological Survey Bulletin* 463: 1–81.

Mount, J. F., Gevirtzman, D. A. and Signor III, P. W. 1983. Precambrian-Cambrian transition problem in western North America: Part 1. Tommotian fauna in the southwestern Great Basin and its implications for the base of the Cambrian system. *Geology* 11: 224–226.

Moussa, M. T. 1970. Nematode fossil trails from the Green River Formation (Eocene) in the Uinta Basin, Utah. *Journal of Paleontology* 44: 304–307.

Muir, M. D. 1974. Microfossils from the Middle Precambrian McArthur Group, Northern Territory, Australia. *Origins of Life* 5: 105–118.

Muir, M. D. 1983. Proterozoic microfossils from the Mara Dolomite Member, Emmerugga Dolomite, McArthur Group, from the Northern Territory, Australia. *Botanical Journal of the Linnean Society* 86: 1–18.

Muir, M. D. and Grant, P. R. 1976. Micropaleontological evidence from the Onverwacht Group, South Africa. In: B. F. Windley (Ed.), *The Early History of the Earth* (John Wiley & Sons: London), pp. 595–604.

Muir, M. D., Bliss, G. M., Grant, P. R. and Fisher, M. J. 1979. Palaeontological evidence for the age of some supposedly Precambrian rocks in Anglesey, North Wales. *Journal of the Geological Society of London* 136: 61–64.

Muir, M. D., Armstrong, K. J. and Jackson, M. J. 1980. Precambrian hydrocarbons in the McArthur Basin, N. T. *BMR Journal of Australian Geology and Geophysics* 5: 301–304.

Müller, K. J. 1977. *Palaeobotryllus* from the Upper Cambrian of Nevada—a probable ascidian. *Lethaia* 10: 107–118.

Müller, K. J. and Miller, J. F. 1976. The problematic microfossil *Utahphospha* from the Upper Cambrian of the western United States. *Lethaia* 9: 391–395.

Müller, K. J. and Walossek, D. 1985. A remarkable arthropod fauna from the Upper Cambrian 'Orsten' of Sweden. *Transactions of the Royal Society of Edinburgh: Earth Sciences* 76: 161–172.

Müller, P. J. and Mangini, A. 1980. Organic carbon decomposition rates in sediments of the Pacific manganese nodule belt by ^{230}Th and ^{231}Pa. *Earth and Planetary Science Letters* 51: 94–114.

Murphy, M. L. 1984. Primary production and grazing in freshwater and intertidal reaches of a coastal stream, Southeast Alaska. *Limnology and Oceanography* 29: 805–815.

Murphy, M. T. J., McCormick, A. and Eglinton, G. 1967. Perhydro-β-carotene in the Green River Shale. *Science* 157: 1040–1042.

Murray, G. E. 1965. Indigenous Precambrian petroleum. *Bulletin of the American Association of Petroleum Geologists* 49(1): 3–31.

Murray, G. E., Kaczor, M. J. and McArthur, R. E. 1980. Indigenous Precambrian petroleum revisited. *American Association of Petroleum Geologists Bulletin* 64: 1681–1700.

Murray, J. and Thomson, A. 1977. Hydrocarbon production in *Anacystis montana* and *Botryococcus braunii*. *Phytochemistry* 16: 465–468.

Muyzer, G., De Koster, S., Van Zul, Y., Boon, J. J. and Westbroek, P. 1966. Immunological studies on microbial mats from Solar Lake (Sinai—a contribution to the organic geochemistry of sediments. *Organic Geochemistry* 10: 697–704.

Mycke, B. and Michaelis, W. 1986. Molecular fossils from chemical degradation of macromolecular organic matter. In: D. Leythaeuser and J. Rullkötter (Eds.), *Advances in Organic Geochemistry 1985* (Pergamon Press: Oxford), pp. 847–858.

Mycke, B., Narjes, F. and Michaelis, W. 1987. Bacteriohopanetetrol from chemical degradation of an oil shale kerogen. *Nature* 326: 179–181.

Myers, J. S. 1984. Archaean tectonics in the Fiskenaesset region of southwest Greenland. In: A. Kröner and R. Greiling (Eds.), *Precambrian Tectonics Illustrated* (E. Schweizerbart'sche Verlagsbuchhandlung: Stuttgart, Germany), pp. 95–112.

Myers, J. S. 1988. Early Archean Narrayer Gneiss Complex, Yilgarn Craton, Western Australia. *Precambrian Research* 38: 297–307.

Myers, J. S. and Williams, I. R. 1985. Early Precambrian crustal evolution at Mount Narryer, Western Australia. *Precambrian Research* 27: 153–163.

Nadal, N. G. M. 1971. Sterols of *Spirulina maxima*. *Phytochemistry* 10: 2537–2538.

Naiman, R. J. 1976. Primary production, standing stock, and export of organic matter in a Mohave Desert thermal stream. *Limnology and Oceanography* 21: 60–73.

Nanney, D. L., Meyer, E. B., Simon, E. M. and Preparata, R.-M. 1989. Comparison of ribosomal and isozymic phylogenies of tetrahymenine ciliates. *Journal of Protozoology* 36: 1–8.

Naqvi, S. M. and Rogers, J. J. W. 1987. *Precambrian Geology of India* (Oxford University Press: New York), 223 pp.

Narbonne, G. M. and Hofmann, H. J. 1987. Ediacaran biota of the Wernecke Mountains, Yukon, Canada. *Palaeontology* 30: 647–676.

Narbonne, G. M., Myrow, P. M., Landing, E. and Anderson, M. M. 1987. a candidate stratotype for the Precambrian-Cambrian boundary, Fortune Head, Burin Peninsula, southeastern Newfoundland. *Canadian Journal of Earth Sciences* 24: 1277–1293.

Nathorst, A. G. 1879. En egendomlig strukturvarietet af lerhaltig kalksten fran Grennatrakten. *Geologiska Föreningens i Stockholm Förhandlingar* 4(8): 216.

Naumova, S. N. 1968. Zonal'nye kompleksy rastitel'nykh mikrofossilij dokembriya i nizhnego kembriya Evrazii i ikh stratigraficheskoe zhachenie [Zonal assemblages of Precambrian and Lower Cambrian plant microfossils of Eurasia and their stratigraphic importance]. *Stratigrafiya nizhnego paleozoya tsentral'noj Evropy* [Stratigraphy of the Lower Paleozoic of central Europe]. *Mezhdunarodyj Geologicheskij Kongress, XXIII Sessiya, Doklady Sovetskikh Geologov Problema* 9: 30–39.

Nautiyal, A. C. 1983. Algonkian (Upper to Middle) micro-organisms from the Semri Group of Son Valley (Mirzapur Distt.), India. *Geoscience Journal* 4: 169–198.

Nazarov, B. B. 1973. Rasiolyarii iz nizhnikh gorizontov kembriya Batenevskogo kryasha [Radiolarions from the Lower Cambrian beds of Bateny Hills]. In: I. T. Zhuravleva (Ed.), *Problemy Paleontologii i Biostratigrafii Nizhnego Kembriya Sibiri i Dal'nego vostoka. Trudy Institut geologii i geofiziki Sibirskoe Otdolenie, Akademiya Nauk SSSR* 49: 5–13.

Nazarov, B. B. 1974. Problematichnye kremnistye obrazovaniya iz nizhnego paleozoya Kazakhstana [Problematic siliceous structures from the lower Paleozoic of Kazakhstan]. In: I. T. Zhuravleva and A. Y. Rozanov (Eds.), *Biostratigrafiya i Paleontologiya Nizhnego Kembriya Evropy i Severnoj Azii* (Nauka: Moscow), pp. 110–112.

Nazarov, B. B. 1975. *Radiolyarii Nizhnego-Srendnego Paleozoya Kazakhstana*. *Akademiya Nauk SSSR, Trudy Institut Geologicheskii*, 275 (Nauka: Moscow), 203 pp.

Nazarov, B. B. and Ormiston, A. R. 1985. Evolution of the Radiolaria in the Paleozoic and its correlation with the development of other marine fossil groups. *Senckenbergiana Lethaea* 66: 203–215.

Needham, R. S., Stuart-Smith, P. G. and Page, R. W. 1988. Tectonic evolution of the Pine Creek Inlier, Northern Territory. *Precambrian Research* 40/41: 543–564.

Needleman, S. B. and Wunsch, A. 1970. A general method applicable to the search for similarities in the amino acid sequence of two proteins. *Journal of Molecular Biology* 48: 443–453.

Nei, M. 1987. *Molecular Evolutionary Genetics* (Columbia University Press: New York), 512 pp.

Neilsen, T. H. and McLaughlin, R. J. 1985. Comparison of tectonic framework and depositional patterns of the Hornelen strike-slip basin of Norway and the Ridge Little Sulphur Creek strike-slip basins of California. In: K. T. Biddle and N. Christie-Blick (Eds.), *Strike-slip Deformation, Basin Formation, and Sedimentation. Society of Economic Paleontologists and Mineralogists Special Publication* 37: 79–104.

Nelson, B. K. and DePaolo, D. J. 1985. Rapid production of continental crust 1.7–1.9 b.y. ago: Nd isotopic evidence from the basement of the North American midcontinent. *Geological Society of America Bulletin* 96: 746–754.

Nelson, D. C. and Castenholz, R. W. 1982. Light responses of *Beggiatoa*. *Archives of Microbiology* 131: 146–155.

Nelson, D. C., Jørgensen, B. B. and Revsbech, N. P. 1986. Growth pattern and yield of a chemoautotrophic *Beggiatoa* sp. in oxygen-sulfide microgradients. *Applied and Environmental Microbiology* 52: 225–233.

Nelson, D. C., Wirsen, C. O. and Jannasch, H. W. In Prep. Massive occurrence of large, autotrophic *Beggiatoa* at hydrothermal vents of the Guaymas Basin.

Nelson, D. R. 1978. Long-Chain methyl-branched hydrocarbons: Occurrence, biosynthesis, and function. *Advances in Insect Physiology* 13: 1–33.

Nes, W. R. 1974. Role of sterols in membranes. *Lipids* 9: 596–612.

Nes, W. R. and McKean, M. L. 1977. *Biochemistry of Steroids and Isopentenoids* (University Press: Baltimore).

Nes, W. R. and Nes, W. D. 1980. *Lipids in Evolution* (Plenum Press: New York), 244 pp.

Netskaya, A. I. and Ivanova, V. A. 1956. Pervaya nakhodka ostrakod v nizhnem kembrii vostochnoj Sibiri. [The first find of ostracodes in the Cambrian of eastern Siberia]. *Doklady Akademii Nauk SSSR* 111: 1095–1097.

Neumann, A. C., Gebelein, C. D. and Scoffin, T. P. 1970. The composition, structure and erodability of subtidal mats, Aboco, Bahamas. *Journal of Sedimentary Petrology* 40: 274–297.

Newell, N. D. 1952. Periodicity in invertebrate evolution. *Journal of Paleontology* 26: 371–385.

Newell, N. D. 1967. Revolutions in the history of life. *Geological Society of America Special Paper* 89: 63–91.

Newell, S. Y., Fallon, R. D. and Tabor, P. S. 1986. Direct microscopy of natural assemblages. In: N. G. Carr and B. A. Whitton (Eds.), *Bacteria in Nature (Vol. 2): Methods and Special Applications in Bacterial Ecology* (Plenum Press: New York), pp. 1–48.

Newman, M. J. and Rood, R. T. 1977. Implications of solar evolution for the earth's early atmosphere. *Science* 198: 1035–1037.

Newman, S. M., Derocher, J. and Cattolico, R. A. 1989. Analysis of chromophytic and rhodophytic ribulose-1,5-Biphosphate carboxylase indicates extensive structural and functional similarities among evolutionary diverse algae. *Plant Physiology* 91: 939–946.

Nichols, B. W. 1973. Lipids composition and metabolism. In: N. G. Carr and B. A. Whitton (Eds.), *The Biology of Blue-Green Algae* (Blackwell Scientific Publishers: Oxford), pp. 144–161.

Nichols, P. D. and Johns, R. B. 1986. The lipid chemistry of sediments from the St. Lawrence estuary. Acyclic unsaturated long chain ketones, diols and ketone alchohols. *Organic Geochemistry* 9: 25–30.

Nicholson, H. A. and Etheridge, R. 1878. *A Monograph of the Silurian Fossils of the Girvan District in Ayrshire with Special Reference to Those Contained in the "Gray Collection". Fasciculus I. (Rhipzoda, Actinozoa, Trilobita)*, 341 pp.

Nicholson, J. M., Stolz, J. F. and Pierson, B. K. 1987. Structure of a microbial mat at Great Sippewissett Marsh, Cape Cod, Massachusetts. *FEMS Microbiology Ecology* 45: 343–364.

Nicol, D. 1966. Cope's Rule and Precambrian and Cambrian invertebrates. *Journal of Paleontology* 40: 1397–1399.

Nicol, D. 1977. The number of living animal species likely to be fossilized. *Florida Scientist* 40: 135–139.

Nikitin, I. F., Gnilovskaya, M. B., Zhuravleva, I. T., Luchinina, V. A. and Myagkova, E. I. 1974. Anderkenskaya biogermnaya gryada i istoriya ee obrazovaniya [Bioherm banks of Anderken and the history of their development]. In: O. A. Betekhtina and I. T. Zhuravleva (Eds.), *Sreda i Zhizn' v Geologicheskom Proshlom* (Nauka: Novosibirsk), pp. 122–159.

Niklas, K. J., Tiffney, B. H. and Knoll, A. H. 1980. Apparent changes in the diversity of fossil plants. *Evolutionary Biology* 12: 1–89.

Niklas, K. J., Tiffney, B. H. and Knoll, A. H. 1983. Patterns in vascular land plant diversification. *Nature* 303: 614–616.

Nikolaeva, I. V., Borodaevskaya, Z. V., Perozio, G. N. and Beloborodova, G. V. 1987. Nizhnij paleozoj Yugo-Vostoka Sibirskoj platformy. Porody nizhnekembrijskikh otlozhenij i ikh genezis [Lower Paleozoic of the South-East Siberian Platform. Lower Cambrian deposits and their origins]. *Trudy Institut geologii i geofiziki Sibirskoe Otdolenie, Akademiya Nauk SSSR* 710: 1–104.

Nisbet, E. G. 1987. *The Young Earth* (Allen and Unwin: London, Sydney, Wellington), 402 pp.

Nishimura, M. and Koyama, T. 1977. The occurrence of stanols in various living organisms and the behavior of sterols in contemporary sediments. *Geochimica et Cosmochimica Acta* 41: 379–385.

Nitecki, M. H. and Debrenne, F. 1979. The nature of radiocyathids and their relationship to receptaculitids and archaeocyathids. *Geobios* 12(1): 5–27.

Nitecki, M. N. (Ed.). 1979. *Mazon Creek Fossils* (Academic Press: New York), 581 pp.

Nordeng, S. C. 1963. Precambrian stromatolites as indicators of polar shifts. In: A. C. Munyan (Ed.), *Polar Wandering and Continental Drift*. Society of Economic Paleontologists and Mineralogists Special Publication 10: 131–139.

Nordli, E. 1957. Experimental studies on the ecology of Ceratia. *Oikos* 8: 200–265.

Norris, E. 1980. Prasinophytes. In: E. R. Cox (Ed.), *Phytoflagellates* (Elsevier North Holland: New York), pp. 85–145.

Norris, J. R. and Swain, H. 1971. Staining bacteria. *Methods of Microbiology* 5A: 105–134.

North, G. R. 1975. Theory of energy-balance climate models. *Journal of Atmospheric Science* 22: 2033–2043.

Novak, O. 1886. Zur Kinntnis der Fauna der Etage F-f in der paläozoischen Schichtengruppe Böhmens. *Sitzung sber. d.k. Böhm. Gesellschaft der Wissenschaften, Prag*: 660–685.

Nowlan, G. S., Narbonne, G. M. and Fritz, W. H. 1985. Small shelly fossils and trace fossils near the Precambrian-Cambrian boundary in the Yukon Territory, Canada. *Lethaia* 18(3): 233–256.

Nursall, J. R. 1959. Oxygen as a prerequisite to the origin of the metazoa. *Nature* 183: 1170–1172.

Nursall, J. R. 1962. On the origins of the major groups of animals. *Evolution* 16: 118–123.

Nutman, A. P., Allaart, J. H., Bridgwater, D., Dimroth, E. and Rosing, M. 1984. Stratigraphic and geochemical evidence for the depositional environment of the early Archean Isua supracrustal belt, Southern West Greenland. *Precambrian Research* 25: 365–396.

Nye, O. B., Dean, D. A. and Hinds, R. 1972. Improved thin section techniques for fossil and recent organisms. *Journal of Paleontology* 46: 271–275.

Nyers, A. 1984. Fauna of the basal conglomerate of the Vassbo lead mine (L. Cambrian; NW Dalecarlia, Sweden). *Neues Jahrbuch für Geologie und Paläontologie, Monatshefte* 1984: 291–299.

Nystuen, J. P. and Siedlecka, A. 1988. The "Sparagmites" of Norway. In: J. A. Winchester (Ed.), *Later Proterozoic Stratigraphy of the Northern Atlantic Regions* (Chapman and Hall: New York), pp. 237–252.

Oberlies, F. and Prashnowsky, A. A. 1968. Biogeochemische und elektronenmikroskopische Untersuchung prakambrischer Gestein. *Naturwissenschaften* 55: 22–28.

O'Brien, C. F. 1970. Eozoon canadense "the dawn animal of Canada". *Isis* 61: 206–223.

O'Brien, S. J., Wardle, R. J. and King, A. F. 1983. The Avalon Zone: A pan-African terrane in the Appalachian Orogen of Canada. *Geological Journal* 18: 195–222.

Odin, G. S., Gale, N. H., Auvray, B., Bielski, M., Doré, F., Lancelot, J.-R. and Pasteels, P. 1983. Numerical dating of Precambrian-Cambrian boundary. *Nature* 301: 21–23.

Odin, G. S., Gale, N. N. and Doré, F. 1985. Radiometric dating of late Precambrian times. *Geological Society of London Memoir* 10: 65–72.

Oehler, D. Z. 1977. Pyrenoid-like structures in Late Precambrian algae from the Bitter Springs Formation of Australia. *Journal of Paleontology* 51: 885–901.

Oehler, D. Z. 1978. Microflora of the Middle Proterozoic Balbirini Dolomite (McArthur Group) of Australia. *Alcheringa* 2: 269–309.

Oehler, J. H. 1976. Experimental studies in Precambrian paleontology: structural and chemical changes in blue-green algae during simulated fossilization in synthetic chert. *Geological Society of America Bulletin* 87: 117–129.

Oehler, J. H. 1977. Microflora of the H.Y.C. Pyritic Shale Member of the Barney Creek Formation (McArthur Group), Middle Proterozoic of northern Australia. *Alcheringa* 1: 315–349.

Oehler, J. H., Oehler, D. Z. and Muir, M. D. 1976. On the significance of tetrahedral tetrads of Precambrian algal cells. *Origins of Life* 7: 259–267.

Oelze, J. 1985. Analysis of bacteriochlorophylls. *Methods of Microbiology* 18: 257–284.

Ogurtsova, R. N. 1975. Nakhodki lontovaskikh akritarkh v otlozheniyakh tommotskogo yarusa Olenekskogo podnyatiya [Lontovan acritarchs of the Tommotian Stage on the Olenek uplift]. *Izvestiya Akademiya Nauk SSSR, seriia Geologicheskaya* 1975(11): 84–89.

Ogurtsova, R. N. 1985. *Rastiten'nye Mikrofossilii Opornogo Razreza Venda-Nizhnego Kembriya Malogo Karatau* [Plant Microfossils of the Reference Section of the Vendian-Lower Cambrian of the Malyj Karatau] (Ilim: Frunze), 137 pp.

Ogurtsova, R. N. and Sergeev, V. N. 1987. Mikrobiota chichkanskoj svity verkhnego dokembriya Malogo Karatau (Yuzhnyj Kazakhstan) [Microbiota of the Chichkan Formation, Little Karatau Range (South Kazakhstan)]. *Paleontologicheskij Zhurnal* 1987(2): 107–116.

Ohki, K., Rueter, J. G. and Fujita, Y. 1986. Cultures of the pelagic Cyanophytes *Trichodesmium erythraeum* and *T. thiebautic* in synthetic medium. *Marine Biology* 91(Berlin): 9–13.

Ohmoto, H. and Felder, R. P. 1987. Bacterial activity in the warmer, sulphate-bearing, Archean oceans. *Nature* 328: 244–246.

Ojakangas, R. W. 1985. Review of Archean clastic sedimentation, Canadian Shield: major felsic volcanic contributions to turbidite and alluvial fan-fluvial facies associations. In: L. D. Ayres, P. C. Thuston, K. D. Card and W. Weber (Eds.), *Evolution of Archean Supracrustal Sequences*. Geological Association of Canada Special Paper 28: 23–48.

Ojakangas, R. W. 1988. Environments of deposition for lower Proterozoic Lake Superior type iron-formation: Biwabik Ironwood and Negaunee iron-formations, western Lake Superior region. *Geological Society of America Abstracts with Programs* 20: 383.

Okamoto, Y., Minami, Y., Matsubara, H. and Sugimura, Y. 1987. Studies on algal cytochromes VI. Some properties and amino acid sequence of cytochrome C_6 from a green alga, *Bryopsis maxima*. *Journal of Biochemistry* 102: 1251–1260.

Okulitch, V. J. 1937. Some changes in nomenclature of Archaeocyathi (Cyathospongia). *Journal of Paleontology* 62: 172–180.

Okulitch, V. J. 1960. The Lower Cambrian fauna. In: T. W. M. Cameron (Ed.), *Evolution, Its Science and Doctrine*. Royal Society of Canada Studia Varia Series, 4 (University of Toronto Press: Toronto), pp. 12–21.

O'Leary, M. H. 1981. Carbon isotope fractionation in plants. *Phytochemistry* 20: 553–567.

O'Leary, M. H. 1984. Measurement of the isotopic fractionation associated with diffusion of carbon dioxide in aqueous solution. *Journal of Physical Chemistry* 88: 823–825.

Olive, P. J. W. 1985. Covariability of reproductive traits in marine invertebrates: implications for the phylogeny of the lower invertebrates. In: S. Conway Morris, J. D. George, R. Gibson and H. M. Platt (Eds.), *The Origin and Relationships of Lower Invertebrates*. Systematics Association Special Volume, 28 (Clarendon Press: Oxford), pp. 42–59.

Oliver, W. A. 1984. *Conchopeltis*: its affinities and significance. In: W. A. Oliver Jr., W. J. Sando, S. D. Cairns, A. G. Coates et al. (Eds.), *Recent Advances in the Paleobiology and Geology of the Cnidaria*. *Palaeontographica Americana* 54: 141–147.

Olsen, G. J. 1987. Earliest phylogenetic branchings: Comparing rRNA-based evolutionary trees inferred with various techniques. *Cold Spring Harbor Symposia on Quantitative Biology* 52: 825–837.

Olsen, G. J. and Woese, C. R. 1989. A brief note concerning archaebacterial phylogeny. *Canadian Journal of Microbiology* 35: 119–123.

Olsen, G. J., Pace, N. R., Nuell, M., Kaine, B. P., Gupta, R. and Woese, C. R. 1985. Sequence of the 16S rRNA gene from the thermoacidophilic Archaebacterium *Sulfolobus solfataricus* and its evolutionary implications. *Journal of Molecular Evolution* 22: 301–307.

O'Neill, P. L. 1981. Polycrystalline echinoderm calcite and its fracture mechanics. *Science* 213: 646–648.

O'Nions, K. R., Oxburgh, R., Hawkesworth, C. J. and Macintyre, R. M. 1973. New isotopic and stratigraphical evidence on the age of Ingletonian: probable Cambrian of northern England. *Journal of the Geological Society of London* 129: 445–452.

O'Nions, R. K. and Pankhurst, R. J. 1978. Early Archean rocks and geochemical evolution of the earth's crust. *Earth and Planetary Science Letters* 38: 211–236.

Onken, B. R. and Signor, P. W. 1988a. *Lidaconus palmettoensis* n.gen. and sp.: an enigmatic Early Cambrian fossil from western Nevada. *Journal of Paleontology* 62(2): 172–180.

Onken, B. R. and Signor, P. W. 1988b. Lower Cambrian stratigraphic paleontology of the southwestern Great Basin (White-Inyo Mountains of eastern California and Esmeralda County, Nevada). *Bulletin of the Southern California Paleontological Society* 20: 131–150.

Oparin, A. I. 1938. *Origin of Life* (Macmillan: New York), 270 pp.

Opdyke, N. D., Jones, D. S., MacFadden, B. J., Smith, D. L., Mueller, P. A. and Shuster, R. D. 1987. Florida as an exotic terrane: Paleomagnetic and geochronologic investigation of lower Paleozoic rocks from the subsurface of Florida. *Geology* 15: 900–903.

Öpik, A. A. 1968. Ordian (Cambrian) Crustacea Bradoriida of Australia. *Australian Bureau of Mineral Resources, Bulletin* 103: 1–45.

Öpik, A. A. 1975. Cymbric Vale fauna of New South Wales and Early Cambrian biostratigraphy. *Geology and Geophysics*. *Australian Bureau of Mineral Resources, Bulletin* 159: 1–74.

Oremland, R. S. 1983. Hydrogen metabolism by decomposing cyanobacterial aggregates in Big Soda Lake, Nevada. *Applied and Environmental Microbiology* 45: 1519–1525.

Oren, A. 1989. Photosynthetic and heterotrophic benthic bacterial communities of a hypersaline sulfur spring on the shore of the Dead Sea (Hamei Mazor). Chapter 6. In: Y. Cohen and E. Rosenberg (Eds.), *Microbial Mats: Physiological Ecology of Benthic Microbial Communities* (American Society for Microbiology: Washington, D.C.), pp. 64–76.

Oren, A. and Shilo, M. 1979. Anaerobic heterotrophic dark metabolism in the cyanobacterium *Oscillatoria limnetica*: Sulfur respiration and lactate fermentation. *Archives of Microbiology* 122: 77–84.

Oren, A., Padan, E. and Malkin, S. 1979. Sulfide inhibition of photosystem II in cyanobacteria (blue-green algae) and tobacco chloroplasts. *Biochimica et Biophysica Acta* 546: 270–279.

Oren, A., Fattom, A., Padan, E. and Tietz, A. 1985. Unsaturated fatty acid composition and biosynthesis in *Oscillatoria limnetica* and other cyanobacteria. *Archives of Microbiology* 141: 138–142.

Orgel, L. E. 1986a. Did template-directed nucleation precede molecular replication? *Origins of Life* 17: 27–34.

Orgel, L. E. 1986b. RNA catalysis and the origins of life. *Journal of Theoretical Biology* 123: 127–149.

Orleanskii, V. K. and Gerasimenko, L. M. 1982. Laboratory modeling of a thermophilic cyanobacterial community. *Mikrobiologiya* 51: 538–542.

Orlowski, S. and Radwanski, A. 1986. Middle Devonian sea-anemone burrows, *Alpertia santacrucensis* ichnogen. et ichnosp. n., from the Holy Cross Mountains. *Acta Geologica Polonica* 36: 233–249.

Oró, J. 1961. Comets and the formation of biochemical compounds on the primitive earth. *Nature* 190: 389–390.

Oró, J., Nooner, D., Zlatkis, A., Wilkstrom, S. A. and Barghoorn, E. S. 1965. Hydrocarbons of biological origin in sediments about two billion years old. *Science* 148: 77–79.

Oró, J., Tornabene, T. G., Nooner, D. W. and Gelpi, E. 1967. Aliphatic hydrocarbons and fatty acids of some marine and freshwater microorganisms. *Journal of Bacteriology* 93: 1811–1818.

Osborne, R. H., Licari, G. R. and Link, M. H. 1982. Modern lacustrine stromatolites, Walker Lake, Nevada. *Sedimentary Geology* 32: 39–61.

Osgood, R. G. 1970. Trace fossils of the Cincinatti area. *Palaeontographica Americana* 41: 281–444.

Ourisson, G., Albrecht, P. and Rohmer, M. 1979. The hopanoids. Palaeochemistry and biochemistry of a group of natural products. *Pure and Applied Chemistry* 51: 709–729.

Ourisson, G., Albrecht, P. and Rohmer, M. 1982. Predictive microbial biochemistry: from molecular fossils to procaryotic membranes. *Trends Biochemical Science* 7: 223–239.

Ourisson, G., Albrecht, P. and Rohmer, M. 1984. The microbial origin of fossil fuels. *Scientific American* 251: 34–41.

Ourisson, G., Rohmer, M. and Poralla, K. 1987. Prokaryotic hopanoids and other polyterpenoid sterol surrogates. *Annual Review of Microbiology* 41: 301–333.

Owen, T., Cess, R. D. and Ramanathan, V. 1979. Early earth: An enhanced carbon dioxide greenhouse to compensate for reduced solar luminosity. *Nature* 277: 640–642.

Oyaizu, H., Debrunner-Vossbrinck, B., Mandelco, L., Studier, J. A. and Woese, C. R. 1987. The green non-sulfur bacteria: deep branching in the eubacterial line of descent. *Systematics and Applied Microbiology* 9: 47–53.

Pace, N. R., Olsen, G. J. and Woese, C. R. 1986a. Ribosomal RNA phylogeny and the primary lines of evolutionary descent. *Cell* 45: 325–326.

Pace, N. R., Stahl, D. A., Lane, D. J. and Olsen, G. J. 1986b. The analysis of natural microbial populations by ribosomal RNA sequences. *Advances in Microbial Ecology* 9: 1–55.

Pachmayr, F. 1960. Vorkommen und Bestimmung von Schwefelverbindungen im Mineralwasser. Ph.D. thesis, Ludwig-Maximillians Universität, Munich, 63 pp.

Paczesna, J. 1986. Upper Vendian and Lower Cambrian ichnocoeneses of Lublin region. *Geology of Poland vol. 7. Biuletyn Instytutu Geologicznego* 355: 31–47.

Padan, E. 1979. Impact of facultatively anaerobic photoautotrophic

References

metabolism on ecology of cyanobacteria. *Advances in Microbial Ecology* 3: 1–48.

Paerl, H. W. and Bebout, B. M. 1988. Direct measurement of O_2-depleted microzones in marine *Oscillatoria*: Relation to N_2 fixation. *Science* 241: 442–445.

Page, R. W. 1988. Geochronology of early to middle Proterozoic fold belts in northern Australia: a review. *Precambrian Research* 40/41: 1–19.

Paine, R. T. 1966. Food web complexity and species diversity. *American Naturalist* 100: 65–75.

Pak, K. L. and Terleev, A. A. 1986. O nakhodke izvestkovykh vodorslej v "rifejskikh" otlozheniyakh loga podtemnogo (Batenevskij krayzh) [On the find of calcareous algae from "Riphean" deposits of the Podtemnyj ravine (Batenevsk ridge)]. In: V. Khomentovskij V and V. Y. Shenfil' (Eds.), *Pozdnij Dokembrij i Rannij Paleozoj Sibiri. Stratigrafiya i Paleontologiya. Inst. Geol. Geofiz. Sibirsk. Otd. Akad. Nauk SSSR* (Nauka: Novosibirsk), pp. 67–74.

Palacios, T. 1987. Microfossiles de pared organica del proterozoica superior (reion central de la Peninsula Iberica). Ph.D. thesis, University of Zaragoza, 131 pp.

Palij, V. M. 1969. O novom vide tsiklomeduz iz Venda Podolii [On a new species of cyclomedusae from the Vendian of Podolia]. *Paleontologicheskiy sbornik Lvovskogo Universiteta* 6: 114–117.

Palij, V. M. 1974a. Podviyni slidi (bilobiti) u vidkladakh baltiyskoi serii pridnistrovya [Double traces (bilobites) in the deposits of the Baltic series in the Dniester area]. *Akademii Nauk Ukrainiskoi SSR, Dopvidi Geologia, Geofizika, Khimiya i Biologia* 36: 499–503.

Palij, V. M. 1974b. Pro znakhidku slidy zhittediyalnosti v rifeyskikh vidkladakh Ovrutskogo kryazhu [On finding a trace fossil in the Riphean deposits of the Ovruch ridge]. *Akademii Nauk Ukrainiskoi SSR, Dopvidi Geologia, Geofizika, Khimiya i Biologia* 36: 34–37.

Palij, V. M. 1976. Ostatki besskeletnoi fauny i sledy zhiznedeyatelnosti iz otlozheniy verkhnego dokembriya i nizhnego kembriya Podolii [Remains of soft-bodied animals and trace fossils from the Late Precambrian and Early Cambrian of Podolia]. In: *Paleontologiya i Stratigrafiya Verkhnego Kembriya i Nizhnego Paleozoya Yugo-Zapada Vostochno-Evropeiskoi Platformy*, pp. 63–77.

Palij, V. M., Posti, E. and Fedonkin, M. A. 1979. Myagkotelye Metazoa i iskopaemye sledy zhivotnykh venda i rannego kembriya. In: B. M. Keller and A. Y. Rosanov (Eds.), *Paleontologiya verkhnedokembriyskikh i kembriyskikh otlozheniy Vostochno-Evropeyskoy platformy* (Nauka: Moscow), pp. 49–82. [Republished in 1983 as: Soft-bodied Metazoa and animal trace fossils in the Vendian and Early Cambrian. In: A. Urbanek and A. Yu. Rozanov (Eds.), Upper Precambrian and Cambrian Palaeontology of the East-European Platform (Wydawnictwa Geologiczne Publishing House, Warsaw), 56–94.]

Palmer, A. R. 1965. Trilobites of the late Cambrian Pterocephaliid biomere in the Great Basin, United States. *U.S. Geological Survey Professional Paper* 493: 1–105.

Palmer, A. R. 1982. Biomere boundaries: a possible test for extraterrestrial perturbation of the biosphere. In: L. T. Silver and P. H. Schultz (Eds.), *Geological Implications of Impacts of Large Asteroids and Comets on the Earth. Geological Society of America Special Publication* 190: 469–476.

Palmer, A. R. and Gatehouse, C. G. 1972. Early and Middle Cambrian Trilobites from Antarctica. *U.S. Geological Survey Professional Paper* 456-D: 1–37.

Palmer, A. R., Borrello, A. V., Cowie, J. W., Lochman-Balk, C. and North, F. K. 1971. The Cambrian of the Great Basin and adjacent areas, western United States. In: C. H. Holland (Ed.), *Cambrian of the New World* (Wiley-Interscience: London), pp. 1–78.

Palmisano, A. C., Cronin, S. E. and Des Marais, D. J. 1988. Analysis of lipophilic pigments from a phototrophic microbial mat community by high performance liquid chromatography. *Journal of Microbiological Methods* 8: 209–217.

Palmisano, A. C., Cronin, S. E., D'Amelio, E. D., Munoz, E. and Des Marais, D. J. 1989a. Distribution and survival of lipophilic pigments in a laminated microbial mat community near Guerrero Negro, Mexico. Chapter 12. In: Y. Cohen and E. Rosenberg (Eds.), *Microbial Mats: Physiological Ecology of Benthic Microbial Communities* (American Society for Microbiology: Washington, D.C.), pp. 138–152.

Palmisano, A. C., Summons, R. E., Cronin, S. E. and Des Marais, D. J. 1989b. Lipophilic pigments from cyanobacterial and diatom mats in Hamelin Pool, Shark Bay, Western Australia. *Journal of Phycology* 25: 655–662.

Pamilo, P. and Nei, M. 1988. Relationships between gene trees and species trees. *Molecular Biology and Evolution* 5: 568–583.

Pantin, H. M. 1955. A probable organic structure from the Dalradian of Ben Vrackie, Perthshire. *Geological Magazine* 92(6): 481–486.

Paoletti, C., Pushpraj, B., Florenzano, G., Capella, P. and Lercker, G. 1976a. Unsaponifiable matter of green and blue-green algal lipids as a factor of biochemical differentiation of their biomasses: II. Terpenic alcohol and sterol fractions. *Lipids* 11: 266–271.

Paoletti, C., Pushpraj, B., Florenzano, G., Capella, P. and Lercker, G. 1976b. Unsaponifiable matter of green and blue-green algal lipids as factor of biochemical differentiation of their biomasses: I. Total unsaponifiable and hydrocarbon fraction. *Lipids* 11: 258–265.

Papanastassiou, D. A. and Wasserburg, G. J. 1971. Lunar chronology and evolution from Rb-Sr studies of Apollo 11 and 12 samples. *Earth and Planetary Science Letters* 11: 37–62.

Park, J. K. and Aitken, J. D. 1986. Paleomagnetism of the Katherine Group in the Mackenzie Mountains: implications for post-Grenville (Hadrynian) apparent polar wander. *Canadian Journal of Earth Sciences* 23: 308–323.

Park, R. and Epstein, S. 1960. Carbon isotope fractionation during photosynthesis. *Geochimica et Cosmochimica Acta* 21: 110–126.

Park, R. K. 1976. A note on the significance of lamination in stromatolites. *Sedimentology* 23: 379–393.

Park, R. K. 1977. The preservation potential of some recent stromatolites. *Sedimentology* 24: 485–506.

Parker, B. C. and Wharton, R. A., Jr. 1985. Physiological ecology of blue-green algal mats (modern stromatolites) in Antarctic oasis lakes. *Arkiv für Hydrobiologie, Supplementband* 71: 331–348.

Parkes, R. J. and Taylor, J. 1986. The relationships between fatty acid distributions and bacterial respiratory types in contemporary marine sediments. *Estuarine, Coastal and Shelf Science* 3: 311–319.

Parkin, L. W. (Ed.). 1969. *Handbook of South Australian Geology* (Geological Survey of South Australia: Adelaide), 268 pp.

Parsley, R. L. 1988. Feeding and respiratory structures in Stylophora. In: C. R. C. Paul and A. B. Smith (Eds.), *Echinoderm Phylogeny and Evolutionary Biology* (Clarendon Press: Oxford), pp. 347–361.

Pashkyavichene, L. T. 1980. *Akritarkhi Pogranichnykh Otlozhenij Venda i Kembriya Zapada Vostochno-Evropejskoj Platformy [Acritarchs from Deposits Near the Vendian-Cambrian Boundary in the Western East European Platform]* (Nauka: Moscow), 76 pp.

Patrick, R. 1977. Ecology of freshwater diatoms and diatom communities. In: D. Werner (Ed.), *The Biology of Diatoms* (University of California Press: Berkeley), pp. 284–332.

Patt, T. E. and Hanson, R. S. 1978. Intracytoplasmic membrane, phospholipid, and sterol content of *Methylobacterium organiphilum* cells grown under different conditions. *Journal of Bacteriology* 134: 636–644.

Patterson, C. 1989. Phylogenetic relations of major groups: conclusions and prospects. In: B. Fernholm, K. Bremer and H. Jörnvall (Eds.), *The Hierarchy of Life* (Elsevier: Amsterdam), pp. 471–488.

Patterson, C. 1990. Reassessing relationships. *Nature* 344: 199–200.

Patterson, C. C. 1956. Age of meteorites and the earth. *Geochimica et Cosmochimica Acta* 10: 230–237.

Patterson, G. W. 1971. The distribution of sterols in algae. *Lipids* 6: 120–127.

Paul, A. Z., Thorndike, E. M., Sullivan, L. G., Heezen, B. C. and Gerard, R. D. 1978. Observations of the deep-sea floor from 202 days of time-lapse photography. *Nature* 272: 812–814.

Paul, C. R. C. 1979. Early echinoderm radiation. In: M. R. House (Ed.), *The Origin of Major Invertebrate Groups. The Systematics Association Special Volume* 12 (Clarendon Press: Oxford), pp. 415–434.

Paul, C. R. C. and Smith, A. B. 1984. The early radiation and phylogeny of echinoderms. *Biological Reviews of the Cambridge Philosophical Society* 59(4): 443–481.

Payzant, J. D., Montgomery, D. S. and Strausz, O. P. 1986. Sulfides in petroleum. *Organic Geochemistry* 10: 357–369.

Peat, C. J. 1984. Precambrian microfossils from the Longmyndian of Shropshire. *Proceedings of the Geologists Association* 95(1): 17–22.

Peat, C. J., Muir, M. D., Plumb, K. A., McKirdy, D. M. and Norvick, M. S. 1978. Proterozoic microfossils from the Roper Group, Northern Territory, Australia. *Bureau of Mineral Resources Journal of Australian Geology and Geophysics* 3: 1–17.

Peel, J. S. 1979. *Anatolepis* from the Early Ordovician of East Greenland—not a fishy tail. *Rapport Grønlands Geologiste Undersoegelse* 91: 111–115.

Peel, J. S. 1988. *Spirellus* and related helically coiled microfossils (cyanobacteria) from the Lower Cambrian of North Greenland. *Rapport Grønlands Geologiste Undersoegelse* 137: 5–32.

Pel'man, Y. L. 1976. Ranne-srednekembrijskije stenotekoidy i novye sakeletnye ostatki neyasnogo sistematicheskogo polozheniya stratotipicheskogo rajona rek Aldana i Leny [Early-Middle Cambrian stenothecoids and new skeletal remains of uncertain systematic position from the type section of the river Aldana and Lena]. In: I. T. Zhuravleva (Ed.), *Stratigrafiya i Paleontologuya Nizhnego i Srednego Kembriya SSSR. Trudy Inst. Geol. Geofiz. Sibirsk. Otd. Akad. Nauk SSSR* 296 (Nauka: Novosibirsk), pp. 176–179.

Pel'man, Y. L. 1977. Ranne- i srednekembrijskie bezzamkovye brakhiopody Sibirskoh platformy [Early and Middle Cambrian inarticulate brachiopods from the Siberian Platform]. *Trudy Institut Geologii i Geofiziki Sibirskoe Otdolenie, Akademiya Nauk SSSR* 316: 1–168.

Pel'man, Y. L. 1985. Novye stenotekoidy iz nizhnego kembriya zapadnoj Mongolii [New stenothecoids from the Lower Cambrian of western Mongolia]. In: B. S. Sokolov and I. T. Zhuravleva (Eds.), *Problematiki Pozdnego Dokembriya i Paleozoya. Trudy Inst. Geol. Geofiz. Sibirsk. Otd. Akad. Nauk SSSR*, 632 (Nauka: Novosibirsk), pp. 103–114.

Pemberton, S. G. and Frey, R. W. 1982. Trace fossil nomenclature and the *Planolites-Palaeophycus* dilemna. *Journal of Paleontology* 56: 843–881.

Peng Lihong 1984. The age and tectonic significance of ophiolites of the Undorsum Group, Nei Monggol Autonomous Region. *Kexue Tongbao* 29(7): 936–939.

Penny, D. 1989. What, if anything is, *Prochloron*? *Nature* 337: 304–305.

Pentecost, A. 1978. Blue-green algae and freshwater carbonate deposits. *Proceedings of the Royal Society of London, Series B* 200: 43–61.

Pentecost, A. 1984. Effects of sedimentation and light intensity on mat-forming Oscillatoriacea with particular reference to *Microcoleus lyngbyaceus* Gomont. *Journal of General Microbiology* 130: 983–990.

Pentecost, A. 1985. Investigation of variation in heterocyst numbers, sheath development and false-branching in natural populations of Scytonemataceae (Cyanobacteria). *Applied and Environmental Microbiology* 102(3): 343–353.

Pentecost, A. 1987. Growth of the freshwater cyanobacterium *Rivularia haematites*. *Proceedings of the Royal Society of London* 232(Series B): 125–136.

Pentecost, A. and Bauld, J. 1988. Nucleation of calcite on the sheaths of cyanobacteria using simple diffusion cell. *Geomicrobiology Journal* 6: 129–135.

Pentecost, A. and Riding, R. 1986. Calcification in cyanobacteria. In: B. S. C. Leadbeater and R. Riding (Eds.), *Biomineralization in Lower Plants and Animals. The Systematics Association Special Volume 30* (Clarendon Press: Oxford), pp. 73–90.

Perasso, R., Baroin, A., Qu, L. H., Bachellerie, J. P. and Adoutte, A. 1989. Origin of the algae. *Nature* 339: 142–144.

Percival, J. A. and Williams, H. R. 1989. Late Archean Quetico accretionary complex, Superior province, Canada. *Geology* 17: 23–25.

Perrin, M., Elston, D. P. and Moussine-Pouchkine, A. 1988. Paleomagnetism of Proterozoic and Cambrian strata, Adrar de Mauritane, Cratonic West Africa. *Journal of Geophysical Research* 93: 2159–2178.

Perroud, H., Van der Voo, R. and Bonhommet, N. 1984. Paleozoic evolution of the Armorica plate on the basis of paleomagnetic data. *Geology* 12: 579–582.

Perry, E. A., Ahmad, S. N. and Swullius, T. M. 1978. The oxygen isotope composition of 3800 m.y. old metamorphosed chert and iron formation from Isua, West Greenland. *Journal of Geology* 86: 223–239.

Perry, E. C. and Ahmad, S. N. 1983. Oxygen isotope geochemistry of Proterozoic chemical sediments. *Geological Society of America Memoir* 161: 253–264.

Perry, E. C. and Tan, F. C. 1972. Significance of oxygen and carbon isotope variations in Early Precambrian cherts and carbonate rocks of southern Africa. *Geological Society of America Bulletin* 83: 647–664.

Perry, E. C., Tan, F. C. and Morey, G. B. 1973. Geology and stable isotope geochemistry of the Biwabik iron formation, northern Minnesota. *Economic Geology* 68: 1110–1125.

Perry, G. J., Gillan, F. T. and Johns, R. B. 1978. Lipid composition of a prochlorophyte. *Journal of Phycology* 14: 369–371.

Peryt, T. M. 1974. Spirorbid-algal stromatolites. *Nature* 249: 239–240.

Pesonen, L. J., Torsvik, T. H., Elming, S.-A. and Bylund, G. 1989. Crustal evolution of Fennoscandia–palaeomagnetic constraints. *Tectonophysics* 162: 27–49.

Petrov, A., V, Mokshantsev, K. B., Fradkin, G. S. and Barykin, S. F. 1977. Tectonics and oil-gas prospects of the Upper Precambrian sediments of the Siberian Platform. *Journal of Petroleum Geology* 18: 388–390.

Petruschevskaya, M. G. 1977. O proiskhozhdenii radioyariy. *Zoologiskii Zhurnal* 56: 1448–1458.

Pfeil, R. W. and Read, J. F. 1980. Cambrian carbonate platform margin facies, Shady Dolomite, southwestern Virginia, U.S.A. *Journal of Sedimentary Petrology* 50(1): 91–116.

Pfennig, N. and Trüper, H. G. 1981. Isolation of members of the families Chromataceae and Chlorobiaceae. In: M. P. Starr, H. Stolp, H. G. Trüper, A. Balows and H. G. Schlegel (Eds.), *The Prokaryotes* (Springer-Verlag: Berlin, Heidelberg, New York), pp. 279–289.

Pfennig, N. and Wagener, S. 1986. An improved method of preparing wet mounts for photomicrographs of microorganisms. *Journal of Microbiological Methods* 4: 303–306.

Pflug, H. D. 1966a. Neue Fossilreste aus den Nama-Schichten in Südwest-Afrika. *Paläontologische Zeitschrift* 40: 14–25.

Pflug, H. D. 1966b. Structured organic remains from the Fig Tree Series of the Barberton Mountain Land. *University of the Witwatersrand Economic Geology Research Unit Information Circular* 28: 1–14.

Pflug, H. D. 1970a. Zur Fauna der Nama-Schichten in Südwest-Afrika. I. Pteridinia, Bau und systematische Zugehörigkeit. *Palaeontographica* A135: 198–231.

Pflug, H. D. 1970b. Zur Fauna der Nama-Schichten in Südwest-Afrika. II. Rangeidae, Bau und systematische Zugehörigkeit. *Palaeontographica* A134: 226–262.

Pflug, H. D. 1972a. The Phanerozoic-Cryptozoic boundary and the origin of the Metazoa. In: A. M. Goodwin (Ed.), *Proceedings of the 24th International Geological Congress, Montreal, section 1* (Hapell's Press Co-operative: Gardenvale, Quebec), pp. 58–67.

Pflug, H. D. 1972b. Zur Fauna der Nama-Schichten in Südwest-Afrika. III. Erniettomorpha, Bau und systematik. *Palaeontographica* A139: 134–170.

Pflug, H. D. 1973. Zur Fauna der Nama-Schichten in Südwest-Afrika. IV. Mikroskopische Anatomie der Petalo-organismen Palaeontographica. *Palaeontographica* A144: 166–202.

Philip, G. M. 1979. Carpoids—echinoderms or chordates? *Biological Reviews* 54: 439–471.

Philip, R. P. and Lewis, C. A. 1987. Organic geochemistry of biomarkers. *Annual Reviews of Earth and Planetary Sciences* 15: 363–395.

Phillips, G. N. 1987. Anomalous gold in the Witwatersrand shales.

References

Economic Geology 82: 2179–2186.

Phillips, J. 1860. *Life on the Earth: Its Origin and Succession* (MacMillan: Cambridge).

Philp, R. P. 1980. Comparative organic geochemical studies of recent algal mats and sediments of algal origin. In: P. A. Trudinger and M. R. Walter (Eds.), *Biogeochemistry of Ancient and Modern Environments* (Australian Academy of Science: Canberra), pp. 173–185.

Philp, R. P. and Calvin, M. 1976. Kerogen structures in recently-deposited algal mats at Laguna Mormona, Baja California: A mocel system for the determination of kerogen structures in ancient sediments. In: J. O. Nriagu (Ed.), *Environmental Biogeochemistry*, vol. 1 (Ann Arbor Science Publishers, Inc.: Ann Arbor), pp. 131–148.

Philp, R. P., Brown, S., Calvin, M., Brassell, S. and Eglinton, G. 1978. Hydrocarbon and fatty acid distributions in recently deposited algal mats at Laguna Guerrero, Baja California. In: W. E. Krumbein (Ed.), *Environmental Biogeochemistry and Geomicrobiology*, vol. 1 (Ann Arbor Science Publishers, Inc.: Ann Arbor), pp. 255–270.

Pickett, J. and Jell, P. A. 1983. Middle Cambrian Sphinctozoa (Porifera) from New South Wales. *Memoirs of the Association of Australasian Palaeontologists* 1: 85–92.

Pickett-Heaps, J. D. 1975. *Green Algae: Structure, Reproduction and Evolution in Selected Genera* (Sinauer Associates: Sunderland, MA), 606 pp.

Pidgeon, R. T. 1984. Geochronological constraints on early volcanic evolution of the Pilbara Block, Western Australia. *Australian Journal of Earth Sciences* 31: 237–242.

Pierce, D. and Cloud, P. 1979. New microbial fossils from ~1.3 billion-year-old rocks of eastern California. *Geomicrobiology Journal* 1: 295–309.

Pierson, B. K. and Castenholz, R. W. 1974. A phototrophic gliding bacterium of hot springs, *Chloroflexus aurantiacus*, gen. and sp. nov. *Archives of Microbiology* 100: 5–24.

Pierson, B. K. and Castenholz, R. W. In Prep. Ecology of microbial mat communities of spouting hot springs in Yellowstone National Park.

Pierson, B. K. and Howard, H. M. 1972. Detection of bacteriochlorophyll-containing microorganisms by infrared flourescence photomicrography. *Journal of General Microbiology* 73: 359–363.

Pierson, B. K., Giovannoni, S. J. and Castenholz, R. W. 1984. Physiological ecology of a gliding bacterium containing bacteriochlorophyll *a*. *Applied and Environmental Microbiology* 47: 576–584.

Pierson, B. K., Giovannoni, S. J., Stahl, D. A. and Castenholz, R. W. 1985. *Heliothrix oregonensis* gen. nov., sp. nov., a phototrophic filamentous gliding bacterium containing bacteriochlorophyll *a*. *Archives of Microbiology* 142: 164–167.

Pierson, B. K., Oesterle, A. and Murphy, G. L. 1987a. Pigments, light penetration and photosynthetic activity in the multi-layered microbial mats of Great Sippewissett Salt Marsh, Massachusetts. *FEMS Microbiology Ecology* 45: 365–376.

Pierson, B. K., Oesterle, A. and Murphy, G. L. 1987b. Pigments, light penetration, and photosynthetic activity in the multi-layered microbial mats of Great Sippewissett Salt (cyanobacteria): morphological biochemical and genetic characterization and effects of water stress on ultrastructure. *Archives of Microbiology* 135: 81–90.

Pinto, J. P. and Holland, H. D. 1988. Paleosols and the evolution of the atmosphere; Part II. *Geological Society of America Special Paper* 216: 21–34.

Pinto, J. P., Gladstone, C. R. and Yung, Y. L. 1980. Photochemical production of formaldehyde in the earth's primitive atmosphere. *Science* 210: 183–185.

Piper, J. D. A. 1976. Paleomagnetic evidence for a Proterozoic supercontinent. *Philosophical Transactions of the Royal Society of London A* 280: 469–490.

Piper, J. D. A. 1982. The Precambrian palaeomagnetic record: the case for the Proterozoic supercontinent. *Earth and Planetary Science Letters* 59: 61–89.

Piper, J. D. A. 1983. Proterozoic palaeomagnetism and single continent plate tectonics. *Geophysical Journal of the Royal Astronomical Society* 74: 163–197.

Piper, J. D. A. 1986. *Paleomagnetism and the Continental Crust* (John Wiley & Sons: New York), 434 pp.

Plumb, K. A. 1985. Subdivision and correlation of late Precambrian sequences in Australia. *Precambrian Research* 29: 303–329.

Plumb, K. A., Derrick, G. M. and Wilson, I. H. 1980. Precambrian Geology of the McArthur River-Mount Isa Region northern Australia. In: R. A. Henderson and P. J. Stephenson (Eds.), *The Geology and Geophysics of Northern Australia* (Geological Society of Australia: Brisbane), pp. 205–307.

Plumb, K. A., Derrick, G. M., Needham, R. S. and Shaw, R. D. 1981. The Proterozoic of northern Australia. In: D. R. Hunter (Ed.), *Precambrian of the Southern Hemisphere* (Elsevier: Amsterdam), pp. 205–307.

Plumb, K. L. and James, H. L. 1986. Subdivision of Precambrian time: recommendations and suggestions by Subcommission on Precambrian Stratigraphy. *Precambrian Research* 32: 65–92.

Plummer, P. S. 1980. Circular structures in a late Precambrian sandstone: fossil medusoids or evidence of fluidization? *Transactions of the Royal Society of South Australia* 104: 13–16.

Pocock, K. J. 1974. A unique case of teratology in trilobite segmentation. *Lethaia* 7: 63–66.

Pojeta, J. and Runnegar, B. 1976. The paleontology of rostroconch mollusks and the early history of the phylum Mollusca. *U.S. Geological Survey Professional Paper* 968: 1–88.

Pojeta, J., Jr. 1975. *Fordilla troyensis* Barrande and early pelecypod phylogeny. *Bulletins of American Paleontology* 67(287): 363–384.

Pompeckj, J. F. 1896. Die Fauna des Cambrium von Tejrovic und Skrej in Böhmen. *Jahrbuch K.-K. geol. Reichsanst. Wien* 45: 495–614.

Pompeckj, J. F. 1927. Ein neues Zeugnis uralten Lebens. *Paläontologische Zeitschrift* 9: 287–313.

Popov, Y. N. 1967. Novaya kembrijskaya stsifomeduza [New Cambrian scyphomedusa]. *Paleontologicheskij Zhurnal* 1967(2): 122–123.

Popov, Y. N. 1968. Stsifomedusa iz opornogo razreza verkhnego kembriya reki Kulyumbe [Scyphomedusa from the type section of the Upper Cambrian of the Kulyumbe river]. *Trudy Nauchno issledovatelskogo Instituta Geologii Arktiki* 155: 211–213.

Popp, B. N., Anderson, T. F. and Sandberg, P. A. 1986. Brachiopods as indicators of original isotopic compositions in some Paleozoic limestones. *Geological Society of America Bulletin* 97: 1262–1269.

Popp, B. N., Takigiku, R., Hayes, J. M., Louda, J. W. and Baker, E. W. 1989. The post-Paleozoic chronology and mechanism of ^{13}C depletion in primary marine organic matter. *American Journal of Science* 289: 436–454.

Porada, H. and Wittig, R. 1983. Turbidites and their significance for the evolution of the Damara Orogen, South West Africa/Namibia. In: R. Miller (Ed.), *Evolution of the Damara Orogen of South West Africa/Namibia. Geological Society of South Africa Special Publication* 11: 21–36.

Pospelov, A. G. 1973. K metodike izucheniya vodoroslej poda *Epiphyton* Bornemann [On the methods of study of the algal genus *Epiphyton* Bornemann]. In: I. T. Zhuravleva (Ed.), *Problemy Paleontologii i Biostratigrafii Nizhnego Kembriya Sibiri i Dal'nego Vos*, pp. 85—89.

Potts, M. and Whitton, B. A. 1977. Nitrogen fixztion by blue-green algal communities on the intertidal zone of the lagoon of Aldabra Atoll. *Oecologia* 27: 275–283.

Potts, M. and Whitton, B. A. 1979. pH and Eh on Aldabra Atoll 2. Intertidal photosynthetic microbial communities showing zonation. *Hydrobiologia* 67: 99–105.

Potts, M. and Whitton, B. A. 1980. Vegetation of the intertidal zone of the lagoon of Aldabra, with particular reference to the photosynthetic prokaryotic communities. *Proceedings of the Royal Society of London, Series B* 208: 13–55.

Pouliot, G. and Hofmann, H. J. 1981. Florencite: a first occurrence in Canada. *Canadian Mineralogist* 19: 535–540.

Poulsen, C. 1967. Fossils from the Lower Cambrian of Bornholm. *Matematisk–Fysiske Meddelelser Kobenhaven: Det Kongelige Domske Videnskabernes Selskab* 36: 1–48.

Poulsen, C. 1969. The Lower Cambrian from Slagelse no. 1, Western Sealand. *Geological Survey of Denmark* 93(Ser II): 1–27.

Poulsen, V. 1963. Notes on *Hyolithellus* Billings, 1871, class Pogonophora Johannson, 1937. *Biol. Meddelelser, Kongelige Danske Videnskabernes Selskab* 23(12): 1–15.

Powell, C. McA., Johnson, B. D. and Veevers, J. J. 1980. A revised fit of east and west Gondwanaland. *Tectonophysics* 63: 13–29.

Powell, C. McA., Johnson, B. D. and Veevers, J. J. 1988. Pre-breakup continental extension in East Gondwanaland and the early opening of the eastern Indian Ocean. *Tectonophysics* 155: 261–283.

Powell, J. W. 1876. *Report on the Geology of the Eastern Portion of the Uinta Mountains and a Region of County Adjacent Thereto* (U.S. Geological Survey: Washington D.C.), 218 pp.

Powell, T. G., Jackson, M. J., Swett, I. P., Crick, I. H. and Summons, R. E. 1987. Petroleum geology and geochemistry, Middle Proterozoic McArthur Basin. *Bureau of Mineral Resources Record* 1987(48): 286.

Powell, T. G., Boreham, C. J., McKirdy, D. M., Michaelsen, B. H. and Summons, R. E. In Press. Petroleum geochemistry of the Murta Member (Mooga Formation) and associated oils, Eromanga Basin, Australia. *Australian Petroleum Exploration Association Journal*.

Pratt, B. R. 1984. *Epiphyton* and *Renalcis*–diagenetic microfossils from calcification of coccoid blue-green algae. *Journal of Sedimentary Petrology* 54: 948–971.

Pratt, L. M., Summons, R. E., Hieshima, G. B. and Hayes, J. M. In Press. Lithofacies and biomarkers in the Precambrian Nonesuch Formation: Petroleum source potential of the Midcontinent Rift System, North America. *Proceedings of the 28th International Geological Congress*.

Preiss, W. V. 1987. The Adelaide Geosyncline—Late Proterozic stratigraphy, sedimentation, palaeontology and tectonics. *Bulletin of the Geological Survey of South Australia* 53: 1–438.

Preiss, W., V 1976a. Basic field and laboratory methods for the study of stromatolites. In: M. R. Walter (Ed.), *Stromatolites: Developments in Sedimentology*, vol. 20 (Elsevier: Amsterdam), pp. 5–13.

Preiss, W., V 1976b. Intercontinental Correlations. In: M. R. Walter (Ed.), *Stromatolites: Developments in Sedimentology*, vol. 20 (Elsevier: Amsterdam), pp. 359–370.

Preiss, W., V and Forbes, B. G. 1981. Stratigraphy, correlation and sedimentary history of Adelaidean (late Proterozoic) Basins in Australia. *Precambrian Research* 15: 255–304.

Preparata, R.-M., Meyer, E. B., Preparata, F. P., Simon, E. M., Vossbrinck, C. R. and Nanney, D. L. 1989. Ciliate evolution: the ribosomal phylogenies of tetrahymenine ciliates. *Journal of Molecular Evolution* 28: 427–441.

Prescott, C. W. 1954. *How to Know the Fresh-Water Algae* (W. C. Brown Co.: Dubuque, Iowa), 211 pp.

Prescott, G. W. 1962. *Algae of the Western Great Lakes Region* (W. C. Brown Co.: Dubuque, Iowa), 977 pp.

Pretorius, D. A. 1976. The nature of the Witwatersrand gold-uranium deposits. In: K. H. Wolf (Ed.), *Handbook of Stratabound and Stratiform Ore Deposits*, vol. 7 (Elsevier: Amsterdam), pp. 28–88.

Pretorius, D. A. 1989. The sources of Witwatersrand gold and uranium: a continued difference of opinion. *Economic Geology Research Unit. University of Witwatersrand Information Circular* 206: 1–43.

Price, P. L., O'Sullivan, T. O. and Alexander, R. 1988. The nature and occurrence of oil in Seram, Indonesia. *Proceeds of the 16th Annual Convention of the Indonesian Petroleum Association, Jakarta, 1987*: 141–173.

Priestly, R. E. and David, T. W. E. 1912. Geological notes of the British Antarctic Expedition, 1907–09. *International Geological Congress* 11: 767–777.

Pyanovskaya, I. A. 1974. Sravnitel'naya kharakteristika nekotorkyh rannei srednekembrijskikh form vodoroslevogo proiskhozhdeniya [Comparative characteristics of some Early and Middle Cambrian forms of algal origin]. In. A. Y. R. I. T. Zhuravleva (Ed.), *Biostratigrafiya i Paleontologiya Nizhnego Kembriya Evropy i Severnoj Azii* (Nauka: Moscow), pp. 229–241.

Pyanovskaya, I. A. 1985. Problematicheskie luchistye organizmy *Radiaxialia* Pjanovskaya, gen. nov. iz kembriya yuzhnogo Tyan'-Shanya [The problematic radiose organisms *Radiaxialia* Pjanovskaya, gen. nov., from the Cambrian of southern Tien-Shan]. In: B. S. Sokolov and I. T. Zhuravleva (Eds.), *Problematiki Pozdnego Dokembriya i Paleozoya. Trudy Inst. Geol. Geofiz. Sibirsk. Otd. Akad. Nauk SSSR*, 632 (Nauka: Novosibirsk), pp. 133–144.

Pyatiletov, V. G. 1976. Mikrofossilii (akritarkhi) iz dokembrijskikh i nizhnekembrijskikh otlozhenij Manskogo progiba [Microfossils (acritarchs) from Precambrian and Lower Cambrian deposits of the Mansk trough]. In: I. T. Zhuravleva (Ed.), *Stratigrafiya i Paleontologiya Niznego i Srednego Kembriya SSSR [Stratigraphy and Paleontology of the Lower and Middle Cambrian of the USSR]. Trudy Inst. Geol. Geofiz. Sibirsk. Otd. Akad. Nauk SSSR* 296 (Nauka: Novosibirsk), pp. 186–8, 240–5.

Pyatiletov, V. G. 1978. Mikrofossilii Manskogo progiba [Microfossils of the Mansk trough]. *Trudy Institut geologii i geofiziki Sibirskoe Otdolenie, Akademiya Nauk SSSR* 400: 175–184, 198–211.

Pyatiletov, V. G. 1980a. O nakhodkakh mikrofossilij roda *Navifusa* v lakhandinskoj svite [On finds of the genus *Navifusa* in the Lakhanda Formation]. *Paleontologicheskij Zhurnal* 1980(3): 143–145.

Pyatiletov, V. G. 1980b. Yudomskij kompleks mikrofossilij Yuzhnoj Yakutii [The Yudomian microfossil assemblage of Southern Yakutia]. *Geologiya i Geofizika* 1980(7): 8–20.

Pyatiletov, V. G. 1986. Mikrofitofossilii pozdnego dokembriya Katangskoj sedloviny i sopredel'nykh territorij (zapadnaya chast' Sibirskoj platformy) [Precambrian microphytofossils from the Katanga saddle and adjacent territories (western part of the Siberian platform)]. In: V. Khomentovskij V and V. Y. Shenfil' (Eds.), *Pozdnij Dokembrij i Rannij Paleozoj Sibiri. Stratigrafiya i Paleontologiya. Inst. Geol. Geofiz. Sibirsk Otd. Akad. Nauk SSSR* (Nauka: Novosibirsk), pp. 129–164.

Pyatiletov, V. G. and Karlova, G. A. 1980. Verkhnerifejskij kompleks rastitel'nykh mikrofossilij Eniseskogo kryazha [Upper Riphean complex of plant microfossils of the Yenisei Ridge]. In: V. Khomentovskij V (Ed.), *Novye Dannye po Stratigrafii Pozdnego Dokembriya Zapada Sibirskoj Platformy i ee Skladchatogo Obramleniya [New Data on the Stratigraphy of the Late Precambrian of the Siberian Platform and its Folded Frame]. Inst. Geol. Geofiz. Sibirsk. Otd. Akad. Nauk SSSR* (Nauka Novosibirsk), pp. 56–135.

Pyatiletov, V. G. and Rudavskaya, V. A. 1985. Akritarkhi yudomskogo kompleksa [Acritarchs of the Yudoma complex]. In: B. S. Sokolov and A. B. Ivanovskij (Eds.), *Vendskaya Sistema 1, Paleontologiya [Vendian System 1, Paleontology]* (Nauka: Moscow), pp. 151–158.

Pyatiletov, V. G., Luchinina, V. A., Shenfil', V. Y. and Yakshin, M. S. 1981. Novye dannye o drevnikh vodorosliyakh Sibiri [New data on Precambrian fossil algae of Siberia]. *Doklady Akademii Nauk SSSR* 261(4): 982–984 (1983, *Doklady Earth Science Sections* 261: 209–211).

Qian, J.-X. and Xiao, B. 1984. [An early Cambrian small shelly fauna from Aksu-Wushi Region, Xinjiang]. *Professional Papers in Stratigraphy and Paleontology* 13: 65–90.

Qian, Y. 1977. [Hyolitha and some problematica from the Lower Cambrian Meishucunian Stage in central and southwestern China]. *Acta Palaeontologica Sinica* 16: 255–275.

Qian, Y. 1978. [The Early Cambrian hyoliths in central and southwest China and their stratigraphical significance]. *Memoirs of the Nanjing Institute of Geology and Palaeontology* 11: 1–38.

Qian, Y. 1984. Several groups of bizarre sclerte fossils from the earliest Cambrian in eastern Yunnan. *Bulletin of the Nanjing Institute of Geology and Palaeontology* 1983(6): 85–99.

Qian, Y. and Bengtson, S. 1989. Palaeontology and biostratigraphy of the Early Cambrian Meishucunian Stage in Yunnan Province,

References

Qian, Y. and Yin, G.-Z. 1984a. Small shelly fossils from the lowermost Cambrian in Guizhou. *Professional Papers in Stratigraphy and Paleontology* 13: 91–123.

Qian, Y. and Yin, G.-Z. 1984b. Zhijinitidae and its stratigraphical significance. *Acta Palaeontologica Sinica* 23(2): 215–223.

Qian, Y. and Zhang, S. 1983. Small shelly fossils from the Xihaoping member of the Tongying Formation in Fangxian County of Hubei Province and their stratigraphical significance. *Acta Palaeontologica Sinica* 22(1): 82–94.

Qian, Y. and Zhang, S. 1985. On the systematic position of small orthoconic fossils with bulbuous initial part from the early Lower Cambrian. *Acta Micropalaeontologica Sinica* 2(1): 1–13.

Qian, Y., Chen, M. and Chen, Y. 1979. Hyolithids and other small shelly fossils from the Lower Cambrian Huangshandong Formation in the eastern part of the Yangtze Gorge. *Acta Palaeontologica Sinica* 18(3): 207–230.

Qin, H.-B. and Ding, L.-F. 1988. [Occurrence of microfossils in the Yangjiagou Member of Tongying Formation, southern Shaanxi]. *Acta Micropalaeontologica Sinica* 5: 171–178.

Qiu, S.-Y. and Liu, H.-F. 1982. Stromatolitic assemblages and their biostratigraphic significance from the upper Precambrian rocks on the minor Qinling Range in Shaanxi Province. *Precambrian Geologic Monograph. Northwest University Scientific Reports*: 1–33.

Qu, L.-H., Perasso, R., Baroin, A., Brugerolle, G., Bachellerie, J.-P. and Adoutte, A. 1988. Molecular evolution of the 5'-terminal domain of large-subunit rRNA from lower eukaryotes. A broad phylogeny covering photosynthetic and nonphotosynthetic protists. *Biosystems* 21: 203–208.

Quesnel, L. B. 1971. Microscopy and micrometry. *Methods of Microbiology* 5A: 1–103.

Raaben, M. E. (Ed.). 1981. *The Tommotian Stage and the Cambrian Lower Boundary Problem* (Amerin Publishing Company, Pvt., Limited: New Dehli), 359 pp.

Radke, M. and Welte, D. H. 1983. The Methylphenanthrene Index (MPI): a maturity parameter based on aromatic hydrocarbons. In: M. Bjorøy, C. Albercht, K. Cornford, K. de Groot, G. Eglinton, E. Galimov, D. Leythaeuser, R. Pelet, J. Rullkotter and G. Speers (Eds.), *Advances in Organic Geochemistry 1981* (John Wiley & Sons: New York), pp. 504–512.

Radugin, K. V. and Stepanova, M. V. 1964. O nitchatykh vodoroslyakh dokembriya iz chasti Vostochnogo Sayana [On filamentous Precambrian algae from the northwest part of East Sayan]. In: L. L. Khalfin (Ed.), *Materialy po Geologii i Poleznym Iskopaemym Zapadnoj Sibiri* (Tomskij Institut: Tomsk), pp. 60–64.

Radugin, K., V 1937. O sootmoshenii kembriya i dokembriya v Gornoj Shorii [On the relation between the Cambrian and Precambrian in Gornaya Shoriya]. *Problemy Soveskoj Geologii* 7: 301.

Radugin, K., V 1962. O rannikh formakh arkheotsiat. [On early types of archaeocyathans]. *Materialy po geologii zapadnoj Sibiri* 63: 7–10.

Radugin, K., V 1964. O novoy gruppe drevnejshikh zhivotnykh [On a new group of ancient animals]. *Trudy Institut geologii i geofiziki Sibirskoe Otdolenie, Akademiya Nauk SSSR* 1: 145–149.

Raff, R. A. and Kaufman, T. C. 1983. *Embryos, Genes, and Evolution* (Macmillan: New York), 395 pp.

Raff, R. A., Field, K. G., Olsen, G. J., Giovannoni, S. J., Lane, D. J., Ghiselin, M. T., Pace, N. R. and Raff, E. C. 1989. Metazoan phylogeny based on analysis of 18S ribosomal RNA. In: B. Fernholm, K. Bremer and H. Jörnvall (Eds.), *The Hierarchy of Life* (Elsevier: Amsterdam), pp. 247–260.

Ragan, M. A. 1988. Ribosomal RNA and the major lines of evolution: a perspective. *Biosystems* 25: 177–188.

Ragan, M. A. 1989. Biochemical pathways and the phylogeny of the eukaryotes. In: B. Fernholm, K. Bremer and H. Jörvall (Eds.), *The Hierarchy of Life* (Elsevier: Amsterdam), pp. 133–143.

Ragan, M. A. and Chapman, D. J. 1977. *A Biochemical Phylogeny of the Protists* (Academic Press: New York), 317 pp.

Raha, P. K. and Sastry, M. V. A. 1982. Stromatolites and Precambrian stratigraphy in India. *Precambrian Research* 18(4): 293–318.

Raiswell, R. and Berner, R. A. 1985. Pyrite formation in euxinic and semi-euxinic sediments. *American Journal of Science* 285: 710–724.

Raiswell, R. and Berner, R. A. 1986. Pyrite and organic matter in Phanerozoic normal marine shales. *Geochimica et Cosmochimica Acta* 50: 1967–1976.

Ramaekers, P. 1981. Hudsonian and Helikian basins of the Athabasca region, northern Saskatchewan. In: F. H. A. Campbell (Ed.), *Proterozoic Basins of Canada. Geological Survey of Canada, Paper 81-10*: 219–234.

Rambler, M. B. and Margulis, L. 1980. Bacterial resistance to ultraviolet irradiation under anaerobiosis: implications for pre-Phanerozoic evolution. *Science* 210: 638–640.

Rao, M. R. S. 1943. Algal structures from the Cuddapah limestones (Pre-Cambrian), south India. *Current Science* 12: 207–208.

Rao, M. R. S. 1944. Algal structures from the Cuddapah Limestones (Precambrian), S. India. *Current Science* 13(3): 75.

Rao, S. R. N. and Mohan, K. 1954. Microfossils from the Dogra slates (Pre-Cambrian) of Kashmir. *Current Science* 23(1): 11–12.

Ratner, M. I. and Walker, J. C. G. 1972. Atmospheric ozone and the history of life. *Journal of Atmospheric Science* 29: 803–808.

Raup, D. M. 1972. Taxonomic diversity during the Phanerozoic. *Science* 177: 1065–1071.

Raup, D. M. 1975. Taxonomic diversity estimation using rarefaction. *Paleobiology* 1: 333–342.

Raup, D. M. 1976. Species diversity in the Phanerozoic: an interpretation. *Paleobiology* 2: 289–297.

Raup, D. M. 1979. Biases in the fossil record of species and genera. *Bulletin of the Carnegie Museum of Natural History* 13: 85–91.

Raup, D. M. 1985. Mathematical models of cladogenesis. *Paleobiology* 11: 42–52.

Raup, D. M. and Valentine, J. W. 1983. Multiple origins of life. *Proceedings of the National Academy of Science USA* 80: 2981–2984.

Raup, D. M., Gould, S. J., Schopf, T. J. M. and Simberloff, D. S. 1973. Stochastic models of phylogeny and the evolution of diversity. *Journal of Geology* 81(5): 525–542.

Raup, R. A. and Kaufman, T. C. 1983. On the early origin of major biologic groups. *Paleobiology* 9: 107–115.

Raven, J. A. 1987. Biochemistry, biophysics and physiology of chlorophyll–b containing algae: implications for taxonomy and phylogeny. *Progress in Phycological Research* 5: 1–122.

Raven, P. H. 1970. A multiple origin for plastids and mitochondria. *Science* 169: 641–646.

Raymond, P. E. 1935. Pre-Cambrian life. *Geological Society of America Bulletin* 46: 375–391.

Rayner, D. H. 1957. A problematical structure from the Ingletonian rocks, Yorkshire. *Transactions of the Leeds Geological Association* 7: 34–42.

Rayner, D. H. 1981. *The Stratigraphy of the British Isles*, 2nd ed. (Cambridge University Press: Cambridge), 460 pp.

Read, J. F. and Pfeil, R. W. 1983. Fabrics of allochthonous reefal blocks, Shady Dolomite (Lower to Middle Cambrian), Virginia Appalachians. *Journal of Sedimentary Petrology* 53(3): 761–778.

Redfield, A. C., Ketchum, B. H. and Richards, F. A. 1963. The influence of organisms on the composition of seawater. In: M. N. Hill (Ed.), *The Sea*, vol. II (Interscience: New York), pp. 26–77.

Reed, J. H., Illich, H. A. and Horsfield, B. 1986. Biochemical evolutionary significance of Ordovician oils and their sources. In: D. Leythaeuser and J. Rullkötter (Eds.), *Advances in Organic Geochemistry 1985* (Pergamon Press: Oxford), pp. 347–358.

Rees, M. N., Pratt, B. R. and Rowell, A. J. 1989. Early Cambrian reefs, reef complexes, and associated lithofacies of the Shackleton Limestone, Transantarctic Mountains. *Sedimentology* 36: 341–361.

Reichenbach, H. 1984. Myxobacteria: a most peculiar group of social prokaryotes. In: E. Rosenberg (Ed.), *Myxobacteria: Development and Cell Interactions* (Springer-Verlag: New York), pp. 1–301.

Reichenbach, H., Ludwig, W. and Stackebrandt, E. 1988. Lack of relationship between gliding cyanobacteria and filamentous gliding heterotrophic eubacteria: comparison of 16S rRNA catalogues of *Spirulina, Saprospira, Vitreoscilla, Leucothrix*, and *Herpetosiphon*. *Archives of Microbiology* 145: 391–395.

Reif, W.-E. 1968. Schwammreste aus dem oberen Ordovizium von Estland und Schweden. *Neues Jahrbuch für Geologie und Paläontologie Monatshefte* 1968: 733–744.

Reith, M. and Cattolico, R. A. 1986. Inverted repeat of *Olisthodiscus luteus* chloroplast DNA contains genes for both subunits of ribulose-1,5-biophosphate carboxylase and the 32,000-dalton Q_B protein: phylogenetic implications. *Proceedings of the National Academy of Science USA* 83: 8599–8603.

Reitz, R. C. and Hamilton, J. G. 1968. The isolation and identification of two sterols from two species of blue-green algae. *Comparative Biochemistry and Physiology* 25: 401–415.

Rejtlinger, E. A. 1948. Kembrijskie foraminifery Yakutii [Cambrian foraminera of Yakutsk]. *Byulleten' Moskovskogo Obshchestva Ispytatelej Prirody, Otdel Geologicheskij* 23(2): 77–81.

Rejtlinger, E. A. 1959. Atlas mikroskpicheskikh organicheskikh ostatkov i problematiki drevnikh tolshch Sibiri [Atlas of microscopic organic remains and problematica of ancient deposits of Siberia]. *Trudy geologicheskii institut, Akademiya Nauk SSSR* 25: 1–62 (translated by Associated Technical Services, Inc., New Jersey).

Renoux, J. M. and Rohmer, M. 1986. Enzymatic cyclization of all-trans pentaprenyl methyl ethers by a cell-free system from the protozoon *Tetrahymena pyriformis*. *European Journal of Biochemistry* 55: 125–132.

Repetski, J. E. 1978. A fish from the Upper Cambrian of North America. *Science* 200: 529–531.

Repetski, J. E. 1981. An Ordovician occurrence of *Utahphospha* Müller & Mille. *Journal of Paleontology* 55(2): 395–400.

Resig, J. M., Lowenstam, H. A., Echols, R. J. and Weiner, S. 1980. An extant opaline foraminifer: test ultrastructure, mineralogy, and taxonomy. *Cushman Foundation Special Publication* 19: 205–214.

Resser, C. E. 1938. Fourth contribution to nomeclature of Cambrian fossils. *Smithsonian Miscellaneous Collections* 97(10): 1–44.

Retallack, G. J. 1986. Reappraisal of 2200 Ma-old paleosol near Waterval Onder, South Africa. *Precambrian Research* 32: 195–232.

Revsbech, N. P. In Press. Diffusion characteristics of microbial communities determined by use of oxygen microsensors. *Journal of Microbiological Methods*.

Revsbech, N. P. and Jørgensen, B. B. 1983. Photosynthesis of benthic microflora measured with high spatial resolution by the oxygen microprofile method: Capabilities and limitations of the method. *Limnology and Oceanography* 28: 749–756.

Revsbech, N. P. and Jørgensen, B. B. 1986. Microelectrodes: their use in microbial ecology. In: K. C. Marshall (Ed.), *Advances in Microbial Ecology* (Plenum Press: New York), pp. 293–352.

Revsbech, N. P. and Ward, D. M. 1983. Oxygen microelectrode that is insensitive to medium chemical composition: use in an acid microbial mat dominated by *Cyanidium caldarium*. *Applied and Environmental Microbiology* 45: 755–759.

Revsbech, N. P. and Ward, D. M. 1984a. Microelectrode studies of interstitial water chemistry and photosynthetic activity in a hot spring microbial mat. *Applied and Environmental Microbiology* 48: 270–275.

Revsbech, N. P. and Ward, D. M. 1984b. Microprofiles of dissolved substances and photosynthesis in microbial mats measured with microelectrodes. In: Y. Cohen, R. W. Castenholz and H. O. Halvorson (Eds.), *Microbial Mat: Stromatolites* (Alan R. Liss, Inc: New York), pp. 171–188.

Revsbech, N. P., Jørgensen, B. B. and Brix, O. 1981. Primary production of microalgae in sediments measured by oxygen microprofile, $H^{14}CO_3^-$ fixation and oxygen exchange methods. *Limnology and Oceanography* 26: 717–730.

Revsbech, N. P., Jørgensen, B. B., Blackburn, T. H. and Cohen, Y. 1983. Microelectrode studies of the photosynthesis and O_2, H_2S, and pH profiles of a microbial mat. *Limnology and Oceanography* 28: 1062–1074.

Revsbech, N. P., Madsen, B. and Jørgensen, B. B. 1986. Oxygen production and consumption in sediments determined at high spatial resolution by computer simulation of oxygen microelectrode data. *Limnology and Oceanography* 31: 293–304.

Revsbech, N. P., Nielsen, L. P., Christensen, P. B. and Sørenson, J. 1988. Combined oxygen and nitrous oxide microsensor from denitrification studies. *Applied and Environmental Microbiology* 54: 2245–2249.

Revsbech, N. P., Nielsen, L. P., Christensen, P. B. and Sørensen, J. In Prep. A combined oxygen and nitrous oxide microsensor for studies of dentrification.

Reymer, A. P. S. and Schubert, G. 1984. Phanerozoic addition rates to the continental crust and crustal growth. *Tectonics* 3: 63–77.

Reymer, A. P. S. and Schubert, G. 1987. Phanerozoic and Precambrian crustal growth rates. In: A. Kröner (Ed.), *Proterozoic Lithospheric Evolution*. American Geophysical Union Geodynamics Series 17: 1–10.

Reynolds, C. S. 1984. *The Ecology of Freshwater Phytoplankton* (Cambridge University Press: Cambridge), 384 pp.

Reynolds, E. S. 1963. Use of lead citrate at high pH as an opaque stain in electron microscopy. *Journal of Cell Biology* 17: 208–212.

Rezanka, T., Zahradnik, J. and Podojil, M. 1982. Hydrocarbons in green and blue-green algae. *Folia Microbiologia* 27: 450–454.

Rhoads, D. C. and Morse, J. W. 1971. Evolutionary and ecologic significance of oxygen-deficient marine basins. *Lethaia* 4: 413–428.

Rhodes, F. H. T. and Bloxam, T. W. 1971. Phosphatic organisms in the Paleozoic and their evolutionary significance. In: E. L. Yochelson (Ed.), *Phosphate in Fossils* (Proceedings of the North American Paleontology Convention K), pp. 1485–1513.

Richards, J. R., Fletcher, I. R. and Blockley, J. G. 1981. Pilbara Galenas: Precise isotopic assay of the oldest Australian leads; model ages and growth-curve implications. *Mineralium Deposita* 16: 7–30.

Richardson, J. B. 1984. Mid-Palaeozoic palynology, facies and correlation. *Stratigraphy. International Geological Congress* 27(1): 341–365.

Richardson, L. L. and Castenholz, R. W. 1987a. Diel vertical movements of the cyanobacterium *Oscillatoria terebriformis* in a sulfide-rich hot spring microbial mat. *Applied and Environmental Microbiology* 53: 2142–2150.

Richardson, L. L. and Castenholz, R. W. 1987b. Enhanced survival of the cyanobacterium *Oscillatoria terebriformis* in darkness under anaerobic conditions. *Applied and Environmental Microbiology* 53: 2151–2158.

Richardson, L. L., Aguilar, C. and Nealson, K. H. 1988. Manganese oxidation in pH and O_2 microenvironments produced by phytoplankton. *Limnology and Oceanography* 33: 352–363.

Richter, D. K. 1985. Die Dolomite der evaporit- und der dolcrete-Playasequenz im mittleren Keuper bei Coburg (NE-Bayern). *Neues Jahrbuch für Geologie und Paläontologie, Abhandlungen* 170: 87–128.

Richter, R. 1955. Die ältesten Fossilien Süd-Afrikas. *Senckenbergiana Lethaea* 36: 243–389.

Ricketts, B. and Donaldson, J. A. 1981. Sedimentary history of the Belcher Group of Hudson Bay. In: F. H. A. Cambell (Ed.), *Proterozoic Basins of Canada. Geological Survey of Canada, Paper* 81-10: 235–254.

Ricketts, B. and Donaldson, J. A. 1989. Stromatolite reef development on a mud-dominated platform in the Middle Precambrian Belcher Group of Hudson Bay. In: H. Gedsetzer, N. P. James and G. Tebbutt (Eds.), *Reefs, Canada and Adjacent Areas. Canadian Society of Petroleum Geologists Memoirs* 13.

Riding, R. 1975. *Girvanella* and other algae as depth indicators. *Lethaia* 8: 173–179.

Riding, R. 1977. Calcified *Plectonema* (blue-green algae), a recent example of *Girvanella* from Aldabra Atoll. *Palaeontology* 20: 33–46.

Riding, R. 1982. Cyanophyte calcification and changes in ocean chemistry. *Nature* 299: 814–815.

Riding, R. and Brasier, M. 1975. Earliest calcareous foraminifera.

Nature 257: 208–210.
Riding, R. and Voronova, L. 1982a. Affinity of the Cambrian alga *Tubomorphophyton* and its significance for the Epiphytaceae. *Palaeontology* 25: 869–878.
Riding, R. and Voronova, L. 1982b. Recent freshwater oscillatoriacean analogue of the Lower Paleozoic calcareous alga *Angulocellularia*. *Lethaia* 15: 105–114.
Riding, R. and Vorona, L. 1984. Assemblages of calcareous algae near the Precambrian/Cambrian boundary in Siberia and Mongolia. *Geological Magazine* 121: 205–210.
Riding, R. and Voronova, L. 1985. Morphological groups and series in Cambrian calcareous algae. In: D. F. Toomey and M. H. Nitecki (Eds.), *Paleoalgology. Contemporary Research and Applications* (Springer-Verlag: Berlin), pp. 56–78.
Riding, R. and Voronova, L. G. 1982. Calcified cyanophytes and the Precambrian-Cambrian transition. *Naturwissenschaften* 69: 498–499.
Riding, R. and Wray, J. L. 1972. Note on the ?algal genera *Epiphyton, Paraepiphyton, Tharama,* and *Chabakovia. Journal of Paleontology* 46(6): 918–919.
Rieger, R. M. 1980. A new group of interstitial worms, Lobatocerbridae nov. fam. (Annelida) and its significance for metazoan phylogeny. *Zoomorphologie* 95: 41–84.
Rieger, R. M. and Sterrer, W. 1975. New spicular skeletons in Turbellaria, and the occurrence of spicules in marine meiofauna. *Zeitschrift für Zoologische Systematik und Evolutionforschung* 13: 207–278.
Rigby, J. K. 1986. Sponges of the Burgess Shale (Middle Cambrian), British Columbia. *Palaeontographica Canadiana* 2: 1–105.
Rigby, J. K. and Nitecki, M. H. 1975. An unusually well preserved heteractinid sponge from the Pennsylvanian of Illinois and a possible classification and evolutionary scheme for the Heteractinida. *Journal of Paleontology* 49(2): 329–339.
Riley, J. P. and Chester, R. 1971. *Introduction to Marine Chemistry* (Academic Press: New York), 465 pp.
Ringwood, A. E. 1979. *Origin of the Earth and Moon* (Springer-Verlag: New York), 295 pp.
Ripley, E. M. and Nicol, D. L. 1981. Sulfur isotopic studies of Archaean slate and graywacke from northern Minnesota: evidence for the existence of sulfate reducing bacteria. *Geochimica et Cosmochimica Acta* 45: 839–846.
Ripperdan, R. L. and Kirschvink, J. L. 1989a. Magnetostratigraphic constraints for reconstructing the Austral-Asian margin of early Paleozoic Gondwanaland. *International Geological Congress* 28(2): 701 (Abstract).
Ripperdan, R. L. and Kirschvink, J. L. 1989b. Paleomagnetic constraints on the Cambrian positions of the North and South China Blocks. *Geological Association of Canada Mineralogical Association of Canada Annual Meeting, Program with Abstracts* V(14): A99.
Rippka, R. 1972. Photoheterotrophy and chemoheterotrophy among unicellular blue-green algae. *Archives of Microbiology* 87: 93–98.
Rippka, R. 1988. Isolation and purification of cyanobacteria. In: L. Packer and A. N. Glazer (Eds.), *Cyanobacteria, Vol. 167 Methods in Enzymology* (Academic Press: London, New York), pp. 3–27.
Rippka, R., Deruelles, J., Waterbury, J. B., Herdman, M. and Stanier, R. Y. 1979. Generic assignments, strain histories and properties of pure cultures of cyanobacteria. *Journal of General Microbiology* 111: 1–61.
Rippka, R., Waterbury, J. B. and Stanier, R. Y. 1981. Isolation and purification of cyanobacteria: some general principles. In: M. P. Starr, H. Stolp, H. G. Trüper, A. Balows and H. G. Schlegel (Eds.), *The Prokaryotes* (Springer-Verlag: Berlin, Heidelberg, New York), pp. 212–220.
Rixhter, D. K. 1985. Die Dolomite der evaporit- und der dolcrete-Playasequenz im mittleren Keuper bei Coburg (NE-Bayern). *Neues Jahrbuch für Geologie und Paläontologie, Abhandlungen* 170: 87–128.
Robbins, E. I., Porter, K. G. and Haberyan, K. A. 1985. Pellet microfossils: possible evidence for metazoan life in early Proterozoic time. *Proceedings of the National Academy of Science USA* 82: 5809–5813.
Roberts, J. D. 1971. Late Precambrian glaciation: an antigreenhouse effect? *Nature* 234: 216–217.
Roberts, J. D. 1976. Late Precambrian dolomites, Vendian glaciation and synchronicity of Vendian glaciations. *Journal of Geology* 84: 47–63.
Robertson, I. D. M. and du Toit, M. C. 1981. The Limpopo Belt. In: D. R. Hunter (Ed.), *Precambrian of the Southern Hemisphere* (Elsevier: Amsterdam), pp. 641–671.
Robertson, I. D. M., du Toit, M. C., Joubert, P., Matthews, P. E., Lockett, N. H., Mendelsohn, F., Broderick, T. J., Bloomfield, K. and Mason, R. 1981. Mobile belts. In: R. R. Hunter (Ed.), *Precambrian of the Southern Hemisphere* (Elsevier: Amsterdam), pp. 641–802.
Robertson, J. A. 1978. Uranium deposits in Ontario. In: M. M. Kimberley (Ed.), *Uranium Deposits: Their Mineralogy and Origin. Mineralogical Association of Canada Short Course Handbook*, 3 (University of Toronto Press: Toronto), pp. 224–280.
Robertson, W. A. 1962. Umbrella-shaped fossils (?) from the Lower Proterozoic of the Northern Territory of Australia. *Journal of the Geological Society of Australia* 9: 87–90.
Robinson, N. and Eglinton, E. 1990. Lipid chemistry of Icelandic hot spring microbial mats. *Organic Geochemistry* 15: 291–298.
Robles, C. D. and Cubit, J. 1981. Influence of biotic factors in an upper intertidal community: Dipteran larvae grazing on algae. *Ecology* 62(6): 1536–1547.
Roblot, M. M. 1964. Sporomorphes du Precambrien Amoricain. *Analytical Chemistry* 50: 105–110.
Roedel, H. 1926. Ein kambrisches Geschiebe mit problematischen Spuren. *Zeitschrift für Geschiebeforschung* 11: 22–26.
Roemer, S. C., Hoagland, K. S. and Rosowski, J. R. 1984. Development of a freshwater periphyton community as influence by diatom mucilages. *Canadian Journal of Botany* 62: 1799–1813.
Roeske, C. A. and O'Leary, M. 1984. Carbon isotope effects on the enzyme-catalyzed carboxylation of ribulose bisphosphate. *Biochemistry* 1984: 6275–6284.
Roeske, C. A. and O'Leary, M. H. 1985. Carbon isotope effect on carboxylation of ribulose bisphosphate catalyzed by ribulose bisphosphate carboxylase from *Rhodospirillum rubrum. Biochemistry* 24: 1603–1607.
Rogers, J. J. W., Dabbagh, M. E., Olsewski, W. J., Jr., Gaudette, H. E., Greenberg, J. K. and Brown, B. A. 1984. Early poststabilization sedimentation and later growth of shields. *Geology* 12: 607–609.
Rohmer, M. and Ourisson, G. 1976. Methylhopanes d'*Acetobacter xylinum* et d'*Acetobacter rancens*: une nouvelle famille de composes triterpenique. *Tetrahedron Letters* 1976(5): 3641–3644.
Rohmer, M., Bouvier, P. and Ourisson, G. 1979. Molecular evolution of biomembranes: Structural equivalents and phylogenetic precursors of sterols. *Proceedings of the National Academy of Science USA* 76: 847–851.
Rohmer, M., Bouvier, P. and Ourisson, G. 1980. Non-specific lanosterol and hopanoid biosynthesis by a cell-free system from the bacterium *Methylococcus capsulatus. European Journal of Biochemistry* 112: 557–560.
Rohmer, M., Bouvier-Nave, P. and Ourisson, P. 1984. Distribution of hopanoid triterpenes in prokaryotes. *Journal of General Microbiology* 130: 1137–1150.
Romanenko, E. V. 1968. Kembrijskie gobki otryada Heteractinellida Altaya [Cambrian sponges of the order Heteractinellida from Altaj]. *Paleontologicheskij Zhurnal* 1968(2): 134–137.
Romano, A. H. and Peloquin, J. P. 1963. Composition of the sheath of *Sphaerotilus natans. Journal of Bacteriology* 86: 252–258.
Ronov, A., Khain, V. and Seslavinsky, K. 1984. *Atlas of Lithological-Paleogeographical Maps of the World. Late Precambrian and Paleozoic of Continents* (USSR Academy of Sciences: Leningrad), 70 pp.
Ronov, A. B. 1958. Organic carbon in sedimentary rocks (in relation to

the presence of petroleum). *Geochemistry* 1958: 510–536.
Ronov, A. B. 1964. Common tendencies in the chemical evolution of the earth's crust, ocean and atmosphere. *Geochemistry International* 1: 713–737.
Ronov, A. B. 1980. *Osadochnaja Oblochka Zemli (Sedimentary Layer of the Earth)* (Nauka: Moscow), 78 pp.
Ronov, A. B. 1982. The earth's sedimentary shell (quantitative patterns of its structure, compositions, and evolution). *International Geological Review* 24: 1313–1388.
Roper, H. 1956. The manganese deposits at Otjosondu, South West Africa. *20th International Geological Congress Symposium on Manganese* 2: 115–122.
Roscoe, S. M. 1968. Huronian rocks and Uraniferrous conglomerates on the Canadian Shield. *Geological Survey of Canada, Paper* 68-40: 1–205.
Roscoe, S. M. 1973. The Huronian Supergroup, a Palaeoaphebian succession showing evidence of atmospheric evolution. In: G. M. Young (Ed.), *Huronian Stratigraphy and Sedimentation. Geological Association of Canada Special Paper* 12: 31–48.
Rosen, B. R. 1979. Modules, members and communes: A postscript introduction to social organisms. In: G. Larwood and B. R. Rosen (Eds.), *Biology and Systematics of Colonial Organisms. Systematics Association Special Volume*, 11 (Clarendon Press: Oxford), pp. 13–35.
Rosen, B. R. 1989. Coloniality. In: D. E. Briggs and P. R. Crowther (Eds.), *Palaeobiology: A Synthesis* (Blackwell Scientific Publications: Oxford), pp. 330–335.
Rosengarten, R., Klein-Struckmeier, A. and Kirchhoff, H. 1988. Rheotactic behavior of a gliding mycoplasma. *Journal of Bacteriology* 170: 989–990.
Rossow, W. B., Henderson-Sellers, A. and Weinrich, S. K. 1982. Cloud feed-back: A stabilizing effect for the early earth? *Science* 217: 1245–1247.
Rothpletz, A. 1908. Ueber Algen und Hydrozoen im Silur von Gotland und Oesel [On algae and hydrozoans in the Silurian of Gotland and Ösel]. *Kungliga Svenska Vetenskapsakademiens Handlingar* 43: 1–25.
Rothpletz, A. 1913. Über Kalkalgen, Spongiostromen und einege andere Fossilien [On calcareous algae, sponges and some other fossils]. *Sveriges Geologiska Undersökning* 10(Series C): 1–57.
Rothschild, L. J. and Heywood, P. 1987. Protistan phylogeny and chloroplast evolution: conflicts and congruence. *Progress in Protistology* 2: 1–68.
Rothschild, L. J., Ragan, M. A., Coleman, A. W., Heywood, P. and Gerbi, S. A. 1986. Are rRNA sequence comparisons the rosetta stone of phylogenetics? *Cell* 47: 640.
Rottem, S. and Markowitz, O. 1979. Carotenoids as reinforcers of the *Acholeplasma laidlawii* lipid bilayer. *Journal of Bacteriology* 140: 944–948.
Round, F. E. 1961. Some algae from the Ennedi Mountains of French equatorial Africa. *Journal of the Royal Microscopical Society* 80: 71–82.
Round, F. E. 1981. Morphology and phyletic relationships of the silicified algae and the archetypal diatom—monophyly or polyphyly. In: T. L. Simpson and B. E. Volcani (Eds.), *Silicon and Siliceous Structures in Biological Systems* (Springer-Verlag: New York), pp. 97–128.
Round, F. E. 1984. The systematics of the Chlorophyta: an historical review leading to some modern concepts [Taxonomy of the Chlorophyta III]. In: D. E. G. Irvine and D. M. John (Eds.), *Systematics of the Green Algae* (Academic Press: London), pp. 1–27.
Roux, A. 1985. Intoduction à l'étude des Algues fossiles paléozoïques (de la Bactérie à la tectonique des plaques) [An introduction to the study of Paleozoic fossil algae (from bacteria to plate tectonics)]. *Centre de Recherche Exploration-Production Elf-Aquitaine, Bulletin* 9(2): 465–699.
Rovnina, L. V. 1981. Palynological method to determine the level of katagensis of organic matter by using Jurassic deposits of Western Siberia. In: J. Brooks (Ed.), *Organic Maturation Studies and Fossil Fuel Exploration* (Academic Press: New York), pp. 427–432.
Rowell, A. J. 1962. The genera of the brachiopod superfamilies Obollelacea and Siphonotretacea. *Journal of Paleontology* 36(1): 136–152.
Rowell, A. J. 1966. Revision of some Cambrian and Ordovician inarticulate brachiopods. *University of Kansas Paleontological Contribution* 7: 1–36.
Rowell, A. J. 1971. Supposed Pre-Cambrian brachiopods. *Smithsonian Contributions to Paleobiology* 3: 71–79.
Rowell, A. J. 1977. Early Cambrian brachiopods from the southwestern Great Basin of California and Nevada. *Journal of Paleontology* 51(1): 68–85.
Rowell, A. J. 1980. Inarticulate brachiopods of the Lower and Middle Cambrian Piocke Shale of the Pioche District, Nevada. *University of Kansas Paleontological Contribution* 98: 1–26.
Rowland, S. M. 1978. Environmental stratigraphy of the lower member of the Poleta Formation (Lower Cambrian), Esmeralda County, Nevada. Ph.D. thesis, University of California, Santa Cruz, 116 pp.
Rowland, S. M. 1981. Archaeocyathid reefs of the southern Great Basin, western United States. In: M. E. Taylor (Ed.), *Short Papers for the Second International Symposium on the Cambrian System 1981. U.S. Geological Survey Open-File Report 81—743* (U.S. Geological Survey: Washington D.C.), pp. 193–197.
Rowland, S. M. and Gangloff, R. A. 1988. Structure and paleoecology of Lower Cambrian reefs. *Palaios* 3: 111–135.
Rowlands, N. J., Blight, P. G., Jarvis, D. M. and Von der Borch, C. C. 1980. Sabkhas and playa environments in late Proterozoic grabens, Willouran Ranges, South Australia. *Journal of the Geological Society of Australia* 27: 55–68.
Roy, S. 1980. Genesis of sedimentary manganese formations: processes and products in recent and older geological ages. In: I. M. Varentsov and G. Y. Grassely (Eds.), *Geology and Geochemistry of Manganese, II* (E. Schweizerbart'sche Verlag: Stuttgart), pp. 13–44.
Roy, S. 1981. *Manganese Deposits* (Academic Press: London), 458 pp.
Rozanov, A. Y. 1979. *Playtisolentii*. In: B. M. Keller and A. Y. Rozanov (Eds.), *Paleontologiya Verkhnedokembriiskikh i Kembrijskikh Otlozhenij Vostochno-Evropejskoj Platformy* (Nauka: Moscow), pp. 83–87.
Rozanov, A. Y. 1983. *Platysolenites*. In: A. Urbanek and A. Y. Rozanov (Eds.), *Upper Precambrian and Cambrian Palaeontology of the East-European Platform* (Wydawnichtwa Geologiczne: Warszawa), pp. 94–100.
Rozanov, A. Y. 1984. The Precambrian-Cambrian boundary in Siberia. *Episodes* 7: 20–24.
Rozanov, A. Y. and Missarzhevskij, V., V 1966. Biostratigrafiya i fauna nizhnikh gorizontov kembriya [Biostratigraphy and fauna of the lower horizons of the Cambrian]. *Trudy Geologicheskii Institut, Akademiya Nauk SSSR* 148: 1–125.
Rozanov, A. Y. and Zhuralev, A. Y. In Press. The Lower Cambrian fossil record in the USSR. In: J. H. Lipps and P. W. Signor (Eds.), *The Origin and Early Evolution of Metazoa* (Plenum: New York).
Rozanov, A. Y., Missarzhevskij, V. V., Volkova, N. A., Voronova, L. G., Krylov, I. N., Keller, B. M., Korolyuk, I. K., Lendzion, K., Mikhnyak, Pykhova and Sidorov, A. D. 1969. *Tommotskij yarus problema nizhnej granitsy kembriya [The Tommotian Stage and the Cambrian lower boundary problem]*. In: M. E. Raaben (Ed.), *Trudy Geol. Inst. Akad. Nauk SSSR 206* (Nauka: Moscow), 380 pp.
Rozen, M., Arad, H., Schonfeld, M. and Tel-Or, E. 1986. Fructose supports glycogen accumulation, heterocyst differentiation, N_2 fixation and growth of the isolated cyanobiont *Anabaena azollae*. *Archives of Microbiology* 145: 187–190.
Rozendaal, A. 1980. The Gamsberg zinc deposit, South Africa: a banded stratiform base metal sulfide ore deposit. In: J. D. Ridge (Ed.), *Proceedings of the 5th IAGOD Symposium* (E. Schweizerbart'sche Verlag: Stuttgart), pp. 619–633.
Rubey, W. W. 1951. Geological history of seawater. An attempt to state

References

the problem. *Geological Society of America Bulletin* 62(Part 1): 1111–1148.

Rubey, W. W. 1955. Development of the hydrosphere and atmosphere, with special reference to probable composition of the early atmosphere. In: A. Poldervaart (Ed.), *Crust of the Earth* (Geological Society of America: New York), pp. 631–650.

Rudavskaya, V. A. 1965. Gistrikhosfery kembriya yuzhnoj chasti Sibiriskoj platformy [Hystrichospheres from the Cambrian of the southern part of the Siberian Platform]. *Palaeofitologicheskij Sbornik. Trudy Vsesoyuznogo Neftyanogo Naucho Issledovatel'skogo Geologorazvedochnogo Instituta (VNIGRI)* 239: 95–107, 388–91.

Rudavskaya, V. A. 1973. Akritarkhi pogranichnykh otlozhenij rifeya i kembriya yuga Vostochnoj Sibiri [Acritarchs from the Riphean-Cambrian boundary deposits in the south of East Siberia]. In: T. F. Vozzhennikova and B. Timofeev V (Eds.), *Miskrofossilii Drevnejshikh Otlozhenij* [*Microfossils of the Oldest Deposits*] (Nauka: Moscow), pp. 17–21.

Rudavskaya, V. A. and Timofeev, B., V 1963. K stratigrafii kembrijskikh otlozhenij Predbajkal'ya [Toward the stratigraphy of the Cambrian deposits of the Prebaikal]. *Geologicheskij Sbornik 8. Trudy Vsesoyuznogo Neftyanogo Naucho Issledovatel'skogo Geologorazvedochnogo Instituta (VNIGRI)* 220: 136–151.

Rudd, J. W. M. and Taylor, C. D. 1980. Methane cycling in aquatic environments. *Advances in Aquatic Microbiology* 2: 77–150.

Rudd, T., Sterritt, R. M. and Lester, J. N. 1984. Formation and conditional stability constants of complexes formed between heavy metals and bacteria extracellular polymers. *Water Research* 18: 379–384.

Rudwick, M. J. S. 1964. The infra-Cambrian glaciation and the origin of the Cambrian fauna. In: A. E. M. Nairn (Ed.), *Problems in Paleoclimatology* (Interscience Publishers: London), pp. 150–155.

Ruedemann, R. 1918. The paleontology of arrested evolution. *New York State Museum Bulletin* 196: 107–134.

Ruedemann, R. 1922a. Additional studies of arrested evolution. *Proceedings of the National Academy of Science USA* 8: 54–55.

Ruedemann, R. 1922b. Further notes on the paleontology of arrested evolution. *American Naturalist* 56: 256–272.

Ruedemann, R. 1933. *Camptostroma*, a Lower Cambrian floating hydrozoan. *Proceedings of the United States National Museum* 82(13): 1–8.

Ruelle, J. C. L. 1982. Depositional environments and genesis of stratiform copper deposits of the Redstone copper belt, Mackenzie Mountains, N.W.T. In: R. W. Hutchinson, C. D. Spence and J. M. Franklin (Eds.), *Precambrian Sulfide Deposits. Geological Association of Canada Special Paper* 25: 701–738.

Rullkötter, J., Meyers, P. A., Schaefer, R. G. and Dunham, K. W. 1986. Oil generation in the Michigan Basin. A biological marker and carbon isotope approach. *Organic Geochemistry* 10: 359–375.

Runnegar, B. 1978. Origin and evolution of the Class Rostroconchia. *Philosophical Transactions of the Royal Society of London* B 284: 319–333.

Runnegar, B. 1981. Muscle scars, shell form and torsion in Cambrian and Ordovician univalved molluscs. *Lethaia* 14(4): 311–322.

Runnegar, B. 1982a. The Cambrian explosion: animals or fossils? *Journal of the Geological Society of Australia* 29: 395–411.

Runnegar, B. 1982b. A molecular-clock date for the origin of the animal phyla. *Lethaia* 15: 199–205.

Runnegar, B. 1982c. Oxygen requirements, a biology and phylogenetic significance of the late Precambrian worm *Dickinsonia* and the evolution of the burrowing habit. *Alcheringa* 6: 223–239.

Runnegar, B. 1983. Molluscan phylogeny revisited. *Memoirs of the Association of Australasian Palaeontologists* 1: 121–144.

Runnegar, B. 1985. Shell microstructures of Cambrian molluscs replicated by phosphate. *Alcheringa* 9: 245–257.

Runnegar, B. In Press—a. *Dickinsonia elongata. Journal of Paleontology*.

Runnegar, B. In Press—b. *Gordia. Ichnos*.

Runnegar, B. In Press—c. Oxygen and the early evolution of the Metazoa. In: C. Bryant (Ed.), *Metazoan Life Without Oxygen* (Chapman and Hall: New York).

Runnegar, B. In Press—d. *Wigwamiella. Precambrian Research*.

Runnegar, B. and Bently, C. 1983. Anatomy, ecology and affinities of the Australian Early Cambrian bivalve *Pojetaia runnegari* Jell. *Journal of Paleontology* 57(1): 73–92.

Runnegar, B. and Jell, P. 1976. Australian Middle Cambrian molluscs and their bearing on early molluscan evolution. *Alcheringa* 1: 109–138.

Runnegar, B. and Jell, P. A. 1980. Australian Middle Cambrian molluscs: corrections and additions. *Alcheringa* 4: 111–113.

Runnegar, B. and Pojeta, J. 1974. Molluscan phylogeny: the paleontological viewpoint. *Science* 186: 311–317.

Runnegar, B. and Pojeta, J., Jr. 1985. Origin and diversification of the Mollusca. *The Mollusca* 10: 1–57.

Runnegar, B. and Strait, B. In Press. *Brooksella canyonensis*, supposed mid-Proterozoic metazoan trace fossil, is a pseudofossil. *Journal of Paleontology*.

Runnegar, B., Pojeta, J., Jr. Morris, N. J., Taylor, J. D., Taylor, M. E. and McClung, G. 1975. Biology of the Hyolitha. *Lethaia* 8: 181–191.

Rutten, M. G. 1971. *The Origin of Life by Natural Causes* (Elsevier: Amsterdam), 420 pp.

Sabrodin, W. 1971. Leben im Präkambrium. *Ideen des exakten Wissens* 12: 835–842.

Safronov, V. S. 1972. *Evolution of the Protoplanetary Cloud and Formation of the Earth and the Planets* (NASA Technical Translations: Washington, D.C.), 206 pp.

Sagan, C. 1961. On the origin and planetary distribution of life. *Radiation Research* 15: 174–192.

Sagan, C. 1973. Ultraviolet selection pressure on the earliest organisms. *Journal of Theoretical Biology* 39: 195–200.

Sagan, C. and Mullen, G. 1972. Earth and Mars: Evolution of atmospheres and surface temperatures. *Science* 177: 52–56.

Sagan, L. 1967. On the origin of mitosing cells. *Journal of Theoretical Biology* 14: 225–274.

Sahni, M. R. 1936. *Fermoria minima*: a revised classification of the organic remains from the Vindhyans of India. *Records of the Geological Survey of India* 69: 458–468.

Sahni, M. R. and Shrivastava, R. N. 1954. New organic remains from the Vindhyan System and the probable systematic position of *Fermoria*, Chapman. *Current Science* 23: 39–41.

Sailland, A., Amiri, I. and Freysinnet, G. 1986. Amino acid sequences of the ribulose-1,5-biophosphate carboxylase/oxygenase small subunit from *Euglena*. *Plant Molecular Biology* 7: 213–218.

Saito, K. 1936. Older Cambrian Brachiopoda, Gastropoda, etc. from North-western Korea. *Journal of the Faculty of Science, Imperial University, Tokyo, Section II, Geology, Mineralogy, Geography, Seismology* 4: 345–367.

Salop, L. J. 1977. *Precambrian of the Northern Hemisphere* (Elsevier: Amsterdam), 378 pp.

Salter, J. W. 1864. On some new fossils from the *Lingula*-flags of Wales. *Quarterly Journal of the Geological Society of London* 20: 233–241.

Salter, J. W. 1866. On the fossils of North Wales. Appendix. *Great Britain Geological Survey Memoir* 3: 240–381.

Saltovskaya, V. D. 1975. Pod *Epiphyton* Bornemann (ego veroyatnye sinonimy i stratigraficheskoe znachenie) [The genus *Epiphyton* Bornemann (its probable synonyms and stratigraphic significance)]. In: M. R. Dzhalilov (Ed.), *Voprosy Paleontologii Tadzhikistana* [*Problems in Paleontology of Tadzhikistan*] (Donish: Dushnabe), pp. 70–88.

Salujha, S. K., Rehman, K. and Arora, C. M. 1970. Microplankton from the Bhimas. *Journal of the Palaeontological Society of India* 15: 10–16.

Salvini-Plawen, L., V 1978. On the origin and evolution of the lower Metazoa. *Zeitschrift für Zoologische Systematik und Evolutionforschung* 16: 40–88.

Sandbeck, K. A. and Ward, D. M. 1981. Fate of immediate methane precursors in low-sulfate, hot-spring algal-bacterial mats. *Applied*

and Environmental Microbiology 41: 775–782.
Sandbeck, K. A. and Ward, D. M. 1982. Temperature adaptations in the terminal processes of anaerobic decomposition of Yellowstone National Park and Icelandic hot spring microbial mats. *Applied and Environmental Microbiology* 44: 844–851.
Santchi, P. B., Bower, P., Nyffeler, U. P., Azevedo, A. and Broeker, W. S. 1983. Estimates of the resistance to chemical transport posed by the deep-sea boundary layer. *Limnology and Oceanography* 28: 899–912.
Sarich, V. and Wilson, A. C. 1973. Generation time-genomic evolution in primates. *Science* 179: 1144–1147.
Saunders, G. W. 1972. Potential heterotrophy in a natural population of *Oscillatoria agardhii* var. *isothrix* Skuja. *Limnology and Oceanography* 17: 704–711.
Savrda, C. E. and Bottjer, D. J. 1986. Trace-fossil model for reconstruction of paleo-oxygenation in bottom waters. *Geology* 14: 3–6.
Savrda, C. E., Bottjer, D. J. and Gorsline, D. S. 1984. Towards development of a comprehensive euxinic biofacies model: evidence from Santa Monica, San Pedro, and Santa Barbara Basins, California Continental Borderland. *American Association of Petroleum Geologists Bulletin* 68: 1179–1192.
Sawkins, F. J. 1976. Widespread continental rifting: Some considerations of timing and mechanism. *Geology* 4: 427–430.
Sayutina, T. A. 1980. Rannekembrijskoe semejstvo Khasaktiidae fam. nov. -vozmozhnye stromatoporaty [The Early Cambrian family Khasaktiidae fam. nov. -possible stromatoporatans]. *Paleontologicheskij Zhurnal* 1980(4): 13–28.
Schaefle, J., Ludwig, B., Albrecht, P. and Ourisson, G. 1977. Hydrocarbures aromatiques d'origine géologique II. *Tetrahedron Letters* 41: 3673–3676.
Schare, M. N., Collings, J. C. and Gray, M. W. 1986. Structure and evolution of the small subunit ribosomal RNA gene of *Crithidia fasciculata*. *Current Genetics* 10: 405–410.
Schidlowski, M. 1979. Antiquity and evolutionary status of bacterial sulfate reduction: sulfur isotope evidence. *Origins of Life* 9: 299–311.
Schidlowski, M. 1982. Content and isotopic composition of reduced carbon in sediments. In: H. D. Holland and M. Schidlowski (Eds.), *Mineral Deposits and the Evolution of the Biosphere* (Springer-Verlag: New York), pp. 103–122.
Schidlowski, M. 1988. A 3,800-million-year isotopic record of life from carbon in sedimentary rocks. *Nature* 333: 313–318.
Schidlowski, M., Hayes, J. M. and Kaplan, I. R. 1983. Isotopic inferences of ancient biochemistries: carbon, sulfur, hydrogen, and nitrogen. Chapter 7. In: J. W. Schopf (Ed.), *Earth's Earliest Biosphere: Its Origin and Evolution* (Princeton University Press: Princeton), pp. 149–187.
Schidlowski, M., Matzigkeit, U. and Krumbein, W. E. 1984. Superheavy organic carbon from hypersaline microbial mats. *Naturwissenschaften* 71: 303–308.
Schidlowski, M., Matzigkeit, U., Mook, W. G. and Krumbein, W. E. 1985. Carbon isotope geochemistry and ^{14}C ages of microbial mats from the Gavish Sabkha and the Solar Lake. In: G. Friedman and W. Krumbein (Eds.), *Hypersaline Ecosystems. Ecological Studies*, 53 (Springer-Verlag: Berlin), pp. 381–401.
Schieber, J. 1988. Redistribution of rare-earth elements (REE) during diagenesis of carbonate rocks from the mid-Proterozoic Newland Formation, Montana, U.S.A. *Chemical Geology* 69: 111–126.
Schindewolf, O. H. 1955. Über die Faunenwende vom Paläozoikum zum Mesozoikum. *Zeitschrift der Deutschen Geologischen Gesellschaft* 105: 153–182.
Schindewolf, O. H. 1956. Über präkambrische Fossilien. In: F. Lotze (Ed.), *Geotektonisches Symposium zu Ehren von Hans Stille* (Kommissionsverlag von Ferdinand Enke: Stuttgart), pp. 455–480.
Schink, B. and Zeikus, J. G. 1983. *Clostridium thermosulfurogenes* sp. nov., a new thermophile that produces elemental sulphur from thiosulphate. *Journal of General Microbiology* 129: 1149–1158.
Schlanger, S. O. and Jenkyns, H. C. 1976. Cretaceous oceanic anoxic events: causes and consequences. *Geologie en Mijnbouw* 55: 179–184.
Schmid, J. C., Connan, J. and Albrecht, P. 1987. Identification of long-chain dialkylthiacyclopentanes in petroleum. *Nature* 329: 54–56.
Schmidt, P. W. and Embleton, B. J. J. 1981. A geotectonic paradox: has the earth expanded? *Journal of Geophysics* 49: 20–25.
Schmitz, F. J. 1983. Uncommon marine steroids. In: P. J. Scheuer (Ed.), *Marine Natural Products. Chemical and Biological Perspectives* (Academic Press: London), pp. 241–297.
Schneider, S. H. and Londer, R. 1984. *The Coevolution of Climate and Life* (Sierra Club Books: San Francisco), 563 pp.
Schnepf, E. and Elbrächter, M. 1988. Cryptophycean-like double membrane-bound chloroplast in the dinoflagellate, *Dinophysis* Ehrenb.: Evolutionary, phylogenetic and toxicological implications. *Botanica Acta* 101: 196–203.
Schoell, M. 1984. Stable isotope studies in petroleum exploration. In: J. D. Brooks and D. H. Welte (Eds.), *Advances in Petroleum Geochemistry*, vol. 1 (Academic Press: London), pp. 215–225.
Schopf, J. W. 1968a. Microflora of the Bitter Springs Formation, Late Precambrian, central Australia. *Journal of Paleontology* 42: 651–688.
Schopf, J. W. 1968b. Precambrian microorganisms from central and South Australia. *American Journal of Botany* 55: 722–723.
Schopf, J. W. 1970. Precambrian micro-organisms and evolutionary events prior to the origin of vascular plants. *Biological Reviews* 45: 319–352.
Schopf, J. W. 1974a. The development and evolution of Precambrian life. *Origins of Life* 5: 119–135.
Schopf, J. W. 1974b. Paleobiology of the Precambrian: the age of blue-green algae. In: M. K. Hecht and W. C. Steere (Eds.), *Evolutionary Biology, Volume 7* (Plenum Press: New York), pp. 1–43.
Schopf, J. W. 1975. Precambrian paleobiology: problems and perspectives. *Annual Review of Earth and Planetary Sciences* 3: 213–249.
Schopf, J. W. 1976. Are the oldest "fossils", fossils? *Origins of Life* 7: 19–36.
Schopf, J. W. 1977. Biostratigraphic usefulness of stromatolitic Precambrian microbiotas: a preliminary analysis. *Precambrian Research* 5: 143–173.
Schopf, J. W. 1978. The evolution of the earliest cells. *Scientific American* 239(3): 110–134.
Schopf, J. W. (Ed.). 1983a. *Earth's Earliest Biosphere: Its Origin and Evolution* (Princeton University Press: Princeton, N.J.), 543 pp.
Schopf, J. W. 1983b. Glossary I: Technical terms. In: J. W. Schopf (Ed.), *Earth's Earliest Biosphere: Its Origin and Evolution* (Princeton University Press: Princeton), pp. 443–458.
Schopf, J. W. 1987. "Hypobradytely": comparison of rates of Precambrian and Phanerozoic Evolution. *Journal of Vertebrate Paleontology* 7(Supplement to Number 3): 25A (Abstract).
Schopf, J. W. 1989. Diversification and extinction in the Proterozoic biosphere. In: *Abstracts, The Origin of Life: Sixth ISSOL Meeting and Ninth International Conference* (Czechoslovak Acad. Sci.: Prague), pp. 234–235.
Schopf, J. W. 1991. Collapse of the Late Proterozoic ecosystem. *South African Journal of Geology* 93.
Schopf, J. W. and Barghoorn, E. S. 1969. Microorganisms from the Late Precambrian of South Australia. *Journal of Paleontology* 43: 111–118.
Schopf, J. W. and Blacic, J. M. 1971. New microorganisms from the Bitter Springs Formation (Late Precambrian) of the north-central Amadeus Basin, Australia. *Journal of Paleontology* 45(6): 925–961.
Schopf, J. W. and Oehler, D. Z. 1976. How old are the eukaryotes? *Science* 193: 47–49.
Schopf, J. W. and Packer, B. M. 1986. Newly discovered Early Archean (3.4–3.5 Ga-old) microorganisms from the Warrawoona Group of Western Australia. In: *Abstracts of the 5th Meeting of the International Society for the Study of the Origins of Life and the 8th International Conference on the Origin of Life* (ISSOL), pp. 163–164.

References

Schopf, J. W. and Packer, B. M. 1987. Early Archean (3.3-billion to 3.5-billion-year-old) Microfossils from Warrawoona Group, Australia. *Science* 237: 70–73.

Schopf, J. W. and Walter, M. R. 1980. Archean microfossils and "microfossil-like" objects—a critical appraisal. In: J. E. Glover and D. I. Groves (Eds.), *Extended Abstracts, Second International Archean Symposium* (Geological Society of Australia and International Geological Correlation Project: Perth, Australia), pp. 23–34.

Schopf, J. W. and Walter, M. R. 1983. Archean microfossils: new evidence of ancient microbes. Chapter 9. In: J. W. Schopf (Ed.), *Earth's Earliest Biosphere: Its Origin and Evolution* (Princeton University Press: Princeton), pp. 214–239.

Schopf, J. W., Barghoorn, E. S., Maser, M. D. and Gordon, R. O. 1965. Electron microscopy of fossil bacteria two billion years old. *Science* 149: 1365–1367.

Schopf, J. W., Kvenvolden, K. A. and Barghoorn, E. S. 1968. Amino acids in Precambrian sediments: an assay. *Proceedings of the National Academy of Science USA* 59(2): 639–646.

Schopf, J. W., Ford, T. D. and Breed, W. J. 1973a. Microorganisms from the Late Precambrian of the Grand Canyon, Arizona. *Science* 179: 1319–1321.

Schopf, J. W., Haugh, B. N., Molnar, R. E. and Satterthwait, D. F. 1973b. On the development of metaphytes and metazoans. *Journal of Paleontology* 47: 1–9.

Schopf, J. W., Hayes, J. M. and Walter, M. R. 1983. Evolution of earth's earliest ecosystems: recent progress and unsolved problems. Chapter 15. In: J. W. Schopf (Ed.), *Earth's Earliest Biosphere: Its Origin and Evolution* (Princeton University Press: Princeton), pp. 361–384.

Schopf, J. W., Zhu, W.-Q., Xu, Z.-L. and Hsu, J. 1984. Proterozoic stromatolitic microbiotas of the 1400–1500 Ma-old Gaoyuzhuang Formation near Jixian, northern China. *Precambrian Research* 24: 335–349.

Schopf, T. J. M. 1978. Fossilization potential of an intertidal fauna: Friday Harbor, Washington. *Paleobiology* 4: 261–270.

Schrader, M., Drews, G., Goleki, J. R. and Weckesser, J. 1982. Isolation and characterization of the sheath from the cyanobacterium *Chlorogloeopsis* PCC 6912. *Journal of General Microbiology* 128: 267–272.

Schubert, K., Rose, G., Wachtel, H., Horhold, C. and Ikekawa, N. 1968. Zum vorkommen von sterinen in bakterien. *European Journal of Biochemistry* 5: 246–251.

Schubert, W., Giani, D., Rongen, P., Krumbein, W. E. and Schmidt, W. 1980. Photoacoustic in-vivo spectra of recent stromatolites. *Naturwissenschaften* 67: 129–132.

Schwartz, R. M. and Dayhoff, M.-O. 1981. Chloroplast origins: inferences from protein and nucleic acid sequences. *Annals of the New York Academy of Sciences* 361: 260–268.

Schwartz, H.-U., Einselle, G. and Herm, D. 1975. Quartz-sandy, grazing contoured stromatolites from coastal embayments of Mauritania, West Africa. *Sedimentology* 22: 529–561.

Schwarz, Z. and Kössel, H. 1980. The primary structure of 16S rDNA from *Zea mays* chloroplast is homologous to *E. coli* 16S rRNA. *Nature* 283: 739–742.

Scmidt, F. 1888. Uber eine neuentdeckte untercambrische Fauna in Estland. *Mémoires de l'Academie impériale des sciences de St. Pétersbourg* 36 (ser. 7): 1–27.

Scoffin, T. P. 1970. The trapping and binding of subtidal carbonate sediments by marine vegetation in Bimini Lagoon, Bahamas. *Journal of Sedimentary Petrology* 40: 249–273.

Scotese, C. R. 1984. Paleozoic paleomagnetism and the assembly of Pangea. In: R. Van der Voo, C. R. Scotese and N. Bonhommet Eds.), *Plate Reconstruction from Paleozoic Paleomagnetism. Geodynamics series*, V. 12 (American Geophysical Union: Washington, D.C.), pp. 1–10.

Sdzuy, K. 1969. Unter- und mittelkambrische Porifera (Chancelloriida und Hexactinellida). *Paläontologische Zeitschrift* 43: 115–147.

Sdzuy, K. 1978. The Precambrian-Cambrian Boundary Beds in Morocco (Preliminary Report). *Geological Magazine* 115: 83–94.

Searcy, D. G. 1986. Some features of thermo-acidophilic archaebacteria preadaptive for the evolution of eukaryotic cells. *Systematics and Applied Microbiology* 7: 198–201.

Searcy, D. G. 1987. Phylogenetic and phenotypic relationships between the eukaryotic nucleocytoplasm and thermophilic archaebacteria. *Annals of the New York Academy of Sciences* 503: 168–179.

Sears, J. W. and Price, R. A. 1978. The Siberian connection: A case for Precambrian separation of the North American and Siberian Cratons. *Geology* 6: 267–270.

Sechback, J. and Kaplan, I. R. 1973. Growth pattern and $^{13}C/^{12}C$ isotope fractionation of *Cyanidium caldarium* and hot spring algal mats. *Chemical Geology* 12: 161–169.

Sedlak, W. 1980. Cambrian megascopic alga-like forms accompanying corallicyathida in quartzite beds of Lysa Góra. *Acta Palaeontologica Polonica* 25: 669–670.

Seely, G. R. and Jensen, R. G. 1965. Effect of solvent on the spectrum of chlorophyll. *Spectrochimica Acta* 21: 1835–1845.

Seewaldt, E. and Stackebrandt, E. 1982. Partial sequence of 16S ribosomal RNA and the phylogeny of *Prochloron*. *Nature* 295: 618–620.

Seifert, W. K. and Moldowan, J. M. 1978. Applications of steranes, terpanes and monoaromatics to the maturation, migration, and source of crude oils. *Geochimica et Cosmochimica Acta* 42: 77–95.

Seilacher, A. 1955. Spuren und Fazies im Unterkambrium. In: O. H. Schindewolf and A. Seilacher (Eds.), *Beiträge zur Kenntnis des Kambriums in der Salt Range (Pakistan)* (Jahrbuch Akademie Wissenschaften und der Literatur: Mainz), pp. 373–396.

Seilacher, A. 1956. Der Beginn des Kambriums als biologische Wende. *Neues Jahrbuch für Geologie und Paläontologie, Abhandlungen* 103: 155–180.

Seilacher, A. 1960. Lebensspuren als Leitfossilien. *Geologische Rundschau* 49: 41–50.

Seilacher, A. 1964a. Biogenic sedimentary structures. In: J. Imbrie and N. D. Newell (Eds.), *Approaches to Paleoecology* (John Wiley & Sons: New York), pp. 296–316.

Seilacher, A. 1964b. Sedimentological classification and nomenclature of trace fossils. *Sedimentology* 3: 253–256.

Seilacher, A. 1983. Precambrian metazoan extinctions. *Geological Society of America Abstracts with Programs* 15(6): 683.

Seilacher, A. 1984. Late Precambrian and Early Cambrian Metazoa: preservational or real extinctions? In: H. D. Holland and A. F. Trendal (Eds.), *Patterns of Change in Earth Evolution* (Springer-Verlag: Berlin), pp. 159–168.

Seilacher, A. 1985. Discussion of Precambrian metazoans. *Philosophical Transactions of the Royal Society of London* B 311: 47–48.

Seilacher, A. 1989. Vendozoa: Organismic construction in the Proterozoic biosphere. *Lethaia* 22: 229–239.

Selg, M. 1986. Algen als Faziesindikatoren: Bioherme und Biostrome im Unter-Kambrium von SW-Sardinien [Algae as facies indicators: bioherms and biostromes in the Lower Cambrian from SW Sardinia]. *Geologische Rundschau* 75(3): 693–702.

Semikhatov, M. A. 1976. Experience in stromatolite studies in the USSR. In. M. R. Walter (Ed.), *Stromatolites: Developments in Sedimentology*, vol. 20 (Elsevier: Amsterdam), pp. 337–358.

Semikhatov, M. A. 1980. On the Upper Precambrian stromatolites standard of North Eurasia. *Earth-Science Reviews* 16(5): 992–1015.

Semikhatov, M. A., Komar, V. A. and Serebryakov, S. N. 1970. Yudomskij kompleks stratotipicheskaj mestnosti. *Adademiya Nauk SSSR, Trudy* 210: 210.

Semikhatov, M. A., Gebelein, C. D., Cloud, P., Awramik, S. M. and Benmore, W. C. 1979. Stromatolite morphogenesis—progress and problems. *Canadian Journal of Earth Sciences* 16: 992–1015.

Sepkoski, J. J., Jr. 1978. A kinetic model of Phanerozoic taxonomic diversity I. Analysis of marine orders. *Paleobiology* 4(3): 223–251.

Sepkoski, J. J., Jr. 1979. A kinetic model of Phanerozoic taxonomic diversity, II. Early Phanerozoic families and multiple equilibria. *Paleobiology* 5: 222–252.

Sepkoski, J. J., Jr. 1981. A factor analytic description of the Phanerozoic

marine fossil record. *Paleobiology* 7: 36–53.
Sepkoski, J. J., Jr. 1982. A compendium of fossil marine families. *Milwaukee Public Museum Contributions in Biology and Geology* 51: 1–125.
Sepkoski, J. J., Jr. 1984. A kinetic model of Phanerozoic taxonomic diversity. III. Post-Paleozoic families and mass extinction. *Paleobiology* 10(2): 246–267.
Sepkoski, J. J., Jr. 1986. Phanerozoic overview of mass extinction. In: D. M. Raup and D. Jablonski (Eds.), *Patterns and Processes in the History of Life* (Springer-Verlag: Berlin), pp. 277–295.
Sepkoski, J. J., Jr. 1988. Alpha, beta, or gamma: where does all the diversity go? *Paleobiology* 14(3): 221–234.
Sepkoski, J. J., Jr. 1989. Periodicity in extinction and the problem of catastrophism in the history of life. *Journal of the Geological Society of London* 146: 7–19.
Sepkoski, J. J., Jr. 1990. Evolutionary faunas. In: D. E. G. Briggs and P. R. Crowther (Eds.), *Palaeobiology. A Synthesis* (Blackwell Scientific Publishers: Oxford), pp. 37–41.
Sepkoski, J. J., Jr., Bambach, R. K., Raup, D. M. and Valentine, J. W. 1981. Phanerozoic marine diversity and the fossil record. *Nature* 293: 435–437.
Serebryakov, S. N. and Semikhatov, M. A. 1974. Riphean and Recent stromatolites: a comparison. *American Journal of Science* 274: 556–574.
Setchell, W. A. 1924. *Ruppia* and its environmental factors. *Proceedings of the National Academy of Science USA* 10: 280–288.
Seward, A. C. 1931. *Plant Life Through the Ages* (Cambridge University Press: Cambridge), 607 pp.
Shabanov, YuY., Astashkin, V. A., Pegel', T. B., Egorova, L. I., Zhuravleva, I. T., Pel'man, YuL., Sundukov, V. M., Sttepanova, M. B., Sukhov, C. C., et al. 1987. *Nizhnij Paleozoj Yugo-Zapadnogo Sklona Anabarskoj anteklizy* (Nauka: Novosibirsk), 207 pp.
Shaler, N. S. and Foerste, A. F. 1888. Preliminary description of North Attleborough fossils. *Bulletin of the Museum Comparative Zoology* 16: 27–41.
Shapiro, J. A. 1988. Bacteria as multicellular organisms. *Scientific American* 258: 82–89.
Shapiro, R. 1986. *Origins–A Skeptics Guide to the Creation of Life on Earth* (Summit Books: New York), 332 pp.
Shaw, D. M. 1976. Development of the early continental crust. Part 2. Prearchean, Protoarchean, and later eras. In: B. F. Windley (Ed.), *The Early History of the Earth* (John Wiley & Sons: New York), pp. 33–54.
Shegelski, R. J. 1980. Archean cratonization, emergence and red bed development, Lake Shebandowan area, Canada. *Precambrian Research* 12: 331–347.
Shegelski, R. J. 1982. The Gunflint Formation in the Thunder Bay area. In: J. M. Franklin (Ed.), *Proterozoic Geology of the Northern Lake Superior Area. Field Trip 4* (Geological Association of Canada, Winnipeg Section: Winnipeg), pp. 15–31.
Shegelski, R. J. 1985. Sulfur isotope study of the Aphebian Gunflint Formation, Ontario. *Geological Association of Canada Mineralogical Association of Canada Annual Meeting, Program with Abstracts* 10: A54.
Shenfil', V. Y. 1980. Obruchevelly v rifejskikh otlozheniyakh enisejskogo dryazha [*Obruchevella* representatives in the Riphean deposits of the Yenisey Ridge Region]. *Doklady Akademii Nauk SSSR* 254(4): 993–994.
Shenfil', V. Y. 1983. Vodorosli v dokembrijskikh otlozheniyakh Vostochnoj Sibiri [Algae in the Precambrian deposits of eastern Siberia]. *Doklady Akademii Nauk SSSR* 269: 471–473 (1984, Doklady Earth Science Sections 269: 177–181).
Shergold, J. H. 1988. Review of trilobite biofaces distributions at the Cambrian-Ordovician Boundary. *Geological Magazine* 125: 363–380.
Sherwood, B. A., Sager, S. L. and Holland, H. D. 1987. Phosphorus in foraminiferal sediments from North Atlantic Ridge cores and in pure limestones. *Geochimica et Cosmochimica Acta* 51: 1861–1866.
Shevyrev, A. A. 1964. The problem of the origin of the Early Cambrian fauna. *International Geology Reviews* 6: 1617–1629.
Shi, Ji-Y., Mackenzie, A. S., Alexander, R., Eglinton, G., Gowar, A. P., Wolff, G. A. and Maxwell, J. R. 1982. A biological marker investigation of petroleums and shales from the Shengli oilfield. *Chemical Geology* 35: 1–31.
Shiea, J., Brassell, S. C. and Ward, D. M. 1990. Mid-chain branched mono- and dimethylalkanes in hot spring cyanobacterial mats: a biogenic source for branched alkanes in petroleums? *Organic Geochemistry* 15: 223–231.
Shiea, J., Brassell, S. C. and Ward, D. M. Submitted. Comparative analysis of free lipids in hot spring cyanobacterial and photosynthetic bacterial mats and photosynthetic bacteria.
Shields, L. M. and Drouet, F. 1962. Distribution of terrestrial algae within the Nevada Test Site. *American Journal of Botany* 49: 547–554.
Shields, L. M., Mitchell, C. and Drouet, F. 1957. Alga- and lichen-stabilized surface crusts as soil nitrogen sources. *American Journal of Botany* 44: 489–498.
Shinn, E. A. 1972. Worm and algal-built columnar stromatolites in the Persian Gulf. *Journal of Sedimentary Petrology* 42: 837–840.
Shiskin, B. B. 1974. Rakovinnaya fauna v nemakit-daldynskoj svite (severo-zapad Anabarskogo podnyatiya) [A shelly fauna in the Nemakit-Daldyn Formation (northwestern Anabar Uplift)]. *Geologiya i Geofizika* 1974(4): 111–114.
Shoemaker, E. M. 1982. The collision of solid bodies. In: J. K. Beatty, B. O'Leary and A. Chaikin (Eds.), *The New Solar System* (Cambridge University Press: Cambridge), pp. 33–44.
Shoemaker, E. M. 1984. Large body impacts through geologic time. In: H. D. Holland and A. F. Trendall (Eds.), *Patterns of Change in Earth Evolution* (Springer-Verlag: Berlin), pp. 15–40.
Shu Degan and Chen Ling 1988. [Discovery of Early Cambrian Radiolaria and its significance]. *Scientia Sinica*: 100–134.
Shukla, M. 1984. Microstromatolites from the "Calc-Zone of Pithoragarh" Kumaun Himalaya. *Geophytology* 14(2): 240–241.
Siever, R. 1988. Evolution of the silica cycle. In: *Abstracts, V. M. Goldschmidt Conference* (The Geochemical Society of Baltimore: Baltimore), p. 74.
Signor, P. W., Mount, J. F. and Onken, B. R. 1987. A Pre-trilobite shelly fauna from the White-Inyo region of eastern California and western Nevada. *Journal of Paleontology* 61: 425–438.
Silver, L. T. and Anderson, T. H. 1974. Possible left-lateral Early to Middle Mesozoic disruption of the southwestern North American Craton Margin. *Geological Society of America Abstracts with Programs* 6: 955–956.
Silvester-Bradley, P. C. 1976. Evolutionary oscillations in prebiology: Igneous activity and the origins of life. *Origins of Life* 7: 9–18.
Simmons, G. M., Hall, G., Mikell, A. and Love, F. 1985. A comparison of biogeological properties of a deep-water marine stromatolite analog with those from ice-covered Antarctic freshwater lakes. *Geomicrobiology Journal* 4(3): 269–283.
Simon, W. 1939. Archaeocyathacea. I. Kritische Sichtung der Superfamilie. II. Die Fauna im Kambrium der Sierra Morena (Spanien). *Abhandlugen Senckenberg Naturforsch. Gesellsch.* 448: 1–87.
Simonetta, A. 1970. Studies on non-trilobite arthropods of the Burgess Shale (Middle Cambrian). *Palaeontographia Italiana* 66: 35–45.
Simonson, B. 1985. Sedimentary constraints on the origin of Precambrian iron-formations. *Geological Society of America Bulletin* 96(2): 244–252.
Simonson, B. M. 1985. Sedimentology of chert in the Early Proterozoic Wishart Formation, Quebec-Newfoundland, Canada. *Sedimentology* 32: 23–40.
Simonson, B. M. and Goode, A. D. T. 1989. First discovery of ferruginous chert arenites in the early Precambrian Hamersley Group of Western Australia. *Geology* 17: 269–272.
Simpson, G. G. 1944. *Tempo and Mode of Evolution* (Columbia University Press: New York), 237 pp.
Simpson, G. G. 1950. *The Meaning of Evolution. A Study of the History of Life and of its Significance for Man* (Oxford University Press:

References

Simpson, G. G. 1953. *The Major Features of Evolution* (Columbia University Press: New York), 434 pp.

Simpson, P. R. and Bowles, J. F. W. 1977. Uranium mineralisation of the Witwatersrand and Dominion Reef systems. *Philosophical Transactions of the Royal Society of London* 286: 527–548.

Simpson, T. L. and Volcani, B. E. 1981. Introduction. In: T. L. Simpson and B. E. Volcani (Eds.), *Silicon and Siliceous Structures in Biological Systems* (Springer-Verlag: New York), pp. 3–12.

Singh, P. and Shukla, S. D. 1981. Fossils from the Lower Tal, their age and its bearing on the stratigraphy of the Lesser Himalaya. *Geoscience Journal* 2(2): 157–176.

Sinha, A. K. 1972. U-Th-Pb systematics and the age of the Onverwacht Series, South Africa. *Earth and Planetary Science Letters* 16: 219–227.

Sinninghe Damste, J. S., De Leeuw, J. W., Kock-van Dalen, A. C., De Zeeuw, M. A., De Lange, F., Rijpstra, W. I. C. and Schenck, P. A. 1987. The occurrence and identification of series of organic sulfur compounds in oils and sediment extracts. I. A study of the Rozel Point Oil (USA). *Geochimica et Cosmochimica Acta* 51: 2369–2391.

Sirevåg, R., Buchanan, B. B., Berry, J. A. and Troughton, J. H. 1977. Mechanisms of CO_2 fixation in bacterial photosynthesis studied by the carbon isotope fractionation technique. *Archives of Microbiology* 112: 35–38.

Sisodya, D. S. 1982. Fossil impressions of jelly-fish in the Nimbahera Limestone, Semri Group of the Vindhyan Supergroup of rocks. *Current Science* 51: 1070–1071.

Sistrom, W. R. and Clayton, R. K. 1964. Studies on a mutant of *Rhodopseudomonas spheroides* unable to grow photosynthetically. *Biochimica et Biophysica Acta* 88: 61–73.

Skyring, G. W. 1984. Sulfate reduction in marine sediments associated with cyanobacterial mats in Australia. In: Y. Cohen, R. W. Castenholz and H. O. Halvorson (Eds.), *Microbial Mats: Stromatolites* (Alan R. Liss Inc.: New York), pp. 265–275.

Skyring, G. W. 1987. Sulfate reduction in coastal ecosystems. *Geomicrobiology Journal* 5: 295–374.

Skyring, G. W. and Bauld, J. 1990. Microbial mats in Australian coastal environments. *Advances in Microbial Ecology* 11: 461–498.

Skyring, G. W., Chambers, L. A. and Bauld, J. 1983. Sulfate reduction in sediments colonized by cyanobacteria, Spencer Gulf, South Australia. *Australian Journal of Marine and Freshwater Research* 34: 359–374.

Skyring, G. W., Lynch, R. M. and Smith, G. D. 1988. Acetylene reduction and hydrogen metabolism by a cyanobacterial/sulfate-reducing bacterial mat ecosystem. *Geomicrobiology Journal* 6: 25–31.

Sleep, N. H., Zahnle, K. J., Kasting, J. F. and Morowitz, H. J. 1989. Annihilation of ecosystems by large asteroid impacts on the early earth. *Nature* 342: 139–142.

Smayda, T. J. 1970. The suspension and sinking of phytoplankton in the sea. *Oceanography and Marine Biology Annual Review* 8: 353–414.

Smith, A. B. 1989. RNA sequence data in phylogenetic reconstruction: Testing the limits of its resolution. *Cladistics* 5: 321–344.

Smith, A. G. and Hallam, A. 1970. The fit of the southern continents. *Nature* 255: 139–144.

Smith, A. G., Hurley, A. M. and Briden, J. C. 1981. *Phanerozoic Paleocontinental World Maps* (Cambridge University Press: Cambridge), 102 pp.

Smith, A. J. 1983. Modes of cyanobacterial carbon metabolism. *Annales Microbiologie (Inst. Paster)* 134 B: 93–113.

Smith, A. J., London, J. and Stanier, R. Y. 1967. Biochemical basis of obligate autotrophy in blue-green algae and thiobacilli. *Journal of Bacteriology* 94: 972–983.

Smith, D. G. 1977. Lower Cambrian palynomorphs from Howth, Co. Dublin. *Geological Journal* 12: 159–168.

Smith, J. W., Schopf, J. W. and Kaplan, I. R. 1970. Extractable organic matter in Precambrian cherts. *Geochimica et Cosmochimica Acta* 34: 659–675.

Smith, J. 1979. Mineralogy of the planets: a voyage in space and time. *Mineralogical Magazine* 43: 1–89.

Smith, R. L. and Bold, H. C. 1966. Phycological Studies VI. Investigations of the Algal Genera *Eremosphaera* and *Oocystis*. *University of Texas Publ. No. 6612* (University of Texas: Austin), 121 pp.

Snajdr, M. 1975. *Konicekion* nov. gen. from the Middle Cambrian of Bohemia (Ostracoda). *Věstnik Ústřeed Ustavu Geologickeho* 50: 153–156.

Sneath, P. H. A. and Sokal, R. R. 1973. *Numerical Taxonomy* (W. H. Freeman & Co.: San Francisco), 573 pp.

Snezhko, A. M. 1986. Novye vidy izvestkobykh vodoroslej iz rannedokembrijskikh otlozhenij Ukrainy [New species of calcareous algae in late Precambrian deposits of the Ukraine]. In: B. S. Sokolov (Ed.), *Aktual'nye Voprosy Sovremennoj Paleoal' gologii* [Current Questions in Contemporary Paleoalgology] (Naukova Dumka: Kiev), pp. 102–105.

Snyder, S. H. 1985. The molecular basis of communication between cells. In: S. H. Snyder (Ed.), *Readings from Scientific American: The Molecules of Life* (Freeman: San Francisco), pp. 84–93.

Soegaard, K. and Eriksson, K. A. 1985. Evidence of tide, storm, and wave interaction on a Precambrian siliciclastic shelf: the 1,700 M.Y. Ortego Group, New Mexico. *Journal of Sedimentary Petrology* 55: 672–684.

Sogin, M. L. and Gunderson, J. H. 1987. Structural diversity of eukaryotic small subunit ribosomal RNAs: evolutionary implications. *Annals of the New York Academy of Sciences* 503: 125–129.

Sogin, M. L., Elwood, H. L. and Gunderson, J. H. 1986a. Evolutionary diversity of eukaryotic small sub-unit rRNA genes. *Proceedings of the National Academy of Science USA* 83: 1383–1387.

Sogin, M. L., Ingold, A., Karlok, M., Nielsen, H. and Engberg, J. 1986b. Phylogenetic evidence for the acquisition of ribosomal RNA introns subsequent to the divergence of some of the major *Tetrahymena* groups. *The EMBO Journal* 5: 3625–3630.

Sogin, M. L., Swanton, M. T., Gunderson, J. H. and Elwood, H. J. 1986c. Sequence of the small subunit ribosomal RNA gene from the hypotrichous aliate *Euplotes aediculatus*. *Journal of Protozoology* 33: 26–29.

Sogin, M. L., Edman, U. and Elwood, H. 1989a. A single kingdom of eukaryotes. In: B. Fernholm, K. Bremer and H. Jörnvall (Eds.), *The Hierarchy of Life* (Elsevier: Amsterdam), p. 133.

Sogin, M. L., Gunderson, J. H., Elwood, H. J., Alonso, R. A. and Peattie, D. A. 1989b. Phylogenetic meaning of the kingdom concept: An unusual ribosomal RNA from *Giardia lamlia*. *Science* 243: 75–77.

Sokolov, B. S. 1965. Drevnelshie otlosheniya rannego kembriya i sabelliditidi' [The oldest Early Cambrian deposits and sabelliditids]. In: *Vsesoyuzn'i Simpozium po Paleontologii Dokembriya i Rannego Kembriya 25—30 Oktyabrya 1965* (Akad. Nauk SSSR Sibirskoe Otdelenie: Novosibirsk), pp. 78–91.

Sokolov, B. S. 1967. Drevneyshiye pognofory [The oldest Pogonophora]. *Akademiya Nauk SSSR, Doklady* 177: 201–204 (English translation, p. 252–255).

Sokolov, B. S. 1968. Sabelliditidy (Pognophora) venda i rannego kembriya SSSR [Vendian and Early Cambrian Sabellitida (Pogonophora) of the USSR]. In: *Proceedings of the International Palaeontological Union, 23rd International Geological Congress*, pp. 79–86.

Sokolov, B. S. 1972. Vendskiy etap v istorii Zemli [The Vendian Period in Earth history]. *Paleontologiya, Doklady Sovetskikh Geologov, Akademiya Nauk SSSR* 7: 114–124.

Sokolov, B. S. 1973. Vendian of northern Eurasia. *American Association of Petroleum Geologists Memoir* 19: 204–218.

Sokolov, B. S. 1975. O paleontologicheskikh nakhodkakh v dousol'skikh otlozheniyakh Irkutskogo amfiteatra [Paleontological discoveries in pre-Usolje deposits of Irkutsk Ampitheatre]. In: B. D. Sokolov and V. V. Khomentovskiy (Eds.), *Analogi Vendskogo Kompleksav Sibiri* (Nauka: Moscow), pp. 112–117.

Sokolov, B. S. 1976a. Organicheskiy mir zemli na puti k fanerozoyskoy differentziatzii [The organic world on the path of Phanerozoic differentiation]. *Vestnik Akademii Nauk SSSR* 1976(1): 126–143.

Sokolov, B. S. 1976b. Precambrian Metazoa and the Vendian-

Cambrian boundary. *Paleontological Journal* 10: 1–13.
Sokolov, B. S. 1977. Organicheskij mir Zemli na puti k fanerozojskoj differntsiatsii [Earth's organic world on the road to Phanerozoic differentiation]. In: *250 Let Akademii Nauk SSSR* (Nauka: Moscow), pp. 423–444.
Sokolov, B. S. (Ed.). 1979. *Paleontologiya Dokembriya i Rannego Kembriya* (Nauka: Leningrad), 304 pp.
Sokolov, B. S. 1984. Vendskiy period i istopii Zemli [The Vendian period of Earth History]. *Priroda* 12: 3–18.
Sokolov, B. S. 1985. Vendskie polikhety [Vendian polychaetes]. In: B. S. Sokolov and A. B. Ivanovsky (Eds.), *Vendskaya Sistema 1* (Nauka: Moscow), pp. 198–200.
Sokolov, B. S. and Fendonkin, M. A. 1984. The Vendian as the terminal system of the Precambrian. *Episodes* 7: 12–19.
Sokolov, B. S. and Fedonkin, M. A. 1986. Global biological events in the late Precambrian. In: O. H. Walliser (Ed.), *Global Bio-Events. A Critical Approach* (Springer-Verlag: Berlin), pp. 105–108.
Sokolov, B. S. and Ivanovsky, A. B. (Eds.), 1985. *Vendskaya Sistema 1* (Nauka: Moscow), 224 pp.
Sokolov, B. S. and Zhuravleva, I. T. 1983. Lower Cambrian Stage subdivision of the Siberia. Atlas of Fossils. *Transaction of the Institute of Geology and Geophysics* 558: 1–216.
Solomon, M., Rafter, T. A. and Durham, K. C. 1971. Sulphur and oxygen isotope studies in the northern Pennines in relation to ore genesis. *Transactions of the Institute of Mining and Metallurgy* 80: 259–275.
Song, X., Liu, G. and Xing, Y. 1984. Characteristics of the Dengyingxia'an, Meishucunian and Qiongzhusian microfloras. In: H. Luo, Z. Jiang, X. Wu, X. Song, L. Ouyang, Y. Xing, G. Liu, S. Zhang and Y. Tao (Eds.), *Sinian-Cambrian Boundary Stratotype Section at Meishucun, Jinning, Yunnan, China* (People's Publishing House: Yunnan Province), pp. 112–115.
Song, Xueliang 1984. *Obruchevella* from the Early Cambrian Meishucun Stage of the Meishucun section, Jinning, Yunnan, China. *Geological Magazine* 121(3): 179–183.
Sorensen, L. O. and Conover, J. T. 1962. Algal mat communities of *Lyngbya confervoides* (Agardh) Gomont. *Publications of the Institute for Marine Science, Austin* 8: 61–74.
Sosnovskaya, O. V. 1980. Novyy rod truvchatykh okamenelostey iz dokembriya Kuznetzkogo Alatau [A new genus of tubular fossils from the Precambrian of the Kuznetz Alatau]. *Paleontologicheskij Zhurnal* 1980(3): 145–147.
Sosnovskaya, O. V. 1981. Newlandievaya problematika v otlozheniyakh verkhnego proterozoya severo-zapadnoy chasti Batenvskogo Kryazha [*Newlandia problematica* in Upper Proterozoic deposits of the northwestern part of the Batenev Ridge]. *Aktual'nye voprosy geologii dekembriya Sibiry. Novsibirsk. Trudy Sibirskogo Nanchno-Isskdovatel'skogo Instituta Geologii, Geofiziki i Mineral'nogo Syr'ya* 290: 73–83.
Sosnovskaya, O. V. and Shiptizyn, V. A. 1984. O nekotoryakh problematichnykh okamenelostyakh iz dokembriya Kuznetzkogo Alatau [Some problematic fossils from the Precambrian sequence of the Kurnetz Alatau]. *Paleontologicheskij Zhurnal* 1984(3): 128–131.
Sournia, A. 1976. Ecologie et productivité d'une Cyanophycée en milieu corallien: *Oscillatoria limosa* Agardh. *Phycologia* 15: 363–366.
Southcott, R. V. 1958. South Australian jellyfish. *South Australian Naturalist* 32: 53–61.
Southgate, P. N. 1986. Depositional environment and mechanism of preservation of microfossils, Upper Proterozoic Bitter Springs Formation, Australia. *Geology* 14: 683–686.
Southgate, P. N. 1989. Relationships between cyclicity and stromatolite form in the Late Proterozoic Bitter Springs Formation, Australia. *Sedimentology* 36: 323–346.
Spamer, E. E. 1988. Geology of the Grand Canyon, v. 3. Part III. An annotated bibliography of the world literature on the Grand Canyon typefossil *Chuaria circularis* Walcott, 1899, an index fossil for the Late Proterozoic. *Geological Society of America, Microfilm Publication* 17: 4 cards.

Sperling, J. A. 1976. Algal ecology of southern Icelandic hot springs in winter. *Ecology* 56: 183–190.
Spizharskij, T. N., Ergaliev, G. K., Zhuravleva, I. T., Repina, L. N., Rozanov, A. Y. and Chernysheva, N. E. 1983. Yarusnaya shkala kembrijskoj sistemy [A scale of stages for the Cambrian System]. *Sovetskaya Geologiya* 1983(8): 57–72.
Spjeldnæs, N. 1963. A new fossil (*Papillomembrana* sp.) from the Upper Precambrian of Norway. *Nature* 200: 63–64.
Spjeldnæs, N. 1967. Fossils from pebbles in the Biskopåsen Formation in southern Norway. *Norges Geologiske Undersoekelse* 251: 53–76.
Sprigg, R. C. 1947. Early Cambrian(?) jellyfishes from the Flinders Ranges, South Australia. *Transactions of the Royal Society of South Australia* 71(2): 212–224.
Sprigg, R. C. 1949. Early Cambrian "jellyfishes" of Ediacara, South Australia and Mount John, Kimberly District, Western Australia. *Transactions of the Royal Society of South Australia* 73(1): 72–99.
Sprigg, R. C. 1988. On the 1946 discovery of the Precambrian Ediacaran fossil fauna in South Australia. *Earth Sciences History* 7: 46–51.
Sprinkle, J. 1973. *Morphology and Evolution of Blastozoan Echinoderms*. Mus. Comp. Zool., Spec. Publ. (Harvard University Press: Boston), 248 pp.
Sprinkle, J. 1976. Classification and phylogeny of 'pelmatzoan' echinoderms. *Systematic Zoology* 25: 83–91.
Spurr, A. R. 1969. A low viscosity epoxy resin embedding medium for electron microscopy. *Journal of Ultrastructal Research* 26: 31–43.
Squire, A. D. 1973. Discovery of Late Precambrian trace fossils in Jersey, Channel Islands. *Geological Magazine* 110: 223–226.
Srikantappa, C., Hormann, P. K. and Raith, M. 1984. Petrology and geochemistry of layered ultramafic to mafic complexes from the Archean craton of Karnataka, southern India. In: A Kröner, G. N. Hanson and A. M. Goodwin (Eds.), *Archaean Geochemistry* (Springer-Verlag: Berlin), pp. 138–160.
Stacey, F. D. 1981. Cooling of the earth—A constraint of paleotectonic hypotheses. In: R. J. O'Conner and S. S. Fyfe (Eds.), *Evolution of the Earth. American Geophysical Union Geodynamics Series* 5: 272–276.
Stackebrandt, E. 1983. A phylogenetic analysis of *Prochloron*. In: H. E. A. Schenk and W. Schwemmler (Eds.), *Endocytobiology II* (W. de Gruyter: Berlin), pp. 921–932.
Stackebrandt, E. and Woese, C. R. 1981. The evolution of prokaryotes. *Symposium of the Society of General Microbiology* 32: 1–32.
Stackebrandt, E., Seewaldt, E., Fowler, V. J. and Schleifer, H.-J. 1982. The relatedness of *Prochloron* sp. isolated from different didemnid ascidian hosts. *Archives of Microbiology* 132: 216–217.
Stahl, D. A., Lane, D. J., Olsen, G. J. and Pace, N. R. 1985. Characterization of a Yellowstone hot spring microbial community by 5S rRNA sequences. *Applied and Environmental Microbiology* 49: 1379–1384.
Stahl, W. J. 1978. Source rock-crude oil correlation by isotopic type curves. *Geochimica et Cosmochimica Acta* 42: 1573–1577.
Stal, L. J., van Gemerden, H. and Krumbein, W. E. 1984. The simultaneous assay of chlorophyll and bacteriochlorophyll in natural microbial communities. *Journal of Microbiological Methods* 2: 295–306.
Stal, L. J., van Gemerden, H. and Krumbein, W. E. 1985. Structure and development of a benthic marine microbial mat. *FEMS Microbiology Ecology* 31: 111–125.
Stanevich, A. M. 1986. Novaya mikrobiota iz chenchinskoj svity Patomskogo nagor'ya [A new microbiota from the Chenchin Formation of the Patom Highland]. In: V. Khomentovskij V and V. Y. Shenfil' (Eds.), *Pozdnij Dokembrij i Rannij Paleozoj Sibiri. Stratigrafiya i Paleontologiya* [*Late Precambrian and Early Paleozoic of Siberia. Stratigraphy and Paleotology*]. *Inst. Geol. Geofiz. Sibirsk. Otd. Akad. Nauk SSSR* (Nauka: Novosibirsk), pp. 115–129.
Stanier, R. Y. and Smith, J. H. C. 1960. The chlorophylls of green bacteria. *Biochimica et Biophysica Acta* 41: 478–484.
Stanier, R. Y., Ingraham, J. L., Wheelis, M. L. and Painter, P. 1986. *The*

References

Microbial World, 5th ed. (Prentice-Hall: New York), 689 pp.

Stanistreet, I. G., De Wit, M. J. and Fripp, R. E. P. 1981. Do graded units of accretionary spheroids in the Barberton Greenstone Belt indicate Archean deep water environment? *Nature* 293: 280–284.

Stanley, G. D. and Yancey, T. E. 1986. A new Late Paleozoic chondrophorine (Hydrozoa, Velellidae) by-the-wind sailor from Malaysia. *Journal of Paleontology* 60: 76–83.

Stanley, G. D., Jr. 1986. Chondrophorine hydrozoans as problematic fossils. In: A. Hoffman and M. H. Nitecki (Eds.), *Problematic Fossil Taxa. Oxford Monographs on Geology and Geophysics*, 5 (Oxford University Press: New York), pp. 68–86.

Stanley, S. M. 1973a. An ecological theory for the sudden origin of multicellular life in the Late Precambrian. *Proceedings of the National Academy of Science USA* 70: 1486–1489.

Stanley, S. M. 1973b. An explanation for Cope's rule. *Evolution* 27: 1–26.

Stanley, S. M. 1975a. Clades versus clones in evolution: why we have sex. *Science* 190: 382–383.

Stanley, S. M. 1975b. A theory of evolution above the species level. *Proceedings of the National Academy of Science USA* 72: 646–650.

Stanley, S. M. 1976a. Fossil data and the Precambrian-Cambrian evolutionary transition. *American Journal of Science* 276: 56–76.

Stanley, S. M. 1976b. Ideas on the timing of metazoan diversification. *Paleobiology* 2: 209–219.

Stanley, S. M. 1976c. Reply [to Towe 1976]. *American Journal of Science* 276: 1180–1181.

Stanley, S. M. 1979. *Macroevolution: Pattern and Process* (W. H. Freeman & Co.: San Francisco), 332 pp.

Stanley, S. M. 1984a. Does bradytely exist? In: N. Eldrige and S. M. Stanley (Eds.), *Living Fossils* (Springer-Verlag: New York), pp. 278–281.

Stanley, S. M. 1984b. Marine mass extinctions: a dominant role for temperature. In: N. H. Nitecki (Ed.), *Extinctions* (University of Chicago Press: Chicago), pp. 69–118.

Stanley, S. M. 1984c. Temperature and biotic crises in the marine realm. *Geology* 12: 205–208.

Star, W. M., Marijnissen, J. P. A. and van Gemert, M. J. C. 1987. New trends in photobiology. Light dosimetry: Status and prospects. *Biology 1. Photochemistry and Photobiology* B: 149–167.

Starr, M. P., Stolp, H., Trüper, H. G. Balows, A. and Schlegel, H. G. (Eds.), 1981. *The Prokaryotes*, vols. 1 & 2 (Springer-Verlag: New York), 2284 pp.

Stasinska, A. 1960. *Velumbrella czarnockii* n. gen., n.sp.—meduse du cambrien inferieur des Monts de Sainte-Croix. *Acta Palaeontologica Polonica* 5(3): 337–346.

Steacy, H. R. 1953. An occurence of uraninite in a black sand. *American Mineralogist* 38: 549–550.

Stebbins, G. L., Jr. 1974. Adaptive radiation and the origin of form in the earliest multicellular organisms. *Systematic Zoology* 22: 478–485.

Steckler, M. S. and Watts, A. B. 1982. Subsidence history and tectonic evolution of Atlantic-type continental margins. In: R. A. Scrutton (Ed.), *Dynamics of Passive Margins. American Geophysical Union Geodynamics Series* 6: 184–196.

Steemann Nielsen, E. 1975. *Marine Photosynthesis. Elsevier Oceanography Series*, 13 (Elsevier Scientific Publ. Co.: Amsterdam), 100 pp.

Steinmuller, K., Kaling, M. and Zetsche, K. 1983. In-vitro synthesis of phycobiliproteids and ribulose-1,5-biphosphate carboxylase by non-poly-adenylated-RNA of *Cyanidium caldarium* and *Porphyridium aerugineum*. *Planta* 159: 308–313.

Steneck, R. S. and Adey, W. H. 1976. The role of environment in control of morphology in *Lithophyllum congestum*, a Caribbean algal ridge builder. *Botanica Marina* 19: 197–215.

Stepanova, M. V. 1974. Novye vidy kembrijskikh vodoroslij Altae-Sayanskoj skladchatoj oblasti [New species of Cambrian algae from the folded region of the Altae-Sayan]. In: A. Y. R. I. T. Zhuravleva (Ed.). *Biostratigrafiya i Paleontologiya Nizhnego Kembriya Evropy i Severnoj Azii* (Nauka: Moscow), pp. 219—222.

Stetter, K. O. 1986. Diversity of extremely thermophilic archaebacteria. In: T. D. Brock (Ed.), *Thermophiles: General, Molecular, and Applied Microbiology* (Wiley-Interscience: New York), pp. 39–74.

Stevens, B. P. J., Barnes, R. G., Brown, R. E., Stroud, W. J. and Willis, I. L. 1988. The Willyama Supergroup in the Broken Hill and Euriowe Blocks, New South Wales. *Precambrian Research* 40/41: 297—327.

Stevens, S. E., Jr. 1988. Cell stability and storage. In: L. Packer and A. N. Glazer (Eds.), *Cyanobacteria, Vol. 167 Methods in Enzymology* (Academic Press: London, New York), pp. 122–124.

Stevenson, D. J. 1983. The nature of the earth prior to the oldest known rock record: The Hadean earth. Chapter 2. In: J. W. Schopf (Ed.), *Earth's Earliest Biosphere: Its Origin and Evolution* (Princeton University Press: Princeton), pp. 32–40.

Stewart, A. J. 1979. A barred-basin marine evaporite in the Upper Proterozoic of the Amadeus Basin, central Australia. *Sedimentology* 26: 33–62.

Stewart, J. H. 1972. Initial deposits in the Cordilleran Geosyncline: Evidence of a Late Precambrian (< 850 m.y.) continental separation. *Geological Society of America Bulletin* 83: 1345–1360.

Stewart, J. H., McMenamin, M. A. S. and Morales-Ramirez, J. M. 1984. Upper Proterozoic and Cambrian rocks in the Caborca region, Sonora, Mexico—Physical stratigraphy, Biostratigraphy, paleocurrent studies, and regional ralations. *U.S. Geological Survey Professional Paper* 1039: 1–36.

Stewart, K. D. and Mattox, K. R. 1980. Phylogeny of phytoflagellates. In: E. R. Cox (Ed.), *Phytoflagellates* (Elsevier: Amsterdam), pp. 433–462.

Stewart, K. D. and Mattox, K. R. 1984. The case for a polyphyletic origin of mitochondria: morphological and molecular comparisons. *Journal of Molecular Evolution* 21: 54–57.

St Jean, J. 1973. A new Cambrian trilobite from the Piedmont of North Carolina. *American Journal of Science* 273-A: 196–216.

Stockner, J. G. 1968. Algal growth and primary productivity in a thermal stream. *Journal of the Fisheries Research Board of Canada* 25: 2037–2058.

Stockwell, C. H. 1961. Structural provinces, orogenies, and time classification of rocks of the Canadian Precambrian Shield. In: J. A. Lowden (Ed.), *Age Determinations by the Geological Survey of Canada. Geological Survey of Canada, Paper* 61-17: 108–118.

Stockwell, C. H., McGlynn, J. C., Emslie, R. F., Stanford, B. V., Norris, A. W. and Donaldson, J. A. 1970. Geology of the Canadian Shield. In: R. J. W. Douglas (Ed.), *Geology and Economic Minerals of Canada, Dept. Mines, Energy, Resour., Econ. Geol. Rept 1* (Geological Survey of Canada: Ottawa), pp. 45–150.

Stöcklein, L., Ludwig, W., Schleifer, K. H. and Stackebrandt, E. 1983. Comparative oligonucleotide cataloguing of 18S ribosomal RNA in phylogenetic studies of eukaryotes. In: U. Jensen and D. E. Fairbrothers (Eds.), *Proteins and Nucleic Acids in Plant Systematics* (Springer-Verlag: Berlin), pp. 58–62.

Störmer, L. 1956. A Lower Cambrian merostome from Sweden. *Arckiv für Zoologi* 9: 507–514.

Stoeser, D. B. and Camp, V. E. 1985. Pan-African microplate accretion of the Arabian Shield. *Geological Society of America Bulletin* 96: 817–826.

Stolz, J. F. 1984. Fine structure of the stratified microbial community at Laguna Figueroa, Baja California, Mexico: II. Transmission electron microscopy as a diagnostic tool in studying microbial communities *in situ*. In: Y. Cohen, R. W. Castenholz and H. O. Halvorson (Eds.), *Microbial Mats: Stromatolites* (Alan R. Liss Inc.: New York), pp. 23–38.

Stolz, J. F. and Margulis, L. 1984. The stratified microbial community at Laguna Figueroa, Baja California, Mexico: a possible model for Pre-phanerozoic laminated microbial communities preserved in cherts. *Origins of Life* 14: 671–679.

St-Onge, M. R., Lucas, M. R., Scott, D. J. and Begin, N. J. 1988. Thin-skinned imbrication and subsequent thick-skinned folding of rift-fill, transitional crust, and ophiolite suites in the 1.9 Ga Cape Smith Belt, northern Quebec. *Current research, Part A. Geological Survey*

of Canada, Paper 88-1C: 1–18.
Stookey, L. L. 1970. Ferrozine—A new spectrophotometric reagent for iron. *Analytical Chemistry* 42: 779–781.
Stowe, C. W. 1984. The early Archaean Selukwe nappe, Zimbabwe. In: A. Kröner and R. Greiling (Eds.), *Precambrian Tectonics Illustrated* (E. Schweizerbart'sche Verlagsbuchhandlung: Stuttgart, Germany), pp. 41–56.
Strathmann, R. R. 1978. Progressive vacating of adaptive types during the Phanerozoic. *Evolution* 32: 907–914.
Strathmann, R. R. and Slatkin, M. 1983. The improbability of animal phyla with few species. *Paleobiology* 9: 97–106.
Stribling, R. and Miller, S. L. 1987. Energy yields for hydrogen cyanide and formaldehyde syntheses: The HCN and amino acid concentrations in the primitive ocean. *Origins of Life* 17: 261–273.
Stringer, S. L., Hudson, K., Blaise, M. A., Walzer, P. D., Cushion, M. T. and Stringer, J. R. 1989. Sequence from ribosomal RNA of *Pneumocystis carinii* compared to those of four fungi suggests an ascomycetous affinity. *Journal of Protozoology* 36: 14s–16s.
Stupavsky, M., Symons, D. T. A. and Gravenor, C. P. 1982. Caledonian remagnetization of the Dalradian tillite, Scotland: Evidence against world-wide late Precambrian glaciation. *EOS* 63: 306.
Summons, R. 1988. Biomarkers: molecular fossils. *Short Courses in Paleontology* 1: 98–113.
Summons, R. E. 1987. Branched alkanes from ancient and modern sediments: isomer discrimination by GC/MS with multiple reaction monitoring. *Organic Geochemistry* 11: 281–290.
Summons, R. E. and Capon, R. J. 1988. Fossil steranes with unprecedented methylation in ring-A. *Geochimica et Cosmochimica Acta* 52: 2733–2736.
Summons, R. E. and Powell, T. G. 1986. *Chlorobiaceae* in Palaeozoic seas revealed by biological markers, isotopes, and geology. *Nature* 319: 763–765.
Summons, R. E. and Powell, T. G. 1987. Identification of aryl isoprenoids in source rocks and crude oils: Biological markers for the green sulphur bacteria. *Geochimica et Cosmochimica Acta* 51: 557–566.
Summons, R. E. and Powell, T. G. In Press-a. Hydrocarbon composition and the depositional environment of source for the Late Proterozoic oils of the Siberian Platform. In: *Early Organic Evolution and Mineral and Energy Resources* (Springer-Verlag: Berlin).
Summons, R. E. and Powell, T. G. In Press-b. Petroleum source rocks of the Amadeus Basin. In: R. J. Korsch (Ed.), *Geological and Geophysical Studies in the Amadeus Basin, Central Australia* (Bureau Mineral Resources Bulletin: Canberra).
Summons, R. E., Volkman, J. K. and Boreham, C. J. 1987. Dinosterane and other steroidal hydrocarbons of dinoflagellate origin in sediments and petroleum. *Geochimica et Cosmochimica Acta* 51: 3075–3082.
Summons, R. E., Brassel, S. C., Eglinton, G., Evans, E., Horodyski, R. J., Robinson, N. and Ward, D. M. 1988a. Distinctive hydrocarbon biomarkers from fossiliferous sediment of the Late Proterozoic Walcott Member, Chuar Group, Grand Canyon, Arizona. *Geochimica et Cosmochimica Acta* 52: 2625–2637.
Summons, R. E., Powell, T. G. and Boreham, C. J. 1988b. Petroleum geology and geochemistry of the Middle Proterozoic McArthur Basin, Northern Australia. III Composition of extractable hydrocarbons. *Geochimica et Cosmochimica Acta* 52: 1747–1763.
Sumner, D. Y., Kirschvink, J. L. and Runnegar, B. N. 1987. Soft-sediment paleomagnetic field tests of Late Precambrian Glaciogenic Sediments. *EOS* 68: 1251.
Sun, W.-G. 1986a. Are there Pre-Ediacarian metazoans? *Precambrian Research* 31: 409.
Sun, W.-G. 1986b. Late Precambrian pennatulids (sea pens) from the eastern Yangtze Gorge, China: *Paracharnia* gen. nov. *Precambrian Research* 31: 361–375.
Sun, W.-G. 1986c. Late Precambrian scyphozoan medusa *Mawsonites randallensis* sp. nov. and its significance in the Ediacara metazoan assemblage, South Australia. *Alcheringa* 10: 169–181.
Sun, W.-G. 1986d. Precambrian medusoids: the *Cyclomedusa* plexus and *Cyclomedusa*-like pseudofossils. *Precambrian Research* 31: 325–360.
Sun, W.-G. 1987a. Discussions on the age of the Liulaobei Formation. *Precambrian Research* 36: 349–352.
Sun, W.-G. 1987b. Palaeontology and biostratigraphy of Late Precambrian macroscopic colonial algae: *Chuaria* Walcott and *Tawuia* Hofmann. *Palaeontographica* 203: 109–134.
Sun, W.-G. and Hou, X.-G. 1987a. Early Cambrian medusae from Chengjiang, Yunnan, China. *Acta Palaeontologica Sinica* 26: 257–270.
Sun, W.-G. and Hou, X.-G. 1987b. Early Cambrian worms from Chengjiang, Yunnan, China: *Maotianshania* gen. nov. *Acta Palaeontologica Sinica* 26(3): 299–305.
Sun, W.-G., Wang, G.-X. and Zjou, B.-H. 1986. Macroscopic worm-like body fossils from the Upper Precambrian (900–700 Ma), Hunainan District, Anhui, China and their stratigraphic and evolutionary significance. *Precambrian Research* 31: 377–403.
Sundquist, E. T. and Broecker, W. S. 1985. *The Carbon Cycle and Atmospheric CO_2: Natural Variations Archean to Present* (American Geophysical Union: Washington), 627 pp.
Suzuki, K., Saito, K., Kawaguchi, A., Okuda, S. and Komagata, K. 1981. Occurrence of ω-cyclohexyl fatty acids in *Curtobacterium pusillum* strains. *Journal of General and Applied Microbiology* 27: 261–266.
Sweeney, R. E. and Kaplan, I. R. 1980. Stable isotope composition of dissolved sulfate and hydrogen sulfide in the Black Sea. *Marine Chemistry* 9: 145–152.
Sysoev, V. A. 1958. Nadotryad Hyolithoidea [Superorder Hyolithoidea]. *Osnovy Paleontologii, Mollyuski Golovonogie* II: 184–190.
Sysoev, V. A. 1965. *Brevilabiatus*—novyj rod khiolitov. [*Brevilabiatus*—a new genus of hyoliths]. In: V. F. Vozin (Ed.), *Paleontologiya i Biostratigrafiya Paleozojskikh i Triasovykh Otlozhenij Yakutii* (Nauka: Moscow), pp. 28–30.
Sysoev, V. A. 1966. O khiolitakh Yudomskoj svity severo-vostochnoj chasti Aldanskoj anteklizy [On the hyoliths of the Yudoma Formation in the northeastern part of the Aldan anteclise]. *Doklady Akademii Nauk SSSR* 166(4): 951–954.
Sysoev, V. A. 1968. *Stratigrafiya i Khiolity Drevnejshikh Sloev Nizhnego Kembriya Sibirskoj Platformy* [Stratigraphy and Hyoliths of the Oldest Lower Cambrian Strata on the Siberian Platform], 78 pp.
Sysoev, V. A. 1970. O novom semijstve nizhnekembrijskikh khiolitov. [On a new family of Lower Cambrian hyoliths]. In: A. K. Bobrov (Ed.), *Stratigrafiya i Paleontolgiya Proterozoya i Kembriya Vostoka Sibirskoj Platformy* (Yak. Knizhnoe Izdat.: Yakutsk), pp. 109–115.
Sysoev, V. A. 1972. Biostratigrafiya i Khiolity Ortotetsimorfi Nizhnego Kembriyya Sibirskoj Platformy [Lower Cambrian Biostratigraphy and Orthothecimorph Hyoliths from the Siberian Platform]: 1–152.
Sysoev, V. A. 1973. Khiolity nizhnej chasti lenskogo yarusa severnogo sklona Aldanskoj anteklizy [Hyoliths from the lower parts of the Lenian Stage of the Aldan anticline]. In: I. T. Zhuravleva (Ed.), *Problemy Paleontologii i Biostratigrafii Nizhnego Kembriya Sibiri i Dal'nego vos. Trudy Inst. Geol. Geofiz. Sibirsk. Otd. Akad. Nauk SSSR* 49 (Nauka: Moscow), pp. 57–68.
Sysoev, V. A. 1974. O pozdnelenskikh khiolitakh Aldanskoj anteklizy [On Late Lenian hyoliths of the Aldan anteclise]. In: I. T. Zhuravleva and A. Y. Rozanov (Eds.), *Biostratigrafiya i Paleontologiya Nizhnego Kembriya Evropy i Severnoj Azii* (Nauka: Moscow), pp. 242–248.
Szaniawski, H. 1982. Chaetognath grasping spines recognized among Cambrian protoconodonts. *Journal of Paleontology* 56: 806–810.
Szaniawski, H. 1983. Structure of protoconodont elements. *Fossils and Strata* 15: 21–27.
Szaniawski, H. 1987. Preliminary structural comparisons of protoconodont, paraconodont, and euconodont elements. In: R. J. Aldridge (Ed.), *Palaeobiology of Conodonts* (Ellis Horwood: Chichester), pp. 35–47.

References

Tabatabai, M. A. 1974. A rapid method for determination of sulfate in water samples. *Environmental Letters* 7: 237–243.

Tabuse, Y., Nishiwaki, K. and Miwa, J. 1989. Mutations in a protein kinase C homolog confer phorbel ester resistance on *Caenorhabditis elegans*. *Science* 243: 1713–1716.

Takahashi, T., Broecker, W. S., Li, Y. H. and Thurber, D. 1968. Chemical and isotopic balances for a meromictic lake. *Limnology and Oceanography* 13: 272–292.

Tanczyk, E. I., Lapointe, P., Morris, W. A. and Schimdt, P. W. 1987. A paleomagnetic study of the layered mafic intrusion at Sept.—Iles, Quebec. *Canadian Journal of Earth Sciences* 24: 1431–1438.

Tankard, A. J., Jackson, M. P. A., Eriksson, K. A., Hobday, D. K., Hunter, D. R. and Minter, W. E. L. 1982. *Crustal Evolution of Southern Africa* (Springer-Verlag: Berlin), 523 pp.

Tappan, H. 1980. *The Paleobiology of Plant Protists* (W. H. Freeman and Co.: San Francisco), 1028 pp.

Tappan, H. and Loeblich, A. R., Jr. 1971. Surface sculpture of the wall in Lower Paleozoic acritarchs. *Micropaleontology* 17: 385–410.

Tarling, D. H. 1972. Another Gondwanaland. *Nature* 238: 92–93.

Tarling, D. H. 1974. A paleomagnetic study of Eocambrian tillites in Scotland. *Journal of the Geological Society of London* 130: 163–177.

Taylor, F. J. R. 1978. Problems in the development of an explicit hypothetical phylogeny of the lower eukaryotes. *Biosystems* 10: 67–89.

Taylor, F. J. R. 1979. Symbionticism revisited: a discussion of the evolutionary impact of intracellular symbioses. *Proceedings of the Royal Society of London, Series B* 202: 267–286.

Taylor, F. J. R. 1980a. Basic biological features of phytoplankton cells. In: I. Morris (Ed.), *The Physiological Ecology of Phytoplankton* (Blackwell Scientific: Oxford), pp. 3–55.

Taylor, F. J. R. 1980b. The stimulation of cell research by endosymbiotic hypotheses for the origin of eukaryote. In: W. Schwemmler and H. E. A. Schenk (Eds.), *Endocytobiology* (W. de Gruyter: Berlin), pp. 917–942.

Taylor, F. J. R. 1987. An overview of the status of evolutionary cell symbiosis theories. *Annals of the New York Academy of Sciences* 503: 1–16.

Taylor, M. E. 1966. Precambrian mollusc-like fossils from Inyo County, California. *Science* 153: 198–201.

Taylor, R. F. 1984. Bacterial triterpenoids. *Microbiological Reviews* 48: 181–198.

Taylor, S. R. and McLennan, S. M. 1985. *The Continental Crust: Its Composition and Evolution* (Blackwell Scientific Publications: Oxford), 312 pp.

Taylor, S. R., Rudnick, R. L., McLennan, S. M. and Eriksson, K. A. 1986. Rare earth element patterns in Archean high-grade metasediments and their tectonic significance. *Geochimica et Cosmochimica Acta* 50: 2267–2279.

Taylor, W. R. 1957. *Marine Algae of the Northeastern Coast of North America* (University of Michigan Press: Ann Arbor), 509 pp.

Tegtmeyer, A. R. and Kroner, A. 1987. U-Pb Zircon ages bearing on the nature of early Archaean greenstone belt evolution, Barberton Mountainland, southern Africa. *Precambrian Research* 36: 1–20.

ten Haven, H. L., de Leeuw, J. W. and Schenck, P. A. 1985. Organic geochemical studies of a Messinian evaporite basin, Northern Appennines (Italy). I: Hydrocarbon biological markers for a hypersaline environment. *Geochimica et Cosmochimica Acta* 49: 2181–2191.

ten Haven, H. L., de Leeuw, J. W., Peakman, T. M. and Maxwell, J. R. 1986. Anomalies in steroid and hopnoid maturity indices. *Geochimica et Cosmochimica Acta* 50: 853–855.

ten Haven, H. L., de Leeuw, J. W., Sinninghe Damste, J. S., Schenck, P. A., Palmer, S. E. and Zumberge, J. E. 1988. Application of biological markers in a recognition of palaeo-hypersaline environments. In: K. Kelts, A. Fleet and M. Talbot (Eds.), *Lacustrine Petroleum Source Rocks* (Blackwell: Oxford), pp. 123–140.

Termier, H. and Termier, G. 1960. L'Ediacarien, premier etage paleontologique. *Revue Generale des Science Pures et Bulletin de l'Association Francaise pour l'Avancement des Sciences* 67(3–4): 79–87.

Teshima, S. and Kanazawa, A. 1972. Occurrence of sterols in the blue-green alga, *Anabaena cylindrica*. *Bulletin of Japan Society Scientific Fisheries* 38: 1197–1202.

Tester, M. and Morris, C. 1987. The penetration of light through soil. *Plant, Cell and Environment* 10: 281–286.

Thaxton, C. B., Bradley, W. L. and Olsen, R. 1984. *The Mystery of Life Origin: Reassessing Current Theories* (Philosophical Library: New York), 228 pp.

Thode, H. G. and Monster, J. 1965. Sulfur-isotope geochemistry of petroleum, evaporites, and ancient seas. *American Association of Petroleum Geologists Review* 4: 367–377.

Thode, H. G., Monster, J. and Dunford, H. B. 1961. Sulphur isotope geochemistry. *Geochimica et Cosmochimica Acta* 25: 159–174.

Thom, J. H. 1975a. Kimberley region. *Geology of Western Australia. Geological Survey of Western Australia Memoirs* 2: 160–193.

Thom, J. H. 1975b. Proterozoic galciogene rocks. *Geology of Western Australia. Geological Survey of Western Australia Memoirs* 2: 211–216.

Thomas, R. C. 1978. *Ion-sensitive Intracellular Microelectrodes. How to Make and Use Them* (Academic Press: London), 110 pp.

Thompson, J. B., Mullins, H. T., Newton, C. R. and Vercoutere, T. L. 1985. Alternative biofacies model for dysaerobic communities. *Lethaia* 18: 167–179.

Thornber, J. P. 1986. Biochemical characterization and structure of pigment-proteins of photosynthetic organisms. In: L. A. Staehelin and C. J. Arntzen (Eds.), *Encyclopedia of Plant Physiology, New Series. Photosynthesis III*, vol. 19 (Springer-Verlag: Berlin), pp. 98–142.

Thurston, P. C., Ayres, L. D., Edwards, G. R., Gelinas, L., Ludden, J. N. and Verpaelst, P. 1985. Archean biomodal volcanism. In: L. D. Ayres, P. C. Thurston, K. D. Card and W. Weber (Eds.), *Evolution of Archean Supracrustal Sequence. Geological Association of Canada Special Paper* 28: 7–22.

Tibbetts, P. J. C., Maxwell, J. R. and Golubić, S. 1978. Carotenoids in an algal mat and ooze from Laguna Mormona, Baja California. In: W. E. Krumbein (Ed.), *Environmental Biogeochemistry and Geomicrobiology*, vol. 1 (Ann Arbor Science Publ. Inc.: Ann Arbor), pp. 271–284.

Tietjen, J. H. 1967. Observations of the ecology of the marine nematode *Monhystera filicandata* Allgen, 1929. *Transactions of the American Microscopical Society* 86: 304–306.

Tietjen, J. H. and Lee, J. J. 1973. Life history and feeding habits of the marine nematode, *Chromadora macrolaimoides* Steiner. *Oecologia* 12: 303–314.

Tiffney, B. H. 1981. Diversity and major events in the evolution of land plants. In: K. J. Niklas (Ed.), *Paleobotany, Paleoecology, and Evolution*, vol. 2 (Praeger: New York), pp. 193–230.

Timofeev, B. V. 1957. O novoj gruppe iskopaemykh spor [On a new group of fossil spores]. *Ezhegodnik Vsesoyuznogo Paleontologicheskogo Obshchestva* 16: 281–285.

Timofeev, B. V. 1958. Spori Proterozoiskikh i rannego stratigraficheskoe znachenie [Spores of the Proterozoic and early Paleozoic deposits of East Siberia and their stratigraphic significance]. *Trudy Mezdu. Sovesc. Raz. Unific. Strat. Sibu*: 226–230 (In Russian).

Timofeev, B. V. 1959. Drevnejshaya flora Pribaltiki i ee stratigraficheskoe znachenie [The most ancient Baltic flora and its significance]. *Trudy Vsesoyuznogo Neftyanogo Naucho Issledovatel'skogo Geologorazvedochnogo Instituta (VNIGRI)* 129: 1–320.

Timofeev, B. V. 1960. Spori dokembriya [Precambrian spores]. In: *Precambrian Stratigraphy and Correlations. 21st Internat. Geol. Cong., Reports of Soviet Geologists, Problem IX* (Izdat. Akad. Nauk SSSR: Moscow), pp. 138–147 (In Russian with English summary).

Timofeev, B. V. 1966. *Mikropaleofitologicheskoe Issledovanie Drevnikh Svit [Micropaleontological Investigations of Ancient Formations]* (Nauka: Moscow), 147 pp.

Timofeev, B. V. 1969. *Sferomorfidy Proterozoya* [*Proterozoic Sphaeromorphida*] (Nauka: Leningrad), 146 pp.

Timofeev, B. V. 1973. *Mikrofitofossilii Dokembriya Ukrainy* [*Microphytofossils from the Precambrian of the Ukraine*] (Nauka: Leningrad), 100 pp.

Timofeev, B. V. and German, T. N. 1976. Verkhnerifejskaya flora r. Mai [Upper Riphean flora of the Maya River]. In: B. Timofeev V, T. N. German and N. S. Mikhajlova (Eds.), *Mikrofitofossilii Dokembriya, Kembriya i Ordovika* [*Microphytofossils of the Precambrian, Cambrian, and Ordovician*] (Nauka: Lenigrad), pp. 44–53.

Timofeev, B. V. and German, T. N. 1979. Dokembriyskaya mikrobiota Lakhandinsloy svity [Precambrian microbiota of the Lakhanda Formation]. In: B. S. Sokolov (Ed.), *Paleontolgiya Dokembriya i Rannego Kembriya* (Nauk: Lenigrad), pp. 137–147.

Timofeev, B., V 1970. Sphaeromorphida géants dans le Precambrien avancé. *Review of Palaeobotany and Palynology* 10: 157–160.

Tissot, B. P. and Welte, D. H. 1984. *Petroleum Formation and Occurrence*, 2nd ed. (Springer-Verlag: Berlin), 699 pp.

Titorenko, T. N. 1974. Novye vidy vodoroslei poda *Renalcis* Vologdin iz nizhnego kembriya Yuga Sibirskoj platformy [New species of algae of the genus *Renalcis* Vologdin from the Lower Cambrian of the south Siberian Platform]. In: A. Y. R. I. T. Zhuravleva (Ed.), *Biostratigrafiya i Paleontologiya Nizhnego Kembriya Evropy i Severnoj Azii* (Nauka: Moscow), pp. 216–218.

Titorenko, T. N. 1986. K. voprosy o sistematike vodoroslej poda *Renalcis* Vologdin, 1932 [Problems and systematics of the algal genus *Renalcis* Vologdin, 1932]. In: B. S. Sokolov (Ed.), *Aktual'nye Voprosy Sovremenoj Paleoal'gologii* [*Current Questions in Contemporary Paleoalgology*] (Naukova Dumka: Kiev), pp. 109–111.

Titorenko, T. N. and Drobkova, E. L. 1974. Biostratigrafiya venda i nizhnego kembriya vostochnoj chasti Irkutskogo amfiteatra po vodoroslyam i mikrofitolitam [Biostratigraphy of the Vendian and Lower Cambrian in the eastern part of the Irkutsk Ampitheater according to the algae and microphytoliths]. *Geologiya i Geofizika* 15(2): 30–39.

Tomas, R. N. and Cox, E. R. 1973. Observations on the symbiosis of *Peridinium balticum* and its intracellular alga. *Journal of Phycology* 9: 304–323.

Tong H.-W. 1987. [Fossil Phosphatocopida from Lower Cambrian of China]. *Acta Micropalaeontologica Sinica* 4: 427–437.

Torell, O. M. 1870. Pertifacta suecana formationis Cambricae. *Lunds Universitat Årsskrift* 6(2): 1–14.

Towe, K. M. 1970. Oxygen-collagen priority and the early metazoan fossil record. *Proceedings of the National Academy of Science USA* 65: 781–788.

Towe, K. M. 1981. Biochemical keys to the emergence of complex life. In: J. Billingham (Ed.), *Life in the Universe* (Massachusetts Institute of Technology Press: Cambridge), pp. 297–306.

Towe, K. M. 1983. Precambrian atmospheric oxygen and banded iron-formations: a delayed ocean model. *Precambrian Research* 20: 161–170.

Treibs, A. 1934. Chlorophyll and heavier derivatives in bituminous rocks, petroleum, mineral waxes, and asphalts. *Annales der Chemie* 510: 42.

Treloar, P. J. 1988. The geological evolution of the Magondi Mobile Belt, Zimbabwe. *Precambrian Research* 38: 55–73.

Trench, R. K. 1982. Physiology, ultrastructure and biochemistry of cyanelles. *Progress in Phycological Research* 1: 257–287.

Trendall, A. F. 1973. Time-distribution and type-distribution of Precambrian iron-formations in Australia. In: *Genesis of Precambrian Iron and Manganese Deposits. Proceedings of the Kiev Symposium, 1970* (UNESCO: Paris), pp. 49–57.

Trendall, A. F. 1983a. The Hamersley Basin. In: R. C. Morris (Ed.), *Iron Formations: Facts and Problems* (Elsevier: Amsterdam), pp. 69–123.

Trendall, A. F. 1983b. Introduction. In: R. C. Morris (Ed.), *Iron Formations: Facts and Problems* (Elsevier: Amsterdam), pp. 1–11.

Trendall, A. F. and Blockey, J. G. 1970. The iron-formations of the Precambrian Hamersley Group, Western Australia. *Geological Survey of Western Australia Bulletin* 119: 1–366.

Trendall, A. F. and Pepper, R. S. 1977. Chemical composition of the Brockman Iron Formation. *Australian Geological Survey Records* 1976/25 (Unpublished).

Treshchetonkova, A. A., Fajzulina, Z. K. and Shirobokov, I. M. 1982. Rastitel'nye mikrofossilii ushakovskoj svity yugozapadnogo Pribajkal'ya [Plant microfossils of the Ushakovka Formation of southwestern Pribaikal]. *Izvestiya Akademiya Nauk SSSR, seriia Geologicheskaya* 1982(5): 116–121.

Troelsen, H. and Jørgensen, B. B. 1982. Seasonal dynamics of elemental sulfur in two coastal sediments. *Estuarine, Coastal and Shelf Science* 15: 255–266.

Trueman, E. R. 1975. *The Locomotion of Soft-bodied Animals* (Edward Arnold: Bristol), 200 pp.

Trüper, H. G. and Imhoff, J. F. 1981. The Genus *Ectothiorhodospira*. In: M. P. Starr, H. Stolp, H. G. Trüper, A. Balows and H. G. Schlegel (Eds.), *The Prokaryotes: a Handbook on Habitats, Isolation and Identification of Bacteria* (Springer-Verlag: Berlin), pp. 274–276.

Trüper, H. G. and Pfennig, N. 1981. Characterization and identification of the anoxygenic phototrophic bacteria. In: M. P. Starr, H. Stolp, H. G. Trüper, A. Balows and H. G. Schlegel (Eds.), *The Prokaryotes: a Handbook on Habitats, Isolation and Identification of Bacteria* (Springer-Verlag: Berlin), pp. 299–312.

Tucker, M. E. 1982. Precambrian dolomites: petrographic and isotopic evidence that they differ from Phanerozoic dolomites. *Geology* 10: 7–12.

Tucker, M. E. 1986. Carbon isotope excursions in Precambrian/Cambrian boundary beds, Morocco. *Nature* 319: 48–50.

Tulliez, J., Bories, G., Boudène, C. and Février, C. 1975. Les hydrocarbures des agules *Spirulines*: nature, étude du devenir de l'heptadécane chez le rat et le porc. *Annales de la Nutrition et de l'Alimentation* 29: 563–572.

Turner, S., Burger-Wiersma, T., Giovannoni, S. J., Mur, L. R. and Pace, N. R. 1989. The relationship of a prochlorophyte *Prochlorothrix hollandica* to green chloroplasts. *Nature* 337: 380–382.

Twist, D. and Cheney, E. S. 1986. Evidence for the transition to an oxygen-rich atmosphere in the Rooiberg Group, South Africa—a note. *Precambrian Research* 33: 255–264.

Tyler, S. A. and Barghoorn, E. S. 1954. Occurrence of structurally preserved plants in Pre-Cambrian rocks of the Canadian Shield. *Science* 119: 606–608.

Tyler, S. A., Barghoorn, E. S. and Barret, L. P. 1957. Anthracitic coal from Precambrian Upper Huronian black shale of the Iron River dist4ict, northern Michigan. *Geological Society of America Bulletin* 68: 1293–1304.

Tynan, M. C. 1983. Coral-like microfossils from the Lower Cambrian of California. *Journal of Paleontology* 57(6): 1188–1211.

Tynni, R. 1978. Lower Cambrian fossils and acritarchs in the sedimentary rocks of Söderfjärden, western Finland. *Geological Survey of Finland Bulletin* 297: 39–81.

Tynni, R. and Donner, J. 1980. A microfossil sedimentation study of the Late Precambrian formation of Hailuoto, Finland. *Bulletin of the Geological Survey of Finland* 311: 1–27.

Tynni, R. and Hokkanen, K. 1982. Annelidien ryömimisjälkiä Lauhanvuoren hiekkakivessä. *Geologi* 7: 129–134.

Uchida, A., Ebata, S., Wada, K., Matsubara, H. and Ishida, Y. 1988. Complete amino acid sequence of ferredoxin from *Peridinium bipes* (Dinophyceae). *Journal of Biochemistry* 104: 700–705.

Ulrich, E. O. and Bassler, R. S. 1931. Cambrian bivalved Crustacea of the order Conchostraca. *United States National Museum Proceedings* 78: 1–130.

Upfold, R. L. 1984. Tufted microbial (cyanobacterial) mats from the Proterozoic Stoer Group, Scotland. *Geology Magazine* 121: 351–355.

Urbanek, A. 1986. The enigma of graptolite ancestry: lesson from a phylogenetic debate. In: A. Hoffman and M. H. Nitecki (Eds.), *Problematic Fossil Taxa* (Oxford University Press: New York), pp. 184–226.

Urbanek, A. and Mierzejewska, G. 1977. The fine structure of zooidal

References

tubes in Sabelliditida and Pogonophora with reference to their affinity. *Acta Palaeontologica Polonica* 22: 223–240.

Urbanek, A. and Mierzejewska, G. 1979. [The fine structure of zooidal tubes in Sabellida and Pogonophora]. In: B. M. Keller and A. Y. Rozanov (Eds.), *Paleotologia Verknedokembriyskikh i Kembriyskikh Otlozheniy Vostochno-Evropeyskoi Platformy* (Nauka: Moscow). [Republished in 1983 as: Soft-Bodied Metazoa and animal trace fossils in the Vendian and Early Cambrian. In: A. Urbanek and A. Yu. Rosanov (Eds.), *Upper Precambrian and Cambrian Palaeontology of the East-European Platform*, (Wydawnictwa Geologiczne Publishing House, Warsaw), 100–111].

Urbanek, A. and Mierzejewska, G. 1983. The fine structure of zooidal tubes in Sabellida and Pogonophora. In: A. Urbanek and A. Y. Rozanov (Eds.), *Upper Precambrian and Cambrian Palaeontology of the East European Platform* (Wydawnictwa Geologiczne: Warsaw), pp. 100–111.

Urbanek, A. and Rozanov, A. Y. (Eds.). 1983. *Upper Precambrian and Cambrian Palaeontology of the East-European Platform* (Publishing House Wydawnictwa Geologiczne: Warsaw), 157 pp.

Urey, H. C. 1952. *The Planets: Their Origin and Development* (Yale University Press: New Haven, Conn.), 245 pp.

Ushatinskaya, G. T. 1987. Unusual inarticulate brachiopods from the Lower Cambrian of Mongolia [Neobychmye bezzamkovye brakhiopody iz nizhnego kenbriya Mongolii]. *Paleontologicheskij Zhurnal* 1987: 62–68.

Vacelet, J. 1985. Coralline sponges and the evolution of the Porifera. In: S. Conway Morris, J. D. George, R. Gibson and H. M. Platt (Eds.), *The Origins and Relationships of Lower Invertebrates. The Systematics Association Special Volume 28* (Clarendon Press: Oxford), pp. 1–13.

Vail, P. R., Mitchum, R. M. and Thompson, S., III 1977. Seismic stratigraphy and global changes of sea level, Part 3: Relative changes of sea level from coastal onlap. *Seismic Stratigraphy–Applications to Hydrocarbon Exploration. American Association of Petroleum Geologists Memoir* 26: 63–81.

Valentin, K. and Zetsche, K. 1989. The genes of both subunit of ribulose-1,5-biophosphate carboxylase constitute an operon on the plastome of a red alga. *Current Genetics* 16: 203–209.

Valentine, J. W. 1969. Patterns of taxonomic and ecological structure of the shelf benthos during Phanerozoic time. *Palaeontology* 12: 684–709.

Valentine, J. W. 1973. *Evolutionary Paleoecology of the Marine Biosphere* (Prentice-Hall: Englewood Cliffs, New Jersey), 511 pp.

Valentine, J. W. 1977. General patterns of metazoan evolution. In: A. Hallam (Ed.), *Patterns of Evolution* (Elsevier: Amsterdam), pp. 27–57.

Valentine, J. W. 1980. Determinants of diversity in higher taxonomic categories. *Paleobiology* 6: 444–450.

Valentine, J. W. 1986. Fossil record of the origin of Bauplans and its implications. In: D. M. Raup and D. Jablonski (Eds.), *Patterns and Processes in the History of Life* (Springer-Verlag: Berlin), pp. 209–222.

Valentine, J. W. 1989a. Bilaterians of the Precambrian-Cambrian transition and the annelid-arthropod relationship. *Proceedings of the National Academy of Science USA* 86: 2272–2275.

Valentine, J. W. 1989b. How good was the fossil record? Clues from the California Pleistocene. *Paleobiology* 15: 83–94.

Valentine, J. W. and Campbell, C. A. 1975. Genetic regulation and the fossil record. *American Scientist* 63: 673–680.

Valentine, J. W. and Erwin, D. H. 1987. Interpreting great developmental experiments: The fossil record. In: R. A. Raff and E. C. Raff (Eds.), *Development as an Evolutionary Process* (Alan R. Liss, Inc.: New York), pp. 71–107.

Valentine, J. W. and Moores, E. M. 1972. Global tectonics and the fossil record. *Journal of Geology* 80: 167–184.

Val'kov, A. K. 1970. Hyolithida amginskogo yarusa srednego kembriya severo-vostoka Sibirskoj platformy [Hyolithida of the Middle Cambrian Amginian Stage of the northeastern part of the Siberian Platform]. In: A. K. Bobrov (Ed.), *Stratigrafiya i Paleontolgiya Proterozoya i Kembriya Vostoka Sibirskoj Platform* (Yak. Knizhnoe Izdat.: Yakutsk), pp. 71–90.

Val'kov, A. K. 1975. *Biostratigrafiya i Khiolity Kembriya Severo-Vostoka Sibirskojplatformy* (Nauka: Moscow), 139 pp.

Val'kov, A. K. 1982. *Biostratigrafiya Nizhnego Kembriya Vostoka Sibirskoj Platformy* (Nauka: Moscow), 91 pp.

Val'kov, A. K. 1983. Rasprostranenie drevnejshikh skeletnykh organizmov i korrelyatsiya nizhnej granitsy kembriya v yugo-vostochnoj chasti Sibirskoj platformy [Distribution of the oldest skeletal organisms and correlation of the lower boundary of the Cambrian in the Southeastern part of the Siberian Platform]. In: V. Khomentovskij V (Ed.), *Pozdnij Dokembrij i Rannij Paleozoj Sibiri–Vendskie Otlozhenya–Sb. nau. tr* (Inst. Geol. Geofiz. SOAN SSSR: Novosibirsk), pp. 37–48.

Val'kov, A. K. 1987. *Biostratigrafiya Nizhnego Kembriya Vostoka Sibirskoj Platformy (Yud.-Ol. region)* (Nauka: Moskva), 136 pp.

Val'kov, A. K. and Karlova, G. A. 1984. Fauna iz perikhodnykh vendsko-kembrijskikh sloev nizhnego techeniya r. Gonam [The fauna of the transitional Vendian-Cambrian beds in the lower reaches of the River Gonam]. In: V. Khomentovskij V (Ed.), *Stratigrafiya Pozdnego Dokembriya i Rannego Paleozoya. Srednyaya Sibir'* (Inst. Geol. Geofiz. SOAN SSSR: Novosibirsk), pp. 12–41.

Val'kov, A. K. and Sysoev, V. A. 1970. Angustiokreidy kembriya Sibiri [Cambrian angustiochreids from Siberia]. In: A. K. Bobrov (Ed.), *Stratigrafiya i Paleontologiya Proterozoya i Kembriya Vostoka Sibirskoj Platform* (Yak. Knizhnoe. Izdat.: Yakutsk), pp. 94–100.

Valley, J. W., Taylor, H. P., Jr. and O'Neil, J. R. 1986. *Stable in High Temperature Geological Processes*. In: P. H. Ribbe (Ed.), *Reviews in Mineralogy*, vol. 16 (Mineralogical Society of America: Washington), 570 pp.

Van Alstine, M. E. and Gillett, S. L. 1979. Paleomagnetism of Upper Precambrian-Cambrian sedimentary rocks from the Desert Range, Nevada. *Journal of Geophysical Research* 84: 4490–4500.

Van Baalen, C., Hoare, D. S. and Brandt, E. 1971. Heterotrophic growth of blue-green algae in dim light. *Journal of Bacteriology* 105: 685–689.

Van den Eynde, H., De Baere, R., De Roeck, E., Van de Peer, Y., Vandenberghe, A., Willerkens, P. and De Wachter, R. 1988. The 5S ribosomal RNA sequences of a red algal rhodoplast and a gymnosperm chloroplast. Implications for the evolution of plastids and cyanobacteria. *Journal of Molecular Evolution* 27: 126–132.

van der Land, J. and Nørrevang, A. 1985. Affinities and intraphyletic relationships of the priapulids. In: S. Conway Morris, J. D. George, R. Gibson and H. M. Platt (Eds.), *The Origins and Relationships of Lower Invertebrates. The Systematics Association Special Volume 28* (Clarendon Press: Oxford), pp. 261–273.

Van der Voo, R. 1988. Paleozoic paleogeography of North America, Gondwana, and interventing displaced terranes: Comparisons of paleomagnetism with paleoclimatology and biogeographical patterns. *Geological Society of America Bulletin* 100: 311–324.

Van der Voo, R., McCabe, C. and Scotese, C. R. 1984. Was Laurentia part of an Eocambrian Supercontinent? In: R. Van der Voo, C. R. Scotese and N. Bonhommet (Eds.), *Plate Reconstruction from Paleozoic Paleomagnetism. Geodynamics Series*, V. 12 (American Geophysical Union: Washington, D.C.), pp. 131–136.

Van Dorsselaer, A., Ensminger, A., Spyckerelle, C., Dastillung, M., Sieskind, O., Arpino, P., Albrecht, P., Ourisson, G., Brooks, P. W., Gaskell, S. J., Kimble, B. J., Philip, R. P., Maxwell, J. R. and Eglinton, G. 1974. Degraded and extended hopane derivatives (C_{27}–C_{35}) as ubiquitous geochemical markers. *Tetrahedron Letters* 1974(14): 1349–1352.

van Gemerden, H., Tughan, C. S., de Wit, R. and Herbert, R. A. 1989. Laminated microbial ecosystems on sheltered beaches in Scapa Flow, Orkney Islands. *FEMS Microbiology Ecology* 62: 87–102.

Vanguestaine, M. 1974. Espèces zonales d'Acritarches du Cambro-Trémadocien de Belgique et de l'Ardenne francaise. *Review of Palaeobotany and Palynology* 18: 63–82.

Van Houten, F. B. 1973. Origin of red beds. A review, 1961–1972.

Annual Review of Earth and Planetary Sciences 1: 39–61.
Van Schmus, R. V. 1965. The geochronology of the Blind River—Bruce Mines area, Ontario, Canada. *Journal of Geology* 73(5): 755–780.
Van Schmus, W. R. 1980. Chronology of igneous rocks associated with the Penokean Orogeny in Wisconsin. *Geological Society of America Special Paper* 182: 159–168.
Van Schmus, W. R. and Hinze, W. J. 1985. The mid-continent rift system. *Annual Review of Earth and Planetary Sciences* 13: 345–384.
Van Schmus, W. R., Bickford, M. E. and Zietz, I. 1987. Early and Middle Proterozoic provinces in the central United States. In: A. Kröner (Ed.), *Proterozoic Lithospheric Evolution. American Geophysical Union Geodynamics Series* 17: 43–68.
Van Valen, L. M. and Maiorana, V. C. 1980. The Archaebacteria and eukaryotic origins. *Nature* 287: 248–250.
Vanyo, J. and Awramik, S. 1982. Length of day and obliquity of the ecliptic 850 MA ago: Preliminary results of a stromatolite growth model. *Geophysics Research Letters* 9: 1124–1128.
Vanyo, J. and Awramik, S. 1985. Stromatolites and earth-sun-moon dynamics. *Precambrian Research* 29: 121–142.
Vanyo, J. P., Hutchinson, R. A. and Awramik, S. M. 1986. Heliotropism in microbial stromatolitic growths at Yellowstone National Park: Geophysical inferences. *EOS* 67: 153–156.
Vasil'ev, B. D., Ivankin, G. A., Koptev, I. I., Nomkonov, V. E., Radugin, K. B. Shipitzyn, V. A. 1968. K probleme raschleneniya verkhnedokembriyskikh otlozheniy Sayan-Altayskoy oblasti i ikh korrelyatzii po ostakam *Newlandia* i drugikh okhamenelostey. In: *Itogi Issledovanity po Gelogii i Georgraffi za 50 Let 1917—1967* (Tomsk University: Tomsk), pp. 84–93.
Vasil'eva, N. I. 1985. K sistematike otryada chancelloriida Walcott, 1920 (incertae sedis) iz nizhnekembrijskijh otlozhenij vostoka Sibirskoj platformy [On the systematics of the order Chancelloriida Walcott, 1920 (incertae sedis) from the Lower Cambrian deposits of the western Siberian Platform]. In: B. S. Sokolov and I. T. Zhuravleva (Eds.), *Problematiki Pozdnego Dokembriya i Paleozoya. Trudy Inst. Geol. Geofiz. Sibirsk. Otd. Akad. Nauk. SSSR*, 632 (Nauka: Novosibirsk), pp. 115–126.
Vasil'eva, N. I. 1986. Novyj rod anabaritid iz nizhnego kembriya Sibirskoj platformy [A new genus of anabaritids from the lower Cambrian of the Siberian Platform]. *Paleontologicheskij Zhurnal* 1986(2): 103–104.
Vasil'eva, N. I. and Sayutina, T. A. 1988. Morfologicheskie raznoobrazie skleritov khantsellorij. [The morphological diversity of chancelloriid sclerites]. In: I, T. Zhuravleva and L. N. Repina (Eds.), *Kembrij Sibiri i Srednej Azii. Trudy Inst. Geol. Geofiz. Sibirsk. Otd. Akad. Nauk SSSR*, 720, vol. 720 (Nauka: Novosibirsk), pp. 190–198.
Veizer, J. 1976a. Evolution of ores of sedimentary affiliation through geologic history: relations to the general tendencies in evolution of the crust. In: K. H. Wolf (Ed.), *Handbook of Stratabound and Stratiform Ore Deposits*, vol. 3 (Elsevier: Amsterdam), pp. 1–41.
Veizer, J. 1976b. Sr_{87}/Sr_{86} evolution of seawater during geologic history and its significance as an index of crustal evolution. In: B. F. Windley (Ed.), *The Early History of the Earth* (John Wiley & Sons: New York), pp. 569–578.
Veizer, J. 1983. Geologic Evolution of the Archean-Early Proterozoic Earth. Chapter 10. In: J. W. Schopf (Ed.), *Earth's Earliest Biosphere: Its Origin and Evolution* (Princeton University Press: Princeton), pp. 240–259.
Veizer, J. 1988a. Continental growth: Comments on "The Archean-Proterozoic transition: evidence from Guyana and Montana" by A. K. Gibbs, C. W. Montgomery, P. A. O'Day and E. A. Erslev. *Geochimica et Cosmochimica Acta* 52: 789–792.
Veizer, J. 1988b. The Earth and its life: systems perspective. *Origins of Life* 18: 13–39.
Veizer, J. 1988c. The evolving exogenic cycle. In: C. B. Gregor, R. M. Garrels, F. T. MacKenzie and J. B. Maynard (Eds.), *Chemical Cycles in the Evolution of the Earth* (John Wiley & Sons: New York), pp. 175–219.
Veizer, J. and Hoefs, J. 1976. The nature of O^{18}/O^{16} and C^{13}/C^{12} secular trends in sedimentary carbonate rocks. *Geochimica et Cosmochimica Acta* 40: 1387–1397.
Veizer, J. and Jansen, S. L. 1979. Basement and sedimentary recycling and continental evolution. *Journal of Geology* 87: 341–370.
Veizer, J. and Jansen, S. L. 1985. Basement and sedimentary recycling—2: time dimension to global tectonics. *Journal of Geology* 93: 625–643.
Veizer, J., Holser, W. T. and Wilgus, C. K. 1980. Correlation of $^{13}C/^{12}C$ and $^{34}S/^{32}S$ secular variations. *Geochimica et Cosmochimica Acta* 44: 579–587.
Veizer, J., Compston, W., Hoefs, J. and Nielson, H. 1982. Mantle buffering of the early oceans. *Naturwissenschaften* 69: 173–180.
Veizer, J., Compston, W., Clauer, N. and Schidlowski, M. 1983. $^{87}Sr/^{86}Sr$ in late proterozoic carbonates: evidence for a "mantle" event at ~900 Ma ago. *Geochimica et Cosmochimica Acta* 47: 295–302.
Veizer, J., Fritz, P. and Jones, B. 1986. Geochemistry of brachiopods: oxygen and carbon isotopic records of Paleozoic oceans. *Geochimica et Cosmochimica Acta* 50: 1679–1696.
Velikanov, V. A., Aseeva, E. A. and Fedonkin, M. A. 1983. *Vend Ukrainy* [*Vendian of the Ukraine*] (Naukova Dumka: Kiev), 164 pp.
Venkatakrishnan, R. and Culver, S. J. 1988. Plate boundaries in West Africa and their implications for Pangean continental fit. *Geology* 16: 322–325.
Vermeij, G. J. 1987. *Evolution and Escalation. An Ecological History of Life* (Princeton University Press: Princeton, N.J.), 527 pp.
Vialov, O. S., Gavrilishin, V. I. and Danyi, V., V 1977. O sleadakh meduz i sposobe ikh obrazovaniya [Jellyfish tracks and their mode of formation]. *Paleontologicheskij Zhurnal* 1977(4): 123–124.
Vidal, G. 1976a. Late Precambrian acritarchs from the Eleonore Bay Group and Tillite Group in East Greenland. *Grønlands Geologiske Undersøgeise Rapport* 78: 1–19.
Vidal, G. 1976b. Late Precambrian microfossils from the Visingsö Beds in southern Sweden. *Fossils and Strata* 9: 1–57.
Vidal, G. 1979a. Acritarchs and the correlation of the Upper Proterozoic. *Publications from the Institutes of Mineralogy, Paleontology and Quaternary Geology, University of Lund, Sweden* 219: 1–22.
Vidal, G. 1979b. Acritarchs from the Upper Proterozoic and Lower Cambrian of East Greenland. *Grønlands Geologiske Undersøgeise Rapport* 134: 1–55.
Vidal, G. 1981a. Aspects of problematic acid-resistant, organic-walled microfossils (acritarchs) in the Upper Proterozoic of the North Atlantic region. *Precambrian Research* 15: 9–23.
Vidal, G. 1981b. Lower Cambrian acritarch stratigraphy in Scandinavia. *Geologiska Föreningens i Stockholm Förhandlingar* 103: 183–192.
Vidal, G. 1981c. Micropaleontology and biostratigraphy of the Upper Proterozoic and Lower Cambrian Sequence in East Finnmark, northern Norway. *Norges Geologiske Undersoekelse* 362: 1–53.
Vidal, G. 1989. Are late Proterozoic carbonaceous megafossils metaphytic algae or bacteria? *Lethaia* 22: 375–379.
Vidal, G. and Ford, T. 1985. Microbiotas from the Late Proterozoic Chuar Group (northern Arizona) and Unita Mountain Group (Utah) and their chronostratigraphic implications. *Precambrian Research* 28: 349–389.
Vidal, G. and Knoll, A. H. 1982. Radiations and extinctions of plankton in the Late Precambrian and Early Cambrian. *Nature* 297: 57–60.
Vidal, G. and Knoll, A. H. 1983. Proterozoic plankton. *Geological Society of America Memoir* 161: 265–277.
Vidal, G. and Moczydłowska, M. 1987. Further reflections on metazoan evolution. *Precambrian Research* 36(3–4): 345–348.
Vidal, G. and Siedlecka, A. 1983. Planktonic, acid-resistant microfossils from the Upper Proterozoic strata of the Barents Sea Region of Varanger Peninsula, East Finnmark, northern Norway. *Norges Geologiske Undersoekelse* 382: 45–79.
Viljoen, M. J. and Viljoen, R. P. 1969. The geology and geochemistry of the lower ultramafic unit of the Onverwacht Group and a proposed new class of igneous rocks. *Geological Society of South Africa Special Publication* 2: 55–86.

References

Viljoen, R. P. and Viljoen, M. J. 1969a. The geological and geochemical significance of the upper formations of the Onverwacht Group. *Geological Society of South Africa Special Publication* 2: 113–152.

Viljoen, R. P. and Viljoen, M. J. 1969b. An Introduction to the geology of the Barberton granite-greenstone terrane. *Geological Society of South Africa Special Publication* 2: 8–28.

Villanueva, E., Luehrsen, K., Gibson, J., Delihas, N. and Fox, G. E. 1985. Phylogenetic origins of the plant mitochondrion based on a comparative analysis of 5S ribosomal RNA sequences. *Journal of Molecular Evolution* 22: 46–52.

Vincent, W. F. and Howard-Williams, C. 1986. Antarctic stream ecosystems: physiological ecology of a blue-green algal epilithon. *Freshwater Biology* 16: 219–233.

Visser, J. N. J. 1971. The deposition of the Griquatown glacial member in the Transvaal Supergroup. *Geological Society of South Africa Transactions* 74: 187–199.

Vodanyuk, S. A. 1989. Ostaki besskeletiykh metazoa iz Khatyspytskoy svity Oleneskogo podnyatiya [Fossils of soft-bodied Metazoa from the Khatspyt Suite of the Olenek Uplift]. In: V. Khomentovskiy V and Y. K. Sovetov (Eds.), *Pozdiiy Dokembriy i Ranniy Paleozoy Sibiri. Aktualnye Voprosy Stratigrafiy* [Late Precambrian and Early Paleozoic of Siberia. Current Problems in Stratigraphy]. *Inst. Geol. Geofiz. Sibirsk. Otd. Akad. Nauk. SSSR* (Nauka: Novosibirsk), pp. 61–74.

Vogel, D. E. 1975. Precambrian weathering in acid metavolcanic rocks from the Superior Province, Villebon Township, south-central Quebec. *Canadian Journal of Earth Sciences* 12: 2080–2085.

Vogel, S. 1981. *Life in Moving Fluids* (Princeton University Press: Princeton), 352 pp.

Vogelmann, T. C. and Björn, L. O. 1984. Measurement of light gradients and spectral regime in plant tissue with a fiber optic probe. *Physiologia Plantarum* 60: 361–368.

Vogelmann, T. C., Knapp, A. K., McClean, T. M. and Smith, W. K. 1988. Measurement of light within thin tissues with fiber optic microprobes. *Physiologia Plantarum* 72: 623–630.

Volkman, J. K. 1986. A review of sterol markers for marine and terrigenous organic matter. *Organic Geochemistry* 9: 83–99.

Volkman, J. K. and Maxwell, J. R. 1986. Acrylic isoprenoids as biological markers. In: R. B. Johns (Ed.), *Biological Markers in the Sedimentary Record* (Elsevier: Amsterdam), pp. 1–42.

Volkman, J. K., Alexander, R., Kagi, R. I. and Rullkötter, J. 1983. GC-MS characterisation of C_{27} and C_{28} triterpanes in sediments and petroleum. *Geochimica et Cosmochimica Acta* 47: 1033–1040.

Volkman, J. K., Burton, H. R., Everitt, D. A. and Allen, D. I. 1987a. Pigment and lipid compositions of algal and bacterial communities in Ace Lake, Vestfold Hills, Antarctica. *Hydrobiologia* 165: 41–57.

Volkman, J. K., Farrington, J. W. and Gagosian, R. B. 1987b. Marine and terrigenous lipids in coastal sediments from the Peru upwelling region at 15°S: Sterols and triterpene alcohols. *Organic Geochemistry* 11: 463–467.

Volkova, N. A. 1968. Akritarkhi dokembrijskikh i nizhnekembrijskikh otlozhenij Ehstonii [Acritarchs of the Precambrian and Lower Cambrian deposits of Estonia]. In: B. M. Keller (Ed.), *Problematiki Pogranichnykh Sloev Rifeya i Kembriya Russkoj Platformy, Urala i Kazakhstana* [Problematics of Riphean and Cambrian Boundary Layers of the Russian Platform, Urals and Kazakhstan]. *Trudy Geol. Inst. Akad. Nauk SSSR* 188 (Nauka: Moscow), pp. 8–36.

Volkova, N. A. 1969. Akritarkhi severo-zapada Russkoj platformy [Acritarchs of the northwest Russian Platform]. In: M. E. Raaben (Ed.), *Tommotskij Yarus i Problema Nizhnej Granitsy Kembriya* [The Tommotian Stage and the Cambrian Lower Boundary Problem]. *Trudy Geol. Inst. Akad. Nauk SSSR* 206 (Nauka: Moscow), pp. 224–236.

Volkova, N. A. 1981. Akritarkhi verkhnego dokembriya Yugo-Vostochnoj Sibiri (ust'kirbinskaya svita) [Acritarchs of the Upper Precambrian of southeastern Siberia (Ust'kirba Formation)]. *Byulleten' Moskovskogo Obshchestva Ispytatelej Prirody, Otdel Geologicheskij* 56(4): 66–75.

Volkova, N. A., Kir'yanov, V. V., Piskun, L. V., Pashkyavichene, L. T. and Yankauskas, T. V. 1979. Rastitel'nye mikrofossilii [Plant microfossils]. In: B. M. Keller and A. Y. Rozanov (Eds.), *Paleontologiya Verkhnedokembrijskikh i Kembrijskikh Otlozhenij Vostochno-Evropejskoj Platformy* [Upper Precambrian and Cambrian Paleontology of the East European Platform] (Nauka: Moscow), pp. 4–38.

Vologdin, A. G. 1931. *Arkheotsiaty Sibiri. Vyupusk 1. Fauna i Flora Izvestnyakov Rajona d. Kameshki i ul. Bej-Buluk Minusinsko-Khakasskogo Kraya i Okamanelosti Izvestnyakov s r. Nizhnej Tersi Kuznetskogo Okruga* [Archaeocyathids of Siberia. Part 1. Fauna and Flora from Limestones near Kameshki Village and Ulus Bej-Buluk, Minusinsk-Khakassia Region and from Fossiliferous Limestone from the Nizhne Ters R., Kuznetsk District] (Geologicheskoe Izdatel'stvo: Moscow), 119 pp.

Vologdin, A. G. 1932. *Arkheotsiaty Sibiri. Vyupusk 2. Fauna Kembrijskikh Izvestnyakov Altaya* [The Archaeocyathinae of Siberia. Part 2. Fauna of the Cambrian Limestones of the Altai Mountains] (Gos Nauchno-Tekh Geol-Raz Izd: Moscow), 106 pp.

Vologdin, A. G. 1937. Arkheotsiaty i vodorosli yuzhnogo sklona Anabarskogo massiva [Archaeocyathids and algae from the southern slopes of the Anabar Massif]. *Trudy Vsesoyuz. Arkticheskij Inst.* 91: 9–66.

Vologdin, A. G. 1939. Arkheotsiaty i vodorosli srednego kembriya Yuzhnogo Urala [Archaeocyathids and algae from the Middle Cambrian of the southern Urals]. *Problemy Paleontologii* 5: 209–276.

Vologdin, A. G. 1940. *Arkheotsiaty i Vodorosli Kemvrijskikh Izvestnyakov Mongolii i Tuvy* [Cambrian Archaeocyathids and Algae from Limestones of Mongolia and Tuva]. *Trudy Mongol'. Komissii, Akad. Nauk SSSR* (Akademiya Nauk SSSR: Moscow), 268 pp.

Vologdin, A. G. 1955a. Kembrijskie *Solenopora* i mollyuski Severnogo Tyan'-Shanya [Cambrian *Solenopora* and molluscs from northern Tien-Shan]. *Doklady Akademii Nauk SSSR* 105(2): 354–356.

Vologdin, A. G. 1955b. Puzyrchataya vodorosli iz verkhneproterozojskikh otlozhenij nizov'ev reki Angary [Vesicular algae from Upper Proterozoic deposits of the upper reaches of the Angar River]. *Doklady Akademii Nauk SSSR* 102(2): 355–356.

Vologdin, A. G. 1958a. Neskol'ko vidov vodoroslej iz sinijskikh i kembrijskikh otlozhenij Kitaya [Some algal species from the Sinian and Cambrian deposits of China]. *Memoirs of the Nanjing Institute of Geology and Palaeontology* 1: 1–32.

Vologdin, A. G. 1958b. Nizhnekembrijskie foraminifery Tuvy [Lower Cambrian foraminifera of Tuva]. *Doklady Akademii Nauk SSSR* 120(2): 405–408.

Vologdin, A. G. 1960. Onekotorykh rezul'tatakh izucheniya sinijskikh vodoroslej i metodike issledovaniya [On some results of studies of Sinian algae and methods of investigation]. *Acta Palaeontologica Sinica* 8(1): 1–26.

Vologdin, A. G. 1962a. *Arkheotsiaty i vodorosli kembriya Bajkal'skogo nagor'ya* [Archaeocyathids and Algae from the Cambrian from the Bajkal Highland]. In: K. B. Kordeh (Ed.), *Trudy Paleontol Inst. Akad. Nauk SSSR* 93 (Akademia Nauk SSSR: Moscow), 119 pp.

Vologdin, A. G. 1962b. Diaomoobrazani'e organizmi' kembriya khrebta Tanny-ola v Tyve. *Doklady Akademii Nauk SSSR* 146(4): 909–912.

Vologdin, A. G. 1962c. *Drevnejshie Vodorosli SSSR* [Ancient Algae of the USSR] (Akademii Nauk SSSR: Moscow), 657 pp.

Vologdin, A. G. 1964. Kribritsiaty—novyj klass arkheotsiat [Cribrycyathans—a new class of archaeocyathans]. *Doklady Akademii Nauk SSSR* 157(6): 1391–1394.

Vologdin, A. G. 1965. K otkrytiyu ostatkov vodoroslej v murandavskoj svite proterozoya Malogo Khingana (DVK) [Discovery of algal remains in the Proterozoic Murandav Formation of the Lesser Khingans (Far East Region)]. *Doklady Akademii Nauk SSSR* 164(3): 677–680 (1966, *Doklady Earth Science Sections* 164: 210–214).

Vologdin, A. G. 1966a. Kribritsiaty kembriya SSSR [Lower Cambrian cribricyathans of the USSR]. *Trudy Paleontologicheskogo Instityuta, Akademiya Nauk SSSR* 109: 1–62.

Vologdin, A. G. 1966b. Ostaki protomeduz iz nizov karagasskoy svity

Vostochnogo Sayan [Remains of protomedusae from the lower part of the Karagassk Suite, eastern Sayan]. *Doklady Akademii Nauk SSSR* 167: 434–436.

Vologdin, A. G. 1966c. Ov otkrytii ostatkov gigantskikh sifonej v drevnikh sloyakh Timanskogo krayzha [Discovery of the remains of gigantic siphonaceous algae in the ancient strata a Timan Ridge]. *Doklady Akademii Nauk SSSR* 169(3): 672–675 (1967, *Doklady Earth Science Sections* 169: 209–213).

Vologdin, A. G. 1967a. K otkrytiyu ostatkov proterozojskikh vodoroslej na Dal'nem Vostoke i na Urale [Discovery of relics of Proterozoic algae in the Far East and in the Urals]. *Doklady Akademii Nauk SSSR* 175(4): 926–928 (1968, *Doklady Earth Science Sections* 175: 214–217).

Vologdin, A. G. 1967b. Ob ostatkakh sifonej iz nizhnego kembriya khrebta Tannu-Ola (Tuva) [Relics of Syphoneae in Lower Cambrian of the Tannu-Ola Mountains, Tuva]. *Doklady Akademii Nauk SSSR* 174(4): 952–955 (1967. *Doklady Earth Science Sections* 174: 202–205).

Vologdin, A. G. and Drozdova, N. A. 1964. Iskopaemaya sinezelenaya vodorosl v pozdnedokembrijskikh otlozheniyakh Dalnego Vostoka [Cyanophycean fossils in Upper Precambrian sediments of the Far East]. *Doklady Akademii Nauk SSSR* 159(3): 576–578.

Vologdin, A. G. and Drozdova, N. A. 1969a. K otkrytiyu vodoroslej semejstva Rivulariaceae v pozdnem dokembrii [A discovery of algae of the family Rivulariaceae in the Upper Precambrian]. *Doklady Akademii Nauk SSSR* 187(5): 1162–1163 (1970, *Doklady Earth Science Sections* 187: 220–222).

Vologdin, A. G. and Drozdova, N. A. 1969b. Novye sinezelenye vodorosli dokembrijskogo vozrasta iz Batenevskogo kryazha [New Precambrian cyanophyceans from Batenev Ridge]. *Doklady Akademii Nauk SSSR* 187(2): 440–442 (1970, *Doklady Earth Science Sections* 187: 211–213).

Vologdin, A. G. and Drozdova, N. A. 1970. Novaya nakhodka drevnejshej fauny [A new find of ancient fauna]. *Doklady Akademii Nauk SSSR* 190(1): 195–197.

Vologdin, A. G. and Kordeh, K. B. 1965. Neskol'ko vidov drevnikh Cyanophyta i ikh tsenozy [Several species of ancient Cyanophyta and their coenoses]. *Doklady Akademii Nauk SSSR* 164(2): 429–432 (1966, *Doklady Earth Science Sections* 164: 207–210).

Vologdin, A. G. and Maslov, A. B. 1960. O novoj gruppe iskopaemykh organizmov iz nizov Yudomdkoj svity Sibirskoj platformy [A new group of fossil organisms from the bottom of the Yudoma Series of the Siberian Platform]. *Doklady Akademii Nauk SSSR* 134(3): 691–693 (1961, *Doklady Earth Science Sections* 134: 1031–1034).

Vologdin, A. G. and Yankauskas, T. V. 1968. Novye kribritsiaty kembriya Sibiri [New cribricyathans from the Cambrian of Siberia]. *Doklady Akademii Nauk SSSR* 183(1): 200–203.

Von Brunn, V. and Mason, T. R. 1977. Silcoclastic-carbonate tidal deposits from the 3000 M.Y. Pongola Supergroup, South Africa. *Sedimentary Geology* 18: 245–255.

Von Damm, K. L. and Bischoff, J. L. 1987. Chemistry of hydrothermal solutions from the southern Juan de Fuca Ridge. *Journal of Geophysical Research* 92: 11, 334–11,346.

Von Damm, K. L., Edmond, J. M., Grant, B., Measures, C. I., Walden, B. and Weiss, R. F. 1985. Chemistry of submarine hydrothermal solutions at 21°N, East Pacific Rise. *Geochimica et Cosmochimica Acta* 49: 2197–2220.

von der Borch, C. C., Christie-Blick, N. and Grady, A. E. 1988. Depositional sequence analysis applied to late Proterozoic Wilpena Group, Adelaide Geosyncline, South Australia. *Australian Journal of Earth Sciences* 35: 59–71.

von Toll, E. 1899. Beiträge zur kenntniss des sibirschen Cambrium [Contributions to the knowledge of the Cambrian of Siberia]. *Mémoires de l'Academie impériale des sciences de St. Pétersbourg. Série 8. Classe des sciences physiques et mathematiques* 8: 1–57.

Voronin, Y. I., Voronova, L. G., Grigor'eva, N. V., Drozdova, N. A., Zhegallo, E. A., Zhuravlev, A. Y., Sayutina, T. A., Sysoev, V. A. and Fodin, V. D. 1982. *Granitsa Dokembriya i Kembriya v Geosinklinal'nykh Oblastyakh* (*Opornyj Razrez Salany-Gol, MNR*) [*The Precambrian-Cambrian Boundary in Geosynclinal Areas* (*Reference Section of Salany-Gol, MPR*)]. *Sovmestnaya Sovetsko-Mongol'skaya Paleontologicheskaya Ehkspeditsiya Trudy*, 18 (Nauka: Moscow), 152 pp.

Voronin, Y. I., Voronova, L. G. and Drozdova, N. A. 1983. Arkheotsiaty i vodorosli nizhnego kembriya bassejna r. Ehgijn-Gol (Severo-Zapadnaya Mongoliya) [Archaeocyathids and algae from the Lower Cambrian of the Ehgijn-Gol River (northwest Mongolia)], *Novye vidy iskopaemykh bespozvonochnykh Mongolii*. *Trudy Sovmestnaia Sovetsko-Mongol'skaia Paleontologicheskaia Ehkspeditsiia* 20: 7–11.

Voronova, L. G. 1969. [Algae]. In: M. E. Raaben (Ed.), *Tommotski Yarus i Problema Nizhnei Granitsy Kembriya* (Nauka: Moscow), pp. 189–195 (1981, translation available: U.S. Dept. of Commerce, National Technical Information Service, Springfield, Virginia; 1981, Amerind Publishing Company, New Delhi).

Voronova, L. G. 1973. Morfologiya vodorslevykh postoek iz nizhnekembrijskikh otlozhenij nizov'eb r. Leny [Morphology of algal structures in Lower Cambrian deposits along the lower reaches of the Lena River]. In: I. T. Zhuravleva (Ed.), *Problemy Paleotologii i Biostratigrafii Nizhnego Kembriya Sibiri i Dal'nego Vos* (Nauka, Sibirskoe Otdeleni: Novosibirsk), pp. 80–84.

Voronova, L. G. 1974. Voprosy morfologii i sistematiki rannekembrijskikh izvestkovykh vodoroslej [Problems in the morphology and systematics of Early Cambrian calcareous algae]. In: I. T. Zhuravleva and A. Y. Rozanov (Eds.), *Biostratigrafiya i Paleontologiya Nizhnego Kembriya Evropy i Severnoj Azii* (Nauka: Moscow), pp. 199–215.

Voronova, L. G. 1976. Izvestkovye vodorosli pogranichnykh sloev dokembriya i kembriya Sibirskoj platformy [Calcareous algae of the Precambrian-Cambrian boundary deposits of the Siberian Platform]. In: L. G. Voronova and E. P. Radionova (Eds.), *Vodorosli i Mikrofitolity Paleozoya* [*Algae and Microphytoliths of the Paleozoic*]. *Trudy Geol. Inst. Akad. Nauk SSSR* 294 (Nauka: Moscow), pp. 1–85.

Voronova, L. G. 1979. Calcitized algae of the Precambrian and the Early Cambrian. *Centre de Recherche Exploration-Production Elf-Aquitaine, Bulletin* 3(2): 867–871.

Voronova, L. G. and Drozdova, N. A. 1986a. K voprosy o biogeografii rannego kembriya [On the problem of Early Cambrian biogeography]. In: A. Y. Rozanov (Ed.), *Problemy Paleobiogeograffi Azii* [*Problems of Paleobiogeography of Asia*]. *Trudy Sovmest. Sovet.-Mongo* 29 (Nauka: Moscow), pp. 34–55.

Voronova, L. G. and Drozdova, N. A. 1986b. O pode *Renalcis* i renal'tsisopodobnykh vodoroslyakh [On the genus *Renalcis* and renalcid-like algae]. In: B. S. Sokolov (Ed.), *Aktual'nye Voprosy Sovremennoj Paleoal'gologii* [*Current Questions in Contermporary Paleoalgology*] (Naukova Dumka: Kiev), pp. 111–114.

Voronova, L. G. and Luchinina, V. A. 1985. Izvestkovye vodorosli Nemakit-Daldynskogo Gorizonta [Calcareous algae from the Nemakit-Daldyn Horizon]. In: B. S. Sokolov (Ed.), *Vendskaya Sistema* (Nauka: Moscow), pp. 162–169.

Voronova, L. G. and Missarzhevskij, V. V. 1969. Nakhodki vodoroslej i trubok chervej v pogranichnykh sloyakh kembriya i dokembriya na severe Sibirskoj platformy [Finds of algae and worm tubes in Precambrian-Cambrian boundary beds in the north of the Siberian Platform]. *Doklady Akademii Nauk SSSR* 184(1): 207–210 (1969, *Doklady Earth Science Sections* 184: 206–209).

Voronova, L. G. and Rajding, R. 1986. Priroda nekotorykh kembrijskikh vodoroslej [Nature of some Cambrian algae]. In: B. S. Sokolov (Ed.), *Aktual'nye Voprosy Sovremennoj Paleoal'gologii* [*Current Questions in Contemporary Paleoalgology*] (Naukova Dumka: Kiev), pp. 114–116.

Voronova, L. G., Drozdova, N. A., Esakova, N. V., Zhegallo, E. A., Zhuravlev, A. Y., Rozanov, A. Y., Sayutina, T. A. and Ushatinskaya, G. T. 1987. Iskopaemye nizhnego kembriya Gor Makkenzi (Kanada) [Lower Cambrian fossils from the Mackenzie Mountains (Canada)]. *Akademiya Nauk SSSR, Trudy* 224: 1–88.

Vossbrinck, C. R., Maddox, J., V, Friedman, S., Debrunner-Vossbrinck,

References

B. A. and Woese, C. R. 1987. Ribosomal RNA sequence suggests microsporicida are extremely ancient eukaryotes. *Nature* 326: 411–414.

Vostokova, V. A. 1962. Kembrijskie gastropody Sibirskoj platformy i Tajmyra [Cambrian gastropods from the Siberian Platform and Tajmyr]. *Sbornik Stat. Paleontol. Biostr., NIIGA* 28: 51–74.

Waagen, W. H. 1885. Salt Range Fossils. Part 4(2). Brachiopoda. *Palaeont. India, Mem.* 1(Ser. 13): 729–770.

Wachter, E. and Hayes, J. M. 1985. Exchange of oxygen isotopes in carbon dioxide—phosphoric acid systems. *Chemical Geology* 52: 365–374.

Wade, M. 1968. Preservation of soft-bodied animals in Precambrian sandstones at Ediacara, South Australia. *Lethaia* 1: 238–267.

Wade, M. 1969. Medusae from uppermost Precambrian or Cambrian sandstones, central Australia. *Palaeontology* 12: 351–365.

Wade, M. 1970. The stratigraphic distribution of the Ediacara fuana in Australia. *Transactions of the Royal Society of South Australia* 94: 87–104.

Wade, M. 1971. Bilateral Precambrian chondrophores from the Ediacara fauna, South Australia. *Proceedings of the Royal Society of Victoria* 84: 183–188.

Wade, M. 1972a. *Dickinsonia*: polychaete worms from the late Precambrian Ediacara fauna, South Australia. *Memoirs of the Queensland Museum* 16: 171–190.

Wade, M. 1972b. Hydrozoa and Scyphozoa and other medusoids from the Precambrian Ediacara fauna, South Australia. *Palaeontology* 15: 197–225.

Wahlstrom, E. E. 1948. Pre-Fountain and recent weathering on Flagstaff Mountain near Boulder, Colorado. *Geological Society of America Bulletin* 59: 1173–1190.

Walcott, C. D. 1883. Pre-Carboniferous strata of the Grand Canyon of the Colorado. *American Journal of Science* 26: 437–442.

Walcott, C. D. 1885. Palaeontologic notes. *American Journal of Science* 29(Ser. 3): 114–117.

Walcott, C. D. 1886. Second contribution to the studies on the Cambrian faunas of North America. *U.S. Geological Survey Bulletin* 30: 1–369.

Walcott, C. D. 1890. Descriptive notes of new genera and species from the Lower Cambrian or *Olenellus* Zone of North America. *Proceedings of the United States National Museum* 12: 33–46.

Walcott, C. D. 1891. The fauna of the Lower Cambrian or *Olenellus* Zone. *U.S. Geological Survey Annual Report* 10: 509–774.

Walcott, C. D. 1895. Algonkian rocks of the Grand Canyon of the Colorado. *Journal of Geology* 3: 312–330.

Walcott, C. D. 1897. Cambrian Brachiopoda: Genera *Iphidea* and *Yorkia* with descriptions of new species of each and of the genus *Acrothele*. *Proceedings of the United States National Museum* 19: 707–718.

Walcott, C. D. 1899. Pre-Cambrian fossiliferous formations. *Geological Society of America Bulletin* 10: 199–244.

Walcott, C. D. 1905. Cambrian Brachiopoda with descriptions of new genera and species. *Proceedings of the United States National Museum* 28: 227–337.

Walcott, C. D. 1910. Cambrian geology and paleontology II. 1. Abrupt appearance of the Cambrian fauna of the North American Continent. *Smithsonian Miscellaneous Collections* 57: 1–16.

Walcott, C. D. 1911a. Cambrian Geology and Paleontology II. 3. Middle Cambrian holothurians and medusae. *Smithsonian Miscellaneous Collections* 57: 41–68.

Walcott, C. D. 1911b. Cambrian geology and paleontology II. Middle Cambrian annelids. *Smithsonian Miscellaneous Collections* 67: 261–364.

Walcott, C. D. 1912. Middle Cambrian Brachiopoda, Malacostraca, Trilobita, and Merostomata. *Smithsonian Miscellaneous Collections* 57: 145–228.

Walcott, C. D. 1914. Cambrian geology and paleontology III, no. 2, Pre-Cambrian Algonkian algal flora. *Smithsonian Miscellaneous Collections* 67(2): 77–156.

Walcott, C. D. 1915. Discovery of Algonkian bacteria. *Proceedings of the National Academy of Science USA* 1: 256–257.

Walcott, C. D. 1917. Cambrian Geology and Paleontology. 4.3. Fauna of the Mount Whyte Formation. *Smithsonian Miscellaneous Collections* 67: 61–113.

Walcott, C. D. 1919. Cambrian geology and paleontology IV. No. 5—Middle Cambrian algae. *Smithsonian Miscellaneous Collections* 67(5): 217–260.

Walcott, C. D. 1920. Cambrian geology and paleontology IV: 6—Middle Cambrian Spongiae. *Smithsonian Miscellaneous Collections* 67(6): 261–364.

Walcott, C. D. 1931. Addenda to descriptions of Burgess Shale fossils. *Smithsonian Miscellaneous Collections* 85: 1–46.

Walde, D. H. G., Gierth, E. and Leonardos, O. H. 1981. Stratigraphy and mineralogy of the manganese ores of Urucum, Mato Grosso, Brazil. *Geologische Rundschau* 70: 1077–1085.

Walker, J. C. G. 1977. *Evolution of the Atmosphere* (MacMillan: New York), 318 pp.

Walker, J. C. G. 1982. Climatic factors on the Archean Earth. *Palaeogeography, Palaeoclimatology, Palaeoecology* 40: 1–11.

Walker, J. C. G. 1984. Suboxic diagenesis in banded iron-formations. *Nature* 309: 340–342.

Walker, J. C. G. 1985. Carbon dioxide on the early Earth. *Origins of Life* 16: 117–127.

Walker, J. C. G. 1987. Was the Archean biosphere upside down? *Nature* 329: 710–712.

Walker, J. C. G. and Brimblecombe, P. 1985. Iron and sulfur in the Prebiologic ocean. *Precambrian Research* 28: 205–222.

Walker, J. C. G., Hays, P. B. and Kasting, J. F. 1981. A negative feedback mechanism for the long-term stabilization of earth's surface temperature. *Journal of Geophysical Research* 86: 9776–9782.

Walker, J. C. G., Klein, C., Schidlowski, M., Schopf, J. W., Stevenson, D. J. and Walter, M. R. 1983. Environmental evolution of the Archean—Early Proterozoic Earth. Chapter 11. In: J. W. Schopf (Ed.), *Earth's Earliest Biosphere: Its Origin and Evolution* (Princeton University Press: Princeton), pp. 260–290.

Walker, T. R. 1967. Formation of red beds in modern and ancient deserts. *Geological Society of America Bulletin* 78: 353–368.

Walker, T. R. 1974. Formation of red beds in moist tropical climates: a hypothesis. *Geological Society of America Bulletin* 85: 633–638.

Walker, T. R., Waugh, B. and Grone, A. J. 1978. Diagenesis in first-cycle desert alluvium of Cenozoic age, southwestern United States and northwestern Mexico. *Geological Society of America Bulletin* 89: 19–32.

Walker, W. F. 1984a. 5S rRNA sequences from Atractiellales and basidomycetous yeasts and fungi imperfecti. *Systematics and Applied Microbiology* 5: 352–359.

Walker, W. F. 1984b. 5S rRNA sequences from Zygomycotina and evolutionary implications. *Systematics and Applied Microbiology* 5: 448–456.

Walker, W. F. 1985a. 5S and 5.8S ribosomal RNA sequences and protist phylogenetics. *Biosystems* 18: 269–278.

Walker, W. F. 1985b. 5S ribosomal RNA sequences from Ascomycetes and evolutionary implications. *Systematics and Applied Microbiology* 6: 48–53.

Walker, W. F. and Doolittle, W. F. 1982a. Nucleotide sequences of ribosomal RNA from four oomycete and chytrid water moulds. *Nucleic Acids Research* 10: 5715–5721.

Walker, W. F. and Doolittle, W. F. 1982b. Redividing the basidiomycetes on the basis of 5D rRNA sequences. *Nature* 299: 723–724.

Walker, W. F. and Doolittle, W. F. 1983. 5S rRNA sequences from 8 basidiomycetes and fungi imperfecti. *Nucleic Acids Research* 11: 7625–7630.

Walker, W. F., Bode, H. R. and Steele, R. E. 1989. Phylogeny and molecular data. *Science* 243: 548–549.

Wall, D. 1962. Evidence from recent plankton regarding the biological affinities of *Tasmanites* Newton 1875 and *Leiosphaeridia* Eisenack 1958. *Geological Magazine* 99: 353–362.

Walliser, O. H. 1958. *Rhombocorniculum comleyense* n.gen., n.sp. (Incertae sedis, Unterkambrium, Shropshire). *Paläontologische Zeitschrift* 32: 176–180.

Walraven, F., Armstrong, R. A. and Kruger, F. J. In Press. A chronological framework for the middle to late Precambrian stratigraphy of the Transvaal, South Africa. *Tectonophysics*.

Walsby, A. E. 1981. Cyanobacteria: planktonic gas vacuolate forms. In: M. P. Starr, H. Stolp, H. G. Trüper, A. Balows and H. G. Schlegel (Eds.), *The Prokaryotes* (Springer-Verlag: Berlin, Heidelberg, New York), pp. 224–235.

Walsh, M. M. and Lowe, D. R. 1985. Filamentous microfossils from the 3,500-Myr-old Onverwacht Group, Barberton Mountain Land, South Africa. *Nature* 314: 530–532.

Walter, M. 1979. Precambrian glaciation. *American Scientist* 67: 142.

Walter, M. R. 1967. Archaeocyatha and the biostratigraphy of the Lower Cambrian Hawker Group, South Australia. *Journal of the Geological Society of Australia* 14(1): 139–152.

Walter, M. R. 1972a. Stromatolites and the biostratigraphy of the Australian Precambrian and Cambrian. *Paleontological Association of London Special Paper* 11: 1–190.

Walter, M. R. 1972b. Tectonically deformed sand volcanoes in a Precambrian greywacke, Northern Territory of Australia. *Journal of the Geological Society of Australia* 18: 395–399.

Walter, M. R. 1976a. Introduction. In: M. R. Walter (Ed.), *Stromatolites: Developments in Sedimentology*, vol. 20 (Elsevier: Amsterdam), pp. 1–3.

Walter, M. R. (Ed.). 1976b. *Stromatolites: Developments in Sedimentology*, vol. 20 (Elsevier: Amsterdam), 790 pp.

Walter, M. R. 1977. Interpreting stromatolites. *American Scientist* 65: 563–571.

Walter, M. R. 1980. Adelaidean and early Cambrian stratigraphy of the southwestern Georgina Basin: Correlation chart and explanatory notes. *Bureau of Mineral Resources Microform. Bureau of Mineral Resources Report* 214: MF 92.

Walter, M. R. 1983. Archean stromatolites: evidence of the earth's earliest benthos. Chapter 8. In: J. W. Schopf (Ed.), *Earth's Earliest Biosphere: Its Origin and Evolution* (Princeton University Press: Princeton), pp. 187–213.

Walter, M. R. and Bauld, J. 1986. Subtidal stromatolites of Shark Bay. *International Sedimentological Congress Abstracts* 12: 315.

Walter, M. R. and Heys, G. R. 1985. Links between the rise of Metazoa and the decline of stromatolites. *Precambrian Research* 29: 149–174.

Walter, M. R. and Hoffman, H. J. 1983. The paleontology and paleoecology of Precambrian iron-formations. In: A. F. Trendall and R. C. Morris (Eds.), *Iron-Formations: Facts and Problems* (Elsevier: Amsterdam), pp. 373–400.

Walter, M. R., Bauld, J. and Brock, T. D. 1972. Siliceous algal and bacterial stromatolites in hot spring and geyser effluents of Yellowstone National Park. *Science* 178: 402–405.

Walter, M. R., Golubić, S. and Priess, W., V 1973. Recent stromatolites from hydromagnesite and aragonite depositing lakes near the Coorong Lagoon, South Australia. *Journal of Sedimentary Petrology* 43: 1021–1030.

Walter, M. R., Bauld, J. and Brock, T. D. 1976a. Microbiology and morphogenesis of columnar stromatolites (*Conophyton, Vacerrilla*) from hot springs in Yellowstone National Park. In: M. R. Walter (Ed.), *Stromatolites: Developments in Sedimentology*, vol. 20 (Elsevier: Amsterdam), pp. 273–310.

Walter, M. R., Goode, A. D. T. and Hall, W. D. M. 1976b. Microfossils from a newly discovered Precambrian stromatolitic iron formation in Western Australia. *Nature* 261: 221–223.

Walter, M. R., Oehler, J. H. and Oehler, D. Z. 1976c. Megascopic algae 1300 million years old from the Belt Supergroup, Montana: a reinterpretation of Walcott's *Helminthoidichnites*. *Journal of Paleontology* 50(5): 872–881.

Walter, M. R., Buick, R. and Dunlop, J. S. R. 1980. Stromatolites 3,400-3,500 Myr old from the North Pole area, Western Australia. *Nature* 284: 443–445.

Walter, M. R., Elphinstone, R. and Heys, G. R. 1989. Proterozoic and Early Cambrian trace fossils from the Amadeus and Georgina Basins, central Australia. *Alcheringa* 13: 209–256.

Walter, M. R., Rulin, D. and Horodyski, R. J. In Press. Coiled carbonaceous megafossils from the Middle Proterozoic of Jixian (Tianjin) and Montana. *American Journal of Science*.

Wang, F. 1981. Sinian microfossils from south-west China. *Nature* 294: 74–76.

Wang, F. 1985. Middle-Upper Proterozoic and lowest Phanerozoic microfossil assemblages from SW China and contiguous areas. *Precambrian Research* 29: 33–43.

Wang, F. and Chen, Q. 1987. Spiniferous acritarchs from the lowest Cambrian, Emei, Sichuan, southwestern China. *Review of Palaeobotany and Palynology* 52: 161–177.

Wang, F., Zhang, X. and Guo, R. 1983. The Sinian microfossils from Jinning, Yunnan, south west China. *Precambrian Research* 23: 133–175.

Wang, F., Chen, Q. and Zhao, X. 1984. New information on Sinian acritarch from SW China and its significance. *English edition. Kexue Tongbao* 29: 656–659.

Wang, G.-X. 1982. Late Precambrian Annelida and Pognophora from the Huanian of Anhui Province. *Bulletin of the Tianjin Institute of Geology and Mineral Resources* 6: 9–22 (In Chinese).

Wang, H.-Z. (Ed.). 1985. *Atlas of the Palaeogeography of China* (Cartographic Publishing House: Beijing), 107 pp.

Wang, R., Lium, R. and Zhao, D. 1987. Precambrian stromatolite oxygen isotopes from Sichuan-Yunnan area and its significance. *Bulletin of the Chengdu Institute of Geological and Mineral Research* 8: 61–68.

Wang, W. C. and Stone, P. H. 1980. Effect of ice-albedo feedback on global sensitivity in a one-dimensional radiative-convective climate model. *Journal of Atmospheric Science* 37: 545–552.

Wang, Y., Lu, Z., Xing, Y., Gao, Z., Lin, W., Ma, G., Zhang, L. and Lu, S. 1980. Subdivision and correlation of the Upper Precambrian in China. In. *Research on Precambrian Geology. Sinian Suberathem in China* (Tianjin Science and Technology Press: Tianjin, China), pp. 1–30.

Waples, D. W. 1977. C/N ratios in source rock studies. *Mineral Industries Bulletin, Colorado School of Mines* 20(5): 1–7.

Waples, D. W., Haug, P. and Welte, D. H. 1974. Occurrence of a regular C_{25} isoprenoid hydrocarbon in Tertiary sediments representing a lagoonal, saline environment. *Geochimica et Cosmochimica Acta* 38: 381–387.

Warburton, G. A. and Zumberge, J. E. 1983. Determination of petroleum sterane distributions by mass spectrometry with selective metastable ion monitoring. *Analytical Chemistry* 55: 123–126.

Ward, B. B. and Carlucci, A. F. 1985. Marine ammonia- and nitrite-oxidizing bacteria: serological diversity determined by immuno-fluorescence in culture and in the environment. *Applied and Environmental Microbiology* 50: 194–201.

Ward, D. M. 1978. Thermophilic methanogenesis in a hot spring algal bacterial mat (71° to 30°C). *Applied and Environmental Microbiology* 35: 1019–1026.

Ward, D. M. 1989. Molecular probes for analysis of microbial communities. In: W. G. Characklis and P. A. Wilderer (Eds.), *Structure and Function of Biofilms. Life Sciences Research Reports 46* (John Wiley & Sons: Chichester), pp. 145–163.

Ward, D. M. and Olsen, G. J. 1980. Terminal processes in the anaerobic degradation of an algal-bacterial mat in a high-sulfate hot spring. *Applied and Environmental Microbiology* 40: 67–74.

Ward, D. M. and Winfrey, M. R. 1985. Interactions between methanogenic and sulfate-reducing bacteria in sediments. *Advances in Aquatic Microbiology* 3: 141–179.

Ward, D. M., Beck, E., Revsbech, N. P., Sandbeck, K. A. and Winfrey, M. R. 1984. Decomposition of hot spring microbial mats. In: Y. Cohen, R. W. Castenholz and H. O. Halvorson (Eds.), *Microbial Mats: Stromatolites* (Alan R. Liss Inc.: New York), pp. 191–214.

Ward, D. M., Brassell, S. C. and Eglinton, G. 1985. Archaebacterial lipids in hot-spring microbial mats. *Nature* 318: 656–659.

References

Ward, D. M., Tayne, T. A., Anderson, K. L. and Bateson, M. M. 1987. Community structure and interactions among community members in hot spring cyanobacterial mats. *Symposium of the Society of General Microbiology* 41: 179–210.

Ward, D. M., Shiea, J., Zeng, Y. B., Dobson, G., Brassell, S. and Eglinton, G. 1989a. Lipids biochemical markers and the composition of microbial mats. In: Y. Cohen and E. Rosenberg (Eds.), *Microbial Mats: Physiological Ecology of Benthic Microbial Communities* (American Society for Microbiology: Washington, D.C.), pp. 439–454.

Ward, D. M., Weller, R., Shiea, J., Castenholz, R. W. and Cohen, Y. 1989b. Hot spring microbial mats: anoxygenic mats of possible evolutionary significance. Chapter 1. In: Y. Cohen and E. Rosenberg (Eds.), *Microbial Mats: Physiological Ecology of Benthic Microbial Communities* (American Society for Microbiology: Washington, D.C.), pp. 3–15.

Ward, D. M., Weller, R. and Bateson, M. M. 1990a. 16S rRNA sequences reveal numerous uncultured microorganisms in a natural community. *Nature* 344: 63–65.

Ward, D. M., Weller, R. and Bateson, M. M. 1990b. 16S rRNA sequences reveal uncultured inhabitants of a well-studied thermal community. *FEMS Microbiology Reviews* 75: 105–116.

Wardle, R. J. and Bailey, D. G. 1981. Early Proterozoic sequences in Labrador. In: F. H. A. Campbell (Ed.), *Proterozoic Basins of Canada. Geological Survey of Canada*, Paper 81–10: 331–358.

Wassman, C. C., Löffelhardt, W. and Bohnert, H. J. 1988. Cyanelles: Organisation and molecular biology. In: P. Fay and C. van Baalen (Eds.), *Cyanobacteria. A Comprehensive Review* (Elsevier: Amsterdam), pp. 303–324.

Watanabe, M. M., Takeda, Y., Sasa, T., Inouye, I., Suda, S., Sawaguchi, T. and Chihara, M. 1987. A green dinoflagellate with chlorophylls *a* and *b*: morphology, fine structure of the chloroplast and chlorophyll composition. *Journal of Phycology* 23: 382–389.

Waterbury, J. B. and Stanier, R. Y. 1981. Isolation and growth of cyanobacteria from marine and hypersaline environments. In: M. P. Starr, H. Stolp, H. G. Trüper, A. Balows and H. G. Schlegel (Eds.), *The Prokaryotes* (Springer-Verlag: Berlin, Heidelberg, New York), pp. 220–223.

Waterbury, J. B. and Willey, J. M. 1988. Isolation and growth of marine planktonic cyanobacteria. In: L. Packer and A. N. Glazer (Eds.), *Cyanobacteria, Vol. 167 Methods in Enzymology* (Academic Press: London, New York), pp. 100–105.

Waterbury, J. B., Watson, S. W., Valois, F. W. and Franks, D. G. 1987. Biological and ecological characterization of the marine unicellular cyanobacterium, *Synechococcus*. In: T. Platt and W. K. W. Li (Eds.), *Photosynthetic Picoplankton* (Canadian Department of Fisheries and Oceans: Ottawa), pp. 71–120.

Watts, A. B., Karner, G. D. and Steckler, M. S. 1982. Lithospheric flexure and the evolution of sedimentary basins. *Philosophical Transactions of the Royal Society of London A* 305: 249–281.

Watts, C. D., Maxwell, J. R. and Kjøsen, H. 1977. The potential of carotenoids as environmental indicators. In: R. Campos and J. Goni (Eds.), *Advances in Organic Geochemistry, 1975* (Empresa Nacianal Adaro de Investigaciones Mineras: Madrid), pp. 391–413.

Watts, D. R., Vand der Voo, R. and French, R. B. 1980a. Palaeomagnetic investigations of the Cambrian Waynesboro and Rome Formations of the Valley and Ridge Province of the Appalachian Mountains. *Journal of Geophysical Research* 85: 5331–5343.

Watts, D. R., Van der Voo, R. and Reeve, S. C. 1980b. Cambrian paleomagnetism of the Llano Uplift, Texas. *Journal of Geophysical Research* 85: 5316–5330.

Wazynska, H. 1967. Wstepne badania mikroflorystyczne osadów sinianu i kambru z obszaru Bialowiezy [Preliminary microflorisitic examinations of the Sinian and Cambrian deposits from the Bialowieza area]. *Kwartalnik Geologiczny* 11: 10–20 (in Polish with English abstract).

Webby, B. D. 1970a. Late Precambrian trace fossils from New South Wales. *Lethaia* 3: 79–109.

Webby, B. D. 1970b. Problematical disk-like structure from the late Precambrian of western New South Wales. *Proceedings of the Linnean Society of New South Wales* 95: 191–193.

Weber, A. L. 1987. The triose model: Glyceraldehyde as a source of energy and monomers for prebiotic condensation reactions. *Origins of Life* 17: 107–120.

Weckesser, J., Hofmann, K., Jürgens, U. J., Whitton, B. A. and Raffelsberger, B. 1988. Isolation and chemical analysis of the sheaths of the filamentous cyanobacteria *Calothrix parietina* and *C. scopulorum*. *Journal of General Microbiology* 134: 629–634.

Wedeking, K. W. and Hayes, J. M. 1983. Carbonization of Precambrian kerogens. In: M. Bjorøy, C. Albercht, K. Cornford, K. de Groot, G. Eglinton, E. Galimov, D. Leythaeuser, R. Pelet, J. Rullkotter and G. Speers (Eds.), *Advances in Organic Geochemistry 1981* (Wiley Heydon Ltd.: Chichester), pp. 546–553.

Weeks, O. B. and Francesconi, M. D. 1978. Occurrence of squalene and sterols in *Cellulomonas dehydrogenans* (Arnaudi 1942) comb. nov. Hester 1971. *Journal of Bacteriology* 136: 614–624.

Weigl, J. W. 1953. Concerning the absorption spectrum of bacteriochlorophyll. *Journal of the American Chemical Society* 75: 999–1000.

Welhan, J. A., Craig, H. and Kim, K. 1984. Hydrothermal gases at 11°N and 13°N on the East Pacific Rise. *EOS* 65: 973–974.

Wells, A. T., Ranford, L. C., Stewart, A. J., Cook, P. J. and Shaw, R. D. 1967. Geology of the north-eastern part of the Amadeus Basin, Northern Territory. *Bureau of Mineral Resources Journal of Australian Geology and Geophysics Report* 113: 93.

Wells, A. T., Forman, D. J., Ranford, L. C. and Cook, R. J. 1970. Geology of the Amadeus Basin, central Australia. *Australia Bureau of Mineral Resources, Geology and Geophysics Bulletin* 100: 1–222.

Wenz, W. 1938. Gastropoda. Allgemeiner Teil und Prosobranchia. In: O. H. Schindewolf (Ed.), *Handbuch der Paläozoologie*, vol. 6 (Bonträger: Berlin), pp. 1–240.

West, G. S. and Fritsch, F. E. 1927. *A Treatise on the British Freshwater Algae* (Cambridge University Press: Cambridge England), 535 pp.

Westergård, A. H. 1936. *Paradoxides oelandicus* beds of Öland, with the account of a diamond boring through the Cambrian at Mossberga. *Sveriges Geologiska Undersökning* 394(Ser. C): 1–66.

Westheide, W. 1984. The concept of reproduction in polychaetes with small body size adaptations in interstitial species. In: A. Fischer and H. D. Pfannenstiel (Eds.), *Polychaete Reproduction. Fortschritte der Zoologie*, 29 (G. Fischer: Stuttgart; New York), pp. 265–287.

Westheide, W. 1985. The systematic position of the Dinophilidae and the archiannelid problem. In: S. Conway Morris, J. D. George, R. Gibson and H. M. Platt (Eds.), *The Origins and Relationships of Lower Invertebrates. Systematics Association Special Volume* 28 (Clarendon Press: Oxford), pp. 310–326.

Wetherill, G. 1985. Occurence of giant impacts during the growth of the terrestrial planets. *Science* 228: 877–879.

Wetherill, G. W. 1972. The beginning of continental evolution. *Tectonophysics* 13: 31–45.

Wetherill, G. W. 1976. The role of large bodies in the formation of the earth and moon. In: R. B. Merrill (Ed.), *Proceedings of the Seventh Lunar Science Conference* (Pergammon Press: New York), pp. 3245–3257.

Wetzel, R. G. 1983. *Limnology*, Second Edition (Saunders College Publishing: Philadelphia, New York), 767 pp.

Whale, G. F. and Walsby, A. E. 1984. Motility of the cyanobacterium *Microcoleus chthonoplastes* in mud. *British Phycological Journal* 19: 117–123.

Wharton, R. A., Jr., Parker, B. C. and Simmons, G. M., Jr. 1983. Distribution, species composition and morphology of algal mats in Antarctic dry valley lakes. *Phycologia* 22: 355–365.

Whatley, J. M. 1981. Chloroplast evolution—ancient and modern. *Annals of the New York Academy of Sciences* 361: 154–164.

Whatley, J. M. and Whatley, P. R. 1981. Chloroplast evolution. *New Phytology* 87: 233–247.

White, R. D. 1986. Cambrian radiolaria from Utah. *Journal of Paleon-*

tology 60: 778–780.

Whiteaves, J. F. 1892. Description of a new genus and species of phyllocarid crustacean from the Middle Cambrian of Mount Stephen. *Canadian Rec. Science* 5: 205–208.

Whiticar, M. J., Faber, E. and Schoell, M. 1986. Biogenic methane formation in marine and freshwater environments: carbon dioxide reduction versus acetate fermentation—isotopic evidence. *Geochimica et Cosmochimica Acta* 50: 693–709.

Whittaker, R. H. 1959. On the broad classification of organisms. *Quarterly Review of Biology* 34: 210–226.

Whittard, W. F. 1953. Palaeoscolex piscatorum gen. et spl nov., a worm from the Remadocian of Shropshire. *Quarterly Journal of the Geological Society of London* 109: 125–135.

Whittington, H. B. 1985. *The Burgess Shale* (Yale University Press: New Haven), 151 pp.

Whittington, H. B. and Briggs, D. E. G. 1985. The largest Cambrian animal, *Anomalocaris*, Burgess Shale, British Columbia. *Philosophical Transactions of the Royal Society of London B* 309: 569–618.

Wickstrom, C. E. and Castenholz, R. W. 1973. Thermophilic ostracod: Aquatic metazoan with the highest known temperature tolerance. *Science* 181: 1063–1064.

Wickstrom, C. E. and Castenholz, R. W. 1978. Association of *Pleurocapsa* and *Calothrix* (cyanophyta) in thermal streams. *Journal of Phycology* 14: 84–88.

Wickstrom, C. E. and Castenholz, R. W. 1985. Dynamics of cyanobacteria-ostracod interactions in an Oregon hot spring. *Ecology* 66: 1024–1041.

Wickstrom, C. E. and Wiegert, R. E. 1980. Response of thermal algal-bacterial mat to grazing by brine flies. *Microbial Ecology* 6: 313–315.

Wiegel, J. and Ljungdahl, L. G. 1981. *Thermoanaerobacter ethanolicus* gen. nov., spec. nov., a new, extreme thermophilic, anaerobic bacterium. *Archives of Microbiology* 128: 343–348.

Wiegel, J., Ljungdahl, L. G. and Rawson, J. R. 1979. Isolation from soil and properties of the extreme thermophile *Clostridium thermohydrosulfuricum*. *Journal of Bacteriology* 139: 800–810.

Wiegel, J., Braun, M. and Gottschalk, G. 1981. *Clostridium thermoautotrophicum* species novum, a thermophile producing acetate from molecular hydrogen and carbon dioxide. *Current Microbiology* 5: 255–260.

Wiegert, R. G. and Mitchell, R. 1973. Ecology of Yellowstone thermal effluent systems: Intersects of blue-green algae, grazing flies (*Paracoenia*, Ephydridae) and water mites (*Parnuniella*, Hydrachnellae). *Hydrobiologia* 41(2): 251–271.

Wilcox, L. W. and Wedemayer, G. L. 1984. *Gymnodinium acidotum* Nygaard (Pyrrophyta), a dinoflagellate with an endosymbiotic cryptomonad. *Journal of Phycology* 20: 236–242.

Wilcox, L. W. and Wedemayer, G. L. 1985. Dinoflagellate with blue-green chloroplasts derived from endosymbiotic eukaryote. *Science* 227: 192–194.

Wilde, P. 1987. Model of progressive ventilation of the late Precambrian-Early Paleozoic ocean. *American Journal of Science* 287: 442–459.

Wilhelms, D. E. 1987. *The Geologic History of the Moon. Geological Survey Professional Paper*, 1348 (United States Government Printing Office: Washington), 302 pp.

Wilkins, A. S. 1986. *Genetic Analysis of Animal Development* (John Wiley & Sons: New York), 583 pp.

Williams, A. and Rowell, A. J. 1965. *Treatise on Invertebrate Paleontology. H. Brachiopoda* (Geological Society of America and University of Kansas Press: Lawrence), 927 pp.

Williams, G. E. 1968. Torridonian weathering and its bearing on Torridonian paleoclimate and source. *Scottish Journal of Geology* 4: 164–184.

Williams, G. E. 1975. Late Precambrian glacial climate and the earth's obliquity. *Geology Magazine* 112: 441–465.

Williams, G. E. 1981. Sunspot periods in the Late Precambrian glacial climate and solar-planetary relations. *Nature* 291: 624–628.

Williams, G. E. 1986. Precambrian permafrost horizons as indicators of paleoclimate. *Precambrian Research* 32: 233–242.

Williams, G. E. 1987. Cosmic signals laid down in stone. *New Scientist* 114: 63–66.

Williams, G. E. and Tonkin, D. G. 1985. Periglacial structures and paleoclimatic significance of a late Precambrian block field in the Cattle Grid Copper Mine, Mt. Gunson, South Australia. *Australian Journal of Earth Sciences* 32: 287–300.

Williams, L. A. and Reimers, C. 1983. Role of bacterial mats in oxygen-deficient marine basins and coastal upwelling regimes: preliminary report. *Geology* 11: 267–269.

Wilmot, N., V and Fallick, A. E. 1989. Original mineralogy of trilobite exoskeletons. *Palaeontology* 32: 297–304.

Wilson, J. F., Bickle, M. J., Hawkesworth, R. J., Martin, A., Nisbet, E. G. and Orphen, J. L. 1978. The granite-greenstone terrains of the Rhodesian Archaean craton. *Nature* 271: 23–27.

Wilson, M. R., Hamilton, P. J., Fallick, A. E., Aftalion, M. and Michard, A. 1985. Sm-Nd, U-Pb and O isotope systematics of granites and Proterozoic crustal evolution in Sweden. *Earth and Planetary Science Letters* 72: 376–388.

Wiman, C. 1894. Paleotologische Notizen, 1 und 2. *Bulletin of the Geological Institutions of the University of Uppsala* 2: 109–117.

Wiman, C. 1903. Studien über das Nordbaltische Silurgebiet. I. Olenellussandstein, Obolussandstein und Ceratopygeschiefer. *Bulletin of the Geological Institutions of the University of Uppsala* 6: 12–76.

Wimbush, M. 1976. The physics of the benthic boundary layer. In: I. N. McCave (Ed.), *The Benthic Boundary Layer* (Plenum Press: New York), pp. 3–10.

Winchester, J. A. (Ed.). 1988. *Later Proterozoic Stratigraphy of the Northern Atlantic Regions* (Blackie: Glasgow), 279 pp.

Windley, B. F. 1977. *The Evolving Continents* (John Wiley & Sons: London), 386 pp.

Windley, B. F. 1981. Precambrian rocks in the light of the plate-tectonic concept. In: A. Kröner (Ed.), *Precambrian Plate Tectonics* (Elsevier: Amsterdam), pp. 1–20.

Windley, B. F. 1983. A tectonic review of the Proterozoic. In: L. G. Medaris Jr., C. W. Byers, D. M. Mickleson and W. C. Shanks (Eds.), *Proterozoic Geology: Selected Papers from an International Proterozoic Symposium. Geological Society of America Memoir* 161: 1–10.

Windley, B. F. 1986. Comparative tectonics of the western Grenville and the western Himalaya. In: J. M. Moore, A. Davidson and A. J. Baer (Eds.), *The Grenville Province. Geological Association of Canada Special Paper* 31: 341–348.

Winsborough, B. M. and Golubić, S. 1987. The role of diatoms in stromatolite growth: two examples from modern freshwater settings. *Journal of Phycology* 23: 195–201.

Winston, D. 1986. Sedimentology of the Ravalli Group, middle Belt carbonate and Missoula Group, Middle Proterozoic Belt Supergroup, Montana, Idaho and Washington. In: S. M. Roberts (Ed.), *Belt Supergroup: A Guide to Proterozoic Rocks of Western Montana and Adjacent Areas. Montana Bureau of Mines and Geology Special Publication* 94: 85–124.

Winston, D. and Wallace, C. A. 1983. The Helena Formation and the Missoula Group at Flint Creek Hill, near Georgetown Lake, Western Montana. Field Guide for Trip No, 3. In: S. W. Hobbs (Ed.), *Guide to field trips, Belt Symposium II* (Geology Department, University of Montana: Missoula), pp. 66–81.

Winter, H. de la R. 1987. A cratonic foreland model for Witwatersrand Basin development in a continental back-arc plate-tectonic setting. *South African Journal of Geology* 90: 409–427.

Winters, K., Parker, P. L. and van Baalen, C. 1969. Hydrocarbons of blue-green algae: geochemical significance. *Science* 163: 467–468.

Witkowski, A. 1986. Microbial mat with an incomplete vertical structure, from brackish water environment, The Puck Bay, Poland. A Possible analog of an "advanced anaerobic ecosystem"? In: *Abstracts of the 5th Meeting of the International Society for the Study of the Origins of Life and the 8th International Conference on the Origin of Life (ISSOL)*, pp. 161–162.

Witt, D. and Stackebrandt, E. 1988. Disproving the hypothesis of a common ancestry for the *Ochromonas danica* chryoplast and *Heliobacterium chlorum*. *Archives of Microbiology* 150: 244–248.

References

Woelters, J. and Erdmann, V. A. 1986. Cladistic analyses of 5S rRNA and 16S rRNA and primary structure—the evolution of eukaryotes and their relation to archaebacteria. *Journal of Molecular Evolution* 24: 152–166.

Woese, C. R. 1982. Archaebacteria and cellular origins: an overview. *Zentralblatt für Bakteriologie, Mikrobiologie und Hygiene 1. Abteilung Originale C* 3: 1–17.

Woese, C. R. 1987. Bacterial evolution. *Microbiological Reviews* 51: 221–271.

Woese, C. R. 1989. Archaebacteria and the nature of their evolution. In: B. Fernholm, K. Bremer and H. Jörvall (Eds.), *The Hierarchy of Life* (Elsevier: Amsterdam), pp. 119–130.

Woese, C. R. and Fox, G. E. 1977. The concept of cellular evolution. *Journal of Molecular Evolution* 10: 1–6.

Woese, C. R. and Olsen, G. J. 1986. Archaebacterial phylogeny: perspectives on the urkingdoms. *Systematics and Applied Microbiology* 7: 161–177.

Woese, C. R., Kandler, O. and Wheelis, M. L. 1990. Towards a natural system of organisms: Proposal for the domains Archaea, Bacteria, and Eucarya. *Proceedings of the National Academy of Science USA* 87: 4576–4579.

Wolk, C. P. 1973. Physiology and cytological chemistry of blue-green algae. *Bacteriological Reviews* 37: 31–101.

Wolk, C. P. 1988. Purification and storage of nitrogen-fixing filamentous cyanobacteria. In: L. Packer and A. N. Glazer (Eds.), *Cyanobacteria, Vol. 167 Methods in Enzymology* (Academic Press: London, New York), pp. 93–95.

Wolters, J. and Erdmann, V. A. 1986. Cladistic analyses of 5S rRNA and 16S rRNA secondary and primary structure—the evolution of eukaryotes and their relation to archaebacteria. *Journal of Molecular Evolution* 24: 152–166.

Wolters, J. and Erdmann, V. A. 1988. Cladistic analyses of ribosomal RNAs—the phylogeny of eukaryotes with respect to the endosymbiotic theory. *Biosystems* 21: 209–214.

Wolters, J. and Erdmann, V. A. 1989. The structure and evolution of archaebacterial ribosomal RNAs. *Canadian Journal of Microbiology* 35: 43–51.

Wood, A. 1948. "*Sphaerocodium,*" a misinterpreted fossil from the Wenlock Limestone. *Proceedings of the Geologists Association* 59: 9–22.

Wood, A. 1957. The type-species of the genus *Girvanella* (calcareous algae). *Palaeontology* 1(1): 22–28.

Wood, G. and Clendening, J. A. 1982. Acritarchs from the Lower Cambrian Murray Shale, Chilhowee Group, of Tennessee, U.S.A. *Palynology* 6: 255–265.

Wood, H. G., Ragsdale, S. W. and Pezacka, E. 1986. The acetyl-CoA pathway of autotrophic growth. *FEMS Microbiology Reviews* 39: 345–362.

Wood, W. B. (Ed.). 1988. *The Nematode* Caenorhabditis elegans (Cold Spring Harbor: New York), 667 pp.

Wray, J. L. 1977. *Calcareous Algae* (Elsevier: Amsterdam), 185 pp.

Wright, J. B., Hastings, D. A., Jones, W. B. and Williams, H. R. 1985. *Geology and Mineral Resources of West Africa* (Allen and Unwin: Boston), 188 pp.

Wright, V. P. and Wright, J. M. 1985. A Stromatolite built by a *Phormidium*-like alga from the Lower Carboniferous of South Wales. In: D. F. Toomey and M. H. Nitecki (Eds.), *Paleoalgology: Contemporary Research and Applications* (Springer-Verlag: Berlin), pp. 40–54.

Wrona, R. 1982. Early Cambrian phosphatic microfossils from southern Spitsbergen (Hornsund Region). *Palaeontologica Polonica* 43: 9–16.

Wrona, R. 1987. Cambrian microfossil *Hadimopanella* Gedik from glacial erratics in West Antarctica. *Palaeontologica Polonica* 49: 37–48.

Wu, F., Van der Voo, R. and Liang, Q. Z. 1989. Reconnaissance magnetostratigraphy of the Precambrian-Cambrian boundary section at Meishucun, southwest China. *Cuadernos de Geologica Iberica* 12: 205–222.

Wyborn, L. A. I., Page, R. W. and Parker, A. J. 1987. Geochemical and geochronological signatures in Australian Proterozoic igneous rocks. *Geological Society of Australia Special Publication* 33: 377–394.

Wyborn, L. A. I., Page, R. W. and McCulloch, M. T. 1988. Petrology, geochronology and isotope geochemistry of the post-1820 Ma granites of the Mount Isa Inlier: mechanism of generation of Proterozoic anorogenic granites. *Precambrian Research* 40/41: 509–541.

Xiao, L.-G. and Zhou, B.-H. 1984. [Early Cambrian Hyolitha from Huainan and Huoquiu Counties in Anhui Province]. *Professional Papers in Stratigraphy and Paleontology* 13: 141–151.

Xing, X.-H., Ding, Q.-X., Luo, H.-L., He, T.-G. and Wang, Y.-G. 1984. The Sinian-Cambrian boundary of China and its related problems. *Geological Magazine* 121(3): 155–169.

Xing, Y. 1982. [Microflora of the Sinian System and Lower Cambrian near Kunming, Yunnan and its stratigraphic significance]. *Acta Geologica Sinica* 56: 42–50 (in Chinese with English abstract).

Xing, Y. and Liu, H. 1978. [*Sinian Microflora*] (Geological Publishing House: Beijing).

Xing, Y. and Liu, K. 1976. *Microapalaeoflora from the Sinian Subera of W. Hupeh and its Stratigraphic Significance* (Publ. Inst. Geol. Min. Resour., Chinese Acad. Geol. Sci.: Peking), 23 pp.

Xing, Y., Ding, Q., Luo, H., He, T., Wang, Y., et al. 1983. [*The Sinian-Cambrian Boundary of China*]. Bull. Inst. Geol., Chinese Acad. Geol. Sci. 10 (Geological Publishing House: Beijing), 262 pp.

Xing, Y.-H., Duan, C.-H., Liang, Y.-Z. and Cao, R.-G. 1985. Late Precambrian palaeontology of China. *Geol. Mem. Ministry Geol. Min. Res.* 2(2): 1–243.

Xing, Y.-S. and Liu, G.-Z. 1979. Coelenterate fossils from the Sinian System of southern Liaoning and its stratigraphical significance. *Acta Geologica Sinica* 53: 167–172 (In Chinese).

Xing, Y.-S., Chen, Y.-Y., Zhang, S.-S., Liu, G.-Z., Xiong, X.-W., Chen, P., Luo, H.-L., Jiang, Z.-W., Wu, X.-C., et al. 1984. [The Sinian-Cambrian boundary of China]. *Bulletin of the Institute of Geology, Chinese Academy of Geological Sciences, Special Issue* 10: 1–260.

Xing, Y.-S., Liu, G.-Z., Yin, C.-G., Yue, Z. and Gao, L.-Z. 1989. Evolution of the terminal Precambrian biotas and its geological significance. *International Geological Congress* 28(3): 392 (Abstract).

Xing, Y.-S. H. 1984. Description of a new worm family—Huaiyuanellidae Xing from the Upper Sinian of North Anhui, China. *Bulletin of the Institute of Geology, Chinese Academy of Geological Sciences* 9: 151–154 (In Chinese).

Xing, Y.-S. 1984. The Sinian and its position in the geological time scale. *International Geological Congress* 27(1): 212 (Abstract).

Xu, H.-K., Rong, J.-Y. and Liu, D.-Y. 1974. [Brachiopoda (Ordovician)]. In: *Handbook of the Stratigraphy and Palaeontology in Southwest China*, pp. 144–154.

Xu, Z. L. 1984a. [Investigation on the procaryotic microfossils from the Gaoyuzhuang Formation, Jixian, North China]. *Acta Botanica Sinica* 26: 216–222 (In Chinese with English abstract).

Xu, Z.-L. 1984b. [Investigation on the procaryotic microfossils from the Gaoyuzhuang Formation, Jixian, North China (cont.)]. *Acta Botanica Sinica* 26: 312–319 (In Chinese).

Yabe, H. and Ozaki, K. 1930. *Girvanella* in the Lower Cambrian of South Manchuria. *Science Report of Tohoku Imperial University, Sendai, Japan, Series 2 (Geology)* 14: 79–85.

Yakshin, M. S. and Luchinina, V. A. 1981. Novye dannye po iskopaemym vodoroslyam semejstva Oscillatoriaceae (Kirchn.) Elenkin [New data on fossil algae of the family Oscillatoriaceae (Kirchn.) Elenkin]. *Trudy Institut Geologii i Geofiziki Sibirskoe Otdolenie, Akademiya Nauk SSSR*: 28.

Yang, D., Oyaizu, Y., Oyaizu, H., Olsen, G. J. and Woese, C. R. 1985. Mitochondrial origins. *Proceedings of the National Academy of Science USA* 82: 4443–4447.

Yang, X.-H., He, Y.-X. and Deng, S.-H. 1983. On the Sinian-Cambrian boundary and the small shelly fossil assemblages in Nanjiang area, Sichuan. *Bulletin of the Chengdu Institute of Geological and Mineral Research* 1983(4): 91–110.

Yang, X.-H., He, Y.-X. and Deng, S.-H. 1984. [New small shelly fossils from Lower Cambrian Meishucun Stage of Nanjiang Area, northern Sichuan]. *Professional Papers in Stratigraphy and Paleontology* 13: 35–47.

Yang, Z., Cheng, Y. and Wang, H. 1986. *The Geology of China* (Clarendon Press: Oxford), 218 pp.

Yanishevskij, M. E. 1926. Ob ostatkakh trubchatykh chervej iz kembrijskoj Sinej Gliny [On remains of tube-dwelling worms from the Cambrian Blue Clay]. *Ezhegodnik Vsesoyuznogo Paleontologicheskogo Obshchestva* 4: 99–111.

Yanishevskij, M. E. 1950. Drevnejshij trilobit iz nizhnekembrijskoj sinej gliny—*Gdowia assatkini* gen. et sp. nov. [The oldest trilobite from the Lower Cambrian Blue Clay—*Gdowia assatkini* gen. et sp. nov.]. *Voprosy Paleontologii* 1: 32–40.

Yankauskas, T. V. 1975. Novye akritarkhi nizhnego kembriya Pribaltiki [New acritarchs from the Lower Cambrian of the Baltic region]. *Paleontologicheskij Zhurnal* 1975(1): 94–104.

Yankauskas, T. V. 1976. Novye vidy akritarkhi iz nizhnego kembriya Pribaltiki [New acritarch species from the Lower Cambrian of the Baltic Region]. In: I. T. Zhuravleva (Ed.), *Stratigrafiya i Paleontologiya Nizhnego i Srednego Kembriya SSSR* [*Stratigraphy and Paleotology of the Lower and Middle Cambrian of the USSR*]. *Trudy Inst. Geol. Geofiz. Sibirsk. Otd. Akad. Nauk SSSR* 296 (Nauka: Novosibirsk), pp. 187–192, 246–247.

Yankauskas, T. V. 1980a. Novye vodorosli iz verkhnego rifeya Yuzhnogo Urala i priural'ya [New algae from the upper Riphean of the southern Urals and adjacent areas]. *Paleontologicheskij Zhurnal* 1980(4): 107–113.

Yankauskas, T. V. 1980b. Shishenyakskaya mikrobiota verkhnego rifeya yuzhnogo Urala [The Upper Riphean Shishenyak microbiota from the southern Urals]. *Doklady Akademii Nauk SSSR* 251: 190–192.

Yankauskas, T. V. 1982. Mikrofossilii rifeya Yuzhnogo Urala [Microfossils of the Riphean in the southern Uras]. In: B. M. Keller (Ed.), *Stratotip Rifeya. Paleontologiya. Paleomagnetizm* [*Stratotype of the Riphean. Paleontology. Paleomagnetism*]. *Trudy Geol. Inst. Akad. Nauk SSSR*, 368 (Nauka: Moscow), pp. 84–120.

Yankauskas, T. V. and Posti, E. 1976. Novye vidy akritarkh kembriya Pribaltiki [New acritarch species from the Cambrian of the Baltic region]. *Eesti NSV Teaduste Akadeemia Toimetised* 25: 145–151.

Yankauskas, T. V., Mikhajlova, N. S. and German, T. N. 1987. V Vsesoyuznyj kollokvium po mikrofossiliyam dokembriya SSSR [The fifth all-union colloquium on Precambrian microfossils of the USSR]. *Izvestiya Akademiya Nauk SSSR, seriia Geologicheskaya* 1987(9): 137–139.

Yeh, L.-S., Sun, S., Chen, Q. and Guo, S.-Z. 1986. Proterozoic and Cambrian phosporites—deposits: Hunyang, Yunnan, China. In: P. J. Cook and J. H. Shergold (Eds.), *Phosphate Deposits of the World. 1. Proterozoic and Cambrian Phosphorites* (Cambridge University Press: Cambridge), pp. 149–154.

Yeo, G. H. 1981. The Late Proterozoic glaciation in the northern Cordillera. In: F. H. A. Campbell (Ed.), *Proterozoic Basins of Canada. Geological Survey of Canada, Paper* 81–10: 25–46.

Yeo, G. M. 1986. Iron-formation in the late Proterozoic Rapitan Group, Yukon and Northwest Territories. In: J. A. Morin (Ed.), *Mineral Deposits of the Northern Cordillera. Canadian Institute of Mining and Metallurgy, Special Volume* 37: 142–153.

Yin, G.-Z. 1978. In: *Handbook of Palaeontology of Southwest China. Guizhou, Part* 1: 383–384.

Yin, J., Ding, L., He, T., Li, S. and Shen, L. 1980. *The Palaeontology and Sedimentary Environment of the Sinian System in Emei-Ganluo Area, Sichuan*, 210 pp.

Yin, L. 1985. Microfossils of the Doushantuo Formation in the Yangtze Gorge district, western Hubei. *Palaeontologia Cathayana* 2: 229–249.

Yin, L. 1987. Microbiotas of latest Precambrian sequences in China. In: Nanjing Institute of Geology and Palaeontology (Ed.), *Stratigraphy and Paleontology of Systemic Boundaries in China, Precambrian-Cambrian Boundary*, vol. 1 (Nanjing University Publishing House: Nanjing), pp. 415–494.

Yochelson, E. L. 1961. The operculum and mode of life of *Hyolithes*. *Journal of Paleontology* 35: 152–161.

Yochelson, E. L. 1974. Redescription of the Early Cambrian *Helenia bella* Walcott, an appendage of *Hyolithes*. *Journal of Research of the U.S. Geological Survey* 2(6): 717–722.

Yochelson, E. L. 1977. Agmata, a proposed extinct phylum of Early Cambrian age. *Journal of Paleontology* 51(3): 437–454.

Yochelson, E. L. 1979. Early radiation of Mollusca and mollusc-like groups. In: M. R. House (Ed.), *The Origin of Major Invertebrate Groups. The Systematics Association Special Volume* 12 (Clarendon Press: Oxford), pp. 323–358.

Yochelson, E. L. and Herrera, H. E. 1974. Un fosil enigmatica del Cambrico inferior de Argentina. *De 'Ameghiniana'* 11(3): 283–294.

Yochelson, E. L. and Stanley, G. D. 1981. An early Ordovician patelliform gastropod, *Palaelophacmaea*, reinterpreted as a coelenterate. *Lethaia* 15: 323–330.

Yochelson, E. L., Stürmer, W. and Stanley, G. D. 1983. *Plectodiscus discoideus* (Rauff): a redescription of a chondrophorine from the Early Devonian Hunsrück Slate, West Germany. *Paläontologische Zeitschrift* 57: 39–68.

Yoshizaki, F., Fukazawa, T., Mishina, Y. and Sugimura, Y. 1989. Some properties and amino acid sequence of plastocyanin from a green alga, *Ulva arasakii*. *Journal of Biochemistry* 106: 282–288.

Young, F. G. 1972. Early Cambrian and older trace fossils from the southern Cordillera. *Canadian Journal of Earth Sciences* 9: 1–17.

Young, G. M. 1970. An extensive Early Proterozoic glaciation in North America. *Palaeogeography, Palaeoclimatology, Palaeoecology* 7: 85–101.

Young, G. M. 1976. Iron-formation and glaciogenic rocks of the Rapitan Group, Northwest Territories, Canada. *Precambrian Research* 3: 137–158.

Young, G. M. 1983. Tectono-sedimentary history of early Proterozoic rocks of the northern Great Lakes region. In: L. G. Medaris Jr. (Ed.), *Early Proterozoic Geology of the Great Lakes Region. Geological Society of America Memoir* 160: 15–32.

Young, G. M., Jefferson, C. W., Delaney, G. D. and Yeo, G. M. 1979. Middle and late Proterozoic evolution of the northern Canadian Cordillera and Shield. *Geology* 7: 125–128.

Yu, W. 1984. Early Cambrian molluscan faunas of Meishucan Stage with special reference to Precambrian-Cambrian boundary. *Acad. Sin. Dev. Geosci. 27th International Geological Congress, 1984, Moscow*: 21–35.

Yu, W. 1986. Lower Cambrian univalved mollusc from Kuruktag, Xinjiang. *Acta Palaeontologica Sinica* 25: 10–16.

Yu, W. 1987. Yangtze micromolluscan fauna in Yangtze region of China with notes on Precambrian-Cambrian boundary. In: *Stratigraphy and Paleontology of Systemic Boundaries in China–PC–C Boundary*, vol. 1 (Nanjing University Publishing House: Nanjing), pp. 19–344.

Yu, W. 1988. [*Jinkenites*–a strange fossil from earliest Cambrian in W. Hubei]. *Acta Palaeontologica Sinica* 27: 303–308.

Yü, W. 1979. Earliest Cambrian monoplacophorans and gastropods from western Hubei with their biostratigraphical significance. *Acta Palaeontologica Sinica* 18(3): 233–266.

Yü, W. 1981. New earliest Cambrian monoplacophorans and gastropods from W. Hubei and E. Yunnan. *Acta Palaeontologica Sinica* 20(6): 552–556.

Yü, W. 1984. On merismoconchids. *Acta Palaeontologica Sinica* 23(4): 432–446.

Yü, W. 1985. *Yangtzedonta*–a problematic Bivalvia from the Meishucunian Stage, China. *Acta Micropalaeontologica Sinica* 2(4): 401–408.

Yue, Z. and He, S.-C. 1989. [Early Cambrian conodonts and bradoriids]. *Acta Micropalaeontologica Sinica* 6: 289–300.

Zablen, L. B., Kissel, M. S., Woese, C. R. and Beutow, D. E. 1975. Phylogenetic origin of the chloroplast and procaryotic nature of its ribosomal RNA. *Proceedings of the National Academy of Science USA* 72: 2418–2422.

Zahnle, K. J. 1986. Photochemistry of methane and the formation of

References

hydrocyanic acid (HCN) in the earth's early atmosphere. *Journal of Geophysical Research* 91: 2819–2834.

Zahnle, K. J., Kasting, J. F. and Pollack, J. B. 1988. Evolution of a steam atmosphere during earth's accretion. *Icarus* 74: 62–97.

Zaika-Novatsky, V. S. 1965. Novye problematicheskie otpechatki iz verkhnego dokembriya Pridestrovya [New problematical imprints from the upper Precambrian of the Dniester region]. *Tezisy dokladov Novosibirskoi* 1965: 98–99.

Zaika-Novatsky, V. S., Velikanov, V. A. and Koval, A. P. 1968. Pervyi predstavitel ediakarskoi fauny v vende Russkoi platformy (verkhnii dokembrii) [The first member of the Ediacara fauna in the Vendian (upper Precambrian) of the Russian Platform]. *Paleontologicheskij Zhurnal* 1968(2): 132–134.

Zaine, M. F. and Fairchild, T. R. 1985. Comparison of *Aulophycus lucianoi* Beurlen & Sommer from Ladario(MS) and the genus *Cloudina* Germs, Ediacaran of Namibia. *Anais de Academia Brasileira de Ciências* 57: 130.

Zaine, M. F. and Fairchild, T. R. 1987. Novas considerações sobre os fósseis de Formação Tamengo, Grupo Corumbá, SW do Brasil. *Anias do 10th Congresso Brasileiro de Paleontologia, Rio de Janeiro, 19 25 Juhlo* 1987: 797–807.

Zang, W. L. and Walter, M. R. 1989. Latest Proterozoic plankton from the Amadeus Basin in central Australia. *Nature* 337: 642–645.

Zbinden, E. A., Holland, H. D., Feakes, C. R. and Dobos, S. K. 1988. The Sturgeon Falls paleosol and the composition of the atmosphere 1.1 b.y.b.p. *Precambrian Research* 42: 141–163.

Zedler, J. B. 1980. Algal mat productivity: comparisons in a salt marsh. *Estuaries* 3: 122–131.

Zeikus, J. G., Hegge, P. W. and Anderson, M. A. 1979. *Thermoanaerobium brockii* gen. nov. and sp. nov., a new chemoorganotrophic, caldoactive, anaerobic bacterium. *Archives of Microbiology* 122: 41–48.

Zeikus, J. G., Ben-Bassat, A. and Hegge, P. W. 1980. Microbiology of methanogenesis in thermal, volcanic environments. *Journal of Bacteriology* 143: 432–440.

Zeikus, J. G., Dawson, M. A., Thompson, T. E., Ingvorsen, K. and Hatchikian, E. C. 1983. Microbial ecology of volcanic sulphidogenesis: isolation and characterization of *Thermodesulfobacterium commune* gen. nov. and sp. nov. *Journal of General Microbiology* 129: 1159–1169.

Zempolich, W. G., Wilkinson, B. H. and Lohmann, K. C. 1988. Diagenesis of Late Proterozoic carbonates: the Beck Spring Dolomite of eastern California. *Journal of Sedimentary Petrology* 58: 656–672.

Zeschke, G. 1961. Transportation of uraninite in the Indus River, Pakistan. *Geological Society of South Africa Transactions* 63: 87–94.

Zhang, R.-J., Feng, S.-N., Xu, G.-H., Yang, D.-L., Yan, D.-P., Li, Z., Jiang, D.-H. and Wu, W. 1990. Discovery of *Chuaria-Tawuia* assemblage in Shilu Group, Hainan Island and its significance. *Science in China* (Series B) 33: 211–220.

Zhang, W.-T. 1987. Early Cambrian Chengjiang fauna and its trilobites. *Acta Palaeontologica Sinica* 26(3): 223–235.

Zhang, W.-T. and Hou, X.-G. 1985. Preliminary notes on the occurrence of the unusual trilobite *Naraoia* in Asia. *Acta Palaeontologica Sinica* 24(6): 591–595.

Zhang, X., Liu, M., Liu, Q., Wang, H. and Li, S. 1982. Preliminary separation and characteristics of phycocyanin in blue-green *Gymnodinium*. *Kexue Tongbao* 27: 1000–1003.

Zhang, X.-G. 1987. Moult stages and dimorphism of Early Cambrian bradoriids from Xichuan, Henan, China. *Alcheringa* 11(1–2): 1–19.

Zhang, Y. 1988. Proterozoic stromatolitic micro-organisms from Hebei, North China: cell preservation and cell division. *Precambrian Research* 38: 165–175.

Zhang, Y. 1989. Multicellular thallophytes with differentiated tissues from Late Proterozoic phosphate rocks of South China. *Lethaia* 22: 113–132.

Zhang, Y. and Hofmann, H. J. 1982. Precambrian stromatolites: image analysis of lamina shape. *Journal of Geology* 90: 253–268.

Zhang, Z. 1984. [Microflora of the Late Sinian Doushantuo Formation, Hubei Province, China]. *Collections Internat. Communication Geol. Sci., Selected Papers, 27th International Geological Congress*: 129–140 (In Chinese, with English abstract).

Zhang, Z.-Y. 1988. *Longfengshania* Du emend.: an earliest record of bryophyte-like fossils. *Acta Palaeontologica Sinica* 27: 416–426.

Zhao, Z.-Q., Xing, Y.-S., Ma, G.-G. and Chen, Y.-Y. (Eds.). 1985. *Biostratigraphy of the Yangtze Gorge area; (1) Sinian* (Geological Publishing House: Beijing), 165 pp.

Zheng, W.-W. 1980. A new occurence of fossil group of *Chuaria* from the Sinian System in north Anhui and its geological meaning. *Bulletin of the Tianjin Institute of Geology and Mineral Resources* 1: 49–69.

Zhong, H. 1977. Preliminary study of the ancient fauna of south China and its stratigraphic significance. *Scientia Geologica Sinica* 1977(2): 118–128.

Zhou, B.-H. and Ziao, L.-G. 1984. [Early Cambrian monoplacophorans and gastropods from Huaninan and Huoqiu Counties, Anhui Province]. *Professional Papers in Stratigraphy and Paleontology* 13: 125–140.

Zhu, S. 1982. An Outline of studies on the Precambrian stromatolites of China. *Precambrian Research* 18(4): 367–396.

Zhu, S., Xu, C. and Gao, J. 1987. Early Proterozoic stromatolites from Wutai Mt. and its adjacent regions. *Bulletin of the Tianjin Institute of Geology and Mineral Resources* 17: 1–221 (In Chinese, with English abstract).

Zhuravlev, A. Y. 1986. Radiocyathids. In: A. Hoffman and M. H. Nitecki (Eds.), *Problematic Fossil Taxa. Oxford Monographys on Geology and Geophysics*, 5 (Oxford University Press: New York), pp. 35–44.

Zhuravlev, A. Y. 1989. Poriferan aspects of archaeocyathan skeletal function. *Memoirs of the Association of Australasian Palaeontologists* 8: 387–399.

Zhuravleva, Yu, A. and Nitecki, M. N. 1985. On the comparative morphology of archaeocyathids and receptaculitids. *Paleontological Journal* 19: 134–136.

Zhuravleva, I. T. 1960. *Arkheotsiaty Sibirskoj Platformy* (Izdanii Akademiia Nauk SSSR: Moscow), 344 pp.

Zhuravleva, I. T. 1983. Tip Gubki [Phylum sponges]. In: B. S. Sokolov and I. T. Zhuraleva (Eds.), *Yarusnoe Raschlenenie Nizhnego Kembriya Sibiri* (Nauka: Moscow), p. 53.

Zhuravleva, I. T. and Myagkova, E. I. 1981. Materialy k izucheniyu Archaeata [Material for the study of Archaeta]. In: B. S. Sokolov (Ed.), *Problematiki Fanerozoya. Trudy Institut Geologii i Geofiziki Sibirskoe Otdolenie, Akademiya Nauk SSSR* 481: 41–74.

Zhuravleva, I. T. and Okuneva, O. G. 1981. O prirode kribritsiat [On the nature of cribricyathans]. In: B. S. Sokolov (Ed.), *Problematiki Fanerozoya. Trudy Institut Geologii i Geofiziki Sibirskoe Otdolenie, Akademiya Nauk SSSR* 481: 23–30.

Zhuravleva, I. T., Zadorozhnaya, N. M., Osadchaya, D. B., Pokrovskaya, N. V., Rodionova, N. M. and Fonin, V 1967. *Fauna Nizhnego Kembriya Tuvy (Opornyj Razrez r. Shivelig-Khem)* (Nauka: Moscow), 176 pp.

Ziegler, A. M., Scotese, C. R., McKerrow, W. S., Johnson, M. E. and Bambach, R. K. 1979. Paleozoic Paleogeography. *Annual Review of Earth and Planetary Sciences* 7: 473–502.

Ziegler, A. M., Hulver, M. L., Lottes, A. L. and Schumachtenberg, W. F. 1984. Uniformitarianism and Palaeoclimates: Inferences from the distribution of carbonate rocks. In: P. Brenchley (Ed.), *Fossils and Climate* (John Wiley & Sons: New York), pp. 3–25.

Zillig, W. 1987. Eukaryotic traits in archaebacteria: Could the eukaryote cytoplase have arisen from archaebacterial origin? *Annals of the New York Academy of Sciences* 503: 78–82.

Zillig, W., Schnabel, R. and Stetter, K. O. 1985. Archaebacteria and the origin of the eukaryotic cytoplasm. *Current Topics in Microbiology and Immunology* 114: 1–18.

Zillig, W., Klenk, H.-P., Palm, P., Puhler, G., Gropp, F., Garrett, R. A. and Leffers, H. 1989. The phylogenetic relations of DNA—dependent RNA polymerase of archaebacteria, eukaryotes, and eubacteria. *Canadian Journal of Microbiology* 35: 73–80.

Zinder, S. H., Doemel, W. N. and Brock, T. D. 1977. Production of volatile sulfur compounds during the decomposition of algal mats. *Applied and Environmental Microbiology* 34: 859–860.

Zuckerkandl, E. and Pauling, L. 1962. Molecular disease, evolution and genetic heterogeneity. In: M. Kasha and B. Pullman (Eds.), *Horizons in Biochemistry* (Academic Press: New York), pp. 189–225.

Zumberge, J. E. 1983. Tricyclic diterpane distributions in the correlation of Paleozoic crude oils from the Williston Basin. In: M. Bjorøy, C. Albercht, K. Cornford, K. de Groot, G. Eglinton, E. Galimov, D. Leythaeuser, R. Pelet, J. Rullkotter and G. Speers (Eds.), *Advances in Organic Geochemistry 1981* (Wiley Heyden Ltd.: Chichester), pp. 738–745.

Zumberge, J. E. 1984. Source rocks of the La Luna Formation (Upper Cretaceous) in the Middle Magdalena Valley, Columbia. In: J. G. Palacas (Ed.), *Petroleum Geochemistry and Source Rock Potential of Carbonate Rocks. American Association of Petroleum Geologists Studies in Geology* 18: 127–134.

Zundel, M. and Rohmer, M. 1985a. Prokaryotic triterpenoids. 1. 3β-methyhopanoids from *Acetobacter* sp. and *Methylococcus capsulatus*. *European Journal of Biochemistry* 150: 23–27.

Zundel, M. and Rohmer, M. 1985b. Prokaryotic triterpenoids. 3. The biosynthesis of 2β-methylhopanoids and 3β-methylhopanoids of *Methylobacterium organophilium* and *Acetobacter pasteurianus* subsp. *pasteurianus*. *European Journal of Biochemistry* 150: 35–39.

Subject Index

Acanthomorph acritarchs, 205, 219, 222, 227–228, 230–231, 514, 533, 557, 591, 883, 902, 904, 907–909, 912, 914, 916, 920, 922–923, 926–927, 931, 933, 935, 938–942, 944–951, 1152, 1163
Accretion: planetary, 9, 13
Accretionary complex, 73; tract, 75
Acritarchs, 219, 433, 441, 453, 455–456, 557, 959, 961, 963, 965, 967, 969, 971, 975, 1121; acanthomorph, 205, 222, 227–228, 230–231, 514, 533, 557, 591, 883, 902, 904, 907–909, 912, 914, 916, 920, 922–923, 926–927, 931, 933, 935, 938–942, 944, 951, 1152, 1163; biostratigraphic use of, 244, 497–498, 513–516; complex, 219, 227, 526, 533–535, 540–548, 561, 1124–1125, 1151–1152, 1157–1158, 1163, 1167–1168; disphaeromorph, 927; diversity of, 521–552; excystment structures of, 228–230; "giant" acanthomorph, 177, 230–231, 498; herkomorph, 205, 219, 222, 225, 227–228, 923, 935, 944–945, 947, 949–951; Late Proterozoic extinction of, 230–232, 244, 455, 498, 557, 859; "megasphaeromorph", 177, 180, 219–222, 230–231, 277, 336, 355–356, 497–498, 516, 533–535, 540–548, 557, 1150–1151, 1157, 1159–1160, 1167–1168; "mesosphaeromorph", 219–222, 227, 230–231, 336, 497–498, 533–535, 540–548, 592, 1149–1150, 1157, 1159–1160, 1167; mid-Proterozoic increase in diversity of, 230–231; netromorph, 205, 219, 222, 227–228, 945, 948, 950–951; nonsphaermorph, 219, 222–232, 498, 1124–1125, 1151–1152, 1157–1158, 1163, 1167–1168; oömorph, 205, 222, 227–228; ornamentation of, 229–230; paleobiogeographic distribution of, 516; polygonomorph, 205, 222, 225, 227–228, 883, 930, 949, 951; prismatomorph, 205, 219, 222, 227–228, 877–878, 883, 922, 939, 941, 948; processes of, 228, 230; pteromorph, 205, 219, 222, 225, 227–228, 880, 906, 910, 920, 949–951; pterospermopsimorphid, 946, 948; pylomes of, 230; range charts showing temporal distribution of, 535–536; sphaeromorph, 205, 219–220, 230–231, 592, 914–915, 1124–1125, 1149–1151, 1157, 1159–1161
Actinians, 434
Adelaide Geosyncline, Australia, 54, 56; PPRG samples from, 629–630; stratigraphy of, 628
Albany-Tracer Province, 54, 56
Alcyonarians, 982, 987
Algae, 871, 876, 879, 886–887, 916, 923, 938, 944, 959, 961, 963, 965, 967, 969, 971, 973, 975, 977, 979, 1036–1037, 1039, 1041, 1043, 1045, 1047, 1049, 1051, 1053; biomarkers of, 85, 87, 89, 114; calcareous, 257, 359–367, 400, 423, 433, 448, 477, 514, 557, 560–561, 565, 860; brown, 419, 477, 479, 481, 961, 977, 979; green, 189, 191, 202, 219, 321–325, 355, 357, 359–361, 419, 479–480, 483, 929, 967, 984, 1047, 1121, 1159–1160; receptaculitid, 404; red, 188, 202, 479, 481–482; yellow-brown, 478–479
Alkaline igneous rocks, 61
Alluvial deposits, 49
Amadeus Basin, Australia, 57; organic matter from, 108; PPRG samples from, 631–632; stratigraphy of, 630

Amazonian Craton, 55, 71, 73
Amino acids, 9
Ammonia, 9, 166; as a greenhouse gas, 166
Amoebas, 431, 483, 592
Anabarites-Protohertzina biozone, 505
Anemones, 392–393
Anhydrite, 22, 59, 62, 165, 169; as an indicator of paleoclimate, 165
Anorthosite, 60–61
Anoxygenic microbial mats, 309–312
Anoxygenic photosynthesis, 280, 313, 341
"Anti-greenhouse" effect, 167
Apparent polar wander (APW) paths, 569
APW path: *See* Apparent polar wander paths.
Arabian-Nubian Shield, 57, 68, 74
Arc accretion, 73
Archaebacteria, 85–87, 133, 185, 188, 243, 287, 291, 314, 316, 320, 471–475, 477; biomarkers from, 85
Archaeooids, 400
Archean: arc magmatism, 18; atmosphere, 9; back arc basins, 18; chemical weathering, 21; convection cells, 18; craton cover, 15; crustal evolution, 16; environment, 40; evaporites, 22; fore arc basins, 18; fossil record, 25, 40; gneiss terranes, 14, 16; granite-greenstones, 14; greenstone belts, 14, 17; iron-formations, 15–16; late bombardment during, 10; liquid water in, 21; mantle circulation, 19; metallogeny, 59; microbiota, 37; microfossils, 25–39, 177, 208; paleoenvironment, 11, 165; plate tectonics, 18, 21; sedimentary rocks, 15; stromatolites, 11–12, 24–27, 41–42, 148–149; terranes, 14; "upside-down" biosphere, 159; volcanic rocks, 15
Arrested evolution, 595. *See also* Hypobradytely.
Ascidians, 402
Athapuscow Aulacogen (East Arm Thrust Belt), Canada: PPRG samples from, 655; stratigraphy of, 655
Atlas of representative Proterozoic microfossils, 1055–1117; listing of type specimens illustrated in, 1059–1061
Atmosphere, 9; box models for evolution of, 159–161, 1185–1186; carbon dioxide concentrations in, 165–166, 1187; composition of, 153–154; formation of, 9; hydrogen peroxide in Archean, 23; lightning in, 10; oxygen:carbon dioxide ratio in, 154; oxygen content of, 23, 153, 159, 160–161, 169; oxidation state of Archean, 9; photochemical products in, 153; Proterozoic, 51, 137, 147, 153, 159, 165; reducing, 9, 150; rise of ozone concentrations in, 162; steam, 9; sulfate reduction and its effect on, 161–162
Avalon Zone, Canada: PPRG samples from, 655–657; stratigraphy of, 655

Bacteria, 1121, 1123–1124, 1126, 1129, 1131, 1137–1138, 1140, 1144, 1159, 1169; acetogenic, 291; aerobic, 591; ammonia-oxidizing, 126; anaerobic, 37, 41, 126, 827; biochemically new class of, 108; chemoautotrophic, 126, 328; coccoid, 203–204;

1287

1288 Subject Index

colorless sulfur, 291; extraction of pigments from, 829–831; facultative, 590–591, 600; fermenting, 587, 827, 841; flavobacteria, 472, 474; fossil, 869–870, 876, 884–885, 890–891, 893–898, 901–903, 905, 907, 909–910, 912–914, 916–917, 919, 921–922, 925, 927–932, 934, 938, 940–941, 943, 949; gliding, 25, 1131; Gram-negative, 86, 185; Gram-positive, 86, 185, 472, 474; green non-sulfur, 320, 472, 474; green photosynthetic, 87, 89, 126; green sulfur, 104, 280–281, 283, 304, 319, 329, 827; halophilic, 84, 87–88, 472; hydrogen, 291; iron, 291, 869, 876; isolation and culture of, 827; methanogenic, 41, 84, 87–88, 125, 289–291, 318, 320, 336, 472, 474–475, 587, 827, 841; methylotrophic, 41, 86, 125, 474; nitrifying, 291; nitrogen-fixing, 827; photosynthetic, 41, 89, 248, 255, 473, 587, 590, 827, 829; purple non-sulfur, 86, 283–284, 304, 319, 472, 474, 480, 827; purple photosynthetic, 126, 272; purple sulfur, 86, 248, 250, 274, 277, 279, 280–281, 283, 301, 319, 326–330, 336, 472, 474, 827; sulfate-reducing, 41, 84, 90–93, 131, 161–162, 336, 587, 827, 841; sulfide-oxidizing, 126, 827; sulfur-metabolizing, 1138, 1144; sulfur-oxidizing, 84, 90–93, 250, 336; thermoacidophilic, 87; thermophilic, 472, 474; "thread cell," 212–213, 1123
Bacteriochlorophylls, 272, 279–281, 283–284, 304, 321, 329–330, 823, 829–831
Baltic Platform, 51
Banded iron-formation (BIF): See Iron-formation.
Barberton Cycle, 15, 17, 68, 70, 75
Barberton Greenstone Belt, South Africa, 15, 70; PPRG samples from, 614–616; stratigraphy of, 614
Barghoorn, E. S., 181
Barite, 22, 59
Basalt: komatiitic, 15; tholeiitic, 15
Basement: distribution of terranes of, 14, 68; half-life of, 63
Bashkiria Anticlinorium, USSR: PPRG samples from, 621–622; stratigraphy of, 620–621
Basin architecture, 49–50
Basin and Range Province, eastern California and southern Nevada, USA: PPRG samples from, 658; stratigraphy of, 658
Basin and Range Province, eastern California, USA: PPRG samples from, 658; stratigraphy of, 658
Basin and Range Province, Utah, USA: PPRG samples from, 658; stratigraphy of, 658
Basin and Range Province, western region, USA: PPRG samples from, 659–660; stratigraphy of, 658–659
Belcher Island Fold Belt, 55
Belt Basin, USA, 48, 660; PPRG samples from, 660–663; stratigraphy of, 660
Beltinids, 354, 357, 501–503, 561, 961
Benchmarks: in acritarch evolution, 228–231; in Precambrian biotic evolution, 244, 587–593
BIF (banded iron-formation): See Iron-formation.
Biogeochemistry: isotopic and molecular, 83. See also Biomarkers; Chemical fossils; Extractable organic matter; Geochemical processing procedures; Kerogen elemental and isotopic compositions; Organic carbon abundance and isotopic composition; Organic carbon isotopic compositions; and Organic geochemical processing procedures.
Biogeography: of calcareous algae, 359–360, 516; of carbonaceous films, 350–353, 438–439, 516; of megascopic dubiofossils, 413, 516: of "megasphaeromorph" acritarchs, 516; of metazoan body fossils, 369, 371, 439, 441, 504, 516; of microfossils, 516; of possible metaphytes, 516; problems in the study of, 515–516; significant geologic sections for, 516; of small shelly fossils, 400–401, 441, 516; of trace fossils, 516
Biomarkers, 84–88; algal, 85, 87, 89, 114; acyclic polyisoprenoids as, 101, 104; alkanes as, 101–102, 104; and biostratigraphy, 489, 511–514, 519; carotenoids as, 101, 104; gammacerane as, 111, 235, 244, 588, 592; hopanoids as, 106; at Laguna Guerrero Negro, 328; in microbial mats, 303–308, 322, 324, 328, 337, 339; and molecular phylogeny, 465, 473–474; monomethyl alkanes as, 102–104; porphyrins as, 84, 87; of prokaryotic cell walls, 303–304; of prokaryotic sheaths, 303–304; of protozoans, 592; steroids as, 85–86, 105. See also Extractable organic matter.
Biostratigraphy: use of acritarchs for, 230–232, 244, 497–498, 513–516; and biozones, 514; use of calcareous algae for, 514; use of camasid dubiofossils for, 506; use of carbon isotopes for, 511–512, 514, 519; use of carbonaceous films for, 501–504, 514; contributors to, 492–493; current status of, 513–514, 519; definition of, 489; historical development of, 489; use of molecular fossils for, 489, 511–514, 519; problems in analyses of, 489, 491–492, 495, 517; use of prokaryotic microfossils for, 497, 513; significant geologic sections for, 494; use of skeletonized metazoan fossils for, 399–400, 505–506, 514; use of skeletonized protozoan fossils for, 499; use of soft-bodied metazoan fossils for, 504, 514; use of stromatolites for, 507–509, 513–516; use of sulfur isotopes for, 511–512; use of trace fossils for, 504–505, 513–514
Biotic diversity—*Phanerozoic fossils*, 523, 527
Biotic diversity—*Proterozoic and Early Cambrian metazoans and metaphytes*, 553–561; and archaeocyathids, 556; and the Burgess Shale fauna, 555–556; and the Cambrian fauna, 559; current status of studies of, 665; data used in the analysis of, 1171–1184; definition of, 523; of the Ediacaran fauna, 554, 556–558; exponential increase of, 553, 559; of higher taxa, 554–556; and Late Proterozoic extinction, 557; of metaphytes, 559–560; of non-metazoan carbonaceous megafossils, 560–561; problems in the analysis of, 523, 525–527, 554; Q-mode factor analysis of, 557–559, 1179–1181; sources of data for analysis of, 553–554, 1171–1184; and stratigraphic units analyzed, 554; of the Tommotian fauna, 556. See also Described taxa.
Biotic diversity—*Proterozoic microfossils*, 230–232, 242, 521–552; and atmospheric carbon dioxide, 552; of benthos, 529–530, 532, 534–535, 537–548, 1125, 1160–1161, 1164–1165, 1167–1168; beta, 530, 532, 536, 538, 542; and coelacanth taxa, 532, 534, 549; current status of studies of, 550–552, 565; data used in the analysis of, 1160–1165, 1167–1168; definition of, 523; of eukaryotes, 538–548; evaluation of, 536–538, 541, 548–550; and geochronology, 526, 529–530, 535, 537; global, 527, 529, 535–538; and global biomass, 538, 540; and "grab-sampling", 544–545; and hypobradytelic evolution, 534; informal classification used in analyses of, 242, 529, 533–535, 1119–1168; Late Proterozoic decline of, 531, 538, 540–548, 551–552; and Lazarus taxa, 529, 532, 534–536, 540–549; of maximally diverse assemblages, 1164–1165; maximum alpha, 529, 532, 536, 538–548; and monographs, 538–543; in "near-shore" stromatolitic carbonates/cherts, 193, 526–527, 529, 531–532; in "off-shore" shales/siltstones, 193, 526–527, 529, 531–532; and paleoenvironments, 192–193, 527, 529–533, 537; of plankton, 133, 529–530, 532, 534–535, 537–548, 1125, 1162–1165, 1167–1168; and preservation, 525–526, 530–531, 533; problems in studies of, 523, 525–527, 529–537, 563–564, 1121; of prokaryotes, 538–548; and publications, 526, 529–531; and the "pull of the Phanerozoic", 531, 538; and sampling, 526–527, 529–532, 537–548; and stromatolite abundance/diversity, 541; taxa omitted from analyses of, 534; and taxonomy, 525–526, 529–530, 533–534, 537. See also Described taxa; and Microfossils.
Biozones: *Anabarites-Protohertzina*, 505; *Grypania*, 501, 519; *Harlaniella podolica*, 392–393; *Pycodes pedum*, 392; *Rusophycus palonensis*, 392; stromatolitic, 513, 519; *Tawuia*, 501, 519; *Vendotaenia*, 504, 519
Birrindudu Basin, Australia: PPRG samples from, 633–634; stratigraphy of, 632–633
Bitumen, 101; in Chuar Group sediments, 109–110; in McArthur Group sediments, 113–114; in Nonesuch Formation sediments, 111–112; in Officer Basin sediments, 107; in Roper Group sediments, 113–114
Bizarre micro-problematica, 195, 233–234, 534
Black smokers, 23
Blue Ridge Province, USA: PPRG samples from, 663; stratigraphy of, 663
Body fossils: current status of studies of, 423; of the Ediacara fauna, 369–390, 433–434, 999–1007; first described, 345; geographic distribution of, 369, 371, 439, 441, 516; history of studies of, 369, 371; oldest biomineralized, 371, 593; preservation of, 372–373, 379, 434–435; pre-Vendian, 369, 1006–1007; regarded as dubiofossils, 369, 1006–1007, 1035–1053; regarded as pseudofossils, 369–370, 1006–1007, 1035–1053; relation to Cambrian strata, 371; taxonomy of, 379, 381; temporal distribution of, 371–372, 593. See also Skeletal metazoan fossils.
Box models for atmosphere/ocean system, 159–161, 1185–1186
Bradytely, 595–597

Subject Index

Broken Hill Block, 55

Calcareous algae, 257, 359–367, 400, 423, 433, 448, 860, 981–998, 1041, 1049; bona fide, 981 993; calcification of, 257, 359; classification of, 359–361; described taxa of, 981–998; diversity of, 361–363, 557, 560–561, 565; earliest known, 447; evaluation of, 981–998; geographic distribution of, 359–360, 516; history of studies of, 359, 361; morphology of, 362; not evaluated, 993–995; paleoecology of, 363; preservation of, 360; regarded as dubiofossils, 359, 361, 996; regarded as pseudofossils, 359, 361, 996–998; stratigraphic ranges of, 363, 366, 400, 514; theories for the origin of, 453
Camasids, 419, 506, 1041, 1043, 1045, 1047, 1049, 1051, 1053
Cape Smith Belt, Canada, 48, 55; PPRG samples from, 663; stratigraphy of, 663
Carbon cycle: burial rate of carbon in, 118, 127; isotopic fractionation in, 89–90, 125–127; linkage of sulfur cycle to, 130, 132
Carbon dioxide, 11, 21, 52, 165–167; concentration by autotrophs of, 126; in early atmosphere, 11, 21, 166–167; in Middle Proterozoic atmosphere, 125–127; and Precambrian climate models, 165–166, 1187
Carbon isotopic composition, 712–762, 791–798: and atmospheric carbon dioxide, 125–127, 133, 167, 174, 338, 511; and biostratigraphy, 511–512, 514, 519; of carbonates, 88, 117–118, 122–123, 125, 721, 748–758; in iron-formations, 150; in Late Proterozoic carbonates, 160; and mass balance, 88–89; of microbial mats, 299–303, 328–330, 332, 337–338, 340; notation for, 88; and organic carbon burial rates, 160; of Precambrian kerogens, 511, 759–762, 791–798; of tawuids, 356. *See also* Organic carbon abundance and isotopic composition; and Organic carbon isotopic compositions.
Carbonaceous films, 423, 433–434, 957–979; biostratigraphic use of, 501–504, 514; definition of, 349; described taxa of, 957–979, 1035–1053; evaluation of, 957–979, 1035–1053; first described, 346, 349; geographic distribution of, 350–353, 438–439, 516; informal categories of, 349; poorly characterized forms of, 354; regarded as dubiofossils or pseudofossils, 1035–1053
Carbonate isotopic compositions, 88, 117–118, 122–123, 125, 721, 748–758
Carbonate-silicate cycle, 166
Carpoids, 412
Catagenesis, 102
Catagraphs, 419, 507–508, 860, 981, 984
Cayeux, M. L., 180
Chaetetids, 985
Chemical fossils, 489, 511–514, 519, 592. *See also* Biomarkers.
Chemical weathering: Archean, 21
Chemoautotrophy: as source of organic carbon in sediments, 126
Chitaldrug Schist Belt, India: PPRG samples from, 622; stratigraphy of, 622
Chitons, 448, 485, 1043
Chondrophores, 381, 434
Chromite deposits, 60–61
Chuar Group: bitumen in sediments of, 109–110
Chuarids, 349, 354–356, 501–503, 961, 965, 967, 971, 973, 975
Ciliates, 267, 295, 430–431, 472, 474–475, 482
Classification of Proterozoic microfossils, 242, 529, 533–535, 1119–1168. *See also* Informal Classification of Proterozoic microfossils.
Climate: global, 51; models of, 159–161, 166, 1185–1186
Cloud, P., 182
"Coal" (shungite), 62
Coccolithophorids, 359
Coelacanth taxa, 532, 534, 549
Coelenterates, 485, 983, 1041, 1043, 1045, 1047. *See also* CNIDARIA (Taxonomic Index).
Colorado Cycle, 68, 71, 74–75
Colorado Plateau Province, USA: PPRG samples from, 664–667; stratigraphy of, 664
Conglomerates, 49, 57, 60
Congo Craton, 73
Continental configurations: Late Proterozoic, 580
Continental crust, 67;
Cycles of formation of, 70; formation of, 69–70
Copepods, 295–296

Copper deposits, 60–62
Coppermine Homocline, Canada: PPRG samples from, 667; stratigraphy of, 667
Corals, 987, 1051
Core formation, 9
Correlation chart for Vendian-Early Cambrian strata, 374–378
Cratons: ages of formation of, 67, 69; deposits of, 53; development of, 72; pre-, early-, late-, and post-stages of, 72; rock associations of, 53–54; stages in formation of, 60
Crocodilians, 595
Crustaceans, 263, 267, 296–297, 381, 408, 416, 485, 595, 961, 969, 1051
Crustal formation, 13, 72; episodes of, 70; on the moon, 13
Crustal recycling, 47
Crustal rifting, 15
Crust-forming episodes, 70
Cryptarchs, 967
Cryptomonads, 479, 482
Cuddapah Basin, India: PPRG samples from, 622; stratigraphy of, 622
Cuvier, G., 489
Cyanelles, 478–479, 481
Cyanobacteria, 84–87, 89, 106, 114, 177, 180, 185–189, 195–202, 206–222, 227, 359, 430, 456, 472–474, 478–479, 533–534, 587, 590, 592, 595–598, 600, 827, 831, 838, 971, 977, 985–986, 1022, 1043, 1049, 1094, 1096–1098, 1100, 1102, 1121, 1123–1124, 1126–1139, 1141–1147, 1159, 1161–1162; ecologic versatility of, 187–188, 597–598; fossil, 876–880, 882–887, 889–951; origin of, 37–39, 150, 218, 473–474; possible Archean presence of, 26–39, 218
Cyphomegachritarchs, 349, 967, 975

Damara Orogen, 56–57, 74
Darwin, C. R., 180, 427, 525
Databases: construction and use of, 855–863
Dawson, J. W., 180
Degassing, 9, 14; history of earth, 21
Described taxa: of Proterozoic and earliest Cambrian carbonaceous remains, trace and body fossils, 953–1053; of Proterozoic and Early Cambrian microfossils and microfossil-like objects, 865–951
Deuterostomes, 485
Dharwars Craton, 15
Diagenesis: effect on carbon isotopic composition of carbonates, 118, 122
Diamictites, 51, 57, 60, 157
Diamonds, 48
Diatoms, 191, 235, 238–239, 250, 262, 268, 283–284, 295–296, 300–301, 304–307, 321–332, 335–336, 449, 479, 831, 838, 872
Dim early sun problem, 165–166
Dinoflagellates, 109, 189–191, 219, 231, 242, 449, 479, 481
Diplomonads, 481
Disphaeromorph acritarchs, 927
Diversity, 521–561. *See also* Biotic diversity.
d'Orbigny, A., 489
Dropstones, 51
Drozdova, N. A., 181

Early Cambrian biota, 433–434, 1035–1053; biogeography of, 400–401, 441; biomineralization of, 441, 447–451; feeding modes of, 440; trophic interactions within, 437–440, 461. *See also* Skeletal metazoan fossils.
Early cratonization stage, 72
Earth: accretion of, 9, 13; crust of, 13; formation of, 13; mantle of, 13
East-European Craton, 73
Echiuroids, 393
Ediacara fauna, 369–390, 423, 433–434, 525, 860, 999–1007; biogeography of, 369, 371, 439, 441, 504, 516; biostratigraphic use of, 504; described taxa of, 999–1007; extinction of, 372; preservation of, 372–373, 379, 434–435, 525–526; taxonomy of, 379, 381, 525–526, 999–1007; trophic interactions within, 437
Ellipsophysids, 349, 356
Endolithic microbial mats, 324, 328
Endolithic microorganisms, 869, 880, 926

Endosymbiosis: and the origin of chloroplasts, 471, 478–480; and the origin of flagella, 480; and the origin of mitochondria, 471–472, 480; and phylogeny, 471–472; and the Precambrian fossil record, 186, 202, 473
Enteropneusts, 393
Eocytes, 471–473, 477
Eoholynids, 354, 357, 502–503, 561, 959
Et Then Basin, 50
Eukaryotic microbial mats, 321–324
Eukaryotic microfossils, 1124–1125, 1149–1152, 1157–1158; earliest known, 186, 209–210, 242, 244, 257, 473; identification of, 186, 202, 209–210, 473, 591–592, 1159–1162; Proterozoic non-acritarch, 1163–1164
Eukaryotic nonmineralized protists: biochemistry of, 189–190; modern distribution of, 190–191; phylogeny of, 480–483; preservation of, 189–190, 192–193
Eukaryotic sexuality: origin of, 244
Eurypterids, 357, 1043, 1045
Evaporites, 47, 51; Archean, 22
Evolutionary conservatism, 595–600. See also Hypobradytely.
Extension: lithospheric, 48–50
Extensional regime, 48
Extinctions: and hypobradytely, 598; Late Proterozoic, 191, 230–232, 244, 372, 455, 498, 557; Phanerozoic, 527, 549
Extractable organic matter, 101, 811–819; composition of analyzed samples of, 813–819; organic geochemical data for, 815–819; rock-eval data for, 815

Facultative photosynthesis, 280, 288–289, 312–313
Faint early sun problem, 165–166
Fans: submarine, 47
Faunal provinces: Gondwanan, 581; Laurentian, 581
Fecal pellets, 105, 392–393, 395, 419, 508
Fermentation in sedimentary environments, 125
Ferns, 86, 871
Fish, 263, 267, 477
Flagellates, 189, 472, 474
Flin Flon Belt, 53, 61
Flow chart and processing procedures for rock samples, 695–698
Formaldehyde, 10
Fucoids, 958, 1038
Fungi, 314, 324, 430, 472–475, 477, 482–483, 869–874, 878, 881–882, 885–886, 902–903, 929, 941, 1152

Gammacerane, 111, 235, 244, 588, 592
Gascoyne Province, Australia: PPRG samples from, 634; stratigraphy of, 634
Gases: volcanic, 9
Gastropods, 263, 267–268, 296, 355, 405, 410
Geochemical processing procedures, 695–708; for carbonate carbon analyses, 702; for elemental (C, H, N) analyses of kerogen, 707; for kerogen, 707–708; for kerogen color, 707–708; for kerogen isolation, 707; for organic carbon analyses, 701; for rock samples, 695–698; for rock-eval analyses, 701–702, for sulfate sulfur analyses, 703, 705; for sulfide sulfur analyses, 703, 705
Geographic and geologic data for PPRG rock samples, 604–693; ages of geologic units represented, 688–693; from Adelaide Geosyncline, Australia, 628–630; from Africa, 614–620; from Amadeus Basin, Australia, 630–632; from Asia, 620–627; from Athapuscow Aulacogen (East Arm Thrust Belt), Canada, 655; from Australia, 628–652; from Avalon Zone, Canada, 655–657; from Barberton Greenstone Belt, South Africa, 614–616; from Bashkiria Anticlinorium, USSR, 620–622; from Basin and Range Province, eastern California and southern Nevada, USA, 658; from Basin and Range Province, eastern California, USA, 658; from Basin and Range Province, Utah, USA, 658; from Basin and Range Province, western region, USA, 658–660; from Belt Basin, USA, 660–663; from Birrindudu Basin, Australia, 632–634; from Blue Ridge Province, USA, 663; from Cape Smith Belt, Canada, 663; from Chitaldrug Schist Belt, India, 622; from Colorado Plateau Province, USA, 664–667; from Coppermine Homocline, Canada, 667; from Cuddapah Basin, India, 622; from Europe, 652–655; from Gascoyne Province, Australia, 634; from Hamersley Basin, Australia, 634–642; from Kaapvaal Craton, South Africa, 617–620; from Karoo Basin, South Africa, 620; from Khantay-Rybninsk Uplift, USSR, 652; from Kimberley Basin (East Kimberley), Australia, 642; from Kimberley Basin (Halls Creek Mobile Zone), Australia, 643–645; from Lake Superior Basin, northern peninsular Michigan, USA, 667–669; from Lake Vättern Graben, Sweden, 652–653; from Lawn Hill Platform, Australia, 645–646; from Limmen Geosyncline, Australia, 646; from Mackenzie Fold Belt, Canada, 669–676; from Malyj Karatau, USSR, 622–623; from Marquette Range, USA, 677; from McArthur Basin, Australia, 646–647; from Medicine Bow Range, USA, 677; for modern samples, South Africa, 620; from Mount Isa Inlier, Australia, 647; from North America, 655–687; from North China Platform, eastern border region, China, 623; from North China Platform, interior region, China, 623–624; from North China Platform, northern border region, China, 624–627; numerical listing of geologic samples, 605–614; from Olenyok Uplift, USSR, 627; from Osen-Røa Nappe Complex, Norway, 653; from Penokean Fold Belt, Canada, 677; from Pilbara Block, Australia, 648–652; from Port Arthur Homocline, Canada, 678–681; from Russian Platform, USSR, 627; from São Francisco Basin, Brazil, 687; from Sierra de Cordoba, Spain, 653; from Sierra de Guadalupe, Spain, 654; from South America, 687; from Superior Province, Canada, 682; from Torridonian Succession, Great Britain, 654–655; from Uchur-Maya Block, USSR, 655; from Uinta Uplift, USA, 682; from Unaka Province, USA, 682; from Van Horn Mobile Belt, USA, 682; from West-Central Brazil, 687; from Wit Mfolozi Inlier, South Africa, 620; from Wopmay Orogen (Foreland Thrust-Fold Belt), Canada, 683–687
Geographic distribution: of basement terranes, 68
Glacial: deposits, 157; dropstones, 51; effects on global temperature, 51, 165; episodes, 51, 57, 157, 165, 173, 526; Huronian occurrence, 165; Late Proterozoic occurrences, 51–52, 151, 165–167, 231, 526; Precambrian occurrences, 1187; rock sequences, 51, 57, 165; varves, 51. See also Diamictites.
Global climate, 51
Glossary of technical terms, 1189–1203
Gold deposits: Archean vein style, 59–60; in iron-formations, 59; Witwatersrand style, 60
Gondwana, 51, 569, 574
Granite-greenstone terranes, 59, 68, 72–75
Graptolites, 371
Grawler Craton, 55
Grazing fauna: effects on microbial mats of, 248, 263, 267, 290, 295–296, 313, 325; effects on stromatolites of, 260, 295–296, 336, 342
Great Bear Magmatic Zone, 53, 55
Greenhouse effect, 9, 11, 166; "anti-greenhouse" effect, 167; and atmospheric gases, 52; Late Proterozoic, 127
Greenstone belts, 14, 17, 53, 59; stage of formation of, 59
Greenstones, 57, 70, 74
Grenville: Orogeny, 71; Province, 56, 58
Grypania biozone, 501, 519
Grypanids, 349, 356, 421, 501–503
Guiana Shield, 53, 55, 71
Gypsum, 22, 59, 62, 165, 169; as an indicator of paleoclimate, 165

Half life: of active margin basins, 63; of basement, 63; of oceanic crust, 63; of orogenic belts, 63; of passive margin basins, 63; of sediments, 63
Hamersley Basin, Australia: PPRG samples from, 634–642; stratigraphy of, 634
Harlaniella podolica biozone, 392–393
Heat balance, 51
Herkomorph acritarchs, 205, 219, 222, 225, 227–228, 923, 935, 944–945, 947, 949–951
Highlands: lunar, 13
Historical development: of Precambrian paleobiology, 179–183, 244, 489; of Proterozoic biostratigraphy, 489; of Proterozoic megapaleontology, 359, 361, 369, 371; of Proterozoic micropaleontology, 179–183, 489
Horotely, 595–596
Horseshoe crabs, 595
Hydrogen peroxide: in Archean atmosphere, 23
Hypobradytely, 243–244, 595: and biochemistry, 596–597; caveats regarding, 596; and cyanobacteria, 595–600; definition of, 596;

Subject Index

and ecologic versatility, 597–598; and extinctions, 598; and genetics, 596, 598; and physiology, 596–597

Ice albedo feedback, 168
Icehouse climate: late Precambrian, 167
Ichnofossils, 369, 372, 381, 416, 419, 423, 433, 437, 1009–1015; Cambrian, 450, 505; definition of, 391; described taxa of, 1009–1015; first described, 345; geographic distribution of, 514; pre-Vendian, 391–392; regarded as dubiofossils, 391, 396, 1010–1015; regarded as pseudofossils, 391, 396, 1010–1015, 1035–1053; stratigraphic ranges of, 392–393, 397, 504–505, 513–514, 1010–1015, 1035–1053; Vendian, 392–393, 433, 437, 454, 1010–1015
Ilmenite deposits, 60–61
Impacts, 9–11, 24; ocean vaporizing, 8
Informal classification of Proterozoic microfossils, 242, 529, 533–535, 1119–1168; for benthos, 1125, 1160–1161, 1164–1165, 1167–1168; for colonial coccoidal and ellipsoidal cells, 1124–1125, 1140–1148, 1155–1156, 1161–1162; for eukaryotes, 1124–1125, 1149–1152, 1157–1158; for maximally diverse Proterozoic assemblages, 1164–1165; for miscellaneous eukaryotes, 1124–1125, 1152, 1158, 1163, 1166; for miscellaneous prokaryotes, 1124–1125, 1148–1149, 1157, 1162–1163; for non-septate filaments, 1123–1129, 1153; for non-sphaeromorph acritarchs, 1124–1125, 1151–1152, 1157–1158; for ordered colonial coccoidal cells, 1124–1125, 1145–1148, 1156; for plankton, 1125, 1162–1165, 1167–1168; for prokaryotic coccoidal or ellipsoidal cells, 1124–1125, 1137–1148, 1155–1157, 1159–1161; for prokaryotic filaments, 1124–1137, 1153–1155; for septate unbranched filaments, 1124–1125, 1129–1137, 1153–1155, 1158–1159; in shales and cherts, 1121; for solitary coccoidal cells and sphaeromorph acritarchs, 1124–1125, 1137–1140, 1149–1151, 1155, 1157, 1159–1161; for unordered colonial coccoidal cells, 1124–1125, 1140–1144, 1155–1156; for unordered colonial ellipsoidal cells, 1124–1125, 1144–1145, 1156
Insects, 263, 267, 295–297, 314, 431, 435, 477, 485, 873, 875, 879
Iridium: in sediments, 24; in spherules, 24
Iron: concentration in oceans, 23; hydrothermal input of, 23, 52
Iron-formations, 15–16, 51–54, 57, 59, 139, 157, 159; abundance of, 140; Archean, 15–16, 140; bulk chemistry of, 141–142; carbon-Al_2O_3 relations in, 149; carbon isotopic composition of, 149; carbon-sulfur relations in, 149; cesium anomalies in, 145; deep marine origin of iron in, 150; definition of, 139; disappearance of, 140, 158; distribution in time of, 140; europium anomalies in, 145; formation in density-stratified ocean system, 149–150; glacial aspects of, 141, 150; hydrothermal input into, 145–146; implications for atmospheric oxygen of, 159, 456, 1185–1187; organic carbon content of, 142–143, 149; origins of, 147–150; oxidation sate of, 141; paleoenvironmental interpretation of, 147; rare earth element (REE) chemistry of, 144; sedimentology of, 139; South African, 148; stratigraphy of, 139–140; trace element chemistry of, 144; transgressive sequence in, 148
Irumide Belt, 68, 71
Isotope effects, 88–92
Isotopes: mass balance calculations using, 88–89; notation used for, 88
Isotopic compositions: of carbonates, 721, 748–758. *See also* Kerogen elemental and carbon isotopic compositions; Organic carbon isotopic compositions; and Sulfur isotopic compositions.
Isua Cycle, 68, 70
Isua rocks, 13–14; greenstone sequence of, 70; metasediments of, 21–22
Kaapvaal Craton, South Africa, 15, 48, 54–55, 59, 71, 148, 158; PPRG samples from, 618–620; stratigraphy of, 617
Karelian Craton, 54
Karoo Basin, South Africa: PPRG samples from, 620; stratigraphy of, 620
Karyotes, 471
Katanjan Belt, 74
Kerogen: formation of, 84; maturation of, 101
Kerogen elemental and carbon isotopic compositions, 95–97, 101, 118, 759–762, 791–798; analysis of, 101; in Archean deposits, 762, 794–796, 798; in Early Proterozoic deposits, 761–762, 793–794, 797–798; in Late Proterozoic deposits, 759–760,

1291

791–792, 796; in Middle Proterozoic deposits, 760–761, 792–793, 797
Key to "probable affinities" of Proterozoic microfossils, 204–205, 243, 1159–1160
Khantay-Rybninsk Uplift, USSR: PPRG samples from, 652; stratigraphy of, 652
Kibaran Orogeny, 55, 68, 71
Kilohigok Basin, 55
Kimberley Basin (East Kimberley), Australia, 55; PPRG samples from, 642; stratigraphy of, 642
Kimberley Basin (Halls Creek Mobile Zone), Australia: PPRG samples from, 643–645; stratigraphy of, 643
Kimberlites, 60, 62
Komatiites, 15, 24, 53
Kordeh (Korde), K. B., 181

Labrador Trough, 55
Laguna Guerrero Negro: biomarkers at, 328; carbon isotopic studies at, 328–329, 332; compared with Solar Lake, 331–332; dissolved inorganic carbon (DIC) at, 328–329; dissolved organic carbon (DOC) at, 328–329; environment of, 325–328; meiofauna at, 839; methods of study of microbial mats at, 822–843; microbial mats at, 248–249, 267, 273, 289, 295–297, 300, 302, 304–306, 325–333; particulate organic matter (POM) at, 325–327; photosynthesis at, 838; salinity gradients at, 325–328; trends in mats at, 328–333
Lake Superior Basin, northern penninsular Michigan, USA: PPRG samples from, 667–669; stratigraphy of, 667
Lake Vättern Graben, Sweden: PPRG samples from, 652–653; stratigraphy of, 652
Late-cratonization stage, 72
Lawn Hill Platform, Australia: PPRG samples from, 646; stratigraphy of, 645–646
Lazarus taxa, 529, 532, 595–596
Lead-zinc deposits: stratiform, 61
Leiospheres, 190, 942, 948, 963
Life: at mid-ocean ridges, 10; origin of, 10
Lightning discharges, 10
Limmen Geosyncline, Australia: PPRG samples from, 646; stratigraphy of, 646
Limpopo Belt, 73
Limulids, 595
Lithification: of microbial mats, 261, 263, 268, 335; of stromatolites, 253, 257–258
Lithosphere: dynamics of, 67; extension of, 48
Lonfengshanids, 349, 356–357, 421, 437, 502–503, 561, 971
Lunar: cratering, 10, 23; cratering record, 10; highlands, 13; maria, 13

Mackenzie Fold Belt, Canada: PPRG samples from, 669–676; stratigraphy of, 669
Magma ocean, 9
Magnesium: concentration in oceans of, 23
Magodi Mobile Belt, 55
Malyj Karatau, USSR: PPRG samples from, 623; stratigraphy of, 622
Manganese deposits, 60–61
Mantle: circulation in Archean, 19
Maria: lunar, 13
Marquette Range, USA: PPRG samples from, 677; stratigraphy of, 677
Maslov, V. P., 181
Mass extinctions, 230–232, 244, 372, 455, 498, 527, 549, 557, 598. *See also* Extinctions.
Massive sulfide deposits, 59–60
McArthur Basin, Australia, 54–55, 71, 102; PPRG samples from, 647; stratigraphy of, 646–647
McArthur Group: bitumen in sediments of, 113–114; total organic carbon (TOC) content of, 102–103
Medicine Bow Range, USA: PPRG samples from, 677; stratigraphy of, 677
Medusoids, 369–370, 372, 379–382, 388, 391, 407, 413, 419, 433, 437, 454, 556, 1002, 1011, 1036–1039, 1043, 1045–1047, 1051, 1053
Megadubiofossils, 347, 414–419, 423, 1035–1053; camasid, 419; discoidal, 416–417; geographic distribution of, 413, 516; horizontal spindle-shaped, 414, 416; vertical cylindrical, 416

Megafossils: biogenicity of, 347, 1035–1053; described taxa of, 953–1053; syngenicity of, 347
Megapseudofossils, 347, 359, 361, 413–415, 417, 419, 423, 1006–1007, 1035–1053
"Megasphaeromorph" acritarchs, 180, 219–222, 227, 230–231, 336, 355–356, 497–498, 516, 533–535, 540–548, 557, 592, 1150–1151, 1157, 1159–1160, 1167–1168
Melanocyrillids, 177, 219, 229, 234–235, 237, 244, 419, 499, 514, 519, 591
"Mesosphaeromorph" acritarchs, 219–222, 227, 230–231, 336, 497–498, 533–535, 540–548, 592, 1149–1150, 1157, 1159–1160, 1167
Metabacteria, 471
Metagenesis, 102
Metallogeny: Archean, 59; Proterozoic, 59
Metamonads, 481, 483
Metazoan body fossils: biostratigraphic use of, 504, 514; current status of studies of, 423; of the Ediacara fauna, 369–390, 433–434, 999–1007; first described, 345; geographic distribution of, 369, 371, 439–441, 516; history of studies of, 369, 371; oldest biomineralized, 371, 505; preservation of, 372–373, 379, 434–435; pre-Vendian, 369; regarded as dubiofossils, 369, 1006–1007, 1035–1053; regarded as pseudofossils, 369–370, 1006–1007, 1035–1053; relation to Cambrian strata, 371; taxonomy of, 379, 381; temporal distribution of, 371–372. *See also* Skeletal metazoan fossils.
Meteorites: Archean impacts of, 19, 23
Methane, 9, 10; biological-oxidation of, 125; concentration in atmosphere of, 163; as a greenhouse gas, 166
Methanogenesis, 125
Methods of investigation of microbial mats, 822–843; for anaerobic processes in hot springs, 841; autofluorescence-using, 823; for culture of microorganisms, 827; electron microscopic, 825; for inorganic chemistry, 843; light microscopic, 823; for light penetration, 833–835; for living material and wet mounts, 823; for meiofaunal field studies, 839; microelectrode-using, 834–835, 837–838; for pigments, 823, 829–831; for preserved and sectioned samples, 823
Microbial mats: anoxygenic, 309–312; biogeochemistry of, 299–308, 322, 324, 328, 331, 337–338; biomarkers in, 303–308, 322, 324, 328, 337, 339; carbon isotope composition of, 299–303, 328–330, 337–338, 340; chemolithotrophic processes in, 291–293; compared with Proterozoic stromatolites, 332–333, 335–342; definition of, 248, 261; diel growth in, 266–267, 271, 273–275, 277–278, 287, 292, 297, 320–321, 340–341; effects of grazing on, 248, 263, 267, 290, 295–297, 313, 325; endolithic, 324, 328; environmental factors controlling formation of, 266–269, 313, 325–328, 332–333, 341; eukaryotic, 321–324; geographic distribution of, 248–250; "green *Chloroflexus*", 309–311; heliotropism in, 266, 278, 287–291, 318; intermittently oxygenic/anoxygenic, 312–313; "inverted", 313–314, 320–321, 342; at Laguna Guerrero Negro, 248–249, 267, 273, 289, 295–297, 300, 302, 304–306, 325–333; laminated fabrics of, 267–268, 313, 332–333, 336–337; light penetration into, 271–273, 277, 287–289, 328, 341, 833–835; lithificatiion of, 261, 263, 268, 335; methods for investigation, 822–843; microbial movements in, 274–278, 313, 340–341; microelectrode studies of, 287–293, 312, 327, 336, 341, 834–835, 837–838; microenvironment in, 271–278, 319–320, 325–333, 341; morphology of, 261–269; oxygen gradients in, 273–274, 277, 287–289, 320–321, 329–330, 332, 341; photosynthesis in, 279–285, 328, 332; preservation of microorganisms in, 319, 330–333, 337; primary productivity of, 285, 340, 823, 837–838; radiotracer studies of, 290–291; terrestrial, 324; thermal, 313–321
Microbial mimicry, 596
Microcontinents, 73
Microdubiofossils, 177, 235, 867, 876–884
Microfossil assemblages, 884–951, 1057–1117; from Archean deposits, 27–39; from Early Cambrian deposits, 945–951; from Early Proterozoic deposits, 884–892, 1059, 1062–1065; from Late Proterozoic deposits, 911–945, 1060–1061, 1081–1117; from Middle Proterozoic deposits, 892–911, 1059–1060, 1066–1080. *See also* Microfossils.
Microfossils: from the Apex Basalt, Warrawoona Group, 28; Archean, 25, 27–39; "assured" eukaryotic, 209–210, 534, 1124–1125, 1149–1152, 1157–1158; benthic, 1125, 1160–1161, 1164–1165, 1167–1168; biostratigraphic use of, 177, 230–232, 244, 497–498, 513–514; bizarre, 534, 1148–1149, 1157, 1162; bona fide, 177, 529, 884–951; coccoidal, 206, 208–210, 219, 255, 534–535, 540–548, 1124–1125, 1137–1142, 1155, 1159–1160, 1167; colonial, 1124–1125, 1140–1148, 1155–1156, 1161–1162; diversity of, 242, 260, 521–552, 884–951; ellipsoidal, 210–211, 1124–1125, 1144, 1156; eukaryotic, 1124–1125, 1149–1152, 1157–1158; evaluation of, 867–951; filamentous eukaryote-like, 534, 591, 1124–1125, 1152, 1158, 1163, 1166; geographic distribution of, 516; illustrations of, 1055–1117; informal classification of, 242, 529, 533–535, 1119–1168; key to the "probable affinities" of, 204–205, 243, 1159–1160; maximally diverse Proterozoic assemblages of, 1164–1165; morphometric analyses of, 195–202, 206–211, 214–222, 227, 529; non-septate filamentous, 211–212, 214–215, 255, 534, 540–548, 1123–1129, 1153; ordered colonial, 534, 540–548, 1124–1125, 1145–1148, 1156; planktonic, 1125, 1162–1165, 1167–1168; possibly eukaryotic, 1163–1164; preservation in cherts vs. shales, 202–206, 210–213, 215–218, 243, 525, 529–530, 1121; problematical, 233–235, 869–884; prokaryotic, 1124–1148, 1153–1156; questionable reports of, 529, 869–884; range charts showing temporal distribution of, 535–536, 1153–1157; septate filamentous, 215–218, 255, 534, 540–548, 1124–1125, 1129–1137, 1153–1155, 1158–1159; unordered colonial, 534, 540–548, 1124–1125, 1140–1145, 1155–1156. *See also* Microfossil assemblages.
Micrononfossils, 177, 179, 867–875
Micropseudofossils, 177, 179, 237, 239, 867–875
Mid-continent rift, 48
Mid-ocean ridge basalt (MORB), 48
Milestones: in acritarch evolution, 228–231; in Precambrian biotic evolution, 244, 587–593
Miller, S. A., 181
Millipedes, 485
Mineral deposits: Proterozoic, 59
Mobergellans, 400–401, 408, 1022
Mobile zones, 68
Molecular fossils, 489, 511–514, 519, 592. *See also* Biomarkers.
Molecular phylogeny: alignment of homologous amino acids and nucleotides for, 467; and biogeochemistry, 465, 473–474; and biologic kingdoms, 472–475; distance procedures for, 468–469; and endosymbioses, 471–472; and metazoan lineages, 485; and molecular clocks, 469, 483, 485, 593; and the origin of animal phyla, 443, 453–454, 459, 593; and the origin of metaphytes, 453, 459; parsimony procedures for, 468–469; and the Precambrian fossil record, 42, 202, 472–474, 483; of prokaryotic lineages, 471–475; of protists, 477–483; techniques of, 467–469
Moon: crustal formation in, 13
Moranids, 349, 357, 501–503, 561
MORB (mid-ocean ridge basalt), 48
Morphological evolutionary conservatism, 595–600. *See also* Hypobradytely.
Morphometric analyses: of Archean microfossils, 32–33; of coccoid microfossils, 206–210, 219–222, 227; of coccoid microorganisms, 206–207; of ellipsoidal microfossils, 211; of ellipsoidal microorganisms, 210–211; of eukaryotic micro-algae, 529; of filamentous microfossils, 214–218; of filamentous microorganisms, 211–213, 215–218; and hypobradytely, 595; of living prokaryotes, 529; of microfossils preserved in cherts vs. shales, 195–202, 243, 529
Mount Isa Inlier, Australia, 55; PPRG samples from, 647; stratigraphy of, 647
Mount Painter Province, 55
Multicellular algae: theories for the origin of, 427, 453
Multicellular animals: early body plans of, 443–446; early phyla of, 443–444; theories for the origin of, 399–400, 427–428
Multicellularity: characteristics of, 429–431; concept of, 429; neobiological evidence for the origin of, 431; theories for the origin of, 399–400, 427–428, 453–457
Musgrave Province, 56, 71
Mycoplasmas, 86

Nain Province, 16
Namaqua Province, 55
Narryer Complex, 70
Nautiloids, 987, 1038–1039
Nematodes, 263, 267, 296–297, 323, 326, 393, 431, 435

Subject Index

Netromorph acritarchs, 205, 219, 222, 227–228, 945, 948, 950–951
New Quebec Belt, 55
Nickel deposits, 60–61
Nonacho Basin, 50
Nonesuch Formation: bitumen in sediments of, 111–112
North China Platform, eastern border region, China: PPRG samples from, 623; stratigraphy of, 623
North China Platform, interior region, China: PPRG samples from, 623–624, 687; stratigraphy of, 623
North China Platform, northern border region, China: PPRG samples from, 625–627, 687; stratigraphy of, 624–625
Nubian Cycle, 68, 71, 73–74
Nudibranchs, 485

"Oblivion age", 63
Oceanic crust: half life of, 63
Oceans: Archean chemical composition of, 21; Archean temperature of, 22; bicarbonate concentration in, 22; calcite-saturation in, 22; carbon isotopic composition of, 170; circulation of, 171; dolomite-saturation in, 22; evidence of from marine evaporites, 169; formation of, 10, 21; iron concentrations in 171; photosynthesis in, 177; Proterozoic chemical composition of, 169, 171; "soda"-type, 22, 169; strontium isotopic ratios in, 171; sulfur isotopic composition of, 170
Officer Basin: bitumen in sediments of, 107
Oil: from Oman, 107; Proterozoic, 62; Siberian, 104
Olary Block, 55
Oldest known: abundant stromatolites, 588; acritarchs with pylomes, 588, 591; "autotrophic" organic carbon-isotopic ratios, 587–588, 590; banded iron-formations, 588, 590; "biogenic" sulfur-isotopic ratios, 587–588; body fossils of metazoans, 588–589, 592–593; calcareous metaphytic algae, 588–589, 592; carbonaceous (kerogenous) organic matter, 590; chroococcaleans, 208, 590; "complete eukaryotes", 592; cyanobacteria, 208, 588–590; detrital uraninites, 588; entophysalidaceans, 590; eukaryotic microfossils, 588, 591–592; fenestrate stromatolites, 587; gammaceranes, 588, 592; "giant" acanthomorph acritarchs, 588; glaciation, 588; invertebrates with exoskeletons, 588, 592–593; metazoan body fossils, 588–589, 592–593; metazoan trace fossils, 588–589, 592–593; methanogenic bacteria, 587; millimetric (eukaryotic) microfossils, 588, 591–592; oscillatoriaceans, 590; oxidized paleosols, 588, 590; protozoans, 588–589, 592; red beds, 588, 590; steranes, 588, 590–591; stromatolites, 587–588, 590; terrestrial rocks, 588; trace fossils of metazoans, 588–589, 592–593; true branching, 591; UV-absorbing ozone, 589; widespread cratonic sediments, 588
Olenyok Uplift, USSR: PPRG samples from, 627; stratigraphy of, 627
Oligochaetes, 485
Oman: oils from, 107
Oömorph acritarchs, 205, 222, 227–228
Oppel, A., 489
Organic carbon abundance and isotopic composition, 95–97, 117–122, 127; 712–758; in Archean whole rock samples, 719–720, 740–747, 756–758; in early Paleozoic whole rock samples, 712, 721, 748; in Early Proterozoic whole rock samples, 717–719, 734–740, 754–756; in Late Proterozoic whole rock samples, 712–715, 721–723, 730, 748–752; in Middle Proterozoic whole rock samples, 715–717, 730–734, 752–754
Organic carbon isotopic compositions, 117–122, 127, 712–758, 791–798; in Archean kerogens, 762, 794–796, 798; in Archean whole rock samples, 719–720, 740–747, 756–758; in early Paleozoic whole rock samples, 712, 721, 748; in Early Proterozoic kerogens, 761–762, 793–794, 797–798; in Early Proterozoic whole rock samples, 717–719, 734–740, 754–756; in Late Proterozoic kerogens, 759–760, 791–792, 796; in Late Proterozoic whole rock samples, 712–715, 721–730, 748–752; in Middle Proterozoic kerogens, 760–761, 792–793, 797; in Middle Proterozoic whole rock samples, 715–717, 730–734, 752–754
Organic geochemical processing procedures, 695–708, 799–809; for analyses of extracts, 805; for carbonate carbon analyses, 702; for elemental (C, H, N) analyses of kerogen, 707; for extraction, 803; for gas chromatography-mass spectrometry, 805; for kerogen, 707–708; for kerogen color, 707–708; for kerogen hydrous pyrolysis, 807; for kerogen isolation, 707; for kerogen pyrolysis-gas chromatography, 807; for organic carbon analyses, 701; for rock samples, 695–698; for rock-eval analyses, 701–702; for sample selection and handling, 801; for sulfate sulfur analyses, 703, 705; for sulfide sulfur analyses, 703, 705; for urea adduction, 809
Organic matter: See Extractable organic matter; and Kerogen elemental and isotopic compositions.
Origin of: aerobic bacteria, 591; anaerobic bacteria, 587; cyanobacteria, 590; eukaryotes, 591–592; eukaryotic sexuality, 244, 588, 591; facultative bacteria, 590–591; life, 10, 40; metaphytes, 453; metazoans, 399–400, 427–457, 592–593; multicelled algae, 453; oxygenic photosynthesis, 41, 150, 208, 590; photosynthetic bacteria, 587, 590; prokaryotes, 587; protozoans, 592. See also Origin of multicellular organisms.
Origin of life, 10, 40; at mid-ocean ridges, 10
Origin of multicellular organisms: and abiotic factors, 427–428, 456–457; and biogeography, 438–441; and biomineralization, 441, 447–451; and biotic factors, 427–431, 454–455; and body plans, 444–446; and calcium, 428, 441, 447–451, 456; and cellular differentiation, 430; and cosmic causes, 428, 457; and diversity, 553–561; and early biotas, 433–435, 437–441, 443–446, 556–559; and ecology, 428, 437–439, 455, 461, 561; and extrinsic factors, 427–428, 456–457; and genealogies, 443–444; and genetics, 428, 431, 455; and glaciation, 428, 457; and intercellular coordination, 430; and intrinsic factors, 427–431, 454–455; and land/sea patterns, 428, 457; and large size, 428–430; and microenvironments, 430–431; and migration, 428, 455; and molecular clocks, 485; and molecular phylogeny, 443, 453–454, 459, 485, 593; and morphogenesis, 431; and opal, 449; and oxygen, 428, 456; and phosphorus, 428, 441, 448–451, 456; and skeletons, 428, 441, 447–451, 454–455; theories for, 399–400, 427–428, 431, 453–459
Orogenic belts: half life of, 63
Orogenic zones, 67
Osen-Røa Nappe Complex, Norway: PPRG samples from, 653; stratigraphy of, 653
Ostracods, 263, 267–268, 296–297, 314, 318–319, 400, 406, 448–450, 558
Outgassing, 9, 14, history of earth, 21
Oxygen: box models of the Proterozoic increase of atmospheric, 159, 1185–1187; controls on atmospheric, 159–160; effect on carbon fixation, 127; isotopic compositions of carbonates, 721; "oases", 159, 590; paleoclimatic implications of isotopes of, 165; sinks, 590–591
Oxygenic/anoxygenic microbial mats, 312–313
Oxygenic photosynthesis, 37–39, 280, 313, 341, 473

Paleobiogeography, 350–353, 359–360, 400–401, 515–516; of the Vendian-Cambrian earth, 569–581. See also Biogeography.
Paleoclimates, 50–51, 157, 168
Paleomagnetic: data, 51, 569; poles, 569
Paleosols, 153–155; identification of, 153; implications for atmospheric oxygen content, 160; oxidation sate of, 154; permeability of, 153; R values of, 164; summary of studies of, 154; transport of gases in, 153
Pan-African Belt, 53, 57, 68
Parkaryotes, 471
Peake and Denison Ranges, 55
Pelecypods, 363, 410, 1036, 1040–1041
Pennatulids, 369, 372, 381, 1045, 1047
Penokean Fold Belt, Canada: PPRG-sample from, 677; stratigraphy of, 677
Petroleum, 62; Proterozoic, 84
Phanerozoic biotic evolution, 595; compared with Proterozoic evolution, 599–600; mode of, 599; "normal rules" of, 597; tempo of, 595, 599
Phillips, J., 527
Phosphate: effect of removal from ocean on atmospheric oxygen content, 162
Phosphorites, 62
Photosynthesis: affect on atmosphere of, 161; anoxygenic, 149, 188, 280, 313, 341, 600; C_3 pathway of, 88–89, 125–126; carbon isotopic fractionation during, 126; facultative, 38, 280, 288–289, 312–313, 597, 600; mechanisms of, 279–280; origin of, 37–39, 41, 321, 342; oxygenic, 149, 280, 313, 341, 590, 600; and photosystem I, 161; and photosystem II, 161; pigments of, 272–273, 279–280, 283–285, 304, 328–330, 332; rates of, 170, 328

Photosynthetic bacteria, 41, 89, 248, 473, 587, 590, 827, 829; carbon isotopic fractionation by, 126; isotopically distinct biomarkers from, 126; as source of organic carbon in sediments, 126
Phycodes pedum biozone, 392
Phylogenetic trees: of biologic kingdoms, 472, 474; of metazoans, 485; of prokaryotes, 472, 474; of protists, 482
Pilbara Block, Australia, 15, 54, 71; age of deposits in, 28; PPRG samples from, 648–652; stratigraphy of, 28, 648
Pine Creek Geosyncline, 48, 55
Planetary accretion, 9, 13
Planetesimals, 9, 13
Plate motions: Vendian-Cambrian, 569–581
Plate tectonics: Archean, 18, 21; Vendian-Cambrian, 569–581
Platinoid deposits, 60–61
Platysolenitids, 238–239, 401
Polychaetes, 268, 406–408, 485
Polygonomorph acritarchs, 205, 222, 225, 227–228, 883, 930, 949, 951
Port Arthur Homocline, Canada: PPRG samples from, 678–681; stratigraphy of, 678
Posibacteria, 477
Post-cratonization stage, 72
Pre-Archean (Hadean) earth, 13
Precambrian basement: distribution of, 14
Precambrian-Cambrian boundary, 5, 569, 580; correlation charts for, 374–378
Precambrian paleobiology: historical development of, 179–183, 489
Precambrian Paleobiology Research Group (PPRG): comparison of PPRG 1979–80 (Archean) and PPRG 1987–88 (Proterozoic), xxiii; history of, xxi-xxiv; members of, xix-xx
Precambrian stratigraphy: subdivisions of, 7; Vendian correlation charts for, 374–378
Pre-cratonization stage, 72
Pre-Ediacaran biota, 433–434; biogeography of, 350–353, 438–439; preservation of, 434–435; trophic interactions within, 437, 461
Preservation: of cell membranes, 84–88; of cell walls, 84, 186–188; of cytoplasm, 186; of lipids, 84–88, 105, 133; of microbial pigments, 84–88; of microfossils in cherts vs. shales, 195–206, 210–213, 215–218, 525; of nonmineralized protists, 189–190, 192–193; probability of sediments, 63–65; of sheaths, 84, 186–188
Pre-Vendian biota, 433–434, biogeography of, 350–353, 438–439; preservation of, 434–435; trophic interactions within, 437, 461
Primary productivity: of microbial mats, 285, 340, 823, 837–838; of stromatolites, 285
Prismatomorph acritarchs, 205, 219, 222, 227–228, 877–878, 883, 922, 939, 941, 948
"Probable affinities" of Proterozoic microfossils, 204–205, 243, 1159–1160
Prokaryotes: fossil, 26–39, 195–218, 497–498, 525, 1124–1148, 1153–1156; modern ecological distribution of, 187. *See also* Microfossils.
Prokaryotic cell walls: biochemistry of, 185; biomarkers of, 303–304; preservation of, 186–188
Prokaryotic degraded cytoplasm: preservation of, 186
Prokaryotic sheath (glycocalyx): biochemistry of, 185; biomarkers of, 303–304; preservation of, 186–188
Proterostomes, 485
Proterozoic biogeography, 350–353, 359–360, 400–401, 515–516. *See also* Biogeography.
Proterozoic biostratigraphy, 489–519. *See also* Biostratigraphy.
Proterozoic biotic evolution, 595; benchmarks in, 587–593; compared with Phanerozoic evolution, 599–600; mode of, 597–600; "primitive rules" of, 599–600; tempo of, 595–597
Proterozoic biozones, 501, 504–505, 513, 519. *See also* Biozones.
Proterozoic-Cambrian boundary, 5, 569, 580; correlation charts for, 374–378
Proterozoic megadubiofossils, 347, 414–419, 423, 1035–1053; camasid, 419, 506; discoidal, 416–417; geographic distribution of, 413; horizontal spindles, 414, 416; vertical cylindrical, 416
Proterozoic megafossils: biogenicity of, 347; described taxa of, 953–1053; syngenicity of, 347
Proterozoic megapseudofossils, 347, 359, 361, 413–415, 417, 419, 423, 1035–1053
Proterozoic metallogeny, 59
Proterozoic microdubiofossils, 235, 867, 876–884
Proterozoic microfossils: evaluation of, 865–951; illustrations of, 1055–1117; informal classification of, 242, 529, 533–535, 1119–1168; key for identification of, 204–205, 243; time-ranges of, 596, 1153–1158. *See also* Microfossils; and Microfossil assemblages.
Proterozoic micrononfossils, 179, 867–875
Proterozoic micropaleontology: current status of studies of, 243–244; historical development of, 179–183, 489
Proterozoic micropseudofossils, 179, 237, 239, 867–875
Proterozoic mineral deposits, 57
Proterozoic oil, 62, 104, 107
Protists: biostratigraphic use of, 499; definition of, 477; evolutionary radiation of, 480–483; origin of, 477–480; preservation of, 84, 189–190, 192–193. *See also* PROTISTA (Taxonomic Index).
Protoconodonts, 449
Pteromorph acritarchs, 205, 219–222, 225, 227–228, 880, 906, 910, 920, 949–951
Pterospermopsimorphid acritarchs, 946, 948

Radiant energy, 51; measurement of in microbial mats, 833–835
Receptaculitids, 984, 986
Recycling rate, 51
Red beds, 157
Redfield ratio, 299
References cited, 1205–1286
Rift assemblage, 48
Rifting: crustal, 15, 54, 56, 71; sediments resulting from, 48
Roper Group: bitumen in sediments of, 113–114; total organic carbon (TOC) content of sediments in, 103
Rotifers, 267
Rubey, W. W., 182
Rubisco: evolution and carbon isotopic effects of, 125
Ruedemann, R., 595
Rusophycus palonensis biozone, 392
Russian Platform, USSR: PPRG samples from, 627; stratigraphy of, 627

Sabelliditids, 354, 357, 371, 401, 407, 434, 502–504, 558, 958–959, 969, 973, 975, 1021
São Francisco Basin, Brazil: PPRG samples from, 687; stratigraphy of, 687
Scelerosponges, 448
Scolecodonts, 371, 406
Seagrass, 325
Sea-pens, 381–382
Sedimentary basins: foreland, 49; passive margin, 48; rift-formed, 47; strike-slip, 49
Seward, A. C., 180
Shungite "coal", 62
Siberian Craton, 73; oil from, 104
Siderite: carbon isotope depletion in iron-formation, 150
Sierra de Cordoba, Spain: PPRG samples from, 653; stratigraphy of, 653
Sierra de Guadalupe, Spain: PPRG samples from, 654; stratigraphy of, 654
Silicate weathering, 52, 166
Simpson, G. G., 595–596
Sinosabelliditids, 354, 357, 421, 433, 437, 502–504, 969
Skeletal metazoan fossils, 423, 860, 1017–1033; biomineralization of, 441, 447–451; characteristics of phyla of, 404–412; definition of, 399; described taxa of, 1017–1033; event resulting in appearance of, 399–400, 428, 453–457; first appearance of types of, 401–404, 505, geographic distribution of, 400–401, 441, 516; history of studies of, 399–400; mineralogical types of, 400–401; representativity of, 399–400; stratigraphic distribution of, 399–400, 505–506, 514, 593, 1017–1033. *See also* Early Cambrian biota; and Metazoan body fossils.
Skeletal protozoan fossils, 592
Slave Craton (Province), 16–17, 55, 70, 72
Slime molds, 431, 483
Small shelly fossils, 423, 860, 1017–1033. *See also* Skeletal metazoan fossils.
Smith, W., 489
"Snowball earth", 51–52
"Soda ocean", 22, 169
Solar: luminosity, 11; system, 13
Solenoporids, 359, 996
Sphaeromorph acritarchs, 177, 180, 205, 219–222, 227–228, 230–

Subject Index

231, 336, 355–356, 497–498, 516, 533–535, 540–548, 557, 592, 876–877, 879–882, 884–912, 914–951, 958–959, 961, 963, 965, 967, 971, 1124–1125, 1149–1151, 1157, 1159–1160
Spirochaetes, 480, 877
Sponges, 237, 268, 402, 437, 448–449, 556, 994, 1038–1041, 1043, 1045, 1047, 1049, 1053
Sprigg, R. C., 181
Stable craton-rifting stage, 61
Steam atmosphere, 9
Stellar evolution, 165
Steranes, 244, 588, 590–591
Stratified ocean: biogeochemical-evidence for, 126
Stromatolites, 24, 50, 981, 984, 994, 996, 1038–1039, 1043, 1045, 1047, 1049; Archean, 11–12, 24–27, 41–42, 148–149; biogeochemical studies of, 337–340; biostratigraphic use of, 507–509, 513–516; biozones of, 513, 519; compared with microbial mats, 332–333, 335–342; conical-columnar, 255–256, 259, 269, 335–336, 508–509, 513–516; definition of, 247–248; effects of grazing on, 260, 267, 295–297, 342, 437, 454; eukaryotic contributions to, 335–336; fabrics of, 253, 507; forms of, 253, groups of, 253; heliotropic growth of, 51; Late Proterozoic decline of, 258–259, 267, 297, 338, 240–241, 437, 507–508; lithification of, 257–258; lithologies of, 253; and microbial evolution, 260, 336, 507; microbiotas of, 255, 335, 337–338; microdigitate, 255–258, 269; morphology of, 253–255; paleoecology of, 255–257, 338; primary productivity of, 285; reef-forming, 256–257, 335; taxonomy of, 253, 255; temporal distribution of, 258, 259, 508–509
Stromatoporids, 1043, 1047
Submarine fans, 47
Sudbury irruptive, 61
Sulfur: cycle, 130–131; isotopic evidence for biological derivation from sulfate, 130–131; isotopic fractionation of, 90–91, 162; isotopic implications for atmospheric oxygen content, 162; mobility of in relation to oceanic ferrous iron, 131–132; vapor as UV screen, 11. *See also* Sulfur abundance and isotopic compositions; and Sulfur isotopic compositions.
Sulfur abundance and isotopic compositions, 129–130, 762–763; in Archean sulfides, 764; in Early Proterozoic sulfides, 763–764; in Late Proterozoic sulfides, 762–763; in Middle Proterozoic sulfides, 763; in Proterozoic and Archean sulfates, 764. *See also* sulfur isotopic compositions.
Sulfur cycle: linkage to carbon cycle, 130–131
Sulfur isotopic compositions, 90–92, 129–130, 762–790; in Archean sulfides and sulfates, 764, 780–786, 790; and biostratigraphy, 511–512; in Early Proterozoic sulfides and sulfates, 763–764, 770–780, 788–790; in Late Proterozoic sulfides and sulfates, 762–766, 787; in Middle Proterozoic sulfides and sulfates, 763, 766–770, 787–788. *See also* sulfur abundance and isotopic compositions.
Superior Cycle, 15–17, 68, 71, 74–75
Superior Province, Canada, 54, 59, 72; PPRG samples from, 682; stratigraphy of, 682
Svecofennian Province, 53, 55, 71, 73

Tachytely, 595–596
Tasmanitids, 229–230
Tawuia biozone, 501, 519
Tawuids, 349, 356, 437, 501–503, 561, 967, 971, 973, 975, 977
Techniques for the study of: mat-building microbial communities, 822–823; molecular phylogeny, 467–469
Tectonic reconstructions: Vendian-Cambrian, 574–578
Temporal distribution of: aerobic bacteria, 589, 591; anaerobic bacteria, 587, 589; banded iron-formations, 588, 591; benchmarks in biotic evolution, 587–593; crustal growth cycles, 588, 595; cyanobacteria, 589–590; detrital uraninites, 588, 591; eukaryotes, 589, 591–593; facultative bacteria, 589–590; glacial sediments, 588; inferred ranges of atmospheric carbon dioxide, 589, 595; inferred ranges of atmospheric oxygen, 589–591, 595; metazoans, 589, 592–593; multicelled algae, 589, 592; organic carbon-isotopic ratios, 589, 595; oxidized paleosols, 588, 591; paleoenvironmental indicators, 588, 590–591; photosynthetic bacteria, 587, 589–590; prokaryotes, 587–591; protozoans, 589, 592; red beds, 588, 591; single-celled algae, 589, 591–592; unoxidized paleosols, 588, 591
Terrestrial microbial mats, 324

Testate amoebae, 592
Tetrahymenol, 86–87, 89, 111, 235, 592
Thecamoebaeans, 592
Thelon Orogen, 55
Thermal microbial mats, 313–321
Thermal subsidence, 48
Tholeites, 15
Thrombolites, 248, 250, 253, 263, 268; temporal distribution of, 258–259, 335–336, 339, 508
Thrust nappe, 49
Timofeev, B. V., 181
Tintinnids, 235, 237
Tonalites, 14, 16–17, 54, 67, 72
Torridonian Succession, Great Britain: PPRG samples from, 654–655; stratigraphy of, 654
Total organic carbon (TOC) abundance, 712–758; in Archean whole rock samples, 719–720, 740–747, 756–758; in early Paleozoic whole rock samples, 712, 721, 748; in Early Proterozoic whole rock samples, 717–719, 734–740, 754–756; in Late Proterozoic whole rock samples, 712–715, 721–730, 748–752; in Middle Proterozoic whole rock samples, 715–717, 730–734, 752–754; in Proterozoic rock samples yielding extractable organic matter, 815–819
Touareg Shield, 68
Trace fossils, 369, 372, 381, 416, 419, 423, 433, 437, 1009–1015, 1035–1053; Cambrian, 450, 505; definition of, 391; described taxa of, 1010–1015; first described, 345; geographic distribution of, 516; pre-Yendian, 391–392; regarded as dubiofossils, 391, 396, 1010–1015, 1035–1053; regarded as pseudofossils, 391, 396, 1010–1015, 1035–1053; stratigraphic ranges of, 392–393, 397, 504–505, 513–514, 593, 1010–1015, 1035–1053; Vendian, 392–393, 433, 437, 454, 593, 1009–1015
Trans-Hudson Orogen, 53, 55, 73–74
Trondhjemite, 14, 16–17, 54, 67, 72
Tunicates, 401, 412
Tyler, S. A., 181

Uchur-Maya Block, USSR: PPRG samples from, 655; stratigraphy of, 655
Uinta Uplift, USA: PPRG samples from, 682; stratigraphy of, 682
Ultraviolet radiation, 11, 162–163; biological protection mechanisms from, 11; sulfur UV screen for, 11
Unaka Province, USA: PPRG samples from, 682; stratigraphy of, 682
Uraninite: detrital, 158
Uranium deposits: related to alkaline igneous rocks, 60–61; related to unconformities, 61; Witwatersrand type, 60

Van Horn Mobile Belt, USA: PPRG samples from, 682; stratigraphy of, 682
Varves: glacial, 51
Vascular plants, 85–87, 105, 153, 160, 599
Vase-shaped microfossils, 177, 219, 229, 234–235, 237, 244, 419, 499, 514, 519, 591, 914, 924, 926–927, 938
Vendotaenia biozone, 504, 519
Vendotaenids, 349, 354, 357, 501–504, 561, 959, 961, 967, 969, 973, 975, 977, 979
Volcanic gases, 9; composition of, 169
Vologdin, A. G., 181

Walcott, C. D., 180, 427
Weathering: of igneous rocks, 153; of silicates, 52
West African Craton, 55, 57, 71
West-Central Brazil: PPRG samples from, 687; stratigraphy of, 687
Wilson cycle, 75
Wit Mfolozi Inlier, South Africa: PPRG samples from, 620; stratigraphy of, 620
Wopmay Orogen (Foreland Thrust-Fold Belt), Canada, 53, 55; PPRG samples from, 683–687; stratigraphy of, 683
Worms, 263, 267–268, 296, 323, 357, 369, 371, 381, 387, 393, 406–407, 413, 416, 431, 434–435, 438, 504, 885, 959, 969, 973, 975, 1014, 1039–1041, 1043, 1051

Yilgarn Block, 15, 59, 73

Zimbabwe Craton, 15, 55, 71
Zircon, 14, 21

Index to Geologic Units

Aaron Formation, 377
ABC Range Quartzite, 374, 628
Abitibi Paleosol, 154
Abner Sandstone, 646
Abu Mahara Formation, 107
Adelaidean Supergroup, 930
Adeyton Group, 944
Aga Formation, 942
Aj Formation, 892
Ajax Limestone, 374, 402, 403, 406, 409
Ajabgarh Group, 879, 917
Ajibik Formation, 416, 1011, 1013, 1037
Akademikerbreen Group, 224, 924
Akaitcho Group, 55, 683
Akaitcho River Formation, 1039
Akberdin Formation, 918, 1061, 1103
Aksumbinsk Formation, 1001
Aktugaj Formation, 622
Aladin Formation, 996
Albanel Formation, 96, 737, 755, 778, 789, 794, 798
Albemarle Group, 1005, 1011–1015
Albert Edward Group, 642
Albinia Formation, 922
Aldridge Formation, 961
Algonkian Shale, 874
Allamoore Formation, 98, 120, 682, 688, 715, 730, 752, 760, 797, 901
Almodovar del Rio Group, 130, 653, 688, 762, 787
Altacha Formation, 950
Altyn Formation, 120, 660–662, 688, 716, 753, 894, 961, 971, 1037, 1039, 1049
Alwar Quartzite, 959, 1037
Amasa Formation, 668
Ambarka Member, 621
Amelia Dolomite, 255, 534, 647, 734, 754, 770, 788, 877, 891, 1059
Amga Formation, 989
Amisk Group, 734, 735, 754, 775, 788
Amos Formation, 647
Amri Unit, 881, 932
Amronia Quartzite, 1051
Anakeesta Formation, 663, 688
Anastas'ino Formation, 988, 989, 994
Andersby Formation, 921, 963
Angara Formation, 946
Angepena Formation, 628
Animikie Group, 494, 678, 778, 789, 1043, 1047, 1053
Annijokka Member, 963
Aorfêrneq Dolomite Member, 931, 932
Apache Group, 1053
Apex Basalt, 28–35, 122, 590, 640, 648–652, 688, 720, 758
Appekunny Argillite, 396, 418, 660, 895

Appekunny Formation, 419, 1011, 1039
Appila Tillite, 628
Applecross Formation, 654, 872
Aralka Formation, 130
Aramon Formation, 506, 1053
Arangi Formation, 876, 902, 959
Aravalli Supergroup, 738, 756, 779, 789, 878, 902
Archaeocyath Limestone, 987, 988, 992
Arcoona Quartzite Member, 883, 941, 961, 979
Areyonga Formation, 97, 119, 630–632, 688, 714, 750
Arisaig Paleosol, 154
Arkaroola Subgroup, 628
Arumbera 3 and 4 Formation, 1011, 1013
Arumbera 3 Formation, 1011–1015
Arumbera II Formation, 1001
Arumbera Quartzite, 1041
Arumbera Sandstone, 630
Arunta Igneous Complex, 631
Asbesheuwels Subgroup, 617
Asbestos Hills Formation, 737, 756
Asha Group, 620
Ashburton Formation, 634
Asiak Formation, 683, 686, 688
Atar Formation, 879, 911, 965
Atar Group, 494, 879, 911
Atdabanian Horizon, 988, 989, 911–994
Attikamagen Formation, 355, 778, 789, 876
Auburn Dolomite, 119, 628, 629, 688, 714, 751, 1061, 1105
Autbea Formation, 654, 911
Avzyan Formation, 120, 621, 753, 900
Awatubi Member, 98, 109, 119, 130, 223, 226, 664, 665, 667, 688, 715, 752, 760, 763, 787, 796, 815, 912, 963, 965
Ayan Formation, 887
Azorejo Sandstone Formation, 654

Backbone Ranges Formation, 378
Backlundtoppen Formation, 130, 725, 749
Badaowan Member, 375
Badami Group, 732, 753, 1041
Badwater Greenstone, 668
Bagrad Horizon, 987–996
Baiyanshao Member, 375
Bairenkonda Subgroup, 622
Bakal Formation, 534, 892, 1059, 1066
Bakeevo Formation, 621, 933
Bakoye Group, 129, 787
Balbirini Dolomite, 96, 97, 419, 534, 646, 770, 788, 792, 797, 817, 893, 1059
Ballachulish Slate, 764, 787
Bambu Group, 880
Bambui Group, 256, 494, 729, 751, 873, 917, 996

1297

Bandak Group, 873, 1053
Banganapalle Formation, 622
Bangemall Group, 731, 752
Bar River Formation, 121, 415, 677, 688, 718, 756, 1037, 1039, 1041, 1051
Baraga Group, 668, 876
Baraboo Quartzite, 55
Barents Sea Group, 229, 929, 930, 963, 965
Barma Formation, 937
Barney Creek Formation, 84, 96, 97, 99, 101, 103, 111, 114, 115, 120, 233, 534, 647, 733, 753, 770, 788, 793, 797, 817, 818, 893, 1059
Barrandium System, 725, 749
Barreiro Formation, 736, 755
Basa Formation, 620
Båsnåring Formation, 930, 965
Bass Limestone, 664, 1037, 1039, 1041
Bastion Formation, 947
Batatal Formation, 739, 756
Båtsfjord Formation, 229, 930, 963
Batten Subgroup, 647
Baviaanskop Formation, 614
Bavly Series, 932
Bayangol Formation, 375
Bazaikhe Horizon, 987–989, 995
Beaumont Dolomite, 628, 629, 688, 1043
Beck Spring Dolomite, 98, 118, 119, 235, 256, 534, 658, 688, 714, 729, 751, 759, 796, 879, 916, 1060, 1082, 1083
Bed 7, 728, 751
Bed 8, 728, 751
Bed 9, 728, 750
Bed 10, 728, 750
Bed 11, 728, 750
Bed 12, 726, 750
Bed 13, 726, 750
Bed 14, 726, 750
Bed 15, 726, 750
Bed 17, 726, 749
Bed 18, 726, 749
Bed 19, 725, 749
Bed 19/20, 725, 749
Beda Formation, 930
Bederysh Member, 120, 621, 688, 715, 752, 905, 906
Belair Subgroup, 628
Belaya Formation, 875, 883, 947
Belcher Group, 234, 473, 494, 507, 869, 876, 884, 885
Belingwe Greenstone Belt, 743, 757
Belingwe Group, 494
Belkin Horizon, 376
Beloj Formation, 992, 995
Belt Supergroup, 54, 129, 130, 237, 494, 495, 660, 732, 753, 769, 788, 871, 877, 894–896, 959, 961, 969, 971, 973
Belvue-Road Formation, 614
Besovets Formation, 889
Bessie Creek Sandstone, 646
Bezymenskaya Formation, 965
Bezymyamyj Formation, 652
Bhander Group, 874, 881, 931
Bhander Limestone, 1041, 1043, 1049, 1051, 1053
Bhima Group, 725, 749, 882, 963
Bhima Series, 932, 994
Bianka Member, 621, 688, 713
Biegaizi Formation, 623, 688
Bijaigarh Formation, 766, 787
Bijawar Group, 878, 902, 1051
Billyakh Group, 897
Billy Creek Formation, 374
Bingerbreek Member, 979
Bird Conglomerate, 617
Biri Formation, 98, 119, 653, 688, 713, 726, 749, 759, 796, 930
Biryan Member, 621
Biscay Formation, 643
Biskipøsen Formation, 653, 935
Bitter Springs Formation, 96, 97, 103, 108, 119, 129, 183, 186, 255, 494, 507, 534, 539, 543, 548, 550, 630–632, 688, 714, 729, 751, 766, 787, 792, 796, 817, 818, 873, 912–914, 993, 1060, 1091–1094

Biwabik Iron Formation, 121, 143, 668, 669, 688, 718, 736, 755, 815, 869, 876, 885
Bjoranes Formation, 727, 750
Black Mudstone, 791, 796
Black Reef Formation, 780, 790
Black Reef Quartzite, 1037
Blackaberget Formation, 725, 749
Blackbrook Group, 377
Blårevbreen Member, 377
Blon Formation, 377
Blondeau Formation, 682, 688
Blueflower Formation, 378, 1005, 1011–1015, 1053
Blyth Dolomite "Member", 628, 922
Bobrovka Formation, 905
Bodajbo Subgroup, 918
Bol'shaya Erba Horizon, 988, 989
Bonanza King Formation, 658
Bonavista Formation, 655, 657, 688, 994, 995
Bonavista Group, 378
Bonney Sandstone, 374, 628
Boolgeeda Iron Formation, 634
Boonall Dolomite, 98, 119, 375, 642, 643, 688, 712, 749, 759, 796
Booysens Shale, 617
Borodino Formation, 961, 967
Bothaville Formation, 617
Bothnia Formation, 871, 890
Bowers Supergroup, 967
Brachina Formation, 96, 628, 724, 749, 791, 796, 1001, 1011
Brachina Subgroup, 374
Brand Group, 377, 1041
Breakaway Shale, 647
Breivik Formation, 376, 1013
Brighton Limestone, 628, 726, 750, 1051
Brigus Formation, 378, 655, 657, 688
Briscal Formation, 378
Brioverian Formation, 937, 1013, 1061
Brockman Iron Formation, 144, 146, 634, 743, 757, 869, 1047
Broken Hill Deposit, 771, 788
Broken Hill Formation, 770
Brøttum Formation, 653
Bruce Formation, 121, 677, 678, 688, 718, 756, 779, 789
Brusno Formation, 890
Buelna Limestone, 990
Bulajskaya Formation, 995
Bulawayan Group, 494, 743, 757, 794, 798
Bulawayan/Shamvaian Group, 782, 790
Bullwhacker Member, 658
Bungaree Formation, 628
Bungle Bungle Dolomite, 35, 96, 98, 120, 494, 633–635, 688, 716, 732, 733, 753, 760, 792, 797, 896, 961, 967, 1060
Bunkers Sandstone, 374
Bunyeroo Formation, 374, 628
Burajskij Horizon, 996
Burgess Shale, 364, 371, 399, 445, 556, 557
Burovaya Formation, 652, 909
Burra Group, 96, 628, 729, 751, 791, 796, 1043
Burzyan Group, 621
Buschmann Formation, 1007
Bushimale System, 994
Bushimay Supergroup, 872, 879, 904
Bushveld Igneous Complex, 617
Butler Paleosol, 154
Butun Formation, 887, 1043
Bystraya Formation, 951

Cadiz Formation, 658
Cailleach Head Formation, 654
Calc Zone, 736, 755
Calcarie Inferieur Formation, 574
Calcarie Supérieur Formation, 574, 722, 748
Caldecote Volcanic Formation, 377
Callanna Group, 628
Cambrian Formation, 722, 748, 1013
Càmbrico Inferior Formation, 1007
Cambridge Argillite, 934
Campbellrand Subgroup, 617
Campito Formation, 378, 659, 1011

Index to Geologic Units

Cango Formation, 724, 749
Canyon Formation, 723, 748
Cap Dolostone, 375
Cape Wrath Paleosol, 154
Cappahayden Formation, 944
Carabini Member, 817
Carawine Dolomite, 634, 740, 756
Carbon Butte Member, 664, 667, 688
Carbon Canyon Member, 98, 109, 119, 664, 666, 667, 688, 715, 752, 760, 763, 797
Carbon Leader Member, 121, 130, 617, 618, 720, 764, 790, 815
Cardenas Lavas, 664
Carpentarian Supergroup, 628
Carrizo Mountain Formation, 682
Carters Bore Rhyolite, 646
Castner Marble, 1039
Catlin Member, 658
Caue Itabirito Formation, 737, 756
Cayetano Formation, 1013
Celsiusberget Group, 967
Central Mount Stuart Formation, 883, 939, 963, 1003, 1005, 1011
Central Rand Group, 617
Chamberlain Shale, 120, 660, 662, 688, 717, 753, 763, 894, 959
Chambers River Formation, 646
Chambless Limestone, 658, 688
Changcheng Group, 494, 625, 870, 871, 876, 877, 889–891, 894, 959, 967, 975, 977, 979, 1037
Changlingzi Formation, 965, 975, 1001, 1003, 1037, 1043–1047
Changlongshan Formation, 354, 912, 961–979
Changzhougou Formation, 625, 626, 688, 1037
Chapel Island 1 Formation, 1011, 1015
Chapel Island 2 Formation, 1011, 1013
Chapel Island Formation, 118, 119, 122, 130, 238, 355, 378, 405, 494, 655–657, 688, 712, 721, 748, 762, 787, 977
Chapoghlu Shale, 959, 963, 967, 969
Chapoma Formation, 925
Chara Formation, 950
Charteris Basalt, 648
Chartorysk Formation, 1061, 1109
Chatkaragaj Formation, 880
Chattisgarh System, 730, 751
Chelmsford Formation, 735, 754, 776, 788, 1039
Chenchinsk Formation, 994, 995
Chengjiajiang Formation, 375, 623, 624, 688
Chernakamenka Formation, 1007
Cheshire Formation, 745, 757
Chesnokovaya Horizon, 992
Chhera Member, 874, 881, 929
Chichkan Formation, 119, 255, 534, 550, 622, 623, 688, 724, 749, 882, 934, 1061, 1110–1112
Chilhowee Group, 682
Chitkanda Formation, 925
Chocolay Group, 668
Chopan Porcellanite, 971
Christie Bay Group, 655
Chuanlinggou Formation, 98, 120, 534, 625, 626, 688, 717, 754, 761, 763, 797, 870, 876, 889, 967, 979
Chuar Group, 96, 97, 99, 109, 111, 189, 223, 226, 229, 234, 235, 494, 592, 664, 817, 906, 909, 910, 912, 914, 959, 961, 963, 965, 971
Chulaktau Formation, 119, 127, 622, 623, 688, 748, 883
Chuniespoort, Group, 617, 884
Churochnaya Formation, 935
Cid Formation, 377
Cijara Formation, 119, 130, 654, 688, 712, 762, 787
Clarke Series, 732, 752
Cleaverville Formation, 122, 640, 648, 649, 688, 720, 758
Clemente Formation, 1037, 1039
Closepet Granite, 622
Clutha Formation, 614
Cobalt Group, 677
Conasauga Group, 682
Conception Group, 494, 726, 750, 942, 1001, 1005, 1049
Connecting Point Group, 942
Conococheague Limestone, 682
Cooley Dolomite Member, 647, 893
Copper Harbor Conglomerate, 668

Coppercap Formation, 57, 119, 130, 669, 672–674, 688, 713, 750, 763, 787
Coppermine River Group, 667
Corbanbirini Formation, 646
Corboy Formation, 648
Corcoran Formation, 97, 646
Coronation Supergroup, 683
Corumbá Group, 494, 1007
Cow Head Group, 239
Cowles Lake Formation, 683
Coxco Dolomite Member, 647
Craighead Limestone, 990
Crawford Formation, 646
Crocodile River Formation, 737, 756
Crystal Spring Formation, 658
Cuddapah Group, 622, 732, 753, 890, 1041
Cuddupah Limestone, 996
Cumbum Subgroup, 622
Curdimurka Subgroup, 628
Cyclops Member, 630

Dabis Formation, 374, 379, 1001–1007, 1011
Dahai Member, 375
Dahongyu Formation, 186, 625, 626, 688, 871, 891
Dakkovarre Formation, 936, 941, 963
Dales Gorge Member, 118, 122, 130, 142, 634, 642, 689, 721, 740, 756, 764, 790
Damara Supergroup, 727, 750, 874
Danilovka Formation, 376
Danilovskaya Oil, 103
Dashka Formation, 234, 925, 1061, 1103, 1104
Daspoort Formation, 778, 789
Daspoort Quartzite, 617
Datanpo Formation, 130, 725, 749
Dazhuang Formation, 623
Deep Lake Group, 55
Deep Spring Formation, 378, 659, 660
Deer Lake Complex, 782, 790
Delhi Supergroup, 879, 917, 959, 1037
Dengying Dolomite, 375
Dengying Formation, 130, 397, 722, 748, 943, 959, 967, 977, 979, 996–998, 1005, 1011, 1013, 1037, 1049, 1053
Denison Paleosol, 154
Deoban Formation, 994, 995
Deoban Limestone, 910
Dereviniskaya Formation, 965
Derevnya Formation, 652, 906, 1060, 1066
Derlo Forrnation, 883, 939, 1061, 1114
Dharwar Group, 622
Diabaig Formation, 654, 727, 750
Diaoyutai Formation, 975
Ding Dong Downs Volcanics, 643
Dismal Lakes Group, 98, 120, 188, 189, 255, 494, 667, 689, 716, 753, 760, 797, 895, 1059, 1070
Dividal Group, 1003
Dogra Slates, 996
Dolomie Inferieur Formation, 574, 722, 748
Dolomite Group, 1039
Dolomite Series, 235
Dominion Group, 617,
Dominion Paleosol, 154
Donglongtan Formation, 375
Donnegan Member, 817
Dornala Formation, 622
Doublet Group, 55
Doushantuo Formation, 130, 231, 375, 534, 550, 724, 749, 883, 940, 941, 967, 990, 994, 1041
Dox Sandstone, 664
Dracoisen Formation, 377, 722, 748, 942
Draken Conglomerate, 234, 534, 550, 924
Draken Formation, 726, 750
Drakenstein Paleosol, 154
Dripping Springs Formation, 732, 753
Drook Formation, 378, 942
Duck Creek Dolomite, 96, 233, 255, 494, 736, 755, 793, 798, 887, 1059
Duerdin Group, 642

Duffer Formation, 28, 648
Duguan Formation, 120, 623, 624, 689, 715, 752
Duliominskaya Oil, 103
Dundas Formation, 965
Dunderberg Formation, 658
Dungaminnie Formation, 646
Duolbasgaissa Formation, 376
Duppa Member, 664, 666, 667, 689
Durnoj Mys Formation, 934
Dwaal Heuvel Quartzite, 617, 618, 689
Dzhetymtau Formation, 883
Dzhide Formation, 996
Dzhur Formation, 996

Earaheedy Group, 96, 494, 735, 754, 793, 797
Early Cambrian Formation, 1011
Easdale Slate, 882, 932
Ecca Group, 620
Echo Bay Formation, 1037
Ediacara Member, 393, 395, 628
Egan Formation, 119, 642–645, 689, 712, 724, 749
Egan Tillite, 375
Ehjno Formation, 926
Ehsehlehkh Formation, 872, 879
Eilean Dubh Formation, 994, 995
Einasleigh Metamorphics, 734, 754
Eino Formation, 926
Ekkerøy Formation, 931, 963
Ekre Formation, 653
Ekre Shale, 376
Ekwi Supergroup, 494, 669–671, 689
Elanka Formation, 987, 989
Elankoe Formation, 376
Elatina Formation, 374, 570, 571, 628, 1003
Elbobreen Formation, 725, 749
Eleonore Bay Group, 494, 727, 750, 880, 926, 959, 963, 1053
Elkera Formation, 1013
Ella Island Formation, 947
Elliot Lake Group, 677
Elvire Formation, 375, 633, 642, 689
Emeroo Subgroup, 628
Emmerugga Dolomite, 647, 891
Empire Formation, 660
Enchantment Lake Formation, 668
Enisej Formation, 991, 995, 996
Epworth Group, 48, 55, 494, 683, 735, 754, 888
Erdaohe Formation, 623, 624, 689
Erkeket Formation, 375
Esfordi Formation, 1005
Espagnola Formation, 121
Espanola Formation, 416, 417, 677, 678, 689, 718, 737, 756, 779, 789
Esperanza Formation, 645
Estenilla Formation, 119, 654, 689, 712
Etcheminian Group, 1011
Etina Limestone, 724, 749
Eurelia Beds, 731, 752, 768, 788
Euro Basalt Formation, 28, 648, 649, 651, 652, 689

Fargoo Tillite Formation, 375, 642, 645, 689
Farina Subgroup, 628
Farnell Group, 1011–1015
Fecho do Funil Formation, 737, 755
Feishui Group, 494, 1007
Fengjiawan Formation, 120, 623, 624, 689, 715, 752
Ferapontyevo Formation, 977
Fermeuse Formation, 944
Fig Tree Group, 15, 25–27, 37, 98, 122, 494, 588, 614–616, 689, 719, 746, 758, 762, 786, 790, 795, 798
Flagstaff Mountain Paleosol, 154
Flat Rock Formation, 375, 642
Flin Flon Paleosol, 154
Flora Formation, 729, 751, 967
Fontano Formation, 98, 120, 683, 684, 686, 689, 715, 754, 761, 763, 797
Forteau Formation, 987, 988, 990, 992, 993
Foselv Formation, 870

Fortescue Group, 15, 28, 39, 494, 588, 634, 745, 758, 795, 798
Fowlers Gap Formation, 1001, 1013, 1037
Franklin Mountain Formation, 675, 689
Franklinsundet Group, 963, 975
Frank River Sandstone, 642, 644, 689
Freda Sandstone, 668
Frere Formation, 233, 234, 255, 494, 735, 754, 888, 1059
Frood Formation, 780, 789
Fuentes Formation, 119, 654, 689, 712

Galeros Formation, 109, 110, 255, 664, 816, 906, 909, 963, 965, 1060, 1079
Gametrail Formation, 378
Gamohaan Formation, 98, 121, 130, 590, 617–620, 689, 719, 740, 756, 762, 764, 790, 798, 884, 993, 1059, 1062
Gandarella Formation, 737, 756
Gangolihat Dolomites, 874, 881, 929
Gangurthi Shale, 963
Gaojiashan Member, 959, 967, 1049
Gaoshanhe Group, 623
Gaoyuzhuang Formation, 35, 98, 120, 255, 355, 501, 534, 625, 626, 689, 716, 733, 753, 761, 797, 877, 894, 959, 975, 1059, 1067–1069
Gap Well Formation, 769, 788
Gaskiers Formation, 378
Gawler Range Volcanics, 628
Gdov Beds, 937
Geluk Subgroup, 614
Gerbikan Horizon, 988
Ghaap Group, 617, 884
Gibbett Hill Formation, 944, 1041
Gibraltar Formation, 736, 755
Giddalur Formation, 622
Gilbert Range Quartzite, 628
Gillen Member, 108, 630, 914
Glasgowbreen Formation, 229, 751, 916
Gleevatsk Formation, 996
Glen Osmond Slate, 628
Golden Dyke Formation, 96, 735, 754, 793, 797
Golneselv Formation, 921, 963
Goloustnaya Formation, 900
Gonam Formation, 881, 1007
Goodrich Quartzite, 668
Gordon Lake Formation, 121, 677, 689, 719, 756, 779, 789, 1037
Gorge Creek Group, 15, 122, 648, 650, 689, 720, 746, 758, 798, 799
Gorynskoy Formation, 979
Gouhou Formation, 623
Government Formation, 783, 790
Government Subgroup, 121, 617, 619, 689, 720
Gowganda Formation, 121, 677, 678, 689, 719, 737, 756, 779, 789, 869, 1039, 1047
Graensesø Formation, 876, 1053
Grand Canyon Supergroup, 664
Grant Bluff Formation, 939
Grasdal Formation, 941, 963
Great Slave Supergroup, 494, 655
Green Head Group, 1043, 1045
Green River Formation, 87
Grenville Metasediments, 767, 787
Grenville Supergroup, 1037, 1039, 1041, 1045, 1051
Greyson Shale, 98, 120, 130, 354, 355, 416, 418, 660, 662, 663, 689, 716, 753, 760, 763, 788, 797, 959, 961, 969, 971, 973, 1051
Grinnell Argillite, 660
Griquatown Formation, 617, 1045
Griquatown Iron Formation, 140, 142
Gross Aub Formation, 374
Grusdievbreen Formation, 728, 750
Grushka Formation, 376
Grythyttan Slate, 735, 755
Guddadarangavanahalli Formation, 622, 689, 717
Gulcheru Formation, 622
Gunflint Iron Formation, 96, 98, 121, 130, 181, 182, 234, 244, 255, 447, 473, 494, 511, 534, 678–681, 689, 718, 737, 755, 760, 764, 778, 789, 793, 798, 815, 869, 870, 876, 885–887, 1059, 1063
Gunpowder Creek Formation, 645

Index to Geologic Units

Hailuoto Sequence, 882, 935, 963
Hakatai Shale, 664, 1037, 1039, 1041
Halls Creek Group, 643
Halls Town Formation, 943
Hamburg Formation, 658
Hamersley Group, 15, 28, 60, 118, 139, 140, 142, 146, 147, 161, 634, 1047
Hank Supergroup, 494
Harbour Main Group, 378
Hardeberga Sand Stone, 226
Hardy Sandstone, 634, 635, 640–642, 652, 689
Harkless Formation, 378, 659
Hartshill Formation, 377
Hastings Group, 1051
Hautes Chutes Formation, 959
Hawker Group, 628
Hay Creek Group, 669
Hazel Formation, 682
Hearne Formation, 736, 755
Heavitree Quartzite, 630
Hector Formation, 726, 750, 931, 961, 967, 1061, 1109
Hedmark Group, 653, 874, 930, 935, 991
Hekpoort Andesite, 617, 618, 620, 689
Hekpoort Paleosol, 154
Helena Formation, 120, 660, 663, 689, 716, 753
Helicopter Siltstone, 633
Hellroaring Member, 1049
Hemlock Formation, 668
Heron Bay Group, 782
Heysen Supergroup, 628
Hinlopenstretet Supergroup, 170
Hiunpani Member, 929
Hoadley Formation, 877
Hodgewater Group, 943
Honeyeater Basalt, 648
Hongshuizhuang Formation, 98, 120, 625, 627, 689, 753, 760, 763, 797, 871, 877, 899
Hooggenoeg Formation, 25, 98, 122, 614–616, 689, 720, 747, 758, 762, 795, 798
Hornby Bay Group, 667, 877, 890
Horsethief Creek Group, 1039, 1049
Hospital Hill Formation, 782, 790
Hotazel Formation, 118, 122, 617, 619, 689, 721, 756
Hough Lake Group, 677
Huainan Group, 494, 623, 918, 961–977, 1007
Huangshandong Member, 375
Hunnberg Formation, 223, 224, 728, 751, 922, 963
Huqf Formation, 723, 748
Huqf Group, 84, 107, 127
Huqf Group Oils, 107
Huronian Supergroup, 54, 71, 157, 158, 677, 789, 870, 1039, 1043, 1047, 1051
Hutuo Group, 494, 887
HYC Pyritic Shale Member, 96, 647, 733, 753, 793, 797, 893
Hyco Formation, 377, 1007
Hyperborean Formation, 918

Ikabijsk Formation, 884
Il'yushkana Formation, 900
Imsdalen Formation, 653
Ingletonian Formation, 1013
Ingta Formation, 378, 1013
Insuzi Group, 121, 588, 620, 689, 720, 746, 758
Inzer Formation, 621, 912
Ironwood Iron Formation, 869
Islay Quartzite Series, 933
Isua Iron Formation, 139, 140, 142, 144, 146
Isua Supergroup, 747, 758, 786, 790, 796, 798
Isua Supracrustal Belt, 13, 14, 22, 494, 587, 588, 590
Iswara Kuppam Formation, 622
Izluchinsk Formation, 933

Jacadigo Group, 930
Jacobina, 782, 790
Jammalamadugu Group, 959, 969
Jarrad Sandstone Member, 119, 633, 642–645, 689, 713, 749
Jatulian Group, 54, 870, 876, 890, 1043

Jatulian Shungite Formation, 735, 754
Jbeliat Group, 374
Jeerinah Formation, 98, 121, 130, 634, 635, 637, 690, 719, 744, 757, 762, 764, 790, 794, 798
Jeppestown Shale, 121, 617, 619, 690, 720, 783, 790
Jingeryu Formation, 625, 873, 916, 963–973
Jinshanzhai Formation, 623, 961
Jiucheng Member, 975–979
Jiudingshan Formation, 119, 623, 690, 714, 751, 918, 991
Jiuliqiao Formation, 501, 504, 959, 965, 969, 973, 975, 977, 1007
Jixian Group, 494, 625, 871, 872, 877, 878, 896, 898, 899, 901
Jodhpur Siltstone, 1043
Joe's Luck Formation, 614
Joffre Member, 142
Johannesburg Subgroup, 617
Johnie Formation, 573, 658
Johnny Cake Shale Member, 642
Johnstone Conglomerate, 617
Jonesboro Limestone, 679, 682, 690, 712
Julie Member, 119, 630, 632, 689, 712, 748
Jupiter Member, 98, 109, 119, 120, 130, 664–666, 690, 715, 752, 760, 763, 787, 797, 909, 965
Jurassic Formation, 1013

Kacha Formation, 947
Kachergat Formation, 901
Kada Formation, 944
Kadali Group, 904
Kadunga Slate, 628
Kahochella Group, 655, 736, 755
Kaimur Group, 1045, 1047, 1049, 1051
Kairovo Formation, 939, 959, 961
Kajrahat Formation, 902
Kakadoon Granite, 646, 647
Kaladgi Group, 734, 754
Kalmarsund Formation, 405, 1013
Kamarovsk Formation, 197, 1061
Kamovsk Formation, 933, 1109
Kameeldoorns Formation, 121, 130, 617, 618, 690, 719, 757, 764, 790
Kameshki Formation, 376, 987, 989, 993, 995, 996
Kamo Formation, 965
Kan'onsk Formation, 927
Kandyk Formation, 655, 930, 931, 965
Kangjia Group, 873, 880, 920
Kaniapiskau Supergroup, 959
Kanilov Formation, 355, 969, 977, 979, 1037, 1043
Kanilov Group, 969
Kanilovka Formation, 883, 939, 945, 1061, 1114
Kaplonosy Formation, 377
Kapp Lord Formation, 729, 751, 963, 975
Karagas Formation, 1005, 1043, 1045, 1051
Karagas Formation Horizon II, 1045
Karakatty Formation, 995
Karatau Group, 621
Karauli Quartzite, 1043
Karelia Complex, 885
Karelian Dolomite, 736, 755
Kasegalik Formation, 255, 534, 736, 755, 884, 1059, 1062
Katanga System, 767, 787
Kataskin Member, 621, 690, 715
Katav Formation, 120, 621, 690, 715, 752, 909
Katherine Group, 669, 1037
Keele Formation, 119, 378, 669–671, 690, 713, 750
Keewatin Group, 121, 679, 690, 719, 740, 741, 756, 757, 780, 781, 790
Keewatin Iron Formation, 740
Keewatin Volcanics, 678
Kennedy Siltstone, 647
Kenyade Horizon, 992
Kessyuse Formation, 375, 991
Kessyusinsk Formation, 995
Ketema Horizon, 996
Keteme Formation, 376, 991
Keweenawan Group, 668
Khairkhan Formation, 375, 411

Khatyspyt Formation, 98, 103, 104, 106, 119, 127, 375, 382, 627, 690, 712, 748, 759, 796, 816, 1001–1003
Kheinjua Formation, 878, 902, 971, 1047
Khmelnitski Formation, 376, 1011–1015
Khopychsk Formation, 994
Khuzhir Formation, 884, 951
Kil'din Series, 900
Kimaj Complex, 652
Kimberley Shale, 121, 619, 690, 720, 758
Kimerot Group, 48
Kingbreen Formation, 130, 729, 751, 917
Kingston Peak Formation, 98, 119, 658, 690, 714, 730, 751, 759, 796, 916
Kintla Member, 1037
Kipalu Iron Formation, 870
Kirgitej Formation, 550, 918
Kirtonryggen Formation, 721, 748
Kiya Formation, 376
Klackaberget Formation, 725, 749
Klipfonteinheuwel Formation, 98, 121, 617–619, 690, 719, 762, 798
Klipriviersberg Group, 617
Klubbnes Formation, 229, 917, 963
Knob Lake Group, 55, 959
Knox Group, 682
Koekas Subgroup, 617
Kokdzhot Formation, 878
Koksu Formation, 622
Komarovo Formation, 1011
Komati Formation, 614, 747, 758
Kona Dolomite, 120, 668, 677, 690, 718, 755, 778, 789
Kongsfjord Formation, 930
Koolpin Formation, 98, 120, 646, 690, 717, 735, 754, 761, 797
Kortbreen Formation, 130, 730, 752, 917
Kostov Formation, 627, 690
Kotlin Formation, 505, 875, 943, 973, 975–979, 1061, 1115
Kotlin Series, 377
Kotujkan Formation, 534, 897
Krivoj Rog Iron Formation, 140, 736, 755, 876
Krivoj Rog Series, 888, 967
Krol Formation, 723, 748
Krol C Formation, 723, 748
Krol D Formation, 723, 748
Krol E Formation, 722, 748
Kromberg Formation, 25, 27, 98, 122, 614–616, 690, 720, 746, 758, 762, 795, 798
Krugersdorp Formation, 617
Krushanovka Formation, 376
Kudash Formation, 933
Kuibis Formation, 723, 748, 883, 979, 1037, 1039
Kuibis Supergroup, 940, 965
Kuktur Member, 621, 622, 690, 716
Kulele Creek Limestone, 735, 754
Kulindinsk Formation, 927
Kulpara Limestone, 411
Kundair Formation, 622
Kundelungu Group, 730, 752, 873, 910, 919
Kuniandi Group, 642
Kuonamka Formation, 375
Kureninsk Formation, 990
Kurgan Formation, 622
Kurnool Group, 622, 959, 969
Kurnool System, 730, 752
Kurtun Formation, 1001
Kuruman Iron Formation, 121, 140, 142, 144, 146–148, 151, 155, 169, 173, 617, 618, 690, 719
Kuruna Siltstone, 121, 130, 634–636, 641, 690, 720, 757, 764, 790
Kushalgarh Formation, 97
Kutorgina Formation, 376, 987, 990, 991, 993
Kwagunt Formation, 103, 109, 110, 111, 223, 226, 229, 255, 664, 730, 751, 815, 816, 912, 914, 959–965, 971, 1060, 1090, 1091
Kylena Basalt, 121, 634, 635, 637, 641, 690, 720, 758
Kyrshabaktin Formation, 622, 883

Labine Group, 1037
Ladoga Formation, 887
Lady Loretta Formation, 645
Lakhanda Formation, 197, 198, 222, 227, 228, 231, 534, 655, 872, 879, 907, 908, 965, 967, 971, 977, 1060, 1072–1078
Lalla Rookh Sandstone, 648, 651, 690
Landrigan Tillite, 642, 727, 750
Langlaagte Quartzite, 617
Lansen Creek Shale Member, 96, 97, 103, 113, 732, 753, 792, 797, 817, 818
Lapichy Formation, 377
Lapponian Group, 71
Late Archean BIF (Ontario), 783, 790
Latham Shale, 658
Lauhanvuoren Formation, 1011
Lausitzer Grauwacken Formation, 943
Lawn Hill Formation, 646
Lecice Member, 882
Leila Sandstone, 647
Lemeza Member, 621
Lena Formation, 993
Leningrad Blue Clay, 534
Lewin Shale, 634, 742, 757
Lewisian Basement, 654
Liantuo Formation, 375
Libby Creek Group, 55, 1013, 1039
Libby Formation, 871, 877
Limbla Member, 630
Limmen Sandstone, 646
Limpopo Belt, 746, 758
Lingulid Sandstone, 226, 947
Linok Formation, 652
Lintiss Vale Formation, 1011–1015
Little Dal Group, 119, 130, 354, 355, 357, 494, 501, 550, 669–672, 674, 676, 690, 714, 751, 763, 787, 919, 961–979, 1037, 1043
Liuchapo Formation, 1039
Liulaobei Formation, 918, 959–977, 1007
Liumao Formation, 1003, 1043, 1045, 1047
Livingstone Conglomerate, 617
Lomagundi Group, 739, 756
Lomfjorden Supergroup, 170
Lone Land Formation, 959
Longarm Quartzite, 1039
Longjiayuan Formation, 623, 624, 690
Longmyndian Supergroup, 921
Longwangmiao Formation, 375
Lontova Formation, 377, 393
Looking Glass Formation, 647
Lookout Shist, 120, 677, 690, 718, 755
Lopatinskij Formation, 919
Lorrain Formation, 121, 677, 690, 719, 756, 779, 789, 1039, 1051
Los Parrales Shale, 654, 690
Los Villares Formation, 653
Louden Volcanics, 648, 651, 690
Louisa Downs Group, 642, 644, 645
Love's Creek Member, 98, 630–632, 690, 715, 759, 769
Lubbock Formation, 642
Lublin Formation, 130, 377, 977
Luipaardsvlei Quartzite, 617
Luonan Group, 623
Luoquan Formation, 98, 119, 130, 623, 624, 690, 713, 750, 759, 763, 787, 796
Luotuoling Formation, 625
Lynott-Carabirini Member, 97
Lynott-Donnegan Member, 97, 797
Lynott Formation, 96, 647, 817
Lynott-Hot Spring Member, 97
Lyubim Formation, 377, 959

Maastakh Formation, 375
Mackenzie Mountains Supergroup, 57, 669
Maddina Basalt, 121, 634, 637, 638, 690, 720, 757
Magaliesberg Quartzite, 617
Magazine Shale, 647
Main Conglomerate, 617, 815
Mainoru Formation, 97, 646
Maiwok Subgroup, 646
Makaryevo Formation, 973, 975, 977
Makawassi Quartz Porphyry, 617
Malga Formation, 655

Malgin Formation, 899
Malka Formation, 919
Mall Bay Formation, 378, 942
Mallapunyah Formation, 647, 734, 754, 771
Malmani Dolomite, 96, 789, 869, 1037, 1041
Malmani Subgroup, 121, 617, 618, 690, 719, 739, 756, 780, 789, 794, 798
Malmesbury Formation, 724, 749
Maloinzer Member, 621, 622, 690, 716
Manganese Group, 419, 1039
Manitounuk Group, 737, 755, 959
Manjeri Formation, 744, 757, 794, 798
Manykaj Formation, 942
Manzurska Formation, 987
Maplewell Group, 377
Maplewood Group, 1001, 1005
Mara Dolomite Member, 647, 891
Maraisburg Quartzite, 617
Marimo Slate, 96, 733, 753, 793, 797
Marystown Group, 378
Marquette Range Supergroup, 668, 876
Marra Mamba Iron Formation, 142, 634, 742, 757, 869
Martyukhe Formation, 881, 930
Martyukhinsk Formation, 991, 994, 996
Marwar Group, 1043
Mashak Formation, 621
Mashan Group, 1003, 1043, 1045, 1047, 1049
Masterton Sandstone, 647
Matinenda Formation, 121, 677, 678, 690, 719, 756, 780, 790
Matzatzal Quartzite, 55
Mazowsze Formation, 225, 377
Maya Formation, 655
Mbuji Mayi Supergroup, 872, 879, 904
McAlly Shale, 642
McArthur Group, 54, 84, 102, 111, 113, 114, 494, 646, 647, 690, 717, 734, 754, 774, 788, 815, 877, 891, 893, 1041
McKim Formation, 121, 677, 678, 690, 719, 756, 780, 790
McLeary Formation, 96, 255, 737, 755, 794, 798, 885, 109
McManus Formation, 1005, 1011–1015
McMinn Formation, 84, 96, 97–99, 103, 113, 114, 120, 228, 646, 690, 716, 732, 753, 760, 792, 797, 815, 817, 818, 897
McPhee Formation, 28, 648, 649, 651, 652, 690
Medicine Peak Formation, 391, 416, 417, 1013
Medicine Peak Quartzite, 1039
Meentheena Carbonate Member, 634, 635, 640, 641, 690, 720, 762, 798
Meishucun Formation, 945, 991
Menihek Formation, 736, 755, 778, 789
Menominee Group, 668
Menominee Iron Formation, 1041
Mescal Limestone, 732, 752, 1053
Mesnard Quartzite, 668
Messel Shale, 126
Michigamme Slate, 668, 959
Michipicoten Iron Formation, 131, 745, 758, 784, 790
Mickwitzia Formation, 225, 226, 1011
Middle Marker Member, 614, 615, 616, 690, 720
Miette Group, 418, 881, 931, 961, 967, 1011
Miller Peak Formation, 1037
Millingport Formation, 377
Minas Series, 870
Mine Series, 994
Min'yar Formation, 119, 255, 534, 550, 621, 690, 713, 750, 928, 929, 1061, 1106, 1107
Mineral Fork Formation, 729, 931
Mineral Fork Tillite, 751
Mingah Tuff Member, 634, 640, 691
Mintaro Shale, 628
Miroedikha Formation, 196–198, 231, 534, 550, 652, 873, 879, 914–916, 965, 967, 969, 977, 1060, 1082–1089
Mississagi Formation, 121, 677, 678, 691, 719, 756, 779, 789, 1043, 1047
Missoula Group, 660, 871, 877, 896
Mistaken Point Formation, 378, 379, 383, 1001, 1005
Mitcham Range Quartzite, 628
Mitchell Yard Dolomite Member, 647
Moelv Formation, 653

Moelv Tillite, 376
Mogamureru Subgroup, 622
Mogilev Formation, 376, 874, 937, 959, 961, 971, 973, 1001–1007, 1011, 1013, 1061, 1112
Mogobane Series, 869
Molar Formation, 967
Mons Cupri Volcanics, 648
Montacute Dolomite, 628, 629, 691
Monte Christo Formation, 884
Montenegro Member, 659, 660, 691
Monterrubio Formation, 654
Montgomery Formation, 782, 790
Moodarra Siltstone, 647
Moodies Group, 15, 16, 614, 616, 746, 758
Moonlight Valley Tillite, 633, 634, 642, 644, 691, 713, 725, 749
Moontana Porphyry, 628
Moorillah Formation, 1001
Mortensnes Tillite Formation, 376, 941
Mosquito Creek Formation, 648, 651, 691
Moty Formation, 875, 883, 947, 979, 995, 1061, 1117
Mount Ada Basalt, 28, 648, 649, 651, 691
Mount Bertram Sandstone, 642
Mount Bruce Supergroup, 28, 634
Mount Forster Sandstone, 375, 633, 634, 642–644, 691
Mount Isa Group, 647, 772, 788, 891
Mount Isa Inlier, 770, 788
Mount John Shale Member, 633, 871
Mount McRae Shale, 96, 634, 742, 757, 794, 798
Mount Parker Sandstone, 633
Mount Partridge Group, 49
Mount Roe Basalt, 634
Mount Rogers Volcanic Group, 682, 1039
Mount Shields Formation, 120, 660, 663, 691, 716, 753
Mount Sylvia Formation, 634, 742, 757
Mount Watson Formation, 908
Mozaan Group, 620, 746, 758, 795, 798
Mozambique Belt, 731, 752
Msauli Chert Member, 614, 616, 691
Muhos Formation, 534, 550, 732, 752, 874, 882, 936
Mule Spring Formation, 378
Mundallio Subgroup, 628
Murandav Formation, 872
Murandavsk Formation, 996
Murchinsonfjorden Supergroup, 494
Murdama Group, 882, 938
Musgravetown Group, 943, 1041
Mwashya Group, 730, 752, 872, 906, 994
Myrtle Shale, 647
Myrtle Springs Formation, 628, 1061, 1104

N. Rhodesian Copperbelt, 787
Nabibis Formation, 374
Naeringselva Member, 965
Nagoriany Formation, 376, 973, 1005
Nallamalai Group, 622
Nama Group, 57, 367, 372, 396, 401, 447, 494, 559, 574, 875, 883, 940, 1001–1007, 1011, 1013, 1039, 1041
Nanfen Formation, 873, 880, 921, 963, 965
Nanguanling Formation, 1043, 1045, 1047
Nankoweap Formation, 370, 664, 1001, 1007, 1043
Nankoweap Group, 391, 418, 494, 1011, 1039, 1043, 1045
Nanshan Formation, 1047
Nantuo Formation, 375
Narryer Gneiss Complex, 16
Narssarssuk Formation, 931
Nasep Formation, 374, 1001, 1005, 1013
Nastapoka Group, 959
Natalevka Formation, 376
Nathan Group, 646, 893
Native Bee Siltstone, 647
Nebida Formation, 988, 990, 992
Negaunee Iron Formation, 870
Negri Volcanics, 648
Nehlehgehr Formation, 872, 878
Neihart Quartzite, 660
Nelidovo Formation, 967
Nel'kan Formation, 919

Nemakit Daldyn Horizon, 494, 505, 506, 558–560, 977, 990, 995, 1001, 1007, 1011, 1013
Nemchany Formation, 934
Nerchinskij Zavod Formation, 944
Neryuen Formation, 965, 967
Newland Formation, 959, 1043–1049
Newland Limestone, 98, 120, 129, 130, 506, 660, 662, 663, 691, 716, 753, 761, 763, 769, 788, 797, 877, 894, 969
Nibel' Formation, 937
Nigrit Series, 943
Nikitinsk Formation, 996
Nimbahera Formation, 1003
Niutitang Formation, 946
Niyuan Formation, 623
Nizhnie Horizon, 991
Nomtsas Formation, 374
Nonesuch Formation, 96, 97, 111, 112, 115, 182, 494
Nonesuch Seep Oil, 103
Nonesuch Shale, 98, 103, 120, 668, 669, 691, 715, 730, 752, 760, 768, 788, 792, 797, 817, 818, 872, 878
Noonday Dolomite, 658, 1039
Nopah Formation, 658
North Leader Member, 121, 617–619, 691, 720
North Star Basalt, 28, 648
Norvik Formation, 751
Nosib Group, 574
Nova Lima Group, 746, 758
Nuccaleena Formation, 374, 628, 724, 749
Nudaus Formation, 374, 979
Nugash Member, 621
Nyborg Formation, 941, 963
Nymerina Basalt, 121, 634, 636, 691, 719, 757
Nyuless Sandstone, 375, 642

Oakway Formation, 730, 752
Obruchev Horizon, 987, 989, 991, 993
Ocoee Supergroup, 48, 56
Odjick Formation, 120, 683, 684, 686, 691, 717, 888
Ogishke Conglomerate, 870
Okhonojsk Formation, 948
Olekma Formation, 948, 992
Olitantshoek Group, 617
Oligicene Formation, 1011
Olistroma del Membrillar, 940
Olkha Formation, 255, 929, 969, 1061, 1106
Olympic Formation, 374
Olympic Member, 630
Olympio Formation, 643, 644, 691
Omakhtinsk Formation, 893, 995, 1059, 1066
Onega Formation, 888
Ongeluk Lava, 617
Onguren Formation, 919
Onon Formation, 940
Onverwacht Group, 25, 27, 37, 39, 494, 590, 614, 747, 758, 786, 790, 795, 798
Onwatin Formation, 735, 754, 776, 789, 1037
Oraparinna Shale, 374
Orbuchev Formation, 376
Ordovician Formation, 1011
Ore Horizon, 1015
Ortega Quartzite, 55
Oselochnaya Formation, 948
Ostrog Series, 936
Ouplaas Member, 144
Outokumpu Ore Deposit, 776, 789
Ovruch Formation, 890, 1013, 1051
Owk Shale, 959, 969
Oxfordbreen Formation, 728, 751, 917

Pachelma Series, 881
Pahrump Group, 48, 256, 367, 494, 658, 879, 916, 990
Paint River Group, 668
Panem Formation, 622
Panorama Formation, 28, 648
Papaghni Subgroup, 622
Parachilna Formation, 374, 628, 1011

Paradise Creek Formation, 96, 120, 255, 645, 646, 691, 717, 734, 754, 793, 891, 1059, 1065
Paraopeba Formation, 687, 691, 873
Parara Limestone, 374, 403, 405, 406, 409, 411
Patherwa Formation, 878, 903
Paun Formation, 919
Pϵ/ϵ Boundary Beds, 722, 748
Pechenga Series, 877, 892
Pecors Formation, 121, 677, 678, 691, 719, 756, 779, 789
Pedroche Formation, 653, 691
Pekanatui Point Formation, 735, 754
Pendjari Formation, 934, 965
Pendkari Group, 934
Penganga Formation, 929
Penge Iron Formation, 779, 789
Perekhod Formation, 376
Persberget Formation, 730, 752
Pertaoorrta Group, 630, 1001
Pertatataka Formation, 96, 97, 103, 107, 108, 119, 224, 231, 630, 631, 691, 712, 723, 748, 791, 796, 817
Pestrotsvet Formation, 376, 655, 721, 748, 948, 987, 988, 992
Pethei Group, 655
Petrovskaya Formation, 967
Piegan Group, 660
Pigeon Formation, 663, 691
Pilbara Supergroup, 28, 648
Pillingani Tuff, 743, 757
Platberg Group, 617
Platonovka Formation, 977, 1007
Pletenyov Formation, 377
Pniel Group, 617
Podinzer Formation, 534, 909, 910, 1060, 1080
Podobruchev Horizon, 989
Pokegama Quartzite, 121, 668, 691, 718, 755, 870, 887
Polarisbreen Group, 938, 942
Poleta Formation, 378, 659, 660, 691, 988, 989, 992
Polese Series, 910
Polesie Formation, 377
Poleste Group, 377
Polo Ground Quartzite, 617
Poludennaya Group, 506, 1043, 1047–1053
Pondinzer Formation, 229, 621
Pongola Supergroup, 15, 71, 494, 617, 620
Porokhtakh Formation, 991, 995
Portage Lake Lavas, 668
Portfjeld Formation, 937
Post-Spilitic Group, 882, 938
Postmasburg Group, 617
Pound Subgroup, 628
Povarovo Group, 959, 973–979
Pre-Ravelli Group, 660
Pretoria Group, 617, 870, 1045
Pribram Formation, 1041
Prichard Formation, 120, 130, 660, 663, 691, 717, 753, 763, 788
Priozersk Formation, 932
Pronto Paleosol, 154
Pulivendla Formation, 622
Pullampet Formation, 622
Puncoviscana Formation, 1043
Purcell Supergroup, 494, 961
Pusa Formation, 119, 654, 691, 712, 961, 963
Puyu Formation, 965

Qianxi Group, 625
Qiaotou Formation, 880, 921
Qingbaikou Group, 494, 623, 625, 873, 879, 908, 912, 916, 961, 963–979
Quirke Lake Group, 677

Radzyn Formation, 377
Rae Group, 667
Rainstorm Member, 573
Ramah Group, 1037, 1039, 1041
Ramsay Lake Formation, 121, 677, 678, 691, 719, 756, 779, 789
Randfontein Quartzite, 619
Random Formation, 378, 655, 657, 691, 944
Randville Dolomite, 668, 736, 755

Index to Geologic Units

Ranford Formation, 375, 642–644, 645, 691
Rapitan Group, 57, 141, 160, 378, 572, 669, 675, 676, 691
Rapitan Iron Formation, 139, 141, 143–145
Ratcliffe Brook Formation, 1011, 1013
Raudstop/Salodd Formation, 751
Ravalli Group, 660
Rawnsley Formation, 1001–1007, 1011–1015
Rawnsley Quartzite, 369, 370, 374, 380, 382, 383–387, 393, 394, 628
Rechka Formation, 931
Recluse Group, 683
Red Pine Shale, 98, 120, 229, 682, 691, 715, 730, 752, 760, 797, 908, 961–965
Red Rock Beds Formation, 643, 644, 691
Reddick Bight Formation, 1037, 1041
Redkino Formation, 197, 198, 377, 967, 1061
Redkino Group, 505, 875, 883, 959, 961, 967, 971, 973, 1001, 1116
Redstone River Formation, 57, 129, 669, 672, 674, 676, 691, 764, 787
Reed Dolomite, 118, 122, 378, 659, 691, 721, 748
Reivilo Formation, 98, 121, 617, 619, 620, 691, 719, 761, 798
Relief Sandstone, 107
Rencontre Formation, 378, 657, 691
Renews Head Formation, 944
Rennes Shale, 1041, 1047, 1049
Reshma Formation, 376
Revet Member, 120, 621, 691, 716, 753
Rewah Group, 1043
Reward Dolomite, 97, 647, 817
Rhynie Formation, 96
Rhynie Sandstone, 119, 628, 629, 691, 714, 751
Rietgat Formation, 121, 617, 618, 691, 719, 744, 757
Ring Formation, 653
Ringwood Member, 630
Riphean Group, 1013
Risky Formation, 378
River Wakefield Subgroup, 98, 119, 628, 629, 691, 714, 728, 751, 759, 796, 922, 1061, 1095–1099
Riversleigh Siltstone, 645
Roaldtoppen Group, 223, 224, 963
Roan Group, 730, 752, 769, 788, 878, 901, 1015
Robertson Bay Group, 967
Rocknest Formation, 98, 120, 122, 494, 683–687, 691, 717, 721, 735, 754, 761, 797
Rodda Beds, 97, 103, 107, 108, 111, 817
Rohtas Formation, 878, 903, 959, 971
Rome Formation, 682
Rooiberg Group, 157
Rooinekke Formation, 121, 617, 619, 692, 719
Roopena Volcanics, 628
Roper Group, 84, 102, 111, 113, 228, 494, 646, 897
Roraima Formation, 55
Rossport Formation, 120, 679, 680, 692, 716, 732, 753
Rove Formation, 96, 98, 120, 678, 679, 680, 692, 717, 735, 754, 761, 778, 789, 793, 797
Rovno Horizon, 377
Rowsell Harbour Formation, 1039
Roy Hill Member, 634
Roznichi Formation, 875, 940
Rushall Slate, 648, 651, 692
Rusty Shale, 961, 971, 975, 977
Ruth Shale, 789
Ryssö Formation, 726, 750, 927, 963

Sabara Formation, 736, 755
Saddleworth Formation, 628
Sakharovsk Formation, 987
Sakukan Formation, 888
Salanygol Formation, 375
Salgash Subgroup, 648
Saline Valley Formation, 378
Samets Formation, 376
Samra Formation, 930
San Vito Formation, 1003, 1043, 1045
Sanashtygol Formation, 376, 988, 996
Sanashtygol Horizon, 987, 989, 990, 991–993, 996
Sandspruit Formation, 614

Sandy Dolomite, 682
Santa Cruz Formation, 930
Santa Domingo Formation, 653
Sarnyere Formation, 990
Satka Formation, 534, 892, 1059, 1065
Saunders Creek Formation, 643
Sayunei Formation, 119, 669, 672–674, 676, 692, 714, 751
Schist-Graywacke Complex, 1043
Schisto-calcarie Group, 996
Schmidtsdrif Subgroup, 617
Schoongezicht Formation, 614
Schwarzkalk Formation, 1013, 1039
Schwarzrand Subgroup, 371, 722, 748, 940, 965, 979
Sebakwian Group, 746, 758, 785, 790
Sediol' Formation, 937
Sekwi Formation, 378
Seliger Formation, 979
Semri Group, 227, 872, 878, 902–904, 959, 963, 967–979, 1041, 1043, 1047, 1051
Serebriya Formation, 875
Serebryanka Formation, 927
Seregin Member, 621
Série de Nara, 873
Série Lie de Vin Formation, 574, 722, 748
Série Schisto Calcaire Formation, 574, 721, 748
Serpent Formation, 121, 677, 678, 692, 719, 756, 779, 789
Sete Lagoas Formation, 880
Seward Group, 55
Shabakty Formation, 409, 411, 622
Shady Bore Quartzite, 645
Shady Dolomite, 363–365, 987, 988, 990, 992, 994
Shaler Group, 764, 787
Shamvaian Group, 744, 757
Shchekur'insk Formation, 996
Sheba Formation, 614
Sheepbed Formation, 98, 119, 130, 378, 669, 670, 671, 692, 713, 749, 759, 762, 787, 796
Shepard Formation, 660, 692, 1039
Shezal Formation, 119, 669, 670, 675, 676, 692, 714, 751
Shibatan Member, 977, 979
Shichang Member, 1043
Shihchiao Formation, 993
Shijia Formation, 623, 969
Shilungdong Formation, 375
Shinumo Quartzite, 664
Shipai Formation, 375
Shiveligsk Formation, 993
Shorikha Formation, 119, 255, 652, 692, 715, 752, 911, 1060, 1080
Shortor Series, 883, 884, 948
Shtandin Formation, 879, 911, 1060, 1080–1082
Shuijintuo Formation, 239, 375
Siberian Platform Oil, 817
Sibley Group, 678, 732, 753
Siehtichan Formation, 872, 878
Siemiatycze-Bialopole Formation, 377
Sierra Bayas Formation, 967
Signal Hill Group, 378, 944, 1041
Sillery Formation, 994
Siltstone Unit 1 Formation, 1005, 1011, 1037
Siltstone Units 1-2 Formation, 1011, 1037
Silverton Formation, 779, 789
Sim Formation, 621, 909, 910, 1060, 1080
Simla Slate Group, 874, 882, 932
Sinian Group, 623, 1001–1003
Sinian Supergroup, 623
Sinsk Formation, 376
Sioux Formation, 1047, 1051
Sioux Quartzite, 55
Sirbu Shale, 881, 931, 1039
Sixtymile Formation, 664
Siyeh Formation, 660–662, 692, 1037, 1041
Skellefte Field, 778, 789
Skillogalee Dolomite, 96, 119, 255, 534, 628, 629, 692, 714, 728, 750, 791, 796, 923, 1061, 1099–1102
Slawatycze Formation, 377
Smalfjord Formation, 963
Snider Formation, 780, 790

Snofield Lake Carbonate, 494
Snowbird Group, 1039
Snows Pond Formation, 943
Snowslip Formation, 660, 663, 692, 896
Soanesville Subgroup, 648
Society Cliffs Formation, 817
Sokoletzkaya Formation, 979
Sokoman Iron Formation, 139, 143, 144, 146, 151, 888
Sortis Group, 876
Sosan Group, 416, 417, 655, 1039
Sotillo Group, 653, 692
Soudan Iron Formation, 121, 668, 669, 692, 719, 744, 757
South Nicolson Group, 645
Spartan Group, 120, 663, 692, 717, 754
Spear Siltstone, 647
Spilitic Group, 882
Spiral Creek Formation, 723, 748
Spokane Formation, 660, 732, 753, 769, 788
Squaw Creek Paleosol, 154
St. John's Group, 378, 418, 494, 944, 1001, 1041, 1043, 1053
Stangenes Formation, 933
Stappogiedde Formation, 376, 943
Staraya Rechka Formation, 937
Steeprock Group, 494, 744, 757, 794, 798
Stein Formation, 642
Stelkuz Formation, 1011, 1013
Stirling Quartzite, 658
Stockdale Formation, 374
Stoer Bay Formation, 730, 752
Stoer Formation, 901
Stoer Group, 654
Stonyfell Quartzite, 628
Strel'nye Gory Formation, 534, 898
Stretton Sandstone, 647
Stretton Series, 921
Striped Peak Formation, 871, 877
Strubenkoop Formation, 779, 789
Studientsa Group, 376, 1011
Studenitza Formation, 969, 979
Sturgeon Falls Paleosol, 154
Sturtian, 727, 750
Suirovo Formation, 937, 1061, 1112
Sujsari Complex, 888
Suket Shale, 534, 872, 878, 903, 959, 963, 967–979, 1041, 1043, 1051
Sukhaya Tunguska Formation, 120, 255, 534, 652, 692, 715, 752, 905, 1060, 1071
Sumian-Sariolan Group, 54, 71
Suncho Formation, 1043
Surprise Creek Formation, 646, 647
Suxian Group, 623
Svanbergfjellet Formation, 130, 224, 227, 231, 727, 750
Svecofennian, 737, 755
Svisloch Formation, 377
Swartkoppie Formation, 614, 747, 758, 795, 798
Swaziland Supergroup, 494, 614

Tabooes Formation, 737, 755
Tadpatri Formation, 622
Taghdout Limestone, 1041
Takiyuak Formation, 683
Tal Formation, 722, 748
Talga Talga Subgroup, 648
Taltheilei Formation, 256, 736, 755
Tamdy Formation, 622, 884, 948
Tamengo Formation, 687, 692, 1007
Tanafjord Group, 933, 936, 941, 963
Tankha Formation, 989
Tanner Member, 98, 109, 120, 130, 664, 665, 692, 715, 752, 760, 763, 787, 797, 906, 965
Taoudenni Basin, 730, 752
Tapeats Sandstone, 580
Tapley Hill Formation, 95–97, 119, 130, 628, 629, 692, 713, 727, 750, 764, 787, 791, 796, 815
Tarcoola Formation, 130
Tatoola Sandstone, 647

Tawallah Group, 647
Tayvallich Limestone, 947
Tean Formation, 642, 692
Teatree Gully Quartzite, 1051
Teena Dolomite, 97, 647
Telemark Formation, 873, 1053
Temiscamie Iron Formation, 778, 789, 870
Tent Hill Formation, 883, 941, 961, 979
Termite Range Formation, 645
Terskij Formation, 927
Tertiary weathering zone, 642
Theespruit Formation, 98, 122, 614, 615, 692, 720, 747, 758, 762, 795, 798
Thule Group, 494, 931, 965, 1041
Tibbit Hill Volcanics, 580
Tieling Formation, 120, 625, 627, 692, 715, 752, 763, 878, 901
Tienheban Formation, 375
Tigonankweine Formation, 1037
Tillery Formation, 377
Tillite Formation, 723, 748, 841
Tillite Group, 941, 963
Timeball Hill Formation, 737, 756, 779, 789
Timeball Hill Quartzite, 617
Timperley Shale, 119, 375, 633, 634, 642, 643, 645, 692, 712, 749
Tindelpina Shale Member, 98, 628, 629, 692, 713, 728, 750, 759, 796
Tindir Group, 237, 494, 931, 991, 1007
Tinnov Formation, 991, 995
Tiskre Formation, 377
Titary Formation, 376
Tjakastad Subgroup, 614
Tjurim Formation, 506
Tokammane Formation, 225, 226, 229, 377, 721, 748, 884, 948
Tokko Formation, 995
Tolbachan Horizon, 992
Tommotian Formation, 1011
Tooganinie Formation, 647, 734, 754, 793, 797
Torgashino Formation, 993
Torpedo Creek Quartzite, 645
Torrearboles Formation, 653
Torridon Group, 119, 654, 655, 692, 714, 751, 872, 911
Torrowangee Group, 1001, 1037
Totem Formation, 374
Tottinsk Formation, 877, 965, 1060, 1066
Toushantou Formation, 991, 995
Towers Formation, 27, 28, 32, 36, 37, 122, 129, 648–651, 692, 720, 746, 758, 764, 790, 795, 798
Transition Beds, 722, 748
Transvaal Dolomite, 869
Transvaal Supergroup, 15, 137, 140, 147–150, 157, 173, 363, 494, 511, 617, 737, 756, 869, 870, 884, 1037–1041, 1045
Tree River Formation, 683
Trenton Limestone, 993
Trezona Formation, 374, 724, 749
Trinny Cove Formation, 943
Troy Quartzite, 1037
Tsaganolom Formation, 375
Tsanglangpu Formation, 375
Tsezotene Formation, 1037
Tsineng Member, 884
Tsipanda Formation, 655, 901, 994
Tsyp-Navolok Shales, 934
Tuanshanzi Formation, 98, 120, 625, 626, 692, 717, 754, 761, 797, 877, 967, 977
Tukan Formation, 120
Tukan Member, 621, 692, 716, 753
Tumbiana Formation, 98, 121, 590, 634–641, 692, 720, 745, 758, 762, 795, 798
Tumen Group, 965
Turee Creek Group, 15, 743, 757
Turffontein Subgroup, 617
Turkutut Formation, 375
Twitya Formation, 98, 119, 130, 378, 669–671, 672, 675, 692, 750, 759, 763, 787, 796
Tyler Formation, 98, 120, 234, 668, 692, 718, 755, 761, 798, 887, 1059, 1064

Index to Geologic Units

Tyurim Formation, 1041-1045, 1049-1053

Uchi Greenstone Belt, 784, 790
Uchur Formation, 655, 881
Udokan Group, 1043
Udokan Series, 778, 789
Ui Group, 965
Uinta Mountain Group, 229, 682, 908, 909, 961, 963, 965
Uk Formation, 933, 1061, 1109
Uk-Kudash Formation, 621
Ulunytuj Formation, 902, 994, 996
Umberatana Formation, 96
Umbertana Group, 628, 725, 749, 791, 796, 1003
Umbolooga Subgroup, 647
Umkondo Group, 735, 754
Uncompahgre Formation, 733, 753
Undalya Quartzite, 628
Ungut Formation, 988, 989
Union Island Group, 98, 120, 655, 692, 717, 754, 761, 797
Unkar Group, 664, 691, 1037, 1039, 1041
Uratanna Formation, 374, 628
Urquhart Shale, 96, 98, 120, 130, 647, 692, 717, 734, 754, 761, 763, 788, 793, 797, 891
Urucum Formation, 930
Urusis Formation, 374, 987
Uryuk Formation, 620
Ushakov Member, 621, 622, 692, 716
Ushakovka Formation, 884, 949, 951
Ushitsa Series, 550, 938
Usol'e Formation, 949, 987, 991, 992, 994, 995
Ust'Botoma Formation, 988
Ustkundat Formation, 376
Ust'Kut Formation, 991
Ust'Kyrbinsk Formation, 965
Ust-Pinega Formation, 377, 380, 382, 384, 386-388, 395, 396, 627, 692, 1001-1007, 1011-1015, 1037
Ust-Sylvitsa Formation, 1043
Ust'Yudoma Formation, 965
Utsingi Formation, 735, 754
Uwharrie Formation, 377

Vadsø Group, 229, 917, 921, 931, 963
Valdai Group, 395, 396, 494, 1001-1007, 1011-1015
Valdai Series, 875, 942, 959, 979
Vallen Group, 870, 876, 889, 1041, 1047, 1053
Valyukhta Formation, 904, 991
Vampire Formation, 378, 505, 875, 949, 950
Vanavarsk Formation, 991
Vangsås Formation, 653
Van Horn Sandstone, 682
V. Chonskaya Oil, 103
Velkerri Formation, 84, 96, 97, 103, 113, 646, 732, 753, 792, 797, 817, 818
Vempalle and G. R. Formations, 120, 754
Vempalle Formation, 255, 622, 692, 717, 890, 1059, 1064
Ventersdorp Supergroup, 15, 121, 494, 617, 619, 692, 719, 744, 757, 794, 798, 869
Venterspost Carbon Reef Member, 617, 619, 692, 719, 757, 790
Venterspost Formation, 130, 619, 618, 719, 757, 764
Vermillion Limestone, 734, 754
Vermillion Member, 1037
Veslyana Formation, 927
Vestertana Group, 941, 943, 963
Veteranen Group, 192, 229, 916, 917, 959
Vetlanda Series, 870
Victor Bay Formation, 97, 817
Vilchitsy Formation, 377
Vindhyan Group, 753, 872, 878
Vindhyan Supergroup, 227, 732, 881, 902-904, 911, 931, 1003
Virgiliana Formation, 377
Virginia Formation, 778, 789
Visingsö Group, 97, 119, 228, 229, 494, 534, 550, 652, 653, 692, 714, 751, 817, 818, 874, 880, 922, 923, 959, 963, 967, 969, 971, 977
Volokov Series, 934
Voron Formation, 881

Vryburg Formation, 617
Vryheid Formation, 620
Vtorokamensk Formation, 901

W-Fold Shale Member, 647
Wade Creek Sandstone, 633
Walcott Member, 98, 109-111, 119, 130, 665-667, 693, 715, 752, 759, 763, 787, 796, 815, 914, 965
Waldo Pedlar Member, 630
Wallace Formation, 871, 877
Walsh Tillite, 724, 749
Wanshan Formation, 623
Warambie Basalt, 648
Warrawoona Group, 22, 26-32, 34, 36-39, 494, 588, 590, 648, 746, 758, 786, 790
Warrie Member, 634, 641, 693
Warrina Park Quartzite, 647
Waterberg Group, 15, 617
Waterton Formation, 877
Weeli Wolli Iron Formation, 634
Weiji Formation, 623
West Rand Group, 617
Westmanbukta Formation, 730, 752, 963
Wewe Slate, 668
Whaleback Shale Member, 634
Whim Creek Group, 28, 71, 648, 649, 651, 652, 693
Whiteout Conglomerate, 988
Whitewater Group, 1037, 1039
Wildbread Formation, 256
Willkawillina Limestone, 374, 628
Willochra Subgroup, 628, 724, 749
Wills Creek Granite, 646, 647
Willyama Complex, 771, 788
Wilmington Formation, 628
Wilpenna Group, 628, 883, 941, 1001-1007, 1011-1015
Wilsonbreen Formation, 377, 724, 749, 938
Windermere Group, 494, 495, 881, 931, 979, 1005, 1011-1015, 1037, 1039, 1049
Windfall Formation, 658, 693, 712
Windidda Formation, 735, 754, 1059
Wirara Formation, 642
Wirrealpa Limestone, 374
Wittenoom Dolomite, 96, 634, 743, 757, 794, 798
Witwatersrand Supergroup, 15, 60, 121, 158, 173, 617, 619, 693, 720, 745, 757, 783, 790, 794, 798
Wlodawa Formation, 130, 377
Wollogorang Formation, 774, 788
Wolstenholme Formation, 1041
Woman Lake Marble Formation, 746, 758
Woman River Iron Formation, 785, 790
Wonoka Formation, 96, 374, 628, 791, 796
Wonokaan Formation, 724, 749
Woocalla Dolomite Member, 98, 99, 119, 628-630, 693, 713, 750, 759, 796, 815
Wood Canyon Formation, 658
Woodhouse Formation, 1001, 1005, 1037
Woodiana Sandstone Member, 634
Woolshed Flat Shale, 628
Wooltana Volcanics, 628
Woongarra Volcanics, 634
Wostenholme Formation, 965
Wuhangshan Group, 965, 1037, 1045
Wumishan Formation, 98, 120, 255, 625-627, 693, 716, 753, 760, 763, 797, 877, 898, 995, 1053
Wyloo Group, 494
Wyman Formation (AUS), 28, 119, 648, 649, 651, 693, 712, 748
Wyman Formation (USA), 378, 659, 693, 712, 748, 1015
Wynniatt Formation, 231

Xiamaling Formation, 625, 873, 879, 908, 973, 975
Xihe Group, 873, 880, 921, 963, 965
Xingii Formation, 624, 693
Xingxiung Formation, 965
Xuhuai Group, 623

Xunjiansi Formation, 120, 623, 624, 693, 716, 753

Yakut Complex, 652
Yalco Formation, 84, 96, 97, 103, 113, 647, 733, 753, 792, 797, 817, 818
Yangjiaping Formation, 946, 992
Yangud Formation, 987, 994, 995
Yangzhuang Formation, 98, 120, 625, 626, 693, 716, 753, 797, 896
Yaryshev Formation, 234, 376, 875, 882, 938, 940, 959, 961, 971, 1001–1007, 1015, 1061, 1113
Yellowknife Supergroup, 743, 757
Yelma Formation, 735, 754
Yilgarn Block, 783, 790
Ymer Formation, 1053
Yuanshan Formation, 375
Yudoma Formation, 119, 231, 234, 255, 376, 427, 534, 655, 693, 712, 723, 748, 941, 965, 991, 993, 1003, 1005, 1115
Yuedej Formation, 990, 995
Yuhucun Formation, 975, 977, 979, 991
Yukandinsk Formation, 995
Yurabi Formation, 642–645, 693, 712, 749
Yurmata Group, 621

Zabit Formation, 991
Zabriskie Quartzite, 658
Zaris Formation, 374, 388, 1007, 1011, 1013
Zhangqu Formation, 119, 623, 693, 715, 752
Zhanatus Series, 622
Zhaowei Formation, 623
Zharnov Formation, 883, 939, 1043
Zharnovki Formation, 376, 969, 1005, 1007
Zherbinsk Formation, 937
Zhongsanyang Formation, 1043, 1045, 1047
Zhongyicun Formation, 375, 1011, 1013
Zhongyicun Member, 397
Zigal'ga Formation, 621, 894
Zigan Formation, 937
Zigazino-Komarovsk Formation, 228, 534, 621, 877, 896
Zizagland Formation, 870, 876, 889, 1041, 1047
Zil'merdak Formation, 229, 255, 419, 534, 621, 905, 906, 1071, 1059
Zin'kov Formation, 942, 1061
Zubovsk Formation, 928
Zubruch Formation, 376

Taxonomic Index

Suprageneric categories are denoted by UPPER CASE type; genera by **Bold Italics**; species by *Normal Italics*. Taxa are illustrated on ***pages listed in Bold Italics***.

Aataenia, 354, 502, 504, 958
    Aataenia reticularis, 958
Abadiella, 1024
ABADIELLIDAE, 1175
Abakania, 1023
Abakanicyathus, 1018
Abicyatbus, 1021
Absidaticonus, 1027
Abundacapsa, 916
    Abundacapsa impages, 916
Acanthamoeba, 482–483
    Acanthamoeba castellanii, 482
Acanthichus, 1040
Acanthina, *362–363*, 366, 981, 984, 986, 996
    Acanthina multiformis, 981, 996
ACANTHINOCYATHIDAE, 1172
Acanthocassis, 1031
ACANTHOCEPHALA, 410
Acanthoclava, 1032
Acanthodiacrodium, 940, 944, 1052
ACANTHOGRAPTIDAE, 1178
Acantholigotriletum, 222, 899, 908, 916, 948, 1151
    Acantholigotriletum primigenum, 908, 916, 948, 1151
Acanthomeridion, 1022
ACANTHOMORPH ACRITARCHS (*See* Subject Index)
ACANTHOPYRGIDAE, 1173
Acanthopyrgus, 1018
Acantosphaera, 1031
ACAROCERATIDAE, 1175
Acetabularia, *323*
Acetobacter, 86
    Acetobacter xylinum, 86
Acetobacterium, 291
    Acetobacterium woodii, 291
Achlya, 482–483
    Achlya bisexualis, 482
ACHOROCYATHIDAE, 1174
Achorocyathus, 1021
Achroonema, 35, 1131
    Achroonema inequale, 35, 1131
Aciconularia, 1022
Acidiscus, 1022
Acidocharacus, 1031
Acidotocarena, 1031
Acimetopus, 1023
ACRANIA, 1178
ACRITARCHS (*See* Subject Index)
Acrosquama, 1026
Acrothele, 1030
ACROTHELIDAE, 1177
Acrothya, 1030

Acrotreta, 1030
ACROTRETIDA, 1030, 1177
ACROTRETIDAE, 1177
Acrum, 227
Actaeus, 1175, 1180–1184
ACTINARIA, 1173
ACTINIANS (*See* Subject Index)
Actinoconus, 1027
Actinoites, 1017
ACTINOMYCETES, 477, 869
Actinophycus, 981, 986, 994, 996, 998
    Actinophycus divaricatus, 996
    Actinophycus liangshanensis, 996
    Actinophycus nanjiangensis, 996
    Actinophycus obrutschevi, 981, 994
    Actinophycus quadricella, 986, 996, 998
Actinotheca, *403*, 1029
    Actinotheca holocyclata, ***403***
Aculeochrea, 1021
Aculochileus, 1027
Acuminachites, 1026
Acus, 981, 996–997
    Acus concentricus, 996
    Acus fasciatus, 997
    Acus muricatus, 997
    Acus platypluteus, 981, 997
Acuticloudina, 1006, 1032
    Acuticloudina borrelloi, 1006
Acutirostriconus, 1027
Acutitheca, 1032
Adoncholaimus, 296
    Adoncholaimus okyris, 296
Adversella, 1026
Adyshevitheca, 1029
Aegides, 1027
AEGUOREIDAE, 1173
Aetholicopalla, ***406***, 1032
    Aetholicopalla adnata, ***406***
Affinovendia, 1000
    Affinovendia arctosa 1000
Afiacyathus, 1018
Agastrocyathus, 1018
AGLASPIDA, 1177
AGLASPIDIDAE, 1177
AGMATA, 401, 407, 1021, 1174
Agmenellum, 850
    Agmenellum quadruplicatum, 850
AGNOSTIDA, 1022, 1175
AGNOSTIDAE, 1175
AGRAULIDAE, 1176
Agraulos, 1023

Aguaraya, 1023
Agyrekia, 1030
AGYREKOCYATHIDAE, 1172
Agyrekocyathus, 1019
Aimia, 907, 1060, **1077**
    Aimia delicata, 907, 1060, **1077**
AIMITIDAE, 1175
Aimitus, 1028
Aimophyton, 872, 1060, **1078**
    Aimophyton varium, 872, 1060, **1078**
Ajacicyathellus, 1018
AJACICYATHIDA, 1018, 1172
AJACICYATHIDAE, 1172
Ajacicyathus, 1018
AKADEMIOPHYLLIDAE, 1174
Akademiophyllum, 1021
Aksuglobulus, 1032
Aksumbensis, 1000
    Aksumbensis aksumbensis, 1000
Aktugaia, 1022
Alalcomenaeus, 1180
Alalcomenaeus, 1025
Alanisia, 1024
ALATAUCYATHIDAE, 1172
Alataucyathus, 1019
Alataurus, 1024
Albumares, 372–373, 381, **386**, 407, 1000
    Albumares brunsae, **386**, 1000
ALBUMARESIDAE, 373, 1173
Alconerscyathus, 1018
ALCYONARIANS (*See* Subject Index)
ALCYONIDA, 1173
Aldanella, **405**, 1026
    Aldanella attleborensis, **405**
ALDANELLIDAE, 1026, 1174
Aldania, 1025
Aldanocyathus, 1018
Aldanolina, 1032
Aldanospina, 1025
Aldanospina, 1029
Aldanotreta, 1030
Aldonaia, 1024
ALGAE (*See* Subject Index)
Algomella, 1027
Alisina, 1030
Allatheca, 1029, 1040
ALLATHECIDAE, 1175
Alliumella, 950
    Alliumella baltica, 950
Allonnia, 1025
ALOKISTOCARIDAE, 1176
Alpertia, 393–**394**
    Alpertia santacrucensis, 393–**394**
Alphacyathus, 1020
ALTAICORNIDAE, 1175
Altaicornus, 1028
Altarmilla, 1032
Altitudella, 1023
Aluta, 1022
ALUTIDAE, 1177
Amanlisia, 1040
    Amanlisia simplex, 1040
Ambarchaeooides, 1032
Ambiguaspora, 943, 1152
    Ambiguaspora parvula, 943, 1152
Ambistapis, 1018
Ambrolinevitus, 1028
Ameliaphycus, 891
    Ameliaphycus croxfordii, 891
Amgaina, 981
    Amgaina compacta, 981
Amganella, 981, 985, 987, 995
    Amganella glabra, 987, 995
AMGASPIDAE, 1176
AMISKWIIDAE, 1174
AMMODISCACEA, 238–239
AMMODISCIDAE, 1171

AMOEBAS (*See* Subject Index)
Amoebinella, 1031
Amphidinium, 479
Amphigeisina, 1031
AMPHIGEISINIDAE, 1031, 1178
Amphipleura, 322
    Amphipleura rutilans, 322
Amphithrix, 322
    Amphithrix janthina, 322
Amphora, 322
    Amphora coffeaeformis, 322
    Amphora ventricosa, 322
Amurella, 872
    Amurella pilosa, 872
AMYDAICORNIDAE, 1175
Anabaena, 281, 322, 848, 852
    Anabaena azollae, 281
    Anabaena cylindrica, 852
    Anabaena solitaria, 852
    Anabaena variabilis, 848
    Anabaena viquieri, 852
Anabaenidium, 886, 894, 912, 935, 1130
    Anabaenidium barghoornii, 886, 1130
    Anabaenidium hailuotoense, 935
    Anabaenidium johnsonii, 912
    Anabaenidium sophoroides, 894
Anabarella, 1026
Anabaritellus, 1021
Anabarites, 373, 381, **403**, 412, 441, 505, 554, 1021
    Anabarites sexalox, **403**
    Anabarites trisulcatus, **403**, 505
ANABARITIDAE, 372–373, 400–401, 403, 407, 434, 447, 556, 559, 1020–1021, 1173
Anabylia, 1000
    Anabylia improvisa, 1000
Anacaena, 296
Anacystis, 848, 850, 852
    Anacystis cyanea, 848
    Anacystis montana, 850
    Anacystis nidulans, 848, 850, 852
    Anacystis variabilis, 852
Anadoxides, 1024
Anafesta, 372–373, 381, **386**, 407, 1000
    Anafesta stankovskii, **386**, 1000
Analox, 1023
Anaptychocyathus, 1019
ANAPTYCTOCYATHIDAE, 1172
Anatolepis, 449
ANCALAGONIDAE, 1173
Ancheilotheca, 1029
Anconochilus, 1027
ANDRARINIDAE, 1176
ANEMONES (*See* Subject Index)
Angaricyathus, 1018
Angaronema, 918, 1136
    Angaronema septatum, 918, 1136
Angulisphaerina, 878–879, 948
    Angulisphaerina keminica, 948
    Angulisphaerina korolevii, 948
    Angulisphaerina perfosoclathrata, 878–879
Angulocellularia, 359, **362**, 366, 981, 983, 987, 991, 994
    Angulocellularia anisotoma, 981, 994
    Angulocellularia mansurkaensis, 987
Angusteva, 1024
Angusticellularia, 981, 994
    Angusticellularia anisotoma, 981, 994
ANGUSTICORNIDAE, 1175
Angusticornus, 1028
Angustiochrea, 1021
Anhuiconus, 1027
Anhuiella, 354, 502, 958, 1021
    Anhuiella sinensis, 958
ANIMALIA, 473
Animikiea, 879, 886–888, 910, 918, 1059, **1063–1064**, 1127
    Animikiea septata, 879, 886, 888, 910, 1059, **1063**, 1127
    Animikiea cf. *septata*, 918
Ankistrodesmus, 885

Taxonomic Index

ANNAMITIIDAE, 1176
ANNELIDA, 381, 406–408, 412, 416, 431, 447, 485, 969, 973, 975, 1010, 1020, 1036–1037, 1039, 1051, 1175
Annularia, 879, 1060, **1076**
  Annularia annulata, 879, 1060, **1076**
Annulocyathella, 1018
ANNULOCYATHIDAE, 1172
Annulocyathus, 1018
Annulofungia, 1018
ANOMALOCARIDAE, 1174, 1180–1184
Anomalocaris, 371, 437, 1025
Anomas, 982, 994
  Anomas ovisimilis, 982, 994
ANOMOCARIDAE, 1176
Antagmus, 1023
Antatlasia, 1024
ANTHASPIDELLIDAE, 1171
Antholithina, 1040
  Antholithina rosacea, 1040
Anthomorpha, 1018
ANTHOMORPHIDAE, 1172
ANTHOZOA, 373, 393, 408, 434, 1020, 1173, 1179, 1181–1184
Antihipponicharion, 1022
Antiquatheca, 1029
Antiquus, 982, 994
  Antiquus cusarandicus, 982, 994
Anzalia, 1032
Aphanocapsa, **195**, 597, 846, 1138–1139, 1141–1143
  Aphanocapsa crassa, 1139
  Aphanocapsa delicatissima, 1141
  Aphanocapsa littoralis var. *macrococca*, 1142–1143
  Aphanocapsa roeseana, 1138
Aphanocapsaopsis, 902–903
  Aphanocapsaopsis ramapuraensis, 903
  Aphanocapsaopsis sitholeyi, 902–903
Aphanothece, **195**, 325, 327–328, 846, 1144
  Aphanothece bullosa, 1144
  Aphanothece halophytica, 325
  Aphanothece pallida, 1144
  Aphanothece saxicola, 1144
Aphetospora, 940–941
  Aphetospora euthenia, 940–941
Apistoconcha, **406**, 1031
  Apistoconcha apheles, **406**
Apocyathus, 1021
Aporosocyathus, 1018
Aptocyathella, 1020
APTOCYATHIDAE, 1172
Aptocyathus, 1020
Aqualisquama, 1032
ARACHNIDA, 1053
Aranidium, 949, 951
  Aranidium aff. *pycnacantum*, 949
  Aranidium sparsum, 951
Arborea, 373, 1000, 1036–1037
  Arborea arborea, 1000, 1037
ARCHAEA, 473–474
Archaeagnostus, 1022
ARCHAEASPIDIDAE, 1175
Archaeaspis, 1024
ARCHAEBACTERIA (*See* Subject Index)
Archaeichnium, 373, 1000, 1037
  Archaeichnium haughtoni, 1000
ARCHAEOCOPIDA, 1022, 1177
ARCHAEOCYATHA, 359, 400–401, 404, 407, 437, 441, 448, 554–556, 558–560, 564, 571–572, 860, 984, 1017–1018, 1020, 1172
ARCHAEOCYATHIDA, 983, 986, 1018, 1037, 1047, 1172
Archaeocyathus, 1018, 1036
ARCHAEODENDRIDA, 1178
ARCHAEODENDRIDAE, 1178
Archaeodiscina, **225**, 947–951
  Archaeodiscina biocostata, 950
  Archaeodiscina minor, 951
  Archaeodiscina nova, 947
  Archaeodiscina prima, 951
  Archaeodiscina umbonulata, **225**, 948, 950

Archaeoellipsoides, 894–895, 898, 1059, **1070**, 1152
  Archaeoellipsoides conjunctivus, 898, 1152
  Archaeoellipsoides grandis, 894–895, 1059, **1070**
  Archaeoellipsoides obesus, 898
Archaeofavosina, 873, 880, 903–904, 932
  Archaeofavosina compta, 932
  Archaeofavosina naumovae, 904
  Archaeofavosina reticulata, 903
  Archaeofavosina cf. *simplex*, 873
  Archaeofavosina sinuta, 904
Archaeofungia, 1018
ARCHAEOFUNGIIDAE, 1172
ARCHAEOGASTROPODA, 1026, 1174
Archaeohystrichospheridium, 883, 937, 944–947, 949, 951, 1152
  Archaeohystrichospheridium acanthaceum, 944, 946–947, 1152
  Archaeohystrichospheridium acer, 944, 1152
  Archaeohystrichospheridium anatropum, 883
  Archaeohystrichospheridium cellulare, 944, 947, 1152
  Archaeohystrichospheridium cuneidentatum, 951
  Archaeohystrichospheridium dasyacanthum, 949
  Archaeohystrichospheridium dorofeevii, 945, 1152
  Archaeohystrichospheridium glebosum, 944
  Archaeohystrichospheridium ignotum, 937
  Archaeohystrichospheridium innominatum, 945, 1152
  Archaeohystrichospheridium janischewskyi, 947
  Archaeohystrichospheridium operculatum, 947
  Archaeohystrichospheridium stipiforme, 945, 1152
  Archaeohystrichospheridium vastum, 883
  Archaeohystrichospheridium vologdaense, 946, 951
Archaeolynthus, 1020
Archaeonema, 888, 984, 912, 923, 1132
  Archaeonema longicellulare, 894, 912, 923, 1132
Archaeooides, **406**, 1032
ARCHAEOOIDS (*See* Subject Index)
Archaeopetasus, **405**, 1032
  Archaeopetasus excavatus, **405**
Archaeopharetra, 1018
ARCHAEOPHARETRIDAE, 1172
ARCHAEOPHYLLIDA, 1173
ARCHAEOPHYLLIDAE, 1173
Archaeophyllum, 1021
Archaeophyton, 1040
  Archaeophyton newberryanum, 1040
Archaeoprotospongia, 1040
Archaeops, 1024
Archaeopsophosphaera, 883, 947
  Archaeopsophosphaera adischevii, 883
  Archaeopsophosphaera cf. *asperata*, 947
Archaeorestis, 205, 233–234, 886, 929, 1148
  Archaeorestis magna, 886, 1148
  Archaeorestis minuta, 929
  Archaeorestis schreiberensis, 886, 1148
Archaeosacculina, 951
  Archaeosacculina atava, 951
  Archaeosacculina salebrosa, 951
  Archaeosacculina torosa, 951
Archaeospherina, 414, 1040
Archaeospira, **405**, 1026
  Archaeospira ornata, **405**
Archaeospongia, 1040
  Archaeospongia radiata, 1040
ARCHAEOSTRACA, 1022
Archaeosycon, 1018
ARCHAEOSYCONIDAE, 1172
Archaeotremaria, 1032
Archaeotrichion, 28, 33, 884, 889, 894, 896, 910, 912, 918–919, 949, 1123, 1126
  Archaeotrichion contortum, 28, 33, 912, 918, 1126
Archaeoxylon, 1040
  Archaeoxylon krasseri, 1040
Archangelia, 1000
  Archangelia valdaica, 1000
Archeoxybaphon, 1032
ARCHEZOA, 481
Archiasterella, **409**, 1025–1026
  Archiasterella hirundo, **409**

Archicladium, 1025
Arcitheca, 1029
Arctacellularia, **198**, 905–906, 911, 914, 919, 933, 1060, **1082**, **1085**, **1087**, 1143–1144
    Arctacellularia doliiformis, **198**, 914, 1060, **1087**, 1144
    Arctacellularia ellipsoidea, 905–906, 914, 919, 933, 1060, **1085**, 1143
ARDOSSACYATHIDAE, 1172
Ardrossania, 1032
Ardrossocyathus, 1018
ARENICOLIDAE, 969, 973
Arenicolites, 392, 501, 505, 1010, 1037, 1040–1041, 1053
    Arenicolites spiralis, 505, 1040, 1053
Argatheca, 1029
Argunaspis, 1024
Arkarua, 372–373, 1000
    Arkarua adami, 1000
Armelia, 1040
    Armelia barrandei, 1040
Armillifera, 1000
    Armillifera parva, 1000
Aroonia, **406**, **439**, 1031
    Aroonia seposita, **406**, **439**
Artemia, 482
    Artemia salina, 842
Arthricocephalites, 1023
Arthricocephalus, 1023
Arthrochites, 1022
Arthrophycus, 1036
ARTHROPODA, 371–373, 381, 397, 401–402, 408, 434, 438, 485, 558, 599, 938, 1017, 1022, 1025, 1037, 1043, 1051, 1175
Arthrosiphon, 914, 1060, **1083**
    Arthrosiphon cornutus, 914, 1060, **1083**
    Arthrosiphon typicus, 914, 1060, **1083**
Arthrospira, 913, 918, 934, 1129
    Arthrospira jenneri, 1129
    Arthrospira khannae, 1129
    Arthrospira platensis var. *tenuis*, 1129
    Arthrospira spirulinoides f. *tenuis*, 1129
Arthurocyathus, 1018
ARTICULATA, 1030, 1177, 1180–1184
Artimyctella, 1031
Arumberia, 1000, 1037, 1040, 1043, 1045, 1047
    Arumberia banksi, 1000, 1040, 1042
ASAPHIDAE, 1176
ASAPHISCIDAE, 1176
ASCIDIANS (*See* Subject Index)
Asiatella, 1024
Asijatheca, 1029
Askepasma, 1030
Asperatopsophosphaera, 875, 895, 898, 908, 947, 1139, 1152
    Asperatopsophosphaera bavlensis, 908
    Asperatopsophosphaera partialis, 947, 1152
    Asperatopsophosphaera umishanensis, 895, 898, 1139
    Asperatopsophosphaera cf. *magna*, 875
    Asperatopsophosphaera cf. *medialis*, 875
    Asperatopsophosphaera unishanensis var. *minor*, 898
Asperoconus, 1027
Aspidella, 372, 416, **418**, 1000–1001, 1042, 1049
    Aspidella costata, 1000
    Aspidella hatyspytia, 372, 1000–1001
    Aspidella terranovica, **418**, 1000, 1042
Astercapsoides, 918
    Astercapsoides borealis, 918
Asteriacites, 1010
Asteriradiatus, 1042
    Asteriradiatus karauliensis, 1042
Asterocythus, 1019
Asterosoma, 419, 1042–1043
Asterosphaeroides, 873
    Asterosphaeroides darsii, 873
    Asterosphaeroides monodii, 873
    Asterosphaeroides richatensis, 873
Asterotumulus, 1019
Astropolichnus, 392
ASTRORHIZACEA, 238–239
Atdabanella, 1023

Atdabanithes, 1028
Atops, 1023
ATREMATA, 967, 969, 973
AULACODIGMATIDAE, 1176
Aulichnites, 1010
Auriculaspira, 1026
Auricultella, 1022
Aurisella, 1031
AURITAMIDAE, 1176
Ausia, 373, 1000
    Ausia fenestrata, 1000
Avalonia, 1023
Avictuspirulina, 931, 1129
    Avictuspirulina minuta, 931, 1129
Aviculocephaloconus, 1032
AVONINIDAE, 1176
Axiculifungia, 1019
Aysheaia, 1177
Azotobacter, 853
    Azotobacter chroococcum, 853
Azyricyathus, 1021
Azyrtalia, 881, 982, 994, 996, 1017, 1172
    Azyrtalia fasciculata, 881, 994
    Azyrtalia globosa, 994
    Azyrtalia telmenica, 996
    Azyrtalia zonulata, 881, 982, 994

Babakovia, 1023
Babetosphaera, 873
    Babetosphaera africana, 873
    Babetosphaera africana var. *africana*, 873
BACATOCYATHIDAE, 1172
Bacatocyathus, 1018
BACILLARIOPHYCEAE, 283
BACTERIA (*See* Subject Index)
Bactrophycus, 898
    Bactrophycus dolichum, 898
    Bactrophycus oblongum, 898
Bagenovia, 1027
Bagenoviella, 1027
Bagongshanella, 349, 502, 958
    Bagongshanella striolata, 958
Baicalia, 508, 519, 627, 661, 666–667
Baikalina, 373, 1000
    Baikalina sessilis, 1000
Baikalocyanthus, 1018
BAIKALOPECTINIDAE, 1172
Bajanophyton, **362**, 366, 982, 987
    Bajanophyton egiingolicum, 987
    Bajanophyton mucosum, 982, 987
Bajiella, 1022
Balangia, 1023
BALANGIIDAE, 1176
Balbiriniella, 893
    Balbiriniella praestans, 893
BALKOCERATIDAE, 1175
Baltisphaeridium, **226**, 227, 229, 881, 902, 904, 938–940, 945, 947–948, 950, **1116**, 1151, 1061
    Baltisphaeridium acerosum, 950
    Baltisphaeridium bohemicum, 938
    Baltisphaeridium brachyspinosum, 950
    Baltisphaeridium cerinum, **226**, 948, 950
    Baltisphaeridium ciliosum, 229, 950
    Baltisphaeridium compressum, 947, 950
    Baltisphaeridium dubium, 950
    Baltisphaeridium gangolihatense, 881
    Baltisphaeridium implicatum, 948, 950
    Baltisphaeridium insigne, 950
    Baltisphaeridium orbiculare, 947, 950
    Baltisphaeridium ornatum, 950
    Baltisphaeridium papillosum, 950
    Baltisphaeridium perrarum, 939, **1116**, 1151, 1061
    Baltisphaeridium pilosiusculum, 950
    Baltisphaeridium primarium, 950
    Baltisphaeridium strigosum, 947, 950
    Baltisphaeridium varium, 950
Bambuites, 880

Taxonomic Index

Bambuites erichsenii, 880
Banffia, 1174
BANGIOPHYCEAE, 189
Barbitositheca, 1022
Barskovia, 1026
Batenevia, 1018
Bathydiscus, 1022
BATHYNOTIDAE, 1175
Bathynotus, 1024
Bathysiphon, 238
    Bathysiphon filliformis, 238
BATHYSIPOHONIDAE, 1171
Bathyuriscellus, 1023–1024
Batillaria, 267
    Batillaria minima, 267
Batinevia, **362**, **364**, 366, 982, 984, 987
    Batinevia bayankolica, 987
    Batinevia nodosa, 987
    Baltinevia ramosa, 982, 987
Batophora, 322
Batschikicyathus, 1020
Bavlinella, 916–917, 922, 930–932, 935, 937–938, 940–943, 948–949, 1061, ***1115***, 1142
    Bavlinella faveolata, 916–917, 922, 930–932, 935, 937–938, 940–943, 948, 1061, ***1115***, 1142
    Bavlinella cf. *faveolata*, 934
Beaumontia, 1042–1043
    Beaumontia eckersleyi, 1042
Beckspringia, 916, 918, 1060, ***1082***, 1130
    Beckspringia communis, 916, 1060, ***1082***, 1130
Beggiatoa, **265**, 267, 271, 273–274, 276–277, 288, 291–293, 296, 314, 326–327, 330–332, 336, 341, 1129
    Beggiatoa alba, 327
    Beggiatoa minima, 1129
BEGGIATOACEAE, 1129
BELLEROPHONTIDA, 1026, 1174
BELLEROPHONTIDAE, 1174
Belliceps, 1024
BELTANACYANTHIDAE, 1172
Beltanacyathus, 1018
Beltanella, 1000
    Beltanella gilesi, 1000
    Beltanella podolica, 1000
Beltanelliformis, 356, 373, **387**, 392–394, 959, 961, 963, 967, 1000–1001, 1003, 1005, 1010–1011, 1037, 1042, 1173
    Beltanelliformis brunsae, **387**, 393–394, 1000–1001, 1003, 1005, 1010–1011
Beltanelloides, 356, **387**, 393, 502, 958, 960, 1000, 1043
    Beltanelloides sorichevae, **387**, 393, 958, 960, 1000
    Beltanelloides sorichevae major, 960
    Beltanelloides sorichevae minor, 960
Beltania, 1033
Beltina, 354, **355**, 357, 503, 960, 973
    Beltina danai, **355**, 960
    Beltina cf. *danai*, 960
BELTINIDS (*See* Subject Index)
Bemella, 1026
Bengtsonia, 1029
Bercutia, 1030
Bergaueria, 356, 372, 392–393, 416, 1010, 1037, 1042
    Bergaueria tumulus, 1010
Bergeroniaspis, 1024
Bergeroniellus, 1024
Beshtashella, 1026
Bestjachica, 982, 994
    Bestjachica rara, 982, 994
BESTJUBELLIDAE, 1176
Bestricophyton, 1042
Beticocyathus, 1018
Beyrichona, 1022
BEYRICHONIDAE, 1177
Bicella, 1023
Biceratops, 1024
Bicia, 1030
Bicorniculum, 1032
Bicuspidata, 883, 1061, ***1114***
    Bicuspidata fusiformis, 883, 1061, ***1114***

BICYATHIDAE, 1172
Bicyathus, 1018
Bidjinella, 1024
Bigeminococcus, 893, 912, 918, 1060, ***1094***, 1146
    Bigeminococcus lamellosus, 912, 918, 1060, ***1094***, 1146
    Bigeminococcus mucidus, 893, 912
Bigotina, 1024
Bigotinops, 1024
Bija, **362**, 366, 982, 987, 1032
    Bija canadensis, 987
    Bija grandis, 987
    Bija sibirica, 982, 987
    Bija cf. *sibirica*, 987
Bilinichnus, 557, 1010
    Bilinichnus simplex, 1010
Billingsella, 1030
BILLINGSELLIDAE, 1177
Binodaspis, 1023
Biocatenoides, 884–885, 888, 893, 895–896, 901, 910, 912, 918, 928, ***1062***, 1129
    Biocatenoides ferrata, 918
    Biocatenoides incrustata, 888, 893, 1129
    Biocatenoides pertenuis, 893
    Biocatenoides rhabdos, 893, 912, 1129
    Biocatenoides sphaerula, 884–885, 895, 912, 1129
Bioistodina, 1031
Bipatinella, 349, 503, 960
    Bipatinella cervicalis, 960
Birrimarnoldi, 237
    Birrimarnoldi antiqua, 237
Bisacculoides, 891, 893, 1146
    Bisacculoides grandis, 891, 893, 1146
    Bisacculoides tabeoviscus, 893
    Bisacculoides vacua, 893
BIVALVIA, 485, 1026, 1175, 1179, 1181–1184
Blastasteria, 1020
Blastulospongia, 239
    Blastulospongia mindyallica, 239
    Blastulospongia monothalamos, 239
    Blastulospongia polytreta, 239
Blayacina, 1024
Bledius, 295
    Bledius capra, 295
Bogutschanophycus, 982
    Bogutschanophycus mariae, 982
Bohaimedusa, 1042
    Bohaimedusa fuxianensis, 1042
Bohemipora, 874
    Bohemipora pragensis, 874
Bojarinovia, 1030
BOLASPIDIDAE, 1176
Bolboparia, 1023
BOMAKELLIDAE, 1173
Bomakiella, 373, 1000
    Bomakiella kelleri, 1000
Bonata, 372–373, **380**, 1000
    Bonata septata, **380**, 1000
BONATIIDAE, 1173
Bondonella, 1024
Bonnaria, 1023
Bonnaspis, 1023
Bonnia, 1023
Bonniella, 1023
Bonniopsis, 1023
Borealicornus, 1029
BOSCECULCYANTHIDAE, 1172
Bostricophyton, 1042–1043
    Bostricophyton bankuiyanensis, 1042
Bothroligotriletum, 888
    Bothroligotriletum exasperatum, 888
Botomaella, **362**, 366, 982–983, 987
    Botomaella aequalis, 987
    Botomaella anabarica, 987
    Botomaella crassa, 987
    Botomaella dubia, 987
    Botomaella mitis, 987
    Botomaella siberica, 987

Botomaella zelenovi, 982, 987
Botomella, 1023
Botominella, **362**, 366, 982, 984–987, 994
　　Botominella limeata, 982, 987
　　Botominella lineata var. *elanskensis*, 987
　　Botominella aff. *lineata*, 987
BOTOMOCYANTHIDAE, 1172
Botomocyathus, 1018
BOTRYOCOCCACEAE, 869, 938, 1152
Botryococcus, 84
Botsfordia, 1030
BOTSFORDIIDAE, 1177
Botswanella, 869, 1053
Bottonaecyathus, 1018
Boxonia, 509
Brabbinthes, 1006, 1017
　　Brabbinthes churkini, 1006
Brachina, 1000
　　Brachina delicata, 1000
BRACHIOPODA, 355, 400–401, 406, 412, 437, 448–450, 485, 558, 595, 959, 967, 969, 971, 973, 979, 1030–1031, 1037–1039, 1047, 1177
Brachypleganon, 941, 1144
　　Brachypleganon khandanum, 941, 1144
Brachysira, 322, **323**, 324
　　Brachysira aponina, 322, **323**, 324
Bracteacoccus, 1140
　　Bracteacoccus grandis, 1140
Bractocyathus, 1019
Bradoria, 1022
BRADORIIDA, 400, 448, 1177
BRADORIIDAE, 1177
Bradorona, 1022
Bradyfallotaspis, 1024
Branchiocaris, 1025, 1177, 1180–1184
Brastadella, 1022
Bredocaris, 1177
Brevilabiatus, 1028
Breviredlichia, 1024
Brevitrichoides, 888, 892, 906, 911, 919–920, 933, 1059–1060, **1066**, **1071**, 1128, 1144
　　Brevitrichoides bashkiricus, 906, 911, 919–920, 933, 1060, **1071**, 1128
　　Brevitrichoides burzjanicus, 892, 1059, **1066**, 1144
　　Brevitrichoides karatavicus, 906, 1060, **1071**
Bristolia, 1024
Brochopsophosphaera, 947
　　Brochopsophosphaera simplex, 947
BRONCHOCYATHIDAE, 1172
Bronicella, 1000–1001
Bronicella podolica, 1000–1001
Brooksella, **369–370**, 391, 418, 1000, 1006, 1010, 1039, 1042–1043, 1045
　　Brooksella canyonensis, **369–370**, 391, 418, 1000, 1006, 1010, 1039, 1042, 1045
BROOKSELLIDAE, 1051
Brushenodus, 1031
BRYOPHYTA, 357
BRYOZOA, 400
Bucania, 1027
Bucanotheca, 1029
Buchholzbrunnichnus, 1000, 1006, 1010
　　Buchholzbrunnichnus kroeneri, 1000, 1006, 1010
Bulaiaspis, 1024
Bullasphaera, 235, 879
　　Bullasphaera variegata, 235, 879
BULMANIDENDRIDAE, 1178
Buluniella, 1026
Bunyerichnus, 1000, 1010
　　Bunyerichnus dalgarnoi, 1000, 1010
Burchalaella, 982, 994
　　Burchalaella parvula, 994
　　Burchalaella pulchella, 982, 994
Burgessia, 1175
BURGESSOCHAETIDAE, 1175
Burithes, 1028

BURLINGIIDAE, 1176
Buschmania, 1006
　　Buschmania roeringi, 1006
Butakovicyathus, 1020
Butovia, 1021
Butunia, 1042
　　Butunia enigmatica, 1042
Byronia, 1021, 1174
BYRONIIDA, 400

Cadniacyathus, 1018
Caenorhabditis, 431
　　Caenorhabditis elegans, 431
CALCAEREA, 400, 1017, 1171, 1179, 1181–1184
Calcihexactina, 1017
Calcitheca, 1029
Callavia, 1024
Callosicoccus, 898
　　Callosicoccus crauros, 898
Calodiscus, 1022
Calothrix, 268, 296–297, 314, 316, **317–318**, 848, 852, 985, 1081
　　Calothrix thermalis, 268, 297
CALYPTOCOSCINIDAE, 1172
Calyptocoscinus, 1019
Calyptothrix, 906, 909, 912, 914, 920, 925, 933, 1060–1061, **1081**, **1083**, **1103**, 1131, 1133
　　Calyptothrix alternata, 906, 909, 920, 933, 1060, **1081**, **1083**, **1103**
　　Calyptothrix annulata, 912, 1131
　　Calyptothrix geminata, 906, 909, 920, 1060, **1081**, 1133
　　Calyptothrix obsoletus, 925, 1061, **1103**
　　Calyptothrix perfecta, 914, 1060, **1083**
Calyptrina, 354, 502, 960, 1006, 1021
　　Calyptrina partita, 1006
　　Calyptrina aff. *striata*, 960
Camasia, 419, 506, 877, 1042–1043, 1053
　　Camasia fruticulata, 1042
　　Camasia spongiosa, 877, 1042
CAMBRASTERIDAE, 1178
Cambria, 1022
Cambricodium, 366, 982, 994
　　Cambricodium capilloides, 994
　　Cambricodium capilloodes, 982, 994
CAMBRIDIIDAE, 1027, 1174
Cambridium, 1027
Cambrina, **362**–363, 366, 982, 986–987
　　Cambrina composita, 987
　　Cambrina fruticulosa, 982, 987
Cambrocassis, 1032
CAMBROCLAVA, 1174
CAMBROCLAVES, 400, 403, 409–410, 438, 1026, 1179, 1181–1184
Cambroclaves, **409**
　　Cambroclaves undulatus, **409**
Cambroclavus, 1026
Cambroconus, 1027
Cambrocrinus, 1178
Cambrocyathellus, 1018
Cambrocyathus, 1018
Cambromedusa, 1173
Cambronanus, 1018
Cambropodus, 1177
Cambroporella, 366, 982, 987
　　Cambroporella tuvensis, 982, 987
Cambroscutum, 1027
Cambrospira, 1027
Cambrothyra, 1032
Cambrotrypa, 1020
Cambrotubulites, 1022
Cambrotubulus, 1021
Camena, 1030
Camenella, **411**, 1030
　　Camenella parilobata, **411**
Camerotheca, 1029
CAMEROTHECIDA, 1029, 1175
CAMEROTHECIDAE, 1175

Taxonomic Index

CAMESELLIDAE, 1176
Campitius, 1021
Camptostroma, 1031
CAMPTOSTROMATIDAE, 1031, 1178
CAMPTOSTROMATOIDEA, 1031, 1178, 1180–1184
CANADASPIDIDA, 1022, 1177
CANADASPIDIDAE, 1177
CANADIIDAE, 1175
Canadiophycus, *362*, 366, 982, 987
    Canadiophycus fibrosus, 982, 987
Canopoconus, 1027, 1174
CAPILLICYATHIDAE, 1174
Capillicyathus, 1021
CAPSOLYNTHIDAE, 1172
Capsolynthus, 1020
Capsulocyathus, 1020
Caragassia, 1037, 1042
    Caragassia karassevi, 1042
Cardiophyllum, 1021
Carelozoon, 1042
    Carelozoon jatulicum, 1042
Carinachites, 1022
CARINACHITIDAE, 1022, 1174
CARINACYATHIDAE, 1172
Carinacyathus, 1018
Carinitheca, 1029
Carinolithes, 1028
Carnarvonia, 1177
Carpicyathus, 1018
CARPOIDS (*See* Subject Index)
Carubacgutes, 1033
Caryosphaeroides, 885, 912, 941, 1060, *1094*
    Caryosphaeroides pristina, 912, 1060, *1094*
    Caryosphaeroides tetras, 912, *1094*
Caryschia, 1042
    Caryschia cyathiformis, 1042
    Caryschia magna, 1042
Cassidina, 1027
Cassubia, 1025
Catinella, 871
    Catinella polymorpha, 871
CATTILLICEPHALIDAE, 1176
Caudiculophycus, 879, 909, 912, 928, 935, 1060, *1086*, *1091*, 1132–1133
    Caudiculophycus acuminatus, 909, 912, 1060, *1091*
    Caudiculophycus curvata, 935
    Caudiculophycus micronulatus, 879, 1060, *1086*
    Caudiculophycus rivularioides, 912, 1060, *1091*, 1132
Caudina, 354, 357, 502, 960
    Caudina cauda, 960
Cavifera, 366, 982, 987
    Cavifera concinna, 982, 987
    Cavifera cf. *concinna*, 987
Cellulomonas, 853
    Cellulomonas dehydrogens, 853
Celtiberium, 948
Centriconus, 1027
CEPHALOCARDIA, 1177, 1180–1184
CEPHALOCHORDATA, 1178
Cephalonyx, 879, 1060, *1086*
    Cephalonyx sibiricus, 879, 1060, *1086*
Cephalophytarion, 895, 898, 906, 912, 914, 925, 939, 1060–1061, *1086–1087*, *1092–1093*, *1104*, 1130–1131, 1133–1134
    Cephalophytarion constrictum, 895, 1133
    Cephalophytarion delicatulum, 912
    Cephalophytarion grande, 912, 1060, *1093*, 1131
    Cephalophytarion laticellulosum, 912, 1060, *1092*
    Cephalophytarion minutum, 912, 1060, *1092*, 1130
    Cephalophytarion piliformis, 925, 1061, *1104*, 1131
    Cephalophytarion taenia, 895, 898
    Cephalophytarion turukhanicum, 914, 1060, *1086–1087*
    Cephalophytarion variabile, 912, 1060, *1093*
CEPHALOPODA, 1175, 1179, 1181–1184
Cerabonusoides, 1032
Ceratium, 191
Ceratoconus, 1027

CERATOCYSTIDAE, 1178
Ceratophyton, 234, 884, 945, 1148
    Ceratophyton vernicosum, 884, 945, 1148
CERATOPYGIDAE, 1176
Cerithidea, 267–268
    Cerithidea californica, 267–268, 296
Cetraria, 850
    Cetraria crispa, 850
    Cetraria nivalis, 850
Chabakovia, *362–363*, 366, 982–987, 992
    Chabakovia cavitata, 987
    Chabakovia chabakoviformis, 987, 992
    Chabakovia flabellata, 987
    Chabakovia fungiformis, 987
    Chabakovia monstrata, 987, 992
    Chabakovia nana, 987
    Chabakovia nodosa, 987
    Chabakovia ramosa, 982, 987
    Chabakovia subglobosa, 988
    Chabakovia tuberosa, 982, 988
Chabakovicyathus, 1020
Chabiosphaera, 922
    Chabiosphaera bohemica, 922
CHAETETIDS, 985
CHAETOGNATHA, 400, 402, 404, 412, 449, 1031, 1178
Chaetomorpha, 322–323
Chakassicyathus, 1018
Chakasskia, 1023
CHAMAESIPHONACEAE, 1096–1097
Chancelloria, *409*, 1025
CHANCELLORIIDA, 408, 1025–1026, 1173
CHANCELLORIIDAE, 1025, 1173
Changaspis, 1023
Changshabaella, 1022
CHANKACYATHIDAE, 1172
Chankacyathus, 1018
Chaoaspis, 1024
Chara, 190
Characodictyon, 1032
Charaussaia, 366, 982, 987
    Charaussaia camptotaenia, 982, 987
Charnia, 373, 381–*382*, 387, 504, 1000, 1003, 1005
    Charnia grandis, 1000, 1003, 1005
    Charnia masoni, *382*, 387, 1000
    Charnia sibirica, 1000, 1005
CHARNIIDAE, 373, 1173
Charniodiscus, 373, 381–*382*, 504, 1000–1001, 1005
    Charniodiscus arboreus, *382*, 1000–1001
    Charniodiscus concentricus, 1000
    Charniodiscus longa, 1000, 1005
    Charniodiscus oppositus, 1000
    Charniodiscus planus, 1000
CHAROPHYCEAE, 190
CHAUNOGRAPTIDAE, 1178
CHEILOCEPHALIDAE, 1176
CHEIRURIDAE, 1176
Cheiruroides, 1023
Chelediscus, 1022
CHELICERATA, 381, 408, 485, 1022
Chengjiangoconus, 1027
Chengkoucyathus, 1019
Chengkouia, 1023
CHENGKOUIIDAE, 1175
Chiella, 1025
Chilodictyon, 1032
CHITINOZOA, 914, 926, 938
CHITONS (*See* Subject Index)
Chlamydomonas, 478–480, 482–483
    Chlamydomonas reinhardtii, 480, 482
Chlamydomonopsis, 886
    Chlamydomonopsis primordialis, 886
CHLORARACHNIOPHYTA, 479
Chlorella, 189, 430, 1138
    Chlorella fusca, 189
    Chlorella pyrenoidsa, 1138
    Chlorella vulgaris, 1138

CHLORELLACEAE, 1138
CHLOROBACTERIA, 473
CHLOROBIACEAE, 86, 104, 126, 276, 282–283, 827
Chlorobium, 87, 266, 274, 301, 304, 307, 310–312, 314, 336, 827, 844
CHLOROCOCCACEAE, 1138–1140, 1149–1150
CHLOROCOCCALES, 189–190
Chlorococcum, 1138–1140
    Chlorococcum aerenosum, 1139
    Chlorococcum diploibionticoideum, 1140
    Chlorococcum endozoicum, 1139
    Chlorococcum humicolum, 1138
    Chlorococcum infusionum, 1139
    Chlorococcum infusorium, 1139
CHLOROFLEXACEAE, 276, 282–283
Chloroflexus, 89, 126, 262, 266–267, 274, 276–278, 280–282, 287, 289, 297, 300, 304–305, 307, 309–314, **316**, **318**–321, 326–331, 336–337, 342, 474, 590, 827, 831, 844, 846
    Chloroflexus aurantiacus, 262, 266, 276–277, 282, 310, 316, **318**, 320, 827, 831
Chlorogloea, 848, 852, 1138
    Chlorogloea fritschii, 848, 852, 1138
    Chlorogloea turgidus, 848
Chlorogloeaopsis, 904
    Chlorogloeaopsis zairensis, 904
Chlorogloeopsis, 300–301, 303, 313–314, 320–321, 844
Chloroherpeton, 276, 827
    Chloroherpeton thalassium, 276
CHLOROPHYCEAE, 104, 205–207, 220, 283, 478, 592, 912–912, 934, 944, 1001, 1053, 1152, 1159–1160
CHLOROPHYTA, 188, 190, 219, 321–325, 355, 357, 359–361, 419, 928
Choia, 1017
CHOIIDAE, 1171
Chomatichnus, 1037, 1042
    Chomatichnus loevcichnus, 1042
Chomustachia, 366, 982, 984, 988
    Chomustachia diadroma, 988
    Chomustachia tuberosa, 988
Chondrastaulina, 1023
Chondrinouvina, 1023
CHONDROPHORES (See Subject Index)
Chondroplon, 1000
    Chondroplon bilobatum, 1000
Chonodrites, 416
CHORDATA, 412, 449, 485, 1178
Chordoichnus, 1042
    Chordoichnus latouchei, 1042
Choubertella, 1024
Chouberticyathus, 1018
CHROMATIACEAE, 126, 276, 282–283, 827
Chromatium, 276–277, 292, 301, 310–311, 314, 327–328, 336, 844, 846
    Chromatium tepidum, 301, 310–311
CHROOCOCCACEAE, 187, 195, 205–209, 212, 218, 220, 255, 590, 595, 597, 879–880, 884–885, 889–890, 891–905, 908, 910–916, 918–950, 1071, 1102, 1137–1139, 1141–1147, 1159
CHROOCOCCALES, 37–38, 195, 206, 255, 218, 243, 911, 926, 934–935, 985, 1159, 1161
Chroococcidiopsis, 927
Chroococcus, 37, **195**, 205–206, 208, 216, 324, 597, 885, 892, 905, 913, 1071, 1124, 1145–1146
    Chroococcus cohaerens, 1145
    Chroococcus giganteus, 206
    Chroococcus macrococcus, 206
    Chroococcus minutus, 1146
    Chroococcus tenax, 1146
    Chroococus turgidus var. *maximus*, 206
CHRYSOPHYCEAE, 189, 478
CHRYSOPHYTA, 237, 359, 449, 483
Chuaria, 180, 349, 353, **354**, 355–357, 433, 439, 503, 514–516, 652–653, 664, 876, 889–890, 903, 906, 908, 911–912, 914, 916–919, 921–923, 926–927, 931–932, 934–935, 940–941, 958–960, 962–964, 966–967, 969, 971, 973, 977, 1042–1043, **1075**, 1150–1151
    Chuaria annularis, 912, 960
    Chuaria circularis, **354**–355, 889, 908, 911–912, 914, 916–919, 921–923, 926–927, 931–932, 934–935, 940–941, 958–960, 962, 964, 966, 969, 977, 1150
    Chuaria aff. *circularis*, 906, 909, 931, 941
    Chuaria cf. *circularis*, 960, 966
    Chuaria fermorei, 966
    Chuaria jacutia, **1075**
    Chuaria minima, 966
    Chuaria multirugosa, 966, 1151
    Chuaria nerjuenica, 966
    Chuaria olavarriensis, 966
    Chuaria wimani, 922, 966
CHUARIACEAE, 349, 963, 965, 967, 975, 977
CHUARIAMORPHIDA, 349, 969, 971
CHUARIDAE, 349, 961, 963, 967, 971, 973
CHUARIDS (See Subject Index)
CILIATES (See Subject Index)
CINCTA, 1178
Circotheca, 1029
CIRCOTHECIDA, 1029, 1175
CIRCOTHECIDAE, 1175
Circulinema, 945, 1136
    Circulinema jinningense, 945, 1136
Circumiella, 937, 1061, **1112**
    Circumiella mogilevica, 937, 1061, **1112**
CIRRIPEDIA, 1177, 1180–1184
Cladophora, 322–**323**
    Cladophora glomerata, 322
Cladophoropsis, 322
    Cladophoropsis membranacea, 322
Clariondia, 1024
CLARUSCOSCINIDAE, 1172
Claruscoscinus, 1018
Claruscyathus, 1018
CLATHRICOSCINIDAE, 1172
Clathricoscinus, 1019
Clathricyathus, 1018
CLAVAGNOSTIDAE, 1175
CLAVOHAMULIDAE, 1178
Cletocampus, 296
    Cletocampus dietersi, 296
CLISOSPIRIDAE, 1174
Clonophycus, 891, 893, 898, 943
    Clonophycus biattina, 893
    Clonophycus elegans, 891, 893
    Clonophycus laceyi, 891
    Clonophycus ostiolum, 893
    Clonophycus cf. *ostiolum*, 893
    Clonophycus refringens, 893
    Clonophycus vulgaris, 893
Clostridium, 291, 320
    Clostridium aceticum, 291
    Clostridium thermoautotrophicum, 320
    Clostridium thermohydrosulphuricum, 320
    Clostridium thermosulfurogenes, 320
Cloudina, 371, 373, **388**, 401, 447–448, 505, 554, 559–560, 564, 593, 1006–1007, 1032
    Cloudina borrelloi, 1006–1007
    Cloudina hartmannae, 371, **388**, 401, 1006–1007
    Cloudina lucianoi, 1006–1007
    Cloudina riemkeae, 1006
    Cloudina waldei, 1006
CLOUDINIDAE, 1174
CNIDARIA, 372, 381, 393, 401, 405, 407, 418, 434, 448–450, 485, 1020, 1043, 1045, 1047, 1173
Cobboldiella, 1029
Coccochloris, 850
    Coccochloris elabens, 850
COCCOLITHOPHORIDS (See Subject Index)
Cocheatina, **1114**
    Cocheatina canilovica, **1114**
Cochlichnus, 373, 392, 1010, 1047
    Cochlichnus serpens, 393, 1010
Codonoconus, 1027
Coelenteratella, 1020
COELOSCLERITOPHORA, 400, 403, 408–409, 438, 448, 1025, 1173, 1179, 1181–1184

Taxonomic Index

COLENTERATES (*See* Cnidaria and Subject Index)
Coleobacter, 893, 1135
    Coleobacter primus, 893, 1135
Coleochaete, 190
Coleogleba, 931
    Coleogleba auctifica, 931
Coleolella, 1029, 1042
    Coleolella billingsi, 1042
COLEOLIDA, 1174, 1179, 1181–1184
COLEOLIDAE, 400–401, 407, 1021, 1174
Coleoloides, 1021
Coleolus, 1021
Collenia, 622, 625, 631, 660, 668–669
    Collenia kona, 668
Colleniella, 635
Collinsia, 1042
    Collinsia mississagiense, 1042
Collyrolenus, 1024
Colonnella, 624
    Colonnella cormosa, 624
    Colonnella heishanensis, 624
Comasphaeridium, **226**, 883, 940, 946, 949
    Comasphaeridium brachyspinosum, **226**
    Comasphaeridium magnum, 940
    Comasphaeridium strigosum, 949
Combinivalvula, 1022, 1177
COMPOSITOCYATHIDAE, 1172
Compositocyathus, 1018
Compsocephalus, 1023
Comptaluta, 1022
COMPTALUTIDAE, 1177
Conannulogungia, 1018
CONCHOPELTIDAE, 1173
Conchopeltis, 408
CONDYLOPYGIDAE, 1175
Confervites, 982, 988
    Confervites primordialis, 982, 988
    Confervites thoreaeformis, 982
Confinisquama, 1032
Conglobocella, 916, 1147
    Conglobocella troxelii, 916, 1147
Conichnus, 392
Conicina, 349, 502, 966
    Conicocina obtusa, 966
Coniunctiophycus, 895, 927
    Coniunctiophycus conglobatum, 895
    Coniunctiophycus gaoyuzhuangense, 895
CONOCARDIOIDA, 1175
CONOCORYPHIDAE, 1176
CONODONTA, 400, 412, 449, 1178, 1180–1184
Conodonta, 871
CONODONTOPHORIDA, 1178
CONOIDOCYATHIDA, 1021
CONOIDOCYATHIDAE, 1174
Conoidocyathus, 1021
Conomedusites, 1000
    Conomedusites lobatus, 1000
Conophyllum, 1042
    Conophyllum minor, 1042
Conophyton, 256, 259, 261, 266, 268–269, 278, 313, 335–336, 508–509, 515, 618–619, 623, 625–626, 661, 686–687
    Conophyton cylindricum, 508, 625–626
    Conophyton garganticum, 626
    Conophyton gaubitza, 623
Conoredlichia, 1024
Conotheca, **403**, 1029
    Conotheca australiensis, **403**
Constrictosphaerina, 948
    Constrictosphaerina alaica, 948
Contitheca, 1029
Contortothrix, 35, 882, 913, 1131
    Contortothrix vermiformis, 35, 913, 1131
Contrahofilum, 878, 929, 1127
    Contrahofilum minutum, 878
    Contrahofilum schopfii, 878, 929, 1127
CONULARIELLIDAE, 1022, 1174

CONULARIIDA, 400–401, 406, 408, 448, 450, 1022, 1174, 1179, 1181–1184
CONULARIIDAE, 1174
COOSELLIDAE, 1176
COPEPODS (*See* Subject Index)
COPLEICYATHIDAE, 1172
Copleicyathus, 1018
Copperia, 419, 506, 1042
    Copperia tubiformis, 1042
CORALS (*See* Subject Index)
Corbularia, 982, 994, 996
    Corbularia conglutinata, 982, 996
Cordilleracyathus, 1018
Cordobicyathus, 1020
CORDYLODODONTIDAE, 1178
Cordyolotus, 580
    Cordyolotus proavis, 580
Coreospira, 1026
COREOSPIRIDAE, 1026
CORIXIDAE, 296
CORNUTA, 1178
Cornutosphaera, 871
    Cornutosphaera polycornuta, 871
Cornutula, 982, 996
    Cornutula kaltatica, 982, 996
    Cornutula kyzassica, 996
CORRALIOIDAE, 1171
Corumbella, 371, 1006, 1032
    Corumbella werneri, 371, 1006
CORUMBELLIDAE, 1173
Corycium, 354, 503, 871, 876, 966
    Corycium enigmaticum, 871, 876, 966
    Corycium oligomerum, 876, 966
Corymbococcus, 876, 886
    Corymbococcus hodgkissii, 886
CORYNEXOCHIDA, 1023, 1176
CORYNEXOCHIDAE, 1176
CORYPYGIDAE, 1176
COSCINOCYATHELLIDAE, 1172
Coscinocyathellus, 1019
COSCINOCYATHIDA, 1019
COSCINOCYATHIDAE, 1172
Coscinocyathus, 1019
Coscinoptycta, 1019
Costatheca, 1029
Costatosphaerina, 872, 948
    Costatosphaerina ramosa, 872
    Costatosphaerina septata, 948
COTHONIIDAE, 1173
COTHURNOCYSTIDAE, 1178
Cowiella, 1031, 1174
CRASSICOSCINIDAE, 1172
Crassifimbra, 1023
Cremnodinotus, 1027
CRENARCHAEOTA, 473–475
CREPICEPHALIDAE, 1176
Crestijachites, 1028
CRESTJAHITIDAE, 1175
Crestoconus, 1027
CRIBIRICYATHEA, 1021
CRIBRICYATHA, 400–401, 407, 983, 1021
CRIBRICYATHIDA, 1021, 1174, 1179, 1181–1184
CRIBRICYATHIDAE, 1174
Cribricyathus, 1021
Cricocosmia, 1020, 1173
Cricopectinus, 1018
CRINOIDEA, 1178, 1180–1184
Crispus, 1021
Crithidia, 481, 483
CROCODILIANS (*See* Subject Index)
Crossbitheca, 1029
CRUSTACEA, 1022
CRUSTACEANS (*See* Subject Index)
Cruziana, 392
CRYPTARCHS (*See* Subject Index)
Crypthecodinium, 191

Crypthecodinium cohnii, 191
CRYPTOMONADS (*See* Subject Index)
CRYPTOPHYCEAE, 481
CRYPTOPOROCYATHIDAE, 1172
Cryptoporocyathus, 1020
Cryptotreta, 1030
CRYPTOTRETIDAE, 1177
Cryptozoon, 180
Ctenichnites, 1042
CTENOCYSTOIDEA, 1178, 1180–1184
CTENOPHORA, 1173, 1179, 1181–1184
Cucumiforma, **197**, 933, 1061, **1109**
    Cucumiforma vanavaria, **197**, 933, 1061, **1109**
Cucurbitella, 592
Cumulosphaera, 870
    Cumulosphaera lamellosa, 870
Cunicularius, 926
    Cunicularius halleri, 926
Cupitheca, 1029
Curticia, 1030
CURTICIIDAE, 1177
Curtitheca, 1029
Curvolithus, 392, 1010–1011, 1013
    Curvolithus aequus, 1010–1011
Cyanidium, 300–301, 304, 307, 314, 324, 479, 848
    Cyanidium caldarium, 301, 304, 307, 324
CYANOBACTERIA (*See* Subject Index)
Cyanonema, 893, 895, 910, 913, 934, 945, **1110**, 1130–1131
    Cyanonema attenuatum, 913, 1130
    Cyanonema disjuncta, 934, **1110**
    Cyanonema inflatum, 893, 945, 1131
    Cyanonema ligamen, 895, 1131
    Cyanonema minor, 893, 1130
Cyanophora, 478–479
CYANOPHYCEAE, 180, 872, 874
Cyanothrixoides, 919
    Cyanothrixoides mirabilis, 919
Cyathocricus, 1018
Cyathospongia, 1042
    Cyathospongia eozoica, 1042
Cycloconchoides, 1027
Cyclocyathella, 1018
CYCLOCYATHELLIDAE, 1172
CYCLOCYSTOIDIDAE, 1178
Cyclomedusa, **380**, 418, 1000, 1036–1037, 1042
    Cyclomedusa annulata, 1042
    Cyclomedusa davidi, 1000
    Cyclomedusa aff. *davidi*, 1042
    Cyclomedusa cf. *davidi*, 1042
    Cyclomedusa delicata, **380**, 1000
    Cyclomedusa gigantea, 1000
    Cyclomedusa gracilis, 1000, 1044
    Cyclomedusa minus, 1000, 1044
    Cyclomedusa minuta, 1000
    Cyclomedusa plana, 1000
    Cyclomedusa radiata, 1000
    Cyclomedusa serebrina, 1000
    Cyclomedusa simplicis, 1044
    Cyclomedusa simplicus, 1000
CYCLOMEDUSIDAE, 373, 1173
CYCLOZOA, 373, 558, 1173, 1179, 1181–1184
Cylindrichnus, **395**, 1010
Cylindrocraterion, 1044
    Cylindrocraterion heroni, 1044
Cymatiogalea, 938
Cymatiosphaera, **225**, 930, 935, 948–951, 1152
    Cymatiosphaera capsulara, 950
    Cymatiosphaera cristata, 950
    Cymatiosphaera favosa, 950
    Cymatiosphaera lavrovii, 948
    Cymatiosphaera membranacea, 950
    Cymatiosphaera minuta, 950
    Cymatiosphaera nerisica, 950
    Cymatiosphaera postii, **225**, 950
    Cymatiosphaera precambrica, 935, 1152
Cymatiosphaeroides, **223**, 912, 922, 1151
    Cymatiosphaeroides kullingii, **223**, 922, 1151
    Cymatiosphaeroides cf. *kullingii*, 912
Cymbia, 1025
Cymbionites, 1178
CYPHOMEGACHRITARCHS (*See* Subject Index)
Cyprideis, 296
    Cyprideis torosa, 296
Cyrtochites, **402**, 412, **439**, 1031, 1174
    Cyrtochites pinnoides, **402**, **439**
CYRTONELLIDA, 1026, 1174
Cytophaga, 1126
    Cytophaga johnsonae, 1126
    Cytophaga krzemieniewskae, 1126
CYTOPHAGACEAE, 1126

Dabanitheca, 1029
Dabashanella, 1022
Dabashanites, 1026
Dactylaria, 314, 324
    Dactylaria gallopava, 314, 324
Dactyloidites, 419, 1044
    Dactyloidites canyonensis, 1044
Dactyosachites, 1026
DAGUINASPIDIDAE, 1176
Daguinaspis, 1024
Dahaiella, 1022
Dailyatia, 1030
Dailycyathus, 1018
Dala, 1177
Daltaenia, 354–**355**, 357, 502, 966
    Daltaenia mackenziensis, **355**, 966
DAOLISHANIIDAE, 1176
Dasycirriphycus, 982, 996
    Dasycirriphycus frutuculosis, 982, 996
DASYCLADACEAE, 361, 860, 878, 932, 982, 984, 986–987, 994, 996, 1053
Dasyconus, 1020
DASYRICYATHIDAE, 1172
Debrennecyathus, 1020
Decoritheca, 1029
Degeletiicyathus, 1018
Degeletticyathellus, 1018
Deiradoclavus, 1026
Deltaclavus, 1026
DEMOSPONGEA, 239, 400–402, 404, 449, 1017, 1171, 1179, 1181–1184
Denaecyathus, 1018
Dendrocyathus, 1018
DENDROCYSTITIDAE, 1178
DENDROGRAPTIDAE, 1178
DENDROIDEA, 1178
Dengyingoconus, 1027
Densocyathus, 1018
Dentachites, 1032
DERMOCARPACEAE, 926
Desmochitina, 873
    Desmochitina grandis, 873
Despujolsia, 1024
DESPUJOLSIIDAE, 1176
Desulfovibrio, 291
    Desulfovibrio barsii, 291
Deunffia, 951
    Deunffia dentrifera, 951
DEUTEROSTOMES (*See* Subject Index)
Diandongia, 1030
Diandongoconus, 1028
DIATOMS (*See* Subject Index)
DICERATOCEPHALIDAE, 1176
Dicerodiscus, 1022
Dickinsonia, 161, 372–373, 379, 381, **383–384**, **386**, 456, 504, 1000–1001, 1005
    Dickinsonia brachina, 1000
    Dickinsonia costata, **383–384**, **386**, 1000–1001, 1005
    Dickinsonia elongata, **384**, 1000
    Dickinsonia lissa, 1000
    Dickinsonia minima, 1000
    Dickinsonia spriggi, 1000
    Dickinsonia tenuis, **384**, 1000

DICKINSONIDAE, 373, 1173
Dictyocoscinus, 1018
DICTYOCYATHIDAE, 1172
Dictyocyathus, 1018
Dictyonina, 1030
Dictyopsophosphaera, 948
　　Dictyopsophosphaera perrara, 948
DICTYOSCINIDAE, 1172
Dictyosphaera, 889, 1141
　　Dictyosphaera macroreticulata, 889, 1141
　　Dictyosphaera sinica, 889
Dictyosphaeridium, 879, 898, 900, 914, 930, 1060, **1085**
　　Dictyosphaeridium eniseicum, 900, 914
　　Dictyosphaeridium tungusum, 898, 1060, **1085**
　　Dictyosphaeridium vittaforme, 879
Dictyostelium, 482–483
　　Dictyostelium discoideum, 482–483
Dictyotidium, 945, 950
　　Dictyotidium birvetense, 945
　　Dictyotidium priscum, 950
Didymaulichnus, 373, 392, 397, 1010–1011
　　Didymaulichnus lyelli, 1010–1011
　　Didymaulichnus meanderiformis, 1010
　　Didymaulichnus miettensis, 1010
　　Didymaulichnus tirasensis, 1010
Didymocyathus, 1018
Dielymella, 1022
DIERESPONGIIDAE, 1172
Dignus, 1027
DIKELOCEPHALIDAE, 1176
Dimidia, 1025
Dimorphichnus, 392
DINESIDAE, 1176
DINOFLAGELLATES (*See* Subject Index)
DINOMISCHIDAE, 1174, 1180–1184
Dinomischus, 1033
Dinophysis, 479
Dioxycaris, 1025, 1177
Dipharus, 1023
DIPLAGNOSTIDAE, 1175
Diplichnites, 392, 1010
Diplococcus, 941
Diplocraterion, 371–372, 392, 397, 1010
　　Diplocraterion parallelum, 1010
Diplocyathellus, 1018
DIPLOMONADS (*See* Subject Index)
Diplopichnus, 416
Diplospinella, 1026
DIPLOTHCIDAE, 1175
Diplotheca, 1028
DISCAGNOSTIDAE, 1175
Discinella, 1022
DISPHAEROMORPH ACRITARCHS (*See* Subject Index)
Distichococcus, 918, 1147
　　Distichococcus minutus, 918, 1147
DITHECODENDRIDAE, 1178
DITHECOIDEA, 1178
Dodecaactinella, **402**, 1017
　　Dodecaactinella cynodontota, **402**
Dokidocyathella, 1020
DOKIDOCYATHIDAE, 1172
Dokidocyathus, 1020
DOKIMOCEPHALIDAE, 1176
DOLEROLENIDAE, 1176
Dolerolenus, 1024
Dolichocyathus, 1021
DOLICHOMETOPIDAE, 1176
Dolichometopsis, 1023
Dolichomocelypha, 1028
Doliutus, 1028
Dominopolia, 950
　　Dominopolia lata, 950
　　Dominopolia longispinosa, 950
DORSOJUGATIDAE, 1175
Dorsojugatus, 1028
Dorsolinevitus, 1028
Doushantuonema, 940

Doushantuonema peatii, 940
***Drepanochites*, 409**, 1026
　　Drepanochites dilatatus, **409**
Drepanopyge, 1026
Drepanuroides, 1024
DRILOMORPHA, 1175
Drosdovia, 1017
Drosophila, 431
　　Drosophilia melanogaster, 431
Dubius, 1021
Dunaliella, 87, 325
Duotingia, 1023
Dupliporocyathus, 1018
Dvinia, 504
Dysnoetopla, 1028
Dysoristus, 1030

Ebianella, 1028
Ebianotheca, 1029
***Eccentrotheca*, 411**, 1030
　　Eccentrotheca guano, **411**, 1030
Echaninia, 881
　　Echaninia mucosa, 881
Echidnina, 239
　　Echidnina runnegari, 239
ECHINODERMATA, 372–373, 400–401, 412, 431, 448–449, 485, 558, 1031, 1041, 1178
ECHIUROIDS (*See* Subject Index)
ECHMATOCRINIDA, 1178
ECHMATOCRINIDAE, 1178
Ectothiorhodospira, 827
EDELSTEINASPIDAE, 1176
Edelsteinia, 982, 994, 1018
　　Edelsteinia cylindrica, 994
　　Edelsteinia mongolica, 994
Ediacaria, 373, 379, **380**, 381, 1000
　　Ediacaria flindersi, 379, **380**, 381, 1000
Edreja, 1030
EDRIOASTERIDA, 1178
EDRIOASTERIDAE, 1178
EDRIOASTEROIDEA, 1178, 1180–1184
Edriodiscus, 1031
EDRIOSTEROIDEA, 1031
Egdetheca, 1029
***Eiffelia*, 402**, 1017
　　Eiffelia araniformis, **402**
EIFFELIIDAE, 1017, 1171
Eladicyathus, 1018
Elasenia, 1000
　　Elasenia aseevae, 1000
Eldonia, 372, 1033, 1174
Elganellus, 1024
ELKANIIDAE, 1177
Elkanospina, 1025
ELLESMEROCERATIDAE, 1175
ELLESMEROCERIDA, 1175
ELLIPSOCEPHALIDAE, 1176
Ellipsocephalus, 1024
Ellipsophysa, 349, 502, 966
　　Ellipsophysa axicula, 966
　　Ellipsophysa proceriaxis, 966
ELLIPSOPHYSIDS (*See* Subject Index)
Ellipsostrenua, 1024
Elliptocephala, 1024
ELVINIIDAE, 1176
Emarginoconus, 1028
Emeiconus, 1028
Emeidus, 1031
Emeiella, 1025
Emeitheca, 1029
Emeithella, 1031
EMERALDELLIDA, 1176
EMERALLIDAE, 1176
Emmensaspis, 1033
EMMRICHELLIDAE, 1176
EMUELLIDAE, 1176
Enammocephalus, 1024

ENDOCERIDA, 1175
Endoconchia, 1032
Enoplus, 296
  Enoplus communis, 296
Ensitheca, 1029
ENTACTINIIDAE, 1171
Enteromorpha, 296, 322–**323**, 325
  Enteromorpha prolifera, 322
Enteromorphites, 354, 357, 502, 966
  Enteromorphites siniansis, 966
ENTEROPNEUSTA, 1178
ENTEROPNEUSTS (*See* Subject Index)
ENTOMASPIDIDAE, 1176
ENTOPHYSALIDACEAE, 207, 218, 255, 590, 595, 597, 884–885, 895–896, 902, 904–905, 914, 918, 923, 926, 941, 1138, 1159
Entophysalis, 187, 262, **265**–266, 269, 300, 325, 597, 846, 850
  Entophysalis major, 262, **265**
Entosphaeroides, 870, 886, 904, 906, 928, 1059, **1063**, 1143
  Entosphaeroides amplus, 870, 886, 1059, **1063**
  Entosphaeroides bilinearis, 904, 1143
  Entosphaeroides irregularis, 904, 1143
  Entosphaeroides aff. *irregularis*, 906
EOACIDASPIDIDAE, 1176
Eoagnostus, 1023
Eoaphanocapsa, 928, 1061, **1108**
  Eoaphanocapsa oparinii, 928, 1061, **1108**
Eoaphanothece, 895
  Eoaphanothece zhuiana, 895
Eoastrion, 205, 233–234, 886–888, 893, **1063**, **1065**, 1148, 1162
  Eoastrion bifurcatum, 886, **1063**, 1148
  Eoastrion simplex, 886, 888, **1063**, 1148
Eoconcha, 1030
Eoconcharium, 1029
Eocoryne, 1017
EOCRINIDAE, 1178
EOCRINOIDEA, 1031, 1178, 1180–1184
Eocucumaria, 1033
Eocyrtolites, 1027
Eocystites, 1178
EOCYTES (*See* Subject Index)
EODISCIDAE, 1175
Eodiscus, 1023
Eodontopleura, 1023
EODONTOPLEURIDAE, 1176
Eoentophysalis, 186–187, 269, 884–885, 893, 895–896, 902, 905, 911, 913, 918, 926, 931, 934, 941, 1060, **1070**–**1071**, 1144, 1148
  Eoentophysalis arcata, 905, 1060, **1071**, 1148
  Eoentophysalis belcherensis, 884–885, 893, 895, 902, 918
  Eoentophysalis cf. *belcherensis*, 931
  Eoentophysalis cumulus, 913, 1144
  Eoentophysalis dismallakesensis, 896, 926, 934, **1070**, 1144
  Eoentophysalis cf. *dismallakensis*, 896
  Eoentophysalis magna, 902
  Eoentophysalis yudomatica, 941, 1144
Eoepiphyton, 982, 994
  Eoepiphyton jalgamicum, 983, 994
  Eoepiphyton jatulicum, 994
Eoescharopora, 1033
Eogloborilus, 1021
Eogloeocapsa, 897
  Eogloeocapsa bella, 897
Eohalobia, 1027–1028
Eoholynia, 354, 357, 502, 504, 966
  Eoholynia mosquensis, 966
EOHOLYNIDS (*See* Subject Index)
Eohyella, 186, 891, 926, 1143
  Eohyella campbellii, 186, 891
  Eohyella dichotoma, 926
  Eohyella endoatracta, 926, 1143
  Eohyella rectoclada, 926, 1143
Eoleptonema, 28, 32
  Eoleptonema australicum, 28, 32
EOLIMULIDAE, 1175
Eolimulus, 1022
Eomarginata, 892, 1059, **1065**, 1151
  Eomarginata striata, 892, 1059, **1065**, 1151
Eomedusa, 372

Eomicrhystridium, 233–234, 882, 886, 888, 928–929, 936, 942–943
  Eomicrhystridium aremoricanum, 928
  Eomicrhystridium barghoornii, 886
Eomicrocoleus, 896, 1130
  Eomicrocoleus crassus, 896, 1130
Eomicrocystis, 897
  Eomicrocystis elegans, 987
  Eomicrocystis irregularis, 897
Eomycetopsis, 881, 884–885, 888–889, 891–895, 897, 902–907, 909–913, 918–920, 922, 925–929, 931–933, 936–937, 940–943, 945, 949, 1060–**1062**, **1068**, **1071**, **1074**, **1080**, **1082**, **1091**, **1093**, **1101**, **1107**, **1110**, **1115**, 1126–1128, 1152
  Eomycetopsis campylomitus, 941, 1126
  Eomycetopsis crassiusculum, 912
  Eomycetopsis cylindrica, 904, 1126
  Eomycetopsis filiformis, 884–885, 891, 895, 902–903, 913, 929, 1126
  Eomycetopsis lata, 918, 1128
  Eomycetopsis pflugii, 903, 1127
  Eomycetopsis psilata, 892, 902–903, 906, 909, 920, 933
  Eomycetopsis reticulata, 903
  Eomycetopsis riberiensis, 919
  eomycetopsis rimata, 906, 909, 920, 933, 1060, **1082**
  Eomycetopsis robusta, 891, 895, 910–911, 918, 922, 925–928, 931, 940, 945, **1080**, **1091**, **1093**, **1107**, 1126
  Eomycetopsis rugosa, 904, 1127
  Eomycetopsis aff. *rugosa*, 909, 920, 933
  Eomycetopsis schopfii, 929, 1152
  Eomycetopsis septata, 902, 904, 932, 1126
  Eomycetopsis siberiensis, 902, 941, 1061, **1115**, 1127
Eonovitatus, 1029
EOORTHIDAE, 1177
Eophormidium, 895, 1135
  Eophormidium capitatum, 895, 1135
  Eophormidium liangii, 895, 1135
  Eophormidium semicirculare, 895
Eophyton, 1044
Eoplectonema, 918
  Eoplectonema minimum, 918
Eopleurocapsa, 918
  Eopleurocapsa cf. *sinica*, 918
Eoporpita, 373, 1000, 1173
  Eoporpita medusa, 1000
Eops, 1024
EOPTERIIDAE, 1175
Eoptychoparia, 1023
Eoredlichia, 1024
Eosaccharomyces, 879, 1060, **1077**
  Eosaccharomyces ramosus, 879, 1060, **1077**
Eosfroma, 1044
  Eosfroma prima, 1044
Eosoconus, 1027
Eosomedusa, 1044
Eosphaera, 205, 233–234, 869, 884, 886, 936, 941–942, 1059, **1063**, 1148
  Eosphaera tyleri, 886, 1059, **1063**, 1148
  Eosphaera aff. *tyleri*, 937
Eospicula, 1044
  Eospicula cayeuxi, 1044
EOSTROPHIIDAE, 1178
Eosynechococcus, 884–885, 897, 902, 905, 911, 913, 916–917, 919, 925, 928–929, 931–932, 936, **1071**, 1144
  Eosynechococcus amadeus, 913, 928, 931, 1144
  Eosynechococcus cf. *amadeus*, 932
  Eosynechococcus brevis, 925, 1144
  Eosynechococcus crassus, 897
  Eosynechococcus depressus, 925, 1144
  Eosynechococcus elongatus, 897
  Eosynechococcus giganteus, 897, 1144
  Eosynechococcus grandis, 885, 897, 911, 919, 1144
  Eosynechococcus isolatus, 902, 1144
  Eosynechococcus major, 897, 1144
  Eosynechococcus medius, 884–885, 905, 925, **1071**, 1144
  Eosynechococcus minutus, 929
  Eosynechococcus moorei, 884–885, 936, 1144
  Eosynechococcus thuleensis, 932, 1144
Eotebenna, 1026

Taxonomic Index

Eotetrahedrion, 913, 1061, **1094**, 1146
    Eotetrahedrion princeps, 913, 1061, **1094**, 1146
Eothele, 1030
EOTOMARIIDAE, 1174
Eovolvox, 233
Eozoon, 413–**414**, 1041, 1044
    Eozoon canadense, **414**, 1044
Eozygion, 885, 903, 913, 929, 942, 1061, **1094**, 1146
    Eozygion grande, 903, 913, 1061, **1094**
    Eozygion minutum, 885, 913, 929, 1061, **1094**, 1146
Epactridion, **406**, 1022
    Epactridion portax, **406**
Ephydra, 297
Epiphyton, 359, 361, **362**, 363–364, **365**, 366, 423, 982–983, 985–986, 988, 993–994, 996
    Epiphyton absimilis, 988
    Epiphyton achoricum, 988
    Epiphyton amplificatum, 988
    Epiphyton anquinum, 988
    Epiphyton baicalicum, 994
    Epiphyton benignum, 986, 988, 993
    Epiphyton bifidum, 988
    Epiphyton bisporangium, 988
    Epiphyton botomense, 988, 993
    Epiphyton cf. *botomense*, 988
    Epiphyton breviramosum, 988
    Epiphyton buguldeicum, 994
    Epiphyton carptum, 988
    Epiphyton celsum, 988
    Epiphyton complexum f. *semenica*, 988
    Epiphyton complexum f. *ungutica*, 988
    Epiphyton condensum, 994
    Epiphyton confractum, 988
    Epiphyton corporatum, 994
    Epiphyton crassum, 988
    Epiphyton crebrum, 988
    Epiphyton crinitum, 983, 988, 991
    Epiphyton crispum, 988, 991
    Epiphyton cristatum, 988, 993
    Epiphyton cudi, 988
    Epiphyton curvatum, 988
    Epiphyton decumanum, 988
    Epiphyton decumanum f. *anastasica*, 988
    Epiphyton decumanum f. *kolbaica*, 988
    Epiphyton decumanum f. *zherzhulica*, 988
    Epiphyton dembovi, 988, 990
    Epiphyton durum, 988, 990
    Epiphyton elegans, 988
    Epiphyton evolutum, 988
    Epiphyton ezhimicum, 988
    Epiphyton falcifruticosum, 988
    Epiphyton fasciatum, 988
    Epiphyton fasciculatum, 989, 994
    Epiphyton fibratum, 988
    Epiphyton fibratus, 994
    Epiphyton flabellatum, 359, 982, 988
    Epiphyton frequens, 989
    Epiphyton frondosum, 989
    Epiphyton fruticosum, 989
    Epiphyton aff. *fruticosum*, 994
    Epiphyton furcatum, 989
    Epiphyton geniculatum, 989
    Epiphyton gigam, 989
    Epiphyton grande, 989, 990
    Epiphyton improcerum, 989
    Epiphyton induratum, 989
    Epiphyton inexpectatum, 989, 990
    Epiphyton inobservabile, 989
    Epiphyton inopinatum, 983, 991
    Epiphyton inopinatus, 989
    Epiphyton cf. *inopinatus*, 994
    Epiphyton intergerinum, 989
    Epiphyton jacutii, 989, 991
    Epiphyton kiyanicum, 989
    Epiphyton longum, 989
    Epiphyton manaense, 989
    Epiphyton manaense f. *giganta*, 989
    Epiphyton mirabile, 989
    Epiphyton naturale, 989
    Epiphyton neodensum, 994
    Epiphyton novum, 989
    Epiphyton nubilum, 989, 993
    Epiphyton ordonatum, 989
    Epiphyton ornatum, 989
    Epiphyton parapusillum, 989
    Epiphyton pencillatum, 989
    Epiphyton cf. *pencillatum*, 989
    Epiphyton periodicum, 989
    Epiphyton plumosum, 989
    Epiphyton pretiosum, 989
    Epiphyton procerum, 989
    Epiphyton pseudoflexuosum, 989
    Epiphyton pusillum, 989
    Epiphyton racemosum, 989
    Epiphyton ramosum, 989
    Epiphyton rectum, 989
    Epiphyton rosulare, 989
    Epiphyton saturum, 989
    Epiphyton scapulum, 989
    Epiphyton scoparium, 989
    Epiphyton simplex, 989, 990
    Epiphyton spissum, 989
    Epiphyton subfruticosum, 989
    Epiphyton suvoravae, 989
    Epiphyton tenue, 989
    Epiphyton tuberculosum, 989
    Epiphyton ulinicum, 989
    Epiphyton varium, 989
    Epiphyton vulgare, 989
    Epiphyton zhuravlevae, 989
Epiphytonoides, **362**–363, 366, 983, 989, 994
    Epiphytonoides affinis, 989
    Epiphytonoides fasciculatus, 989, 994
    Epiphytonoides nurmogoicus, 989
    Epiphytonoides ornatus, 989
    Epiphytonoides roselatus, 989
    Epiphytonoides sanashtykgolicus, 983, 989
    Epiphytonoides shevelicus, 990
    Epiphytonoides tenuiramosus, 990
Erbiella, 1023
Erbina, 983, 994, 996
    Erbina aristata, 983, 994
Erbiopsidella, 1023
Erbiopsis, 1023
ERBOCYATHIDAE, 1172
Erbocyathus, 1018
Eremactis, **409**, 1025
    Eremactis conara, **409**
Eremosphaera, 1149–1150
    Eremosphaera gigas, 1149
    Eremosphaera viridus, 1150
ERIXANIIDAE, 1176
Erniaster, 1000
    Erniaster apertus, 1000
    Erniaster patellus, 1000
Ernietta, 373, 381, 1000–1003, 1005
    Ernietta aarensis, 1000
    Ernietta plateauensis, 1000–1001, 1003, 1005
    Ernietta tsachanabis, 1002
ERNIETTAMORPHA, 373, 1173
ERNIETTIDAE, 373, 1173
Erniobaris, 373, 1002
    Erniobaris baroides, 1002
    Erniobaris epistula, 1002
    Erniobaris gula, 1002
    Erniobaris parietalis, 1002
Erniobeta, 373, 1002
    Erniobeta forensis, 1002
    Erniobeta scapulosa, 1002
Erniocarpus, 373, 1002
    Erniocarpus carpoides, 1002
    Erniocarpus sermo, 1002
Erniocentris, 373, 1002
    Erniocentris centriformis, 1002

Erniocoris, 373, 1002
    Erniocoris orbiformis, 1002
Erniodiscus, 373, 1002
    Erniodiscus clipeus, 1002
    Erniodiscus rutilis, 1002
Erniofossa, 373, 1002
    Erniofossa prognatha, 1002
Erniograndis, 373, 1002
    Erniograndis paraglossa, 1002
    Erniograndis sandalix, 1002
Ernionorma, 373, 1002
    Ernionorma abyssoides, 1002
    Ernionorma clausula, 1002
    Ernionorma corrector, 1002
    Ernionorma peltis, 1002
    Ernionorma rector, 1002
    Ernionorma tribunalis, 1002
Erniopelta, 373, 1002
    Erniopelta scrupula, 1002
Erniotaxis, 373, 1002
    Erniotaxis segmentrix, 1002
Ernogia, 1031
Erphyllum, 1021
Erraticornus, 1028
Erugatocyathus, 1019
Erythrobacter, 281, 284
Escasona, 1022
Escherichia, 471, 483, 853
    Escherichia coli, 471, 483, 853
Esmeraldina, 1024
Estiastra, **225**, 949, 951
    Estiastra minima, **225**, 949, 951
Estrangia, 1024
ETHMOCOSCINIDAE, 1172
Ethmocoscinus, 1019
ETHMOCYATHIDAE, 1172
Ethmocyathus, 1018
ETHMOLYNTHIDAE, 1172
ETHMOPECTINIDAE, 1172
Ethmopectinus, 1018
ETHMOPHYLLIDAE, 1172
Ethmophyllum, 1018
Ethmosphaeridium, 939
EUBACTERIA, 86, 133, 185–189, 342, 471–474, 477–478
Eucapsis, 891, **1065**, 1148
Eucapsomorpha, 919, 1148
    Eucapsomorpha rara, 919, 1148
EUCARYA, 473–474
Euglena, 478–479, 481–483
    Euglena gracilis, 482
EUGLENOPHYCEAE, 478–479
EUKARYOTA, 220, 471, 473
EUKARYOTES (*See* Subject Index)
EUREKIIDAE, 1176
EURYARCHAEOTA, 473–475
Euryaulidion, 942
    Euryaulidion cylindratum, 942
Eurycyphus, 349, 502, 966
    Eurycyphus altilis, 966
    Eurycyphus lycotropus, 966
EURYPTERIDS (*See* Subject Index)
Eustypocystis, 1178
Evittia, 949
    Evittia cf. *irregulare*, 949
Evmiaksia, 1002
    Evmiaksia aksionovi, 1002
Exilitheca 1029
EXILITHECIDA, 1175
EXILITHECIDAE, 1175
Exochobrachium, 233–234, 886, 1148
    Exochobrachium triangulum, 886, 1148
Exoculatus, 1044
    Exoculatus arschaniensis, 1044
    Exoculatus selgineikensis, 1044
Extentitheca, 1029

Facivermis, 1033

FALITIDAE, 1177
FALLOCYATHIDAE, 1172
Fallocyathus, 1018
Fallotaspis, 1024
Fangxianites, 1025
FANSYCYATHIDAE, 1172
Fansycyathus, 1018
Fasciculella, 354, 502, 966
    Fasciculella bagungshanensis, 966
Fasciculus, 1173
Favilynthus, 1020
Favososphaeridium, 887, 890, 898–901, 910, 916, 922, 925, 927, 932, 935–936, 940, 942, 947, 949, 1142–1143
    Favososphaeridium bothnicum, 890, 898, 900, 927, 1142
    Favososphaeridium cf. *bothnicum*, 900
    Favososphaeridium favosum, 887, 899–900, 922, 927, 942
    Favososphaeridium scandicum, 949, 1142
    Favososphaeridium variabilis, 901
Favosphaera, 938, 1142
    Favosphaera conglobata, 938, 1142
    Favosphaera sola, 938
Feilongshania, 1023
Fenestrocyathus, 1018
Fengyangella, 349, 502, 966
    Fengyangella doedica, 966
FEOCYATHIDAE, 1172
Fermoria, 349, 503, 911, 966
    Fermoria capsella, 966
    Fermoria granulosa, 968
    Fermoria minima, 911, 968
FERMORIIDAE, 349
Ferralsia, 1024
Ferrimonilis, 893, 1130
    Ferrimonilis variabile, 893, 1130
Fibularix, 871–873
    Fibularix funicula, 871
    Fibularix garlickii, 873
    Fibularix mendelsohnii, 872–873
    Fibularix porulosa, 871
    Fibularix spinosa, 871
    Fibularix verrucosa, 871
    Fibularix zambiana, 873
Fieldia, 1177
FIELDIIDAE, 1173
Filamentella, 871, 877
    Filamentella plurima, 877
Filaria, 363, 366, 983, 990, 994
    Filaria calcarata, 983, 990
    Filaria mira, 994
    Filaria seriata, 983, 990
    Filaria sporifera, 990
Filiconstrictosus, 35–36, 896, 913–914, 1060–1061, **1086**, **1091–1092**, 1132–1133
    Filiconstrictosus diminutus, 35, 913–914, 1061, **1091–1092**, 1132
    Filiconstrictosus eniseicum, 914, 1060, **1086**, 1133
    Filiconstrictosus majusculus, 913, 1061, **1091**
FINKELNBURGIIDAE, 1177
Firmicornus, 1029
Fistulella, 366, 983, 990
    Fistulella decipiens, 983, 990
    Fistulella sanashtykgolica, 983, 990
Flabellaforma, **1077**
    Flabellaforma compacta, **1077**
Flabellina, 983, 996
    Flabellina multiformis, 983, 996
Flabetheca, 1032
FLAGELLATES (*See* Subject Index)
Flagellis, 939, 1061, **1114**, 1127
    Flagellis tenuis, 939, 1061, **1114**, 1127
Flexannulus, 1018
Flexicyathus, 1019
Flindersicoscinus, 1018
FLINDERSICYATHIDAE, 1173
FLORIDEOPHYCEAE, 189
FLORIOSOCYATHIDAE, 1172
Floritheca, 936

Taxonomic Index

Floritheca muhosensis, 936
Foliaceria, 983, 994
   Foliaceria polymorpha, 983, 994
Fomitchella, 404, 1031, 1174, 1180–1184
Foninia, 366, 983, 996
   Foninia fasciculata, 983, 996
FORAMINIFERIDA, 237–239, 263, 355, 359, 431, 433, 451, 454, 499, 592, 877, 967, 1041, 1045, 1171
Fordaspis, 1023
Fordilla, 1026
FORDILLIDAE, 1026, 1175
FORMOSOCYATHIDAE, 1172
Formosocyathus, 1018
Fragilicyathus, 1018
FRANSUASAECYATHIDAE, 1172
Fransuasaecyathus, 1020
Fremontella, 1024
Fremyella, 852
   Fremyella disposiphon, 852
Frinalicyathus, 1018
Frutexites, 886, 994
   Frutexites microstroma, 886
FRYXELLODONTIDAE, 1178
FUNGI (*See* Subject Index)
FURNISHINIDAE, 1178
Fusosquamula, 354, 502, 968
   Fusosquamula vlasovi, 968
Fusuconharium, 1029
Fuxianhuia, 1025, 1177

Gagarinicyathus, 1018
Gakarusia, 1044
   Gakarusia addisoni, 1044
Galaxiopsis, 886
   Galaxiopsis melanocentra, 886
GALICORNIDAE, 1175
Galicornus, 1028
Galinaecyathus, 1020
Gallatinia, 1044
   Gallatinia pertexa, 1044
   Gallatinia pretexa, 1044
   Gallatinia scalariformis, 1044
Gallionella, 291–292
   Gallionella ferruginea, 291
Ganloudina, 1031
Gaoqiaoella, 1025
Gaparella, 1032
Gastreochrea, 1021
GASTROCONIDAE, 1173
Gastroconus, 1020
GASTROPODA, 1026, 1174, 1179, 1181–1184
GASTROPODS (*See* Subject Index)
Gdowia, 1006, 1025, 1177
   Gdowia assatkini, 1006
Gelasene, 1023
Gemma, 362, 366, 983, 990
   Gemma inclusa, 983, 990
   Gemma maculosa, 990
Geniculicyathus, 1019
Geocyathus, 1018
Georyssus, 296
GERBICANICYATHIDAE, 1172
Gerbikanaecyathus, 1020
Geresia, 1030
Germinosphaera, 233–234, 926, 1061, **1103**, 1148
   Germinosphaera bispinosa, 926, 1061, **1103**, 1148
   Germinosphaera unispinosa, 926, 1061, **1103**, 1148
Ghoshia, 879
   Ghoshia bifurcata, 879
Giardia, 472, 481, 483
Gibbaspira, 1027
GIGANTOPYGIDAE, 1176
Gigantopygus, 1024
Gigoutella, 1024
Ginella, 1026
Ginospina, 1025–1026
Giordanella, 1023

Girphanovella, 1020
Girvanella, 359, 361–**362**, **364**, 366, 423, 514, 916, 982–984, 986–987, 990–991, 994
   Girvanella antiqua, 990, 994
   Girvanella antiquoformis, 994
   Girvanella iyangensis, 990
   Girvanella manchurica, 990
   Girvanella mexicana, 990
   Girvanella ocellatus, 994
   Girvanella problematica, 359, 983, 990
   Girvanella recta, 994
   Girvanella roberti, 994
   Girvanella sibirica, 990
   Girvanella sinensis, 990
   Girvanella staminae, 990
Glaessneria, 1002
   Glaessneria imperfecta, 1002
Glaessnericyathus, 1018
Glaessnerina, 373, 1002, 1044
   Glaessnerina grandis, 1002
Glaucocystis, 478
Glauderia, 1021
Glenobotrydion, 976, 885, 891, 895–896, 902, 904–906, 910, 913, 922–923, 926–929, 933, 1060, **1071**, **1093–1094**, **1108**, 1143–1145
   Glenobotrydion aenigmatis, 876, 902, 905–906, 910, 913, 919, 922–923, 927, 933, **1093–1094**, 1143
   Glenobotrydion compressus, 919
   Glenobotrydion granulosum, 891, 1145
   Glenobotrydion kanshiensis, 904
   Glenobotrydion majorinum, 885, 902, 910, 913, 928–929, **1108**
   Glenobotrydion solutum, 906, 1060, **1071**
   Glenobotrydion tetragonalum, 904, 1144
   Glenobotrydion varioforme, 895
Gleotheceopsis, 891
   Gleotheceopsis aggregata, 891
Globifructus, 1032
Globigerina, 1041
Globophycus, 885, 893, 897, 903, 905, 910, 913, 934, 1061, 1146
   Globophycus minor, 893
   Globophycus rugosum, 913, 934, 1061, 1146
GLOBORILIDA, 1029, 1175
GLOBORILIDAE, 1175
Globorilus, 1029
Globoritubulus, 1029
GLOBOSOCYATHIDAE, 1172
Globosocyathus, 1020
Globuloella, 366, 983, 994, 996
   Globuloella botomensis, 983, 994
   Globuloella incompacta, 994
   Globuloella notabila, 994
   Globuloella pellucida, 996
Globulus, 366, 983, 990
   Globulus gregalis, 983, 990
Gloeocapsa, 37, **195**, 205–206, 208, 216, 597, 905, 913, 928, 940, 1124, 1145
   Gloeocapsa punctata, 1145
   Gloeocapsa repestris, 1145
Gloeocapsoides, 918
   Gloeocapsoides media, 918
Gloeocapsomorpha, 878–879, 889–890, 892, 899, 901, 911, 916, 918, 924–925, 930, 933, 935–938, 944, 946–947, 1141
   Gloeocapsomorpha hebeica, 889, 930
   Gloeocapsomorpha karauliensis, 879
   Gloeocapsomorpha makrocysta, 889, 899, 901–902, 924, 947, 1141
   Gloeocapsomorpha prisca, 879, 889, 899–900, 911, 916, 924–925, 937–938, 946–947
   Gloeocapsomorpha priscata, 892, 918
Gloeocystis, 1139
   Gloeocystis gigas, 1139
Gloeodiniopsis, 891–892, 895, 897, 905, 913, 925, 928, 932, 934, 1059, 1061, **1065**, **1071**, **1093**, **1106**, 1142, 1145–1146
   Gloeodiniopsis dilutus, 934, 1146
   Gloeodiniopsis grandis, 928, 1061, **1106**
   Gloeodiniopsis gregaria, 913, 925, 1142
   Gloeodiniopsis hebeiensis, 895

1324 Taxonomic Index

Gloeodiniopsis lamellosa, 905, 913, 928, **1071**, **1093**, **1106**, 1145
  Gloeodiniopsis cf. *lamellosa*, 932
  Gloeodiniopsis magna, 928, **1006**, 1145
  Gloeodiniopsis mikros, 925, 1145
  Gloeodiniopsis pangjapuensis, 895, 1145
  Gloeodiniopsis uralicus, 892, 1059, **1065**, 1145
Gloeotrichia, 850, 852
  Gloeotrichia echinulata, 850, 852
Glomovertella, 359, 366, 983, 990
  Glomovertella firma, 983, 990
Gloriosocyathus, 1018
Glossophyton, 349, **354**, 356–357, 502, 968
  Glossophyton foliformis, 968
  Glossophyton hailaiensis, 968
  Glossophyton mucronatus, **354**, 968
Glottimorpha, 870–871, 874
  Glottimorpha asiatica, 870, 874
  Glottimorpha ordinata, 871
Glycine, 482
  Glycine max, 482
Glyptias, 1030
Glyptoria, 1030
GNALTACYATHIDAE, 1172
Gnaltacyathus, 1018
Gogia, 1031
Goldfieldia, 1023
Gonamella, 1032
Gonamispongia, 1020
Gonamophyton, 881
  Gonamophyton ovale, 881
Gongrosira, 322, 324
Goniocornus, 1028
Goniosphaeridium, **226**, 949
  Goniosphaeridium implicatum, **226**
  Goniosphaeridium primarium, 949
Gordia, 373, 392, **395**, 505, 1010, 1013
  Gordia antiqua, 1010
  Gordia arcuata, 1010
  Gordia marina, 1010, 1013
Gordonicyathellus, 1018
Gordonicyathus, 1019
Gordonifungia, 1019
Gordonophyton, 361, **362**, 363, **365**–366, 983, 990
  Gordonophyton axillare, 990
  Gordonophyton dembovi, 988, 990
  Gordonophyton distinctum, 983, 990
  Gordonophyton durum, 988, 990
  Gordonophyton grandis, 989–990
  Gordonophyton inexpectatum, 989–990
  Gordonophyton nodosum, 990
  Gordonophyton parvulum, 990
  Gordonophyton simplex, 989–990
Gorgonisphaeridium, 231
Gorskinocyathus, 1019
Gracilaria, 322
  Gracilaria blodgettii, 322
Gracilicyathus, 1021
Gracilitheca, 1029
GRACILITHECIDAE, 1175
Grammatophora, 325
Granoconus, 1028
Granomarginata, **225**, 514, 878–879, 881, 902–904, 932, 935–936, 938, 946–947, 949
  Granomarginata dhalii, 932
  Granomarginata exquisita, 932
  Granomarginata minuta, 904
  Granomarginata prima, 946, 949
  Granomarginata primitiva, 881, 932
  Granomarginata regia, 902–903
  Granomarginata rotata, 903
  Granomarginata simlaensis, 932
  Granomarginata squamacea, **225**, 514, 947, 949
  Granomarginata cf. *squamacea*, 938
  Granomarginata typica, 878–879
  Granomarginata vetula, 902
Granularia, 1023

GRANULARIIDAE, 1176
GRAPHOSCYPHIIDAE, 1173
GRAPTOLITES (See Subject Index)
GRAPTOLITHINA, 1178, 1180–1184
Graviglomus, 926, 1144
  Graviglomus incrustus, 926, 1144
Greysonia, 506, 1043–1044
  Greysonia basaltica, 1044
Grypania, 349, **354**–**355**, 356–357, 433–434, 473–474, 501, 503, 514, 519, 588, 591, 959, 968, 975
  Grypania spiralis, **354**, 473–474, 968
GRYPANIDS (See Subject Index)
Guangyuanaspis, 1024
Guangyuanella, 1022
Guizhoudiscus, 1023
Gumbycyathus, 1019
Gunflintia, 877, 885–888, 891, 893–895, 901–902, 904, 910, 929, 940, 944, **1063**–**1064**, **1068**, 1130–1131, 1133, 1152
  Gunflintia barghoornii, 904, 1152
  Gunflintia brueckneri, 944
  Gunflintia grandis, 886, 929, 1059, **1063**, 1131
  Gunflintia cf. *grandis*, 887, **1064**
  Gunflintia magna, 904, 1133
  Gunflintia minuta, 885–888, 891, 895, 910, 929, 1059, **1063**, 1130
  Gunflintia cf. *minuta*, 891, 902, 940
  Gunflintia oehlerae, 877
  Gunflintia septata, 893, 1130
Gutticonus, 1027
Gyalosphaera, 932
  Gyalosphaera fluitans, 932
  Gyalosphaera cf. *fluitans*, 932
Gymnodinium, 479
Gymnosolen, 509, 519, 631
Gyratosphaerina, 948
  Gyratosphaerina aspera, 948
GYROCYSTIDAE, 1178

Habrocephalus, 1024
Hadimopanella, **411**, 1031
  Hadimopanella apicata, **411**
Hadrotreta, 1030
Hagenetta, 373, 393, 1002, 1010
  Hagenetta aarensis, 393, 1002, 1010
Halichondrites, 1044
  Halichondrites graphitiferus, 1044
HALICHONDRITIDAE, 1171
Halkieria, 1026
HALKIERIIDAE, 408, 1026, 1173
Hallidaya, 1002
  Hallidaya brueri, 1002
Hallucigenia, 410
HALLUCIGENIIDAE, 1174
HALOBACTERIA, 472–473, 475
Halophiloscia, 296
Halosphaera, 189, 191
Halysicyathus, 1019
Halythrix, 36, 885, 895, 913, 927, 1061, **1092**, 1132, 1134
  Halythrix leningradica, 927, 1134
  Halythrix nodosa, 36, 913, 1061, **1092**, 1132
Hamatoconus, 1028
Hamatolenus, 1024
Hampilina, 1027
HAMPTONIIDAE, 1171
Hamusella, 1027
HANBURIIDAE, 1176
Hanchiangella, 1022
Hanchungella, 1022
Hanshuiella, 1031
Harlaniella, 373, 392–393, 505, 557, 657, 1010
  Harlaniella podolica, 392, 505, 1010
HARPIDIDAE, 1176
Hastina, 1031
Haupiria, 1030
HAWKEICYATHIDAE, 1173
Hazelia, 1017
HAZELIIDAE, 1171

Taxonomic Index

Hebediscus, 1023
Hecamede, 296
  Hecamede grisescens, 296
Heckericyathus, 1019
Hedstroemia, **362**, 366, 982–983, 985, 990, 993, 995
  Hedstroemia borealis, 995
  Hedstroemia flabellata, 990, 993
  Hedstroemia halimedoidea, 983
  Hedstroemia igarcaensis, 933
  Hedstroemia series, 990
Helcionella, 1027
HELCIONELLIDAE, 1174
Helenia, 1028
Heliconema, 882, 895, 909, 913–914, 944, 1060–1061, **1080**, **1083**, **1092**, 1129
  Heliconema australiense, 895, 913, 1129
  Heliconema funiculum, 913, 1061, **1092**, 1129
  Heliconema randomense, 944
  Heliconema turukhania, 914, 1060, **1083**
  Heliconema uralense, 909, 1060, **1080**
HELICOPLACIDAE, 1178
HELICOPLACOIDEA, 412, 1031, 1178, 1180–1184
Helicoplacus, 1031
Heliobacillus, 827
HELIOBACTERIA, 473
Heliobacterium, 238, 478, 827
  Heliobacterium chlorium, 238
Heliomedusa, 1030
Heliothrix, 281–283, 314, 321, 827
  Heliothrix oregonensis, 282–283, 321, 827
Helmetia, 1177
Helminthoida, 395, 437, 1010
Helminthoidichnites, 349, **354–355**, 356, 373, 392–393, 501, 503, 505, 968, 1010–1011, 1013, 1015
  Helminthoidichnites meeki, **355**–356, 968
  Helminthoidichnites neihartensis, 968
  Helminthoidichnites spiralis, 505, 968
  Helminthoidichnites tenuis, 393, 1010–1011, 1013, 1015
Helminthopsis, 392, 397, 1010
HEMICHORDATA, 1178
Hemisphaerammina, 239
Henaniodus, 1031
Heosomocelypha, 1028
Heraultia, 1026
Heraultipegma, 1026
HERKOMORPH ACRITARCHS (*See* Subject Index)
HESSLANDONIDAE, 1177
HETERACTINIDA, 1017
HETERACTINIDAE, 1171
Heterocyathus, 1020
Heterosculpotheca, 1032
Heterostella, 1017
HEXACTINELLIDA, 400–402, 404, 449, 1017
HEXACTINELLIDAE, 1172, 1179, 1181–1184
Hexangulaconularia, **406**, 1022
  Hexangulaconularia formosa, **406**
Hicksia, 1024
HICKSIIDAE, 1176
Hiemalora, 373, **382**, 1002, 1005
  Hiemalora pleiomorphus, **382**, 1002
  Hiemalora stellaris, 1002, 1005
HIEMALORIIDAE, 1173
Hindermeyeria, 1024
HINTZESPONGIIDAE, 1172
Hipponicharion, 1022
Hippopharangites, **409**, 1026
  Hippopharangites dailyi, **409**
Hoffetella, 1024
Holmia, 1024
Holmiella, 1024
Holmitheca, 1029
Homeothrix, 322
HOMOIOSTELEA, 1178, 1180–1184
HOMOSTELEA, 1178, 1180–1184
Homotreta, 1030
Honanella, 983, 991
  Honanella densa, 983, 991

Hormosiroidea, 1010, 1039
  Hormosiroidea arumbera, 1010
  Hormosiroidea canadensis, 1010
Houlongdongella, 1022
HOUSIIDAE, 1176
Hsuaspis, 1024
HUAIHECERATIDAE, 1175
Huainanella, 354, 502, 968, 1006, 1021
  Huainanella cylindrica, 968, 1006
  Huainanella striata, 968
Huainania, 349, 502, 968
  Huainania comma, 968
Huaiyuanella, 354, 502, 968, 1021
  Huaiyuanella baiguashanensis, 968
  Huaiyuanella marginata, 968
  Huaiyuanella aff. *marginata*, 968
  Huaiyuanella minuta, 968
  Huaiyuanella striata, 968
HUAIYUANELLIDAE, 349, 969
Huanglingella, 1027
Huangshandongella, 1033
Huangshandongoconus, 1027
Huanospongia, 1017
Hubeinella, 1028
Hubeispira, 1027
HUENELLIDAE, 1178
Huizenodus, 1031
Hujiagouella, 1027
Humboldtochaeta, 1032
HUNGAIIDAE, 1176
Hunnanocephalus, 1023
HUPECYATHELLIDAE, 1172
Hupecyathellus, 1019
Hupecyathus, 1018
Hupeia, 1024
Hupeidiscus, 1023
Hurdia, 1177
Huroniospora, 233, 876, 879, 884–888, 891–893, 919, 929, 940, 942, 1059, **1062–1063**, **1115**, 1137–1138
  Huroniospora macroreticulata, 886, 929, 1059, **1063**
  Huroniospora microreticulata, 876, 886, 891, 929, 1059, **1063**
  Huroniospora ornata, 891
  Huroniospora psilata, 886, 891–892, 929, 1059, **1063**
  Huroniospora rimosa, 919
Hyaloxybaphon, 1032
HYDNODICTYIDAE, 1172
Hydrocoleum, **196**, 1128
  Hydrocoleum heterotrichum, 1128
HYDROCONIDAE, 1173
HYDROCONOZOA, 400, 1173, 1179, 1181–1184
Hydroconus, 1020
HYDROIDA, 1173
HYDROZOA, 373, 434, 983, 1020, 1051, 1173, 1179, 1181–1184
HYDRYCONIDA, 1020
Hyella, 926
  Hyella gigas, 926
HYELLACEAE, 218, 926
HYMENOCARIDIDAE, 1177
Hymenocaris, 1022
Hymenophacoides, 880
HYMENOSTRACA, 1022, 1177
HYOLITHA, 355, 400–403, 410, 438, 447, 556, 558, 1028–1029, 1175
HYOLITHELLIDAE, 1174
Hyolithellus, **403**, 408, 1021
  Hyolithellus filiformis, **403**
  Hyolithellus cf. *micans*, **403**
HYOLITHELMINTHES, 400–401, 403, 408, 448, 1021, 1174, 1179, 1181–1184
Hyolithes, 1028
HYOLITHIDA, 1028, 1175
HYOLITHIDAE, 1175
HYOLITHOMORPHA, 1175, 1179, 1181–1184
Hyperammina, 238
HYPSELOCONIDAE, 1174
Hyptiotheca, **403**, 1028
  Hyptiotheca karraculum, **403**

HYPTOCYATHIDAE, 1172
Hyptocyathus, 1019
Hyypiana, 870
   Hyypiana jatulica, 870

Ichangia, 1024
Ichnusa, 1002, 1044–1045
   Ichnusa cocozzi, 1002, 1044
Ichnusina, 1044–1045
Ichnusocyathus, 1019
IDAHOIIDAE, 1176
IDIOTUBIDAE, 1178
Igorella, 1027
Igorellina, 1027
Ijinicyathus, 1019
Ikeyia, 1044
   Ikeyia tumida, 1044
ILLAENURIDAE, 1176
Ilsanella, 1027
IMBRICATA, 1178
Inactis, 870
Inacyathella, 1019
Inaria, **372**–373, 393, 1002
   Inaria karli, **372**, 1002
INARTICULATA, 558, 1030, 1177, 1180–1184
Incurvocyathus, 1020
Indiana, 1022
INDIANIDAE, 1177
Indianites, 1022
Indocera, 1031
Indota, 1022
Inessocyathus, 1019
Inflexiostella, 1017
Inkrylovia, 373, 1002
   Inkrylovia lata, 1002
INSECTS (*See* Subject Index)
Insignicornus, 1028
INSILICORYPHIDAE, 1175
Intrites, 393, **396**, 505, 1010
   Intrites punctatus, 392–393, **396**, 1010
Inuoyina, 1023
Invaginatibalteus, 1032
Inzeria, 519, 632
   Inzeria intia, 632
Iphidea, 1030
IRINAECYATHIDAE, 1172
Irinaecythus, 1019
IRREGULARES, 554, 1018, 1172, 1179, 1183–1184
Irridinitus, 418, 1002
   Irridinitus multiradiatus, 1002
ISCHYRINIIDAE, 1174
Isitella, 1027
Isiticyathus, 1019
Isitiella, 1027
Isititheca, 1029
ISITITHECIDAE, 1175
Isoclavus, 1026
ISOROPHIDA, 1178
Isophaera, 287, 320, 827
   Isophaera pallida, 287, 320, 827
Isoxys, 1025, 1177
Israelaria, 1030
Ivshinella, 1030
Iyaia, 1044
   Iyaia sayanica, 1044

Jacutianema, 907, 1060, **1074**, 1127
   Jacutianema solubila, 907, 1060, **1074**, 1127
Jacuticornus, 1028
Jakobina, 1028
JAKUTIDAE, 1176
Jakutiochrea, 1021
Jakutocarinus, 1019
Jakutus, 1023
Janguadacyathus, 1019
Jangudaspis, 1023
Japhanicyathus, 1019

Jaraktina, 1032
Jatuliana, 876, 983, 995
   Jatuliana furcata, 876, 983, 995
Jawonya, 239, 1017
Jebiletticoscinus, 1019
Jianfengia, 1025
Jiangispirellus, 1032
Jingshanodus, 1031
Jinkenites, 1028
Jixiella, 1002, 1044
   Jixiella capistratus, 1002, 1044
Judaiella, 1023
Judomia, 1024
Judomiella, 1024
JUGALICYATHIDAE, 1173
Jussenia, 1044, 1046
   Jussenia edelsteini, 1046
   Jussenia cf. *edeldsteini*, 1044

Kadvoya, 983, 991
   Kadvoya mirabilis, 983, 991
Kadyella, 1024
Kaimuria, 1046
   Kaimuria chambalensis, 1046
KAINELLIDAE, 1176
Kaisalia, **380**, 1002
   Kaisalia levis, 1002
   Kaisalia mensae, **380**, 1002
Kaiyangites, 1031
Kakabekia, 205, 233–234, 872, 878, 885–886, 888, 1059, **1063**, 1149, 1162
   Kakabekia flabelliformis, 872
   Kakabekia rarea, 872
   Kakabekia umbellata, 886, 888, 1059, **1063**, 1149
   Kakabekia cf. *umbellata*, 888
Kalbyella, 1027
KALTATOCYATHIDAE, 1172
Kaltatocyathus, 1020
Kalusina, 354, 502
Kameschkoviella, 1024
Kandatocyathus, 1019
Kanilovia, 354, 357, 502, 968
   Kanilovia insolita, 968
Karatubulus, 1017
Kareliana, 870, 983, 995–996
   Kareliana ukrainica, 996
   Kareliana zonata, 870, 983, 995
KARYOTES (*See* Subject Index)
Kaschkadakia, 1027
Katnia, 349, 354, 503
Katunioides, 1027
Kaunchuanella, 1028
KAZAKHSTANICYATHIDA, 1173
KAZAKHSTANICYATHIDAE, 1173
Kazakhstanicyathus, 1018
Kazakovichyathus, 1020
Kazyrycyathus, 1019
Keeleaspis, 1023
Kelanella, 1029–1030
KELANELLIDAE, 1029, 1174
Kellericyathus, 1019
Kempia, 1039, 1046
   Kempia huronense, 1039, 1046
Kenella, 366, 983, 991
   Kenella ornata, 983, 991
Kennardia, 1030
Ketemecornus, 1028
Ketemella, 983, 991
   Ketemella lenaica, 983, 991
Khairkhania, 1027
Khasagtina, 1030
Khasaktia, 1018
KHASAKTIDAE, 1171
Khatyspytia, 1002
   Khatyspytia grandis, 1002
Kheinjuasphaera, 902
   Kheinjuasphaera vulgaris, 902

Khetatheca, 1029
KIDRJASOCYATHIDAE, 1172
Kidrjassocyathus, 1020
Kijacus, 1031
KIJACYATHIDAE, 1172
Kijacyathus, 1019
Kijanella, 1024
Kildinella, **198**, 223, 229, 349, 503, 892, 894, 896, 898–907, 909, 911–912, 914–915, 917–935, 937, 939, 941–946, 948, 951, 968, 970, 1061, **1075**, **1103**, 1140, 1149–1150
    Kildinella exsculpta, 925
    Kildinella hyperboreica, 892, 896, 898–901, 905, 907, 909, 912, 917–935, 939, 941–944, 948, 951, 1149
    Kildinella jacutica, 907, 909, 914, 925, 942, 968, 1149
    Kildinella cf. *jacutica*, 922
    Kildinella kulgunica, 906, 909
    Kildinella lophostriata, **198**, 223, 229, 896, 1061, **1103**
    Kildinella magna, 914, 922, 968, 970, 1150
    Kildinella minuta, 902–903
    Kildinella miroedichia, 914
    Kildinella nordia, 896, 900, 906, 909, 920, 933–934, 1149
    Kildinella rifeica, 896, 898, 905, 909, 1140
    Kildinella sinica, 896, 898–901, 905, 907, 909, 911, 915, 917, 921–923, 925–927, 930–931, 934, 941, 945–946
    Kildinella cf. *sinica*, 922, 935
    Kildinella suketensis, 903
    Kildinella timanica, 898, 900, 925, 927
    Kildinella timofeevii, 904, 1149
    Kildinella tschapomica, 892, 896, 898, 900, 905–906, 909, 915, 925, 927, 933
    Kildinella vesljanica, 905, 907, 909, 918, 927
    Kildinella cf. *vesljanica*, 923
Kildinosphaera, **223**, 228–229, 889, 908–912, 914, 917, 929–930
    Kildinosphaera chagrinata, **223**, 228–229, 908–909, 911–912, 917, 929
    Kildinosphaera granulata, 917, 930
    Kildinosphaera lophostriata, **223**, 909, 914, 930
    Kildinosphaera verrucata, 911–912, 930
Kimberella, 1002–1003
    Kimberella quadrata, 1002–1003
KIMBERELLIDAE, 1173
Kimberia, 1002
    Kimberia quadrata, 1002
Kingaspis, 1024
KINGSTONIIDAE, 1176
Kinneyia, 1046
    Kinneyia simulans, 1046
Kinzercystis, 1031
Kisasaecyathus, 1019
Kistasella, 1027
Kjerulfia, 1024
KNOYRIIDAE, 1172
Koksodus, 1031
Koksuja, 1021
KOLBICYATHIDAE, 1172
Kolbinella, 1023
KOMASPIDIDAE, 1176
Konderia, 877, 1060, **1066**
    Konderia elliptica, 877, 1060, **1066**
Konicekion, 1022
Kootenia, 1023
KORDECYATHIDAE, 1172
Kordecyathus, 1019
Kordephyton, 361–**362**, **365**–366, 983, 988, 991
    Kordephyton conglutinatum, 991
    Kordephyton crinitum, 988, 991
    Kordephyton crispum, 988, 991
Korilacus, 1031
Korilithes, 1028
Korilophyton, **362**, 366, 983, 991, 995
    Korilophyton angustum, 983, 991
    Korilophyton inopinatum, 983, 991
Korovinella, 1018
KOROVINELLIDAE, 1173
Kotuites, 1021
Kotujella, 1030
Kotuyicathellus, 1019

Kotuyicyathus, 1019
Kotuyitheca, 1029
Kotyikanites, 1021
Krasnopeevacyathus, 1019
Krishnania, 349, 356, 503, 911, 970
    Krishnania acuminata, 911, 970
Krolina, 1024
Kuamaia, 1025, 1177
Kuanyangia, 1023–1024
Kueichowia, 1029
KUEICHOWIIDAE, 1176
Kugdatheca, 1023–1024
Kuibisia, 1002
    Kuibisia glabra, 1002
Kullingia, 1002
    Kullingia concentrica, 1002
Kulparina, 1030
Kundatella, 1030
Kundatia, 366, 983, 991
    Kundatia composita, 983, 991
Kunmingella, 1022
Kunyangella, 1022
Kunyangotheca, 1029
Kuonamkicornus, 1028
Kuraya, 1020
Kussiella, 519
Kutorgina, 1030
KUTORGINIDA, 1030, 1177
KUTORGINIDAE, 1177
Kyarocyathus, 1020
Kyzassia, 363–366, 983, 991
    Kyzassia elegans, 991
    Kyzassia formosa, 983, 991

Labradoria, 1023
Labyrinthus, 1033
Lacerathus, 1021
Ladaecyathus, 1019
Ladatheca, 1029
Lagenaconularia, 1022
Lagenicyathus, 1021
Lakhandinia, 349, 353, 356, 503, 907, 970, 1060, **1076**
    Lakhandinia prolata, 907, 970, 1060, **1076**
Laminarites, 349, 354, 501–502, 504, 873–875
    Laminarites antiquissimus, 501, 873
Lancastria, 1023
Lanceoforma, **354**, 357, 503, 970
    Lanceoforma striata, **354**, 970
Lanicyathus, 1019
Lapworthella, **411**, 1030
    Lapworthella fasciculata, **411**
LAPWORTHELLIDAE, 410, 1030, 1174
Laratheca, 1029
Lathamella, 1031
Laticephalus, 1023
Laticonus, 1028
Laticornus, 1028
Latimeria, 595
    Latimeria chalumnae, 595
Latiredlichia, 1024
Latirostratus, 1027
Latisphaera, 916, 1142
    Latisphaera wrightii, 916, 1142
Latoporata, 883
    Latoporata naumovae, 883
Latouchella, 1027
Laudonia, 1024
Leanchoilia, 1025
LEANCHOILIIDA, 1175
LEANCHOILIIDAE, 1175
Lebediscyathus, 1019
LECANOPYGIDAE, 1176
Lecythioscopa, 1173
Leguminella, 1031
Leibaella, 1021
LEIBAELLIDAE, 1174
Leibotheca, 1029

Leiofusa, 877–879, 948
   Leiofusa actinomorpha, 878–879
   Leiofusa bicornuta, 877
   Leiofusa digitata, 877
Leiofusidium, 910, **1080**
   Leiofusidium dubium, 910, **1080**
Leioligotriletum, 887, 890, 898, 938, 943
   Leioligotriletum compactum, 898
   Leioligotriletum crassum, 887, 938
   Leioligotriletum cf. *crassum*, 943
   Leioligotriletum glumaceum, 898
   Leioligotriletum minutissimum, 890
   Leioligotriletum nitidum, 887
Leiomarginata, 946
   Leiomarginata simplex, 946
Leiominuscula, 876, 889, 892, 896, 900, 906, 910, 951, 1138
   Leiominuscula incrassata, 876
   Leiominuscula minuta, 892, 896, 900, 906, 951
   Leiominuscula aff. *minuta*, 889, 910
   Leiominuscula orientalis, 889
   Leiominuscula pellucentis, 889, 1138
Leiopsophosphaera, 878, 883, 889, 908, 918, 920–921, 933, 947
   Leiopsophosphaera aspertus, 889, 908, 920–921
   Leiopsophosphaera convexiplicata, 933
   Leiopsophosphaera effusa, 947
   Leiopsophosphaera aff. *effusus*, 889, 921
   Leiopsophosphaera microrugosa, 933
   Leiopsophosphaera minor, 889, 908, 920–921
   Leiopsophosphaera mugodjarica, 878
   Leiopsophosphaera pelucidus, 889, 918, 920–921
   Leiopsophosphaera rotunda, 883
Leioria, 1030
LEIORIIDAE, 1177
Leiosphaeridia, 189, **198**, **223**, 228–229, 877–878, 889, 896, 901–906, 908–912, 915, 917, 925, 930–931, 935–945, 947, 949–950, 965, 1060–1061, **1075**, **1080**, **1114**, **1116**, 1149–1150
   Leiosphaeridia asperata, 906, 908–909, 911, 915, 917, 930
   Leiosphaeridia bicrura, 228, 896, 947
   Leiosphaeridia bituminosa, 877, 901, 905, 915, 938, 940, 1149
   Leiosphaeridia dehisca, 229, 945
   Leiosphaeridia densum, 902
   Leiosphaeridia eisenackia, 937–938
   Leiosphaeridia incrassatula, **198**, 939, 1061, **1116**
   Leiosphaeridia kanshiensis, 904
   Leiosphaeridia kulgunica, 229, 910, 1060, **1080**
   Leiosphaeridia microgranulosa, 902
   Leiosphaeridia ochroleuca, 944, 947, 1150
   Leiosphaeridia porcellanitensis, 902
   Leiosphaeridia pylomifera, 945
   Leiosphaeridia sarjeantii, 878
   Leiosphaeridia subgranulata, 950
   Leiosphaeridia undulata, 942, 1061, **1114**
LEIOSTEGIIDAE, 1176
Leiothrichoides, **197**, 504, 906–907, 915, 920, 933, 1060, **1072**, **1088–1089**, 1127
   Leiothrichoides maculatus, 915, 1060, **1089**, 1127
   Leiothrichoides tenuitunicatus, **197**, 907, 1060, **1072**
   Leiothrichoides typicus, 906, 915, 920, 933, **1088**, 1127
Leiovalia, 936, 951
   Leiovalia tenera, 951
Lenadiscus, 1023
Lenaella, 983, 991, 1020
   Lenaella longa, 991
   Lenaella reticulata, 983, 991
Lenalituus, 1028
Lenargyrion, 1031
LENARGYRIONIDAE, 1174, 1180–1184
Lenaspis, 1023
Lenastella, 1017
Lenatheca, 1029
Lenica, 1017, 1171
LENOCYATHIDAE, 1172
Lenocyathus, 1019
Lentitheca, 1029
Lepidites, 1028
LEPIDOCYSTIDAE, 1178
Lepidocystis, 1031

LEPIDOCYSTOIDEA, 1031, 1178
Lepochites, 1026
Leptochilodiscus, 1023
LEPTOMIDAE, 1171
Leptomitus, 1017
Leptosocyathus, 1019
Leptostega, 1027
Leptoteichos, 844, 886–887, 922
   Leptoteichos golubicii, 886–887
Leptothrix, 304
   Leptothrix discophora, 304
Leshanella, 1022
LEUCOTHRICACEAE, 35, 1131
Liangshanella, 1022
Liantuoconus, 1028
Liaoningia, 1002, 1046
   Liaoningia fuxianensis, 1002, 1046
   Liaoningia cf. *fuxianensis*, 1046
LICHENOIDIDAE, 1178
Licmorpha, 325
Lidaconus, 1032
Liepaina, 946
Lignum, 871
   Lignum nematoideum, 871
   Lignum punctulosum, 871
   Lignum striatum, 871
Ligyrokala, 1028
LUMULAVIDA, 1176
LUMULIDS (*See* Subject Index)
Linella, 509
Linevitus, 1028
Linguiformis, 349, 502, 970
   Linguiformis loeris, 970
Lingula, 1046
   Lingula calumet, 1046
Lingulella, 503, 970, 1030, 1046
   Lingulella montana, 1046
LINGULIDA, 400, 595, 1030, 1177
Linnarssonia, 1030
Liorichita, 1028
LIOSTRACINIDAE, 1176
LIPOPORIDAE, 1173
Lithapium, 1032
LITHISTIDA, 1171
Lithophyllum, 361
   Lithophyllum congestum, 361
Liulaobeia, 349, 502, 970
   Liulaobeia mesacosta, 970
Livia, 1025
Ljadlovites, 349, 502, 970
   Ljadlovites reticulatus, 970
Ljadovia, 875, 1061, **1113**
   Ljadovia perforata, 875, 1061, **1113**
Lobiochrea, 1021
Loculicyathellus, 1019
Loculicyathus, 1019
Loculitheca, 1029
LOGANELLIDAE, 1176
Lomasulcavichites, 1026
Lomasulsacachites, 1026
Lomatiocyathus, 1021
Lomentunella, 907, 1060, **1076**, 1136
   Lomentunella vaginata, 907, 1060, **1076**, 1136
Lomosovis, 1002
   Lomosovis malus, 1002
LONCHOCEPHALIDAE, 1176
Longaevus, 1021
Longduia, 1024
LONGDUIIDAE, 1176
Longfengshania, 349, **354–355**, 356–357, 433, 501, 503, 514, 970–971
   Longfengshania elongata, 970
   Longfengshania gemmiforma, 970
   Longfengshania longipetiolata, 970
   Longfengshania ovalis, **354**, 970
   Longfengshania spheria, 970
   Longfengshania stipitata, **355**, 970

Taxonomic Index

LONGFENGSHANIDS (*See* Subject Index)
Longianda, 1024
Longicyathus, 1021
Longiochrea, 1021
Lopholigotriletum, 901–902, 924, 938, 944, 946–947
    Lopholigotriletum crispum, 924, 938, 944
    Lopholigotriletum grumosum, 946
    Lopholigotriletum spathaeforme, 901–902, 924, 947
Lophominuscula, 951
    Lophominuscula rugosa, 951
Lophorytidodiacrodium, 945
    Lophorytidodiacrodium tosnaense, 945
Lophosphaeridium, 903, 932, 937–938, 948–951
    Lophosphaeridium bellus, 932
    Lophosphaeridium granulatum, 904
    Lophosphaeridium rarum, 948
    Lophosphaeridium tentativum, 950
    Lophosphaeridium truncatum, 949, 951
    Lophosphaeridium vetulum, 903
Lophotheca, 1021
Lophyridia, 296
    Lophyridia aulica, 296
Lopochites, 1026
Lorenzinites, 373, 1002
    Lorenzinites rarus, 1002
Loriforma, 349, 354, 502, 970
    Loriforma closta, 970
Luaspis, 1024
Luella, 1022
Luguviella, 1030
Lunachites, 1031
Lunolenus, 1024
Lunulacyathus, 1019
Lunulidia, 936
    Lunulidia nana, 936
Lusatiops, 1024
Luxella, 1023
Luyanhaochiton, 1026
Lyngbya, 188, **196**, 214, 262, **265**–266, 268, 296, 300, 304, 597, 846, 850, 894, 929, 934, 940, 1067, 1126–1128, 1136–1137
    Lyngbya aestuarii, 188, 262, **265**–266, 304, 850
    Lyngbya bipunctata, 1126
    Lyngbya borgerti, 1127
    Lyngbya calcifera, 1128
    Lyngbya chaetomorphae, 1126
    Lyngbya cryptovaginata, 1127
    Lyngbya digueti, 1126
    Lyngbya epiphytica, 1126
    Lyngbya kuetzingii, 1126
    Lyngbya lagerhaimii, 850
    Lyngbya limnetica, 1126
    Lyngbya loriae, 1126
    Lyngbya magnifica, 1128
    Lyngbya majuscula, 850
    Lyngbya martensiana, 1128
    Lyngbya mesotricha, 1127
    Lyngbya perelegans, 1126
    Lyngbya polysiphoniae, 1127
    Lyngbya rubida, 1127
    Lyngbya semiplena, 1128
    Lyngbya sordida, 1136
    Lyngbya spiralis, 1127
    Lyngbya stagnina, 1137

Mabiania, 1032
Mabianoconullus, 1022
Macanopsis, 416
MACHAERIDIA, 412
Mackenzia, 434, 1173
    Mackenzia costalis, 434
Mackenziecyathus, 1019
Mackenziella, 1028
Mackenziephycus, **362**, 366, 984, 991
    Mackenziephycus medullaris, 984, 991
Mackinnonia, 1027
MACLURITIDAE, 1174
Macromonas, 1144
Macromonas mobilis, 1144
Macroptycha, 196–**197**, 199, 915, 939, 1060, **1085**
    Macroptycha biplicata, **197**, 199, 915, 939, 1060, **1085**
    Macroptycha multiplicata, 915
    Macroptycha triplicata, 199, 915
    Macroptycha uniplicata, **197**, 199, 915, 1060, **1085**
Macrostomum, 296
Maculosphaera, 916
    Maculosphaera kingstonensis, 916
Madigania, 373, 1002
    Madigania annulata, 1002
Magnicanalis, 1030
Maidipingoconus, 1027
Maikhanella, **405**, 1027–1028, 1174
    Maikhanella cambrica, **405**
    Maikhanella pristinis, **405**
Maishucunconus, 1028
Majaella, 984, 986, 991, 1002
    Majaella verkhojanica, 984, 991, 1002
Majaphyton, 872, 1060, **1078**
    Majaphyton antiquaum, 872, 1060, **1078**
    Majaphyton ceratum, 872
    Majaphyton cyatum, 872
Majasphaeridium, 872, 1060, **1078**
    Majasphaeridium carpogenum, 872, 1060, **1078**
Majatheca, 1029
Makarakia, 1027
MALACOSTRACA, 1022, 1177, 1180–1184
Maldeotaia, 1031
Mallagnostus, 1023
Malongella, 1022
Malungia, 1024
Malykanotheca, 1029
MANACYATHIDAE, 1174
Manacyathus, 1021
Manchuriophycus, 414, 1037, 1041, 1046, 1051
    Manchuriophycus inexpectans, 1046
    Manchuriophycus sawadai, 1046
    Manchuriophycus yamamotoi, 1046
Manicosiphonia, 984, 997
    Manicosiphonia bambusa, 984, 997
    Manicosiphonia conica, 997
    Manicosiphonia conserta, 997
    Manicosiphonia fissilis, 997
    Manicosiphonia furcata, 997
    Manicosiphonia hanyuanensis, 997
MANICOSIPONIACEAE, 361, 981, 984–985
Maotianshania, 1020, 1173
Marenita, 366, 984, 996
    Marenita bayankolica, 996
    Marenita kundatica, 984, 996
Margaretia, 1033, 1173
Margaritichnus, **395**, 1010
    Margaritichnus linearis, 1010
Margominuscula, 884, 889, 892, 896, 920–921, 951, 1059, **1066**, 1144
    Margominuscula antiqua, 889, 951, 1144
    Margominuscula prima, 951
    Margominuscula regularis, 884
    Margominuscula rugosa, 889, 892, 896, 920–921, 951, 1059, **1066**
    Margominuscula aff. *tennela*, 889
    Margominuscula tremata, 884, 951
Margoporata, 946
    Margoporata conflata, 946
Mariochrea, 1021
MARJUMIIDAE, 1176
Markuelia, 1032
MARRELOMORPHA, 1175, 1180–1184
MARRIOCARIDA, 1177
Martinssonia, 1177
Marywadea, 1002
    Marywadea ovata, 1002
Mashania, 1002, 1046
    Mashania angusta, 1002, 1046
    Mashania annulata, 1002, 1046
    Mashania deformata, 1002, 1046

Mashania longshanensis, 1002, 1046
Mashania minuta, 1046
Mashania sinensis, 1046
Mastakhella, 1027
Mastiocladus, 274
Mastogloia, 322, **323**, 324
Mastogloia halophila, 322, 324
Mastogloia cf. *halophila*, **323**
Mastogloia reimeri, 322
Mattajacyathus, 1019
MATTEWCYATHIDAE, 1173
MATTHEVIIDAE, 1174
Matutella, 1030
Mawsonella, 984, 997, 1046
Mawsonella wooltanensis, 984, 997, 1046
Mawsonites, 381, 1002
Mawsonites randallensis, 1002
Mawsonites spriggi, 381, 1002
Mayiella, 1024
MAYIELLIDAE, 1176
MAYLISORIIDAE, 1171
Medusichnites, 1046, 1053
Medusichnites form gamma, 1046, 1053
Medusina, 1002
Madusina asteroides, 1002
Medusina filamentus, 1002
Medusina mawsoni, 1002
Medusinites, 1002–1004, 1046
Medusinites asteroides, 1002–1003
Medusinites paliji, 1002
Medusinites patellaris, 1002
Medusinites simplex, 1002, 1046
Medusinites sokolovi, 1004
MEDUSINITIDAE, 1173
MEDUSOIDS (*See* Subject Index)
Medvezhichnus, 505, 1010
Medvezhichnus pudicum, 1010
Megagrapton, 1046
Magegrapton regulare, 1046
Magalytrum, 886
Megalytrum diacenum, 886
Megapalaeolenus, 1024
Megasphaera, 1032
"MEGASPHAEROMORPH" ACRITARCHS (*See* Subject Index)
MEGASPHAEROMORPHIDA, 349, 969, 971, 977
Meghystrichosphaeridium, 1032
Meishucunchiton, 1028
Meishucunella, 1022
Meitanovitus, 1028
MELANOCYRILLIDS (*See* Subject Index)
Melanocyrillium, 191, 234–235, 237, 244, 592, 914, 926, 1060, **1090**, 1151
Melanocyrillium fimbriatum, 914, **1090**, 1151
Melanocyrillium hexodiadema, 914, 1060, **1090**, 1151
Melanocyrillium horodyskii, 914, 1151
Melasmatosphaera, 885, 888, 910, 918, 1138
Melasmatosphaera magna, 885, 888, 918
Melasmatosphaera media, 885, 910
Melasmatosphaera parva, 885, 1138
Melkanicyathus, 1020
Mellopegma, 1027
Membranacyathus, 1019
Menghinella, 1023
Menneria, 876, 878, 880–883
Menneria adischevii, 883
Menneria foraminis, 880–882
Menneria granosa, 882–883
Menneria levis, 876, 878, 880–881
Menneria roblotae, 881
Mennericyathus, 1019
MENOMONIIDAE, 1176
Merismoconcha, 1028
Merismopedia, 923, 1102, 1147
Merismopedia glauca, 1147
MEROSTOMATA, 961, 1022, 1043, 1045, 1175, 1180–1184
MEROSTROMOIDEA, 1176, 1180–1184

Mesetaia, 1024
Mesodema, 1024
Mesodinium, 479
"MESOSPHAEROMORPH" ACRITARCHS (*See* Subject Index)
METABACTERIA (*See* Subject Index)
METACOSCINIDAE, 1173
Metacoscinus, 1018
METACYATHIDAE, 1173
Metadoxides, 1024
Metafungia, 1018
Metaldetes, 1018
Metallogenium, 233–234, 873, 888, **1065**, 1148
Metallogenium personatum, 873
METAMONADS (*See* Subject Index)
METAPHYTA, 959, 961, 967, 973, 975, 977, 979
Metaredlichioides, 1024
Metethmophyllum, 1018
METHANOBACTERIA, 473, 475
METHANOBACTERIACEAE, 1140
METHANOBACTERIALES, 474–475
Methanobacterium, 287, 291, 320
Methanobacterium thermoautotrophicum, 287, 320
METHANOCOCCALES, 474–475
Methanococcus, 291, 1140
Methanococcus vannielii, 1140
METHANOMICROBIALES, 474–475
Methanosarcina, 88
Methanosarcina barkerii, 88
Methanospirillum, 304
Methanospirillum hungatei, 304
Methylobacterium, 853
Methylobacterium organophilum, 853
Methylococcus, 86, 853
Methylococcus capsulatus, 86, 853
Metisaspina, 1023
Mezenia, 349, 353, 502, 970
Mezenia kossovoyi, 970
Mialsemia, 1004
Mialsemia semichatovi, 1004
Micatheca, 1029
Michniakia, 1027
Mickwitzia, 1030
Micmacca, 1024
Micmaccopsis, 1023
Micrasterias, 886
Micrhystridium, **225**, 228, 882, 892, 938, 942, 945–947, 949–951, 1152
Micrhystridium ampliatum, 946
Micrhystridium brevicornum, 950
Micrhystridium dissimilare, **225**, 949, 951
Micrhystridium lanatum, 949, 951
Micrhystridium lubomlense, 951
Micrhystridium cf. *minutum*, 949
Micrhystridium nannacanthum, 938
Micrhystridium notatum, 951
Micrhystridium obscurum, 951
Micrhystridium oligum, 951
Micrhystridium ordense, **225**
Micrhystridium pallidum, 950
Micrhystridium radzynicum, 951
Micrhystridium spinosum, 951
Micrhystridium tornatum, 942, 945, 949, 1152
Micrhystridium villosum, 951
Micrina, 1030
MICROCOCCACEAE, 1140
Microcodium, 361, 984, 991, 995
Microcodium elegans, 984
Microcodium laxus, 361, 984, 991, 995
Microcoleus, 188, **196**, 262, **264–265**, 267–268, 273–276, 280–283, 290, 296, 299, 300, 301, 306, 312–313, 324–328, 330–332, 336, 597–598, 844, 846, 850, 870, 894, 906, 933, 1128
Microcoleus chthonoplastes, 188, 262, **264–265**, 267, 273–276, 280–281, 290, 306, 312–313, 324–328, 330–332, 336, 850
Microcoleus lacustris f. *minor*, 1128
Microcoleus lyngbyaceus, 276
Microcoleus subtorulosus, 1128
Microcoleus vaginatus, 598

Taxonomic Index

Microconcentrica, 908, 915, 920, 1141
    Microconcentrica induplicata, 908, 920, 1141
    Microconcentrica aff. *induplicata*, 920
Microcornus, 1028
Microcoryne, **402**, 1020
    Microcoryne cephalata, **402**
Microcystis, **195**, 430, 597, 850, 852, 931, 1137–1138, 1141
    Microcystis aeruginosa, 850, 852, 1137, 1141
    Microcystis flos-aquae, 1137
    Microcystis pseudofilamentosa, 1137
    Microcystis ramosa, 1138
    Microcystis robusta, 1138
    Microcystis scripta, 1138
    Microcystis viridis, 1137, 1141
    Microcystis wesenbergii, 850
Microcystopsis, 895
    Microcystopsis yaoi, 895
MICRODICTYON, 1029
Microdictyon, 404, 410–**411**, 448, 1029, 1180–1184
    Microdictyon rhomboidale, **411**
    Microdictyon sinicum, **411**
MICRODICTYONIDAE, 1174
Microdiscus, 1023
Microlaimus, 296
Micromitra, **406**, 1030
    Micromitra undosa, **406**
Micromonospora, 853
Micropylepora, 1033
Microsachites, 1026
Microsphaera, 879
    Microsphaera foveolata, 879
MICROSPORIDIA, 472, 474, 481–483
Microtaenia, 872
    Microtaenia granulosa, 872
Microvalia, 936
    Microvalia spinosa, 936
Millaria, 871
    Millaria implexa, 871
MILLIPEDES (*See* Subject Index)
MIMETASTERIDAE, 1175
MINERVAECYSTIDAE, 1178
Minitheca, 1029
Minjaria, 509, 519
Minusella, 1024
Minutiafilum, 881
    Minutiafilum minutum, 881
Minymerisma, 1028
Mirabella, 1027
Mirabichitina, 1032
Mirabifolliculus, 1032
Mirandocyathus, 1020
Miranella, 1023
Miratheca, 1032
Miriella, 1031
MISKOIIDAE, 1173
Misracyathus, 1046
    Misracyathus vindhicanus, 1046
Misraea, 354, 502, 970
    Misraea psilata, 970
    Misraea vindhyanensis, 970
MISSISQUOIIDAE, 1176
Mitroporata, 883
    Mitroporata armata, 883
MITROSAGOPHORA, 410, 1174
Mobergella, 404–**405**, **439**, 1022
    Mobergella bella, **405**
    Mobergella holsti, **405**, **439**
    Mobergella turgida, **405**
MOBERGELLANS (*See* Subject Index)
MODIOMORPHOIDA, 1026, 1175
Molaria, 1177
MOLLISONIIDAE, 1177
MOLLUSCA, 373, 381, 393, 400–401, 405, 408, 410, 416, 438, 447–448, 485, 969, 1026–1027, 1174
MONASTERIIDAE, 1177
Monasterium, 1022
MONAXONIDA, 1171
Monglodus, **402**, 1031
    Monglodus cf. *rostriformis*, **402**
Mongoliacus, 1032
Mongolitubulus, **411**, 1031, 1174, 1180–1184
Monhystera, 296–297
Monoconvexa, 1030
Monocraterion, 392, 397, 416, 1010, 1045
    Monocraterion tentaculatum, 1010
MONOCYATHIDA, 1020, 1172
MONOCYATHIDAE, 1172
Monodites, 873
    Monodites princeps, 873
Monomorphichnus, 392, 1010, 1012
    Monomorphichnus bilinearis, 1010
    Monomorphichnus lineatus, 1010, 1012
Mononotella, 1022
MONOPLACOPHORA, 405, 408, 410, 558, 1026, 1174, 1179, 1181–1184
Monospinites, 1026
Monostroma, 357
Monotrematum, 943
Montanella, 871
    Montanella beltensis, 871
Montfortia, 357, 501, 505
    Montfortia filiformis, 505
Mooritheca, 1029
Mootwingeecyathus, 1019
Morania, 349, **355**, 357, 503, 959, 970
    Morania antiqua, 970
MORANIDS (*See* Subject Index)
Motina, 1025
MRASSUCYATHIDAE, 1172
Mrassucyathus, 1019
Mucchatocyathus, 1019
Mucilina, 994
    Mucilina fossilis, 994
Mucorites, 872, 1060, **1077–1078**
    Mucorites rifeicus, 872, 1060, **1077–1078**
Muensteria, 1037
Muhosspora, 936
    Muhosspora reticulata, 936
MULTIFARIIDAE, 1174
Multiplicisphaeridium, **226**, 951
    Multiplicisphaeridium dactilum, **226**
    Multiplicisphaeridium dendroideum, 951
    Multiplicisphaeridium vilnense, 951
Multisiphonia, 984, 997
    Multisiphonia hemicirculis, 997
    Multisiphonia nanshanensis, 984, 997
MULTIVASCULATIDAE, 1172
Mundocephalina, 1024
Muniaichnus, 1046
Murandavia, 872, 881, 930, 984, 996
    Murandavia amurica, 872, 984
    Murandavia granulosa, 881, 996
    Murandavia magna, 996
    Murandavia marginata, 996
    Murandavia rotunda, 872, 996
    Murandavia aff. *rotunda*, 930, 996
Musca, 296
    Musca drassirostris, 296
MYCOPLASMAS (*See* Subject Index)
Mycosphaeroides, 872–873, 1060, **1077–1078**
    Mycosphaeroides aggregatus, 872, 1060, **1077**
    Mycosphaeroides caudatus, 873, 1060, **1078**
Mycteroligotriletum, 888
    Mycteroligotriletum marmoratum, 888
Myopsolenus, 1024
Myoscolex, 1020
MYRIAPODA, 1177, 1180–1184
Myriasporella, 893, 1141
    Myriasporella pyriformis, 893, 1141
MYXOBACTERIA, 430
Myxococcoides, 879, 881, 885, 891, 893–894, 896–898, 902–904, 910, 913, 919, 922–923, 925–927, 929, 932, 940, 942, 945–946, 1059, 1061, **1093**, **1115**, 1138, 1140–1143, 1146
    Myxococcoides bansensis, 929

Myxococcoides cantabrigiensis, 922, 925–926, 1146
Myxococcoides compactus, 879
Myxococcoides congoensis, 904, 1142
Myxococcoides cracens, 894, 1143
Myxococcoides globosa, 903, 1143
Myxococcoides grandis, 896, 945, 1059
Myxococcoides cf. *grandis*, 898
Myxococcoides indicus, 929
Myxococcoides inornata, 879, 885, 902, 913, 919, 922, 940
Myxococcoides kingii, 891, 893, 946
Myxococcoides konzalovae, 891
Myxococcoides magnus, 903
Myxococcoides minor, 879, 881, 885, 891, 903, 910, 919, 925, 1061, **1093**, 1141
Myxococcoides minuta, 891, 894, 1141
Myxococcoides muricata, 923, 1138
Myxococcoides ovata, 925
Myxococcoides ramapuraensis, 903
Myxococcoides reniformis, 891
Myxococcoides reticulata, 913, 1141
Myxococcoides staphylidion, 942, 1061, **1115**, 1141
Myxococcoides verrucosa, 904
Myxococcus, 430
Myxococcus xanthus, 430
Myxomorpha, 891, 894, 1141
Myxomorpha janecekii, 891, 894, 1141
Myxosarcina, 923, 1100, 1102, 1124, 1146

Nadalina, 1004
Nadalina yukonensis, 1004
Naegleria, 483
Nalivkinicyathus, 1019
Namalia, 373, 1004
Namalia villiersiensis, 1004
NAMANOIIDAE, 1176
Nanamanicosiphonia, 984, 997
Nanamanicosiphonia lepradosa, 997
Nanamanicosiphonia liangshanensis, 997
Nanamanicosiphonia minuta, 984, 997
Nanamanicosiphonia ninglangensis, 997
Nanamanicosiphonia yunnanensis, 997
Nanchengella, 1022
Nanchengella jinningensis, 1022
Nanjiangochitina, 1032
Nanjiangofolliculus, 1032
Nannococcus, 891, 893–895, 940, 1141
Nannococcus vulgaris, 891, 893–895, 940, 1141
Nannocystis, 86, 307, 853
Nannocystis exedens, 86, 307, 853
Naraoia, 1025
NARAOIIDAE, 1176
Narynella, 1030
Nasepia, 1004
Nasepia altae, 1004
NAUTILOIDS (*See* Subject Index)
Navicula, 322, 325, 327
Neantia, 1046, 1048
Neantia deformata, 1046
Neantia reticulata, 1048
Neantia rhedonensis, 1048
Neantia verrucosa, 1048
Nebela, 592
Nectocaris, 1174
Nehanniaspis, 1023
NEKTASPIDA, 1176
Nelcanella, 881
Nelcanella annularia, 881
Nelcanella radians, 881
Nelcanella solaris, 881
Nelcanella stellata, 881
NELEGEROCORNIDAE, 1175
Nelegerocornus, 1028
Neltneria, 1024
NEMATODES (*See* Subject Index)
Nemiana, 356, 373, 393, 1004, 1043
Nemiana simplex, 393, 1004

Nenoxites, 393, **395**, 437, 505, 557, 1012
Nenoxites curvus, **395**, 1012
Neobolus, 1030
Neochloris, 1140
Neochloris pseudostigmatica, 1140
Neocobboldia, 1023
Neogloborilus, 1029
Neokunmingella, 1022
Neoloculicyathus, 1019
Neonereites, 373, 392–393, **395–396**, **418**–419, 437, 1011–1012, 1039
Neonereites biserialis, 393, **395**, 1012
Neonereites renarius, **396**, 1012
Neonereites uniserialis, 393, 419, 1011–1012
Neopagetina, 1023
Neophrooides, 1032
Neoredlichia, 1024
NEOREDLICHIIDAE, 1176
NEPEIDAE, 1176
Nephelostroma, 984, 991
Nephelostroma lecomtei, 984, 991
Nephroformia, 349, 356–357, 502, 972
Nephroformia liulaobeiensis, 972
Nereites, 1012
NETROMORPH ACRITARCHS (*See* Subject Index)
Neurospora, 482–483
Neurospora crassa, 482
Nevadatubulus, 1032
Nevadella, 1024
Nevadia, 1024
Nevidia, 929, 1143
Nevidia multicellaria, 929, 1143
Nevidia sphaerocellaria, 929
Newlandia, 419, 506, 1048–1049
Newlandia concentrica, 1048
Newlandia cristata, 1048
Newlandia frondosa, 1048–1049
Newlandia cf. *frondosa*, 1048
Newlandia lamellosa, 1048
Newlandia major, 1048
Newlandia obrutchevi, 1048
Newlandia prava, 1048
Newlandia sarcinula, 1048
Newlandia tchurakovi, 1048
Newlandia usovi, 1048
Nicholsonia, **362**, **364**, 366, 984, 991
Nicholsonia composita, 991
Nicholsonia glomerata, 984, 991
Nicholsonia grandis, 991
Nicholsonia involuntans, 991
Nicotiana, 479
Nikatheca, 1029
NILEIDAE, 1176
Nimbia, 1004
Nimbia dniesteri, 1004
Nimbia occlusa, 1004
Nimbia paula, 1004
Ninella, 1030
Ningqiangsclerites, 1031
Niphadus, 1017
Nisusia, 1030
NISUSIIDAE, 1177
Nitella, 190
Nitocra, 296
Nitoricornus, 1028
Nitrobacter, 291
Nitrosomonas, 291
Nitzschia, 322, 327–328
Nitzschia closterium, 322
Nochoroicyathellus, 1019
NOCHOROICYATHIDAE, 1172
Nochoroicyathus, 1019
Nochoroiella, 1030
Nodularites, 945
Nodularites maslovii, 945
Nolichuckia, 1178

Taxonomic Index

Nomgoliella, 1026
NORWOODIIDAE, 1176
Nostoc, 220, 324, 355, 357, 416, 591, 597–598, 850, 852, 876, 959, 965, 971, 975, 977
   Nostoc carneum, 852
   Nostoc commune, 598, 850, 852
   Nostoc harveyana, 852
   Nostoc muscorum, 850, 852
   Nostoc planctonicum, 597
NOSTOCACEAE, 914
NOSTOCALES, 195–196, 218, 220, 243, 255, 884, 914, 943, 949, 971, 1124
Nostocites, 873
   Nostocites vesiculosa, 873
Nostocomorpha, 895, 943, 949
   Nostocomorpha prisca, 895
NOTABILITIDAE, 1175
Notabilitus, 1028
Nothamusium, 1027
Nothozoe, 1025, 1177
NOVITATIDAE, 1175
Novitatus, 1029
Nubecularities, 984, 991–992
   Nubecularities polymorphus, 984, 991–992
Nucellohystrichosphaera, 907, 1060, **1075**, 1151
   Nucellohystrichosphaera megalea, 907, 1060, **1075**, 1151
Nucellosphaeridium, 889, 894, 899, 903–904, 907, 915, 919, 926, 936, 939, 946, 1061, **1103**, 1150
   Nucellosphaeridium bellum, 907, 915, 1150
   Nucellosphaeridium deminatum, 907, 915, 1150
   Nucellosphaeridium magnum, 904, 1150
   Nucellosphaeridium minimum, 903, 946
   Nucellosphaeridium minutum, 889, 894
   Nucellosphaeridium cf. *minutum*, 939
   Nucellosphaeridium spumosum, 926, 1061, **1103**
   Nucellosphaeridium triangulatum, 904
   Nucellosphaeridium zonale, 899
   Nucellosphaeridium zonatum, 903–904
Nuchacyathus, 1019
NUCULOIDA, 1026, 1175
NUDIBRANCHS (*See* Subject Index)
Nuia, 982, 984, 991
   Nuia siberca, 984, 991

Obconicophycus, 913, 1061, **1092**, 1134
   Obconicophycus amadeus, 913, 1061, **1092**, 1134
Obliquatheca, 1029
OBLIQUATHECIDAE, 1175
Oblisicornus, 1028
Obolella, 503, 970, 1030
OBOLELLIDA, 1030, 1177
OBOLELLIDAE, 1177
OBOLIDAE, 1177
Obolopsis, 1030
Obolus, 1030
Obruchevella, 359, **362**, **364**, 366, 434, 448, 514, 877, 918, 934, 940, 946, 982–984, 986, 990–991
   Obruchevella blandita, 991
   Obruchevella condensata, 918, 991
   Obruchevella delicata, 984, 991
   Obruchevella delicata var. *elongata*, 991
   Obruchevella delicata f. *semeikini*, 991
   Obruchevella ditissimus, 991
   Obruchevella meishucunensis, 991
   Obruchevella minor, 940, 991
   Obruchevella parva, 946, 991
   Obruchevella parvissima, 991
   Obruchevella pusilla, 991
   Obruchevella sibirica, 991
   Obruchevella tungusica, 991
Obtusoconus, 1027
Ochromonas, 237, 478–479, 482–483
   Ochromonas danica, 482
Ochthebius, 296
   Ochthebius auratus, 296
Ocridosphaeridium, 918

Ocruranus, **405**, 1027–1028, 1174
   Ocruranus trulliformis, **405**
OCTOCORALLIA, 400–402, 434, 448, 1020
Octoedryxium, 514, 907, 922, 930, 936, 939, 941, 948, 1061, **1114**, **1117**, 1151
   Octoedryxium simmetricum, 939, 1061, **1114**, 1151
   Octoedryxium truncatum, 922, 930, 939, 941, 948, 1061, **1114**, **1117**, 1151
ODARAIDAE, 1177, 1180–1184
ODONTOGRIPHIDAE, 1174
ODONTOPLEURIDA, 1023, 1176
ODONTOPLEURIDAE, 1176
Oelandia, 1027
Oelandiella, 1027
Oelandocaris, 1177
OEPIKALUTIDAE, 1177
Oesia, 1174
OGYGOPSIDAE, 1176
Ogygopsis, 1023
Okulitchicyathus, 1018
Oldhamia, 1048
Olekmaspis, 1024
OLENELLIDAE, 1176
Olenellus, 1024
Olenichnus, 1012
   Olenichnus irregularis, 1012
OLENIDAE, 1176
Olenoides, 1023
Olgaecyathus, 1019
OLIGOCHAETES (*See* Subject Index)
Olisthodiscus, 479
Olivooides, 1032, 1048
   Olivooides papillatus, 1048
Omachtenia, 508, 519
Omalenlina, 1028
Omalophyma, 1127
   Omalophyma gracilia, 1127
Oncchocephalus, 1023
Oncholaimus, 296
Onega, 373, **386**, 434, 1004
   Onega stepanovi, **386**, 434, 1004
Onegia, 1004
   Onegia nenoxa, 1004
ONEOTODONTIDAE, 1178
Oniscoichnus, 1048
Onuphionella, 401, 1032
Onychia, 1025
ONYCHOCHILIDAE, 1026, 1174
ONYCHOPHORA, 371, 1177, 1180–1184
OOCYSTACEAE, 1139–1140, 1149–1150
Ooidium, 822, 932
   Ooidium aspertum, 822
OÖMORPH ACRITARCHS (*See* Subject Index)
OOMYCETES, 483
OPHABINIIDAE, 1174
OPHELIIDAE, 1175
Ophiosema, 1022
Orbiasterocyathus, 1019
Orbicoscinus, 1020
Orbicyathellus, 1019
Orbicyathus, 1019
Orbisiana, 354, 972
   Orbisiana simplex, 972
Orienticyathus, 1020
ORSTENOCARDIA, 1177
ORTHIDA, 1030, 1177
ORTHIDAE, 1177
Orthogonium, 1004
   Orthogonium parallelum, 1004
Orthotheca, 1029
ORTHOTHECIDA, 410, 558, 1029, 1175
ORTHOTHECIDAE, 1175
ORTHOTHECIMORPHA, 1175, 1179, 1181–1184
Ortonella, 984, 991
   Ortonella furcata, 984, 981
ORYCTOCEPHALIDAE, 1176

Oryctocephalus, 1023
Oryctolagus, 482
    Oryctolagus caniculus, 482
Orygmatosphaeridium, 877, 890, 899, 901, 903–904, 916, 918, 929, 939
    Orygmatosphaeridium exile, 877
    Orygmatosphaeridium plicatum, 903, 929
    Orygmatosphaeridium rubiginosum, 916
    Orygmatosphaeridium trizonatum, 904
    Orygmatosphaeridium vulgarum, 904
Oryza, 482
    Oryza sativa, 482
Oryzoconcha, 1026
Oscillatoria, 36, **196**, 267, 276–277, 281, 301, 311–314, 322–**323**, 326–327, 332, 597–598, 846, 850, 929, 934, 1130–1135
    Oscillatoria acuta, 36, 1133
    Oscillatoria agardhii, 281, 850
    Oscillatoria amoena, 1131
    Oscillatoria amphigranulata, 281, 301, 311
    Oscillatoria animalis, 1131
    Oscillatoria antillarum, 1135
    Oscillatoria boryana, 276, 281, 312–313
    Oscillatoria chalybea, 1134
    Oscillatoria chlorina, 1132
    Oscillatoria curviceps, 1134
    Oscillatoria grunowiana, 1132, 1159
    Oscillatoria limnetica, 281, 597
    Oscillatoria margaritifera, 276
    Oscillatoria miniata, 1135
    Oscillatoria minima, 598
    Oscillatoria nigra, 1133
    Oscillatoria princeps, 276, 598, 913
    Oscillatoria pseudogeminata, 1130
    Oscillatoria rubescens, 1133
    Oscillatoria salina f. *major*, 1133
    Oscillatoria schultzii, 1131
    Oscillatoria subtilissima, 598
    Oscillatoria tenuis, 1133
    Oscillatoria terebriformis, 267, 276–277, 281, 301, 1132
    Oscillatoria trichoides, 1130
    Oscillatoria williamsii, 850
    Oscillatoria woronichinii, 850
OSCILLATORIACEAE, 36, 195–196, 201–202, 212–218, 255, 590, 595–597, 869, 880, 884, 886–887, 890, 893–898, 901, 903–907, 909–917, 920–921, 923, 925–929, 932–935, 938–941, 944–945, 949, 1093, 1101, 1107, 1110, 1123–1124, 1126–1137, 1158
Oscillatoriopsis, 35, 879, 887, 893, 895–898, 903, 905–906, 910, 913, 915, 920, 922, 932, 935, 938, 949, 1060–1061, **1068**, **1070–1071**, **1089**, **1092**, 1131–1135, 1137
    Oscillatoriopsis acuminata, 895, 1131
    Oscillatoriopsis bacilaris, 915, 1060, **1089**
    Oscillatoriopsis bothnica, 935, 1134
    Oscillatoriopsis breviconvexa, 913, 1061, **1092**, 1133
    Oscillatoriopsis cf. *breviconvexa*, 895
    Oscillatoriopsis constricta, 935
    Oscillatoriopsis cuboides, 887
    Oscillatoriopsis curta, 896, **1070**, 1132
    Oscillatoriopsis disciformis, 895
    Oscillatoriopsis glabra, 895
    Oscillatoriopsis hemisphaerica, 895
    Oscillatoriopsis magna, 935, 1135
    Oscillatoriopsis majuscula, 887, 1137
    Oscillatoriopsis maxima, 895, 940, 1135
    Oscillatoriopsis media, 905, 910, 922, 935, 1060, **1071**, 1134
    Oscillatoriopsis obtusa, 879, 913, 1061, **1092**, 1132
    Oscillatoriopsis psilata, 903, 949
    Oscillatoriopsis robusta, 896, 1135
    Oscillatoriopsis schopfii, 893
    Oscillatoriopsis taimirica, 927, 1134
    Oscillatoriopsis tuberculata, 895
    Oscillatoriopsis variabilis, 932, 1134
    Oscillatoriopsis wernadskii, 915, 938
Oscillatorites, 938–939, 945, **1115**, 1134
    Oscillatorites wernadskii, 945, **1115**, 1134
Ostiana, 879, 1060, **1084**
    Ostiana microcystis, 879, 1060, **1084**

OSTRACODA, 1022, 1177, 1180–1184
OSTRACODS (*See* Subject Index)
Otekmaspis, 1023
Ottoia, 1020
OTTOIIDAE, 1173
Ouijjana, 1024
Ovalitheca, 1029
Ovaluta, 1022
Ovatoscutum, 372, 381, **383–384**, 1004, 1173
    Ovatoscutum concentricum, **383–384**, 1004
Ovidiscina, 349, **354**, 503, 912, 972
    Ovidiscina bagongshanica, 912, 972
    Ovidiscina longa, **354**, 912, 972
    Ovidiscina pakungshan(n)i(c)a, 972
Ovolites, 1048
    Ovolites primus, 1048
Ovulum, 951
    Ovulum lanceolatum, 951
    Ovulum saccatum, 951
Oxytus, 1029

Pachyaspis, 1024
Pachyredlichia, 1024
Pachysphaera, 189, 191
Padina, 361
    Padina jamaicensis, 361
Pagetides, 1023
Pagetiellus, 1023
PAGETIIDAE, 1175
PAGODIIDAE, 1176
Pahvantia, 1177
PAIUTIIDA, 400–401, 408, 1021, 1174
Paiutitubulites, 1022
PAIUTITUBULITIDAE, 1021, 1174
Pakria, 878
    Pakria kheinjuaensis, 878
Palaeacmaea, 1027
PALAEACMAEIDAE, 1173
Palaeoanacystis, 873, 876, 880, 885, 888, 891, 894–895, 902–904, 913, 936, 1141
    Palaeoanacystis irregularis, 876
    Palaeoanacystis plumbii, 891, 894
    Palaeoanacystis psilata, 904
    Palaeoanacystis punctatus, 903
    Palaeoanacystis reticulatus, 903
    Palaeoanacystis suketensis, 902–903
    Palaeoanacystis verrucosus, 903
    Palaeoanacystis vulgaris, 873, 880, 885, 891, 894–895, 902–903, 913, 1141
Palaeoaphanizomenon, 926, 1061, **1103**
    Palaeoaphanizomenon scabratus, 926, 1061, **1103**
Palaeobotryllus, 449
Palaeocalothrix, 873, 877, 1060, **1087**
    Palaeocalothrix divaricatus, 873, 1060, **1087**
    Palaeocalothrix xui, 877
Palaeoconularia, 1020
PALAEOCOPIDA, 1022, 1177
Palaeocryptidium, 874, 880, 928
    Palaeocryptidium cayeuxii, 880, 928
Palaeocuniculichnites, 1048
    Palaeocuniculichnites osangustus, 1048
Palaeogirvanella, 984, 995
    Palaeogirvanella erebiensis, 984, 995
    Palaeogirvanella sajanica, 984
Palaeolenella, 1024
Palaeolenides, 1024
Palaeolentus, 1024
Palaeolenus, 1024
PALAEOLORICATA, 1174
Palaeolyngbya, **198**, 880, 894–895, 897, 907, 910–911, 913, 915, 917, 920, 929, 933, 935, 940, 1060–1061, **1072**, **1083**, **1089**, **1091**, **1093**, **1109**, 1136
    Palaeolyngbya baraudensis, 880, 1061, **1091**, **1093**
    Palaeolyngbya barghoorniana, 895, 913
    Palaeolyngbya cf. *barghoorniana*, 913
    Palaeolyngbya catenata, 915, 1136

Palaeolyngbya distincta, 880
Palaeolyngbya elongata, 917, 1136
Palaeolyngbya helva, 907, 1060, **1072**, 1136
Palaeolyngbya lata, 935, 1136
Palaeolyngbya maxima, 895
Palaeolyngbya minor, 910, 913, 920, 1061, **1091**
Palaeolyngbya cf. *minor*, 911
Palaeolyngbya sphaerocephala, **198**, 915, 1060, **1083**, **1089**
Palaeolyngbya zilimica, 933, 1061, **1109**
Palaeomicrocoleus, 869–870
Palaeomicrocoleus gruneri, 870
Palaeomicrocystis, 904, 984, 996
Palaeomicrocystis cambrica, 984, 996
Palaeomicrocystis schopfii, 904
Palaeonostoc, 929
Palaeonostoc barghoornii, 929
Palaeopascichnus, 373, 392–393, **395**, 437, 505, 1012, 1015
Palaeopascichnus delicatus, **395**, 1012, 1015
Palaeopascichnus sinuosus, 1012
Palaeophycus, 416, 1012–1013
Palaeophycus alternatus, 1012
Palaeophycus tubularis, 1012
Palaeoplatoda, 1004, 1048
Palaeoplatoda segmentata, 1004, 1048
Palaeopleurocapsa, 892, 902–903, 911, 923, 926, 929, 935–936, 1059, 1061, **1065**, **1106**, **1111**, 1146–1147
Palaeopleurocapsa fusiforma, 935, 1061, **1111**, 1147
Palaeopleurocapsa kamaelgensis, 929, 1061, **1106**, 1147
Palaeopleurocapsa kelleri, 892, 1059, **1065**
Palaeopleurocapsa reniforma, 935, 1061, **1111**
Palaeopleurocapsa wopfneri, 902–903, 923, 1146
Palaeopleurocapsa cf. *wopfneri*, 911
Palaeopleurocapsa anhuiensis, 972, 1006
Palaeoschmidtites, 1030
PALAEOSCOLECIDA, 1175, 1179, 1181–1184
PALAEOSCOLECIDAE, 1175
Palaeoscolex, 1020
Palaeoscytonema, 880, 886, 904, 1136
Palaeoscytonema indicum, 880
Palaeoscytonema intermingla, 880
Palaeoscytonema misrae, 880
Palaeoscytonema moorhousei, 886
Palaeoscytonema srivastavae, 904, 1136
Palaeosiphonella, 205, 916, 923, 1060, **1083**, **1100**, 1129
Palaeosiphonella cloudii, 916, 1060, **1083**, 1129
Palaeospiralis, 886
Palaeospiralis canadensis, 886
Palaeospirulina, 886
Palaeospirulina arcuata, 886
Palaeospirulina minuta, 886
Palaeosulcachites, 1026
Palaeotrochis, 1048, 1050–1051
Palaeotrochis major, 1050
Palaeotrochis minor, 1050
Palaeovaucheria, 907, 1060, **1072**, **1073**, **1074**
Palaeovaucheria clavata, 907, 1060, **1072**, **1073**, **1074**
Paleamorpha, 943, 949
Paleobotryllus, 1174
Paleocenosphaera, 1032
Paleocrassilimbus, 1032
Paleohexadictyon, 1032
Paleoisocystis, 895
Paleoisocystis disporata, 895
Paleoisocystis monosporata, 895
Paleolina, 354, 371, 502, 972, 1006–1007, 1021
Paleolina evenkiana, 972, 1006
Paleolina tortuosa, 972, 1006
Paleomegasquama, 1032
PALEOMERIDAE, 1177
Paleomerus, 1022
Paleonites, 984, 989, 991
Paleonites jacutii, 984, 989, 991
Paleonostochopsis, 878
Paleonostochopsis vindhyanensis, 878
Paleorhyncus, 354, 357, 502, 972, 1006, 1021
Paleoxiphosphaera, 1032

Paliella, 1004
Paliella patelliformis, 1004
PALMELLACEAE, 1139
Palmellococcus, 1139
Palmellococcus marinus, 1139
Palmericyathellus, 1018
Palmericyathus, 1019
Panomninella, 363, 366, 984, 986, 991, 995
Panomninella copiosa, 995
Panomninella floribunda, 995
Panomninella laxa, 991, 995
Panomninella lecta, 995
Panomninella ornata, 984, 991
Panomninella petrosa, 991
Panomninella rotunda, 995
Panomninella silicea, 995
Paokannia, 1024
Papillionata, 1004
Papillionata eyrei, 1004
Papillocyathus, 1020
Papillomembrana, 984, 991, 1053
Papillomembrana compta, 984, 991
PAPYRIASPIDIDAE, 1176
Paraaldanella, 1026
Parabadiella, 1024
PARABOLINOIDIDAE, 1176
Paracanthodus, 1031
Paracarinachites, 1026
PARACARINACHITIDAE, 403, 409–410, 1026, 1174
Paraceratoconus, 1028
Parachabakovia, 366, 982, 984, 991
Parachabakovia dura, 984, 991
Paracharinachites, **409**
Paracharinachites parabolicus, **409**
Paracharinachites sinensis, **409**
Paracharinachites spinus, **409**
Paracharnia, 369, 373, 1004, 1053
Paracharnia dengyingensis, 1004
Paracoccus, 480
Paracoccus denitrificans, 480
Paracoenia, 297
Paracoenia turbida, 297
PARACONODONTIDA, 1178
Paraconularoides, 1022
Paraconus, 1027
Paracoscinus, 1018
Paracrassosphaera, 948, 1061, **1117**, 1152
Paracrassosphaera dedalea, 948, 1061, **1117**, 1152
Paracymatiosphaera, 946
Paracymatiosphaera annularis, 946
Paracymatiosphaera hunnanensis, 946
Paracymatiosphaera irregularis, 946
Paracymatiosphaera regularis, 946
Paradoxides, 1050
Paradoxides barberi, 1050
PARADOXIDIDAE, 1176
Parafallotaspis, 1024
Paraformichella, 1028
Paragloborilus, **403**, 1021
Paragloborilus subglobosus, **403**
Paragraulos, 1023
Parahyolithes, 1029
Parailsanella, 1027
Parakorilithes, 1028
Parakunmingella, 1022
Paraleptomitella, 1017
Paralongfengshania, 349, **354**, 357, 503, 972
Paralongfengshania sicyoides, **354**, 972
Paramalungia, 1024
Paramecium, 482
Paramecium tetraurelia, 482
Paramedusium, 379, 1004
Paramedusium africanum, 1004
Paramobergella, 1032
Paranabarites, 1022
Paranacyathus, 1018

Paraphaseolella, 1022
Parapunctella, 1025
Pararedlichia, 1025
Pararenicola, 354, 357, 371, 433–434, 501–502, 972, 1006–1007, 1021
    Pararenicola huaiyuanensis, 371, 972, 1006–1007
Parasabellidites, 1021
Parascenella, 1028
Parasolenopora, 984, 997
    Parasolenopora irregularis, 997
    Parasolenopora subradiata, 984, 997
Paratermierella, 1025
Paratetraphycus, 940, 1147
    Paratetraphycus giganteus, 940, 1147
PARATRILOBITA, 1175, 1180–1184
Paratungusella, 1025
Parazhijinites, 1031
Parenchymodiscus, 926
    Parenchymodiscus endolithicus, 926
Pareocrinus, 1178
Pareops, 1025
PARKARYOTES (See Subject Index)
Parkula, **403**, 1028
    Parkula bounites, **403**
Parmorphorella, 1027
Partitiofilum, 906, 913, 920, 1060–1061, **1066**, **1091**, 1132, 1134, 1158
    Partitiofilum gongyloides, 913, 1061, **1091**, 1132, 1158
    Partitiofilum aff. *gongyloides*, 920
    Partitiofilum tungusum, 906, 1060, **1066**, 1134
Parvancorina, 373, 1004
    Parvancorina minchami, 1004
Paterimitra, 1030
Paterina, 1030
PATERINIDA, 1030, 1177
PATERINIDAE, 1177
Paterisquama, 1032
Patinisquama, 1032
PAUXILLITIDAE, 1175
Peachella, 1025
Pectenocyathus, 1019
Pediastrum, 190
Pegmatreta, 1031
Pelagiella, 1026–1027
PELAGIELLIDAE, 1174
PELECYPODS (See Subject Index)
Pellicularia, 873, 879, 1060, **1082**
    Pellicularia tenera, 873, 879, 1060, **1082**
Peltigera, 850
    Peltigera canina, 850
PENNATULIDS (See Subject Index)
PENTAMERIDA, 1178
PEPTOCOCCACEAE, 1137
PEREGRINICYATHIDAE, 1172
Peregrinicyathus, 1019
Peridionites, 1178
Periomma, 1024
Periommella, 1024
Peripteratocyathus, 1021
PERONOCHAETIDAE, 1175
Persicitheca, 1029
Persimedusites, 1004
    Persimedusites chahgazensis, 1004
PERSPICARIDIDAE, 1177
Perspicaris, 1022
Perulagranum, 926
    Perulagranum obovatum, 926
PETALONAMAE, 373, 381, 558, 1173, 1179, 1181–1184
Petalostroma, 1004
    Petalostroma kuibis, 1004
Petasisquama, 1032
Peteinosphaeridium, 923, 1152
    Peteinosphaeridium reticulatum, 923, 1152
Petraphera, 869
    Petraphera vivescenticula, 869
Peytoia, 1025
Phacelofimbria, 984, 997

    Phacelofimbria emeishanensis, 984, 997
    Phacelofimbria minor, 997
PHACOPIDA, 1176
PHAEOPHYTA, 359, 419, 477, 961, 975, 977, 979
PHALACROMIDAE, 1175
Phanerosphaerops, 897, 913, 922, 927
    Phanerosphaerops capitaneus, 913, 922, 927
Phascolites, 349, 356, 503, 972
    Phascolites symmetricus, 972
Phaseolella, 1022
Philoxenella, 1026
Phobetractinia, 1017
PHOBETRACTINIDS, 1172
Phormidium, 261–262, **265**–267, 280, 300–301, 313–314, **318**, 320, 327–328, 597, 844, 846, 850, 852, 904, 1126–1127, 1135–1136
    Phormidium bigranulatum, 1126
    Phormidium bohneri, 1135
    Phormidium favosum, 1136
    Phormidium hendersonii, 262, 266
    Phormidium incrustatum, 1127, 1136
    Phormidium luridum, 850, 852
    Phormidium molle, 267
    Phormidium mucosum, 1127
    Phormidium purpurascens, 1126
    Phormidium retzii f. *major*, 1127
    Phormidium subincrustatum, 1127
    Phormidium tenue, 313, **318**
    Phormidium tenue var. *granuliferum*, 262
    Phormidium unicinatum, 1136
PHOSPHATOCOPIDA, 1177
Phycodes, 392, 397, 505, 1012
    Phycodes pedum, 392, 397, 505, 1012
Phycomycetes, 357, 882, 907, 914
Phyllochites, 1031
Phyllochiton, 1031
PHYLLOCYSTIDAE, 1178
Phyllozoon, 373, 1004
    Phyllozoon hanseni, 1004
Phymatocyathus, 1020
Physarum, 482–483
    Physarum polycephalum, 483
Piamaecyathellus, 1019
PIAMAECYATHIDAE, 1172
Piamaecyathus, 1019
Piaziella, 1024
Pikaia, 1178
Pila, 938
Pilavia, 894, 1137
    Pilavia maculata, 894, 1137
Pilbaria, 508
Pileconus, 1028
Pilitella, 354, 502, 972
    Pilitella composita, 972
Pinacocyathus, 1018
Pinegia, 1004
    Pinegia stellaris, 1004
Pinnulina, 985, 996
    Pinnulina cambrica, 985, 996
PIRANIIDAE, 1171
Pirenella, 267, 296
    Pirenella conica, 267, 296
PLACOSEMATIDAE, 1176
Plagiogmus, 371–372, 392, 397, 1012
    Plagiogmus arcuatus, 1012
Planaspis, 1025
Planispiralichnus, 1012
    Planispiralichnus grandis, 1012
Planktosphaeria, 1150
    Planktosphaeria maxima, 1150
Planolites, 373, 392–393, 416, **418**, 505, 1012, 1036, 1050
    Planolites ballandus, 393, 1012
    Planolites beverleyensis, 416, 1012
    Planolites corrugatus, 416, **418**, 1050
    Planolites montanus, 1012
    Planolites superbus, 416, **418**, 1050
Planomedusites, 1003–1004
    Planomedusites grandis, 1004

Taxonomic Index

Planomedusites patellaris, 1003–1004
Planotheca, 1028
PLANTAE, 473, 1038–1039
Planuspira, 1027
Plasmodium, 482
    Plasmodium berghei, 482
PLATYHELMINTHES, 408, 1049
PLATYOSIPOHONIDAE, 1171
Platypholinia, 850, 1004
    Platypholinia pholiata, 850, 1004
Platysolenites, **238**–239, 1032
    Platysolenites antiquissimus, **238**
PLATYSOLENITIDS (*See* Subject Index)
Platyspinites, 1026
Plectodiscus, 372
Plectonema, 850
    Plectonema terebrans, 850
PLECTONOCERATIDAE, 1175
PLECTONOCERIDA, 1175
PLECTRIFERIDAE, 1176
Plenocaris, 1177
PLETHOPELTIDAE, 1176
Pleurocapsa, 268, 297, 314, 316, **317**–**318**, 597, 885, 1124, 1146
    Pleurocapsa minor, 268, 297
PLEUROCAPSACEAE, 218, 597, 912, 923, 1096–1098, 1100, 1102
PLEUROCAPSALES, 892, 926–927
Plicatidium, 357, 873, 920, 1060, **1082**
    Plicatidium latum, 873, 1060, **1082**
Plicatolingula, 1031
Plicitheca, 1029
Plicocyathus, 1019
Plinthoconion, 1032
Pluralicoscinus, 1020
Pneumocystis, 482
Podoliella, 882, 939, 1061, **1114**
    Podoliella irregulare, 939, 1061, **1114**
    Podoliella prismatica, 882
    Podoliella regulare, 939
Podolimirus, 1004
    Podolimirus mirus, 1004
Podolina, **223**, 514, 883, 930, 1152
    Podolina angulata, 883
    Podolina echinata, 883
    Podolina minuta, **223**, 883, 930, 1152
Podospora, 482–483
    Podospora anserina, 482
POGONOPHORA, 357, 371, 407–408, 485, 504, 959, 961, 969, 973, 975
Pojetaia, 1026, 1175
Poletaevella, 1025
Poliella, 1023
Poliellaspis, 1023
Poliellina, 1023
Polliaxis, 1023
Pollingeria, 1174
Pollofructus, 1032
Pollukia, 1001, 1004
    Pollukia serebrina, 1001, 1004
    Pollukia shulgae, 1004
Polyactinella, 1017
Polybessurus, 195, 205, 923, 926, 940, **1096**, **1097**, **1098**, **1099**, 1124, 1148, 1162
    Polybesurus bipartitus, 926, 1148
    Polybessurus cf. *bipartitus*, **1096**, **1097**, **1098**, **1099**
Polycavita, 937, 1061, **1109**, **1112**
    Polycavita bullata, 937, 1061, **1109**, **1112**
    Polycavita frillata, **1109**, **1112**
Polycellaria, 871, 874
    Polycellaria bonnerensis, 871
    Polycellaria indistincta, 871
    Polycellaria levis, 871
    Polycellaria longata, 871
POLYCHAETA, 1175, 1179, 1181–1184
POLYCHAETES (*See* Subject Index)
Polychlamydum, 940
    Polychlamydum insigne, 940

Polycladium, 1025
POLYCOSCINIDAE, 1172
Polycoscinus, 1020
POLYCYRTASPIDIDAE, 1176
    Polyedrosphaeridium, 884, 899, 934, 939, 942, 944, 947–948, 951, 1141
    Polyedrosphaeridium bullatum, 884, 899, 934, 939, 942, 944, 947–948, 1141
Polyedryxium, 882, 948, 1061, **1117**
    Polyedryxium neftelenicum, 948, 1061, **1117**
POLYGONOMORPH ACRITARCHS (*See* Subject Index)
Polynucella, 899
    Polynucella biconcentrica, 899
Polyphyma, 1022
POLYPLACOPHORA, 1174, 179, 1181–1184
Polyplacus, 1031
Polyporata, 870–873, 875, 947
    Polyporata nidia, 875, 947
    Polyporata obsoleta, 870–873
    Polyporata verrucosa, 875, 947
Polysphaeroides, 915, 926, 1060–1061, **1084**–**1085**, **1104**, 1132
    Polysphaeroides contextus, 915, 1060, **1085**, 1132
    Polysphaeroides filliformis, 915, 1060, **1084**
    Polysphaeroides lineatus, 915
    Polysphaeroides nuclearis, 926, 1061, **1104**
Polysphaerula, 873
    Polysphaerula globulosa, 873
Polystillicidocyathus, 1020
Polythrichoides, 906, 910, 933, **1088**, 1137
    Polythrichoides lineatus, 906, 910, 933, **1088**, 1137
Pomerania, 1025, 1177
Pomeria, 1004
    Pomoria corolliformis, 1004
POMORIIDAE, 1173
Pompeckium, 1031
Pontigulasia, 592
Poratites, 1033
Porcaconus, 1027
Porcauricula, 1030
PORIFERA, 404, 1017, 1171
POROCOSCINIDAE, 1172
Porocoscinus, 1020
Porocyathellus, 1019
POROCYATHIDAE, 1172
Porocyathus, 1019
Porphyra, 357
Porphyridium, 233, 479
PORPITIDAE, 1020, 1173
Portalia, 1174
POSIBACTERIA (*See* Subject Index)
Postacanthella, 1028
Postestephaconus, 1028
Potamocypris, 268, 297
Potentillina, 363, 366, 985, 991
    Potentillina companulata, 985, 991
    Potentillina monstrata, 991
Poulsenia, 1024
Praecambridium, 373, **384**, 1004
    Praecambridium sigillum, **384**, 1004
PRAECANTHOCHITONIDAE, 1174
Praelamellodonta, 1028
PRAENUCULIDADAE, 1026, 1175
Praesolenopora, 985, 997
    Praesolenopora fascicularis, 997
    Praesolenopora flabella, 985, 997
    Praesolenopora formosa, 997
    Praesolenopora furcata, 997
    Praesolenopora hanyuanensis, 997
    Praesolenopora liaoningensis, 997
    Praesolenopora magniflabella, 997
PRASINOPHYCEAE, 189, 191, 219–220, 225, 230, 241
PRASINOPHYTA, 219, 961
Prethmophyllum, 1019
PRETIOSOCYATHIDAE, 1172
Pretiosocyathus, 1019
PRIAPULIDA, 405, 1020, 1173, 1179, 1181–1184
Primaconulariella, 1031

Primevifilum, 28, **30**, **31**, 32–33, 35
   Primaevifilum amoenum, **31**, 32, 35
   Primaevifilum conicoterminatum, **30**, 33, 35
   Primaevifilum delicatulum, **30**, 32, 35
   Primaevifilum septatum, 28, 32, 35
Primoflagella, 972
   Primoflagella speciosa, 972
Primorivularia, 880, 886, 915, 1060, **1083**
   Primorivularia dissimilara, 915, 1060, **1083**
   Primorivularia robusta, 880
   Primorivularia thunderbayensis, 886
Primorphlagella, 357
Priscansermarinus, 1177
PRISMATOMORPH ACRITARCHS (*See* Subject Index)
PRISMOCYATHIDAE, 1173
Prismocyathus, 1018
Proaulopora, 359, **362**, **364**, 366, 981, 984–987, 989, 992–993, 995
   Prosaulopora composita, 992
   Proaulopora crassa, 992
   Proaulopora extincta, 992
   Proaulopora flexuosa, 992
   Proaulopora glabra, 981, 986–987, 989, 993, 995
   Proaulopora longa, 992
   Proaulopora microspora, 992
   Proaulopora rarissima, 981, 985, 995
   Proaulopora sajanica, 992
Proboscicaris, 1177
PROCHELICERATA, 1045
PROCHLOROBACTERIA, 473–474
Prochloron, 474, 478, 480–481, 852
PROCHLOROPHYTA, 474, 478–479, 590
Prochlorothrix, 474, 478, 480
PROCONODONTIDAE, 1178
Proecceylipterus, 1027
Proerbia, 1023
Profallotaspis, 1025
Proichangia, 1025
PROKARYOTA, 35
PROKARYOTES (*See* Subject Index)
Prokootenia, 1023
Proliostracus, 1024
Proplina, **405**, 1027
PROPRIOLYNTHIDAE, 1172
Propriolynthus, 1020
Prorocentrum, 482
   Prorocentrum micans, 482
Prosinuites, 1027
PROTACTINOCERATIDAE, 1175
PROTACTINOCERIDA, 1175
Protadelaidea, 1050
   Protadelaidea browni, 1050
   Protadelaidea howchini, 1050
Protagraulos, 1025
Protarchaeosacculina, 882
   Protarchaeosacculina atava, 882
Protegocista, 235, 879
PROTEROCAMEROCERATIDAE, 1175
PROTEROSTOMES (*See* Subject Index)
Proterotainia, 354, 503, 972
   Proterotainia montana, 972
   Proterotainia neihartensis, 972
Proteus, 430
   Proteus mirabilis, 430
PROTISTA, 400, 431, 449, 473, 477, 924
Protoarenicola, 354, 357, 433–434, 501–502, 514, 972, 1006, 1021
   Protoarenicola baiguashanensis, 972, 1006
Protobolella, 349, 503, 972
   Protobolella jonesi, 972
   Protobolella minima, 972
Protobolus, 1031
PROTOCARIDIDAE, 1177
Protocaris, 1025
PROTOCONODONTIDA, 1178, 1180–1184
PROTOCONODONTS (*See* Subject Index)
Protoconus, 1028
PROTOCYCLOCYATHIDAE, 1173
Protocyclocyathus, 1018

Protodipleurosoma, 1004
   Protodipleurosoma rugulosum, 1004
   Protodipleurosoma wardi, 1004
Protoechiurus, 1004
   Protoechiurus edmondsi, 1004
Protohertzina, **402**, 412, 441, 505, 554, 1031
   Protohertzina anabarica, **402**, 505
   Protohertzina unguliformis, **402**, 505
Protohyalostelia, 1017, 1171
Protoleiosphaeridium, 867, 873, 880, 888–889, 899, 904, 908, 920–921, 1141
   Protoleiosphaeridium bullatum, 880
   Protoleiosphaeridium conglutinatum, 888, 1141
   Protoleiosphaeridium densum, 904
   Protoleiosphaeridium aff. *faveolatum*, 908
   Protoleiosphaeridium infriatum, 908, 920–921
   Protoleiosphaeridium laevigatum, 904
   Protoleiosphaeridium problematicum, 874
   Protoleiosphaeridium pusillum, 920
   Protoleiosphaeridium solidum, 899, 908
Protolenella, 1025
PROTOLENIDAE, 1176
Protolenus, 1025
Protolyellia, 1020
Protoniobia, 1050
   Protoniobia wadea, 1050
Protoortonella, 983, 985, 992–993
   Protoortonella flabellata, 992–993
   Protoortonella igarcaensis, 992–993
Protopharetra, 1018
PROTOPHARETRIDAE, 1173
Protopterygotheca, 1026
Protorivularia, 985, 995
   Protorivularia onega, 985, 995
PROTORTHIDAE, 1177
Protosolenopora, 946, 985, 992
   Protosolenopora distincta, 946, 985, 992
Protosphaeridium, 867, 876, 882, 884–885, 887–890, 892, 894, 896, 898–900, 904–905, 907–912, 915–916, 918–925, 927–940, 942–951, 1138–1140
   Protosphaeridium acis, 888–890, 892, 898–899, 908, 916, 918–919, 923–924, 927, 932, 935, 937–938, 942–944, 949
   Protosphaeridium cf. *acis*, 901
   Protosphaeridium asaphum, 894, 899, 908, 919, 924, 932, 937, 942, 944–945
   Protosphaeridium crispum, 900
   Protosphaeridium densum, 884, 887, 889–890, 899–902, 905, 907–908, 911, 918–921, 923–925, 927–928, 930–931, 934, 936–938, 943–946, 948–949
   Protosphaeridium discum, 892, 918, 925
   Protosphaeridium flexuosum, 887–890, 892, 896, 898, 900–902, 909–911, 918, 920, 923, 931–933, 937, 939, 942, 945
   Protosphaeridium cf. *flexuosum*, 911, 922–923
   Protosphaeridium gibberosum, 900, 907, 916, 924, 937
   Protosphaeridium cf. *gibberosum*, 911
   Protosphaeridium laccatum, 889–890, 892, 898–900, 902, 907, 909, 923–924, 934, 937–938, 940, 942, 944, 947
   Protosphaeridium cf. *laccatumn*, 911
   Protosphaeridium ostiolatum, 916
   Protosphaeridium paleaceum, 889, 898, 901, 915, 931, 935, 937, 944, 1139
   Protosphaeridium papyraceum, 900–901, 919, 923, 927, 932–934, 1140
   Protosphaeridium parvulum, 887, 898, 900–901, 909, 918, 925, 932, 947, 1139
   Protosphaeridium cf. *parvulum*, 901, 911
   Protosphaeridium patelliforme, 889–890, 899, 901, 907, 916, 919, 923–924, 935, 943, 945–946, 948–950, 1139
   Protosphaeridium pusillum, 890, 899, 901, 905, 918, 925, 931–932, 937, 940, 944
   Protosphaeridium rigidulum, 887, 889, 892, 898, 901, 907, 918–919, 923, 927, 931, 937, 942–943, 1139
   Protosphaeridium scabridum, 889–890, 892, 899, 905, 909, 918–919, 927, 934, 1139
   Protosphaeridium torulosum, 899, 905, 918, 927, 932, 937, 939, 942, 947
   Protosphaeridium tuberculiferum, 890, 892, 899–902, 905, 909,

Taxonomic Index

911, 915, 918, 924, 929, 931, 933–934, 937–939, 944–946, 950–951, 1140
Protosphaeridium vermium, 909, 915, 918, 928
Protosphaeridium volkovae, 882, 904
Protosphaeridium wimanii, 923
Protosphaerites, 1032
Protospiralichnus, 1012
Protospiralichnus circularis, 1012
Protospongia, 1017
PROTOSPONGIIDAE, 1172
Prototreta, 1030
Protovirgularia, 1050
Protowenella, 1026
PROTOZOA, 407, 431, 477, 592, 1171
Protuberantia, 985, 995
Protuberantia vesicularis, 985, 995
Protypus, 1023
Prozacanthoide, 1023
Pruvostina, 1025
Pruvostinoides, 1025
PRYMNESIOPHYCEAE, 479
PRYMNESIOPHYTA, 237
Psamathopalass, 1031
Psammichnites, 1012
Psammichnites gigas, 1012
Psammosphaera, 239
Pseudanabaena, 314
Pseudaptos, 1024
Pseudoacus, 985, 997
Pseudoacus renalis, 985, 997
Pseudoanthus, 994
Pseudoanthus cambricum, 994
Pseudoarctolepis, 1177
Pseudochara, 879
Pseudoclavus, 1026
Pseudogymnosolen, 508, 625–627
Pseudogymnosolen mopangyuensis, 625–626
Pseudoichangia, 1025
Pseudojussenia, 1050
Pseudojussenia parva, 1050
Pseudokadyella, 1025
Pseudokunmingella, 1022
Pseudolancastria, 1023
Pseudolenus, 1025
Pseudomonas, 430
Pseudooides, 1032
Pseudopollicina, 1028
Pseudoredlichia, 1025
Pseudoresserops, 1025
Pseudorhizostomites, 373, 379, **382**, 1004–1005
Pseudorhizostomites howchini, 373, 379, **382**, 1004–1005
Pseudorhopilema, 1004
Pseudorhopilema chapmani, 1004
Pseudorthotheca, 1021
Pseudosaukianda, 1025
Pseudosyringocnema, 1018
PSEUDOSYRINGOCNEMIDAE, 1173
Pseudotasmanites, 950
Pseudotasmanites parvus, 950
Pseudovendia, 1004
Pseudovendia charnwoodensis, 1004
Pseudozonosphaera, 880, 896, 898, 908, 921
Pseudozonosphaera sinica, 908, 921
Pseudozonosphaera cf. *sinica*, 880
Pseudozonosphaera verrucosa, 896, 898
Psiloria, 1030
Psophosphaera, 947
Psophosphaera obscura, 947
Psophosphaera selebrosa, 947
PTERIDINIIDAE, 373, 1173
Pteridinium, 371–373, 379, 381, **386**, 393, 434, 504, 1001, 1003–1005, 1007, 1047
Pteridinium carolinaense, 379, **386**, 1004–1005
Pteridinium latum, 1003–1004, 1007
Pteridinium nenoxa, 1004
Pteridinium simplex, 1004–1005
PTEROBRANCHIA, 1178

PTEROCEPHALIDAE, 1176
Pterocyathus, 1021
PTEROMORPH ACRITARCHS (*See* Subject Index)
Pteroredlichia, 1025
Pterospermella, **225**, 228, 910, 936, 949–951, **1080**, 1151
Pterospermella simica, 228, 910, 936, **1080**, 1151
Pterospermella solida, 951
Pterospermella cf. *solida*, **225**, 949
Pterospermella vitalis, 950
Pterospermella vitrea, 951
Pterospermopsimorpha, **197**, 200, 514, 871, 877, 880, 889, 896, 899–900, 906–907, 910–911, 915, 920, 923–924, 927–928, 931, 936, 941, 948, 950, 1060, **1076**, **1085**, 1151
Pterospermopsimorpha annulare, 900, 1151
Pterospermopsimorpha binata, 889
Pterospermopsimorpha capsulata, 896, 906, 910, 920, 1060 1151
Pterospermopsimorpha concentrica, 880
Pterospermopsimorpha densicoronata, 923
Pterospermopsimorpha insolita, 900, 915, 928, 1060, **1085**, 1151
Pterospermopsimorpha mogilevica, 931, 941
Pterospermopsimorpha ornata, 936
Pterospermopsimorpha pileiformis, 899, 907, 915
Pterospermopsimorpha wolynica, 950
PTEROSPERMOPSIMORPHID ACRITARCHS (*See* Subject Index)
Pterospermopsis, 228, 899, 908, 910, 1060, **1080**, 1151
Pterospermopsis concentricus, 899
Pterospermopsis dubius, 910, 1060, **1080**, 1151
Pterospermopsis oculatus, 899, 908
Pterospermopsis simicus, 228, 910, 1060, **1080**, 1151
PTYCHASPIDIDAE, 1176
PTYCHOPARIIDA, 1023, 1176
PTYCHOPARIIDAE, 1176
Ptychoparopsis, 1024
Pubericyathus, 1021
Pulvinomorpha, 876–877
Pulvinomorpha angulata, 877
Pulvinomorpha mitis, 876
Pumilibaxa, 349, 356–357, 503, 969, 972
Pumilibaxa huaiheiana, 972
Pumilibaxa cf. *huaiheiana*, 972
Punctatus, 1032
Puratanichnus, 1050
Puratanichnus bijawarensis, 1050
Purella, 1028, 1174
Pustularia, 985, 996
Pustularia taeniata, 985, 996
Pustularia vetusta, 996
PUTAPACYATHIDA, 1172
PUTAPACYATHIDAE, 1172
PYCHNOIDOCOSINIDAE, 1173
Pycnoidocyathus, 1018
PYRAMIMONADALES, 961
Pyrgites, 1032, 1174
Pyrodictium, 188, 291, 474
Pyrodictium brockii, 291
Pyrodictium occultum, 188
PYRROPHYTA, 189
PYXIDOCYATHIDAE, 1174
Pyxidocyathus, 1021

Qiaotingaspis, 1025
Quadratapora, 1029
Quadratimorpha, 871, 877–878
Quadratimorpha florentis, 871
Quadratimorpha ordinata, 877
Quadratimorpha simplisis, 878
Quadratimorpha tenera, 877
Quadrochites, 1032
Quadrorites, 1031
Quadrosiphogonuchites, 1022
Quadrotheca, 1028
Quebecia, 1031
Quinquelithes, 1028

Rachovskia, 1018
Radiaxialia, 1033
Radicicyathus, 1021
Radicula, 354, 357, 503, 972
    Radicula podolicina, 972
Radiocerniculum, 1032
RADIOCYATHA, 361, 400, 404, 860, 981, 1020
Radiocyathus, 1020
RADIOLARIA, 237–239, 433, 449, 451, 454, 499, 876, 1171, 1179, 1181–1184
Radiophycus, 946
    Radiophycus yangjiapingensis, 946
Ramacia, 893
    Ramacia carpentariana, 893
Ramapuraea, 1050
    Ramapuraea vindhyanensis, 1050
Ramellina, 1004
    Ramellina pennata, 1004
Ramentia, 1028, 1174
Ramifer, 1021
Ramivaginalis, 929, 1061, **1107**, 1129
    Ramivaginalis uralensis, 929, 1061, **1107**, 1129
Randomia, 1027
Rangea, 345, 371, **372**, 373, 381, 504, 1004–1005
    Rangea arborea, 1004
    Rangea brevior, 1004
    Rangea grandis, 1004
    Rangea longa, 1004
    Rangea schneiderhoehni, 345, **372**, 1004–1005
    Rangea sibirica, 1004
RANGEIDAE, 373, 1173
RANGEOMORPHA, 373, 1173
Rarissimetus, 1028
Rarocyathus, 1021
Raropectinus, 1019
RASETTASPIIDAE, 1176
Rasetticyathus, 1019
Rattus, 482
    Rattus norvegicus, 482
RAYMONDINIDAE, 1176
Razumovskya, 361–**362**, **364**, 366, 982, 985, 990, 992, 995
    Razumovskya alta, 995
    Razumovskya cf. *alta*, 985, 995
    Razumovskya ethmoidale, 992
    Razumovskya fibrosa, 992
    Razumovskya gracilis, 995
    Razumovskya grandis, 992
    Razumovskya hispida, 992
    Razumovskya kiyanica, 992
    Razumovskya lata, 992
    Razumovskya multispora, 992
    Razumovskya seriata, 992
    Razumovskya uralica, 985, 992
RECEPTACULITIDS (*See* Subject Index)
Rectannulus, 1019
Redkina, 233–234, 371, 373, **388**, 938, 999, 1006, 1032, 1061, **1113**, 1149, 1174
    Redkina fedonkinispis, 938, 1061, **1113**, 1149
    Redkina spinosa, 371, **388**, 1006
Redlichaspis, 1025
Redlichia, 1025
REDLICHIIDA, 556, 1024, 1175
REDLICHIIDAE, 1176
Redlichina, 1025
Redlichops, 1025
Redoubtia, 1174
REGULARES, 554, 1018, 1172, 1179, 1183–1184
Rehbachiella, 1177
REIBEIRIIDAE, 1175
REMOPLEURIDIDAE, 1176
Renalcis, 359–361, **362–363**, 366, 423, 514, 982–987, 992, 995–996
    Renalcis chabakoviformis, 987, 992
    Renalcis cibus, 992
    Renalcis compositus, 992
    Renalcis conchaeformis, 992
    Renalcis densum, 992
    Renalcis elegans, 992
    Renalcis erbinatus, 992
    Renalcis gelatinosus, 986, 992
    Renalcis granosus, 985, 992
    Renalcis granosus var. *fequratus*, 992
    Renalcis granosus var. *plenus*, 992
    Renalcis granulatus, 992
    Renalcis halisiteformis, 995
    Renalcis jacuticus, 992
    Renalcis aff. *jacuticus*, 992
    Renalcis ex gr. *jacuticus*, 992
    Renalcis lenaicum, 992
    Renalcis levis, 992
    Renalcis minutus, 992
    Renalcis nodularis, 992
    Renalcis novum, 992
    Renalcis pectunculus, 992
    Renalcis polymorphus, 984, 991–992
    Renalcis aff. *polymorphus*, 993
    Renalcis pseudoradiatus, 995
    Renalcis rotundus, 993, 995
    Renalcis seriata, 993
    Renalcis simplex, 995
    Renalcis textularites, 995
    Renalcis tuberculatus, 993
Renitheca, 1029
Resegia, 1032
Resimopsis, 1025
Resserops, 1025
Retecoscinus, 1020
Retetumulus, 1020
RETICULOSA, 1172
Retifacies, 1025
Retiforma, 937, 1061, **1112**
    Retiforma tolparica, 937, 1061, **1112**
Retisphaeridium, **225**, 904, 945, 1150
    Retisphaeridium densum, 945
    Retisphaeridium dichamerum, **225**
    Retisphaeridium vindhyanense, 904, 1150
Rewardocyathus, 1019
Reynella, 1050
    Reyenella howchini, 1050
Rhabdochites, 1031
Rhabdocyathella, 1020
RHABDOCYATHELLIDAE, 1172
RHABDOPLEURIDA, 371, 1178
Rhabdotubus, 1178
Rhicnonema, 885, 898, 918, 929, 941, 1135
    Rhicnonema antiquum, 885, 929, 941
    Rhicnonema crassivaginatum, 918, 1135
RHIZACYATHIDA, 1172
RHIZOPODEA, 237, 483, 592, 1171, 1179, 1181–1184
RHODOBACTERIA, 472
RHODOPHYCEAE, 205–207, 220, 481, 592, 934, 944, 1159
RHODOPHYTA, 110, 189, 357, 359–360, 367, 928, 981, 986, 994
RHODOSPIRILLACEAE, 283, 827
Rhombicalvaria, 1025, 1177
Rhombocorniculum, 412, 1031, 1174, 1180–1184
Rhysonetron, 414–**415**, 1039, 1041, 1050–1051
    Rhysonetron byei, 1050
    Rhysonetron lahtii, 1050
    Rhysonetron ramapuraensis, 1050
RHYSSOMETOPIDAE, 1176
RIBEIRIIDAE, 1026
RIBEIRIOIDA, 1026, 1174
Rimouskia, 1024
Rinconia, 1025
Ringifungia, 1019
Rivularia, 250
RIVULARIACEAE, 869, 909, 912
ROBERTOCYATHIDAE, 1172
Robertocyathus, 1019
Robertsonia, 296
    Robertsonia salsa, 296
Robustocyathellus, 1019
ROBUSTOCYATHIDAE, 1172
Robustocyathus, 1019
Rondocephalus, 1023

Taxonomic Index

Rosella, 1026
Rosellatona, 1020, 1173
Rossocyathella, 1019
ROSTROCONCHIA, 410, 1026, 1174, 1179, 1181–1184
Rostroconus, 1028
Rotadiscus, 372, 1020
Rothpletzella, 982–983, 985, 992–993
    Rothpletzella flabellata, 990, 992–993
    Rothpletzella igarcaensis, 990, 992–993
ROTIFERS (*See* subject Index)
Rotundocyathus, 1019
Roualtia, 1050
    Roualtia rewaensis, 1050
Rozanovicoscinus, 1020
ROZANOVICYATHIDAE, 1172
Rozanovicyathus, 1020
Rozanoviella, 1028
Rudanalus, 1020
Ruedemannella, 354, 502, 972, 974, 1006
    Ruedemannella minuta, 972, 974, 1006
Rugatotheca, 1032
Rugoconites, 1004
    Rugoconites enigmaticus, 1004
    Rugoconites tenuirugosus, 1004
Rugoinfractus, 391, 1012, 1050
    Rugoinfractus ovruchensis, 391, 1012, 1050
RUGOSA, 1173
Rugosoopsis, 907, 1060, **1074**, 1128
    Rugosoopsis tenuis, 907, 1060, **1074**, 1128
Ruminococcus, 1137
    Ruminococcus albus, 1137
Runnegarochiton, 1028
Ruppia, 325
    Ruppia maritima, 325
Rushtonia, 1021
Rushtonites, 1031, 1174, 1180–1184
Rusophycus, 371–372, 392, **439**
    Rusophycus dispar, **439**
    Rusophycus palonensis, 392
Russocyathus, 1019
Rustella, 1031

Saarina, 354, 371, **388**, 502, 1006, 1021
SAARINIDAE, 371, 959, 961, 1174
Sabellarites, 416, 1037, 1050
Sabellidites, 354–**355**, 357, 371, 502, 974, 1021
    Sabellidites cambriensis, 974
SABELLIDITIDA, 1021, 1174, 1179, 1181–1184
SABELLIDITIDAE, 1174
SABELLIDITIDS (*See* Subject Index)
Saccharomyces, 482
    Saccharomyces cerevisiae, 482
Sacciconus, 1028
Sacciella, 1022
Saccocaris, 1177
Sachites, 1026
SACHITIDA, 408, 1026, 1173
SACHITIDAE, 1026, 1173
Sagacyathus, 1019
Sagittitheca, 1032
Sailycaspis, 1023
Sajanacyathus, 1019
Sajanaspis, 1025
Sajanella, 1004, 1050
    Sajanella arshanica, 1004, 1050
Sajania, 362, 366, 985–986, 993, 995
    Sajania fasciculata, 993
    Sajania frondosa, 985, 993
    Sajania pennata, 993
SAJANOCYATHIDAE, 1172
Sajanolynthus, 1020
Salairocyathus, 1020
Salanacus, 1031
Salanyella, 1027
Salanygolina, 1020
Salanytheca, 1021
Salome, 925, 941, 1137

Salome hubeiensis, 941, 1137
Salome svalbardensis, 925, 941, 1137
Salopiella, 1032
Salterella, 1021
SALTERELLIDAE, 1174
SANARKOCYATHIDAE, 1172
Sanarkocyathus, 1019
Sanaschtykgolia, 1024
Sanctacaris, 1175
Sangshuania, 349, **354**, 356, 503, 969, 974
    Sangshuania linearis, **354**, 356, 974
    Sangshuania sangshuanensis, **354**, 356, 974
Saralinskia, 419, 1050
    Saralinskia boulinnikovi, 1050
    Saralinskia glomeria, 1050
    Saralinskia multiangulata, 1050
    Saralinskia plana, 1050
    Saralinskia radiata, 1050
    Saralinskia ramosa, 1050
    Saralinskia serrata, 1050
    Saralinskia stellata, 1050
SARCODINA, 592
Sardospongia, 1017
Sargaella, 1027
Sarmenta, 354, 357, 502, 974
    Sarmenta capitata, 974
Sarotrocercus, 1175
Satka, **197**, 200, 877, 892–893, 897, 914, 917, 933, 939, 1059, 1061, **1066**, **1103**, **1109**, **1116**, 1143
    Satka colonialica, 877, 914, 917, 1061, **1103**, 1143
    Satka elongata, 897, 1061, **1103**
    Satka favosa, 892–893, 897, 1059, **1066**
    Satka granulosa, **197**, 200, 227, 939, 1061, **1116**
    Satka squamifera, 933, **1109**
Satpulispora, 874, 882, 902
    Satpulispora major, 882, 902
    Satpulispora microreticulata, 882, 902
    Satpulispora minuta, 874, 902
    Satpulispora psilata, 882, 902
Saukianda, 1025
SAUKIANDIDAE, 1176
Saukiandiops, 1025
SAUKIIDAE, 1176
Scambocris, 1031
Scamboscamna, 1028
Scaniella, 237
Scaphita, 915
    Scaphita eniseica, 915
Scenedesmus, 190
Scenella, 1020
SCENELLIDAE, 1020, 1173
Scenellopsis, 1027
Schiderticyathellus, 1019
SCHIDERTYCYATHIDAE, 1172
Schismatosphaeridium, 881
    Schismatosphaeridium kumauni, 881
Schizopholis, 1030
SCHIZORAMIA, 381
Schizothrix, 262, 300, 322, 324, 885
    Schizothrix atavia, 885
    Schizothrix calcicola, 262, 324
Schizothropsis, 895
    Schizothropsis caudata, 895
Schlecofura, 1173
Schmidtiellus, 1025
Schumnycyathus, 1020
Scintilla, 871
    Scintilla perforata, 871
Scissilisphaera, 927, 1143
    Scissilisphaera regularis, 927, 1143
Scissotheca, 1032
SCLEROSPONGES (*See* Subject Index)
SCOLECODONTS (*See* Subject Index)
Scolicia, 1012
Scoponodus, **409**, 1026
    Scoponodus renustus, **409**
Scutatestomaconus, 1027

SCYPHOZOA, 381, 407-408, 558, 1173, 1179, 1181-1184
Scytonema, 261-262
   Scytonema mychrous, 262
SCYTONEMATACEAE, 880
SEA-PENS (*See* Subject Index)
SEAGRASS (*See* Subject Index)
SEBARGASIIDAE, 1171
Securiconus, 1028
Sekwia, 1004
   Sekwia excentrica, 1004
   Sekwia kaptarenkoe, 1004
Sekwiaspis, 1023
Sekwicyathus, 1020
Selindeochrea, 1021
Selkirkia, 1020
SELKIRKIIDAE, 1173
Sellaulichnus, 373, 392, **395**, 1011-1012
   Sellaulichnus meishucunensis, **395**, 1011-1012
Sellula, 1022
Semielliptotheca, 1029
Serebrina, 354, 502
Serioides, 1027
Serligia, **362**, 985, 993, 995
   Serligia fragilis, 985, 993
   Serligia cf. *fragilis*, 985
   Serligia gracilis, 993
   Serligia cf. *gracilis*, 993
Serracaris, 1025, 1177
Serrodiscus, 1023
Shaanxilithes, 354, 502
Shabaella, 1023
Shabaktiella, 1027
Shafferia, 1177
Shatania, 1025
SHELBYOCERIDAE, 1174
Shensiella, 1022
Shifangia, 1025
Shipaiella, 1025
SHIRAKIELLIDAE, 1176
Shivelicus, 1023
Shiveligocyathus, 1018
Shizudiscus, 1023
Shouhsienia, 349, **354**, 356-357, 502, 908, 912, 969, 973-974
   Shouhsienia longa, 908, 912, 974
   Shouhsienia magna, 974
   Shouhsienia multirugosa, 912, 974
   Shouhsienia shouhsienensis, **354**, 908, 912, 974
Shujana, 985, 996
   Shujana praefulgida, 996
   Shujana shulgini, 985, 996
SHUMARDIIDAE, 1176
Shuntaria, 919
   Shuntaria evidens, 919
SIBERIOGRAPTIDAE, 1178
Sibirecyathus, 1019
Sibiria, 1030
Sibiriafilum, 873, 1060, **1086-1087**
   Sibiriafilum tunicum, 873, 1060, **1086-1087**
Sibiriaspis, 1025
Sichotecyathus, 1019
Sichuanolenus, 1025
Sichuanospira, 1027
Sicyus, 349, 503
Siderocapsa, 37
Sidneyia, 1052
   Sidneyia groenlandica, 1052
SIDNEYIIDAE, 1176
SIGMOCOSCINIDAE, 1172
Sigmocoscinus, 1020
SIGMOCYATHIDAE, 1172
Sigmocyathus, 1019
Sigmofungia, 1018
SIGMOFUNGIIDAE, 1173
Silesicaris, 1022
Silurovella, 372
Silviacoscinus, 1020
Simplotubus, 1029

Sinaiicoccus, 930, 1139
   Sinaiicoccus avnimelechii, 930
   Sinaiicoccus delicatus, 930
   Sinaiicoccus minutus, 930, 1139
Sinijanella, 1023
Sinocapsa, 985, 933
   Sinocapsa honanica, 985, 993
Sinoclavus, 1026
Sinodiscus, 1023
Sinolenus, 1025
Sinosabellidites, 354, 356-357, 433-434, 501, 503, 514, 525, 973-974, 1006-1007, 1021
   Sinosabellidites huainanensis, 974, 1006-1007
SINOSABELLIDITIDS (*See* Subject Index)
Sinosachites, 1026
Sinotaenia, 354, 503, 974
   Sinotaenia liulaobeiensis, 974
Sinotubulites, 1032, 1174
Sinskia, 1025
Sinuconus, 1028
SINUITIDAE, 1174
SINUOPEIDAE, 1174
Siphogonuchites, **409**, 1026
   Siphogonuchites triangularis, **409**
SIPHOGONUCHITIDAE, 408, 1026, 1173
Siphonia, 985, 997-998
   Siphonia columella, 997
   Siphonia decussa, 997
   Siphonia florisglobosa, 997
   Siphonia herbacea, 985, 997
   Siphonia songlinensis, 998
SIPHONOPHORIDA, 391, 1020, 1173
Siphonophycus, 28, 33, 363, 884, 887-889, 894-895, 898, 904, 906, 910, 912-913, 916, 918-920, 922, 925, 927, 929, 932-933, 936, 941, 943-944, 949, 985, 933, 1060-**1062**, **1080-1081**, **1092**, **1098**, **1101**, **1107**, **1109**, 1123, 1126-1128
   Siphonophycus antiquus, 28, 33
   Siphonophycus beltense, 894, 912, 1127
   Siphonophycus capitaneum, 929, 1061, **1107**
   Siphonophycus costatus, 906, 920, 933, 1060, **1080-1081**, **1109**, 1128
   Siphonophycus crassiusculum, 894, 1126
   Siphonophycus hughesii, 944, 1127
   Siphonophycus indicum, 929
   Siphonophycus inornatum, 895, 898, 925, 941, 1127
   Siphonophycus kestron, 895, 910, 913, 927, 944, 985, 933, **1092**, 1128
   Siphonophycus lamellosum, 918, 1128
   Siphonophycus punctatus, 904, 1128
   Siphonophycus sinense, 941
   Siphonophycus transvaalense, 884, 993, 1059, **1062**, 1128
SIPHONOTRETIDAE, 1177
Siphula, 850
   Siphula ceratites, 850
SIPUNCULIDA, 410, 485
Sissospina, 1025
Sivaglicania, 893, 1059, **1066**
   Sivaglicania tadasii, 893, 1059, **1066**
Skania, 1177
Skara, 1177
SKARACARIDA, 1177
Skiagia, **226**, 228, 231, 949
   Skiagia ciliosa, 949
   Skiagia compressa, **226**, 949
   Skiagia orbiculare, 949
   Skiagia ornata, 949
   Skiagia scottica, 949
Skinnera, 372-373, 381, 407, 1004
   Skinnera brooksi, 1004
Skolithos, **396**, 397, 416-**417**, 1012, 1038, 1045, 1049, 1052
   Skolithos delinatus, **396**, 1012
   Skolithos miaoheensis, 1052
   Skolithos ramosus, 1012
SLIME MOLDS (*See* Subject Index)
SOANICYATHIDAE, 1172
Soanicyathus, 1020
Sokolovina, 1021

Taxonomic Index

Sokolovitheca, 1029
Solenopleurella, 1024
SOLENOPLEURIDAE, 1176
Solenopora, 359, **362**, 366, 982, 985, 993, 995–996
    Solenopora compacta, 993
    Solenopora spongioides, 985
    Solenopora tjanshanica, 995
SOLENOPORIDS (See Subject Index)
Solenotia, 1031
SOLUTA, 1178
Somphocyathus, 1020
Sonella, 1030
Songlinella, 1022
Songlingella, 1027
Sonskia, 1052
    Sonskia prava, 1052
Spatangopsis, 1020
Spatuloconus, 1028
Sphaerocodium, 985
    Sphaerocodium gotlandica, 985
Sphaerocongregus, 930–931, **1109**, 1142
    Sphaerocongregus variabilis, 930–931, **1109**, 1142
Sphaerophycus, 876, 885, 888, 891, 893–894, 896, 898, 910, 913, 925, 929, 936, 1061, **1070**, **1094**, 1142, 1145, 1147
    Sphaerophycus densus, 879
    Sphaerophycus gigas, 876
    Sphaerophycus medium, 896, 898, 925, 929, **1070**, 1145
    Sphaerophycus cf. *medium*, 896
    Sphaerophycus parvum, 885, 891, 893–894, 896, 910, 913, 925, 1061, **1094**, 1145
    Sphaerophycus aff. *parvum*, 936, **1070**
    Sphaerophycus reticulatum, 891, 894, 1142
    Sphaerophycus tetragonale, 891, 1147
    Sphaerophycus wilsonii, 925
Sphaerotilus, 304
    Sphaerotilus natans, 304
Sphenoderia, 237
Sphinctocyathus, 1018
SPHINCTOZOA, 239, 448, 556, 1017, 1171
Spinicerniculum, 1032
Spinocera, 1031
Spinulitheca, 1029
SPINULITHECIDAE, 1175
Spinulothele, 1030
SPIOMORPHA, 1175
SPIONIDAE, 1175
Spirellus, 434, 1032
SPIRILLACEAE, 1129
SPIRILLICYATHIDAE, 1173
Spirillicyathus, 1018
Spirillinema, 930, 1129
    Spirillinema bentorii, 930, 1129
Spirillum, 1129
    Spirillum volutans, 1129
SPIROCHAETES (See Subject Index)
Spiroichnus, 505, 969
Spirosaccus, 874
    Spirosaccus punctata, 874
Spiroscolex, 1052
Spirosolenites, 238–239, 1032
Spirulina, 195, 276, 309, 327–328, 850, 852, 913–914, 918, 934, 1123–1124, 1129
    Spirulina labyrinthiformis, 309
    Spirulina laxissima, 1129
    Spirulina maxima, 852
    Spirulina meneghiniana, 1129
    Spirulina platensis, 850, 852
    Spirulina subsalsa, 276
SPONGES (See Subject Index)
Spongiochloris, 1140, 1149
    Spongiochloris excentrica, 1140
    Spongiochloris incrassata, 1140
    Spongiochloris spongiosa, 1149
Spongoides, 1052
    Spongoides grandis, 1052
Sporinula, 363, 366, 985–986, 993
    Sporinula palmata, 985, 993

Spriggia, 372–373, **382**, 1003–1004, 1173
    Spriggia annulata, **382**, 1003–1004
    Spriggia wadea, 1004
Spriggina, 372–373, 381, **385**, 434, 1003–1005
    Spriggina borealis, 1004
    Spriggina floundersi, **385**, 1003–1005
    Spriggina ovata, 1004
SPRIGGINIDAE, 373, 1175
SPUMELLARIA, 1171
Spumiosina, 883–884, 1061, **1116**
    Spumiosina cineraria, 883
    Spumiosina rubiginosa, 883, 1061, **1116**
    Spumiosina solida, 884
Spumosata, 883, 948, 1152
    Spumosata minor, 948
    Spumosata nova, 883, 948, 1152
    Spumosata simplex, 948
Squamodictyon, 1052
SQUAMOSOCYATHIDAE, 1172
Squamosocyathus, 1019
Staphylococcus, 853, 1140
    Staphylococcus aureus, 853
    Staphylococcus saprophyticus, 1140
Stapicyathus, 1019
Statanulocyathus, 1020
STAURINIDAE, 1173
Staurinidia, 1004
    Staurinidia crucicula, 1004
Stefania, 1031, 1174
Stellaria, 1026
Stellinium, 946
Stelloglyphus, 505
Stellostomites, 372
Stellostromites, 1033
Stenocyphus, 349, 503, 974
    Stenocyphus subtilis, 974
Stenotheca, 1027
Stenothecella, 1027
STENOTHECIDAE, 1174
STENOTHECOIDEA, 1027, 1174, 1179, 1181–1184
Stenothecoides, 1027
Stenozonoligotriletum, 887, 899, 901, 935, 938–939, 943, 945
    Stenozonoligotriletum sokolovii, 887, 935, 939, 943, 945
    Stenozonoligotriletum validum, 887, 899, 901, 938
Stephanoconus, 1028
STEPHENOSCOLECIDAE, 1175
Stephochetus, 994
    Stephochetus ocellatus, 994
Stictoconus, 1032
Stictosphaeridium, 884, 887, 889–892, 898–902, 905, 907–911, 915, 917–918, 921–928, 931, 933–934, 937–938, 941–942, 948–949, 951, 1140, 1149
    Stictosphaeridium implexum, 889–890, 892, 899–901, 905, 908, 910–911, 918, 924–925, 934, 937–938, 946, 949, 1149
    Stictosphaeridium pectinale, 884, 899–902, 907, 909, 915, 926, 928, 933–934, 951
    Stictosphaeridium sibiricum, 899
    Stictosphaeridium sinapticuliferum, 887, 889, 891, 899–902, 907–908, 915, 922, 924–926, 933–934, 937–938, 942, 945–946
    Stictosphaeridium cf. *sinapticuliferum*, 911, 923, 926
    Stictosphaeridium tortulosum, 899–900, 915, 945, 948, 1140
    Stictosphaeridium verrucatum, 923, 1140
STIGONEMATALES, 879
STILLICIDOCYATHIDAE, 1172
Stillicidocyathus, 1020
Stimulitheca, 1029
Stoibostrombus, **411**, 1032
    Stoibostrombus crenulatus, **411**,
Stoliconus, 1028
STRABOPIDAE, 1177
Strabops, 1022
Stratifera, 625
Strenuaeva, 1025
Strenuella, 1025
Streptomyces, 853
    Streptomyces olivaceous, 853

Strettonia, 1023
Striatella, *198*, 325, 875, 1061, *1112*, *1116*
    Striatella coriacea, *198*, 875, 1061, *1112*, *1116*
Striatocyathus, 1021
Stromatocystites, 1031
STROMATOCYSTITIDA, 1178
STROMATOCYSTITIDAE, 1178
Stromatopora, 993
    Stromatopora compacta, 993
STROMATOPORIDS (See Subject Index)
STROMATOPOROIDEA, 1171
Studenicia, 883, 1061, *1114*
    Studenicia bacotica, 883, 1061, *1114*
Stylonychia, 482
    Stylonychia pustulata, 482
STYLOPHORA, 1178, 1180–1184
Subphyllochorda, 1012
Subtiflora, *362*, 366, 982, 985–987, 993
    Subtiflora delicata, 985, 993
    Subtiflora mazasia, 993
    Subtiflora lineata, 987
Subtilocyathus, 1020
Sugaites, 1026
Sulcagloborilus, 1029
SULCAVITIDAE, 1175
Sulcavitus, 1028
Sulcocarina, 1027
SULFOBACTERIA, 473–475
SULFOLOBALES, 477
Sulfolobus, 291
Sunella, 1022
Sunicyathus, 1021
Sunnaginia, 1030
SUNNAGINIDAE, 410, 1030, 1174
Surindia, 1031
Suvorovella, 984, 986, 993, 1004
    Suvorovella aldanica, 986, 993, 1004
Suzmites, 505, 1006, 1012, 1052
    Suzmites tenuis, 1006
    Suzmites volutatus, 1006, 1012
SVEALUTIDAE, 1177
Svetlanocyathus, 1019
Swantonia, 1031
Symplassosphaeridium, 882, 889, 892–893, 897, 904, 915, 920, 936, 938–939, 942–943, 946, 948, 1059, *1066*, 1141, 1143
    Symplassosphaeridium bulbosum, 904
    Symplassosphaeridium bushimayense, 904
    Symplassosphaeridium incrustatum, 936, 938, 942–943, 1141
    Symplassosphaeridium parvum, 936
    Symplassosphaeridium subcoalitum, 948, 1143
    Symplassosphaeridium tumidulum, 946, 1143
    Symplassosphaeridium undosum, 892–893, 897, 920, 1059, *1066*
Symploca, *196*, 904, 1127
    Symploca muscorum, *1127*
Syndianella, 1025
Synechococcus, *195*, 209, 277, 287, 297, 300, 313, 314, 316, *318*–320, 328, 331, 337, 844, 850, 1144
    Synechococcus aeruginosus, 1144
    Synechococcus bacillaris, 850
    Synechococcus lividus, 277, 287, 313, 316, *318*–320
Synechocystis, 301
Syniella, 982–983, 986, 991, 993
    Syniella invenusta, 986, 993
Synodophycus, 925
    Synodophycus euthemos, 925
Synsphaeridium, 867, 879, 882, 885, 893, 897–901, 905–908, 910, 915, 917, 920–921, 923, 928, 930–933, 935–936, 939–940, 944, 951
    Synsphaeridium conglutinatum, 885, 898–900, 907–908, 920–921, 928, 932, 936, 940, 944
    Synsphaeridium sorediforme, 901, 905, 907, 931, 939
    Synsphaeridium switjasium, 951
Syringocnema, 1018
SYRINGOCNEMIDIDA, 1173
SYRINGOCNEMIDIDAE, 1173
SYRINGOCOSCINIDAE, 1173

Syringocoscinus, 1018
Syringomorpha, 393, 1012
    Syringomorpha nilssoni, 393, 1012
Syringsella, 1018
Syspacephalus, 1024
Szechaunaspis, 1023

Tabatopygellina, 1023
TABELLAECYATHIDAE, 1173
Tabellaecyathus, 1018
TABULACONIDAE, 1173
Tabulaconus, 1020
TABULACYATHIDAE, 1173
Tabulacyathus, 1018
TABULATA, 400, 1173
Taeniatum, 878, 880, 919, 929, 942–944
    Taeniatum crassum, 878, 880
    Taeniatum simplex, 880
Taenitrichoides, 882, 1061, *1112*
    Taenitrichoides jaryschevicus, 882, 1061, *1112*
TAKAKKAWIIDAE, 1171
Tanbaoites, 1026
Taninia, 366, 986, 993
    Taninia tomentosa, 986, 993
Tannuella, 1027
TANNUOLACYATHIDAE, 1172
Tannuolaia, 986, 993
    Tannuolaia fonini, 986, 993
Tannuolina, *411*, 1030
    Tannuolina zhangwentangi, *411*
TANNUOLINIDAE, 1030, 1174
Tannuspira, 1027
Taonichnites, 1052
Taphrhelminthoida, 392, 397, 1012
    Taphrhelminthoida dailyi, 1012
Taphrhelminthopsis, 392, 396–397, 1011–1012, 1015
    Taphrhelminthopsis circularis, 396, 1012, 1015
    Taphrhelminthopsis dailyi, 1011
Taraxaculum, *402*, 1017
    Taraxaculum volans, *402*
Tarthinia, *362*, 366, 986, 992–993, 995–996
    Tarthinia rotunda, 986, 993, 995
    Tarthinia zachirica, 995
Tarynaspis, 1023
Tasmanadia, 1052
    Tasmanadia dassii, 1052
Tasmanites, 189, *225*, 229, 503, 904, 908–910, 917–918, 924, 932, 946–947, 949–951, 974, 1060, *1080*, 1149–1150, 1159
    Tasmanites bobrowskii, 949, 951, 1150
    Tasmanites piritaensis, 950
    Tasmanites refejicus, 229, 908–910, 917–918, 1060, *1080*
    Tasmanites tenellus, 946, 949, 1150
    Tasmanites variabilis, 947, 1150
    Tasmanites vindhyanensis, 229, 904, 974, 1150
    Tasmanites volkovae, *225*, 949, 951, 1149
Tateana, 372–373, *380*, 1001, 1006
    Tateana inflata, *380*, 1001, 1006
Tatijanaecyathus, 1020
Tawuia, 349, 353, *354–355*, 356–357, 433–434, 439, 453, 461, 501–502, 514, 516, 519, 525, 561, 592, 959, 961, 967, 969, 971, 973–976, 979
    Tawuia dalensis, *354–355*, 974, 976–977
    Tawuia cf. *dalensis*, 974
    Tawuia fusiformis, 976
    Tawuia hippocrepica, 976
    Tawuia rampuraensis, 976
    Tawuia sinensis, 976
    Tawuia striatia, 976
    Tawuia striatis, 976
    Tawuia suketensis, 976
TAWUIDS (See Subject Index)
Taylorcyathus, 1019
Tchangsichiton, 1031
Tcharatheca, 1029
Tcharella, 1030
Tchuranitheca, 1029
TCHURANOTHECIDAE, 1175

Taxonomic Index

TECHNOPHORIDAE, 1175
TEGANIIDAE, 1172
Tegerocoscinus, 1020
TEGEROCYATHIDAE, 1172
Tegerocyathus, 1019
TEGOPELTIDAE, 1176
Teichichnus, 392
Telemarkites, 873, 1052
    Telemarkites enigmaticus, 873, 1052
Templuma, 366, 986, 995, 1052
    Templuma sinica, 986, 995, 1052
TENNERICYATHIDAE, 1172
Tennericyathus, 1019
Tenuofilum, 914, 925, 927
    Tenuofilum septatum, 914, 925, 927
Teophipolia, 945
    Teophipolia lacerata, 945
TERCYATHIDAE, 1172
Tercyathus, 1019
TEREBELLIDAE, 1175
TEREBELLOMORPHA, 1175
Terechtaspis, 1025
TERIDONTIDAE, 1178
Termieraspis, 1025
Termierella, 1025
Terraecyathus, 1019
Tersia, 986, 993
    Tersia filiforma, 986, 993
Tersicyathus, 1019
Tesella, 1030
TETRADELLIDAE, 1022, 1177
Tetrahymena, 87, 592
    Tetrahymena pyriformis, 87
TETRALOBULIDAE, 1178
Tetraphycus, 894, 897, 903, 918, 925, 932, 935, 942, 1061, **1111**, **1115**, 1147
    Tetraphycus acinulus, 894, 1147
    Tetraphycus amplus, 897
    Tetraphycus bistratosus, 935, 1061, **1111**
    Tetraphycus congregatus, 903, 1147
    Tetraphycus conjunctum, 942, 1061, **1115**
    Tetraphycus diminutivus, 894, 925
    Tetraphycus gregalis, 894
    Tetraphycus hebeiense, 918
    Tetraphycus major, 894, 1147
Tetrasphaera, 907, 1060, **1075**
    Tetrasphaera antiqua, 907, 1060, **1075**
Tetratheca, 1029
TETRATHECIDAE, 1175
Thalamocyathus, 1019
Thalamopectinus, 1019
Thalassinoides, 416
Thalassiothrix, 322
    Thalassiothrix woodii, 322
THALASSOCYATHIDA, 1018, 1172
Thalassocyathus, 1018
Thambetolepis, **409**, 1026
    Thambetolepis delicata, **409**
Thaumatophycus, 986
    Thaumatophycus furcatus, 986
THECAMOEBAEANS (*See* Subject Index)
Theococyathus, 1021
Thermoanaerobacter, 320
    Thermoanaerobacter ethanolicus, 320
Thermoanaerobium, 320
    Thermoanaerobium brockii, 320
Thermobacteroides, 287, 320
    Thermobacteroides acetoethylicus, 287, 320
THERMOCOCCALES, 474
Thermodesulfobacterium, 287, 290, 320
    Thermodesulfobacterium commune, 287, 290, 320
Thermomicrobium, 320
    Thermomicrobium roseum, 320
THERMOPROTEALES, 477
Thermoproteus, 291, 474
    Thermoproteus tenax, 291
Thermothrix, 291–292, 313–314, **316**

Thermothrix thioparus, 292
Thermotoga, 474
    Thermotoga maritima, 474
THERMOTOGALES, 474
Thermus, 287, 314, 320
    Thermus aquaticus, 287, 320
Thiobacillus, 291, 853
    Thiobacillus ferrooxidans, 291
    Thiobacillus thioparus, 853
Thiocapsa, 272
    Thiocapsa pfennigii, 272
Thiocystis, 846
Thiomicrospira, 291
Thioploca, 250, 291–292, 336
Thiothrix, 35, 291–292, 1131
    Thiothrix nivea, 35, 1131
Thiovulum, 205–206, 1138
    Thiovulum majus, 206, 1138
THORACICA, 1177
Thoralaspis, 1025
Thorslundella, 1022
Thylacocausticus, 926
    Thylacocausticus globorum, 926
Thymos, 886
    Thymos halis, 886
Tianshandiscus, 1032
Tianzhushanella, 1031
Tianzhushania, 1026
Tianzhushanospira, 1028
Tichkaella, 1027
Tigillites, 1052
    Tigillites bohmei, 1052
Tiksitheca, 1021
Timanella, 986, 993
    Timanella gigas, 986, 993
TINTINNIDS (*See* Subject Index)
Tirasiana, 373, **380**, 1006
    Tirasiana cocarda, 1006
    Tirasiana concentralis, 1006
    Tirasiana coniformis, 1006
    Tirasiana disciformis, **380**, 1006
TIRASIANIDAE, 1173
Tolliccyathus, 1018
Tomentula, 366, 986, 993
    Tomentula interrupta, 993
    Tomentula villosa, 986, 993
Tommotia, 1030
TOMMOTIIDA, 400, 403, 410–411, 438, 448, 556, 574, 1029, 1174, 1179, 1181–1184
TOMMOTIIDAE, 1030, 1174
Tommotitubulus, 1032
Tomocyathus, 1020
Topolinocyathus, 1021
Torellella, **403**, 1021
TORELLELLIDAE, 1174
Torellelloides, 1021
Tormentella, 871
    Tormentella tubiformis, 871
Toromorpha, 883
Torosocyathus, 1019
Torridoniphycus, 912
    Torridoniphycus lepidus, 912
Torrowangea, 392–393, 1012, 1014–1015
    Torrowangea rosei, 1012, 1014
Tortocyathus, 1012
Tortunema, 906, 910–911, 915, 920, 933, 935, 1060, **1081**, **1087**, **1089**, 1132, 1135
    Tortunema bothica, 935
    Tortunema eniseica, 906, 915, 920, 1060, **1087**, 1135
    Tortunema sibirica, 906, 910, 915, 920, 933, 1060, **1089**, 1132
TOTIGLOBIDAE, 1178
Tracheleuglypha, 237
TRACHELOCRINIDAE, 1178
Trachyhystrichosphaera, **224**, 907, 920, 922, 927, 935, 1060, **1075**, 1151
    Trachyhystrichosphaera aimika, 907, 920, 1060, **1075**, 1151
    Trachyhystrichosphaera bothnica, 935, 1151

Trachyhystrichosphaera vidalii, **224**, 922, 927, 1151
TRACHYLINIDA, 1173
Trachymarginata, 875, 883, 939, 948
   Trachymarginata aberrantis, 948
   Trachymarginata globulosa, 875
   Trachymarginata rustosa, 883
   Trachymarginata speciosa, 939
Trachyoligotriletum, 887–888, 890, 898, 937, 943, 948
   Trachyoligotriletum asperatum, 888
   Trachyoligotriletum gyratum, 948
   Trachyoligotriletum incrassatum, 919, 938
   Trachyoligotriletum cf. *incrassatum*, 943
   Trachyoligotriletum laminaritum, 887–888
   Trachyoligotriletum minutum, 898
   Trachyoligotriletum nevelense, 890
   Trachyoligotriletum obsoletum, 890
   Trachyoligotriletum planum, 890
Trachypsophosphaera, 884
   Trachypsophosphaera exilis, 884
Trachysphaeridium, 190, 229, 349, 503, 879, 902, 907–909, 911–912, 916, 919–923, 925–927, 929, 935–945, 948–949, 976, 1060, **1075**, 1140, 1150–1151
   Trachysphaeridium apertum, 923
   Trachysphaeridium attenuatum, 937–939, 942, 945
   Trachysphaeridium chihsienense, 879
   Trachysphaeridium aff. *chihsienense*, 921
   Trachysphaeridium cultum, 908, 920–921
   Trachysphaeridium decorum, 929
   Trachysphaeridium hyalinum, 889, 908, 920–921, 1140
   Trachysphaeridium incrassatum, 920–921
   Trachysphaeridium lachandinum, 907–908, 916, 976
   Trachysphaeridium laminaritum, 229, 912, 916, 923, 925, 936, 938–939, 945, 948, 1060, **1075**
   Trachysphaeridium aff. *laminaritum*, 908, 920
   Trachysphaeridium laufeldii, 190, 229, 908–909, 912, 923, 926–927, 940, 1151
   Trachysphaeridium levis, 911, 922–923, 927, 935–936
   Trachysphaeridium maicum, 908, 916, 1150
   Trachysphaeridium minor, 908
   Trachysphaeridium patellare, 945
   Trachysphaeridium planum, 920–921
   Trachysphaeridium rugosum, 920
   Trachysphaeridium simplex, 889, 920–921
   Trachysphaeridium stipticum, 908
   Trachysphaeridium timofeevii, 922–923, 941, 949
   Trachysphaeridium uspenskyi, 939
   Trachysphaeridium vetterni, 923, 976, 1150
Trachythichoides, 916, 1060, **1087**, **1089**, 1152
   Trachythichoides ovalis, 916, 1060, **1087**, **1089**, 1152
Trapezochites, 1026
Trapezotheca, 1029
TRAPEZOVITIDAE, 1175
Trapezovitus, 1028
Trematobolus, 1030
Trematosia, 1030
Trematosphaeridium, 229, 880, 887, 890–893, 897–902, 904–906, 908–911, 916, 918, 922–924, 926, 929–934, 942–944, 948, 1140
   Trematosphaeridium bhimaii, 932
   Trematosphaeridium holtedahlii, 229, 887, 890–891, 893, 897, 899–900, 905–906, 909–911, 916, 918, 922–924, 926, 930, 933–934, 942, 948
   Trematosphaeridium sinuatum, 899, 902, 931, 944, 1140
   Trematosphaeridium zairense, 904
Treptichnus, 392, 397, 1014
   Treptichnus birfurcus, 1014
   Treptichnus triplex, 1014
Triangulaspis, 1023
Triangulina, 1023
Triangumorpha, 871
   Triangumorpha striata, 871
TRIBRACHIDIDEA, 373
Tribrachidium, 372–373, 379, 381, **386**, 407, 504, 1006, 1173
   Tribrachidium heraldicum, **386**, 1006
Tricellaria, 871
   Tricellaria delylensis, 871
Trichocorixa, 296
Trichodesmium, 430, 597, 850

Trichodesmium erythaeum, 850
TRICREPICEPHALIDAE, 1176
Tricuspidatia, 1052
   Tricuspidatia trigonata, 1052
Tridia, 1052
   Tridia koptevi, 1052
   Tridia salebrosa, 1052
TRIDIIDAE, 1053
TRILOBITA, 355, 381, 400, 402, 438, 440, 553, 556, 558, 581, 1022, 1037, 1051, 1175, 1180–1184
TRILOBOZOA, 372–373, 381, 401, 407, 407, 434, 559, 1020–1021, 1173, 1179, 1181–1184
Trinema, 237
Trininaecyathus, 1019
TRINODIDAE, 1175
Triplicatella, 1032
TROCHOCYSTITIDAE, 1178
Tropidiana, 1022
Truncatoconus, 1028
TRYBLIDIIDA, 1027, 1174
Trypanosoma, 482–483
   Trypanosoma brucei, 482
TSINANIIDAE, 1176
Tsunydiscus, 1023
Tsunyiella, 1022
TUARANGIIDA, 1174
Tubercularia, 986, 995
   Tubercularia latiuscula, 986, 995
Tubericyathus, 1020
Tuberoconus, 1027
Tubicoscinus, 1020
Tubiphyton, 1052
   Tubiphyton taghdoutensis, 1052
Tubocyathus, 1018
TUBOIDEA, 1178
Tubomorphophyton, 359, **362**, 366, 986, 988–989, 993
   Tubomorphophyton benignum, 988, 993
   Tubomorphophyton botomense, 988, 993
   Tubomorphophyton cristatum, 988, 993
   Tubomorphophyton latum, 993
   Tubomorphophyton limpidum, 993
   Tubomorphophyton nubilum, 989, 993
   Tubomorphophyton pseudofruticosum, 993
Tubophyllum, 984–987, 995
   Tubophyllum victori, 986–987, 995
Tubulella, 1032, 1174
Tubulosa, 940, 1061, **1112**, 1127
   Tubulosa corrugata, 940, 1061, **1112**, 1127
Tulenicornus, 1028
Tumulcoscinus, 1020
Tumulduria, 404, 1031, 1174
Tumilfungia, 1019
TUMULIFUNGIIDAE, 1172
TUMULIOLYNTHIDAE, 1172
Tumuliolynthus, 1020
Tumulocyathellus, 1019
TUMULOCYATHIDAE, 1172
Tumulocyathus, 1019
Tumuloglobosus, 1020
Tungusella, 1025
Tungussia, 519
TUNICATES (*See* Subject Index)
Tuojdachites, 1028
Tuoraconus, 1027
Turcutheca, 1029
Turricyathus, 1021
Turuchanica, **198**–200, 890, 892–893, 897, 901, 905, 907–910, 916, 924, 928, 934, 936, 938, 1060, **1076**
   Turuchanica alara, 900, 905, 907–908, 924, 934, 936, 938
   Turuchanica aff. *kulgunica*, 936
   Turuchanica maculata, 936
   Turuchanica ternata, **198**, 200, 890, 892–893, 897, 900–901, 905, 908–910, 916, 928, 934, 1060, **1076**
Tuserospina, 1026
Tuvaeconus, 1020
Tuvanella, 1025
Tuvinia, 1033

Taxonomic Index

Tuyunaspis, 1023
Tuzoia, 1025
TUZOIIDAE, 1177
Tyloligotriletum, 936
    Tyloligotriletum asper, 936
Tylosphaeridium, 947, 1149
    Tylosphaeridium tallinicum, 947, 1149
Tyrasotaenia, 354–**355**, 357, 503–504, 657, 890, 976, 978
    Tyrasotaenia filiforma, 976
    Tyrasotaenia podolica, **355**, 976
    Tyrasotaenia cf. *podolica*, 890, 976
    Tyrasotaenia tungusica, 976
Tyrkanispongia, 1006
    Tyrkanisponia tenua, 1006

Udzhaites, 1021
Uktaspis, 1023
Ulcundia, 1032
Ulophyton, 879, 1060, **1078**
    Ulophyton rifeicum, 879, 1060, **1078**
Ulothrix, 322–**323**
    Ulothrix cf. *aequalis*, 322
ULOTHRICACEAE, 1001, 1152
Unbellula, 986, 993
    Unbellula minuta, 986, 993
Uncinaspira, 1027
Uniformitheca, 1029
Uniporata, 875, 947–948
    Uniporata nidia, 875, 947
    Uniporata striata, 948
    Uniporata torosa, 875
UNIRAMIA, 381, 408
Uralocyathella, 1020
URALOCYATHELLIDAE, 1172
URALOCYATHIDAE, 1172
Uranosphaera, 1020
Uranovia, 986, 995
    Uranovia granosa, 986, 995
    Uranovia multa, 995
Urcyathella, 1019
Urcyathus, 1019
Urococcus, 1139
    Urococcus foslienanus, 1139
Ushbaspis, 1025
Ushkarella, 1022, 1177
Utahphospha, 412
UTAHPHOSPHIDAE, 400, 403, 411–412, 449, 1031
Utchurella, 984, 986, 995
    Utchurella explicata, 986, 995

Vairimorpha, 481, 483
Vaizitsinia, 1006
    Vaizitsinia sophia, 1006
Valdainia, 1006
    Valdainia plumosa, 1006
Valeria, 918, 920, 1103
    Valeria lophostriata, 918, 920, 1103
Validaspis, 1025
Vallenia, 870, 1052
    Vallenia erlingi, 870, 1052
Vandalosphaeridium, **226**, 228, 514, 914, 931, 933, 941, 1152
    Vandalosphaeridium reticulatum, 228
    Vandalosphaeridium varangeri, 931, 933, 941, 1152
    Vandalosphaeridium walcottii, **226**, 228, 914, 1152
Varicamanicosiphonia, 981, 986, 998
    Varicamanicosiphonia quadricella, 981, 998
    Varicamanicosiphonia segmenta, 998
Variopelta, 1023
Varitheca, 1029
Vaucheria, 907
VAUCHERIACEAE, 1152
Vauxia, 1017
VAUXIIDAE, 1171
Vaveliksia, 1006
    Vaveliksia velikanovi, 1006
Vavososphaeridium, 882, 904, 932, 1149
    Vavososphaeridium bharadwajii, 904
    Vavososphaeridium densum, 904, 1149
    Vavososphaeridium reticulatum, 932
    Vavososphaeridium vindhyanense, 904
Velancorina, 1006
    Velancorina martina, 1006
Velumbrella, 372, 1020, 1173
Vendella, 1005–1006
    Vendella haelenicae, 1006
    Vendella larini, 1006
    Vendella sokolovi, 1005–1006
Vendia, 373, **386**, 1006
    Vendia sokolovi, **386**, 1006
VENDIATA, 373
Vendichnus, 505, 1014
    Vendichnus vendicus, 1014
Vendomia, 373, **386**, 1006
    Vendomia menneri, **386**, 1006
VENDOPHYCEAE, 349
Vendotaenia, 220, 354, 357, 502, 504, 514, 519, 627, 883, 940, 943, 958, 969, 978
    Vendotaenia antiqua, 958, 978
    Vendotaenia cf. *antiqua*, 943
VENDOTAENIDS (*See* Subject Index)
Vendovermites, 1014, 1052
VENDOZOA, 381
Veprina, 1006
    Veprina undosa, 1006
VERMES, 373, 969
Vermiforma, 396, 1006, 1014
    Vermiforma antiqua, 396, 1006, 1014
Veronicacyathus, 1020
VERTEBRATA, 400, 412, 431, 448–451, 599
Veryhachium, 233–234, 887, 1149
Vesicophycus, 880
    Vesicophycus problematicus, 880
Vesicophyton, 930, 1142
    Vesicophyton punctatum, 930, 1142
Vesiculosphaerina, 948
    Vesiculosphaerina singularis, 948
VESTROGOTHIIDAE, 1177
Veteronostocale, 914, 935, 1061, **1091**, 1131
    Veteronostocale amoenum, 914, 1061, **1091**, 1131
    Veteronostocale copiosus, 935
Vetulicola, 1025, 1177
Vetustovermis, 1020
Vimenites, 392–393, 505, 1013–1014
    Vimenites bacillaris, 393, 1013–1014
Vindhyanella, 349, 503, 978
    Vindhyanella jonesi, 978
Vindhyania, 354, 503, 978
    Vindhyania jonesi, 978
Vindhyavasinia, 349, 354, 503
Virgatotheca, 1029
Vittia, 1018
Vladmissa, 1006
    Vladmissa missarzhevskii, 1006
Volborthella, 1021
VOLBORTHELLIDA, 1174, 1179, 1181–1184
Volodia, 1052
    Volodia annulata, 1052
Vologdinella, 986, 993
    Vologdinella fragile, 986, 993
    Vologdinella grandis, 993
    Vologdinella pulchra, 993
VOLOGDINELLIDAE, 1174
Vologdinia, 366, 986, 993, 1052
    Vologdinia concentrica, 1052
    Vologdinia major, 1052
    Vologdinia shorica, 1052
    Vologdinia verticillata, 986, 993
Vologdinocyathellus, 1019
VOLOGDINOCYATHIDAE, 1172
Vologdinocyathus, 1019
VOLOGDINOPHYLLACEAE, 1021
VOLOGDINOPHYLLIDAE, 1174
Vologdinophyllum, 1021
Volvacyathus, 1018

VOLVORTHELLIDAE, 1174
Volvox, 430
Volyniella, 875, 882, 911, 920, 935, 938–939, 1060–1061, **1082**, **1112**, **1114–1115**, 1127, 1129
    Volyniella canilovica, 939, 1061, **1114**, 1127
    Volyniella cylindrica, 935, 1129
    Volyniella glomerata, 911, 920, 1060, **1082**
    Volyniella rara, 875, 1061, **1115**
    Volyniella valdaica, 938, 1061, **1112**
Voroninicyathus, 1019
Voznesenskicyathus, 1018

Walcotella, 1022
Walcottina, 1030
Walossekia, 1177
Wangzishia, 1025
Wanneria, 1025
WAPKIIDAE, 1171
WAPTIIDA, 1177
WAPTIIDAE, 1177
Watsonella, 1026
Waucobella, 1031
WESTERGAARDODINIDAE, 1178
Westgardia, 1174
Westonia, 1030
Wetheredella, 239, 359, **362**, 366, 986, 993
    Wetheredella silurica, 986
Weymouthia, 1023
Wigwamiella, 373, 1006
    Wigwamiella enigmatica, 1006
WIWAXIIDA, 408, 1173
Wollea, 356
WORMS (*See* Subject Index)
Worthenella, 1174
WRIGHTICYATHIDAE, 1172
Wrighticyathus, 1019
Wuchiapingella, 1022
Wushichites, 1032
Wutingaspis, 1025
Wutingella, 1022
Wyattia, 660, 1032

XANTHOPHYCEAE, 189
XENOPODA, 1053
Xenopus, 482
    Xenopus laevis, 482
Xenothrix, 233–234, 887, 1149
    Xenothrix inconcreta, 887, 1149
Xenusion, 1033, 1174
Xestecyathus, 1020
Xiadongoconus, 1028
Xiadongtubulus, 1029
Xianfengella, **405**, 1027, 1177
    Xianfengella prima, **405**
Xianfengoconcha, 1028
Xiangquianaspis, 1024
XIAOSHANOCERATIDAE, 1175
Xilingxia, 1024
Xilingxiaconus, 1027
Xiuqiella, 1023

Yacutolituus, 1028
Yangtzechiton, 1026
Yangtzeconus, 1027
Yangtzedonta, 1025
Yangtzemerisma, 1028
Yangtzesclerites, 1032
Yangtzespira, 1027
YANHECERATIDAE, 1175
YANHECERIDA, 1175
Yanischevskyites, 238, 1032

Yankogovitus, 1028
Yaoyingella, 1022
Yelovichnus, 373, 392–393, **395**, 437, 1011, 1014
    Yelovichnus gracilis, **395**, 1011, 1014
Yentaiia, 986, 993
    Yentaiia liaoyangensis, 986, 993
Yeshanella, 1022
Yiliangella, 1025
Yiliangellina, 1025
Yinites, 1025
YINITIDAE, 1176
Yochelcionella, 1027
YOCHELCIONELLIDAE, 1174
YOHOIIDAE, 1175, 1180–1184
Yorkia, 1030
YORKIIDAE, 1177
Yuanjiapingella, 1031
Yudjaicyathus, 1019
Yuehsienszella, 1024
Yukoncyathus, 1019
Yukonia, 1023
Yukonides, 1023
Yuliunia, 1032
Yunnanaspidella, 1025
Yunnanaspis, 1025
YUNNANOCEPHALIDAE, 1176
Yunnanocephalus, 1025
Yunnanodus, **402**, 1031, 1174
    Yunnanodus dolerus, **402**
Yunnanomedusa, 372, 1033
Yunnanopleura, 1028
Yunnanospira, 1027
Yunnanotheca, 1030
Yuwenia, 1026

ZACANTHOIDIDAE, 1176
Zacanthopsis, 1023
Zaganolomia, 986, 998
    Zaganolomia buralica, 986, 998
Zangerlispongia, 1017
Zea, 482
    Zea mays, 482
Zepaera, 1022
Zeugites, 1028
ZHANATELLIDAE, 1177
Zhenbaspis, 1025
Zhenpingella, 1022
Zhijinella, 1022
Zhijinites, 1026
ZHIJINITIDAE, 1026, 1174
Zhongbaoella, 1022
Zhuravlevaecyathus, 1020
Zigzagella, 1017
Zolotytsia, 1006
    Zolotytsia biserialis, 1006
Zonacoscinus, 1020
Zonacyathus, 1019
Zonooidium, 945
    Zonooidium guttiforme, 945
    Zonooidium mirabile, 945
Zonosphaeridium, 900, 904–905, 916, 920
    Zonosphaeridium crassum, 900
    Zonosphaeridium densum, 904
    Zonosphaeridium foveolatum, 905
    Zonosphaeridium minutum, 920
    Zonosphaeridium punctatum, 904
Zoophycos, 1043
Zoothamnium, 430
    Zoothamnium geniculatum, 430
Zosterosphaera, 885, 914
    Zosterosphaera tripunctata, 914
Zygogonium, 301, 314, 324